BUSINESS WRITER'S QUICK REFERENCE GUIDE

D1710730

Terry R. Bacon, PhD
Lawrence H. Freeman, PhD
Shipley Associates

Editorial Board

Larry Best, PhD
Phillip Bozek, PhD
Jeffrey J. Butler, DA
Richard Callahan, PhD
James DeMoux, PhD
Sidney Jenson, PhD
A. Dennis Mead, PhD
Vicki Mickelsen, PhD
David Pugh, PhD
Mary Jo Roberts, PhD
William C. Stringham, PhD
Barbara von Diether, PhD

Business, Law and General Books Division

John Wiley & Sons, Inc.
New York • Chichester • Brisbane • Toronto • Singapore

Library of Congress Cataloging-in-Publication Data
Main entry under title:

Business writer's quick reference guide.

Includes index.
1. English language—Rhetoric—Handbooks, manuals, etc. 2. English language—Business English—Handbooks, manuals, etc. I. Shipley Associates.
PE1479.B87B87 1986 808′.066651 85-29533
ISBN 0-471-84541-8

Printed in the United States of America

87 10 9 8 7 6 5 4

TABLE OF CONTENTS

Reference Glossary

Model Documents

Letters

Memos

Other

Reports

PREFACE

The *Business Writer's Quick Reference Guide* is for writers, editors, secretaries, reviewers, supervisors, executives, and others who communicate business and technical ideas in English and who need practical answers to questions about writing in the world of work.

We have published this *Business Writer's Guide* because few of the other style guides available today address the daily writing and editing problems that professional men and women face. Most style guides present the rules of grammar, punctuation, and spelling, but they generally do not answer these kinds of questions:

- How should you set up headings?
- How should you use lists?
- How should you use key words?
- What are the principles of organization?
- How do you present ideas that your readers will not like or will not accept?
- How do you write for managers? For technical and scientific peers?
- How can you recognize passive sentences? How do you convert passive sentences into active sentences?
- When should you use visual aids? Which kind of visual should you use, and how do you set it up?
- How do you write good captions?

To provide the answers to these practical questions, we have created an alphabetical glossary of style that includes topics of interest to professionals in the world of work. Following that is a documents section with models of business and technical documents. To help you find information, we have included a detailed table of contents and index and have thoroughly cross-referenced both sections of the *Business Writer's Guide*.

English and the State of Style

Like all languages, English is a set of conventions: sounds and ways to spell these sounds, words and ways to combine them, sentence structures, punctuation symbols, and word meanings that range from the concrete to the abstract.

These conventions change over time. Words are born, grow, and change in meaning; they evolve through usage and die from disuse when writers and speakers no longer need them. Similarly, punctuation, spelling, and stylistic conventions change. They evolve as the language adapts to printing presses, computers, space shuttles, television, new industries, changing social concerns and political issues, new perspectives on history, new economic theories—in short, to everything in a constantly changing world.

English has changed dramatically since eighth-century *Beowulf*, one of the earliest English texts. Today, the original text of *Beowulf* looks as though it is written in a foreign language. English has even changed since Shakespeare was writing—only 400 years ago. The original language in Shakespeare's plays is often incomprehensible to modern readers. And four hundred years from now, readers of English will likely consider today's English just as incomprehensible.

Living languages like English constantly change. (Dead languages like Latin do not change.) If English were static, we could give precise rules for usage. We could ensure that those rules followed logic, were consistent, and had no unruly exceptions. But English is dynamic, and over time its conventions have evolved—often in unpredictable and seemingly nonsensical ways. So its rules are not always logical, they are rarely consistent, and they have many, many exceptions.

However, with a little diligence and the right tools, you can use English well. This *Business Writer's Guide* is one of the right tools. You have to supply the diligence.

In this book, we have recorded the currently accepted stylistic conventions of English. We have labelled those conventions *rules*, but you should understand that these rules merely describe current stylistic conventions, especially as they apply to business and technical documents. The rules are not laws, and over time they will surely change. If you compare this *Business Writer's Guide* to other style guides, you will probably find some disagreements. As we wrote this book, we often had to make decisions about stylistic preferences. Where authorities disagreed with each other, we chose the style most in harmony with the needs of technical and business writers and readers, and where an academic authority's stylistic preference conflicted with common and accepted usage in today's business and technical community, we chose common usage.

We have simplified some discussions and descriptions to make them more useful to writers who are not experts in grammar and punctuation. Our simplifications do not misrepresent the current conventions of English grammar, but they may overlook certain exceptions and complexities (often of value only to university scholars).

Acknowledgments

Before acknowledging those people who actively contributed to this *Business Writer's Guide,* we wish to thank the tens of thousands of engineers, geologists, chemists, technicians, biologists, field foremen, landmen, geophysicists, secretaries, editors, lawyers, accountants, auditors, managers, supervisors, and others who have participated in Shipley Associates workshops. Without their questions, concerns, problems, uncertainties, challenges, and insights, we would not have been able to produce this book. In teaching them, we have learned; in learning from us, they have taught.

Among the many professionals at Shipley Associates who have materially contributed to the *Business Writer's Guide* are those on our editorial board: Doctors Sid Jenson, Vicki Mickelsen, Larry Best, Phil Bozek, Jeff Butler, Richard Callahan, James DeMoux, A. Dennis Mead, David Pugh, Mary Jo Roberts, William Stringham, and Barbara von Diether. These fine teachers and researchers reviewed the concept and design of the *Business Writer's Guide,* critiqued sections as they were prepared, and provided invaluable assistance in editing and proofreading the final text. We are indebted as much to their skills as writers and reviewers as we are to their knowledge of language and their perceptions of the *Business Writer's Guide's* audience.

We also acknowledge the contributions of a fine staff: Barbara Petersen, DeNeil Petersen, Merilee Meyer, and Patti Ferrin. These professionals are responsible for the typing, proofreading, editing, design, and layout of the book. Without their labor, intelligence, and dedication, this book would not exist.

Terry R. Bacon
Lawrence H. Freeman
November 1985

SURVEYING THE PARTS OF THE GUIDE

Use the chart below and the illustrations on the following pages when you have questions about using the *Business Writer's Guide*. As the chart indicates, you can locate information on writing style and the conventions of English either in the reference glossary (pp. 3-328) or in the model documents section (pp. 241-308). The *Business Writer's Guide* also has a table of contents (pp. iii-iv) and a very thorough index (pp. 311-319).

If you have a specific question about punctuation, grammar, or style	→	**Go to the appropriate alphabetical topic in the reference glossary (pp. 3-238).**
To find out how to write a particular type of letter, memo, or other document	→	**Go to the model documents section (pp. 241-308).**
If you have trouble deciding what to look up	↗	**Go to the table of contents (pp. v-vi).**
	↘	**Go to the index (pp. 309-319).**

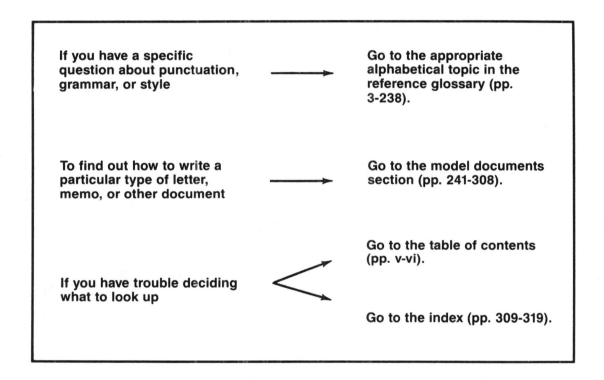

If you have a specific question about punctuation, grammar, or style—

Go to the appropriate alphabetical topic in the reference glossary (pp. 3-238).

Sample Questions

- Should I use a comma or a semicolon to separate items in a series?

- Should I use a pie chart or a bar chart to illustrate my data?

- Should I use headings in a report only three pages long?

You may have to check more than one topic in the reference glossary to answer a specific question. If so, the cross references will direct you to the right topics.

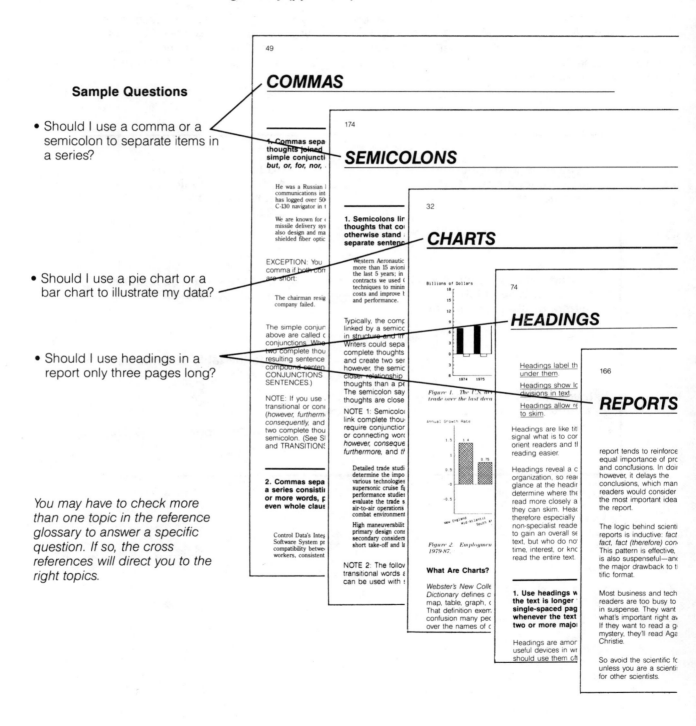

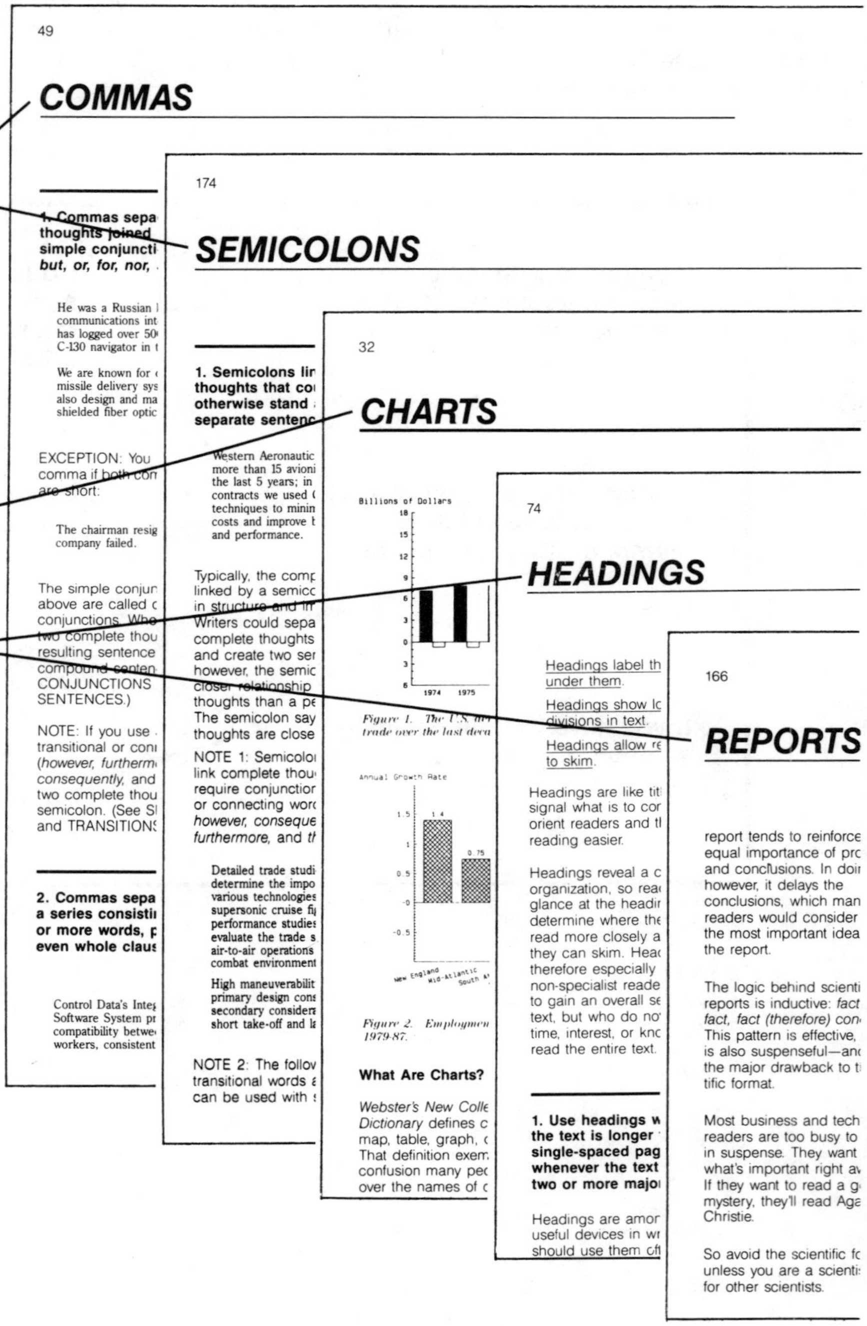

What should you look for once you find a topic that addresses your question?

Sample Question

- Should I use a comma or a semicolon to separate items in a series?

Step 1: Scan the topic for the pertinent rule. (Rules are numbered and printed in blue.)

Step 2: Look through the examples for one like your sentence or paragraph.

Step 3: Check the notes for exceptions or additional examples that might be helpful.

Step 4: Look up the cross references (in all capital letters) if you need further information.

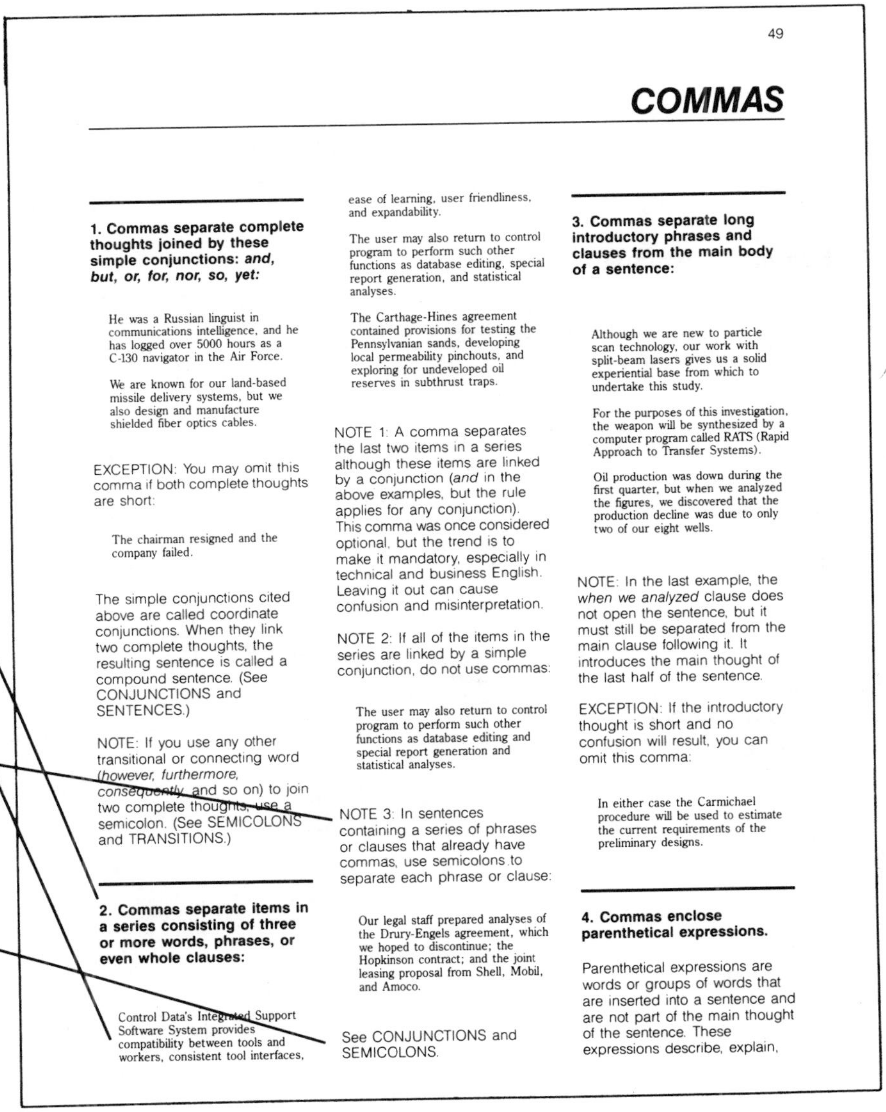

49

COMMAS

1. Commas separate complete thoughts joined by these simple conjunctions: *and, but, or, for, nor, so, yet:*

He was a Russian linguist in communications intelligence, and he has logged over 5000 hours as a C-130 navigator in the Air Force.

We are known for our land-based missile delivery systems, but we also design and manufacture shielded fiber optics cables.

EXCEPTION: You may omit this comma if both complete thoughts are short:

The chairman resigned and the company failed.

The simple conjunctions cited above are called coordinate conjunctions. When they link two complete thoughts, the resulting sentence is called a compound sentence. (See CONJUNCTIONS and SENTENCES.)

NOTE: If you use any other transitional or connecting word (*however, furthermore, consequently*, and so on) to join two complete thoughts, use a semicolon. (See SEMICOLONS and TRANSITIONS.)

2. Commas separate items in a series consisting of three or more words, phrases, or even whole clauses:

Control Data's Integrated Support Software System provides compatibility between tools and workers, consistent tool interfaces, ease of learning, user friendliness, and expandability.

The user may also return to control program to perform such other functions as database editing, special report generation, and statistical analyses.

The Carthage-Hines agreement contained provisions for testing the Pennsylvanian sands, developing local permeability pinchouts, and exploring for undeveloped oil reserves in subthrust traps.

NOTE 1: A comma separates the last two items in a series although these items are linked by a conjunction (*and* in the above examples, but the rule applies for any conjunction). This comma was once considered optional, but the trend is to make it mandatory, especially in technical and business English. Leaving it out can cause confusion and misinterpretation.

NOTE 2: If all of the items in the series are linked by a simple conjunction, do not use commas:

The user may also return to control program to perform such other functions as database editing and special report generation and statistical analyses.

NOTE 3: In sentences containing a series of phrases or clauses that already have commas, use semicolons to separate each phrase or clause:

Our legal staff prepared analyses of the Drury-Engels agreement, which we hoped to discontinue; the Hopkinson contract; and the joint leasing proposal from Shell, Mobil, and Amoco.

See CONJUNCTIONS and SEMICOLONS.

3. Commas separate long introductory phrases and clauses from the main body of a sentence:

Although we are new to particle scan technology, our work with split-beam lasers gives us a solid experiential base from which to undertake this study.

For the purposes of this investigation, the weapon will be synthesized by a computer program called RATS (Rapid Approach to Transfer Systems).

Oil production was down during the first quarter, but when we analyzed the figures, we discovered that the production decline was due to only two of our eight wells.

NOTE: In the last example, the *when we analyzed* clause does not open the sentence, but it must still be separated from the main clause following it. It introduces the main thought of the last half of the sentence.

EXCEPTION: If the introductory thought is short and no confusion will result, you can omit this comma:

In either case the Carmichael procedure will be used to estimate the current requirements of the preliminary designs.

4. Commas enclose parenthetical expressions.

Parenthetical expressions are words or groups of words that are inserted into a sentence and are not part of the main thought of the sentence. These expressions describe, explain,

To find out how to write a particular type of letter, memo, or other document—

Go to the model documents section (pp. 241-308). To see what models are available, check page vi of the table of contents.

Sample Questions

- How should I write a letter complaining about a product?
- How should I summarize information for executive bodies?
- How should I set up and organize a procedure for a procedure manual?

You may find more than one useful model.

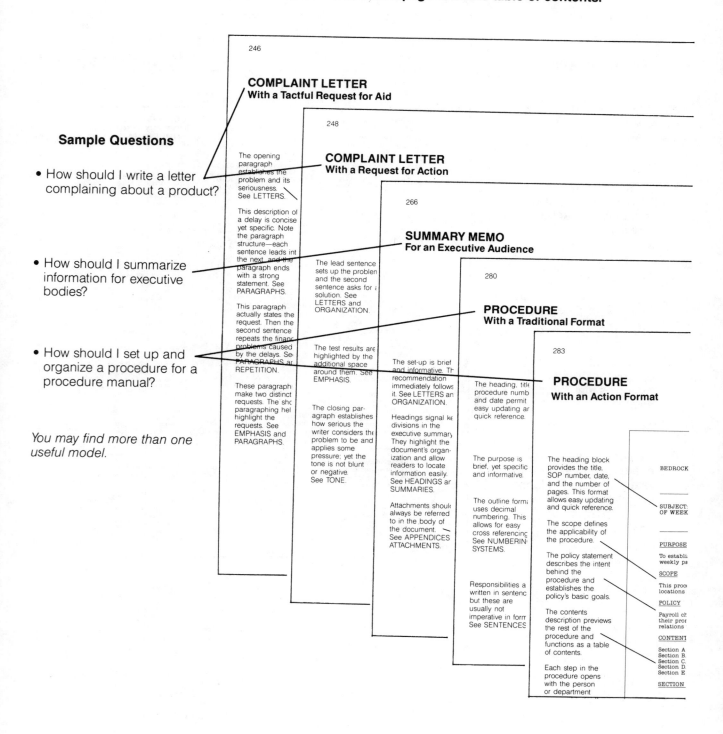

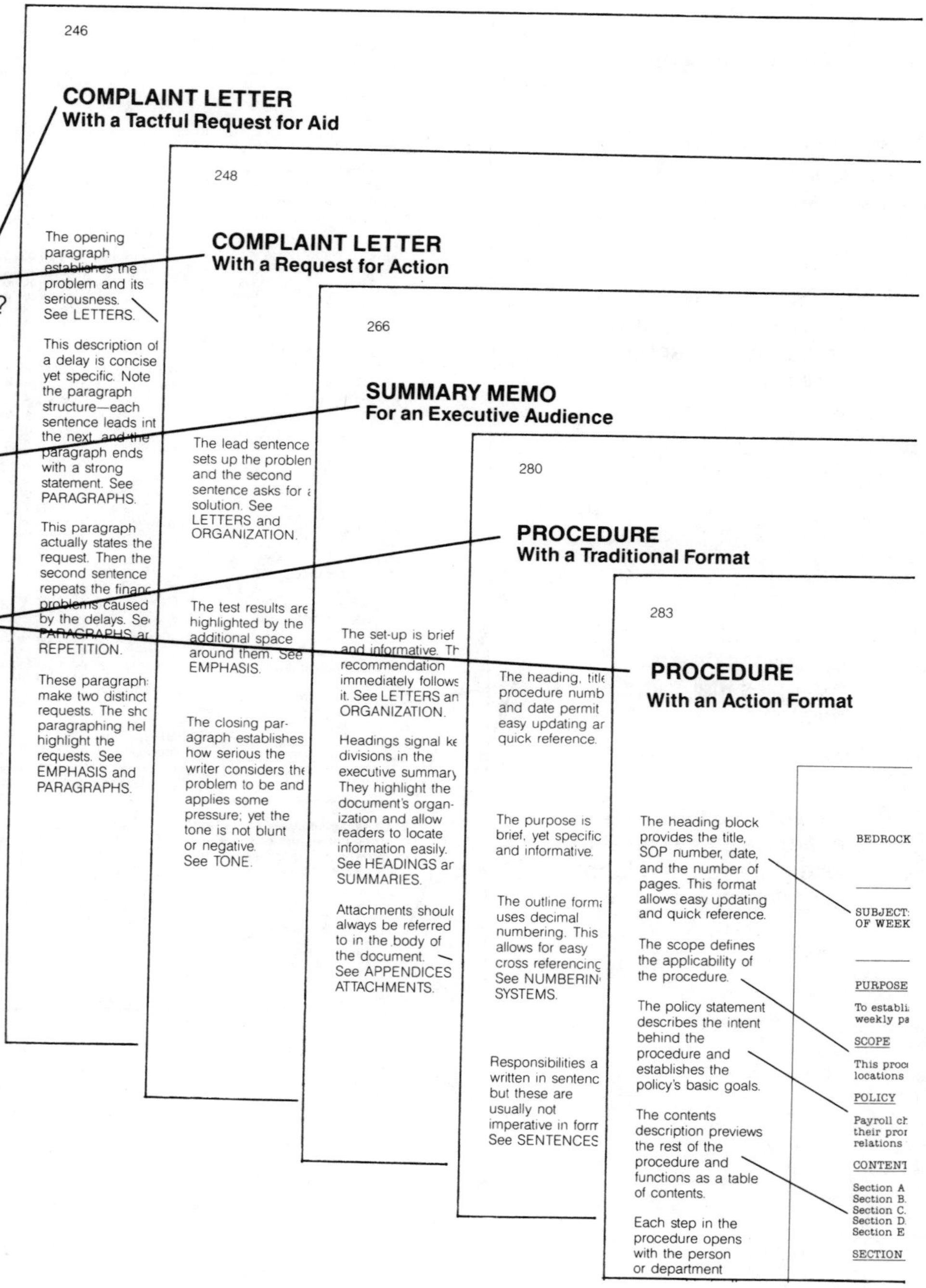

What should you look for once you locate a model similar to the letter, memo, or document you want to write?

Sample Question

- How should I write a letter complaining about a product?

Step 1: Read the model, noting its organization, format, style, and tone.

Step 2: Study the marginal comments for important notes, suggestions, and options.

Step 3: Read the bottom comment for general guidance on writing the letter, memo, or document.

Step 4: For further information or clarification, check one or more of the cross references.

248

COMPLAINT LETTER
With a Request for Action

FARMLAND FROZEN
FAMILY FOODS

August 8, 1983

Ms. Patricia Goodway
Customer Support Manager
The Pacific Baking Company
9924 West Pacific Way
San Francisco, CA 97521

Dear Ms. Goodway:

Recent samples of your breader contain more fines than our specifications allow. Please investigate the problem and let me know as soon as possible what you can do to solve it.

On August 5, our Medford plant inspectors noticed excessive carbon specks on breaded products, so they checked the breader lots on hand (codes 341X and 345X). According to our specification, fines through a U.S. Standard Sieve No. 80 must fall within these limits: 1 percent ± 1 percent (or a maximum of 2 percent). The breader lots on hand gave these results:

341X — 3.30%

345X — 4.10%

As you can see, the fines in these two lots exceeded our specification by 1.3 and 2.1 percent. These levels are not yet serious, but they do cause excessive carbon specks in the frying oil and on the finished product.

We have enjoyed a long relationship with Pacific Baking and hope to continue doing business with you. We are therefore anxious to solve this problem and will appreciate your prompt action.

Sincerely,

James Van Prooven

James Van Prooven
Plant Manager

JVP:qw

303 Blossom Avenue Des Moines, Iowa 50321 (515) 521-4911

The lead sentence sets up the problem, and the second sentence asks for a solution. See LETTERS and ORGANIZATION.

The test results are highlighted by the additional space around them. See EMPHASIS.

The closing paragraph establishes how serious the writer considers the problem to be and applies some pressure; yet the tone is not blunt or negative. See TONE.

Complaint letters, even if written in anger, should not sound angry. The tone should be positive and constructive. The writer should present clear evidence of dissatisfaction, but should strive to balance the complaint with positive solutions. See TONE and LETTERS.

As in this letter, complaint letters usually start with a concise statement of the problem followed by a request for action or resolution. The problem statement should be as brief as possible but long enough to make the request understandable. See the discussion of set-ups in ORGANIZATION.

This letter illustrates the semi-block format with standard punctuation. See LETTERS.

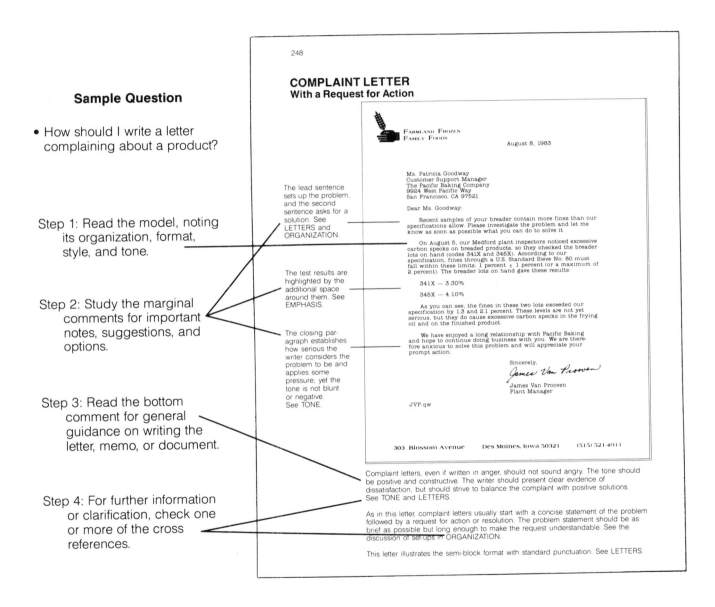

If you have trouble deciding what to look up—

Go to the table of contents (pp. iii-iv) or the index (pp. 309-319).

Sample Question

- How should I begin a response letter?

Step 1: Check the table of contents for topics that might help you with your problem.

Step 2: Turn to the most promising topic in the reference glossary or model in the model documents section.

Step 3: If your first topic or model doesn't help you, go to another one.

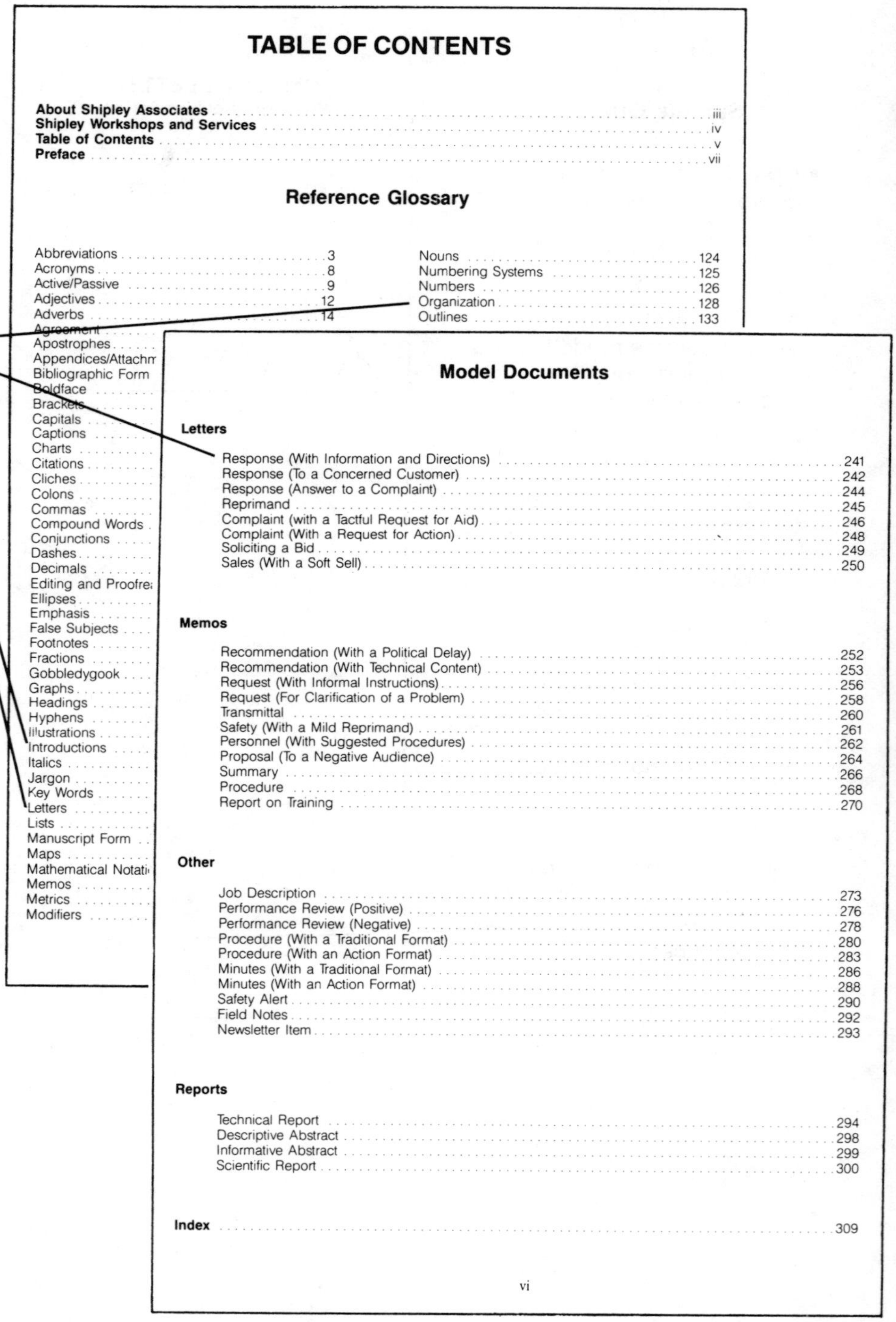

TABLE OF CONTENTS

Reference Glossary

Model Documents

Letters

Memos

Other

Reports

vi

Sample Question

- How should I begin a response letter?

Step 1: Survey the index for topics that might help you with your problem.

Step 2: Check any cross references to other topics in the index.

Step 3: Look up the most promising topic in the reference glossary or model in the model documents section.

Step 4: If your first topic or model doesn't help you, go to another one.

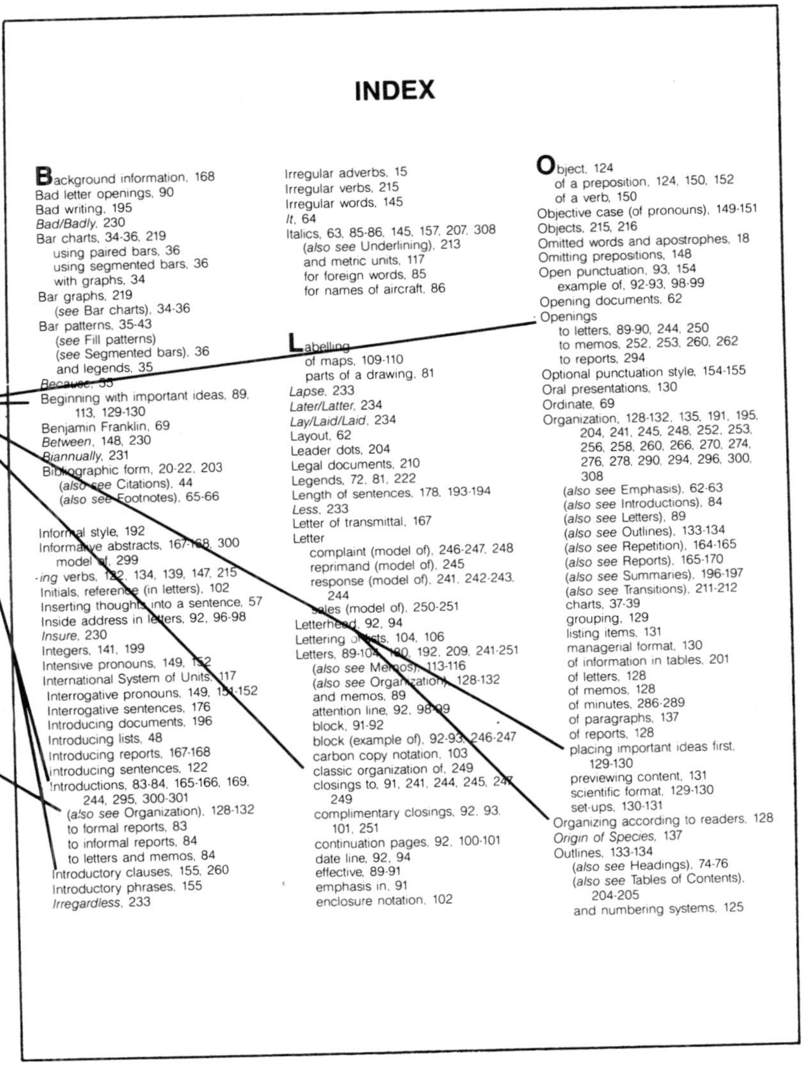

INDEX

Background information, 168
Bad letter openings, 90
Bad writing, 195
Bad/Badly, 230
Bar charts, 34-36, 219
  using paired bars, 36
  using segmented bars, 36
  with graphs, 34
Bar graphs, 219
  (*see* Bar charts), 34-36
Bar patterns, 35-43
  (*see* Fill patterns)
  (*see* Segmented bars), 36
  and legends, 35
Because, 33
Beginning with important ideas, 89, 113, 129-130
Benjamin Franklin, 69
Between, 148, 230
Biannually, 231
Bibliographic form, 20-22, 203
  (*also see* Citations), 44
  (*also see* Footnotes), 65-66

Informal style, 192
Informative abstracts, 167-168, 300
  model of, 299
-ing verbs, 122, 134, 139, 147, 215
Initials, reference (in letters), 102
Inserting thoughts into a sentence, 57
Inside address in letters, 92, 96-98
Insure, 230
Integers, 141, 199
Intensive pronouns, 149, 152
International System of Units, 117
Interrogative pronouns, 149, 151-152
Interrogative sentences, 176
Introducing documents, 196
Introducing lists, 48
Introducing reports, 167-168
Introducing sentences, 122
Introductions, 83-84, 165-166, 169, 244, 295, 300-301
  (*also see* Organization), 128-132
  to formal reports, 83
  to informal reports, 84
  to letters and memos, 84
Introductory clauses, 155, 260
Introductory phrases, 155
Irregardless, 233

Irregular adverbs, 15
Irregular verbs, 215
Irregular words, 145
It, 64
Italics, 63, 85-86, 145, 157, 207, 308
  (*also see* Underlining), 213
  and metric units, 117
  for foreign words, 85
  for names of aircraft, 86

Labelling
  of maps, 109-110
  parts of a drawing, 81
Lapse, 233
Later/Latter, 234
Lay/Laid/Laid, 234
Layout, 62
Leader dots, 204
Legal documents, 210
Legends, 72, 81, 222
Length of sentences, 178, 193-194
Less, 233
Letter of transmittal, 167
Letter
  complaint (model of), 246-247, 248
  reprimand (model of), 245
  response (model of), 241, 242-243, 244
  sales (model of), 250-251
Letterhead, 92, 94
Lettering objects, 104, 106
Letters, 89-104, 180, 192, 209, 241-251
  (*also see* Memos), 113-116
  (*also see* Organization), 128-132
  and memos, 89
  attention line, 92, 98-99
  block, 91-92
  block (example of), 92-93, 246-247
  carbon copy notation, 103
  classic organization of, 249
  closings to, 91, 241, 244, 245, 247, 249
  complimentary closings, 92, 93, 101, 251
  continuation pages, 92, 100-101
  date line, 92, 94
  effective, 89-91
  emphasis in, 91
  enclosure notation, 102

Object, 124
  of a preposition, 124, 150, 152
  of a verb, 150
Objective case (of pronouns), 149-151
Objects, 215, 216
Omitted words and apostrophes, 18
Omitting prepositions, 148
Open punctuation, 93, 154
  example of, 92-93, 98-99
Opening documents, 62
Openings
  to letters, 89-90, 244, 250
  to memos, 252, 253, 260, 262
  to reports, 294
Optional punctuation style, 154-155
Oral presentations, 130
Ordinate, 69
Organization, 128-132, 135, 191, 195, 204, 241, 245, 248, 252, 253, 256, 258, 260, 266, 270, 274, 276, 278, 290, 294, 296, 300, 308
  (*also see* Emphasis), 62-63
  (*also see* Introductions), 84
  (*also see* Letters), 89
  (*also see* Outlines), 133-134
  (*also see* Repetition), 164-165
  (*also see* Reports), 165-170
  (*also see* Summaries), 196-197
  (*also see* Transitions), 211-212
  charts, 37-39
  grouping, 129
  listing items, 131
  managerial format, 130
  of information in tables, 201
  of letters, 128
  of memos, 128
  of minutes, 286-289
  of paragraphs, 137
  of reports, 128
  placing important ideas first, 129-130
  previewing content, 131
  scientific format, 129-130
  set-ups, 130-131
Organizing according to readers, 128
Origin of Species, 137
Outlines, 133-134
  (*also see* Headings), 74-76
  (*also see* Tables of Contents), 204-205
  and numbering systems, 125

REFERENCE GLOSSARY

ABBREVIATIONS

Abbreviations allow writers to avoid cumbersome repetition of lengthy words and phrases. They are a form of shorthand and are appropriate in technical and business writing, particularly in lists, tables, charts, graphs, and other visual aids where space is limited.

See ACRONYMS.

1. Eliminate periods in and after most abbreviations.

Formerly, most abbreviations required periods. Today, the trend is to eliminate periods in and after abbreviations, especially in the abbreviated names of governmental agencies, companies, private organizations, and other groups:

AFL-CIO	AMA	CBS	DOE
FTC	IOOF	NFL	NLRB
OPEC	TVA	TWA	YWCA

Abbreviations for units of measure now appear without the period, and the same form is used for both the singular and plural:

m oz ft cm yd mm lb yr

NOTE 1: By convention some abbreviations still require periods:

a.d.	a.m.	b.c.	Dr.
e.g.	etc.	i.e.	Mr.
Mrs.	Ms.	p.m.	pp.
U.K.	U.S.A.	U.S.S.R.	

Retain the period, too, in abbreviations that spell normal words:

in., inches (*not* in)

no., number (*not* no)

A recent dictionary, such as *Webster's New Collegiate Dictionary*, is the best resource for determining if an abbreviation requires periods. (See REFERENCES.)

NOTE 2: Abbreviations with periods should be typed without spaces between letters and periods:

e.g. (*not* e. g.)

U.K. (*not* U. K.)

2. Clarify an unfamiliar abbreviation by enclosing its unabbreviated form within parentheses following its first use in a document:

The applicant had insurance through CHAMPUS (Civilian Health and Medical Program of the Uniformed Services).

The alloy is hardened with 0.2 percent Np (neptunium). Adding Np before cooling alters the crystalline structure of manganese host alloys.

NOTE 1: Some authorities prefer to cite the unabbreviated form of the word before its abbreviation. We believe that this practice can inhibit, rather than enhance, the reader's comprehension of the abbreviation:

The applicant had insurance through the Civilian Health and Medical Program of the Uniformed Services (CHAMPUS).

The alloy is hardened with 0.2 percent neptunium (Np). Adding Np before cooling alters the crystalline structure of manganese host alloys.

NOTE 2: Do not use an unfamiliar abbreviation unless you plan to use it more than once in the same document.

3. Do not abbreviate a unit of measurement unless it is used in conjunction with a number:

Pipe diameters will be measured in inches.

but

Standard pipe diameter is 3 in.

The dimensions of the property were recorded in both meters and feet.

but

The property is 88 ft by 130 ft.

The southern property line is 45.3 m.

4. Do not abbreviate a title unless it precedes a name:

The cardiac research unit comprises five experienced doctors.

but

Our program director is Dr. Royce Smith.

5. Spell out abbreviations that begin a sentence (except for abbreviated words that, by convention, are never spelled out, like *Mr.* and *Mrs.*):

Oxygen extraction will be accomplished at high temperatures.

not

O extraction will be accomplished at high temperatures.

except

Ms. Jean MacIntyre will be responsible for modifying our subsea sensors.

ABBREVIATIONS

6. Spell out abbreviated words that are connected to other words by hyphens:

6-foot gap (*not* 6-ft)
12-meter cargo bay (*not* 12-m)
3.25-inch pipe (*not* 3.25-in.)

Other Conventions

7. Do not abbreviate the names of months and days within normal text. Use the abbreviations in chronologies, notes, tables, and charts:

The facilities modernization plan is due January 1985. (*not* Jan 1985 *or* 1/85)

8. Avoid the symbol forms of abbreviations except in charts, graphs, illustrations, and other visual aids:

55 percent (*not* 55%)
15 ft (*not* 15′)
32.73 in. (*not* 32.73″)

9. Use a single period when an abbreviation ends a sentence:

To head our laser redesign effort, we have hired the 1980 Nobel prize winner from the U.S.A. (*not* U.S.A..)

NOTE: If the clause or sentence ends with something other than a period, (*e.g.*, commas, semicolons, colons, question marks, exclamation points), then the other mark of punctuation follows the period at the end of the abbreviation:

Have we hired the 1980 Nobel prize winner from the U.S.A.?

If you plan to arrive by 6 p.m., you will not need to guarantee your reservation.

List of Abbreviations

Following is a list of many common abbreviations, both for words and common measurements. In this listing, some abbreviations appear with periods, although the trend is to eliminate the periods (see rule 1 above). For example, *Ph.D.* appears with periods to assist writers and typists who wish to retain the periods, although many writers today prefer the increasingly more common *PhD* without periods.

Also refer to *The Chicago Manual of Style*, 13th Edition, and to *Webster's New Collegiate Dictionary*. (See REFERENCES.)

Abbreviations of Words and Phrases

AA, Alcoholics Anonymous
A.B. or B.A., bachelor of arts
abbr., abbreviation
abs., abstract
acct., account
A.D. (*anno Domini*), in the year of our Lord
ADP, automated data processing
a.k.a., also known as
A.M. (*anno mundi*), in the year of the world
A.M. or M.A., master of arts
a.m. (*ante meridiem*), before noon
approx., approximately
Ave., avenue
a.w.l., absent with leave
a.w.o.l., absent without official leave

B.C., before Christ
bf., boldface
Bldg., building
B.Lit(t). or Lit(t).B., bachelor of literature
Blvd., boulevard
b.o., buyer's option
B.S. or B.Sc., bachelor of science

ca. (*circa*), about
ca, centiare
c. and s.c., caps and small caps
c.b.d., cash before delivery
cf., confer, compare, or see
Co., company
c.o.d., cash on delivery
COLA, cost of living adjustment
con., continued
Conus., continental United States
Corp., corporation
c.p., chemically pure
C.P.A., certified public accountant
cr., credit; creditor
Ct., court

d.b.a., doing business as
D.D., doctor of divinity
D.D.S., doctor of dental surgery
Dist. Ct., District Court
D.Lit(t). or Lit(t).D., doctor of literature
do. (ditto), the same
DP (*no periods*), displaced person
D.P.H., doctor of public health
D.P.Hy., doctor of public hygiene
dr., debit; debtor
Dr., doctor; drive
D.V.M., doctor of veterinary medicine

E., east
e.g. (*exempli gratia*), for example
emcee, master of ceremony
e.o.m., end of month
et al. (*et alii*), and others
et seq. (*et sequentia*), and the following
etc. (*et cetera*), and others

f., ff., and following page, pages
f°, folio
f.o.b., free on board
4°, quarto

GI, general issue; government issue
G.M.&S., general, medical, and surgical
GNP, gross national product
Gov., governor
gr. wt., gross weight

HE (*no periods*), high explosive
HF (*no periods*), high frequency

ibid. (*ibidem*), in the same place
id. (*idem*), the same
i.e. (*id est*), that is
IF (*no periods*), intermediate frequency
Insp. Gen., Inspector General
IOU, I owe you
IQ, intelligence quotient

J.D. (*jurum doctor*), doctor of laws
Jr., junior

ABBREVIATIONS

lat., latitude
LC, Library of Congress
lc., lowercase
liq., liquid
lf., lightface
LF, low frequency
LL.B., bachelor of laws
LL.D., doctor of laws
loc. cit. (*loco citato*), in the place cited
long., longitude
Ltd., limited
Lt. Gov., lieutenant governor

M, money supply: M_1; M_{1B}; M_2
M., monsieur; MM., messieurs
m. (*meridies*), noon
M.D., doctor of medicine
memo, memorandum
MF, medium frequency
MIA, missing in action (*plural*, MIA's)
Mlle., mademoiselle
Mme., madam; Mmes., mesdames
mo., month
Mr., mister (*plural*, Messrs.)
Mrs., mistress
Ms., coined feminine title (*plural*, Mses.)
M.S., master of science
MS., MSS., manuscript, manuscripts
Msgr., monsignor
m.s.l., mean sea level

N., north
NA., not available; not applicable
NE., northeast
n.e.c., not elsewhere classified
n.e.s., not elsewhere specified
net wt., net weight
No., Nos., number, numbers
n.o.i.b.n., not otherwise indexed by name
n.o.p., not otherwise provided (for)
n.o.s., not otherwise specified
n.s.k., not specified by kind
n.s.p.f., not specifically provided for
NW., northwest

OK, OK'd, OK'ing, OK's
op. cit. (*opere citato*), in the work cited

PA, public address system
PAC, political action committee (*plural*, PAC's)
Ph.B. or B.Ph., bachelor of philosophy
Ph.D. or D.Ph., doctor of philosophy
Ph.G., graduate in pharmacy
PIN, personal identification number
Pl., place
p.m. (*post meridiem*), afternoon
P.O. Box (*with number*), *but* post office box (*in general sense*)
POW, prisoner of war (*plural*, POW's)
Prof., professor
pro tem (*pro tempore*), temporarily
P.S. (*post scriptum*), postscript; public school (*with number*)

QT, on the quiet

RAM, random access memory
R&D, research and development
Rd., road
RDT&E, research, development, testing, and evaluation
Rev., reverend
RF, radio frequency
R.F.D., rural free delivery
RIF, reduction(s) in force; RIF'd, RIF'ing, RIF's
R.N., registered nurse
RR., railroad
Rt. Rev., right reverend
Ry., railway

S., south; Senate bill (*with number*)
S&L('s), savings and loan(s)
sc. (*scilicet*), namely (*see also* ss)
s.c., small caps
s.d. (*sine die*), without date
SE., southeast
2d, 3d, second, third
SHF, superhigh frequency
sic, thus
SOP, standard operating procedure
SOS, wireless distress signal
sp. gr., specific gravity
Sq., square (*street*)
Sr., senior
SS, steamship
ss (*scilicet*), namely (*in law*) (*see also* sc.)
St., Ste., SS.; Saint, Sainte, Saints
St., street
STP, standard temperature and pressure
Supt., superintendent
Surg., surgeon
SW., southwest

T., Tps., township, townships
Ter., terrace
t.m., true mean
TV, television

uc., uppercase
UHF, ultrahigh frequency
U.S.A., United States of America
USA, U.S. Army
U.S. 40, U.S. No. 40, U.S. Highway No. 40

v. or vs. (*versus*), against
VAT, value added tax
VCR, video cassette recorder
VHF, very high frequency
VIP, very important person
viz (*videlicet*), namely
VLF, very low frequency
VTR, video tape recording

W., west
w.a.e., when actually employed
wf, wrong font
w.o.p., without pay

ZIP Code, Zone Improvement Plan Code (*Postal Service*)
ZIP+4, 9-digit ZIP Code

Abbreviations of Units of Measurement

A, ampere
Å, angstrom
a, are
a, atto (*prefix*, one-quintillionth)
aA, attoampere
abs, absolute (*temperature and gravity*)
ac, alternating current
AF, audiofrequency
Ah, ampere-hour
A/m, ampere per meter
AM, amplitude modulation
asb, apostilb
At, ampere-turn
at, atmosphere
atm, atmosphere (*infrequently*, As)
at wt, atomic weight
au, astronomical units
avdp, avoirdupois

b, barn
B, bel
b, bit
bbl, barrel
bbl/d, barrel per day
Bd, baud
bd. ft., board foot (*obsolete*); *use* fbm
Bé, Baumé
Bev (*obsolete*); *see* GeV
Bhn, Brinell hardness number
bhp, brake horsepower
bm, board measure
bp, boiling point
Btu, British thermal unit
bu, bushel

c, ¢, ct; cent(s)
c, centi (*prefix*, one-hundredth)
C, coulomb
c, cycle (*radio*)
°C, degree Celsius
cal, calorie (*also*: cal_{IT}, International Table; cal_{th}, thermochemical)
cc. (*obsolete*), *use* cm^3
cd, candela (*obsolete*: candle)
cd/in^2, candela per square inch
cd/m^2, candela per square meter
c.f.m. (*obsolete*), *use* ft^3/min
c.f.s. (*obsolete*), *use* ft^3/s
cg, centigram

ABBREVIATIONS

c·h, candela-hour
Ci, curie
cL, centiliter
cm, centimeter
c/m, cycles per minute
cm^2, square centimeter
cm^3, cubic centimeter
cmil, circular mil
cp, candlepower
cP, centipoise
cSt, centistokes
cu ft (*obsolete*), *use* ft^3
cu in (*obsolete*), *use* in^3
cwt, hundredweight

D, darcy
d, day
d, deci (*prefix*, one-tenth)
d, pence
da, deka (*prefix*, 10)
dag, dekagram
daL, dekaliter
dam, dekameter
dam^2, square dekameter
dam^3, cubic dekameter
dB, decibel
dBu, decibel unit
dc, direct current
dg, decigram
dL, deciliter
dm, decimeter
dm^2, square decimeter
dm^3, cubic decimeter
dol, dollar
doz, dozen
dr, dram
dwt, deadweight tons
dwt, pennyweight
dyn, dyne

EHF, extremely high frequency
emf, electromotive force
emu, electromagnetic unit
erg, erg
esu, electrostatic unit
eV, electronvolt

°F, degree Fahrenheit
F, farad
f, femto (*prefix*, one-quadrillionth)
F, fermi (*obsolete*); *use* fm, femtometer
fbm, board foot; board foot measure
fc, footcandle
fL, footlambert
fm, femtometer
FM, frequency modulation
ft, foot
ft^2, square foot
ft^3, cubic foot
ftH_2O, conventional foot of water
ft·lb, foot-pound
ft·lbf, foot pound-force
ft/min, foot per minute
ft^2/min, square foot per minute
ft^3/min, cubic foot per minute
ft-pdl, foot poundal
ft/s, foot per second
ft^2/s, square foot per second
ft^3/s, cubic foot per second
ft/s^2, foot per second squared
ft/s^3, foot per second cubed

G, gauss
G, giga (*prefix*, one billion)
g, gram; acceleration of gravity
Gal, gal cm/s^2
gal, gallon
gal/min, gallons per minute
gal/s, gallons per second
Gb, gilbert
g/cm^3, gram per cubic centimeter
GeV, gigaelectronvolt
GHz, gigahertz (gigacycle per second)
gr, grain; gross

h, hecto (*prefix*, 100)
H, henry
h, hour
ha, hectare
HF, high frequency
hg, hectogram
hL, hectoliter
hm, hectometer
hm^2, square hectometer
hm^3, cubic hectometer
hp, horsepower
hph, horsepower-hour
Hz, hertz (cycles per second)

id, inside diameter
ihp, indicated horsepower
in., inch
in^2, square inch
in^3, cubic inch
in/h, inch per hour
inH_2O, conventional inch of water
inHg, conventional inch of mercury
in-lb, inch-pound
in/s, inch per second

J, joule
J/K, joule per kelvin

K, kayser
K, kelvin (*degree symbol improper*)
k, kilo (*prefix*, 1,000)
k, thousand (7k = 7,000)
kc, kilocycle; *see also* kHz (kilohertz), kilocycles per second
kcal, kilocalory
keV, kiloelectronvolt
kG, kilogauss
kg, kilogram
kgf, kilogram-force
kHz, kilohertz (kilocycles per second)
kL, kiloliter
klbf, kilopound-force
km, kilometer
km^2, square kilometer
km^3, cubic kilometer
km/h, kilometer per hour
kn, knot (*speed*)
kΩ, kilohm
kt, kiloton; carat
kV, kilovolt
kVa, kilovoltampere
kvar, kilovar
kW, kilowatt
kWh, kilowatthour

L, lambert
L, liter (*also* l)
lb, pound
lb ap, apothecary pound
lb avdp, avoirdupois pound
lbf, pound-force
lbf/ft, pound-force foot
lbf/ft^2, pound-force per square foot
lbf/ft^3, pound-force per cubic foot
lbf/in^2, pound-force per square inch
lb/ft, pound per foot
lb/ft^2, pound per square foot
lb/ft^3, pound per cubic foot
lct, long calcined ton
ldt, long dry ton
LF, low frequency
lin ft, linear foot
l/m, lines per minute
lm, lumen
lm/ft^2, lumen per square foot
lm/m^2, lumen per square meter
lm·s, lumen second
lm/W, lumen per watt
l/s, lines per second
L/s, liter per second
lx, lux

M, mega (*prefix*, 1 million)
M, million (3M = 3 million)
m, meter
m, milli (*prefix*, one-thousandth)
M_1, monetary aggregate
m^2, square meter
m^3, cubic meter
μ, micro (*prefix*, one-millionth)
μ, micron (*obsolete*); *use* μm, micrometer
mA, milliampere
μA, microampere
mbar, millibar
μbar, microbar
Mc, megacycle; *see also* MHz (megahertz), megacycles per second
mc, millicycle; *see also* mHz (millihertz), millicycles per second
mcg, microgram (*obsolete*); *use* μg
mD, millidarcy
meq, milliquivalent
MeV, megaelectronvolts
mF, millifarad
μF, microfarad
mG, milligauss
mg, milligram
μg, microgram
Mgal/d, million gallons per day
mH, millihenry
μH, microhenry
mho, mho (*obsolete*); *use* S, siemens
MHz, megahertz

ABBREVIATIONS

mHz, millihertz
mi, mile (*statute*)
mi^2, square mile
mi/gal, mile(s) per gallon
mi/h, mile per hour
mil, mil
min, minute (*time*)
μin, microinch
mL, milliliter
mm, millimeter
mm^2, square millimeter
mm^3, cubic millimeter
mμ, (*obsolete*); *see* nm, nanometer
μm, micrometer
μm^2, square micrometer
μm^3, cubic micrometer
μμ, micromicron (*use of compound prefixes is obsolete*); *use* pm, picometer
μμf, micromicrofarad (*use of compound prefixes is obsolete*); *use* pF
mmHg, conventional millimeter of mercury
μmho, micromho (*obsolete*); *use* μS, microsiemens
MΩ, megohm
mo, month
mol, mole (*unit of substance*)
mol wt, molecular weight
mp, melting point
ms, millisecond
μs, microsecond
Mt, megaton
mV, millivolt
μV, microvolt
MW, megawatt
mW, milliwatt
μW, microwatt
MWd/t, megawatt-days per ton
Mx, maxwell

n, nano (*prefix*, one-billionth)
N, newton
nA, nanoampere
nF, nanofarad
nm, nanometer (millimicron, *obsolete*)
N·m, newton meter
N/m^2, newton per square meter
nmi, nautical mile
Np, neper
ns, nanosecond
$N{\cdot}s/m^2$, newton second per square meter
nt, nit

od, outside diameter
Oe, oersted (*use of* A/m, amperes per meter, *preferred*)
oz, ounce (*avoirdupois*)

p, pico (*prefix*, one-trillionth)
P, poise
Pa, pascal
pA, picoampere
pct, percent
pdl, poundal
pF, picofarad (micromicrofarad, *obsolete*)
pF, water-holding energy
pH, hydrogen-ion concentration
ph, phot; phase
pk, peck
p/m, parts per million
ps, picosecond
pt, pint
pW, picowatt

qt, quart
quad, quadrillion (10^{15})

°R, rankine
°R, roentgen
R, degree rankine
R, degree reaumur
rad, radian
rd, rad
rem, roentgen equivalent man
r/min, revolutions per minute
rms, root mean square
r/s, revolutions per second

s, second (*time*)
s, shilling
S, siemens
sb, stilb
scp, spherical candlepower
s·ft, second-foot
shp, shaft horsepower
slug, slug
sr, steradian
sSf, standard saybolt fural
sSu, standard saybolt universal
$stdft^3$, standard cubic foot (feet)
Sus, saybolt universal second(s)

T, tera (*prefix*, 1 trillion)
Tft^3, trillion cubic feet
T, tesla
t, tonne (*metric ton*)
tbsp, tablespoonful
thm, therm
ton, ton
tsp, teaspoonful
Twad, twaddell

u, (unified) atomic mass unit
UHF, ultrahigh frequency

V, volt
VA, voltampere
var, var
VHF, very high frequency
V/m, volt per meter

W, watt
Wb, weber
Wh, watthour
W/(m·K), watt per meter kelvin
W/sr, watt per steradian
$W/(sr{\cdot}m^2)$, watt per steradian square meter

x, unknown quantity

yd, yard
yd^2, square yard
yd^3, cubic yard
yr, year

ACRONYMS

Acronyms are abbreviations that are pronounced as words:

ALGOL (ALGOrithmic Language)
Amtrak (American Track)
BASIC (Beginners' All-purpose Symbolic Instruction Code)
Bit (BInary Translation)
COBOL (COmmon Business-Oriented Language)
FORTRAN (FORmula TRANslator)
laser (light amplification by stimulated emission of radiation)
loran (long-range navigation)
NATO (North Atlantic Treaty Organization)
NOW (National Organization for Women)
Pepco (Potomac Electric Power Company)
PERT (Program Evaluation and Review Technique)
radar (radio detecting and ranging)
SALT (Strategic Arms Limitation Treaty)
secant (separation control of aircraft by nonsynchronous techniques)
sonar (sound navigation ranging)
START (Strategic Arms Reduction Treaty)
WAC (Womens Army Corps)
Waves (Women Accepted for Volunteer Emergency Service)
ZIP (Zone Improvement Plan)

Because acronyms are intended to replace the longer expressions they represent, you can use them anywhere in writing. They are, in fact, new words.

Acronyms may be written in ALL CAPITALS if they form proper names. However, some acronyms are conventionally upper and lower case:

Amtrak   Aramco   Pepco

The most common acronyms, those representing generic technical concepts rather than organizations or programs, are typically all lowercase:

laser   radar   sonar

See ABBREVIATIONS.

---

1. When you introduce new or unfamiliar acronyms, use the acronym and then, in parentheses, spell out the name or expression:

> Our program fully complies with the provisions of STEP (the Supplemental Training and Employment Program). To implement STEP, however, we had to modify subcontracting agreements with four components suppliers.

NOTE: Some writers and editors prefer to introduce unfamiliar acronyms by first spelling out the component words and then placing the acronym in parentheses.

---

2. Avoid overusing acronyms, especially if your readers are unlikely to be very familiar with them.

Until readers learn to recognize and instantly comprehend an acronym (like *laser*), the acronym hinders reading. It creates a delay while the reader's mind recalls and absorbs the acronym's meaning. Therefore, you should be cautious about using acronyms, especially unfamiliar ones. Overloading a text with acronyms makes the text unreadable, even if you have previously introduced and explained the acronyms.

Acronyms are good shorthand devices, but use them judiciously.

ACTIVE/PASSIVE

Sentences expressing an action can (and usually do) have three basic elements: the actor (the person or thing performing the action), the action (the verb), and the receiver (the person or thing receiving the action).

When the structure of the sentence has the actor in front of the action, the sentence is in the **active voice**:

> Michigan companies manufacture millions of precision machine tools.

Companies is the actor; *manufacture* is the action; and *tools* receives the action. Because the actor comes before the action, the sentence is active. The subject of the sentence performs the action.

When the structure of the sentence has the receiver in front of the action, the sentence is in the **passive voice**:

> Millions of precision machine tools are manufactured by Michigan companies.

In this sentence, the subject (*tools*) is not doing the manufacturing. The tools are being manufactured. They are being acted upon; they are receiving the action. Therefore, the subject—and the sentence—is passive.

1. Prefer active sentences.

Active sentences are usually shorter and more dynamic than passive sentences. They generally have more impact and seem more "natural" because readers expect (and are accustomed to) the actor-action-receiver pattern. Active writing is more forceful and more self-confident.

Passive writing, on the other hand, can seem weak-willed, indecisive, or evasive. In passive sentences, the reader encounters the action before learning who performed it. In some passive sentences, the reader never discovers who performed the action. So passive sentences seem static. Passive sentences are useful—even preferable—in some circumstances, but you should prefer active sentences.

When to Use Passives

2. Use a passive sentence when you don't know or don't want to mention the actor:

> The failure occurred because metal shavings had been dropped into the worm-gear housing.
>
> Clearly, the site had been inspected, but we found no inspection report and could not identify the inspectors.

In the first example above, a passive sentence is acceptable because we don't know who dropped the metal shavings into the housing. In the second example, we might know who inspected the site but don't want to mention names because the situation could be sensitive or politically charged.

3. Use a passive sentence when the receiver is more important than the actor.

The strongest part of most sentences is the opening. Therefore, the sentence element appearing first will receive greater emphasis than those elements appearing later in the sentence. For this reason, a passive sentence is useful when you wish to emphasize the receiver of the action:

> Cross-sectional analysis techniques—the most important of our innovations—are currently being tested in our Santa Barbara Laboratory.
>
> Minimum material size or thickness requirements will then be established to facilitate recuperator weight, size, and cost estimates.

In both examples, we wish to emphasize the receiver of the action. Note how emphasis changes if we restructure the first example:

> The most important of our innovations (cross-sectional analysis techniques) is currently being tested in our Santa Barbara Laboratory.
>
> Our Santa Barbara Laboratory is currently testing the most important of our innovations—cross-sectional analysis techniques.
>
> Our Santa Barbara Laboratory is currently testing cross-sectional analysis techniques—the most important of our innovations.

The emphasis in each sentence differs, depending upon sentence structure. The first revision emphasizes *innovations*, and it is still a passive sentence. The last two revisions are active, and both stress *our Santa Barbara Laboratory*.

ACTIVE/PASSIVE

The ending of a sentence is also emphatic (although not as emphatic as the beginning), so the sentence ending with *techniques* does place secondary emphasis on *techniques*. However, the best way to emphasize *cross-sectional analysis techniques* is by opening the sentence with that phrase.

4. Use a passive sentence when you need to form a smooth transition from one sentence to the next.

Occasionally, writers must arrange sentence elements so that key words appearing in both sentences are near enough to each other for readers immediately to grasp the connection between the sentences. In the example below, for instance, the writer needs to form a smooth transition between sentences by repeating the key words *work packages*:

> We will develop a simplified matrix of tasks that will include all budgetary and operational work packages. These work packages will be scheduled and monitored by individual program managers.

The second sentence is passive. It would be shorter and stronger as an active sentence:

> Individual program managers will schedule and monitor these work packages.

However, the active version of the second sentence does not connect as well with the previous sentence:

> We will develop a simplified matrix of tasks that will include all budgetary and operational work packages. Individual program managers will schedule and monitor these work packages.

For a brief moment, the second sentence seems to have changed the subject. Not until readers reach the end of the second sentence will they realize that both sentences concern work packages. Therefore, making the second sentence passive creates a smoother transition and actually improves the passage.

Passives and First Person

5. Do not use passive sentences to avoid using first person pronouns.

Some writers use passives to avoid using first person pronouns (*I, me, we,* or *us*). These writers mistakenly believe that first person pronouns are inappropriate in business or technical writing. In fact, the first person is preferable to awkward or ambiguous passive sentences like the examples below:

> It is recommended that a state-of-the-art survey be added to the initial redesign studies.

Who is recommending it? You? The customer? Someone else? And who is supposed to add the survey?

> Cost data will be collected and maintained to provide a detailed history of the employee hours expended during the program. This tracking effort will be accomplished by the use of an established employee-hour accumulating system.

Things seem to be happening in these sentences, but no one seems to be doing them.

Writers who overuse the passive to avoid first person pronouns convey the impression that they don't want to accept the responsibility for their actions. This implication is why passive sentences can seem evasive even when the writer doesn't intend them to be.

Passives allow you to eliminate the actor. In some cases, eliminating the actor is appropriate and desirable. In other cases (as in the previous examples), eliminating the actor creates confusion and doubt. Active versions of these examples, using first person pronouns, are much better:

> We recommend that the initial redesign studies include a state-of-the-art survey.

> Using our employee-hour accumulating system, we will collect and maintain cost data to provide a detailed history of the employee hours expended during the program.

How to Convert Passives

Technical and scientific writers generally use too many passives. They use them unnecessarily, often more from habit than choice. Converting unneeded passives to actives will strengthen the style of the document, making it appear crisper and more confident. Following are some ways to convert passives.

ACTIVE/PASSIVE

6. Make sentences active by turning the clause or sentence around:

These chemical methods are described in more detail in section 6.

Section 6 describes these chemical methods in more detail.

A functional outline of the program is included in the Work Breakdown Structure (figure 1.1-2).

The Work Breakdown Structure (figure 1.1-2) includes a functional outline of the program.

Brakes on both drums are activated as required by the control system to regulate speed and accurately position the launcher.

The control system activates brakes on both drums as required to regulate speed and accurately position the launcher.

After these requirements are identified, we will develop a comprehensive list of applicable technologies.

After identifying these requirements, we will develop a comprehensive list of applicable technologies.

7. Make sentences active by changing the verb:

The solutions were achieved only after extensive development of fabrication techniques.

The solutions occurred only after extensive development of fabrication techniques.

The Gaussian elimination process can be thought of as a means of "decomposing" a matrix into three factors.

The Gaussian elimination process "decomposes" a matrix into three factors.

The Navy recuperator requirements are expected to bring added emphasis to structural integrity.

The Navy recuperator requirements will probably emphasize structural integrity.

Coalescence was always observed to start at the base of the column.

Coalescence always started at the base of the column.

8. Make sentences active by rethinking the sentence:

Special consideration must be given to structural mounting, heat exchanger shape, ducting losses, and ducting loads.

Structural mounting, heat exchanger shape, ducting losses, and ducting loads are especially important.

To ensure that a good alternate design approach is not overlooked, a comparison between plate-fin and tubular designs will be made during the proposed study program.

Comparing plate-fin and tubular designs during the proposed study program will ensure that we thoughtfully consider alternate design approaches.

This study will show what can be done to alleviate technology failure by selectively relaxing requirements.

This study will show how selectively relaxing requirements can alleviate technology failure.

It must be said, however, that while maximum results are gained by a design synthesis approach such as we propose, the area to be covered is so large that it will still be necessary to concentrate on the most important technologies and their regions of interest.

Our proposed design synthesis approach will yield maximum results. Nevertheless, the area of interest is very large. Concentrating on the most important technologies and their regions of interest will still be necessary.

ADJECTIVES

Adjectives describe or modify nouns and other adjectives. They typically precede nouns or follow either verbs of sense (*feel, look, sound, taste, smell*) or linking verbs (*be, seem, appear, become*):

The slow process . . . (*or* The process is slow.)

Warm weather . . . (*or* The weather seems warm.)

The cautious superintendent . . . (*or* The superintendent became cautious.)

Harry felt bad. (*not* badly, *which is an adverb*)

Adjectives also tell which, what kind of, or how many people or things are being discussed.

Adjectives and Adverbs

Adjectives and adverbs are similar. They both describe or modify other words, and they both can compare two or more things. Sometimes they appear in similar positions in sentences:

Harry felt cautious. (*adjective*)

Harry felt cautiously along the bottom of the muddy stream. (*adverb*)

The guard remained calm. (*adjective*)

The guard remained calmly at his post. (*adverb*)

The car came close to me. (*adverb*)

The corporal watched the prisoner closely. (*adverb*)

NOTE: Not all adverbs end in *-ly* (for example, the adverbs *deep, fair, hard, wide*). Some adverbs have two forms: an *-ly* form and another form that is identical to the adjective (*deep/deeply, fair/fairly, hard/hardly, wide/widely*). You can determine whether words are adjectives by trying to put them in front of a noun. In the examples above, *cautious Harry* and *the calm guard* both make sense, so *cautious* and *calm* are adjectives. The *close car* does not mean what the original sentence meant, so *close* is an adverb in this context. (See ADVERBS.)

Comparatives and Superlatives

Adjectives have different forms for comparing two objects (the comparative form) and comparing more than two objects (the superlative form):

Our word processor is slower than the new IBM Displaywriter. (Slower *is the comparative form.*)

The Gemini software package was the slowest one we surveyed. (Slowest *is the superlative form.*)

Stocks are a likelier investment than bonds if long-term growth is the goal. (*or* more likely)

Nissan's likeliest competitor in the suburban wagon market is General Motors. (*or* most likely)

The 1986 budget is more adequate than the 1985 budget.

The cooling provisions are the most adequate feature of the specifications.

NOTE: One-syllable words use *-er/-est* to form comparatives or superlatives. Two-syllable words use either *-er/-est* or *more/most*. Three-syllable words use *more/most*. A few adjectives have irregular comparative forms: *good (well), better, best; bad, worse, worst; many, more, most*.

---

1. Use the comparative (*-er/more*) forms when comparing two people or things and the superlative (*-est/most*) forms when comparing more than two:

Of the two designs, Boeing's seems more efficient.

The Shearson-American Express proposal is the most attractive. (*More than two options are implied, so the superlative is proper.*)

Weekly deductions are the best method for financing the new hospital insurance plan.

Weekly deductions are better than any other method for financing the new hospital insurance plan. (*The comparative* better *is used because the various options are being compared one by one, not as a group.*)

Nouns Used as Adjectives

Nouns often behave like adjectives, especially in complex technical phrases. Here is a typical phrase from an aircraft manual: *C-5A airframe weight calculation error percentage*. The first five words in this phrase are nouns used as adjectives: *C-5A, airframe, weight, calculation*, and *error*. These five nouns are sometimes called a noun string. Such nouns are extremely useful because, as in this case, English often does not have an adjective form with the same meaning as the noun.

Although useful and often necessary, nouns used as adjectives may be clear only to technically knowledgeable people:

aluminum honeycomb edge panels

What is aluminum? The honeycomb, the edges, or the panels? Only a knowledgeable reader can tell for sure. Sometimes, the order of the words suggests an interpretation:

aluminum edge honeycomb panels

From this phrase, we may expect the edges, and not the honeycomb, to be aluminum, but we still can't know for sure if *aluminum edge* and *honeycomb* equally modify *panels* or if *aluminum edge* and *honeycomb* combine to become a single modifier of *panels* or if *aluminum* modifies something called *edge honeycomb*:

(aluminum + edge) + honeycomb panels

or

(aluminum+ edge + honeycomb) panels

or

aluminum + (edge + honeycomb) panels

In alphabetical lists of parts, the main noun being modified must be listed first. Therefore, the modifying words appear afterwards, usually separated by commas. The modifying words are typically listed in reverse order, with the most general modifiers closest to the main noun:

panels, honeycomb, aluminum edge

or

panels, edge, aluminum honeycomb

or

panels, aluminum edge honeycomb

A helpful technique for discovering or clarifying the structure of noun strings is to ask the question, *What kind?* Begin with the main noun being modified and proceed from there to build the string of modifying nouns.

panels

What kind of panels?

honeycomb

What kind of honeycomb?

aluminum edge

In this case, we have assumed that *aluminum edge* describes a particular type of honeycomb. Because *aluminum* and *edge* jointly modify *honeycomb*, they act as one word. We usually show that two or more words are acting together as joint or compound modifiers by hyphenating them:

aluminum-edge honeycomb panels

See HYPHENS.

---

2. Arrange nouns used as adjectives in technical expressions so that the more general nouns are closest to the word they are modifying:

semi-automatic slat worm gear

automatic slat worm gear

semi-automatic strut backoff gear

automatic strut backoff gear

NOTE 1: The structure of such phrases (as well as the logic behind this rule) is revealed in catalogued lists:

gear
- backoff
  - automatic strut
  - semi-automatic strut
- worm
  - automatic slat
  - semi-automatic slat

NOTE 2: Some technical writers and editors rarely use internal punctuation (either hyphens or commas) to separate nouns in noun strings. In many scientific and technical fields, hyphens that would normally connect parts of a unit modifier are eliminated:

methyl bromide solution (*not* methyl-bromide solution)

black peach aphid (*not* black-peach aphid *or* black peach-aphid)

grey willow leaf beetle

swamp black currant seedlings

Hyphens in many technical words are, however, very hard to predict: *horse-nettle* vs. *horseradish* or *devilsclaw* vs. *devils-paintbrush*. In instances where the first word is capitalized, the compound is often hyphenated: *China-laurel, Queen-Annes-lace, Australian-pea*, etc. (See HYPHENS.)

ADJECTIVES

NOTE 3: Commas are not used to separate nouns in noun strings. However, we use commas to separate true adjectives when the adjectives equally modify the same noun:

grey, burnished, elliptical sphere

sloppy, poorly written, inadequate proposal

See NOUNS and COMMAS.

ADVERBS

Adverbs are modifiers that give the how, where, when, and extent of the action within a sentence. Most adverbs end in *-ly*, but some common adverbs do not: *so, now, later, then, well*, etc. Adverbs often modify the main verbs in sentences:

The engineer slowly prepared the design plan. (*How?*)

The supply ship moved close to the drilling platform. (*Where?*)

They later surveyed all participants in the research project. (*When?*)

The crude oil flowed rapidly from the ruptured pipeline. (*Extent?*)

Adverbs can also modify adjectives or other adverbs:

Their proposal was highly entertaining.

Costs were much lower than expected.

The well was so deep that its costs became prohibitive.

The board of directors cut costs more severely and more rapidly than we anticipated.

NOTE: Some adverbs have two forms: *close/closely, deep/deeply, loud/loudly, slow/slowly, wide/widely.* (See ADJECTIVES.)

1. Place adverbs such as *only, almost, nearly, merely,* and *also* as close as possible to the word they modify:

The bank examiners looked at only five accounts. (*not* The bank examiners only looked at five accounts.)

The engineer had almost finished the specifications. (*not* The engineer almost had finished the specifications.)

Adverbs and Adjectives

Adverbs and adjectives are quite similar. They each modify or describe other words, and they often appear in similar positions in sentences, but they have quite different meanings:

The lab technician carefully smelled the sample. (*Adverb*)

The cheese smelled bad. (*Adjective*)

The new motor worked badly the first day. (*Adverb*)

Not knowing the language, they stayed close to the interpreter. (*Adverb*)

We closely studied the blueprints. (*Adverb*)

The election was so close that no one was a clear winner. (*Adjective*)

See ADJECTIVES.

Comparative and Superlative Forms

Adverbs have different forms to show comparison of two things (the comparative form) and comparison of more than two things (the superlative form). The comparative uses an *-er* form or *more*; the superlative uses an *-est* or *most*:

ADVERBS

The counselor left sooner than expected. (*comparative*)

The fluid returned more slowly to its original level. (*comparative*)

They debated most successfully the wisdom of expanding into the West Coast market. (*superlative*)

The most rapidly moving car turned out to be the new Ford high performance model. (*superlative*)

NOTE 1: Some adverbs have irregular comparative and superlative forms: *well, better, best; badly, worse, worst; little, less, least; much, more, most.*

NOTE 2: See ADJECTIVES for a more complete discussion of whether to use *-er* or *more* for comparatives and *-est* or *most* for superlatives. The rules for adverbs are identical to those for adjectives.

AGREEMENT

English nouns and verbs have singular and plural forms. One of the most standard rules of grammar is that subjects of sentences must agree in number with their verbs: singular subjects require singular verbs and plural subjects require plural verbs:

The proposal was finished. (*not* were finished)

The boilers have become corroded. (*not* has become)

The notion of agreement also refers to the singular/plural agreement between pronouns and their antecedents (the words the pronouns stand for) and between types of pronouns (first, second, or third person) when those pronouns have the same or similar types of antecedents:

Jane Swenson submitted her report. (*The pronoun* her *agrees with its antecedent* Jane Swenson.)

See PRONOUNS and VERBS.

1. The subject of a sentence should agree in number with the sentence verb:

The geologist is analyzing the well data.

The employees are discussing the benefit package.

The Elements of Geometry is the basic textbook.

Midwest states normally include Kentucky and Missouri.

A list of Midwest states normally includes Kentucky and Missouri.

Note: A noun ending with an *s* or *es* is usually plural. A verb ending with an *s* or *es* is usually singular. *Employees* is plural. The verbs *is* and *includes* are both singular.

Some verbs do not change their form to reflect singular and plural: *will include, included, had included, will have included*, etc.

Agreement problems sometimes occur because the subject of the sentence is not clearly singular or plural:

None of the crew is going to take leave.

or

None of the crew are going to take leave.

Some writers become confused, too, when the subject is separated from the verb by words or phrases that do not agree in number with the subject:

Only one of the issues we discussed is on the agenda for tomorrow's meeting.

Few aspects of the problem we are now facing are as clear as they should be.

AGREEMENT

The availability of rice, as well as of medical supplies, determines the life expectancy of a typical adult in Hong Kong.

Normal wear and tear, along with planned obsolescence, is the reason most automobiles provide only an average of 6.5 years of service.

The number of the subject does not change if the subject follows the verb:

What are your arguments for tertiary recovery in reef-producing ranges?

There are five new pumps sitting in the warehouse.

Discussed are the basic design flaws in the preliminary specifications and the lack of adequate detail in the drawings.

Finally, some noun subjects look plural because they end in *s* or *ics*, but they are still singular:

Politics has changed drastically with the advent of television.

The news from Algeria continues to be discouraging.

Measles rarely occurs in adults.

2. Subjects connected by *and* require a plural verb:

The fasteners and the ceiling panel have been fabricated.

The regional engineer and the field geologist agree that we should plug and abandon the well.

A personal computer and a photo copier are essential business tools today.

EXCEPTION: Sometimes words connected by *and* become so closely linked that they become singular in meaning, thus requiring a singular verb:

Bacon and eggs is my favorite breakfast.

My name and address is on the inside cover.

3. Singular subjects connected by *either . . . or*, *neither . . . nor*, and *not only . . . but also* require a singular verb:

Either the tail assembly or the wing design is causing excessive fuel consumption.

Neither the district engineer nor the superintendent has approved the plans.

Not only the cost but also the design is a problem.

NOTE: When one of a pair of subjects is plural, the verb agrees with the subject closest to it:

Either the tail assembly or the wing struts are causing excessive fuel consumption.

Either the wing struts or the tail assembly is causing excessive fuel consumption.

4. When used as a subject or as the modifier of the subject, *each, every, either, neither, one, another, much, anybody, anyone, everybody, everyone, somebody, someone, nobody*, and *no one* require singular verbs:

Every proposal has been evaluated.

Each engineer is responsible for the final proofing of engineering proposals.

Everyone has received the pension information.

Somebody was responsible for the drop in production.

No one but the design engineer knows the load factors used in the calculations.

NOTE: Although words ending with *-one* and *-body* require a singular verb, sentences with such words often become awkward when a pronoun refers to the words:

Everyone turns in his report on Monday.

Using the pronoun *his* maintains the agreement with the subject, but if the *everyone* mentioned includes women, the expression is sexist. Some writers and editors argue that male pronouns (*he, his, him, himself*) are generic, that they refer to both males and females. Others maintain that this convention discriminates against women. Writers and editors who share this view prefer to include both men and women in their sentences:

Everyone turns in his or her report on Monday.

Finally, some liberal editors argue that *everyone* implies a plurality, so *their* becomes the acceptable pronoun:

Everyone turns in their reports on Monday.

AGREEMENT

The sexism problem is avoidable in most sentences simply by making the subject plural and eliminating such troublesome words as *everyone*:

All engineers turn in their reports on Monday.

See PRONOUNS and SEXIST LANGUAGE.

5. When used as a subject or as the modifier of a subject, *both, few, several, many,* and *others* require plural verbs:

Both proposals were unsatisfactory.

Several were available earlier this month.

Few pipes were still in service.

6. *All, any, more, most, none, some, one-half of, two-thirds of, a part of,* and *a percentage of* require either a singular or a plural verb, depending upon the noun they refer to:

All of the work has been assigned. (*singular*)

All of the trees have been removed. (*plural*)

Most sugar is now made from sugar beets.

Most errors were caused by carelessness.

Some of the report was written in an overly ornate style.

Some design features were mandatory.

One-half of the project has been completed.

One-half of the pages have been proofed.

A percentage of the room is for storage.

A percentage of the employees belong to the company credit union.

7. Collective nouns and expressions with time, money, and quantities take a singular or a plural verb, depending upon their intended meaning:

The committee votes on pension policy when disputes occur. (Committee, *a collective noun, is considered singular.*)

The committee do not agree on the interpretation of the mandatory retirement clause. (Committee, *a collective noun, is considered plural.*)

The audience was noisy, especially during the final act.

The audience were in their seats by 7:30.

Two years is the usual waiting period. (Two years *is an expression of time considered as a single unit.*)

The two years were each divided into quarters for accounting purposes. (Two years *is an expression of time considered as a plural of* year.)

Six dollars is the fee.

Six dollars were spread out on the counter.

Five liters is all the tank can hold.

Five liters of wine were sold before noon.

NOTE: Sometimes sentences with collective nouns become awkward because they seem both singular and plural. In such cases, rephrasing often helps:

Audience members were in their seats by 7:30.

APOSTROPHES

1. Apostrophes indicate omitted letters or words in a contraction:

It's not going to be easy. (It is not going to be easy.)

It won't be easy. (It will not be easy.)

We will coordinate with the manufacturer who's chosen to supply the semiconductors. (who is chosen)

2. Apostrophes indicate possession:

Boeing's airframe manufacturing capabilities are world renowned.

The unit's most unique capability is its amplification of weak echoes.

- When the possessive word is plural and ends in *s*, the apostrophe follows the *s*:

The suppliers' requests are not unreasonable considering the amount of time required for fabrication.

We consider the states' environmental quality offices to be our partners in reclamation.

NOTE: Irregular plurals that do not end in *s* require an *'s*:

The report on women's status in the executive community is due next Friday.

Materials for children's toys must conform to Federal safety standards.

- When the possessive word is singular, the apostrophe comes before the *s*:

Rockwell International's process for budgeting is one of the most progressive in the industry.

The circuit's most unusual capability is its error detection and correction function.

- When the possessive word is singular and already ends with an *s*, the apostrophe follows the *s* and may itself be followed by another *s* (although most writers prefer the apostrophe alone):

General Dynamics' (*or* Dynamics's) management proposal is very project specific.

Our project manager would be Martin Jones. Dr. Jones' (*or* Jones's) experience with laser refractors has made him a leader in the field.

NOTE: The possessive form of the pronoun *it* is *its*, not *it's*. (*It's* is the contraction of *it is* or *it was*.):

Possessive: Its products have over ten thousand hours of testing behind them.

Contraction: It's (It is) in the interests of economy and efficiency that we pursue atmospheric testing as well.

Similarly, the possessive form of *who* is *whose*, not *who's*. *Who's* is a contraction for *who is* or *who has*.

See POSSESSIVES.

3. Apostrophes indicate the passage of time in certain stock phrases:

a month's pay
an hour's time
4 days' work
3 years' study

4. Apostrophes may precede the s in the plural of letters, signs, symbols, figures, acronyms, and abbreviations, although the trend is to omit the apostrophe unless omitting it would be confusing:

The X's indicate insertable material. (*or* x's *or* Xs *but not* xs)

Our risk management process is designed to eliminate the if's and but's.

All of our senior staff have PhDs. (*or* Ph.D.'s *or* Ph.D.s)

The manufacturer indicates fragile material by placing #'s in any of the last three positions in the transportation code.

Packagers should code all categories, including the A's and I's. (*not* As *or* Is *nor* as *or* is)

General Dynamics began its task definition program in the early 1960s. (*or* 1960's)

The tracer tests will be run on all APOs in Europe. (*or* APO's)

The Bureau of Land Management has prepared three EAs (Environmental Assessments) for those grazing allotments. (*or* EA's)

APPENDICES/ATTACHMENTS

Appendices (often informally referred to as attachments) are more and more common in documents, especially those intended for busy peers, supervisors, and managers who do not have time to or cannot wade through pages of data and analysis. Appendices and attachments are acceptable (often desirable) in letters and memos as well as reports.

The following types of information can and often do appear in appendices or attachments:

Background data
Case studies
Computations
Derivations
Detailed component descriptions
Detailed test results
Excerpts from related research
Histories
Lengthy analyses
Parts lists
Photographs
Raw data
Sources of additional information
Supporting letters and memos
Tables of data

The word *appendix* has two acceptable plurals: *appendices* and *appendixes*. *Appendices* is still widely used by educated speakers and writers, but *appendixes* is growing in popularity because it follows the regular method for making English words plural.

---

1. Use appendices to streamline reports and memos that would otherwise be too lengthy.

In business and technical reports and memos, assess your readers' need to know the background and analysis behind the relevant conclusions and recommendations.

Relevant conclusions and recommendations should appear very early in most business and technical reports, often as part of an executive summary. Busy readers can therefore receive a streamlined report of 8 to 10 pages (instead of the traditional formal report of 30 to 50 pages) with appendices containing appropriate background information, detailed results, and lengthy analyses.

See SUMMARIES.

---

2. Avoid making appendices a dumping ground for unnecessary information.

Because the appendices are not part of the body of the report, some writers believe they have the license to include in the appendices every scrap of information they know about the subject. This practice leads to massive, often confusing appendices that discourage readers.

Would a knowledgeable reader need the information in the appendices to interpret the conclusions and recommendations? If so, then the appendices are justified. In writing your document, determine who the readers will be and ask yourself what additional information these readers will need to better understand your approach, analysis, results, conclusions, and recommendations.

One rule of thumb is that appendices should contain only information prepared expressly for the project in question. Background information from files and tangential reports (general background information) should not appear in appendices. Often, readers know such background information and reports anyway.

To summarize, if a reader needs certain information to understand a report, this information belongs in the body or in the appendices of a report. All other information belongs in backup files.

---

3. Number or letter appendices and attachments sequentially.

Sequential numbering or lettering is essential: Appendix A, Appendix B, etc., or Attachment 1, Attachment 2, etc. Numbers and letters are both correct, so either is acceptable. In longer documents, your choice may depend upon whether you have numbered or lettered the sections or chapters. If your sections or chapters are numbered, then use letters to label appendices. Conversely, if your sections or chapters are lettered, use numbers for the appendices. The system you use to label appendices should indicate a clear distinction between the appendices and the body of the document.

Typically, the appendices are numbered in the order in which the references to them appear in the body of the report. So the first appendix mentioned in the report becomes Appendix A (or Appendix 1), the second one mentioned is Appendix B (or Appendix 2), and so on.

APPENDICES/ATTACHMENTS

NOTE: You should also give each appendix or attachment a title. Referring to appendices or attachments by number is not informative and can be confusing. Number or letter your appendices and attachments and then title them. In the text, refer to the appendix or attachment by number and title.

4. Refer to all appendices and attachments in the body of the document.

Refer to all appendices or attachments in the body of the document so that readers know that the information within them is available.

Your references should be informative rather than cryptic. A cryptic reference (such as *See appendix C*) does not tell readers enough about the appended information. The following references are informative:

Particulate counts from all collection points in the study area appear in appendix C, Particulate Data.

Attachment 5, A Report on Reserve Faulting in the Boling Dome, provides further evidence of the complex faulting that may control production.

See appendix A (Prescription Trends During the 1970s) for further analysis of valium use and abuse since its introduction.

BIBLIOGRAPHIC FORM

Bibliographic forms appear in standard bibliographies at the end of chapters, articles, and books. Whatever the exact form, bibliographic entries include the name of the author, the title, and the full publication history (including the edition, the publisher or press, the city of publication, the date of publication).

The forms of bibliographic entries vary greatly, depending on the professional background of the author, the profession's needs and traditions, the type of publication, and the publisher. The bibliographic form that Shipley Associates recommends represents a standard format useful for a variety of professions and publishers.

However, we advise you to find out the specific format requirements (including bibliographic form) of the publisher to whom you are submitting a document.

NOTE: In the following rules, the titles of publications in bibliographic entries are underlined. In printed documents, the titles appearing in bibliographic entries would be italicized. Underlining replaces italics when documents are typed or when italics is not available. Elsewhere in this *Style Guide*, book titles are italicized.

1. For a book, give the name of the author or authors, the date of publication, the full title, the volume number, the edition, the city of publication, and the publisher:

Book by one author

Dempster, Jacob B. 1982. The Art of Fine Book Publishing. New Haven: The Cottage Press, Inc.

Book by two authors

Gallo, George, and L.J. Lane. 1978. Paper and Paper-Making. Baltimore: The Freedom Press & Co.

BIBLIOGRAPHIC FORM

Book by three authors

Green, H.J., Ellen Jacoby, and James Reed. 1976. The Art of Graphic Illustration. New Orleans: The Creole Community Press.

Book by more than three authors

Groundvik, K., *et al.* 1971. The Evolution of the Printing Press. Los Angeles: The Hispanic Press.

Book by one editor

Hough, R. William, ed. 1968. Fine Lettering. New York: Simon and Schuster.

Book by two editors

Millman, Howie J., and Fred Stein, eds. 1974. Preparing Leather Book Covers. Boston: J.L. Cabot and Sons Publishing.

Two volumes by an organization

Modern Language Association of America. 1974. Scholarly Publishing in North America. 2 vols. New Haven: The Classical Press.

Chapter of a book

Williams, Clive. 1979. "The Opacity of Ink." In The Art of Printing, edited by Jason Farnsworth. New York: Holt, Rinehart & Winston.

NOTE 1: In these entries, the date directly follows the name of the author or authors. This convention complements the author/date style of citations in the text. (See CITATIONS.) In this style, the text of a document contains parenthetical references:

A 1981 study revealed that fleas transmit the virus (Babcock 1981). This study relied on two earlier studies (Duerdun 1976 and Abbott 1973).

or

A 1981 study revealed that fleas transmit the virus (Babcock). This study relied on two earlier studies (Duerdun 1976 and Abbott 1973). [*Because the date of Babcock's study is already in the sentence, including the date in the citation is unnecessary.*]

NOTE 2: Publications in the humanities usually cite the publication date following the name of the publisher:

Smithson, Arthur J. The History of Modern China. New York: Simon and Schuster, 1976.

This bibliographic form complements the footnoting pattern of citations routinely used by most scholars in the humanities. For more information on this style, see *The Chicago Manual of Style*, 13th edition. (See also FOOTNOTES.)

NOTE 3: Bibliographic entries in the physical and biological sciences often capitalize only the first word of the title:

Smithson, Arthur J. 1976. The history of modern China. New York: Simon and Schuster. [China *is capitalized because it is a proper noun.*]

---

2. For a journal and for a magazine article, give the name of the author or authors, the year of publication, the full title of the article, the name of the journal or magazine, the volume, the month or quarter of publication, and the pages:

Article by one author

Broward, Charles Evans. 1981. "Traveling the Southern California Desert." UCLA Chronicle 15 (Spring): 45-54.

Article by two authors

Calleston, Dwight R., and James Buchanan. 1976. "The Desert Tortoise: Its Vanishing Habitat." The Californian 7 (April): 23-28. [*Follow the book format above for articles with more than two authors.*]

Article appearing in more than one issue

Stevens, Harold, and Jason Drew. 1976. "The Family of Bighorn Sheep." The Bighorn Sheep Newsletter 8 (Fall and Winter): 34-35, 28-31.

Article from a popular magazine

Trump, Josiah. 1969. "The Desert Indians." Time, December 12, 45-49.

Review of a published book

Williams, Ellen. 1980. Review of Prospecting in the Southern Desert by Amy Van Pol and James Freeman. The Californian 10 (July): 24-31.

NOTE 1: As with books, these entries cite the year of publication immediately after the name of the author or authors. In publications for the humanities, the date appears (with the month) after the volume of the journal or magazine:

Stillman, Wendy. "Photographing Desert Sunsets." UCLA Chronicle 15 (Spring 1981): 4-8.

NOTE 2: Some editors, especially in the biological and physical sciences, prefer to omit the quotation marks around the title of the article and to capitalize only the first word of the title:

Stillman, Wendy. 1981. Photographing desert sunsets. UCLA Chronicle 15 (Spring): 4-8.

BIBLIOGRAPHIC FORM

3. For unpublished material, give the author or authors, the title, and as much of its history as available:

Dissertation or thesis

Johnson, Dugdale. 1983. "The Habitat of the Desert Tortoise: Its Inter-Relationship with Man." D.Sc. diss., University of Southern California.

Professional paper

Rusk, Joan, and Elaine Yardley. 1980. "The Diseases of the Bighorn Sheep." Paper presented at the annual meeting of the Bighorn Sheep Society, Los Angeles, 24-26 May.

Personal communication

Turgott, Edward. 1983. Letter to the author, 31 May.

NOTE: The formats for other unpublished documents (television shows, radio shows, interviews, duplicated material, diaries, etc.) should supply as much bibliographic information as possible. The bibliographic form should allow readers to locate the document easily.

4. For public documents, give the country, state, county, or other government division, the full title, and complete publication information:

Iowa. State Assembly. Committee on Farm Commodities. 1974. Report to the Farm Bureau on Corn Subsidies. 45th Assembly, 2nd. sess.

U.S. Congress. House. Committee on Ways and Means. 1945. Hearings on Import Duties on Shellfish. 79th Cong., 1st sess.

U.S. Bureau of the Census. 1984. Gross and Net Fishing Revenues 1980. Prepared by the Commerce Division in cooperation with the Commodity Division. Washington, D.C.: United States Government Printing Office.

Alphabetizing Bibliographic Entries

5. Alphabetize bibliographic entries by the author's last name:

Adam
Adams
Berg
Berger
Bergerson
Michael
Michaels
Michaelsen
Michaelson
Mickael
Zucker

If two or more authors have the same last name, alphabetize according to first names or initials. A set of initials always precedes a first name beginning with the same letter:

Brown, A.W.
Brown, Andrew
Brown, J.B.
Brown, Jane
Brown, John

If single- and multiple-author entries begin with the same last name, list the single authors first:

Davis, Jeanne
Davis, Jeanne and Kristen Cooper
Davis, Jeanne, Kristen Cooper, and Ellen James

Treat all names beginning with *Mc* and *Mac* as though they begin with *Mac*. Alphabetize them letter by letter, as you would with other words:

Mabrey
McDonald
MacDougal
McHenry
MacMillian

If you attribute the document or item to an institution or agency, the first word in the institution's or agency's name becomes the key word for alphabetizing:

Atomic Energy Commission
Boston Globe
MacMillian Institute
Manchester Chronicle
Merrimack Morning News
U.S. Department of Commerce (*not* Department of Commerce)
U.S. Geological Survey

NOTE: In names beginning with articles (*a*, *an*, and *the*), alphabetize by the second word in the name, but list the article if the article is part of the legal name:

Albany State College
The American University
Antioch College
Brown University
The Johns Hopkins University
Syracuse University

BOLDFACE

Boldface type features thicker and darker letters than normal type: **boldface type** vs. normal type.

Until recently, boldface type was available only in printed material. Typewriters could not provide boldface type unless they had changeable typewheels or type elements (*e.g.*, IBM Selectrics).

Modern word processors and computers can easily provide boldface, so more and more boldface is appearing in business letters, memos, and technical reports.

---

1. Use boldface to highlight headings or key words and phrases requiring emphasis.

The above guideline, for instance, appears in boldface, both because it is similar to a heading (despite being a complete sentence) and because it provides key information. Boldface type is emphatic because it draws attention to itself on a page otherwise filled with normal type.

Boldface type is particularly effective in distinguishing between different levels of headings. First-level headings are frequently centered and appear in boldface type. Second-level headings may appear at the left-hand margin, but they can also be centered. If a second-level heading is centered, it should appear in normal type—the contrast between boldface and normal type thus distinguishes between first- and second-level headings.

Be careful about overusing boldface type. If you use too much boldface type for emphasis, the effect will be lost. In fact, boldface type creates more strain on the eyes than normal type, so too much of it in a text is disturbing, not emphatic. Try especially to avoid long boldface passages.

See EMPHASIS and HEADINGS.

---

2. Use boldface for index section titles and for the titles of tables or illustrations.

In a complicated index, boldface type could indicate major sections while normal type indicates subsections. Similarly, boldface type can highlight the titles of tables and illustrations.

Boldface type is emphatic because of the contrast between it and surrounding print. So you can use boldface type effectively whenever you need a word, phrase, sentence, heading, or title to stand out.

BRACKETS

1. Use brackets to insert comments or corrections in quoted material:

"Your quoted price [$3750] is far more than our budget allows."

"Our engineers surveyed the cite [site] for its suitability as a waste disposal cite [site]."

NOTE 1: In these examples, the brackets indicate that the material quoted did not have the information included within the brackets.

NOTE 2: A common use of brackets, especially in published articles, is to insert *sic* in brackets following an error:

"We studied the affect [*sic*] of the new design on production outputs."

Sic, borrowed from Latin, means "thus" or "so." It tells readers that the text quoted appears exactly as it did in the original, including the error. In the example above, the word preceding [*sic*] should have been *effect*.

2. Use brackets to enclose parenthetical or explanatory material that occurs within material that is already enclosed within parentheses:

We decided to reject the bid from Gulf Industries International. (Actually the bid [$58,000] was tempting because it was far below our estimate and because Gulf Industries usually does good work.)

See PARENTHESES.

NOTE: You can sometimes use dashes instead of the outer parentheses and then replace the brackets with parentheses:

this

The Board of Directors—or more accurately, a committee of the actual owners (Hyatt, Burke, and Drake)—are answerable to no one but themselves.

not this

The Board of Directors (or more accurately, a committee of the actual owners [Hyatt, Burke, and Drake]) are answerable to no one but themselves.

NOTE: Some writers and editors consider the version without brackets preferable because having both parentheses and brackets in the same sentence can look clumsy and can be confusing. (See DASHES.)

3. For mathematical expressions, place parentheses inside brackets inside braces inside parentheses:

({ [( )] }).

See MATHEMATICAL NOTATION.

4. No other marks of punctuation need to come before or after brackets unless the bracketed material has its own mark of punctuation or the overall sentence needs punctuation:

this

The procedure was likely to be costly. (Actually, the cost [$38 per unit] included some of the research and development expenses.)

not this

The procedure was likely to be costly. (Actually, the cost, [$38 per unit], included some of the research and development expenses.)

CAPITALS

Capitalization follows two basic rules—the first two rules cited below. Unfortunately, these two rules cannot begin to account for the number of exceptions and options facing writers who have to decide whether a word should be capitalized.

Because of the number of exceptions and options, this section includes many minor rules that supplement the two basic rules. Together, the basic and supplementary rules provide guidance, but you should also check an up-to-date dictionary for additional guidance if the proper choice is still not clear. (See REFERENCES.)

1. Capitalize proper names—that is, those specific, one-of-a-kind names for a person, place, university or school, organization, religion, race, month or holiday, historic event, trade name, and title of a person or a document:

John F. Kennedy
Gail Sawyer
David Lewis
Jeanne Kirkpatrick
Henry Ford
Nancy Kassebaum
Sally

the Far East
China
the Eastern Shore (Maryland)
Massachusetts
Grove County
Baltimore City (*or* Baltimore)
United States of America
Lake Michigan
the Missouri River

the University of Utah
Western High School
the Golden Daycare Center
Shell Oil Company
the Prudential Life Insurance Co.
the American Legion
the Elks
the United Mine Workers
the Republican Party

Baptists
Judaism
Japanese
Hindus

May
September
Fourth of July
New Year's Day

the Reformation
World War I
Battle of Bull Run
the Crucifixion

Cyclone (fence)
Xerox copier
Band-Aids
Kodak
Coca-Cola

Mrs. Louise Brantly
Mr. Wing Phillips
Dr. Georgia Burke
Professor Robert Borson
Lieutenant Jeb Stuart

Handbook of Chemical Terms
The New York Times
The American Heritage Dictionary
"Time-Sharing" in *Training Magazine*

See TITLES.

NOTE 1: As the many instances of lowercase *the* above indicate, *the* is usually not capitalized unless it has become part of the full official name: *The Hague, The Johns Hopkins University*, etc.

NOTE 2: In proper names several words long, conjunctions, short prepositions, and articles (*a, an*, and *the*) are not capitalized:

the Federal Republic of Germany
Johnson and Sons, Inc.
"Recovery of Oil in Plugged and Abandoned Wells"

NOTE 3: Capitalize an individual's title only when it precedes the individual's proper name:

Professor George Stevens (*but* George Stevens, who is a professor . . .)

Captain Ellen Dobbs (*but* Ellen Dobbs, who is the captain of our company . . .)

President Henry Johnson (*but* Henry Johnson, president of Johnson Motors . . .)

NOTE 4: Adjectives derived from proper names are capitalized only when the original sense is maintained:

a French word (*but* french fries)
Venetian art (*but* venetian blinds)
Siamese cuisine (*but* siamese twins)

Even in these cases, dictionaries often differ; for instance, the current *American Heritage Dictionary* recommends *Siamese twins* rather than the form preferred above. So you often have to use your judgment. However, be consistent throughout a document.

2. Do not capitalize common nouns—that is, those nouns that are general or generic:

a geologist
my accountant
the engineers
your secretary

a country
a planet
a river
north
the city

a trade school
college
high school

CAPITALS

a holiday
the swing shift

a copier
the facial tissue

a foreman
my mentor
my supervisor
the doctor

spring
summer
fall
winter

twentieth century
the thirties (*however,* the Gay Nineties)

NOTE 1: One useful test to determine whether a noun is common is to ask if *a* or *an* does or can precede it in your context. If *a* or *an* makes sense before the noun, then the noun is common:

a pope (*but* the Pope)
an attorney
a U.S. senator

but

a President (*referring to any President of the United States*)

Because of special deference, the word *President* is always capitalized when it refers to any or all of the Presidents of the United States. This violates rule 2 above.

NOTE 2: Titles that follow a noun rather than precede it are not capitalized:

Theo Jones, who is our comptroller . . .
Betty Stevens, my secretary . . .
Rene Leon, who is a staff geophysicist . . .

NOTE 3: Common nouns separated from their proper nouns (or names) can occasionally be capitalized:

— Titles of high company officials, when the titles take the place of the officials' names:

We spoke to the President about the new labor policy.

The State Director has to sign before the plan goes into effect.

— Names of departments when they replace the whole name of the department:

We sent the letter to Accounting.

According to Maintenance, the pump had been replaced just last month.

— Names of countries, national divisions, governmental groups when the common noun replaces the full name (often in internal government correspondence):

From the beginning of the Republic, a balance of powers was necessary.

The State submitted a brief as a friend of the court.

The Department has a policy against overtime for employees at professional levels.

The House sent a bill to the Conference Committee.

— Names of close family members used in place of their proper names, especially in direct address:

Please understand, Mother, that I intend to pay my fair share.

Before leaving I spoke to Mother, Father, and Uncle George.

NOTE 4: Capitalize plural common nouns following two or more proper nouns unless the common nouns represent topographical features (such as lakes, rivers, mountains, oceans, and so on):

West and South High Schools
the Korean and Vietnam Wars
the State and Defense Departments

but

the Mississippi and Missouri rivers
the Wasatch and Uinta mountains
the Pacific and Indian oceans

3. Capitalize the first word of sentences, quotations, and listed items (either phrases or sentences):

Researchers propose to complete eight projects this year.

The technical specifications stated: "All wing strut pins should have a 150 percent load factor."

The accountant discussed the following issues:

—Budget design
—Cost overruns
—Entry postings

NOTE: Words following a colon or a dash are often capitalized, although some editors prefer not to. A good rule of thumb is that full sentences and long quotations (usually sentences) begin with a capital letter after a colon or dash:

We followed one principle: Short-term investments must be consistent with long-term goals.

or

We followed one principle—short-

CAPITALS

term investments must be consistent with long-term goals.

The Bible states: "The race is not to the swift."

4. Capitalize the names of directions when they indicate specific geographical areas. Do not capitalize them when they merely indicate a direction or a general or unspecified portion of a larger geographical area:

the Deep South
the Midwest
the Near East
the North
the Northwest

blowing from the southeast
eastern Missouri
southern Italy
the northern Midwest
toward the south
traveling north

NOTE: Sometimes titles are not clearly a geographical area—for instance, *East Texas*. If local custom identifies *East Texas* as including a particular number of counties, then the capital *E* is correct. If, however, *east Texas* means merely the general eastern portion of the state, then the lowercase *e* is correct. Of course, *eastern Texas* (rather than *east Texas*) would be a clearer means of indicating direction rather than geographical area.

5. Capitalize names for the Deity, names for the Bible and other sacred writings, names of religious bodies and their adherents, and names denoting the Devil:

Christ
God
He, Him
Lord
Messiah
Son of Man
the Almighty
Thee

God's Word
the Good Book
the Old Testament
the Word

a Lutheran
an Episcopalian
Episcopal Church
Lutheran Church

His Satanic Majesty
Satan
the Great Malevolence

6. Capitalize the first word and all main words of headings and subheadings and of titles of books, articles, and other documents. Do not capitalize the articles (*a*, *an*, and *the*), the coordinate conjunctions (*and, but, or, nor, for, so, yet*), or the short prepositions (*to*, *from*, *of*, etc.) unless they appear as the first word:

"An Examination of Church-State Relations"
Declaration of Independence
Oil and Gas Journal
The Geology of East Texas
"The Greening of Panama" in *Scientific American*

See HEADINGS.

7. Capitalize the second word in hyphenated titles and headings except when the compound is normally hyphenated (that is, when the compound has not been temporarily formed as a compound modifier):

A Report on Tin-Lined Acid Converters
Power-Driven Extraction Methodologies
High-Pressure Drilling of Diamond-Bearing Ores

but

"Shut-off Techniques for Nuclear Reactors"
Up-to-date Visual Aids
Follow-up Analyses of Tertiary Recovery Techniques

NOTE: In hyphenated numbers, the second word is never capitalized:

Ninety-six
One-fourth
Seventy-five
Three-quarters
Twenty-three

See COMPOUNDS, HYPHENS, and NUMBERS.

8. Capitalize the geological names of eras, periods, systems, series, epochs, and ages:

Jurassic Period
Late Cretaceous
Little Willow
Paleozoic Era
Upper Triassic

NOTE 1: Do not capitalize structural terms such as *arch, basin, formation, zone, field, pay, pool, dome, uplift, anticline, reservoir,* or *trend* when they combine with geological names:

Cincinnati arch
Delaware basin
East Texas field
Ozark uplift

CAPITALS

NOTE 2: Experts disagree about the capitalization of *upper, middle, lower* and *late, middle, early*. The following list from the *United States Government Printing Office Style Manual* (1984) provides the best summary of the difficult capitalization conventions in this technical area. Note that both *upper Oligocene* and *Upper Devonian* are correct, although the capitalization of *upper* is inconsistent.

Alexandrian
Animikie
Atoka
Belt
Cambrian:
  Upper, Late
  Middle, Middle
  Lower, Early
Carboniferous Systems
Cayuga
Cenozoic
Cincinnatian
Chester
Coahuila
Comanche
Cretaceous:
  Upper, Late
  Lower, Early
Des Moines
Devonian:
  Upper, Late
  Middle, Middle
  Lower, Early
Eocene:
  upper, late
  middle, middle
  lower, early
glacial:
  interglacial
  postglacial
  preglacial
Glenarm
Grand Canyon
Grenville
Guadalupe
Gulf
Gunnison River
Holocene
Jurassic:
  Upper, Late
  Middle, Middle
  Lower, Early
Keweenawan
Kinderhook
Leonard
Little Willow
Llano
Meramec
Mesozoic:
  pre-Mesozoic
  post-Mesozoic
Miocene:
  upper, late
  middle, middle
  lower, early
Mississippian:
  Upper, Late
  Lower, Early
Missouri
Mohawkian
Morrow
Niagara
Ochoa
Ocoee
Oligocene:
  upper, late
  middle, middle
  lower, early
Osage
Ordovician:
  Upper, Late
  Middle, Middle
  Lower, Early
Pahrump
Paleocene:
  upper, late
  middle, middle
  lower, early
Paleozoic
Pennsylvanian:
  Upper, Late
  Middle, Middle
  Lower, Early
Permian:
  Upper, Late
  Lower, Early
Pleistocene
Pliocene:
  upper, late
  middle, middle
  lower, early
Precambrian:
  upper
  middle
  lower
Quaternary
red beds
Shasta
Silurian:
  Upper, Late
  Middle, Middle
  Lower, Early
St. Croixan
Tertiary
Triassic:
  Upper, Late
  Middle, Middle
  Lower, Early
Virgil
Wolfcamp
Yavapai

NOTE 3: Topographical terms are usually capitalized, but the general terms *province* and *section* are not:

Hudson Valley
Interior Highlands
Middle Rocky Mountains
Ozark Plateau
Uinta Basin

Navajo section
Pacific Border province

---

9. In text, do not capitalize a common noun used with a date, number, or letter merely to denote time or sequence:

appendix A
collection 3
drawing 8
figure 5
page 45
paragraph 2
plate VI
section c
volume III

NOTE 1: These common nouns should be capitalized if they appear in headings, titles, or captions: *Appendix A* or *Figure 5*.

NOTE 2: In these cases, *no.*, *#*, or *No.* (for *Number*) is unnecessary:

Appendix A (*not* Appendix No. A)
page 45 (*not* page no. 45)
site 5 (*not* site #5)

CAPITALS

NOTE 3: Some technical and scientific fields do capitalize a common noun used with a date, number, or letter. For example, The Society of Petroleum Engineers recommends in its *Style Guide* that writers observe the following style of capitalization:

Method 3
Sample 2
Table 4
Wells A22 and B7

10. Capitalize proper nouns combined with common nouns, as in the names of plants, animals, diseases, and scientific laws or principles:

Boyle's law
Brittany spaniel
Cooper's hawk
Down's syndrome
Fremont silktassel
Gunn effect
Hodgkin's disease
Virginia clematis

CAPTIONS

Captions are the titles or explanatory labels accompanying visual aids. A good caption indicates what the visual aid is about and also provides enough information about the visual aid for readers to interpret it accurately.

See VISUAL AIDS. See also CHARTS, GRAPHS, ILLUSTRATIONS, MAPS, PHOTOGRAPHS, and TABLES.

1. Use action captions whenever possible.

Some captions are **telegraphic** (not action) in style and provide a minimum amount of information:

Figure 23. Glass Containers: 1984 Profile

Table 2. Quartz Hill Project Area Soils

Table 14. Existing Water Quality Characterization

Such captions are appropriate when the visual aids are essentially self-explanatory, as with tables of raw data. If the purpose of the table or figure is to present encyclopedic information that requires little or no interpretation and if the visual has a clear point of view, then telegraphic captions are acceptable. However, if the table or figure is not clear in and of itself, if it is subject to varying interpretations or could be misunderstood, then use action captions.

Action captions are usually longer than telegraphic captions and provide enough information to help readers understand the central points that the visual conveys. Action captions are usually complete sentences:

Figure 4. The cabin temperature control system features a modulated cabin sensor.

Figure 23(b). The check valve permits air flow in one direction only.

Table 17. Air intersect data: Only stations 23 and 45 experienced significant increases in CO levels during the study period.

The caption for figure 4 tells readers that this figure illustrates the cabin temperature control system, but it also focuses the reader's attention on the modulated cabin sensor.

CAPTIONS

The caption helps provide a clear perspective on the visual by drawing the reader's attention to the most important part of the drawing.

The caption for figure 23(b) indicates that the figure shows the check valve. Further, it establishes how the check valve operates, so readers looking at the illustration will immediately grasp the central point of the figure and understand how they should view it.

The caption for table 17 presents an alternative form of the action caption: A telegraphic title followed by an explanation. This table presents air intersect data. After establishing the subject matter, the caption highlights significant data and, in effect, tells readers how to read the table. By pointing out what is most important in the table, the caption gives direction to the reader's experience and reinforces the major points that the writer will make in the text.

Here are some additional examples of action captions:

Figure 1. The axial-flow design has the greatest performance potential.

Figure 2. General's design-to-cost strategy will guarantee low life-cycle costs.

Figure 3. Our project management system ensures maximum responsiveness to program requirements at all levels.

Table 1. Population impacts. If population redistribution trends continue, the Southwest will exceed baseline figures by 1996.

Table 2. The socioeconomic mitigation measures proposed are effective but very costly.

Figure 4. Seasonal streamflow patterns: Keta River and White Creek peak in mid-October during the snow goose migration.

Table 3. Contrary to media opinion, the Thunder Basin Project will create—not destroy—jobs in the Sequaw Valley: at least 500 within the next 10 years.

Figure 5. Flame-envelope thermal model 2 results in lower ambient temperatures but produces diffuse radiation.

Figure 6. The analog input multiplexer design limits components that are not included in self-testing to a few passive components.

NOTE: You can create action captions without using complete sentences if the captions establish a point of view that tells readers how to read and interpret the visual aid:

Figure 14. Declining production through the 1970s.

A telegraphic or title-like caption for this same figure would be *Production through the 1970s* or simply *Production (1970-79)*. Neither caption provides enough information. If the point of the visual aid is that production declined through the 1970s, then the caption should indicate that. If the point is that production declined steadily, then the action caption might be as follows:

Figure 14. The steady decline in production through the 1970s.

or

Figure 14. Steadily declining production through the 1970s.

or

Figure 14. Production declined steadily through the 1970s.

Whether or not an action caption is a complete sentence, it provides a point of view on the visual. It tells readers not only what the visual is about but also what the visual means. It tells readers how to interpret what they're looking at.

2. Number figures and tables sequentially throughout the document, and place the number before the caption.

Figures and tables should be numbered sequentially as they appear in the document. If you present an important figure or table twice, treat it as two separate visuals and number each according to its position in the sequence. (See VISUAL AIDS.)

Figure and table numbers should be whole numbers: *Figure 1, Figure 2, Figure 3*, etc. If you are numbering visuals by chapter, then use a hyphen to separate chapter number from visual number: *Figure 14-2* (chapter 14, figure 2), *Table 2-8* (chapter 2, table 8). You can also use the decimal numbering system with hyphens to designate visual numbers within a section of a report if the report's sections have been numbered decimally: *Figure 23.2-1* (section 23.2, figure 1), *Table 7.4-13* (section 7.4, table 13). If the visual aid has several parts and you need to identify all parts, use parentheses and lowercase alphabetical characters to designate subparts: *Figure 34(a), Figure 34(b), Figure 34(c)*, etc.

CAPTIONS

The figure or table number should precede the caption. Use a period following the number, and then leave two spaces before the caption:

> Figure 89. The U.S. aerospace industry has maintained a favorable balance of trade over the last decade and continues to dominate the U.S. market.
>
> Figure 17.6-2. Telephone equipment trade balances
>
> Table 14-2(b). Federal shipbuilding and repair budget

NOTE 1: The style of punctuation varies. Some editors prefer a colon or a period and a dash following the number. Some editors prefer no punctuation and leave three or four spaces between the number and the caption:

> Figure 14-2: Federal shipbuilding and repair budget
>
> Figure 14-2.—Federal shipbuilding and repair budget
>
> Figure 14-2 Federal shipbuilding and repair budget

NOTE 2: In captions, the words *figure* and *table* should be capitalized. However, when you are referring to visuals in the text, even if you refer to a specific visual, do not capitalize *figure* or *table* unless it begins a sentence:

> As shown in figure 33,
>
> According to table 14.2-4,
>
> *however*
>
> Figure 33 shows
>
> Table 14.2-4 presents

See CAPITALS.

3. Use periods following action captions but no punctuation following telegraphic captions.

Action captions are usually complete sentences and should therefore end with a period. Telegraphic captions, on the other hand, are like titles or headings and are normally not complete sentences, so they require no punctuation.

If you mix action and telegraphic captions, end all of them with periods.

4. Captions may appear above or below visuals, but be consistent.

Some authorities argue that captions should always appear above their visuals because the captions are like titles. Others argue that captions should always appear below their visuals because the visuals are more important than the captions and placing captions below their visuals is more aesthetic. Still others argue that captions for tables should appear above tables but captions for figures should appear below figures.

We see no reason to treat tables differently than figures. However, you should be consistent. If you're going to place captions ahead of (or below) visuals, then do so throughout your document. Deciding whether to place captions before or after visuals will require some judgment, but the mechanics of reading suggest the placement principles indicated in the next two rules.

5. Place short or telegraphic captions above their visuals.

Short captions form a quick introduction to the visual and should be seen first as readers are reading from the top of the page down. Short captions are very much like titles and would function as titles.

6. Place long captions, especially action captions, below their visuals.

Placing a lengthy caption above a visual would make the visual look top heavy and therefore not aesthetically appealing. Furthermore, a lengthy caption above a visual would slow readers, and they would therefore be apt to skip the caption. So if you are using action captions and some of them become lengthy (more than two lines), then place all of the captions in the document below their visuals.

The key, again, is to be consistent throughout the document.

CHARTS

What Are Charts?

Webster's New Collegiate Dictionary defines *chart* as a map, table, graph, or diagram. That definition exemplifies the confusion many people have over the names of different types of visual aids.

Originally, *chart* meant a document, although most charts were maps. When maps were combined with tabular data (*e.g.*, the mileage charts on modern road maps), *chart* came to mean a tabular or matrix display. When the tabular data was plotted on a coordinate graph, the word *chart* became synonymous with *graph*, especially if the data was displayed in bars or circles. Bar and circle/pie graphs (or charts) thus became associated with geometric shapes and were also called diagrams.

Ultimately, the nomenclature of the different types of visual aids is unimportant. For the purposes of this *Guide*, however, we will define charts as those visual aids that fall into these loose categories: **bar charts, surface charts, pie charts, flowcharts, organization charts, Gantt charts,** and **combination charts** (or hybrids). Note, however, that some writers and illustrators refer to bar and surface charts as graphs and to pie charts as circular diagrams.

Many charts do not fall into neat categories, especially those involving combinations of displays. Such combinations are frequently the most inventive, dramatic, and effective visual displays of information, so we urge you not to become overly concerned with nomenclature.

The best visuals are those that rapidly and effectively communicate their central ideas, regardless of what one might call them.

See VISUAL AIDS and GRAPHS.

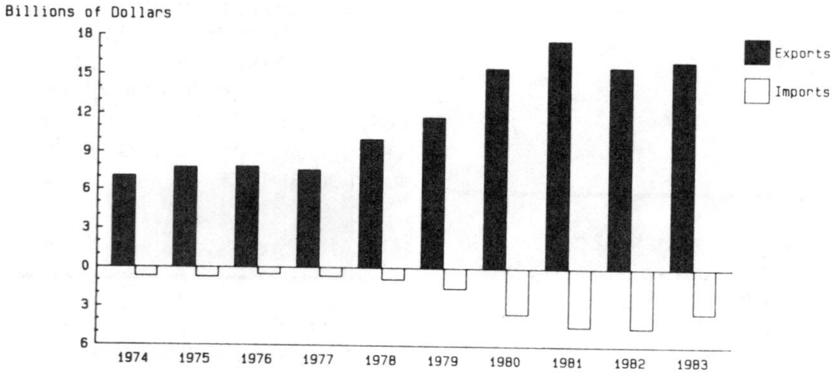

Figure 1. The U.S. aerospace industry has maintained a favorable balance of trade over the last decade and still dominates the U.S. market.

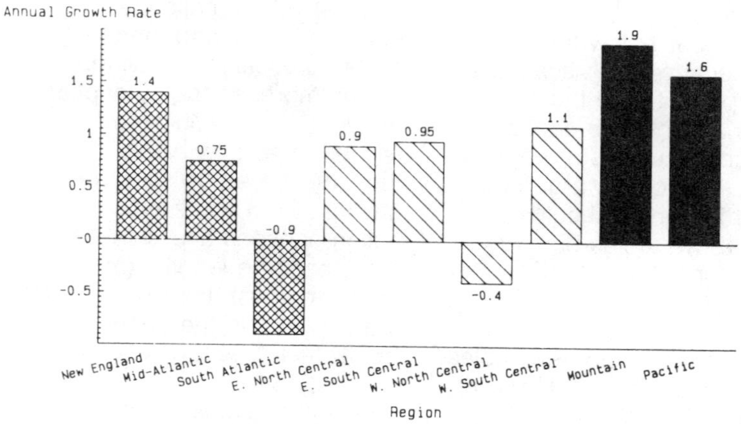

Figure 2. Employment growth rates for all selected industries: annual change 1979-87.

General Rules for Constructing Charts

1. If appropriate, use scales to indicate the quantity, magnitude, or range of each axis.

If the horizontal (*x*) axis or the vertical (*y*) axis indicates quantities, magnitudes, or ranges, use a scale that shows the axis minimum and maximum, as well as the numeric intervals. (See the vertical scales on figures 1, 2, 3, and 7, and the horizontal scales on figures 4, 6, and 8.)

Label the minimums and maximums. Also label a sequence of intervals along the scale. If the minimum is 0 and the maximum is 1000, for instance, you might label the scale in steps of 100. These interval labels indicate the scale and allow readers to interpret data.

Do not use so many labels that the scale becomes crowded; however, do not use so few that readers cannot easily interpret data. Generally, try to leave one to two spaces between interval labels. (See GRAPHS.)

2. If appropriate, use tick marks to help readers interpret data.

Tick marks are short lines on and perpendicular to an axis that indicate the intervals along a scale. Use longer tick marks beside interval labels; use shorter tick marks between labels. Generally, try to use twice as many tick marks as labels so that you have tick marks at the midpoints between labels.

Do not use too many tick marks. The more tick marks you use, the more crowded the scale becomes. Try to make the scale no more detailed than it has to be for readers to interpret the data represented on the chart. In other words, do not create finer distinctions than necessary for the data being shown. The tick marks on figure 1 are appropriate. The tick marks on figure 2 are far more detailed than necessary, especially since the data labels above or below each bar already show the size of the bar. Detailed interpretation of figure 2 is unnecessary, so the tick marks (spaced only 0.05 increments apart) are also unnecessary. (See GRAPHS.)

3. Orient horizontally all letters, numbers, words, and phrases in headings, legends, and labels.

All of the letters, numbers, and words on a chart should be readable from one reading perspective. Readers should not have to reorient the page to read one part of a chart.

EXCEPTION: In some cases, the vertical axis label must be oriented vertically, usually because of space restrictions.

4. Place footnotes and source information below the chart.

Footnotes typically explain or clarify the information appearing in the entire chart or in one small part of it. Often, footnotes

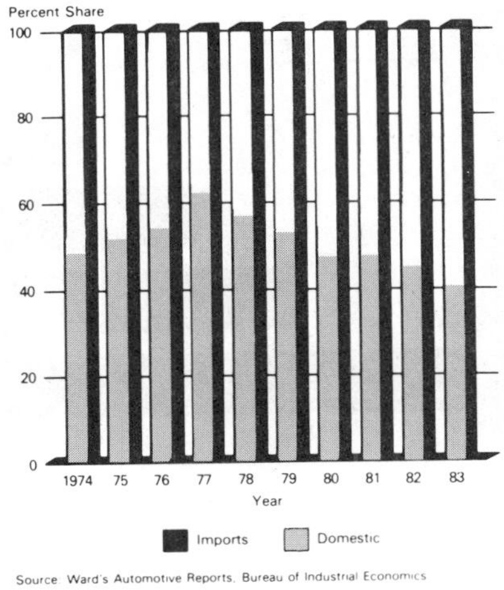

Figure 3. Subcompact car sales.

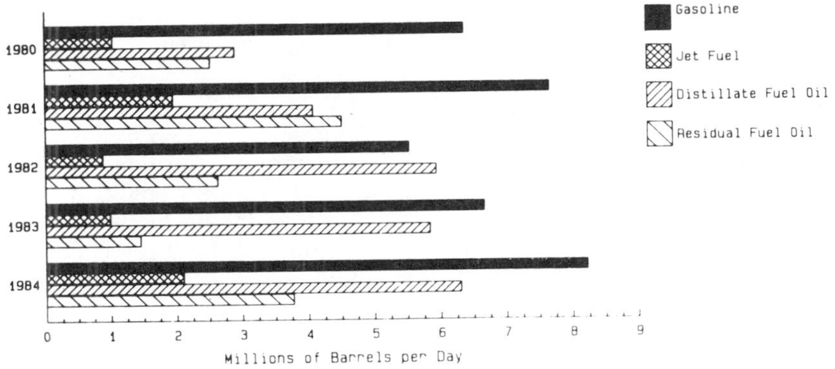

Figure 4. Petroleum products supplied (1980-84).

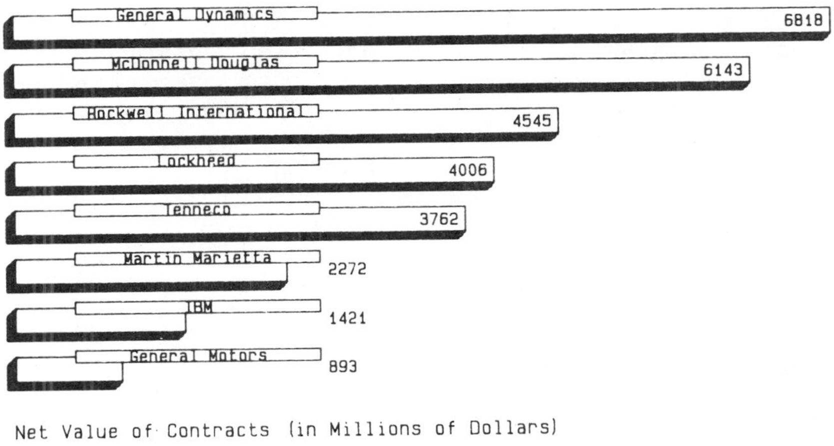

Figure 5. DoD contracts in 1983 for eight prime contractors.

CHARTS

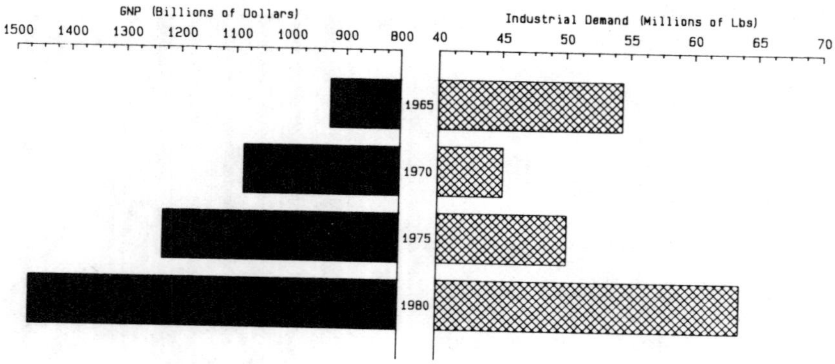

Source: U.S. Department of Agriculture

Figure 6. U.S. molybdenum demand measured against real GNP.

tell what the data applies to, where it came from, or how accurate it is:

[1] All data is in 1977 dollars.

[2] For major industrial groups only.

[3] According to the National Stockmen's Council. The AABP estimates that production has declined only 3.55% since 1981.

The footnotes for each chart are numbered independently from footnotes in the text and from footnotes in other visuals. Begin with footnote 1 and proceed sequentially. Within the body of the chart, use superscripted footnote numbers. Place the footnote explanations (in numerical order) immediately below the chart and flush with the left margin. Repeat the superscripted footnote number and then provide the appropriate explanation, followed by a period.

If the chart covers more than one page, place the appropriate footnotes with each page. If caption, footnotes, and text appear below a chart, place the footnotes ahead of the text but below the caption.

If footnote numbers would be confusing in the body of the chart, use letters [a], [b], [c], [d], etc.), asterisks (*, **, ***), or other symbols.

Source information may appear in footnotes if the referenced source provided only that data indicated by the footnote and not the data for the rest of the chart:

[4] From *The Wall Street Journal*, May 14, 1984.

[5] Source: U.S. Department of the Interior.

Source information may also appear within parentheses in the caption (regardless of where the caption appears) or within brackets under the caption if the caption appears ahead of the chart:

Figure 1. U.S. aerospace balance of trade: 1974-83 (U.S. Department of Commerce)

Figure 1. U.S. aerospace balance of trade: 1974-83

[U.S. Department of Commerce]

See FOOTNOTES.

5. Ensure that the visual characteristics of the chart reflect the magnitude and importance of the data being represented.

The value of charts is their visual impact. Consequently, writers can mislead readers by producing visuals that give more or less prominence to an idea or piece of data the writer wishes to emphasize or de-emphasize.

Distorting bar lengths or sizes or pie slice areas can mislead readers into thinking that something is larger or smaller than it really is. Similarly, using bright colors for insignificant data and dull colors for significant data can confuse readers and lead some to think that the insignificant is really more important. To ensure that you have presented a truthful and accurate picture of the situation being depicted, strive to make the visual impression created by the chart consistent with reality. (See VISUAL AIDS.)

Bar Charts

Bar charts depict the relationship between two or more variables, one of which is usually time. These charts typically show how the other variables change over time. Consequently, bar charts are useful for depicting trends (see figure 1). Because bar charts can show multiple variables, they can also depict how several variables change relative to one another over time (see figure 6).

CHARTS

Bar charts are not useful if the quantities depicted do not differ significantly. And if you expand or distort the scales to dramatize slight differences, the bar chart will look suspicious to alert readers and may damage your credibility.

Bar charts may be horizontal or vertical. In vertical bar charts, time is usually plotted along the horizontal axis (see figures 1 and 3). In horizontal bar charts, time is usually plotted along the vertical axis (see figures 4 and 6).

Some writers and illustrators argue that vertical bar charts are better for showing trends (figure 1) and that horizontal bar charts are better for comparisons (figures 4 and 5) and for showing magnitude changes (figure 6). Certainly, readers are more used to seeing trends shown along a horizontal axis. However, comparisons and magnitude changes are usually clear in either orientation. Use your judgment.

Bar charts may be used with other visual forms, such as line or coordinate graphs (see GRAPHS), maps (see MAPS), or pie charts (see figure 11).

6. Clearly label each bar.

Ensure that readers understand what each bar represents. See figures 1 through 9 for examples.

7. Make the bars wider than the spaces between them.

See figures 1 through 9. A bar chart with the spaces wider than the bars would seem "airy." The dominant visual effect in a bar chart should be the bars.

8. Use different bar patterns to indicate differences in types of data.

Bar patterns allow you to distinguish areas, regions, groups, and parts. Patterns also allow you to focus readers' attention on areas of the chart you consider most important. Bar patterns include but are not limited to the following:

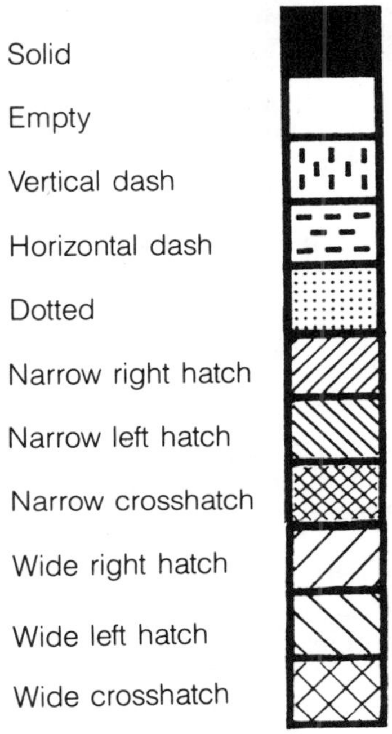

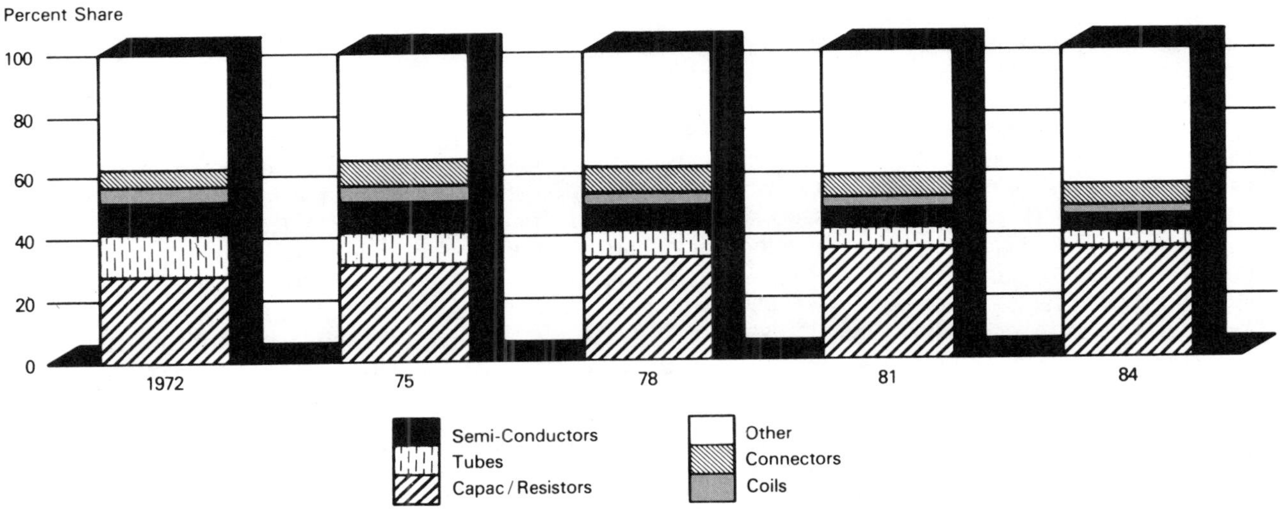

Source: Bureau of the Census, Bureau of Industrial Economics.

Figure 7. U.S. electronic component shipments.

CHARTS

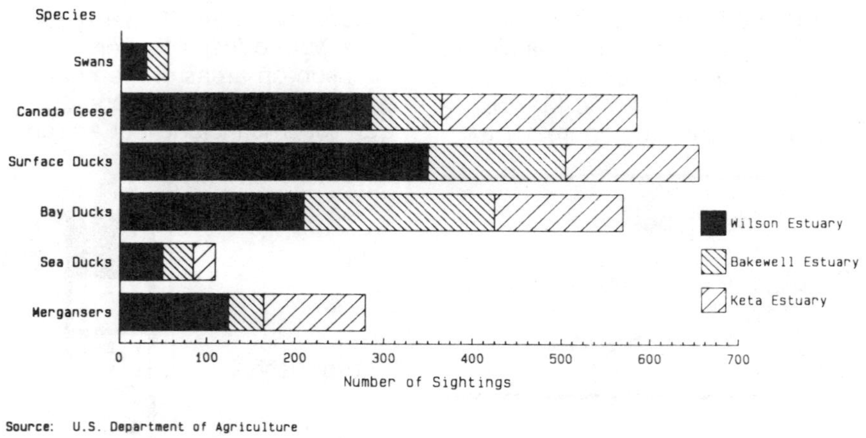

Figure 8. Cumulative annual waterfowl sightings by species in three estuaries (1982).

Figure 1 uses a simple black/white contrast to dramatize the difference between imports and exports. To further emphasize those differences, figure 1 also places exports above the zero dividing line and imports below the line. Note that the vertical axis reflects positive numbers on both sides of the zero dividing line.

Figure 2 uses a narrow crosshatch pattern to indicate eastern regions, a wide right hatch pattern to indicate mid-region states, and solid bars to indicate western states. The solid bars dominate the chart, causing readers to focus on the growth rates of the western regions, which are the highest in the country. The writer might have wished to emphasize the declining growth rates in the South Atlantic and West North Central regions, and could have done so by using empty bars for other regions and solid bars for the two declining regions.

Figure 3 uses combination solid and empty bar patterns within the same bars to indicate percent shares.

Figure 4 uses four bar patterns to distinguish between four types of petroleum products.

Figure 6 uses two bar patterns to isolate the bars that reflect two very different scales.

9. If you use different bar patterns, provide a legend that identifies the bars.

In figures 2 and 6, the bar patterns are clear without a legend. In figures 1, 3, 4, 7, 8, and 9, however, a legend is necessary. Legends typically appear on the bottom or right-hand side of the chart. They give the bar patterns in small boxes along with labels indicating what each pattern represents.

10. Use paired bars to depict sets of data having different scales.

Figure 6 shows a paired bar chart. That kind of chart is necessary when the data being depicted can be divided into two sets of data that operate on different numerical scales. In figure 6, GNP is measured in billions of dollars while industrial demand for molybdenum is measured in millions of pounds. Depicting these relationships would be difficult unless the chart allowed for two distinct *x*-axis scales.

11. Use segmented bars to depict three or more variables.

Segmented charts (figures 7 through 9) allow you to show at least three variables. Figure 7 depicts time, percent share, and types of electronic components. Figure 8 shows species, estuaries, and numbers of sightings.

Figure 9 is a three-dimensional bar chart depicting four variables: ownership, time, units (in Million $), and, through bar segmentation, where semiconductor imports come from (Europe, Japan, SE Asia, and Other).

Pie Charts

Pie charts are circles (or pies) divided into sectors (or slices) to show the relationship of parts to a whole. Naturally, the parts must add up to 100 percent.

Pie charts are useful for general comparisons of relative size. However, they are not useful if accuracy is important. Further, pie charts are not useful for showing a large number of items.

The eye can measure linear distances far more easily than radial distances or areas.

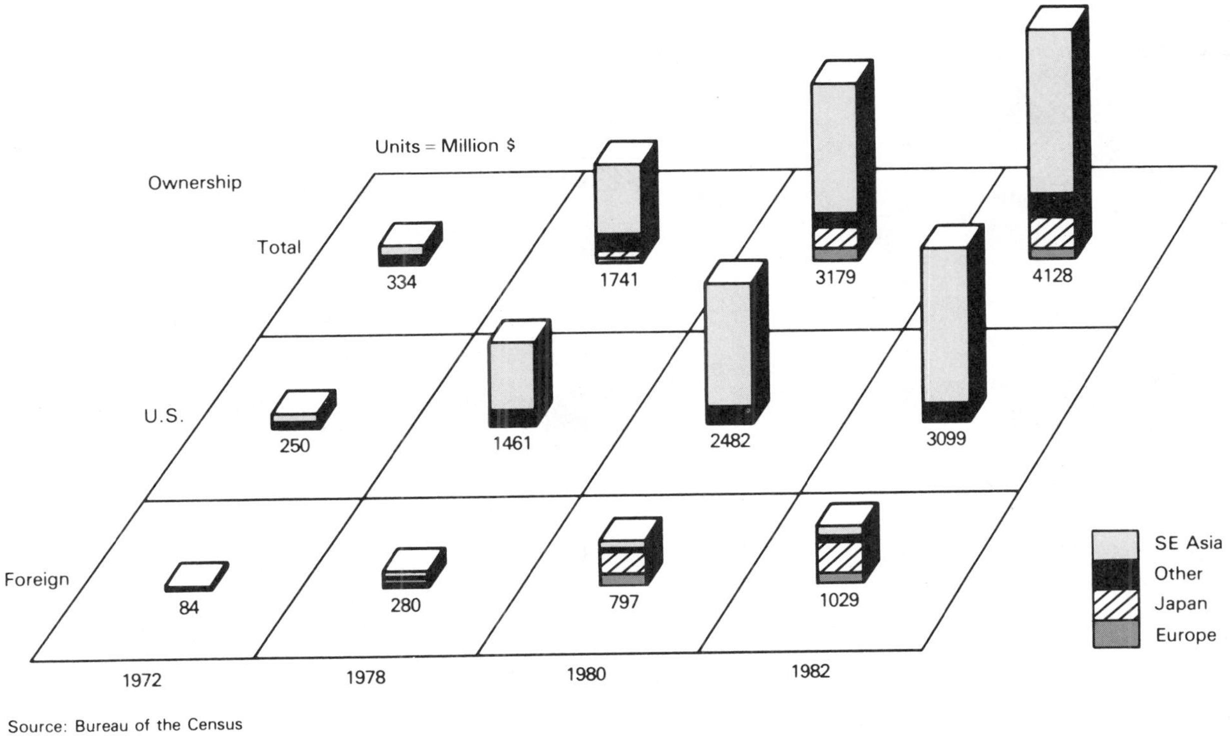

Figure 9. Origins of U.S. semiconductor imports.

Therefore, visual comparisons of bars on a bar chart are much easier than visual comparisons of sectors in a pie chart. Moreover, readers can usually make more accurate judgments about data relationships expressed in a linear fashion. So if you need accuracy, use a bar chart. If you need to show how parts relate to one another and to a whole—and if precise numbers are not important—use a pie chart.

12. Identify each sector of the pie and, if appropriate, the percentage it represents.

Pie charts do not have axes and therefore cannot be very precise, so if percentages are important, identify them (see figure 12).

Always identify each sector of the pie. You may do this by using labels (figures 10 and 11) or fill patterns and a legend (figure 12). For further information on fill patterns, see the preceding discussion of bar charts.

13. Differentiate adjacent pie sectors by using alternating fill patterns or colors.

To help readers distinguish the sectors, use alternating fill patterns or colors (see figures 10 through 12). Reserve the solid (or black) fill pattern for the prominent sectors, the ones you wish to emphasize. Never use the same fill pattern for adjacent sectors.

14. Ensure that the size of each sector reflects the data it represents.

The size of the sector conveys a powerful visual message. Readers often "grasp" a pie chart simply by perceiving the relative size of the sectors, even if accompanying data labels clearly indicate what percentage each sector is supposed to reflect.

15. Group small percentage items under a general label, such as "Other."

Pie charts should have no more than 8 to 10 sectors, depending upon the size of the pie. The larger the pie, the more sectors you can safely divide it into. However, beyond some

CHARTS

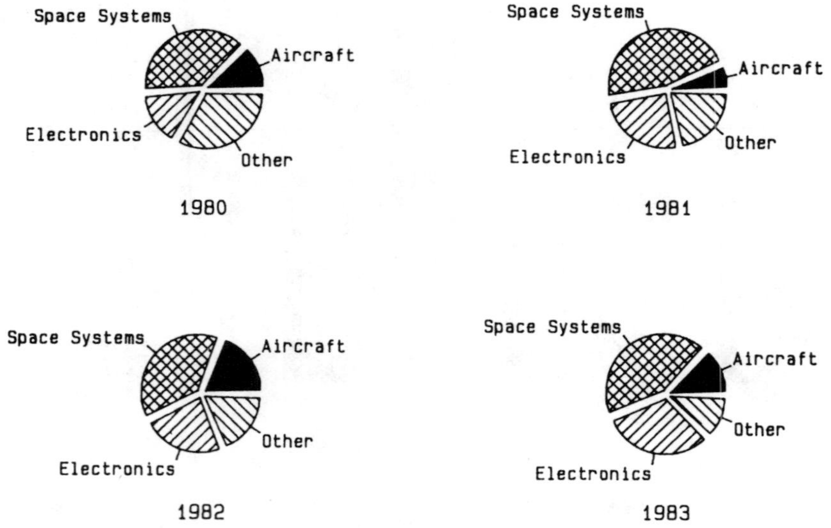

Source: U.S. Department of Defense

Figure 10. DoD prime contract awards for RDT&E

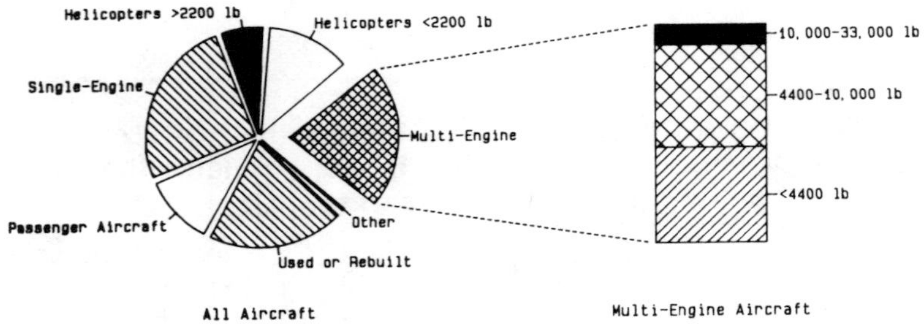

Figure 11. Sales of multi-engine aircraft accounted for 22 percent of total civil aircraft exports in 1983.

reasonable number of sectors (10), a pie chart becomes too busy and therefore difficult to read.

If you have a number of small percentage items, you should group them and give them a common label, such as "Other Parts," "Other Exports," or simply "Other" if the context of the chart and the names of the other labels indicate what "Other" refers to. If you have items of moderate size (say 10 to 20 percent) that are unimportant and would distract readers, you might group them under a common label. (See figure 10, where the "Other" category occupies over 25 percent of the pie labelled "1980.")

16. Use pie-bar combination charts to show the composition of an important sector of the pie.

Figure 11 shows a combination pie-bar chart. Here, the focus of the chart is on multi-engine exports, and the bar associated with that pie sector is a segmented bar indicating the types of multi-engine aircraft being exported and their percentage of the whole.

Each sector of a pie can potentially be expanded into a segmented bar. In this way, you are adding one more variable to the pie chart.

17. Use a series of pie charts to add time as a variable.

Another way of adding a variable is to use a series of pie charts (see figures 10 and 12). The most common reason for using a series is to add time as a variable.

Organization Charts

Organization charts depict the structure of an organization. These charts typically show the divisions and subdivisions of the organization, the hierarchy and relationship of the groups to one another, lines of control (responsibility and authority), and lines of communication and coordination. Organization charts help readers visualize the structure of an organization and the relationships within it.

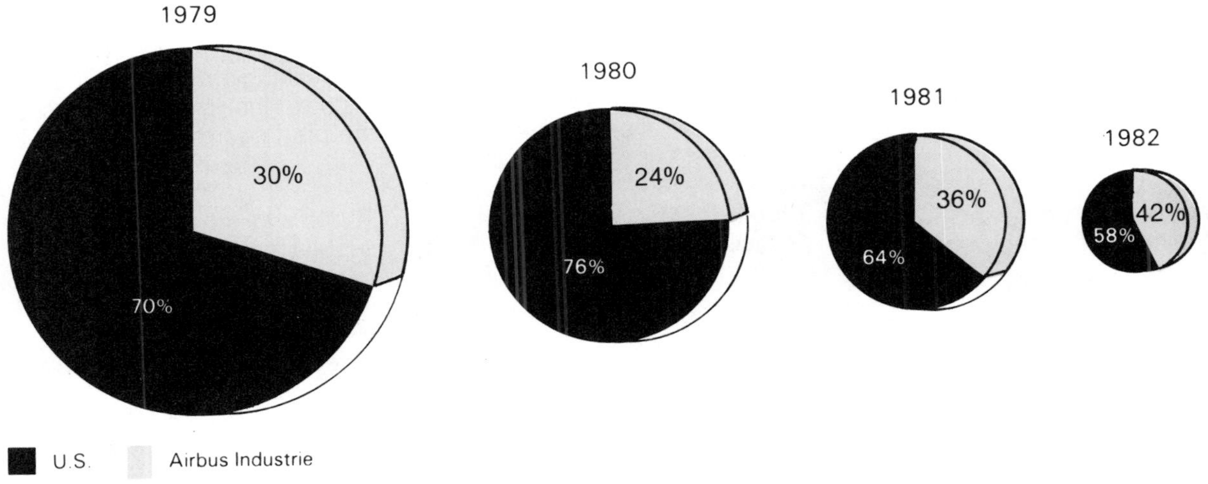

Pie diagrams depicted in proportion to the constant dollar value of widebody transport orders minus cancellations taken in years shown.
Source: Bureau of Industrial Economics.

Figure 12. Free world widebody transport market shares based on value of firm orders.

18. Use squares or rectangles to indicate divisions and subdivisions within the organization.

The conventional means of indicating divisions, subdivisions, groups, project teams, functional areas, etc., is to enclose the name of the organizational unit within a square or rectangle.

You might distinguish between higher and lower units by changing the size of the rectangle or by changing its border (from boldface or thick lines representing upper-level units to thinner, normal lines representing lower-level units).

19. Structure an organization chart from the top down.

This rule is important for two reasons: it reflects the way we read, and it reflects our perception and understanding of organizational structure.

Readers of English read from left to right and from top to bottom. Therefore, an organizational chart should be structured from left to right and top to bottom. The left to right progression may or may not be useful, depending upon the type of organization you are depicting. However, the top to bottom progression is always useful, simply because almost all organizations are based upon a hierarchy.

Accordingly, you should display the structure of the organization in descending order of authority, with the highest authority or level at the top of the chart and the lowest authorities or levels at the bottom of the chart. This structure reflects the metaphor of top-down management and thus reinforces the readers' expectations about organizational structure.

If the organization you are describing does not operate on a top-down basis, be inventive and create an organizational display that does reflect the organization's operational style and structure.

20. Use solid lines to indicate direct relationships and dotted or dashed lines to indicate indirect relationships.

Solid lines usually show direct lines of control. Dashed or dotted lines usually indicate lines of communication or coordination.

In figure 13, the dashed lines forming the rectangles along the right-hand side of the chart indicate that the Vice President for Engineering coordinates with the Division Liaison office and has lines of communication and coordination down through the Division Liaison organization, but the Vice President's direct authority extends through Systems Engineering, Product Engineering, and Electronics.

CHARTS

Flowcharts

Flowcharts depict a process. They show readers the parts of a process and how those parts are related.

Flowcharts use a symbol system to indicate the types of activities and control or transfer points being depicted.

Squares and rectangles typically indicate activities in the process. In figure 14, for instance, the upper left-hand rectangle represents "Ore Crushing," the first activity taking place in this ore processing system. The arrow linking this rectangle to the rectangle directly to the right indicates that, after being crushed, the ore undergoes a chemical bath.

The arrows indicate the sequence of activity in the process and show a chronological (and sometimes cause-and-effect) relationship between linked activities or control points.

Circles typically indicate control or transfer points. Control points are those points in the process where the activities are monitored, started, stopped, or in some other way controlled. Transfer points are those points where the sequence of activity leaves one flowchart and continues on to another. In figure 14, the two right-most circles indicate that the ore has been processed and is ready for packaging. To continue following the process, the reader must go to the packaging flowchart (which is shown in figure 15).

The "Pack" circles in figure 14 represent all of figure 15. Consequently, figure 14 is the more general flowchart. If "Ore Crushing" involved a series of steps, the writer could have turned the "Ore Crushing" symbol in figure 14 into a circle and then constructed another subordinate flowchart (like figure 15) that represented all of the activities involved in crushing the ore.

Note that by putting all of the packaging activities into one subordinate flowchart, the writer has avoided significant repetition in figure 14.

Diamonds typically represent decision points. Often, as in figures 14 and 15, these

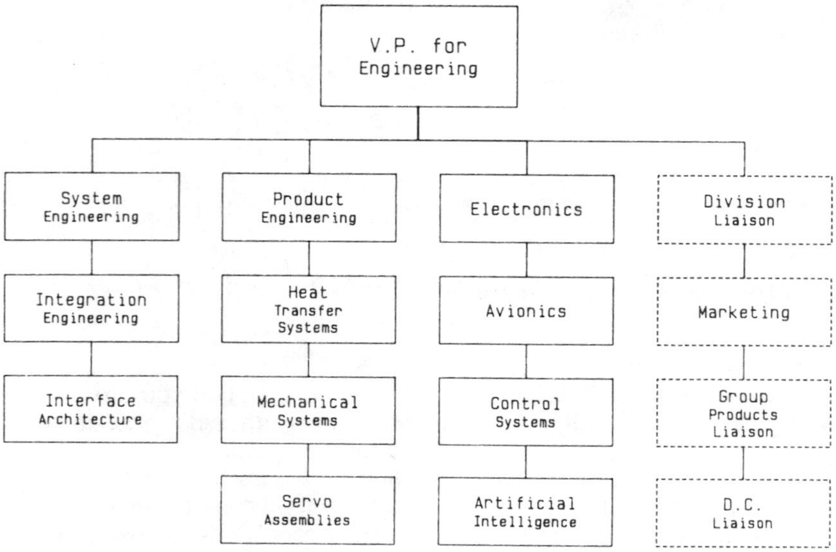

Figure 13. Engineering department organization.

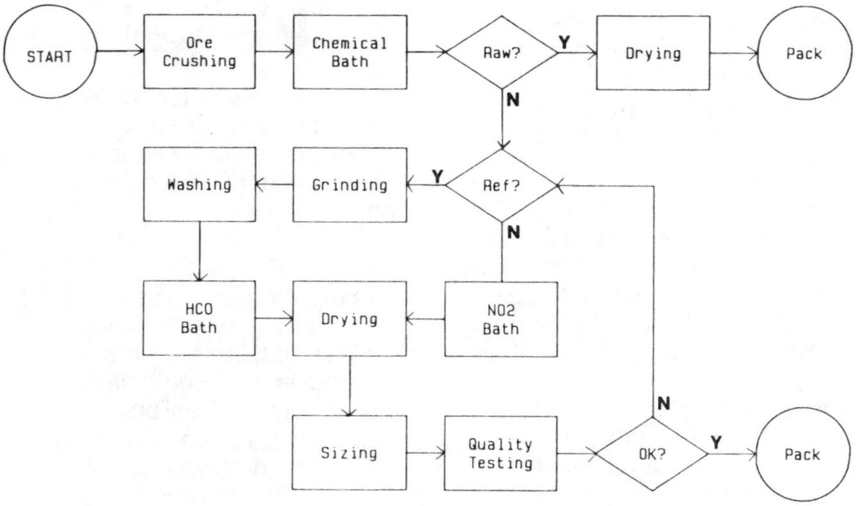

Figure 14. Ore processing flowchart.

CHARTS

decisions are represented by simple yes/no questions. Diamonds normally have three lines linking them to other symbols: one incoming line indicating what precedes the decision, one outgoing line indicating "yes," and the other outgoing line indicating "no."

Other symbols are possible, particularly in data processing and other specialized fields, such as architecture and electrical engineering. These symbols often have very specific meanings and have become traditional means of expression in particular scientific and technical applications. If you need to create flowcharts for these specialized areas, consult appropriate trade journals and textbooks.

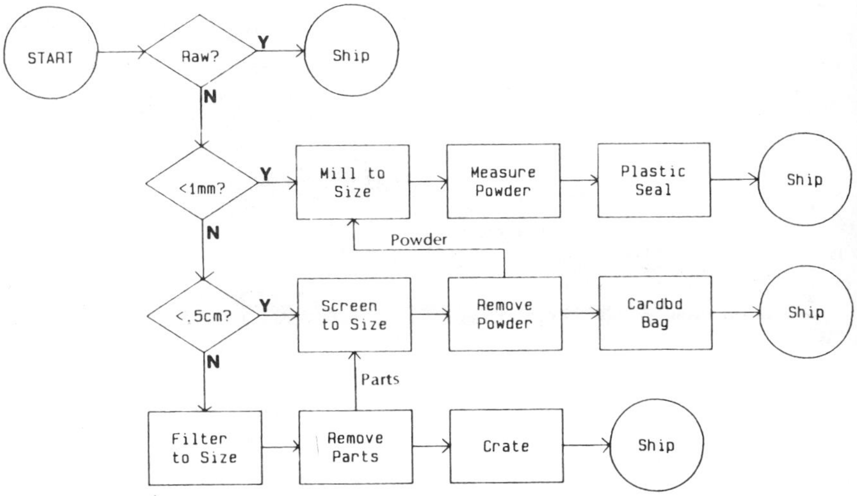

Figure 15. Processed ore packaging flowchart.

21. Place the starting activity in the upper left-hand corner of the chart and proceed to the right and down. Place the ending activity in the lower right-hand corner of the chart.

Readers of English read from the left to the right and from the top to the bottom. Therefore, readers will expect the flowchart to begin in the upper left-hand corner. Don't disappoint them.

If possible, try to end the flowchart in the lower right-hand corner for the reasons cited above.

22. Break large or complicated flowcharts into smaller, simpler flowcharts.

Flowcharts that become too large or too complicated are unreadable as well as intimidating. To avoid repetition and to keep flowcharts from becoming too long, break them into general or overview charts and subordinate or component charts.

23. Use arrows to show the sequence or direction of flow within the flowchart.

Flowcharts with activities or control points linked only by lines are often confusing. Place an arrow on the end of the line to indicate the sequence or direction of flow (see figures 14 and 15).

24. Use footnotes to explain symbols, abbreviations, and connections with other flowcharts.

As in figures 14 and 15, footnotes help explain abbreviations and activities or control points. Footnotes can also help readers understand the structure and sequence of related flowcharts. The final note on figure 14, for instance, indicates that "Pack" refers to a subordinate flowchart on packaging.

Surface Charts

Surface charts (figure 16) show the effect of cumulative additions on a range of data. These charts resemble multiple line graphs (see GRAPHS), but their purpose is not to allow for accurate interpretation as much as it is to display cumulative changes; therefore, the surface chart is considered a chart and not a graph.

Surface charts are created by plotting data accumulations (usually two or more variables over a period of time) and then coloring or shading the area between successive lines to demonstrate both the effect of accumulation and the relationships between the variables plotted.

CHARTS

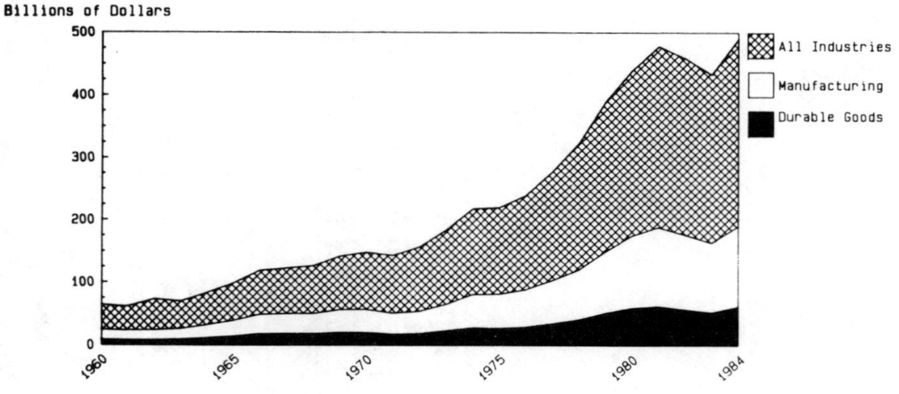

Figure 16. New plant and equipment expenditures (1960-84).

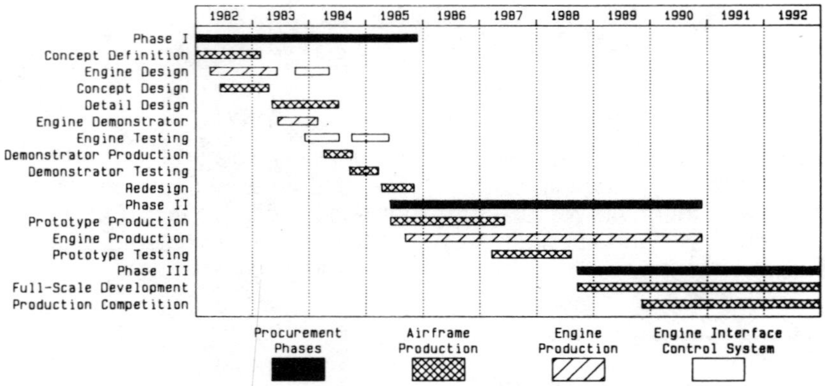

Figure 17. F25AF production schedule.

Surface charts are effective only when they depict gradual changes. Further, you cannot use them if any of the curves overlap. Otherwise, the general principles outlined above for charts apply to surface charts. Refer also to the principles cited in GRAPHS and VISUAL AIDS.

25. Use different patterns to shade areas beneath lines. If possible, use color as well.

The effect of a surface chart comes from the visual impact of the shaded areas, so use patterns to illustrate the area between successive lines. (For an illustration of possible patterns, see Bar Charts above.) If possible, color the shaded areas. Use the brightest color for the area you wish to highlight or the area that is most important.

Gantt Charts

Gantt charts are horizontal bar charts used to schedule tasks, projects, and programs. Gantt charts help readers visualize a sequence of activity occurring over a long period. They help readers see how sequential and concurrent activities are related to each other in time and how activities depend upon one another for completion on schedule.

The horizontal x-axis is always time. Time may be represented in decades, years, quarters, months, weeks, days, and hours. If necessary or helpful to readers, you can use more than one x-axis (see figure 18). To make Gantt charts easier to read, use vertical dotted or dashed lines running the length of the chart to mark major time periods (see figure 17).

26. List activities in chronological order beginning with the first event in the sequence.

Gantt charts suggest strict chronology. Do not violate the reader's expectation that the events listed from top to bottom along the vertical y-axis will appear in chronological order.

27. Clearly label or identify the bars.

Always identify what each bar represents, either with a bar label along the left margin or with a legend. Traditionally, the bar labels appear on the left-hand side of the chart.

Note, however, that in figure 17 the engine interface control system activities that follow certain engine production activities are indicated only by the empty bars, which are explained in the legend at the bottom of the chart. If these interface activities were more important as stand-alone items, they would require their own rows.

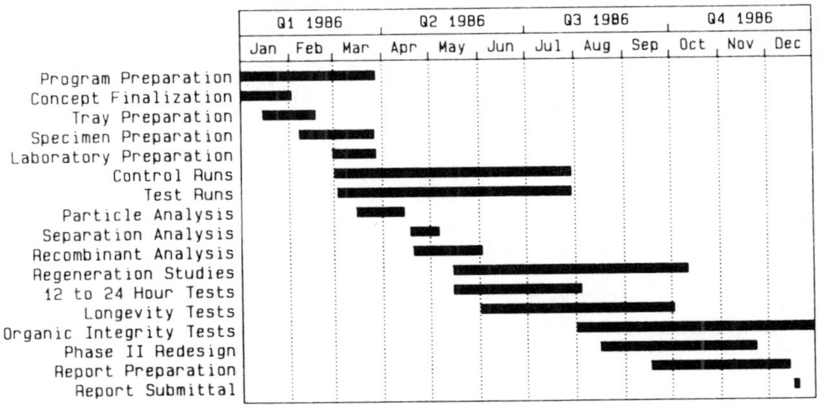

Figure 18. DeLorge recombinant test schedule.

28. If appropriate, indicate milestones on the chart.

Milestones may be indicated with small circles, dots, or triangles. If you have reporting, control, or performance events or deadlines that constitute milestones for tracking and monitoring progress, then indicate them on the chart.

29. Use bar patterns to identify groups of related activities. If you do so, also include a legend explaining what the patterns represent.

Bar patterns (see Bar Charts above) can be used to indicate similar or identical activities that occur at different times. In figure 17, for instance, all airframe production activities are shown with narrow crosshatch bars.

The bars representing engine production activities have a wide right hatch pattern. These patterns enable readers to see how related activities fit within the whole sequence.

If you use bar patterns, always include a legend to the right or at the bottom of the chart.

Combination Charts

Combination charts are not distinct categories of charts, but writers and graphic artists often combine chart types in such unique and creative ways that the result seems to constitute a new category.

Remember that a chart is only as good as the effect it creates. Charts should communicate quickly and simply. They should be integrated with the text and should convey information more forcefully or dramatically than is possible in text.

If you can combine chart types to convey a rich, unusual, interesting, or dramatic message, then do so. You do not gain extra points for adhering strictly to some predetermined form. Let your purpose, readers, medium, data, and ideas dictate the form of chart (or other visual aid) that would be most effective.

CITATIONS

Citations enable writers to identify in the text itself the sources of their information. The methods of citation vary, depending upon the technical field and its traditions, the type of publication, and the publisher. Many professional societies and journals also have their own method of citation.

The method of citation that Shipley Associates recommends represents the standard convention in the physical sciences and engineering disciplines. However, if you are writing for a particular professional society or technical journal, you should follow its method of citation.

1. Enclose the author's name and the date of the publication in parentheses following the material quoted or the ideas referred to. Attach at the end of the text an alphabetical list of the cited works:

> One critic called the whole dispute a "galaxy of confusion" (Jameson 1976). In reply, the spokesman for the conservative wing rebutted Jameson point by point (S. Clarke 1977).

This system is the briefest, most efficient system for citing sources. The information in parentheses is so brief that it does not interrupt the text, and the author's name indicates the source.

NOTE 1: This system, or some variation of it, is favored by physical and biological scientists, as well as many social scientists. Scholars in the humanities still prefer to use footnotes. (See FOOTNOTES.)

NOTE 2: An alternative method uses only numbers in the text, not the author's last name and date:

> One critic called the whole dispute a "galaxy of confusion" (1). In reply, the spokesman for the conservative wing rebutted Jameson point by point (2).

or

> One critic called the whole dispute a "galaxy of confusion" [1]. In reply, the spokesman for the conservative wing rebutted Jameson point by point [2].

The numbers appearing within parentheses or brackets are keyed to a list of sources that appears at the end of the text. The list is not alphabetized because it follows the order in which the sources were cited in the text. If a single citation changes early in the document, then all later citations and numbers must be changed. Consequently, this alternative method of citation seems less efficient.

2. Use a consistent format for citing the name of the author and the date of publication:

> (Jakobson 1981)
> (Bains and Eveslong 1984)
> (Federal Science Committee 1979)
> (Smithson, Haarke, and Bruppe 1982)
> (U.S. Department of Agriculture 1978)

NOTE: Some authors and journals prefer to place a comma between the author and the date:

> (Jakobson, 1981)

Another common variation is to include the page number or volume and page number following the date:

> (Jakobson 1981, 43-48)

The abbreviations *p.* and *pp.* (for *page* and *pages*) are unnecessary:

> (Bains and Eveslong 1984, 156)
> (Federal Science Committee 1979, 2:34-36)

In the last example, the number *2* is the volume number.

Whatever the format used, authors should be consistent in their method of citation within the same document.

3. Include a full alphabetized list of cited sources following the article or chapter:

Book with a single author

Bricke, Larry N. 1984. Canadian Political Parties. Toronto: New Country Press.

Book with two authors

Campbell, Josiah, and Wallace Daughterly. 1976. Conflict in the Provinces. Edmonton, Alberta: The Royal Penny Press.

Journal or magazine article

Mahoney, Edward G. 1978. "A Dissident's View of Canadian Politics." The Political Review 2:56-59.

NOTE 1: In these bibliographic entries, the date of publication comes immediately after the name of the author. In the more common bibliographic form, the date appears after the name of the publisher:

Bricke, Larry N. Canadian Political Parties. Toronto: New Country Press, 1984.

See BIBLIOGRAPHIC FORM.

NOTE 2: The titles are underlined rather than italicized because italics is often not available on standard typewriters and on some word processors.

CLICHES

A cliche is a worn-out phrase that was originally effective, even vivid:

innocent bystander
irony of fate
too funny for words
cool as a cucumber
moot point
far and wide

Such phrases are so common that writers and speakers use them habitually, without thinking. Their familiarity makes cliches convenient. So, when writers are struggling to express an idea, using a cliche becomes a tempting alternative to serious thought. (See WORDY PHRASES.)

The Origin of Cliches

The word *cliche* comes from French and is often still written with the French accent: *cliché*. Writers of English typically omit the accent mark because the word has been anglicized. (See SPELLING.)

Many cliches were originally metaphors and were therefore vivid. Their wittiness and sparkle made them memorable, so they were repeated often. However, any effect diminishes with repetition, so over time, the vividness of the original metaphor is dulled by repetition, and the expression becomes a cliche.

The first person in medieval Europe to associate the concept of avoidance with the Black Death must have created a vivid image in listeners' minds. But today's users of the expression *to avoid like the plague* experience little if any of the original effect.

Today, the cliche means little. We've heard it too often.

Other cliches developed and survived because they sounded good:

bag and baggage
rack and ruin
not wisely but too well
snug as a bug in a rug
willynilly

The alliteration (repetition of initial consonants) and repetition of *bag* in the expression *bag and baggage* likely ensured the phrase's survival. Similarly, *willynilly* has survived so long that its original meaning has been lost: "whether you are willing or unwilling." Now *willynilly* seems to imply haphazard or weak actions. The logical choice in its original meaning has faded. Now the cliche has more sound than substance.

---

1. Use cliches sparingly, if at all.

Writers would find eliminating all cliches a hard row to hoe. Reasonable cliches are fine in certain contexts, but the one in the preceding sentence is clearly inappropriate. The context of this discussion makes a cliche based on farm chores ridiculous.

However, some cliches—in the right context—are valuable:

Following the testimony, the judge had to sift through 1,000 pages of unreadable and often contradictory testimony.

Here, the cliche of *sifting through* many pages is not objectionable. It is, in fact, a fine metaphor given the circumstance to which it applies. In this case, the cliche does not clash with the context.

So one test of a cliche's acceptability is the degree to which it is relevant to the context in which it occurs and the extent to which it goes unnoticed. If the cliche does not call attention to itself, it is probably acceptable. The moment a writer (or reader) knows the expression is a cliche, it is unacceptable in most contexts.

NOTE: Sometimes a cliche can be used to advantage if its meaning or phrasing allows the writer to play against the cliche itself, as in the following quote from Oscar Wilde:

"Truth is never pure, and rarely simple."

By rephrasing the cliche, Wilde asks readers to reexamine the cliche about *pure and simple truth*.

Some Common Cliches

English contains hundreds of cliches. The following list includes some of the more common ones currently in use:

a bad scene
a can of worms
acid test
active consideration
add insult to injury
agree to disagree
all things considered
all too soon
along these lines
among those present

CLICHES

ample opportunity
an end run
armed to the teeth
a roundhouse punch
as a matter of fact
as the crow flies
at a loss for words
at one fell swoop
attached hereto
auspicious moment
avoid like the plague
awaiting further orders

back at the ranch
back to the drawing board
bag and baggage
bated breath
beat a hasty retreat
be at loggerheads
beginning of the end
benefit of the doubt
best-laid plans
better late than never
better left unsaid
beyond the shadow of a doubt
bite the bullet
bitter end
blissful ignorance
block out
bloody but unbowed
bolt from the blue
bone of contention
bottom line
bright and shining faces
broad daylight
brook no delay
brute force
budding genius
built-in safeguards
burning question
burning the midnight oil
busy as a bee
by leaps and bounds
by the same token

calm before the storm
capacity crowd
cast a pall
casual encounter
chain reaction
charged with emotion
checkered career/past
cherished belief
chief cook and bottle washer
circumstances beyond my control
city fathers
civic wrath
clear as crystal/day
colorful display
come full circle
common/garden variety
confirming our conversation
conservative estimate
considered opinion
consigned to oblivion
conspicuous by its absence
contents noted
controlling factor
cool as a cucumber
crying need
curiously enough
cut a long story short
cut down in his prime

dark horse
date with destiny
days are numbered
dazed condition
dead as a doornail
deadly earnest
deafening crash
deficits mount
deliberate falsehood
depths of despair
diamond in the rough
dig in your heels
discreet silence
do not hesitate to
doomed to disappointment
doom is sealed
dotted on the landscape
dramatic new move
drastic action
due consideration
dynamic personality

each and every
easier said than done
eat, drink, and be merry
eloquent silence
eminently successful
enclosed herewith
engage in conversation
enjoyable occasion
entertaining high hopes of
epic struggle
equal to the occasion
errand of mercy
even tenor
exception that proves the rule
existing conditions
express one's appreciation
eyeball to eyeball

failed to dampen spirits
fair sex
fall on bad times
fall on deaf ears
far and wide
far be it from me
far cry
fateful day
fate worse than death
feedback loop
feel free to
feel vulnerable
festive occasion
few and far between
few well-chosen words
fickle finger of fate
final analysis
fine-tune one's plans
finishing touches
fit as a fiddle
floral tribute
food for thought
fools rush in
foregone conclusion
foul play
from the sublime to the ridiculous

gala occasion
generation gap
generous to a fault
gild the lily
give the green light to
glowing cheeks
go down the drain
goes without saying
goodly number
good team player
grateful acknowledgement
grave concern
green with envy
grim reaper
grind to a halt

hale and hearty
hands across the sea
happy pair
hastily summoned
have the privilege
heartfelt thanks/appreciation
heart of the matter
heart's desire
heated argument
heave a sigh of relief
height of absurdity
herculean efforts
hook, line, and sinker
hook or crook
hope for the future
hope springs eternal
hot pursuit
how does that grab you?
hunker down
hurriedly retraced his steps

ignominious retreat
ignorance is bliss
ill-fated
immaculately attired
immeasurably superior
impenetrable mystery
in close proximity
inextricably linked
infinite capacity
inflationary spiral
innocent bystander
in no uncertain terms
in our midst
in reference/regard to
in short supply
internecine strife
in the limelight
in the nick of time
in the same boat with

CLICHES

in the twinkling of an eye
in this day and age
into full swing
iron out the difficulty
irony of fate
irreducible minimum
irreparable/irreplaceable loss
it dawned on me

just deserts
just for openers

keep options open

labor of love
lashed out at
last analysis
last but not least
last-ditch effort
leaps and bounds
leave no stone unturned
leaves much to be desired
leave up in the air
lend a helping hand
let well enough alone
like a bolt from the blue
limped into port
line of least resistance
little woman
lit up like a Christmas tree
lock, stock, and barrel
logic of events
long arm of the law
long-felt need

make good one's escape
man the barricades
marked contrast
masterpiece of understatement
matter of life and death
mecca for travelers
method to/in his madness
milk of human kindness
miraculous escape
moment of truth
momentous decision/occasion
monumental traffic jam
moot point
more in sorrow than in anger
more sinned against than sinning
more than meets the eye
more the merrier
motley crew

narrow escape
nearest and dearest
needs no introduction
never a dull moment
never before in the history of
nipped in the bud
none the worse for wear
no sooner said than done
not wisely but too well

one and the same
ongoing dialogue
on more than one occasion
on unimpeachable authority
open secret
order out of chaos
other things being equal
outer directed
overwhelming odds
own worst enemy

pales into insignificance
paralyzed with fright
paramount importance
part and parcel
patience of Job
pay the piper
peer group
pet peeve
pick and choose
pie in the sky
pinpoint the cause
pipe dream
place in the sun
play hard ball
play it by ear
point with pride
poor but honest
powder keg
powers that be
pretty kettle of fish
pros and cons
proud heritage
proud possessor
pull one's weight

rack and ruin
ravishing beauty
red-letter day
regrettable incident
reigns supreme
reliable source
remedy the situation
right on
riot-torn area
ripe old age
round of applause
rude habitation

sadder but wiser
saw the light of day
scathing sarcasm
sea of faces
seat of learning
second to none
seething mass of humanity
select few
selling like hotcakes
shattering effect
shift into high gear
shot in the arm
sigh of relief
silence broken only by
silhouetted against the sky
simple life
skeleton in the closet
snug as a bug in a rug
social amenities
something hitting the fan
spectacular event
spirited debate
steaming jungle
stick out like a sore thumb
stick to one's guns
straight and narrow path
structure one's day
such is life
sum and substance
superhuman effort
supreme sacrifice
sweat of his brow
sweeping changes
sweet sixteen

take the bull by the horns
take up the cudgels
telling effect
tender mercies
terror stricken
thanking you in advance
there's the rub
this day and age
those present
throw a monkey wrench
throw a party
throw caution to the winds
thrust of your report
thunderous applause
tie that binds
time immemorial
time of one's life
tongue in cheek
too funny for words
too numerous to mention
tough it out/through
tower of strength
trials and tribulations
trust implicitly
tumultuous applause

uncharted seas
unprecendented situation
untimely end
untiring efforts
uptight

vale of tears
vanish into thin air
viable alternative

watery grave
wax eloquent/poetic
weaker sex
wear and tear
wend one's way
whirlwind tour
wide open spaces
words fail to express
word to the wise
work one's wiles
worse for wear
wrought havoc

COLONS

1. Colons link related thoughts, one of which must be capable of standing alone as a sentence.

Colons emphasize the second thought (unlike semicolons, which emphasize both thoughts equally, and dashes, which emphasize the break in the sentence and can emphasize the first thought).

Colons tend to throw emphasis forward: they tend to make the second thought the most important part of the sentence. When such is the case, the colon indicates that explanation or elaboration follows:

> The Franklin Shipyard needed one thing to remain solvent: to win the Navy's supercarrier contract.

> The Franklin Shipyard needed one thing to remain solvent: it had to win the Navy's supercarrier contract.

NOTE: The two complete thoughts in the second example could also appear as two sentences:

> The Franklin Shipyard needed one thing to remain solvent. It had to win the Navy's supercarrier contract.

However, linking these thoughts with a colon emphasizes their close connection. Writing them as two sentences is less emphatic if the writer wishes to stress that the one thing Franklin needs is to win the contract.

2. Colons introduce lists or examples:

> Our management development study revealed the need for greater monitoring during these crucial phases:
>
> 1. Initial organization
> 2. Design and development
> 3. Fabrication and quality control

> The Mars Division's audit of field service personnel centers revealed the following general deficiencies:
>
> 1. Service personnel do not fully understand the new rebate policy.
> 2. Parts inventories are inadequate.
> 3. The centralized customer records are not operational, although the computer terminals have all been installed.

NOTE: The items listed do not require periods unless they are complete sentences. (See LISTS.)

See PARALLELISM.

3. Colons separate hours from minutes, volumes from pages, and the first part of a ratio from the second:

> The deadline is 3:30 p.m. on Friday.

> See *Government Architecture* 15:233.

> The ratio of direct to indirect costs is 1:1.45.

4. Colons follow the salutation in a formal letter:

> Dear Ms. Labordean:

> Ladies and Gentlemen:

See LETTERS.

5. Colons separate titles from subtitles:

> *Government Architecture: Managing Interface Specifications*

COMMAS

1. Commas separate complete thoughts joined by these simple conjunctions: *and, but, or, for, nor, so, yet:*

He was a Russian linguist in communications intelligence, and he has logged over 5000 hours as a C-130 navigator in the Air Force.

We are known for our land-based missile delivery systems, but we also design and manufacture shielded fiber optics cables.

EXCEPTION: You may omit this comma if both complete thoughts are short:

The chairman resigned and the company failed.

The simple conjunctions cited above are called coordinate conjunctions. When they link two complete thoughts, the resulting sentence is called a compound sentence. (See CONJUNCTIONS and SENTENCES.)

NOTE: If you use any other transitional or connecting word (*however, furthermore, consequently,* and so on) to join two complete thoughts, use a semicolon. (See SEMICOLONS and TRANSITIONS.)

2. Commas separate items in a series consisting of three or more words, phrases, or even whole clauses:

Control Data's Integrated Support Software System provides compatibility between tools and workers, consistent tool interfaces, ease of learning, user friendliness, and expandability.

The user may also return to control program to perform such other functions as database editing, special report generation, and statistical analyses.

The Carthage-Hines agreement contained provisions for testing the Pennsylvanian sands, developing local permeability pinchouts, and exploring for undeveloped oil reserves in subthrust traps.

NOTE 1: A comma separates the last two items in a series although these items are linked by a conjunction (*and* in the above examples, but the rule applies for any conjunction). This comma was once considered optional, but the trend is to make it mandatory, especially in technical and business English. Leaving it out can cause confusion and misinterpretation.

NOTE 2: If all of the items in the series are linked by a simple conjunction, do not use commas:

The user may also return to control program to perform such other functions as database editing and special report generation and statistical analyses.

NOTE 3: In sentences containing a series of phrases or clauses that already have commas, use semicolons to separate each phrase or clause:

Our legal staff prepared analyses of the Drury-Engels agreement, which we hoped to discontinue; the Hopkinson contract; and the joint leasing proposal from Shell, Mobil, and Amoco.

See CONJUNCTIONS and SEMICOLONS.

3. Commas separate long introductory phrases and clauses from the main body of a sentence:

Although we are new to particle scan technology, our work with split-beam lasers gives us a solid experiential base from which to undertake this study.

For the purposes of this investigation, the weapon will be synthesized by a computer program called RATS (Rapid Approach to Transfer Systems).

Oil production was down during the first quarter, but when we analyzed the figures, we discovered that the production decline was due to only two of our eight wells.

NOTE: In the last example, the *when we analyzed* clause does not open the sentence, but it must still be separated from the main clause following it. It introduces the main thought of the last half of the sentence.

EXCEPTION: If the introductory thought is short and no confusion will result, you can omit this comma:

In either case the Carmichael procedure will be used to estimate the current requirements of the preliminary designs.

4. Commas enclose parenthetical expressions.

Parenthetical expressions are words or groups of words that are inserted into a sentence and are not part of the main thought of the sentence. These expressions describe, explain,

COMMAS

or comment upon something in the sentence, typically the word or phrase preceding the parenthetical expression:

> The transport will, according to our calculations, require only 10,000 feet of runway.
>
> The survey results, though not what we had predicted, confirm that the rate of manufacturer acceptance will exceed 60 percent.

Parentheses and dashes may also enclose parenthetical expressions. Use commas most of the time, but when you want to make the expression stand out, enclose it with parentheses (which are more emphatic than commas) or dashes—which are more emphatic than parentheses. (See PARENTHESES and DASHES.)

5. Commas separate nonessential modifying and descriptive phrases and clauses from a sentence, especially those clauses beginning with *who, which,* or *that*:

> These biocybernetic approaches, which merit further investigation, will improve performance of the man/machine interface.

In this sentence, *which merit further investigation* is not essential because the reader will already know which biocybernetic approaches the sentence refers to. The clause beginning with *which* is non-essential and could be left out:

> These biocybernetic approaches will improve performance of the man/machine interface.

If there were several biocybernetic approaches, however, and the writer needed to identify only those meriting further investigation, the clause would be essential, could not be left out, and would not take commas:

> Improving the performance of the man/machine interface meant identifying those biocybernetic approaches that merit further investigation.

The *that* in the preceding example commonly introduces essential clauses although *which* sometimes appears.

Modifying or descriptive clauses should always follow the words they modify. If they cannot be removed from the sentence without changing the meaning, they are essential and must not be separated by commas from the word they modify. If they can be removed, they are nonessential and must be separated by commas from the main thought in the sentence.

> *Essential*: She is the Dr. Gruber who developed analytical engine compressor stability models for NASA.

She is the Dr. Gruber does not make sense as an independent statement. The descriptive clause beginning with *who* is essential and therefore cannot be separated by a comma from *Gruber.*

> *Nonessential*: Our Design Team Leader will be Dr. Janet Gruber, who developed analytical engine compressor stability models for NASA.

Our Design Team Leader will be Dr. Janet Gruber does stand alone as a complete and independent thought. In this case, the descriptive clause beginning with *who* is nonessential. Separating it from *Gruber* with a comma shows that it is additional and nonessential information. Note that a comma would follow *NASA* if the sentence continued.

See PRONOUNS for a discussion of relative pronouns.

6. Commas separate two or more adjectives that equally modify the same noun:

> This configuration features an advanced, multimission payload capacity.

NOTE: If two or more adjectives precede a noun, however, and one adjective modifies another adjective—and together they modify the noun—you must use a hyphen.

A good test for determining whether two or more adjectives equally modify a noun is to insert *and* between them. If the resulting phrase makes sense, then the adjectives are equal, and you should use commas to replace the *and*'s:

> old and rusty pipe (*therefore* old, rusty pipe)
>
> *however*
>
> old and rusty and steam pipe (*The* and *between* rusty *and* steam *makes no sense. Therefore, the phrase should be:* old, rusty steam pipe.)

See HYPHENS and ADJECTIVES.

COMMAS

7. Commas separate items in dates and addresses:

The proposal was signed on March 15, 1985.

Contact Benson Aerodynamics, Lindsay, Indiana, for further information.

NOTE: A comma follows the year when the month and day precede the year. However, when the date consists only of month and year, a comma is not necessary:

The final report will be due January 15, 1985, just a month before the board meeting.

but

The final report will be due in January 1985.

When the date appears in the day-month-year sequence, no commas are necessary:

The report is due 15 January 1985.

8. Commas separate titles and degrees from names:

The chief liaison will be Roger Hillyard, Project Review Board Chairman.

Mary Sarkalion, PhD, will coordinate modeling and simulation studies.

Modeling and simulation studies will be the responsibility of Mary Sarkalion, PhD.

NOTE: When the degree or title appears in the middle of a sentence, commas must appear before and after it.

9. Commas follow the salutation in informal letters and the complimentary closing in all letters:

Dear Joan,

Sincerely,

See COLONS and LETTERS.

10. Commas set off the names of people addressed:

So, Bob, if you'll check your records, we'll be able to adjust the purchase order to your satisfaction.

11. Commas set off the following transitional words and expressions when they introduce sentences or when they link two complete thoughts: *accordingly, consequently, for example, for instance, further, furthermore, however, indeed, nevertheless, nonetheless, on the contrary, on the other hand, then, thus:*

Consequently, the primary difference between CDSP and other synthesis programs is development philosophy.

Synthesis programs are now common in industry; however, CDSP has several features that make it especially suitable for this type of study.

or

Synthesis programs are now common in industry; CDSP has, however, several features that make it especially suitable for this type of study.

NOTE: A few of these transitional words (*however, thus, then, indeed*) are occasionally part of the main thought of the sentence and do not form an actual transition. When such is the case, omit the punctuation before and after the words:

However unreliable cross-section analysis may be, it is still the most efficient means of scaling mathematical models.

Thus translated, the decoded message can be used to diagram nonlinear relationships.

12. Commas, like periods, always go inside of closing quotation marks. Commas go outside of parentheses or brackets:

The specifications contained many instances of the phrase "or equal," which is an attempt to avoid actually specifying significant features of a required product.

Thanks to the USGS (United States Geological Survey), we have an up-to-date water resources survey for Dade County.

See SPACING.

COMPOUND WORDS

Compound words are words formed when two or more words act together. The compound may be written as a single word (with no space between the joined words), with a hyphen between the joined words, or with spaces between the joined words:

footnote
ourselves
right-of-way
3-minute break
delayed-reaction switch
land bank loan
parcel post delivery

The form of the compound varies with custom and usage, as well as with the length of time the compound has existed.

Compound words usually begin as two or more separate, often unrelated words. When writers and speakers begin using the words together as nouns, verbs, adjectives, or adverbs, the compound generally has a hyphen or a space between words, depending on custom and usage. As the new compound becomes more common, the hyphen and space might drop, and the compound might be written as one word:

on-site *has become* onsite
co-operate *has become* cooperate
rail road *has become* railroad
auto body *has become* autobody

However, because of custom or usage, some compounds retain the hyphen or space between words:

all-inclusive
deep-rooted
living room
middle-sized
re-cover (*to cover again*)
re-create (*to create again*)
rough-coat (*used as a verb*)
sand-cast (*used as a verb*)
satin-lined
steam-driven

sugar water
summer time
terra firma
throw line
under secretary

Because new compound words are continually appearing in the language and because even familiar compounds may appear in different forms, depending upon how they are used in a sentence, writers may have difficulty deciding which form of a compound to use. Recent dictionaries can often help by indicating how a word or compound has appeared previously.

However, for new compounds and for compounds not covered in dictionaries, use the principles of clarity and consistency, as well as the following guidelines, to select the form of the compound.

1. Write compounds as two words when the compounds appear with the words in their customary order and when the meaning is clear:

test case	report card
sick leave	barn door
flood control	social security
real estate	civil rights

NOTE 1: Many such combinations are so common that we rarely think of them as compounds (especially because they do not have hyphens and are written with spaces between words). In many cases, writing them as a single word would be ridiculous: *floodcontrol, realestate.*

NOTE 2: We continue to pronounce such compounds with fairly equal stress on the joined words, especially when one or more of the words has two or more syllables (as in *social security*).

2. Write compounds as single words (no spaces between joined words) when the first word of the compound receives the major stress in pronunciation:

airplane
cupboard
doorstop
dragonfly
footnote
nightclerk
seaward
warehouse

NOTE 1: The stress often shifts to the first word when that word has only one syllable, as in the preceding examples.

NOTE 2: Words beginning with the following prefixes are not true compounds. Such words are usually written without a space or a hyphen:

*after*birth
*Anglo*mania
*ante*date
*bi*weekly
*by*law
*circum*navigation
*co*operate
*contra*position
*counter*case
*de*energize
*demi*tasse
*ex*communicate
*extra*curricular
*fore*tell
*hyper*sensitive
*hypo*acid
*in*bound
*infra*red
*inter*view
*intra*spinal
*intro*vert
*iso*metric
*macro*analysis
*meso*thorax

COMPOUND WORDS

*meta*genesis
*micro*phone
*mis*spelling
*mono*gram
*multi*color
*neo*phyte
*non*neutral
*off*set
*out*bake
*over*active
*over*flow
*pan*cosmic
*para*centric
*parti*coated
*peri*patetic
*plano*convex
*poly*nodal
*post*script
*pre*exist
*pro*consul
*pseudo*scientific
*re*enact
*retro*spect
*semi*official
*step*father
*sub*secretary
*super*market
*thermo*couple
*tran*sonic
*trans*ship
*tri*color
*ultra*violet
*un*necessary
*under*flow

NOTE 3: Words ending with the following suffixes are not true compounds. Such words are usually written without a space or hyphen:

port*able*
cover*age*
oper*ate*
plebis*cite*
twenty*fold*
spoon*ful*
kilo*gram*
geo*graphy*
man*hood*
self*ish*
meat*less*
out*let*
wave*like*
procure*ment*
partner*ship*
lone*some*
home*stead*
north*ward*
clock*wise*

3. Hyphenate compounds that modify or describe other words:

rear-engine bracket
tool-and-die shop
two-phase engine-replacement program
down-to-cost model

See HYPHENS and ADJECTIVES.

NOTE 1: Such compounds are hyphenated only when they come before the word they modify. If the words forming the compound appear after the word they are describing, leave out the hyphens:

bracket for the rear engine (*but* rear-engine bracket)
a shop making tools and dies (*but* tool-and-die shop)
a program with two phases (*but* two-phase program)

NOTE 2: When the meaning is clear, such compound modifiers may not need hyphens:

land management plan
life insurance company
per capita cost
production credit clause
speech improvement class

NOTE 3: Do not hyphenate if the first word of the compound modifier ends with *-ly*:

barely known problem
eminently qualified researcher
highly developed tests
gently sloping range

however

well-developed tests
well-known problem
well-qualified researcher

4. Treat compounds used as verbs as separate words:

to break down
to check out
to follow up
to get together
to go ahead
to know how
to run through
to shut down
to shut off
to stand by
to start up
to take off
to trade in

The parallel compound nouns are usually either written as one word or hyphenated:

breakdown
checkout
follow-up
get-together
go-ahead
know-how
run-through
shutdown
shutoff
standby
start-up
takeoff
trade-in

However, some verb phrases are identical to the compound noun form:

cross-reference (*both a noun and verb*)

When in doubt, check your dictionary.

CONJUNCTIONS

Conjunctions connect words, phrases, or clauses and at the same time indicate the relationship between them. Conjunctions include the simple coordinate conjunctions (*and, but, or, for, nor, so, yet*), the subordinate conjunctions (*because, since, although, when, if, so that,* etc.), the correlative conjunctions (*either . . . or, neither . . . nor, both . . . and*), and the conjunctive adverbs (*however, thus, furthermore,* etc.).

Coordinate Conjunctions

The simple coordinate conjunctions are *and, but, or, for, nor, so,* and *yet*. They often connect two independent clauses (or complete thoughts):

> The geologist analyzed the drill cores, and the engineer planned the future drilling operations.
>
> Our proposal was a day late, but we were not eliminated from competition.
>
> The pump will have to be replaced, or we will continue to suffer daily breakdowns.
>
> We rejected his budget, yet he continued to argue that all contested items were justified.

See SENTENCES.

These simple connectors establish the relationship between the thoughts being coordinated:

—*And* shows addition
—*Or* shows alternative
—*Nor* shows negative alternative
—*But* and *yet* show contrast
—*For* and *so* show causality

NOTE 1: When you use a coordinate conjunction to connect two independent clauses or complete thoughts, place a comma before the conjunction, as in the sentences above. (See COMMAS.) However, you may omit the comma when the two clauses are short and closely related. Also, a semicolon can replace both the comma and the conjunction. (See SEMICOLONS.)

NOTE 2: The conjunctions *and* and *or* (preceded by a comma) also connect the last two items in a series:

> The engineer designed an emergency exit door, a narrow outside stairway, and a concrete support pad.
>
> She requested full written disclosure, an apology, or financial compensation.

See COMMAS.

1. Ensure that in choosing *and* and *or* you select the conjunction that conveys exactly what you mean.

At first glance, these conjunctions merely join two or more items, but they can and often do imply much more:

<u>And</u>

In the following sentences *and* does more than merely connect the ideas. What *and* implies is stated in parentheses following each example:

> He saw the accident, and he called the police. (*therefore*)
>
> My boss is competent, and David is not. (*contrast*)
>
> He changed the tire, and he replaced the hub cap. (*then*)
>
> Explain the cost savings, and I'll approve your proposal. (*condition*)

<u>Or</u>

The conjunction *or* usually means one of two possibilities (*I want either a Ford or an Oldsmobile*). However, *or* sometimes has other, occasionally confusing, implications:

> The faulty part or the worm gear seemed to be causing our problem. (*Are the faulty part and the worm gear the same? Only knowledgeable readers would know for sure.*)
>
> Add to the bid, or I'll reject your offer. (*negative condition*)
>
> He began doing the schematics, or at least he appeared to be doing them. (*correction*)

2. Occasionally, sentences may begin with a coordinate conjunction.

This advice contradicts the rule that many of us learned in school: "Never begin a sentence with *and*." Some writers and editors still offer this advice, but most have now recognized that this so-called rule has no basis. Even Shakespeare began some of his sentences with coordinate conjunctions.

A coordinate conjunction at the beginning of a sentence links the sentence to the preceding sentence or paragraph. Sometimes, the linking is unnecessary:

We objected to the proposal because of its length. And others felt that it had errors in fact.

The *and* at the beginning of the second sentence is simply unnecessary. It adds nothing to the thought and may easily be omitted:

We objected to the proposal because of its length. Others felt that it had errors in fact.

Using a conjunction to begin a sentence is not grammatically incorrect. Sometimes, it is good stylistic variation. But it tends to look and sound informal, so avoid this practice in formal documents.

3. Do not use *and* or *but* before *which* (or *that*, *who*, *whose*, *whom*, *where*) unless there is a preceding parallel *which* (or *that*, *who*, *whose*, *whom*, *where*):

We explored the DeMarcus itinerary, which you explained in your letter but which you failed to mention in Saturday's meeting.

The meetings should take place where we met last year or where we can arrange for equally good facilities.

The following sentence violates this principle. Consequently, it is grammatically incorrect:

The plans called for a number of innovative features, especially regarding extra insulation, and which should save us much in fuel costs. (*Deleting the* and *would solve the lack of parallelism in this sentence.*)

See PARALLELISM.

Subordinate Conjunctions

In contrast to the limited set of coordinate conjunctions, subordinate conjuctions are a varied and diverse group:

after, although, as, because, before, if, once, since, that, though, until, when, where, while

in that, so that, such that, except that, in order that, now (that), provided (that), supposing (that), considering (that), as far as, as long as, so long as, sooner than, rather than, as if, as though, in case

if . . . (then)
although . . . yet/nevertheless
as . . . so
more/-er/less . . . than
as . . . as
so . . . (that)
such . . . as
such . . . (that)
no sooner . . . than
whether . . . or (not)
the . . . the

Subordinate conjunctions introduce subordinate or dependent clauses (clauses that do not convey complete thoughts and are therefore not independent):

After the engineer gave her talk
Because of the voltage loss
When the test results come in
While still producing fluids
In that you had already made the request
Except that the procedure was costly
Provided that you calculate the results
As though it hadn't rained enough
If we fail
As aware as he is, so
So expensive that it was prohibitive
Whether or not you submit the report

These phrases or dependent clauses must be attached to independent clauses (or complete thoughts) to form sentences:

After the engineer gave her talk, several colleagues had questions.

In that you had already made the request, we decided to omit the formal interview.

If we fail, the project stops. (*or* If we fail, then the project stops.)

As aware as he is, so he must be sensitive to the personnel problems.

See SENTENCES.

NOTE 1: A phrase or subordinate clause that opens a sentence should be followed by a comma. (See COMMAS.) The preceding sentences illustrate this rule.

NOTE 2: When the phrase or subordinate clause follows the independent clause or main thought of the sentence, no commas are necessary:

The experiment failed because of the voltage loss.

We would have denied the request except that the procedure was so costly.

We wondered whether you would turn in your report.

NOTE 3: Occasionally, the phrase or dependent clause interrupts the main clause and must have commas on both sides of it to indicate where the phrase or dependent clause appears:

CONJUNCTIONS

The President and the Joint Chiefs of Staff, after receiving the latest aerial reconnaissance photos of the area, decided on a naval blockade of all ports.

Our budgetary problems, regardless of the Madiera Project expense, would have taken care of themselves if the prime rate hadn't gone up three points.

4. Subordinate conjunctions can begin sentences:

When the test results come in, we'll have to analyze them carefully.

Because the project manager was unfamiliar with the budget codes, we failed to expense the costs of fabrication.

NOTE: The old school rule, "Never begin a sentence with *because*," was and remains a bad rule. You may begin a sentence with *because* as long as the dependent clause it introduces is followed by an independent clause or complete thought.

5. Distinguish between some subordinate conjunctions that have overlapping or multiple meanings (especially *because/since/as* and *while/although/as*).

Avoid using *since* and *as* to mean "because":

Because the Leiper Project failed, several engineers were reassigned to electro-optics. (*not* Since the project failed . . .)

Because we had ample supplies, no new batteries were ordered. (*not* As we had ample supplies . . .)

(Avoid using *while* and *as* to mean "although":

Although many employees begin work at 8 a.m., others begin at 7 a.m. (*not* While many employees begin work at 8 a.m. . . .)

Although the value of the test results declined, we still felt we could meet the deadline. (*not* As the value of the test results declined . . .)

Correlative Conjunctions

Correlative conjunctions are pairs of coordinate conjunctions:

both . . . and
either . . . or
neither . . . nor
not only . . . but also

6. Make the constructions following each coordinate conjunction parallel:

The committee was interested in both real estate holdings and stock investments. (*not* . . . both in real estate holdings and the stock investments.)

The investigation revealed that either the budget was inaccurate or our records had gaps. (*not* The investigation revealed either that the budget was inaccurate or our records had gaps.)

NOTE: Faulty parallelism problems occur when the same phrase structure or word patterns do not occur after each coordinate conjunction:

He was aware that not only was the pipe too small but also that the pipe supports were made of aluminum instead of stainless steel.

This sentence is confused because the two *that*'s are not parallel. The first *that* comes before *not only* while the second *that* comes after *but also*.

A parallel version of the sentence is much smoother:

He was aware not only that the pipe was too small but also that the pipe supports were made of aluminum instead of stainless steel.

See PARALLELISM.

Conjunctive Adverbs

Conjunctive adverbs are adverbs that function as conjunctions, typically by connecting independent clauses or complete thoughts. The most common conjunctive adverbs are *accordingly, also, besides, consequently, further, furthermore, hence, however, moreover, nevertheless, otherwise, then, therefore, thus*, and *too*.

See TRANSITIONS.

7. Conjunctive adverbs used to join two complete thoughts must be preceded by a semicolon and followed by a comma:

Aircraft assembly is a lengthy production process; however, the individual assembly steps must still be tightly controlled.

Increasing pressure in the T-valves is potentially dangerous; nevertheless, we will not be able to monitor effluent discharge without doing so.

See SEMICOLONS.

NOTE: You can omit the comma following the conjunctive adverb if the sentence is short:

I think; therefore I am.

CONJUNCTIONS

8. Conjunctive adverbs at the beginning of a sentence are usually followed by a comma:

> Therefore, I am recommending that Osage abandon plans to build another coal-fired generator.

> However, sulfur compounds may not be the answer either.

NOTE 1: You may omit this comma if the sentence is short:

> Thus the plan failed.

NOTE 2: If the adverb appears at the beginning of the sentence but does not behave as a conjunction, it is part of the sentence and cannot be followed by a comma:

> However we examined the problem, we could not resolve the fundamental dispute between offshore drilling companies and the leaseholders' association.

> Then the seam split at the forward discharge valve, and the boiler lost pressure rapidly.

DASHES

Dashes are excellent devices for emphasizing key material and for setting off explanatory information in a sentence. They can also be used to indicate where each item in a list begins and to separate paragraph headings from succeeding text. (See HEADINGS, LISTS, and PUNCTUATION.)

You can create dashes on a typewriter by typing two unspaced hyphens. When you use a dash between two words, leave no space on either side of the dash. (See SPACING.)

1. Dashes link introductory or concluding thoughts to the rest of the sentence.

Dashes linking thoughts emphasize the break in the sentence. Dashes often make the first thought the most important part of the sentence:

> Winning the Navy's supercarrier contract—that's what the Franklin Shipyard needed to remain solvent.

Dashes can act like colons, however, and throw emphasis forward:

> We subjected the design to rigorous testing—but to no avail because stress, we discovered, was not the problem.

Often, the information following the dash clarifies, explains, or reinforces what came before the dash:

> We consider our plan bold and unusual—bold because no one has tried to approach the problem from this angle, unusual because it's not how one might expect to use laser technology.

Dashes can also link otherwise complete sentences:

> The technical problem was not the design of the filter—the problem was poor quality assurance.

2. Dashes interrupt a sentence for insertion of thoughts related to, but not part of, the main idea of the sentence:

> The F-18 had been in the design phase—airfoil studies were being done by Barnett Industries—for 6 years before the Air Force canceled its contract.

In this case parentheses could replace the dashes; with parentheses, the sentence becomes slightly less emphatic. (See PARENTHESES.)

3. Dashes emphasize explanatory information enclosed in a sentence:

> Two of Barnett's primary field divisions—Industrial Manufacturing and Product Field Testing—will supervise the construction and implementation of the prototype.

In this case, commas or parentheses could replace the dashes. The commas would not be as emphatic as dashes; the parentheses would be more emphatic than commas, but less emphatic than dashes. (See PARENTHESES and COMMAS.)

4. Dashes link particulars to a following summary statement:

> Reliability and trust—this is what Bendix has to offer.

> Developing products that become the industry standard, minimizing the risk of failure, and controlling costs through aggressive management—these have become the hallmarks of our reputation.

DECIMALS

Decimal numbers are a linear way to represent fractions based on multiples of 10. The decimal *0.45* represents the following fraction:

45/100

See FRACTIONS.

The decimal point (or period) is the mark dividing the whole number on the left from the decimal fraction on the right:

504.678

In many foreign countries, writers use a comma for the decimal point:

504,678

---

1. Use figures for all decimals and do not write the equivalent fractions:

4.5 (*not* 4 5/10)
0.356 (*not* 356/1000)
0.5 (*not* 5/10)
0.4690 (*not* 4690/10000)

---

2. If the decimal does not have a whole number, insert a zero before the decimal point:

0.578 (*not*.578)
0.2 (*not*.2)

NOTE: This rule has a few exceptions, including:

Colt .45
A batting average of .345
A probability of p =.07

---

3. Retain the zero after the decimal point or at the end of the decimal number only if the zero represents exact measurement (or a significant digit):

0.45 *or* 0.450
28.303 *or* 28.3030

NOTE: Also retain the final zero in a decimal if the zero results from the rounding off of the decimal:

23.180 *for* 23.1789 (*if the decimal number is supposed to be rounded to three digits in the decimal fraction*)

---

4. Use spaces but not commas to separate groups of three digits in the decimal fraction.

In the metric system, the decimals may be broken into groups of three digits by inserting spaces:

56.321 677 90
707.004 766 321

but 567.4572 (*not* 567.457 2)

You can use commas to separate groups of three digits that appear in the whole number part of the decimal:

56,894.65
500,067.453 467

However, do not use commas to separate groups of three digits in the decimal fraction:

4.67234 (*not* 4.672,34)
2344.000 567 (*not* 2344.000,567)

See METRIC SYSTEM.

---

5. In columns, line up the decimal points:

```
 56
  0.004
115.9
 56.24445
  0.6
```

NOTE: Whole numbers without decimals (*e.g., 56* above) do not require a decimal point.

---

6. Do not begin a sentence with a decimal number:

this

The timer interrupts the processor 14.73 times a second.

not this

14.73 times a second the timer interrupts the processor.

See NUMBERS.

EDITING AND PROOFREADING SYMBOLS

Writers, editors, secretaries, reviewers, proofreaders, typesetters, printers—those people who work with documents must have a system for indicating changes to a text.

The standard system of editing and proofreading symbols (listed in most dictionaries) is far more complex than most of us need because it includes symbols that were developed to allow copyeditors, typesetters, printers, and others involved in publishing to make minute corrections to a text about to be printed.

The editing and proofreading symbols listed below constitute a simplified set of the standard symbols. This simplified set addresses the needs of most business and technical writers who must communicate suggestions and editorial corrections to writers, reviewers, or secretaries in a business environment. If you are interested in the complete set of proofreading symbols, see *The Chicago Manual of Style*, 13th Edition, p. 94, or the *United States Government Printing Office Style Manual* (March 1984), p. 5.

The example in the box illustrates the simplified method of editing and proofreading a short piece of text. This example also follows the rules cited below.

1. Use consistent proofreading symbols to indicate changes or corrections to text:

- Delete or take out.
- ∧ Insert a phrase, word, or punctuation mark.
- ∾ Transpose letters, words, or phrases.
- ⊐ Move to the right.
- ⊏ Move to the left.
- ≡ Use capital letter(s).
- / Use lowercase letter(s).
- ⁀‿ Close up a space.
- # Add a space.
- ¶ Make a new paragraph.

NOTE 1: Professional proofreaders sometimes use a different symbol in the margin than they use in the text. For instance, the # sign in the margin indicates that a space should be added. In text a slash mark indicates where the space should be added:

The incorrect/proposal

NOTE 2: In addition to the proofreading symbols shown above, some reviewers also use the symbol *sp* to indicate a spelling error. Most editors would correct a spelling error by inserting, deleting, or transposing letters, but if the reviewer wants the author to make the corrections, then indicating spelling errors with *sp* is helpful.

Original

Writers and Secrtaries of word pro cessing specailists have to agree on what to use when editingand proofreading drat materials. without such an agreement and a consistent convetion, erros kreep in and quality writing is impossible.

Corrected

Writers and secretaries or word processing specialists have to agree on what symbols to use when editing and proofreading draft materials.

Without such an agreement, errors creep in and quality writing is impossible.

EDITING AND PROOFREADING SYMBOLS

2. Use marginal marks to indicate corrections made within lines.

Changes to a text are sometimes difficult to see, particularly those changes made in pencil or black ink, which readers may have trouble distinguishing from surrounding print. To highlight changes or corrections, you should use a red or green pencil for changes. Even the change in color is sometimes difficult to see, however, particularly for color-blind reviewers or secretaries.

So indicate changes by marking the change within the line but also using marginal marks to show that a change appears on the line beside the mark. Professional proofreaders use marginal proofreading symbols to highlight changes, but even a simple check mark or an *X* can be effective in catching the reader's eye.

Be consistent, whether you use standard marginal proofreading symbols or simply a checkmark to indicate that a change has occurred in the line beside the check mark.

3. Use different colors of ink for different proofreadings (either by the same person or several people).

A text going through multiple revisions can become difficult to decipher if the different revisions are not somehow indicated. A very good system is to change the color of the reviewer's or proofreader's pencil (as in the example below).

The first reviewer might indicate changes with a red pencil, the second reviewer with a green pencil, the third with a blue pencil, and so on. The color of the suggestion thus indicates when and by whom the suggestion was made. This system is particularly effective during peer review.

An Example of Multiple Revisions

The Grayson plant operated by Mogo recovers almost all of the propane, butane and gaoline, but no ethane and the rsidue gas is sold to TransState Pipeline Co. TPS processes the the residue stream and recovers most the ethane and remaining NGL'S. TPS purchases th egas at the Grayson Plant outlet and then transmits it some 6 miles to its procesing plant near Abilene, TX. Once there, the residue steam is processed within some thirty-six hours and the resulting products are sold both to other companies although TPS does ships some of the products to its chcmical nearby subsidary. The TPS operation clearly compliments the Mogo operation at Grayson, so we should consider bidding on the TPS facilities (if, of course, the price is reasonable. Actually, we've heared rumors that TPS is interested in selling.).

ELLIPSES

An ellipsis, which consists of three spaced periods (. . .), indicates omissions, primarily in quoted material. Ellipses are the opposite of brackets, which indicate insertions in quoted material. (See BRACKETS.)

1. Use an ellipsis within quoted material to indicate omissions of words, sentences, or paragraphs:

"Labor costs . . . caused an operating loss for January of nearly $10,000."

Original: Labor costs, which our executive committee has been studying, caused an operating loss for January of nearly $10,000.

"No tax increases for 1986 . . . will occur."

Original: No tax increases for 1986 in personal withholding will occur.

NOTE 1: If omitted material comes at the beginning of a sentence, the quoted material opens with an ellipsis, especially if the material appears to be a complete sentence:

" . . . the printed budget will remain unchanged."

Original: Despite a few inconsistencies, the printed budget will remain unchanged.

NOTE 2: If the omitted material comes at the end of a sentence, the quoted material ends with an ellipsis plus the ending punctuation of the sentence:

"The Department of Energy denied our request for an energy subsidy . . . ."

Original: The Department of Energy denied our request for an energy subsidy even though we felt our request would be cost effective.

NOTE 3: Some authorities prefer to have no space following the last word in the sentence: *"energy subsidy. . . . "* (See SPACING.)

NOTE 4: Instead of spaced periods (. . .), the *United States Government Printing Office Style Manual* recommends asterisks (* * *). However, this practice rarely occurs outside of printed federal government materials. Even there, use of asterisks is not consistent.

2. Do NOT use an ellipsis to omit words if such omissions change the meaning or intent of the original quotation:

Chairman James Aubrey indicated that financing the debt load would . . . seriously undermine efforts to recover delinquent loans.

Original: Chairman James Aubrey indicated that financing the debt load would not detract from or seriously undermine efforts to recover delinquent loans.

3. Do not use an ellipsis to open or close a quotation if the quotation is clearly only part of an original sentence:

this

He kept returning to "the unexplored option," as the report termed it.

not this

He kept returning to " . . . the unexplored option . . . ," as the report termed it.

---

this

We discussed the "three legal loopholes" mentioned in the last Supreme Court decision on school busing.

not this

We discussed the " . . . three legal loopholes . . . " mentioned in the last Supreme Court decision on school busing.

4. Use a line of spaced periods to indicate that one or more entire lines of text are omitted:

Friends, Romans, countrymen, lend me your ears;
I come to bury Caesar, not to praise him.
. . . .
He was my friend, faithful and just to me:

NOTE 1: The line of periods does not tell a reader how much was omitted. The writer is responsible for retaining the intent and meaning of the original material.

NOTE 2: Poems and other long quotations do not require quotation marks. Instead, indentation and extra lines above and below the quoted material indicate that it is a quotation. (See QUOTATION MARKS.)

ELLIPSES

5. Use an ellipsis to indicate omitted material in mathematical expressions:

$a_1, a_2, \ldots, a_n$

$1 + 2 + \ldots + n$

See MATHEMATICAL NOTATION.

6. Use an ellipsis to indicate faltering speech:

I protest . . . or maybe I should only suggest that you have made a mistake.

I wonder . . . perhaps . . . if . . . that is a wise choice.

EMPHASIS

Effective writers **emphasize** important ideas by ensuring that they are more prominent than ideas, data, or details of lesser importance. Effective writers control their readers' eyes and minds. Effective writers don't just write documents—they design them.

Effective business and technical writing is always emphatic.

Emphasis applies to all levels of writing—from the layout and structure of an entire document to individual sentences. On any level, you emphasize words and ideas by manipulating **position** and **appearance.**

Position refers to the placement of words within a sentence, paragraph, or section. Appearance refers to the layout of ideas on a page and to the physical character of the words or ideas: spacing, indentation, boldface type, underlining, type size, type style, etc.

See BOLDFACE and SPACING.

1. Open and close with your important ideas.

The beginning and the ending of writing structures are more prominent than the middle.

The first and last words in a sentence receive the greatest emphasis. The opening and closing sentences in a paragraph are more prominent, as are the opening and closing paragraphs in a section or subsection.

Therefore, emphasize your important ideas by placing them in these stronger positions. Because readers encounter the opening first, always try to begin with the most important ideas.

Your details—the data, explanation, support, elaboration—belong in the middle.

In longer paragraphs, in sections, and in documents, end with important ideas, even if you repeat something stated earlier.

See ORGANIZATION.

2. Subordinate minor ideas.

Your major ideas deserve more attention. Therefore, you should subordinate minor information by using less space to discuss it or by placing the minor information in an appendix.

The minor information can include raw data; lengthy but relatively unimportant discussions of systems, techniques, or processes; and routine explanations of matters related to but less important than your central ideas.

Do not spend inappropriate amounts of time discussing unimportant matters. If you do, you throw the document out of focus.

See APPENDICES/ATTACHMENTS.

EMPHASIS

3. Repeat important ideas.

Repetition is emphatic because it reinforces the idea in the reader's mind. However, repetition may be either ineffective or effective.

Ineffective repetition occurs when writers repeat something too quickly—without sufficient intervening discussion—and when they use the same combination of words to express an idea.

Effective repetition occurs when an important idea appears in several different forms—typically at the beginning and ending of the section or paragraph concerning the idea. The first occurrence of the idea is introductory; the next occurrence is summary. Between them is explanation or elaboration, example, description, definition, or proof.

See REPETITION and KEY WORDS.

4. Use space to isolate important ideas.

Leaving more space around important ideas makes the page look less cluttered and also draws attention to the ideas. Instead of double spacing between paragraphs, leave three spaces. Or center an important idea in the middle of a page and leave extra spaces on both sides of it.

5. Use headings to highlight information.

Headings stand out and are therefore always emphatic. Use standard headings or headlines—longer headings that summarize and announce conclusions, directions, accomplishments, and startling or unusual facts. Theme headings, which announce the major thrust of a section, are especially emphatic.

Headings always draw attention to the information that follows them. They allow readers to be selective in reading, and they allow them to set the document aside and return later to a specific section.

See HEADINGS and CAPTIONS.

6. Use lists to highlight serial information.

Lists are visually more emphatic than paragraphs. Typically, lists are indented, have space between each item, and use numbers, letters, bullets, or dashes to indicate where each item begins. Readers pay more attention to displayed (or vertical) lists than they do to standard text.

See LISTS.

7. Use visual aids to emphasize important ideas.

Visual aids are naturally emphatic. They draw the reader's eye simply because they are different from text. One of the best ways to emphasize information is to make it visual. Create charts, graphs, drawings, schematic diagrams, flowcharts, tree diagrams, illustrations, etc.

See VISUAL AIDS, CHARTS, GRAPHS, ILLUSTRATIONS, MAPS, PHOTOGRAPHS, and TABLES.

8. Use single-sentence paragraphs to emphasize ideas.

Single-sentence paragraphs are more emphatic because they are shorter. They demand less effort to read, so readers tend to pay more attention to them. Use them judiciously, however. Too many single-sentence paragraphs make a document look as if it has no paragraphs.

See PARAGRAPHS.

9. Use typographical features to emphasize words or sentences.

CAPITAL LETTERS, <u>underlining</u>, **boldface type**, *italics*, and other typographical features are more emphatic than normal text. Be careful not to overuse them. Too much typographical variation makes the text look chaotic.

See UNDERLINING, BOLDFACE, and ITALICS.

10. Use color to emphasize.

If the document is to be run on an offset press, use color to highlight. Printing all of the thematic headings or theme statements in dark blue, for instance, would make them stand out. Readers thumbing through the document and reading only the blue theme statements would glean most of the important information.

See VISUAL AIDS.

FALSE SUBJECTS

False subjects are pronouns like *it* and *there* that have no concrete antecedents; that is, the pronouns do not refer to anything real. They are abstractions.

False subjects often occur at the beginning of a sentence and displace the true subject:

> It is this phase that is important.

In this sentence, *it* seems to stand for *phase*, but replacing the pronoun with its apparent antecedent creates nonsense:

> This phase is this phase that is important.

The true subject of the sentence is *phase*. Beginning with the true subject creates a shorter, much crisper sentence:

> This phase is important.

False subjects can also appear within sentences:

> We decided that it was important for the costs to be explained.

The false subject *it* weakens the middle of the sentence and adds unnecessary additional words. The sentence is far stronger without the false subject:

> We decided that explaining the costs was important.
>
> *or*
>
> We decided to explain the costs.

1. Eliminate false subjects.

Whenever possible, eliminate pronouns that lack concrete antecedents. Getting rid of these false subjects makes your writing more concise and often clearer.

Sometimes, however, the false subject is necessary, as in the expression *it is raining*. The word *it* is an abstraction, but what could you say in its place? Ask yourself, what is raining? No other word or words will be as functional or as expedient as the false subject. In such limited circumstances as *it is raining* or *it is noon*, false subjects are acceptable. Otherwise, try to eliminate them.

Below are additional before-and-after examples of false subjects. Note how much simpler, more direct, and shorter the sentences are when we eliminate the false subjects along with any accompanying words:

> In designing a thermal protection system, it is possible to meet the 1,100 degree fire requirement yet not be reliable. (*20 words*)
>
> Thermal protection system designs can meet the 1,100 degree fire requirement without being reliable. (*14 words*)

> Within the family of hydrates, there are several solid metal oxides that both chemically combine with water (hydration) and mechanically retain water (capillary condensation). (*24 words*)
>
> Within the family of hydrates, several solid metal oxides both chemically combine with water (hydration) and mechanically retain water (capillary condensation). (*21 words*)

> It will also be possible, as the study proceeds, to identify and extract important performance degradations resulting from failure to improve a given technology. (*24 words*)
>
> As the study proceeds, we can also identify and extract important performance degradations resulting from failure to improve a given technology. (*21 words*)
>
> *or*
>
> As the study proceeds, identifying and extracting important performance degradations resulting from failure to improve a given technology will also be possible. (*22 words*)

> From table 5.3, it appears that the use of relatively compact heat-transfer surfaces in the 10- to 20-fins/in. range will provide the compactness necessary to achieve the Navy goal of reduced size and weight. (*34 words*)
>
> As table 5.3 suggests, using relatively compact heat-transfer surfaces in the 10- to 20-fins/in. range will apparently provide the compactness necessary to achieve the Navy goal of reduced size and weight. (*31 words*)

> There are five factors that influenced our decision to repartition the system. (*12 words*)
>
> Five factors influenced our decision to repartition the system. (*9 words*)

> It will be the responsibility of the team manager to ensure that the total required time-phased quantity and skill mix can be supplied from the onsite pool. (*27 words*)
>
> The team manager will be responsible for ensuring that the total required time-phased quantity and skill mix can be supplied from the onsite pool. (*24 words*)

FOOTNOTES

Footnotes are the most common method of citing sources, especially in the humanities. As their name implies, footnotes originally appeared at the bottom of a page, but now they usually appear at the end of a chapter or article, making typing and page layout easier.

Instead of using footnotes, writers in the physical and biological sciences usually cite the author and the date of publication by enclosing them within parentheses in the text. These citations are developed fully in bibliographies that appear at the end of the text. (See CITATIONS and BIBLIOGRAPHIC FORM.)

Underlining replaces italics in the following footnote examples. This underlining is often necessary because typewriters and some word processors cannot italicize words.

1. Use raised Arabic numerals immediately following a quotation or paraphrase to indicate that the quotation or paraphrase has a footnote:

Within your text you may have a quotation from a published book or article: "Writers should always use quotation marks for exact quotations."[1] Sometimes you may be paraphrasing someone's ideas.[2] In these cases, your footnote number should come as close to the idea as possible,[3] even if the particular sentence goes on to discuss a second source.[4] Naturally, in a normal document, you should avoid having footnotes after every sentence or phrase.

NOTE: The footnote number comes after all punctuation, except for a dash. Footnotes are numbered sequentially within a chapter of a book and within an article.

2. In the first footnote to a source, include the author or authors, the full title, complete publishing information, and the pages being referred to:

Book by one author

1. David G. Peters, The Energetic West (Los Angeles: The Peter Pauper Press, 1982), 90.

Book by two authors

2. Frank S. Sloan and Jane Seymour, The Road West in the 1850's (Salt Lake City: The Popular Press, 1976), 34-35.

Book by more than three authors; information from several pages

3. Ralph Davidson, et al., The Western Fault System (Omaha: University of Nebraska Press, 1968), 126-127, 175, and 189.

Journal article

4. Janice Wesley, "Metal Matrix Alignment in Fiber Production," Massachusetts Institute of Technology Journal, 16 (October 1981): 45-46.

Public document

5. U.S. Congress. Senate. Foreign Affairs Committee. Report on Two Chinas in the Coming Decade, 91st Cong., 2d sess., 1981 (Washington, D.C.: United States Government Printing Office, 1983), 187-188.

Dissertation or thesis

6. O. X. Jones, "The Influence of Congressional Resolutions on Trade with China: A Study of Inconsistencies" (Ph.D. dissertation, University of Maryland, 1982), 87-88.

Personal letter

7. Senator Frank Church of Idaho to O. X. Jones, 23 November 1975. Personal files of O. X. Jones, Portland, Oregon.

Interview

8. Sidney Sung, interview during the annual meeting of the American Political Association, Seattle, Washington, November 1983.

NOTE 1: Footnotes are similar in form to paragraphs. The first line is indented, and all items are punctuated as if the information were the first sentence in the paragraph. The items are usually separated by commas.

NOTE 2: Writers can add a comment or additional facts to a typical footnote:

8. Sidney Sung, interview during the annual meeting of the American Political Association. Seattle, Washington, November 1983. Mr. Sung, the cultural attache in San Francisco for the People's Republic of China, granted this interview with the understanding that all of his comments would be off-the-record until after the 1984 U.S. Presidential election.

NOTE 3: Superscripts are a common way to number footnotes presented either at the bottom of the page or at the end of the chapter or article:

[9]Jason K. Bacon, The Two-China Policy (New York: Columbia University Press, 1976), 85.

FOOTNOTES

3. In second and subsequent references, make footnotes brief. Generally, include only the author's last name and the page number of the material referred to:

Second or subsequent footnote for one author

10. Bacon, 56-57.

Second or subsequent footnote for two authors

11. Sloan and Seymour, 18-21.

Second or subsequent footnote in which two or more authors by the same last name have been mentioned in first footnotes

12. Bacon, Two-China Policy, 76.

NOTE: This convention for second and subsequent footnotes eliminates the need for such Latin abbreviations as *ibid.*, *op. cit.*, and *loc. cit.*, which make footnotes difficult to read.

FRACTIONS

Fractions are mathematical expressions for the quotient of two quantities: *5/4* or *1/2*. In these fractions, the slash mark means *5* divided by *4* and *1* divided by *2*. Strictly speaking, decimals are also fractions. (See DECIMALS.)

1. Spell out and hyphenate fractions appearing by themselves in ordinary text, especially if they are followed by *of a* or *of an*:

two-thirds of an inch (*not* 2/3 of an inch)
. . . decreased by one-third
one-half foot
one-fourth inch
one-tenth
one-hundreth of a mile
eighty-four one-thousandths
(better 0.084)

NOTE 1: The longer a fractional expression becomes, especially if whole numbers are involved, the more desirable it is to express the fraction in figures (or a decimal):

56/64
98/100 (*or* 0.98)
2 1/2 times (*or 2½ times)*
6 3/4 (*or* 6.75)
29 1/3

NOTE 2: Measurements, especially in scientific and technical documents, require figures:

1/3-foot step
1/2-inch pipe (*or ½-inch pipe*)
2/3-inch-diameter pipe
7 1/2 meters (*or 7½ meters)*
8 1/2-by-11-inch paper

See NUMBERS and HYPHENS.

NOTE 3: Express fractions in figures when they are combined with abbreviations or symbols:

5 1/4 V
34 1/3 km
8 1/2 hr
5 1/2" x 6 2/3"

See NUMBERS.

GOBBLEDYGOOK

Gobbledygook is the imaginative name for language that is so pompous, long-winded, and abstract that it is unintelligible. Some dictionaries trace the term to the verb *gobble*, describing the sounds made by turkeys, and it is tempting to believe that writers of gobbledygook resemble this Thanksgiving favorite. Actually, such writers are usually well intentioned. They might even take pride in writing what they consider to be sophisticated and complex language.

Perhaps the best way to appreciate gobbledygook is to read a couple of samples:

> This office's activities during the year were primarily continuing their primary functions of education of the people to acquaint them of their needs, problems, and alternate problem solutions, in order that they can make wise decisions in planning and implementing a total program that will best meet the needs of the people, now and in the future.
>
> Because the heavy mistletoe infestation in the Cattle Creek drainage area has rendered the residual timber resources useless for timber production, the ultimate goal is to establish a healthy, viable new stand of Douglas fir.

The average reader has to read these passages several times before beginning to decipher such nonsense. Why are the passages so difficult?

—Words and phrases are abstract. What does *alternate problem solutions* mean? Similarly, are *residual timber resources* the same thing as *trees*? If so, the writer should say *trees*. (See WORDY PHRASES and REDUNDANT WORDS.)

—Words and phrases are pompous sounding. Are the office's activities primarily their primary functions? Is a healthy timber stand different from a viable timber stand? If not, then the writer should stick with the simpler word: *healthy*.

—Sentences are long and clumsy. By themselves, the 57 words in the first paragraph would make reading difficult, but the clumsy phrasing makes the reading impossible. The 35 words in the second paragraph are closer to a reasonable number, but the writer delays the major thought in the sentence with a massive introductory clause (beginning with *because*). As written, the sentence demands that readers remember a long opening condition while they try to absorb the main thought. The sentence would be clearer if the main and introductory clauses were reversed.

See STRONG VERBS, ACTIVE/PASSIVE, and PARALLELISM.

How To Avoid Gobbledygook

1. Use concrete and specific words and phrases whenever possible:

not this

> The environmental effects, although extremely important, are often so subtle and so confounded with and perhaps complicated by other environmental effects, which are no less important, that we neither gain a recognition of nor fully learn to appreciate the climatic effects that in fact exist and the resulting advantages of properly recognizing the environmental conditions that are the result of the aforementioned environmental effects.

this

> The environmental effects are often so slight and so hard to distinguish from other effects that we fail to appreciate their impact on the climate. We may even fail to appreciate the importance of properly assessing their environmental impact.

---

not this

> In order to bring the proposed recreational plan to completion, to evaluate existing recreation site appurtenances and facilities, and to include applicable facilities such as tables, fireplaces, etc., in the proposed new recreational plan, it will be necessary to receive photographs of all current appurtenances and facilities located within the state park area.

this

> To complete the recreational plan, we will need pictures of all tables, fireplaces, and other existing camping facilities in the state park.

See WORDY PHRASES and REDUNDANT WORDS.

2. Avoid pompous words and phrases.

The word *appurtenance* in the preceding example is an excellent example of a pompous word. Most readers will not understand *appurtenance,* and forcing them to look up the word in a dictionary may not clarify the passage. Two recent desk dictionaries define *appurtenance* quite differently: "something added to another, more important thing; accessory" (*The American*

GOBBLEDYGOOK

Heritage Dictionary) and "an incidental right (as a right-of-way) attached to a principal property right and passing in possession with it" (*Webster's New Collegiate Dictionary*). Which meaning should readers choose? More to the point, why make them choose?

If a word is not in common usage, avoid it or use it in such a way that the context provides a definition. In the example above, *appurtenance* surely fails the test. Here are some other pompous words and phrases with possible substitutes in parentheses:

accordingly (so)
acquaint (inform, tell)
activate (start)
additional (more)
adhere (stick)
ameliorate (improve)
apprise (tell, inform)

cognizant (aware)
commence (begin)
compensation (pay)
component (part)
concur (agree)
configuration (shape, design)
conflagration (fire)
curtail (slow, shorten)

demonstrate (show)
descend (fall, climb down)
donate (give)

encounter (meet)
evacuate (leave, empty, clear)
exhibit (show)

fabricate (make)
factor (fact)
feasible (likely, possible)
fracture (break)
function (work, act)

implicate (involve)
impotent (weak)
incinerate (burn)
increment (amount, bit)
indubitably (doubtless, undoubtedly)
inform (tell)
in isolation (alone, by itself)
initiate (begin)

locality (place)
locate (find)

major (chief, main)
manifest (show)
manipulate (operate)
manufacture (make)
modification (change)
moreover (besides)

necessitate (compel)
necessity (need)

paramount (main, chief)
perspective (view)
phenomenal (unusual)
philosophy (belief, idea)
potent (strong)
practically (nearly, most, all but)
proceed (go)
purchase (buy)

ramification (result)
render (make)
request (ask)
reside (live)
residence (home)

sophisticated (complex)
spotlight (stress)
state (say)
stimulate (excite)
succor (help)
sufficient (enough)

thoroughfare (aisle, street)
terminate (end, fire)
transmit (send)

vacillate (waver)
veracious (true)
visualize (imagine, picture)

---

3. Make sentences direct and clear.

Sentence length is only one sign of complexity. A 10-word sentence can be unclear because it is poorly structured or contains abstract and pompous words. A 50-word sentence, on the other hand, may be clear and easily understandable. Generally, however, the longer a sentence becomes, the more complex its structure is likely to be and the more difficult it will be to read:

not this

A tax deduction can be claimed in respect to any person whom the individual maintains at his own expense and who is (1) a relative of his, or of his wife, and is incapacitated by old age or infirmity from maintaining himself or herself, or (2) his or his wife's widowed mother, whether incapacitated or not, or (3) his daughter who is resident with him and upon whose services he is compelled to depend by reason of old age or infirmity.

this

If you support a relative who is unable to work because of old age or poor health, you can claim a deduction of $XXXX. You can claim this deduction if you support your widowed mother or your spouse's widowed mother, whether or not she is able to work. If you support a daughter who lives with you because you or your spouse is old or in poor health, you can claim a deduction of $XXXX.

See STRONG VERBS, ACTIVE/PASSIVE, and PARALLELISM.

Gobbledygook and Jargon

Gobbledygook and jargon can both make reading difficult, but they are not the same.

Jargon chiefly includes terms known and used by a specific technical or professional group. Carpenters, for instance, have a number of jargon terms: *stud, joist, sill plate, header, cap plate, trip-L-grip,* etc. All such fields use a specialized vocabulary. (See JARGON.)

Gobbledygook can include terms that constitute technical jargon, but gobbledygook generally also includes

GOBBLEDYGOOK

nontechnical words that are simply unfamiliar, unnecessary, or too large. Using jargon is inappropriate only if readers will not comprehend it. Gobbledygook offends everyone. Good writing can include some jargon, particularly if the words are defined or understandable within the context, but good writing never includes gobbledygook.

A Final Illustration

Some 200 years ago, opponents of Benjamin Franklin argued that to vote a man had to own property. Franklin's supporters disagreed and stated their case as follows:

> It cannot be adhered to with any reasonable degree of intellectual or moral certainty that the inalienable right man possesses to exercise his political preferences by employing his vote in referendums is rooted in anything other than man's own nature, and is, therefore, properly called a natural right. To hold, for instance, that this natural right can be limited externally by making its exercise dependent on a prior condition of ownership of property, is to wrongly suppose that man's natural right to vote is somehow more inherent in and more dependent on the property of man than it is on the nature of man. It is obvious that such belief is unreasonable, for it reverses the order of rights intended by nature.

Franklin agreed with this argument but knew that people wouldn't be moved by such pompous oratory. So he explained his position as follows:

> To require property of voters leads us to this dilemma: I own a jackass; I can vote. The jackass dies; I cannot vote. Therefore the vote represents not me but the jackass.

GRAPHS

Graphs depict numerical data and are useful for showing trends, cycles, cumulative changes, relationships between variables, and distributions. They are not as effective as tables in providing precise data, but readers should be able to extract relatively accurate numerical data from the lines plotted.

Graphs are normally plotted on grid lines, with a horizontal axis (*x*-axis or abscissa) and a vertical axis (*y*-axis or ordinate). The axes may also be diagonal or radial. The grid lines are usually equally spaced in horizontal and vertical directions and reflect the numerical scales along each axis. However, grids may be irregular, reflecting logarithmic scales, for instance, or probability distribution curves (see figure 6).

Including grid lines on graphs was the status quo for many years. Recently, however, graphs have begun to appear without grid lines, principally because many graphs are now produced by computer (see figures 1 through 5). Omitting grid lines is now acceptable in many graphs. However, if readers will be expected to extract precise data from graphs, you should not omit grid lines unless you combine the graph with a table (see figure 5).

The most common graphs are line or coordinate graphs and logarithmic graphs. In this *Style Guide*, graphs are distinguished from charts, including bar, pie, and surface charts. For information on those visual forms, see CHARTS.

For general information on visual aids, see VISUAL AIDS.

See also ILLUSTRATIONS, MAPS, PHOTOGRAPHS, and TABLES.

For information on captions, see CAPTIONS.

1. Make graphs simple and easy to read.

Ensure that each graph has a single important point to make

GRAPHS

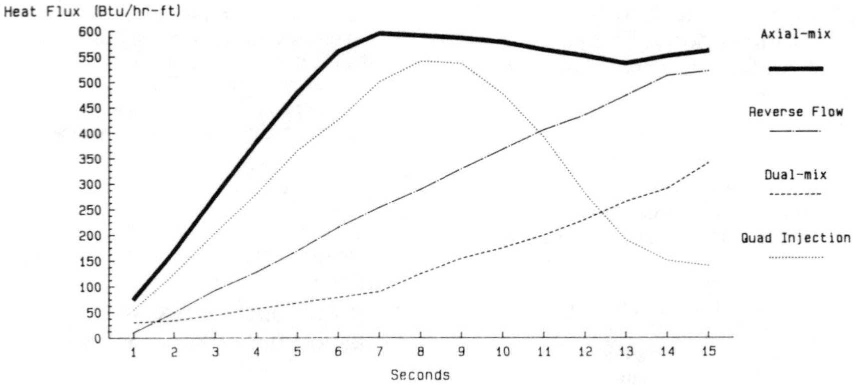

Figure 1. Of the four alternative combustor designs considered, the axial-transverse mix is the most efficient.

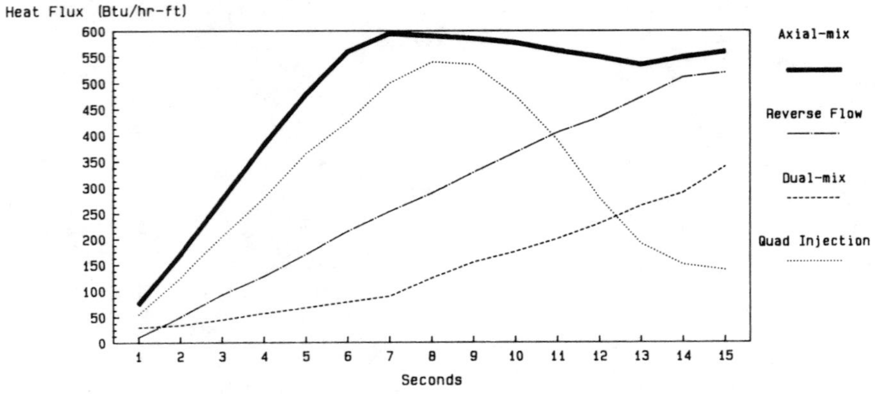

Figure 2. Of the four alternative combustor designs considered, the axial-transverse mix is the most efficient.

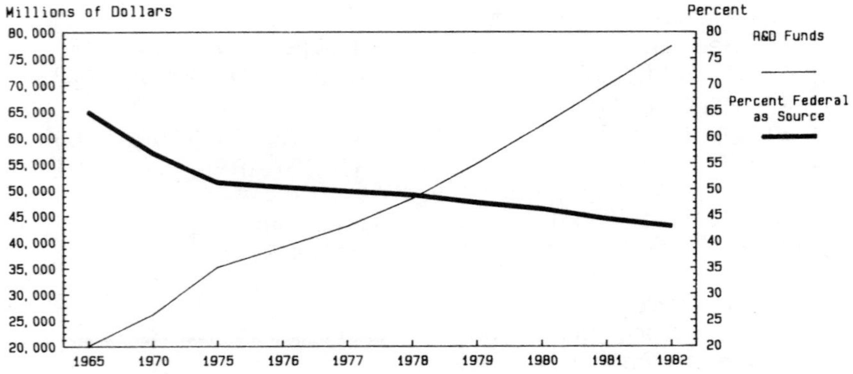

Source: Statistical Abstract of the United States (1982–83).

Figure 3. Federal support of research and development programs has steadily declined since 1965.

or a single relationship to show. Don't try to do too much with each graph. Complicated graphs are often confusing, even to technically competent readers.

Simplify graphs by eliminating anything that does not contribute to the central visual message of the graph, and do not use more labels, numbers, tick marks, or grid lines than necessary to do the job. Too much clutter makes graphs difficult to read.

2. Use accurate captions to identify the content and purpose of each graph.

Give each graph a figure number, and use accurate, specific captions to tell readers what the graph is about and how to read it. If at all possible, use action captions (see figures 1, 2, and 3). (See CAPTIONS.)

3. Label each axis and provide the appropriate scale for the data being plotted.

Use axis labels to identify each axis. Indicate the units of measurement in the axis label or by placing the units of measurement within parentheses after or below the axis label:

Wheat Shipments in Metric Tons

Wheat Shipments (Metric Tons)

Wheat Shipments
(Metric Tons)

See figures 1 through 4.

GRAPHS

If *time* is one variable, plot it on the abscissa (see figures 1, 2, and 3).

Scales should increase from bottom to top along the ordinate axis and left to right along the abscissa. Scale maximums and minimums should appear on the farthest grid lines along each axis. Scale labels should appear at appropriate intervals to facilitate data interpretation. See the next rule.

Generally, scales are indicated along the left and bottom axes; however, if you enclose the graph in a box (see figures 2 and 3) and the graph is unusually wide, you may label the ordinate scales on both left and right axes.

Occasionally, you may want to plot two or more lines that have different scales. If so, show the different scales on left and right ordinate axes (see figure 3). Use legends, axis labels, and different line patterns to show readers which line pertains to which scale. Avoid putting the different scales on the same left or right axis. Try to put different ordinate scales on opposite ordinate axes so that you emphasize their difference.

4. Ensure that scales accurately reflect the data being presented.

Do not use scales that exaggerate or distort the numerical relationships that actually exist. You can make small and insignificant differences look important by using a minute scale, and you can hide critical differences by using an overly large scale.

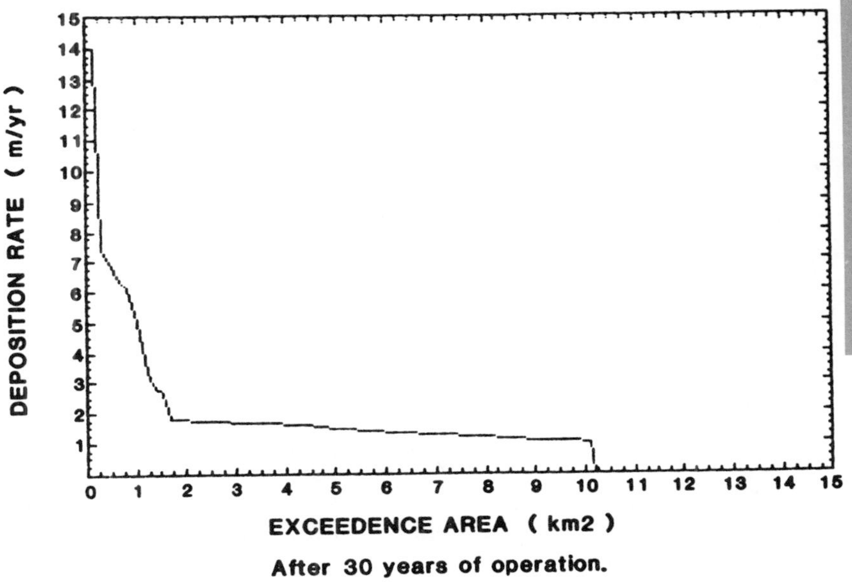

Figure 4. Deposition rates and exceedence areas for Boca de Quadra.

Your graphs should reflect the reality of the data being plotted. Therefore, your choice of scale is critically important.

5. Use tick marks to aid data interpretation.

Tick marks are short lines on and perpendicular to an axis that indicate the numerical interval along the axis. Tick marks reflect the axis scale.

Generally, you should have twice as many tick marks as scale labels (the numbers indicating the scale interval).

However, tick marks may be more numerous (four per scale label is acceptable; eight per label is less so). Avoid using more tick marks than necessary.

The scales (and tick marks) should be sufficiently detailed to allow readers to extract data values by determining where any data point on the plotted line falls on each axis. Most readers interpret data by placing a straight edge on the plotted line parallel to one axis and then reading the scale labels and tick marks along the perpendicular axis to determine the *x* or *y* value of points on the line.

The tick marks corresponding to scale labels should be twice as long as the tick marks that do not correspond to scale labels.

6. Make your key data lines heavier than axis and grid lines and less important data lines.

Axis and grid lines are less important than data lines and should therefore be thinner and lighter. The visual emphasis in the graph should be on the plotted (or data) lines, not on

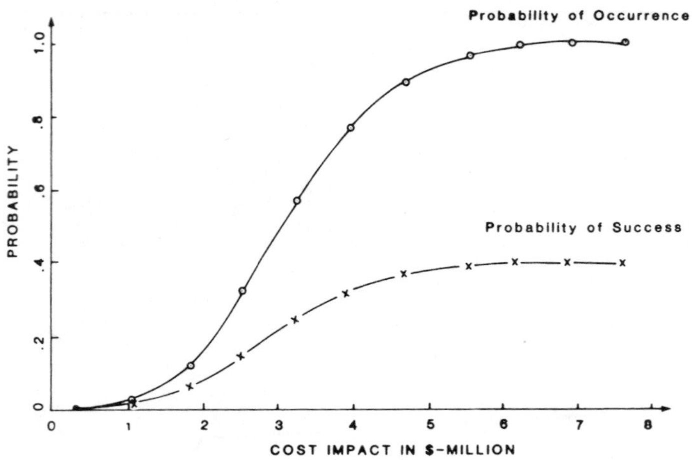

COST (K$)	PROBABILITY OCCURRENCE	CUMULATIVE PROBABILITY	PROBABILITY SUCCESS	CUMULATIVE PROBABILITY
0	0.000	0.000	0.000	0.000
724	0.004	0.004	0.001	0.001
1,445	0.022	0.026	0.014	0.015
2,172	0.094	0.120	0.045	0.060
2,896	0.202	0.322	0.086	0.146
3,620	0.243	0.565	0.097	0.243
4,344	0.200	0.765	0.076	0.319
5,068	0.128	0.893	0.046	0.365
5,792	0.066	0.959	0.019	0.384
6,516	0.027	0.986	0.007	0.391
7,239	0.010	0.996	0.002	0.393
7,963	0.003	0.999	0.000	0.393
8,687	0.000	0.999	0.000	0.393

Figure 5. Cost impact distribution.

the grid lines or axes. Equally, your important data lines should be more emphatic than less important data lines.

Note that in figures 1 and 2, the data line for the axial-mix configuration is much heavier than any of the other lines on the graph. Because the axial-mix configuration is the focus of the graph, its line thickness is greater. Similarly, in figure 3, the line representing declining federal support of R&D is shown in boldface, while the line representing total R&D funds is of a standard width.

7. Use multiple lines, if necessary, to show the relationships between three or more variables, and use different line patterns to depict different variables.

Lines of different patterns are useful for plotting more than two variables. Line patterns include solid, thin solid, wide solid, thin dashes, wide dashes, small dots, large dots, hollow (two thin lines together), and mixed dots and dashes of varying sizes. If you use different line patterns, include a legend to the right of or below the graph that identifies the lines. The legend should include an example of each line (see figures 1, 2, and 3).

Whether or not you use a legend, always clearly identify each line, as well as each axis and scale (see figures 1, 2, and 5).

8. Do not smooth out data lines.

Make graphs as realistic as possible, even if the data lines are "jerky" or erratic. The small but unsightly dips that make lines look ragged also make them look real. Data lines that are too smooth and polished look unreal.

9. Label important values on a data line.

If you wish to highlight or discuss certain important values on data lines, label them within the grid system. Readers will pay more attention to labeled data values and will not be forced to interpret data values (see figure 6).

To help readers interpret the data, you may also use markers on the plotted line to indicate plotted points (see figure 5). The types of markers include circles, dots, x's, dashes, and asterisks. You can also help readers interpret data by combining the graph with a table, as in figure 5.

10. If labels or numbers appear within the grid, blank out the grid area beneath the labels or numbers.

Grid lines can "mask" letters and numbers. So if your labels or numbers will appear within the grid, blank out the grid lines beneath the labels or numbers

GRAPHS

by creating a white rectangular area (see figure 6). Make these blank areas no larger than necessary and try to minimize your use of them within the same graph. Too many blank areas makes the graph look choppy and erratic.

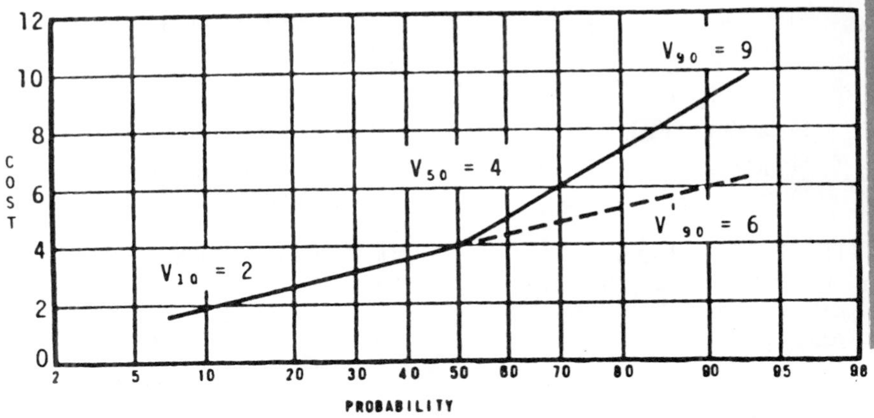

Figure 6. Element probability curve.

11. Orient all labels, numbers, and letters so that they are parallel with the horizontal axis.

Placing all lettering horizontal on the page makes graphs easier to read. The labels in figures 1, 2, and 6 are preferable to those shown in figures 4 and 5.

The only exception to this rule is long vertical labels. If your ordinate axis label is too long to write horizontally, then write it vertically with the base of the letters parallel and adjacent to the ordinate axis. Readers should be able to read long vertical labels by turning the page clockwise. (See CHARTS.)

12. Place source and explanatory information below the graph and flush left.

If you have taken the information for the graph from another source, indicate that you have by providing a source line.

Source and explanatory information belong below the grid area. If necessary, use superscripted numbers, letters, or symbols (such as *) to key data points, labels, or numbers to a footnote. Repeat the number or symbol in the footnote area (below and left of the graph) before providing sources or explanation. (See CHARTS and TABLES.)

HEADINGS

Headings label the text under them.

Headings show logical divisions in text.

Headings allow readers to skim.

Headings are like titles. They signal what is to come. They orient readers and thus make reading easier.

Headings reveal a document's organization, so readers can glance at the headings to determine where they need to read more closely and where they can skim. Headings are therefore especially useful for non-specialist readers who need to gain an overall sense of the text, but who do not have the time, interest, or knowledge to read the entire text.

---

1. Use headings whenever the text is longer than one single-spaced page or whenever the text conveys two or more major ideas.

Headings are among the most useful devices in writing. You should use them often enough (without overdoing them) to make the text as easy to read as possible.

---

2. Choose different levels of headings to indicate logical divisions and groupings in the text.

A chapter title, for instance, may be a major (or first-level) heading. Section titles in the chapter may be minor (or second-level) headings. Subsection titles could then be subheadings (third-level headings), and so on.

The level of the heading indicates its logical relation to other headings as well as to the whole. The levels are most apparent in an outline:

3.4 Component Descriptions

3.4.1 Gearbox Assembly
3.4.2 Brakes
3.4.3 Hydraulic Motors
3.4.4 AIU Interfaces

3.4.4.1 Controller
3.4.4.2 Encoder
3.4.4.3 Tachometer

3.4.5 Servo Valves

In both text and outlines (such as a table of contents), the levels may be indicated by a numbering system (like the one above) or by the placement, size, or appearance of the headings.

See OUTLINES and TABLES OF CONTENTS.

---

3. To show the level of the heading, use the three types of typographical variation: the placement of the heading on the page, the size of the type, and the appearance of the type.

Placement variations include centering, placing the heading flush left, indenting, and using a run-in heading (placing the heading on the same line as the text following it):

A CENTERED HEADING

A FLUSH LEFT HEADING

An Indented Heading

A Run-in Heading. The text begins following the period (or dash) and two spaces. Sometimes, a run-in heading has only three spaces between it and the succeeding text:

A Run-in Heading When you do not use punctuation (as in this case), you should make the heading visually distinct from the text in the rest of the paragraph. You might use boldface type or a larger typeface.

Size variations are possible when you can vary the point size of the lettering in the heading. The larger the point size, the higher the level of the heading. A 24-point heading is on a higher level than a 12-point heading:

A 24-point He

An 18-point Headi

A 14-point Heading

A 12-point Heading

Appearance variations include ALL CAPITAL LETTERS, underlining, **boldface type**, different type faces, and *italics*. (See EMPHASIS.)

Heading Levels

Use the following placement and appearance lists to create different levels of headings. The variables are listed in decreasing order of importance:

Placement

1. Centered
2. Flush left
3. Indented
4. Run-in (on the same line as text)

Appearance

1. ALL CAPITALS
2. Underlining
3. **Boldface**
4. *Italics*

As this first list shows, a centered heading is on a higher level than a flush left heading. A flush left heading, on the other hand, is on a higher level than an indented heading.

The appearance variations may be added singly or in combination. Thus, a heading may be in all capital letters, or it may be underlined, or both. Appearance variations used in combination create higher level headings than headings with a single appearance feature. So an underlined, all-capital-letter heading is on a higher level than a heading featuring only all capital letters.

Appearance, size, and placement variations used together allow writers many heading types—and consequently many heading levels.

The Number of Levels

How many levels should you use? Dozens are possible, but readers could not comprehend that many levels of subordination. Practically speaking, you should use no more than four to five levels, depending upon your readers. The less technical or less educated your readers, the fewer levels you should use. Experienced and well-educated scientific readers are more used to reading text with multiple levels of subordination. If you suspect that your readers will have trouble remembering the heading levels, use fewer levels.

Numbers with Headings

4. Use a numbering system with your headings if you have more than four levels of headings or if your document is lengthy and you or your readers will need to refer to sections of it by number.

Report writers normally construct their tables of contents by listing the headings and subheadings in their reports. If you use a numbering system with your headings, include the numbers in the table of contents.

See NUMBERING SYSTEMS, TABLES OF CONTENTS, and OUTLINES.

Effective Headings

5. Make your headings informative, specific, and inclusive.

Informative, specific headings allow readers to determine immediately the contents of a section. Unfortunately, many standard headings are neither specific nor informative: *Introduction, Discussion, Results*, and so on. Better versions of these would be as follows: *Purpose of the Drilling Proposal, Implications of Three Proposed Tests, Valid Data from the Third Test.*

Inclusive headings signal that only material mentioned in the heading will actually be covered in the section. If the heading is *Valid Data from the Third Test*, then no data from the second or first test should appear following the heading.

6. Make headings parallel in structure.

Parallel structure means that all headings at the same level have the same basic grammatical structure. So if one heading opens with an *-ing* word, the other headings will also:

this

Developing the Appropriate Tests
Sending Out for Bids
Selecting the Winner
Agreeing on Preliminary Contract Talks

not this

Developing Appropriate Tests
Sending Out for Bids
Selection of the Winner
Preliminary Contract Talks

See PARALLELISM.

Question Headings

Most headings are declarative. They state or announce a topic:

Facilities in the Local Impact Area
Summerhill Treatment Plant
Complex Traps in Gulf Coast Fields
The Geomorphology of Exploration

HEADINGS

An alternative heading form is the question heading:

How Great Is the Avalanche Danger?
Will Surface Water Quality Be Degraded?
What Are the Alternatives?
Should Exxon Proceed with the Project?
Will the Public Accept Our Position?
How Likely Is a Major Bank Default?

As long as you don't overdo them, question headings offer interesting possibilities and can be very effective. Question headings not only announce the topic but also stimulate interest because they pose a question that curious readers will want to see answered.

Question headings are generally more engaging than declarative headings because they seem to speak directly to the reader. However, be careful not to pose obvious or condescending questions:

Doesn't Everyone Know about Anticlines?

7. Use the same type of heading at each heading level.

Be consistent. If you decide to use question headings, use them for all headings at that heading level. Mixing question and declarative headings at the same level is confusing. However, you can use question headings at a major level and then use declarative headings at subordinate levels, or vice versa.

HYPHENS

Hyphenation is one of the trickier aspects of English. There are many rules of hyphenation—including some that apply only in limited circumstances—and all of the rules have exceptions. Below are the most common conventions of hyphen usage. For a more thorough discussion of hyphenation, refer to *The Chicago Manual of Style, the Gregg Reference Manual,* or the *United States Government Printing Office Style Manual.* (See REFERENCES.)

Hyphens Connect

Fundamentally, hyphens show a connection. Typically, the connection is between two words or between a prefix and a word. The connected words (known as compounds) can function as nouns, verbs, or adjectives:

Connected words as nouns

brother-in-law
ex-mayor
follow-up
foot-pound
know-how
run-through
self-consciousness
time-saver
two-thirds

Connected words as verbs

to blue-pencil
to double-space
to spot-check
to tape-record

Connected words as adjectives

all-around person
black-and-white print
coarse-grained wood
decision-making authority
even-handed person
half-hearted attempts
high-grade ore
high-pressure lines
interest-bearing notes
little-known program
long-range plans
low-lying plains
matter-of-fact approach
off-the-record comment
old-fashioned system
part-time employees
30-fold increase
three-fourths majority
twenty-odd inspections
up-to-date methods
well-known researcher

Unfortunately, not all connected (or compound) nouns, verbs, and adjectives require hyphens.

HYPHENS

Here are a few of the exceptions:

Connected but unhyphenated nouns

ball of fire
breakdown
fellow employee
goodwill
problem solving
quasi contract
takeoff
trademark
trade name

Connected but unhyphenated verbs

to downgrade
to handpick
to highlight
to proofread
to waterproof

Connected but unhyphenated adjectives

barely known researcher
bright red building
crossbred plants
halfhearted attempts
highly complex task
10 percent increase
twofold increase
unselfconscious person
very well known researcher
worldwide problem

As the above examples illustrate, connected words have three possible forms. They can appear as two separate words (*highly motivated*), as one word formed by connecting the two original words with a hyphen (*high-pressure*), and as one word formed by joining the original two words (*highbrow*). (See COMPOUND WORDS.)

Convention and tradition often dictate which form the connected words will take. If you are not sure which form is correct, refer to a recent dictionary.

Rules of Hyphenation

1. Hyphenate two or more words that act together to create a new meaning:

a counterflow plate-fin
the V-space between units
the Grumman F-14A airplane
one-half of the annular ring
to double-check the tests

This rule indicates a potential use of the hyphen, not a mandatory one. In some instances the two words become a single word, without a hyphen: *highlight, bumblebee, barrelhead*. In other instances, the two words remain separate: *base line, any one* (one item from a group), *amino acid*. The words sometimes remain separate because combining them would produce strange-looking forms: *aminoacid, beautyshop, breakfastroom*. Because the presence or absence of a hyphen is often a matter of convention, check a recent dictionary if you are not sure how the compound word should be written.

See CAPITALS for the proper capitalization of hyphenated words in titles.

2. Hyphenate two or more words that act together to modify another word:

brazed-and-welded construction
cross-counterflow unit
engine-to-recuperator mountings
full-scale testing
no-flow heat exchanger
pressure-drop decrease
3-year, multimillion-dollar program
12-foot-wide embayment
up-to-scale modeling
U-tube arrangement
well-documented success

This rule applies only when the connected or compound modifier occurs before the word it modifies. (See rule 4 below.)

3. Hyphenate compound numbers from twenty-one to ninety-nine and compound adjectives with a numerical first part:

thirty-four
eighty-one
five-volume proposal
13-phase plan
10-dollar fee
24-inch tape
500-amp circuit

4. Do not hyphenate connected words that function as adjectives if they occur after the word they modify:

The boiler was brazed and welded.
The compartment is 32 feet wide.
The program is well documented.

but

The brazed-and-welded boiler
The 32-foot-wide compartment
The well-documented program

5. Do not hyphenate connected words that act as adjectives if the first word ends in *-ly*:

highly motivated engineer
poorly conceived design
vastly different approach
completely revised program

NOTE: The words ending in *-ly* are actually adverbs. The *-ly* form indicates the structure of the modifying phrase, so a hyphen is unnecessary.

HYPHENS

6. Prefixes generally do not require hyphens:

counterblow
midpoint
nonperson
progovernment
supercar
undersea

NOTE 1: Hyphens do appear when the prefix precedes a capitalized word:

un-American
mid-August
non-Soviet

NOTE 2: Hyphens do connect some prefixes (especially those ending in a vowel) to words: *anti-inflationary, ultra-conservative*. If you are not sure whether a prefix requires a hyphen, refer to a recent dictionary or to the 1984 edition of the *United States Government Printing Office Style Manual*. (See COMPOUND WORDS.)

7. Hyphenate words that must be divided at the end of a line.

Words are always divided between syllables, and hyphens should appear at the end of the line where the word division has occurred. Try not to divide a word on the last line of a page.

Hyphens and Technical Terminology

The use of hyphens in technical expressions varies considerably. Some writers try to adhere strictly to the rules outlined above. But many, because of tradition, convention, or local preference, violate the rules of hyphenation when they believe that the technical expression will be clear:

We will need a high pressure hose.

In this sentence, *high* obviously modifies *pressure*. The sentence refers to a hose that is capable of withstanding high pressures. It is not a pressure hose that happens to be high (off the ground). Yet if we follow the rules of hyphenation strictly, the sentence should be:

We will need a high-pressure hose.

Hyphens are often omitted in technical expressions because the expressions are clear to technical readers without the hyphens. The context clarifies the expression. In many cases, however, missing hyphens can cause confusion or a complete lack of comprehension, as in this sentence from an aircraft maintenance manual:

Before removing the retaining pin, refer to the wing gear truck positioning actuator assembly schematic.

Nontechnical (or technical but unknowledgeable) readers can only guess which words are associated with which other words. Does *truck* link with *wing gear*, or does *truck* modify *positioning*? Hyphens would help clarify the modifier relationships:

wing-gear truck-positioning actuator assembly

If convention or tradition allows you to eliminate hyphens where they would normally appear in compound words, do so unless eliminating them will mystify some readers. If you are going to err, err on the side of caution. Proper use of hyphens will not baffle knowledgeable technical readers, and it will help those readers who are not familiar with a technical expression.

See ADJECTIVES.

ILLUSTRATIONS

Illustrations, diagrams, and drawings include a wide range of visuals whose purpose is to depict parts, functions, relationships, activities, and processes that would be difficult to describe in text. These visuals range from schematic drawings of electrical systems to assembly diagrams, and from pictorial schematics to equipment illustrations with exploded views.

Producing good illustrations almost always requires a professional graphic artist. This *Guide* will not discuss either the art or mechanics of creating effective illustrations, diagrams, and drawings. Instead, it will focus on how writers should conceive of and use these visuals and what writers can do to assist graphic artists.

The visual possibilities of illustrations are enormous. The examples shown in this section represent a very small part of what is possible. In designing illustrations, diagrams, and drawings, you should use your imagination and seek advice early in the writing process from a graphic artist. Consider the broad range of alternatives before settling on final designs or options.

For general information on visual aids, see VISUAL AIDS. See also CHARTS, GRAPHS, MAPS, PHOTOGRAPHS, and TABLES.

For information on captions, see CAPTIONS.

1. Use illustrations, diagrams, and drawings to visualize a system, process, or piece of equipment that would be difficult to describe in text.

Illustrations are very effective at showing views of objects or systems that do not exist (a drawing of a proposed tool), that are abstractions (organizational or functional systems), or that would be impossible to show otherwise (exploded views or cutaways).

Illustrations allow readers to see inside something that is sealed, to see opposite or hidden sides of something simultaneously, and to see, in close-up, the details of a small assembly that would otherwise not be visible while looking at the larger object that contains the small assembly (see figure 1).

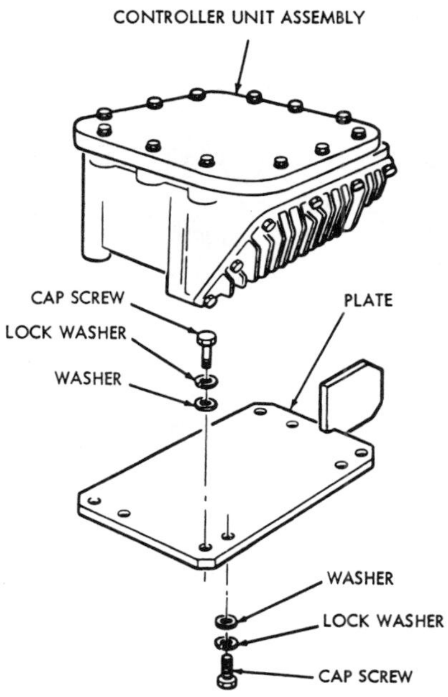

Figure 1. Attaching the controller unit to the plate is simple.

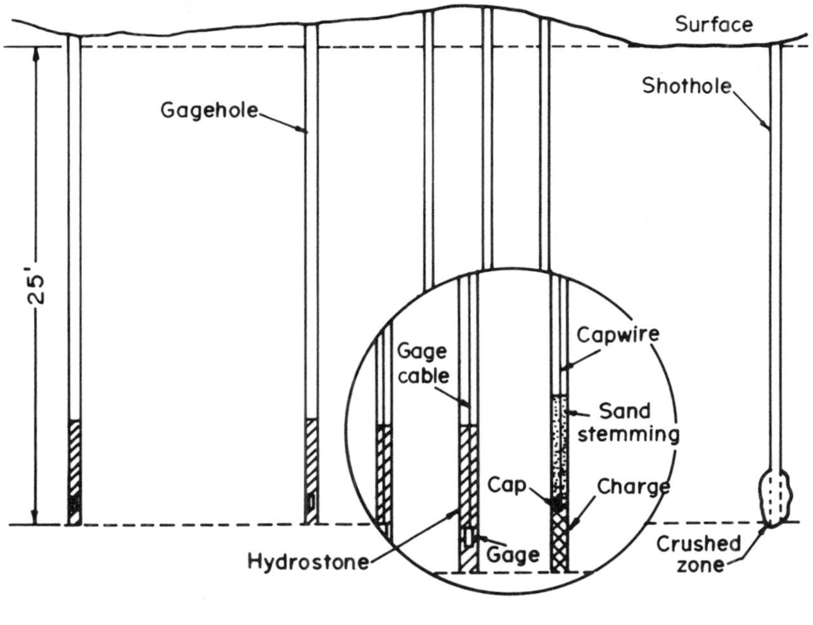

Figure 2. Gage and charge placement.

ILLUSTRATIONS

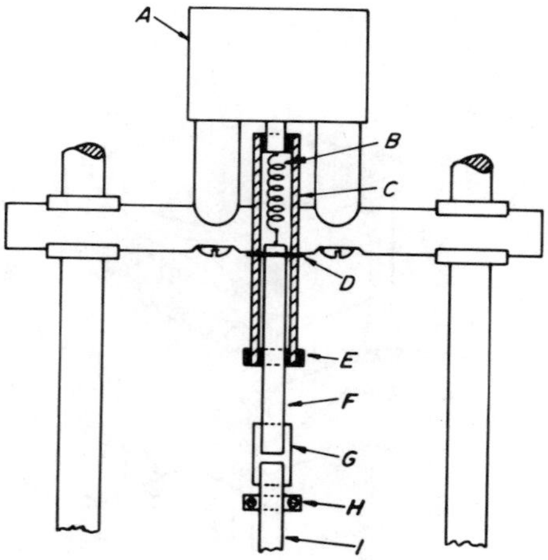

KEY

A Motor
B Spring
C Slotted tube
D Pin
E Collar
F Copper tubing
G Tygon tubing
H Ball bearing
I Stainless steel tubing

Figure 3. The activator assembly.

Exploded views allow you to exaggerate (or magnify) certain parts of an object to show the details of the exaggerated part while keeping the rest of the object in its correct perspective (see figure 2). Typically, exploded views accomplish two objectives simultaneously: They show how and where the magnified part fits in with the larger object, and they show how the magnified part is constructed. Cutaways are like an orange sliced in half: They show the internal structure (or assembly) of an object that is normally sealed (see figure 3).

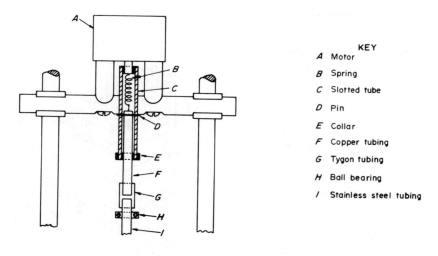

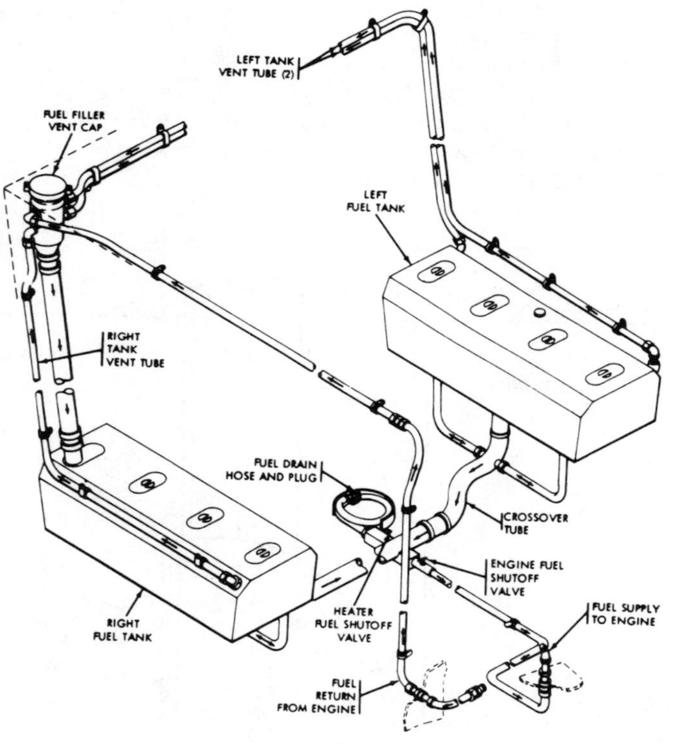

Figure 4. Our dual-tank fuel system features a fuel return line and redundant vent tubes.

2. Keep illustrations simple and give each one a perspective that enables readers to understand it.

Like other visual aids, illustrations and drawings should be focused. That is, they should clearly present a single central concept. Therefore, they should be clean and uncluttered. Everything not pertaining to the central concept should be eliminated. No detail should be present that does not contribute to the presentation of that single central idea.

As well as being simple, a good illustration has a clear perspective. Illustrations allow you to distort reality, so you must ensure that readers understand the perspective from which the illustration presents its subject. Illustrations almost always show their subjects out of context. Therefore, you may need to establish what the reader is viewing and how that thing relates to other things in its real environment.

ILLUSTRATIONS

Scales may be necessary if the size and relationship of the object depicted to other things in its environment are not clear. You can also use labels to indicate size, direction, orientation, and nomenclature (see figures 1 through 6). If you do not indicate size and distance relationships, readers may not be able to determine the correct proportions of the object shown or its correct orientation in the world outside of the illustration or drawing.

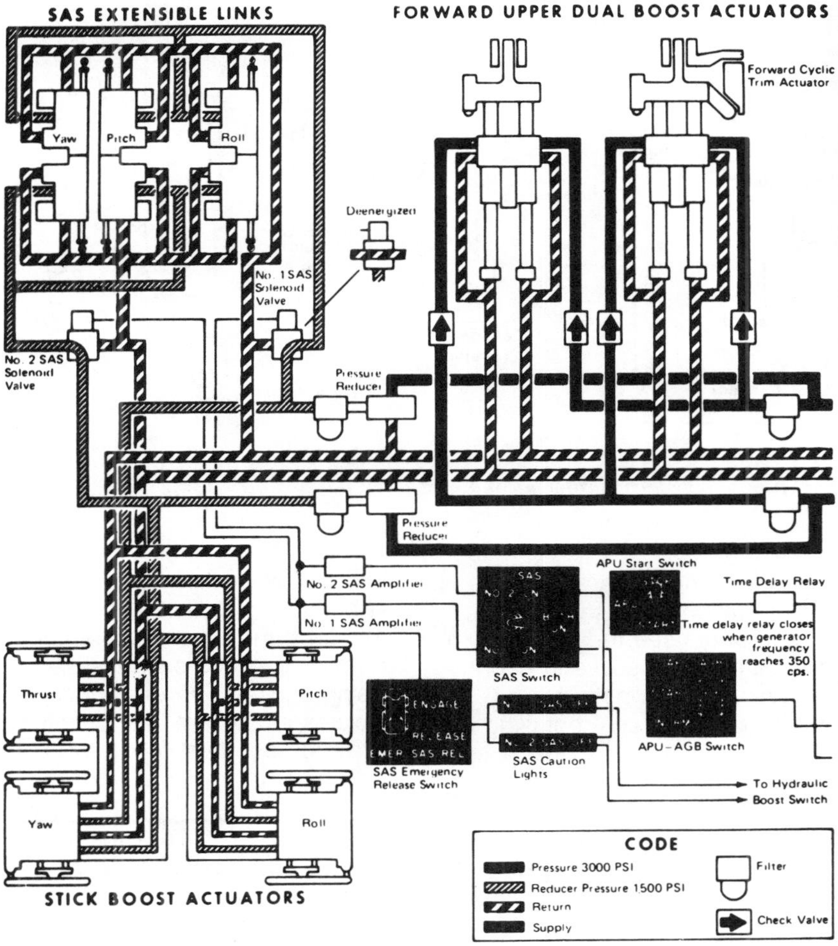

Figure 5. Flight control system actuators.

3. Label each illustration clearly and, if necessary, label the parts of the object shown.

Figure 1 shows a typical nomenclature illustration. The controller unit assembly is shown as one unit because it is not the focus of the illustration. The cap screws, lock washers, washers, and plate are the reason this illustration exists, so each is labelled separately. Lead lines drawn through the axis of each part show how the parts fit together.

Labelling of the significant parts of a drawing is crucial for reader comprehension. You may use word labels and arrows, as in figures 1, 2, and 4, or you may use numbers, letters, or symbols in the drawing itself (figure 3) with a legend or key.

4. Ensure that all letters, numbers, and labels are horizontally oriented on the drawing or illustration.

The language appearing on any part of a drawing or illustration should never be vertically oriented unless the bases of the individual letters or numbers are horizontal:

E
X
A
M
P
L
E

The lettering and numbering on an illustration should be oriented so that readers can read it without reorienting the illustration. If you run out of space, use arrows and move the labels away from the busy area of the illustration. If necessary, omit the labels and use letters, numbers, or symbols and a legend or key (see rule 3 above).

5. In a series of illustrations, make the viewing angle consistent.

If you are showing the same object in a series of illustrations, and the point of the series is to

ILLUSTRATIONS

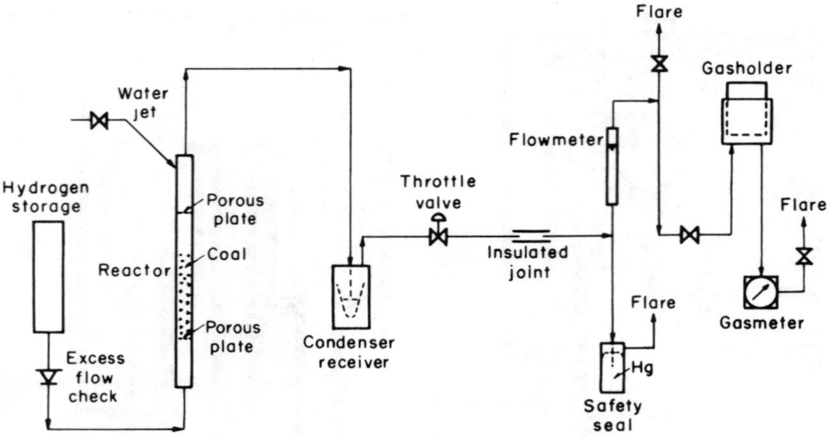

Figure 6. A reactor water jet and a series of flares control temperature during processing.

show assembly/disassembly steps or operational phases, ensure that readers see the object from the same perspective in each illustration. Changing the perspective is very confusing.

6. If necessary for clarity, remove surrounding detail from illustrations.

Figure 4 shows a schematic drawing or illustration in which surrounding but irrelevant detail has been removed. You often see this sort of illustration in subsystem pictorials. Removing the surrounding detail allows readers to focus on the system being shown. The drawing isolates its subject and therefore provides excellent focus.

7. Use line patterns in an illustration to show how different subsystems interact within a system.

If you are showing how different subsystems fit together and function, you may need to use different line patterns, as in figure 5. The line patterns allow readers to isolate subsystems while viewing the whole system. For an example of the line patterns possible, see GRAPHS.

If you use line patterns, provide a legend explaining what the different patterns represent.

8. If your drawing shows a process, structure the process from top to bottom and left to right.

Figure 6 is a schematic flowsheet. It shows a process and the equipment used in that process. These drawings must be oriented from top to bottom and left to right so that readers "read" them as they would read text. In all process drawings, readers will expect the process to start at the left and end at the right. Don't disappoint them. (See CHARTS.)

INTRODUCTIONS

As the name suggests, an introduction introduces. It conducts the reader into a document, usually by establishing the reason for the document's existence, its relation to other documents or projects, and any special circumstances, facts, conditions, or decisions that help the reader understand the body of the document.

The body of the document is more project or product oriented—it focuses on the situation or thing being described or analyzed. The introduction (as well as the conclusion) is more reader oriented—it orients the reader by providing a context for the reading. In short, the introduction prepares the reader for what will follow.

When To Write an Introduction

Although introductions usually come first in a document, you should write them last. The introduction is often the hardest part of the document to write because it sets up information that appears throughout the rest of the document. So skip the introduction until you have written most, if not all, of the body of the document.

The Different Types of Introductions

Introductions differ, depending upon the type of document in which they appear. Introductions for formal reports are often fairly long, especially if they summarize technical or scientific literature. These introductions often provide lengthy background or historical information that allows uninformed readers to develop a perspective on the text that follows. These introductions also sometimes define special terms or establish the assumptions upon which the succeeding analysis was based.

Introductions to informal reports and to letters and short memos often are quite short. These introductions are usually not called introductions, nor are they separated from the rest of the text by headings. These introductions usually set up the major points that follow, and they can refer to previous documents, meetings, or conversations.

Finally, introductions can function as executive summaries in both technical and nontechnical reports. (See SUMMARIES.)

Introductions to Formal Reports

Formal reports are likely to have the most developed and structured introductions. No two introductions are alike, but most include some of the following information:

—The **problem** or **opportunity** prompting the project or investigation

—The **goals** or **purposes** of the project or investigation

—The likely **audience** for the project or investigation

—The **scope** of the project or investigation

—The **sources** of relevant information

—The **methods** used in the project or investigation

—This project's or investigation's **relation** to previous or concurrent projects or investigations

—Any useful working **definitions**

The problem or opportunity that the document addresses arises from the historical background. What work was done that stimulated this project? What other work has been done in this or related fields? Often, this section will survey relevant literature. (See CITATIONS.)

If this project is part of a larger or related project, state their relationship briefly. Brevity is important because readers rarely want or need to learn the point-by-point history of a project or related projects.

The purpose of the project explains why the project was undertaken and what it is expected to achieve. The key objectives might read as follows:

1. To evaluate the ground-water resources of the alluvial aquifer of the Carmel Valley, California, ground-water basin.

2. To develop a two-dimensional, digital, ground-water flow model of the aquifer that will aid in the understanding of the geohydrology of the aquifer.

3. To identify data inadequacies that might be needed for future studies of this aquifer.

INTRODUCTIONS

NOTE: Several objectives can combine to form the single overall purpose of the project.

The scope refers to the limits of the project and the document itself. Provide the scope by stating what the document will cover and what it will not cover.

The methods explain how a project was conducted: how the investigator developed the experimental design, constructed or designed the apparatus, collected the data, analyzed the results, and developed the conclusions. If the methods were routine, provide no more than a brief summary. If the methods were unusual or original, explain them thoroughly in the introduction or consider discussing methods in a separate section or subsection of the document. (See REPORTS.)

Introductions to Informal Reports

Introductions to informal reports often function as executive summaries, which include conclusions and recommendations along with the necessary background information.

Readers of informal reports are often busy managers or supervisors. They are usually aware of the general details of an investigation or project but trust subordinates to evaluate the problem properly and to solve it efficiently. Such readers will become impatient with lengthy digressions and unnecessary explanation, support, and justification. Consequently, they typically want a succinct executive summary. Only if questions arise will they go beyond the summary and read the background information and analysis.

The key feature of a combined introduction and executive summary is a statement of major conclusions and recommendations. A full list of conclusions and recommendations often appears later in the report, but you should never force managers or supervisors to search for them.

Introductions to Letters and Memorandums

Introductions to letters and memos establish the writer's tone and approach as well as set the stage for the ideas and supporting details that follow. (See LETTERS and MEMOS.)

<u>Establishing a Tone</u>

Letters and memorandums are intended to be more personal and less formal than reports. To convey this more human dimension, some writers choose to use personal pronouns: *I*, *you*, and *we*. Others open with personal remarks or social greetings, much as we often do in personal conversations. Even writers whose purpose is avowedly serious may "break the ice" by calling the reader by name:

> Please let us know, Jim, if our proposal for the replacement pumps begins to answer your needs.

See TONE.

<u>Setting the Stage</u>

Setting the stage may mean no more than a brief phrase: *As we discussed yesterday* . . . or *According to our records* . . . . In other instances, writers may decide to provide the background for the document before stating the point. This background or set-up is quite common in letters conveying bad news, such as enforcement of a financial penalty or a personnel reprimand.

However, be cautious about spending too much time setting the stage. Almost always, the best strategy is to get to the point quickly.

See ORGANIZATION.

ITALICS

Italics (slanted typeface) is available only in printed material or on word processing systems capable of printing italics. In handwritten or typed material, underlining replaces italics. (See UNDERLINING.)

Because italics is unavailable to most writers, some conventions concern only printers. For example, in printed mathematical expressions, letters are italicized while numerals are set in normal type. This distinction is not normally made in handwritten or typed material. The following standard rules apply in most technical and business documents.

---

1. Use italics for words used as words:

> In all offshore contracts, *consolidation* does not mean what it normally means.
>
> The Anaguae reservoir study was confusing because the author kept referring to the anomolous formations as *anonymous* formations.

NOTE 1: Use italics to emphasize words, phrases, and even letters when discussing them as examples of language. This use of italics (or underlining) sets the words or phrases apart from the other words in the sentence:

> We have traditionally used the symbol *M* to mean million. In data processing, however, *M* means thousand, so one indicates million with the symbol *MM*.
>
> One should avoid opening letters with the phrase *in reference to.* Similarly, according to our corporate guidelines, one should never end a letter by saying *very truly yours.* Both phrases are too wordy.

NOTE 2: Quotation marks sometimes replace italics (underlining) as a way of highlighting words and phrases, especially in handwritten texts, where even underlining is less easy for a reader to see:

> After some discussion, we decided to order "A Dictionary of Mining, Minerals and Related Terms" and the "Society of Petroleum Engineers Publication Style Guide." The two volumes should help us prepare articles for the "Journal of Petroleum Technology" and "Petroleum Transactions."

---

2. Use italics for foreign words and phrases that have not yet been absorbed into English:

> The initial concept of the United Nations captured a certain *Weltanschauung.*
>
> The *couturier* insisted on keeping the new dress designs secret.
>
> The staple crop in South Africa is *kaffir,* which is a form of sorghum raised for cattle fodder.

NOTE: Some foreign words and phrases have become so common in English that they are not italicized:

> ad hoc
> habeas corpus
> per annum
> rendezvous
> vice versa

Some recent dictionaries indicate if words are still considered foreign, but others do not. If your dictionary does not, use your judgment to determine if a word is sufficiently foreign to be italicized. Foreign words usually retain their foreign pronunciations and meanings.

---

3. Use italics for titles of books, magazines, newspapers, movies, plays, and other works individually produced or published:

> To remain current on advances in space technology, we subscribed to *Aviation Week.*
>
> The documentary *Before Their Time* showed what is possible when companies wisely invest IR&D funds.

See TITLES.

NOTE 1: Sections of these published works are not italicized. So chapters, magazine articles, acts within a play, and editorials in a newspaper require quotation marks, not italics:

> Last week's *Time* had an article entitled "The Roots of International Terrorism."
>
> The final chapter of the annual report is entitled "Prospects for Growth in the 1990s."

See QUOTATION MARKS.

NOTE 2: In some typed documents, especially in the publishing business, the titles of books, magazines, and newspapers are in all capital letters and are not underlined:

> DELTAIC OSCILLATION
> FORBES MAGAZINE
> GEOTIMES
> THE WALL STREET JOURNAL

ITALICS

4. Use italics for the names of aircraft, vessels, and spacecraft:

Discovery
Friendship 6
H.M.S. *Intrepid*
NS *Savannah*
U.S.S. *Iowa*
U.S.S. *Nautilus*

NOTE: In these examples only the names are italicized, not the abbreviations or numerals associated with the names.

5. Use italics for names of genera, subgenera, species, and subspecies. Names of higher groups (phyla, classes, orders, families, tribes) are not italicized:

the genera *Quercus* and *Liriodendron*
the family Leguminosae

See the *Council of Biology Editors Style Manual* for additional information and examples. (See REFERENCES.)

JARGON

Jargon has two meanings, both negative. First, it means using familiar words in unfamiliar ways (using *hot* to mean "crucial" or "exciting") or using excessive "shorthand" to describe something that would normally require more words (a police officer saying, "That guy was really a ninety-nine," which is police jargon for code 99, the radio code indicating an unbalanced person).

Second, *jargon* means technical or specialized language unfamiliar to a particular reader or listener. Thus, one person's technical or specialized vocabulary becomes another person's jargon. The following discussion focuses on this second, more common meaning of the word *jargon*.

Every technical discipline needs and has its own vocabulary.

Medical doctors have innumerable special terms, often derived from Latin: *amebic dysentery, uvula, gastric hernia*.

Lawyers also use a number of common terms that have developed special meanings: *property, liability, consideration, conveyance. Consideration*, for instance, means a payment of some kind as a sign of agreement on a contract. This special legal meaning is not obvious to the uninitiated, who might not even know the word has a special meaning. In some contexts, readers may not know whether the ordinary meaning or the technical legal meaning is intended.

The cooks and waitresses in restaurants might develop special slang terms and abbreviations, so when a patron picks up a check, the items written on it are not immediately clear. This is a form of jargon, one very local and limited, yet still very useful.

Carpenters and architects also have their own language: *joist, rafter, gambrel roof, header, sole plate, cross bridging*, etc. These jargon words have special meanings and are useful terms, but someone not familiar with carpentry is not likely to understand them.

See STYLE, SCIENTIFIC/TECHNICAL STYLE, and TONE.

1. Do not use jargon unless your readers will understand it. If they will not understand—and you must use a term—then define it.

The doctor who gives a diagnosis only in medical terminology has failed to communicate with most patients. Similarly, the engineer who speaks only through coordinate graphs and equations will baffle, and perhaps alienate, the general reader who wants an overall sense of the proposed engineering project.

Here are two technical examples, both of which may use jargon unfamiliar to nontechnical readers:

JARGON

> Wastewater treatment that employs fixed-film biological BOD removals has been shown to be more efficient than was predicted in our pilot studies. This result may be due to product mix, concentration, primary treatment, media type, wall effects, etc.

> Dry rubble stones shall consist of trap rock, granite, gneiss or other approved hard, durable, tough rock. They should be sound, free from weathered or decomposed pieces, shattered ends, and structural defects, and shall be approved by the Contracting Officer.

In both of these examples, a general or nontechnical reader would encounter unfamiliar terms and abbreviations: *BOD, primary treatment, media type, trap rock, gneiss, shattered ends, structural defects*, etc. To the right reader, these terms have perfectly legitimate meanings, but to the uninitiated reader, the words might be confusing or nonsensical.

Jargon and the Social Sciences

Writers in the social sciences—especially in psychology and sociology—have often been accused of using excessive jargon. Writers in both fields use many common English words with special, often stipulated meanings: *response, learning, training, feeling, concept, idea, group, class, family*, etc.

A psychologist discussing a *tertiary mediated response* is referring to a response coming through an intermediate person and delivered third hand. The concept and its expression are valuable in a limited context and to a limited audience. Otherwise, they are meaningless.

Similarly, a sociologist in talking about families might need to define where the family ends—perhaps at second cousins twice removed. So the sociologist begins to use the term *extended family* for all relatives, including second cousins twice removed. To an uninitiated reader, the term will likely be confusing or awkward.

Here is an example of how difficult a jargon-filled passage from the social sciences can become:

> Another very common psychological use of the analysis of variance is seen in test development techniques and procedures where the measurement or test specialist has designed a new test instrument and administered it to a large normative sample of subjects, including students.

Rewritten, this passage can be shorter and clearer:

> An analysis of variance allows test specialists to analyze the questions on a newly developed test by comparing a large number of student responses.

Jargon-filled writing is always difficult to uninitiated readers, but making the writing clear and concise does aid comprehension. Often, the context in which jargon appears helps readers understand what the writer intended.

Jargon and Gobbledygook

Jargon and gobbledygook are not quite the same. Gobbledygook is the use of abstract or pompous words and long, convoluted sentences. It is clearly bad writing. Jargon, by contrast, is a specialized vocabulary for a particular technical field and is often a useful shorthand. (See GOBBLEDYGOOK.)

KEY WORDS

Key words are like flags—they rise above the rest of the text and signify what is most important. In a paragraph, subsection, or section, the key words are those that give the text meaning. Key words impart the central message.

You can deliberately repeat key words and phrases to reinforce your message. Key words ensure that readers who are not reading carefully will still get the point of what you are saying and remember the central or most important ideas.

The example below is from a short section on condenser operation. Note how the writer drives home the message by using repeated key words as variations on an important theme:

> This highly effective water-separation process is possible because the condenser design positively prevents two potential **icing** problems: (1) blockage of the low-pressure side by **snow-laden** **cold** air, and (2) **freezing** of condensate on the **cold** metal surfaces. As noted earlier, the entering **cold-side** air is below the **freezing** point of water and, although the condenser **heats** it, the outlet **temperature** is still **below 32 degrees F** Consequently, much of the entrained **snow** is not evaporated and must pass completely through the condenser without blocking the flow passages. Because **cold-side** air **temperatures** are consistently below the **freezing** point, the condenser must be carefully designed so that the metal surface **temperatures** remain above the **freezing** point.

The key words concerning temperature represent one important line of thought in this paragraph. An equally important line of thought concerns the design, mechanics, and operation of the condenser:

> This highly effective water-separation process is possible because the **condenser design** positively prevents two potential icing problems: (1) **blockage** of the low-pressure side by snow-laden cold air, and (2) freezing of **condensate** on the cold **metal surfaces** As noted earlier, the **entering** cold-side air is below the freezing point of water and, although the **condenser** heats it, the **outlet** temperature is still below 32 degrees F Consequently, much of the entrained snow is **not evaporated** and must **pass** completely through the **condenser** without **blocking** the **flow passages** Because cold-side air temperatures are consistently below the freezing point, the **condenser** must be carefully **designed** so that the **metal surface** temperatures remain above the freezing point.

Note how the two sets of key words work together to create the overall effect and to establish both a primary problem (icing) and a primary need (a condenser design that will prevent it).

See PARAGRAPHS and TRANSITIONS.

The key words in the next paragraph provide both a solution and a sharp contrast:

> Figure 2-1 shows how the **condenser design prevents** these **icing** problems. A special **hot** section on the **cold-side** face **prevents ice blockage** A small **flow** of **hot** air from the compressor outlet **passes** through the **hot** section tubes, **raising** the metal **temperatures above freezing** and allowing the **snow or ice** to be **evaporated** in the main core. The **hot** air then reenters the high-pressure **air-flow** at the turbine inlet.

Headings and Captions

Headings and captions should contain the most important of the text's key words. In fact, they should be composed almost entirely of key words:

> Condenser Design Prevents Icing

See HEADINGS, CAPTIONS, and EMPHASIS.

LETTERS

Letters are one of the principal forms of business communication. Good letters do more than convey information, actions, and decisions. They establish the personal style of the sender and the image of the sender's organization, and they act as surrogate conversations between parties.

Well-written letters are clear and concise. The important points appear early in the letter, usually in the opening sentence. The writer emphasizes crucial data and ideas, wastes no words, and includes nothing that is not relevant to the central theme of the letter. Readers understand clearly why the letter was written and what they should do after reading it.

Letter writing styles have changed over the years. Decades ago, lengthy, rambling letters were acceptable. Today, writers and readers prefer a simplified, concise letter, probably in response to the enormous increase in written communication that has occurred in the last 20 years. Today, busy readers do not have time to spend on lengthy digressions and explanations. They want the writer to get to the point quickly.

Letters and memos are similar in many respects. The principal distinction is that letters are written to persons outside the sender's organization and memos are written to persons inside the sender's organization. (See MEMOS.)

This section begins with a discussion of the principles of good letter writing. Then it describes the format styles of business letters, including the block, modified block, semiblock, and simplified styles, and the two most common styles of punctuation, open and standard. (See PUNCTUATION.)

Effective Letter Writing

1. Begin most letters with the most important point.

If you can, try to open the letter with the most important idea in the letter:

> Dear Mr. Smith:
>
> On March 1, we will meet with General's attorneys to discuss the Bellocq acquisition, and we would like you to be present.
>
> *or*
>
> Could you meet with us and General's attorneys on March 1 to discuss the Bellocq acquisition?

> Dear Ms. Atkins:
>
> MOGO recommends plugging and abandoning the C.C. Baker 12. Production continues to decline and is now below economically feasible levels.
>
> *or*
>
> Because production on the C.C. Baker 12 continues to decline and is now below economically feasible levels, MOGO recommends plugging and abandoning this well.

> Dear Dr. Jones:
>
> The test results from the Hampstead facility indicate that exposure levels do not currently exceed EPA safety levels.

> Dear Mr. Johnson:
>
> For the reasons cited below, we have decided not to adopt your suggestion to delay platform renovations until the 4th Quarter. However, we will modify the containment area according to the revised specifications that you submitted.

The opening is the strongest part of a letter. Putting your important ideas somewhere other than the opening is unfair to busy readers. Furthermore, if you begin the letter with information that your readers know is relatively unimportant, they may begin to skim, and in skimming, they may miss important points.

Beginning with your important points establishes your purpose right away and gives readers a perspective for understanding the rest of what you have to say. (See ORGANIZATION.)

Sometimes, you cannot open with your most important idea because readers either will not understand it or will not accept it. If that is the case, then set up the major ideas with a brief explanation:

> Dear Ms. Atkins:
>
> Our 8-month effort to stimulate the C.C. Baker 12 has not succeeded, and production continues to decline. If we keep operating this well at current and projected production levels, production costs will outweigh revenues. Therefore, MOGO recommends plugging and abandoning this well.

Set-ups, like the one shown above, should always lead directly to your major point.

LETTERS

Never open a letter with

— references:

This is in reference to your letter of February 15. *(See rule 2 below.)*

— unnecessary social statements:

Here's hoping that the weather in Tampa is fine and that your family is doing well in the new year.

— mechanical enclosure indications:

Enclosed is . . .
Attached is . . .
Enclosed herewith . . .
Attached hereto . . .
Enclosed please find . . .

— statements that suggest your topic but do not indicate your position:

Delaying platform renovation until the 4th Quarter is an idea worthy of consideration.

— information that supports, explains, or illustrates your major points (unless that information legitimately sets up your major points):

Delaying platform renovation until the 4th Quarter would mean rebudgeting funds that have already been allocated during 2nd Quarter. *(This is justification for your decision, but you should state the decision first, then justify it.)*

---

2. Subordinate references to previous documents, conversations, and meetings.

As noted above, you should never open a letter with references. References are never the major point of a letter. Therefore, they belong after the major point or before the text (in a subject or reference block). Here are some ways to work in references:

this

Dear Mr. Smith:

Thank you for agreeing to be present on March 1 during our meeting with General's attorneys to discuss the Bellocq acquisition. As Mary Evans indicated when she phoned you yesterday with the invitation, we are especially concerned about the Forbish property.

not this

Dear Mr. Smith:

This confirms Mary Evans' telephone conversation with you on February 15 in regard to our March 1 meeting with General's attorneys to discuss the Bellocq acquisition.

---

this

Dear Dr. Jones:

The test results from the Hampstead facility indicate that exposures do not currently exceed EPA safety levels. Your letter of July 23 expressed concern over the large amounts of radiation present.

not this

Dear Dr. Jones:

This is in reference to your letter of July 23 in which you expressed concern over the large amount of radiation present.

Subordinate references by placing them after the major point or by putting them in reference or subject blocks ahead of the text:

RE: Your Letter of July 23 Concerning Radiation Levels at Hampstead

SUBJECT: OUR NOVEMBER 9 TELEPHONE CONVERSATION ON PRODUCTION QUOTAS

RE: Yesterday's Meeting on Supervisory Policy

SUBJECT: Recommendation to P&A the C.C. Baker 12 (Re: your letter of April 5)

Avoid the following types of reference statements, especially as letter openings. They have all become cliches:

This is in reference to your letter of . . .

In reference to our telephone conversation concerning . . .

This confirms our telephone conversation of . . .

Reference is made to our recent meeting in which we . . .

This is in response to your inquiry regarding . . .

NOTE: As stated above under rule 1, also avoid the following cliched references to enclosures:

Enclosed herewith
Attached hereto

You should state that you have enclosed or attached something, but do so later in the letter. Don't open your letter with *enclosed* or *attached*, and never add *herewith* or *hereto*, which sound legalistic (their first offense) and are also redundant (the *coup de grace*). Perhaps the worst reference to enclosures is the following:

Enclosed please find

LETTERS

If you enclose it, they'll find it. Don't ever use this ridiculous statement.

Enclosures typically appear in letters because whatever accompanies the letter is enclosed within the envelope. Attachments typically appear in memos because memos generally do not come in envelopes. Whatever accompanies the memo must be attached (via rubber band, staple, or paper clip).

Always subordinate references to enclosures or attachments:

> Dear Dr. Jones:
>
> The test results from the Hampstead facility indicate that exposures do not currently exceed EPA safety levels. In your letter of July 23 (enclosure 1), you expressed concern over the large amounts of radiation present.
>
> On June 17, EPA representatives inspected the Hampstead facility (see enclosure 2). They did find traces of arsenic in the vent system but could not . . .

3. Ensure that your letters are clearly and logically organized.

Logical organization is crucial to effective letters. Readers should understand from the early paragraphs what the letter is about and how the writer has organized his or her thoughts.

Letter organization does not differ significantly from the logical organization of memos, manuals, specifications, proposals, and reports. However, the audience for a business letter is usually different than for those other types of documents, and letter readers typically read with a different purpose.

So you must organize your thoughts carefully, keeping your readers in mind. Unless you have a compelling reason to do otherwise, follow the organizational principles discussed under ORGANIZATION.

4. Throughout letters, emphasize key data and ideas.

Use headings, lists, numbering systems, visual aids, white space, single-sentence paragraphs, repetition, and other emphatic devices to highlight major points.

Letter writers sometimes mistakenly assume that the emphatic devices listed in the previous paragraph are inappropriate in letters. In fact, these devices help break up large blocks of text and make letters more readable. Do not fail to use them if the opportunity arises.

See EMPHASIS.

5. Avoid cliched letter closings.

Letter closings should be as simple and direct as letter openings. The closing may reiterate an important point stated earlier in the letter, or it may provide useful information, such as a due date, a response deadline, or the name and telephone number of a person to contact for assistance.

Avoid these kinds of statements (which have become cliches):

> Thanking you in advance . . .
>
> Should you have any further questions or be in need of further assistance . . .
>
> Do not hesitate to contact me . . .
>
> Feel free to contact me . . .

Instead, make an offer of assistance sound natural and direct:

> If you have further questions, please call me at 123-4567.

Also, avoid complimentary closings that sound exaggerated:

> Very truly yours,
>
> Truly yours,
>
> Deepest regards,
>
> In sincerest appreciation,
>
> Thanking you for everything,

The best complimentary closings are simple and brief:

> Sincerely,
>
> Respectfully,
>
> Thank you,

See CLICHES.

Letter Format and Punctuation Styles

Format Styles

On the following pages are four letter models: **block**, **modified**

LETTERS

Block Letter, Open Punctuation

OSAGE
GAS AND ELECTRIC
COMPANY, INC.
511 Franklin Road
Suite 8
Lincoln, Nebraska 68506
(402) 355-7070

May 31, 1985

In reply to: Invoice 5068

American Gas Company
Engineering Department
3498 Anyplace Drive
Alameda, OH 87543

Attention Mrs. Joyce Johnson

Ladies and Gentlemen

Subject: The Form of the Block Letter, with Open Punctuation

The block format means that every line is flush with the left-hand margin, even the date and the complimentary closing.

Open punctuation means that no punctuation follows the salutation and complimentary closing. Block letters may also follow the standard punctuation style (see the model of the modified block letter).

The date appears 2 to 6 lines below the letterhead, depending upon the length of the letter. Following the date is the reference line, which may include an invoice number, letter date, telephone date, file number, account number, or other pertinent information. The reference line is optional.

The inside address follows the spelling, format, and abbreviation style used in the letterhead of the organization receiving the letter. If you are writing to a specific person within the addressed organization, then begin the inside address with that person's name. Use an attention line only if the inside address does not contain a person's name (as in this example) but you want to route the letter to a specific person. Using an attention line indicates that the letter concerns a business matter and may be handled by anyone in the department receiving the letter. The attention line may contain the name of a department or group.

block, **semi-block**, and **simplified**. These models illustrate the use, placement, and punctuation of the following elements of letters:

letterhead/return address
date line
reference line or block
special notations
inside address
attention line
salutation
subject line or block
text or body
headings for continuation page
complimentary closing
signature line
reference initials
enclosure notation
carbon copy notation
blind carbon copy notation
postscript

The elements in boldface type are standard; the others are optional. (NOTE: In the the simplified letter, the salutation and complimentary closing are omitted.)

In the **block** style, all of these elements appear flush with the left-hand margin, and the paragraphs are not indented.

In the **modified block** style, the date, reference line, complimentary closing, and signature block appear right of center; everything else is flush left. The paragraphs are not indented.

The **semi-block** style, is similar to the modified style except that the paragraphs are indented (usually five spaces).

The **simplified** style is similar to the block style except that the salutation and complimentary closing are omitted. Paragraphs are not indented.

For many years, the semi-block style was standard. Today's letter writers favor the block or simplified style for several reasons. First, moving every element flush left is easier for typists. Second, indenting paragraphs is no longer necessary because writers routinely leave one blank line between paragraphs. When paragraphs were not separated by blank lines, indentation was essential for showing where new paragraphs began.

Note, however, that the semi-block letter has a more casual feel to it. Even the modified block style, with some information moved right of center, seems less formal than the full block style.

Your organization may dictate a stylistic preference. If not, you should choose a style that is consistent with the image you wish to project. If you wish to appear formal and businesslike, use the block or simplified styles. If you want to be more casual, then select the modified or semi-block styles.

The chief distinction of the simplified style is the absence of salutation and complimentary closing. These omissions solve a problem unique to today's letter writers: how to write to women without offending them by using traditional but sexist language.

Salutations and complimentary closings are traditional but non-essential elements of letters. They are forms of social address that were mandatory when letter writing had a different social purpose than it does today. In a purely business climate, the traditional greeting *Dear* and the traditional closings, such as *Yours truly*, are not necessary. Therefore, if you wish, you may omit them and follow the simplified letter style.

Punctuation Styles

In the **open** style, the writer omits all non-essential punctuation, including a colon or comma after the salutation and a comma after the complimentary closing.

In the **standard** style, the writer uses minimal punctuation in the letter but does include a colon (or a comma) after the salutation and a comma after the complimentary closing. The standard style is almost always used in letters with indented paragraphs (the semi-block style).

The **closed** (or **close**) style has all but vanished from business letters written in the United States, although some European firms still use it. In the closed style, writers retain all of the punctuation marks used in standard punctuation and add others:

Block Letter, Open Punctuation

American Gas Company
Engineering Department
May 31, 1985
Invoice 5068
Page 2

The preferred salutation in a business letter not addressed to an individual is "Ladies and Gentlemen." "Gentlemen" shows sexual bias and may be offensive (see SEXIST LANGUAGE), and "To Whom It May Concern" is obnoxiously formal. You may avoid this problem by beginning the inside address with "Mrs. Joyce Johnson," and then writing "Dear Mrs. Johnson" as the salutation.

The optional subject line is underlined for emphasis, but some writers and editors prefer all capital letters. The subject line should be as specific and informative as possible, even if it requires more words. The subject line should tell readers specifically what this letter is all about.

The text has block paragraphs (no first-line indentation) that have one line between them. If the text extends beyond the first page, ensure that you have at least four lines of text on the second page. On the second page, the heading begins approximately six lines down from the top of the page (depending upon the length of the material on the second page). The text begins at least two spaces below the heading.

The complimentary closing is "Sincerely," which is a good choice for both informal and formal business letters. To leave space for the signature, the author's name and title appear four lines beneath the complimentary closing.

The author's and the typist's initials appear two lines below the author's title and are flush left. Immediately beneath the initials comes the enclosure line, with the number of enclosures indicated (in parentheses). The letter ends with the list of carbon copies (cc) and, if appropriate, blind carbon copies (bcc).

Sincerely

Marion R. Garvey

Marion R. Garvey
Chief Engineer

MRG:st
Enclosures (3)

cc Mrs. Florence Lynch
Edward Jenkinson
bcc Joseph Franks

— A period after the date

— Commas after each line of the address, with a period after the final line:

Mr. Edwin Jones,
Wellhead Oil Company,
1359 Fifth Avenue,
Bellevue, MI 65431

— Commas after each line in the signature block, with a period after the last line.

The **block** and **simplified** letter models illustrate the open punctuation style. The **modified** and **semi-block** letter models show standard punctuation.

Margins and Spacing

The left-hand and right-hand margins should be roughly equal, but the exact spacing will vary depending upon the length of the letter. In long letters, the margins are usually at least an inch and a quarter wide. In short

LETTERS

Modified Block Letter, Standard Punctuation

FARMLAND FROZEN
FAMILY FOODS

October 31, 1985

File SD 87/6
Your letter September 5, 1985

Mr. George Freed, Jr.
Assistant Manager
Stevenson Retail Mart
349 Highland Boulevard
Miami Beach, FL 96502

Dear Mr. Freed:

In modified block letters, the date, reference block (optional), and complimentary closing are right of center (sometimes flush right); everything else is flush left. Paragraphs are not indented. In standard punctuation, a comma follows the complimentary close and a colon (formal) or comma (informal) follows the salutation.

The reference block, which is optional, can contain this letter's file number, as well as references to previous documents, meetings, or conversations. Often, reference blocks include the file number, subject, and date of relevant documents preceding this letter or to which this letter is responding.

The inside address includes the full name and title of the person addressed along with the company name and address (written as they appear in that company's letterhead). The inside address is conventionally 3 or 4 lines below the date, but you can leave additional blank lines if your letter is short.

The salutation follows those conventions described in the block and semi-block letter models.

Do not indent paragraphs. To indicate where paragraphs begin and end, leave a blank line between paragraphs.

The second page continuation heading includes the name of the addressee, the date of the letter, and the file number. If the letter has a subject line, then the continuation heading should include the subject as well (file references are also optinal). Place continuation headings six lines from the top of the page. Start the text four lines (or more if the continued text is short) from the heading.

The complilmentary closing can appear (1) directly beneath the date in the heading, (2) flush right, or (3) five spaces right of center. The

303 Blossom Avenue   Des Moines, Iowa 50321   (515) 521-4911

letters, the margins can be wider; in extremely short letters, wide margins can be combined with double or triple spacing. The typist's goal should be to center the letter on the page, so the typist may have to reset the margins for letters of unusual size.

Letterhead/Return Address

All business letters should have either a printed letterhead or typed return address. Letterheads should contain the following:

— Logo (optional)

— Full legal name of the organization

— Full legal address—including post office box number, suite number, city, state, and the full ZIP code

— Area code and telephone number(s)

— Telex or cable instructions (optional)

In addition, departments, branch offices, and company officers may have their own letterheads, which may include titles, building numbers, and other specific identifiers:

E. G. Walters, Jr.
Office Manager
Viewmont Branch Office
High Fidelity Savings & Loan
1234 S. Main Street
Logan, IN 44444-4444
(123) 456-7890

If you are not using stationery with a printed letterhead, then you must type a return address that contains the same information that a letterhead would contain. The semi-block model illustrates the format for a typed return address; in block and simplified letters, such an address would appear flush with the left-hand margin.

Date Line

Placement of the date line varies depending upon the length of the letter and the style of the letterhead.

The date line should never extend into the right or left margins, but any other placement is possible, depending upon the letterhead.

The date line usually appears 2 or 3 lines below the letterhead, but you can leave as much as 6 or 7 lines if the letter is short. If your stationery does not have a letterhead; then place the date on the line below the typed return address (as in the semi-block model).

LETTERS

(See the four model letters for further information on placement of the date.)

The standard date line in the United States is month, day, and year: *March 15, 1985.* Do not use abbreviations.

Writers in the U.S. Government, including the military, and in many foreign countries prefer to list day, month, and year: *15 March 1985.*

Reference Line or Block

Reference lines appear beneath the date line, usually two lines down, but some companies prefer only a single line. Reference lines are typically aligned with the date or are flush with the left- or right-hand margin, depending upon the format style of the letter:

this

March 15, 1985

Invoice SD-4576A

or this

15 March 1985

Invoice 45890

Reference lines are optional, but you should seriously consider using them, especially if the letter refers to several invoices, files, letters, or telephone conversations. References are easy to see under the date line, and they eliminate the need for writers to include such references in the crucial opening paragraph of the letter. Reference lines under the date line rarely need a lead-in, but one is optional if it can clarify the reference:

Modified Block Letter, Standard Punctuation

Mr. Freed - 2 - October 31, 1985
File SD 87/6
Your letter September 5, 1985

author's name and title should appear at least four lines below and flush with the left side of the complimentary closing.

The reference initials should appear flush left, two lines below the author's title. If you include the author's initials, they appear first and in all capitals. The typist's initials appear in lowercase letters following the author's initials. If someone other than the author will sign the letter and you want to include that person's initials, place them in all capitals to the left of the author's initials. Separate all initials with a colon.

Sincerely,

Frank Jefferson

Frank Jefferson
Sales Manager

FJ:CC:vb

Avoid postscripts if at all possible. If you must use one, place it at the end of the last page. Postscripts follow the paragraph style established in the text. They need not begin with the word *postscript* or with the initials *PS* (but if they do, they must be followed by a colon). Following the postscript, you do not need another signature but may add one.

15 March 1985
In reply to: Your telephone call
of 7 March 1985

Sometimes, writers include the reference line or block below the inside address, either flush left or right of center. These reference blocks can begin with *RE:* and may run several lines. In the style adopted by the Department of Defense, writers list all references by number or letter in a reference block.

References:

A. DOD Directive 5202.43-1

B. SECNAVINST 452.1

C. COMCINCPAC Ltr 85-00064-5, dtd 23 April 1985

D. DOD Manual 34.2

Then when writers have to refer to those references in the body of the letter, they refer to them by number or letter:

LETTERS

MIDLAND OIL AND GAS OPERATIONS, INC.
7000 Jalepeno Boulevard
Dallas, Texas 75234
(214) 735-9600

434 Fish Lake Road
Salmon, ID 43287
June 8, 1985

In reply to: Invoice 45/765
May 6, 1985

Mrs. Joanne G. Kelsey
Executive Vice President
Year-Long Heating Company, Inc.
4376 Grand View Avenue
Anchorage, AK 98754

Dear Mrs. Kelsey:

Semi-Block Letter, Standard Punctuation

In semi-block letters, the return address, date, and complimentary closing appear right of center. The inside address, salutation, and headings are flush left. All paragraphs are indented (usually five spaces). In standard punctuation, a comma follows the complimentary closing and a colon (formal) or a comma (informal) follows the salutation.

If the paper has no letterhead, then use a return address as shown above. If the paper has a letterhead, the date and optional reference block appear three or four lines below the letterhead.

The inside address opens with the name of the individual receiving the letter, followed by the person's title, the exact name of the company (spelled and punctuated as on that company's letterhead), and the address.

Salutations commonly begin with *Dear*. Formal salutations use the receiver's title and last name: *Dear Mrs. Kelsey*. Less formal salutations can use only a first name: *Dear Joanne*. Use a colon after the name in formal salutations, and a comma after the name in informal salutations. Abbreviate *Doctor* (Dr.) and all gender titles: *Messrs., Mr., Mrs.,* and *Ms.* Do not abbreviate other titles: *Senator, Mayor, General, Professor*, etc.

The subject line (optional) may be centered on the page two lines below the salutation (as shown above).

> Reference E suggests that formulation of a new European policy is imminent. However, reference F cites an EEC memorandum stating that economic goals set last year would not be changed until 1988.

This practice seems sensible if you are going to be discussing a number of references and don't want to repeat the name, subject, and date of the reference every time you refer to it.

When letters are longer than one page, reference lines sometimes appear under the date in the heading on all continued pages of the letter. (See the modified block model for an example.)

Special Notations

Special notations appear between the date (or reference line) and the inside address. Such notations include the following: SPECIAL DELIVERY, REGISTERED MAIL, CERTIFIED MAIL, CONFIDENTIAL, PERSONAL. These notations usually appear above the inside address and are usually typed in all capital letters (for visibility). (See the simplified letter model for an example.)

If two or more of these notations apply to a letter, the second and additional notations appear directly beneath the first one. Leave no blank lines between them. These notations also appear on the envelope, usually above the receiver's address, but placement varies according to the size of the envelope and the appearance of the address. Always ensure that such notations are clearly visible.

Inside Address

The inside address includes the name and address of the organization receiving the letter. The spelling, format, and punctuation of the receiving organization's name and address should be consistent with the spelling, format, and punctuation shown on that organization's letterhead or typed return address.

The spacing of the inside address below the date (and reference line) will vary depending on the length of the letter.

The inside address usually includes (1) the addressee's courtesy title—*Mr., Mrs., Ms., Miss*, or *Dr.*; (2) the addressee's business title; (3) the name of the organization; (4) the street address and, if appropriate, the post office box, suite number, mail drop, or other mailing information; (5) the city and state, and (6) the full ZIP code.

Semi-Block Letter, Standard Punctuation

Mrs. Joanne G. Kelsey -2- June 6, 1985

Headings

Headings are excellent ways to highlight the organization of the letter and to emphasize key points or sections, especially if the letter is more than one page long. Set off the heading by leaving a blank line above and below it and by underlining the heading or typing it in boldface type.

The continuation heading for the second and additional pages should appear as shown above. Continued pages should have at least three lines of text.

Displayed Lists

Displayed lists are effective in business letters, especially those running more than one page and having a number of paragraphs. Optional formats for lists include:

• Bulleted lists. They are perhaps the most emphatic lists because bullets are so dark. On word processors and typewriters that do not have bullets, use a lowercase *o* followed by two spaces. Indent the listed items on both the right and the left, as illustrated here.

-- Lists introduced by a dash (two hyphens). They are a little less emphatic than bulleted lists. The dash usually appears without a space between it and the text. Indent the listed items on both the right and the left.

1. Numbered or alphabetical lists. They help readers cross-reference items and are valuable if the items are listed in descending order of importance. Leave two spaces after the period. Indent the listed items on both the right and the left.

The complimentary closing should be *Respectfully* (formal) or *Sincerely* (less formal). The writer's name and title appear four lines below the closing.

The typist's initials appear flush left, followed by notations for enclosures and carbon copies. (See the other letter models for examples of these items).

Sincerely,

Ellen G. Sanderson

Ellen G. Sanderson

rgt

Ms. Louise H. Hansen
Director of Manual Preparation
The Locklear Company, Inc.
Suite 3546, First National Building
456 Second Street
Houston, TX 82398

Dr. Edwin B. Roberts
Chief, Psychiatric Services
Saint Benedict's Hospital
P.O. Box 67
North Medford, OR 76598

NOTE 1: Proper titles are a complex issue, so if you are writing to national political figures, royalty, or foreign officials, check with a standard reference such as Lois Hutchinson's *Standard Handbook for Secretaries* (1977) or *Webster's Secretarial Handbook* (1976); these and other resources are listed in REFERENCES.

NOTE 2: If you are writing to a woman, try to determine her title preference: *Miss, Mrs.*, or *Ms.* If you cannot determine a preference, then omit the title entirely. *Ms.* used to be the preference in such cases, but a recent survey indicated that many married and divorced women did not wish to be addressed as *Ms.* under any circumstance (see SEXIST LANGUAGE). If you omit the title, do so in both the inside address and the salutation:

Carolyn D. Faust
Personnel Manager
Osage Power and Light
1212 Circuit Street
Omaha, NE 55532

Carolyn D. Faust: (*salutation*)

or

Dear Carolyn D. Faust:

NOTE 3: If you don't know who will read your letter or if you are writing to an organization rather than to an individual within the organization, then address either a position title or the name of the organization (or a department within it):

District Engineer
Andrews District Office
MOGO Oil Company
901 West Street
Andrews, OK 55555

Department of Geophysical Research
New Orleans Regional Office
MOGO Oil Company, Inc.
657 Basin Street
New Orleans, LA 22222

NOTE 4: Do not use abbreviations in the inside address except for the standard U.S. Postal Service abbreviations for states. Use the following two-character state abbreviations both in the inside address and on envelopes:

Alabama	AL
Alaska	AK
Arizona	AZ
Arkansas	AR
California	CA
Canal Zone	CZ
Colorado	CO
Connecticut	CT
Delaware	DE

LETTERS

Simplified Letter, Open Punctuation

SKY AVIATION
822 Ocean View Drive
Long Beach, California 90802
(714) 332-3978

March 15, 1985

CONFIDENTIAL

Ms. Susan Willey
Finance Officer
G.L. Findley and Company
345 Anchor Street
Portland, OR 76209

THE SIMPLIFIED LETTER

A simplified letter, Ms. Workman, follows the format and style developed by the Administrative Management Society. Its chief features are a full block format (everything flush left), open punctuation, and the omission of both the salutation and the complimentary closing.

As illustrated above, both the date and inside address appear as they would in a block letter (see the model of the block letter). Notations such as CONFIDENTIAL and PERSONAL are optional.

A subject line (all capitalized) replaces the salutation.

The reader's name usually appears somewhere early in the first paragraph. Such a reference is a nice personal touch. The paragraphs are not indented.

Lists

Lists, especially numbered lists, are usually flush left with double spacing between items to set them off:

1. Listed item 1

2. Listed item 2, and if the item has more than one line, subsequent lines are flush left. The idea is for typists to type as few extra spaces and punctuation marks as possible.

3. Listed item 3

District of Columbia	DC
Florida	FL
Georgia	GA
Guam	GU
Hawaii	HI
Idaho	ID
Illinois	IL
Indiana	IN
Iowa	IA
Kansas	KS
Kentucky	KY
Louisiana	LA
Maine	ME
Maryland	MD
Massachusetts	MA
Michigan	MI
Minnesota	MN
Mississippi	MS
Missouri	MO
Montana	MT
Nebraska	NE
Nevada	NV
New Hampshire	NH
New Jersey	NJ
New Mexico	NM
New York	NY
North Carolina	NC
North Dakota	ND
Ohio	OH
Oklahoma	OK
Oregon	OR
Pennsylvania	PA
Puerto Rico	PR
Rhode Island	RI
South Carolina	SC
South Dakota	SD
Tennessee	TN
Texas	TX
Utah	UT
Vermont	VT
Virgin Islands	VI
Virginia	VA
Washington	WA
West Virginia	WV
Wisconsin	WI
Wyoming	WY

NOTE 5: Address formats for Canadian and other foreign addresses vary slightly from American address formats. The biggest difference is that the name of the country appears on a separate line and is usually typed in all capital letters:

134 Western Province Boulevard
Edmonton, Alberta
T5J 2H7
CANADA

Unter den Eichen 56
Heidelberg 3886
WEST GERMANY (*or* FEDERAL REPUBLIC OF GERMANY)

Attention Line

An attention line is necessary when the inside address does not contain either the name of an individual or the name of a department. In these cases, an attention line appears two lines below the inside address and is flush with the left-hand margin:

H. Allen and Sons Insurance
Company
Suite 3409, Valley Bank Building
408 Pico Boulevard
Long Beach, CA 88888

Attention Miss Georgia Banks

or

Denver Regional Office
Midland Oil and Gas Company
4509 Western Avenue
Denver, CO 77777

ATTENTION EXPLORATION DEPARTMENT

LETTERS

As these examples illustrate, you do not need a colon following *Attention*. Note that *attention* may be typed with an initial capital letter or with all capital letters. If you use all capitals, then type the name following *ATTENTION* in all capitals, too. See the block letter model.

Simplified Letter, Open Punctuation

Ms. Susan Willey
Page 2
March 15, 1985

Other types of lists using either bullets or dashes are usually indented at least 5 spaces:

- Bullets are made by typing lowercase *o*'s and filling in the centers with a pen. The text in such a list is indented from both the right and left margins, as in this example.

-- A dash, actually two unspaced hyphens, is less emphatic than a bullet. Dashes usually appear with no space between them and the text following them. The text is aligned in a block.

The heading of continued pages is printed block fashion, as illustrated above. At least three lines of text should appear on continued pages.

No complimentary close appears in a simplified letter. However, the closing line of the letter can be a courteous closing: "We welcome the opportunity to work with you," or "please call us at 123-456-7890 if we can assist you further." The writer's name and title (both in all capitals) appear five lines below the final paragraph; the name and title can be separated by a spaced hyphen or a comma.

Reference initials and notations about enclosures and carbon copies follow the name and title of the writer, as shown below.

Kirk Youngsblood, Jr

KIRK YOUNGSBLOOD, JR.—CHIEF ACCOUNTANT

KY:lgh
Enclosures (4)

cc Alvin G. Harris

Salutation

The salutation usually begins with the conventional greeting *Dear* and is followed by the title and name of the addressee. In the open punctuation style, nothing follows the salutation. In standard punctuation, use a colon (for formal letters) or a comma (for informal letters) after the salutation.

Here are some sample salutations when the writer knows the addressee's name and title:

Dear Mr. Neal:
Dear Frank: (*or* Dear Frank,)
Dear Mrs. Skoal:
Dear Miss Anderson:
Dear Ms. Branch:
Dear Cheryl: (*or* Dear Cheryl,)
Dear Dr. Burns:
Dear Professor Bettridge:
Dear President Maloney:
Dear Miss Dearden and Mr. Wu:
Dear Mrs. Anderson and Ms. Blaine:

When you don't know the addressee's name, you have several options:

Ladies and Gentlemen:
Gentlemen and Ladies:
Dear Sir or Madam:
Dear Madam or Sir:
Ladies: (*all women*)
Gentlemen: (*all men*)

or

Dear Colleagues:
Dear Friends:
Dear Members of the Council:
Dear Landowners:

Traditional salutations include such forms as *Gentlemen*, *Dear Sirs*, and *Dear Mr.* __________ . Avoiding these sexist greetings is sometimes problematic. Many people consider *Ladies and Gentlemen* and *Dear Sir or Madam* to be overly formal and old fashioned. If you know the gender of the person you are addressing, then *Dear Mr.* or *Dear Ms.* (plus the name) is acceptable. However, if you don't know that person's gender, then using a greeting that identifies gender could be a problem: *Dear Mr. Smith* (what if Smith is a woman?) or *Dear Ms. A. B. Cooper* (what if A. B. Cooper is a man?). Sometimes, a person's name suggests gender: John Smith, Mary Jones, George Hayes, Linda Meyers. However, you cannot always be certain: Actors Michael Learned and Glenn Close are women. (See SEXIST LANGUAGE.)

Many people today suggest that gender should not be an identifier

LETTERS

in the business and technical world. Some companies insist that their employees be addressed by first and middle initials and last name: *C. H. Hardy, B. W. Richmond*, etc. Many other companies insist that letter writers not use traditional but sexist addresses, such as *Gentlemen* and *Dear Sirs*. The dilemma occurs when you don't want to use a sexist salutation but don't like any of the alternatives. One solution is the simplified letter, which omits the salutation altogether. Another solution is simply to omit the gender title: *Dear A. B. Cooper.*

NOTE 1: As noted above, when you don't know a woman's title, your best option may be to use the woman's name without a title:

> Dear Helen Brown:

NOTE 2: Some writers prefer a more formal letter style, especially in the salutation:

> My dear Mr. Devon:
> My dear Susan:

These salutations sound too stiff to be acceptable today. Many people would also find them condescending. You should also avoid the following previously acceptable salutations:

> Dear Messrs. Franks and Harris *(for two men)*
> Dear Mesdames Long and Minor *(for two women)*

Even very educated readers in the U.S. would have difficulty pronouncing these French forms and would likely consider the writer odd.

Subject Line

Subject lines are useful ways of establishing the letter's subject. Subject lines allow readers to file letters by subject and retrieve them fairly easily from files. If you use a subject line, make it as specific as possible so that readers know instantly what the letter is about. (See HEADINGS.)

Insert subject lines two lines below the salutation and two lines above the first line of the text. (See the block and semi-block letter models for examples. See the simplified letter model for slightly different spacing.)

Highlight the subject line by choosing underlining, all capital letters, or boldface type.

Text or Body

The text or body begins two lines below the salutation (or optional subject line). In the simplified letter, the text begins three lines below the subject line.

The text of most letters is single spaced, although double spacing is acceptable if the letter is very short. Leave a blank line between paragraphs regardless of line spacing and no matter which letter format style you follow.

In block and simplified letters, do not indent paragraphs. In modified and semi-block letters, indent paragraphs (usually 5 spaces). Some organizations indent paragraphs up to 10 spaces.

Reversed indentation (sometimes called "hanging indentation") is a format option, especially in advertising letters. In these letters, the overall format can follow one of the four models of common business letters, but the paragraphs look like this:

> A hanging-indented paragraph begins flush with the left-hand margin, but subsequent lines in the paragraph are indented, usually 5 spaces.

Long quotations within the text of a letter are indented 5 to 10 spaces on both the left and right margins. Double space before and after such quotations, so that the quotations are framed by white space.

Use similar right and left indentation for lists. (See the semi-block and simplified letter models for examples of how to set up lists; also, see LISTS.)

If the text continues beyond the first page, then ensure that at least three lines of text appear on the second page. If necessary, adjust margins and line spacing so that the first page is not too crowded and the second page and additional pages have enough text to justify a continued page.

Headings for Continuation Pages

Continuation pages should begin with a heading containing the name of the person receiving

LETTERS

the letter, the page number, and the date. Two patterns are common. See the models of the block and simplified letters for a block pattern. See the model of the semi-block letter for an alternative pattern.

The continuation heading can also repeat information mentioned on the first page in the reference line: invoice number, file number, date of a previous letter or memo, etc.

Complimentary Closing

The complimentary closing appears two lines below the closing line of the text. The block, modified block, and semi-block letters require complimentary closings. The simplified letter omits the complimentary closing.

Alignment of the complimentary closing varies according to the format style of the letter. As the models indicate, the complimentary closing in block letters appears flush with the left-hand margin. In modified and semi-block letters, the complimentary closing appears right of center and is sometimes flush with the right-hand margin.

Your choice of a closing is one of the clearest ways you convey the level of formality and the degree of personal feeling you have toward the reader.

In most business letters—those that are relatively formal without being stiff or distant—choose one of the following closings:

Sincerely,
Sincerely yours,
Thank you,

NOTE: In ordinary business letters, avoid these often preferred but excessively formal closings:

Yours truly,
Very truly yours,
Yours very truly,

Truly has become a cliche in letter closings, so avoid it.

In informal, friendly letters, you might use these closings:

Best wishes,
Regards,
Best regards,
Kindest regards,
Cordially,

In highly formal letters, such as those addressed to dignitaries and high government or ecclesiastical officials, you might use one of the following closings:

Yours sincerely,
Respectfully yours,
Respectfully,

These formal closings usually match similarly formal salutations. If your letters demand more formality, check with Lois Hutchinson's *Standard Handbook for Secretaries* (1977), with *Webster's Secretarial Handbook* (1976), or with one of the other resources listed in REFERENCES.

Signature Block

The signature block follows the complimentary closing. In simplified letters, the signature block appears four or five lines below the last line of the text.

Alignment of the signature block varies with letter format styles. In block and simplified letters, the signature block is flush with the left-hand margin. In the modified and semi-block styles, the signature block is usually right of center (see those models).

The signature block consists of the following:

Company name (*optional*)
Handwritten signature of the writer
Full typed name of the writer
Title of the writer

The following three signature blocks (along with the complimentary closings) are typical. The first illustrates a company name:

Yours,

D. & L. DRILLING EQUIPMENT

Dwight G. Edwards
Sales Manager

Sincerely,

Ivan G. Nostromo, Jr.
Chief Engineer

Best wishes,

Howard G. Balock, PhD
Personnel Manager
Engineering Division

NOTE 1: The company name is necessary only when the letter represents a company policy, position, or decision, especially in legal matters. Note that the signature of the company official plus the typed name and title of the official also appear in the signature block.

LETTERS

NOTE 2: Sign formal and official letters with your full legal name. In informal and friendly letters, you need to sign only your first name (as in the third example above). Do not include courtesy titles such as *Mr., Miss, Mrs., Ms.,* or *Dr.* in your written signature.

NOTE 3: Women's signatures generally include the woman's given and family names, with no courtesy title, such as *Ms., Miss,* or *Mrs.* These titles, if appropriate, would appear, either with or without parentheses, in the typed version of the name following the signature:

Sincerely,

(Mrs.) Elaine Raddison
Treasurer

or

Mrs. Elaine Raddison

If a woman prefers to use her husband's full name, then the husband's name is typed below the signature:

Sincerely yours,

Mrs. Thomas Raddison
Treasurer

NOTE 4: Academic titles and professional titles are not part of the signature. If you use them, they appear following the typed name. If used, these academic titles and professional titles replace *Dr.* or other courtesy titles preceding the name:

not this

Dr. Grace Babbitt, M.D.

this

Grace Babbitt, M.D.

NOTE 5: Secretaries who sign letters for an author should sign the author's name and then add their own initials either in the middle or on the right-hand side under the signature:

Sincerely,

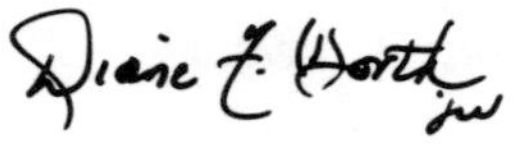

Diane F. Worth
Benefits Specialist

If secretaries or others sign their own names, rather than the author's name, then they should sign *for* the author:

Sincerely,

For Frank Procter, P.E.
District Engineer

Reference Initials

The reference initials consist of the secretary's or typist's initials and often the writer's initials.

These initials appear two lines below the last line of the signature block and are always flush with the left-hand margin.

If only the typist's initials appear, they are usually lowercase:

goj

If you include writer's initials, type them in all capitals, followed by a slash mark or a colon, followed by the typist's initials in lower case:

LHF/goj
LHF:goj

In some instances, the writer is different from the person sending the letter. In these cases, the signer's initials come first, the writer's initials come next, and the typist's initials come last:

LHF/TK/goj
LHF:TK:goj

Enclosure Notation

Enclosure notations remind readers that one or more items were enclosed with the original letter. Such notations usually come directly under the reference initials. (See the model letters for examples of their placement.)

Enclosure notations differ greatly in their forms. Here are some of the commonly accepted and correct forms:

Enclosure
Enclosure (4)
4 Enclosures
Enclosures 4
Enc.

Sometimes the types of enclosures are indicated:

Enclosures
1. Invoice 5487/87
2. File 54A-R333
3. Map 28g

NOTE: If the items "enclosed" were sent separately, indicate that as follows:

Enclosures
1. Invoice 5487/87

Sent separately
2. Map 28g
3. FFFF Price list

Carbon Copy Notation

Carbon Copy (and blind carbon copy) notations show the distribution of the letter. This notation comes two lines below enclosure notations and is flush with the left-hand margin.

Carbon copy notations may appear as follows:

cc
cc:
Copy to
Copies to

The usual practice is to list all people receiving the letter besides the person addressed in the inside address or attention line:

cc G. L. Lane
H. D. Fisk
N. O. Pope

If some copies circulate to people without the addressee's knowledge, then these people's initials appear only on the internal carbon following the abbreviation *bcc* (for blind carbon copies):

bcc V. N. Hoopes
W. X. Salvatore

Postscript

Postscripts are for additions to the letter after it has been typed or for items needing emphasis.

Postscripts appear two lines below the last line of the carbon copy notation (or reference initials).

Postscripts may or may not start with initials: *PS* or *PPS*. See the postscript in the model of the modified block letter for an example without such initials.

Envelopes

All business envelopes, regardless of size, have two mandatory features:

—The addressee's full name and address. These should be centered vertically on the envelope and should be centered horizontally between the return address and the right edge of the envelope.

—The sender's full name and full address. This return address is usually printed or typed in the upper left-hand corner, two or three lines below the top edge and five spaces from the left edge of the envelope.

Besides the address and return address, envelopes may have the following:

—Special mailing notations (*SPECIAL DELIVERY, CERTIFIED MAIL, REGISTERED MAIL*) come beneath the stamp in the upper right-hand corner.

—Other miscellaneous notations (*Personal, Confidential, Please Forward*, and *Hold for Arrival*) appear above the receiver's address.

NOTE 1: The names and addresses should be consistent with those in the letterhead, the inside address, and the signature block of the letter.

NOTE 2: The address should contain no abbreviations except those in the legal name of an organization and in the Postal Service's two-character abbreviations for states. The name and address should be typed in block style:

Mr. Hank Stephenson
Financial Officer
G.H. Vogel and Company, Inc.
Mail Drop 567-3
650 First Avenue
Los Angeles, CA 90000

NOTE 3: Abbreviations are permissible in mass mailouts using addresses from computers. The U.S. Postal Service has provided standard sets of abbreviations for long names of cities and towns, as well as more general terms like *road* and *university*.

NOTE 4: Carefully fold letters before inserting them into envelopes. The two common methods of folding letters are as follows:

Folds for Long Business Envelopes (No. 10)

—Fold the bottom third of the letter up and crease. Next fold the top third of the letter down and crease. (Caution: The top fold should not come far enough down to bend or crease the third of the paper folded up from the bottom.)

LETTERS

Folds for Regular Business Letters (No. 6¾)

— Fold horizontally almost in half, with about one-half inch of the top of the paper visible above the folded portion.

— Then fold the paper in the vertical direction. This time, fold the paper into thirds: the right third over the middle third, and then the left third over the other thirds. (If folded properly, the upper left corner of the letter is on the top of all of the folds.)

LISTS

Lists include a series of items embedded within a paragraph (called **paragraph lists**) and a series displayed vertically (called **displayed lists**):

1. Listed item a
2. Listed item b
3. Listed item c
4. Listed item d

Paragraph Lists

1. Use a list within a paragraph whenever the list is short (fewer than six items) and you do not wish to emphasize the list:

Five collective protection countermeasures were identified: (1) simple activated-carbon absorption filters, (2) regenerative filters, (3) closed-loop or recirculation (4) environmental control systems, pyrolytic destruction of agents, and (5) corona discharge and other molecular disruption techniques.

Displayed Lists

2. Use a displayed list for a long series of items and for any series you wish to emphasize:

Five collective protection countermeasures were identified:

1. Simple activated-carbon absorption filters
2. Regenerative filters
3. Closed-loop or recirculation environmental control systems
4. Pyrolytic destruction of agents
5. Corona discharge and other molecular disruption techniques

Lists and Numbers or Letters

3. Use numbers or letters to identify each item in a paragraph series. Enclose the number or letter within parentheses:

The HCF memory is in three sections: (1) program, (2) nonvolatile RAM, and (3) scratch-pad RAM.

or

The HCF memory is in three sections: (a) program, (b) nonvolatile RAM, and (c) scratch-pad RAM.

4. Use numbers, letters, bullets, or dashes to identify each item in a displayed list.

Use numbers or letters whenever the list is lengthy, whenever the text must refer to items in the list, or whenever the items are listed in decreasing order of importance. The numbers or letters should not be enclosed by parentheses, but they should be followed by a period:

a. Definition of systems and subsystems

b. Progressive apportionment of figure-of-merit requirements to subsystems

c. Progressive definition of functional requirements for subsystems to meet mission requirements

d. Definition of subsystem design and interface constraints as dictated by the chosen deployment strategy

See NUMBERING SYSTEMS.

LISTS

Lists and Bullets or Dashes

5. Use bullets or dashes to identify each item in a displayed list when the list contains items of equal importance and those items will not have to be referred to by number or letter:

We selected these means of accomplishing the scope of work for the following reasons:

- They respond to the program needs as defined in the RFP.
- They reflect the approach to project definition steps that experience indicates is effective in other isotope separation projects.
- They reflect our experience in assisting R&D personnel in transferring the requisite technology to documentation rapidly enough to support tight project schedules.

Create bullets by typing a lowercase *o* and using a black ink pen to fill in the center.

Lists and Capitalization

6. Capitalize the first word of each item in a displayed list:

- Item a
- Item b
- Item c

NOTE: The exception to this rule occurs whenever the listed items complete the thought begun in the introductory sentence (see rule 12 below).

7. Capitalize the first word of each item in a paragraph list only if each item is a complete sentence or if an item begins with a proper noun:

We propose that the qualification program include (1) documentation of tests conducted on similar equipment, (2) service history of similar equipment, and (3) Bell Laboratories' testing of the equipment.

Lists and Colons

8. Use a colon to introduce a list when the sentence preceding the list contains such anticipatory words or phrases as *the following, as follows, thus,* and *these*:

The 1553 interface will be programmed to respond to the following mode commands:

a. Synchronize (without data word)

b. Synchronize (with data word)

c. Transmit status word

d. Reset terminal

9. Do not end the introductory sentence with a colon if the sentence is lengthy and the anticipatory word or phrase occurs very early in the sentence or if another sentence comes between the introductory sentence and the list:

The following steps are required to process the message received by the bit processor. Note that the sequence parallels both subsystem interface architectures.

1. Recognize a valid RT enable flag.
2. Read the 16-bit word from the bit processor.
3. Check the T/R bit.
4. Decode the subaddress field where incoming data will be stored.

10. In paragraph lists, do not precede the list with a colon if the list follows a preposition or a verb:

The 1553 interface will be programmed to respond to (a) synchronize (without data word), (b) synchronize (with data word), (c) transmit status word, and (d) reset terminal.

NOTE 1: If this list becomes a displayed list, however, use the colon even though the list follows a verb:

The 1553 interface commands are:

a. Synchronize (without data word)

b. Synchronize (with data word)

c. Transmit status word

d. Reset terminal

NOTE 2: This example (where a colon follows a verb introducing a displayed list) is more and more widely accepted. However, some editors would still revise the lead-in sentence to read as follows:

The 1553 interface has these commands:

LISTS

Lists and Periods

11. Do not end items in a displayed list with periods unless one or more of the items is a complete sentence.

See the example under rule 8, where periods have been omitted. In the example below, each item listed is a complete sentence, so each requires a period:

> Western Aeronautics has a wide range of related experience:
>
> - We built and tested crash-proof recorders for the T41 and T7G engine data analyzers.
> - We designed, developed, tested, and manufactured the Central Air Flight Data Computer (CAFDC) systems for the Air Force.
> - We produced the full range of flight data computers with 18 different designs flown on 35 different aircraft.

12. End the last item in a paragraph list with a period. End the last item in a displayed list with a period only if the list completes the sentence begun with the introductory statement and each item in the list is separated by a comma or semicolon:

> The life of a counterflow plate-fin recuperator may be prolonged by
>
> 1. increasing the fin thickness,
> 2. providing airflow passages inside the manifold hoops, and
> 3. providing for gas flow through the hollow side bars.

NOTE: Lists with continued punctuation (as in this example) are now rare, probably because writers, influenced by advertising, are using capitals and spacing to make lists more visually emphatic. Such continued lists usually are not introduced with a colon.

Lists Within Lists

13. Whenever one list occurs inside another list, use numbers for the outer list and letters for the inner or nested list:

> 1. The physical characteristics of the regenerator include (a) ferritic stainless steel construction, compatible with a moist, coastal salt-air environment; (b) an internally insulated turbine exhaust duct; (c) a horizontal configuration; and (d) high performance rectangular fins.
> 2. The performance data includes (a) 88 percent thermal effectiveness, (b) over 4,600 hours of operating time, and (c) no evidence of corrosion or fouling.

NOTE: For a third level of nested lists, use lowercase Roman numerals (*i, ii, iii,* etc.). Also use caution. Lists within lists within lists become confusing and irritating.

Lists and Parallelism

14. Ensure that items in lists are parallel in structure. Furthermore, begin each item with the same type of word (noun, verb, adjective, etc.):

> This programming language interface will allow the simulation model programs to access the data base in the following ways:
>
> 1. Retrieve records with specific key values
> 2. Retrieve records in data base sequences
> 3. Insert records into the data base
> 4. Delete records from the data base
> 5. Modify and replace records in the data base

Each item begins not only with a verb, but with the same kind of verb. The list would not be parallel if the verb or sentence forms were changed:

> This programming language interface will allow the simulation model programs to access the data base in the following ways:
>
> 1. Retrieve records with specific key values
> 2. Records in data base sequences can also be retrieved
> 3. Insertion of records into the data base
> 4. Deleting records from the data base
> 5. Modification and replacement of records in the data base

Lists, whether in paragraph or display form, must always be parallel. The items listed must be consistent in form and structure. (See PARALLELISM.)

MANUSCRIPT FORM

Manuscript originally referred to handwritten copy. It now means the typed or word-processed copy from which a final copy is prepared. The final copy may be typeset and printed, or it may be photocopied from the manuscript and then circulated within an organization.

The following rules apply to most manuscripts, whether formally printed or not. If you are writing for a particular publication, however, consult the editors for specific guidelines on the form your manuscript should take.

---

1. Type manuscripts on good quality paper, usually 8½-by-11-inch bond.

Do not use erasable paper, onion-skin paper, or odd-sized paper. Odd-sized paper can create difficulties in photocopying. Corrections made on erasable and onion-skin paper are difficult to read.

---

2. Type manuscripts using a standard type size and style, either elite or pica.

Interchangeable typing elements and print wheels give writers a choice of typefaces and sizes. Nevertheless, in manuscripts you should avoid those typefaces that present too radical an image for sustained reading. Do not use typefaces that are too fat or too thin, and avoid italics, script, and gothic typefaces for the bulk of your text. However, you can use exotic typefaces for effect if you use them sparingly. (See EMPHASIS.)

Standard typesizes are 10- or 12-pitch. Some typewriters and most computer printers allow for larger or smaller typesizes. Use larger typesizes for headings only. Avoid smaller typesizes except in such printed matter as forms and contracts and in tables, charts, maps, and other visual aids where space is limited.

---

3. Double- or triple-space manuscripts and leave generous margins at the top, bottom, and sides of each page.

Double or triple spacing and generous margins allow for editorial insertions and corrections. Usually the top and the left-hand margins should be at least 1½ inches, and the bottom and the right-hand margins should be at least 1 inch. Journals and presses often have their own requirements. Some provide paper with a ruled box to ensure that writers and typists leave proper margins.

Footnotes, bibliographies, and inserted material (such as extensive quotations) should also be double-spaced, especially if such items are being prepared for a final printed copy. Footnotes should be listed separately, chapter by chapter, rather than inserted at the bottom of each page. (See FOOTNOTES and BIBLIOGRAPHIC FORM.)

If your final copy will be prepared directly from manuscript, and if your final copy will be single-spaced, resist having the manuscript single-spaced too early. Double-spaced manuscripts are much easier to revise and edit.

On word processors, turning a draft double-spaced manuscript into a single-spaced final copy is especially easy.

---

4. Number manuscript pages consecutively, beginning at the first page of text or, if your text has chapters, at the chapter divider for the first chapter.

Page numbers should appear at the top of each page, centered or flush right.

Consecutive numbering throughout the manuscript is advisable, but in longer texts, especially technical publications, numbering by chapter is advisable: 4-65 for Chapter 4, page 65.

Chapter dividers and the first pages of chapters should have odd page numbers, although those page numbers are usually not printed in the text. These pages should appear on the right-hand or facing pages of double-sided manuscripts. If you use a chapter divider, the divider page is p. 1, its reverse side is an unnumbered p. 2, and the first page of text is p. 3.

Front material, including the table of contents, title page, preface, and list of illustrations, are usually numbered with small Roman numerals: *v, vi, vii*.

Avoid inserted pages if you can. If you cannot, number them as follows: *36a, 36b, 36c,* etc. The inserted pages in this example would follow page 36.

Numbers for figures, tables, sections, and chapters should be Arabic. The numbers of figures and tables often reflect the chapter numbers as well: *Figure 6-8* for Chapter 6, Figure 8.

End material (appendices, glossaries, notes or footnotes, and bibliographies) should follow the sequential numbering begun in the text. If the text is numbered chapter by chapter (*4-18, 4-19,* etc.), then end material should receive its own section numbers: *A-15* for page 15 of Appendix A.

5. Print corrections above the text, preferably in ink, and make them as legible as possible.

If corrections will not fit above the line in question, cut the text apart and insert a newly printed or typed version into the space left in the original copy. Avoid making elaborate insertions in the margins, and never use the reverse side of a page for corrections or comments. If inserted material is lengthy, use inserted pages and clearly number them. See rule 4 above.

Formal proofreading symbols are necessary only when you are dealing with professional typesetters and printers. (See EDITING AND PROOFREADING SYMBOLS.) If you use your own system for noting changes, however, ensure that typists understand the correction symbols.

Always keep a complete copy of your manuscript, including all corrections. This precaution is especially important if you are submitting your manuscript to a publisher.

6. Develop and maintain consistent headings within a manuscript.

Modern word processors allow headings to be more varied than they could be with typewriters. If you have access to word processing, explore the options. If your manuscript will be printed, you have an even greater range of possibilities. Typeset and printed headings can be set in larger type sizes or different type styles, and they can be printed in boldface or color. (See HEADINGS.)

7. Plan your tables, figures, and other visual aids as early as possible.

Plan your visual aids before writing much of the text, especially in technical documents. Many publications have strict guidelines for the size, style, and quality of visual aids. You should be aware of the visual aid opportunities and limitations before devoting too much effort to the text, and you should ensure that the visuals you produce will be consistent with a publisher's guidelines. Also, visual aids often take longer to produce than text. (See VISUAL AIDS.)

8. Avoid extensive cross references throughout a manuscript, especially references to page numbers.

Page references may change every time a text has to be repaginated. If cross references are necessary, use sections or chapter headings. Also, remember that extensive cross references demand extra proofreading and checking for internal consistency.

Publishing a Book or Journal Article

If you wish to publish a book or journal article, ask the editors for a copy of any necessary editorial guidelines. In addition, you may wish to refer to *The Chicago Manual of Style*, 13th Edition. It has an excellent chapter on manuscript preparation and copyediting.

MAPS

Maps show the geographic features of an area and indicate spatial relationships, locations, and distances. Maps can also show geographic distributions of people, housing, manufacturing sites, wells, geologic features, crops, mineral occurrences, watersheds, etc.

Maps are a type of figure and should receive figure numbers unless you are using a great many maps. If the number of maps is large, you can label them *Map 1, Map 2, Map 3,* etc.

For general information on using visual aids, see VISUAL AIDS. See also CHARTS, GRAPHS, ILLUSTRATIONS, PHOTOGRAPHS, and TABLES.

For information on captions, see CAPTIONS.

1. Ensure that maps are scaled correctly to show what you want to show.

All maps must have a scale. The maps you choose should depict the features you wish to emphasize in enough detail for readers to grasp the geographic relationships shown. If you use too small a scale, you may not be able to provide a useful geographic perspective on what you're showing and your maps may become too large. If you use too large a scale, your features may become too small and the surrounding area too large—the map may overwhelm what you're trying to show.

Choosing the proper scale may be difficult if your range of scales is limited. Try, however, to choose a scale that properly focuses readers' attention on the geographic features you wish to show. If too large an area of the map has no features of interest or is not relevant to your subject, then you are using the wrong scale.

Once you have chosen a scale, indicate distances clearly and use familiar units of measure. Label distances in feet or miles rather than in meters and kilometers. If the document in which the map appears will have international readers, then select either English or metric units and provide a conversion table or alternate scale.

See UNITS OF MEASUREMENT.

2. Clearly label the map and its parts.

Ensure that readers understand all of the features on your maps. Orient all of the letters, numbers,

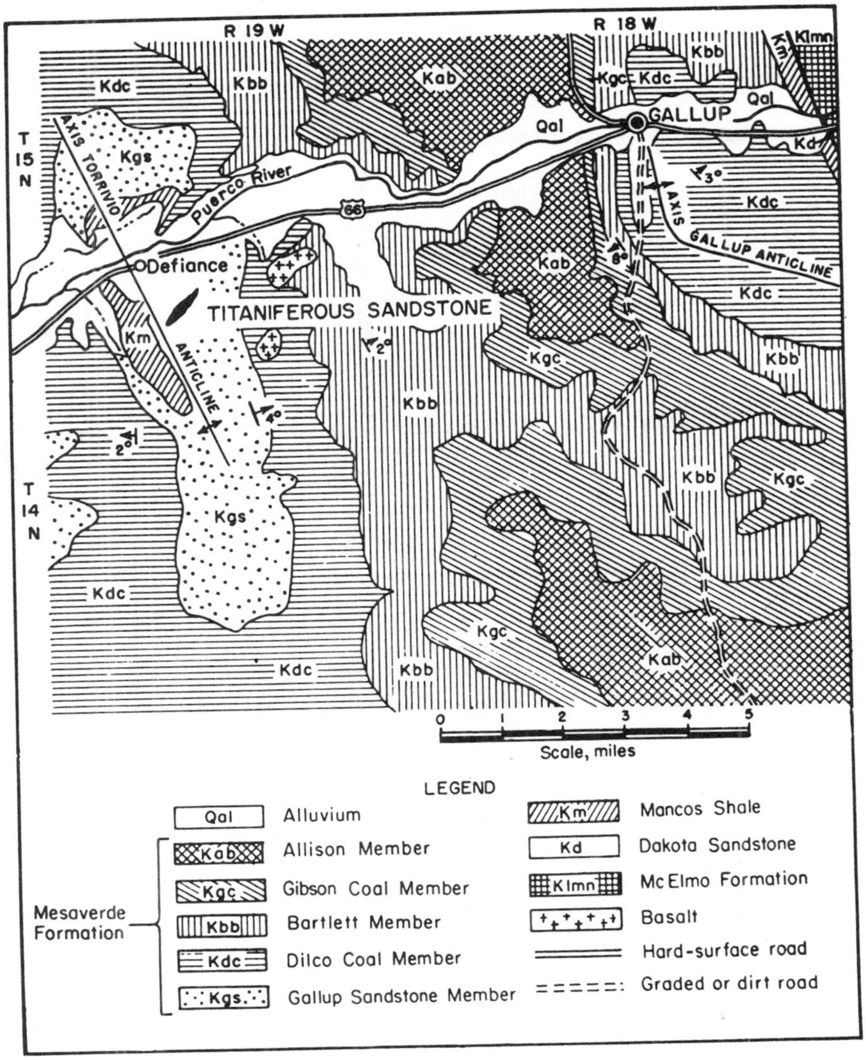

Figure 1. Titaniferons sandstone deposits near Gallup, New Mexico.

MAPS

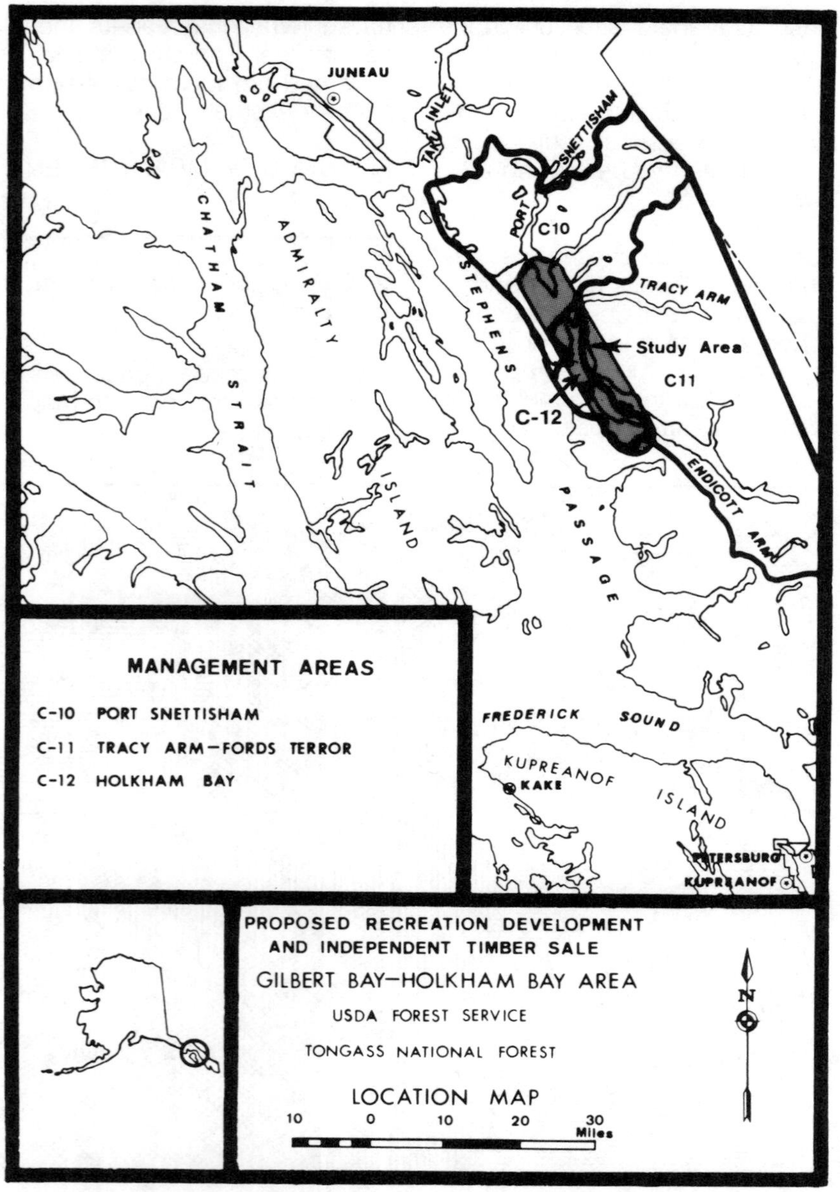

Figure 2. Location map for the Gilbert Bay - Holkam Bay study area.

and labels horizontally on the page, and use a legend if necessary (see figure 1).

3. Simplify maps by eliminating unnecessary detail.

Like other visual aids, maps must have a single central concept or idea. They should present one important idea and should be focused on that idea. Anything on the map that is extraneous to that central concept—such as contours, unrelated roads, creeks, and trails—is clutter. So don't make maps more detailed than necessary, or readers will waste time worrying over features that have nothing to do with the central concept of the map.

4. If necessary, establish the larger geographical perspective of the map by using an inset map of the larger geographical area.

Figure 2 is a map of a study area in southeast Alaska. Readers unfamiliar with this area will not recognize the study area or know where in Alaska it is located. Therefore, to provide geographical perspective, an inset map of Alaska is used. The study area depicted on the map is circled so readers can determine where the area is located.

5. Use shading, color, and fill patterns to emphasize the features you want readers to focus on.

Shading, color, and fill patterns help distinguish features while calling attention to them. As long as you don't overdo them, these devices help readers see what you want them to see. For an example of different fill patterns, see CHARTS. You can also use overlays to identify and distinguish particular features.

Figure 1 shows a shaded map. The legend at the bottom indicates what the different fill patterns represent. However, because this map presents so many different formations, the fill patterns are supplemented by formation initials. Without these initials, the map would be harder to grasp.

Note that Route 66 and the city of Gallup provide just enough

MAPS

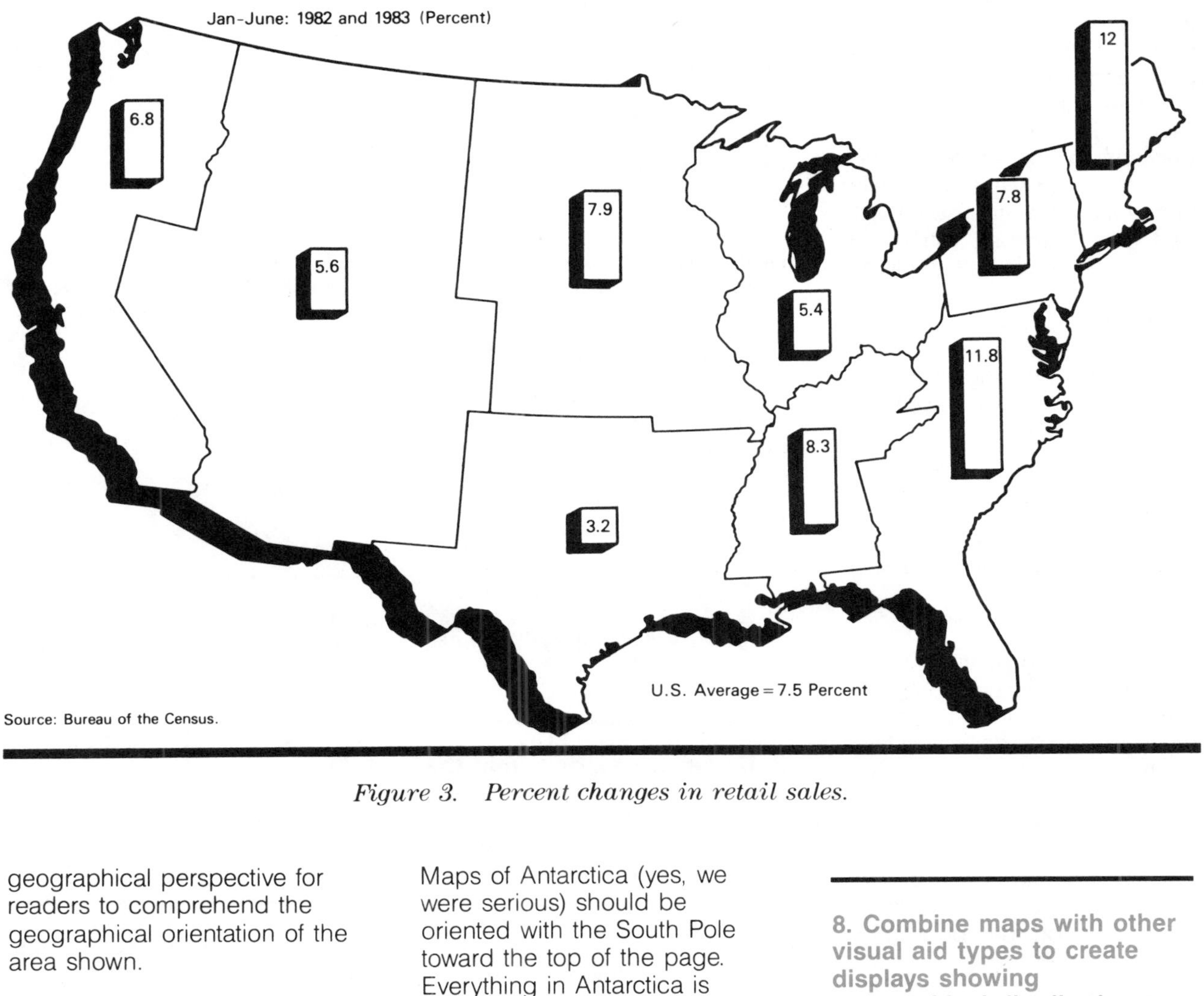

Figure 3. Percent changes in retail sales.

geographical perspective for readers to comprehend the geographical orientation of the area shown.

6. Always indicate compass direction and always orient north toward the top of the page.

Unless you're using a map of Antarctica, you should orient the map so that north is toward the top of the page and south is toward the bottom. In any case, always include compass directions.

Maps of Antarctica (yes, we were serious) should be oriented with the South Pole toward the top of the page. Everything in Antarctica is upside down.

7. Use larger lettering to label the principal points of interest on maps.

In figure 1, the principal point of interest is the titaniferous sandstone deposit, so that deposit is labelled with larger letters. Note that a white rectangular background over the fill patterns makes the lettering stand out.

8. Combine maps with other visual aid types to create displays showing geographical distribution.

Figure 3 is an excellent example of a combination visual. The map of the United States provides a geographical perspective, and the bar features indicate distribution by states. Combination visuals give you many opportunities for visualizing complex relationships. Most of the time, you are limited only by your imagination.

MATHEMATICAL NOTATION

The style of mathematical notation varies from one journal to another and from one publisher to another. So before preparing manuscripts for professional submission, determine which style the editor or publisher prefers. Adhering to the preferred style is especially important if the manuscript will be typeset because setting mathematical symbols is often the most costly phase of typesetting.

The following principles of mathematical notation apply to most publications.

1. Be consistent in writing mathematical signs, symbols, and units.

Because the conventions for writing mathematical signs, symbols, and units vary, you must establish a consistent methodology and adhere to it. If you are writing the Greek letters sigma and theta, for instance, you can use either σ or ς for lowercase sigma and θ or ϑ for lowercase theta. Either form is acceptable, but be consistent in the one you choose.

A related principle is to be consistent in your choice of units from one equation to the next. If in one equation a figure is expressed in meters, then related equations should express equivalent figures in meters, not yards.

See SIGNS AND SYMBOLS.

NOTE: Consider keeping a list of the conventions you prefer for writing signs, symbols, and units. Referring to the list as necessary will help you maintain consistency throughout your text.

2. Use displayed (separate-line) expressions for lengthy equations or for special equations.

Displayed expressions are separated from the text and are usually centered on the page with two or more spaces above and below the equation. Major equations and those equations too lengthy or too complicated to place within the text should be displayed. You can also display equations you wish to emphasize:

The initial form of the equation was

$$\int_a^b f(x)dx = \frac{b-a}{6}\left[f(a)+4f\left(\frac{a+b}{2}\right)+f(b)\right]$$

3. Use expressions within the text when the equations are minor or routine, when they are short, and when they are not important enough to highlight.

Expressions within text are more difficult for readers to see. However, in many circumstances, the writer does not need to highlight an equation by displaying it. Expressions written within text must be simple enough for readers to comprehend easily:

The expression $1/(x+y)$ becomes increasingly smaller as the values of x and y increase.

When you convert displayed expressions into textual expressions, add parentheses or brackets as necessary to clarify the mathematical relationships:

$\frac{3}{x-y}ab$	*becomes*	$[3/(x-y)]ab$
$6+\frac{x+y}{3}+18$	*becomes*	$6+(x+y)/3+18$

Similarly, when converting a textual expression into a displayed expression, remove unnecessary parentheses and brackets:

$l/(x/y)$	*becomes*	$\frac{l}{x/y}$
$(a+3)/(b+6)$	*becomes*	$\frac{a+3}{b+6}$

4. Do not punctuate displayed expressions that continue a sentence in the text:

The revised equation is

$$y = A+B(x-x_1)+C(x-x_1)^2$$

5. Divide displayed expressions that extend more than one line before the equals sign or the sign of operation:

$$\int_a^b f(x)dx = \frac{1}{3}\Big[y_o+y_n+4(y_1+y_2+\ldots+y_{n-1})$$
$$+2(y_2+y_4+\ldots+y_{n-2})\Big]\triangle x$$

NOTE 1: The opposite is true for expressions within text:

After conversion, the alternate version has $\pi(y_{i-1}+y_i)(\triangle x_i^2+\triangle y_i^2) = \pi(y_{i-1}+y_i)[1+(\triangle y_i/\triangle x_i)^2]^{1/2}\triangle x_i$

MATHEMATICAL NOTATION

NOTE 2: Do not divide short expressions. If possible, avoid dividing any expressions.

6. For reference, number displayed expressions in parentheses to the right of the expression:

$$V^5 2\pi \int_a^b xf(x)dx \qquad (15)$$

NOTE: Unless you refer to the expressions elsewhere in the text, numbering of displayed expressions is unnecessary.

7. For grouped expressions, place parentheses inside brackets inside braces inside parentheses:

$$X = a\left(b\left\{c + \left[d + 2(e+7)\right]\right\}\right)$$

NOTE: Sometimes these symbols appear by themselves or the sequence does not apply:

Sets

$$\{a, b, c\}$$

Expressions with Functions

$$f(g(x))$$

Expressions with Upper and Lower Limits

$$\left[\tan \theta\right]_o^\pi$$

See BRACKETS.

MEMOS

The term *memos* is a shortened version of *memoranda* or *memorandums*. The longer, formal words are correct, but the more convenient *memos* is now widely used and is also acceptable.

Memos are essentially letters written to persons within the writer's organization. Hence, memos are often referred to as "interoffice correspondence." Memos may also function as informal technical reports. Some oil companies, for instance, publish internal documents called Geologic Memorandums. In form, these documents are memos; in content, they are technical reports. So the distinction between letters, memos, and reports is often inexact. Nevertheless, memos are almost always intended for an internal audience, and they tend to be less formal than letters and reports.

Memo length varies, depending upon the organization and the purpose of the memo. Some organizations insist that memos be no longer than one page. They argue that memos are for transmitting and storing day-to-day internal messages (memos are a more permanent record than telephone calls), and that messages requiring more than one page have enough content to justify a report.

Other organizations allow memos to be as long as necessary, and they are often as much as 20 pages long. In longer memos, the writer must use summaries, headings, lists, and other emphatic devices to break up the content and make the memo readable (see EMPHASIS).

Memo content also varies considerably—from brief notices of meetings to full-fledged analyses of alternatives and recommendations for action. The fact is that memos are useful devices for transmitting any type of information to other persons within your organization.

Effective Memo Writing

Because memos are essentially letters that stay within an organization, the principles of good letter writing apply equally to memos:

—Begin memos with the most important ideas.

—Subordinate references to previous documents, conversations, and meetings.

MEMOS

Memo (Printed Form)

To: M.L. Abrams Date: March 15, 1985
From: Joan Abercrombie
Subject: Memorandum Format on Printed Forms Invoice 45897 / A

On memorandum forms with To, From and Subject printed, the names and subject title should be aligned with the left margin of the body of the memo (as shown here). If the printed items are flush left (as above), then the names (corresponding to To and From) should be aligned with the subject title. If the printed items are aligned flush right, then the names and subject title appear two spaces after the items (or colons).

The addressee's name appears without a gender title (Mr., Mrs., Ms., or Miss). If the memo is addressed to two or more people, list other primary readers after or beneath the first addressee. If the form does not have enough space to list all addressees, then write "Distribution" in the To line and list all addressees in a distribution list at the end of the memo. List all secondary readers in a cc line at the end of the memo.

The date is usually written with the month spelled out: March 15, 1985 (not 3/15/1985).

Brief reference lines can appear two lines beneath the date. (See the discussion of reference lines in LETTERS.) Extensive references to previous documents, meetings, or conversations must follow the subject line because of space limitations on most printed forms; such references might include the names and dates of previous memos that have a bearing on this memo. If several items appear in the reference list, number them for easy reference in the body of the memo (see LETTERS).

Some organizations omit the From line and place an author's signature block at the end of the memo. Other organizations retain the From line but add the author's initials at the end of the memo. Titles following the author's name or initials are usually unnecessary but can be used.

The content of the subject line should be specific enough to tell readers exactly what the memo is about. The subject line may therefore include dates, invoice numbers, project information, loan agreement numbers, and other similar information. (A specific enough subject line may make a reference line unnecessary.) If the subject line extends beyond one line, the information should be single spaced and centered:

Subject: Revisions of the Payment Clause
Loan Agreement 5676-34

—Ensure that your memos are clearly and logically organized.

—Throughout memos, emphasize key data and ideas.

—Avoid cliched closings.

For a thorough discussion of these principles, see LETTERS. See also ORGANIZATION, EMPHASIS, REPETITION, KEY WORDS, PARAGRAPHS, and VISUAL AIDS.

Memo Format

Memo format varies considerably from organization to organization. However, memos often have these components: **heading, body, signature line, reference initials, attachment notation,** and **carbon copy notation.** See the model memo in this section for an illustration of these components.

The Heading

Memo headings, whether printed or typed, usually contain these elements:

To:
From:
Subject:
Date:

The order of these elements, their spacing and punctuation, and their placement on the page vary considerably. In printed memo forms, the heading elements often do not have colons. Typed headings usually have colons.

Some memos open with *To* and then give the subject line. Others place the date after the *To* line. Still others arrange the items in two parallel lists:

To	From
Department	Department
Subject	Date

Some memos omit the *From* line, opting instead for a typed name and signature at the end of the memo.

Two optional elements of the heading are a distribution list and a reference line or block.

The distribution list may appear in a box that follows or includes *To*. If used in the heading, the distribution list replaces the cc (carbon copy) list at the end of the memo.

Short reference lines or blocks can appear two lines below the date if the date appears by itself just right of the center of the page. Extensive references need

MEMOS

to have their own lines, usually before or after the *Subject* line:

> 1. F.H. Howell, "Testing of the Wing Plate Assembly," May 18, 1984
>
> 2. J.K. Jameson, "Design Options in the Wing Plate Assembly," March 22, 1984

Numbering references helps writers refer to the references later in the text:

> Reference 1 notes that all wing plate assemblies have passed inspection this year. However, reference 2 indicates that design modifications must be undertaken to improve reliability.

The names of both the sender and the receiver do not require courtesy titles (*Mr., Mrs., Ms.,* or *Miss*), but *Dr.* is sometimes used. Names should be as complete as possible even if the sender and receiver are close friends. Long after the memo has been filed, future readers will probably not know whom Hank or Sue refers to, and the names could be important.

The subject line should be as specific as possible (see HEADINGS):

> *this*
>
> Subject: Recommendation to Test Two Methods of Lowering Salt Content
>
> *not this*
>
> Subject: Salt Content Tests
>
> ---
>
> *this*
>
> Subject: The Sales Decline in the Northern Region
>
> *not this*
>
> Subject: Northern Region Sales

Memo (Printed Form)

M.L. Abrams
Page 2
March 15, 1985

The body of the memo has single-spaced paragraphs with one blank line between paragraphs. Paragraphs may appear in block form (no indentation) or with indentation (usually five spaces).

Headings and Lists

Headings are always valuable, but they are mandatory if the memo becomes two or more pages long. Lists can help even a one-page memo. See LISTS for a discussion of the different kinds of lists.

The heading for continued pages should contain the addressee's name, the date, and the page number. An optional form to the block pattern shown above is:

M.J. Abrams   -2-   March 15, 1985

The reference initials appear flush with the left-hand margin and are two lines below the writer's initials or signature or two lines below the text (if the writer's initials or signature are omitted).

An attachment notation appears immediately below and flush with the reference initials. The number of attachments appears within parentheses, as shown below.

The memo ends with the cc list of those secondary readers receiving copies. A bcc (blind carbon copy) list can also appear if the primary readers are not supposed to know or do not need to know the complete circulation list. The bcc list appears only on the copies, not on the original (see LETTERS).

J.A.

bop
Attachments (2)

cc Sidney White
   Blake James
   Sharon Billig

The Body

Paragraphs in the body of the memo are usually single-spaced with a double space between them. These paragraphs may or may not be indented (5 to 10 spaces). Both the indented and block forms are correct and usually acceptable; however, your organization may have a preferred style.

Headings and lists are important techniques, especially when a memo is more than a page or two long (see HEADINGS and LISTS).

The Signature Line

Traditional memos have no signature line. The author's name appears after *From* in the heading.

Recently, however, many writers have begun signing their initials or their whole names two lines

MEMOS

below the final line of the text. The name or initials may be typed, but they may also be handwritten.

Rarely do such signature lines contain titles, probably because the people within an organization already know job titles or can easily look them up in a directory.

Reference Initials

Reference initials in memos usually contain only the typist's initials. These initials appear either two lines below the signature line or two lines below the final line of text (if the memo has no signature line). The initials are usually in lowercase letters:

dor

If the reference initials also contain the author's initials, then they would precede those of the typist and would follow one of these forms:

GLK/dor   GLK:dor   glk:dor

When someone other than the sender writes the memo, the sender's initials come first, then the writer's initials, and then the typist's initials:

GLK/TER/dor   GLK:TER:dor

The Attachment Notation

Attachment notations are not very common in memos. If used, they appear on the line immediately below the reference notation:

GLK/dor
Attachments (3)

In some very technical memos with a number of attachments (such as maps or charts), the attachments may be listed at the bottom of the memo following the attachment notation.

Carbon Copy Notations

If used, *cc* (carbon copy or courtesy copy) notations appear two lines below the reference or attachment notations. The form varies:

cc
cc:
Copy to
Copies to

Memos sent to a large number of readers often have a distribution list instead of a carbon copy list. The word *Distribution* appears in the heading following *To. Distribution* also appears instead of *cc* in the carbon copy notation, and following *Distribution* is a list of the names and (if appropriate) departments of those people who should receive copies of the memo.

Only occasionally do *bcc* (blind carbon copy) lists appear on memos. Blind carbon copy lists appear only on the copies and not on the original memo.

If used, the *bcc* list appears two lines below the *cc* list.

See SPACING.

METRICS

The metric system is now used worldwide by scientists in the physical and biological sciences. The most precise version of metric system is the International System of Units or SI (from the *Système international d'unités*).

Despite its widespread acceptance, SI has not been adopted by all U.S. firms and government agencies. Retooling to metric standards has been a slow and costly process, and redrafting existing maps and design layouts has been unfeasible.

Nevertheless, SI is accepted internationally and uses unambiguous symbols. It is, therefore, the preferred system of measurement for all sciences and many areas of engineering.

1. Use the following base units and their SI symbols:

length	meter (m)
mass	kilogram (kg)
time	second (s)
current	ampere (A)
thermodynamic temperature	kelvin (K)
amount of substance	mole (mol)
luminous intensity	candela (cd)
plane angle	radian (rad)
solid angle	steradian (sr)

2. Do not capitalize or italicize SI symbols, except those derived from proper names (*e.g.*, A and K). The symbols do not change in the plural and are never followed by a period:

46 m (*not* 46m)	6 K
1 kg	6 kg
15 s	22.5 cd

Table 1. SI-Derived units with special names

QUANTITY	NAME	SYMBOL	EXPRESSED IN SI UNITS
absorbed dose of ionizing radiation	gray	Gy	J/kg
activity of radionuclides	becquerel	Bq	s^{-1}
electric capacitance	farad	F	C/V
electric conductance	siemens	S	A/V
electric potential, potential difference, electromotive force	volt	V	W/A
electric resistance	ohm	Ω	V/A
energy, work, quantity of heat	joule	J	N·m
force	newton	N	$m \cdot kg/s^2$
frequency	hertz	Hz	s^{-1}
illuminance	lux	lx	lm/m^2
inductance	henry	H	Wb/A
luminous flux	lumen	lm	cd·sr
magnetic flux	weber	Wb	V·s
magnetic flux density	tesla	T	Wb/m^2
power, radiant flux	watt	W	J/s
pressure, stress	pascal	Pa	N/m^2
quantity of electricity, electric charge	coulomb	C	s·A

NOTE 1: Use the metric abbreviations only when the metric unit follows a number. If the metric unit appears without a number, spell it out:

We measured 2 kg of salt.

but

We had several kilograms of salt.

NOTE 2: Wherever possible, choose SI units so that the numerical values will be between 0.1 and 1000:

54 m (*not* 54 000 mm)
3.6 mm (*not* 0.0036 m)

3. Use a point or period as the decimal marker, and use spaces to separate long numbers into easily readable groups of three:

45 671.378 34
0.634 701

METRICS

Table 2. SI-Derived units with no special names

QUANTITY	DESCRIPTION	EXPRESSED IN SI UNITS
acceleration—linear	meter per second squared	m/s^2
—angular	radian per second squared	rad/s^2
area	square meter	m^2
concentration (of amount of substance)	mole per cubic meter	mol/m^3
current density	ampere per square meter	A/m^2
density, mass density	kilogram per cubic meter	kg/m^3
dynamic viscosity	pascal second	Pa·s
electric charge density	coulomb per cubic meter	C/m^3
electric field strength	volt per meter	V/m
energy density	joule per cubic meter	J/m^3
heat capacity, entropy	joule per kelvin	J/K
heat flux density, irradiance	watt per square meter	W/m^2
luminance	candela per square meter	cd/m^2
magnetic field strength	ampere per meter	A/m
molar energy	joule per mole	J/mol
molar entropy, molar heat capacity	joule per mole kelvin	J/(mol·K)
moment of force	newton meter	N·m
permeability	henry per meter	H/m
permittivity	farad per meter	F/m
specific energy	joule per kilogram	J/kg
specific heat capacity, specific entropy	joule per kilogram kelvin	J/(kg·K)
specific volume	cubic meter per kilogram	m^3/kg
speed—linear	meter per second	m/s
—angular	radian per second	rad/s
surface density of charge, flux density	coulomb per square meter	C/m^2
surface tension	newton per meter	N/m
thermal conductivity	watt per meter kelvin	W/(m·K)
volume	cubic meter	m^3
wave number	1 per meter	m^{-1}

NOTE 1: When only four numbers appear on one side of the decimal, the space is optional but not preferred:

5.7634 *or* 5.763 4
8764 *or* 8 764

NOTE 2: In many foreign countries, writers use a comma as the decimal marker. If you are writing for a foreign journal or publisher, you may need to use a comma as the decimal marker:

5,763
0,634
3,1415

See DECIMALS and PERIODS.

4. Some SI-derived units have special names:

See table 1 on the previous page.

5. Some SI-derived units have no special names:

See table 2.

METRICS

6. Use the table of prefixes to form the names and symbols of multiples and submultiples of SI units:

See table 3.

NOTE 1: Without using a space or a hyphen, attach the prefixes directly to the SI base unit: *kilogram, millisecond, gigameter,* etc. Similarly, the abbreviations for the prefixes attach directly to the abbreviation for the SI units: *cm, Mg, mK,* etc.

NOTE 2: Do not use two or more of the prefixes to make compounds of the SI units. Write *ns* (nanosecond), not *mμs* (millimicrosecond).

NOTE 3: Although kilogram is the base unit for mass, the prefixes are added to gram (g), not kilogram (kg).

7. Some non-SI units are still permissible within SI:

See table 4.

8. Avoid certain metric units that have been replaced by SI units:

See table 5.

9. Use table 6 to convert SI units into the common units of measure still widely used in the United States.

Table 3. Prefixes and their symbols for SI units

MULTIPLYING FACTOR	PREFIX	SYMBOL
1 000 000 000 000 000 000 = 10^{18}	exa	E
1 000 000 000 000 000 = 10^{15}	peta	P
1 000 000 000 000 = 10^{12}	tera	T
1 000 000 000 = 10^{9}	giga	G
1 000 000 = 10^{6}	mega	M
1 000 = 10^{3}	kilo	k
100 = 10^{2}	hecto	h
10 = 10^{1}	deca	da
0.1 = 10^{-1}	deci	d
0.01 = 10^{-2}	centi	c
0.001 = 10^{-3}	milli	m
0.000 001 = 10^{-6}	micro	μ
0.000 000 001 = 10^{-9}	nano	n
0.000 000 000 001 = 10^{-12}	pico	p
0.000 000 000 000 001 = 10^{-15}	femto	f
0.000 000 000 000 000 001 = 10^{-18}	atto	a

Table 4. Non-SI units permissible within SI

QUANTITY	NAME	SYMBOL	DEFINITION
area	hectare	ha	1 ha = 1 hm^2 = 10 000 m^2
mass	ton, tonne	t	1 t = 1 000 kg = 1 Mg
plane angle	degree	°	1° = (π/180) rad
	minute	′	1′ = (π/10 800) rad
	second	″	1″ = (π/648 000) rad
temperature	degree Celsius	°C	0°C = 273.15 K However, for temperature intervals 1°C = 1 K
time	minute	min	1 min = 60 s
	hour	h	1 h = 3600 s
	day	d	1 d = 86 400 s
	year	a	
volume	liter	l or L	1 l = 1 dm^3

METRICS

Table 5. Metric units replaced by SI units

QUANTITY	NAME	SYMBOL	DEFINITION
absorbed dose of ionizing radiation	rad	rad	1 rad = 10 mGy = 10 mJ/kg
activity	curie	Ci	1 Ci = 37 GBq = 37 ns^{-1}
area	are	a	1 a = 100 m^2
	barn	b	1 b = 100 fm^2
conductance	mho	mho	1 mho = 1 S
energy	calorie	cal	1 cal = 4.1868 J
	erg	erg	1 erg = 0.1 μj
force	kilogram-force	kgf	1 kgf = 9.806 65 N
	kilopond	kp	1 kp = 9.806 65 N
	dyne	dyn	1 dyn = 10 μN
illuminance	phot	ph	1 ph = 10 klx
length	angstrom	Å	1 Å = 0.1 nm
	micron	μ	1 μ = 1 μm
	fermi	fm	1 fermi = 1 femtometer = 1 fm
	X unit	—	1 X unit = 100.2 fm
luminance	stilb	sb	1 sb = 1 cd/cm^2
magnetic field strength	oersted	Oe	1 Oe corresponds to $\frac{1000}{4\pi}$ A/m
magnetic flux	maxwell	Mx	1 Mx corresponds to 0.01·μWb
magnetic flux density	gauss	Gs. G	1 Gs corresponds to 0.1 mT
magnetic induction	gamma	γ	1 γ = 1 nT
mass	metric carat	—	1 metric carat = 200 mg
	gamma	γ	1 γ = 1 μg
pressure	torr	torr, Torr	1 torr = 1.333 22 × 10^2 Pa
viscosity —dynamic	poise	P	1 P = 1 dyn·s/cm^2 = 0.1 Pa·s
—kinematic	stokes	St	1 St = 1 cm^2/s
volume	stere	st	1 st = 1 m^3
	lambda	λ	1 λ = 1 μl = 1 mm^3

METRICS

Table 6. Metric values and their equivalents

LENGTH

Myriameter (obs.)	10,000 meters	6.2137 miles
Kilometer	1,000 meters	0.62137 mile
Hectometer	100 meters	328 feet 1 inch
Dekameter	10 meters	393.7 inches
Meter	1 meter	39.37 inches
Decimeter	0.1 meter	3.937 inches
Centimeter	0.01 meter	0.3937 inch
Millimeter	0.001 meter	0.0394 inch

AREA

Hectare	10,000 square meters	2.471 acres
Are	100 square meters	119.6 square yards
Centiare	1 square meter	1,550 square inches

WEIGHT

Name	Number of grams	Volume of water corresponding to weight	Avoirdupois weight of water
Metric ton, millier or tonneau	1,000,000	1 cubic meter	2,204.6 pounds
Kilogram or kilo	1,000	1 liter	2.2046 pounds
Hectogram	100	1 deciliter	3.5274 ounces
Dekagram	10	10 cubic centimeters	0.3527 ounce
Gram	1	1 cubic centimeter	15.432 grains
Decigram	.1	0.1 cubic centimeter	1.5432 grains
Centigram	.01	10 cubic millimeters	0.1543 grain
Milligram	.001	1 cubic millimeter	0.0154 grain

CAPACITY

Name	Number of liters	Metric cubic measure	United States measure	British measure
Kiloliter or stere	1,000	1 cubic meter	1.308 cubic yards	1.308 cubic yards
Hectoliter	100	0.1 cubic meter	2.838 bushels 26.417 gallons	2.75 bushels 22.00 gallons
Dekaliter	10	10 cubic decimeters	1.135 pecks 2.6417 gallons	8.80 quarts 2.200 gallons
Liter	1	1 cubic decimeter	0.908 dry quart 1.0567 liquid quarts	0.880 quart
Deciliter	.1	0.1 cubic decimeter	6.1023 cubic inches 0.845 gill	0.704 gill
Centiliter	.01	10 cubic centimeters	0.6102 cubic inch 0.338 fluid ounce	0.352 fluid ounce
Milliliter	.001	1 cubic centimeter	0.061 cubic inch 0.271 fluid dram	0.284 fluid dram

COMMON MEASURES AND THEIR METRIC EQUIVALENTS

Common measure	Equivalent
Inch	2.54 centimeters
Foot	0.3048 meter
Yard	0.9144 meter
Rod	5.029 meters
Mile	1.6093 kilometers
Square inch	6.452 square centimeters
Square foot	0.0929 square meter
Square yard	0.836 square meter
Square rod	25.29 square meters
Acre	0.4047 hectare
Square mile	259 hectares
Cubic inch	16.39 cubic centimeters
Cubic foot	0.0283 cubic meter
Cubic yard	0.7646 cubic meter
Cord	3.625 steres
Liquid quart, United States	0.9463 liter
Dry quart, United States	1.101 liters
Quart, imperial	1.136 liters
Gallon, United States	3.785 liters
Gallon, imperial	4.546 liters
Peck, United States	8.810 liters
Peck, imperial	9.092 liters
Bushel, United States	35.24 liters
Bushel, imperial	36.37 liters
Ounce, avoirdupois	28.35 grams
Pound, avoirdupois	0.4536 kilogram
Ton, long	1.0160 metric tons
Ton, short	0.9072 metric ton
Grain	0.0648 gram
Ounce, troy	31.103 grams
Pound, troy	0.3732 kilogram

MODIFIERS

Modifiers are words or groups of words that describe or limit other words. Modifiers include adjectives, adverbs, prepositional phrases, nouns used as adjectives, and clauses that function as adjectives or adverbs:

> The entire proposal had excellent graphics. (*adjectives*)
>
> The manager eventually explained the reasons for his disapproval. (*adverb*)
>
> The pump next to the intake line was serviced last month. (*prepositional phrase*)
>
> The Sky Aviation proposal, which scored second in technical merit, had some interesting innovations. (*adjectival clause*)
>
> The ventilation fan was replaced because its peak circulation volume fell short of our needs. (*adverbial clause*)

See ADJECTIVES, ADVERBS, NOUNS, PREPOSITIONS, and CONJUNCTIONS.

Writers and editors usually depend upon their ears to tell them where a modifier should appear in a sentence. Essentially, however, modifiers should be as close as possible to the words they modify. If they aren't, readers might misinterpret the sentence. The most common sentence problems associated with modifiers result from dangling or misplaced modifiers.

For information on compound modifiers, see HYPHENS.

Dangling Modifiers

Modifiers dangle if they do not seem to be related to anything in the sentence or if they are not placed near enough to the words they modify to seem attached to those words. Modifiers dangle when they float, unattached, in a sentence.

Dangling modifiers can be adjectives, adverbs, prepositional phrases, infinitive verbs, appositives, or clauses. Quite often, dangling modifiers are participial phrases, usually beginning with a present participle (such as *knowing*):

> Knowing that standard 3/4-inch pipe was too small, the specifications included provisions for larger pipe.

The phrase beginning with *knowing* seems to modify the noun *specifications*, but, clearly, specifications cannot know anything. The phrase must modify a human being to make logical sense, but no humans are mentioned in the sentence, so the modifier dangles.

Whenever you open a sentence with an action stated by an *-ing* verb (present participle) or *-ed* verb (past participle) and do not follow it with the name of the person doing the action, you will have a dangling modifier (sometimes called a dangling participle):

> *not this*
>
> After discussing interest rate trends, the decision was made to refinance our present loan.
>
> *this*
>
> After discussing interest rate trends, we decided to refinance our present loan.

> *not this*
>
> Having analyzed the technical problems, the recommendation was to route the feed-forward signal through a broadband transmitter.
>
> *this*
>
> Having analyzed the technical problems, she recommended routing the feed-forward signal through a broadband transmitter.
>
> *or this*
>
> An analysis of the technical problems led researchers to suggest routing the feed-forward signal through a broadband transmitter.

---

1. Ensure that modifiers, particularly those expressing action, have a clear noun to modify and are placed as close as possible to that noun (preferably just before it):

> *not this*
>
> Having missed our connecting flight, no flights later that day were going to Albuquerque. (*Who missed the flight?*)
>
> *this*
>
> Having missed our connecting flight, we discovered that no later flights were going to Albuquerque.

> *not this*
>
> While reviewing the figures, many errors became apparent. (*Who reviewed the figures?*)
>
> *nor this*
>
> Many errors became apparent while reviewing the figures.
>
> *this*
>
> While reviewing the figures, we discovered many errors.

NOTE 1: Dangling modifiers do not necessarily introduce the sentence; they can appear anywhere:

MODIFIERS

The report was inaccurate, comparing it with the prior ones. (*Who compared it?*)

NOTE 2: Some introductory participles (usually ending in *-ing*) have become so common that they do not require clear words to modify:

Considering your reluctance, you should not represent us before the Texas Railway Commission.

Judging from the revised figures, the report will never be approved.

Misplaced Modifiers

Modifiers are misplaced when they do not appear in their customary place in a sentence. Readers often misread sentences in which the modifiers are misplaced:

not this

Hughes was told that he was no longer needed by the personnel manager.

this

The personnel manager told Hughes that the company no longer needed him.

---

not this

Your letter regarding the workover of March 15 reached me today.

this

Your March 15 letter regarding the workover reached me today.

or this

Your letter regarding the March 15 workover reached me today.

In both of these examples, the writer might not have intended what the "better" versions say, but the "better" versions are much clearer. Your goal as a writer should be to write so that you cannot be misunderstood. One way to achieve this goal is to ensure that modifiers appear where they should.

---

2. Ensure that modifiers appear either next to or as close as possible to the word or words modified:

not this

The manager only was interested in production data. (*Does* only *modify* manager *or* data?)

this

The manager was interested only in production data.

See ADVERBS.

The book on the shelf with all the samples is our only copy. (*Does the book or the shelf contain the samples?*)

The report on geological formations in southern Utah that our manager studied was as up-to-date as possible. (*Did the manager study the report or the formations? The clause* that our manager studied *should appear immediately after* report *or immediately after* Utah. *The placement of the modifying clause conveys its meaning.*)

---

not this

A computer program has been written for calculating estimates of the gradients on the mainframe IBM in the Production Department.

this

A computer program for estimating gradients has been written for the mainframe IBM in the Production Department.

---

not this

We are shipping the circuit board that failed under separate cover. (*Was it okay until you shipped it separately?*)

this

We are separately shipping the faulty circuit board.

or this

The circuit board that failed is being shipped separately.

NOTE 1: If a modifier refers to two nouns, it should appear with the first noun mentioned:

not this

The land is rocky on the west side and somewhat less rocky on the east side of the allotment.

this

The land is rocky on the west side of the allotment and somewhat less rocky on the east side.

NOTE 2: Unmodified nouns might need an article (*a, an,* or *the*) or an adjective to clarify their meaning:

not this

The secretary and treasurer attended our meeting. (*Is the* secretary and treasurer *one person or are they two people?*)

this

The secretary and the treasurer attended our meeting.

NOUNS

Nouns signify persons, places, things, and ideas. Even more significant, perhaps, nouns are the main words in a variety of noun phrases:

a bottle
the comprehensive report
a slowly changing pattern
some tomatoes for lunch
the young engineer who works next door

Noun phrases, in turn, become key building blocks in the English sentence. Within a sentence, a noun phrase can be a subject, an object, or a complement:

The proposed electrical changes will be expensive. (*subject*)

The engineer designed two holding ponds. (*object*)

The applicant was the person who was busily filling out forms. (*complement*)

Noun phrases can also complete a prepositional phrase by becoming the object of the preposition:

near the fuel storage tank
beyond the property line
at the amount we requested

1. Distinguish between singular and plural nouns and those nouns that are neither.

Nouns that can be singular and plural signify things that can be counted:

a bottle/two bottles
every desk/six desks
neither proposal/three proposals
each pump/30 pumps
either ox/five oxen

Nouns that are neither singular nor plural signify things that cannot be counted:

furniture/some furniture
meat/most meat
wheat/less wheat
hospitality/more hospitality
warmth/some warmth

See PLURALS and AGREEMENT.

The difference between these two types of nouns is important because each type will accept only certain modifying words:

a bottle (*not* much bottle)
three pumps (*not* less pumps)

some warmth (*not* each warmth)
more hospitality (*not* three hospitality)

Native speakers of English usually choose the proper modifying words unconsciously. Only occasionally do they make mistakes: *The zoo had less animals than we expected.* Because *animals* can be counted, the proper modifier is *fewer*: *The zoo had fewer animals than we expected.*

Many nouns can belong to either type, but their meanings change, depending upon the context:

She's had many odd experiences.
This job requires experience.

We bought an evening paper.
Wrap the parcel in butcher paper.

The talks will take place in Cairo.
He dislikes idle talk.

2. Use collective nouns as either singular or plural.

Collective nouns are nouns that signify groups of people or things: *staff, team, family, committee, majority, crew, squad*, etc.:

The committee has met, and it has rejected the amendment. (*singular*)

The committee have met, and they have rejected the amendment. (*plural*)

The majority has made its viewpoint clear to the candidate. (*singular*)

The majority have made their viewpoint clear to the candidate. (*plural*)

See AGREEMENT.

3. Distinguish between common and proper nouns:

Common Nouns

a company	the professor
three lines	the avenue
some paper	a river
an idea	our dentist

Proper Nouns

Acme Glass Company
Professor Thomas Miles
Second Avenue
the Mississippi River
Dr. John Wray

NOTE: Proper nouns are capitalized, while common nouns are not. (For more information, see CAPITALS, TITLES, and ADJECTIVES.)

NUMBERING SYSTEMS

Numbering systems are used with outlines, tables of contents, and headings to display a document's organization and allow readers easy access to parts of the document. The two basic numbering systems are the traditional outline system and the decimal system. (See OUTLINES.)

Traditional System

Traditional outlines use the following numbering and lettering conventions:

1. Uppercase Roman numeral
2. Capital letter
3. Numeral
4. Lowercase letter
5. Numeral in parentheses
6. Lowercase letter in parentheses
7. Numeral with right parenthesis
8. Lowercase letter with right parenthesis

Here, along with the standard indentations, is the traditional system:

```
I.
   A.
      1.
         a.
            (1)
               (a)
                  1)
                     a)
```

Some authorities (*e.g.*, *The Chicago Manual of Style*, 13th Edition, p. 247) prefer a different, but similar, system of subordination.

Decimal System

In the decimal system, successive decimal points indicate levels of subordination:

```
1.0
   1.1
      1.1.1
      1.1.2
      1.1.3
         1.1.3.1
         1.1.3.2
2.0
   2.1
   2.2
      2.2.1
      2.2.2
```

1. Use the traditional system in most cases where you want to show multiple subordination levels, but use the decimal system for very lengthy documents.

The decimal system is preferable in very lengthy documents with a multitude of numbered subsections and in any document with so many major headings that the Roman numerals would become large enough to create confusion among those readers unfamiliar with Roman numerals.

However, with more than four or five levels of subordination, the decimal system is less desirable because readers cannot easily comprehend the text's logical structure. (See TABLES OF CONTENTS and OUTLINES.)

Numbering Systems and Punctuation

Numbers and letters in the traditional outline system always require punctuation. The higher level subdivisions take periods; the lower level subdivisions take either parentheses or a single right parenthesis.

In the decimal system, the numbers can be followed by several spaces instead of punctuation:

2.1   Testing Procedures

See HEADINGS and LISTS.

NUMBERS

Whether you write out a number or use a figure depends upon the size of the number, what it stands for, and how exact it is. The stylistic conventions for number usage vary, so you may find conflicting suggestions from one dictionary or style guide to the next. The recommendations that follow are based on the current standard practice for technical and scientific writing.

1. Use figures for any number expressing time, measurement, or money:

3 a.m.
$15
45 ft
1 in.
8 cm
34.17 m

Measurement includes length, weight, volume, velocity, etc.

Because figures are easier to see, you should prefer them. However, this convention has exceptions, the most important of which appear in rules 2 through 12.

2. Write out numbers expressing quantity if the numbers are below 10; otherwise, use figures:

five systems
three mission capabilities
14 mission capabilities
57 technicians
four copies

NOTE: In nontechnical writing, writers often write out numbers less than 100. Numbers written out are less emphatic than numbers expressed as figures. Consequently, writing out numbers less than 100 avoids overemphasizing double-digit numbers in nontechnical documents, which typically contain few numbers.

3. Write out numbers that begin a sentence:

Twelve inches from the centerline are two slots for plate fins.

Four years ago, we initiated an IR&D study of argon-atmosphere braze furnaces.

4. Rewrite sentences beginning with a very large number:

this

Every second, the oscillator receives 363 signals from the bit generator.

not this

363 times a second the oscillator receives a signal from the bit generator.

nor this

Three hundred sixty-three times a second the oscillator receives a signal from the bit generator.

5. Write out round numbers expressing approximations:

about three thousand
approximately sixty applicants
over three million signals
around five hundred transmissions
five or six hundred transmissions

NOTE: Some authorities (notably the *United States Government Printing Office Style Manual*) prefer numerals with such words as *nearly, about, around*, and *approximately*. Use your judgment. Figures convey a greater sense of precision than words. Thus figures may seem to contradict the idea of approximating.

6. Use a combination of letters and figures for very large round numbers (1 million or greater):

We have invested over $45 million on laser research in the last 5 years.

Our annual IR&D budget exceeds $16 million.

7. Be consistent.

Treat numbers of the same type equally within a sentence, paragraph, or section. (However, <u>never</u> begin a sentence with a figure.):

this

Unit A will require 5 outlets; Unit B, 17 outlets; Unit C, 9 outlets; and Unit D, 14 outlets.

not this

Unit A will require five outlets; Unit B, 17 outlets; Unit C, nine outlets; and Unit D, 14 outlets.

---

this

Seven of the stations carry 39 spare controllers. The other 14 stations carry only 8 spares.

not this

7 of the stations carry 39 spare controllers. The other 14 stations carry only eight spares.

NUMBERS

nor this

Seven of the stations carry thirty-nine spare controllers. The other fourteen stations carry only eight spares.

nor this

Seven of the stations carry 39 spare controllers. The other fourteen stations carry only 8 spares.

The sentence cannot begin with a figure, so *seven* must be written out. The *14 stations* uses figures because 14 is greater than 9; so the two references to *stations* cannot be consistent. The number *39* is too large to write out, so both of the numbers referring to spare controllers are written as figures, although *8* expresses a quantity and is less than 10.

8. Use figures for quantities containing both whole numbers and fractions:

The proposal calls for 8½-by-11-inch paper.

See FRACTIONS.

9. Always use figures for percentages and decimal fractions:

The rectangular fins are 0.07 in. high.

The maximum core diameter is 2.54 mm.

The tests require an 8 percent solution.

NOTE: In the last example, *8%* would also be acceptable, although many style guides prefer that writers use the percent sign only in tables and visual aids. In accounting and other financial documents, the percent sign is common in text. (See SIGNS AND SYMBOLS.)

10. Always use figures for dates.

June 14, 1985
14 June 1985
the 14th of June 1985
June 1985

NOTE: If you use the preferred style (month-day-year, as shown in the first example above), always separate the day and year with a comma. The second example shows the alternate style: day-month-year.

If you write only month and year (as in the last example above), use no punctuation. Separating the month and the year is unnecessary. (See COMMAS.)

11. Form the plural of a number expressed as a figure by adding a lowercase -*s*:

before the 1970s
temperatures well into the 200s
the 5s represent actual strikes

NOTE: Plurals of numbers written out are formed like the plurals of other words:

in the twenties
groups of threes or fours

See PLURALS.

12. Use a comma to separate groups of three digits:

55,344,500
10,001
9,999
678

NOTE 1: In some technical fields, the preferred style is to omit the comma separating digits in numbers only four digits long:

5600
9999

NOTE 2: A common practice outside of the United States is to use a space instead of a comma to separate groups of three digits:

98 072.1
7 143

See METRICS.

ORGANIZATION

The ideas presented in a document should be structured in a natural but emphatic sequence that conveys the most important information to readers at the most critical times.

The principles of organization differ slightly from document to document depending upon the type of document, the readers, the content, and the writer's purpose. Nevertheless, logic and common sense dictate that a well-organized document must have certain features:

- The ideas in the document must be clear and sensible, given the subject, and comprehensible, given the readers.
- The document should conform to the readers' sense of what the most important points are and of how these points are arranged.
- The document should announce its organizational scheme and then stick to it.

Letters, memos, and reports differ somewhat in their organizational patterns, mostly because their readers differ.

Readers of letters are typically outside of the organization sending the letter. Their relationship to the writer is therefore more distant, and consequently more formal, than the relationship between the writer and others within the writer's company. (See LETTERS.)

Readers of memos, on the other hand, are typically from within the writer's organization and share various assumptions, experiences, and knowledge—all of which tend to make memos less formal than letters. (See MEMOS.)

The distance and formality between writer and reader affect organization in several ways. The greater the distance, the more the need to set up (introduce and perhaps explain) the ideas in the document. The greater the distance, the greater the need to substantiate information that might be subject to differing interpretations. The more formal the document, the more the writer must consider format traditions and reader expectations in organizing material.

Reports, technical or otherwise, often have prescribed formats (or organizations). Scientific report format is based on a long tradition in the sciences. The format of such reports is strictly prescribed, and writers have very few options in varying that format. Technical (but nonscientific) reports offer somewhat more latitude, but even there some companies have strict guidelines on organizing technical reports.

Within the limitations imposed by tradition, logic, and audience, writers must carefully consider how to arrange their ideas and supporting data so that the document serves its purpose and satisfies the readers' needs. The principles listed below suggest how you can accomplish these tasks.

See LETTERS, MEMOS, and REPORTS.

---

1. Organize information according to your readers' needs.

How you organize information often depends upon your readers. You may organize the same information differently for different readers depending upon their needs and your purpose in writing to them. Here, for instance, is the text of a short letter written to the test director of a laboratory:

> We wish to request the following tests on the dry field cement samples that we shipped on July 20 to Mr. J. F. Springer of your laboratory:
>
> - Thickening time
> - Rheology
> - High temperature-high pressure fluid loss
> - 12- and 24-hr compressive strength
>
> Davidson-Warner, a cementing company, has been using this cement in our Mt. Hogan Field. On July 17, they experienced a cementing failure while setting a string of 3½-in. casing at 11,332 ft. in our Hogan BB-62 well. They pumped 688 barrels of cement and 78 barrels of displacement fluid before halting displacement when the pressure increased to 5000 psi.
>
> To facilitate your testing, we have attached pertinent well logs, cement data, and a copy of Davidson-Warner's laboratory blend test results. Please submit your findings to me at your earliest convenience.

This letter begins, appropriately enough, with a request. The writer wants something of the reader. Establishing what the writer wants makes sense as an opening statement. The specific detail concerning the cementing failure does not appear until the middle paragraph because this particular reader will not need to know this information except as background for conducting the tests. The details of the cementing failure are less important than a list of the tests the writer is requesting.

ORGANIZATION

However, if the document had been written to the production engineer who will now be responsible for this well, it might have begun like this:

> The Hogan BB-62 is currently shut in because of a cementing failure that occurred on July 17. The regional office would like us to return this well to production by July 28.
>
> On July 15, this well was shut in to allow Davidson-Warner to set a new string of 3½-in. casing from 10,500 ft. to 11,890 ft. While setting the string at 11,332 ft., they halted displacement when the pressure increased to 5000 psi. Before stopping, they had pumped 688 barrels of cement and 78 barrels of displacement fluid. They left approximately 35 barrels of cement in the casing (with a cement top at 8,992 ft.).
>
> Wiley Laboratories has been asked to test dry field samples of the cement. In the meantime, AGF Cement has been contracted to finish setting the string. They will be onsite no later than July 25. You should plan to be present.
>
> Mt. Hogan Field production figures are down 4.3 percent in July, primarily due to this cementing failure. The regional production manager has asked that we resume full production by July 28. If you need assistance, call me at 555-6666.

This memo is written from supervisor to subordinate. Its tone is obviously different (more forceful, more directive) than the letter written to the laboratory. The organization of ideas is also very different.

The memo to the engineer begins with a statement of fact (a set-up), followed by a deadline. As in the first letter, the details of the cementing failure appear in the middle, but in this second example, the details lead to an amplification of the implied directive that appears in the opening paragraph. The memo closes with a compelling reason for action (production figures down) and a reminder of the deadline.

As you organize a document, always consider what information your readers need from you. In the examples above, the test director at Wiley Laboratories will not care that Davidson-Warner left 35 barrels of cement in the casing. The engineer will not care that the dry field samples were shipped to Mr. Springer. Each document above reflects those concerns that its readers will care most about.

The data and ideas that you include in a document and the way you organize these data and ideas depend upon (1) whom you are writing to and (2) why you are writing to them.

2. Group similar ideas.

Separating similar ideas creates chaos. In the examples above, the details concerning the cementing failure appear in the same place. If they had been scattered, the effect could have been devastating for readers:

> The Hogan BB-62 is currently shut in because of a cementing failure that occurred on July 17. Wiley Laboratories has been asked to test dry field samples of the cement.
>
> On July 15, this well was shut in to allow Davidson-Warner to set a new string of 3½″ casing from 10,500′ to 11,890′. Please try to return this well to production by July 28. AGF Cement has been contracted to finish setting the string. Before stopping, Davidson-Warner had pumped 688 barrels of cement and 78 barrels of displacement fluid. AGF Cement will be onsite no later than July 25.

As this demonstration shows, separating related ideas creates confusion and jars readers.

3. Place your most important ideas first.

A frequent problem with business and technical writing is the tendency to lead to, rather than from, major ideas. Many writers believe that they have to build their case, that skeptical readers will not agree with their conclusions unless they first demonstrate how they arrived at those conclusions. This tendency results in documents that are unemphatic, difficult to follow, and filled with unnecessary detail.

The strongest part of a document is its beginning. Readers typically pay more attention at the beginning because they are discovering what the document is about. The beginning, then, is the most emphatic part of document by virtue of its position. Because the beginning is so strong, you should begin with the most important ideas in the document—and then support those ideas by presenting your evidence afterwards.

The Scientific Format. Many of those writers who tend to lead down to their major ideas have been schooled in the

ORGANIZATION

scientific method. According to the scientific method, one presents the facts, observations, and data that lead to and support a conclusion. The strength of this method is that it presents a series of steps that culminates in an inevitable conclusion. Anyone should be able to repeat the steps and reach the same conclusion. Therefore, the steps are as important as the conclusion.

In some scientific reports (notably those written from one scientist to another), an organizational scheme based on the scientific method is desirable:

Summary (or Abstract)

Introduction

Materials and Methods

Results and Discussion

Fact 1
Fact 2
Fact 3
Fact 4

(therefore)

Conclusions

Recommendations (optional)

Summary (optional)

This format is acceptable only if readers will be as interested in the process of arriving at the conclusions as they are in the conclusions themselves. When readers are more interested in the conclusions, follow the managerial format.

The Managerial Format. The managerial format is the reverse of the scientific format. Managers (and most other nonscientific readers) are far more interested in the conclusions than they are in the steps leading to them. This is not to say that these readers will not want to see the conclusions supported—only that they will want the conclusions before the results and discussion:

Summary/Executive Summary

Introduction

Conclusions (and Recommendations)

(because of)

Results

Fact 1
Fact 2
Fact 3
Fact 4

Having the conclusions early in the report facilitates reading because the reader is given a perspective from which to understand the facts and data being presented. Furthermore, busy managers often know the background and tests that have led to the conclusions.

You should follow the managerial format in all documents except scientific documents written for scientific peers.

See REPORTS.

NOTE 1: The principle of emphasis through placement extends to all documents and sections of documents. Your most important ideas should appear at the beginning of your documents and of individual sections. The most important idea in most paragraphs should appear in the opening sentence. The most important words in a sentence typically come at the beginning of the sentence. (See PARAGRAPHS and SENTENCES.)

NOTE 2: A corollary to note 1 is that you should always subordinate detail. Place it in the middle of sentences, paragraphs, sections, and documents. Detail includes data, explanation, elaboration, description, analyses, results, etc.

NOTE 3: In lengthy documents, begin and end with important ideas.

The lengthier a document becomes, the more crucial this rule is. Readers of long passages need to be introduced to the subject, learn the most important points early, receive the supporting detail and explanation, and then have it all wrapped up in a tidy closing statement that reiterates the important points.

An old adage regarding oral presentations (but applicable to writing) is that you should tell 'em what you're gonna tell 'em, tell 'em, and then tell 'em what you told 'em. (See REPETITION.)

See REPORTS and EMPHASIS.

4. Keep your set-ups short.

Sometimes you cannot begin by stating your most important idea because the reader either will not understand it or will not accept it. If such is the case, you need to set up the most important idea by providing introductory information meant either to inform readers or to persuade them.

A fundamental of organization in business and technical writing is to keep your set-ups short. Do not delay your major ideas any longer than necessary.

When you give people positive information (i.e., when you say "yes" to them), you should give them the positive information right away. They want to hear it, and hearing it will make them more receptive toward you and the rest of the information you provide.

However, when you give readers negative information (when you say "no" to them), giving them the negative information first will put them off, and they will not be receptive to what follows. Moreover, they may become antagonistic toward you and may believe that you have made the negative decision precipitously.

Therefore, you should say "no" to readers only after you have set them up for it. Be careful, however, not to delay the "no" too long. Keep your set-ups short, as in the following example:

> I have been asked to reply to your request for additional compensation following approval of your Engineering Change Order dated March 3.
>
> As you know, a Health Department inspector ordered the design changes, and our contract states that all design changes required for safety reasons are warranted under the contractor's bond. Therefore, additional compensation would be inappropriate at this time.

The first sentence sets the stage. The second provides brief rationale for the decision. The third states the decision. The two-sentence set-up in this example makes the decision more palatable than if the writer had begun by saying: *"We will not be providing the additional compensation you requested."*

See INTRODUCTIONS.

---

5. List items in descending order of importance.

Readers typically assume that information in lists appears in descending order of importance: most important listed item first, least important item last.

Numbering or lettering systems reinforce this assumption. We all know that being number 1 is better than being number 6. We know from school that an A is better than an F. Rightly or wrongly, we assume a natural ranking of items. Therefore, writers should list items in descending order of importance.

If you wish to create a list in which items are equally important, use bullets or dashes instead of numbers or letters, and state that the listed items are equal.

See LISTS.

---

6. In long or complex documents, preview your most important ideas and your major content areas.

In longer documents, you must establish the structural framework of the document. If you don't, readers may become overwhelmed by the document's size or complexity and never develop a good understanding of its content.

Summaries and introductions are ideal devices for previewing content, but you can also preview content in opening paragraphs.

Your preview should sound natural and should be unobtrusive. Generally, when the preview refers to itself as a preview, it is obtrusive, as the example below illustrates:

not this

> This report discusses the results of the Hamerling Study (March-October 1979), which found that predators have played only a minor role in the recent population decline of the cutthroat trout. The first section concerns the quality of the watershed, which has declined significantly since 1965.
>
> Following that section is a discussion of the role of climate changes, particularly a 2-degree increase in temperature throughout the study area. In section 3, the report notes the effect of deforestation in one part of the study area. In its concluding section, the report discusses the combined impact of watershed degradation, climate changes, and deforestation. As the report notes, these changes have reshaped the cutthroat trout's habitat, perhaps beyond the species' ability to adapt.

this

> The Hamerling Study (March-October 1979) found that predators have played only a minor role in the recent population decline of the cutthroat trout. Far more serious impacts on this species are a degraded watershed, temperature increases, and deforestation.

ORGANIZATION

> Together, these environmental changes have reshaped the cutthroat trout's habitat, perhaps beyond the species' ability to adapt.

The last example amounts to a summary of the report. It could actually appear in the summary, become part of an abstract, or open the introduction. It could even appear in all three places.

See INTRODUCTIONS and SUMMARIES.

7. Discuss items in the same order in which you introduce them.

When you introduce items, you should discuss them in the same order later. Saying that you are going to talk about A, B, and C, and then beginning by discussing B violates the readers' sense of order. Follow these examples:

> The three greatest influences on Cutthroat Trout population are *a degraded watershed, temperature changes*, and *deforestation.*
>
> *The watershed* has been declining in quality since 1965 when . . .
>
> *Temperature changes* over the last 5 years have resulted in a 2 degree . . .
>
> *Deforestation* through the study area has also affected . . .

---

this

> The acquisition improved our *cash flow* while providing significant tax advantages and allowing us to capitalize expenses. Prior to the takeover, we had negative *cash flow* on several . . .

not this

> The acquisition improved our *cash flow* while providing significant tax advantages and allowing us to capitalize expenses. Prior to the takeover, our *expenses* were not capitalized . . .

The second example demonstrates a subtle but important use of a series. The writer introduces three ideas: cash flow, taxes, and expenses. To be consistent with the order in which these ideas were introduced, the writer must follow the introductory statement with cash flow, not taxes or expenses, as occurs in the final version.

8. Use headings, transitions, key words, and paragraph leads to provide cues to the document's organization.

Throughout documents, you should signal organizational shifts or changes in direction by using headings, transitions, repeated key words, and opening or closing statements in paragraphs.

Headings are especially useful when you need to signal abrupt changes in direction, such as the transition from one topic to another (unrelated) topic. If the shifts are too radical, you cannot easily indicate them in text.

Transitions and repeated key words provide for smoother changes in direction and are useful between sentences and paragraphs, as in the example below. Note how the underlined words indicate organizational patterns and shifts in direction:

> The coal seam trends northwesterly for approximately 9,500 meters before pinching out on a fault line. <u>However</u>, seismic evidence suggests that <u>another</u> seam of coal extends from a point 75 meters downdip of the pinchout. This <u>second</u> seam appears to trend northerly for another 5,000 meters. <u>Together</u>, these seams represent a sizeable reserve of recoverable coal, <u>but</u> initiating mining operations will still be extremely <u>difficult</u>.
>
> The biggest <u>difficulty</u> is landowner resistance to strip mining . . .

See HEADINGS, KEY WORDS, OUTLINES, PARAGRAPHS, and TRANSITIONS.

OUTLINES

Outlines are convenient tools for the schematic organization of material. (See ORGANIZATION.)

Preliminary or draft outlines help writers determine early in the writing process, usually before the document is written, whether the content is logical and complete. Preliminary outlines do not have to be neat or accurately numbered.

Final outlines (which usually form the table of contents) display the overall structure of the content. Final outlines may or may not be numbered. If they are numbered, they typically use either the traditional format (I/A/1/a, etc.) or the decimal format (1/1.1/1.1.1, etc.). (See NUMBERING SYSTEMS.)

Traditional Outlines

Traditional outlines are those using the following numbering and lettering system:

TITLE

I. First-level division
  A. Second-level division
    1. Third-level division
      a. Fourth-level division
        (1) Fifth-level division
          (a) Sixth-level division
II. First-level division
  A. Second-level division
  B. Second-level division

NOTE 1: Some writers and editors prefer *a)* instead of *a.* to indicate a fourth-level division.

NOTE 2: Because Roman numerals vary in length, they are customarily aligned according to the period, not the length of the numeral:

  I.
 II.
III.

Decimal Outlines

TITLE

1. First-level division
  1.1 Second-level division
    1.1.1 Third-level division
      1.1.1.1 Fourth-level division
  1.2 Second-level division
    1.2.1 Third-level division
      1.2.1.1 Fourth-level division
2. First-level division

NOTE: A variation of the decimal format uses hundreds and tens. This format is not widely used, perhaps because it is less flexible than the decimal and traditional formats:

TITLE

100 First-level division
  110 Second-level division
    111 Third-level division
    112 Third-level division
  120 Second-level division
200 First-level division

---

1. Avoid outlines with more than four or five levels of subordination.

Subordination refers to the number of successive subdivisions. The more intricate and involved the subdivisions become, the harder it is for readers to grasp and remember the organizational scheme. When the level of subordination reaches six or seven levels, readers will not be able to comprehend the successive subdivisions. Even in outlines that you create only for yourself, too many subdivisions will cause confusion.

---

2. Do not make your outlines too brief.

Brevity may be the soul of wit, but too brief an outline will not allow you to explore the content sufficiently, and it will not assist readers in understanding your organizational scheme:

I. Intro
II. The Problem
III. Options
IV. Analysis
V. Conclusions
VI. Further Work

This outline has sparse (indeed, cryptic) titles and no subdivisions. In effect, it provides nothing beyond a vague sense of general direction. Fleshing out this outline will help the writer think much more carefully about the exact content needed and about the logic and the arrangement of ideas.

OUTLINES

See HEADINGS for a discussion of informative headings and CAPTIONS for a description of action captions. Also, see ORGANIZATION.

3. Maintain parallel structure in formal outlines.

Each division heading and each subheading should be grammatically parallel with items at the same level of heading. So if you use *-ing* verb forms for some headings at one level, use *-ing* verb forms for all headings at that level:

not this

2.1.1 Preparing for the trial run
2.1.2 Checking safety procedures
2.1.3 The trial run itself
2.1.4 Preliminary findings

this

2.1.1 Preparing for the trial run
2.1.2 Checking safety procedures
2.1.3 Conducting the trial run
2.1.4 Determining preliminary findings

In many cases, parallelism allows writers to avoid repeating words that occur within all headings at the same level:

not this

2.8.1 DOD test results
2.8.2 Test results from the Acme Testing Laboratory
2.8.3 Test results from Conair Environmental Labs

this

2.8.1 Test results

  2.8.1.1 DOD
  2.8.1.2 Acme Testing Laboratory
  2.8.1.3 Conair Environmental Labs

4. If possible, design your outline so that each subdivision has at least two points.

If a subdivision has only one subpoint in it, then the subpoint should probably become the subdivision heading:

not this

3. Cost analysis

   a. Overhead rates

4. Labor issues

this

3. Overhead rates

4. Labor issues

In the first example above, the subdivision for *overhead rates* has but a single point. If the cost analysis consists of nothing more than overhead rates, why list cost analysis as an activity? The second example properly recognizes that the cost analysis is nothing more than a determination of overhead rates.

5. Use an outline to check the logical consistency and basic organization of a piece of writing.

If an outline is not parallel and is not logical, then the document based on the outline is likely to be chaotic. (See ORGANIZATION and TABLES OF CONTENTS.)

PARAGRAPHS

Paragraphs are visual and logical signals to readers.

As visual signals, paragraphs help readers to perceive divisions within a document. The typographical devices used to show the divisions include indentation, blank lines above and below paragraphs, and paragraph numbers. Without visual paragraphing, texts would be a mass of undifferentiated sentences.

As logical signals, paragraphs reflect the major divisions and subdivisions in content within a document. In good writing, the transition from topic to topic is reflected in the transition from paragraph to paragraph. Each major topic has its own paragraph (and sometimes more than one), and each paragraph concerns only one topic.

A Case Study in Paragraphing

Some writers dump everything they can think of about a topic into a single paragraph. Then when a paragraph becomes long enough (by whatever standard), they pause, indent, and start another paragraph. This sort of paragraphing is neither logical nor effective, as the following example illustrates:

> Oxides of nitrogen include nitrogen dioxide (NO_2) and nitric oxide (NO). NO_2 is a pungent gas that causes nose and eye irritation and pulmonary discomfort. NO is converted to NO_2 by atmospheric chemical reaction. Both NO and NO_2 participate in photochemical reactions leading to smog. Sulfur dioxide (SO_2) is a colorless and pungent gas that causes irritation to the respiratory tract and eyes and causes bronchoconstriction at high concentrations. Hydrocarbons react with NO or NO_2 and sunlight to form photochemical oxidants or smog. Health effects include irritation of the eye, nose, and throat. Extended periods of high levels of oxidants produce headaches and cause difficulty in breathing in patients suffering from emphysema.

What did the writer of this paragraph want to accomplish? Is the first sentence on NO and NO_2 an accurate reflection of the rest of the content? How do the other facts and points in the paragraph fit together? Can readers see a definite pattern or structure to the facts?

These and similar questions suggest several remedies:

- Shorten the paragraph and focus on only one topic.
- State this topic in the opening sentence.
- Supply organizational cues in the opening sentence and, as appropriate, in later sentences.

By applying these remedies, we can revise the paragraph as follows:

> Nitrogen oxides, hydrocarbons, and sulfur dioxide—these constituents of smog can cause health problems. Nitrogen dioxide (NO_2) is a pungent gas that causes nose and eye irritation and pulmonary discomfort. Hydrocarbons that react with NO_2 or with nitric oxide (NO) and sunlight form photochemical oxidants that can irritate the eyes, nose, and throat. Extended exposure to high levels of oxidants can produce headaches and cause persons with emphysema to have trouble breathing. Sulfur dioxide (SO_2) is a colorless and pungent gas that irritates the eyes and respiratory tract and, at high concentrations, can cause bronchoconstriction.

The paragraph might also be revised to focus on a single health problem, such as eye irritation. If so, eye irritation would become the focus of the opening sentence. Then all succeeding sentences would relate to eye irritation.

The original paragraph might also be broken into separate paragraphs that discuss each type of pollutant. We can't know which approach is correct unless we know the context in which the revised paragraph will appear and the purpose of the document as a whole. (See ORGANIZATION.)

---

1. Limit paragraphs to a single topic or major idea.

Ensure that your paragraphs focus on a single topic or idea. When you go to a new topic, start a new paragraph. If your paragraph on a single topic becomes too long, start a new paragraph at a logical point and have two (or more) paragraphs dealing with the same topic. When such is the case, you normally focus each paragraph on a subtopic related to the overall topic.

---

2. Do not allow paragraphs to become too long.

Quantifying paragraph length is difficult, but in business and technical writing, paragraphs exceeding 150 to 175 words should be rare. Most paragraphs will consist of three to six sentences. If a single-spaced paragraph goes beyond

PARAGRAPHS

one-third of a page, it is probably too long. A double-spaced paragraph should not exceed half a page in length.

The document's format should influence paragraph length. If a document has narrow columns (two or three to the page), then paragraphs should be shorter, perhaps on the average no more than 125 words long. If a document uses a full page format (one column), then average paragraph length can reach 175 words.

Length is therefore a function of appearance and visual relief. Almost all readers have difficulty with dense pages of print, no matter how well written and logically organized the text may be. Remember that paragraphs are visual devices meant to make reading easier, so keep them shorter rather than longer.

3. Vary the length of your paragraphs.

A document containing paragraphs of uniform length would be dull and difficult to read. For the sake of variety and to stimulate reader interest, you should vary the length of your paragraphs, especially in documents over one page long.

The length of successive paragraphs will of course depend upon content. The logic of the material will dictate, at least to some extent, where paragraphs can logically begin and end, but you still have a great deal of latitude.

A particularly involved point may require lengthy explanation and two or three examples. If so, you might state the point and explain it in one or two paragraphs and then make each example a separate paragraph. Dividing the topic in this fashion is especially desirable in a double- or triple-column page format.

Your paragraph stating the main point could be relatively short. Short paragraphs usually draw attention to themselves, so they are useful for stating major ideas. The explanatory paragraph should be much longer. The paragraphs providing the examples should vary in length, with the most important example appearing in the longest paragraph.

Are single-sentence paragraphs acceptable?

Yes. A common misconception about paragraphing is that single-sentence paragraphs somehow violate the principles of writing. In fact, single-sentence paragraphs are very emphatic, especially if they are surrounded by longer paragraphs. You should take care not to use single-sentence paragraphs too often, however. Too many single-sentence paragraphs and you have no paragraphs. So use them judiciously. (See EMPHASIS.)

4. Ensure that your opening sentence accurately reflects the content of the paragraph.

At the very least, your opening sentence should establish a key word or phrase that indicates the paragraph's topic. If your paragraph will focus on health problems, then the opening sentence should contain at least two key words: *health* and *problems*. These key words help establish the paragraph's viewpoint, which is often called its thesis.

If possible, the paragraph should open with a sentence that clearly states its thesis. Such sentences are often called topic sentences:

> Timber sales along the Graveny ridge have substantially increased erosion. From 1972 to 1983, the Forest Service conducted four timber sales that . . .
>
> MOGO's reservoir study of May 1984 indicates that remaining recoverable reserves exceed previous estimates by over 66 percent. Seismic data gathered in conjunction with the study . . .
>
> The Packaging Department examined the problem and recommends replacing our standard cardboard containers with molded plastic wrap. The plastic is applied from a hot roller after the cases . . .

However, in business and technical writing, many paragraphs cannot begin with topic sentences. In a technical report discussing a series of tests, for instance, the results section of the report might have several paragraphs opening as follows:

> Test 1, series 1, involved decreasing eluants by .4 cc/hr and noting pH changes occurring as the solution was heated to 250 degrees F. . . .
>
> During test 1, series 2, eluants were removed altogether, and the solution was subjected to pressure variations during heating . . .

While these paragraphs do not open with typical topic sentences, they do have opening sentences that clearly establish the paragraph's content and general direction. As readers, we should expect the first paragraph to focus entirely on test 1, series 1. Any information in that paragraph that is not related to test 1, series 1, does not belong there. Similarly, the second paragraph should focus on test 1, series 2.

Try to open every paragraph with a topic sentence. If you cannot, then open the paragraph with a sentence that clearly establishes the paragraph's subject, content, or general direction.

See ORGANIZATION.

5. Organize paragraphs logically.

The structure of the ideas within a paragraph should be logical. Furthermore, the paragraph structure should be obvious to readers.

Sometimes, this structure follows a classic organizational pattern: chronological, whole to parts, problem to solution, cause to effect, most important to least important, general to specific, and so on. Sometimes, the structure follows some logic that is inherent to the subject. A paragraph on drilling rig problems, for instance, might be organized according to a series of problems that relate to each other in some way that uninformed readers would not perceive.

The paragraph below is paraphrased from Charles Darwin's *Origin of Species*. It demonstrates a classic organizational pattern: general to specific. Note that the opening sentence is a topic (or thesis) sentence and that the three succeeding sentences substantiate Darwin's thesis:

> Without exception, every species naturally **reproduces** at so high a **rate** that, if not destroyed, the earth would soon be covered by the progeny of a single pair. Even **slow-breeding** man has **doubled** in 25 years, and at this **rate**, in less than 1,000 years, there would literally not be standing room for his progeny. Linneaus has calculated that if an annual plant **produced** only two seeds—and no plant is so **unproductive**—and their seedlings next year **produced** two, and so on, then in 20 years there should be 1,000,000 plants. The elephant is reckoned the slowest **breeder** of all animals, and I have taken some pains to estimate its minimum **reproductive rate**; it will be safest to assume that it begins **breeding** when 30 years old, and goes on **breeding** till 90 years old, bringing forth six young in the interval, and surviving till 100 years old; if this be so, after 750 years there would be nearly 19,000,000 elephants alive, descended from the first pair.

6. Use key words and other devices to ensure that paragraphs are coherent.

Coherence refers to the cohesiveness of a paragraph's sentences. In a coherent paragraph, the sentences seem to "stick together"—they all clearly belong in the paragraph and are logically connected to one another.

In the preceding paragraph example, the boldfaced key words indicate one of the most common methods of achieving coherence: repeating key words. The key words in the Darwin paragraph form a clear link between sentences. Consequently, as you read the paragraph, every sentence seems to belong. In the paragraph below, that sense of coherence is lacking:

> A great number of apparatus is available today for field work in the broad sense, including gas-chromatographs, as well as infrared, electrochemical, and other analyzers. In connection with the early prediction of possible pollutants and the assessment of natural discharge prior to the development of geothermal resources, however, hydrogen sulfide and volatiles such as ammonia, mercury, and arsenic are the major concern. Under the conditions prevailing before industrial development, preliminary evaluation and prediction of the discharge of such chemicals depends to a sizeable extent on water analyses. Surveying mercury content in air might deserve consideration; however, it is not discussed here, as mercury determination in soil is likely to be a valid substitute for it.

The opening sentence to this paragraph suggests that the paragraph will discuss the apparatus available for field work, particularly those apparatus listed. However, this equipment is never again discussed. The second, third, and fourth sentences seem loosely connected, but the paragraph never "gels," it never seems to be focused on a single topic. In short, the paragraph is incoherent.

You can achieve coherence by opening with a topic sentence, by using a clear organizational scheme, by repeating key words, by using transitional words (such as *however, furthermore, consequently, next, then, additionally,* etc.), and by using pronouns to link

PARAGRAPHS

sentences to the major idea or theme of the paragraph. (See ORGANIZATION, KEY WORDS, TRANSITIONS, and PRONOUNS.)

7. Emphasize the important ideas within a paragraph.

The opening and closing sentences of a paragraph tend to be the most emphatic sentences in the paragraph simply by virtue of their position. Readers pay more attention to those sentences than to the sentences that fall in the middle of the paragraph. Therefore, you should try to place your most important ideas in those sentences.

The Darwin paragraph above opens with a clear statement of Darwin's thesis. The opening sentence is the strongest sentence in the paragraph, and Darwin has wisely used it to state his most important idea. The closing sentence is also strong, and Darwin uses it to give his best example. If his thesis is true of the elephant, which is the slowest breeder of all animals, then it must also be true of every other species.

8. Provide transitions between paragraphs.

In most well-written documents, the information flows from paragraph to paragraph. To achieve this effect, you must provide smooth transitions between paragraphs.

Writers can set up transitions by previewing content. If you announce, for instance, that you will be discussing five topics and then list those topics, you have set up a progression that the reader will expect. As you move from topic to topic, the transitions will be automatic:

> The first topic concerns . . .
>
> Likewise, the second topic . . .
>
> The third topic . . .
>
> However, the fourth topic . . .
>
> Finally, the fifth topic . . .

As you can see, these paragraph openings also use some transitional words to effect the transition, but merely moving to and announcing the next topic is sufficient.

Sometimes, you can create the transition between paragraphs by using key words to connect the closing sentence of one paragraph and the opening sentence of the next:

> . . . because deep **salt domes** usually occur as a result of normal **faulting**
>
> Thrust or reverse **faults**, on the other hand, are normally responsible for piercement **salt domes** . . .

In this excerpt, the first paragraph closes with a key word (*faulting*) that is repeated in the opening sentence of the next paragraph. The first paragraph concerns deep salt domes; the second concerns piercement salt domes. Repeating a variation of *faulting* helps makes the transition.

In this example, the opening sentence of the succeeding paragraph makes the transition. However, the transition could also be made by the closing sentence of the preceding paragraph:

> . . . because deep **salt domes** usually occur as a result of normal **faulting**. Thrust or reverse **faults** on the other hand, are normally responsible for **piercement salt domes**
>
> **Piercement domes**, which are common along the Texas and Louisiana Gulf Coast, produce from traps caused when complex **faulting** forces a salt core upward through overlying sediments . . .

In some documents, the information cannot easily flow from paragraph to paragraph because the paragraph topics are too disjointed. When such is the case, use headings, lists, and numbering systems to indicate the transition from one topic to another:

> . . . Reef-producing areas may or may not be obvious from overlying sediments.
>
> Piercement Domes
>
> Piercement domes, which are common along the Texas and Louisiana Gulf Coast, produce from traps caused when complex faulting forces a salt core upward through overlying sediments . . .

See HEADINGS, LISTS, NUMBERING SYSTEMS, and TRANSITIONS.

9. If appropriate, break up or replace paragraphs with lists.

If a paragraph consists of a long series of items, or if a paragraph contains such a series, consider replacing the paragraph with a displayed list. Lists are more emphatic than paragraphs, so if you want to emphasize the series of items, display it. (See LISTS.)

PARALLELISM

Parallelism is essentially a convention of sentence construction. The principle behind it is that similar ideas should be expressed in a similar fashion, thereby demonstrating their similarity and making reading easier. The following sentence is not parallel:

> The analysis will include organizing, dividing, and assessment of turnaround functions.

The sentence verb *include* is followed by three key words: *organizING*, *dividING*, and *assessMENT*. These three words appear in series. They are equal in purpose and use in the sentence. Therefore, they should have the same grammatical form:

> The analysis will include organizing, dividing, and assessing turnaround functions.

1. Ensure that two or more parts of speech behaving similarly in a sentence or coordinated (connected) in some way are parallel in construction.

Parallelism applies not only to verbs, but also to nouns, adjectives, phrases, and every other part of a sentence:

> The Interface Team will be responsible <u>for integrating</u> the functional units developed by the OR Team and <u>for executing</u> the model test matrix.

> Applying <u>abstraction</u>, <u>partition</u>, and <u>projection</u> to the system development process results in the traditional top-down view of the software engineering process.

> Figure 2.2-1 shows the documentation relationships: <u>where things happen</u>, <u>why things happen</u>, and <u>how things can be changed</u>.

> Multi-level training was necessary to meet the needs of <u>managers</u>, <u>designers</u>, and <u>programmers</u>.

> A final report was prepared, <u>describing</u> the case study process and <u>referencing</u> the documents containing the code.

2. Make items in lists parallel.

Parallelism is especially important in lists. A list, whether displayed vertically on the page or embedded within a paragraph, is a series. To make it parallel, each item should be constructed similarly and should begin with the same kind of word (noun, verb, etc.):

> This file will include the following items:
>
> 1. Problem headings
> 2. Data base specifications
> 3. Reporting intervals
> 4. Restart options
> 5. Level of detail options
> 6. Links to report macros

The following list is also parallel (each item completes the sentence started by the introductory statement). Note that each item begins with the same kind of verb:

> The study concluded that the ATAC fighter must:
>
> 1. Have a long-range, high-payload capability.
> 2. Be flexible in mission and payload design.
> 3. Be survivable against A-A and S-A threats.
> 4. Be maneuverable in the F-15/F-16 class.

NOTE: This example has several variations. Many authorities would insist on a different lead-in sentence (one with a complete grammatical structure):

> A study concluded that the ATAC fighter must have these features:

See COLONS, CONJUNCTIONS, GOBBLEDYGOOK, and LISTS.

PARENTHESES

1. Parentheses enclose explanatory sentences within a paragraph:

> Only the total systems approach can deal with the tradeoff in performance between the weapon and the aircraft platform. Existing beyond-visual-range air-to-air missiles are inhibited, for instance, by the lack of an effective IFF system. The total systems approach, with its full range of analysis tools, may be the only acceptable means of evaluating tradeoffs prior to the detail design phase. (The discussion of IFF design under Targeting Systems on p. 89 reveals how we solved the problem cited above.)

2. Parentheses enclose references, examples, ideas, and citations that are not part of the main thought of a sentence:

> Our Level 6 analysis (see figure 9.4) illustrates how a single multi-mission destroyer can contribute to task force operations.

> Our design accounts for all environmental factors that may affect sensitivity (smoke, terrain, weather, and physical damage).

> Affordability (cited in the RFP as a primary concern) was the guiding principle behind our application of new technologies.

> Our previous state-of-the-art survey (conducted over a 3-month period in 1982) suggested that RDF SOP's were not current.

> The most recent research (Smithson 1983) revealed pollution problems from nearby gasoline storage tanks.

See CITATIONS.

Parentheses, Commas, and Dashes

Commas and dashes also enclose explanatory ideas. Commas are less emphatic than parentheses; dashes are more emphatic. Note how emphasis progressively increases in the following examples:

> Cost analyses using both parametric and detail O&S cost methodologies helped us determine the right support systems.

> Cost analyses, using both parametric and detail O&S cost methodologies, helped us determine the right support systems.

> Cost analyses (using both parametric and detail O&S cost methodologies) helped us determine the right support systems.

> Cost analyses—using both parametric and detail O&S cost methodologies—helped us determine the right support systems.

See COMMAS and DASHES.

3. Parentheses enclose numbers in a paragraph list:

> The operational characteristics we will discuss below are (1) manning, (2) training, and (3) providing required support.

4. Parentheses enclose acronyms, abbreviations, and figures that have been written out:

> The CARP (Capital Area Renovation Project) is adequately funded as long as the contractor trims costs by using off-the-shelf materials wherever possible.

> United's South Fork Mine can deliver over 20,000 dwt (deadweight tons) of ore every month.

> By the project deadline date, Northrop will deliver fifty (50) centrifugal pump assemblies to the San Diego facility.

NOTE: The practice of writing out numbers and enclosing the figure in parentheses is not necessary except in legal, contractual, or requisition documents. Do it only when you need to protect against unauthorized alteration of numbers in a document.

See NUMBERS.

Parentheses and Brackets

Brackets are, in effect, parentheses. They enclose incidental or explanatory words and phrases within parentheses or within quoted material:

> The environment and activities of opposing forces may change the capabilities of a particular sensor (see appendix 4, Battlefield Adaptability Requirements, for a fuller discussion of RGS [Remote Ground Sensing] and ground-based sensor limitations).

> Your original letter stated: "Our on-site project coordinator [Walt Petersen] will be responsible for maintaining the Schedule of Deliverables."

See BRACKETS.

Parentheses and Periods

If the entire sentence is enclosed by parentheses, the period at the end of the sentence goes inside the closing parenthesis:

> (See appendix 2 for the complete test results.)

If only part of a sentence is enclosed by parentheses and the closing parenthesis occurs at the end of the sentence, the period goes outside the closing parenthesis:

> Hydrostatic and thermostatic monitors ensure system equilibrium (see figure 5-15 for monitor locations).

See PERIODS and QUESTION MARKS.

PERIODS

1. Periods follow statements, commands, indirect questions, and questions intended as suggestions:

Statements

The workover plan was finished.

Tomorrow we will visit the mine site.

Mr. Smythe owes OP&L $75.

Commands

Stop working on the project now.

Please help us tomorrow.

Redesign the pump housing to accommodate the larger intake pipe.

Indirect Questions

I wonder how he managed the project.

Jane Greer asked whether we would approve the budget.

Questions Intended as Suggestions

Will you please return the forms by a week from Monday.

Would you let me know if you have any questions.

2. Periods follow numerals or letters marking a list, but periods need not follow the items listed unless they are full sentences (see rule 1 above):

a. A larger pump
b. An extra ventilation fan
c. A heavy duty circuit breaker

---

1. The cost is 50 percent greater than was budgeted.
2. Materials were not equal to those specified.
3. Installation procedures were violated.

NOTE: You may need a period to end a list that continues the syntax established in its lead-in sentence:

We tested the procedure by

1. increasing the flow,
2. decreasing the temperature, and
3. contaminating the water.

This pattern of continued syntax and punctuation is much rarer than it used to be. (See LISTS.)

3. Periods separate integers from decimals:

4.567
327.5
1,456.25

NOTE: In some foreign countries, a comma separates integers from decimals and spaces separate groups of three numerals in longer numerals:

4,567
56 764,534 45

See METRICS.

4. Use a period with run-in headings, but not with displayed headings:

Two Options. The first option is to discontinue the testing until safety procedures are developed. The second option . . .

This same heading would have no following period if it appeared on its own line:

Two Options

The first option is to discontinue the testing until safety procedures are developed. The second option . . .

NOTE: You can also use run-in headings with dashes, colons, or no punctuation. But be consistent. Once you've established a pattern, use it throughout a document.

See HEADINGS.

5. Periods follow some abbreviations:

10 a.m.	6 p.m.
A.D. 1910	225 B.C.
U.S.A.	Mr./Mrs./Ms.
e.g.	S. Pugh
Dr. William Lange	U.K.
	i.e.

NOTE: Many abbreviations no longer require periods, especially names of fraternal organizations, government agencies, corporations, colleges and universities:

BPOE	BLM
DOE	GM
UCLA	LSU

See ABBREVIATIONS, PARENTHESES, and QUESTION MARKS.

PHOTOGRAPHS

Figure 1. An offshore rig in the Barents Sea.

Figure 2. The scaffolding outlines the new building while providing masons with a platform upon which to work.

Figure 3. The scaffolding outlines the new building while providing masons with a platform upon which to work.

Photographs convey realism and authenticity. They are the most persuasive kind of visual aid because readers tend to trust the exactness of what is presented. Therefore, photos are useful for presenting evidence.

If you want to show readers what is, use a photograph; if you want to show readers what is possible, use an illustration.

For general information on using visual aids, see VISUAL AIDS. See also CHARTS, GRAPHS, ILLUSTRATIONS, MAPS, and TABLES.

For information on captions, see CAPTIONS.

1. Ensure that each photograph has a central point of interest.

Like other visual aids, photographs should focus on a single central concept or idea. The photo should have what photographers call a principal point of interest. Ideally, the photograph is "composed" so that the reader's eye is led naturally and immediately to that principal point of interest. To accomplish this feat, the photo must be simple and uncluttered (see figure 1).

To establish the central point of interest, ensure that the photograph is taken from an angle that maximizes coverage of the object that should be of greatest interest. The same object may be photographed from a variety of angles, so choose the angle that best focuses on the object of interest and that best shows what you want the photograph to show.

Figure 1 is effective at focusing the reader's attention on the principal point of interest. Figures 2, 3, and 4 are not effective.

Figure 2 attempts to show the scaffolding that brick masons erect in constructing a new building. However, the angle at which the photograph was taken distorts the view of the scaffolding and does not present enough of the superstructure for readers to appreciate what they are supposed to be seeing.

Figure 3 also does not present enough of the superstructure for readers to comprehend the scaffolding and is too cluttered (see rule 2 below).

Figure 4 presents the scaffolding from an edge-on angle, which does not provide adequate perspective, and confuses the principal point of interest by including two workmen who are staring at the camera.

Figure 5 is a much better photograph. It presents enough of the superstructure for readers to appreciate the scaffolding, and it includes no extraneous items.

2. Ensure that photographs are simple and uncluttered. If necessary, crop photos to eliminate distracting detail.

Each photograph should have a single central subject. Try to eliminate everything in the picture that does not contribute or relate to that subject (for good examples, see figures 1 and 5; for a bad example, see figure 4).

PHOTOGRAPHS

Take photographs from an angle that maximizes the focus on the principal point of interest and minimizes surrounding detail. Often, the photographer must crop the photograph in the darkroom to eliminate detail that detracts from the principal point of interest or that is simply irrelevant to the point of the photo.

Figure 6 is an example of a photograph that needs to be cropped. The principal item of interest, the building under construction, occupies only a small part of the frame. The low building in the background is extraneous. Its presence in the frame detracts from the principal point of interest and creates a confused photograph.

The principles of photographic composition are beyond the scope of this discussion, but in a photograph you should not allow details to remain that compete with the object you want readers to notice first and to focus on. Sometimes, to eliminate distractions or nonessential detail, you may have to take the photograph from another angle.

So, when the photographs are taken, the photographer must consider how the photo will be used, what the photo is intended to show, what the reader should be focusing on, and whether the objects visible through the lens from each angle are appropriate.

3. Ensure that photos have sufficient contrast and are in focus.

Proper separation of light and shadow (contrast) makes for a good black and white photograph. If the contrast is not sufficient (figure 7), readers may not be able to distinguish the shapes and details of the objects photographed. If the photo has too much contrast (figure 8), detail will be lost and the photo will look sharp and unrealistic.

Proper contrast occurs when the photograph shows a clear range of tones: crisp whites, distinct shades of gray, and deep blacks (figure 9).

In addition, the photo should be sharply focused.

4. Establish the size and proportion of the object photographed by using scales or by including objects in the photo that permit scale comparisons.

Photos taken extremely close up or extremely far away may be difficult for readers to grasp because they lack perspective.

Figure 4. The scaffolding outlines the new building while providing masons with a platform upon which to work.

Figure 5. The scaffolding outlines the new building while providing masons with a platform upon which to work.

Figure 6. Trusses are weight-bearing members that also establish and maintain the super structure's frame and angles.

Figure 7. Though unfinished, the building already shows its strong vertical orientation.

PHOTOGRAPHS

Figure 8. Though unfinished, the building already shows its strong vertical orientation.

Figure 9. Though unfinished, the building already shows its strong vertical orientation.

Figure 10. Atlas oversized bricks are sturdier and more durable than normal bricks.

Without perspective, readers may interpret a close-up of the hair on a man's arm as a bizarre forest and a close-up of fabric as a net made of thick ropes. Photos taken outside of the normal range of our visual experience must be placed into perspective.

You could print a scale beside a photo or place a familiar object on the thing being photographed. A close-up of bricks, for instance, might include a pencil (see figure 10). Readers must still be told that they're looking at the texture of a brick, but the pencil will give them a sense of size and proportion.

Photographs of geologic features are especially difficult for readers to interpret unless the photograph includes a vehicle, a person, or some other familiar object that permits a scale comparison.

PLURALS

Nouns (and some pronouns) usually signal whether they are singular or plural by changes in spelling.

Such changes are no problem if they follow the regular pattern: an *-s* or an *-es* added to the singular form makes the plural form:

report + s	=	reports
book + s	=	books
church + es	=	churches
tax + es	=	taxes

Problems arise when the plural does not follow the regular pattern:

mouse	mice
datum	data
chassis	chassis
fungus	fungi
matrix	matrices
I	we
he, she, it	they

The following discussion covers these and other irregular plurals. The best guide, however, is a good modern dictionary. (See REFERENCES.)

See NOUNS and SPELLING.

1. Use the following list to determine the plurals of many irregular forms, especially those technical terms borrowed from Latin or other languages:

addendum, addenda
agendum, agenda
alga, algae
alumnus, alumni (*masc.*)
alumna, alumnae (*fem.*)
antenna, antennas (antennae, *zoology*)
appendix, appendixes (*or* appendices)
axis, axes
basis, bases
cactus, cactuses
calix, calices
cicatrix, cicatrices
Co., Cos.
coccus, cocci
consortium, consortia
crisis, crises
criterion, criteria
curriculum, curriculums (*or* curricula)
datum, data
desideratum, desiderata
ellipsis, ellipses
equilibrium, equilibriums (equilibria, *scientific*)
erratum, errata
executrix, executrices
focus, focuses
folium, folia
formula, formulas
fungus, fungi
genus, genera
gladiolus (*singular and plural*)
helix, helices
hypothesis, hypotheses
index, indexes (indices, *scientific*)
lacuna, lacunae
larva, larvae
larynx, larynxes
lens, lenses
locus, loci
madam, mesdames
matrix, matrices
medium, mediums (*or* media)
memorandum, memorandums (*better* memo, memos)
minutia, minutiae
nucleus, nuclei
oasis, oases
octopus, octopuses
opus, opera
parenthesis, parentheses
phylum, phyla
plateau, plateaus
radius, radii
radix, radixes
referendum, referendums
septum, septa
seta, setae
stimulus, stimuli
stratum, strata
stylus, styluses
syllabus, syllabuses (*or* syllabi)
symposium, symposia
synopsis, synopses
terminus, termini
testatrix, testatrices
thesaurus, thesauri
thesis, theses
thorax, thoraxes
vertebra, vertebras (vertebrae, *zoology*)
virtuoso, virtuosos
vortex, vortexes

NOTE 1: Many of the above forms now have regular plurals (*appendix, appendixes* or *memorandum, memorandums*). However, some editors still prefer the irregular forms (usually based on the word's origin in Latin or another language: *appendices, memoranda*). The longer a word is in English, the stronger the tendency is to make the plural conform to the regular English pattern (adding an *s* or *es* to the singular form).

NOTE 2: Though many of these words come from other languages, they are now sufficiently English and do not need underlining or italics. (See UNDERLINING and ITALICS.)

2. In compound terms add the plural ending (usually *s* or *es*) to the most significant word:

attorneys at law
bills of fare
brothers-in-law
comptrollers general
daughters-in-law
goings-on
grants in aid
lookers-on
reductions in force
surgeons general

assistant chiefs of staff
assistant surgeons general

assistant attorneys
deputy judges
lieutenant colonels
trade unions

hand-me-downs
higher-ups
pick-me-ups

3. Nouns ending in *o* preceded by a consonant usually add *es* for the plural:

PLURALS

echo, echoes
veto, vetoes
potato, potatoes

EXCEPTIONS: This rule has many exceptions, so if in doubt, check a good dictionary. (See REFERENCES.) Here are some of the common exceptions:

dynamo, dynamos
Eskimo, Eskimos
ghetto, ghettos
halo, halos
indigo, indigos
magneto, magnetos
octavo, octavos
piano, pianos
sirocco, siroccos
two, twos
zero, zeros

4. The coined plurals of abbreviations, titles, figures, letters, and symbols require an *s* and sometimes an apostrophe plus an *s*:

OK's
ABC's
CODs or COD's
the three Rs
SOS's
g's
1 by 4's

NOTE: As the above examples illustrate, editors do not agree about the need for an apostrophe. The trend, however, is for the apostrophe to vanish, leaving the simple -*s* signal that the item is a plural. (See APOSTROPHES.)

5. Plurals of pronouns, when they exist, are likely to be very irregular:

I	we
you	you
he, she, it	they

See PRONOUNS.

POSSESSIVES

Possessives are those forms of nouns and pronouns that show ownership or, in some cases, other close relationships:

Ownership

IBM's service booklet
Mr. Vaughan's store
his store
Susan's desk
her desk
Lewis' report
the engineer's schedule
the Lewises' house

Other Relationships

the book's cover
its cover
the corporation's support
his lawyer's consent
the captain's story
a summer's day
a day's absence
a doctor's degree

NOTE: Noun possessive forms routinely require an apostrophe or an apostrophe plus an *s*. (See APOSTROPHES.)

1. Distinguish between true possessives and descriptive terms:

Possessives (whose . . .)

Exxon's reply
the employee's record
Oregon's laws
the Smiths' house

Descriptive Terms (what kind of . . .)

an Exxon reply
the employee record
Oregon laws
the Smith house

NOTE 1: Either form of the above phrases is correct, so decide which form you prefer and then be consistent within the same document.

NOTE 2: The names of countries, organized bodies, and groups ending in *s* usually do not require apostrophes:

United States plan
technicians guide
Massachusetts statutes
editors handbook
United Nations publication

POSSESSIVES

2. For singular nouns not ending in *s* and for plural nouns not ending in *s*, form the possessive by adding an apostrophe plus an *s*:

the cat's paw
Anne's statement
a man's coat
men's coats
an accountant's books
the children's payments

3. For both plural nouns and singular nouns ending in *s* or an *s* sound, the possessive form requires only an apostrophe:

General Dynamics' proposal
Sears' 4th Quarter Report
Penneys' reaction
the boss' idea
James' speech

NOTE: Some editors and writers prefer to add both an apostrophe and an *s* to singular nouns ending in *s*, especially if the new word has an extra syllable:

General Dynamics's proposal
the actress's script
the boss's idea
James's speech

4. Add an apostrophe plus an *s* to the end of personal and organizational names showing possession:

Sears & Roebuck's policy

Charles F. Shook's decision

the Odd Fellows's initiation (*or* Odd Fellows' initiation)

E.F. Hutton & Company's merger with Smith Barney

NOTE: Corporate and organizational practices vary, so, if possible, check the letterhead or other correspondence for exceptional cases:

American Bankers Association
Steelworkers Union
Investors Profit Sharing

5. Possessive forms of personal pronouns and of the relative pronoun *who* do not require an apostrophe:

My secretary had <u>mine</u>.

Your supervisor had <u>hers</u>.

The company lost <u>its</u> comptroller.

We refused to pay for <u>ours</u>.

They lost <u>theirs</u> when the market fell.

He mentioned <u>his</u>.

<u>Whose</u> report is this?

NOTE 1: These are the possessive forms of personal pronouns: *my/mine, your/yours, his, her/hers, its, our/ours, their/theirs*. In cases with two forms, the first form must come before the noun it possesses while the second form comes after the verb, with its noun implied:

They were her ideas.
The ideas were hers.

NOTE 2: Distinguish between possessive forms without apostrophes and contractions with apostrophes:

The pump had lost its cover.

We decided that it's (*it is*) time to redrill.

He had changed his job recently.

He's changing his job.

Whose idea was this?

Who's going to be at the meeting?

6. Possessive forms of indefinite pronouns require apostrophes:

anyone else's task
one's ideas
the other's notion
the others' schedules
anybody's recommendation
someone else's job

7. Possessives sometimes occur without a following noun:

My ideas are like Sue's.

His nose is like a bloodhound's.

I'll be at Jim's.

She was at the doctor's.

IBM's is good but Apple's is better.

NOTE: Large, established companies sometimes violate the usual rule (singular form plus an apostrophe and an *s*):

We conducted a survey at Macys. (*or* Macys')

Harrods is splitting its stock.

8. A possessive modifies an *-ing* form of a verb used as a noun:

Bill's speaking to my boss helped.

I objected to his working on the rig overnight.

I admired Sue's planning.

PREPOSITIONS

Prepositions are words that connect or relate nouns and pronouns to preceding words and phrases:

> The engineer moved from his desk.
>
> The plans for the new substation have yet to be completed.
>
> The case against her became even more convincing.
>
> There is truth in what you say.
>
> The firm submitted a summary of the specifications.

The simple prepositions are *at, by, in, on, down, from, off, out, through, to, up, for, of,* and *with*. More complex, even phrasal, prepositions also exist: *against, beneath, in front of, on top of, on board (of), outside of, according to, on account of, by means of,* etc. Although they number less than a hundred, prepositions are essential words in English: most normal sentences contain one or more prepositions.

1. Do not be overly concerned if you end a sentence with a preposition.

For over a thousand years English has had normal sentences that ended with prepositions:

> That's something we can't put up with.
>
> The Universal acquisition is the most difficult deal we've gotten into.
>
> This is the report I've been telling you about.
>
> She's the accountant you spoke to?
>
> This project is the one that you objected to.

Winston Churchill was once corrected for ending a sentence with a preposition, and he replied, "That is the sort of [nonsense] up with which I shall not put."

As Churchill's witty reply indicates, many English sentences sound awkward if you try to avoid ending them with prepositions:

> *not*
>
> The Universal acquisition is the most difficult deal into which we've gotten.
>
> This is the report about which I told you.
>
> She is the accountant to whom you spoke.
>
> This project is the one to which you objected.

Make your sentences as simple, smooth, and direct as possible, and don't worry about such misconceptions as not ending a sentence with a preposition. The fact is, a preposition is a fine word to end a sentence with.

2. Distinguish between the prepositions *between* and *among*.

Between usually refers to two things, while *among* refers to more than two things. However, *between* can also refer to more than two things if each of the things is compared to all the others as a group:

> The judge divided the land between the two parties. (*not* among)
>
> The judge divided the land among a dozen parties. (*not* between)
>
> We had trouble deciding between the two pumps. (*not* among)
>
> We carefully analyzed the differences between the six alternatives. (*or* among)
>
> We had trouble deciding among all the possible small trucks. (*or* between)

3. You can sometimes omit prepositions without changing the meaning of the sentence:

> All (of) the engineers visited the site.
>
> We moved the pipe off (of) the loading dock.
>
> The filing cabinet is too near (to) the door.
>
> We met (at) about 8 p.m.
>
> Where are they (at)?

In each of the above sentences, the preposition within parentheses is unnecessary. In the last sentence, *at* is both unnecessary and incorrect. Generally, native speakers of English can be guided by their innate sense of what constitutes smooth and clear uses of prepositions.

See MODIFIERS.

PRONOUNS

Pronouns are words that take the place of more specific nouns or noun phrases:

George completed the drawings.
He completed the drawings.

The young engineer spoke up.
She spoke up.

John's survey was efficient.
His survey was efficient.

First State Bank went bankrupt.
It went bankrupt.

John, Sue, Esther, and George left the party early.
They left the party early.

Who completed the drawings?
(Who = *somebody unknown*)

The man hit himself.
We ourselves paid for the damage.

The bid and the interest were issues.
Those were issues.

John, Sue, Ester, George, etc. left the party early.
Everyone left the party early.

As the above examples indicate, pronouns include a variety of types of words: **personal** pronouns (*he, she, his, they, etc.*), **interrogative** and **relative** pronouns (*who, that,* and *which*), **reflexive** and **intensive** pronouns (*himself, ourselves, etc.*), **demonstrative** pronouns (*this, that, these,* and *those*), and **indefinite** pronouns (*everyone, anybody, someone, etc.*).

Personal Pronouns

			SUBJECTIVE CASE	OBJECTIVE CASE	POSSESSIVE CASE
1st Person	Singular		I	me	my/mine
	Plural		we	us	our/ours
2nd Person	Singular		you	you	your/yours
	Plural		you	you	your/yours
3rd Person	Singular	Masculine	he	him	his
		Feminine	she	her	hers
		Non-human	it	it	its
	Plural		they	them	their/theirs

Personal Pronouns

Personal pronouns are those that commonly replace the names of individuals or objects: *I, we, you, he, she, it,* and *they*:

I and *we* are first person pronouns—that is, they are used for the person(s) speaking.

You (singular and plural) is the second person pronoun—that is, it is used for person(s) spoken to.

He, she, it, and *they* are third person pronouns—that is, they are used for persons or objects spoken about.

Most personal pronouns also have different forms for the singular and the plural, for different uses in sentences, and for possessives. The table on this page summarizes these different forms of the personal pronouns.

While the differences between the singular and plural forms are obvious, the differences between the cases are often confusing. Different cases (or pronoun forms) have different uses in sentences:

I surveyed the site. (I *is the subjective case.*)

The committee quizzed me. (Me *is the objective case.*)

My report was too long. (My *is the possessive case—the one that comes before its noun.*)

The report is mine. (Mine *is the possessive case—the one that follows its noun.*)

See PLURALS.

1. Use subjective case pronouns for the subject of a verb or when the pronoun follows a form of *be* (*am, is are, was, were, be, been*):

Subject of a Verb

Jan and I analyzed the blueprints.
We discussed the design options.
You were omitted from the roll.
He called two colleagues.
She and her employee came to the 11 a.m. meeting.
It was a poor choice.
They and Harold met about the legal problem.

PRONOUNS

NOTE: In cases where a pronoun and a noun are both subjects of the same verb, you can test the pronoun by removing the noun. In the sentence *She and her employee came to the 11 a.m. meeting*, no one would read it as follows: *Her came to the 11 a.m. meeting*. So *she* is the correct pronoun.

Following a Form of *Be*

The engineer chosen was she.
The contractor who won is he.
Was it they who called?
It could have been I.

NOTE: Pronouns following forms of *be* often sound strange:

It is I.
This is she.

Normally, objective case pronouns follow verbs: *It is me* or *This is her.* These sentences are acceptable in informal speech, which is where they ordinarily would appear anyway. In formal speech and in most writing, the subjective case forms are correct. (Of course, if your sentence sounds stiff or strange, rephrase it to avoid the problem. You could say *I am here* instead of *It is I*.)

2. Use objective case pronouns when the pronoun is the object of a verb or the object of a preposition:

Object of a Verb

The committee chose her.
Carol did not include Jane and him.
The pollution affected them.
The President contacted Harry and me.

NOTE: When a noun and a pronoun both follow a verb, you can check the pronoun by omitting the noun. So *The President contacted Harry and me* becomes *The President contacted me*. Few people would be comfortable with *I*, which is the wrong pronoun, but many speakers often use *I* when a noun comes before it: *The President contacted Harry and I*. This sentence is incorrect.

Object of a Preposition

Frank walked by her.
The team studied with me.
On account of me the project ended.
The proposed schedule depended on Jane and her.
Acme Inc. worked against Jason and me.

NOTE: In the last two sentences, you can test for the correct pronoun by removing the noun that comes between the preposition and the pronoun. (*Wrong*: The proposed schedule depended on she. Acme Inc. worked against I.)

3. Use possessive cases correctly:

That was her report.
The report was hers.

The problems were ours, not yours.
They were our problems, not your problems.

I rejected your arguments.
The arguments were yours, not mine.

NOTE 1: In cases where two possessive forms exist for a single pronoun, the simple form (without an *s*) comes before the noun; the other form (with an *s*) follows a form of *be* (*am, is, are, was, were, be, been*). *His* and *its* have only a single form.

NOTE 2: Do not confuse the possessive pronouns with simple contractions. Possessive pronouns do not have apostrophes; contractions have apostrophes:

Possessive Pronoun	Contraction
its	it's
his	he's
their/theirs	they're
your/yours	you're

4. Choose pronouns so that they agree with their antecedents in number, in gender, and in case:

Agreement in Number

The men worked on the design plan all night, but they were unable to finish. (men [*antecedent*] = *plural*; they = *plural*)

Cheryl Higgins prepared a revised safety procedure, but she failed to get managerial approval. (Cheryl Higgins [*antecedent*] = *singular*; she = *singular*)

Agreement in Gender

Sidney Brown was eager to take his first actuarial exam. (Sidney [*antecedent*] *can be either male or female, so the pronoun is a key sign of which is intended. Often, of course, the female spelling is* Sydney.)

NOTE: When the antecedent is an indefinite pronoun, writers often don't know whether to use a singular or plural pronoun, much less a male or female pronoun:

PRONOUNS

Everyone should arrange their (his *or* her?) desk before leaving.
Everybody is responsible for their (his *or* her?) own time cards.

The forms with *their* have become more acceptable, especially in speech, even though *everyone* and *everybody* are usually considered singular pronouns. To avoid any sexist language and to eliminate the clumsy *his or her*, the best option would be to rewrite the sentences as plurals:

All employees should arrange their desks before leaving.
All clerks and secretaries are responsible for their own time cards.

See AGREEMENT, SEXIST LANGUAGE, and the discussion below of indefinite pronouns.

Agreement in Case

Betty argued with them—especially Susan and him. (them = *objective case*; him = *objective case*)

Interrogative and Relative Pronouns

Interrogative and relative pronouns are similar in their uses, and some interrogative pronouns are identical to the main relative pronouns: *who, whom, whose,* and *which*. Relative pronouns also include *that, whoever, whomever, whatever, why,* and *where*. Interrogatives also include *what*, which is not a relative form.

Many interrogative and relative pronouns have only one form, but *who* changes its form just as the personal pronouns do:

Subjective case: who, whoever
Objective case: whom, whomever
Possessive case: whose

That and the other relative and interrogative forms do not change to reflect the different cases.

5. For interrogative and relative *who*, use the subjective case for subjects and following a form of *be*; use the objective case for objects of verbs and prepositions:

Subjective Case

Who is the project engineer?
Whoever writes the report gets all the credit.
The agent who spoke to us was courteous.
The supervisor determined who would be the representative.
The representative is whoever has the most years of service.

Objective Case

Whom did you nominate?
Whom did you wish to speak to?
Or: To whom did you wish to speak?
Whomever you nominate, we'll find someone to balance the ticket.
The man whom you spoke to is our president.

NOTE 1: The first two examples under the objective case sound almost too formal, even stiff. Most speakers of English are more comfortable if a subjective *who* begins the sentence or question. For this reason, two other versions of these questions are acceptable in informal, spoken English:

Who did you nominate?
Who did you wish to speak to?

Similarly, even with non-questions, subjective *who* sometimes replaces more correct *whom*:

Whoever you nominate, we'll find someone to balance the ticket.

These forms with *who* or *whoever* are incorrect in writing even though acceptable in informal speech.

NOTE 2: Deciding on the correct case is especially difficult when the relative pronoun is part of a complex sentence. The trick is to isolate only the clause containing the relative pronoun:

who wants to study chemistry
whom the proposal team chose
whoever is the top candidate

These are correct uses of *who* and *whom*, so you can insert them into any sentence, and they'll be correct:

John, who wants to study chemistry, is our lab technician.

We left with Sandra, whom the proposal team chose.

Whoever is the top candidate will be our speaker.

We planned to interview whoever is the top candidate.

So to decide on the case of a troublesome use of *who* or *whom*, try to isolate the clause containing the relative pronoun (underlined below):

We decided to abandon the direction established by the geologist who/whom we first talked to.

PRONOUNS

The proper form then is *whom*, used as the object of the preposition *to*.

NOTE 3: Adjective clauses introduced by a relative pronoun are often difficult to punctuate. The simplest rule is to enclose such clauses with commas when they are non-essential:

Jack Craven, who is our project coordinator, has been with the company for 15 years.

We presented the proposal to the team from AirFlo Inc., which is a firm based in Denver.

When the clauses are essential, no commas are required:

The proposal that we sent to the Department of Transportation has been cancelled. (*Without the clause* that we sent to the Department of Transportation, *the sentence would not be clear to most readers, especially if more than one proposal is possible.*)

The Governor proposed banning all autos that did not have a safety inspection. (*The clause* that did not have a safety inspection *is essential to the meaning.*)

See COMMAS.

Reflexive and Intensive Pronouns

Reflexive and intensive pronouns are identical in appearance: both end in *-self* (singular) and *-selves* (plural). Personal pronouns have reflexive and intensive forms: *myself, ourselves, yourself, yourselves, himself, herself, itself,* and *themselves*.

Reflexive pronouns "reflect" back to a noun mentioned earlier in the same sentence:

George injured himself.
The team quarreled among themselves.
Amy cut herself.
The dog licked itself.

Intensive pronouns usually follow the nouns they intensify:

George himself was injured.
The team themselves were arguing.
Amy herself ran the errand.
I myself checked on the data.

6. Avoid the unnecessary use of intensive/reflexive forms when an ordinary personal pronoun would suffice:

With our permission, Jeff became the spokesman for Jim and myself. (*better*: Jim and me)

John and myself are both uncomfortable with the proposal. (*better*: John and I)

Demonstrative Pronouns

Only four demonstrative pronouns exist:

	Singular	Plural
"near"	this	these
"far"	that	those

Demonstrative pronouns sometimes appear before a noun and sometimes they can replace the noun or noun phrase entirely:

This design plan has problems.
This has problems. (*This sentence assumes that the reader or listener knows what the* this *refers to.*)

That proposal for the DOE demonstrates effective graphics.
That demonstrates effective graphics.

7. Avoid vague uses of demonstrative pronouns (usually when the pronoun is used without the noun it is describing):

This is something to consider. (*better*: This shortfall in payments is something to consider.)

These are difficult. (*better*: These exercises are difficult.)

NOTE: Such vague sentences are particularly annoying when, for example, a *this* refers back to an entire sentence or several things in a sentence:

not this

The travel plans were a jumble and even turned out to be beyond our budget. This meant that we could not make connections as planned.

this

The travel plans were a jumble and even turned out to be beyond our budget. The jumble (*or* This jumble) meant that we could not make connections as planned.

Indefinite Pronouns

Indefinite pronouns include a number of words that have unspecified or vague meanings, usually because the writer or speaker does not want to identify or can't identify the person(s) or thing(s) referred to. These sentences illustrate some of the common indefinite pronouns:

PRONOUNS

Everyone should proofread the report before we send it out.

We cleaned out everything before vacating the office.

Both were involved in the design tests last summer.

We determined that somebody had changed the entry code in the computer.

The procedures required that anyone leaving the security area sign and note the date and time.

Nobody had a pass to be on the construction site after normal work hours.

A List of Indefinite Pronouns

everyone everybody each both	everything every all
someone somebody some one	something somewhere another
anyone anybody either	anything anywhere any
no one nobody none	nothing nowhere neither
many several more	few others most

NOTE: Many of these indefinite pronouns can act very much like adjectives:

Several books were stolen from our library.

Most participants would want to attend the final session.

or

Several were stolen from our library. Most would want to attend the final session.

8. Indefinite pronouns used as subjects should agree in number with their verbs:

Anyone likes to receive a positive performance review. (Anyone = *singular*; likes = *singular*)

Several were contacted before we chose a final candidate. (Several = *plural*; were = *plural*)

All of the sugar was tainted. (All of the sugar = *singular*; was = *singular*)

All of the employees were notified of the new vacation policy. (All of the employees = *plural*; were = *plural*)

NOTE: Sometimes the indefinite pronoun, while considered grammatically singular, requires a plural pronoun:

Everyone should abandon their attempt to discover the error.

Strict editors would argue that *everyone* is singular while *their* is plural, so traditionally the correct form has been this version:

Everyone should abandon his attempt to discover the error.

Feminists and others have criticized this version as being sexist because females are ignored. One option would be to replace *his* with *his or her*. A more reasonable option is to rewrite the sentence so that it is clearly plural:

All engineers should abandon their attempts to discover the error.

See AGREEMENT, COMMAS, SEXIST LANGUAGE, STYLE, and WORD PROBLEMS. Also see the discussion above of personal pronouns.

PUNCTUATION

The marks of punctuation are as much a part of language as words. Like words, punctuation is a code. Knowing and following the code allows writers to communicate with others who know the code.

Yet the rules of punctuation are not entirely firm, fixed, and unchanging. Like words, the marks of punctuation are evolutionary. The rules of punctuation have changed slightly in the last 25 years and can be expected to continue changing.

Furthermore, punctuation is subject to tradition, convention, and local usage. Conventions differ from industry to industry and publication to publication. The punctuation style that one publisher prefers may differ from the style another publisher prefers.

Despite such variations, many punctuation rules are firm and universal, as, for instance, the comma in *June 15, 1984*. Other "rules" are optional.

Writers who use every permissible mark of punctuation follow a **mandatory style**. Writers who prefer to omit optional punctuation follow an **optional style**. Other terms can be and have been used to describe these two styles of punctuation: formal vs. informal, conservative vs. liberal, and closed (or close) vs. open.

How we label these styles of punctuation is not as important as the fact that punctuation is a code and that 95 percent of the rules are not subject to debate. Even those writers who prefer the optional style will agree with most of the punctuation decisions made by writers opting for the mandatory style. Without such agreement, writing would be as chaotic and difficult to read as the following example:

> Four drilling crew's would drill the shot holes; using truck, mounted drills' with, port'able pits? each hole would be-drilled: to no more than a/maximum' depth, of 200 feet, Drilling, would require, approximately/200 gallons-of water; for each hole, and it! would be supplied by: trucks, the water-would be acquired. from "the nearest, available" source.

The following two examples illustrate the differences between the mandatory and optional styles.

Example 1: Commas in a Series

Writers following the optional style usually omit the comma before the *and* joining the last two items in a series:

> The report analyzed the possible market in 1985-1986, the projected labor costs and the supply of raw materials.
>
> We requested employment figures for 1981, 1982 and 1983.

Writers following the mandatory style always include the comma before the *and* joining the last two items in a series.

In most sentences the comma before the *and* adds nothing to the sentence, but on occasion, omitting the comma can create ambiguity:

> The maintenance people replaced the rocker arm bracket, hinge pin and wheel assembly.

Are the hinge pin and the wheel part of the same assembly? A comma after *pin* clearly signals that the three items are separate:

> The maintenance people replaced the rocker arm bracket, hinge pin, and wheel assembly.

Because clarity and precision are so important in technical and scientific writing, we recommend that writers adopt the mandatory style and retain the comma before the *and* in a series.

Example 2: Commas in Dates

Punctuating the day, month, and year is usually simple and straightforward. A comma separates the day and the year:

> June 15, 1985

What happens, however, when the date appears in the middle of a sentence?

> We moved on June 15, 1985 into our new office.

Writers following the optional style would not insert a comma after *1985*. Writers following the mandatory style would add a comma after *1985*:

> We moved on June 15, 1985, into our new office.

Both versions are "correct." Both are acceptable because the presence or absence of the comma after *1985* does not affect clarity.

PUNCTUATION

Which style should you follow? The mandatory style allows for more precision and can help you avoid ambiguity, so we recommend it for technical, scientific, and legal documents. We also recommend it for any formal or critical documents. In letters and memos to familiar readers, however, you might wish to follow the optional, less formal style.

Mandatory / Optional Punctuation

The following rules are those that allow for optional punctuation. Writers following the optional style might omit punctuation in the cases cited below except when clarity would suffer. If omitting punctuation would make the writing less clear, more ambiguous, or at all confusing, then use the appropriate mark of punctuation.

—Use an apostrophe and an *s* to form the singular possessive of words of more than one syllable ending in *s*:

Harris's report
Davis's plan

OPTIONAL STYLE: These possessive forms would be *Harris'* and *Davis'*. Note that the apostrophe is not optional.

—Use a comma before a coordinate conjunction to separate two independent thoughts, even when they are quite short and simple:

The plan was finished, and the budget calculated.

OPTIONAL STYLE: You would omit the comma after *finished*.

—Use a comma before the *and* that joins the last two items in a series:

We ordered two water pumps, a fan, and three replacement belts.

OPTIONAL STYLE: You would omit the comma after *fan*.

—Use a comma to enclose parenthetical expressions, even very short ones such as *thus*:

He was, thus, surprised by the answer.

OPTIONAL STYLE: You would omit both commas around *thus*.

—Use a comma after introductory phrases and clauses, even when they are very short and simple:

After we wrote the report, we submitted it.

OPTIONAL STYLE: You would omit the comma after *report*.

—Use a comma after the year in a date and the state in an address when either appears in the middle of a sentence:

The time sheets for July 6, 1984, show no overtime.

His speech in Joplin, Missouri, was most forgettable.

OPTIONAL STYLE: You would omit the commas after *1984* and *Missouri*.

—Use hyphens to form compound words that modify other words:

We will install a high-tension line above the Bradley overpass.

OPTIONAL STYLE: You would omit the hyphen between *high* and *tension*.

—Use periods after all abbreviations, including the names of government agencies, colleges and universities, and private organizations:

I.B.M.
S.U.N.Y. (State University of New York)
N.A.A.C.P.

OPTIONAL STYLE: You would omit the periods in these and most other abbreviations.

—Use a semicolon along with a coordinate conjunction to join independent clauses or complete thoughts that already contain a comma, even if they are clear without the semicolon:

Although new rain gauges helped us monitor total precipitation, we could not have anticipated the heavy spring runoff; and the resulting floods caused considerable damage to the watershed.

OPTIONAL STYLE: You would use a comma after *runoff*, not a semicolon.

For further information on commas, periods, semicolons, colons, dashes, and other marks of punctuation, see the alphabetical entries elsewhere in this *Guide*.

QUESTION MARKS

1. Use question marks to indicate direct questions:

Will analysis modeling be required?

NOTE 1: Use a separate question mark for each in a series of incomplete or elliptical questions:

When will the preliminary targeting studies be due? In 30 days? 60? 90?

NOTE 2: Use question marks at the end of statements written as declarations but intended as questions:

These are the final figures?

Regulating the supply pressure was the only realistic solution?

NOTE 3: Do *not* use question marks for indirect questions or for statements written as questions but intended as courteous requests:

During our preliminary design studies, we ask <u>whether the weapon parameters would be an integrated part of the overall system requirements</u>. (*The underlined clause is an indirect question. Its phrasing is not identical to a direct question.*)

Will you please forward five copies of DD Form 1425.

See PERIODS.

2. Use question marks to indicate questions within a sentence:

Your project managers have the authority, don't they, to reallocate resources based on changing needs?

NOTE 1: When the question follows an introductory statement, capitalize the first word of the question and use a question mark:

The remaining question is, Can shortfalls in determining performance results be quantified with the Phase 2 synthesis procedure?

Today's manufacturer asks, How can Q&A expenses be distributed across project and functional lines?

NOTE 2: If the question precedes a concluding sentence remark, the question mark goes after the question, and the sentence remark begins with a lowercase word:

Which mission would prove most productive? was the remaining question.

3. Use question marks within parentheses to indicate doubt:

CAD/CAM was first investigated at MIT in 1964(?).

Question Marks with other Punctuation

4. Place question marks inside quotation marks, parentheses, or brackets only when they are part of the quoted or parenthetical material:

During the preproposal conference, an Allied representative asked, "Does the Statement of Work represent a minimal subset of requirements?"

What did the contracting officer mean when she said, "The Statement of Work <u>is</u> the minimal subset of requirements"?

When did she ask, "Who is the project manager?"

The engineering program includes analysis, design, testing (fabrication?), and integration of the unit into the helicopter.

Doesn't the Program Plan call for Task F completion no later than 75 days ARO (August 17)?

See PARENTHESES, BRACKETS, and QUOTATION MARKS.

QUOTATION MARKS

1. Use quotation marks to enclose direct quotations:

The RFP says, "All pages in the proposal must be numbered."

Direct quotations include the actual words and phrases from a document or from a person speaking.

Indirect quotations do not take quotation marks:

The RFP says that all pages in the proposal must be numbered.

Indirect quotations do not give every word and phrase in the direct quotation. Often an indirect quotation is only a paraphrase:

The RFP says that we should number all pages in the proposal.

See QUOTATIONS.

NOTE 1: Long quotations have two equally acceptable conventions: (1) They may be enclosed by quotation marks, or (2) they may be indented from both the left and right margins (in which case they do not require quotation marks). If quotation marks are used and the quotation extends for more than one paragraph, quotation marks should appear at the beginning and ending of the entire quotation and at the beginning of each new paragraph within the quotation.

NOTE 2: Single quotation marks are used to indicate a quotation within a quotation:

According to the NASA report, "National Aerodynamics argued for a 'differential scale of evolution' "

2. Use quotation marks to indicate the title of an article, section, volume, and other parts of a longer document:

The contracting officer must approve all items specified under "Special Equipment" in the cost proposal.

This volume must include a completed and signed Standard Form 33, "Solicitation, Offer, and Award."

See TITLES.

3. Use quotation marks to indicate that a word is used in a special or abnormal sense:

The 1978 study suggested that NASA's definition of "suitability" contradicts the goals of the program.

Only in English do "thin chance" and "fat chance" mean the same.

NOTE: Italics (or underlining) can replace quotation marks when you want to refer to a word as a word:

In the Smythington contract, *boundaries* refers only to those property lines surveyed after July 1982.

See ITALICS and UNDERLINING.

Quotation Marks with Other Punctuation

4. Always place periods and commas inside closing quotation marks:

We have completed the section entitled "Representations, Certifications, and Acknowledgements."

The logarithmic decrease of the differential threshold is sometimes mistakenly called "Fechner's law."

NOTE: This relatively recent convention seems illogical, but it is now standard. The convention developed because printers wanted to put the smaller periods and commas inside of closing quotation marks for a cleaner appearance on the page.

5. Always place semicolons and colons outside closing quotation marks:

The corporation's experience belongs under "Related Experience"; the project manager's experience belongs under "Resumes of Key Personnel."

Include the following under "Manhours and Materials":

a. Work statement tasks
b. Estimated completion schedule
c. Materials/equipment required

6. Place dashes and question marks inside of quotation marks if they are part of the quotation; otherwise, place them outside quotation marks:

The section entitled "Personnel Qualifications"—the only part of the proposal where we can address the team's APL experience—is restricted to five pages.

He said, "Can we improve the unload utilities without losing the language interface?"

See QUESTION MARKS and SPACING.

QUOTATIONS

Quotations are an effective tool when you have to refer persuasively to data and conclusions from another document.

When you quote from another document, be careful not to misrepresent the content and the intent of the original passage. The following rules are suggestions. You must judge for yourself if you are accurately representing the original document.

1. Quote only the key or relevant passages:

> Your letter made an excellent case for the "procedural lapse" that caused the double billing to your account.
>
> In the analysis of the data, Jameson (1983) argued for three "equally persuasive hypotheses."

NOTE 1: Sometimes only a two- or three-word phrase is sufficient to capture the flavor of the original document. Rarely should you need to quote whole sentences or paragraphs from the original document.

Long quotes distract the reader and often signal that the writers have not done the work necessary to boil the quoted document down to its essentials.

NOTE 2: As illustrated in the two examples above, all quoted words, phrases, and sentences should normally be enclosed by quotation marks. (See QUOTATION MARKS.)

NOTE 3: In cases where words are omitted from the middle of quoted material, an ellipsis signals that material has been omitted:

> Your letter made an excellent case for the "procedural lapse . . . and the sloppy record keeping" that caused the double billing of your account. (*The original read:* ". . .the procedural lapse when my account was opened and the sloppy record keeping ever since.")

See ELLIPSES.

2. Cite the sources for any quoted material—from words and phrases to whole sentences and paragraphs. (See CITATIONS.)

Inexperienced writers sometimes are careless in their citations.

No material from another source, especially copyrighted material, should ever appear in another document without full and adequate credit being given to the original author(s).

No reader should ever have to guess what is original and what the writer is borrowing from someone else. Both accurate citations and accurate use of quotation marks and ellipses can remove such uncertainties.

3. Distinguish carefully between direct and indirect quotations.

Direct quotations contain only the original words and phrases of the document being quoted.

Indirect quotations are a writer's summary of someone else's words. Some minor words may come from the original, but the writer has made significant changes. Even then, key words should appear within quotation marks.

Here is an original document. Following it are examples of a direct quotation and an indirect quotation:

Original Passage

> Hank Stevens was over 30 minutes late three times during the week of November 3. He called in on one of these mornings, but we received no calls on the other two mornings. Although it is now November 15, he has given no satisfactory excuse or explanation of his lateness.

Direct Quotation

> According to Hank Stevens' supervisor, Hank has failed to give a "satisfactory excuse or explanation of his lateness" on three mornings during the week of November 3. Hank did call in one of the mornings, but he did not call in the other two mornings.

Indirect Quotation

> According to Hank Stevens' supervisor, Hank has not explained adequately his three instances of lateness during the week of November 3. Hank was over 30 minutes late in each case, and he only called in one time.

NOTE 1: As this example illustrates, the direct quotation is often embedded in a passage that contains some indirect quotations. As rule 1 above indicates, you should quote only the pertinent words and phrases.

QUOTATIONS

NOTE 2: Often the change from direct to indirect quotations involves a change in syntax and wording (especially in the pronouns):

Direct Quotations

> "We need to receive your response by no later than January 15, 1985."

> "I have been unable to locate the original data collected in 1976. The files seem not to have been moved when we moved in 1980 from the old building to our present building."

Indirect Quotations

> MOGO Oil said that they must have our written response by January 15, 1985.

> Gene Sayers has been unable to find the original 1976 data. He suspects that the files were not moved in 1980 when we moved into our present building.

REDUNDANT WORDS

A redundant word or phrase unnecessarily qualifies another word.

For instance, in the expression *basic fundamentals*, the word *basic* is unnecessary because, by definition, all fundamentals are basic.

Past experience is redundant because for experience to be experience, it must have been acquired in the past. *Past history* is redundant for the same reason. You can't have present or future history. All history is past. (However, *ancient history* and *recent history* are acceptable because they both refer to specific parts of the past.)

---

1. Eliminate redundant words.

A fundamental of good writing style is to eliminate unnecessary words. Do not say *basic fundamentals*. Just say *fundamentals*. Do not speak of your *past experience*. It is simply your *experience*. Documents are not *attached together*. They are simply *attached*.

See WORDY PHRASES.

Redundancies and Emphasis

---

2. Use redundant words to emphasize or dramatize a situation or condition:

> Mailing the ramjet study by Friday is absolutely essential.

The only proper use of redundancies in writing is to emphasize. The expression *absolutely essential* is redundant because nothing can be more essential than *essential*. If you need to heighten the sense of urgency in a situation by exaggerating, then redundant words are acceptable. However, do NOT overuse redundancies for emphasis. The effect diminishes quickly. You can become so exaggerated in your style that readers pay less attention to your ideas. Too much emphasis becomes no emphasis.

See EMPHASIS and GOBBLEDYGOOK.

A List of Redundancies

The following list of redundancies will help you identify those you habitually use. The redundant expression appears in the left column; in the right column are possible substitutes:

absolutely complete	complete
absolutely essential	essential
absolutely nothing	nothing
accidentally stumbled	stumbled
a.c. current	a.c./ alternating current
actual experience	experience
adequate enough	adequate/ enough
advance forward	advance

REDUNDANT WORDS

advance planning	planning
aluminum metal	aluminum
and etc.	etc.
any and all	any/all
arrive on the scene	arrive
ask the question	ask
assembled together	assembled
attached hereto	attached
attach together	attach
basic fundamentals	fundamentals
before in the past	before/in the past
betwixt and between	between
brief in duration	brief/quick/fast
check up on	check
circle around	circle
close proximity	proximity
collect together	collect
combine together	combine
completely destroyed	destroyed
completely opposite	opposite
connect together	connect
consensus of opinion	consensus
consequent results	results
consolidate together	consolidate
continue on	continue
continue to remain	remain
contributing factor	factor
cooperate together	cooperate
couple together	couple
desirable benefits	benefits
diametrically opposite	opposite
disappear from sight	disappear
disregard altogether	disregard
each and every	each/every
early beginnings	beginnings
empty cavity	cavity
enclosed herewith	enclosed
endorse on the back	endorse
end product	product
end result	result
entirely destroyed	destroyed
equally as good	as good/equally good
exactly identical	identical
expired and terminated	expired/terminated
extremely immoderate	immoderate
fast in action	fast
few in number	few
filled to capacity	filled
final completion	completion
final conclusion	conclusion
finally ended	ended
first beginnings	beginnings
following after	following/after
funeral obsequies	obsequies
fused together	fused
heat up	heat
hidden pitfall	pitfall
hopeful optimism	hope/optimism
important essentials	essentials
joint cooperation	cooperation
join together	join
joint partnership	partnership
just exactly	just/exactly
large in size	large
large-sized	large
lift up	lift
living incarnation	incarnation
main essentials	essentials
melt down	melt
mingle together	mingle
mix together	mix
more preferable	preferable
mutual cooperation	cooperation
necessary requisite	requisite
new innovation	innovation
one and the same	the same
one definite reason	one reason
one particular example	one example
one specific case	one case
part and parcel	part
past experience	experience
period of time	period
personal friend	friend
personal opinion	opinion
pervade the whole	pervade
plan ahead	plan
plan for the future	plan
plan in advance	plan
postponed until later	postponed
presently planned	planned
prolong the duration	prolong
qualified expert	expert
really and truly	really
reason is because	reason is that/because
recur again	recur
red in color	red
reduce down	reduce
regress back	regress
remand back	remand
repeat again	repeat
resultant effect	effect
same identical	same
seems apparent	seems/is apparent
separate and distinct	separate/distinct
shuttle back and forth	shuttle
single unit	unit
skirt around	skirt
small in size	small
small-sized	small
specific example	example
still continue	continue
still remains	remains
suddenly collapsed	collapsed
summer months	summer
surprising upset	upset
surrounding circumstances	circumstances
surround on all sides	surround
ten miles distant from	ten miles from
three hours of time	three hours
throughout the entire	throughout
throughout the whole	throughout
total of ten	ten
to the northward	north/northward
traverse across	traverse
true fact	fact
ultimate end	end
universal the world over	universal
unsolved problem	problem
visit with	visit
ways and means	ways/means

REFERENCES

The *Business Writer's Quick Reference Guide* will answer most of your stylistic questions and will help you to become a more effective writer. However, for further information, including far more exhaustive treatments of some stylistic issues, refer to the books listed below. These references are among the finest available in their special areas.

Dictionaries

An up-to-date dictionary is an essential writer's tool. Besides providing correct spellings, a dictionary gives definitions, pronunciations, word origins, synonyms, and guidance on word usage.

Most of the dictionaries listed below are abridged dictionaries written for general readers as well as readers in the colleges and universities. The exception is *Webster's Third New International Dictionary of the English Language,* unabridged; this dictionary was completed in 1961 and has been reprinted many times since. Although it is over 20 years old, this dictionary is still the best single source of information about American English.

The American Heritage Dictionary. 1982. 2nd college ed. New York: Houghton Mifflin & Company.

Funk & Wagnalls Standard College Dictionary. 1977. New York: T.Y. Crowell & Company.

Random House College Dictionary. 1984. New York: Random House. Inc.

Webster's New Collegiate Dictionary. 1984. 9th ed. Springfield, Massachusetts: G. & C. Merriam.

Webster's New-World Dictionary. 1983. 2nd college ed. New York: World Publishing Company.

Webster's Third New International Dictionary of the English Language, Unabridged. 1961. Springfield, Mass: G. & C. Merriam.

Specialized Dictionaries

In addition to those dictionaries listed above, many professional groups or disciplines have specialized dictionaries that list the technical terms and jargon particular to their professional area. Below are just a few of these specialized dictionaries.

A Dictionary of Mining, Mineral, and Related Terms. 1968. Ed. by Paul W. Thrush and the Staff of the Bureau of Mines. Washington: U.S. Department of the Interior, Bureau of Mines.

Dictionary of Geological Terms. 1976. Garden City, New York: Anchor Press.

Glossary of Geology. 1980. 2nd ed. Ed. by Robert L. Bates and Julia A. Jackson. Falls Church, Virginia: American Geological Institute.

The Illustrated Petroleum Reference Dictionary. 1982. 2nd ed. Ed. by R.D. Langenkamp. Tulsa, Oklahoma: PennWell Publishing Co.

Langenkamp, R.D. 1981. *Handbook of Oil Industry Terms and Phrases.* 3rd ed. Tulsa, Oklahoma: PennWell Publishing Company.

Schwarz, Charles F., Edward C. Thor, and Gary H. Elsner. 1976. *Wildland Planning Glossary.* USDA Forest Serv. General Technical Report, PSW-13. Pacific Southwest Forest and Range Experimental Station. Berkeley, California.

Spelling Guides

Spelling guides are alphabetized lists of words meant primarily for persons who need to check spelling or word division.

10,000 Medical Words: Spelled and Divided for Quick Reference. 1972. New York: McGraw-Hill Book Company.

20,000 Words: Spelled and Divided for Quick Reference. 1977. 7th ed. New York: McGraw-Hill Book Company.

Webster's Instant Word Guide. 1980. Springfield, Mass: Merriam-Webster, Inc.

Thesauruses

A thesaurus provides synonyms and often antonyms. Use a thesaurus when you need to find optional ways of expressing an idea, when you can't think of a word but know the word exists, or when the only word you can think of is not exactly correct and you need to find a more precise

REFERENCES

alternative. Do not use a thesaurus to find bigger words than the ones you can think of. In other words, don't try to sound impressive by finding and using big words (see GOBBLEDYGOOK).

The New Roget's Thesaurus in Dictionary Form. 1978. Ed. by Norman Lewis. New York: G. P. Putnam's Sons.

Roget's II: The New Thesaurus. 1983. By the editors of the *American Heritage Dictionary.* New York: Houghton Mifflin and Company.

Webster's New Dictionary of Synonyms. 1984. 2nd ed. Springfield, Massachusetts: Merriam-Webster, Inc.

Webster's New World Thesaurus. 1971. Ed. by Charlton Laird. New York: Simon and Schuster.

General Style Guides

The *Business Writer's Quick Reference Guide* is a general style guide. However, it differs from those listed below in that it provides much more information about effective writing techniques and the writing process, and it provides models of effective letters, memos, and other documents.

General style guides are, next to an up-to-date dictionary, the best writer's resources. They usually cover punctuation, abbreviations, spelling problems, capitalization, and special signs and symbols. In addition, each of those listed below has special features. *The Chicago Manual of Style*, for instance, has a long discussion of printing and binding. The *United States Government Printing Office Style Manual* has special guidelines for Congressional publications and a fine chapter on word compounds.

The Chicago Manual of Style. 1982. 13th ed. Chicago: The University of Chicago Press.

The Gregg Reference Manual. 1984. 6th ed. Ed. by William A. Sabin. New York: Gregg Division, McGraw-Hill Book Co.

Hutchinson, Lois. 1977. *Standard Handbook for Secretaries.* New 8th ed. New York: McGraw-Hill Book Company.

The McGraw-Hill Style Manual: A Concise Guide for Writers and Editors. 1983. Ed. by Marie Longyear. New York: McGraw-Hill Book Company.

Pickens, Judy E. 1985. *The Copy to Press Handbook: Preparing Words and Art for Print.* New York: John Wiley & Sons, Inc.

United States Government Printing Office. 1984. *Style Manual.* Washington, D.C.: United States Government Printing Office.

Webster's Secretarial Handbook. 1983. 2nd ed. Springfield, Massachusetts: Merriam-Webster, Inc.

Words Into Type. 1974. 3rd ed. Englewood Cliffs, New Jersey: Prentice-Hall, Inc.

Specialized Style Guides

Specialized style guides are those intended for a particular professional group. Despite their special audience, they usually also cover basic punctuation and other items of general interest.

American Chemical Society. 1978. *Handbook for Authors of Papers in American Chemical Society Publications.* Washington, DC: American Chemical Society.

American Institute of Physics. 1978. *Style Manual for Guidance in the Preparation of Papers.* 3rd rev. ed. New York: American Institute of Physics.

American Medical Association, Scientific Publications Division. 1976. *Style Book and Editorial Manual.* 6th ed. Chicago: American Medical Association.

American Psychological Association. 1983. *Publication Manual.* 3rd ed. Washington, DC: American Psychological Association.

Council of Biology Editors. 1983. *CBE Style Manual.* 5th ed. Bethesda, Maryland: Council of Biology Editors, Inc.

Golen, Steven P., Glenn Pearce, and Ross Figgins. 1985. *Report Writing for Business and Industry.* New York: John Wiley & Sons, Inc.

Swanson, Ellen. 1982. *Mathematics into Type: Copyediting and Proofreading of Mathematics for Editorial Assistants and Authors.* Rev. ed. Providence, Rhode Island: American Mathematical Society.

U.S. Geological Survey. 1978. *Suggestions to Authors of the Reports of the United States Geological Survey.* 6th ed. Washington, DC: Government Printing Office.

REFERENCES

Zweifel, Frances W. 1961. *A Handbook of Biological Illustration*. Chicago: University of Chicago Press, Phoenix Books.

Grammar and Usage Handbooks

Grammar handbooks, especially those intended for college students, survey the principles of English grammar and provide many dos and don'ts for writers. The Harbrace and Random House grammars listed below are only two of the many such current grammar books available in book stores.

Usage handbooks provide rules for the proper use of words. Such rules quite often reflect the writers' prejudices, so keep that in mind as you use such handbooks.

Baker, Sheridan Baker. 1976. *The Complete Stylist and Handbook*. New York: Crowell.

Bernstein, Theodore M. 1977. *The Careful Writer: a Modern Guide to English Usage*. New York: Atheneum.

Bernstein, Theodore M. 1971. *Miss Thistlebottom's Hobgoblins: the Careful Writer's Guide to the Taboos, Bugbears, and Outmoded Rules of English Usage*. New York: Simon and Schuster, Inc.

Bernstein, Theodore M. 1958. *Watch Your Language: A Lively, Informal Guide to Better Writing*. Great Neck, New York: Channel Press.

Copperud, Roy H. 1980. *American Usage and Style: the Consensus*. New York: Van Nostrand Reinhold & Company.

Evans, Bergan, and Cornelia Evans. 1957. *A Dictionary of Contemporary American Usage*. New York: Random House.

Follett, Wilson. 1966. *Modern American Usage: A Guide*. Edited and completed by Jacques Barzun and others. New York: Hill and Wang.

Fowler, Henry Watson. 1965. *Dictionary of Modern English Usage*. 2nd rev. ed. by Sir Ernest Gowers. Oxford: Clarendon Press.

Harbrace College Handbook. 1984. 9th ed. Ed. by John C. Hodges and Mary E. Whitten. New York: Harbrace Jovanovich, Inc.

Markgraf, Carl. 1979. *Punctuation*. New York: John Wiley & Sons, Inc.

Nicholson, Margaret. 1957. *A Dictionary of American-English Usage, Based on Fowler's Modern English Usage*. New York: Oxford University Press.

Random House Handbook. 1983. 4th ed. Ed. by Frederick Crews. New York: Random House, Inc.

Romine, Jack. 1975. *Vocabulary for Adults*. New York: John Wiley & Sons, Inc.

Smith, Leila. 1985. *Basic English for Business and Technical Careers*. New York: John Wiley & Sons, Inc.

Books on Writing

The following books cover writing, both as a process and as a final product. Strunk and White's book is perhaps the most famous and the most readable book of those listed, but the others are also valuable.

Bell, P. 1985. *High-Tech Writing: How to Write for the Electronics Industry*. New York: John Wiley & Sons, Inc.

Ewing, David. 1979. *Writing for Results: In Business, Government, the Sciences and the Professions*. 2nd ed. New York: John Wiley & Sons, Inc.

Flesch, Rudolf. 1972. *Say What You Mean*. New York: Harper & Row Publishers.

Lanham, Richard A. 1981. *Revising Business Prose*. New York: Charles Scribner's Sons.

Strunk, W., Jr., and E. B. White. 1979. *The Elements of Style*. 3d ed. New York: Macmillan.

Tichy, H. J. 1966. *Effective Writing for Engineers, Managers, Scientists*. New York: John Wiley & Sons.

Williams, Joseph M. 1981. *Style: Ten Lessons in Clarity and Grace*. Glenview, Illinois: Scott, Foresman and Company.

Zinsser, William Knowlton. 1985. *On Writing Well: An Informal Guide to Writing Nonfiction*. 3d ed. New York: Harper and Row.

REPETITION

Repetition is sometimes considered a trait of poor writing. In fact, repetition can be a valuable emphasis technique. It can enhance the impact and readability of a document. But you must be careful not to repeat an idea too soon after stating it the first time, and when you repeat ideas, you should state them in a different way than you did originally:

> *not this*
>
> In my opinion, the Rothskeller algorithm will not alleviate error detection inaccuracies. The mathematical model proposed does not account for errors introduced by migrant electrons. Therefore, I do not think that the Rothskeller algorithm will alleviate error detection inaccuracies.

This is poor repetition because the repeated idea (first and last sentences) is repeated too quickly, and the author uses virtually the same combination of words to state the idea both times. Here is an example of good repetition:

> *this*
>
> In my opinion, the Rothskeller algorithm will not alleviate error detection inaccuracies. The mathematical model proposed does not adequately account for errors introduced by migrant electrons. These electrons interact randomly with bits and can change bit patterns in fundamental and disastrous ways. Superior error detection techniques not only must minimize the mushrooming effect of errors compounded by repetition of a bit sequence (which Rothskeller does very effectively) but also must account for random errors that may inadvertently be replicated hundreds of times.
>
> The Rothskeller algorithm for random error detection identifies logical bit pattern inconsistencies and applies logic correction. However, its identification strategies assume that errors will be of predictable types. Non-predictable logic errors, caused by random migrant electrons, can go undetected. Therefore, I do not believe that the Rothskeller algorithm will effectively alleviate error detection inaccuracies.

The repeated idea (first and last sentences) is separated by lengthy discussion. Furthermore, the last sentence states the idea somewhat differently than it was stated originally. Someone reading the revised passage will be less likely to find the repetition obtrusive.

Effective repetition of a fact or idea reinforces it in the reader's mind. The reinforcement emphasizes the idea and helps readers remember it.

See ORGANIZATION, EMPHASIS, and KEY WORDS.

1. Design documents with clean and deliberate repetition in mind.

As you design a document and organize your ideas, build in effective repetition. Consider repeating (1) key requests or recommendations, (2) major conclusions, and (3) most important or most convincing facts.

Most professionals in the world of work have more documents crossing their desks than they have time to read. Research shows that the majority of intended readers skim through the documents they receive. Rarely do they read the documents in depth, and almost never will they read something more than once.

Consequently, you must convey your important ideas quickly and emphatically. Repeating key requests, recommendations, conclusions, and facts helps to ensure that skimming readers will not miss the most important ideas in your documents.

2. Use the inherent repetition in formal report structure to reinforce major ideas and to strengthen logic and impact.

Technical and other formal reports are structured for deliberate repetition. The writer's conclusions, for instance, typically appear in these sections of the report: abstract, executive summary, conclusions, and discussion. The recommendations will appear one way or the other in the abstract, executive summary, and perhaps the discussion. They will also appear in their own section.

Repetition in reports is deliberate. It allows readers to read selectively. Some readers will read the abstract and will not need to read further. Others will read the executive summary and perhaps the conclusions and recommendations sections. Still others will glance at the results section and then read through the discussion.

Because reports are deliberately repetitive, all of these readers will have encountered the major conclusions and recommendations. Those who read the major conclusions and recommendations more than

REPETITION

once will have had those ideas reinforced. (See REPORTS.)

3. Use repetition to emphasize the logic behind a discussion, especially when details are parallel in their intent.

Repetition of a key word or phrase indicates that a sequence of ideas is parallel:

> Skillful writers have a clear sense of purpose. Skillful writers are aware of their readers' wants, needs, and concerns. Skillful writers can choose between many stylistic options. In short, skillful writers focus their task and use the right techniques to create the right effect.

This kind of repetition is effective in emphasizing key points and limiting sentence length. (Imagine how difficult this passage would be to read if it were a single sentence.) Beware of using this kind of repetition too often, however. With much repetition, the effect diminishes and the writing begins to sound unnatural.

See PARALLELISM, KEY WORDS, and EMPHASIS.

REPORTS

Reports cover a broad range of business and technical documents, including formal reports, scientific reports (often published), corporate technical reports (following internal guidelines), progress reports, trip reports, laboratory and research reports, accident reports, memorandum reports, and financial reports.

Many of these reports are periodic documents that convey the status of a program, project, task, study, or other organizational effort. Other reports are written in response to specific needs and situations.

Some reports are more informal than others, but readers of reports generally expect to find certain information (summaries, conclusions, recommendations, analyses, supporting facts, etc.), and they expect the report's tone to be businesslike—not officious or bureaucratic, but objective, factual, and honest (see TONE).

Formal reports have traditional components, which are discussed below. For further information relevant to reports, see MEMOS, ORGANIZATION, and VISUAL AIDS. See also the models located in the second part of this *Guide*.

Scientific Reports

Scientific reports are tightly controlled by scientific convention and tradition. Many scientific reports are published by professional groups and conferences, and many appear in technical or scientific journals. These professional organizations usually have explicit editorial guidelines that authors must follow.

The tradition of the scientific method dictates the format of most scientific reports:

Abstract

Introduction

Materials and Methods

Results and Discussion

Fact 1
Fact 2
Fact 3

(therefore)

Conclusions

Recommendations (if any)

Summary (optional)

This format is roughly chronological—moving from the problem to be solved, to the test design (including materials as well as methods), to the test results, to an analysis of the results, and finally to the conclusions.

This logical pattern roughly duplicates the process the scientist used while conducting the study. Ideally, readers should be able to duplicate the process themselves and reach the same conclusions.

In scientific reports, the process of arriving at the conclusions is generally as important as the conclusions themselves. And the pattern of the scientific

REPORTS

report tends to reinforce the equal importance of process and conclusions. In doing so, however, it delays the conclusions, which many readers would consider the most important ideas in the report.

The logic behind scientific reports is inductive: *fact, fact, fact, fact (therefore) conclusion*. This pattern is effective, but it is also suspenseful—and that's the major drawback to the scientific format.

Most business and technical readers are too busy to be held in suspense. They want to know what's important right away. If they want to read a good mystery, they'll read Agatha Christie.

So avoid the scientific format unless you are a scientist writing for other scientists.

In fairness, we should note that reports of all kinds are typically divided into clearly marked sections, and most readers never read the entire report anyway. They read selected sections, depending upon their needs. Scientific reports do allow for conclusions to appear early: in the abstract, for instance, and often in the introduction. So if readers of scientific reports want to know the conclusions first, they go to the section entitled "Conclusions."

The inductive mode of thought behind the scientific format is contrary to the way most readers want to encounter information. For most readers you should use the far more common format found in standard technical reports.

Technical Reports

Technical and scientific reports often share many features, but technical reports usually differ from scientific reports in several crucial respects:

—They are distributed within an organization and are generally not formally published.

—They are intended for internal use only (many are even proprietary) or have a very limited distribution outside of the parent organization.

—Their readers are decisionmakers and others who need to have all of the important information (key findings, conclusions, and recommendations) right away and who may not be at all interested in how the writer arrived at that important information.

Technical report formats usually follow a managerial format, which emphasizes the conclusions and recommendations by placing them at the beginning of the report and subordinates the results and discussion:

<u>Executive Summary</u>

<u>Introduction</u>

<u>Conclusions and Recommendations</u>

(because of)

<u>Results and Discussion</u>

Fact 1
Fact 2
Fact 3

<u>Summary (optional)</u>

The managerial format opens with a summary (often called an executive summary) that presents a distillation of the report's most important ideas. Following the summary, writers often provide a list of conclusions and recommendations so that busy managers and supervisors have to read no further to discover the essence of the report.

The managerial format follows an inverted logic: *conclusion (based on) fact, fact, fact, fact*. If readers wish, they can read the facts to determine how the writer arrived at the conclusion. But they don't have to. They can read the conclusion alone and then go on to something else.

Decisionmakers are almost always part of the audience for technical reports, and they are usually the primary readers. So most technical reports should follow the managerial format.

Memorandum Reports

Memorandum reports are less formal and usually shorter versions of technical reports, although both types of reports may be very similar in content. Memorandum reports almost always follow the managerial format, but they usually do not include all of the components of a standard technical report. They do not, for instance, have

REPORTS

covers, title pages, tables of contents, lists of figures, abstracts, and other formal sections required for either scientific or technical reports. These formal sections are necessary only when reports are widely circulated or published.

See MEMOS.

Parts of Reports

Scientific, technical, and some memorandum reports might include the following:

Letter of Transmittal
Cover
Abstract
Title Page
Preface or Foreword
Table of Contents
List of Figures
Body
Bibliography or List of
  References
Appendix

Letter of Transmittal

A letter of transmittal accompanies and introduces a report. It might explain what the report is about, why it was written, how it relates to previous reports or projects, what problems the writer encountered, why the report includes or excludes particular data, and what certain readers may find of interest.

A letter of transmittal can provide information that would not be appropriate in the report itself, especially sensitive or confidential information. Hence, different letters of transmittal may be written for different readers of the same report.

Letters of transmittal are usually brief. They tend to be less formal than the reports they transmit. However, the more formal the report, the more formal the letter of transmittal is likely to be.

Cover

Covers are appropriate on formal reports and on those intended for widespread or public distribution. The information on the cover is usually similar to the information found on the title page (see below). However, covers are usually well designed and often include artwork.

Abstract

An abstract is a very brief distillation of a report's content. It is intended to describe the report's content and sometimes to provide information about key findings, conclusions, and recommendations.

Some abstracts, especially those for published scientific reports, are primarily useful in data banks and library catalogues. Such abstracts will be printed in catalogues or bibliographies. Prospective readers should be able to determine from reading the abstract whether they would profit from reading the entire report.

Less formal abstracts often function as one-page summaries of corporate technical reports. Actual summaries may be longer and include more information. In addition, summaries are part of the report and should not be separated from it. Abstracts, on the other hand, are not considered part of the report and should always be understandable in and of themselves.

Abstracts are usually either **descriptive** or **informative**. In either case, they present the key information in a brief paragraph or two (usually no more than about 250 words).

Descriptive Abstracts

Descriptive abstracts describe the content of the report but do not include interpretive statements, conclusions, or recommendations:

> The report analyzes the effects of caffeine on three groups of heart patients: (1) those with diagnosed hypertension and initial signs of heart trouble, (2) those using blood pressure medication but who have not had surgery, and (3) those having had heart surgery. The report discusses the correlation between caffeine and variations in blood pressure for these three groups.

This abstract describes the general scope of the research, but does not provide results or conclusions.

Informative Abstracts

Informative abstracts are generally longer and more comprehensive than descriptive abstracts. Typically, they describe the research or project and summarize key results and conclusions:

> Caffeine, in moderate amounts (no more than two cups of coffee per

REPORTS

> day), has no significant impact on patients with heart problems (ranging from those with diagnosed hypertension to those having had actual heart surgery). Beyond two cups, however, the impacts become increasingly severe. Patients with recent heart surgery showed the most effects, including very high blood pressure and chest pains. Patients on blood pressure medication could cancel the effects of the medication by drinking more than two cups of coffee. Patients with diagnosed hypertension showed elevated blood pressures for up to 3 hours after drinking over two cups of coffee. In conclusion, the effects of caffeine increased substantially with every cup of coffee beyond the two-cup threshold.

In this (fictitious) informative abstract, readers interested in the subject can determine if they would want to read the full report. Informative abstracts provide key results and conclusions. Consequently, if the research techniques are obvious, knowledgeable readers may need no more than the abstract.

If you have a choice, always write an informative abstract.

Title Page

The title page can contain the following information:

- The title of the report
- The name of the person(s) writing the report
- The name of the person(s) for whom the report is prepared
- The date of submission
- The name of the division, group, or department, as well as the name of the organization
- A research number or other documentation aid
- A copyright notice and other special notations (such as *SECRET* or *PROPRIETARY INFORMATION*)

Preface or Foreword

Prefaces or forewords (they are the same) generally appear only in formal or published reports—and often not even there. Informal and memorandum reports rarely include a preface or foreword.

If used, a preface or foreword can include the following:

- References to other researchers or reports to which the author is indebted
- Background information regarding the origin of the report—such as who requested it, who funded it, what the goals were, and so on
- Acknowledgment of contributors, including other researchers, managers, technicians, reviewers, editors, proofreaders, and so on
- Financial implications
- Observations regarding unusual conclusions or recommendations
- Miscellaneous personal comments about the contents, including areas for future study

Table of Contents

The table of contents is an outline of the report.

It helps readers understand the structure of the report and locate particular sections. It helps writers organize their thoughts (or check their organization).

A table of contents should contain enough second- and third-level headings to capture the actual content and approach of the chapters. Chapter headings by themselves are often too cryptic:

not this

I. Introduction
II. Preliminary Conditions
III. Governmental Controls

this

I. Introduction
  A. Corporate policy on experiments with animals
  B. Precedents for this research
  C. Guidelines and goals of this research
II. Preliminary Conditions
  A. Physiological profiles of the test animals
  B. Structure of control and experimental groups
  C. Checks and balances in the research procedures
III. Governmental Controls
  A. Documentation needed for report to the FDA
  B. External verification of results
  C. Legal penalties for failure to report

See TABLES OF CONTENTS.

List of Figures (or Tables or Maps)

A list of figures (or tables or maps) is necessary only if the report is extensive and contains many of these or other visual aids. (See VISUAL AIDS.)

If used, the list of figures appears following the table of contents and on a separate page. (NOTE: The table of contents should include the list of figures and its page number. See TABLES OF CONTENTS.)

Tables and figures are usually listed separately, so if you list tables, do so in a List of Tables.

Figures include charts, graphs, maps, photographs, and diagrams. If you have a large number of any particular type of figure, you can list them as separate types of visuals:

List of Maps
List of Charts
List of Photographs

Number tables and figures (and other specific types of visuals) separately, and number them sequentially as they appear in the report. (See CAPTIONS.)

NOTE: If your report has large separate sections, you can number visuals sequentially within each section (see CAPTIONS).

Body

The body of a report can follow either the scientific format or the managerial format. (See the opening section of this discussion of reports and also see ORGANIZATION.)

<u>Scientific Organization</u>

Abstract
Introduction
Materials and Methods
Results and Discussion
Conclusions
Recommendations (if any)
Summary (optional)

<u>Managerial Organization</u>

Executive Summary
Introduction
Conclusions
Recommendations
Materials and Methods (optional)
Results and Discussion
Summary (optional)

See SUMMARIES.

<u>Introduction</u>

The introduction sets the stage. It normally includes the historical background of the report (and the project or program being reported) and establishes the scope of the report. The introduction may also define special terms and discuss the report's relation to other reports or research efforts.

Introductions also discuss the content and organization of the report. In other words, the introduction tells readers what the report contains and where to find it. In essence, the introduction is a roadmap.

If the report does not contain a preface, some of the items covered in the preface may also appear in the introduction:

—Person or group authorizing the research

—Contributors, especially other researchers

—Financial implications

—Noteworthy points about the conclusions and recommendations

—Other special items of interest

One major difference between an introduction and a summary is that the introduction does not contain the conclusions and recommendations. Another major difference is that a summary does not provide background information or lay out the structure of the report. (See INTRODUCTIONS.)

<u>Materials and Methods</u>

This section includes the materials and methods used during the experiment, study, or project. Limit this section to those materials and methods unfamiliar to knowledgeable readers. If the materials and methods are standard, you can mention them briefly in the introduction and then omit this section.

<u>Results and Discussion</u>

This section presents relevant data, discusses the meaning and significance of the data, makes inferences, and states the conclusions. If you have a lot of raw data to present, place it in an appendix and extract only the most important data to present in this section.

In some reports, especially formal scientific reports, the results are separate from the discussion.

REPORTS

Conclusions

This section brings together everything in the report and states your convictions. Every conclusion should grow out of information elsewhere in the report. Without such logical support, readers will justifiably feel that you have failed to accomplish the goals of the research.

Recommendations

Recommendations are suggestions for future actions—either managerial action or future research. Recommendations are almost always present in reports directed to corporate managers and supervisors.

In some cases, conclusions and recommendations are presented in a single section.

Summary and Executive Summary

Traditionally, the summary appeared at the end of the report. In the scientific format, the summary still appears at the end (if it appears at all). Its purpose at the end of a report is to "sum up" the major ideas presented, to remind readers what was important about what they read—the key findings, the conclusions, and any recommendations.

Summaries at the beginning of a report are becoming more common. They are highly desirable in reports directed to managers and supervisors.

Quite often, this opening summary is called an "executive summary." The title indicates clearly how this summary is meant to be read and who is meant to read it. It opens the body of a report written in the managerial format. If well done, the executive summary includes everything a busy manager or supervisor needs to know to make a decision. The detailed results and data often appear only as an appendix. Consequently, a good executive summary in effect makes the rest of the report superfluous—and it should. If you are writing well, readers should not have to read beyond your summaries unless they have a particular need for the detail that follows.

The information in a summary must be consistent with information appearing throughout the body of the report. Furthermore, you should have nothing in the summary that does not also appear elsewhere in the report.

Summaries are always part of the body of a report. Abstracts, on the other hand, should always be able to stand by themselves.

See SUMMARIES.

Bibliography or List of References

A bibliography or list of references is necessary only in more formal reports or in reports with a number of references.

If you have only two or three references, cover them fully in the text:

> As George Stevens established in his report entitled "The Life Cycle of the Toad" (*Animal Physiology,* X [March 1983] 234-237), toads have very low metabolic levels.

See CITATIONS and BIBLIOGRAPHIC FORM for full information on the use of parenthetical citations and for different ways to list bibliographic entries.

Appendix

The appendix is for information that is not properly part of the text or is too lengthy to be included in the text (voluminous data, computer programs, lengthy descriptions of methods, etc.).

If information in the appendix is of more than one kind, use two or more appendices, each identified by letter and title:

> Appendix A—Graphs
> Appendix B—Photographs
> Appendix C—Programs

Ensure that you always mention the appendices in the text. Where reference to an appendix would help readers, identify which appendix is appropriate and what the reader can expect to find there:

> Appendix E presents raw distillation data.
>
> For further information regarding these formulas, see appendix B1.
>
> The names and addresses of all of those who responded are listed in appendix H.

See APPENDICES/ ATTACHMENTS.

SCIENTIFIC/TECHNICAL STYLE

Scientists and technical specialists write and speak a different language. They do use many of the words and sentence patterns that non-technical writers use, but their language is sufficiently different to be called a scientific or technical style. (See STYLE.)

Most obviously, scientists and technical specialists use technical terms: *joule, volt, electron, ion, protozoa, electroencephalogram, uterine tube, ethyl ethers, hypochlorous acid*, etc. Such terms have specific meanings and are generally foreign to lay readers. Often, lay readers consider such technical terms to be jargon because they do not readily know and perhaps cannot easily understand them. (See JARGON.)

Less obviously, scientists and technical specialists sometimes use common words in uncommon ways. Often, such uses are more confusing to lay readers than technical words, which lay readers expect to be foreign to their experience. Here are two examples of uncommon uses of common words:

> A continuous function in an *n*-dimensional vector space is considered smooth when the function has certain well-defined features.

> We were unable to calculate the work because the granite outcrop would not budge, even after a standard charge exploded in a hole drilled into its base produced a hairline fracture evident throughout the circumference of the neck.

In the first example, mathematicians have borrowed the intuitive notion of smoothness to describe an abstract mathematical concept.

In the second example, physicists define *work* to be the effect when a force moves an object a certain distance. Without movement, work (according to the definition) has not taken place. Equally, *neck* describes that portion of an outcropping where the base of the outcropping most clearly resembles the border between the outcropping and the mass of rock from which the outcropping protrudes. The resemblance to a human neck is clear, but lay readers may not make the connection as readily as geologists.

Both technical terms and common words used in uncommon ways complicate scientific and technical writing. In both cases, lay readers may become lost, even when the writing is clear, concise, and logical. Therefore, if you are writing a scientific or technical document that is intended, at least in part, for lay readers, try to define technical terms (or don't use them), and avoid using common terms in uncommon ways.

If you are writing for scientific or technical readers, you must still try to make the writing as clear, concise, and logical as you can. Technical word usage may not hinder technical readers, but writing that is poorly organized, clumsily phrased, and inaccurately conceived will still be difficult if not impossible to read.

---

1. Make your use of technical and scientific terms appropriate for your readers.

Avoid unusual or overly technical terms if possible. If you cannot avoid them and your readers will not understand them, then define the terms. Do not use unfamiliar abbreviations unless your readers will understand them or unless you carefully explain the abbreviations first (see ABBREVIATIONS and ACRONYMS). Once you have established the terms and abbreviations, be consistent in your usage throughout the document.

Do not assume that your readers (all of your readers) know and understand your terms as you are using them. This is a trap.

First, rarely do all readers of a document have the same kind and level of technical background. You should always write for the lowest common denominator—that is, for the least technically sophisticated reader.

Second, even well-established technical terms, concepts, and abbreviations may be subject to dispute, so you may have to stipulate definitions so that your readers know exactly what you intended when you used a particular word.

One of the easiest ways to handle difficult or unfamiliar technical terms is to provide an informal definition when the terms first appear:

> A further health problem in many African countries results from filariasis, which means that the blood contains small thread-like worms.

or

SCIENTIFIC/TECHNICAL STYLE

The presence of small thread-like worms in the blood (called filariasis) is a further health problem in many African countries.

Third, scientists sometimes use different words to describe essentially the same thing, and this creates substantial confusion, even among scientific readers. Do not speak of *aspects* of a procedure in one paragraph, and then refer to the same concept as *elements* or *features* later.

2. Use concise, direct sentences.

Sentences are the building blocks of effective writing of any kind. However, they are especially important in writing that is inherently complicated and includes jargon. So strive to make your sentences clear, direct, and concise. (See SENTENCES.)

The principles of writing direct, concise sentences are summarized below. Elsewhere in this *Guide*, you will find each topic discussed in depth.

- **Choose active rather than passive sentences:**

this

We determined that the coefficient of friction varied most at extremely cold temperatures.

not this

It was determined that the coefficient of friction varied most at extremely cold temperatures.

See ACTIVE/PASSIVE.

- **Avoid wordiness:**

this

Our revised proposal presented further justification for development costs while offering to reduce our profit fee by 20 percent.

not this

Further justification for development costs, along with an offer for the 20 percent reduction of our profit fee, was presented in our revised proposal.

See WORDY PHRASES.

- **Use strong verbs:**

this

After lengthy study, we adjusted the effluent guidelines to accommodate projected economic hardships.

not this

After lengthy study, we made an adjustment of the effluent guidelines to effect greater accommodation with projected economic hardships.

See STRONG VERBS.

- **Avoid false subjects:**

this

Five-spot pattern recoveries are probably less efficient in sand trap reservoirs that have been depleted to within 15 percent of recoverable reserves.

not this

It is probable that five-spot pattern recoveries are less efficient in sand trap reservoirs that have been depleted to within 15 percent of recoverable reserves.

See FALSE SUBJECTS.

- **Use pronouns to make your writing more personal and direct:**

—Pronouns are appropriate when a scientific or technical writer is recommending something, drawing conclusions, or conveying deliberate decisions or choices:

The evidence suggests that this species is in fact indigenous to the Everglades. Therefore, I recommend broadening the scope of the USDA's habitat study before it writes the Los Puertos EIS.

The globule had a specific density of 6.78, more than twice what the OCS recovery team had estimated, so we concluded that zinc, not iron, was its major constituent.

Once we had analyzed the data from the pilot tests, we determined how to control temperature variations in succeeding tests.

—Some technical and scientific writing does not lend itself to first person pronouns. The historical description of a laboratory procedure, for instance, should not mention the person who performed the procedure unless the focus of the discussion is on that person and not the procedure. If the procedure itself is more important, do not use personal pronouns:

The ore sample is washed in a weak solution of hydrochloric acid. Next, the ore is cleansed thoroughly in a water bath and then dried in a heat chamber.

NOTE: If this passage appeared in a set of instructions, second person pronouns and imperative sentences would be appropriate:

Wash the ore sample in a weak solution (no more than 5 percent) of hydrochloric acid. Then you should clean the sample thoroughly in a water bath and dry it in a heat chamber.

SCIENTIFIC/TECHNICAL STYLE

or

Wash the ore sample in a weak solution (no more than 5 percent) of hydrochloric acid. Then clean the sample thoroughly in a water bath and dry it in a heat chamber.

The use of pronouns in technical writing has become more common in the last several decades, but don't overuse pronouns. Most scientific and technical writing is meant to convey information objectively, not to establish the writer's personality. If you are writing for the general public, however, use more pronouns and try to convey both clarity and warmth. For an example, read anything by Carl Sagan, Issac Asimov, or Arthur C. Clarke, three successful scientific writers whose readers are primarily non-technical. (See PRONOUNS.)

3. Design scientific and technical documents for visual impact and readability.

Visual appearance has become more and more important as a feature of good scientific and technical style. Well-conceived graphs, tables, and illustrations are essential if a document is to look fully professional.

In essence, the text supports the visuals, not the reverse. So time spent planning and designing visuals is critical for successful scientific and technical documents.

Below are some key principles of using visual aids. For a much more complete discussion, see VISUAL AIDS, CAPTIONS, CHARTS, GRAPHS, ILLUSTRATIONS, TABLES, MAPS, and PHOTOGRAPHS.

- Create your visuals before you write your text.
- Select visuals that are appropriate for your readers.
- Focus your visuals on key points and keep them simple and uncluttered.
- Introduce visuals before they appear in the text.
- Use clear, active captions on your visuals.
- Use emphatic devices (underlining, italics, boldface type, shading, etc.) to emphasize important ideas in your visuals as well as your text.

4. Above all, be as accurate as possible.

Good scientific and technical writing is precise and accurate—in its facts, its references, its procedures, its analyses.

Check every detail. If your document mentions that the site of the project is southwest of Dry Gulch, then the map should not show the site to be northwest of Dry Gulch. Similarly, if you mention production data in the text, the same figures (rounded to the same significant figures) should appear in accompanying tables. Every detail in the document should be as accurate as the writer can humanly guarantee.

A single inaccurate figure or fact can destroy the credibility of a document and its writer.

NOTE: Some writers defend weasel wording or hedging by stating that we can't know anything for sure; therefore, we can make no absolute statements. According to them, we can never state a fact because we can't be sure that anything is irrevocably true. The words typically used to hedge are: *generally, usually, likely, possibly, notion, surmise, speculation, conjecture, indicate, suggest, appear, seem, believe,* etc.

If necessary, use such words to indicate honest doubt or uncertainty, but do not overdo them. If you are reasonably certain of something, say so. Weasel wording weakens your document and your image.

not this

We believe that the evidence favors an interpretation that gas reserves in the Mellencamp zone are likely to be potentially significant.

this

Gas reserves in the Mellencamp zone are probably significant.

SEMICOLONS

1. Semicolons link complete thoughts that could otherwise stand alone as separate sentences:

> Western Aeronautics has completed more than 15 avionics contracts in the last 5 years; in 11 of those contracts we used CAD/CAM techniques to minimize development costs and improve both reliability and performance.

Typically, the complete thoughts linked by a semicolon are equal in structure and importance. Writers could separate the complete thoughts with a period and create two sentences; however, the semicolon shows a closer relationship between the thoughts than a period does. The semicolon says, "These thoughts are closely related."

NOTE 1: Semicolons used to link complete thoughts do not require conjunctions (transitional or connecting words like *however, consequently, furthermore,* and *thus*):

> Detailed trade studies helped us determine the importance of the various technologies in designing a supersonic cruise fighter; similar performance studies helped us evaluate the trade study findings in air-to-air operations in a simulated combat environment.

> High maneuverability was our primary design consideration; our secondary considerations included short take-off and large payloads.

NOTE 2: The following transitional words and phrases can be used with semicolons if the writer needs to indicate or clarify the relationship between the thoughts before and after the semicolon: *accordingly, consequently, for example, for instance, further, furthermore, however, indeed, moreover, nevertheless, nonetheless, on the contrary, on the other hand, therefore, thus*:

> We use aircraft geometry to generate survivability parameters such as radar signature; furthermore, we add performance, weapons, and avionics characteristics from the developed aircraft design to evaluate military effectiveness. (*Note the comma following* furthermore.)

> The SDC is a multi-terminal facility dedicated to a variety of projects; however, its size and flexibility, coupled with strict management controls, guard against fragmentation and crossover.

NOTE 3: Do NOT use the simpler conjunctions (*and, but, or, for, nor, so,* and *yet*) with semicolons; use these simple conjunctions and an accompanying comma when linking two complete thoughts.

However, if the complete thoughts are lengthy and already contain commas, then use a semicolon with these simple conjunctions to join the two complete thoughts:

> The committee began its scrutiny of the preliminary design study, which General Avionics submitted upon just two weeks' notice; and they found that the transformational analyses, although obviously hurried, were still superior to those of other designers who'd had much more time.

See COMMAS, CONJUNCTIONS, and TRANSITIONS.

2. Semicolons clarify items in series when one or more of the items has a comma:

> Our cost breakdown demonstrates our cost consciousness; our commitment to low overhead, especially through our direct contract-costing procedures; and our desire to minimize risk by using proven resources.

Without the semicolons to separate each major item in the series, readers may not understand how many items the series contains. The need for clarification varies from sentence to sentence. Sometimes it is not crucial; sometimes it is:

> Our avionics designs incorporate state-of-the-art subcomponents, including dual Intel 86000 microprocessors; Marquette frequency stabilizers, the most advanced anti-ECM devices currently available; and Barnett Industries' redesigned RFX regulators.

If the three items in this series had been separated with commas, the purpose of the phrase *the most advanced anti-ECM devices currently available* would not be clear. It would seem to be a fourth item in the series, although its true purpose is to describe the Marquette frequency stabilizers. The semicolons clarify its purpose in the sentence.

SENTENCES

Sentences are the building blocks of thought. Without sentences and the context in which they appear, communication would not be possible.

Sentences are fundamental to language, yet they are hard to define. *Webster's New Collegiate Dictionary* says that a sentence is a "grammatically self-contained speech unit." Many teachers would call it "a complete thought," and perhaps add that it usually contains, as a minimum, a subject and a verb. That definition would seem to rule out the following, all of which are sentences:

Yes.
No.
Maybe.
Now.
Stop.
Hello.
Me?
When?
Where?
Why?
How?
Ok.
Oh!

Most sentences are longer than these and do state both the subject and the verb:

The invoice was late.

Because the invoice was late, we could not include it in accounts payable for November.

When did the invoice arrive?

It arrived late.

The invoice, which was late, missed the deadline for November accounts payable.

The invoice was late, so we could not include it in November accounts payable.

Next time, send us the invoice promptly.

What happened with the late invoice was that we could not include it in November accounts payable.

In some cases, parts of the sentence are understood but not stated:

Too late. (The invoice was too late.)

Late again! (The invoice was late again!)

These examples reveal several things about sentences. First, they are self-contained, although they may rely heavily on something said earlier (or later) for readers to fully comprehend them. Second, they consist of a meaningful word or group of words. A word constituting a sentence does not have to be a particular kind of word, nor does it have to be meaningful in normal contexts. *Ok* and *Yes* are clearly self-contained expressions, but *String* could also be a sentence:

What did you use to secure the box?

String. (I used string. *The* I used *is understood.*)

Single verbs can also function as sentences:

What do you suggest I do during lunch?

Run. (I suggest that you run.)

So, must sentences express complete thoughts to be sentences? Yes, although the completeness of the thought usually depends upon the context in which the sentences appear, and some sentences may not state words that are understood. For an utterance to be a sentence, it must either state or imply a complete thought, given its context.

Must sentences contain a subject and a verb to be sentences? Yes and no. Sentences do have a grammatical structure, including a subject and a verb, but either or both can be understood:

When should I leave?

Now. (You should leave now. *In this sentence, the subject* [you] *and the verb* [should leave] *are both understood.*)

Sometimes, the subject and or verb are complicated and don't convey the primary meaning or central thought in the sentence:

What happened with the late invoice was that we could not include it in November accounts payable.

The subject is *What happened with the late invoice*. The verb is *was*. The subject is a complicated noun clause, and the main meaning is not in the main verb, but in the final clause: *that we could not include it in November accounts payable.*

Still, in most business and technical writing, sentences do have a subject and verb, even if these grammatical slots are filled with many words and have complex grammatical relationships.

Finally, all written sentences begin with a capital letter and end with a mark of punctuation, usually a period but sometimes a question mark or exclamation point. This convention applies to all sentences, even those with only one word and those in which words that are understood have been left out.

SENTENCES

Perhaps because sentences are difficult to define, most grammar handbooks settle for two fairly simple, yet practical, systems for cataloguing sentences:

Purpose or Intent

- Declarative Sentences
- Interrogative Sentences
- Exclamatory Sentences
- Imperative Sentences

Grammatical Structures

- Simple Sentences
- Compound Sentences
- Complex Sentences
- Compound-Complex Sentences

Purpose or Intent of Sentences

1. Use declarative sentences to make statements of fact and opinion. Usually such sentences follow the subject-verb word order, and they end with a period:

We reviewed the report.

Because of the detailed analyses involved, our review of the report is likely to take several days.

The report is only five pages long.

The report, which is only five pages long, will still take several days to review because the analyses are lengthy.

2. Use interrogative sentences to ask questions. Interrogative sentences usually begin with a question word (*who, which, where, when, why,* and *how*) or with a verb:

Who is the engineer in charge?

Which plan is likely to be approved?

When will the construction project end?

How often do they propose to inspect the site?

Have you filled in all the necessary forms?

Were the construction specifications adequate?

3. Use exclamatory sentences to make strong assertions or surprising observations. Exclamatory sentences usually end with an exclamation point:

What a surprising conclusion!

That's wrong!

What a field day for the lawyers that will be!

Oh, I doubt that!

NOTE: Exclamatory sentences often have grammatical structures very different from normal declarative sentences. In fact, they may be only a word or two long:

No!
How surprising!
A shame!

4. Use imperative sentences to give directions or commands. Imperative sentences usually begin with a verb and end with a period (although an exclamation point is also occasionally possible):

Move the recycling pump to the second floor.

Adjust the flange on the steam connection to prevent leakage.

Do not submit the pink copy of this form!

Stop. (*or* Stop!)

Grammatical Structures of Sentences

Simple Sentences. Simple sentences are sentences that express one complete thought. Essentially, they contain a single subject and a single verb, although both subject and verb may be compound:

The pump failed. (*single subject and verb*)

The new steam pump failed after only three weeks of service. (*single subject and verb*)

We analyzed the blueprints. (*single subject and verb*)

James Hawkins and I analyzed the blueprints for the new maintenance facility. (*compound subject, single verb*)

James Hawkins and I analyzed and revised the blueprints for the new maintenance facility. (*compound subject and verb*)

NOTE 1: As indicated above, simple sentences can contain compound subjects, such as *James Hawkins and I*, or compound verbs, such as *analyzed and revised*. Although compound, such subjects and verbs form a single unit, at least for the purpose of the sentence in question, so the sentence is still considered simple.

NOTE 2: A quick test for a simple sentence is that you cannot logically break the sentence at any point and come up with two other simple sentences. For example, the following simple sentence, even with compound subject and compound verb, cannot be broken into two simple sentences, so the sentence is a single, simple unit:

> My supervisor and I joked about the assignment and then worked on ways to accomplish it.

Compound Sentences. Compound sentences are essentially a union of two or more simple sentences. These simple sentences are usually linked by one of the simple coordinating conjunctions: *and, but, or, nor, for, yet*, and *so*. (See COMMAS and CONJUNCTIONS.) Sometimes, the simple sentences are linked by a semicolon or by a semicolon and a conjunctive adverb. (See SEMICOLONS and CONJUNCTIONS.)

> The project was expensive, and management still hadn't decided to proceed with it.

> Our supervisor wanted to increase office productivity, but the turnover in personnel made such an increase unlikely.

> The data confirmed our initial assumptions about the problems in prototype production; with these problems, the project will almost certainly exceed the budget.

> The surveying was to have been completed by October 15; however, construction must start on or before November 15.

NOTE 1: A quick test for a compound sentence is to see if you can divide the sentence into two or more simple sentences:

> The surveying was to have been completed by October 15.
> Construction must start on or before November 15.

NOTE 2: Sometimes three or more simple sentences can combine into a compound sentence:

> The site was ready, the construction crew was ready, and the materials were ready, but the weather was not cooperative.

Complex Sentences. A complex sentence is a simple sentence with a dependent or subordinate clause attached to it. The dependent or subordinate clause can appear in front of the main clause (the otherwise simple sentence), in the middle of the main clause, or behind the main clause.

Here are some dependent or subordinate clauses:

> Although our bid was the lowest . . .

> . . . who was the most expensive candidate . . .

> . . . because the tailings pile is virtually inert.

Adding a simple sentence to each of these dependent clauses forms a complex sentence:

> Although our bid was the lowest, another contractor had more experience.

> Cameron Blake, who was the most expensive candidate, did have the most impressive credentials.

> Reclamation of the mine site will be difficult because the tailings pile is virtually inert.

NOTE 1: A test for a complex sentence is to separate the dependent clause and the main clause (or simple sentence). This test will work for the sentences above, but not for this sentence:

> What Jack wanted to discuss with us became clear once the meeting got under way.

What Jack wanted to discuss with us is a noun clause that functions as the subject of the sentence. Trying to separate it from the rest of the sentence would result in two sentence fragments, neither of which can stand alone. Although this sentence fails the test, it is still a complex sentence.

NOTE 2: Dependent clauses are usually introduced by these words:

<u>Subordinate Conjunctions</u>

> Because, since, although, even though, after, before, so that, while, when, etc. (See CONJUNCTIONS.)

<u>Relative Pronouns</u>

> Who, whom, whose, which, that, whoever, whomever, why, when, where, etc. (See PRONOUNS.)

Compound-Complex Sentences. Compound-complex sentences are a combination of the two previous sentence types. They are <u>both</u> compound and complex. Therefore, a compound-complex sentence has two attached independent clauses—a compound sentence (two simple sentences attached to each

SENTENCES

other)—and at least one dependent clause.

> Because the firm's manufacturing capacity could not be increased rapidly enough, they were unable to fill their orders; consequently, competitors gained a significant foothold on the market.

NOTE: This sentence is compound because the semicolon separates two independent clauses, each of which could stand alone as a complete thought. The introductory dependent clause beginning with *because* makes the sentence complex.

Sentence Length and Readability

All readability formulas include sentence length as one measure of the readability of a piece of writing. A formula usually asks you to determine the average sentence length in the document or passage. Some authorities argue that average sentence length for any level of reader should be kept below some maximum (15 to 20 words), and that in writing for younger or less sophisticated readers, the average sentence length should be even less (8 to 12 words, depending upon how young the readers are).

Sentence length is one important factor in readability. Readers have to read, comprehend, remember, and interpret information sentence by sentence. The longer a sentence, the more they have to hold in their minds to comprehend, remember, and interpret the thought being expressed. Long sentences place an unnecessary burden upon readers.

Another factor in readability is sentence syntax (or structure). Long sentences may still be easily readable if they are constructed so that the sentence is easy to follow and easy to understand:

> We accepted the bid from the Cranston Construction Company—although it was the highest bidder—because (1) it has the manpower and equipment to start the project immediately, (2) its personnel are experienced in this type of construction, (3) it has the necessary permits in hand, (4) it is a local company using local resources, and (5) its management was willing to post a substantial performance bond.

This 67-word complex sentence is long but readable because it states its central point immediately: We accepted the bid from Cranston. The dependent clause enclosed by dashes clearly states a fact contrary to our expectations: Cranston was the highest bidder. We don't expect the contract to go to the highest bidder, and we wonder why Cranston received the award. So the numbered list provides the rationale for the decision. The numbering of the list helps clarify the syntax.

However, long sentences must be written this clearly to be readable. If your average sentence length becomes too high and you do not have the time to structure each sentence with great clarity, then your writing will be difficult, if not impossible, to read. (See STYLE.)

5. Keep your average sentence length below 20 words for typical business and technical writing.

The 20-word average is based on readability experiments and studies that reinforce the discussion of sentence length presented above. Obviously, in arriving at a 20-word per sentence average, you will have some longer sentences and some shorter sentences.

Longer sentences were more acceptable many years ago. Writers in the 19th century produced some mammoth sentences. Today, the trend is toward short, concise sentences. Busy readers don't have the time for or the interest in longwinded prose. The short sentence is the better sentence.

6. Use a variety of sentence types and sentence lengths.

Writing that is uniform is tedious. Make some sentences long and some short. Make most of them moderate in length. Remember that long sentences are useful for presenting involved concepts and for elaborating upon a point that requires some thoughtfulness. Short sentences are useful for stating clear, crisp thoughts.

Short sentences are naturally emphatic. Long sentences are not.

Create variety, too, in your choice of sentence types. Use mostly simple sentences, but do

SENTENCES

not avoid the compound, complex, and compound-complex sentences. Compound and complex sentences are very useful for expressing related ideas. If you use only simple sentences, you will be producing Dick-and-Jane writing. Compound and complex sentences lend themselves to the expression of related, connected, contrasting, and sequential thoughts.

7. Strive to make all sentences direct.

Keep the syntax (or structure) of the sentence as uncomplicated as possible. You can do this by keeping the subject as close as possible to the verb, by keeping modifiers as close as possible to the words they modify, and by using conjunctions and transitions to show progress, sequence, connection, and contrast.

See TRANSITIONS, PRONOUNS, MODIFIERS, ADJECTIVES, ADVERBS, NOUNS, VERBS, and CONJUNCTIONS.

As you write or revise sentences, ask yourself, "What is the single, central concept I am trying to express in this sentence?" In other words, "What is the point?" State that point clearly and directly, regardless of the sentence type or length.

SEXIST LANGUAGE

English from its earliest history has often marked words as either male or female (and even sometimes neuter). Pronouns are the commonest examples: *he, him, his* vs. *she, her, hers* vs. *it, its*. A number of nouns also have different male and female forms: *waiter/waitress, stewardess/steward, heir/heiress, countess/count, host/hostess, actress/actor, usher/usherette.* And some words used for everyone seem to include only males: *mankind, layman, manpower*, and so on.

Many such distinctions, called gender distinctions, have become objectionable, especially in recent years with the debate about equal rights for women. So many publishing firms and most writers routinely remove unnecessary and often objectionable gender distinctions from published writing.

This trend is the basis for the following rules, most of which require little effort from writers.

1. Do not use words that unnecessarily distinguish between male and female:

These	Not these
flight attendant	stewardess
people, humans	mankind
work force	manpower
layperson	layman
employee	workman
heir	heiress
serving person	waitress

NOTE 1: The use of female forms such as *waitress* and *heiress* has declined. *Heir* now includes both male and female; *waiter* still has male echoes, but these may fade soon. The best advice is to be sensitive to this issue and then to use female forms only when you have a definite need to signal a gender difference.

NOTE 2: Historically the word *man* (especially used in compound words like *layman*) could include both males and females; its closest modern equivalent would be, for instance, the indefinite pronoun *one* or *person*. This historical meaning has however been forgotten, so much so that many women now argue that they are silently being left out when compounds with *man* are used.

2. Avoid unnecessary uses of *he, him, or his* to refer to such indefinite pronouns as *everyone, everybody, someone,* and *somebody.*

SEXIST LANGUAGE

The problem sentences are often ones where the indefinite pronouns introduce a single person and then a later pronoun refers to that person:

> Everyone should take (his? her?) coat.
> Someone left (his? her?) report.

Unless we clearly know who *everyone* and *someone* refer to, we cannot pick the proper singular pronoun. We thus have to choose among several options:

—Make the sentences plural, if possible:

> All employees should take their coats.

—Remove the pronoun entirely:

> Someone left a (*or* this) report.

—Use both the male and female pronouns:

> Each employee should take his or her coat.
> Someone left his or her report.

—Use the plural pronoun *their* (or maybe *they* or *theirs*):

> Each employee should take their coat.
> Someone left their report.

NOTE: This last option is fine for informal or colloquial speech, but most editors and writers would object to the use of the plural pronouns to refer to the singular *everyone* and *someone*.

See PRONOUNS and AGREEMENT.

---

3. Avoid unnecessary uses of *he, him, his* or *she, her, hers* when the word refers to both males and females:

not these

> A secretary should set her (his?) priorities each day.

> The engineer opened her (his?) presentation with an overhead transparency.

> A writer should begin his (her?) outline with the main point.

As with rule 2 above, writers have several options:

—Change the sentences to plurals:

> Secretaries should set their priorities each day.

> Writers should begin their outlines with the main point.

—Remove the pronouns:

> The engineer began the presentation with an overhead transparency.

> A secretary should set firm priorities each day.

NOTE: A third option is to use the phrase *his or her,* but this becomes clumsy in a text of any length, so use one of the two options above.

---

4. Avoid the traditional salutation *Gentlemen* if the organization receiving the letter includes men and women.

The best option is to use one of the following:

> Ladies and Gentlemen:
> Gentlemen and Ladies:

If the letter is going to a single person whose name you do not know, then use these forms:

> Dear Sir or Madam:
> Dear Director:
> Dear Personnel Manager:

NOTE: In recent years a number of unusual salutations have appeared, but you should avoid them:

> Dear Gentlepersons:
> Dear Gentlepeople:
> Dear People:
> Dear Folks:

See LETTERS.

---

5. Do not substitute *s/he, he/she, hisorher,* or other such hybrid forms for standard personal pronouns.

These hybrid forms are unpronounceable and are not universally accepted by English users, so avoid them. Instead, either remove pronouns or change the sentences to plurals, as suggested under rule 3. Where you must use singular personal pronouns, use *he and she, his or her,* or *him and her.*

---

6. Do not call adult females *girls*, especially in a business or technical situation.

Referring to adult females as *girls* is no longer acceptable to most people except in contexts (typically humorous) in which it would also be appropriate to refer to adult males as *boys*. Labelling women as *girls* in serious (or even casual) conversation indicates a bias (intentional or otherwise) that is inappropriate in the business and technical community.

SIGNS AND SYMBOLS

Signs and symbols are increasingly important in scientific and technical writing. So the signs and symbols used in scientific and technical documents should be standard, and you should ensure that your signs and symbols are consistent within a document.

1. Choose standard signs and symbols, and ensure that readers understand them.

Some symbols are so well known that you don't need to explain them, regardless of where they appear: =, +, −, $, %, ×, ±, and ÷.

Other symbols require an explanation, either in notes at the bottom of a table or in a separate list of signs and symbols for a specific text.

2. Limit signs and symbols to tables, figures, and other visual aids; avoid them in the text itself.

EXCEPTION: Some very common symbols, such as the percent sign (%), may be used in texts written for a specific group of readers. Accounting documents, for instance, routinely refer to percentages. Writing out *percent* rather than using the symbol (%) is both time consuming and unnecessary. Similarly, chemical symbols are appropriate in documents written for people who know and understand the symbols, but such symbols are inappropriate in business documents and technical documents intended for nonchemists.

3. No space appears on either side of the signs +, −, ±, ×, and ÷ :

C+D	245±5
675−41	84÷12
7×12	

EXCEPTION: When the × is used to mean "crossed with" (as in plant or animal breeding) or to indicate magnification, a space appears on each side of the symbol:

Early Roma × Big Girl
× 20 (magnification)

NOTE: A space does appear on either side of the equals sign:

$x+y = 4$

See MATHEMATICAL NOTATION.

Common Signs and Symbols

Chemical

Element	Symbol	Atomic number	Atomic weight[1]
Actinium	Ac	89	227.0278
Aluminium	Al	13	26.98154
Americium	Am	95	(243)
Antimony (Stibium).	Sb	51	121.75
Argon	Ar	18	39.948
Arsenic	As	33	74.9216
Astatine	At	85	(210)
Barium	Ba	56	137.33
Berkelium	Bk	97	(247)
Beryllium	Be	4	9.01218
Bismuth	Bi	83	208.9804
Boron	B	5	10.81
Bromine	Br	35	79.904
Cadmium	Cd	48	112.41
Caesium	Cs	55	132.9054
Calcium	Ca	20	40.08
Californium	Cf	98	(251)
Carbon	C	6	12.011
Cerium	Ce	58	140.12
Chlorine	Cl	17	35.453
Chromium	Cr	24	51.996
Cobalt	Co	27	58.9332
Copper	Cu	29	63.546
Curium	Cm	96	(247)
Dysprosium	Dy	66	162.50
Einsteinium	Es	99	(252)
Erbium	Er	68	167.26
Europium	Eu	63	151.96
Fermium	Fm	100	(257)
Fluorine	F	9	18.998403
Francium	Fr	87	(223)
Gadolinium	Gd	64	157.25
Gallium	Ga	31	69.72
Germanium	Ge	32	72.59
Gold	Au	79	196.9665
Hafnium	Hf	72	178.49
Helium	He	2	4.00260
Holmium	Ho	67	164.9304
Hydrogen	H	1	1.00794
Indium	In	49	114.82
Iodine	I	53	126.9045
Iridium	Ir	77	192.22
Iron	Fe	26	55.847
Krypton	Kr	36	83.80
Lanthanum	La	57	138.9055
Lawrencium	Lr	103	(260)
Lead	Pb	82	207.2
Lithium	Li	3	6.941
Lutetium	Lu	71	174.967
Magnesium	Mg	12	24.305
Manganese	Mn	25	54.9380
Mendelveium	Md	101	(258)
Mercury	Hg	80	200.59
Molybdenum	Mo	42	95.94
Neodymium	Nd	60	144.24
Neon	Ne	10	20.179
Neptunium	Np	93	237.0482
Nickel	Ni	28	58.69
Niobium	Nb	41	92.9064
Nitrogen	N	7	14.0067
Nobelium	No	102	(259)
Osmium	Os	76	190.2
Oxygen	O	8	15.9994
Palladium	Pd	46	106.42
Phosphorus	P	15	30.97376
Platinum	Pt	78	195.08
Plutonium	Pu	94	(244)
Polonium	Po	84	(209)
Potassium (Kalium).	K	19	39.0983
Praseo-dymium	Pr	59	140.9077
Promethium	Pm	61	(145)
Protactinium	Pa	91	231.0359
Radium	Ra	88	226.0254
Radon	Rn	86	(222)
Rhenium	Re	75	186.207
Rhodium	Rh	45	102.9055
Rubidium	Rb	37	85.4678
Ruthenium	Ru	44	101.07

SIGNS AND SYMBOLS

Element	Symbol	Atomic number	Atomic weight[1]
Samarium	Sm	62	150.36
Scendium	Sc	21	44.9559
Selenium	Se	34	78.96
Silicon	Si	14	28.0855
Silver	Ag	47	107.8682
Sodium (Natrium).	Na	11	22.98977
Strontium	Sr	38	87.62
Sulfur	S	16	32.06
Tantalum	Ta	73	180.9479
Technetium	Tc	43	(98)
Tellurium	Te	52	127.60
Terbium	Tb	65	158.9254
Thallium	Tl	81	204.383
Thorium	Th	90	232.0381
Thulium	Tm	69	168.9342
Tin	Sn	50	118.69
Titanium	Ti	22	47.88
Tungsten (Wolfram).	W	74	183.85
(Unnil-hexium)	(Unh)	106	(263)
(Unnil-pentium)	(Unp)	105	(262)
(Unnil-quadium)	(Unq)	104	(261)
Uranium	U	92	238.0289
Vanadium	V	23	50.9415
Xenon	Xe	54	131.29
Ytterbium	Yb	70	173.04
Yttrium	Y	39	88.9059
Zinc	Zn	30	65.38
Zirconium	Zr	40	91.22

[1]These atomic weights apply to elements as they exist naturally on Earth and to certain artificial elements. Values in parentheses are for radioactive elements whose atomic weights cannot be quoted precisely without knowledge of the origin of the elements. The value given is the atomic mass number of the isotope of that element of longest known half life.

Electrical

ℜ reluctance
↔ reaction goes both right and left
↕ reaction goes both up and down
↕ reversible
→ direction of flow; yields
→ direct current
⇆ electrical current
⇆ reversible reaction
⇄ reversible reaction
⇆ alternating current
⇄ alternating current
⇌ reversible reaction beginning at left
⇋ reversible reaction beginning at right
Ω ohm; omega
MΩ megohm; omega
μΩ microohm, mu omega
ω angular frequency, solid angle; omega
Φ magnetic flux; phi
Ψ dielectric flux; electrostatic flux; psi
γ conductivity; gamma
ρ resistivity; rho
Λ equivalent conductivity
HP horsepower

Geologic Systems

J Jurassic
T̄R Triassic
P Permian
P Pennsylvanian
M Mississippian
D Devonian
S Silurian
O Ordovician
Є Cambrian
pЄ Precambrian
C Carboniferous
Q Quaternary
T Tertiary
K Cretaceous

NOTE: These standard letter symbols are used by the Geological Survey on geologic maps. A capital letter indicates the system, and one or more lowercased letters designate the formation and member where used.

Mathematical

— vinculum (above letters)
∺ geometrical proportion
∹ difference, excess
∥ parallel
∥s parallels
∦ not parallels
| | absolute value
· multiplied by
: is to; ratio
÷ divided by
∴ therefore; hence
∵ because
∷ proportion; as
≪ is dominated by
> greater than
⊏ greater than
≥ greater than or equal to
≧ greater than or equal to
≷ greater than or less than
≯ is not greater than
< less than
⊐ less than
≶ less than or greater than
≮ is not less than
⋖ smaller than
≤ less than or equal to
≦ less than or equal to
⋝ *or* ≥ greater than or equal to
⋜ equal to or less than
⋜ equal to or less than
⋠ is not greater than equal to or less than
⋝ equal to or greater than
⋡ is not less than equal to or greater than
≟ equilateral
⊥ perpendicular to
⊢ assertion sign
≐ approaches
≑ approaches a limit
≚ equal angles
≠ not equal to
≡ identical with
≢ not identical with
𝍸 score
≈ *or* ≒ nearly equal to
= equal to
∼ difference
≃ perspective to
≅ congruent to approximately equal
≏ difference between
≎ geometrically equivalent to
(included in
) excluded from
⊂ is contained in
∪ logical sum or union
∩ logical product or intersection
√ radical
√ root
∛ square root
∛ cube root
∜ fourth root
⁵√ fifth root
⁶√ sixth root
π pi
ε base (2.718) of natural system of logarithms; epsilon
∈ is a member of; dielectric constant; mean error; epsilon
+ plus
+ bold plus
− minus
− bold minus
/ shill(ing); slash; virgule
± plus or minus
∓ minus or plus
× multiplied by
= bold equal
number
℔ per
% percent
∫ integral
| single bond
╲ single bond
╱ single bond
‖ double bond
⑊ double bond
⫽ double bond
⌬ benzene ring
∂ *or* δ differential; variation
∂ Italian differential
→ approaches limit of
∼ cycle sine

SIGNS AND SYMBOLS

∽ horizontal integral
∮ contour integral
∝ variation; varies as
Π product
Σ summation of; sum; sigma
! or ∟ factorial product

See MATHEMATICAL NOTATION.

Measurement

℔ pound
ʒ dram
fʒ fluid dram
℥ ounce
f℥ fluid ounce
O pint

Miscellaneous

§ section
† dagger
‡ double dagger
℀ account of
℅ care of
𝍸 score
¶ paragraph
þ Anglo-Saxon

℄ center line
☌ conjunction
⊥ perpendicular to
〃 or " ditto
∝ variation
℞ recipe
⊐ move right
⊏ move left
○ or ⊙ or ① annual
⊙⊙ or ② biennial
∈ element of
℈ scruple
ƒ function
! exclamation mark
⊞ plus in square
♃ perennial
ϕ diameter
c̄ mean value of c
∪ mathmodifier
⊂ mathmodifier
⊡ dot in square
◬ dot in triangle
⊠ station mark
@ at

Money

¢ cent
¥ yen
£ pound sterling
₥ mills

Sex

♂ or ♁ male
□ male, in charts
♀ female
○ female, in charts
⚥ hermaphrodite

Weather

T thunder
⩼ thunderstorm; sheet lightning
< sheet lightning
↓ precipitate
◍ rain
⟵ floating ice crystals
↔ ice needles
▲ hail
⊗ sleet
∞ glazed frost
⊔ hoarfrost
V frostwork
✳ snow or sextile
⊠ snow on ground
+ drifting snow (low)
≡ fog
∞ haze
⌂ Aurora

SLASHES

A slash (/) is also called a solidus or a virgule.

1. Use a slash when a season or a time period extends beyond a single year:

winter 1980/1981
fiscal year 1984/85
425/424 B.C.

NOTE: A hyphen sometimes replaces the slash:

Winter 1980-1981
425-424 B.C.

2. A slash can replace *per*:

yards/mile
feet/second *or* ft/sec

3. A slash is used in mathematical expressions written on a single line:

$x/a - y/c = 1$ *for* $\frac{x}{a} - \frac{y}{c} = 1$

$(C/D)/(C+2D)^3$ *for* $\frac{\frac{C}{D}}{(C+2D)^3}$

4. A slash separates lines of quoted poetry presented in prose form:

As Shakespeare so aptly noted: "There is a tide in the affairs of men, / Which, taken at the flood, leads on to fortune."

SLASHES

5. A slash appears in *and/or*:

We asked for a rebate and/or an explanation.

NOTE: Some writers and editors object to all uses of *and/or*, arguing that the expression is ambiguous. In the above example, the meaning includes *a rebate and an explanation* as well as *a rebate or an explanation*. Many editors would prefer to rewrite the sentence:

We asked for a rebate or an explanation or both.

SPACING

Proper, even creative, spacing can help guarantee that a final document looks professional and is easy to read. Spacing decisions range from the arrangement of text and visuals on the page to the amount of space left after each paragraph and the number of spaces following the final period in a sentence.

Page Formats

Decide early in the writing process how you want your pages to look. Options, especially with word processing, include variable margins, indentations, a variety of type faces, and variable line spacing. For important writing projects, you might consult professional designers before writing the text. Sometimes, design considerations affect how much and what you write.

See MANUSCRIPT FORM and EMPHASIS.

1. Leave ample white space on your pages, especially around important ideas or data.

Writers (and typists) often cram too much writing onto a page, trying to stay within page limits or adhering to custom, an arbitrary format, or just plain myth about how pages ought to look. Some writers ignore the appearance of the document because they have never had to think about it.

Without careful and early attention to the desired page layout, the text may not fit within prescribed limits, and the tables, figures, and other visuals may not complement the text.

Although every page layout is different, here are some principles to follow as you work with available space:

—Adjust the margins in letters and memos so that the top and bottom margins are roughly equal and the left-hand and right-hand margins are also equal. Your goal is to center the letter or memo on the page. (See LETTERS and MEMOS.)

—In single-spaced text, double-space between paragraphs. Even triple spacing may be desirable for extra space on a page or for highlighting key paragraphs.

—Avoid excessively long paragraphs or a series of short, choppy paragraphs. (See PARAGRAPHS.)

—Add lists, tables, figures, or other visual aids to break up long stretches of text. (See LISTS and TABLES.)

—Design a system of headings that allows you to divide the text frequently and to highlight key ideas. (See HEADINGS and EMPHASIS.)

NOTE: A rough mock-up of your document is an excellent planning aid. A mock-up typically consists of a series of blank pages, one for each page in the proposed document, with titles, headings, lists, visual aids, and perhaps paragraphs sketched in.

SPACING

The mock-up suggests where and how long each of the major textual units will be. You should produce the mock-up well before writing the actual text.

Spacing and Punctuation

One sign of an inexperienced typist is erratic spacing before and after punctuation marks. The following list covers the basics of spacing around punctuation:

—Leave two spaces after any mark of punctuation that ends a sentence.

—Always place commas and periods inside of quotation marks. Place colons, semicolons, and dashes outside of quotation marks. Place question marks and exclamation marks inside or outside of quotation marks, depending upon whether they are part of the quotation. (See QUOTATION MARKS.)

—Leave no space before a semicolon and one space after it.

—Leave no space before or after dashes. On a standard typewriter, create a dash by typing two hyphens with no space between them:

The plan--a method for extracting iron ore--is cost effective.

See DASHES.

—Leave no space before a colon and two spaces after it within a sentence.

—Leave one space before an opening or left parenthesis within a sentence. Leave two spaces when the opening parenthesis follows another sentence; if it is a complete sentence, the parenthetical material opens with a capital and the final punctuation comes before the final parenthesis. (See PARENTHESES.)

—Leave no space before a closing or right parenthesis and one space after it within a sentence. When an entire sentence is enclosed within parentheses, the final parenthesis goes outside of the closing punctuation and two spaces follow the right parenthesis.

—Leave one space before and after each of the three periods in an ellipsis. If an ellipsis concludes a sentence, the ellipsis has four, rather than three, periods. The first period has one space between it and the last word in the sentence. Two spaces follow the last period. (See ELLIPSES.)

NOTE: Each of the punctuation marks discussed above has its own entry in this *Guide*. Refer to those entries if you have questions or wish to see additional examples.

SPELLING

Correct spelling is the final ingredient in any professional document—from formal report to everyday letter or memo.

Spelling is one of those details that are important for the sake of both clarity and credibility. Most misspellings do not cause readers to misinterpret the sentence in which the misspelling occurs. But the misspelled word draws attention to itself, which slows down readers and diverts their attention away from the ideas being expressed.

Language is the medium. When the medium draws attention to itself, it detracts from the message.

Misspelled words may also cause readers to question the writer's competence, intelligence, and credibility. How much confidence would you have in this writer's engineering abilities?

> Raw seewater has been considured as a posible alternet sorce for the consentrater principle water supply. However, bench scale tests indacate that the high consentration of disolved salts in seewater interfeer in the eficenct recovary of minaral from the ore.

Misspellings in a document make the writer and the writer's organization look incompetent, sloppy, careless, and potentially untrustworthy.

However, spelling in English is far from simple. Roughly 90 percent of the words in English are regular, but the other 10 percent are demons. Which words in the following pairs are correct?

accomodate/accommodate
committment/commitment
concientious/conscientious
changable/changeable
imperceptable/imperceptible
indispensible/indispensable
inevitible/inevitable
irresistable/irresistible
occurence/occurrence
offerred/offered
preceed/precede
prefered/preferred
prevalant/prevalent
privlege/privilege
seperate/separate
similiar/similar
transfered/transferred
truely/truly

If you are like most people, you had to pause on at least two or three of the above pairs. Perhaps you still aren't sure. Did you look up any of them in a dictionary?

English is a hybrid language. It evolved over centuries of influence from the languages of the armies that invaded England: the Romans, the Saxons, the Vikings, the Normans, and so on. English is "impure" in this regard and consequently has an inconsistent base system of words. That's why English pronunciation and spelling are inconsistent.

1. Challenge the spelling of every word in your document, especially those words you have difficulty with.

The best proofreaders and editors challenge every word, especially those that are known to be difficult (such as those listed above).

If the word is common enough, you can trust yourself to recognize correct spelling. If you are unsure, however, check a dictionary or spelling dictionary. (See REFERENCES.)

2. Use those spelling rules that you find helpful.

The spelling rules are difficult to remember, and most have many exceptions. If you take the time to memorize the rules, you should probably also memorize the exceptions. At some point, the exercise becomes tedious, and the rewards are questionable.

However, you should use those rules that you have found helpful and that you remember well enough to apply.

Probably the most well-known and most useful rule is "*i* before *e* except after *c*." Here are some of the exceptions:

counterfeit
foreign
freight
height
heir
neighbor
sleigh
vein
weigh
weight

Some of the other common rules are briefly summarized below:

- Change a final *y* to *i* before adding a suffix to a word, but keep the *y* before *-ing*:

activity	activities
deny	denies, denying
happy	happily, happier, happiest, happiness
likely	likelihood
study	studies, studied, studying

SPELLING

- Drop a silent final *e* before suffixes beginning with a vowel but not before suffixes beginning with a consonant:

age	aging
desire	desirable
mobile	mobility
notice	noticing
scarce	scarcity
care	careful
manage	management
safe	safety
wife	wifely

Exceptions

acreage
argument
changeable
courageous
judgment
lineage
mileage
ninth
truly
wholly

- Double a final consonant before a suffix beginning with a vowel (1) if the consonant ends a stressed syllable (or a single-syllable word) and (2) if the consonant is preceded by a single vowel:

bag	bagged
brag	bragged
gun	gunned
shop	shopped, shopper
stop	stopped
begin	beginning
occur	occurred
prefer	preferred
regret	regretting, regretted

3. Form plurals carefully. Many irregular forms exist:

man	men
ox	oxen
analysis	analyses
matrix	matrices
potato	potatoes
piano	pianos

See PLURALS for a discussion of these irregular forms as well as a list of the most common irregular plurals.

4. Keep a list of the words you have trouble spelling.

Remembering spelling rules and their exceptions is difficult. A simpler and nearly foolproof method for improving your spelling is to keep a list of the words you commonly misspell. Look up the correct spellings and list the words alphabetically.

When you see that you have misspelled a word, add it to your list. Then refer to the list when you need to use one of those words. Over time, your mind will come to recognize the look of the word as it is spelled correctly, and you will no longer need the list.

Until you no longer need it, keep the list in a convenient place: tucked inside your dictionary, on the wall in front of your desk or writing area, under the glass on top of your desk, or taped inside your notebook. Keep it where you can see it easily as you write.

A List of Common Spelling Demons

Below is a list of some of the most common spelling demons. These words, interestingly enough, are not technical ones because most of us learn to spell technical words as we learn our technical subjects.

Common words are the problem because their irregularities are often difficult to predict and almost impossible to remember. Also, we often see common words misspelled, so we remember the look of the misspelling, not the look of the correct spelling.

absorb
acetic (*acid*)
acceptable
accessible
accommodate
accompanied
accuracy
accustomed
achievement
acoustic
acquire
acreage
adapter
adsorb
aegis
affect (*usually a verb*)
affected
aggression
aging
aid (*help*)
aide (*assistant, helper*)
aisles
all ready (*all prepared*)
all right
all together (*all those in group*)
all ways (*by every means*)
a lot of
already (*previously*)
altogether (*entirely*)
aluminum
always (*all the time*)
amateur
analogous
announcement
anonymous
antibiotics
any one (*any specific person or object*)
anyone (*any person*)
appall, appalled
apparent
appearance
appraise (*estimate value*)
apprise (*inform*)
appropriate
aquatic
archaeology
artisan
ascetic (*austere*)
aspirin
athletics
attendance

SPELLING

authentic
a while (*noun*)
awhile (*adverb*)

bargain
basically
beside (*next to*)
besides (*in addition*)
beveled
beneficial
benefited
biased
breath (*noun*)
breathe (*verb*)
bulletin
bureaucracy
business

caffeine
calendar
caliber
caliper
category
calk
calorie
canceled, canceling
cancellation
candor
canvas (*cloth*)
canvass (*solicit*)
capital (*city*)
capitol (*building*)
carat (*gem weight*)
caret (*arrow mark*)
cemetery
census
challenge
changeable
channel
characteristic
chisel, chiseled
choose (*present tense*)
chose (*past tense*)
coarsely
commitment
committee
competent
competition
complement (*complete*)
compliment (*praise*)
conceited
conceive
condemn
confidant (*person*)
confident (*sure*)
conscience
conscientious
consensus
consistent
continuous
controlled
controversial
councilor (*of council*)
counselor (*advisor*)
courteous

criticism
criticize
curiosity
curious

deceive
decision
definitely
descend
descendant
description
desirable
despair
desperate
despicable
device (*noun*)
devise (*verb*)
dietitian
disappoint
disapprove
disastrous
discipline
discreet (*prudent*)
discrete (*distinct or separate*)
disease
distill, distilled
distinct
doctor
dyeing (*coloring*)
dying (*death*)

easily
ecstasy
effect (*usually a noun*)
efficient
eighth
elaborately
elicit (*to draw*)
embarrass
emigrate (*go from*)
employee
enroll, enrolled
ensure (*guarantee*)
entirely
envelop (*verb*)
envelope (*noun*)
environment
equipment
equipped
especially
every day (*each day*)
everyday (*ordinary*)
evidently
exaggerate
except
exhaust
existence
experiment
explanation
eying

familiar
farther (*distance*)
fascinate
favorite

February
fiber
finally
financially
flammable (*not* inflammable)
fluorescent
fluorine
foreign
foresee
foretell
forgo (*relinquish*)
  forego (*precede*)
forty
forward (*ahead*)
  foreword (*preface*)
fulfill, fulfilled
further (*degree*)
fuselage

gauge
generally
glamour
government
governor
grammar
guaranteed
guerrilla

happened
harass
heard
height
heroes
hindrance
hoping
humane
humorous
hurriedly
hypocrisy
hypocrite

ideally
idiosyncrasy
ignorant
illicit (*illegal*)
illogical
imaginary
imagine
imitate
immediately
immensely
immigrate (*go into*)
incalculable
incidentally
incredible
indispensable
inequity
influential
initiative
innocuous
insurance
insure (*guarantee financially*)
intelligent
interference
integrate

SPELLING

interrupt
irrelevant
irresistible
irritated

jealousy
jewelry
judgment

kilogram
knowledge

laboratory
laid
lath (*wood*)
lathe (*machine*)
led
leisure
length
lenient
leukemia
liable
library
license
lightning
likelihood
liquefy
liveliest
logistics
loose (*adjective*)
lose (*verb*)
luxury
lying

magazine
magnificent
maintenance
manageable
management
maneuver
mantel (*shelf*)
mantle (*cloak*)
margarine
marijuana
marriage
material (*goods*)
materiel (*military goods*)
mathematics
meant
medicine
meteorology
mileage
miniature
minor
mirror
mischievous
missile
morale
mortgage
mucus (*noun*)
mucous (*adjective*)
muscle
mysterious

naturally
necessary
nevertheless
nickel
niece
nineteen
ninety
ninth
noticeable
nowadays
nuclear
nuisance
numerous

occasion
occasionally
occurred
occurrence
occurring
off
offense
official
omission
omitted
omitting
oneself
opponent
opportunity
opposite
oppression
optimism
ordinance (*law*)
ordnance (*military*)
ordinarily
originally

pamphlet
parallel
paralleled
parole
particle
particularly
pastime
peaceable
peculiar
penetrate
perceive
performance
perquisite (*privilege*)
perhaps
permanent
personal (*individual*)
personnel (*employees*)
perspective (*viewpoint*)
persuade
pertain
phosphorus (*noun*)
phosphorous (*adjective*)
physical
picnicking
pigeon
poison
politician
pollute
possession
possibly
practical
practically
precede
precedence (*priority*)
precedents (*prior instances*)
predominant
preferred
prejudice
prerequisite (*requirement*)
prevail
prevalent
preventive (*not* preventative)
principal (*chief or main*)
principle (*theory or idea*)
prisoner
privilege
probably
procedure
proceed
processes
professor
programmed, programmer,
programming
prominent
pronounce
pronunciation
propaganda
prophecy (*noun*)
prophesy (*verb*)
prospective (*expected*)
psychology
publicly
pursue
pursuing
pursuit

quandary
quarreled
quarreling
questionnaire
quiet
quite
quizzes

rarefy
rarity
rebel
receipt
receive
recession
recipe
reconnaissance
reconnoiter
recommend
recyclable
referring
regular
regulate
rehearsal
reinforce
relief (*noun*)
relieve (*verb*)
religious
remembrance
reminisce
repellant (*noun*)
repellent (*adjective*)

SPELLING

repetition
resemblance
resistance
restaurant
rhythm
ridiculous

sacrifice
salvage (*save*)
safety
satellite
scarcity
scenery
schedule
secede
secretary
seismology
seize
selvage (*edging*)
separate
sergeant
sheriff
shining
shrubbery
signaled, signaling
significant
similar
sincerely
sizable
some time (*some time ago*)
sometime (*formerly*)
sometimes (*at times*)
souvenir
spacious (*space*)
specious (*deceptive*)
sponsor
stationary (*fixed*)
stationery (*paper*)
statistics
stepped
stopped
straight
strategy
strength
strenuous
stretch
studies
studying
subpoena, subpoenaed
subtlety
suburban
succeed
succession
suicide
sulfur
superintendent
supersede
suppress
surely
surreptitious
surround
surveillance
suspicious
susceptible
synonymous

technical
technique
temperature
temporary
their (*pronoun*)
themselves
there (*adverb*)
therefore
thorough
though
through
tie, tied, tying
till
tobacco
totaled, totaling
too
tragedy
transferable
transferred
traveled, traveling
tremendous
truly
twelfth
typical
tyranny

unanimous
unconscious
undoubtedly
until
usage
usually

vacuum
various
vengeance
villain
violence
visible
vitamins

warrant
warring
weather
Wednesday
where
wherever
whether
whichever
whiskey
wholly
woman (*singular*)
women (*plural*)
writing
written

yield

STRONG VERBS

A common stylistic problem is using weak, rather than strong, verbs.

Weak verbs are those simple verbs that occur so frequently in our language that they have little impact: *is, are, was, were, can, could, has, had, have, do, did, done, make, use, come.*

Obviously, these verbs are essential to English. Using them, either as primary or auxiliary sentence verbs, is inescapable. However, writers often use them unnecessarily to create wordy, weak sentences:

The system has wide applicability for a variety of industrial cogeneration situations. (*12 words*)

STRONG VERBS

In this sentence, the writer has transformed *apply*, a much stronger verb than *has*, into an awkward, bureaucratic noun, *applicability*, and used the weaker verb as the sentence verb. Using *apply* as the sentence verb creates a shorter, stronger sentence:

> The system applies to a variety of industrial cogeneration situations. (*10 words*)

See SCIENTIFIC/TECHNICAL STYLE.

---

1. Use Strong Verbs.

Strong verbs are less common; therefore, readers tend to pay more attention to them. They have more impact in a sentence, and they help writers avoid big, bureaucratic nouns:

not this

Membranes for the separation and enrichment of gas mixtures have been under study by our Basic Development Department since 1965.

this

Since 1965, our Basic Development Department has studied membranes for the separation and enrichment of gas mixtures.

---

not this

We have made vast improvements in our reaction mechanisms.

this

We have vastly improved our reaction mechanisms.

---

not this

We would like to conduct further investigations into the pilot program before giving a proposed factory location.

this

We would like to investigate the pilot program further before proposing a factory location.

---

not this

We will give special emphasis to the evaluation of plating techniques for the deposition of amorphous or glassy metal coatings.

this

We will especially emphasize evaluating plating techniques for depositing amorphous or glassy metal coatings.

---

not this

Before proceeding with catalyst development, we would make technical and economic assessments of the advantages of the various source materials available.

this

Before developing the catalyst, we would assess the technical and economic advantages of the various source materials.

STYLE

Style is the sum of the choices, both conscious and unconscious, that writers make while planning, designing, writing, and editing documents.

These choices include the type of document, the words chosen, the structure and length of sentences, the length and type of paragraphs, the document's organization, the use of emphatic devices (headings, lists, white space), the use and kind of visual aids, the typeface and type size, the paper, and so on.

Such choices give each document a unique style, tone, and feeling. Each telex, each field note, each business letter, each quick memo, each formal report—they are all unique because they differ in tone, attitude, perspective, and style from every other telex, note, letter, memo, or report.

STYLE

Style and Tone

Style and *tone* are often confused. The terms are similar, and some speakers use them interchangeably. However, style is the cause and tone is the effect:

> *Tone* refers to the feeling or impression a document conveys to its readers. (See TONE.)
>
> *Style* refers to those choices writers make that create the tone conveyed to readers.

Style is often categorized as being either **formal** or **informal**. These distinctions typically revolve around the type of document being written, the intended readers, the writer's relationship to the readers, the document's purpose, and the message being conveyed.

Formal documents, such as legal agreements, technical reports, and many letters, are written for readers with whom the writer has no personal relationship. Therefore, familiarity or levity is generally unacceptable. It would seem inappropriate to the circumstance and would interfere with the message. Formal documents are usually written to convey information objectively to readers who may not know (or care to know) the author.

The tone of formal documents tends to be impersonal, objective, restrained, deliberate, and factual. Formal documents that are intended to convince readers to do or accept something are often also forceful, dynamic, and perhaps intensive. Most documents that convey negative or unpleasant information are formal.

Informal documents, such as personal letters, newsletters, trip reports, and most memos, are generally written for people the writer knows or feels comfortable with, perhaps only through employment in the same organization. Familiarity, levity, and wit are often acceptable. The informality creates a more relaxed atmosphere, which makes the message warmer, more easily acceptable. Informal documents are often informative, friendly, and subjective.

The tone of informal documents is generally relaxed, informative, helpful, casual, personal, positive, nonthreatening, and perhaps cheerful. Informal documents may be persuasive, but they are rarely forceful or aggressive. When writers need to be aggressive and when they need to convey negative or unpleasant information, they generally become more formal in their approach. (See TONE.)

Style also refers to styles associated with particular disciplines: geologic style, geophysical style, legal style, medical style, engineering style, auditing style, academic style, scientific style, social science style, bureaucratic style, and so on. (See SCIENTIFIC/ TECHNICAL STYLE.)

In each of these styles, the writer uses jargon particular to the discipline and writes according to a long tradition of document preparation and appearance. We are perhaps most familiar with the legal and bureaucratic styles, but all disciplines have a set of standards and traditions that affect the way people communicate in writing.

Word Choice and Style

Each word you choose helps establish the style of your writing. Words are one of the most visible traits of style.

You may, for instance, wish to discuss the effects of a decision. The word *effects* is a fairly neutral choice. Instead of *effects*, you might speak of *impacts*, *consequences*, or *results*. *Impacts* suggests some negative connotations, as does *consequences*, which is more formal sounding (perhaps because of its length) than *impacts*. *Results*, on the other hand, has a positive feeling to it: *We're going to get results*.

You have other less common choices: *aftermath, corollary, end product, eventuality, outcome, sequel, upshot*. *Aftermath* has definite negative implications, besides being almost dramatic in its tone. *Corollary* has limited usefulness, if for no other reason than its mathematical echoes (and hence almost too educated a tone). *End product* seems plain, yet still wordy when compared to *effects* or *results*. *Eventuality* implies some final or ultimate result; again, its length makes it sound more formal. *Outcome* is about as neutral as any of the words in this list, but it may have negative connotations, as in the outcome of a medical test. *Sequel* implies a second follow-up event, not a real effect. And *upshot* suggests surprise, even chaos. Even more words are possible: *development, fruit, outgrowth, ramification, repercussion, conclusion*.

STYLE

Part of the richness of English is its large vocabulary, which offers multiple possibilities for expressing any idea. Yet no two words mean exactly the same thing, so when you choose one word rather than another, you change the style of the document, if only slightly. Within a few lines, then, you will make dozens of choices, all of which combine to establish the style (and resulting tone) of the document.

Obviously, some word choices make a bigger difference than others. Selecting *effects* rather than *results* changes the style very little. But if you use *impacts*, *consequences*, or *aftermath*, the document may shift radically in tone and effect:

> What is the effect of altering course in mid-flight?
>
> What are the consequences of altering course in mid-flight?

> The pipeline may affect the salmon population in the Little Middle River.
>
> The pipeline will impact the salmon population in the Little Middle River.

As you can see, some word choices are important stylistic signals. Others are inconsequential:

> What is the effect of altering course in mid-flight?
>
> What is the result of altering course in mid-flight?

Writers in various technical professions have a body of technical words that affect the style:

Legal

> tort, legatee, real property, contract, conveyance, *amicus curiae*, party, sue, brief, witness, jurisdiction, plaintiff, etc.

Medical

> curette, mamillary, uvulae, amoebic, gastric hernia, leucoplast, dermatitis, proboscis, etc.

Computer

> batch processing, cursor, default, field, file, logon, real time, sign off, etc.

Construction

> sill, head, transom, mullion, fascia, neoprene spaces, support mullion, jamb, seat board, glazing gasket, butt glazed, hopper sash, soffit, rowlock

Such specialized scientific and technical terms are unavoidable given today's complex technologies (see JARGON). The presence of this jargon is the most visible sign to readers that a document reflects a particular style.

Sentences and Style

Writers have virtually an infinite number of ways to express ideas. Even in an ordinary 15- to 30-word sentence, the possibilities run into the tens of thousands, both in terms of word choice and sentence structure. The sheer range of possibility means that every sentence except the shortest and most trivial is potentially unique (never having been written or said before).

Sentence options—**length** and **structure**—are the most important features of a writer's style. However, these options are often less obvious to readers than the choice of a particular word or technical term. Readers notice the style of sentences only when something goes wrong, as in an awkward sentence or one where something is deliberately unusual:

> That is something up with which I shall not put.
>
> *or*
>
> Turning into the wrong driveway, a tree was hit by me which I don't have.

Sentence Length

Sentence length by itself normally will not establish a definite style but it will contribute to style. Most sentences average anywhere from 12 to 25 words in length. Readers are accustomed to those lengths, so they are likely to notice only those sentences that are either extremely short or extremely long:

> We refuse.
>
> *or*
>
> Science is nothing but trained and organized common sense, differing from the latter only as a veteran may differ from a raw recruit: and its methods differ from those of common sense only as far as the guardsman's cut and thrust differ from the manner in which a savage wields his club. (from Thomas Huxley's *Collected Essays*)

An individual sentence, even if potentially noteworthy, won't be noteworthy unless it stands out from the sentences surrounding it. So **average sentence length** is probably more of a direct indication of style than the

STYLE

length of a single sentence. Readability formulas always include average sentence length as a measure of readability because length is a good indicator of the difficulty of a document. (See SENTENCES.)

Long sentences can reflect different styles (and tones) depending upon other features within them. Long, well-structured sentences with a sophisticated vocabulary usually convey an educated or thoughtful quality. But if the sentence seems longer than necessary for the ideas being expressed, and if the vocabulary is more sophisticated than necessary, then the sentence may seem stuffy, extravagant, or pompous:

> If biological populations or habitats that may require additional protection are identified by the Deputy Conservation Manager, Offshore Field Operations (DCMOFO) in the leasing area, the DCMOFO will require the lessee to conduct environmental surveys or studies, including sampling, as approved by the DCMOFO, to determine existing environmental conditions, the extent and composition of biological populations or habitats, and the effects of proposed or existing operations on the populations or habitats that might require additional protective measures.

This sentence's length—some 77 words—is surely excessive, but other features contribute to its bureaucratic, stiff, faintly legal style:

- The use of the unfamiliar acronym *DCMOFO* gives the sentence a bureaucratic touch, especially with its repetitions of the acronym.
- The repetition within the sentence, especially of the phrase *population or habitats*, reinforces the bureaucratic, even stuffy, tone.
- The delay of the main subject and verb (*the DCMOFO will require*) until after the long introductory *if* clause forces readers to absorb, comprehend, and remember too much information at once. Consequently, the sentence is more difficult to read than it should be.
- Some of the phrasing is clumsy and ill-placed. The phrase *in the leasing area* comes so late in the opening clause that its meaning is fuzzy. Is the DCMOFO in the leasing area? Are the populations and habitats in the leasing area? Have any or all of these been identified in the leasing area?

For an example of a well-structured long sentence, see SENTENCES.

Sentence Structure

Sentence structure—including grammatical structure, the sequence of ideas, and the various repeated word patterns—all contribute to the style of a sentence or passage. The following versions of the same basic sentence say much the same thing, but their different structures create different effects:

(1) We considered how best to present the conflicting data and our interpretations of these conflicts.

(2) How to present the conflicting data, as well as our interpretations of these conflicts, was under consideration.

(3) Because of conflicting data and differing interpretations, we were considering different presentation strategies.

(4) We were considering different strategies for presenting the conflicting data and our interpretations of the data.

(5) It was difficult to decide on strategies for presenting the conflicting data and the differing interpretations of the data.

Sentences 1 and 4 are the most direct (and they happen to be the ones that follow most closely normal English word order). Sentence 2 is formal, even stuffy, because its opening clause is so long that the verb *was* is almost lost. Sentence 3 is fairly ordinary, even though it opens with the conditional *because* clause. Sentence 5 is perhaps the most stuffy; it opens with a false subject (see FALSE SUBJECTS), and it avoids all pronouns (see PRONOUNS).

The structural patterns for a single sentence present a broad range of possibilities. Putting sentences together increases the possibilities exponentially.

A string of formal, oddly structured sentences not only slows down readers but conveys a tone of formality or stuffiness. A string of short, direct sentences can sound clean and efficient (or abrupt and efficient, depending on the context).

Other Stylistic Choices

Many other features in a document besides words and sentences can convey a particular style.

The basic format of a document is usually significant. A document with narrow margins, single-spaced text, and long paragraphs conveys a dense, information-packed but potentially dull image. Readers may consider the language heavy and ponderous. A document with generous margins and lots of open space makes readers feel that the writing is open and inviting, easier to read.

Besides format choices, many other features influence a reader's perception of a document:

- The typeface used for the text
- The type of paper—both weight and texture
- The number and quality of the visual aids
- The care with which the proofreading and editing has been done
- The professionalism of the binding and the quality of the printing
- The presence or absence of color

Individual Style

We all have styles of speaking and writing that are unique to us, regardless of the circumstances in which we speak or write. This fact reflects the basic and pervasive nature of style.

In speech, an individual's style is easy to identify. Most of us can recognize a close friend, not only from the sound of the friend's voice, but also from the structure and content of the speech, from the words chosen, and from the sentence patterns used. We know our friends as talkative, quiet, abrupt, cheerful, depressed, thoughtful, humorous, or tactful. In writing, an individual's style may be harder to identify, and yet it exists in each choice the individual has made to produce a document.

Style and Ineffective Writing

Ultimately, style is the writer. We can't describe a universally preferable style because the decisions writers make depend on the context: the subject or content, the purpose of the document, the readers, previous or related documents, and the situation or climate in which the document is produced.

Nevertheless, good writing is distinguishable from bad writing, and you should never confuse bad writing with style.

Good writing is clear, emphatic, well organized, and concise. Bad writing is often vague or confusing, unemphatic, chaotic, and wordy.

Good writers obey the principles of effective writing—regardless of subject matter, purpose, readers, context, style, or tone.

For a review of those principles, see this *Guide,* particularly the sections on SENTENCES, PARAGRAPHS, ORGANIZATION, EMPHASIS, ACTIVE/PASSIVE, STRONG VERBS, FALSE SUBJECTS, and KEY WORDS.

For information on writing specific types of documents, see LETTERS, MEMOS, REPORTS, and SUMMARIES. See also the demonstration models in the last part of this *Guide*. These models will not reflect all stylistic choices or all possible styles, but they do embody the principles of effective business and technical writing.

SUMMARIES

Summaries are abridgements or compendiums of the important points in a document. Summaries are essential for readers who don't have time to read the entire document, are not interested in reading the entire document, or need to review the important points without reading the entire document.

Traditional summaries appear at the end of documents, especially those organized scientifically (see ORGANIZATION). These summaries present the main points from the preceding discussion.

Recently, **executive summaries** have begun to appear at the beginning of documents, especially in documents organized according to the managerial format (see ORGANIZATION). These summaries preview the main points that will appear in the discussion that follows.

Traditional Summaries

These summaries briefly repeat the major ideas, especially conclusions and recommendations. They include little, if any, background information and no supporting data or detail of any kind. They are typically shorter than executive summaries, which may include some background and supporting information.

Traditional summaries should not be separated from the rest of the document. They are a final summation, so they depend on information presented earlier in the document. Everything in them must already have been stated earlier.

The writer of a traditional summary often assumes that readers will use the summary after having read the rest of the document.

Executive Summaries

Executive summaries—which are also known simply as summaries—appear at the beginning of documents. They usually include the following:

- Background/introduction to the document. Although very brief, this section gives readers enough information so that they'll understand the reason for the document, the key problems addressed, and any special conditions or situations that the reader should be aware of. (See ORGANIZATION and INTRODUCTIONS.)

- Main conclusions. These may be a little longer than in traditional summaries because they are not repeated from conclusions presented earlier. Therefore, in addition to the conclusion itself, you may need a little explanation or elaboration. Just keep it short.

- Recommendations, if any. Again, you are not repeating recommendations presented earlier, so each recommendation may require some explanation. If appropriate, tie each recommendation to the specific conclusion or conclusions that prompted it.

- A review of data (optional). This section is limited to pertinent items, not all the data. The complete data usually appear in an appendix. However, you may need to present key data in an executive summary so that readers are aware of the key supporting information that your document presents. Remember that readers may read only the executive summary, not the entire document. Give them enough detail to substantiate your conclusions, but not so much that the rest of the document becomes unnecessary.

In some instances, business and technical documents may consist only of an executive summary, with supporting or explanatory material located in appendices or attachments. Such documents reflect the attempt of many businesses to limit documentation to the essentials.

Summaries vs. Abstracts

Scientific and technical writers often include abstracts with their reports. Abstracts are condensations of a document, usually written so that readers can preview the content of the document to determine whether they are interested in reading the entire document. In a sense, abstracts are extended titles.

Abstracts may contain much of the same information found in summaries, but abstracts are meant to be detached from their

SUMMARIES

documents. Frequently, abstracts are published separately, in a catalog or list of abstracts. Sometimes, journals request that potential authors submit abstracts of their articles. The editors read the abstracts to determine which articles they want to read in full.

Summaries, on the other hand, are an integral part of the documents they belong to, and they are not normally detached.

By convention, abstracts are rarely longer than one paragraph (about 200 words). Summaries have no conventional length restrictions. They should be concise, of course, but summaries may range from 50 to 5000 words, depending upon the length of the parent document. An extensive report or multi-volume proposal could very well have an executive summary of 20 pages.

See REPORTS and ORGANIZATION. See also the demonstration models located at the back of this *Guide*.

How To Write Summaries

You may wish to produce a preliminary summary of a long or complicated document. A preliminary summary will help you organize your thoughts and determine your most significant points. However, the final summary (the one appearing in your document) should be written last. Here is a procedure for writing good summaries (especially executive summaries):

1. Read through the entire document. Ensure that you have a firm grasp of the document's purpose, scope, point of view, and major ideas.

2. Identify the major ideas and data. You can underline or circle major ideas, or put stars beside them, or highlight them with a highlighting pen. Identify all of the major statements: observations, conclusions, recommendations, key supporting data, key facts, etc.

3. Pull the major ideas together and note how they are developed sequentially through the document. If necessary, refer to details in the text to clarify any points not absolutely clear.

4. Condense by combining sentences, generalizing, eliminating unnecessary supporting information, and eliminating unnecessary words and phrases. Use simple, direct sentences, and eliminate all unnecessary jargon and big words. Keep your writing simple and straightforward.

5. Use transitional words and phrases to link ideas and provide a smooth flow of thought from one sentence to the next.

6. Test the result by challenging every word, phrase, sentence, and idea. If something isn't pulling its weight, get rid of it.

7. Challenge the overall summary. Does it accurately reflect the content of the whole document? Have you left out major ideas? Does the summary distort any facts, relationships, conclusions, or recommendations? Does the summary provide ample information for readers to comprehend the ideas without reading the whole document? Can the summary stand by itself? If readers read nothing but the summary, will they be adequately informed?

Remember that the purpose of the summary is not to convey <u>everything</u>. Summaries convey the essentials. If readers want additional support or proof, they should read the rest of the document. But make sure that the summary provides readers with a firm grasp of the major ideas.

TABLES

Tables are information displays organized by rows and columns.

Tables allow writers to present precise data: *3.1415, 9.8690, 31.0035,* etc. Such data cannot be presented as precisely in any other type of visual aid (see VISUAL AIDS). Tables allow for quick and accurate comparisons and can also depict trends and relationships, although not as well as charts and graphs (see CHARTS and GRAPHS).

Tables are useful primarily because they provide for very organized displays of precise data. Tabular information is usually more understandable than the same information presented in text.

Use tables when you must present a large amount of information, when you need to give readers the exact figures, or when you want readers to be able to compare figures or other information prèsented in different rows and columns.

For general information on using visual aids, see VISUAL AIDS. See also CHARTS, GRAPHS, ILLUSTRATIONS, MAPS, and PHOTOGRAPHS. For information on captions, see CAPTIONS.

For a much more elaborate discussion of tables, including printing considerations, see the *United States Government Printing Office Style Manual.* (See REFERENCES.)

Parts of Tables

The standard parts of tables are the following: the **table number** and **caption**, the **boxhead** (containing column headings), the **stub** (containing row headings), the **field** or **body, rules**, **footnotes**, and the **source line**. See figure 1 for a sample table layout.

Table Number and Caption

Number tables sequentially as they appear in the document. Number them separately from figures.

Figure 1. Sample table layout

Table Number — NO. **201.** PERCENT OF POPULATION ENGAGED IN PHYSICAL EXERCISE, BY TYPE OF EXERCISE, SEX, AND AGE: 1975 — *Ca*

SEX AND AGE	Total population (1,000)	Percent exercising regularly [1]	PERCENT, BY TYPE OF EXERCISE [2]						
			Ride bicycle	Calisthenics	Jog	Lift weights	Swim	Walk	All other
Total, 20 years and over	**135,655**	**48.6**	**10.9**	**13.5**	**4.8**	**3.4**	**11.8**	**33.8**	**6.8**
20–44 years	71,084	53.7	16.1	17.3	7.3	5.4	16.9	33.8	6.9
45–64 years	43,145	43.4	6.5	10.8	2.7	1.5	8.0	32.9	6.5
65 years and over	21,426	42.3	2.9	6.1	1.2	(B)	2.8	35.7	6.9
Male, 20 years and over	**63,665**	**48.5**	**10.8**	**13.5**	**7.2**	**6.3**	**13.3**	**32.5**	**6.4**
20–44 years	34,268	52.7	14.9	17.5	10.6	10.1	18.8	31.4	6.2
45–64 years	20,567	42.0	6.7	10.1	3.8	2.6	8.1	31.4	5.9
65 years and over	8,830	47.3	4.3	5.9	2.1	(B)	4.1	39.4	8.1
Female, 20 years and over	**71,990**	**48.7**	**11.1**	**13.5**	**2.7**	**.8**	**10.5**	**35.0**	**7.1**
20–44 years	36,816	54.6	17.2	17.1	4.1	1.1	15.0	36.0	7.5
45–64 years	22,579	44.6	6.4	11.4	1.6	(B)	7.8	34.2	7.1
65 years and over	12,595	38.7	1.8	6.3	(B)	(B)	1.9	33.0	6.0

Stub — *Be* — *Fi*

B Base less than minimum required for reliability. [1] Regular exercise is any exercise done on a weekly basis. [2] More than one type of exercise can be reported per person. — *Footnotes*

Source: U.S. National Center for Health Statistics, *Health, United States, 1976–1977.* — *Source line*

TABLES

Unless you are presenting raw data, try to make the table caption an action caption. Use telegraphic, title-like captions, however, for tables listing large amounts of raw information.

The table caption should clearly identify the table and tell readers how to read or interpret the table:

Table 1. The industrial gas shipment decline since 1980.

Table 2. Wood panel products (1975-84): Production has more than doubled in 10 years.

Table 3. Increasing arsenic concentrations in groundwater, Sonoma County, California, 1985.

These captions indicate both what the tables are about and how readers should interpret them. If you are presenting a large amount of raw data, however, and your purpose is to present, not interpret, raw data, then use shorter, title-like captions:

Table 1. Industrial gas shipments (1976-84)

Table 2. Wood panel product production (1975-84)

Table 3. Arsenic concentrations in groundwater, Sonoma County, California, 1985

Unless the caption is too long (more than two full lines), place it above the table, just after or below the table number (see table 1). If you use lengthy captions for tables, place them below their tables. However, if you place captions below tables, use a short title with the table number and center the table number and title above each table.

If a unit of measurement applies throughout the table, you may state the unit of measurement in the caption or within parentheses beneath the caption. If the unit of measurement appears below the caption, you can subordinate it by printing it in a smaller typesize.

See CAPTIONS.

Boxhead

The boxhead contains the stub heading and the column headings (see figure 1). Place a rule above and beneath the boxhead to separate it from the table number and caption and from the field. Make column and stub headings as concise as possible. Use more than one line, if necessary, to complete a column heading, but try not to exceed three lines for any heading. Orient the headings horizontally.

If appropriate, include units of measurement in the headings or enclose the units of measurement within parentheses below the headings. If the column headings require more than one line and you place units of measurement below the headings, separate the headings and units with a thin rule (see tables 2 and 3). Use thicker rules around the boxhead itself.

Stub

The stub is the left-hand column. Use it to label the rows. Make the row headings as concise as possible. As necessary, put row units of measurement either within parentheses following the row heading (see table 4) or in another column beside the stub.

If the rows consist of major and subordinate items, place the major items flush left and indent the subordinate items (see table 1).

Field

The field consists of the data rows and columns below the boxhead and to the right of the stub. If necessary for clarity, use leaders (rows of periods) between rows to facilitate reading (see tables 2, 3, or 4). Align the data presented in each column by placing words left flush within columns, integers (*e.g., 40*) right flush by digit, and real numbers (*e.g., 40.0* or *40.068*) vertically by decimal point.

Rules

Rules are horizontal or vertical lines that separate parts of the table. Always place rules between the boxhead and the field. Also, place a rule above the boxhead to separate the boxhead from the table number and caption. If you use footnotes or a source line, place a rule between them and the bottom of the field.

If your table is large, you may need to place rules between groups of rows. Typically, the rules appear between groups of five. These rules help readers follow information down large tables.

The rules surrounding the boxhead should be thicker than those appearing within the field.

NOTE 1: Instead of placing rules after every fifth row, you may leave an extra line to separate groups of rows.

NOTE 2: Some authorities also use vertical rules to separate the stub from the field and groups

TABLES

Table 1. U.S. machine tool consumption, selected numerically controlled machines, 1979-82, (in millions of current dollars.)

Item	1979	1980	1981	1982
Machining Centers				
Production	356.5	413.0	482.6	339.5
Exports	45.4	56.7	53.4	34.0
Imports	39.6	93.4	195.7	188.9
Consumption	350.7	449.7	624.9	494.4
Imports as share of consumption	11.3%	20.8%	31.3%	38.2%
Horizontal spindle turning machines				
Production	284.4	321.7	346.6	238.6
Exports	n.a.	20.5	20.4	30.4
Imports	n.a.	159.8	278.0	194.1
Consumption	n.a.	461.0	604.2	402.2
Imports as share of consumption	n.a.	34.7%	46.0%	48.2%
Punching and shearing machines				
Production	82.5	110.6	97.1	59.0
Exports	n.a.	13.2	28.3	8.3
Imports	n.a.	16.5	32.5	25.5
Consumption	n.a.	113.9	101.3	76.2
Imports as share of consumption	n.a.	14.5%	32.1%	33.4%

Note: Data in this table differ from those in other tables due to different source of production data and exclusion of data on machine tool parts.

Sources: Bureau of the Census, "Current Industrial Report for Metalworking Machinery," MQ-35W; "U.S. Imports for Consumption," IM146; and "U.S. Exports," EM522.

of columns from each other. The trend today is to eliminate all vertical rules.

Footnotes

Use footnotes to clarify the headings, identify unfamiliar abbreviations or units of measurement, or explain the data appearing within the field.

Use superscripted footnote numbers or letters (¹, ², ³) or (ᵃ, ᵇ, ᶜ) or symbols (*, **, ***) to link the footnoted information in the table and its explanation (see tables 2 through 4). If the footnote applies to the entire table, you do not need a footnote symbol (see the note beneath table 1).

The footnotes should appear below the table, beginning flush left. If a footnote extends across the table and you have many footnotes, break the footnote references into two columns and print them in a smaller typesize, if possible (see table 4).

Source Line

The source line identifies the source of the information presented in the table. Source lines always appear below footnotes and should be aligned with the footnote references (see tables 1 through 4).

Rules for Using Tables

1. Keep tables as simple as possible.

Tables can become complicated quickly, so simplify them as much as possible. If the information you are trying to present becomes too complicated, break it up into two or more tables.

2. Place important tables in the body of the document; place unimportant tables in appendices or attachments.

Tables conveying critical information must appear in the text. However, if you are presenting tables of raw data, consider placing them in an appendix or attachment. Do not force readers to ponder raw data unless they want to. (See EMPHASIS and APPENDICES/ ATTACHMENTS.)

3. Use table numbers and captions to identify tables and to help readers understand the information presented.

Table numbers help readers track tables through a document and help them find important tabular information.

Captions label the information found in a table and can tell readers how to read the table. The caption is your opportunity to influence the reader's perception of the information you present. You should take advantage of that opportunity and write an action caption. (See CAPTIONS.)

4. Orient tables horizontally on the page.

Try to orient tables horizontally, so that readers do not have to reorient the page to read the table. This means that most tables can be longer rather than wider (that is, they can have more rows than columns). If you need to have more columns

TABLES

than rows, consider redesigning your table or breaking it into two tables. As a last resort, orient the table sideways with the boxhead toward the left-hand side of the page.

5. Clearly label tables that extend beyond one page, and use continuation headings on continued pages.

If you must continue a table from one page to another, write *continued* at the bottom right-hand corner of the table on each page to be continued. Then on the top of each new page where the table has been continued, write, for instance, *Table 4, continued*.

Finally, repeat the boxhead on each continued page.

Table 2. Production and imports of crude oil, natural gas, and natural gas plant liquids, 1960-84.

Year	Production				Gross Imports	
	Crude oil[1] (billions of barrels)	Natural gas[2] (trillions of cu. ft.)	Natural gas plant liquids (billions of barrels)	Total (quadrillions of Btu's)	Crude oil (billions of barrels)	Natural gas (trillions of cu. ft.)
1960	2.575	12.23	0.340	29.05	0.373	0.16
1970	3.517	21.01	0.606	44.58	0.482	0.82
1972	3.455	21.62	0.638	44.85	0.813	1.02
1973	3.361	21.73	0.634	44.25	1.183	1.03
1974	3.203	20.71	0.616	42.25	1.270	0.96
1975	3.057	19.24	0.596	39.74	1.497	0.95
1976	2.976	19.10	0.587	39.07	1.936	0.96
1977	3.009	19.16	0.590	39.35	2.413	1.01
1978	3.178	19.12	0.572	40.17	2.321	0.97
1979	3.121	19.66	0.578	40.47	2.380	1.25
1980	3.146	19.60	0.576	40.42	1.925	0.98
1981	3.129	19.40	0.587	40.15	1.605	0.90
1982	3.157	17.75	0.566	38.55	1.273	0.93
1983[3]	3.161	16.28	0.562	37.26	1.278	0.86
1984[4]	3.126	17.41	0.567	38.23	1.636	0.90

[1] Includes lease condensate.
[2] Net dry natural gas, including nonhydrocarbon gases.
[3] Estimated.
[4] Forecast.
Source: Energy Information Administration.

6. Use white space, boxes, or lines to separate tables from surrounding text.

Tables are visual aids and should be placed as thoughtfully as any other visual aids. Separate them from surrounding text by leaving three lines above and below tables (more white space on the page), by placing the tables within boxes (or frames), or by using thin rules above and below the tables to separate them from the text. (See EMPHASIS.)

7. Organize rows and columns logically so that they reflect the purpose of the table and make the table easy to read.

The arrangement of rows and columns will depend upon the information being presented and upon the purpose of the table.

Use a chronological order for information presented by time or date or sequence. Use a whole-to-parts pattern for major and subordinate items. Use logical grouping for items that should appear together because of type, size, or relationship. (See ORGANIZATION.)

Place information that you want readers to compare in adjacent rows or columns. As much as possible, avoid forcing readers to compare or contrast information separated by other rows or columns.

Place information that must be numerically tallied or compared in columns. Performing mathematical operations across

TABLES

Table 3. Prices and price indexes for U.S. imported and domestically produced crude oil, 1972-84. (Price in dollars per barrel: index 1972-100.)

Year	Refiner acquisition cost of U.S. imported crude		Refiner acquisition cost of U.S. produced crude		Wellhead price of U.S. produced crude	
	Price	Index	Price	Index	Price	Index
1972	3.22	100.0	3.67	100.0	3.39	100.0
1973	4.08	126.7	4.17	113.6	3.89	114.7
1974	12.52	388.8	7.18	195.6	6.87	202.7
1975	13.93	432.6	8.39	228.6	7.67	226.3
1976	13.48	418.6	8.84	240.9	8.19	241.6
1977	14.53	451.2	9.55	260.2	8.57	252.8
1978	14.57	452.5	10.61	289.1	9.00	265.5
1979	21.67	673.0	14.27	388.8	12.64	372.9
1980	33.89	1,052.5	24.23	660.2	21.59	636.9
1981	37.05	1,150.6	34.33	935.4	31.77	937.2
1982	33.55	1,041.9	31.22	850.7	28.52	841.3
1983[1]	29.13	904.7	28.97	789.4	26.37	777.9
1984[2]	29.00	900.6	29.00	790.2	NA	NA

[1] Estimated.
[2] Forecast.
NA = not available
Source: Energy Information Administration.

rows is much more difficult than performing the same operations down columns.

When time is one of the variables, you can arrange it along rows or columns. However, time is traditionally displayed on the horizontal or x-axis of charts, so you should try to display time across columns (see table 1). (Note, however, that tables 2 and 3 are also easy to read. Time is displayed in rows in these tables because displaying time in columns would create wider, rather than longer, tables. See rule 4 above.)

8. Use row headings (in the stub) and column headings (in the boxhead) to identify the information listed in each row and column.

Tables without adequate headings are often incomprehensible. Ensure that your headings are concise and yet descriptive. If necessary, use footnotes to explain or fully describe the headings.

9. Use rules to separate the boxhead from the field, and, as necessary, use rules to separate groups of data columns or rows from each other.

Complex tables are much easier to read and to follow when the information is separated by rules. Use thicker rules to separate the boxhead and the stub from the field and thinner rules to separate groups of rows and columns.

10. Where space is limited, use abbreviations.

Abbreviations are appropriate in tables (as well as other visual aids). Use them where space is limited. If the abbreviation is uncommon, explain it in a footnote. (See ABBREVIATIONS.)

11. Base comparable numerical amounts on the same unit of measurement, and convert all fractions to decimals.

Numbers that readers will want or need to compare must be presented in the same unit of measurement. Do not present some numbers in meters and similar numbers in centimeters, or some data in feet and comparable data in yards.

Further, convert all fractions to decimals and use real numbers (numbers with decimals) as necessary for accuracy. However, do not make numbers more precise than accuracy allows. In other words, do not give more than the significant digits in a real number. If you do, you will convey a false sense of accuracy.

See DECIMALS and FRACTIONS.

TABLES

12. Align decimals vertically within columns.

Except where you are mixing types of numbers, ensure that decimals are aligned vertically. The exception occurs in columns containing different types of real numbers, as in table 1. The percentage figures are distinct from the production figures and should not be aligned, although both sets of data contain decimals. Note that the percent signs clearly distinguish the percentages from the production figures.

Table 4. Brick and structural clay tile (SIC 3251): trends and projections 1972-84, (in millions of dollars except as noted.)

Item	1972	1977	1979	1981	1982[1]	1983[2]	Compound annual rate of growth 1972–83	1984[3]	Percent change 1983–84
Industry data									
Value of shipments[4]	513.0	777.7	915.9	738.6	659.0	987.0	6.1	—	—
Value of shipments (1972 $)[4]	513.0	469.9	433.7	306.1	263.2	363.0	−3.1	390.0	7.4
Total employment (000)	24.1	20.5	22.9	17.0	14.1	13.9	−4.9	17.0	22.3
Production workers (000)	20.4	17.0	18.5	13.5	10.8	10.8	−5.6	14.0	29.6
Average hourly earnings of production workers ($)	3.05	4.39	5.37	6.16	6.58	6.65	7.4	—	—
Capital expenditures	38.6	49.6	78.4	68.5	—	—	—	—	—
Product data									
Value of shipments[5]	471.3	715.3	853.0	666.1	595.0	863.0	5.7	—	—
Value of shipments (1972 $)[5]	471.3	432.2	403.9	276.0	245.0	338.0	−3.0	363.0	7.4
Product price index (1972 = 100)	100.0	164.8	210.9	241.8	251.1	272.5	9.5	—	—
Quantity shipped (billion bricks)	8.4	8.7	7.7	5.1	4.4	6.1	−2.9	—	—
Trade									
Value of exports	2.0	5.3	4.5	6.8	4.9	6.8	11.8	—	—
Value of imports	7.2	22.8	23.5	18.0	16.7	20.0	9.7	—	—
Export/shipments ratio	0.004	0.007	0.005	0.010	0.008	0.008	—	—	—
Import/new supply ratio[6]	0.011	0.024	0.022	0.022	0.027	0.023	—	—	—

[1] Estimated except for product price index, exports, and imports.
[2] Estimated.
[3] Forecast.
[4] Value of all products and services sold by industry SIC 3251.
[5] Value of shipments of brick and structural clay tile products produced by all industries.
[6] New supply is the sum of product shipments plus imports.
Source: Bureau of the Census and Bureau of Industrial Economics. Estimates and forecasts by the Bureau of Industrial Economics.

13. Use zeros, dashes, ellipses, or *NA* to indicate that information is missing or not applicable.

Do not leave entries in rows or columns blank where the information is unavailable or not applicable. Readers will not know whether the blank entry was intentional.

Instead, indicate missing or inapplicable information by entering a dash, a zero, an ellipsis mark, or the abbreviation *NA*.

NOTE: Use the abbreviation *do* (for *ditto*) where a data entry in a column is the same as the data entry directly above it.

14. Use footnotes to explain or clarify table headings and entries.

Tabular information is governed as much by space limitations as it is by content necessity, and you may not be able to explain fully a table heading or entry in the space available. So use footnotes where necessary for clarification. (See VISUAL AIDS.)

Footnotes belong below the table and are usually flush left. Footnotes usually have the following format: superscripted reference number or letter indented, second or additional lines flush with the left-hand margin, etc. See tables 2, 3, and 4.

If you are referencing numerical entries located within the field, use footnote letters or symbols rather than numbers, which some readers may confuse with the number being referenced. However, use a consistent system of reference. If you establish a system of footnote letters, then use that system throughout all of your tables.

See the example tables with this section. (See also FOOTNOTES.)

15. Identify all data sources.

For the sake of clarity, as well as substantiation, identify all data sources. The source line may follow the standard bibliographic format (see table 1) or may identify the source by name alone (tables 2 through 4). (See BIBLIOGRAPHIC FORM and CITATIONS.)

Whether you provide full source information depends upon the nature of the source documents, the extent to which you have summarized from the original, the purpose of your document, and the purpose of the table. If readers will want to examine the original sources, then provide full source information. However, if you have gathered data from a variety of sources or documents, you may not be able to identify specific documents.

If you have examined a number of EPA documents, for instance, and collated data from many of those documents, your source line should not list every single source. Instead, you should simply indicate:

> Source: Environmental Protection Agency

Begin source lines with the word *source* or *sources*, followed by a colon. End the source line with a period.

TABLES OF CONTENTS

A table of contents helps readers in two ways: (1) It outlines the structure of the document and thus provides insight into the document's organization, and (2) it provides the page numbers for all sections and subsections, thus helping readers to locate parts of the document.

Creating a preliminary table of contents is a useful writing technique because writers have to think carefully about the document's organization. Gaps, illogical order, and misplaced emphasis become more apparent when the writer is forced to clarify and complete a preliminary table of contents.

The final table of contents cannot be prepared until the document is finished and the pages numbered.

1. Use a table of contents for any report longer than 10 pages.

The 10-page figure is arbitrary, but remember that a table of contents helps readers to see the overall organization of the document as well as to find key sections. Documents under 10 pages are generally short and uncomplicated enough for readers to determine the structure by skimming through the document before reading. However, skimming through longer documents may not provide an adequate sense of structure because the reader's mind is being asked to comprehend too much information spread over too great a distance.

2. Include major divisions (often chapter headings) and the next level of subdivisions in the table of contents.

Major divisions and the next level subdivisions are essential if readers are to comprehend the document's structure. Make the division titles as specific as possible:

this

II. Testing for Flammability

Temperatures
Duration
Flash Points

not this

II. Testing

NOTE: Ensure that all of the divisions and subdivisions that appear in the table of contents also appear in the body of the document.

3. Use letters and numbers with division and subdivision titles in the table of contents only if you also use them in the text.

A table of contents can resemble an outline with page numbers, but such an outline structure (*I, A, 1, a,* etc.) is not necessary unless the chapter or section titles and the headings in the document reflect the same numbering system. Tables of contents often have a decimal numbering system (*1.1.1, 1.1.2, 1.1.3,* etc.) in place of the standard outline system. Either system is acceptable.

See OUTLINES and NUMBERING SYSTEMS.

4. Use blank lines, indentation, and leader dots to lay out the table of contents and help readers find page numbers.

Leave **blank lines** between major entries, and ensure that the spacing reflects the logical structure of the document. For instance, you may want to leave two lines between major divisions (first-level headings) and one line between major subdivisions (second-level headings). Readers should be able to tell where major divisions occur simply by noting the number of lines between entries.

Use **indentation** to show levels of subordination. Major divisions (first-level headings) should be flush left. Major subdivisions (second-level headings) should be indented five spaces. Minor subdivisions (third-level headings) should be indented 10 spaces, and so on. Note the table of contents example accompanying this section.

Finally, consider using **leader dots** (rows of spaced periods) to connect entries with their page numbers. Leader dots allow readers to trace the connection across the page.

To emphasize major divisions or subdivisions, you might also use all capital letters, boldface type, and other emphatic techniques. (See EMPHASIS.)

The accompanying table of contents example shows one method of displaying the organizational structure of a document and of providing page numbers to help readers

TABLES OF CONTENTS

locate the document's parts. For a further example, see the table of contents to the *Business Writer's Quick Reference Guide*.

5. Include preliminary material and appendices or attachments in the table of contents.

Some writers ignore the preliminary material and attachments when they construct their tables of contents. However, a table of contents should reflect the structure of the entire document, so include all of the document's parts.

TABLE OF CONTENTS

TELEX

Telex and TWX are just two of the automated communication systems available today. These services, both offered by Western Union, are the most widely used telegraph services in American business. Other kinds of electronic mail systems and services are now available, including rapid transmission of information over computer networks and via satellite transmission. The discussion below pertains to Telex and TWX, but the principles apply to most other electronic mail systems.

Telex is a low-cost international system, with a network covering over 120 nations. Its terminals are teletypewriters with speed limitations that change as equipment improves. Businesses use the telex for orders, confirmations, and other information transmissions where the message is relatively brief.

TWX is a Western Union service that supplements Telex. TWX is faster and its keyboard is larger, offering more characters than on the Telex keyboard. Because of its speed and flexibility, TWX is better suited to the transmission of long documents.

---

1. Clearly identify the subject, any code or transmittal numbers, the addressee(s), and the sender(s).

Use the proper coding system (usually present on most preprinted forms). Always double check any code or transmittal numbers for accuracy.

---

2. Determine and then work within any limitations in the available keyboards and typefaces.

Telex messages, for instance, require all capitals, so distinctions based on capital letters (vs. lowercase letters) are impossible to make. A message written in all capital letters is often difficult to read, so keep your sentences and paragraphs short and use numbered lists whenever possible.

Some teletypewriters, especially those used for Telex messages, have only a few symbols. Therefore, you will have to write out most symbols: *percent* for %, *dollars* for $, and so on.

---

3. Repeat important words to ensure clarity and accuracy. Also, write out important numbers:

> Our firm does not, repeat not, want to subscribe to the proposed leasing project.
>
> We are sending you our first payment of twenty-one thousand five hundred and twenty-one dollars.

Data transmission errors on Telex and TWX are rare, but they still occur often enough to warrant concern for accuracy. You can't transmit *$10,000* and be entirely certain that the receiver won't get *$100,000*. So when accuracy matters (as in sums of money or numbers of items ordered), write out the numbers.

---

4. Clearly indicate deadlines, including whether and when you expect a response.

Both sender and receiver should be sensitive to the need for prompt processing. Messages sent electronically are normally those messages requiring prompt action. So, whenever appropriate, the sender should indicate deadlines, and the addressee should have office procedures that guarantee an answer within a reasonable period (usually under 24 hours).

As you set deadlines, be sure to account for time zone differences.

TITLES

Titles of people, organizations, governments, and publications often require special capitalization, punctuation, and other format conventions. (See NOUNS.)

1. Capitalize titles when they immediately precede personal names, but do not capitalize them when they follow personal names:

Mrs. Robert T. Evans
Mr. Edward Johnson
Miss Sylvia Smead
Ms. Josephine Kukor
President Amy Kaufmann
Assistant Professor Ned Davis
Mayor-elect Boon Hollenbeck
General Laswell Hopkins
Lieutenant Cynthia Wagner
the Reverend John Tyler
Rabbi Tochterman

Amy Kaufmann, the president of Union College, spoke to the press.

Ned Davis is an assistant professor at Columbia University.

We voted for Boon Hollenbeck, who is now mayor-elect.

Laswell Hopkins was our general for only two months.

Our lieutenant was Cynthia Wagner. (*Here* was *separates the title from the name.*)

NOTE 1: The titles of high-ranking international, national, and state officials often retain their capitalization, even when the name of the individual is either absent or does not follow the title:

The President spoke before the Congress.

We wrote the Vice President.

The Pope toured South America.

The Governor still had two years to serve.

The Prime Minister of Ghana was invited to the White House.

NOTE 2: Titles of company or corporation executives, as well as titles of lesser federal and state officials are sometimes capitalized. Such capitalization is unnecessary, but you should follow company or agency practice:

The Mayor announced an end to the New York transit strike. (*or* The mayor)

The Vice President for Finance is resigning Monday, September 18.

The Superintendent refused to approve our budget request.

NOTE 3: Titles used in a general sense are not capitalized:

a U.S. representative
a king
a prime minister
an ambassador

2. Capitalize the names of companies, schools, organizations, and religious bodies:

the Johnson Wax Company
the University of Oregon
the Young Men's Christian Association
the Urban League of Detroit
the Republican Party
St. John's Lutheran Church

NOTE: The words capitalized are those normally capitalized in any title (see CAPITALS). The initial *the* in most such titles is not capitalized unless the company, school, organization, or religious body has established the initial *the* as part of its legal name: *The Johns Hopkins University, The Travelers Insurance Company.*

3. Capitalize the names of government bodies:

the United Nations
the Cabinet
the Bureau of the Budget
the California Legislature
the Ohio Board of Education
the Davis County Commission

NOTE: Except for international and national bodies, shortened forms of these government bodies or common terms are not capitalized:

the House (*for* House of Representatives)
the Department (*for* Department of Agriculture)
the Court (*for* the U.S. Supreme Court)
the police department
the county council
the board of education

4. Capitalize and italicize (or underline) the titles of books, magazines, newspapers, plays, movies, television series, and other separately published works:

Oliver Twist (*book or movie*)
Newsweek
the *New York Times*
West Side Story
Superman
NOVA

See UNDERLINING and ITALICS.

TITLES

5. Capitalize and use quotation marks for chapters of books, articles in magazines, news stories or editorials, acts within a play, episodes of a television series, or other sections of something separately produced or published:

The last chapter was called "The Final Irony."

"The Colombian Connection" was the lead article in last week's *Time*.

We supported his editorial, "A Streamlined Election System."

We watched "The Fatal Circle" last night on *Gunsmoke*.

See QUOTATION MARKS, UNDERLINING, and ITALICS.

TONE

Your tone reflects your attitude toward your subject and your readers. Your writing may strike your readers as personal or impersonal, friendly or distant. You may sound warm and engaging or cold and abrupt.

Your style reflects your disposition as a writer and the choices you make while writing: the words you choose, the way you structure your sentences, whether you feel comfortable using personal pronouns, whether you lecture to readers or invite them to join you in considering ideas.

See STYLE.

Style and tone are often confused. Some people use the terms interchangeably, but one is the cause and the other the effect:

> **Style** refers to those writers' choices that create the tone readers perceive. Style is the writer's "manner of speaking," the way the writer uses language to express ideas.
>
> **Tone** refers to the feeling or impression a document conveys to its readers. It is one of the products of the writer's style.

Tone, then, is the impression readers receive from your writing, the attitude conveyed in your treatment of the subject. We usually describe the tone of a piece of writing with these types of words:

abrasive	formal
aggressive	forthright
assertive	friendly
authoritative	impersonal
blunt	informal
bureaucratic	informative
casual	objective
cold	officious
condescending	personal
courteous	polite
demanding	sincere
discourteous	stiff
distant	subjective
earnest	threatening
engaging	warm

Desirable Business Tone

Most of the time, your business documents should be:

courteous	informative
forthright	personal
friendly	polite
helpful	sincere
informal	warm

The extent to which your documents are personal will depend upon your relationship with the reader. But never fail to be courteous, polite, informative, sincere, and helpful—especially when you don't know the reader.

If you need to convey negative information, as in a poor performance appraisal or reprimand, or in a document threatening legal action, your document may need to be:

assertive	impersonal
formal	objective

However, you should never write documents that are:

abrasive	condescending
blunt	discourteous
bureaucratic	officious
cold	

Good business documents—no matter how tough or adversarial

TONE

they are—should never be discourteous.

Below are the stylistic choices that will help you write business letters and memos that have an effective tone. Not coincidentally, these are the same rules that make writing clear, concise, and easy to read.

1. Use pronouns to establish a personal, human tone in letters and memos.

Probably no single language choice is as effective in making business documents sound human and personal as well-chosen pronouns. Of the pronouns possible, *you* is the most important. You should always be aware of your readers and address them directly:

this

During the discussion of your April bill, you mentioned that you had called your local service representative at least three times during the month. Do you remember the representative's name and the dates when you called?

not this

Concerning the April bill, the local service representative may have been called, but these calls cannot be verified unless the representative's name and the dates when the representative was called are provided to this office.

The ineffective version has no personal pronouns; consequently, the reader is ignored. Omitting personal pronouns makes the letter cold and informal. The passive verbs contribute to the impersonal tone and help make the letter sound unfriendly at best. (See ACTIVE/PASSIVE.)

Next to *you*, the pronouns *I* and *we* are essential for effective letters and memos. Some people argue that the writer should not be mentioned in documents. They argue that documents should not reflect personal opinions or the personality of the author. This argument fails to distinguish between personal opinions and personal responsibility for one's actions. Contrast these two examples:

this

Based on the data, I (*or* we) conclude that MOGO should plug and abandon the Gilbert Ray Well 3.

not this

Based on the data, it is concluded that MOGO should plug and abandon the Gilbert Ray Well 3.

The second version is mechanical, almost robotic. No person seems to have acted. The conclusion simply occurred, like something out of the Twilight Zone. The result is a faceless, anonymous tone, one calculated to avoid responsibility and perhaps to confuse readers or keep them deliberately in the dark.

2. Make your letters and memos sound very much like the language you would choose if you actually talked with the readers.

Read your document aloud. Would you be comfortable saying those words to someone in person? If you delivered the message to readers orally, would you express yourself this way?

The tone of a good business document is a natural one. It isn't full of slang or homey conversational expressions (*Well shucks, I reckon we ought to drill anyways*), but it should sound natural, not forced or contrived. If the document does not sound natural, if it is stiff and complex, if it is formal and faceless, you should rethink your tone.

A business document is not a transcription of your actual words, pauses, corrections, and other verbal lapses (*Well, uh, I think that, uh, if we, uh . . .*). It should, however, be similar to the way you talk. The words, phrases, and sentences you use should be simple and direct, even though you have edited and revised them:

this

I recommend that we immediately replace the roof on the Bradley building.

not this

It is recommended that the roof on the Bradley building be replaced forthwith.

this

Before leaving the room, turn out the lights.

not this

Prior to evacuating the premises, ensure that the illumination has been terminated.

The simplest remedy for overly stiff, bureaucratic writing is to write like a human being. Don't write like an officious, faceless

TONE

bureaucrat. Just be yourself. Imagine that you're talking to other people in person. Try to sound human, not mechanical.

Using personal pronouns will help considerably. Here are some other suggestions:

- Keep your sentences short and direct. Challenge any sentence that is longer than 30 words, and try to keep your average sentence length down to about 20 words. You can often break a long sentence into two shorter sentences. Ensure that each sentence, whatever its length, is as clear and direct as possible. (See SENTENCES.)

- Avoid long, unnecessarily complex words and unnecessary technical terms. (See GOBBLEDYGOOK and JARGON.) Never use words because you think those words are impressive. The writer who struggles to sound intelligent and educated (especially by using a thesaurus) often winds up sounding silly:

  this

  We think your water pipes have corroded so much that only a trickle of water can flow through them.

  not this

  Our hypothesis is that your water supply system has undergone severe corrosion and reached the debilitating point where water normally available is unavailable in the quantity and at the pressure provided for in the original specifications for your domicile.

Leaving aside the laughable words (*debilitating* and *domicile*), the second version still suffers from terminal wordiness. Isn't *pipes* better than *water supply system*? And *corroded so much* better than *undergone severe corrosion*? What do the inexact references to quantity and pressure accomplish that the word *trickle* doesn't do more vividly? The idea being expressed does not require technical terms, especially if the primary reader is a homeowner, not an engineer.

Legal documents often suffer from the same kind of wordiness and unnecessary complexity:

not this

In accordance with the provisions of the aforementioned procedure, the attached conveyance should be executed by you in triplicate, with the signature duly witnessed and attested to by a Notary Public, and the executed set of conveyance forms should then be returned to this office on or before, and no later than, Friday, May 23.

this

According to the procedure outlined above, please sign all three copies of the conveyance and have your signature notarized. Then return the completed forms to this office by Friday, May 23.

---

3. Choose sentence structures that reflect a friendly, conversational tone.

- Avoid passive sentences:

  this

  Our review of your claim indicates that you should receive a refund of $72. This refund would apply to 1982 charges.

  not this

  Your claim has been reviewed and it has been determined that $72 should be refunded to you for the period January to December 1982.

  ---

  this

  We analyzed the drilling reports for the source of the discrepancies. Over 90 percent of the discrepancies turned out to be simple errors in daily recording.

  not this

  The drilling reports have been analyzed to determine the source of the discrepancies. It is concluded that over 90 percent of the discrepancies were caused by simple errors in daily recording.

  See ACTIVE/PASSIVE.

- Avoid false subjects:

  this

  Unless we address these issues during this quarter, they will distort our financial report for the entire year.

  not this

  There are certain issues that we should address during this quarter that will distort our financial report for the entire year.

  ---

  this

  Under MOGO'S policies, we will not acquire additional drilling pipe until after the beginning of the new fiscal year.

  not this

  It is likely that, given MOGO'S policies, it will be impossible to acquire additional drilling pipe until after the beginning of the new fiscal year.

  See FALSE SUBJECTS.

TONE

4. Include personal information and personal references.

Readers like to know that you have addressed their needs. So, if appropriate, include information from previous letters, memos, or discussions in your document. Or include information they have either requested or will need:

> I recommend that you file a complaint with the Federal Trade Commission. Your review of the relevant correspondence with the company persuaded me that you have a case.

A mechanical, yet easy, way to make a letter or memo personal is to include the reader's name in the body of the letter:

> If we proceed, Cal, you should ask Product Research for a copy of their assessment form. I think you'd find it helpful.
>
> *or*
>
> So, Beth, if you have any more suggestions, please call me at ext. 3578.

5. Choose your paper, typeface, and format to reflect a personal, friendly tone.

Even physical concerns can affect the tone of a letter or memo.

Choose good quality paper (usually 20-pound rag paper) and a pleasing typeface. Now that most firms have word processors and computers, the typeface options are growing. Know your options. Your goal is to capture the personal, friendly quality of the content of your letter (consider Helvetica or Optima). Some typefaces are too rigid and stark (American Typewriter). Others are too informal (Script, which attempts to look like cursive writing).

Next, design your document so that it has a lean and open look, one conveying a personal, friendly tone. This usually means using generous margins, short paragraphs (on the average), headings, lists, and lots of white space. (See EMPHASIS.)

TRANSITIONS

Transitions are words or phrases that connect ideas and show how they are related. Occurring between two sentences or paragraphs, a transition shows how the sentences or paragraphs are connected, thus making the writing smoother and more logical. A transition creates a point of reference for readers, allowing them to see how the writing is organized and where it's heading.

Following is a list of transitions and their functions.

A List of Transitions

Addition

additionally, again, also, besides, further, furthermore, in addition, likewise, moreover, next, too, what is more

Comparison or Contrast

by contrast, by the same token, conversely, however, in contrast, in spite of, instead, in such a manner, likewise, nevertheless, otherwise, on the contrary, on the one hand, on the other hand, rather, similarly, still, yet

Concession

anyway, at any rate, be that as it may, even so, however, in any case, in any event, nevertheless, of course, still

Consequence

accordingly, as a result, consequently, hence, otherwise, so, then, therefore, thus

Diversion

by the way, incidentally

Generalization

as a rule, as usual, for the most part, generally, in general, ordinarily, usually

TRANSITIONS

Illustration

for example, for instance

Place

here, there, near, nearby, close

Restatement

in essence, in other words, namely, that is

Summary

after all, all in all, briefly, by and large, finally, in any case, in any event, in brief, in conclusion, in short, in summary, on balance, on the whole, ultimately

Time and Sequence

after a while, afterward, at first, at last, at the same time, currently, finally, first (second, third, etc.), first of all, for now, for the time being, immediately, instantly, in conclusion, in the first place, in the meantime, in time, in turn, later, meanwhile, next, presently, previously, simultaneously, soon, subsequently, then, to begin with

See CONJUNCTIONS, ORGANIZATION, and PARAGRAPHS.

Punctuation and Transitions

1. Use commas to separate transitions from the main body of a sentence:

However, uncontrolled R&D efforts that do not design quality into the product, process, or service may be of dubious value.

Consequently, development areas will include high performance ceramic seals and ceramic-plate material characterization.

The regulator poppet is, however, normally held open by the regulating spring.

Transitions interrupt and are generally not part of the main thought of the sentence; therefore, you should separate them from the rest of the sentence. When a transitional word occurs where two complete thoughts are joined, as in the following sentence, use a semicolon where the two complete thoughts join (usually in front of the transition) and a comma after the transition:

The scope of this study will not permit a review of all technologies; however, we believe our experience in advanced ship design will allow us to maximize the study of those areas with greater operation potential.

See SEMICOLONS and COMMAS.

2. If the transition interrupts the flow of a sentence, place commas on both sides of the transitional word or phrase:

A lightweight airframe, for instance, partially offsets poor propulsion technology.

The sensor probe, on the other hand, contains thermistor sensing elements.

3. If a transitional word occurs at the beginning of a sentence and is essential to the meaning of the sentence, do NOT separate it with a comma:

However warm air enters the cabin, the conditioned temperature will not rise above the nominal range.

NOTE: Such sentences can be confusing, so you should consider rephrasing the sentence:

Regardless of how warm air enters the cabin, the conditioned temperature will not rise above the nominal range.

UNDERLINING

In typed material, underlining replaces italics as a tool for highlighting certain unusual words and phrases. (See ITALICS.) You can also underline words, phrases, and sentences to emphasize them.

1. Underline words used as words:

> The words affect and effect are often confused.
>
> The contract stipulated that monitoring must be continuous, but during negotiation they stated that periodic monitoring would suffice. Should the contract read continual or periodic rather than continuous?

NOTE: Single letters, words, and even phrases should be underlined to separate them from the ordinary words within a sentence:

> The phrase come hell or high water has a long and interesting history.

Some editors prefer to use quotation marks for words and phrases used unusually within a sentence. However, quotation marks clutter up a sentence if several words are being talked about:

> The forms "am," "is," "are," "was," and "were" don't even resemble "be," which is the principal form of the verb.

2. Underline foreign words and phrases that have not been absorbed into English:

> After a coup d'oeil, the detective was ready to question the suspect. (Coup d'oeil means "a quick survey.")

NOTE: Contrast coup d'oeil, which is clearly not part of English, with coup d'état, which is now so familiar that no underlining is necessary. Many modern dictionaries fail to specify whether a word should be considered foreign or not, so you may have to use your own judgment.

3. Underline the titles of separate publications:

> I bought a copy of the Wall Street Journal.
>
> Have you read the novel All Quiet on the Western Front?
>
> The Sunshine Patriot is a pamphlet being circulated by the Republican Party.
>
> We attended a preview of The Story of the Bell System.
>
> We have tickets for the opening night of Aida.

NOTE: Underline separately published or produced items, but use quotation marks for book chapters, articles in magazines, or sections of the separate works:

> The latest issue of the Oil and Gas Journal contained an article entitled "Shearing Problems with Sucker Rods."
>
> The Salt Lake Tribune had an editorial entitled "A Bold Proposal."

See QUOTATION MARKS and TITLES.

4. Underline the names of aircraft, vessels, and spacecraft:

> U.S.S. Constitution
> H.M.S. Bounty
> Gemini 4

5. Underline words and phrases for emphasis:

> Please send two copies to us by Friday, October 13 at the latest.
>
> Unscrew the fitting by turning it clockwise, not counter-clockwise.

NOTE: Use underlining for emphasis sparingly. Underlined text is difficult too read, and the effect diminishes quickly if you overuse it. (See EMPHASIS.)

UNITS OF MEASUREMENT

Units of measurement include either English units (also called the U.S. customary system) or metric units. Many U.S. firms still favor English units, but metric units, especially the SI system, are now widely used by scientists and engineers. (See METRICS.)

1. Use the following common English units to measure length and area.

NOTE: The abbreviations for these units of measurement usually appear only in tables, charts, graphs, and other visual aids. (See ABBREVIATIONS.)

2. Use the following common English units for volume or capacity and weight.

NOTE 1: Although the United States and the British Commonwealth both use the same names for units and their

LENGTH

English Unit	U.S. Equivalents	Metric Equivalents
inch	0.083 foot	2.540 centimeters
foot	⅓ yard, 12 inches	30.480 centimeters
yard	3 feet, 36 inches	0.914 meter
rod	5½ yards, 16½ feet	5.029 meters
mile (statute, land)	1,760 yards, 5,280 feet	1.609 kilometers
mile (nautical, international)	1.151 statute miles	1.852 kilometers

AREA

English Unit	U.S. Equivalents	Metric Equivalents
square inch	0.007 square foot	6.452 square centimeters
square foot	144 square inches	929.030 square centimeters
square yard	1,296 square inches, 9 square feet	0.836 square meter
acre	43,560 square feet, 4,840 square yards	4,047 square meters
square mile	640 acres	2.590 square kilometers

VOLUME OR CAPACITY

English Unit	U.S. Equivalents	Metric Equivalents
cubic inch	0.00058 cubic foot	16.387 cubic centimeters
cubic foot	1,728 cubic inches	0.028 cubic meter
cubic yard	27 cubic feet	0.765 cubic meter

English Liquid Measure	U.S. Equivalents	Metric Equivalents
fluid ounce	8 fluid drams, 1,804 cubic inches	29.573 milliliters
pint	16 fluid ounces, 28.875 cubic inches	0.473 liter
quart	2 pints, 57.75 cubic inches	0.946 liter
gallon	4 quarts, 231 cubic inches	3.785 liters
barrel	varies from 31 to 42 gallons, established by law or usage	

English Dry Measure	U.S. Equivalents	Metric Equivalents
pint	½ quart, 33.6 cubic inches	0.551 liter
quart	2 pints, 67.2 cubic inches	1.101 liters
peck	8 quarts, 537.605 cubic inches	8.810 liters
bushel	4 pecks, 2,150.420 cubic inches	35.239 liters

WEIGHT

English Avoirdupois Unit	U.S. Equivalents	Metric Equivalents
grain	0.036 dram, 0.002285 ounce	64.798 milligrams
dram	27.344 grains, 0.0625 ounce	1,772 grams
ounce	16 drams, 437.5 grains	28.350 grams
pound	16 ounces, 7,000 grains	453.592 grams
ton (short)	2,000 pounds	0.907 metric ton (1,000 kilograms)
ton (long)	1.12 short tons, 2,240 pounds	1.016 metric tons

Apothecary Weight Unit	U.S. Equivalents	Metric Equivalents
scruple	20 grains	1,296 grams
dram	60 grains	3.888 grams
ounce	480 grains, 1.097 avoirdupois ounces	31.103 grams
pound	5,760 grains, 0.823 avoirdupois pound	373.242 grams

UNITS OF MEASUREMENT

abbreviations, the two systems do differ, so be cautious in interpreting publications using these units. Here, for example, are the equivalents between the British Imperial units and the U.S. English units:

British Imperial Liquid and Dry Measure	U.S. English Equivalents	Metric Equivalents
fluid ounce	0.961 U.S. fluid ounce, 1.734 cubic inches	28.413 milliliters
pint	1.032 U.S. dry pints, 1.201 U.S. liquid pints, 34.678 cubic inches	568.245 milliliters
quart	1.032 U.S. dry quarts, 1.201 U.S. liquid quarts, 69.354 cubic inches	1.136 liters
gallon	1.201 U.S. gallons, 277.420 cubic inches	4.546 liters
peck	554.84 cubic inches	0.009 cubic meter
bushel	1.032 U.S. bushels, 2,219.36 cubic inches	0.036 cubic meter

VERBS

Verbs are the key action words in most sentences. They tell what the subject has done, is doing, or will be doing, and they indicate the subject's relationship to the object or complement. Because verbs also signal time through their different tenses (forms), they are potentially the most important words in a sentence. For instance, varying only the verb in a sentence produces major shifts in the meaning:

She shows us her report.
She showed us her report.
She will show us her report.

She has shown us her report.
She had shown us her report.
She will have shown us her report.

She is showing us her report.
She was showing us her report.
She will be showing us her report.

These nine sentences only begin to illustrate all the possible verb forms. If we include more complex verb phrases, the possibilities multiply:

She is going to be showing us her report.

She must have been showing us her report.

Principal Verb Forms

Verbs commonly have several standard forms (often called principal parts) from which all the other verb forms are built:

base form	call eat cut
-s form (3rd person singular present—he, she, it)	calls eats cuts
past form	called ate cut
-ed participle (past participle)	called eaten cut
-ing participle (present participle)	calling eating cutting

Regular verbs, like *call*, routinely require only an *-s*, *-ed*, or *-ing* to change the base form. If a dictionary does not supply any forms except the base form, the verb is regular, like *call*.

Irregular verbs, like *eat* and *cut*, are unpredictable, so writers have to know the different forms, not just follow the regular pattern. Dictionaries include these irregular forms in their entries for these verbs.

NOTE: Unfortunately, not all verbs are clearly regular or irregular. The following verbs,

VERBS

for example, have two different forms of the past participle, one regular, the other irregular:

mow, mowed, mown (*or* mowed)
show, showed, showed (*or* shown)
swell, swelled, swollen (*or* swelled)

---

1. Check a recent dictionary to determine the correct forms for any verb you are unsure of.

Here, for instance, are the main forms for some of the common irregular verbs:

buy, bought, bought
cost, cost, cost
drink, drank, drunk
freeze, froze, frozen
keep, kept, kept
lead, led, led
lie, lay, lain
light, lighted/lit, lighted/lit
rise, rose, risen
sell, sold, sold
sit, sat, sat
speak, spoke, spoken
spoil, spoilt/spoiled, spoilt/spoiled
take, took, taken
tear, tore, torn
think, thought, thought
wet, wet/wetted, wet/wetted
write, wrote, written

NOTE: Where two forms exist, the regular forms (with *-ed*) are becoming more common. Over time, many irregular verbs have changed and are changing into regular verbs.

Verb Tenses

Verbs have the following basic tenses or times:

Basic Verb Tenses

Present:	They study.
Past:	They studied.
Future:	They will study.
Present Perfect:	They have studied.
Past Perfect:	They had studied.
Future Perfect:	They will have studied.

Then a parallel set of progressive forms exists, which indicates that the action is continuing:

Progressive Verb Tenses

Present:	They are studying.
Past:	They were studying.
Future:	They will be studying.
Present Perfect:	They have been studying.
Past Perfect:	They had been studying.
Future Perfect:	They will have been studying.

Finally, a parallel set of passive verb tenses also exists:

Passive Verb Tenses

Present:	The report is studied.
Past:	The report was studied.
Future:	The report will be studied.
Present Perfect:	The report has been studied.
Past Perfect:	The report had been studied.
Future Perfect:	The report will have been studied.

NOTE: As in the above sentences, passive verb sentences highlight the object or thing receiving the action, not the person or thing performing the action. The passive sentences do not identify the person who is, was, or will be studying the report. In most cases, you should prefer the active voice and avoid the passive. (See ACTIVE/PASSIVE.)

---

2. Vary your verb tenses to reflect the often complex data in your writing.

This rule contradicts what you may have learned in grammar school: "Don't mix your tenses." Actually, you can and should vary your tenses to reflect the often complicated time relationships of your subject:

Yesterday, we <u>analyzed</u> (*past*) the samples for any traces of zinc ore. We <u>found</u> (*past*) none. Today, however, we <u>were reexamining</u> (*past progressive*) the sample when we <u>found</u> (*past*) two promising pieces of rock. They <u>have</u> (*present*) veins like zinc ore, although their color <u>is</u> (*present*) not quite right. Our report <u>will</u> therefore <u>show</u> (*future*) the potential presence of zinc.

Most writers choose their tenses unconsciously, but several basic conventions exist for selecting tenses in technical writing:

—Record in the past tense experiments and tests performed in the past:

The second run produced flawed data because the heating unit failed. We did not detect the failure until the run was almost over.

—Use the present tense for scientific facts and truths:

Water freezes at 32°F., unless a chemical in the water changes its freezing point.

Newton discovered that every action has an equal and opposite reaction.

VERBS

—Use the present tense to discuss data within a published report:

> The slope of the temperature curve decreases sharply at 20 minutes. The figures in table 3-14 document this decrease.

—Shift from present to past tense as necessary to refer to research studies and prior papers. When you are discussing an author and his or her research, use the past tense:

> Jones (1976) studied a limited dose of the drug. He concluded that no harmful side effects occurred.

—When you are discussing different current theories, use the present tense:

> Jones (1976) argues that limited doses of the drug produce no harmful side effects. His data, however, is flawed because he failed to distinguish between the natural and synthetic versions of the drug.

Auxiliary Verbs

Auxiliary verbs are the most common verbs in the English Language: *is, are, was, were, be, been, can, could, do, did, has, have, had, may, might, shall, should, will, would, must, ought to,* and *used to.*

Auxiliaries are crucial to many of the tenses presented above, but auxiliaries also can function by themselves as main sentence verbs:

> The tests are complete.
> He did the primary drawings.
> They have no budget.

See STRONG VERBS.

Verbs and Agreement

The verb should agree in number with its subject. So a plural subject requires a plural verb, and a singular subject requires a singular verb:

> The geologist has completed the tests. (*singular*)
> The geologists have completed the tests. (*plural*)
>
> A test was completed last week.
> Several tests were completed last week.
>
> The report analyzes the impact.
> The reports analyze the impact.

See AGREEMENT.

---

3. Ensure that your verbs agree with your subjects.

The only circumstance in which verbs change their forms to adjust to different numbers is in the third person forms of the present tense:

> She works every day. (*singular*)
> They work every day. (*plural*)
>
> She is the candidate. (*singular*)
> They are the candidates. (*plural*)
>
> He has the answer. (*singular*)
> They have the answer. (*plural*)
>
> It is broken. (*singular*)
> They are broken. (*plural*)

NOTE 1: Third person singular verbs have an *-s* ending, as in *works* above. The third person plural verbs have no *-s*, as in *work*. So the rule for verbs is the opposite of nouns: the forms with *-s* endings are the singular forms.

NOTE 2: The verb *be* is exceptional because it changes in the present tense to agree with different pronouns:

> I am studying.
> You are studying. (*singular*)
> He, she, it *is* studying.
>
> We are studying.
> You are studying. (*plural*)
> They are studying.

Subjunctive Verbs

Subjunctive verbs are special verb forms that signal recommendations or conditions contrary to fact. Centuries ago, subjunctives were very common verb forms, but today they are limited to the instances covered in the following two rules.

---

4. Use a subjunctive verb in *if* clauses to state a situation that is untrue, impossible, or highly unlikely:

> If I were (*not* was) the candidate, I would not agree to a debate.
>
> If it were (*not* was) raining, we couldn't conduct the experiment.
>
> If I were (*not* was) you, I would change banks.

NOTE 1: The above sentences require *were* rather than normal *was*, which would appear to agree with the subjects. This use amounts to an historical survival, so it doesn't fit our modern expectations. (Actually, other verbs in *if* clauses are subjunctive, but

VERBS

only the *were/was* pattern looks or sounds exceptional.)

NOTE 2: If the *if* clause states something that is possible or likely, then do not use a subjunctive:

> If he leaves this job, he'll get $500 in severance pay.
>
> If it was an error and I suspect it was, then we'll have to pay you damages.

5. Use subjunctive verbs in sentences making strong recommendations or demands, or indicating necessity:

> I recommend that the case be settled by Tuesday.
>
> He demands that the money be refunded.
>
> The court has resolved that the witness be found in contempt of court.
>
> It is essential that he leave by noon.
>
> They urge that she return the money.
>
> They resolved that Dan write the termination letter.

NOTE 1: As the first three sentences show, if the verb in the *that* clause would normally be *am, is,* or *are*, then its subjunctive form is *be*.

NOTE 2: As the last three sentences show, if the verb in the *that* clause is normally a third person singular verb, then its subjunctive form does not take the usual *-s* ending.

VISUAL AIDS

Visual aids are one of the writer's best devices for emphasizing information. Because they are visual rather than verbal (as writing is), visual aids are much more emphatic than the written text around them.

They stimulate the reader's interest in the topic, they focus the reader's attention, and they aid the reader's understanding of the information being presented.

Visuals can emphasize important data and ideas in ways that text cannot. Visuals can be dramatic, revealing, stimulating, even surprising. To achieve these same effects, text would have to be written far better than most writers are capable of writing.

Visuals show data in a concise and effective manner (in tables), they show how data compare or contrast (in charts), they show changing data relationships (in graphs), and they show configurations that would be difficult, if not impossible, to describe (in illustrations). Thus visuals enhance the reader's ability to understand, interpret, and remember the data visualized. (See CHARTS, GRAPHS, ILLUSTRATIONS, and TABLES.)

Because visuals are so strong, writers sometimes overuse them.

A text with inadequate visuals suffers, but so does a text with too many visuals. In an ideal document, the text and the visuals are balanced. They complement one another. The complete story is not told in either form; instead, the two forms work together in harmony to convey and emphasize the highpoints of the message being delivered.

Properly balancing visuals and the accompanying text is not difficult if you obey a few general principles.

VISUAL AIDS

1. Use visual aids to emphasize your important ideas and data, and place the visuals for maximum impact.

Visuals should capture the key recommendation, the surprising trend, the unexpected financial problem, the most convincing data—in short, use visuals for the highlights of your message.

If your purpose is to recommend a new high volume pump, then consider contrastive bar graphs showing volumes for the old and the new pump. If your purpose is to revise prior estimates of the effects of acid rain on New England lakes, design your visuals with clear before and after contrasts (perhaps a bar chart, a pie chart, or even contrastive photographs). You control your readers' minds by using appropriate visuals to control their eyes.

Beware, however, of using visuals for mere impact. By their very nature, visuals are highly emphatic, so reserve their use for important information. If you waste visuals on unimportant or unnecessary information, you will be wasting a valuable opportunity, and you may confuse or perplex your readers.

Place visuals as strategically and for as much impact as possible. The right visual should appear at the right moment. Don't let visuals come too early (before the reader can properly appreciate or comprehend them), and don't let visuals come too late (after the reader has already spent time reading and absorbing the information being visualized).

Ideally, readers should encounter the visuals just after the ideas being visualized have been introduced and stated concisely but before lengthy explanation or elaboration.

If possible, try to place visuals so that the visuals and the text concerning them are on the same page.

2. Create your visuals before you write the text.

Visuals should never be an afterthought. They are more emphatic than text and should therefore receive greater attention early in the writing process.

Create your visuals first; write your text last.

As you generate ideas and begin to focus your message, list the important ideas that you will be conveying to your readers. Then ask yourself whether and how you could visualize those ideas. Do some rough sketches. For longer documents, you might even do a mock-up of the document, which is simply a collection of projected pages, with potential visuals and accompanying text sketched in (often with mere squiggles, empty boxes, etc.).

Later, as you write the draft, return to your notes or the mock-up to check up on how and where the reader will be encountering your visuals.

3. Eliminate unnecessary visuals.

- If the visual aid duplicates information already in the text and if the visual will not significantly enhance the reader's comprehension, eliminate it.
- If the visual aid cannot present information more effectively than text, eliminate it.
- If the visual aid presents unnecessary, irrelevant, or unimportant information, eliminate it.
- If the visual aid was created for another document and does not exactly fit the information and circumstances of your document, eliminate it.
- If you are submitting a document to a journal or a publisher, eliminate all visuals that do not contribute substantially to your message, especially those visuals involving color, which is expensive to reproduce.

4. Select visuals that are appropriate for your readers.

The visuals you select should depend in part on the orientation and skill of your readers. For instance, you should not use logarithmic graphs with nontechnical readers (who will probably not understand them); conversely, you should not use simple charts to present complex technical data to highly technical readers.

VISUAL AIDS

Documents become difficult, however, when they need to be read by many readers, all of whom have different technical backgrounds and different uses for the information.

Remember that your visuals (and your text) are indirectly controlled by the least technical of your projected readers. Keep these readers in mind as you generate your visuals and your text. You may, of course, sometimes make a particular visual or a part of your text a little too difficult for such readers, but if you do this, you'll need to build in nontechnical explanations somewhere else in the text.

Some of the best writers and editors deliberately vary the readability of their text and visuals; they are assuming that different readers will read and interpret different sections of a document. Readers with much technical knowledge may survey only the key table or the most technical graph. Readers with less technical knowledge may stop after reading the introduction and the summary (or abstract).

So, writers need to analyze carefully their prospective readers—all of their readers. Once they know just who their readers are and what their technical backgrounds are, they can begin to adapt their document to these readers.

5. Select visuals that are appropriate for your topic.

Visual aids come in many forms: charts, graphs, illustrations, maps, photographs, and tables. These forms—and all of the variations within them—give you many alternatives for transforming ideas and data into visual representations.

Charts depict relationships between two or more variables, possibly at distinct points in time (bar charts). Charts can also display organizational relationships (organization charts), identify the relationship of parts to a whole (pie charts), and illustrate the flow and relationship of steps in a process (flowcharts). (See CHARTS.)

Graphs depict the relationship of two or more variables and show how those variables change. Graphs allow for comparisons and show trends. (See GRAPHS.)

Illustrations and diagrams show conceptual objects or assemblies and provide perspectives on existing objects or assemblies that photographs cannot capture. Illustrations can show exploded views, which focus attention on smaller parts of a larger object or assembly. (See ILLUSTRATIONS.)

Maps show topographical relationships and indicate scale and distance. (See MAPS.)

Photographs convey realism. They allow you to show readers exactly what something looks like. With photographs comes authenticity. (See PHOTOGRAPHS.)

Tables display data in rows and columns. They allow for quick comparisons of precise data. (See TABLES.)

The visual aid you select will depend upon your reader, but it will also depend upon what you are trying to achieve with the visual. The table on the next page shows how various visual aids can be used to accomplish a writer's purposes.

6. Keep your visuals focused and keep them simple and uncluttered.

When you are designing and constructing a visual aid, ask yourself, What is the point? What is the most important idea that I am trying to convey visually? What is my central concept?

Focus on your key purpose or concept and then build your visual aids around it.

Do not ask the visual to do two or more things. Keep it simple. Ineffective visuals typically fail because writers have not designed them with a clear concept in mind (which makes the visuals unfocused) or because writers have tried to make the visuals do too much (which makes them cluttered or too complex). Like a good paragraph, a good visual focuses on one idea—and conveys it sharply and purposefully.

Keep visuals uncluttered by eliminating all extraneous information. Your preliminary coordinate graph or the complete table of readings contains a lot of valuable information, but neither of these is likely to be effective unless you pare it down to its essentials.

VISUAL AIDS

DOCUMENT PURPOSES

Type of Visual	Costs	Causes Effects	Trends	Organizational Relationships	Policies Procedures	Decisions Alternatives	Work Flow	Chronology	Design Parts Apparatus	Comparison Contrast	Advantages Disadvantages
Tables	X	X	X	X	X	X		X		X	X
Line Graphs	X		X					X		X	
Bar Charts	X		X			X		X		X	X
Pie Charts	X		X			X				X	X
Schematic Diagrams									X		
Flowcharts		X		X			X	X			
Maps/Site Plans		X				X					
Photographs		X							X	X	X
Tree Diagrams				X	X	X	X				
Illustrations									X		
Blueprints									X		
Combination		X	X	X	X	X	X	X	X	X	X

A good visual contains nothing that is not directly related to its central concept.

7. Introduce visuals in the text before the visuals appear.

Always introduce visuals in the text. A visual that suddenly appears without introduction or explanation generally confuses readers.

The introduction should come *before* the visual. Furthermore, the introduction should be informative and specific:

informative and specific

As figure 3 shows, produced water from the 2nd Langley is much more acidic than produced water from the 1st Langley.

This project is estimated to cost $356,200. The cost breakdown in table 15 shows that hardware costs account for nearly 65 percent of total costs, while labor costs constitute only 12 percent of the total.

not informative

See figure 3.

The total cost of this project is estimated to be $356,200 (Table 15).

not specific

Figure 3 shows the chemical analysis of produced water from two wells.

The total cost of this project is estimated to be $356,200. Table 15 provides a cost breakdown.

As these examples illustrate, a good introduction indicates not only what the visual aid is about but also, at least in part, what the reader should get from it.

A good introduction tells the reader how to interpret the visual aid.

See CAPTIONS.

EXCEPTION: Sometimes, a visual aid must appear in the text before you can introduce it. This would occur, for instance, when a small table or chart appears at the top of a new page and is followed by a column of text. Rather than break up the text, you might keep the table or chart at the top of the page. The

VISUAL AIDS

introduction would then have to appear after the visual. See rule 12 below.

8. Number the visuals in the order of their appearance and use clear, active captions.

Figures include all visuals that are not tables, including graphs, charts, diagrams, illustrations, photographs, and maps. By convention, tables and figures are numbered separately. Therefore, table 5 could appear in a document after figure 20.

In a short document or a document with few visuals, number the visuals sequentially through the entire text. In a lengthy document with chapters or numbered sections, give the visuals hyphenated numbers, such as *figure 3-2*. The first number indicates the chapter or section, and the second number indicates the number of the visual within that chapter or section.

In longer documents, especially formal reports and publications (pamphlets, books, research studies, etc.), include in the table of contents a list of figures and, if appropriate, a list of tables.

A list of figures (and a list of tables) should provide, in sequence, each figure's number, caption (or title), and page number. If the document has a large number of specialized figures, then create a separate list of each special type. For instance, a lengthy report containing a number of maps might include a list of maps. Another report, in which photographs play an important role, might include a list of photographs. If the document does not warrant such special listings, however, don't create them.

Lists of tables and figures appear immediately after the table of contents (usually on separate pages), and should themselves be listed in the table of contents. If you create both a list of tables and a list of figures, either list may appear first. (See TABLES OF CONTENTS.)

The captions for your visual aids should be informative and specific. Indicative titles (those that merely indicate what the visual is about) are not as helpful:

> Figure 17. Pronghorn Antelope Trends in Montana
>
> Table 2. Particulates in Sonoma Valley, California, from May to October 1984

Action captions are informative. They state not only what the visuals are about but also what readers should learn from them:

> Figure 17. The pronghorn antelope population in Montana has declined steadily since 1971.
>
> Table 2. From May to October 1984, particulates in Sonoma Valley, California, have remained well below EPA emission standards.

Action captions generally contain a subject and a verb. They should be informative without becoming too lengthy. If necessary, you may want to shorten the caption when you list it in the list of tables or figures. (See CAPTIONS for further examples.)

9. Design visuals that are easy to read.

Design the visual so that the central concept is immediately apparent. Leave out extraneous elements that tend to draw attention away from the central concept. One good test of a visual aid's effectiveness is to ask an uninformed reader to look at the visual and identify its central concept. If this reader cannot tell you what the point is within 5 or 10 seconds, the visual is not effective.

Choose a simple typeface for the lettering on visuals, and use capitals and lowercase letters for all headings and labels. (Lettering all in capitals is hard to read, especially if it extends beyond two or three words.) Ensure also that the lettering is large enough so that if the visual is reduced, the lettering will still be readable.

Make the lines and lettering that indicate scales, axes, notes, and legends lighter than the lines and lettering indicating data points, curves, areas, or bars. Ensure that the visual aid's features do not hide or detract from the central concept.

10. Use emphatic devices to emphasize important ideas in visuals.

Emphatic devices include underlining, italics, boldface type, larger type sizes, shading, line patterns, and color. Use these devices to highlight important words and data in visuals. Furthermore, use the same emphatic device each

VISUAL AIDS

time to emphasize the same kind of word or data. Establish a typographical system and be consistent throughout all of your visuals.

Color is especially useful in helping readers understand the visual, locate information, and distinguish between different phases, parts, or configurations. Color can also highlight special features, such as cautions (usually yellow) and warnings (usually red). Using color to highlight a line, row, column, slice, area, circle, or data point can focus the reader's attention on that item.

As you select colors, try to establish a color scheme that makes sense: green for things that are prospering, yellow for things in transition, red for things that are failing. Or blues to indicate coolness, reds to indicate warmth. Use contrasting colors to show contrasting concepts or major changes; use variations of one color to show minor variations. Use the brightest colors to emphasize the most important ideas, the dullest colors to subdue the less important ideas. Finally, beware of using too much color at once and making the visual look kaleidoscopic.

11. Orient visuals horizontally on the page.

Text has a horizontal orientation, which means that it is comprehensible from the left to the right across the page (i.e., horizontally). If possible, visuals should be oriented the same way. Readers should be able to "read" visuals without turning the page sideways.

The standard page layouts for visual aids are full-page, half-page, and quarter-page formats (see the models at the end of this section).

Larger visuals (such as maps) may be printed on larger paper, folded, and inserted into a flap at the end of the document or bound into the document. Visuals that are bound into the document and that fold out for readers to view are called foldouts. Typically, foldout pages are 11 inches high and 17 inches wide (for a one-fold) or 25.5 inches wide (for a two-fold). (The widths may be adjusted to make room for the binding.)

Full- and quarter-page visuals are higher than they are wide. Half-page visuals and foldouts are wider than they are high. Separate visuals (such as those inserted into a flap) may have any dimensions.

As you design visual aids, you must consider whether the information you wish to present in the visual is consistent with the size and the dimensions of the visual aid format that is most appropriate for the information.

For instance, if you are designing a table and have twice as many columns as rows, you must use a format that allows for more width than height. Typically, you would select a half-page format. However, a half-page format may occupy too much space on a page. If the information is not that critical, consider reorienting the table by switching the columns and the rows, thereby creating a table that is higher than it is wide, which may allow you to use a quarter-page visual.

On the other hand, if your tabular information requires more columns than rows, and if you are tabulating a substantial amount of information, you may be forced to use a format that is wider than it is high. If a half-page format does not allow enough space, then you may have to orient the visual sideways and use a full-page for the table. If a full-page (oriented sideways) is still insufficient, you may have to break up the table into smaller tables or use a foldout.

If you can, avoid orienting visuals sideways. If a visual will not fit within an acceptable format, redesign the visual or divide it into workable parts.

12. In double-sided text, avoid unnecessary breaks in the text, and if possible, balance your pages so that the left-hand pages complement the facing right-hand pages.

The model page layouts shown in this section illustrate possible placements of visual aids on a page. These layouts reflect the following guidelines:

—Avoid unnecessary breaks in the text. This guideline is easy to follow if only one visual appears on a page (models 2, 10, 11, 16, and 17). If more than one visual appears on a page, then text should not be broken (models 4, 5, 7, 8, 15, and 18). Model 14 is an exception; in it, the writer wants to separate the visuals in the first column from the single visual in the second column.

VISUAL AIDS

PAGE LAYOUT MODELS

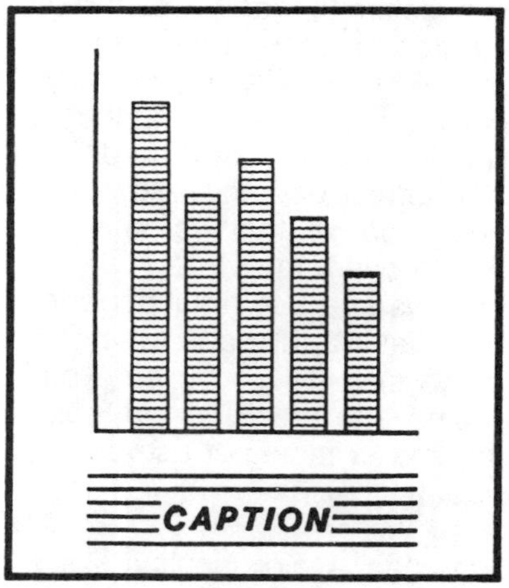

Model 1

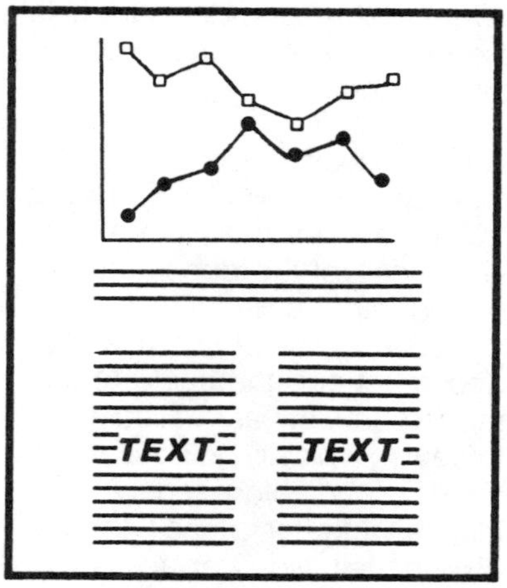

Model 2

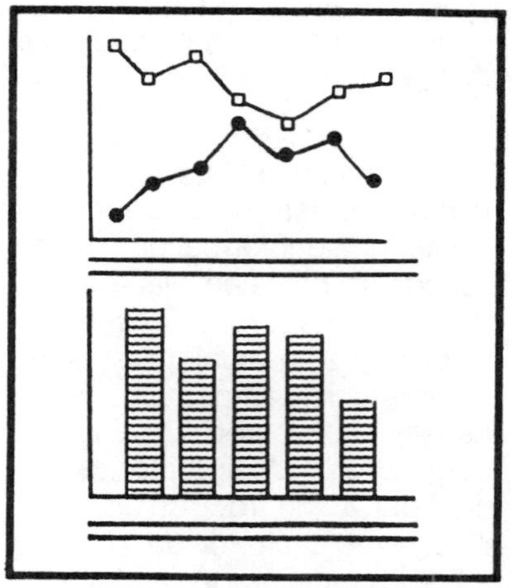

Model 3

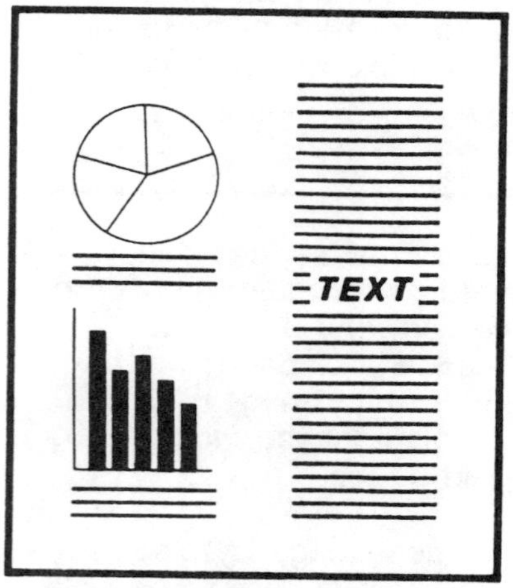

Model 4 Left-Facing

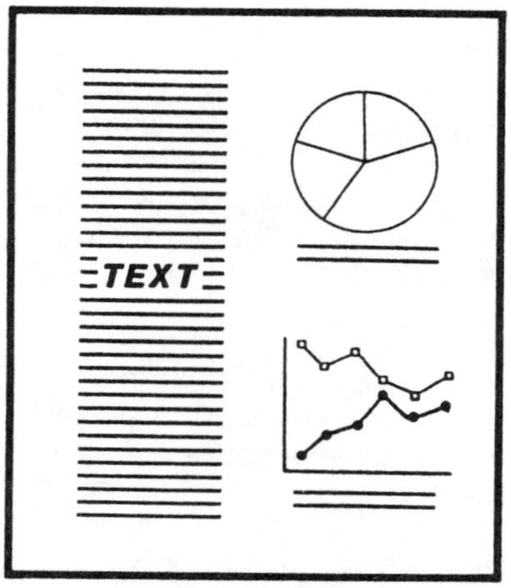

Model 5 Right-Facing

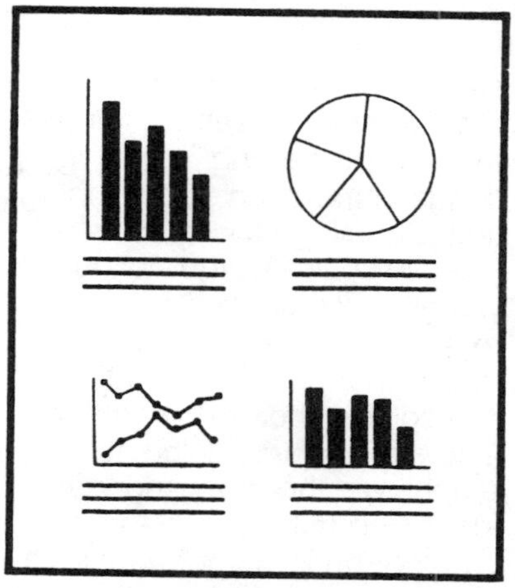

Model 6

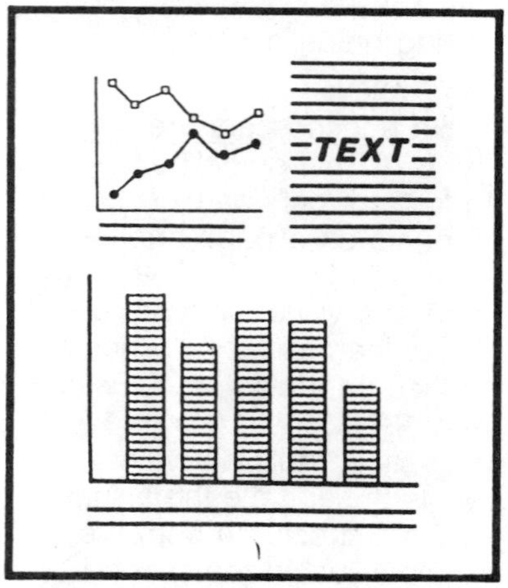

Model 7 Left-Facing

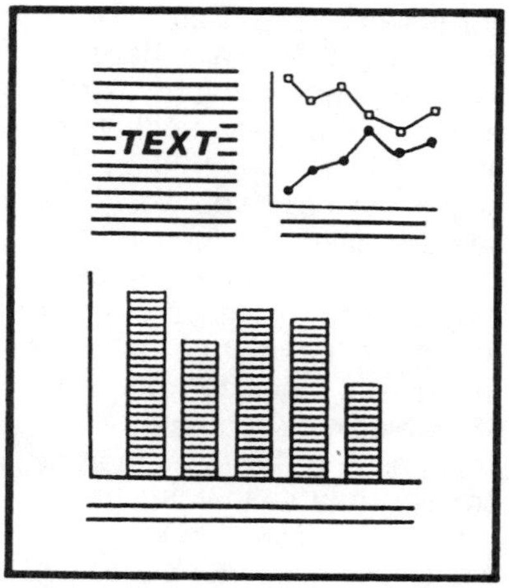

Model 8 Right-Facing

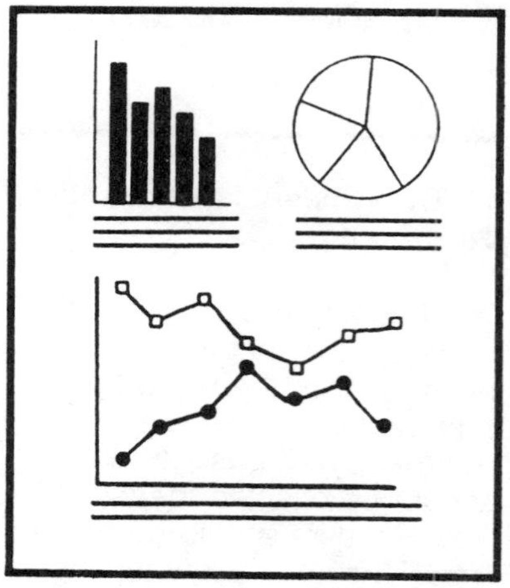

Model 9

VISUAL AIDS

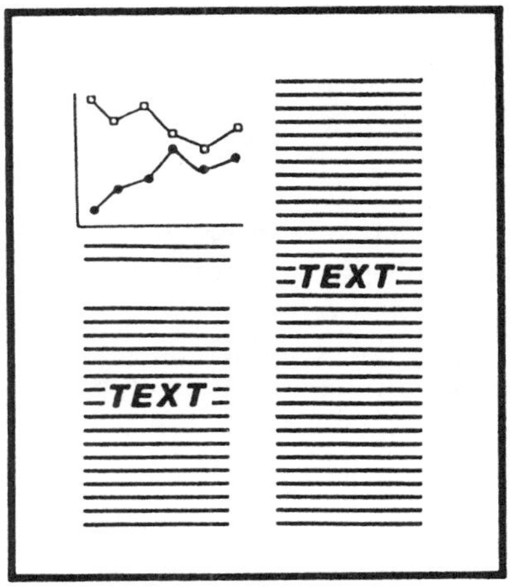

Model 10 Left-Facing

Model 11 Right-Facing

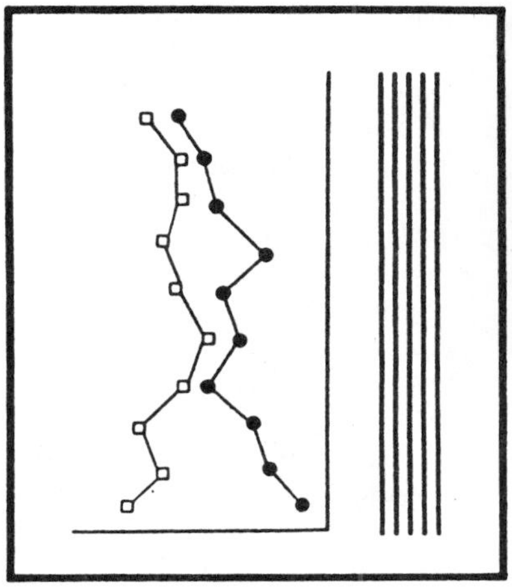

Model 12

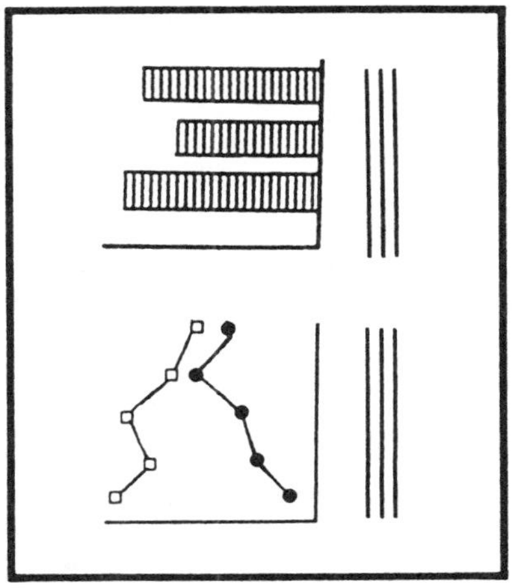

Model 13

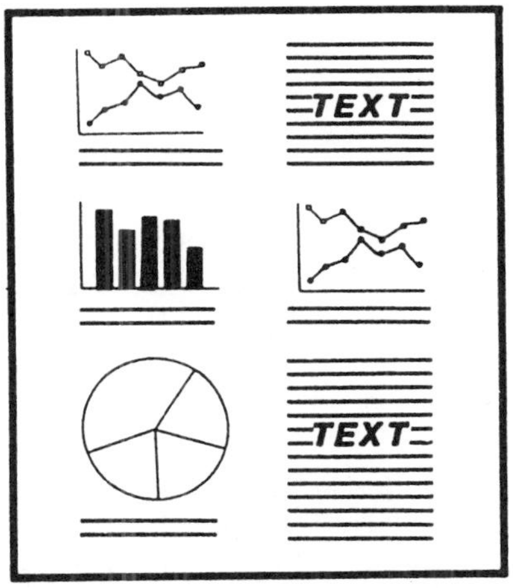

Model 14 Left-Facing

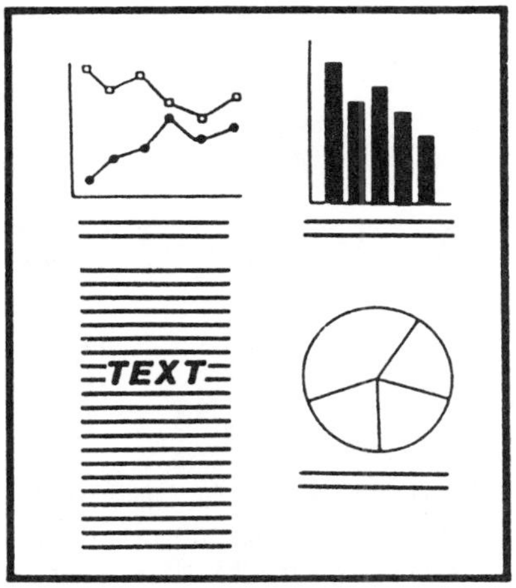

Model 15 Right-Facing

Model 16

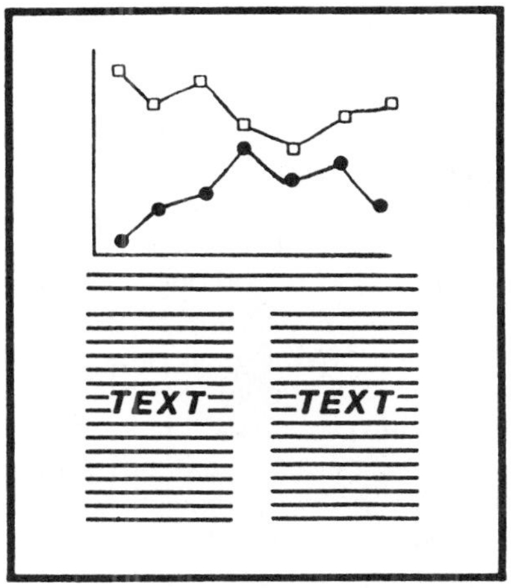

Model 17

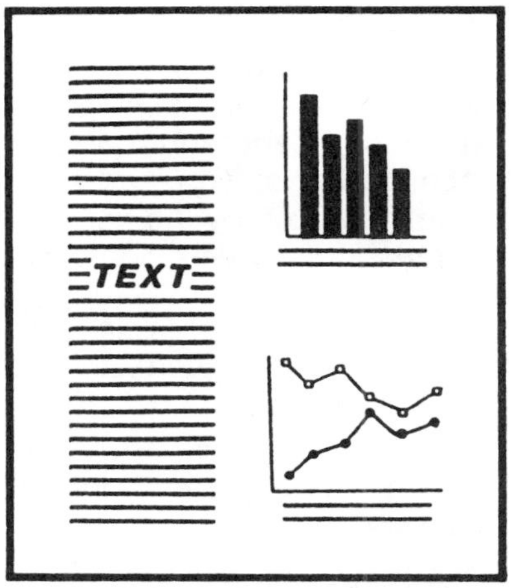

Model 18 Right-Facing

VISUAL AIDS

—Move visuals to the top and outside edges of the page (the left side of left-hand pages and the right side of right-hand pages). Models 2, 4, 5, 7, 8, 10, 11, 15, and 18 illustrate this guideline. Model 16, used as a left-hand page, would be an exception; but the writer might choose this option if the reference to the visual came in the middle of the second column of text. The visual needs to appear as soon after its introduction as possible.

— Balance visuals on facing left- and right-hand pages. Models 4 and 5, 7, and 8, and 10 and 11 are the best examples of balanced facing pages. Balanced pages mirror each other, so if a left-hand page opens with a visual in the upper left-hand corner, the facing right-hand page will have a visual in the upper right-hand corner. Such balanced pages are difficult to achieve, especially given the normal variations in text as well as visuals of different sizes and types. Models 14 and 15, for instance, partially balance each other even though model 14 has an additional visual on it.

NOTE: These guidelines do not apply to single-sided pages. On single-sided pages, you should place visuals as close as possible to their introductions. See rule 7 above.

WORD PROBLEMS

Words are symbols representing persons, places, things, actions, qualities, characteristics, states of being, and abstract ideas. Words are the substance of language.

To use words well, you must know what the words mean (both their **denotations** and **connotations**); you must select words that convey the right impression to your readers; you must use specific, concrete words whenever possible; and you must use the correct words for the context.

Denotations and Connotations

The denotation of a word is what that word signifies or stands for, what it explicitly represents. The connotation of a word is something suggested by the word, something implied.

The word *pig*, for instance, denotes an animal of certain characteristics, specifically, a swine. *Pig* also denotes an earthenware crock, a device used to clean out pipes (a *piggot*), and an oblong mass of metal (as in *pig iron*). *Pig* even came in the 1960s to denote a person in authority, particularly a police officer, but this sense has now become less common. As with the word *pig*, many words have multiple, often changing, denotations. *Pig* connotes any of those attributes commonly associated with swine that may be applied to persons: sloppiness, filthiness, gluttony, sloth, immorality, or unwholesomeness.

What a word explicitly stands for may constitute only a small part of the word's meaning. Thus connotations play an important role in English. They help our language expand, adapt to changing needs, and exercise the flexibility that makes it such an expressive language.

However, you must be aware that connotations affect word usage possibilities in ways that are sometimes difficult to predict. Words that meant (denoted) one thing years ago may today mean (connote) something different enough to be undesirable in your context. *Chauvinism*, for instance, denotes excessive patriotism and devotion to duty, which may or may not be favorable, depending upon the prevailing political climate. The term originated with Nicholas Chauvin of Rochefort, whose devotion to Napoleon was at first celebrated and later criticized. Whether pro or con, the term meant excessive patriotism.

Today, however, *chauvinism* has been applied to persons, usually male, who exhibit favoritism toward men and a bias against women. This connotation has been used so widely that, today, labelling someone a chauvinist is to suggest sexual bias, not excessive patriotism, so the term is entirely negative.

If you are to use language well, you must be aware of the connotations of words, particularly in the minds of your readers. If your readers will assume that *effects* are favorable and that *impacts* are unfavorable, then your word choice will depend upon which impression you wish your readers to have. The right word then is the precise word that conveys your intended meaning. (See TONE.)

Concrete vs. Abstract Words

The distinction between concrete and abstract words is similar to the distinction between specific and general. Concrete words are those words representing specific people, places, objects, and actions—all things that we can experience through our senses. So concrete words create a vivid mental image. Abstract words are usually more general. They represent ideas, qualities, actions, conditions, and other things that are removed from direct sensory experience. Here are some examples of concrete and abstract words:

Concrete	Abstract
Apple	Food
M-60 tank	Transportation
Arthur Smith	Co-worker
Teledyne	Corporation
Los Angeles	Urban area
IBM Selectric	Office equipment

Using specific, concrete terms is important because in doing so you narrow the range of possible interpretation. If you use an abstraction like *food* when you mean *apples*, the reader will probably not receive the message you intended to send. Here, from the *Guide for Air Force Writing*, is an example of an abstraction ladder:

Weapon System
Hardware
Aircraft
Bomber
B-52

The higher you go on the ladder, the more abstract the words become. The problem many writers have is that they use more general, abstract words than they should. Instead of *B-52*, some writers might say *strategic bomber*, *aircraft*, or even *weapon system*. A B-52 is certainly a strategic bomber, but so is a B-1. Therefore, *strategic bomber* is more abstract than *B-52*. Saying *bomber* rather than *B-52* will not create the image of a B-52 in readers' minds, and they may not understand precisely what you mean. *Weapon system* is even further removed from the concrete reality of a B-52 bomber, so a writer mentioning a weapon system allows readers to imagine everything from a B-52 to the latest attack submarine.

Avoid this problem by using the most specific, concrete words you can. If you need abstractions, use them. But be as specific as you can with every word choice.

The Correct Word for the Context

A major writing problem can occur if you use the wrong word for the context. Which of the words within parentheses in the following sentences is correct?

(You're, Your) supposed to call the manager before leaving.

The (principal, principle) problem with this alternative is its reliance on a second wash cycle between chemical baths.

The plan will (affect, effect) residents living south of the city.

In all three cases, the first word is correct. Using the wrong word may not prevent readers from understanding the sentences. But many readers might begin to wonder if you are literate. And if you seem

WORD PROBLEMS

illiterate, then even the facts in a letter may appear questionable. So using correct words is important.

You should be aware, however, that using incorrect or imprecise words in legal documents (or in any document that can potentially be subject to interpretation in a courtroom) is very dangerous. Courts tend to support the word on the page, not what the author claims to have intended.

Sometimes, using the incorrect word causes serious confusion and potential misinterpretation:

> Please ensure that the third production cycle is stopped continuously for quality assurance testing.

Continuously means "uninterrupted or constant." *Continually* means "recurring often." If the production cycle is stopped continuously, then it never runs. If the cycle is stopped continually, it is stopped frequently, but not constantly. (Some readers may not make this distinction. For the sake of clarity, the writer would be better off issuing a more specific order: *Please ensure that the third production cycle is stopped every half hour for quality assurance testing*.)

Some words are clearly correct or incorrect. You can't write *it's* when you mean *its*, and vice versa. However, some distinctions are more difficult to make. Even the dictionaries do not offer clear-cut distinctions in some cases:

> (Since, because) we had not received payment, we decided to close the account.
>
> They proposed to publish a (bimonthly, semimonthly) newsletter for all employees.
>
> The proposal (that, which) you prepared has turned out to be very profitable.

In the first sentence, both choices do include the notion of causality. However, strict editors prefer *because*. It suggests causality and nothing else, while *since* primarily means before now or from some time in the past until now. Only secondarily does *since* suggest causality. Thus, strict editors consider *because* less ambiguous.

In the second sentence, *bimonthly* is unclear because the ambiguous *bi-* can mean either twice within a time period (twice a month) or every other time period (every two months). *Semimonthly*, which has always meant twice a month, has suffered because of the confusion over *bimonthly*. Given the confusion between the two terms, you should perhaps avoid both terms and just say *twice a month* or *every two months*.

In the third sentence, both words are acceptable to some writers and editors. Strictly speaking, *that* is more correct because the clause it introduces is essential to the sentence and cannot be removed; if the clause were non-essential, *which* would be the more correct word and the non-essential clause would be separated from the main clause by commas. However, many educated writers and speakers violate this distinction, some using *that* and *which* almost interchangeably. So insisting that *that* is the only correct choice for the third sentence is difficult.

Language and usage rules cannot be and never have been independent from the living language they describe. The spoken and written language that people use is always the final arbiter in disputes over word correctness. And a living language is always changing, so today's correct choice may be tomorrow's error. Good writers must remain alert to changes in a word's meaning and their acceptability.

The following list of problem word choices reflects some of the main choices facing today's writers. You should supplement this list and update it as necessary to remain current.

Word Problems

Accent/Ascent/Assent. *Accent* means "to emphasize or stress." *Ascent* means "a going up or rising movement." *Assent* means "to agree":

> His accent on the word *demotion* betrayed his true feelings.
>
> His ascent up the corporate ladder has been rapid.
>
> The trustee's assent is necessary before we sign the agreement.

Accept/Except. *Accept* is a verb meaning "to receive." *Except* is a preposition meaning "to the exclusion of":

WORD PROBLEMS

Have you accepted our explanation for the overpayment?

We completed everything except the two proposals for Acme, Inc.

Adapt/Adept/Adopt. *Adapt* means "to adjust to a situation." *Adept* means "skillful." *Adopt* means "to put into practice or to borrow":

Within a week she adapted to the new billing procedure.

She won the promotion because she was so adept at her job.

Just last year we adopted a new method for maintaining inventory.

Adjacent/Contiguous/Conterminous. *Adjacent* is the most general word, usually meaning "close to and nearby" and only sometimes "sharing the same boundary":

Burger King is adjacent to the Cottonwood Mall.

The adjacent lots were both owned by the same construction company.

Contiguous usually means "sharing the same boundary" even though it includes the notion of "adjacent" in most of its uses:

The two mining claims turned out to be contiguous once the survey was completed; the owners had originally believed that a strip of state land separated the claims.

Conterminous (also *coterminous*) is the most specific of the terms and also the rarest. Its most distinctive meaning is "contained within one boundary" even though it also includes the senses of "sharing the same boundary" and quite rarely of being "adjacent." Its most distinctive use, however, is as follows:

The conterminous United States includes only 48 of the 50 states.

These three words are a problem because they share a common meaning: "close to or nearby each other." At the same time, they each have more specific meanings, as illustrated. As with any confused words, writers should choose other phrasing if they wish to be as precise as possible:

not

Our two lots were adjacent. (*neither* contiguous, *nor* conterminous)

better

Our two lots shared a common boundary on the north.

or with a different meaning

Our two lots fell entirely within the city boundary.

Adverse/Averse. *Adverse* is an adjective meaning "unfavorable." *Averse* is adjective meaning "having a dislike or a distaste for something." The two also contrast in how they are used in sentences. *Averse* appears only after the verb *be* or occasionally *feel*:

We studied the adverse data before making our decision to plug and abandon the well.

An adverse comment destroyed the negotiations.

The President was averse to cutting the Defense budget.

We felt averse to signing for such a large loan given the adverse economic forecasts.

Advice/Advise. *Advice* is a noun meaning "recommendations." *Advise* is a verb meaning "to make a recommendation":

My advice was to meet with the client about the service problem.

Did someone advise you to hire a lawyer?

Affect/Effect. *Affect* is usually a verb meaning "to change or influence." *Effect* is usually a noun meaning "a result or consequence":

Temperature variations will affect the test results.

The technician analyzed the effects of the new sample on the data.

NOTE: *Affect* can also be a noun meaning "the subjective impression of feeling or emotion," and *effect* can also be a verb meaning "to bring about or cause":

His strange affect (*noun*) caused the psychiatrist to sign the committal order.

The general manager's directive effected (*verb*) an immediate restructuring of all senior staff operations.

All right/Alright. *All right* is the standard spelling; *alright* is an informal or nonstandard spelling and is not considered correct. Never use *alright*.

Allusion/Illusion/Delusion. *Allusion* means "a reference to something." *Illusion* means "a mistaken impression." *Delusion* means "a false belief":

His allusion to Japanese management techniques was not well received, but he made his point.

WORD PROBLEMS

> Like the old magician's illusion of the floating lady, the bank manager created an illusion of solvency that fooled even seasoned investors.
>
> The patient had delusions about being watched by the FBI.

Alternate/Alternative. Confusion comes from competing adjective uses. *Alternate* as an adjective means "occurring in turns" or "every other one." *Alternative* as an adjective means "allowing for a choice between two or more options":

> Winners were chosen from alternate lines rather than from a single line.
>
> The alternative candidate was a clear compromise between the two parties.

Sometimes, these two adjective meanings almost merge, especially when an alternative plan is viewed as a plan which replaces another:

> An alternate/alternative plan provided for supplementary bank financing. (*Either is correct.*)

NOTE: Strict editors attempt to restrict *alternative* (in its noun and adjective uses) to only two options:

> Life is the alternative to death.

Actual usage, however, has broadened *alternative* to include any number of options:

> The planning commission analyzed five alternative sites. (*or* The planning commission analyzed five alternatives.)

Altogether/All together. *Altogether* means "completely or entirely." *All together* means "in a group":

> We had altogether too much trouble getting a simple answer to our question.
>
> The spare parts lists are all together now and can be combined.

Among/Between. *Among* refers to more than two choices. *Between* usually refers to two choices only, but it can refer to more than two:

> We had difficulty deciding among the many options—over 200 colors.
>
> The contracting officer has eliminated three bidders, so the Source Selection Authority must choose between us and Universal Data.

Strict editors do try to restrict *between* to two choices only. But occasionally, you can use *between* instead of *among*, especially where *among* would not sound right:

> The research group analyzed the differences between the five alternatives.

Anyone/Any one. *Anyone* means "any person." *Any one* means "a specific person or object":

> Anyone who wants a copy of the Camdus report should receive one.
>
> We were supposed to eliminate any one of the potential mine sites.

Assure/Insure/Ensure. All three words mean "to make certain or to guarantee." *Assure* is limited to references with people:

> The doctor assured him that the growth was non-malignant.

Insure is used in discussing financial guarantees:

> His life was insured for $150,000.
>
> The company failed to insure the leased automobile.

In many common instances, *insure* and *ensure* are interchangeable:

> His plan ensured (*or* insured) that the company would not end the year with a deficit.

Bad/Badly. Originally, *bad* was the adjective form, and *badly* was the adverb form. Now, however, *badly* has begun to function in sentences the same way *bad* has. This overlap is a problem when you use the verbs of the senses (*feel, look, smell*, etc.):

<u>Originally</u>

> Harold felt bad (*adjective*) all day from the blow on his head.
>
> The machine worked badly (*adverb*) despite the overhaul.

<u>Currently Acceptable</u>

> Harold felt badly all day from the blow on his head.

NOTE: Only the verb *feel* allows for either *bad* or *badly*, as in the preceding sentence. Other confusions between the adjective and the adverb are not acceptable, especially in written English:

<u>Not Acceptable</u>

> He looked badly after the bachelor party. (*correct*: bad)

WORD PROBLEMS

The lab smelled badly after the drainage samples arrived. (*correct*: bad)

Biannually/Biennially. *Biannually* means "two times a year." *Biennially* means "every two years":

Because we meet biannually, we will have 10 meetings over the next five years.

Our long-range planning committee meets biennially—on even numbered years.

Bimonthly/Semimonthly. *Bimonthly* can mean either "every two months" or "twice a month." *Semimonthly* means "twice a month." Because of the potential confusion surrounding *bimonthly*, you should avoid the word and write *every two months* or *twice a month*. *Semimonthly* has only one meaning and should not be confusing. Still, it has suffered from the ambiguity of *bimonthly*:

Our bimonthly newsletter appears in January, March, May, July, September, and November. (*better*: Our newsletter appears every two months, beginning in January.)

We proposed semimonthly meetings of the legislative committee.

Can/May. *Can* means (1) "ability," (2) "permission," and (3) "theoretical possibility." *May* means (1) "permission" and (2) "possibility":

Ability

She can speak German, but she can't write it very well.

Permission

Can I help you with your project?

May I help you with your project?

NOTE: *May* sounds more formal than *can*. If you wish to sound formal, use *may*. Otherwise, use *can*.

Possibility

George can make mistakes if he's rushed. (*or* may)

The project can be stopped if necessary. (*or* may)

The trail may be blocked, but we won't know until later.

Your objection may be reasonable, but we still don't agree.

Capital/Capitol. *Capital* means "the central city or site of government," "invested money," and "an upper case letter." *Capitol* means "the main government building":

Paris is the capital of France.

The necessary capital for such an elegant restaurant is $1.5 million, but I doubt that investors will put up that much.

THIS SENTENCE IS WRITTEN IN CAPITALS.

The legislature authorized a complete renovation of the capitol dome.

Carat/Caret/Karat. *Carat* means "the weight of a gem." *Caret* means "a mark showing an insertion." *Karat* means "a unit for the purity of gold":

The ring had a 2.2 carat diamond.

I've used a caret to indicate where to insert the new paragraph.

The ring is made of 18-karat gold.

Cite/Sight/Site. *Cite* means "to quote." *Sight* means "vision." *Site* means "a location":

During the trial, our attorney cited earlier testimony.

Most of the tunnel was out of sight, so we could not estimate the extent of the damage.

The contractor prepared the site by bulldozing all the brush off to the side.

Complement/Compliment. *Complement* means "completing or supplementing something." *Compliment* means "an expression of praise":

The report's recommendations complement those made by the executive committee last year.

The manager passed on a compliment from the vice president, who was impressed with the proposal team's efforts.

Comprise/Compose. Strict editors carefully distinguish between these two words—that is, *comprise* means "to include or contain" and *compose* means "to make up from many parts":

The U.S. Congress comprises the House of Representatives and the Senate.

The House of Representatives and the Senate compose the U.S. Congress.

Such sentences, especially those with *comprise*, are beginning to sound stiff and overly formal, and passive alternatives are more and more common, although not accepted by all editors:

WORD PROBLEMS

The U.S. Congress is comprised of the House of Representatives and the Senate.

As with other disputed word uses, choose an alternate version whenever possible:

The House of Representative and the Senate constitute the U.S. Congress.

Continual/Continuous. *Continual* means "intermittent, but frequently repeated." *Continuous* means "without interruption":

Because the pipes are so old, continual leaks appear despite our repair efforts.

Because of a short in the wiring, the horn sounded continuously for 10 minutes.

Council/Counsel/Consul. *Council* means "a group of people." *Counsel* means "to advise" (verb), "advice" (noun) or "an attorney." *Consul* means "a foreign representative":

The safety council passed a motion to ban smoking in shaft elevators.

The consultant counseled us in ways to improve our management of ID team efforts.

His counsel was to rewrite the proposal.

MOGO's counsel made the opening statement.

The French consul helped us obtain an import license.

Councilor/Counselor. *Councilor* means "a member of council." *Counselor* is "an advisor or lawyer":

The councilors decided to table the motion until the next meeting.

Our staff medical counselor has a PhD in clinical psychology.

Credible/Creditable/Credulous. *Credible means "believable." Creditable* means "praiseworthy." *Credulous* means "gullible":

His revised report was more credible, chiefly because the manpower estimates were scaled to match the price.

In spite of some short cuts, the proposal team came up with a creditable proposal; in fact, they won the contract.

He was so credulous that anyone could fool him.

Data. *Data* (the plural of *datum*) is now often used as both the singular and plural forms of the word. In some technical and scientific writing, however, *data* is still traditionally plural only. If the convention in your discipline or organization is to use *data* as a plural, then be sure that your sentences reflect correct agreement of subject and verb:

Our production data are being examined by the EPA because a citizen complained about excessive emissions from our plant.

The data have been difficult to analyze, chiefly because of sloppy record keeping.

These sentences may sound strange or awkward to many readers. Consequently, some technical writers avoid phrasing that calls attention to the plural meaning of *data*. The two sentences above, for instance, could read as follows:

EPA is examining our production data because a citizen complained about excessive emissions from our plant.

We found the data difficult to analyze, chiefly because of the sloppy record keeping.

Different from/Different than. These two forms can and should be used interchangeably. Strict editors, however, may insist that *different from* is somehow better than *different than*. Actually, well-educated writers and many editors have used both forms for well over three hundred years. The argument over these two forms is an example of a preference being mistaken for a rule. In fact, no clear distinction between the two forms has ever existed:

The results were far different from those we expected. (*or* different than)

The study is different than we had been led to expect. (*or* different from the one we had been led to expect.)

Discreet/Discrete. *Discreet* means "tactful or prudent." *Discrete* means "separate or individual":

The counselor was so discreet that no one learned we had been meeting with him.

The testing included three discrete samples of the ore body.

Disinterested/Uninterested. Originally, *disinterested* meant "neutral or unbiased," and *uninterested* meant "without interest." Careful writers and editors still maintain this distinction:

The judge was appointed because he was clearly disinterested in the dispute.

WORD PROBLEMS

The President was so uninterested in the problem that he failed to act.

Disburse/Disperse. *Disburse* is the verb meaning "to pay out." *Disperse* is the verb meaning "to scatter":

The payroll clerk disburses the petty cash funds as needed.

The reserved top soil was dispersed over the site after the project was completed.

Elapse/Lapse. *Elapse* is a verb meaning "to pass by or to slip"; it usually refers to time. *Lapse* is a verb with many meanings, most derived from sense of "to drift, to discontinue, or to terminate":

Two weeks elapsed before we heard from the Internal Revenue Service.

The speaker lapsed into silence after the embarrassing question.

Our contract lapsed before we could negotiate its renewal.

Eminent/Imminent. *Eminent* means "outstanding or prestigious." *Imminent* means "very near or impending":

Only eminent researchers will win Nobel prizes.

The dam's collapse was imminent, so we evacuated downstream communities.

Envelop/Envelope. *Envelop* is the verb meaning "to enclose or to encase." *Envelope* is the noun meaning "something that contains or encloses":

They proposed to envelop the storage tank with the fire-retardant foam.

We placed in the envelope both the final report and the backup surveys.

Farther/Further. As far back as Shakespeare these words have been confused, and in some contexts they are clearly interchangeable—*e.g., farther/further from the truth*.

Strict editors still maintain that *farther* should be restricted to senses involving distance, while *further* includes other senses:

The assembly site was farther from testing area than we wished.

A further consideration was the inflation during those years.

Fewer/Less. *Fewer*, the comparative form of *few*, usually refers to things that can be counted. *Less*, one comparative form of *little*, refers to mass items, such as sugar or salt, which cannot be counted, and to abstractions:

We analyzed fewer well sites than the government wanted us to analyze.

Less sodium chloride in the water meant that we had fewer problems with corrosion.

Fewer teachers, less education.

See NOUNS and ADJECTIVES.

Forward/Foreword. *Forward* is an adjective and an adverb, both generally meaning "at or near the front." *Foreword* is the noun meaning "the introduction to a book":

The hopper moves forward when the drying phase is nearly finished.

The foreword to the book was only two pages long.

Imply/Infer. *Imply* means "to suggest or hint." *Infer* means "to draw a conclusion or to deduce":

The report implies that the break-even point may be difficult to reach, but it fails to give supporting data.

Based on our comments, she inferred that we would not give our wholehearted support to her project.

In regard to/As regards/In regards to. The first two forms are acceptable. By convention, the third form (*in regards to*) is unacceptable. Do not use *in regards to*:

In regard to your report, our firm is still busy analyzing it.

As regards your second question, we have taken all of the steps necessary to acquire mineral rights on the Bruneau lease.

A better choice in most sentences is to use the shorter and simpler *regarding*:

Regarding your second question, we have taken all of the steps necessary to acquire mineral rights on the Bruneau lease.

Irregardless/Regardless. *Irregardless* is an unacceptable version of *regardless*. Do not use *irregardless* in either speech or writing.

We decided to fund the project regardless of the cash flow problems we were having.

Its/It's. *Its* is the possessive pronoun. *It's* is the contraction for *it is*:

Its last section was unclear and probably inaccurate.

It's time for the annual turnaround maintenance check.

See PRONOUNS.

WORD PROBLEMS

Later/Latter. *Later*, the comparative form of *late*, means "coming after something else." *Latter* is an adjective meaning "the second of two objects or persons":

> Later in the evening a fire broke out.
>
> In our analysis of the Lankford and Nipon sites, we finally decided that the latter site was preferable.

NOTE: *Latter* (and its parallel *former*) are sometimes confusing, so rewrite to avoid them:

> We finally decided that the Nipon site was preferable to the Lankford site.

Lay/Laid/Laid. These three words are the principal parts of the verb *lay*. The verb itself means "to put or to place." It must have an object:

> The contractor promised to lay the sod before the fall rains began. (*The object is* sod.)
>
> The manager laid his plan before his colleagues.
>
> Our recent talks with the Russians have laid the groundwork for control of nuclear weapons in space.

Lie/Lay/Lain. These three forms are the principal parts of the verb *lie*. *Lie* means "to rest or recline." In contrast with *lay*, *lie* cannot have an object:

> The main plant entrance lies south of the personnel building.
>
> The new access road lay on the bench above the floodplain.
>
> That supply has lain there for over a decade.

Maybe/May be. *Maybe* is the adverb meaning "perhaps." *May be* is a verb form meaning "possibility":

> Maybe we should analyze the impacts before going ahead with the project.
>
> Whatever happens may be beyond our control, especially if inflation is unchecked.

Practical/Practicable. These two words mean much the same thing, and dictionaries disagree on their distinctions. Given this confusion, writers should stay with the common form *practical* and avoid *practicable*.

Strict editors do maintain that the difference between *practical* and *practicable* is similar to the difference between *useful* and *possible*. *Practical* means "not theoretical; useful, proven through practice." *Practicable* means "capable of being practiced or put into action; feasible":

> Despite the uniqueness of the problem, the contractor developed a practical method for shoring up the foundation.
>
> Although technically practicable, the solution was not practical because it would have put us over budget.

The last sentence says that we were technically capable of achieving the solution but that this solution was not feasible because of financial constraints. Similarly, building a house on top of Mt. St. Helens is probably possible (it is practicable), but doing so is not feasible (it is not practical) for obvious reasons.

Precedence/Precedents. *Precedence* is the noun meaning "an established priority." *Precedents* is the plural form of the noun meaning "an example or instance, as in a legal case":

> The Robbins account should take precedence over the Jackson account; after all, Robbins gives us over 50 percent of our business.
>
> The Brown decision was the precedent for many later decisions involving racial issues and education.

Principal/Principle. *Principal* is a noun or adjective meaning "main or chief." *Principle* is a noun meaning "belief, moral standard, or law governing the operation of something":

> The principal technical problems we faced were simply beyond current technologies.
>
> The principles of electricity explain the voltage drop in lines.
>
> If we act according to our principles, we will not allow the transaction to proceed.

NOTE: Some writers are confused by *principal* because it refers to the head of a school and to money borrowed from a bank. These are noun forms of a word that used to be only an adjective. The noun forms of *principal* come from noun phrases: *the principal teacher* and *the principal amount*. Over time, the nouns *teacher* and *amount* were dropped, and *principal* assumed the full meaning of the original phrases. Now, *principal* is a noun as well as an adjective.

Raise/Raised/Raised. These three forms are the principal parts of the verb *raise*. *Raise* means "to move (something) upward." *Raise* always requires an object:

> They will raise the funds by January 1, 1986.
>
> We raised the water level some 20 feet to accommodate the changing use patterns.
>
> They had raised the amount to cover the travel costs.

Respectfully/Respectively. *Respectfully* means "with respect." *Respectively* means "in the sequence named":

> Our representatives were not treated very respectfully.
>
> According to production data for March, May, July, and September, the number of cases were, respectively, 868, 799, 589, and 803.

NOTE: *Respectively* often makes sentences difficult to interpret, so avoid *respectively* whenever you can.

Rise/Rose/Risen. These three forms are the principal parts of the verb *rise. Rise* means "to stand up or move upward." *Rise* does not take an object:

> The balloon rises/will rise once the air heats up.
>
> Because the water rose, we had to evacuate the ground floor.
>
> The moisture level in the gas has risen substantially over the last week.

Said. The word *said* often becomes a shorthand term for a document or item previously mentioned:

> We have examined said plans and can find no provisions for the clay soils on the site.

Such uses of *said* are not appropriate in normal business and technical writing. Only in legal writing (and maybe not even there) should writers ever use *said* in this way. The above sentence could be rewritten as follows:

> We have examined the plans and can find no provisions for the clay soils on the site. (*Readers will usually know from the context what plans the writer is referring to.*)

Set/Set/Set. These three forms are the principal parts of the verb *set. Set means "to put or to place (something)." Set* must have an object:

> They set the surveying equipment in the back of the truck. (equipment *is the object*)
>
> Yesterday we set up the derrick so that drilling could begin at the beginning of today's shift. (derrick *is the object*)
>
> After we had set the flow, we began to monitor fluctuations from changes in pressure.

Sit/Sat/Sat. These three forms are the principal parts of the verb *sit. Sit* means "to rest or to recline." *Sit* does not take an object:

> The well sits at the foot of a steep cliff.
>
> The committee sat through the long session with very few complaints.
>
> The oil drums must have sat on the loading dock all weekend.

Stationary/Stationery. *Stationary* is an adjective meaning "fixed in one spot, unmoving." *Stationery* is a noun meaning "paper for writing on":

> Because the boiler was bolted to the floor, it remained stationary despite the vibration.
>
> The new letterhead on our stationery made our company seem more up-to-date.

Than/Then. *Than* is used in comparisons. *Then* is an adverb meaning "at that time":

> George's report was shorter than Mary's.
>
> We then decided to analyze the trace minerals in the water samples.

Their/There/They're. *Their* is a possessive pronoun. *There* is an adverb meaning "at that place." *They're* is the contraction for *they are*:

> The engineers turned in their reports for printing.
>
> The well site was there along the base of the plateau.
>
> They're likely to object if we try to include those extra expenses in the invoice.

To/Too/Two. *To* is the preposition. *Too* is both an adverb meaning "excessively" and a conjunctive adverb meaning "also." *Two* is the numeral:

> The proposal went to the Department of Interior for approval.
>
> The design was too costly considering our budget. (too = *excessively*)
>
> The issue, too, was that technology is only now beginning to cope with these low-temperature problems. (too = *also*)

Toward/Towards. *Toward* and *towards* are merely different forms of the same word. *Toward*

WORD PROBLEMS

is the preferred form in American English. In British English, *towards* is more common than *toward*.

Who/Whom. See the discussion of interrogative and relative pronouns in PRONOUNS.

Who's/Whose. *Who's* is the contraction for *who is*. *Whose* is the possessive form of the pronoun *who*:

> Who's the contractor for the site preparation work?
>
> She was the supervisor whose workers had all that overtime.

Your/You're. *Your* is the possessive form of *you*. *You're* is the contraction for *you are*:

> Your letter arrived too late for us to adjust the original invoice.
>
> If you're interested, we can survey the production history of that sand over the last decade or so.

WORDY PHRASES

As their name indicates, wordy phrases are phrases that use too many words to express an idea. Many of them are similar to or incorporate redundancies, and many wordy phrases have been overused so much they've become cliches:

> all of a sudden
> at a later date
> beyond a shadow of a doubt
> in light of the fact that
> in the environment of
> in the neighborhood of
> kept under surveillance
> on two different occasions
> reported to the effect
> the fullest possible extent

None of these phrases is necessary, but many writers use them habitually so they seem "natural" somehow. In informal speech and writing, these phrases may in fact be preferable, depending on your audience. In technical and business writing, however, you should avoid them.

See REDUNDANT WORDS, CLICHES, GOBBLEDYGOOK, and SCIENTIFIC/TECHNICAL STYLE.

---

1. Eliminate Wordy Phrases.

A fundamental of good writing style is to eliminate unnecessary words. Do not say *all of a sudden*. Just say *suddenly*. Instead of *at a later date*, say *later*. *In light of the fact that* simply means *because of*.

The shorter, simpler words or expressions make your writing more concise and, consequently, make it look and sound more professional.

A List of Wordy Phrases

The following list of wordy phrases will help you identify those you habitually use. The wordy phrase appears in the left column; beside it, in the right column, are possible substitutes:

according to the law	legally
add an additional	add
add the point that	add that
afford an opportunity	permit/allow
a great deal of	much
a great number of	more
a great number of times	often/frequently
a large number of	many
a little less than	almost
all of a sudden	suddenly
along the lines of	like
a majority of	most
a number of	several/many/some
any one of the two	either
a period of several weeks	several weeks
as a general rule	usually, generally
as a matter of fact	in fact
a small number of	few
as of now	now
as of this date	today
as regards	about
as related to	for/about
assuming that	if
as to	about
a sufficient number	enough
at a later date	later
at all times	always
at an early date	soon
at hand	here
at present	now
at regular intervals of time	regularly
at that time	then
at the conclusion of	after
at the present time	now

WORDY PHRASES

at the rear of	behind
at the same instant	simultaneously
at this time	now
at which time	then
based on the fact that	due to/ because
beyond a shadow of a doubt	doubtless
brought to a sudden halt	halted
by means of	by
by the time that	when
by the use of	by
by way of illustration	for example
called attention to the fact	reminded
came to a stop	stopped
cannot be possible	impossible
come to an end	end
cost the sum of	cost
despite the fact that	although
detailed information	details
draw to a close	end
due to the fact that	because
during the course of	during
during the time that	when/while
during which time	while
estimated at about	estimated at
estimated roughly at	estimated at
exactly alike	identical
except in a small number of cases	usually
exhibit a tendency to	tend to
expose to elevated temperature	heat
few in number	few
for a short space of time	for a short time
for the purpose of	for/to
for the reason that	because/since
for this reason	so
from now on	in the future
from the point of view of	for
from time to time	occasionally
having reference to this	for/about
if and when	if/when
if at all possible	if possible
if that were the case	if so
in accordance with	by/under
in addition (to)	also/besides
in a number of cases	many/some
in a position to	can/may
in a satisfactory manner	satisfactorily
inasmuch as	because/since/ as
in back of	behind
in case of	if
in close proximity	near
in conjunction with	with
in consideration of the fact	because
in excess of	more than
in favor of	for
in few cases	seldom
in few instances	seldom
in lieu of	instead of/ in place of
in light of the fact that	because
in many cases	often
in most cases	usually
in order to	to
in other words	or/that is
in rare cases	rarely
in (with) reference to	about/ concerning
in regard to	about/ concerning
in relation to	with
in respect to	about/ concerning
in short supply	scarce
in terms of	according to
in the absence of	without
in the amount of	of/for
in the case of	for/by/in/if
in the course of	during
in the environment of	around/near
in the event of	if
in the event that	should
in the first place	first/primarily
in the instance of	for
in the majority of cases	usually
in the matter of	about
in the nature of	like
in the near future	soon
in the neighborhood of	about/near
in the proximity of	near/nearly/ about
in the vicinity of	around/near
introduced a new	introduced
in view of the fact that	considering
involve the necessity of	requires
is/are in the possession of	has/have
is defined as	is
is in the process of making	is making
is of the opinionthat	believes
is representative of	typifies
it is apparent that	apparently
it is clear that	clearly
is is evident that	evidently
it is obvious that	obviously
it is often the case that	frequently
it is plain that	plainly
it is unquestionable that	unquestionably
it would appear that	it seems
kept an eye on	watched
kept under surveillance	watched
last of all	last
leaving out of consideration	disregarding
made an investigation of	investigated
major portion of	most of
make application to	apply
make a purchase	buy
make contact with	meet/contact
more and more	increasingly
notwithstanding the fact that	although
of considerable magnitude	big/large/great
off of	off
of no mean ability	capable
of small diameter	fine/thin
of very minor importance	unimportant
on account of	because
on a few occasions	occasionally
on a stretch of road	on a road
on behalf of	for
once in a great while	seldom/rarely
one after another	alternately
one by one	singly
one part in a hundred	1 percent
on the basis of	by
on the grounds that	because
on the part of	by
on two different occasions	twice
ought to	should
outside of	except
owing to the fact that	since/because
period of time	interval/period
pertaining to	about
possibly might	might
prior to	before
probed into	probed
proceed to investigate, control, study,	analyze/study
provide a continuous indication of	show continuously

WORDY PHRASES

pursuant to	following
range all the way from	range from
reduced to basic essentials	simplified
refer back	refer
relative to	about
repeat again	repeat
reported to the effect	reported
revise downward	lower
seldom if ever	rarely
separate into two equal parts	halve
since the time when	since
started off with	started with
subsequent to	after/following
take appropriate measures	act
taking this factor into consideration	therefore
that is to say	that is
the foregoing	the/this/that/these/those
the fullest possible extent	mostly/fully/completely
the only difference being that	except that
there is no doubt that	doubtless/no doubt
there is no question that	unquestionably
through the use of	by
to be cognizant of	to know
to summarize the above	in summary
total operating costs	operating costs
turn up	turn
two by two	paired/in pairs
until and unless	unless
until such time as	until
up to now	formerly
went on to say	added/continued
when and if	if
with a view to	intending to
with full approval	approved
within the realm of possibility	possibly/possible
without variation	constant/stable
with reference to	about
with regard to	about/regarding
with respect to	about/respecting
with the exception of	except
with the object of	to
with the result that	so that

MODEL DOCUMENTS

RESPONSE LETTER
With Information and Directions

BCC
BOUNTIFUL CHEMICAL COOPERATIVE, INC.
139 Sequoia Park Way
Suite 303
Oakland, California 90022
(213) 451-7561

February 16, 1984

Mrs. Louise Lantham
12039 Plaza Drive
Tallahassee, FL 32303

Dear Mrs. Lantham:

John Smythe asked me to write to you to explain how you can add your husband as a dependent under the group insurance plan underwritten by the Savannah Life Insurance Company.

To enroll your husband, simply complete the enclosed enrollment card and medical information sheet and submit them, along with a medical report from your physician, to me at P.O. Box 3443, Durango, Colorado 81301. The medical report should be up-to-date and complete and should include the following information:

- Current illnesses or injuries, including your physician's prognosis
- Illnesses and injuries within the past 5 years
- Diagnoses and dates of treatments for all illnesses or injuries listed above
- Types and dosages of medications, including whether he is still taking them, and, if not, when he stopped
- Any other information your physician believes might help us to evaluate your husband's physical condition

As soon as we receive this information, we will be able to process your application. If you have further questions, please call me at 305-123-4567.

Sincerely,
Walt Cavanaugh

Walt Cavanaugh
Assistant Manager
Personnel Department

WC/ght
Enclosures

The set-up is short but essential. It explains why John Smythe isn't writing. See ORGANIZATION.

The response comes early in the letter and is clear and straightforward.

Listing the requirements emphasizes them. Using a bulleted list makes every item equally important. See COLONS, LISTS, EMPHASIS, and PARALLELISM.

The listed items do not end in periods because none of them is a complete sentence. See LISTS.

The closing is social and informative. See LETTERS.

A brief set-up is essential but the response itself must appear as early as possible. The list enhances readability and emphasizes the requirements. See LISTS.

The tone of the letter is businesslike but courteous. The writer tries to establish warmth and cooperation in the response. See TONE.

This letter illustrates the semi-block format with standard punctuation. See LETTERS.

RESPONSE LETTER
To a Concerned Customer

The opening paragraph establishes the context of the response and is a courteous acknowledgment of the customer's concern.

This paragraph attempts to develop rapport with the reader. The tone is informal and personal. See TONE and LETTERS.

This paragraph states FFFF's position and leads into the summary of the testing—which provides evidence validating FFFF's position.

A displayed list highlights the test results, which are presented in everyday language. See SCIENTIFIC/ TECHNICAL STYLE and TONE.

FARMLAND FROZEN
FAMILY FOODS

January 16, 1984

Ms. Josephine Lambert
1667 Willow Avenue
Seattle, WA 96508

Dear Ms. Lambert:

Thank you for your recent inquiry about the use of mozzarella cheese substitute in our FFFF lasagna.

I too am worried about the increasing presence of artificial products and additives in our food, so I can sympathize with your concern over finding a mozzarella cheese substitute as an ingredient in FFFF lasagna.

Let me assure you, however, that the mozzarella cheese substitute and all other ingredients in our lasagna are as wholesome and safe as we can make them. We decided to use the mozzarella cheese substitute only after extensive testing. Here are the results:

—Mozzarella cheese substitute has no cholesterol but does have all of the vitamins, minerals, and protein found in natural mozzarella. Many of our customers have asked for lower cholesterol levels in our products.

—Our consumer test panel, in extensive blind testing, could detect no difference in taste, smell, or appearance between the mozzarella cheese substitute and natural mozzarella.

—Ingredients in natural mozzarella and the substitute are almost identical: water, milk protein, and fat. Natural cheese has animal fat, which contributes the cholesterol, while the substitute has soybean oil. Both products have similar minor additives to enhance the flavor, prevent spoilage, and guarantee consistent quality.

303 Blossom Avenue   Des Moines, Iowa 50321   (515) 521-4911

Letters responding to customer complaints should be as personal and direct as possible. The writer must address all questions or grievances in the reader's earlier letter and must do so in a friendly, sympathetic, and courteous manner.

In responses to complaints or negative inquiries (as in this letter), beware of repeating the negative events or details verbatim. Instead, summarize the complaints. Don't force readers to relive the circumstances that made them angry enough to write in the first place.

RESPONSE LETTER
To a Concerned Customer

Josephine Lambert - 2 - January 16, 1984

—Using mozzarella cheese substitute helps us lower the cost of FFFF lasagna, especially given the recent increase in the price of natural cheese. This economic advantage also allows us to put more mozzarella into our lasagna—something many of our customers want.

Even with the mozzarella cheese substitute, we believe that our lasagna is a very high quality product. It is both delicious and nutritious.

I hope this letter has answered your questions about our use of a cheese substitute and will make you feel comfortable about purchasing our lasagna. You mentioned that you enjoy a variety of FFFF products, so I am sending you several complimentary coupons. Thank you for your concern and for your interest in FFFF products.

Sincerely,

Martha Frampton

Martha Frampton
Public Relations

MF:sj
Enclosures

The opening sentence begins with a dependent clause and delays its major point (*our lasagna is high quality*); however, the introductory clause mentions *cheese substitute*, so it forms an effective transition from previous thoughts. See TRANSITIONS.

The personal pronoun *I* is appropriate even though the writer is speaking for FFFF. She shifts to the collective *we* when she speaks for the company as a whole. Such shifts in pronouns are natural and acceptable. See TONE, LETTERS, and PRONOUNS.

This letter illustrates the semi-block format with standard punctuation. See LETTERS.

RESPONSE LETTER
Answer to a Complaint

FARMLAND FROZEN
FAMILY FOODS

June 25, 1984

Mr. Wiley G. Elkins
1456 Jackson Avenue
Jordan, MO 64833

Dear Mr. Elkins:

The opening is courteous, personal, and reassuring. See INTRODUCTIONS and TONE.

Thank you for your recent letter about Potato Ripples. The blackened material you found is not harmful although I can appreciate your dissatisfaction at finding it.

The explanation is simple and non-technical. See JARGON and GOBBLEDYGOOK.

What you found was a small chunk of burned potato. These blackened chunks occur when particles break loose from potatoes and collect in our deep-fat frying system. To prevent these chunks from being packaged, we continuously filter the oil in our fryers, and we periodically clean the fryers. In addition, our inspectors visually check all Potato Ripples before packaging. Despite these precautions, a chunk of blackened potato occasionally finds its way into a package of Potato Ripples.

This single-sentence paragraph establishes a key idea: quality.

We do value the quality of our products and will continue to make every effort to prevent chunks of burned potato from being packaged with our Potato Ripples.

The closing (and the coupons) further customer relations. The last sentence repeats the opening. See REPETITION.

I am enclosing a coupon for a free package of Potato Ripples, Potato Crisps, Zucchini Fritters, or Eggplant Gems. We hope you will use the coupon to try Potato Ripples again or to try one of our other products. Again, thank you for your letter and for your interest in FFFF's products.

Sincerely yours,

Jackson Blaine

Jackson Blaine
Quality Control

JB:mm
Enclosure

303 Blossom Avenue **Des Moines, Iowa 50321** **(515) 521-4911**

Tact and a personal touch are crucial in letters to disgruntled customers. Personal pronouns help establish a tactful tone, and the details include specific references to Mr. Elkins' letter. See TONE.

The writer avoids technical complexity and jargon (he might have called the chunk of potato a "carbonized carbohydrate particle"). See JARGON, GOBBLEDYGOOK, and SCIENTIFIC/TECHNICAL STYLE.

The letter follows the modified block format with standard punctuation. See LETTERS.

REPRIMAND LETTER

SKY AVIATION
822 Ocean View Drive
Long Beach, California 90802
(714) 332-3978

March 28, 1984

Mr. Fred Benson
7654 Laguna Boulevard
Long Beach, CA 94986

Dear Fred

I regretted having to talk to you last Wednesday about your failure to call in before being absent from work on March 22. Based on our discussion and your prior attendance problems, the company has decided to give you 2 days off without pay on April 1 and 2.

I have enrolled you in the Long Beach Alcohol and Drug Abuse program, which is a confidential service open to all of our employees. The program coordinator, John Hughes, has made an appointment at 10:30 a.m. on April 1 for you to see Roberta Crenshaw, an alcoholism counselor. Roberta is located in the main office at our Alameda facility. John also requested that you begin attending AA meetings each Thursday evening at the Paloma Community Church.

We hope, Fred, that this counseling will help you solve your problems. If further unannounced absences occur, the company will have to take more stringent disciplinary actions, including possible termination.

Sincerely

G. Terry Nielsen

G. Terry Nielsen
Director of Personnel

GTN:fg

The first sentence sets up the major idea in the second sentence. See ORGANIZATION, LETTERS and TONE.

This paragraph presents details. Listing these details is optional, but the list might create too formal an impression in this letter. See LISTS and PARAGRAPHS.

The closing paragraph is personal and hopeful while reiterating the seriousness of the situation.

Reprimand letters are hard to write, especially when the reader is a valuable employee. The best strategy is to soften the tone by being friendly and direct. Using the reader's name and using the pronouns *I* and *you* help make the tone friendly and personal. Consequently, this letter will be easier for the reader to accept. See TONE.

As in any effective letter, each of the paragraphs has a clear and definite purpose. Clear organization helps readers read and understand the content. See ORGANIZATION and PARAGRAPHS.

This letter illustrates the modified block format with open punctuation. See LETTERS.

COMPLAINT LETTER
With a Tactful Request for Aid

The opening paragraph establishes the problem and its seriousness. See LETTERS.

This description of a delay is concise yet specific. Note the paragraph structure—each sentence leads into the next, and the paragraph ends with a strong statement. See PARAGRAPHS.

This paragraph actually states the request. Then the second sentence repeats the financial problems caused by the delays. See PARAGRAPHS and REPETITION.

These paragraphs make two distinct requests. The short paragraphing helps highlight the requests. See EMPHASIS and PARAGRAPHS.

MIDLAND OIL AND GAS OPERATIONS, INC.
7000 Jalepeno Boulevard
Dallas, Texas 75234
(214) 735-9600

November 15, 1984

District Director
U.S. Immigration and Naturalization Service
2645 St. Anne Street
Houston, TX 77004

Dear Sir:

During the past year, MOGO ships delivering crude oil to our Houston refinery complex have been unnecessarily delayed by immigration officials who have not arrived as scheduled for boarding. These delays cost MOGO Oil (and ultimately consumers) an average of $2,745/hr.

The most recent delay occurred on September 28, 1984, when the MV Seaworthy docked in Gulfview. Boarding was scheduled for 0400 that morning, but immigration officials did not arrive until 0930. When the officials finally arrived, they apologized for being late by saying, "We forgot." As you know, dock workers cannot unload a foreign vessel until immigration officials have cleared it, so unloading was delayed for over 5 hours. That delay cost MOGO over $15,000.00.

We would appreciate whatever you can do to help eliminate or at least reduce these delays. Occasional delays are inevitable, but we cannot sustain financial losses like the one above and remain competitive unless we increase the price we charge consumers for our oil products.

Please clarify for us the regulations regarding unloading of a vessel prior to boarding by immigration officials. In particular, can dock personnel unload a vessel that has cleared Customs but has yet to clear Immigration?

We would also appreciate having an after-hours telephone number to call when immigration officials do not arrive on schedule.

Complaint letters have to be tactful yet firm. They are usually based on one or more incidents, which must be established before you can request a remedy. The first two paragraphs of this letter provide background information that establishes the nature and seriousness of the problem. The request in the third paragraph is therefore understandable and reasonable. See ORGANIZATION and LETTERS.

Tone in complaint letters is important. Such letters should not sound shrill, angry, or unreasonable. In this letter, the author writes calmly, but directly. He presents a specific case and then makes several understandable requests. The letter ends with a plea for cooperation. See TONE.

COMPLAINT LETTER
With a Tactful Request for Aid

U.S. Immigration and Naturalization Service
November 15, 1984
Page Two

The telephone number strengthens the pleasant, yet direct closing. See LETTERS.

Please call Jack Severenson or me at 886-2233 to discuss solutions to this problem. We want to do everything we can to cooperate with the Immigration and Naturalization Service.

Respectfully,

Frank W. Whitaker

Frank W. Whitaker
Shipping

FWW:dfe

This letter illustrates block format with standard punctuation. See LETTERS.

COMPLAINT LETTER
With a Request for Action

FARMLAND FROZEN
FAMILY FOODS

August 8, 1983

Ms. Patricia Goodway
Customer Support Manager
The Pacific Baking Company
9924 West Pacific Way
San Francisco, CA 97521

Dear Ms. Goodway:

The lead sentence sets up the problem, and the second sentence asks for a solution. See LETTERS and ORGANIZATION.

Recent samples of your breader contain more fines than our specifications allow. Please investigate the problem and let me know as soon as possible what you can do to solve it.

On August 5, our Medford plant inspectors noticed excessive carbon specks on breaded products, so they checked the breader lots on hand (codes 341X and 345X). According to our specification, fines through a U.S. Standard Sieve No. 80 must fall within these limits: 1 percent ± 1 percent (or a maximum of 2 percent). The breader lots on hand gave these results:

The test results are highlighted by the additional space around them. See EMPHASIS.

341X — 3.30%

345X — 4.10%

As you can see, the fines in these two lots exceeded our specification by 1.3 and 2.1 percent. These levels are not yet serious, but they do cause excessive carbon specks in the frying oil and on the finished product.

The closing paragraph establishes how serious the writer considers the problem to be and applies some pressure; yet the tone is not blunt or negative. See TONE.

We have enjoyed a long relationship with Pacific Baking and hope to continue doing business with you. We are therefore anxious to solve this problem and will appreciate your prompt action.

Sincerely,

James Van Prooven

James Van Prooven
Plant Manager

JVP:qw

303 Blossom Avenue   Des Moines, Iowa 50321   (515) 521-4911

Complaint letters, even if written in anger, should not sound angry. The tone should be positive and constructive. The writer should present clear evidence of dissatisfaction, but should strive to balance the complaint with positive solutions. See TONE and LETTERS.

As in this letter, complaint letters usually start with a concise statement of the problem followed by a request for action or resolution. The problem statement should be as brief as possible but long enough to make the request understandable. See the discussion of set-ups in ORGANIZATION.

This letter illustrates the semi-block format with standard punctuation. See LETTERS.

LETTER SOLICITING A BID

SKY AVIATION
822 Ocean View Drive
Long Beach, California 90802
(714) 332-3978

April 24, 1985

Mr. James Quirk
Wizard Machine Tools, Inc.
568 Flatbush Boulevard
Montrose Island, IL 44572

Dear Mr. Quirk

This single-sentence opening paragraph concisely states the letter's purpose. It states the "What's new?" portion of the message. See PARAGRAPHS.

Sky Aviation is soliciting bids for delta Q indicators for our Foxx 175 business jets.

The second paragraph provides essential detail and indicates that the letter has enclosures.

After examining the enclosed drawing and specifications, you are invited to submit a quote for producing 1,980 of these indicators—delivered at a rate of 55 per month for 36 months beginning November 1, 1985.

This paragraph is an effective—though not entirely subtle—appeal to the reader to keep the bid price low.

I understand that the indicators you already provide for our Windstream 88 fixed wing aircraft are very similar to the proposed delta Q indicator. This similarity should make design and fabrication of the new indicators relatively easy and should therefore reduce development and manufacturing costs.

The closing paragraph provides final details. It constitutes the "What's next?" portion of the message.

Please send your quote to Mrs. Ernestine Gonzales in our Purchasing Department. The bids should arrive no later than Friday, May 14. If you have any questions, please call me or Jim Booth at 343-4545.

Sincerely

Arnie Madsen

Arnold Madsen
Manager, Engineering

AM:mm
Enclosures

This letter features classic letter design: A concise opening that clearly indicates the letter's purpose (to solicit bids); a crisp middle with details appearing in short, well-designed paragraphs; and a courteous closing that provides final administrative details and ends with an offer of assistance. See LETTERS and ORGANIZATION.

This letter illustrates the block format with open punctuation. See LETTERS.

SALES LETTER
With a Soft Sell

SKY AVIATION
822 Ocean View Drive
Long Beach, California 90802
(714) 332-3978

June 18, 1984

International Aeronautics
P.O. Box 1149
Galveston, TX 41504

Attention: Mr. Boon Hollenbeck

Subject: Cabin Pressurization System for the K-38

Gentlemen:

We were delighted to hear that International Aeronautics won the contract to develop the K-38 Heavy Cargo Helicopter.

As you know, Sky Aviation has done much of the pioneering work in cabin pressurization. Our pressurization systems are the state of the art in aircraft cabin pressurization, largely because of our patented flowback valve, the VA-321-E.

As you initiate detail design studies for the K-38, we hope you will consider basing cabin pressurization on our VA-321-E and allowing us to assist in cabin pressurization design.

The RFP indicates that the K-38 will require an 8-lb/min. valve capable of maintaining a delta P of 0.3 to 0.5 psi. Valve VA-321-E meets these requirements and is more efficient than any of the other valves currently used in aircraft pressurization systems.

The enclosed drawing (DRA-321-E) shows the standard Sky Aviation configuration for the VA-321-E valve, including the outflow, check, and solenoid subsystems. Here are the control modes within this configuration:

> Minimum Differential Pressure Mode. In a 2-psi vacuum with the solenoid valve energized, the outflow valve will be fully open. With an 8 lb/min. through-flow, this will result in maximum differential pressure of 0.75 inches of water.
>
> Positive Differential Pressure Control. With the solenoid valve de-energized, the outflow valve will move toward the closed position and regulate the differential pressure to 0.6 psi. The regulation point is achieved by the correct sizing of the main poppet return spring.

The opening is positive, yet low-keyed, and the writer establishes the purpose of the letter within the first three short paragraphs.

RFP (Request for Proposal) is an abbreviation that will be familiar to the readers. Therefore, the writer does not have to spell it out. See ABBREVIATIONS.

The run-in headings highlight key technical points. See HEADINGS and LETTERS.

A sales letter is a blend of fact and sales pitch. Thinking positively is essential, but sincerity and reality are also necessary ingredients of successful sales letters. This letter opens and closes with a soft sales pitch. The technical information in the middle is emphasized by headings and underlining. The writer's goal is to make the technical information seem substantial and convincing. See EMPHASIS, LETTERS, TONE, and TECHNICAL/SCIENTIFIC STYLE.

This letter illustrates the modified block format with standard punctuation. See LETTERS.

SALES LETTER
With A Soft Sell

Mr. Boon Hollenbeck
June 18, 1984
Page Two

Negative Differential Pressure Control. As presently configured, the proposed system would incorporate negative differential pressure control, i.e., if the cabin pressure should become less than the ambient air pressure, the outflow valve will open and admit ambient air into the cabin at approximately 0.4 psi. If you do not want this feature in your application, we can eliminate it through a minor change to the valve.

Leaving an extra blank line here helps distinguish the control mode discussion from the continuation of the text.

Valve VA-321-E is currently not in production. Before initiating production, we would have to size the outflow valve return spring. This design and test phase should take no more than 4 weeks.

If you decide to use our valve, you will have all of the technical resources of Sky Aviation at your command, including the engineers who have designed the pressurization systems for over 8,000 aircraft flying today. We now manufacture 25 percent of the pressurization systems for commercial aircraft and 15 percent of the systems for military aircraft.

This paragraph presents relevant past performance information—justification for selecting Sky Aviation.

For further information, please call Fred Huber or me at (216) 777-8787.

Sincerely,

Howard C. Patterson

Howard C. Patterson
Vice President

Sincerely is a good complimentary closing to a business letter. *Very truly yours* is too effusive. See LETTERS.

HCP:cv
Enclosure

RECOMMENDATION MEMO
With a Political Delay

TO: Bob Conners

FROM: Denise Van Horn

SUBJECT: Recommendation to Lower the Toggle Testing Requirement on the Nose Landing Gear Transducer

DATE: June 10, 1985

We feel confident that our production of nose landing gear transducers has improved to the point where 100 percent toggle testing is no longer necessary. As I told Tom Rogers on the phone yesterday, 31 of 32 transducers in the second lot passed the test, and the only reject was just barely out of specification. Consequently, we request that you lower the toggle testing requirement to 20 percent.

Toggle testing delays production considerably and is both time consuming and extremely expensive. We hope to eliminate the testing procedure and rely on quality assurance checks, which is our standard practice. For now, however, we believe that 20 percent testing will guarantee quality assurance.

If I can provide further information, please let me know.

SKY AVIATION

The subject block is very specific. Even though the reader will be averse to the request, the subject block should not be coy, misleading, or vague.

The opening paragraph sets up the request by providing sound support for it. The reader is being asked to do something he does not want to do; therefore, the writer <u>must</u> validate the request before actually stating it. See LETTERS and ORGANIZATION.

Note how the sentences lead from one to the other. See PARAGRAPHS and TRANSITIONS.

Repeating the 20 percent figure is helpful because it reinforces one of the writer's major points. See REPETITION and EMPHASIS.

Normally, writers should open their memos by clearly stating their request. Sometimes, however, politics forbids such directness. In this case, Bob Conners has been adamant about 100 percent toggle testing, so Denise knows that she cannot open the memo with a blunt request for him to change his mind. She has to <u>prepare</u> him to change his mind, so she leads off with a two-sentence set-up that introduces the idea and provides solid evidence to support it. See MEMOS, LETTERS, and ORGANIZATION.

This organization is most effective when, for political reasons, writers cannot open with a direct statement of the request. If politics are not at issue, however, writers should open with the request itself.

RECOMMENDATION MEMO
With Technical Content

The references are complete enough to facilitate document retrieval and are numbered to facilitate citation in the text of the memo. See MEMOS and LETTERS.

The opening paragraph states the author's position and lists three key reasons for that position.

The summaries of the references are optional, but they should appear somewhere in the memo. They inform readers not familiar with the references and remind readers who are familiar with the references. See MEMOS.

This paragraph establishes the major topics to be discussed in the memo. It is a preview of key content. Note that the three topics listed become major headings. See REPETITION and ORGANIZATION.

To: F. Winters Date: December 7, 1984

Subject: Priming the Windstream Fuselage with Rivlin 780

Reference: (a) Memo 5870-5-113 (10-25-84), R. Trent to D. Birdwell, "Long Beach Paint Hangar Requirements To Support Windstream Airplane Programs."

(b) Memo 3281-6-98 (11-18-84), S. Blaine to R. Trent, "Painting the Windstream Fuselage in the Long Beach Paint Hangar."

We continue to believe that Rivlin 780 should not be used on the Windstream airplanes painted at the Foxx facility in Long Beach. Rivlin 680 is more cost effective, safer to use, and easier to apply.

Reference (a) stated that Long Beach would have the capacity to apply exterior paint to the Windstream by early 1985. However, Long Beach capability will depend upon the ground rules attached to reference (a).

Reference (b) stated that the Windstream airplanes will require Rivlin 780 instead of Rivlin 680. The memo requested adjusting the applicable rules to accommodate Rivlin 780.

Though the authors of both memos argued persuasively for Rivlin 780, we continue to support Rivlin 680 as the primer for the Windstream. Rivlin 780 is prohibitively expensive, particularly in pollution control costs; it requires costly additional manufacturing steps; and it is riskier to use because of the danger of staining.

Pollution Control Costs

Rivlin 780 contains approximately six times as much chromium as Rivlin 680 and must therefore be disposed of via a tank truck instead of the sanitary sewer. Disposal costs for a single Windstream would be approximately $650.

SKY AVIATION

Recommendations should either open with the recommendation (as above) or with a brief set-up that prepares readers for the recommendation. Only in exceptional cases should a memo lead up to final sentence or paragraph that reveals the recommendation. See ORGANIZATION and MEMOS.

The references are often extensive, especially in technical discussions. They should follow the order they are mentioned in the memo, and all references listed at the beginning of the memo should be mentioned in the text.

RECOMMENDATION MEMO
With Technical Content

Numbering this list emphasizes the extra steps necessary. Using a displayed list (rather than a paragraph list) adds white space, which also emphasizes the extra steps. See EMPHASIS and LISTS.

Each item is a sentence, so each ends with a period. Further, each item is the same kind of sentence (imperative), so the list is parallel. See PARALLELISM, COLONS, and LISTS.

The single-sentence paragraphs emphasize separate but important points. See EMPHASIS.

F. Winters -2- December 7, 1984

Using Rivlin 780 in the areas to be painted would not eliminate the need for Rivlin 680 on unpainted areas of the fuselage. Therefore, we would need to perform five extra steps.

Additional Manufacturing Steps

These are the additional steps necessary when using Rivlin 780:

1. Mask the skin area to remain unpainted, using metal-foil tape to prevent contact with Rivlin 780.

2. Apply Rivlin 780.

3. Rinse.

4. Allow to dry.

5. Remove metal-foil masking tape.

After taking these steps, we would still need to apply Rivlin 680 to unpainted areas in the conventional manner.

Labor and production delay costs for these extra steps would be the equivalent of adding at least one 8-hour shift for each Windstream painted.

Risk of Staining

Even with metal-foil tape, Rivlin 780 is impossible to mask. In the past, Rivlin 780 has seeped under the tape and stained adjacent skin surfaces due to leaks under the masked areas. Removing these stains requires hand or mechanical polishing, especially in critical areas. Often, such polishing is unsatisfactory and has led customers to reject an airplane.

Deliberate repetition is a valuable technique. The opening paragraph establishes three key reasons for adopting the writer's recommendation. These reasons are discussed and restated in the fourth paragraph and in the closing summary. See REPETITION.

The initial *we* and the *our* in the summary paragraph indicate that H. Roterman and the staff in Long Beach Production are making the recommendation. Using personal pronouns is much better than a common approach (*It is recommended that* . . .) using false subjects and passive verbs. See FALSE SUBJECTS, ACTIVE/PASSIVE, and TONE.

RECOMMENDATION MEMO
With Technical Content

The summary reinforces major points made earlier in the memo. Closing summaries like this one should contain no information that has not already appeared in the memo.

F. Winters -3- December 7, 1984

Summary

In our judgment, the performance benefits of Rivlin 780—durability and corrosion protection—do not warrant its use. Because of increased manufacturing costs, we should consider staying with a uniform application of Rivlin 680.

Harry Roterman

H. Roterman, Manager
Long Beach Production
Supervisor

The writer's name is necessary at the end because it does not appear in the opening of the memo. The title and department are optional. See MEMOS.

dfg

cc: G. Yeatman
R. Trent
W. Samson

REQUEST MEMO
With Informal Instructions

A set-up is unnecessary, so the writer opens with the major idea, a request. The word *please* is courteous, and it adds a personal touch. See ORGANIZATION and LETTERS.

The opening sentence captures the gist of the paragraph. See PARAGRAPHS.

This brief lead-in sentence could also be a heading: **Duties of the Monitor.** See HEADINGS.

The NOTE (all in capitals) emphasizes a key point. See EMPHASIS.

Each item in the list begins with an infinitive verb; therefore, the list is parallel. See VERBS, LISTS, and PARALLELISM.

To:	All Managers and Supervisors
From:	Jacqueline Burrows
Date:	April 7, 1984
Subject:	Monitors for the Air Conditioning Survey

Please appoint someone from your department to monitor the effectiveness of the air conditioning system between now and May 1. During this period, Constant Air, Inc., will be reviewing and balancing the systems in Buildings 3 and 4 to guarantee that all departments can independently maintain the temperatures appropriate for their areas.

Prior to Constant Air's visit, ask all employees to remove cardboard, styrofoam, tape, and other material from the ventilation grills. Many employees have tried to control the temperature in their offices by blocking the vents. However, Constant Air will not be able to draw valid conclusions about the current system unless the vents are clear.

The person you appoint to monitor the system will have the following duties:

1. To contact Constant Air with questions and reports. Constant Air's number is 976-3421. NOTE: Neither the monitor nor any other employee should call Building Services with requests related to the air conditioning.

2. To keep an hourly log of temperatures at selected points in your department. Constant Air will identify these points for you during its initial visit to your department on April 9 or 10.

3. To survey employees at least once a day to determine their satisfaction with room temperatures.

BCC

BOUNTIFUL CHEMICAL COOPERATIVE, INC.

Although a request, this memo is very neutral both in content and in tone. The writer minimizes the use of *I, my,* and *me*. Instead, she follows the *you* approach. Note how often and how directly she addresses the readers. See TONE.

The paragraphs and sentences are short and direct, and the list allows readers to scan the memo for key points. See EMPHASIS, LISTS, and SENTENCES.

REQUEST MEMO
With Informal Instructions

This single-sentence paragraph emphasizes and repeats a key point. See REPETITION, PARAGRAPHS, and EMPHASIS.

All Managers and Supervisors -2- April 7, 1984

4. To prepare a weekly report summarizing survey results. Constant Air will bring sample forms for this report to your department when they visit on April 9 or 10.

Only Constant Air, not Building Services, should adjust the air conditioning system during this survey.

Please call me (Ext. 456) or Ned Trent (Ext. 459) if you or your monitor have any questions about this survey.

J.B.

asd

REQUEST MEMO
For Clarification of a Problem

The specific subject line simplifies cross-referencing and filing. See HEADINGS.

The opening paragraph establishes the topic and conveys the purpose of the memo.

The second paragraph sets up the list and leads into the details. See ORGANIZATION.

Displayed lists are naturally emphatic, so the BW recommendations are highlighted. Note that both listed items begin the same way. See PARALLELISM, LISTS, and EMPHASIS.

Compound modifiers, such as *1-inch,* must be hyphenated. See NUMBERS, HYPHENS, and MODIFIERS.

Apparently and *instead* are conjunctive adverbs that introduce a sentence; therefore, they are followed by commas. See TRANSITIONS and COMMAS.

To: Howard Deedy — Date: November 24, 1984

From: Charlotte Smart

Subject: Improper Repairs on the Barrett-Woodward Rotary Car Dumper and Positioner

I understand that the recent repairs on the BW (Barrett-Woodward) rotary car dumper and positioner did not follow the recommended procedure and are therefore inadequate. I would appreciate your investigating and clarifying this situation.

Here are the facts as I understand them. Please inform me if these facts are incorrect.

Early in November, we discovered many problems with the BW rotary car dumper and positioner and called the BW Customer Service Engineering Group. They analyzed the problems and recommended the following:

1. Replacing the 3/8-inch steel rails with 1-inch rails. The positioner rides on these rails. An AR (abrasion resistant) plate caps the steel, and both rest on a concrete pedestal. The concrete is spalling, and the rail welds are beginning to crack.
2. Installing the 1-inch rails by grouting under them and continuously welding the AR plate to the rails.

Management authorized the repairs on November 9. Before they were completed, the BW consulting engineer discovered that the AR plate had been stitch welded, not continuously welded. He claims that he reported this to Howard Beale. However, I learned of the improper welding only after John Sturgees conducted a safety inspection and discovered it more or less by accident.

Apparently, the improper welding was not replaced. Instead, the repairs were covered with grout. BW believes that the AR plate will eventually curl up and become loose because of the improper welding.

We seem to have ignored BW's advice and potentially wasted BCC maintenance funds, especially if we have to redo these repairs a year or two from now.

BCC
BOUNTIFUL CHEMICAL COOPERATIVE, INC.

While basically a request for information, this memo is potentially critical of the reader or people working for the reader. In such politically or humanly sensitive matters, tone is important. The writer must be forceful without being aggressive, direct but not blunt, businesslike but not inhuman.

Clearly, this memo is written to someone who already knows a lot about the subject. The abbreviations *BW* and *AR* are explained, but some terms, such as *spalling,* are not defined. The writer has assumed a knowledgeable audience. See SCIENTIFIC/TECHNICAL STYLE.

REQUEST MEMO
For Clarification of a Problem

The closing is direct but not harsh. *Please* makes the closing more courteous, and *shed whatever light,* which is a cliche, makes it more colloquial and therefore personal. See TONE and CLICHES.

Howard Deedy -2- November 24, 1984

I would like to know why the repairs were not made according to BW's recommendations and who was responsible. Please shed whatever light you can on this issue.

C.S.

rw

TRANSMITTAL MEMO
With Attachment

Both the subject line and the opening line of the first paragraph focus on the key point—a new field test. The word *attached* appears after the key point. See LETTERS and ORGANIZATION.

The parenthetical comment separated by commas could have been separated by parentheses or dashes. See DASHES and PARENTHESES.

Long introductory clauses should be separated from the sentences they introduce. See COMMAS.

Courteous writers give their telephone or extension numbers when they offer to provide more information. See LETTERS.

To: Production Supervisors Date: February 28, 1984

From: Clarence Hough

Subject: A New Field Test of Effluents for Sulfides

A new field test for determining the presence of sulfides in the effluents we release into the county water system is attached. This test is simpler and faster than our current tests and is the one that the Dade County Sanitation District currently uses.

The new test, commonly called the Alka-Seltzer test, uses half of an Alka-Seltzer tablet in one cup of the sample. The liberated gases pass through a test paper, which turns brown if sulfides are present (a positive reaction).

While simple and fast, the test will not reveal the concentration of the sulfides present. Samples with a positive reaction must go to the laboratory for further testing.

If the Alka-Seltzer test is positive, then the Hamilton test will usually give a positive reading. When the Hamilton test and the Alka-Seltzer test disagree, use the Alka-Seltzer results to determine whether to release effluents into the county water system.

Please call Steve Hankin (Ext. 589) or me (Ext. 591) if you need further information about this new test.

C. H.

nm
Attachment

MIDLAND OIL AND GAS OPERATIONS, INC.

This routine transmittal memo properly starts with its point—the new field test. Thus the memo follows a managerial format (see ORGANIZATION); the writer wisely avoids a recital of the history of sulfides testing.

The paragraphs are short and direct. These, plus the short sentences, contribute to the readability of the memo. See PARAGRAPHS and SENTENCES.

SAFETY MEMO
With a Mild Reprimand

The specific subject captures both issues raised in the memo. See HEADINGS.

The first paragraph combines good and bad news—thus softening the impact of the reprimand. The contractions make the memo sound less formal, more personal.

This detail paragraph presents information simply and directly. The logical pattern is cause to effect. See PARAGRAPHS.

The tone of the reprimand is professional, not personal or harsh. The last sentence, with its inclusive *we*, softens the impact of the memo while reinforcing its message.

To: Jerry Falmouth Date: March 14, 1984

From: Wally Velder

Subject: Your Accident on December 23 and the Prevention of Accidents

Jerry, I'm pleased that you weren't seriously injured in your accident on December 23, but I'm also disappointed that this accident happened to one of our foremen. Managers and supervisors must be especially safety conscious in our business.

I understand that you were working as a substitute foreman on the Acid Line and that you stepped on a steel meter cover, slipped, and twisted your knee. It had been raining earlier in the afternoon, and the plate was wet. You were wearing safety shoes, but apparently your shoes are old and the tread is worn smooth.

As a foreman, you should be especially alert to hazardous conditions, such as wet flooring and worn treads on safety shoes. You set an example that those people working for you should follow. If we, as management, are to prevent accidents, we must anticipate them.

Please be more cautious in the future.

W. V.

mm

BCC

BOUNTIFUL CHEMICAL COOPERATIVE, INC.

Personnel memos (or letters) are often difficult to write, especially as in this case, when the letter contains even a mild reprimand. A proper tone, therefore, is essential. As in the above example, tone comes from several things: (1) positive information—as in the opening comment about Jerry's not being seriously injured, (2) a personal tone—as in the pronouns and use of Jerry's name in the opening, and (3) a firm but courteous closing—both with reasonable and honest requests for improvements and with the use of *please*. See TONE, MEMOS, LETTERS, and PRONOUNS.

PERSONNEL MEMO
With Suggested Procedures

Although more humorous than informative, the subject does fit the light tone of the memo.

To: All Employees of the Sales Department

From: James Haworth

Date: May 30, 1984

Subject: Your "Home Away from Home"

The opening sentence states the point of the memo. Beginning with a contraction creates an informal, conversational tone. See TONE, MEMOS, and LETTERS.

Let's all try to do everything we can to make our offices as pleasant, comfortable, and well maintained as possible.

Offices should reflect our personalities, but they also should convey a professional impression to visitors. So please take time to consider ways to ensure that your office reflects our excellent products.

Suggestions

The bulleted list is a good way to highlight the suggestions. Note that each listed item is parallel in structure. See LISTS and PARALLELISM.

The use of the run-in headings (in all capitals) visually highlights each suggestion.

- NAILS. Please use low-impact nails for hanging pictures. They will support most pictures, and they do less damage to the walls. My secretary has these if you need them.
- SCOTCH TAPE. Please do not use it. Tape is a quick and easy way to get that poster, map, or note up on the wall, but when it is removed, it removes paint and sometimes plaster.
- POSTERS, MAPS, PICTURES. Please frame any that you intend to hang. Several inexpensive frame shops around town can make a $2.98 print look like a $200 lithograph. Family pictures likewise get special attention if attractively framed.
- MACRAME AND OTHER PLANT HANGERS. The secret to hanging things from ceilings is a little gadget called the "handi-hook," which is designed for suspended ceilings. It hooks over the metal separators and is easy to install or move. However, it can't support very heavy plants. My secretary has a supply of handi-hooks.

Personnel notices often require a light touch. Employees rightly object to personnel notices that are too serious, impersonal, or even critical in tone (and content). Managers and supervisors sometimes have to send reminders and requests that are potentially negative (as this notice could be). The lighter tone helps writers convey negative information in a manner that readers will not find objectionable. See MEMOS, LETTERS, and TONE.

PERSONNEL MEMO
With Suggested Procedures

The social closing (beginning with *please*) is also conversational in tone.

Sales Department Employees -2- May 30, 1984

- GENERAL HOUSEKEEPING. Neatness and organization take a lot of effort, but they are worth it. We'll all be able to walk the halls without glancing into offices and wincing. Even more importantly, we'll be able to locate information more easily if the information is carefully labelled and filed.

Please give some thought to making your office as attractive as possible. We have a beautiful facility, and with a little effort from us all, it will remain so.

J.H.

mm

PROPOSAL MEMO
To a Negative Audience

The arrangement of these items varies from memo to memo. See MEMOS.

To: Jack Ladda and Harvey Smith

From: Charles Percival

Date: August 23, 1984

Subject: Upgrading Substation Testing and Maintenance

Because the readers are antagonistic, the memo proposes a meeting, not a solution, and the language is deliberately conciliatory. Note the inclusive *we*. See TONE.

Let's explore ways to upgrade substation testing and maintenance. As we found out in Tuesday's meeting, we can agree on a number of the current problems; now we need to discover ways to solve these problems. If you are available, let's meet at 2 p.m. Wednesday, September 6 in my office to discuss our ideas and solutions.

To highlight a paragraph list, the numbers are enclosed within parentheses. See PARENTHESES and LISTS.

During Tuesday's meeting we identified many specific test and maintenance areas needing improvement: (1) maintenance intervals, (2) tests between maintenance, and (3) adherence to maintenance procedures. Even more important is our overall commitment to maintenance as a priority responsibility, especially given the complex and expensive equipment in the OG&E electrical system.

The opening sentence to this paragraph is a topic sentence: it states the point of the paragraph. See PARAGRAPHS and COLONS. The sentences are short and active. See ACTIVE/PASSIVE and SENTENCES.

Yet the signals are clear: Maintenance is not a high priority in many substations. Many substations have inadequate maintenance personnel, and many of the personnel they do have are not properly trained. Some maintenance crews spend too much time on ordinary construction. Finally, maintenance personnel who leave are not being replaced. Obviously, management needs to address these problems if we are going to improve our maintenance situation.

As a starting point, here are a few preliminary suggestions for improving our test and maintenance procedures:

1. Use contract or construction personnel, not maintenance personnel, for all construction work.

2. Make comprehensive manpower and equipment forecasts based on present maintenance needs. Use these to hire additional maintenance personnel.

OSAGE
GAS AND ELECTRIC
COMPANY, INC.

Because of the antagonistic audience, the writer deliberately makes his memo more tentative than it would ordinarily need to be. The opening paragraph is conciliatory, and the six listed items are called suggestions, not proposals; the six are also listed well into the memo rather than at the beginning. See ORGANIZATION, LETTERS, AND MEMOS.

PROPOSAL MEMO
To a Negative Audience

The displayed list highlights the suggestions. Note that the items listed are parallel and begin with imperative verbs. See PARALLELISM, SENTENCES, and VERBS.

Repeating the meeting time reinforces the central message of the memo and solicits the reader's cooperation. See REPETITION and EMPHASIS.

Jack Ladda and Harvey Smith -2- August 23, 1984

3. Train existing personnel to meet the above forecasts.
4. Contract with Noble Engineering to perform key testing and to inspect and maintain power circuit breakers. Noble personnel should be on call 24 hours a day but should not be used only to catch up during crises.
5. Establish a trouble-shooting maintenance crew in the Central Office. This crew would have special responsibility for extra high voltage equipment.
6. Make Engineering, not Operations, responsible for maintenance and testing.

Can you meet at 2 p.m. on Wednesday, September 6 in my office to discuss these and other suggestions? Let's do what we must to arrive at constructive solutions to our test and maintenance problems.

C.P.

uio

SUMMARY MEMO
For an Executive Audience

The set-up is brief and informative. The recommendation immediately follows it. See LETTERS and ORGANIZATION.

Headings signal key divisions in the executive summary. They highlight the document's organization and allow readers to locate information easily. See HEADINGS and SUMMARIES.

Attachments should always be referred to in the body of the document. See APPENDICES/ATTACHMENTS.

To: Steward Pollack Date: June 12, 1984

From: Brenda Hamilton

Subject: Osage Gas Plant
Compressor Engine and Cylinder Overhaul
Contract MR-789-65

The following mechanical overhaul contractors have submitted time and material bids to perform the compressor engine and cylinder overhauls at the Osage Gas Plant:

Martin Energy Services
J & L Incorporated
Efficiency Production Services
White River Maintenance, Inc.

Recommendation

Based on the attached bid summary and the related analysis, I recommend contracting with White River Maintenance, Inc., to perform this work. All bidders, except for J & L Incorporated, appear to be equally qualified, but White River was the lowest bidder by an estimated $50,000.

Bid Calculation

Bids included only time and material, so we had to analyze the bids by developing a hypothetical crew size and calculating a composite manhour rate. From these calculations, we were able to determine the total labor costs for each contractor. We also calculated probable transportation costs as an add-on amount to labor costs. Finally, we built in an 85 percent contingency amount based on the uncertainty of the parts needed to complete the overhaul. For further details, see the attached bid calculations.

MIDLAND OIL AND GAS OPERATIONS, INC.

An executive summary can open a report or, as in this case, provide a general summary for detail appearing in attachments. Readers are usually busy managers and supervisors who already know something about the subject. Consequently, they may not need more information than that appearing in a brief summary to make a decision. See SUMMARIES and REPORTS.

SUMMARY MEMO
For an Executive Audience

The paragraph divisions parallel the headings. See HEADINGS and PARAGRAPHS.

One-sentence paragraphs emphasize key points. See PARAGRAPHS and EMPHASIS.

Steward Pollack
Page Two
June 12, 1984

Scope of Work and Specifications

The scope of the work includes the overhaul of 5 Hall-Burke HB-76 gas engines and 12 compressor cylinders. White River will unload all equipment and will handle all procurement through their office. They will also pay for all transportation costs and for the complete installation of necessary replacement parts. Finally, they will be responsible for aligning and grouting the crankshaft.

A copy of the specifications is available from Sven Nordstrom of Engineering Services.

Implementation

Either J. K. Barnes or I will submit a weekly progress report and budget update. J. K. Barnes will be on-site at the White River shop to guarantee the quality of the work.

B. H.

gsp
Attachments

After a brief set-up that orients readers on the subject, an executive summary should state its major point, which may be a conclusion, recommendation, request, or announcement. A limited amount of supporting information or detail should follow. A summary must include all of the major ideas, facts, or data that appear in the attachments or in the larger report being summarized. See ORGANIZATION.

PROCEDURE MEMO

The opening sentence establishes the purpose of the memo. See LETTERS and ORGANIZATION.

Parentheses enclose supplemental information. See PARENTHESES.

Each procedure opens with an imperative verb. See PARALLELISM, SENTENCES, and VERBS.

Each item in the subordinate list begins with a dash. The format and indentation differences between the major (or outer) list and the subordinate (or inner) list help readers tell them apart. See LISTS.

The reason for the procedures is less important than the procedures themselves, so the rationale comes last.

The paragraphs are short, which aids readability. See PARAGRAPHS.

To: All Department Heads

From: Susan Hall

Date: December 28, 1984

Subject: Records Stored at Trolley Street Warehouse

Please comply with the following procedures when you need to store records at the Trolley Street Warehouse.

1. Use a standard records transmittal box and a records transmittal slip. (Call 287-9009 to order these boxes. Each empty box will contain a blank records transmittal slip.)

2. Include the following information on each records transmittal slip:

   —Description of the contents

   —Destruction date according to policy R-23 (Duplicate copies of this policy are available from Susan Jameson in Corporate Services.)

   —Name of responsible supervisor

   —Department number (and extension)

3. Enclose the white and yellow copies of the records transmittal slip inside each box of records. At the Trolley Warehouse, the number and location of each box will be recorded on the yellow copy, which will be returned to the department sending the records. Each department should maintain a file of the yellow copies.

These procedures result from the recent reorganization of the records retention area. This reorganization was necessary because too many records were being lost and too many out-of-date records were being retained.

SKY AVIATION

A clear format is essential in procedures. In this memo, the numbered and dashed lists are crucial if the procedure is to be clear and readable. See EMPHASIS and LISTS for options; also see the more formal procedures illustrated elsewhere in these model documents.

The imperative verbs are also essential in procedures. They highlight actions the reader should take, and they allow writers to condense their directions. For the sake of parallelism, state <u>all</u> directions or steps using imperative verbs. See SENTENCES.

PROCEDURE MEMO

Giving the extension number is courteous. Writers should do so even when they know the readers well.

The writer's initials (either typed or signed) are optional.

All Department Heads -2- December 28, 1984

Under this new procedure, all departments should know what is and is not stored. Periodically reviewing the records transmittal files should enable departments to determine what stored materials can be destroyed.

If you have any questions about these procedures, please call me at extension 2344.

S. H.

blp

MEMO REPORTING ON TRAINING

The subject is specific. See HEADINGS and MEMOS.

To: James MacWhortle

From: Seymour Hirschfield

Date: December 18, 1984

Subject: Assessment of the Applications Engineer School, Dallas, November 5-27

The first paragraph summarizes the assessment and makes a recommendation. See ORGANIZATION, MEMOS, and LETTERS.

The Dallas school for applications engineers was both effective and worthwhile. I recommend that all of our applications engineers attend this school. The training should more than pay for itself in reduced maintenance expenses.

Based on what I learned, here are some general recommendations that apply to our situation at Gulfport:

The displayed list highlights what the writer learned from the school, including potential changes in Gulfport procedures. This is more important than the course content. See ORGANIZATION and LISTS.

1. We should develop documentation standards for applications engineers. We spend far too much time tracing and flow charting existing programs. These inconsistent and ad hoc documentation efforts lead to excessive maintenance time, needless repetition of maintenance, and frequent errors during installations.
2. All Gulfport applications engineers should be given a master copy of the standard logic flow. The standard logic flow lists steps in the proper order, thus helping to ensure efficient adjustments to the program. Working from the standard logic would minimize omissions and mistakes, especially those resulting from ad hoc decisions. This standard logic would also help to improve our documentation. (See the attachment for an example of a standard logic flow.)
3. All Gulfport applications engineers should attend the Dallas 3-week school. This school is just as valuable as the basic school for control engineers. The optimum arrangement would make the Control Engineer School a prerequisite for the Applications Engineer School.

The course content should be no longer than necessary. If readers want detail, give it to them; if not, summarize.

Course Content

The school provides the applications engineer with enough background knowledge for understanding the computer system

MIDLAND OIL AND GAS OPERATIONS, INC.

This memo deliberately opens with the "bottom line," so that busy supervisors or managers do not have to read beyond the first paragraph unless they want more information. The key applications follow immediately. Everything else is optional (depending upon whether readers wish to know more of the details). See ORGANIZATION.

MEMO REPORTING ON TRAINING

Parentheses enclose additional but non-essential information. See PARENTHESES.

The headings divide content into identifiable areas, making it easier for readers to understand how the document is organized and where to find particular information. See HEADINGS.

James MacWhortle -2- December 18, 1984

and utility programs, as well as the basics of computer programming. We also reviewed the fundamentals of process dynamics.

During the first week (the lecture week), the instructors gave us reading assignments. Each day's instruction ended with a question-and-answer period, with plenty of time to review essential concepts.

During the second and third weeks (the laboratory weeks), we used a computer simulation of a real process at our East Chicago plant. The goal was to build, tune, and document a control scheme to control the process. Inherent in the simulation were process noise, non-linear valve characteristics, and dead time. To test the process, we subjected it to large load upsets.

The first assignment was to build all variables and to write supervisory logic for the control scheme. Trial-and-error tuning was not effective, so we used SLEUTH to obtain tuning constants. When documentation was complete, we subjected the loop to large load upsets and then observed and recorded the responses. We learned that ordinary supervisory control was not fast enough for an acceptable response.

Next, we used DDC programming with two feed-forward algorithms. The first was a simple ration controller, while the second included dynamic lead/lag terms. East Chicago's Shut Blocks (pre-programmed, shared, or common general blocks) were used for the feed-forward algorithm. We determined the feed-forward constants using SLEUTH and PFTUNE. This control scheme caused significantly less deviation in the control variable, and response time was significantly faster.

<u>Course Presentation</u>

Both the instructors and the materials were excellent. The laboratory simulations were the high point of the school. Such practical experience guarantees that students retain the skills and concepts learned and know how to apply them in actual plants and facilities.

The recommendation about other engineers attending the school is the memo's most important idea; therefore, it appears very early in the memo. The recommendation about rental cars is less important and would detract from the memo's purpose if it came first, so the author subordinates it by putting it at the end of the memo.

MODEL DOCUMENTS

MEMO REPORTING ON TRAINING

James MacWhortle -3- December 18, 1984

The instructors had all had field experience as applications engineers, so they were well versed in the content. Moreover, they were excellent presenters and knew how to involve students throughout the 3 weeks of training.

Final Recommendation

Some of us attending the school felt isolated during the weekends, so I would recommend either making several rental cars available or holding the course at a less isolated site. The training took place at the downtown Holiday Inn, where much happens during the week. But over the weekend, everyone leaves for the suburbs. The town seemed deserted, and walking around after dark was not advisable.

S. H.

mm
Attachment

The tone of the memo is conversational. The writer's use of the pronouns *I* and *we* is appropriate because he is relating a personal experience. See ACTIVE/PASSIVE and TONE.

JOB DESCRIPTION

These initial facts will vary from company to company, but job codes and the date of the last revision are usually desirable pieces of information.

The opening section (and its heading) highlights those general personnel traits usually valued in any employee (not just an entry-level auditor).

The traits are listed beginning with a verb. This eliminates much needed repetition—for instance, "the employee will . . . ." In essence, this approach is an action one. See PARALLELISM.

The numbering system (and the lists) will allow for easy references to a particular duty or responsibility. See LISTS and NUMBERING SYSTEMS.

JOB DESCRIPTION

JOB TITLE Internal Auditor	
JOB LEVEL 101-Au	PAY RANGE 5
DEPARTMENT Internal Auditing	GROUP OR AREA Production

General Requirements

A. Works effectively as a member of the Audit Department

1. Maintains the highest standards of professionalism
2. Cooperates with co-workers to improve both the quality of the work and the morale of the department
3. Works diligently to improve the department's audit processes and procedures
4. Assists in making audit reports and other departmental documents as professional as possible

B. Maintains good job behavior

1. Is punctual and conscientious about work hours, and makes appropriate arrangements for vacations and special leave
2. Is properly groomed at all times
3. Performs tasks cheerfully and without complaint
4. Accepts responsibility for all properly assigned tasks
5. Accepts criticism well and makes sincere efforts to improve

LAST REVISED August 1983 PAGE 1 OF 3

Job descriptions are difficult to write because they should be as specific as possible, and writers (often the employees themselves) don't want to be very specific about the duties. Also, many employees think job descriptions are a waste of time and effort. (Actually, they may be a waste if they are only written for the files or if they are misused by supervisors.)

MODEL DOCUMENTS

JOB DESCRIPTION

The specific duties and responsibilities are organized almost chronologically—from the beginning of the audit to the end. Other organizations would be possible. See ORGANIZATION.

The parentheses set off an extra piece of information. See PARENTHESES.

The numbering of the main subheadings (A, B, C, etc.) continues sequentially throughout the job description (despite different major headings). This sequential numbering also permits good cross referencing.

Job Title Internal Auditor

Specific Duties and Responsibilities

C. Analyzes upcoming audit situations

1. Reviews relevant prior audits (including working papers)
2. Solicits help (if appropriate) from auditors who worked on prior audits
3. Prepares the individuals being audited—either in person or through correspondence
4. Coordinates upcoming tasks with other auditors participating in the audit

D. Conducts efficient and professional audits

1. Maintains an objective and unbiased attitude toward those being audited; is conspicuously fair
2. Briefs those being audited as to goals, procedures, and purposes
3. Creatively plans and projects necessary audit tasks and procedures
4. Devises appropriate strategies for tracking key tasks
5. Prepares orderly working papers and other records of the facts discovered during the audit
6. Consults with other auditors (including the audit supervisor) to guarantee that the audit is proceeding properly and according to SOP
7. Maintains good personal relations with those being audited
8. Conducts (or participates in) an effective exit interview

Page 2 of 3

The above description covers two areas: (1) general requirements and (2) specific duties and responsibilities. General requirements by themselves may be too general—hence, the need for some specific job-related duties and responsibilities.

Job Title Internal Auditor

E. Prepares a high quality, professional audit report

1. Submits draft materials by assigned deadlines
2. Assists colleagues during reviews of the overall audit report
3. Writes clear and effective audit summaries, findings, and recommendations
4. Coordinates the preparation of audit reports with secretaries and word processing personnel

F. Assists in transmitting the audit report to appropriate managers and supervisors

1. Prepares individual oral briefings, if necessary
2. Arranges for followup investigations and reports

Page 3 of 3

Some organizations would prefer, of course, to have a proposed rating scale as part of the job description. So each duty or task would have have some sort of scale: *Exceeds expectations, Meets expectations,* etc. Some personnel specialists even argue for evaluations tied to percentages or production: *Has fewer than 5 percent errors* or *Produces 185 widgets per day.* Such scales are not applicable to all types of jobs, so the above job description does not attempt to incorporate them.

PERFORMANCE REVIEW
Positive

The initial sentence (and the following paragraph) summarize the positive nature of the review. See ORGANIZATION.

The review uses third person (*she*) because it goes into the personnel file and not directly to Judith (although she would see this statement). See PRONOUNS.

The discussion opens with strengths—a good strategy in any review.

The run-in headings are keyed to specific duties listed in Judith's job description. See the model job description given earlier in this section.

Each strength is illustrated by an example, even proof.

These suggestions should be as specific as possible, but they are often based only on general impressions.

PERFORMANCE REVIEW

NAME Judith Hirsch	
JOB LEVEL 101-Au	DEPARTMENT Internal Auditing
EVALUATED BY Denise Mead	DATE February 11, 1985

WRITTEN COMMENTARY

Judith Hirsch has been a credit to the Auditing Department during her first six months with us. She has learned a lot about our company, she has learned to work well with her colleagues in the Auditing Department, and, most importantly, she has become an excellent auditor.

The following comments support the numerical ratings in the attached department performance checklist.

Strengths

Duty D2. Judith did an excellent job preparing the Shipping Department for the audit we conducted for them in February. She initially met with their department manager, who commented that she made a good impression on everyone, especially those who were intimidated at the thought of an audit.

Duty D5. Judith's working papers are excellent. She takes accurate notes, and her marginal commentary will aid future auditors in tracking the basis for her findings.

Duty E3. Judith promises to be one of the most able writers in our department. While she is still learning to use our format and organization, her writing is clear and concise. I found it satisfying not to have to rewrite her sections of the final audit report.

Areas Needing Improvement

Duty D3. Judith needs to be more aggressive in challenging questionable practices discovered during an audit. Given her

Denise Mead
EVALUATOR SIGNATURE

PAGE 1 OF 2

The positive performance review should be almost a pleasure to write. Resist, however, the temptation to skip the positive review because "everything's going fine." Even the most eager of workers needs the praise of an honestly earned review.

Reviews are usually part of a checklist, where actual numerical ratings have to be recorded. These ratings are not sufficient, however, so always plan to add written comments. While you should always comment on low ratings, don't forget to mention the positive examples.

PERFORMANCE REVIEW
Positive

Name Judith Hirsch

Job Title Internal Auditor Date 2-11-85

newness to the company, she was concerned with establishing good relationships with personnel from other departments, but she needs to be personable while challenging questionable practices.

Duty D6. Judith needs to work more closely with the other auditors in the department. She seems reluctant to reveal uncertainties, but newly hired auditors must ask questions about procedures and about departments scheduled for upcoming audits.

The summary repeats in different language the points made in paragraph 1. See REPETITION and ORGANIZATION.

Summary

Except for the above duties, where Judith's evaluations were either lower or higher than expected, her evaluations were typical of promising new hires. She is, therefore, an asset to the Auditing Department. I look forward to following her career with us.

Page 2 of 2

Written comments should be as specific as your time and knowledge allow. Someone reading your comments should believe that you had a firm basis for the different ratings you have given. Also, specific comments (even proof) are the most memorable features of a review.

PERFORMANCE REVIEW
Negative

The initial sentence opens with something positive despite the negative nature of the review. The positive point should be an honest one.

The promotion issue is also valuable background for what is to follow. See LETTERS and ORGANIZATION.

The initial points should explain Judith's strengths in very specific terms. These points help to put the negative comments into perspective.

The run-in headings correspond to the duties listed in the job description. See the model job description given earlier in this section.

The direct quotation is an excellent piece of evidence (positive or negative). See QUOTATIONS and QUOTATION MARKS.

PERFORMANCE REVIEW

NAME Judith Hirsch	
JOB LEVEL 101-Au	DEPARTMENT Internal Auditing
EVALUATED BY Denise Mead	DATE February 11, 1985

WRITTEN COMMENTARY

Judith Hirsch continues to improve as an auditor, but she still has some serious weaknesses. Unless these weaknesses are corrected, Judith will not be eligible for promotion within the Auditing Department.

The following comments support the numerical ratings in the attached department performance checklist, which is based on the job description for an internal auditor.

Strengths

Duties B3 and B4. Judith accepts assignments cheerfully and attacks them with enthusiasm. When I asked her to fill in for Steve Broom (who suddenly quit), she immediately agreed and energetically began to assume his role in the auditing team for the Production Department.

Duties D2, D7 and D8. Judith seems to relate very well with the field personnel we often deal with. Last month, for instance, she participated in an audit of the Maintenance Department. According to the senior auditor who managed that audit, Judith was a "strong, likable representative of the Auditing Department." I suspect this is due to Judith's mechanical abilities, as well as her solid one-on-one skills.

Areas Needing Improvement

Duty D5. Judith's working papers are not acceptable. They need to be carefully planned and cross referenced so that all findings have clear supporting data. This is a skill related to general writing ability. I suggest that Judith prepare a detailed written outline or mock-up of future working papers and that she check this outline with either a senior auditor or me.

Denise Mead
EVALUATOR SIGNATURE

PAGE 1 OF 2

Negative reviews are always a problem. Face-to-face meetings are bad enough, but having to put negative judgments into writing is even worse. To make such written reviews as painless as possible, bring in some positive traits and keep your discussion as specific as possible. After all, the employee needs to know exactly why you are dissatisfied with the job performance.

Positive points are essential so that the employee can't argue that you have not even tried to be fair. Managers should also reinforce positive behaviors whenever possible.

PERFORMANCE REVIEW
Negative

Besides being specific, the negative comments should, if possible, suggest ways to improve. Here Judith is asked to develop a "plan of attack."

The summary is as upbeat as possible, but still as honest as necessary. It does partially repeat paragraph 1. See REPETITION.

Name Judith Hirsch

Job Title Internal Auditor Date 2-11-85

Duty E3. Judith's writing skills are still below average. She works hard to make her writing correct, but correct writing is not necessarily adequate. Judith's writing is usually "thin," often unsupported. She seems to have trouble expanding on or supporting her findings. I suggest that she develop a plan of attack to work on this deficiency; we can involve several of our more seasoned auditors in the plan. Perhaps they could review draft materials and give specific feedback on a variety of writing tasks.

Duties C1 and D6. Judith still needs to be more willing to ask for help within the Auditing Department. Her good one-on-one skills need to extend to areas of professional activity. Judith needs to avail herself of department resources if she intends to continue to improve.

Summary

Judith could have a bright future with our department, but as I have explained to her, her advancement will be slow if she is unable to grow in the areas discussed above. The three deficiencies mentioned above are critical to this growth, so these will be the basis for the next performance review in six months.

Page 2 of 2

Candor, while difficult, is essential, especially if some negative personnel action (firing, demotion, or a transfer) is likely. Without candor, the employee has no idea of how to improve, and ultimately you have no written record on which to base the firing, demotion, or transfer.

Despite being part of a pre-printed form (usually including a checklist), the review should still use the basics of good writing: emphasis, a clear organization, specific examples, clear and concise sentences, etc. See EMPHASIS, ORGANIZATION, and STYLE.

MODEL DOCUMENTS

PROCEDURE
With a Traditional Format

The heading, title, procedure number, and date permit easy updating and quick reference.

The purpose is brief, yet specific and informative.

The outline format uses decimal numbering. This allows for easy cross referencing. See NUMBERING SYSTEMS.

Responsibilities are written in sentences, but these are usually not imperative in form. See SENTENCES.

China Lake Refinery — Refinery Instruction MOGO-111

EQUIPMENT NUMBERING PROCEDURE

August 5, 1984 — Page 1 of 3

1. Purpose

   This instruction establishes the procedure for numbering new equipment, replacement equipment, equipment transferred between plants, and equipment transferred to a plant from reclamation. The equipment of primary concern includes pumps, pump drivers, turbines, blowers, fans, and compressors.

2. Responsibilities

   The responsibilities and operation of the equipment numbering system are as follows:

   2.1 The Engineering Division is responsible for assigning plant equipment numbers. Equipment should be numbered as follows:

   2.11 Replacement Equipment

   Equipment purchased to replace existing equipment receives the existing equipment's number. For example, if a new pump replaces existing pump G-89, the new pump becomes pump G-89.

   2.12 New Capital Installation

   Equipment purchased for new installation or expansion receives the next sequential plant number according to the Drafting Department's records.

   2.13 Used Equipment Transferred Between Plants or Equipment Transferred to a Plant from Reclamation

   This equipment receives an existing equipment number if it replaces equipment (under paragraph

MIDLAND OIL AND GAS OPERATIONS, INC.

Traditional procedures use some techniques to make reading easier, but they also more closely follow the patterns of ordinary business and technical language. Procedures using an action format (as in the following example) are usually more telegraphic or skeletal in their format and their language. The action format is often easier for readers to read and to use.

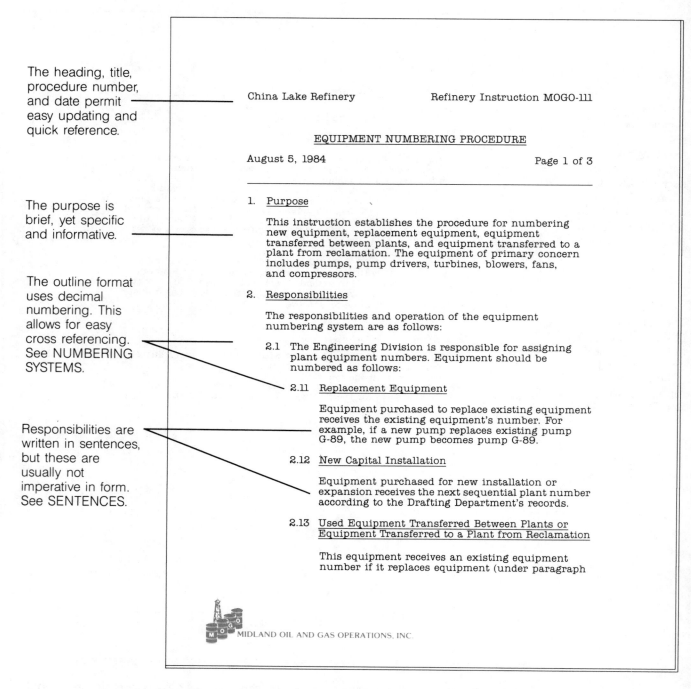

PROCEDURE
With a Traditional Format

Headings on all pages repeat the key information from the first page. This allows quick identification of the procedure if the pages become separated.

Examples are essential for clarity.

The responsibilities under this section are phrased in parallel *-ing* forms. See PARALLELISM.

China Lake Refinery — Refinery Instruction MOGO-111

EQUIPMENT NUMBERING PROCEDURE

August 5, 1984 — Page 2 of 3

2.11) or the next sequential number if it is for new installation or expansion (under paragraph 2.12).

2.2 For Engineering filing and record purposes only, the following modification to the above numbering system is necessary:

2.21 Equipment that replaces existing equipment receives the existing equipment's plant number, as described above.

To indicate the number of replacements occurring in a particular service, the equipment number will receive a suffix. For example, a pump replacing G-89 would be numbered G-89-1. A pump replacing G-89-1 would be numbered G-89-2. Due to present computer system limitations, maintenance equipment records will reflect only the plant number, as discussed in paragraph 2.1.

2.3 The engineer assigned to monitor equipment numbers is responsible for the following:

2.13 Ensuring that new equipment receives proper numbers and that Engineering has recorded the correct numbers, according to paragraph 2.2 above.

2.32 Ensuring that drawings, data sheets, performance curves, installation, operating and maintenance procedures, and other associated information on equipment being replaced are removed from the plant file and photostat books and are transferred either to the reclamation or idle equipment files.

MIDLAND OIL AND GAS OPERATIONS, INC.

Other organizations of this procedure would be possible. For instance, the general subsection entitled "Responsibilities" is unnecessary. So a writer could convert subsection 2.1 to 2; subsection 2.2 to 3; and subsection 2.3 to 4. Of course, all the other numbering would change.

MODEL DOCUMENTS

PROCEDURE
With a Traditional Format

Comments in parentheses often emphasize points. Boldface type is another option. See EMPHASIS and BOLDFACE.

Paragraph lists are less visible than displayed lists, but they still visually emphasize points. See EMPHASIS and LISTS.

China Lake Refinery — Refinery Instruction MOGO-111

EQUIPMENT NUMBERING PROCEDURE

August 5, 1984 — Page 3 of 3

Further, ensuring that new equipment information is placed in the appropriate plant file. (Plot plans do not require updating because the plant number does not change.)

2.33 Providing maintenance coordinators with the following information: (1) appropriate equipment numbers; (2) installation, operation, and maintenance instructions; (3) complete parts lists including suggested spare parts; (4) performance curves; (5) data sheets; (6) wiring diagrams and electrical information; and (7) outline drawings. Maintenance coordinators are responsible for updating their records to reflect the new equipment.

2.34 For equipment transfers between plants or from reclamation, ensuring that equipment drawings and other equipment information are transferred to the appropriate plant files and photostat books, and that those files and books reflect the new equipment numbers.

MIDLAND OIL AND GAS OPERATIONS, INC.

PROCEDURE
With an Action Format

The heading block provides the title, SOP number, date, and the number of pages. This format allows easy updating and quick reference.

The scope defines the applicability of the procedure.

The policy statement describes the intent behind the procedure and establishes the policy's basic goals.

The contents description previews the rest of the procedure and functions as a table of contents.

Each step in the procedure opens with the person or department responsible for the step. Next comes the action associated with the step. Note that the action is stated in imperative sentences. See SENTENCES.

BEDROCK MINING & MILLING CORPORATION

Administrative
Standard Operating Procedure

SUBJECT: DISTRIBUTION OF WEEKLY PAYCHECKS	DATE: 8 SEP 84	S.O.P. 24 page 1 of 5

PURPOSE

To establish a procedure for the security and distribution of weekly paychecks.

SCOPE

This procedure applies to Bedrock Mining & Milling Corporation locations in the Billings, MT and Rock Springs, WY areas.

POLICY

Payroll checks will be distributed in such manner as to ensure their prompt delivery to maintain good management/employee relations while providing sound internal controls.

CONTENTS

SECTION A. CHECK CUSTODIANS

RESPONSIBILITY	ACTION
Each Department Head	A-1. Appoints check custodian for his or her department by signed memo to the CDS (Check Distribution Section), giving name, dept. number, and signature of check custodian.

BMM

Procedures using an action format are as schematic as possible, with headings, lists, and imperative statements. This format enhances readability and allows readers to find their particular responsibilities and actions.

The most common problem with procedures is the passive voice. Writers list actions passively, and readers often don't know who is supposed to do what. A procedure that says "Checks must be examined before delivery" does not indicate who is

PROCEDURE
With an Action Format

SUBJECT: DISTRIBUTION OF WEEKLY PAYCHECKS — DATE: 8 SEP 84 — S.O.P. 24

page 2 of 5

		Informs CDS by memo of changes in check custodians. In no instance appoints a supervisor/manager who approves time cards as the check custodian.
Check Distribution Office	A-2.	Assigns a specified time for each check custodian to pick up checks at the Check Distribution Office.
	A-3.	Forwards checks to outlying check custodians via BMM courier, using locked bags.
Check Custodian	A-4.	Picks up checks at the assigned time, signing the pick-up log for the checks. If he or she misses the assigned time, picks up checks as soon as possible.
	A-5.	Prepares Form C82-974, "Notification of Personnel Transfer" (See attachment A), whenever employees are transferred to other departments or locations, and forwards it to the CDS no later than noon on the Wednesday before payday.
	A-6.	Delivers paychecks to payees only. UNDER NO CIRCUMSTANCES GIVES PAYCHECK TO ANYONE OTHER THAN THE EMPLOYEE WHOSE NAME APPEARS ON THE CHECK.

Quotation marks enclose titles. See QUOTATION MARKS.

Underlining and capital letters emphasize key warnings and cautions. See EMPHASIS and UNDERLINING.

supposed to do the examining. Therefore, all steps in procedures must identify not only the action but also the person or department responsible. See ACTIVE/PASSIVE.

Inevitably, procedures must be updated, so the numbering system, date of creation, and any revision date need to appear on each page.

PROCEDURE
With an Action Format

SUBJECT: DISTRIBUTION OF WEEKLY PAYCHECKS | DATE: 8 SEP 84 | S.O.P. 24 page 3 of 5

SECTION B. DISTRIBUTION OF FIRST SHIFT PAY CHECKS

RESPONSIBILITY	ACTION
First Shift Check Custodian	B-1. Picks up checks each Friday at the assigned time.
	B-2. Delivers checks to employees after pickup and safeguards undelivered checks. Returns checks not picked up by employees by 2:30 p.m. to the CDS by 2:45 p.m. DOES NOT HOLD CHECKS UNTIL THE FOLLOWING DAY.
First Shift Employee Who Did Not Receive a Paycheck	B-3. Picks up undelivered check at the Check Distribution Section on the following work day.

NOTE: This procedure would have two more pages. These other pages would add little to the model, so we have omitted them.

MINUTES
With a Traditional Format

The title or subject is specific. See HEADINGS.

The displayed list highlights the criteria. Each item is parallel with the others. See LISTS and PARALLELISM.

Short paragraphs enhance readability. See PARAGRAPHS.

September 20, 1984

MINUTES OF THE SEPTEMBER 15 MEETING ON THE SHUTTLE SYSTEM FROM GREEN BRIAR TO LAKEVIEW CENTER.

Attending: G.L. Benson, Frank Houck, Martha Memmert, and Jeanne Skorut. Absent: Fred Householder.

AN OVERVIEW OF THE SHUTTLE SYSTEM CRITERIA

G. L. Benson reviewed the working criteria for the proposed shuttle system:

1. One round-trip every two hours (two buses in rotation), with trips starting at 7 a.m. and ending at 6 p.m. (Travel time is estimated to be about 50 minutes one way.)

2. Buses must be comfortable and attractive; they must convey a good image for our company.

3. Buses must carry at least 12 passengers.

4. Buses must be economical to operate and maintain.

MLT BUS SYSTEM'S PRELIMINARY BID

Frank Houck reviewed the preliminary bid from MLT System. They proposed using two vehicles at $37.00/hour (including buses, drivers, and maintenance).

If we run the buses 11 hours a day, our weekly, monthly, and yearly fees would be as follows:

Weekly	$ 2,035
Monthly	$ 8,954
Yearly	$107,448

G. L. Benson suggested that we look at other arrangements, given the costs of MLT Bus System. We should also develop comparable figures on costs should we decide to lease our own vans and hire drivers. Martha Memmert volunteered to get competitive bids from several other companies. Jeanne Skorut is going to develop an estimate of what it would cost us to lease our own vans and hire drivers. Both Martha and Jeanne will report on their findings at the October 1 meeting with Fred Householder.

Traditional minutes usually attempt to record a good deal of what happened during the meeting, and they usually follow the events chronologically. The above minutes begin with a review of the proposed criteria and end with the report to Fred Householder. See ORGANIZATION.

MINUTES
With a Traditional Format

Each paragraph summarizes a different motion, with some details about the discussion. The writer is following a chronological pattern: The results of the motions come late in the paragraphs.

Minutes of the September 15 Meeting | Page 2

POSSIBLE CHANGES TO THE CRITERIA

At Frank's suggestion, we discussed where and how our criteria could be more flexible. Martha made a motion that we use a single bus for most round trips, with double buses on the trips at 7 a.m. and at noon. The motion passed, so Martha and Jeanne will get estimates as they investigate the financial arrangements.

The standard paint for MLT vans is "school bus" yellow, and the initial criteria called for the buses to be painted blue and white, in keeping with our corporate colors. However, MLT estimated that painting the vans would add a one-time cost of $1,000 per van. After some discussion, we decided that the costs of painting exceeded the value of blue vans.

Frank suggested eliminating air conditioning on the buses. MLT's proposal also recommended eliminating air conditioning, which costs $2,500 per bus in initial charges and $250 per bus per week in additional operating expenses (because of fuel and maintenance costs). Jeanne pointed out that a 50-minute bus ride in the August heat and humidity would be detrimental to anyone's sanity, but the costs are astronomical. Frank suggested studying the problem further, perhaps by contacting local bus companies and the weather bureau to try to determine the feasibility of eliminating the air conditioning. The motion passed, and Frank was elected to study the problem. He will report on his findings by October 5.

George Benson will discuss our suggestions for revising the criteria with Fred Householder, who developed the initial criteria. George will inform the rest of us of Fred's decisions regarding changes.

Respectfully submitted,

Jeanne Skorut

Jeanne Skorut

The headings, lists, short paragraphs, and brief sentences all contribute to the overall readability, but motions and action items are potentially lost. For this reason, we recommend the action format for minutes (see the following example). See also EMPHASIS.

MODEL DOCUMENTS

MINUTES
With an Action Format

The subject line is quite specific. See HEADINGS and MEMOS.

An accurate record of attendees is essential.

Action items emphasize the people (with names boldfaced), their responsibilities, and the due dates. See BOLDFACE, EMPHASIS, and ORGANIZATION.

This overview supplies necessary background information (much as corrections to prior minutes would do).

The numbered list helps to highlight key information. See LISTS.

September 20, 1984

MINUTES OF THE SEPTEMBER 15 MEETING ON THE SHUTTLE SYSTEM FROM GREEN BRIAR TO LAKEVIEW CENTER

Attending: G.L. Benson, Frank Houck, Martha Memmert, and Jeanne Skorut
Absent: Fred Householder

ACTION ITEMS

Martha Memmert	Obtain bids from several other transportation companies. (Due October 1)
Jeanne Skorut	Develop an estimate of what it would cost for us to lease our own vans and hire drivers. (Due October 1)
Frank Houck	Study the feasibility of eliminating air conditioning from the buses. (Due October 5)
George Benson	Discuss our work with the criteria with Fred Householder, who originally drafted the criteria. (Due October 1)

AN OVERVIEW OF THE SHUTTLE SYSTEM CRITERIA

G.L. Benson opened the meeting with a review of the criteria:

1. Buses must make one round trip every 2 hours (two buses to operate simultaneously), with trips starting at 7 a.m. and ending at 6 p.m. (Travel time is estimated to be about 50 minutes one way.)
2. Buses must be comfortable and attractive—ones that convey a good image for our company.
3. Buses must carry at least 12 passengers.
4. Buses must be economical to operate and maintain.

BCC
BOUNTIFUL CHEMICAL COOPERATIVE, INC.

Action minutes highlight (1) actions during the meeting being recorded and (2) actions needed in the future (usually before the next meeting). Action minutes do not attempt to capture everything that was discussed, and they deliberately do not record the meeting in strict chronological order. See ORGANIZATION.

Some repetition is inevitable, especially if the meeting is long and the issues complex. As above, the action items duplicate items mentioned elsewhere in the minutes. See REPETITION.

MINUTES
With an Action Format

The bid information is necessary background to the following motions.

This motion highlights the name of the person making the motion, but the heading could identify the issue: Other Busing Systems. See HEADINGS.

These headings highlight the content of the motion, but the person's name is attached, in parentheses.

Minutes of the September 15 Meeting Page 2

MLT BUS SYSTEM'S PRELIMINARY BID

Fred Houck reviewed the preliminary bid from MLT System. They proposed using two vehicles at a cost per hour of $37.00 (including the buses, their drivers, and all maintenance).

If we maintained the proposed schedule of 11 hours a day, our weekly, monthly, and yearly rates would be as follows:

Weekly Fee	$2,035
Monthly Fee	$8,954
Yearly Fee	$107,448

G. L. Benson's motion: That we investigate other arrangements, given the costs of the MLT Bus System. Martha volunteered to investigate other companies, and Jeanne will develop an estimate of costs for leasing buses and hiring drivers.

POSSIBLE CHANGES TO THE CRITERIA

Number of Trips (motion by Martha Memmert): That we use a single bus for most trips, with double buses on the trips at 7 a.m. and at noon. The motion passed, so Martha and Jeanne will include this new criterion in their reports.

Painting of the Buses (informal motion): That we not paint the yellow MLT buses blue (BCC's corporate color). Motion passed, saving a one-time cost per van of $1,000.

Elimination of Air Conditioning (motion by Frank Houck): That we eliminate air conditioning (also recommended in the MLT proposal). Initial installation costs would be $2,500 per bus, with weekly costs of $250 per bus (because of fuel, maintenance, etc.). Frank will investigate the problem and report no later than October 5.

Report to Fred Householder: George Benson agreed to report recommended criteria revisions to Fred, who originally developed the criteria. George will report on Fred's response by October 1.

Respectfully submitted,

Jeanne Skorut

Jeanne Skorut

The format above, while not mandatory, does allow for visual openness and impact, thus making the minutes easy for readers to review. See EMPHASIS.

Action minutes still record enough information for them to be useful as a record of what the participants discussed. This information is important in case someone besides an attendee has to review the minutes.

MODEL DOCUMENTS

SAFETY ALERT

The subject line is specific enough to inform readers of the problem.

The actual alert is the point, so it opens the document. See ORGANIZATION.

The investigation revealed certain findings: these appear in short paragraphs, but a numbered list would also be possible. See LISTS and PARAGRAPHS.

Note that similar ideas are grouped together. See ORGANIZATION.

The parentheses enclose additional but non-essential information. See PARENTHESES.

This opening sentence is deliberately passive in order to highlight *alert*. See ACTIVE/PASSIVE.

ALERT RESPONSE

Alert No.:	P5-G-81	Alert Response:	TM-678-BB
Date Received:	July 11, 1984	Response Date:	August 1, 1984
Subject:	Asbestos in Rivlin RK 786 Adhesive		

<u>Alert</u>

All personnel should use extreme caution when working with Rivlin TK 786 adhesive. These precautions include wearing dust respirators and wet sanding areas bonded with the adhesive (see Safety Procedure SP-56). For further information, see the attached safety notice.

<u>Investigation</u>

Rivlin TK 786 adhesive is used by Sky Aviation and our suppliers. This epoxy-based adhesive contains asbestos as an inert filler.

Our Production group uses the adhesive to bond flight hardware (both electrical and nonstructural) and to conduct some test operations. Our suppliers use the adhesive to bond fiberglass epoxy cable tray supports.

Although sanding and abrading is not always required, situations such as spillage can demand sanding and abrading. Sanding releases asbestos in the dust, so wet sanding is a standard precaution.

<u>Actions Taken</u>

This alert (and the accompanying safety notice) has been issued to all Production floor personnel. Procurement will circulate the alert and safety notice to our suppliers so they can warn their employees. Future procurement packages will contain the safety notice.

SKY AVIATION

This document—called an "Alert Response"—is typical of many special documents developed to notify readers of a problem or to document an investigation. Documents such as this generally have a prescribed format that the writer must follow. The headings in the document reflect that format and indicate where readers can find particular types of information (*e.g.*, investigative findings under "Investigation"). In keeping with the nature of the subject, the tone is formal and factual. See TONE.

SAFETY ALERT

The final warning note uses all capital letters to emphasize the point. See EMPHASIS and HEADINGS.

Alert Response TM-678-BB -2- August 1, 1984

Materials Engineering is consulting with NASA to determine whether we can replace TK 786 with TK 780, which is a non-asbestos version of the adhesive.

Actions Pending

None

NOTE: THE ALERT ON TK 786 IS A PERSONNEL SAFETY PROBLEM AND DOES NOT AFFECT SKY HARDWARE.

Kitzi Gelb
RELIABILITY MANAGER

Attachment

SKY AVIATION

FIELD NOTES

The information in these blanks should be as complete as possible.

Some data, such as the date, may need to be added to printed forms.

While telegraphic, comments should not be too brief and too cryptic.

Avoid symbols—for instance, the " for *inches* and ' for *feet*.

Be sure items in one column line up with the parallel information in other columns.

DRILLING AND SAMPLING LOG

HOLE NO. S-4
ELEV. 4900 ±
FEATURE Irrigation Well — DEPTH 75 ft
LOCATION Sec. 18T 17N R15 E — COUNTY Jasper
LOGGED BY L.H. Freyman — DATE DRILLED 7/18/84 and 7/21/84 — WATER LEVEL 16 ft (approx.)
DRILLING CO. Acme — DRILLER J. Stevens — DRILL RIG Glaser HD300 Air Rotary Drill Master
8 in. hammer bit — Starting time 3:52

	ELEV. (DEPTH)	CLASS.	DESCRIPTION FIELD IDENTIFICATION	SAMPLE NUMBER	MODE	REMARKS
7/18	0-15		Glacial debris & colluvium		Started up noon	Set conductor 5-6 ft. Dampness at start but no free water.
	15-25		Added Argo foam at 25 ft	Sample #1 at 25 ft	Water at 15 ft 1:23 p.m.	
	25-47		Chips of Olivine basalt			35 ft — stopped to add Argo foam — to hold back caving
	47 ft.		Drill rate 1 ft/min	Sample #2 47 ft	Down 2:48-3:10	Resume drilling 3:10
	47-50		Drilling going well			
	50-60		" " "	Sample #3 50 ft		
	60-75				Down at 4:45	Broken air valve
			Caving - attempted to clean out — high use of Argo foam		Shut down 6:30 p.m.	
7/21			Set 40' of slotted casing		Started up at 8:30 a.m.	
			Water level 8 ft 4 in.		Completed casing at 9:45 a.m.	

SHEET 1 OF 1

Field notes or logs contain essential technical (even legal) information. As such, they need to be carefully written and systematically filed. Writers should use ink (or, at least, a dark lead pencil), and they should attempt to print or write as clearly and neatly as possible. If possible, writers should review their notes at least once a day and correct any obvious errors or sloppy writing (such as dotting *i*'s or connecting loops on *o*'s).

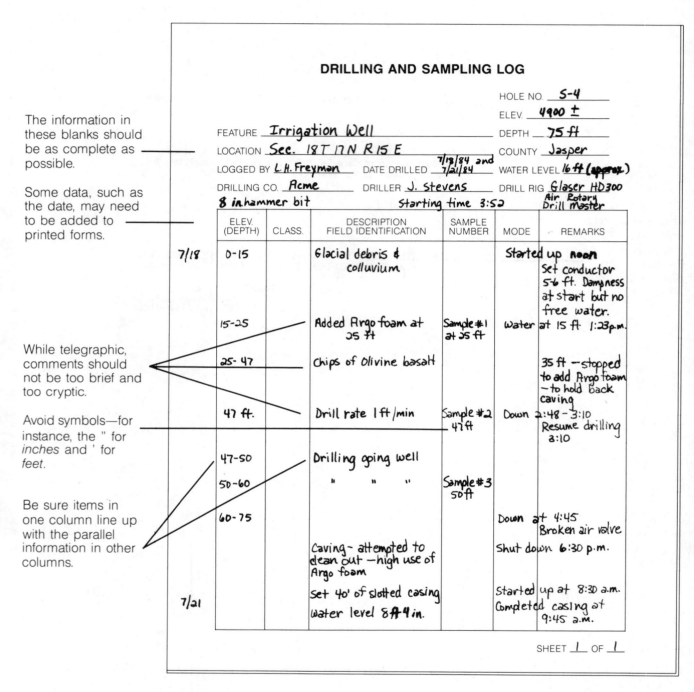

NEWSLETTER ITEM

The headline captures the key idea.

The opening paragraph summarizes the who, what, where, when, how, and why.

The paragraphs are short because newsletters usually appear in narrow columns. See PARAGRAPHS.

Actual quotations help make the story more vivid and readable. See QUOTATIONS.

The dash emphasizes Joan's final summary comment. See DASHES.

Highland Wins 1985 Safety Award; Jason is Runner-up

The Highland Plant in Rimrock, Texas, has won the 1985 BCC Safety Award. This is the third straight win for Highland, which posted no lost-time accidents during 1985.

As winners, the Highland workers will all receive their choice of the following: a Coleman camp stove, a set of Oneida stainless steel dinnerware, or a Black and Decker shop vacuum, plus attachments.

According to Joan Tyree, Highland plant manager, "Everyone at Highland contributed to the success—it was a real team effort."

Highland supervisors conducted weekly safety meetings, and the plant offered its own $50 safety award. Jack Henderson won the award for his suggestion that the badly worn non-slip flooring in the men's shower room be replaced.

The Jason Plant in Yarrow, Oklahoma, was the 1985 runner-up for the BCC Safety Award. Yarrow employees will each receive their choice of a Norbest frozen turkey or a Swift Premium ham. The Jason Plant had only one lost-time accident during 1985.

Newsletters should be as specific and informative as possible. Thus writers should use direct quotations as well as other specific facts (such as the award prizes in the above example). Newsletters should follow the inverted organization, with the main point in the opening paragraph; this is similar to the managerial organization used in many letters and memos. See ORGANIZATION.

Newsletters should be as readable as possible, with short sentences and simple, direct words. See STYLE.

TECHNICAL REPORT

The title and associated data allow for careful cross-referencing and document storage and retrieval. The page notation (*1 of 4*) helps readers keep track of pages.

The one-page summary, although not asking for executive action, is almost an executive summary. See REPORTS and SUMMARIES.

The first paragraph establishes the purpose as well as the original line of investigation. See MEMOS, LETTERS, and ORGANIZATION.

The listed findings are concise. The sentences are deliberately short. See SENTENCES and SCIENTIFIC/TECHNICAL STYLE.

The recommendations and their potential cost savings conclude the summary.

RESEARCH AND DEVELOPMENT DEPARTMENT

TITLE:			RDD NO
DRILLING OF THE G-175 STRUT FITTING WITH AN ACME 570			8795-3
ROUTING STATUS	CHARGE NUMBER	MODEL NUMBER(s)	PAGE NO.
Routine	3-T3743-8042-286444	G-175	1 of 4

SUMMARY

The purpose of this investigation was to determine the cause of the hole elongation in the drilling of the G-175 strut fitting. Three Acme Model 570 drills are currently used for this operation. We originally thought that improper sequencing of the feed and clamp-up system caused the elongation problem. In investigating this problem, we did the following:

1. Laboratory tests showed that one drill motor was unclamping with the drill still in the hole. We rebuilt the feed and clamp-up system to remedy this problem.

2. Further tests showed that the area of the clamp foot was too small to prevent the motor from rocking on its axis during drilling, so we designed a larger clamp foot.

3. Tests indicated that the drill motor feed rates were excessive. We corrected the feed rates on all three Acme drills.

4. The drill motors tended to stall right at the breakout of the drill, so we fabricated a Skylube application system. This modification prevented stalling.

We issued Memo 5698-4-76 recommending the larger clamp foot (CF-8765-54-A) and the new Skylube application system (LA-5767-87) for use with all Acme 570 drills now being used. We estimate that these changes will save approximately $6,000 per year by eliminating the need to rework poorly drilled holes.

Diane Metcalf	5/30/85
PREPARED BY	DATE
Wallace Petersen	6/10/85
APPROVED BY	DATE
Robert Hogge	6/12/85
APPROVED BY	DATE

SKY AVIATION

The summary (likely limited to one page) is a powerful technique for limiting documentation and for making technical reports more accessible. Many readers will not want or need to read more than the summary. If the summary contains a few of the important details, these readers will be able to determine whether they need to read further. See REPORTS, SUMMARIES, and ORGANIZATION.

TECHNICAL REPORT

The brief introduction establishes the reason for and purpose of the investigation. As appropriate, relevant prior work and other background information might also appear in the introduction. See INTRODUCTIONS and REPORTS.

The displayed list highlights the proposed test and redesign program. This list actually repeats information covered under Tests. See REPETITION and LISTS.

The headings are not specific, but they are probably standard. All research/investigation reports in this company have the same headings. The consistent format helps readers of many similar reports find information easily. See HEADINGS and ORGANIZATION.

Introduction

The Acme 570 drills were drilling many unsatisfactory holes in the strut fittings for the G-175 airplanes. The holes were often elongated or bell-mouthed, requiring reworking of the holes to an oversized diameter. Production Research initiated a program to determine the cause of these unsatisfactory holes.

Program Approach

To determine the cause of the problem, we developed a test and redesign program as follows:

1. Observe the Acme 570 drills in actual operation on a G-175.
2. Observe the clamp-drill cycle with a high-speed TV camera.
3. Redesign the clamp foot to increase the clamp area and reduce flexing.
4. Prepare a recommendation for Production, including a revised drilling procedure and accompanying drawings.

Tests

To determine the cause of the hole elongation, we brought one of the Acme 570 drills from the Production line to the laboratory for testing. We used a high-speed TV camera to observe the clamp-drill cycle of the drill motor. The camera showed that this motor was unclamping with the drill still in the hole. We sent this drill to Small Tool Repair for the overhaul of the clamp feed system.

The repaired Acme 570 was again tested and observed with the high-speed camera. The drill functioned well this time, but the clamp foot seemed to cover an insufficient area. The drill motor was able to move slightly during drilling. Apparently, vibration causes the drill to "migrate" during high speed drilling, even though the bit tends to hold the drill in place. Pressure on the inside of the drill hole causes minute imperfections in the drilling circumference, which become exaggerated when the drill bites into one of these imperfections and causes it to elongate.

No. 8795-3
Page 2 of 4

The body of the report follows a scientific format rather than a managerial format. Thus the conclusions and recommendations appear at the end rather than at the beginning. In most cases, the sequence and specificity of the headings in the body are almost irrelevant. Most readers will not read carefully beyond the summary, and readers familiar with the report format will already understand the content and organization of ideas appearing in the body.

TECHNICAL REPORT

The tests are explained in the chronological order in which they were performed. The chronological pattern helps readers to follow the test sequence and therefore its logic. Note that each step ends with a conclusion or recommendation for further study.

Most paragraphs are organized chronologically, from problem to result or finding. This pattern is common in technical and scientific reports. See PARAGRAPHS.

To solve this problem, we designed a new clamp foot (CF-8765-54-A) with 57 percent greater surface area. Then we subjected the new clamp to 140 drill tests. During these tests, the drill did not "migrate" as before, nor did the hole elongate significantly, although some imperfections in the drilling holes were observable to the naked eye. (See attached figure 1.)

We next checked the feed rate on the drill motor for drilling the .309 holes. The standard rate (4,300 rpm) produces a uniform drilling shaft so long as the drill bit is aligned precisely in the drill. But if the bit is not aligned precisely, the bit produces excessive vibration and hole elongation.

To determine an optimal drill speed, we tested the drill at five speed ranges: 2,000; 2,500; 3,000; 3,500; and 4,000 rpm. Rates below 3,000 rpm were unsatisfactory because the reduced speed created more friction and thus more heat. Rates above 3,500 rpm produced excessive vibration and hole elongation with drill bits not precisely aligned. So we repeated this test using another four speed ranges: 3,100; 3,200; 3,300; and 3,400 rpm.

Of these ranges, 3,400 rpm proved to be optimal. Further adjusting revealed that 3,460 rpm (± 30 rpm) is the best compromise rate. Accordingly, we adjusted the drilling speed to 3,460 rpm. (See Ref. 5Y114-87, Drilling and Reaming Feeds and Speeds).

Finally, we noticed that the drill tended to stall just before the drill finished the hole. We designed a Skylube application system (LA-5767-87) to lubricate the bit (see attached figure 2). This system eliminated stalling problems. For further information on this application system, see RDD NO. 8799-6.

<u>Conclusions and Recommendations</u>

The Acme 570 drill was returned to the Production line in mid September. It had a new clamp foot, a correct feed rate, and the

The conclusions and recommendations can be so brief because they have already been covered—first in the summary and later in the Tests section. See ORGANIZATION and REPORTS.

Skylube application system. We are continuing to monitor the performance of this drill, but preliminary results are promising.

We sent a recommendation memo (5698-4-76) to B. Worth recommending that all Acme 570 drills be modified with a larger clamp foot (CF-8765-54-A) and the new Skylube application system (LA-5767-87).

No. 8795-3
Page 4 of 4

DESCRIPTIVE ABSTRACT

The opening sentence identifies the what, where, and why of the study. No actual results should appear.

The pronoun *we* softens the impersonal tone and removes the need for passive sentences. See ACTIVE/PASSIVE.

The closing mention of the "guidelines" does not summarize the content of the guidelines.

After inconclusive laboratory tests, we conducted field tests in the Lubbock, Texas, area to determine if Stimuflo (from Fluid Engineering Company) is cost effective in enhancing acid stimulation. These tests, conducted from April 1983 to October 1984, contrasted oil and gas production from 14 wells stimulated with Stimuflo with production from 11 control wells stimulated using conventional acid techniques. We analyzed the results in light of Stimuflo production costs as well as differences in the average payout for all wells tested. Based on this analysis, we developed guidelines for the potential use of Stimuflo in future stimulations.

MIDLAND OIL AND GAS OPERATIONS, INC.

Descriptive abstracts describe the general content of a study or report, but they do not get into the actual results. As such, they are primarily useful for bibliographic cross referencing, where someone wants to know the what, the why, and the how, but not the actual conclusions or recommendations. A descriptive abstract will, of course, mention enough specific key words to flag its content during a computer search. Key words in the above abstract would be *acid stimulation, Stimuflo, oil and gas, production,* and so on.

INFORMATIVE ABSTRACT

The opening sentence gives the key result—the cost effectiveness of Stimuflo, under certain conditions.

The scope of the study is briefly summarized.

The results are specific enough that most readers would not need to read the actual report.

The dashes highlight a key qualification—Stimuflo's lack of success on mature wells. See DASHES.

Field tests indicate that Fluid Engineering Company's Stimuflo is a cost effective method of enhancing acid stimulation of wells with attractive recoverable reserves. Despite inconclusive laboratory tests, field tests in the Mountain View Field (Lubbock, Texas) have demonstrated this cost effectiveness. Over a 5-month period, 14 Mountain View wells were stimulated using Stimuflo. Another 11 wells were stimulated using conventional acid techniques. The results indicate an average recovery increase of 17.65 BOPD with Stimuflo rather than with conventional acid techniques. We predicted average payout with Stimuflo to be 14.3 months, depending on a well's production history and reservoir type. Mature wells—ones with prior stimulations—did not respond well enough to Stimuflo to warrant its use, especially given the 30 to 50 percent additional costs of using Stimuflo.

MIDLAND OIL AND GAS OPERATIONS, INC.

Informative abstracts give the actual information discovered—the results and any pertinent conclusions or recommendations. Primary readers for informative abstracts would be those already very familiar with the subject field and thus able to use the content of the abstract in place of ordering the whole report. As with a descriptive abstract, informative abstracts do contain many of the key words used in computer searches. Informative abstracts mention research methods and other techniques only if they are likely to be unknown to knowledgeable readers.

SCIENTIFIC REPORT

The introduction states the problem that generated the research. See INTRODUCTIONS and REPORTS.

This abstract is informative because it presents actual conclusions. See ABSTRACTS, REPORTS, and SUMMARIES.

The purpose comes as early in the introduction as possible.

The scope of the tests—both their sites and dates—follows the statement of purpose. The chemicals tested could be listed here, but they already appear in the Abstract, and they will appear under Methods and Materials.

The keywords identify other words useful for computer searches based on the keywords.

The first footnote, including the CAUTION, is a standard disclaimer used by many Federal and State agencies as well as some private research groups.

TERMITE CONTROL STUDIES IN PANAMA

ABSTRACT

Subterranean termite control studies in a tropical area (Panama) are described. Testing began in 1943 on Barro Colorado Island, which was formed when the Panama Canal was completed in the early 1900s.

Materials tested included DDT (various concentrations and formulations), BHC (benzene hexachloride), trichlorobenzene, sodium arsenite, pentachlorophenol, sodium flurosilicate, copper ammonium fluoride, aldrin, chlordane, dieldrin, and heptachlor. Dieldrin (1.0 percent), applied to the soil as a water emulsion, was still 100 percent effective after 27 years, when the tests were terminated. Tests with concentrations of 0.25 percent of aldrin, chlordane, and heptachlor were initiated in 1963, and all three chemicals were still 100 percent effective after 16 years.[1]

Additional keywords: Field studies, soil treatments, test procedures, tropics.

1. This publication reports research involving insectides. It does not contain recommendations for their use, nor does it imply that the uses discussed here have been registered. All uses of pesticides must be registered by appropriate State and Federal agencies before they can be recommended.

INTRODUCTION

Termites have been damaging facilities of the Panama Canal ever since it was constructed in the early 1900s. Termites have also severely damaged nearby military facilities in the Canal Zone.

To evaluate various chemicals for their effectiveness in preventing termite damage in a tropical environment, a series of long-term field evaluations began in 1943 and were continued in 1946.

These initial tests were conducted on Barro Colorado Island, Panama. The island, formed by canal construction, is in the Canal about 18 miles from the Atlantic outlet. Since 1924 it has been under the jurisdiction of the Smithsonian Institute.

From 1951 to 1953, termite control studies were considerably expanded in an area known as the Curundu Jungle Test Site at

CAUTION: Pesticides can be injurious to humans, domestic animals, desirable plants, and fish or other wildlife—if they are not handled or applied properly. Use all pesticides selectively and carefully. Follow recommended practices for the disposal of surplus pesticides and pesticide containers.

This scientific report is based on a technical paper published by the U.S. Department of Agriculture.

The report follows the scientific pattern—the discussion leads down to the conclusions. Note, however, that the abstract summarizes key conclusions. See ORGANIZATION and REPORTS.

SCIENTIFIC REPORT

The figures use action captions and are clearly labelled. See VISUAL AIDS, FIGURES, and CAPTIONS.

The sponsor of the research is optional if the title, the publication name, or footnotes do not clearly identify the sponsor.

The introduction ends with a brief note suggesting value or applications.

Methods and Materials should be no longer than necessary. The actual test methods should be described, but some of the detail about the chemicals tested might be left for tables or appendices. See REPORTS and APPENDICES/ATTACHMENTS.

Fort Clayton on the Pacific side of the Isthmus. In 1963, more tests were initiated at a site adjacent to Curundu.

The U.S. Tropical Entomology Laboratory of the United States War and Navy Departments originally sponsored these tests; in 1951, however, sponsorship passed to the Pesticide Service of the Department of Agriculture.

All tests ceased in 1979 when jurisdiction of the test areas reverted to the Republic of Panama.

Although many of the test chemicals did not satisfactorily prevent termites from damaging test materials, the test results should provide baseline data for interpreting the results from future termite control tests in tropical environments.

METHODS AND MATERIALS

Two standard test methods were used during these studies. The test methods were altered slightly in later tests because of lessons learned in the early tests.

Standard Test Methods

Method 1. The stake test, illustrated in figure 1, consisted of digging a hole 38 cm in diameter and 48 cm deep, removing approximately 0.057 m^3 of soil, and then treating the soil before replacing it. A wooden stake (5 x 10 x 46 cm) was driven 31 cm deep in the center of this treated soil to serve as bait for the termites.

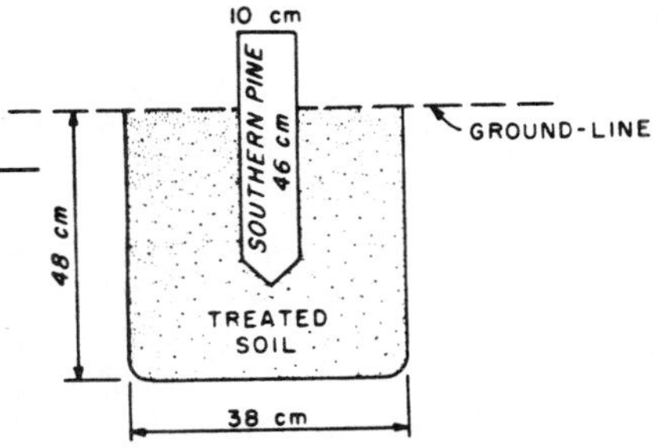

Figure 1.—*Stake test method for evaluating insecticides.*

This bait stake was either southern pine or some other termite-susceptible wood that indicated at annual inspections whether termites had penetrated the treated soil around the stake.

Each stake site was placed a minimum of 1.5 m away from other sites. Each concentration of each chemical was repeated 10 times in a randomized test block. When termites had penetrated the soil in 5 of the 10 identical sites, the chemical treatment was considered a failure.

The 1943 Barro Colorado stake tests, which used 39 different chemical treatments, were the first termite studies conducted under tropical conditions. Treatments included sodium arsenite as a dry powder and as a 10 percent solution in water; creosote in various oils; 5 percent pentachlorophenol; orthodichlorobenzene in oil, in creosote, and in creosote plus diesel oil; and diesel oil controls.

The tone of the report is formal and generally impersonal. Few pronouns appear, and some sentences are deliberately passive to avoid using personal pronouns. See SCIENTIFIC/TECHNICAL STYLE.

The amount of detail in reports is always open to debate. This report might well have been little more than a research note—with findings limited to the final studies and their results. Such abbreviated reports would probably reduce or eliminate the

MODEL DOCUMENTS

SCIENTIFIC REPORT

The reference to the figure is specific and falls on the same page as the figure. See FIGURES, CAPTIONS, and VISUAL AIDS.

The hyphen is necessary because *43-cm²* is a compound adjective modifying *area*. See HYPHENS.

The paragraphs giving the different test materials open with dates and sites because the dates and sites are used as subheadings in the Results section. This logical tracking helps to improve readability. See HEADINGS, PARAGRAPHS, and KEY WORDS.

The 1946 Barro Colorado stake tests included different dosages of 16 chemicals and methods for a total of 54 treatments. Dosages were 1.69, 3.38, and 6.76 liter/m^3 of soil for plots with the standard 38-cm diameter x 48-cm deep hole (0.057 m^3). Some of the more recognizable formulations included 5.0 percent DDT in water; 5.0 percent DDT in acetylenetetrachloride; 0.8 percent benzene hexachloride (BHC) in kerosene; copper naphthanate (2 percent copper in kerosene); lead arsenate, as dry powder at 227-g dosage and in water, 227 g in 0.94 liter at a 0.94-liter dosage; 5 percent monochloronaphthalene in kerosene; kerosene controls; and untreated controls.

In 1951-52, a new series of standardized stake tests was installed at Curundu in a new test area. The 1952 group included the following emulsions and fuel oil solutions:

— 5.0 percent DDT in oil

— 5.0 percent DDT plus 2.0 percent chlordane in oil

— 20.0 percent DDT in Xylene

— 0.50, 1.0, and 2.0 percent dieldrin in water

— 0.4 percent gamma BHC in oil

— Trichlorobenzene in diesel oil (3:1 ratio)

— Untreated controls, both oil and water.

Dosages were 6.76 and 10.1 liters/m^3 for the emulsions and oils, and 0.94 and 1.88 liters for the DDT concentrate.

Method 2. The ground-board test, as illustrated in figure 2, was designed to test conditions when wooden military equipment had to be laid on the jungle floor. However, the test also evaluated wooden construction materials used beneath slab-type houses.

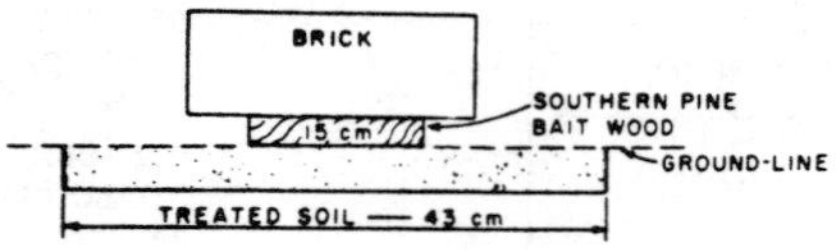

Figure 2.—*Standard ground-board method for evaluating insecticides.*

The method consisted of removing the duff and debris from a 43-cm^2 area of soil to expose the surface soil. The chemical to be evaluated was then sprinkled on the soil and a wooden pine board (30 x 30 x 2.5 cm) was placed in the center of the treated area. A rock or brick was then laid on the board to hold it in place.

The board was examined annually for termite damage, which, if found, indicated that termites had penetrated the treated soil. Ten duplicate sites for each test material were also used in this method, and randomizing was complete within blocks. Again, when termites had penetrated 5 of the 10 identical sites, the treatment was considered a failure.

In 1946, a series of ground-board treatments were established on Barro Colorado Island not only to determine the chemical effectiveness in controlling termites but also to compare the results with earlier stake test results (method 1).

Methods and Materials discussion and omit discussion of those chemicals that were ineffective. However, this data may be very significant because it tells other researchers what not to consider in future studies.

The two-column format is standard for many research reports. For this format to be effective, paragraphs must be shorter than normal because long single-column paragraphs become excessive in a two-column format. See SPACING, PARAGRAPHS, and VISUAL AIDS.

The numbers are not written out because some of them are greater than nine. See NUMBERS.

The displayed list, introduced by dashes, helps make the dense technical names and amounts more readable. See LISTS.

Test boards were placed on 10 tilled sites and on 10 untilled sites in this heavily shaded jungle area. The 43 materials tested included 3 dosages each of 13 chemical formulations and 4 untreated controls. The formulations consisted of the following:

— Acetylenetetrachloride

— 5.0 percent DDT in diesel oil

— 5.0 percent DDT in waste motor oil

— 5.0 percent DDT in kerosene

— 5.0 percent DDT in gasoline

— 10.0 percent sodium arsenite in water

— Diesel oil, waste motor oil, kerosene, and gasoline

Dosages were 2.5, 5.1, and 10.1 liters of formulation/m^2 area of soil surface.

In 1951-52 and again in 1953, ground-board surface tests, using 0.37 m^2 of treated soil and 2 x 15 x 15-cm bait boards, were initiated at the Curundu Jungle Test Site. The 1951-52 tests used the following materials:

— 0.5 percent pentachlorophenol in fuel oil and water

— 0.5 percent gamma BHC in oil

— 0.4 percent gamma BHC in water

— 5.0 percent DDT in oil and water

— 2.0 percent chlordane in oil and water

The oil formulations were put in at dosages of 0.47 and 0.94 liters; the water emulsions were at 0.94 and 1.41 liters. Additional formulations included:

— 5.0 percent pentachlorophenol plus 0.5 percent gamma BHC in oil

— 5.0 percent DDT in oil

— 5.0 percent sodium flurosilicate dry powder (113- and 170-g dosages)

— 0.5 percent copper ammonium fluoride in water

— Tetrachlorobenzene mixture with fuel oil

— Trichlorobenzene mixture with fuel oil

— Orthodichlorobenzene mixture with fuel oil

In 1953, tests of 10 dieldrin emulsion treatments of 1.0 and 2.0 percent, and 3 oil solutions of 0.5, 1.0, and 2.0 percent were initiated at 0.47 and 0.94 liters each.

In 1963, tests with a much wider range of previously proven insecticides were initiated in a jungle site close to the Curundu Jungle Test Site. This site was carefully selected to include as many species of subterranean termites as possible. The purposes were (1) to determine the lowest effective rate and dosage of emulsifiable concentrate of aldrin, chlordane, dieldrin, and heptachlor in a tropical exposure, and (2) to determine the

SCIENTIFIC REPORT

The Results section presents information based on site and test data. This pattern reflects the pattern established in Materials and Methods. See REPETITION and ORGANIZATION.

The boldfaced subheadings enhance readability and allow for quick cross referencing and skimming. See HEADINGS and EMPHASIS.

The reference to table 2 becomes part of the actual discussion of results. See VISUAL AIDS.

efficacy of granular forms of insecticides when applied to the soil surface.

The 48 separate treatments included chlordane and dieldrin at 0.03, 0.06, 0.12, 0.25, 0.50, 1.0, and 2.0 percent and aldrin and heptachlor at 0.06 and 0.25 percent. The application rates were 946 and 1982 ml/929 cm^2. Aldrin, dieldrin, and heptachlor granules were applied to the soil to give equivalencies of 0.12, 0.25, 0.50, and 1.0 percent applied at the rate of 10.12 liters/m^2. Chlordane granules were applied at only 0.25 and 0.50 percent.

RESULTS AND DISCUSSION

Barro Colorado - 1943 - Stake Tests

By 1952 (9 years), all treatments except those that included sodium arsenite had failed to prevent termites from attacking the wood bait stakes. When the tests ended in 1954, the arsenite treatments of 810, 1620, and 2430 g/m^3 of dry chemical and 4.73 ml of 10 percent solution in water/m^3 were providing termite protection under the severe tropical exposure.

Barro Colorado - 1946 - Stake Tests

By 1952 (6 years), all treatments except 2 DDT formulations, 2 BHC, 4 monochloronaphthalene, 4 copper naphthanate, and 1 lead arsenate had failed to prevent termite attacks. The 1954 (8-year) inspection showed that only the 8.0 percent DDT in acetylenetetrachloride treatment still had limited effectiveness. In the DDT treatment, soil in 6 of the 10 sites had not been penetrated by termites—60 percent protection.

Barro Colorado - 1946 - Ground-board Tests

When the test was closed in 1954 (8 years), only 2 formulations continued to provide control: 5.0 percent DDT in diesel oil and 10 percent sodium arsenate in water. They were giving 80 percent protection, which, by today's standards, would not be acceptable for recommendation as subterranean termite control. The results of the new method were so similar to those from the more difficult and time-consuming stake tests that the new ground-board method was selected for future studies.

Curundu - 1952-53 - Stake Tests

As table 1 indicates, dieldrin and chlordane were the most effective test materials; DDT was the next most effective, and BHC was the least effective.

The recorded attacks during the second and third test years to the test sites with the highest concentrations of dieldrin and chlordane may have been anomalies because few or no further attacks occurred at sites with these concentrations.

In soil treated with BHC, attacks were noted at the end of the fifth year. By the end of the ninth year, multiple attacks throughout the treated soil had occurred.

Curundu - 1951-53 - Ground-board Tests

As table 2 indicates, the only chemical concentrations that remained 100 percent effective for the entire 26 years were 1.0 percent dieldrin at 15.18 liters/m^2 and 2.0 percent dieldrin at 10.12 liters/m^2.

SCIENTIFIC REPORT

The table is clear and easy to read despite the mass of data presented. See TABLES.

This first table provides actual and essential results; earlier tests are summarized in the text, not in fully developed and unnecessary tables. See TABLES.

The table is oriented horizontally so readers do not have to turn the page. See TABLES.

The table caption cannot be an action caption, given the amount of data presented. See CAPTIONS and TABLES.

Table 1.—*Evaluation of insecticides applied as soil treatments in standard stake tests in 1952 in the Panama Canal Zone*

Treatment designation and material	Rate of application liters/m^3	% of stakes undamaged by termites after exposure for indicated years															
		1	2	3	4	5	7	9	11	13	15	17	19	21	23	25	27
5.0% DDT in fuel oil	6.76	100	100	100	100	100	100	100	80	70	70	50	....	...	...	...	....
5.0% DDT in fuel oil	10.1	100	100	100	100	100	100	90	70	50	....	....	....	...	...	...	....
5.0% DDT in water	6.76	100	90	80	80	80	60	50	....	....	....	....	....	...	...	...	....
5.0% DDT in water	10.1	100	100	100	100	100	100	80	80	80	80	80	70	70	70	60	50
2.0% chlordane in fuel oil	10.1	100	100	100	100	100	100	100	100	100	100	100	100	90[1]	90	90	90
2.0% chlordane in water	10.1	100	90	90	90	90	90	90	90	90	90	90	90	90	90	90	90
5.0% DDT + 2% chlordane in fuel oil	6.76	100	100	100	100	100	100	100	80	80	80	80	80	80	80	80	80
5.0% DDT + 2% chlordane in fuel oil	10.1	100	90	90	90	90	90	90	80	80	80	80	70	70	70	70	70
5.0% penta in fuel oil	6.76	100	100	100	70	70	70	70	30	....	....	....	....	...	...	...	....
5.0% penta in fuel oil	10.1	100	100	100	100	100	90	70	50	....	....	....	....	...	...	...	....
0.5% gamma BHC in fuel oil	6.76	100	100	100	100	100	80	70	60	50	....	....	....	...	...	...	....
0.5% gamma BHC in fuel oil	10.1	100	100	100	100	90	70	50	....	....	....	....	....	...	...	...	....
0.4% gamma BHC emulsion	6.76	100	100	90	90	80	70	50	....	....	....	....	....	...	...	...	....
0.4% gamma BHC emulsion	10.1	100	100	100	100	90	90	70	60	60	60	60	50	...	...	...	....
Fuel oil control	10.1	100	100	80	60	40	20	....	....	....	....	....	....	...	...	...	....
20% DDT concentration in Xylene	3.38	100	100	100	100	90	80	80	70	60	60	60	60	60	60	60	60
20% DDT concentration in Xylene	6.76	100	100	100	100	90	90	80	80	80	80	70	60	60	50[1]	...	....
Untreated control	....	50	20	0	0	0	80	30	40	30	40	40	20	30[1]	20	10	0
25% trichlorobenzene in fuel oil 1:3	6.76	100	90	90	80	70	60	50	....	....	....	....	....	...	...	...	....
0.5% dieldrin in fuel oil	6.76	100	100	100	100	100	100	100	100	100	100	100	90	90	70[1]	60	50
0.5% dieldrin in fuel oil	10.1	100	100	100	100	100	100	100	90	90	90	90	80	80	70[1]	40	....
1.0% dieldrin in fuel oil	6.76	100	100	100	100	100	100	100	80	80	80	80	80	80	80	80	80
1.0% dieldrin in fuel oil	10.1	100	100	100	100	100	100	100	90	90	90	90	90	90	90	90	90
2.0% dieldrin in fuel oil	6.76	100	100	100	100	100	100	100	100	90	90	90	90	90	90	90	90
2.0% dieldrin in fuel oil	10.1	100	100	100	100	100	100	100	100	100	100	100	100	100	100	100	100
1.0% dieldrin emulsion	10.1	100	100	100	100	100	100	100	100	100	100	100	100	100	100	100	100
2.0% dieldrin emulsion	10.1	100	100	90	90	90	90	80	80	80	80	80	80	80	80	80	80

[1] These attacks were made by *Heterotermes* sp.

Table 2.—*Evaluation of insecticides applied as soil treatments in ground-board tests in 1953 in the Panama Canal Zone*

Treatment designation and material	Rate of application liters/m^2	% of ground boards undamaged by termites after exposure for indicated years															
		1	2	3	4	5	6	8	10	12	14	16	18	20	22	24	26
MOGO Oil Co. #1 Termicide Oil	5.06	100	50	....	....	....	....	....	....	....	....	....	....	...	...	....	...
Same as above plus 5% penta.	5.06	90	50	40	....	....	....	....	....	....	....	....	....	...	...	....	...
Same as above plus 2% copper naphthanate	5.06	90	50	50	....	....	....	....	....	....	....	....	....	...	...	....	...
MOGO Oil Co. #6 Weed-Death	5.06	100	70	60	60	50	....	....	....	....	....	....	....	...	...	....	...
Same as above plus 5% penta	5.06	90	50	....	....	....	....	....	....	....	....	....	....	...	...	....	...
Same as above plus 2% copper naphthanate	5.06	90	60	50	....	....	....	....	....	....	....	....	....	...	...	....	...
1.0% Dieldrin in water	5.06	100	100	100	100	100	100	100	100	100	89	89	89	89	89	75	75
1.0% Dieldrin in water	7.59	100	100	100	100	100	100	100	100	100	100	100	100	100	100	100	100
2.0% Dieldrin in water	5.06	100	100	100	100	100	100	100	100	100	100	100	100	100	100	100	100
2.0% Dieldrin in water	7.59	100	100	100	100	100	100	100	100	89	89	89	89	89	89	89	89
0.5% Dieldrin in fuel oil	2.53	100	100	100	100	100	100	100	100	100	90	90	90	70	50[1]	....	...
0.5% Dieldrin in fuel oil	5.06	100	100	100	100	100	100	100	100	100	100	80	80	80	30[1]	....	...
1.0% Dieldrin in fuel oil	2.53	100	100	100	100	100	100	100	90	90	90	90	90	80	50[1]	....	...
1.0% Dieldrin in fuel oil	5.06	100	100	100	100	100	100	100	80	80	80	80	80	70[1]	70	70	70
2.0% Dieldrin in fuel oil	2.53	100	100	100	100	100	100	100	100	100	100	100	100	100	100	100	75[1]
2.0% Dieldrin in fuel oil	5.06	100	100	100	100	100	100	100	100	100	100	100	100	89	89	89	89
Untreated control	....	50	60	50	70	50	50	40	50	50	70	70	30	20	40	40	40[1]

[1] These attacks were made by *Heterotermes* sp.

SCIENTIFIC REPORT

In fuel oil mixtures, the 2.0 dieldrin sustained some attacks after 26 years, the 1.0 dieldrin after 20 years. Both concentrations gave excellent protection, but since oil is no longer used as a carrier for termiticides except in special cases, these formulations are not suggested for use.

Dieldrin at both 0.5 percent and 1.0 percent at 5.06 liters/m^2 of soil surface area failed (less than 50 percent effective) at 22 years. Even though this is excellent long-term protection, it did not protect as long as dieldrin installed in tests in Mississippi.[2]

---

2. Jason B. Kline and Josephine Everett, "Termite-Resistant Woods," *Southern Agricultural Studies*, 11 (February 1972), 35-37.

Table 3.—*Insecticides evaluated against subterranean termites* (Coptotermes *sp. and* Heterotermes *sp.) in the Panama Canal Zone in 1963*

Formulation (approx. % by wt.)	Rate of application liters/m^2	% of ground boards undamaged by termites after exposure for indicated years															
		1	2	3	4	5	6	7	8	9	10	11	12	13	14	15	16
Aldrin (actual)																	
0.067	5.06	100	100	100	100	100	100	100	100	90	80	80	70	60	60	30	....
	10.12	100	100	100	100	100	100	100	100	100	100	100	100	90	70	70	60
0.25	5.06	100	100	100	100	100	100	100	100	100	100	100	100	100	100	100	100
	10.12	100	100	100	100	100	100	100	100	100	100	100	100	100	100	100	100
Dieldrin (actual)																	
0.033	5.06	100	100	100	90	90	90	80	70	50	....	....	....	....	....	....	....
	10.12	100	100	90	90	80	80	80	80	50	....	....	....	....	....	....	....
0.067	5.06	100	100	100	90	90	90	90	80	60	50	....	....	....	....	....	....
	10.12	100	100	100	100	100	100	90	80	80	80	50	....	....	....	....	....
0.125	5.06	100	100	100	100	100	100	100	100	100	100	90	70	50	....	....	....
	10.12	100	100	100	100	100	100	100	100	100	100	100	100	100	70	60	60
0.25	5.06	100	100	100	100	100	100	100	100	100	100	100	100	100	100	100	90
	10.12	100	100	100	100	100	100	100	100	100	100	100	100	100	100	100	100
0.50	5.06	100	100	100	100	100	100	100	100	100	100	100	100	100	100	100	100
	10.12	100	100	100	100	100	100	100	100	100	100	100	100	100	100	100	100
1.0	5.06	100	100	100	100	100	100	100	100	100	100	100	100	100	100	100	100
	10.12	100	100	100	100	100	100	100	100	100	100	100	100	100	100	100	100
Heptachlor (actual)																	
0.067	5.06	100	100	100	100	100	100	100	100	100	100	90	70	60	....	....	....
	10.12	100	100	100	100	100	100	100	100	100	100	100	100	80	80	50	....
0.25	5.06	100	100	100	100	100	100	100	100	100	100	100	100	100	100	100	90
	10.12	100	100	100	100	100	100	100	100	100	100	100	100	100	100	100	100
Chlordane (technical)																	
0.033	5.06	100	100	100	80	70	60	60	60	50	....	....	....	....	....	....	....
	10.12	100	100	100	70	70	70	40	....	....	....	....	....	....	....	....	....
0.067	5.06	100	90	90	90	80	80	80	70	40	....	....	....	....	....	....	....
	10.12	100	100	100	100	100	90	90	80	70	70	50	....	....	....	....	....
0.125	5.06	100	100	100	100	100	100	90	90	80	80	70	50	....	....	....	....
	10.12	100	100	100	100	100	90	90	90	80	80	80	70	50	....	....	....
0.25	5.06	100	100	100	100	100	100	100	100	90	60	60	40	....	....	....	....
	10.12	100	100	100	100	100	100	100	100	100	100	100	100	100	100	100	100
0.5	5.06	100	100	100	100	100	100	100	100	100	90	80	80	80	80	80	80
	10.12	100	100	100	100	100	100	100	100	100	100	100	100	100	100	100	100
1.0	5.06	100	100	100	100	100	100	100	100	100	100	100	100	100	100	100	100
	10.12	100	100	100	100	100	100	100	100	90	90	90	90	90	90	90	90
2.0	5.06	100	100	100	100	100	100	100	100	100	100	100	100	100	100	100	100
	10.12	100	100	100	100	100	100	100	100	100	100	100	100	100	100	100	100
Untreated control	0	50	50	70	40	60	50	40	50	10	0	10	0	0	40	10	10

Paragraphs (for the 1963 ground-board results) are each focused on a single type of result. See PARAGRAPHS.

Besides dieldrin (the only true pesticide), the tested chemicals included mainly different oils, which were ineffective. As table 2 indicates, only a MOGO Oil Company[3] weed killer gave 50 percent control for more than 4 years.

The results of these 1951-53 tests apply especially to *Heterotermes convexinotatus* Snyder and *Heterotermes tenuis* Hagen because these were the predominant species found in the study areas.

Ground-board Series - 1963

At concentrations of 0.25 percent or more, aldrin, heptachlor, chlordane, and dieldrin all provided excellent protection until the tests ended in the sixteenth year (1979). As table 3 indicates, aldrin was slightly more effective than the others, all of which had at least some attacks by the sixteenth year.

At concentrations of 0.067 percent (see table 3), aldrin was also the most resistant, although some attacks did occur as early as the ninth year, with failure in the fifteenth year. Heptachlor was the next most resistant, with attacks occurring in the eleventh year and failure in the thirteenth year. At this concentration, dieldrin and chlordane both failed in the ninth year. (Concentrations were considered failures when termites attacked over 50 percent of the test sites with those concentrations.)

Table 4 presents the results of the granular insecticides applied directly to the soil. Only 5 treatments (0.50 percent and 1.0 percent aldrin, 0.50 percent and 1.0 percent dieldrin, and 1.0 percent heptachlor) remained 100

---

3. MOGO has neither financed nor sponsored this research. All results and conclusions are the responsibility of the authors.

Table 4.—*Granular insecticides applied in ground-board tests in 1963 in the Panama Canal Zone*

Formulation (approx. % by weight)	Weight of toxicant applied[1] (g/932 cm^2)	% of ground boards undamaged by termites after indicated years															
		1	2	3	4	5	6	7	8	9	10	11	12	13	14	15	16
Aldrin (actual)																	
0.125	1.19	100	100	100	100	100	90	90	90	90	90	90	80	70	70	30	....
0.25	2.37	100	100	100	100	100	100	100	100	100	100	100	100	90	90	90	80
0.50	4.73	100	100	100	100	100	100	100	100	100	100	100	100	100	100	100	100
1.0	9.46	100	100	100	100	100	100	100	100	100	100	100	100	100	100	100	100
Dieldrin (actual)																	
0.125	1.19	100	100	100	100	100	100	100	100	100	100	90	80	80	80	80	70
0.25	2.37	100	100	100	100	100	100	100	100	100	100	100	100	100	100	100	90
0.50	4.73	100	100	100	100	100	100	100	100	100	100	100	100	100	100	100	100
1.0	9.46	100	100	100	100	100	100	100	100	100	100	100	100	100	100	100	100
Chlordane (technical)																	
0.25	2.37	100	100	100	100	100	100	100	100	80	80	80	50	....	....	....	....
0.50	4.73	100	100	100	100	100	100	100	100	100	100	100	100	100	90	90	90
Heptachlor (actual)																	
0.125	1.19	100	100	100	100	90	90	90	90	80	80	70	60	60	40	....	....
0.25	2.37	100	100	100	100	90	90	90	90	90	80	80	70	60	50	....	....
0.50	4.73	100	100	100	90	90	90	90	90	90	90	90	90	90	90	90	90
1.0	9.46	100	100	100	100	100	100	100	100	100	100	100	100	100	100	100	100
Untreated control	....	50	50	70	40	60	50	40	50	10	0	10	0	0	40	10	10

[1] The amounts shown in this column are equivalent to amounts of toxicant that are applied for each percentage of 946 ml/932 cm^2 in water emulsion.

SCIENTIFIC REPORT

The numbered list of conclusions emphasizes their importance even though they come last in the report. See LISTS, ORGANIZATION, and REPORTS.

Italics is used for the names of the different species of termites. See ITALICS.

percent effective for the duration (16 years) of the study. The earliest attack occurred in the fifth year on 0.125 percent and 0.25 percent heptachlor. Generally, these granular treatments did not perform as well as the emulsions, but this was expected because the granular materials were more subject to washing by rainfall than emulsions.

The species of termites penetrating the soil (either *Coptotermes* sp. or *Heterotermes* sp.) were recorded, but in many cases, the wooden monitoring baits were destroyed and no termites were present. Based on the termites that could be identified, the predominant termites in the study area were *Coptotermes niger* Snyder, *H. convexinotatus* and *H. tenuis*. *Nasutitermes corniger* Motsch. and *Microcerotermes arboreus* Emerson were also found in the area.

CONCLUSIONS

1. Aldrin is the best chemical to use. No attacks occurred on any soil treated with 0.25 percent solution or higher.

2. Heptachlor is slightly more effective than dieldrin. Both were 90 percent effective at 0.25 percent after 16 years, but 0.067 percent dieldrin was attacked earlier than 0.067 heptachlor.

3. All granular materials at 0.50 percent appear equally effective; however, use labels are not available for granules at this time.

INDEX

BUSINESS
WRITER'S
QUICK
REFERENCE
GUIDE

INDEX

ABOUT SHIPLEY ASSOCIATES

The Shipley Concept

Customized Programs

Serving clients with excellence has been the Shipley watchword since the early 1970s when our consultants began teaching professionals in the world of work how to communicate better. We offer our international clientele an array of professional training programs, products, and services.

Shipley Associates training is unique in that it is highly **tailored**. Instead of the usual "off-the-shelf" or "canned" programs, **we customize our programs and manuals for every client we serve**. Our extensive tailoring makes our programs especially relevant to our clients' specific training needs.

We have a training staff of over 40 consultants, each with a **doctorate**, each an **expert** in his or her field.

In addition to our training programs, we offer a broad range of **consulting services** from assistance with major writing projects to copy-editing final drafts, from training the trainers to providing communications audits, training needs assessment, and diagnostic services. Our proposal consultation services range from proposal team building to RFP analysis, from strategy formulation to red teaming. We also perform audits of proposal production procedures and processes.

Shipley Training Works

Sound Communication Skills

We build **professional competence**. Most of the people in the world of work whom we teach are very good in their areas of expertise, but they do not always have effective communication skills.

We teach them to be more efficient and precise in the way they conceptualize, organize, compose, edit, and present their ideas. Our training for **executives** and **managers** is reinforced in the training we give to **technicians** and **first-line supervisors**. **Secretaries** learn writing, editing, and proofreading techniques that complement the writing and presenting skills that we teach the **engineers, geologists,** and **scientists** for whom they work.

Because we have a **unified and coordinated training program** to assist professionals at all levels in an organization, we are able to build skills, competence, and confidence among individuals in a way that benefits the entire organization.

Shipley Benefits for Clients

Needs, Analysis, Applicability

- The Shipley approach to training is **practical** and **applicable** to the real-world problems of the people we teach.
- We help our clients **identify and assess needs**.
- We provide a thorough **diagnostic analysis**.
- Where the need exists, we can conduct a thorough **communications audit** of the client organization.
- We **tailor** our training and materials for every program.
- Our **PhD consultants** have broad experience with a multitude of firms and agencies in a variety of settings.
- Our clients can schedule one workshop or dozens of workshops on a range of topics.
- Our decade of steady growth and our concentration on **communication** and **communication management** guarantee our clients a **reliable training and consulting resource**.

SHIPLEY WORKSHOPS AND SERVICES

Shipley Training Programs

A Variety of Offerings

Technical Writing teaches engineers, scientists, chemists, geologists, researchers, physicists, computer scientists, technical writers, technical editors, and other data-oriented professionals a practical, repeatable strategy for producing effective technical documents.

Effective Business Writing gives participants the skills necessary for writing efficiently and for producing effective documents.

How to Write Winning Proposals introduces participants to source selection practices and takes them through proposal development from beginning to end. Features intensive hands-on exercises, including three proposal simulations. Includes a 300-page reference manual.

How to Manage Winning Proposals helps functional managers, program managers, proposal managers, volume leaders, and marketing professionals understand the intricacies of managing proposal teams.

Executive Writing provides a specially designed improvement program for busy executives and managers who want to improve their efficiency and excellence in writing and reviewing written communication.

Secretarial Proofreading, Editing, and Writing strengthens the writing and editing skills of secretaries, clerks, administrative assistants, and others in key support staff positions.

Audit Report Writing is designed specifically for internal auditors who write sensitive audit reports, findings, memorandums, and summary letters—documents that must be clear, accurate, and objective.

Making Professional Presentations is an exciting workshop for professionals who need to be more self-confident, polished, persuasive, and informative in making presentations before groups. (This program is tailored for technical presentations, briefings, and sales presentations.)

Environmental Documents Writing I and II have been presented nationwide to professionals from BIA, Forest Service, BLM, Corps of Engineers, Geological Survey, Fish and Wildlife Service, tribal organizations, and many state agencies. These workshops focus on NEPA, CEQ Regulations, and agency guidelines as they apply to environmental impact statements, management plans, decision documents, and other environmental documents.

Managing the NEPA Process promotes planning, organizing, and managing for greater efficiency. Participants learn to take advantage of scoping, tiering, adoption, categorical exclusion, and incorporation by reference.

Construction Specifications Writing develops practical skills using current specifications produced by the sponsoring agency or organization and includes detailed information on 12 essential topics, including CSI form, arrangement, and sources.

Equipment Specifications Writing teaches professionals involved in equipment and supply contracting how to write accurate and legally defensible specifications and bidding packages.

Managing Written Communication includes videotape, workbook, and instruction designed to teach executives, managers, and supervisors how to manage writers to increase productivity, improve document quality, and enhance professionalism among those who write, as well as among those who review and edit writing.

Rapid Reading for the Professional develops efficient skills in participants who desire more information in less time, who must process large amounts of written information quickly, and who want more time for their important tasks both on the job and at home.

Shipley Professional Services

Dedicated Onsite Assistance

Communications Audits are thorough, fact-based analyses of the communications systems, procedures, and effectiveness within an organization. Includes a lengthy, computer-analyzed questionnaire, onsite documents and systems analysis, interviews with key personnel at all levels, and expert observation of communications processes and systems. Culminates in a lengthy report of findings (with recommendations) and onsite debriefings of key managers and other personnel.

Proposal Consultation Services include a range of onsite proposal assistance activities: proposal planning, competitive assessments, strategy formulation, RFP analysis, proposal outlining and storyboarding, proposal draft reviews, and red teaming.

Proposal Process Audits analyze an organization's proposal production system and offer specific recommendations for improving proposal management, team performance, and production systems.

Document Management and Review activities include planning, preparing, and reviewing of large-scale documents such as site preparation plans, environmental impact statements, computer user manuals, and other lengthy reports.

BUSINESS WRITER'S QUICK REFERENCE GUIDE

Please help us provide you with the best possible products in the future by filling out and returning this questionnaire. Your comments will be greatly appreciated. If you desire further information as stated below, one of our representatives will contact you either by phone or by mail.

NAME: ______________________________

TITLE: ______________________________

COMPANY OR ORGANIZATION: ______________________________

ADDRESS: ______________________________

CITY, STATE, ZIP: ______________________________

1. How did you receive the book?

______ A. Distributed by my company

______ B. Through a Shipley Associates seminar

______ C. Ordered personally

If C, how did you hear about the book? ______________________________

2. What are your general comments about the *Guide*? ______________________________

3. Are there any topics, sections, or features you find particularly helpful or beneficial? ______________________________

4. Are there any specific topics that you would like to see added, or covered more fully in the *Guide*? ______________________________

****

I would like to learn more about:

- ☐ discount prices for bulk orders.
- ☐ in-house training seminars conducted by Shipley Associates, in business writing and communication.
- ☐ materials for running my own in-house seminar.
- ☐ customizing the *Guide* to my company's specific needs.

☐ Please contact me directly:

☐ by mail ☐ by phone; My number is (_____)________________.

NO POSTAGE
NECESSARY
IF MAILED
IN THE
UNITED STATES

BUSINESS REPLY MAIL

FIRST CLASS PERMIT NO. 2277 NEW YORK, NY

Postage will be paid by addressee:

JOHN WILEY & SONS, INC.
605 Third Ave.
New York, N.Y. 10158

Attn: David Sobel

Your time is valuable. Make the most of your time *inside and outside* the lab.

To help you manage your time inside and outside the A&P lab classroom, this best-selling manual works hand-in-hand with **Mastering A&P**, the leading online homework and learning program for A&P. This edition features dozens of new, full-color figures and photos, revamped Clinical Application questions, an expanded set of pre-lab videos, dissection videos, and more.

EXERCISE 9

The Axial Skeleton

Learning Outcomes

- Name the three parts of the axial skeleton.
- Identify the bones of the axial skeleton, either by examining disarticulated bones or by pointing them out on an articulated skeleton or skull, and name the important bone markings on each.
- Name and describe the different types of vertebrae.
- Discuss the importance of intervertebral discs and spinal curvatures.
- Identify three abnormal spinal curvatures.
- List the components of the thoracic cage.
- Identify the bones of the fetal skull by examining an articulated skull or image.
- Define *fontanelle*, and discuss the function and fate of fontanelles.
- Discuss important differences between the fetal and adult skulls.

Pre-Lab Quiz

Instructors may assign these and other Pre-Lab Quiz questions using Mastering A&P™

1. The axial skeleton can be divided into the skull, the vertebral column, and the:
   a. thoracic cage **c.** hip bones
   b. femur **d.** humerus
2. Eight bones make up the ____________, which encloses and protects the brain.
   a. cranium **b.** face **c.** skull
3. The ____________ vertebrae articulate with the corresponding ribs.
   a. cervical **c.** spinal
   b. lumbar **d.** thoracic
4. The ____________, commonly referred to as the breastbone, is a flat bone formed by the fusion of three bones: the manubrium, the body, and the xiphoid process.
   a. coccyx **b.** sacrum **c.** sternum
5. A fontanelle:
   a. is found only in the fetal skull
   b. is a fibrous membrane
   c. allows for compression of the skull during birth
   d. all of the above

Go to Mastering A&P™ > Study Area to improve your performance in A&P Lab.

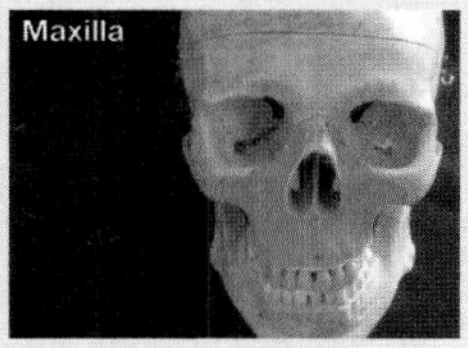

> Lab Tools > Bone & Dissection Videos

Instructors may assign new Building Vocabulary coaching activities, Pre-Lab Quiz questions, Art Labeling activities, related bone videos and coaching activities, Practice Anatomy Lab Practical questions (PAL), and more using the Mastering A&P™ Item Library.

Materials

- Intact skull and Beauchene skull
- X-ray images of individuals with scoliosis, lordosis, and kyphosis (if available)
- Articulated skeleton, articulated vertebral column, removable intervertebral discs
- Isolated cervical, thoracic, and lumbar vertebrae, sacrum, and coccyx
- Isolated fetal skull

The **axial skeleton** (the green portion of Figure 8.1 on p. 104) can be divided into three parts: the skull, the vertebral column, and the thoracic cage. This division of the skeleton forms the longitudinal axis of the body and protects the brain, spinal cord, heart, and lungs.

115

NEW! Mastering A&P study tools are highlighted on the first page of each lab exercise, along with a photo preview of a related pre-lab video, image from Practice Anatomy Lab 3.1 (PAL), or animation.

NEW! Mastering A&P assignments, including NEW Building Vocabulary Coaching Activities, are signaled at appropriate points throughout the manual to help you connect the exercises to relevant assignments that can be auto-graded in Mastering A&P.

See p. 115

Be Prepared: Learning in A&P Lab is an *Active* Process.

Before going into the lab, read the **background information** for the exercise, connect your reading to the figures and photos, complete the **pre-lab quiz**, and preview the questions in the tear-out **Exercise Review Sheet**. After lab, review your lab notes to remember important concepts. To improve your performance on lab practical exams, log into **Mastering A&P**, where you can watch related videos, practice with customized flashcards, and more.

NEW! Dozens of full-color figures and photos have been added to the Exercise Review Sheets, replacing black-and-white line drawings. Selected labeling questions are available as **new Art-Labeling assignments in Mastering A&P**.

Muscles of the Head and Neck

3. Using choices from the key at the right, correctly identify muscles provided with leader lines on the illustration.

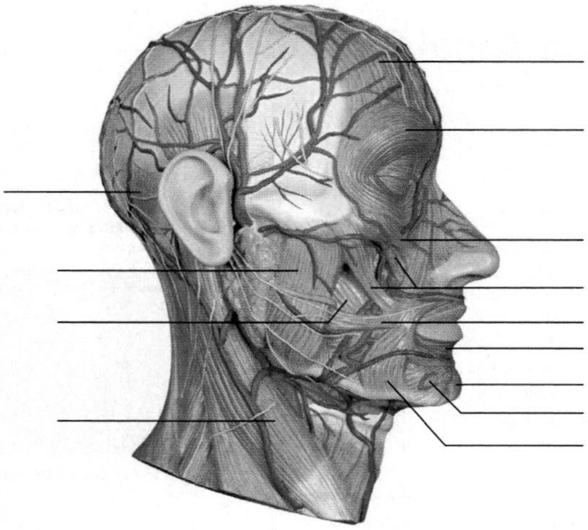

Key:

a. buccinator
b. depressor anguli oris
c. depressor labii inferioris
d. frontal belly of the epicranius
e. levator labii superioris
f. masseter
g. mentalis
h. occipital belly of the epicranius
i. orbicularis oculi
j. orbicularis oris
k. risorius
l. sternocleidomastoid
m. zygomaticus minor and major

See p. 224

Compare to Previous Edition

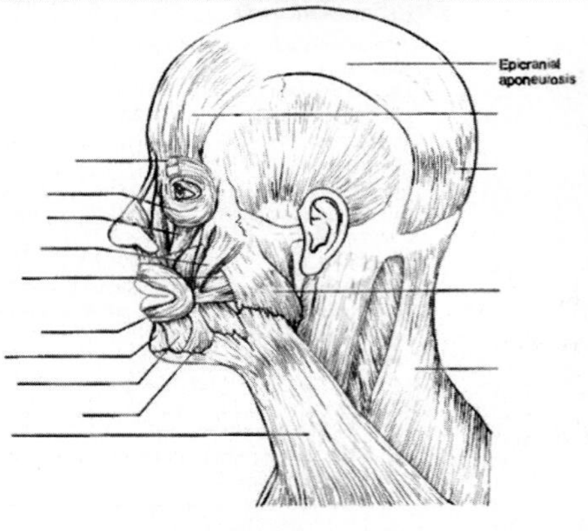

NEW! Clinical Application Questions have been added to the Exercise Review Sheets to help you connect lab concepts with real-world clinical scenarios.

27. ➕ As we age, we often become shorter. Explain why this might occur.

See p. 141

NEW! Building Vocabulary Coaching Activities are a fun way to learn word roots and A&P terminology while building and practicing important language skills.

Get 24/7 videos, coaching, and practice with Mastering A&P.

EXPANDED! 8 new Pre-Lab Video Coaching Activities in Mastering A&P (for a total of 18) focus on key concepts in the lab activity and walk you through important procedures. New pre-lab video topics include Preparing and Observing a Wet Mount, Examining a Long Bone, Initiating Pupillary Reflexes, Palpating Superficial Pulse Points, Auscultating Heart Sounds, and more.

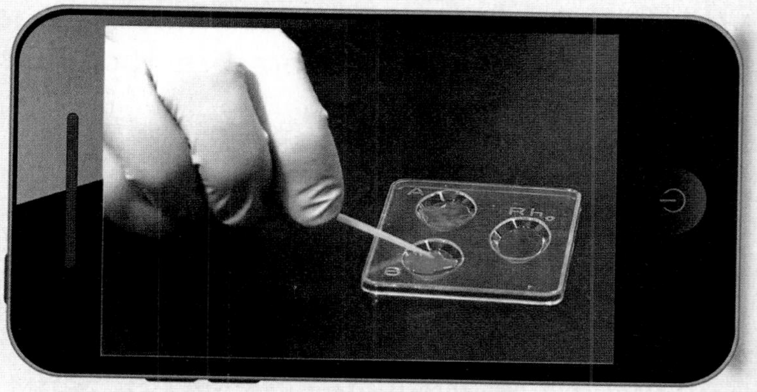

NEW! Cat and Fetal Pig Dissection Video Coaching Activities help you prepare for dissection by previewing key anatomical structures. Each video includes one to two comparisons to human structures.

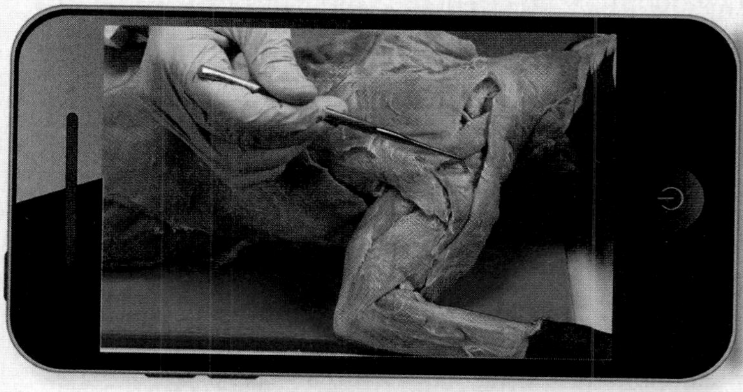

NEW! Customizable Practice Anatomy Lab (PAL) Flashcards allow you to create a personalized, mobile-friendly deck of flashcards and quizzes using images from PAL 3.1. You can generate flashcards using only the structures that your instructor has emphasized in lecture or lab.

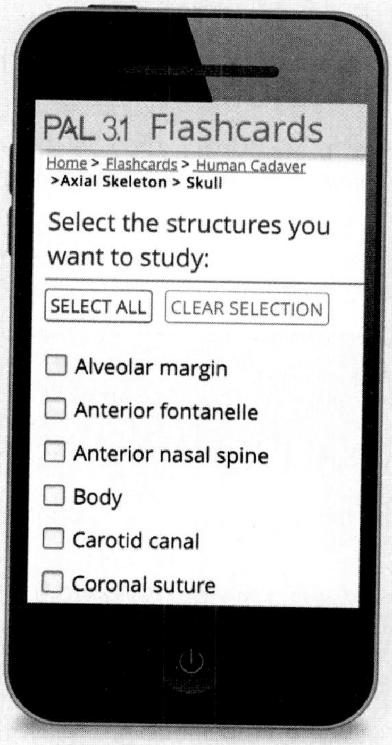

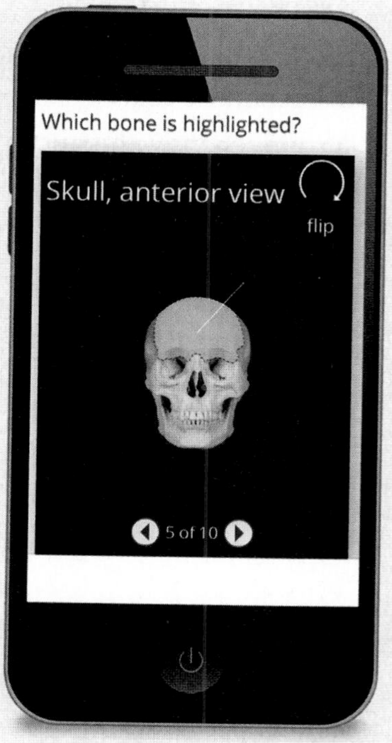

IMPROVED! The Pearson eText mobile app allows you to access the complete lab manual online or offline, along with all of the videos described above.

Additional Support for Students & Instructors

Mastering A&P offers thousands of tutorials, activities, and questions that can be assigned for homework and practice. Highlights of popular assignment options include the following:

PhysioEx™ 9.1 is an easy-to-use lab simulation program that consists of 12 exercises containing 63 physiology lab activities that can be used to supplement or substitute for wet labs.

IMPROVED! Practice Anatomy Lab 3.1 is now accessible on all mobile devices to give students 24/7 access to the most widely used lab specimens, including human cadaver, anatomical models, histology slides, cat, and fetal pig.

Dynamic Study Modules are manageable, mobile-friendly sets of questions with extensive feedback for students to test, learn, and retest until they master basic concepts.

- **NEW!** Instructors can select or deselect specific questions to customize assignments.
- **EXPANDED!** The Lab Manual Mastering A&P course now offers over 3,000 Dynamic Study Module questions, shared with the Marieb/Hoehn texbook ***Human Anatomy & Physiology* 11th Edition**.

The Mastering A&P Instructor Resources Area includes the following downloadable tools:

- **Customizable PowerPoint® lecture outlines** include customizable images and provide a springboard for lab prep.
- **All of the figures, photos, and tables from the manual** are available in JPEG and PowerPoint® formats, in labeled and unlabeled versions, and with customizable labels and leader lines.
- **Test bank** provides thousands of customizable questions across Bloom's taxonomy levels and includes all lab practical and quiz questions from Practice Anatomy Lab 3.1. Each question is tagged to chapter learning outcomes that can also be tracked within Mastering A&P assessments. Available in Microsoft® Word and TestGen® formats.
- **Animations and videos** bring A&P concepts to life and include pre-lab videos, bone videos, and dissection videos.
- **A comprehensive Instructor's Guide**, co-authored by Elaine Marieb and Lori Smith, includes prep instructions for each exercise, along with answer keys for all of the Exercise Review Sheets.

A Photographic Atlas for Anatomy & Physiology
By Nora Hebert, Ruth E. Heisler, et al.
ISBN 9780321869258

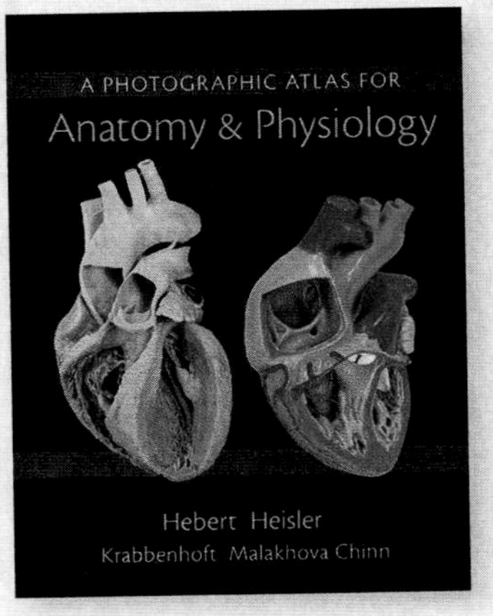

Instructor Resource DVD with PowerPoint Lecture Outlines
ISBN 9780134777092

Instructor's Guide for Human Anatomy & Physiology Lab Manual 13/e
ISBN 9780134778839

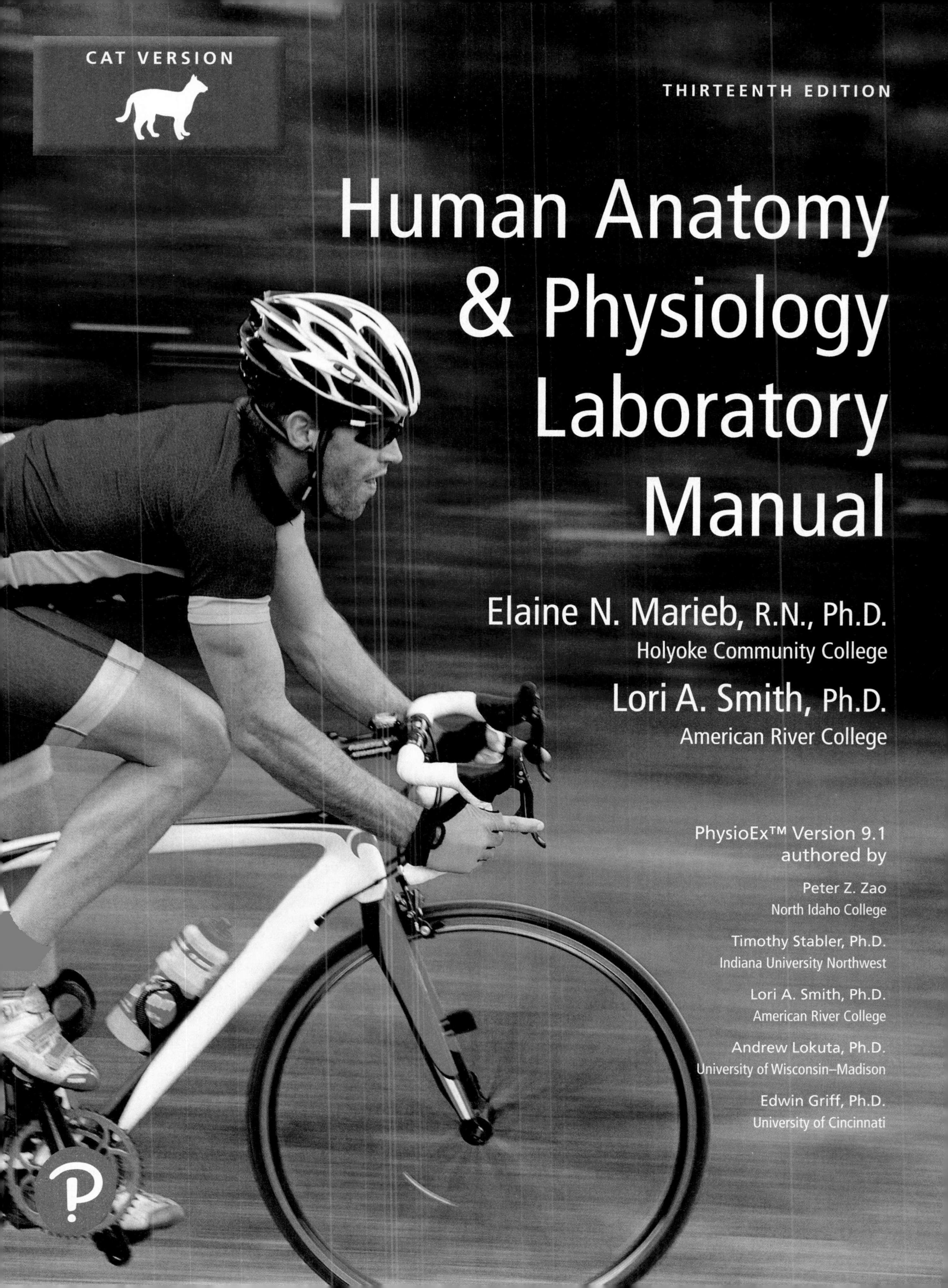

CAT VERSION

THIRTEENTH EDITION

Human Anatomy & Physiology Laboratory Manual

Elaine N. Marieb, R.N., Ph.D.
Holyoke Community College

Lori A. Smith, Ph.D.
American River College

PhysioEx™ Version 9.1
authored by

Peter Z. Zao
North Idaho College

Timothy Stabler, Ph.D.
Indiana University Northwest

Lori A. Smith, Ph.D.
American River College

Andrew Lokuta, Ph.D.
University of Wisconsin–Madison

Edwin Griff, Ph.D.
University of Cincinnati

Editor-in-Chief: Serina Beauparlant
Senior Courseware Portfolio Manager: Lauren Harp
Managing Producer: Nancy Tabor
Content & Design Manager: Michele Mangelli, Mangelli Productions, LLC
Courseware Editorial Assistant: Dapinder Dosanjh
Rich Media Content Producer: Kimberly Twardochleb
Mastering Content Developer: Cheryl Chi
Production Supervisor: Janet Vail
Copyeditor: Sally Peyrefitte
Proofreader: Betsy Dietrich

Compositor: iEnergizer Aptara®, Ltd.
Art Coordinator: David Novak
Interior & Cover Designer: Hespenheide Design
Illustrator: Imagineering STA Media Services, Inc.
Rights & Permissions Management: Ben Ferrini
Rights & Permissions Project Manager: Cenveo © Publishing Services, Matt Perry
Photo Researcher: Kristin Piljay
Manufacturing Buyer: Stacey Weinberger
Director of Product Marketing: Allison Rona
Senior Anatomy & Physiology Specialist: Derek Perrigo

Cover Photo Credit: technotr/Getty Images

Acknowledgments of third-party content appear on page C-1, which constitutes an extension of this copyright page.

Pearson www.pearson.com

ISBN 10: 0-13-463233-8 (Student edition)
ISBN 13: 978-0-13-463233-9 (Student edition)
ISBN 10: 0-13-477708-5 (Instructor's Review Copy)
ISBN 13: 978-0-13-477708-5 (Instructor's Review Copy)

13 2022

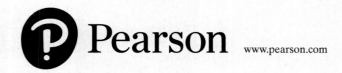

About the Authors

Elaine N. Marieb

After receiving her Ph.D. in zoology from the University of Massachusetts at Amherst, Elaine N. Marieb joined the faculty of the Biological Science Division of Holyoke Community College. While teaching at Holyoke Community College, where many of her students were pursuing nursing degrees, she developed a desire to better understand the relationship between the scientific study of the human body and the clinical aspects of the nursing practice. To that end, while continuing to teach full time, Dr. Marieb pursued her nursing education, which culminated in a Master of Science degree with a clinical specialization in gerontology from the University of Massachusetts. It is this experience that has informed the development of the unique perspective and accessibility for which her publications are known.

Dr. Marieb has given generously to provide opportunities for students to further their education. She funds the E. N. Marieb Science Research Awards at Mount Holyoke College, which promotes research by undergraduate science majors, and has underwritten renovation of the biology labs in Clapp Laboratory at that college. Dr. Marieb also contributes to the University of Massachusetts at Amherst, where she provided funding for reconstruction and instrumentation of a cutting-edge cytology research laboratory. Recognizing the severe national shortage of nursing faculty, she underwrites the Nursing Scholars of the Future Grant Program at the university.

In 2012 and 2017, Dr. Marieb gave generous philanthropic support to Florida Gulf Coast University as a long-term investment in education, research, and training for healthcare and human services professionals in the local community. In honor of her contributions, the university is now home to the Elaine Nicpon Marieb College of Health and Human Services.

Lori A. Smith

Lori A. Smith received her Ph.D. in biochemistry from the University of California at Davis. Before discovering her passion for teaching, she worked as a research scientist and project leader in the medical diagnostics industry. In 1999, she joined the faculty at American River College in the Biology Department, where she teaches anatomy and physiology and microbiology to students preparing for nursing or other allied health careers. Since 2005, she has coauthored Pearson's PhysioEx™: Laboratory Simulations in Physiology and has continued to coauthor several Pearson lab manuals. Dr. Smith has been named Instructor of the Year by the American River College Associated Student Body, and she is a member of the Human Anatomy and Physiology Society (HAPS) and California Academy of Sciences. When not teaching or writing, she enjoys spending time with her family: hiking, cycling, and kayaking.

Preface to the Instructor

The philosophy behind the revision of this manual mirrors that of all earlier editions. It reflects a still developing sensibility for the way teachers teach and students learn, informed by years of teaching the subject and by collecting suggestions from other instructors as well as from students enrolled in multifaceted healthcare programs. *Human Anatomy & Physiology Laboratory Manual* was originally developed to facilitate and enrich the laboratory experience for both teachers and students. This edition retains those same goals.

This manual, intended for students in introductory human anatomy and physiology courses, presents a wide range of laboratory experiences for students concentrating in nursing, physical therapy, pharmacology, respiratory therapy, and exercise science, as well as biology and premedical programs. The manual's coverage is intentionally broad, allowing it to serve both one- and two-semester courses, and it is available in versions that contain detailed guidelines for dissecting a cat or fetal pig laboratory specimen.

Basic Approach and Features

The generous variety of experiments in this manual provides flexibility that enables instructors to gear their laboratory approach to specific academic programs or to their own teaching preferences. The manual remains independent of any textbook, so it contains the background discussions and terminology necessary to perform all experiments. Such a self-contained learning aid eliminates the need for students to bring a textbook into the laboratory.

Each of the 46 exercises leads students toward a coherent understanding of the structure and function of the human body. The manual begins with anatomical terminology and an orientation to the body, which together provide the necessary tools for studying the various body systems. The exercises that follow reflect the dual focus of the manual—both anatomical and physiological aspects receive considerable attention. As the various organ systems of the body are introduced, the initial exercises focus on organization, from the cellular to the organ system level. As indicated by the table of contents, the anatomical exercises are usually followed by physiological experiments that familiarize students with various aspects of body functioning and promote the critical understanding that function follows structure. The numerous physiological experiments for each organ system range from simple experiments that can be performed without specialized tools to more complex experiments using laboratory equipment, computers, and instrumentation techniques.

Features

The dissection scissors icon appears at the beginning of activities that entail the dissection of isolated animal organs. In addition to the figures, isolated animal organs, such as the sheep heart and pig kidney, are employed to study anatomy because of their exceptional similarity to human organs.

Homeostasis is continually emphasized as a requirement for optimal health. Pathological conditions are viewed as a loss of homeostasis; these discussions can be recognized by the homeostatic imbalance logo within the descriptive material of each exercise. This holistic approach encourages an integrated understanding of the human body. The homeostatic imbalance icon directs the student's attention to conditions representing a loss of homeostasis.

A safety icon notifies students that specific safety precautions must be observed when using certain equipment or conducting particular lab procedures. For example, when working with ether, students are to use a hood; and when handling body fluids such as blood, urine, or saliva, students are to wear gloves. All exercises involving body fluids (blood, urine, saliva) incorporate current Centers for Disease Control and Prevention (CDC) guidelines for handling human body fluids. Because it is important that nursing students in particular learn how to safely handle bloodstained articles, the manual has retained the option to use human blood in the laboratory. However, the decision to allow testing of human (student) blood or to use animal blood in the laboratory is left to the discretion of the instructor in accordance with institutional guidelines. The CDC guidelines for handling body fluids are reinforced by the laboratory safety procedures described on the inside front cover of this text, in Exercise 29: Blood, and in the *Instructor's Guide*. You can photocopy the inside front cover and post it in the lab to help students become well versed in laboratory safety.

Group Challenge activities are designed to enhance collaborative group learning and to challenge students to think critically, identify relationships between anatomical structures and physiological functions, and achieve a deeper understanding of anatomy and physiology concepts.

BIOPAC® The BIOPAC® icon in a relevant exercise materials list signals the use of the BIOPAC® Student Lab System and alerts you to the equipment needed. BIOPAC® is used in Exercises 14, 18, 20, 21, 31, 33, 34, and 37. The instructions in the lab manual are for use with the BIOPAC® MP36/35 and MP45 data acquisition unit. Note that some exercises are not compatible with the MP45 data acquisition unit. For those exercises, the MP45 will not be listed in the Materials section. In this edition, the lab manual instructions are for use with BSL software 4.0.1 and above for Windows 10/8.x/7 or Mac OS X10.9–10.12. Refer to the Materials section in each exercise for the applicable software version. The Instructor Resources area of Mastering A&P provides the following additional support for alternative data acquisitions systems, including exercises that can be distributed to students:

- *BIOPAC® Instructions for the MP36 (or MP35/30) data acquisition unit* using BSL software versions earlier than 4.0.1 (for Windows and Mac) for Exercises 14, 18, 20, 21, 31, and 34
- *Powerlab®* Instructions for Exercises 14, 21, 31, 33, 34, and 37
- *iWorx®* Instructions for Exercises 14, 18, 21, 31, 33, 34, and 37
- *Intelitool®* Instructions for Exercises 14i, 21i, 31i, and 37i

- **Exercise Review Sheets** follow each laboratory exercise and provide space for recording and interpreting experimental results and require students to label diagrams and answer matching and short-answer questions. Selected questions can be assigned and automatically graded in Mastering A&P.
- **PhysioEx™ 9.1 Exercises**, located in the back of the lab manual and accessible through a subscription to Mastering A&P, are easy-to-use computer simulations that supplement or take the place of traditional wet labs safely and cost-effectively. These 12 exercises contain a total of 63 physiology laboratory activities that allow learners to change variables and test out various hypotheses for the experiments. PhysioEx™ allows students to repeat labs as often as they like, perform experiments without harming live animals, and conduct experiments that are difficult to perform because of time, cost, or safety concerns.

Updated Content in This Edition of the Lab Manual

Throughout the manual, the narrative text has been streamlined and updated to make the language more understandable and to better meet the needs of today's students. Additional highlights include the following:

- **Dozens of new full-color figures and photos** replace black-and-white line drawings in the Exercise Review Sheets. Selected labeling questions in the manual can be assigned in Mastering A&P.
- **New Clinical Application questions** have been added to the Exercise Review Sheets and challenge students to apply lab concepts and critical-thinking skills to real-world clinical scenarios.
- **Updated BIOPAC® procedures** are included in the manual for eight lab exercises for the BIOPAC® 4.0 software upgrade. Procedures for Intelitool®, PowerLab®, and iWorx® remain available in the Instructor Resources area of Mastering A&P.
- **New Mastering A&P visual previews** appear on the first page of each lab exercise, highlighting a recommended pre-lab video, a related image from Practice Anatomy Lab 3.1 (PAL 3.1), or a helpful animation.
- **New Mastering A&P assignment recommendations** are signaled at appropriate points throughout the manual to help instructors assign related auto-graded activities and assessments.
- **Extensive updates and improvements** have been made to each of the 46 laboratory exercises in the manual to increase clarity and reduce ambiguity for students. Art within the exercises, the narrative, as well as the questions and figures within the Review Sheets have been updated. For a complete list of content updates, please refer to the *Instructor's Guide for Human Anatomy & Physiology Laboratory Manual* 13/e (ISBN 9780134778839 or in the Instructor Resources area of Mastering A&P).

Highlights of Updated Content in Mastering A&P

Mastering A&P, the leading online homework, tutorial, and assessment system is designed to engage students and improve results by helping them stay on track in the course and quickly master challenging anatomy and physiology concepts. Mastering A&P assignments support interactive features in the lab manual, including pre-lab video coaching activities; bone, muscle, and dissection videos; Dynamic Study Modules; *Get Ready for A&P*; plus a variety of Art Labeling questions, Clinical Application questions, and more. Highlights for this edition include the following:

- **8 new Pre-Lab Video Coaching Activities in Mastering A&P** (for a total of 18) focus on key concepts in the lab activity and walk students through important procedures. New pre-lab video titles include Preparing and Observing a Wet Mount, Examining a Long Bone, Initiating Pupillary Reflexes, Palpating Superficial Pulse Points, Auscultating Heart Sounds, and more.
- **New Cat and Fetal Pig Dissection Video Coaching Activities** help students prepare for dissection by previewing key anatomical structures. Each video includes one to two comparisons to human structures.
- **IMPROVED! Practice Anatomy Lab™ (PAL™ 3.1) is now fully accessible on all mobile devices**, including smartphones, tablets, and laptops. PAL is an indispensable virtual anatomy study and practice tool that gives students 24/7 access to the most widely used lab specimens, including human cadaver; anatomical models from leading manufacturers such as 3B Scientific, SOMSO, Denoyer-Geppert, Frey Scientific/Nystrom, Altay Scientific, and Ward's; histology; cat; and fetal pig. PAL 3.1 is easy to use and includes built-in audio pronunciations, rotatable bones, and simulated fill-in-the-blank lab practical exams.
- **New Customizable Practice Anatomy Lab (PAL) Flashcards** enable students to create a personalized, mobile-friendly deck of flashcards and quizzes using images from PAL 3.1. Students can generate flashcards using only the structures that their instructor emphasizes in lecture or lab.
- **New Building Vocabulary Coaching Activities** are a fun way for students to learn word roots and A&P terminology while building and practicing important language skills.
- **Expanded Dynamic Study Modules** help students study effectively on their own by continuously assessing their activity and performance in real time. Students complete a set of questions and indicate their level of confidence in their answer. Questions repeat until the student can answer them all correctly and confidently. These are available as graded assignments prior to class and are accessible on smartphones, tablets, and computers.
  - The Lab Manual Mastering A&P course now offers over 3000 Dynamic Study Module questions, shared with Marieb/Hoehn *Human Anatomy & Physiology*, 11th Edition.
  - Instructors can now remove questions from Dynamic Study Modules to better fit their course.
- **Expanded Drag-and-Drop Art Labeling Questions** allow students to assess their knowledge of terms and structures in the lab manual. Selected Exercise Review Sheet labeling activities in the manual are now assignable.

Please refer to the preceding pages for additional information about Mastering A&P and other resources for instructors and students.

Acknowledgments

Continued thanks to our colleagues and friends at Pearson who collaborated with us on this edition, especially Editor-in-Chief Serina Beauparlant, Acquisitions Editor Lauren Harp, Editorial Assistant Dapinder Dosanjh, and Rich Content Media Producers Kimberly Twardochleb and Lauren Chen. We also thank the Pearson Sales and Marketing team for their work in supporting instructors and students, especially Senior A&P Specialist Derek Perrigo and Director of Product Marketing Allison Rona.

Special thanks go out to Amanda Kaufmann for her leadership and expertise in producing the 18 pre-lab videos that support this edition, and to Mike Mullins of BIOPAC®, who helped us update the instructions for consistency with the upgraded software.

We're also grateful to Michele Mangelli and her superb production team, who continue to cross every hurdle with uncommon grace and skill, including Janet Vail, production coordinator; David Novak, art and photo coordinator; Kristin Piljay, photo researcher; Gary Hespenheide, interior and cover designer; and Sally Peyrefitte, copyeditor.

Last but not least, we wish to extend our sincere thanks to the many A&P students who have circulated through our lab classrooms and have used this lab manual over the years—you continue to inspire us every day! As always, we welcome your feedback and suggestions for future editions.

Elaine N. Marieb & Lori A. Smith

THIRTEENTH EDITION REVIEWERS

We wish to thank the following reviewers, who provided thoughtful feedback and helped us make informed decisions for this edition of both the lab manual and Mastering A&P resources:

Matthew Abbott, *Des Moines Area Community College*
Lynne Anderson, *Meridian Community College*
Penny Antley, *University of Louisiana, Lafayette*
Marianne Baricevic, *Raritan Valley Community College*
Christopher W. Brooks, *Central Piedmont Community College*
Jocelyn Cash, *Central Piedmont Community College*
Christopher D'Arcy, *Cayuga Community College*
Mary E. Dawson, *Kingsborough Community College*
Karen Eastman, *Chattanooga State Community College*
Jamal Fakhoury, *College of Central Florida*
Lisa Flick, *Monroe Community College*
Michele Finn, *Monroe Community College*
Juanita Forrester, *Chattahoochee Technical College*
Larry Frolich, *Miami Dade College*
Michelle Gaston, *Northern Virginia Community College, Alexandria*
Tejendra Gill, *University of Houston*
Abigail M. Goosie, *Walters State Community College*
Karen Gordon, *Rowan Cabarrus Community College*
Jennifer Hatchel, *College of Coastal Georgia*
Clare Hays, *Metropolitan State University*
Nathanael Heyman, *California Baptist University*
Samuel Hirt, *Auburn University*
Alexander Ibe, *Weatherford College*
Shahdi Jalilvand, *Tarrant County College–Southeast*
Marian Leal, *Sacred Heart University*
Geoffrey Lee, *Milwaukee Area Technical College*
Tara Leszczewiz, *College of Dupage*
Mary Katherine Lockwood, *University of New Hampshire*
Francisco J. Martinez, *Hunter College of CUNY*
Bruce Maring, *Daytona State College*
Geri Mayer, *Florida Atlantic University*
Tiffany B. McFalls-Smith, *Elizabethtown Community & Technical College*
Melinda A. Miller, *Pearl River Community College*
Todd Miller, *Hunter College of CUNY*
Susan Mitchell, *Onondaga Community College*
Erin Morrey, *Georgia Perimeter College*
Jill O'Malley, *Erie Community College*
Suzanne Oppenheimer, *College of Western Idaho*
Lori Paul, *University of Missouri - St. Louis*
Stacy Pugh-Towe, *Crowder College*
Suzanne Pundt, *The University of Texas at Tyler*
Jackie Reynolds, *Richland College*
Anthony Rizzo, *Polk State College*
Jo Rogers, *University of Cincinnati*
James Royston, *Pearl River Community College*
Connie E. Rye, *East Mississippi Community College*
Mark Schmidt, *Clark State Community College*
Jennifer Showalter, *Waubonsee Community College*
Teresa Stegall-Faulk, *Middle Tennessee State University*
Melissa Ann Storm, *University of South Carolina–Upstate*
Bonnie J. Tarricone, *Ivy Tech Community College*
Raymond Thompson, *University of South Carolina*
Anna Tiffany Tindall-McKee, *East Mississippi Community College*
Allen Tratt, *Cayuga Community College*
Khursheed Wankadiya, *Central Piedmont Community College*
Diane L. Wood, *Southeast Missouri State University*

Contents

A Pre-Lab video is available in Mastering **A&P™** for selected activities.

THE SKELETAL SYSTEM

THE MUSCULAR SYSTEM

THE NERVOUS SYSTEM

Instructors may download two additional lab activities: Inhibiting the Nerve and Visualizing the Compound Action Potential with an Oscilloscope. Instructors, please go to Mastering A&P™ > Instructor Resources > Additional Resources > Additional Exercises

THE ENDOCRINE SYSTEM

THE CIRCULATORY SYSTEM

EXERCISE 1

The Language of Anatomy

Learning Outcomes

▶ Describe the anatomical position, and explain its importance.
▶ Use proper anatomical terminology to describe body regions, orientation and direction, and body planes.
▶ Name the body cavities, and indicate the important organs in each.
▶ Name and describe the serous membranes of the ventral body cavities.
▶ Identify the abdominopelvic quadrants and regions on a torso model or image.

Pre-Lab Quiz
 Instructors may assign these and other Pre-Lab Quiz questions using Mastering A&P™

1. Circle True or False. In anatomical position, the body is lying down.
2. Circle the correct underlined term. With regard to surface anatomy, <u>abdominal</u> / <u>axial</u> refers to the structures along the center line of the body.
3. The term *superficial* refers to a structure that is:
   a. attached near the trunk of the body
   b. toward or at the body surface
   c. toward the head
   d. toward the midline
4. The _____ plane runs longitudinally and divides the body into right and left sides.
   a. frontal
   b. sagittal
   c. transverse
   d. ventral
5. Circle the correct underlined terms. The dorsal body cavity can be divided into the <u>cranial</u> / <u>thoracic</u> cavity, which contains the brain, and the <u>sural</u> / <u>vertebral</u> cavity, which contains the spinal cord.

Go to Mastering A&P™ > Study Area to improve your performance in A&P Lab.

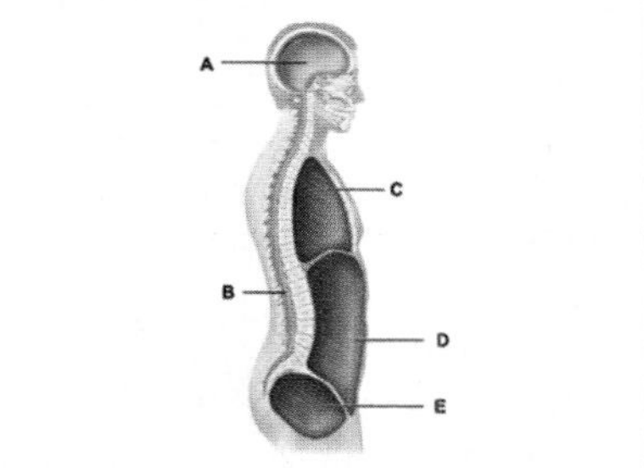

 Instructors may assign new Building Vocabulary coaching activities, Pre-Lab Quiz questions, Art Labeling activities, and more using the Mastering A&P™ Item Library.

Materials
▶ Human torso model (dissectible)
▶ Human skeleton
▶ Demonstration: sectioned and labeled kidneys (three separate kidneys uncut or cut so that [a] entire, [b] transverse sectional, and [c] longitudinal sectional views are visible)
▶ Gelatin-spaghetti molds
▶ Scalpel

A student new to any science is often overwhelmed at first by the terminology used in that subject. The study of anatomy is no exception. But without specialized terminology, confusion is inevitable. For example, what do *over*, *on top of*, *above*, and *behind* mean in reference to the human body? Anatomists have an accepted set of reference terms that are universally understood. These allow body structures to be located and identified precisely with a minimum of words.

This exercise presents some of the most important anatomical terminology used to describe the body and introduces you to basic concepts of **gross anatomy**, the study of body structures visible to the naked eye.

Anatomical Position

When anatomists or doctors refer to specific areas of the human body, the picture they keep in mind is a universally accepted standard position called the **anatomical position**. In the anatomical position, the human body is erect, with the feet only slightly apart, head and toes pointed forward, and arms hanging at the sides with palms facing forward (**Figure 1.1a**). It is also important to remember that "left" and "right" refer to the sides of the individual, not the observer.

☐ Assume the anatomical position. The hands are held unnaturally forward rather than hanging with palms toward the thighs.

Check the box when you have completed this task.

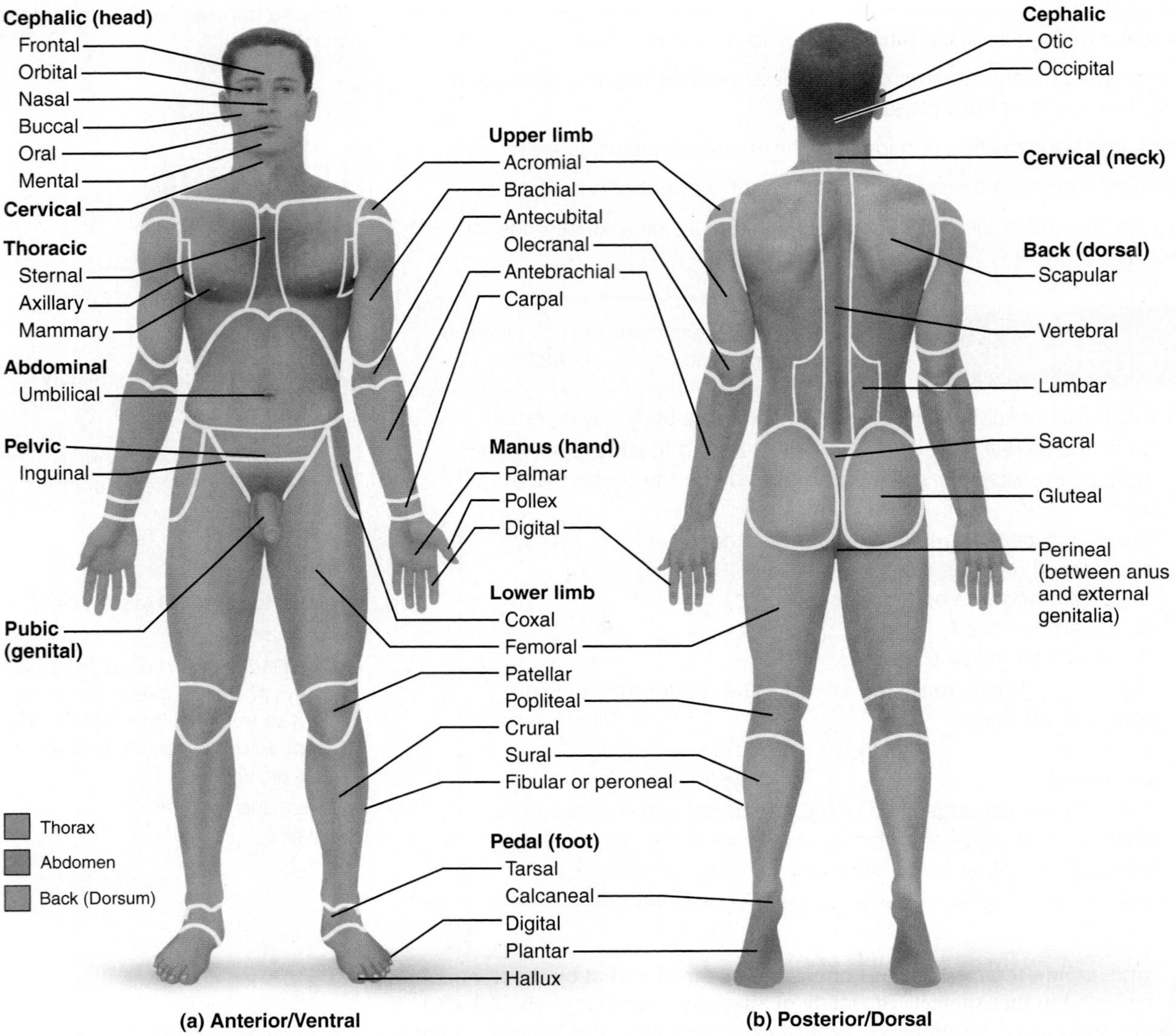

Figure 1.1 Anatomical position and regional terms. Heels are raised to illustrate the plantar surface of the foot, which is actually on the inferior surface of the body.

Instructors may assign this figure as an Art Labeling Activity using Mastering A&P™

Regional Anatomy

The body is divided into two main regions, the axial and appendicular regions. The **axial region** includes the head, neck, and trunk; it runs along the vertical axis of the body. The **appendicular region** includes the limbs, which are also called the appendages or extremities. The body is also divided up into smaller regions within those two main divisions. **Table 1.1** summarizes the body regions that are illustrated in Figure 1.1.

1

Table 1.1 Regions of the Human Body (Figure 1.1)

Region	Description	Region	Description
Abdominal	Located below the ribs and above the hips	Nasal	Nose
Acromial	Point of the shoulder	Occipital	Back of the head
Antebrachial	Forearm	Olecranal	Back of the elbow
Antecubital	Anterior surface of the elbow	Oral	Mouth
Axillary	Armpit	Orbital	Bony eye socket
Brachial	Arm (upper portion of the upper limb)	Otic	Ear
Buccal	Cheek	Palmar	Palm of the hand
Calcaneal	Heel of the foot	Patellar	Kneecap
Carpal	Wrist	Pedal	Foot
Cephalic	Head	Pelvic	Pelvis
Cervical	Neck	Perineal	Between the anus and the external genitalia
Coxal	Hip	Plantar	Sole of the foot
Crural	Leg	Pollex	Thumb
Digital	Fingers or toes	Popliteal	Back of the knee
Femoral	Thigh	Pubic	Genital
Fibular (peroneal)	Side of the leg	Sacral	Posterior region between the hip bones
Frontal	Forehead	Scapular	Shoulder blade
Gluteal	Buttocks	Sternal	Breastbone
Hallux	Great toe	Sural	Calf
Inguinal	Groin	Tarsal	Ankle
Lumbar	Lower back	Thoracic	Chest
Mammary	Breast	Umbilical	Navel
Manus	Hand	Vertebral	Spine
Mental	Chin		

Activity 1

Locating Body Regions

Locate the anterior and posterior body regions on yourself, your lab partner, and a human torso model.

Directional Terms

Study the terms below, referring to Figure 1.2 for a visual aid. Notice that certain terms have different meanings, depending on whether they refer to a four-legged animal (quadruped) or to a human (biped).

Superior/inferior *(above/below):* These terms refer to placement of a structure along the long axis of the body. The nose, for example, is superior to the mouth, and the abdomen is inferior to the chest.

Anterior/posterior *(front/back):* In humans, the most anterior structures are those that are most forward—the face, chest, and abdomen. Posterior structures are those toward the backside of the body. For instance, the spine is posterior to the heart.

Medial/lateral *(toward the midline/away from the midline or median plane):* The sternum (breastbone) is medial to the ribs; the ear is lateral to the nose.

The terms of position just described assume the person is in the anatomical position. The next four term pairs are more absolute. They apply in any body position, and they consistently have the same meaning in all vertebrate animals.

1

Cephalad (cranial)/caudal *(toward the head/toward the tail):* In humans, these terms are used interchangeably with *superior* and *inferior*, but in four-legged animals they are synonymous with *anterior* and *posterior*, respectively.

Ventral/dorsal *(belly side/backside):* These terms are used chiefly in discussing the comparative anatomy of animals, assuming the animal is standing. In humans, the terms *ventral* and *dorsal* are used interchangeably with the terms *anterior* and *posterior*, but in four-legged animals, *ventral* and *dorsal* are synonymous with *inferior* and *superior*, respectively.

Proximal/distal *(nearer the trunk or attached end/farther from the trunk or point of attachment):* These terms are used primarily to locate various areas of the body limbs. For example, the fingers are distal to the elbow; the knee is proximal to the toes. However, these terms may also be used to indicate regions (closer to or farther from the head) of internal tubular organs.

Superficial (external)/deep (internal) *(toward or at the body surface/away from the body surface):* For example, the skin is superficial to the skeletal muscles, and the lungs are deep to the rib cage.

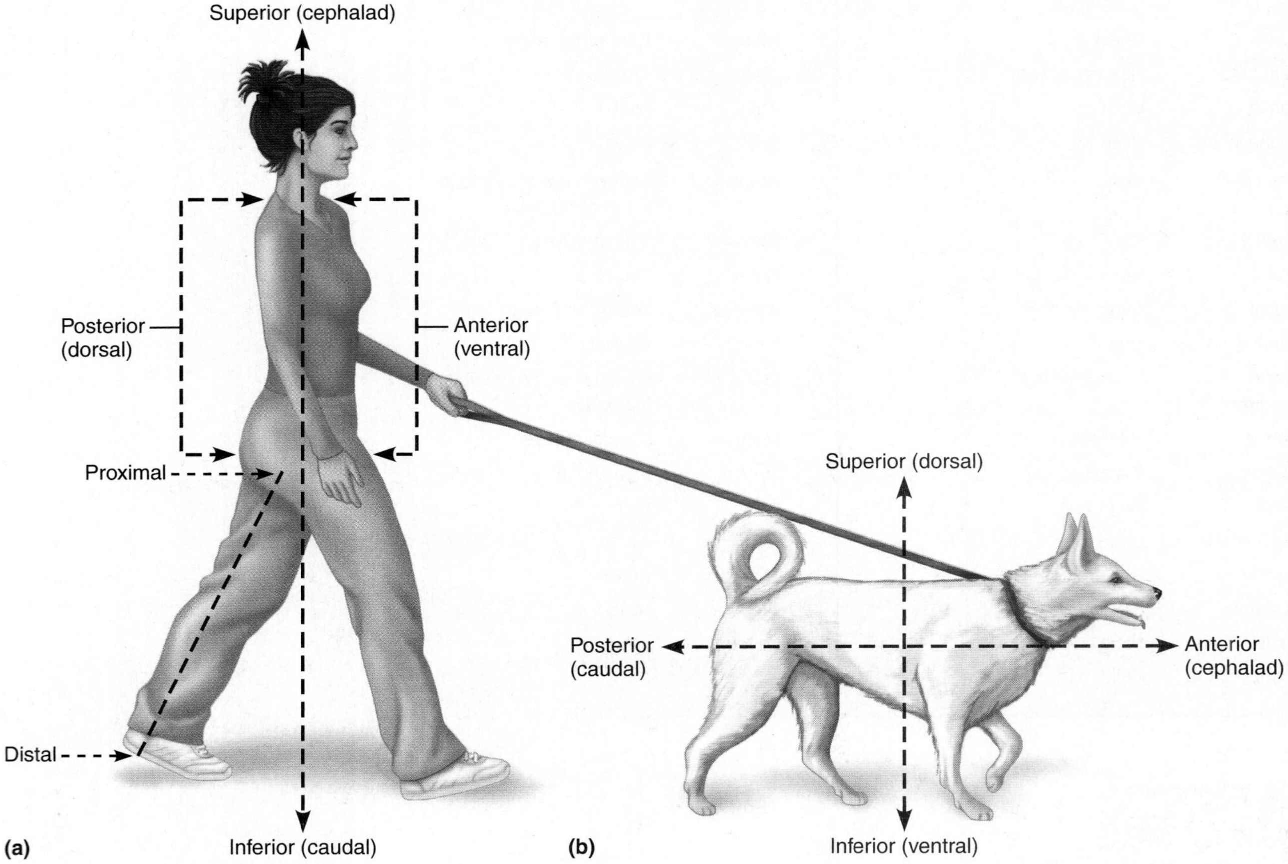

Figure 1.2 Directional terms. (a) With reference to a human. **(b)** With reference to a four-legged animal.

Instructors may assign this figure as an Art Labeling Activity using Mastering A&P™

Activity 2

Practicing Using Correct Anatomical Terminology

Use a human torso model, a human skeleton, or your own body to practice using the regional and directional terminology.

1. The popliteal region is ________________. (anterior or posterior)

2. The acromial region is ________________ to the otic region. (medial or lateral)

3. The femoral region is ________________ to the tarsal region. (proximal or distal)

4. The bones are ________________ to the skin. (superficial or deep)

Body Planes and Sections

The body is three-dimensional, and in order to observe its internal structures, it is often necessary to make a **section**, or cut. When the section is made through the body wall or through an organ, it is made along an imaginary surface or line called a **plane**. A section is named for the plane along which it is cut. Anatomists commonly refer to three planes (**Figure 1.3**), or sections, that lie at right angles to one another.

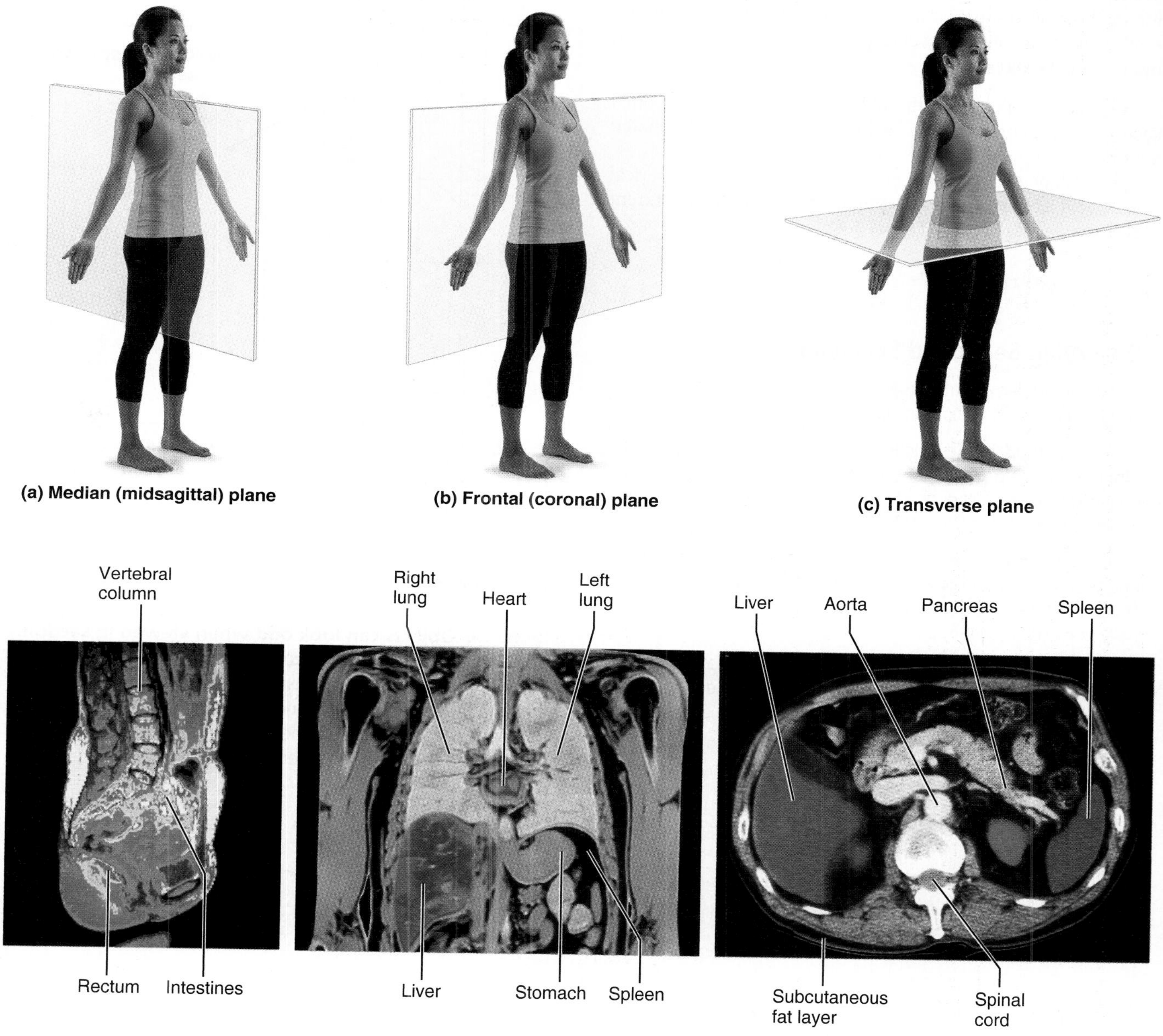

Figure 1.3 Planes of the body with corresponding magnetic resonance imaging (MRI) scans. Note the transverse section is an inferior view.

Instructors may assign this figure as an Art Labeling Activity using Mastering **A&P™**

Sagittal plane: A sagittal plane runs longitudinally and divides the body into right and left parts. If it divides the body into equal parts, right down the midline of the body, it is called a **median**, or **midsagittal, plane**.

Frontal plane: Sometimes called a **coronal plane**, the frontal plane is a longitudinal plane that divides the body (or an organ) into anterior and posterior parts.

Transverse plane: A transverse plane runs horizontally, dividing the body into superior and inferior parts. When organs are sectioned along the transverse plane, the sections are commonly called **cross sections**.

On microscope slides, the abbreviation for a longitudinal section (sagittal or frontal) is l.s. Cross sections are abbreviated x.s. or c.s.

A median or frontal plane section of any nonspherical object, be it a banana or a body organ, provides quite a different view from a cross section (**Figure 1.4**).

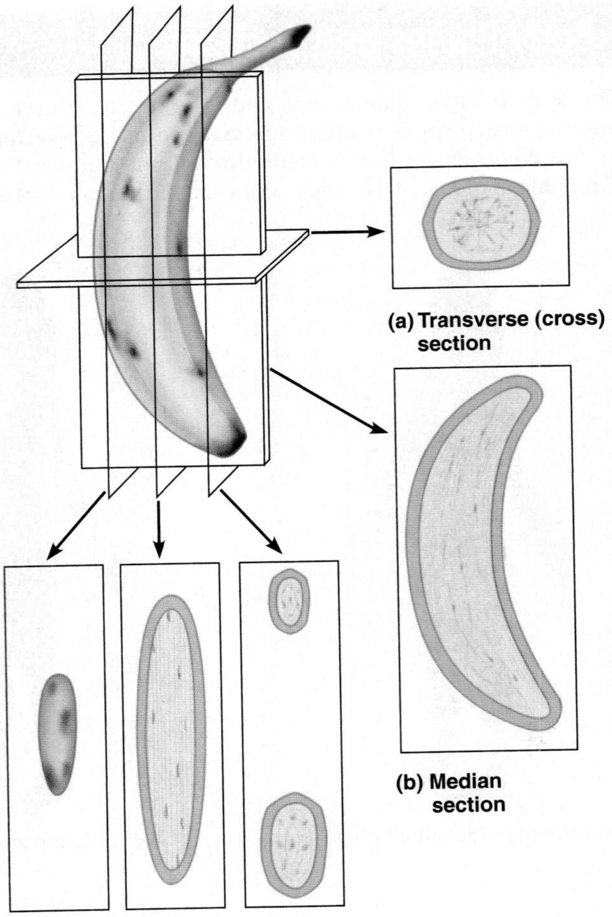

Figure 1.4 Objects can look odd when viewed in section. This banana has been sectioned in three different planes **(a–c)**, and only in one of these planes **(b)** is it easily recognized as a banana. If one cannot recognize a sectioned organ, it is possible to reconstruct its shape from a series of successive cuts, as from the three serial sections in **(c)**.

Activity 3

Observing Sectioned Specimens

1. Go to the demonstration area and observe the transversely and longitudinally cut organ specimens (kidneys).
2. After completing instruction 1, obtain a gelatin-spaghetti mold and a scalpel, and take them to your laboratory bench. (Essentially, this is just cooked spaghetti added to warm gelatin, which is then allowed to gel.)
3. Cut through the gelatin-spaghetti mold along any plane, and examine the cut surfaces. You should see spaghetti strands that have been cut transversely (x.s.) and some cut longitudinally (a median section).
4. Draw the appearance of each of these spaghetti sections below, and verify the accuracy of your section identifications with your instructor.

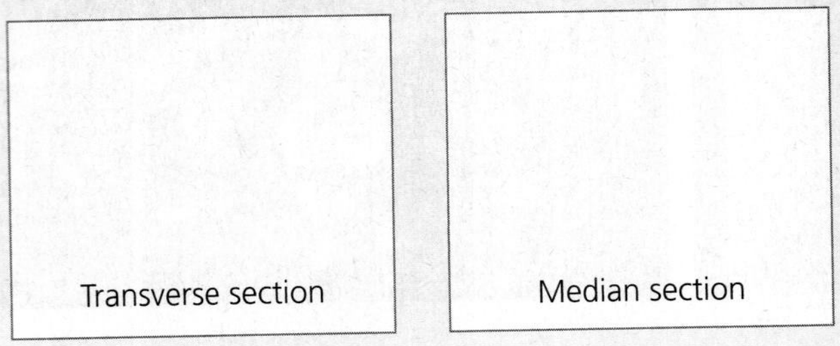

Body Cavities

The axial region of the body has two large cavities that provide different degrees of protection to the organs within them (**Figure 1.5**).

Dorsal Body Cavity

The dorsal body cavity can be subdivided into the **cranial cavity**, which lies within the rigid skull and encases the brain, and the **vertebral** (or **spinal**) **cavity**, which runs through the bony vertebral column to enclose the delicate spinal cord.

Ventral Body Cavity

Like the dorsal cavity, the ventral body cavity is subdivided. The superior **thoracic cavity** is separated from the rest of the ventral cavity by the dome-shaped diaphragm. The heart and lungs, located in the thoracic cavity, are protected by the bony rib cage. The cavity inferior to the diaphragm is referred to as the **abdominopelvic cavity**. Although there is no further physical separation of the ventral cavity, some describe the abdominopelvic cavity as two areas: a superior **abdominal cavity**, the area that houses the stomach, intestines, liver, and other organs, and an inferior **pelvic cavity**, the region that is partially enclosed by the bony pelvis and contains the reproductive organs, bladder, and rectum.

Serous Membranes of the Ventral Body Cavity

The walls of the ventral body cavity and the outer surfaces of the organs it contains are covered with a very thin, double-layered membrane called the **serosa**, or **serous membrane**. The part of the membrane lining the cavity walls is referred to as the **parietal serosa**, and it is continuous with a similar membrane, the **visceral serosa**, covering the external surface of the organs within the cavity. These membranes produce a thin lubricating fluid that allows the visceral organs to slide over one another or to rub against the body wall with minimal

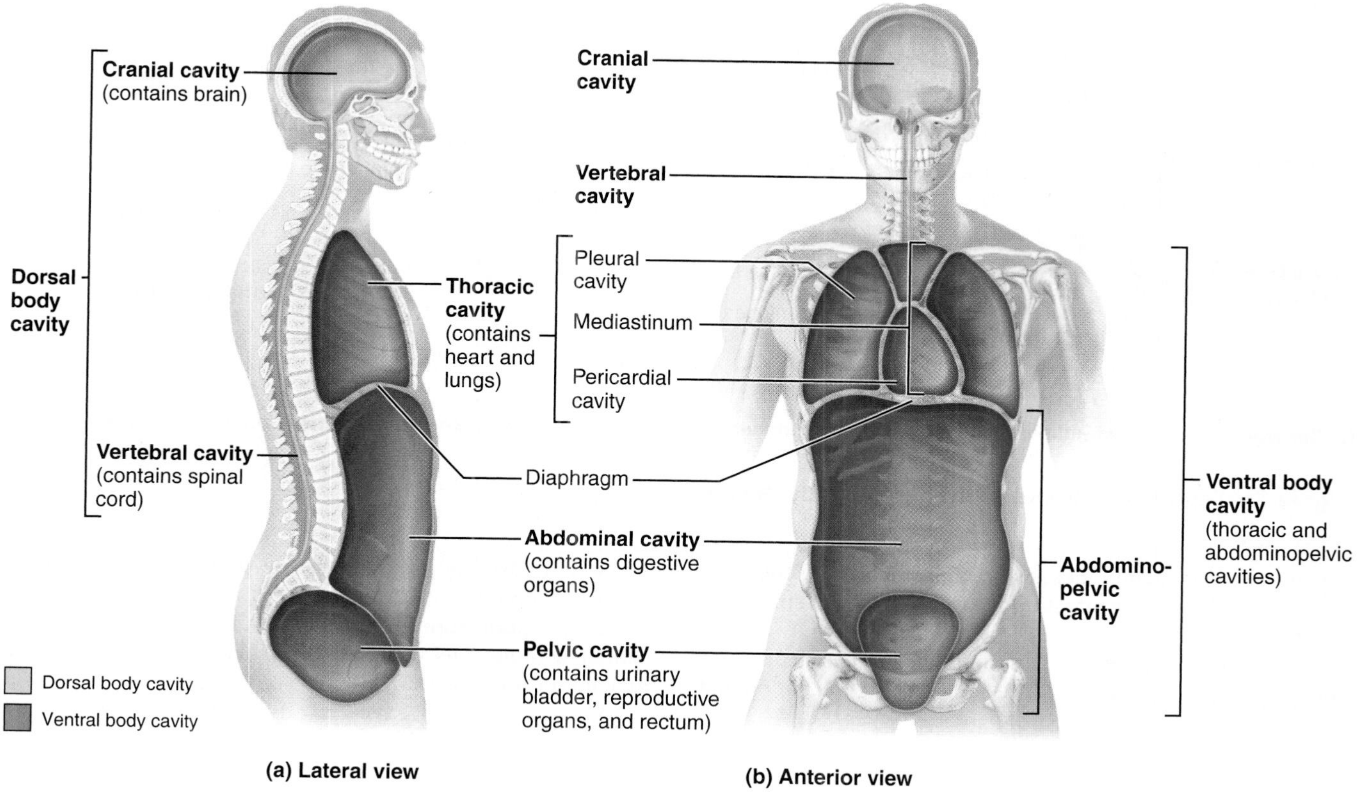

Figure 1.5 Dorsal and ventral body cavities and their subdivisions.

Instructors may assign this figure as an Art Labeling Activity using Mastering A&P™

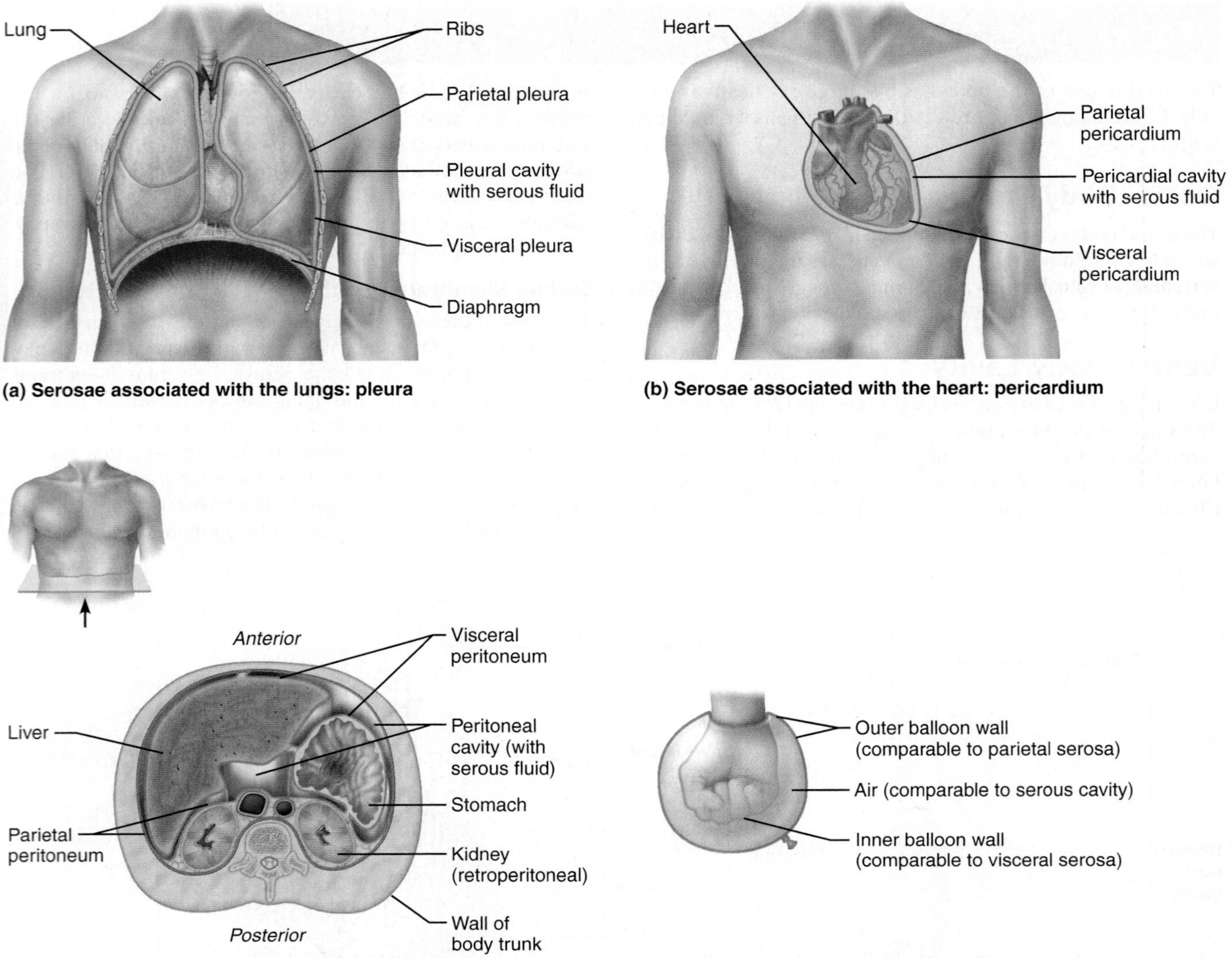

(a) Serosae associated with the lungs: pleura

(b) Serosae associated with the heart: pericardium

(c) Serosae associated with the abdominal viscera: peritoneum

(d) Model of the serous membranes and serous cavity

Figure 1.6 Serous membranes of the ventral body cavities.

friction. Serous membranes also compartmentalize the various organs to prevent infection in one organ from spreading to others.

The specific names of the serous membranes depend on the structures they surround. The serosa lining the abdominal cavity and covering its organs is the **peritoneum**, the serosa enclosing the lungs is the **pleura**, and the serosa around the heart is the **pericardium** (**Figure 1.6**). A fist pushed into a limp balloon demonstrates the relationship between the visceral and parietal serosae (Figure 1.6d).

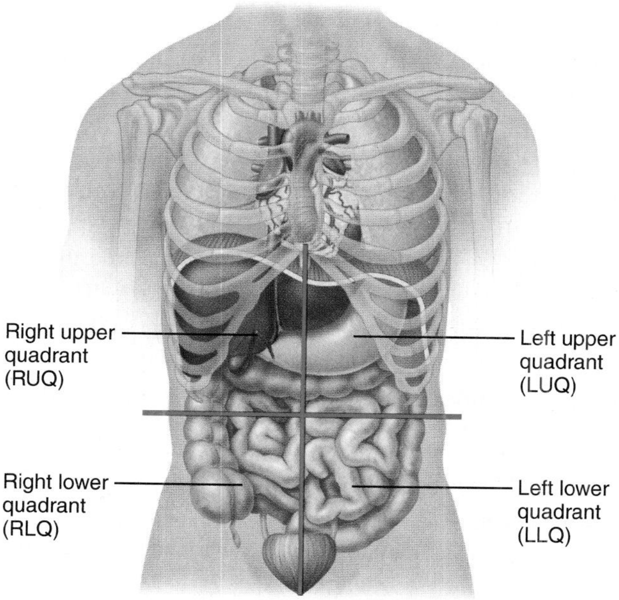

Figure 1.7 Abdominopelvic quadrants. Superficial organs are shown in each quadrant.

Activity 4

Identifying Organs in the Abdominopelvic Cavity

Examine the human torso model to respond to the following questions.

Name two organs found in the left upper quadrant.

_______________ and _______________

Name two organs found in the right lower quadrant.

_______________ and _______________

What organ (Figure 1.7) is divided into identical halves by the median plane? _______________

Abdominopelvic Quadrants and Regions

Because the abdominopelvic cavity is quite large and contains many organs, it is helpful to divide it up into smaller areas for discussion or study.

Most physicians and nurses use a scheme that divides the abdominal surface and the abdominopelvic cavity into four approximately equal regions called **quadrants**. These quadrants are named according to their relative position—that is, *right upper quadrant*, *right lower quadrant*, *left upper quadrant*, and *left lower quadrant* (**Figure 1.7**). Note that the terms *left* and *right* refer to the left and right side of the body in the figure, not the left and right side of the art on the page.

A different scheme commonly used by anatomists divides the abdominal surface and abdominopelvic cavity into nine separate regions by four planes (**Figure 1.8**). As you read

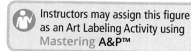

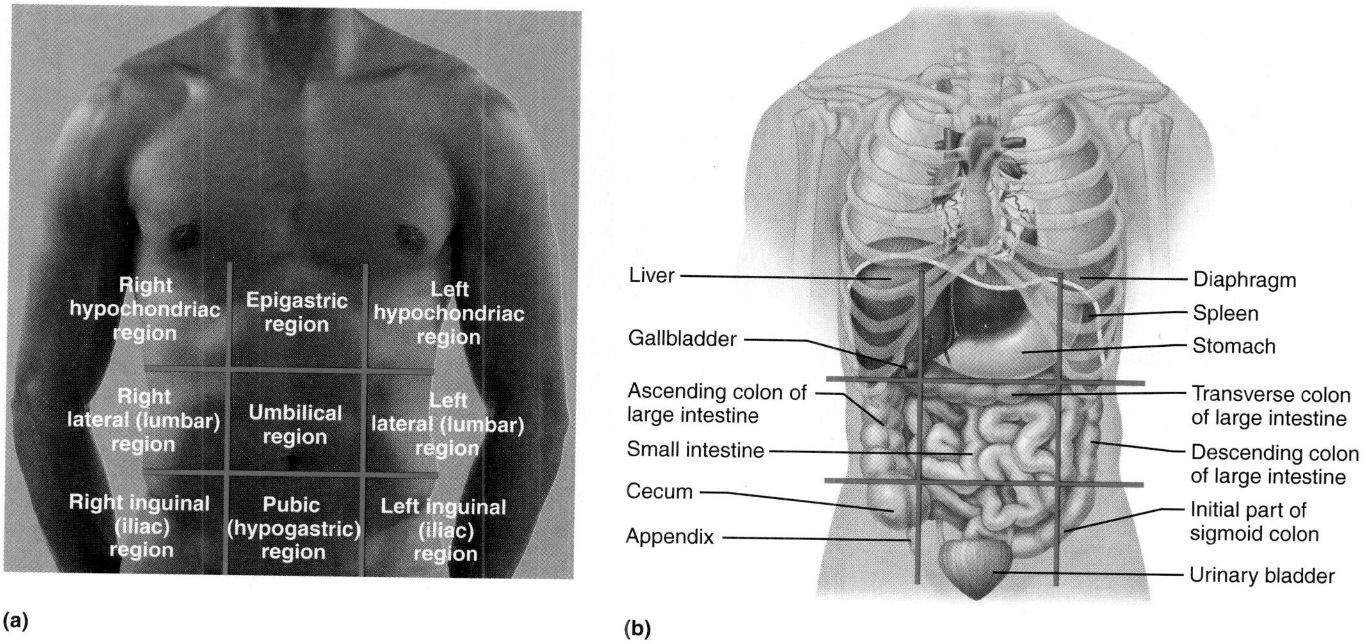

Figure 1.8 Abdominopelvic regions. Nine regions delineated by four planes. **(a)** The superior horizontal plane is just inferior to the ribs; the inferior horizontal plane is at the superior aspect of the hip bones. The vertical planes are just medial to the nipples. **(b)** Superficial organs are shown in each region.

Instructors may assign this figure as an Art Labeling Activity using Mastering **A&P™**

through the descriptions of these nine regions, locate them in Figure 1.8, and note the organs contained in each region.

Umbilical region: The centermost region, which includes the umbilicus (navel)

Epigastric region: Immediately superior to the umbilical region; overlies most of the stomach

Pubic (hypogastric) region: Immediately inferior to the umbilical region; encompasses the pubic area

Inguinal, or iliac, regions: Lateral to the hypogastric region and overlying the superior parts of the hip bones

Lateral (lumbar) regions: Between the ribs and the flaring portions of the hip bones; lateral to the umbilical region

Hypochondriac regions: Flanking the epigastric region laterally and overlying the lower ribs

Activity 5

Locating Abdominopelvic Surface Regions

Locate the regions of the abdominopelvic surface on a human torso model.

Other Body Cavities

Besides the large, closed body cavities, there are several types of smaller body cavities (**Figure 1.9**). Many of these are in the head, and most open to the body exterior.

Oral cavity: The oral cavity, commonly called the *mouth*, contains the tongue and teeth. It is continuous with the rest of the digestive tube, which opens to the exterior at the anus.

Nasal cavity: Located within and posterior to the nose, the nasal cavity is part of the passages of the respiratory system.

Orbital cavities: The orbital cavities (orbits) in the skull house the eyes and present them in an anterior position.

Middle ear cavities: Each middle ear cavity lies just medial to an eardrum and is carved into the bony skull. These cavities contain tiny bones that transmit sound vibrations to the hearing receptors in the inner ears.

Synovial cavities: Synovial cavities are joint cavities—they are enclosed within fibrous capsules that surround the freely movable joints of the body, such as those between the vertebrae and the knee and hip joints. Like the serous membranes of the ventral body cavity, membranes lining the synovial cavities secrete a lubricating fluid that reduces friction as the enclosed structures move across one another.

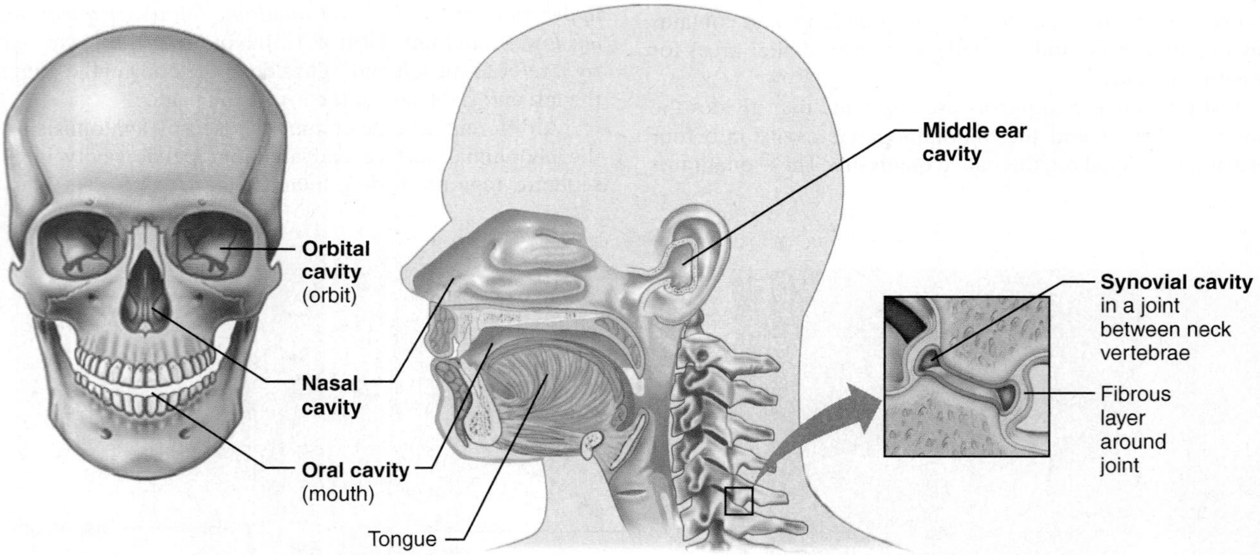

Figure 1.9 Other body cavities. The oral, nasal, orbital, and middle ear cavities are located in the head and open to the body exterior. Synovial cavities are found in joints between bones, such as the vertebrae of the spine, and at the knee, shoulder, and hip.

EXERCISE 1

REVIEW SHEET
The Language of Anatomy

Instructors may assign a portion of the Review Sheet questions using Mastering A&P™

Name ______________________ Lab Time/Date ______________________

Regional Terms

1. Describe completely the standard human anatomical position. ______________________

____

2. Use the regional terms to correctly label the body regions indicated on the figures below.

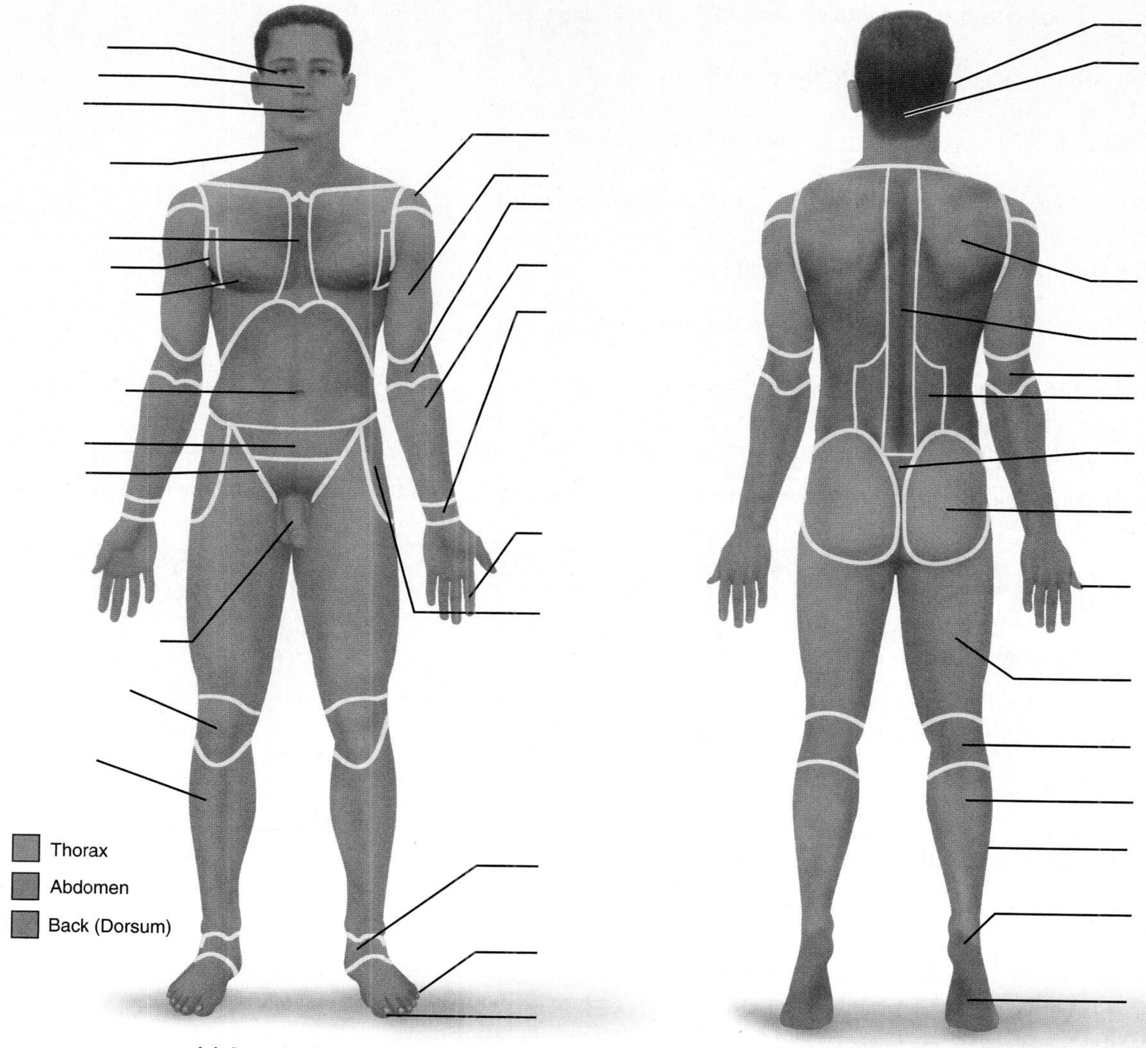

(a) Anterior/Ventral **(b) Posterior/Dorsal**

Directional Terms, Planes, and Sections

3. Define *plane*. ______________________________

4. Several incomplete statements appear below. Correctly complete each statement by choosing the appropriate anatomical term from the choices. Use each term only once.

anterior	inferior	posterior	superior
distal	lateral	proximal	transverse
frontal	medial	sagittal	

1. The thoracic cavity is ______________ to the abdominopelvic cavity.
2. The trachea (windpipe) is ______________ to the vertebral column.
3. The wrist is ______________ to the hand.
4. If an incision cuts the heart into left and right parts, a ______________ plane of section was used.
5. The nose is ______________ to the cheekbones.
6. The thumb is ______________ to the ring finger.
7. The vertebral cavity is ______________ to the cranial cavity.
8. The knee is ______________ to the thigh.
9. The plane that separates the head from the neck is the ______________ plane.
10. The popliteal region is ______________ to the patellar region.
11. The plane that separates the anterior body surface from the posterior body surface is the ______________ plane.

5. Correctly identify each of the body planes by writing the appropriate term on the answer line below the drawing.

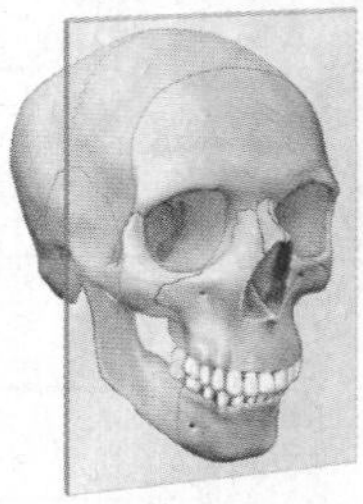

(a) ______________

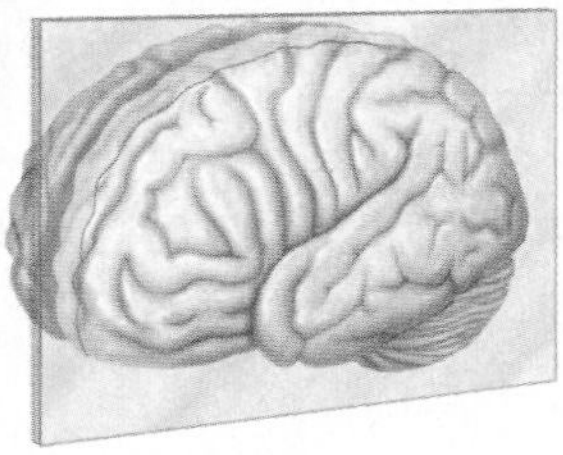

(b) ______________

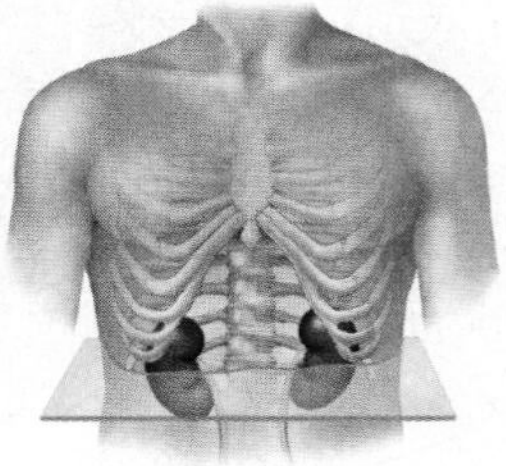

(c) ______________

Body Cavities

6. Name the muscle that subdivides the ventral body cavity. ______________________________

7. Which body cavity provides the least protection to its internal structures? ______________________________

8. For the body cavities listed, name one organ located in each cavity.

1. cranial cavity ______________________________.
2. vertebral cavity ______________________________.

3. thoracic cavity ______________________________.

4. abdominal cavity ______________________________.

5. pelvic cavity ______________________________.

6. mediastinum ______________________________.

9. Name the abdominopelvic region where each of the listed organs is located.

1. spleen ______________________________

2. urinary bladder ______________________________

3. stomach (largest portion) ______________________________

4. cecum ______________________________

10. Explain how serous membranes protect organs from infection. ______________________________

____

11. Which serous membrane(s) is/are found in the thoracic cavity? ______________________________

____

12. Which serous membrane(s) is/are found in the abdominopelvic cavity? ______________________________

____

13. Using the key choices, identify the small body cavities described below.

Key: a. middle ear cavity
b. nasal cavity
e. oral cavity
d. orbital cavity
e. synovial cavity

__________ 1. holds the eyes in an anterior-facing position

__________ 2. houses three tiny bones involved in hearing

__________ 3. contained within the nose

__________ 4. contains the tongue

__________ 5. surrounds a joint

14. ✚ Name the body region that blood is usually drawn from. ______________________________

15. ✚ A patient has been diagnosed with appendicitis. Use anatomical terminology to describe the location of the person's pain.

Assume that the pain is referred to the surface of the body above the organ. ______________________________

16. ✚ Which body cavity would be opened to perform a hysterectomy? ______________________________

____

17. ✚ Which smaller body cavity would be opened to perform a total knee joint replacement? ______________________________

____

18. ✚ An abdominal hernia results when weakened muscles allow the protrusion of abdominal structures. In the case of an umbilical hernia, parts of a serous membrane and the small intestine form the bulge. Which serous membrane is involved?

____

EXERCISE 2 Organ Systems Overview

Learning Outcomes

- Name the human organ systems, and indicate the major functions of each.
- List several major organs of each system, and identify them in a dissected rat, human cadaver or cadaver image, or a dissectible human torso model.
- Name the correct organ system for each organ when presented with a list of organs.

Go to Mastering A&P™ > Study Area to improve your performance in A&P Lab.

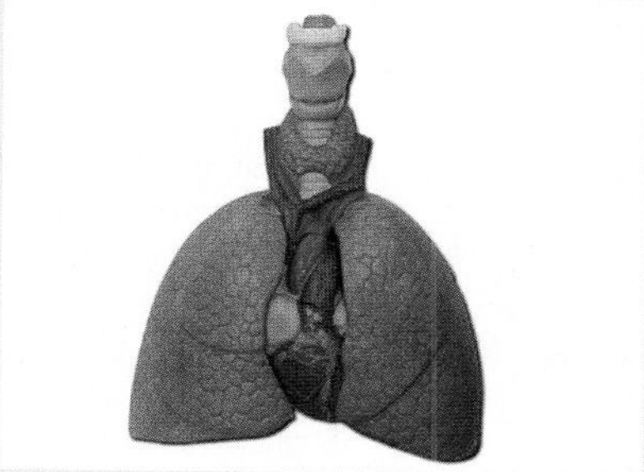

> Lab Tools > Practice Anatomy Lab > Anatomical Models

Instructors may assign new Building Vocabulary coaching activities, Pre-Lab Quiz questions, Art Labeling activities, Practice Anatomy Lab Practical questions (PAL), and more using the Mastering A&P™ Item Library.

Pre-Lab Quiz

Instructors may assign these and other Pre-Lab Quiz questions using Mastering A&P™

1. Name the structural and functional unit of all living things. ________
2. The small intestine is an example of a(n) ________, because it is composed of two or more tissue types that perform a particular function for the body.
   a. epithelial tissue
   b. muscular tissue
   c. organ
   d. organ system
3. The ____________ system is responsible for maintaining homeostasis of the body via rapid transmission of electrical signals.
4. The kidneys are part of the ____________ system.
5. The thin muscle that separates the thoracic and abdominal cavities is the ____________.

Materials

- Freshly killed or preserved rat (predissected by instructor as a demonstration or for student dissection [one rat for every two to four students]) or predissected human cadaver
- Dissection trays
- Twine or large dissecting pins
- Scissors
- Probes
- Forceps
- Disposable gloves
- Human torso model (dissectible)

The basic unit of life is the **cell**. Cells fall into four different categories according to their structures and functions. These categories correspond to the four primary tissue types: epithelial, muscular, nervous, and connective. A **tissue** is a group of cells that are similar in structure and function. An **organ** is a structure composed of two or more tissue types that performs a specific function for the body.

An **organ system** is a group of organs that act together to perform a particular body function. For example, the organs of the digestive system work together to break down foods and absorb the end products into the bloodstream in order to provide nutrients and fuel for all the body's cells. In all, there are 11 organ systems, described in **Table 2.1** on p. 16.

Read through this summary of the body's organ systems (Table 2.1) before beginning your rat dissection or examination of the predissected human cadaver. If a human cadaver is not available, Figures 2.3 to 2.6 will serve as a partial replacement.

Table 2.1 Overview of Organ Systems of the Body

Organ system	Major component organs	Function
Integumentary	Skin, hair, and nails; cutaneous sense organs and glands	• Protects deeper organs from mechanical, chemical, and bacterial injury, and from drying out • Excretes salts and urea • Aids in regulation of body temperature • Produces vitamin D
Skeletal	Bones, cartilages, tendons, ligaments, and joints	• Body support and protection of internal organs • Provides levers for muscular action • Cavities provide a site for blood cell formation • Bones store minerals
Muscular	Muscles attached to the skeleton	• Primary function is to contract or shorten; in doing so, skeletal muscles allow locomotion (running, walking, etc.), grasping and manipulation of the environment, and facial expression • Generates heat
Nervous	Brain, spinal cord, nerves, and sensory receptors	• Allows body to detect changes in its internal and external environment and to respond to such information by activating appropriate muscles or glands • Helps maintain homeostasis of the body via rapid transmission of electrical signals
Endocrine	Pituitary, thymus, thyroid, parathyroid, adrenal, and pineal glands; ovaries, testes, and pancreas	• Helps maintain body homeostasis, promotes growth and development; produces chemical messengers called hormones that travel in the blood to exert their effect(s) on various target organs of the body
Cardiovascular	Heart and blood vessels	• Primarily a transport system that carries blood containing oxygen, carbon dioxide, nutrients, wastes, ions, hormones, and other substances to and from the tissue cells where exchanges are made; blood is propelled through the blood vessels by the pumping action of the heart • Antibodies and other protein molecules in the blood protect the body
Lymphatic	Lymphatic vessels, lymph nodes, spleen, and thymus	• Picks up fluid leaked from the blood vessels and returns it to the blood • Cleanses blood of pathogens and other debris • Houses lymphocytes that act via the immune response to protect the body from foreign substances
Respiratory	Nasal cavity, pharynx, larynx, trachea, bronchi, and lungs	• Keeps the blood continuously supplied with oxygen while removing carbon dioxide • Contributes to the acid-base balance of the blood
Digestive	Oral cavity, pharynx, esophagus, stomach, small and large intestines, and accessory structures including teeth, salivary glands, liver, and pancreas	• Breaks down ingested foods to smaller particles, which can be absorbed into the blood for delivery to the body cells • Undigested residue removed from the body as feces
Urinary	Kidneys, ureters, bladder, and urethra	• Rids the body of nitrogen-containing wastes including urea, uric acid, and ammonia, which result from the breakdown of proteins and nucleic acids • Maintains water, electrolyte, and acid-base balance of blood
Reproductive	Male: testes, prostate gland, scrotum, penis, and duct system, which carries sperm to the body exterior	• Provides gametes called sperm for perpetuation of the species
	Female: ovaries, uterine tubes, uterus, mammary glands, and vagina	• Provides gametes called eggs; the uterus houses the developing fetus until birth; mammary glands provide nutrition for the infant

DISSECTION AND IDENTIFICATION

The Organ Systems of the Rat

Many of the external and internal structures of the rat are quite similar in structure and function to those of the human. So, a study of the gross anatomy of the rat should help you understand our anatomy. The following instructions include directions for dissecting and observing a rat. In addition, the descriptions of the organs (**Activity 4, Examining the Ventral Body Cavity**, which begins on p. 18) also apply to superficial observations of a previously dissected human cadaver. The general instructions for observing external structures also apply to human cadaver observations. The photographs in Figures 2.3 to 2.6 will provide visual aids.

Note that four organ systems (integumentary, skeletal, muscular, and nervous) will not be studied at this time, because they require microscopic study or more detailed dissection.

Activity 1

Observing External Structures

1. If your instructor has provided a predissected rat, go to the demonstration area to make your observations. Alternatively, if you and/or members of your group will be dissecting the specimen, obtain a preserved or freshly killed rat, a dissecting tray, dissecting pins or twine, scissors, probe, forceps, and disposable gloves, and bring them to your laboratory bench.

If a predissected human cadaver is available, obtain a probe, forceps, and disposable gloves before going to the demonstration area.

2. Don the gloves before beginning your observations. This precaution is particularly important when handling freshly killed animals, which may harbor pathogens.

3. Observe the major divisions of the body—head, trunk, and extremities. If you are examining a rat, compare these divisions to those of humans.

Activity 2

Examining the Oral Cavity

Examine the structures of the oral cavity. Identify the teeth and tongue. Observe the extent of the hard palate (the portion underlain by bone) and the soft palate (immediately posterior to the hard palate, with no bony support). Notice that the posterior end of the oral cavity leads into the throat, or pharynx, a passageway used by both the digestive and respiratory systems.

Activity 3

Opening the Ventral Body Cavity

1. Pin the animal to the wax of the dissecting tray by placing its dorsal side down and securing its extremities to the wax with large dissecting pins as shown in **Figure 2.1a**.

Text continues on next page →

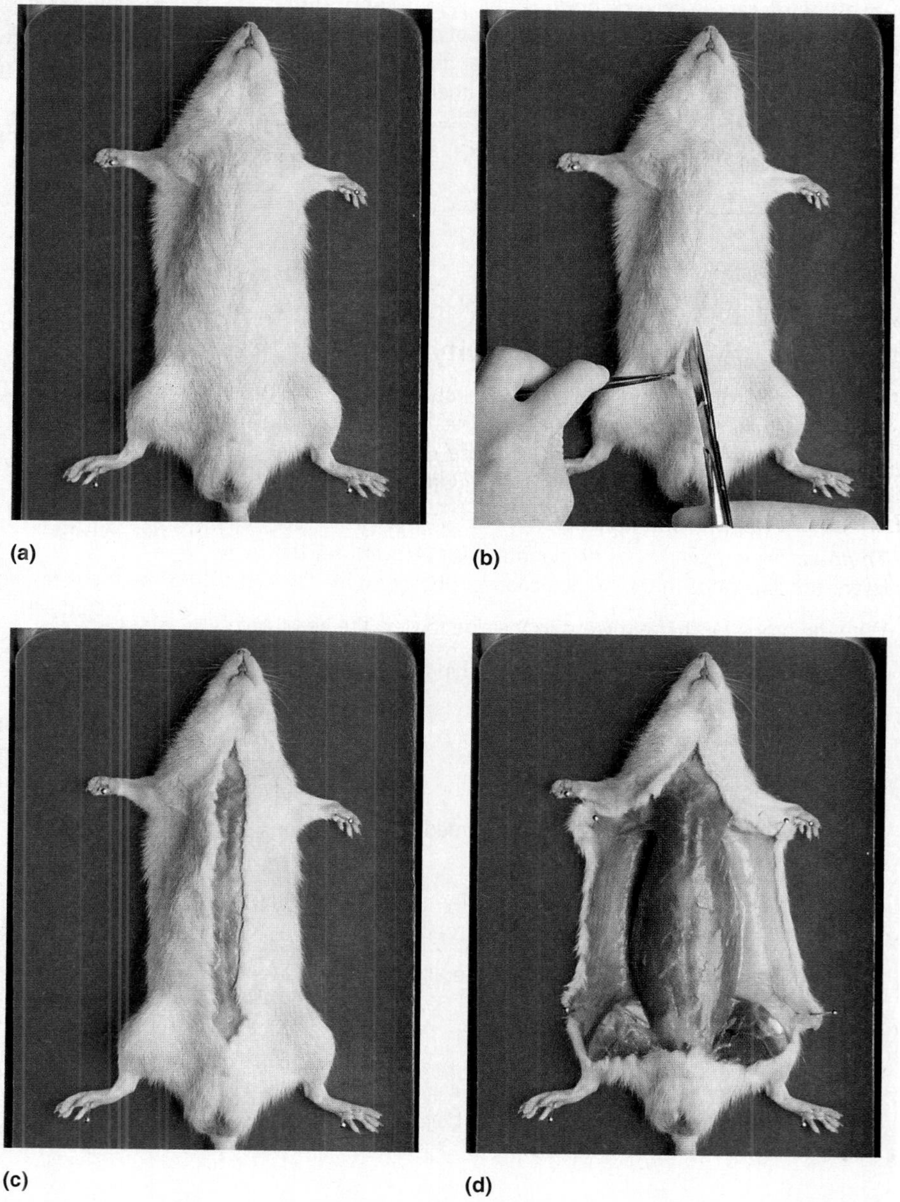

Figure 2.1 Rat dissection: Securing for dissection and the initial incision. **(a)** Securing the rat to the dissection tray with dissecting pins. **(b)** Using scissors to make the incision on the median line of the abdominal region. **(c)** Completed incision from the pelvic region to the lower jaw. **(d)** Reflection (folding back) of the skin to expose the underlying muscles.

2

2. Lift the abdominal skin with a forceps, and cut through it with the scissors (Figure 2.1b). Close the scissor blades, and insert them flat under the cut skin. Moving in a cephalad direction, open and close the blades to loosen the skin from the underlying connective tissue and muscle. Now, cut the skin along the body midline, from the pubic region to the lower jaw (Figure 2.1c). Finally, make a lateral cut about halfway down the ventral surface of each limb. Complete the job of freeing the skin with the scissor tips, and pin the flaps to the tray (Figure 2.1d). The underlying tissue that is now exposed is the skeletal musculature of the body wall and limbs. Notice that the muscles are packaged in sheets of pearly white connective tissue (fascia), which protect the muscles and bind them together.

3. Carefully cut through the muscles of the abdominal wall in the pubic region, avoiding the underlying organs. Now, hold and lift the muscle layer with a forceps and cut through the muscle layer from the pubic region to the bottom of the rib cage. Make two lateral cuts at the base of the rib cage (**Figure 2.2**). A thin membrane attached to the inferior boundary of the rib cage should be obvious; this is the **diaphragm**, which separates the thoracic and abdominal cavities. Cut the diaphragm where it attaches to the ventral ribs to loosen the rib cage. Cut through the rib cage on either side. You can now lift the ribs to view the contents of the thoracic cavity. Cut across the flap, at the level of the neck, and remove the rib cage.

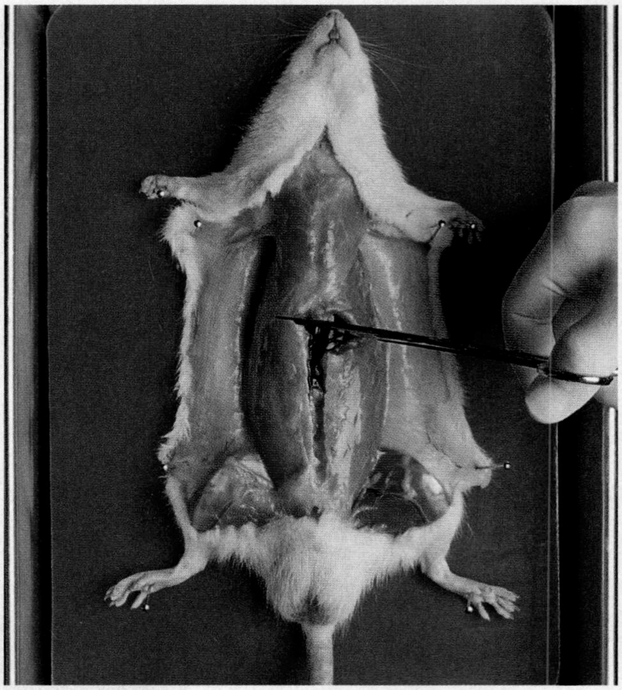

Figure 2.2 **Rat dissection.** Making lateral cuts at the base of the rib cage.

Activity 4

Examining the Ventral Body Cavity

1. Starting with the most superficial structures and working deeper, examine the structures of the thoracic cavity. Refer to **Figure 2.3** as you work. Choose the appropriate view depending on whether you are examining a rat (a) or a human cadaver (b).

Thymus: An irregular mass of glandular tissue overlying the heart (not illustrated in the human cadaver photograph).

With the probe, push the thymus to the side to view the heart.

Heart: Medial oval structure enclosed within the pericardium (serous membrane).

Lungs: Lateral to the heart on either side.

Now observe the throat region to identify the trachea.

Trachea: Tubelike "windpipe" running medially down the throat; part of the respiratory system.

Follow the trachea into the thoracic cavity; notice where it divides into two branches. These are the bronchi.

Bronchi: Two passageways that plunge laterally into the tissue of the two lungs.

To expose the esophagus, push the trachea to one side.

Esophagus: A food chute; the part of the digestive system that transports food from the pharynx (throat) to the stomach.

Diaphragm: A thin muscle attached to the inferior boundary of the rib cage.

Follow the esophagus through the diaphragm to its junction with the stomach.

Stomach: A curved organ important in food digestion and temporary food storage.

2. Examine the superficial structures of the abdominopelvic cavity. Lift the **greater omentum**, an extension of the peritoneum (serous membrane) that covers the abdominal viscera. Continuing from the stomach, trace the rest of the digestive tract (**Figure 2.4**, p. 20).

Small intestine: Connected to the stomach and ending just before the saclike cecum.

Large intestine: A large muscular tube connected to the small intestine and ending at the anus.

Text continues on page 20. →

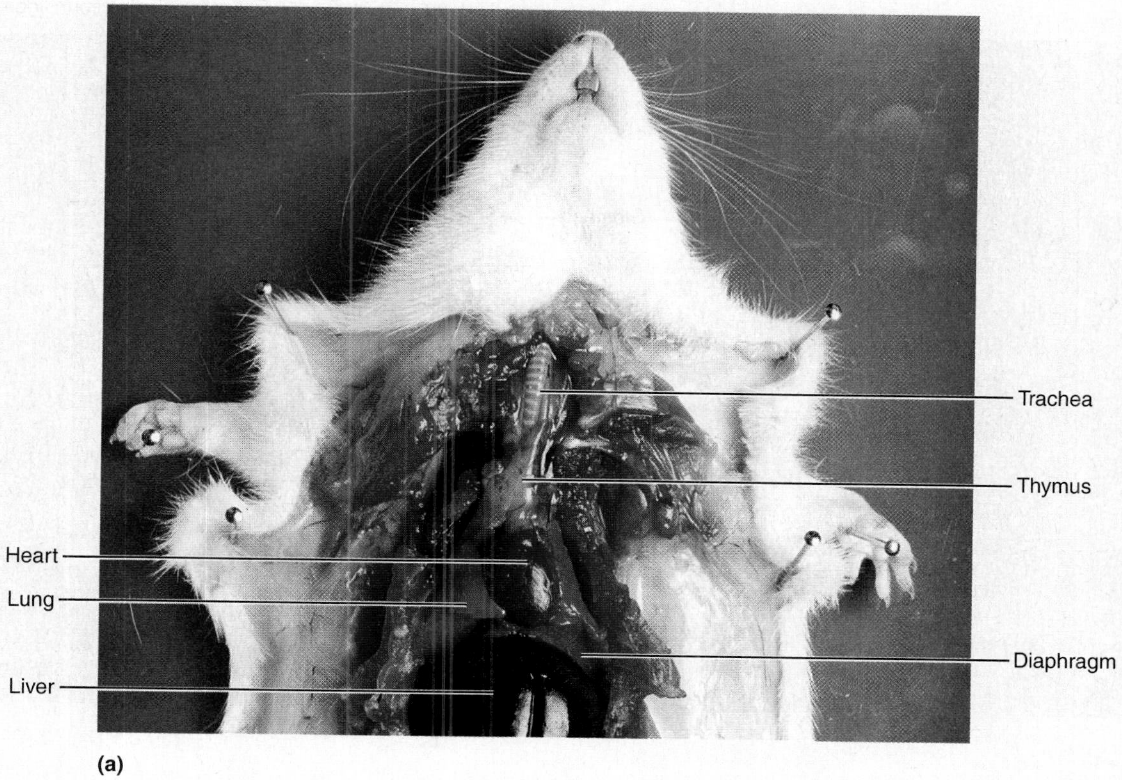

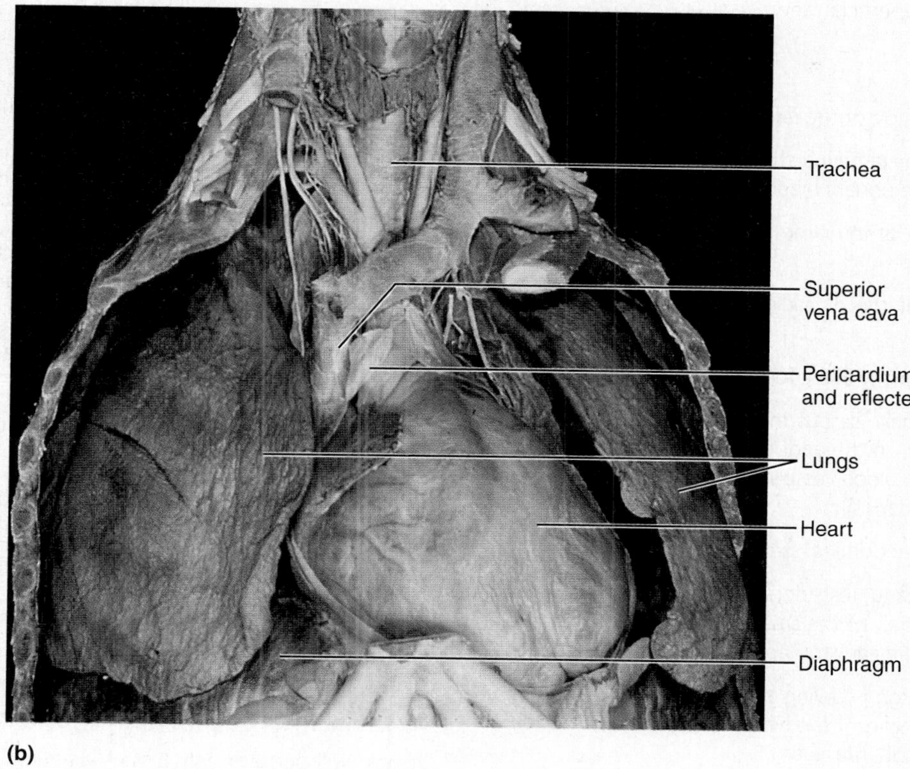

Figure 2.3 Superficial organs of the thoracic cavity. (a) Dissected rat. **(b)** Human cadaver.

Instructors may assign this figure as an Art Labeling Activity using Mastering **A&P™**

2

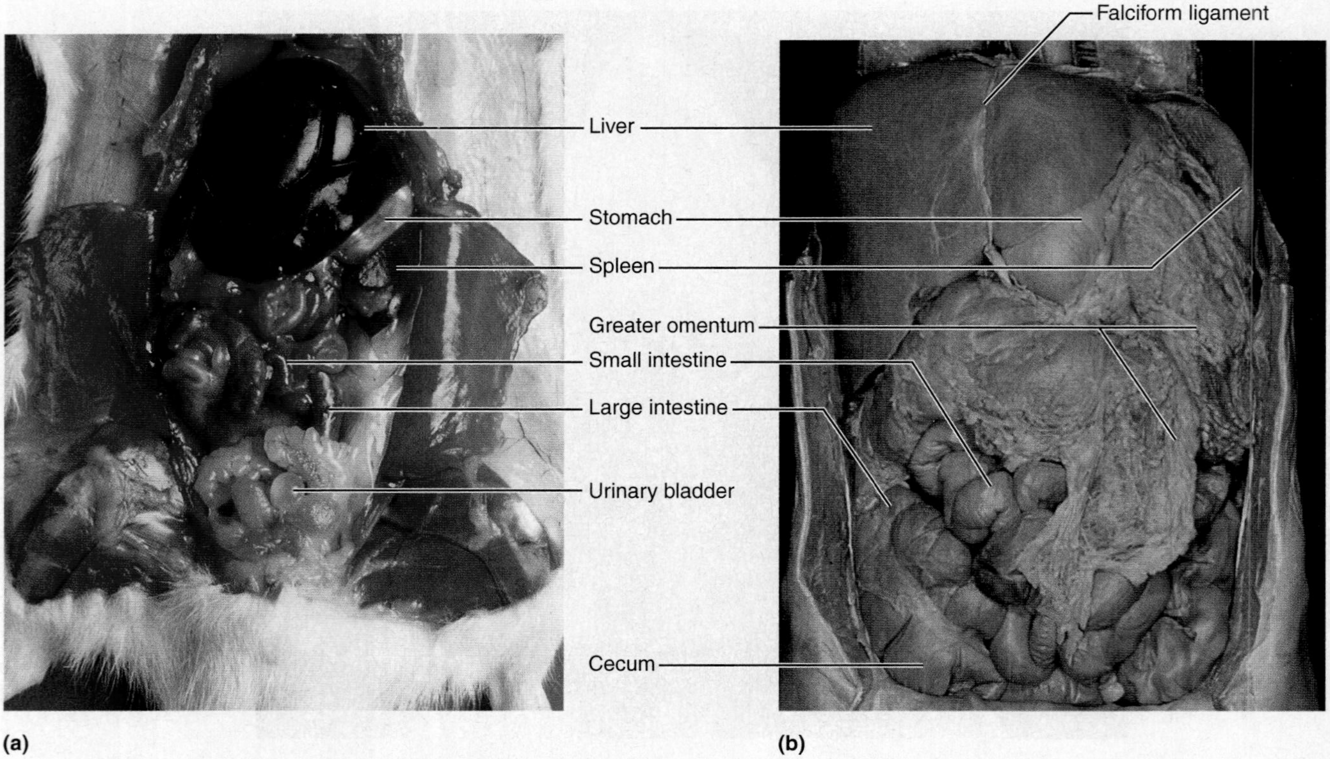

Figure 2.4 Abdominal organs. (a) Dissected rat, superficial view. **(b)** Human cadaver, superficial view.

Instructors may assign this figure as an Art Labeling Activity using Mastering A&P™

Cecum: The initial portion of the large intestine.

Follow the course of the large intestine to the rectum, which is partially covered by the urinary bladder (**Figure 2.5**).

Rectum: Terminal part of the large intestine; continuous with the anal canal.

Anus: The opening of the digestive tract (through the anal canal) to the exterior.

Now lift the small intestine with the forceps to view the mesentery.

Mesentery: An apronlike serous membrane; suspends many of the digestive organs in the abdominal cavity. Notice that it is heavily invested with blood vessels and, more likely than not, riddled with large fat deposits.

Locate the remaining abdominal structures.

Pancreas: A diffuse gland; rests dorsal to and in the mesentery between the first portion of the small intestine and the stomach. You will need to lift the stomach to view the pancreas.

Spleen: A dark red organ curving around the left lateral side of the stomach; an organ of the lymphatic system, it is often called the red blood cell "graveyard."

Liver: Large and brownish red; the most superior organ in the abdominal cavity, directly beneath the diaphragm.

3. To locate the deeper structures of the abdominopelvic cavity, move the stomach and the intestines to one side with the probe.

Examine the posterior wall of the abdominal cavity to locate the two kidneys (Figure 2.5).

Kidneys: Bean-shaped organs; retroperitoneal (behind the peritoneum).

Adrenal glands: Large endocrine glands that sit on top of each kidney; considered part of the endocrine system.

Carefully strip away part of the peritoneum with forceps and attempt to follow the course of one of the ureters to the bladder.

Ureter: Tube running from the indented region of a kidney to the urinary bladder.

Urinary bladder: The sac that serves as a reservoir for urine.

4. In the midline of the body cavity lying between the kidneys are the two principal abdominal blood vessels:

Inferior vena cava: The large vein that returns blood to the heart from the lower body regions.

Descending aorta: Deep to the inferior vena cava; the largest artery of the body; carries blood away from the heart.

5. You will perform only a brief examination of reproductive organs. If you are working with a rat, first determine if the animal is a male or female. Observe the ventral body surface beneath the tail. If a saclike scrotum and an opening for the anus are visible, the animal is a male. If three body openings—urethral, vaginal, and anal—are present, it is a female.

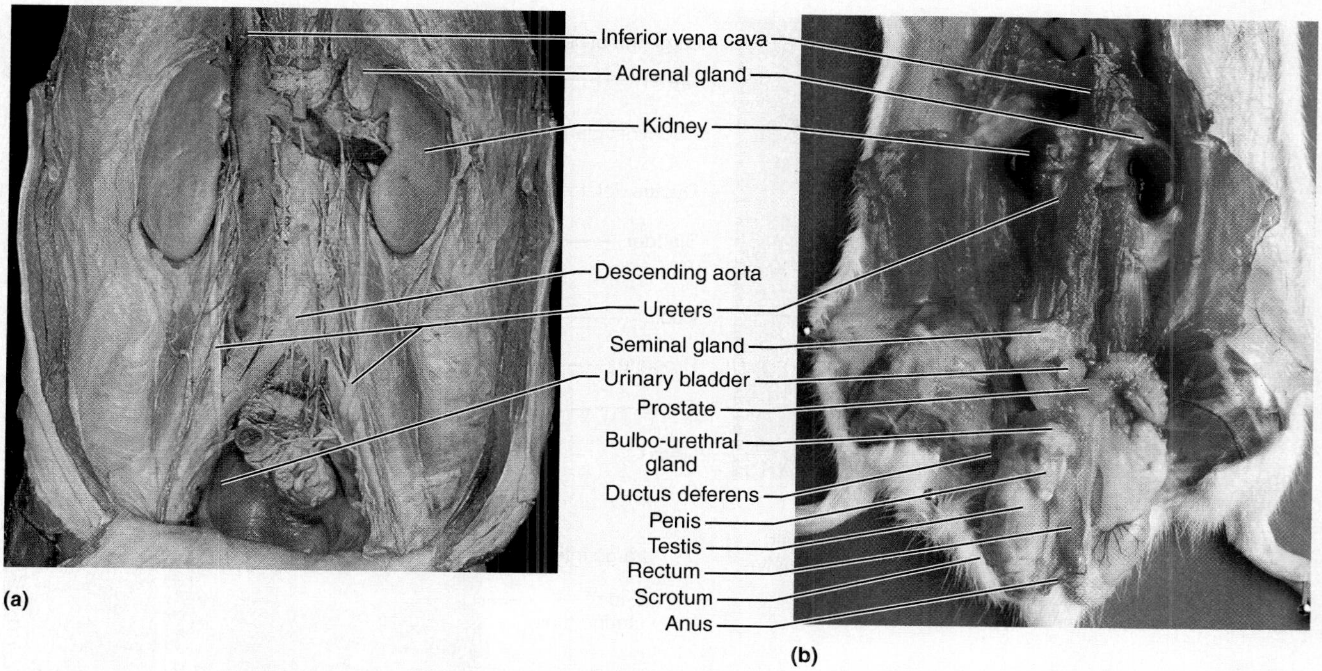

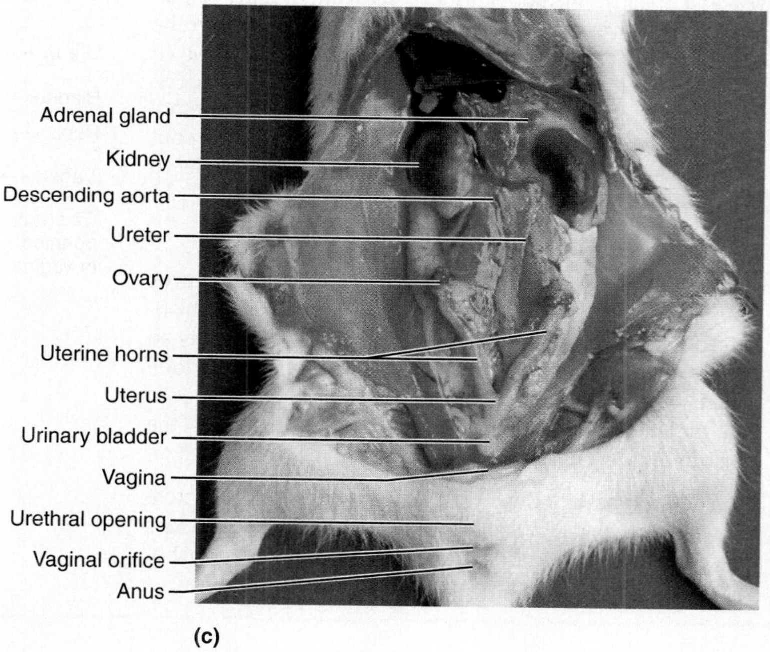

Figure 2.5 Deep structures of the abdominopelvic cavity. (a) Human cadaver. **(b)** Dissected male rat. (Some reproductive structures also shown.) **(c)** Dissected female rat. (Some reproductive structures also shown.)

Male Rat

Make a shallow incision into the **scrotum**. Loosen and lift out one oval **testis**. Exert a gentle pull on the testis to identify the slender **ductus deferens**, or **vas deferens**, which carries sperm from the testis superiorly into the abdominal cavity and joins with the urethra. The urethra runs through the penis and carries both urine and sperm out of the body. Identify the **penis**, extending from the bladder to the ventral body wall. Figure 2.5b indicates other glands of the male rat's reproductive system, but they need not be identified at this time.

Female Rat

Inspect the pelvic cavity to identify the Y-shaped **uterus** lying against the dorsal body wall and superior to the bladder (Figure 2.5c). Follow one of the uterine horns superiorly to identify an **ovary**, a small oval structure at the end of the uterine horn. (The rat uterus is quite different from the uterus of a human female, which is a single-chambered organ about the size and shape of a pear.) The inferior undivided part of the rat uterus is continuous with the **vagina**, which leads to the body exterior. Identify the **vaginal orifice** (external vaginal opening).

Text continues on next page →

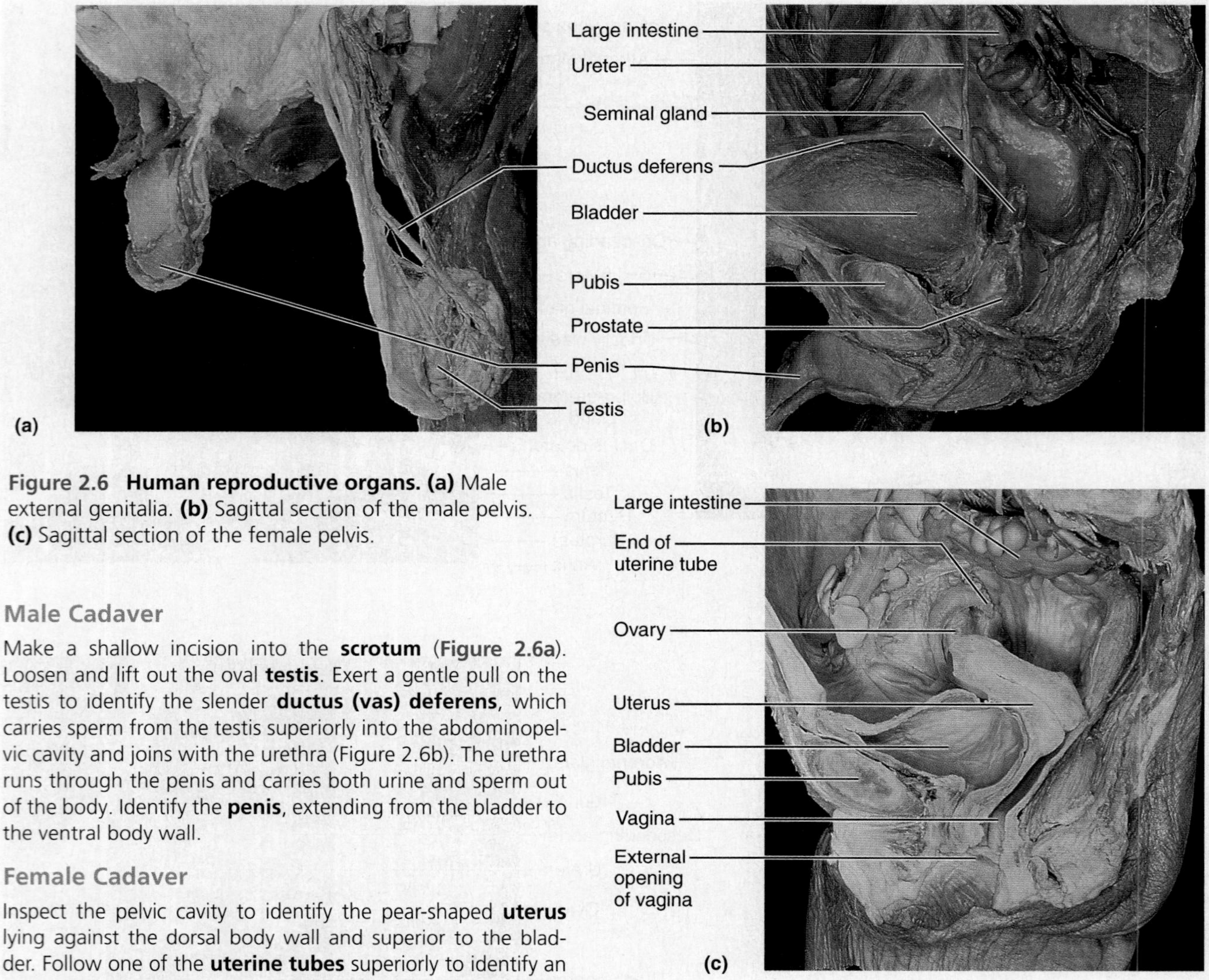

Figure 2.6 Human reproductive organs. (a) Male external genitalia. **(b)** Sagittal section of the male pelvis. **(c)** Sagittal section of the female pelvis.

Male Cadaver

Make a shallow incision into the **scrotum** (**Figure 2.6a**). Loosen and lift out the oval **testis**. Exert a gentle pull on the testis to identify the slender **ductus (vas) deferens**, which carries sperm from the testis superiorly into the abdominopelvic cavity and joins with the urethra (Figure 2.6b). The urethra runs through the penis and carries both urine and sperm out of the body. Identify the **penis**, extending from the bladder to the ventral body wall.

Female Cadaver

Inspect the pelvic cavity to identify the pear-shaped **uterus** lying against the dorsal body wall and superior to the bladder. Follow one of the **uterine tubes** superiorly to identify an **ovary**, a small oval structure at the end of the uterine tube (Figure 2.6c). The inferior part of the uterus is continuous with the **vagina**, which leads to the body exterior. Identify the **vaginal orifice** (external vaginal opening).

6. When you have finished your observations, rewrap or store the dissection animal or cadaver according to your instructor's directions. Wash the dissecting tools and equipment with laboratory detergent. Dispose of the gloves as instructed.

Activity 5

Examining the Human Torso Model

Examine a human torso model to identify the organs listed. Check off the boxes as you locate the organs. Some model organs will have to be removed to see the deeper organs.

- ☐ Adrenal gland
- ☐ Aortic arch
- ☐ Brain
- ☐ Diaphragm
- ☐ Esophagus
- ☐ Heart
- ☐ Inferior vena cava
- ☐ Kidneys
- ☐ Large intestine
- ☐ Liver
- ☐ Lungs
- ☐ Mesentery
- ☐ Pancreas
- ☐ Small intestine
- ☐ Spleen
- ☐ Stomach
- ☐ Thyroid gland
- ☐ Trachea
- ☐ Ureters
- ☐ Urinary bladder

Instructors may assign a portion of the Review Sheet questions using Mastering A&P™

2 REVIEW SHEET

EXERCISE

Organ Systems Overview

Name ______________________________ Lab Time/Date ______________________

1. Label each of the organs at the end of the supplied leader lines.

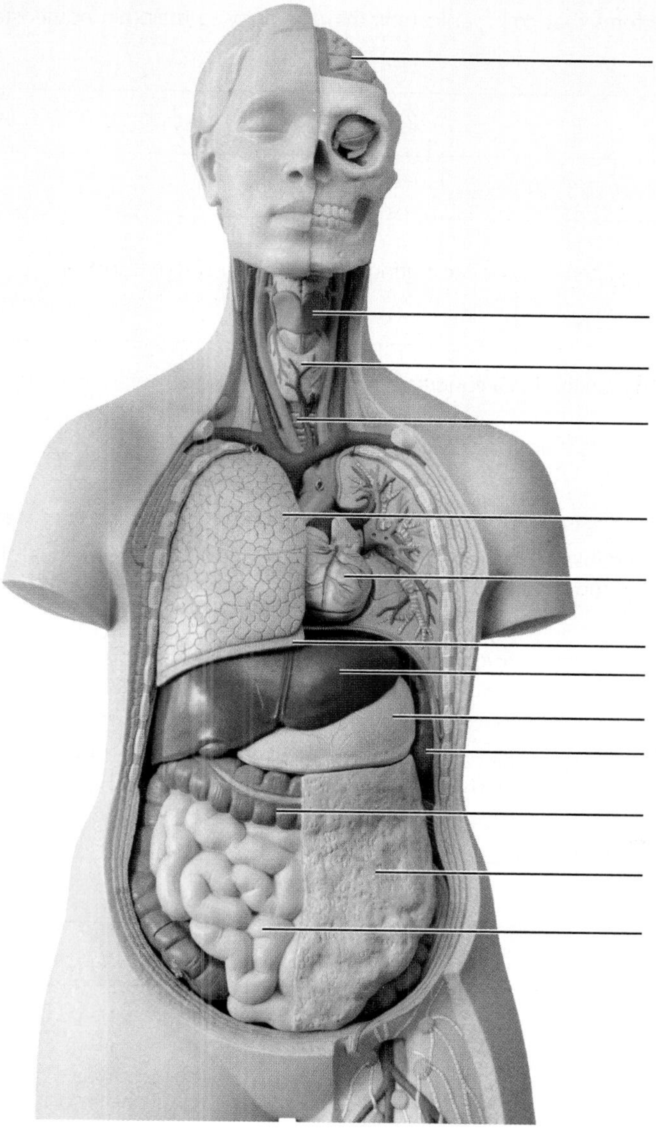

2. Name the *organ system* to which each of the following sets of organs or body structures belongs.

____________	1. thymus, spleen, lymphatic vessels	____________	5. epidermis, dermis, cutaneous sense organs
____________	2. bones, cartilages, tendons	____________	6. testis, prostate
____________	3. pancreas, pituitary gland	____________	7. liver, large intestine, rectum
____________	4. trachea, bronchi, lungs	____________	8. kidneys, ureter, urethra

3. Name the cells that are produced by the testes and ovaries. ______________________________

____

4. List the four primary tissue types. ______________________________

____

5. Explain why an artery is an organ. ______________________________

____

6. Name the two main organ systems that communicate within the body to maintain homeostasis. Briefly explain their different control mechanisms. ______________________________

____

____

7. Explain the role that the skeletal system plays in facilitating cardiovascular system function. ______________________________

____

8. Untreated diabetes mellitus can lead to a condition in which the blood is more acidic than normal. Name two organ systems that play the largest role in compensating for acid-base imbalances. ______________________________

____

9. The mother of a child scheduled to receive a thymectomy (removal of the thymus gland) asks you whether there will be any side effects from the removal of the gland. Which two organ systems would you mention in your explanation?

____

____

10. Individuals with asplenia are missing their spleen or have a spleen that doesn't function well. It is recommended that these patients talk to their doctor about vaccines that are indicated for their health condition. Explain how this recommendation correlates to their chronic health condition. ______________________________

____

____

EXERCISE 3

The Microscope

Learning Outcomes

- Identify the parts of the microscope, and list the function of each.
- Describe and demonstrate the proper techniques for care of the microscope.
- Demonstrate proper focusing technique.
- Define *total magnification*, *resolution*, *parfocal*, *field*, *depth of field*, and *working distance*.
- Measure the field diameter for one objective lens, calculate it for all the other objective lenses, and estimate the size of objects in each field.
- Discuss the general relationships between magnification, working distance, and field diameter.

Go to Mastering A&P™ > Study Area to improve your performance in A&P Lab.

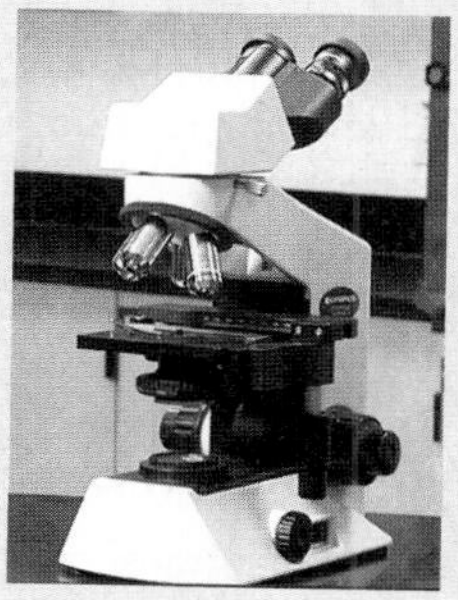

> Lab Tools > Pre-Lab Videos > Compound Microscope

Instructors may assign new Building Vocabulary coaching activities, Pre-Lab Quiz questions, Art Labeling activities, Pre-Lab Video Coaching Activities for The Compound Microscope, and more using the Mastering A&P™ Item Library.

Pre-Lab Quiz

Instructors may assign these and other Pre-Lab Quiz questions using Mastering A&P™

1. The microscope slide rests on the ____________ while being viewed.
   a. base **c.** iris
   b. condenser **d.** stage
2. Your lab microscope is *parfocal*. What does this mean?
   a. The specimen is clearly in focus at this depth.
   b. The slide should be almost in focus when changing to higher magnifications.
   c. You can easily discriminate two close objects as separate.
3. If the ocular lens magnifies a specimen 10×, and the objective lens used magnifies the specimen 35×, what is the total magnification being used to observe the specimen? ____________
4. How do you clean the lenses of your microscope?
   a. with a paper towel
   b. with soap and water
   c. with special lens paper and cleaner
5. Circle True or False. You should always begin observation of specimens with the oil immersion lens.

With the invention of the microscope, biologists gained a valuable tool to observe and study structures, such as cells, that are too small to be seen by the unaided eye. This exercise will familiarize you with the workhorse of microscopes—the compound microscope—and provide you with the necessary instructions for its proper use.

***Note to the Instructor:** The slides and coverslips used for viewing cheek cells are to be soaked for 2 hours (or longer) in 10% bleach solution and then drained. The slides and disposable autoclave bag containing coverslips, lens paper, and used toothpicks are to be autoclaved for 15 min at 121°C and 15 pounds pressure to ensure sterility. After autoclaving, the disposable autoclave bag may be discarded in any disposal facility, and the slides and glassware washed with laboratory detergent and prepared for use. These instructions apply as well to any bloodstained glassware or disposable items used in other experimental procedures.*

Materials*

- Compound microscope
- Millimeter ruler
- Prepared slides of the letter e or newsprint
- Immersion oil
- Lens paper
- Prepared slide of grid ruled in millimeters
- Prepared slide of three crossed colored threads
- Clean microscope slide and coverslip
- Toothpicks (flat-tipped)
- Physiological saline in a dropper bottle
- Iodine or dilute methylene blue stain in a dropper bottle
- Filter paper or paper towels
- Beaker containing fresh 10% household bleach solution for wet mount disposal
- Disposable autoclave bag
- Prepared slide of cheek epithelial cells

Care and Structure of the Compound Microscope

The **compound microscope** is a precision instrument and should always be handled with care. *At all times you must observe the following rules for its transport, cleaning, use, and storage:*

- When transporting the microscope, hold it in an upright position, with one hand on its arm and the other supporting its base. Do not swing the instrument during its transport or jar the instrument when setting it down.
- Use only special grit-free lens paper to clean the lenses. Use a circular motion to wipe the lenses, and clean all lenses before and after use.
- Always begin the focusing process with the scanning objective lens in position, changing to the higher-power lenses as necessary.
- Use the coarse adjustment knob only with the scanning objective lens.
- Always use a coverslip with wet mount preparations.
- Before putting the microscope in the storage cabinet, remove the slide from the stage, rotate the scanning objective lens into position, wrap the cord as directed, and replace the dust cover or return the microscope to the appropriate storage area.
- Never remove any parts from the microscope; inform your instructor of any mechanical problems that arise.

Activity 1

Identifying the Parts of a Microscope

1. Using the proper transport technique, obtain a microscope and bring it to the laboratory bench.

☐ Record the number of your microscope in the **Summary chart** (p. 28).

Compare your microscope with **Figure 3.1**, and identify the microscope parts described in **Table 3.1** on p. 29.

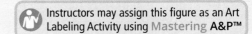

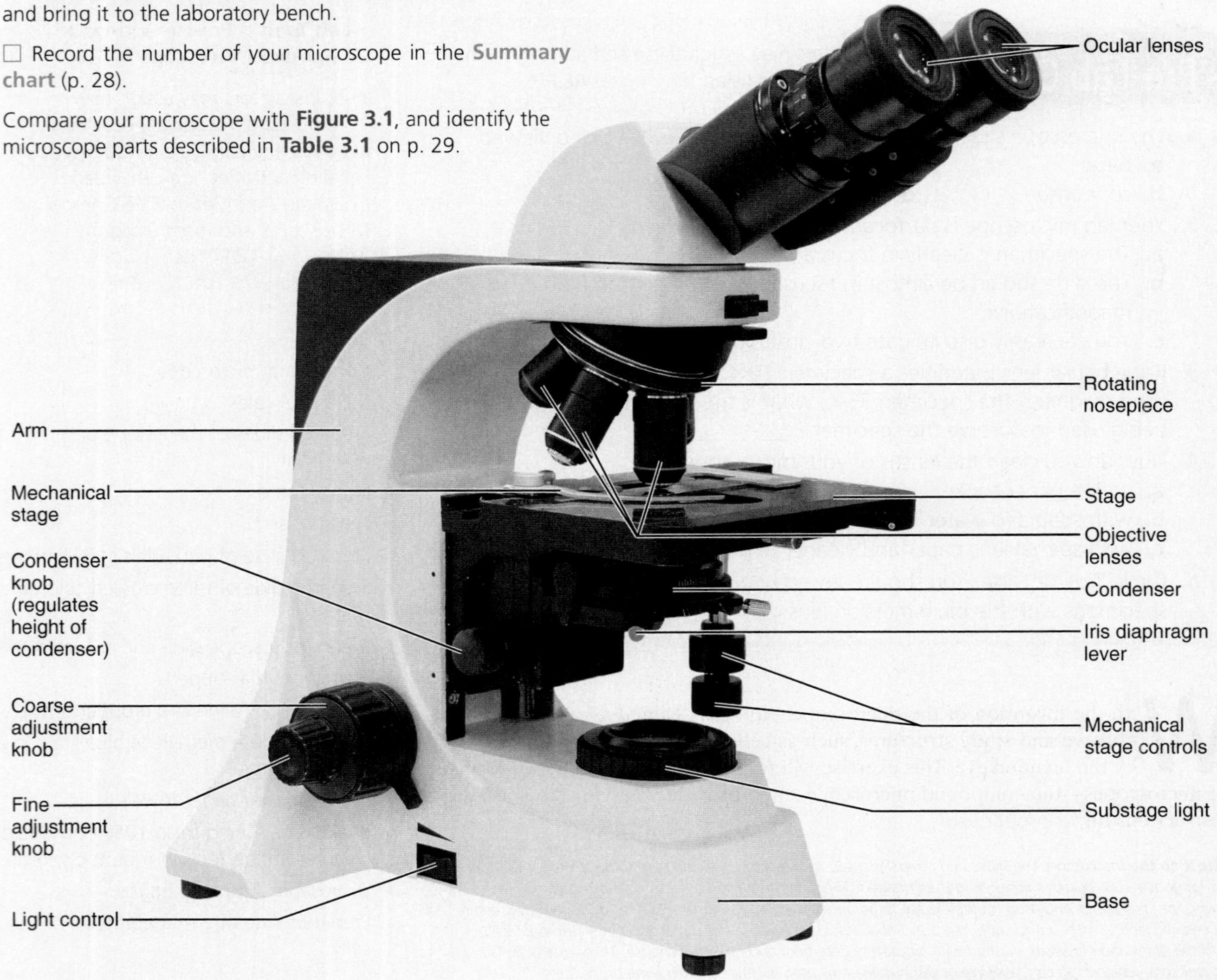

Figure 3.1 Compound microscope and its parts.

Instructors may assign this figure as an Art Labeling Activity using Mastering A&P™

2. Examine the objective lenses carefully; note their relative lengths and the numbers inscribed on their sides. On many microscopes, the scanning lens, with a magnification of 4×, is the shortest lens. The low-power objective lens typically has a magnification of 10×. The high-power objective lens is of intermediate length and has a magnification range from 40× to 50×. The oil immersion objective lens is usually the longest of the objective lenses and has a magnifying power of 100×. Some microscopes lack the oil immersion lens.

☐ Record the magnification of each objective lens of your microscope in the first row of the Summary chart (p. 28). Also, cross out any column relating to a lens that your microscope does not have.

Magnification and Resolution

The microscope is an instrument of magnification. With the compound microscope, magnification is achieved through the interplay of two lenses—the ocular lens and the objective lens. The objective lens magnifies the specimen to produce a **real image** that is projected to the ocular. This real image is magnified by the ocular lens to produce the **virtual image** that your eye sees (**Figure 3.2**).

The **total magnification** (TM) of any specimen being viewed is equal to the power of the ocular lens multiplied by the power of the objective lens used. For example, if the ocular lens magnifies 10× and the objective lens being used magnifies 45×, the total magnification is 450× (or 10 × 45).

- Determine the total magnification for each of the objectives on your microscope, and record the figures on the third row of the Summary chart.

The compound light microscope has certain limitations. Although the level of magnification is almost limitless, the **resolution** (or resolving power), that is, the ability to discriminate two close objects as separate, is not. The human eye can resolve objects about 100 μm apart, but the compound microscope has a resolution of 0.2 μm under ideal conditions. Objects closer than 0.2 μm are seen as a single fused image.

Resolution is determined by the amount and physical properties of the visible light that enters the microscope. In general, the more light delivered to the objective lens, the greater the resolution. The size of the objective lens aperture (opening) decreases with increasing magnification, allowing less light to enter the objective. Thus, you will probably find it necessary to increase the light intensity at the higher magnifications.

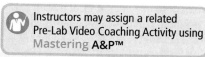

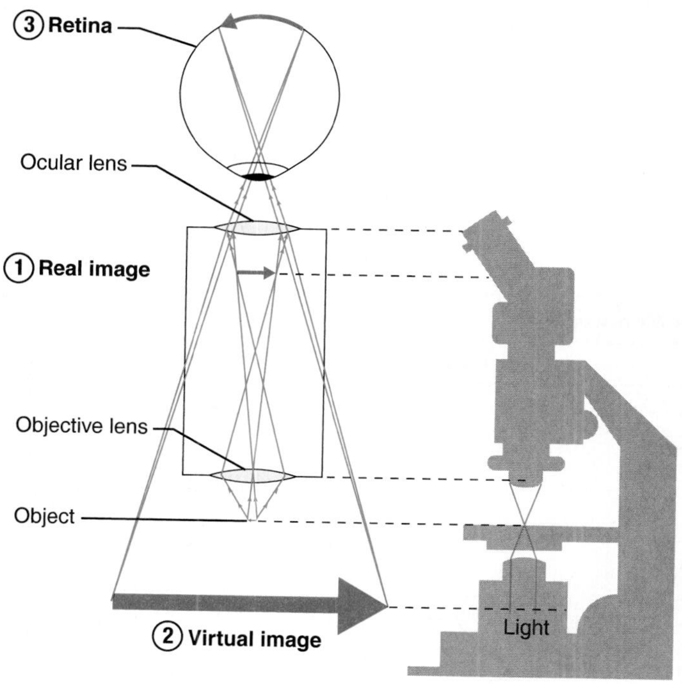

Figure 3.2 Image formation in light microscopy. **Step** ① The objective lens magnifies the object, forming the real image. **Step** ② The ocular lens magnifies the real image, forming the virtual image. **Step** ③ The virtual image passes through the lens of the eye and is focused on the retina.

Activity 2

Instructors may assign a related Pre-Lab Video Coaching Activity using Mastering **A&P™**

Viewing Objects Through the Microscope

1. Obtain a millimeter ruler, a prepared slide of the letter *e* or newsprint, a dropper bottle of immersion oil, and some lens paper. Adjust the condenser to its highest position, and switch on the light source of your microscope.

2. Secure the slide on the stage so that you can read the slide label and the letter *e* is centered over the light beam passing through the stage. On the mechanical stage of your microscope, open the jaws of its slide holder by using the control lever, typically located at the rear left corner of the mechanical stage. Insert the slide squarely within the confines of the slide holder.

3. With your scanning objective lens in position over the stage, use the coarse adjustment knob to bring the objective lens and stage as close together as possible.

4. Look through the ocular lens and adjust the light for comfort using the iris diaphragm lever. Now use the coarse adjustment knob to focus slowly away from the *e* until it is as clearly focused as possible. Complete the focusing with the fine adjustment knob.

5. Sketch the letter *e* in the circle on the Summary chart (p. 28) just as it appears in the **field**—the area you see through the microscope.

How far is the bottom of the objective lens from the surface of the slide? In other words, what is the **working distance**? (See Figure 3.3.) Use a millimeter ruler to measure the working distance.

Record the working distance in the Summary chart.

Text continues on next page. →

How has the apparent orientation of the e changed top to bottom, right to left, and so on?

___

___

6. Move the slide slowly away from you on the stage as you view it through the ocular lens. In what direction does the image move?

___

Move the slide to the left. In what direction does the image move?

___

7. Today, most good laboratory microscopes are **parfocal**; that is, the slide should be in focus (or nearly so) at the higher magnifications once you have properly focused at the lower magnification. *Without touching the focusing knobs*, increase the magnification by rotating the next higher magnification lens into position over the stage. Make sure it clicks into position. Using the fine adjustment only, sharpen the focus. If you are unable to focus with a new lens, your microscope is not parfocal. Do not try to force the lens into position. Consult your instructor. Note the decrease in working distance. As you can see, focusing with the coarse adjustment knob could drive the objective lens through the slide, breaking the slide and possibly damaging the lens. Sketch the letter e in the Summary chart. What new details become clear?

___

___

As best you can, measure the distance between the objective and the slide.

Record the working distance in the Summary chart.

Is the image larger or smaller? ___

Approximately how much of the letter e is visible now?

___

Is the field diameter larger or smaller? ___

Why is it necessary to center your object (or the portion of the slide you wish to view) before changing to a higher power?

___

Move the iris diaphragm lever while observing the field. What happens?

___

Is it better to increase *or* to decrease the light when changing to a higher magnification?

___ Why? ___

8. If you have just been using the low-power objective, repeat the steps given in direction 7 using the high-power objective lens. What new details become clear?

___

___

Record the working distance in the Summary chart.

Summary Chart for Microscope #___

	Scanning	Low power	High power	Oil immersion
Magnification of objective lens	___ ×	___ ×	___ ×	___ ×
Magnification of ocular lens	10 ×	10 ×	10 ×	10 ×
Total magnification	___ ×	___ ×	___ ×	___ ×
Working distance	___ mm	___ mm	___ mm	___ mm
Detail observed letter e	◯	◯	◯	◯
Field diameter	___ mm ___ µm	___ mm ___ µm	___ mm ___ µm	___ mm ___ µm

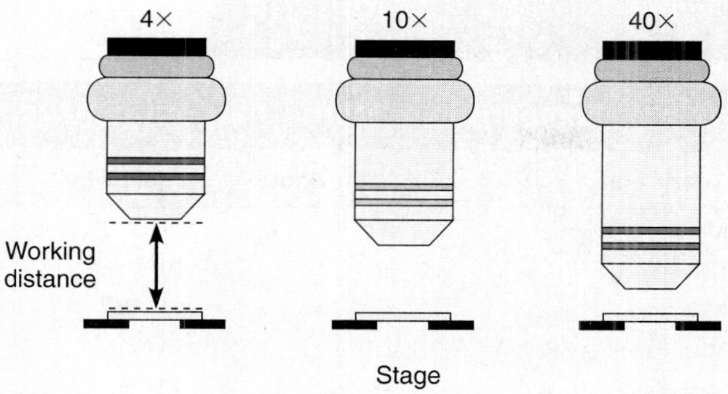

Figure 3.3 Relative working distances of the 4×, 10×, and 40× objectives.

9. Without touching the focusing knob, rotate the high-power lens out of position so that the area of the slide over the opening in the stage is unobstructed. Place a drop of immersion oil over the e on the slide and rotate the oil immersion lens into position. Set the condenser at its highest point (closest to the stage), and open the diaphragm fully. Adjust the fine focus and fine-tune the light for the best possible resolution.

Note: *If for some reason the specimen does not come into view after adjusting the fine focus, do not go back to the 40× lens to recenter. You do not want oil from the oil immersion lens to cloud the 40× lens. Turn the revolving nosepiece in the other direction to the low-power lens, and recenter and refocus the object. Then move the immersion lens back into position, again avoiding the 40× lens. Sketch the letter* e *in the Summary chart. What new details become clear?*

____

____

Is the field diameter again decreased in size? __________

As best you can, estimate the working distance, and record it in the Summary chart. Is the working distance less *or* greater than it was when the high-power lens was focused?

____

Compare your observations on the relative working distances of the objective lenses with the illustration in **Figure 3.3**. Explain why it is desirable to begin the focusing process at the lowest power.

____

____

10. Rotate the oil immersion lens slightly to the side, and remove the slide. Clean the oil immersion lens carefully with lens paper, and then clean the slide in the same manner with a fresh piece of lens paper.

Table 3.1 Parts of the Microscope

Microscope part	Description and function
Base	The bottom of the microscope. Provides a sturdy flat surface to support and steady the microscope.
Substage light	Located in the base. The light from the lamp passes directly upward through the microscope.
Light control	Located on the base or arm. This dial allows you to adjust the intensity of the light passing through the specimen.
Stage	The platform that the slide rests on while being viewed. The stage has a hole in it to allow light to pass through the stage and through the specimen.
Mechanical stage	Holds the slide in position for viewing and has two adjustable knobs that control the precise movement of the slide.
Condenser	Small nonmagnifying lens located beneath the stage that concentrates the light on the specimen. The condenser may have a knob that raises and lowers the condenser to vary the light delivery. Generally, the best position is close to the inferior surface of the stage.
Iris diaphragm lever	The iris diaphragm is a shutter within the condenser that can be controlled by a lever to adjust the amount of light passing through the condenser. The lever can be moved to close the diaphragm and improve contrast. If your field of view is too dark, you can open the diaphragm to let in more light.
Coarse adjustment knob	This knob allows you to make large adjustments to the height of the stage to initially focus your specimen.
Fine adjustment knob	This knob is used for precise focusing once the initial coarse focusing has been completed.
Head	Attaches to the nosepiece to support the objective lens system. It also provides for attachment of the eyepieces which house the ocular lenses.
Arm	Vertical portion of the microscope that connects the base and the head.
Nosepiece	Rotating mechanism connected to the head. Generally, it carries three or four objective lenses and permits positioning of these lenses over the hole in the stage.
Objective lenses	These lenses are attached to the nosepiece. Usually, a compound microscope has four objective lenses: scanning (4×), low-power (10×), high-power (40×), and oil immersion (100×) lenses. Typical magnifying powers for the objectives are listed in parentheses.
Ocular lens(es)	Binocular microscopes will have two lenses located in the eyepieces at the superior end of the head. Most ocular lenses have a magnification power of 10×. Some microscopes will have a pointer and/or reticle (micrometer), which can be positioned by rotating the ocular lens.

The Microscope Field

The microscope field decreases with increasing magnification. Measuring the diameter of each of the microscope fields will allow you to estimate the size of the objects you view in any field. For example, if you have calculated the field diameter to be 4 mm and the object being observed extends across half this diameter, you can estimate that the length of the object is approximately 2 mm.

Microscopic specimens are usually measured in micrometers and millimeters, both units of the metric system. You can get an idea of the relationship and meaning of these units from **Table 3.2**. A more detailed treatment appears in the appendix.

Table 3.2 Comparison of Metric Units of Length

Metric unit	Abbreviation	Equivalent
Meter	m	(about 39.37 in.)
Centimeter	cm	10^{-2} m
Millimeter	mm	10^{-3} m
Micrometer (or micron)	μm (μ)	10^{-6} m
Nanometer	nm (mμ)	10^{-9} m

(Refer to the Getting Started exercise on MasteringA&P for tips on metric conversions.)

Activity 3

Instructors may assign a related Pre-Lab Video Coaching Activity using Mastering A&P™

Estimating the Diameter of the Microscope Field

1. Obtain a grid slide, which is a slide prepared with graph paper ruled in millimeters. Each of the squares in the grid is 1 mm on each side. Use your scanning objective lens to bring the grid lines into focus.

2. Move the slide so that one grid line touches the edge of the field on one side, and then count the number of squares you can see across the diameter of the field. If you can see only part of a square, as in the accompanying diagram, estimate the part of a millimeter that the partial square represents.

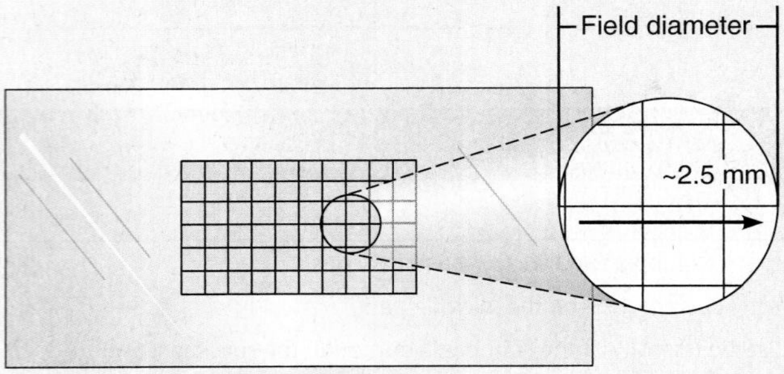

Record this figure in the appropriate space marked "field diameter" on the Summary chart (p. 28). (If you have been using the scanning lens, repeat the procedure with the low-power objective lens.)

Complete the chart by computing the approximate diameter of the high-power and oil immersion fields. The general formula for calculating the unknown field diameter is:

Diameter of field A × total magnification of field A = diameter of field B × total magnification of field B

where A represents the known or measured field and B represents the unknown field. This can be simplified to

$$\text{Diameter of field } B = \frac{\text{diameter of field } A \times \text{total magnification of field } A}{\text{total magnification of field } B}$$

For example, if the diameter of the low-power field (field A) is 2 mm and the total magnification is 50×, you would compute the diameter of the high-power field (field B) with a total magnification of 100× as follows:

Field diameter B = (2 mm × 50)/100

Field diameter B = 1 mm

3. Estimate the length (longest dimension) of the following drawings of microscopic objects. *Base your calculations on the field diameters you have determined for your microscope and the approximate percentage of the diameter that the object occupies.* The first one is done for you.

Fat cell seen in 400× (total magnification, TM) field:

Field diameter = 0.4 mm = 400 μm

Portion of the field diameter occupied by the object = 1/3

Approximate length = 133 μm

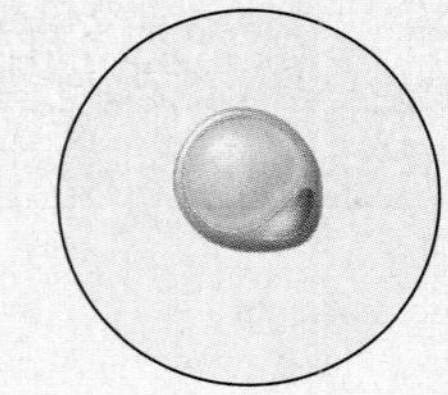

Smooth muscle cell seen in 400× (TM) field:

approximate length:

______________ mm

or ______________ μm

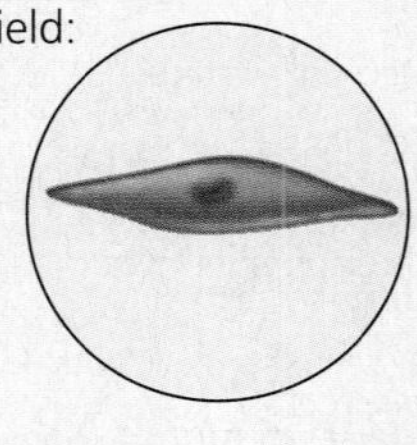

Cheek cell seen in oil immersion field:

approximate length:

______________ μm

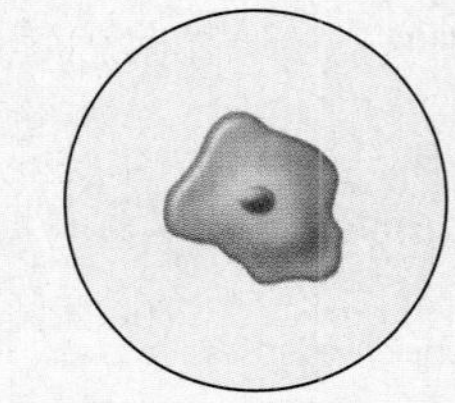

Perceiving Depth

Any microscopic specimen has depth as well as length and width; it is rare indeed to view a tissue slide with just one layer of cells. Normally you can see two or three cell thicknesses. Therefore, it is important to learn how to determine relative depth with your microscope. In microscope work, the **depth of field** (the thickness of the plane that is clearly in focus) is greater at lower magnifications. As magnification increases, depth of field decreases.

Activity 4

Perceiving Depth

1. Obtain a slide with colored crossed threads. Focusing at low magnification, locate the point where the three threads cross each other.

2. Use the iris diaphragm lever to greatly reduce the light, thus increasing the contrast. Focus down with the coarse adjustment until the threads are out of focus, then slowly focus upward again, noting which thread comes into clear focus first. Observe: As you rotate the adjustment knob forward (away from you), does the stage rise or fall? If the stage rises, then the first clearly focused thread is the top one; the last clearly focused thread is the bottom one.

If the stage descends, how is the order affected?____________

____

Record your observations, relative to which color of thread is uppermost, middle, or lowest:

Top thread ________________________________

Middle thread ______________________________

Bottom thread ______________________________

Viewing Cells Under the Microscope

There are various ways to prepare cells for viewing under a microscope. One method is to mix the cells in physiological saline (called a *wet mount*) and stain them.

If you are not instructed to prepare your own wet mount, obtain a prepared slide of epithelial cells to make the observations in step 10 of Activity 5.

Activity 5

Instructors may assign a related Pre-Lab Video Coaching Activity using Mastering A&P™

Preparing and Observing a Wet Mount

1. Obtain the following: a clean microscope slide and coverslip, two flat-tipped toothpicks, a dropper bottle of physiological saline, a dropper bottle of iodine or methylene blue stain, and filter paper (or paper towels). Handle only your own slides throughout the procedure.

2. Place a drop of physiological saline in the center of the slide. Using the flat end of the toothpick, *gently* scrape the inner lining of your cheek. Transfer your cheek scrapings to the slide by agitating the end of the toothpick in the drop of saline (**Figure 3.4a** on p. 32).

! *Immediately* discard the used toothpick in the disposable autoclave bag provided.

3. Add a tiny drop of the iodine or methylene blue stain to the preparation. (These epithelial cells are nearly transparent and thus difficult to see without the stain, which colors the nuclei of the cells.) Stir again, using a second toothpick.

! *Immediately* discard the used toothpicks in the disposable autoclave bag provided.

4. Hold the coverslip with your fingertips so that its bottom edge touches one side of the drop (Figure 3.4b), then *slowly* lower the coverslip onto the preparation (Figure 3.4c). *Do not just drop the coverslip*, or you will trap large air bubbles under it, which will obscure the cells. *Always use a coverslip with a wet mount* to protect the lens.

5. Examine your preparation carefully. The coverslip should be tight against the slide. If there is excess fluid around its edges, you will need to remove it. Obtain a piece of filter paper, fold it in half, and use the folded edge to absorb the excess fluid.

Text continues on next page. →

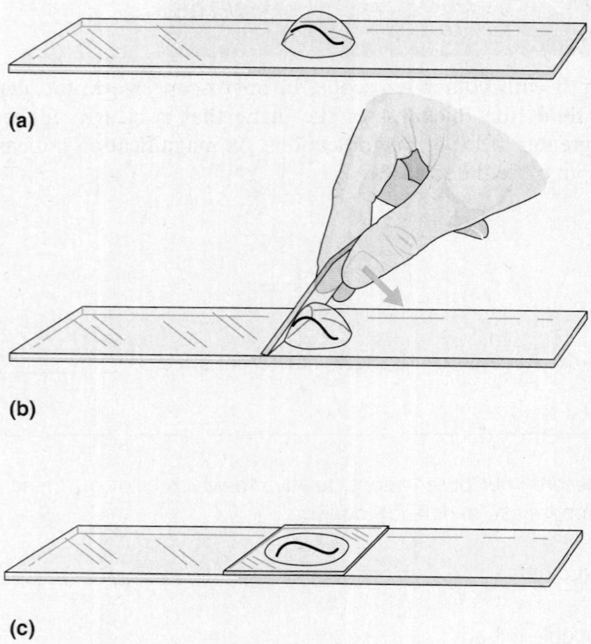

Figure 3.4 Procedure for preparation of a wet mount. **(a)** Place the object in a drop of water (or saline) on a clean slide; **(b)** hold a coverslip at a 45° angle with the fingertips; and **(c)** lower the coverslip slowly.

⚠ Before continuing, discard the filter paper or paper towel in the disposable autoclave bag.

6. Place the slide on the stage, and locate the cells at the lowest power. You will probably want to dim the light to provide more contrast for viewing the lightly stained cells.

7. Cheek epithelial cells are very thin, flat cells. In the cheek, they provide a smooth, tilelike lining (**Figure 3.5**). Move to high power to examine the cells more closely.

8. Make a sketch of the epithelial cells that you observe.

Use information on your Summary chart (p. 28) to estimate the diameter of cheek epithelial cells. Record the total magnification (TM) used.

__________ μm __________ × (TM)

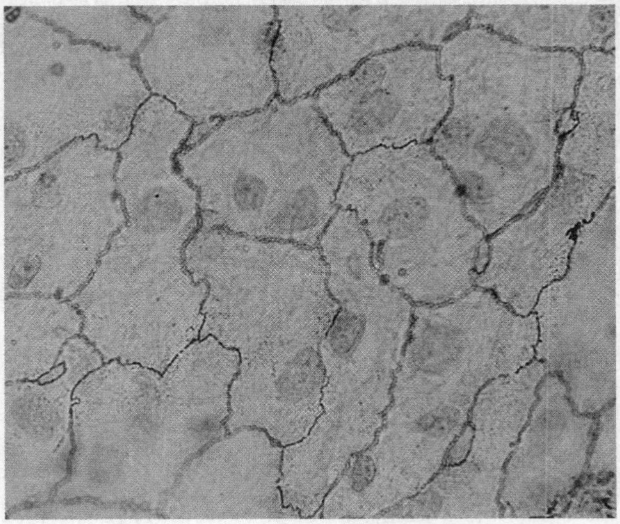

Figure 3.5 Epithelial cells of the cheek cavity (surface view, 600×).

Why do *your* cheek cells look different from those in Figure 3.5? (Hint: What did you have to *do* to your cheek to obtain them?)

____

____

⚠ 9. When you complete your observations of the wet mount, dispose of your wet mount preparation in the beaker of bleach solution, and put the coverslips in an autoclave bag.

10. Obtain a prepared slide of cheek epithelial cells, and view them under the microscope.

Estimate the diameter of one of these cheek epithelial cells using information from the Summary chart (p. 28).

__________ μm __________ × (TM)

Why are these cells more similar to those in Figure 3.5 and easier to measure than those of the wet mount?

____

____

____

11. Before leaving the laboratory, make sure all other materials are properly discarded or returned to the appropriate laboratory station. Clean the microscope lenses, and return the microscope to the storage cabinet.

Instructors may assign a portion of the Review Sheet questions using Mastering A&P™

REVIEW SHEET
EXERCISE 3 The Microscope

Name ______________________ Lab Time/Date ______________________

Care and Structure of the Compound Microscope

1. Label all indicated parts of the microscope.

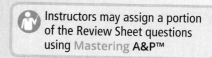

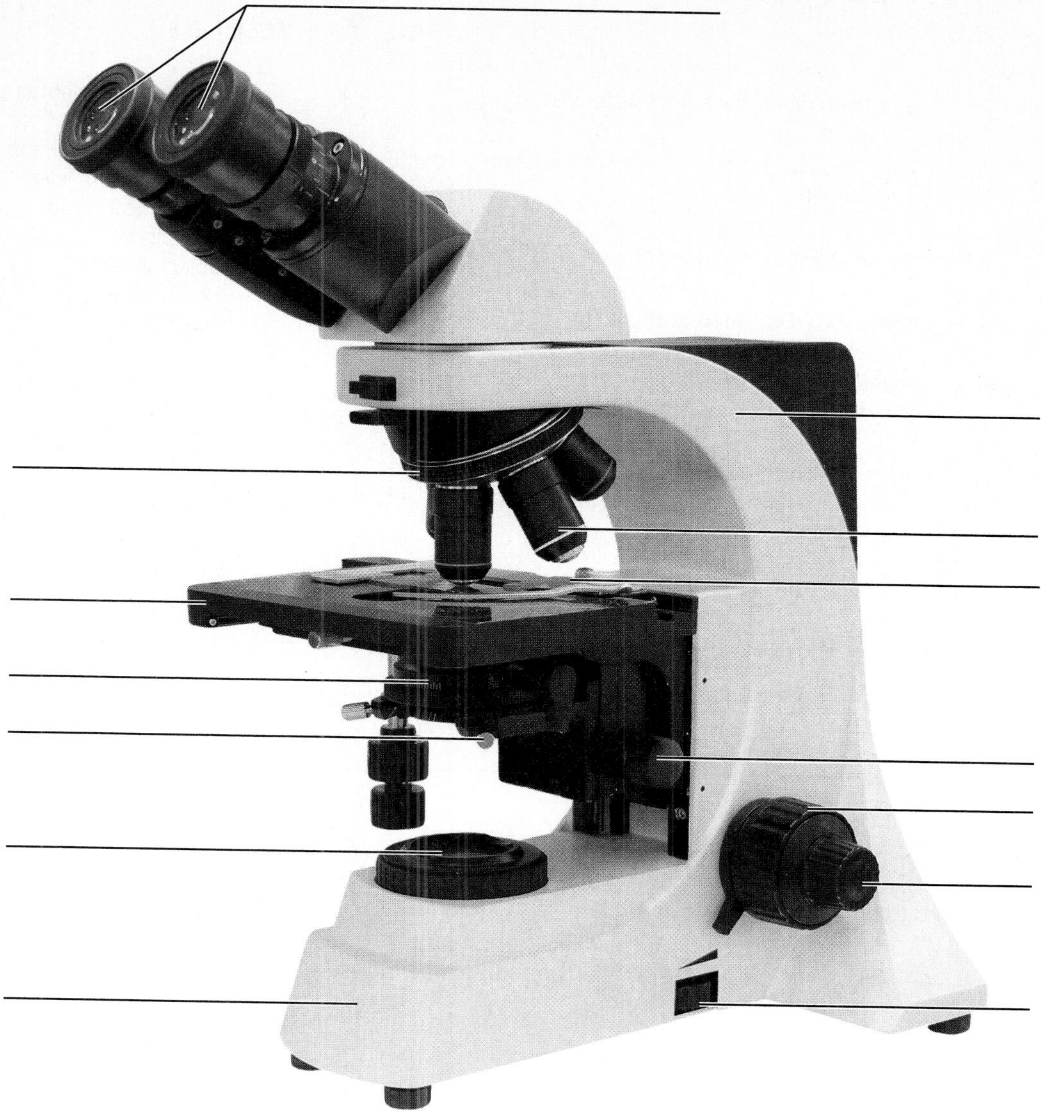

2. Explain the proper technique for transporting the microscope.

____

3. Each of the following statements is either true or false. If true, write *T* on the answer blank. If false, correct the statement by writing on the blank the proper word or phrase to replace the one that is underlined.

________________ 1. The microscope lens may be cleaned <u>with any soft tissue</u>.

________________ 2. The microscope should be stored with the <u>oil immersion</u> lens in position over the stage.

________________ 3. When beginning to focus, use the <u>scanning objective</u> lens.

________________ 4. When focusing on high power, always use the <u>coarse</u> adjustment knob to focus.

________________ 5. A coverslip should always be used <u>with wet mounts</u>.

4. Match the microscope structures in column B with the statements in column A that identify or describe them.

Column A

________ 1. platform on which the slide rests for viewing

________ 2. used to adjust the amount of light passing through the specimen

________ 3. controls the movement of the slide on the stage

________ 4. delivers a concentrated beam of light to the specimen

________ 5. used for precise focusing once initial focusing has been done

________ 6. carries the objective lenses; rotates so that the different objective lenses can be brought into position over the specimena.

Column B

a. coarse adjustment knob
b. condenser
c. fine adjustment knob
d. iris diaphragm lever
e. mechanical stage
f. nosepiece
g. objective lenses
h. ocular lens
i. stage

5. Define the following terms.

total magnification: ________________________________

____

resolution: ________________________________

Viewing Objects Through the Microscope

6. Complete, or respond to, the following statements:

________________ 1. The distance from the bottom of the objective lens to the surface of the slide is called the __________.

________________ 2. Assume there is an object on the left side of the field that you want to bring to the center (that is, toward the apparent right). In what direction would you move your slide? __________.

________________ 3. The area of the slide seen when looking through the microscope is the __________.

________________ 4. If a microscope has a 10× ocular lens and the total magnification is 950×, the objective lens in use at that time is __________×.

_______________ 5. Why should the light be dimmed when looking at living (nearly transparent) cells?

_______________ 6. If, after focusing in low power, you need to use only the fine adjustment to focus the specimen at the higher powers, the microscope is said to be _______________.

_______________ 7. You are using a 10× ocular and a 15× objective, and the field diameter is 1.5 mm. The approximate field size with a 30× objective is _______________ mm.

_______________ 8. If the diameter of the low-power field is 1.5 mm, an object that occupies approximately a third of that field has an estimated diameter of _______________ mm.

7. You have been asked to prepare a slide with the letter *F* on it (as shown below). In the circle below, draw the *F* as seen in the low-power field.

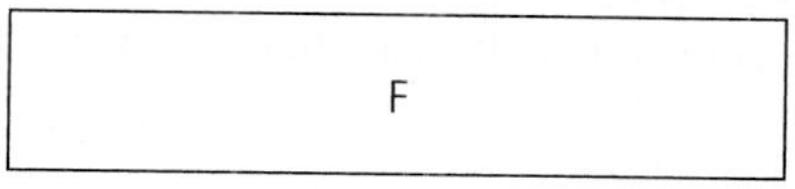

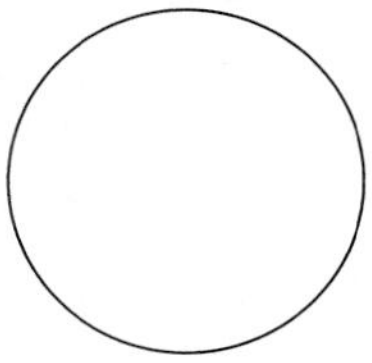

8. Estimate the length (longest dimension) of the object in μm:

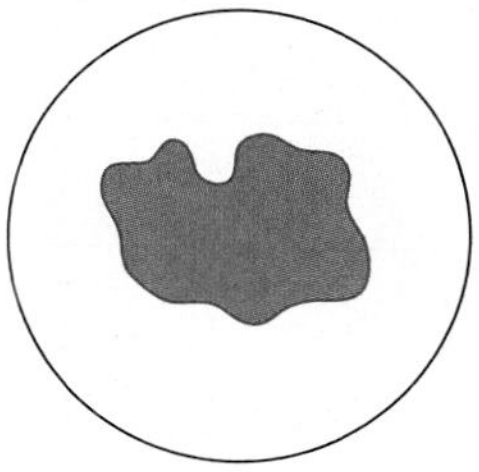

Total magnification = 100×

Field diameter = 1.6 mm

Length of object = _______________ μm

9. Say you are observing an object in the low-power field. When you switch to high power, it is no longer in your field of view.

Why might this occur? ____

____

What should you do initially to prevent this from happening? ____

____

10. Do the following factors increase or decrease as one moves to higher magnifications with the microscope?

resolution: _______________ amount of light needed: _______________

working distance: _______________ depth of field: _______________

11. A student has the high-power lens in position and appears to be intently observing the specimen. The instructor, noting a working distance of about 1 cm, knows the student isn't actually seeing the specimen.

How so? ____

12. Describe the proper procedure for preparing a wet mount.

____

____

____

13. Indicate the probable cause of the following situations during use of a microscope.

a. Only half of the field is illuminated: _______________

____

b. The visible field does not change as the mechanical stage is moved: _______________

____

14. + A blood smear is used to diagnose malaria. In patients with malaria, the protozoa can be found near and inside red blood cells. Explain why a microscope capable of high magnification and high resolution would be needed to diagnose malaria.

____

____

15. + Histopathology is the use of microscopes to view tissues to diagnose and track the progression of diseases. Why are thin slices of tissue ideal for this procedure? _______________

____

EXERCISE 4

The Cell: Anatomy and Division

Learning Outcomes

- Define *cell*, *organelle*, and *inclusion*.
- Identify on a cell model or diagram the following cellular regions and list the major function of each: nucleus, cytoplasm, and plasma membrane.
- Identify the cytoplasmic organelles and discuss their structure and function.
- Compare and contrast specialized cells with the concept of the "generalized cell."
- Define *interphase*, *mitosis*, and *cytokinesis*.
- List the stages of mitosis, and describe the key events of each stage.
- Identify the mitotic phases on slides or appropriate diagrams.
- Explain the importance of mitotic cell division, and describe its product.

Pre-Lab Quiz

Instructors may assign these and other Pre-Lab Quiz questions using Mastering A&P™

1. When a cell is not dividing, the DNA is loosely spread throughout the nucleus in a threadlike form called:
   a. chromatin
   b. chromosomes
   c. cytosol
   d. ribosomes
2. The plasma membrane not only provides a protective boundary for the cell but also determines which substances enter or exit the cell. We call this characteristic:
   a. diffusion
   b. membrane potential
   c. osmosis
   d. selective permeability
3. Because these organelles are responsible for providing most of the ATP that the cell needs, they are often referred to as the "powerhouses" of the cell. They are the:
   a. centrioles
   b. lysosomes
   c. mitochondria
   d. ribosomes
4. Circle True or False. The end product of mitosis is four genetically identical daughter nuclei.
5. DNA replication occurs during:
   a. cytokinesis
   b. interphase
   c. metaphase
   d. prophase

Go to Mastering A&P™ > Study Area to improve your performance in A&P Lab.

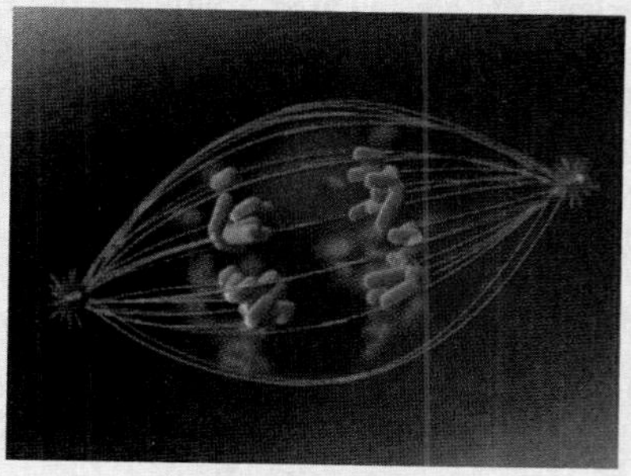

> Animations & Videos > A&P Flix > Mitosis

Instructors may assign new Building Vocabulary coaching activities, Pre-Lab Quiz questions, Art Labeling activities, and more using the Mastering A&P™ Item Library.

Materials

- Three-dimensional model of the "composite" animal cell or laboratory chart of cell anatomy
- Compound microscope
- Prepared slides of simple squamous epithelium, teased smooth muscle (l.s.), human blood smear, and sperm
- Animation/video of mitosis
- Three-dimensional models of mitotic stages
- Prepared slides of whitefish blastulas
- Chenille sticks (pipe cleaners), two different colors cut into 3-inch pieces, 8 pieces per group

Note to the Instructor: *See directions for handling wet mount preparations and disposable supplies (p. 25, Exercise 3). For suggestions on the animation/video of mitosis, see the Instructor's Guide.*

The **cell** is the structural and functional unit of all living things. The cells of the human body are highly diverse, and their differences in size, shape, and internal composition reflect their specific roles in the body. Still, cells do have many common anatomical features, and all cells must carry out certain functions to sustain life. For example, all cells can maintain their boundaries, metabolize, digest nutrients and dispose of wastes, grow and reproduce, move, and respond to a stimulus. This exercise begins by describing the structural similarities found in many cells, illustrated by a "composite," or "generalized," cell (**Figure 4.1a**), and then considers the function of cell reproduction (cell division).

Anatomy of the Composite Cell

In general, all animal cells have three major regions, or parts, that can readily be identified with a light microscope: the **nucleus**, the **plasma membrane**, and the **cytoplasm**. The nucleus is near the center of the cell. It is surrounded by cytoplasm, which in turn is enclosed by the plasma membrane. Figure 4.1a is a diagram representing the fine structure of the composite cell. An electron micrograph (Figure 4.1b) reveals the cellular structure, particularly of the nucleus.

Nucleus

The nucleus contains the genetic material, DNA, sections of which are called *genes*. Often described as the control center of the cell, the nucleus is necessary for cell reproduction. A cell that has lost or ejected its nucleus is programmed to stop dividing.

When the cell is not dividing, the genetic material is loosely dispersed throughout the nucleus in a threadlike form called **chromatin**. When the cell is in the process of dividing to form daughter cells, the chromatin coils and condenses, forming dense, rodlike bodies called **chromosomes**—much in the way a stretched spring becomes shorter and thicker when it is released.

The nucleus also contains one or more small spherical bodies, called **nucleoli**, composed primarily of proteins and ribonucleic acid (RNA). The nucleoli are assembly sites for ribosomes that are particularly abundant in the cytoplasm.

The nucleus is bound by a double-layered porous membrane, the **nuclear envelope**. The nuclear envelope is similar in composition to other cellular membranes, but it is distinguished by its large **nuclear pores**. They are spanned by protein complexes that regulate what passes through, and they permit easy passage of protein and RNA molecules.

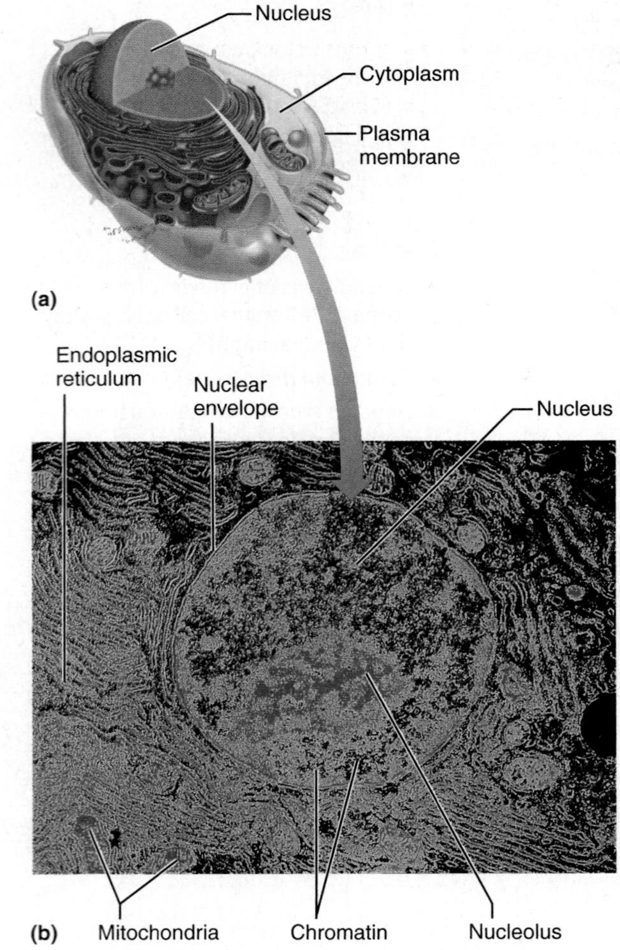

Figure 4.1 Anatomy of the composite animal cell. (a) Diagram. **(b)** Transmission electron micrograph (5000×).

Activity 1

Identifying Parts of a Cell

Identify the nuclear envelope, chromatin, nucleolus, and the nuclear pores in Figure 4.1a and b and Figure 4.3.

Plasma Membrane

The **plasma membrane** separates cell contents from the surrounding environment, providing a protective barrier. Its main structural building blocks are phospholipids (fats) and globular protein molecules. Some of the externally facing proteins and lipids have sugar (carbohydrate) side chains attached to them that are important in cellular interactions (**Figure 4.2**). As described by the fluid mosaic model, the membrane is a bilayer of phospholipid molecules in which the protein molecules float. Occasional cholesterol molecules dispersed in the bilayer help stabilize it.

Because of its molecular composition, the plasma membrane is selective about what passes through it. It allows nutrients to enter the cell but keeps out undesirable substances. By the same token, valuable cell proteins and other substances are kept within the cell, and excreta, or wastes, pass to the exterior. This property is known as **selective permeability**.

Additionally, the plasma membrane maintains a resting potential that is essential to normal functioning of excitable cells, such as neurons and muscle cells, and plays a vital role

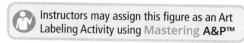

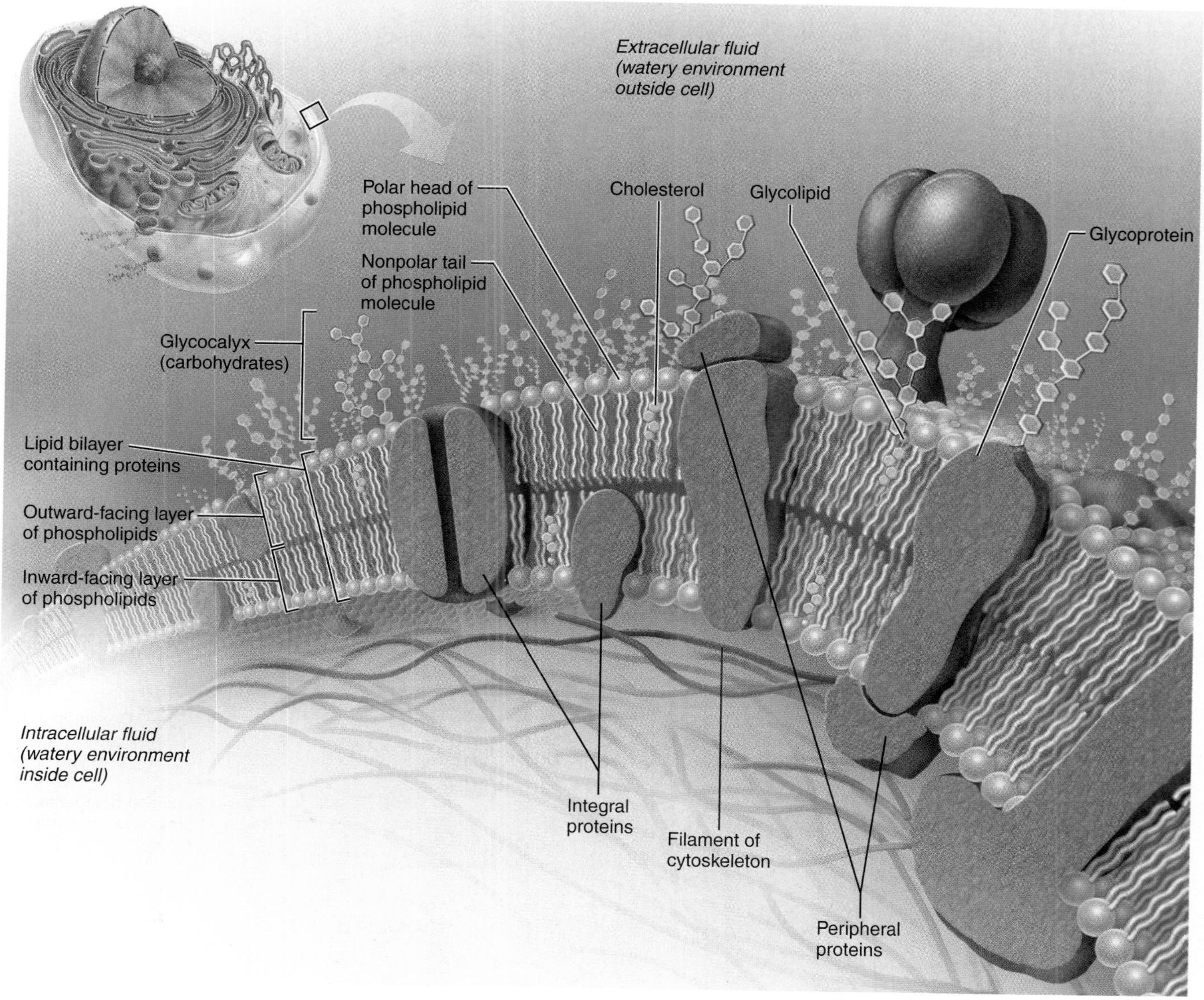

Figure 4.2 Structural details of the plasma membrane.

Instructors may assign this figure as an Art Labeling Activity using Mastering A&P™

in cell signaling and cell-to-cell interactions. In some cells, the membrane is thrown into tiny fingerlike projections or folds called **microvilli** (Figure 4.3, p. 40). Microvilli greatly increase the surface area of the cell available for absorption or passage of materials and for the binding of signaling molecules.

Activity 2

Identifying Components of a Plasma Membrane

Identify the phospholipid and protein portions of the plasma membrane in Figure 4.2. Also locate the sugar (*glyco* = carbohydrate) side chains and cholesterol molecules. Identify the microvilli in the generalized cell diagram (Figure 4.3).

Cytoplasm and Organelles

The cytoplasm consists of the cell contents between the nucleus and plasma membrane. Suspended in the **cytosol**, the fluid cytoplasmic material, are many small structures called **organelles** (literally, "small organs"). The organelles are the metabolic machinery of the cell, and they are highly organized to carry out specific functions for the cell as a whole. The cytoplasmic organelles include the ribosomes, smooth and rough endoplasmic reticulum, Golgi apparatus, lysosomes, peroxisomes, mitochondria, cytoskeletal elements, and centrioles.

4

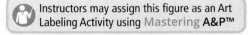

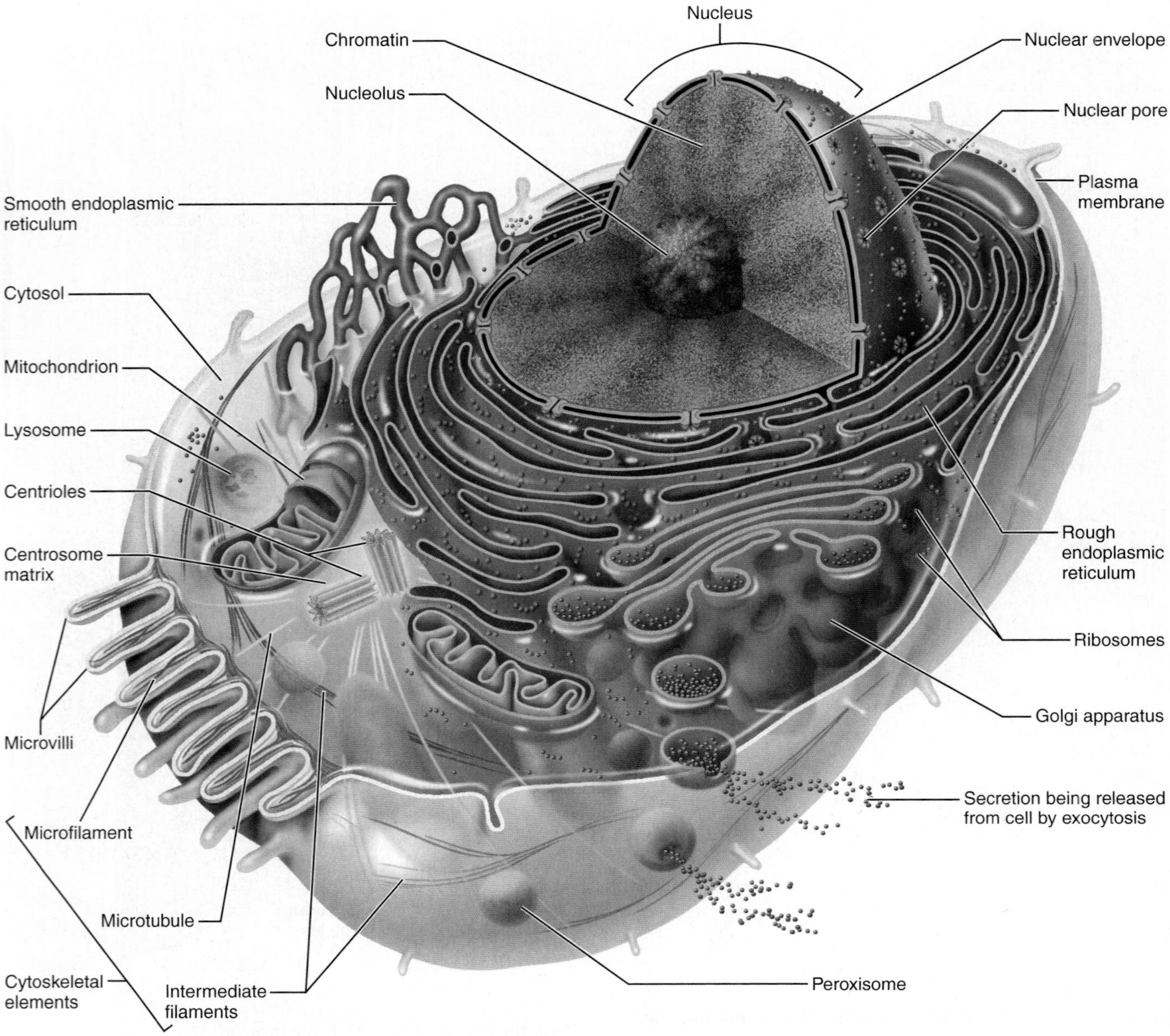

Figure 4.3 Structure of the generalized cell. No cell is exactly like this one, but this composite illustrates features common to many human cells. Not all organelles are drawn to the same scale in this illustration.

Instructors may assign this figure as an Art Labeling Activity using Mastering A&P™

Activity 3

Locating Organelles

Each organelle type is described in **Table 4.1**. Read through the table, and then, as best you can, locate the organelles in Figures 4.1b and 4.3.

Activity 4

Examining the Cell Model

Once you have located all of these structures in the art (Figures 4.1b and 4.3), examine the cell model (or cell chart) to repeat and reinforce your identifications.

The cell cytoplasm may or may not contain **inclusions**. Examples of inclusions are stored foods (glycogen granules and lipid droplets), pigment granules, crystals of various types, water vacuoles, and ingested foreign materials.

Table 4.1 Summary of Structure and Function of Cytoplasmic Organelles

Organelle		Location and function
Ribosomes		Tiny spherical bodies composed of RNA and protein; floating free or attached to a membranous structure (the rough ER) in the cytoplasm. Actual sites of protein synthesis.
Endoplasmic reticulum (ER)		Membranous system of tubules that extends throughout the cytoplasm; two varieties: rough and smooth. Rough ER is studded with ribosomes; tubules of the rough ER provide an area for storage and transport of the proteins made on the ribosomes to other cell areas. Smooth ER, which has no function in protein synthesis, is a site of steroid and lipid synthesis, lipid metabolism, and drug detoxification.
Golgi apparatus	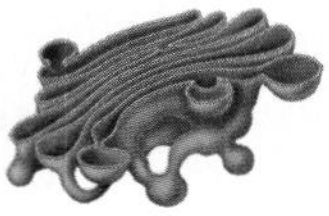	Stack of flattened sacs with bulbous ends and associated small vesicles; found close to the nucleus. Plays a role in packaging proteins or other substances for export from the cell or incorporation into the plasma membrane and in packaging lysosomal enzymes.
Lysosomes	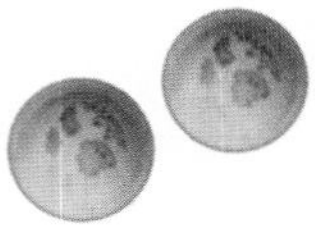	Various-sized membranous sacs containing digestive enzymes including acid hydrolases; function to digest worn-out cell organelles and foreign substances that enter the cell. Have the capacity of total cell destruction if ruptured and are for this reason referred to as "suicide sacs."
Peroxisomes	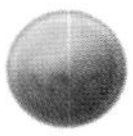	Small lysosome-like membranous sacs containing oxidase enzymes that detoxify alcohol, free radicals, and other harmful chemicals. They are particularly abundant in liver and kidney cells.
Mitochondria		Generally rod-shaped bodies with a double-membrane wall; inner membrane is shaped into folds, or cristae; contain enzymes that oxidize foodstuffs to produce cellular energy (ATP); often referred to as "powerhouses of the cell."
Centrioles		Paired, cylindrical bodies that lie at right angles to each other, close to the nucleus. Internally, each centriole is composed of nine triplets of microtubules. As part of the centrosome, they direct the formation of the mitotic spindle during cell division; form the bases of cilia and flagella and in that role are called *basal bodies*.
Cytoskeletal elements: microfilaments, intermediate filaments, and microtubules	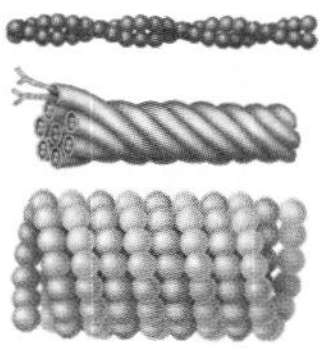	Form an internal scaffolding called the *cytoskeleton*. Provide cellular support; function in intracellular transport. Microfilaments are formed largely of actin, a contractile protein, and thus are important in cell mobility, particularly in muscle cells. Intermediate filaments are stable elements composed of a variety of proteins and resist mechanical forces acting on cells. Microtubules form the internal structure of the centrioles and help determine cell shape.

Differences and Similarities in Cell Structure

Activity 5

Observing Various Cell Structures

1. Obtain a compound microscope and prepared slides of simple squamous epithelium, smooth muscle cells (teased), human blood, and sperm.

2. Observe each slide under the microscope, carefully noting similarities and differences in the cells. See photomicrographs for simple squamous epithelium (Figure 3.5 in Exercise 3) and teased smooth muscle (Figure 6.7c in Exercise 6). The oil immersion lens will be needed to observe blood and sperm. Distinguish the boundaries of the individual cells, and notice the shape and position of the nucleus in each case. When you

Text continues on next page. →

4

Simple squamous epithelium
Diameter ___________

Sperm cells
Length ___________

Human red blood cells
Diameter ___________

Teased smooth muscle cells
Length ___________

look at the human blood smear, direct your attention to the red blood cells, the pink-stained cells that are most numerous. The color photomicrographs illustrating a blood smear (Figure 29.3 in Exercise 29) and sperm (Figure 43.3 in Exercise 43) may be helpful in this cell structure study. Sketch your observations in the circles provided above.

3. Measure the length or diameter of each cell, and record below the appropriate sketch.

4. How do these four cell types differ in shape and size?

____

____

____

____

How might cell shape affect cell function?

____

____

____

____

____

Which cells have visible projections? ___________

____

How do these projections relate to the function of these cells?

____

____

Do any of these cells lack a plasma membrane? ___________

A nucleus? ___________

In the cells with a nucleus, can you discern nucleoli?

____

Cell Division

The cell cycle is the series of changes that a cell goes through from the time it is formed until it reproduces. The outer ring of **Figure 4.4** shows the two main periods of the cell cycle, interphase (in green) and the mitotic phase (in yellow). **Interphase** is the longer period, during which the cell grows and carries out its usual activities. **Cell division**, or the **mitotic phase**, is the period when the cell reproduces itself by dividing. In an interphase cell about to divide, the genetic material (DNA) is copied exactly via DNA replication. Once this important event has occurred, cell division ensues.

Cell division is essential for growth and repair. Cell division, which is also called the **M (mitotic) phase** of the cell cycle, consists of two events called *mitosis* and *cytokinesis*. **Mitosis** is the division of the copied DNA of the mother cell to two daughter nuclei. **Cytokinesis** is the division of the cytoplasm, which begins when mitosis is nearly complete. Although mitosis is usually accompanied by cytokinesis, in some instances cytoplasmic division does not occur, leading to the formation of binucleate or multinucleate cells.

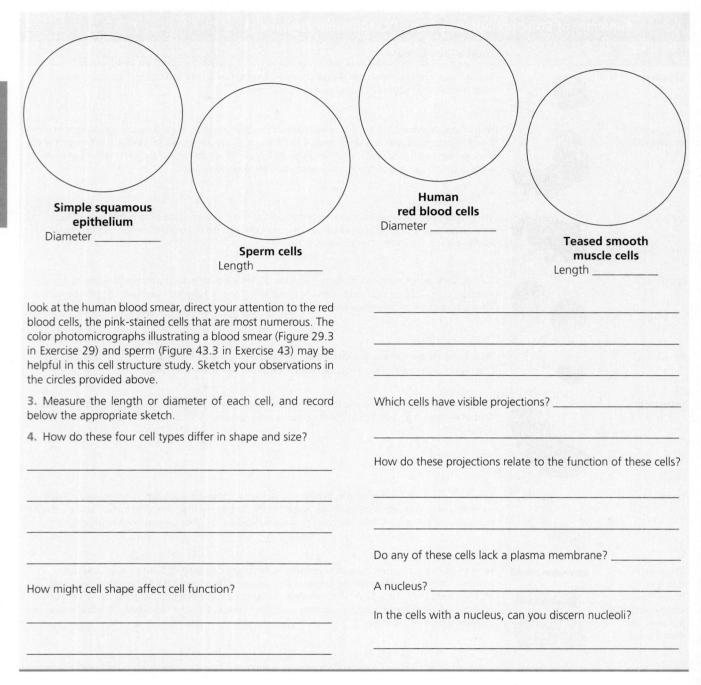

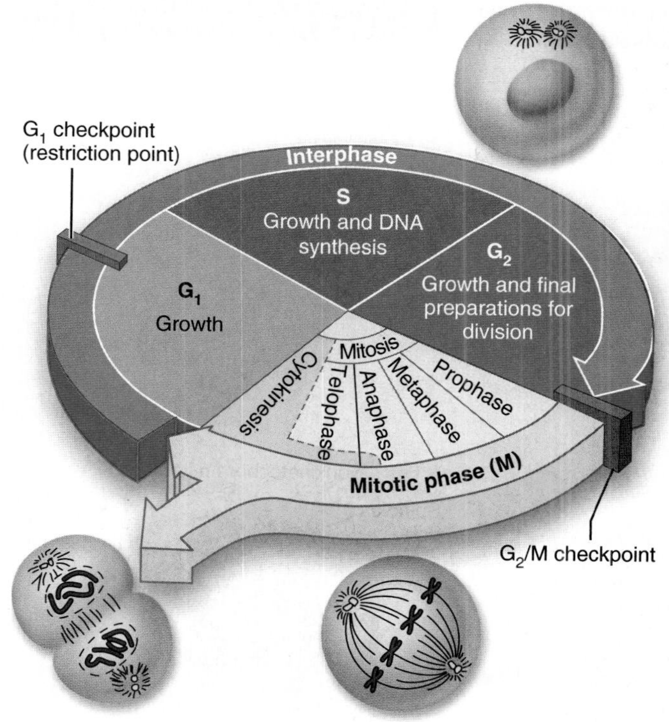

Figure 4.4 The cell cycle. The two main phases are interphase (green in the outer ring) and the mitotic phase (yellow in the outer ring).

The product of **mitosis** is two daughter nuclei that are genetically identical to the mother nucleus. **Meiosis**, which yields four daughter nuclei that differ genetically in composition from the mother nucleus, is used only for the production of gametes (eggs and sperm) for sexual reproduction.

The phases of mitosis include **prophase**, **metaphase**, **anaphase**, and **telophase**. The detailed events of interphase, mitosis, and cytokinesis are described and illustrated in **Figure 4.5** on pp. 44–45.

Mitosis is essentially the same in all animal cells, but depending on the tissue, it takes from 5 minutes to several hours to complete. In most cells, centriole replication occurs during interphase of the next cell cycle.

At the end of cell division, two daughter cells exist—each with a smaller cytoplasmic mass than the mother cell but genetically identical to it. The daughter cells grow and carry out the normal spectrum of metabolic processes until it is their turn to divide.

Cell division is extremely important during the body's growth period. Most cells divide until puberty, when adult body size is achieved and overall body growth ceases. After this time in life, only certain cells carry out cell division routinely—for example, cells subjected to abrasion (epithelium of the skin and lining of the gut). Other cell populations—such as liver cells—stop dividing but retain this ability should some of them be removed or damaged. Skeletal muscle, cardiac muscle, and most mature neurons almost completely lose this ability to divide and thus are severely handicapped by injury.

Activity 6

Identifying the Mitotic Stages

1. Watch an animation or video presentation of mitosis (if available).

2. Using the three-dimensional models of dividing cells provided, identify each of the mitotic phases illustrated and described in Figure 4.5.

3. Obtain a prepared slide of whitefish blastulas to study the stages of mitosis. The cells of each *blastula* (a stage of embryonic development consisting of a hollow ball of cells) are at approximately the same mitotic stage, so it may be necessary to observe more than one blastula to view all the mitotic stages. Examine the slide carefully, identifying the four mitotic phases and the process of cytokinesis. Compare your observations with the photomicrographs (Figure 4.5), and verify your identifications with your instructor.

Activity 7

"Chenille Stick" Mitosis

1. Obtain a total of eight 3-inch pieces of chenille stick, four of one color and four of another color (e.g., four green and four purple).

2. Assemble the chenille sticks into a total of four chromosomes (each with two sister chromatids) by twisting two sticks of the same color together at the center with a single twist.

What does the twist at the center represent? ____________

____

3. Arrange the chromosomes as they appear in early prophase.

Name the structure that assembles during this phase.

____

Draw early prophase in the space provided in the Review Sheet (question 6, p. 49).

4. Arrange the chromosomes as they appear in late prophase.

What structure on the chromosome centromere do the growing spindle microtubules attach to? ____________

What structure is now present as fragments? ____________

____

Text continues on page 46. →

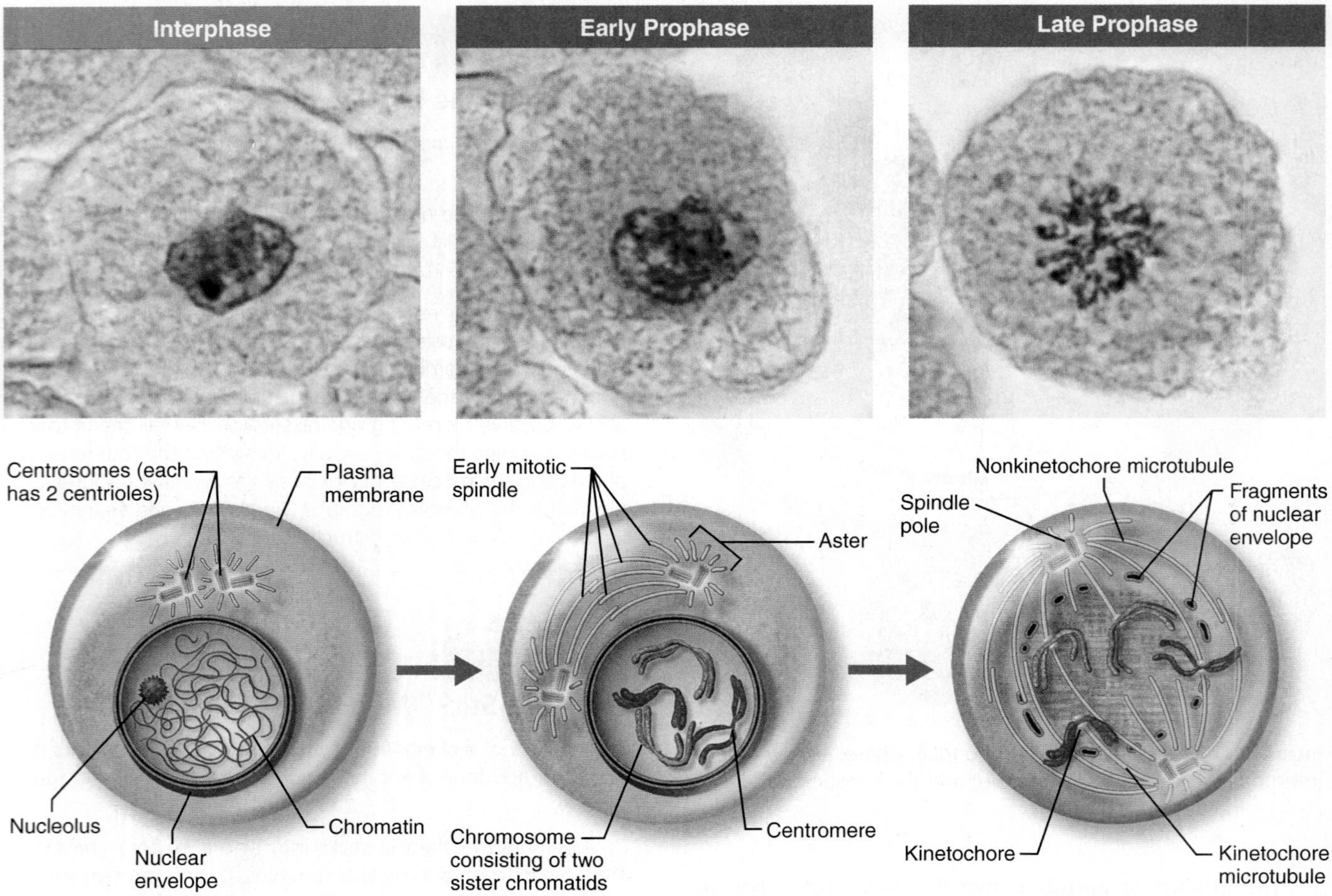

Interphase

Interphase is the period when the cell carries out its normal metabolic activities and grows. Interphase is not part of mitosis.

- During interphase, the DNA-containing material is in the form of chromatin. The nuclear envelope and one or more nucleoli are intact and visible.
- There are three distinct periods of interphase:
  G_1: The centrioles begin replicating.
  S: DNA is replicated.
  G_2: Final preparations for mitosis are completed, and centrioles finish replicating.

Prophase—first phase of mitosis

Early Prophase

- The chromatin condenses, forming barlike chromosomes.
- Each duplicated chromosome consists of two identical threads, called **sister chromatids**, held together at the **centromere**. (Later when the chromatids separate, each will be a new chromosome.)
- As the chromosomes appear, the nucleoli disappear, and the two centrosomes separate from one another.
- The centrosomes act as focal points for growth of a microtubule assembly called the **mitotic spindle**. As the microtubules lengthen, they propel the centrosomes toward opposite ends (poles) of the cell.
- Microtubule arrays called **asters** ("stars") extend from the centrosome matrix.

Late Prophase

- The nuclear envelope breaks up, allowing the spindle to interact with the chromosomes.
- Some of the growing spindle microtubules attach to **kinetochores**, special protein structures at each chromosome's centromere. Such microtubules are called **kinetochore microtubules**.
- The remaining spindle microtubules (not attached to any chromosomes) are called **nonkinetochore microtubules**. The microtubules slide past each other, forcing the poles apart.
- The kinetochore microtubules pull on each chromosome from both poles in a tug-of-war that ultimately draws the chromosomes to the center, or equator, of the cell.

Figure 4.5 The interphase cell and the events of cell division. The cells shown are from an early embryo of a whitefish. Photomicrographs are above; corresponding diagrams are below. (Micrographs approximately 1600×.)

Instructors may assign this figure as an Art Labeling Activity using Mastering A&P™

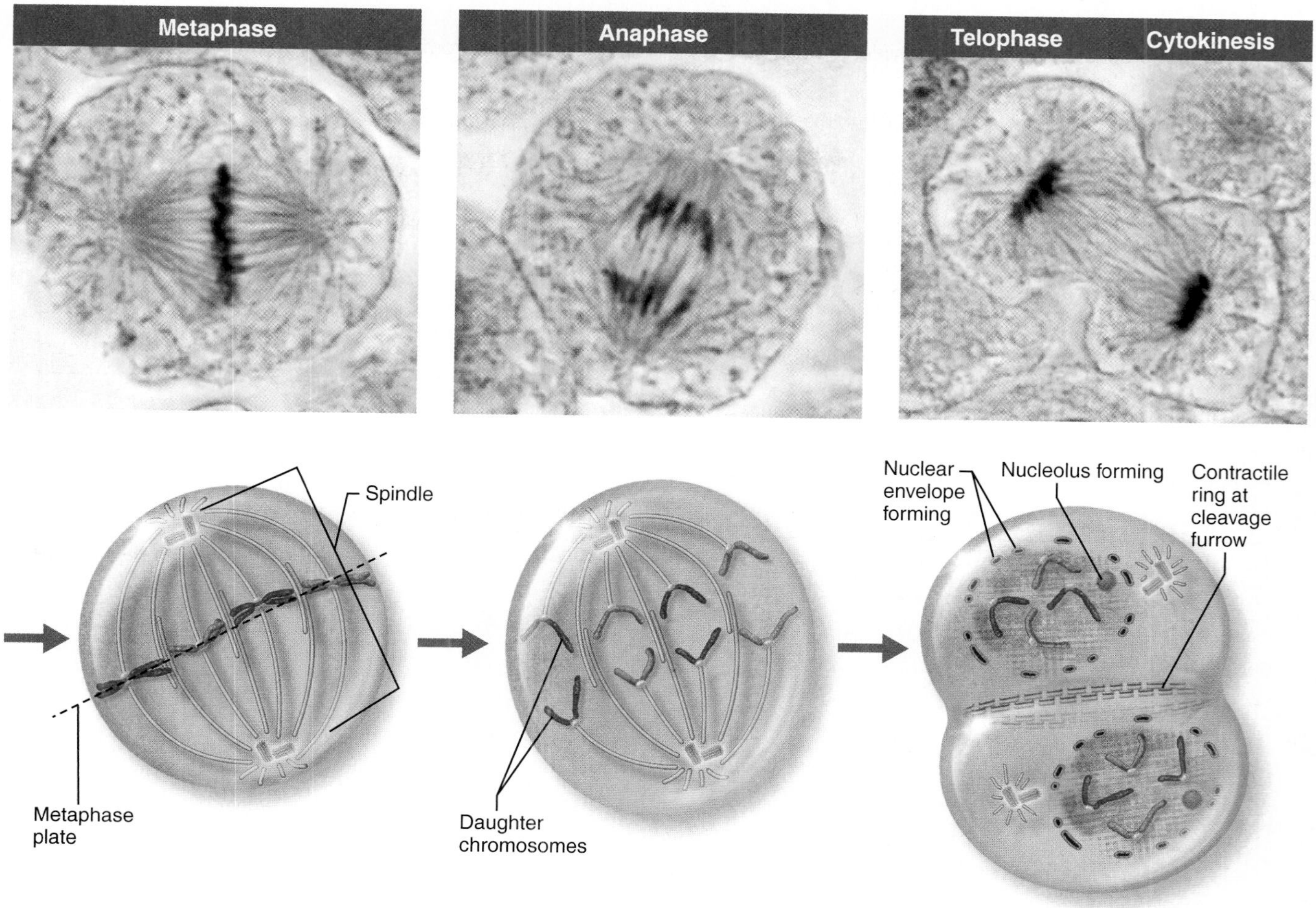

Metaphase—second phase of mitosis

- The two centrosomes are at opposite poles of the cell.
- The chromosomes cluster at the midline of the cell, with their centromeres precisely aligned at the **equator** of the spindle. This imaginary plane midway between the poles is called the **metaphase plate**.
- Enzymes act to separate the chromatids from each other.

Anaphase—third phase of mitosis

The shortest phase of mitosis, anaphase begins abruptly as the centromeres of the chromosomes split simultaneously. Each chromatid now becomes a chromosome in its own right.

- The kinetochore microtubules, moved along by motor proteins in the kinetochores, gradually pull each chromosome toward the pole it faces.
- At the same time, the microtubules slide past each other, lengthen, and push the two poles of the cell apart.
- The moving chromosomes look V shaped. The centromeres lead the way, and the chromosomal "arms" dangle behind them.
- The fact that the chromosomes are short, compact bodies makes it easier for them to move and separate. Diffuse threads of chromatin would trail, tangle, and break, resulting in imprecise "parceling out" to the daughter cells.

Telophase—final phase of mitosis

Telophase begins as soon as chromosomal movement stops. This final phase is like prophase in reverse.

- The identical sets of chromosomes at the opposite poles of the cell uncoil and resume their threadlike chromatin form.
- A new nuclear envelope forms around each chromatin mass, nucleoli reappear within the nuclei, and the spindle breaks down and disappears.
- Mitosis is now ended. The cell, for just a brief period, is binucleate (has two nuclei), and each new nucleus is identical to the original parent nucleus.

Cytokinesis—division of cytoplasm

Cytokinesis begins during late anaphase and continues through and beyond telophase. A contractile ring of actin microfilaments forms the **cleavage furrow** and pinches the cell apart.

Figure 4.5 *(continued)*

4

Draw late prophase in the space provided on the Review Sheet (question 6, p. 49).

5. Arrange the chromosomes as they appear in metaphase.

What is the name of the imaginary plane that the chromosomes align along? ______________________

Draw metaphase in the space provided on the Review Sheet (question 6, p. 49).

6. Arrange the chromosomes as they appear in anaphase.

What does untwisting of the chenille sticks represent?

____

____

Each sister chromatid has now become a ____________.

Draw anaphase in the space provided on the Review Sheet (question 6, p. 49).

7. Arrange the chromosomes as they appear in telophase.

Briefly list four reasons why telophase is like the reverse of prophase.

____

____

____

Draw telophase in the space provided on the Review Sheet (question 6, p. 49).

Instructors may assign a portion of the Review Sheet questions using Mastering A&P™

EXERCISE 4 REVIEW SHEET
The Cell: Anatomy and Division

Name ______________________ Lab Time/Date ______________________

Anatomy of the Composite Cell

1. Label the cell structures using the leader lines provided.

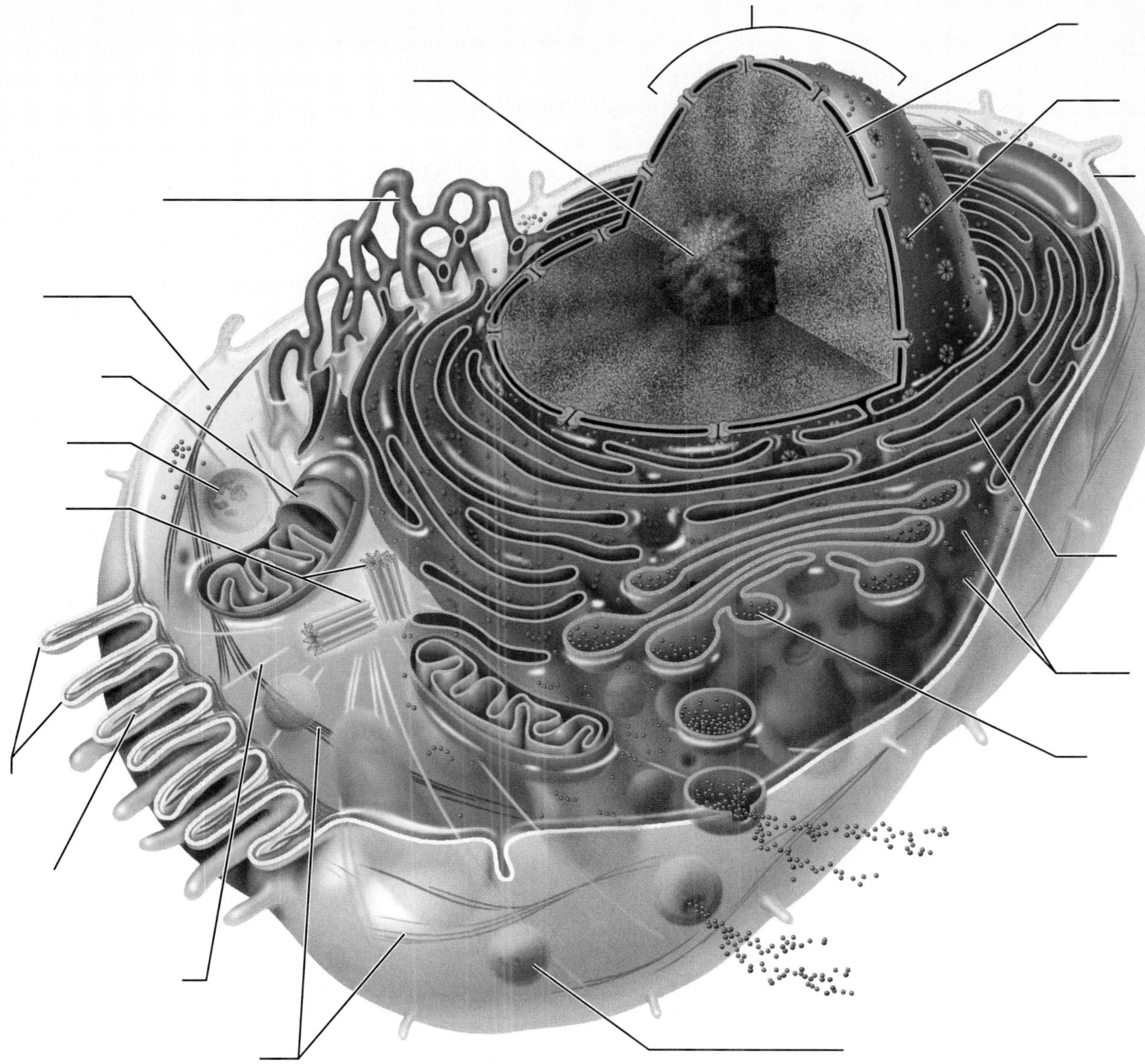

2. Match each cell structure listed on the left with the correct description on the right.

________	1. ribosome	a.	main site of ATP synthesis
________	2. smooth ER	b.	encloses the chromatin
________	3. mitochondrion	c.	sac of digestive enzymes
________	4. nucleus	d.	examples include glycogen granules and ingested foreign materials
________	5. Golgi apparatus	e.	forms basal bodies and helps direct mitotic spindle formation
________	6. lysosome	f.	site of protein synthesis
________	7. centriole	g.	forms the external boundary of the cell
________	8. cytoskeleton	h.	site of lipid synthesis
________	9. inclusion	i.	packaging site for ribosomes
________	10. plasma membrane	j.	packages proteins for transportation
________	11. nucleolus	k.	internal cellular network of rodlike structures

Differences and Similarities in Cell Structure

3. Choose the specimen observed in Activity 5 (squamous epithelium, sperm cells, smooth muscle, or human red blood cells) that fits the description below.

1. ____________________ cell has a flagellum for movement
2. ____________________ cells have an elongated shape (tapered at each end)
3. ____________________ cells are close together
4. ____________________ cells are circular
5. ____________________ cells are thin and flat, with irregular borders
6. ____________________ cells are anucleate (without a nucleus)
7. ____________________ longest cell

Cell Division

4. What is the function of mitotic cell division? ____

____

5. Identify the four phases of mitosis shown in the following photomicrographs, and select the events from the key that correctly identify each phase. On the appropriate answer line, write the letters that correspond to these events.

Key:

a. The nuclear envelope re-forms.
b. Chromosomes line up in the center of the cell.
c. Chromatin coils and condenses, forming chromosomes.
d. Chromosomes stop moving toward the poles.
e. The chromosomes are V shaped.
f. The nuclear envelope breaks down.
g. Chromosomes attach to the spindle fibers.
h. The mitotic spindle begins to form.

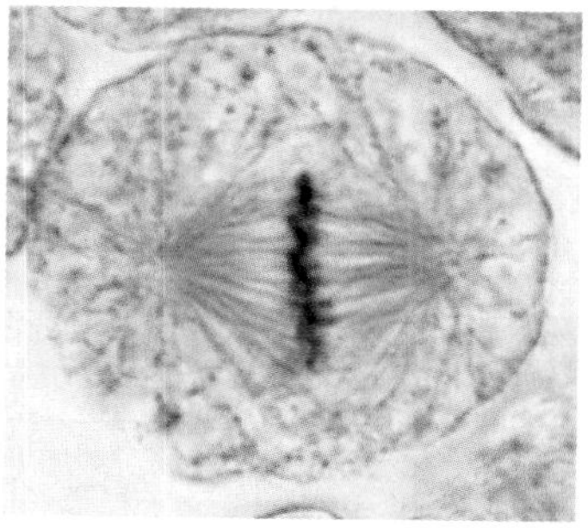

1. Phase: ______________________

Events: ______________________

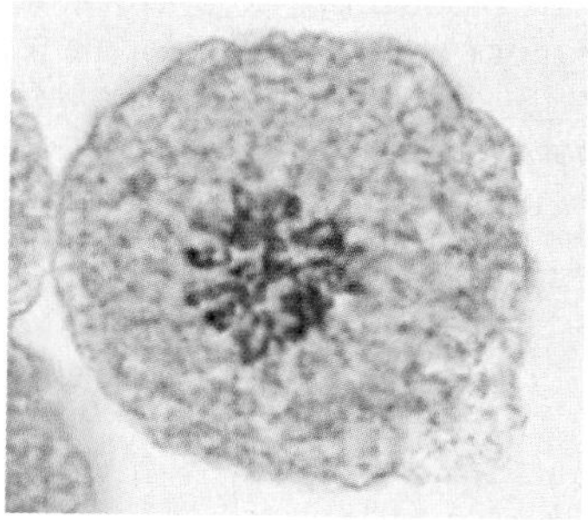

2. Phase: ______________________

Events: ______________________

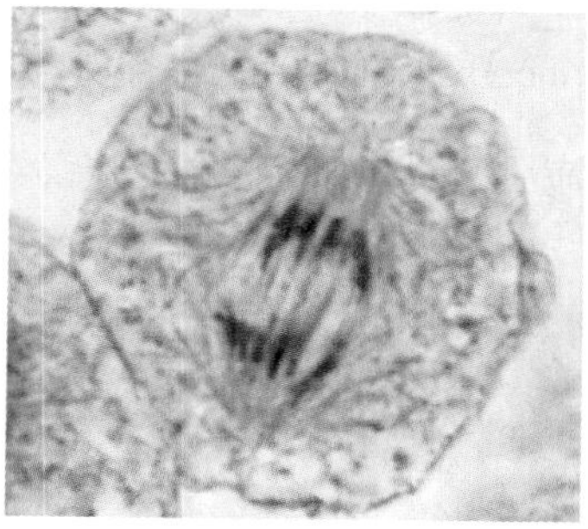

3. Phase: ______________________

Events: ______________________

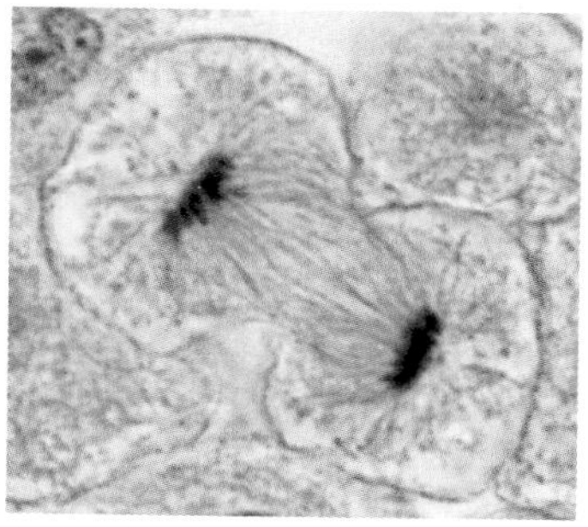

4. Phase: ______________________

Events: ______________________

6. Draw the phases of mitosis for a cell that contains four chromosomes as its diploid, or $2n$, number.

7. Describe the events that occur during interphase.

____

____

8. Complete or respond to the following statements:

Division of the __1__ is referred to as mitosis. Cytokinesis is division of the __2__. The major structural difference between chromatin and chromosomes is that the latter are __3__. Chromosomes attach to the spindle fibers by undivided structures called __4__. If a cell undergoes mitosis but not cytokinesis, the product is __5__. The structure that acts as a scaffolding for chromosomal attachment and movement is called the __6__. __7__ is the period of cell life when the cell is not involved in division. Three cell populations in the body that do not routinely undergo cell division are __8__, __9__, and __10__.

1. ______________________
2. ______________________
3. ______________________
4. ______________________
5. ______________________
6. ______________________
7. ______________________
8. ______________________
9. ______________________
10. ______________________

9. + Plasma cells are key to the immune response because they secrete antibodies. Given that antibodies are made of protein, which membrane-enclosed cell organelle would you expect the plasma cells to have in abundance? Why? ______________

____

10. + Name which organelle you would expect to play the largest role in decomposition of the human body. Why? __________

____

11. + Some antifungal medications work by blocking DNA synthesis in the fungal cell. Describe where in the cell cycle such a medication would halt the fungal cell and the consequences of this early termination of the cycle. ______________

____

EXERCISE 5

The Cell: Transport Mechanisms and Cell Permeability

Learning Outcomes

- Define *selective permeability*, and explain the difference between active and passive transport processes.
- Define *diffusion*, and explain how simple diffusion and facilitated diffusion differ.
- Define *osmosis*, and explain the difference between isotonic, hypotonic, and hypertonic solutions.
- Define *filtration*, and discuss where it occurs in the body.
- Define *vesicular transport*, and describe phagocytosis, pinocytosis, receptor-mediated endocytosis, and exocytosis.
- List the processes that account for the movement of substances across the plasma membrane, and indicate the driving force for each.
- Determine which way substances will move passively through a selectively permeable membrane when given appropriate information about their concentration gradients.

Go to Mastering A&P™ > Study Area to improve your performance in A&P Lab.

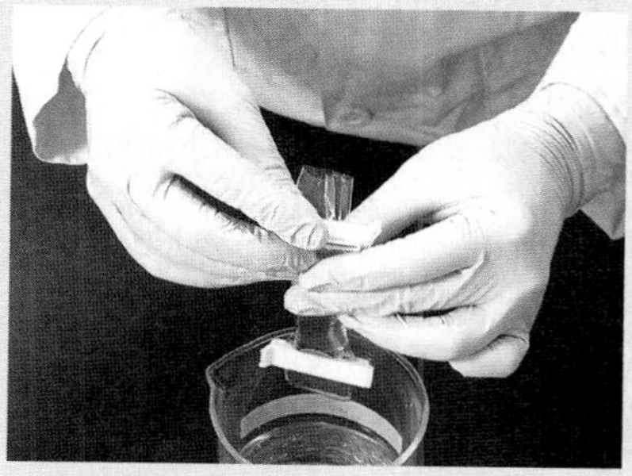

> Lab Tools > Pre-Lab Videos > Diffusion and Osmosis

Instructors may assign new Building Vocabulary coaching activities, Pre-Lab Quiz questions, Art Labeling activities, Pre-Lab Video Coaching Activities for Diffusion and Osmosis, PhysioEx activities, and more using the Mastering A&P™ Item Library.

Pre-Lab Quiz

Instructors may assign these and other Pre-Lab Quiz questions using Mastering A&P™

1. Circle the correct underlined term. A passive process, diffusion / osmosis is the movement of solute molecules from an area of greater concentration to an area of lesser concentration.
2. A solution surrounding a cell is *hypertonic* if:
   a. it contains fewer nonpenetrating solute particles than the interior of the cell
   b. it contains more nonpenetrating solute particles than the interior of the cell
   c. it contains the same amount of nonpenetrating solute particles as the interior of the cell
3. Which of the following would require an input of energy?
   a. diffusion
   b. filtration
   c. osmosis
   d. vesicular transport
4. Circle the correct underlined term. In pinocytosis / phagocytosis, parts of the plasma membrane and cytoplasm extend and engulf a relatively large or solid material.
5. Circle the correct underlined term. In active / passive processes, the cell provides energy in the form of ATP to power the transport process.

Materials

Passive Processes

Diffusion of Dye Through Agar Gel

- Petri dish containing 12 ml of 1.5% agar-agar
- Millimeter-ruled graph paper
- Wax marking pencil
- 3.5% methylene blue solution (approximately 0.1 *M*) in dropper bottles
- 1.6% potassium permanganate solution (approximately 0.1 *M*) in dropper bottles
- Medicine dropper

Text continues on next page. →

Diffusion and Osmosis Through Nonliving Membranes

- Four dialysis sacs
- Small funnel
- 25-ml graduated cylinder
- Wax marking pencil
- Fine twine or dialysis tubing clamps
- 250-ml beakers
- Distilled water
- 40% glucose solution
- 10% sodium chloride (NaCl) solution
- 40% sucrose solution
- Laboratory balance
- Paper towels
- Hot plate and large beaker for hot water bath
- Benedict's solution in dropper bottle
- Silver nitrate ($AgNO_3$) in dropper bottle
- Test tubes in rack, test tube holder

Experiment 1

- Deshelled eggs
- 400-ml beakers
- Wax marking pencil
- Distilled water
- 30% sucrose solution
- Laboratory balance
- Paper towels
- Graph paper
- Weigh boat

Experiment 2

- Clean microscope slides and coverslips
- Medicine dropper
- Compound microscope
- Vials of mammalian blood obtained from a biological supply house or veterinarian—at option of instructor
- Freshly prepared physiological (mammalian) saline solution in dropper bottle
- 5% sodium chloride solution in dropper bottle
- Distilled water
- Filter paper
- Disposable gloves
- Basin and wash bottles containing 10% household bleach solution
- Disposable autoclave bag
- Paper towels

Diffusion Demonstrations

1. Diffusion of a dye through water

Prepared the morning of the laboratory session with setup time noted. Potassium permanganate crystals are placed in a 1000-ml graduated cylinder, and distilled water is added slowly and with as little turbulence as possible to fill to the 1000-ml mark.

2. Osmometer

Just before the laboratory begins, the broad end of a thistle tube is closed with a selectively permeable dialysis membrane, and the tube is secured to a ring stand. Molasses is added to approximately 5 cm above the thistle tube bulb, and the bulb is immersed in a beaker of distilled water. At the beginning of the lab session, the level of the molasses in the tube is marked with a wax pencil.

Filtration

- Ring stand, ring, clamp
- Filter paper, funnel
- Solution containing a mixture of uncooked starch, powdered charcoal, and copper sulfate ($CuSO_4$)
- 10-ml graduated cylinder
- 100-ml beaker
- Lugol's iodine in a dropper bottle

Active Processes

- Video/animation showing phagocytosis (if available)

Note to the Instructor: *See directions for handling wet mount preparations and disposable supplies (p. 25, Exercise 3).*

PEx PhysioEx™ 9.1 Computer Simulation Ex.1 on p. PEx-3.

Because of its molecular composition, the plasma membrane is selective about what passes through it. It allows nutrients to enter the cell but keeps out undesirable substances. By the same token, valuable cell proteins and other substances are kept within the cell, and excreta or wastes pass to the exterior. This property is known as **selective**, or **differential**, **permeability**. Transport through the plasma membrane occurs in two basic ways. In **passive processes**, concentration or pressure differences drive the movement. In **active processes**, the cell provides energy (ATP) to power the transport process.

Passive Processes

The two important passive processes of membrane transport are *diffusion* and *filtration*. Diffusion is an important transport process for every cell in the body. By contrast, filtration usually occurs only across capillary walls.

Diffusion

Molecules possess **kinetic energy** and are in constant motion. As molecules move about randomly at high speeds, they collide and ricochet off one another, changing direction with each collision (**Figure 5.1**). A **concentration gradient** is present when molecules are unevenly distributed, resulting in an area of higher concentration and an area of lower concentration. **Diffusion** is the movement of molecules from a region of their higher concentration to a region of their lower concentration. Because the driving force for diffusion is the kinetic energy of the molecules, the speed of diffusion depends on molecular size and temperature. Smaller molecules move faster, and molecules move faster as temperature increases.

There are many examples of diffusion in nonliving systems. For example, if a bottle of ether were uncorked at the front of the laboratory, very shortly thereafter you would be nodding off as the molecules became distributed throughout the room.

In general, molecules diffuse passively through the plasma membrane if they can dissolve in the lipid portion of the membrane, as CO_2 and O_2 can. The unassisted diffusion of solutes (dissolved substances) through a selectively permeable membrane is called **simple diffusion**.

Certain molecules, glucose for example, are transported across the plasma membrane with the assistance of a protein carrier molecule. The substances move by a passive transport process called **facilitated diffusion**. As with simple diffusion, the substances move from an area of higher concentration to one of lower concentration, that is, down their concentration gradients.

Osmosis

The flow of water across a selectively permeable membrane is called **osmosis**. During osmosis, water moves down its concentration gradient. The concentration of water is inversely related to the concentration of solutes. If the solutes can diffuse across the membrane, both water and solutes will move down their concentration gradients through the membrane. If the particles in solution are nonpenetrating solutes (prevented from crossing the membrane), water alone will move by osmosis and in doing so will cause changes in the volume of the compartments on either side of the membrane.

Diffusion of Dye Through Agar Gel and Water

The relationship between molecular weight and the rate of diffusion can be examined easily by observing the diffusion of two different types of dye molecules through an agar gel. The dyes used in this experiment are methylene blue, which has a molecular weight of 320 and is deep blue in color, and potassium permanganate, a purple dye with a molecular weight of 158. Although the agar gel appears quite solid, it is primarily (98.5%) water and allows the dye molecules to move freely through it.

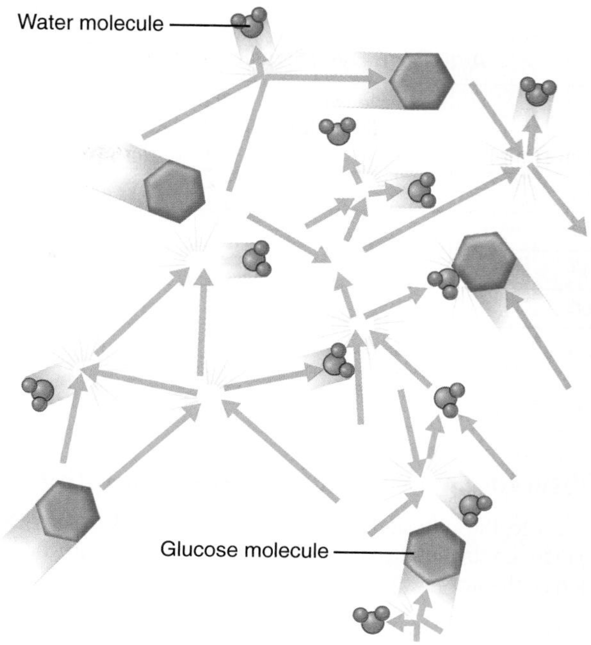

Figure 5.1 Random movement and numerous collisions cause molecules to become evenly distributed.

Activity 1

Observing Diffusion of Dye Through Agar Gel

1. Work with members of your group to formulate a hypothesis about the rates of diffusion of methylene blue and potassium permanganate through the agar gel. Justify your hypothesis.

2. Obtain a petri dish containing agar gel, a piece of millimeter-ruled graph paper, a wax marking pencil, dropper bottles of methylene blue and potassium permanganate, and a medicine dropper.

3. Using the wax marking pencil, draw a line on the bottom of the petri dish dividing it into two sections. Place the petri dish on the ruled graph paper.

4. Create a well in the center of each section using the medicine dropper. To do this, squeeze the bulb of the medicine dropper, and push it down into the agar. Release the bulb as you slowly pull the dropper vertically out of the agar. This should remove an agar plug, leaving a well in the agar. (See **Figure 5.2a**.)

5. Carefully fill one well with the methylene blue solution and the other well with the potassium permanganate solution (Figure 5.2b).

Record the time. ______________________

6. At 15-minute intervals, measure the distance the dye has diffused from each well by measuring the diameter of the dye. Continue these observations for 1 hour, and record the results in the **Activity 1 chart** (p. 54).

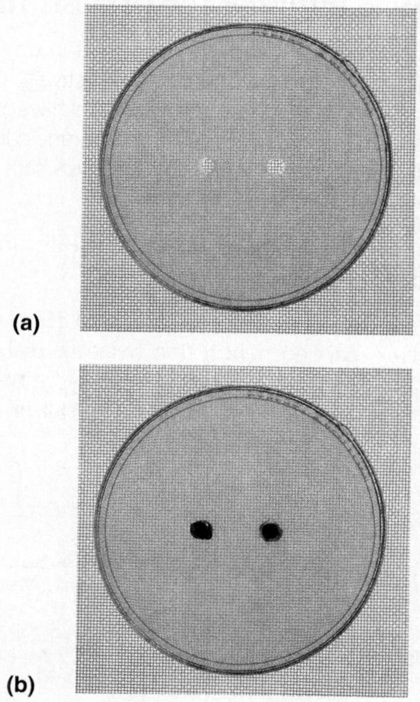

Figure 5.2 Comparing diffusion rates. Agar-plated petri dish as it appears after the placement of 0.1 *M* methylene blue in one well and 0.1 *M* potassium permanganate in another.

Text continues on next page. →

Activity 1: Dye Diffusion Results		
Time (min)	**Diameter of methylene blue (mm)**	**Diameter of potassium permanganate (mm)**
15		
30		
45		
60		

Which dye diffused more rapidly? ______________

What is the relationship between molecular weight and rate of molecular movement (diffusion)?

____

Why did the dye molecules move? ______________

____

Activity 2

Observing Diffusion of Dye Through Water

1. Go to the diffusion demonstration area, and observe the cylinder containing dye crystals and water set up at the beginning of the lab.

2. Measure the number of millimeters the dye has diffused from the bottom of the graduated cylinder, and record.

______________ mm

3. Record the time the demonstration was set up and the time of your observation. Then compute the rate of the dye's diffusion through water and record below.

Time of setup ______________

Time of observation ______________

Rate of diffusion ______________ mm/min

Activity 3

Instructors may assign a related Pre-Lab Video Coaching Activity using Mastering A&P™

Investigating Diffusion and Osmosis Through Nonliving Membranes

The following experiment provides information on the movement of water and solutes through selectively permeable membranes called dialysis sacs. Dialysis sacs have pores of a particular size. The selectivity of living membranes depends on more than just pore size, but using the dialysis sacs will allow you to examine selectivity due to this factor.

1. Read through the experiments in this activity, and develop a hypothesis for each part.

2. Obtain four dialysis sacs, a small funnel, a 25-ml graduated cylinder, a wax marking pencil, fine twine or dialysis tubing clamps, and four beakers (250 ml). Number the beakers 1 to 4 with the wax marking pencil, and half fill all of them with distilled water except beaker 2, to which you should add 125 ml of the 40% glucose solution.

3. Prepare the dialysis sacs one at a time. Using the funnel, half fill each with 20 ml of the specified liquid (see Activity 3 chart). Press out the air, fold over the open end of the sac, and tie it securely with fine twine or clamp it. Before proceeding to the next sac, rinse it under the tap, and quickly and carefully blot the sac dry by rolling it on a paper towel. Weigh it with a laboratory balance. Record the weight in the **Activity 3 chart**, and then drop the sac into the corresponding beaker. Be sure the sac is completely covered by the beaker solution, adding more solution if necessary. **Figure 5.3** illustrates the configuration of the beakers with the contents of the dialysis sacs and the beaker solutions.

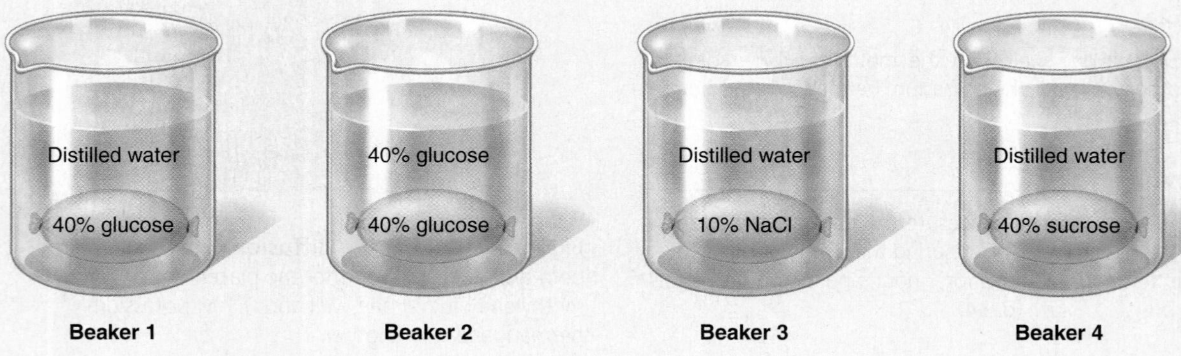

Figure 5.3 Setup for observing diffusion through nonliving membranes.

Activity 3: Experimental Data on Diffusion and Osmosis Through Nonliving Membranes						
Beaker	**Contents of sac**	**Initial weight**	**Final weight**	**Weight change**	**Tests—beaker fluid**	**Tests—sac fluid**
Beaker 1 ½ filled with distilled water	Sac 1, 20 ml of 40% glucose solution				Benedict's test:	Benedict's test:
Beaker 2 ½ filled with 40% glucose solution	Sac 2, 20 ml of 40% glucose solution					
Beaker 3 ½ filled with distilled water	Sac 3, 20 ml of 10% NaCl solution				$AgNO_3$ test:	
Beaker 4 ½ filled with distilled water	Sac 4, 20 ml of 40% sucrose solution				Benedict's test:	

5

- Sac 1: 40% glucose solution
- Sac 2: 40% glucose solution
- Sac 3: 10% NaCl solution
- Sac 4: 40% sucrose solution

Allow the sacs to remain undisturbed in the beakers for 1 hour. Use this time to continue with other experiments.

4. After an hour, boil a beaker of water on the hot plate. Obtain the supplies you will need to determine your experimental results: dropper bottles of Benedict's solution and silver nitrate solution, a test tube rack, four test tubes, and a test tube holder.

5. Quickly and gently blot sac 1 dry and weigh it. (**Note:** Do not squeeze the sac during the blotting process.) Record the weight in the data chart.

Was there any change in weight? ____________

Conclusions: ____________

____

____

Place 5 drops of Benedict's solution in each of two test tubes. Put 4 ml of the beaker fluid into one test tube and 4 ml of the sac fluid into the other. Mark the tubes for identification, and then place them in a beaker containing boiling water. Boil 2 minutes. Cool slowly. If a green, yellow, or rusty red precipitate forms, the test is positive, meaning that glucose is present. If the solution remains the original blue color, the test is negative. Record results in the data chart.

Was glucose still present in the sac? ____________

Was glucose present in the beaker? ____________

Conclusions: ____________

____

____

6. Blot gently and weigh sac 2. Record the weight in the data chart.

Was there an *increase* or *decrease* in weight? ____________

With 40% glucose in the sac and 40% glucose in the beaker, would you expect to see any net movement of water (osmosis) or of glucose molecules (simple diffusion)?

____________ Why or why not? ____________

____

7. Blot gently and weigh sac 3. Record the weight in the data chart.

Was there any change in weight? ____________

Conclusions: ____________

____

Take a 5-ml sample of beaker 3 solution and put it in a clean test tube. Add a drop of silver nitrate ($AgNO_3$). The appearance of a white precipitate or cloudiness indicates the presence of silver chloride ($AgCl$), which is formed by the reaction of $AgNO_3$ with NaCl (sodium chloride). Record results in the data chart.

Text continues on next page. →

5

Results: ___________________________

Conclusions: ___________________________

8. Blot gently and weigh sac 4. Record the weight in the data chart.

Was there any change in weight? ___________________________

Conclusions: ___________________________

____

Take a 1-ml sample of beaker 4 solution and put the test tube in boiling water in a hot water bath. Add 5 drops of Benedict's solution to the tube and boil for 5 minutes. The presence of glucose (one of the hydrolysis products of sucrose) in the test tube is indicated by the presence of a green, yellow, or rusty colored precipitate.

Did sucrose diffuse from the sac into the water in the small beaker? ___________________________

Conclusions: ___________________________

____

9. In which of the test situations did net osmosis occur?

____

In which of the test situations did net simple diffusion occur?

____

What conclusions can you make about the relative size of glucose, sucrose, NaCl, and water molecules?

____

____

____

With what cell structure can the dialysis sac be compared?

____

10. Prepare a lab report for the experiment. (See Getting Started, on MasteringA&P.) Be sure to include in your discussion the answers to the questions proposed in this activity.

Activity 4

Observing Osmometer Results

Before leaving the laboratory, observe the *osmometer demonstration* set up before the laboratory session to follow the movement of water through a membrane (osmosis). Measure the distance the water column has moved during the laboratory period, and record below. (The position of the meniscus [the surface of the water column] in the thistle tube at the beginning of the laboratory period is marked with wax pencil.)

Distance the meniscus has moved: ______________ mm

Did net osmosis occur? Why or why not?

____

____

____

Activity 5

Investigating Diffusion and Osmosis Through Living Membranes

To examine permeability properties of plasma membranes, conduct the following experiments. As you read through the experiments in this activity, develop a hypothesis for each part.

Experiment 1

1. Obtain two deshelled eggs and two 400-ml beakers. Note that the relative concentration of solutes in deshelled eggs is about 14%. Number the beakers 1 and 2 with the wax marking pencil. Half fill beaker 1 with distilled water and half fill beaker 2 with 30% sucrose.

2. Carefully blot each egg by rolling it gently on a paper towel. Place a weigh boat on a laboratory balance and tare the balance (that is, make sure the scale reads 0.0 with the weigh boat on the scale). Weigh egg 1 in the weigh boat, record the initial weight in the **Activity 5 chart**, and gently place it into beaker 1. Repeat for egg 2, placing it in beaker 2.

3. After 20 minutes, remove egg 1 and gently blot it and weigh it. Record the weight, and replace it into beaker 1. Repeat for egg 2, placing it into beaker 2. Repeat this procedure at 40 minutes and 60 minutes.

4. Calculate the change in weight of each egg at each time period, and enter that number in the data chart. Also calculate the percent change in weight for each time period and enter that number in the data chart.

Activity 5: Experiment 1 Data from Diffusion and Osmosis Through Living Membranes						
Time	**Egg 1 (in distilled H_2O)**	**Weight change**	**% Change**	**Egg 2 (in 30% sucrose)**	**Weight change**	**% Change**
Initial weight (g)		—	—		—	—
20 min						
40 min						
60 min						

How has the weight of each egg changed?

Egg 1 ______________________________

Egg 2 ______________________________

Make a graph of your data by plotting the percent change in weight for each egg versus time.

How has the appearance of each egg changed?

Egg 1 ______________________________

Egg 2 ______________________________

A solution surrounding a cell is **hypertonic** if it contains more nonpenetrating solute particles than the interior of the cell. Water moves from the interior of the cell into a surrounding hypertonic solution by osmosis. A solution surrounding a cell is **hypotonic** if it contains fewer nonpenetrating solute particles than the interior of the cell. Water moves from a hypotonic solution into the cell by osmosis. In both cases, water moved down its concentration gradient. Indicate in your conclusions whether distilled water was a hypotonic or hypertonic solution and whether 30% sucrose was hypotonic or hypertonic.

Conclusions: ______________________________

____

____

Experiment 2

Now you will conduct a microscopic study of red blood cells suspended in solutions of varying tonicities. The objective is to determine whether these solutions have any effect on cell shape by promoting net osmosis.

1. The following supplies should be available at your laboratory bench to conduct this experimental series: two clean slides and coverslips, a vial of mammalian blood, a medicine dropper, physiological saline, 5% sodium chloride solution, distilled water, filter paper, and disposable gloves.

! Wear disposable gloves at all times when handling blood (steps 2–5).

2. Place a very small drop of physiological saline on a slide. Using the medicine dropper, add a small drop of the blood to the saline on the slide. Tilt the slide to mix, cover with a coverslip, and immediately examine the preparation under the high-power lens. Notice that the red blood cells retain their normal smooth disclike shape (**Figure 5.4a**, p. 58). This is because the physiological saline is **isotonic** to the cells. That is, it contains a concentration of nonpenetrating solutes (e.g., proteins and some ions) equal to that in the cells (same solute/water concentration). Consequently, the cells neither gain nor lose water by osmosis. Set this slide aside.

3. Prepare another wet mount of the blood, but this time use 5% sodium chloride (saline) solution as the suspending medium. Carefully observe the red blood cells under high power. What is happening to the normally smooth disc shape of the red blood cells?

____

____

____

This crinkling-up process, called **crenation**, is due to the fact that the 5% sodium chloride solution is hypertonic to the cytosol of the red blood cell. Under these circumstances, water leaves the cells by osmosis. Compare your observations to the figure above (Figure 5.4b).

4. Add a drop of distilled water to the edge of the coverslip. Fold a piece of filter paper in half and place its folded edge at the opposite edge of the coverslip; it will absorb the saline solution and draw the distilled water across the cells. Watch the red blood cells as they float across the field. Describe the change in their appearance.

____

____

Distilled water contains *no* solutes (it is 100% water). Distilled water and *very* dilute solutions (that is, those containing less than 0.9% nonpenetrating solutes) are hypotonic to the cell. In a hypotonic solution, the red blood cells first "plump up"

Text continues on next page. →

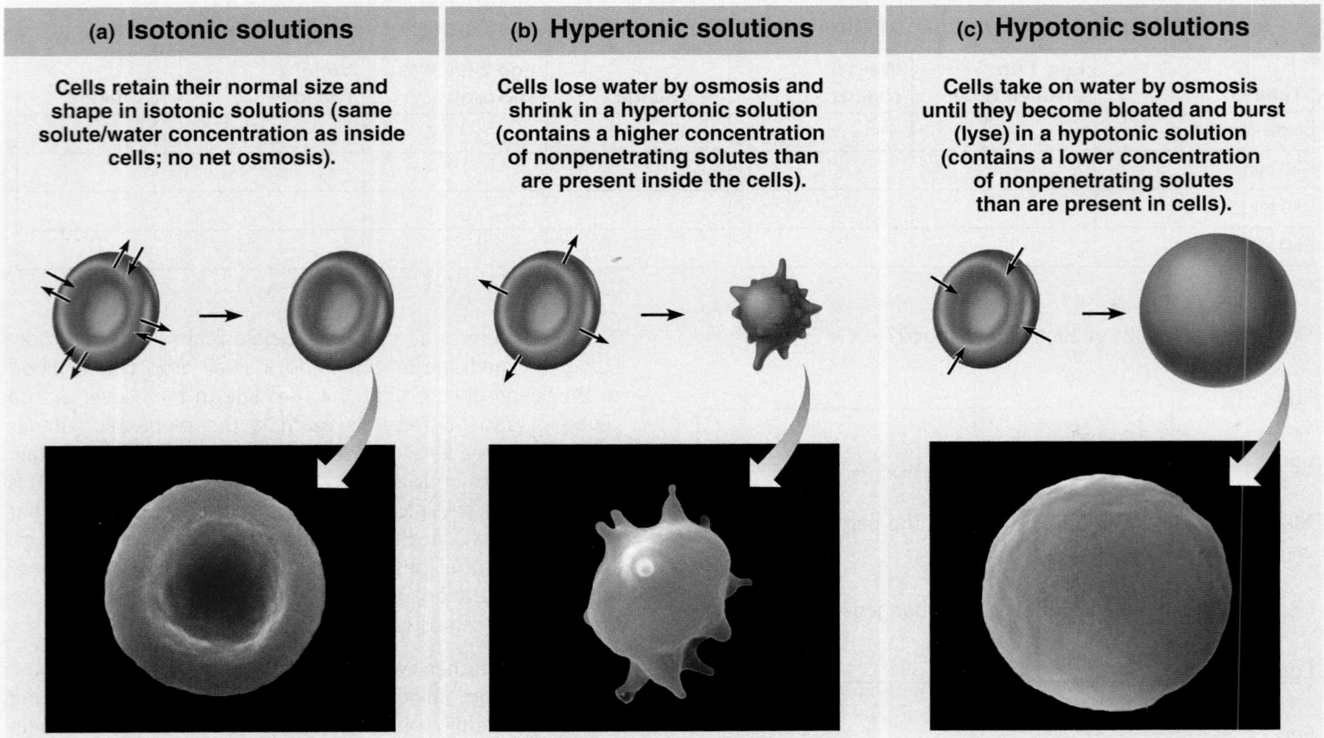

Cells retain their normal size and shape in isotonic solutions (same solute/water concentration as inside cells; no net osmosis).

(b) Hypertonic solutions

Cells lose water by osmosis and shrink in a hypertonic solution (contains a higher concentration of nonpenetrating solutes than are present inside the cells).

(c) Hypotonic solutions

Cells take on water by osmosis until they become bloated and burst (lyse) in a hypotonic solution (contains a lower concentration of nonpenetrating solutes than are present in cells).

Figure 5.4 Influence of isotonic, hypertonic, and hypotonic solutions on red blood cells.

Instructors may assign this figure as an Art Labeling Activity using Mastering A&P™

(Figure 5.4c), but then they suddenly start to disappear. The red blood cells burst as the water floods into them, leaving "ghosts" in their wake—a phenomenon called **hemolysis**.

5. Place the blood-soiled slides and test tube in the bleach-containing basin. Put the coverslips you used into the disposable autoclave bag. Obtain a wash (squirt) bottle containing 10% bleach solution, and squirt the bleach liberally over the bench area where blood was handled. Wipe the bench down with a paper towel wet with the bleach solution, and allow it to dry before continuing. Remove gloves, and discard in the autoclave bag.

6. Prepare a lab report for experiments 1 and 2. (See Getting Started, on MasteringA&P.) Be sure to include in the discussion answers to the questions proposed in this activity.

Filtration

Filtration is a passive process in which water and solutes are forced through a membrane by hydrostatic (fluid) pressure. For example, fluids and solutes filter out of the capillaries in the kidneys and into the kidney tubules because the blood pressure in the capillaries is greater than the fluid pressure in the tubules. Filtration is not selective. The amount of filtrate (fluids and solutes) formed depends almost entirely on the pressure gradient (difference in pressure on the two sides of the membrane) and on the size of the membrane pores.

Activity 6

Observing the Process of Filtration

1. Obtain the following equipment: a ring stand, ring, and ring clamp; a funnel; a piece of filter paper; a beaker; a 10-ml graduated cylinder; a solution containing uncooked starch, powdered charcoal, and copper sulfate; and a dropper bottle of Lugol's iodine. Attach the ring to the ring stand with the clamp.

2. Fold the filter paper in half twice, open it into a cone, and place it in a funnel. Place the funnel in the ring of the ring stand and place a beaker under the funnel. Shake the starch solution, and fill the funnel with it to just below the top of the filter paper. When the steady stream of filtrate changes to countable filtrate drops, count the number of drops formed in 10 seconds and record.

______________ drops

When the funnel is half empty, again count the number of drops formed in 10 seconds, and record the count.

______________ drops

3. After all the fluid has passed through the filter, check the filtrate and paper to see which materials were retained by the paper. If the filtrate is blue, the copper sulfate passed. Check both the paper and filtrate for black particles to see whether the charcoal passed. Finally, using a 10-ml graduated cylinder, put a 2-ml filtrate sample into a test tube. Add several drops of Lugol's iodine. If the sample turns blue/black when iodine is added, starch is present in the filtrate.

Passed: ______________________________

Retained: ______________________________

What does the filter paper represent? ______________

During which counting interval was the filtration rate greatest? ______________________________

Explain: ______________________________

____

____

What characteristic of the three solutes determined whether or not they passed through the filter paper?

____

Active Processes

Whenever a cell uses the bond energy of ATP to move substances across its boundaries, the process is an *active process*. Substances moved by active means are generally unable to pass by diffusion. They may not be lipid soluble; they may be too large to pass through the membrane channels; or they may have to move against rather than with a concentration gradient. There are two types of active processes: *active transport* and *vesicular transport*.

Active Transport

Like carrier-mediated facilitated diffusion, **active transport** requires carrier proteins that combine specifically with the transported substance. Active transport may be primary, driven directly by hydrolysis of ATP, or secondary, driven indirectly by energy stored in ionic gradients. In most cases, the substances move against concentration or electrochemical gradients or both. These substances are insoluble in lipid and too large to pass through membrane channels but are necessary for cell life.

Vesicular Transport

In **vesicular transport**, fluids containing large particles and macromolecules are transported across cellular membranes inside membranous sacs called *vesicles*. Like active transport, vesicular transport moves substances into the cell **(endocytosis)** and out of the cell **(exocytosis)**. Vesicular transport requires energy, usually in the form of ATP, and all forms of vesicular transport involve protein-coated vesicles to some extent.

There are three types of endocytosis: phagocytosis, pinocytosis, and receptor-mediated endocytosis. In **phagocytosis** ("cell eating"), the cell engulfs some relatively large or solid material such as a clump of bacteria, cell debris, or inanimate particles (**Figure 5.5a**, p. 60). When a particle binds to receptors on the cell's surface, cytoplasmic extensions called pseudopods form and flow around the particle. This produces a vesicle called a *phagosome*. In most cases, the phagosome then fuses with a lysosome and its contents are digested. Indigestible contents are ejected from the cell by exocytosis.

In **pinocytosis** ("cell drinking"), also called **fluid-phase endocytosis**, the cell "gulps" a drop of extracellular fluid containing dissolved molecules (Figure 5.5b). Since no receptors are involved, the process is nonspecific. Unlike phagocytosis, pinocytosis is a routine activity of most cells, allowing them a way of sampling the extracellular fluid. It is particularly important in cells that absorb nutrients, such as cells that line the intestines.

The main mechanism for *specific* endocytosis of most macromolecules is **receptor-mediated endocytosis** (Figure 5.5c). The receptors for this process are plasma membrane proteins that bind only certain substances. This exquisitely selective mechanism allows cells to concentrate material that is present only in small amounts in the extracellular fluid. The ingested vesicle may fuse with a lysosome that either digests or releases

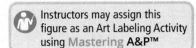

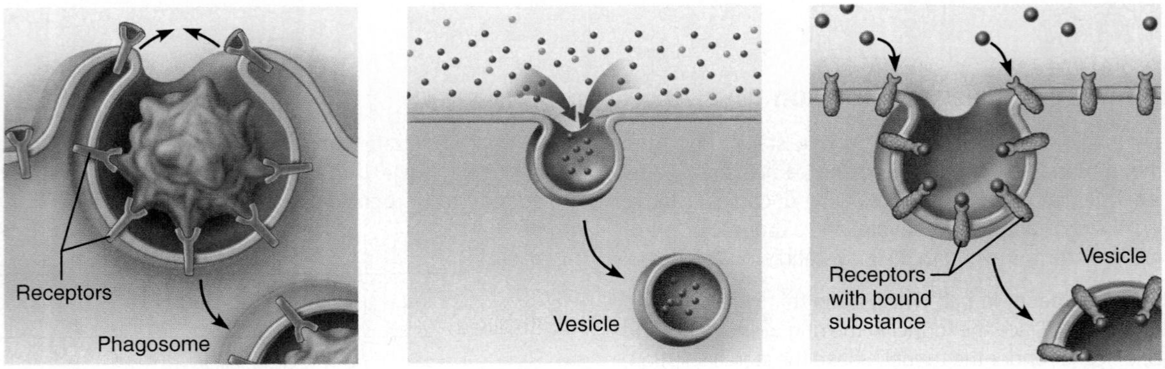

(a) Phagocytosis (b) Pinocytosis (c) Receptor-mediated endocytosis

Figure 5.5 Three types of endocytosis. (a) In phagocytosis, cellular extensions flow around the external particle and enclose it within a phagosome. **(b)** In pinocytosis, fluid and dissolved solutes enter the cell in a tiny vesicle. **(c)** In receptor-mediated endocytosis, specific substances attach to cell-surface receptors and enter the cell in protein-coated vesicles.

Instructors may assign this figure as an Art Labeling Activity using Mastering A&P™

its contents, or it may be transported across the cell to release its contents by exocytosis. The latter case is common in endothelial cells lining blood vessels because it provides a quick means to get substances from blood to extracellular fluid. Substances taken up by receptor-mediated endocytosis include enzymes, insulin and some other hormones, cholesterol (attached to a transport protein), and iron.

Exocytosis is a vesicular transport process that ejects substances from the cell into the extracellular fluid. The substance to be removed from the cell is first enclosed in a protein-coated vesicle called a **secretory vesicle**. In most cases, the vesicle migrates to the plasma membrane, fuses with it, and then ruptures, spilling its contents out of the cell. Exocytosis is used for hormone secretion, neurotransmitter release, mucus secretion, and ejection of wastes.

Activity 7

Observing Phagocytosis

Go to the video viewing area and watch the video demonstration of phagocytosis (if available).

Note: *If you have not already done so, complete Activity 2 (Observing Diffusion of Dye Through Water, p. 54) and Activity 4 (Observing Osmometer Results, p. 56).*

REVIEW SHEET

EXERCISE 5

The Cell: Transport Mechanisms and Permeability

Instructors may assign a portion of the Review Sheet questions using Mastering A&P™

Name ______________________ Lab Time/Date ______________________

Choose all answers that apply to questions 1 and 2, and place their letters on the response blanks to the right.

1. The movement of molecules ______________________

a. reflects the kinetic energy of molecules
b. reflects the potential energy of molecules
c. is ordered and predictable
d. is random and erratic

2. Speed of molecular movement ______________________.

a. is higher in larger molecules
b. is lower in larger molecules
c. increases with increasing temperature
d. decreases with increasing temperature
e. reflects kinetic energy

3. Summarize below the results of Activity 3, Investigating Diffusion and Osmosis Through Nonliving Membranes.

Sac 1: 40% glucose suspended in distilled water

Did glucose diffuse out of the sac? __________ Did the sac weight change? ______________________

Explanation: ______________________

____

Sac 2: 40% glucose suspended in 40% glucose

Was there net movement of glucose into or out of the sac? ______________________

Explanation: ______________________

Did the sac weight change? ______________________

Explanation: ______________________

____

Sac 3: 10% NaCl suspended in distilled water

Was there net movement of NaCl out of the sac? ______________________

Direction of net osmosis: ______________________

Sac 4: 40% sucrose suspended in distilled water

Was there net movement of sucrose out of the sac? ______________________

Explanation: ______________________

Direction of net osmosis: ______________________

4. What single characteristic of the selectively permeable membranes *used in the laboratory* determines the substances that can pass through them? ___

In addition to this characteristic, what other factors influence the passage of substances through living membranes?

___

___

5. A semipermeable sac filled with a solution containing 4% NaCl, 9% glucose, and 10% albumin is suspended in a solution with the following composition: 10% NaCl, 10% glucose, and 40% albumin. The diagram below illustrates the solutes inside and outside of the sac. Assume that the sac is permeable to all substances except albumin. With respect to net movement, state whether each of the following will (a) move into the sac, (b) move out of the sac, or (c) not move.

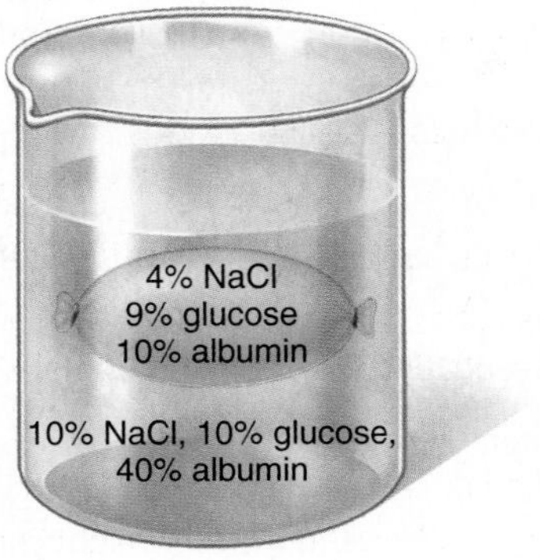

glucose: ___ albumin: ___

water: ___ NaCl: ___

6. Summarize below the results of Activity 5, Experiment 1 (Investigating Diffusion and Osmosis Through Living Membranes—the egg). List and explain your observations.

Egg 1 in distilled water: ___

___

___

Egg 2 in 30% sucrose: ___

___

___

7. The diagrams below represent three microscope fields containing red blood cells. Which field contains a hypertonic solution? ____________ The cells in this field are said to be ____________. Which field contains an isotonic bathing solution? ____________ Which field contains a hypotonic solution? ____________ What is happening to the cells in this field? ______________________________

____

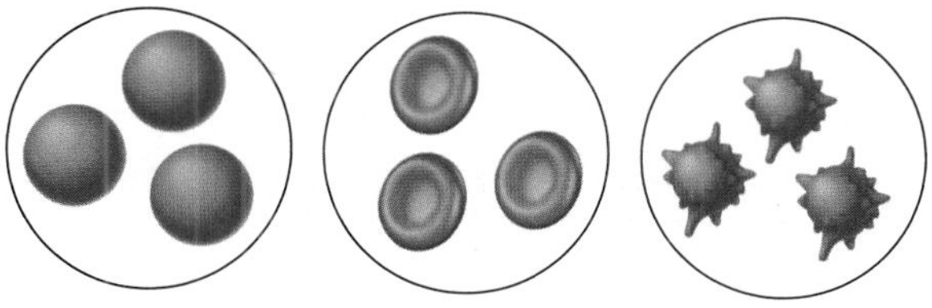

(a) (b) (c)

8. What determines whether a transport process is active or passive?

____

9. Characterize membrane transport as fully as possible by choosing all the phrases that apply and inserting their letters on the answer blanks.

Passive processes: ______________________ Active processes: ______________________

a. account for the movement of fats and respiratory gases through the plasma membrane
b. include phagocytosis and pinocytosis
c. include osmosis, simple diffusion, and filtration
d. occur against concentration and/or electrical gradients
e. use hydrostatic pressure or molecular energy as the driving force

10. For the osmometer demonstration (Activity 4), explain why the level of the water column rose during the laboratory session.

____

____

11. Name one similarity and one difference between simple diffusion and osmosis.

____

____

12. Name one similarity and one difference between simple diffusion and facilitated diffusion.

____

____

13. Name one similarity and one difference between pinocytosis and receptor-mediated endocytosis. ______________________

____

14. Many classroom protocols for extracting DNA from cheek cells instruct students to swish an isotonic sports drink in their mouths as they gently scrape the inside of the mouth with the teeth. Why would it be better to use an isotonic sports drink than plain water? ____

____

15. ➕ Drinking too much plain water in a short period of time can result in water intoxication. As a result, blood plasma will become hypotonic. What effect do you think this would have on cells, and why? ______________________________

____

16. ➕ Receptor-mediated endocytosis is used to remove low-density lipoproteins (LDLs) from circulating in the blood. Explain the effect that defective LDL receptors would have on a patient's cholesterol levels and overall risk for heart disease. (Hint: LDLs are the "bad cholesterol.")

____

____

EXERCISE 6

Classification of Tissues

Learning Outcomes

- ▶ Name the four primary tissue types in the human body, and state a general function of each.
- ▶ Name the major subcategories of the primary tissue types, and identify the tissues of each subcategory microscopically or in an appropriate image.
- ▶ State the locations of the various tissues in the body.
- ▶ List the general function and structural characteristics of each of the tissues studied.

Pre-Lab Quiz

Instructors may assign these and other Pre-Lab Quiz questions using Mastering A&P™

1. Epithelial tissues can be classified according to cell shape. _________________ epithelial cells are scalelike and flattened.
   - **a.** Columnar
   - **b.** Cuboidal
   - **c.** Squamous
   - **d.** Transitional
2. All connective tissue is derived from an embryonic tissue known as:
   - **a.** cartilage
   - **b.** ground substance
   - **c.** mesenchyme
   - **d.** reticular
3. All the following are examples of connective tissue *except:*
   - **a.** bones
   - **b.** ligaments
   - **c.** neurons
   - **d.** tendons
4. Circle the correct underlined term. Of the two major cell types found in nervous tissue, neurons / neuroglial cells are highly specialized to generate and conduct electrical signals.
5. How many basic types of muscle tissue are there? _________________

The human body is organized into structural levels of organization. The simplest level is the chemical level, where atoms combine to form molecules. Molecules form organelles, the functional units of cells. The cellular level is the functional unit of life. In humans and other multicellular organisms, cells function together to maintain homeostasis in the body.

Groups of cells that are similar in structure and function are called **tissues**. The four primary tissue types—epithelium, connective tissue, nervous tissue, and muscle—have distinctive structures, patterns, and functions. The four primary tissues are further divided into subcategories, as described shortly.

To perform specific body functions, the tissues are organized into **organs** such as the heart, kidneys, and lungs. Most organs contain several representatives of the

Go to Mastering A&P™ > Study Area to improve your performance in A&P Lab.

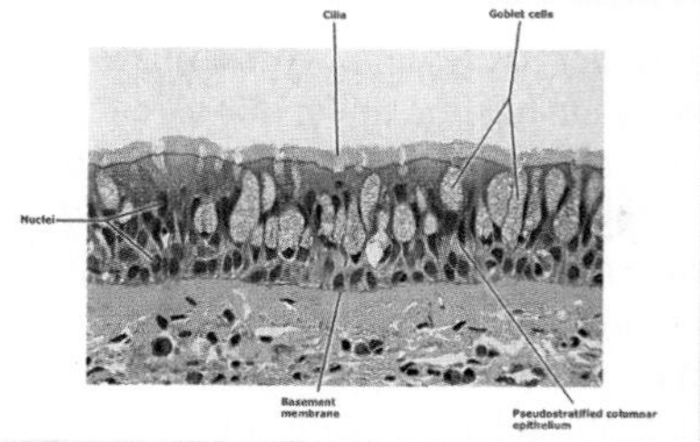

> Lab Tools > Practice Anatomy Laboratory > Histology

Instructors may assign new Building Vocabulary coaching activities, Pre-Lab Quiz questions, Art Labeling activities, Practice Anatomy Lab Practical questions (PAL), and more using the Mastering A&P™ Item Library.

Materials

- ▶ Compound microscope
- ▶ Immersion oil
- ▶ Prepared slides of simple squamous, simple cuboidal, simple columnar, stratified squamous (nonkeratinized), stratified cuboidal, stratified columnar, pseudostratified ciliated columnar, and transitional epithelium
- ▶ Prepared slides of mesenchyme; of adipose, areolar, reticular, and dense (both regular and irregular connective tissues); of hyaline and elastic cartilage; of fibrocartilage; of bone (x.s.); and of blood
- ▶ Prepared slide of nervous tissue (spinal cord smear)
- ▶ Prepared slides of skeletal, cardiac, and smooth muscle (l.s.)

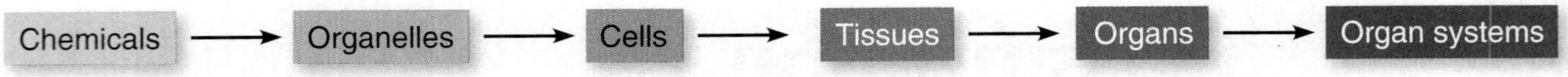

Figure 6.1 Levels of structural organization.

primary tissues, and the arrangement of these tissues determines the organ's structure and function. Thus **histology**, the study of tissues, complements a study of gross anatomy and provides the structural basis for a study of organ physiology.

The next level of organization is the organ system level, where organs work together. **Figure 6.1** summarizes the structural level of organization in the body from the simplest to the most complex.

Epithelial Tissue

Epithelial tissue, or an **epithelium**, is a sheet of cells that covers a body surface or lines a body cavity. It occurs in the body as (1) covering and lining epithelium and (2) glandular epithelium.

Epithelial functions include protection, absorption, filtration, excretion, secretion, and sensory reception. For example, the epithelium covering the body surface protects against bacterial invasion and chemical damage. Epithelium specialized to absorb substances lines the stomach and small intestine. In the kidney tubules, the epithelium absorbs, secretes, and filters. Secretion is a specialty of glandular epithelium.

The following characteristics distinguish epithelial tissues from other types:

- Polarity. The membranes always have one free surface, called the *apical surface*, and typically that surface is significantly different from the *basal surface.*
- Specialized contacts. Cells fit closely together to form membranes, or sheets of cells, and are bound together by specialized junctions.
- Supported by connective tissue. The cells are attached to and supported by an adhesive **basement membrane**, which is an acellular material secreted partly by the epithelial cells (*basal lamina*) and connective tissue cells (*reticular lamina*) that lie next to each other.
- Avascular but innervated. Epithelial tissues are supplied by nerves but have no blood supply of their own (are avascular). Instead they depend on diffusion of nutrients from the underlying connective tissue.
- Regeneration. If well nourished, epithelial cells can easily divide to regenerate the tissue. This is an important characteristic because many epithelia are subjected to a good deal of abrasion.

The covering and lining epithelia are classified according to two criteria—arrangement or relative number of layers and cell shape (**Figure 6.2**). On the basis of arrangement, epithelia are classified as follows:

- **Simple** epithelia consist of one layer of cells attached to the basement membrane.
- **Stratified** epithelia consist of two or more layers of cells.

Based on cell shape, epithelia are classified into three categories:

- **Squamous** (scalelike)
- **Cuboidal** (cubelike)
- **Columnar** (column-shaped)

The terms denoting shape and arrangement of the epithelial cells are combined to describe the epithelium fully. *Stratified epithelia are named according to the cells at the apical surface of the epithelial sheet*, not those resting on the basement membrane.

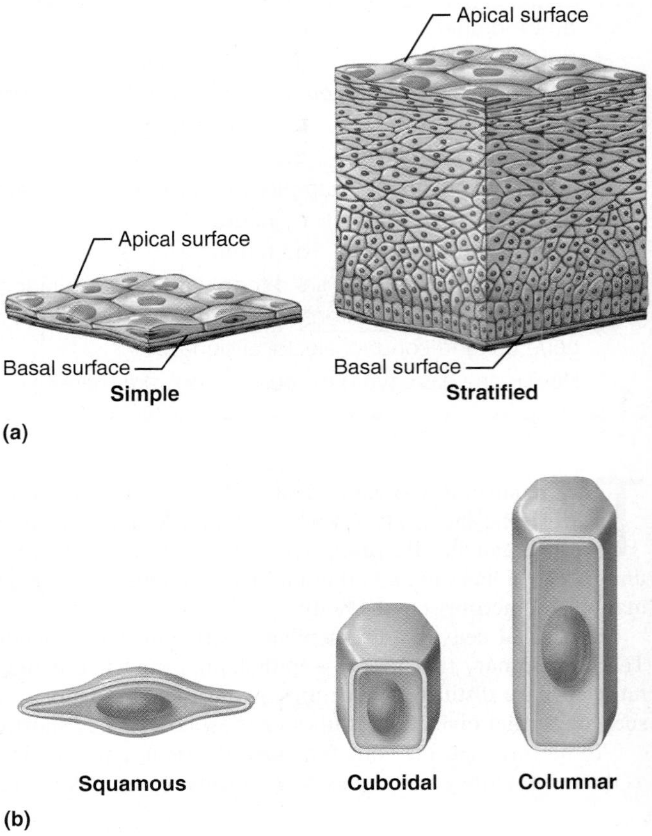

Figure 6.2 Classification of epithelia. (a) Classification based on number of cell layers. **(b)** Classification based on cell shape.

There are, in addition, two less easily categorized types of epithelia.

- **Pseudostratified epithelium** is actually a simple columnar epithelium (one layer of cells), but because its cells vary in height and the nuclei lie at different levels above the basement membrane, it gives the false appearance of being stratified. This epithelium is often ciliated.
- **Transitional epithelium** is a rather peculiar stratified squamous epithelium formed of rounded, or "plump," cells with the ability to slide over one another to allow the organ to be stretched. Transitional epithelium is found only in urinary system organs subjected to stretch, such as the bladder. The superficial cells are flattened (like true squamous cells) when the organ is full and rounded when the organ is empty.

Epithelial cells forming glands are highly specialized to remove materials from the blood and to manufacture them into new materials, which they then secrete. There are two types of glands, *endocrine* and *exocrine*. **Endocrine glands** lose their surface connection (duct) as they develop; thus they are referred to as ductless glands. They secrete hormones into the extracellular fluid, and from there the hormones enter the blood or the lymphatic vessels that weave through the glands. **Exocrine glands** retain their ducts, and their secretions empty through these ducts either to the body surface or into body cavities. The exocrine glands include the sweat and oil glands, liver, and pancreas.

The most common types of epithelia, their characteristic locations in the body, and their functions are described in **Figure 6.3**.

Activity 1

Examining Epithelial Tissue Under the Microscope

Obtain slides of simple squamous, simple cuboidal, simple columnar, stratified squamous (nonkeratinized), pseudostratified ciliated columnar, stratified cuboidal, stratified columnar, and transitional epithelia. Examine each carefully, and notice how the epithelial cells fit closely together to form intact sheets of cells, a necessity for a tissue that forms linings or the coverings of membranes. Scan each epithelial type for modifications for specific functions, such as cilia (motile cell projections that help to move substances along the cell surface), and microvilli, which increase the surface area for absorption. Also be alert for goblet cells, which secrete lubricating mucus. Compare your observations with the descriptions and photomicrographs in Figure 6.3.

While working, check the questions in the Review Sheet at the end of this exercise. A number of the questions there refer to some of the observations you are asked to make during your microscopic study.

(a) Simple squamous epithelium

Description: Single layer of flattened cells with disc-shaped central nuclei and sparse cytoplasm; the simplest of the epithelia.

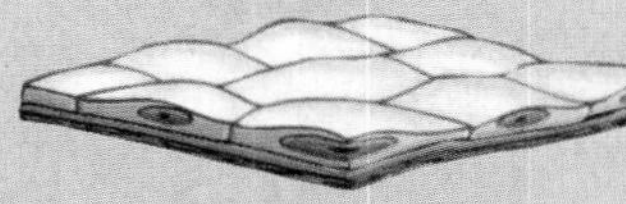

Function: Allows materials to pass by diffusion and filtration in sites where protection is not important; secretes lubricating substances in serosae.

Location: Kidney glomeruli; air sacs of lungs; lining of heart, blood vessels, and lymphatic vessels; lining of ventral body cavity (serosae).

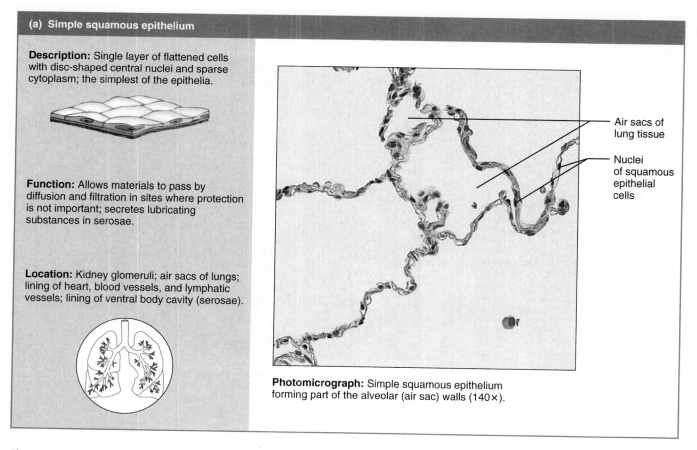

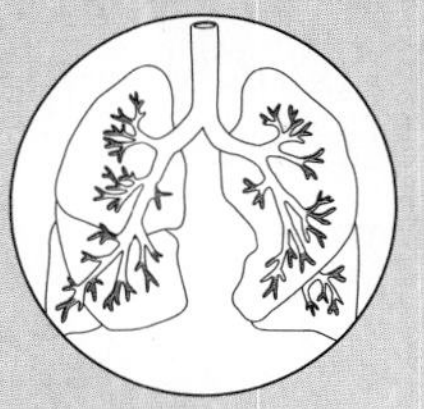

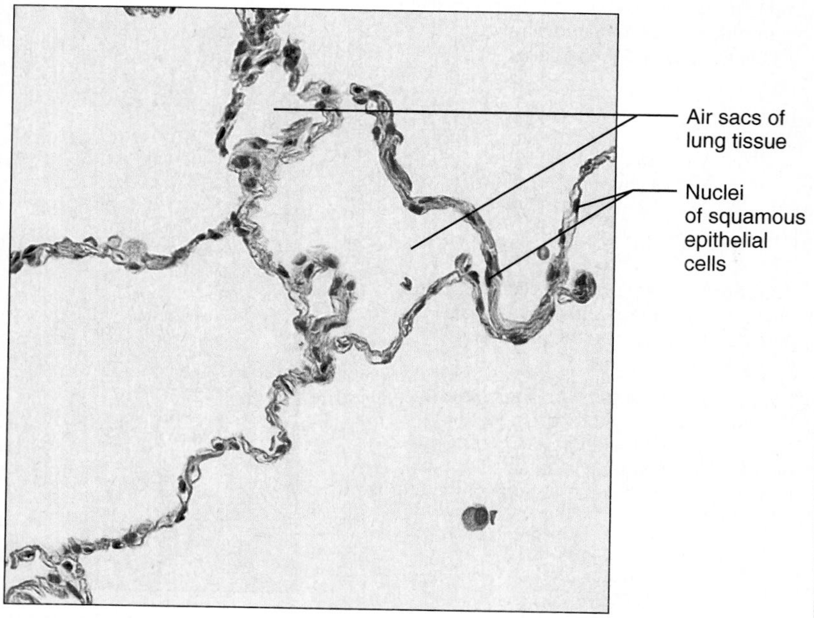

Photomicrograph: Simple squamous epithelium forming part of the alveolar (air sac) walls (140×).

Figure 6.3 Epithelial tissues. Simple epithelia **(a)**.

6

(b) Simple cuboidal epithelium

Description: Single layer of cubelike cells with large, spherical central nuclei.

Function: Secretion and absorption.

Location: Kidney tubules; ducts and secretory portions of small glands; ovary surface.

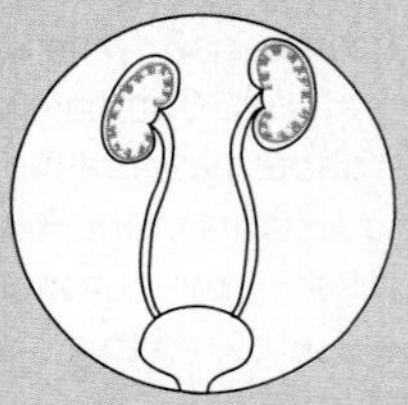

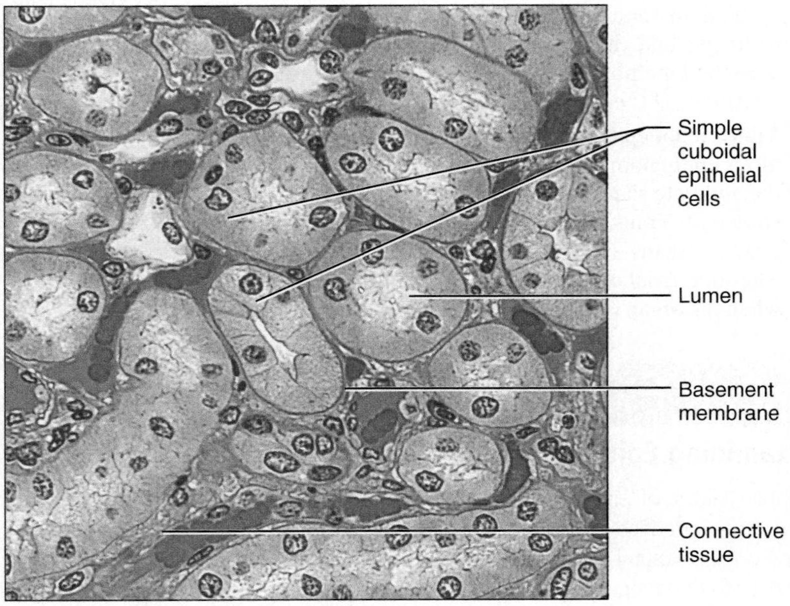

Photomicrograph: Simple cuboidal epithelium in kidney tubules (430×).

(c) Simple columnar epithelium

Description: Single layer of tall cells with *round* to *oval* nuclei; some cells bear cilia; layer may contain mucus-secreting unicellular glands (goblet cells).

Function: Absorption; secretion of mucus, enzymes, and other substances; ciliated type propels mucus (or reproductive cells) by ciliary action.

Location: Nonciliated type lines most of the digestive tract (stomach to rectum), gallbladder, and excretory ducts of some glands; ciliated variety lines small bronchi, uterine tubes, and some regions of the uterus.

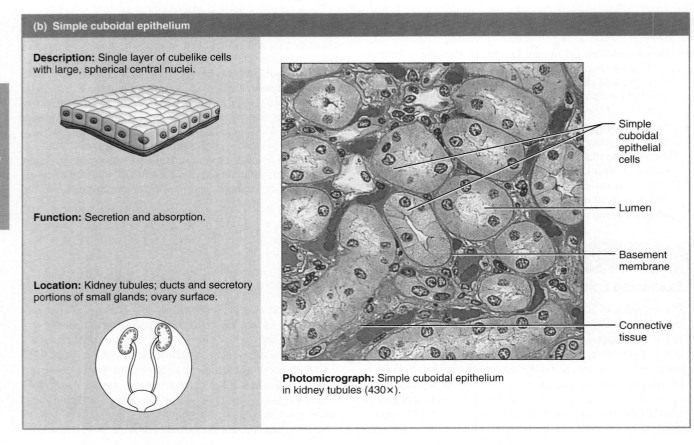

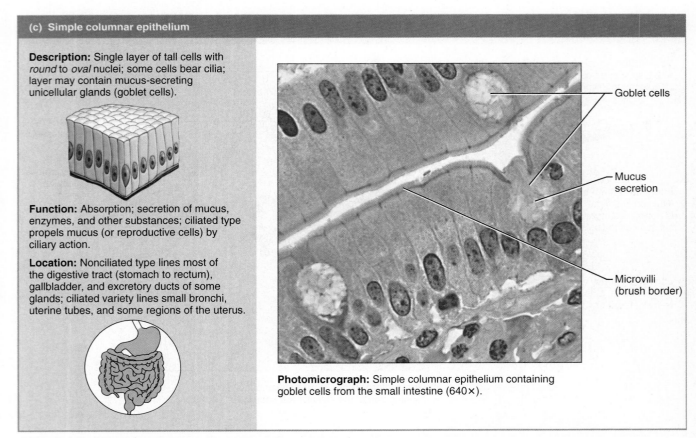

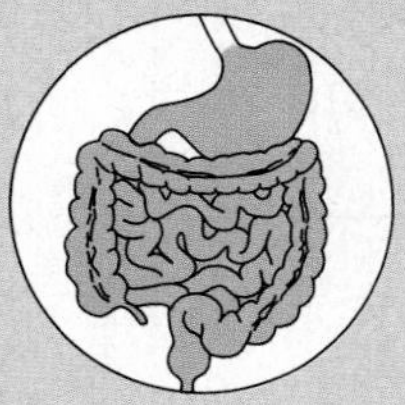

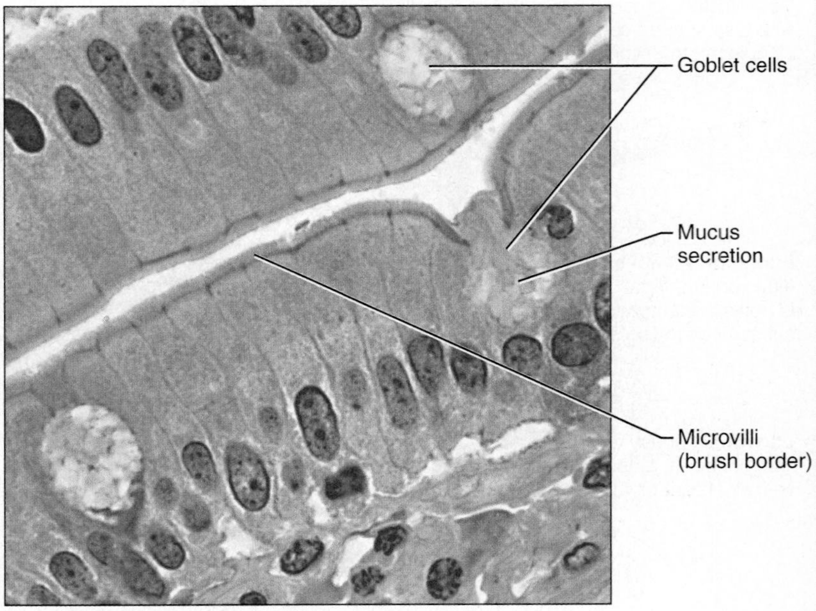

Photomicrograph: Simple columnar epithelium containing goblet cells from the small intestine (640×).

Figure 6.3 *(continued)* **Epithelial tissues.** Simple epithelia **(b)** and **(c)**.

(d) Pseudostratified columnar epithelium

Description: Single layer of cells of differing heights, some not reaching the free surface; nuclei seen at different levels; may contain mucus-secreting goblet cells and bear cilia.

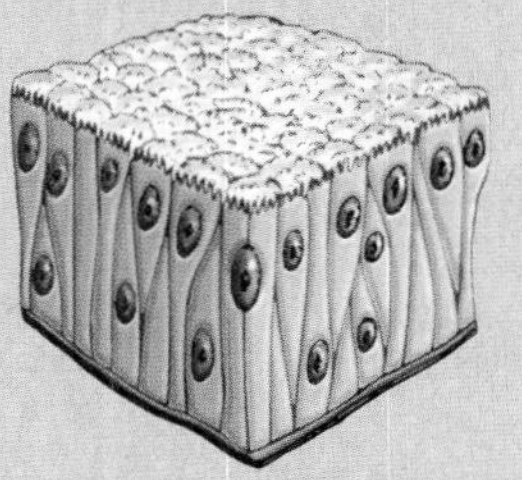

Function: Secretes substances, particularly mucus; propulsion of mucus by ciliary action.

Location: Nonciliated type in male's sperm-carrying ducts and ducts of large glands; ciliated variety lines the trachea, most of the upper respiratory tract.

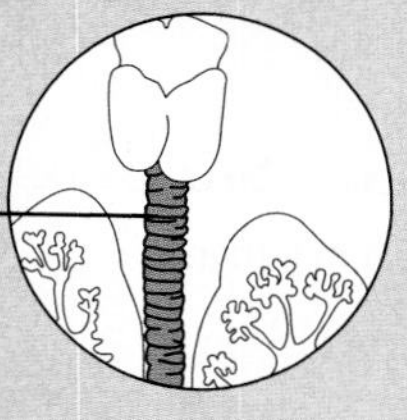

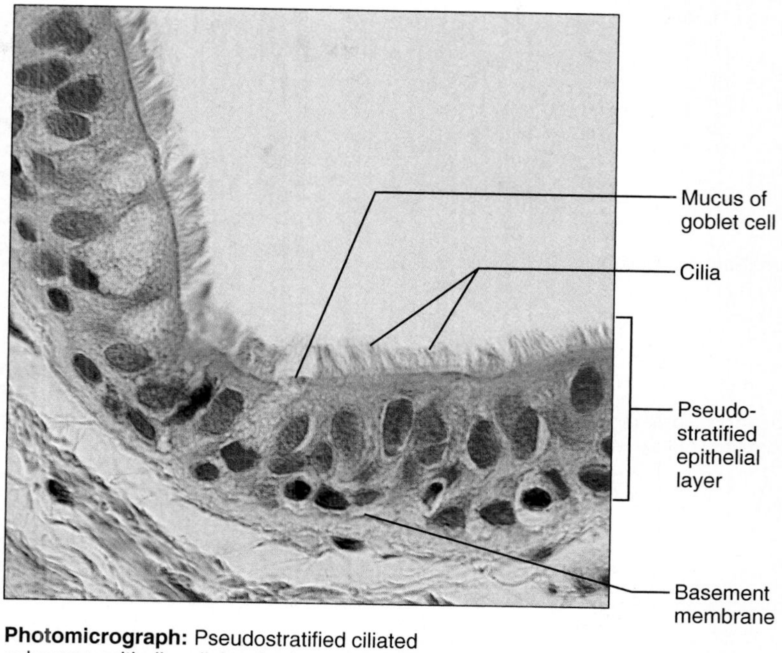

Photomicrograph: Pseudostratified ciliated columnar epithelium lining the human trachea (530×).

6

(e) Stratified squamous epithelium

Description: Thick membrane composed of several cell layers; basal cells are cuboidal or columnar and metabolically active; surface cells are flattened (squamous); in the keratinized type, the surface cells are full of keratin and dead; basal cells are active in mitosis and produce the cells of the more superficial layers.

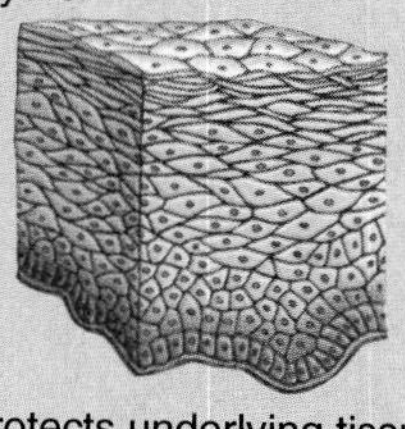

Function: Protects underlying tissues in areas subjected to abrasion.

Location: Nonkeratinized type forms the moist linings of the esophagus, mouth, and vagina; keratinized variety forms the epidermis of the skin, a dry membrane.

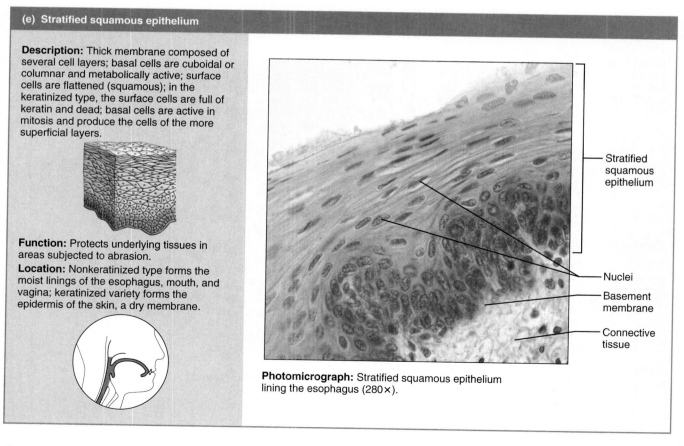

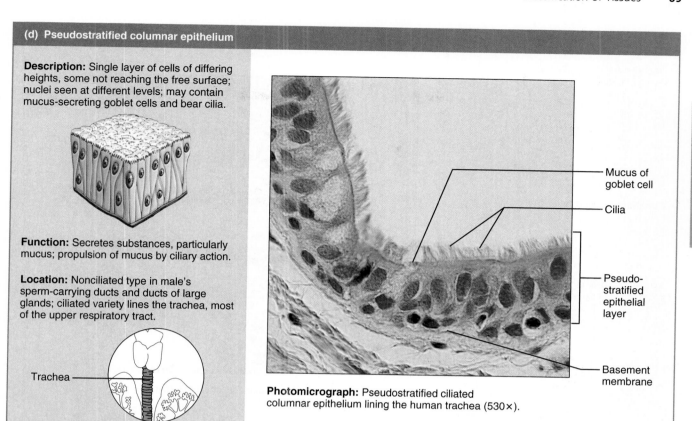

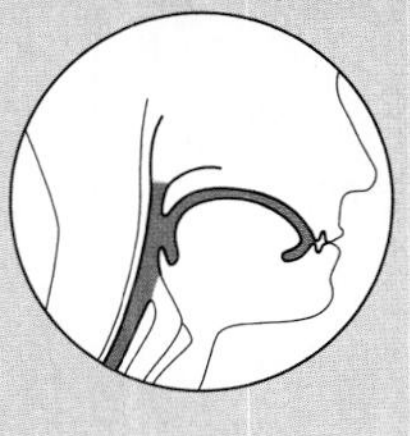

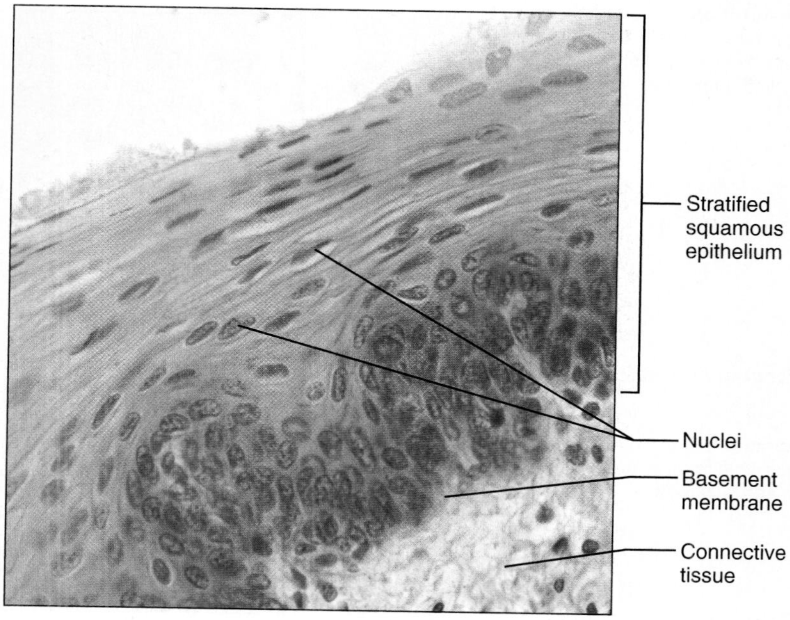

Photomicrograph: Stratified squamous epithelium lining the esophagus (280×).

Figure 6.3 *(continued)* Stratified epithelia **(d)** and **(e)**.

➤

(f) Stratified cuboidal epithelium

Description: Generally two layers of cubelike cells.

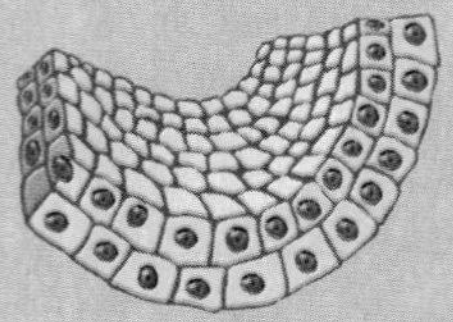

Function: Protection.

Location: Largest ducts of sweat glands, mammary glands, and salivary glands.

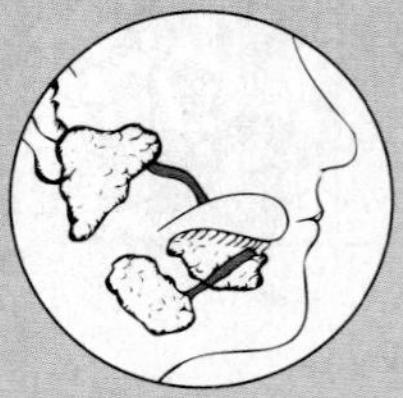

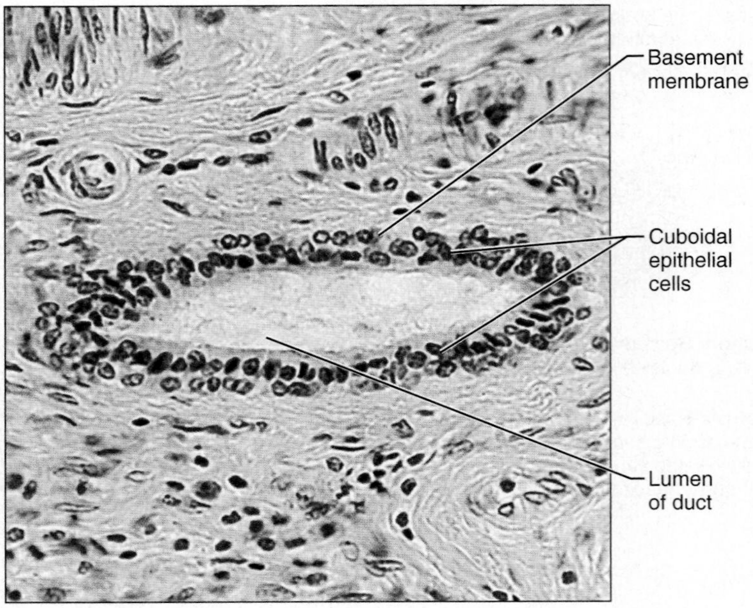

Photomicrograph: Stratified cuboidal epithelium forming a salivary gland duct (290×).

(g) Stratified columnar epithelium

Description: Several cell layers; basal cells usually cuboidal; superficial cells elongated and columnar.

Function: Protection; secretion.

Location: Rare in the body; small amounts in male urethra and in large ducts of some glands.

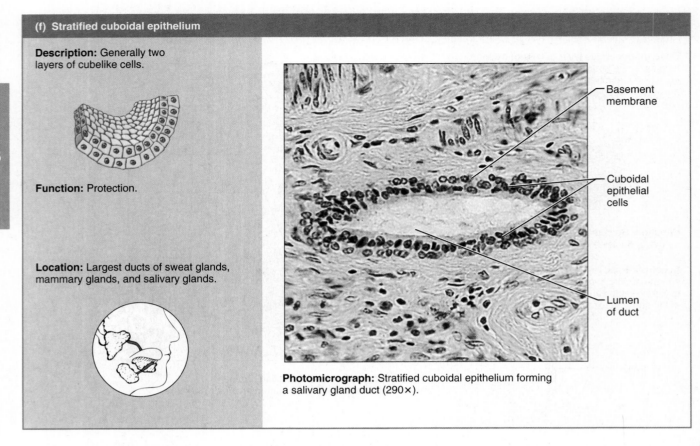

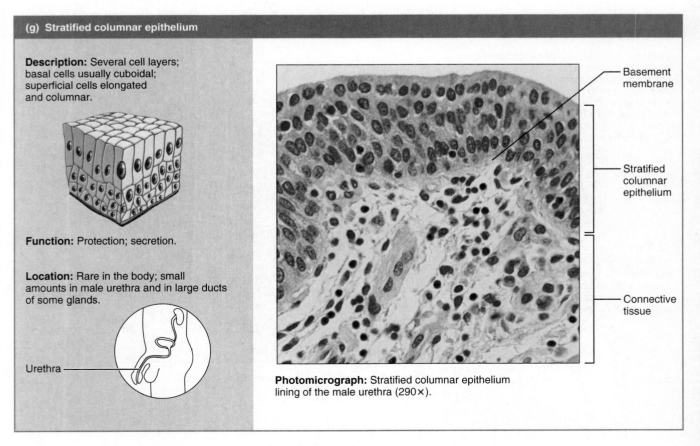

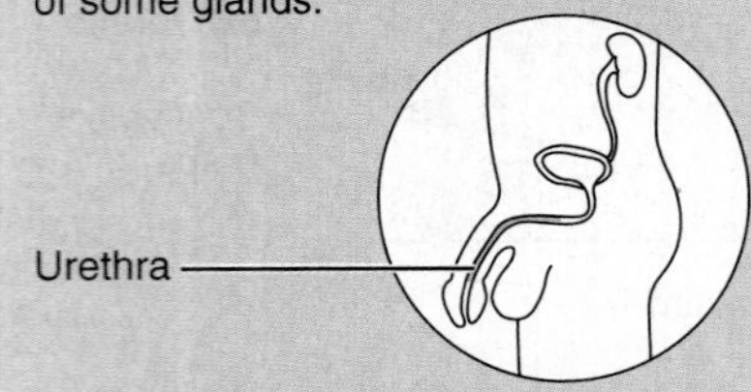

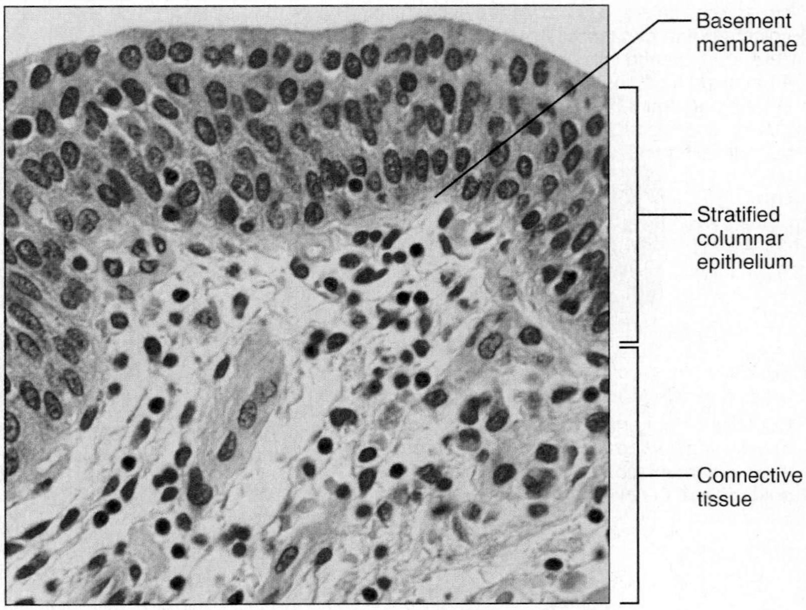

Photomicrograph: Stratified columnar epithelium lining of the male urethra (290×).

Figure 6.3 *(continued)* **Epithelial tissues.** Stratified epithelia **(f)** and **(g)**.

(h) Transitional epithelium

Description: Resembles both stratified squamous and stratified cuboidal; basal cells cuboidal or columnar; surface cells dome shaped or squamouslike, depending on degree of organ stretch.

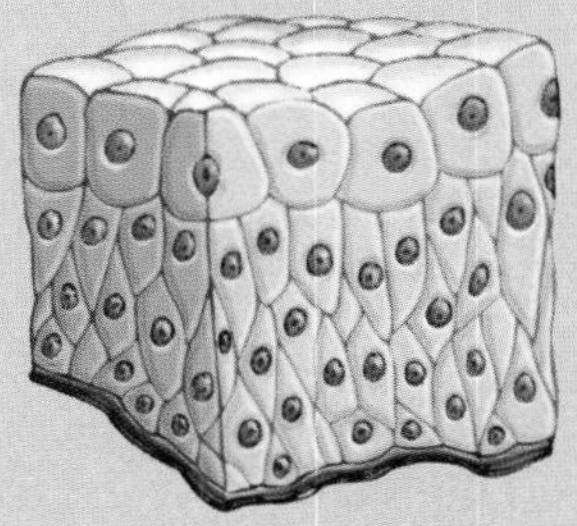

Function: Stretches readily and permits distension of urinary organ by contained urine.

Location: Lines the ureters, urinary bladder, and part of the urethra.

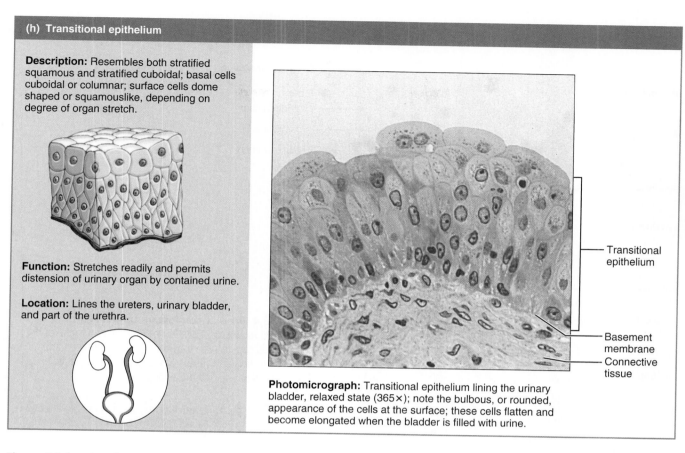

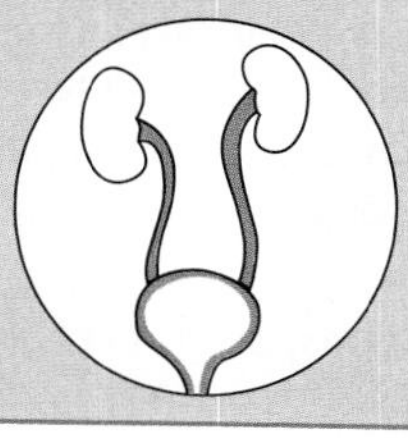

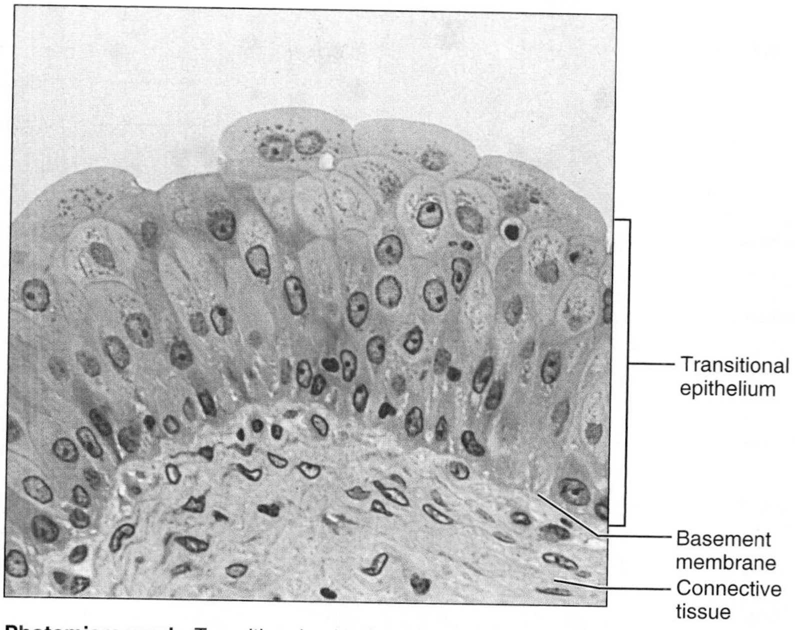

Photomicrograph: Transitional epithelium lining the urinary bladder, relaxed state (365×); note the bulbous, or rounded, appearance of the cells at the surface; these cells flatten and become elongated when the bladder is filled with urine.

Figure 6.3 *(continued)* Stratified epithelia **(h)**.

Connective Tissue

Connective tissue is found in all parts of the body as discrete structures or as part of various body organs. It is the most abundant and widely distributed of the tissue types.

There are four main types of adult connective tissue. These are **connective tissue proper**, **cartilage**, **bone**, and **blood**. Connective tissue proper has two subclasses: **loose connective tissues** (areolar, adipose, and reticular) and **dense connective tissues** (dense regular, dense irregular, and elastic). **Connective tissues** perform a variety of functions, but they primarily protect, support, insulate, and bind together other tissues of the body. For example, bones are composed of connective tissue (**bone**, or **osseous tissue**), and they protect and support other body tissues and organs. The ligaments and tendons **(dense regular connective tissue)** bind the bones together or connect skeletal muscles to bones.

Areolar connective tissue (Figure 6.4, p. 72) is a soft packaging material that cushions and protects body organs. **Adipose** (fat) tissue provides insulation for the body tissues and a source of stored energy.

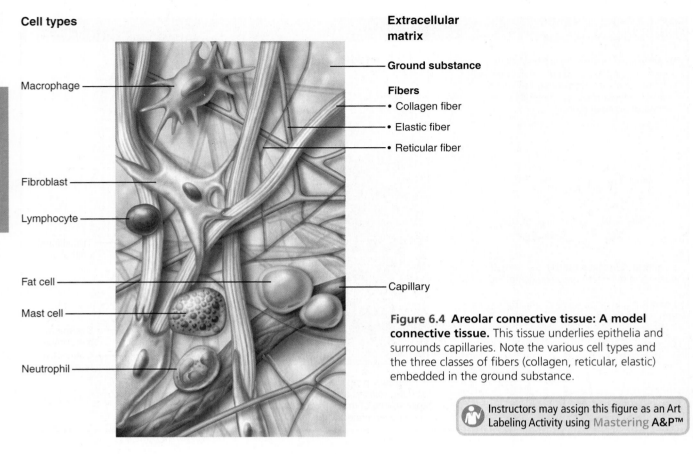

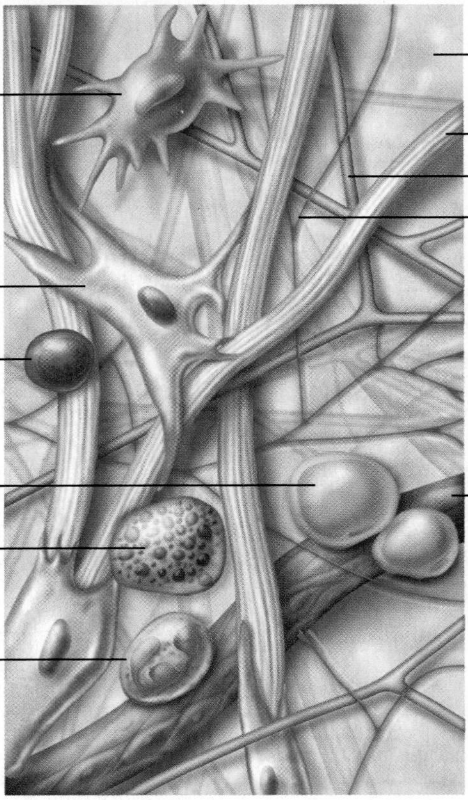

Figure 6.4 Areolar connective tissue: A model connective tissue. This tissue underlies epithelia and surrounds capillaries. Note the various cell types and the three classes of fibers (collagen, reticular, elastic) embedded in the ground substance.

Instructors may assign this figure as an Art Labeling Activity using Mastering A&P™

The characteristics of connective tissue include the following:

- **Common origin.** All connective tissues are derived from embryonic tissue (*mesenchyme*).
- **Degrees of vascularity.** Many types of connective tissue have a rich blood supply. Exceptions include cartilage, which is avascular, and dense connective tissue, which is poorly vascularized.
- **Extracellular matrix.** There is a great deal of noncellular, nonliving material (matrix) between the cells of connective tissue. The composition and amount of matrix vary for connective tissues.

The extracellular matrix has two components—ground substance and fibers. The **ground substance** is composed chiefly of interstitial fluid, cell adhesion proteins, and proteoglycans. Depending on its specific composition, the ground substance may be liquid, semisolid, gel-like, or very hard. When the matrix is firm, as in cartilage and bone, the connective tissue cells reside in cavities in the matrix called *lacunae.* The fibers, which provide support, include **collagen** (white) **fibers**, **elastic** (yellow) **fibers**, and **reticular** (fine collagen) **fibers**. Of these, the collagen fibers are most abundant.

The connective tissues have a common structural plan seen best in *areolar connective tissue* (Figure 6.4). Since all other connective tissues are variations of areolar, it is considered the model, or prototype, of the connective tissues. Notice that areolar tissue has all three varieties of fibers, but they are sparsely arranged in its transparent gel-like ground substance (Figure 6.4). The cell type that secretes its matrix is the *fibroblast*, but a wide variety of other cells (including phagocytic cells, such as macrophages, and certain white blood cells and mast cells that act in the inflammatory response) are present as well. The more durable connective tissues, such as bone, cartilage, and the dense connective tissues, characteristically have a firm ground substance and many more fibers.

Figure 6.5 lists the general characteristics, location, and function of some of the connective tissues found in the body.

Activity 2

Examining Connective Tissue Under the Microscope

Obtain prepared slides of mesenchyme; of adipose, areolar, reticular, dense regular, elastic, and dense irregular connective tissue; of hyaline and elastic cartilage and fibrocartilage; of osseous connective tissue (bone); and of blood. Compare your observations with the views illustrated in Figure 6.5.

Distinguish the living cells from the matrix. Pay particular attention to the denseness and arrangement of the matrix. For example, notice how the matrix of the dense regular and dense irregular connective tissues, respectively making up tendons and the dermis of the skin, is packed with collagen fibers. Note also that in the *regular* variety (tendon), the fibers are all running in the same direction, whereas in the dermis they appear to be running in many directions.

While examining the areolar connective tissue, notice how much empty space there appears to be (*areol* = small empty space), and distinguish the collagen fibers from the coiled elastic fibers. Identify the starlike fibroblasts. Also, try to locate a **mast cell**, which has large, darkly staining granules in its cytoplasm (*mast* = stuffed full of granules). This cell type releases histamine, which makes capillaries more permeable during inflammation and allergies.

In adipose tissue, locate a "signet ring" cell, a fat cell in which the nucleus can be seen pushed to one side by the large, fat-filled vacuole that appears to be a large empty space. Also notice how little matrix there is in adipose (fat) tissue. Distinguish the living cells from the matrix in the dense connective tissue, bone, and hyaline cartilage preparations.

Scan the blood slide at low and then high power to examine the general shape of the red blood cells. Then, switch to the oil immersion lens for a closer look at the various types of white blood cells. How does the matrix of blood differ from all other connective tissues?

___

___

(a) Embryonic connective tissue: mesenchyme

Description: Embryonic connective tissue; gel-like ground substance containing fibers; star-shaped mesenchymal cells.

Function: Gives rise to all other connective tissue types.

Location: Primarily in embryo.

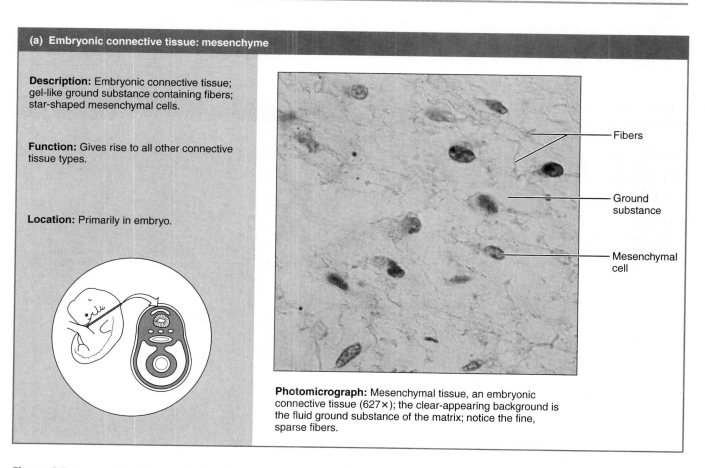

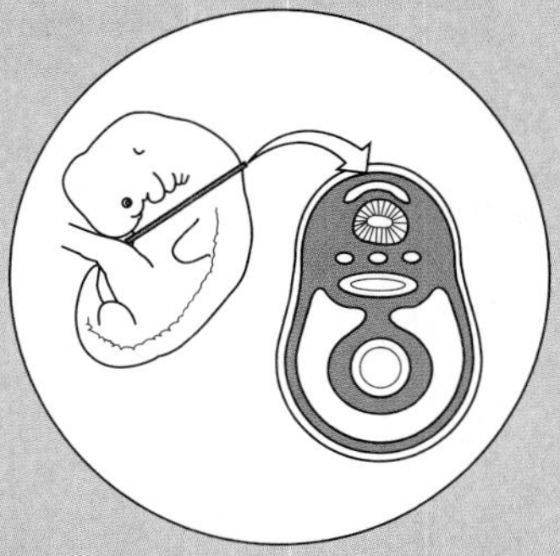

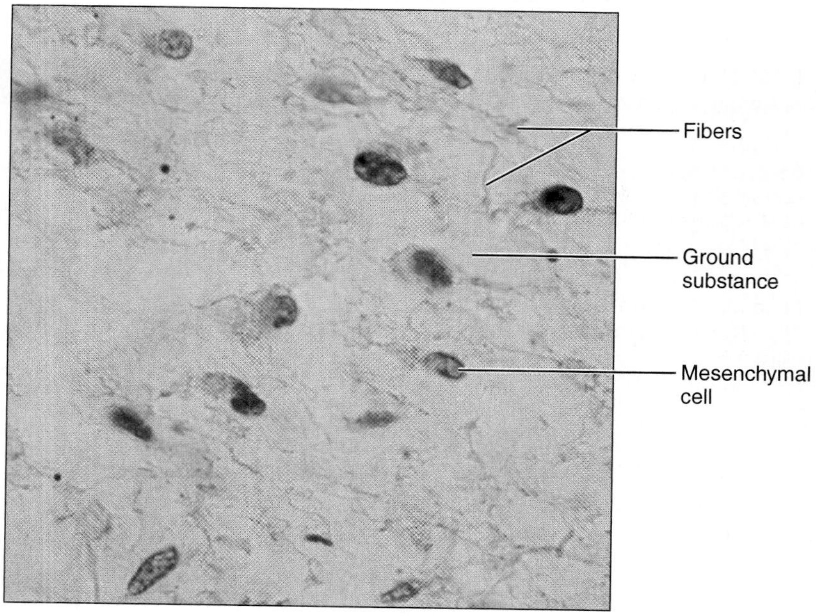

Photomicrograph: Mesenchymal tissue, an embryonic connective tissue (627×); the clear-appearing background is the fluid ground substance of the matrix; notice the fine, sparse fibers.

Figure 6.5 Connective tissues. Embryonic connective tissue **(a)**.

➤

(b) Connective tissue proper: loose connective tissue, areolar

Description: Gel-like matrix with all three fiber types; cells: fibroblasts, macrophages, mast cells, and some white blood cells.

Function: Wraps and cushions organs; its macrophages phagocytize bacteria; plays important role in inflammation; holds and conveys tissue fluid.

Location: Widely distributed under epithelia of body, e.g., forms lamina propria of mucous membranes; packages organs; surrounds capillaries.

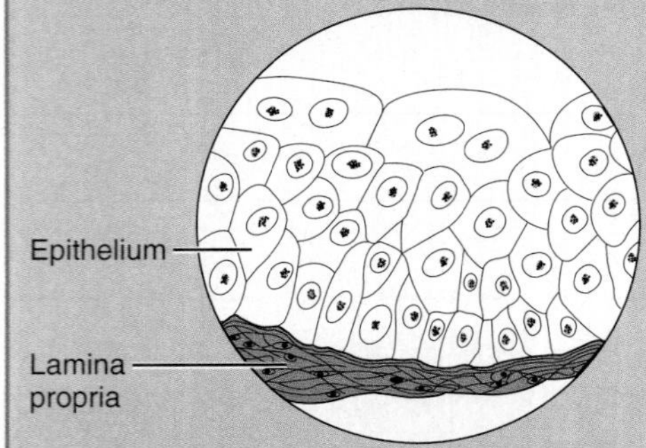

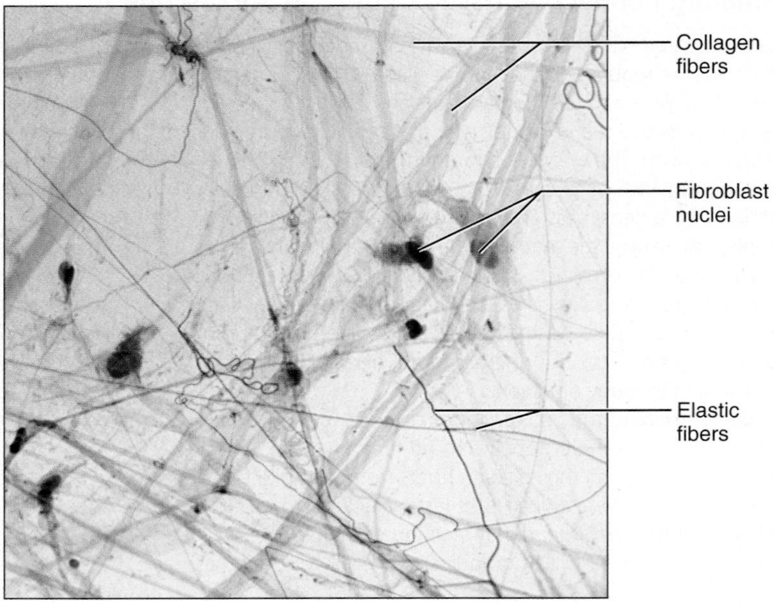

Photomicrograph: Areolar connective tissue, a soft packaging tissue of the body (365×).

(c) Connective tissue proper: loose connective tissue, adipose

Description: Matrix as in areolar, but very sparse; closely packed adipocytes, or fat cells, have nucleus pushed to the side by large fat droplet.

Function: Provides reserve fuel; insulates against heat loss; supports and protects organs.

Location: Under skin; around kidneys and eyeballs; within abdomen; in breasts.

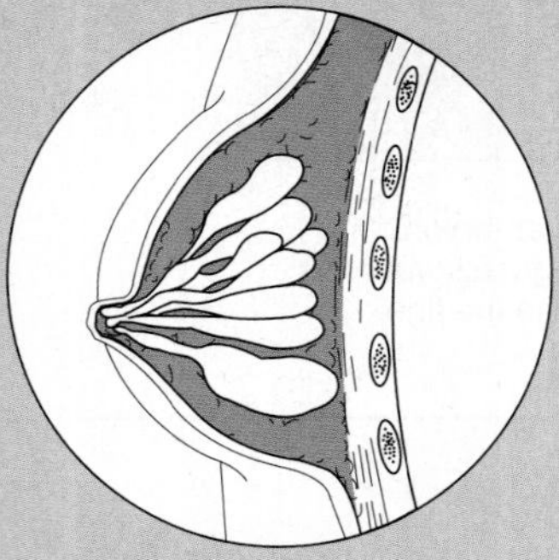

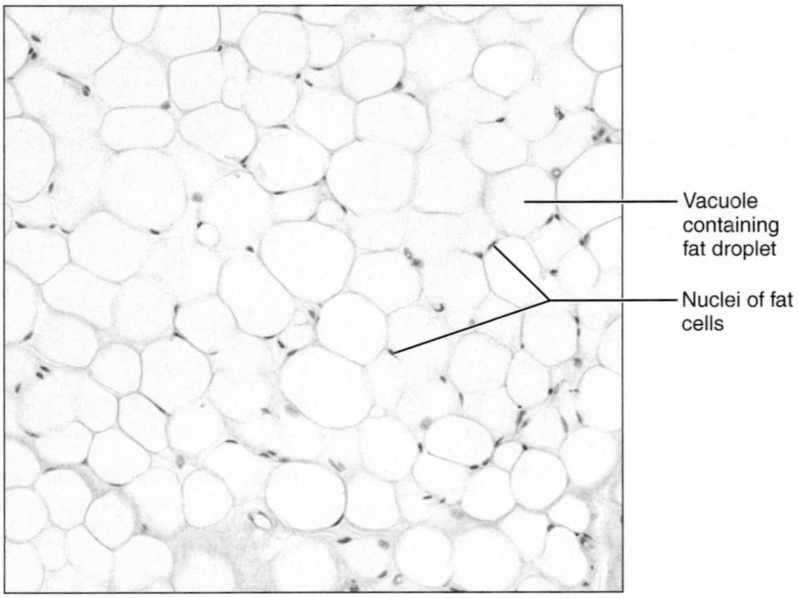

Photomicrograph: Adipose tissue from the subcutaneous layer under the skin (110×).

Figure 6.5 *(continued)* **Connective tissues.** Connective tissue proper **(b)** and **(c)**.

(d) Connective tissue proper: loose connective tissue, reticular

Description: Network of reticular fibers in a typical loose ground substance; reticular cells lie on the network.

Function: Fibers form a soft internal skeleton (stroma) that supports other cell types, including white blood cells, mast cells, and macrophages.

Location: Lymphoid organs (lymph nodes, bone marrow, and spleen).

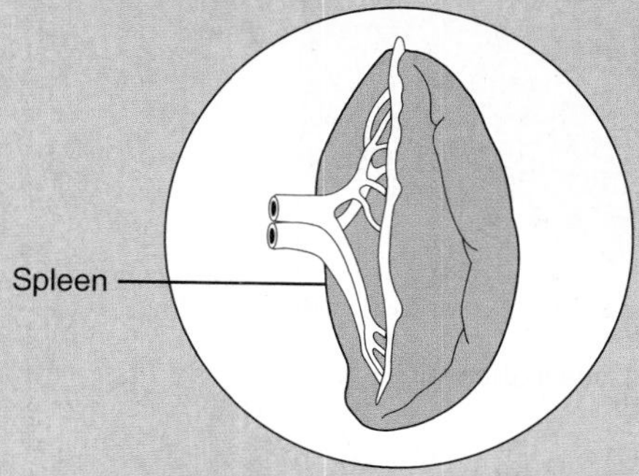

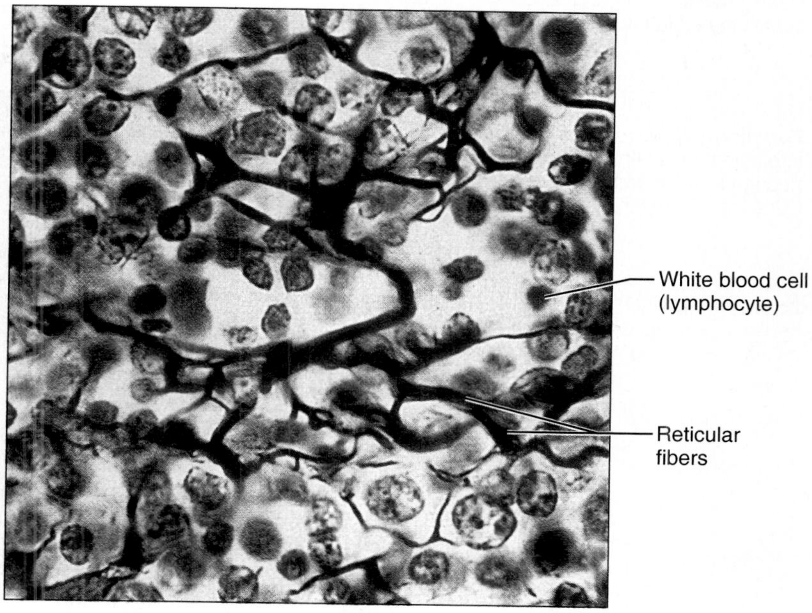

Photomicrograph: Dark-staining network of reticular connective tissue fibers forming the internal skeleton of the spleen (350×).

(e) Connective tissue proper: dense regular connective tissue

Description: Primarily parallel collagen fibers; a few elastic fibers; major cell type is the fibroblast.

Function: Attaches muscles to bones or to other muscles; attaches bones to bones; withstands great tensile stress when pulling force is applied in one direction.

Location: Tendons, most ligaments, aponeuroses.

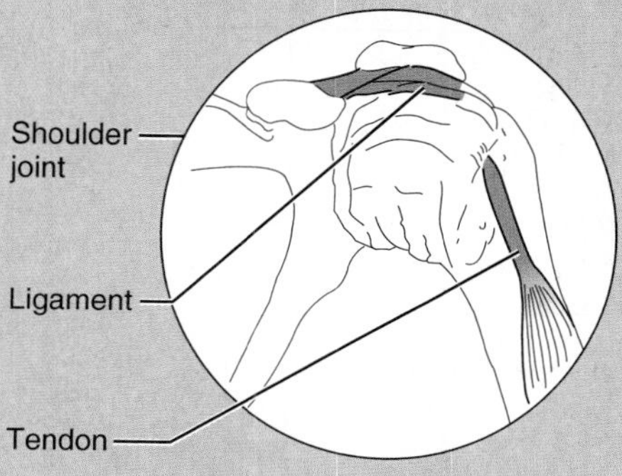

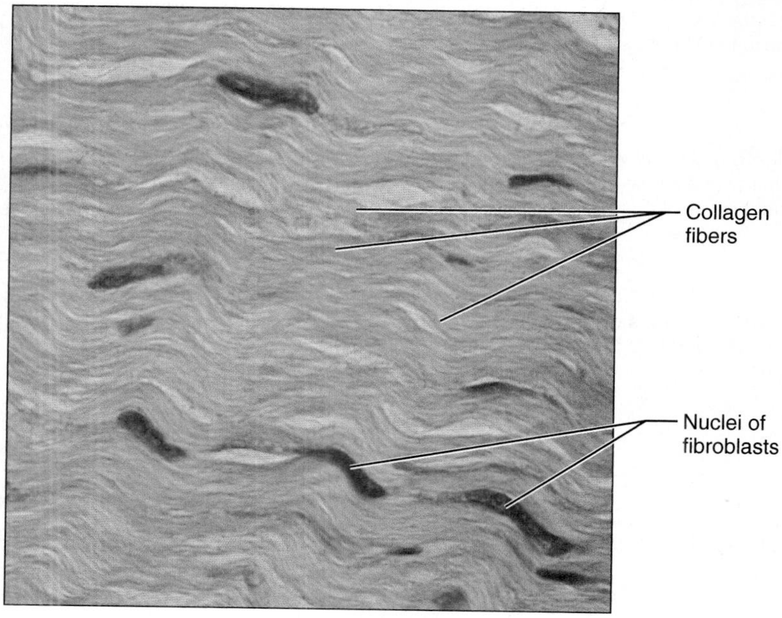

Photomicrograph: Dense regular connective tissue from a tendon (590×).

Figure 6.5 *(continued)* Connective tissue proper **(d)** and **(e)**.

➤

(f) Connective tissue proper: elastic connective tissue

Description: Dense regular connective tissue containing a high proportion of elastic fibers.

Function: Allows recoil of tissue following stretching; maintains pulsatile flow of blood through arteries; aids passive recoil of lungs following inspiration.

Location: Walls of large arteries; within certain ligaments associated with the vertebral column; within the walls of the bronchial tubes.

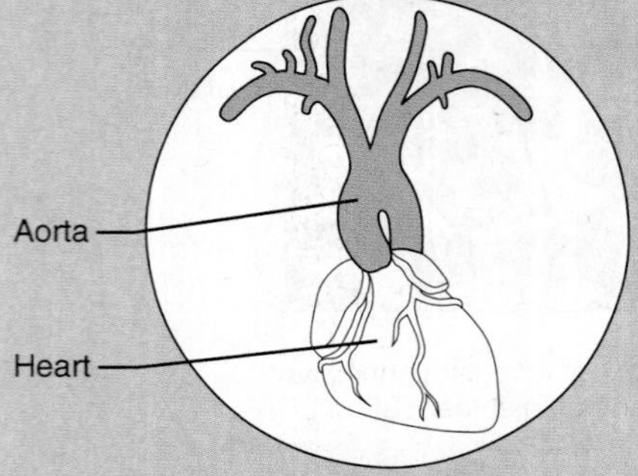

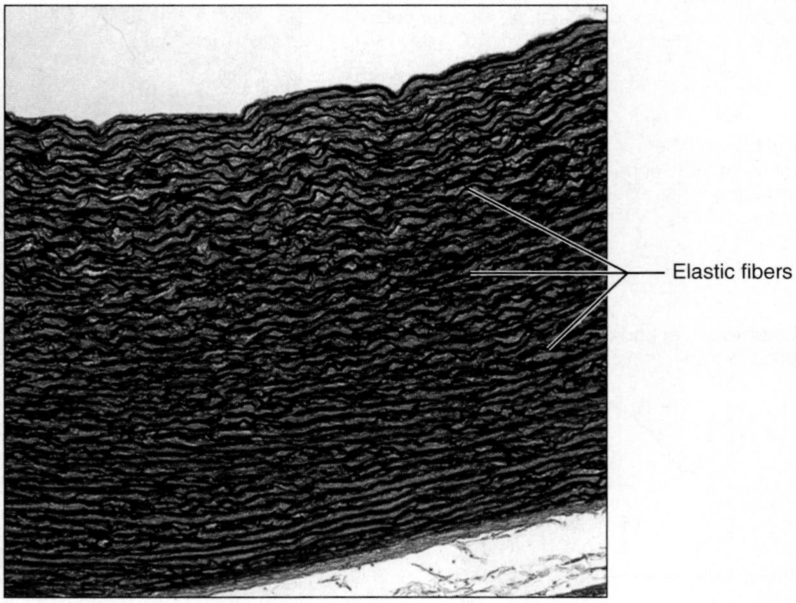

Photomicrograph: Elastic connective tissue in the wall of the aorta (250×).

(g) Connective tissue proper: dense irregular connective tissue

Description: Primarily irregularly arranged collagen fibers; some elastic fibers; major cell type is the fibroblast.

Function: Able to withstand tension exerted in many directions; provides structural strength.

Location: Fibrous capsules of organs and of joints; dermis of the skin; submucosa of digestive tract.

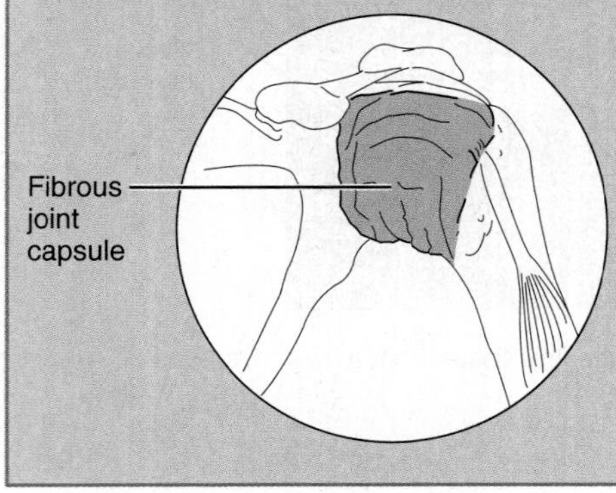

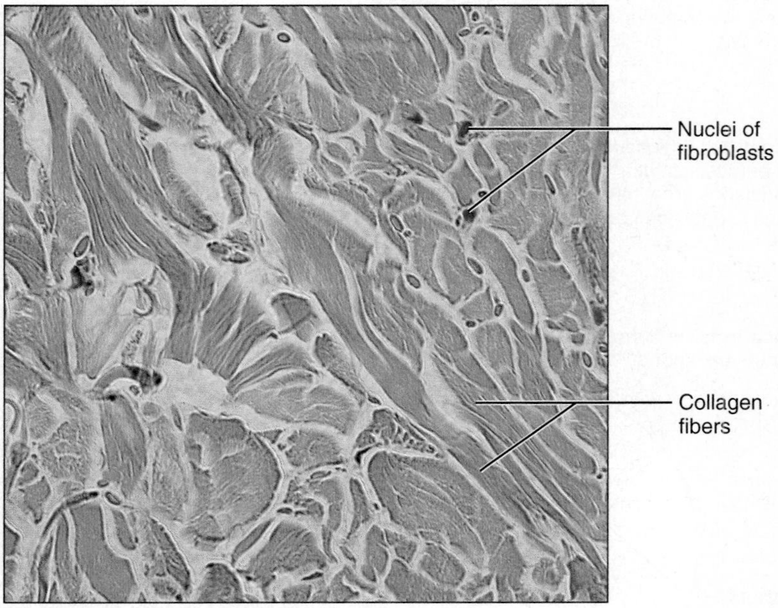

Photomicrograph: Dense irregular connective tissue from the dermis of the skin (210×).

Figure 6.5 *(continued)* **Connective tissues.** Connective tissue proper **(f)** and **(g)**.

(h) Cartilage: hyaline

Description: Amorphous but firm matrix; collagen fibers form an imperceptible network; chondroblasts produce the matrix and, when mature (chondrocytes), lie in lacunae.

Function: Supports and reinforces; serves as resilient cushion; resists compressive stress.

Location: Forms most of the embryonic skeleton; covers the ends of long bones in joint cavities; forms costal cartilages of the ribs; cartilages of the nose, trachea, and larynx.

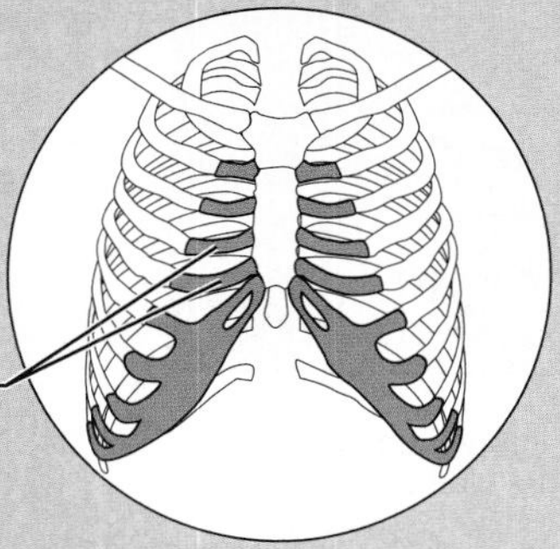

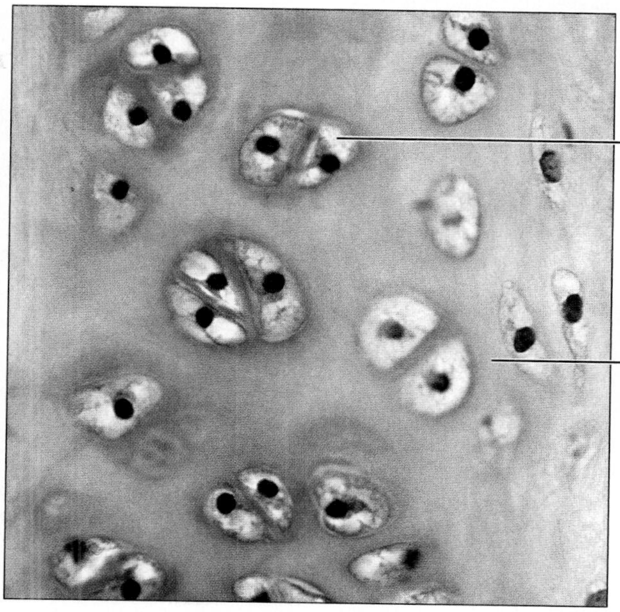

Photomicrograph: Hyaline cartilage from a costal cartilage of a rib (470×).

(i) Cartilage: elastic

Description: Similar to hyaline cartilage, but more elastic fibers in matrix.

Function: Maintains the shape of a structure while allowing great flexibility.

Location: Supports the external ear (auricle); epiglottis.

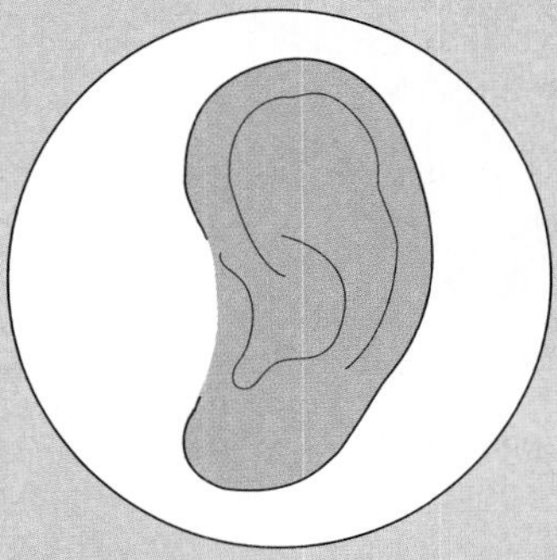

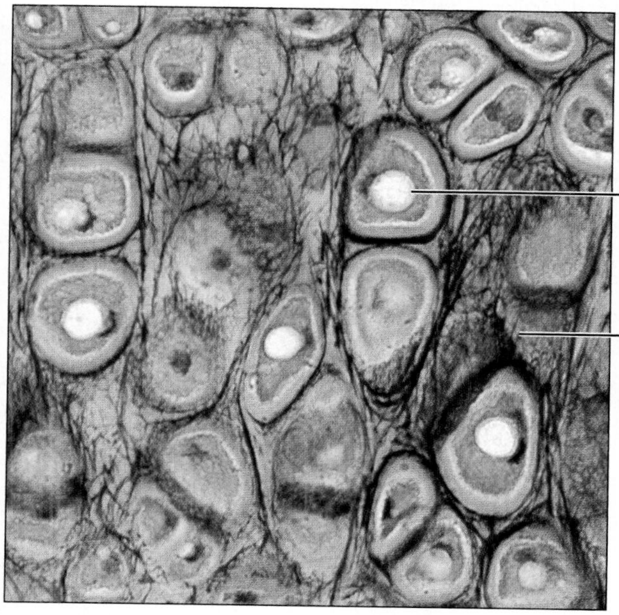

Photomicrograph: Elastic cartilage from the human ear auricle; forms the flexible skeleton of the ear (510×).

Figure 6.5 *(continued)* Cartilage **(h)** and **(i)**.

(j) Cartilage: fibrocartilage

Description: Matrix similar to but less firm than matrix in hyaline cartilage; thick collagen fibers predominate.

Function: Tensile strength with the ability to absorb compressive shock.

Location: Intervertebral discs; pubic symphysis; discs of knee joint.

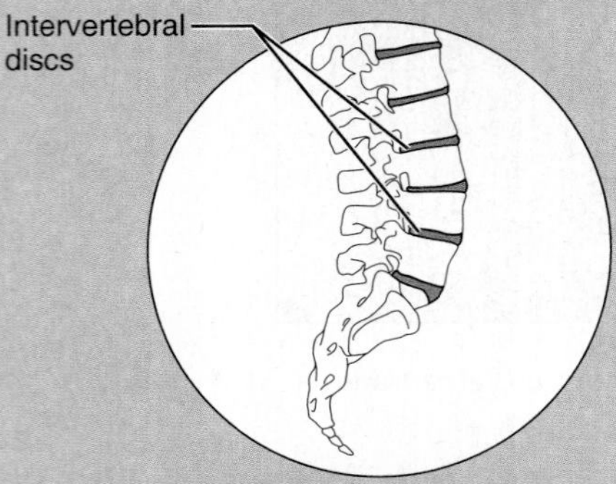

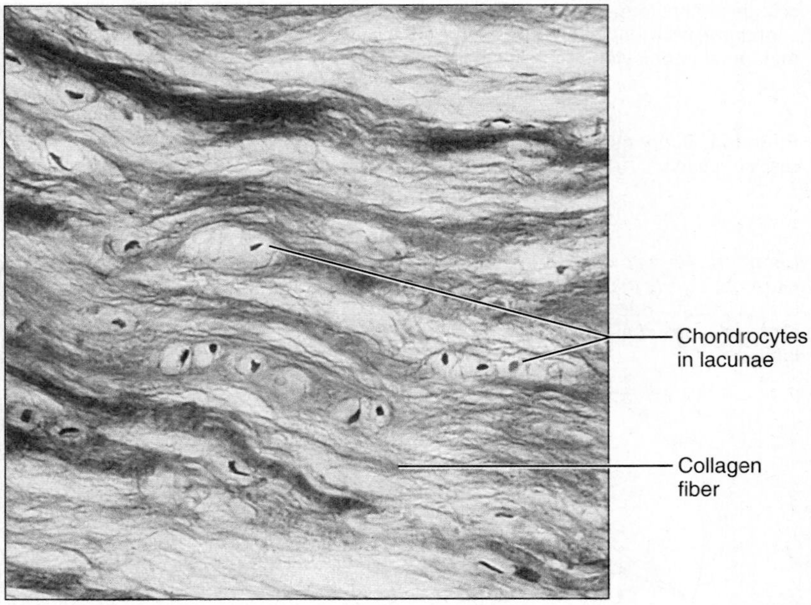

Photomicrograph: Fibrocartilage of an intervertebral disc (160×). Special staining produced the blue color.

(k) Bones (osseous tissue)

Description: Hard, calcified matrix containing many collagen fibers; osteocytes lie in lacunae. Very well vascularized.

Function: Bone supports and protects (by enclosing); provides levers for the muscles to act on; stores calcium and other minerals and fat; marrow inside bones is the site for blood cell formation (hematopoiesis).

Location: Bones.

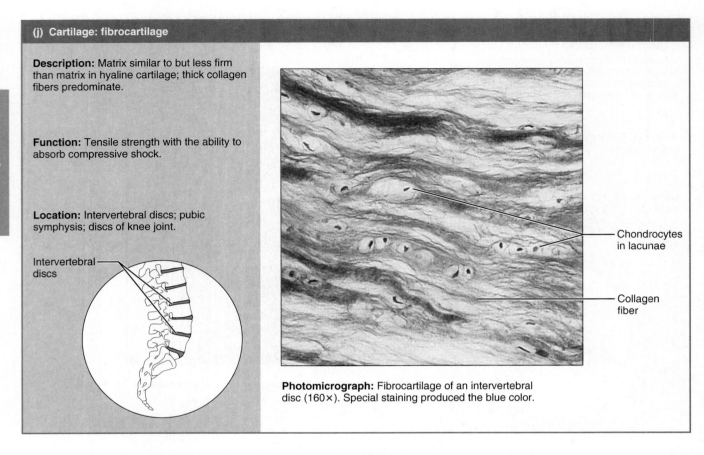

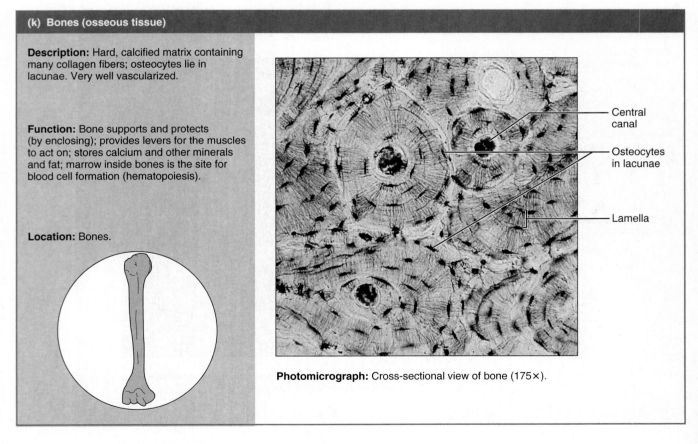

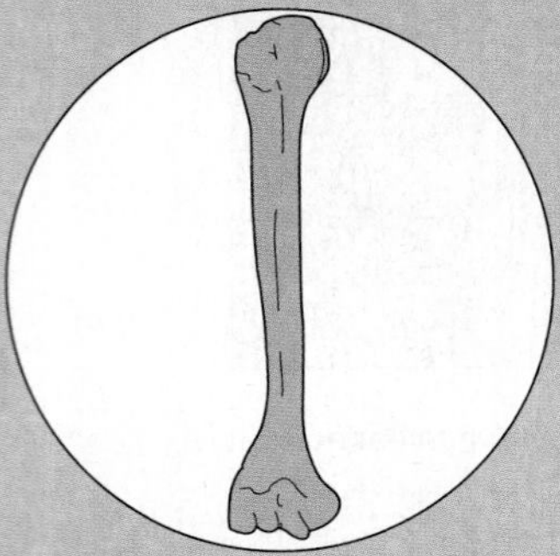

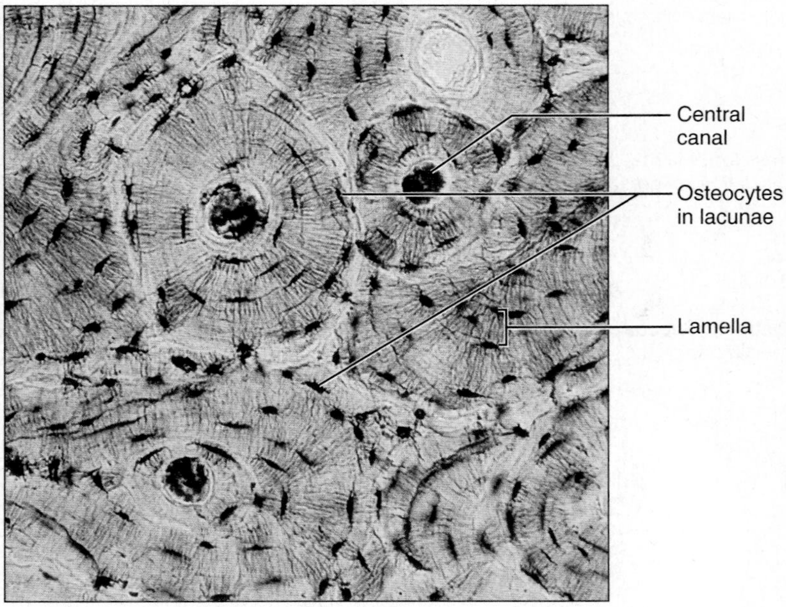

Photomicrograph: Cross-sectional view of bone (175×).

Figure 6.5 *(continued)* **Connective tissues.** Cartilage **(j)** and bone **(k)**.

(l) Blood

Description: Red and white blood cells in a fluid matrix (plasma).

Function: Transport of respiratory gases, nutrients, wastes, and other substances.

Location: Contained within blood vessels.

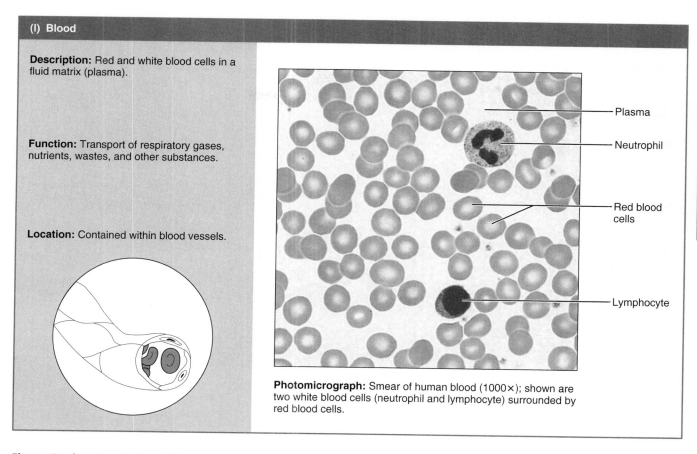

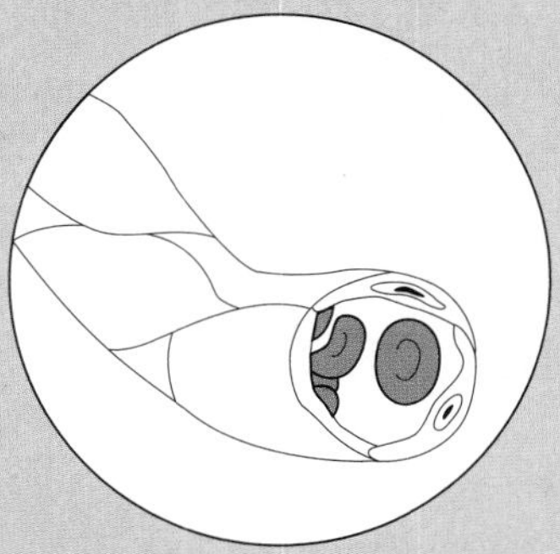

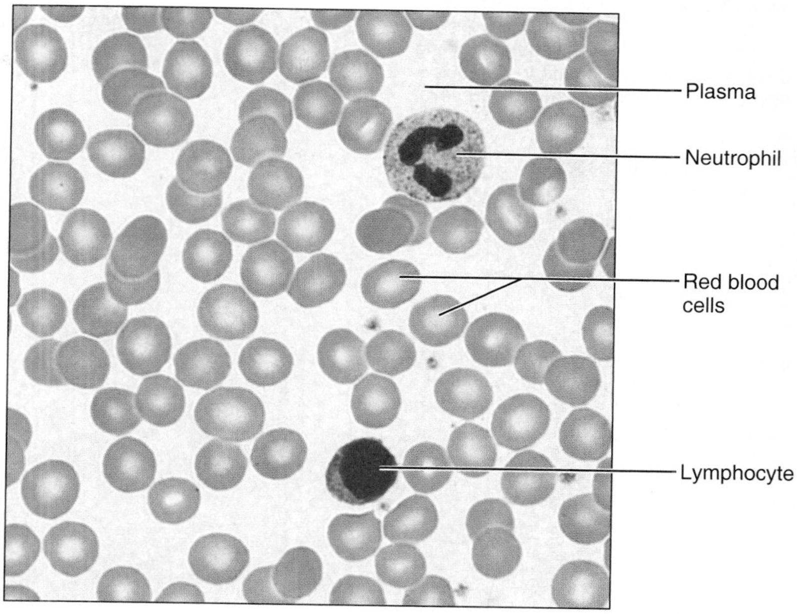

Photomicrograph: Smear of human blood (1000×); shown are two white blood cells (neutrophil and lymphocyte) surrounded by red blood cells.

Figure 6.5 *(continued)* Blood **(l)**.

Nervous Tissue

Nervous tissue is made up of two major cell populations. The **neuroglia** are special supporting cells that protect, support, and insulate the more delicate neurons. The **neurons** are highly specialized to receive stimuli (excitability) and to generate electrical signals that may be sent to all parts of the body (conductivity).

The structure of neurons is markedly different from that of all other body cells. They have a nucleus-containing cell body, and their cytoplasm is drawn out into long extensions (cell processes)—sometimes as long as 1 m (about 3 feet), which allows a single neuron to conduct an electrical signal over relatively long distances. (More detail about the anatomy of the different classes of neurons and neuroglia appears in Exercise 15.)

Activity 3

Examining Nervous Tissue Under the Microscope

Obtain a prepared slide of a spinal cord smear. Locate a neuron and compare it to **Figure 6.6**, p. 80. Keep the light dim—this will help you see the cellular extensions of the neurons. (See also Figure 15.2 in Exercise 15.)

Nervous tissue

Description: Neurons are branching cells; cell processes that may be quite long extend from the nucleus-containing cell body; also contributing to nervous tissue are nonexcitable supporting cells.

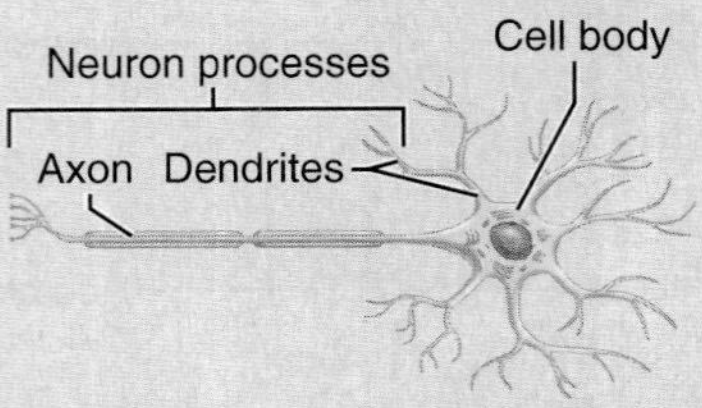

Function: Neurons transmit electrical signals from sensory receptors and to effectors (muscles and glands); supporting cells support and protect neurons.

Location: Brain, spinal cord, and nerves.

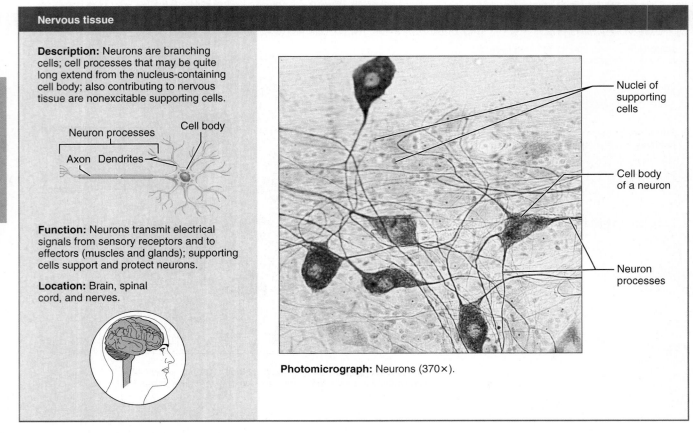

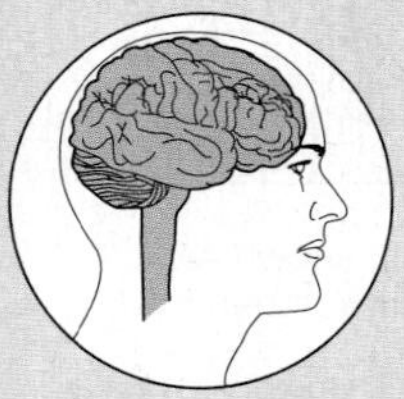

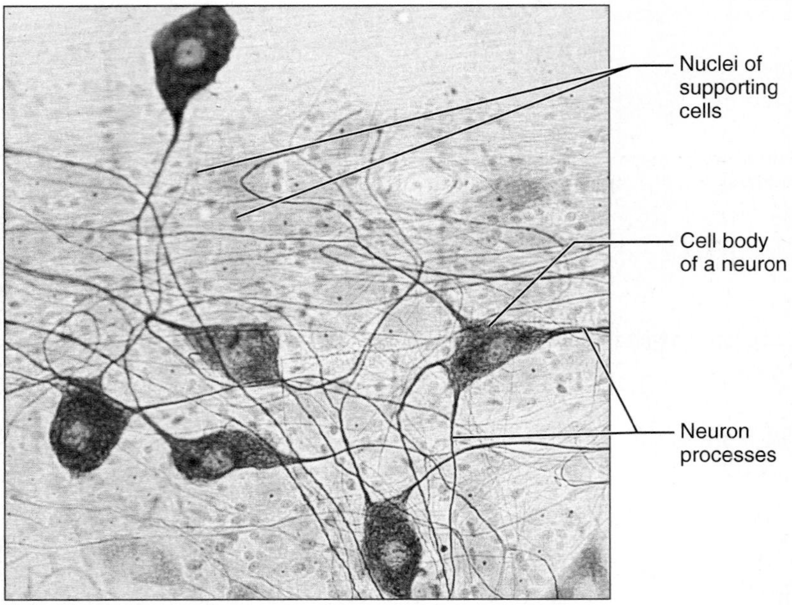

Photomicrograph: Neurons (370×).

Figure 6.6 **Nervous tissue.**

Muscle Tissue

Muscle tissue (Figure 6.7) is highly specialized to contract and produces most types of body movement. The three basic types of muscle tissue are described briefly here.

Skeletal muscle, the flesh of the body, is attached to the skeleton. It is under voluntary control (consciously controlled), and its contraction moves the limbs and other external body parts. The cells of skeletal muscles are long, cylindrical, nonbranching, and multinucleate (several nuclei per cell), with the nuclei pushed to the periphery of the cells; they have obvious *striations* (stripes).

Cardiac muscle is found only in the heart. As it contracts, the heart acts as a pump, propelling the blood into the blood vessels. Cardiac muscle, like skeletal muscle, has striations, but cardiac cells are branching uninucleate cells that interdigitate (fit together) at junctions called **intercalated discs**. These structural modifications allow the cardiac muscle to act as a unit. Cardiac muscle is under involuntary control, which means that we cannot voluntarily or consciously control the operation of the heart.

Smooth muscle is found mainly in the walls of hollow organs (digestive and urinary tract organs, uterus, blood vessels). Typically it has two layers that run at right angles to each other; consequently its contraction can constrict or dilate the lumen (cavity) of an organ and propel substances. Smooth muscle cells are quite different in appearance from those of skeletal or cardiac muscle. No striations are visible, and the uninucleate smooth muscle cells are spindle-shaped (tapered at the ends, like a candle). Like cardiac muscle, smooth muscle is under involuntary control.

Activity 4

Examining Muscle Tissue Under the Microscope

Obtain and examine prepared slides of skeletal, cardiac, and smooth muscle. Notice their similarities and dissimilarities in your observations and in the illustrations and photomicrographs in Figure 6.7.

(a) Skeletal muscle

Description: Long, cylindrical, multinucleate cells; obvious striations.

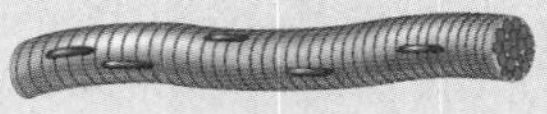

Function: Voluntary movement; locomotion; manipulation of the environment; facial expression; voluntary control.

Location: In skeletal muscles attached to bones or occasionally to skin.

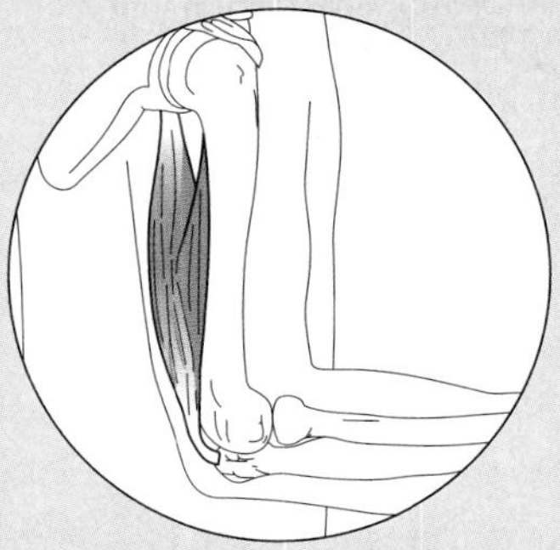

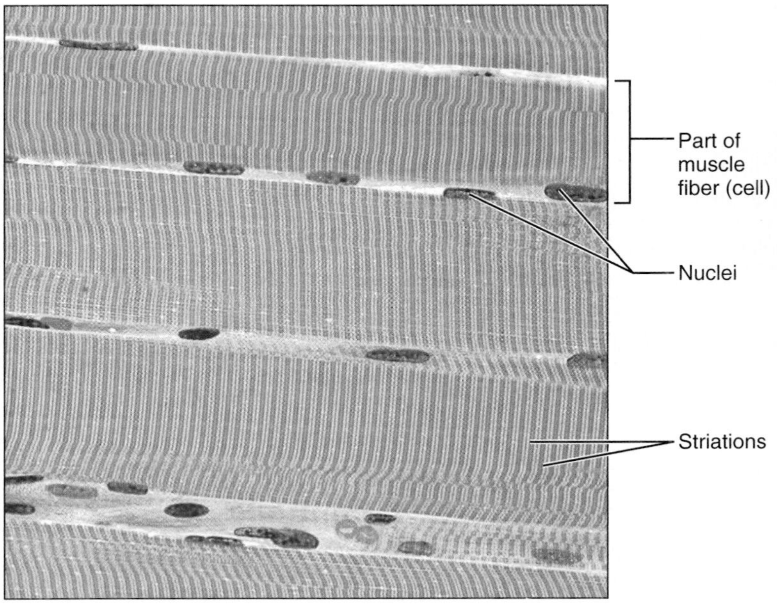

Photomicrograph: Skeletal muscle (approx. 550×). Notice the obvious banding pattern and the fact that these large cells are multinucleate.

(b) Cardiac muscle

Description: Branching, striated, generally uninucleate cells that interdigitate at specialized junctions called intercalated discs.

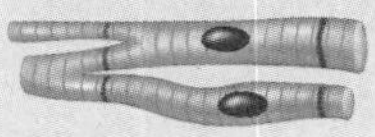

Function: As it contracts, cardiac muscle propels blood into the circulation; involuntary control.

Location: The walls of the heart.

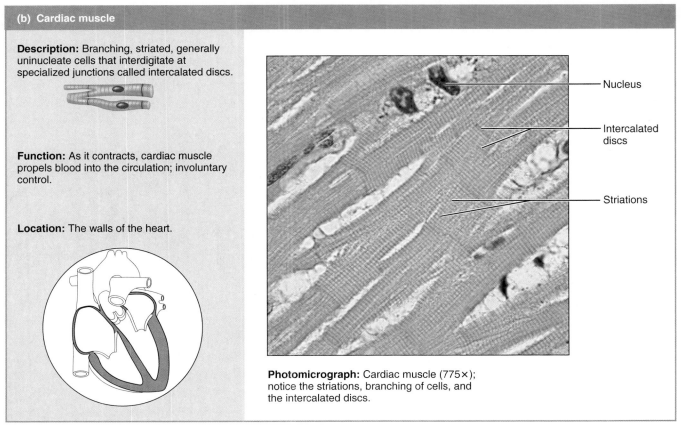

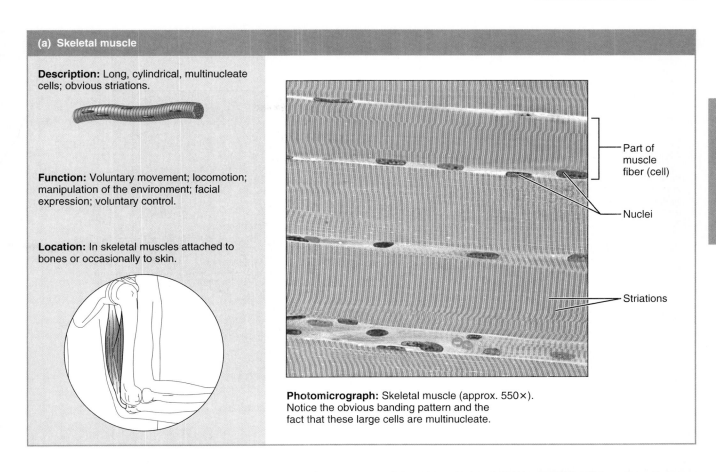

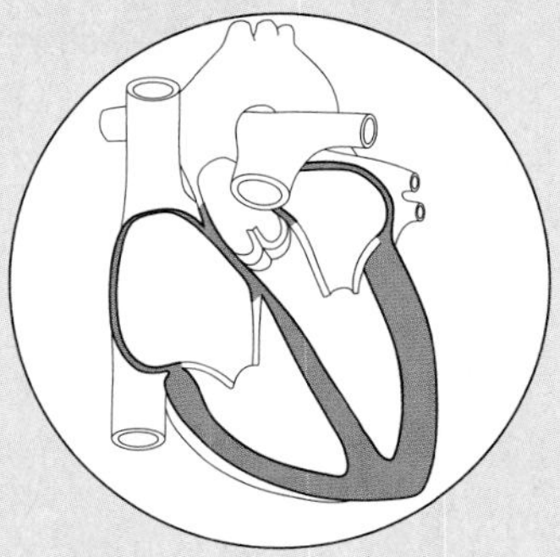

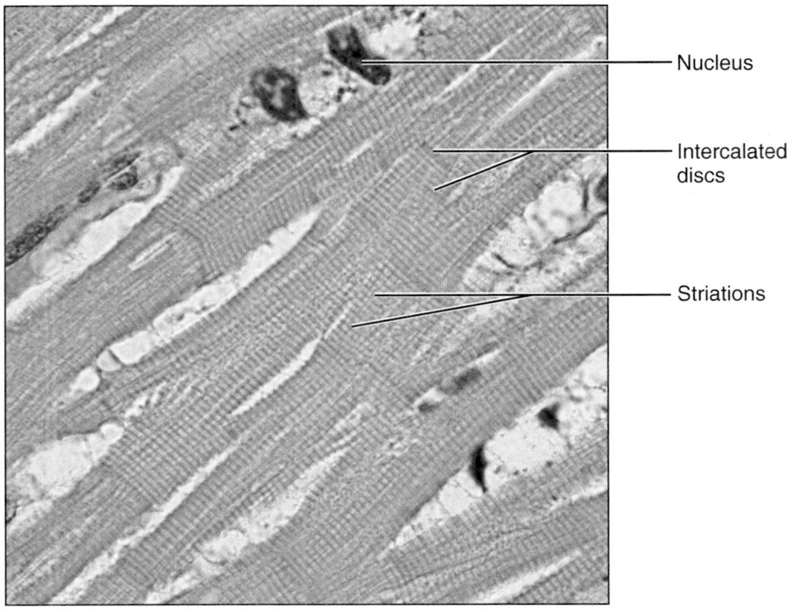

Photomicrograph: Cardiac muscle (775×); notice the striations, branching of cells, and the intercalated discs.

Figure 6.7 **Muscle tissues.** Skeletal muscle **(a)** and cardiac muscle **(b)**.

(c) Smooth muscle

Description: Spindle-shaped cells with central nuclei; no striations; cells arranged closely to form sheets.

Function: Propels substances (foodstuffs, urine) or a baby along internal passageways; involuntary control.

Location: Mostly in the walls of hollow organs.

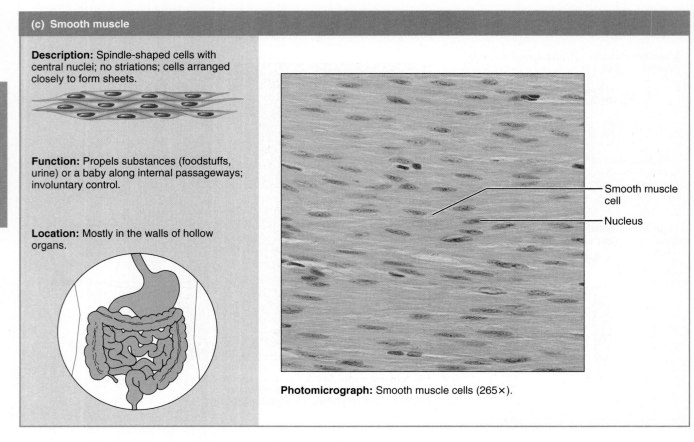

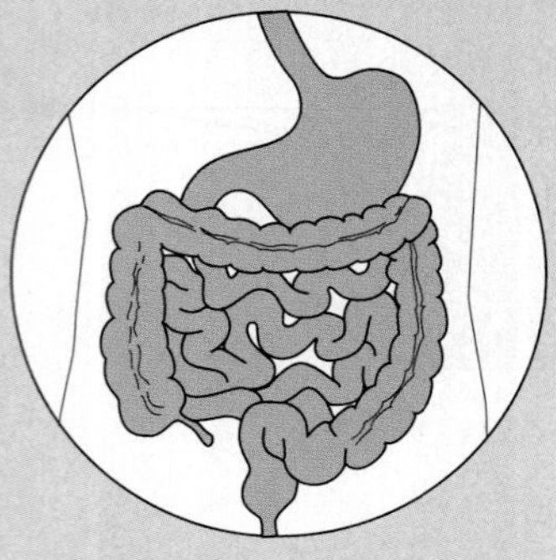

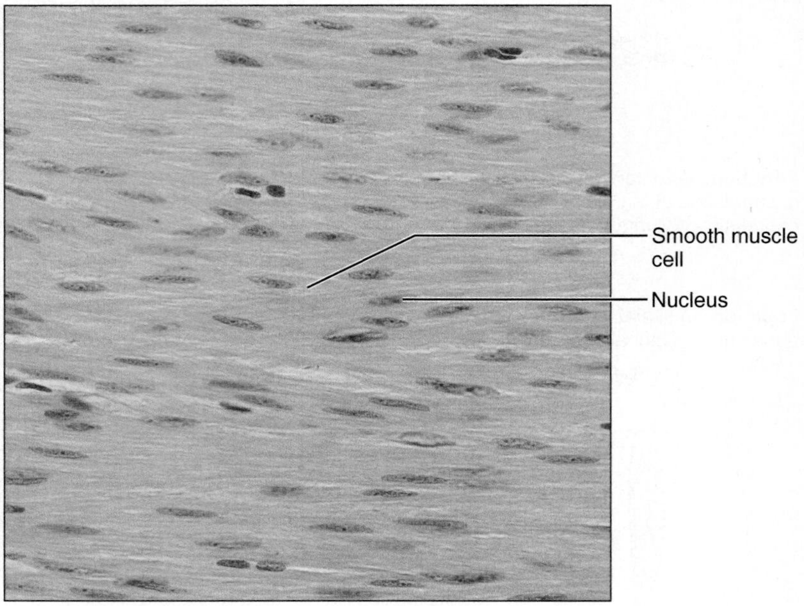

Photomicrograph: Smooth muscle cells (265×).

Figure 6.7 *(continued)* **Muscle Tissues.** Smooth muscle **(c)**.

Instructors may assign a portion of the Review Sheet questions using Mastering A&P™

EXERCISE 6 REVIEW SHEET
Classification of Tissues

Name ______________________ Lab Time/Date ______________________

Tissue Structure and Function—General Review

1. List the following in order from least complex to most complex: organ, cell, tissue, and organ system. ______________________

____

2. Define *histology*. ______________________

____

3. Use the key choices to identify the major tissue types described below.

Key: a. connective tissue b. epithelium c. muscle d. nervous tissue

______________ 1. lines body cavities and covers the body's external surface

______________ 2. pumps blood, flushes urine out of the body, allows one to swing a bat

______________ 3. forms endocrine and exocrine glands

______________ 4. anchors, packages, and supports body organs

______________ 5. classified based on the shape and arrangement of the cells

______________ 6. derived from mesenchyme

______________ 7. major function is to contract

______________ 8. transmits electrical signals

______________ 9. consists of cells within an extracellular matrix

______________ 10. most widespread tissue in the body

______________ 11. forms nerves and the brain

Epithelial Tissue

4. Describe five general characteristics of epithelial tissue. ______________________

____

____

____

5. For each function listed, name one type of epithelium and an organ that provides for that function.

Function 1: Protection ______________________________

Function 2: Diffusion ______________________________

Function 3: Secretion ______________________________

Function 4: Filtration ______________________________

Function 5: Absorption ______________________________

6. What structural feature do epithelia that provide for protection have in common?

____

7. Describe the following cell-surface modifications using the table below.

Cell-surface modification	Type(s) of epithelia with the modification	Function (include a specific organ)
Cilia		
Goblet cells		
Microvilli		

8. Transitional epithelium is actually stratified squamous epithelium with special characteristics.

How does it differ structurally from other stratified squamous epithelia? ______________________________

____

How does the structural difference support its function? ______________________________

____

____

9. How do the endocrine and exocrine glands differ in structure and function? ______________________________

____

____

____

Connective Tissue

10. What are three general characteristics of connective tissues? ______

______

______

11. What functions are performed by connective tissue? ______

______

12. How are the functions of connective tissue reflected in its structure? ______

______

______

13. Using the key, choose the best response to identify the connective tissues described below.

Key:
a. adipose connective tissue
b. areolar connective tissue
c. dense irregular connective tissue
d. dense regular connective tissue
e. elastic cartilage
f. elastic connective tissue
g. fibrocartilage
h. hyaline cartilage
i. osseous tissue

______ 1. attaches bones to bones and muscles to bones

______ 2. insulates against heat loss

______ 3. forms the fibrous joint capsule

______ 4. makes up the intervertebral discs

______ 5. composes basement membranes; a soft packaging tissue with a jellylike matrix

______ 6. forms the larynx, the costal cartilages of the ribs, and the embryonic skeleton

______ 7. provides a flexible framework for the external ear

______ 8. provides levers for muscles to act on

______ 9. forms the walls of large arteries

Nervous Tissue

14. What two physiological characteristics are highly developed in neurons? ______

______

15. In what ways are neurons similar to other cells? ______

______

How are they structurally different? ______

16. Describe how the unique structure of a neuron relates to its function in the body.

______

______

Muscle Tissue

17. The terms and phrases in the key relate to the muscle tissues. For each of the three muscle tissues, select the terms or phrases that characterize it, and write the corresponding letter of each term on the answer line.

Key:					
a.	striated	e.	voluntary	i.	attached to bones
b.	branching cells	f.	involuntary	j.	intercalated discs
c.	spindle-shaped cells	g.	one nucleus	k.	in wall of bladder and stomach
d.	cylindrical cells	h.	many nuclei	l.	forms heart walls

Skeletal muscle: ______________________ Cardiac muscle: ______________________ Smooth muscle: ______________________

18. ✚ Orthopedic surgeons are fond of saying, "It is better to break a bone than it is to tear a tendon or ligament." Based on your understanding of these two types of connective tissue, explain why that would be true. ______________________

____

____

19. ✚ A buccal swab procedure removes stratified squamous cells to obtain the DNA profile of an individual. Explain why a buccal swab shouldn't cause bleeding. ______________________

____

20. ✚ When cardiac muscle tissue dies in adults, it is replaced with scar tissue composed of dense connective tissue. Explain how the function of the scar tissue would differ from the function of the cardiac muscle tissue. ______________________

____

21. ✚ Smoking impairs cilia because the toxins paralyze and can destroy the cilia. Based on this loss of function, explain which types of infections smokers would be more susceptible to. ______________________

____

For Review

22. Label the tissue types illustrated here and on the next page, and identify all structures provided with leaders.

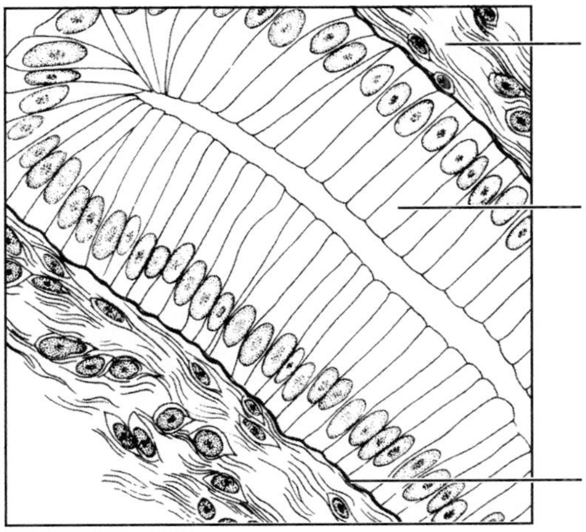

(a) ______________________________

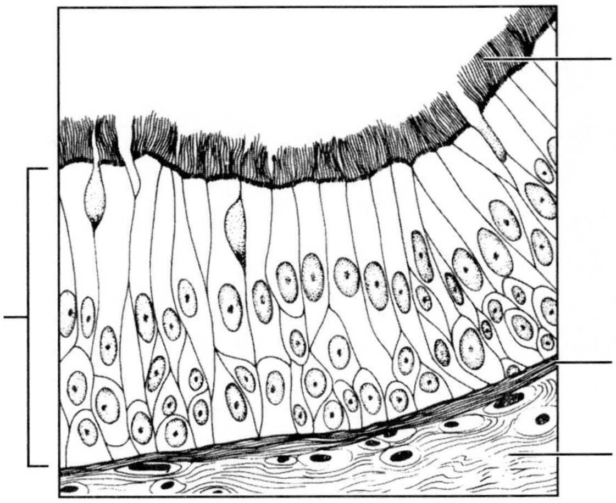

(b) ______________________________

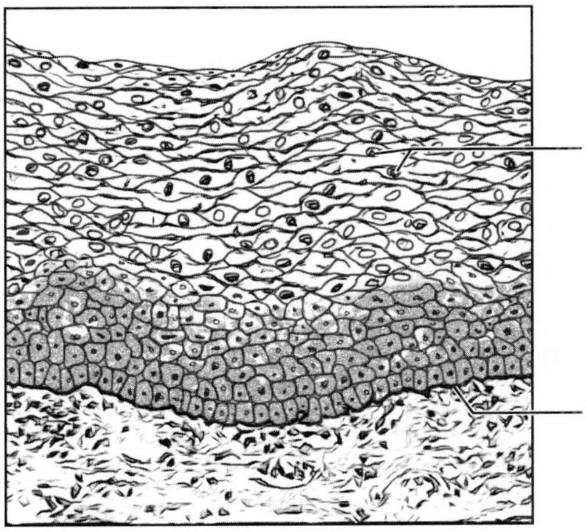

(c) ______________________________

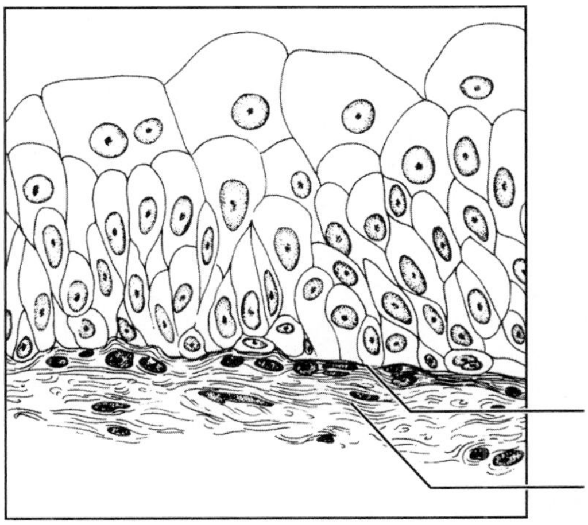

(d) ______________________________

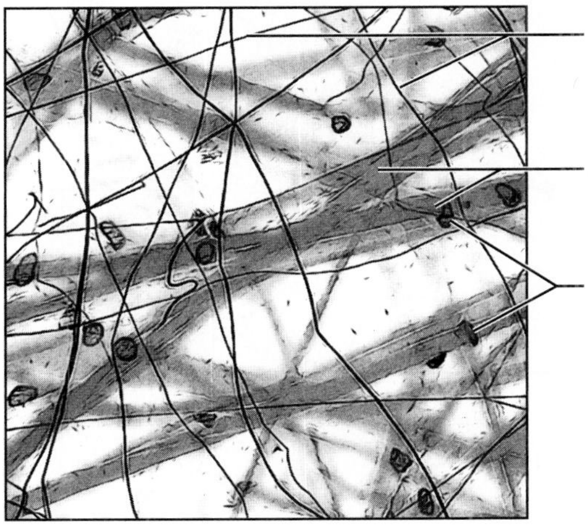

(e) ______________________________

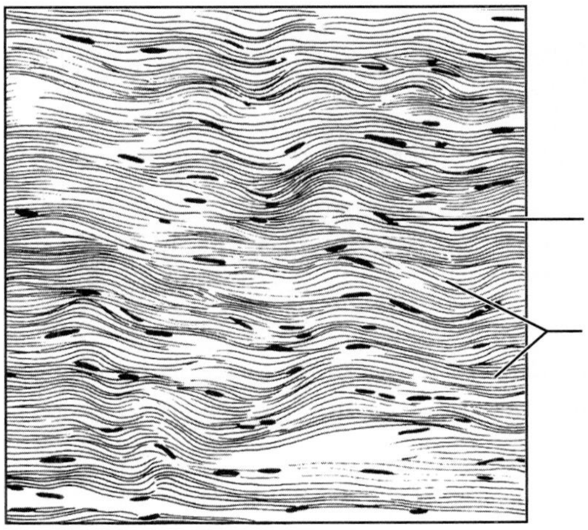

(f) ______________________________

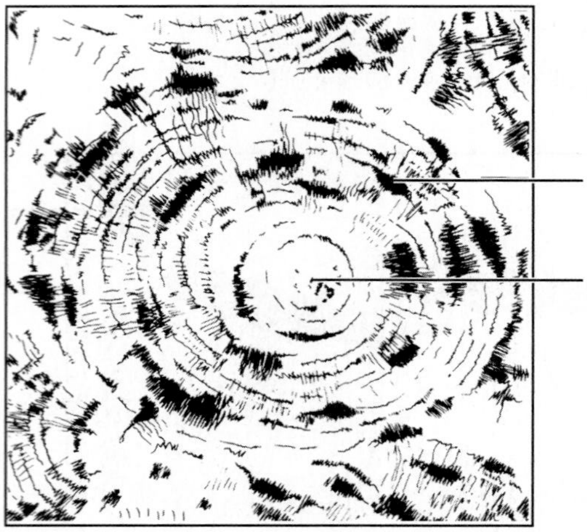

(g) ______________________________

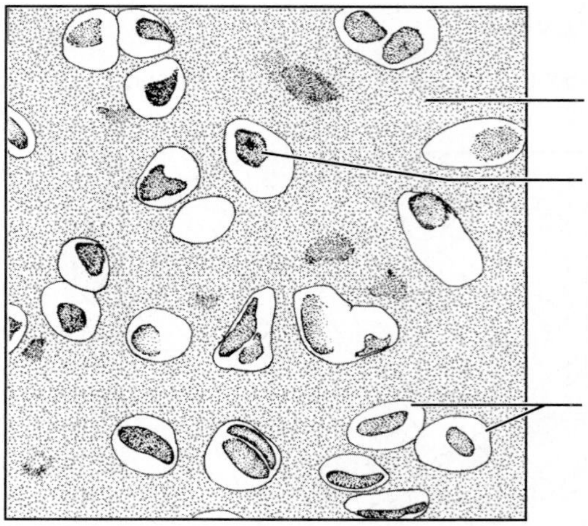

(h) ______________________________

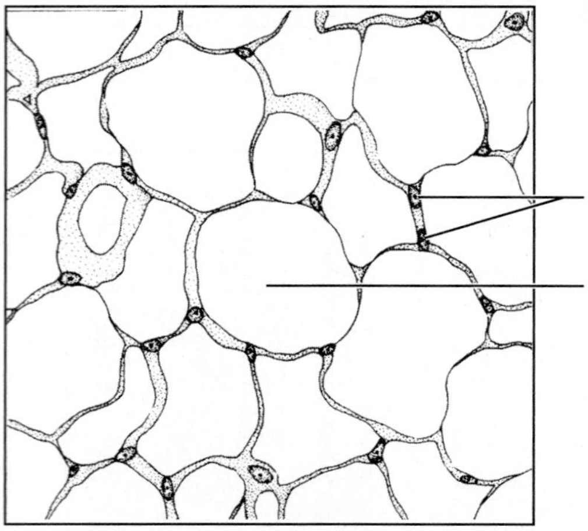

(i) ______________________________

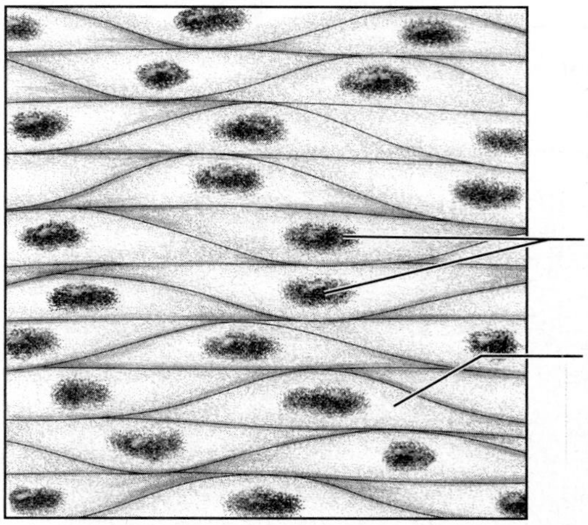

(j) ______________________________

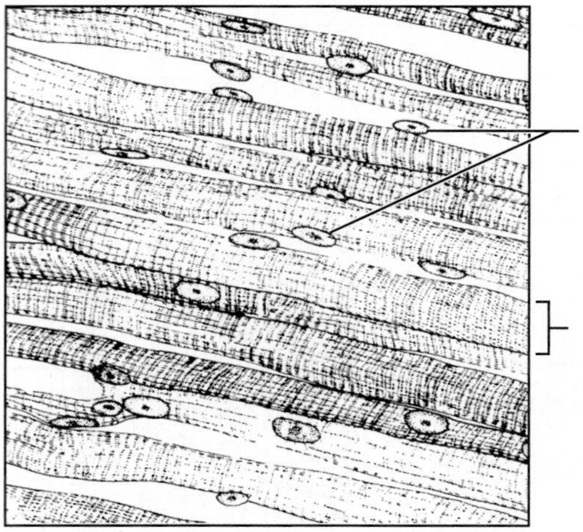

(k) ______________________________

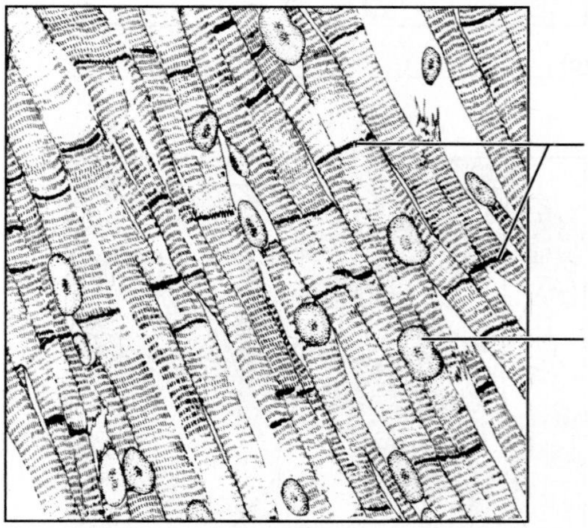

(l) ______________________________

EXERCISE 7

The Integumentary System

Learning Outcomes

- List several important functions of the skin, or integumentary system.
- Identify the following skin structures on a model, image, or microscope slide: epidermis, dermis (papillary and reticular layers), hair follicles and hair, sebaceous glands, and sweat glands.
- Name and describe the layers of the epidermis.
- List the factors that determine skin color, and describe the function of melanin.
- Identify the major regions of nails.
- Describe the distribution and function of hairs, sebaceous glands, and sweat glands.
- Discuss the difference between eccrine and apocrine sweat glands.
- Compare and contrast the structure and functions of the epidermis and the dermis.

Go to Mastering A&P™ > Study Area to improve your performance in A&P Lab.

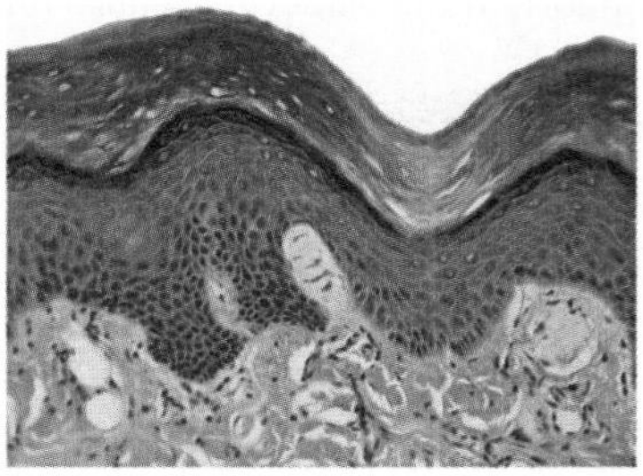

> Lab Tools > Practice Anatomy Lab > Histology

Instructors may assign new Building Vocabulary coaching activities, Pre-Lab Quiz questions, Art Labeling activities, Practice Anatomy Lab Practical questions (PAL) for Histology of the Integumentary System, and more using the Mastering **A&P™** Item Library.

Pre-Lab Quiz

Instructors may assign these and other Pre-Lab Quiz questions using Mastering **A&P™**

1. All the following are functions of the skin *except*:
   a. excretion of body wastes
   b. insulation
   c. protection from mechanical damage
   d. site of vitamin A synthesis
2. The most superficial layer of the epidermis is the:
   a. stratum basale **c.** stratum granulosum
   b. stratum spinosum **d.** stratum corneum
3. These cells produce a brown-to-black pigment that colors the skin and protects DNA from ultraviolet radiation damage. The cells are:
   a. dendritic cells **c.** melanocytes
   b. keratinocytes **d.** tactile cells
4. The portion of a hair that projects from the surface of the skin is known as the:
   a. bulb **b.** matrix **c.** root **d.** shaft
5. Circle the correct underlined term. The ducts of sebaceous / sweat glands usually empty into a hair follicle but may also open directly on the skin surface.

Materials

- Skin model (three-dimensional, if available)
- Compound microscope
- Prepared slide of human scalp
- Prepared slide of skin of palm or sole
- Sheet of 20# bond paper ruled to mark off cm^2 areas
- Scissors
- Povidone-iodine swabs, or Lugol's iodine and cotton swabs
- Adhesive tape
- Disposable gloves
- Data collection sheet for plotting distribution of sweat glands
- Porelon® fingerprint pad or portable inking foils
- Ink cleaner towelettes
- Index cards (4 in. × 6 in.)
- Magnifying glasses

The **integument** is considered an organ system because it consists of multiple organs, the **skin** and its accessory organs. It is much more than an external body covering; architecturally, the skin is a marvel. It is tough yet pliable, a characteristic that enables it to withstand constant insult from outside agents.

The skin has many functions, most concerned with protection. It insulates and cushions the underlying body tissues and protects the entire body from abrasion, exposure to harmful chemicals, temperature extremes, and bacterial invasion. The hardened uppermost layer of the skin prevents water loss from the body surface.

The skin's abundant capillary network plays an important role in temperature regulation by regulating heat loss from the body surface.

The skin has other functions as well. For example, it acts as an excretory system; urea, salts, and water are lost through the skin pores in sweat. The skin also has important metabolic duties. For example, it is the site of vitamin D synthesis for the body. Vitamin D plays a role in calcium absorption in the digestive system. Finally, the sense organs for touch, pressure, pain, and temperature are located here.

Basic Structure of the Skin

The skin has two distinct regions—the superficial *epidermis* composed of epithelium and an underlying connective tissue, the *dermis* (**Figure 7.1**). These layers are firmly "cemented" together along a wavy border. But friction, such as the rubbing of a poorly fitting shoe, may cause them to separate, resulting in a blister. Immediately deep to the dermis is the **subcutaneous layer**, or **hypodermis**, which is not considered part of the skin. It consists primarily of adipose tissue. The main skin areas and structures are described below.

Activity 1

Locating Structures on a Skin Model

As you read, locate the following structures in the diagram (Figure 7.1) and on a skin model.

Epidermis

Structurally, the avascular epidermis is a keratinized stratified squamous epithelium consisting of four distinct cell types and four or five distinct layers.

Cells of the Epidermis

- **Keratinocytes** (literally, keratin cells): The most abundant epidermal cells, their main function is to produce keratin

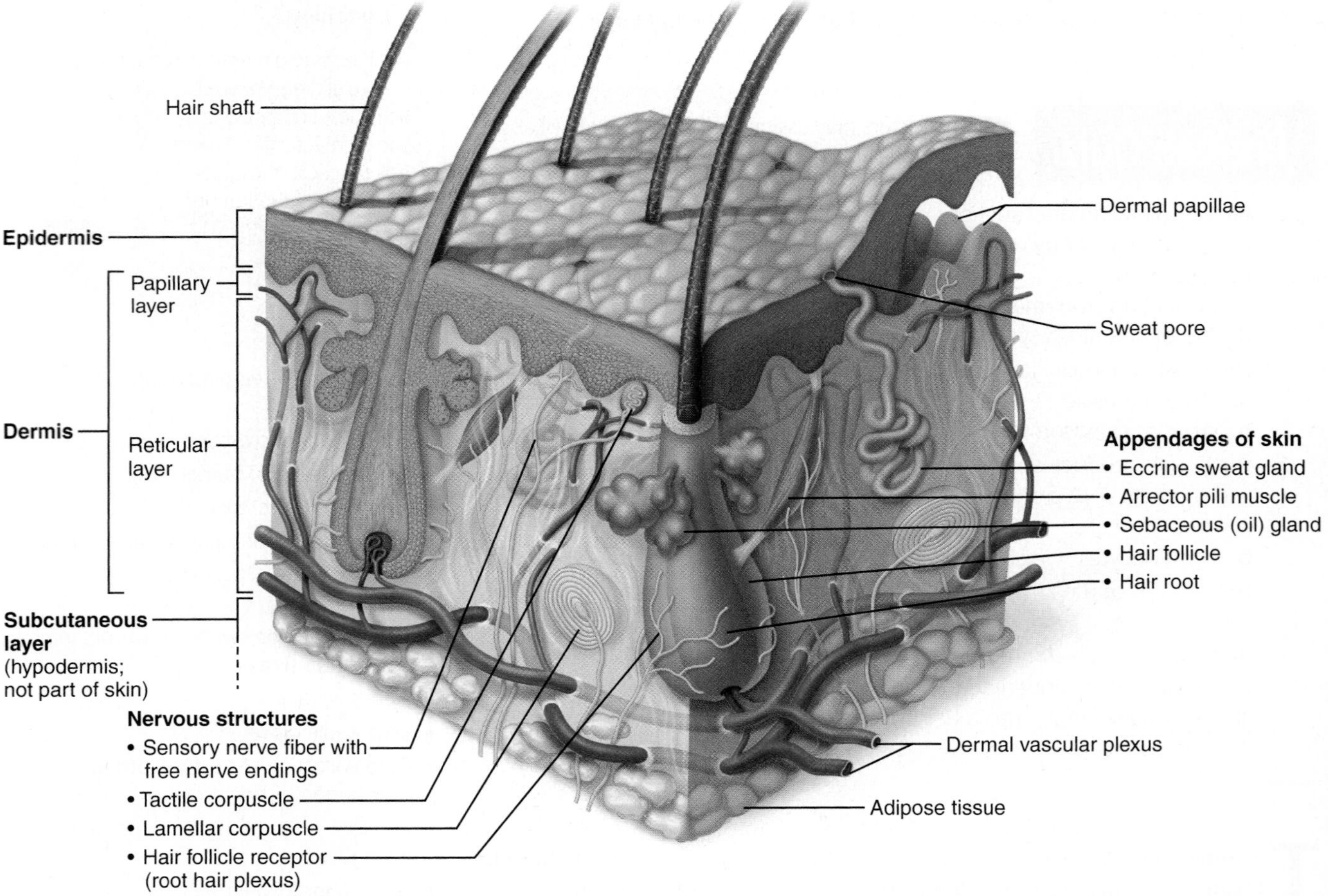

Figure 7.1 Skin structure. Three-dimensional view of the skin and the underlying tissue. The epidermis and dermis have been pulled apart at the right corner to reveal the dermal papillae. Tactile corpuscles are not common in hairy skin but are included here for illustrative purposes.

Instructors may assign this figure as an Art Labeling Activity using Mastering A&P™

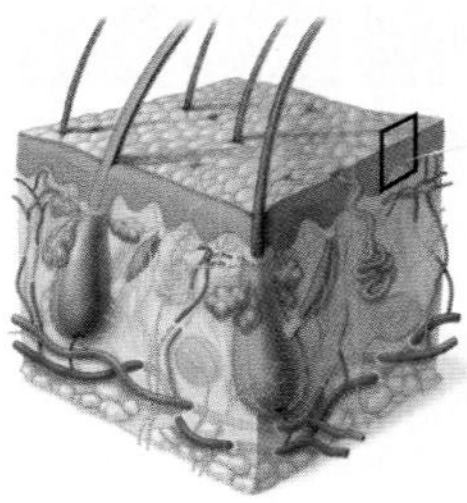

Figure 7.2 The main structural features in the epidermis of thin skin. (a) Photomicrograph depicting the four major epidermal layers (430×). **(b)** Diagram showing the layers and relative distribution of the different cell types. Notice that the keratinocytes are joined by numerous desmosomes. The stratum lucidum, present in thick skin, is not illustrated here.

Instructors may assign this figure as an Art Labeling Activity using Mastering **A&P™**

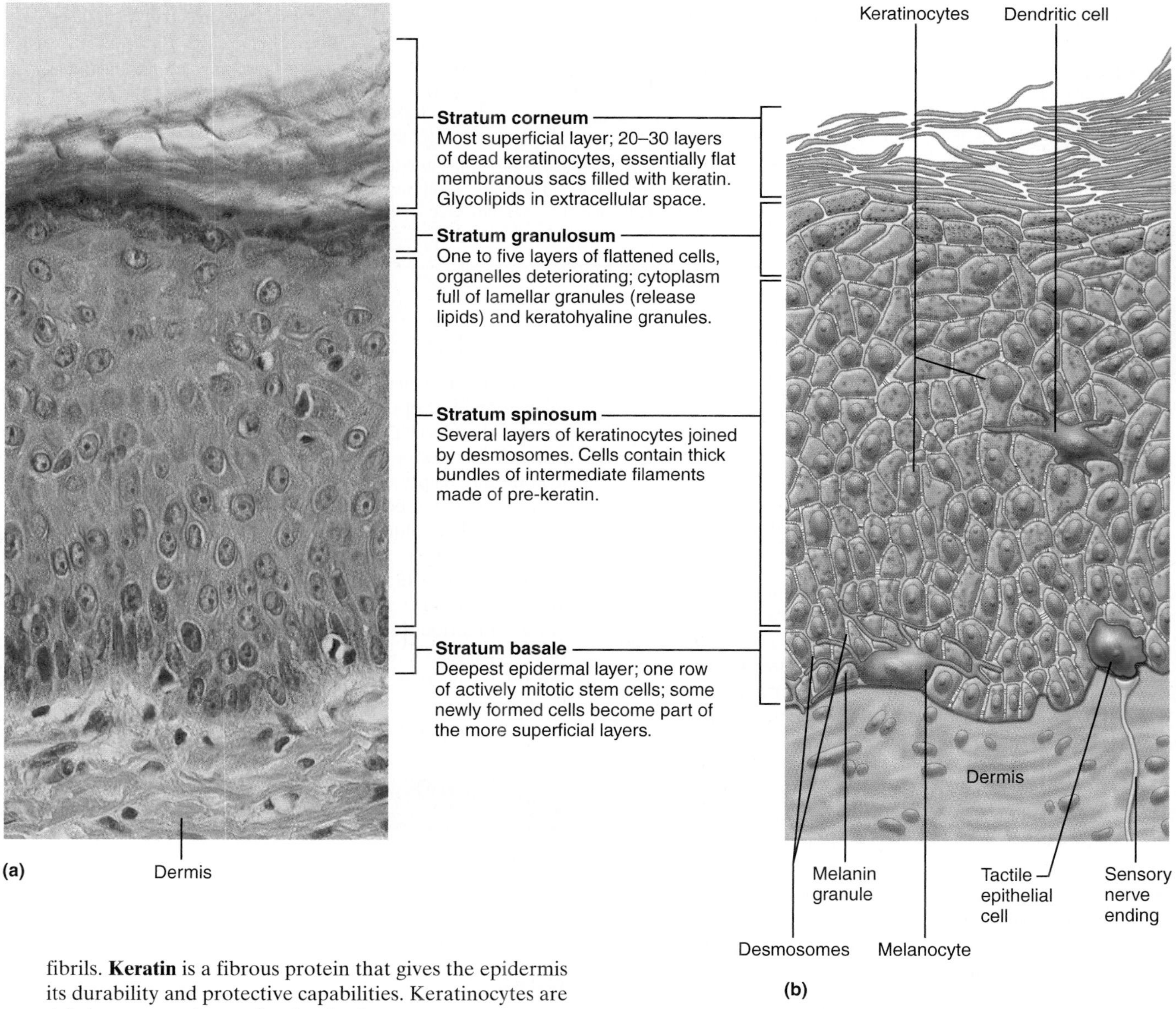
Stratum corneum
Most superficial layer; 20–30 layers of dead keratinocytes, essentially flat membranous sacs filled with keratin. Glycolipids in extracellular space.
Stratum granulosum
One to five layers of flattened cells, organelles deteriorating; cytoplasm full of lamellar granules (release lipids) and keratohyaline granules.
Stratum spinosum
Several layers of keratinocytes joined by desmosomes. Cells contain thick bundles of intermediate filaments made of pre-keratin.
Stratum basale
Deepest epidermal layer; one row of actively mitotic stem cells; some newly formed cells become part of the more superficial layers.
(a)
Dermis
Keratinocytes
Dendritic cell
Dermis
Melanin granule
Tactile epithelial cell
Sensory nerve ending
Desmosomes
Melanocyte
(b)

7

fibrils. **Keratin** is a fibrous protein that gives the epidermis its durability and protective capabilities. Keratinocytes are tightly connected to each other by desmosomes.

Far less numerous are the following types of epidermal cells (**Figure 7.2**):

- **Melanocytes:** Spidery black cells that produce the brown-to-black pigment called **melanin**. The skin tans because melanin production increases when the skin is exposed to ultraviolet radiation (UVR) in sunlight. The melanin provides a protective pigment umbrella over the nuclei of the cells in the deeper epidermal layers, thus shielding their genetic material (DNA) from the damaging effects of ultraviolet radiation. *Freckles* and *moles (nevi)* are areas where melanin is more concentrated.
- **Dendritic cells:** Also called *Langerhans cells*, these cells arise from the bone marrow and migrate to the epidermis. They ingest foreign substances and play a key role in activating the immune response.
- **Tactile epithelial cells:** Occasional spiky hemispheres that, in combination with sensory nerve endings, form sensitive touch receptors located at the epidermal-dermal junction.

Layers of the Epidermis

The epidermis consists of four layers in thin skin, which covers most of the body. Thick skin, found on the palms of the hands and soles of the feet, contains an additional layer, the stratum

Table 7.1 Layers of the Epidermis (from superficial to deep)

Epidermal layer	Description
Stratum corneum (horny layer)	The outermost layer consisting of 20–30 layers of dead, scalelike keratinocytes. They are constantly being exfoliated and replaced by the division of the deeper cells.
Stratum lucidum (clear layer)	Present only in thick skin. A very thin transparent band of flattened, dead keratinocytes with indistinct boundaries.
Stratum granulosum (granular layer)	A thin layer named for the abundant granules its cells contain. These granules are (1) *lamellar granules*, which contain a waterproofing glycolipid that is secreted into the extracellular space; and (2) *keratohyaline granules*, which help to form keratin in the more superficial layers. At the upper border of this layer, the cells are beginning to die.
Stratum spinosum (spiny layer)	Several layers of cells that contain thick, weblike bundles of intermediate filaments made of a pre-keratin protein. The cells in this layer appear spiky because when the tissue is prepared, the cells shrink, but their desmosomes hold tight to adjacent cells. Cells in this layer and the basal layer are the only ones to receive adequate nourishment from diffusion of nutrients from the dermis.
Stratum basale (basal layer)	A single row of cells immediately above the dermis. Its cells are constantly undergoing mitosis to form new cells, hence its alternate name, *stratum germinativum*. Some 10–25% of the cells in this layer are melanocytes, which thread their processes through this and adjacent layers of keratinocytes. Occasional tactile epithelial cells are also present in this layer.

lucidum. From deep to superficial, the layers of the epidermis are the stratum basale, stratum spinosum, stratum granulosum, stratum lucidum, and stratum corneum (Figure 7.2). The layers of the epidermis are summarized in **Table 7.1**.

Dermis

The connective tissue proper making up the dermis consists of two principal regions—the papillary and reticular areas (Figure 7.1). Like the epidermis, the dermis varies in thickness.

- **Papillary dermis:** The more superficial dermal region composed of areolar connective tissue. It is very uneven and has fingerlike projections from its superior surface, the **dermal papillae**, which attach it to the epidermis above. These projections lie on top of the larger dermal ridges. In the palms of the hands and soles of the feet, they produce the *fingerprints*, unique patterns of *epidermal ridges* that remain unchanged throughout life. Abundant capillary networks in the papillary layer furnish nutrients for the epidermal layers and allow heat to radiate to the skin surface. The pain receptors (free nerve endings) and touch receptors (*tactile corpuscles* in hairless skin) are also found here.
- **Reticular dermis:** The deepest skin layer. It is composed of dense irregular connective tissue and contains many arteries and veins, sweat and sebaceous glands, and pressure receptors (*lamellar corpuscles*).

The abundant dermal blood supply allows the skin to play a role in the regulation of body temperature. When body temperature is high, the arterioles serving the skin dilate, and the capillary network of the dermis becomes engorged with the heated blood. Thus body heat is allowed to radiate from the skin surface.

The dermis is also richly provided with lymphatic vessels and nerve fibers. Many of the nerve endings bear highly specialized receptor organs that, when stimulated by environmental changes, transmit messages to the central nervous system for interpretation. Some of these receptors—free nerve endings (pain receptors), a lamellar corpuscle, a tactile corpuscle, and a hair follicle receptor (also called a *root hair plexus*)—are shown in Figure 7.1. (These receptors are discussed in depth in Exercise 22.)

Skin Color

Skin color is a result of the relative amount of melanin in skin, the relative amount of carotene in skin, and the degree of oxygenation of the blood. *Carotene* is a yellow-orange pigment present primarily in the stratum corneum and in the adipose tissue of the hypodermis. Its presence is most noticeable when large amounts of carotene-rich foods (carrots, for instance) are eaten.

Skin color may be an important diagnostic tool. For example, flushed skin may indicate hypertension or fever, whereas pale skin is typically seen in anemic individuals. When the blood is inadequately oxygenated, as during asphyxiation and serious lung disease, both the blood and the skin take on a bluish cast, a condition called **cyanosis**. **Jaundice**, in which the tissues become yellowed, is almost always diagnostic for liver disease, whereas a bronzing of the skin hints that a person's adrenal cortex is hypoactive **(Addison's disease)**. +

Accessory Organs of the Skin

The accessory organs of the skin—cutaneous glands, hair, and nails—are all derivatives of the epidermis, but they reside primarily in the dermis. They originate from the stratum basale and extend into the dermis.

Nails

Nails are hornlike derivatives of the epidermis (**Figure 7.3**). Their named parts are:

- **Nail plate:** The visible attached portion.
- **Free edge:** The portion of the nail that grows out away from the body.
- **Hyponychium:** The region beneath the free edge of the nail.
- **Nail root:** The part that is embedded in the skin and adheres to an epithelial nail bed.
- **Nail folds:** Skin folds that overlap the borders of the nail.
- **Eponychium:** Projection of the thick proximal nail fold commonly called the cuticle.

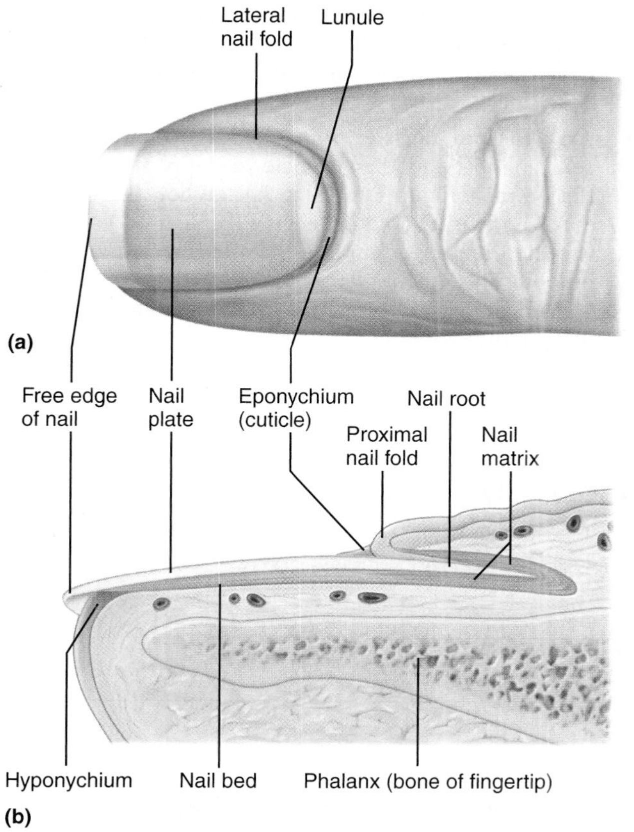

Figure 7.3 **Structure of a nail. (a)** Surface view of the distal part of a finger. The nail matrix that forms the nail lies beneath the lunule; the epidermis of the nail bed underlies the nail. **(b)** Sagittal section of the fingertip.

- **Nail bed:** Extension of the stratum basale beneath the nail.
- **Nail matrix:** The thickened proximal part of the nail bed containing germinal cells responsible for nail growth. As the matrix produces the nail cells, they become heavily keratinized and die. Thus nails, like hairs, are mostly nonliving material.
- **Lunule:** The proximal region of the thickened nail matrix, which appears as a white crescent moon. Everywhere else, nails are transparent and nearly colorless, but they appear pink because of the blood supply in the underlying dermis. When someone is cyanotic because of a lack of oxygen in the blood, the nail beds take on a blue cast.

Activity 2

Identifying Nail Structures

Identify the parts of a nail (as shown in Figure 7.3) on yourself or your lab partner.

Hairs and Associated Structures

Hairs, enclosed in hair follicles, are found all over the entire body surface, except for thick-skinned areas (the palms of the hands and the soles of the feet), parts of the external genitalia, the nipples, and the lips.

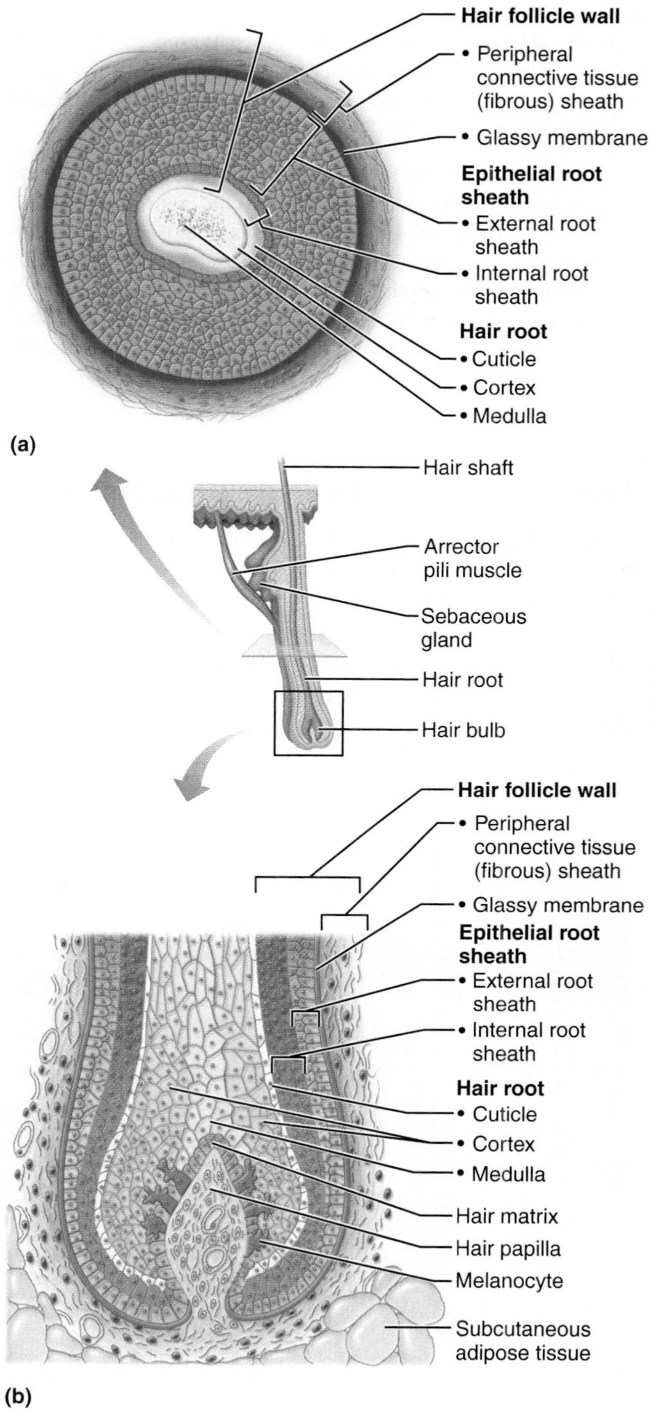

Figure 7.4 **Structure of a hair and hair follicle.**
(a) Diagram of a cross section of a hair within its follicle.
(b) Diagram of a longitudinal view of the hair bulb of the hair follicle, which encloses the hair matrix, the actively dividing epithelial cells that produce the hair.

Hair consists of two primary regions: the **hair shaft**, the region projecting from the surface of the skin, and the **hair root**, which is beneath the surface of the skin and is embedded within the **hair follicle**. The **hair bulb** is a collection of well-nourished epithelial cells at the base of the hair follicle (Figure 7.4).

The hair shaft and the hair root have three layers of keratinized cells: the *medulla* in the center, surrounded by the *cortex*, and the protective *cuticle*. Abrasion of the cuticle at the tip of the hair shaft results in split ends. Hair color depends on the amount and type of melanin pigment found in the hair cortex.

- **Hair follicle:** A structure formed from both epidermal and dermal cells (Figure 7.4). Its **epithelial root sheath**, with two parts (internal and external), is enclosed by a thickened basement membrane, the glassy membrane, and by a **peripheral connective tissue** (or fibrous) **sheath**, which is essentially dermal tissue. A small nipple of dermal tissue protrudes into the hair bulb from the peripheral connective tissue sheath and provides nutrition to the growing hair. It is called the **hair papilla**. A layer of actively dividing epithelial cells called the **hair matrix** is located on top of the hair papilla.
- **Arrector pili muscle:** Small bands of smooth muscle cells connect each hair follicle to the papillary layer of the dermis (Figures 7.1 and 7.4). When these muscles contract (during cold or fright), the slanted hair follicle is pulled upright, dimpling the skin surface with goose bumps. This phenomenon is especially dramatic in a scared cat, whose fur actually stands on end to increase its apparent size.

7

Activity 3

Comparing Hairy and Relatively Hair-Free Skin Microscopically

Whereas thick skin has no hair follicles or sebaceous (oil) glands, thin skin usually has both. The scalp, of course, has the highest density of hair follicles.

1. Obtain a prepared slide of the human scalp, and study it carefully under the microscope. Compare your tissue slide to **Figure 7.5a**, and identify as many as possible of the diagrammed structures in Figure 7.1.

How is this stratified squamous epithelium different from that observed in the esophagus (Exercise 6)?

____

____

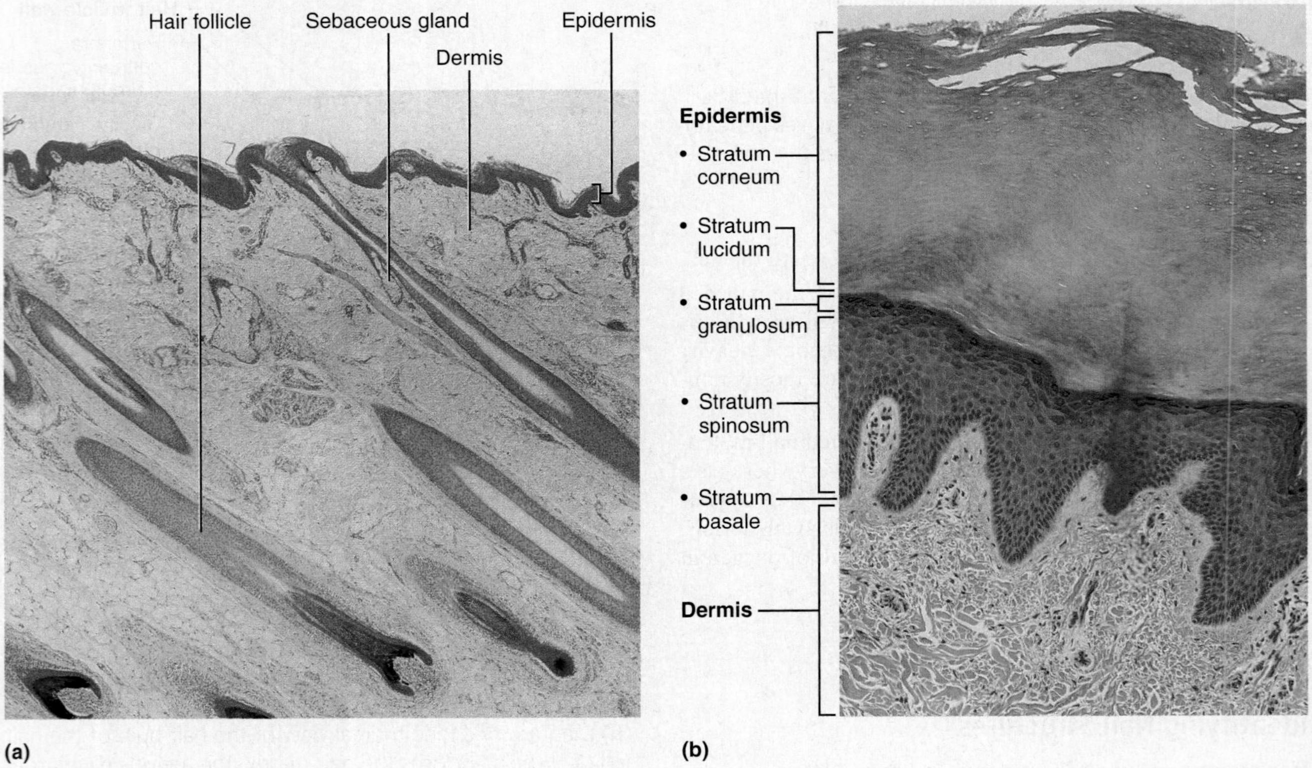

Figure 7.5 Photomicrographs of skin. (a) Thin skin with hairs (40×). **(b)** Thick skin (75×).

How do these differences relate to the functions of these two similar epithelia?

____

____

____

2. Obtain a prepared slide of hairless skin of the palm or sole (Figure 7.5b). Compare the slide to the photomicrograph in Figure 7.5a. In what ways does the thick skin of the palm or sole differ from the thin skin of the scalp?

____

____

Cutaneous Glands

The cutaneous glands fall primarily into two categories: the sebaceous glands and the sweat glands (Figure 7.1 and **Figure 7.6**).

Sebaceous (Oil) Glands

The sebaceous glands are found nearly all over the skin, except for the palms of the hands and the soles of the feet. Their ducts usually empty into a hair follicle, but some open directly on the skin surface.

Sebum is the product of sebaceous glands. It is a mixture of oily substances and fragmented cells that acts as a lubricant to keep the skin soft and moist (a natural skin cream) and keeps the hair from becoming brittle. The sebaceous glands become particularly active during puberty, when more male hormones (androgens) begin to be produced for both genders; thus the skin tends to become oilier during this period of life.

Blackheads are accumulations of dried sebum, bacteria, and melanin from epithelial cells in the oil duct. **Acne** is an active infection of the sebaceous glands.

Sweat (Sudoriferous) Glands

Sweat, or sudoriferous, glands are exocrine glands that are widely distributed all over the skin. Outlets for the glands are epithelial openings called *pores*. Sweat glands are categorized by the composition of their secretions.

- **Eccrine sweat glands:** Also called **merocrine sweat glands**, these glands are distributed all over the body. They produce a clear secretion consisting primarily of water, salts (mostly NaCl), and urea. Eccrine sweat glands, under the control of the nervous system, are an important part of the body's heat-regulating apparatus. They secrete sweat when the external temperature or body temperature is high. When sweat evaporates, it carries excess body heat with it.
- **Apocrine sweat glands:** Found predominantly in the axillary and genital areas, these glands secrete the basic components of eccrine sweat plus proteins and fat-rich substances. Apocrine sweat is an excellent nutrient medium for the microorganisms typically found on the skin. This sweat is initially odorless, but when bacteria break down its organic components, it begins to smell unpleasant.

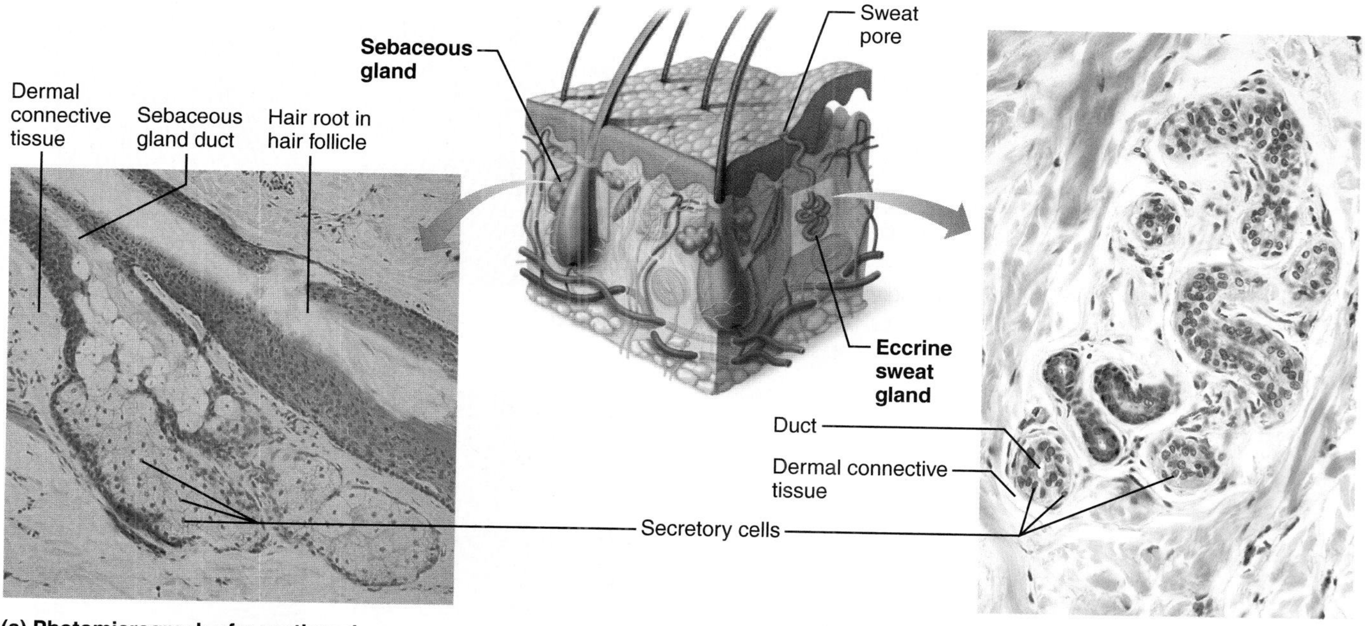

(a) Photomicrograph of a sectioned sebaceous gland (100×)

(b) Photomicrograph of a sectioned eccrine sweat gland (145×)

Figure 7.6 **Cutaneous glands.**

Activity 4

Differentiating Sebaceous and Sweat Glands Microscopically

Using the slide *thin skin with hairs* and the photomicrographs of cutaneous glands (Figure 7.6) as a guide, identify sebaceous and eccrine sweat glands. What characteristics relating to location or gland structure allow you to differentiate these glands?

____

____

____

7

Activity 5

Plotting the Distribution of Sweat Glands

1. Form a hypothesis about the relative distribution of sweat glands on the palm and forearm. Justify your hypothesis.

2. The bond paper for this simple experiment has been preruled in cm^2—put on disposable gloves, and cut along the lines to obtain the required squares. You will need two squares of bond paper (each 1 cm × 1 cm), adhesive tape, and a povidone-iodine swab *or* Lugol's iodine and a cotton-tipped swab.

3. Paint an area of the medial aspect of your left palm (avoid the deep crease lines) and a region of your left forearm with the iodine solution, and allow it to dry thoroughly. The painted area in each case should be slightly larger than the paper squares to be used.

4. Have your lab partner *securely* tape a square of bond paper over each iodine-painted area, and leave the paper squares in place for 20 minutes. (If it is very warm in the laboratory while this test is being conducted, you can obtain good results within 10 to 15 minutes.)

5. After 20 minutes, remove the paper squares, and count the number of blue-black dots on each square. The presence of a blue-black dot on the paper indicates an active sweat gland. The iodine in the pore is dissolved in the sweat and reacts chemically with the starch in the bond paper to produce the blue-black color. You have produced "sweat maps" for the two skin areas.

6. Which skin area tested has the greater density of sweat glands?

____

____

____

7. Tape your results (bond paper squares) to a data collection sheet labeled "palm" and "forearm" at the front of the lab. Be sure to put your paper squares in the correct columns on the data sheet.

8. Once all the data have been collected, review the class results.

9. Prepare a lab report for the experiment. (See Getting Started, on MasteringA&P).

Dermography: Fingerprinting

Each of us has a unique genetically determined set of fingerprints. Because fingerprinting is useful for identifying and apprehending criminals, most people associate this craft solely with criminal investigations. However, fingerprints are also invaluable for quick identification of amnesia victims, missing persons, and unknown deceased, such as people killed in major disasters.

The friction ridges responsible for fingerprints appear in several patterns, which are clearest when the fingertips are inked and then pressed against white paper. Impressions are also made when perspiration or any foreign material such as blood, dirt, or grease adheres to the ridges and the fingers are then pressed against a smooth, nonabsorbent surface. The three most common patterns are *arches*, *loops*, and *whorls* (**Figure 7.7**).

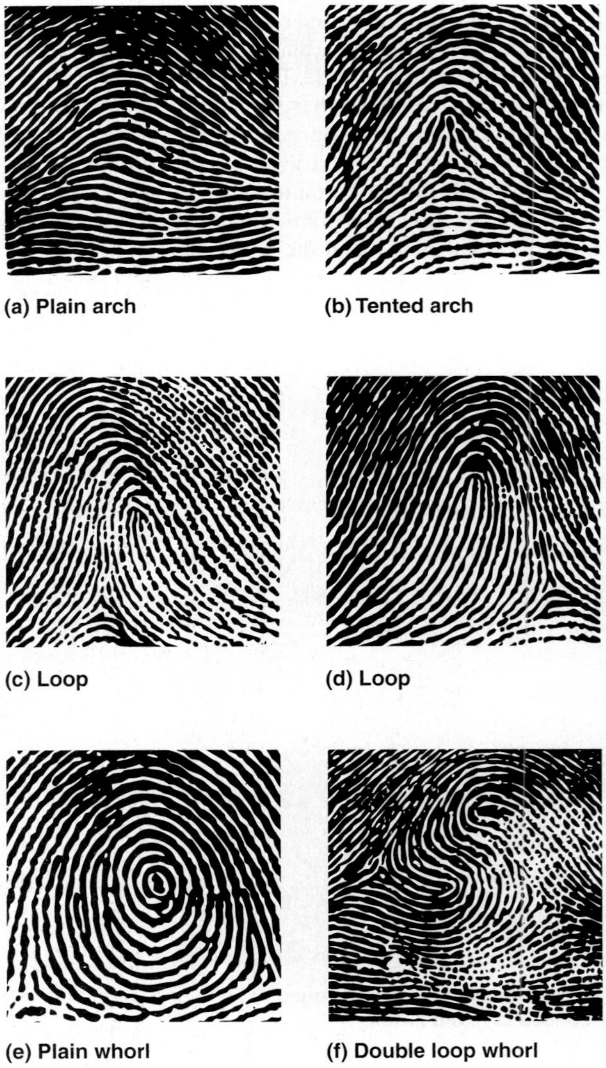

(a) Plain arch (b) Tented arch (c) Loop (d) Loop (e) Plain whorl (f) Double loop whorl

Figure 7.7 Main types of fingerprint patterns. **(a, b)** Arches. **(c, d)** Loops. **(e, f)** Whorls.

Activity 6

Taking and Identifying Inked Fingerprints

For this activity, you will be working as a group with your lab partners. Though the equipment for professional fingerprinting is fairly basic, consisting of a glass or metal inking plate, printer's ink (a heavy black paste), ink roller, and standard 8 in. × 8 in. cards, you will be using supplies that are even easier to handle. Each student will prepare two index cards, each bearing his or her thumbprint and index fingerprint of the right hand.

1. Obtain the following supplies and bring them to your bench: two 4 in. × 6 in. index cards per student, Porelon® fingerprint pad or portable inking foils, ink cleaner towelettes, and a magnifying glass.

2. The subject should wash and dry the hands. Open the ink pad or peel back the covering over the ink foil, and position it close to the edge of the laboratory bench. The subject should position himself or herself at arm's length from the bench edge and inking object.

3. A second student, called the *operator*, stands to the left of the subject and with two hands holds and directs movement of the subject's fingertip. During this process, the subject should look away, try to relax, and refrain from trying to help the operator.

4. The thumbprint is to be placed on the left side of the index card, the index fingerprint on the right. The operator should position the subject's right thumb or index finger on the side of the bulb of the finger in such a way that the area to be inked spans the distance from the fingertip to just beyond the first joint, and then roll the finger lightly across the inked surface until its bulb faces in the opposite direction. To prevent smearing, the thumb is rolled away from the body midline (from left to right as the subject sees it; see **Figure 7.8**), and the index finger is rolled toward the body midline (from right to left). The same ink foil can be reused for all the students at the bench; the ink pad is good for thousands of prints. Repeat the procedure (still using the subject's right hand) on the second index card.

5. If the prints are too light, too dark, or smeary, repeat the procedure.

6. While other students in the group are making clear prints of their thumb and index finger, those who have completed that activity should clean their inked fingers with a towelette and attempt to classify their own prints as arches, loops, or whorls. Use the magnifying glass as necessary to see ridge details.

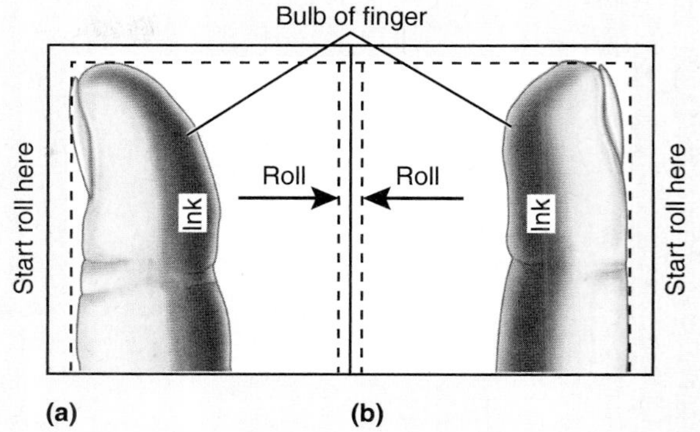

Figure 7.8 Fingerprinting. Method of inking and printing **(a)** the thumb and **(b)** the index finger of the right hand.

7. When all group members at a bench have completed the above steps, they are to write their names on the backs of their index cards. The students combine their cards, shuffle them, and transfer them to the opposite bench. Finally, students classify the patterns and identify which prints were made by the same individuals.

How difficult was it to classify the prints into one of the three categories given?

____

Why do you think this is so?

____

Was it easy or difficult to identify the prints made by the same individual?

____

Why do you think this was so?

____

Instructors may assign a portion of the Review Sheet questions using Mastering A&P™

EXERCISE 7

REVIEW SHEET
The Integumentary System

Name ______________________ Lab Time/Date ______________________

Basic Structure of the Skin

1. Complete the following statements by writing the appropriate word or phrase on the blank line:

1. The superficial region of the skin is the__________, composed of __________ __________ __________ (3 words) tissue.

2. The deeper region tissue is the __________, composed of connective tissue.

3. The most numerous cell of the epidermis is the __________.

4. The two primary layers of the dermis are the __________ dermis, composed of areolar connective tissue, and the __________ dermis, composed of dense irregular connective tissue.

2. Four protective functions of the skin are:

a. __________ c. __________

b. __________ d. __________

3. Using the key choices, choose all responses that apply to the following descriptions. Some terms are used more than once.

Key: a. stratum basale
b. stratum corneum
c. stratum granulosum
d. stratum lucidum
e. stratum spinosum
f. papillary dermis
g. reticular dermis

__________ 1. layer of translucent cells in thick skin containing dead keratinocytes

__________ 2. two layers containing dead cells

__________ 3. dermal layer responsible for fingerprints

__________ 4. epidermal layer exhibiting the most rapid cell division

__________ 5. layer including scalelike dead cells, full of keratin, that constantly slough off

__________ 6. layer named for the numerous granules present

__________ 7. location of melanocytes and tactile epithelial cells

__________ 8. area where weblike pre-keratin filaments first appear

__________ 9. deep layer of the dermis

__________ 10. layer that secretes a glycolipid that prevents water loss from the skin

4. Label the integumentary structures and areas indicated in the diagram.

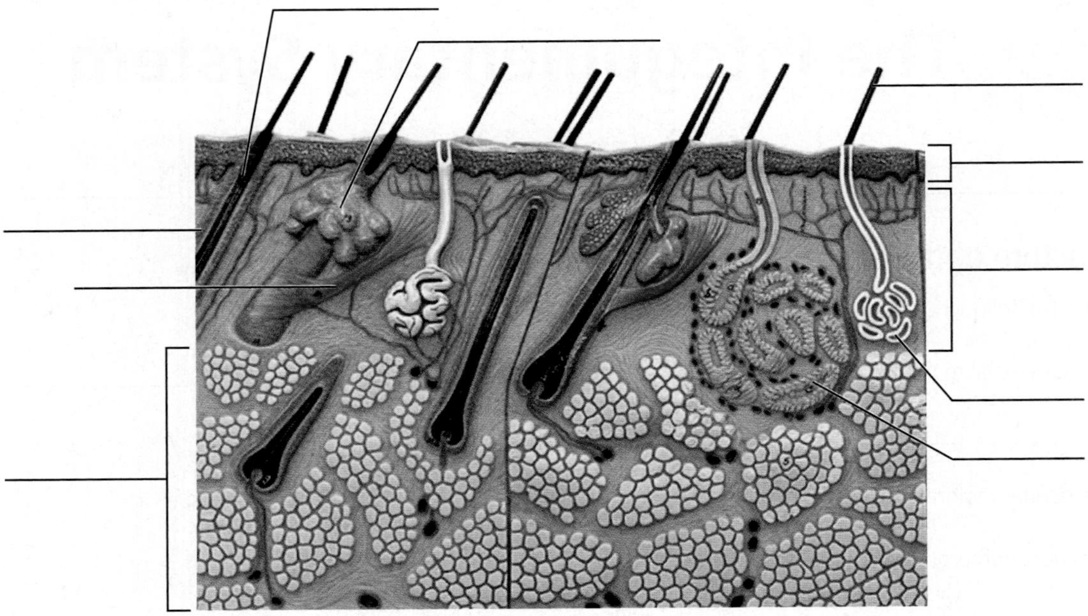

5. Label the layers of the epidermis in thick skin. Then, complete the statements that follow.

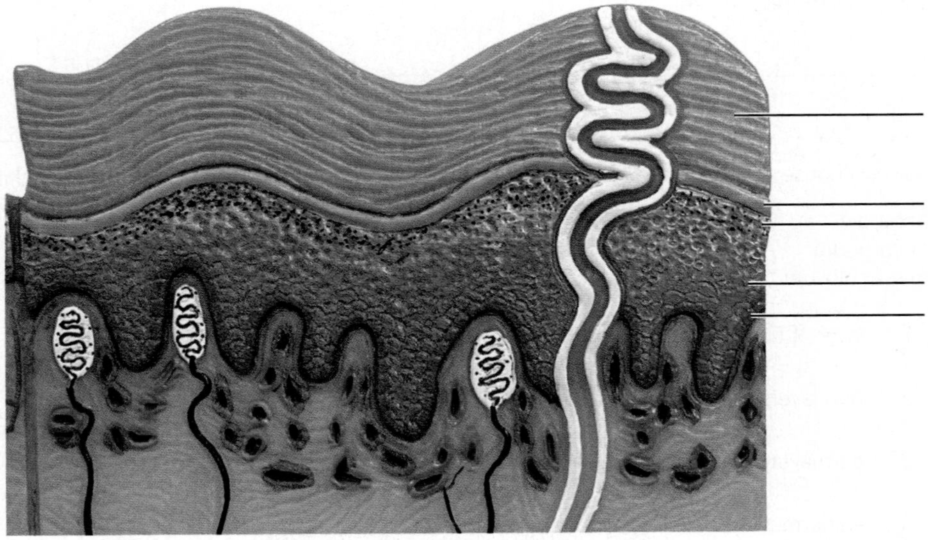

a. Glands that respond to rising androgen levels are the ______________________ glands.

b. ________________ ________________ are epidermal cells that play a role in the immune response.

c. Tactile corpuscles are located in the ________________________________ ________________________________.

d. ________________________ corpuscles are located deep in the dermis.

6. What substance is manufactured in the skin and plays a role in calcium absorption elsewhere in the body?

____

7. List the sensory receptors found in the dermis of the skin. ______________________________

____

____

Accessory Organs of the Skin

8. Match the key choices with the appropriate descriptions. Some terms are used more than once.

Key: a. arrector pili
b. cutaneous receptors
c. hair
d. hair follicle
e. nail
f. sebaceous glands
g. sweat gland—apocrine
h. sweat gland—eccrine

__________ 1. tiny muscles, attached to hair follicles, that pull the hair upright during fright or cold

__________ 2. sweat gland with a role in temperature control

__________ 3. sheath formed of both epithelial and connective tissues

__________ 4. less numerous type of sweat-producing gland; found mainly in the pubic and axillary regions

__________ 5. primarily dead/keratinized cells (two responses from key)

__________ 6. specialized nerve endings that respond to temperature, touch, etc.

__________ 7. secretes a lubricant for hair and skin

9. Describe two integumentary system mechanisms that help regulate body temperature. ______________________________

____

____

10. Several structures of the hair are listed below. Identify each by matching its letter with the appropriate area on the photomicrograph.

a. cortex

b. cuticle

c. hair matrix

d. hair papilla

e. medulla

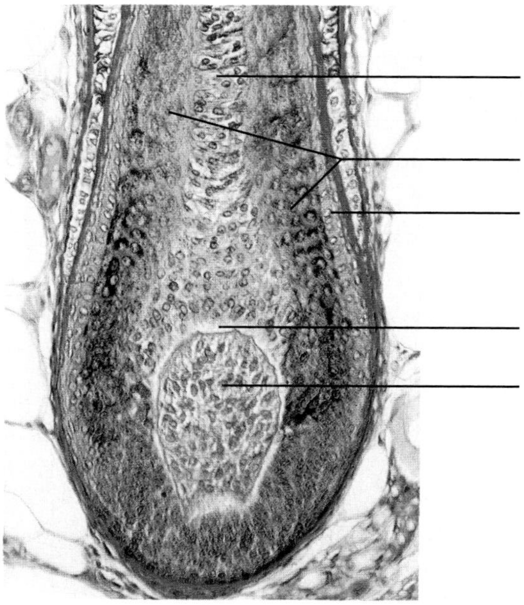

Plotting the Distribution of Sweat Glands

11. With what substance in the bond paper does the iodine painted on the skin react? ____________

12. Based on class data, which skin area—the forearm or palm of hand—has more sweat glands? ____________

Was this an expected result? ____________ Explain. ____________

____

Which other body areas would, if tested, prove to have a high density of sweat glands? ____________

____

13. What organ system controls the activity of the eccrine sweat glands? ____________

Dermography: Fingerprinting

14. Why can fingerprints be used to identify individuals? ____________

____

15. Name the three common fingerprint patterns.

____________, ____________ and ____________

16. Henna tattoos are temporary tattoos that last about 2 weeks. Hypothesize why henna tattoos do not last as long as permanent tattoos. ____________

____

____

17. ✚ Vitiligo is a disorder in which the pigmentation of the skin is uneven, resulting in white patches. Recent research suggests that vitiligo might be an autoimmune disorder. Which cells would you expect to be most affected, and why?

____

____

18. ✚ Keratinase is an enzyme produced by dermatophytes. Which organs in the body would these pathogenic fungi tend to proliferate in, and why? ____________

____

EXERCISE 8

Overview of the Skeleton: Classification and Structure of Bones and Cartilages

Learning Outcomes

- Name the two primary tissue types that form the skeleton.
- List the functions of the skeletal system.
- Locate and identify the three major types of skeletal cartilages.
- Name the four main groups of bones based on shape.
- Identify surface bone markings and list their functions.
- Identify the major anatomical areas on a longitudinally cut long bone or on an appropriate image.
- Explain the role of inorganic salts and organic matrix in providing flexibility and hardness to bone.
- Locate and identify the major parts of an osteon microscopically, or on a histological model or appropriate image of compact bone.

Pre-Lab Quiz

Instructors may assign these and other Pre-Lab Quiz questions using Mastering A&P™

1. All the following are functions of the skeleton *except*:
   a. attachment for muscles
   b. production of melanin
   c. site of red blood cell formation
   d. storage of lipids
2. Circle the correct underlined term. The axial / appendicular skeleton consists of bones that surround the body's center of gravity.
3. The type of cartilage that has the greatest strength and is found in the knee joint and intervertebral discs is:
   a. elastic **b.** fibrocartilage **c.** hyaline
4. Circle the correct underlined term. The shaft of a long bone is known as the epiphysis / diaphysis.
5. The structural unit of compact bone is the:
   a. osteon **b.** canaliculus **c.** lacuna

Go to Mastering A&P™ > Study Area to improve your performance in A&P Lab.

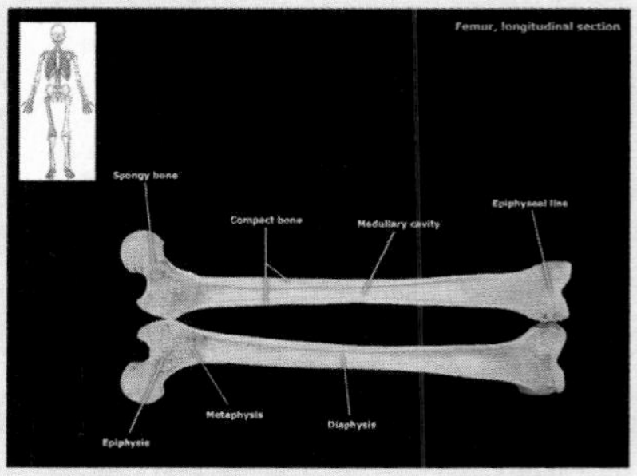

> Lab Tools > Practice Anatomy Lab > Human Cadaver: Appendicular Skeleton

Instructors may assign new Building Vocabulary coaching activities, Pre-Lab Quiz questions, Art Labeling activities, Practice Anatomy Lab Practical questions (PAL), and more using the Mastering A&P™ Item Library.

Materials

- Long bone sawed longitudinally (beef bone from a slaughterhouse, if possible, or prepared laboratory specimen)
- Disposable gloves
- Long bone soaked in 10% hydrochloric acid (HCl) (or vinegar) until flexible
- Long bone baked at 250°F for more than 2 hours
- Compound microscope
- Prepared slide of ground bone (x.s.)
- Three-dimensional model of microscopic structure of compact bone
- Prepared slide of a developing long bone undergoing endochondral ossification
- Articulated skeleton

Besides supporting and protecting the body as an internal framework, the **skeleton** provides a system of levers with which the skeletal muscles work to move the body. In addition, the bones store lipids and many minerals (the most important of which is calcium). Finally, the red marrow of bones provides a site for blood cell formation.

The skeleton is made up of bones that are connected at *joints*, or *articulations*. The skeleton is subdivided into two divisions: the **axial skeleton** (the bones that lie around the body's center of gravity) and the **appendicular skeleton** (bones of the limbs, or appendages) (**Figure 8.1**, p. 104).

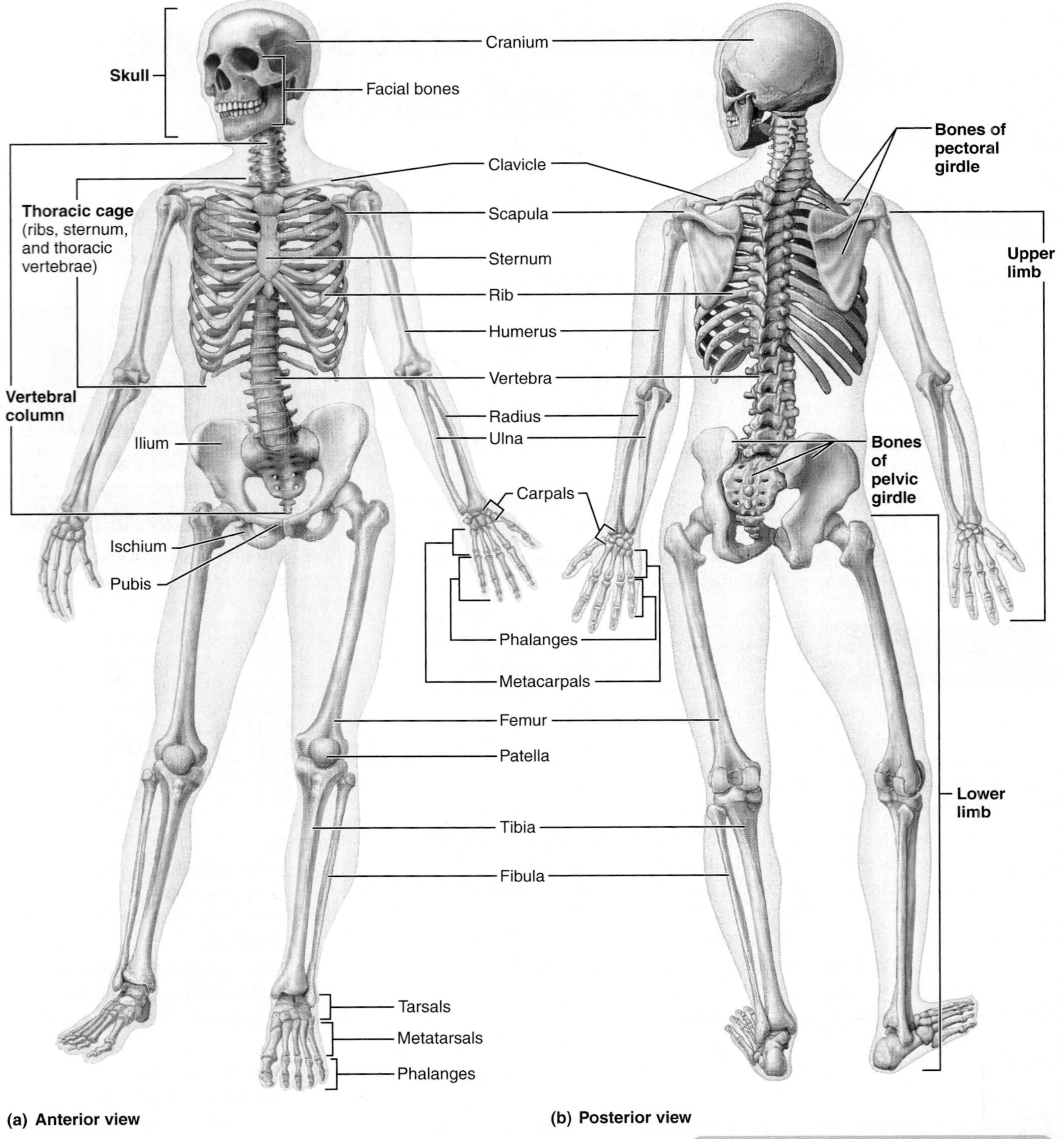

Figure 8.1 The human skeleton. The bones of the axial skeleton are colored green to distinguish them from the bones of the appendicular skeleton.

Instructors may assign this figure as an Art Labeling Activity using Mastering A&P™

Cartilages of the Skeleton

The three types of cartilage found in the body are hyaline cartilage, elastic cartilage, and fibrocartilage. In embryos, the skeleton is predominantly made up of hyaline cartilage, but in adults, most of the cartilage has been replaced by more rigid bone.

Cartilage contains no nerves or blood vessels and is surrounded by a covering of dense irregular connective tissue, called a *perichondrium*, which acts to resist distortion of the cartilage. You will see from the examples that hyaline cartilage is the most common type found in the body, whereas elastic cartilage is known for its flexibility. Fibrocartilage provides strength and shock absorption.

- **Hyaline cartilage:**
  - *Articular cartilages*—cover the ends of most bones at movable joints
  - *Costal cartilages*—connect the ribs to the sternum
  - *Respiratory cartilages*—found in the larynx and other respiratory structures
  - *Nasal cartilages*—support the external nose
- **Elastic cartilage:**
  - Found in the external ear and the epiglottis (the guardian of the airway)
- **Fibrocartilage:**
  - *Intervertebral discs*—pads located between the vertebrae
  - *Menisci*—pads located in the knee joint
  - *Pubic symphysis*—located where the hip bones join anteriorly

Bone Classification and Bone Markings

The 206 bones of the adult skeleton are composed of two basic kinds of osseous tissue that differ in their texture. **Compact bone** is dense and made up of organizational units called *osteons*. **Spongy** (or *cancellous*) **bone** is composed of small *trabeculae* (columns) of bone and lots of open space.

Bones may be classified further on the basis of their gross anatomy into four groups: long, short, flat, and irregular bones (**Figure 8.2**).

Long bones, such as the femur and phalanges (singular: phalanx), are much longer than they are wide, generally consisting of a shaft with heads at either end. Long bones are composed mostly of compact bone. **Short bones** are typically cube shaped, and they contain more spongy bone than compact bone. The tarsals and carpals are examples.

Flat bones are generally thin, with two waferlike layers of compact bone sandwiching a thicker layer of spongy bone between them. Although the name "flat bone" implies a structure that is straight, many flat bones are curved (for example, the bones of the cranium). Bones that do not fall into one of the preceding categories are classified as **irregular bones**. The vertebrae are irregular bones.

8

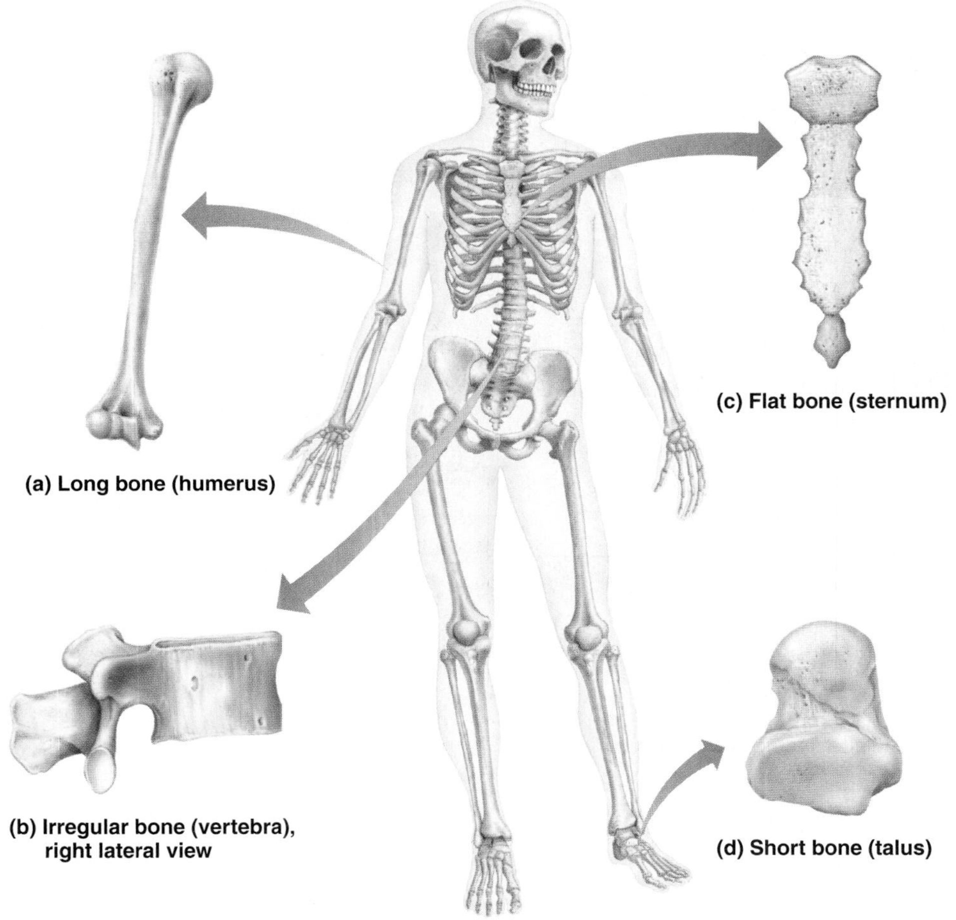

Figure 8.2 Classification of bones on the basis of shape.

Some anatomists also recognize two other subcategories of bones. **Sesamoid bones** are special types of short bones formed within tendons. The patellas (kneecaps) are sesamoid bones. **Sutural bones** are tiny bones between cranial bones.

Even a casual observation of the bones will reveal that bone surfaces are not featureless smooth areas. They have an array of bumps, holes, and ridges. These **bone markings** reveal where bones form joints with other bones, where muscles, tendons, and ligaments were attached, and where blood vessels and nerves passed. Bone markings fall into two main categories: projections that grow out from the bone and serve as sites of muscle attachment or help form joints; and depressions or openings in the bone that often serve as conduits for nerves and blood vessels. The bone markings are summarized in **Table 8.1**.

Table 8.1 Bone Markings

Name of bone marking	Description
Projections That Are Sites of Muscle and Ligament Attachment	
Tuberosity	Large rounded projection; may be roughened
Crest	Narrow ridge of bone; usually prominent
Trochanter	Very large, blunt, irregularly shaped process (the only examples are on the femur)
Line	Narrow ridge of bone; less prominent than a crest
Tubercle	Small rounded projection or process
Epicondyle	Raised area on or above a condyle
Spine	Sharp, slender, often pointed projection
Process	Bony prominence

Illustration:

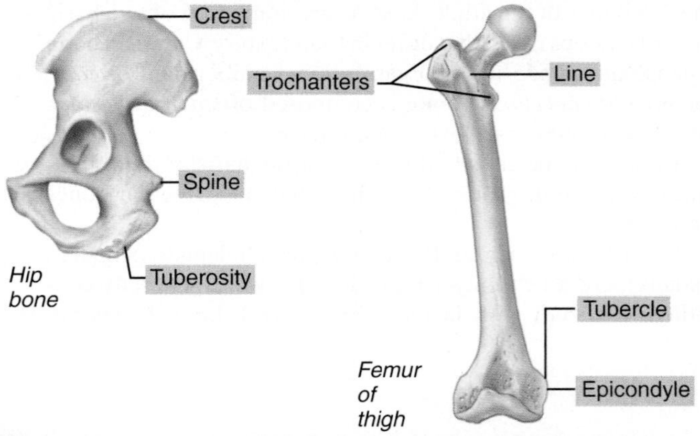

Name of bone marking	Description
Projections That Help Form Joints	
Head	Bony expansion carried on a narrow neck
Facet	Smooth, nearly flat articular surface
Condyle	Rounded articular projection
Ramus	Armlike bar of bone

Illustration:

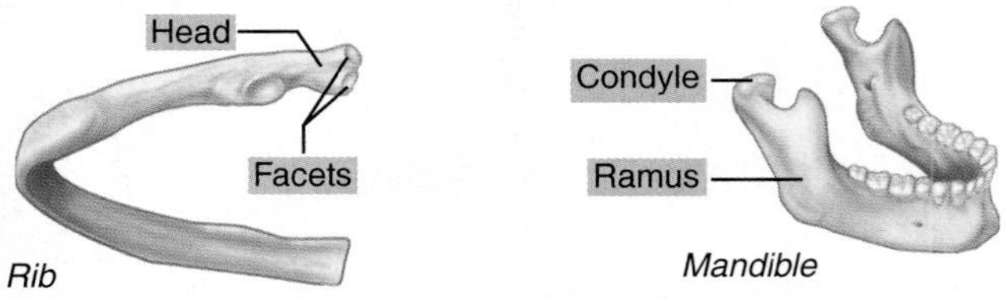

Name of bone marking	Description
Depressions and Openings	
For Passage of Vessels and Nerves	
Fissure	Narrow, slitlike opening
Foramen	Round or oval opening through a bone
Notch	Indentation at the edge of a structure
Others	
Meatus	Canal-like passageway
Sinus	Bone cavity, filled with air and lined with mucous membrane
Fossa	Shallow basinlike depression in a bone, often serving as an articular surface

Illustration:

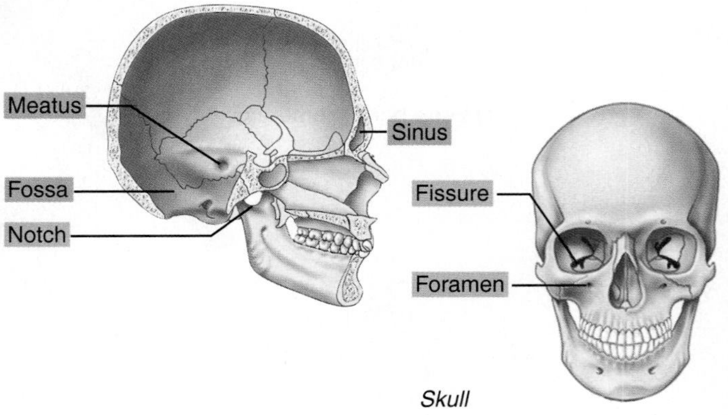

Gross Anatomy of the Typical Long Bone

Activity 1

Instructors may assign a related video and coaching activity for Examining a Long Bone using Mastering A&P™

Examining a Long Bone

1. Obtain a long bone that has been sawed along its longitudinal axis. If a cleaned dry bone is provided, no special preparations need be made.

Note: If the bone supplied is a fresh beef bone, don disposable gloves before beginning your observations.

Identify the **diaphysis** or shaft (**Figure 8.3** may help). Observe its smooth surface, which is composed of compact bone. If you are using a fresh specimen, carefully pull away the **periosteum**, a fibrous membrane covering made up of dense irregular connective tissue, to view the bone surface. Notice that many fibers of the periosteum penetrate into the bone. These collagen fibers are called **perforating fibers**. Blood vessels and nerves travel through the periosteum and invade the bone. *Osteoblasts* (bone-forming cells) and *osteoprogenitor cells* are found on the inner, or osteogenic, layer of the periosteum.

2. Now inspect the **epiphysis**, the end of the long bone. Notice that it is composed of a thin layer of compact bone that encloses spongy bone.

Text continues on next page. →

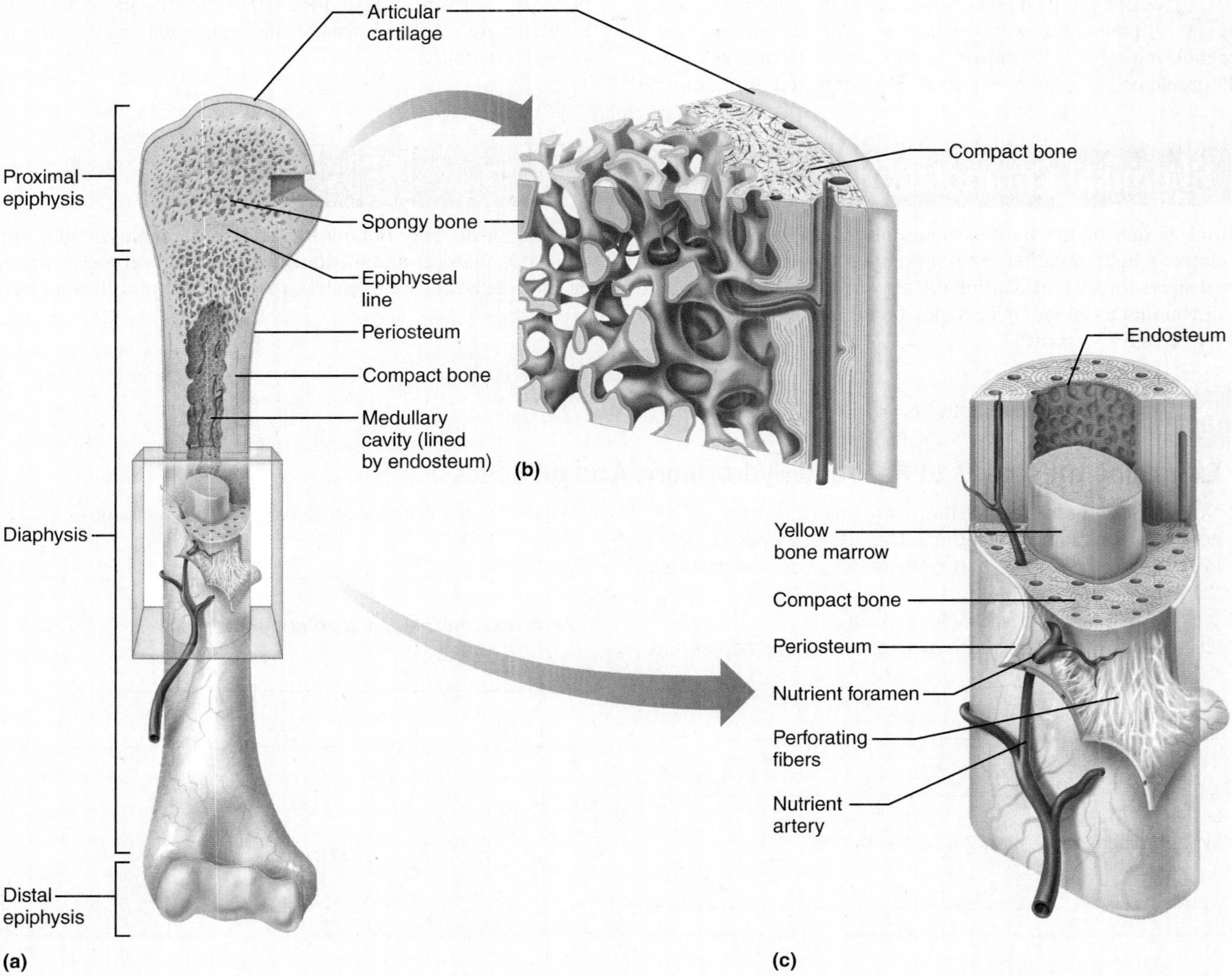

Figure 8.3 The structure of a long bone (humerus of the arm). (a) Anterior view with longitudinal section cut away at the proximal end. **(b)** Pie-shaped, three-dimensional view of spongy bone and compact bone of the epiphysis. **(c)** Cross section of diaphysis (shaft). Note that the external surface of the diaphysis is covered by a periosteum but that the articular surface of the epiphysis is covered with hyaline cartilage.

Instructors may assign this figure as an Art Labeling Activity using Mastering A&P™

3. Identify the **articular cartilage**, which covers the epiphysis in place of the periosteum. The glassy hyaline cartilage provides a smooth surface to minimize friction at joints.

4. If the animal was still young and growing, you will be able to see the **epiphyseal plate**, a thin area of hyaline cartilage that provides for longitudinal growth of the bone during youth. Once the long bone has stopped growing, these areas are replaced with bone and appear as thin, barely discernible remnants—the **epiphyseal lines**.

5. In an adult animal, the central cavity of the shaft (*medullary cavity*) is essentially a storage region for adipose, or **yellow bone marrow**. In the infant, this area is involved in forming blood cells, and so **red bone marrow** is found in the marrow cavities. In adult bones, the red bone marrow is confined to the interior of the epiphyses, where it occupies the spaces between the trabeculae of spongy bone.

6. If you are examining a fresh bone, look carefully to see if you can distinguish the delicate **endosteum** lining the shaft. The endosteum also covers the trabeculae of spongy bone and lines the central and perforating canals of compact bone. Like the periosteum, the endosteum contains osteoprogenitor cells that differentiate into osteoblasts. As the bone grows in diameter on its external surface, it is constantly being broken down on its inner surface. Thus the thickness of the compact bone layer composing the shaft remains relatively constant.

! 7. If you have been working with a fresh bone specimen, return it to the appropriate area and properly dispose of your gloves, as designated by your instructor.

Longitudinal bone growth at epiphyseal plates (growth plates) follows a predictable sequence and provides a reliable indicator of the age of children exhibiting normal growth. If problems of long-bone growth are suspected (for example, pituitary dwarfism), X-ray films are taken to view the width of the growth plates. An abnormally thin epiphyseal plate indicates growth retardation. +

Chemical Composition of Bone

Bone is one of the hardest materials in the body. Although relatively light, bone has a remarkable ability to resist tension and shear forces that continually act on it. An engineer would tell you that a cylinder (like a long bone) is one of the strongest structures for its mass.

The hardness of bone is due to the inorganic calcium salts deposited in its ground substance. Its flexibility comes from the organic elements of the matrix, particularly the collagen fibers.

Activity 2

Examining the Effects of Heat and Hydrochloric Acid on Bones

Obtain a bone sample that has been soaked in hydrochloric acid (HCl) (or in vinegar) and one that has been baked. Heating removes the organic part of bone, whereas acid dissolves the minerals.

Gently apply pressure to each bone sample. What happens to the heated bone?

____

____

What happens to the bone treated with acid?

____

What does the acid appear to remove from the bone?

____

What does baking appear to do to the bone?

____

____

Microscopic Structure of Compact Bone

Spongy bone has a spiky, open-work appearance, resulting from the arrangement of the **trabeculae** that compose it, whereas compact bone appears to be dense and homogeneous on the outer surface. However, microscopic examination of compact bone reveals that it is riddled with passageways carrying blood vessels, nerves, and lymphatic vessels that provide the living bone cells with needed substances and a way to eliminate wastes.

Activity 3

Examining the Microscopic Structure of Compact Bone

1. Obtain a prepared slide of ground bone, and examine it under low power. Focus on a central canal (**Figure 8.4**). The **central canal** (*Haversian canal*) runs parallel to the long axis of the bone and carries blood vessels, nerves, and lymphatic vessels through the bony matrix. Identify the **osteocytes** (mature bone cells) in **lacunae** (chambers), which are arranged in concentric circles called **concentric lamellae** around the central canal. Because bone remodeling is going on all the time, you will also

Text continues on next page. →

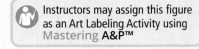

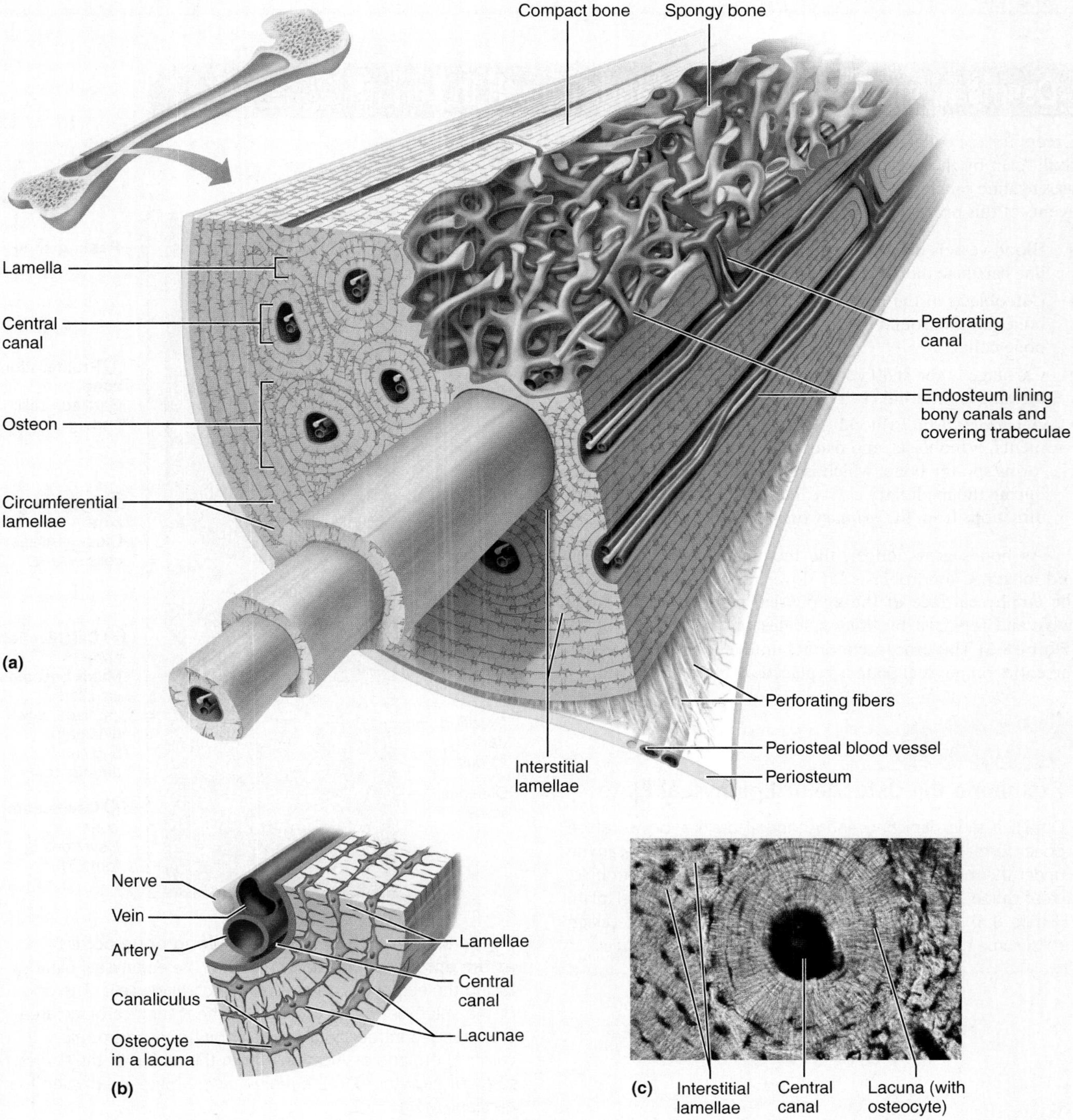

Figure 8.4 Microscopic structure of compact bone. (a) Diagram of a pie-shaped segment of compact bone, illustrating its structural units (osteons). **(b)** A portion of one osteon. **(c)** Photomicrograph of a cross-sectional view of an osteon (320×).

Instructors may assign this figure as an Art Labeling Activity using Mastering **A&P™**

8

see some *interstitial lamellae*, remnants of osteons that have been broken down (Figure 8.4c).

A central canal and all the concentric lamellae surrounding it are referred to as an **osteon**, or *Haversian system*. Also identify **canaliculi**, tiny canals radiating outward from a central canal to the lacunae of the first lamella and then from lamella to lamella. The canaliculi form a dense transportation network through the hard bone matrix, connecting all the living cells of the osteon to the nutrient supply. You may need a higher-power magnification to see the fine canaliculi.

2. Also note the **perforating canals** (*Volkmann's canals*) (Figure 8.4). These canals run at right angles to the shaft and connect the blood and nerve supply of the medullary cavity to the central canals.

3. If a model of bone histology is available, identify the same structures on the model.

Ossification: Bone Formation and Growth in Length

Except for the collarbones, all bones of the body inferior to the skull form by the process of **endochondral ossification**, which uses hyaline cartilage as a model for bone formation. The major events of this process are as follows:

- Blood vessels invade the perichondrium covering the hyaline cartilage model and convert it to a periosteum.
- Osteoblasts at the inner surface of the periosteum secrete bone matrix around the hyaline cartilage model, forming a bone collar.
- Cartilage in the shaft center calcifies and then hollows out, forming an internal cavity.
- A *periosteal bud* (blood vessels, nerves, red marrow elements, osteoblasts, and osteoclasts) invades the cavity and forms spongy bone, which is removed by osteoclasts, producing the medullary cavity. This process proceeds in both directions from the *primary ossification center*.

As bones grow longer, the medullary cavity gets larger and longer. Chondroblasts lay down new cartilage matrix on the epiphyseal face of the epiphyseal plate, and it is eroded away and replaced by bony spicules on the diaphyseal face (**Figure 8.5**). This process continues until late adolescence when the entire epiphyseal plate is replaced by bone.

Activity 4

Examining the Osteogenic Epiphyseal Plate

Obtain a slide depicting endochondral ossification (cartilage bone formation), and bring it to your bench to examine under the microscope. Identify the proliferation, hypertrophic, calcification, and ossification zones of the epiphyseal plate (Figure 8.5). Then, also identify the area of resting cartilage cells, some hypertrophied chondrocytes, and bony spicules.

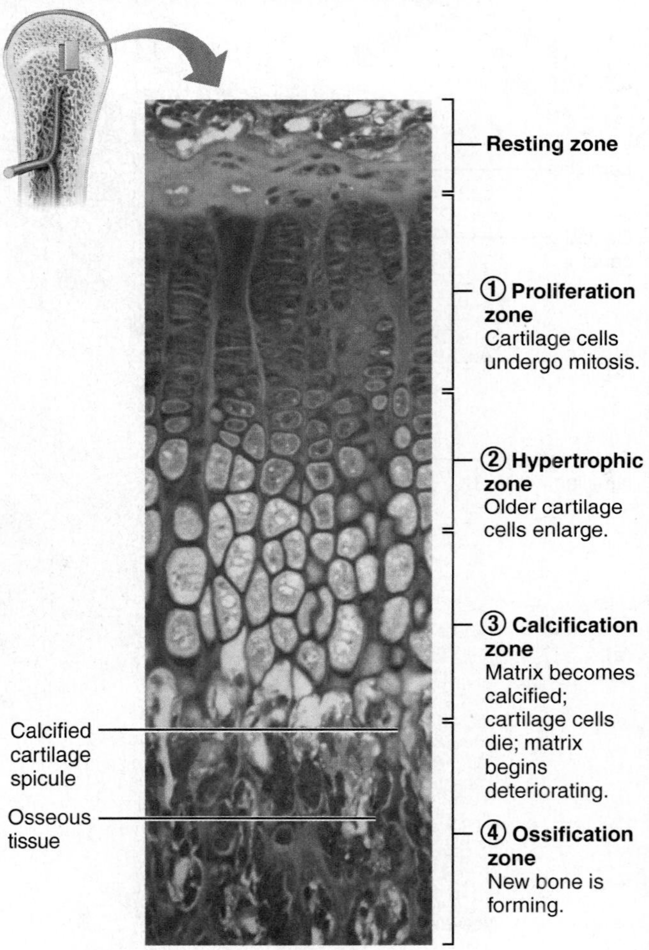

Figure 8.5 Growth in length of a long bone occurs at the epiphyseal plate. The side of the epiphyseal plate facing the epiphysis contains resting cartilage cells. The cells of the epiphyseal plate proximal to the resting cartilage area are arranged in four zones—proliferation, hypertrophic, calcification, and ossification—from the region of the earliest stage of growth ① to the region where bone is replacing the cartilage ④ (125×).

EXERCISE 8

REVIEW SHEET

Overview of the Skeleton: Classification and Structure of Bones and Cartilages

Instructors may assign a portion of the Review Sheet questions using Mastering A&P™

Name ______________________ Lab Time/Date ______________________

Cartilages of the Skeleton

1. Using the key choices, identify each type of cartilage described (in terms of its body location or function) below.

Key: a. elastic b. fibrocartilage c. hyaline

_______ 1. supports the external ear

_______ 2. between the vertebrae

_______ 3. forms the walls of the voice box (larynx)

_______ 4. the epiglottis

_______ 5. articular cartilages

_______ 6. meniscus in a knee joint

_______ 7. connects the ribs to the sternum

_______ 8. most effective at resisting compression

_______ 9. most springy and flexible

_______ 10. most abundant

Classification of Bones

2. The four major anatomical classifications of bones are long, short, flat, and irregular. Which category has the least amount of spongy bone relative to its total volume? ______________________

3. Classify each of the bones in the chart below as either long, short, flat, or irregular by placing a check mark in the appropriate column. Also use a check mark to indicate whether the bone is a part of the axial or the appendicular skeleton. Use Figure 8.1 as a guide.

	Long	Short	Flat	Irregular	Axial skeleton	Appendicular skeleton
Sternum						
Radius						
Calcaneus (tarsal bone)						
Parietal bone (cranial bone)						
Phalanx (single bone of a digit)						
Vertebra						

Bone Markings

4. Match the terms in column B with the appropriate description in column A.

Column A		Column B
____________	1. sharp, slender process*	a. condyle
____________	2. small rounded projection*	b. crest
____________	3. narrow ridge of bone*	c. epicondyle
____________	4. large rounded projection*	d. facet
____________	5. structure supported on neck†	e. fissure
____________	6. armlike projection†	f. foramen
____________	7. rounded, articular projection†	g. fossa
____________	8. narrow slitlike opening‡	h. head
____________	9. canal-like structure	i. meatus
____________	10. round or oval opening through a bone‡	j. process
____________	11. shallow depression	k. ramus
____________	12. air-filled cavity	l. sinus
____________	13. large, irregularly shaped projection*	m. spine
____________	14. raised area on or above a condyle*	n. trochanter
____________	15. bony projection	o. tubercle
____________	16. smooth, nearly flat articular surface†	p. tuberosity

*a site of muscle and ligament attachment
†takes part in joint formation
‡a passageway for nerves or blood vessels

Gross Anatomy of the Typical Long Bone

5. Match the key terms with the descriptions.

Key:
a. articular cartilage
b. diaphysis
c. endosteum
d. epiphyseal line
e. epiphysis
f. medullary cavity
g. periosteum
h. red marrow

____________ 1. end portion of a long bone

____________ 2. helps reduce friction at joints

____________ 3. site of blood cell formation

____________ 4. two membranous sites of osteoprogenitor cells

____________ 5. scientific term for bone shaft

____________ 6. contains yellow marrow in adult bones

____________ 7. growth plate remnant

6. Use the key terms to identify the structures marked by leader lines in the diagrams. (Some terms are used more than once.)

Key:
- a. articular cartilage
- b. compact bone
- c. diaphysis
- d. endosteum
- e. epiphyseal line
- f. epiphysis
- g. medullary cavity
- h. nutrient artery
- i. periosteum
- j. spongy bone
- k. yellow bone marrow

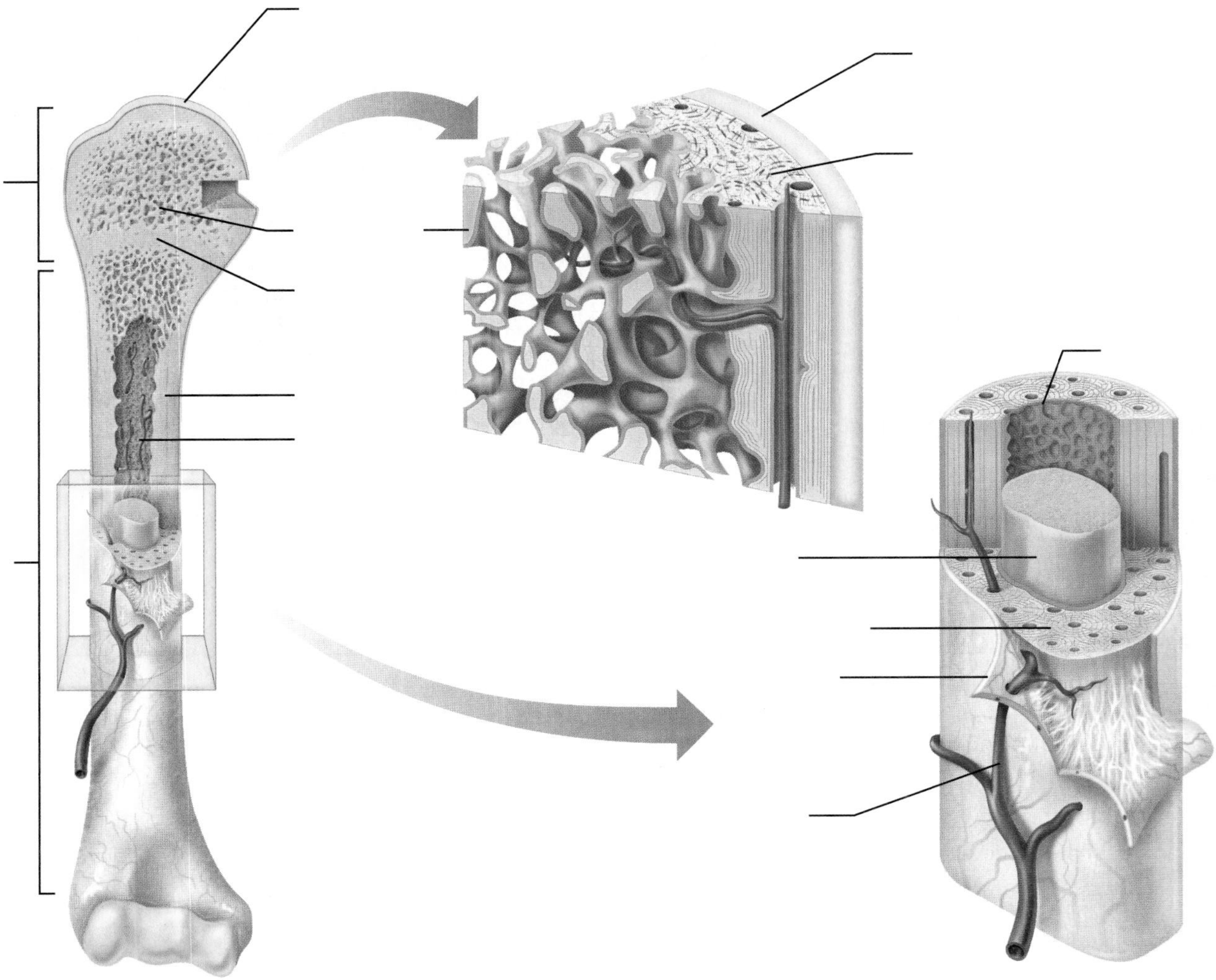

Chemical Composition of Bone

7. What is the function of the organic matrix in bone?

8. Name the important organic bone components. ______________________________

9. Calcium salts form the bulk of the inorganic material in bone. What is the function of the calcium salts?

____

10. Baking removes ______________________ from bone. Soaking bone in acid removes ______________________.

Microscopic Structure of Compact Bone

11. Several descriptions of bone structure are given below. Identify the structure involved by choosing the appropriate term from the key and placing its letter in the blank. Then, on the photomicrograph of bone on the right (175×), identify all structures named in the key and draw a bracket enclosing a single osteon.

Key: a. canaliculi b. central canal c. lacuna d. lamella

__________ 1. layer of bony matrix around a central canal

__________ 2. site of osteocytes

__________ 3. longitudinal canal carrying blood vessels, lymphatics, and nerves

__________ 4. tiny canals connecting osteocytes of an osteon

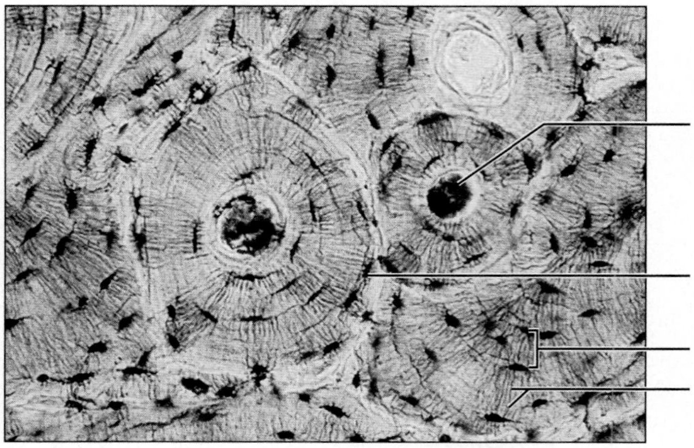

Ossification: Bone Formation and Growth in Length

12. Compare and contrast events occurring on the epiphyseal and diaphyseal faces of the epiphyseal plate.

Epiphyseal face: __________

Diaphyseal face: __________

13. + The pain in the leg that is referred to as "shin splints" is often caused by microtears in the periosteum and perforating fibers. These tears lead to inflammation of the periosteum. Considering the type of tissue found in the periosteum, which cells do you think would be most involved in the repair process?

____

14. + In a child with rickets, the bones are not properly calcified. Which treated bone in Activity 2 most closely resembles the bones of a child with rickets? Why? __________

____

____

15. + Achondroplasia is a type of dwarfism in which the long bones stop growing during childhood, resulting in limbs that are disproportionately shorter than the torso. This genetic disorder is characterized by deficiencies in the epiphyseal plate that include a low number of chondrocytes and inability of chondrocytes to enlarge. Which zones do you think would be most affected by this disorder, and why? __________

____

____

EXERCISE 9 The Axial Skeleton

Learning Outcomes

- Name the three parts of the axial skeleton.
- Identify the bones of the axial skeleton, either by examining disarticulated bones or by pointing them out on an articulated skeleton or skull, and name the important bone markings on each.
- Name and describe the different types of vertebrae.
- Discuss the importance of intervertebral discs and spinal curvatures.
- Identify three abnormal spinal curvatures.
- List the components of the thoracic cage.
- Identify the bones of the fetal skull by examining an articulated skull or image.
- Define *fontanelle*, and discuss the function and fate of fontanelles.
- Discuss important differences between the fetal and adult skulls.

Go to Mastering A&P™ > Study Area to improve your performance in A&P Lab.

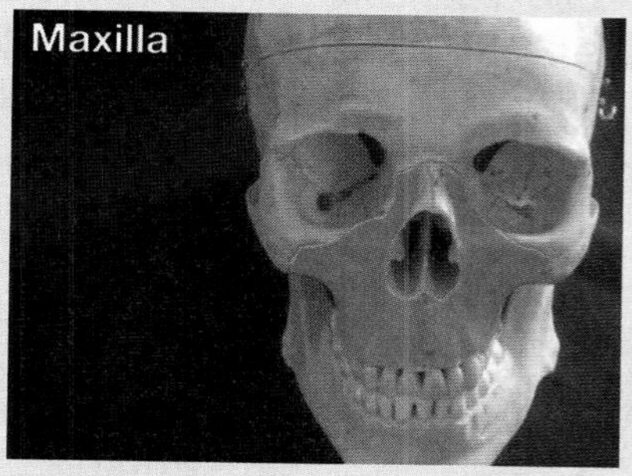

> Lab Tools > Bone & Dissection Videos

Instructors may assign new Building Vocabulary coaching activities, Pre-Lab Quiz questions, Art Labeling activities, related bone videos and coaching activities, Practice Anatomy Lab Practical questions (PAL), and more using the Mastering **A&P™** Item Library.

Materials

- Intact skull and Beauchene skull
- X-ray images of individuals with scoliosis, lordosis, and kyphosis (if available)
- Articulated skeleton, articulated vertebral column, removable intervertebral discs
- Isolated cervical, thoracic, and lumbar vertebrae, sacrum, and coccyx
- Isolated fetal skull

Pre-Lab Quiz

Instructors may assign these and other Pre-Lab Quiz questions using Mastering **A&P™**

1. The axial skeleton can be divided into the skull, the vertebral column, and the:
   a. thoracic cage **c.** hip bones
   b. femur **d.** humerus
2. Eight bones make up the ____________, which encloses and protects the brain.
   a. cranium **b.** face **c.** skull
3. The ____________ vertebrae articulate with the corresponding ribs.
   a. cervical **c.** spinal
   b. lumbar **d.** thoracic
4. The ____________, commonly referred to as the breastbone, is a flat bone formed by the fusion of three bones: the manubrium, the body, and the xiphoid process.
   a. coccyx **b.** sacrum **c.** sternum
5. A fontanelle:
   a. is found only in the fetal skull
   b. is a fibrous membrane
   c. allows for compression of the skull during birth
   d. all of the above

The **axial skeleton** (the green portion of Figure 8.1 on p. 104) can be divided into three parts: the skull, the vertebral column, and the thoracic cage. This division of the skeleton forms the longitudinal axis of the body and protects the brain, spinal cord, heart, and lungs.

The Skull

The **skull** is composed of two sets of bones. Those of the **cranium** (8 bones) enclose and protect the fragile brain tissue. The **facial bones** (14 bones) support the eyes and position them anteriorly. They also provide attachment sites for facial muscles. All but one of the bones of the skull are joined by interlocking fibrous joints called *sutures*. The mandible is attached to the rest of the skull by a freely movable joint.

Activity 1

Identifying the Bones of the Skull

The bones of the skull (**Figures 9.1–9.10**, pp. 117–125) are described in **Tables 9.1** and **9.2** on p. 116 and pp.120–123. As you read through this material, identify each bone on an intact and/or Beauchene skull (see Figure 9.10).

Note: *Important bone markings are listed in the tables for the bones on which they appear, and each bone name is colored to correspond to the bone color in the figures.*

The Cranium

The cranium may be divided into two major areas for study—the **cranial** vault, or **calvaria**, forming the superior, lateral, and posterior walls of the skull; and the **cranial base**, forming the

(Text continues on p. 122.) →

9

Table 9.1A The Axial Skeleton: Cranial Bones and Important Bone Markings

Cranial bone	Important markings	Description
Frontal (1) Figures 9.1, 9.3, 9.7, 9.9, and 9.10	N/A	Forms the forehead, superior part of the orbit, and the floor of the anterior cranial fossa.
	Supraorbital margin	Thick margin of the eye socket that lies beneath the eyebrows.
	Supraorbital foramen (notch)	Opening above each orbit allowing blood vessels and nerves to pass.
	Glabella	Smooth area between the eyes.
Parietal (2) Figures 9.1, 9.3, 9.6, 9.7, and 9.10	N/A	Form the superior and lateral aspects of the skull.
Temporal (2) Figures 9.1, 9.2, 9.3, 9.6, 9.7, and 9.10	N/A	Form the inferolateral aspects of the skull and contribute to the middle cranial fossa; each has squamous, tympanic, and petrous parts.
	Squamous part	Located inferior to the squamous suture. The next two markings are located in this part.
	Zygomatic process	A bridgelike projection that articulates with the zygomatic bone to form the zygomatic arch.
	Mandibular fossa	Located on the inferior surface of the zygomatic process; receives the condylar process of the mandible to form the temporomandibular joint.
	Tympanic part	Surrounds the external ear opening. The next two markings are located in this part.
	External acoustic meatus	Canal leading to the middle ear and eardrum.
	Styloid process	Needlelike projection that serves as an attachment point for ligaments and muscles of the neck. (This process is often missing from demonstration skulls because it has broken off.)
	Petrous part	Forms a bony wedge between the sphenoid and occipital bones and contributes to the cranial base. The remaining temporal markings are located in this part.
	Jugular foramen	Located where the petrous part of the temporal bone joins the occipital bone. Forms an opening which the internal jugular vein and cranial nerves IX, X, and XI pass.
	Carotid canal	Opening through which the internal carotid artery passes into the cranial cavity.
	Foramen lacerum	Almost completely closed by cartilage in the living person but forms a jagged opening in dried skulls.
	Stylomastoid foramen	Tiny opening between the mastoid and styloid processes through which cranial nerve VII leaves the cranium.
	Mastoid process	Located posterior to the external acoustic meatus; serves as an attachment point for neck muscles.

Table continues on p. 120. ➤

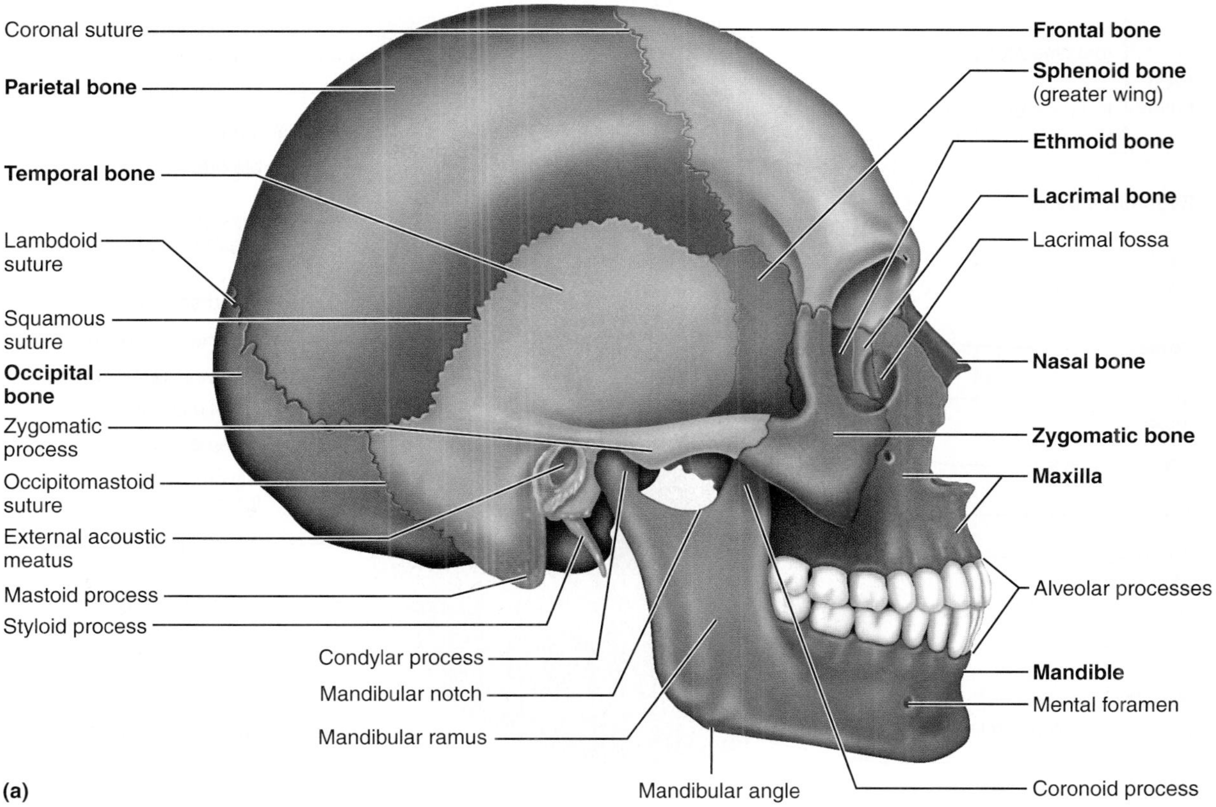

(a)

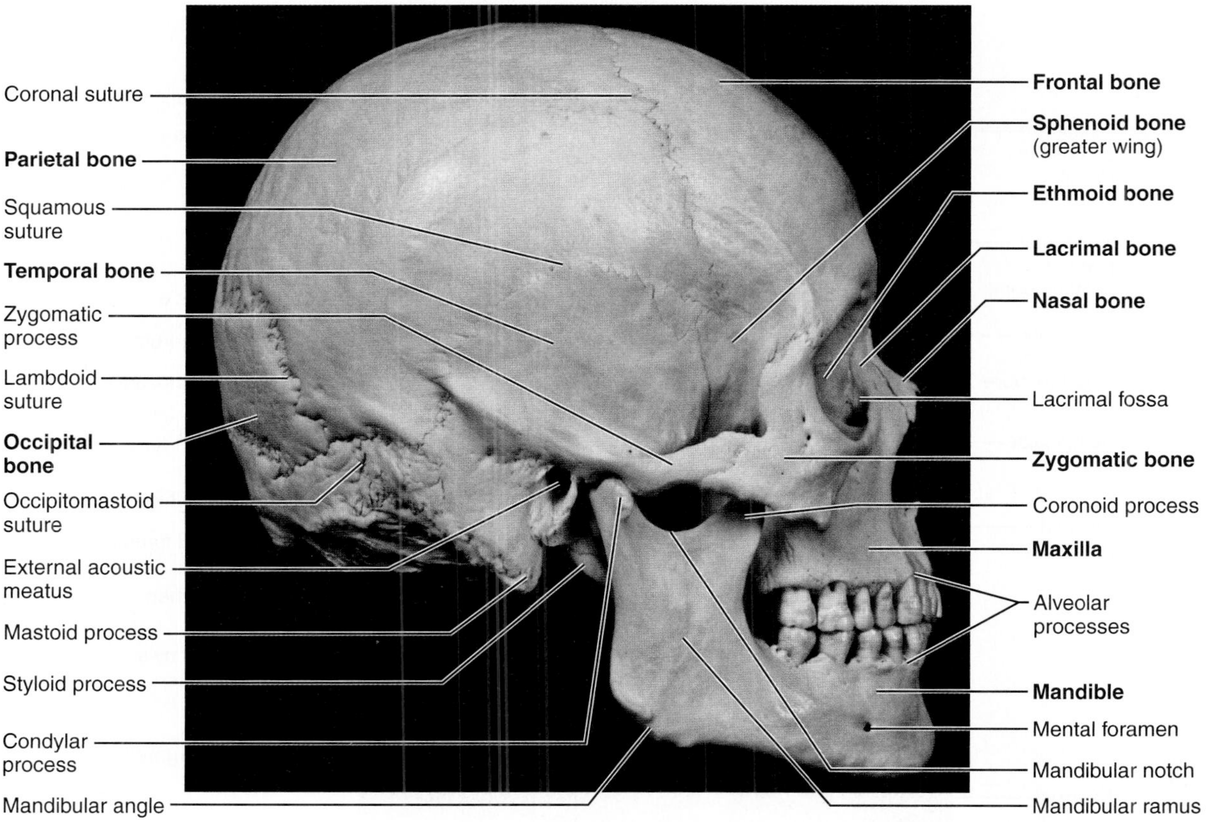

(b)

Figure 9.1 External anatomy of the right lateral aspect of the skull. **(a)** Diagram. **(b)** Photograph.

Instructors may assign related videos and coaching activities for the Cranium and Temporal Bone using Mastering A&P™

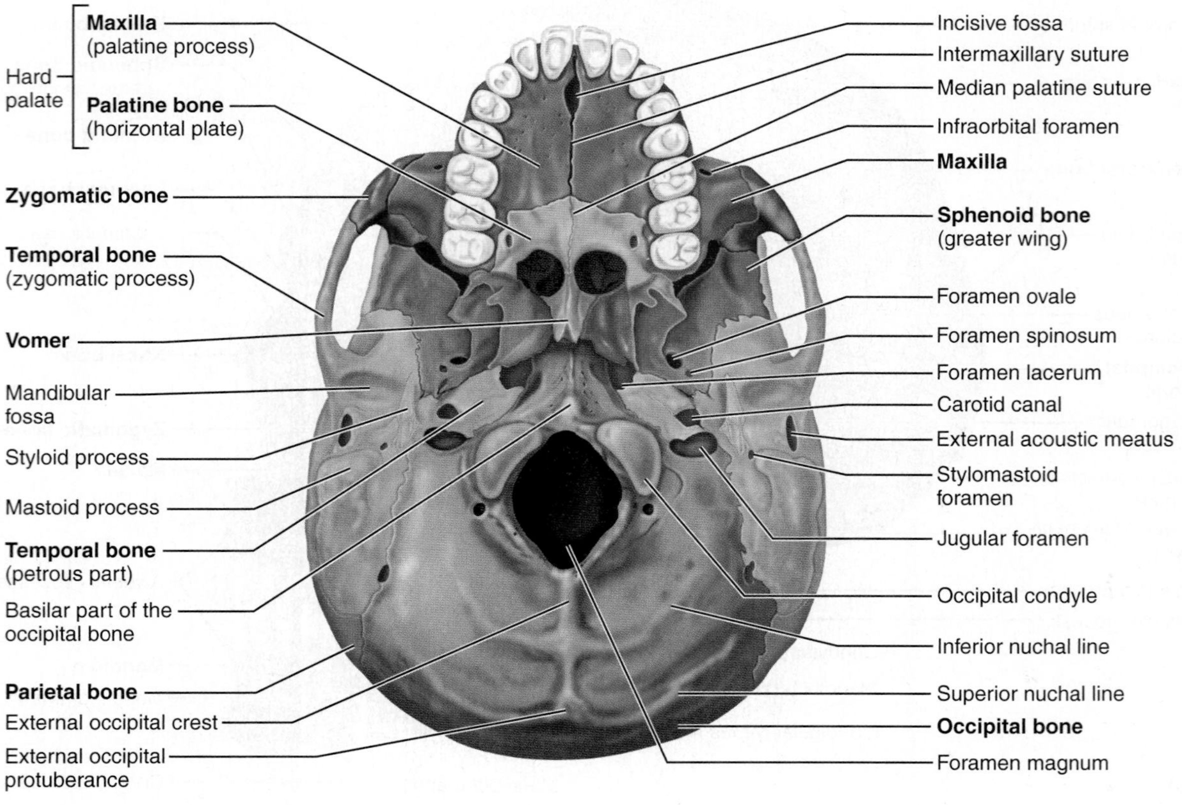

(a)

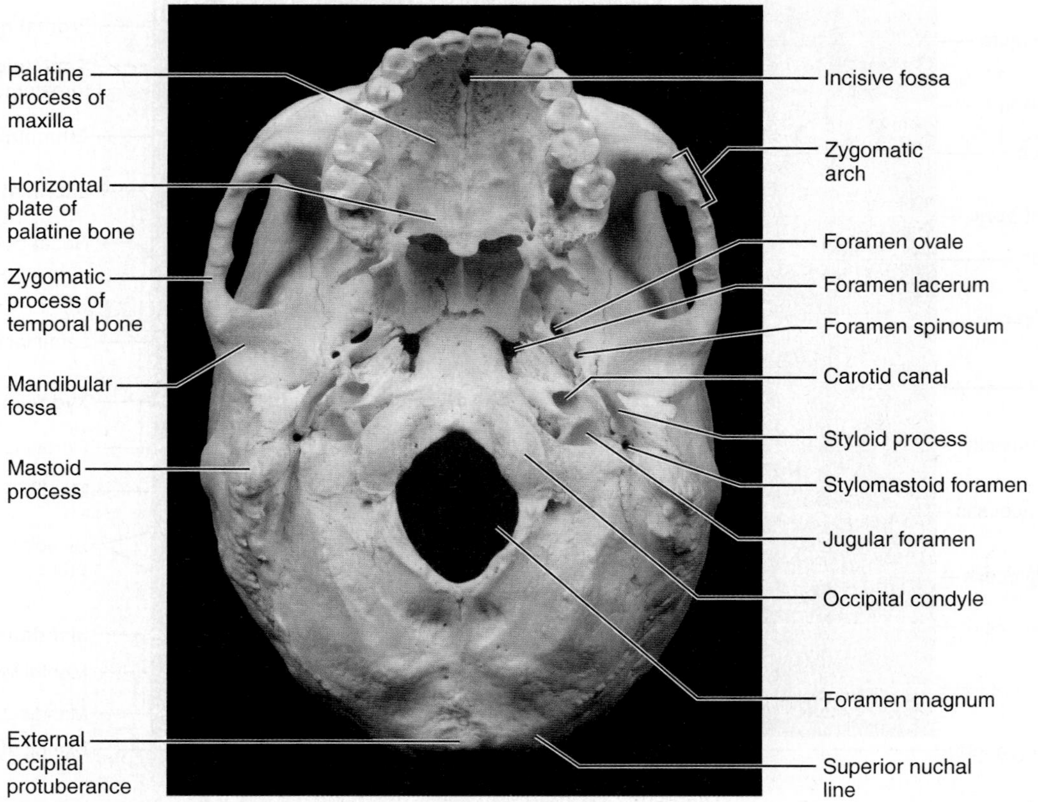

(b)

Figure 9.2 Inferior view of the skull, mandible removed.

Instructors may assign a related video and coaching activity for the Zygomatic Arch using Mastering A&P™

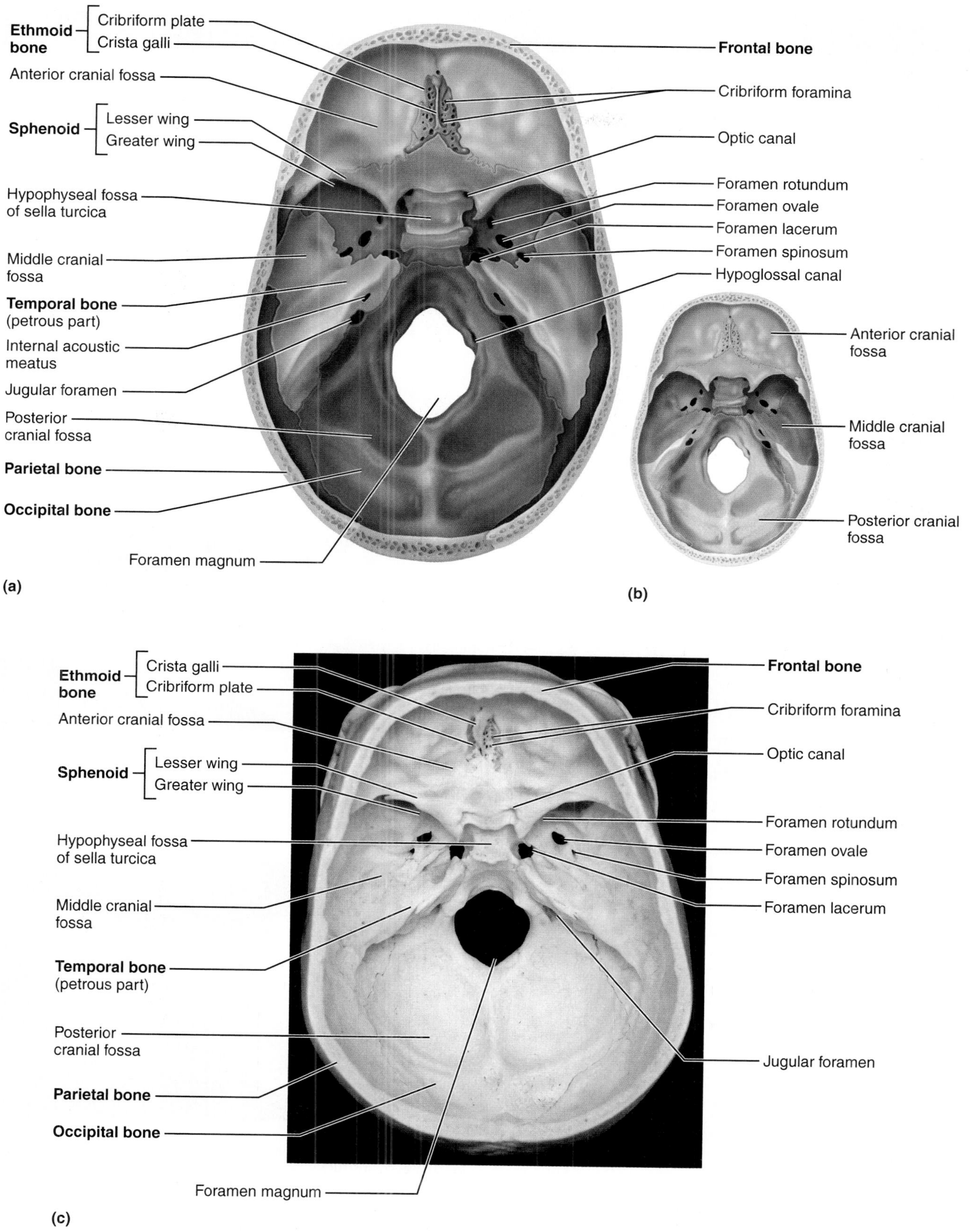

Figure 9.3 Internal anatomy of the inferior portion of the skull. (a) Superior view of the base of the cranial cavity, calvaria removed. **(b)** Diagram of the cranial base showing the extent of its major fossae. **(c)** Photograph of superior view of the base of the cranial cavity, calvaria removed.

Instructors may assign a related video and coaching activity for the Occipital Bone using Mastering **A&P™**

9

Table 9.1A The Axial Skeleton: Cranial Bones and Important Bone Markings *(continued)*

Cranial bone	Important markings	Description
Occipital (1) Figures 9.1, 9.2, 9.3, and 9.6	N/A	Forms the posterior aspect and most of the base of the skull.
	Foramen magnum	Large opening in the base of the bone, which allows the spinal cord to join with the brain stem.
	Occipital condyles	Rounded projections lateral to the foramen magnum that articulate with the first cervical vertebra (atlas).
	Hypoglossal canal	Opening medial and superior to the occipital condyle through which cranial nerve XII (the hypoglossal nerve) passes.
	External occipital protuberance	Midline prominence posterior to the foramen magnum.

The number in parentheses ( ) following the bone name indicates the total number of such bones in the body.

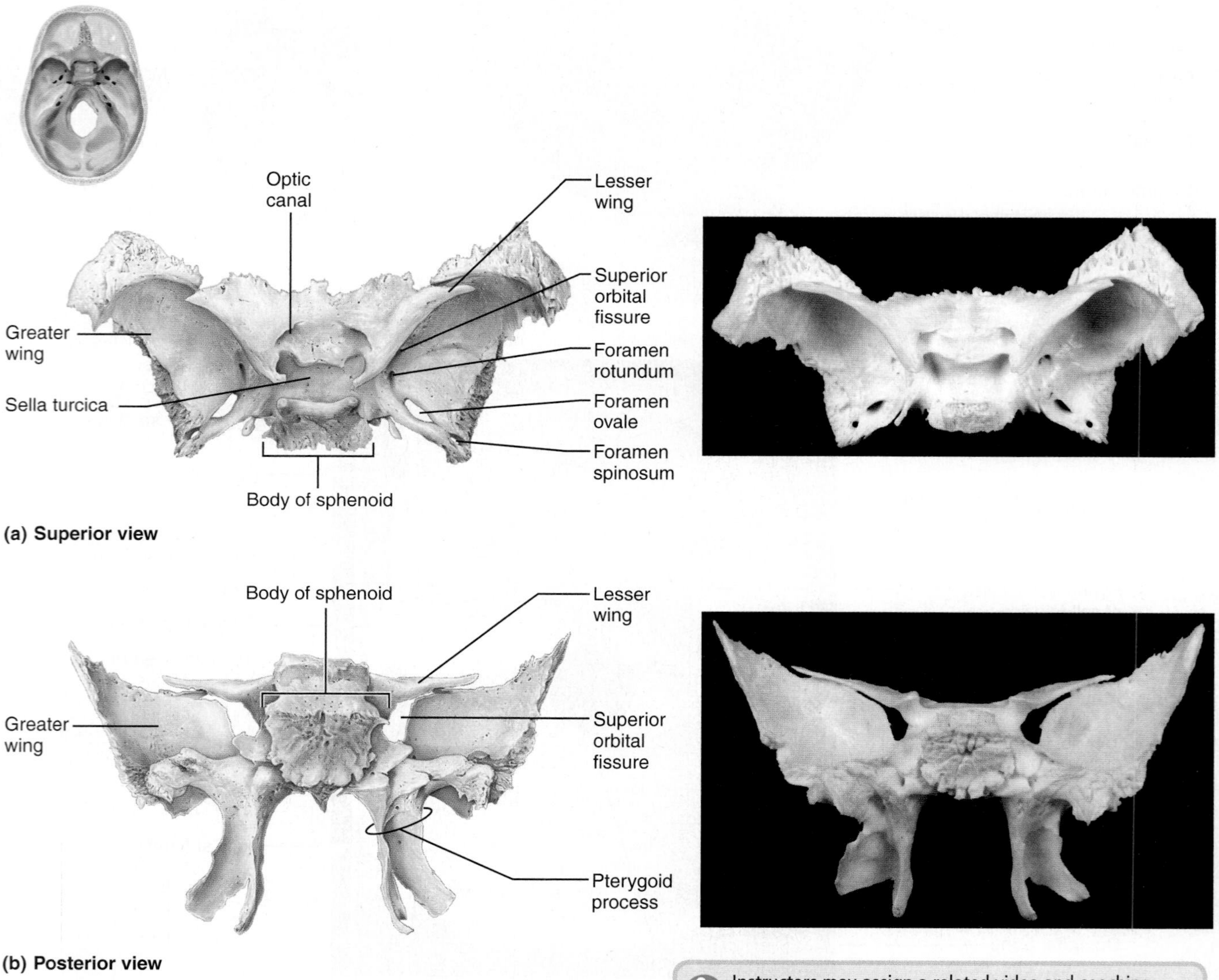

Figure 9.4 The sphenoid bone.

Instructors may assign a related video and coaching activity for the Sphenoid Bone using Mastering A&P™

Table 9.1B The Axial Skeleton: Cranial Bones and Important Bone Markings

Cranial bone	Important markings	Description
Sphenoid bone (1) Figures 9.1, 9.2, 9.3, 9.4, 9.7, and 9.10	N/A	Bat-shaped bone that is described as the keystone bone of the cranium because it articulates with all other cranial bones.
	Greater wings	Project laterally from the sphenoid body, forming parts of the middle cranial fossa and the orbits.
	Pterygoid processes	Project inferiorly from the greater wings; attachment site for chewing muscles (pterygoid muscles).
	Superior orbital fissures	Slits in the orbits providing passage of cranial nerves that control eye movements (III, IV, VI, and the ophthalmic division of V).
	Sella turcica	"Turkish saddle" located on the superior surface of the body; the seat of the saddle, called the *hypophyseal fossa*, holds the pituitary gland.
	Lesser wings	Form part of the floor of the anterior cranial fossa and part of the orbit.
	Optic canals	Openings in the base of the lesser wings; cranial nerve II (optic nerve) passes through to serve the eye.
	Foramen rotundum	Openings located in the medial part of the greater wing; a branch of cranial nerve V (maxillary division) passes through.
	Foramen ovale	Openings located posterolateral to the foramen rotundum; a branch of cranial nerve V (mandibular division) passes through.
	Foramen spinosum	Openings located posterolateral to the foramen ovale; provides passageway for the middle meningeal artery.

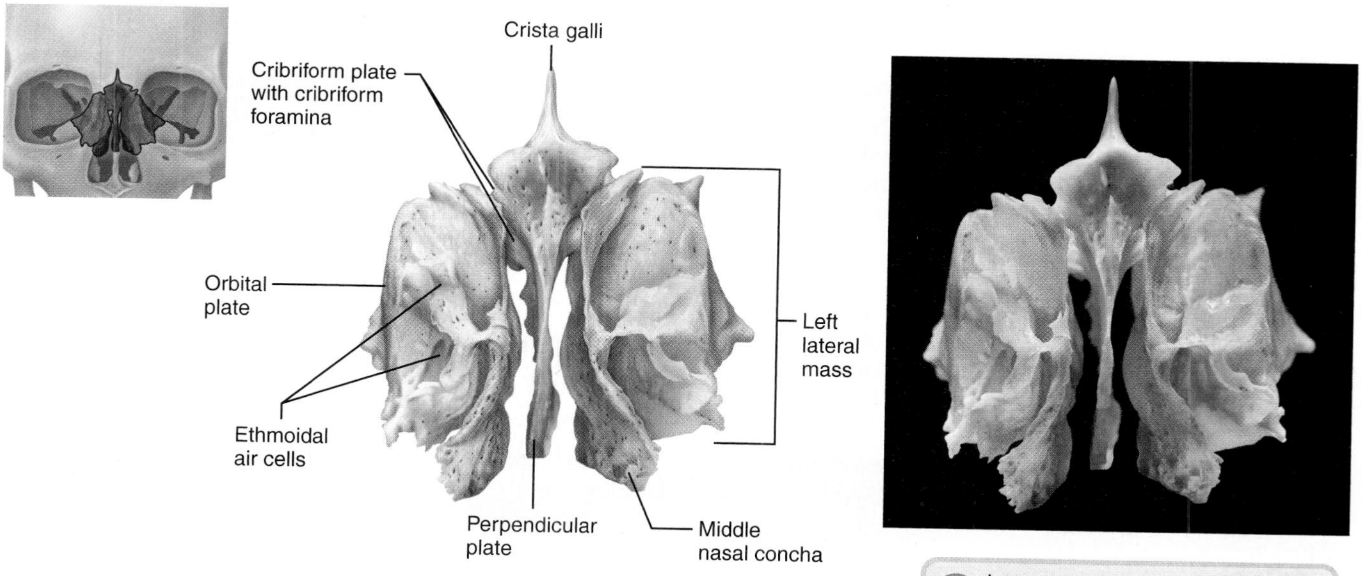

Figure 9.5 The ethmoid bone. Anterior view. The superior nasal conchae are located posteriorly and are therefore not visible in the anterior view.

Instructors may assign a related video and coaching activity for the Ethmoid Bone using Mastering A&P™

Table 9.1C The Axial Skeleton: Cranial Bones and Important Bone Markings

Cranial bone	Important markings	Description
Ethmoid (1) Figures 9.1, 9.3, 9.5, 9.7, and 9.10	N/A	Contributes to the anterior cranial fossa; forms part of the nasal septum and the nasal cavity; contributes to the medial wall of the orbit.
	Crista galli	"Rooster's comb"; a superior projection that attaches to the dura mater, helping to secure the brain within the skull.
	Cribriform plates	Located lateral to the crista galli; form a portion of the roof of the nasal cavity and the floor of the anterior cranial fossa.

Table 9.1C The Axial Skeleton: Cranial Bones and Important Bone Markings *(continued)*

Cranial bone	Important markings	Description
Ethmoid (1) *(continued)*	Cribriform foramina	Tiny holes in the cribriform plates that allow for the passage of filaments of cranial nerve I (olfactory nerve).
	Perpendicular plate	Inferior projection that forms the superior portion of the nasal septum.
	Lateral masses	Flank the perpendicular plate on each side and are filled with sinuses called *ethmoidal air cells*.
	Orbital plates	Lateral surface of the lateral masses that contribute to the medial wall of the orbits.
	Superior and middle nasal conchae	Extend medially from the lateral masses; act as turbinates to improve airflow through the nasal cavity.

skull bottom. Internally, the cranial base has three distinct depressions: the **anterior**, **middle**, and **posterior cranial fossae** (see Figure 9.3). The brain sits in these fossae, completely enclosed by the cranial vault. Overall, the brain occupies the *cranial cavity*.

9

Major Sutures

The four largest sutures are located where the parietal bones articulate with each other and where the parietal bones articulate with other cranial bones:

- **Sagittal suture:** Occurs where the left and right parietal bones meet superiorly in the midline of the cranium (Figure 9.6).
- **Coronal suture:** Running in the frontal plane, where the parietal bones meet the frontal bone anteriorly (Figure 9.1).
- **Squamous suture:** Occurs where a parietal bone and temporal bone meet on the lateral aspect of the skull (Figure 9.1).
- **Lambdoid suture:** Occurs where the parietal bones meet the occipital bone posteriorly (Figure 9.6).

Facial Bones

Of the 14 bones composing the face, 12 are paired. *Only the mandible and vomer are single bones.* An additional bone, the hyoid bone, although not a facial bone, is considered here because of its location.

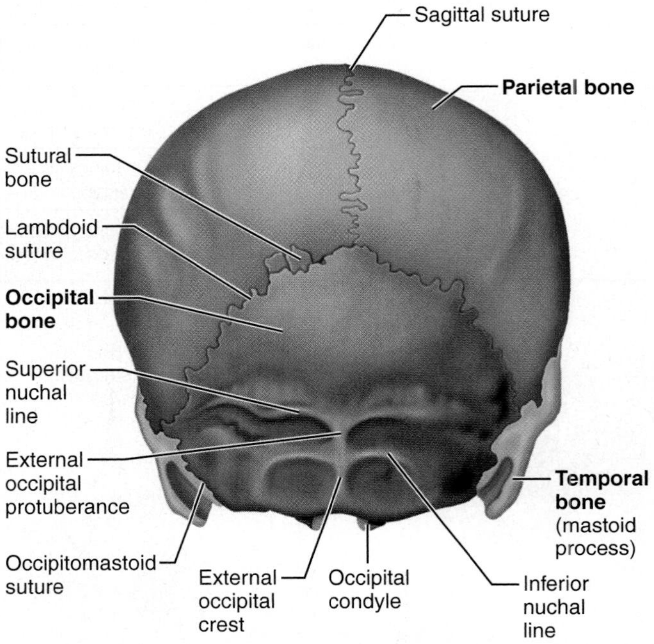

Figure 9.6 Posterior view of the skull.

Instructors may assign this figure as an Art Labeling Activity using Mastering A&P™

Table 9.2 The Axial Skeleton: Facial Bones and Important Bone Markings (Figures 9.1, 9.7, 9.9, and 9.10, with additional figures listed for specific bones)

Facial bone	Important markings	Description
Nasal (2)	N/A	Small rectangular bones forming the bridge of the nose.
Lacrimal (2)	N/A	Each forms part of the medial orbit in between the maxilla and ethmoid bone.
	Lacrimal fossa	Houses the lacrimal sac, which helps to drain tears from the nasal cavity.
Zygomatic (2) (also Figure 9.2)	N/A	Commonly called the cheekbones; each forms part of the lateral orbit
Inferior nasal concha (2)	N/A	Inferior turbinate; each forms part of the lateral walls of the nasal cavities; improves the airflow through the nasal cavity
Palatine (2) (also Figure 9.2)	N/A	Forms the posterior hard palate, a small part of the nasal cavity, and part of the orbit.
	Horizontal plate	Forms the posterior portion of the hard palate.
	Median palatine suture	Median fusion point of the horizontal plates of the palatine bones.
Vomer (1)	N/A	Thin, blade-shaped bone that forms the inferior nasal septum.
Maxilla (2) (also Figures 9.2 and 9.8)	N/A	Keystone facial bones because they articulate with all other facial bones except the mandible; form the upper jaw and parts of the hard palate, orbits, and nasal cavity.
	Frontal process	Forms part of the lateral aspect of the bridge of the nose.
	Infraorbital foramen	Opening under the orbit that forms a passageway for the infraorbital artery and nerve.

Table 9.2 (continued)

Facial bone	Important markings	Description
Maxilla (2) *(continued)*	Palatine process	Forms the anterior hard palate; meet anteriorly in the intermaxillary suture (Note: Seen in inferior view).
	Zygomatic process	Articulation process for zygomatic bone.
	Alveolar process	Inferior margin of the maxilla; contains sockets in which the teeth lie
Mandible (1) (also Figures 9.2 and 9.8)	N/A	The lower jawbone, which articulates with the temporal bone to form the only freely movable joints in the skull (the temporomandibular joint).
	Condylar processes	Articulate with the mandibular fossae of the temporal bones.
	Coronoid processes	"Crown-shaped" portion of the ramus for muscle attachment.
	Mandibular notches	Separate the condylar process and the coronoid process.
	Body	Horizontal portion that forms the chin.
	Ramus	Vertical extension of the body.
	Mandibular angles	Posterior points where the ramus meets the body.
	Mental foramina	Paired openings on the body (lateral to the midline); transmit blood vessels and nerves to the lower lip and skin of the chin.
	Alveolar process	Superior margin of the mandible; contains sockets in which the teeth lie.
	Mandibular foramina	Located on the medial surface of each ramus; passageway for the nerve involved in tooth sensation. (Dentists inject anesthetic into this foramen before working on the lower teeth.)

9

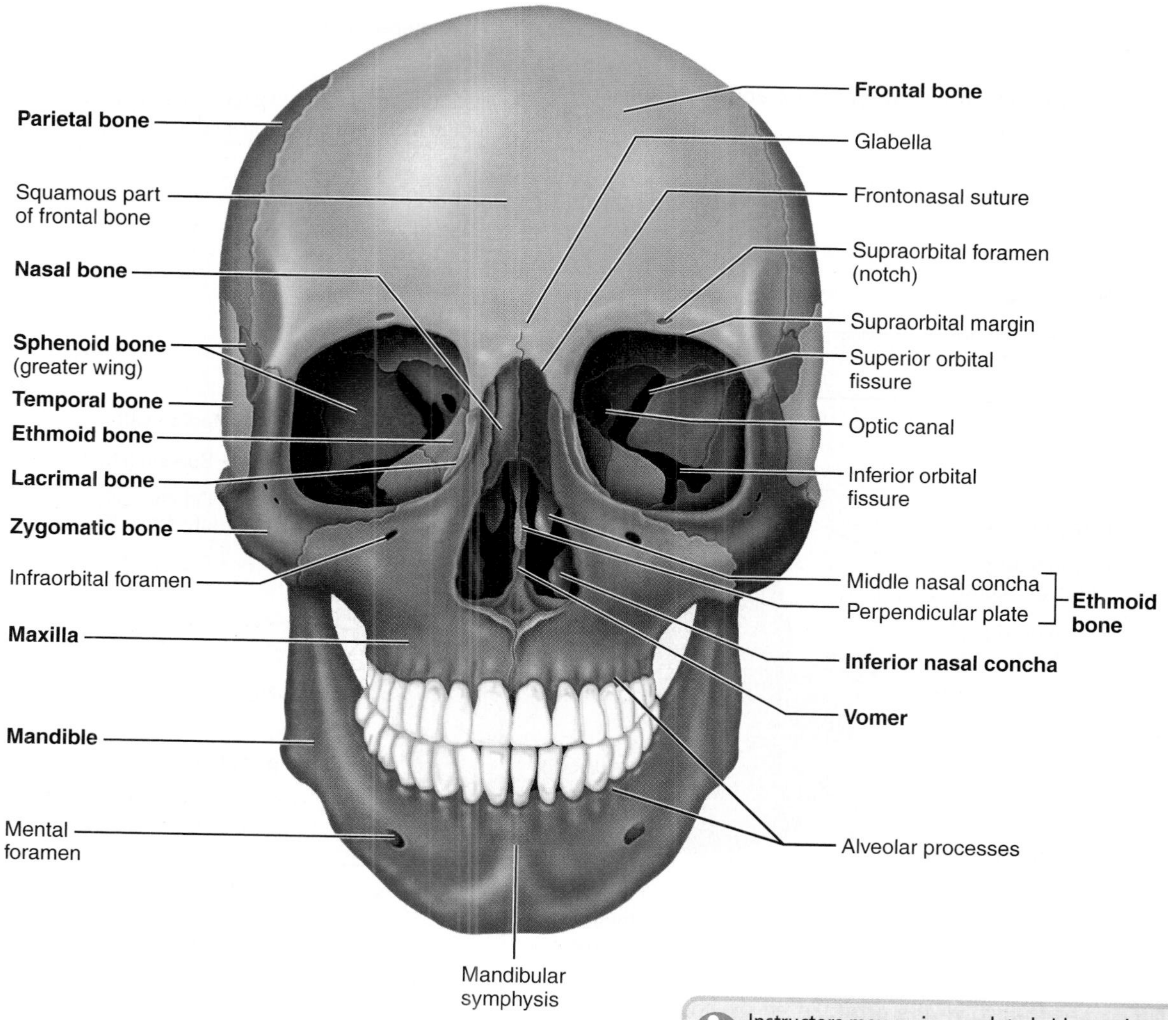

Figure 9.7 Anterior view of the skull.

Instructors may assign a related video and coaching activity for the Facial Bones using Mastering A&P™

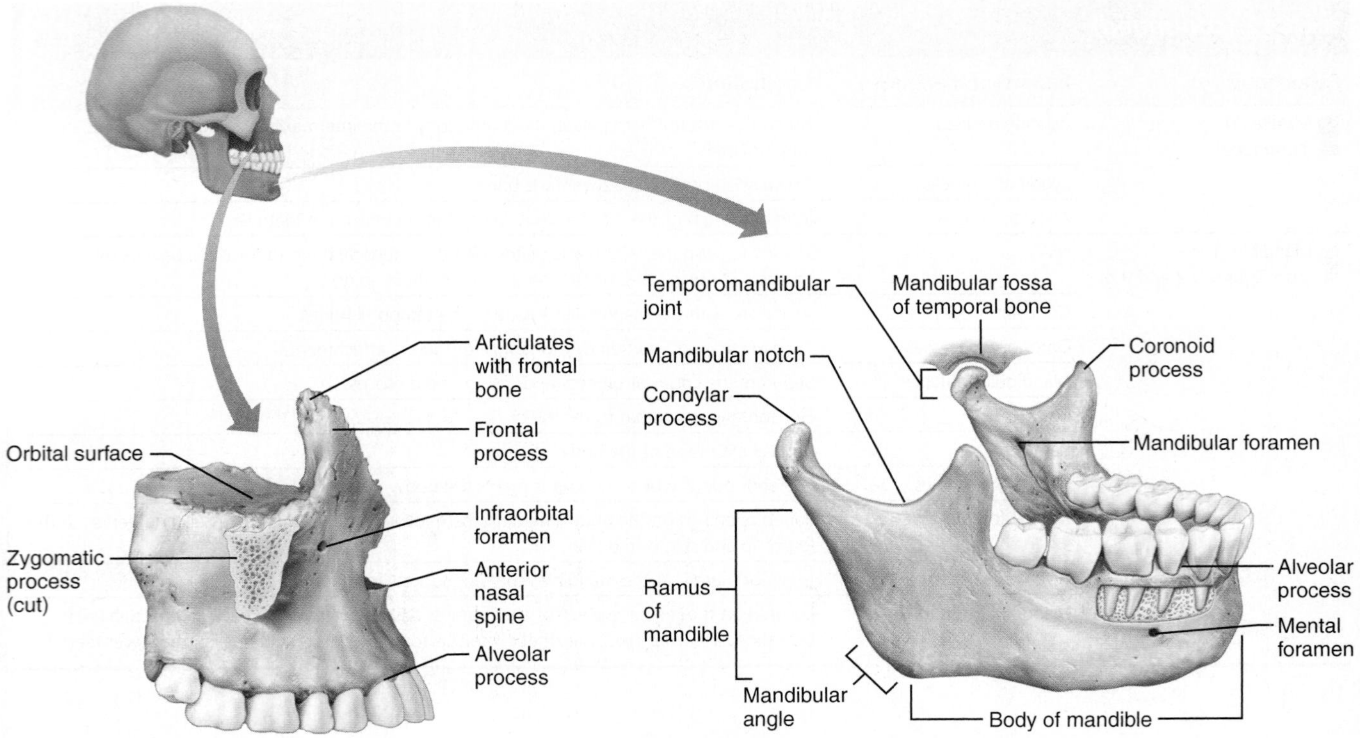

(a) Maxilla, right lateral view

(b) Mandible, right lateral view

Figure 9.8 Detailed anatomy of the maxilla and mandible.

Instructors may assign related videos and coaching activities for the Maxilla and Mandible using Mastering A&P™

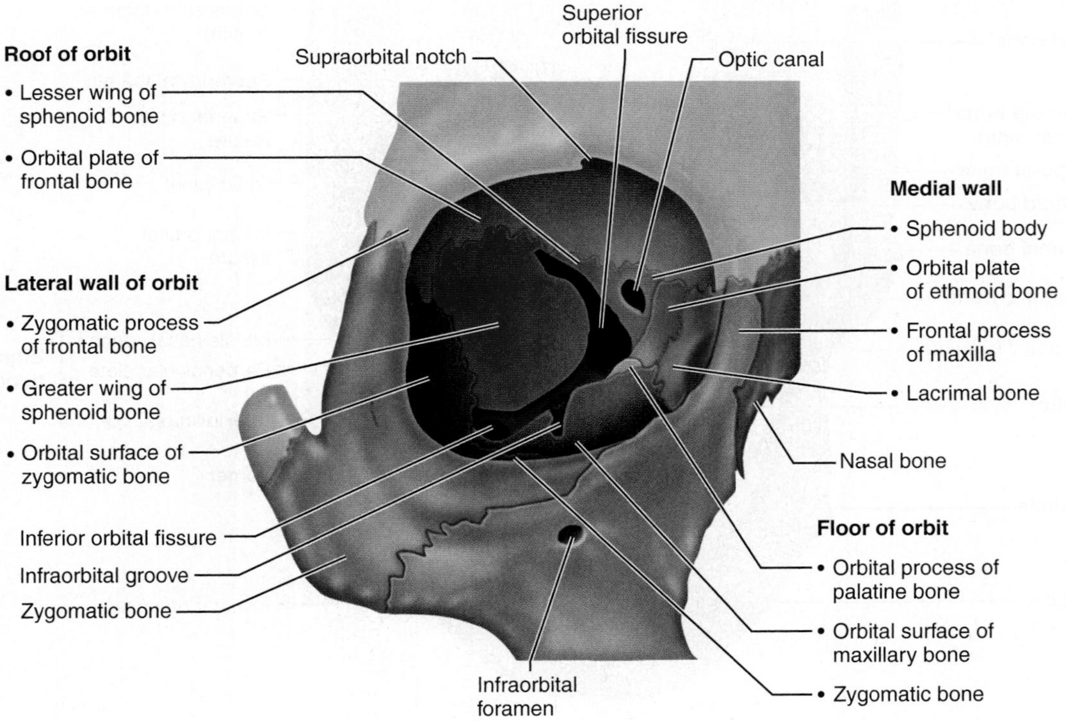

Figure 9.9 Bones that form the orbit. Seven skull bones form the orbit, the bony cavity that surrounds the eye: the frontal, sphenoid, ethmoid, lacrimal, maxilla, palatine, and zygomatic.

Instructors may assign this figure as an Art Labeling Activity using Mastering A&P™

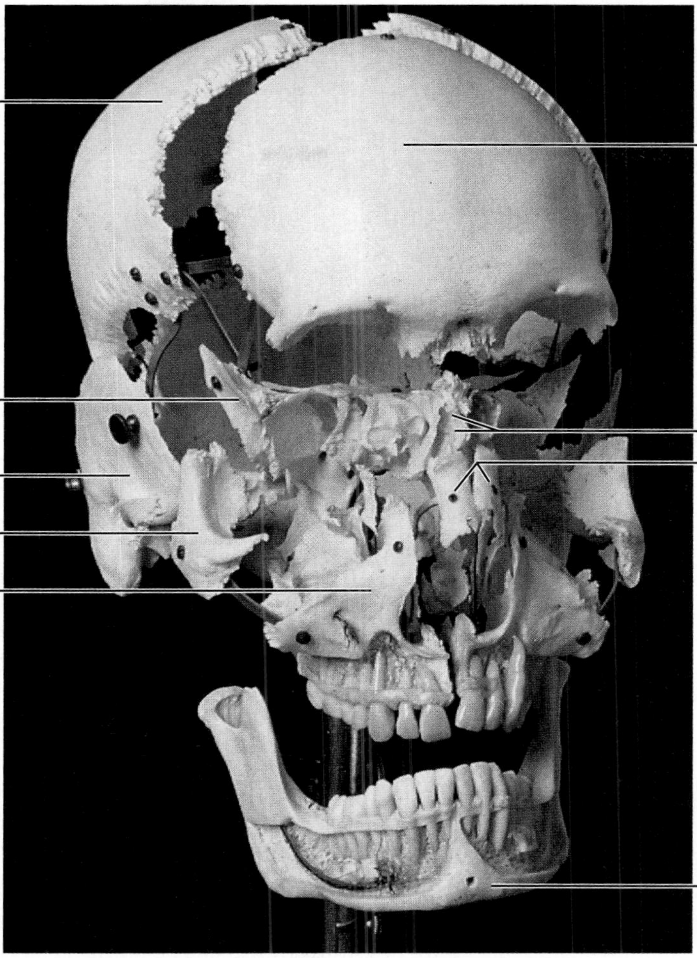

Figure 9.10 Anterior view of the Beauchene skull.

9

Group Challenge

Odd Bone Out

Each of the following sets contains four bones. One of the listed bones does not share a characteristic that the other three do. Work in groups of three, and discuss the characteristics of the bones in each group. On a separate piece of paper, one student will record the characteristics of each bone. For each set of bones, discuss the possible candidates for the "odd bone out" and which characteristic it lacks, based on your notes. Once your group has come to a consensus, circle the bone that doesn't belong with the others, and explain why it is singled out. What characteristic is it missing? Include as many characteristics as you can think of, but make sure your choice does not have the key characteristic. Use an articulated skull, disarticulated skull bones, and the pictures in your lab manual to help you select and justify your answer.

1. Which is the "odd bone"?	Why is it the odd one out?
Zygomatic bone Maxilla Vomer Nasal bone	

2. Which is the "odd bone"?	Why is it the odd one out?
Parietal bone Sphenoid bone Frontal bone Occipital bone	

3. Which is the "odd bone"?	Why is it the odd one out?
Lacrimal bone Nasal bone Zygomatic bone Maxilla	

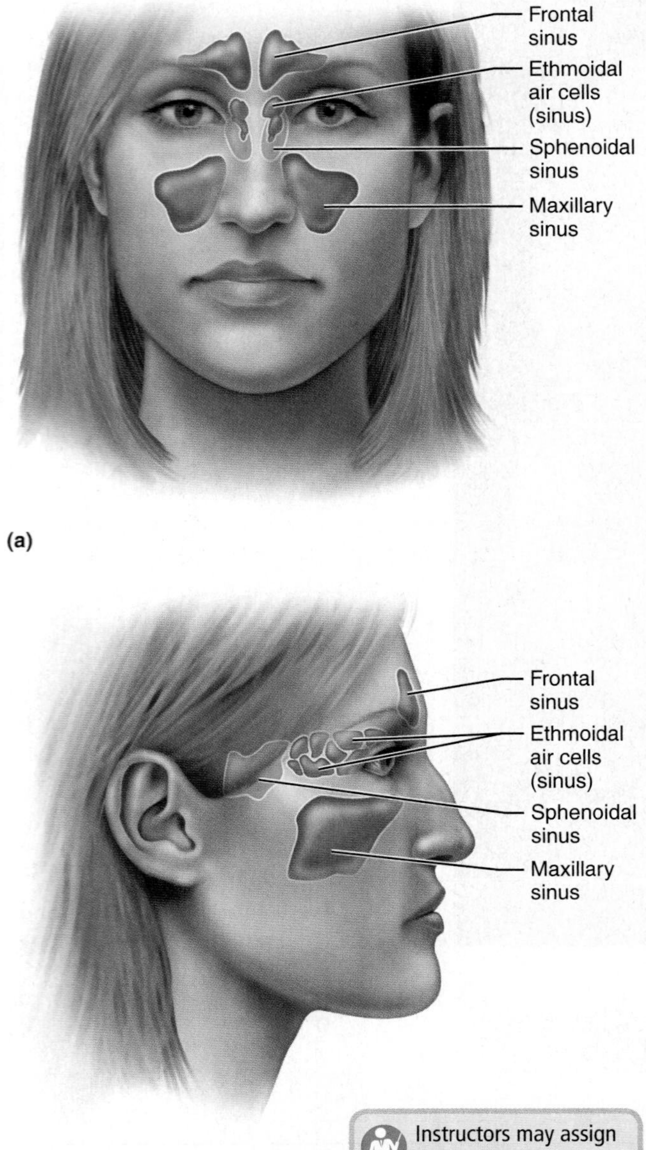

Instructors may assign this figure as an Art Labeling Activity using Mastering A&P™

Figure 9.11 Paranasal sinuses. **(a)** Anterior aspect. **(b)** Medial aspect.

Paranasal Sinuses

Five skull bones—the frontal, sphenoid, ethmoid, and paired maxillary bones—contain mucus-lined, air-filled cavities called **paranasal sinuses** (see Figure 9.5 and **Figure 9.11**). These paranasal sinuses lighten the skull and act as resonance chambers for speech. The maxillary sinus is the largest of the sinuses found in the skull.

Sinusitis, or inflammation of the sinuses, sometimes occurs as a result of an allergy or bacterial invasion of the sinus cavities. In such cases, some of the connecting passageways between the sinuses and nasal passages may become blocked with thick mucus or infectious material. Then, as the air in the sinus cavities is absorbed, a partial vacuum forms. The result is a sinus headache localized over the inflamed sinus area. Severe sinus infections may require surgical drainage to relieve this painful condition. +

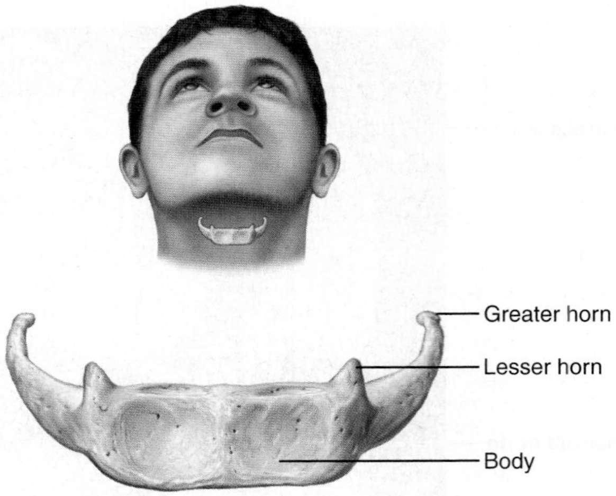

Figure 9.12 Hyoid bone.

Hyoid Bone

Not really considered or counted as a skull bone, the hyoid bone is located in the throat above the larynx. It serves as a point of attachment for many tongue and neck muscles. It does not articulate with any other bone and is thus unique. It is horseshoe shaped with a body and two pairs of **horns**, or **cornua** (**Figure 9.12**).

Activity 2

Palpating Skull Markings

Palpate the following areas on yourself. Place a check mark in the boxes as you locate the skull markings. Ask your instructor for help with any markings that you are unable to locate.

☐ Zygomatic bone and arch. (The most prominent part of your cheek is your zygomatic bone. Follow the posterior course of the zygomatic arch to its junction with your temporal bone.)

☐ Mastoid process (the rough area behind your ear).

☐ Temporomandibular joints. (Open and close your jaws to locate these.)

☐ Greater wing of sphenoid. (Find the indentation posterior to the orbit and superior to the zygomatic arch on your lateral skull.)

☐ Supraorbital foramen. (Apply firm pressure along the superior orbital margin to find the indentation resulting from this foramen.)

☐ Infraorbital foramen. (Apply firm pressure just inferior to the inferomedial border of the orbit to locate this large foramen.)

☐ Mandibular angle (most inferior and posterior aspect of the mandible).

☐ Mandibular symphysis (midline of chin).

☐ Nasal bones. (Run your index finger and thumb along opposite sides of the bridge of your nose until they "slip" medially at the inferior end of the nasal bones.)

☐ External occipital protuberance. (This midline projection is easily felt by running your fingers up the furrow at the back of your neck to the skull.)

☐ Hyoid bone. (Place a thumb and index finger beneath the chin just anterior to the mandibular angles, and squeeze gently. Exert pressure with the thumb, and feel the horn of the hyoid with the index finger.)

The Vertebral Column

The **vertebral column**, extending from the skull to the pelvis, forms the body's major axial support. Additionally, it surrounds and protects the delicate spinal cord while allowing the spinal nerves to emerge from the cord via openings between adjacent vertebrae. The term *vertebral column* might suggest a rigid supporting rod, but this is far from the truth. The vertebral column consists of 24 single bones called **vertebrae** and two composite, or fused, bones (the sacrum and coccyx) that are connected in such a way as to provide a flexible curved structure (**Figure 9.13**). Of the 24 single vertebrae, the seven bones of the neck are called *cervical vertebrae*; the next 12 are *thoracic vertebrae*; and the 5 supporting the lower back are *lumbar vertebrae*. Remembering common mealtimes for breakfast, lunch, and dinner (7 A.M., 12 noon, and 5 P.M.) may help you to remember the number of bones in each region.

The vertebrae are separated by pads of fibrocartilage, **intervertebral discs**, that cushion the vertebrae and absorb shocks. Each disc has two major regions, a central gelatinous *nucleus pulposus* that behaves like a rubber ball, and the *anulus fibrosus* that stabilizes the disc and contains the nucleus pulposus. The anulus fibrosus is composed of an outer ring of collagen fibers and an inner ring of fibrocartilage.

As a person ages, the water content of the discs decreases, and the discs become thinner and less compressible. This situation, along with other degenerative changes such as weakening of the ligaments and tendons of the vertebral column, predisposes older people to a ruptured disc, called a **herniated disc**. In a herniated disc, the anulus fibrosus commonly ruptures and the nucleus pulposus protrudes (herniates) through it. This event typically compresses adjacent nerves, causing pain. +

The thoracic and sacral curvatures of the spine are referred to as *primary curvatures* because they are present and well developed at birth. Later the *secondary curvatures* are formed. The cervical curvature becomes prominent when the baby begins to hold its head up independently, and the lumbar curvature develops when the baby begins to walk.

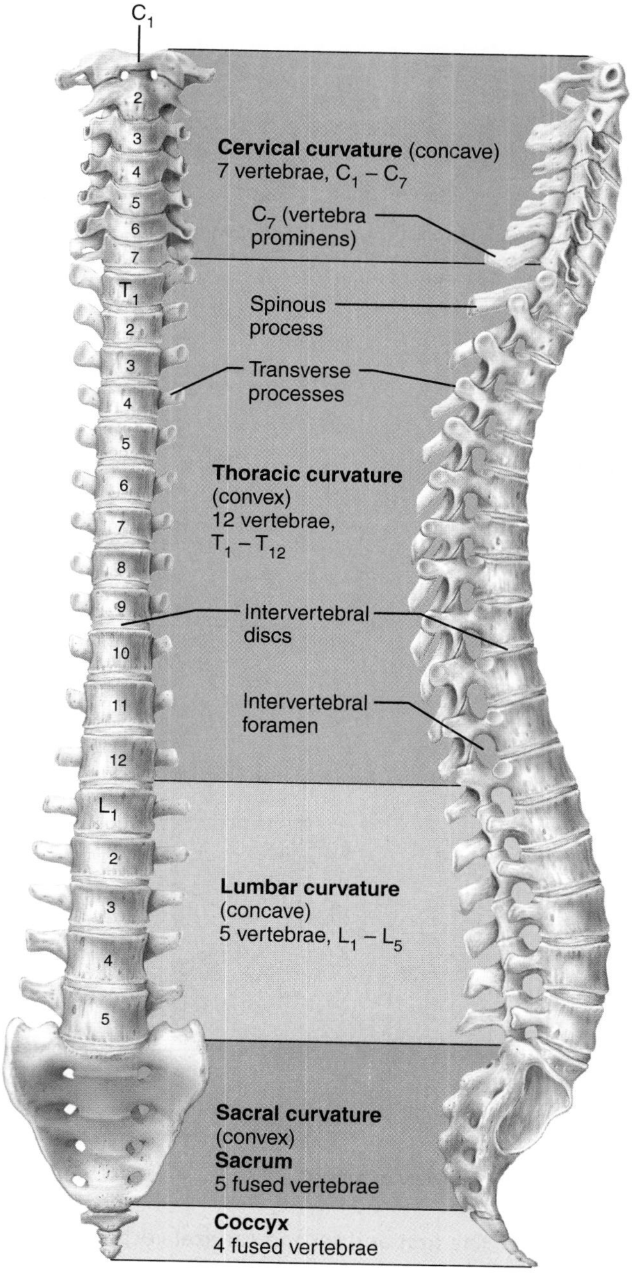

Figure 9.13 The vertebral column. Notice the curvatures in the lateral view. (The terms *convex* and *concave* refer to the curvature of the posterior aspect of the vertebral column.)

Activity 3

Examining Spinal Curvatures

1. Observe the normal curvature of the vertebral column in the articulated vertebral column or laboratory skeleton and compare it to Figure 9.13. Note the differences between normal curvature and three abnormal spinal curvatures seen in the figure-*scoliosis*, *kyphosis*, and *lordosis* (**Figure 9.14**). These abnormalities may result from disease or poor posture. Also examine X-ray images, if they are available, showing these same conditions in a living patient.

Text continues on next page →

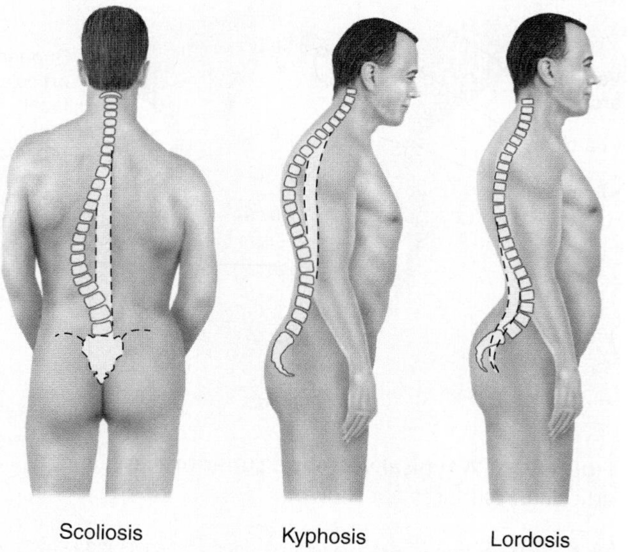

Figure 9.14 Abnormal spinal curvatures.

9

2. Then, using the articulated vertebral column (or an articulated skeleton), examine the freedom of movement between two lumbar vertebrae separated by an intervertebral disc.

When the fibrous disc is properly positioned, are the spinal cord or peripheral nerves impaired in any way?

____

Remove the disc, and put the two vertebrae back together. What happens to the nerve?

____

What would happen to the spinal nerves in areas of malpositioned or "slipped" discs?

____

Structure of a Typical Vertebra

Although they differ in size and specific features, all vertebrae have some features in common (**Figure 9.15**).

- **Body (centrum):** Rounded central weight-bearing portion of the vertebra, which faces anteriorly in the human vertebral column.
- **Vertebral arch:** Composed of two pedicles and two laminae.
- **Vertebral foramen:** Opening enclosed by the body and vertebral arch; a passageway for the spinal cord.
- **Transverse processes:** Two lateral projections from the vertebral arch.
- **Spinous process:** Single medial and posterior projection formed at the junction of the two laminae.

The transverse and spinous processes are attachment sites for muscles.

- **Superior and inferior articular processes:** Paired projections lateral to the vertebral foramen that enable articulation with adjacent vertebrae. The superior articular processes typically face toward the spinous process (posteriorly), whereas the inferior articular processes face (anteriorly) away from the spinous process.

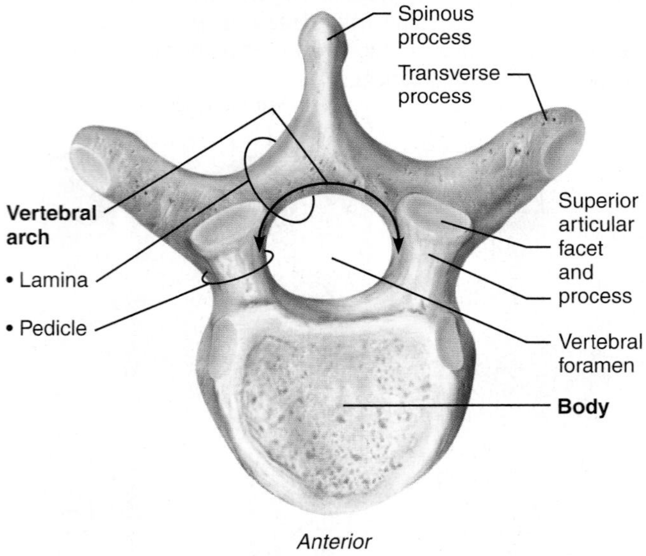

Figure 9.15 A typical vertebra, superior view. Inferior articulating surfaces not shown.

Instructors may assign a related video and coaching activity for the Typical Vertebra using Mastering A&P™

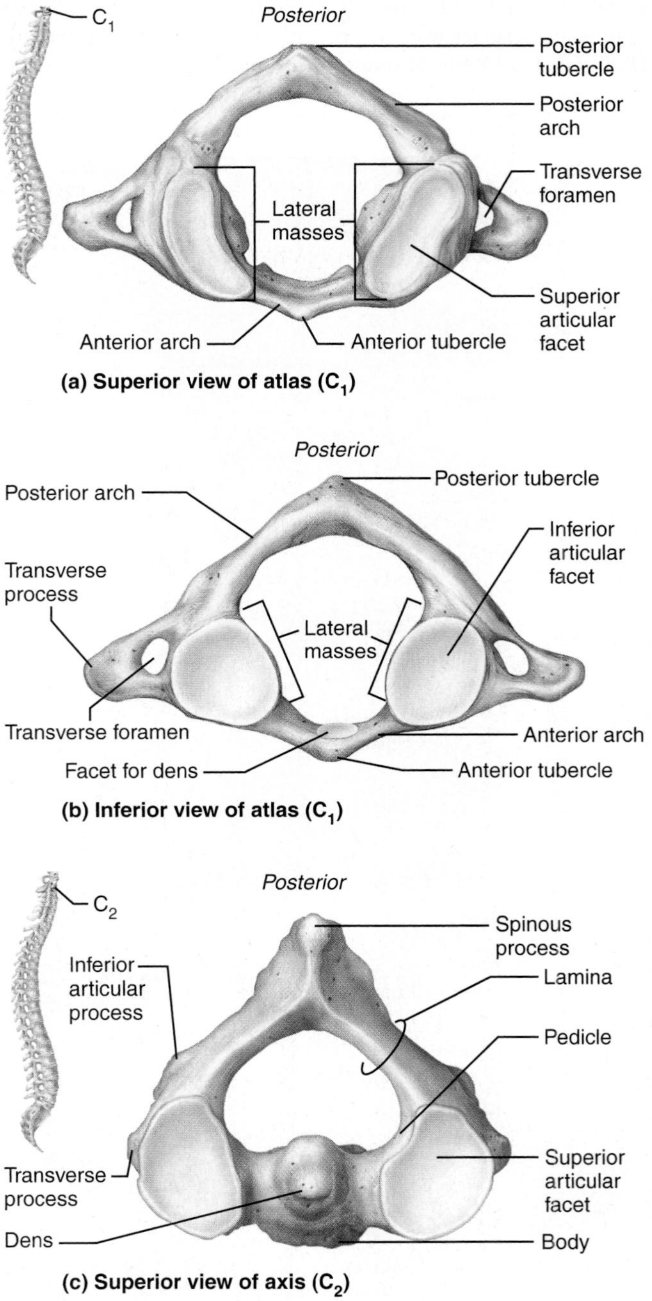

Figure 9.16 The first and second cervical vertebrae.

Instructors may assign a related video and coaching activity for the Atlas and Axis using Mastering A&P™

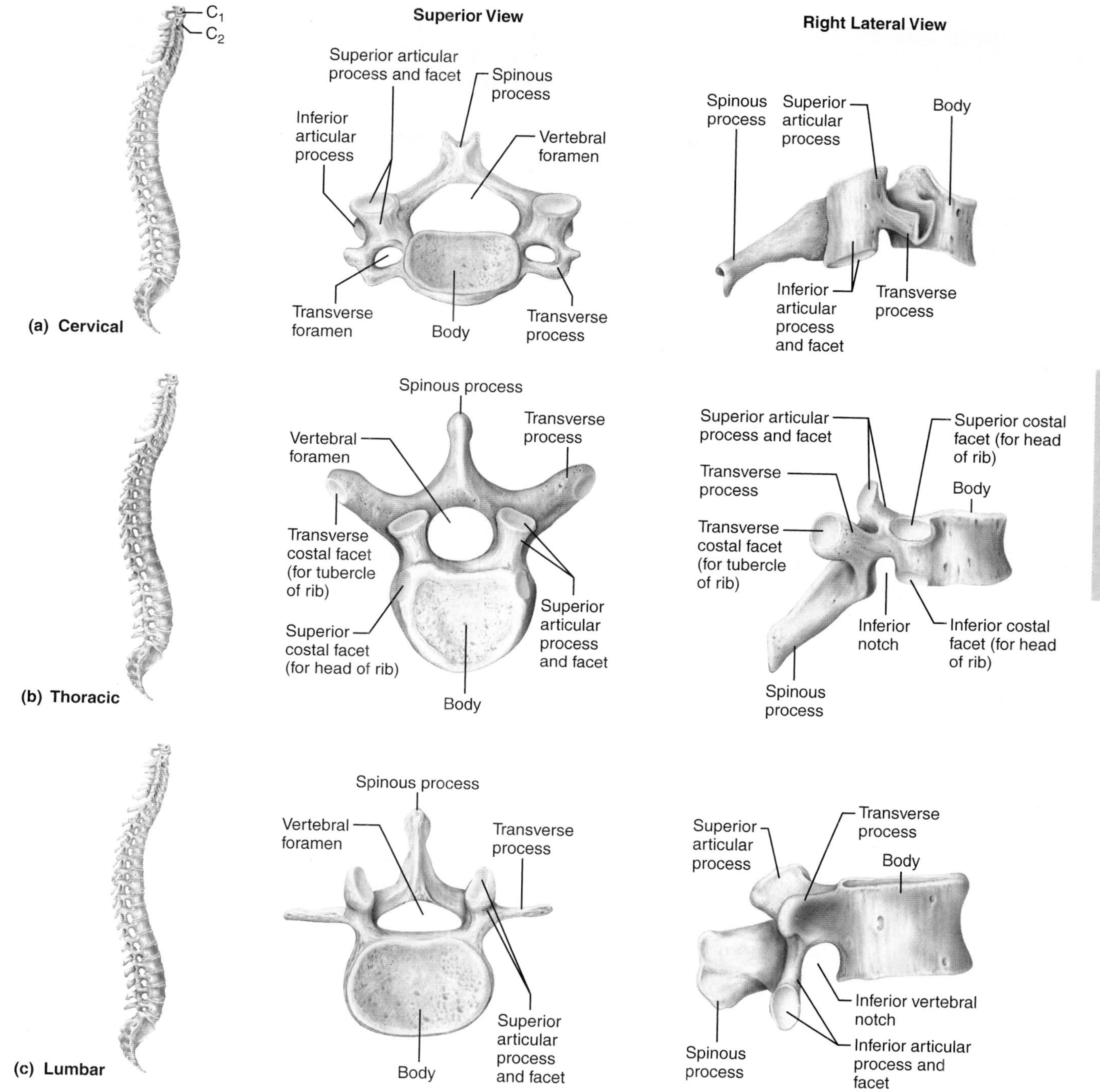

Figure 9.17 Superior and right lateral views of typical vertebrae.

- **Intervertebral foramina:** The right and left pedicles have notches (see Figure 9.17) on their inferior and superior surfaces that create openings, the intervertebral foramina, for spinal nerves to leave the spinal cord between adjacent vertebrae.

Figures 9.16–9.18 and **Table 9.3** on p. 130 show how specific vertebrae differ; refer to them as you read the next sections.

Cervical Vertebrae

The seven cervical vertebrae (referred to as C_1 through C_7) form the neck portion of the vertebral column. The first two cervical vertebrae (atlas and axis) are highly modified to perform special functions (**Figure 9.16**). The **atlas** (C_1) lacks a body, and its lateral processes contain large concave depressions on their superior surfaces that receive the occipital condyles of the skull. This joint enables you to nod "yes." The **axis** (C_2) acts as a pivot for the rotation of the atlas (and skull) above. It bears a large vertical process, the **dens**, that serves as the pivot point. The articulation between C_1 and C_2 allows you to rotate your head from side to side to indicate "no."

The more typical cervical vertebrae (C_3 through C_7) are distinguished from the thoracic and lumbar vertebrae by several features (see Table 9.3 and **Figure 9.17**). They are the smallest, lightest vertebrae, and the vertebral foramen is triangular. The

Table 9.3 Regional Characteristics of Cervical, Thoracic, and Lumbar Vertebrae

Characteristic	(a) Cervical (C_3–C_7)	(b) Thoracic	(c) Lumbar
Body	Small, wide side to side	Larger than cervical; heart shaped; bears costal facets	Massive; kidney shaped
Spinous process	Short; bifid; projects directly posteriorly	Long; sharp; projects inferiorly	Short; blunt; projects directly posteriorly
Vertebral foramen	Triangular	Circular	Triangular
Transverse processes	Contain foramina	Have costal facets for ribs (except T_{11} and T_{12})	Thin and tapered
Superior and inferior articulating processes	Superior facets directed superoposteriorly	Superior facets directed posteriorly	Superior facets directed posteromedially (or medially)
	Inferior facets directed inferoanteriorly	Inferior facets directed anteriorly	Inferior facets directed anterolaterally (or laterally)
Movements allowed	Flexion and extension; lateral flexion; rotation; the spine region with the greatest range of movement	Rotation; lateral flexion possible but restricted by ribs; flexion and extension limited	Flexion and extension; some lateral flexion; rotation prevented

9

spinous process is short and often bifurcated (divided into two branches). The spinous process of C_7 is not branched, however, and is substantially longer than that of the other cervical vertebrae. Because the spinous process of C_7 is visible through the skin at the base of the neck, it is called the *vertebra prominens* (Figure 9.13) and is used as a landmark for counting the vertebrae. Transverse processes of the cervical vertebrae are wide, and they contain foramina through which the vertebral arteries pass superiorly on their way to the brain. Any time you see these foramina in a vertebra, you can be sure that it is a cervical vertebra.

☐ Palpate your vertebra prominens. Place a check mark in the box when you locate the structure.

Thoracic Vertebrae

The 12 thoracic vertebrae (referred to as T_1 through T_{12}) may be recognized by the following structural characteristics. They have a larger body than the cervical vertebrae (see Figure 9.17). The body is somewhat heart shaped, with two small articulating surfaces, or **costal facets**, on each side (one superior, the other inferior) close to the origin of the vertebral arch. These facets articulate with the heads of the corresponding ribs. The vertebral foramen is oval or round, and the spinous process is long, with a sharp downward hook. The closer the thoracic vertebra is to the lumbar region, the less sharp and shorter the spinous process. Articular facets on the transverse processes articulate with the tubercles of the ribs. Besides forming the thoracic part of the spine, these vertebrae form the posterior aspect of the thoracic cage. Indeed, they are the only vertebrae that articulate with the ribs.

Lumbar Vertebrae

The five lumbar vertebrae (L_1 through L_5) have massive blocklike bodies and short, thick, hatchet-shaped spinous processes extending directly backward (see Table 9.3 and Figure 9.17). The superior articular facets face posteromedially; the inferior ones are directed anterolaterally. These structural features reduce the mobility of the lumbar region of the spine. Since most stress on the vertebral column occurs in the lumbar region, these are also the sturdiest of the vertebrae.

The spinal cord ends at the superior edge of L_2, but the outer covering of the cord, filled with cerebrospinal fluid, extends an appreciable distance beyond. Thus a *lumbar puncture* (for examination of the cerebrospinal fluid) or the administration of "saddle block" anesthesia for childbirth is normally done between L_3 and L_4 or L_4 and L_5, where there is little or no chance of injuring the delicate spinal cord.

The Sacrum

The **sacrum** (**Figure 9.18**) is a composite bone formed from the fusion of five vertebrae. Superiorly it articulates with L_5, and inferiorly it connects with the coccyx. The **median sacral crest** is a remnant of the spinous processes of the fused vertebrae. The winglike **alae**, formed by fusion of the transverse processes, articulate laterally with the hip bones. The sacrum is concave anteriorly and forms the posterior border of the pelvis. Four ridges (lines of fusion) cross the anterior part of the sacrum, and **sacral foramina** are located at either end of these ridges. These foramina allow blood vessels and nerves to pass. The vertebral canal continues inside the sacrum as the **sacral canal** and terminates near the coccyx via an enlarged opening called the **sacral hiatus**. The **sacral promontory** (anterior border of the body of S_1) is an important anatomical landmark for obstetricians.

☐ Attempt to palpate the median sacral crest of your sacrum. (This is more easily done by thin people and obviously in privacy.) Place a check mark in the box when you locate the structure.

The Coccyx

The **coccyx** (see Figure 9.18) is formed from the fusion of three to five small irregularly shaped vertebrae. It is literally the human tailbone, a vestige of the tail that other vertebrates have. The coccyx is attached to the sacrum by ligaments.

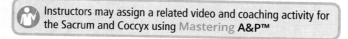

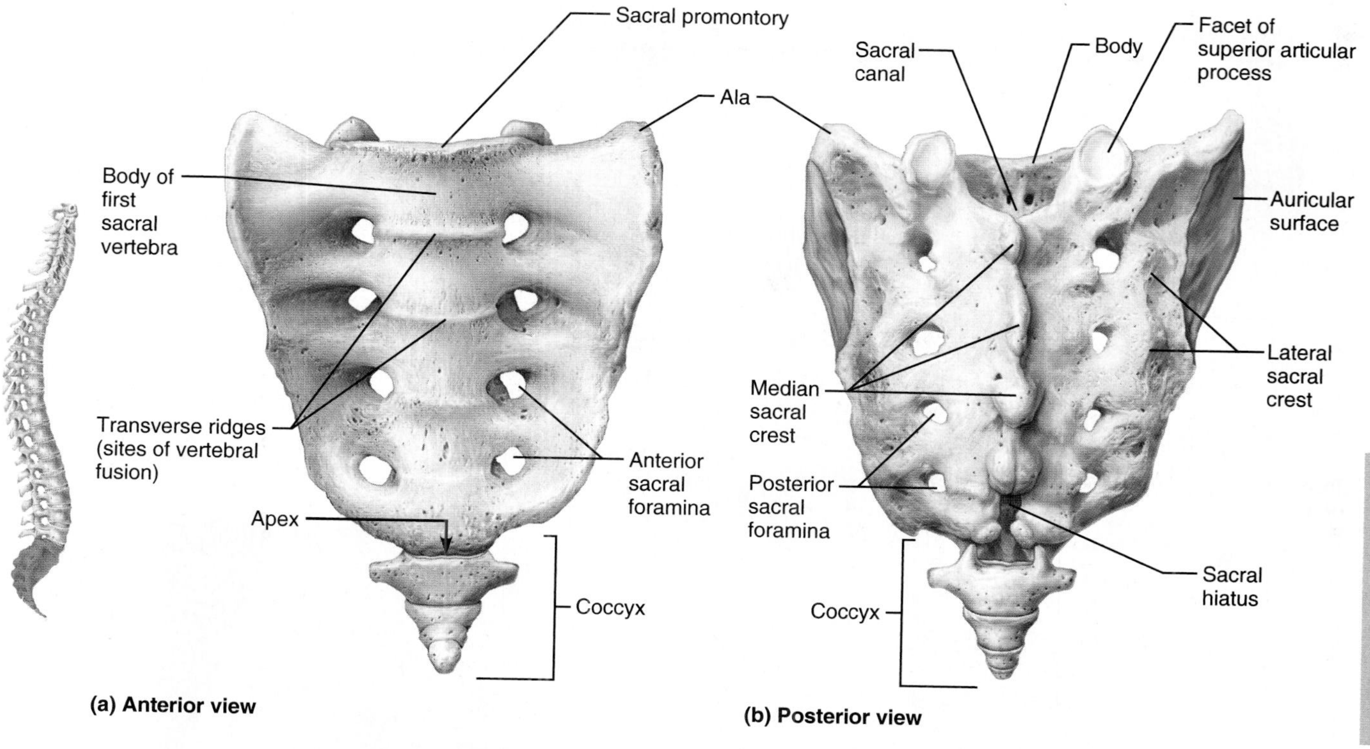

Figure 9.18 Sacrum and coccyx.

Instructors may assign a related video and coaching activity for the Sacrum and Coccyx using Mastering A&P™

Activity 4

Examining Vertebral Structure

Obtain examples of each type of vertebra, and examine them carefully, comparing them to Figures 9.16, 9.17, 9.18, and Table 9.3 and to each other.

The Thoracic Cage

The **thoracic cage** consists of the bony thorax, which is composed of the sternum, ribs, and thoracic vertebrae, plus the costal cartilages (**Figure 9.19**). Its cone-shaped, cagelike structure protects the organs of the thoracic cavity, including the critically important heart and lungs.

The Sternum

The **sternum** (breastbone), a typical flat bone, is a result of the fusion of three bones—the manubrium, body, and xiphoid process. It is attached to the first seven pairs of ribs.

The superiormost **manubrium** looks like the knot of a tie; it articulates with the clavicle (collarbone) laterally. The **body** forms the bulk of the sternum. The **xiphoid process** constructs the inferior end of the sternum and lies at the level of the fifth intercostal space. Although it is made of hyaline cartilage in children, it is usually ossified in adults over the age of 40.

In some people, the xiphoid process projects dorsally. This may present a problem because physical trauma to the chest can push such a xiphoid into the underlying heart or liver, causing massive hemorrhage. +

The sternum has three important bony landmarks—the jugular notch, the sternal angle, and the xiphisternal joint. The **jugular notch** (concave upper border of the manubrium) can be palpated easily; generally it is at the level of the disc in between the second and third thoracic vertebrae. The **sternal angle** is a result of the manubrium and body meeting at a slight angle to each other, so that a transverse ridge is formed at the level of the second ribs. It provides a handy reference point for counting ribs to locate the second intercostal space for listening to certain heart valves, and it is an important anatomical landmark for thoracic surgery. The **xiphisternal joint**, the point

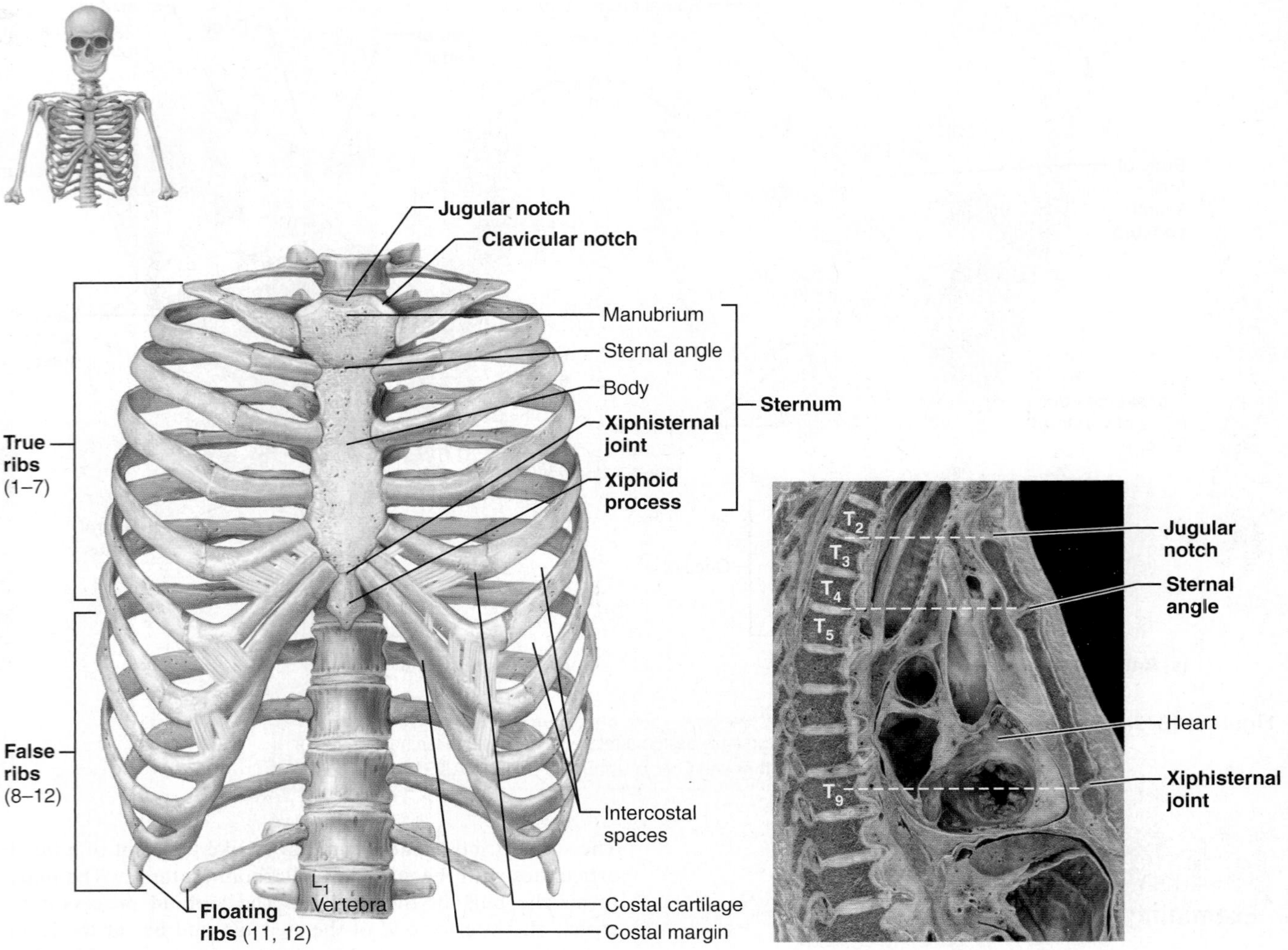

(a) (b)

Figure 9.19 The thoracic cage. (a) Anterior view with costal cartilages shown in blue. **(b)** Median section of the thorax, illustrating the relationship of the surface anatomical landmarks of the thorax to the thoracic portion of the vertebral column.

Instructors may assign a related video and coaching activity for the Sternum using Mastering **A&P™**

where the sternal body and xiphoid process fuse, lies at the level of the ninth thoracic vertebra.

☐ Palpate your sternal angle and jugular notch. Place a check mark in the box when you locate the structures.

Because of its accessibility, the sternum is a favored site for obtaining samples of blood-forming (hematopoietic) tissue for the diagnosis of suspected blood diseases. A needle is inserted into the marrow of the sternum, and the sample is withdrawn (sternal puncture).

The Ribs

The 12 pairs of **ribs** form the walls of the thoracic cage (see Figure 9.19 and **Figure 9.20**). All of the ribs articulate posteriorly with the vertebral column via their heads and tubercles and then curve downward and toward the anterior body surface. The first seven pairs, called the *true*, or *vertebrosternal*, ribs, attach directly to the sternum by their "own" costal cartilages. The next five pairs are called *false ribs*; they attach indirectly to the sternum or entirely lack a sternal attachment. Of these, rib pairs 8–10, which are also called *vertebrochondral ribs*, have indirect cartilage attachments to the sternum via the costal cartilage of rib 7. The last two pairs, called *floating*, or *vertebral*, *ribs*, have no sternal attachment.

Activity 5

Examining the Relationship Between Ribs and Vertebrae

First take a deep breath to expand your chest. Notice how your ribs seem to move outward and how your sternum rises. Then examine an articulated skeleton to observe the relationship between the ribs and the vertebrae.

(Refer to Activity 3, Palpating Landmarks of the Trunk, section on The Thorax: Bones, steps 1 and 3, in Exercise 46, Surface Anatomy Roundup.)

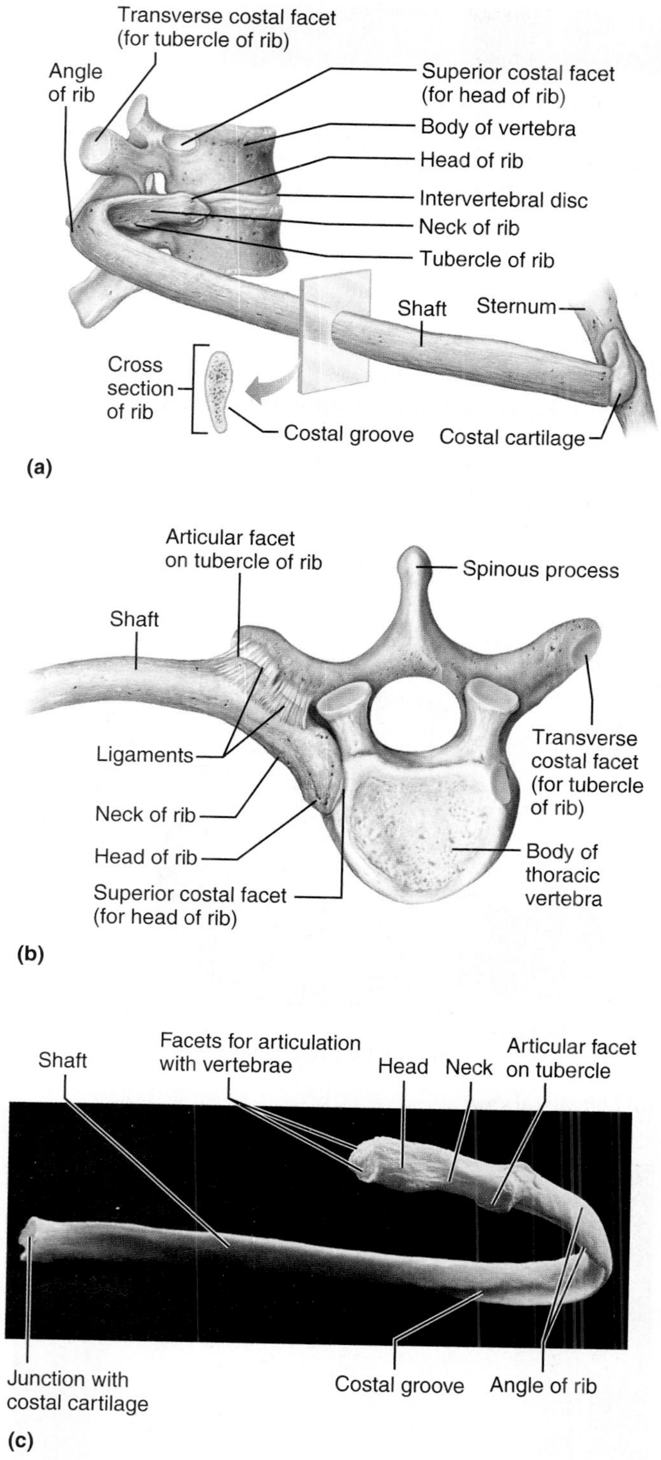

Figure 9.20 Structure of a typical true rib and its articulations. (a) Vertebral and sternal articulations of a typical true rib. **(b)** Superior view of the articulation between a rib and a thoracic vertebra, with costovertebral ligaments. **(c)** Right rib 6, posterior view.

Instructors may assign a related video and coaching activity for the Rib using Mastering **A&P™**

The Fetal Skull

One of the most obvious differences between fetal and adult skeletons is the huge size of the fetal skull relative to the rest of the skeleton. Skull bones are incompletely formed at birth and connected by fibrous membranes called **fontanelles**. The fontanelles allow the fetal skull to be compressed slightly during birth and also allow for brain growth in the fetus and infant. They ossify (become bone) as the infant ages, completing the process by the time the child is 1½ to 2 years old.

Activity 6

Examining a Fetal Skull

1. Obtain a fetal skull, and study it carefully.

- Does it have the same bones as the adult skull?
- How does the size of the fetal face relate to the cranium?
- How does this compare to what is seen in the adult?

2. Locate the following fontanelles on the fetal skull (refer to **Figure 9.21**): *anterior* (or *frontal*) *fontanelle, mastoid fontanelle, sphenoidal fontanelle,* and *posterior* (or *occipital*) *fontanelle*.

3. Notice that some of the cranial bones have conical protrusions. These are **ossification (growth) centers**. Notice also that the frontal bone is still in two parts and that the temporal bone is incompletely ossified, little more than a ring of bone.

4. Before completing this study, check the questions on the Review Sheet at the end of this exercise to ensure that you have made all of the necessary observations.

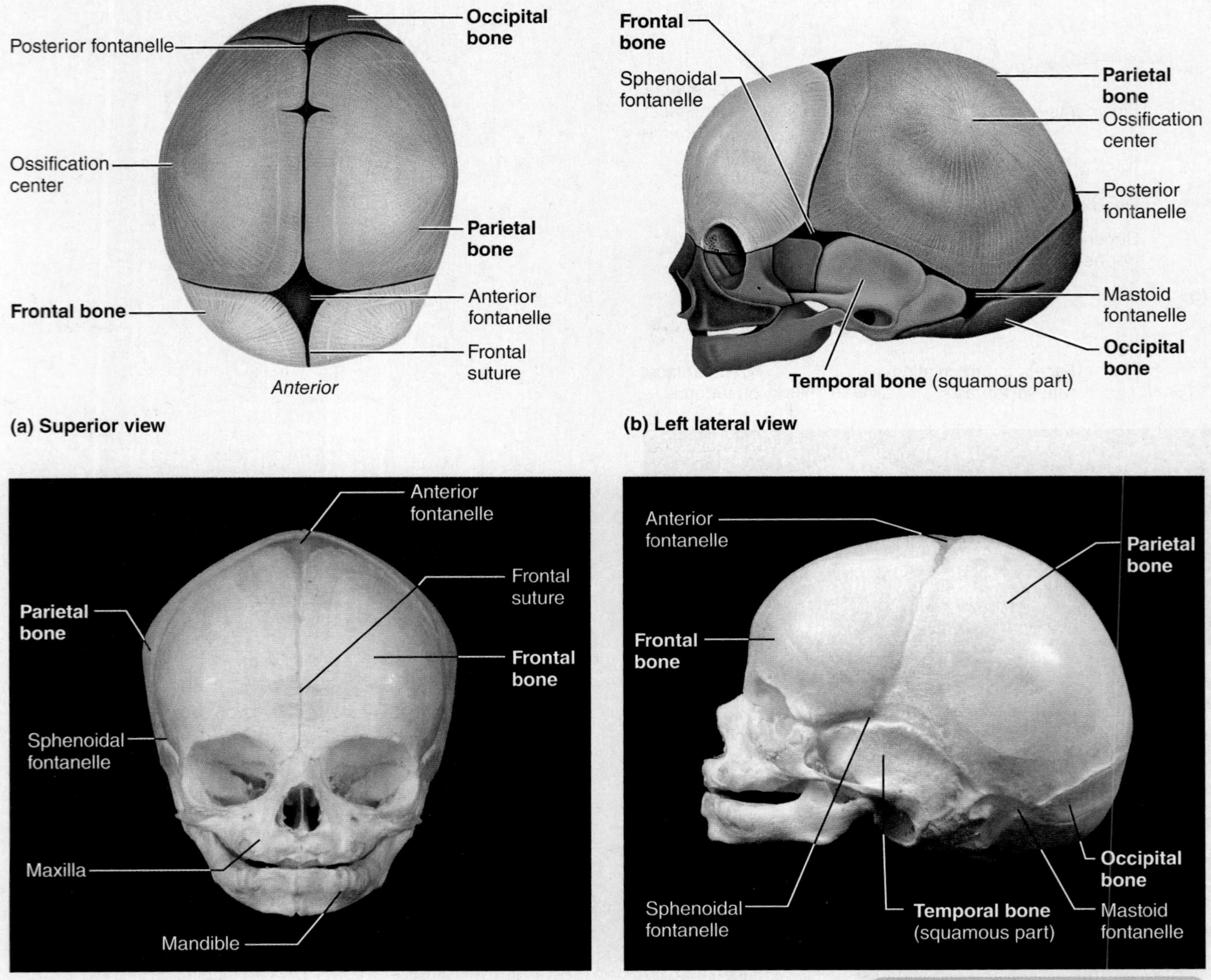

Figure 9.21 Skull of a newborn.

Instructors may assign this figure as an Art Labeling Activity using Mastering A&P™

Instructors may assign a portion of the Review Sheet questions using Mastering A&P™

REVIEW SHEET

EXERCISE 9 The Axial Skeleton

Name ______________________________ Lab Time/Date ______________________

The Skull

1. First, match the bone names in column B with the descriptions in column A (the items in column B may be used more than once). Then, circle the bones in column B that are cranial bones.

Column A

__________ 1. forms the anterior cranium

__________ 2. cheekbone

__________ 3. bridge of nose

__________ 4. posterior bones of the hard palate

__________ 5. much of the lateral and superior cranium

__________ 6. single, irregular, bat-shaped bone forming part of the cranial base

__________ 7. tiny bones bearing tear ducts

__________ 8. anterior part of hard palate

__________ 9. superior and middle nasal conchae form from its projections

__________ 10. site of mastoid process

__________ 11. has condyles that articulate with the atlas

__________ 12. small U-shaped bone in neck, where many tongue muscles attach

__________ 13. organ of hearing found here

_____, _____ 14. two bones that form the nasal septum

__________ 15. forms the most inferior turbinate

Column B

a. ethmoid
b. frontal
c. hyoid
d. inferior nasal concha
e. lacrimal
f. mandible
g. maxilla
h. nasal
i. occipital
j. palatine
k. parietal
l. sphenoid
m. temporal
n. vomer
o. zygomatic

2. Using choices from the numbered key to the right, identify all bones and bone markings provided with various leader lines in the two following photographs. A colored dot at the end of a leader line indicates a bone. Leader lines without a colored dot indicate bone markings. Note that vomer, sphenoid bone, and zygomatic bone will each be labeled twice.

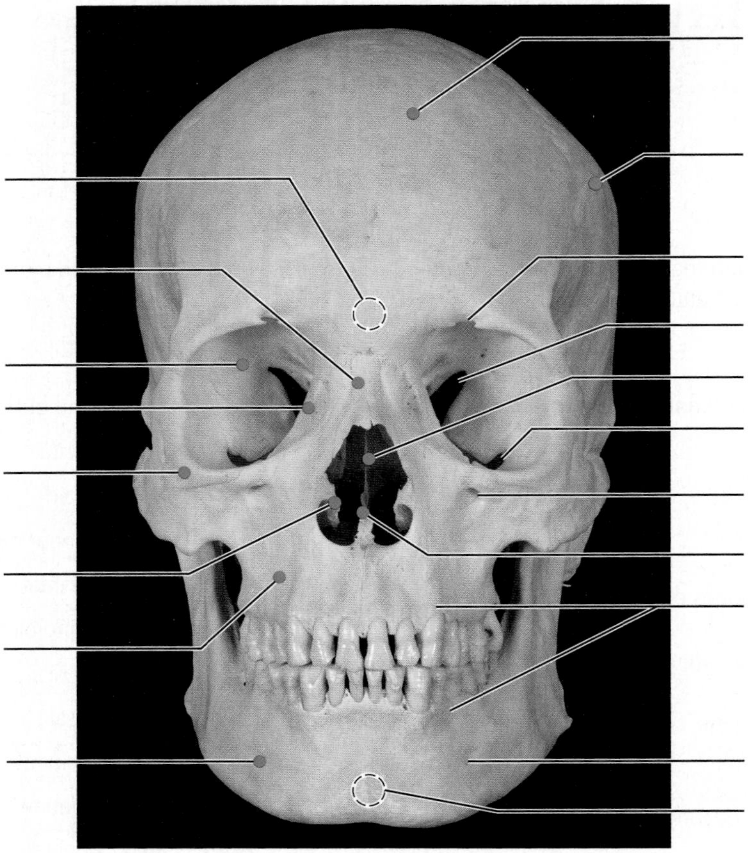

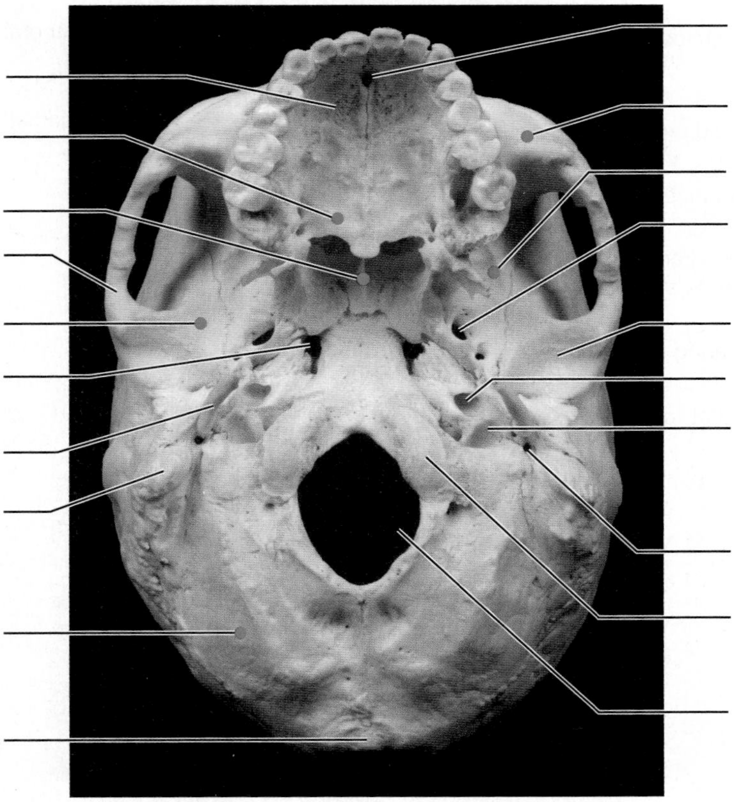

Key:
1. alvelolar processes
2. carotid canal
3. ethmoid bone (perpendicular plate)
4. external occipital protuberance
5. foramen lacerum
6. foramen magnum
7. foramen ovale
8. frontal bone
9. glabella
10. incisive fossa
11. inferior nasal concha
12. inferior orbital fissure
13. infraorbital foramen
14. jugular foramen
15. lacrimal bone
16. mandible
17. mandibular fossa
18. mandibular symphysis
19. mastoid process
20. maxilla
21. mental foramen
22. nasal bone
23. occipital bone
24. occipital condyle
25. palatine bone
26. palatine process of maxilla
27. parietal bone
28. sphenoid bone
29. styloid process
30. stylomastoid foramen
31. superior orbital fissure
32. supraorbital foramen
33. temporal bone
34. vomer
35. zygomatic bone
36. zygomatic process

3. Define *suture.* ______

4. With one exception, the skull bones are joined by sutures. Name the exception.

______

5. What bones are connected by the lambdoid suture?

______

What bones are connected by the squamous suture?

______

6. Name the eight bones of the cranium. (Remember to include left and right.)

______ ______ ______ ______

______ ______ ______ ______

7. List the bones that have sinuses, and give two possible functions of the sinuses.

______

______

______

8. What is the bony orbit? ______

What bones contribute to the formation of the orbit? ______

______

9. Why can the sphenoid bone be called the keystone bone of the cranium? ______

______

The Vertebral Column

10. The distinguishing characteristics of the vertebrae composing the vertebral column are noted below. Correctly identify each described structure by choosing a response from the key.

Key: a. atlas
b. axis
c. cervical vertebra—typical
d. coccyx
e. lumbar vertebra
f. sacrum
g. thoracic vertebra

__________ 1. vertebra type containing foramina in the transverse processes, through which the vertebral arteries ascend to reach the brain

__________ 2. dens here provides a pivot for rotation of the first cervical vertebra (C_1)

__________ 3. transverse processes faceted for articulation with ribs; spinous process pointing sharply downward

__________ 4. composite bone; articulates with the hip bone laterally

__________ 5. massive vertebra; weight-sustaining

__________ 6. "tail bone" fused vertebrae

__________ 7. supports the head; allows a rocking motion in conjunction with the occipital condyles

11. Using the key, correctly identify the vertebral parts/areas described below. (More than one choice may apply in some cases.) Also use the key letters to correctly identify the vertebral areas in the diagram.

Key: a. body
b. intervertebral foramina
c. lamina
d. pedicle
e. spinous process
f. superior articular facet
g. transverse process
h. vertebral arch
i. vertebral foramen

__________ 1. cavity enclosing the spinal cord

__________ 2. weight-bearing portion of the vertebra

_____, _____ 3. provide levers against which muscles pull

_____, _____ 4. provide an articulation point for the ribs

__________ 5. openings providing for exit of spinal nerves

_____, _____ 6. structures that form an enclosure for the spinal cord

_____, _____ 7. structures that form the vertebral arch

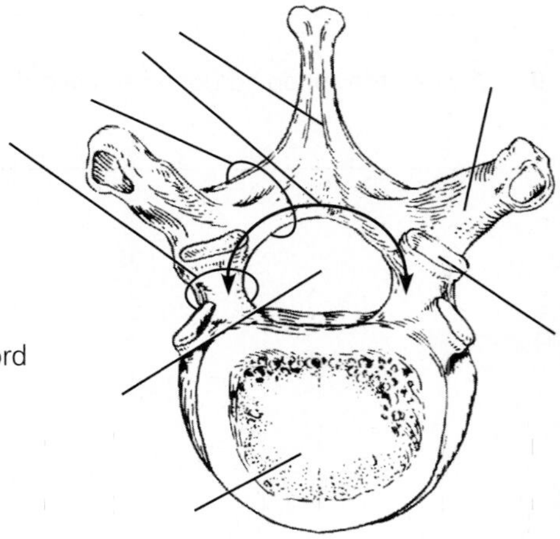

12. Describe how a spinal nerve exits from the vertebral column. ______________________________

____

13. Name two factors/structures that permit flexibility of the vertebral column.

______________________________ and ______________________________

14. What kind of tissue makes up the intervertebral discs? ______________________________

15. What is a herniated disc? ______________________________

What problems might it cause? ______________________________

____

16. Which two spinal curvatures are obvious at birth? ______________ and ______________

Under what conditions do the secondary curvatures develop? ______________________________

____

____

17. Use the key to label the structures on the thoracic region of the vertebral column.

Key: a. intervertebral discs
b. intervertebral foramina
c. spinous prosesses
d. thoracic vertebrae
e. transverse processes

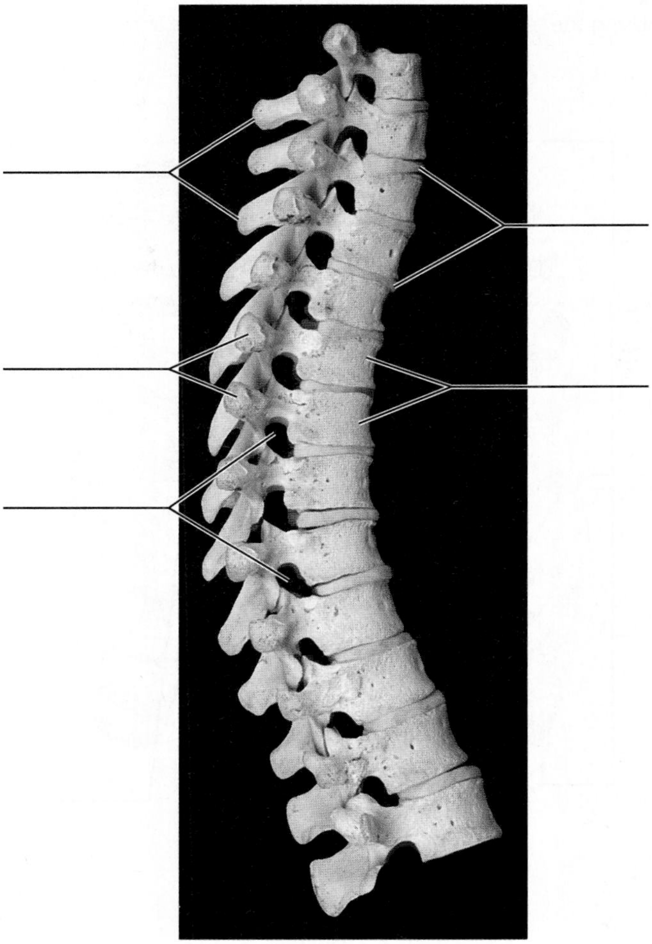

The Thoracic Cage

18. The major bony components of the thorax (excluding the vertebral column) are the ______________________

and the ______________________

19. Differentiate between a true rib and a false rib. ______________________

____

____

Is a floating rib a true or a false rib? ______________________.

20. What is the general shape of the thoracic cage? ______________________

21. Using the terms in the key, identify the regions and landmarks of the thoracic cage.

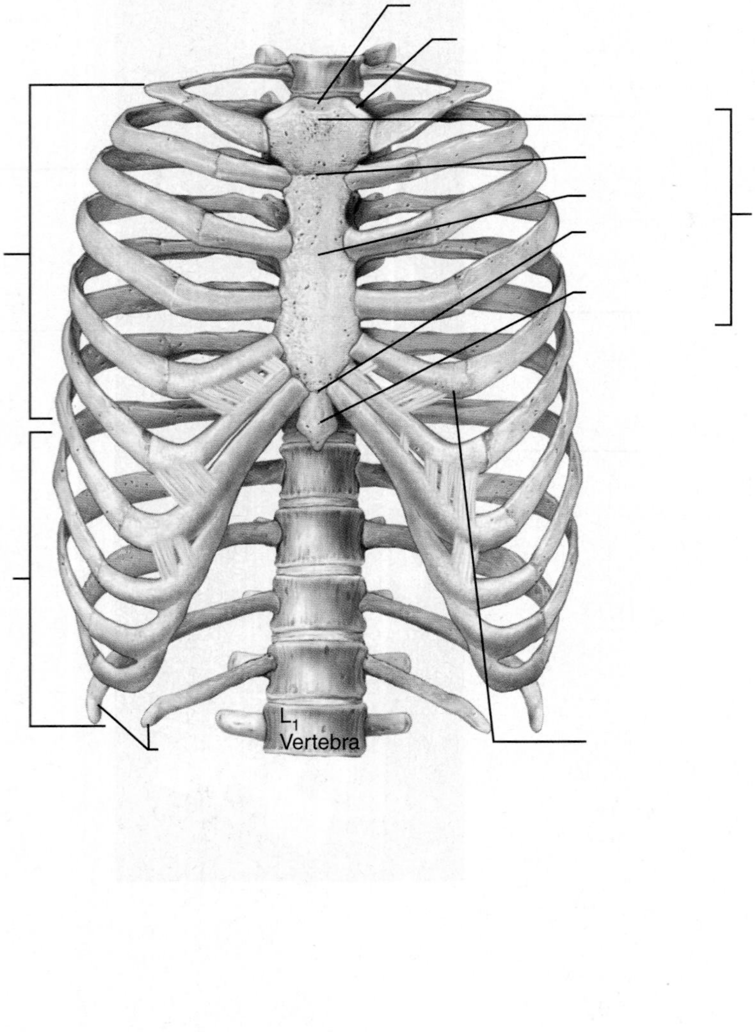

Key:
a. body
b. clavicular notch
c. costal cartilage
d. false ribs
e. floating ribs
f. jugular notch
g. manubrium
h. sternal angle
i. sternum
j. true ribs
k. xiphisternal joint
l. xiphoid process

The Fetal Skull

22. Are the same skull bones seen in the adult also found in the fetal skull? ______

23. How does the size of the fetal face compare to its cranium? ______

______

How does this compare to the adult skull? ______

______

______

24. What are the outward conical projections on some of the fetal cranial bones? ______

25. What is a fontanelle? ______

What is its fate? ______

What is the function of the fontanelles in the fetal skull? ______

______

26. ✚ Craniosynostosis is a condition in which one or more of the fontanelles is replaced by bone prematurely. Discuss the ramifications of this early closure.

______

______

27. ✚ As we age, we often become shorter. Explain why this might occur. ______

______

______

28. ✚ The xiphoid process is often missing from the sternum in bone collections. Hypothesize why it might be missing. ______

______

EXERCISE 10

The Appendicular Skeleton

Learning Outcomes

- Identify the bones of the pectoral and pelvic girdles and their attached limbs by examining isolated bones or an articulated skeleton, and name the important bone markings on each.
- Describe the differences between a male and a female pelvis, and explain why these differences are important.
- Compare the features of the human pectoral and pelvic girdles, and discuss how their structures relate to their specialized functions.
- Arrange unmarked, disarticulated bones in their proper places to form an entire skeleton.

Go to Mastering A&P™ > Study Area to improve your performance in A&P Lab.

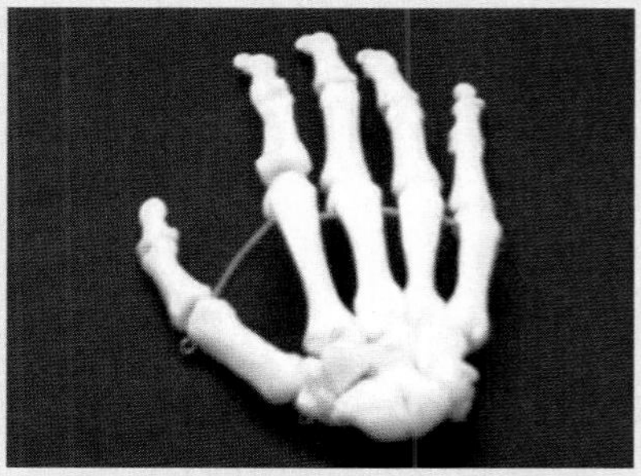

> Lab Tools > Bone & Dissection Videos

Instructors may assign new Building Vocabulary coaching activities, Pre-Lab Quiz questions, Art Labeling activities, related bone videos and coaching activities, Practice Anatomy Lab Practical questions (PAL), and more using the Mastering **A&P™** Item Library.

Pre-Lab Quiz

Instructors may assign these and other Pre-Lab Quiz questions using Mastering **A&P™**

1. Circle the correct underlined term. The pectoral / pelvic girdle attaches the upper limb to the axial skeleton.
2. The arm consists of one long bone, the:
   a. femur **c.** tibia
   b. humerus **d.** ulna
3. The strongest, heaviest bone of the body is in the thigh. It is the
   a. femur **b.** fibula **c.** tibia
4. The ________________, or "knee cap," is a sesamoid bone that is found within the quadriceps tendon.
5. Each foot has a total of ________________ bones.

Materials

- Articulated skeletons
- Disarticulated skeletons (complete)
- Articulated pelves (male and female for comparative study)
- X-ray images of bones of the appendicular skeleton

The **appendicular skeleton** (the gold-colored portion of Figure 8.1) is composed of the 126 bones of the appendages and the pectoral and pelvic girdles, which attach the limbs to the axial skeleton. Although the upper and lower limbs differ in their functions and mobility, they have the same fundamental plan, with each limb made up of three major segments connected by freely movable joints.

Activity 1

Examining and Identifying Bones of the Appendicular Skeleton

Carefully examine each of the bones described in this exercise, and identify the characteristic bone markings of each (see Tables 10.1–10.5). The markings aid in determining whether a bone is the right or left member of its pair; for example, the glenoid cavity is on the lateral aspect of the scapula, and the spine is on its posterior aspect. *This is a very important instruction because you will be constructing your own skeleton to finish this laboratory exercise.* Additionally, when corresponding X-ray images are available, compare the actual bone specimen to its X-ray image.

Table 10.1 The Appendicular Skeleton: The Pectoral (Shoulder) Girdle (Figures 10.1 and 10.2)

Bone	Important markings	Description
Clavicle ("collarbone") (Figures 10.1 and 10.2)	Acromial (lateral) end	Flattened lateral end that articulates with the acromion of the scapula to form the acromioclavicular (AC) joint
	Sternal (medial) end	Oval or triangular medial end that articulates with the sternum to form the lateral walls of the jugular notch (see Figure 9.19, p. 131)
	Conoid tubercle	A small, cone-shaped projection located on the lateral, inferior end of the bone; serves to anchor ligaments
Scapula ("shoulder blade") (Figures 10.1 and 10.2)	Superior border	Short, sharp border located superiorly
	Medial (vertebral) border	Thin, long border that runs roughly parallel to the vertebral column
	Lateral (axillary) border	Thick border that is closest to the armpit and ends superiorly with the glenoid cavity
	Glenoid cavity	A shallow socket that articulates with the head of the humerus
	Spine	A ridge of bone on the posterior surface that is easily felt through the skin
	Acromion	The lateral end of the spine of the scapula that articulates with the clavicle to form the AC joint
	Coracoid process	Projects above the glenoid cavity as a hooklike process; helps attach the biceps brachii muscle
	Suprascapular notch	Small notch located medial to the coracoid process that allows for the passage of blood vessels and a nerve
	Subscapular fossa	A large shallow depression that forms the anterior surface of the scapula
	Supraspinous fossa	A depression located superior to the spine of the scapula
	Infraspinous fossa	A broad depression located inferior to the spine of the scapula

Bones of the Pectoral Girdle and Upper Limb

The Pectoral (Shoulder) Girdle

The paired **pectoral**, or **shoulder, girdles** (**Figures 10.1** and **10.2** and **Table 10.1**) each consist of two bones—the anterior clavicle and the posterior scapula. The clavicle articulates with the axial skeleton via the sternum. The scapula does not articulate with the axial skeleton. The shoulder girdles attach the upper limbs to the axial skeleton and provide attachment points for many trunk and neck muscles.

The shoulder girdle is exceptionally light and allows the upper limb a degree of mobility not seen anywhere else in the body. This is due to the following factors:

- The sternoclavicular joints are the *only* site of attachment of the shoulder girdles to the axial skeleton.
- The relative looseness of the scapular attachment allows it to slide back and forth against the thorax with muscular activity.
- The glenoid cavity is shallow and does little to stabilize the shoulder joint.

However, this exceptional flexibility exacts a price: the arm bone (humerus) is very susceptible to dislocation, and fracture of the clavicle disables the entire upper limb. This is because the clavicle acts as a brace to hold the scapula and arm away from the top of the thoracic cage.

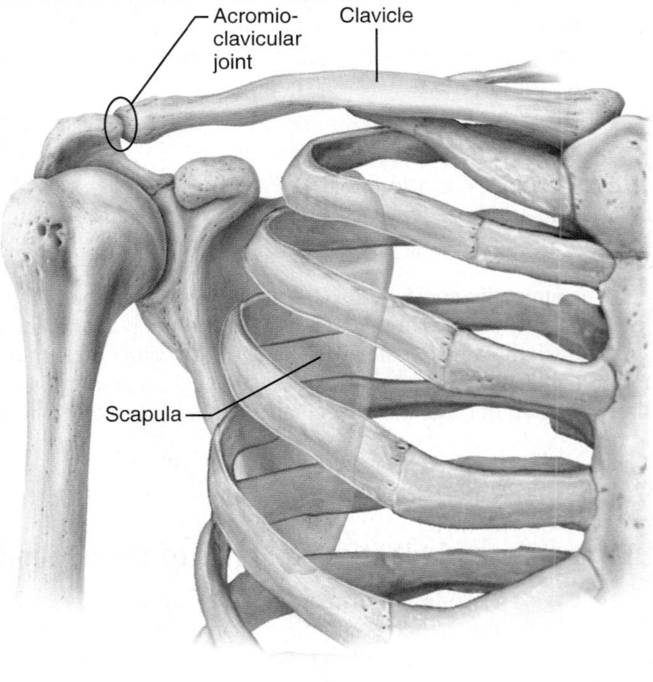

Figure 10.1 Articulated bones of the pectoral (shoulder) girdle. The right pectoral girdle is articulated to show the relationship of the girdle to the bones of the thorax and arm.

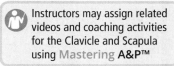

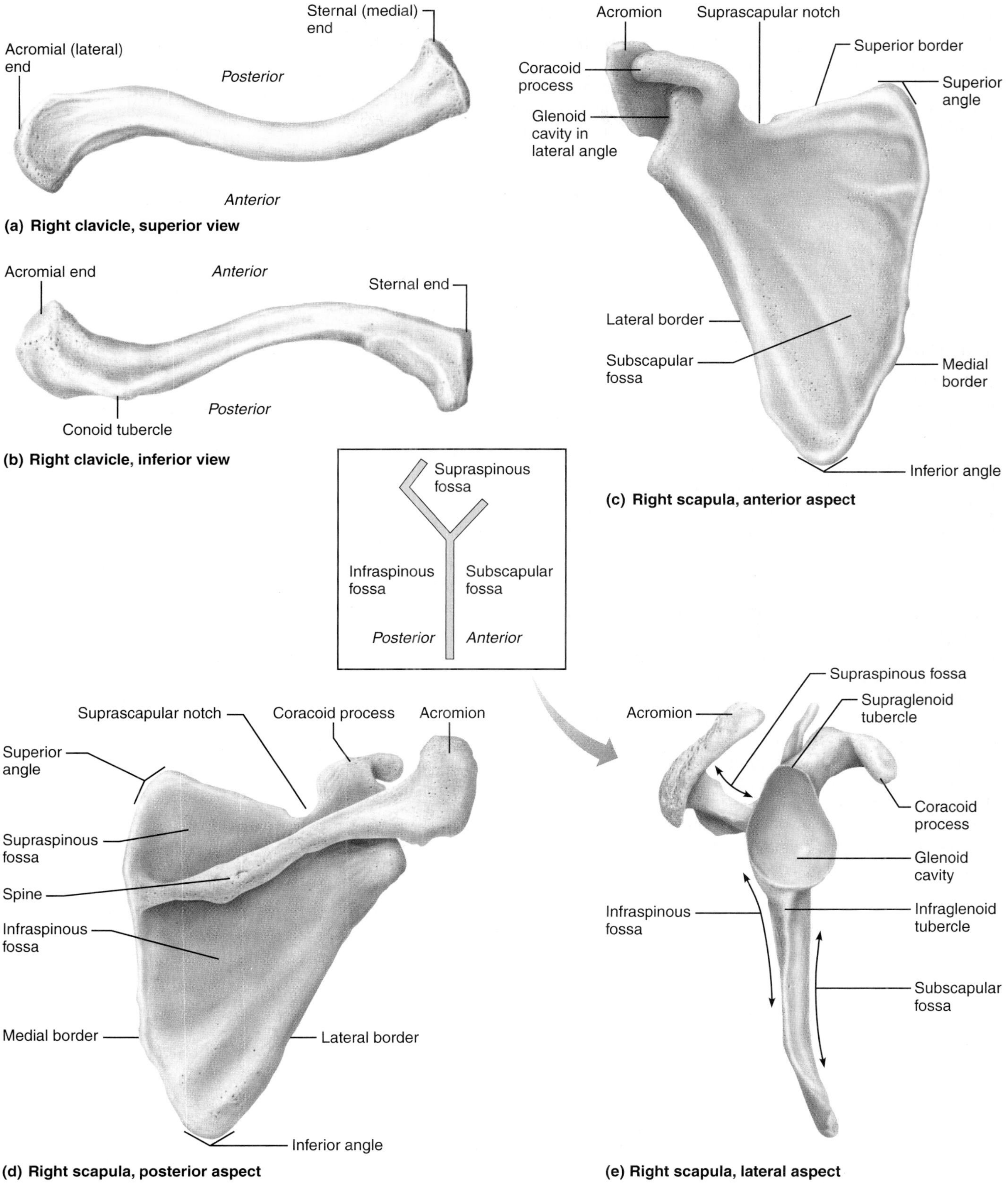

Figure 10.2 Individual bones of the pectoral (shoulder) girdle. View **(e)** is accompanied by a schematic representation of its orientation.

Instructors may assign related videos and coaching activities for the Clavicle and Scapula using Mastering **A&P™**

The Arm

The arm, or brachium (**Figure 10.3** and **Table 10.2A**), contains a single bone—the **humerus**. Proximally its rounded *head* fits into the shallow glenoid cavity of the scapula. The head is separated from the shaft by the *anatomical neck* and the more constricted *surgical neck*, which is a common site of fracture.

The Forearm

Two bones, the radius and the ulna, compose the skeleton of the forearm, or antebrachium (**Figure 10.4**, p. 148 and **Table 10.2B**, p. 148). When the body is in the anatomical position, the **radius** is lateral, the **ulna** is medial, and the radius and ulna are parallel.

The Hand

10

The skeleton of the hand, or manus (**Figure 10.5**, p. 149), includes three groups of bones, those of the carpus, the metacarpals, and the phalanges.

The wrist is the proximal portion of the hand. It is referred to anatomically as the **carpus**; the eight bones composing it are the **carpals**. The carpals are arranged in two irregular rows of four bones each, illustrated in Figure 10.5. In the proximal row (lateral to medial) are the *scaphoid*, *lunate*, *triquetrum*, and *pisiform bones*; the scaphoid and lunate articulate with the distal end of the radius. In the distal row (lateral to medial) are the *trapezium*, *trapezoid*, *capitate*, and *hamate*.

The **metacarpals**, numbered I to V from the thumb side of the hand toward the little finger, radiate out from the wrist like spokes to form the palm of the hand. The *bases* of the metacarpals articulate with the carpals; their more bulbous *heads* articulate with the phalanges distally. When the fist is clenched, the heads of the metacarpals become prominent as the knuckles.

The fingers are numbered from 1 to 5, beginning from the thumb (*pollex*) side of the hand. The 14 bones of the fingers, or digits, are miniature long bones, called **phalanges** (singular: *phalanx*). Each finger contains three phalanges (proximal, middle, and distal) except the thumb, which has only two (proximal and distal).

Activity 2

Palpating the Surface Anatomy of the Pectoral Girdle and the Upper Limb

Identify the following bone markings on the skin surface of the upper limb. It is usually preferable to palpate the bone markings on your lab partner since many of these markings can only be seen from the posterior aspect. Place a check mark in the boxes as you locate the bone markings. For any markings that you are unable to locate, ask your instructor for help.

☐ Clavicle: Palpate the clavicle along its entire length from sternum to shoulder.

☐ Acromioclavicular (AC) joint: The high point of the shoulder, which represents the junction point between the clavicle and the acromion of the scapula.

☐ Spine of the scapula: Extend your arm at the shoulder so that your scapula moves posteriorly. As you do this, your scapular spine will be seen as a winglike protrusion on your posterior thorax and can be easily palpated by your lab partner.

☐ Lateral epicondyle of the humerus: After you have located the epicondyle, run your finger posteriorly into the hollow immediately dorsal to the epicondyle. This is the site where the extensor muscles of the forearm are attached and is a common site of the excruciating pain of tennis elbow.

☐ Medial epicondyle of the humerus: Feel this medial projection at the distal end of the humerus. The ulnar nerve, which runs behind the medial epicondyle, is responsible for the tingling, painful sensation felt when you hit your "funnybone."

☐ Olecranon of the ulna: Work your elbow—flexing and extending—as you palpate its posterior aspect to feel the olecranon moving into and out of the olecranon fossa on the posterior aspect of the humerus.

☐ Ulnar styloid process: With the hand in the anatomical position, feel this small inferior projection on the medial aspect of the distal end of the ulna.

☐ Radial styloid process: Find this projection at the distal end of the radius (lateral aspect). It is most easily located by moving the hand medially at the wrist. Next, move your fingers (not your thumb) just medially onto the anterior wrist. You should be able to feel your pulse at this pressure point, which lies over the radial artery (radial pulse).

☐ Pisiform: Just distal to the ulnar styloid process, feel the rounded, pealike pisiform bone.

☐ Metacarpophalangeal joints (knuckles): Clench your fist and locate your knuckles—these are your metacarpophalangeal joints.

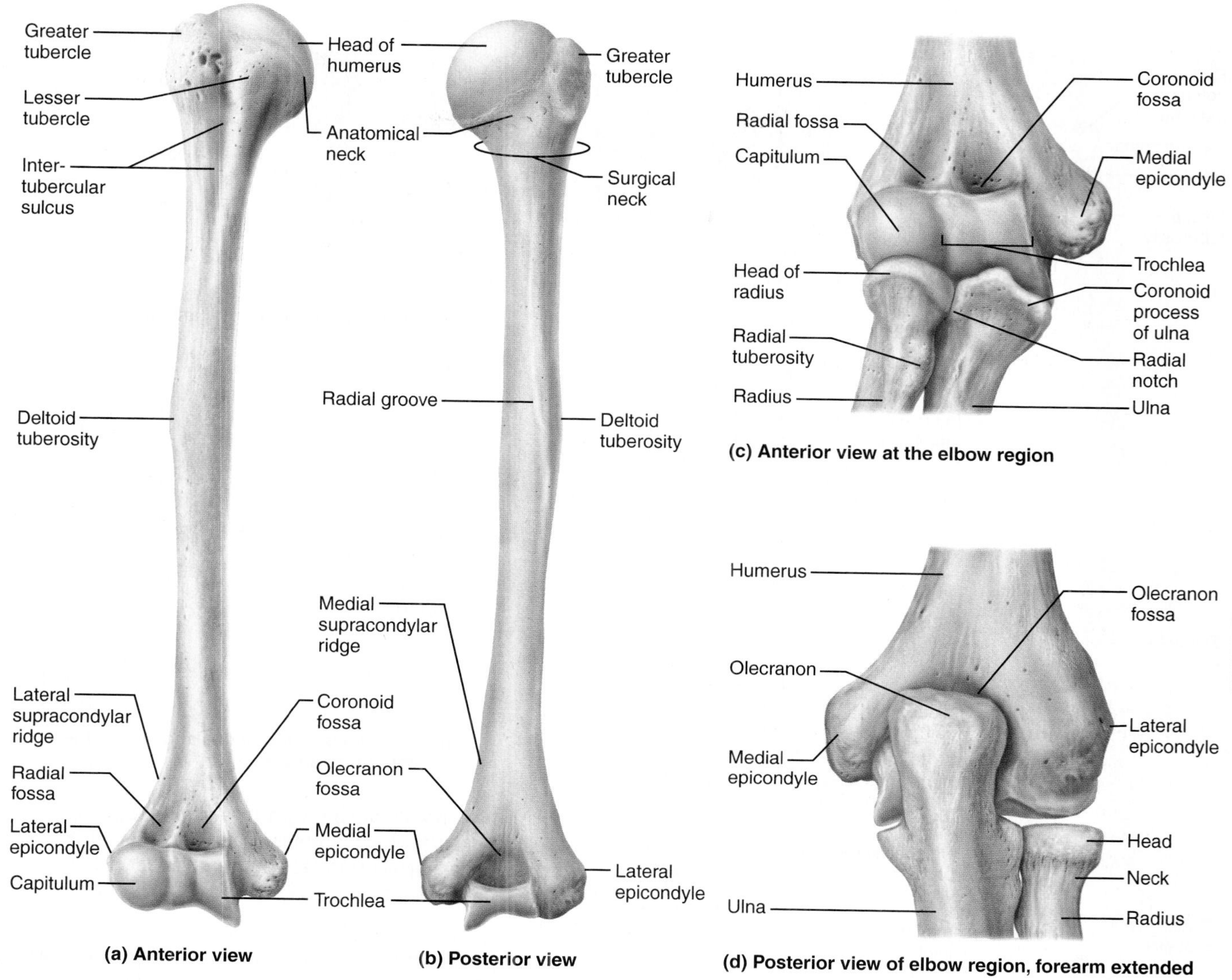

Figure 10.3 Bone of the right arm. (a, b) Humerus. **(c, d)** Detailed views of elbow region.

Instructors may assign a related video and coaching activity for the Humerus using Mastering **A&P™**

Table 10.2A The Appendicular Skeleton: The Upper Limb (Figure 10.3)

Bone	Important markings	Description
Humerus (Figure 10.3) (only bone of the arm)	Greater tubercle	Large lateral prominence; site of the attachment of rotator cuff muscles
	Lesser tubercle	Small medial prominence; site of attachment of rotator cuff muscles
	Intertubercular sulcus	A groove separating the greater and lesser tubercles; the tendon of the biceps brachii lies in this groove
	Deltoid tuberosity	A roughened area about midway down the shaft of the lateral humerus; site of attachment of the deltoid muscle
	Radial fossa	Small lateral depression; receives the head of the radius when the forearm is flexed
	Coronoid fossa	Small medial anterior depression; receives the coronoid process of the ulna when the forearm is flexed
	Capitulum	A rounded lateral condyle that articulates with the radius
	Trochlea	A flared medial condyle that articulates with the ulna
	Lateral epicondyle	Small condyle proximal to the capitulum
	Medial epicondyle	Rough condyle proximal to the trochlea
	Radial groove	Small posterior groove, marks the course of the radial nerve
	Olecranon fossa	Large distal posterior depression that accommodates the olecranon of the ulna

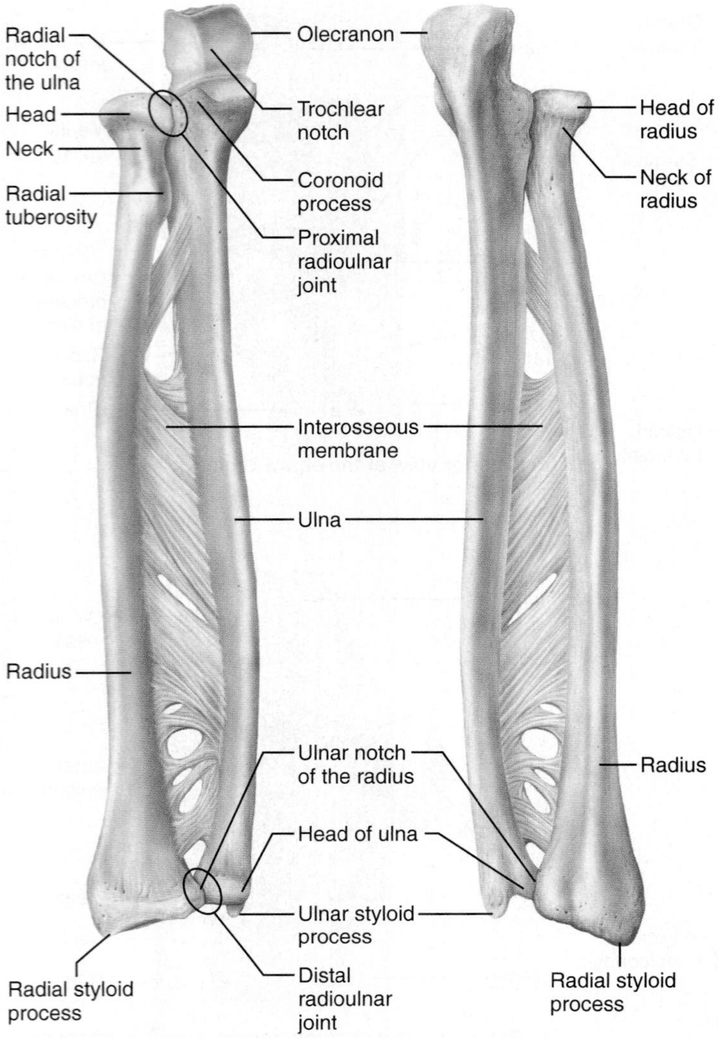

(a) Anterior view

(b) Posterior view

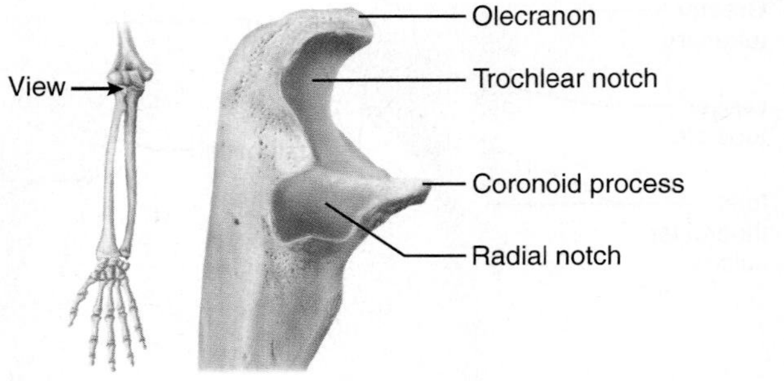

(c) Proximal portion of ulna, lateral view

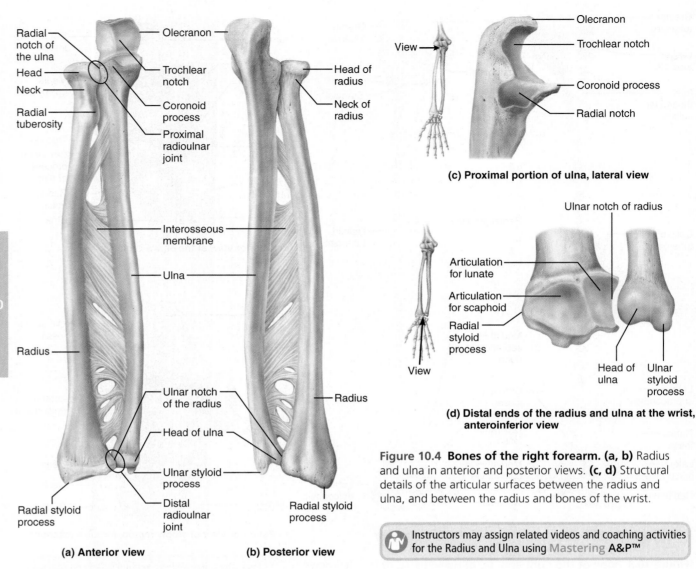

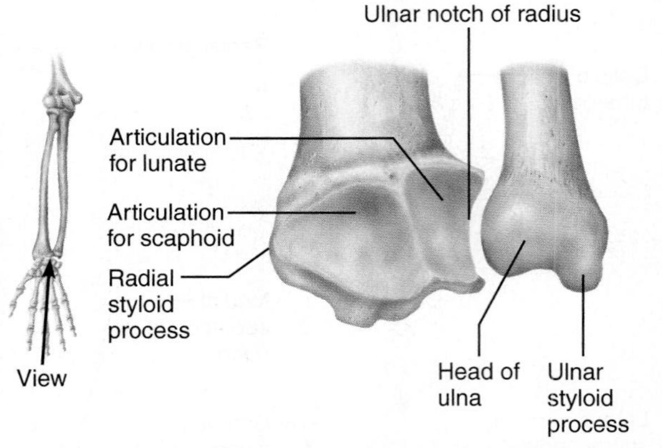

(d) Distal ends of the radius and ulna at the wrist, anteroinferior view

Figure 10.4 Bones of the right forearm. (a, b) Radius and ulna in anterior and posterior views. **(c, d)** Structural details of the articular surfaces between the radius and ulna, and between the radius and bones of the wrist.

Instructors may assign related videos and coaching activities for the Radius and Ulna using Mastering A&P™

10

Table 10.2B The Appendicular Skeleton: The Upper Limb (Figures 10.3 and 10.4)

Bone	Important markings	Description
Radius (lateral bone of the forearm in the anatomical position) (Figures 10.4 and 10.3c, d)	Head	Proximal end of the radius that forms part of the proximal radioulnar joint and articulates with the capitulum of the humerus
	Radial tuberosity	Medial prominence just below the head of the radius; site of attachment of the biceps brachii
	Radial styloid process	Distal prominence; site of attachment for ligaments that travel to the wrist
	Ulnar notch	Small distal depression that accommodates the head of the ulna, forming the distal radioulnar joint
Ulna (medial bone of the forearm in the anatomical position) (Figures 10.4 and 10.3c, d)	Olecranon	Prominent process on the posterior proximal ulna; articulates with the olecranon fossa of the humerus when the forearm is extended
	Trochlear notch	Deep notch that separates the olecranon and the coronoid process; articulates with the trochlea of the humerus
	Coronoid process	Shaped like a point on a crown; articulates with the trochlea of the humerus
	Radial notch	Small proximal lateral notch that articulates with the head of the radius; forms part of the proximal radioulnar joint
	Head	Slim distal end of the ulna; forms part of the distal radioulnar joint
	Ulnar styloid process	Distal pointed projection; located medial to the head of the ulna

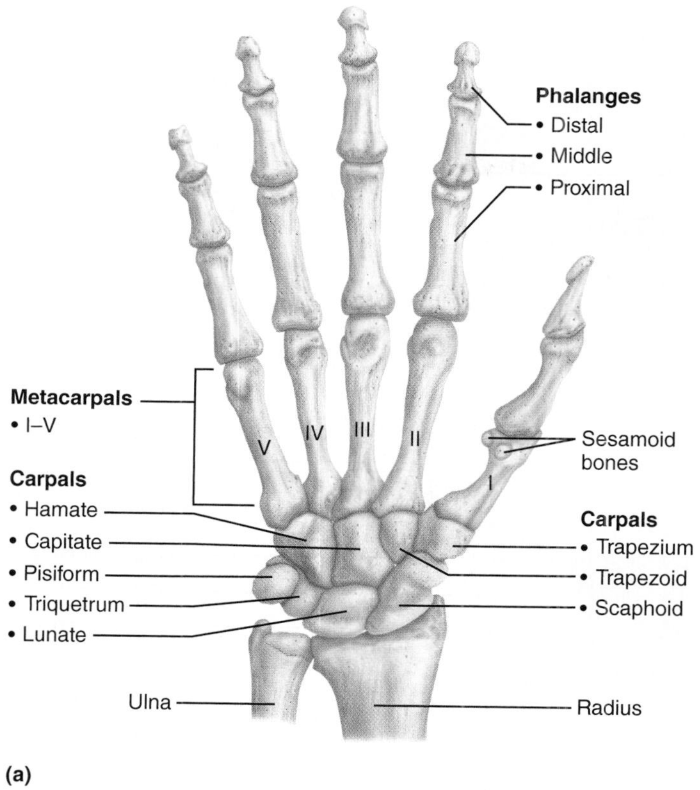

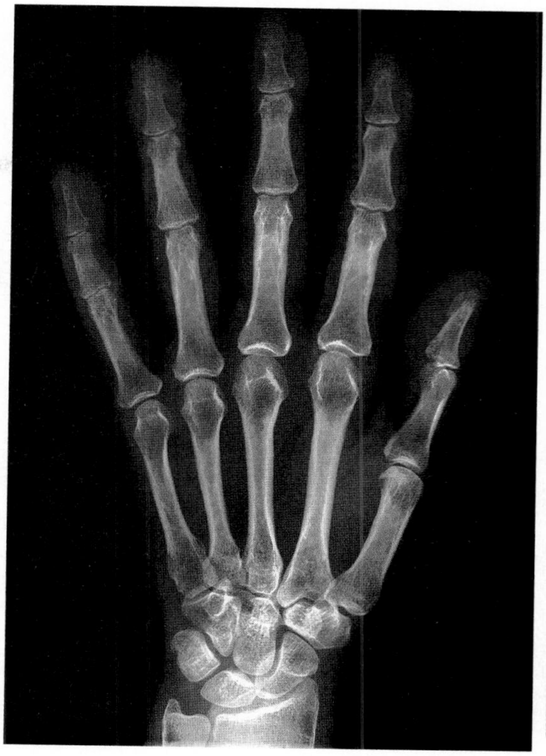

Figure 10.5 Bones of the right hand. (a) Anterior view showing the relationships of the carpals, metacarpals, and phalanges. **(b)** X-ray image of the right hand in the anterior view.

Instructors may assign a related video and coaching activity for the Hand using Mastering A&P™

10

Bones of the Pelvic Girdle and Lower Limb

The Pelvic (Hip) Girdle

As with the bones of the pectoral girdle and upper limb, pay particular attention to bone markings needed to identify right and left bones.

The **pelvic girdle**, or **hip girdle** (**Figure 10.6**, p. 150 and **Table 10.3**, p. 151), is formed by the two **hip bones** (also called the **ossa coxae**, or *coxal bones*) and the sacrum. The deep structure formed by the hip bones, sacrum, and coccyx is called the **pelvis** or *bony pelvis*. In contrast to the bones of the shoulder girdle, those of the pelvic girdle are heavy and massive, and they attach securely to the axial skeleton. The sockets for the heads of the femurs (thigh bones) are deep and heavily reinforced by ligaments to ensure a stable, strong limb attachment. The ability to bear weight is more important here than mobility and flexibility. The combined weight of the upper body rests on the pelvic girdle.

Each hip bone is a result of the fusion of three bones—the **ilium**, **ischium**, and **pubis**—which are distinguishable in the young child.

The ilium, ischium, and pubis fuse at the deep socket called the **acetabulum** (literally, "vinegar cup"), which receives the head of the thigh bone. The rami of the pubis and ischium form a bar of bone enclosing the **obturator foramen**, through which a few blood vessels and nerves pass.

Activity 3

Observing Pelvic Articulations

Take time to examine an articulated pelvis. Notice how each hip bone articulates with the sacrum posteriorly and how the two hip bones join at the pubic symphysis. The sacroiliac joint is a common site of lower back problems because of the pressure it must bear.

Comparison of the Male and Female Pelves

The female pelvis reflects differences for childbearing—it is wider, shallower, lighter, and rounder than that of the male. Her pelvis not only must support the increasing size of a fetus, but also must be large enough to allow the infant's head (its largest dimension) to descend through the birth canal at birth.

The **pelvic brim** is a continuous oval ridge of bone that runs along the pubic symphysis, pubic crests, arcuate lines, sacral alae, and sacral promontory. The **false pelvis** is that portion superior to the pelvic brim; it is bounded by the alae of the ilia laterally and the sacral promontory and lumbar vertebrae posteriorly. The false pelvis supports the abdominal

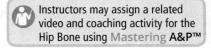

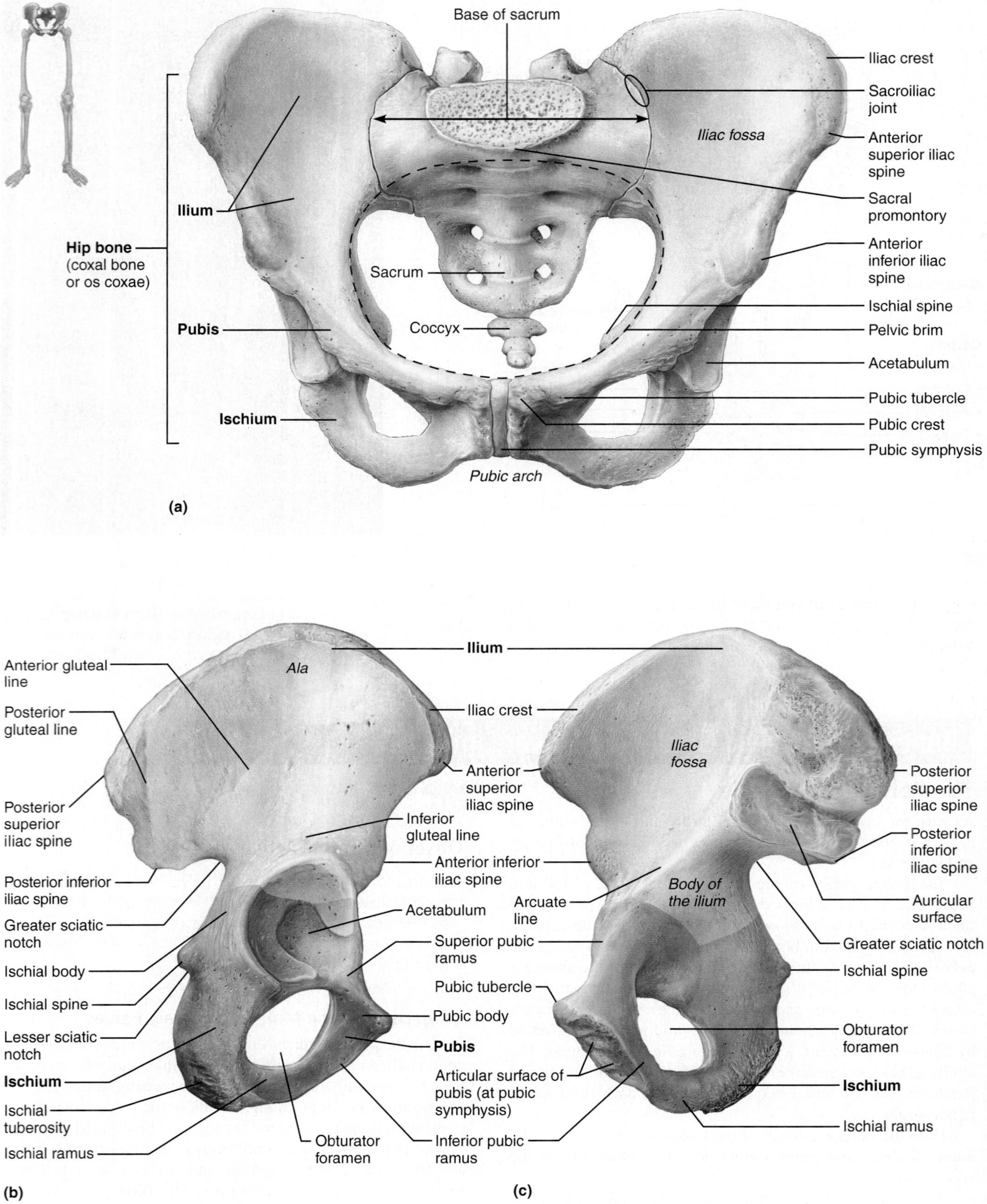

Figure 10.6 Bones of the pelvic girdle. (a) Articulated pelvis, showing the two hip bones (coxal bones), which together with the sacrum comprise the pelvic girdle, and the coccyx. **(b)** Right hip bone, lateral view, showing the point of fusion of the ilium, ischium, and pubis. **(c)** Right hip bone, medial view.

Table 10.3 The Appendicular Skeleton: The Pelvic (Hip) Girdle (Figure 10.6)

Bone	Important markings	Description
Ilium (Figure 10.6)	Iliac crest	Thick superior margin of bone
	Anterior superior iliac spine	The blunt anterior end of the iliac crest
	Posterior superior iliac spine	The sharp posterior end of the iliac crest
	Anterior inferior iliac spine	Small projection located just below the anterior superior iliac spine
	Posterior inferior iliac spine	Small projection located just below the posterior superior iliac spine
	Greater sciatic notch	Deep notch located inferior to the posterior inferior iliac spine; allows the sciatic nerve to enter the thigh
	Iliac fossa	Shallow depression below the iliac crest; forms the internal surface of the ilium
	Auricular surface	Rough medial surface that articulates with the auricular surface of the sacrum, forming the sacroiliac joint
	Arcuate line	A ridge of bone that runs inferiorly and anteriorly from the auricular surface
Ischium ("sit-down" bone) (Figure 10.6)	Ischial tuberosity	Rough projection that receives the weight of our body when we are sitting
	Ischial spine	Located superior to the ischial tuberosity and projects medially into the pelvic cavity
	Lesser sciatic notch	A small notch located inferior to the ischial spine
	Ischial ramus	Narrow portion of the bone that articulates with the pubis
Pubis (Figure 10.6)	Superior pubic ramus	Superior extension of the body of the pubis
	Inferior pubic ramus	Inferior extension of the body of the pubis; articulates with the ischium
	Pubic crest	Thick anterior border
	Pubic tubercle	Lateral end of the pubic crest; lateral attachment for the inguinal ligament
	Articular surface	Surface of each pubis that combines with fibrocartilage to form the pubic symphysis

viscera, but it does not restrict childbirth in any way. The **true pelvis** is the region inferior to the pelvic brim that is almost entirely surrounded by bone. Its posterior boundary is formed by the sacrum. The ilia, ischia, and pubic bones define its limits laterally and anteriorly.

The **pelvic inlet** is the opening delineated by the pelvic brim. The widest dimension of the pelvic inlet is from left to right, that is, along the frontal plane. The **pelvic outlet** is the inferior margin of the true pelvis. It is bounded anteriorly by the pubic arch, laterally by the ischia, and posteriorly by the sacrum and coccyx. Since both the coccyx and the ischial spines protrude into the outlet opening, a sharply angled coccyx or large, sharp ischial spines can dramatically narrow the outlet. The largest dimension of the outlet is the anterior-posterior diameter.

Activity 4

Comparing Male and Female Pelves

Examine male and female pelves for the following differences. For *female* pelves:

- The pelvic inlet is larger and more circular.
- The pelvis as a whole is shallower, and the bones are lighter and thinner.
- The sacrum is broader and less curved, and the pubic arch is more rounded and broader.
- The acetabula are smaller and farther apart, and the ilia flare more laterally.
- The ischial spines are shorter, farther apart, and everted, thus enlarging the pelvic outlet.

The major differences between the male and female pelves are summarized in **Table 10.4** on p. 152.

The Thigh

The **femur**, or thigh bone (**Figure 10.7b**, p. 153 and **Table 10.5A**, p. 153), is the only bone of the thigh. It is the heaviest, strongest bone in the body. The ball-like head of the femur articulates with the hip bone via the deep, secure socket of the acetabulum.

The femur angles medially as it runs downward to the leg bones; this brings the knees in line with the body's center of gravity. The medial course of the femur is more noticeable in females because of the wider female pelvis.

The **patella** (Figure 10.7a), or kneecap, is a triangular sesamoid bone enclosed in the quadriceps tendon that secures the anterior thigh muscles to the tibia.

(Text continues on p. 155.) →

Table 10.4 Comparison of the Male and Female Pelves

Characteristic	Female	Male
General structure and functional modifications	Tilted forward; adapted for childbearing; true pelvis defines the birth canal; cavity of the true pelvis is broad, shallow, and has a greater capacity	Tilted less forward; adapted for support of a male's heavier build and stronger muscles; cavity of the true pelvis is narrow and deep
Bone thickness	Bones lighter, thinner, and smoother	Bones heavier and thicker, and markings are more prominent
Acetabula	Smaller; farther apart	Larger; closer together
Pubic arch	Broader angle (80°–90°); more rounded	Angle is more acute (50°–60°)
Anterior view	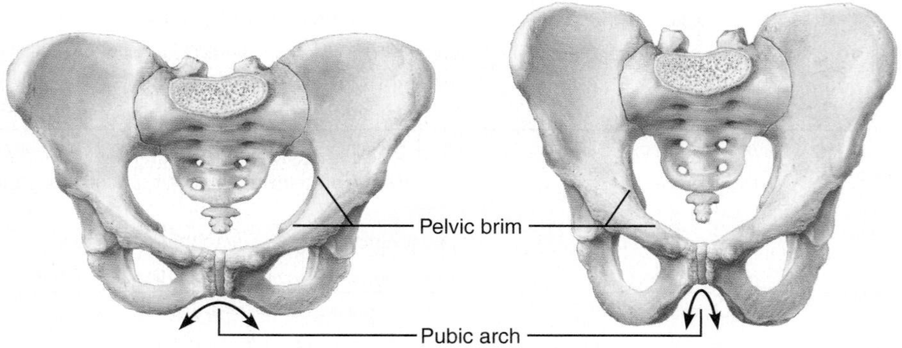	
Sacrum	Wider; shorter; sacrum is less curved	Narrow; longer; sacral promontory projects anteriorly
Coccyx	More movable; straighter; projects inferiorly	Less movable; curves and projects anteriorly
Left lateral view	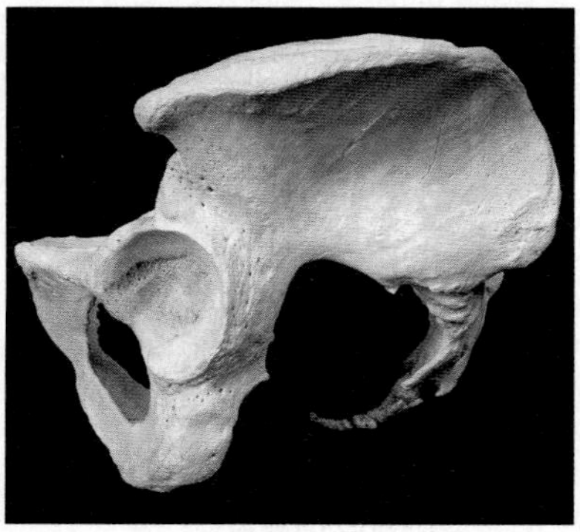	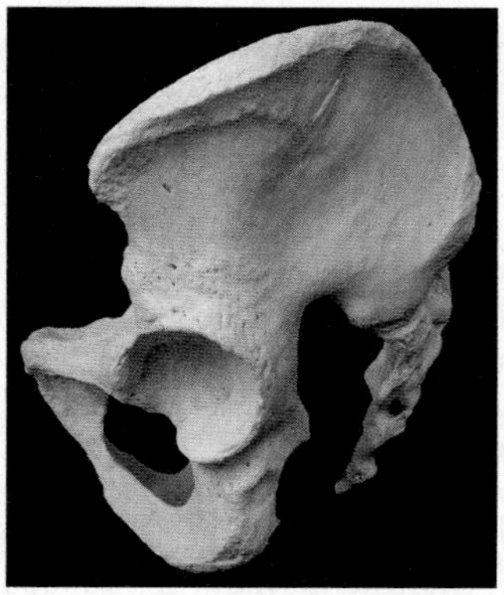
Pelvic inlet	Wider; oval from side to side	Narrow; basically heart shaped
Pelvic outlet	Wider; ischial spines shorter, farther apart, and everted	Narrower; ischial spines longer, sharper, and point more medially
Posteroinferior view	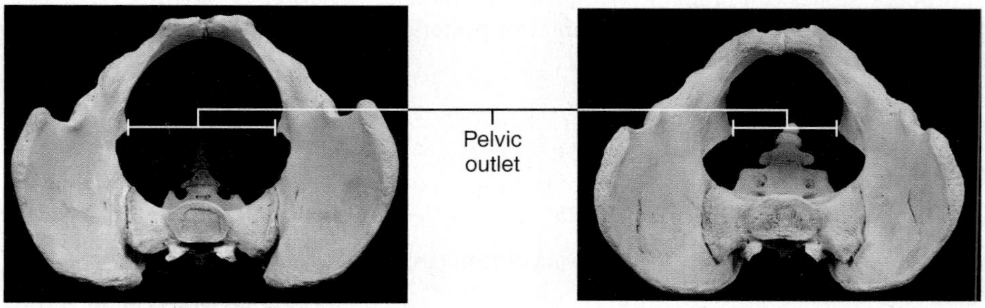	

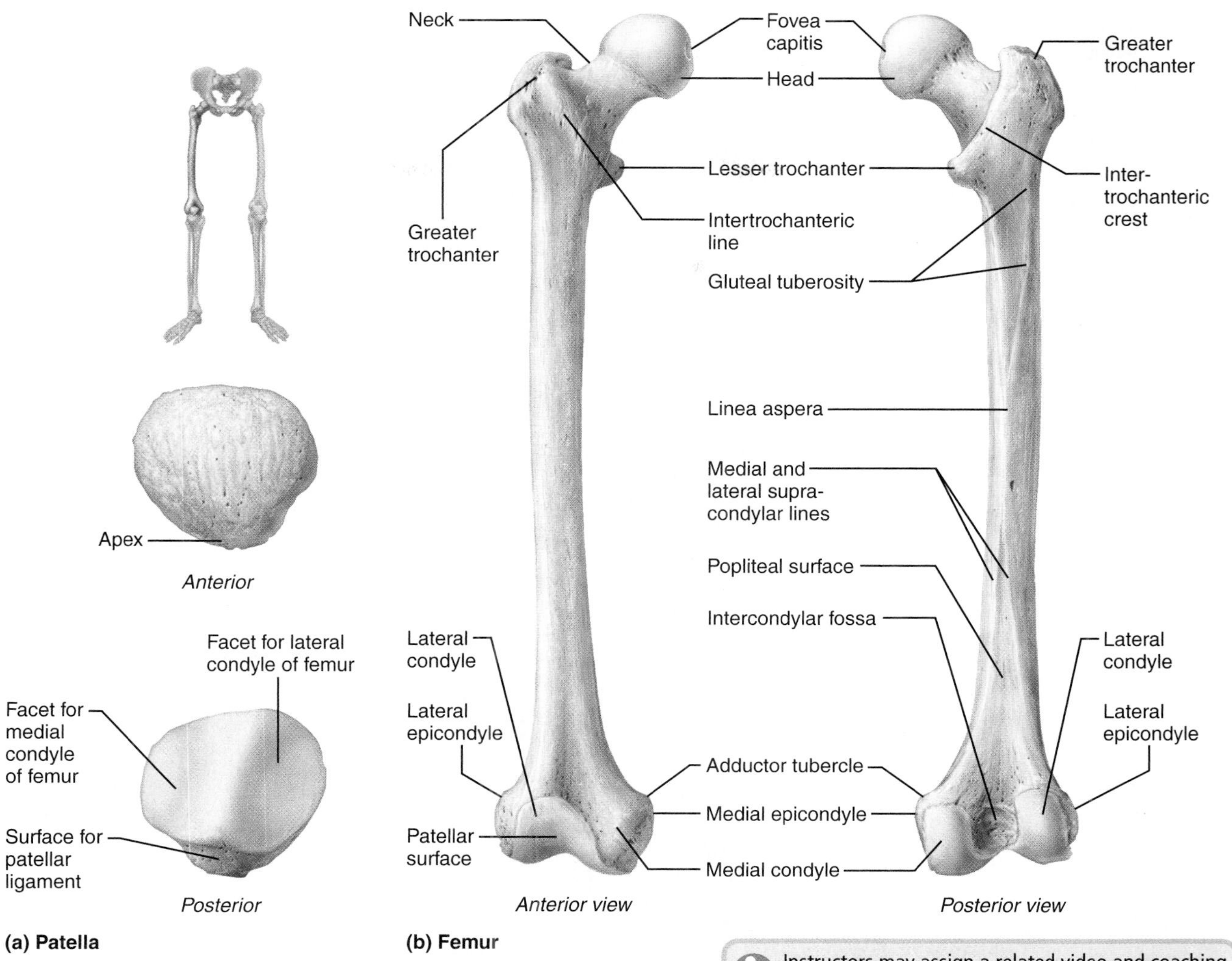

Figure 10.7 Bones of the right knee and thigh.

Instructors may assign a related video and coaching activity for the Femur using Mastering A&P™

Table 10.5A The Appendicular Skeleton: The Lower Limb (Figure 10.7)

Bone	Important markings	Description
Femur (thigh bone) (Figure 10.7)	Fovea capitis	A small pit in the head of the femur for the attachment of a short ligament that runs to the acetabulum
	Neck	Weakest part of the femur, the usual fracture site of a "broken hip"
	Greater trochanter	Large lateral projection; serves as site for muscle attachment on the proximal femur
	Lesser trochanter	Large posteromedial projection; serves as site for muscle attachment on the proximal femur
	Intertrochanteric line	Thin ridge of bone that connects the two trochanters anteriorly
	Intertrochanteric crest	Prominent ridge of bone that connects the two trochanters posteriorly
	Gluteal tuberosity	Thin ridge of bone located posteriorly; serves as a site for muscle attachment on the proximal femur
	Linea aspera	Long vertical ridge of bone on the posterior shaft of the femur
	Medial and lateral supracondylar lines	Two lines that diverge from the linea aspera and travel to their respective condyles
	Medial and lateral condyles	Distal "wheel shaped" projections that articulate with the tibia, each condyle has a corresponding epicondyle
	Intercondylar fossa	Deep depression located between the condyles and beneath the popliteal surface
	Adductor tubercle	A small bump on the superior portion of the medial epicondyle; attachment site for the large adductor magnus muscle
	Patellar surface	Smooth distal anterior surface between the condyles; articulates with the patella

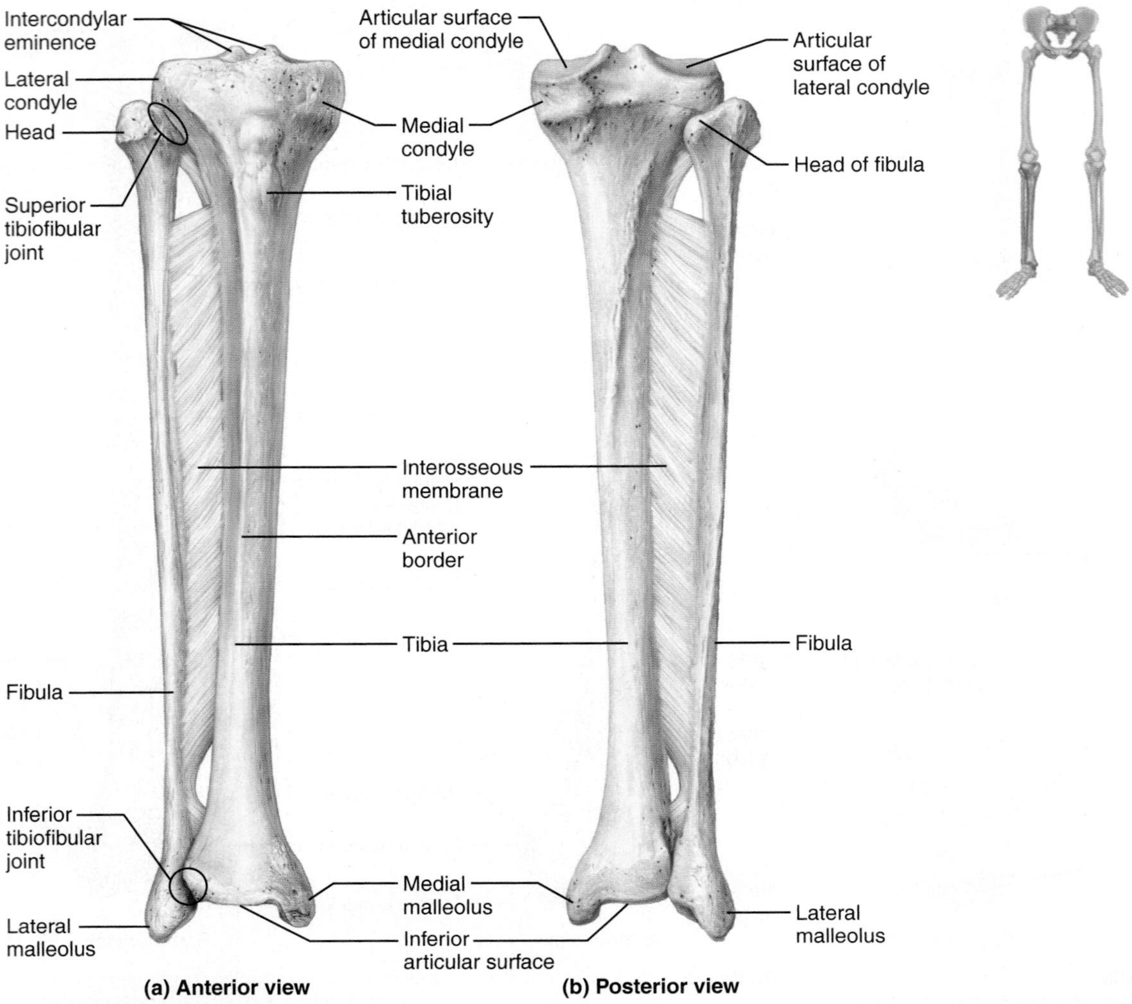

Figure 10.8 Bones of the right leg. Tibia and fibula, anterior and posterior views.

Instructors may assign related videos and coaching activities for the Fibula and Tibia using Mastering A&P™

Table 10.5B The Appendicular Skeleton: The Lower Limb (Figure 10.8)

Bone	Important markings	Description
Tibia (shinbone, medial bone of the leg) (Figure 10.8)	Lateral condyle	Slightly concave surface that articulates with the lateral condyle of the femur; the inferior region of this condyle articulates with the fibula to form the superior tibiofibular joint
	Medial condyle	Slightly concave surface that articulates with the medial condyle of the femur
	Intercondylar eminence	Irregular projection located between the two condyles
	Tibial tuberosity	Roughened anterior surface; site of patellar ligament attachment
	Anterior border	Sharp ridge of bone easily palpated because it is close to the surface
	Medial malleolus	Forms the medial bulge of the ankle
	Inferior articular surface	Distal surface of the tibia that articulates with the talus
	Fibular notch	Smooth lateral surface that articulates with the fibula to form the inferior tibiofibular joint
Fibula (lateral bone of the leg) (Figure 10.8)	Head	Proximal end of the fibula that articulates with the tibia to form the superior tibiofibular joint
	Lateral malleolus	Forms the lateral bulge of the ankle and articulates with the talus

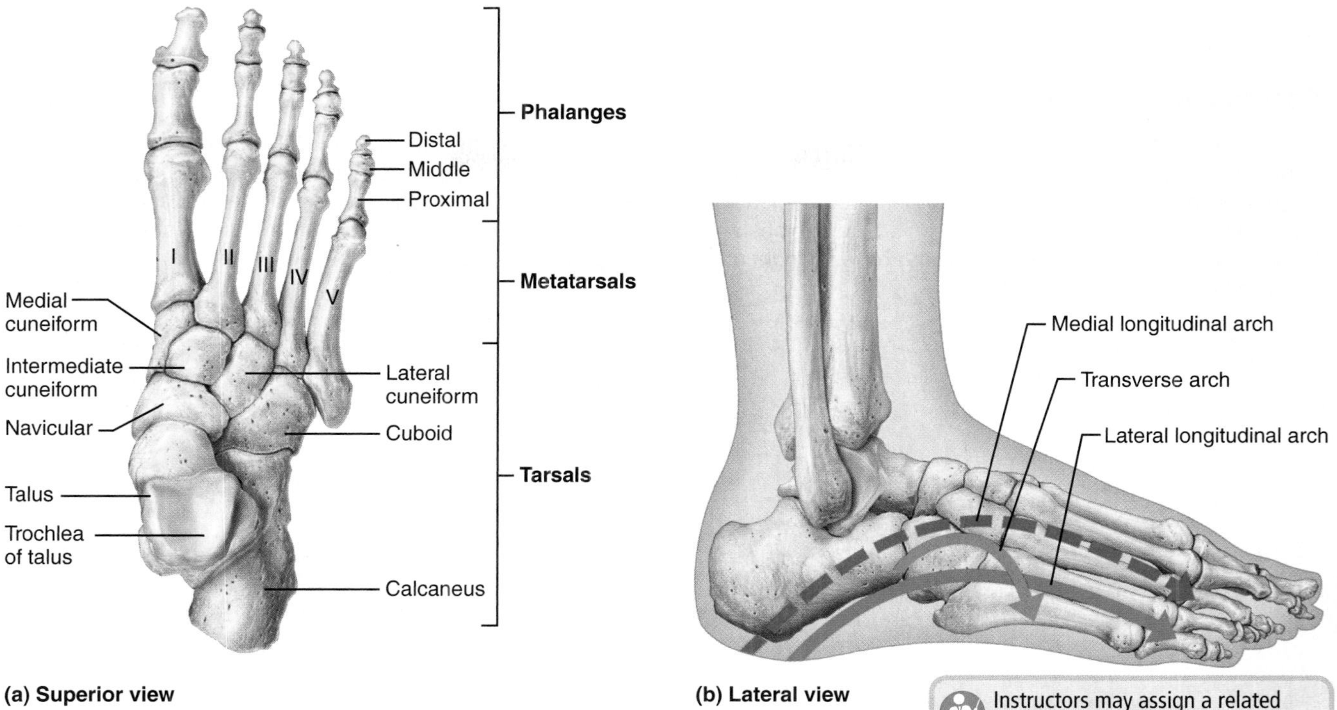

Figure 10.9 Bones of the right foot. Arches of the right foot are diagrammed in **(b)**.

Instructors may assign a related video and coaching activity for the Foot using Mastering **A&P™**

The Leg

Two bones, the tibia and the fibula, form the skeleton of the leg (**Figure 10.8** and **Table 10.5B**). The **tibia**, or *shinbone*, is the larger, medial, weight-bearing bone of the leg.

The **fibula**, which lies parallel to the tibia, takes no part in forming the knee joint. Its proximal head articulates with the lateral condyle of the tibia.

The Foot

The bones of the foot include the 7 **tarsal** bones, 5 **metatarsals**, which form the instep, and 14 **phalanges**, which form the toes (**Figure 10.9**). Body weight is concentrated on the two largest tarsals, which form the posterior aspect of the foot. These are the larger *calcaneus* (heel bone) and the *talus*, which lies between the tibia and the calcaneus. (The other tarsals are named and identified in Figure 10.9). The metatarsals are numbered I through V, medial to lateral. Like the fingers of the hand, each toe has three phalanges except the great toe, which has two.

The bones in the foot are arranged to produce three strong arches—two longitudinal arches (medial and lateral) and one transverse arch (Figure 10.9b). Ligaments bind the foot bones together, and tendons of the foot muscles hold the bones firmly in the arched position but still allow a certain degree of give. Weakened tendons and ligaments supporting the arches are referred to as *fallen arches* or *flat feet*.

Activity 5

Palpating the Surface Anatomy of the Pelvic Girdle and Lower Limb

Locate and palpate the following bone markings on yourself and/or your lab partner. Place a check mark in the boxes as you locate the bone markings. Ask your instructor for help with any markings that you are unable to locate.

☐ Iliac crest and anterior superior iliac spine: Rest your hands on your hips—they will be overlying the iliac crests. Trace the crest as far posteriorly as you can, and then follow it anteriorly to the anterior superior iliac spine. This latter bone marking is clearly visible through the skin of very slim people.

☐ Greater trochanter of the femur: This is easier to locate in females than in males because of the wider female pelvis. Try to locate it on yourself as the most lateral point of the proximal femur. It typically lies about 6 to 8 inches below the iliac crest.

☐ Patella and tibial tuberosity: Feel your kneecap and palpate the ligaments attached to its borders. Follow the inferior patellar ligament to the tibial tuberosity.

☐ Medial and lateral condyles of the femur and tibia: As you move from the patella inferiorly on the medial (and then the lateral) knee surface, you will feel first the femoral and then the tibial condyle.

☐ Medial malleolus: Feel the medial protrusion of your ankle, the medial malleolus of the distal tibia.

☐ Lateral malleolus: Feel the bulge of the lateral aspect of your ankle, the lateral malleolus of the fibula.

☐ Calcaneus: Attempt to follow the extent of your calcaneus, or heel bone.

Activity 6

Constructing a Skeleton

1. When you finish examining yourself and the disarticulated bones of the appendicular skeleton, work with your lab partner to arrange the unlabeled, disarticulated bones on the laboratory bench to form an entire skeleton. Careful observation of bone markings should help you distinguish between right and left members of bone pairs.

2. When you believe that you have accomplished this task correctly, ask the instructor to check your arrangement. If it is not correct, go to the articulated skeleton and check your bone arrangements. Also review the descriptions of the bone markings as necessary to correct your bone arrangement.

Instructors may assign a portion of the Review Sheet questions using Mastering A&P™

REVIEW SHEET

EXERCISE 10

The Appendicular Skeleton

Name ______________________________ Lab Time/Date ______________________________

Bones of the Pectoral Girdle and Upper Limb

1. Fill in the blank to complete the statements below:

a. The bones that form the pectoral girdle are the ______________________ and ______________________.

b. The upper limb is formed by the arm bone, the ______________________, and the two bones of the forearm, the ______________________ and ______________________.

c. The ______________ are the wrist bones. List the proximal row of wrist bones from lateral to medial: ____.

List the distal row of wrist bones from lateral to medial: ____.

d. The ______________ form the palm of the hand, and the heads of these bones form the knuckles.

e. A single finger bone is called a ____________. Each hand has ____________ finger bones, called ____________.

2. Match the bone markings in column B with the descriptions in column A.

Column A		Column B
__________	1. depression in the scapula that articulates with the humerus	a. acromion
__________	2. surface on the radius that receives the head of the ulna	b. capitulum
__________	3. lateral rounded knob on the distal humerus	c. coracoid process
__________	4. posterior depression on the distal humerus	d. coronoid fossa
__________	5. a roughened area on the lateral humerus: deltoid attachment site	e. deltoid tuberosity
__________	6. hooklike process; biceps brachii attachment site	f. glenoid cavity
__________	7. surface on the ulna that receives the head of the radius	g. medial epicondyle
__________	8. medial condyle of the humerus that articulates with the ulna	h. olecranon fossa
__________	9. lateral end of the spine of the scapula; clavicle articulation site	i. radial notch
__________	10. small bump on the humerus, often called the "funny bone"	j. trochlea
__________	11. anterior depression, superior to the trochlea, that receives part of the ulna when bending at the elbow	k. ulnar notch

3. Using items from the list at the right, identify the anatomical landmarks and regions of the scapula.

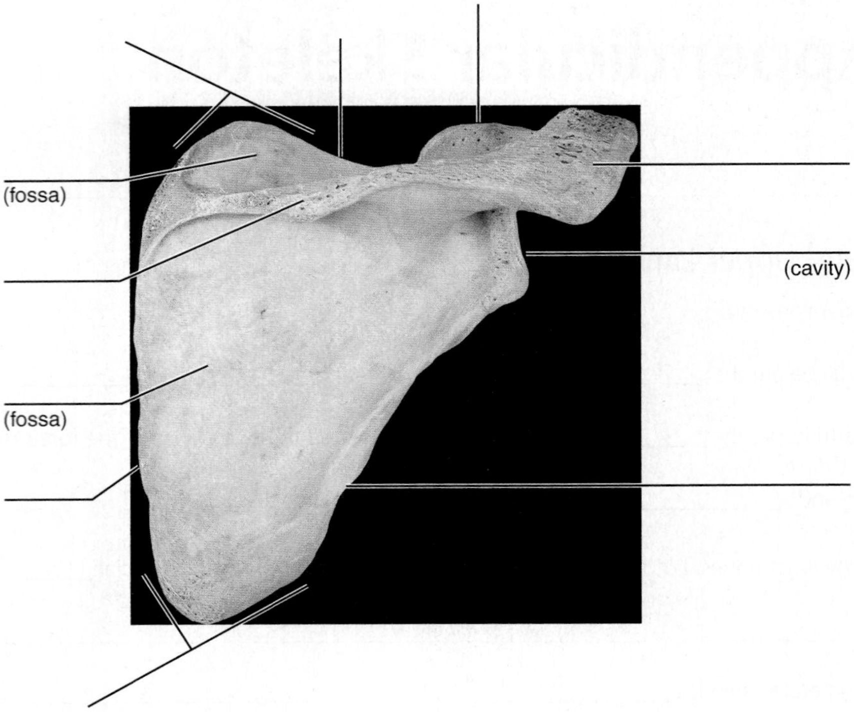

Key:

a. acromion
b. coracoid process
c. glenoid cavity
d. inferior angle
e. infraspinous fossa
f. lateral border
g. medial border
h. spine
i. superior angle
j. superior border
k. supraspinous fossa

4. Match the terms in the key with the appropriate leader lines on the photograph of the humerus.

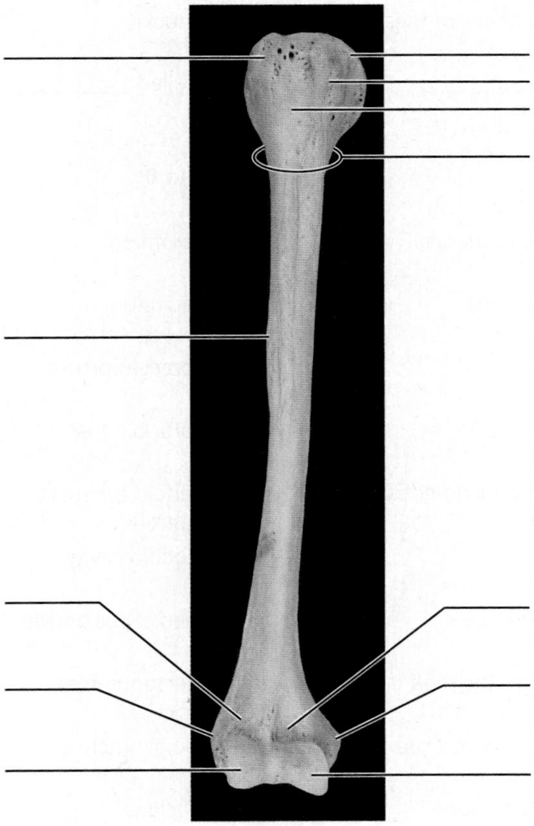

Key:

a. capitulum
b. coronoid fossa
c. deltoid tuberosity
d. greater tubercle
e. head
f. intertubercular sulcus
g. lateral epicondyle
h. lesser tubercle
i. medial epicondyle
j. radial fossa
k. surgical neck
l. trochlea

5. Match the terms in the key with the appropriate leader lines on the photographs of the posterior view of the radius on the left and the lateral view of the ulna on the right.

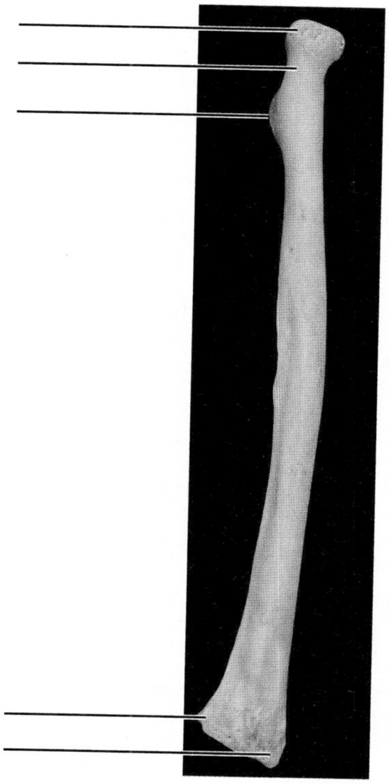

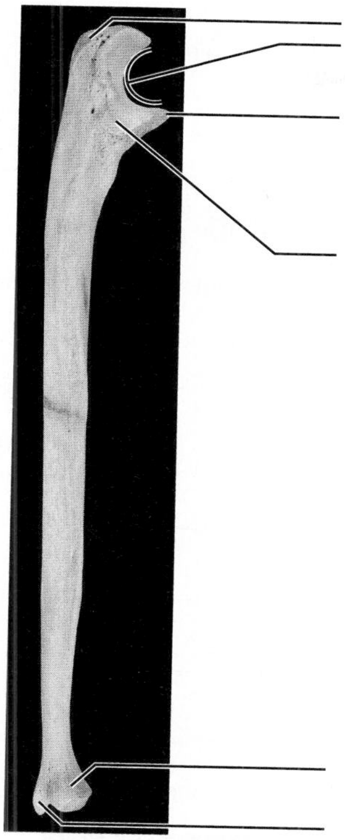

Key:

a. coronoid process
b. head of the radius
c. head of the ulna
d. neck of the radius
e. olecranon
f. radial notch of the ulna
g. radial styloid process
h. radial tuberosity
i. trochlear notch
j. ulnar notch of the radius
k. ulnar styloid process

6. Match the terms in the key with the appropriate leader lines on the photograph of the anterior view of the hand.

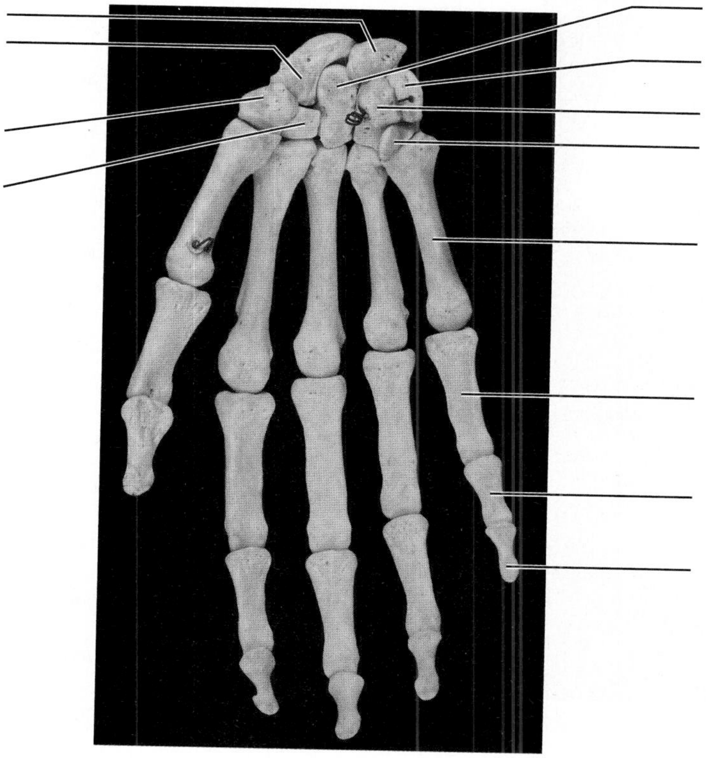

Key:

a. capitate
b. distal phalanx
c. hamate
d. lunate
e. metacarpal
f. middle phalanx
g. pisiform
h. proximal phalanx
i. scaphoid
j. trapezium
k. trapezoid
l. triquetrum

7. Name the two bone markings that form the proximal radioulnar joint.

____

8. Name the two bone markings that form the distal radioulnar joint.

____

Bones of the Pelvic Girdle and Lower Limb

9. Compare the pectoral and pelvic girdles by choosing appropriate descriptive terms from the key.

Key: a. flexibility most important
b. massive
c. lightweight
d. insecure axial and limb attachments
e. secure axial and limb attachments
f. weight-bearing most important

Pectoral: ________, ________, ________ Pelvic: ________, ________, ________

10. Distinguish between the true pelvis and the false pelvis. ____

____

____

11. Match the terms in the key with the appropriate leader lines on the photograph of the lateral view of the hip bone.

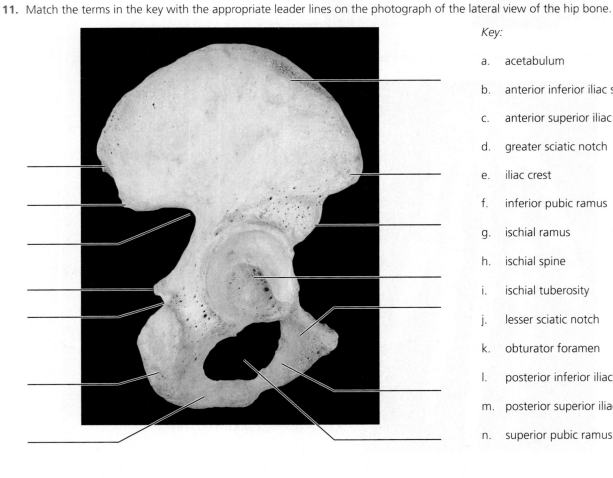

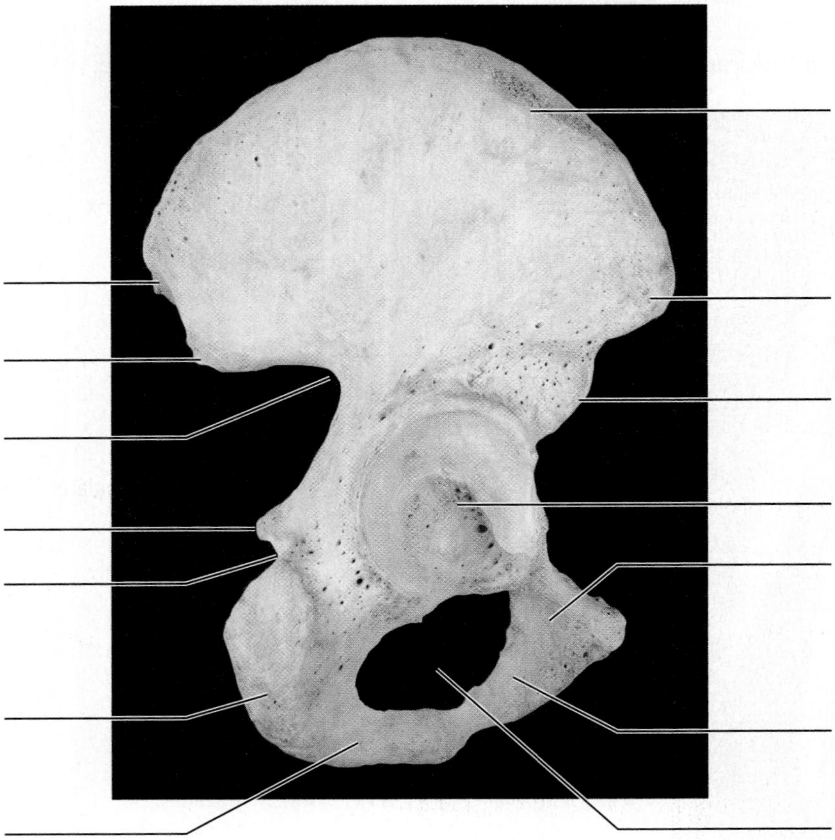

Key:

a. acetabulum
b. anterior inferior iliac spine
c. anterior superior iliac spine
d. greater sciatic notch
e. iliac crest
f. inferior pubic ramus
g. ischial ramus
h. ischial spine
i. ischial tuberosity
j. lesser sciatic notch
k. obturator foramen
l. posterior inferior iliac spine
m. posterior superior iliac spine
n. superior pubic ramus

12. Match the bone names and markings in column B with the descriptions in column A. The items in column B may be used more than once.

Column A	Column B
______, ______, and	a. acetabulum
______ 1. fuse to form the hip bone	b. calcaneus
______ 2. rough projection that supports body weight when sitting	c. femur
______ 3. point where the hip bones join anteriorly	d. fibula
______ 4. superiormost margin of the hip bone	e. gluteal tuberosity
______ 5. deep socket in the hip bone that receives the head of the thigh bone	f. greater and lesser trochanters
______ 6. joint between axial skeleton and pelvic girdle	g. greater sciatic notch
______ 7. longest, strongest bone in body	h. iliac crest
______ 8. thin, lateral leg bone	i. ilium
______ 9. permits passage of the sciatic nerve	j. ischial tuberosity
______ 10. notch located inferior to the ischial spine	k. ischium
______ 11. point where the patellar ligament attaches	l. lateral malleolus
______ 12. kneecap	m. lesser sciatic notch
______ 13. shinbone	n. medial malleolus
______ 14. medial ankle projection	o. metatarsals
______ 15. lateral ankle projection	p. obturator foramen
______ 16. largest tarsal bone	q. patella
______ 17. ankle bones	r. pubic symphysis
______ 18. bones forming the instep of the foot	s. pubis
______ 19. opening in hip bone formed by the pubic and ischial rami	t. sacroiliac joint
______ and ______ 20. sites of muscle attachment on the proximal femur	u. talus
	v. tarsals
______ 21. tarsal bone that "sits" on the calcaneus	w. tibia
______ 22. weight-bearing bone of the leg	x. tibial tuberosity
______ 23. tarsal bone that articulates with the tibia	

13. Match the terms in the key with the appropriate leader lines on the photograph of the anterior view of the femur.

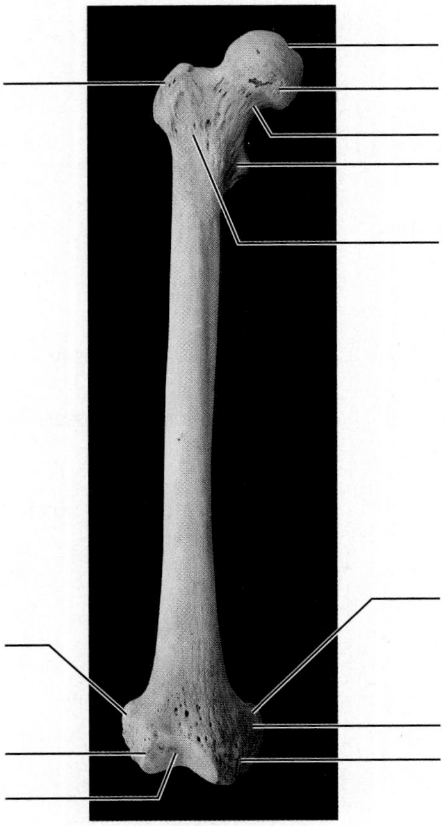

Key:

a. adductor tubercle
b. fovea capitis
c. greater trochanter
d. head
e. intertrochanteric line
f. lateral condyle
g. lateral epicondyle
h. lesser trochanter
i. medial condyle
j. medial epicondyle
k. neck
l. patellar surface

14. Match the terms in the key with the appropriate leader lines on the photograph of the anterior view of the tibia.

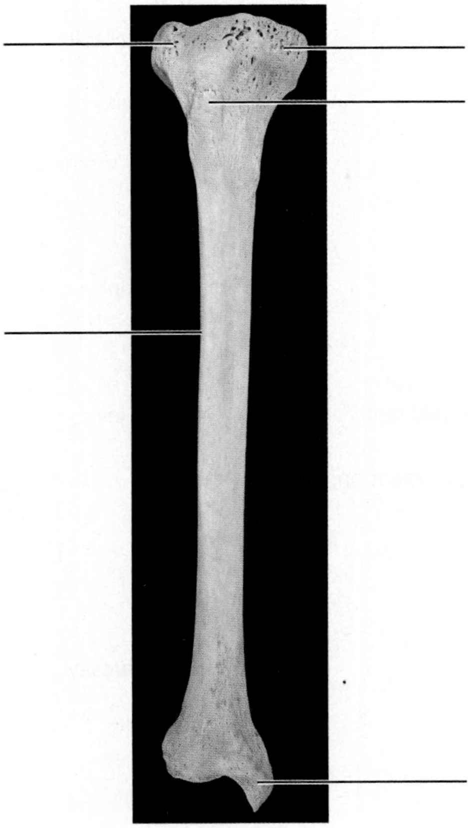

Key:

a. anterior border
b. lateral condyle
c. medial condyle
d. medial malleolus
e. tibial tuberosity

15. Match the terms in the key with the appropriate leader lines on the photograph of the posterior view of the articulated tibia and fibula.

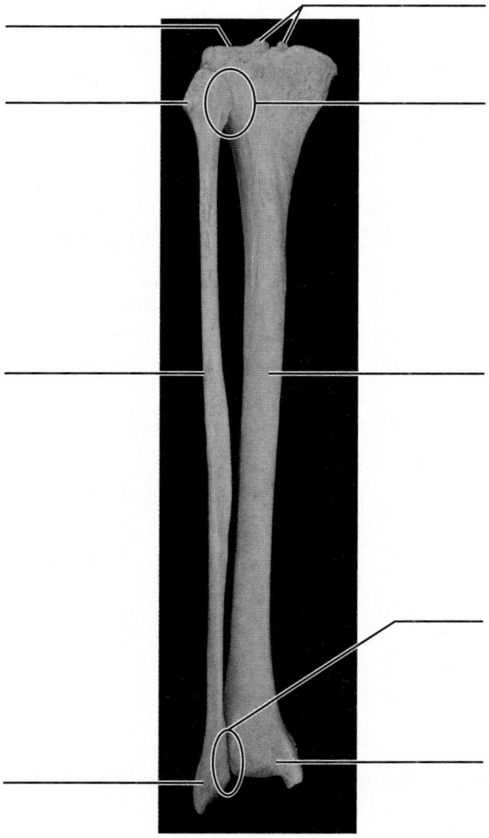

Key:

a. articular surface of the lateral condyle

b. head of the fibula

c. inferior tibiofibular joint

d. intercondylar eminence

e. lateral malleolus

f. medial malleolus

g. shaft of the fibula

h. shaft of the tibia

i. superior tibiofibular joint

16. Are the bones of the leg shown above from the left or from the right leg? ______________________________

Explain how you can tell which side of the body they are from. ______________________________

____

17. Match the terms in the key with the appropriate leader lines on the photograph of the superior view of the articulated foot.

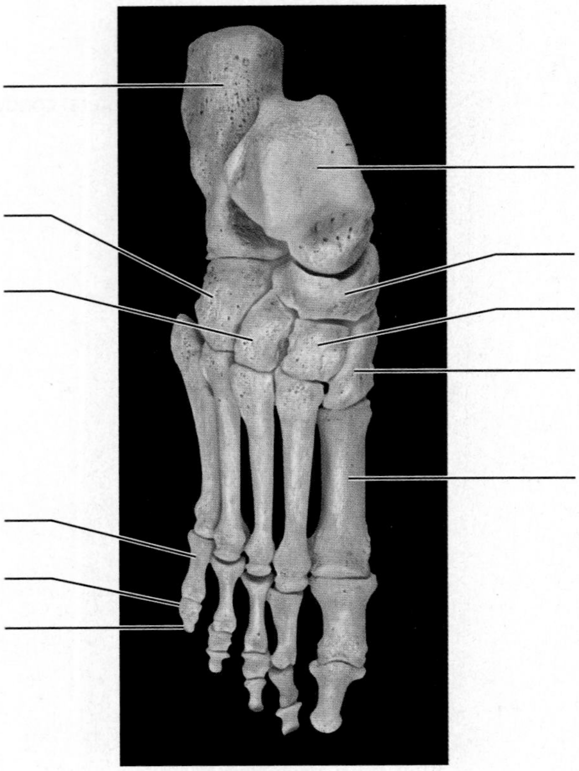

Key:

a. calcaneus

b. cuboid

c. distal phalanx

d. intermediate cuneiform

e. lateral cuneiform

f. medial cuneiform

g. metatarsal

h. middle phalanx

i. navicular

j. proximal phalanx

k. talus

18. FOOSH is an acronym that stands for **F**all **o**n **O**ut**s**tretched **H**and. Discuss possible fractures and dislocations that might occur with an injury of this type.

____

____

19. Describe some of the features of the female pelvis that provide for compatibility with vaginal birth. ____________________

____

____

20. Your X-ray exam reveals that you have fractured your fibula. Your physician remarks, "Well, it's better than breaking your tibia." Explain why a fracture of the tibia would be worse than a fracture of the fibula. ____________________

____

EXERCISE 11

Articulations and Body Movements

Learning Outcomes

- Name and describe the three functional categories of joints.
- Name and describe the three structural categories of joints, and discuss how their structure is related to mobility.
- Identify the types of synovial joints; indicate whether they are nonaxial, uniaxial, biaxial, or multiaxial; and describe the movements each makes.
- Define *origin* and *insertion* of muscles.
- Demonstrate or describe the various body movements.
- Compare and contrast the structure and function of the shoulder and hip joints.
- Describe the structure and function of the knee and temporomandibular joints.

Pre-Lab Quiz

Instructors may assign these and other Pre-Lab Quiz questions using Mastering A&P™

1. The functional classification of joints is based on:
   a. a joint cavity
   b. amount of connective tissue
   c. amount of movement allowed by the joint
2. Circle the correct underlined term. Sutures, which have their irregular edges of bone joined by short fibers of connective tissue, are an example of <u>fibrous</u> / <u>cartilaginous</u> joints.
3. Circle the correct underlined term. Every muscle of the body is attached to a bone or other connective tissue structure at two points. The <u>origin</u> / <u>insertion</u> is the more movable attachment.
4. The hip joint is an example of a _________ synovial joint.
   a. ball-and-socket **c.** pivot
   b. hinge **d.** plane
5. Movement of a limb *away* from the midline or median plane of the body in the frontal plane is known as:
   a. abduction **c.** extension
   b. eversion **d.** rotation

Go to Mastering A&P™ > Study Area to improve your performance in A&P Lab.

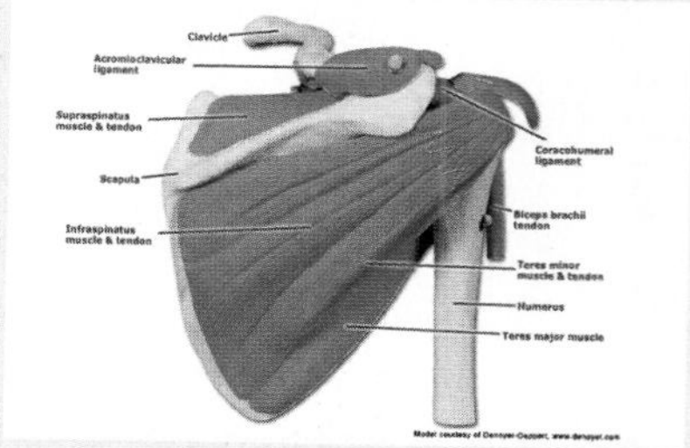

> Lab Tools > Practice Anatomy Lab > Anatomical Models

Instructors may assign new Building Vocabulary coaching activities, Pre-Lab Quiz questions, Art Labeling activities, Practice Anatomy Lab Practical questions (PAL), and more using the Mastering A&P™ Item Library.

Materials

- Skull
- Articulated skeleton
- X-ray image of a child's bone showing the cartilaginous growth plate
- Anatomical chart of joint types
- Diarthrotic joint (fresh or preserved), preferably a beef knee joint sectioned sagittally (Alternatively, pig's feet with phalanges sectioned frontally could be used)
- Disposable gloves
- Water balloons and clamps
- Functional models of hip, knee, and shoulder joints
- X-ray images of normal and arthritic joints

With rare exceptions, every bone in the body is connected to, or forms a joint with, at least one other bone. **Articulations**, or joints, perform two functions for the body. They (1) hold the bones together and (2) allow the rigid skeletal system some flexibility so that gross body movements can occur.

Classification of Joints

Joints are classified structurally and functionally. The structural classification is based on the type of connective tissue between the bones and the presence or absence of a joint cavity. Structurally, there are *fibrous*, *cartilaginous*, and *synovial joints*.

The functional classification focuses on the amount of movement allowed at the joint. On this basis, there are **synarthroses**, or immovable joints; **amphiarthroses**, or slightly movable joints; and **diarthroses**, or freely movable joints.

The structural and functional classifications of joints are summarized in **Table 11.1**. A sample of the types of joints can be found in **Figure 11.1**.

Activity 1

Identifying Fibrous Joints

Examine a human skull. Notice that adjacent bone surfaces do not actually touch but are separated by fibrous connective tissue. Also examine a skeleton and anatomical chart of joint types and Table 11.3, pp. 177–178, for examples of fibrous joints.

Activity 2

Identifying Cartilaginous Joints

Identify the cartilaginous joints on a human skeleton, Table 11.3, and an anatomical chart of joint types. View an X-ray image of the cartilaginous growth plate (epiphyseal plate) of a child's bone if one is available.

11

Table 11.1 Summary of Joint Classifications (Figure 11.1)

Structural class	Structural characteristics	Structural types	Examples	Functional classification
Fibrous	Adjoining bones connected by dense fibrous connective tissue; no joint cavity	Suture (short fibers)	Squamous suture between the parietal and temporal bones	Synarthrosis (immovable)
		Syndesmosis (longer fibers)	Between the tibia and fibula	Amphiarthrosis (slightly movable and immovable)
		Gomphosis (periodontal ligament)	Tooth in a bony socket (Figure 38.12)	Synarthrosis (immovable)
Cartilaginous	Adjoining bones united by cartilage; no joint cavity	Synchondrosis (hyaline cartilage)	Between the costal cartilage of rib 1 and the sternum and the epiphyseal plate in growing long bones (Figure 8.5)	Synarthrosis (immovable)
		Symphysis (fibrocartilage)	Intervertebral discs between adjacent vertebrae and the anterior connection between the pubic bones	Amphiarthrosis (slightly movable)
Synovial	Adjoining bones covered in articular cartilage; separated by a joint cavity and enclosed in an articular capsule lined with a synovial membrane	Plane joint	Between the carpals of the wrist	Diarthrosis (freely movable)
		Hinge joint	Elbow joint	
		Pivot joint	Proximal radioulnar joint	
		Condylar joint	Between the metacarpals and the proximal phalanx	
		Saddle joint	Between the trapezium (carpal) and metatarsal I	
		Ball-and-socket joint	Shoulder joint	

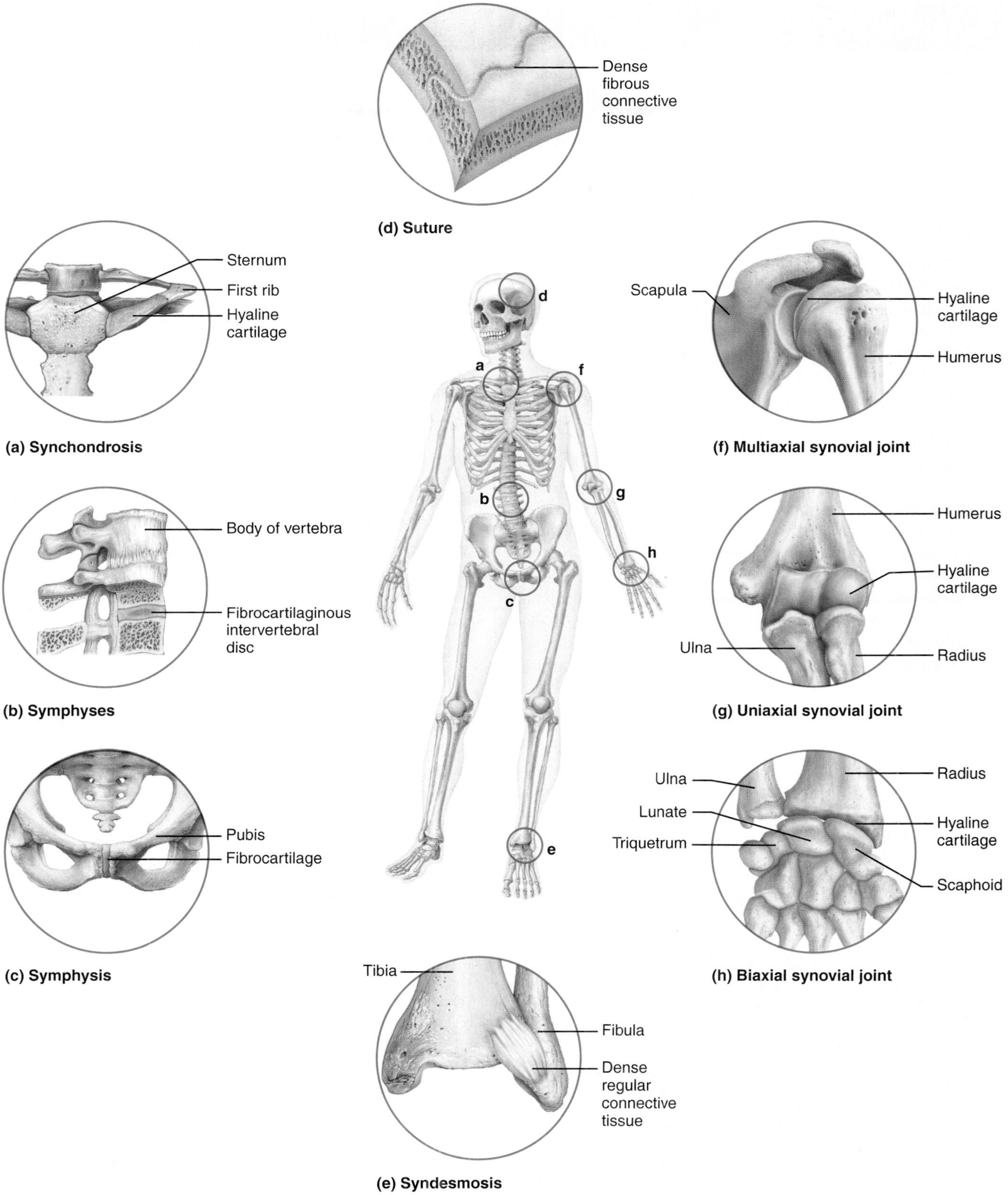

Figure 11.1 Types of joints. Joints to the left of the skeleton are cartilaginous joints; joints above and below the skeleton are fibrous joints; joints to the right of the skeleton are synovial joints. **(a)** Joint between costal cartilage of rib 1 and the sternum. **(b)** Intervertebral discs of fibrocartilage connecting adjacent vertebrae. **(c)** Fibrocartilaginous pubic symphysis connecting the pubic bones anteriorly. **(d)** Dense fibrous connective tissue connecting interlocking skull bones. **(e)** Ligament of dense regular connective tissue connecting the inferior ends of the tibia and fibula. **(f)** Shoulder joint. **(g)** Elbow joint. **(h)** Wrist joint.

Instructors may assign this figure as an Art Labeling Activity using Mastering A&P™

11

Synovial Joints

Synovial joints are those in which the articulating bone ends are separated by a joint cavity containing synovial fluid (see Figure 11.1f–h). All synovial joints are diarthroses, or freely movable joints. Most joints in the body are synovial joints.

Synovial joints typically have the following structural characteristics (**Figure 11.2**):

- **Joint (articular) cavity:** A space between the articulating bones that contains a small amount of synovial fluid.
- **Articular cartilage:** Hyaline cartilage that covers the surfaces of the bones forming the joint.
- **Articular capsule:** Two layers that enclose the joint cavity. The tough external layer is the *fibrous layer* composed of dense irregular connective tissue. The inner layer is the *synovial membrane* composed of loose connective tissue.
- **Synovial fluid:** A viscous fluid, the consistency of egg whites, located in the joint cavity and the articular cartilage. This fluid acts as a lubricant, reducing friction.
- **Reinforcing ligaments:** Some joints are reinforced by ligaments. Most ligaments are **capsular ligaments**, which are thickenings of the fibrous layer of the articular capsule. Some ligaments are **extracapsular**, found outside the articular capsule. **Intracapsular ligaments** are found within the articular capsule.
- **Nerves and blood vessels:** Sensory nerve fibers detect pain and joint stretching. Most of the blood vessels supply the synovial membrane.
- **Articular discs:** Pads composed of fibrocartilage (menisci) may also be present to minimize wear and tear on the bone surfaces.
- **Bursa and tendon sheath:** Sac filled with synovial fluid that reduces friction where ligaments, muscles, skin, tendons, and bones rub together. A tendon sheath is an elongated bursa that wraps around a tendon like a bun around a hot dog.

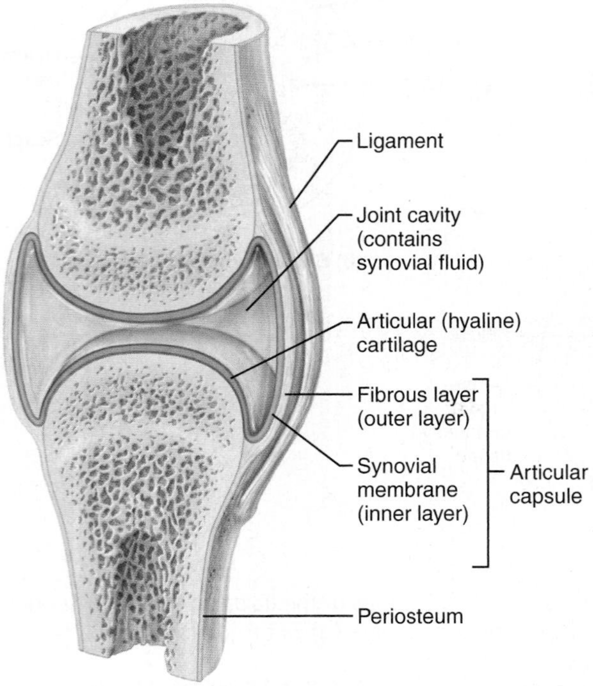

Figure 11.2 General structure of a synovial joint.

Activity 3

Examining Synovial Joint Structure

Examine a beef or pig joint to identify the general structural features of diarthrotic joints as listed above.

! If the joint is freshly obtained from the slaughterhouse and you will be handling it, don disposable gloves before beginning your observations.

Activity 4

Demonstrating the Importance of Friction-Reducing Structures

1. Obtain a small water balloon and clamp. Partially fill the balloon with water (it should still be flaccid), and clamp it closed.

2. Position the balloon atop one of your fists and press down on its top surface with the other fist. Push on the balloon until your two fists touch and move your fists back and forth over one another. Assess the amount of friction generated.

3. Unclamp the balloon and add more water. The goal is to get just enough water in the balloon so that your fists cannot come into contact with one another, but instead remain separated by a thin water layer when pressure is applied to the balloon.

4. Repeat the movements in step 2 to assess the amount of friction generated.

How does the presence of a cavity containing fluid influence the amount of friction generated?

____

What anatomical structure(s) does the water-containing balloon mimic?

____

Types of Synovial Joints

The many types of synovial joints can be subdivided according to their function and structure. The shapes of the articular surfaces determine the types of movements that can occur at the joint, and they also determine the structural classification of the joints (**Tables 11.2** and **11.3**, p. 177 and **Figure 11.3**).

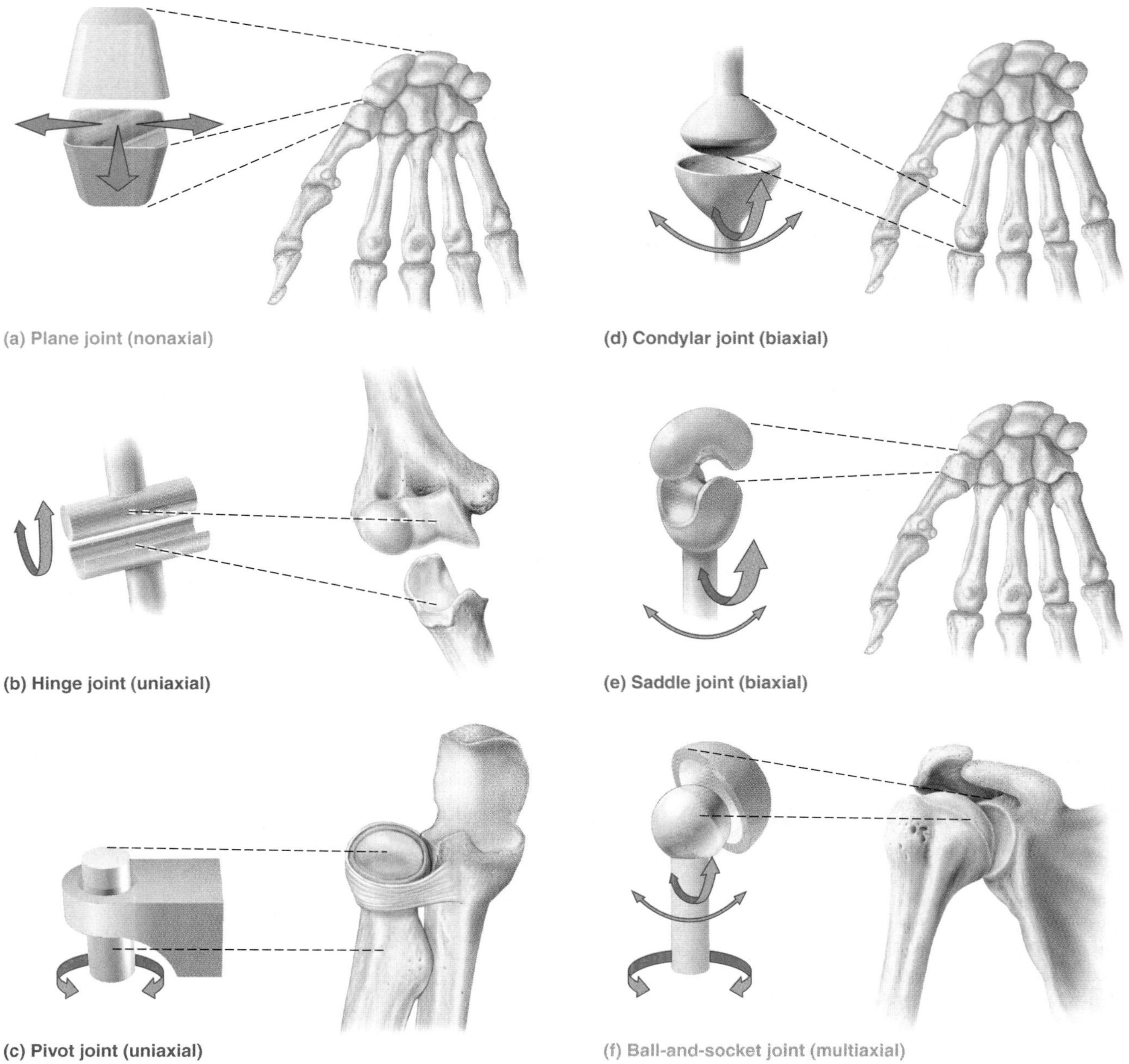

Figure 11.3 Types of synovial joints. Dashed lines indicate the articulating bones. **(a)** Intercarpal joint. **(b)** Elbow. **(c)** Proximal radioulnar joint. **(d)** Metacarpophalangeal joint. **(e)** Carpometacarpal joint of the thumb. **(f)** Shoulder.

Instructors may assign this figure as an Art Labeling Activity using Mastering A&P™

11

Table 11.2 Types of Synovial Joints (Figure 11.3)

Synovial joint	Description of articulating surfaces	Movement	Examples
Plane	Flat or slightly curved bones	Nonaxial: gliding	Intertarsal, intercarpal joints
Hinge	A rounded or cylindrical bone fits into a concave surface on the other bone	Uniaxial: flexion and extension	Elbow, interphalangeal joints
Pivot	A rounded bone fits into a sleeve (a concave bone plus a ligament)	Uniaxial: rotation	Proximal radioulnar, atlantoaxial joints
Condylar	An oval condyle fits into an oval depression on the other bone	Biaxial: flexion, extension, adduction, and abduction	Metacarpophalangeal (knuckle) and radiocarpal joints
Saddle	Articulating surfaces are saddle shaped; one surface is concave, the other surface is convex	Biaxial: flexion, extension, adduction, and abduction	Carpometacarpal joint of the thumb
Ball-and-socket	The ball-shaped head of one bone fits into the cuplike depression of the other bone	Multiaxial: flexion, extension, adduction, abduction, and rotation	Shoulder, hip joints

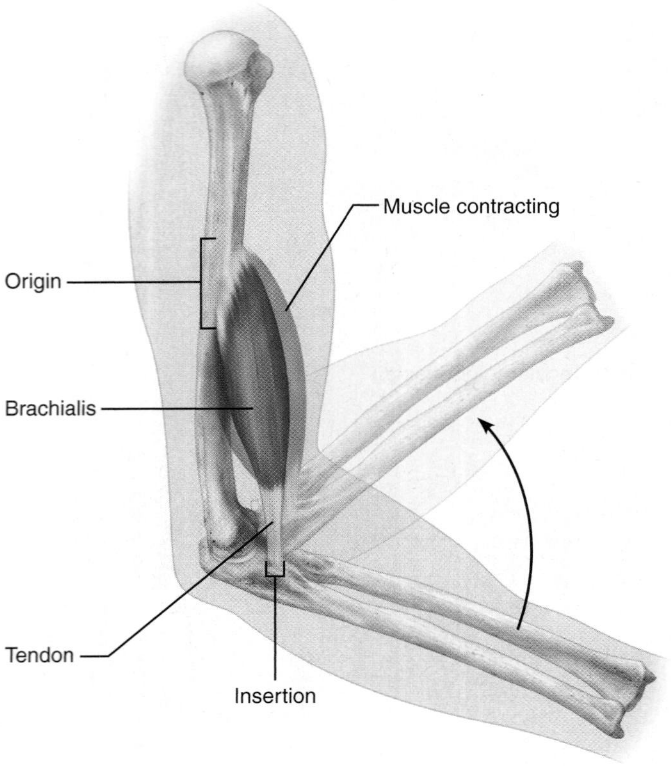

Figure 11.4 Muscle attachments (origin and insertion). When a skeletal muscle contracts, its insertion is pulled toward its origin.

Movements Allowed by Synovial Joints

Every muscle of the body is attached to bone (or other connective tissue structures) at two points. The **origin** is the stationary, immovable, or less movable bone, and the **insertion** is the more movable bone. Body movement occurs when muscles contract across diarthrotic synovial joints (**Figure 11.4**). When the muscle contracts and its fibers shorten, the insertion moves toward the origin. The type of movement depends on the construction of the joint and on the placement of the muscle relative to the joint. The most common types of body movements are described below (and illustrated in **Figure 11.5**).

Activity 5

Demonstrating Movements of Synovial Joints

Try to demonstrate each movement as you read through the following material:

Flexion (Figure 11.5a–c): A movement, generally in the sagittal plane, that decreases the angle of the joint and reduces the distance between the two bones. Flexion is typical of hinge joints (bending the knee or elbow) but is also common at ball-and-socket joints (bending forward at the hip).

Extension (Figure 11.5a–c): A movement that increases the angle of a joint and the distance between two bones or parts of the body; the opposite of flexion. If extension proceeds beyond anatomical position (bends the trunk backward), it is termed *hyperextension*.

Abduction (Figure 11.5d): Movement of a limb away from the midline of the body, along the frontal plane, or the fanning movement of fingers or toes when they are spread apart.

Adduction (Figure 11.5d): Movement of a limb toward the midline of the body or drawing the fingers or toes together; the opposite of abduction.

Rotation (Figure 11.5e): Movement of a bone around its longitudinal axis without lateral or medial displacement. Rotation, a common movement of ball-and-socket joints, also describes the movement of the atlas around the dens of the axis.

Circumduction (Figure 11.5d): A combination of flexion, extension, abduction, and adduction commonly observed in ball-and-socket joints like the shoulder. The limb as a whole outlines a cone.

Text continues on page 172. →

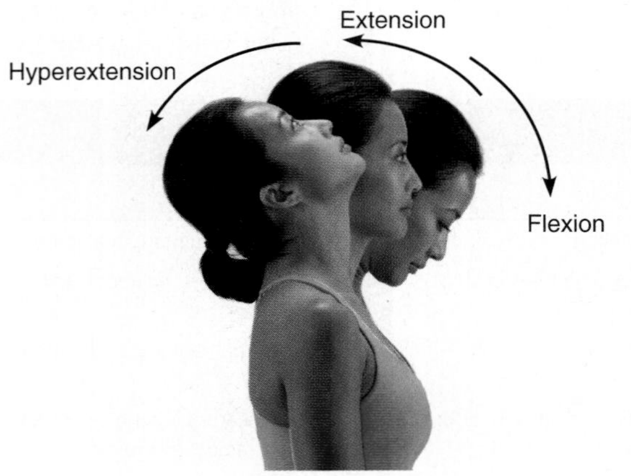

(a) Flexion, extension, and hyperextension of the neck

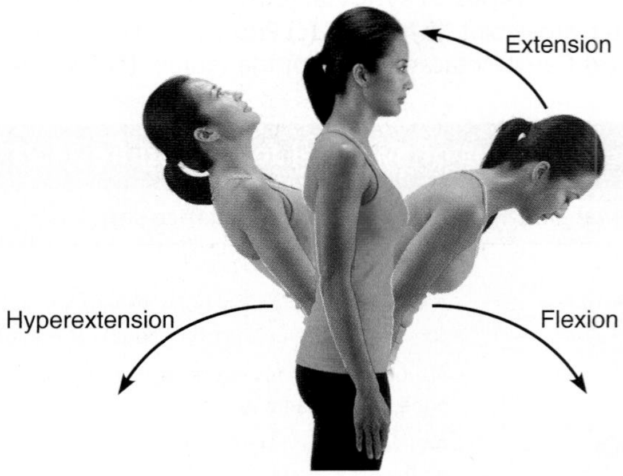

(b) Flexion, extension, and hyperextension of the vertebral column

Figure 11.5 Movements occurring at synovial joints of the body.

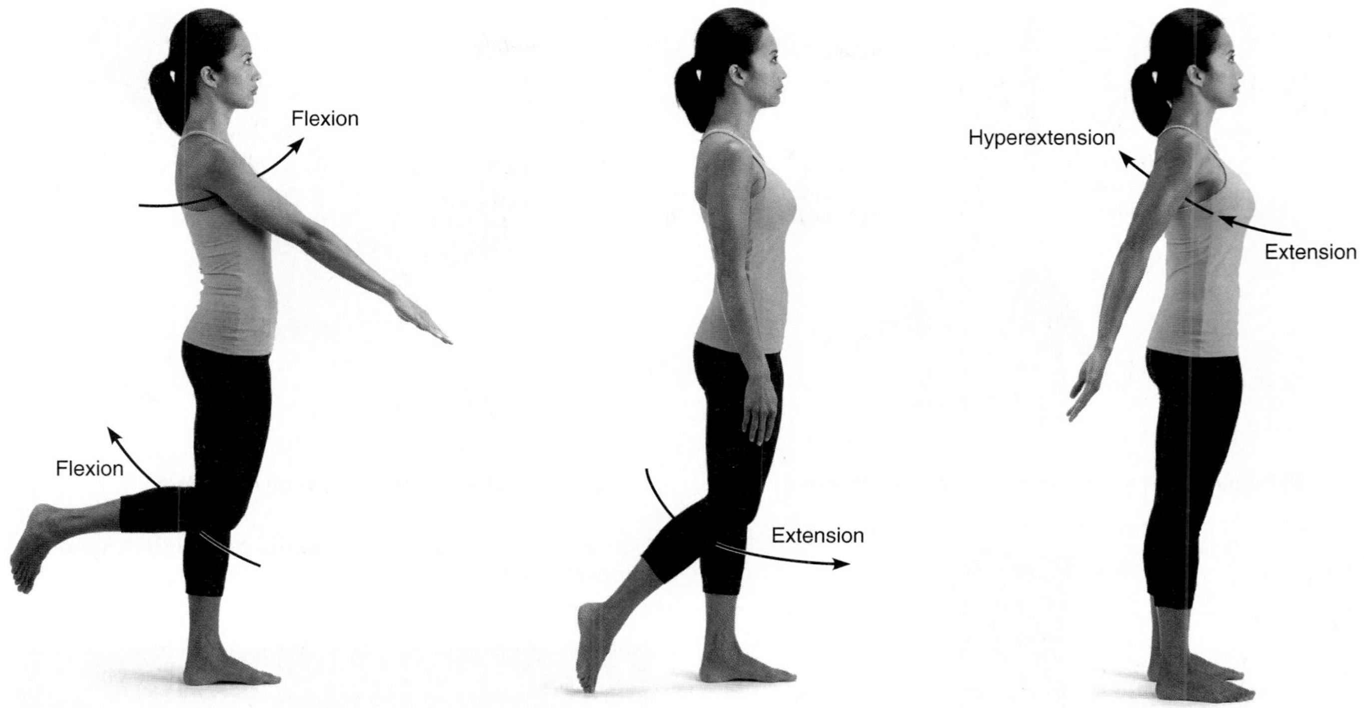

(c) Flexion and extension at the shoulder and knee, and hyperextension of the shoulder

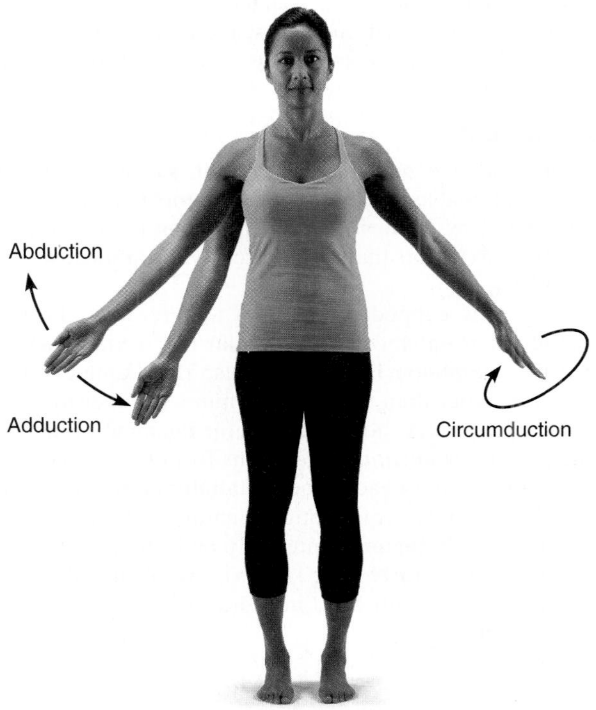

(d) Abduction, adduction, and circumduction of the upper limb at the shoulder

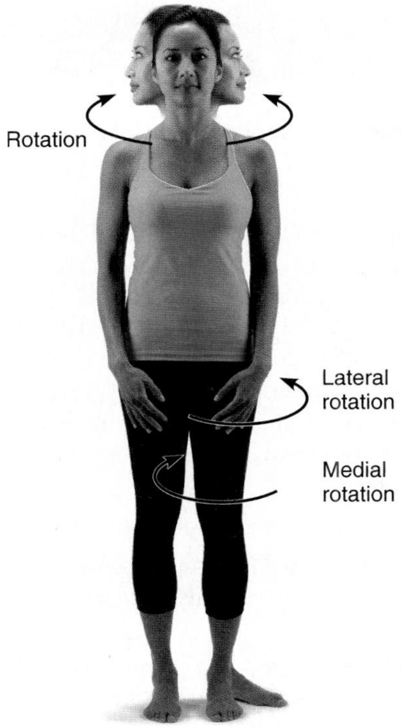

(e) Rotation of the head and lower limb

Figure 11.5 *(continued)*

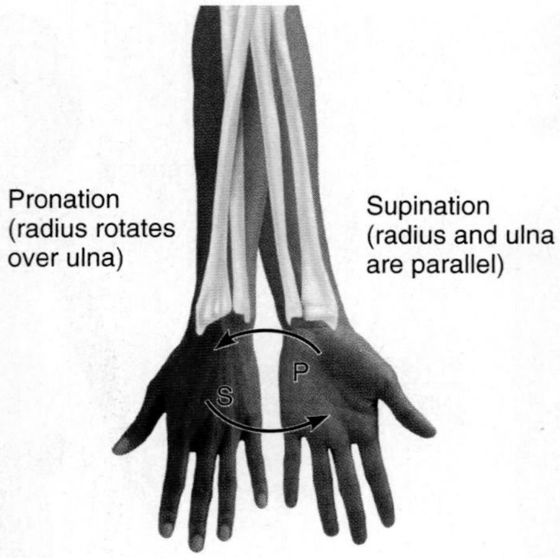

(f) Pronation (P) and supination (S) of the forearm

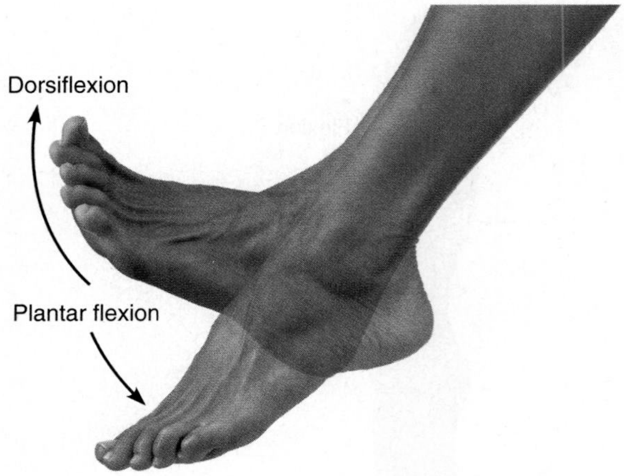

(g) Dorsiflexion and plantar flexion of the foot

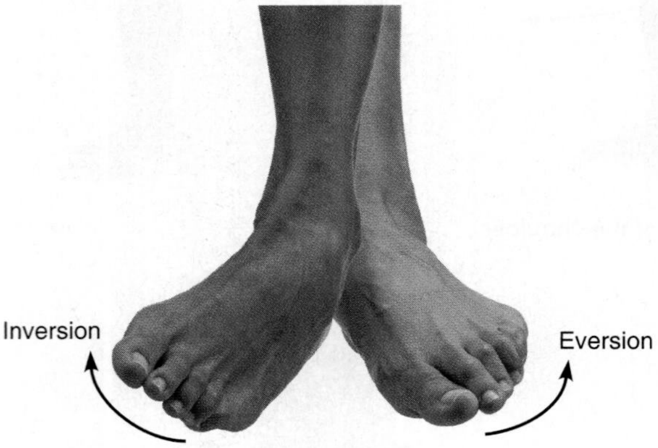

(h) Inversion and eversion of the foot

Figure 11.5 *(continued)* **Movements occurring at synovial joints of the body.**

11

Pronation (Figure 11.5f): Movement of the palm of the hand from an anterior or upward-facing position to a posterior or downward-facing position. The distal end of the radius rotates over the ulna so that the bones form an X with pronation of the forearm.

Supination (Figure 11.5f): Movement of the palm from a posterior position to an anterior position (the anatomical position); the opposite of pronation. During supination, the radius and ulna are parallel.

The last four terms refer to movements of the foot:

Dorsiflexion (Figure 11.5g): A movement of the ankle joint that lifts the foot so that its superior surface approaches the shin.

Plantar flexion (Figure 11.5g): A movement of the ankle joint in which the foot is flexed downward as if standing on one's toes or pointing the toes.

Inversion (Figure 11.5h): A movement that turns the sole of the foot medially.

Eversion (Figure 11.5h): A movement that turns the sole of the foot laterally; the opposite of inversion.

Selected Synovial Joints

The Hip and Knee Joints

Both of these joints are large weight-bearing joints of the lower limb, but they differ substantially in their security. Read through the brief descriptive material below, and look at the questions in the review sheet at the end of this exercise before beginning your comparison.

The Hip Joint

The hip joint is a ball-and-socket joint, so movements can occur in all possible planes. However, its movements are definitely limited by its deep socket and strong reinforcing ligaments, the two factors that account for its exceptional stability (**Figure 11.6**).

The deeply cupped acetabulum that receives the head of the femur is enhanced by a circular rim of fibrocartilage called the **acetabular labrum**. Because the diameter of the labrum is smaller than that of the femur's head, dislocations of the hip are rare. A short ligament, the **ligament of the head of the femur** (*ligamentum teres*) runs from the pitlike **fovea capitis** on the femur head to the acetabulum where it helps to secure the femur. Several strong ligaments, including the **iliofemoral** and **pubofemoral** anteriorly and the **ischiofemoral** that spirals posteriorly (not shown), are arranged so that they "screw" the femur head into the socket when a person stands upright.

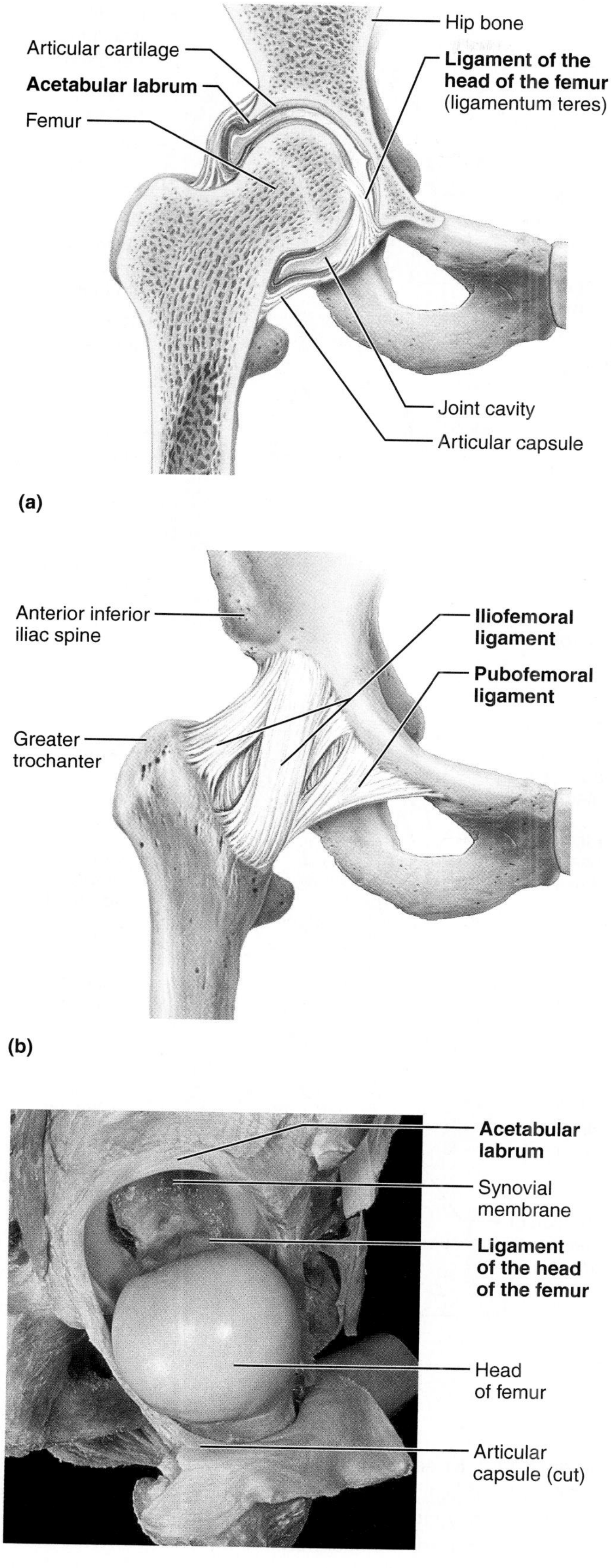

Figure 11.6 Hip joint relationships. (a) Frontal section through the right hip joint. **(b)** Anterior superficial view of the right hip joint. **(c)** Photograph of the interior of the hip joint, lateral view.

Activity 6

Demonstrating Actions at the Hip Joint

If a functional hip joint model is available, identify the joint parts and manipulate it to demonstrate the following movements: flexion, extension, abduction, and medial and lateral rotation that can occur at this joint.

Reread the information on what movements the associated ligaments restrict, and verify that information during your joint manipulations.

The Knee Joint

The knee is the largest and most complex joint in the body. Three joints in one (**Figure 11.7**, p. 174), it allows extension, flexion, and a little rotation. The **tibiofemoral joint**, actually a bicondyloid joint between the femoral condyles above and the **menisci** (semilunar cartilages) of the tibia below, is functionally a hinge joint, a very unstable one made slightly more secure by the menisci (Figure 11.7b and d). Some rotation occurs when the knee is partly flexed, but during extension, the menisci and ligaments counteract rotation and side-to-side movements. The other joint is the **femoropatellar joint**, the intermediate joint anteriorly (Figure 11.7a and c).

The knee is unique in that it is only partly enclosed by an articular capsule. Anteriorly, where the capsule is absent, are three broad ligaments, the **patellar ligament** and the **medial** and **lateral patellar retinacula** (retainers), which run from the patella to the tibia below and merge with the capsule on either side.

Extracapsular ligaments including the **fibular** and **tibial collateral ligaments** (which prevent rotation during extension) and the **oblique popliteal** and **arcuate popliteal ligaments** are crucial in reinforcing the knee. The knees have a built-in locking device that must be "unlocked" by the popliteus muscles (Figure 11.7e) before the knees can be flexed again. The **anterior** and **posterior cruciate ligaments** are intracapsular ligaments that cross (*cruci* = cross) in the notch between the femoral condyles. They help to prevent anterior-posterior displacement of the joint and overflexion and hyperextension of the joint.

Activity 7

Demonstrating Actions at the Knee Joint

If a functional model of a knee joint is available, identify the joint parts and manipulate it to illustrate the following movements: flexion, extension, and medial and lateral rotation.

Reread the information on what movements the various associated ligaments restrict, and verify that information during your joint manipulations.

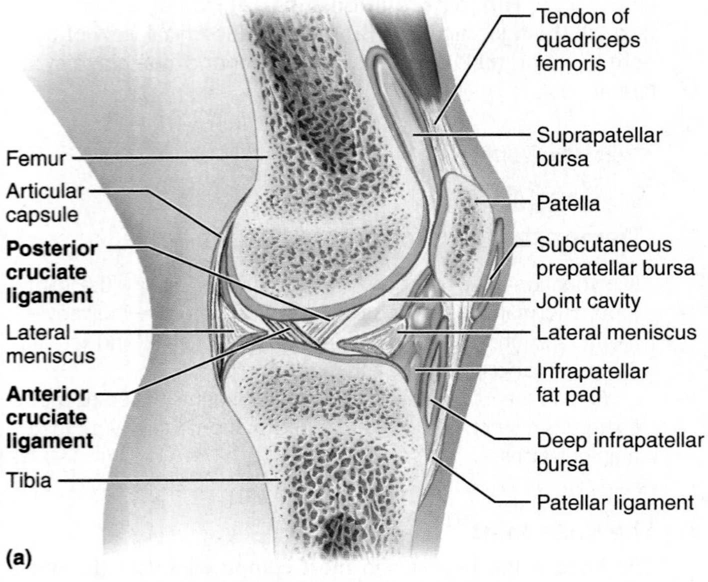

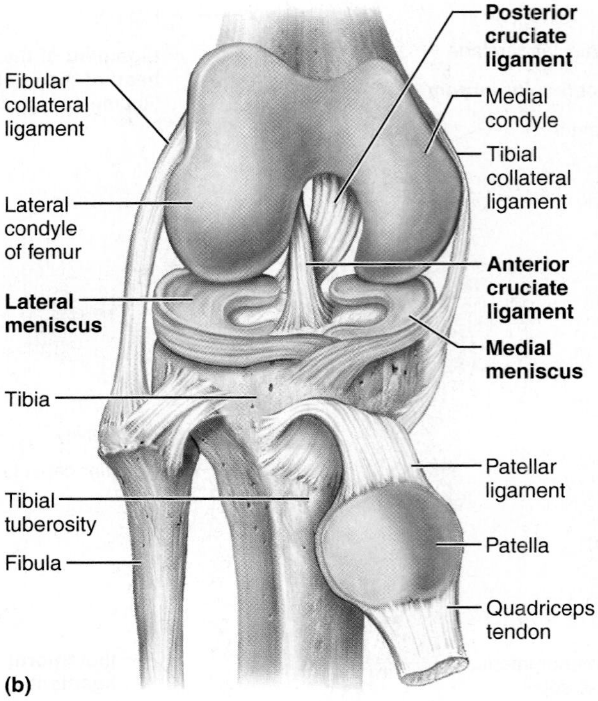

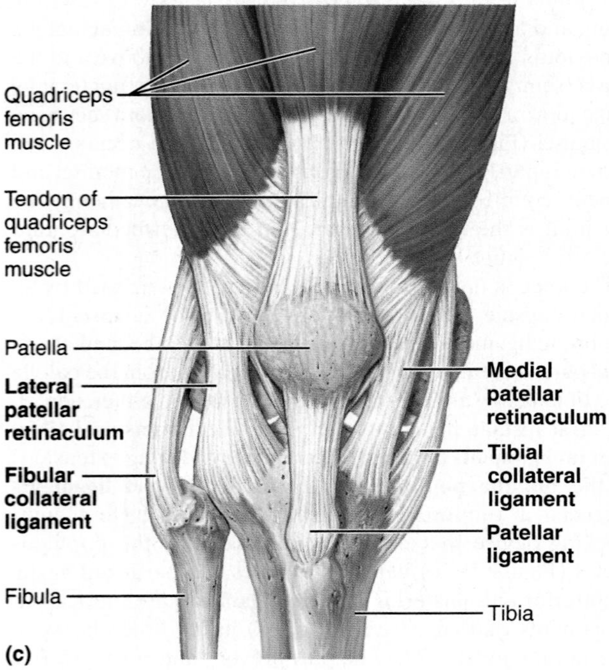

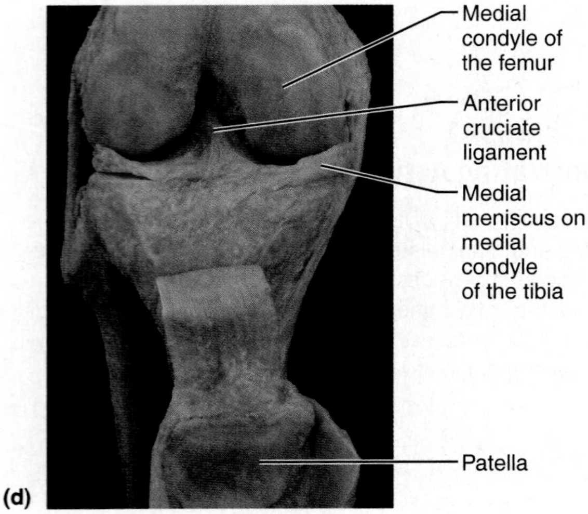

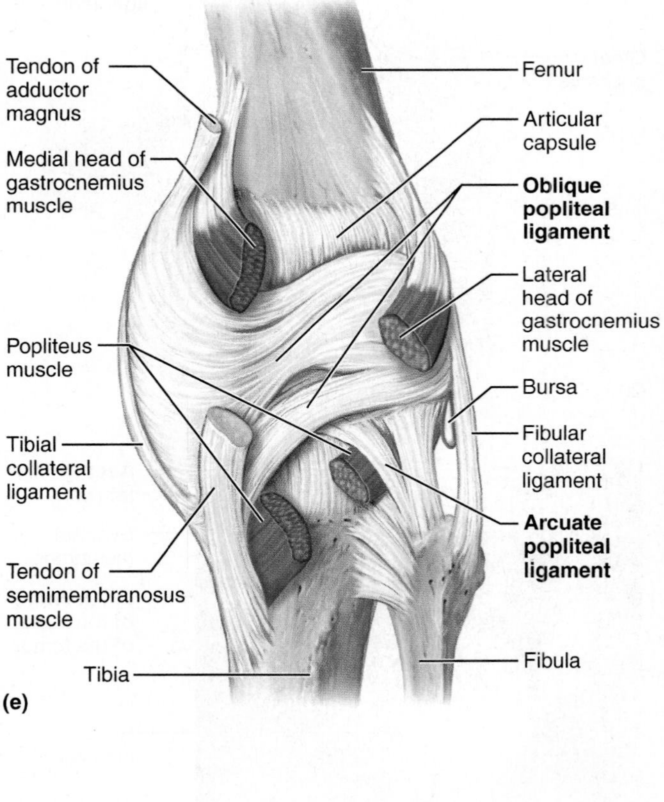

Figure 11.7 Knee joint relationships. (a) Median section of right knee joint. **(b)** Anterior view of slightly flexed right knee joint showing the cruciate ligaments. Articular capsule has been removed; the quadriceps tendon has been cut and reflected distally. **(c)** Anterior superficial view of the right knee. **(d)** Photograph of an opened knee joint corresponds to view in (b). **(e)** Posterior superficial view of the ligaments clothing the knee joint.

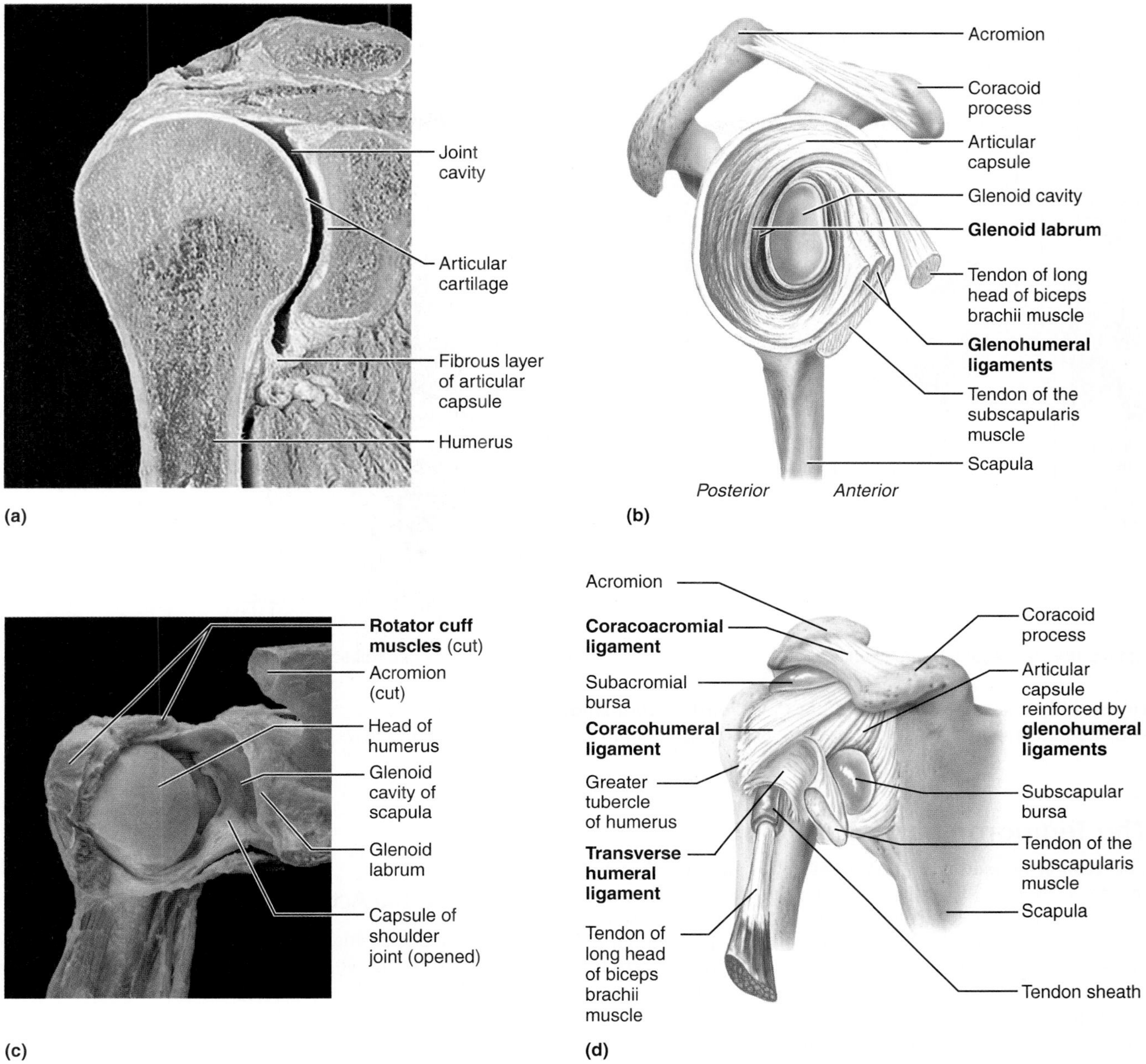

Figure 11.8 Shoulder joint relationships. (a) Frontal section through the shoulder. **(b)** Right shoulder joint, cut open and viewed from the lateral aspect; humerus removed. **(c)** Photograph of the interior of the shoulder joint, anterior view. **(d)** Anterior superficial view of the right shoulder.

The Shoulder Joint

The shoulder joint, or **glenohumeral joint**, is the most freely moving joint of the body. The rounded head of the humerus fits the shallow glenoid cavity of the scapula (**Figure 11.8**). A rim of fibrocartilage, the **glenoid labrum**, deepens the cavity slightly.

The articular capsule enclosing the joint is thin and loose, contributing to ease of movement. The **coracohumeral ligament** helps support the weight of the upper limb, and three weak **glenohumeral ligaments** strengthen the front of the capsule. Muscle tendons from the biceps brachii and **rotator cuff** muscles contribute most to shoulder stability.

Activity 8

Demonstrating Actions at the Shoulder Joint

If a functional shoulder joint model is available, identify the joint parts and manipulate the model to demonstrate the following movements: flexion, extension, abduction, adduction, circumduction, and medial and lateral rotation.

Note where the joint is weakest, and verify the most common direction of a dislocated humerus.

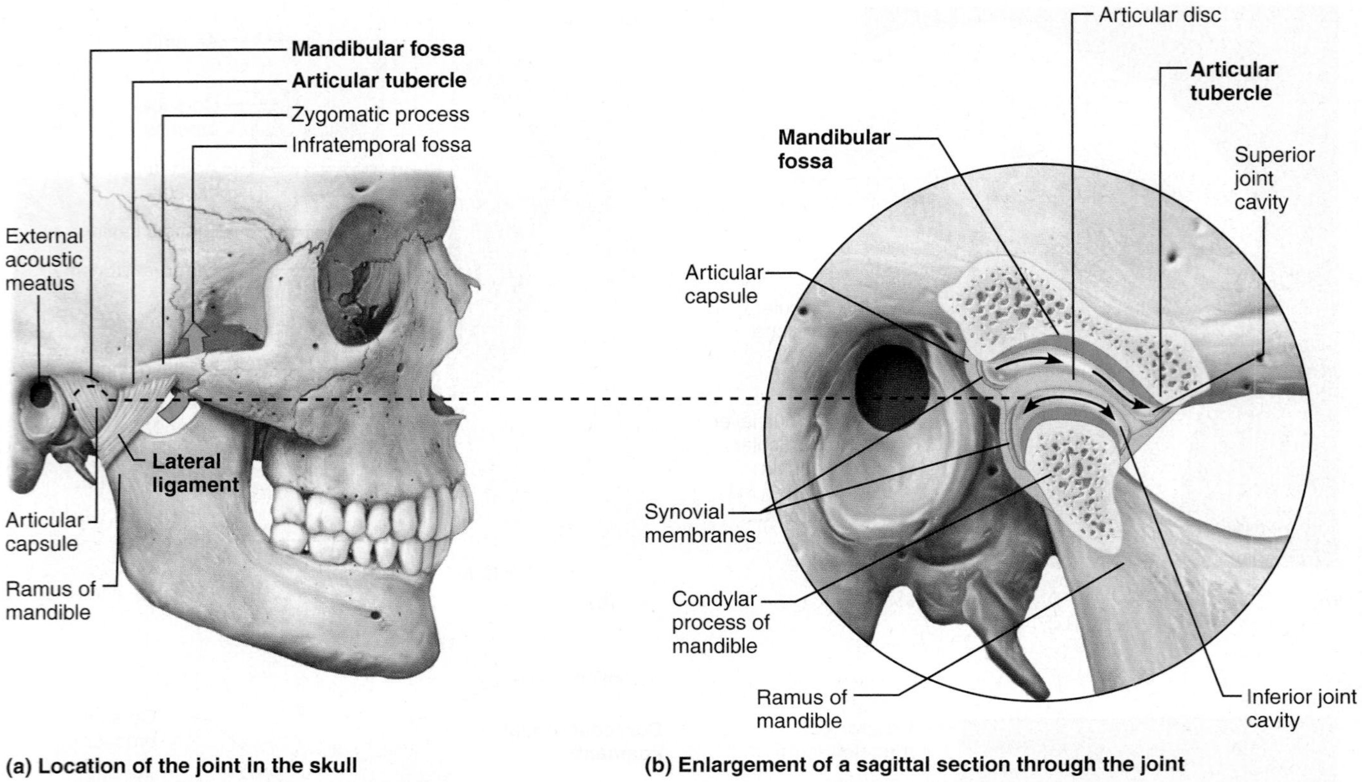

Figure 11.9 The temporomandibular (jaw) joint relationships. Note that the superior and inferior compartments of the joint cavity allow different movements indicated by arrows.

The Temporomandibular Joint

The **temporomandibular joint (TMJ)** lies just anterior to the ear (**Figure 11.9**), where the egg-shaped condylar process of the mandible articulates with the inferior surface of the squamous part of the temporal bone. The temporal bone joint surface has a complicated shape: posteriorly is the **mandibular fossa** and anteriorly is a bony knob called the **articular tubercle**. The joint's articular capsule, though strengthened by the **lateral ligament**, is loose; an articular disc divides the joint cavity into superior and inferior compartments. Typically, the condylar process–mandibular fossa connection allows the familiar hingelike movements of elevating and depressing the mandible to open and close the mouth. However, when the mouth is opened wide, the condylar process glides anteriorly and is braced against the dense bone of the articular tubercle so that the mandible is not forced superiorly through the thin mandibular fossa when we bite hard foods.

Activity 9

Examining the Action at the TMJ

While placing your fingers over the area just anterior to the ear, open and close your mouth to feel the hinge action at the TMJ. Then, keeping your fingers on the TMJ, yawn to demonstrate the anterior gliding of the condylar process of the mandible.

Table 11.3 Structural and Functional Characteristics of Body Joints

Illustration	Joint	Articulating bones	Structural type*	Functional type; movements allowed
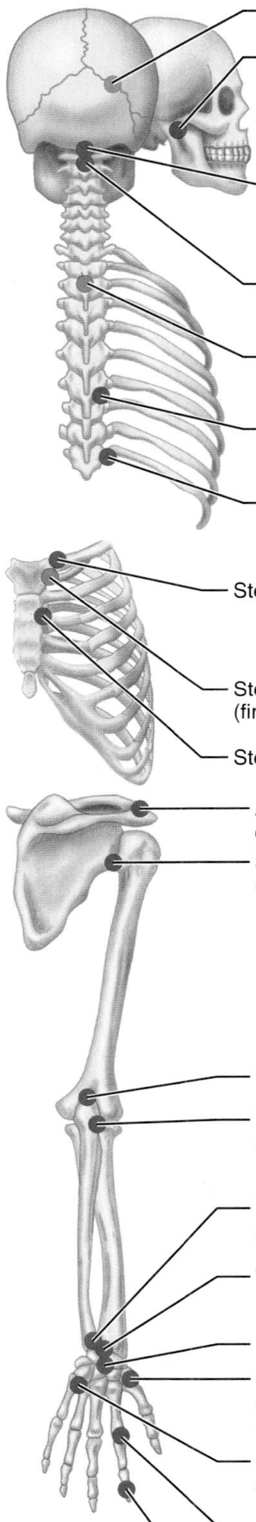	Skull	Cranial and facial bones	Fibrous; suture	Synarthrotic; no movement
	Temporo-mandibular	Temporal bone of skull and mandible	Synovial; modified hinge† (contains articular disc)	Diarthrotic; gliding and uniaxial rotation; slight lateral movement, elevation, depression, protraction, and retraction of mandible
	Atlanto-occipital	Occipital bone of skull and atlas	Synovial; condylar	Diarthrotic; biaxial; flexion, extension, lateral flexion, circumduction of head on neck
	Atlantoaxial	Atlas (C_1) and axis (C_2)	Synovial; pivot	Diarthrotic; uniaxial; rotation of the head
	Intervertebral	Between adjacent vertebral bodies	Cartilaginous; symphysis	Amphiarthrotic; slight movement
	Intervertebral	Between articular processes	Synovial; plane	Diarthrotic; gliding
	Costovertebral	Vertebrae (transverse processes or bodies) and ribs	Synovial; plane	Diarthrotic; gliding of ribs
	Sternoclavicular	Sternum and clavicle	Synovial; shallow saddle (contains articular disc)	Diarthrotic; multiaxial (allows clavicle to move in all axes)
	Sternocostal (first)	Sternum and rib 1	Cartilaginous; synchondrosis	Synarthrotic; no movement
	Sternocostal	Sternum and ribs 2–7	Synovial; double plane	Diarthrotic; gliding
	Acromio-clavicular	Acromion of scapula and clavicle	Synovial; plane (contains articular disc)	Diarthrotic; gliding and rotation of scapula on clavicle
	Shoulder (glenohumeral)	Scapula and humerus	Synovial; ball and socket	Diarthrotic; multiaxial; flexion, extension, abduction, adduction, circumduction, rotation of humerus
	Elbow	Ulna (and radius) with humerus	Synovial; hinge	Diarthrotic; uniaxial; flexion, extension of forearm
	Proximal radioulnar	Radius and ulna	Synovial; pivot	Diarthrotic; uniaxial; pivot (head of radius rotates in radial notch of ulna)
	Distal radioulnar	Radius and ulna	Synovial; pivot (contains articular disc)	Diarthrotic; uniaxial; rotation of radius to allow pronation and supination
	Wrist	Radius and proximal carpals	Synovial; condylar	Diarthrotic; biaxial; flexion, extension, abduction, adduction, circumduction of hand
	Intercarpal	Adjacent carpals	Synovial; plane	Diarthrotic; gliding
	Carpometacarpal of digit I (thumb)	Carpal (trapezium) and metacarpal I	Synovial; saddle	Diarthrotic; biaxial; flexion, extension, abduction, adduction, circumduction, opposition of metacarpal I
	Carpometacarpal of digits II–V	Carpal(s) and metacarpal(s)	Synovial; plane	Diarthrotic; gliding of metacarpals
	Metacarpo-phalangeal (knuckle)	Metacarpal and proximal phalanx	Synovial; condylar	Diarthrotic; biaxial; flexion, extension, abduction, adduction, circumduction of fingers
	Interphalangeal (finger)	Adjacent phalanges	Synovial; hinge	Diarthrotic; uniaxial; flexion, extension of fingers

Table 11.3 Structural and Functional Characteristics of Body Joints *(continued)*

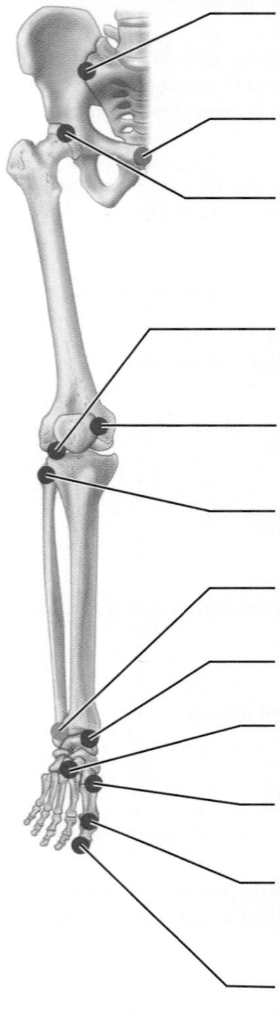

Illustration	Joint	Articulating bones	Structural type*	Functional type; movements allowed
	Sacroiliac	Sacrum and hip bone	Synovial; plane	Diarthrotic; little movement, slight gliding possible (more during pregnancy)
	Pubic symphysis	Pubic bones	Cartilaginous; symphysis	Amphiarthrotic; slight movement (enhanced during pregnancy)
	Hip (coxal)	Hip bone and femur	Synovial; ball and socket	Diarthrotic; multiaxial; flexion, extension, abduction, adduction, rotation, circumduction of thigh
	Knee (tibiofemoral)	Femur and tibia	Synovial; modified hinge† (contains articular discs)	Diarthrotic; biaxial; flexion, extension of leg, some rotation allowed
	Knee (femoropatellar)	Femur and patella	Synovial; plane	Diarthrotic; gliding of patella
	Superior tibiofibular	Tibia and fibula (proximally)	Synovial; plane	Diarthrotic; gliding of fibula
	Inferior tibiofibular	Tibia and fibula (distally)	Fibrous; syndesmosis	Synarthrotic; slight "give" during dorsiflexion of foot
	Ankle	Tibia and fibula with talus	Synovial; hinge	Diarthrotic; uniaxial; dorsiflexion, and plantar flexion of foot
	Intertarsal	Adjacent tarsals	Synovial; plane	Diarthrotic; gliding; inversion and eversion of foot
	Tarsometatarsal	Tarsal(s) and metatarsal(s)	Synovial; plane	Diarthrotic; gliding of metatarsals
	Metatarso-phalangeal	Metatarsal and proximal phalanx	Synovial; condylar	Diarthrotic; biaxial; flexion, extension, abduction, adduction, circumduction of great toe
	Interphalangeal (toe)	Adjacent phalanges	Synovial; hinge	Diarthrotic; uniaxial; flexion, extension of toes

***Fibrous joint** indicated by orange circles; **cartilaginous joints**, by blue circles; **synovial joints**, by purple circles.
†These modified hinge joints are structurally bicondylar.

Joint Disorders

Most of us don't think about our joints until something goes wrong with them. Joint pains and malfunctions have a variety of causes. For example, a hard blow to the knee can cause a painful bursitis, known as "water on the knee," due to damage to, or inflammation of, the patellar bursa.

Sprains and dislocations are other types of joint problems. In a **sprain**, the ligaments reinforcing a joint are damaged by overstretching or are torn away from the bony attachment. Since both ligaments and tendons are cords of dense regular connective tissue with a poor blood supply, sprains heal slowly and are quite painful.

Dislocations occur when bones are forced out of their normal position in the joint cavity. They are normally accompanied by torn or stressed ligaments and considerable inflammation. The process of returning the bone to its proper position, called reduction, should be done only by a physician. Attempts by the untrained person to "snap the bone back into its socket" are often more harmful than helpful.

Advancing years also take their toll on joints. Weight-bearing joints in particular eventually begin to degenerate. *Adhesions* (fibrous bands) may form between the surfaces where bones join, and extraneous bone tissue (*spurs*) may grow along the joint edges. Such degenerative changes lead to the complaint so often heard from the elderly: "My joints are getting so stiff. . . ."

- If possible, compare an X-ray image of an arthritic joint to one of a normal joint. ✚

REVIEW SHEET

11 Articulations and Body Movements

EXERCISE

Instructors may assign a portion of the Review Sheet questions using Mastering A&P™

Name ______________________________ Lab Time/Date ______________________

Fibrous, Cartilaginous, and Synovial Joints

1. Use key responses to identify the joint types described below.

Key: a. cartilaginous b. fibrous c. synovial

______________ 1. includes shoulder, elbow, and wrist joints

______________ 2. includes joints between the vertebral bodies and the pubic symphysis

______________ 3. sutures are memorable examples

______________ 4. found in the epiphyseal plate

______________ 5. found in a gomphosis

______________ 6. all are freely movable or diarthrotic

2. Label the diagram of a typical synovial joint using the terms provided in the key and the appropriate leader lines.

Key: a. articular capsule

b. articular cartilage

c. fibrous layer

d. joint cavity

e. ligament

f. periosteum

g. synovial membrane

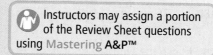

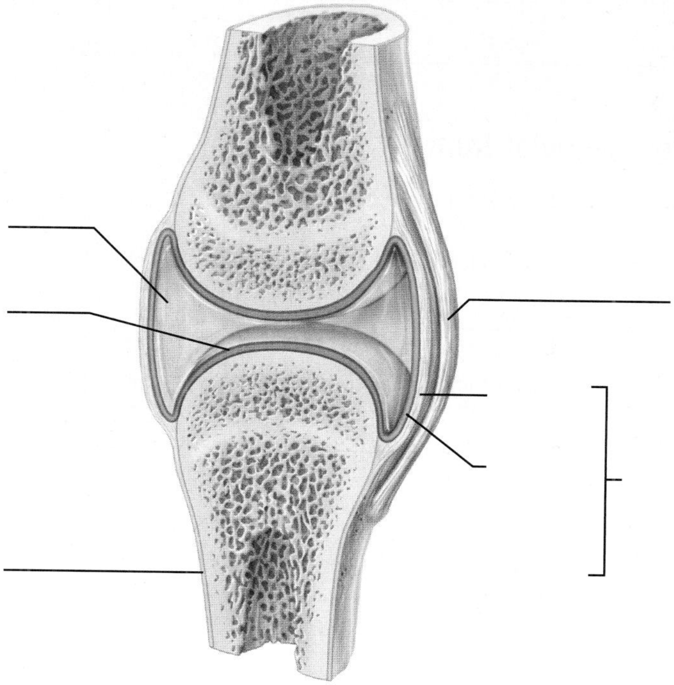

3. How does a tendon sheath differ from a bursa? ______________________________

____

4. Which structure in the synovial joint produces synovial fluid? ______________________

5. Match the synovial joint categories in column B with their descriptions in column A. Some terms may be used more than once.

Column A

_______________ 1. joint between the axis and atlas

_______________ 2. hip joint

_______________ 3. intervertebral joints (between articular processes)

_______________ 4. joint between forearm bones and wrist

_______________ 5. elbow

_______________ 6. interphalangeal joints

_______________ 7. intercarpal joints

_______________ 8. joint between talus and tibia/fibula

_______________ 9. joint between skull and vertebral column

_______________ 10. joint between jaw and skull

_______________ 11. joints between proximal phalanges and metacarpal bones

_______________ 12. a multiaxial joint

_______________, _______________ 13. biaxial joints

_______________, _______________ 14. uniaxial joints

Column B

a. ball-and-socket

b. condylar

c. hinge

d. pivot

e. plane

f. saddle

Selected Synovial Joints

6. Which joint, the hip or the knee, is more stable? _______________

Name two important factors that contribute to the stability of the hip joint.

_______________ and _______________

Name two important factors that contribute to the stability of the knee.

_______________ and _______________

7. Label the photograph of a knee joint model using the terms provided in the key and the appropriate leader lines.

Key:

a. anterior cruciate ligament

b. fibula

c. fibular collateral ligament

d. lateral condyle of the femur

e. lateral meniscus

f. medial meniscus

g. patella

h. patellar ligament

i. tibia

j. tibial collateral ligament

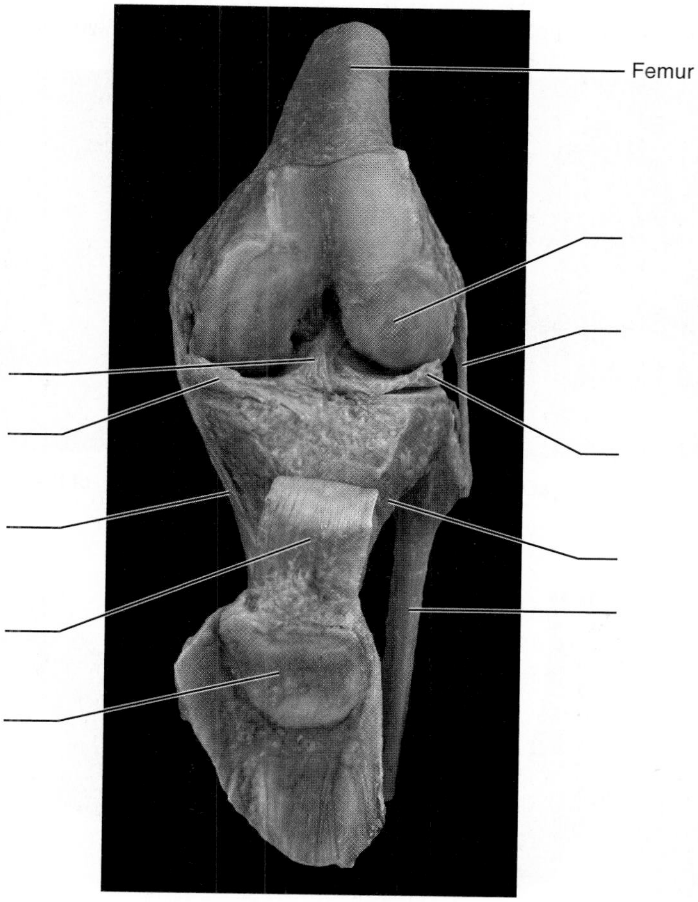

8. The shoulder joint is built for mobility. List three factors that contribute to the large range of motion at the shoulder:

1. ______________________________

2. ______________________________

3. ______________________________

Movements Allowed by Synovial Joints

9. Complete the descriptions below the diagrams by inserting the type of movement in each answer blank.

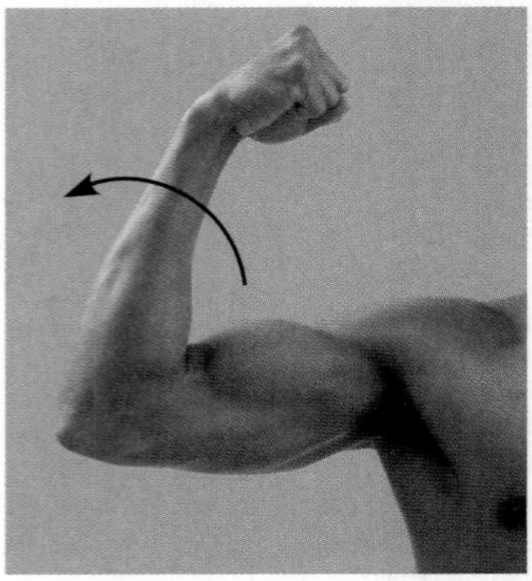

(a) ________________ **at the elbow**

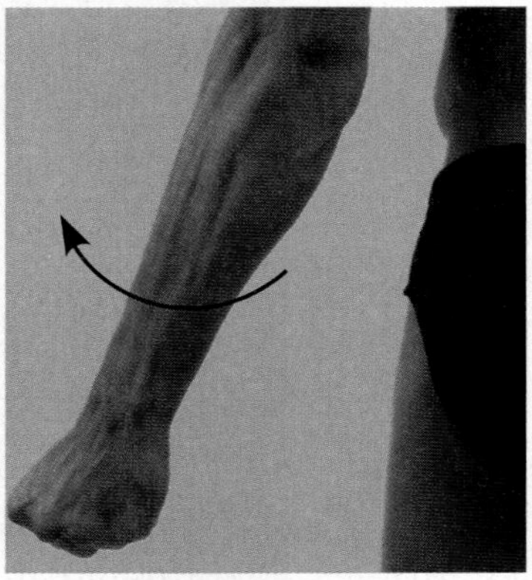

(b) ________________ **of the upper limb**

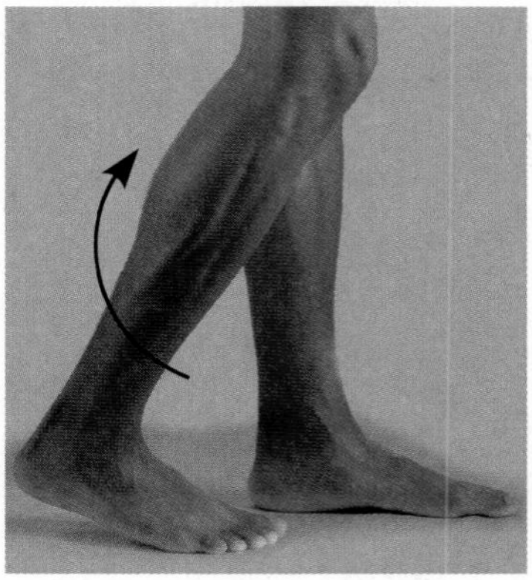

(c) ________________ **at the knee**

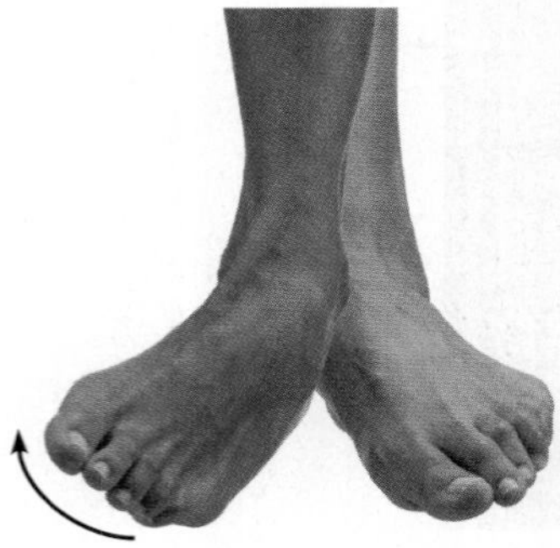

(d) ________________ **of the foot**

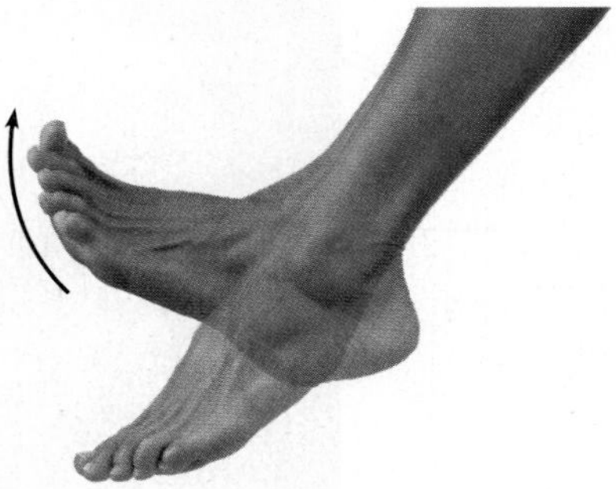

(e) ________________ **of the foot**

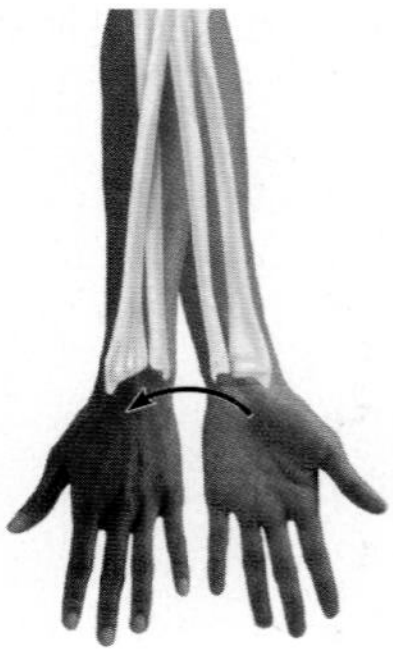

(f) ________________ **of the forearm**

10. ✚ The glenoid labrum can become torn from overuse or direct injury to the shoulder joint. Considering the function of the glenoid labrum, describe some of the consequences of a large tear in the glenoid labrum. ________________________________

____

____

11. ✚ A physician diagnoses you with "olecranon bursitis." Predict the location and cause of the swelling that you are experiencing.

____

____

12. ✚ The menisci in the knee joint can be torn for a variety of reasons. Considering the structure of the menisci, would you expect these tears to heal on their own?________________________________

Why or why not? ____

____

EXERCISE 12 Microscopic Anatomy and Organization of Skeletal Muscle

Learning Outcomes

- Define *muscle fiber*, *myofibril*, and *myofilament*, and describe the structural relationships among them.
- Describe thick (myosin) and thin (actin) filaments and their relationship to the sarcomere.
- Discuss the structure and location of T tubules and terminal cisterns.
- Define *endomysium*, *perimysium*, and *epimysium*, and relate them to muscle fibers, fascicles, and entire muscles.
- Define *tendon* and *aponeurosis*, and describe the difference between them.
- Describe the structure of skeletal muscle from gross to microscopic levels.
- Explain the connection between motor neurons and skeletal muscle, and discuss the structure and function of the neuromuscular junction.

Go to Mastering A&P™ > Study Area to improve your performance in A&P Lab.

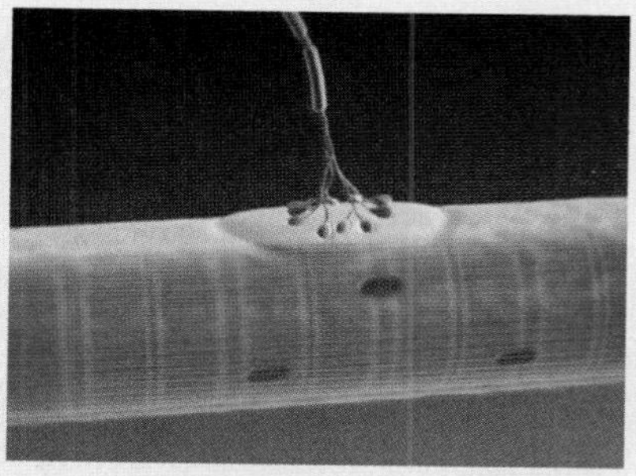

> Animations & Videos > A&P Flix > Events at the NMJ

Instructors may assign new Building Vocabulary coaching activities, Pre-Lab Quiz questions, Art Labeling activities, Practice Anatomy Lab Practical questions (PAL), and more using the Mastering **A&P™** Item Library.

Pre-Lab Quiz

Instructors may assign these and other Pre-Lab Quiz questions using Mastering **A&P™**

1. Circle the correct underlined term. Because the cells of skeletal muscle are relatively large and cylindrical in shape, they are also known as fibers / tubules.
2. Each muscle fiber is surrounded by thin connective tissue called the:
   a. aponeurosis
   b. epimysium
   c. endomysium
   d. perimysium
3. A cordlike structure that connects a muscle to another muscle or bone is:
   a. a fascicle
   b. a tendon
   c. deep fascia
4. The junction between an axon and a muscle fiber is called a ______________________.
5. Circle the correct underlined term. The contractile unit of muscle is the sarcolemma / sarcomere.

Materials

- Three-dimensional model of skeletal muscle fibers (if available)
- Forceps
- Dissecting needles
- Clean microscope slides and coverslips
- 0.9% saline solution in dropper bottles
- Chicken breast or thigh muscle (fresh from the market)
- Compound microscope
- Prepared slides of skeletal muscle (l.s. and x.s. views) and skeletal muscle showing neuromuscular junctions
- Three-dimensional model of skeletal muscle showing neuromuscular junction (if available)

Most of the muscle tissue in the body is **skeletal muscle**, which attaches to the skeleton or associated connective tissue. Skeletal muscle shapes the body and gives you the ability to move—to walk, run, jump, and dance; to draw, paint, and play a musical instrument; and to smile and frown. The remaining muscle tissue of the body consists of smooth muscle that forms the walls of hollow organs and cardiac muscle that forms the walls of the heart. Smooth and cardiac muscle move materials within the body. For example, smooth muscle moves digesting food through the digestive system, and urine from the kidneys to the exterior of the body. Cardiac muscle moves blood through the blood vessels.

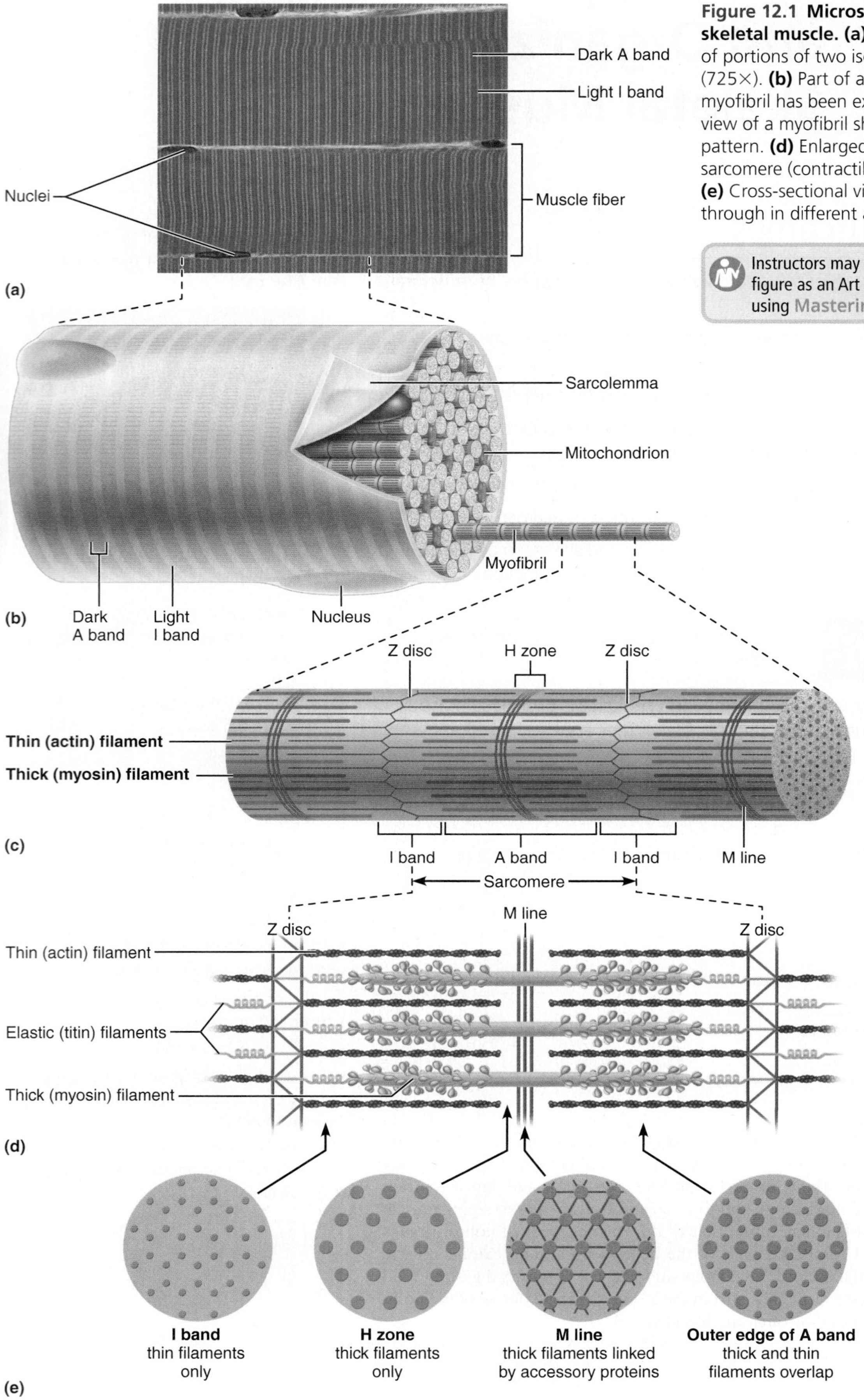

Figure 12.1 Microscopic anatomy of skeletal muscle. (a) Photomicrograph of portions of two isolated muscle fibers (725×). **(b)** Part of a muscle fiber. One myofibril has been extended. **(c)** Enlarged view of a myofibril showing its banding pattern. **(d)** Enlarged view of one sarcomere (contractile unit) of a myofibril. **(e)** Cross-sectional view of a sarcomere cut through in different areas.

Instructors may assign this figure as an Art Labeling Activity using Mastering A&P™

12

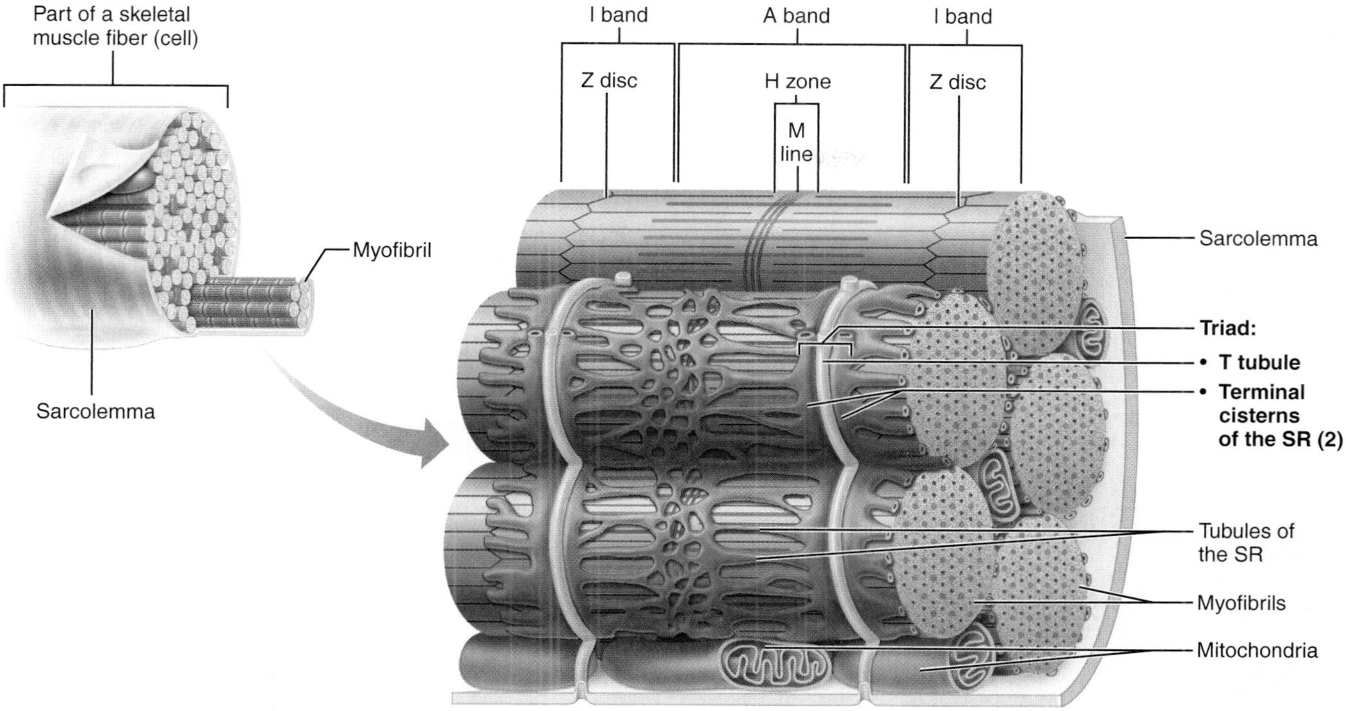

Figure 12.2 Relationship of the sarcoplasmic reticulum and T tubules to the myofibrils of skeletal muscle.

Instructors may assign this figure as an Art Labeling Activity using Mastering A&P™

Each of the three muscle types has a structure and function uniquely suited to its function in the body. Our focus here is to investigate the structure of skeletal muscle.

Skeletal muscle is also known as *voluntary muscle* because it can be consciously controlled, and as striated muscle because it appears to be striped.

The Cells of Skeletal Muscle

Skeletal muscle is made up of relatively large, long cylindrical cells, called **muscle fibers**. These cells range from 10 to 100 μm in diameter. Some are up to 30 cm long.

Since hundreds of embryonic cells fuse to produce each muscle fiber, the cells (**Figure 12.1a** and **b**) are multinucleate; multiple oval nuclei can be seen just beneath the plasma membrane (called the *sarcolemma* in these cells). The nuclei are pushed peripherally by the longitudinally arranged **myofibrils**, long, rod-shaped organelles that nearly fill the sarcoplasm, the cytoplasm of the muscle cell. Alternating light (I) and dark (A) bands along the length of the perfectly aligned myofibrils give the muscle fiber its striped appearance.

Electron microscope studies have revealed that the myofibrils are made up of even smaller threadlike structures called **myofilaments** (Figure 12.1d). The myofilaments are composed largely of two varieties of contractile proteins—**actin** and **myosin**—which slide past each other during muscle activity to bring about shortening or contraction of the muscle cells. The actual contractile units of muscle, called **sarcomeres**, extend from the middle of one I band (its Z disc) to the middle of the next along the length of the myofibrils (Figure 12.1c and d.) Cross sections of the sarcomere in areas where **thick** and **thin filaments** overlap show that each thick filament is surrounded by six thin filaments; each thin filament is surrounded by three thick filaments (Figure 12.1e).

At each junction of the A and I bands, the sarcolemma indents into the muscle cell, forming a **transverse tubule (T tubule)**. These tubules run deep into the muscle fiber between cross channels, or **terminal cisterns**, of the elaborate smooth endoplasmic reticulum called the **sarcoplasmic reticulum (SR)** (**Figure 12.2**). Regions where the SR terminal cisterns border a T tubule on each side are called **triads**.

Activity 1

Examining Skeletal Muscle Cell Anatomy

1. Look at the three-dimensional model of skeletal muscle fibers, noting the relative shape and size of the cells. Identify the nuclei, myofibrils, and light and dark bands.

2. Obtain forceps, two dissecting needles, slide and coverslip, and a dropper bottle of saline solution. With forceps, remove a very small piece of muscle (about 1 mm diameter) from a fresh chicken breast or thigh. Place the tissue on a clean microscope slide, and add a drop of the saline solution.

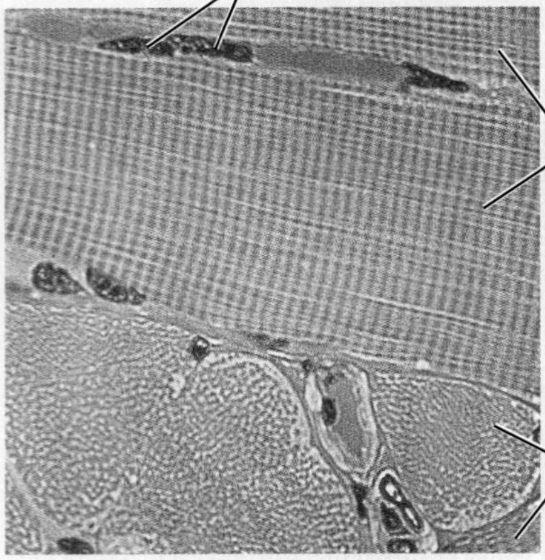

Figure 12.3 Photomicrograph of muscle fibers, longitudinal and cross sections (800×).

3. Pull the muscle fibers apart (tease them) with the dissecting needles or forceps until you have a fluffy-looking mass of tissue. Cover the teased tissue with a coverslip, and observe under the high-power lens of a compound microscope. Look for the banding pattern by examining muscle fibers isolated at the edge of the tissue mass. Regulate the light carefully to obtain the highest possible contrast.

4. Now compare your observations with the photomicrograph (**Figure 12.3**) and with what can be seen in professionally prepared muscle tissue. Obtain a slide of skeletal muscle (longitudinal section), and view it under high power. From your observations, draw a small section of a muscle fiber in the space provided below. Label the nuclei, sarcolemma, and A and I bands.

What structural details become apparent with the prepared slide?

____

____

Organization of Skeletal Muscle Cells into Muscles

Muscle fibers are soft and surprisingly fragile. Thousands of muscle fibers are bundled together with connective tissue to form the organs we refer to as skeletal muscles (**Figure 12.4**). Each muscle fiber is enclosed in a delicate, areolar connective tissue sheath called the **endomysium**. Several sheathed muscle fibers are wrapped by a collagenic membrane called the **perimysium**, forming a bundle of muscle fibers called a **fascicle**. A large number of fascicles are bound together by a much coarser "overcoat" of dense irregular connective tissue called the **epimysium**, which sheathes the entire muscle. All three sheaths converge to form strong cordlike **tendons** or sheetlike **aponeuroses**, which attach muscles to each other or indirectly to bones. A muscle's more movable attachment is called its *insertion*, whereas its fixed (or immovable) attachment is the *origin* (Exercise 11).

Tendons perform several functions; two of the most important are to provide durability and to conserve space. Because tendons are tough dense regular connective tissue, they can span rough bony projections that would destroy the more delicate muscle tissues. Because of their relatively small size, more tendons than fleshy muscles can pass over a joint.

In addition to supporting and binding the muscle fibers, and providing strength to the muscle as a whole, the connective tissue wrappings provide a route for the entry and exit of nerves and blood vessels that serve the muscle fibers. The larger, more powerful muscles have relatively more connective tissue than muscles involved in fine or delicate movements.

As we age, the mass of the muscle fibers decreases, and the amount of connective tissue increases; thus the skeletal muscles gradually become more sinewy, or "stringier." +

Activity 2

Observing the Histological Structure of a Skeletal Muscle

Identify the muscle fibers, their peripherally located nuclei, and their connective tissue wrappings—the endomysium, perimysium, and epimysium, if visible (use Figure 12.4 as a reference).

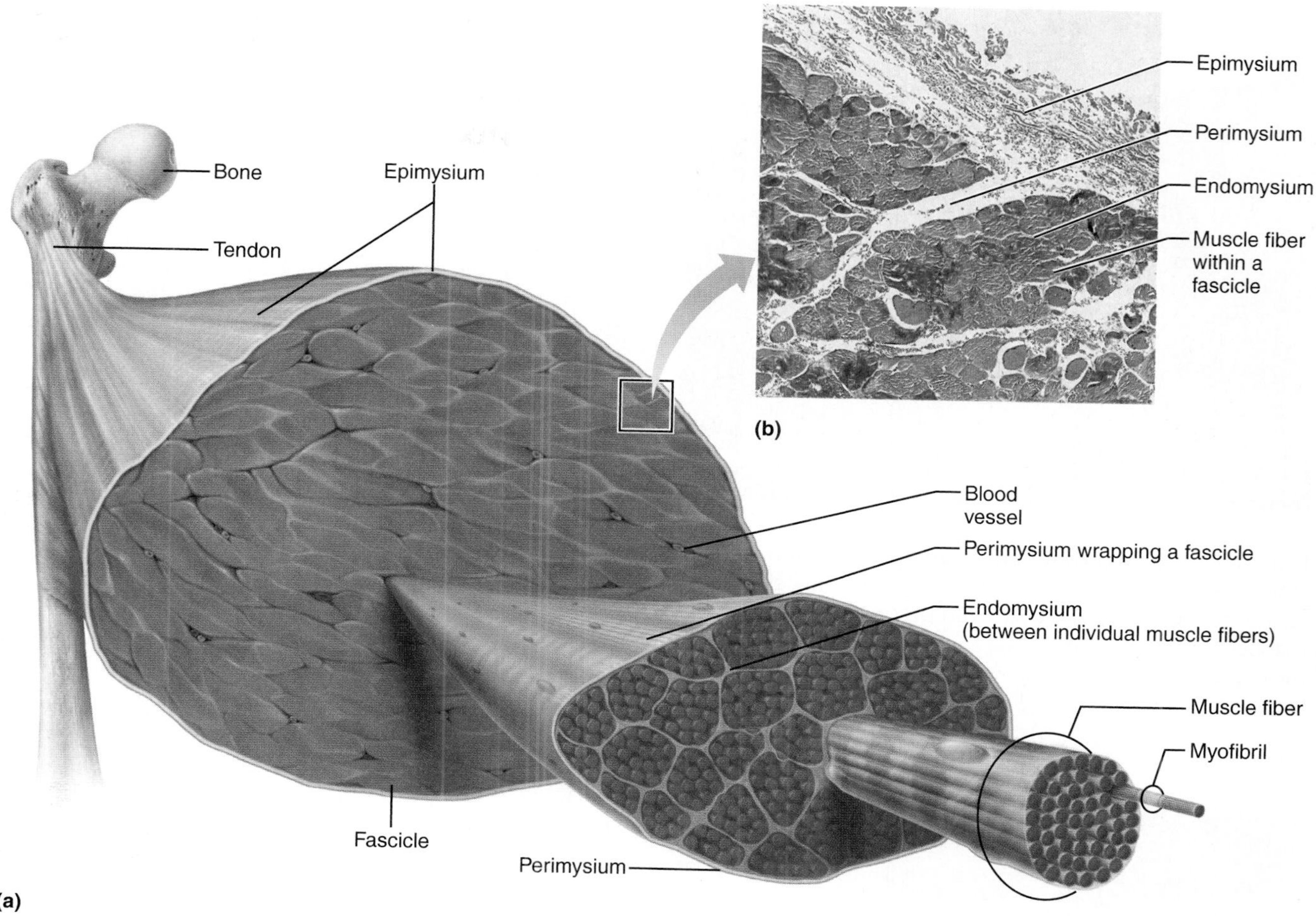

Figure 12.4 Connective tissue sheaths of skeletal muscle. (a) Diagram. **(b)** Photomicrograph of a cross section of skeletal muscle (40×).

The Neuromuscular Junction

The voluntary skeletal muscle cells must be stimulated by motor neurons via nerve impulses. The junction between an axon of a motor neuron and a muscle fiber is called a **neuromuscular junction** (Figure 12.5, p. 188).

Each axon of the motor neuron usually divides into many branches called *terminal branches* as it approaches the muscle. Each of these branches ends in an axon terminal that participates in forming a neuromuscular junction with a single muscle fiber. Thus a single neuron may stimulate many muscle fibers. Together, a neuron and all the muscle fibers it stimulates make up the functional structure called the **motor unit**. Part of a motor unit, showing two neuromuscular junctions, is shown in Figure 12.6, p. 188. The neuron and muscle fiber membranes, close as they are, do not actually touch. They are separated by a small fluid-filled gap called the **synaptic cleft** (see Figure 12.5).

Within the axon terminals are many mitochondria and vesicles containing a neurotransmitter called acetylcholine (ACh). When an action potential reaches the axon terminal, voltage-gated Ca^{2+} channels open. Ca^{2+} enters the axon terminal and causes ACh to be released by exocytosis. The ACh rapidly diffuses across the synaptic cleft and combines with the receptors on the sarcolemma. When receptors bind ACh, a change in the permeability of the sarcolemma occurs. Ion channels open briefly, depolarizing the sarcolemma, and subsequent contraction of the muscle fiber occurs.

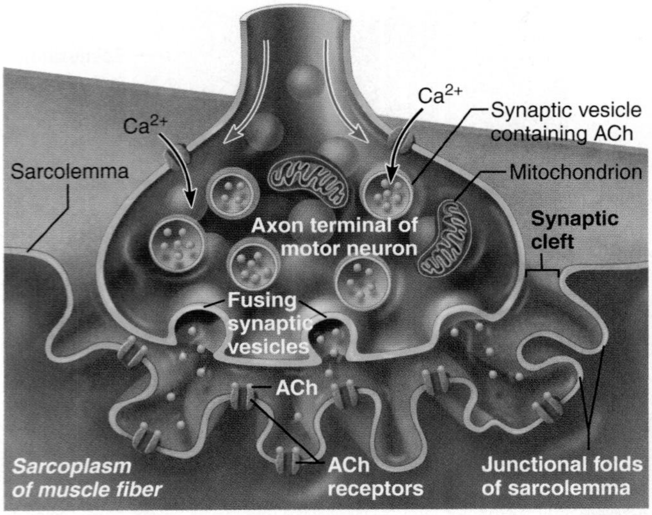

Figure 12.5 The neuromuscular junction. Pink arrows indicate arrival of the action potential, which ultimately causes vesicles to release ACh. The ACh receptor is part of the ion channel that opens briefly, causing depolarization of the sarcolemma.

Instructors may assign this figure as an Art Labeling Activity using Mastering A&P™

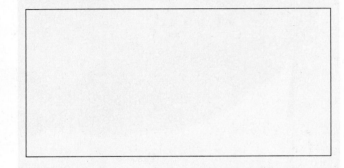

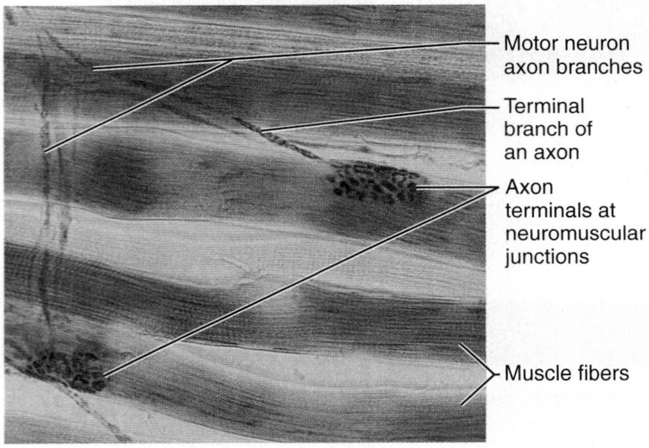

Figure 12.6 Photomicrograph of neuromuscular junctions (750×).

Activity 3

Studying the Structure of a Neuromuscular Junction

1. If possible, examine a three-dimensional model of skeletal muscle fibers that illustrates the neuromuscular junction. Identify the structures just described.

2. Obtain a slide of skeletal muscle stained to show a portion of a motor unit. Examine the slide under high power to identify the axon branches that extend like a leash to the muscle fibers.

Follow one of the axons to its terminal branch to identify the oval-shaped axon terminal at the end of the branch. Compare your observations to the photomicrograph (Figure 12.6). Sketch a small section in the space provided below. Label the axon of the motor neuron, its terminal branches, axon terminals, and muscle fibers.

Instructors may assign a portion of the Review Sheet questions using Mastering A&P™

REVIEW SHEET

EXERCISE 12

Microscopic Anatomy and Organization of Skeletal Muscle

Name ______________________ Lab Time/Date ______________________

Skeletal Muscle Cells and Their Organization into Muscles

1. Use the items in the key to correctly identify the structures described below.

__________ 1. connective tissue covering a bundle of muscle fibers

__________ 2. bundle of muscle fibers

__________ 3. contractile unit of muscle

__________ 4. superficial sheath that covers the entire muscle

__________ 5. thin areolar connective tissue surrounding each muscle fiber

__________ 6. plasma membrane of the muscle fiber

__________ 7. a long organelle with a banded appearance found within muscle fibers

__________ 8. actin- or myosin-containing structure

__________ 9. cord of collagen fibers that attaches a muscle to a bone

Key:

a. endomysium
b. epimysium
c. fascicle
d. myofibril
e. myofilament
f. perimysium
g. sarcolemma
h. sarcomere
i. tendon

2. List three reasons why the connective tissue sheaths of skeletal muscle are important.

____

____

____

3. Why are there more indirect—that is, tendinous—muscle attachments to bone than there are direct attachments?

____

____

____

4. How does an aponeurosis differ from a tendon structurally? ______________________

____

How is an aponeurosis functionally similar to a tendon? ______________________

____

5. The drawing and photomicrograph below show a relaxed sarcomere. Using the terms from the key, identify each structure indicated by a leader line or bracket. The number 2 in parentheses indicates that the structure will be labeled twice.

Key:

a. actin filament

b. A band

c. H zone

d. I band (2)

e. M line

f. myosin filament

g. Z disc (2)

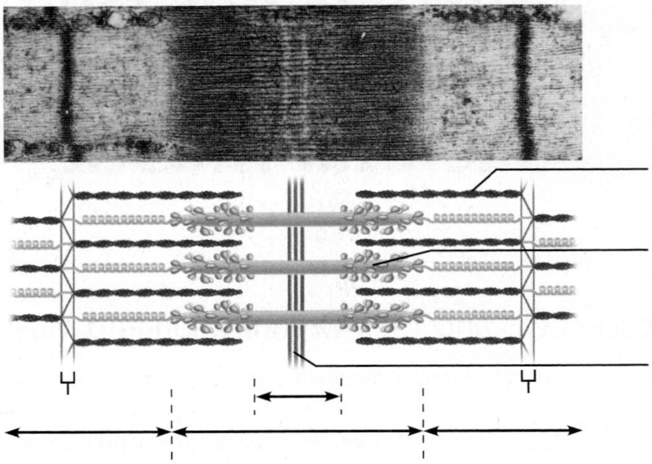

6. On the following figure, label the endomysium, epimysium, a fascicle, a muscle fiber, a myofibril, perimysium, and the tendon.

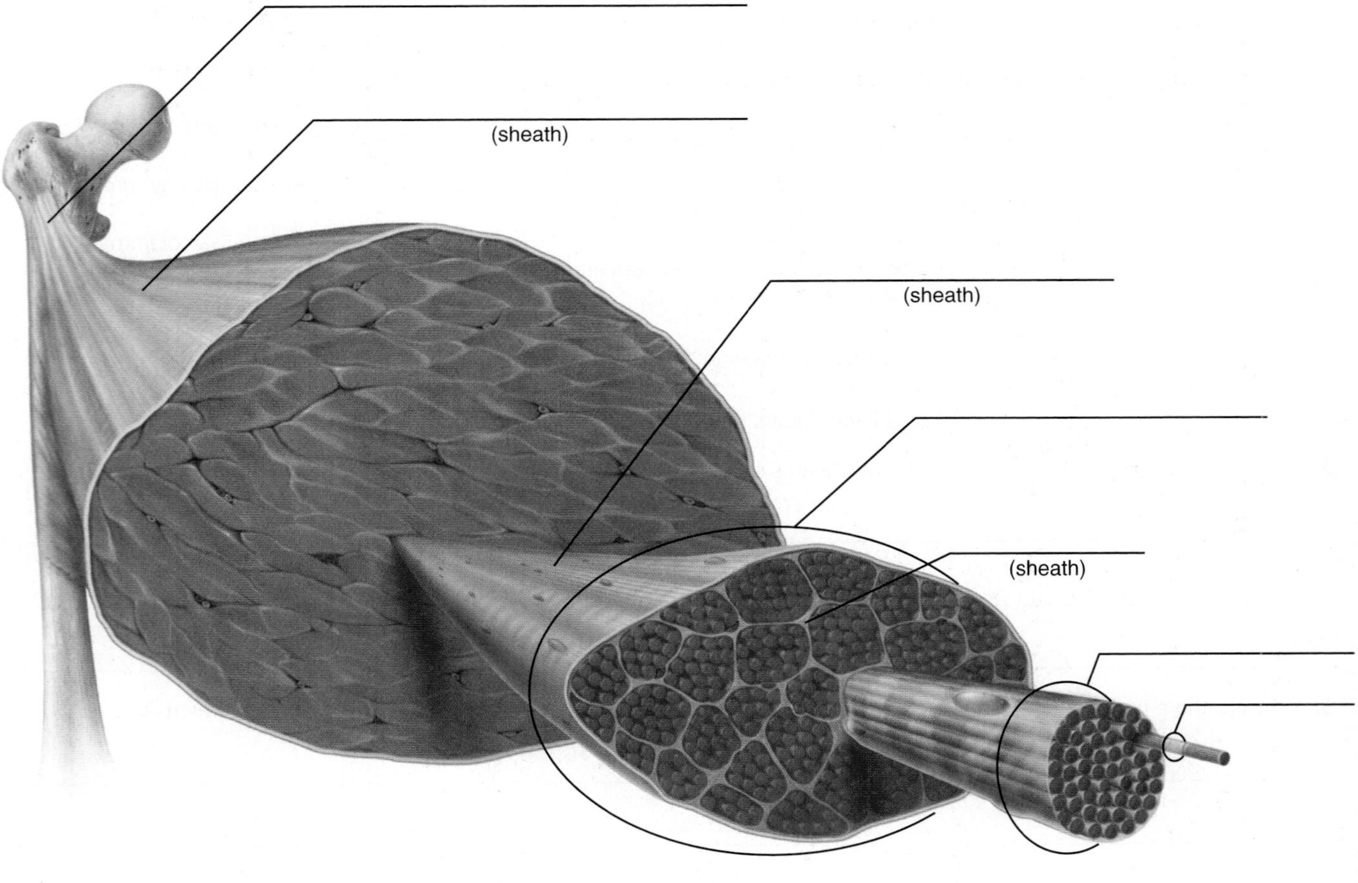

The Neuromuscular Junction

7. Complete the following statements:

The junction between a motor neuron's axon and the muscle fiber plasma membrane is called a __1__ junction. A motor neuron and all of the skeletal muscle fibers it stimulates is called a __2__. The actual gap between the axon terminal and the muscle fiber is called a __3__. Within the axon terminal are many small vesicles containing a neurotransmitter substance called __4__. When the __5__ reaches the ends of the axon, the neurotransmitter is released and diffuses to the muscle cell membrane to combine with receptors there. The combining of the neurotransmitter with the muscle membrane receptors causes a change in permeability of the membrane, resulting in __6__ of the sarcolemma. Then, contraction of the muscle fiber occurs.

1. ______________________
2. ______________________
3. ______________________
4. ______________________
5. ______________________
6. ______________________

8. The events that occur at a neuromuscular junction are depicted below. Identify every structure provided with a leader line.

Note: The pink arrows depict the propagation of the action potential.

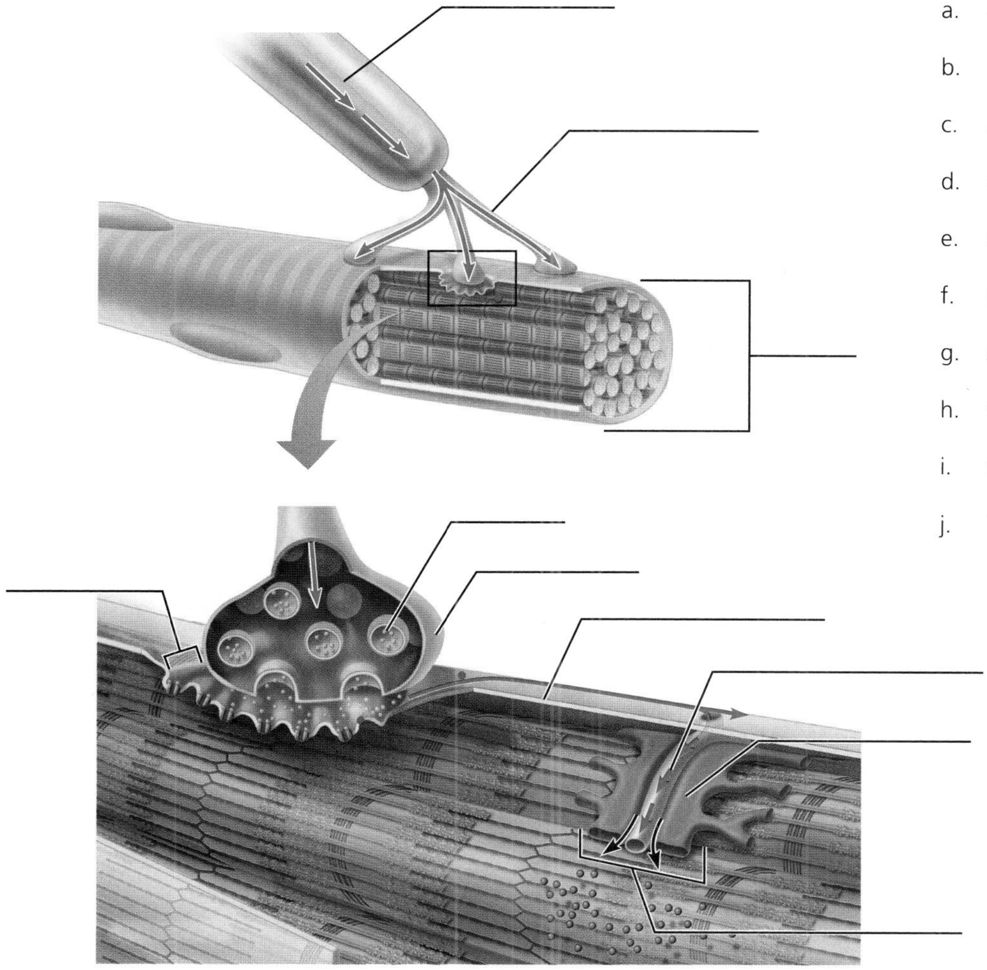

Key:

a. axon terminal of motor neuron
b. motor neuron axon branch
c. myelinated axon of motor neuron
d. muscle fiber
e. sarcolemma of muscle fiber
f. synaptic cleft
g. synaptic vesicle containing ACh
h. terminal cistern of the SR
i. triad
j. T tubule

9. Necrotizing fasciitis is a serious bacterial infection. Necrosis is death of tissues in the body. Considering the organization of the connective tissue sheaths of skeletal muscle, explain how this infection could spread rapidly throughout the body.

____

____

10. The bacterium *Clostridium botulinum* secretes botulinum toxin, a neurotoxin. The toxin blocks the release of acetylcholine from the axon terminal of a motor neuron. Explain how the toxin binding would change the normal sequence of events at the neuromuscular junction. ____

____

____

EXERCISE 13 Gross Anatomy of the Muscular System

Learning Outcomes

- Define *prime mover (agonist)*, *antagonist*, *synergist*, and *fixator*.
- List the criteria used in naming skeletal muscles.
- Identify the major muscles of the human body on a torso model, a human cadaver, lab chart, or image, and state the action of each.
- Name muscle origins and insertions as required by the instructor.
- Explain how muscle actions are related to their location.
- List antagonists for the prime movers listed.

Go to Mastering A&P™ > Study Area to improve your performance in A&P Lab.

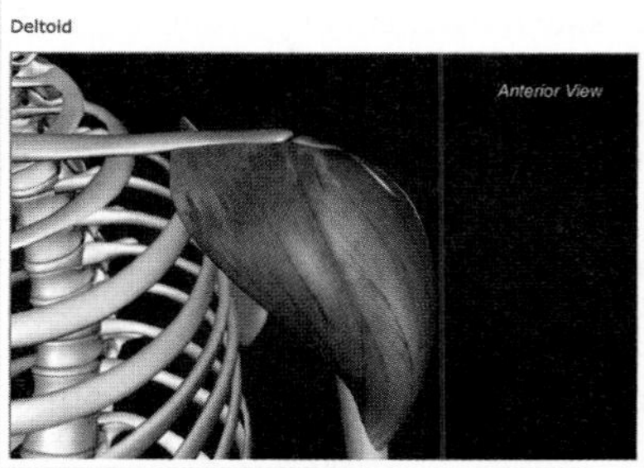

> Animations & Videos > A&P Flix > Group Muscle Actions & Joints

Instructors may assign new Building Vocabulary coaching activities, Pre-Lab Quiz questions, Art Labeling activities, related A&P Flix activities, Practice Anatomy Lab Practical questions (PAL), and more using the Mastering **A&P™** Item Library.

Pre-Lab Quiz

Instructors may assign these and other Pre-Lab Quiz questions using Mastering **A&P™**

1. A prime mover, or ___________, produces a particular type of movement.
   a. agonist **b.** antagonist **c.** fixator **d.** synergist
2. Circle True or False. Muscles of facial expression differ from most skeletal muscles because they usually do not insert into a bone.
3. Muscles that act on the ___________ cause movement at the hip, knee, and foot joints.
   a. lower limb **b.** trunk **c.** upper limb
4. This two-headed muscle bulges when the forearm is flexed. It is the most familiar muscle of the anterior humerus. It is the:
   a. biceps brachii **b.** flexor carpii radialis **c.** extensor digitorum **d.** triceps brachii
5. These abdominal muscles are responsible for giving me my "six-pack." They also stabilize my pelvis when walking. They are the ___________ muscles.
   a. internal intercostals **b.** rectus abdominis **c.** quadriceps **d.** triceps femoris

Materials

- Human torso model or large anatomical chart showing human musculature
- Human cadaver for demonstration (if available)
- Disposable gloves
- Tubes of body (or face) paint
- 1" wide artist's brushes

For instructions on animal dissections, see the dissection exercises (starting on p. 695) in the cat and fetal pig editions of this manual.

Skeletal muscles enable movement. Smiling, frowning, speaking, singing, breathing, dancing, running, and playing a musical instrument are just a few examples. Most often, purposeful movements require the coordinated action of several skeletal muscles.

Classification of Skeletal Muscles

Types of Muscles

Muscles that are most responsible for producing a particular movement are called **prime movers**, or **agonists**. Muscles that oppose or reverse a movement are called **antagonists**. When a prime mover is active, the fibers of the antagonist are stretched and in the relaxed state. Antagonists can be prime movers in their own right. For example, the biceps muscle of the arm (a prime mover of flexion at the elbow) is antagonized by the triceps (a prime mover of extension at the elbow).

Synergists aid the action of agonists either by assisting with the same movement or by reducing undesirable or unnecessary movement. Contraction of a muscle crossing two or more joints would cause movement at all joints spanned if the synergists were not there to stabilize them. For example, the muscles that flex the fingers cross both the wrist and finger joints, but you can make a fist without bending at the wrist because synergist muscles stabilize the wrist joint.

13

Fixators, or fixation muscles, are specialized synergists. They immobilize the origin of a prime mover so that all the tension is exerted at the insertion. Muscles that help maintain posture are fixators; so too are muscles of the back that stabilize or "fix" the scapula during arm movements.

Naming Skeletal Muscles

Remembering the names of the skeletal muscles is a monumental task, but certain clues help. Muscles are named on the basis of the following criteria:

- **Direction of muscle fibers:** Some muscles are named for the direction in which their muscle fibers run with reference to some imaginary line, usually the midline of the body. A muscle with fibers running parallel to that imaginary line will have the term *rectus* (straight) in its name. For example, the rectus abdominis is the straight muscle of the abdomen. Likewise, the terms *transverse* and *oblique* indicate that the muscle fibers run at right angles and obliquely (respectively) to the imaginary line. Muscle structure is determined by fascicle arrangement. The most common patterns of fascicle arrangement are circular, convergent, parallel, and pennate (**Figure 13.1**).
- **Muscle size:** Terms such as *maximus* (largest), *minimus* (smallest), *longus* (long), and *brevis* (short) are often used in naming muscles—as in gluteus maximus and gluteus minimus.
- **Muscle location:** Some muscles are named for the bone with which they are associated. For example, the temporalis muscle overlies the temporal bone.
- **Number of origins:** When the term *biceps*, *triceps*, or *quadriceps* forms part of a muscle name, you can generally assume that the muscle has two, three, or four origins (respectively). For example, the biceps brachii has two origins.
- **Location of the attachments:** For example, the sternocleidomastoid muscle has its origin on the sternum (*sterno*) and clavicle (*cleido*), and inserts on the mastoid process of the temporal bone.
- **Muscle shape:** For example, the deltoid muscle is roughly triangular (*deltoid* = triangle), and the trapezius muscle resembles a trapezoid.
- **Muscle action:** For example, all the adductor muscles of the anterior thigh bring about its adduction, and all the extensor muscles of the wrist extend the hand.

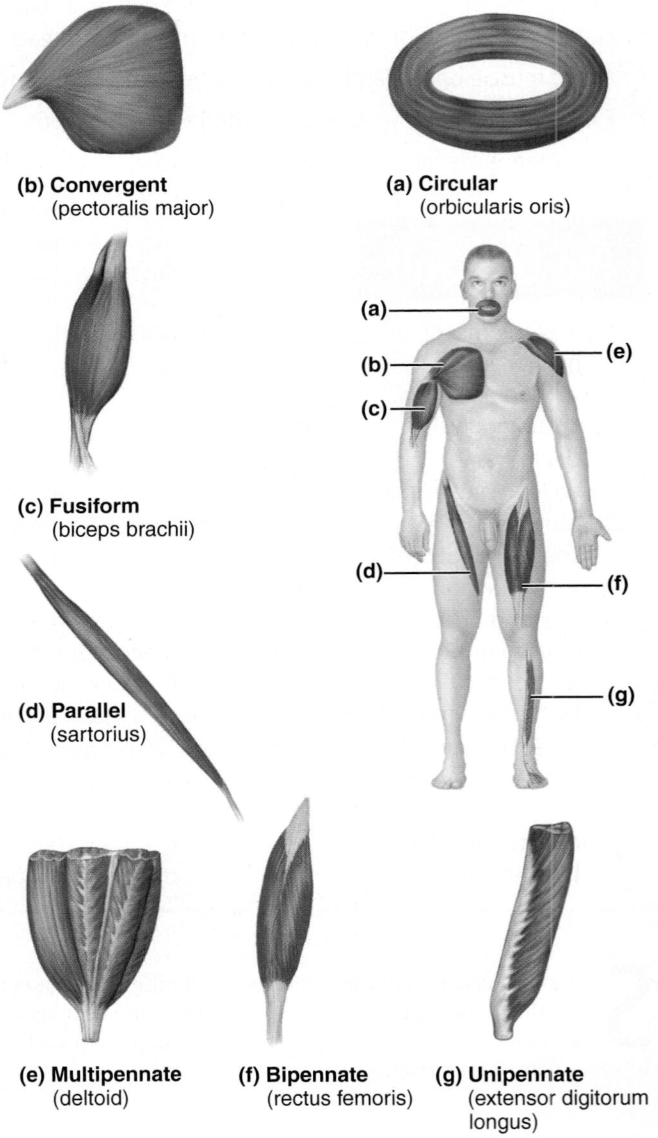

Figure 13.1 Patterns of fascicle arrangement in muscles.

Identification of Human Muscles

While reading the tables and identifying the various human muscles in the figures, try to visualize what happens when the muscle contracts. Since muscles often have many actions, we have indicated the primary action of each muscle in blue type in the tables. Then, use a torso model or an anatomical chart to identify as many muscles as possible. If a human cadaver is available, your instructor will provide specific instructions. Then, carry out the instructions for demonstrating and palpating muscles. **Figures 13.2** and **13.3**, pp. 196–197, are summary figures illustrating the superficial musculature of the body.

Muscles of the Head and Neck

The muscles of the head serve many specific functions. For instance, the muscles of facial expression differ from most skeletal muscles because they insert into the skin or other muscles rather than into bone. As a result, they move the facial skin, allowing a wide range of emotions to be expressed. Other muscles of the head are the muscles of mastication, which move the mandible during chewing, and the six extrinsic eye muscles located within the orbit, which aim the eye. (Orbital muscles are studied in Exercise 23.) Neck muscles provide for the movement of the head and shoulder girdle.

Activity 1

Identifying Head and Neck Muscles

Read the descriptions of specific head and neck muscles and identify the various muscles in the figures (**Tables 13.1** and **13.2**, pp. 198–201, and **Figures 13.4** and **13.5**, pp. 199–201), trying to visualize their action when they contract. Then identify them on a torso model or anatomical chart.

Demonstrating Operations of Head Muscles

1. Raise your eyebrow to wrinkle your forehead. You are using the *frontal belly* of the *epicranius* muscle.
2. Blink your eyes; wink. You are contracting *orbicularis oculi*.
3. Close your lips and pucker up. This requires contraction of *orbicularis oris*.
4. Smile. You are using *zygomaticus*.
5. To demonstrate the *temporalis*, place your hands on your temples and clench your teeth. The *masseter* can also be palpated now at the angle of the jaw.

Muscles of the Trunk

The trunk musculature includes muscles that move the vertebral column; anterior thorax muscles that act to move ribs, head, and arms; and muscles of the abdominal wall that play a role in the movement of the vertebral column but more importantly form the "natural girdle," or the major portion of the abdominal body wall.

Activity 2

Identifying Muscles of the Trunk

Read the descriptions of specific trunk muscles and identify them in the figures (**Table 13.3**, p. 202, and **Table 13.4**, pp. 204–208, and **Figures 13.6–13.10**, pp. 202–208), visualizing their action when they contract. Then identify them on a torso model or anatomical chart.

Demonstrating Operations of Trunk Muscles

Now, work with a partner to demonstrate the operation of the following muscles. One of you can demonstrate the movement; the following steps are addressed to this partner. The other can supply resistance and palpate the muscle being tested.

1. Fully abduct the arm and extend at the elbow. Now adduct the arm against resistance. You are using the *latissimus dorsi*.
2. To observe the action of the *deltoid*, try to abduct your arm against resistance. Now attempt to elevate your shoulder against resistance; you are contracting the upper portion of the *trapezius*.
3. The *pectoralis major* is used when you press your hands together at chest level with your elbows widely abducted.

13

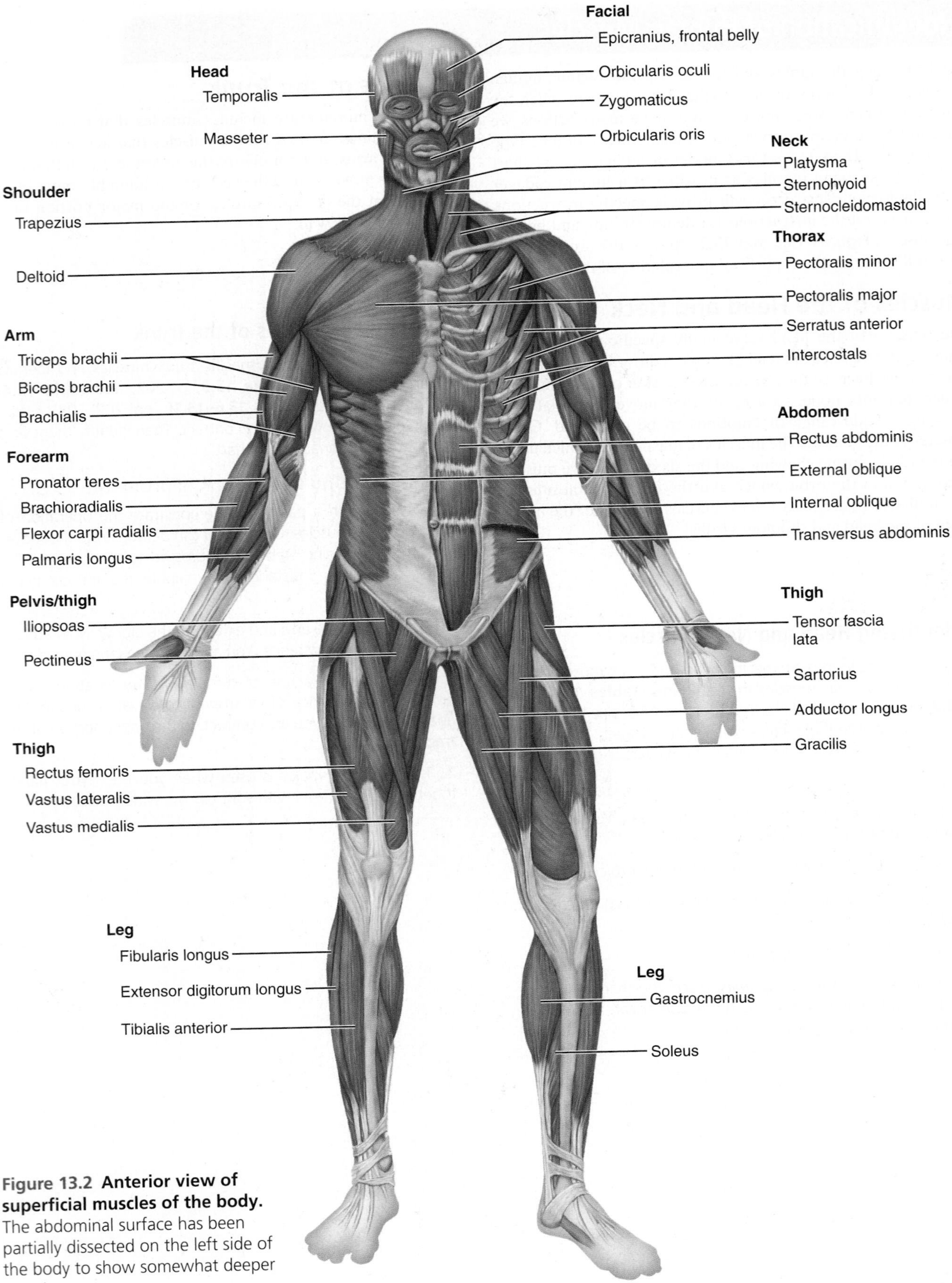

Figure 13.2 Anterior view of superficial muscles of the body. The abdominal surface has been partially dissected on the left side of the body to show somewhat deeper muscles.

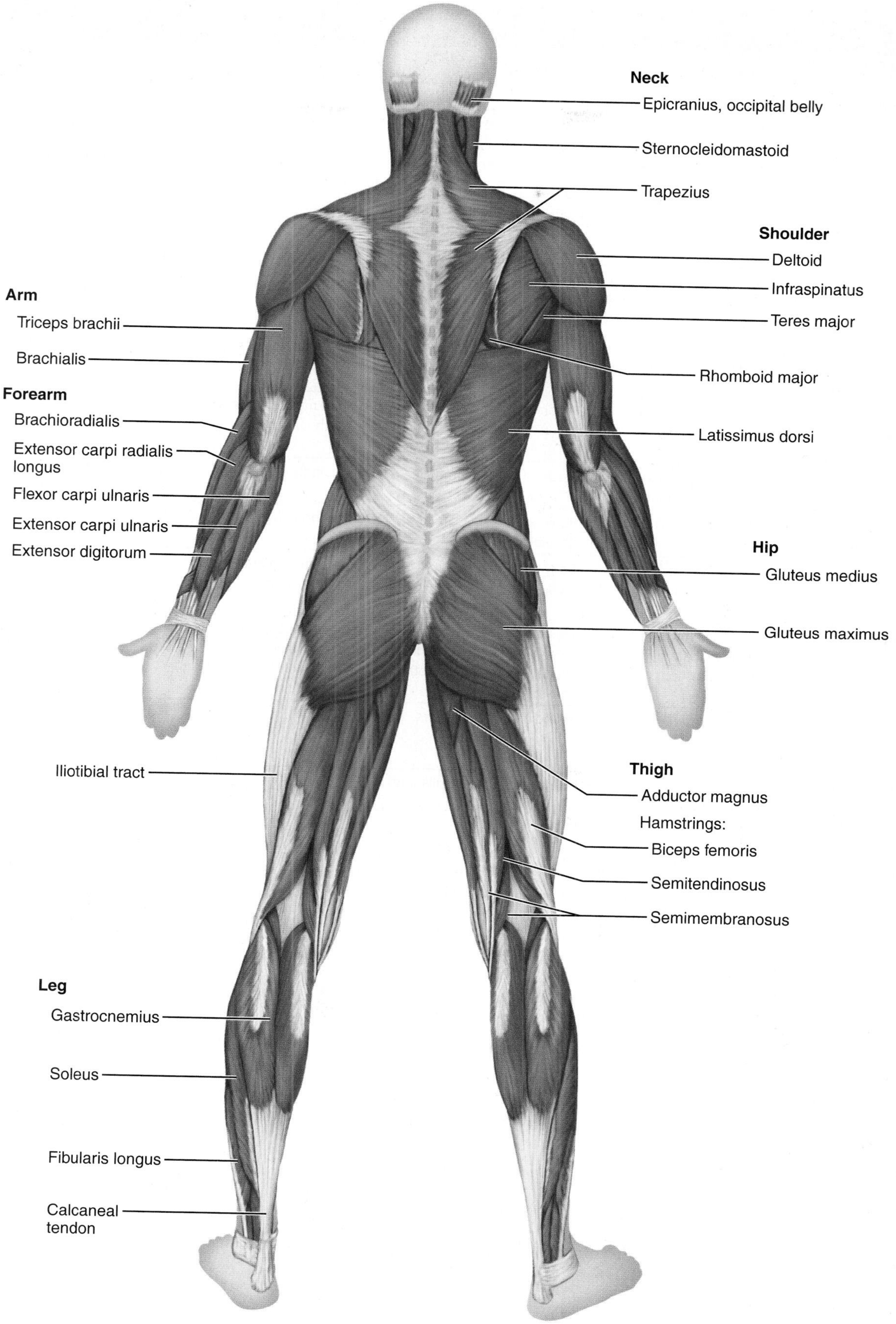

Figure 13.3 Posterior view of superficial muscles of the body.

13

Table 13.1 Major Muscles of the Head (Figure 13.4)

Muscle	Comments	Origin	Insertion	Action
Facial Expression (Figure 13.4a, b)				
Epicranius—frontal and occipital bellies	Bipartite muscle consisting of frontal and occipital bellies, which covers dome of skull	Frontal belly—epicranial aponeurosis; occipital belly—occipital and temporal bones	Frontal belly—skin of eyebrows and root of nose; occipital belly—epicranial aponeurosis	With aponeurosis fixed, frontal belly raises eyebrows; occipital belly fixes aponeurosis and pulls scalp posteriorly
Orbicularis oculi	Tripartite sphincter muscle of eyelids	Frontal and maxillary bones and ligaments around orbit	Tissue of eyelid	Closes eye, produces blinking, squinting, and draws eyebrows inferiorly
Corrugator supercilii	Small muscle; acts with orbicularis oculi	Arch of frontal bone above nasal bone	Skin of eyebrow	Draws eyebrows medially and inferiorly; wrinkles skin of forehead vertically
Levator labii superioris	Thin muscle between orbicularis oris and inferior eye margin	Zygomatic bone and infraorbital margin of maxilla	Skin and muscle of upper lip	Raises and furrows upper lip; opens lips
Zygomaticus—major and minor	Extends diagonally from corner of mouth to cheekbone	Zygomatic bone	Skin and muscle at corner of mouth	Raises lateral corners of mouth upward (smiling muscle)
Risorius	Slender muscle; runs inferior and lateral to zygomaticus	Fascia of masseter muscle	Skin at angle of mouth	Draws corner of lip laterally; tenses lip; zygomaticus synergist
Depressor labii inferioris	Small muscle running from lower lip to mandible	Body of mandible lateral to its midline	Skin and muscle of lower lip	Draws lower lip inferiorly
Depressor anguli oris	Small muscle lateral to depressor labii inferioris	Body of mandible below incisors	Skin and muscle at angle of mouth below insertion of zygomaticus	Zygomaticus antagonist; draws corners of mouth downward and laterally
Orbicularis oris	Multilayered muscle of lips with fibers that run in many different directions; most run circularly	Arises indirectly from maxilla and mandible; fibers blended with fibers of other muscles associated with lips	Encircles mouth; inserts into muscle and skin at angles of mouth	Closes lips; purses and protrudes lips (kissing and whistling muscle)
Mentalis	One of muscle pair forming V-shaped muscle mass on chin	Mandible below incisors	Skin of chin	Protrudes lower lip; wrinkles chin
Buccinator	Principal muscle of cheek; runs horizontally, deep to the masseter	Molar region of maxilla and mandible	Orbicularis oris	Draws corner of mouth laterally; compresses cheek (as in whistling); holds food between teeth during chewing
Platysma	Unpaired, thin, sheetlike superficial neck muscle, plays role in facial expression	Fascia of chest (over pectoral muscles and deltoid)	Lower margin of mandible, skin, and muscle at corner of mouth	Tenses skin of neck; depresses mandible; pulls lower lip back and down

➤

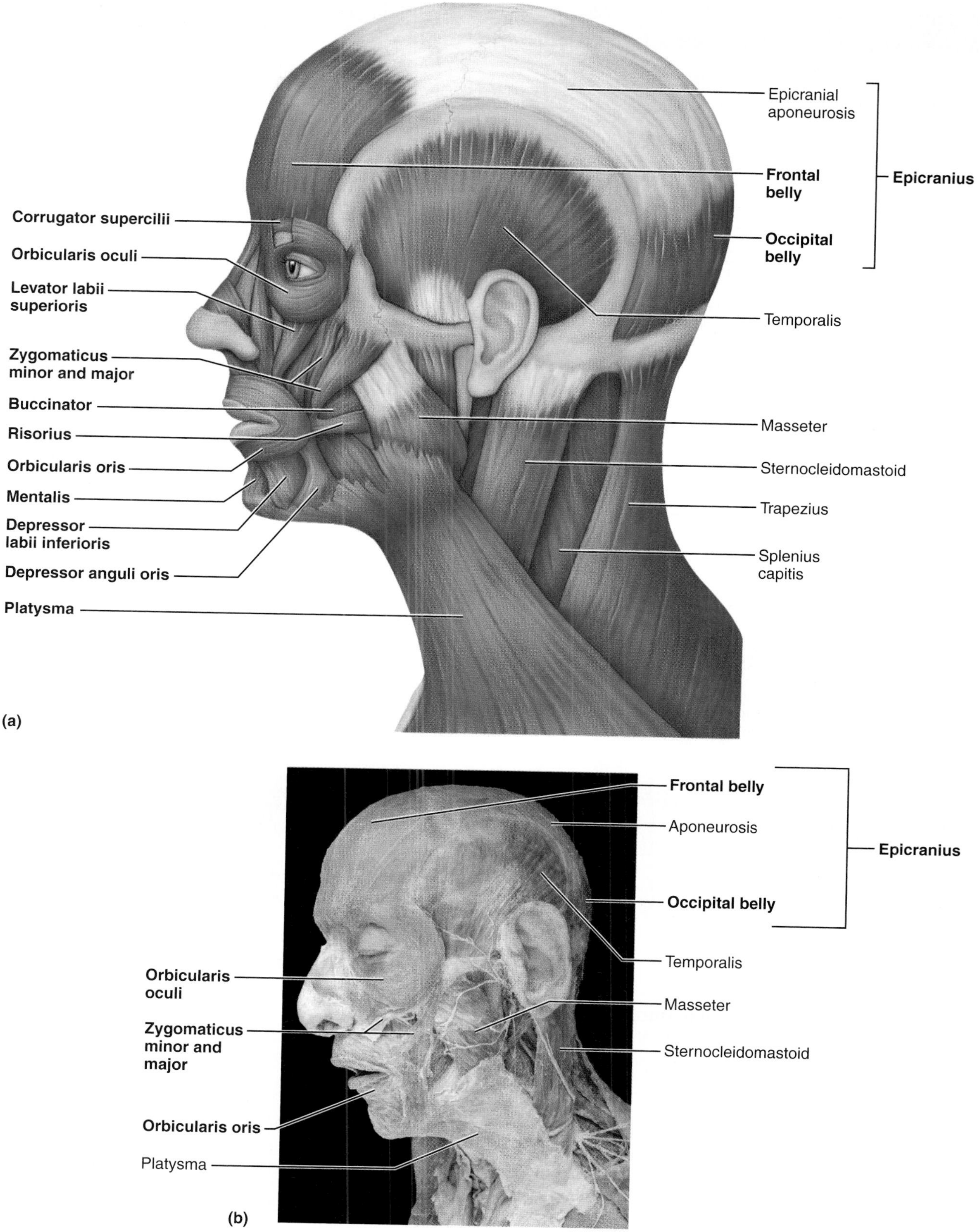

Figure 13.4 Muscles of the scalp, face, and neck (left lateral view). (a) Superficial muscles. **(b)** Photo of superficial structures of head and neck.

Table 13.1 Major Muscles of the Head *(continued)*

Muscle	Comments	Origin	Insertion	Action
Muscles of Mastication (Figure 13.4c, d)				
Masseter	Covers lateral aspect of mandibular ramus; can be palpated on forcible closure of jaws	Zygomatic arch and zygomatic bone	Angle and ramus of mandible	Prime mover of jaw closure; elevates mandible
Temporalis	Fan-shaped muscle lying over parts of frontal, parietal, and temporal bones	Temporal fossa	Coronoid process of mandible	Closes jaw; elevates and retracts mandible
Buccinator	(See muscles of facial expression.)			
Medial pterygoid	Runs along internal (medial) surface of mandible (thus largely concealed by that bone)	Sphenoid, palatine, and maxillary bones	Medial surface of mandible, near its angle	Synergist of temporalis and masseter; elevates mandible; in conjunction with lateral pterygoid, aids in grinding movements of teeth
Lateral pterygoid	Superior to medial pterygoid	Greater wing of sphenoid bone	Condylar process of mandible	Protracts jaw (moves it anteriorly); in conjunction with medial pterygoid, aids in grinding movements of teeth

13

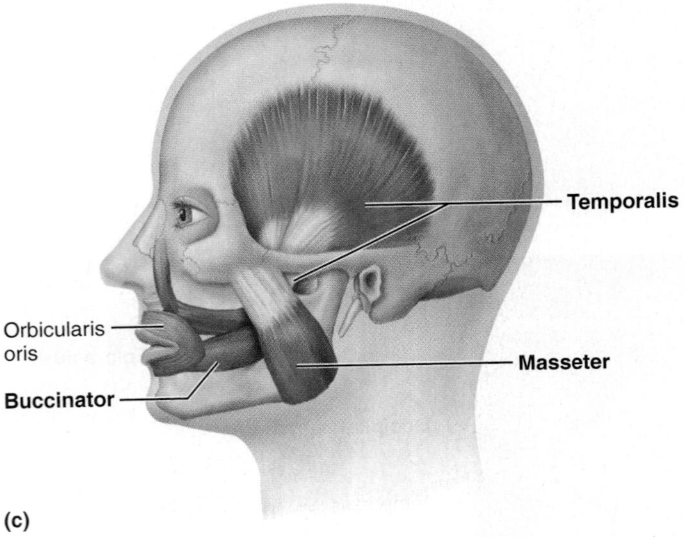

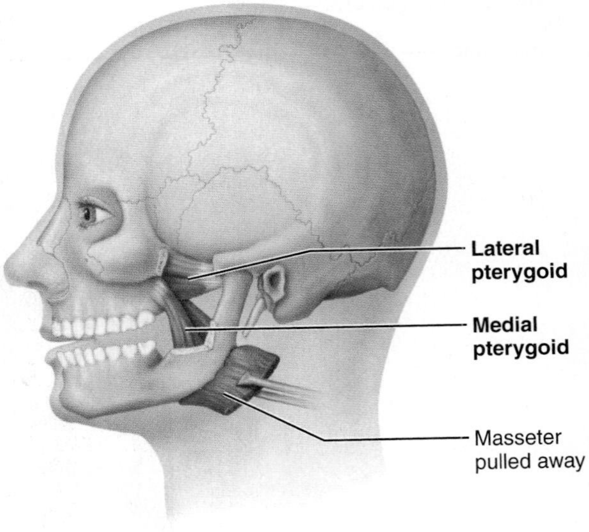

Figure 13.4 *(continued)* **Muscles of mastication. (c)** Lateral view of the temporalis, masseter, and buccinator muscles. **(d)** Lateral view of the deep chewing muscles, the medial and lateral pterygoid muscles.

Table 13.2 Muscles of the Neck and Throat (Figure 13.5)

Muscle	Comments	Origin	Insertion	Action
Suprahyoid Muscles				
Digastric	Consists of two bellies united by an intermediate tendon; forms a V-shape under chin	Lower margin of mandible (anterior belly) and mastoid process (posterior belly)	By a connective tissue loop to hyoid bone	Acting together, elevate hyoid bone; open mouth and depress mandible
Stylohyoid	Slender muscle parallels posterior border of digastric; below angle of jaw	Styloid process of temporal bone	Hyoid bone	Elevates and retracts hyoid bone
Mylohyoid	Just deep to digastric; forms floor of mouth	Medial surface of mandible	Hyoid bone and median raphe	Elevates hyoid bone and base of tongue during swallowing
Infrahyoid Muscles				
Sternohyoid	Runs most medially along neck; straplike	Manubrium and medial end of clavicle	Lower margin of hyoid bone	Acting with sternothyroid and omohyoid, depresses larynx and hyoid bone if mandible is fixed; may also flex skull
Sternothyroid	Lateral and deep to sternohyoid	Posterior surface of manubrium	Thyroid cartilage of larynx	(See Sternohyoid above)
Omohyoid	Straplike with two bellies; lateral to sternohyoid	Superior surface of scapula	Hyoid bone; inferior border	(See Sternohyoid above)
Thyrohyoid	Appears as a superior continuation of sternothyroid muscle	Thyroid cartilage	Hyoid bone	Depresses hyoid bone; elevates larynx if hyoid is fixed

13

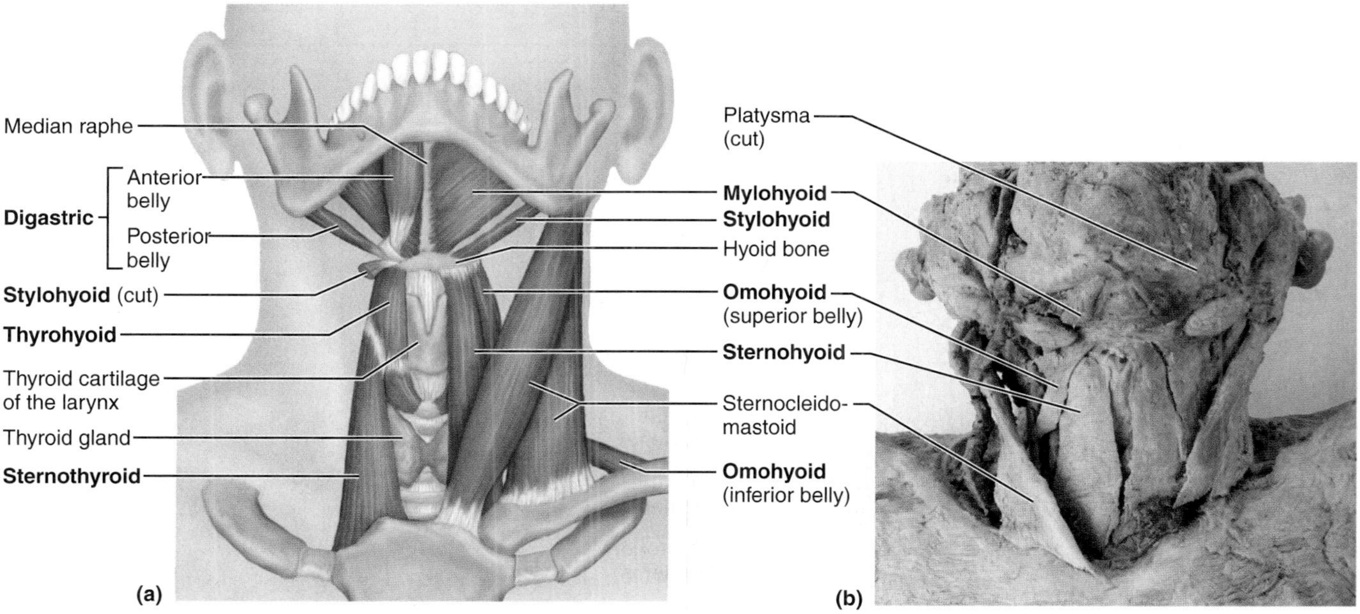

Figure 13.5 Anterior muscles of the neck and throat. (a) Anterior view of suprahyoid and infrahyoid muscles. **(b)** Cadaver photo of suprahyoid and infrahyoid muscles.

13

Table 13.3 Muscles of the Neck and Vertebral Column: Head Movements and Trunk Extension (Figure 13.6)

Muscle	Comments	Origin	Insertion	Action
Sternocleidomastoid	Two-headed muscle located deep to platysma on anterolateral surface of neck; indicate limits of anterior and posterior triangles of neck	Manubrium of sternum and medial portion of clavicle	Mastoid process of temporal bone and superior nuchal line of occipital bone	Simultaneous contraction of both muscles of pair causes flexion of neck, acting independently, rotate head toward shoulder on opposite side
Scalenes—anterior, middle, and posterior	Located more on lateral than anterior neck; deep to platysma and sternocleidomastoid	Transverse processes of cervical vertebrae	Anterolaterally on ribs 1–2	Flex and slightly rotate neck; elevate ribs 1–2 (aid in inspiration)
Splenius	Superficial muscle (capitis and cervicis parts) extending from upper thoracic region to skull	Ligamentum nuchae and spinous processes of C_7–T_6	Mastoid process, occipital bone, and transverse processes of C_2–C_4	As a group, extend or hyperextend head; when only one side is active, head is rotated and bent toward the same side
Erector spinae	A long tripartite muscle composed of iliocostalis (lateral), longissimus, and spinalis (medial) muscle columns; superficial to semispinalis muscles; extends from pelvis to head	Iliac crest, transverse processes of lumbar, thoracic, and cervical vertebrae, and/or ribs 3–6 depending on specific part	Ribs and transverse processes of vertebrae about six segments above origin; longissimus also inserts into mastoid process	Extend and bend the vertebral column laterally; fibers of the longissimus also extend and rotate head
Semispinalis	Deep composite muscle of the back—thoracis, cervicis, and capitis portions	Transverse processes of C_7–T_{12}	Occipital bone and spinous processes of cervical vertebrae and T_1–T_4	Acting together, extend head and vertebral column; acting independently (right vs. left), causes rotation toward the opposite side
Quadratus lumborum	Forms greater portion of posterior abdominal wall	Iliac crest and lumbar fascia	Inferior border of rib 12; transverse processes of lumbar vertebrae	Each flexes vertebral column laterally; together extend the lumbar spine and fix rib 12; maintains upright posture

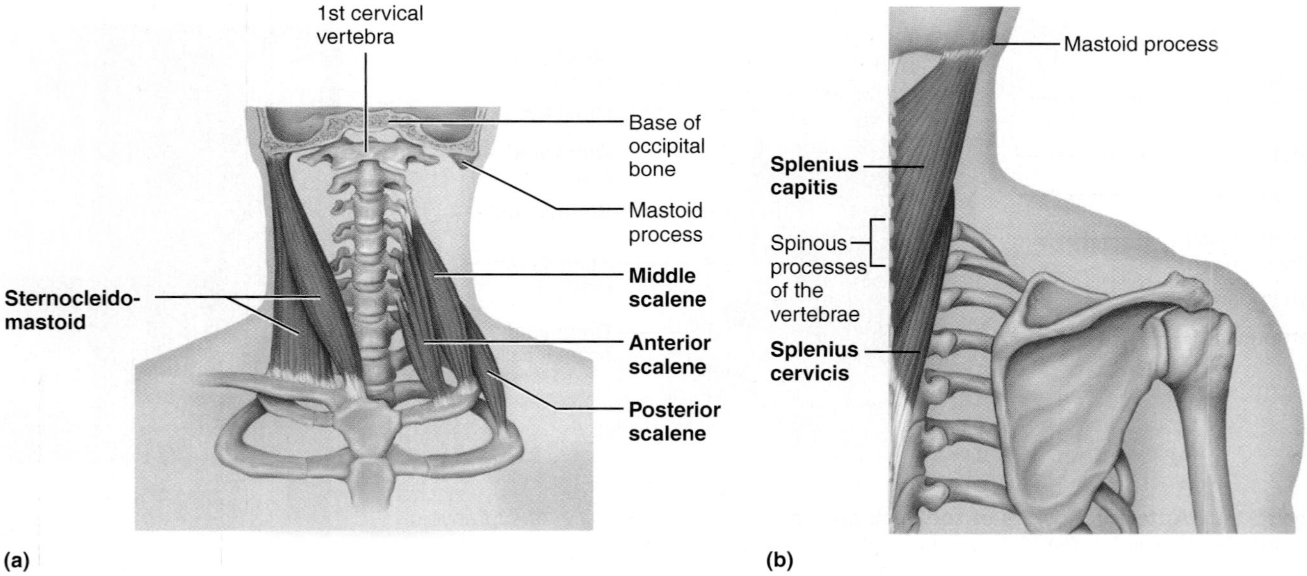

Figure 13.6 Muscles of the neck and vertebral column that move the head and trunk. **(a)** Muscles of the anterolateral neck; superficial platysma muscle and deeper neck muscles removed. **(b)** Deep muscles of the posterior neck; superficial muscles removed.

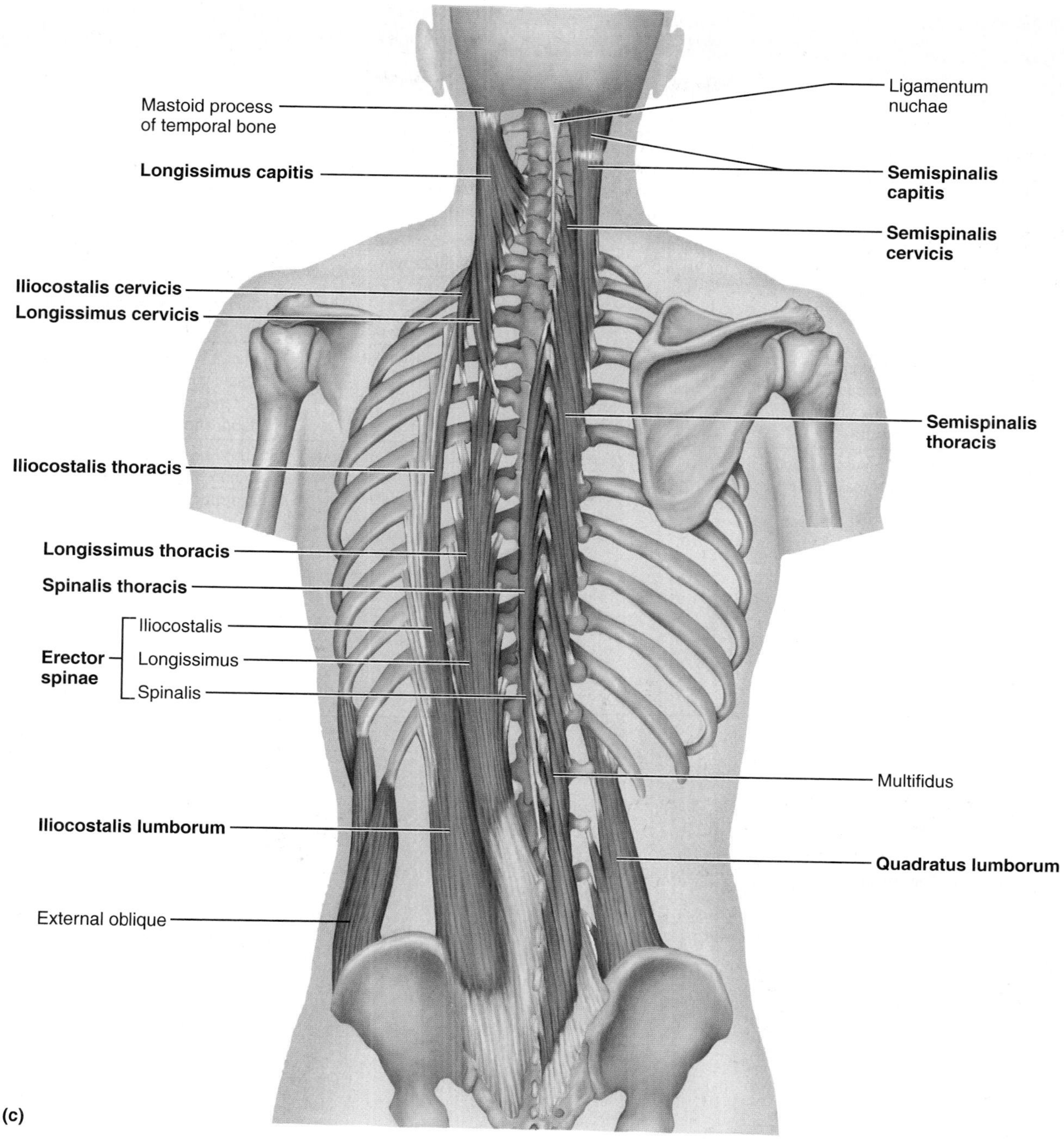

Figure 13.6 *(continued)* **(c)** Deep muscles of the back. The superficial and splenius muscles have been removed. The three muscle columns (iliocostalis, longissimus, and spinalis) forming the erector spinae are shown on the left, and deeper semispinalis and quadratus lumborum are shown on the right.

Table 13.4 Muscles of the Thorax, Shoulder, and Abdominal Wall (Figures 13.7, 13.8, 13.9, and 13.10)

Muscle	Comments	Origin	Insertion	Action
Thorax and Shoulder, Superficial (Figure 13.7)				
Pectoralis major	Large fan-shaped muscle covering upper portion of chest	Clavicle, sternum, cartilage of ribs 1–6 (or 7), and aponeurosis of external oblique muscle	Fibers converge to insert by short tendon into intertubercular sulcus of humerus	Prime mover of arm flexion; adducts, medially rotates arm; with arm fixed, pulls chest upward (thus also acts in forced inspiration)
Serratus anterior	Fan-shaped muscle deep to scapula; deep and inferior to pectoral muscles on lateral rib cage	Lateral aspect of ribs 1–8 (or 9)	Vertebral border of anterior surface of scapula	Prime mover to protract and hold scapula against chest wall; rotates scapula, causing inferior angle to move laterally and upward; essential to raising arm; fixes scapula for arm abduction
Deltoid (see also Figure 13.10)	Fleshy triangular muscle forming shoulder muscle mass; intramuscular injection site	Lateral 1/3 of clavicle; acromion and spine of scapula	Deltoid tuberosity of humerus	Acting as a whole, prime mover of arm abduction; when only specific fibers are active, can act as a synergist in flexion, extension, and rotation of arm
Pectoralis minor	Flat, thin muscle deep to pectoralis major	Anterior surface of ribs 3–5, near their costal cartilages	Coracoid process of scapula	With ribs fixed, draws scapula forward and inferiorly; with scapula fixed, draws rib cage superiorly

13

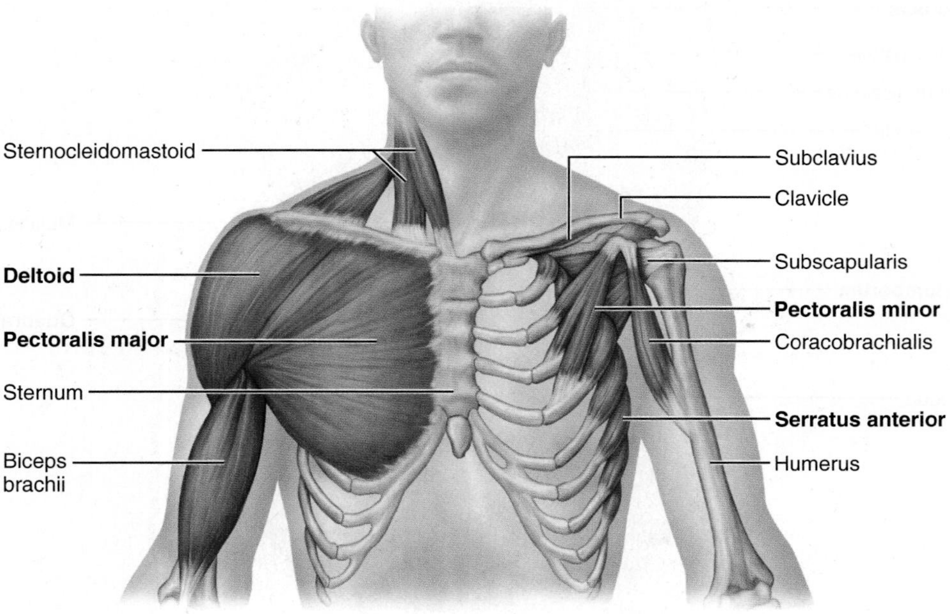

Figure 13.7 Muscles of the thorax and shoulder acting on the scapula and arm (anterior view). The superficial muscles, which effect arm movements, are shown on the left side of the figure. These muscles have been removed on the right side of the figure to show the muscles that stabilize or move the pectoral girdle.

Table 13.4 ***(continued)***

Muscle	Comments	Origin	Insertion	Action
Thorax, Deep: Muscles of Respiration (Figure 13.8)				
External intercostals	11 pairs lie between ribs; fibers run obliquely downward and forward toward sternum	Inferior border of rib above (not shown in figure)	Superior border of rib below	Pull ribs toward one another to elevate rib cage; aid in inspiration
Internal intercostals	11 pairs lie between ribs; fibers run deep and at right angles to those of external intercostals	Superior border of rib below	Inferior border of rib above (not shown in figure)	Draw ribs together to depress rib cage; aid in forced expiration; antagonistic to external intercostals
Diaphragm	Broad muscle; forms floor of thoracic cavity; dome-shaped in relaxed state; fibers converge from margins of thoracic cage toward a central tendon	Inferior border of rib and sternum, costal cartilages of last six ribs and lumbar vertebrae	Central tendon	Prime mover of inspiration flattens on contraction, increasing vertical dimensions of thorax; increases intra-abdominal pressure

➤

13

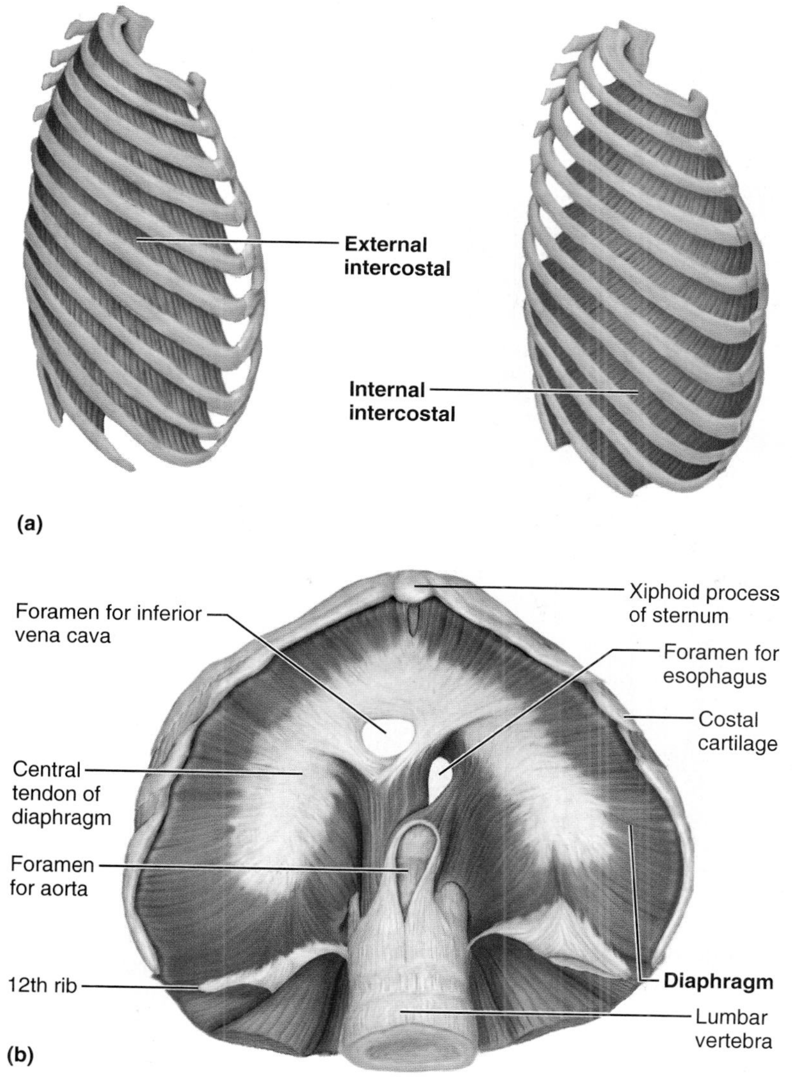

Figure 13.8 Deep muscles of the thorax: muscles of respiration. (a) The external intercostals (inspiratory muscles) are shown on the left, and the internal intercostals (expiratory muscles) are shown on the right. These two muscle layers run obliquely and at right angles to each other. **(b)** Inferior view of the diaphragm, the prime mover of inspiration. Notice that its muscle fibers converge toward a central tendon, an arrangement that causes the diaphragm to flatten and move inferiorly as it contracts. The diaphragm and its tendon are pierced by the great vessels (aorta and inferior vena cava) and the esophagus.

Table 13.4 Muscles of the Thorax, Shoulder, and Abdominal Wall *(continued)*

Muscle	Comments	Origin	Insertion	Action
Abdominal Wall (Figure 13.9)				
Rectus abdominis	Medial superficial muscle, extends from pubis to rib cage; ensheathed by aponeuroses of oblique muscles; segmented	Pubic crest and symphysis	Xiphoid process and costal cartilages of ribs 5–7	Flexes the vertebral column; increases abdominal pressure; fixes and depresses ribs; stabilizes pelvis during walking; used in sit-ups and curls
External oblique	Most superficial lateral muscle; fibers run downward and medially; ensheathed by an aponeurosis	Anterior surface of lower eight ribs	Linea alba,* pubic crest and tubercles, and iliac crest	See rectus abdominis, above; compresses abdominal wall; also aids muscles of back in trunk rotation and lateral flexion; used in oblique curls
Internal oblique	Most fibers run at right angles to those of external oblique, which it underlies	Lumbar fascia, iliac crest, and inguinal ligament	Linea alba, pubic crest, and costal cartilages of last three ribs	As for external oblique
Transversus abdominis	Deepest muscle of abdominal wall; fibers run horizontally	Inguinal ligament, iliac crest, cartilages of last five or six ribs, and lumbar fascia	Linea alba and pubic crest	Compresses abdominal contents

*The linea alba (white line) is a narrow, tendinous sheath that runs along the middle of the abdomen from the sternum to the pubic symphysis. It is formed by the fusion of the aponeurosis of the external oblique and transversus muscles.

13

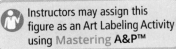

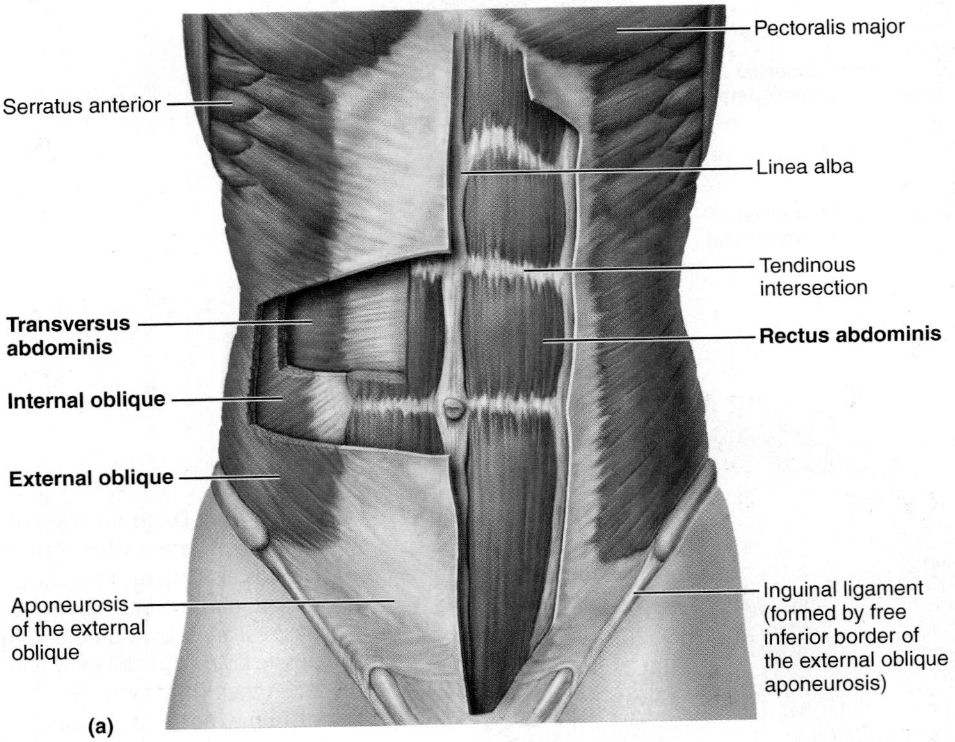

Figure 13.9 Anterior view of the muscles forming the anterolateral abdominal wall. **(a)** The superficial muscles have been partially cut away on the left side of the diagram to reveal the deeper internal oblique and transversus abdominis muscles.

Instructors may assign this figure as an Art Labeling Activity using Mastering **A&P™**

Table 13.4 *(continued)*

Muscle	Comments	Origin	Insertion	Action
Muscles of the Shoulder and Thorax (Figure 13.10)				
Trapezius	Most superficial muscle of posterior thorax; very broad origin and insertion	Occipital bone; ligamentum nuchae; spines of C_7 and all thoracic vertebrae	Acromion and spinous process of scapula; lateral third of clavicle	Extends head; raises, rotates, and retracts (adducts) scapula and stabilizes it; superior fibers elevate scapula (as in shrugging the shoulders); inferior fibers depress scapula
Latissimus dorsi	Broad flat muscle of lower back (lumbar region); extensive superficial origins	Indirect attachment to spinous processes of lower six thoracic vertebrae, lumbar vertebrae, last three to four ribs, and iliac crest	Floor of intertubercular sulcus of humerus	Prime mover of arm extension; adducts and medially rotates arm; brings arm down in power stroke, as in striking a blow
Infraspinatus	A rotator cuff muscle; partially covered by deltoid and trapezius	Infraspinous fossa of scapula	Greater tubercle of humerus	Lateral rotation of arm; helps hold head of humerus in glenoid cavity; stabilizes shoulder
Teres minor	A rotator cuff muscle; small muscle inferior to infraspinatus	Lateral margin of posterior scapula	Greater tubercle of humerus	Same as for infraspinatus

➤

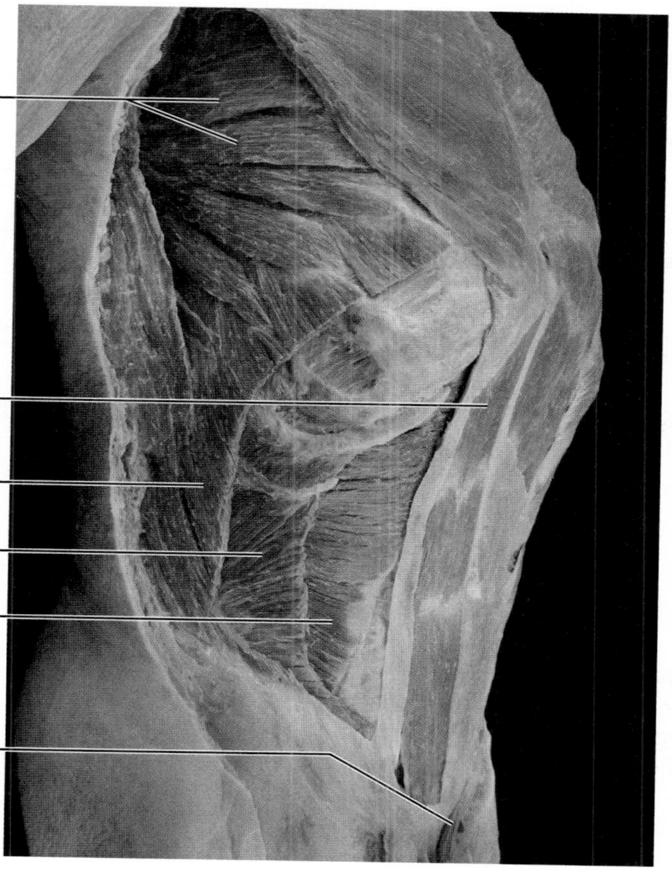

Figure 13.9 *(continued)* **(b)** Photo of the anterolateral abdominal wall.

Table 13.4 Muscles of the Thorax, Shoulder, and Abdominal Wall *(continued)*

Muscle	Comments	Origin	Insertion	Action
Teres major	Located inferiorly to teres minor	Posterior surface at inferior angle of scapula	Intertubercular sulcus of humerus	Extends, medially rotates, and adducts arm; synergist of latissimus dorsi
Supraspinatus	A rotator cuff muscle; obscured by trapezius	Supraspinous fossa of scapula	Greater tubercle of humerus	Initiates abduction of arm; stabilizes shoulder joint
Levator scapulae	Located at back and side of neck, deep to trapezius	Transverse processes of C_1–C_4	Medial border of scapula superior to spine	Elevates and adducts scapula; with fixed scapula, laterally flexes neck to the same side
Rhomboids—major and minor	Beneath trapezius and inferior to levator scapulae; rhomboid minor is the more superior muscle	Spinous processes of C_7 and T_1–T_5	Medial border of scapula	Pull scapulae medially (retraction); stabilize scapulae; rotate glenoid cavity downward

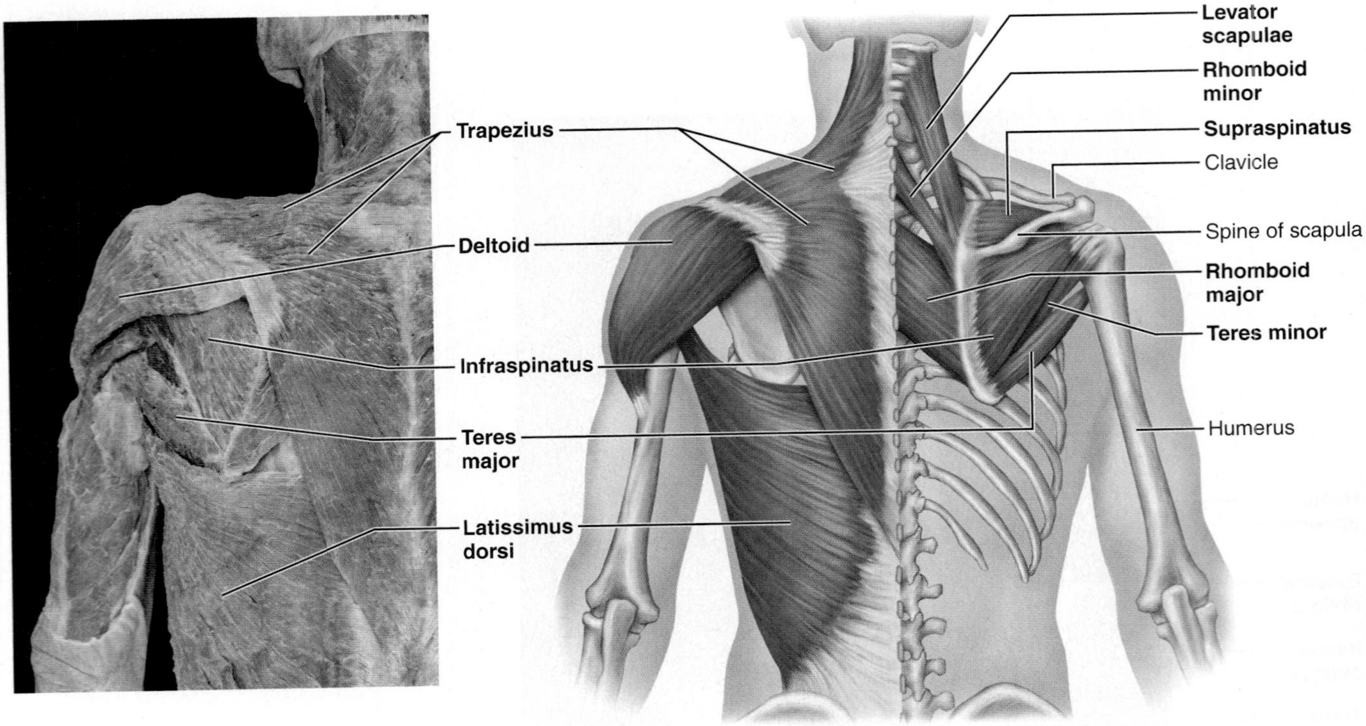

Figure 13.10 Muscles of the shoulder and thorax (posterior view). The superficial muscles of the back are shown for the left side of the body, with a corresponding photograph. The superficial muscles are removed on the right side of the illustration to reveal the deeper muscles acting on the scapula and the rotator cuff muscles that help to stabilize the shoulder joint.

Muscles of the Upper Limb

The muscles that act on the upper limb fall into three groups: those moving the arm, those causing movement of the forearm, and those moving the hand and fingers.

The muscles that cross the shoulder joint to insert on the humerus and move the arm (subscapularis, supraspinatus and infraspinatus, deltoid, and so on) are primarily trunk muscles that originate on the axial skeleton or shoulder girdle. These muscles are included with the trunk muscles.

The second group of muscles, which cross the elbow joint and move the forearm, consists of muscles forming the musculature of the humerus. These muscles arise mainly from the humerus and insert in forearm bones. They are responsible for flexion, extension, pronation, and supination.

The third group forms the musculature of the forearm. For the most part, these muscles cross the wrist to insert on the digits and produce movements of the hand and fingers.

Activity 3

Identifying Muscles of the Upper Limb

Study the origins, insertions, and actions of muscles that move the forearm, and identify them (**Table 13.5**, p. 210, and **Figure 13.11**, p. 210).

Do the same for muscles acting on the wrist and hand (**Table 13.6**, pp. 211–212, and **Figure 13.12**, pp. 211–213). They are more easily identified if you locate their insertion tendons first.

Then see if you can identify the upper limb muscles on a torso model, anatomical chart, or cadaver. Complete this portion of the exercise with palpation demonstrations as outlined next.

Demonstrating Operations of Upper Limb Muscles

1. To observe the *biceps brachii*, attempt to flex your forearm (hand supinated) against resistance. The insertion tendon of this biceps muscle can also be felt in the lateral aspect of the cubital fossa (where it runs toward the radius to attach).

2. If you acutely flex at your elbow and then try to extend the forearm against resistance, you can demonstrate the action of your *triceps brachii*.

3. Strongly flex your hand, and make a fist. Palpate your contracting wrist flexor muscles' origins at the medial epicondyle of the humerus and their insertion tendons, at the anterior aspect of the wrist.

4. Flare your fingers to identify the tendons of the *extensor digitorum* muscle on the dorsum of your hand.

Table 13.5 Muscles of the Humerus That Move the Forearm (Figure 13.11)

Muscle	Comments	Origin	Insertion	Action
Triceps brachii	Large fleshy muscle of posterior humerus; three-headed origin	Long head—inferior margin of glenoid cavity; lateral head—posterior humerus; medial head—distal radial groove on posterior humerus	Olecranon of ulna	Powerful forearm extensor; antagonist of forearm flexors (brachialis and biceps brachii)
Anconeus	Short triangular muscle blended with triceps	Lateral epicondyle of humerus	Lateral aspect of olecranon of ulna	Abducts ulna during forearm pronation; synergist of triceps brachii in forearm extension
Biceps brachii	Most familiar muscle of anterior humerus because this two-headed muscle bulges when forearm is flexed	Short head: coracoid process; long head: supraglenoid tubercle and lip of glenoid cavity; tendon of long head runs in intertubercular sulcus and within capsule of shoulder joint	Radial tuberosity	Flexion (powerful) and supination of forearm; "it turns the corkscrew and pulls the cork"; weak arm flexor
Brachioradialis	Superficial muscle of lateral forearm; forms lateral boundary of cubital fossa	Lateral ridge at distal end of humerus	Base of radial styloid process	Synergist in forearm flexion
Brachialis	Immediately deep to biceps brachii	Distal portion of anterior humerus	Coronoid process of ulna	Flexor of forearm

13

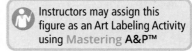

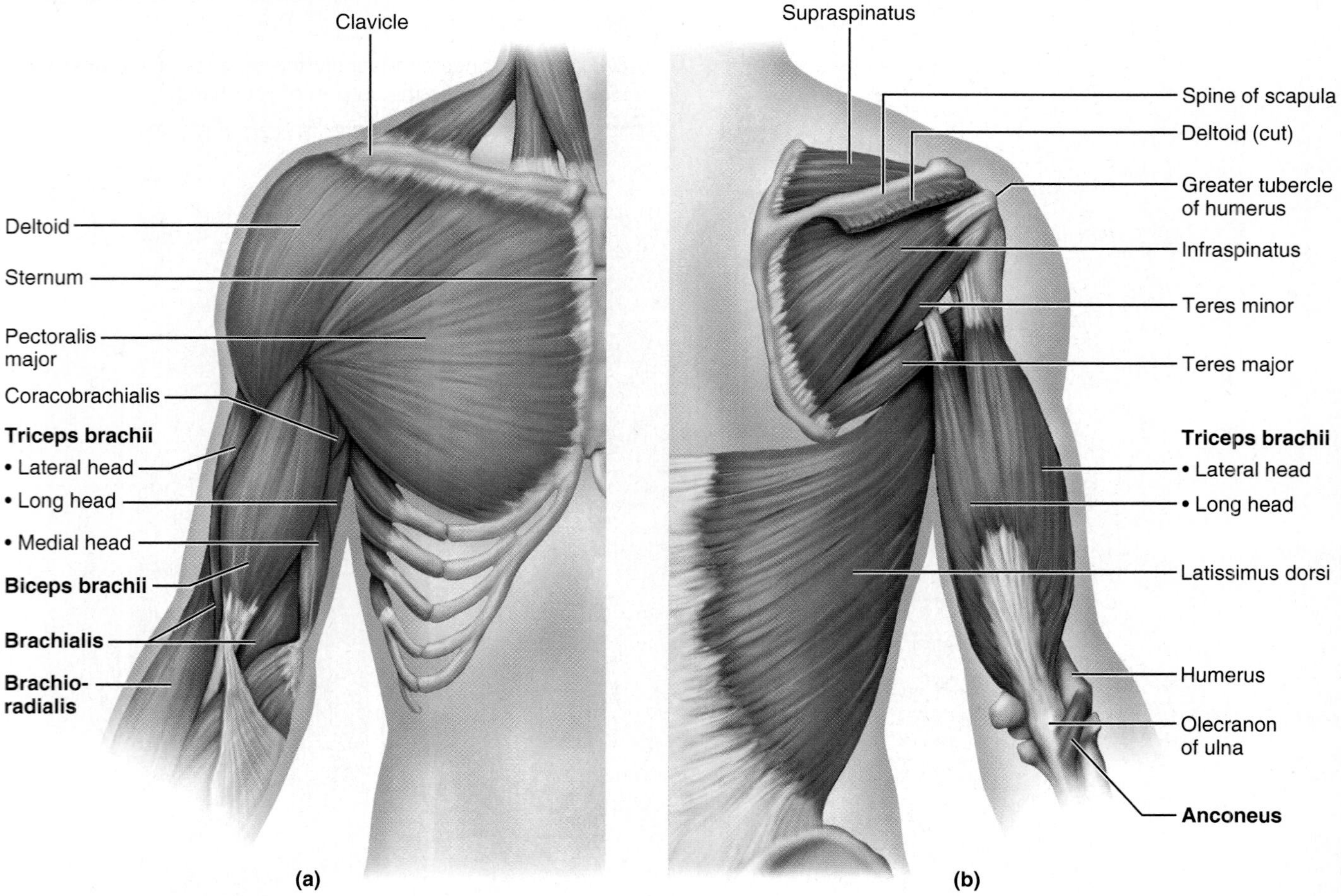

Figure 13.11 Muscles acting on the arm and forearm. (a) Superficial muscles of the anterior thorax, shoulder, and arm, anterior view. **(b)** Posterior aspect of the arm showing the lateral and long heads of the triceps brachii muscle.

Instructors may assign this figure as an Art Labeling Activity using Mastering A&P™

Table 13.6 Muscles of the Forearm That Move the Hand and Fingers (Figure 13.12)

Muscle	Comments	Origin	Insertion	Action
Anterior Compartment, Superficial (Figure 13.12a, b, c)				
Pronator teres	Seen in a superficial view between proximal margins of brachioradialis and flexor carpi radialis	Medial epicondyle of humerus and coronoid process of ulna	Midshaft of radius	Acts synergistically with pronator quadratus to pronate forearm; weak forearm flexor
Flexor carpi radialis	Superficial; runs diagonally across forearm	Medial epicondyle of humerus	Base of metacarpals II and III	Powerful flexor and abductor of the hand
Palmaris longus	Small fleshy muscle with a long tendon; medial to flexor carpi radialis	Medial epicondyle of humerus	Palmar aponeurosis; skin and fascia of palm	Flexes hand (weak); tenses skin and fascia of palm

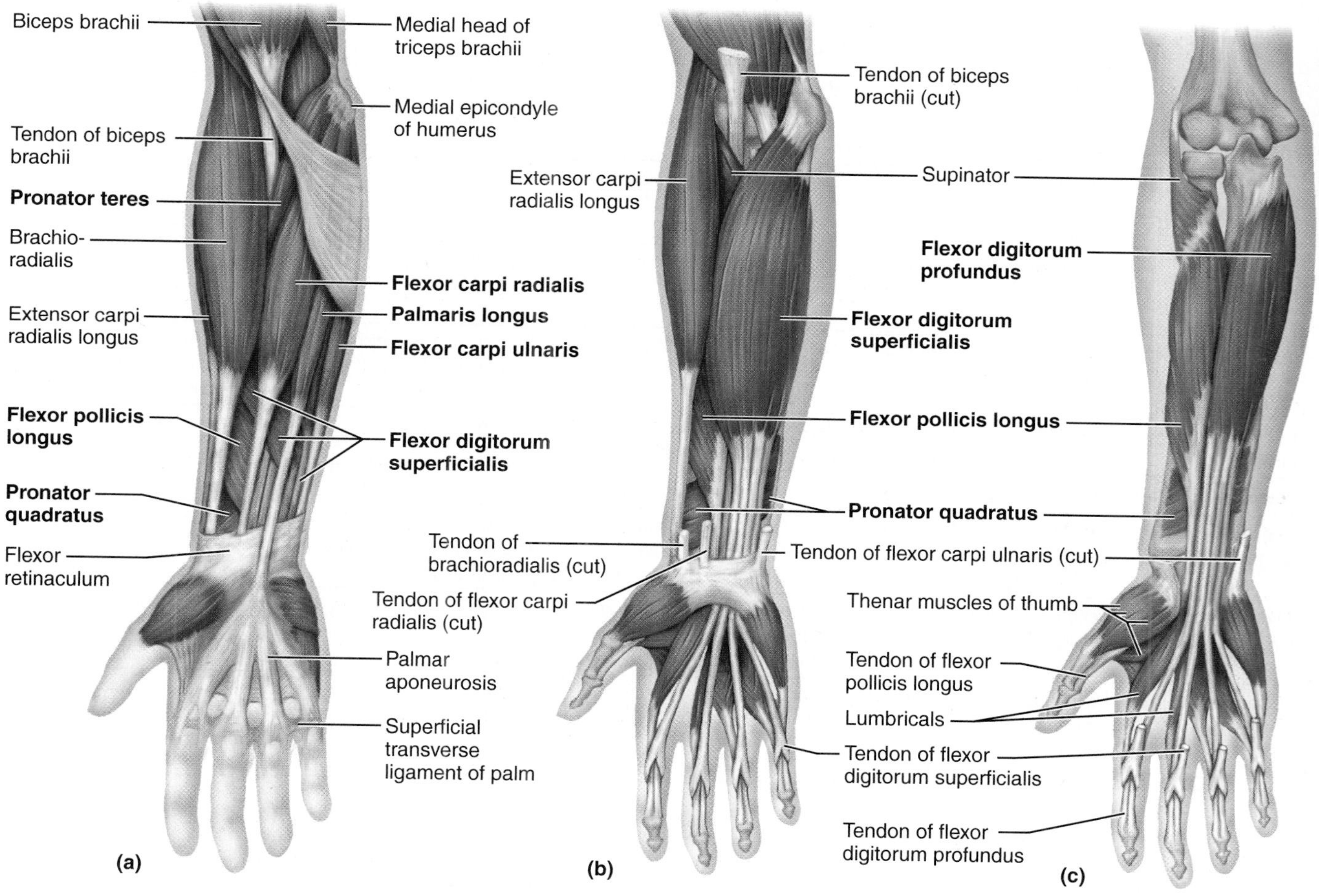

Figure 13.12 Muscles of the forearm and wrist. (a) Superficial anterior view of right forearm and hand. **(b)** The brachioradialis, flexors carpi radialis and ulnaris, and palmaris longus muscles have been removed to reveal the position of the somewhat deeper flexor digitorum superficialis. **(c)** Deep muscles of the anterior compartment. Superficial muscles have been removed. (*Note:* The thenar muscles of the thumb and the lumbricals that help move the fingers are illustrated here but are not described in Table 13.6.)

13

Table 13.6 Muscles of the Forearm That Move the Hand and Fingers *(continued)*

Muscle	Comments	Origin	Insertion	Action
Anterior Compartment, Superficial *(continued)*				
Flexor carpi ulnaris	Superficial; medial to palmaris longus	Medial epicondyle of humerus and olecranon and posterior surface of ulna	Base of metacarpal V; pisiform and hamate bones	Flexes and adducts hand
Flexor digitorum superficialis	Deeper muscle (deep to muscles named above); visible at distal end of forearm	Medial epicondyle of humerus, coronoid process of ulna, and shaft of radius	Middle phalanges of fingers II–V	Flexes hand and middle phalanges of fingers II–V
Anterior Compartment, Deep (Figure 13.12a, b, c)				
Flexor pollicis longus	Deep muscle of anterior forearm; distal to and paralleling lower margin of flexor digitorum superficialis	Anterior surface of radius, and interosseous membrane	Distal phalanx of thumb	Flexes thumb (*pollex* is Latin for "thumb")
Flexor digitorum profundus	Deep muscle; overlain entirely by flexor digitorum superficialis	Anteromedial surface of ulna, interosseous membrane, and coronoid process	Distal phalanges of fingers II–V	Sole muscle that flexes distal phalanges; assists in hand flexion
Pronator quadratus	Deepest muscle of distal forearm	Distal portion of anterior ulnar surface	Anterior surface of radius, distal end	Pronates forearm
Posterior Compartment, Superficial (Figure 13.12d, e, f)				
Extensor carpi radialis longus	Superficial; parallels brachioradialis on lateral forearm	Lateral supracondylar ridge of humerus	Base of metacarpal II	Extends and abducts hand
Extensor carpi radialis brevis	Deep to extensor carpi radialis longus	Lateral epicondyle of humerus	Base of metacarpal III	Extends and abducts hand; steadies wrist during finger flexion
Extensor digitorum	Superficial; medial to extensor carpi radialis brevis	Lateral epicondyle of humerus	By four tendons into distal phalanges of fingers II–V	Prime mover of finger extension; extends hand; can flare (abduct) fingers
Extensor carpi ulnaris	Superficial; medial posterior forearm	Lateral epicondyle of humerus; posterior border of ulna	Base of metacarpal V	Extends and adducts hand
Posterior Compartment, Deep (Figure 13.12d, e, f)				
Extensor pollicis longus and brevis	Muscle pair with a common origin and action; deep to extensor carpi ulnaris	Dorsal shaft of ulna and radius, interosseous membrane	Base of distal phalanx of thumb (longus) and proximal phalanx of thumb (brevis)	Extend thumb
Abductor pollicis longus	Deep muscle; lateral and parallel to extensor pollicis longus	Posterior surface of radius and ulna; interosseous membrane	Metacarpal I and trapezium	Abducts and extends thumb
Supinator	Deep muscle at posterior aspect of elbow	Lateral epicondyle of humerus; proximal ulna	Proximal end of radius	Synergist of biceps brachii to supinate forearm; antagonist of pronator muscles

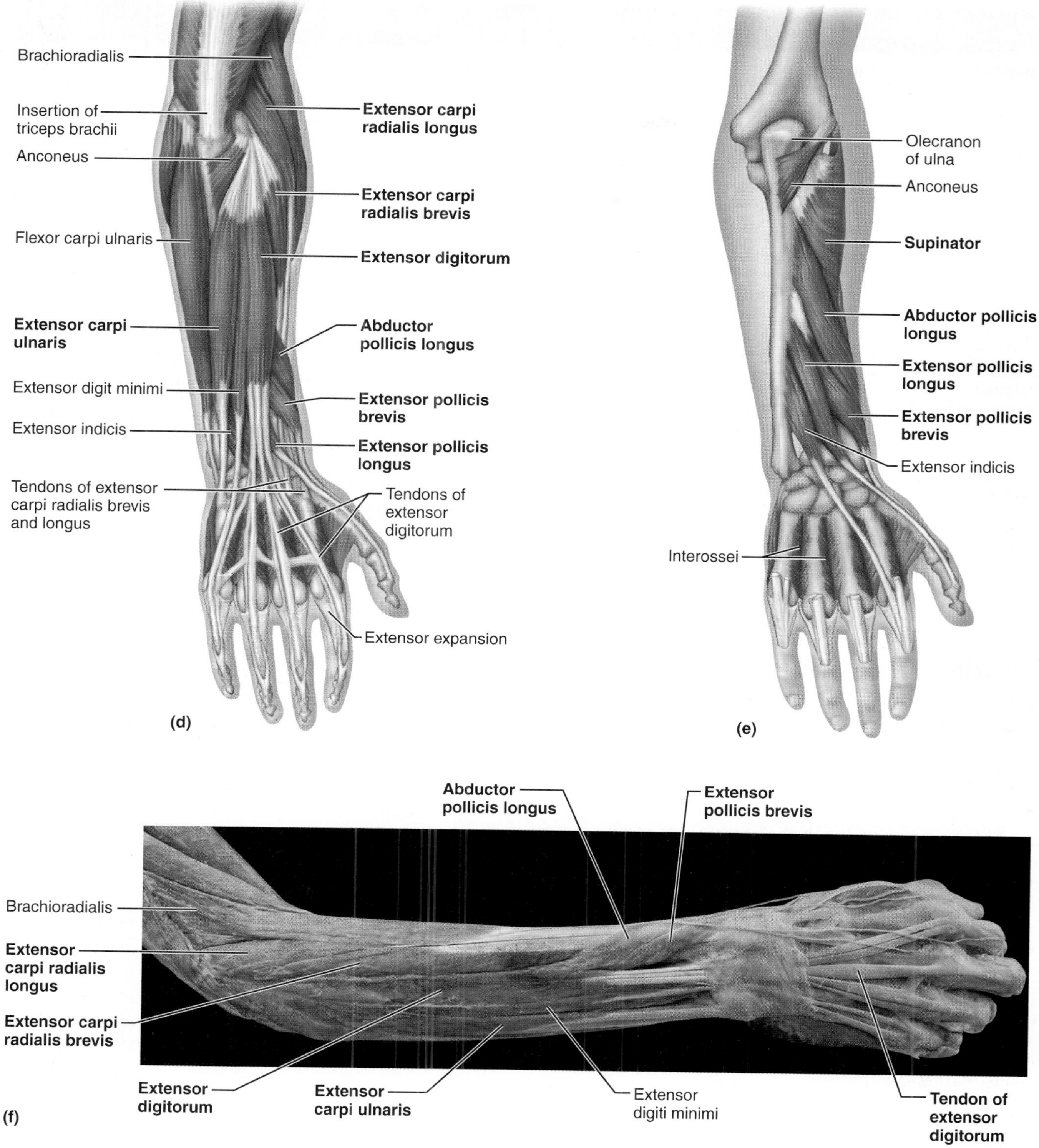

Figure 13.12 *(continued)* **Muscles of the forearm and wrist. (d)** Superficial muscles, posterior view. **(e)** Deep posterior muscles; superficial muscles have been removed. The interossei, the deepest layer of intrinsic hand muscles, are also illustrated. **(f)** Photo of posterior muscles of the right forearm.

Table 13.7 Muscles That Move the Thigh and Leg, Anterior and Medial Aspects (Figure 13.13)

Muscle	Comments	Origin	Insertion	Action
Origin on the Pelvis				
Iliopsoas—iliacus and psoas major	Two closely related muscles; fibers pass under inguinal ligament to insert into femur via a common tendon; iliacus is more lateral	Iliacus—iliac fossa and crest, lateral sacrum; psoas major—transverse processes, bodies, and discs of T_{12} and lumbar vertebrae	On and just below lesser trochanter of femur	Flex trunk at hip joint, flex thigh; lateral flexion of vertebral column (psoas)
Sartorius	Straplike superficial muscle running obliquely across anterior surface of thigh to knee	Anterior superior iliac spine	By an aponeurosis into medial aspect of proximal tibia	Flexes, abducts, and laterally rotates thigh; flexes leg; known as "tailor's muscle" because it helps effect cross-legged position in which tailors are often depicted
Medial Compartment				
Adductors—magnus, longus, and brevis	Large muscle mass forming medial aspect of thigh; arise from front of pelvis and insert at various levels on femur	Magnus—ischial and pubic rami and ischial tuberosity; longus—pubis near pubic symphysis; brevis—body and inferior pubic ramus	Magnus—linea aspera and adductor tubercle of femur; longus and brevis—linea aspera	Adduct and medially rotate and flex thigh; posterior part of magnus is also a synergist in thigh extension
Pectineus	Overlies adductor brevis on proximal thigh	Pectineal line of pubis (and superior pubic ramus)	Inferior from lesser trochanter to linea aspera of femur	Adducts, flexes, and medially rotates thigh
Gracilis	Straplike superficial muscle of medial thigh	Inferior ramus and body of pubis	Medial surface of tibia just inferior to medial condyle	Adducts thigh; flexes and medially rotates leg, especially during walking

➤

13

Muscles of the Lower Limb

Muscles that act on the lower limb cause movement at the hip, knee, and ankle joints. Since the human pelvic girdle is composed of heavy, fused bones that allow very little movement, no special group of muscles is necessary to stabilize it. This is unlike the shoulder girdle, where several muscles (mainly trunk muscles) are needed to stabilize the scapulae.

Muscles acting on the thigh (femur) cause various movements at the multiaxial hip joint. These include the iliopsoas, the adductor group, and others.

Muscles acting on the leg form the major musculature of the thigh. (Anatomically, the term *leg* refers only to that portion between the knee and the ankle.) The thigh muscles cross the knee to allow flexion and extension of the leg. They include the hamstrings and the quadriceps.

The muscles originating on the leg cross the ankle joint and act on the foot and toes.

Activity 4

Identifying Muscles of the Lower Limb

Read the descriptions of specific muscles acting on the thigh and leg and identify them (**Table 13.7**, p. 216, and **Table 13.8**, pp. 216–217, and **Figures 13.13** and **13.14**, pp. 215–217), trying to visualize their action when they contract.

Since some of the muscles acting on the leg also have attachments on the pelvic girdle, they can cause movement at the hip joint.

Do the same for muscles acting on the foot and toes (**Table 13.9**, pp. 218–220, **Figure 13.15**, p. 219, and **Figure 13.16**, pp. 221–222).

Then identify all the muscles on a model or anatomical chart.

Demonstrating Operations of Lower Limb Muscles

1. Go into a deep knee bend and palpate your own *gluteus maximus* muscle as you extend your thigh to resume the upright posture.

2. Demonstrate the contraction of the anterior *quadriceps femoris* by trying to extend your leg against resistance. Do this while seated and note how the quadriceps tendon reacts. The *biceps femoris* of the posterior thigh comes into play when you flex your leg against resistance.

3. Now stand on your toes. Have your partner palpate the lateral and medial heads of the *gastrocnemius* and follow it to its insertion in the calcaneal tendon.

4. Dorsiflex and invert your foot while palpating your *tibialis anterior* muscle (which parallels the sharp anterior crest of the tibia laterally).

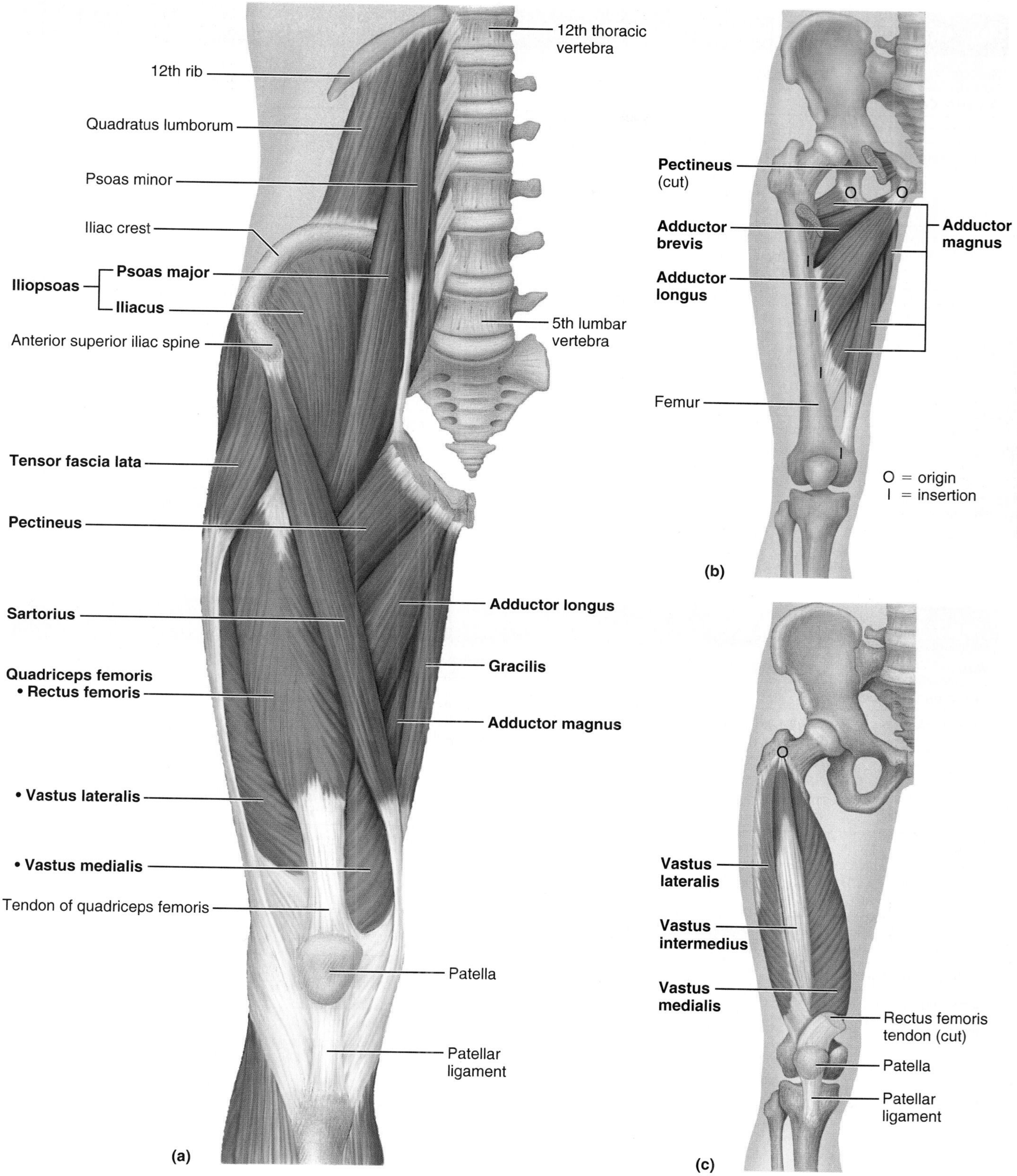

Figure 13.13 Anterior and medial muscles promoting movements of the thigh and leg. (a) Anterior view of the deep muscles of the pelvis and superficial muscles of the right thigh. **(b)** Adductor muscles of the medial compartment of the thigh. **(c)** The vastus muscles (isolated) of the quadriceps group.

Instructors may assign this figure as an Art Labeling Activity using Mastering A&P™

13

13

Table 13.7 Muscles That Move the Thigh and Leg, Anterior and Medial Aspects *(continued)*

Muscle	Comments	Origin	Insertion	Action
Anterior Compartment				
Quadriceps femoris*				
Rectus femoris	Superficial muscle of thigh; runs straight down thigh; only muscle of group to cross hip joint	Anterior inferior iliac spine and superior margin of acetabulum	Tibial tuberosity and patella	Extends leg and flexes thigh
Vastus lateralis	Forms lateral aspect of thigh; intramuscular injection site	Greater trochanter, intertrochanteric line, and linea aspera	Tibial tuberosity and patella	Extends leg and stabilizes knee
Vastus medialis	Forms inferomedial aspect of thigh	Linea aspera and intertrochanteric line	Tibial tuberosity and patella	Extends leg; stabilizes patella
Vastus intermedius	Obscured by rectus femoris; lies between vastus lateralis and vastus medialis on anterior thigh	Anterior and lateral surface of femur	Tibial tuberosity and patella	Extends leg
Tensor fascia lata	Enclosed between fascia layers of thigh	Anterior aspect of iliac crest and anterior superior iliac spine	Iliotibial tract (lateral portion of fascia lata)	Flexes, abducts, and medially rotates thigh; steadies trunk

*The quadriceps form the flesh of the anterior thigh and have a common insertion in the tibial tuberosity via the patellar ligament. They are powerful leg extensors, enabling humans to kick a football, for example.

Table 13.8 Muscles That Move the Human Thigh and Leg, Posterior Aspect (Figure 13.14)

Muscle	Comments	Origin	Insertion	Action
Origin on the Pelvis				
Gluteus maximus	Largest and most superficial of gluteal muscles (which form buttock mass); intramuscular injection site	Dorsal ilium, sacrum, and coccyx	Gluteal tuberosity of femur and iliotibial tract*	Complex, powerful thigh extensor (most effective when thigh is flexed, as in climbing stairs—but not as in walking); antagonist of iliopsoas; laterally rotates and abducts thigh
Gluteus medius	Partially covered by gluteus maximus; intramuscular injection site	Upper lateral surface of ilium	Greater trochanter of femur	Abducts and medially rotates thigh; steadies pelvis during walking
Gluteus minimus (not shown in figure)	Smallest and deepest gluteal muscle	External inferior surface of ilium	Greater trochanter of femur	Abducts and medially rotates thigh; steadies pelvis
Posterior Compartment				
Hamstrings†				
Biceps femoris	Most lateral muscle of group; arises from two heads	Ischial tuberosity (long head); linea aspera and distal femur (short head)	Tendon passes laterally to insert into head of fibula and lateral condyle of tibia	Extends thigh; laterally rotates leg; flexes leg

Table 13.8 *(continued)*

Muscle	Comments	Origin	Insertion	Action
Semitendinosus	Medial to biceps femoris	Ischial tuberosity	Medial aspect of upper tibial shaft	Extends thigh; flexes leg; medially rotates leg
Semimembranosus	Deep to semitendinosus	Ischial tuberosity	Medial condyle of tibia; lateral condyle of femur	Extends thigh; flexes leg; medially rotates leg

*The iliotibial tract, a thickened lateral portion of the fascia lata, ensheathes all the muscles of the thigh. It extends as a tendinous band from the iliac crest to the knee.

†The hamstrings are the fleshy muscles of the posterior thigh. As a group, they are strong extensors of the thigh; they counteract the powerful quadriceps by stabilizing the knee joint when standing.

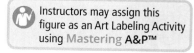

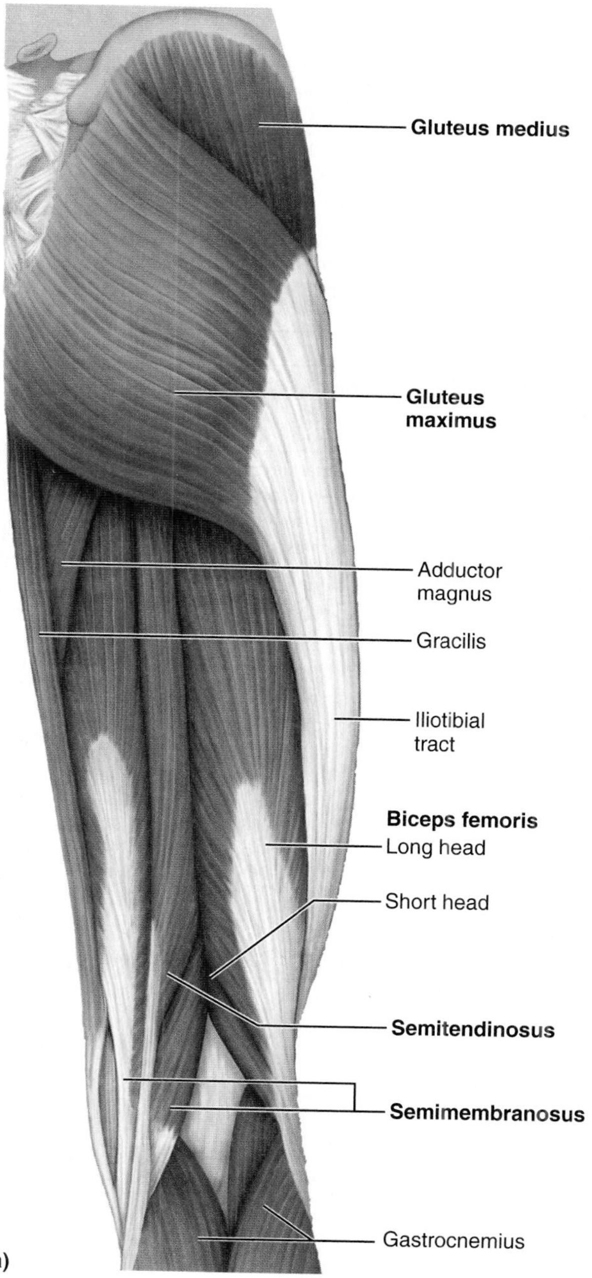

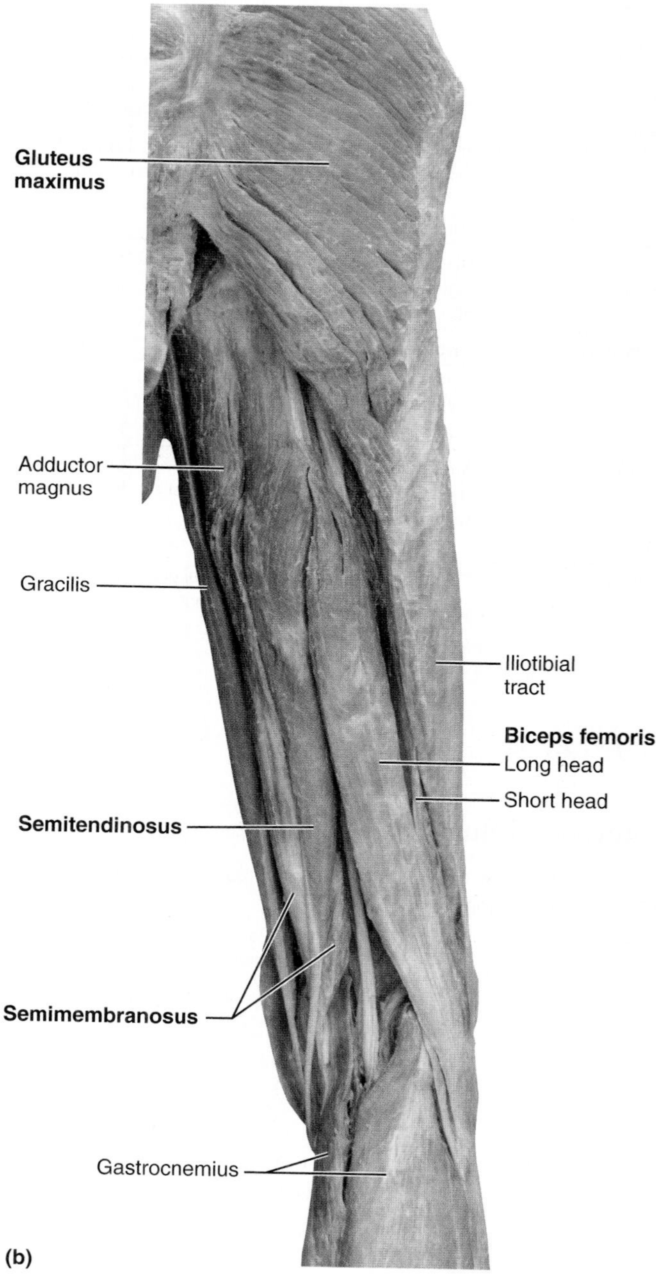

Figure 13.14 Muscles of the posterior aspect of the right hip and thigh. **(a)** Superficial view showing the gluteus muscles of the buttock and hamstring muscles of the thigh. **(b)** Photo of muscles of the posterior thigh.

Instructors may assign this figure as an Art Labeling Activity using Mastering A&P™

Table 13.9 Muscles That Move the Foot and Ankle (Figures 13.15 and 13.16)

Muscle	Comments	Origin	Insertion	Action
Lateral Compartment (Figure 13.15a, b and Figure 13.16b)				
Fibularis (peroneus) longus	Superficial lateral muscle; overlies fibula	Head and upper portion of fibula	By long tendon under foot to metatarsal I and medial cuneiform	Plantar flexes and everts foot
Fibularis (peroneus) brevis	Smaller muscle; deep to fibularis longus	Distal portion of fibula shaft	By tendon running behind lateral malleolus to insert on proximal end of metatarsal V	Plantar flexes and everts foot
Anterior Compartment (Figure 13.15a, b)				
Tibialis anterior	Superficial muscle of anterior leg; parallels sharp anterior margin of tibia	Lateral condyle and upper 2/3 of tibia; interosseous membrane	By tendon into inferior surface of first cuneiform and metatarsal I	Prime mover of dorsiflexion; inverts foot; supports longitudinal arch of foot
Extensor digitorum longus	Anterolateral surface of leg; lateral to tibialis anterior	Lateral condyle of tibia; proximal ¾ of fibula; interosseous membrane	Tendon divides into four parts; inserts into middle and distal phalanges of toes II–V	Prime mover of toe extension; dorsiflexes foot
Fibularis (peroneus) tertius	Small muscle; often fused to distal part of extensor digitorum longus	Distal anterior surface of fibula and interosseous membrane	Tendon inserts on dorsum of metatarsal V	Dorsiflexes and everts foot
Extensor hallucis longus	Deep to extensor digitorum longus and tibialis anterior	Anteromedial shaft of fibula and interosseous membrane	Tendon inserts on distal phalanx of great toe	Extends great toe; dorsiflexes foot
Posterior Compartment, Superficial (Figure 13.16a; also Figure 13.15)				
Triceps surae	Refers to muscle pair below that shapes posterior calf		Via common tendon (calcaneal) into calcaneus of the heel	Plantar flex foot
Gastrocnemius	Superficial muscle of pair; two prominent bellies	By two heads from medial and lateral condyles of femur	Calcaneus via calcaneal tendon	Plantar flexes foot when leg is extended; crosses knee joint; thus can flex leg (when foot is dorsiflexed)
Soleus	Deep to gastrocnemius	Proximal portion of tibia and fibula; interosseous membrane	Calcaneus via calcaneal tendon	Plantar flexion; an important muscle for locomotion

13

Activity 5

Review of Human Musculature

Review the muscles by labeling the figures in the Review Sheet (questions 11 and 12, pages 228–229).

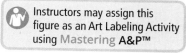

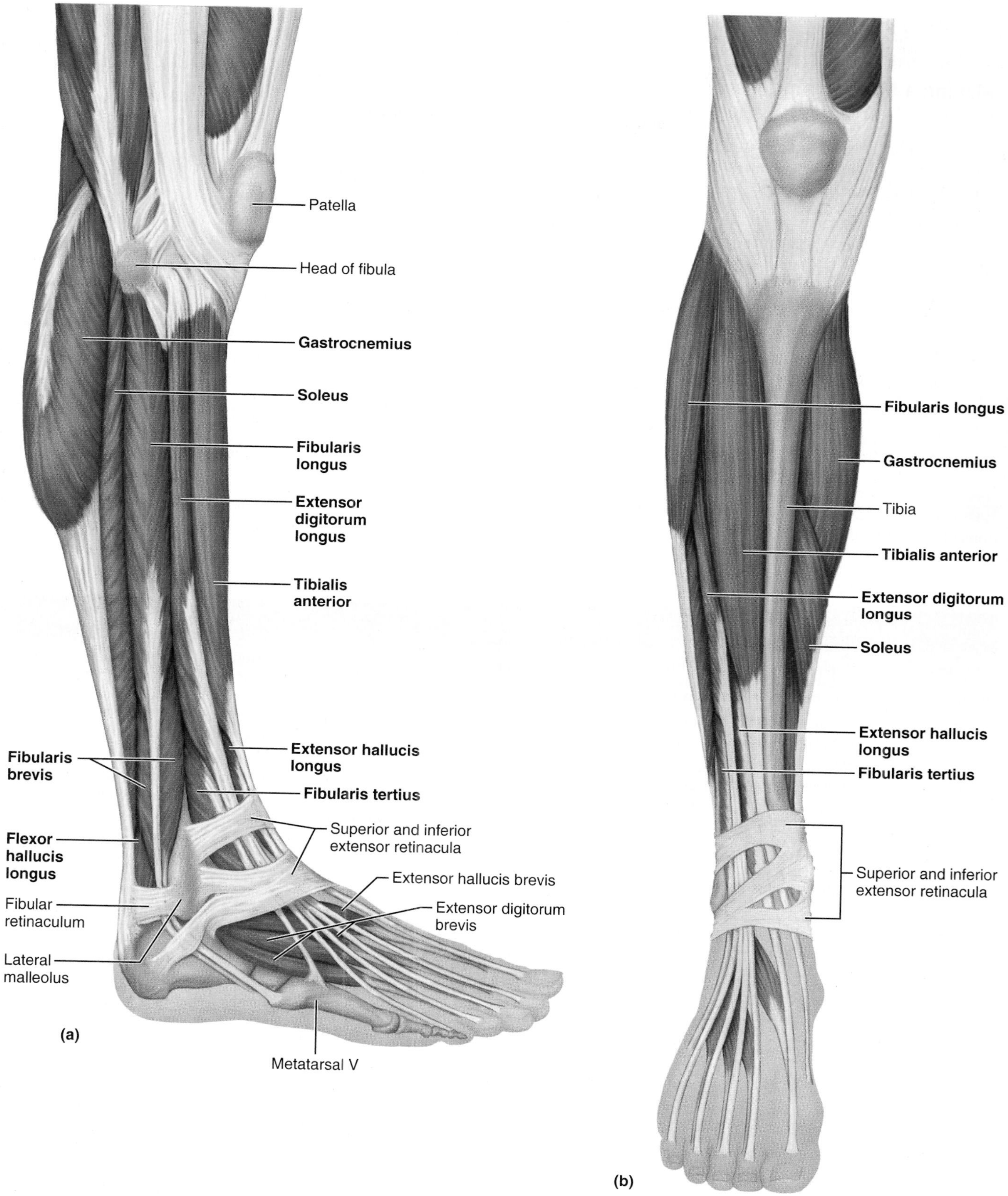

Figure 13.15 Muscles of the anterolateral aspect of the right leg. (a) Superficial view of lateral aspect of the leg, illustrating the positioning of the lateral compartment muscles (fibularis longus and brevis) relative to anterior and posterior leg muscles. **(b)** Superficial view of anterior leg muscles.

Instructors may assign this figure as an Art Labeling Activity using Mastering A&P™

13

Activity 6

Making a Muscle Painting

1. Choose a male student to be "muscle painted."

2. Obtain brushes and water-based paints from the supply area while the "volunteer" removes his shirt and rolls up his pant legs (if necessary).

3. Using different colored paints, identify the muscles listed below by painting his skin. If a muscle covers a large body area, you may opt to paint only its borders.

- biceps brachii
- deltoid
- erector spinae
- pectoralis major
- rectus femoris
- tibialis anterior
- triceps brachii
- vastus lateralis
- biceps femoris
- extensor carpi radialis longus
- latissimus dorsi
- rectus abdominis
- sternocleidomastoid
- trapezius
- triceps surae
- vastus medialis

4. Check your "human painting" with your instructor before cleaning your bench and leaving the laboratory.

For instructions on animal dissections, see the dissection exercises (starting on p. 695) in the cat and fetal pig editions of this manual.

Table 13.9 Muscles That Move the Foot and Ankle *(continued)*

Muscle	Comments	Origin	Insertion	Action
Posterior Compartment, Deep (Figure 13.16b–e)				
Popliteus	Thin muscle at posterior aspect of knee	Lateral condyle of femur and lateral meniscus	Proximal tibia	Flexes and rotates leg medially to "unlock" knee when leg flexion begins
Tibialis posterior	Thick muscle deep to soleus	Superior portion of tibia and fibula and interosseous membrane	Tendon passes obliquely behind medial malleolus and under arch of foot; inserts into several tarsals and metatarsals II–IV	Prime mover of foot inversion; plantar flexes foot; stabilizes longitudinal arch of foot
Flexor digitorum longus	Runs medial to and partially overlies tibialis posterior	Posterior surface of tibia	Distal phalanges of toes II–V	Flexes toes; plantar flexes and inverts foot
Flexor hallucis longus (see also Figure 13.15a)	Lies lateral to inferior aspect of tibialis posterior	Middle portion of fibula shaft; interosseous membrane	Tendon runs under foot to distal phalanx of great toe	Flexes great toe (*hallux* = great toe); plantar flexes and inverts foot; the "push-off muscle" during walking

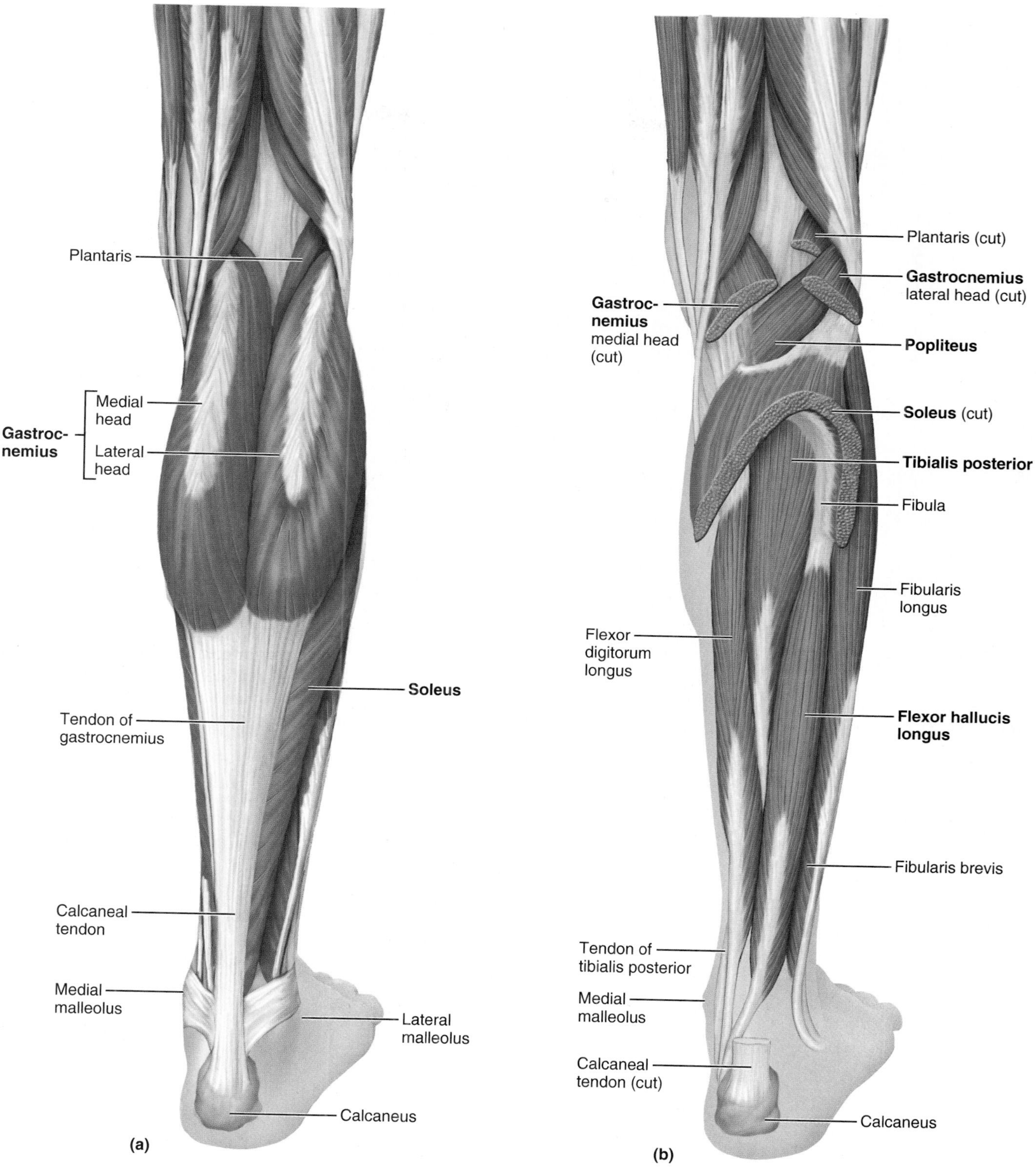

Figure 13.16 Muscles of the posterior aspect of the right leg. (a) Superficial view of the posterior leg. **(b)** The triceps surae has been removed to show the deep muscles of the posterior compartment.

13

➤

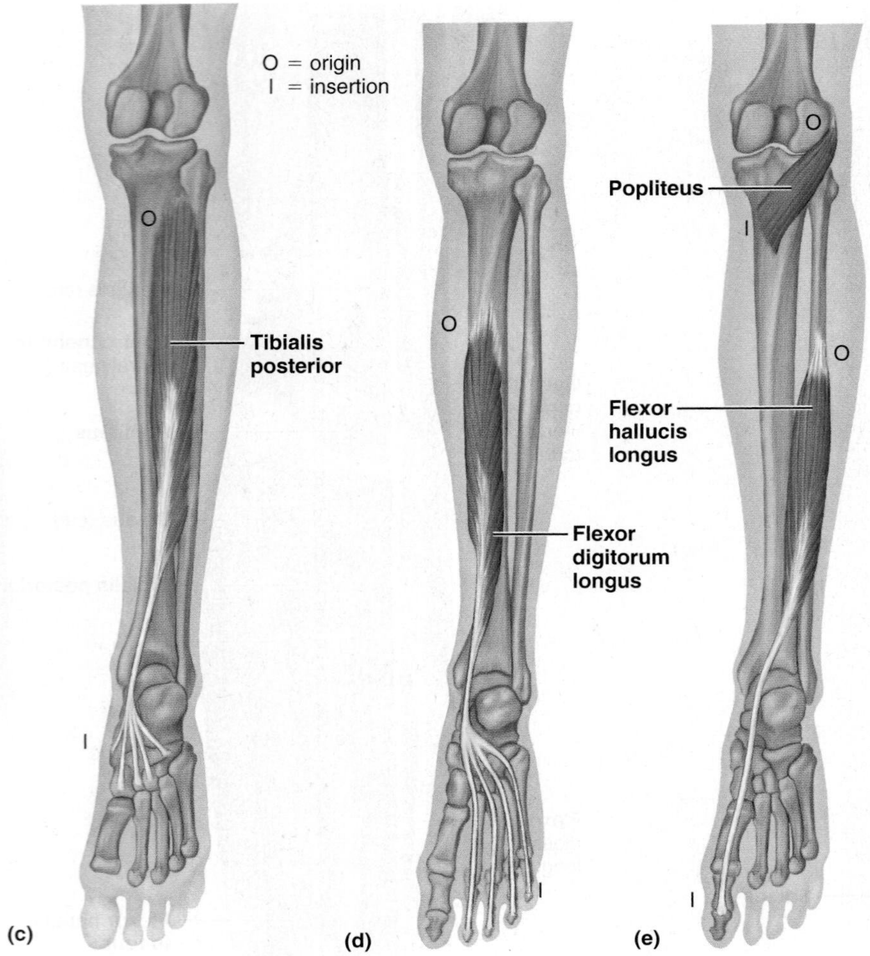

Figure 13. 16 *(continued)* **Muscles of the posterior compartment of the right leg. (c–e)** Individual muscles are shown in isolation so that their origins and insertions may be observed.

REVIEW SHEET

EXERCISE 13

Gross Anatomy of the Muscular System

Instructors may assign a portion of the Review Sheet questions using Mastering A&P™

Name ______________________ Lab Time/Date ______________________

Classification of Skeletal Muscles

1. Several criteria were given for the naming of muscles.

For each of the criteria below, list at least two muscles that are named for the given criterion.

1. Muscle location: ______________________
2. Muscle shape: ______________________
3. Muscle size: ______________________
4. Direction of muscle fibers: ______________________
5. Number of origins: ______________________
6. Location of attachments: ______________________
7. Muscle action: ______________________

2. Match the key terms to the descriptions below.

Key: a. prime mover (agonist) b. antagonist c. synergist d. fixator e. origin f. insertion

______________ 1. term for the biceps brachii during forearm flexion

______________ 2. term that describes the relation of brachioradialis to biceps brachii during forearm flexion

______________ 3. term for the triceps brachii during forearm flexion

______________ 4. term for the more movable muscle attachment

______________ 5. term for the more fixed muscle attachment

______________ 6. term for the rotator cuff muscles and deltoid when the forearm is flexed and the hand grabs a tabletop to lift the table

Muscles of the Head and Neck

3. Using choices from the key at the right, correctly identify muscles provided with leader lines on the illustration.

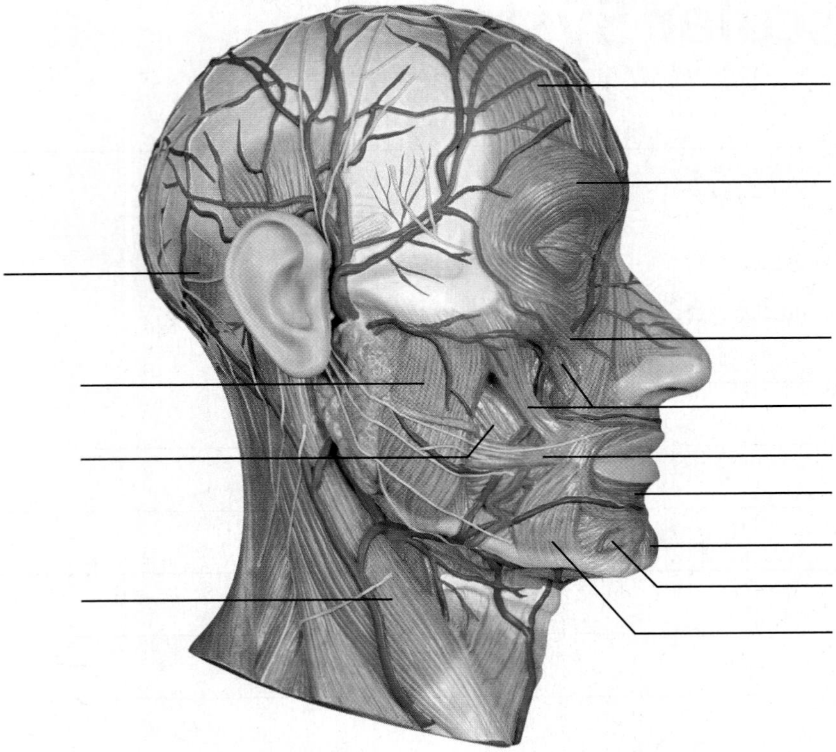

Key:

a. buccinator

b. depressor anguli oris

c. depressor labii inferioris

d. frontal belly of the epicranius

e. levator labii superioris

f. masseter

g. mentalis

h. occipital belly of the epicranius

i. orbicularis oculi

j. orbicularis oris

k. risorius

l. sternocleidomastoid

m. zygomaticus minor and major

4. Using the key provided in question 3, identify the muscles described next.

________ 1. used in smiling

________ 2. used to suck in your cheeks

________ 3. used in blinking and squinting

________ 4. used to pout (pulls the corners of the mouth downward)

________ 5. raises your eyebrows for a questioning expression

________ 6. used to turn and tilt the head toward the shoulder

________ 7. your kissing muscle

________ 8. prime mover of jaw closure

________ 9. draws corners of the lip back (laterally)

Muscles of the Trunk

5. Correctly identify both intact and transected (cut) muscles depicted in the illustration, using the key given at the right.

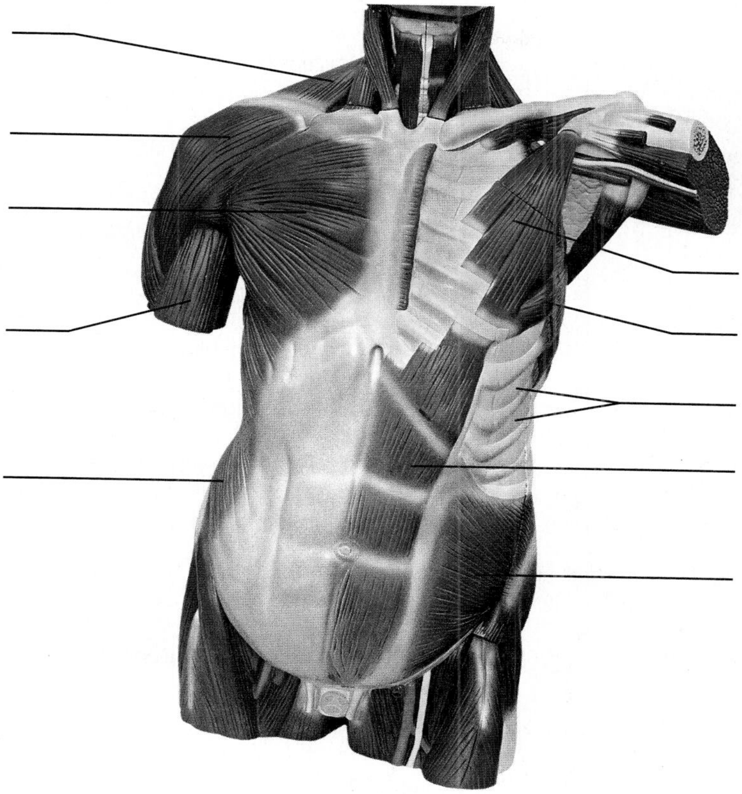

Key:

a. biceps brachii (cut)
b. deltoid
c. external intercostals
d. external oblique
e. internal oblique
f. pectoralis major
g. pectoralis minor
h. rectus abdominis
i. serratus anterior
j. trapezius

6. Using the key provided in question 5 above, identify the major muscles described below.

________ 1. a major flexor of the vertebral column

________ 2. prime mover for forearm flexion

________ 3. prime mover for arm flexion

________ 4. assume major responsibility for forming the abdominal wall (three pairs of muscles)

________ 5. prime mover of arm abduction

________ 6. with ribs fixed, pulls scapula forward and downward

________ 7. moves the scapula forward and rotates scapula upward

________ 8. small, inspiratory muscles between the ribs; elevate the rib cage

________ 9. extends the head

Muscles of the Upper Limb

7. Using terms from the key on the right, correctly identify all muscles provided with leader lines in the illustration.

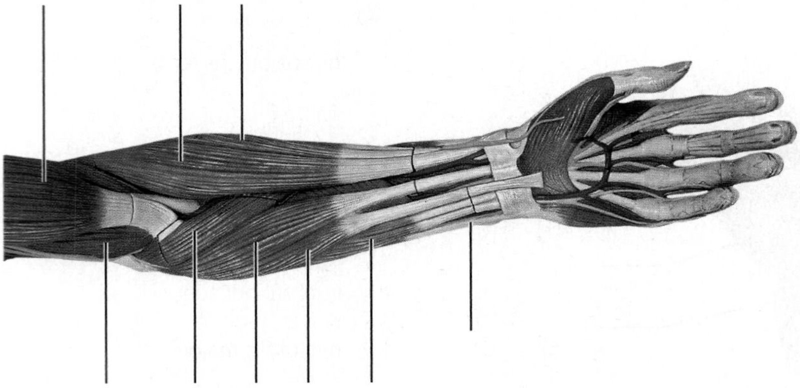

Key:

a. biceps brachii
b. brachialis
c. brachioradialis
d. extensor carpi radialis longus
e. flexor carpi radialis
f. flexor carpi ulnaris
g. flexor digitorum superficialis
h. palmaris longus
i. pronator teres

8. Use the key provided in question 7 to identify the muscles described below. (Some choices from the key will be used more than once.)

________ 1. flexes and supinates the forearm

________ 2. muscle located in the posterior compartment of the forearm

________ 3. forearm flexors; no role in supination (two muscles)

________ 4. muscle located medial to the palmaris longus

________ 5. flexes and abducts the hand

________ 6. flexes the hand and middle phalanges

________ 7. pronates the forearm

________ 8. flexes and adducts the hand

________ 9. extends and abducts the hand

________ 10. flat muscle that is a weak hand flexor, tenses skin of the palm

Muscles of the Lower Limb

9. Using the terms from the key on the right, correctly identify all muscles provided with leader lines in the illustrations below.

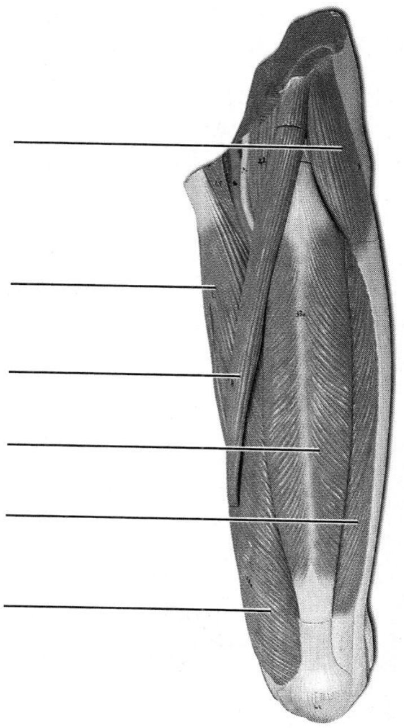

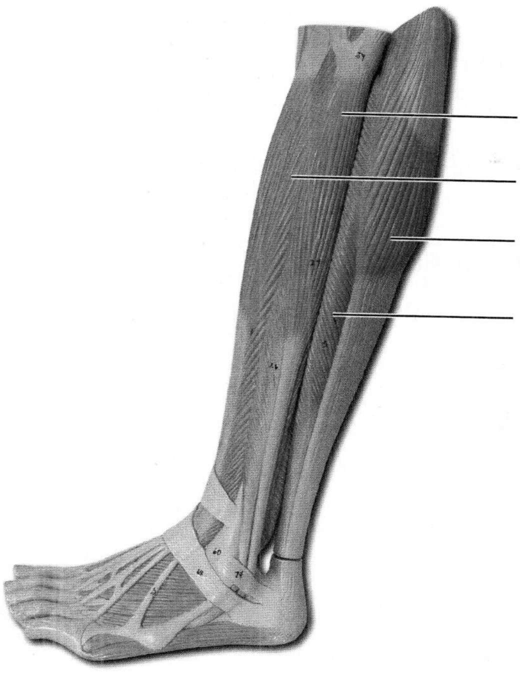

Key:

a. adductor longus

b. extensor digitorum longus

c. fibularis longus

d. gastrocnemius

e. rectus femoris

f. sartorius

g. soleus

h. tensor facia lata

i. vastus lateralis

j. vastus medialis

10. Use the key terms in question 9 to respond to the descriptions below.

________ 1. "tailor's muscle"

________ 2. lateral compartment muscle that plantar flexes and everts the foot

________ 3. abducts the thigh to take the "at ease" stance

________ 4. extend leg and stabilize knee (two muscles)

________ 5. posterior compartment muscles that plantar flex the foot (two muscles)

________ 6. adducts the thigh, as when standing at attention

________ 7. extends the toes

________ 8. extends leg and flexes thigh

General Review: Muscle Recognition

11. Identify each lettered muscle in the illustration of the human anterior superficial musculature by matching its letter with one of the following muscle names:

________ 1. adductor longus
________ 2. biceps brachii
________ 3. brachioradialis
________ 4. deltoid
________ 5. extensor digitorum longus
________ 6. external oblique
________ 7. fibularis longus
________ 8. flexor carpi radialis
________ 9. flexor carpi ulnaris
________ 10. frontal belly of epicranius
________ 11. gastrocnemius
________ 12. gracilis
________ 13. iliopsoas
________ 14. intercostals
________ 15. internal oblique
________ 16. masseter
________ 17. orbicularis oculi
________ 18. orbicularis oris
________ 19. palmaris longus
________ 20. pectineus
________ 21. pectoralis major
________ 22. pectoralis minor
________ 23. platysma
________ 24. pronator teres
________ 25. rectus abdominis
________ 26. rectus femoris
________ 27. sartorius
________ 28. serratus anterior
________ 29. soleus
________ 30. sternocleidomastoid
________ 31. sternohyoid
________ 32. temporalis
________ 33. tensor fascia lata
________ 34. tibialis anterior
________ 35. transversus abdominis
________ 36. trapezius
________ 37. triceps brachii
________ 38. vastus lateralis
________ 39. vastus medialis
________ 40. zygomaticus

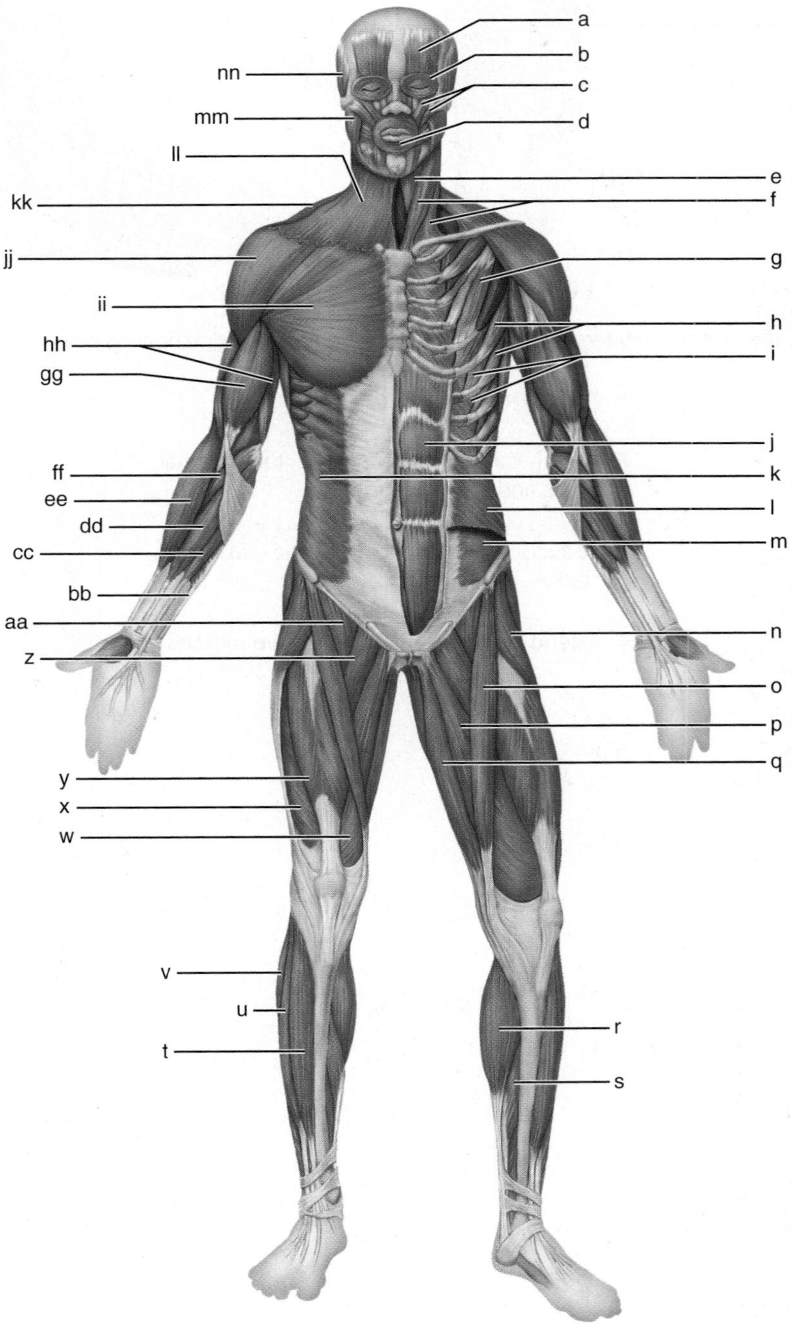

12. Identify each lettered muscle in this illustration of the human posterior superficial musculature by matching its letter with one of the following muscle names:

________ 1. adductor magnus

________ 2. biceps femoris

________ 3. brachialis

________ 4. brachioradialis

________ 5. deltoid

________ 6. extensor carpi radialis longus

________ 7. extensor carpi ulnaris

________ 8. extensor digitorum

________ 9. external oblique

________ 10. flexor carpi ulnaris

________ 11. gastrocnemius

________ 12. gluteus maximus

________ 13. gluteus medius

________ 14. gracilis

________ 15. iliotibial tract (tendon)

________ 16. infraspinatus

________ 17. latissimus dorsi

________ 18. occipital belly of epicranius

________ 19. semimembranosus

________ 20. semitendinosus

________ 21. sternocleidomastoid

________ 22. teres major

________ 23. trapezius

________ 24. triceps brachii

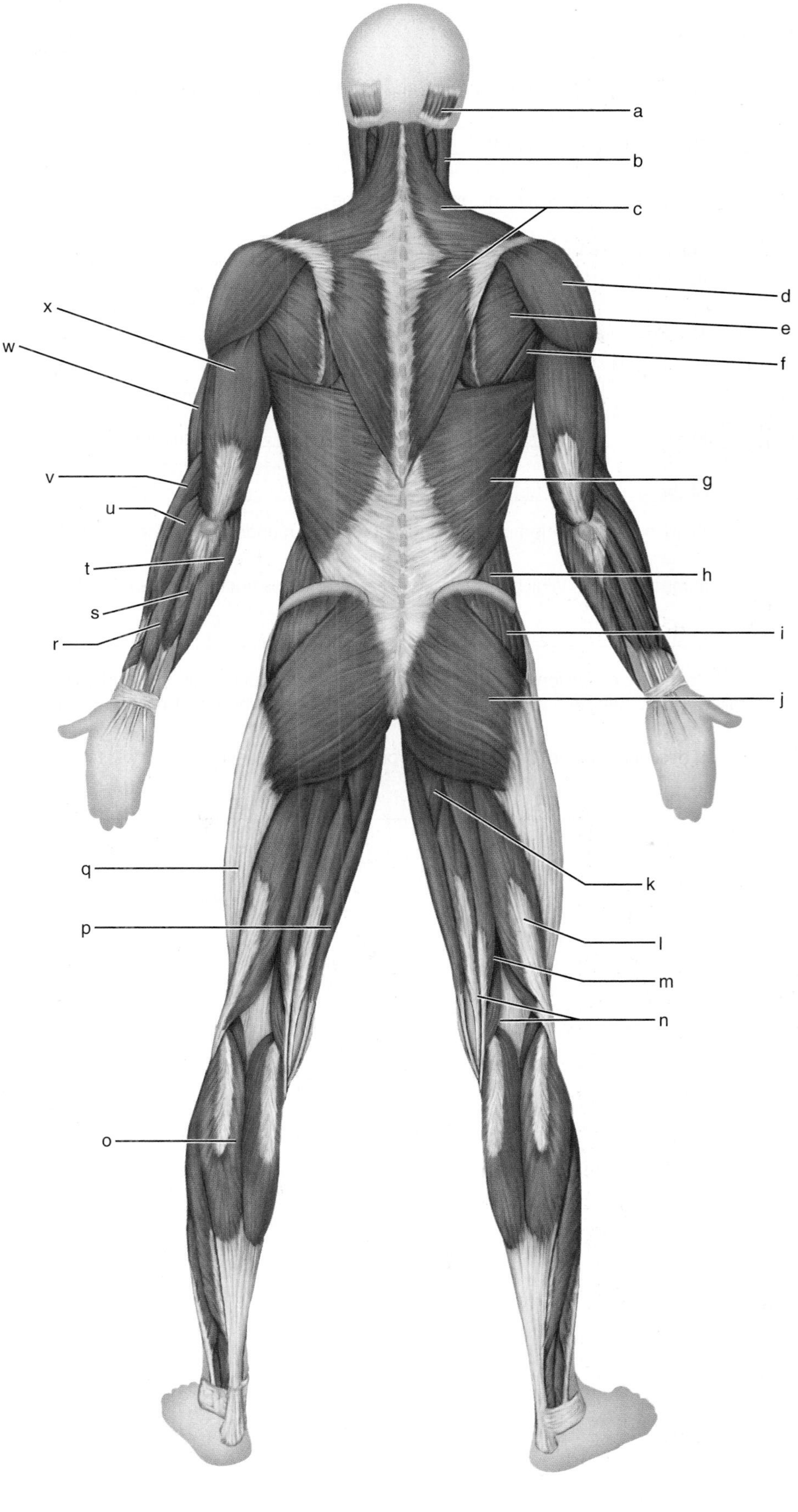

General Review: Muscle Descriptions

13. Identify the muscles described by completing the following statements. Use an appropriate reference as needed.

1. The ______________, ______________, ______________, and ______________ are commonly used for intramuscular injections (four muscles).

2. The insertion tendon of the ______________ group contains a large sesamoid bone, the patella.

3. The triceps surae insert in common into the ______________ tendon.

4. The bulk of the tissue of a muscle tends to lie ______________ to the part of the body it causes to move.

5. The extrinsic muscles of the hand originate on the ______________.

6. Most flexor muscles are located on the ______________ aspect of the body; most extensors are located ______________. An exception to this generalization is the extensor-flexor musculature of the ______________.

14. Bruxism is a condition in which individuals clench and/or grind their teeth. It often occurs as they sleep, leading to jaw pain and damaged teeth. Which muscles contract during this nocturnal event? ______________

____

15. Repetitive extension of the hand at the wrist and abduction of the hand can lead to lateral epicondylitis. Although sometimes called "tennis elbow," it more often affects individuals who don't play tennis. Based on the name *lateral epicondylitis* and the action described above, which muscle would most likely have microscopic tears in the tendon? ______________

____

EXERCISE 14

Skeletal Muscle Physiology: Frogs and Human Subjects

Learning Outcomes

- Define the terms *resting membrane potential*, *depolarization*, *repolarization*, *action potential*, and *refractory period*, and explain the physiological basis of each.
- Observe muscle contraction microscopically, and describe the role of ATP and various ions in muscle contraction.
- Define *muscle twitch*, and describe its three phases.
- Differentiate among subthreshold stimulus, threshold stimulus, and maximal stimulus.
- Define *tetanus* in skeletal muscle, and explain how it comes about.
- Define *muscle fatigue*, and describe the effects of load on muscle fatigue.
- Define *motor unit*, and relate recruitment of motor units and temporal summation to production of a graded contraction.
- Demonstrate how a physiograph or a computer with a data acquisition unit can be used to record skeletal muscle activity.
- Explain the significance of recordings obtained during experimentation.

Pre-Lab Quiz

Instructors may assign these and other Pre-Lab Quiz questions using Mastering A&P™

1. Circle the correct underlined term. The potential difference, or voltage, across the plasma membrane is the result of the difference in membrane permeability to anions / cations, most importantly Na^+ and K^+.
2. The ________ wave follows the depolarization wave across the sarcolemma.
   a. hyperpolarization b. refraction c. repolarization
3. Circle True or False. The voltage at which the first noticeable contractile response is achieved is called the threshold stimulus.
4. Circle True or False. A single muscle is made up of many motor units, and the gradual activation of these motor units results in a graded contraction of the whole muscle.
5. A sustained, smooth, muscle contraction that is a result of high-frequency stimulation is:
   a. tetanus b. tonus c. twitch

The contraction of skeletal and cardiac muscle fibers can be considered in terms of three events—electrical excitation of the muscle fiber, excitation contraction coupling, and shortening of the muscle fiber due to sliding of the myofilaments within it.

At rest, all cells maintain a potential difference, or voltage, across their plasma membrane; the inner face of the membrane is approximately −60 to −90 millivolts (mV) compared with the cell exterior. This potential difference is a result of differences in membrane permeability to cations, most importantly sodium

Go to Mastering A&P™ > Study Area to improve your performance in A&P Lab.

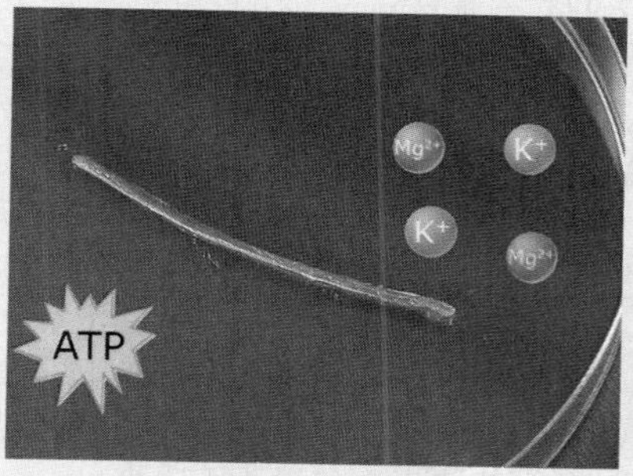

> Lab Tools > Pre-Lab Videos > Muscle Contraction

Instructors may assign new Building Vocabulary coaching activities, Pre-Lab Quiz questions, Art Labeling activities, Pre-Lab Video Coaching Activities for Muscle Contraction, Practice Anatomy Lab Practical questions (PAL), PhysioEx activities, and more using the Mastering A&P™ Item Library.

Materials

- ATP muscle kits (glycerinated rabbit psoas muscle;* ATP and salt solutions obtainable from Carolina Biological Supply)
- Petri dishes
- Microscope slides
- Coverslips
- Millimeter ruler
- Compound microscope
- Stereomicroscope
- Pointed glass probes (teasing needles)
- Small beakers (50 ml)
- Distilled water
- Glass-marking pencil
- Textbooks or other heavy books
- Watch or timer
- Ringer's solution (frog)
- Scissors

Text continues on next page. →

- Metal needle probes
- Medicine dropper
- Cotton thread
- Forceps
- Disposable gloves
- Glass or porcelain plate
- Pithed bullfrog†
- Physiograph or BIOPAC® equipment:‡ Physiograph apparatus, physiograph paper and ink, force transducer, pin electrodes, stimulator, stimulator output extension cable, transducer stand and cable, straight pins, frog board

BIOPAC® BSL System with BSL software version 4.0.1 and above (for Windows 10/8.x/7 or Mac OS X 10.9–10.12), data acquisition unit MP36/35 or MP45, PC or Mac computer, Biopac Student Lab electrode lead set, hand dynamometer, headphones, metric tape measure, disposable vinyl electrodes, and conduction gel.

Instructors using the MP36/35/30 data acquisition unit with BSL software versions earlier than 4.0.1 (for Windows or Mac) will need slightly different channel settings and collection strategies. Instructions for using the older data acquisition unit can be found on MasteringA&P.

Notes to the Instructor: **At the beginning of the lab, the muscle bundle should be removed from the test tube and cut into approximately 2-cm lengths. Both the cut muscle segments and the entubed glycerol should be put into a petri dish. One muscle segment is sufficient for every two to four students making observations.*

†Bullfrogs to be pithed by lab instructor as needed for student experimentation. (If instructor prefers that students pith their own specimens, an instructional diagram on that procedure suitable for copying for student handouts is provided in the Instructor's Guide.)

‡Additionally, instructions for Activity 3 using a kymograph can be found in the Instructor Guide. Instructions for using PowerLab® equipment can be found on MasteringA&P.

PEx PhysioEx™ 9.1 Computer Simulation Ex. 2 on p. PEx-17.

(Na^+) and potassium (K^+) ions. Intracellular potassium concentration is much greater than its extracellular concentration, and intracellular sodium concentration is considerably less than its extracellular concentration. Hence, steep concentration gradients across the membrane exist for both cations. Because the plasma membrane is more permeable to K^+ than to Na^+, the cell's **resting membrane potential** is more negative inside than outside. The resting membrane potential is of particular interest in excitable cells, such as muscle fibers and neurons, because changes in that voltage underlie their ability to do work (to contract and/or issue electrical signals).

Action Potential

When a muscle fiber is stimulated, the sarcolemma becomes temporarily more permeable to Na^+, which enters the cell. This sudden influx of Na^+ alters the membrane potential. That is, the cell interior becomes less negatively charged at that point, an event called **depolarization**. When depolarization reaches a certain level and the sarcolemma momentarily changes its polarity, a depolarization wave travels along the sarcolemma. Even as the influx of Na^+ occurs, the sarcolemma becomes less permeable to Na^+ and more permeable to K^+. Consequently, K^+ ions move out of the cell, restoring the resting membrane potential (but not the original ionic conditions), an event called **repolarization**. The repolarization wave follows the depolarization wave across the sarcolemma. This rapid depolarization and repolarization of the membrane that is propagated along the entire membrane from the point of stimulation is called the **action potential**.

During repolarization, a muscle fiber is said to be in a **refractory period** because the cell cannot be stimulated again until repolarization is complete. Note that repolarization restores only the electrical conditions of the resting state. The ATP-dependent Na^+-K^+ pump restores the ionic conditions of the resting state.

Contraction

Propagation of the action potential down the T tubules of the sarcolemma causes the release of calcium ions (Ca^{2+}) from storage in the sarcoplasmic reticulum within the muscle fiber. When the calcium ions bind to the regulatory protein troponin on the actin myofilaments, they act as an ionic trigger that initiates contraction, and the actin and myosin filaments slide past each other. Once the action potential ends, the calcium ions are almost immediately transported back into the sarcoplasmic reticulum. Instantly the muscle fiber relaxes.

Activity 1

Instructors may assign a related Pre-Lab Video Coaching Activity using Mastering A&P™

Observing Muscle Fiber Contraction

In this simple experiment, you will have the opportunity to review your understanding of muscle fiber anatomy and to watch fibers respond to the presence of ATP and/or a solution of K^+ and magnesium ions (Mg^{2+}).

This experiment uses preparations of glycerinated muscle. The glycerination process denatures troponin and tropomyosin. Consequently, calcium, so critical for contraction in vivo, is not necessary here. The role of magnesium and potassium salts as cofactors in the contraction process is not well understood, but magnesium and potassium salts seem to be required for ATPase activity in this system.

1. Talk with other members of your lab group to develop a hypothesis about requirements for muscle fiber contraction for this experiment. The hypothesis should have three parts: (1) salts and ATP, (2) ATP only, and (3) salts only.

Briefly describe your hypothesis for each part on the space provided. ________

____

____

2. Obtain the following materials from the supply area: two glass teasing needles or forceps; six glass microscope slides and six coverslips; millimeter ruler; dropper bottles containing the following solutions: (a) 0.25% ATP plus 0.05 *M* KCl plus 0.001 *M* $MgCl_2$ in distilled water, (b) 0.25% ATP in triply distilled water, and (c) 0.05 *M* KCl plus 0.001 *M* $MgCl_2$ in distilled water; a petri dish; a beaker of distilled water; a glass-marking pencil; and a small portion of a previously cut muscle bundle segment. While you are at the supply area, place the muscle fibers in the petri dish, and pour a small amount of glycerol (the fluid in the supply petri dish) over your muscle fibers. Also obtain both a compound and a stereomicroscope and bring them to your laboratory bench.

3. Using clean fine glass needles or forceps, tease the muscle segment to separate its fibers. The objective is to isolate *single* muscle fibers for observation. Be patient and work carefully so that the fibers do not get torn during this isolation procedure.

4. Transfer one or more of the fibers (or the thinnest strands you have obtained) onto a clean microscope slide with a glass needle, and cover with a coverslip. Examine the fiber under the compound microscope at low- and then high-power magnifications to observe the striations and the smoothness of the fibers when they are in the relaxed state.

5. Clean three microscope slides well and rinse in distilled water. Label the slides A, B, and C.

6. Transfer three or four fibers to microscope slide A with a glass needle. Using the needle as a prod, carefully position the fibers so that they are parallel to one another and as straight as possible. Place this slide under a *stereomicroscope*, and measure the length of each fiber by holding a millimeter ruler adjacent to it. Alternatively, you can rest the microscope slide *on* the millimeter ruler to make your length determinations. Record the data in the **Activity 1 chart**.

7. Flood the fibers (situated under the stereomicroscope) with several drops of the solution containing ATP, K^+, and Mg^{2+}. Watch the reaction of the fibers after adding the solution. After 30 seconds (or slightly longer), remeasure each fiber and record the observed ending lengths on the chart. Calculate the percentage of contraction by using the simple formula below, and record this data on the chart.

$$\text{Initial length (mm)} - \text{ending length (mm)} = \text{net change (mm)}$$

then:

$$\frac{\text{net change (mm)}}{\text{initial length (mm)}} \times 100 = \underline{\qquad}\text{\% contraction}$$

8. Carefully transfer one of the fibers from step 7 to a clean, unmarked microscope slide, cover with a coverslip, and observe with the compound microscope. Mentally compare your initial observations with the view you are observing now. What differences do you see? (Be specific.)

____

____

____

What zones (or bands) have disappeared?

____

9. Repeat steps 6 through 8 twice more, using clean slides and fresh muscle fibers. On slide B, use the solution of ATP in distilled water (no salts). Then on slide C, use the solution containing only salts (no ATP) for the third series. Record data on the Activity 1 chart.

Text continues on next page. →

14

Activity 1: Observations of Muscle Fiber Contraction

	Muscle fiber 1	Muscle fiber 2	Muscle fiber 3	Average
Salts and ATP, slide A				
Initial length (mm)				
Ending length (mm)				
% Contraction				
ATP only, slide B				
Initial length (mm)				
Ending length (mm)				
% Contraction				
Salts only, slide C				
Initial length (mm)				
Ending length (mm)				
% Contraction				

10. Calculate the averages for each slide. Collect the average data from all the groups in your laboratory and use these data to prepare a lab report. (See Getting Started, on MasteringA&P.) Include in your discussion the following questions:

What % contraction was observed when ATP was applied in the absence of K^+ and Mg^{2+}?

____

What % contraction was observed when the muscle fibers were flooded with a solution containing K^+ and Mg^{2+}, and lacking ATP?

____

What conclusions can you draw about the importance of ATP, K^+, and Mg^{2+} to the contractile process?

____

____

Can you draw exactly the same conclusions from the data provided by each group? List some variables that might have been introduced into the procedure and that might account for any differences.

____

____

14

Activity 2

Inducing Contraction in the Frog Gastrocnemius Muscle

Physiologists have learned a great deal about the way muscles function by isolating muscles from laboratory animals and then stimulating these muscles to observe their responses. Various stimuli—electrical shock, temperature changes, extremes of pH, certain chemicals—elicit muscle activity, but laboratory experiments of this type typically use electrical shock. This is because it is easier to control the onset and cessation of electrical shock, as well as the strength of the stimulus.

Various types of apparatus are used to record muscle contraction. All include a way to mark time intervals, a way to indicate exactly when the stimulus was applied, and a way to measure the magnitude of the contractile response. Instructions are provided here for setting up a physiograph apparatus (**Figure 14.1**).

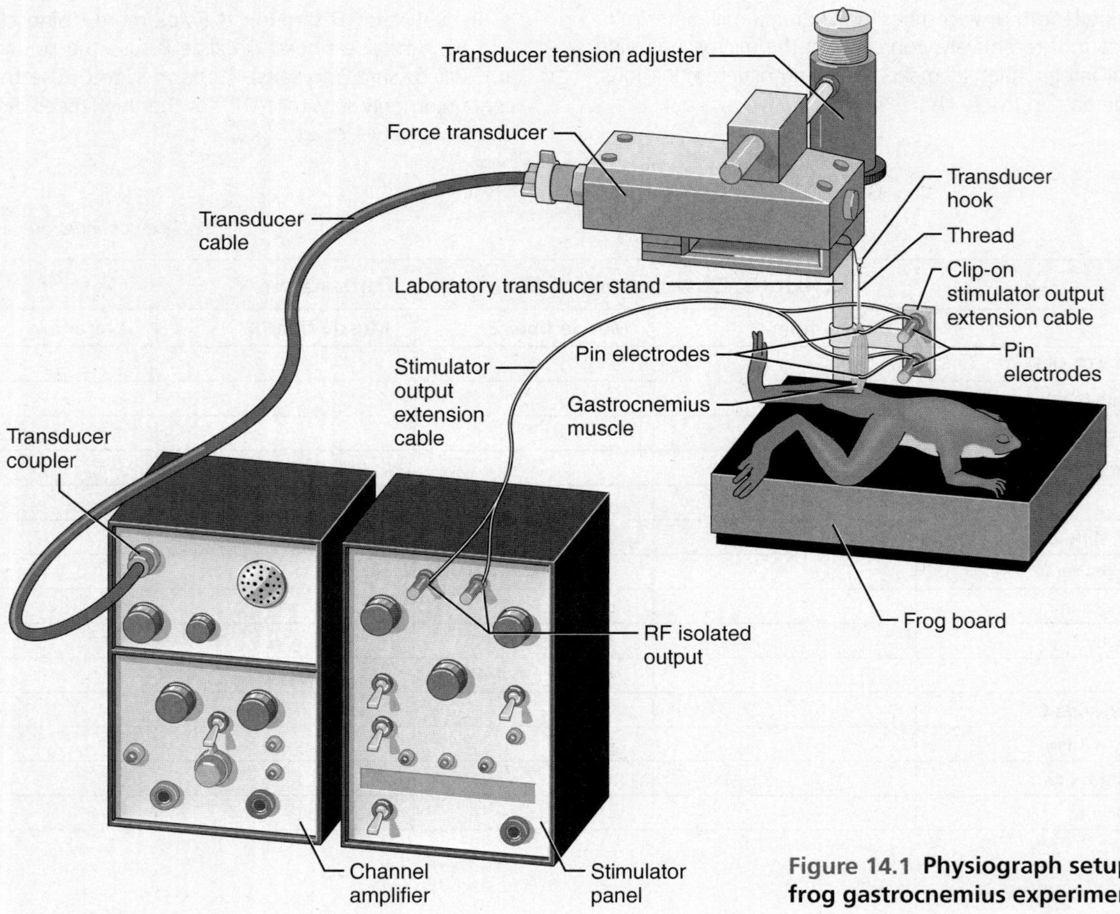

Figure 14.1 Physiograph setup for frog gastrocnemius experiments.

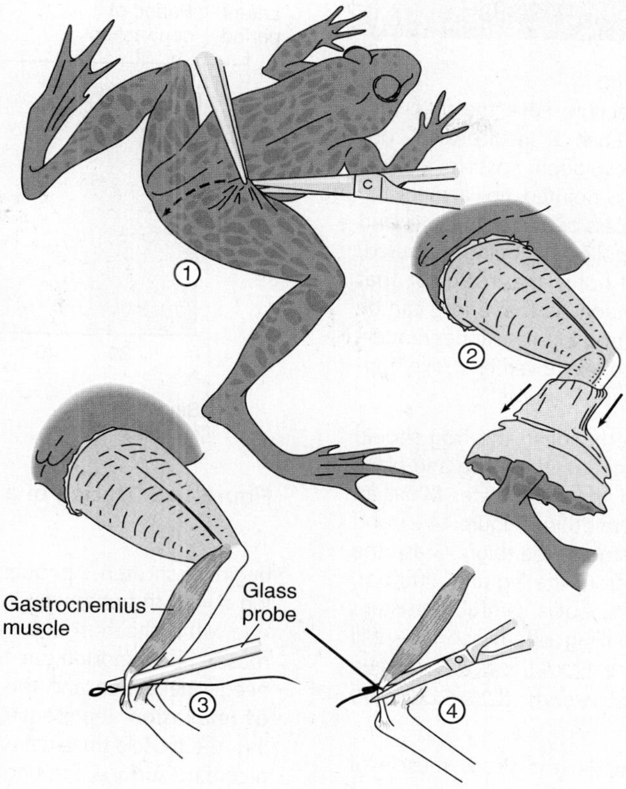

Figure 14.2 Preparation of the frog gastrocnemius muscle. Numbers indicate the sequence of steps for preparing the muscle.

Specific instructions for use of recording apparatus during recording will be provided by your instructor.

Preparing a Muscle for Experimentation

The preparatory work that precedes the recording of muscle activity tends to be quite time-consuming. If you work in teams of two or three, the work can be divided. While one of you is setting up the recording apparatus, one or two students can dissect the frog leg (**Figure 14.2**). Experimentation should begin as soon as the dissection is completed.

Materials

Channel amplifier and transducer cable

Stimulator panel and stimulator output extension cable

Force transducer

Transducer tension adjuster

Transducer stand

Two pin electrodes

Frog board and straight pins

Prepared frog (gastrocnemius muscle freed and calcaneal tendon ligated)

Frog Ringer's solution

Procedure

1. Connect transducer to transducer stand and attach frog board to stand.

2. Attach transducer cable to transducer and to input connection on channel amplifier.

3. Attach stimulator output extension cable to output on stimulator panel (red to red, black to black).

4. Using clip at opposite end of extension cable, attach cable to bottom of transducer stand adjacent to frog board.

5. Attach two pin electrodes securely to electrodes on clip.

6. Place knee of prepared frog in clip-on frog board and secure by inserting a straight pin through tissues of frog. Keep frog muscle moistened with Ringer's solution.

7. Attach thread from the calcaneal tendon of frog to transducer spring hook.

8. Adjust position of transducer on stand to produce a constant tension on thread attached to tendon (taut but not tight). Gastrocnemius muscle should hang vertically directly below hook.

9. Insert free ends of pin electrodes into the muscle, one at proximal end and the other at distal end.

Text continues on next page. →

DISSECTION

Frog Hind Limb

1. Before beginning the frog dissection, have the following supplies ready at your laboratory bench: a small beaker containing 20 to 30 ml of frog Ringer's solution, scissors, a metal needle probe, a glass probe with a pointed tip, a medicine dropper, cotton thread, forceps, a glass or porcelain plate, and disposable gloves. While these supplies are being gathered, one member of your team should notify the instructor that you are ready to begin experimentation, so that a frog can be prepared (pithed). Preparation of a frog in this manner renders it unable to feel pain and prevents reflex movements (like hopping) that would interfere with the experiments.

! 2. All students who will be handling the frog should don disposable gloves. Obtain a pithed frog and place it ventral surface down on the glass plate. Make an incision into the skin approximately midthigh (Figure 14.2), and then continue the cut completely around the thigh. Grasp the skin with the forceps, and strip it from the leg and hindfoot. The skin adheres more at the joints, but a careful, persistent pulling motion—somewhat like pulling off a stocking—will enable you to remove it in one piece. From this point on, keep the exposed muscle tissue moistened with the Ringer's solution to prevent spontaneous twitches.

3. Identify the gastrocnemius muscle (the fleshy muscle of the posterior calf) and the calcaneal tendon that secures it to the heel.

4. Slip a glass probe under the gastrocnemius muscle, and run it along the entire length and under the calcaneal tendon to free them from the underlying tissues.

5. Cut a piece of thread about 10 inches long, and use the glass probe to slide the thread under the calcaneal tendon. Knot the thread firmly around the tendon and then sever the tendon distal to the thread. Alternatively, you can bend a common pin into a Z shape and insert the pin securely into the tendon. The thread is then attached to the opposite end of the pin. Once the tendon has been tied or pinned, the frog is ready for experimentation (see Figure 14.2).

Recording Muscle Activity

Skeletal muscles consist of thousands of muscle fibers and react to stimuli with graded responses. Thus muscle contractions can be weak or vigorous, depending on the requirements of the task. Graded responses (different degrees of shortening) of a skeletal muscle depend on the number of muscle fibers being stimulated. In the intact organism, the number of motor units firing at any one time determines how many muscle fibers will be stimulated. In this experiment, the frequency and strength of an electrical current determines the response.

A single contraction of skeletal muscle is called a **muscle twitch**. A tracing of a muscle twitch (**Figure 14.3**) shows three distinct phases: latent, contraction, and relaxation. The **latent period** is the interval from stimulus application until the muscle begins to shorten. Although no activity is indicated on the tracing during this phase, excitation-contraction coupling is occurring within the muscle. During the **period of contraction**, the muscle fibers shorten; the tracing shows an increasingly higher needle deflection and the tracing peaks. During the **period of relaxation**, represented by a downward curve of the tracing, the muscle fibers relax and lengthen. On a slowly moving recording surface, the single muscle twitch appears as a spike rather than a bell-shaped curve, as in Figure 14.3, but on a rapidly moving recording surface, the three distinct phases just described become recognizable.

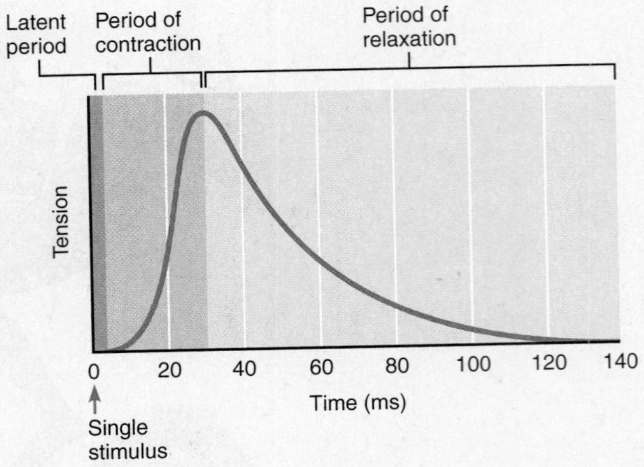

Figure 14.3 Tracing of a muscle twitch.

Determining the Threshold Stimulus

1. Assuming that you have already set up the recording apparatus, set the time marker to deliver one pulse per second and set the paper speed at a slow rate, approximately 0.1 cm per second.

2. Set the duration control on the stimulator between 7 and 15 milliseconds (msec), multiplier ×1. Set the voltage control at 0 V, multiplier ×1. Turn the sensitivity control knob of the stimulator fully clockwise (lowest value, greatest sensitivity).

3. Administer single stimuli to the muscle at 1-minute intervals, beginning with 0.1 V and increasing each successive stimulus by 0.1 V until a contraction is obtained (shown by a spike on the paper).

At what voltage did contraction occur? ________________ V

The voltage at which the first perceptible contractile response is obtained is called the **threshold stimulus**. All stimuli applied prior to this point are termed **subthreshold stimuli**, because at those voltages no response was elicited.

4. Stop the recording and mark the record to indicate the threshold stimulus, voltage, and time. Do not remove the record from the recording surface; continue with the next experiment. Remember: keep the muscle preparation moistened with Ringer's solution at all times.

Observing Graded Muscle Response to Increased Stimulus Intensity

1. Follow the previous setup instructions, but set the voltage control at the threshold voltage (as determined in the first experiment).

2. Deliver single stimuli at 1-minute intervals. Initially increase the voltage between shocks by 0.5 V; then increase the voltage by 1 to 2 V between shocks as the experiment continues, until contraction height increases no further. Stop the recording apparatus.

What voltage produced the highest spike (and thus the maximal strength of contraction)? ____________ V

This voltage, called the **maximal stimulus** (for *your* muscle specimen), is the weakest stimulus at which all muscle fibers are being stimulated.

3. Mark the record *maximal stimulus.* Record the maximal stimulus voltage and the time you completed the experiment.

4. What is happening to the muscle as the voltage is increased?

____

____

What is another name for this phenomenon? (Use an appropriate reference if necessary.)

____

5. Explain why the strength of contraction does not increase once the maximal stimulus is reached.

____

____

Timing the Muscle Twitch

1. Follow the previous setup directions, but set the voltage for the maximal stimulus (as determined in the preceding experiment) and set the paper advance or recording speed at maximum. Record the paper speed setting:

____________________ mm/sec

2. Determine the time required for the paper to advance 1 mm by using the formula

$$\frac{1 \text{ mm}}{\text{mm/sec (paper speed)}}$$

(Thus, if your paper speed is 25 mm/sec, each mm on the chart equals 0.04 sec.) Record the computed value:

1 mm = ____________ sec

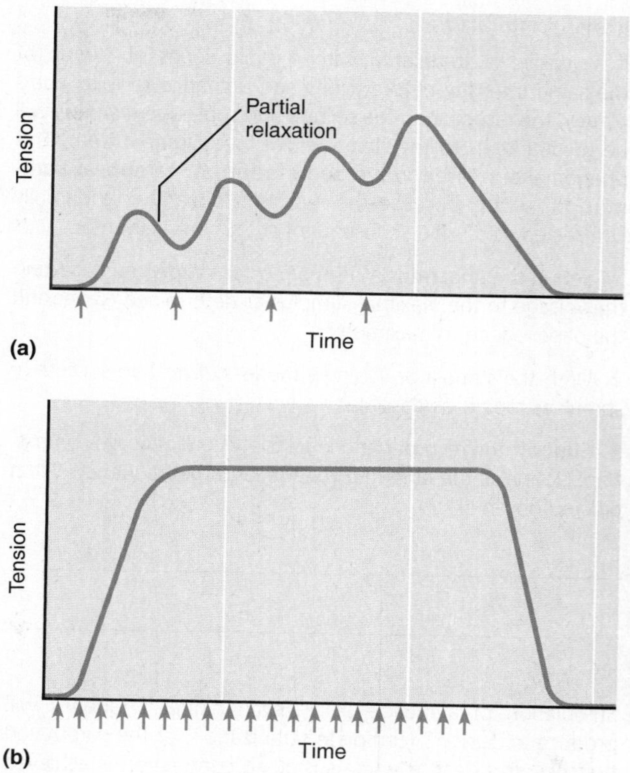

Figure 14.4 Muscle response to stimulation. Arrows represent stimuli. **(a)** Wave summation at low-frequency stimulation. **(b)** Fused tetanus occurs as stimulation rate is increased.

3. Deliver single stimuli at 1-minute intervals to obtain several "twitch" curves. Stop the recording.

4. Determine the duration of the latent, contraction, and relaxation phases of the twitches, and record the data below.

Duration of latent period: ____________ sec

Duration of period of contraction: ____________ sec

Duration of period of relaxation: ____________ sec

5. Label the record to indicate the point of stimulus, the beginning of contraction, the end of contraction, and the end of relaxation.

6. Allow the muscle to rest (but keep it moistened with frog Ringer's solution) before continuing with the next experiment.

Observing Graded Muscle Response to Increased Stimulus Frequency

Muscles subjected to frequent stimulation, without a chance to relax, exhibit two kinds of responses—wave summation and tetanus—depending on the stimulus frequency (**Figure 14.4**).

Text continues on next page. →

14

Wave Summation

If a muscle is stimulated with a rapid series of stimuli of the same intensity before it has had a chance to relax completely, the response to the second and subsequent stimuli will be greater than to the first stimulus (see Figure 14.4a). This phenomenon, called **wave summation**, or **temporal summation**, occurs because the muscle is already in a partially contracted state when subsequent stimuli are delivered.

1. Set up the apparatus as in the previous experiment, setting the voltage to the maximal stimulus as determined earlier and the paper speed to maximum.

2. With the stimulator in single mode, deliver two successive stimuli as rapidly as possible.

3. Shut off the recorder and label the record as *wave summation*. Note also the time, the voltage, and the frequency. What did you observe?

____

____

Tetanus

Stimulation of a muscle at an even higher frequency will produce a "fusion" (complete tetanization) of the summated twitches. In effect, a single sustained contraction is achieved in which no evidence of relaxation can be seen (see Figure 14.4b). **Fused tetanus**, or **complete tetanus**, demonstrates the maximum force generated by a skeletal muscle Outside the laboratory, fused tetanus occurs infrequently.

1. To demonstrate fused tetanus, maintain the conditions used for wave summation except for the frequency of stimulation. Set the stimulator to deliver 60 stimuli per second.

2. As soon as you obtain a single smooth, sustained contraction (with no evidence of relaxation), discontinue stimulation and shut off the recorder.

3. Label the tracing with the conditions of experimentation, the time, and the area of *fused* or *complete tetanus*.

Inducing Muscle Fatigue

Muscle fatigue is a reversible physiological condition in which a muscle is unable to contract even though it is being stimulated. Fatigue can occur with short-duration maximal contraction or long-duration submaximal contraction. Although the phenomenon of muscle fatigue is not completely understood, several factors appear to contribute to it. Most affect excitation-contraction coupling. One theory involves the buildup of inorganic phosphate (P_i) from ATP and creatine phosphate breakdown, which may block calcium release from the sarcoplasmic reticulum (SR). Another theory suggests that potassium accumulation in the T tubules may block calcium release from the SR and alter the membrane potential of the muscle fiber. Lactic acid buildup, long implicated as a cause of fatigue, does not appear to play a role.

1. To demonstrate muscle fatigue, set up an experiment like the tetanus experiment but continue stimulation until the muscle completely relaxes and the contraction curve returns to the baseline.

2. Measure the time interval between the beginning of complete tetanus and the beginning of fatigue (when the tracing begins its downward curve). Mark the record appropriately.

3. Determine the time required for complete fatigue to occur (the time interval from the beginning of fatigue until the return of the curve to the baseline). Mark the record appropriately.

4. Allow the muscle to rest (keeping it moistened with Ringer's solution) for 10 minutes, and then repeat the experiment.

What was the effect of the rest period on the fatigued muscle?

____

____

What might be the physiological basis for this reaction?

____

____

Determining the Effect of Load on Skeletal Muscle

When the fibers of a skeletal muscle are slightly stretched by a weight or tension, the muscle responds by contracting more forcibly and thus is capable of doing more work. When the actin and myosin barely overlap, sliding can occur along nearly the entire length of the actin filaments. If the load is increased beyond the optimum, the latent period becomes longer, contractile force decreases, and relaxation (fatigue) occurs more quickly. With excessive stretching, the muscle is unable to develop any active tension and no contraction occurs. Since the filaments no longer overlap at all with this degree of stretching, the sliding force cannot be generated.

If your equipment allows you to add more weights to the muscle specimen or to increase the tension on the muscle, perform the following experiment to determine the effect of loading on skeletal muscle and to develop a work curve for the frog's gastrocnemius muscle.

1. Set the stimulator to deliver the maximal voltage as previously determined.

2. Stimulate the unweighted muscle with single shocks at 1- to 2- second intervals to achieve three or four muscle twitches.

3. Stop the recording apparatus and add 10 g of weight or tension to the muscle. Restart and advance the recording about 1 cm, and then stimulate again to obtain three or four spikes.

4. Repeat the previous step seven more times, increasing the weight by 10 g each time until the total load on the muscle is 80 g or the muscle fails to respond. If the calcaneal tendon tears, the weight will drop, which ends the trial. In such cases, you will need to prepare another frog's leg to continue the

experiments and the maximal stimulus will have to be determined for the new muscle preparation.

5. When these "loading" experiments are completed, discontinue recording. Mark the curves on the record to indicate the load (in grams).

6. Measure the height of contraction (in millimeters) for each sequence of twitches obtained with each load, and insert this information in the **Activity 2 chart**.

7. Compute the work done by the muscle for each twitch (load) sequence.

Weight of load (g) × distance load lifted (mm) = work done

Enter these calculations into the chart in the column labeled Work done, Trial 1.

8. Allow the muscle to rest for 5 minutes. Then conduct a second trial in the same manner (i.e., repeat steps 2 through 7). Record this second set of measurements and calculations in the columns labeled Trial 2. Be sure to keep the muscle well moistened with Ringer's solution during the resting interval.

9. Using two different colors, plot a line graph of work done against the weight on the grid accompanying the chart for each trial. Label each plot appropriately.

10. Dismantle all apparatus and prepare the equipment for storage. Dispose of the frog remains in the appropriate container. Discard the gloves as instructed and wash and dry your hands.

11. Inspect your records of the experiments, and make sure each is fully labeled with the experimental conditions, the date, and the names of those who conducted the experiments. For future reference, attach a tracing (or a copy of the tracing) for each experiment to this page.

Activity 2: Results for Effect of Load on Skeletal Muscle

	Distance load lifted (mm)		Work done	
Load (g)	**Trial 1**	**Trial 2**	**Trial 1**	**Trial 2**
0				
10				
20				
30				
40				
50				
60				
70				
80				

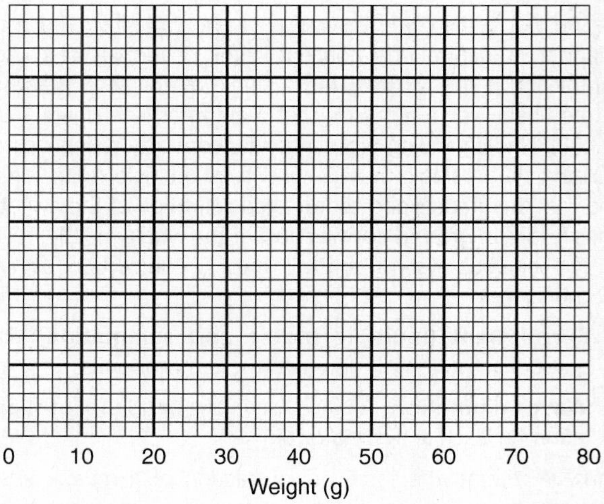

14

Activity 3

Demonstrating Muscle Fatigue in Humans

1. Work in small groups. In each group select a subject, a timer, and a recorder.

2. Obtain a copy of the laboratory manual and a copy of the textbook. Weigh each book separately, and then record the weight of each in the **Activity 3 chart** in step 6.

3. The subject is to extend an upper limb straight out in front of him or her, holding the position until the arm shakes or the muscles begin to ache. Record the time to fatigue on the chart.

4. Allow the subject to rest for several minutes. Now ask the subject to hold the laboratory manual while keeping the arm and forearm in the same position as in step 3 above. Record the time to fatigue on the chart.

5. Allow the subject to rest again for several minutes. Now ask the subject to hold the textbook while keeping the upper limb in the same position as in steps 3 and 4 above. Record the time to fatigue on the chart.

6. Each person in the group should take a turn as the subject, and all data should be recorded in the chart.

Activity 3: Results for Human Muscle Fatigue

		Time elapsed until fatigue		
Load	**Weight of object**	**Subject 1**	**Subject 2**	**Subject 3**
Appendage	N/A			
Lab Manual				
Textbook				

7. What can you conclude about the effect of load on muscle fatigue? Explain.

____

____

____

Activity 4

Electromyography in a Human Subject Using BIOPAC®

Part 1: Temporal and Multiple Motor Unit Summation

This activity is an introduction to a procedure known as **electromyography**, the recording of skin-surface voltage that indicates underlying skeletal muscle contraction. The actual visible recording of the resulting voltage waveforms is called an **electromyogram (EMG)**.

A single skeletal muscle consists of numerous elongated *skeletal muscle fibers* (Exercise 12). These muscle fibers are excited by *motor neurons* of the central nervous system whose *axons* terminate at the muscle. An axon of a motor neuron branches profusely at the muscle. Each branch produces multiple **axon terminals**, each of which innervates a single muscle fiber. The number of muscle fibers controlled by a single motor neuron can vary greatly, from five (for fine control needed in the hand) to 500 (for gross control, such as in the buttocks). The most important organizational concept in the physiology of muscle contraction is the **motor unit**, a single motor neuron and all of the fibers within a muscle that it activates (see Figure 12.6, p. 188). Understanding gross muscular contraction depends on realizing that a single muscle consists of multiple motor units, and that the gradual and coordinated activation of these motor units results in **graded contraction** of the whole muscle.

The nervous system controls muscle contraction by two mechanisms:

- **Recruitment (multiple motor unit summation):** The gradual activation of more and more motor units
- **Wave summation:** An increase in the *frequency* of stimulation for each active motor unit

Thus, increasing the force of contraction of a muscle arises from gradually increasing the number of motor units being activated and increasing the frequency of nerve impulses delivered by those active motor units.

Setting Up the Equipment

1. Connect the BIOPAC® unit to the computer, then turn the computer **ON**.
2. Make sure the BIOPAC® unit is **OFF**.
3. Plug in the equipment (as shown in **Figure 14.5**).
   - Electrode lead set—CH 1
   - Headphones—back of MP36/35 unit or top of MP45 unit
4. Turn the BIOPAC® unit **ON**.
5. Attach three electrodes to the subject's dominant forearm as shown in **Figure 14.6** and attach the electrode leads according to the colors indicated.
6. Start the Biopac Student Lab program by double-clicking the icon on the desktop or by following your instructor's guidance.
7. Select lesson **L01-Electromyography (EMG) I** from the menu and click **OK**.
8. Type in a filename that will save this subject's data on the computer hard drive. You may want to use the subject's last name followed by EMG-1 (for example, SmithEMG-1). Then click **OK**.

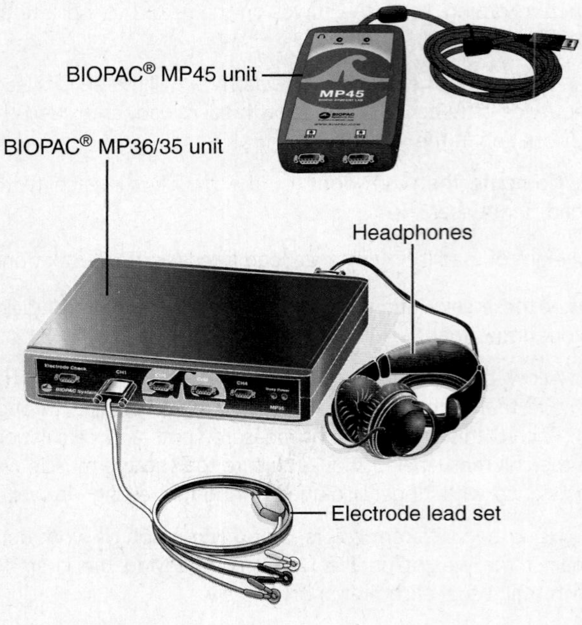

Figure 14.5 Setting up the BIOPAC® equipment to observe recruitment and temporal summation. Plug the headphones into the back of the MP36/35 data acquisition unit or into the top of the MP45 unit, and the electrode lead set into Channel 1. Electrode leads and headphones are shown connected to the MP36/35 unit.

Calibrating the Equipment

1. With the subject in a still position, click **Calibrate**. This initiates the process by which the computer automatically establishes parameters to record the data properly for the subject.

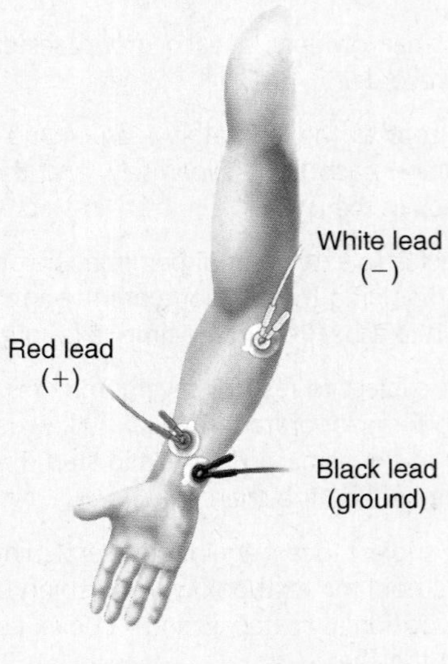

Figure 14.6 Placement of electrodes and the appropriate attachment of electrode leads by color.

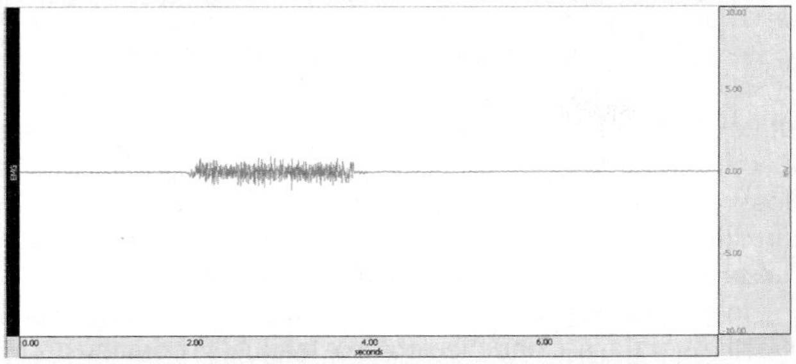

Figure 14.7 Example of waveform during the calibration procedure.

2. After you click **OK**, have the subject wait for 2 seconds, clench the fist tightly for 2 seconds, then release the fist and relax. The computer then automatically stops the recording.

3. Observe the recording of the calibration data, which should look like the waveform in the example (**Figure 14.7**).

- If the data look very different, click **Redo Calibration** and repeat the steps above.
- If the data look similar, proceed to the next section.

Recording the Data

1. Tell the subject that the recording will be of a series of four fist clenches, with the following instructions: First clench the fist softly for 2 seconds, then relax for 2 seconds, then clench harder for 2 seconds, and relax for 2 seconds, then clench even harder for 2 seconds, and relax for 2 seconds, and finally clench with maximum strength, then relax. The result should be a series of four clenches of increasing intensity.

When the subject is ready to do this, click **Record**; then click **Suspend** when the subject is finished.

2. Observe the recording of the data, which should look like the waveforms in the example (**Figure 14.8**).

- If the data look very different, click **Redo** and repeat the steps above.

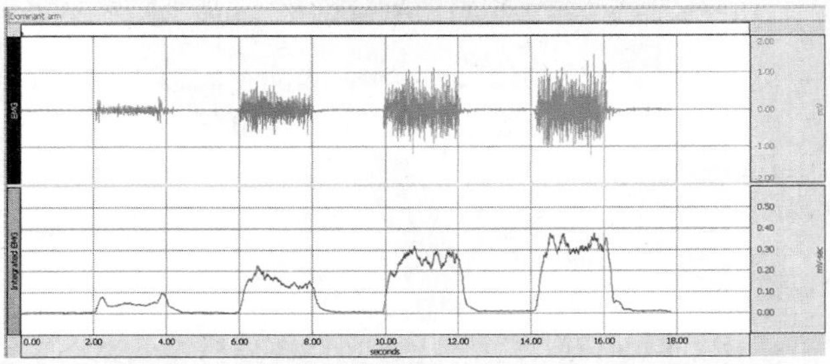

Figure 14.8 Example of waveforms during the recording of data. Note the increased signal strength with the increasing force of the clench.

- If the data look similar, click **STOP**. Click **YES** in response to the question, "Are you finished with both forearm recordings?"

Optional: Anyone can use the headphones to listen to an "auditory version" of the electrical activity of contraction by clicking **Listen** and having the subject clench and relax. Note that the frequency of the auditory signal corresponds with the frequency of action potentials stimulating the muscles. The signal will continue to run until you click **STOP**.

3. Click **Done**, and then remove all electrodes from the forearm.

- If you wish to record from another subject, choose the **Record from another subject** option and return to step 5 under Setting Up the Equipment.
- If you are finished recording, choose **Analyze current data file** and click **OK**. Proceed to Data Analysis, step 2.

Data Analysis

1. If you are just starting the BIOPAC® program to perform data analysis, enter **Review Saved Data** mode and choose the file with the subject's EMG data (for example, SmithEMG-1).

2. Observe the **Raw EMG** recording and computer-calculated **Integrated EMG**. The raw EMG is the actual recording of the voltage (in mV) at each instant in time, while the integrated EMG reflects the absolute intensity of the voltage from baseline at each instant in time.

3. To analyze the data, set up the first four pairs of channel/measurement boxes at the top of the screen by selecting the following channels and measurement types from the drop-down menus.

Channel	Measurement	Data
CH 1	Min	raw EMG
CH 1	Max	raw EMG
CH 1	P-P	raw EMG
CH 40	Mean	integrated EMG

4. Use the arrow cursor and click the I-beam cursor box at the lower right of the screen to activate the "area selection" function. Using the activated I-beam cursor, highlight the first EMG cluster, representing the first fist-clenching (**Figure 14.9**, p. 242).

5. Notice that the computer automatically calculates the **Min**, **Max**, **P-P**, and **Mean** for the selected area. These measurements, calculated from the data by the computer, represent the following:

Min: Displays the *minimum* value in the selected area

Max: Displays the *maximum* value in the selected area

P-P (Peak-to-Peak): Measures the difference in value between the highest and lowest values in the selected area

Mean: Displays the average value in the selected area

Text continues on next page. →

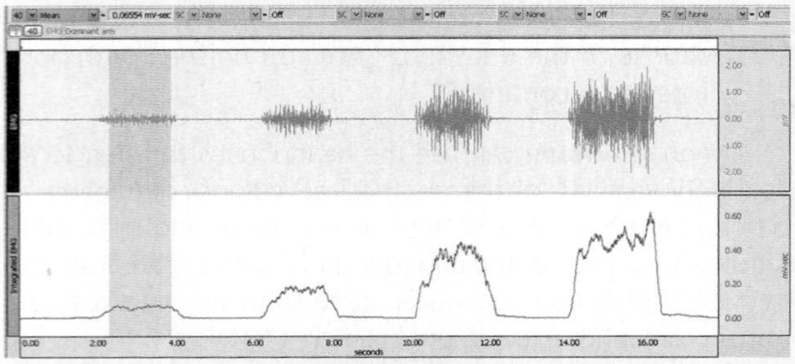

Figure 14.9 Using the I-beam cursor to highlight a cluster of data for analysis.

6. Write down the data for clench 1 in the chart in step 7 (round to the nearest 0.01 mV).

7. Using the I-beam cursor, highlight the clusters for clenches 2, 3, and 4, and record the data in the **EMG Cluster Results chart.**

EMG Cluster Results				
	Min	**Max**	**P-P**	**Mean**
Clench 1				
Clench 2				
Clench 3				
Clench 4				

From the data recorded in the chart, what trend do you observe for each of these measurements as the subject gradually increases the force of muscle contraction?

____

____

What is the relationship between maximum voltage for each clench and the number of motor units in the forearm that are being activated?

____

____

Part 2: Force Measurement and Fatigue

In this set of activities, you will be observing graded muscle contractions and fatigue in a subject. **Graded muscle contractions**, which represent increasing levels of force generated by a muscle, depend on (1) the gradual activation of more motor units, and (2) increasing the frequency of action potentials (stimulation) for each active motor unit. This permits a range of forces to be generated by any given muscle or group of muscles, all the way up to the maximum force.

For example, the biceps muscle will have more active motor units and exert more force when lifting a 10-kg object than when lifting a 2-kg object. In addition, the motor neuron of each active motor unit will increase the frequency of action potentials delivered to the motor units, resulting in tetanus. When all of the motor units of a muscle are activated and in a state of tetanus, the maximum force of that muscle is achieved. Recall that fatigue is a condition in which the muscle gradually loses some or all of its ability to contract after contracting for an extended period of time. Recent experimental evidence suggests that this is mostly due to ionic imbalances.

In this exercise, you will observe and measure graded contractions of the fist, and then observe fatigue, in both the dominant and nondominant arms. To measure the force generated during fist contraction, you will use a **hand dynamometer** (*dynamo* = force; *meter* = measure). The visual recording of force is called a **dynagram**, and the procedure of measuring the force itself is called **dynamometry**.

You will first record data from the subject's **dominant arm** (forearm 1) indicated by his or her "handedness," then repeat the procedures on the subject's **nondominant arm** (forearm 2) for comparison.

Setting Up the Equipment

1. Connect the BIOPAC® unit to the computer and turn the computer **ON**.

2. Make sure the BIOPAC® unit is **OFF**.

3. Plug in the equipment (as shown in **Figure 14.10**).
   - Electrode lead set—CH1
   - Hand dynamometer—CH2
   - Headphones—back of MP36/35 unit or top of MP45 unit

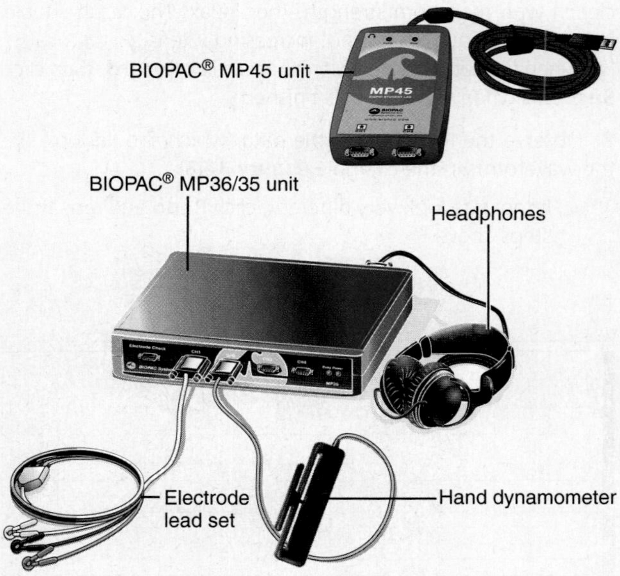

Figure 14.10 Setting up the BIOPAC® equipment to observe graded muscle contractions and muscle fatigue. Plug the headphones into the back of the MP36/35 data acquisition unit or into the top of the MP45 unit, the electrode lead set into Channel 1, and the hand dynamometer into Channel 2. Electrode leads and dynamometer are shown connected to the MP36/35 unit.

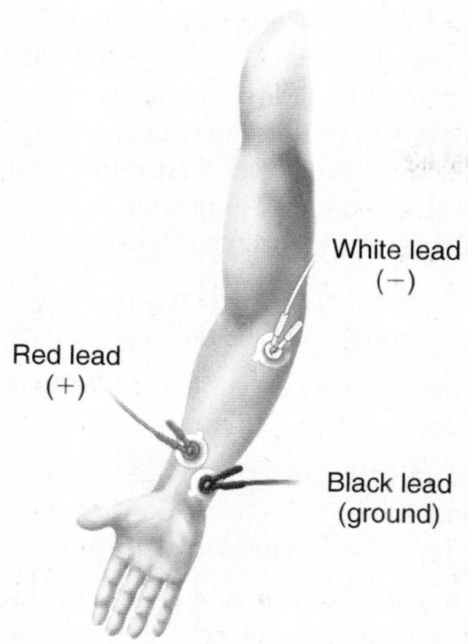

Figure 14.11 Placement of electrodes and the appropriate attachment of electrode leads by color.

4. Turn the BIOPAC® unit **ON**.

5. Attach three electrodes to the subject's *dominant* forearm (forearm 1; see the attachments in **Figure 14.11**), and attach the electrode leads according to the colors indicated.

6. Start the Biopac Student Lab program on the computer by double-clicking the icon on the desktop or by following your instructor's guidance.

7. Select lesson **L02-EMG-2** from the menu and click **OK**.

8. Type in a filename that will save this subject's data on the computer hard drive. You may want to use subject's last name followed by EMG-2 (for example, SmithEMG-2). Then click **OK**.

Calibrating the Equipment

1. With the hand dynamometer at rest on the table, click **Calibrate**. This initiates the process by which the computer automatically establishes parameters to record the data properly for the subject.

2. A pop-up window prompts the subject to remove any grip force. This is to ensure that the dynamometer has been at rest on the table and that no force is being applied. When this is so, click **OK**.

3. As instructed by the pop-up window, the subject is to grasp the hand dynamometer with the dominant hand. With model SS25LA/LB grasp the dynamometer with the palm of the hand against the short grip bar (as shown in **Figure 14.12a**). Hold model SS25LA/LB vertically. With model SS56L, wrap the hand around the bulb (as shown in Figure 14.12b). Do not curl the fingers into the bulb. (The older SS25L hand dynamometer may also be used with any of the data acquisition units.) Then, click **OK**. The instructions that follow apply to this model of dynamometer.

4. Tell the subject that the instructions will be to wait 2 seconds, then squeeze the hand dynamometer as hard as possible for 2 seconds, and then relax. The computer will automatically stop the calibration.

5. When the subject is ready to proceed, click **OK** and follow the instructions in step 4, which are also in the pop-up window. The calibration will stop automatically.

6. Observe the recording of the calibration data, which should look like the waveforms in the calibration example (**Figure 14.13**, p. 244).

- If the data look very different, click **Redo Calibration** and repeat the steps above.
- If the data look similar, proceed to the next section.

Recording Incremental Force Data for the Forearm

1. Using the force data from the calibration procedure, estimate the **maximum force** that the subject generated (kg).

2. Divide that maximum force by four. In the following steps, the subject will gradually increase the force in approximately

Text continues on next page. →

14

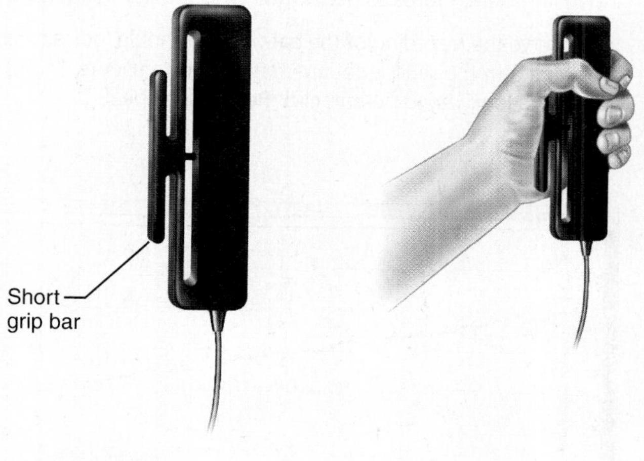

(a) Model SS25LA

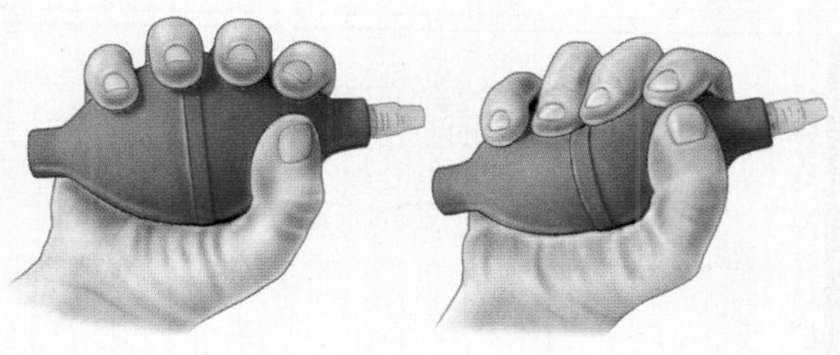

(b) Model SS56L

Figure 14.12 Proper grasp of the hand dynamometer using either model SS25LA/LB or model SS56L.

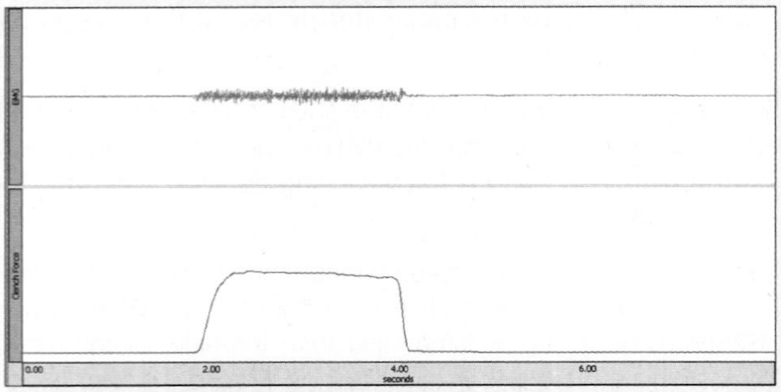

Figure 14.13 Example of calibration data. Electromyography is measured in millivolts at the top, and force is measured in kilograms or kgf/m^2 at the bottom.

these increments. For example, if the maximum force generated was 20 kg, the increment will be 20/4 = 5 kg. The subject will grip at 5 kg, then 10 kg, then 15 kg, and then 20 kg. The subject should watch the tracing on the computer screen and compare it to the scale on the right to determine each target force. Click **Continue**.

3. After you click **Record**, have the subject wait 2 seconds, clench at the first force increment 2 seconds (for example, 5 kg), then relax 2 seconds, clench at the second force increment 2 seconds (10 kg), then relax 2 seconds, clench at the third force increment 2 seconds (15 kg), then relax 2 seconds, then clench with the maximum force 2 seconds (20 kg), and then relax. When the subject relaxes after the maximum clench, click **Suspend** to stop the recording.

4. Observe the recording of the data, which should look similar to the data in the example (**Figure 14.14**).

- If the data look very different, click **Redo** and repeat the steps above.
- If the data look similar, click **Continue** and proceed to observation and recording of muscle fatigue.

Recording Muscle Fatigue Data for the Forearm

Continuing from the end of the incremental force recording, the subject will next record muscle fatigue.

1. After you click **Record**, the recording will continue from where it stopped and the subject will clench the dynamometer with maximum force for as long as possible. A "marker" will appear at the top of the data, denoting the beginning of this recording segment. When the subject's clench force falls below 50% of the maximum (for example, below 10 kg for a subject with 20 kg maximum force), click **Suspend**. The subject should not watch the screen during this procedure; those helping can inform the subject when it is time to relax.

2. Observe the recording of the data, which should look similar to the data in the muscle fatigue data example (**Figure 14.15**).

- If the data look very different, click **Redo** and repeat the steps above.
- If the data look similar, and you want to record from the nondominant arm, click **Continue** and proceed to Recording from the Nondominant Arm.
- If the data look similar, and you do not want to record from the nondominant arm or you have just finished recording the nondominant arm, click **STOP**. A dialog box comes up asking if you are sure you want to stop recording. Click **NO** to return to the **Record** or **Stop** options, providing one last chance to redo the fatigue recording. Click **YES** to end the recording session and automatically save your data.

Optional: Anyone can use the headphones to listen to an "auditory version" of the electrical activity of contraction by clicking **Listen** and having the subject clench and relax. Note that the frequency of the auditory signal corresponds to the frequency of action potentials stimulating the muscles. The signal will continue to run until you click **STOP**.

3. Click **Done**. Choose **Analyze current data file** and proceed to Data Analysis, step 2.

4. Remove all electrodes from the arm of the subject.

Recording from the Nondominant Arm

1. To record from the nondominant forearm, attach three electrodes to the subject's forearm (as shown in Figure 14.11) and attach the electrode leads according to the colors indicated.

2. Click **Record**. Repeat the Clench-Release-Wait cycles with increasing clench force as performed with the dominant arm.

3. Observe the recording of the data, which should look similar to the data in the muscle fatigue data example (Figure 14.15). If the data look very different, click **Redo** and repeat.

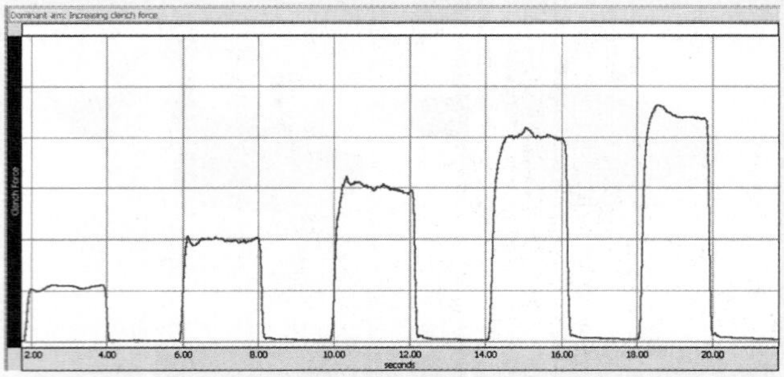

Figure 14.14 Example of incremental force data.

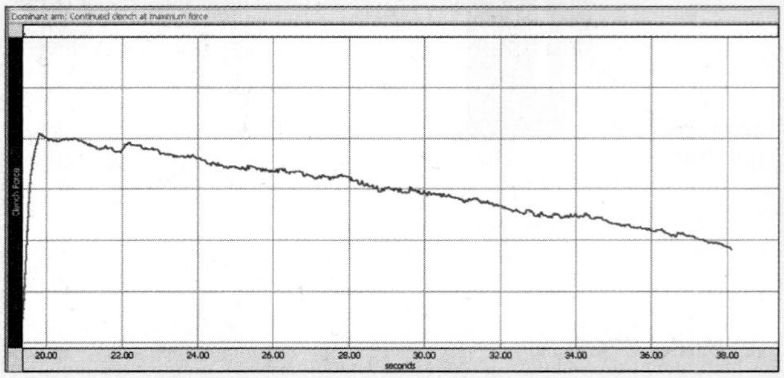

Figure 14.15 Example of muscle fatigue data.

4. End the session by repeating the steps for muscle fatigue. Click **Continue** and repeat steps 1 and 2 of the muscle fatigue section, recording the nondominant arm.

Data Analysis

1. In **Review Saved Data** mode, select the file that is to be analyzed (for example, SmithEMG-2-1-L02).

2. Observe the recordings of the clench **Force** (kg or kgf/m^2), **Raw EMG** (mV), and computer-calculated **Integrated EMG** (mV). The force is the actual measurement of the strength of clench in kilograms at each instant in time. The raw EMG is the actual recording of the voltage (in mV) at each instant in time, and the integrated EMG indicates the absolute intensity of the voltage from baseline at each instant in time.

3. To analyze the data, set up the first three pairs of channel/measurement boxes at the top of the screen. Select the following channels and measurement types:

Channel	Measurement	Data
CH 41	Mean	clench force
CH 40	Mean	integrated EMG

4. Use the arrow cursor and click the I-beam cursor box on the lower right side of the screen to activate the "area selection" function. Using the activated I-beam cursor, highlight the "plateau phase" of the first clench cluster. The plateau should be a relatively flat force in the middle of the cluster (**Figure 14.16**).

5. Observe that the computer automatically calculates the **P-P** and **Mean** values for the selected area. These measurements, calculated from the data by the computer, represent the following:

P-P (Peak-to-Peak): Measures the difference in value between the highest and lowest values in the selected area

Mean: Displays the average value in the selected area

6. In the chart in step 7, record the data for clench 1 (for example, 5-kg clench) to the nearest 0.01.

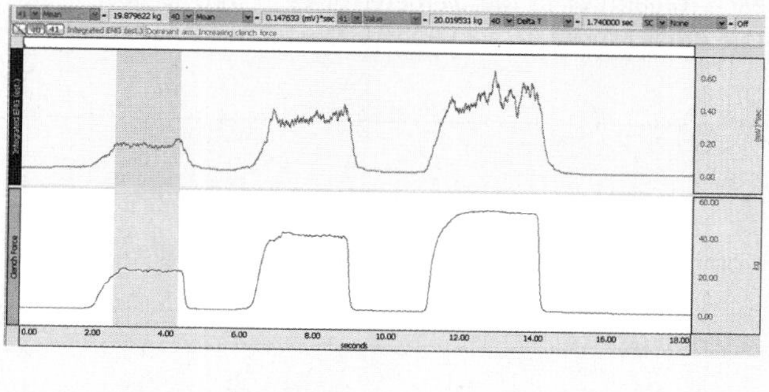

Figure 14.16 Highlighting the "plateau" of clench cluster 1.

7. Using the I-beam cursor, highlight the clusters for the subsequent clenches, and record the data in the **Dominant Forearm Clench Increments chart**.

Dominant Forearm Clench Increments		
	Force at plateau Mean (kg or kgf/m^2)	**Integrated EMG Mean (mV-sec)**
Clench 1		
Clench 2		
Clench 3		
Clench 4		

8. Scroll along the bottom of the data page to the segment that includes the recording of muscle fatigue (it should begin after the "marker" that appears at the top of the data).

9. Change the channel/measurement boxes so that the first two selected appear as follows (the third should be set to "none"):

Channel	Measurement	Data
CH 41	Value	force
CH 40	Delta T	integrated EMG

Value: Measures the highest value in the selected area (force measured in kg with SS25LA/LB or kgf/m^2 with the SS56L.)

Delta T: Measures the time elapsed in the selected area (time measured in seconds.)

10. Use the arrow cursor and click on the I-beam cursor box on the lower right side of the screen to activate the "area selection" function.

11. Using the activated I-beam cursor, select just the single point of maximum clench strength at the start of this data segment as shown in **Figure 14.17**.

12. In the **Dominant Forearm Fatigue Measurement chart** on p. 246, note the maximum force measurement for this point.

Text continues on next page. →

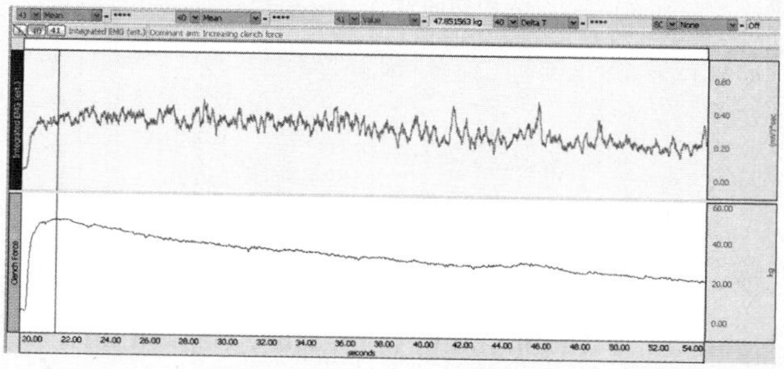

Figure 14.17 Selection of single point of maximum clench.

Dominant Forearm Fatigue Measurement		
Maximum clench force (kg or kgf/m²)	**50% of the maximum clench force (divide maximum clench force by 2)**	**Time to fatigue (seconds)**

13. Calculate the value of 50% of the maximum clench force, and record this in the data chart.

14. Using a metric tape measure, measure the circumference of the subject's dominant forearm at its greatest diameter:

______________ cm

15. Measure the amount of time that elapsed between the initial maximum force and the point at which the subject fatigued to 50% of this level. Using the activated I-beam cursor, highlight the area from the point of 50% clench force back to the point of maximal clench force, as shown in the example (**Figure 14.18**).

16. Note the time it took the subject to reach this point of fatigue (CH 40 Delta T), and record this data in the Dominant Forearm Fatigue Measurement chart.

Repeat Data Analysis for the Nondominant Forearm

1. Return to step 1 of the Data Analysis section, and repeat the same measurements for the nondominant forearm (forearm 2).

2. Record your data in the **Nondominant Forearm Clench Increments chart** and the **Nondominant Forearm Fatigue Measurement chart**. These data will be used for comparison.

3. Using a tape measure, measure the circumference of the subject's nondominant forearm at its greatest:

______________ cm

4. When finished, exit the program by going to the **File** menu at the top of the screen and clicking **Quit**.

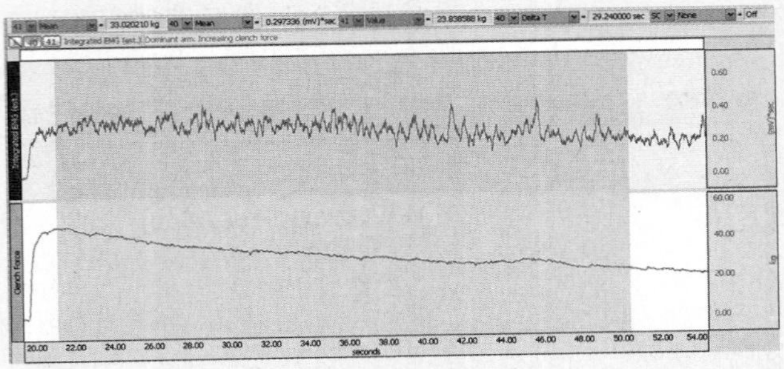

Figure 14.18 Highlighting to measure elapsed time to 50% of the maximum clench force.

Is there a difference in maximal force that was generated between the dominant and nondominant forearms? If so, how much?

____

Calculate the percentage difference in force between the dominant maximal force and nondominant maximal force.

____

____

Nondominant Forearm Clench Increments		
	Force at plateau Mean (kg or kgf/m²)	**Integrated EMG Mean (mV-sec)**
Clench 1		
Clench 2		
Clench 3		
Clench 4		

Nondominant Forearm Fatigue Measurement		
Maximum Clench Force (kg or kgf/m²)	**50% of the maximum clench force (divide maximum clench force by 2)**	**Time to fatigue (seconds)**

Is there a difference in the circumference between the dominant and nondominant forearms? If so, how much?

____

If there is a difference in circumference, is this difference likely to be due to a difference in the *number* of muscle fibers in each forearm or in the *diameter* of each muscle fiber in the forearms? Explain. Use an appropriate reference if needed.

____

____

Compare the time to fatigue between the two forearms.

____

____

Instructors may assign a portion of the Review Sheet questions using Mastering A&P™

REVIEW SHEET

EXERCISE 14

Skeletal Muscle Physiology: Frogs and Human Subjects

Name ______________________ Lab Time/Date ______________________

Muscle Activity

1. The following group of incomplete statements begins with a description of a muscle fiber in the resting state just before stimulation. Complete each statement by choosing the correct response from the key items.

Key:
a. Na^+ diffuses out of the cell
b. K^+ diffuses out of the cell
c. Na^+ diffuses into the cell
d. K^+ diffuses into the cell
e. inside the cell
f. outside the cell
g. relative ionic concentrations on the two sides of the membrane
h. electrical conditions
i. activation of the ATP-dependent sodium-potassium pump

At rest, there is a greater concentration of Na^+ ____________; there is a greater concentration of K^+ ____________.

When the stimulus is delivered, the permeability of the membrane at that point is changed; and ____________, initiating the depolarization of the membrane. Almost as soon as the depolarization wave has begun, a repolarization wave follows it across the membrane. This occurs as ____________. Repolarization restores the ____________ of the resting state. The ____________ is (are) reestablished by ____________.

2. Number the following statements in the proper sequence to describe the contraction mechanism in a skeletal muscle fiber. Number 1 has already been designated.

_____1_____ Depolarization occurs, and the action potential is generated.

____________ The muscle fiber relaxes.

____________ The calcium ion concentrations at the myofilaments increase; the myofilaments slide past one another, and the muscle fiber shortens.

____________ The action potential, carried deep into the muscle fiber by the T tubules, triggers the release of calcium ions from the sarcoplasmic reticulum.

____________ The concentration of the calcium ions at the myofilaments decreases as they are actively transported into the sarcoplasmic reticulum.

3. Refer to your observations of muscle fiber contraction in Activity 1 to answer the following questions.

a. Did your data support your hypothesis? ____________

b. *Explain* your observations fully. ____________

____

____

c. Draw a relaxed and a contracted sarcomere below. Label the Z discs, thick filaments, and thin filaments.

Relaxed	Contracted

Induction of Contraction in the Frog Gastrocnemius Muscle

4. Why is it important to destroy the brain and spinal cord of a frog before conducting physiological experiments on muscle contraction? ______________________________

____

5. What kind of stimulus (electrical or chemical) travels from the motor neuron to skeletal muscle? ______________

What kind of stimulus (electrical or chemical) travels from the axon terminal to the sarcolemma? ______________

6. Give the name and duration of each of the three phases of the muscle twitch, and describe what is happening during each phase.

a. ______________, ______________ msec, ______________________________

____

b. ______________, ______________ msec, ______________________________

c. ______________, ______________ msec, ______________________________

7. Use the items in the key to identify the conditions described.

Key:

a. maximal stimulus
b. recruitment
c. subthreshold stimulus
d. tetanus
e. threshold stimulus
f. wave summation

______________ 1. sustained contraction without any evidence of relaxation

______________ 2. stimulus that results in no perceptible contraction

______________ 3. stimulus at which the muscle first contracts perceptibly

______________ 4. increasingly stronger contractions owing to stimulation at a rapid rate

______________ 5. increasingly stronger contractions owing to increased stimulus strength

______________ 6. weakest stimulus at which all muscle fibers in the muscle are contracting

8. Complete the following statements by writing the appropriate words on the corresponding numbered blanks at the right.

When a weak but smooth muscle contraction is desired, a few motor units are stimulated at a __1__ rate. Within limits, as the load on a muscle is increased, the muscle contracts __2__ (more/less) strongly.

1. ______________________________

2. ______________________________

9. During the frog experiment on muscle fatigue, how did the muscle contraction pattern change as the muscle began to fatigue?

____

____

How long was stimulation continued before fatigue was apparent? ______________________________

If the sciatic nerve that stimulates the living frog's gastrocnemius muscle had been left attached to the muscle and the stimulus had been applied to the nerve rather than the muscle, would fatigue have become apparent sooner, later, or at the same time?

____

10. What will happen to a muscle in the body when its nerve supply is destroyed or badly damaged? ______________________________

____

____

11. Explain the relationship between the load on a muscle and its strength of contraction. ______________________________

____

____

12. The skeletal muscles are maintained in a slightly stretched condition for optimal contraction. How is this accomplished?

____

Why does stretching a muscle beyond its optimal length reduce its ability to contract? (Include an explanation of the events at the level of the myofilaments.) ______________________________

____

____

13. If the length but not the tension of a muscle is changed, the contraction is called an isotonic contraction. In an isometric contraction, the tension is increased but the muscle does not shorten. Which type of contraction did you observe most often during the laboratory experiments?

____

Electromyography in a Human Subject Using BIOPAC®

14. If you were a physical therapist applying a constant voltage to the forearm, what might you observe if you gradually increased the *frequency* of stimulatory impulses, keeping the voltage constant each time?

____

____

15. Describe what is meant by the term *recruitment.* ____

____

16. Describe the physiological processes occurring in the muscle fibers that account for the gradual onset of muscle fatigue.

____

____

____

17. Most subjects use their dominant forearm far more than their nondominant forearm. What does this indicate about degree of activation of motor units and these factors: muscle fiber diameter, maximum muscle fiber force, and time to muscle fatigue? (You may need to use your textbook for help with this one.)

____

____

____

____

18. Define *dynamometry.* ____

____

19. + Carpal tunnel syndrome is a condition in which the median nerve is compressed as it travels through a narrow passage in the wrist. Explain how hand dynamometry might be used to assess a patient's need for surgery to release carpal tunnel compression. ____

____

EXERCISE 15

Histology of Nervous Tissue

Learning Outcomes

- ▶ Discuss the functional differences between neurons and neuroglia.
- ▶ List six types of neuroglia, and indicate where each is found in the nervous system.
- ▶ Identify the important anatomical features of a neuron on an appropriate image.
- ▶ List the functions of dendrites, axons, and axon terminals.
- ▶ Explain how a nerve impulse is transmitted from one neuron to another.
- ▶ State the function of myelin sheaths, and explain how Schwann cells myelinate nerve fibers (axons) in the peripheral nervous system.
- ▶ Classify neurons structurally and functionally.
- ▶ Differentiate between a nerve and a tract, and between a ganglion and a CNS nucleus.
- ▶ Identify an endoneurium, perineurium, and epineurium microscopically or in an appropriate image and cite their functions.

Pre-Lab Quiz

Instructors may assign these and other Pre-Lab Quiz questions using Mastering A&P™

1. Neuroglia of the peripheral nervous system include:
   a. ependymal cells and satellite cells
   b. oligodendrocytes and astrocytes
   c. satellite cells and Schwann cells
2. These branching neuron processes serve as receptive regions and transmit electrical signals toward the cell body. They are:
   a. axons
   b. collaterals
   c. dendrites
   d. neuroglia
3. Circle the correct underlined term. Axons running through the central nervous system form <u>tracts</u> / <u>nerves</u> of white matter.
4. Neurons can be classified according to structure.
   ________________ neurons have many processes that issue from the cell body.
   a. Bipolar **b.** Multipolar **c.** Unipolar
5. Within a nerve, each axon is surrounded by a covering called the:
   a. endoneurium **b.** epineurium **c.** perineurium

Go to Mastering A&P™ > Study Area to improve your performance in A&P Lab.

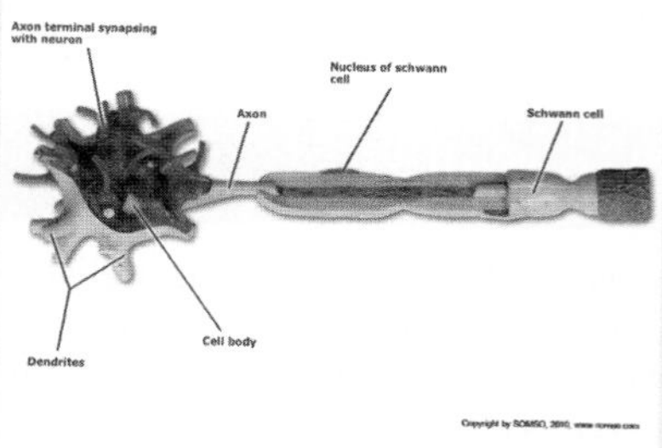

> Lab Tools > Practice Anatomy Lab > Anatomical Models: Nervous System > Peripheral Nervous System

Instructors may assign new Building Vocabulary coaching activities, Pre-Lab Quiz questions, Art Labeling activities, Practice Anatomy Lab Practical questions (PAL), and more using the Mastering A&P™ Item Library.

Materials

- ▶ Model of a "typical" neuron (if available)
- ▶ Compound microscope
- ▶ Immersion oil
- ▶ Prepared slides of an ox spinal cord smear and teased myelinated nerve fibers
- ▶ Prepared slides of Purkinje cells (cerebellum), pyramidal cells (cerebrum), and a dorsal root ganglion
- ▶ Prepared slide of a nerve (x.s.)

The nervous system is the master integrating and coordinating system, continuously monitoring and processing sensory information both from the external environment and from within the body. Every thought, action, and sensation is a reflection of its activity. Like a computer, it processes and integrates new "inputs" with information previously fed into it to produce a response.

Two primary divisions make up the nervous system: the central nervous system, or CNS, consisting of the brain and spinal cord, and the peripheral nervous system, or PNS, which includes all the nervous elements located outside the central nervous system. PNS structures include nerves, sensory receptors, and some clusters of neuron cell bodies.

Despite its complexity, nervous tissue is made up of just two principal cell types: neurons and neuroglia.

Neuroglia

The **neuroglia** ("nerve glue"), or **glial cells**, of the CNS include *astrocytes*, *oligodendrocytes*, *microglial cells*, and *ependymal cells* (**Figure 15.1**). The neuroglia found in the PNS include *Schwann cells*, also called neurolemmocytes, and *satellite cells*.

Neuroglia serve the needs of the delicate neurons by supporting and protecting them. In addition, they act as phagocytes (microglial cells), myelinate the cytoplasmic extensions of the neurons (oligodendrocytes and Schwann cells), play a role in capillary-neuron exchanges, and control the chemical environment around neurons (astrocytes). Although neuroglia resemble neurons in some ways (many have branching cellular extensions), they are not capable of generating and transmitting nerve impulses. In this exercise, we focus on the highly excitable neurons.

15

Neurons

Neurons, or nerve cells, are the basic functional units of nervous tissue. They are highly specialized to transmit messages from one part of the body to another in the form of nerve impulses. Although neurons differ structurally, they have many identifiable features in common (**Figure 15.2a** and **b**). All have a **cell body** from which slender processes extend. The cell body is both the *biosynthetic center* of the neuron and part of its *receptive region*. Neuron cell bodies make up the gray matter of the CNS, and form clusters there that are called **nuclei**. In the PNS, clusters of neuron cell bodies are called **ganglia**.

The neuron cell body contains a large round nucleus surrounded by cytoplasm. Two prominent structures are found in the cytoplasm: One is cytoskeletal elements called **neurofibrils**, which provide support for the cell and a means to transport substances throughout the neuron. The second is darkly staining structures called **chromatophilic substance**, clusters of rough endoplasmic reticulum involved in protein synthesis.

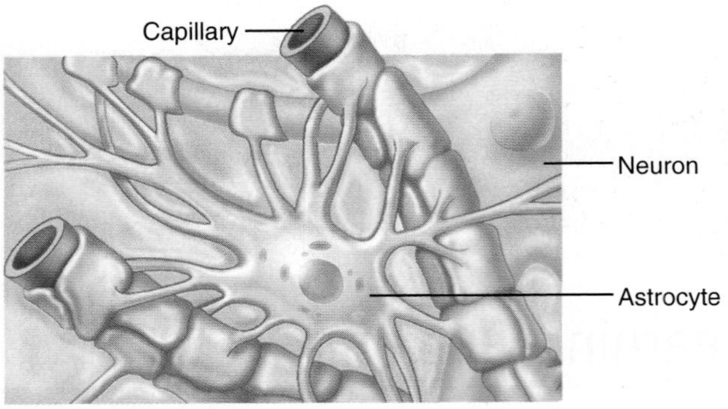

(a) Astrocytes are the most abundant CNS neuroglia.

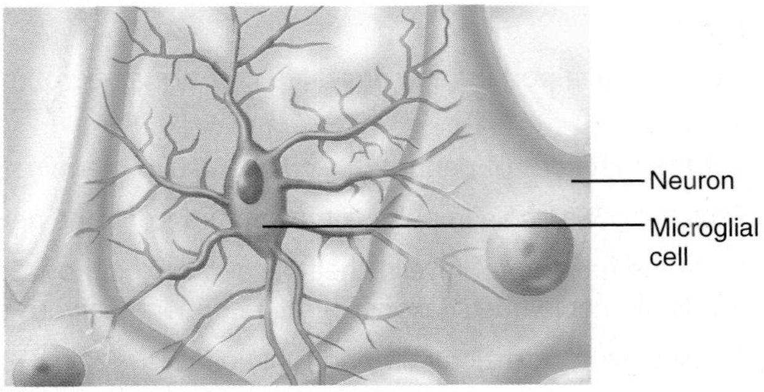

(b) Microglial cells are defensive cells in the CNS.

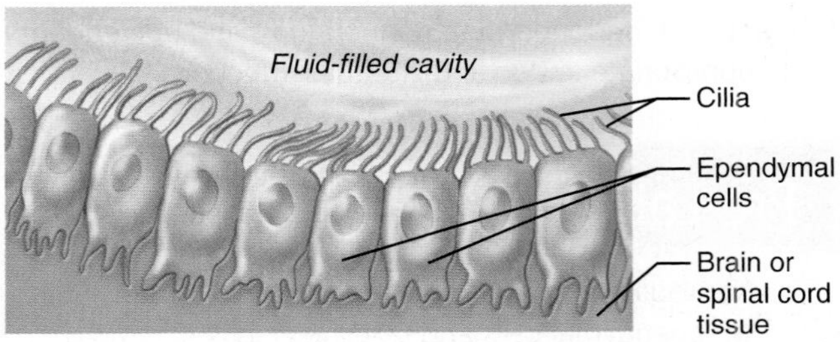

(c) Ependymal cells line cerebrospinal fluid–filled cavities.

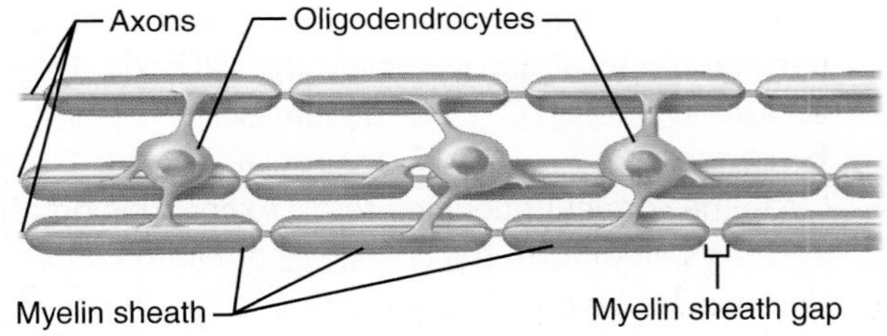

(d) Oligodendrocytes have processes that form myelin sheaths around CNS nerve fibers.

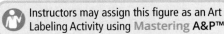

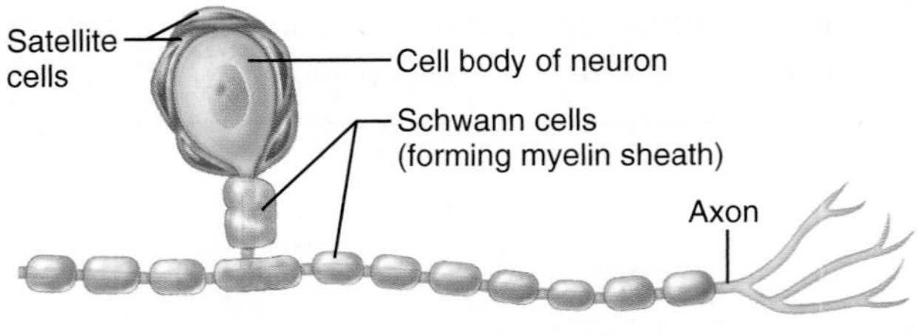

(e) Satellite cells and Schwann cells (which form myelin) surround neurons in the PNS.

Figure 15.1 Neuroglia. (a–d) The four types of neuroglia in the central nervous system. **(e)** Neuroglia of the peripheral nervous system.

Instructors may assign this figure as an Art Labeling Activity using Mastering A&P™

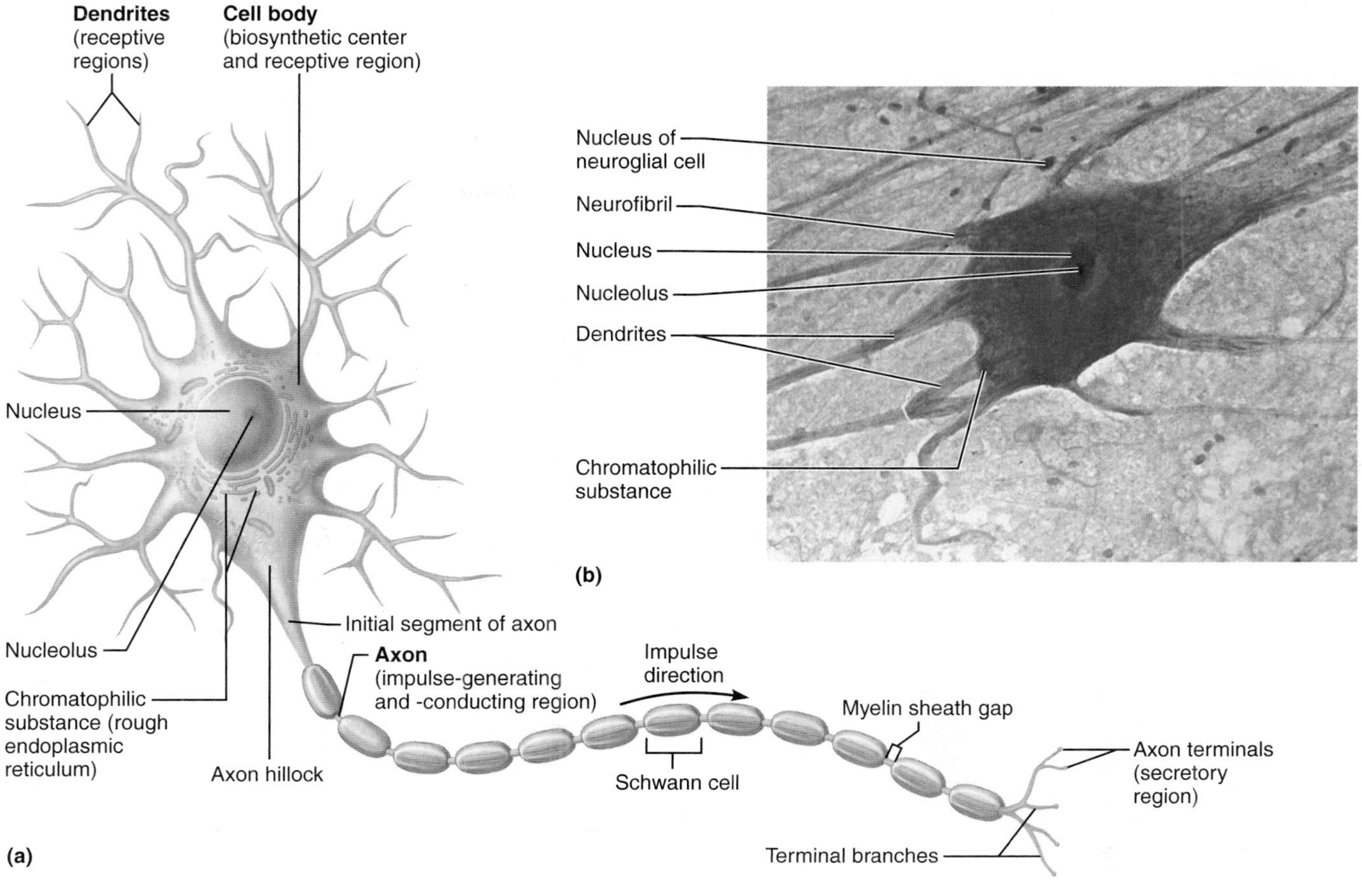

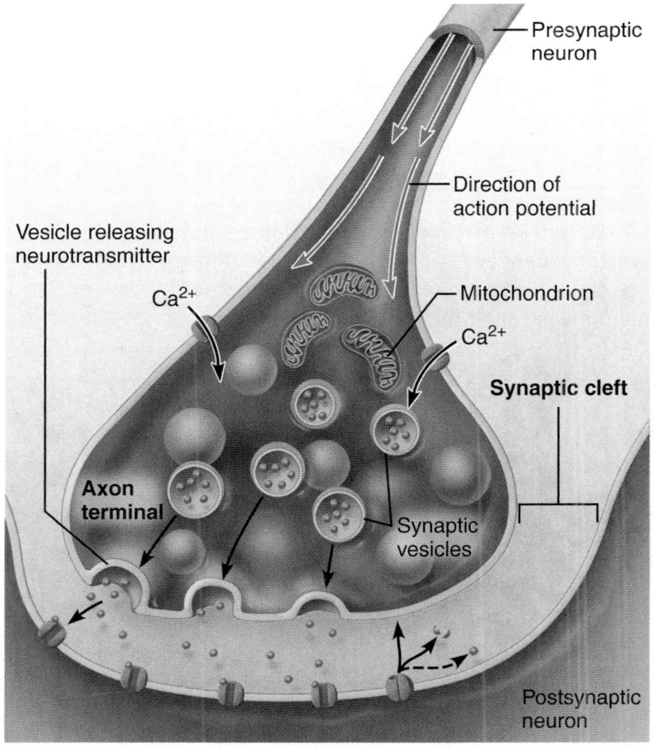

Figure 15.2 Structure of a typical motor neuron. **(a)** Diagram. **(b)** Photomicrograph (450×). **(c)** Enlarged diagram of a synapse.

Instructors may assign this figure as an Art Labeling Activity using Mastering A&P™

Neurons have two types of processes. **Dendrites** are *receptive regions* that bear receptors for neurotransmitters released by the axon terminals of other neurons. **Axons**, also called *nerve fibers* when they are long, form the *impulse generating and conducting region* of the neuron. The white matter of the nervous system is made up of axons. Neurons may have many dendrites, but they have only a single axon. The axon may branch, forming one or more processes called **axon collaterals**.

In general, a neuron is excited by other neurons when their axons release neurotransmitters close to its dendrites or cell body. The electrical signal produced travels across the cell body and if it is great enough, it elicits a regenerative electrical signal, a *nerve impulse* or *action potential*, that travels down the axon. The axon in motor neurons begins just distal to a slightly enlarged cell body structure called the **axon hillock** (Figure 15.2a). The axon ends in many small structures called **axon terminals** which form **synapses** with neurons or effector cells. These terminals store the neurotransmitter chemical in tiny vesicles. Each axon terminal of the neuron is separated from the cell body or dendrites of the next neuron by a tiny gap called the **synaptic cleft** (Figure 15.2c).

Most long nerve fibers are covered with a fatty material called *myelin*, and such fibers are referred to as **myelinated fibers**. Nerve fibers in the peripheral nervous system are typically heavily myelinated by special supporting cells called **Schwann cells**, which wrap themselves tightly around the axon in jelly roll fashion (**Figure 15.3**, p. 254). The wrapping is the **myelin sheath**. Since the myelin sheath is formed by many individual Schwann cells, it is a discontinuous sheath. The gaps, or indentations, in the sheath are called **myelin sheath gaps** or **nodes of Ranvier** (see Figure 15.2a).

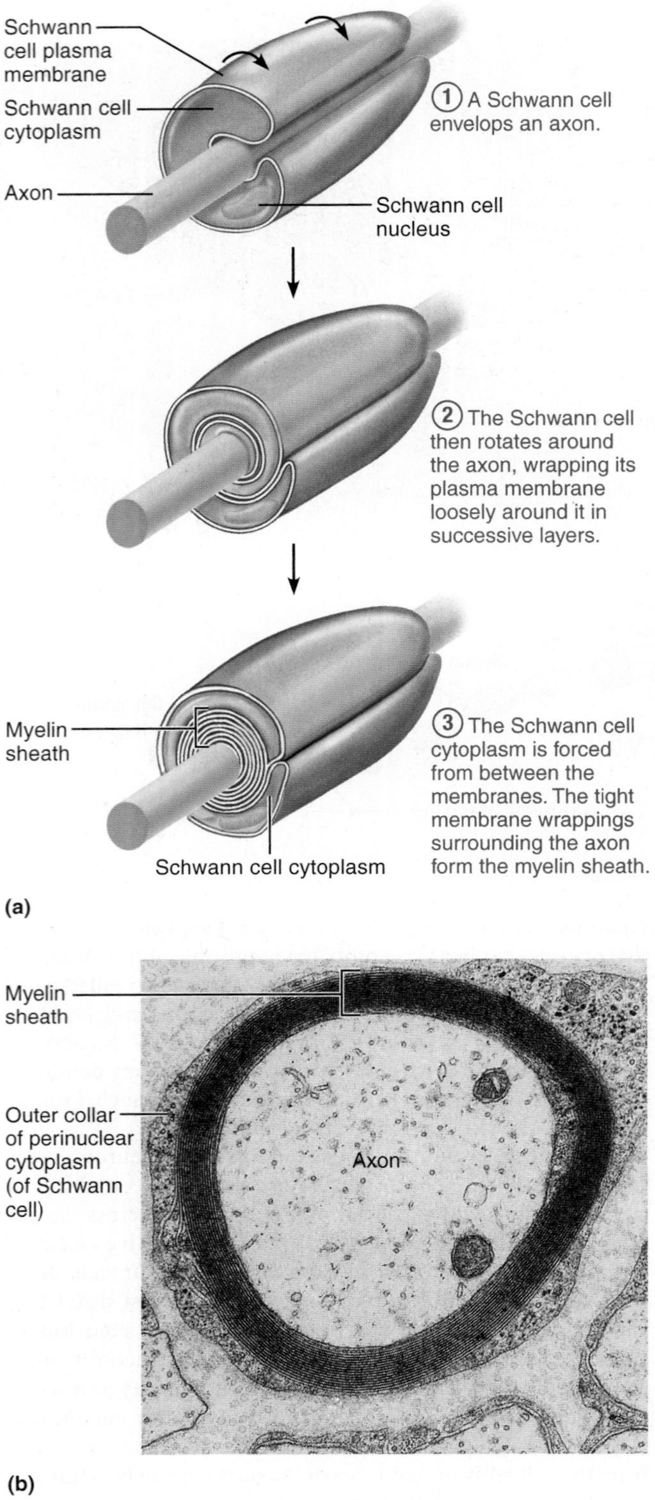

Figure 15.3 Myelination of a nerve fiber (axon) by Schwann cells. (a) Nerve fiber myelination. **(b)** Electron micrograph of cross section through a myelinated axon (7500×).

Within the CNS, myelination is accomplished by neuroglia called **oligodendrocytes** (see Figure 15.1d). Because of its chemical composition, myelin electrically insulates the fibers and greatly increases the transmission speed of nerve impulses.

Activity 1

Identifying Parts of a Neuron

1. Study the illustration of a typical motor neuron (Figure 15.2), noting the structural details described, and then identify these structures on a neuron model.

2. Obtain a prepared slide of the ox spinal cord smear, which has large, easily identifiable neurons. Study one representative neuron under oil immersion and identify the cell body; the nucleus; the large, prominent "owl's eye" nucleolus; and the granular chromatophilic substance. If possible, distinguish the axon from the many dendrites.

Sketch the neuron in the space provided below, and label the important anatomical details you have observed. Compare your sketch to Figure 15.2b.

3. Obtain a prepared slide of teased myelinated nerve fibers. Identify the following (use **Figure 15.4** as a guide): myelin sheath gaps, axon, Schwann cell nuclei, and myelin sheath.

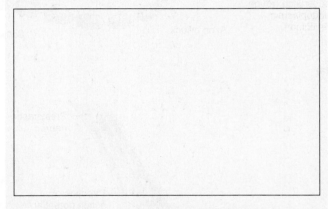

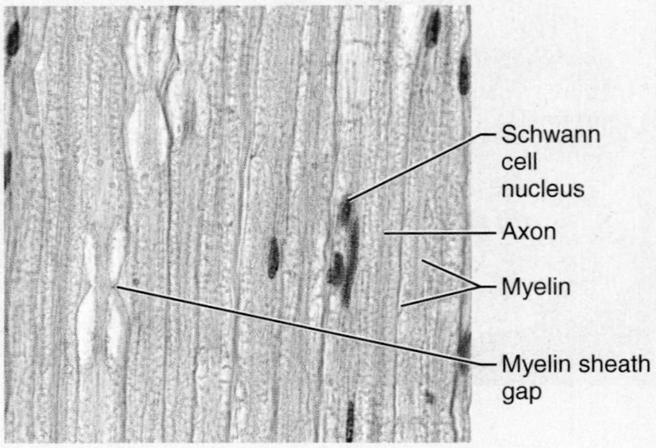

Figure 15.4 Photomicrograph of a small portion of a peripheral nerve in longitudinal section (265×).

Sketch a portion of a myelinated nerve fiber in the space provided below, illustrating a myelin sheath gap. Label the axon, myelin sheath, myelin sheath gap, and Schwann cell nucleus.

Do the gaps seem to occur at consistent intervals, or are they irregularly distributed? ______________________

Explain the functional significance of this finding: __________

____

____

Neuron Classification

Neurons may be classified on the basis of structure or of function.

Classification by Structure

Structurally, neurons may be differentiated by the number of processes attached to the cell body (**Figure 15.5a**). In **unipolar neurons**, one very short process, which divides into *peripheral* and *central processes*, extends from the cell body. Functionally, only the most distal parts of the peripheral process act as receptive endings; the rest acts as an axon along with the central process. Unipolar neurons are more accurately called **pseudounipolar neurons** because they are

15

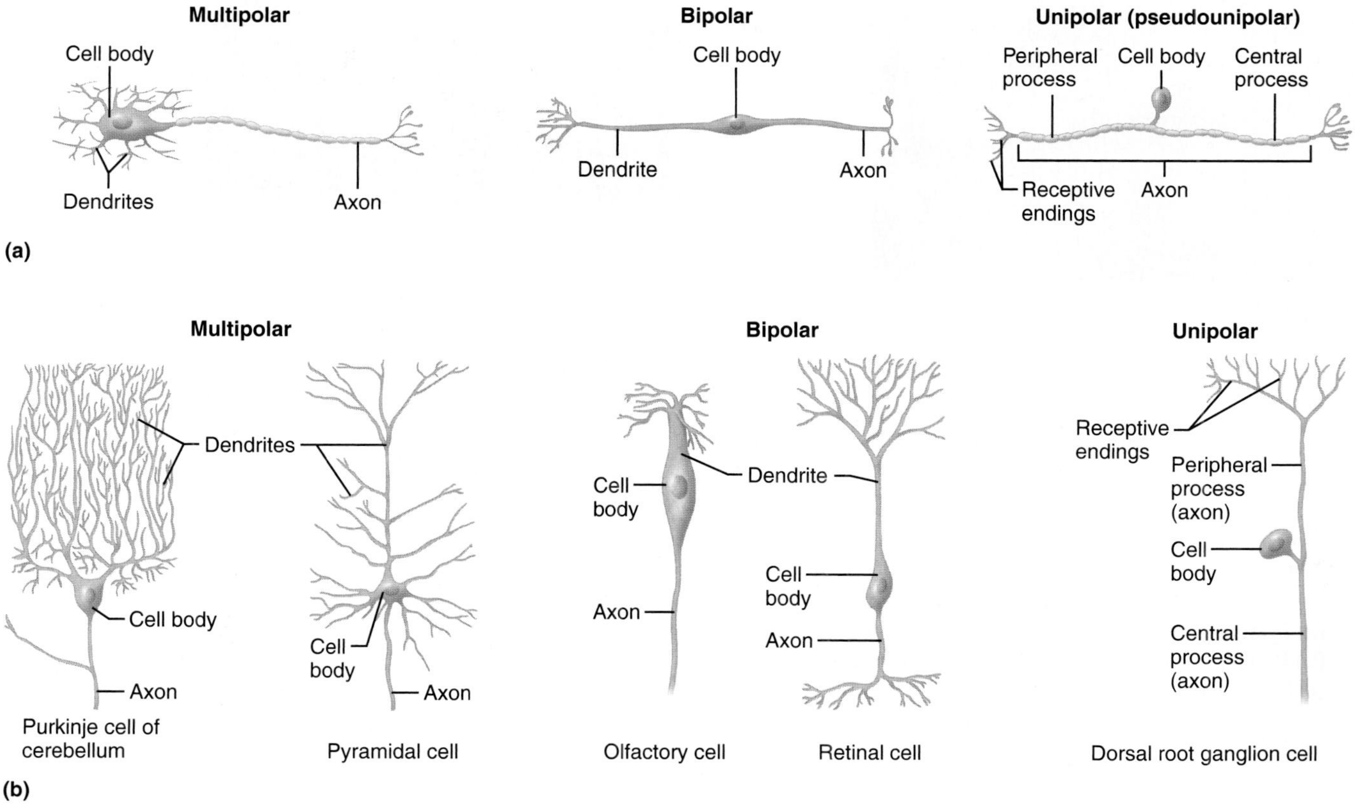

Figure 15.5 Classification of neurons according to structure. (a) Classification of neurons based on structure (number of processes extending from the cell body). **(b)** Structural variations within the classes.

derived from bipolar neurons. Nearly all neurons that conduct impulses toward the CNS are unipolar.

Bipolar neurons have two processes attached to the cell body. This neuron type is quite rare, typically found only as part of the receptor apparatus of the eye, ear, and olfactory mucosa.

Many processes issue from the cell body of **multipolar neurons**, all classified as dendrites except for a single axon. Most neurons in the brain and spinal cord and those whose axons carry impulses away from the CNS fall into this last category.

Activity 2

Studying the Microscopic Structure of Selected Neurons

Obtain prepared slides of pyramidal cells of the cerebral cortex, Purkinje cells of the cerebellar cortex, and a dorsal root ganglion. As you observe them under the microscope, try to pick out the anatomical details and compare the cells to Figure 15.5b and **Figure 15.6**. Notice that the neurons of the cerebral and cerebellar tissues (both brain tissues) are extensively branched; in contrast, the neurons of the dorsal root ganglion are more rounded. The many small nuclei visible surrounding the neurons are those of bordering neuroglia.

Which of these neuron types would be classified as multipolar neurons?

____

As unipolar? ______________________________

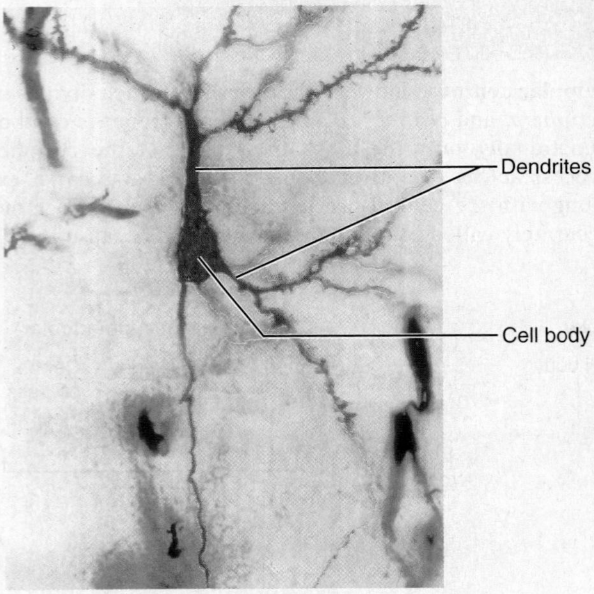

(a)

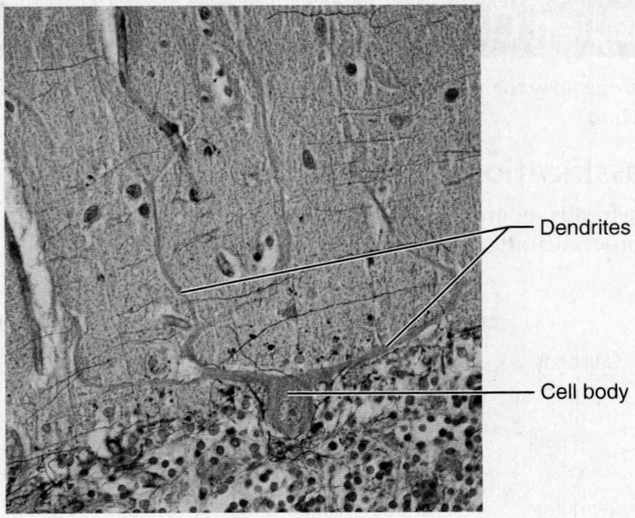

(b)

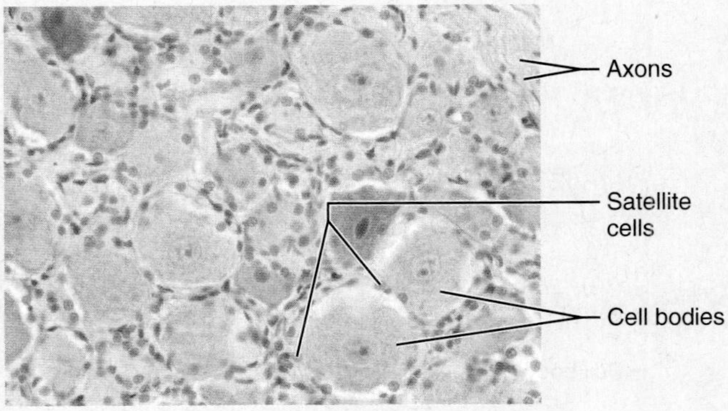

(c)

Figure 15.6 Photomicrographs of neurons.
(a) Pyramidal neuron from the cerebral cortex (600×).
(b) Purkinje cell from the cerebellar cortex (600×).
(c) Dorsal root ganglion cells (245×).

15

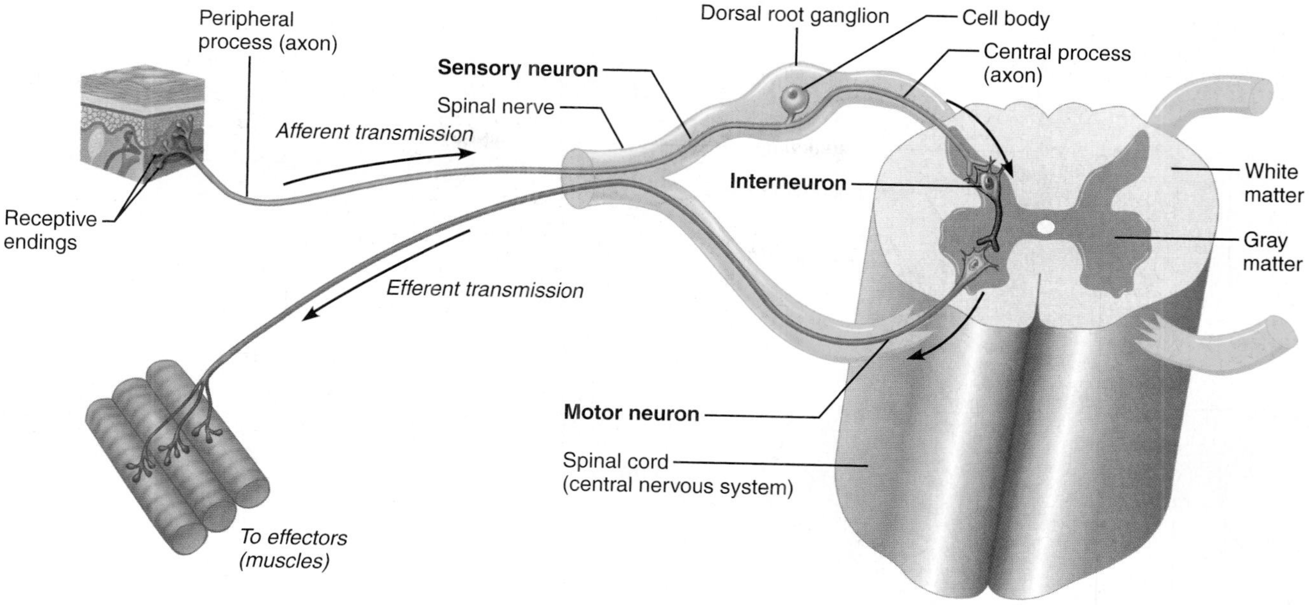

Figure 15.7 Classification of neurons on the basis of function. Sensory (afferent) neurons conduct impulses from the body's sensory receptors to the central nervous system; most are unipolar neurons with their cell bodies in ganglia in the peripheral nervous system (PNS). Motor (efferent) neurons transmit impulses from the CNS to effectors (muscles). Interneurons complete the communication line between sensory and motor neurons. They are typically multipolar, and their cell bodies reside in the CNS.

Classification by Function

In general, neurons carrying impulses from sensory receptors in the internal organs (viscera), the skin, skeletal muscles, joints, or special sensory organs are termed **sensory**, or **afferent, neurons** (Figure 15.7). The receptive endings of sensory neurons are often equipped with specialized receptors that are stimulated by specific changes in their immediate environment. (The structure and function of these receptors are considered separately in Exercise 22, General Sensation.) The cell bodies of sensory neurons are always found in a ganglion outside the CNS, and these neurons are typically unipolar.

Neurons carrying impulses from the CNS to the viscera and/or body muscles and glands are termed **motor**, or **efferent, neurons**. Motor neurons are most often multipolar, and their cell bodies are almost always located in the CNS.

The third functional category of neurons is **interneurons**, which are situated between and contribute to pathways that connect sensory and motor neurons. Their cell bodies are always located within the CNS, and they are multipolar neurons structurally.

Structure of a Nerve

In the CNS, bundles of axons are called **tracts**. In the PNS, bundles of axons are called **nerves**. Wrapped in connective tissue coverings, nerves extend to and/or from the CNS and visceral organs or structures of the body periphery such as skeletal muscles, glands, and skin.

Like neurons, nerves are classified according to the direction in which they transmit impulses. **Sensory (afferent) nerves** conduct impulses only toward the CNS. A few of the cranial nerves are pure sensory nerves. **Motor (efferent) nerves** carry impulses only away from the CNS. The ventral roots of the spinal cord are motor nerves. Nerves carrying both sensory (afferent) and motor (efferent) fibers are called **mixed nerves**; most nerves of the body, including all spinal nerves, are mixed nerves.

Within a nerve, each axon is surrounded by a delicate connective tissue sheath called an **endoneurium**, which insulates it from the other neuron processes adjacent to it. The endoneurium is often mistaken for the myelin sheath; it is instead an additional sheath that surrounds the myelin sheath. Groups of axons are bound by a coarser connective tissue, called the **perineurium**, to form bundles of fibers called **fascicles**. Finally, all the fascicles are bound together by a white, fibrous connective tissue sheath called the **epineurium**, forming the cord-like nerve (Figure 15.8, p. 258). In addition to the connective tissue wrappings, blood vessels and lymphatic vessels serving the fibers also travel within a nerve.

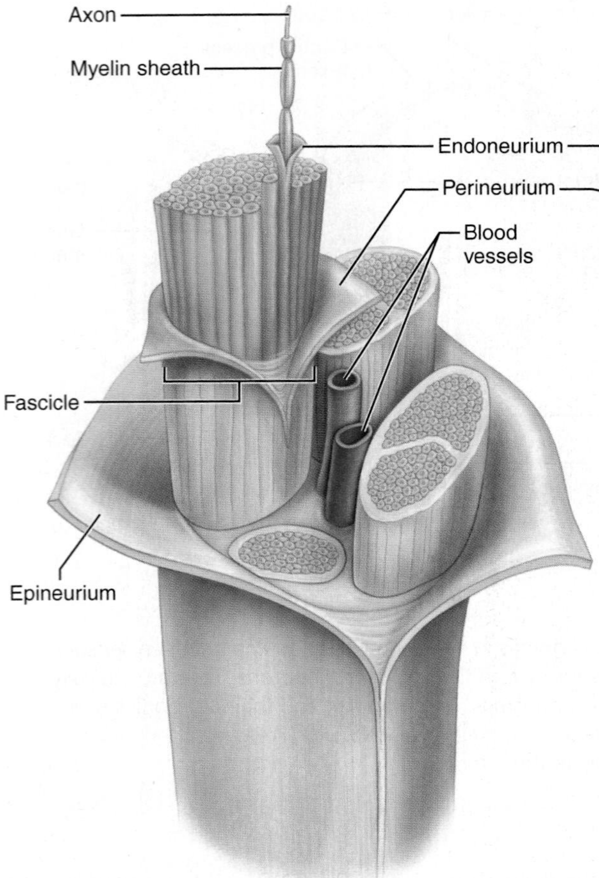

(a)

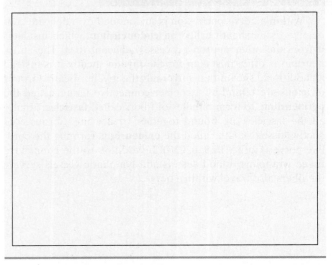

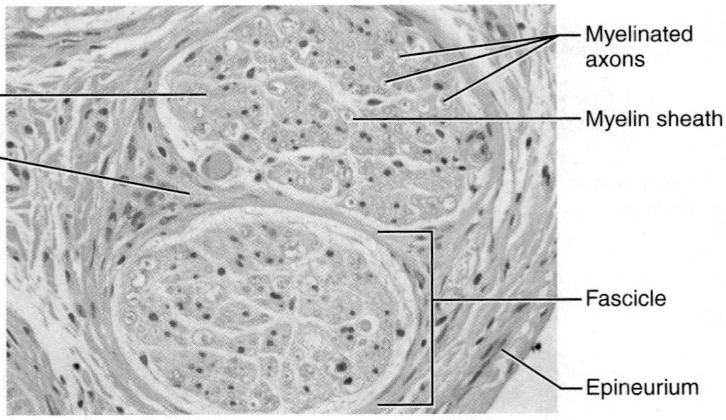

(b)

Figure 15.8 Structure of a nerve showing connective tissue wrappings. (a) Three-dimensional view of a portion of a nerve. **(b)** Photomicrograph of a cross-sectional view of part of a peripheral nerve (200×).

Instructors may assign this figure as an Art Labeling Activity using Mastering **A&P™**

15

Activity 3

Examining the Microscopic Structure of a Nerve

Use the compound microscope to examine a prepared cross section of a peripheral nerve. Using the photomicrograph (Figure 15.8b) as an aid, identify axons, myelin sheaths, fascicles, and endoneurium, perineurium, and epineurium sheaths. If desired, sketch the nerve in the space below.

Instructors may assign a portion of the Review Sheet questions using Mastering A&P™

15 REVIEW SHEET

EXERCISE

Histology of Nervous Tissue

Name ______________________ Lab Time/Date ______________________

1. The basic functional unit of the nervous system is the neuron. What is the major function of this cell type?

____

2. Match each statement with the correct type of neuroglia by filling in the blank.

______________ 1. forms the myelin sheath in the CNS

______________ 2. lines CSF-filled cavities

______________ 3. surrounds the cell body of a neuron found in the PNS

______________ 4. act as a phagocyte in the CNS

______________ 5. forms the myelin sheath in the PNS

______________ 6. controls the chemical environment around neurons in the CNS

3. Match each description with a term from the key.

Key:
a. afferent neuron
b. central nervous system
c. efferent neuron
d. ganglion
e. interneuron
f. neuroglia
g. neurotransmitters
h. nerve
i. nucleus
j. peripheral nervous system
k. synaptic cleft
l. tract

______________ 1. the brain and spinal cord collectively

______________ 2. specialized supporting cells in the nervous system

______________ 3. junction or point of close contact between neurons

______________ 4. a bundle of axons inside the PNS

______________ 5. neuron serving as part of the conduction pathway between sensory and motor neurons

______________ 6. ganglia and spinal and cranial nerves

______________ 7. collection of neuron cell bodies found within the CNS

______________ 8. neuron that conducts impulses away from the CNS to muscles and glands

______________ 9. neuron that conducts impulses toward the CNS from the body periphery

______________ 10. chemicals released by neurons that stimulate or inhibit other neurons or effectors

______________ 11. collection of neuron cell bodies found in the PNS

______________ 12. bundle of axons inside the CNS

Neuron Anatomy

4. Match the following anatomical terms (column B) with the appropriate description or function (column A).

Column A

_____________ 1. region of the cell body from which the axon originates

_____________ 2. secretes neurotransmitters

_____________ 3. receptive regions of a neuron (2 terms)

_____________ 4. insulates the nerve fibers

_____________ 5. site of the nucleus and most important metabolic area

_____________ 6. involved in the transport of substances within the neuron

_____________ 7. essentially rough endoplasmic reticulum, important metabolically

_____________ 8. impulse generator and transmitter

Column B

a. axon

b. axon terminal

c. axon hillock

d. cell body

f. chromatophilic substance

g. dendrite

h. myelin sheath

i. neurofibril

5. Label the following structures on the diagram of a multipolar neuron shown below: cell body, nucleus, nucleolus, chromatophilic substance, dendrites, initial segment of axon, myelin sheath, myelin sheath gaps, and axon terminals.

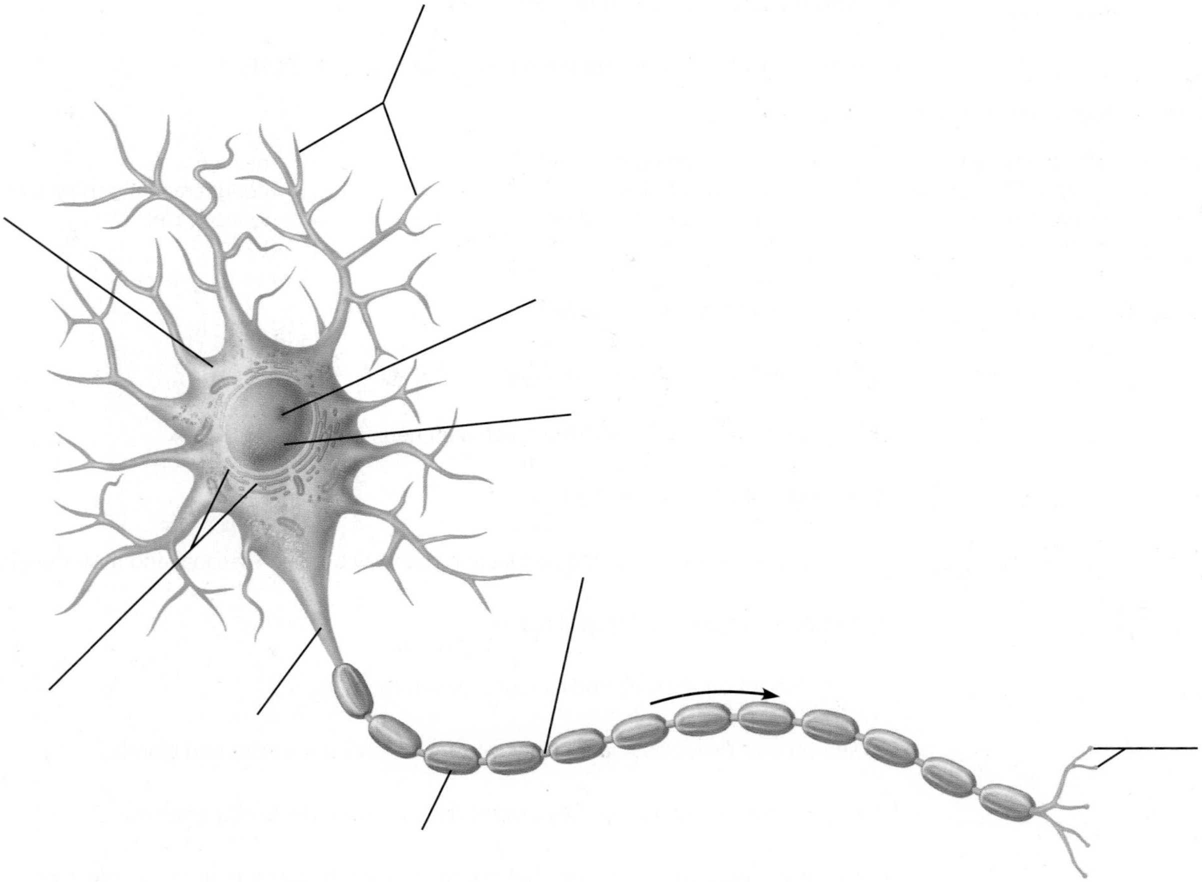

6. What substance is found in synaptic vesicles of the axon terminal? ________________________________

7. What anatomical characteristic determines whether a particular neuron is classified as unipolar, bipolar, or multipolar?

____

Make a simple line drawing of each type here.

Unipolar neuron	Bipolar neuron	Multipolar neuron

8. Correctly identify the sensory (afferent) neuron, interneuron, and motor (efferent) neuron in the figure below.

Which of these neuron types is/are unipolar? ___________________________

Which is/are most likely multipolar? ___________________________

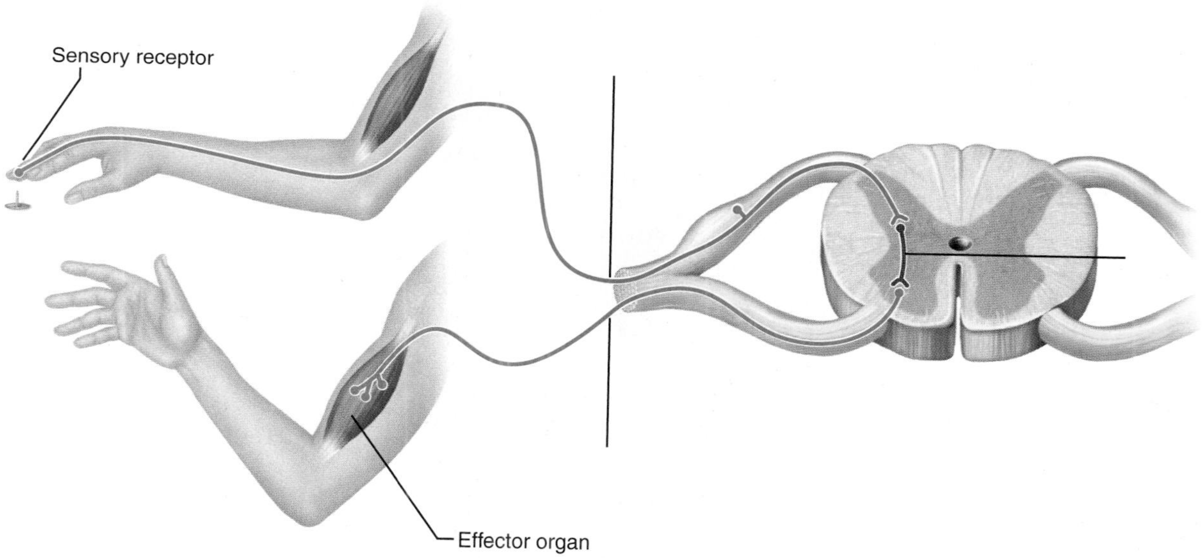

9. Describe how the Schwann cells form the myelin sheath encasing the nerve fibers.

____

____

Structure of a Nerve

10. What is a nerve? ____

____

11. State the location of each of the following connective tissue coverings.

endoneurium: ____________________

perineurium: ____________________

epineurium: ____________________

12. What is the function of the connective tissue wrappings found in a nerve? ____________________

____

13. Define *mixed nerve*. ____________________

____

14. Identify all indicated parts of the nerve section.

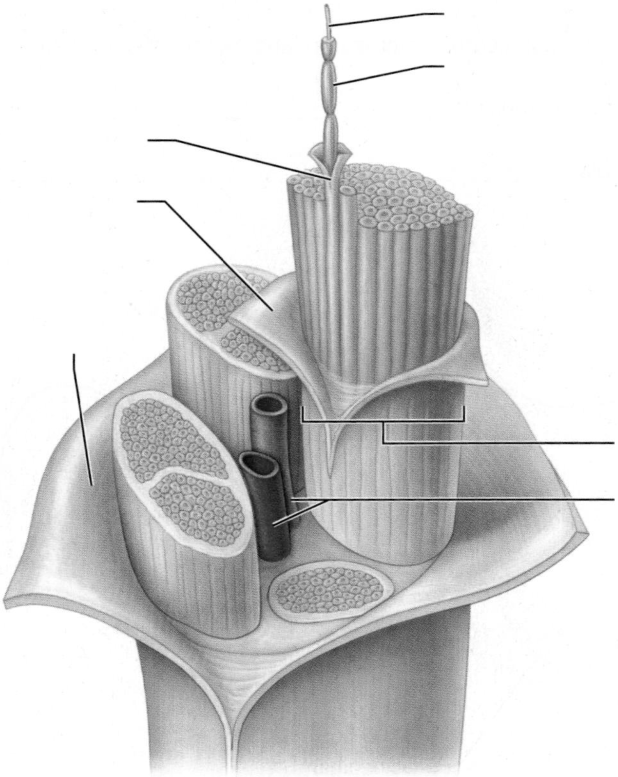

15. Amyotrophic lateral sclerosis is a neurodegenerative disease in which motor neurons are progressively destroyed. Excess levels of the neurotransmitter glutamate have been implicated in this process. Which type of neuroglia would play a role in controlling glutamate levels in the chemical environment of neurons?

____

16. Peripheral neuropathy has a variety of causes. Worldwide, the most common cause is leprosy, also known as Hansen's disease. Would you expect peripheral neuropathy to cause damage to tracts or to nerves? ____________________

Why? ____________________

EXERCISE 16

Neurophysiology of Nerve Impulses: Frog Subjects

Learning Outcomes

- Describe the resting membrane potential in neurons.
- Define *depolarization*, *repolarization*, *action potential*, and *relative refractory period* and *absolute refractory period*.
- Describe the events that lead to the generation and conduction of an action potential.
- Explain briefly how a nerve impulse is transmitted from one neuron to another, and how a neurotransmitter may be either excitatory or inhibitory to the recipient cell.
- Define *compound action potential*, and discuss how it differs from an action potential in a single neuron.
- Describe the preparation used to examine contraction of the frog gastrocnemius muscle.
- List various substances and factors that can stimulate neurons.

Pre-Lab Quiz

Instructors may assign these and other Pre-Lab Quiz questions using Mastering **A&P™**

1. Circle the correct underlined term. Excitability / Conductivity is the ability to transmit nerve impulses to other neurons.
2. When a neuron is stimulated, the membrane becomes more permeable to Na^+ ions, which diffuse into the cell and cause:
   a. depolarization
   b. hyperpolarization
   c. repolarization
3. As an action potential progresses, the permeability to Na^+ decreases, and the permeability to this ion increases:
   a. Ca^{2+} **b.** K^+ **c.** Na^+
4. The period of time when the neuron is totally insensitive to further stimulation and cannot generate another action potential is:
   a. absolute refractory period **c.** repolarization
   b. membrane potential **d.** threshold
5. What muscle and nerve will you need to isolate to study the physiology of nerve fibers?
   a. gastrocnemius and sciatic
   b. sartorius and femoral
   c. triceps brachii and radial

Go to Mastering **A&P™ > Study Area to improve your performance in A&P Lab.**

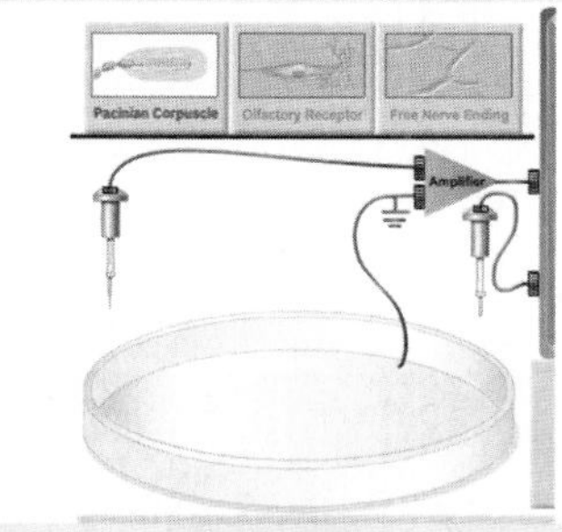

> Lab Tools > PhysioEx 9.1

Instructors may assign new Building Vocabulary coaching activities, Pre-Lab Quiz questions, Art Labeling activities, Practice Anatomy Lab Practical questions (PAL), PhysioEx activities, and more using the Mastering **A&P™** Item Library.

Materials

- *Rana pipiens**
- Dissecting instruments and tray
- Disposable gloves
- Ringer's solution (frog) in dropper bottles at room temperature
- Thread
- Glass rods or probes
- Glass plates or slides
- Ring stand and clamp
- Stimulator; platinum electrodes
- Forceps
- Filter paper
- 0.01% hydrochloric acid (HCl) solution
- Sodium chloride (NaCl) crystals
- Heat-resistant mitts
- Bunsen burner
- Safety goggles

* Instructor to provide freshly pithed frogs (*Rana pipiens*) for student experimentation.

PEx PhysioEx™ 9.1 Computer Simulation Ex. 3 on p. PEx-35.

Neurons are **excitable**; they respond to stimuli by producing an electrical signal. Excited neurons communicate—they transmit electrical signals to neurons, muscles, and glands, a property called **conductivity**. In a resting neuron, the interior of the cell membrane is slightly more negatively charged than the cell exterior (**Figure 16.1**). The difference in electrical charge produces a **resting membrane potential** across the membrane that is measured in millivolts.

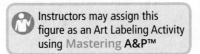

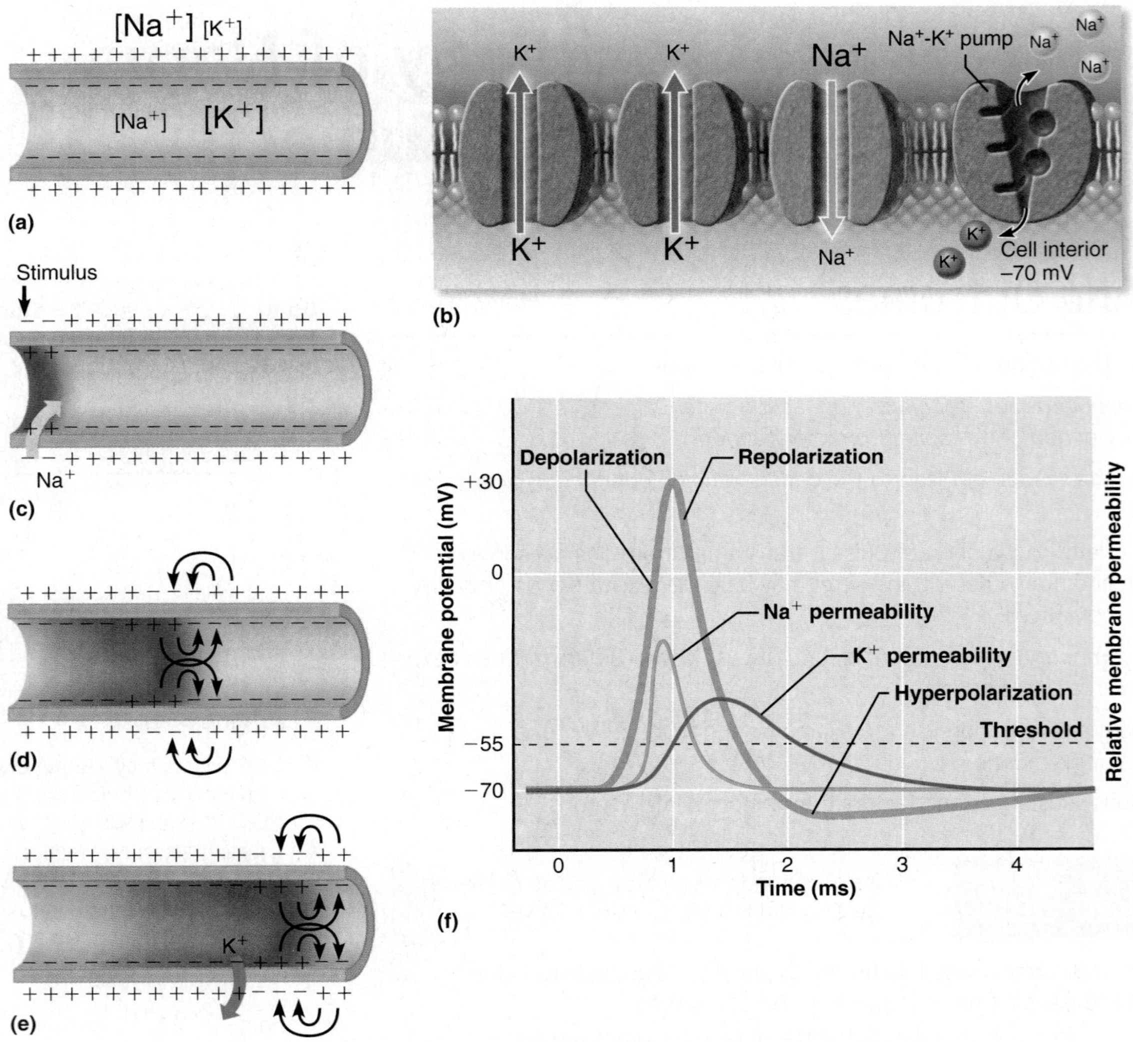

Figure 16.1 The action potential. **(a)** Resting membrane potential (RMP). There is an excess of positive ions at the external cell surface, with Na^+ the predominant extracellular fluid ion and K^+ the predominant intracellular ion. **(b)** Na^+ ions leak into the cell, and K^+ ions leak out. The RMP is maintained by the active sodium-potassium pump. **(c)** Depolarization—reversal of the RMP. Application of a stimulus changes the membrane permeability, and Na^+ ions are allowed to diffuse rapidly into the cell. **(d)** Generation of the action potential. If the stimulus reaches threshold, the depolarization wave spreads rapidly along the entire length of the membrane. **(e)** Repolarization—reestablishment of the RMP. The negative charge on the internal plasma membrane surface and the positive charge on its external surface are reestablished by diffusion of K^+ ions out of the cell, proceeding in the same direction as in depolarization. **(f)** The action potential is caused by permeability changes in the plasma membrane. The purple line shows the changes to the membrane potential over time. The yellow and green lines show the changes to the permeability of Na^+ and K^+, respectively, over time.

Instructors may assign this figure as an Art Labeling Activity using Mastering A&P™

As in most cells, the predominant intracellular cation is K^+; Na^+ is the predominant cation in the extracellular fluid. In a resting neuron, Na^+ leaks into the cell and K^+ leaks out. The resting membrane potential is maintained by the sodium-potassium pump, which transports Na^+ back out of the cell and K^+ back into the cell.

The Action Potential

When a neuron receives an excitatory stimulus, the membrane becomes more permeable to sodium ions, and Na^+ diffuses down its electrochemical gradient into the cell. As a result, the interior of the membrane becomes less negative (Figure 16.1c), an event called **depolarization**. If the stimulus is great enough to depolarize the the axon to **threshold**, an **action potential** is generated.

When the threshold voltage is reached, the membrane permeability to Na^+ increases rapidly (Figure 16.1f). As the neuron depolarizes, the polarity of the membrane reverses: the interior surface now becomes more positive than the exterior (Figure 16.1d). As the membrane permeability to Na^+ falls, the permeability to K^+ increases, and K^+ diffuses down its electrochemical gradient and out of the cell (Figure 16.1e). Once again the interior of the membrane becomes more negative than the exterior. This event is called **repolarization**. As you can see, the action potential is a brief reversal of the neuron's membrane potential.

The period of time when Na^+ permeability is rapidly changing and maximal, and the period immediately following when Na^+ permeability becomes restricted, together correspond to a time when the neuron is insensitive to further stimulation and cannot generate another action potential. This period is called the **absolute refractory period**. As Na^+ permeability is gradually restored to resting levels during repolarization, an especially strong stimulus to the neuron may provoke another action potential. This period of time is the **relative refractory period**. Restoration of the resting membrane potential restores the neuron's normal excitability.

Once generated, the action potential propagates along the entire length of the axon. It is never partially transmitted. Furthermore, it retains a constant amplitude and duration; the action potential is not small when a stimulus is small and large when a stimulus is large. Since the action potential of a given neuron is always the same, it is said to be an all-or-none response. When an action potential reaches the axon terminals, it usually causes neurotransmitter to be released. The neurotransmitter may be excitatory or inhibitory to the next cell in the transmission chain, depending on the receptor types on that cell.

Physiology of Nerves

The sciatic nerve is composed of a bundle of axons that vary in diameter. An electrical signal recorded from a nerve represents the summed electrical activity of all the axons in the nerve. This summed activity is called a **compound action potential**. Unlike an action potential in a single axon, the compound action potential varies in shape according to which axons are producing action potentials. When a nerve is stimulated by external electrodes, as in our experiments, the largest axons reach threshold first and generate action potentials. Higher-intensity stimuli are required to produce action potentials in smaller axons.

In this laboratory session, you will investigate the functioning of a nerve by subjecting the sciatic nerve of a frog to various types of stimuli.

16

DISSECTION

Isolating the Gastrocnemius Muscle and Sciatic Nerve

1. Don gloves and safety glasses. Obtain a pithed frog from your instructor, and bring it to your laboratory bench. Also obtain dissecting instruments, a tray, thread, two glass rods or probes, and frog Ringer's solution from the supply area.

2. Prepare the sciatic nerve as illustrated (**Figure 16.2**). Place the pithed frog on the dissecting tray, dorsal side down. Make a cut through the skin around the circumference of the frog approximately halfway down the trunk, and then pull the skin down over the muscles of the legs. Open the abdominal cavity, and push the abdominal organs to one side to expose the origin of the glistening white sciatic nerve, which arises from

Text continues on next page. →

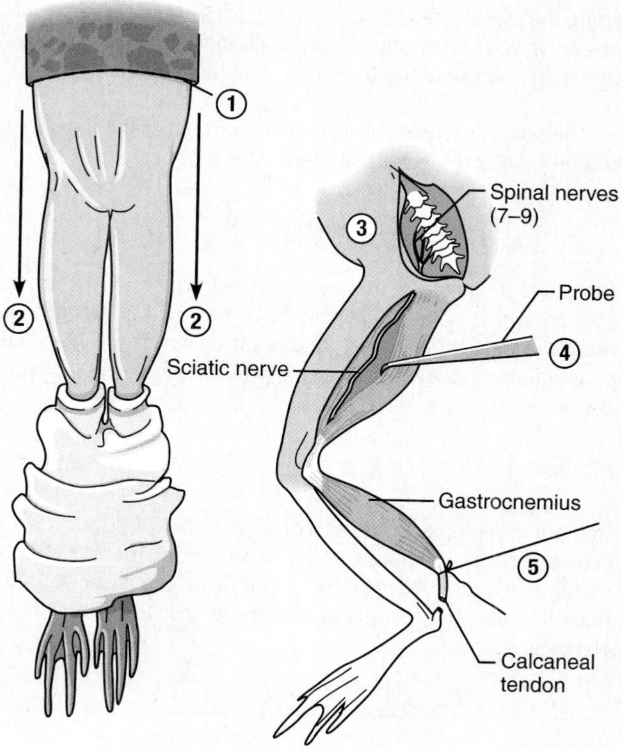

Figure 16.2 Removal of the sciatic nerve and gastrocnemius muscle. (1) Cut through the frog's skin around the circumference of the trunk. (2) Pull the skin down over the trunk and legs. (3) Make a longitudinal cut through the abdominal musculature, and expose the roots of the sciatic nerve (arising from spinal nerves 7–9). Ligate the nerve and cut the roots proximal to the ligature. (4) Use a glass probe to expose the sciatic nerve beneath the posterior thigh muscles. (5) Ligate the calcaneal tendon, and cut it free distal to the ligature. Release the gastrocnemius muscle from the connective tissue of the knee region.

the last three spinal nerves. *Once the sciatic nerve has been exposed, keep it continually moist with room-temperature frog Ringer's solution*.

3. Using a glass probe, slip a piece of thread moistened with frog Ringer's solution under the sciatic nerve close to its origin at the vertebral column. Make a single ligature (tie it firmly with the thread), and then cut through the nerve roots to free the proximal end of the sciatic nerve from its attachments. Using a glass rod or probe, carefully separate the posterior thigh muscles to locate and then free the sciatic nerve, which runs down the posterior aspect of the thigh.

4. Tie a piece of thread around the calcaneal tendon of the gastrocnemius muscle, and then cut through the tendon distal to the ligature to free the gastrocnemius muscle from the heel. Using a scalpel, carefully release the gastrocnemius muscle from the connective tissue in the knee region. At this point you should have completely freed both the gastrocnemius muscle and the sciatic nerve, which innervates it.

Activity 1

To access **Activity 2: Inhibiting the Nerve** and **Activity 3: Visualizing the Compound Action Potential with an Oscilloscope**, go to Mastering **A&P™** > Study Area > Additional Study Tools

Stimulating the Nerve

In this first set of experiments, stimulation of the nerve and generation of the compound action potential will be indicated by contraction of the gastrocnemius muscle. Because you will make no mechanical recording (unless your instructor asks you to), you must keep complete and accurate records of all experimental procedures and results.

1. Obtain a glass slide or plate, ring stand and clamp, stimulator, electrodes, salt (NaCl), forceps, filter paper, 0.01% hydrochloric acid (HCl) solution, Bunsen burner, and heat-resistant mitts. With glass rods, transfer the isolated muscle-nerve preparation to a glass plate or slide, and then attach the slide to a ring stand with a clamp. Allow the end of the sciatic nerve to hang over the free edge of the glass slide, so that it is easily accessible for stimulation. *Remember to keep the nerve moist at all times.*

2. You are now ready to investigate the response of the sciatic nerve to various stimuli, beginning with electrical stimulation. Using the stimulator and platinum electrodes, stimulate the sciatic nerve with single shocks, gradually increasing the intensity of the stimulus until the threshold stimulus is determined.

The muscle as a whole will just barely contract at the threshold stimulus. Record the voltage of this stimulus:

Threshold stimulus: ______________ V

Continue to increase the voltage until you find the point beyond which no further increase occurs in the strength of muscle contraction—that is, the point at which the maximal contraction of the muscle is obtained. Record this voltage below.

Maximal stimulus: ______________ V

Delivering multiple or repeated shocks to the sciatic nerve causes volleys of impulses in the nerve. Shock the nerve with multiple stimuli. Observe the response of the muscle. How does this response compare with the response to the single electrical shocks?

____

____

3. To investigate mechanical stimulation, pinch the free end of the nerve by firmly pressing it between two glass rods or by pinching it with forceps. What is the result?

____

4. Test chemical stimulation by applying a small piece of filter paper saturated with HCl solution to the free end of the nerve. What is the result?

____

Drop a few crystals of salt (NaCl) on the free end of the nerve. What is the result?

____

5. Now test thermal stimulation. Wearing the heat-resistant mitts, heat a glass rod for a few moments over a Bunsen burner. Then touch the rod to the free end of the nerve. What is the result?

____

What do these muscle reactions say about the excitability and conductivity of neurons?

____

____

Most neurons within the body are stimulated to the greatest degree by a particular stimulus (in many cases, a chemical neurotransmitter), but a variety of other stimuli may trigger nerve impulses, as seen in the experimental series just conducted. Generally, no matter what type of stimulus is present, if the affected part responds by becoming activated, it will always react in the same way. Familiar examples are the well-known phenomenon of "seeing stars" when you receive a blow to the head or press on your eyeball, both of which trigger impulses in your optic nerves.

Instructors may assign a portion of the Review Sheet questions using Mastering A&P™

REVIEW SHEET

EXERCISE 16

Neurophysiology of Nerve Impulses: Frog Subjects

Name ______________________ Lab Time/Date ______________________

The Action Potential

1. Match the terms in column B to the appropriate definition in column A.

Column A

_________ 1. period of depolarization of the neuron membrane during which it cannot respond to a second stimulus

_________ 2. reversal of the resting potential due to an influx of sodium ions

_________ 3. period during which potassium ions diffuse out of the neuron because of a change in membrane permeability

_________ 4. period of repolarization when only a strong stimulus will elicit an action potential

_________ 5. mechanism in which ATP is used to move sodium out of the cell and potassium into the cell; restores the resting membrane voltage and intracellular ionic concentrations

Column B

a. absolute refractory period

b. action potential

c. depolarization

d. relative refractory period

e. repolarization

f. sodium-potassium pump

2. Define the term *depolarization*. ______________________

____

How does an action potential differ from simple depolarization? ______________________

____

3. The graph shown below depicts the changes in membrane permeability that result in an action potential. Label the areas on the tracing that represent the following: depolarization, repolarization, hyperpolarization, Na^+ permeability, and K^+ permeability.

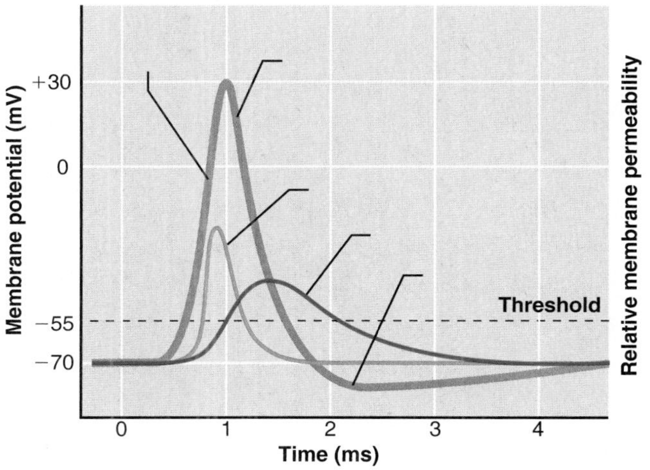

Physiology of Nerves Stimulating and Inhibiting the Nerve

4. Respond appropriately to each question posed below. Insert your responses in the corresponding numbered blanks to the right.

1–4. Name four types of stimuli that resulted in action potential generation in the sciatic nerve of the frog.

5. Which of the stimuli resulted in the most effective nerve stimulation?

6. What is the usual mode of stimulus transfer in neuron-to-neuron interactions?

7. During this set of experiments, how did you recognize that impulses were being transmitted?

1. ____________________
2. ____________________
3. ____________________
4. ____________________
5. ____________________
6. ____________________
7. ____________________

5. + Some local anesthetics that are used to block nerve pain work by decreasing the permeability of sodium in the plasma membrane of the neuron. Explain the effect that this change in permeability would have on the generation of action potentials and why. ____________________

____

____

EXERCISE 17

Gross Anatomy of the Brain and Cranial Nerves

Learning Outcomes

- List the elements of the central and peripheral divisions of the nervous system.
- Discuss the difference between the sensory and motor portions of the nervous system, and name the two divisions of the motor portion.
- Recognize the terms that describe the development of the human brain, and discuss the relationships between the terms.
- As directed by your instructor, identify the bold terms associated with the cerebral hemispheres, diencephalon, brain stem, and cerebellum on a dissected human brain, brain model, or appropriate image. State the functions of these structures.
- State the difference between gyri, fissures, and sulci.
- Describe the composition of gray matter and white matter in the nervous system.
- Name and describe the three meninges that cover the brain, state their functions, and locate the falx cerebri, falx cerebelli, and tentorium cerebelli.
- Discuss the formation, circulation, and drainage of cerebrospinal fluid.
- Identify the cranial nerves by number and name on a model or image, stating the origin and function of each.
- Identify at least four key anatomical differences between the human and sheep brain.

Go to Mastering A&P™ > Study Area to improve your performance in A&P Lab.

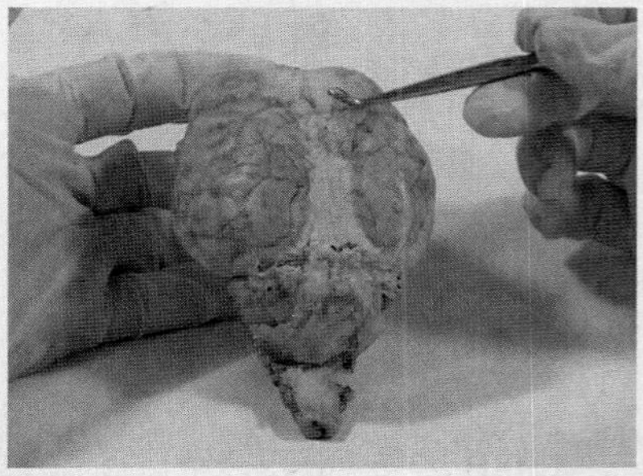

Lab Tools > Bone and Dissection Videos > Sheep Brain

Instructors may assign new Building Vocabulary coaching activities, Pre-Lab Quiz questions, Art Labeling activities, related videos and coaching activities for the Sheep Brain dissection, Practice Anatomy Lab Practical questions (PAL), and more using the Mastering **A&P™** Item Library.

Pre-Lab Quiz

Instructors may assign these and other Pre-Lab Quiz questions using Mastering **A&P™**

1. Circle the correct underlined term. The central nervous system / peripheral nervous system consists of the brain and spinal cord.
2. Circle True or False. Deep grooves within the cerebral hemispheres are known as gyri.
3. On the ventral surface of the brain, you can observe the optic nerves and chiasma, the pituitary gland, and the mammillary bodies. These externally visible structures form the floor of the:
   a. brain stem **b.** diencephalon **c.** frontal lobe **d.** occipital lobe
4. Circle the correct underlined term. The outer cortex of the brain contains the cell bodies of cerebral neurons and is known as white matter / gray matter.
5. The brain and spinal cord are covered and protected by three connective tissue layers called:
   a. lobes **b.** meninges **c.** sulci **d.** ventricles

Materials

- Human brain model (dissectible)
- Preserved human brain (if available)
- Three-dimensional model of ventricles
- Frontally sectioned human brain slice (if available)
- Materials as needed for cranial nerve testing (see Table 17.2): aromatic oils (e.g., vanilla and cloves); eye chart; ophthalmoscope; penlight; safety pin; blunt probe; cotton; solutions of sugar, salt, vinegar, and quinine; ammonia; tuning fork; and tongue depressor
- Preserved sheep brain (meninges and cranial nerves intact)
- Dissecting instruments and tray
- Disposable gloves

When viewed alongside all Earth's animals, humans are indeed unique, and the key to their uniqueness is found in the brain. Each of us is a reflection of our brain's experience. If all past sensory input could mysteriously and suddenly be "erased," we would be unable to walk, talk, or communicate in any manner. Spontaneous movement would occur, as in a fetus, but no voluntary integrated function of any type would be possible. Clearly we would cease to be the same individuals.

For convenience, the nervous system is considered in terms of two principal divisions: the central nervous system and the peripheral nervous system. The **central nervous system (CNS)** consists of the brain and spinal cord, which primarily interpret incoming sensory information and issue instructions based on that information and on past experience. The **peripheral nervous system (PNS)** consists of the cranial and spinal nerves, ganglia, and sensory receptors.

The PNS has two major subdivisions: the **sensory portion**, which consists of nerve fibers that conduct impulses from sensory receptors toward the CNS, and the **motor portion**, which contains nerve fibers that conduct impulses away from the CNS. The motor portion, in turn, consists of the **somatic division** (sometimes called the *voluntary system*), which controls the skeletal muscles, and the **autonomic nervous system (ANS)**, which controls smooth and cardiac muscles and glands.

In this exercise the brain (CNS) and cranial nerves (PNS) will be studied together because of their close anatomical relationship.

The Human Brain

During embryonic development of all vertebrates, the CNS first makes its appearance as a simple tubelike structure, the **neural tube**, that extends down the dorsal median plane. By the fourth week, the human brain begins to form as an expansion of the anterior or rostral end of the neural tube (the end toward the head). Shortly thereafter, constrictions appear, dividing the developing brain into three major regions—**forebrain**, **midbrain**, and **hindbrain** (**Figure 17.1**). The remainder of the neural tube becomes the spinal cord.

The central canal of the neural tube, which remains continuous throughout the brain and cord, enlarges in four regions of the brain, forming chambers called **ventricles** (see Figure 17.8a and b, p. 279).

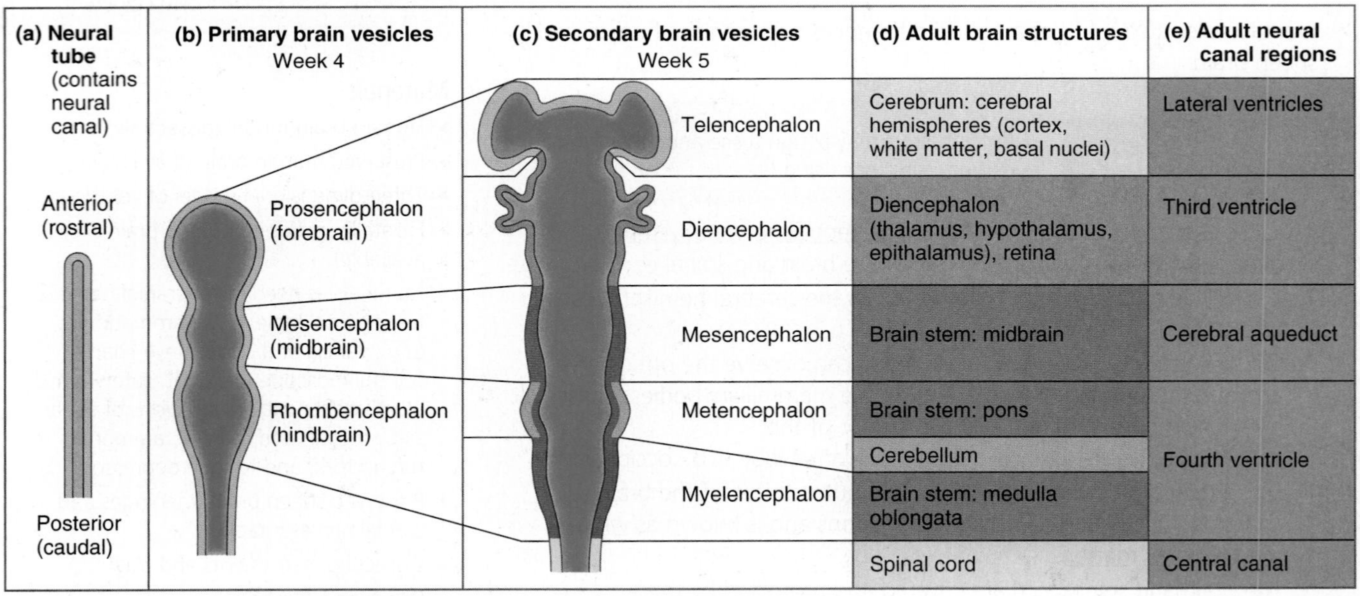

Figure 17.1 Embryonic development of the human brain. (a) The neural tube subdivides into **(b)** the primary brain vesicles, which subsequently form **(c)** the secondary brain vesicles, which differentiate into **(d)** the adult brain structures. **(e)** The adult structures derived from the neural canal.

Activity 1

Identifying External Brain Structures

Identify external brain structures using the figures cited. Also use a model of the human brain and other learning aids as they are mentioned.

Generally, the brain is studied in terms of four major regions: the cerebral hemispheres, diencephalon, brain stem, and cerebellum.

Cerebral Hemispheres

The **cerebral hemispheres** are the most superior portion of the brain (**Figure 17.2**, p. 272). Their entire surface is thrown into elevated ridges of tissue called **gyri** that are separated by shallow grooves called **sulci** or deeper grooves called **fissures**. Many of the fissures and gyri are important anatomical landmarks.

The cerebral hemispheres are divided by a single deep fissure, the **longitudinal fissure**. The **transverse cerebral fissure** separates the cerebral hemispheres from the cerebellum below. The **central sulcus** divides the **frontal lobe** from the **parietal lobe**, and the **lateral sulcus** separates the **temporal lobe** from the parietal lobe. The **parieto-occipital sulcus** on the medial surface of each hemisphere divides the **occipital lobe** from the parietal lobe. A fifth lobe of each cerebral hemisphere, the **insula**, is buried deep within the lateral sulcus, and is covered by portions of the temporal, parietal, and frontal lobes. Notice that most cerebral hemisphere lobes are named for the cranial bones that lie over them.

Some important functional areas of the cerebral hemispheres have also been located (Figure 17.2d).

The cell bodies of cerebral neurons involved in these functions are found only in the outermost gray matter of the cerebrum, the **cerebral cortex**. Most of the balance of cerebral tissue—the deeper **cerebral white matter**—is composed of myelinated fibers bundled into tracts carrying impulses to or from the cortex. Some of these functional areas are summarized in **Table 17.1**.

Using a model of the human brain (and a preserved human brain, if available), identify the areas and structures of the cerebral hemispheres described above and in the table.

Then continue using the model and preserved brain along with the figures as you read about other structures.

Diencephalon

The **diencephalon** is embryologically part of the forebrain, along with the cerebral hemispheres (see Figure 17.1).

Turn the brain model so the ventral surface of the brain can be viewed. Identify the externally visible structures that mark the position of the floor of the diencephalon using **Figure 17.3** on p. 273 as a guide. These are the **olfactory bulbs** (synapse point of cranial nerve I) and **tracts**, **optic nerves** (cranial nerve II), **optic chiasma** (where the medial fibers of the optic nerves cross over), **optic tracts**, **pituitary gland**, and **mammillary bodies**.

Brain Stem

Continue to identify the **brain stem** structures—the **cerebral peduncles** (fiber tracts in the **midbrain** connecting the pons below with cerebrum above), the pons, and the medulla oblongata. *Pons* means "bridge," and the **pons** consists primarily of motor and sensory fiber tracts connecting the brain with lower CNS centers. The lowest brain stem region, the **medulla oblongata** (often shortened to just *medulla*), is also composed primarily of fiber tracts. You can see the **decussation of pyramids**, a crossover point for the major motor tracts (pyramidal tracts) descending from the motor

Text continues on p. 273. →

17

Table 17.1 Important Functional Areas of the Cerebral Cortex

Functional sensory areas	Location	Functions
Primary somatosensory cortex	Postcentral gyrus of the parietal lobe	Receives information from the body's sensory receptors in the skin and from proprioceptors in the skeletal muscles, joints, and tendons
Primary visual cortex	Occipital lobe	Receives visual information that originates in the retina of the eye
Primary auditory cortex	Temporal lobe in the gyrus bordering the lateral sulcus	Receives sound information from the receptors for hearing in the internal ear
Olfactory cortex	Medial surface of the temporal lobe, in a region called the uncus	Receives information from olfactory (smell) receptors in the superior nasal cavity
Functional motor areas	**Location**	**Functions**
Primary motor cortex	Precentral gyrus of the frontal lobe	Conscious control of voluntary movement of skeletal muscles
Broca's area	Anterior to the inferior region of the premotor area in the frontal lobe in only one hemisphere	Controls the muscles involved in speech production and also plays a role in the planning of nonspeech motor functions

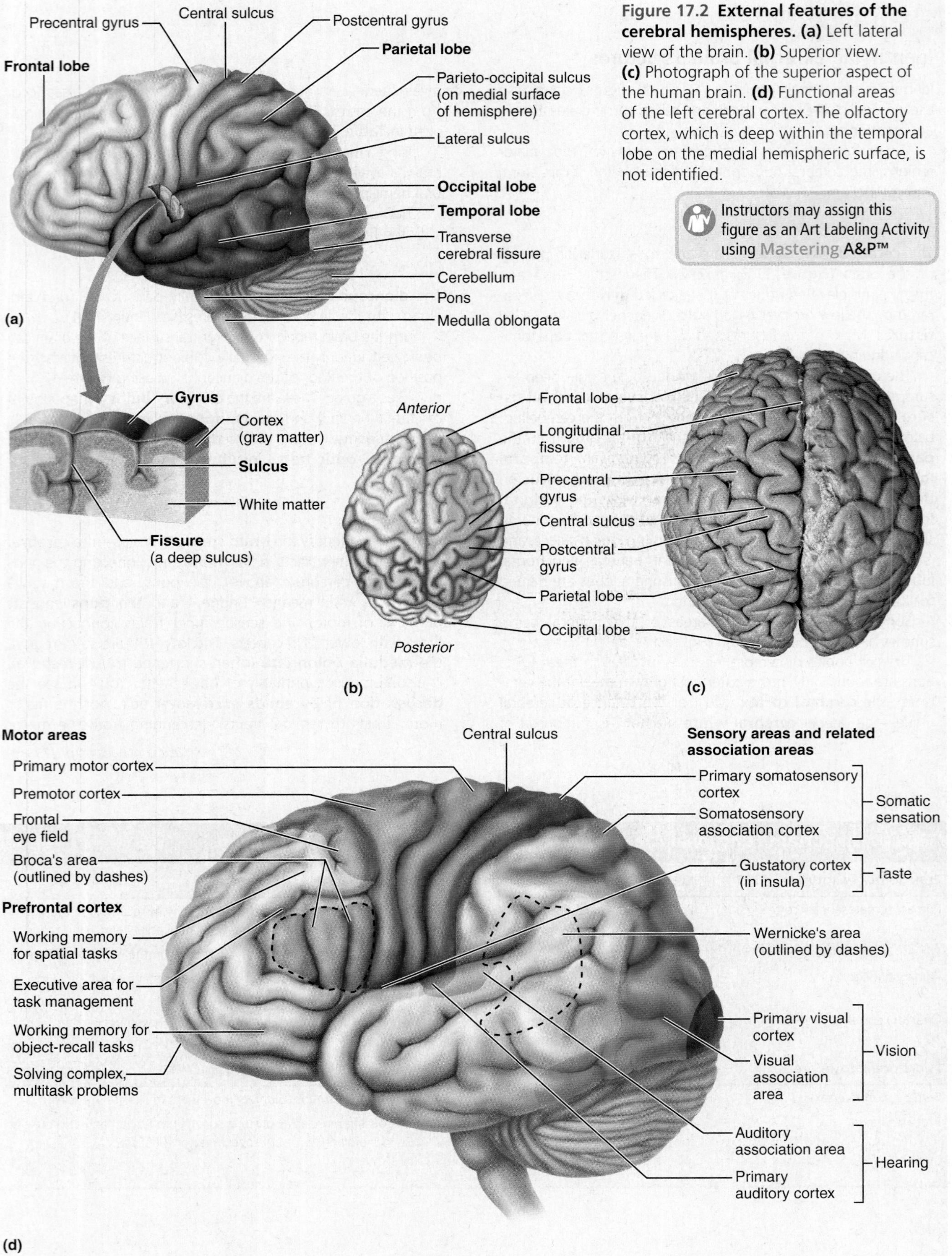

Figure 17.2 External features of the cerebral hemispheres. (a) Left lateral view of the brain. **(b)** Superior view. **(c)** Photograph of the superior aspect of the human brain. **(d)** Functional areas of the left cerebral cortex. The olfactory cortex, which is deep within the temporal lobe on the medial hemispheric surface, is not identified.

Instructors may assign this figure as an Art Labeling Activity using Mastering A&P™

17

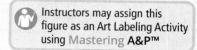

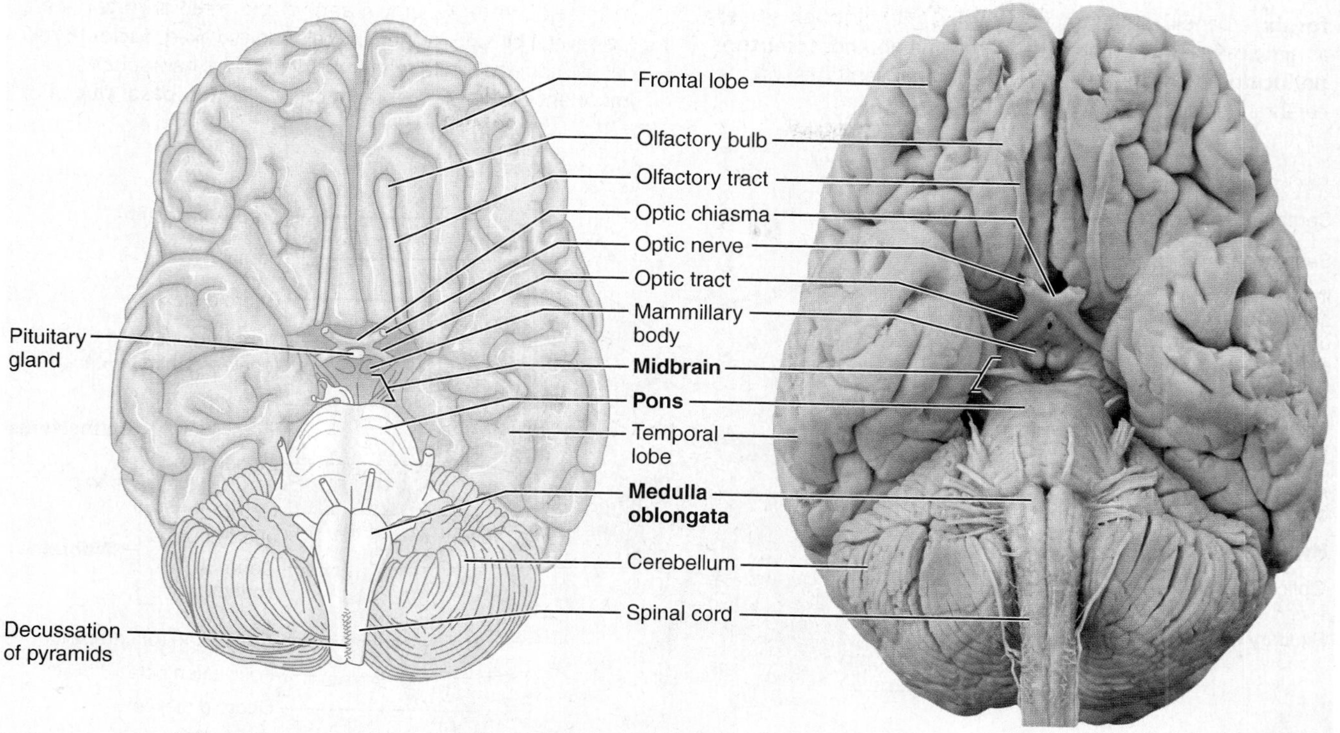

Figure 17.3 Ventral (inferior) aspect of the human brain, showing the three regions of the brain stem. Only a small portion of the midbrain can be seen; the rest is surrounded by other brain regions.

Instructors may assign this figure as an Art Labeling Activity using Mastering A&P™

17

areas of the cerebrum to the spinal cord, on the medulla's surface. The medulla also houses many vital autonomic centers involved in the control of heart rate, respiratory rhythm, and blood pressure as well as involuntary centers involved in vomiting and swallowing.

Cerebellum

1. Turn the brain model so you can see the dorsal aspect. Identify the large cauliflower-shaped **cerebellum**, which projects dorsally from under the occipital lobe of the cerebrum.

Notice that, like the cerebrum, the cerebellum has two major hemispheres and a convoluted surface (see Figure 17.6). It also has an outer cortex made up of gray matter with an inner region of white matter.

2. Remove the cerebellum to view the **corpora quadrigemina** (Figure 17.4), located on the posterior aspect of the midbrain, a brain stem structure. The two superior prominences are the **superior colliculi** (visual reflex centers); the two smaller inferior prominences are the **inferior colliculi** (auditory reflex centers).

Activity 2

Identifying Internal Brain Structures

The deeper structures of the brain have also been well mapped. As the internal brain areas are described, identify them on the figures cited. Also, use the brain model as indicated to help you in this study.

Cerebral Hemispheres

1. Take the brain model apart so you can see a medial view of the internal brain structures (**Figure 17.4**, p. 274). Observe the model closely to see the extent of the outer cortex (gray matter), which contains the cell bodies of cerebral neurons. The pyramidal cells of the cerebral motor cortex (studied in Exercise 15 and Figure 15.5) are representative of the neurons seen in the precentral gyrus.

2. Observe the deeper area of white matter, which is composed of fiber tracts. The fiber tracts found in the cerebral hemisphere white matter are called *association tracts* if they connect two portions of the same hemisphere, *projection tracts* if they run between the cerebral cortex and lower brain structures or spinal cord, and *commissures* if they run from one hemisphere to another. Observe the large **corpus callosum**, the major commissure connecting the cerebral hemispheres. The corpus callosum arches above the structures of the diencephalon and roofs over the lateral ventricles. Notice also the

Text continues on next page. →

fornix, a bandlike fiber tract concerned with olfaction as well as limbic system functions, and the membranous **septum pellucidum**, which separates the lateral ventricles of the cerebral hemispheres.

3. In addition to the gray matter of the cerebral cortex, there are several clusters of neuron cell bodies called **nuclei** buried deep within the white matter of the cerebral hemispheres. One important group of cerebral nuclei, called the **basal nuclei** or

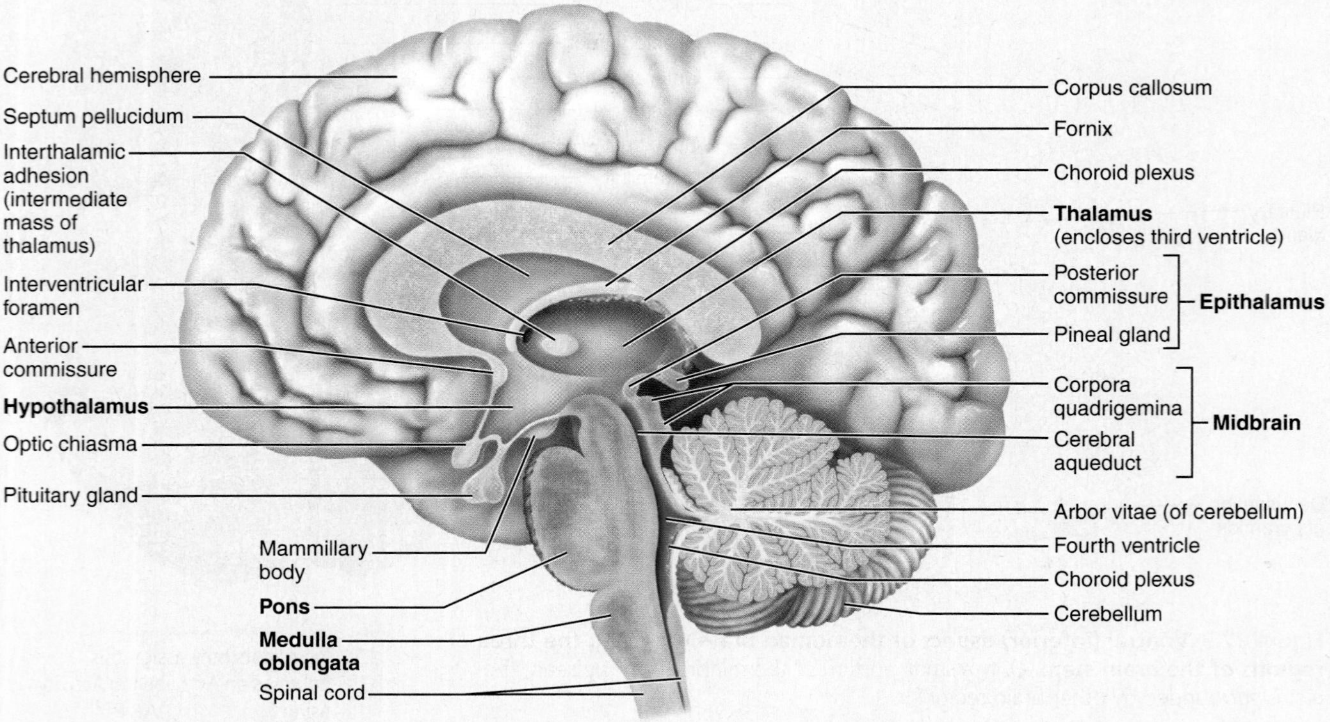

(a)

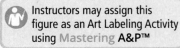

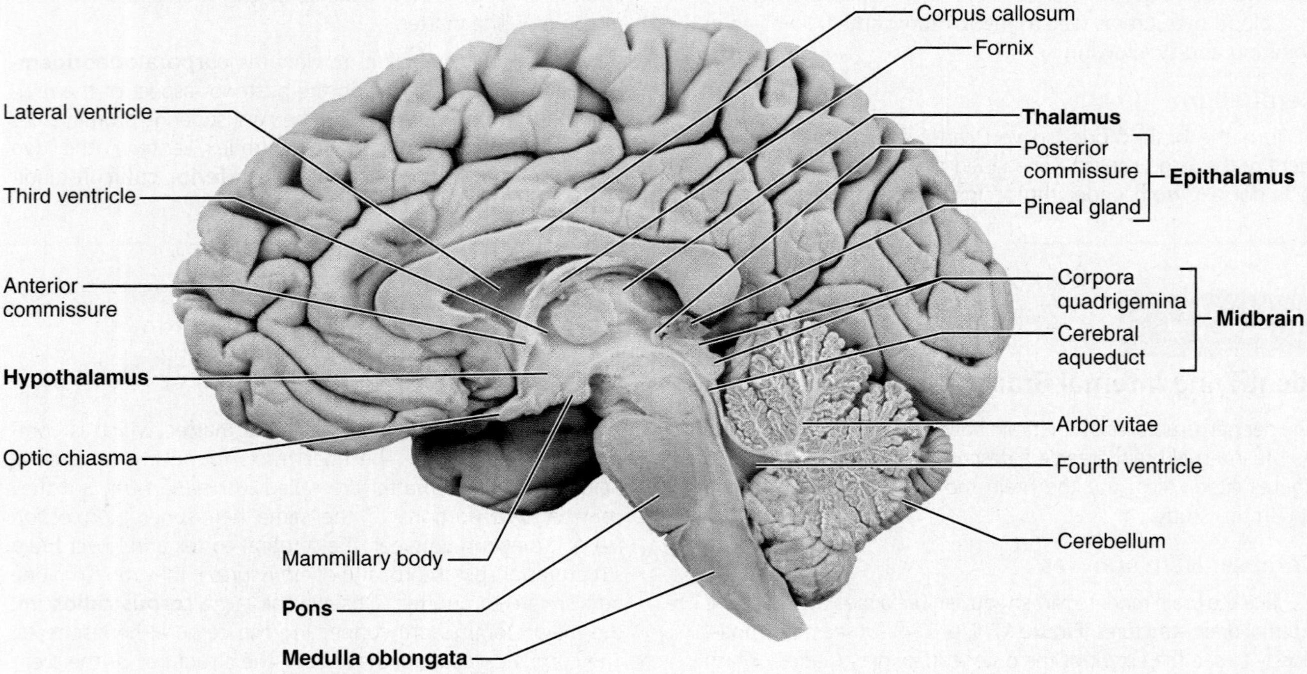

(b)

Figure 17.4 Diencephalon and brain stem structures as seen in a medial section of the brain. (a) Diagram. **(b)** Photograph.

Instructors may assign this figure as an Art Labeling Activity using **Mastering A&P™**

17

basal ganglia,* flank the lateral and third ventricles. You can see these nuclei if you have a dissectible model or a coronally or cross-sectioned human brain slice. (Otherwise, **Figure 17.5** will suffice.)

*We follow the guidelines of *Terminologia Anatomica* and use the term *basal nuclei* in this lab manual. However, the use of the term *basal ganglia* is widespread in clinical settings.

The basal nuclei are involved in regulating voluntary motor activities. The most important of them are the arching, comma-shaped **caudate nucleus**, the **putamen**, and the **globus pallidus**. The closely associated *amygdaloid body* (located at the tip of the tail of the caudate nucleus) is part of the *limbic system*.

Text continues on next page. →

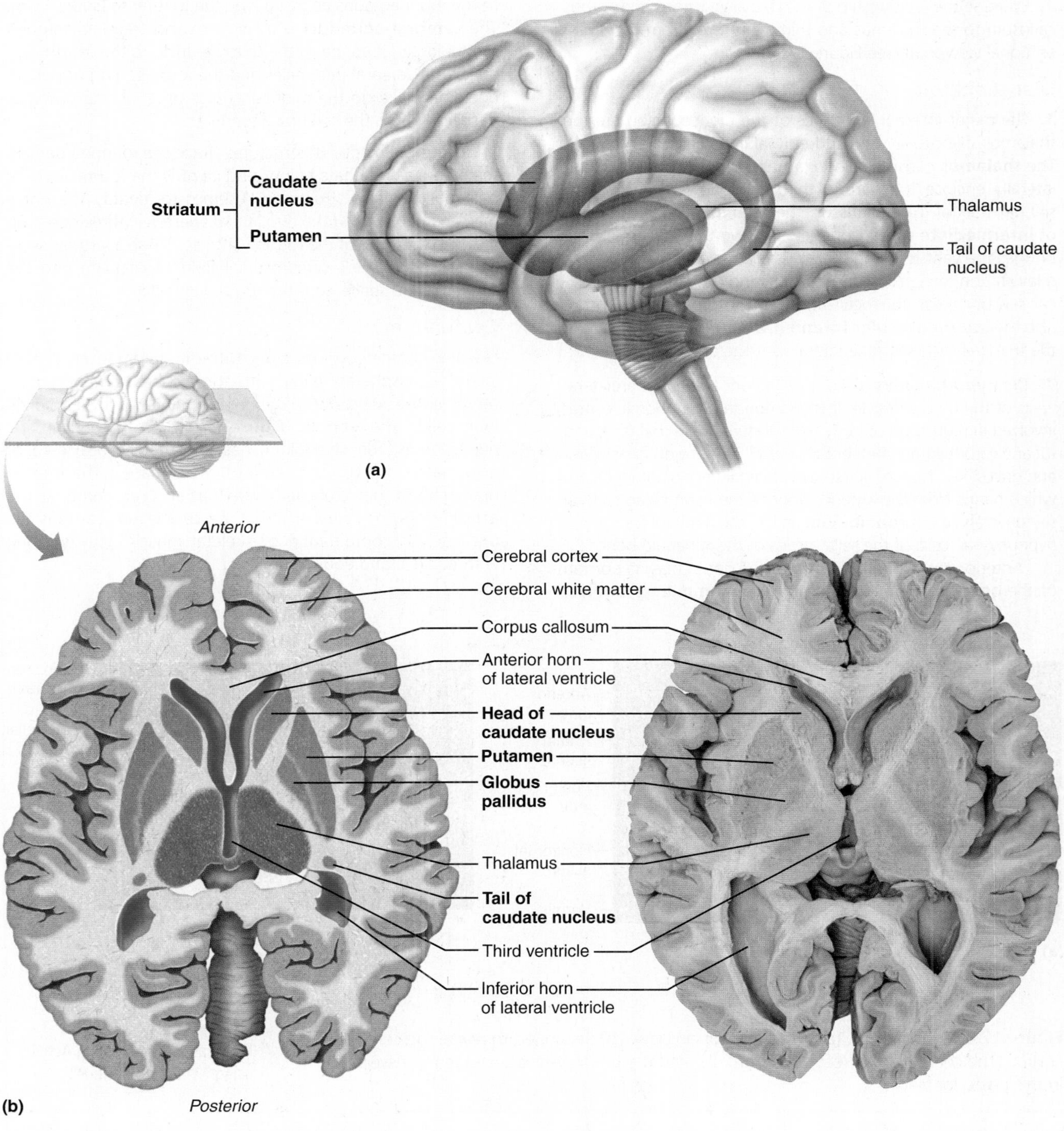

Figure 17.5 Basal nuclei. (a) Three-dimensional view of the basal nuclei showing their positions within the cerebrum. **(b)** A transverse section of the cerebrum and diencephalon showing the relationship of the basal nuclei to the thalamus and the lateral and third ventricles.

17

The **corona radiata**, a spray of projection fibers coursing down from the precentral (motor) gyrus, combines with sensory fibers traveling to the sensory cortex to form a broad band of fibrous material called the **internal capsule**. The internal capsule passes between the thalamus and the basal nuclei and through parts of the basal nuclei, giving them a striped appearance. This is why the caudate nucleus and the putamen are referred to collectively as the **striatum**, or "striped body" (Figure 17.5a).

4. Examine the relationship of the lateral ventricles and corpus callosum to the thalamus and third ventricle—from the cross-sectional viewpoint (see Figure 17.5b).

Diencephalon

1. The major internal structures of the diencephalon are the thalamus, hypothalamus, and epithalamus (see Figure 17.4). The **thalamus** consists of two large lobes of gray matter that laterally enclose the narrow third ventricle of the brain. A slender stalk of thalamic tissue, the **interthalamic adhesion**, or **intermediate mass**, connects the two thalamic lobes and bridges the ventricle. The thalamus is a major integrating and relay station for sensory impulses passing upward to the cortical sensory areas for localization and interpretation. Locate also the **interventricular foramen**, a tiny opening connecting the third ventricle with the lateral ventricle on the same side.

2. The **hypothalamus** makes up the floor and the inferolateral walls of the third ventricle. It is an important autonomic center involved in regulation of body temperature, water balance, and fat and carbohydrate metabolism as well as many other activities and drives (sex, hunger, thirst). Locate again the pituitary gland, which hangs from the anterior floor of the hypothalamus by a slender stalk, the **infundibulum**. In life, the pituitary rests in the hypophyseal fossa of the sella turcica of the sphenoid bone.

Anterior to the pituitary, identify the optic chiasma portion of the optic pathway to the brain. The **mammillary bodies**, relay stations for olfaction, bulge exteriorly from the floor of the hypothalamus just posterior to the pituitary gland.

3. The **epithalamus** forms the roof of the third ventricle and is the most dorsal portion of the diencephalon. Important structures in the epithalamus are the **pineal gland**, and the **choroid plexus** of the third ventricle.

Brain Stem

1. Now trace the short midbrain from the mammillary bodies to the rounded pons below. (Continue to refer to Figure 17.4). The **cerebral aqueduct** is a slender canal traveling through the midbrain; it connects the third ventricle to the fourth ventricle. The cerebral peduncles and the rounded corpora quadrigemina make up the midbrain tissue anterior and posterior (respectively) to the cerebral aqueduct.

2. Locate the hindbrain structures. Trace the rounded pons to the medulla oblongata below, and identify the fourth ventricle posterior to these structures. Attempt to identify the single median aperture and the two lateral apertures, three openings found in the walls of the fourth ventricle. These apertures serve as passageways for cerebrospinal fluid to circulate into the subarachnoid space from the fourth ventricle.

Cerebellum

Examine the cerebellum. Notice that it is composed of two lateral hemispheres, each with three lobes (*anterior*, *posterior*, and a deep *flocculonodular*) connected by a midline lobe called the **vermis** (**Figure 17.6**). As in the cerebral hemispheres, the cerebellum has an outer cortical area of gray matter and an inner area of white matter. The treelike branching of the cerebellar white matter is referred to as the **arbor vitae**, or "tree of life." The cerebellum controls the unconscious coordination of skeletal muscle activity along with balance and equilibrium.

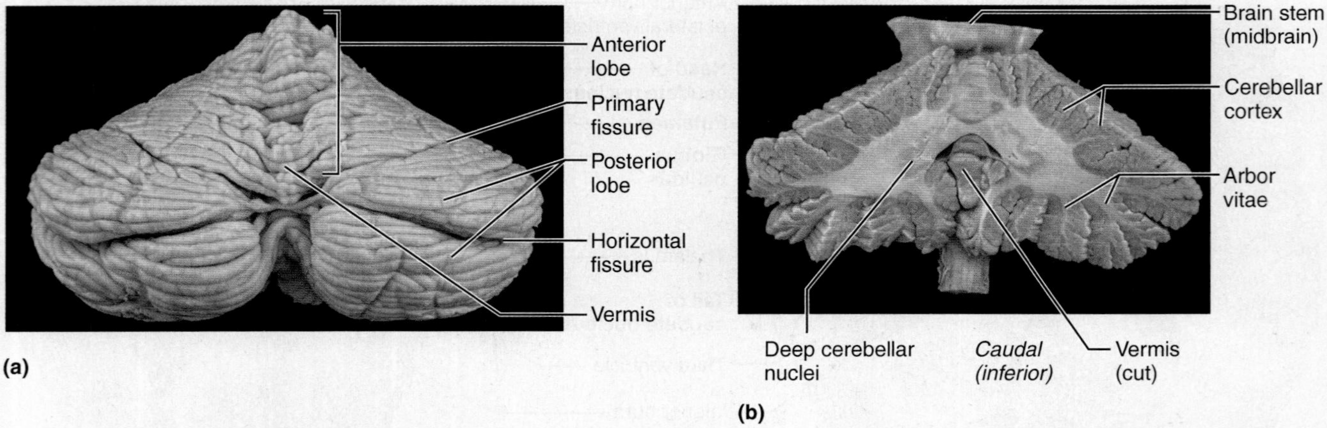

Figure 17.6 Cerebellum. (a) Posterior (dorsal) view. **(b)** Sectioned to reveal the cerebellar cortex. (The cerebellum is sectioned frontally, and the brain stem is sectioned transversely, in this posterior view.)

Instructors may assign this figure as an Art Labeling Activity using Mastering **A&P™**

Meninges of the Brain

The brain and spinal cord are covered and protected by three connective tissue membranes called **meninges** (**Figure 17.7**). The outermost meninx is the leathery **dura mater**, a double-layered membrane. One of its layers (the *periosteal layer*) is attached to the inner surface of the skull, forming the periosteum. The other (the *meningeal layer*) forms the outermost brain covering and is continuous with the dura mater of the spinal cord.

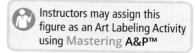

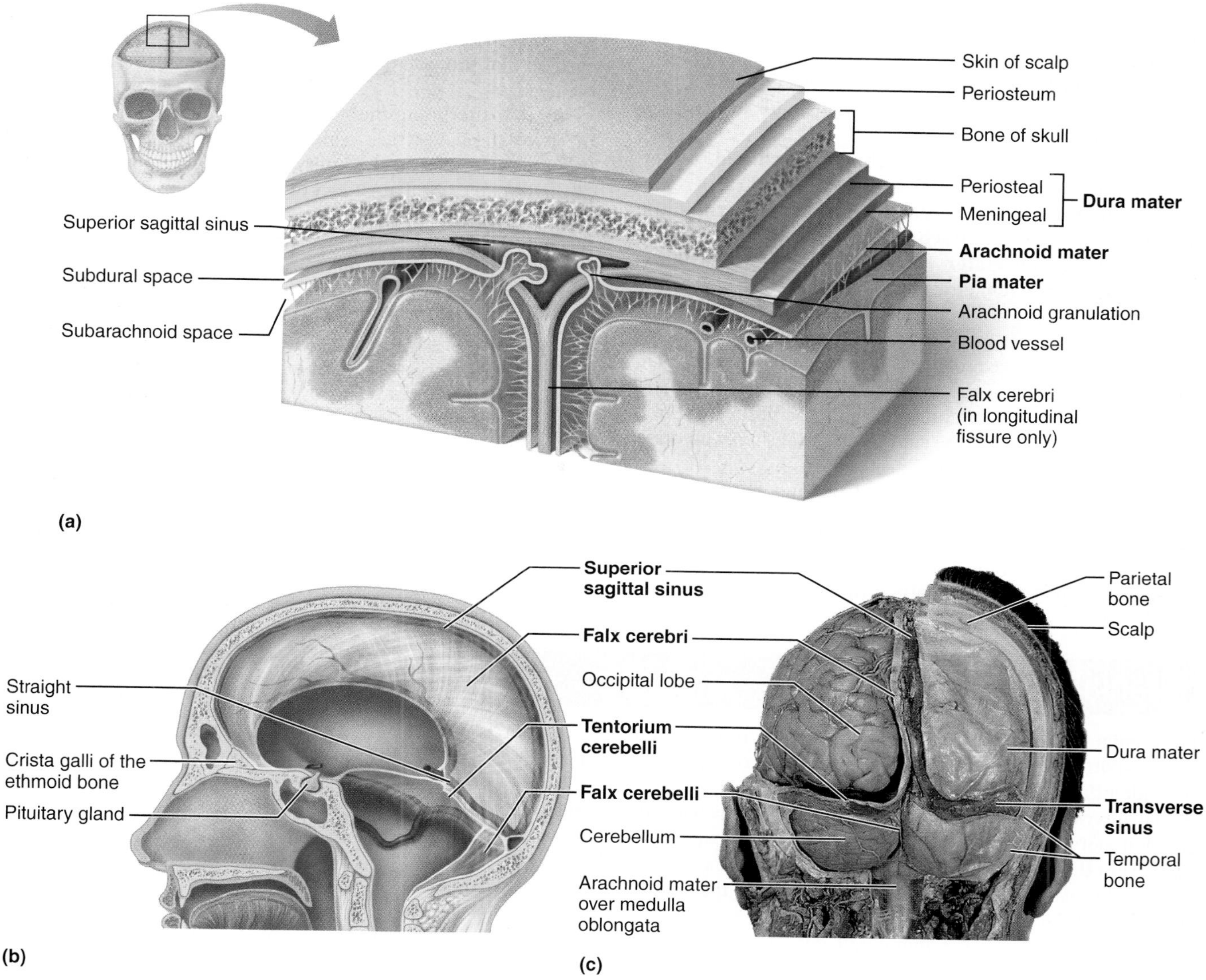

Figure 17.7 Meninges of the brain. **(a)** Three-dimensional frontal section showing the relationship of the dura mater, arachnoid mater, and pia mater. The meningeal dura forms the falx cerebri fold, which extends into the longitudinal fissure and attaches the brain to the ethmoid bone of the skull. The superior sagittal sinus is enclosed by the dural membranes superiorly. Arachnoid granulations, which return cerebrospinal fluid to the dural venous sinus, are also shown. **(b)** Midsagittal view showing the position of the dural folds: the falx cerebri, tentorium cerebelli, and falx cerebelli. **(c)** Posterior view of the brain in place, surrounded by the dura mater. Sinuses between periosteal and meningeal dura contain venous blood.

Instructors may assign this figure as an Art Labeling Activity using Mastering A&P™

The dural layers are fused together except in three places where the inner membrane extends inward to form a septum that secures the brain to structures inside the cranial cavity. One such extension, the **falx cerebri**, dips into the longitudinal fissure between the cerebral hemispheres to attach to the crista galli of the ethmoid bone of the skull (Figure 17.7a). The cavity created at this point is the large **superior sagittal sinus**, which collects blood draining from the brain tissue. The **falx cerebelli**, separating the two cerebellar hemispheres, and the **tentorium cerebelli**, separating the cerebrum from the cerebellum below, are two other important inward folds of the inner dural membrane.

The middle meninx, the weblike **arachnoid mater**, underlies the dura mater and is partially separated from it by the **subdural space**. Threadlike projections bridge the **subarachnoid space** to attach the arachnoid to the innermost meninx, the **pia mater**. The delicate pia mater is highly vascular and clings tenaciously to the surface of the brain, following its gyri.

In life, the subarachnoid space is filled with cerebrospinal fluid. Specialized projections of the arachnoid tissue called **arachnoid granulations** protrude through the dura mater. These granulations allow the cerebrospinal fluid to drain back into the venous circulation via the superior sagittal sinus and other dural venous sinuses.

Meningitis, inflammation of the meninges, is a serious threat to the brain because of the intimate association between the brain and meninges. Should infection spread to the neural tissue of the brain itself, life-threatening **encephalitis** may occur. Meningitis is often diagnosed by taking a sample of cerebrospinal fluid (via a spinal tap) from the subarachnoid space. +

Cerebrospinal Fluid

The cerebrospinal fluid (CSF), much like plasma in composition, is continually formed by the **choroid plexuses**, small capillary knots hanging from the roof of the ventricles of the brain. The cerebrospinal fluid in and around the brain forms a watery cushion that protects the delicate brain tissue against blows to the head.

Within the brain, the cerebrospinal fluid circulates from the two lateral ventricles (in the cerebral hemispheres) into the third ventricle via the **interventricular foramina**, and then through the cerebral aqueduct of the midbrain into the fourth ventricle (**Figure 17.8**). CSF enters the subarachnoid space through the paired **lateral apertures** in the side walls of the fourth ventricle and the **median aperture** in its roof. There it bathes the outer surfaces of the brain and spinal cord. The fluid returns to the blood in the dural venous sinuses via the arachnoid granulations.

Ordinarily, cerebrospinal fluid forms and drains at a constant rate. However, under certain conditions—for example, obstructed drainage or circulation resulting from tumors or anatomical deviations—cerebrospinal fluid accumulates and exerts increasing pressure on the brain which, uncorrected, causes neurological damage in adults. In infants, **hydrocephalus** (literally, "water on the brain") is indicated by a gradually enlarging head. The infant's skull is still flexible and contains fontanelles, so it can expand to accommodate the increasing size of the brain. +

Cranial Nerves

The **cranial nerves** are part of the peripheral nervous system and not part of the brain proper, but they are most appropriately identified while studying brain anatomy. The 12 pairs of cranial nerves primarily serve the head and neck. Only one pair, the vagus nerves, extends into the thoracic and abdominal cavities. All but the first two pairs (olfactory and optic nerves) arise from the brain stem and pass through foramina in the base of the skull to reach their destination.

The cranial nerves are numbered consecutively, and in most cases their names reflect the major structures they control. The cranial nerves are described by name, number (Roman numeral), origin, course, and function in the list (**Table 17.2**, pp. 280–282). This information should be committed to memory. A mnemonic device that might be helpful for remembering the cranial nerves in order is "*O*n *o*ccasion, *o*ur *t*rusty *t*ruck *a*cts *f*unny—*v*ery *g*ood *v*ehicle *a*ny*h*ow." The first letter of each word and the "a" and "h" of the final word "anyhow" will remind you of the first letter of the cranial nerve name.

Most cranial nerves are mixed nerves (containing both motor and sensory fibers). But close scrutiny of the list (Table 17.2) will reveal that two pairs of cranial nerves (optic and olfactory) are purely sensory in function.

Activity 3

Identifying and Testing the Cranial Nerves

1. Observe the ventral surface of the brain model to identify the cranial nerves. (**Figure 17.9** on p. 280 may also aid you in this study.) Notice that the first (olfactory) cranial nerves are not visible on the model because they consist only of short axons that run from the nasal mucosa through the cribriform foramina of the ethmoid bone. (However, the synapse points of the first cranial nerves, the *olfactory bulbs*, are visible on the model.)

Text continues on p. 280 →

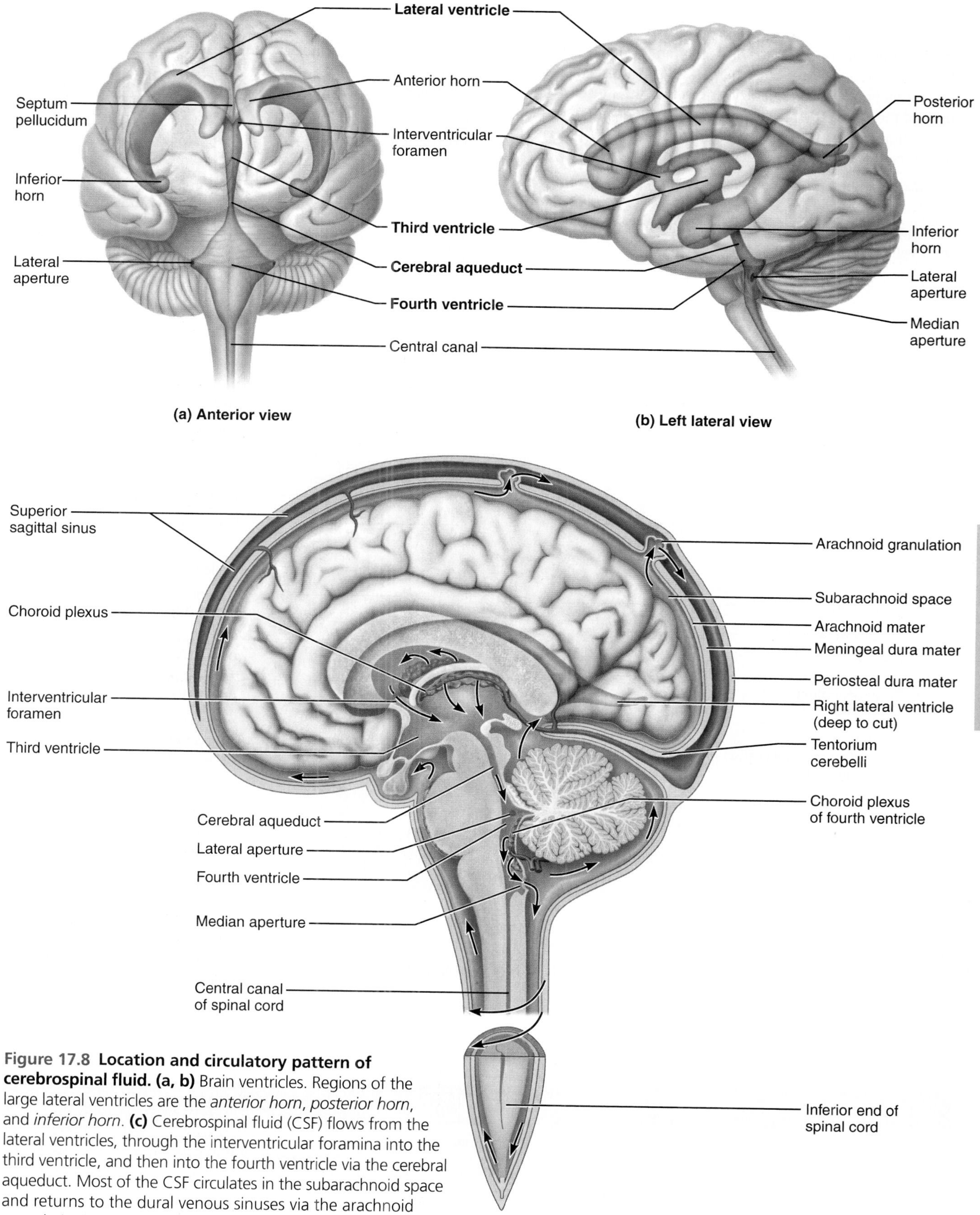

Figure 17.8 Location and circulatory pattern of cerebrospinal fluid. (a, b) Brain ventricles. Regions of the large lateral ventricles are the *anterior horn*, *posterior horn*, and *inferior horn*. **(c)** Cerebrospinal fluid (CSF) flows from the lateral ventricles, through the interventricular foramina into the third ventricle, and then into the fourth ventricle via the cerebral aqueduct. Most of the CSF circulates in the subarachnoid space and returns to the dural venous sinuses via the arachnoid granulations.

Instructors may assign this figure as an Art Labeling Activity using Mastering **A&P™**

2. Testing cranial nerves is an important part of any neurological examination. See the last column of Table 17.2 for techniques you can use for such tests. Conduct tests of cranial nerve function following directions given in the "testing" column of the table. The results may help you understand cranial nerve function, especially as it pertains to some aspects of brain function.

3. Several cranial nerve ganglia are named in the **Activity 3 chart**. *Using your textbook or an appropriate reference*, fill in the chart by naming the cranial nerve the ganglion is associated with and stating the ganglion location.

Activity 3: Cranial Nerve Ganglia

Cranial nerve ganglion	Cranial nerve	Site of ganglion
Trigeminal		
Geniculate		
Inferior		
Superior		
Spiral		
Vestibular		

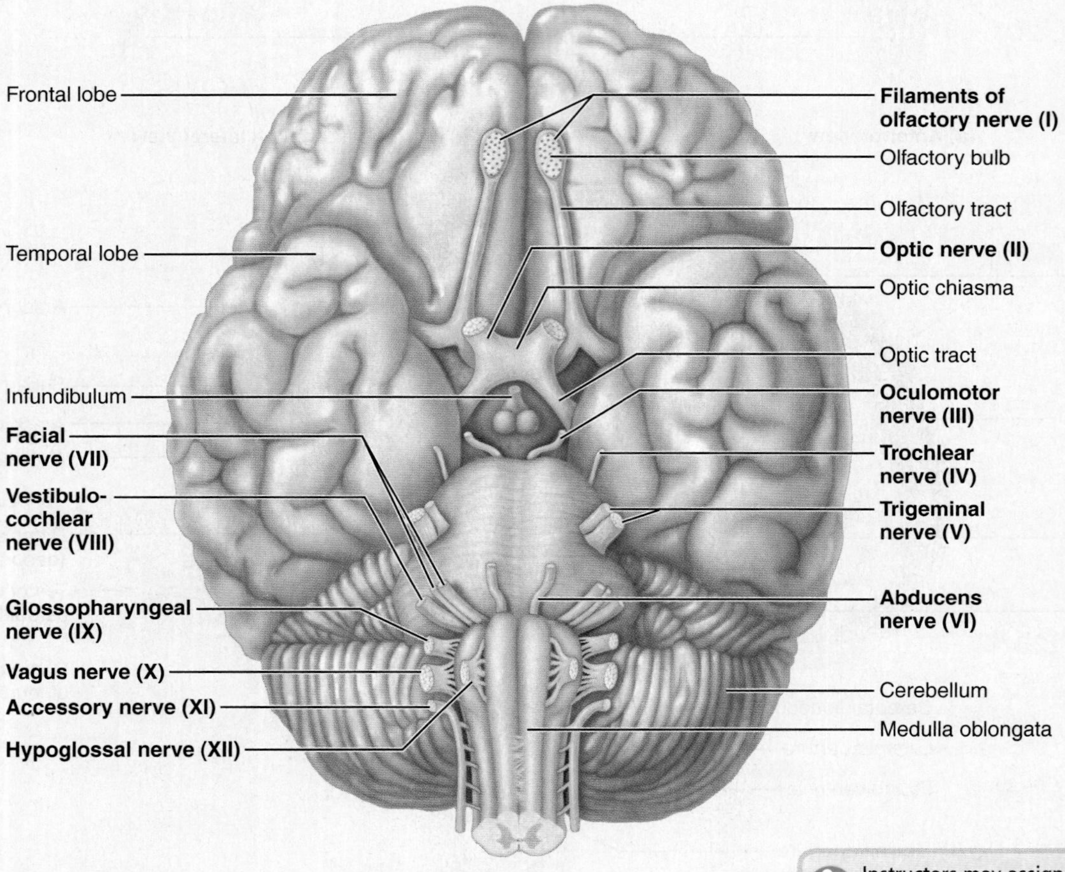

Figure 17.9 Ventral aspect of the human brain, showing the cranial nerves. (See also Figure 17.3.)

Instructors may assign this figure as an Art Labeling Activity using Mastering A&P™

Table 17.2 The Cranial Nerves (Figure 17.9)

Number and name	Origin and course	Function*	Testing
I. Olfactory	Fibers arise from olfactory epithelium and run through cribriform foramina of ethmoid bone to synapse in olfactory bulb.	Purely sensory—carries afferent impulses for sense of smell.	Person is asked to sniff aromatic substances, such as oil of cloves and vanilla, and to identify each.

Table 17.2 ***(continued)***

Number and name	Origin and course	Function*	Testing
II. Optic	Fibers arise from retina of eye and pass through optic canal of sphenoid bone. Fibers partially cross over at the optic chiasma and continue on to the thalamus as the optic tracts. Final fibers of this pathway travel from the thalamus to the primary visual cortex as the optic radiation.	Purely sensory—carries afferent impulses associated with vision.	Vision and visual field are determined with eye chart and by testing the point at which the person first sees an object (finger) moving into the visual field. Fundus of eye viewed with ophthalmoscope to detect papilledema (swelling of optic disc, or point at which optic nerve leaves the eye) and to observe blood vessels.
III. Oculomotor	Fibers emerge from ventral midbrain and course ventrally to enter the orbit. They exit from skull via superior orbital fissure.	Primarily motor—somatic motor fibers to inferior oblique and superior, inferior, and medial rectus muscles, which direct eyeball, and to levator palpebrae muscles of the superior eyelid; parasympathetic fibers to smooth muscle controlling lens shape and pupil size.	Pupils are examined for size, shape, and equality. Pupillary reflex is tested with penlight (pupils should constrict when illuminated). Convergence for near vision is tested, as is subject's ability to follow objects with the eyes.
IV. Trochlear	Fibers emerge from midbrain and exit from skull via superior orbital fissure.	Primarily motor—provides somatic motor fibers to superior oblique muscle that moves the eyeball.	Tested with cranial nerve III.
V. Trigeminal	Fibers run from face to pons and form three divisions: mandibular division fibers pass through foramen ovale in sphenoid bone, maxillary division fibers pass via foramen rotundum in sphenoid bone, and ophthalmic division fibers pass through superior orbital fissure of sphenoid bone.	Mixed—major sensory nerve of face; conducts sensory impulses from skin of face and anterior scalp, from mucosae of mouth and nose, and from surface of eyes; mandibular division also contains motor fibers that innervate muscles of mastication and muscles of floor of mouth.	Sensations of pain, touch, and temperature are tested with safety pin and hot and cold probes. Corneal reflex tested with wisp of cotton. Motor branch assessed by asking person to clench the teeth, open mouth against resistance, and move jaw side to side.
VI. Abducens	Fibers leave inferior pons and exit from skull via superior orbital fissure.	Primarily motor—carries somatic motor fibers to lateral rectus muscle that abducts the eyeball.	Tested with cranial nerve III.
VII. Facial	Fibers leave pons and travel through temporal bone via internal acoustic meatus, exiting via stylomastoid foramen to reach the face.	Mixed—supplies somatic motor fibers to muscles of facial expression and the posterior belly of the digastric muscle; parasympathetic motor fibers to lacrimal and salivary glands; carries sensory fibers from taste receptors of anterior tongue.	Anterior two-thirds of tongue is tested for ability to taste sweet (sugar), salty, sour (vinegar), and bitter (quinine) substances. Symmetry of face is checked. Subject is asked to close eyes, smile, whistle, and so on. Tearing is assessed with ammonia fumes.
VIII. Vestibulocochlear	Fibers run from inner ear equilibrium and hearing apparatus, housed in temporal bone, through internal acoustic meatus to enter pons.	Mostly sensory—vestibular branch transmits impulses associated with sense of equilibrium from vestibular apparatus and semicircular canals; cochlear branch transmits impulses associated with hearing from cochlea. Small motor component adjusts the sensitivity of the sensory receptors.	Hearing is checked by air and bone conduction using tuning fork.
IX. Glossopharyngeal	Fibers emerge from medulla oblongata and leave skull via jugular foramen to run to throat.	Mixed—somatic motor fibers serve pharyngeal muscles, and parasympathetic motor fibers serve salivary glands; sensory fibers carry impulses from pharynx, tonsils, posterior tongue (taste buds), and from chemoreceptors and pressure receptors of carotid artery.	A tongue depressor is used to check the position of the uvula. Gag and swallowing reflexes are checked. Subject is asked to speak and cough. Posterior third of tongue may be tested for taste.

Table 17.2 **The Cranial Nerves *(continued)***

Number and name	Origin and course	Function*	Testing
X. Vagus	Fibers emerge from medulla oblongata and pass through jugular foramen and descend through neck region into thorax and abdomen.	Mixed—fibers carry somatic motor impulses to pharynx and larynx and sensory fibers from same structures; very large portion is composed of parasympathetic motor fibers, which supply heart and smooth muscles of abdominal visceral organs; transmits sensory impulses from viscera.	As for cranial nerve IX (IX and X are tested together, since they both innervate muscles of throat and mouth).
XI. Accessory	Fibers arise from the superior aspect of spinal cord, enter the skull, and then travel through jugular foramen to reach muscles of neck and back.	Mixed (but primarily motor in function)—provides somatic motor fibers to sternocleidomastoid and trapezius muscles.	Sternocleidomastoid and trapezius muscles are checked for strength by asking person to rotate head and shrug shoulders against resistance.
XII. Hypoglossal	Fibers arise from medulla oblongata and exit from skull via hypoglossal canal to travel to tongue.	Mixed (but primarily motor in function)—carries somatic motor fibers to muscles of tongue.	Person is asked to protrude and retract tongue. Any deviations in position are noted.

*Does not include sensory impulses from proprioceptors.

17

DISSECTION

The Sheep Brain

The sheep brain is enough like the human brain to warrant comparison. Obtain a sheep brain, disposable gloves, dissecting tray, and instruments, and bring them to your laboratory bench.

1. Don disposable gloves. If the dura mater is present, remove it as described here. Place the intact sheep brain ventral surface down on the dissecting pan, and observe the dura mater. Feel its consistency and note its toughness. Cut through the dura mater along the line of the longitudinal fissure (which separates the cerebral hemispheres) to enter the superior sagittal sinus. Gently force the cerebral hemispheres apart laterally to expose the corpus callosum deep to the longitudinal fissure.

2. Carefully remove the dura mater and examine the superior surface of the brain. Notice that its surface, like that of the human brain, is thrown into convolutions (fissures and gyri). Locate the arachnoid mater, which appears on the brain surface as a delicate "cottony" material spanning the fissures. In contrast, the innermost meninx, the pia mater, closely follows the cerebral contours.

3. Before beginning the dissection, turn your sheep brain so that you are viewing its left lateral aspect. Compare the various areas of the sheep brain (cerebrum, brain stem, cerebellum) to the photo of the human brain (**Figure 17.10**). Relatively speaking, which of these structures is obviously much larger in the human brain?

Ventral Structures

Turn the brain so that its ventral surface is uppermost. (**Figure 17.11a** and **b** show the important features of the ventral surface of the brain.)

1. Look for the clublike olfactory bulbs anteriorly, on the inferior surface of the frontal lobes of the cerebral hemispheres. Axons of olfactory neurons run from the nasal mucosa through the cribriform foramina of the ethmoid bone to synapse with the olfactory bulbs.

How does the size of these olfactory bulbs compare with those of humans?

____

Is the sense of smell more important for protection and foraging in sheep or in humans?

____

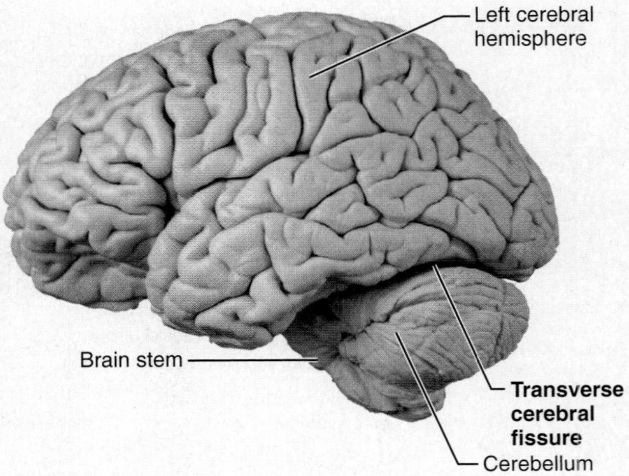

Figure 17.10 Photograph of lateral aspect of the human brain.

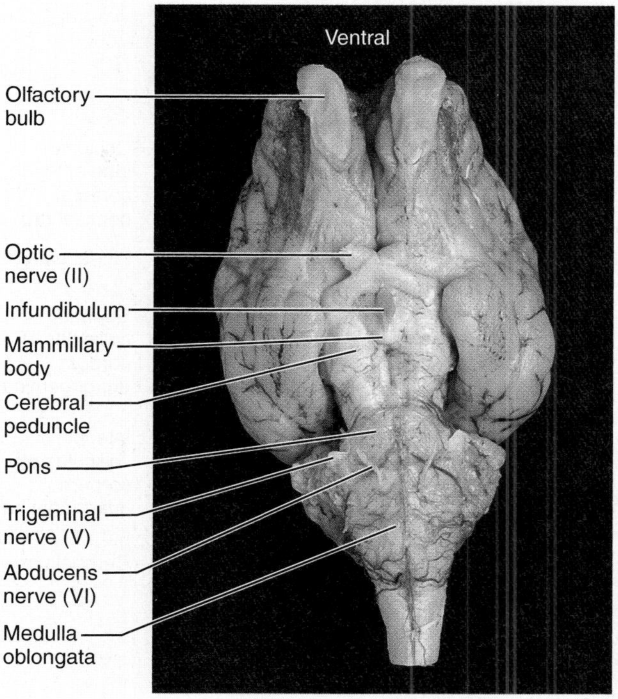

(a)

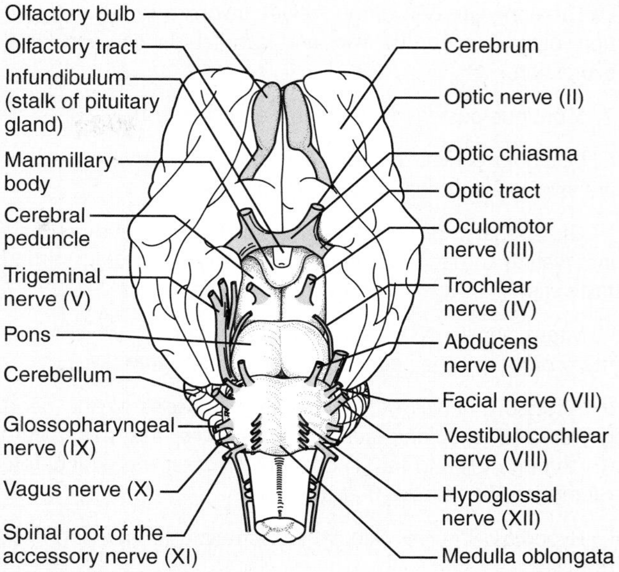

(b)

Figure 17.11 Intact sheep brain. (a) Photograph of ventral view. **(b)** Diagram of ventral view. **(c)** Photograph of the dorsal view.

Instructors may assign a related video and coaching activity for the Sheep Brain Dissection

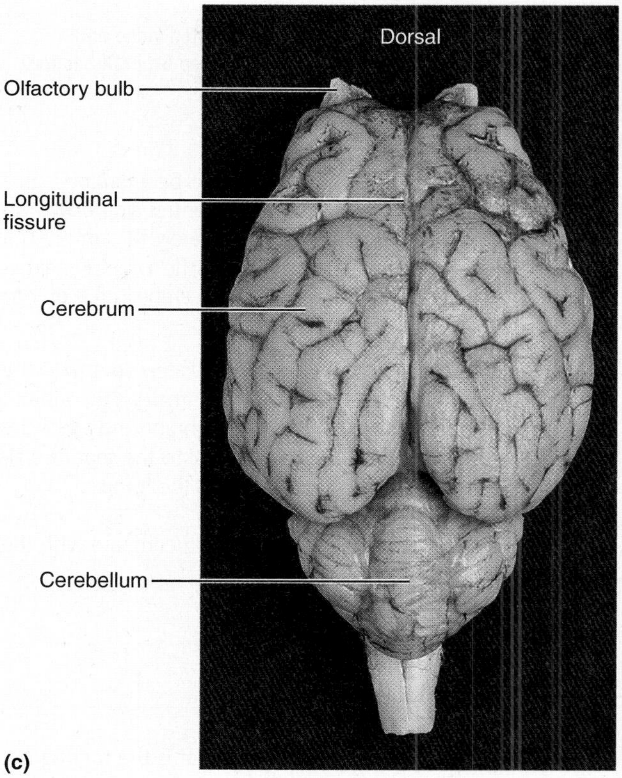

(c)

17

2. The optic nerve (II) carries sensory impulses from the retina of the eye. Thus this cranial nerve is involved in the sense of vision. Identify the optic nerves, optic chiasma, and optic tracts.

3. Posterior to the optic chiasma, two structures protrude from the ventral aspect of the hypothalamus—the infundibulum (stalk of the pituitary gland) immediately posterior to the optic chiasma and the mammillary body. Notice that the sheep's mammillary body is a single rounded eminence. In humans, it is a double structure.

4. Identify the cerebral peduncles on the ventral aspect of the midbrain, just posterior to the mammillary body of the hypothalamus. The cerebral peduncles are fiber tracts connecting the cerebrum and medulla oblongata. Identify the large oculomotor nerves (III), which arise from the ventral midbrain surface, and the tiny trochlear nerves (IV), which can be seen at the junction of the midbrain and pons. Both of these cranial nerves provide motor fibers to extrinsic muscles of the eyeball.

5. Move posteriorly from the midbrain to identify first the pons and then the medulla oblongata, structures composed primarily of ascending and descending fiber tracts.

6. Return to the junction of the pons and midbrain, and proceed posteriorly to identify the following cranial nerves, all arising from the pons. Check them off as you locate them.

☐ Trigeminal nerves (V), which are involved in chewing and sensations of the head and face.

☐ Abducens nerves (VI), which abduct the eye (and thus work in conjunction with cranial nerves III and IV).

Text continues on next page. →

☐ Facial nerves (VII), large nerves involved in taste sensation, gland function (salivary and lacrimal glands), and facial expression.

7. Continue posteriorly to identify and check off:

☐ Vestibulocochlear nerves (VIII), mostly sensory nerves that are involved with hearing and equilibrium.

☐ Glossopharyngeal nerves (IX), which contain motor fibers innervating throat structures and sensory fibers transmitting taste stimuli (in conjunction with cranial nerve VII).

☐ Vagus nerves (X), often called "wanderers," which serve many organs of the head, thorax, and abdominal cavity.

☐ Accessory nerves (XI), which serve muscles of the neck, larynx, and shoulder; actually arise from the spinal cord (C_1 through C_5) and travel superiorly to enter the skull before running to the muscles that they serve.

☐ Hypoglossal nerves (XII), which stimulate tongue and neck muscles.

It is likely that some of the cranial nerves will have been broken off during brain removal. If so, observe sheep brains of other students to identify those missing from your specimen, using your check marks as a guide.

17

Dorsal Structures

1. Refer to the dorsal view photograph (Figure 17.11c) as a guide in identifying the following structures. Reidentify the now exposed cerebral hemispheres. How does the depth of the fissures in the sheep's cerebral hemispheres compare to that of the fissures in the human brain?

____

2. Examine the cerebellum. Notice that, in contrast to the human cerebellum, it is not divided longitudinally, and that its fissures are oriented differently. What dural fold (falx cerebri or falx cerebelli) is missing that is present in humans?

____

3. Locate the three pairs of cerebellar peduncles, fiber tracts that connect the cerebellum to other brain structures, by lifting the cerebellum dorsally away from the brain stem. The most posterior pair, the inferior cerebellar peduncles, connect the cerebellum to the medulla. The middle cerebellar peduncles attach the cerebellum to the pons, and the superior cerebellar peduncles run from the cerebellum to the midbrain.

4. To expose the dorsal surface of the midbrain, gently separate the cerebrum and cerebellum (as shown in **Figure 17.12.**) Identify the corpora quadrigemina, which appear as four rounded prominences on the dorsal midbrain surface.

What is the function of the corpora quadrigemina?

____

____

Also locate the pineal gland, which appears as a small oval protrusion in the midline just anterior to the corpora quadrigemina.

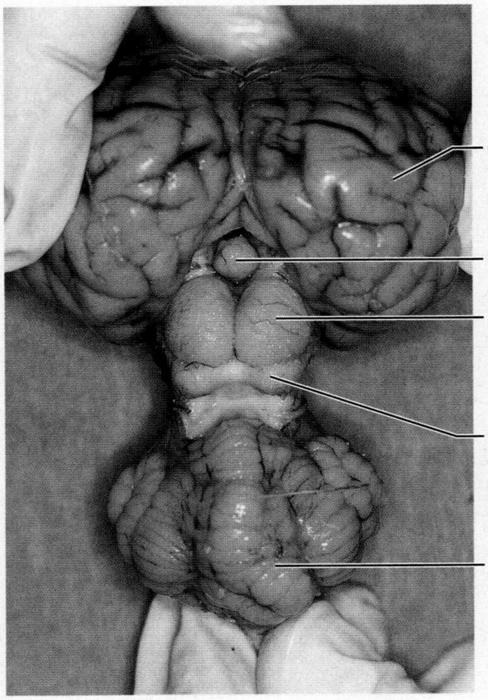

Figure 17.12 Means of exposing the dorsal midbrain structures of the sheep brain.

Instructors may assign a related video and coaching activity for the Sheep Brain Dissection

Internal Structures

1. The internal structure of the brain can be examined only after further dissection. Place the brain ventral side down on the dissecting tray and make a cut completely through it in a superior to inferior direction. Cut through the longitudinal fissure, corpus callosum, and midline of the cerebellum. Refer to **Figure 17.13** as you work.

2. A thin nervous tissue membrane immediately ventral to the corpus callosum that separates the lateral ventricles is the septum pellucidum. If it is still intact, pierce this membrane and probe the lateral ventricle cavity. The fiber tract ventral to the septum pellucidum and anterior to the third ventricle is the fornix.

How does the size of the fornix in this brain compare with the size of the human fornix?

____

____

Why do you suppose this is so? (Hint: What is the function of this band of fibers?)

____

3. Identify the thalamus, which forms the walls of the third ventricle and is located posterior and ventral to the fornix. The interthalamic adhesion appears as an oval protrusion of the thalamic wall.

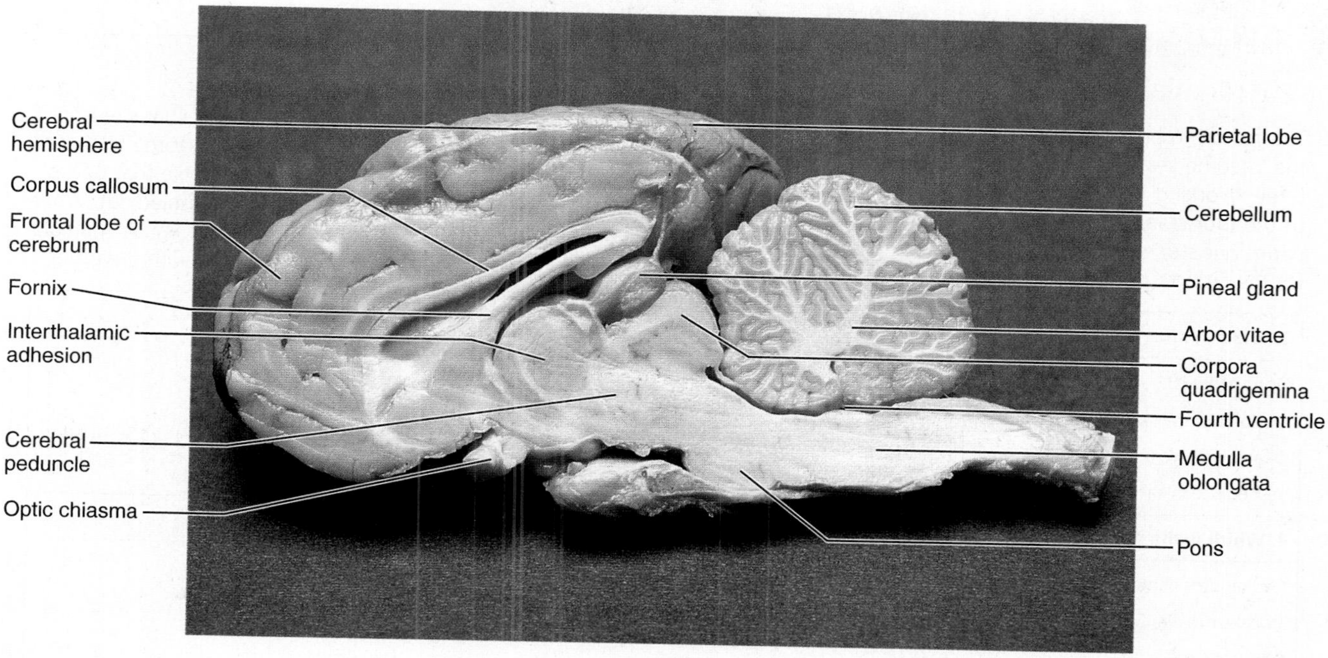

Figure 17.13 Photograph of median section of the sheep brain showing internal structures.

Instructors may assign this figure as an Art Labeling Activity using Mastering A&P™

4. The hypothalamus forms the floor of the third ventricle. Identify the optic chiasma, infundibulum, and mammillary body on its exterior surface. The pineal gland is just beneath the junction of the corpus callosum and fornix.

5. Locate the midbrain by identifying the corpora quadrigemina that form its dorsal roof. Follow the cerebral aqueduct through the midbrain tissue to the fourth ventricle. Identify the cerebral peduncles, which form its anterior walls.

6. Identify the pons and medulla oblongata, which lie anterior to the fourth ventricle. The medulla continues into the spinal cord without any obvious anatomical change, but the point at which the fourth ventricle narrows to a small canal is generally accepted as the beginning of the spinal cord.

7. Identify the cerebellum posterior to the fourth ventricle. Notice its internal treelike arrangement of white matter, the arbor vitae.

8. If time allows, obtain another sheep brain and section it along the frontal plane so that the cut passes through the infundibulum. Compare your specimen with **Figure 17.14**, and attempt to identify all the structures shown in the figure.

9. Check with your instructor to determine whether a small portion of the spinal cord from your brain specimen should be saved for spinal cord studies (Exercise 19). Otherwise, dispose of all the organic debris in the appropriate laboratory containers and clean the laboratory bench, the dissection instruments, and the tray before leaving the laboratory.

17

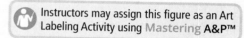

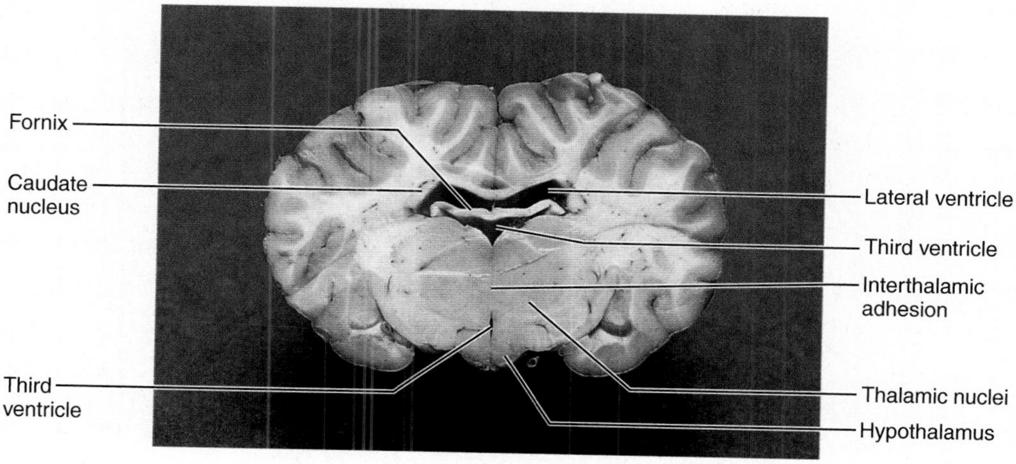

Figure 17.14 Frontal section of a sheep brain. Major structures include the thalamus, hypothalamus, and lateral and third ventricles.

Instructors may assign a related video and coaching activity for the Sheep Brain Dissection

Group Challenge

Odd (Cranial) Nerve Out

The following boxes each contain four cranial nerves. One of the listed nerves does not share a characteristic with the other three. Working in groups of three, discuss the characteristics of the four cranial nerves in each set. On a separate piece of paper, one student will record the characteristics for each nerve for the group. For each set of four nerves, discuss the possible candidates for the "odd nerve" and which characteristic it lacks based upon your notes. Once you have come to a consensus within your group, circle the cranial nerve that doesn't belong with the others, and explain why it is singled out. What characteristic is it missing? Sometimes there may be multiple reasons why the cranial nerve doesn't belong with the others.

1. Which is the "odd" nerve?	**Why is it the odd one out?**
Optic nerve (II) Oculomotor nerve (III) Olfactory nerve (I) Vestibulocochlear nerve (VIII)	

2. Which is the "odd" nerve?	**Why is it the odd one out?**
Oculomotor nerve (III) Trochlear nerve (IV) Abducens nerve (VI) Hypoglossal nerve (XII)	

3. Which is the "odd" nerve?	**Why is it the odd one out?**
Facial nerve (VII) Hypoglossal nerve (XII) Trigeminal nerve (V) Glossopharyngeal Nerve (IX)	

Instructors may assign a portion of the Review Sheet questions using Mastering A&P™

REVIEW SHEET

EXERCISE 17 Gross Anatomy of the Brain and Cranial Nerves

Name ______________________________ Lab Time/Date ______________________________

The Human Brain

1. Using the terms from the key, identify the structures of the brain.

Key:

a. brain stem
b. central sulcus
c. cerebellum
d. frontal lobe
e. lateral sulcus
f. occipital lobe
g. parietal lobe
h. parieto-occipital sulcus
i. postcentral gyrus
j. precentral gyrus
k. temporal lobe
l. transverse cerebral fissure

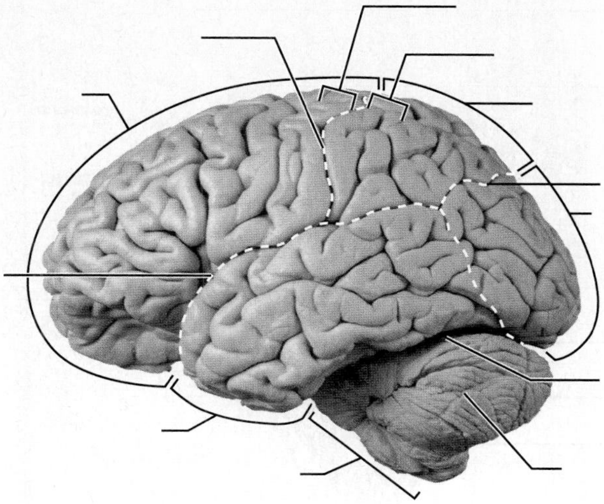

2. In which of the cerebral lobes are the following functional areas found?

primary auditory cortex: ______________________ olfactory cortex: ______________________

primary motor cortex: ______________________ primary visual cortex: ______________________

primary somatosensory cortex: ______________________ Broca's area: ______________________

3. Which of the following structures are not part of the brain stem? (Circle the appropriate response or responses.)

cerebral hemispheres pons midbrain cerebellum medulla oblongata diencephalon

4. Complete the following statements by writing the proper word or phrase on the corresponding blanks at the right.

A(n) __1__ is an elevated ridge of cerebral tissue. The convolutions seen in the cerebrum are important because they increase the __2__. Gray matter is composed of __3__. White matter is composed of __4__. A fiber tract that provides for communication between different parts of the same cerebral hemisphere is called a(n) __5__ tract, whereas one that carries impulses from the cerebrum to lower CNS areas is called a(n) __6__ tract. The caudate nucleus and putamen are collectively called the __7__.

1. ______________________
2. ______________________
3. ______________________
4. ______________________
5. ______________________
6. ______________________
7. ______________________

5. Using the terms from the key, identify the structures on the following midsagittal view of the human brain.

Key:
- a. anterior commissure
- b. cerebellum
- c. cerebral aqueduct
- d. cerebral hemisphere
- e. choroid plexus
- f. corpora quadrigemina
- g. corpus callosum
- h. fornix
- i. fourth ventricle
- j. hypothalamus
- k. interthalamic adhesion
- l. mammillary body
- m. medulla oblongata
- n. midbrain
- o. optic chiasma
- p. pineal gland
- q. pituitary gland
- r. pons
- s. septum pellucidum
- t. thalamus

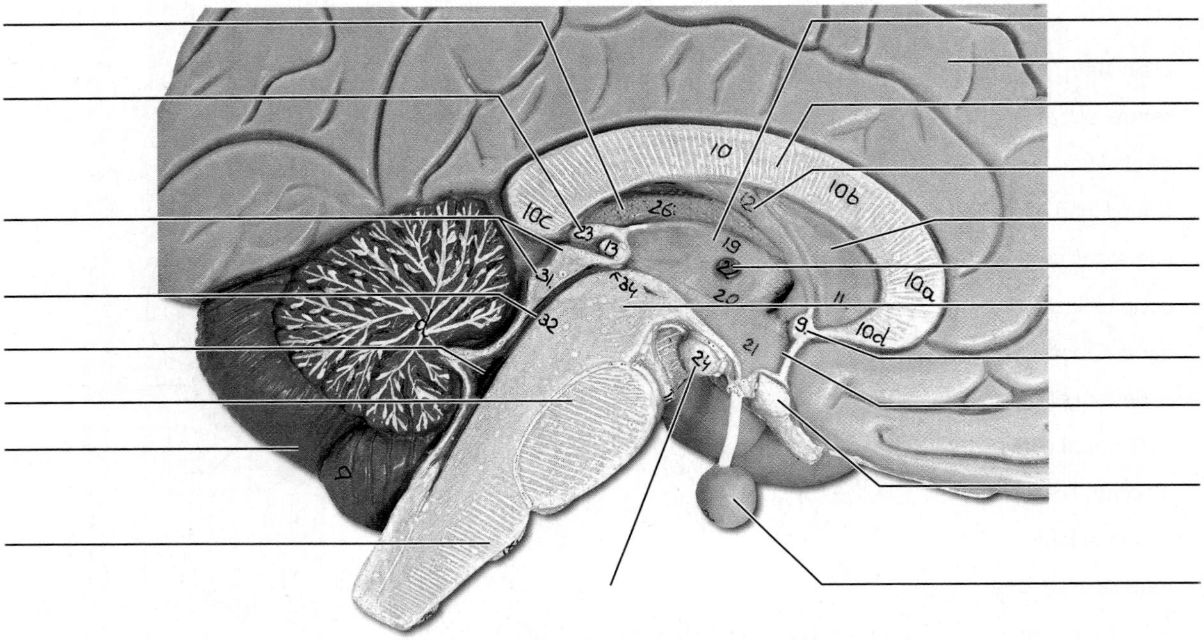

6. Using the key from question 5, match the appropriate structures with the descriptions given below.

______ 1. site of regulation of body temperature and water balance; most important autonomic center

______ 2. site where medial fibers of the optic nerves cross

______ 3. located in the midbrain; contains reflex centers for vision and hearing

______ 4. responsible for regulation of posture and coordination of complex muscular movements

______ 5. important synapse site for afferent fibers traveling to the sensory cortex

______ 6. contains autonomic centers regulating blood pressure, heart rate, and respiratory rhythm, as well as coughing, sneezing, and swallowing centers

______ 7. large fiber tract connecting the cerebral hemispheres

______ 8. relay stations for olfactory pathways

______ 9. canal that connects the third and fourth ventricles

______ 10. portion of the brain stem where the cerebral peduncles are located

7. Embryologically, the brain arises from the rostral end of a tubelike structure that quickly becomes divided into three major regions. Groups of structures that develop from the embryonic brain are listed below. Designate the embryonic origin of each group as the hindbrain, midbrain, or forebrain.

__________ 1. the diencephalon, including the thalamus, optic chiasma, and hypothalamus

__________ 2. the medulla oblongata, pons, and cerebellum

__________ 3. the cerebral hemispheres

8. What is the function of the basal nuclei? ______________________________

____

Meninges of the Brain

9. Identify the meningeal (or associated) structures described below:

______________________________ 1. outermost meninx covering the brain; composed of tough fibrous connective tissue

______________________________ 2. innermost meninx covering the brain; delicate and highly vascular

______________________________ 3. structures instrumental in returning cerebrospinal fluid to the venous blood in the dural venous sinuses

______________________________ 4. structure that produces the cerebrospinal fluid

______________________________ 5. middle meninx; like a cobweb in structure

______________________________ 6. its outer layer forms the periosteum of the skull

______________________________ 7. a dural fold that attaches the cerebrum to the crista galli of the skull

______________________________ 8. a dural fold separating the cerebrum from the cerebellum

Cerebrospinal Fluid

10. Label the structures involved with circulation of cerebrospinal fluid on the accompanying diagram.

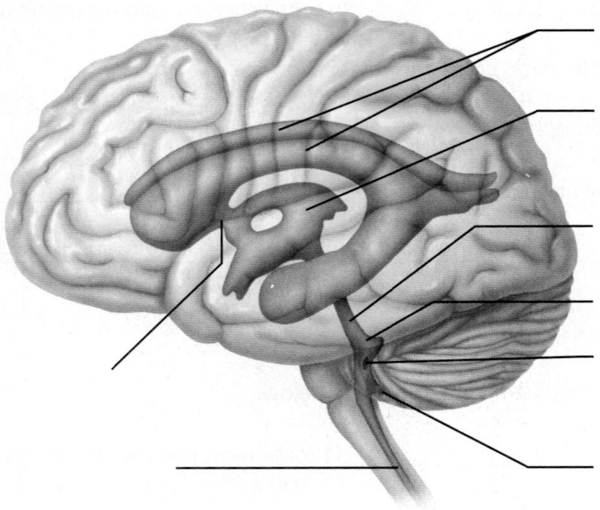

Add arrows to the figure above to indicate the flow of cerebrospinal fluid from its formation in the lateral ventricles to the site of its exit from the fourth ventricle. Then fill in the blanks in the following paragraph.

Cerebrospinal fluid flows from the fourth ventricle into the __1__ space surrounding the brain and spinal cord. From this space it drains through the __2__ into the __3__.

1. ______________________
2. ______________________
3. ______________________

11. Provide the name and number of the cranial nerves involved in each of the following activities, sensations, or disorders.

__________ 1. rotating the head

__________ 2. smelling a flower

__________ 3. raising the eyelids; pupillary constriction

__________ 4. slowing the heart; increasing motility of the digestive tract

__________ 5. involved in Bell's palsy (facial paralysis)

__________ 6. chewing food

__________ 7. listening to music; seasickness

__________ 8. secretion of saliva; tasting well-seasoned food

__________ 9. involved in "rolling" the eyes (three nerves—provide numbers only)

__________ 10. feeling a toothache

__________ 11. reading the newspaper

__________ 12. purely or mostly sensory in function (three nerves—provide numbers only)

Cranial Nerves

12. Using the terms below, correctly identify all structures indicated by leader lines on the diagram.

a. abducens nerve (VI)
b. accessory nerve (XI)
c. facial nerve (VII)
d. glossopharyngeal nerve (IX)
e. hypoglossal nerve (XII)
f. longitudinal fissure
g. mammillary body
h. medulla oblongata
i. oculomotor nerve (III)
j. olfactory bulb
k. olfactory tract
l. optic chiasma
m. optic nerve (II)
n. optic tract
o. pons
p. trigeminal nerve (V)
q. trochlear nerve (IV)
r. vagus nerve (X)
s. vestibulocochlear nerve (VIII)

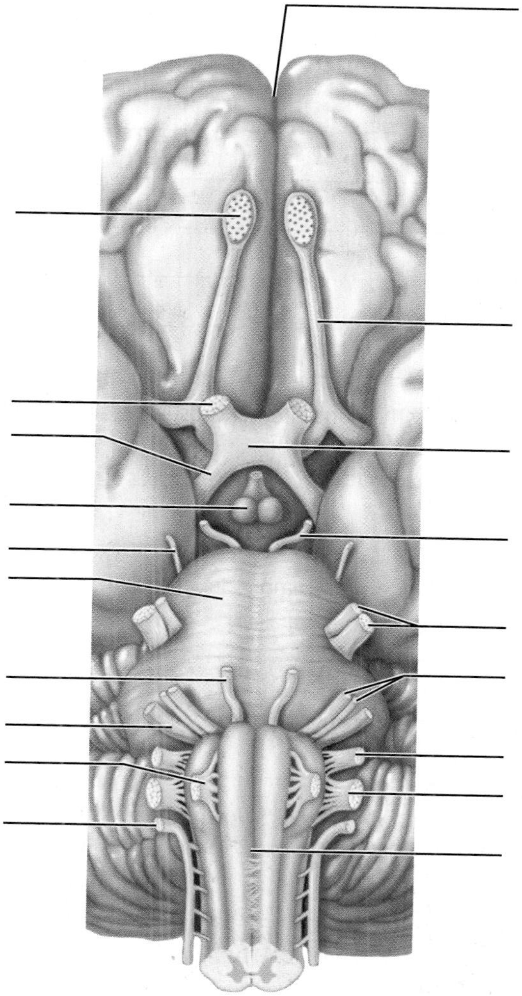

Dissection of the Sheep Brain

13. In your own words, describe the firmness and texture of the sheep brain tissue as observed when you cut into it.

____

Given that formalin hardens all tissue, what conclusions might you draw about the firmness and texture of living brain tissue?

____

14. When comparing human and sheep brains, you observed some profound differences between them. Record your observations in the chart below.

Structure	Human	Sheep
Olfactory bulb		
Pons/medulla relationship		
Location of cranial nerve III		
Mammillary body		
Corpus callosum		
Interthalamic adhesion		
Relative size of superior and inferior colliculi		
Pineal gland		

15. + A brain hemorrhage within the region of the right internal capsule results in paralysis of the left side of the body. Explain why the left side (rather than the right side) is affected. ______________________________

____

16. + Explain why trauma to the brain stem is often much more dangerous than trauma to the frontal lobes.

____

____

____

17. + Patients with unresponsive wakefulness syndrome (UWS) have lost awareness of self and their environment. In many cases, there is no damage to the cerebral cortex or the brain stem. If signal transmission to the cerebral cortex is affected, what part of the brain is most likely to have been damaged? ______________________________

18. + Patients with unresponsive wakefulness syndrome (UWS) often reflexively respond to visual and auditory stimuli. Explain how this phenomenon relates to the unaffected parts of their brain involved in sensory input. ______________________________

____

EXERCISE 18

Electroencephalography

Learning Outcomes

- Define *electroencephalogram (EEG)*, and discuss its clinical significance.
- Describe or recognize typical tracings of alpha, beta, theta, and delta brain waves, and indicate the conditions when each is most likely to occur.
- Indicate the source of brain waves.
- Define *alpha block*.
- Monitor the EEG in a human subject.
- Describe the effect of a sudden sound, mental concentration, and respiratory alkalosis on the EEG.

Go to Mastering A&P™ > Study Area to improve your performance in A&P Lab.

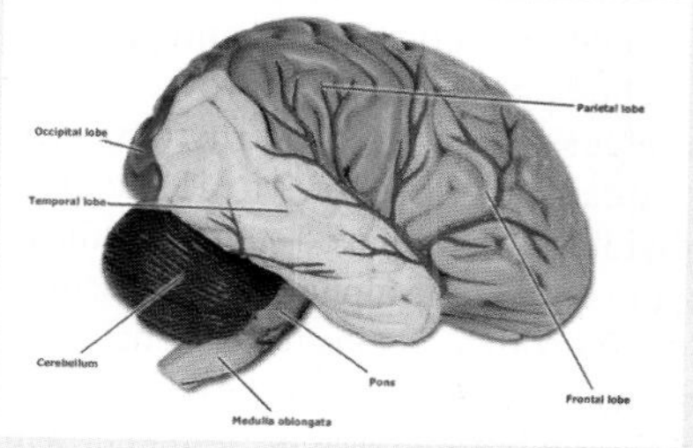

> Lab Tools > Practice Anatomy Lab > Anatomical Models

Instructors may assign new Building Vocabulary coaching activities, Pre-Lab Quiz questions, Art Labeling activities, Practice Anatomy Lab Practical questions (PAL), and more using the Mastering A&P™ Item Library.

Pre-Lab Quiz

Instructors may assign these and other Pre-Lab Quiz questions using Mastering A&P™

1. What does an electroencephalogram (EEG) measure?
   a. electrical activity of the brain
   b. electrical activity of the heart
   c. emotions
   d. physical activity of the subject
2. Circle the correct underlined term. Alpha waves / Beta waves are typical of the attentive or awake state.
3. Circle True or False. Brain waves can change with age, sensory stimuli, and the chemical state of the body.
4. Where will you place the indifferent (ground) electrode on your subject?
   a. the earlobe
   b. the forehead
   c. over the occipital lobe
   d. over the temporal bone
5. During today's activity, students will instruct subjects to *hyperventilate*. What should the subjects do?
   a. breathe in a normal manner
   b. breathe rapidly
   c. breathe very slowly
   d. hold their breath until they almost pass out

Materials

- Oscilloscope and EEG lead-selector box or physiograph and high-gain preamplifier
- Cot (if available) or pillow
- Electrode gel
- EEG electrodes and leads
- Collodion gel or long elastic EEG straps
- Abrasive pad

BIOPAC® BSL System with BSL software version 4.0.1 and above (for Windows 10/8.x/7 or Mac OS X 10.9–10.12), data acquisition unit MP36/35 or MP45, PC or Mac computer, Biopac Student Lab electrode lead set, disposable vinyl electrodes, swim cap or supportive wrap to press electrodes against head for improved contact, and a cot or lab bench and pillow.

Instructors using the MP36/35/30 data acquisition unit with BSL software versions earlier than 4.0.1 (for Windows or Mac) will need slightly different channel settings and collection strategies. Instructors for using the older data acquisition unit can be found on MasteringA&P.

As curious humans we are particularly interested in how the brain thinks, reasons, learns, remembers, and controls consciousness. As students we have learned that the brain accomplishes its tasks through electrical activities of neurons. The remarkable noninvasive technologies of twenty-first-century neuroscience have advanced our understanding of brain functions, as has the long-used technique of electroencephalography—the recording of electrical activity from the surface portions of the brain. It is incredible that the sophisticated equipment used to record an electroencephalogram (EEG) is commonly available in undergraduate laboratories. As you use it, you begin to explore the complex higher functions of the human brain.

Brain Wave Patterns and the Electroencephalogram

The **electroencephalogram (EEG)**, a record of the electrical activity of the brain, can be obtained through electrodes placed at various points on the skin or scalp of the head. This electrical activity, which is recorded as waves (**Figure 18.1**), represents the summed electrical activity of many neurons at the surface of the cerebral cortex.

Certain characteristics of brain waves are known. They have a frequency of 1 to 30 hertz (Hz) or cycles per second, a dominant rhythm of 10 Hz, and an average amplitude (voltage) of 20 to 100 microvolts (μV). They vary in frequency in different brain areas, occipital waves having a lower frequency than those associated with the frontal and parietal lobes.

The first of the brain waves to be described by scientists were the alpha waves (or alpha rhythm). **Alpha waves** have an average frequency range of 8 to 13 Hz and are produced when the individual is in a relaxed state with the eyes closed. **Alpha block**, suppression of the alpha rhythm, occurs if the eyes are opened or if the individual begins to concentrate on some mental problem or visual stimulus. Under these conditions, the waves decrease in amplitude but increase in frequency. Under conditions of fright or excitement, the frequency increases even more.

Beta waves, closely related to alpha waves, are faster (14 to 30 Hz) and have a lower amplitude. They are typical of the attentive or alert state.

Very large (high-amplitude) waves with a frequency of 4 Hz or less that are seen in deep sleep are **delta waves. Theta waves** are large, irregular waves with a frequency of 4 to 7 Hz. Although theta waves are normal in children, they are uncommon in awake adults.

Brain waves change with age, sensory stimuli, brain pathology, and the chemical state of the body. Glucose deprivation, oxygen poisoning, and sedatives all interfere with the rhythmic activity of brain output by disturbing the metabolism of neurons. Sleeping individuals and patients in a coma have EEGs that are slower (lower frequency) than the alpha rhythm of normal adults. Fright, epileptic seizures, and various types of drug intoxication can be associated with comparatively faster cortical activity. As these examples show, impairment of cortical function is indicated by neuronal activity that is either too fast or too slow; unconsciousness occurs at both extremes of the frequency range.

Because spontaneous brain waves are always present, even during unconsciousness and coma, the absence of brain waves (a "flat" EEG) is taken as clinical evidence of brain death. The EEG is used clinically to diagnose and localize many types of brain lesions, including epileptic foci (cortical area responsible for causing the seizure), infections, abscesses, and tumors. ✚

Activity 1

Observing Brain Wave Patterns Using an Oscilloscope or Physiograph

If one electrode (the *active electrode*) is placed over a particular cortical area and another (the *indifferent, or ground, electrode*) is placed over an inactive part of the head, such as the earlobe, all of the activity of the cortex underlying the active electrode will, theoretically, be recorded. The inactive area provides a zero reference point, or a baseline, and the EEG represents the difference between "activities" occurring under the two electrodes.

1. Connect the EEG lead-selector box to the oscilloscope preamplifier, or connect the high-gain preamplifier to the physiograph channel amplifier. Adjust the horizontal sweep and sensitivity according to the directions given in the instrument manual or by your instructor.

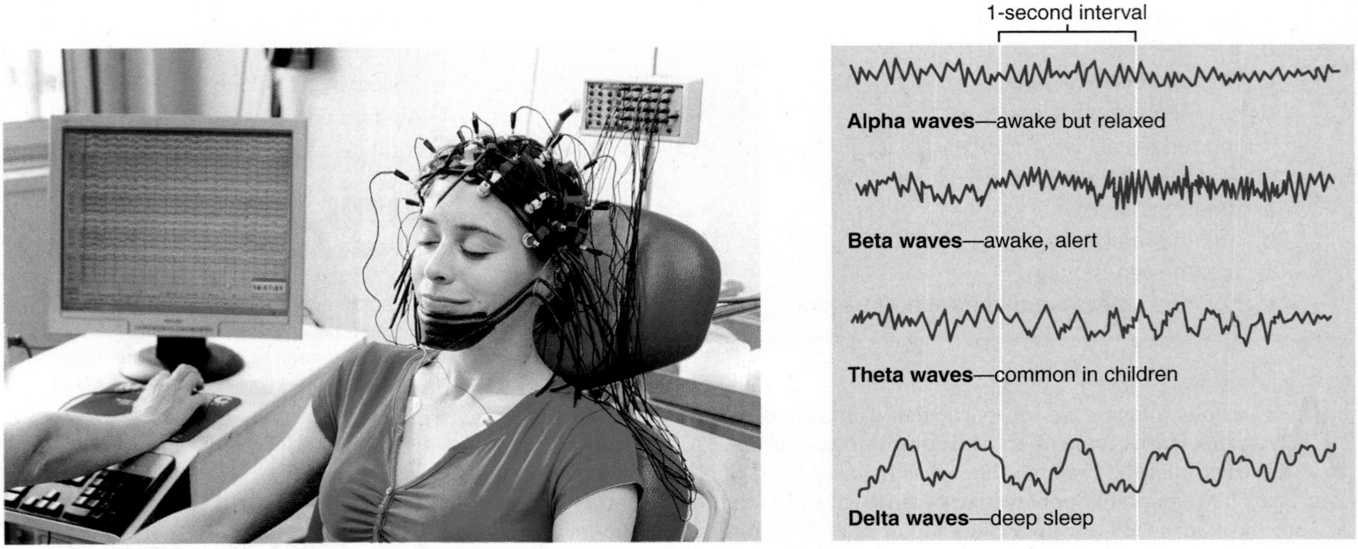

(a) (b)

Figure 18.1 Electroencephalography and brain waves. (a) Scalp electrodes are positioned on the patient to record brain waves. **(b)** Typical EEGs.

2. Prepare the subject. The subject should lie undisturbed on a cot or on the lab bench with eyes closed in a quiet, dimly lit area. (Someone who is able to relax easily makes a good subject.) Apply a small amount of electrode gel to the subject's forehead above the left eye and on the left earlobe. Press an electrode to each prepared area and secure each by (1) applying a film of collodion gel to the electrode surface and the adjacent skin or (2) using a long elastic EEG strap (knot tied at the back of the head). If collodion gel is used, allow it to dry before you continue.

3. Connect the active frontal lead (forehead) to the EEG lead-selector box outlet marked "L Frontal." Connect the lead from the indifferent electrode (earlobe) to the ground outlet (or to the appropriate input terminal on the high-gain preamplifier).

4. Turn the oscilloscope or physiograph on, and observe the EEG pattern of the relaxed subject for a period of 5 minutes. If the subject is truly relaxed, you should see a typical alpha-wave pattern. (If the subject is unable to relax and the alpha-wave pattern does not appear in this time interval, test another subject.) Since the electrical activity of muscles interferes with EEG recordings, discourage all muscle movement during the monitoring period. If 60-cycle "noise" (appearing as fast, regular, low-amplitude waves superimposed on the more irregular brain waves) is present in your record because of the presence of other electronic equipment, consult your instructor to eliminate it.

5. Abruptly and loudly clap your hands. The subject's eyes should open, and alpha block should occur. Observe the immediate brain wave pattern. How do the frequency and amplitude (rhythm) of the brain waves change?

____

____

Would you characterize this as beta rhythm? ____________

Why? ______________________________

____

6. Allow the subject about 5 minutes to achieve relaxation once again, then ask him or her to compute a number problem that requires concentration (for example, add 3 and 36, subtract 7, multiply by 2, add 50, etc.). Observe the brain wave pattern during the period of mental computation.

Observations: ______________________________

____

____

7. Once again allow the subject to relax until alpha rhythm resumes. Then, instruct him or her to hyperventilate (breathe rapidly) for 3 minutes. *Be sure to tell the subject when to stop hyperventilating*. Hyperventilation rapidly flushes carbon dioxide out of the lungs, decreasing carbon dioxide levels in the blood and producing respiratory alkalosis.

Observe the changes in the rhythm and amplitude of the brain waves occurring during the period of hyperventilation.

Observations: ______________________________

____

____

8. Think of other stimuli that might affect brain wave patterns. Test your hypotheses. Describe what stimuli you tested and what responses you observed.

____

____

____

Activity 2

Electroencephalography Using BIOPAC®

In this activity, the EEG of the subject will be recorded during a relaxed state, first with the eyes closed, then with the eyes open while silently counting to ten, and finally with the eyes closed again.

18

Setting Up the Equipment

1. Connect the BIOPAC® unit to the computer and turn the computer **ON**.

2. Make sure the BIOPAC® unit is **OFF**.

3. Plug in the equipment as shown in **Figure 18.2** on p. 296.
   - Electrode lead set—CH 1

4. Turn the BIOPAC® unit **ON**.

5. Attach three electrodes to the subject's scalp and earlobe as shown in **Figure 18.3** on p. 296. Follow these important guidelines to assist in effective electrode placement:
   - Select subjects with the easiest access to the scalp.
   - Move hair apart at electrode site, and gently abrade skin.
   - Apply some gel to the electrode. A fair amount of gel must be used to obtain a good electrode-to-scalp connection.
   - Apply pressure to the electrodes for 1 minute to ensure attachment.
   - Use a swimcap or supportive wrap to maintain attachment. Subject should not press electrodes against scalp.
   - Do not touch the electrodes while recording.
   - The earlobe electrode may be folded under the lobe itself.

Text continues on next page. →

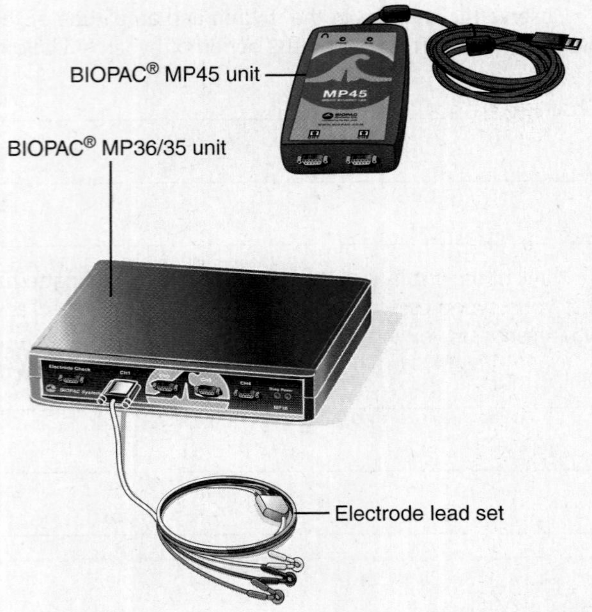

Figure 18.2 Setting up the BIOPAC® equipment. Plug the electrode set into Channel 1. Electrode leads are shown connected to the MP36/35 unit.

6. When the electrodes are attached, the subject should lie down and relax with eyes closed for 5 minutes before recording.

7. Start the Biopac Student Lab program on the computer by double-clicking the icon on the desktop or by following your instructor's guidance.

8. Select lesson **L03-EEG-1** from the menu, and click **OK**.

9. Type in a filename that will save this subject's data on the computer hard drive. You may want to use the subject's last name followed by EEG-1 (for example, SmithEEG-1). Then click **OK**.

10. During this preparation, the subject should be very still and in a relaxed state with eyes closed. Allow the subject to relax with minimal stimuli.

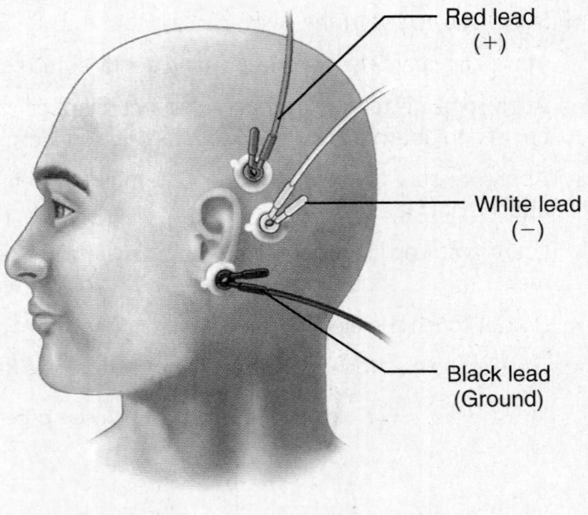

Figure 18.3 Placement of electrodes and the appropriate attachment of electrode leads by color.

Calibrating the Equipment

1. Make sure that the electrodes remain firmly attached to the surface of the scalp and earlobe. The subject should remain absolutely still and try to avoid moving the body or face.

2. With the subject in a relaxed position, click **Calibrate**.

3. You will be prompted to check electrode attachment one final time. When ready, click **OK**; the computer will record for 8 seconds and stop automatically.

4. Observe the recording of the calibration data (it should look like **Figure 18.4** with baseline at zero).

- If the data look very different, click **Redo Calibration** and repeat the steps above.
- Once the data look similar to Figure 18.4, click **Continue** and proceed to the next section. Note that despite your best efforts, electrode adhesion may not be strong enough to record data; try another subject or different electrode placement.

Recording the Data

1. The subject should remain relaxed with eyes closed.

2. After clicking **Record**, the "director" will instruct the subject to keep eyes closed for the first 20 seconds of recording, then open the eyes and *mentally* (not verbally) count to 20, then close the eyes again and relax for 20 seconds. The director will insert a marker by pressing the **F4** key (PC or Mac) when the command to open eyes is given, and another marker by pressing the **F5** key (PC or Mac) when the subject reaches the count of twenty and closes the eyes. Click **Suspend** 20 seconds after the subject recloses the eyes.

3. Observe the recording of the data (it should look similar to the data in **Figure 18.5**).

- If the subject moved too much during the recording, it is likely that artifact spikes will appear in the data. Remind the subject to be very still.
- Look carefully at the **alpha rhythm** band of data. The intensity of the alpha signal should decrease during the "eyes open" phase of the recording. If the data do not demonstrate this change, make sure that the electrodes are firmly attached.
- If the data show artifact spikes or the alpha signal fails to decrease when the eyes are open, click **Redo**.
- If the data look similar to the example (Figure 18.5), click **Continue** and proceed to the next step.

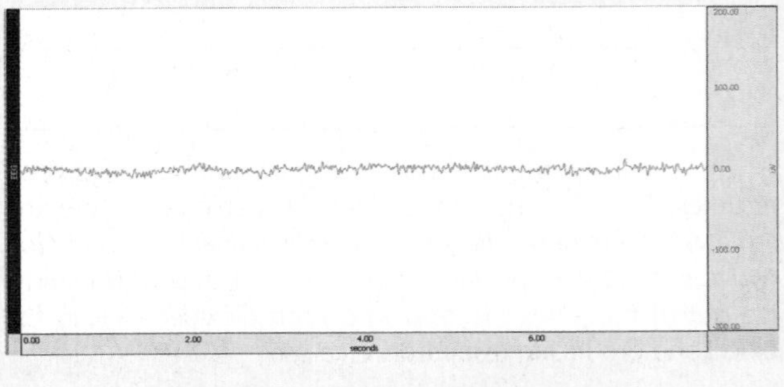

Figure 18.4 Example of calibration data.

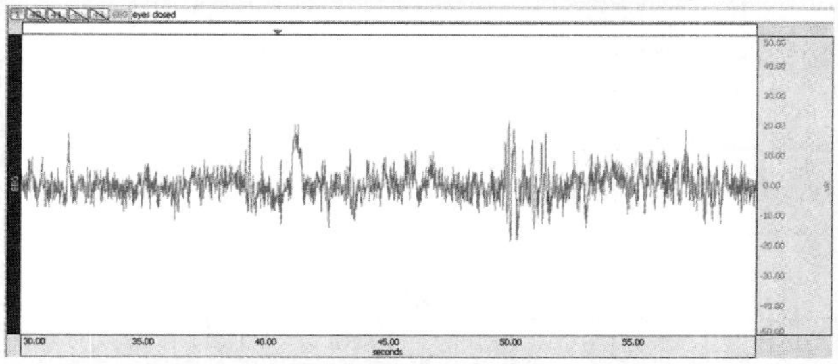

Figure 18.5 Example of EEG data.

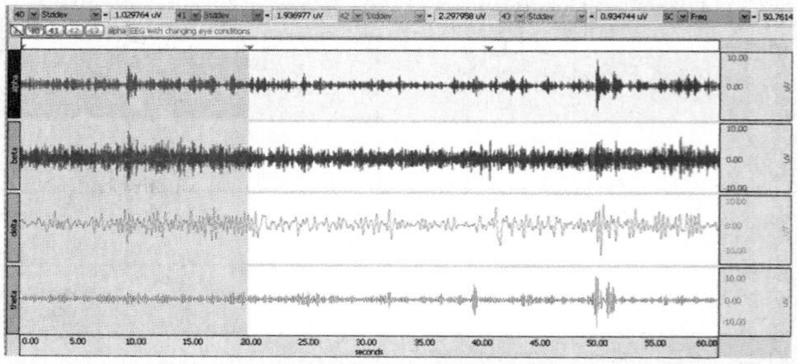

Figure 18.6 Highlighting the first data segment.

4. When finished, click **Done**. If you are certain you want to stop recording, click **YES**. Remove the electrodes from the subject's scalp.

5. A pop-up window will appear. To record from another subject, select **Record from another subject** and return to step 5 under Setting Up the Equipment. If continuing to the Data Analysis section, select **Analyze current data file** and proceed to step 2.

Data Analysis

1. If just starting the BIOPAC® program to perform data analysis, enter **Review Saved Data** mode, and choose the file with the subject's EEG data (for example, SmithEEG-1).

2. Observe the way the channel numbers are designated: CH 40–**alpha**; CH 41–**beta**; CH 42–**delta**; and CH 43–**theta**. CH 1 (raw EEG) is hidden. The software used it to extract and display each frequency band. If you want to see CH 1, hold down the **Alt** key (PC) or **Option** key (Mac) while using the cursor to click channel box 1 (the small box with a 1 at the upper left of the screen).

3. To analyze the data, set up the first four pairs of channel/measurement boxes at the top of the screen by selecting the following channels and measurement types from the drop-down menus:

Channel	Measurement	Data
CH 40	Stddev	alpha
CH 41	Stddev	beta
CH 42	Stddev	delta
CH 43	Stddev	theta

Stddev (Standard deviation): This is a statistical calculation that estimates the variability of the data in the area highlighted by the I-beam cursor. This function minimizes the effects of extreme values and electrical artifacts that may unduly influence interpretation of the data.

4. Use the arrow cursor and click the I-beam cursor box at the lower right of the screen to activate the "area selection" function. Using the activated I-beam cursor, highlight the first 20-second segment of EEG data, which represents the subject at rest with eyes closed (**Figure 18.6**).

5. Observe that the computer automatically calculates the stddev for each of the channels of data (alpha, beta, delta, and theta).

6. Record the data for each rhythm in the **Standard Deviations chart**, rounding to the nearest 0.01 μV.

7. Repeat steps 4–6 to analyze and record the data for the next two segments of data, with eyes open, and with eyes reclosed. The triangular markers inserted at the top of the data should provide guidance for highlighting.

8. To continue the analysis, change the settings in the first four pairs of channel/measurement boxes. Select the following channels and measurement types:

Channel	Measurement	Data
CH 40	Freq	alpha
CH 41	Freq	beta
CH 42	Freq	delta
CH 43	Freq	theta

Freq (frequency): This gives the frequency in hertz (Hz) of an individual wave that is highlighted by the I-beam cursor.

9. To view an individual wave from among the high-frequency waveforms, you must use the zoom function. To activate the

Text continues on next page. →

Standard Deviations (Stddev) of Signals in Each Segment

Rhythm	Channel	Eyes closed Segment 1 Seconds 0–20	Eyes open Segment 2 Seconds 21–40	Eyes reclosed Segment 3 Seconds 41–60
Alpha	CH 40			
Beta	CH 41			
Delta	CH 42			
Theta	CH 43			

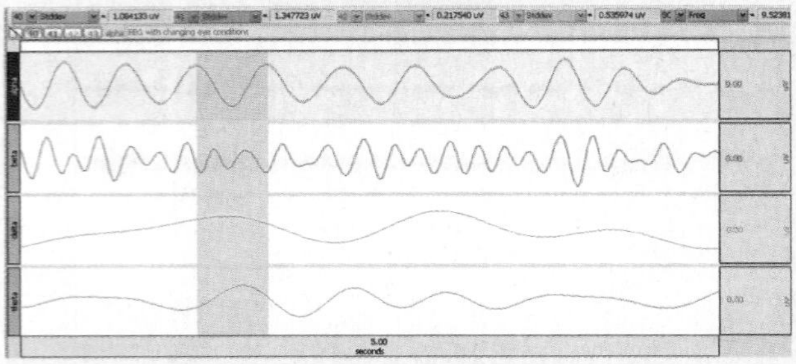

Figure 18.7 Highlighting a single alpha wave.

zoom function, use the cursor to click the magnifying glass at the lower-right corner of the screen (near the I-beam cursor box). The cursor will become a magnifying glass.

10. As the analysis begins, CH 40—the alpha data—will be automatically activated. To examine individual waves within the **alpha** data, click that band with the magnifying glass until it is possible to observe the peaks and troughs of individual waves within **Segment 1**.

- To properly view each of the waveforms, you may have to click the **Display** menu and select **Autoscale Waveforms**. This function rescales the data for the rhythm band that is selected.

11. At this time, focus on alpha waves only. Reactivate the I-beam cursor by clicking its box in the lower-right corner. Highlight a *single* alpha wave from peak to peak (as shown in **Figure 18.7**).

12. Read the calculated frequency (in Hz) in the measurement box for CH 40, and record this as the frequency of Wave 1 for alpha rhythm in the **Frequencies of Waves chart**.

13. Use the I-beam cursor to select two more individual **alpha** waves, and record their frequencies in the chart.

14. You will now perform the same frequency measurements for three waves in each of the **beta** (CH 41), **delta** (CH 42), and **theta** (CH 43) data sets. Record these measurements in the chart.

15. Calculate the average of the three waves measured for each of the brain rhythms, and record the average in the chart.

16. When finished, answer the following questions, and then exit the program by going to the **File** menu at the top of the page and clicking **Quit**.

Look at the waveforms you recorded, and carefully examine all three segments of the alpha rhythm record. Is there a difference in electrical activity in this frequency range when the eyes are open versus closed? Describe your observations.

Carefully examine all three segments of the beta rhythm record. Is there a difference in electrical activity in this frequency range when the eyes are open versus closed? Describe what you observe.

This time, compare the intensity (height) of the alpha and beta waveforms throughout all three segments. Does the intensity of one signal appear more varied than the other in the record? Describe your observations.

Examine the data for the delta and theta rhythms. Is there any change in the waveform as the subject changes states? If so, describe the change observed.

The degree of variation in the intensity of the signal was estimated by calculating the standard deviation of the waves in each segment of data. In which time segment (eyes open, eyes closed, or eyes reclosed) is the difference in the standard deviations the greatest?

Frequencies of Waves for Each Rhythm (Hz)

Rhythm	Channel	Wave 1	Wave 2	Wave 3	Average
Alpha	CH 40				
Beta	CH 41				
Delta	CH 42				
Theta	CH 43				

Instructors may assign a portion of the Review Sheet questions using Mastering A&P™

18 REVIEW SHEET

EXERCISE

Electroencephalography

Name ______________________ Lab Time/Date ______________________

Brain Wave Patterns and the Electroencephalogram

1. Define *EEG*. ______________________

2. Identify the type of brain wave pattern described in each statement below.

   ______________ below 4 Hz; slow, large waves; normally seen during deep sleep

   ______________ rhythm generally apparent when an individual is in a relaxed, nonattentive state with the eyes closed

   ______________ correlated to the alert state; usually about 14 to 30 Hz

3. What is meant by the term *alpha block*?

   ____

4. List at least four types of brain lesions that may be determined by EEG studies. ______________________

   ____

5. What is the common result of hypoactivity or hyperactivity of the brain neurons? ______________________

Observing Brain Wave Patterns

6. How was alpha block demonstrated in the laboratory experiment? ______________________

   ____

7. What was the effect of mental concentration on the brain wave pattern? ______________________

   ____

   ____

8. What effect on the brain wave pattern did hyperventilation have? ______________________

   ____

   ____

Electroencephalography Using BIOPAC®

9. Observe the average frequency of the waves you measured for each rhythm. Did the calculated average for each fall within the specified range indicated in the introduction to encephalograms?

____

10. ➕ Name three different conditions that electroencephalography could be used to diagnose. ______________________

____

11. ➕ Which lobe in the brain would you expect to see an abnormal EEG for children with congenital visual deficiencies?

____

19 EXERCISE

The Spinal Cord and Spinal Nerves

Learning Outcomes

- List two major functions of the spinal cord.
- Define *conus medullaris*, *cauda equina*, and *filum terminale*.
- Name the meninges of the spinal cord, and state their function.
- Indicate two major areas where the spinal cord is enlarged, and explain the reasons for the enlargement.
- Identify important anatomical areas on a model or image of a cross section of the spinal cord, and where applicable, name the neuron type found in these areas.
- Locate on a diagram the fiber tracts in the spinal cord, and state their functions.
- Note the number of pairs of spinal nerves that arise from the spinal cord, describe their division into groups, and identify the number of pairs in each group.
- Describe the origin and fiber composition of the spinal nerves, differentiating between roots, the spinal nerve proper, and rami, and discuss the result of transecting these structures.
- Discuss the distribution of the dorsal and ventral rami of the spinal nerves.
- Identify the four major nerve plexuses on a model or image, name the major nerves of each plexus, and describe the destination and function of each.

Go to Mastering A&P™ > Study Area to improve your performance in A&P Lab.

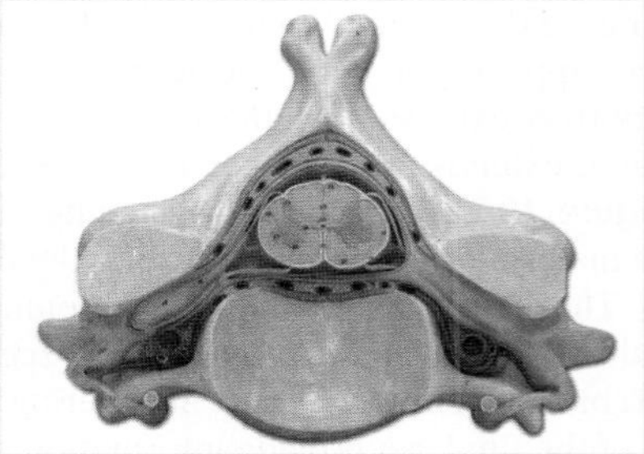

> Lab Tools > Practice Anatomy Lab > Anatomical Models

Instructors may assign new Building Vocabulary coaching activities, Pre-Lab Quiz questions, Art Labeling activities, Practice Anatomy Lab Practical questions (PAL), and more using the Mastering A&P™ Item Library.

Pre-Lab Quiz

Instructors may assign these and other Pre-Lab Quiz questions using Mastering A&P™

1. The spinal cord extends from the foramen magnum of the skull to the first or second lumbar vertebra, where it terminates in the:
   a. conus medullaris **c.** filum terminale
   b. denticulate ligament **d.** gray matter
2. How many pairs of spinal nerves do humans have?
   a. 10 **c.** 31
   b. 12 **d.** 47
3. Circle the correct underlined term. Fiber tracts conducting impulses to the brain are called ascending or <u>sensory</u> / <u>motor</u> tracts.
4. The ventral rami of all spinal nerves except T_2 through T_{12} form complex networks of nerves known as:
   a. fissures **c.** plexuses
   b. ganglia **d.** sulci
5. Circle the correct underlined term. The sciatic nerve divides into the tibial and <u>posterior femoral cutaneous</u> / <u>common fibular</u> nerves.

Materials

- Spinal cord model (cross section)
- Three-dimensional models or laboratory charts of the spinal cord and spinal nerves
- Red and blue pencils
- Preserved cow spinal cord sections with meninges and nerve roots intact (or spinal cord segment saved from the brain dissection in Exercise 17)
- Dissecting instruments and tray
- Disposable gloves
- Dissecting microscope
- Prepared slide of spinal cord (x.s.)
- Compound microscope

For instructions on animal dissections, see the dissection exercises (starting on p. 695) in the cat and fetal pig editions of this manual.

The cylindrical **spinal cord**, a continuation of the brain stem, is a communication center. It plays a major role in spinal reflex activity and provides neural pathways to and from the brain.

Anatomy of the Spinal Cord

Enclosed within the vertebral canal of the spinal column, the spinal cord extends from the foramen magnum of the skull to the first or second lumbar vertebra, where it terminates in the cone-shaped **conus medullaris**. Like the brain, the cord is cushioned and protected by meninges (**Figure 19.1**). The dura mater and arachnoid mater extend beyond the conus medullaris, approximately to the level of S_2. The **filum terminale**, a fibrous extension of the conus medullaris covered by pia mater, extends even farther to attach to the posterior coccyx (**Figure 19.2**). **Denticulate ligaments**, saw-toothed shelves of pia mater, secure the spinal cord to the dura mater.

The cerebrospinal fluid–filled meninges extend beyond the end of the spinal cord, providing an excellent site for removing cerebrospinal fluid without endangering the spinal cord. Analysis of the fluid can provide information about suspected bacterial or viral infections of the meninges. This procedure, called a *lumbar puncture*, is usually performed below L_3.

In humans, 31 pairs of spinal nerves arise from the spinal cord and pass through intervertebral foramina to serve the body area near their level of emergence. The cord is about the circumference of a thumb for most of its length, but there are enlargements in the *cervical* and *lumbar* areas where the nerves serving the upper and lower limbs arise.

Because the spinal cord does not extend to the end of the vertebral column, the lumbar and sacral spinal nerve roots must travel through the vertebral canal before reaching their intervertebral foramina. This collection of spinal nerve roots is called the **cauda equina** (Figure 19.2a and d) because of its similarity to a horse's tail.

Figure 19.3 (p. 304) illustrates the spinal cord in cross section. Notice that the gray matter looks like a butterfly.

Activity 1

Identifying Structures of the Spinal Cord

Obtain a three-dimensional model or laboratory chart of a cross section of a spinal cord, and identify its structures as they are described in **Table 19.1** on p. 304 and shown in Figure 19.3.

19

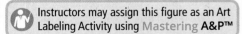

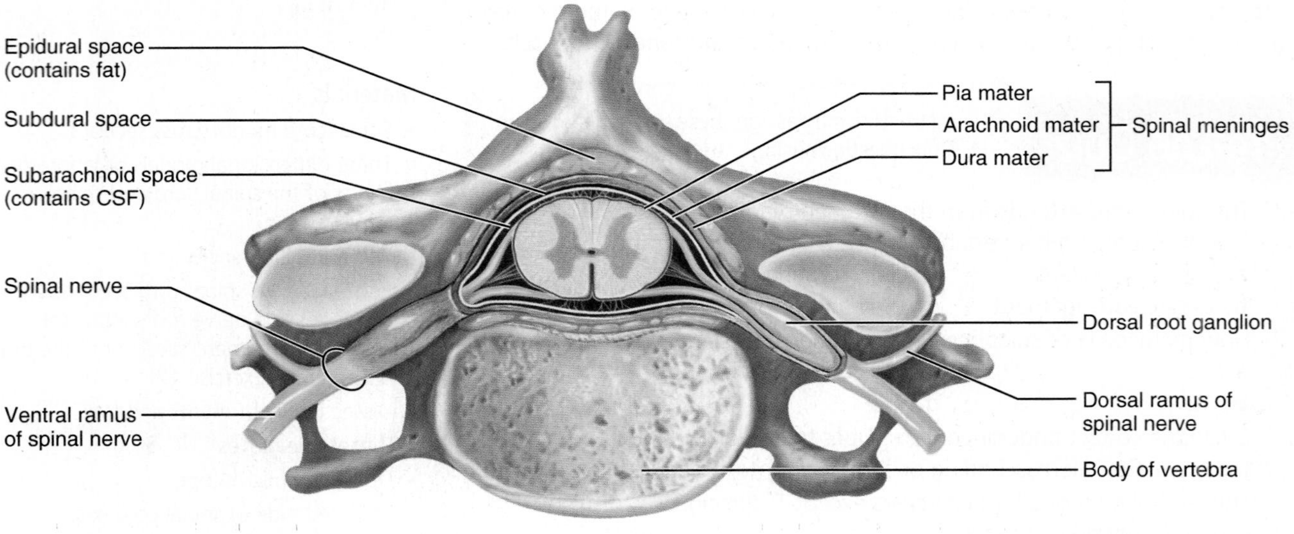

Figure 19.1 Cross section through the spinal cord illustrating its relationship to the surrounding vertebra, cervical region.

Instructors may assign this figure as an Art Labeling Activity using Mastering **A&P™**

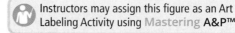

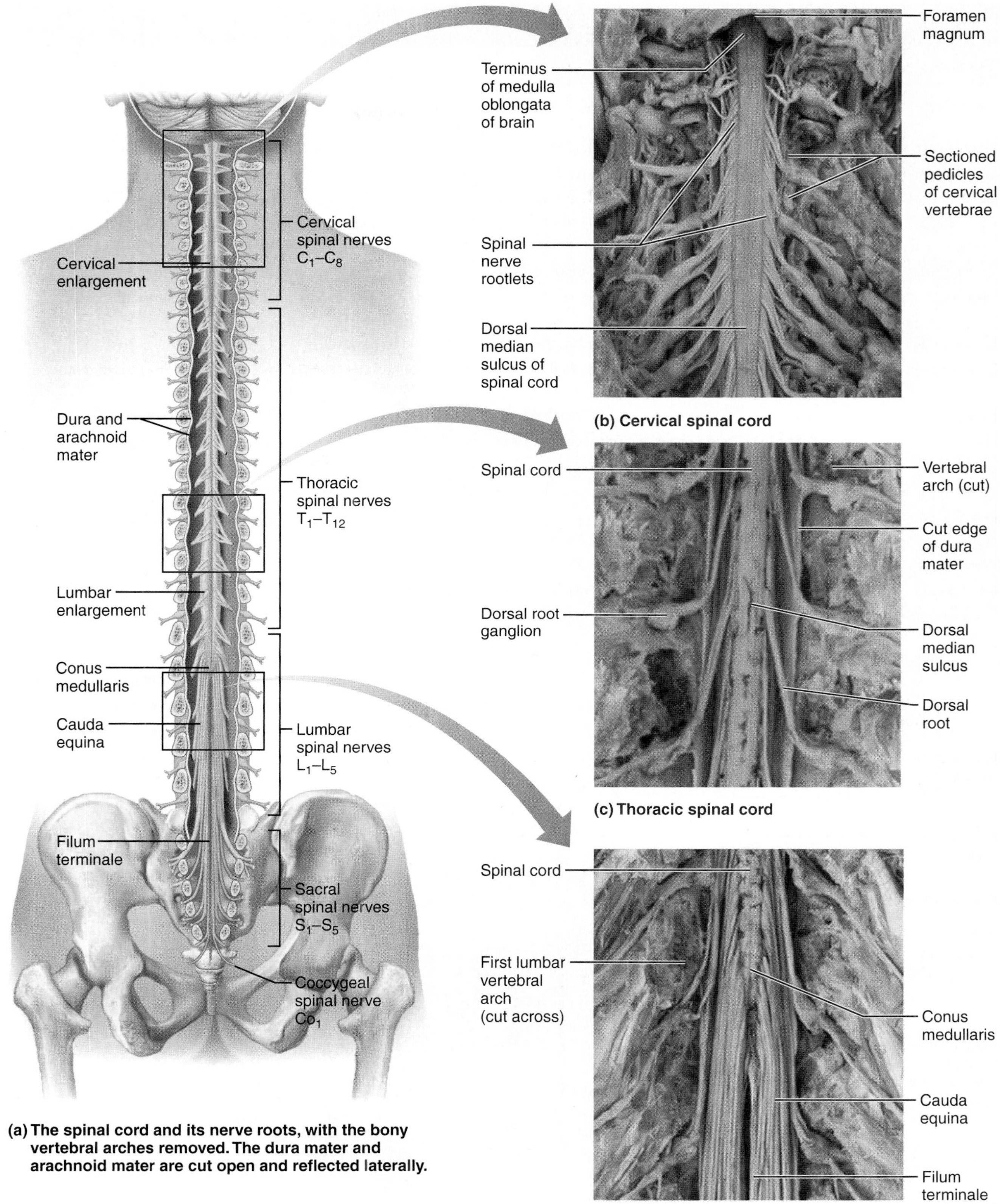

(a) The spinal cord and its nerve roots, with the bony vertebral arches removed. The dura mater and arachnoid mater are cut open and reflected laterally.

(b) Cervical spinal cord

(c) Thoracic spinal cord

(d) Inferior end of spinal cord, showing conus medullaris, cauda equina, and filum terminale

Figure 19.2 **Gross structure of the spinal cord, dorsal view.**

Instructors may assign this figure as an Art Labeling Activity using Mastering **A&P™**

19

Table 19.1 Anatomy of the Spinal Cord in Cross Section (Figure 19.3)

Structure	Description
Gray matter	Located in the center of the spinal cord and shaped like a butterfly or the letter H. Areas of the gray matter are individually named and described below.
Dorsal (posterior) horns	Posterior projections of the gray matter that contain primarily interneurons and the axons of sensory neurons.
Ventral (anterior) horns	Anterior projections of the gray matter that contain the cell bodies of somatic motor neurons and some interneurons.
Lateral horns	Small lateral projections that are present only in the thoracic and lumbar regions of the gray matter. When present, they contain the cell bodies of motor neurons of the autonomic nervous system.
Gray commissure	The cross bar of the H that surrounds the central canal.
Central canal	A narrow central cavity that is continuous with the ventricles of the brain.
Dorsal root	A nerve root that fans out into dorsal rootlets to connect to the posterior spinal cord. It contains the axons of sensory neurons.
Dorsal root ganglion	A bulge on the dorsal root that contains the cell bodies of sensory neurons.
Ventral root	A nerve root that is formed by the ventral rootlets connected to the anterior spinal cord. It contains the axons of motor neurons.
Spinal nerve	Formed by the fusion of dorsal and ventral roots. They are mixed nerves because they contain both sensory and motor fibers.
White matter	Forms the outer region of the spinal cord and is composed of myelinated and nonmyelinated axons organized into tracts.
Ventral median fissure	The anterior, more open of the two grooves that partially divide the spinal cord into left and right halves.
Dorsal median sulcus	The posterior, shallower of the two grooves that partially divide the spinal cord into left and right halves.
White columns	Each side of the spinal cord has three funiculi (columns): dorsal (posterior) funiculus, lateral funiculus, and ventral (anterior) funiculus, which are further divided into tracts.
Ascending (sensory) tracts	Carry sensory information from the sensory neurons to the brain.
Descending (motor) tracts	Carry motor instructions from the brain to the body's muscles and glands.

19

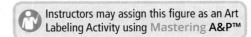

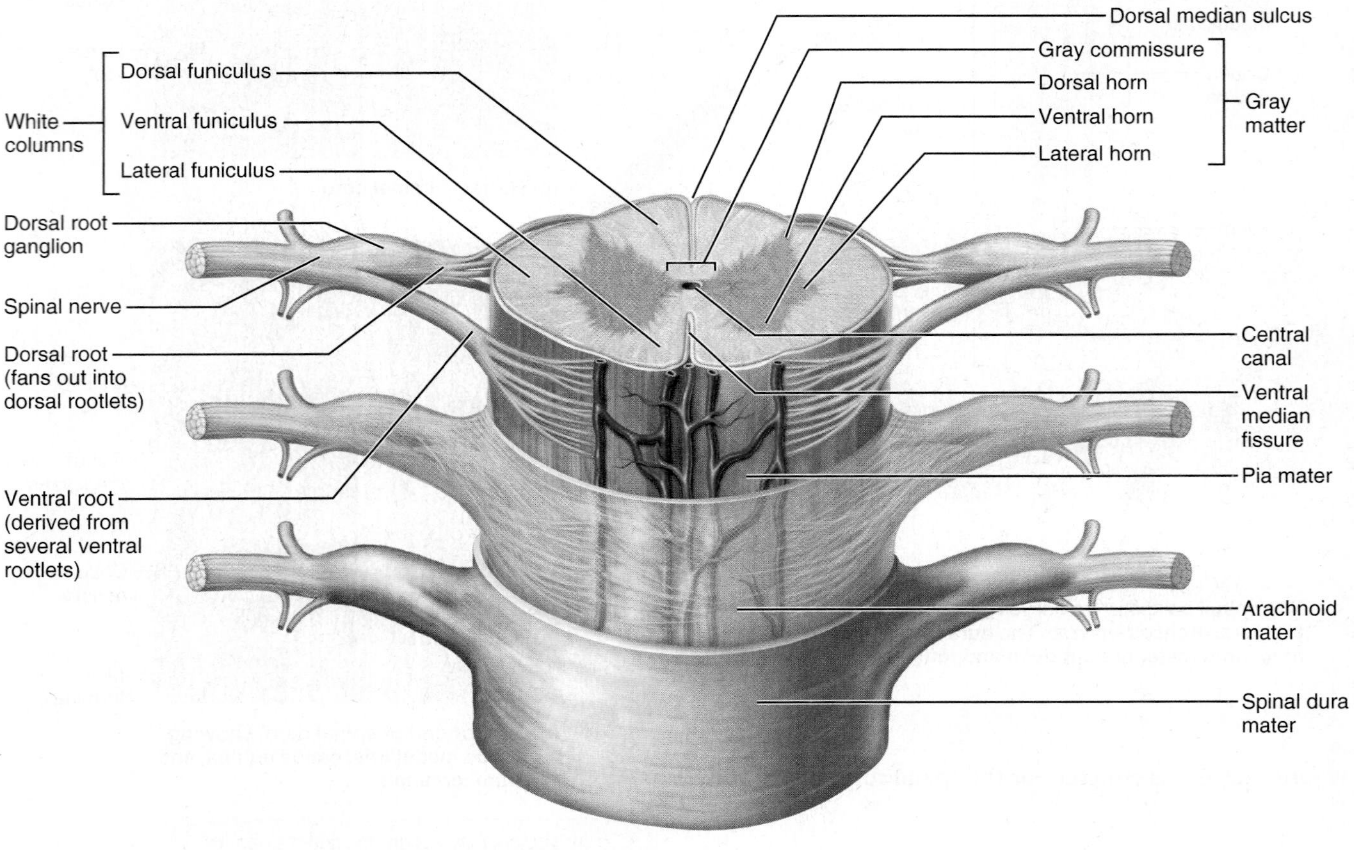

Figure 19.3 Anterior view of the spinal cord in cross section and its meninges, thoracic region.

Instructors may assign this figure as an Art Labeling Activity using Mastering A&P™

Even though the spinal cord is protected by meninges and cerebrospinal fluid in the vertebral canal, it is highly vulnerable to traumatic injuries, such as might occur in an automobile accident.

When the cord is damaged, both motor and sensory functions are lost in body areas normally served by that region and lower regions of the spinal cord. Injury to certain spinal cord areas may even result in a permanent flaccid paralysis of both legs, called **paraplegia**, or of all four limbs, called **quadriplegia**. +

Activity 2

Identifying Spinal Cord Tracts

With the help of your textbook, label the spinal cord diagram (**Figure 19.4**) with the tract names that follow. Each tract is represented on both sides of the cord, but for clarity, label the motor tracts on the right side of the diagram and the sensory tracts on the left side of the diagram. *Color ascending (sensory) tracts blue, and descending (motor) tracts red.* Then fill in the functional importance of each tract beside its name below. As you work, try to be aware of how the naming of the tracts is related to their anatomical distribution.

Dorsal columns

Fasciculus gracilis ____________________

Fasciculus cuneatus ____________________

Dorsal spinocerebellar ____________________

Ventral spinocerebellar ____________________

Lateral spinothalamic ____________________

Ventral spinothalamic ____________________

Lateral corticospinal ____________________

Ventral corticospinal ____________________

Rubrospinal ____________________

Tectospinal ____________________

Vestibulospinal ____________________

Medial reticulospinal ____________________

Lateral reticulospinal ____________________

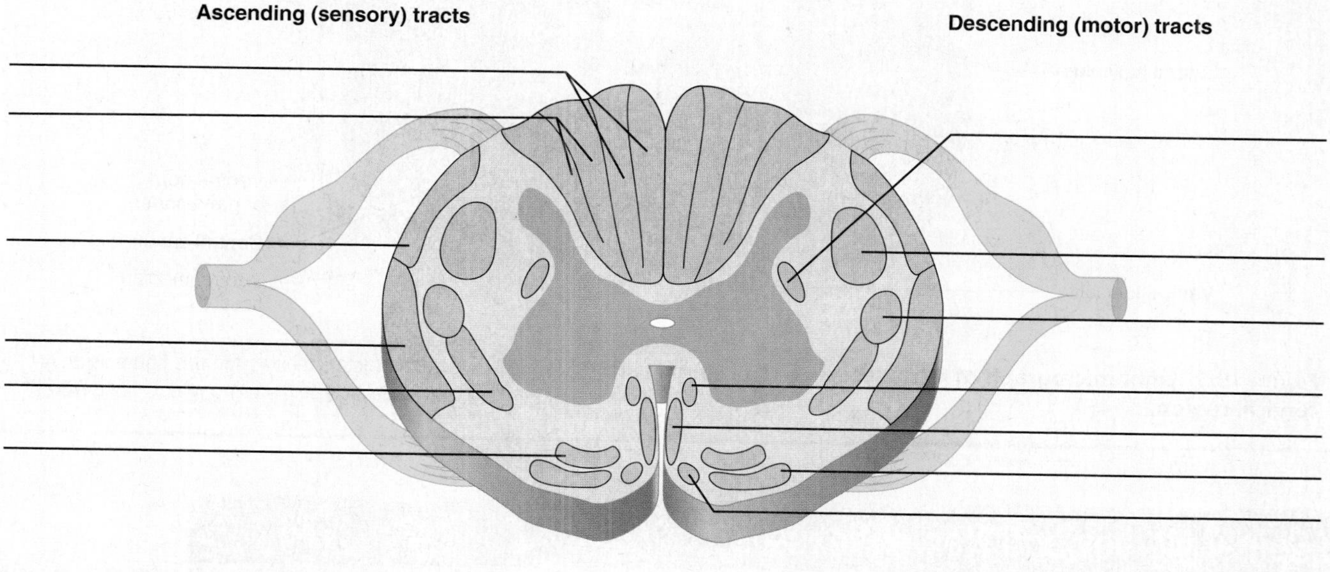

Figure 19.4 Cross section of the spinal cord showing the relative positioning of its major tracts.

DISSECTION

Spinal Cord

1. Obtain a dissecting tray and instruments, disposable gloves, and a segment of preserved spinal cord (from a cow or saved from the brain specimen used in Exercise 17). Identify the dura mater and the weblike arachnoid mater.

What name is given to the third meninx, and where is it found?

____

____

Peel back the dura mater, and observe the fibers making up the dorsal and ventral roots. If possible, identify a dorsal root ganglion.

2. Cut a thin cross section of the cord, and identify the ventral and dorsal horns of the gray matter with the naked eye or with the aid of a dissecting microscope. Notice that the dorsal horns are more tapered than the ventral horns and that they extend closer to the edge of the spinal cord.

Also identify the central canal, white matter, ventral median fissure, dorsal median sulcus, and dorsal, ventral, and lateral funiculi.

3. Obtain a prepared slide of the spinal cord (cross section) and a compound microscope. Examine the slide carefully under low power (refer to **Figure 19.5** to identify spinal cord features). Observe the shape of the central canal.

Is it basically circular or oval? ____________________

Name the type of neuroglia that lines this canal. ____________

____

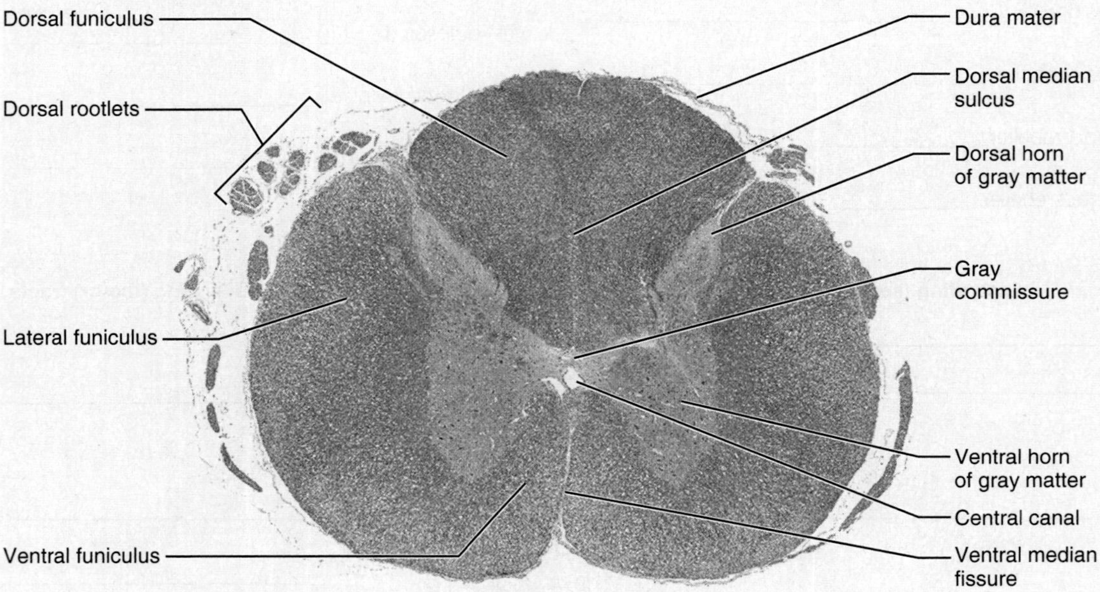

Figure 19.5 Light micrograph of cross section of the spinal cord (8×), cervical region.

Instructors may assign this figure as an Art Labeling Activity using Mastering **A&P™**

Spinal Nerves and Nerve Plexuses

The 31 pairs of human spinal nerves arise from the fusions of the ventral and dorsal roots of the spinal cord and are therefore **mixed nerves** containing both sensory and motor fibers (see Figure 19.1). There are 8 pairs of cervical nerves (C_1–C_8), 12 pairs of thoracic nerves (T_1–T_{12}), 5 pairs of lumbar nerves (L_1–L_5), 5 pairs of sacral nerves (S_1–S_5), and 1 pair of coccygeal nerves (Co_1) (**Figure 19.6a**). The first pair of spinal nerves leaves the vertebral canal between the base of the skull and the atlas, but all the rest exit via the intervertebral foramina. The first through seventh pairs of cervical nerves emerge *above* the vertebra for which they are named; C_8 emerges between C_7 and T_1. (Notice that there are 7 cervical vertebrae, but 8 pairs of cervical nerves.) The remaining spinal nerve pairs emerge from the spinal cord area *below* the same-numbered vertebra.

Almost immediately after emerging, each nerve divides into **dorsal** and **ventral rami**. Thus each spinal nerve is only about 1 or 2 cm long. The rami, like the spinal nerves, contain both motor and sensory fibers. The smaller dorsal rami serve the skin and musculature of the posterior body trunk at their approximate level of emergence. The ventral rami of spinal nerves T_2 through T_{12} pass anteriorly as the **intercostal nerves** to supply the muscles of intercostal spaces, and the skin and

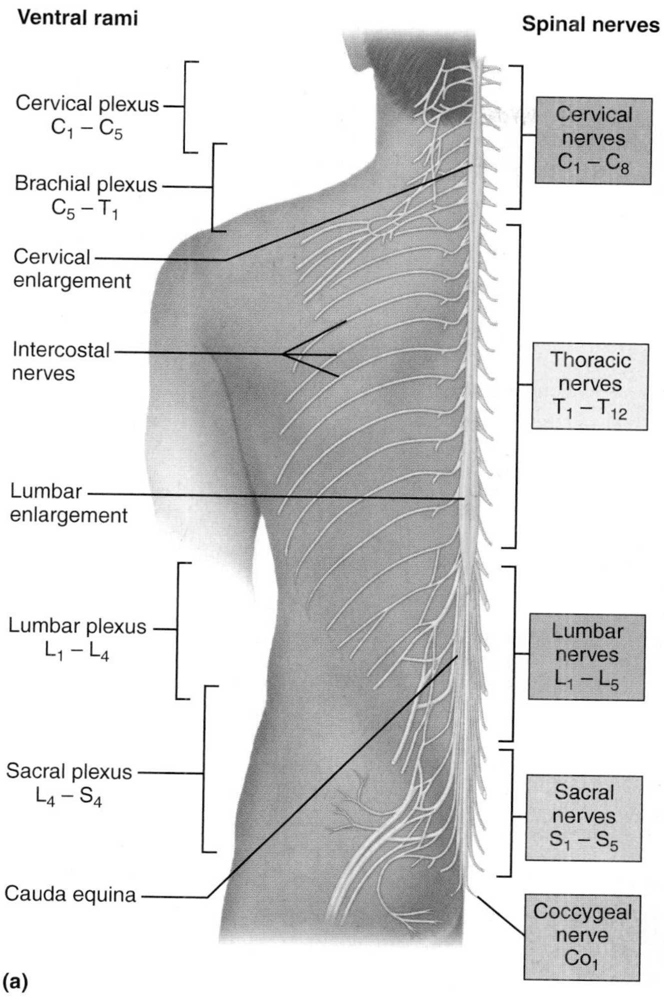

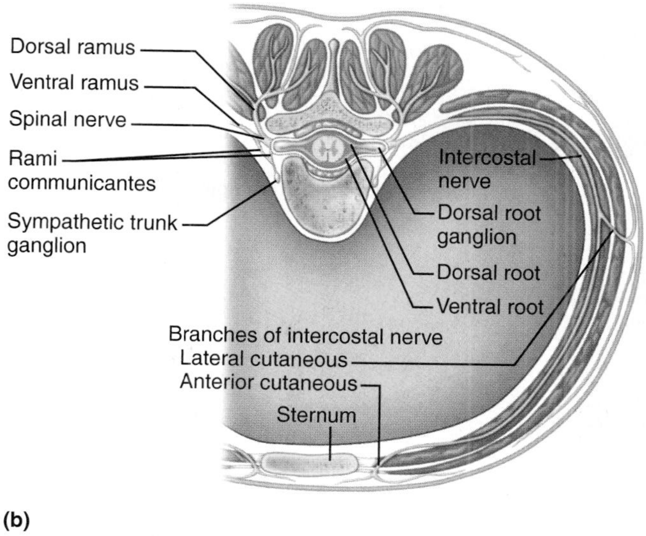

Figure 19.6 Human spinal nerves. (a) Spinal nerves are shown at right; ventral rami and the major nerve plexuses are shown at left. **(b)** The distribution of dorsal and ventral rami. In the thoracic region, each ventral ramus continues as an intercostal nerve. Dorsal rami innervate the intrinsic muscles and skin of the back.

Instructors may assign this figure as an Art Labeling Activity using Mastering A&P™

muscles of the anterior and lateral trunk. The ventral rami of all other spinal nerves form complex networks of nerves called **nerve plexuses**. These plexuses primarily serve the muscles and skin of the limbs. The fibers of the ventral rami unite in the plexuses (with a few rami supplying fibers to more than one plexus). From the plexuses the fibers diverge again to form peripheral nerves, each of which contains fibers from more than one spinal nerve. (The four major nerve plexuses and their chief peripheral nerves are described in Tables 19.2–19.5 and illustrated in Figures 19.7–19.10. Their names and site of origin should be committed to memory). The tiny S_5 and Co_1 spinal nerves contribute to a small plexus that serves part of the pelvic floor.

Cervical Plexus and the Neck

The **cervical plexus** (**Figure 19.7** and **Table 19.2**, p. 309) arises from the ventral rami of C_1 through C_5 to supply muscles of the shoulder and neck. The major motor branch of this plexus is the **phrenic nerve**, which arises from C_3 through C_4 (plus some fibers from C_5) and passes into the thoracic cavity in front of the first rib to innervate the diaphragm. The primary danger of a broken neck is that the phrenic nerve may be severed, leading to paralysis of the diaphragm and cessation of breathing. A rhyme to help you remember the rami (roots) forming the phrenic nerves is "C_3, C_4, C_5 keep the diaphragm alive."

Brachial Plexus and the Upper Limb

The **brachial plexus** is large and complex, arising from the ventral rami of C_5 through C_8 and T_1 (**Table 19.3**, p. 309). The plexus, after being rearranged consecutively into *trunks*, *divisions*, and *cords*, finally becomes subdivided into five major *peripheral nerves* (**Figure 19.8**, p. 308).

19

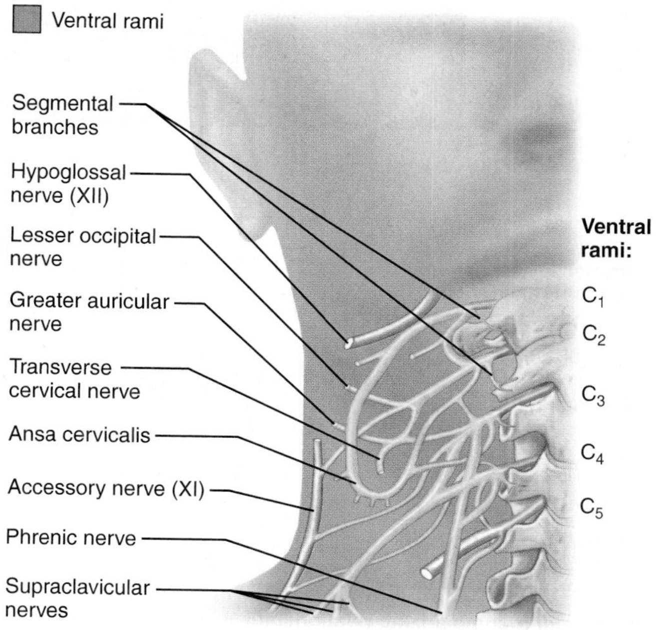

Figure 19.7 The cervical plexus. Note cranial nerves XI and XII do not belong to the cervical plexus. (See Table 19.2.)

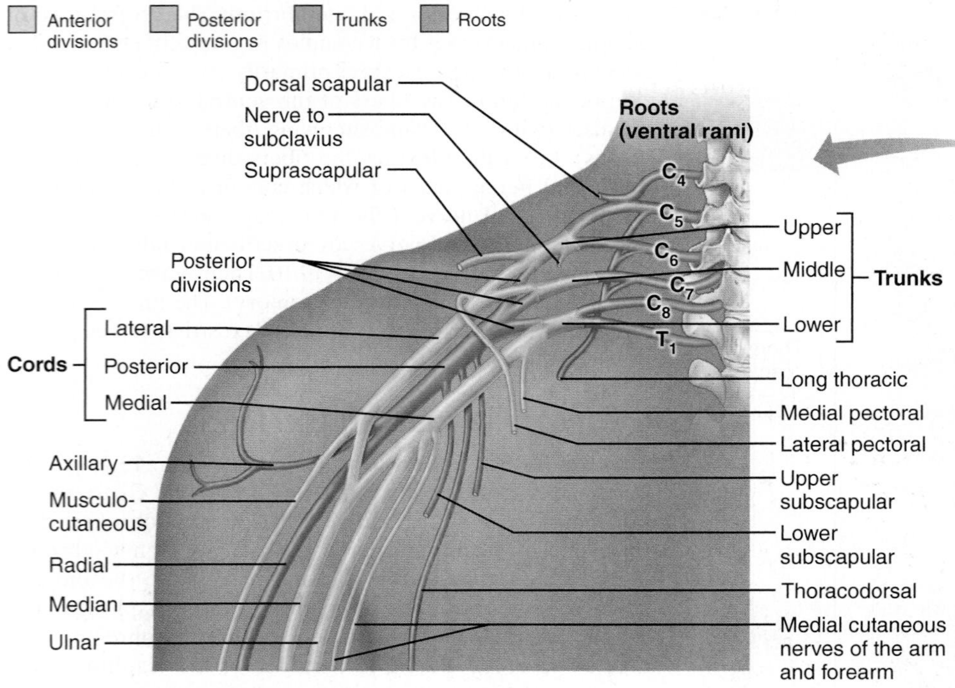

(a) Roots (rami C_5–T_1), trunks, divisions, and cords

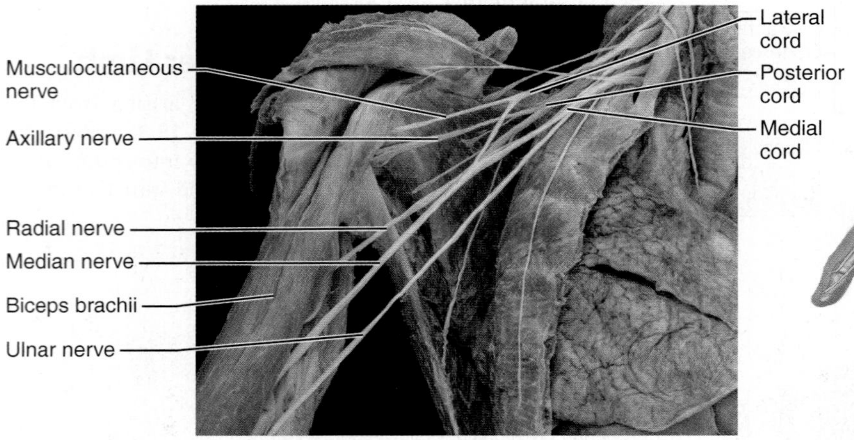

(b) Cadaver photo

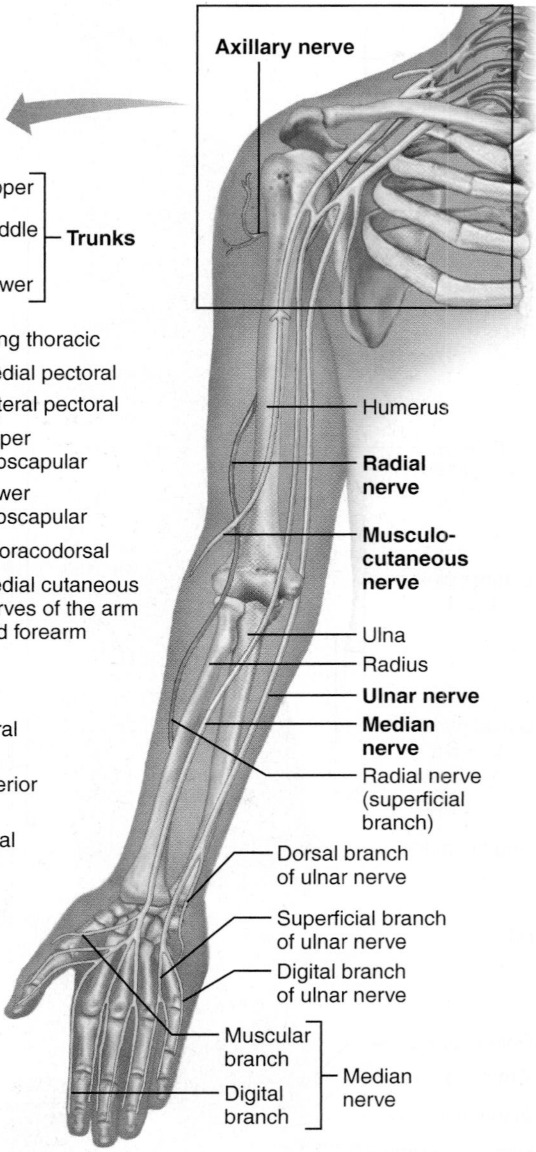

(c) The major nerves of the upper limb

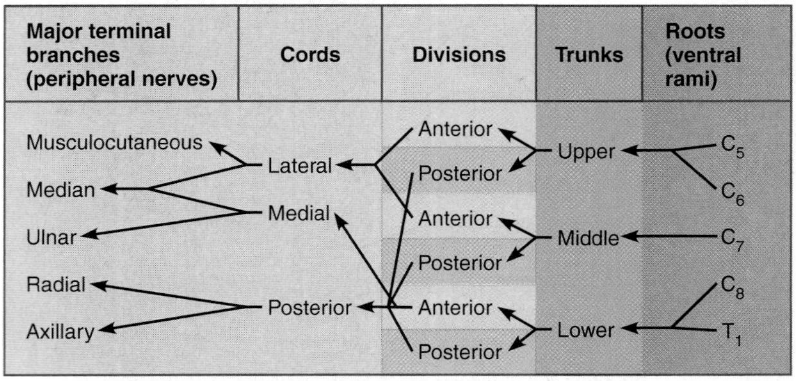

(d) Flowchart summarizing relationships within the brachial plexus

Figure 19.8 The brachial plexus. (See Table 19.3.)

Instructors may assign this figure as an Art Labeling Activity using Mastering A&P™

Table 19.2 Branches of the Cervical Plexus (Figure 19.7)

Nerves	Ventral rami	Structures served
Cutaneous Branches (Superficial)		
Lesser occipital	C_2 (C_3)	Skin on posterolateral aspect of head and neck
Greater auricular	C_2, C_3	Skin of ear, skin over parotid gland
Transverse cervical	C_2, C_3	Skin on anterior and lateral aspect of neck
Supraclavicular (medial, intermediate, and lateral)	C_3, C_4	Skin of shoulder and clavicular region
Motor Branches (Deep)		
Ansa cervicalis (superior and inferior roots)	C_1–C_3	Infrahyoid muscles of neck (omohyoid, sternohyoid, and sternothyroid)
Segmental and other muscular branches	C_1–C_5	Deep muscles of neck (geniohyoid and thyrohyoid) and portions of scalenes, levator scapulae, trapezius, and sternocleidomastoid muscles
Phrenic	C_3–C_5	Diaphragm (sole motor nerve supply)

Table 19.3 Branches of the Brachial Plexus (Figure 19.8)

Nerves	Cord and ventral rami	Structures served
Axillary	Posterior cord (C_5, C_6)	Muscular branches: deltoid and teres minor muscles Cutaneous branches: some skin of shoulder region
Musculocutaneous	Lateral cord (C_5–C_7)	Muscular branches: flexor muscles in anterior arm (biceps brachii, brachialis, coracobrachialis) Cutaneous branches: skin on anterolateral forearm (extremely variable)
Median	By two branches, one from medial cord (C_8, T_1) and one from the lateral cord (C_5–C_7)	Muscular branches to flexor group of anterior forearm (palmaris longus, flexor carpi radialis, flexor digitorum superficialis, flexor pollicis longus, lateral half of flexor digitorum profundus, and pronator muscles); intrinsic muscles of lateral palm and digital branches to the fingers Cutaneous branches: skin of lateral two-thirds of hand on ventral side and dorsum of fingers II and III
Ulnar	Medial cord (C_8, T_1)	Muscular branches: flexor muscles in anterior forearm (flexor carpi ulnaris and medial half of flexor digitorum profundus); most intrinsic muscles of hand Cutaneous branches: skin of medial third of hand, both anterior and posterior aspects
Radial	Posterior cord (C_5–C_8, T_1)	Muscular branches: posterior muscles of arm and forearm (triceps brachii, anconeus, supinator, brachioradialis, extensors carpi radialis longus and brevis, extensor carpi ulnaris, and several muscles that extend the fingers) Cutaneous branches: skin of posterolateral surface of entire limb (except dorsum of fingers II and III)
Dorsal scapular	Branches of C_5 rami	Rhomboid muscles and levator scapulae
Long thoracic	Branches of C_5–C_7 rami	Serratus anterior muscle
Subscapular	Posterior cord; branches of C_5 and C_6 rami	Teres major and subscapularis muscles
Suprascapular	Upper trunk (C_5, C_6)	Shoulder joint; supraspinatus and infraspinatus muscles
Pectoral (lateral and medial)	Branches of lateral and medial cords (C_5–T_1)	Pectoralis major and minor muscles

19

The **axillary nerve**, which serves the muscles and skin of the shoulder, has the most limited distribution. The large **radial nerve** passes down the posterolateral surface of the arm and forearm, supplying all the extensor muscles of the arm, forearm, and hand and the skin along its course. The radial nerve is often injured in the axillary region by the pressure of a crutch or by hanging one's arm over the back of a chair. The **median nerve** passes down the anteromedial surface of the arm to supply most of the flexor muscles in the forearm and several muscles in the hand (plus the skin of the lateral surface of the palm of the hand).

☐ Hyperextend at the wrist to identify the long, obvious tendon of your palmaris longus muscle, which crosses the exact midline of the anterior wrist. Your median nerve lies immediately deep to that tendon, and the radial nerve lies just *lateral* to it.

The **musculocutaneous nerve** supplies the arm muscles that flex the forearm and the skin of the lateral surface of the forearm. The **ulnar nerve** travels down the posteromedial surface of the arm. It courses around the medial epicondyle of the humerus to supply the flexor carpi ulnaris, the ulnar head of the flexor digitorum profundus of the forearm, and all intrinsic muscles of the hand not served by the median nerve. It supplies the skin of the medial third of the hand, both the anterior and posterior surfaces. Trauma to the ulnar nerve, which often occurs when the elbow is hit, is commonly referred to as "hitting the funny bone" because of the odd, painful, tingling sensation it causes.

Severe injuries to the brachial plexus cause weakness or paralysis of the entire upper limb. Such injuries may occur when the upper limb is pulled hard and the plexus is stretched (as when a football tackler yanks the arm of the halfback), and by blows to the shoulder that force the humerus inferiorly (as when a cyclist is pitched headfirst off his motorcycle and grinds his shoulder into the pavement). +

Lumbosacral Plexus and the Lower Limb

The **lumbosacral plexus**, which serves the pelvic region of the trunk and the lower limbs, is actually a complex of two plexuses, the lumbar plexus and the sacral plexus (Figures 19.9 and 19.10). These plexuses interweave considerably and many fibers of the lumbar plexus contribute to the sacral plexus.

The Lumbar Plexus

The **lumbar plexus** arises from ventral rami of L_1 through L_4 (and sometimes T_{12}). Its nerves serve the lower abdominopelvic region and the anterior thigh (**Table 19.4** and **Figure 19.9**). The largest nerve of this plexus is the **femoral nerve**, which passes beneath the inguinal ligament to innervate the anterior thigh muscles. The cutaneous branches of the femoral nerve (median and anterior femoral cutaneous and the saphenous nerves) supply the skin of the anteromedial surface of the entire lower limb.

The Sacral Plexus

Arising from L_4 through S_4, the nerves of the **sacral plexus** supply the buttock, the posterior surface of the thigh, and virtually all sensory and motor fibers of the leg and foot (**Table 19.5**, p. 312, and **Figure 19.10**). The major peripheral nerve of this plexus is the **sciatic nerve**, the largest nerve in the body. The sciatic nerve leaves the pelvis through the greater sciatic notch and travels down the posterior thigh, serving its flexor muscles

(Text continues on p. 312.) →

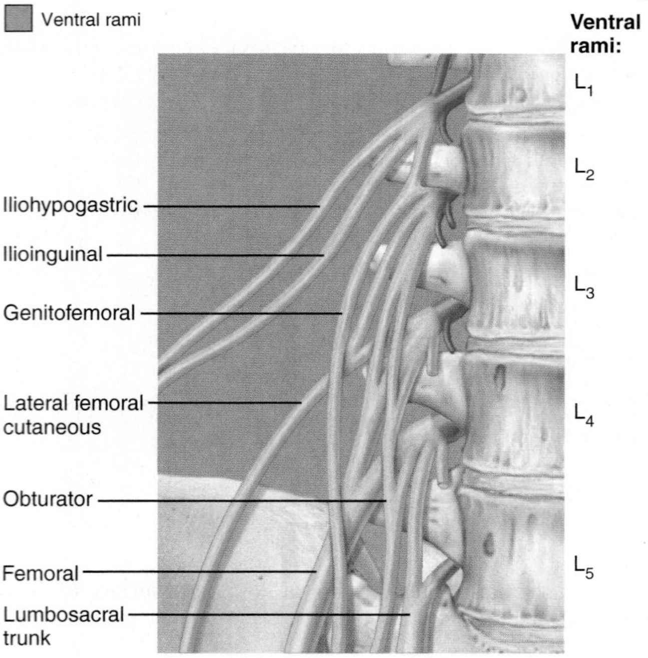

(a) Ventral rami and major branches of the lumbar plexus

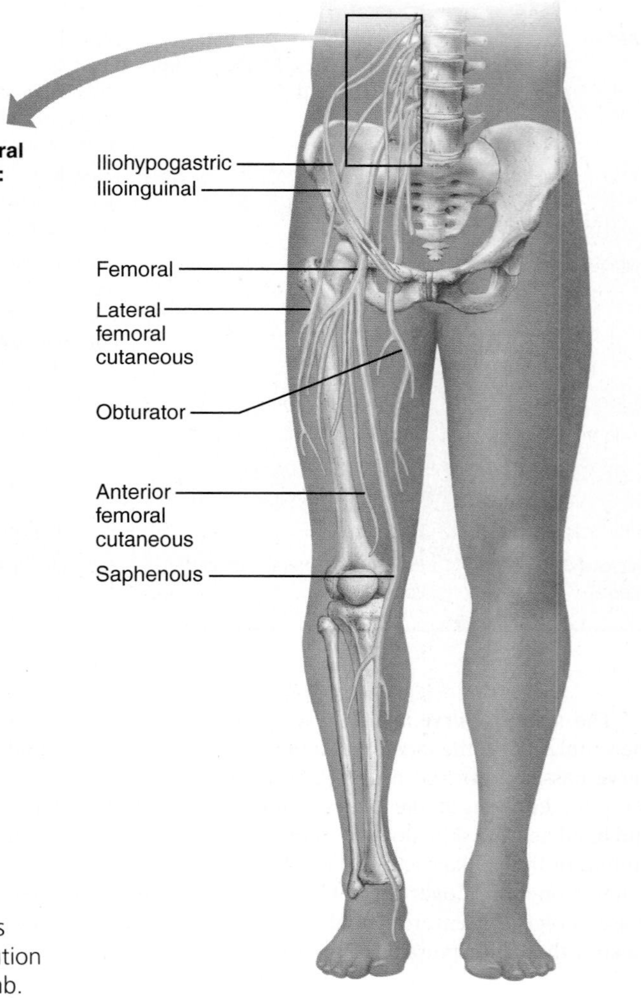

(b)

Figure 19.9 The lumbar plexus (anterior view.) (a) Spinal roots (ventral rami) and major branches of the lumbar plexus. **(b)** Distribution of the major peripheral nerves of the lumbar plexus in the lower limb. (See Table 19.4.)

Instructors may assign this figure as an Art Labeling Activity using Mastering A&P™

Table 19.4 Branches of the Lumbar Plexus (Figure 19.9)

Nerves	Ventral rami	Structures served
Femoral	L_2–L_4	Skin of anterior and medial thigh via *anterior femoral cutaneous* branch; skin of medial leg and foot, hip and knee joints via *saphenous* branch; motor to anterior muscles (quadriceps and sartorius) of thigh and to pectineus, iliacus
Obturator	L_2–L_4	Motor to adductor magnus (part), longus, and brevis muscles, gracilis muscle of medial thigh, obturator externus; sensory for skin of medial thigh and for hip and knee joints
Lateral femoral cutaneous	L_2, L_3	Skin of lateral thigh; some sensory branches to peritoneum
Iliohypogastric	L_1	Skin of lower abdomen and hip; muscles of anterolateral abdominal wall (internal obliques and transversus abdominis)
Ilioinguinal	L_1	Skin of external genitalia and proximal medial aspect of the thigh; inferior abdominal muscles
Genitofemoral	L_1, L_2	Skin of scrotum in males, of labia majora in females, and of anterior thigh inferior to middle portion of inguinal region; cremaster muscle in males

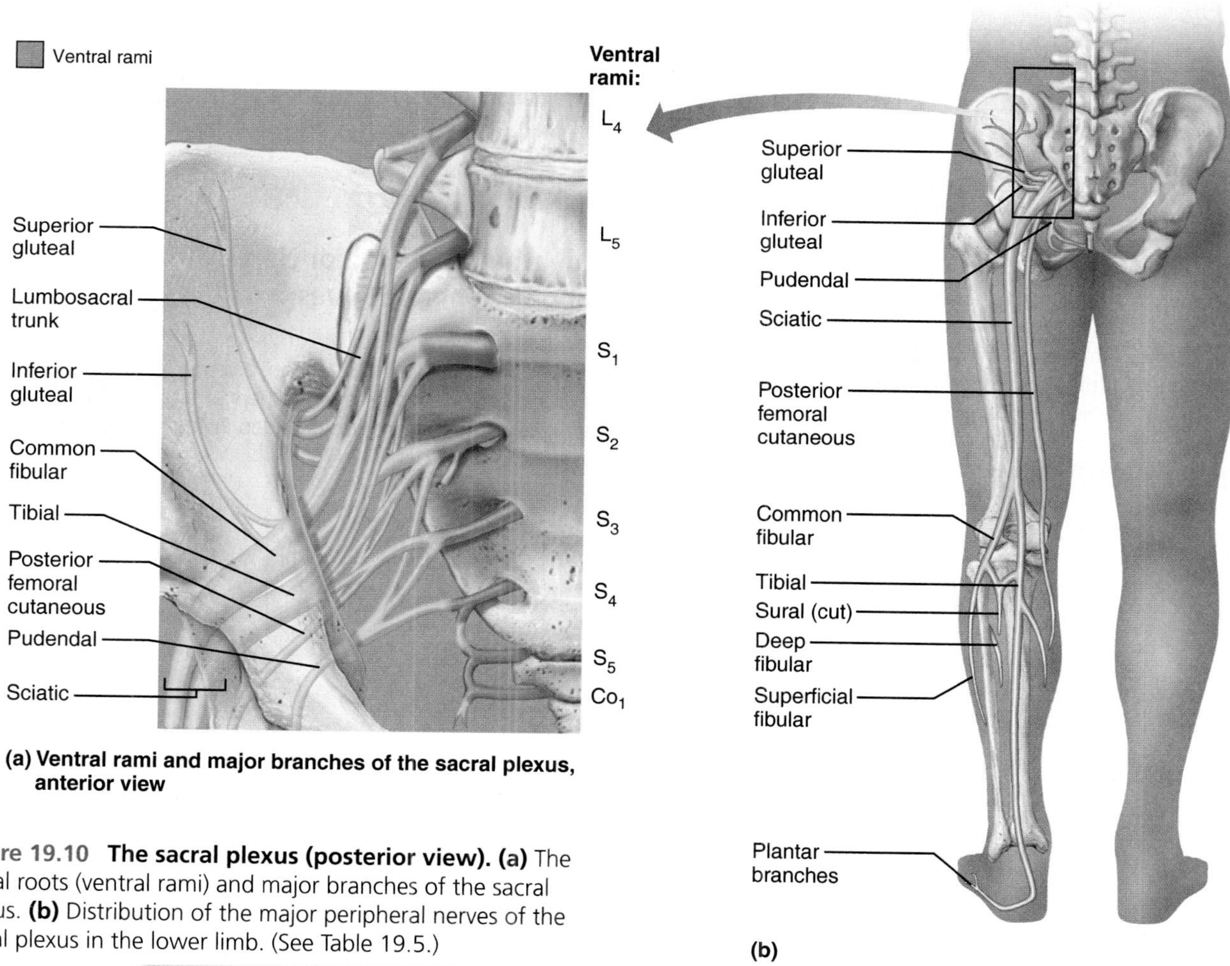

Figure 19.10 The sacral plexus (posterior view). (a) The spinal roots (ventral rami) and major branches of the sacral plexus. **(b)** Distribution of the major peripheral nerves of the sacral plexus in the lower limb. (See Table 19.5.)

Instructors may assign this figure as an Art Labeling Activity using Mastering A&P™

Table 19.5 Branches of the Sacral Plexus (Figure 19.10)

Nerves	Ventral rami	Structures served
Sciatic nerve	L_4–S_3	Composed of two nerves (tibial and common fibular) in a common sheath; they diverge just proximal to the knee
• Tibial (including sural, medial and lateral plantar, and medial calcaneal branches)	L_4–S_3	Cutaneous branches: to skin of posterior surface of leg and sole of foot Motor branches: to muscles of back of thigh, leg, and foot (hamstrings [except short head of biceps femoris], posterior part of adductor magnus, triceps surae, tibialis posterior, popliteus, flexor digitorum longus, flexor hallucis longus, and intrinsic muscles of foot)
• Common fibular (superficial and deep branches)	L_4–S_2	Cutaneous branches: to skin of anterior and lateral surface of leg and dorsum of foot Motor branches: to short head of biceps femoris of thigh, fibularis muscles of lateral leg, tibialis anterior, and extensor muscles of toes (extensor hallucis longus, extensors digitorum longus and brevis)
Superior gluteal	L_4–S_1	Motor branches to gluteus medius and minimus and tensor fascia lata
Inferior gluteal	L_5–S_2	Motor branches to gluteus maximus
Posterior femoral cutaneous	S_1–S_3	Skin of buttock, posterior thigh, and popliteal region; length variable; may also innervate part of skin of calf and heel
Pudendal	S_2–S_4	Supplies most of skin and muscles of perineum (region encompassing external genitalia and anus and including clitoris, labia, and vaginal mucosa in females, and scrotum and penis in males); external anal sphincter

and skin. In the popliteal region, the sciatic nerve divides into the **common fibular nerve** and the **tibial nerve**, which together supply the balance of the leg muscles and skin, both directly and via several branches.

Injury to the proximal part of the sciatic nerve, as might follow a fall or disc herniation, results in a number of lower limb impairments. **Sciatica**, characterized by stabbing pain radiating over the course of the sciatic nerve, is common. When the sciatic nerve is completely severed, the leg is nearly useless. The leg cannot be flexed and the foot drops into plantar flexion (it dangles), a condition called **footdrop**. +

Activity 3

Identifying the Major Nerve Plexuses and Peripheral Nerves

Identify each of the four major nerve plexuses and their major nerves (Figures 19.7–19.10) on a large laboratory chart or model. Trace the courses of the nerves and relate those observations to the information provided in Tables 19.2–19.5.

EXERCISE 19

REVIEW SHEET

The Spinal Cord and Spinal Nerves

Instructors may assign a portion of the Review Sheet questions using Mastering A&P™

Name ______________________ Lab Time/Date ______________________

Anatomy of the Spinal Cord

1. Match each anatomical term in the key to the descriptions given below.

Key: a. cauda equina b. conus medullaris c. filum terminale d. foramen magnum

_______________ 1. most superior boundary of the spinal cord

_______________ 2. meningeal extension beyond the spinal cord terminus

_______________ 3. spinal cord terminus

_______________ 4. collection of spinal nerves traveling in the vertebral canal below the terminus of the spinal cord

2. Match the key letters on the diagram with the following terms.

_______________ 1. central canal

_______________ 2. dorsal horn

_______________ 3. dorsal median sulcus

_______________ 4. dorsal root ganglion

_______________ 5. dorsal root of spinal nerve

_______________ 6. gray commissure

_______________ 7. lateral horn

_______________ 8. spinal nerve

_______________ 9. ventral horn

_______________ 10. ventral median fissure

_______________ 11. ventral root of spinal nerve

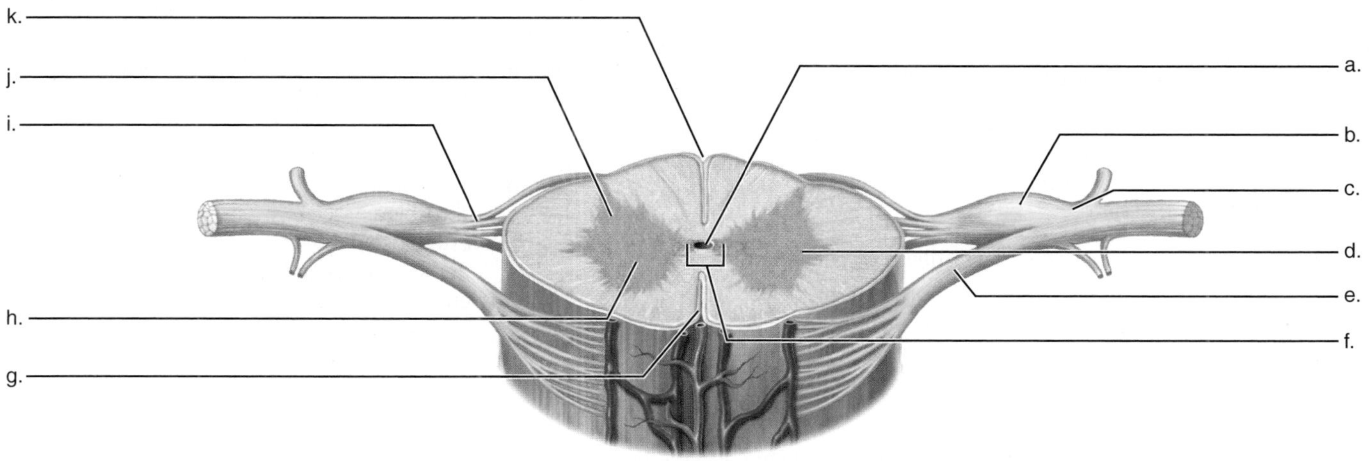

3. Choose the proper answer from the following key to respond to the descriptions relating to spinal cord anatomy. (Some terms are used more than once.)

Key: a. sensory b. motor c. both sensory and motor d. interneurons

_______________ 1. primary neuron type found in dorsal horn

_______________ 2. primary neuron type found in ventral horn

_______________ 3. neuron type in dorsal root ganglion

_______________ 4. fiber type in ventral root

_______________ 5. fiber type in dorsal root

_______________ 6. fiber type in spinal nerve

4. Where in the vertebral column is a lumbar puncture generally done? _______________

Why is this the site of choice? _______________

____

5. The spinal cord is enlarged in two regions, the _______________ and the _______________ regions.

What is the significance of these enlargements? _______________

____

6. How does the position of the gray and white matter differ in the spinal cord and the cerebral hemispheres?

____

____

7. Describe the function of the denticulate ligaments. _______________

____

8. Explain why spinal nerves are mixed nerves. _______________

____

9. From the key, choose the name of the tract that might be damaged when the following conditions are observed. (More than one choice may apply; some terms are used more than once.)

_______________ 1. uncoordinated movement

_______________ 2. lack of voluntary movement

_______________ 3. tremors, jerky movements

_______________ 4. diminished pain perception

_______________ 5. diminished sense of touch

Key: a. dorsal columns (fasciculus cuneatus and fasciculus gracilis)
b. lateral corticospinal tract
c. ventral corticospinal tract
d. tectospinal tract
e. rubrospinal tract
f. vestibulospinal tract
g. lateral spinothalamic tract
h. ventral spinothalamic tract

Dissection of the Spinal Cord

10. Compare and contrast the meninges of the spinal cord and the brain. _______________

____

11. How can you distinguish the dorsal from the ventral horns? ______________________________

____

Spinal Nerves and Nerve Plexuses

12. In the human, there are 31 pairs of spinal nerves, named according to the region of the vertebral column from which they issue. The spinal nerves are named below. Indicate how they are numbered.

cervical nerves ______________________ sacral nerves ______________________

lumbar nerves ______________________ thoracic nerves ______________________

13. The ventral rami of spinal nerves C_1 through T_1 and T_{12} through S_4 take part in forming ______________________, which serve the ______________________ of the body. The ventral rami of T_2 through T_{12} run between the ribs to serve the ______________________. The dorsal rami of the spinal nerves serve ______________________.

14. What would happen if the following structures were damaged or transected? (Use the key choices for responses.)

Key: a. loss of motor function b. loss of sensory function c. loss of both motor and sensory function

__________ 1. dorsal root of a spinal nerve

__________ 2. ventral root of a spinal nerve

__________ 3. ventral ramus of a spinal nerve

15. Define *nerve plexus*. ____

____

16. Name the major nerves that serve the following body areas.

______________________ 1. biceps brachii

______________________ 2. diaphragm

______________________ 3. posterior thigh

______________________ 4. fibularis muscles

______________________ 5. flexor carpi radialis

______________________ 6. deltoid

______________________ 7. gracilis

______________________ 8. anterior thigh

______________________ 9. muscles of the perineum

17. After a person recovers from the chickenpox, the virus can lie dormant in the dorsal root ganglion in multiple levels of the spinal cord and cranial nerves. Later in life, the virus can be reactivated and travel within a sensory neuron to cause shingles. Do you think it is possible to get shingles more than once? ________ Why or why not? ______________________________

____

18. Wrist drop results in an inability to extend the hand at the wrist. Which nerve would most likely be affected in this injury, and why? ____

____

EXERCISE 20 The Autonomic Nervous System

Learning Outcomes

- State how the autonomic nervous system differs from the somatic nervous system.
- Identify the site of origin and the function of the sympathetic and parasympathetic divisions of the autonomic nervous system.
- Identify the neurotransmitters associated with the sympathetic and parasympathetic fibers.
- Record and analyze data associated with the galvanic skin response.

Pre-Lab Quiz

Instructors may assign these and other Pre-Lab Quiz questions using Mastering A&P™

1. The ______________ nervous system is the subdivision of the peripheral nervous system which regulates body activities that are generally not under conscious control.
   a. autonomic b. cephalic
   c. somatic d. vascular
2. Circle the correct underlined term. The parasympathetic division of the autonomic nervous system is also known as the craniosacral / thoracolumbar division.
3. Circle True or False. Cholinergic fibers release epinephrine.
4. The ______________________ division of the autonomic nervous system is responsible for the "fight-or-flight" response because it adapts the body for extreme conditions such as exercise.
5. Circle True or False. The galvanic skin response measures an increase in water and electrolytes at the skin surface.

Go to Mastering A&P™ > Study Area to improve your performance in A&P Lab.

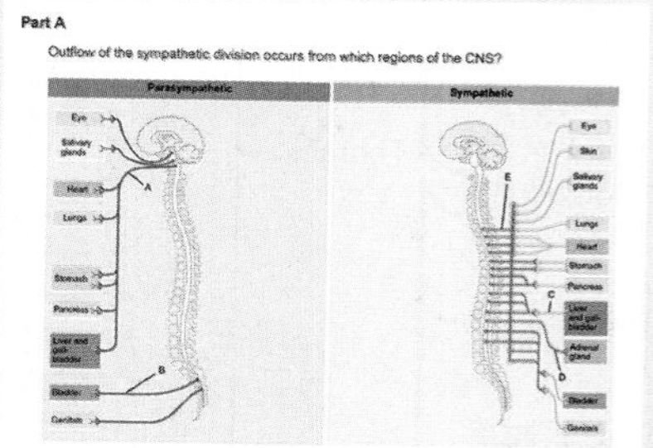

Instructors may assign new Building Vocabulary coaching activities, Pre-Lab Quiz questions, Art Labeling activities, Practice Anatomy Lab Practical questions (PAL), and more using the Mastering A&P™ Item Library.

Materials

- Laboratory chart or three-dimensional model of the sympathetic trunk (chain)

BIOPAC® BIOPAC® BSL System with BSL software version 4.0.1 and above (for Windows 10/8.x/7 or Mac OS X 10.9–10.12), data acquisition unit MP36/35 or MP45, PC or Mac computer, respiratory transducer belt, EDA/GSR finger leads or disposable finger electrodes with EDA pinch leads, Biopac Student Lab electrode lead set, disposable vinyl electrodes, conduction gel, and nine 8½ × 11 inch sheets of paper of different colors (white, black, green, red, blue, yellow, orange, brown, and pink) to be viewed in this sequence.

Instructors using the MP36/35/30 data acquisition unit with BSL software versions earlier than 4.0.1 (for Windows or Mac) will need slightly different channel settings and collection strategies. Instructions for using the older data acquisition unit can be found on MasteringA&P.

For instructions on animal dissections, see the dissection exercises (starting on p. 695) in the cat and fetal pig editions of this manual.

The **autonomic nervous system (ANS)** is the subdivision of the peripheral nervous system (PNS) that regulates body activities that are generally not under conscious control. For this reason, the ANS is also called the *involuntary nervous system*.

The motor pathways of the **somatic** (voluntary) **nervous system**, innervate the skeletal muscles, and the motor pathways of the autonomic nervous system innervate smooth muscle, cardiac muscle, and glands. In the somatic division, the cell bodies of the motor neurons reside in the brain stem or ventral horns of the spinal cord, and their axons extend directly to the skeletal muscles they serve. However, the autonomic nervous system consists of chains of two motor neurons. The first motor neuron of each pair, called the *preganglionic neuron*, resides in the brain stem or the spinal cord. Its axon leaves the central nervous system (CNS) to synapse with the second motor neuron, the *postganglionic neuron*, whose cell body is located in an autonomic ganglion outside the CNS. The axon of the postganglionic neuron then extends to the organ it serves.

The ANS has two major functional subdivisions: the sympathetic and parasympathetic divisions. Both serve most of the same organs but generally cause opposing, or antagonistic, effects. **Table 20.1** on p. 318 compares the parasympathetic and sympathetic divisions. Refer also to **Figure 20.1**, p. 318, and **Figure 20.2**, p. 319.

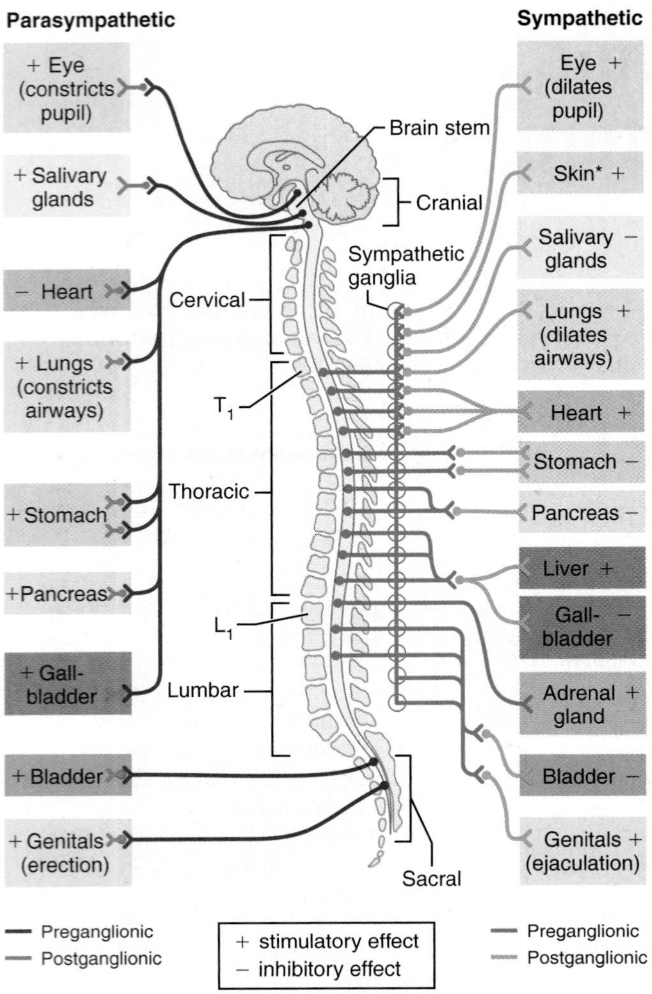

Figure 20.1 Overview of the subdivisions of the autonomic nervous system. Although sympathetic innervation to the skin(*) is shown here mapped to the cervical area, all nerves to the periphery carry postganglionic sympathetic fibers.

Activity 1

Locating the Sympathetic Trunk

Locate the sympathetic trunk (chain) on the spinal nerve chart or three-dimensional model. Notice that the trunk resembles a string of beads. There is a trunk on either side of the vertebral column.

As we grow older, our sympathetic nervous system gradually becomes less and less efficient, particularly in causing vasoconstriction of blood vessels. When elderly people stand up quickly after sitting or lying down, they often become light-headed or faint. The sympathetic nervous system is not able to react quickly enough to counteract the pull of gravity by activating the vasoconstrictor fibers. So, blood pools in the feet. This condition, **orthostatic hypotension**, is a type of low blood pressure resulting from changes in body position as described. Orthostatic hypotension can be prevented to some degree if the person changes position *slowly*. This gives the sympathetic nervous system a little more time to react and adjust. ✚

20

Table 20.1 Anatomical and Physiological Comparison of the Parasympathetic and Sympathetic Divisions (Figure 20.1)

Characteristic	Parasympathetic (craniosacral)	Sympathetic (thoracolumbar)
Origin in the CNS (contains the preganglionic cell body)	Brain stem nuclei of cranial nerves III (oculomotor), VII (facial), IX (glossopharyngeal), and X (vagus); spinal cord segments S_2 through S_4	Lateral horns of the gray matter of the spinal cord T_1 through L_2
Location of ganglia (contains the postganglionic cell body)	Located close to the target organ (**terminal ganglia**) or within the wall of the target organ (**intramural ganglia**)	Located close to the CNS: alongside the vertebral column (**sympathetic trunk ganglia**) or anterior to the vertebral column (**collateral ganglia**)
Axon lengths	Long preganglionic axons	Short preganglionic axons
	Short postganglionic axons	Long postganglionic axons
White and gray rami communicantes (see Figure 20.2)	None	Each white ramus communicans contains myelinated preganglionic axons. Each gray ramus communicans contains nonmyelinated postganglionic axons.
Degree of branching of preganglionic axons	Minimal	Extensive
Functional role	Performs maintenance functions; conserves and stores energy; rest-and-digest response	Prepares the body for emergency situations and vigorous physical activity; fight-or-flight response
Neurotransmitters	Preganglionic axons release acetylcholine (cholinergic fibers)	Preganglionic axons release acetylcholine
	Postganglionic axons release acetylcholine	Most postganglionic axons release norepinephrine (adrenergic fibers); postganglionic axons serving sweat glands and the blood vessels of skeletal muscle release acetylcholine; neurotransmitter activity is supplemented by the release of epinephrine and norepinephrine by the adrenal medulla
Effects of the division	More specific and local	More general and widespread

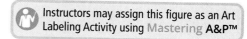

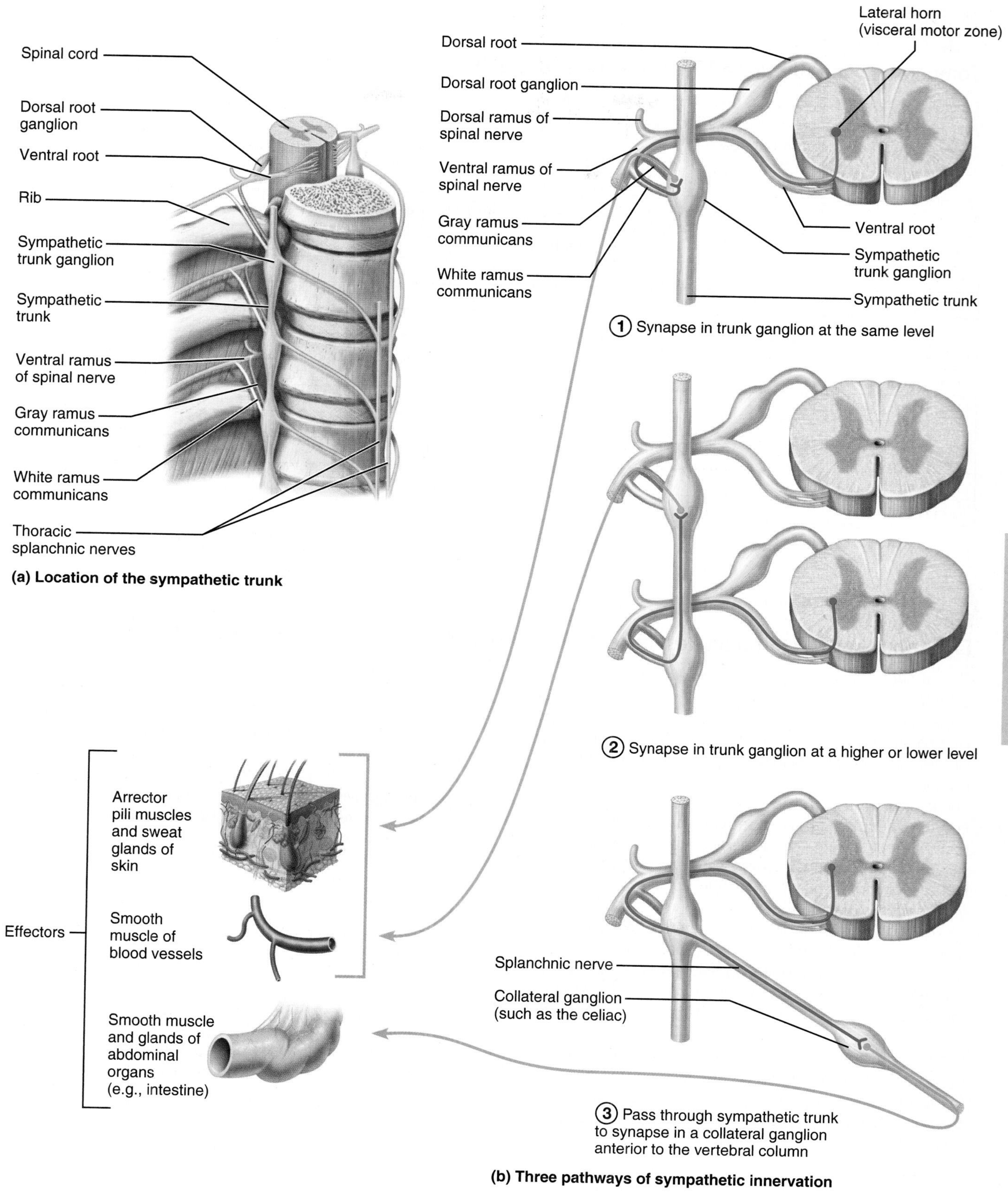

Figure 20.2 Sympathetic trunks and pathways. (a) Diagram of the right sympathetic trunk in the posterior thorax. **(b)** Synapses between preganglionic and postganglionic sympathetic neurons can occur at three different locations.

Instructors may assign this figure as an Art Labeling Activity using Mastering A&P™

20

Activity 2

Comparing Sympathetic and Parasympathetic Effects

Several body organs are listed in the **Activity 2 chart**. Using your textbook as a reference, list the effect of the sympathetic and parasympathetic divisions on each.

Activity 2: Parasympathetic and Sympathetic Effects		
Organ	**Parasympathetic effect**	**Sympathetic effect**
Heart		
Bronchioles of lungs		
Digestive tract		
Urinary bladder		
Iris of the eye		
Blood vessels (most)		
Penis/clitoris		
Sweat glands		
Adrenal medulla		
Pancreas		

Activity 3

Exploring the Galvanic Skin Response (Electrodermal Activity) Within a Polygraph Using BIOPAC®

The autonomic nervous system is closely integrated with the emotions of an individual. A sad event, sharp pain, or simple stress can bring about measurable changes in autonomic regulation of heart rate, respiration, and blood pressure. In addition to these obvious physiological signs, more subtle autonomic changes can occur in the skin. Specifically, changes in autonomic tone in response to external circumstances can influence the rate of sweat gland secretion and blood flow to the skin that may not be readily seen but can be measured. The **galvanic skin response** is an electrophysiological measurement of changes that occur in the skin due to changes in autonomic stimulation.

The galvanic skin response, also referred to as electrodermal activity (EDA), is measured by recording the changes in **galvanic skin resistance (GSR)** and **galvanic skin potential (GSP)**. Resistance, recorded in *ohms* (Ω), is a measure of the opposition to the flow of current from one electrode to another. Increasing resistance results in decreased current. Potential, measured in *volts* (V), is a measure of the amount of charge separation between two points. Increased sympathetic stimulation of sweat glands decreases resistance on the skin because of increased water and electrolytes on the skin surface.

In this experiment you will record heart rate, respiration, and EDA/GSR while the subject is exposed to various conditions. Because "many" variables will be "recorded," this process is often referred to as a **polygraph**. The goal of this exercise is to record and analyze data to observe how this process works. This is not a "lie detector test," as its failure rate is far too high to provide true scientific or legal certainty. However, the polygraph can be used as an investigative tool.

Setting Up the Equipment

1. Connect the BIOPAC® unit to the computer, and turn the computer **ON**.
2. Make sure the BIOPAC® unit is **OFF**.
3. Plug in the equipment (as shown in **Figure 20.3**).
   - Respiratory transducer belt—CH 1
   - Electrode lead set—CH 2
   - EDA/GSR finger leads or disposable finger electrodes and EDA pinch leads—CH 3
4. Turn the BIOPAC® unit **ON**.

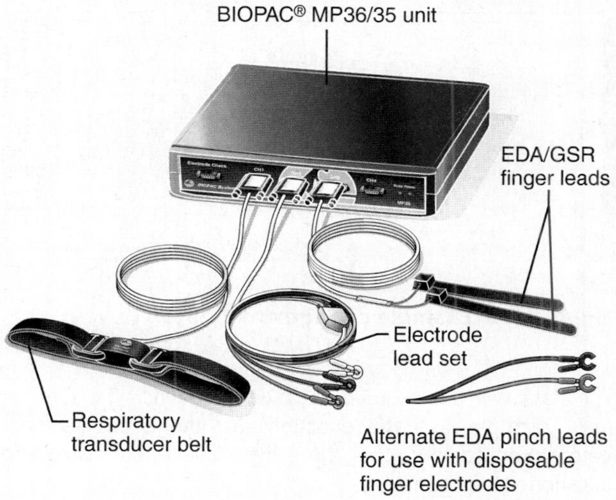

Figure 20.3 Setting up the BIOPAC® equipment. Plug the respiratory transducer belt into Channel 1, the electrode lead set into Channel 2, and the EDA/GSR finger leads into Channel 3.

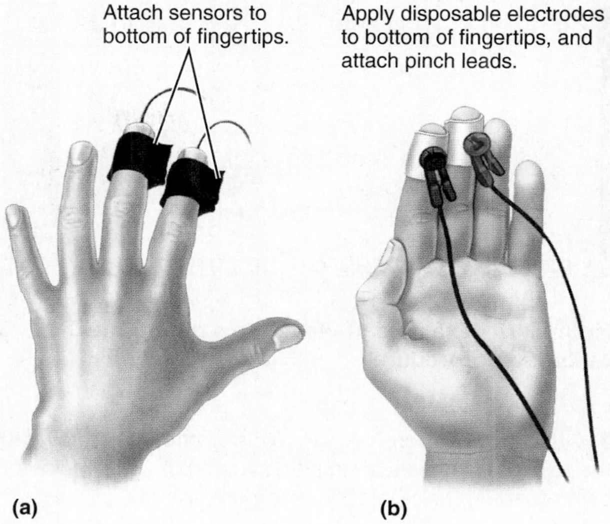

Figure 20.5 Placement of the EDA/GSR finger lead sensors or disposable electrodes on the fingers.

5. Attach the respiratory transducer belt to the subject (as shown in **Figure 20.4**). It should be fastened so that it is slightly tight even at the point of maximal expiration.

6. To pick up a good EDA/GSR signal, it is important that the subject's hand have enough sweat (as it normally would). *The subject should not have freshly washed or cold hands.* Place the electrodes on the middle and index fingers with the sensors on the skin, not the fingernail. They should fit snugly but not be so tight as to cut off circulation. If using EDA/GSR finger lead sensors, fill both cavities of the leads with conduction gel, and attach the sensors to the subject's fingers (as shown in **Figure 20.5a**). If using disposable finger electrodes, apply to subject's fingers and attach pinch leads (as shown in Figure 20.5b). Attach the electrodes at least 5 minutes before recording.

7. In order to record the heart rate, place the electrodes on the subject (as shown in **Figure 20.6**). Place an electrode on the medial surface of each leg, just above the ankle. Place another electrode on the right anterior forearm just above the wrist.

Text continues on next page. ➔

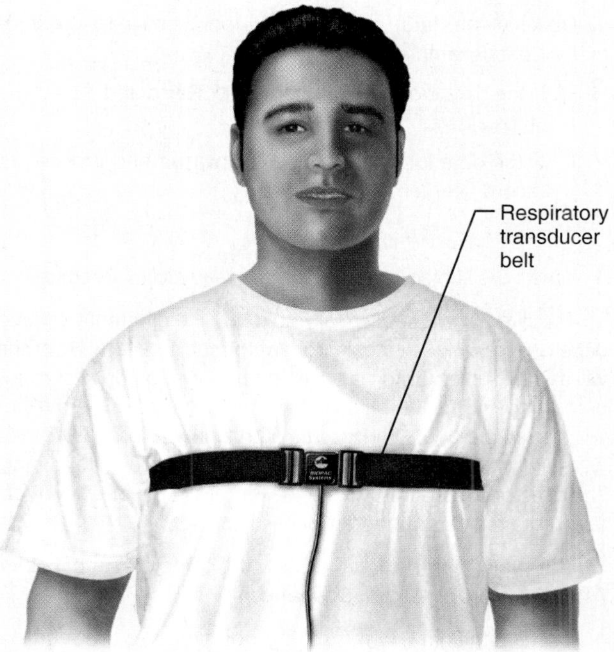

Figure 20.4 Proper placement of the respiratory transducer belt around the subject's thorax.

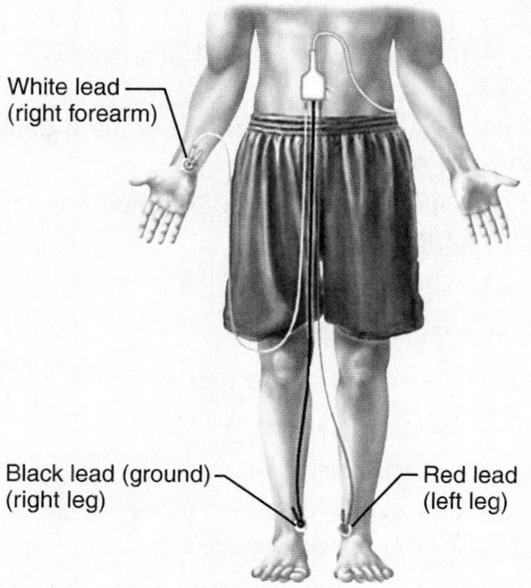

Figure 20.6 Placement of electrodes and the appropriate attachment of electrode leads by color.

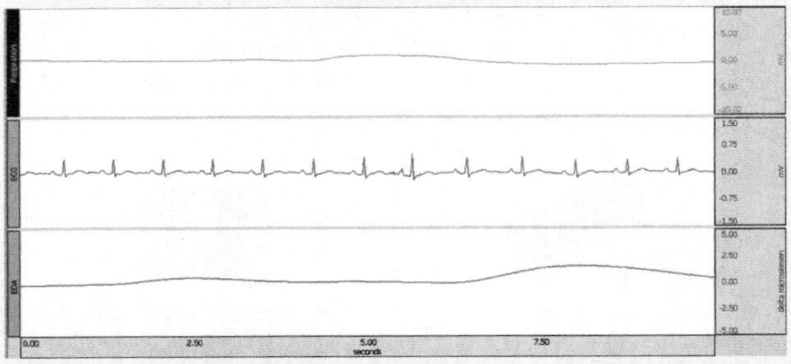

Figure 20.7 Example of waveforms during the calibration procedure.

8. Attach the electrode lead set to the electrodes according to the colors shown in the example (Figure 20.6). Wait 5 minutes before starting the calibration procedure.

9. Start the Biopac Student Lab program on the computer by double-clicking the icon on the desktop or by following your instructor's guidance.

10. Select lesson **L09-Poly-1** from the menu, and click **OK**.

11. Type in a filename that will save this subject's data on the computer hard drive. You may want to use the subject's last name followed by Poly-1 (for example, SmithPoly-1), then click **OK**.

Calibrating the Equipment

1. Have the subject sit facing the director, but do not allow the subject to see the computer screen. The subject should remain immobile but be relaxed with legs and arms in a comfortable position.

2. When the subject is ready, click **Calibrate** and then click **Yes** and **OK** if prompted. After 3 seconds, the subject will hear a beep and should inhale and exhale deeply for one breath.

3. Wait for the calibration to stop automatically after 10 seconds.

4. Observe the data, which should look similar to that in **Figure 20.7**.
   - If the data look very different, click **Redo Calibration** and repeat the steps above.
   - If the data look similar, click **Continue** and proceed to the next section.

Recording the Data

Hints to obtaining the best data:

- Do not let the subject see the data as it is being recorded.
- Conduct the exam in a quiet setting.
- Keep the subject as still as possible.
- Take care to have the subject move the mouth as little as possible when responding to questions.
- Make sure the subject is relaxed at resting heart rate before the exam begins.

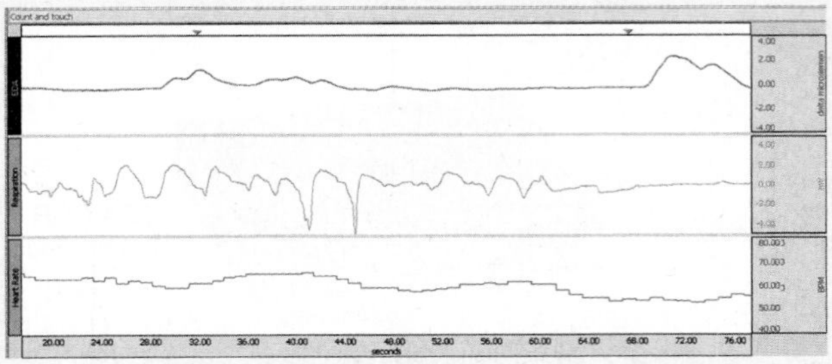

Figure 20.8 Example of Segment 1 data.

The data will be recorded in three segments. The director must read through the directions for the entire segment before proceeding so that the subject can be prompted and questioned appropriately.

Segment 1: Baseline Data

1. When the subject and director are ready, click **Record**.

2. After waiting 5 seconds, the director will ask the subject to respond to the following questions and should remind the subject to minimize mouth movements when answering. Use the **F9** key (PC) or **ESC** key (Mac) to insert a marker after each response. Wait about 5 seconds after each answer.
   - Quietly state your name.
   - Slowly count down from 10 to zero.
   - Count backward from 30 by odd numbers (29, 27, 25, etc.).
   - Finally, the director lightly touches the subject on the cheek.

3. After the final, cheek-touching test, click **Suspend**.

4. Observe the data, which should look similar to the Segment 1 data example (**Figure 20.8**).
   - If the data look very different, click **Redo** and repeat the steps above.
   - If the data look similar, click **Continue** and proceed to record Segment 2.

Segment 2: Response to Different Colors

1. When the subject and director are ready, click **Record**.

2. The director will sequentially hold up nine differently colored paper squares about 2 feet in front of the subject's face. He or she will ask the subject to focus on the particular color for 10 seconds before moving to the next color in the sequence. The director will display the colors and insert a marker in the following order: white, black, green, red, blue, yellow, orange, brown, and pink. The director or assistant will use the **F9** key (PC) or **ESC** key (Mac) to insert a marker at the start of each color.

3. The subject will be asked to view the complete set of colors. After the color pink, click **Suspend**.

4. Observe the data, which should look similar to the Segment 2 data example (**Figure 20.9**).

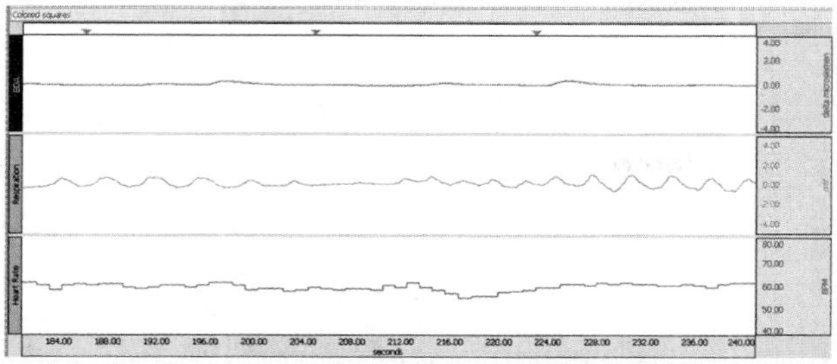

Figure 20.9 Example of Segment 2 data.

- If the data look very different, click **Redo** and repeat the steps above.
- If the data look similar, click **Continue** and proceed to record Segment 3.

Segment 3: Response to Different Questions

1. When the subject and director are ready, click **Record**.

2. The director will ask the subject the 10 questions in step 3 and note if the answer is Yes or No. In this segment, the recorder will use the **F9** key (PC) or **ESC** key (Mac) to insert a marker at the end of each question and the end of each answer. The director will circle the Yes or No response of the subject in the "Response" column of the Segment 3 Measurements chart (p. 325).

3. The following questions are to be asked and answered either Yes or No:
   - Are you currently a student?
   - Are your eyes blue?
   - Do you have any brothers?
   - Did you earn an "A" on the last exam?
   - Do you drive a motorcycle?
   - Are you less than 25 years old?
   - Have you ever traveled to another planet?
   - Have aliens from another planet ever visited you?
   - Do you watch *Sesame Street*?
   - Have you answered all of the preceding questions truthfully?

4. After the last question is answered, click **Suspend**.

5. Observe the data, which should look similar to the Segment 3 data example (**Figure 20.10**).
   - If the data look very different, click **Redo** and repeat the steps above.
   - If the data look similar, click **Done**. Click **Yes** if you are finished recording.

6. Without recording, simply ask the subject to respond once again to all of the questions as honestly as possible. The director circles the Yes or No response of the subject in the "Truth" column of the Segment 3 Measurements chart (p. 325).

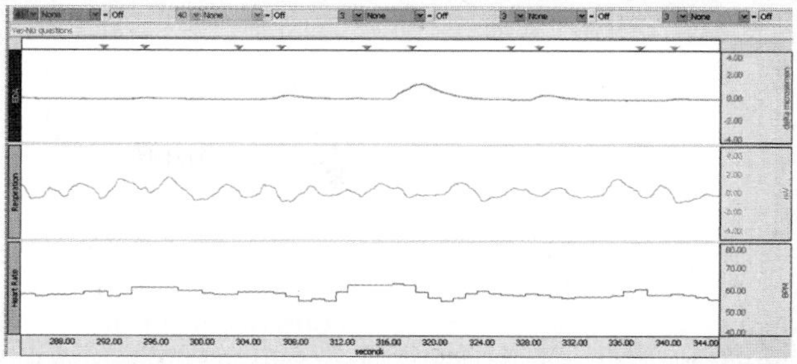

Figure 20.10 Example of Segment 3 data.

7. Remove all of the sensors and equipment from the subject, and continue to Data Analysis.

Data Analysis

1. If you are just starting the BIOPAC® program to perform data analysis, enter **Review Saved Data** mode and choose the file with the subject's EDA/GSR data (for example, Smith-Poly-1). If **Analyze Current Data File** was previously chosen, proceed to analysis.

2. Observe how the channel numbers are designated (as shown in **Figure 20.11**): CH 3—**EDA/GSR**; CH 40—**Respiration**; CH 41—**Heart Rate**.

3. You may need to use the following tools to adjust the data in order to clearly view and analyze the first 5 seconds of the recording.
   - Click the magnifying glass in the lower right corner of the screen (near the I-beam box) to activate the **zoom** function. Use the magnifying glass cursor to click on the very first waveforms until the first 5 seconds of data are represented (see horizontal time scale at the bottom of the screen).
   - Select the **Display** menu at the top of the screen, and click **Autoscale Waveforms** in the drop-down menu. This function will adjust the data for better viewing.

Text continues on next page. →

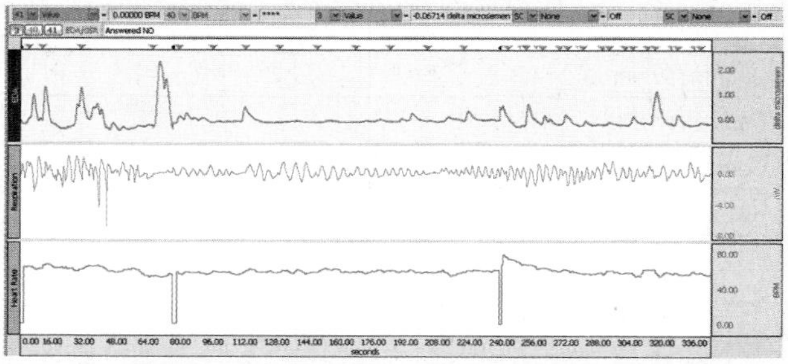

Figure 20.11 Example of polygraph recording with EDA/GSR, respiration, and heart rate.

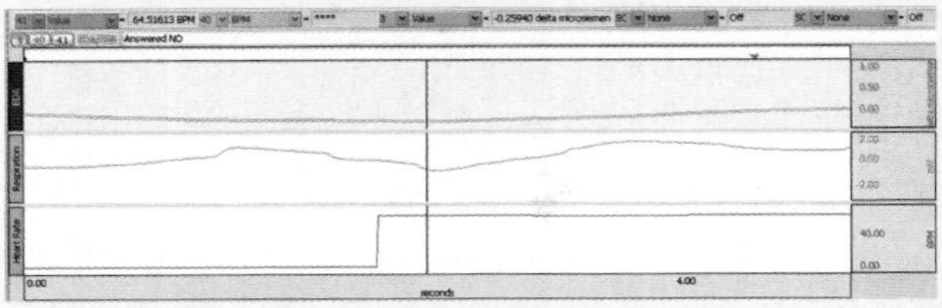

Figure 20.12 Selecting the 2-second point for data analysis.

4. To analyze the data, note the first three pairs of channel/measurement boxes at the top of the screen. (Each box activates a drop-down menu when you click it.) The following channels and measurement types should already be set:

Channel	Measurement	Data
CH 41	Value	heart rate
CH 40	Value	respiration
CH 3	Value	EDA/GSR

Value: Displays the value of the measurement (for example, heart rate or EDA/GSR) at the point in time that is selected.

BPM: In this analysis, the BPM calculates breaths per minute when the area that is highlighted starts at the beginning of one inhalation and ends at the beginning of the next inhalation.

5. Using the arrow cursor, click the I-beam cursor box at the lower right side of the screen to activate the "area selection" function. Using the activated I-beam cursor, select the 2-second point on the data (as shown in **Figure 20.12**). Record the heart rate and EDA/GSR values for Segment 1 data in the **Segment 1 Measurements chart**. This point represents the resting, or baseline, data.

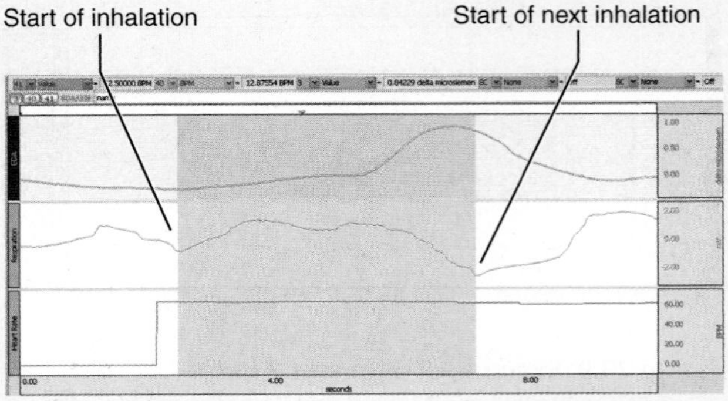

Figure 20.13 Highlighting the waveforms from the start of one inhalation to the start of the next.

6. Using data from the first 5 seconds, use the I-beam cursor tool to highlight an area from the start of one inhalation to the start of the next inhalation (as shown in **Figure 20.13**). The start of an inhalation is indicated by the beginning of the ascension of the waveform. Record this as the baseline respiratory rate in the Segment 1 Measurements chart.

7. Using the markers as guides, scroll along the bottom scroll bar until the data from Segment 1 appears.

8. Analyze all parts of Segment 1. Using the tools described in steps 5 and 6, acquire the measurements for the heart rate, EDA/GSR, and respiration rate soon after each subject response. Use the maximum EDA/GSR value in that time frame as the point of measurement for EDA/GSR and heart rate. Use the beginning of two consecutive inhalations in that same time frame to measure respiration rate. Record these data in the Segment 1 Measurements chart.

9. Repeat these same procedures to measure EDA/GSR, heart rate, and respiration rate for each color in Segment 2. Record these data in the **Segment 2 Measurements chart.**

Segment 1 Measurements

Procedure	Heart rate [CH 41 value]	Respiratory rate [CH 40 BPM]	EDA/GSR [CH 3 value]
Baseline			
Quietly say name			
Count from 10			
Count from 30			
Face is touched			

Segment 2 Measurements			
Color	**Heart rate [CH 41 value]**	**Respiratory rate [CH 40 BPM]**	**EDA/GSR [CH 3 value]**
White			
Black			
Green			
Red			
Blue			
Yellow			
Orange			
Brown			
Pink			

Segment 3 Measurements					
Question	**Response**	**Truth**	**Heart rate [CH 41 value]**	**Resp. rate [CH 40 BPM]**	**EDA/GSR [CH 3 value]**
Student?	Y N	Y N			
Blue eyes?	Y N	Y N			
Brothers?	Y N	Y N			
Earn "A"?	Y N	Y N			
Motorcycle?	Y N	Y N			
Under 25?	Y N	Y N			
Planet?	Y N	Y N			
Aliens?	Y N	Y N			
Sesame?	Y N	Y N			
Truthful?	Y N	Y N			

10. Repeat these same procedures to measure EDA/GSR, heart rate, and respiration rate for responses to each question in Segment 3. Record these data in the **Segment 3 Measurements chart**.

11. Examine EDA/GSR, heart rate, and respiration rate of the baseline data in the Segment 1 Measurements chart.

12. For every condition to which the subject was exposed, write **H** if that value is higher than baseline, write **L** if the value is lower, and write **NC** if there is no significant change. Repeat this analysis for Segments 2 and 3.

Examine the data in the Segment 1 Measurements chart. Is there any noticeable difference between the baseline EDA/GSR, heart rate, and respiration rate after each prompt? Under which prompts is the most significant change noted?

____

____

____

Examine the data in the Segment 2 Measurements chart. Is there any noticeable difference between the baseline EDA/GSR, heart rate, and respiration rate after each color presentation? Under which colors is the most significant change noted?

____

____

____

Examine the data in the Segment 3 Measurements chart. Is there any noticeable difference between the baseline EDA/GSR, heart rate, and respiration rate after each question? After which is the most significant change noted?

____

____

____

Text continues on next page. →

Speculate as to the reasons why a subject may demonstrate a change in EDA/GSR from baseline under different color conditions.

____

____

____

Speculate as to the reasons why a subject may demonstrate a change in EDA/GSR from baseline when a particular question is asked.

____

____

____

Which branch of the autonomic nervous system is dominant during a galvanic skin response?

____

EXERCISE 20

REVIEW SHEET
The Autonomic Nervous System

Instructors may assign a portion of the Review Sheet questions using Mastering A&P™

Name ______________________ Lab Time/Date ______________________

Parasympathetic and Sympathetic Divisions

1. List the names of the two motor neurons of the autonomic nervous system.

____

2. List the names and numbers of the four cranial nerves that the parasympathetic division of the ANS arises from.

____

3. List two types of sympathetic ganglia that contain postganglionic cell bodies.

____

4. List two types of parasympathetic ganglia that contain postganglionic cell bodies.

____

5. Which part of the rami communicantes contains nonmyelinated fibers? ______________________________

____

6. The following chart states a number of characteristics. Use a check mark to show which division of the autonomic nervous system is involved in each.

Sympathetic division	Characteristic	Parasympathetic division
	Postganglionic axons secrete norepinephrine; adrenergic fibers	
	Postganglionic axons secrete acetylcholine; cholinergic fibers	
	Long preganglionic axon; short postganglionic axon	
	Short preganglionic axon; long postganglionic axon	
	Arises from cranial and sacral nerves	
	Arises from spinal nerves T_1 through L_3	
	Normally in control	
	"Fight-or-flight" system	
	Has more specific effects	
	Has rami communicantes	
	Has extensive branching of preganglionic axons	

Galvanic Skin Response (Electrodermal Activity) Within a Polygraph Using BIOPAC®

7. Describe exactly how, from a physiological standpoint, EDA/GSR can be correlated with activity of the autonomic nervous system.

____

____

____

8. Based on this brief exposure to a polygraph, explain why this might not be an exact tool for testing the sincerity and honesty of a subject. Refer to your data to support your conclusions.

____

____

____

9. + Ogilvie syndrome is a condition that mimics a bowel obstruction. The patient experiences abdominal bloating and constipation in the absence of a mechanical blockage. Ogilvie syndrome is usually preceded by surgery and results in extreme loss of motor activity in the bowel. Which division of the autonomic nervous system is underactive in this case, especially with respect to the gastrointestinal system?

____

10. + Neostigmine is a drug that is classified as an acetylcholinesterase inhibitor. Explain how neostigmine could reverse the effects of Ogilvie syndrome.

____

____

EXERCISE 21

Human Reflex Physiology

Learning Outcomes

- Define *reflex* and *reflex arc*.
- Describe the differences between autonomic and somatic reflexes.
- Explain why reflex testing is an important part of every physical examination.
- Name, identify, and describe the function of each element of a reflex arc.
- Describe and discuss several types of reflex activities as observed in the laboratory; indicate the functional or clinical importance of each; and categorize each as a somatic or autonomic reflex action.
- Explain why cord-mediated reflexes are generally much faster than those involving input from the higher brain centers.
- Investigate differences in reaction time between intrinsic and learned reflexes.

Go to Mastering A&P™ > Study Area to improve your performance in A&P Lab.

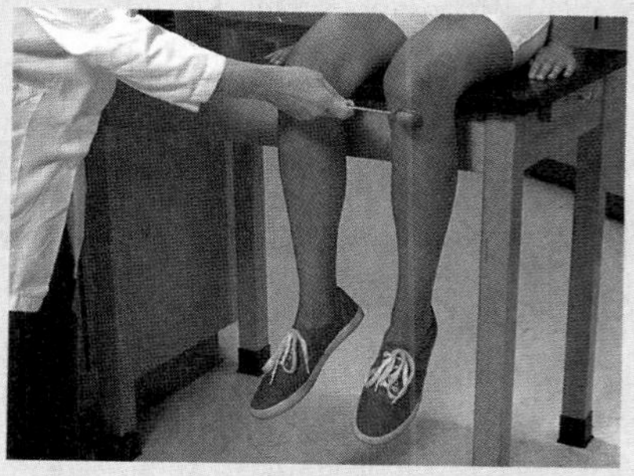

> Lab Tools > Pre-Lab Videos > Stretch Reflexes

Instructors may assign new Building Vocabulary coaching activities, Pre-Lab Quiz questions, Art Labeling activities, Pre-Lab Video Coaching Activities for Stretch Reflexes, Practice Anatomy Lab Practical questions (PAL), and more using the Mastering **A&P™** Item Library.

Pre-Lab Quiz

Instructors may assign these and other Pre-Lab Quiz questions using Mastering **A&P™**

1. Circle the correct underlined term. Autonomic / Somatic reflexes include all those reflexes that involve stimulation of skeletal muscles.
2. In a reflex arc, the ___________ transmits afferent impulses to the central nervous system.
   a. integration center
   b. motor neuron
   c. receptor
   d. sensory neuron
3. Stretch reflexes are initiated by tapping a ___________, which stretches the associated muscle.
   a. bone
   b. muscle
   c. tendon or ligament
4. An example of an autonomic reflex that you will be studying in today's lab is the ___________ reflex.
   a. crossed-extensor
   b. gag
   c. plantar
   d. salivary
5. Circle True or False. A reflex that occurs on the same side of the body that was stimulated is an ipsilateral response.

Materials

- Reflex hammer
- Sharp pencils
- Cot (if available)
- Absorbent cotton (sterile)
- Tongue depressor
- Metric ruler
- Flashlight
- 100- or 250-ml beaker
- 10- or 25-ml graduated cylinder
- Lemon juice in dropper bottle
- Wide-range pH paper
- Large laboratory bucket containing freshly prepared 10% household bleach solution for saliva-containing glassware
- Disposable autoclave bag
- Wash bottle containing 10% bleach solution
- Reaction time ruler (if available)

Text continues on next page. →

BIOPAC® BIOPAC® BSL System with BSL software version 4.0.1 and above (for Windows 10/8.x/7 or Mac OS X 10.9–10.12), data acquisition unit MP36/35, PC or Mac computer, hand switch and headphones.

Instructors using the MP36/35/30 data acquisition unit with BSL software versions earlier than 4.0.1 (for Windows or Mac) will need slightly different channel settings and collection strategies. Instructions for using the older data acquisition unit can be found on MasteringA&P.

Note: *Instructions for using PowerLab® equipment can be found on MasteringA&P.*

Reflexes are rapid, predictable, involuntary motor responses to stimuli; they are mediated over neural pathways called reflex arcs. Many of the body's control systems are reflexes, which can be either inborn (intrinsic) or learned (acquired).

Another way to categorize reflexes is into one of two large groups: autonomic reflexes and somatic reflexes. **Autonomic** (visceral) **reflexes** are mediated through the autonomic nervous system, and we are not usually aware of them. These reflexes activate smooth muscles, cardiac muscle, and the glands of the body, and they regulate body functions such as digestion, elimination, blood pressure, salivation, and sweating. **Somatic reflexes** include all those reflexes that involve stimulation of skeletal muscles by the somatic division of the nervous system.

Reflex testing is an important diagnostic tool for assessing the condition of the nervous system. If the spinal cord is damaged, the easily performed reflex tests can help pinpoint the level of spinal cord injury. Motor nerves above the injured area may be unaffected, whereas those at or below the lesion site may be unable to participate in normal reflex activity. +

Components of a Reflex Arc

Reflex arcs have five basic components (**Figure 21.1**):

1. The *receptor* is the site of stimulus action.
2. The *sensory neuron* transmits afferent impulses to the CNS.
3. The *integration center* may be as simple as a single synapse between a sensory neuron and a motor neuron. A reflex of this type is a *monosynaptic reflex*. *Polysynaptic reflexes* involve one or more interneurons synapsing to form the integration center.
4. The *motor neuron* conducts efferent impulses from the integration center to an effector organ.
5. The *effector*, a muscle fiber or a gland cell, responds to efferent impulses by contracting or secreting, respectively.

The simple patellar, or knee-jerk, reflex (**Figure 21.2a**) is an example of a monosynaptic reflex arc. It will be demonstrated in the laboratory. However, most reflexes are more complex and polysynaptic, involving the participation of one or more interneurons in the reflex arc pathway. An example of a polysynaptic reflex is the flexor reflex (Figure 21.2b). Because the reflex may be delayed or inhibited at the synapses, the more synapses encountered in a reflex pathway, the more time is required for the response.

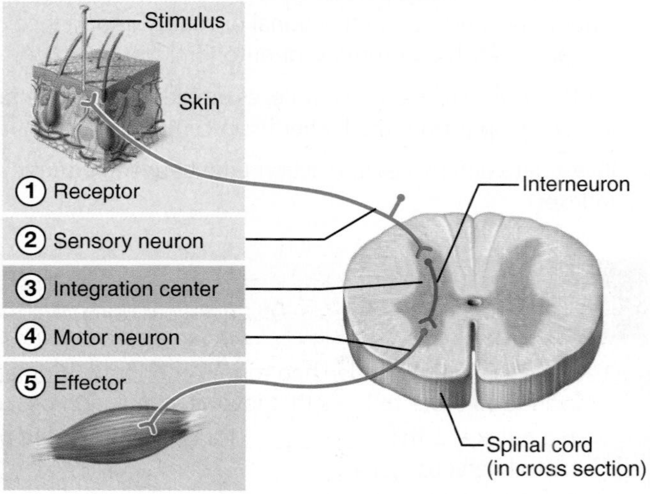

Figure 21.1 The five basic components of reflex arcs. The reflex illustrated is polysynaptic.

Instructors may assign this figure as an Art Labeling Activity using Mastering **A&P™**

Somatic Reflexes

There are several types of somatic reflexes, including several that you will be observing during this laboratory session—the stretch, crossed-extensor, superficial, corneal, and gag reflexes. Spinal reflexes are mediated by the spinal cord without direct involvement of higher brain centers. Although these inborn spinal reflexes don't require brain involvement, the brain is often "advised" of the reflex activity. The reflex activity can be facilitated or inhibited, depending on the circumstances. Some somatic reflexes are mediated by cranial nerves.

Spinal Reflexes

Stretch Reflexes

Stretch reflexes are important for maintaining and adjusting muscle tone for posture, balance, and locomotion. Stretch reflexes are initiated by tapping a tendon or ligament, which stretches the muscle to which the tendon is attached (**Figure 21.3**, p. 332). This stimulates the muscle spindles and causes reflex contraction of the stretched muscle or muscles. Branches of the afferent fibers from the muscle spindles also synapse with interneurons controlling the antagonist muscles. The inhibition of those interneurons and the antagonist muscles, called *reciprocal inhibition*, causes them to relax and prevents them from resisting (or reversing) the contraction of the stretched muscle. Additionally, impulses are relayed to higher brain centers (largely via the dorsal white columns) to advise of muscle length, speed of shortening, and the like—information needed to maintain muscle tone and posture.

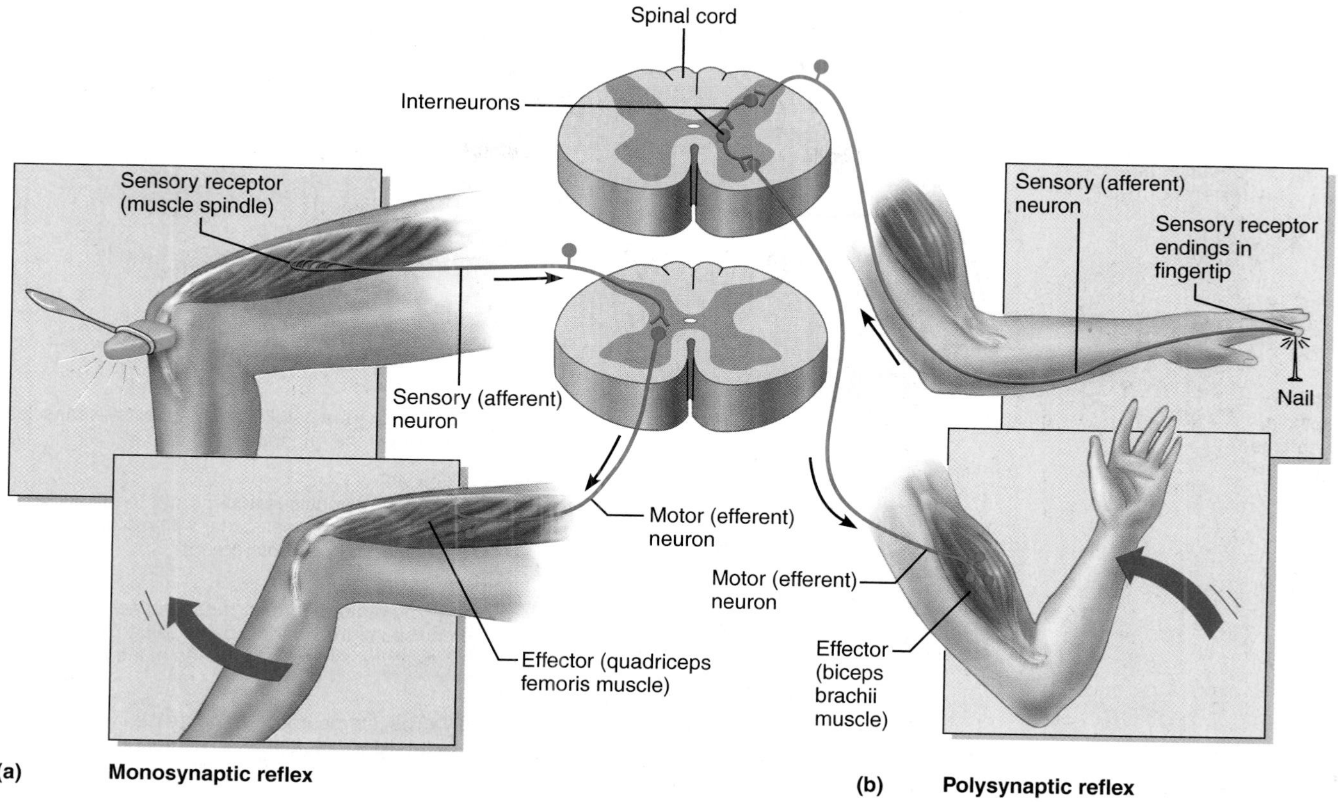

Figure 21.2 Monosynaptic and polysynaptic reflex arcs. The integration center is in the spinal cord, and in each example the receptor and effector are in the same limb. **(a)** The patellar reflex, a two-neuron monosynaptic reflex. **(b)** A flexor reflex, an example of a polysynaptic reflex.

Instructors may assign this figure as an Art Labeling Activity using Mastering A&P™

21

Activity 1

Instructors may assign a related video and coaching activity for the Stretch Reflexes using Mastering A&P™

Initiating Stretch Reflexes

1. Test the **patellar**, or **knee-jerk**, **reflex** by seating a subject on the laboratory bench with legs hanging free (or with knees crossed). Tap the patellar ligament sharply with the broad side of the reflex hammer just below the knee between the patella and the tibial tuberosity, as shown in **Figure 21.4** on p. 332. The knee-jerk response assesses the L_2–L_4 level of the spinal cord. Test both knees and record your observations.

____

____

Which muscles contracted? ____________________

____

What nerve is carrying the afferent and efferent impulses?

____

2. Test the effect of mental distraction on the patellar reflex by having the subject add a column of three-digit numbers while you test the reflex again. Is the response more *or* less vigorous than the first response?

____

What are your conclusions about the effect of mental distraction on reflex activity?

____

3. Now test the effect of muscular activity occurring simultaneously in other areas of the body. Have the subject clasp the edge of the laboratory bench and vigorously attempt to pull it upward with both hands. At the same time, test the patellar reflex again. Is the response more or less vigorous than the first response?

____

4. Fatigue also influences the reflex response. The subject should jog in position until she or he is very fatigued (*really*

Text continues on next page. →

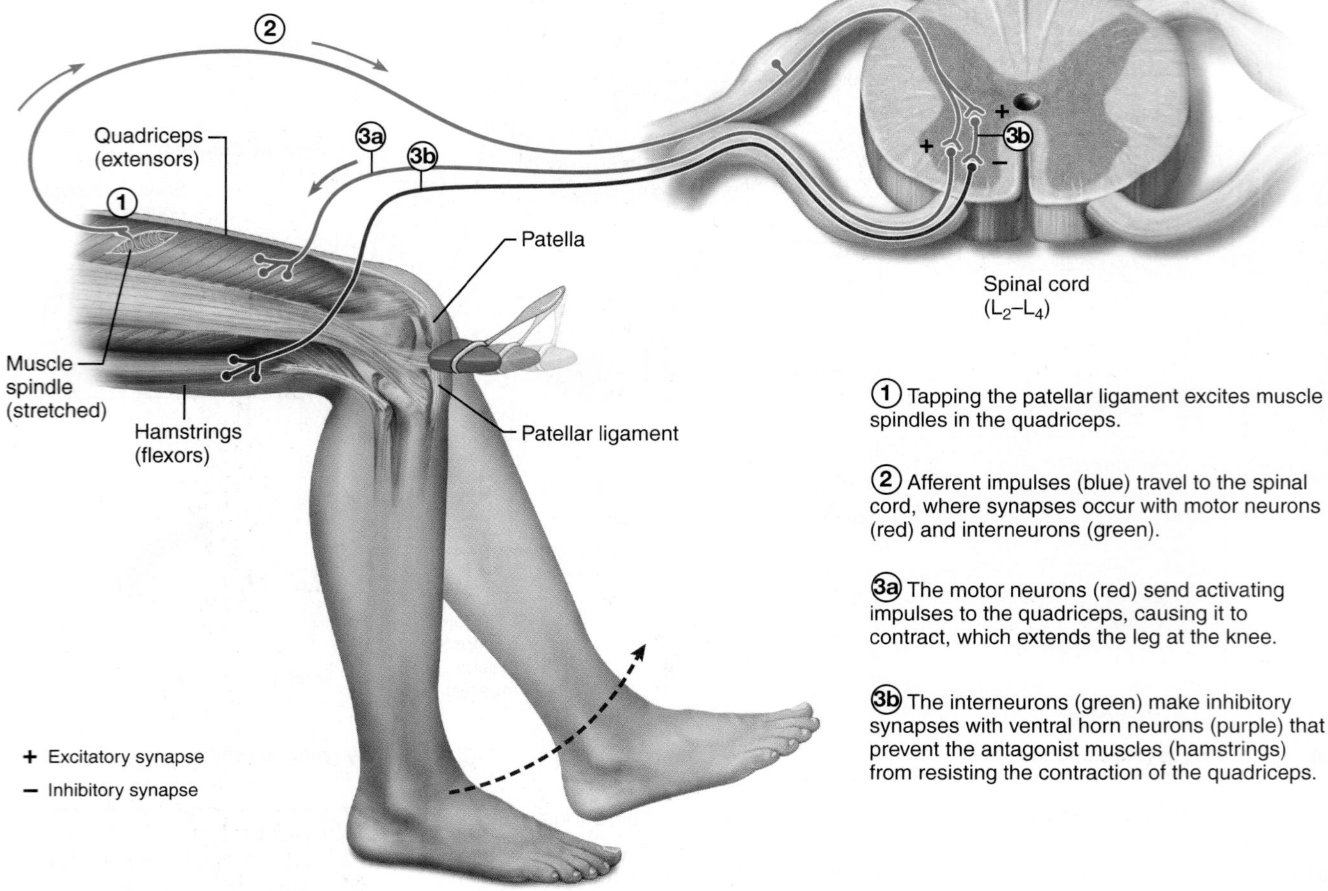

Figure 21.3 The patellar (knee-jerk) reflex—a specific example of a stretch reflex.

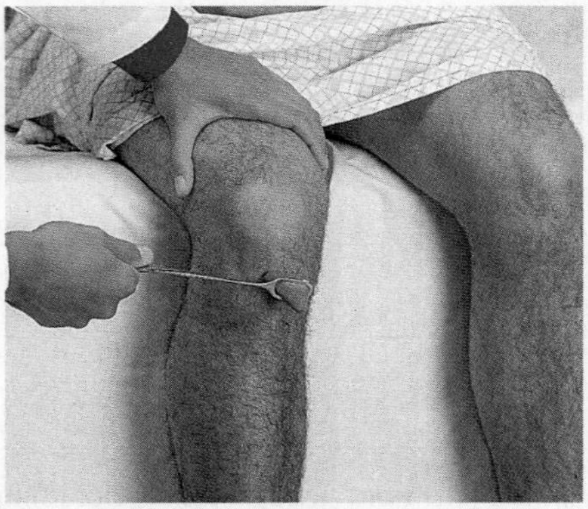

Figure 21.4 Testing the patellar reflex. The examiner supports the subject's knee so that the subject's muscles are relaxed, and then strikes the patellar ligament with the reflex hammer. The proper location may be ascertained by palpation of the patella.

fatigued—no slackers). Test the patellar reflex again, and record whether it is more or less vigorous than the first response.

____

Would you say that nervous system activity *or* muscle function is responsible for the changes you have just observed?

Explain your reasoning. ____

____

____

____

5. The **calcaneal tendon**, or **ankle-jerk**, **reflex** assesses the first two sacral segments of the spinal cord. With your shoe removed and your foot dorsiflexed slightly to increase the tension of the gastrocnemius muscle, have your partner sharply tap your calcaneal tendon with the broad side of the reflex hammer (**Figure 21.5**).

What is the result? ____

____

During walking, what is the action of the gastrocnemius?

____

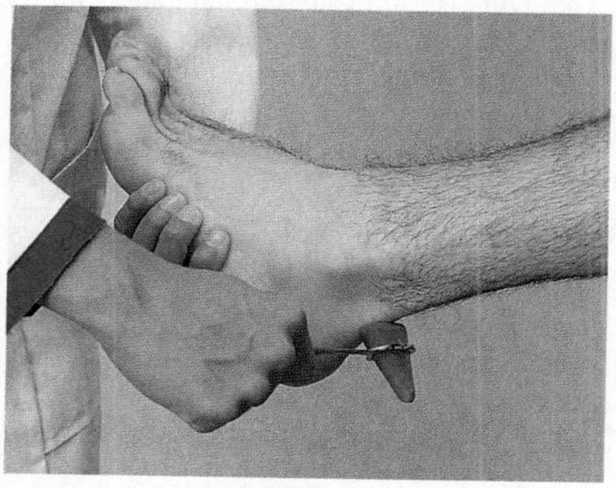

Figure 21.5 Testing the calcaneal tendon reflex. The examiner slightly dorsiflexes the subject's ankle by supporting the foot lightly in the hand, and then taps the calcaneal tendon just above the ankle.

Crossed-Extensor Reflex

The **crossed-extensor reflex** is more complex than the stretch reflex. It consists of a flexor, or withdrawal, reflex followed by extension of the opposite limb.

This reflex is quite obvious when, for example, a stranger suddenly and strongly grips one's arm. The immediate response is to withdraw the clutched arm and push the intruder away with the other arm. The reflex is more difficult to demonstrate in a laboratory because it is anticipated, and under these conditions the extensor part of the reflex may be inhibited.

Activity 2

Initiating the Crossed-Extensor Reflex

The subject should sit with eyes closed and with the back of one hand resting on the laboratory bench. Obtain a sharp pencil, and suddenly prick the subject's index finger. What are the results?

____

____

____

Did the extensor part of this reflex occur simultaneously or more slowly than the other reflexes you have observed?

____

What are the reasons for this? ______________________________________

____

____

The reflexes that have been demonstrated so far—the stretch and crossed-extensor reflexes—are examples of reflexes in which the reflex pathway is mediated at the spinal cord level only.

Superficial Reflexes

The **superficial reflexes** (abdominal, cremasteric, and plantar reflexes) are initiated by stimulation of receptors in the skin and mucosae. The superficial reflexes depend *both* on functional upper-motor pathways and on the spinal cord–level reflex arc. Since only the plantar reflex can be tested conveniently in a laboratory setting, we will use this as our example.

The **plantar reflex**, an important neurological test, is elicited by stimulating the cutaneous receptors in the sole of the foot. In adults, stimulation of these receptors causes the toes to flex and move closer together. Damage to the primary motor cortex or the corticospinal tract, however, produces *Babinski's sign*, an abnormal response in which the great toe moves in an upward direction, and the smaller toes fan out. In a newborn infant, it is normal to see Babinski's sign because myelination of the nervous system is incomplete.

Activity 3

Initiating the Plantar Reflex

Have the subject remove a shoe and sock and lie on the cot or laboratory bench, with knees slightly bent and thighs rotated so that the posterolateral side of the foot rests on the cot. Alternatively, the subject may sit up and rest the lateral surface of the foot on a chair. Draw the handle of the reflex hammer firmly along the lateral side of the exposed sole from the heel to the base of the great toe (**Figure 21.6**). 21

What is the response? ______________________________________

____

Is this a normal plantar reflex or a Babinski's sign?

____

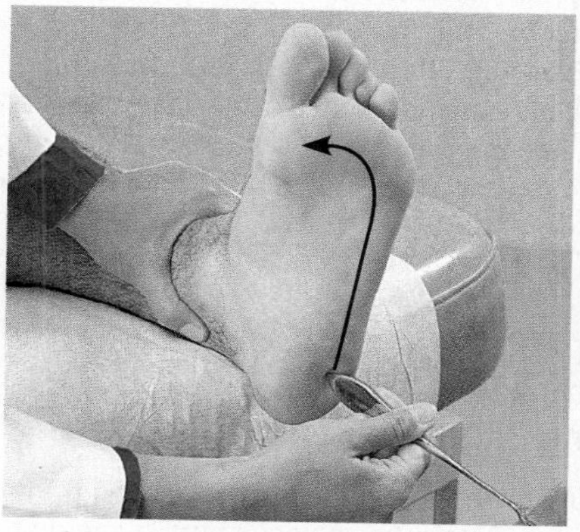

Figure 21.6 Testing the plantar reflex. Using a moderately sharp object, the examiner strokes the lateral border of the subject's sole, starting at the heel and continuing toward the great toe across the ball of the foot.

Cranial Nerve Reflexes

In these experiments, you will be working with your lab partner to illustrate two somatic reflexes mediated by cranial nerves.

Corneal Reflex

The **corneal reflex** is mediated through the trigeminal nerve. The absence of this reflex is an ominous sign because it often indicates damage to the brain stem resulting from compression of the brain or other trauma.

Activity 4

Initiating the Corneal Reflex

Stand to one side of the subject; the subject should look away from you toward the opposite wall. Wait a few seconds and then quickly, *but gently*, touch the subject's cornea (on the side toward you) with a wisp of absorbent cotton. What reflexive reaction occurs when something touches the cornea?

____

What is the function of this reflex?

____

____

Gag Reflex

The **gag reflex** tests the somatic motor responses of cranial nerves IX and X (glossopharyngeal and vagus nerves). When the oral mucosa on the side of the uvula is stroked, each side of the mucosa should rise, and the amount of elevation should be equal. The uvula is the fleshy tab hanging from the roof of the mouth.

Activity 5

Initiating the Gag Reflex

For this experiment, select a subject who does not have a queasy stomach, because regurgitation is a possibility. Gently stroke the oral mucosa on each side of the subject's uvula with a tongue depressor. What happens?

____

____

! Discard the used tongue depressor in the disposable autoclave bag before continuing. *Do not* lay it on the laboratory bench at any time.

Autonomic Reflexes

The autonomic reflexes include the pupillary, ciliospinal, and salivary reflexes, as well as many other reflexes.

Pupillary Reflexes

There are several types of pupillary reflexes. The **pupillary light reflex** and the **consensual reflex** will be examined here. In both of these pupillary reflexes, the retina of the eye is the receptor, the optic nerve contains the afferent fibers, the oculomotor nerve contains the efferent fibers, and the smooth muscle of the iris is the effector. Absence of normal pupillary reflexes is generally a late indication of severe trauma or deterioration of the vital brain stem tissue due to metabolic imbalance.

Activity 6

Instructors may assign a related video and coaching activity for the Pupillary Reflexes using Mastering A&P™

Initiating Pupillary Reflexes

1. Conduct the reflex testing in an area where the lighting is relatively dim. Before beginning, obtain a metric ruler and a flashlight. Measure and record the size of the subject's pupils as best you can.

Right pupil: ___________ mm Left pupil: ___________ mm

2. Stand to the left of the subject to conduct the testing. The subject should shield his or her right eye by holding a hand vertically between the eye and the right side of the nose.

3. Shine a flashlight into the subject's left eye. What is the pupillary response?

____

Measure the size of the left pupil: ___________ mm

4. Without moving the flashlight, observe the right pupil. Has the same type of change (called a *consensual response*) occurred in the right eye?

____

Measure the size of the right pupil: ___________ mm

The consensual response, or any reflex observed on one side of the body when the other side has been stimulated, is called a **contralateral response**. The pupillary light response, or any reflex occurring on the same side stimulated, is referred to as an **ipsilateral response**.

What does the occurrence of a contralateral response indicate about the pathways involved?

____

____

What is the function of these pupillary responses?

____

____

____

Ciliospinal Reflex

The **ciliospinal reflex** is another example of reflex activity in which pupillary responses can be observed. This response may initially seem a little bizarre, especially in view of the consensual reflex just demonstrated.

Activity 7

Initiating the Ciliospinal Reflex

1. While observing the subject's eyes, gently stroke the skin (or just the hairs) on the left side of the back of the subject's neck, close to the hairline.

What is the reaction of the left pupil? _______________

The reaction of the right pupil? _______________

2. If you see no reaction, repeat the test using a gentle pinch in the same area.

The response you should have noted—pupillary dilation—is consistent with the pupillary changes occurring when the sympathetic nervous system is stimulated. Such a response may also be elicited in a single pupil when more impulses from the sympathetic nervous system reach it for any reason. For example, when the left side of the subject's neck was stimulated, sympathetic impulses to the left iris increased, resulting in the ipsilateral reaction of the left pupil.

On the basis of your observations, would you say that the sympathetic innervation of the two irises is closely integrated?

_______________ Why or why not? _______________

____

____

Salivary Reflex

Unlike the other reflexes, in which the effectors were smooth or skeletal muscles, the effectors of the **salivary reflex** are glands. The salivary glands secrete varying amounts of saliva in response to reflex activation.

Activity 8

Initiating the Salivary Reflex

1. Obtain a small beaker, a graduated cylinder, lemon juice, and wide-range pH paper. After refraining from swallowing for 2 minutes, the subject is to expectorate (spit) the accumulated saliva into a small beaker. Using the graduated cylinder, measure the volume of the expectorated saliva and determine its pH.

Volume: _______________ cc pH: _______________

2. Now place 2 or 3 drops of lemon juice on the subject's tongue. Allow the lemon juice to mix with the saliva for 5 to 10 seconds, and then determine the pH of the subject's saliva by touching a piece of pH paper to the tip of the tongue.

pH: _______________

As before, the subject is to refrain from swallowing for 2 minutes. After the 2 minutes is up, again collect and measure the volume of the saliva and determine its pH.

Volume: _______________ cc pH: _______________

3. How does the volume of saliva collected after the application of the lemon juice compare with the volume of the first saliva sample?

____

How does the final saliva pH reading compare to the initial reading?

____

How does the final saliva pH reading compare to that obtained 10 seconds after the application of lemon juice?

____

! Dispose of the saliva-containing beakers and the graduated cylinders in the laboratory bucket that contains bleach, and put the used pH paper into the disposable autoclave bag. Wash the bench down with 10% bleach solution before continuing.

Reaction Time of Intrinsic and Learned Reflexes

The time required for reaction to a stimulus depends on many factors—sensitivity of the receptors, velocity of nerve conduction, the number of neurons and synapses involved, and the speed of effector activation, to name just a few. There is no clear-cut distinction between intrinsic and learned reflexes, as most reflex actions are subject to modification by learning or conscious effort. In general, however, if the response involves a simple reflex arc, the response time is short. Learned reflexes involve a far larger number of neural pathways and many types of higher intellectual activities, including choice and decision making, which lengthens the response time.

There are various ways of testing reaction time of reflexes. The following activities provide an opportunity to demonstrate the major time difference between a simple reflex arc and learned reflexes and to measure response time under various conditions.

Activity 9

Testing Reaction Time for Intrinsic and Learned Reflexes

1. Using a reflex hammer, elicit the patellar reflex in your partner. Note the relative reaction time needed for this intrinsic reflex to occur.

2. Now test the reaction time for learned reflexes. The subject should hold a hand out, with the thumb and index finger extended. Hold a metric ruler so that its end is exactly 3 cm above the subject's outstretched hand. The ruler should be in the vertical position with the numbers reading from the bottom up. When the ruler is dropped, the subject should be able to grasp it between thumb and index finger as it passes, without having to change position. Have the subject catch the ruler five times, varying the time between trials. The relative speed of reaction can be determined by reading the number on the ruler at the point of the subject's fingertips.* (Thus if the number at the fingertips is 15 cm, the subject was unable to catch the ruler until 18 cm of length had passed through the fingers; 15 cm of ruler length plus 3 cm to account for the distance of the ruler above the hand.)† Record the number of centimeters that pass through the subject's fingertips (or the number of seconds required for reaction) for each trial:

Trial 1: ________ cm Trial 4: ________ cm
________ sec ________ sec

Trial 2: ________ cm Trial 5: ________ cm
________ sec ________ sec

Trial 3: ________ cm
________ sec

3. Perform the test again, but this time say a simple word each time you release the ruler. Designate a specific word as a signal for the subject to catch the ruler. On all other words, the subject is to allow the ruler to pass through the fingers. Trials in which the subject erroneously catches the ruler are to be disregarded. Record the distance the ruler travels (or the number of seconds required for reaction) in five *successful* trials:

Trial 1: ________ cm Trial 4: ________ cm
________ sec ________ sec

Trial 2: ________ cm Trial 5: ________ cm
________ sec ________ sec

Trial 3: ________ cm
________ sec

Did the addition of a specific word to the stimulus increase or decrease the reaction time?

____

4. Perform the testing once again to investigate the subject's reaction to word association. As you drop the ruler, say a word—for example, *hot*. The subject is to respond with a word he or she associates with the stimulus word—for example, *cold*—catching the ruler while responding. If unable to make a word association, the subject must allow the ruler to pass through the fingers. Record the distance the ruler travels (or the number of seconds required for reaction) in five successful trials, as well as the number of times the subject does not catch the ruler.

Trial 1: ________ cm Trial 4: ________ cm
________ sec ________ sec

Trial 2: ________ cm Trial 5: ________ cm
________ sec ________ sec

Trial 3: ________ cm
________ sec

Number of times the subject did *not* catch the ruler:

____

*Distance (d) can be converted to time (t) using the simple formula:

$$d \text{ (in cm)} = (1/2)(980 \text{ cm/sec}^2)t^2$$

$$t^2 = (d/490 \text{ cm/sec}^2)$$

$$t = \sqrt{(d/(490 \text{ cm/sec}^2)}$$

†An alternative would be to use a reaction time ruler, which converts distance to time (seconds).

Activity 10

Measuring Reaction Time Using BIOPAC®

Setting Up the Equipment

1. Connect the BIOPAC® unit to the computer, and turn the computer **ON**.

2. Make sure the BIOPAC® unit is **OFF**.

3. Plug in the equipment (as shown in **Figure 21.7**).
   - Hand switch—CH1
   - Headphones—back of MP36/35 unit

4. Turn the BIOPAC® unit **ON**.

5. Start the Biopac Student Lab program on the computer by double-clicking the icon on the desktop or by following your instructor's guidance.

6. Select lesson **L11-Reaction Time I** from the menu and click **OK**.

7. Type in a filename that will save this subject's data on the computer hard drive. You may want to use the subject's last name followed by React-1 (for example, SmithReact-1), then click **OK**.

Calibrating the Equipment

1. Seat the subject comfortably so that he or she cannot see the computer screen and keyboard.

2. Put the headphones on the subject, and give the subject the hand switch to hold.

3. Tell the subject that he or she is to push the hand switch button when a "click" is heard.

4. Click **Calibrate**, and then click **OK** when the subject is ready.

5. Observe the recording of the calibration data, which should look like the waveforms in the calibration example (**Figure 21.8**).

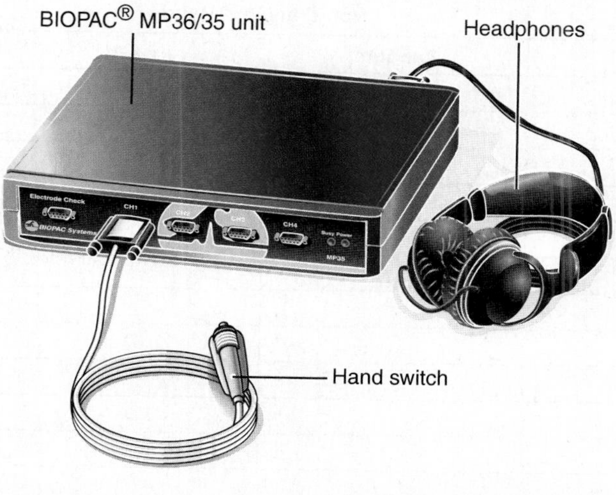

Figure 21.7 Setting up the BIOPAC® equipment. Plug the headphones into the back of the MP36/35 data acquisition unit and the hand switch into Channel 1. Hand switch and headphones are shown connected to the MP36/35 unit.

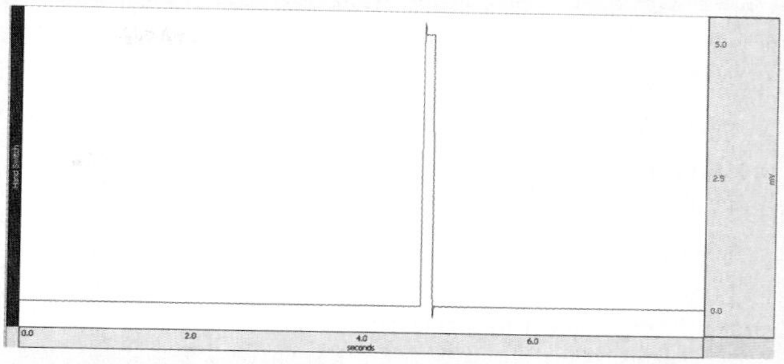

Figure 21.8 Example of waveforms during the calibration procedure.

- If the data look very different, click **Redo Calibration** and repeat the steps above.
- If the data look similar, click **Continue** to proceed to the next section.

Recording the Data

In this experiment, you will record four different segments of data. In Recordings 1 and 3, the subject will respond to random click stimuli. In Recordings 2 and 4, the subject will respond to click stimuli at fixed intervals (about 4 seconds). The director will click **Record** to initiate Recording 1, and **Continue** to initiate Recordings 2, 3, and 4. The subject should focus only on responding to the sound.

Recording 1: Random Interval - dominant hand

1. Each time a sound is heard, the subject should respond by pressing the button on the hand switch with their dominant hand as quickly as possible.

2. When the subject is ready, the director should click **Record** to begin the stimulus-response sequence. The recording will stop automatically after 10 clicks.
   - A triangular marker will be inserted above the data each time a "click" stimulus occurs.
   - An upward-pointing "pulse" will be inserted each time the subject responds to the stimulus.

3. Observe the recording of the data, which should look similar to the data-recording example (**Figure 21.9**).

Text continues on next page. →

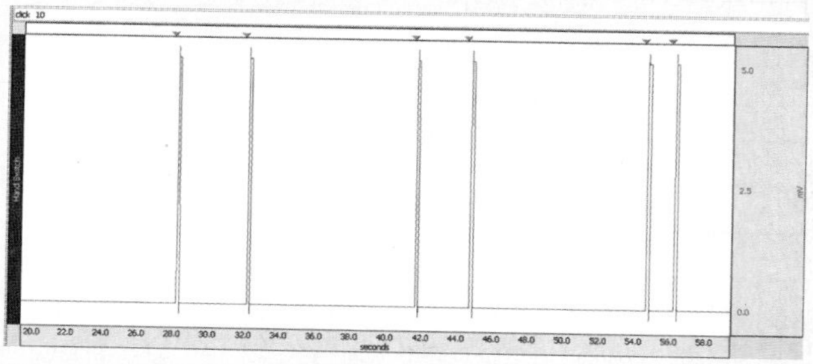

Figure 21.9 Example of waveforms during the recording of data.

- If the data look very different, click **Redo** and repeat the steps above.
- If the data look similar, click **Continue** to move on to the next recording.

Recording 2: Fixed Interval - dominant hand

1. Each time a sound is heard, the subject should respond by pressing the button on the hand switch with their dominant hand as quickly as possible.

2. When the subject is ready, the director should click **Record** to begin the stimulus-response sequence. The recording will stop automatically after 10 clicks.

3. Observe the recording of the data, which should again look similar to the data-recording example (Figure 21.9).

- If the data look very different, click **Redo** and repeat the steps above.
- If the data look similar, click **Continue** to move on to the next recording.

Recording 3: Random Interval - non-dominant hand

1. Repeat the steps for Recording 1 with the non-dominant hand.

Recording 4: Fixed Interval - non-dominant hand

1. Repeat the steps for Recording 2 with the non-dominant hand.

2. If the data after this final recording are fine, click **Done.** A pop-up window will appear; to record from another subject select **Record from another subject**, and return to step 7 under Setting Up the Equipment. If continuing to the Data Analysis section, select **Analyze current data file** and proceed to step 2 in the Data Analysis section.

21

Data Analysis

1. If you are just starting the BIOPAC® program to perform data analysis, enter **Review Saved Data** mode and choose the file with the subject's reaction data (for example, SmithReact-1).

2. Observe that all 10 reaction times are automatically calculated for each segment and are placed in the journal at the bottom of the computer screen.

3. Write the 10 reaction times for each segment in the **Reaction Times chart**.

4. Delete the highest and lowest values of each segment, then calculate and record the average for the remaining eight data points.

5. When you are finished, exit the program by going to the **File** menu at the top of the page and clicking **Quit**.

Do you observe a significant difference between the average response times of Recording 1 and Recording 3? If so, what might account for the difference, even though they are both random trials?

____

____

Likewise, do you observe a significant difference between the average response times of Recording 2 and Recording 4? If so, what might account for the difference, even though they are both fixed interval trials?

____

____

Reaction Times (seconds)

	Dominant Hand		**Non-dominant Hand**	
Stimulus #	**Random**	**Fixed**	**Random**	**Fixed**
1				
2				
3				
4				
5				
6				
7				
8				
9				
10				
Average				

Instructors may assign a portion of the Review Sheet questions using Mastering A&P™

21 REVIEW SHEET
EXERCISE

Human Reflex Physiology

Name ______________________ Lab Time/Date ______________________

The Reflex Arc

1. Label the five components of a reflex arc using the leader lines in the figure below.

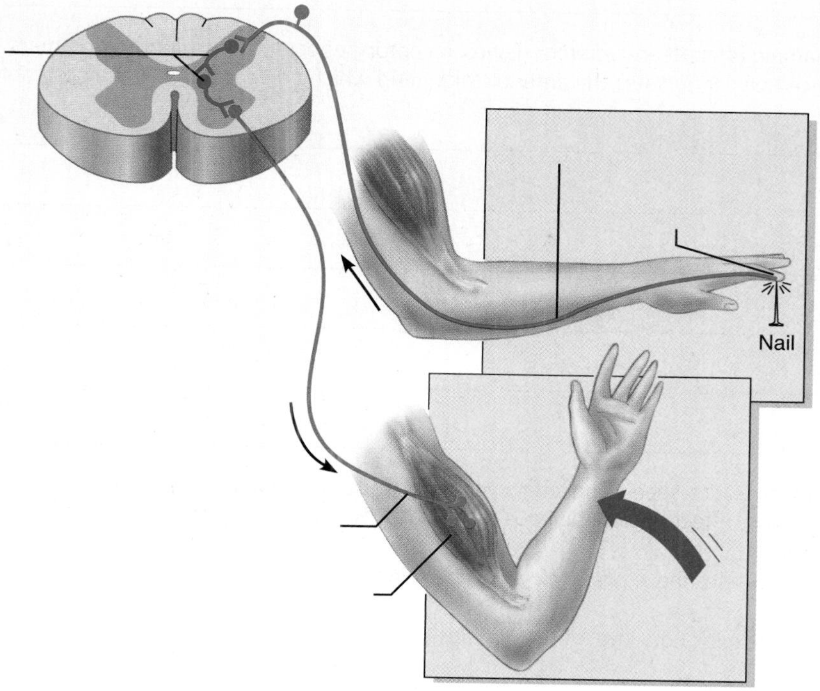

2. In general, what is the importance of reflex testing in a routine physical examination? ______________________

____

____

Somatic and Autonomic Reflexes

3. Use the key terms to complete the statements given below. (Some terms are used more than once.)

Key:
a. abdominal reflex
b. calcaneal tendon reflex
c. ciliospinal reflex
d. corneal reflex
e. crossed-extensor reflex
f. gag reflex
g. patellar reflex
h. plantar reflex
i. pupillary light reflex

Reflexes classified as somatic reflexes include a ________, ________, ________, ________, ________, ________, and ________.

Of these, the stretch reflexes are ______________ and ______________, and the superficial reflexes are ______________ and ______________.

Reflexes classified as autonomic reflexes include ________ and ________.

4. Name three somatic spinal reflexes that are mediated at the spinal cord level. ______

______

5. Name three somatic spinal reflexes in which the higher brain centers participate. ______

______

6. Can the stretch reflex be elicited in a singly pithed animal (that is, an animal in which the brain has been destroyed but the spinal cord is still intact.)? ______

Explain your answer. ______

______

7. Trace the reflex arc, naming efferent and afferent nerves, receptors, effectors, and integration centers, for the two reflexes listed. (Hint: Remember which nerve innervates the anterior thigh, and which nerve innervates the posterior thigh.)

patellar reflex: ______

______

______

calcaneal tendon reflex: ______

______

______

8. Three factors that influence the speed and effectiveness of reflex arcs were investigated in conjunction with patellar reflex testing—mental distraction, effect of simultaneous muscle activity in another body area, and fatigue.

Which of these factors increases the excitatory level of the spinal cord? ______

Which factor decreases the excitatory level of the muscles? ______

When the subject was concentrating on an arithmetic problem, did the change noted in the patellar reflex indicate that brain activity is necessary for the patellar reflex or only that it may modify it? ______

______

9. Name the division of the autonomic nervous system responsible for each of the reflexes listed.

ciliospinal reflex: ______ salivary reflex: ______

pupillary light reflex: ______

10. The pupillary light reflex, the crossed-extensor reflex, and the corneal reflex illustrate the purposeful nature of reflex activity. Describe the protective aspect of each.

pupillary light reflex: ______

corneal reflex: ______

crossed-extensor reflex: ____________________

____

____

____

11. Was the pupillary consensual response contralateral or ipsilateral? ____________________

Why would such a response be of significant value in this particular reflex? ____________________

____

12. Differentiate between the types of activities accomplished by somatic and autonomic reflexes.

____

____

____

13. Several types of reflex activity were not investigated in this exercise. The most important of these are autonomic reflexes, which are difficult to illustrate in a laboratory situation. To rectify this omission, complete the following chart, using references as necessary.

Reflex	Organ involved	Receptors stimulated	Action
Micturition (urination)			
Defecation			
Carotid sinus			

Reaction Time of Intrinsic and Learned Reflexes

14. How do intrinsic and learned reflexes differ? ____________________

____

15. Name at least three factors that may modify reaction time to a stimulus. ____________________

____

16. In general, how did the response time for the learned activity performed in the laboratory compare to that for the simple patellar reflex? ______

17. Did the response time without verbal stimuli decrease with practice? ______ Explain the reason for this.

______

18. Explain, in detail, why response time increased when the subject had to react to a word stimulus.

______

______

19. When you were measuring reaction time in the BIOPAC® activity, was there a difference in reaction time when the stimulus was predictable versus unpredictable? Explain your answer.

______

______

20. ✚ Hyporeflexia occurs when normal reflexes are weak but not absent. Explain how this condition could be due to damage to skeletal muscle, a sensory neuron, or a motor neuron. ______

______

______

21. ✚ With pathological tetanus (also known as "lockjaw"), the prime mover and antagonistic muscle groups contract at the same time because of the effect of the tetanus neurotoxin. Describe the process that the tetanus toxin affects and the role that this process plays in the stretch reflexes. ______

______

______

EXERCISE 22

General Sensation

Learning Outcomes

- List the stimuli that activate general sensory receptors.
- Define *exteroceptor*, *interoceptor*, and *proprioceptor*.
- Recognize and describe the various types of general sensory receptors as studied in the laboratory, and list the function and locations of each.
- Explain the tactile two-point discrimination test, and state its anatomical basis.
- Define *tactile localization*, and describe how this ability varies in different areas of the body.
- Define *adaptation*, and describe how this phenomenon can be demonstrated.
- Discuss *negative afterimages* as they are related to temperature receptors.
- Define *referred pain*, and give an example of it.

Go to Mastering A&P™ > Study Area to improve your performance in A&P Lab.

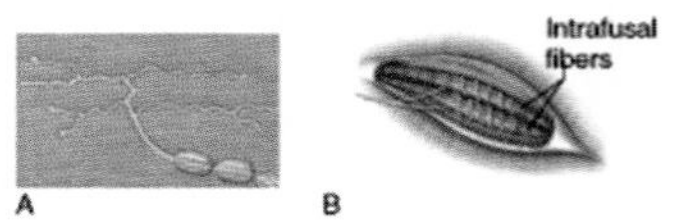

Instructors may assign new Building Vocabulary coaching activities, Pre-Lab Quiz questions, Art Labeling activities, Practice Anatomy Lab Practical questions (PAL), and more using the Mastering **A&P™** Item Library.

Pre-Lab Quiz

Instructors may assign these and other Pre-Lab Quiz questions using Mastering **A&P™**

1. Sensory receptors can be classified according to their source of stimulus. __________ are found close to the body surface and react to stimuli in the external environment.
   a. Exteroceptors
   b. Interoceptors
   c. Proprioceptors
   d. Visceroceptors
2. Lamellar corpuscles respond to:
   a. deep pressure and vibrations
   b. light touch
   c. pain and temperature
3. Circle True or False. A map of the sensory receptors for touch, heat, cold, and pain shows that they are not evenly distributed throughout the body.
4. When a stimulus is applied for a prolonged period, the rate of receptor discharge slows, and conscious awareness of the stimulus declines. This phenomenon is known as:
   a. accommodation
   b. adaptation
   c. adjustment
   d. discernment
5. You will test referred pain in this activity by immersing the subject's:
   a. face in ice water to test the cranial nerve response
   b. elbow in ice water to test the ulnar nerve response
   c. hand in ice water to test the axillary nerve response
   d. leg in ice water to test the sciatic nerve response

Materials

- Compound microscope
- Immersion oil
- Prepared slides (longitudinal sections) of lamellar corpuscles, tactile corpuscles, tendon organs, and muscle spindles
- Calipers or esthesiometer
- Small metric rulers
- Fine-point, felt-tipped markers (black, red, and blue)
- Large beaker of ice water; chipped ice
- Hot water bath set at 45°C; laboratory thermometer
- Towel
- Four coins (nickels or quarters)
- Three large finger bowls or 1000-ml beakers

People are very responsive to *stimuli*, which are changes within a person's environment. Hold a freshly baked apple pie before them, and their mouths water. Tickle them, and they giggle.

The body's **sensory receptors** respond to stimuli. The tiny sensory receptors of the **general senses** are activated by touch, pressure, pain, heat, cold, stretch, vibration, and changes in body position. In contrast to these widely distributed *general sensory receptors*, the receptors of the special senses are large, complex *sense organs*. The **special senses** include vision, hearing, equilibrium, smell, and taste. Only the anatomically simpler **general sensory receptors** will be studied in this exercise. Sensory receptors may be classified by the location of the stimulus.

- **Exteroceptors** are sensitive to stimuli in the external environment, and typically they are found close to the body surface. Exteroceptors include the simple cutaneous receptors in the skin (**Figure 22.1**) and the highly specialized receptors of the special senses.

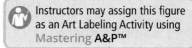

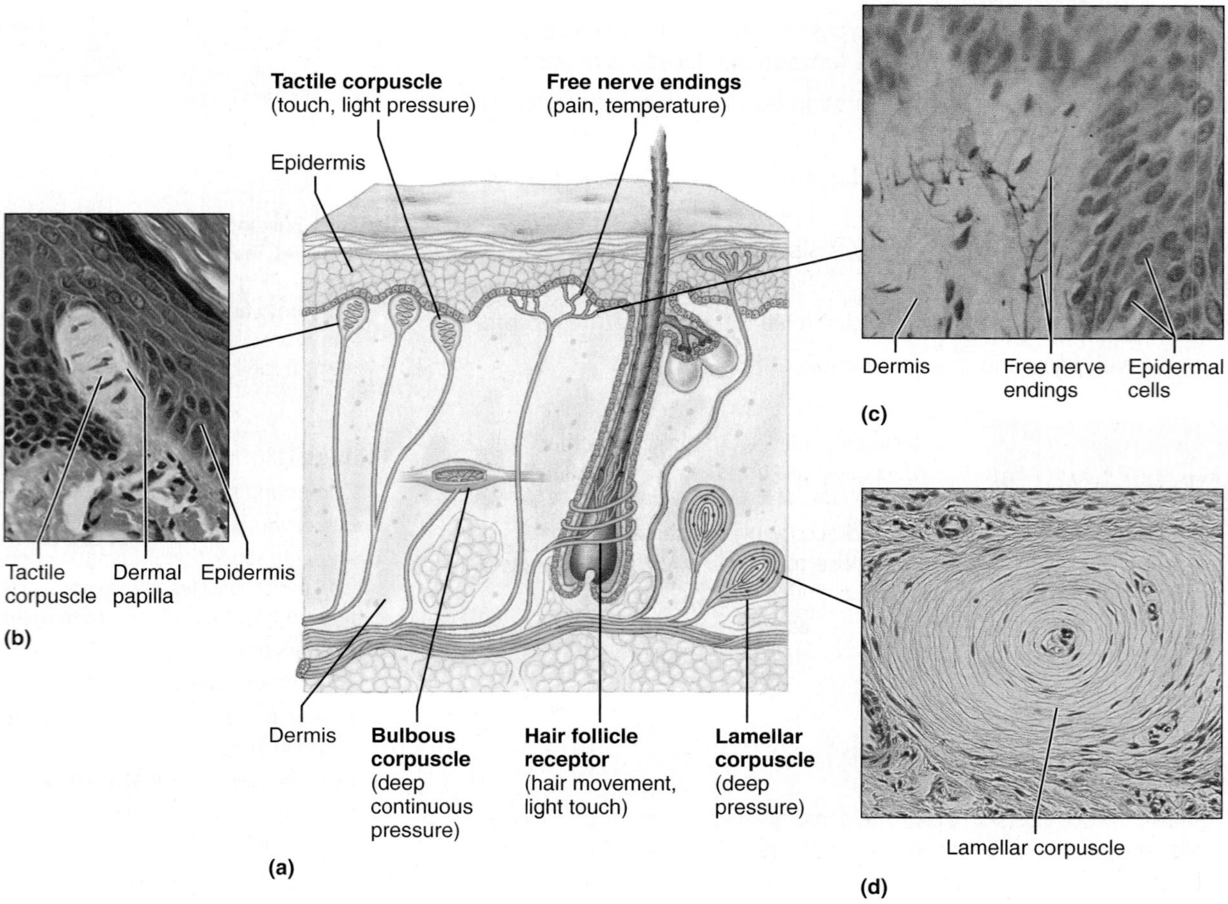

Figure 22.1 Examples of cutaneous receptors. Drawing **(a)** and photomicrographs **(b–d)**. **(a)** Free nerve endings, hair follicle receptor, tactile corpuscles, lamellar corpuscles, and bulbous corpuscle. Epithelial tactile complexes are not illustrated. **(b)** Tactile corpuscle in a dermal papilla (400×). **(c)** Free nerve endings at dermal-epidermal junction (330×). **(d)** Cross section of a lamellar corpuscle in the dermis (220×).

Instructors may assign this figure as an Art Labeling Activity using Mastering **A&P™**

- **Interoceptors**, or *visceroceptors*, respond to stimuli arising within the body. Interoceptors are found in the internal visceral organs and include stretch receptors (in walls of hollow organs), chemoreceptors, and others.
- **Proprioceptors**, like interoceptors, respond to internal stimuli but are restricted to skeletal muscles, tendons, joints, ligaments, and connective tissue coverings of bones and muscles. They provide information about body movements and position by monitoring the amount of stretch of those structures.

Structure of General Sensory Receptors

Anatomically, general sensory receptors are nerve endings that are either nonencapsulated or encapsulated. General sensory receptors are summarized in **Table 22.1** on p. 346.

Activity 1

Studying the Structure of Selected Sensory Receptors

1. Obtain a compound microscope and microscope slides of lamellar and tactile corpuscles. Locate, under low power, a tactile corpuscle in the dermal layer of the skin. As mentioned above, these are usually found in the dermal papillae. Then switch to the oil immersion lens for a detailed study. Notice that the free nerve fibers within the corpuscle are aligned parallel to the skin surface. Compare your observations to the photomicrograph of a tactile corpuscle (Figure 22.1b).

2. Next observe a lamellar corpuscle located much deeper in the dermis. Try to identify the slender naked nerve ending in the center of the receptor and the many layers of connective tissue surrounding it (which looks rather like an onion cut lengthwise). Also, notice how much larger the lamellar corpuscles are than the tactile corpuscles. Compare your observations to the photomicrograph of a lamellar corpuscle (Figure 22.1d).

3. Obtain slides of muscle spindles and tendon organs, the two major types of proprioceptors (**Figure 22.2**). In the slide of **muscle spindles**, note that minute extensions of the nerve endings of the sensory neurons coil around specialized slender skeletal muscle cells called **intrafusal fibers**. The **tendon organs** are composed of nerve endings that ramify through the tendon tissue close to the attachment between muscle and tendon. Stretching of muscles or tendons excites these receptors, which then transmit impulses that ultimately reach the cerebellum for interpretation. Compare your observations to Figure 22.2.

22

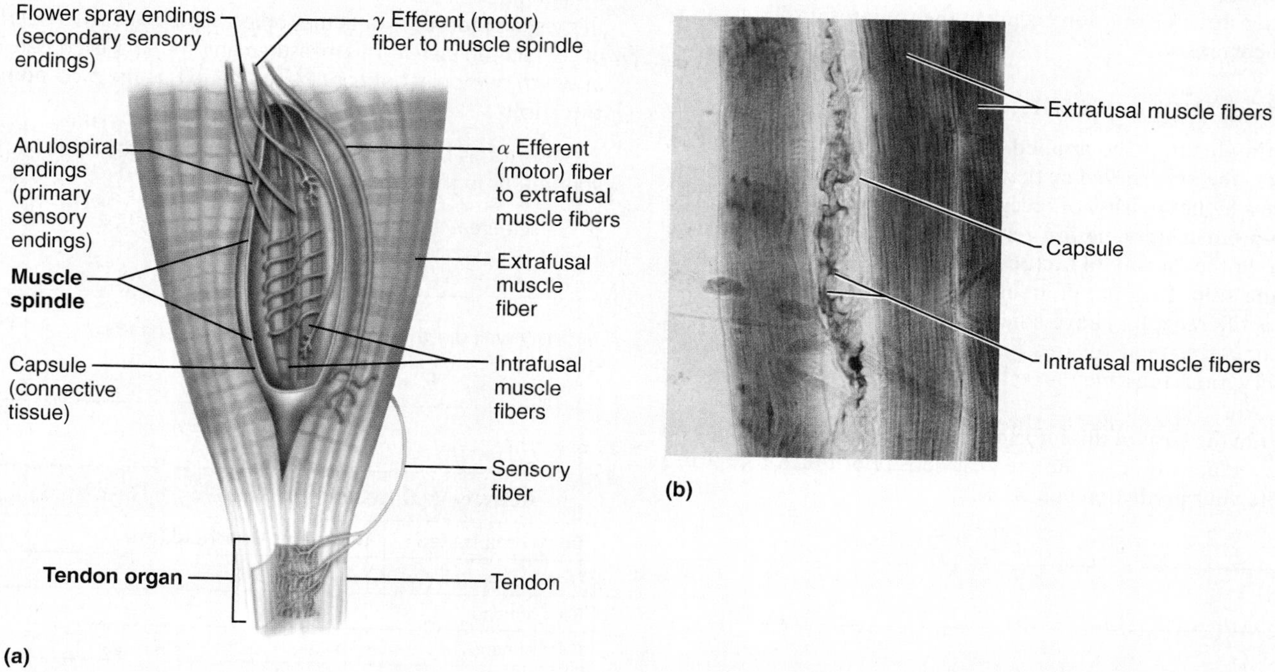

Figure 22.2 Proprioceptors. (a) Diagram of a muscle spindle and tendon organ. Myelin has been omitted from all nerve fibers for clarity. **(b)** Photomicrograph of a muscle spindle (80×).

Instructors may assign this figure as an Art Labeling Activity using Mastering A&P™

Table 22.1 Receptors of the General Senses (Figures 22.1 and 22.2)

Structural class	Body location(s)	Stimulus type
Nonencapsulated		
Free nerve endings	Most body tissues; especially the epithelia and connective tissues	Primarily pain, heat, and cold
Epithelial tactile complexes	Stratum basale of the epidermis	Light pressure
Hair follicle receptors	Wrapped around hair follicles	Bending of hairs
Encapsulated		
Tactile (Meissner's) corpuscles	Dermal papillae of hairless skin	Light pressure, discriminative touch
Bulbous corpuscles	Deep in the dermis, subcutaneous tissue, and joint capsules	Deep pressure and stretch
Lamellar corpuscles	Dermis, subcutaneous tissue, periosteum, tendons, and joint capsules	Deep pressure, stretch, and vibration
Muscle spindles	Skeletal muscles	Muscle stretch (proprioception)
Tendon organs	Tendons	Tendon stretch, tension (proprioception)

Receptor Physiology

Transduction is the process by which environmental *stimuli* change into nerve impulses that are relayed to the CNS. *Sensation* (awareness of the stimulus) and *perception* (conscious interpretation of the stimulus) occur in the brain. Nerve impulses from cutaneous receptors are relayed to the primary somatosensory cortex, where stimuli from different body regions form a body map. Therefore, each location on the body is represented by a specific cortical area. It is this cortical organization that allows us to know exactly where a sensation comes from on the body. Further interpretation of the sensory information occurs in the somatosensory association cortex.

22

Two-Point Discrimination Test

A stimulus must be applied to a sensory neuron's **receptive field**—the area served by that neuron. Some areas in the body have a higher density of receptors in their receptive field. The **two-point discrimination test** is used to determine a crude map of the density of tactile receptors in the various regions of the skin. In general, areas that have the greatest density of tactile receptors have a heightened ability to "feel." These areas correspond to areas that receive the greatest motor innervation; thus they are also typically areas of fine motor control.

On the basis of this information, which areas of the body do you *predict* will have the greatest density of touch receptors? Write your prediction below.

____

____

Activity 2

Determining the Two-Point Threshold

1. Using calipers or an esthesiometer and a metric ruler, test the ability of the subject to differentiate two distinct sensations when the skin is touched simultaneously at two points. Beginning with the face, start with the caliper arms completely together. Gradually increase the distance between the points, testing the subject's skin after each adjustment. Continue with this testing procedure until the subject reports that *two points* of contact can be felt. This measurement, the smallest distance at which two points of contact can be felt, is the **two-point threshold**.

2. Repeat this procedure on the body areas listed and record your results in the **Activity 2 chart**.

3. Which area has the smallest two-point threshold?

____

4. How well did this compare with your prediction?

____

Activity 2: Determining Two-Point Threshold

Body area tested	Two-point threshold (mm)
Face	
Back of hand	
Palm of hand	
Fingertip	
Lips	
Back of neck	
Anterior forearm	

Tactile Localization

Tactile localization is the brain's ability to determine which portion of the skin has been touched. The receptive field of the body periphery has a corresponding "touch" field in the brain's primary somatosensory cortex. Some body areas are well represented with touch receptors, allowing tactile stimuli to be localized with great accuracy, but in other body areas, touch-receptor density allows only crude discrimination. In general, the smaller the receptive field of the sensory neurons serving the area, the greater the brain's ability to detect the location of the stimulus.

Activity 3

Testing Tactile Localization

1. The subject's eyes should be closed during the testing. The experimenter touches the palm of the subject's hand with a pointed black felt-tipped marker. The subject should then try to touch the exact point with his or her own marker, which should be of a different color. Measure the error of localization (the distance between the two marks) in millimeters.

2. Repeat the test in the same spot twice more, recording the error of localization for each test. Average the results of the three determinations, and record it in the **Activity 3 chart**.

Activity 3: Testing Tactile Localization	
Body area tested	**Average error (mm)**
Palm of hand	
Fingertip	
Anterior forearm	
Back of hand	
Back of neck	

3. Repeat the procedure on the body areas listed and record the averaged results in the chart above.

4. Which area has the smallest error of localization?

___

5. Which body area tested has the smallest receptive field?

___

Adaptation of Sensory Receptors

The number of impulses transmitted by sensory receptors often changes both with the intensity of the stimulus and with the length of time the stimulus is applied. In many cases, when a stimulus is applied for a prolonged period without movement, the rate of receptor discharge slows, and conscious awareness of the stimulus declines or is lost until some type of stimulus change occurs. This phenomenon is referred to as **adaptation**. The touch receptors adapt particularly rapidly, which is highly desirable. Who, for instance, would want to be continually aware of the pressure of clothing on their skin?

Activity 4

Demonstrating Adaptation of Touch Receptors

1. The subject's eyes should be closed. Obtain four coins. Place one coin on the anterior surface of the subject's forearm, and determine how long the sensation persists for the subject. Duration of the sensation:

___ sec

2. Repeat the test, placing the coin at a different forearm location. How long does the sensation persist at the second location?

___ sec

3. After awareness of the sensation has been lost at the second site, stack three more coins atop the first one.

Does the pressure sensation return? ___

If so, for how long is the subject aware of the pressure in this instance?

___ sec

Are the same receptors being stimulated when the four coins, rather than the one coin, are used? ___

Explain how perception of the stimulus intensity has changed. ___

___

Activity 5

Demonstrating Adaptation of Temperature Receptors

Adaptation of temperature receptors also can be tested.

1. Obtain three large finger bowls or 1000-ml beakers and fill the first with 45°C water. Have the subject immerse her or his left hand in the water and report the sensation. Keep the left hand immersed for 1 minute, and then also immerse the right hand in the same bowl.

What is the sensation of the left hand when it is first immersed?

___

What is the sensation of the left hand after 1 minute as compared to the sensation in the right hand just immersed?

___

Text continues on next page. ➔

Had adaptation occurred in the left hand? ____________

2. Rinse both hands in tap water, dry them, and wait 5 minutes before conducting the next test. Just before beginning the test, refill the finger bowl with fresh 45°C water, fill a second with ice water, and fill a third with water at room temperature.

3. Place the *left* hand in the ice water and the *right* hand in the 45°C water. What is the sensation in each hand after 2 minutes as compared to the sensation perceived when the hands were first immersed?

____

____

Which hand seemed to adapt more quickly?

____

4. After reporting these observations, the subject should then place both hands simultaneously into the finger bowl containing the water at room temperature. Record the sensation in the

left hand: ____________

The right hand: ____________

Referred Pain

Pain receptors are densely distributed in the skin, and they adapt very little, if at all. This lack of adaptability is due to the protective function of the receptors. The sensation of pain often indicates tissue damage or trauma to body structures. Thus no attempt will be made in this exercise to localize the pain receptors or to prove their nonadaptability, since both would cause needless discomfort.

However, the phenomenon of referred pain is easily demonstrated in the laboratory, and such experiments provide information that may be useful in explaining common examples of this phenomenon. **Referred pain** is a sensory experience in which pain is perceived as arising in one area of the body when in fact another, often quite remote area, is receiving the painful stimulus. Thus the pain is said to be "referred" to a different area.

22

Activity 6

Demonstrating the Phenomenon of Referred Pain

Immerse the subject's elbow in a finger bowl containing ice water. In the **Activity 6 chart**, record the quality of the sensation (such as discomfort, tingling, or pain) and the localization of the sensations he or she reports for the intervals indicated. The elbow should be removed from ice water after the 2-minute reading. The last recording is to occur 3 minutes after removal of the subject's elbow from the ice water.

The ulnar nerve, which serves the medial third of the hand, is involved in the phenomenon of referred pain experienced during this test. How does the localization of this referred pain correspond to the areas served by the ulnar nerve?

____

Activity 6: Demonstrating Referred Pain

Time of observation	Quality of sensation	Localization of sensation
On immersion		
After 1 min		
After 2 min		
3 min after removal		

Instructors may assign a portion of the Review Sheet questions using Mastering A&P™

22 EXERCISE REVIEW SHEET General Sensation

Name ______________________________ Lab Time/Date ______________________

Structure of General Sensory Receptors

1. Differentiate between interoceptors and exteroceptors relative to location and stimulus source.

interoceptor: ____

exteroceptor: ____

2. Label the cutaneous receptors indicated by leader lines on the diagram.

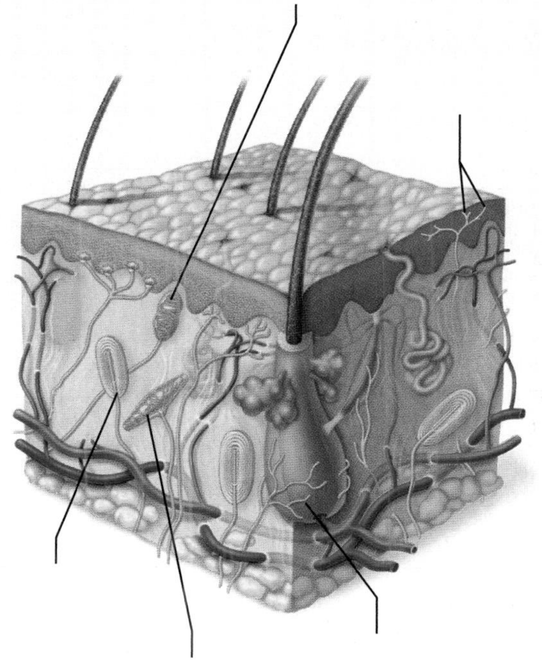

3. Match the general sensory receptors with their descriptions.

______________ 1. muscle spindle

______________ 2. tactile corpuscle

______________ 3. epithelial tactile complex

______________ 4. free nerve endings

______________ 5. lamellar corpuscles

a. touch receptor located in the deepest layer of the epidermis

b. deep pressure receptor located in the dermis

c. pain receptors

d. touch receptor located in the dermal papillae

e. proprioceptor located in skeletal muscle

Receptor Physiology

4. Define *receptive field*. ____

5. Explain how the two-point discrimination test illustrates how the density of touch receptors in a receptive field varies in different areas of the body. ______________________________

____

How well did your results correspond to your predictions? ______________________________

____

6. Explain how the error of localization correlates to the size of the receptive field as demonstrated by the tactile localization experiment. ______________________________

____

7. Several questions regarding general sensation are posed below. Answer each by placing your response in the appropriately numbered blanks to the right.

1. Which cutaneous receptors are the most numerous?	1. ______________
2–3. Which two body areas tested were most sensitive to touch?	2–3. ______________
4–5. Which two body areas tested were least sensitive to touch?	4–5. ______________
	6. ______________
6–8. Where would referred pain appear if the following organs were receiving painful stimuli: (6) gallbladder, (7) kidneys, and (8) appendix? (Use your textbook if necessary.)	7. ______________
	8. ______________
9. Where was referred pain felt when the elbow was immersed in ice water during the laboratory experiment?	9. ______________
10. What region of the cerebrum interprets the kind and intensity of stimuli that cause cutaneous sensations?	10. ______________

8. Define *adaptation of sensory receptors.* ______________________________

____

9. Why is it advantageous to have pain receptors that are sensitive to all vigorous stimuli, whether heat, cold, or pressure?

____

Why is the nonadaptability of pain receptors important? ______________________________

____

10. ✚ Patients with Hansen's disease (also known as leprosy) initially experience loss of cutaneous sensation in their fingertips and toes. Explain the consequences of this loss of sensation, and identify the receptors that become deficient with this disease.

____

____

11. ✚ An individual arrives at the emergency room complaining of pain in the left shoulder and upper extremity. The patient is whisked away into the first available room. Explain why these symptoms could be life-threatening. ______________________________

____

____

EXERCISE 23

Special Senses: Anatomy of the Visual System

Learning Outcomes

- Identify the anatomy of the eye and its accessory anatomical structures on a model or appropriate image, and list the function(s) of each; identify the structural components that are present in a preserved sheep or cow eye (if available).
- Define *conjunctivitis*, *cataract*, and *glaucoma*.
- Describe the cellular makeup of the retina.
- Explain the difference between rods and cones with respect to visual perception and retinal localization.
- Trace the visual pathway to the primary visual cortex, and indicate the effects of damage to various parts of this pathway.

Go to Mastering A&P™ > Study Area to improve your performance in A&P Lab.

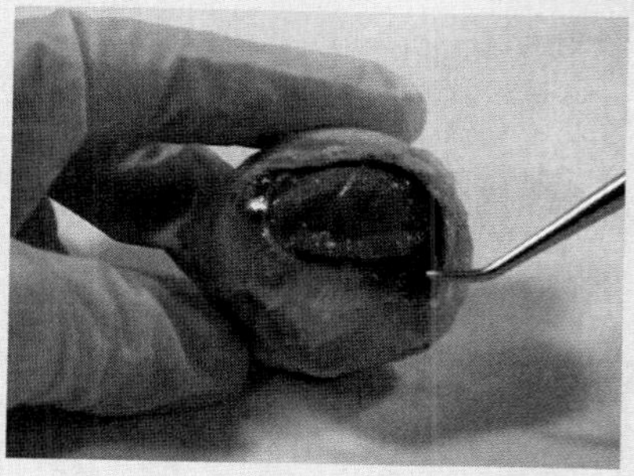

> Lab Tools > Bone and Dissection Videos > Cow Eye

Instructors may assign new Building Vocabulary coaching activities, Pre-Lab Quiz questions, Art Labeling activities, related videos and coaching activities for the Cow Eye dissection, Practice Anatomy Lab Practical questions (PAL), and more using the Mastering **A&P™** Item Library.

Materials

- Chart of eye anatomy
- Dissectible eye model
- Prepared slide of longitudinal section of an eye showing retinal layers
- Compound microscope
- Preserved cow or sheep eye
- Dissecting instruments and tray
- Disposable gloves

Pre-Lab Quiz

Instructors may assign these and other Pre-Lab Quiz questions using Mastering **A&P™**

1. Name the mucous membrane that lines the internal surface of the eyelids and continues over the anterior surface of the eyeball.

   ____

2. How many extrinsic eye muscles are attached to the exterior surface of each eyeball?
   a. three **c.** five
   b. four **d.** six
3. The wall of the eye has three layers. The outermost fibrous layer is made up of the opaque white sclera and the transparent:
   a. choroid **c.** cornea
   b. ciliary gland **d.** lacrima
4. Circle the correct underlined term. The aqueous humor / vitreous humor is a clear, watery fluid that helps to maintain the intraocular pressure of the eye and provides nutrients for the avascular lens and cornea.
5. Circle True or False. At the optic chiasma, the fibers from the medial side of each eye cross over to the opposite side.

Anatomy of the Eye

Accessory Structures

The adult human eye is a sphere measuring about 2.5 cm (1 inch) in diameter. Only about one-sixth of the eye's anterior surface is observable (**Figure 23.1**, p. 352); the remainder is enclosed and protected by a cushion of fat and the walls of the bony orbit.

The accessory structures of the eye include the eyebrows, eyelids, conjunctivae, lacrimal apparatus, and extrinsic eye muscles (**Table 23.1**, p. 352, and **Figure 23.2**, p. 353).

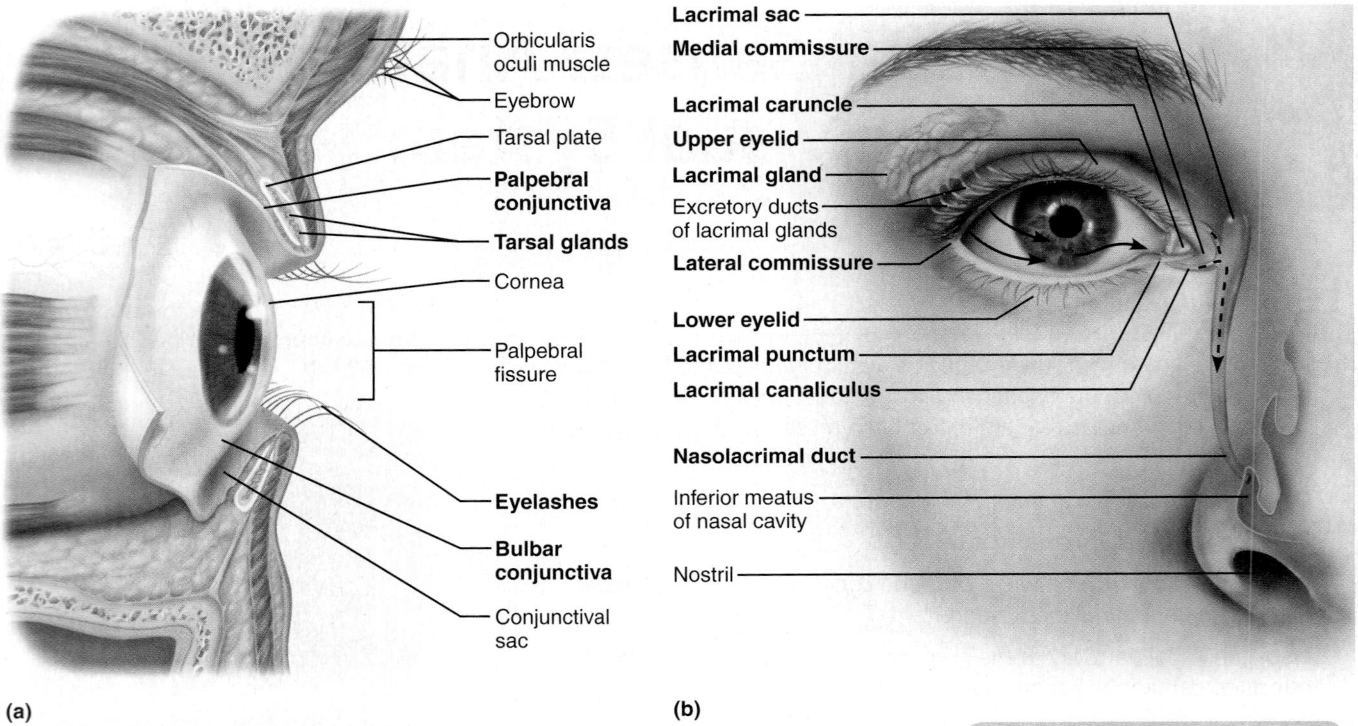

Figure 23.1 The eye and accessory structures. (a) Lateral view. **(b)** Anterior view with lacrimal apparatus. Arrows indicate the direction of the flow of lacrimal fluid.

Instructors may assign this figure as an Art Labeling Activity using Mastering **A&P™**

23

Table 23.1 Accessory Structures of the Eye (Figures 23.1 and 23.2)

Structure	Description	Function
Eyebrows	Short hairs located on the supraorbital margins	Shade and prevent sweat from entering the eyes.
Eyelids (palpebrae)	Skin-covered upper and lower lids, with eyelashes projecting from their free margin	Protect the eyes and spread lacrimal fluid (tears) with blinking.
Tarsal glands	Modified sebaceous glands embedded in the tarsal plate of the eyelid	Secrete an oily secretion that lubricates the surface of the eye.
Ciliary glands	Typical sebaceous and modified sweat glands that lie between the eyelash follicles	Secrete an oily secretion that lubricates the surface of the eye and the eyelashes. An infection of a ciliary gland is called a **sty**.
Conjunctivae	A clear mucous membrane that lines the eyelids (palpebral conjunctivae) and lines the anterior white of the eye (bulbar conjunctiva)	Secrete mucus to lubricate the eye. Inflammation of the conjunctiva results in conjunctivitis, (commonly called "pinkeye").
Medial and lateral commissures	Junctions where the eyelids meet medially and laterally	Form the corners of the eyes. The medial commissure contains the lacrimal caruncle.
Lacrimal caruncle	Fleshy reddish elevation that contains sebaceous and sweat glands	Secretes a whitish oily secretion for lubrication of the eye (can dry and form "eye sand").
Lacrimal apparatus	Includes the lacrimal gland and a series of ducts that drain the lacrimal fluid into the nasal cavity	Protects the eye by keeping it moist. Blinking spreads the lacrimal fluid.
Lacrimal gland	Located in the superior and lateral aspect of the orbit of the eye	Secretes lacrimal fluid, which contains mucus, antibodies, and lysozyme.
Lacrimal puncta	Two tiny openings on the medial margin of each eyelid	Allow lacrimal fluid to drain into the superior and inferiorly located lacrimal canaliculi.
Lacrimal canaliculi	Two tiny canals that are located in the eyelids	Allow lacrimal fluid to drain into the lacrimal sac.
Lacrimal sac	A single pouch located in the medial orbital wall	Allows lacrimal fluid to drain into the nasolacrimal duct.
Nasolacrimal duct	A single tube that empties into the nasal cavity	Allows lacrimal fluid to flow into the nasal cavity.
Extrinsic eye muscles	Six muscles for each eye; four recti and two oblique muscles (see Figure 23.2)	Control the movement of each eyeball and hold the eyes in the orbits.

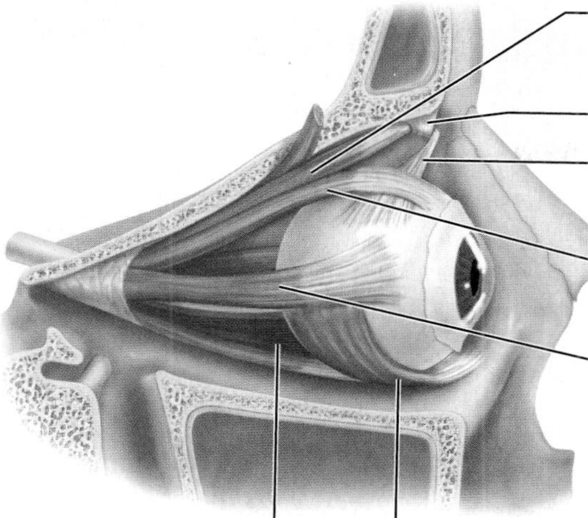

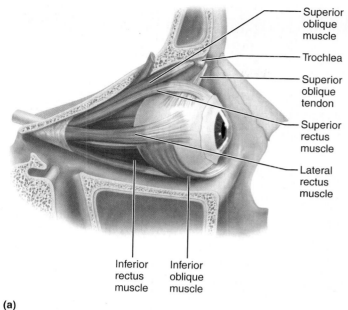

(a)

Muscle	Action
Lateral rectus	Moves eye laterally
Medial rectus	Moves eye medially
Superior rectus	Elevates eye and turns it medially
Inferior rectus	Depresses eye and turns it medially
Inferior oblique	Elevates eye and turns it laterally
Superior oblique	Depresses eye and turns it laterally

(b)

Instructors may assign this figure as an Art Labeling Activity using Mastering A&P™

Figure 23.2 Extrinsic muscles of the eye. (a) Lateral view of the right eye. **(b)** Summary of actions of the extrinsic eye muscles.

Activity 1

Identifying Accessory Eye Structures

Using a chart of eye anatomy or Figure 23.1, observe the eyes of another student, and identify as many of the accessory structures as possible. Ask the student to look to the left. Which extrinsic eye muscles are responsible for this action?

Right eye: ______________________________

Left eye: ______________________________

Internal Anatomy of the Eye

Anatomically, the wall of the eye is constructed of three layers: the **fibrous layer**, the **vascular layer**, and the **inner layer** (**Table 23.2**, p. 354, and **Figure 23.3**, p. 355).

Distribution of Photoreceptors

The photoreceptor cells are distributed over most of the neural retina, except where the optic nerve leaves the eyeball. This site is called the **optic disc**, or *blind spot*, and is located in a weak spot in the **fundus** (posterior wall). Lateral to each blind spot, and directly posterior to the lens, is an area called the **macula lutea** ("yellow spot"), an area of high cone density. In its center is the **fovea centralis**, a tiny pit, which contains only cones and is the area of greatest visual acuity. Focusing for detailed color vision occurs in the fovea centralis.

Internal Chambers and Fluids

The lens divides the eye into two segments: the **anterior segment** anterior to the lens, which contains a clear watery fluid called the **aqueous humor**, and the **posterior segment** behind the lens, filled with a gel-like substance, the **vitreous humor**. The anterior segment is further divided into **anterior** and **posterior chambers**, located before and after the iris, respectively. The aqueous humor is continually formed by the capillaries of the **ciliary processes** of the ciliary body. It helps to maintain the intraocular pressure of the eye and provides nutrients for the avascular lens and cornea. The aqueous humor is drained into the **scleral venous sinus**. The vitreous humor provides the major internal reinforcement of the posterior part of the eyeball, and helps to keep the retina pressed firmly against the wall of the eyeball. It is formed *only* before birth and is not renewed. In addition to the cornea, the light-bending media of the eye include the lens, the aqueous humor, and the vitreous humor.

Anything that interferes with drainage of the aqueous fluid increases intraocular pressure. When intraocular pressure reaches dangerously high levels, the retina and optic nerve are compressed, resulting in pain and possible blindness, a condition called **glaucoma**. +

23

Activity 2

Identifying Internal Structures of the Eye

Obtain a dissectible eye model and identify its internal structures described above. (As you work, also refer to Figure 23.3.)

Table 23.2 Layers of the Eye (Figure 23.3)

Structure	Description	Function
Fibrous Layer (External Layer)		
Sclera	Opaque white connective tissue that forms the "white of the eye."	Helps to maintain the shape of the eyeball and provides an attachment point for the extrinsic eye muscles.
Cornea	Structurally continuous with the sclera; modified to form a transparent layer that bulges anteriorly; contains no blood vessels.	Forms a clear window that is the major light bending (refracting) medium of the eye.
Vascular Layer (Middle Layer)		
Choroid	A blood vessel–rich, dark membrane.	The blood vessels nourish the other layers of the eye, and the melanin helps to absorb excess light.
Cilary body	Modification of the choroid that encircles the lens.	Contains the ciliary muscle and the ciliary process.
Ciliary muscle	Smooth muscle found within the ciliary body.	Alters the shape of the lens with contraction and relaxation.
Ciliary process	Radiating folds of the ciliary muscle.	Capillaries of the ciliary process form the aqueous humor by filtering plasma.
Ciliary zonule (Suspensory ligament)	A halo of fine fibers that extends from the ciliary process around the lens.	Attaches the lens to the ciliary process.
Iris	The anterior portion of the vascular layer that is pigmented. It contains two layers of smooth muscle (sphincter pupillae and dilator pupillae).	Controls the amount of light entering the eye by changing the size of the pupil diameter. The sphincter pupillae contract to constrict the pupil. The dilator pupillae contract to dilate the pupil.
Pupil	The round central opening of the iris.	Allows light to enter the eye.
Inner Layer (Retina)		
Pigmented layer of the retina	The outer layer that is composed of only a single layer of pigment cells (melanocytes).	Absorbs light and prevents it from scattering in the eye. Pigment cells act as phagocytes for cleaning up cell debris and also store vitamin A needed for photoreceptor renewal.
Neural layer of the retina	The thicker inner layer composed of three main types of neurons: photoreceptors (rods and cones), bipolar cells, and ganglion cells.	Photoreceptors respond to light and convert the light energy into action potentials that travel to the primary visual cortex of the brain.

Microscopic Anatomy of the Retina

Cells of the retina include the pigment cells of the outer pigmented layer and the neurons of the neural layer (**Figure 23.4**, p. 356). The inner neural layer is composed of three major populations of neurons. These are, from outer to inner aspect, the **photoreceptors**, the **bipolar cells**, and the **ganglion cells**.

The **rods** are the specialized photoreceptors for dim light. Visual interpretation of their activity is in gray tones. The **cones** are color photoreceptors that permit high levels of visual acuity, but they function only under conditions of high light intensity; thus, for example, no color vision is possible in moonlight. The fovea centralis contains only cones, the macula lutea contains mostly cones, and from the edge of the macula to the retina periphery, cone density declines gradually. By contrast, rods are most numerous in the periphery, and their density decreases as the macula is approached.

Light must pass through the ganglion cell layer and the bipolar cell layer to reach and excite the rods and cones. As a result of a light stimulus, the photoreceptors undergo changes in their membrane potential that influence the bipolar cells. These in turn stimulate the ganglion cells, whose axons leave the retina in the tight bundle of fibers known as the **optic nerve** (Figure 23.3). In addition to these three major types of neurons, the retina also contains horizontal cells and amacrine cells (both are interneurons), which play a role in visual processing.

Activity 3

Studying the Microscopic Anatomy of the Retina

Use a compound microscope to examine a histologic slide of a longitudinal section of the eye. Identify the retinal layers by comparing your view to the photomicrograph (Figure 23.4b).

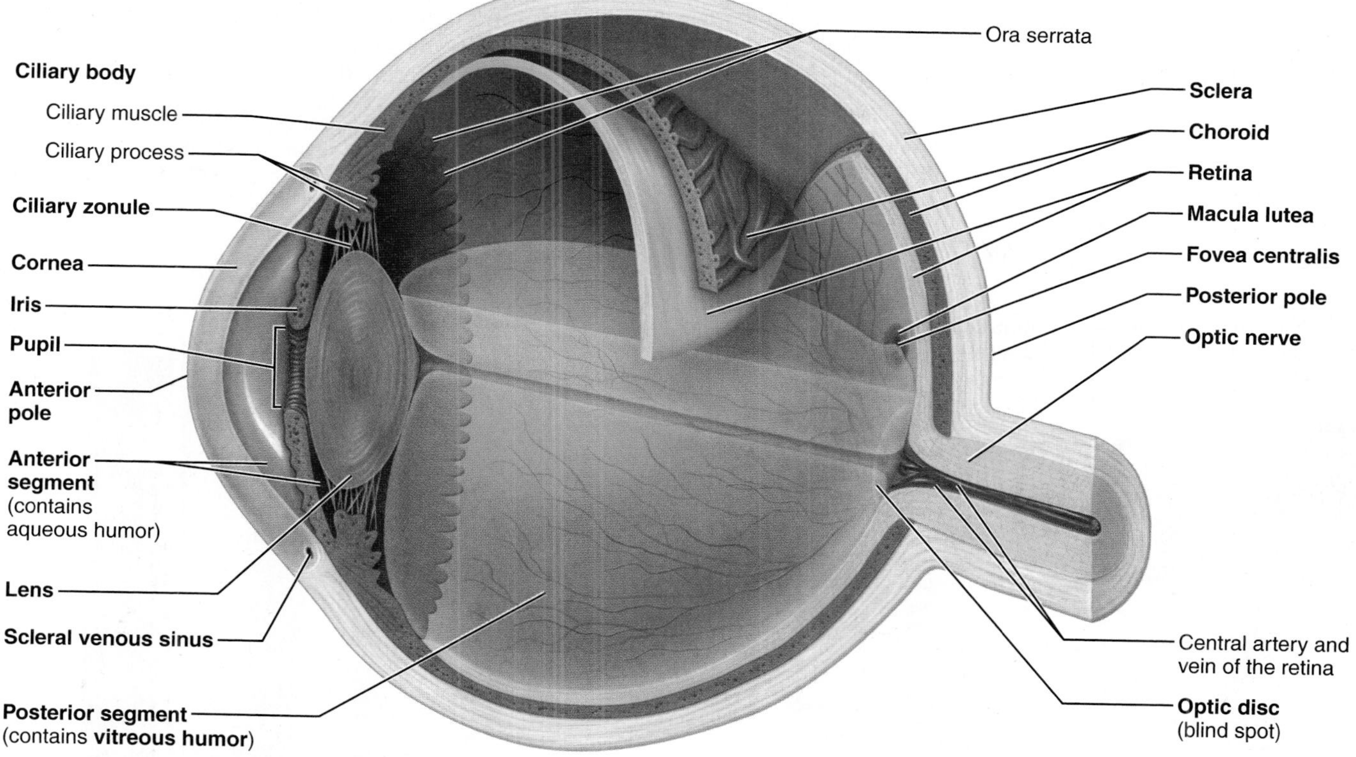

(a) Diagram of sagittal section of the eye. The vitreous humor is illustrated only in the bottom half of the eyeball.

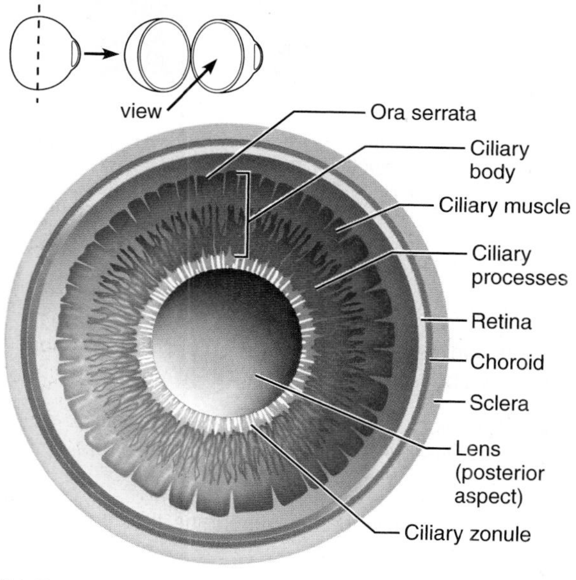

(b) Posterior view of anterior half of the eye

Figure 23.3 **Internal anatomy of the eye.**

DISSECTION

The Cow (Sheep) Eye

1. Obtain a preserved cow or sheep eye, dissecting instruments, and a dissecting tray. Don disposable gloves.

2. Examine the external surface of the eye, noting the thick cushion of adipose tissue. Identify the optic nerve as it leaves the eyeball, the remnants of the extrinsic eye muscles, the conjunctiva, the sclera, and the cornea. (Refer to **Figure 23.5** on p. 357 as you work.)

3. Trim away most of the fat and connective tissue, but leave the optic nerve intact. Holding the eye with the cornea facing downward, carefully make an incision with a sharp scalpel into

Text continues on next page. →

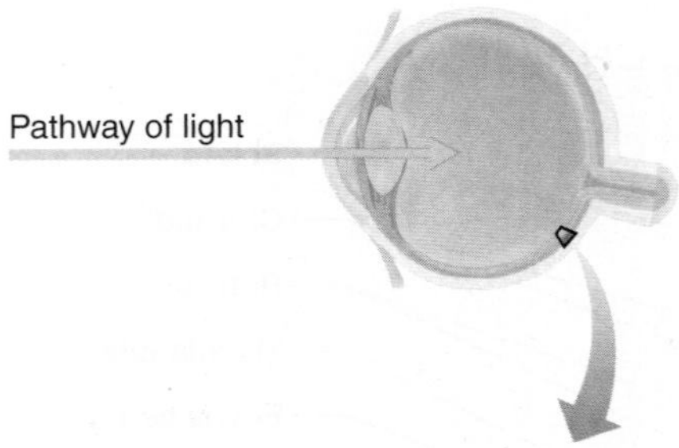

Figure 23.4 Microscopic anatomy of the retina. **(a)** Diagram of cells of the retina. Note the pathway of light through the retina. Neural signals (output of the retina) flow in the opposite direction. **(b)** Photomicrograph of the retina (140×).

Instructors may assign this figure as an Art Labeling Activity using Mastering **A&P™**

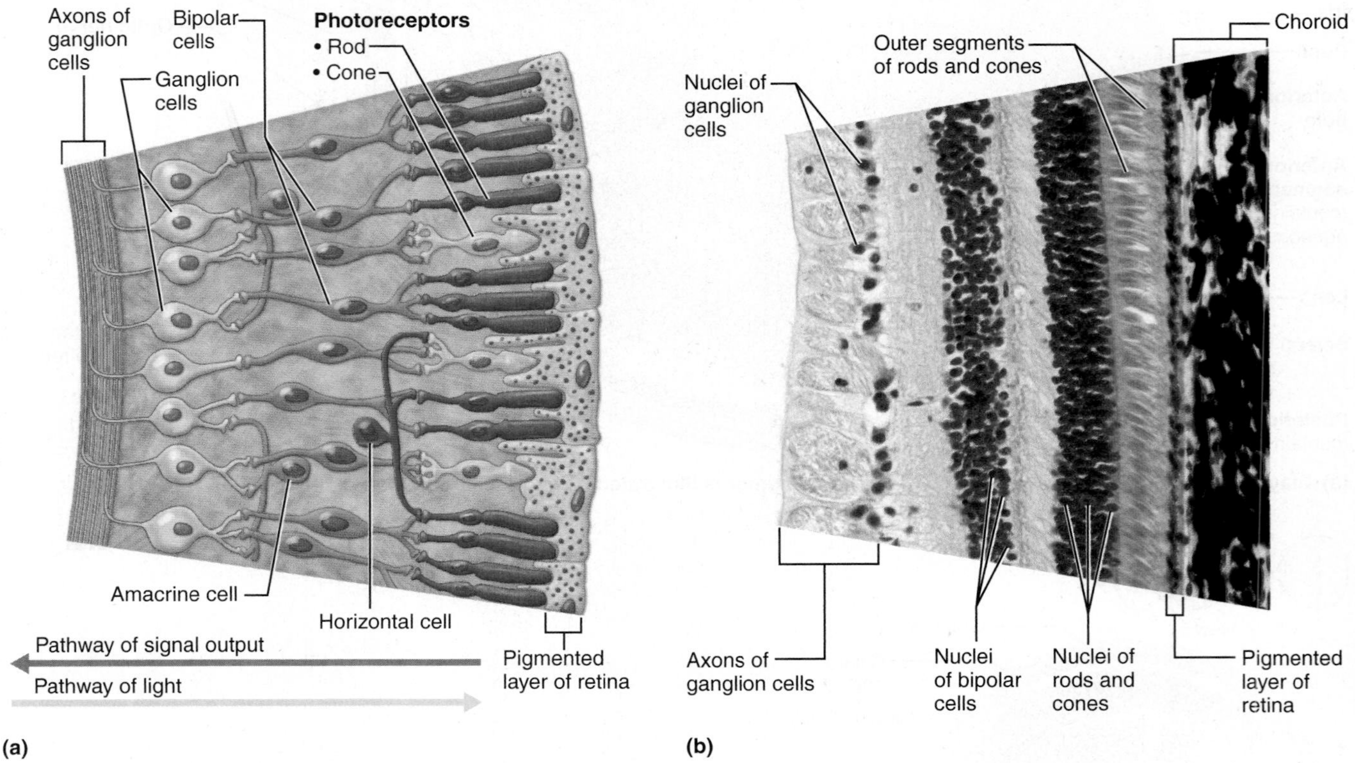

the sclera about 6 mm (¼ inch) above the cornea. Using scissors, complete the incision around the circumference of the eyeball paralleling the corneal edge.

4. Carefully lift the anterior part of the eyeball away from the posterior portion. The vitreous humor should remain with the posterior part of the eyeball.

5. Examine the anterior part of the eye, and identify the following structures:

Ciliary body: Black pigmented body that appears to be a halo encircling the lens.

Lens: Biconvex structure that is opaque in preserved specimens.

Carefully remove the lens and identify the adjacent structures:

Iris: Anterior continuation of the ciliary body penetrated by the pupil.

Cornea: More convex anteriormost portion of the sclera; normally transparent but cloudy in preserved specimens.

6. Examine the posterior portion of the eyeball. Carefully remove the vitreous humor, and identify the following structures:

Retina: Appears as a delicate tan membrane that separates easily from the choroid.

Note its posterior point of attachment. What is this point called?

____

Pigmented choroid coat: Appears iridescent in the cow or sheep eye owing to a modification called the **tapetum lucidum**. This specialized surface reflects the light within the eye and is found in the eyes of animals that live under conditions of dim light. It is not found in humans.

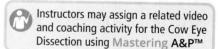

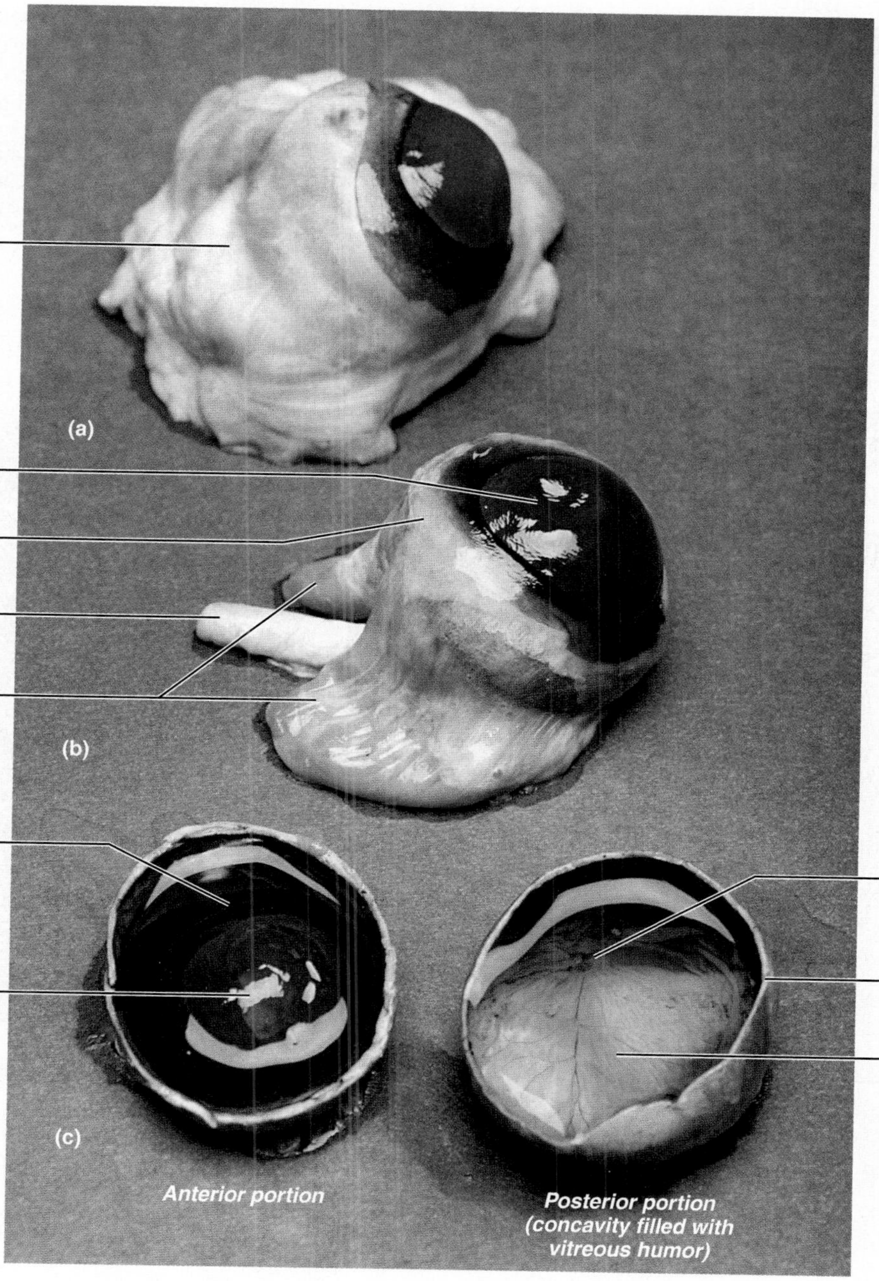

Figure 23.5 Anatomy of the cow eye. **(a)** Cow eye (entire) removed from the bony orbit (notice the large amount of fat cushioning the eyeball). **(b)** Cow eye (entire) with fat removed to show the extrinsic muscle attachments and optic nerve. **(c)** Cow eye cut along the frontal plane to reveal internal structures.

Instructors may assign a related video and coaching activity for the Cow Eye Dissection using Mastering **A&P™**

23

Visual Pathways to the Brain

The axons of the ganglion cells of the retina converge at the posterior aspect of the eyeball and exit from the eye as the optic nerve. At the **optic chiasma**, the fibers from the medial side of each eye cross over to the opposite side (**Figure 23.6**, p. 358). The fiber tracts thus formed are called the **optic tracts**. Each optic tract contains fibers from the lateral side of the eye on the same side and from the medial side of the opposite eye.

The optic tract fibers synapse with neurons in the **lateral geniculate nucleus** of the thalamus, whose axons form the **optic radiation**, terminating in the **primary visual cortex** in the occipital lobe of the brain. Here they synapse with the cortical neurons, and visual interpretation occurs.

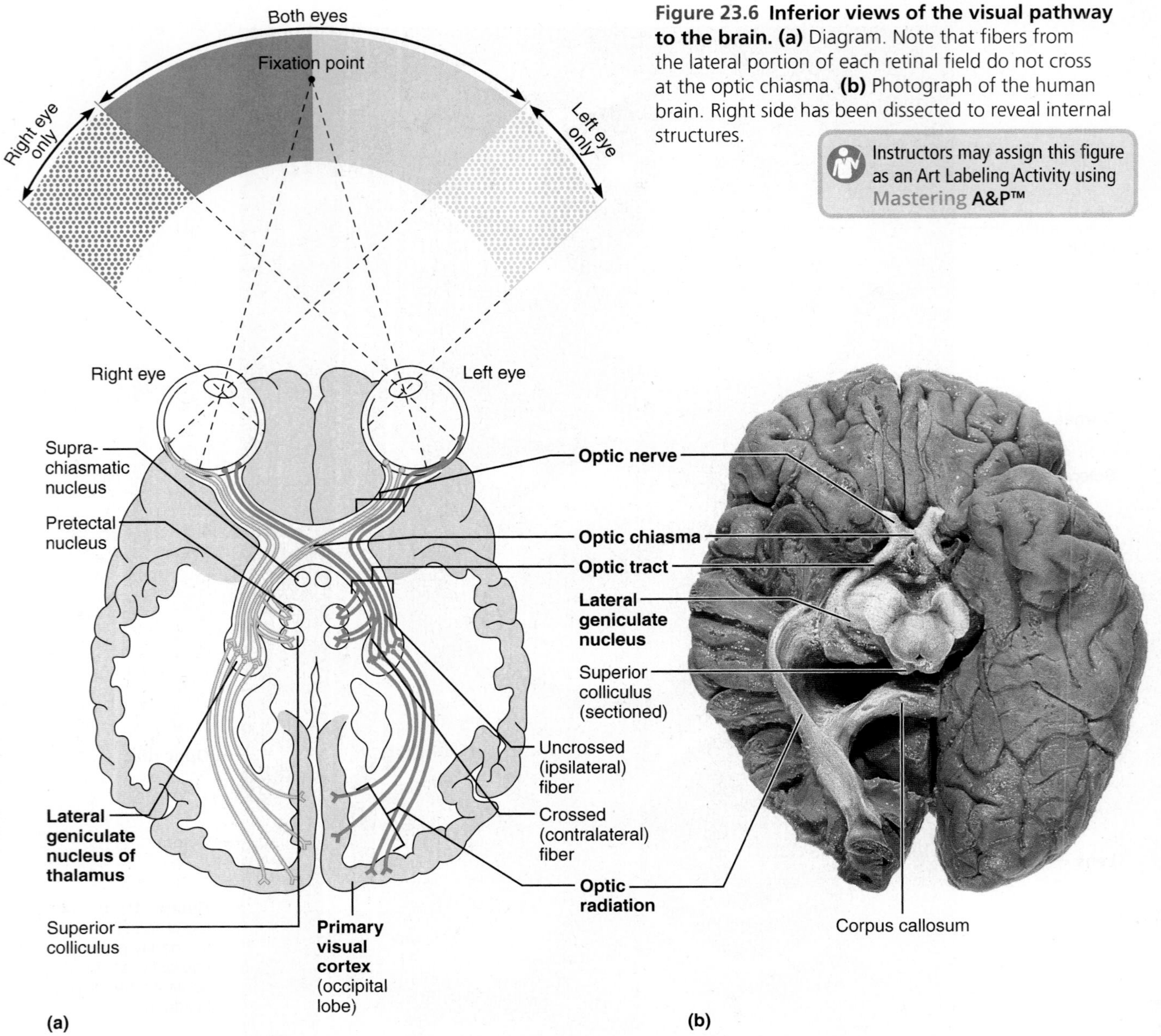

Figure 23.6 Inferior views of the visual pathway to the brain. (a) Diagram. Note that fibers from the lateral portion of each retinal field do not cross at the optic chiasma. **(b)** Photograph of the human brain. Right side has been dissected to reveal internal structures.

Instructors may assign this figure as an Art Labeling Activity using Mastering **A&P™**

23

Activity 4

Predicting the Effects of Visual Pathway Lesions

After examining the visual pathway diagram (Figure 23.6a), determine what effects lesions in the following areas would have on vision:

In the right optic nerve: ______________________________

____

Through the optic chiasma: ______________________________

____

In the left optic tract: ______________________________

____

In the right cerebral cortex (primary visual cortex): __________

____

Instructors may assign a portion of the Review Sheet questions using Mastering A&P™

REVIEW SHEET

EXERCISE 23 Special Senses: Anatomy of the Visual System

Name ______________________ Lab Time/Date ______________________

Anatomy of the Eye

1. Name five accessory eye structures that contribute to the formation of lacrimal fluid (tears) and/or help lubricate the eyeball, and then describe the major secretory product of each.

Accessory structures	Product

2. The eyeball is wrapped in adipose tissue within the bony orbit. What is the function of the adipose tissue?

____

3. Why does one often have to blow one's nose after crying? ______________________

____

4. Identify the extrinsic eye muscle predominantly responsible for each action described below.

______________________ 1. turns the eye laterally

______________________ 2. turns the eye medially

______________________ 3. turns the eye up and laterally

______________________ 4. turns the eye down and medially

______________________ 5. turns the eye up and medially

______________________ 6. turns the eye down and laterally

5. What is a sty? ______________________

____

Conjunctivitis? ______________________

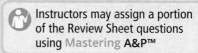

6. Correctly identify each lettered structure in the diagram by writing the letter next to its name in the numbered list. Use an appropriate reference if necessary.

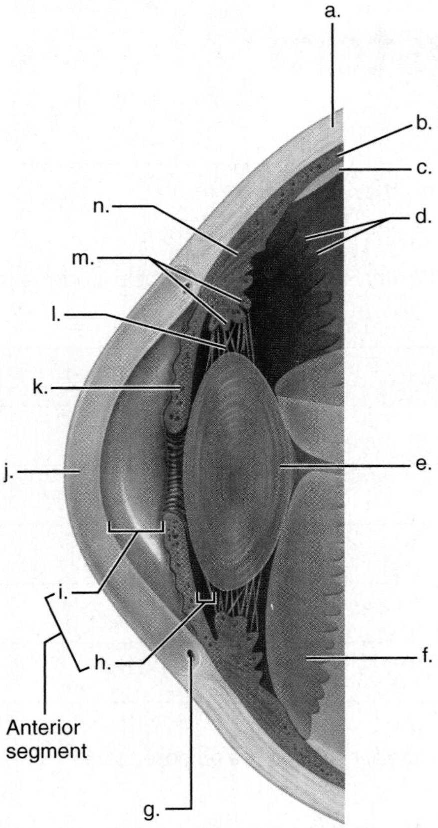

__________ 1. anterior chamber

__________ 2. choroid

__________ 3. ciliary muscle

__________ 4. ciliary processes

__________ 5. ciliary zonule

__________ 6. cornea

__________ 7. iris

__________ 8. lens

__________ 9. ora serrata

__________ 10. posterior chamber

__________ 11. posterior segment

__________ 12. retina

__________ 13. sclera

__________ 14. scleral venous sinus

7. Match the key responses with the descriptive statements that follow. (Some choices will be used more than once.)

Key: a. aqueous humor
b. choroid
c. ciliary body
d. ciliary processes of the ciliary body
e. cornea
f. fovea centralis
g. iris
h. lens
i. optic disc
j. retina
k. sclera
l. scleral venous sinus
m. vitreous humor

__________ 1. fluid filling the anterior segment of the eye

__________ 2. the "white" of the eye

__________ 3. part of the retina that lacks photoreceptors

__________ 4. modification of the choroid that contains the ciliary muscle

__________ 5. drains aqueous humor from the eye

__________ 6. layer containing the rods and cones

__________ 7. substance occupying the posterior segment of the eyeball

__________ 8. forms the bulk of the heavily pigmented vascular layer

__________, __________ 9. composed of smooth muscle structures (2)

________________ 10. area of critical focusing and detailed color vision

________________ 11. form the aqueous humor

________________, ________________ 12. light-bending media of the eye (4)

________________, ________________

________________ 13. anterior continuation of the sclera—your "window on the world"

________________ 14. composed of tough, white, opaque, fibrous connective tissue

Microscopic Anatomy of the Retina

8. The two major layers of the retina are the pigmented and neural layers. In the neural layer, the neuron populations are arranged as follows from the pigmented layer to the vitreous humor. (Circle the proper response.)

bipolar cells, ganglion cells, photoreceptors

photoreceptors, ganglion cells, bipolar cells

ganglion cells, bipolar cells, photoreceptors

photoreceptors, bipolar cells, ganglion cells

9. The axons of the ________________ cells form the optic nerve, which exits from the eyeball.

10. Complete the following statements by writing either *rods* or *cones* on each blank.

The dim light receptors are the ________________. Only ________________ are found in the fovea centralis, whereas mostly ________________ are found in the periphery of the retina. ________________ are the photoreceptors that operate best in bright light and allow for color vision.

Dissection of the Cow (Sheep) Eye

11. What modification of the choroid that is not present in humans is found in the cow eye? ________________

What is its function? ________________

____

12. What does the retina look like? ________________

At what point is it attached to the posterior aspect of the eyeball? ________________

Visual Pathways to the Brain

13. The visual pathway to the occipital lobe of the brain consists most simply of a chain of five cells. Beginning with the photoreceptor cell of the retina, name them, and note their location in the pathway.

1. ________________
2. ________________
3. ________________
4. ________________
5. ________________

14. How is the right optic *tract* anatomically different from the right optic *nerve*? ______________________________

____

____

15. ✚ Visual field tests are done to reveal destruction along the visual pathway from the retina to the optic region of the brain. Note where the lesion is likely to be in the following cases.

Normal vision in left eye visual field; absence of vision in right eye visual field: ______________________________

Normal vision in both eyes for right half of the visual field; absence of vision in both eyes for left half of the visual field:

____

16. ✚ A cornea transplant involves the grafting of a donor cornea into a recipient's anterior eye. The sutures to hold the graft in place must stay in place for a long period of time because the cornea is slow to heal. Explain why the healing process is so slow and also why graft rejection is unlikely with a cornea transplant. ______________________________

____

____

EXERCISE 24

Special Senses: Visual Tests and Experiments

Learning Outcomes

- ▶ Discuss the mechanism of image formation on the retina.
- ▶ Define the following terms: *accommodation*, *astigmatism*, *emmetropic*, *hyperopia*, *myopia*, *refraction*, and *presbyopia*. Describe several simple visual tests to which the terms apply.
- ▶ Discuss the benefits of binocular vision.
- ▶ Define *convergence*, and discuss the importance of the pupillary and convergence reflexes.
- ▶ State the importance of an ophthalmoscopic examination.

Go to Mastering A&P™ > Study Area to improve your performance in A&P Lab.

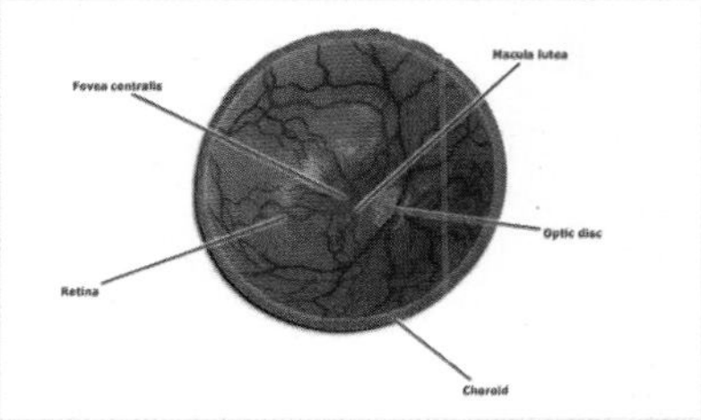

> Lab Tools > Practice Anatomy Lab > Anatomical Models

Instructors may assign new Building Vocabulary coaching activities, Pre-Lab Quiz questions, Art Labeling activities, Practice Anatomy Lab Practical questions (PAL), and more using the Mastering **A&P™** Item Library.

Pre-Lab Quiz

Instructors may assign these and other Pre-Lab Quiz questions using Mastering **A&P™**

1. Circle the correct underlined term. Photoreceptors are distributed over the entire neural retina, except where the optic nerve leaves the eyeball. This site is called the <u>macula lutea</u> / <u>optic disc</u>.
2. Circle True or False. People with difficulty seeing objects at a distance are said to have myopia.
3. A condition that results in the loss of elasticity of the lens and difficulty focusing on a close object is called:
   a. myopia **c.** hyperopia
   b. presbyopia **d.** astigmatism
4. Photoreceptors of the eye include rods and cones. Which one is responsible for interpreting color; which can function only under conditions of high light intensity?

   ____
5. Circle the correct underlined term. <u>Extrinsic</u> / <u>Intrinsic</u> eye muscles are controlled by the autonomic nervous system.

Materials

- ▶ Metric ruler; meter stick
- ▶ Common straight pins
- ▶ Snellen eye chart, floor marked with chalk or masking tape to indicate 20-ft distance from posted Snellen chart
- ▶ Ishihara's color plates
- ▶ Two pencils
- ▶ Test tubes large enough to accommodate a pencil
- ▶ Laboratory lamp or penlight
- ▶ Ophthalmoscope (if available)

The Optic Disc

In this exercise, you will perform several visual tests and experiments focusing on the physiology of vision. The first test involves demonstrating the blind spot (optic disc), the site where the optic nerve exits the eyeball.

Activity 1

Demonstrating the Blind Spot

1. Hold the figure for the blind spot test (**Figure 24.1**) about 46 cm (18 inches) from your eyes. Close your left eye, and focus your right eye on the X, which should be positioned so that it is directly in line with your right eye. Move the figure slowly toward your face, keeping your right eye focused on the X. When the dot focuses on the blind spot, which lacks photoreceptors, it will disappear.

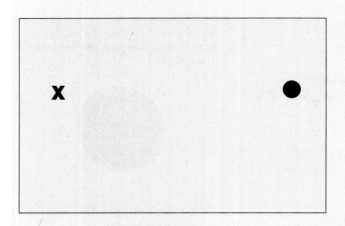

Figure 24.1 Blind spot test figure.

2. Have your laboratory partner record in centimeters the distance at which this occurs. The dot will reappear as the figure is moved closer. Distance at which the dot disappears:

Right eye ______________________________

Repeat the test for the left eye, this time closing the right eye and focusing the left eye on the dot. Record the distance at which the X disappears:

Left eye ______________________________

Refraction, Visual Acuity, and Astigmatism

When light rays pass from one medium to another, their speed changes, and the rays are bent, or **refracted**. Thus the light rays in the visual field are refracted as they encounter the cornea, aqueous humor, lens, and vitreous humor of the eye.

The refractive index (bending power) of the cornea, aqueous humor, and vitreous humor are constant. But the lens's refractive index can be varied by changing the lens's shape. The greater the lens convexity, or bulge, the more the light will be bent. Conversely, the less convex the lens (the flatter it is), the less it bends the light.

In general, light from a distant source (over 6 m, or 20 feet) approaches the eye as parallel rays, and no change in lens convexity is necessary for it to focus properly on the retina. However, light from a close source tends to diverge, and the convexity of the lens must increase to make close vision possible. To achieve this, the ciliary muscle contracts, decreasing the tension on the ciliary zonule attached to the lens and allowing the elastic lens to bulge. Thus, a lens capable of bringing a *close* object into sharp focus is more convex than a lens focusing on a more distant object. The ability of the eye to focus differentially for objects of close vision (less than 6 m, or 20 feet) is called **accommodation**. It should be noted that the image formed on the retina as a result of the refractory activity of the lens (**Figure 24.2**) is a **real image** (reversed from left to right, inverted, and smaller than the object).

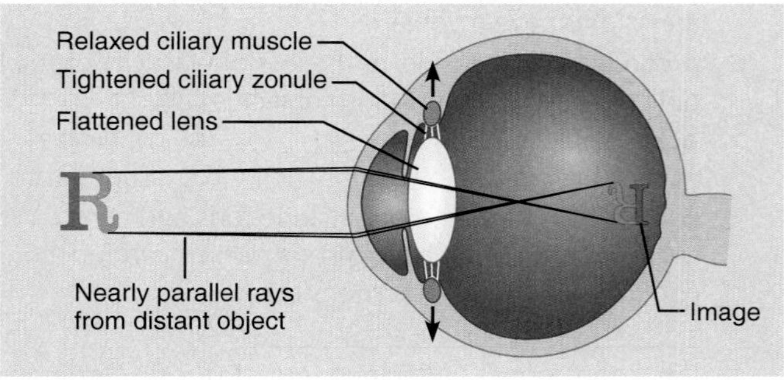

Figure 24.2 Refraction and real images. The refraction of light in the eye produces a real image (reversed, inverted, and reduced) on the retina.

The normal, or **emmetropic**, eye is able to accommodate properly. However, visual problems may result (1) from lenses that are too strong or too "lazy" (overconverging and underconverging, respectively); (2) from structural problems, such as an eyeball that is too long or too short to provide for proper focusing by the lens; or (3) from a cornea or lens with improper curvatures. Problems of refraction are summarized in **Figure 24.3**.

Irregularities in the curvatures of the lens and/or the cornea lead to a blurred vision problem called **astigmatism**. Cylindrically ground lenses are prescribed to correct the condition. ✚

Accommodation

The elasticity of the lens decreases dramatically with age, resulting in difficulty in focusing for near or close vision, especially when reading. This condition is called **presbyopia**—literally, old vision. Lens elasticity can be tested by measuring the **near point of vision**. The near point of vision is about 10 cm from the eye in young adults. It is closer in children and farther in old age.

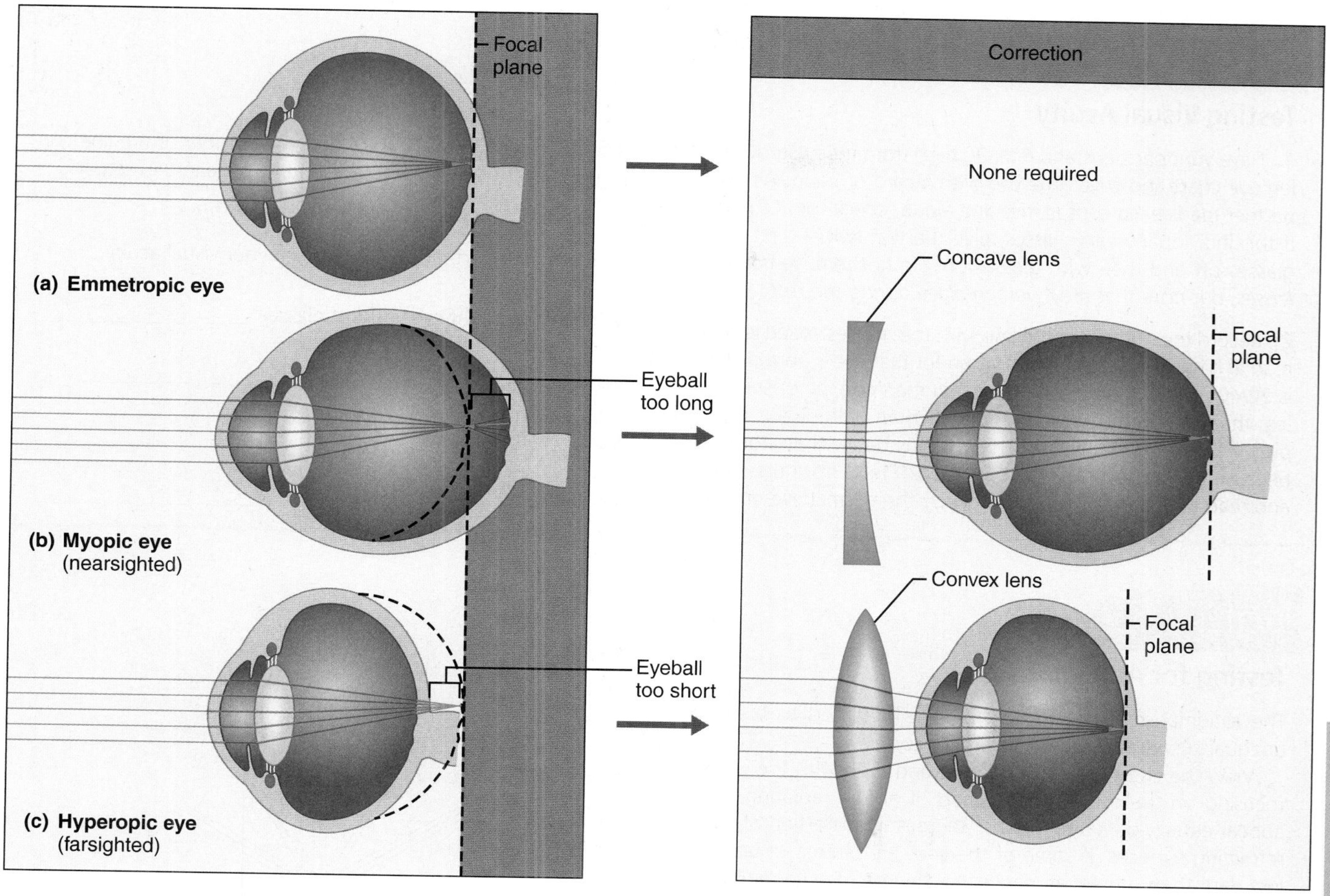

Figure 24.3 Problems of refraction. **(a)** In the emmetropic eye, light from near and far objects is focused properly on the retina. **(b)** In a myopic eye, light from distant objects is brought to a focal point before reaching the retina. Applying a concave lens focuses objects properly on the retina. **(c)** In the hyperopic eye, light from a near object is brought to a focal point behind the retina. Applying a convex lens focuses objects properly on the retina. The refractory effect of the cornea is ignored here.

24

Activity 2

Determining Near Point of Vision

To determine your near point of vision, hold a common straight pin (or other object) at arm's length in front of one eye. Slowly move the pin toward that eye until the pin image becomes distorted. Have your lab partner use a metric ruler to measure the distance in centimeters from your eye to the pin at this point, and record the distance. Repeat the procedure for the other eye.

Near point for right eye: ______________________

Near point for left eye: ______________________

Visual Acuity

Visual acuity, or sharpness of vision, is generally tested with a Snellen eye chart. This test is based on the fact that letters of a certain size can be seen clearly with normal vision at a specific distance. The distance at which the emmetropic eye can read a line of letters is printed at the end of that line.

Activity 3

Testing Visual Acuity

1. Have your partner stand 6 m (20 feet) from the posted Snellen eye chart and cover one eye with a card or hand. As your partner reads each consecutive line aloud, check for accuracy. If this individual wears glasses, give the test twice—first with glasses off and then with glasses on. *Do not remove contact lenses, but note that they were in place during the test.*

2. Record the number of the line with the smallest-sized letters read. If it is 20/20, the person's vision for that eye is normal. If it is 20/40, or any ratio with a value less than one, he or she has less than the normal visual acuity. (Such an individual is myopic.) If the visual acuity is 20/15, vision is better than normal, because this person can stand at 6 m (20 feet) from the chart and read letters that are discernible by the normal eye only at 4.5 m (15 feet). Give your partner the number of the line corresponding to the smallest letters read, to record in step 4.

3. Repeat the process for the other eye.

4. Have your partner test and record your visual acuity.

Visual acuity, right eye without glasses: ____________

Visual acuity, right eye with glasses: ____________

Visual acuity, left eye without glasses: ____________

Visual acuity, left eye with glasses: ____________

Activity 4

Testing for Astigmatism

The astigmatism chart (**Figure 24.4**) is designed to test for unequal curvatures of the lens and/or cornea.

View the chart first with one eye and then with the other, focusing on the center of the chart. If all the radiating lines appear equally dark and distinct, there is no distortion of your refracting surfaces. If some of the lines are blurred or appear less dark than others, at least some degree of astigmatism is present.

Is astigmatism present in your left eye? ____________

Right eye? ____________

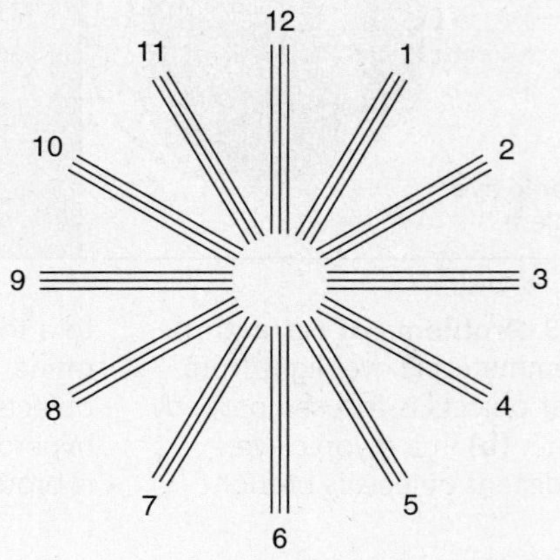

Figure 24.4 Astigmatism testing chart.

Color Blindness

Ishihara's color plates are designed to test for deficiencies in the color photoreceptor cells—the cones. There are three cone types, each containing a different light-absorbing pigment. One type primarily absorbs the red wavelengths of the visible light spectrum, another the blue wavelengths, and a third the green wavelengths. Nerve impulses reaching the brain from these different photoreceptor types are then interpreted (seen) as red, blue, and green, respectively. Overlapping input from more than one cone type leads the brain to interpret the intermediate colors of the visible light spectrum.

Activity 5

Testing for Color Blindness

1. Find the interpretation table that accompanies the Ishihara color plates, and prepare a sheet to record data for the test. Note which plates are patterns rather than numbers.

2. View the color plates in bright light or sunlight while holding them about 0.8 m (30 inches) away and at right angles to your line of vision. Report to your laboratory partner what you see in each plate. Take no more than 3 seconds for each decision.

3. Your partner should record your responses and then check their accuracy with the correct answers provided in the color plate book. Is there any indication that you have some degree of color blindness? _______________ If so, what type?

____

Binocular Vision

Humans, cats, predatory birds, and most primates are endowed with *binocular vision*. Their visual fields, each about 170 degrees, overlap to a considerable extent, and each eye sees a slightly different view (**Figure 24.5**). The primary visual cortex fuses the slightly different images, providing **depth perception** (or **three-dimensional vision**).

In contrast, the eyes of rabbits, pigeons, and many other animals are on the sides of their head. Such animals see in two different directions and thus have a panoramic field of view and *panoramic vision*. A mnemonic device to keep these straight is "Eyes in the front—likes to hunt; eyes to the side—likes to hide."

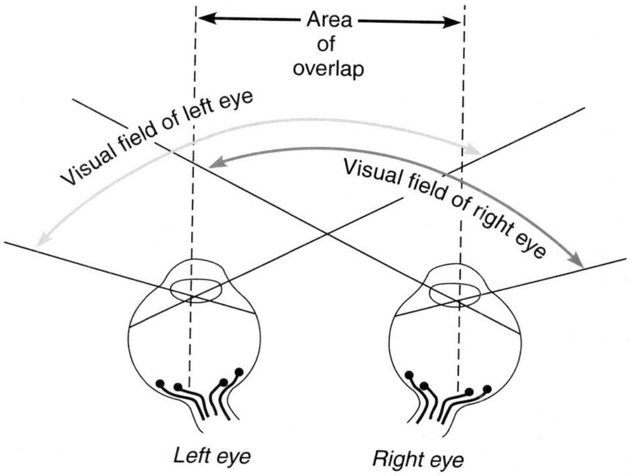

Figure 24.5 Overlapping of the visual fields.

24

Activity 6

Testing for Depth Perception

1. To demonstrate that a slightly different view is seen by each eye, perform the following simple experiment.

Close your left eye. Hold a pencil at arm's length directly in front of your right eye. Position another pencil directly beneath it and then move the lower pencil about half the distance toward you. As you move the lower pencil, make sure it remains in the *same plane* as the stationary pencil, so that the two pencils continually form a straight line. Then, without moving the pencils, close your right eye and open your left eye. Notice that with only the right eye open, the moving pencil stays in the same plane as the fixed pencil, but that when viewed with the left eye, the moving pencil is displaced laterally away from the plane of the fixed pencil.

2. To demonstrate the importance of two-eyed binocular vision for depth perception, perform this second experiment. Have your laboratory partner hold a test tube vertical about arm's length in front of you. With both eyes open, quickly insert a pencil into the test tube. Remove the pencil, bring it back close to your body, close one eye, and quickly insert the pencil into the test tube. *(Do not feel for the test tube with the pencil!)* Repeat with the other eye closed.

Was it as easy to dunk the pencil with one eye closed as with both eyes open?

____

Eye Reflexes

Both intrinsic (internal) and extrinsic (external) muscles are necessary for proper eye functioning. The *intrinsic muscles*, controlled by the autonomic nervous system, are those of the ciliary body (which alters the lens curvature in focusing) and the sphincter pupillae and dilator pupillae muscles of the iris (which control pupil size and thus regulate the amount of light entering the eye). The *extrinsic muscles* are the rectus and oblique muscles, which are attached to the eyeball exterior (see Figure 23.2). These muscles control eye movement and make it possible to keep moving objects focused on the fovea centralis. They are also responsible for **convergence**, or medial eye movements, which is essential for near vision. When convergence occurs, both eyes are directed toward the near object viewed. The extrinsic eye muscles are controlled by the somatic nervous system.

Activity 7

Demonstrating Reflex Activity of Intrinsic and Extrinsic Eye Muscles

Involuntary activity of both the intrinsic and extrinsic muscle types is brought about by reflex actions that can be observed in the following experiments.

Pupillary Light Reflex

Sudden illumination of the retina by a bright light causes the pupil to constrict reflexively in direct proportion to the light intensity. This protective response prevents damage to the delicate photoreceptor cells.

Obtain a laboratory lamp or penlight. Have your laboratory partner sit with eyes closed and hands over the eyes. Turn on the light and position it so that it shines on the subject's right hand. After 1 minute, ask your partner to uncover and open the right eye. Quickly observe the pupil of that eye. What happens to the pupil?

____

Shut off the light, and ask your partner to uncover and open the opposite eye. What are your observations of the pupil?

____

____

Accommodation Pupillary Reflex

Have your partner gaze for approximately 1 minute at a distant object in the lab—*not* toward the windows or another light source. Observe your partner's pupils. Then hold some printed material 15 to 25 cm (6 to 10 inches) from his or her face, and direct him or her to focus on it.

How does pupil size change as your partner focuses on the printed material?

____

Explain the value of this reflex. _______________

____

____

____

Convergence Reflex

Repeat the previous experiment, this time using a pen or pencil as the close object to be focused on. Note the position of your partner's eyeballs while he or she gazes at the distant object, and then at the close object. Do they change position as the object of focus is changed?

_______________ In what way? _______________

____

____

____

Ophthalmoscopic Examination of the Eye (Optional)

The ophthalmoscope is an instrument used to examine the *fundus*, or eyeball interior, to determine visually the condition of the retina, optic disc, and internal blood vessels. Certain pathological conditions such as diabetes mellitus, arteriosclerosis, and degenerative changes of the optic nerve and retina can be detected by such an examination. The ophthalmoscope consists of a set of lenses mounted on a rotating disc (the **lens selection disc**), a light source regulated by a **rheostat control**, and a mirror that reflects the light so that the eye interior can be illuminated.

The lens selection disc is positioned in a small slit in the mirror, and the examiner views the eye interior through this slit, appropriately called the **viewing window**. The focal length of each lens is indicated in diopters preceded by a plus (+) sign if the lens is convex and by a negative (−) sign if the lens is concave. When the zero (0) is seen in the **diopter window**, on the examiner side of the instrument, there is no lens positioned in the slit. The depth of focus for viewing the eye interior is changed by changing the lens.

The light is turned on by depressing the red **rheostat lock button** and then rotating the rheostat control in the clockwise direction. The **aperture selection dial** on the front of the instrument allows the nature of the light beam to be altered. The **filter switch**, also on the front, allows the choice of a green, unfiltered, or polarized light beam. Generally, green light allows for clearest viewing of the blood vessels in the eye interior and is most comfortable for the subject.

Once you have examined the ophthalmoscope and have become familiar with it, you are ready to conduct an eye examination.

Activity 8

Conducting an Ophthalmoscopic Examination

1. Conduct the examination in a dimly lit or darkened room with the subject comfortably seated and gazing straight ahead. To examine the right eye, sit face-to-face with the subject, hold the instrument in your right hand, and use your right eye to view the interior of the eye. You may want to steady yourself by resting your left hand on the subject's shoulder. To view the left eye, use your left eye, hold the instrument in your left hand, and steady yourself with your right hand.

2. Begin the examination with the 0 (no lens) in position. Grasp the instrument so that the lens disc may be rotated with the index finger. Holding the ophthalmoscope about 15 cm (6 inches) from the subject's eye, direct the light into the pupil at a slight angle—through the pupil edge rather than directly through its center. You will see a red circular area that is the illuminated eye interior.

3. Move in as close as possible to the subject's cornea (to within 5 cm, or 2 inches) as you continue to observe the area. Steady your instrument-holding hand on the subject's cheek if necessary. If both your eye and that of the subject are normal, the fundus can be viewed clearly without further adjustment of the ophthalmoscope. If the fundus cannot be focused, slowly rotate the lens disc counterclockwise until the fundus can be clearly seen. When the ophthalmoscope is correctly set, the fundus of the right eye should appear as in the photograph (**Figure 24.6**). (**Note:** If a positive [convex] lens is required and your eyes are normal, the subject has hyperopia. If a negative [concave] lens is necessary to view the fundus and your eyes are normal, the subject is myopic.)

When the examination is proceeding correctly, the subject can often see images of retinal vessels in his own eye that appear rather like cracked glass. If you are unable to achieve a sharp focus or to see the optic disc, move medially or laterally and begin again.

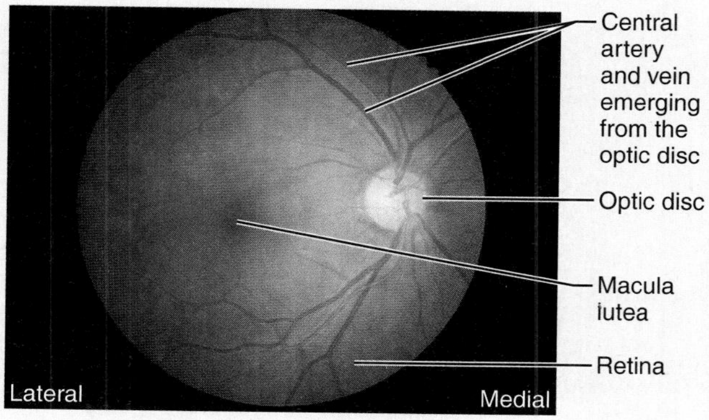

Figure 24.6 Fundus (posterior wall) of right retina.

4. Examine the optic disc for color, elevation, and sharpness of outline, and observe the blood vessels radiating from near its center. Locate the macula lutea, lateral to the optic disc. It is a darker area in which blood vessels are absent, and the fovea centralis appears to be a slightly lighter area in its center. The macula lutea is most easily seen when the subject looks directly into the light of the ophthalmoscope.

! Do not examine the macula lutea for longer than 1 second at a time.

5. When you have finished examining your partner's retina, shut off the ophthalmoscope. Change places with your partner (become the subject), and repeat steps 1–4.

EXERCISE 24

REVIEW SHEET

Special Senses: Visual Tests and Experiments

Instructors may assign a portion of the Review Sheet questions using Mastering A&P™

Name ______________________ Lab Time/Date ______________________

The Optic Disc, Refraction, Visual Acuity, and Astigmatism

1. Explain why vision is lost when light hits the blind spot. ______________________

2. Match the terms in column B with the descriptions in column A.

Column A		Column B
________	1. light bending	a. accommodation
________	2. ability to focus for close (less than 20 feet) vision	b. astigmatism
________	3. normal vision	c. emmetropia
________	4. inability to focus well on close objects (farsightedness)	d. hyperopia
________	5. nearsightedness	e. myopia
________	6. blurred vision due to unequal curvatures of the lens or cornea	f. refraction

3. Complete the following statements:

In farsightedness, the light is focused _1_ the retina. The lens required to treat myopia is a _2_ lens. The near point of vision increases with age because the _3_ of the lens decreases as we get older. A convex lens, like that of the eye, produces an image that is upside down and reversed from left to right. Such an image is called a _4_ image.

1. ______________________
2. ______________________
3. ______________________
4. ______________________

4. Use terms from the key to complete the statements concerning near and distance vision. (Some choices will be used more than once.)

Key: a. contracted b. decreased c. increased d. loose e. relaxed f. taut

During distance vision, the ciliary muscle is __________, the ciliary zonule is __________, the convexity of the lens is __________, and light refraction is __________. During close vision, the ciliary muscle is __________, the ciliary zonule is __________, lens convexity is __________, and light refraction is __________.

5. Using your Snellen eye test results, answer the following questions.

Is your visual acuity normal, less than normal, or better than normal? ______________________

Explain your answer. ______________________

____

Color Blindness

6. How can you explain the fact that we see a great range of colors even though only three cone types exist?

____

____

Binocular Vision

7. What is the advantage of binocular vision? ____

What factor(s) are responsible for binocular vision? ____

____

Eye Reflexes

8. In the experiment on the convergence reflex, what happened to the position of the eyeballs as the object was moved closer to the subject's eyes? ____

Which extrinsic eye muscles control the movement of the eyes during this reflex? ____

Ophthalmoscopic Examination

9. What pathological conditions can be detected with an ophthalmoscopic exam? ____

____

10. + Macular degeneration is an eye disease in which the macula lutea deteriorates. Explain why this would have a more profound effect on vision than deterioration of other parts of the retina. ____

____

11. + The "near triad" refers to the fact that three reflexes are required for near-point vision: accommodation of the lenses, constriction of the pupils, and convergence of the eyeballs. Strabismus is a misalignment of the eyes that can lead to double vision.

Which reflex is affected in this condition? ____

____

Are intrinsic or extrinsic eye muscles affected? ____

EXERCISE 25 Special Senses: Hearing and Equilibrium

Learning Outcomes

- Identify the anatomical structures of the external, middle, and internal ear on a model or appropriate diagram, and explain their functions.
- Describe the anatomy of the organ of hearing (spiral organ in the cochlea), and explain its function in sound reception.
- Discuss how one is able to localize the source of sounds.
- Define *sensorineural deafness* and *conduction deafness*, and relate these conditions to the Weber and Rinne tests.
- Describe the anatomy of the organs of equilibrium in the internal ear, and explain their relative function in maintaining equilibrium.
- State the locations and functions of endolymph and perilymph.
- Discuss the effects of acceleration on the semicircular canals.
- Define *nystagmus*, and relate this event to the balance and Barany tests.
- State the purpose of the Romberg test.
- Explain the role of vision in maintaining equilibrium.

Pre-Lab Quiz

Instructors may assign these and other Pre-Lab Quiz questions using Mastering A&P™

1. Circle the correct underlined term. The ear is divided into <u>three</u> / <u>four</u> major areas.
2. The external ear is composed primarily of the _______________ and the external acoustic meatus.
   a. auricle
   b. cochlea
   c. eardrum
   d. stapes
3. Circle the correct underlined term. Sound waves that enter the external acoustic meatus eventually encounter the <u>tympanic membrane</u> / <u>oval window</u>, which then vibrates at the same frequency as the sound waves hitting it.
4. Three small bones found within the middle ear are the malleus, incus, and _______________.
5. The snail-like _______________, found in the internal ear, contains sensory receptors for hearing.

Go to Mastering A&P™ > Study Area to improve your performance in A&P Lab.

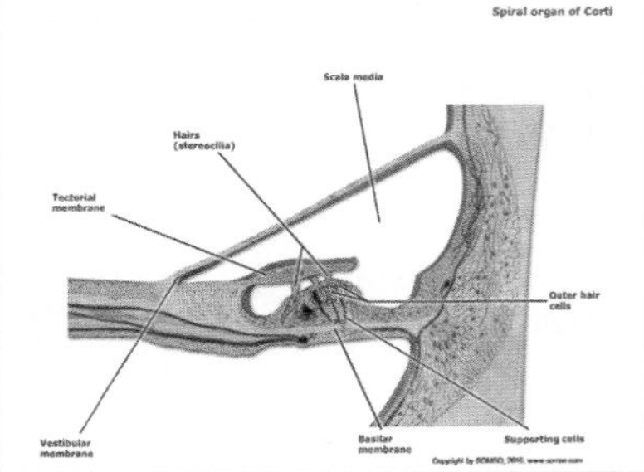

> Lab Tools > Practice Anatomy Lab > Anatomical Models

Instructors may assign new Building Vocabulary coaching activities, Pre-Lab Quiz questions, Art Labeling activities, Practice Anatomy Lab Practical questions (PAL), and more using the Mastering **A&P™** Item Library.

Materials

- Three-dimensional dissectible ear model and/or chart of ear anatomy
- Otoscope (if available)
- Disposable otoscope tips (if available) and autoclave bag
- Alcohol swabs
- Compound microscope
- Prepared slides of the cochlea of the ear
- Absorbent cotton
- Pocket watch or clock that ticks
- Metric ruler
- Tuning forks (range of frequencies)
- Rubber mallet
- Audiometer and earphones
- Red and blue pencils
- Demonstration: Microscope focused on a slide of a crista ampullaris receptor of a semicircular canal
- Three coins of different sizes
- Rotating chair or stool
- Blackboard and chalk or whiteboard and markers

The ear is a complex structure containing sensory receptors for hearing and equilibrium. The ear is divided into three major areas: the *external ear*, the *middle ear*, and the *internal ear* (**Figure 25.1**). The external and middle ear structures serve the needs of the sense of hearing *only*, whereas internal ear structures function both in equilibrium and hearing.

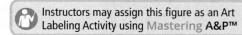

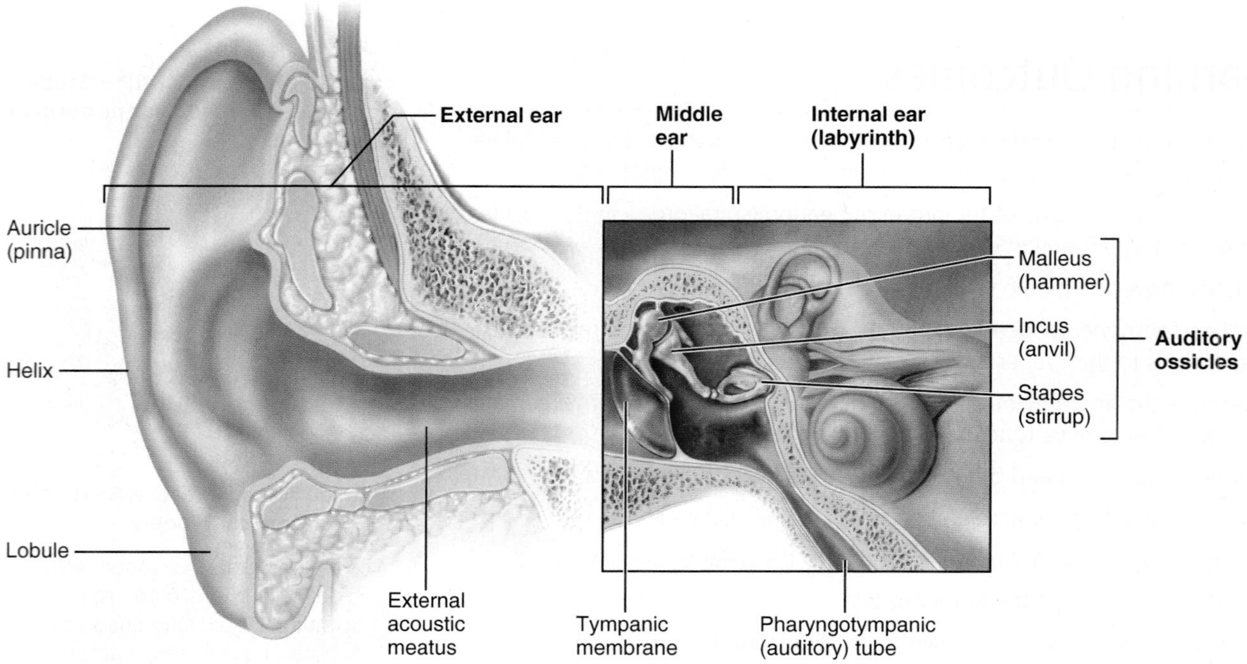

Figure 25.1 Anatomy of the ear.

Anatomy of the Ear

Gross Anatomy

Activity 1

Identifying Structures of the Ear

Obtain a dissectible ear model or chart of ear anatomy, and identify the structures summarized in **Table 25.1**.

Because the mucosal membranes of the middle ear cavity and nasopharynx are continuous through the pharyngotympanic tube, **otitis media**, or inflammation of the middle ear, is a fairly common condition, especially among children prone to sore throats. In cases where large amounts of fluid or pus accumulate in the middle ear cavity, an emergency myringotomy (lancing of the eardrum) may be necessary to relieve the pressure. Frequently, tiny ventilating tubes are put in during the procedure. +

The **internal ear** consists of a system of mazelike chambers and is therefore also referred to as the **labyrinth** (**Figure 25.2**). The internal ear has two main divisions: the bony labyrinth and the membranous labyrinth. The **bony labyrinth** is a cavity in the temporal bone that includes the semicircular canals, the vestibule, and the cochlea. The **membranous labyrinth** is a collection of ducts and sacs including the semicircular ducts, the utricle, the saccule, and the cochlear duct. The bony labyrinth is filled with an aqueous fluid called **perilymph**. The membranous labyrinth is filled with a more viscous fluid called **endolymph**. The ducts and sacs of the membranous labyrinth are suspended in the perilymph. The snail-shaped **cochlea** contains the sensory receptors for hearing. The **vestibule** and **semicircular canals** are involved with equilibrium.

Table 25.1 Structures of the External, Middle, and Internal Ear (Figure 25.1)

External Ear

Structure	Description	Function
Auricle (pinna)	Elastic cartilage covered with skin	Collects and directs sound waves into the external acoustic meatus
Lobule ("earlobe")	Portion of the auricle that is inferior to the external acoustic meatus	Completes the formation of the auricle
External acoustic meatus	Short, narrow canal carved into the temporal bone; lined with ceruminous glands	Transmits sound waves from the auricle to the tympanic membrane
Tympanic membrane (eardrum)	Thin membrane that separates the external ear from the middle ear	Vibrates at exactly the same frequency as the sound wave(s) hitting it and transmits vibrations to the auditory ossicles

Middle Ear

A small air-filled chamber — the tympanic cavity. Contains the auditory ossicles (malleus, incus, and stapes)

Structure	Description	Function
Malleus (hammer)	Tiny bone shaped like a hammer; its "handle" is attached to the eardrum	Transmits and amplifies vibrations from the tympanic membrane to the incus
Incus (anvil)	Tiny bone shaped like an anvil that articulates with the malleus and the stapes	Transmits and amplifies vibrations from the malleus to the stapes
Stapes (stirrup)	Tiny bone shaped like a stirrup; its "base" fits into the oval window	Transmits and amplifies vibrations from the incus to the oval window
Oval window	Oval-shaped membrane located deep to the stapes	Transmits vibrations from the stapes to the perilymph of the scala vestibuli
Pharyngotympanic (auditory) tube	A tube that connects the middle ear to the superior portion of the pharynx (throat)	Equalizes the pressure in the middle ear cavity with the external air pressure so that the tympanic membrane can vibrate properly

Internal Ear

Bony labyrinth	Membranous labyrinth (within the bony labyrinth)	Structure that contains the receptors	Function of the receptors
Cochlea	Cochlea duct	Spiral organ	Hearing
Vestibule	Utricle and saccule	Maculae	Equilibrium: static equilibrium and linear acceleration of the head
Semicircular canals	Semicircular ducts	Ampullae	Equilibrium: rotational acceleration of the head

25

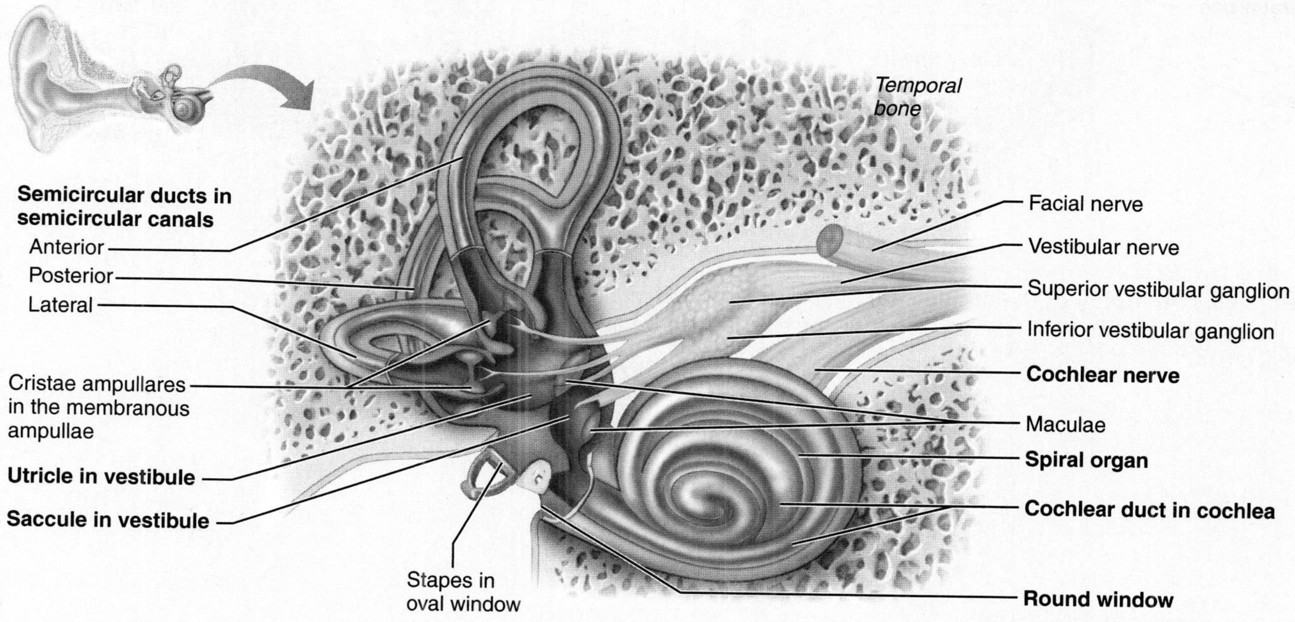

Figure 25.2 Internal ear. Right membranous labyrinth (blue) shown within the bony labyrinth (tan). The locations of sensory organs for hearing and equilibrium are shown in purple.

Instructors may assign this figure as an Art Labeling Activity using Mastering **A&P™**

Activity 2

Examining the Ear with an Otoscope (Optional)

1. Obtain an otoscope and two alcohol swabs. Inspect your partner's external acoustic meatus, and then select the speculum with the largest *diameter* that will fit comfortably into his or her ear to permit full visibility. Clean the speculum thoroughly with an alcohol swab, and then attach the speculum to the battery-containing otoscope handle. Before beginning, check that the otoscope light beam is strong. If not, obtain another otoscope or new batteries. Some otoscopes come with disposable tips. Be sure to use a new tip for each ear examined. Dispose of these tips in an autoclave bag after use.

2. When you are ready to begin the examination, hold the lighted otoscope securely between your thumb and forefinger (like a pencil), and rest the little finger of the otoscope-holding hand against your partner's head. This maneuver forms a brace that allows the speculum to move as your partner moves and prevents the speculum from penetrating too deeply into the external acoustic meatus during unexpected movements.

3. Grasp the ear auricle firmly and pull it up, back, and slightly laterally. If your partner experiences pain or discomfort when the auricle is manipulated, an inflammation or infection of the external ear may be present. If this occurs, do not attempt to examine the ear canal.

4. Carefully insert the speculum of the otoscope into the external acoustic meatus in a downward and forward direction only far enough to permit examination of the tympanic membrane, or eardrum. Note its shape, color, and vascular network. The healthy tympanic membrane is pearly white. During the examination, notice whether there is any discharge or redness in the external acoustic meatus, and identify earwax.

5. After the examination, thoroughly clean the speculum with the second alcohol swab before returning the otoscope to the supply area.

25

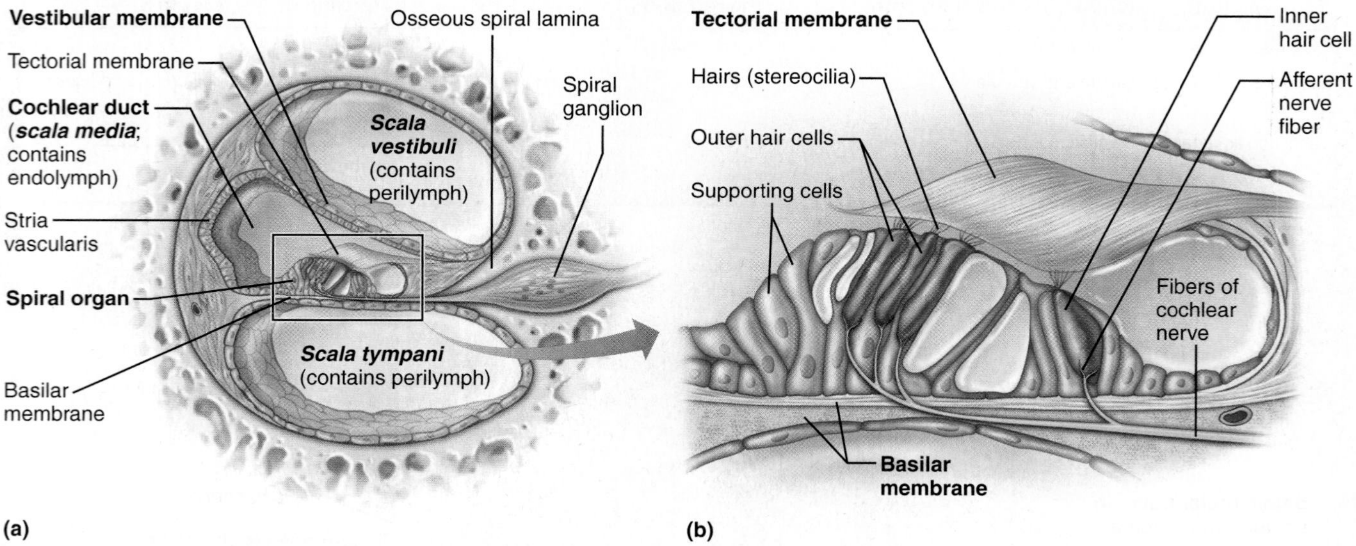

Figure 25.3 Anatomy of the cochlea. (a) Magnified cross-sectional view of one turn of the cochlea, showing the relationship of the three scalae. **(b)** Detailed structure of the spiral organ.

Microscopic Anatomy of the Spiral Organ

The membranous **cochlear duct** is a soft wormlike tube about 3.8 cm long. It winds through the full two and three-quarter turns of the cochlea and separates the perilymph-containing cochlear cavity into upper and lower chambers, the **scala vestibuli** and **scala tympani** (Figure 25.3). The scala vestibuli begins at the oval window, which "seats" the foot plate of the stapes located laterally in the tympanic cavity. The scala tympani is bounded by a membranous area called the **round window**. The round window can bulge into the tympanic cavity and act as a pressure relief valve for the exit of pressure waves. The cochlear duct is the middle **scala media**. It is filled with endolymph and supports the **spiral organ**, which contains the receptors for hearing—the sensory hair cells and nerve endings of the **cochlear nerve**, a division of the vestibulocochlear nerve (VIII).

In the spiral organ, the auditory receptors are hair cells that rest on the **basilar membrane**, which forms the floor of the cochlear duct (Figure 25.3). Their hairs are stereocilia that project into a gelatinous membrane, the **tectorial membrane**, that overlies them. The roof of the cochlear duct is called the **vestibular membrane**.

Activity 3

Examining the Microscopic Structure of the Cochlea

Obtain a compound microscope and a prepared microscope slide of the cochlea, and identify the areas shown in **Figure 25.4**.

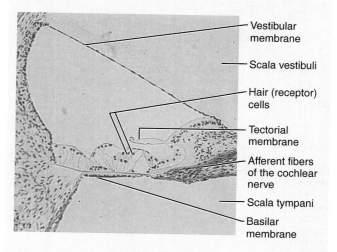

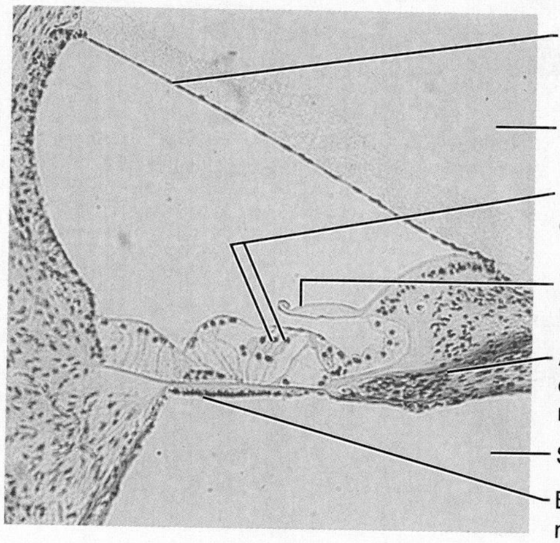

Figure 25.4 Histological image of the spiral organ (100×).

25

The Mechanism of Hearing

The mechanism of hearing begins as sound waves pass through the external acoustic meatus and through the middle ear into the internal ear, where the vibration eventually reaches the spiral organ, which contains the receptors for hearing.

Vibration of the stapes at the oval window creates pressure waves in the perilymph of the scala vestibuli, which are transferred to the endolymph in the cochlear duct (scala media). As the waves travel through the cochlear duct, they displace the basilar membrane and bend the hairs (sterocilia) of the hair cells. Hair cells at any given location on the basilar membrane are stimulated by sounds of a specific frequency (pitch). High-frequency waves (high-pitched sounds) displace the basilar membrane near the base where the fibers are short and stiff. Low-frequency waves (low-pitched sounds) displace the basilar membrane near the apex, where the fibers are long and flexible.

Once the hair cells are stimulated, they depolarize and begin the chain of nervous impulses that travel along the cochlear nerve and eventually reach the auditory centers of the temporal lobe cortex. This series of events results in the phenomenon we call hearing (**Figure 25.5**, p. 378).

Sensorineural deafness results from damage to neural structures anywhere from the cochlear hair cells through neurons of the auditory cortex. **Presbycusis** is a type of sensorineural deafness that occurs commonly in people by the time they are in their sixties. It results from a gradual deterioration and atrophy of the spiral organ and leads to a loss in the ability to hear high tones and speech sounds. +

Although presbycusis is considered to be a disability of old age, early onset is becoming much more common because of headphone abuse. Prolonged or excessive noise tears the cilia from hair cells, and the damage is progressive and cumulative.

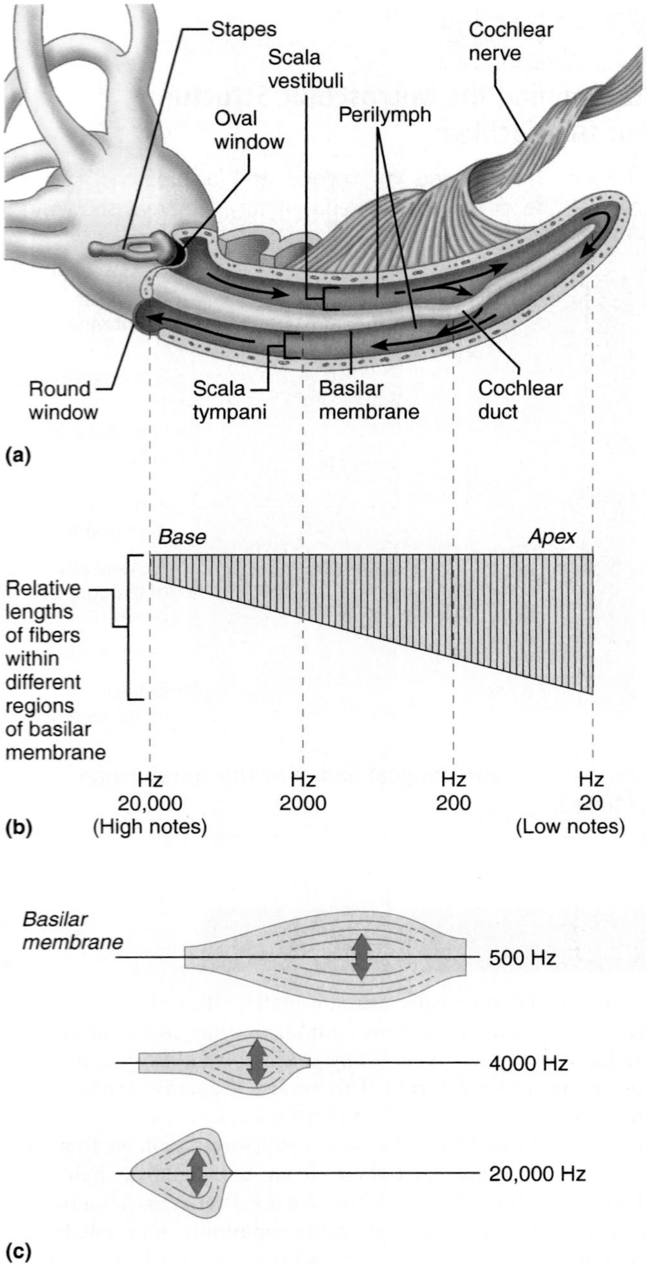

Figure 25.5 Resonance of the basilar membrane. The cochlea is depicted as if it has been uncoiled. **(a)** Perilymph movement in the cochlea following the stapes pushing the oval window. The compressional wave thus created causes the round window to bulge into the middle ear. Pressure waves set up vibrations in the basilar membrane. **(b)** Fibers span the basilar membrane. The stiffness of the fibers "tunes" specific regions to vibrate at specific frequencies. **(c)** Different frequencies of pressure waves in the spiral organ stimulate particular hair cells located in the basilar membrane.

Activity 4

Conducting Laboratory Tests of Hearing

Perform the following hearing tests in a quiet area. Test both the right and left ears.

Acuity Test

Have your lab partner pack one ear with cotton and sit quietly with eyes closed. Obtain a ticking clock or pocket watch, and hold it very close to his or her *unpacked* ear. Then slowly move it away from the ear until your partner signals that the ticking is no longer audible. Record the distance in centimeters at which ticking is inaudible, and then remove the cotton from the packed ear.

Right ear: ______________ Left ear: ______________

Is the threshold of audibility sharp or indefinite?

____

Sound Localization

Ask your partner to close both eyes. Hold the pocket watch at an audible distance (about 15 cm) from the ear, and move it to various locations (front, back, sides, and above the head). Have your partner locate the position by pointing in each instance. Can the sound be localized equally well at all positions?

____

If not, at what position(s) was the sound *less* easily located?

____

____

The ability to localize the source of a sound depends on two factors—the difference in the loudness of the sound reaching each ear and the time of arrival of the sound at each ear. How does this information help to explain your findings?

____

____

Frequency Range of Hearing

Obtain three tuning forks: one with a low frequency (75 to 100 Hz [cps]), one with a frequency of approximately 1000 Hz, and one with a frequency of 4000 to 5000 Hz. Strike the lowest-frequency fork on the heel of your hand or with a rubber mallet, and hold it close to your partner's ear. Repeat with the other two forks.

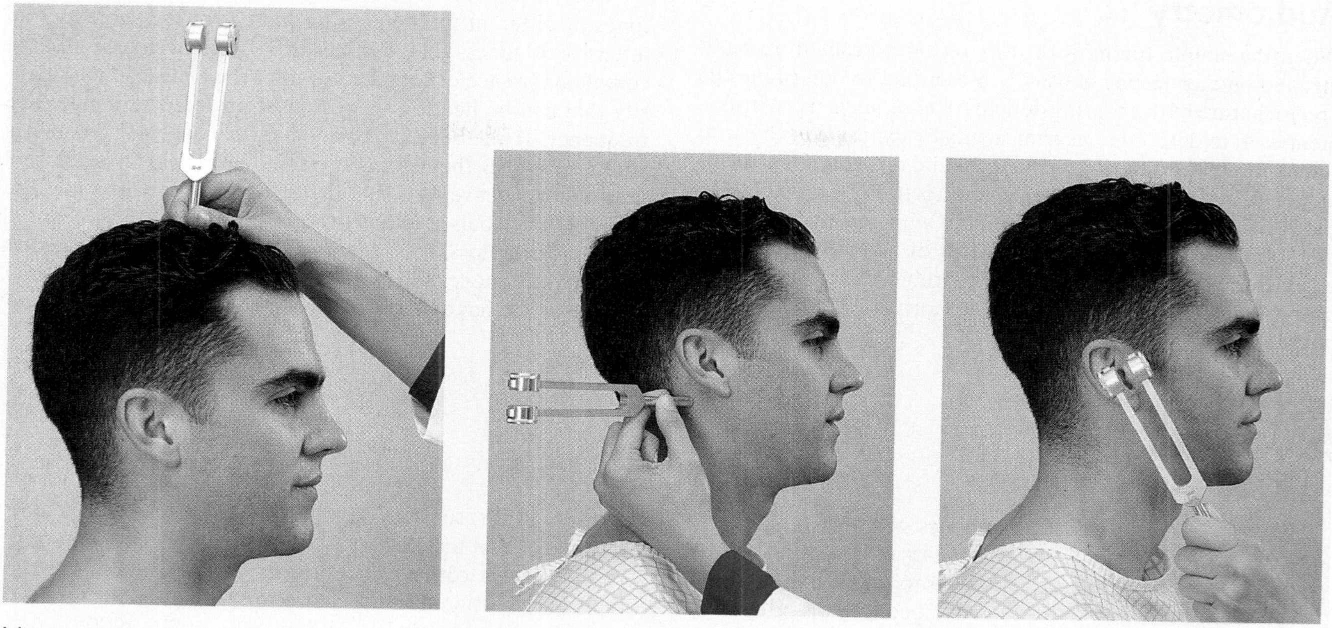

(a) (b) (c)

Figure 25.6 The Weber and Rinne tuning fork tests. (a) The Weber test to evaluate whether the sound remains centralized (normal) or lateralizes to one side or the other (indicative of some degree of conduction or sensorineural deafness). **(b, c)** The Rinne test to compare bone conduction and air conduction.

Which fork was heard *most* clearly and comfortably?

_______________ Hz

Which was heard *least* well? _______________ Hz

Weber Test to Determine Conduction and Sensorineural Deafness

Strike a tuning fork and place the handle of the tuning fork medially on your partner's head (**Figure 25.6a**). Is the tone equally loud in both ears, or is it louder in one ear?

____

If it is equally loud in both ears, your partner has equal hearing or equal loss of hearing in both ears. If sensorineural deafness is present in one ear, the tone will be heard in the unaffected ear but not in the ear with sensorineural deafness.

Conduction deafness occurs when something prevents sound waves from reaching the fluids of the internal ear. Compacted earwax, a perforated eardrum, inflammation of the middle ear (otitis media), and damage to the ossicles are all causes of conduction deafness. If conduction deafness is present, the sound will be heard more strongly in the ear in which there is a hearing loss due to sound conduction by the bone of the skull. Conduction deafness can be simulated by plugging one ear with cotton.

Rinne Test for Comparing Bone- and Air-Conduction Hearing

1. Strike the tuning fork, and place its handle on your partner's mastoid process (Figure 25.6b).

2. When your partner indicates that the sound is no longer audible, hold the still-vibrating prongs close to his or her external acoustic meatus (Figure 25.6c). If your partner hears the fork again (by air conduction) when it is moved to that position, hearing is not impaired, and you record the test result as positive (+). (Record below step 5.)

3. Repeat the test on the same ear, but this time test air-conduction hearing first.

4. After the tone is no longer heard by air conduction, hold the handle of the tuning fork on the bony mastoid process. If the subject hears the tone again by bone conduction after hearing by air conduction is lost, there is some conduction deafness, and the result is recorded as negative (–).

5. Repeat the sequence for the opposite ear.

Right ear: _______________ Left ear: _______________

Does the subject hear better by bone or by air conduction?

____

Audiometry

When the simple tuning fork tests reveal a problem in hearing, audiometer testing is usually prescribed to determine the precise nature of the hearing deficit. An *audiometer* is an instrument used to determine hearing acuity by exposing each ear to sound stimuli of differing *frequencies* and *intensities*. The hearing range of human beings during youth is from 20 to 20,000 Hz, but hearing acuity declines with age, with reception for the high-frequency sounds lost first. Most of us tend to be fairly unconcerned until we begin to have problems hearing sounds in the range of 125 to 8000 Hz, the normal frequency range of speech.

The basic procedure of audiometry is to initially deliver tones of different frequencies to one ear of the subject at an intensity (loudness) of 0 decibels (dB). Zero decibels is not the complete absence of sound, but rather the softest sound intensity that can be heard by a person of normal hearing at each frequency. If the subject cannot hear a particular frequency stimulus of 0 dB, the hearing threshold level control is adjusted until the subject reports that he or she can hear the tone. The number of decibels of intensity required above 0 dB is recorded as the hearing loss. For example, if the subject cannot hear a particular frequency tone until it is delivered at 30 dB intensity, then he or she has a hearing loss of 30 dB for that frequency.

Activity 5

Audiometry Testing

1. Obtain an audiometer and earphones, and a red and a blue pencil. Before beginning the tests, examine the audiometer to identify the two tone controls: one to regulate frequency and a second to regulate the intensity (loudness) of the sound stimulus. Identify the two output control switches that regulate the delivery of sound to one ear or the other (*red* to the right ear, *blue* to the left ear). Also find the *hearing threshold level control*, which is calibrated to deliver a basal tone of 0 dB to the subject's ears.

2. Place the earphones on the subject's head so that the red cord or ear-cushion is over the right ear and the blue cord or ear-cushion is over the left ear. Instruct the subject to raise one hand when he or she hears a tone.

3. Set the frequency control at 125 Hz and the intensity control at 0 dB. Press the red output switch to deliver a tone to the subject's right ear. If the subject does not respond, raise the sound intensity slowly by rotating the hearing level control counterclockwise until the subject reports (by raising a hand) that a tone is heard. Repeat this procedure for frequencies of 250, 500, 1000, 2000, 4000, and 8000.

4. Record the results in the grid (below) for frequency versus hearing loss by marking a small red circle on the grid at each frequency-dB junction at which a tone was heard. Then connect the circles with a red line to produce a hearing acuity graph for the right ear.

5. Repeat steps 3 and 4 for the left (blue) ear, and record the results with blue circles and connecting lines on the grid.

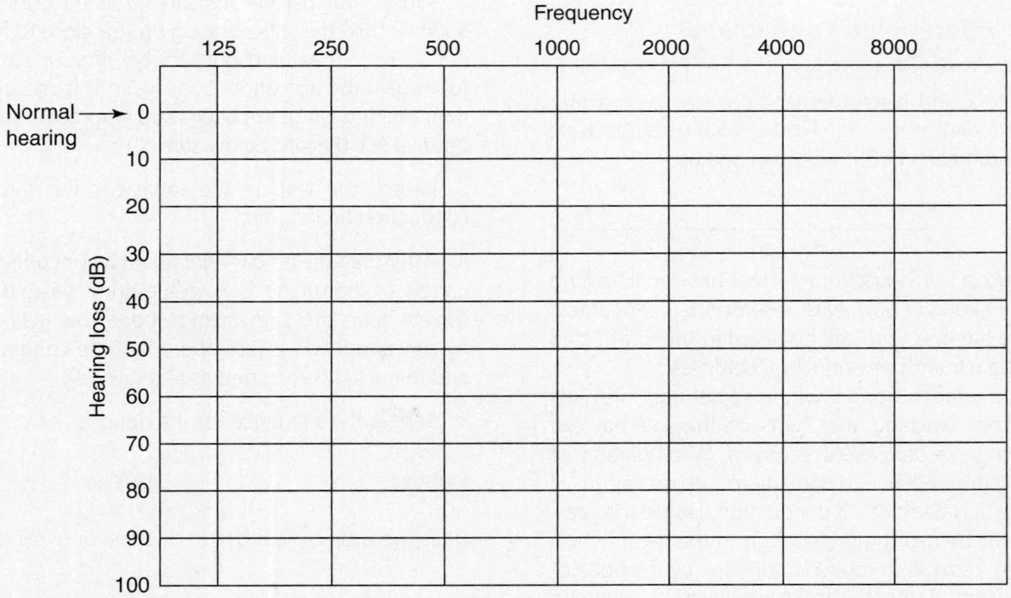

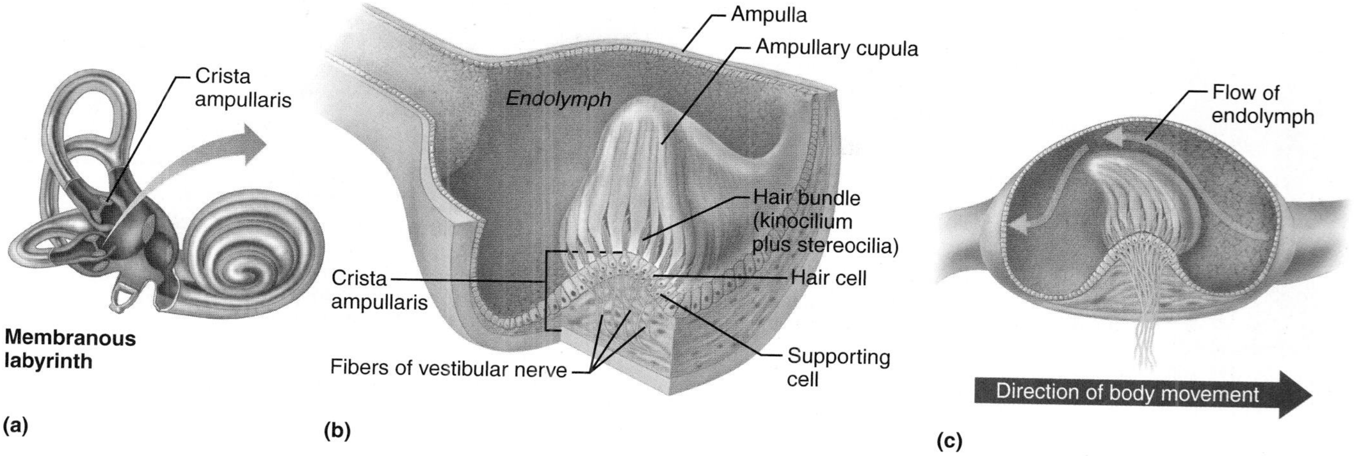

Figure 25.7 Structure and function of the crista ampullaris. (a) Arranged in the three spatial planes, the semicircular ducts in the semicircular canals each have a swelling called an ampulla at their base. **(b)** Each ampulla contains a crista ampullaris, a receptor that is essentially a cluster of hair cells with hairs projecting into a gelatinous cap called the ampullary cupula. **(c)** Movement of the cupula during rotational acceleration of the head.

Microscopic Anatomy of the Equilibrium Apparatus and Mechanisms of Equilibrium

The equilibrium receptors of the internal ear are collectively called the **vestibular apparatus** and are found in the vestibule and semicircular canals of the bony labyrinth. The vestibule contains the saclike **utricle** and **saccule**, and the semicircular chambers contain membranous **semicircular ducts.** These membranes are filled with endolymph and contain receptor cells that are activated by the bending of the hairs on their hair cells.

Semicircular Canals

The semicircular canals monitor rotational acceleration of the head. This process is called **dynamic equilibrium**. The canals are 1.2 cm in circumference and are oriented in three planes—horizontal, frontal, and sagittal. At the base of each semicircular duct is an enlarged region, the **ampulla.** Within each ampulla is a receptor region called a **crista ampullaris**, which consists of a tuft of hair cells covered with a gelatinous cap, or **ampullary cupula** (**Figure 25.7**).

The cristae respond to changes in the velocity of rotational head movements. During acceleration, as when you begin to twirl around, the endolymph in the canal lags behind the head movement because inertia causes the ampullary cupula to push—like a swinging door—in the opposite direction. The head movement depolarizes the hair cells, and as a result, impulse transmission is enhanced in the vestibular division of the eighth cranial nerve to the brain (Figure 25.7c). If the body continues to rotate at a constant rate, the endolymph eventually comes to rest and moves at the same speed as the body. The ampullary cupula returns to its upright position, hair cells are no longer stimulated, and you lose the sensation of spinning. When rotational movement stops suddenly, the endolymph keeps on going in the direction of head movement. This pushes the ampullary cupula in the *same* direction as the previous head movement and hyperpolarizes the hair cells; as a result, fewer impulses are transmitted to the brain. This tells the brain that you have stopped moving and accounts for the reversed motion sensation you feel when you stop twirling suddenly.

25

Activity 6

Examining the Microscopic Structure of the Crista Ampullaris

Go to the demonstration area and examine the slide of a crista ampullaris. Identify the areas depicted in the photomicrograph (**Figure 25.8**) and labeled diagram (Figure 25.7b).

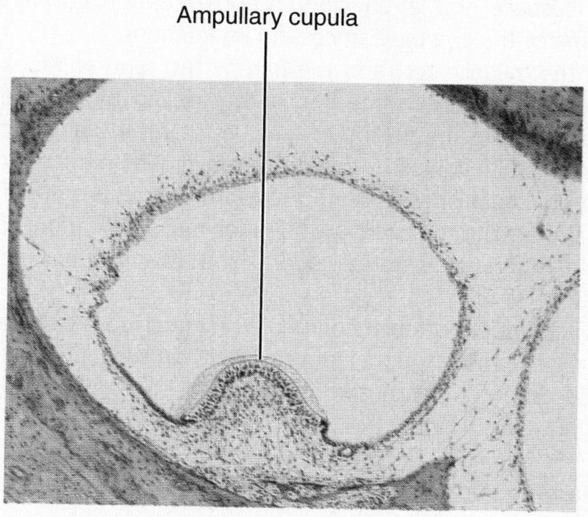

Figure 25.8 Micrograph of a crista ampullaris (42×).

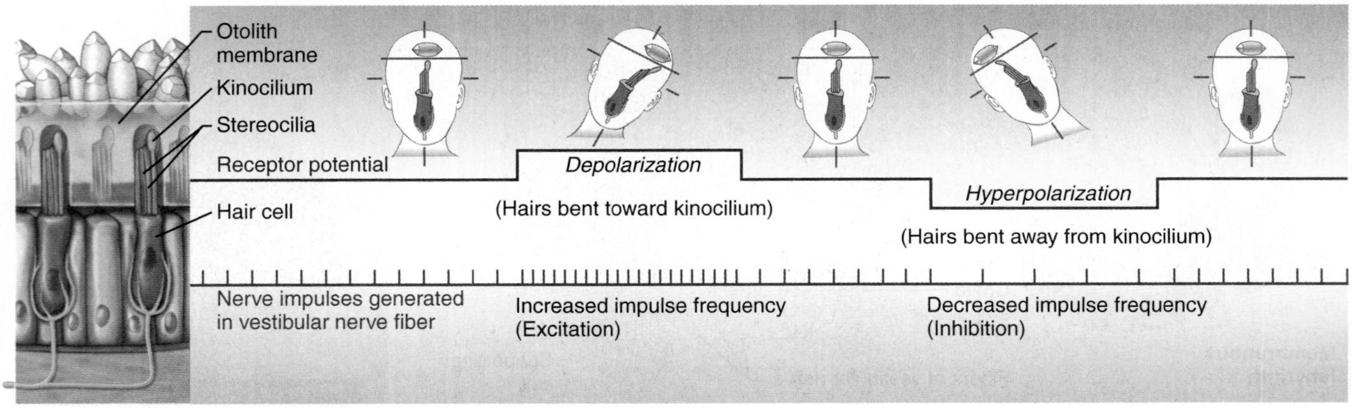

Figure 25.9 The effect of gravitational pull on a macula receptor in the utricle. When movement of the otolith membrane bends the hair cells in the direction of the kinocilium, the hair cells depolarize, exciting the nerve fibers, which generates action potentials more rapidly. When the hairs are bent in the direction away from the kinocilium, the hair cells become hyperpolarized, inhibiting the nerve fibers and decreasing the action potential rate (i.e., below the resting rate of discharge).

Maculae

Maculae in the membranous utricle and saccule contain another set of **hair cells**, receptors that in this case monitor head position and acceleration in a straight line. This monitoring process is called **static equilibrium**. The maculae respond to gravitational pull, thus providing information on which way is up or down as well as changes in linear speed. The hair cells in each macula have **sterocilia** plus one **kinocilium** that are embedded in the **otolith membrane**, a gelatinous material containing small grains of calcium carbonate called **otolith**. When the head moves, the otoliths move in response to variations in gravitational pull. As they deflect different hair cells, they trigger hyperpolarization or depolarization of the hair cells and modify the rate of impulse transmission along the vestibular nerve (**Figure 25.9**).

Although the receptors of the semicircular canals and the vestibule are responsible for dynamic and static equilibrium respectively, they rarely act independently. Complex interaction of many of the receptors is the rule. Processing is also complex and involves the brain stem and cerebellum as well as input from proprioceptors and the eyes.

25

Activity 7

Conducting Laboratory Tests on Equilibrium

The function of the semicircular canals and vestibule are not routinely tested in the laboratory, but the following simple tests illustrate normal equilibrium apparatus function as well as some of the complex processing interactions.

In the balance test and the Barany test, you will look for **nystagmus**, which is the involuntary rolling of the eyes in any direction or the trailing of the eyes slowly in one direction (slow phase), followed by their rapid movement in the opposite direction (rapid phase). During rotation, the slow drift of the eyes is related to the backflow of endolymph in the semicircular ducts. Nystagmus is normal during and after rotation; abnormal otherwise.

Nystagmus is often accompanied by **vertigo**—a sensation of dizziness and rotational movement when such movement is not occurring or has ceased.

Balance Tests

1. Have your partner walk a straight line, placing one foot directly in front of the other.

Is he or she able to walk without significant wobbling from side to side? ______________________

Did he or she experience any dizziness? ______________

The ability to walk with balance and without dizziness, unless subject to rotational forces, indicates normal function of the equilibrium apparatus.

Was nystagmus present? ______________________

2. Place three coins of different sizes on the floor. Ask your lab partner to pick up the coins, and carefully observe his or her muscle activity and coordination.

Did your lab partner have any difficulty locating and picking up the coins? ______________________

Describe your observations and your lab partner's observations during the test.

____

____

What kinds of complex interactions involving balance and coordination must occur for a person to move fluidly during this test?

____

____

3. If a person has a depressed nervous system, mental concentration may result in a loss of balance. Ask your lab partner to stand up and count backward from ten as rapidly as possible.

Did your lab partner lose balance? ______________

Barany Test (Induction of Nystagmus and Vertigo)

This experiment evaluates the semicircular canals and should be conducted as a group effort to protect the test subject(s) from possible injury.

! Read the following precautionary notes before beginning:

- The subject(s) chosen should not be easily inclined to dizziness during rotational or turning movements.
- Rotation should be stopped immediately if the subject feels nauseated.
- Because the subject(s) will experience vertigo and loss of balance as a result of the rotation, several classmates should be prepared to catch, hold, or support the subject(s) as necessary until the symptoms pass.

1. Instruct the subject to sit on a rotating chair or stool, and to hold on to the arms or seat of the chair, feet on stool rungs. The subject's head should be tilted forward approximately 30 degrees (chin almost touching the chest). The horizontal (lateral) semicircular canal is stimulated when the head is in this position. The subject's eyes are to remain *open* during the test.
2. Four classmates should position themselves so that the subject is surrounded on all sides. The classmate posterior to the subject will rotate the chair.
3. Rotate the chair to the subject's right approximately 10 revolutions in 10 seconds, then suddenly stop the rotation.
4. Immediately note the direction of the subject's resultant nystagmus; and ask him or her to describe the feelings of movement, indicating speed and direction sensation. Record this information below.

____

____

____

If the semicircular canals are operating normally, the subject will experience a sensation that the stool is still rotating immediately after it has stopped and *will* demonstrate nystagmus.

When the subject is rotated to the right, the ampullary cupula will be bent to the left, causing nystagmus during rotation in which the eyes initially move slowly to the left and then quickly to the right. Nystagmus will continue until the ampullary cupula has returned to its initial position. Then, when rotation is stopped abruptly, the ampullary cupula will be bent to the right, producing nystagmus with its slow phase to the right and its rapid phase to the left. In many subjects, this will be accompanied by a feeling of vertigo and a tendency to fall to the right.

Romberg Test

The Romberg test determines the integrity of the dorsal white column of the spinal cord, which transmits impulses to the brain from the proprioceptors involved with posture.

1. Have your partner stand with his or her back to the blackboard or whiteboard.
2. Draw one line parallel to each side of your partner's body. He or she should stand erect, with eyes open and staring straight ahead for 2 minutes while you observe any movements. Did you see any gross swaying movements?

____

3. Repeat the test. This time the subject's eyes should be closed. Note and record the degree of side-to-side movement.

____

4. Repeat the test with the subject's eyes first open and then closed. This time, however, the subject should be positioned with his or her left shoulder toward, but not touching, the board so that you may observe and record the degree of front-to-back swaying.

____

Do you think the equilibrium apparatus of the internal ear was operating equally well in all these tests?

____

The proprioceptors? ______________

Why was the observed degree of swaying greater when the eyes were closed?

____

____

Text continues on next page. →

What conclusions can you draw regarding the factors necessary for maintaining body equilibrium and balance?

___

___

Role of Vision in Maintaining Equilibrium

To further demonstrate the role of vision in maintaining equilibrium, perform the following experiment. (Ask your lab partner to record observations and act as a "spotter.") Stand erect, with your eyes open. Raise your left foot approximately 30 cm off the floor, and hold it there for 1 minute.

Record the observations: ___

___

___

Rest for 1 or 2 minutes; and then repeat the experiment with the same foot raised but with your eyes closed. Record the observations:

___

___

REVIEW SHEET

EXERCISE 25

Special Senses: Hearing and Equilibrium

Instructors may assign a portion of the Review Sheet questions using Mastering A&P™

Name ______________________ Lab Time/Date ______________________

Anatomy of the Ear

1. Identify all indicated structures and ear regions in the following photograph.

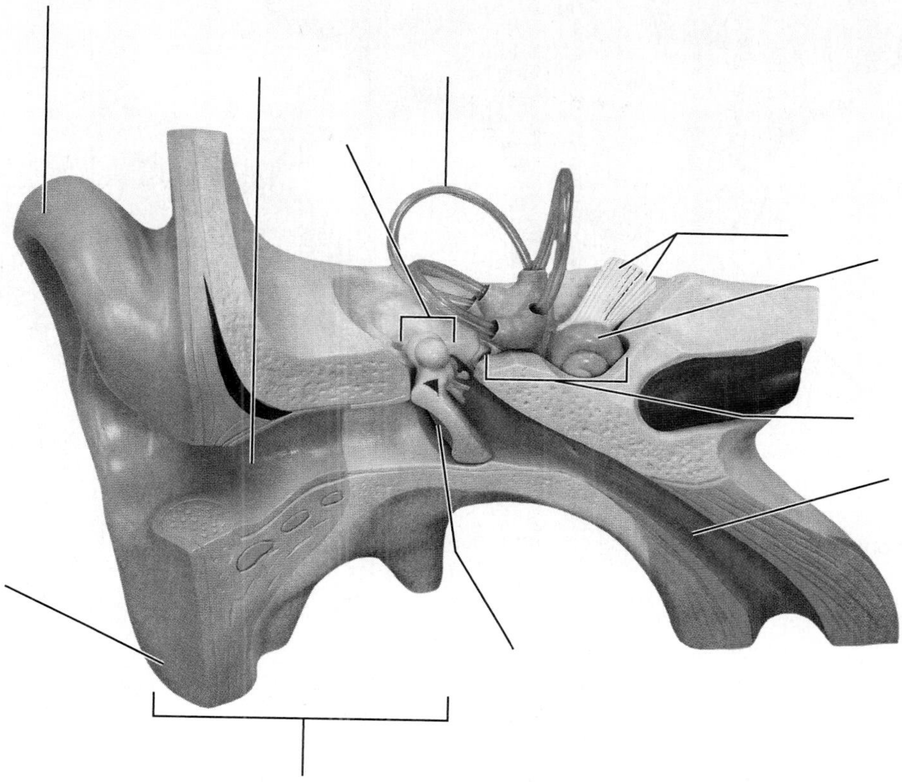

2. Identify the structures of the middle and internal ear indicated in the following photograph.

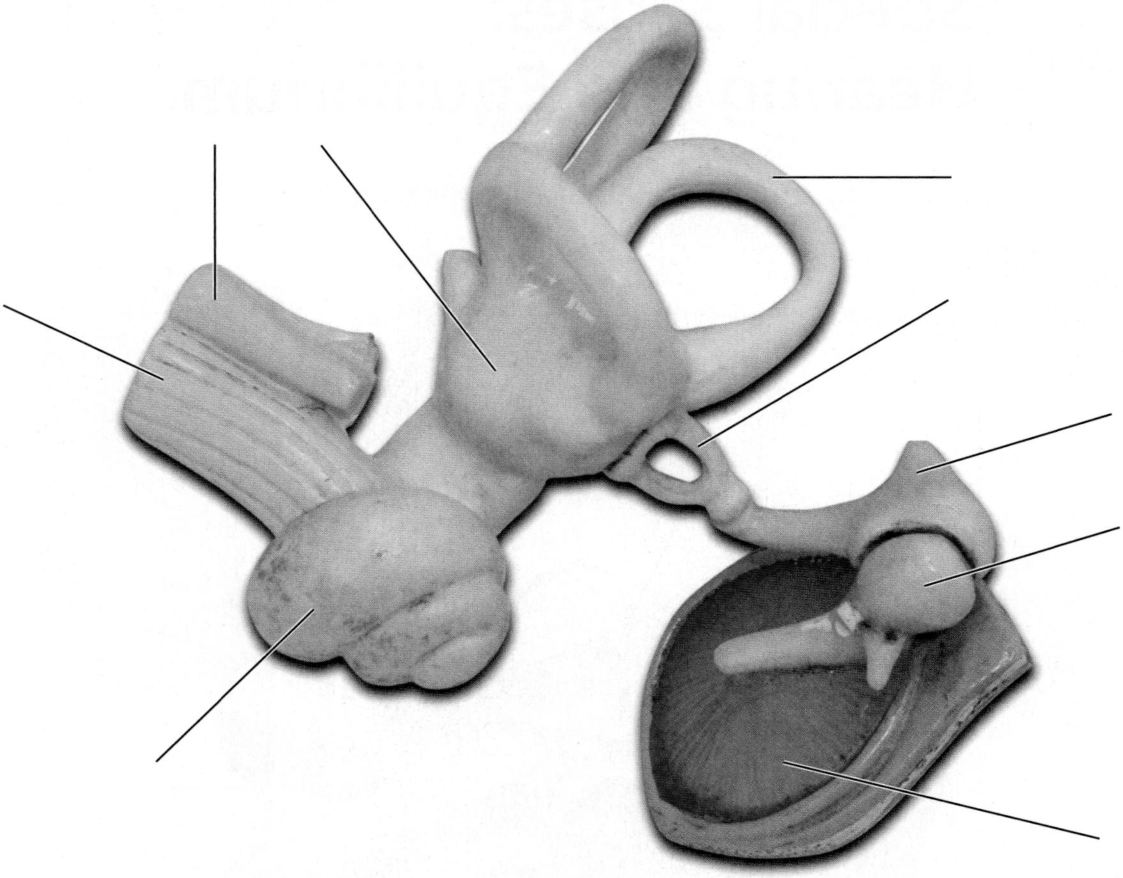

3. Select the terms from column B that match the column A descriptions.

Column A		Column B
_______, _______, _______	1. collectively called the auditory ossicles	a. ampulla
__________, __________	2. sacs found within the vestibule; sites of the maculae	b. ampullary cupula
________________	3. vibrates at the same frequency as the sound waves hitting it; transmits the vibrations to the ossicles	c. basilar membrane
________________	4. fluid contained within the membranous labyrinth	d. cochlear duct
________________	5. fluid contained within the bony labyrinth	e. endolymph
________________	6. grains of calcium carbonate in the maculae	f. incus (anvil)
________________	7. location of the spiral organ	g. malleus (hammer)
________________	8. involved in equalizing the pressure in the middle ear with the external air pressure	h. otoliths
________________	9. positioned in all three special planes	i. oval window
________________	10. gelatinous cap overlying hair cells of the crista ampullaris	j. perilymph
		k. pharyngotympanic tube
		l. round window

______	11. transmits the vibrations of the stapes to the fluid in the scala vestibuli	m. saccule
		n. semicircular ducts
______	12. hair cells of the spiral organ rest on this membrane	o. stapes (stirrup)
______	13. contains the crista ampullaris	p. tectorial membrane
______	14. acts as a pressure relief valve for pressure waves in the scala tympani	q. tympanic membrane
		r. utricle
______	15. gelatinous membrane overlying in the hair cells of the spiral organ	

4. Sound waves hitting the tympanic membrane initiate its vibratory motion. Trace the pathway through which vibrations and fluid currents are transmitted to finally stimulate the hair cells in the spiral organ. (Name the appropriate ear structures in their correct sequence.)

Tympanic membrane → ______________________________

____

____

5. Explain how the basilar membrane allows us to differentiate sounds of different pitch. ______________________________

____

____

____

6. Explain the role of the endolymph of the semicircular ducts in activating the receptors during angular motion.

____

____

____

7. Explain the role of the otoliths in perception of static equilibrium (head position). ______________________________

____

____

____

Laboratory Tests

8. Was the auditory acuity measurement made in Activity 4 (on p. 378) the same or different for both ears?

____

What factors might account for a difference in the acuity of the two ears? ______________________________

____

____

9. During the sound localization experiment in Activity 4 (on p. 378), note the position(s) in which the sound was least easily located.

How can you explain this phenomenon? ______________________________

____

10. In the frequency experiment in Activity 4 (on p. 378), note which tuning fork was the most difficult to hear. ____________

What conclusion can you draw? ______________________________

____

11. When the tuning fork handle was pressed to the forehead during the Weber test, where did the sound seem to originate?

____

Where did it seem to originate when one ear was plugged with cotton? ______________________________

12. Indicate whether the following conditions relate to conduction deafness (C), sensorineural deafness (S), or both (C and S).

____________ 1. can result from the fusion of the ossicles

____________ 2. can result from a lesion on the cochlear nerve

____________ 3. sound heard in one ear but not in the other during bone and air conduction

____________ 4. can result from otitis media

____________ 5. can result from impacted cerumen or a perforated eardrum

____________ 6. can result from a blood clot in the primary auditory cortex

13. The Rinne test evaluates an individual's ability to hear sounds conducted by air or bone. Which is more indicative of normal hearing? ______________________________

14. The Barany test investigated the effect that rotatory acceleration had on the semicircular canals. Explain *why* the subject still had the sensation of rotation immediately after being stopped. ______________________________

____

15. What is the usual reason for conducting the Romberg test? ______________________________

____

16. + Acute labyrinthitis is sudden onset of inflammation of the structures that form the membranous labyrinth. List the structures that could be inflamed with this condition. ______________________________

____

17. + Acute labyrinthitis causes temporary impairment of the receptors located in the membranous labyrinth. Describe the symptoms that might accompany this condition. ______________________________

____

26 EXERCISE Special Senses: Olfaction and Taste

Learning Outcomes

- State the location and cellular composition of the olfactory epithelium.
- Describe the structure of olfactory sensory neurons, and state their function.
- Discuss the locations and cellular composition of taste buds.
- Describe the structure of gustatory epithelial cells, and state their function.
- Identify the cranial nerves that carry the sensations of olfaction and taste.
- Name five basic qualities of taste sensation, and list the chemical substances that elicit them.
- Explain the interdependence between the senses of smell and taste.
- Name two factors other than olfaction that influence taste appreciation of foods.
- Define *olfactory adaptation*.

Go to Mastering A&P™ > Study Area to improve your performance in A&P Lab.

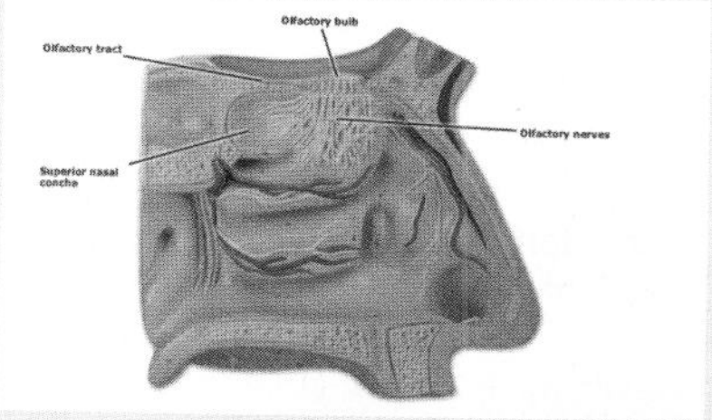

> Lab Tools > Practice Anatomy Lab > Anatomical Models

Instructors may assign new Building Vocabulary coaching activities, Pre-Lab Quiz questions, Art Labeling activities, Practice Anatomy Lab Practical questions (PAL), and more using the Mastering **A&P™** Item Library.

Pre-Lab Quiz

Instructors may assign these and other Pre-Lab Quiz questions using Mastering **A&P™**

1. The organ of smell is the ___________, located in the roof of the nasal cavity.
   a. nares
   b. nostrils
   c. olfactory epithelium
   d. olfactory nerve
2. Circle the correct underlined term. Olfactory receptors are <u>bipolar</u> / <u>unipolar</u> sensory neurons whose olfactory cilia extend outward from the epithelium.
3. Most taste buds are located in ___________, peglike projections of the tongue mucosa.
   a. cilia
   b. concha
   c. papillae
   d. supporting cells
4. Circle the correct underlined term. Vallate papillae are arranged in a V formation on the <u>anterior</u> / <u>posterior</u> surface of the tongue.
5. Circle the correct underlined term. Most taste buds are made of <u>two</u> / <u>three</u> types of modified epithelial cells.

The receptors for olfaction and taste are classified as **chemoreceptors** because they respond to chemicals in solution. Although five relatively specific types of taste receptors have been identified, the olfactory receptors are considered sensitive to a much wider range of chemical sensations.

Materials

- Prepared slides: nasal olfactory epithelium (l.s.); the tongue showing taste buds (x.s.)
- Compound microscope
- Paper towels
- Packets of granulated sugar
- Disposable autoclave bag
- Paper plates
- Equal-sized food cubes of apple, raw potato, dried prunes, banana, and raw carrot (These prepared foods should be in an opaque container; a foil-lined egg carton would work well.)
- Toothpicks
- Disposable gloves
- Cotton-tipped swabs
- Paper cups
- Flask of distilled or tap water
- Prepared vials of oil of cloves, oil of peppermint, and oil of wintergreen or corresponding flavorings found in the condiment section of a supermarket
- Chipped ice

Text continues on next page. →

- Five numbered vials containing common household substances with strong odors (herbs, spices, etc.)
- Nose clips
- Absorbent cotton

Olfactory Epithelium and Olfaction

A pseudostratified epithelium called the **olfactory epithelium** is the organ of smell. It occupies an area lining the roof of the nasal cavity (**Figure 26.1a**). Since most of the air entering the nasal cavity enters the respiratory passages below, the superiorly located nasal epithelium is in a rather poor position for performing its function. This is why sniffing, which brings more air into contact with the receptors, increases your ability to detect odors.

Three cell types are found within the olfactory epithelium:

- **Olfactory sensory neurons:** Specialized receptor cells that are bipolar neurons with nonmotile olfactory cilia.

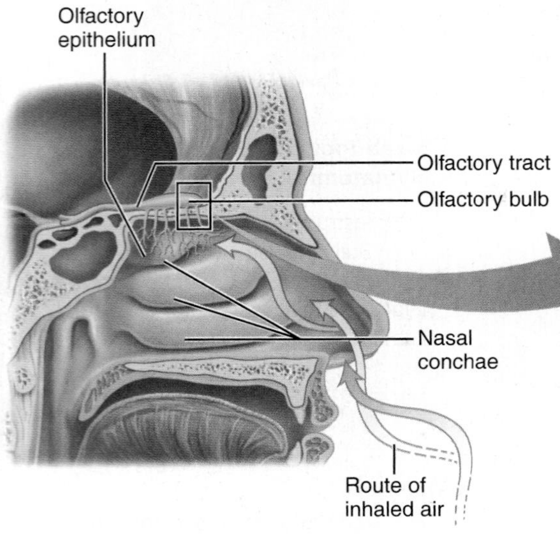

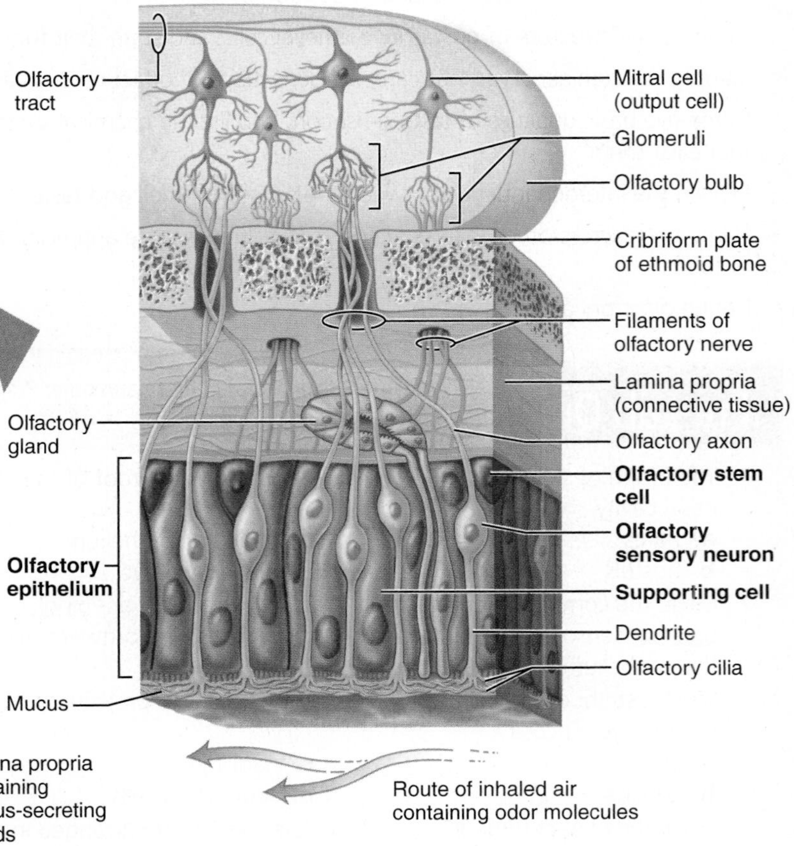

(a)

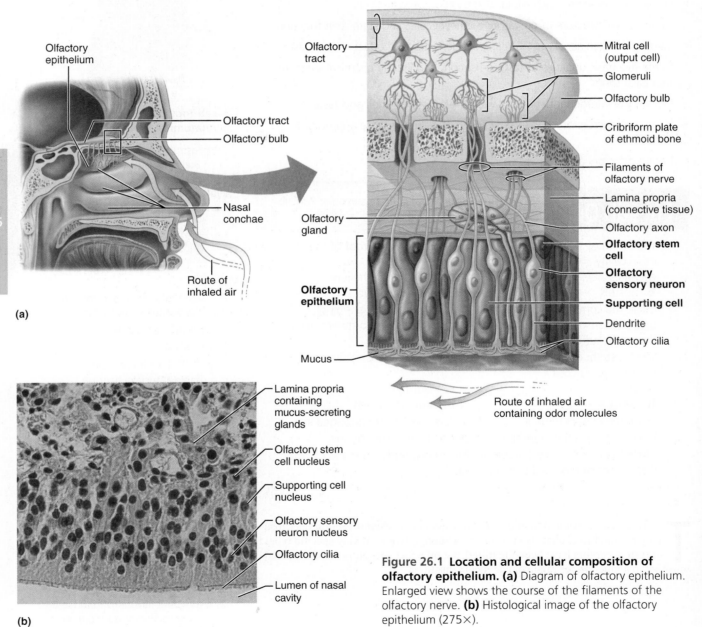

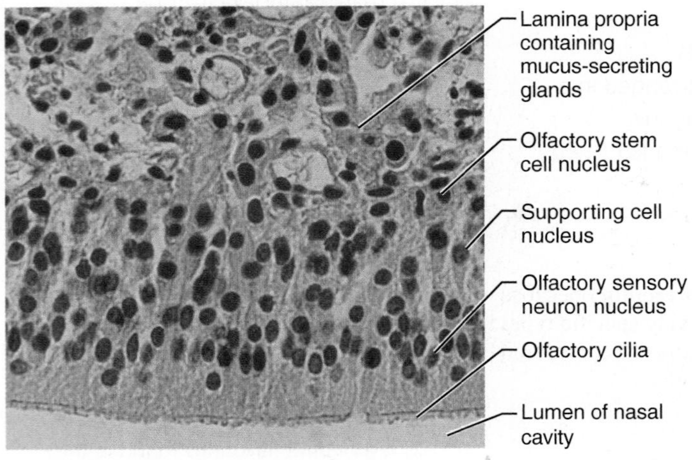

(b)

Figure 26.1 Location and cellular composition of olfactory epithelium. (a) Diagram of olfactory epithelium. Enlarged view shows the course of the filaments of the olfactory nerve. **(b)** Histological image of the olfactory epithelium (275×).

Instructors may assign this figure as an Art Labeling Activity using Mastering A&P™

- **Supporting cells:** Columnar cells that surround and support the olfactory sensory neurons. They form the bulk of the olfactory epithelium.
- **Olfactory stem cells:** Located near the basal surface of the epithelium, they divide to form new olfactory sensory neurons.

The axons of the olfactory sensory neurons form small fascicles called the *filaments of the olfactory nerve* (cranial nerve I), which penetrate the cribriform foramina and synapse in the olfactory bulbs.

Activity 1

Microscopic Examination of the Olfactory Epithelium

Obtain a longitudinal section of olfactory epithelium. Examine it closely using a compound microscope, comparing it to the photomicrograph (Figure 26.1b).

Taste Buds and Taste

The **taste buds**, containing specific receptors for the sense of taste, are widely but not uniformly distributed in the oral cavity. Most are located in **papillae**, peglike projections of the mucosa, on the dorsal surface of the tongue. A few are found on the soft palate, epiglottis, pharynx, and inner surface of the cheeks.

Taste buds are located in the side walls of the large **vallate papillae** (arranged in a V formation on the posterior surface of the tongue); in the side walls of the **foliate papillae**; and on the tops of the more numerous, mushroom-shaped **fungiform papillae** (Figure 26.2).

Each taste bud consists largely of an arrangement of two types of modified epithelial cells:

- **Gustatory epithelial cells:** The receptors for taste; they have long microvilli called **gustatory hairs** that project through the epithelial surface through a **taste pore**.

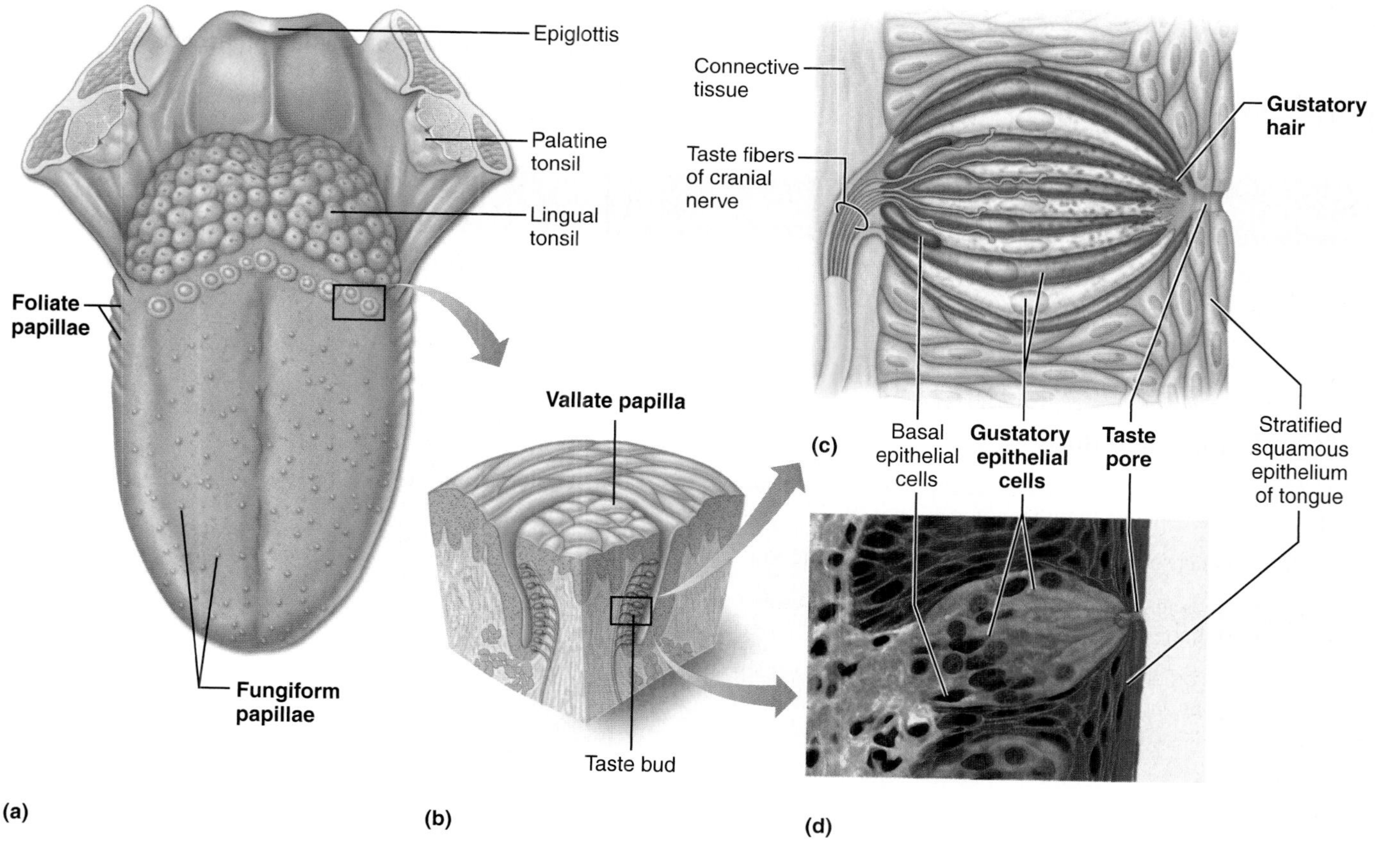

Figure 26.2 Location and structure of taste buds. (a) Taste buds on the tongue are associated with papillae, projections of the tongue mucosa. **(b)** An enlarged section of a vallate papilla shows the position of the taste buds in its lateral walls. **(c)** An enlarged view of a taste bud. **(d)** Photomicrograph of a taste bud (445×).

Instructors may assign this figure as an Art Labeling Activity using Mastering A&P™

- **Basal epithelial cells:** Precursor cells that divide to replace the gustatory epithelial cells.

When the gustatory hairs contact food molecules dissolved in saliva, the gustatory epithelial cells depolarize. The sensory (afferent) neurons that innervate the taste buds are located in three cranial nerves: the *facial nerve (VII)* serves the anterior two-thirds of the tongue; the *glossopharyngeal nerve (IX)* serves the posterior third of the tongue; and the *vagus nerve (X)* carries a few fibers from the pharyngeal region.

When taste is tested with pure chemical compounds, most taste sensations can be grouped into one of five basic qualities—sweet, sour, bitter, salty, or umami (oo-mom'ē; "delicious"). Although all taste buds are believed to respond in some degree to all five classes of chemical stimuli, each type responds optimally to only one.

The *sweet* receptors respond to a number of seemingly unrelated compounds such as sugars (fructose, sucrose, glucose), saccharine, some lead salts, and some amino acids. *Sour* receptors are activated by acids, specifically their hydrogen ions (H^+) in solution. *Salty* taste seems to be due to an influx of metal ions, particularly Na^+, while *umami* is elicited by the amino acids glutamate and aspartate, which are responsible for the "meat taste" of beef and the flavor of monosodium glutamate (MSG). *Bitter* taste is elicited by alkaloids (e.g., caffeine and quinine) and other nonalkaloid substances, such as aspirin.

Activity 2

Microscopic Examination of Taste Buds

Obtain a microscope and a prepared slide of a tongue cross section. Locate the taste buds on the tongue papillae (use Figure 26.2b as a guide). Make a detailed study of one taste bud. Identify the taste pore and gustatory hairs if observed. Compare your observations to the photomicrograph (**Figure 26.3**).

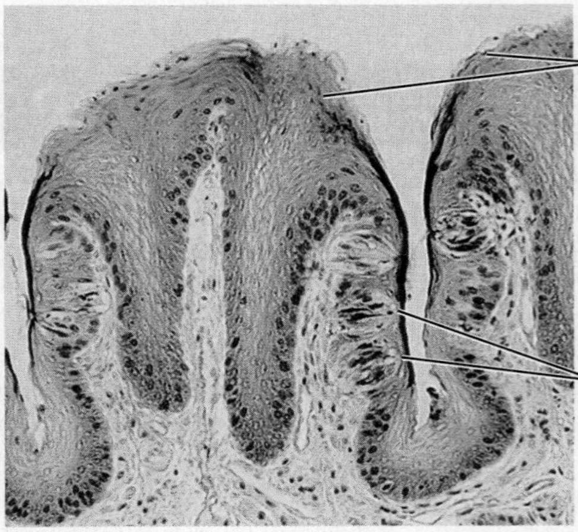

Figure 26.3 Taste buds on the lateral aspects of foliate papillae of the tongue (140×).

26

Laboratory Experiments

⚠ *Notify instructor of any food or scent allergies or restrictions before beginning experiments.*

Activity 3

Stimulating Taste Buds

1. Obtain several paper towels, a sugar packet, and a disposable autoclave bag, and bring them to your bench.

2. With a paper towel, dry the dorsal surface of your tongue.

⚠ Immediately dispose of the paper towel in the autoclave bag.

3. Tear off a corner of the sugar packet, and shake a few sugar crystals on your dried tongue. Do *not* close your mouth.

How long does it take to taste the sugar? ____________ sec

Why couldn't you taste the sugar immediately?

____

Activity 4

Examining the Combined Effects of Smell, Texture, and Temperature on Taste

Effects of Smell and Texture

1. Ask the subject to sit with eyes closed and to pinch the nostrils shut.

2. Using a paper plate, obtain samples of the food items provided by your laboratory instructor. At no time should the subject be allowed to see the foods being tested. Wear disposable gloves, and use toothpicks to handle food.

3. For each test, place a cube of food in the subject's mouth, and ask him or her to identify the food by using the following sequence of activities:

- First, manipulate the food with the tongue.
- Second, chew the food.
- Third, if the subject does not make a positive identification with the first two techniques and the taste sense, ask the subject to release the pinched nostrils and to continue chewing with the nostrils open. This may help the subject make a positive identification.

In the **Activity 4 chart**, record the type of food, and then put a check mark in the appropriate column for the result.

Was the sense of smell equally important in all cases?

____

Where did it seem to be important, and why?

____

____

Discard gloves in autoclave bag.

Effect of Olfactory Stimulation

What is commonly referred to as taste depends heavily on stimulation of the olfactory receptors, particularly in the case of strongly odoriferous substances. The following experiment should illustrate this fact.

1. Obtain vials of oil of wintergreen, peppermint, and cloves, paper cup, flask of water, paper towels, and some fresh cotton-tipped swabs. Ask the subject to sit so that he or she cannot see which vial is being used. Then ask the subject to dry the tongue and pinch the nostrils shut.

2. Use a cotton swab to apply a drop of one of the oils to the subject's tongue. Can he or she distinguish the flavor?

____

! Put the used swab in the autoclave bag. *Do not redip the swab into the oil.*

3. Have the subject open the nostrils, and record the change in sensation he or she reports.

____

4. Have the subject rinse the mouth well and dry the tongue.

5. Prepare two swabs, each with one of the two remaining oils.

6. Hold one swab under the subject's open nostrils while touching the second swab to the tongue.

Record the reported sensations. ______________________________

____

____

! 7. Dispose of the used swabs and paper towels in the autoclave bag before continuing.

Which sense, taste or smell, appears to be more important in the proper identification of a strongly flavored volatile substance?

____

Effect of Temperature

Olfaction and food texture are not the only influences on our taste sensations. Temperature also helps determine whether we appreciate or even taste the foods we eat. To illustrate this, have your partner hold some chipped ice on the tongue for approximately a minute and then close the eyes. Immediately place any of the foods previously identified in the mouth, and ask for an identification.

Results? ______________________________

____

____

Activity 4: Identification by Texture and Smell

Food tested	Texture only	Chewing with nostrils pinched	Chewing with nostrils open	Identification not made

Activity 5

Assessing the Importance of Taste and Olfaction in Odor Identification

1. Go to the designated testing area. Close your nostrils with a nose clip, and breathe through your mouth. Breathing through your mouth only, attempt to identify the odors of common substances in the numbered vials at the testing area. Do not look at the substance in the container. Record your responses on the chart above.

2. Remove the nose clips, and repeat the tests using your nose to sniff the odors. Record your responses in the **Activity 5 chart**.

3. Record any other observations you make as you conduct the tests.

4. Which method gave the best identification results?

____

What can you conclude about the effectiveness of the senses of taste and olfaction in identifying odors?

____

Activity 5: Identification by Mouth and Nasal Inhalation

Vial number	Identification with nose clips	Identification without nose clips	Other observations
1			
2			
3			
4			
5			

Activity 6

Demonstrating Olfactory Adaptation

Obtain some absorbent cotton and two of the following oils: oil of wintergreen, peppermint, or cloves. Place several drops of oil on the absorbent cotton. Press one nostril shut.

Hold the cotton under the open nostril, and exhale through the mouth. Record the time required for the odor to disappear (for olfactory adaptation to occur).

______________ sec

Repeat the procedure with the other nostril.

______________ sec

Immediately test another oil with the nostril that has just experienced olfactory adaptation. What are the results?

____

What conclusions can you draw? ______________________________

____

Instructors may assign a portion of the Review Sheet questions using Mastering A&P™

REVIEW SHEET

EXERCISE 26

Special Senses: Olfaction and Taste

Name ______________________________ Lab Time/Date ______________________

Olfactory Epithelium and Olfaction

1. Match the terms in column B with the appropriate description in column A.

Column A

__________ 1. the organ of smell

__________ 2. cell type that forms most of the olfactory epithelium

__________ 3. cell type that differentiates to replace olfactory sensory neurons

__________ 4. bipolar neurons with nonmotile cilia

__________ 5. axons that pass through the cribriform foramina

Column B

a. filaments of the olfactory nerve
b. olfactory epithelium
c. olfactory sensory neurons
d. olfactory stem cells
e. supporting cells

2. How and why does sniffing increase your ability to detect an odor? ______________________________

____

Taste Buds and Taste

3. Name five sites where receptors for taste are found, and circle the predominant site.

____________________, ____________________, ____________________,

____________________, and ____________________

4. Describe the cellular makeup and arrangement of a taste bud. (Use a diagram, if helpful.) ____________________

____

5. Taste and smell receptors are both classified as ______________________________, because they both

respond to ____

6. Why is it impossible to taste substances if your tongue is dry? ______________________________

____

7. The basic taste sensations are mediated by specific chemical substances or groups. Name them for the following taste modalities.

salt: ____________________ sour: ____________________ umami: ____________________

bitter: ____________________ sweet: ____________________

Laboratory Experiments

8. Name three factors that influence our enjoyment of foods. Substantiate each choice with an example from the laboratory experience.

1. ____________________ Substantiation: ____

____

2. ____________________ Substantiation: ____

____

3. ____________________ Substantiation: ____

____

Which of the factors chosen is most important? ____________________ Substantiate your choice with an example from everyday life. ____

____

Expand on your explanation and choices by explaining why a cold, greasy hamburger is unappetizing to most people.

____

____

9. One symptom of the common cold is loss of appetite. Explain why this occurs.

____

____

10. Invasive dental procedures can permanently or temporarily alter gustation. If taste sensations at the tip of the tongue are absent, which cranial nerve is most likely to be affected? ____

____

EXERCISE 27 Functional Anatomy of the Endocrine Glands

Learning Outcomes

- Identify the major endocrine glands of the body using an appropriate image.
- List the major hormones, and discuss the target and general function of each.
- Explain how hormones contribute to body homeostasis, using appropriate examples.
- Discuss some mechanisms that stimulate release of hormones from endocrine glands.
- Describe the structural and functional relationship between the hypothalamus and the pituitary gland.
- Correctly identify the histology of the thyroid, parathyroid, pancreas, anterior and posterior pituitary, adrenal cortex, and adrenal medulla by microscopic inspection or in an image.
- Name and identify the specialized hormone-secreting cells in the above tissues.
- Describe the pathology of hypersecretion and hyposecretion of several of the hormones studied.

Pre-Lab Quiz

Instructors may assign these and other Pre-Lab Quiz questions using Mastering A&P™

1. Circle the correct underlined term. An endocrine / exocrine gland is a ductless gland that empties its hormone into the extracellular fluid.
2. The pituitary gland, also known as the ______________, is located in the sella turcica of the sphenoid bone.
   a. hypophysis **b.** hypothalamus **c.** thalamus
3. The ______________ gland is composed of two lobes and located in the throat, just inferior to the larynx.
   a. pancreas **c.** thymus
   b. posterior pituitary **d.** thyroid
4. This gland is rather large in an infant, begins to atrophy at puberty, and is relatively inconspicuous by old age. It produces hormones that direct the maturation of T cells. It is the _________ gland.
   a. pineal **c.** thymus
   b. testes **d.** thyroid
5. Circle the correct underlined term. Pancreatic islets / Acinar cells form the endocrine portion of the pancreas.

Go to Mastering A&P™ > Study Area to improve your performance in A&P Lab.

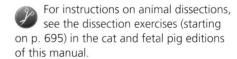

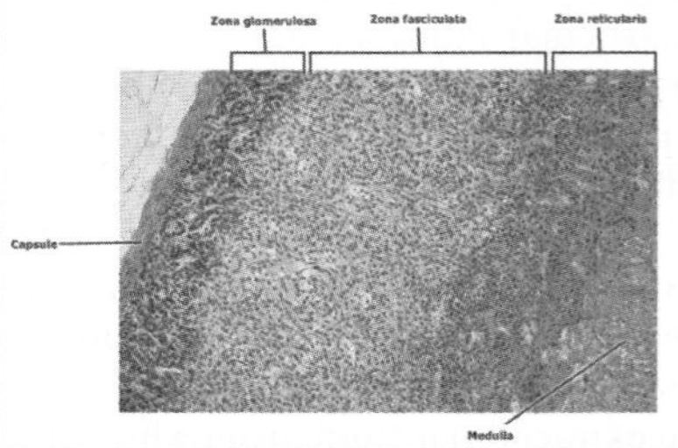

> Lab Tools > Practice Anatomy Lab > Histology

Instructors may assign new Building Vocabulary coaching activities, Pre-Lab Quiz questions, Art Labeling activities, Practice Anatomy Lab Practical questions (PAL), and more using the Mastering A&P™ Item Library.

Materials

- Human torso model
- Anatomical chart of the human endocrine system
- Compound microscope
- Prepared slides of the anterior pituitary and pancreas (with differential staining), posterior pituitary, thyroid gland, parathyroid glands, and adrenal gland

For instructions on animal dissections, see the dissection exercises (starting on p. 695) in the cat and fetal pig editions of this manual.

The **endocrine system** is the second major control system of the body. Acting with the nervous system, it helps coordinate and integrate the activity of the body. The nervous system uses electrochemical impulses to bring about rapid control, whereas the more slowly acting endocrine system uses chemical messengers, or **hormones**.

The term *hormone* comes from a Greek word meaning "to arouse." The body's hormones, which are steroids or amino acid–based molecules, arouse the body's tissues and cells by stimulating changes in their metabolic activity. These changes lead to growth and development and to the physiological homeostasis of many body systems. Although hormones travel through the blood, a given hormone affects only a specific organ or organs. Cells within an organ that respond to a particular hormone are referred to as the **target cells** (also **target**) of that hormone. The ability of the target to respond depends on the ability of the hormone to bind with specific cellular receptors.

Although the function of most hormone-producing glands is purely endocrine, the function of others (the pancreas and gonads) is mixed—both endocrine and exocrine. The endocrine glands release their hormones directly into the extracellular fluid, from which the hormones enter blood or lymph.

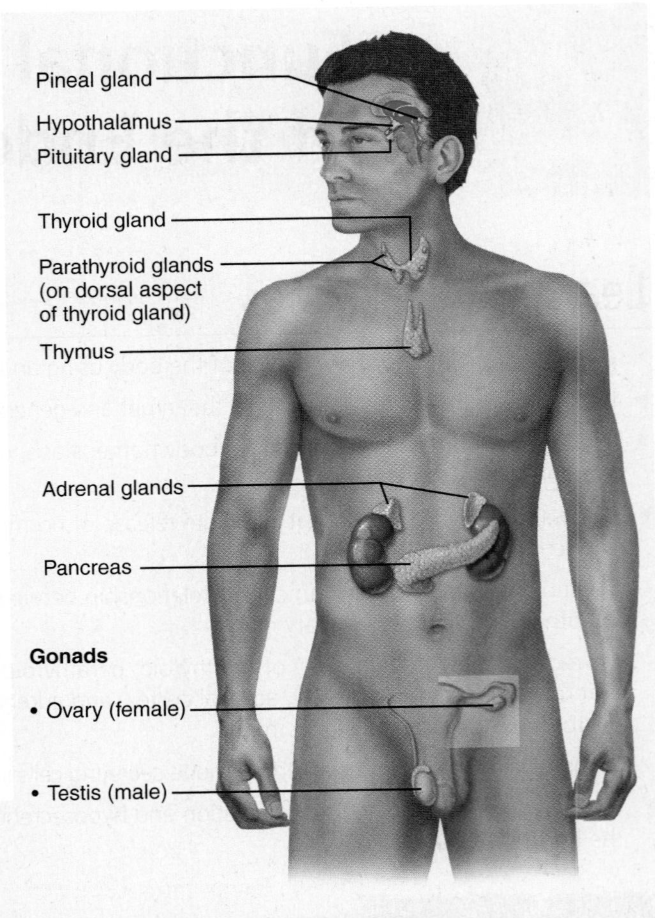

Figure 27.1 Human endocrine organs.

Instructors may assign this figure as an Art Labeling Activity using Mastering A&P™

Activity 1

Identifying the Endocrine Organs

27

Locate the endocrine organs in **Figure 27.1**. Also locate these organs on the anatomical charts or torso model. As you locate the organs, read through Tables 27.1–27.4.

Endocrine Glands

Pituitary Gland (Hypophysis)

The **pituitary gland**, or **hypophysis**, is located in the sella turcica of the sphenoid bone. It consists largely of two functional *lobes*, the **adenohypophysis**, or **anterior pituitary**, and the **neurohypophysis**, consisting of the **posterior pituitary** and the **infundibulum**—the stalk that attaches the pituitary gland to the hypothalamus (**Figure 27.2**).

The anterior pituitary produces and secretes a number of hormones, four of which are **tropic hormones**. The target organ of a tropic hormone (tropin) is another endocrine gland.

Because the anterior pituitary controls the activity of many other endocrine glands, it is sometimes called the *master endocrine gland*. However, because *releasing* or *inhibiting hormones* from neurons of the ventral hypothalamus control anterior pituitary cells, the hypothalamus supersedes the anterior pituitary as the major controller of endocrine glands.

The ventral hypothalamic hormones control production and secretion of the anterior pituitary hormones. The hypothalamic hormones reach the cells of the anterior pituitary through the **hypophyseal portal system** (Figure 27.2), a complex vascular arrangement of two capillary beds that are connected by the hypophyseal portal veins.

The posterior pituitary is not an endocrine gland, because it does not synthesize the hormones it releases. Instead, it acts as a storage area for two *neurohormones* transported to it via the axons of neurons in the paraventricular and supraoptic nuclei of the hypothalamus. Refer to **Table 27.1** on pp. 397–398, which summarizes the hormones released by the pituitary gland.

Pineal Gland

The *pineal gland* is a small cone-shaped gland located in the roof of the third ventricle of the brain. Its major endocrine product is **melatonin**, which exhibits a diurnal (daily) cycle. It peaks at night, making us drowsy, and is lowest around noon. Recent evidence suggests that melatonin has anti-aging properties. Melatonin appears to play a role in the production of antioxidants.

Thyroid Gland

The *thyroid gland* is composed of two lobes joined by a central mass, or isthmus. It is located in the anterior neck, just inferior to the larynx.

Parathyroid Glands

The *parathyroid glands* are found embedded in the posterior surface of the thyroid gland. Typically, there are two small oval

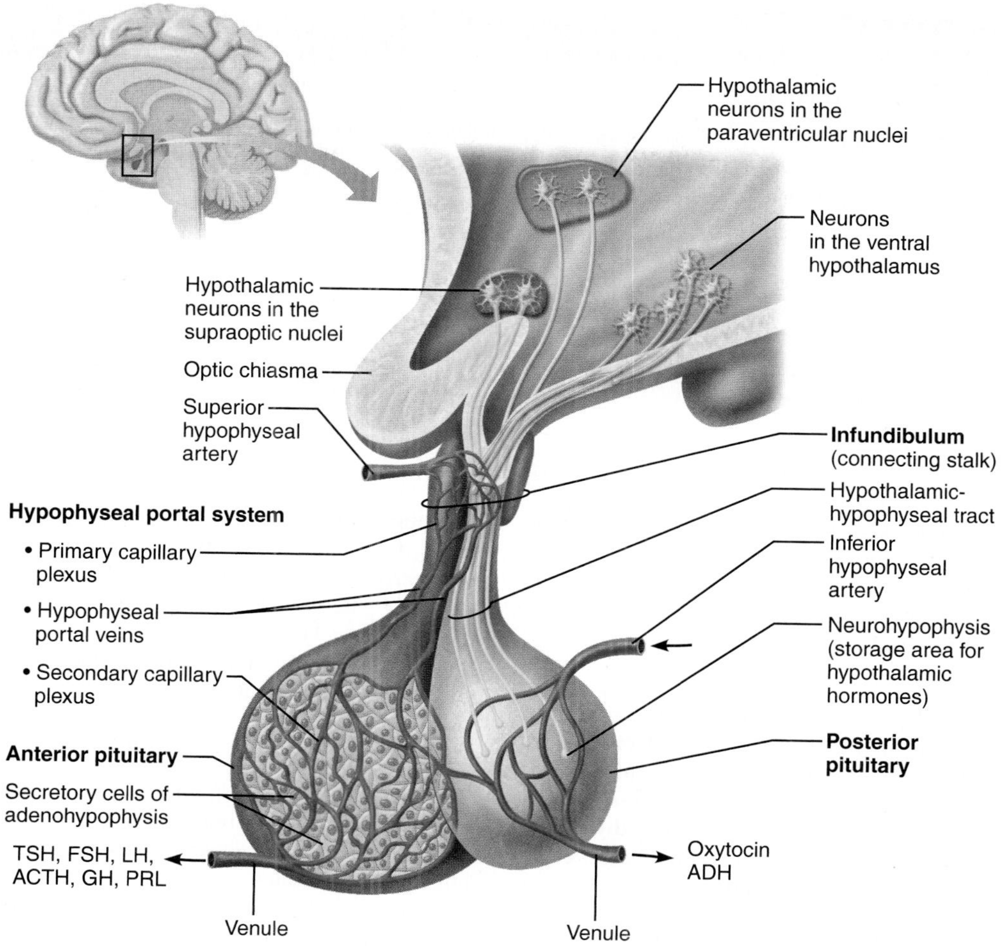

Figure 27.2 Hypothalamus and pituitary gland. Neural and vascular relationships between the hypothalamus and the anterior and posterior lobes of the pituitary are depicted.

Table 27.1A **Pituitary Gland Hormones (Figure 27.2)**

Hormone	Stimulus for release	Target	Effects
Anterior Pituitary Gland: Tropic Hormones			
Thyroid-stimulating hormone (TSH)	Thyrotropin-releasing hormone (TRH)*	Thyroid gland	Stimulates the secretion of thyroid hormones (T_3 and T_4)
Follicle-stimulating hormone (FSH)	Gonadotropin-releasing hormone (GnRH)*	Ovaries and testes (gonads)	**Females**—stimulates ovarian follicle maturation and estrogen production **Males**—stimulates sperm production
Luteinizing hormone (LH)	Gonadotropin–releasing hormone (GnRH)*	Ovaries and testes (gonads)	**Females**—triggers ovulation and stimulates ovarian production of estrogen and progesterone **Males**—stimulates testosterone production
Adrenocorticotropic hormone (ACTH)	Corticotropin-releasing hormone (CRH)*	Adrenal cortex	Stimulates the release of glucocorticoids and androgens (mineralocorticoids to a lesser extent)
Anterior Pituitary Gland: Other Hormones (Not Tropic)			
Growth hormone (GH)	Growth hormone–releasing hormone (GHRH)*	Liver, muscle, bone, and cartilage, mostly	Stimulates body growth and protein synthesis, mobilizes fat and conserves glucose
Prolactin (PRL)	A decrease in the amount of prolactin-inhibiting hormone (PIH)*	Mammary glands in the breasts	Stimulates milk production (lactation)

* Indicates hormones produced by the hypothalamus.

Table 27.1B Pituitary Gland Hormones (Figure 27.2)

Hormone	Stimulus for release	Target	Effects
Posterior Pituitary Gland (Hormones That Are Synthesized by the Hypothalamus and Stored in the Posterior Pituitary)			
Oxytocin*	Nerve impulses from hypothalamic neurons in response to cervical/uterine stretch or suckling of an infant	Uterus and mammary glands	Stimulates powerful uterine contractions during birth and stimulates milk ejection (let-down) in lactating mothers
Antidiuretic hormone (ADH)*	Nerve impulses from hypothalamic neurons in response to increased blood solute concentration or decreased blood volume	Kidneys	Stimulates the kidneys to reabsorb more water, reducing urine output and conserving body water

* Indicates hormones produced by the hypothalamus.

glands on each lobe, but there may be more and some may be located in other regions of the neck.

Table 27.2 summarizes the hormones secreted by the thyroid and parathyroid glands.

Thymus

The *thymus* is a bilobed gland situated in the superior thorax, posterior to the sternum and overlying the heart. Conspicuous in the infant, it begins to atrophy at puberty, and by old age it is relatively inconspicuous. The thymus produces several different families of peptide hormones, including **thymulin**, **thymosins**, and **thymopoietins**. These hormones are thought to be involved in the development of T lymphocytes and the immune response, but their roles are poorly understood. They appear to act locally as paracrines.

Adrenal Glands

The two *adrenal*, or *suprarenal*, *glands* are located atop the kidneys. Anatomically, the **adrenal medulla** develops from neural crest tissue, and it is directly controlled by the sympathetic nervous system. The medullary cells respond to this stimulation by releasing a hormone mix of **epinephrine** (80%) and **norepinephrine** (20%), which act with the sympathetic nervous system to elicit the fight-or-flight response to stressors. Table 27.3 summarizes the hormones secreted by the adrenal glands.

The gonadocorticoids are produced throughout life in relatively insignificant amounts; however, hypersecretion of these hormones produces abnormal hairiness (**hirsutism**) and masculinization. +

Pancreas

The *pancreas*, located behind the stomach and close to the small intestine, functions as both an endocrine and exocrine gland. It produces digestive enzymes as well as insulin and glucagon, important hormones concerned with the regulation of blood sugar levels. Table 27.4 summarizes two of the hormones produced by the pancreas.

The Gonads

The *female gonads*, or *ovaries*, are paired, almond-sized organs located in the pelvic cavity. In addition to producing the female sex cells (ova), the ovaries produce two steroid hormone groups, the **estrogens** and **progesterone**. The endocrine and exocrine functions of the ovaries do not begin until the onset of puberty.

Table 27.2 Thyroid and Parathyroid Gland Hormones

Hormone(s)	Stimulus for release	Target	Effects
Thyroid Gland			
Thyroxine (T_4) and Triiodothyronine (T_3), collectively referred to as thyroid hormone (TH)	Thyroid-stimulating hormone (TSH)	Most cells of the body	Increases basal metabolic rate (BMR); regulates tissue growth and development.
Calcitonin	High levels of calcium in the blood	Bones	No known physiological role in humans. When the hormone is supplemented at doses higher than normally found in humans, it does have some pharmaceutical applications.
Parathyroid Gland (Located on the Posterior Aspect of the Thyroid Gland)			
Parathyroid hormone (PTH)	Low levels of calcium in the blood	Bones and kidneys	Increases blood calcium by stimulating osteoclasts and by stimulating the kidneys to reabsorb more calcium. PTH also stimulates the kidneys to convert vitamin D to calcitriol, which is required for the absorption of calcium in the intestines.

Table 27.3 Adrenal Gland Hormones

Cortical area	Hormone(s)	Stimulus for release	Target	Effects
Adrenal Cortex				
Zona glomerulosa	Mineralocorticoids: mostly aldosterone	Angiotensin II release and increased potassium in the blood (ACTH only in times of severe stress)	Kidneys	Increases the reabsorption of sodium and water by the kidney tubules. Increases the secretion of potassium in the urine.
Zona fasciculata	Glucocorticoids: mostly cortisol	ACTH	Most body cells	Promotes the breakdown of fat and protein, promotes stress resistance, and inhibits the immune response.
Zona reticularis	Gonadocorticoids: androgens (most are converted to testosterone and some to estrogen)	ACTH	Bone, muscle, integument, and other tissues	In females, androgens contribute to body growth, contribute to the development of pubic and axillary hair, and enhance sex drive. They have insignificant effects in males.
Cells	**Hormone(s)**	**Stimulus for release**	**Target**	**Effects**
Adrenal Medulla				
Chromaffin cells	Catecholamines: epinephrine and norepinephrine	Nerve impulses from preganglionic sympathetic fibers	Most body cells	Mimics sympathetic nervous system activation, "fight-or-flight" response.

Table 27.4 Pancreas and Gonad Hormones

Hormone	Stimulus for release	Target(s)	Effects
Pancreas			
Insulin	Increased blood glucose levels, parasympathetic nervous system stimulation	Most cells of the body	Accelerates the transport of glucose into body cells; promotes glycogen, fat, and protein synthesis
Glucagon	Decreased blood glucose levels, sympathetic nervous system stimulation	Primarily the liver and adipose	Accelerates the breakdown of glycogen to glucose, stimulates the conversion of lactic acid into glucose, releases glucose into the blood from the liver
Ovaries (Female Gonads)			
Estrogens	Luteinizing hormone (LH) and follicle-stimulating hormone (FSH)	Most cells of the body	Promote the maturation of the female reproductive organs and the development of secondary sex characteristics
Estrogens and progesterone (together)	LH and FSH	Uterus and mammary glands	Regulate the menstrual cycle and promote breast development
Testes (Male Gonads)			
Testosterone	LH and FSH	Most cells of the body	Promotes the maturation of the male reproductive organs, the development of secondary sex characteristics, sperm production, and sex drive

The paired oval *testes* of the male are suspended in a pouchlike sac, the scrotum, outside the pelvic cavity. In addition to the male sex cells (sperm), the testes produce the male sex hormone, **testosterone**. Both the endocrine and exocrine functions of the testes begin at puberty. For a more detailed discussion of the function and histology of the ovaries and testes, see Exercises 42 and 43. Table 27.4 summarizes the hormones produced by the gonads.

27

Microscopic Anatomy of Selected Endocrine Glands

Activity 2

Examining the Microscopic Structure of Endocrine Glands

Obtain a microscope and one of each assigned slide. Compare your observations with the images (**Figure 27.3a–f**).

Thyroid Gland

1. Scan the thyroid under low power, noting the **follicles**, generally spherical sacs containing a pink-stained material (*colloid*). Stored T_3 and T_4 are attached to the protein colloidal material stored in the follicles as **thyroglobulin** and are released gradually to the blood. Compare the tissue viewed to the photomicrograph of thyroid tissue (Figure 27.3a).

2. Observe the tissue under high power. Notice that the walls of the follicles are formed by simple cuboidal or squamous epithelial cells. The **parafollicular**, or **C, cells** you see between the follicles produce calcitonin.

When the thyroid gland is actively secreting, the follicles appear small. When the thyroid is hypoactive or inactive, the follicles are large and plump, and the follicular epithelium appears to be squamous.

Parathyroid Glands

Observe the parathyroid tissue under low power to view its two major cell types, the parathyroid cells and the oxyphil cells. Compare your observations to the photomicrograph of parathyroid tissue (Figure 27.3b). The **parathyroid cells**, which synthesize parathyroid hormone (PTH), are small and abundant. The function of the scattered, much larger **oxyphil cells** is unknown.

Pancreas

1. Observe pancreas tissue under low power to identify the roughly circular **pancreatic islets** (also called *islets of Langerhans*), the endocrine portions of the pancreas. The islets are scattered amid the more numerous **acinar cells** and stain differently (usually lighter), which makes it possible to identify them. The deeper-staining acinar cells form the major portion of the pancreatic tissue. Acinar cells produce the exocrine secretion of digestive enzymes. Alkaline fluid produced by duct cells accompanies the hydrolytic enzymes. (See Figure 27.3c.)

2. Focus on islet cells under high power. Notice that they are densely packed and have no definite arrangement (Figure 27.3c). In contrast, the cuboidal acinar cells are arranged around secretory ducts. In Figure 27.3c, it is possible to distinguish the **alpha (α) cells**, which stain darker and produce glucagon, from the **beta (β) cells**, which stain lighter and synthesize insulin. If differential staining is used, the beta cells are larger and stain gray-blue, and the alpha cells are smaller and appear bright pink.

Pituitary Gland

1. Observe the general structure of the pituitary gland under low power to differentiate the glandular anterior pituitary from the neural posterior pituitary.

2. Using the high-power lens, focus on the nests of cells of the anterior pituitary. When differential stains are used, it is possible to identify the specialized cell types that secrete the specific hormones. Using Figure 27.3d as a guide, locate the reddish pink–stained **acidophil cells**, which produce growth hormone and prolactin, and the **basophil cells**, stained blue to purple in color, which produce the tropic hormones (TSH, ACTH, FSH, and LH). **Chromophobe cells**, the third cellular population, do not take up the stain and appear colorless. The role of the chromophobe cells is controversial, but they apparently are not directly involved in hormone production.

3. Now focus on the posterior pituitary, where two hormones (oxytocin and ADH) synthesized by hypothalamic neurons are stored. Observe the nerve fibers of hypothalamic neurons. Also note the **pituicytes** (Figure 27.3e).

Adrenal Gland

1. Hold the slide of the adrenal gland up to the light to distinguish the cortex and medulla areas. Then scan the cortex under low power to distinguish the differences in cell appearance and arrangement in the three cortical areas. Refer to Figure 27.3f as you work. In the outermost **zona glomerulosa**, where most mineralocorticoid production occurs, the cells are arranged in spherical clusters. The deeper intermediate **zona fasciculata** produces glucocorticoids. Its cells are arranged in parallel cords. The innermost cortical zone, the **zona reticularis** produces sex hormones and some glucocorticoids. The cells here stain intensely and form a branching network.

2. Switch to higher power to view the large, lightly stained cells of the adrenal medulla, which produce epinephrine and norepinephrine. Notice their clumped arrangement.

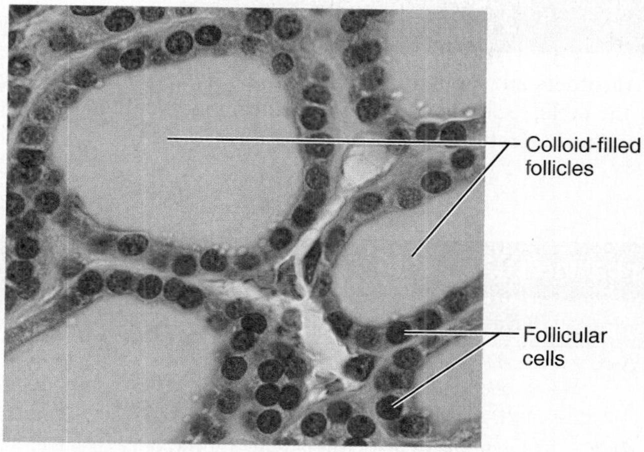

(a) Thyroid gland (360×)

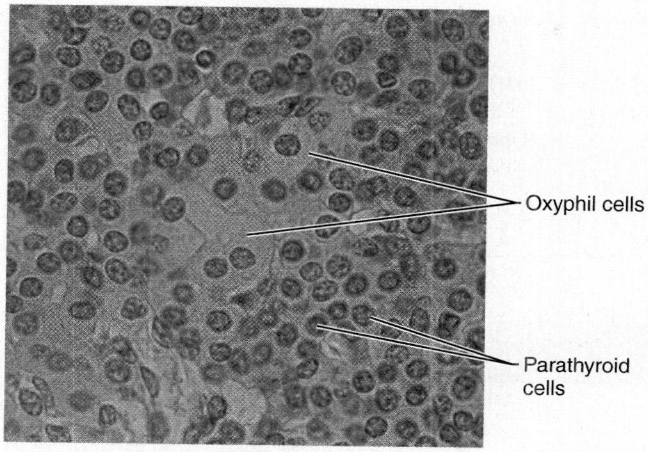

(b) Parathyroid gland (375×)

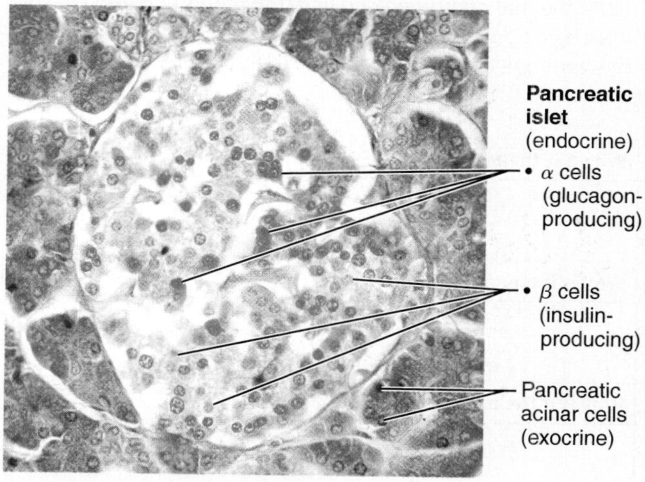

(c) Pancreatic islet (205×)

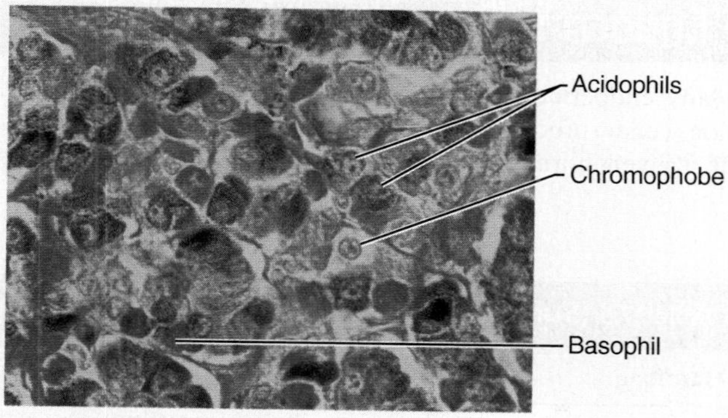

(d) Anterior pituitary (380×)

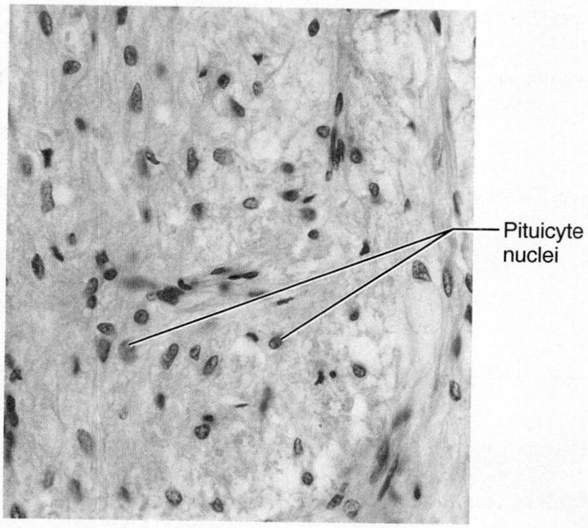

(e) Posterior pituitary (400×)

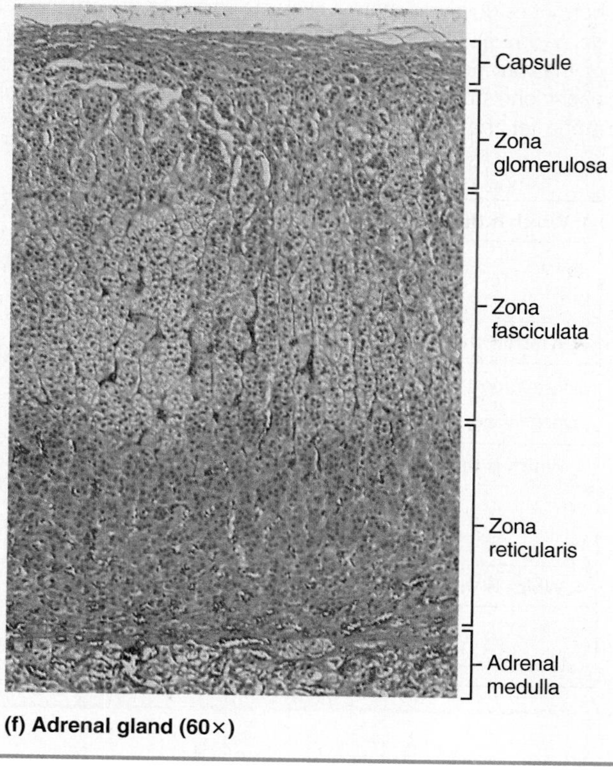

(f) Adrenal gland (60×)

Figure 27.3 **Microscopic anatomy of selected endocrine organs.**

27

Endocrine Disorders

Many endocrine disorders are a result of either hyposecretion (underproduction) or hypersecretion (overproduction) of a given hormone. The characteristics of select endocrine disorders are summarized in **Table 27.5**. As you read through the table, recall the targets for the hormones and the effects of normal secretion levels.

Table 27.5 Summary of Select Endocrine Homeostatic Imbalances

Hormone	Effects of hyposecretion	Effects of hypersecretion
Growth hormone	*In children:* **pituitary dwarfism**, which results in short stature with normal proportions	*In children:* **gigantism**, abnormally tall *In adults:* **acromegaly**, abnormally large bones of the face, feet, and hands
Antidiuretic hormone	**Diabetes insipidus**, a condition characterized by thirst and excessive urine output	**Syndrome of inappropriate ADH secretion**, a condition characterized by fluid retention, headache, and disorientation
Thyroid hormone	*In children:* **cretinism**, intellectual disability with a disproportionately short-sized body *In adults:* **myxedema**, low metabolic rate, edema, physical and mental sluggishness	**Graves' disease**, elevated metabolic rate, sweating, irregular heart rate, weight loss, protrusion of the eyeballs, and nervousness
Parathyroid hormone	**Hypoparathyroidism**, neural excitability with tetany (muscle spasms) and convulsions	**Hyperparathyroidism**, loss of calcium from bones, causing deformation, and spontaneous fractures
Insulin	**Diabetes mellitus**, which results in an inability of cells to take up and utilize glucose and in loss of glucose in the urine (may be due to hyposecretion or hypoactivity of insulin)	**Hypoglycemia**, which results in low blood sugar and is characterized by anxiety, nervousness, tremors, and weakness

27

Group Challenge

Odd Hormone Out

Each box below contains four hormones. One of the listed hormones does not share a characteristic that the other three do. Work in groups of three, and discuss the characteristics of the four hormones in each group. On a separate piece of paper, one student will record the characteristics for each hormone for the group. For each set of four hormones, discuss the possible candidates for the "odd hormone" and which characteristic it lacks based upon your recorded notes. Once you have come to a consensus among your group, circle the hormone that doesn't belong with the others and explain why it is singled out. Sometimes there may be multiple reasons why the hormone doesn't belong with the others.

1. Which is the "odd hormone"?		Why is it the odd one out?
ACTH	oxytocin	
LH	FSH	

2. Which is the "odd hormone"?		Why is it the odd one out?
aldosterone	cortisol	
epinephrine	ADH	

3. Which is the "odd hormone"?		Why is it the odd one out?
PTH	testosterone	
LH	FSH	

4. Which is the "odd hormone"?		Why is it the odd one out?
insulin	cortisol	
calcitonin	glucagon	

REVIEW SHEET

EXERCISE 27

Functional Anatomy of the Endocrine Glands

Instructors may assign a portion of the Review Sheet questions using Mastering A&P™

Name ______________________ Lab Time/Date ______________________

Gross Anatomy and Basic Function of the Endocrine Glands

1. Both the endocrine and nervous systems are major regulating systems of the body; however, the nervous system has been compared to a text message, and the endocrine system to mailing a letter. Briefly explain this comparison.

____

____

____

2. Define *hormone*. ______________________

____

3. Chemically, hormones belong chiefly to two molecular groups, the ______________________

and the ______________________

4. Define *target cell*. ______________________

____

5. If hormones travel in the bloodstream, why don't all tissues respond to all hormones? ______________________

____

6. Identify the endocrine organ described by each of the following statements.

______________ 1. located in the anterior neck; produces key hormones for metabolism

______________ 2. produces the hormones that are stored in the posterior pituitary

______________ 3. a mixed gland, located behind the stomach and close to the small intestine

______________ 4. paired glands suspended in the scrotum

______________ 5. bilobed gland located in the sella turcica

______________ 6. found in the pelvic cavity of the female, concerned with ova and female hormone production

______________ 7. found in the upper thorax overlying the heart; large during youth

______________ 8. found in the roof of the third ventricle of the brain

7. The table below lists the functions of many of the hormones you have studied. From the keys below, fill in the hormones responsible for each function, and the endocrine glands that produce each hormone. Glands may be used more than once.

Hormones Key:

ACTH	FSH	prolactin
ADH	glucagon	PTH
aldosterone	insulin	T_3/T_4
cortisol	LH	testosterone
epinephrine	oxytocin	TSH
estrogens	progesterone	

Glands Key:

adrenal cortex	parathyroid glands
adrenal medulla	posterior pituitary
anterior pituitary	testes
hypothalamus	thyroid gland
ovaries	
pancreas	

Function	Hormone(s)	Synthesizing Gland(s)
Regulate the function of another endocrine gland (tropic)	1.	
	2.	
	3.	
	4.	
Maintain salt and water balance in the extracellular fluid	1.	
	2.	
Directly involved in milk production and ejection	1.	
	2.	
Controls the rate of body metabolism and cellular oxidation	1.	
Regulates blood calcium levels	1.	
Regulate blood glucose levels; produced by the same "mixed" gland	1.	
	2.	
Released in response to stressors	1.	
	2.	
Drive development of secondary sex characteristics in males	1.	
Directly responsible for regulation of the menstrual cycle	1.	
	2.	

8. Although the pituitary gland is sometimes referred to as the master gland of the body, the hypothalamus exerts control over the pituitary gland. How does the hypothalamus control both anterior and posterior pituitary functioning?

___

___

9. Indicate whether the release of the hormones listed below is stimulated by (A) another hormone; (B) the nervous system (neurotransmitters, or neurosecretions); or (C) humoral factors (the concentration of specific nonhormonal substances in the blood or extracellular fluid).

_________ 1. ACTH
_________ 2. calcitonin
_________ 3. estrogens
_________ 4. insulin
_________ 5. norepinephrine
_________ 6. parathyroid hormone
_________ 7. T_4/T_3
_________ 8. testosterone
_________ 9. TSH, FSH

10. Name the hormone(s) produced in *inadequate* amounts that directly result in the following conditions.

_________________________ 1. tetany
_________________________ 2. excessive urine output without high blood glucose levels
_________________________ 3. loss of glucose in the urine
_________________________ 4. abnormally small stature, normal proportions

11. Name the hormone(s) produced in *excessive* amounts that directly result in the following conditions.

_________________________ 1. in the adult: large bones of the hands, feet, and face
_________________________ 2. nervousness, irregular pulse rate, sweating
_________________________ 3. demineralization of bones, spontaneous fractures

Microscopic Anatomy of Selected Endocrine Glands

12. Choose a response from the key below to name the hormone(s) produced by the cell types listed.

Key:
a. calcitonin
b. GH, prolactin
c. glucagon
d. glucocorticoids
e. insulin
f. mineralocorticoids
g. PTH
h. T_4/T_3
i. TSH, ACTH, FSH, LH

_________ 1. parafollicular cells of the thyroid
_________ 2. follicular cells of the thyroid
_________ 3. beta cells of the pancreatic islets
_________ 4. alpha cells of the pancreatic islets
_________ 5. basophil cells of the anterior pituitary
_________ 6. zona fasciculata cells
_________ 7. zona glomerulosa cells
_________ 8. parathyroid cells
_________ 9. acidophil cells of the anterior pituitary

13. + Pituitary gland tumors can secrete excess amounts of growth hormone. Describe the signs and symptoms that these tumors cause in an adult experiencing hypersecretion of the growth hormone. _________________________

____

14. + Tumors of the adrenal medulla, called pheochromocytomas, cause hypersecretion of catecholamines. Describe the expected signs and symptoms of this tumor. _________________________

____

15. Identify the endocrine glands, and name all structures indicated by a leader line.

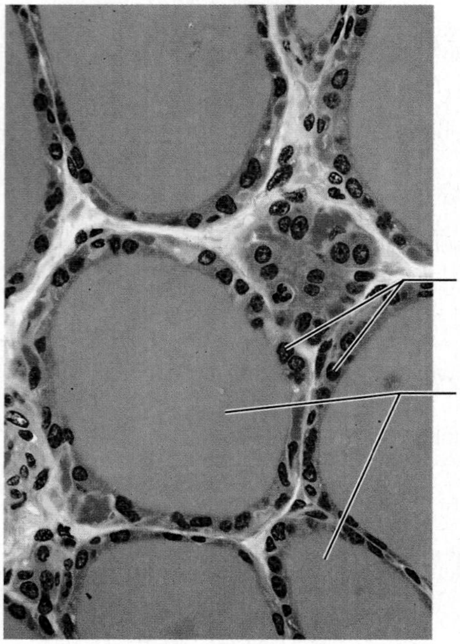

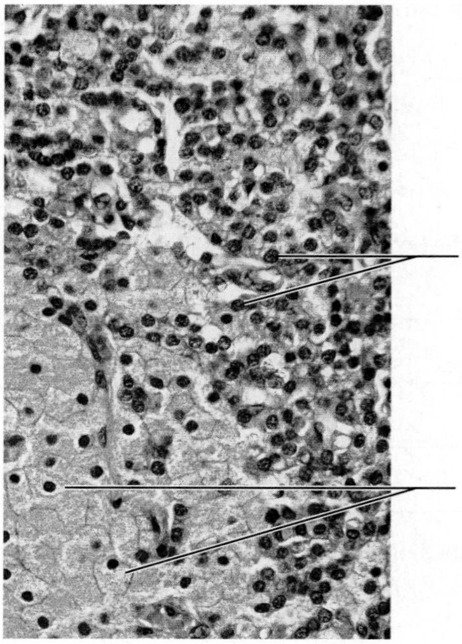

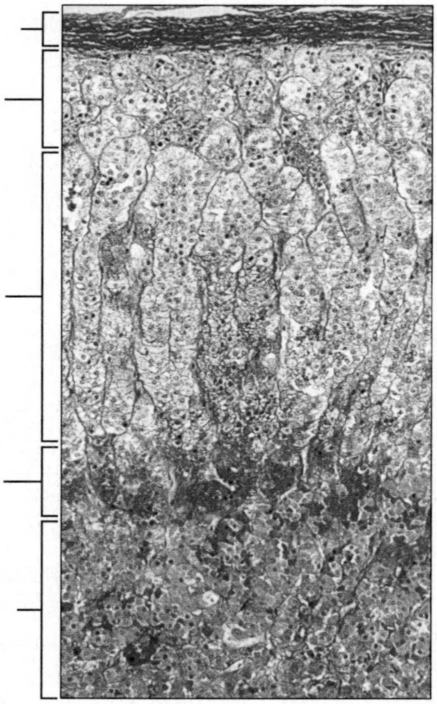

EXERCISE 28

Endocrine Wet Labs and Human Metabolism

Learning Outcomes

- Describe the effects of pituitary extract in the frog, and indicate which hormone(s) is/are responsible for these effects.
- Describe the symptoms of hyperinsulinism in the fish, and explain how these symptoms were reversed.
- Define *metabolism*.
- State the functions of thyroid hormone in the body.
- Explain how negative feedback mechanisms regulate thyroid hormone secretion.
- Describe and explain the various pathologies associated with hypothyroidism and hyperthyroidism.

Pre-Lab Quiz

Instructors may assign these and other Pre-Lab Quiz questions using Mastering A&P™

1. Circle True or False. Gonadotropins are produced by the anterior pituitary gland.
2. Circle the correct underlined term. Many people with diabetes mellitus need injections of insulin / glucagon to maintain glucose homeostasis.
3. Circle the correct underlined term. Catabolism / Anabolism is the process by which substances are broken down into simpler compounds.
4. _________ is the single most important hormone responsible for influencing the rate of cellular metabolism and body heat production.
   a. Calcitonin
   b. Estrogen
   c. Insulin
   d. Thyroid hormone
5. Basal metabolic rate (BMR) is:
   a. decreased in individuals with hyperthyroidism
   b. increased in individuals with hyperthyroidism
   c. increased in obese individuals

The endocrine system performs many complex and interrelated effects on the body as a whole, as well as on specific organs and tissues. Most scientific knowledge about this system is recent, and new information is constantly being reported. Many experiments on the endocrine system require relatively large laboratory animals; are time-consuming (requiring days to weeks of observation); and often involve technically difficult surgical procedures to remove the glands or parts of them, all of which makes it difficult to conduct more general types of laboratory experiments. Nevertheless, the two technically unsophisticated experiments presented here should illustrate how dramatically hormones affect body functioning. (Also, students may perform simulated endocrine wet labs in PhysioEx Exercise 4.)

Go to Mastering A&P™ > Study Area to improve your performance in A&P Lab.

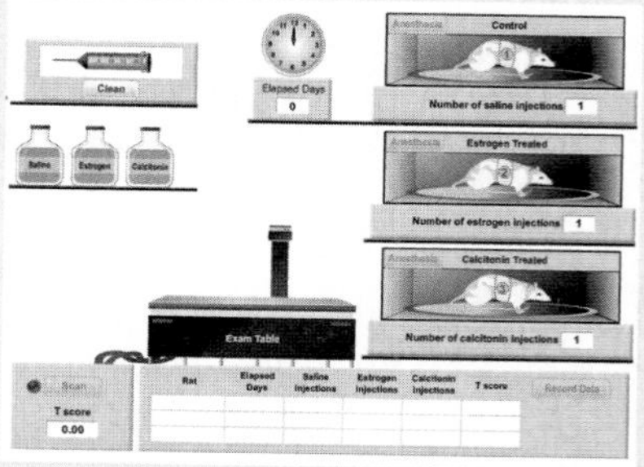

> Lab Tools > PhysioEx 9.1

Instructors may assign new Building Vocabulary coaching activities, Pre-Lab Quiz questions, Art Labeling activities, Practice Anatomy Lab Practical questions (PAL), PhysioEx activities, and more using the Mastering A&P™ Item Library.

Materials

Activity 1: Pituitary hormone and ovary*

- Female frogs *(Rana pipiens)*
- Disposable gloves
- Battery jars
- Syringe (2-ml capacity)
- 20- to 25-gauge needle
- Frog pituitary extract
- Physiological saline
- Spring or pond water
- Wax marking pencils

Activity 2: Hyperinsulinism*

- 500- or 600-ml beakers
- 20% glucose solution
- Commercial insulin solution (400 international units [IU] per 100 ml of H_2O)
- Finger bowls
- Small (4–5 cm, or 1½–2 in.) freshwater fish (guppy, bluegill, or sunfish—listed in order of preference)
- Wax marking pencils

Text continues on next page. →

**The Selected Actions of Hormones and Other Chemical Messengers* video (available to qualified adopters from Pearson Education) may be used in lieu of student participation in Activities 1 and 2.

PEx PhysioEx™ 9.1 Computer Simulation Ex. 4 on p. PEx-59.

28

Endocrine Experiments: Gonadotropins and Insulin

Activity 1

Determining the Effect of Pituitary Hormones on the Ovary

The anterior pituitary gonadotropic hormones—follicle-stimulating hormone (FSH) and luteinizing hormone (LH)—regulate the ovarian cycles of the female (see Exercise 43). Although amphibians normally ovulate seasonally, many can be stimulated to ovulate "on demand" by injecting an extract of pituitary hormones. In the following experiment, you will need to inject the frog the day before the lab session or return to check results the day after the scheduled lab session.

1. Don disposable gloves, and obtain two frogs. Place them in separate battery jars to bring them to your laboratory bench. Also bring back a syringe and needle, a wax marking pencil, pond or spring water, and containers of pituitary extract and physiological saline.

2. Before beginning, examine each frog for the presence of eggs. Hold the frog firmly with one hand and exert pressure on its abdomen toward the cloaca (in the direction of the legs). If ovulation has occurred, any eggs present in the uterine tube will be forced out and will appear at the cloacal opening. If no eggs are present, continue with step 3.

   If eggs are expressed, return the animal to your instructor and obtain another frog for experimentation. Repeat the procedure for determining whether eggs are present until you have obtained two frogs that lack eggs.

3. Draw 1 to 2 ml of the pituitary extract into a syringe. Inject the extract subcutaneously into the anterior abdominal (peritoneal) cavity of the frog you have selected to be the experimental animal. To inject into the peritoneal cavity, hold the frog with its ventral surface superiorly. Insert the needle through the skin and muscles of the abdominal wall in the lower quarter of the abdomen. Do not insert the needle far enough to damage any of the vital organs. With a wax marker, label its large battery jar "experimental," and place the frog in it. Add a small amount of pond or spring water to the battery jar before continuing.

4. Draw 1 to 2 ml of physiological saline into a syringe and inject it into the peritoneal cavity of the second frog—this will be the control animal. (Make sure you inject the same volume of fluid into both frogs.) Place this frog into the second battery jar, marked "control." Add a small amount of pond or spring water to the battery jar. Allow the animals to remain undisturbed for 24 hours.

5. After 24 hours, again check each frog for the presence of eggs in the cloacal opening. (See step 2.) If no eggs are present, make arrangements with your laboratory instructor to return to the lab on the next day (at 48 hours after injection) to check your frogs for the presence of eggs.

6. Return the frogs to the terrarium before leaving or continuing with the lab.

In which of the prepared frogs was ovulation induced?

____

Specifically, what hormone in the pituitary extract causes ovulation to occur?

____

Activity 2

Observing the Effects of Hyperinsulinism

Many people with diabetes mellitus need injections of insulin to maintain normal blood glucose levels. Adequate amounts of blood glucose are essential for proper functioning of the nervous system; thus, the administration of insulin must be carefully controlled. If blood glucose levels fall sharply, the patient will go into insulin shock.

A small fish will be used to demonstrate the effects of hyperinsulinism. Since the action of insulin on the fish parallels that in the human, this experiment should provide valid information concerning its administration to humans.

1. Prepare two finger bowls. Using a wax marking pencil, mark one A and the other B. To finger bowl A, add 100 ml of the commercial insulin solution. To finger bowl B, add 200 ml of 20% glucose solution.

2. Place a small fish in finger bowl A, and observe its actions carefully as the insulin diffuses into its bloodstream through the capillary circulation of its gills.

Approximately how long did it take for the fish to become comatose?

____

What types of activity did you observe in the fish before it became comatose?

____

____

____

____

3. When the fish is comatose, carefully transfer it to finger bowl B and observe its actions. What happens to the fish after it is transferred?

____

____

____

Approximately how long did it take for this recovery?

____

4. After all observations have been made and recorded, carefully return the fish to the aquarium.

Human Metabolism and Thyroid Hormones

Metabolism is a broad term referring to all chemical reactions that are necessary to maintain life. It involves both *catabolism*, enzymatically controlled processes in which substances are broken down to simpler substances, and *anabolism*, processes in which larger molecules or structures are built from smaller ones. Most catabolic reactions in the body are accompanied by a net release of energy. Some of the liberated energy is captured to make ATP, the energy-rich molecule used by body cells to energize all their activities; the balance is lost in the form of thermal energy or heat. Maintaining body temperature is linked to the heat-liberating aspects of metabolism.

Various foods make different contributions to the process of metabolism. For example, carbohydrates, particularly glucose, are generally broken down or oxidized to make ATP, whereas fats are utilized to form cell membranes and myelin sheaths and to insulate the body with a fatty cushion. Fats are used secondarily for producing ATP, particularly when the diet is inadequate in carbohydrates. Proteins and amino acids tend to be conserved by body cells, and understandably so, since most structural elements of the body are built with proteins.

Thyroid hormone (TH, collectively T_3 and T_4), produced by the thyroid gland, is the single most important hormone influencing an individual's basal metabolic rate (BMR) and body heat production. Basal metabolic rate, often called the "energy cost of living," is the energy needed to perform essential activity such as breathing and maintaining organ function. The level of thyroid hormone produced directly affects BMR; the more thyroid hormone produced, the higher the BMR. In addition, thyroid hormone regulates growth and development.

The tropic hormone thyroid-stimulating hormone (TSH), produced by the anterior pituitary, controls the secretory activity of the thyroid gland. The hypothalamic hormone thyrotropin-releasing hormone (TRH) stimulates the release of TSH from cells of the anterior pituitary gland. Rising blood levels of thyroid hormone act on both the anterior pituitary and the hypothalamus to inhibit secretion of TSH. (**Figure 28.1**

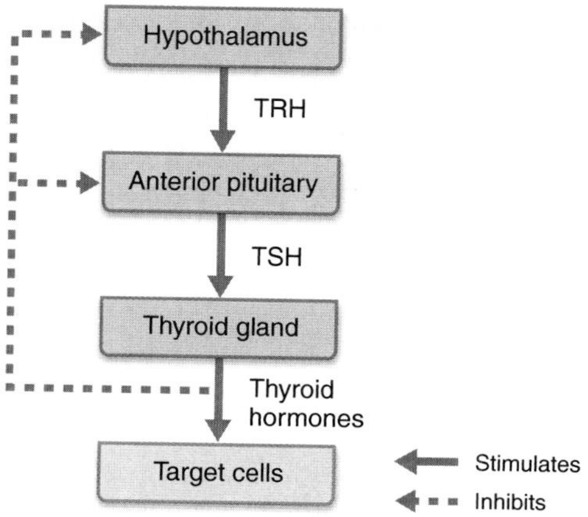

Figure 28.1 Regulation of thyroid hormone secretion.

illustrates the feedback loop that regulates thyroid hormone secretion.)

A **goiter** is an enlargement of the thyroid gland. Both *hypothyroidism* and *hyperthyroidism* can result in production of a goiter. In either case, the goiter is a result of excessive stimulation of the thyroid gland.

Hypothyroidism, also called **myxedema**, produces symptoms including low metabolic rate; feeling chilled; constipation; thick, dry skin and puffy skin ("bags") beneath the eyes; edema; lethargy; and mental sluggishness. A goiter occurs when hypothyroidism is caused by (1) primary failure of the thyroid gland or (2) an iodine-deficient diet that prevents the thyroid gland from producing TH. In both cases, the low levels of TH remove the inhibition for secretion of TSH, and its levels rise. When hypothyroidism is secondary to hypothalamic or anterior pituitary failure, TRH and/or TSH levels fall, and no goiter is produced.

Symptoms of *hyperthyroidism* include elevated metabolism; sweating; a rapid, more forceful heartbeat; nervousness; weight loss; difficulty concentrating; and changes in skin texture. The most common cause of hyperthyroidism is Graves' disease. Protrusion of the eyeballs sometimes occurs in patients with Graves' disease and is a unique symptom of this type of hyperthyroidism. Graves' disease is an autoimmune disorder in which the body makes abnormal antibodies that mimic the action of TSH on follicular cells of the thyroid. Despite low levels of TSH, the thyroid is being powerfully stimulated and produces a large goiter. Hyperthyroidism can also arise secondary to excess hypothalamic or anterior pituitary secretion. In this case, TSH levels are high and a goiter also occurs. A hypersecreting thyroid tumor also causes hyperthyroidism. TSH levels are low when such a tumor is present, and there is no goiter. +

Use the information above to answer the questions associated with the case studies in the following Group Challenge.

Group Challenge

Thyroid Hormone Case Studies

Work in groups of three, and discuss the two cases presented. Record your group's answers to the questions in the space provided.

Case 1: Marty is a 24-year-old male. He has noticed a bulge on his neck that has been increasing in size over the past few months. His physician orders a blood test with the following results:

Component	Results	Normal range	Units
TSH	<0.1	0.1–5.5	μIU/ml
Free T_4	5.3	0.8–1.7	ng/dL

Does Marty have hypothyroidism or hyperthyroidism?

____

Name and briefly describe the most likely cause of his

thyroid disorder. ________________________________

____

What other signs and symptoms might Marty be

experiencing? ________________________________

____

Case 2: Heather is a 60-year-old female. She complains of swelling in her limbs and fatigue. Her physician orders a blood test with the following results:

Component	Results	Normal range	Units
TSH	5.7	0.1–5.5	μIU/ml
Free T_4	0.5	0.8–1.7	ng/dL

Does Heather have hypothyroidism or hyperthyroidism?

____

Name and briefly describe the most likely cause(s) of her

thyroid disorder. ________________________________

____

Does Heather have a goiter? ________________________________

What other signs and symptoms might Heather be

experiencing? ________________________________

____

Instructors may assign a portion of the Review Sheet questions using Mastering A&P™

REVIEW SHEET

EXERCISE 28

Endocrine Wet Labs and Human Metabolism

Name ______________________ Lab Time/Date ______________________

Determining the Effect of Pituitary Hormones on the Ovary

1. In the experiment on the effects of pituitary hormones, two anterior pituitary hormones caused ovulation to occur in the experimental animal. Which of these actually triggered ovulation?

   ______________________ The normal function of the second hormone involved, ______________________,

   is to ______________________.

2. Why was a second frog injected with saline? ______________________

Observing the Effects of Hyperinsulinism

3. Briefly explain why the fish became comatose when it was immersed in the insulin solution.

   ____

   ____

4. What is the mechanism of the recovery process observed? ______________________

   ____

5. What would you do to help a friend who had inadvertently taken an overdose of insulin? ______________________

   ______________________ Why? ______________________

   ____

Human Metabolism and Thyroid Hormones

6. Use an appropriate reference to indicate which of the following would be associated with increased or decreased BMR. Indicate increase by ↑ and decrease by ↓.

   increased exercise __________ aging __________ infection/fever __________

   increased stress __________ obesity __________ sex (♂ or ♀) __________

7. Considering the effects of pituitary hormones on the ovary, name the hormones that could theoretically be supplemented to improve fertility in humans. Correlate the functions of the hormones to their role in improving fertility.

____

____

____

Which hormone might contribute to multiple births, and why?

____

____

8. Congenital hyperinsulinism is a rare condition in which infants experience frequent episodes of hypoglycemia. Based on the functions of the two main hormones that control blood sugar homeostasis, hypothesize how medications might be used to counteract the effects of congenital hyperinsulinism. ______________________________

____

____

____

EXERCISE 29

Blood

Learning Outcomes

- Name the two major components of blood, and state their average percentages in whole blood.
- Describe the composition and functional importance of plasma.
- Define *formed elements*, and list the cell types composing them, state their relative percentages, and describe their major functions.
- Identify erythrocytes, basophils, eosinophils, monocytes, lymphocytes, and neutrophils when provided with a microscopic preparation or appropriate image.
- Provide the normal values for a total white blood cell count and a total red blood cell count, and state the importance of these tests.
- Conduct the following blood tests in the laboratory, and state their norms and the importance of each: differential white blood cell count, hematocrit, hemoglobin determination, clotting time, and plasma cholesterol concentration.
- Define *leukocytosis*, *leukopenia*, *leukemia*, *polycythemia*, and *anemia*; cite a possible cause for each condition.
- Perform an ABO and Rh blood typing test in the laboratory, and discuss the reason for transfusion reactions resulting from the administration of mismatched blood.

Go to Mastering A&P™ > Study Area to improve your performance in A&P Lab.

> Lab Tools > Pre-Lab Videos > Hematocrit

Instructors may assign new Building Vocabulary coaching activities, Pre-Lab Quiz questions, Art Labeling activities, Pre-Lab Video Coaching Activities for Determining the Hematocrit and Blood Typing, Practice Anatomy Lab Practical questions (PAL), PhysioEx activities, and more using the Mastering **A&P™** Item Library.

Pre-Lab Quiz

Instructors may assign these and other Pre-Lab Quiz questions using Mastering **A&P™**

1. Three types of formed elements found in blood include erythrocytes, leukocytes, and:
   a. electrolytes **b.** fibers **c.** platelets **d.** sodium salts
2. The least numerous but largest of all agranulocytes is the:
   a. basophil **b.** lymphocyte **c.** monocyte **d.** neutrophil
3. ________ are the leukocytes responsible for releasing histamine and other mediators of inflammation.
   a. Basophils **b.** Eosinophils **c.** Monocytes **d.** Neutrophils
4. Circle the correct underlined term. The normal hematocrit value for <u>females</u> / <u>males</u> is generally higher than that of the opposite sex.
5. Circle the correct underlined term. Blood typing is based on the presence of proteins known as <u>antigens</u> / <u>antibodies</u> on the outer surface of the red blood cell plasma membrane.

Materials

General supply area:

- Disposable gloves
- Safety glasses (provided by students)
- Bucket or large beaker containing 10% household bleach solution for slide and glassware disposal
- Spray bottles containing 10% bleach solution
- Autoclave bag
- Designated lancet (sharps) disposal container
- Plasma (obtained from an animal hospital or prepared by centrifuging animal [for example, cattle or sheep] blood obtained from a biological supply house)
- Test tubes and test tube racks
- Wide-range pH paper

Text continues on next page. →

- Stained smears of human blood from a biological supply house or, if desired by the instructor, heparinized animal blood obtained from a biological supply house or an animal hospital (for example, dog blood), or EDTA-treated red cells (reference cells*) with blood type labels obscured (available from Immucor, Inc.)

Note to the Instructor: *See directions below for handling of soiled glassware and disposable items.*

**The blood in these kits (each containing four blood cell types—A1, A2, B, and O—individually supplied in 10-ml vials) is used to calibrate cell counters and other automated clinical laboratory equipment. This blood has been carefully screened and can be safely used by students for blood typing and determining hematocrits. It is not usable for hemoglobin determinations or coagulation studies.*

- Clean microscope slides
- Glass stirring rods
- Wright's stain in a dropper bottle
- Distilled water in a dropper bottle
- Sterile lancets
- Absorbent cotton balls
- Alcohol swabs (wipes)
- Paper towels
- Compound microscope
- Immersion oil
- Assorted slides of white blood count pathologies labeled "Unknown Sample ____________"
- Timer

Because many blood tests are to be conducted in this exercise, it is advisable to set up a number of appropriately labeled supply areas for the various tests, as designated below. Some needed supplies are located in the general supply area.

Note: *Artificial blood prepared by Ward's Natural Science can be used for differential counts, hematocrit, and blood typing.*

Activity 4: Hematocrit

- Heparinized capillary tubes
- Microhematocrit centrifuge and reading gauge (if the reading gauge is not available, a millimeter ruler may be used)
- Capillary tube sealer or modeling clay

Activity 5: Hemoglobin determination

- Hemoglobinometer, hemolysis applicator, and lens paper; or Tallquist hemoglobin scale and test paper

Activity 6: Coagulation time

- Capillary tubes (nonheparinized)
- Fine triangular file

Activity 7: Blood typing

- Blood typing sera (anti-A, anti-B, and anti-Rh [anti-D])
- Rh typing box
- Wax marking pencil
- Toothpicks
- Medicine dropper
- Blood test cards or microscope slides

Activity 8: Demonstration

- Microscopes set up with prepared slides demonstrating the following bone (or bone marrow) conditions: macrocytic hypochromic anemia, microcytic hypochromic anemia, sickle cell anemia, lymphocytic leukemia (chronic), and eosinophilia

Activity 9: Cholesterol measurement

- Cholesterol test cards and color scale

PEx PhysioEx™ 9.1 Computer Simulation Ex. 11 on p. PEx-161.

29

In this exercise, you will study plasma and formed elements of blood and conduct various hematologic tests. These tests are useful diagnostic tools for the physician because blood composition (number and types of blood cells, and chemical composition) reflects the status of many body functions and malfunctions.

ALERT: Special precautions when handling blood. This exercise provides information on blood from several sources: human, animal, human treated, and artificial blood. The instructor will decide whether to use animal blood for testing or to have students test their own blood in accordance with the educational goals of the student group. For example, for students in the nursing or laboratory technician curricula, learning how to safely handle human blood or other human wastes is essential. Whenever blood is being handled, special attention must be paid to safety precautions. Instructors who opt to use human blood are responsible for its safe handling. Precautions should be used regardless of the source of the blood. This will both teach good technique and ensure the safety of the students.

Follow exactly the safety precautions listed below.

1. Wear safety gloves at all times. Discard appropriately.
2. Wear safety glasses throughout the exercise.
3. Handle only your own, freshly drawn (human) blood.
4. Be sure you understand the instructions and have all supplies on hand before you begin any part of the exercise.
5. Do not reuse supplies and equipment once they have been exposed to blood.
6. Keep the lab area clean. Do not let anything that has come in contact with blood touch surfaces or other individuals in the lab. Keep track of the location of any supplies and equipment that come into contact with blood.
7. Immediately after use dispose of lancets in a designated disposal container. Do not put them down on the lab bench, even temporarily.
8. Dispose of all used cotton balls, alcohol swabs, blotting paper, and so forth, in autoclave bags; place all soiled glassware in containers of 10% bleach solution.
9. Wipe down the lab bench with 10% bleach solution when you finish.

Composition of Blood

Circulating blood is a rather viscous substance that varies from bright red to a dull brick red, depending on the amount of oxygen it is carrying. Oxygen-rich blood is bright red. The average volume of blood in the body is about 5–6 L in adult males and 4–5 L in adult females.

Blood is classified as a type of connective tissue because it consists of cells within a matrix. The nonliving fluid matrix is the **plasma**, and the cells and cell fragments are the **formed elements**. The fibers typical of a connective tissue matrix become visible in blood only when clotting occurs. They then appear as fibrin threads, which form the structural basis for clot formation.

More than 100 different substances are dissolved or suspended in plasma (**Figure 29.1**), which is over 90% water. These include nutrients, gases, hormones, various wastes and metabolites, many types of proteins, and electrolytes. The composition of plasma varies continuously as cells remove or add substances to the blood.

Three main types of formed elements are present in blood (**Table 29.1**, p. 418). Most numerous are **erythrocytes**, or **red blood cells (RBCs)**, which are literally sacs of hemoglobin molecules that transport the bulk of the oxygen carried in the blood (and a small percentage of the carbon dioxide). **Leukocytes**, or **white blood cells (WBCs)**, are part of the body's nonspecific defenses and the immune system, and **platelets** function in hemostasis (blood clot formation); together they make up <1% of whole blood. Formed elements normally constitute about 45% of whole blood; plasma accounts for the remaining 55%.

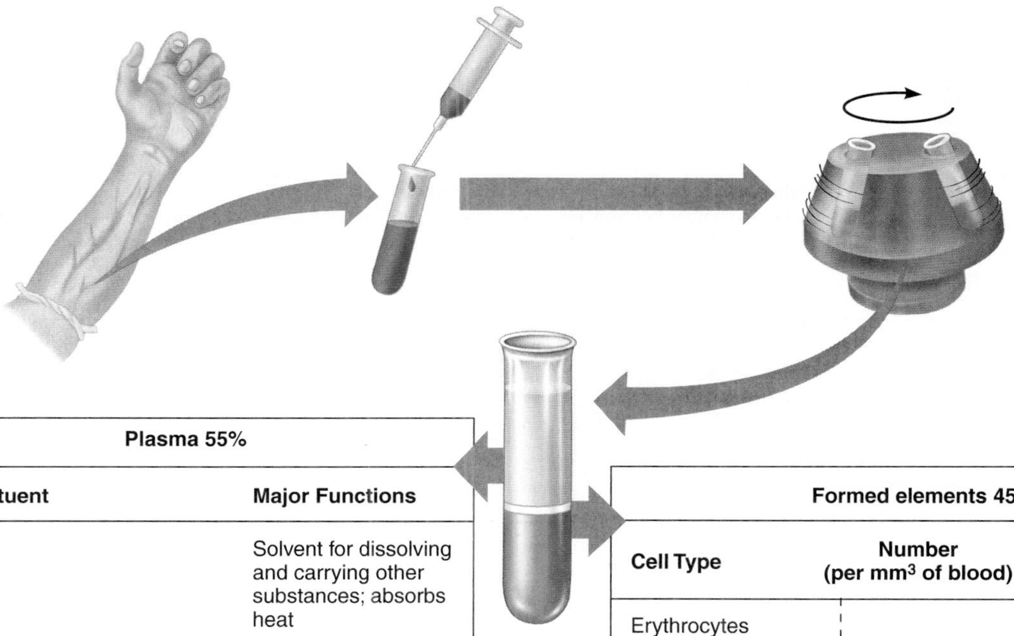

Plasma 55%	
Constituent	**Major Functions**
Water	Solvent for dissolving and carrying other substances; absorbs heat
Salts (electrolytes) Sodium Potassium Calcium Magnesium Chloride Bicarbonate	Osmotic balance, pH buffering
Plasma proteins Albumin Fibrinogen Globulins	 Osmotic balance Clotting of blood Defense (antibodies) and lipid transport
Substances transported by blood Nutrients (glucose, fatty acids, amino acids, vitamins) Waste products of metabolism (urea, uric acid) Respiratory gases (O_2 and CO_2) Hormones	

Formed elements 45%		
Cell Type	**Number (per mm^3 of blood)**	**Functions**
Erythrocytes (red blood cells) 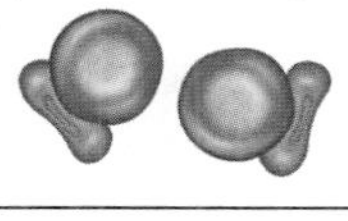	4 – 6 million	Transport oxygen and help transport carbon dioxide
Leukocytes (white blood cells)	4800 – 10,800	Defense and immunity
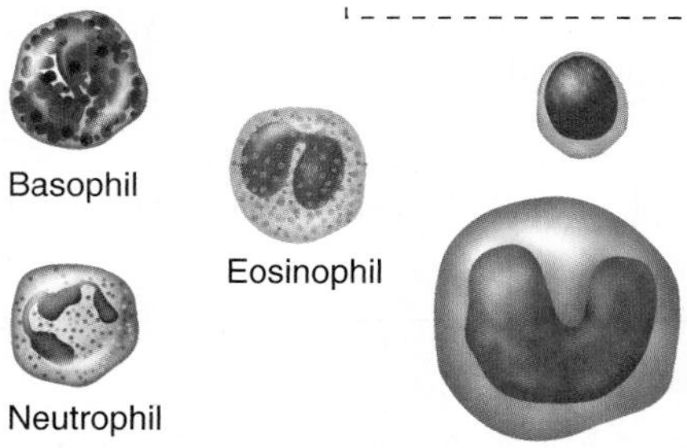 Basophil, Eosinophil, Neutrophil		Lymphocyte, Monocyte
Platelets 	150,000 – 400,000	Blood clotting

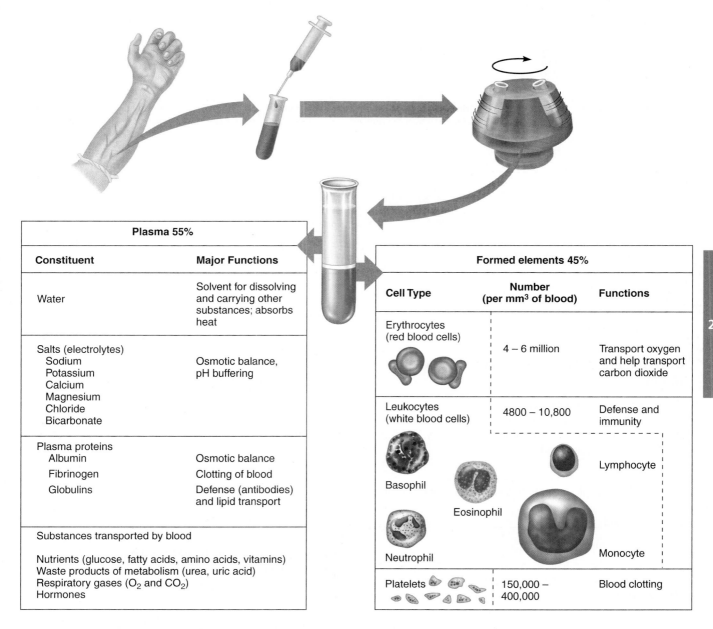

Figure 29.1 The composition of blood. Note that leukocytes and platelets are found in the band between plasma (above) and erythrocytes (below). This band is known as the *buffy coat*.

29

Table 29.1 Summary of Formed Elements of Blood

Cell type	Illustration	Description*	Cells/mm³ (µl) of blood	Function
Erythrocytes (red blood cells, RBCs)	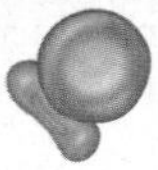	Biconcave, anucleate disc; orange-pink color; diameter 7–8 µm	4–6 million	Transport oxygen and carbon dioxide
Leukocytes (white blood cells, WBCs)		Spherical, nucleated cells	4800–10,800	
Granulocytes Neutrophil	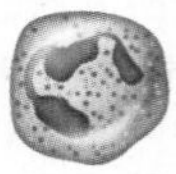	Nucleus multilobed; pale red and blue cytoplasmic granules; diameter 10–12 µm	3000–7000 Differential count: 50–70%	Phagocytize pathogens or debris
Eosinophil	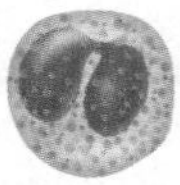	Nucleus bilobed; red cytoplasmic granules; diameter 10–14 µm	100–400 Differential count: 2–4%	Kill parasitic worms; slightly phagocytic; complex role in allergy and asthma
Basophil	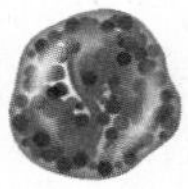	Nucleus lobed; large blue-purple cytoplasmic granules; diameter 10–14 µm	20–50 Differential count: <1%	Release histamine and other mediators of inflammation; contain heparin, an anticoagulant
Agranulocytes Lymphocyte		Nucleus spherical or indented; pale blue cytoplasm; diameter 5–17 µm	1500–3000 Differential count: 25–45%	Mount immune response by direct cell attack or via antibody production
Monocyte	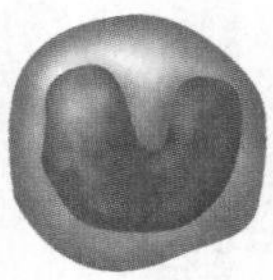	Nucleus U- or kidney-shaped; gray-blue cytoplasm; diameter 14–24 µm	100–700 Differential count: 3–8%	In tissues, develop into macrophages that phagocytize pathogens or debris
Platelets	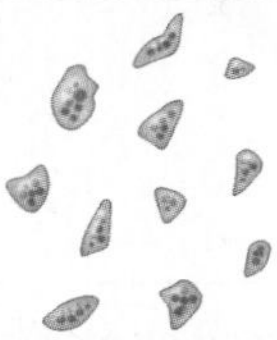	Cytoplasmic fragments containing granules; stain deep purple; diameter 2–4 µm	150,000–400,000	Seal small tears in blood vessels; instrumental in blood clotting

*Appearance when stained with Wright's stain.

Activity 1

Determining the Physical Characteristics of Plasma

Go to the general supply area, and carefully pour a few milliliters of plasma into a test tube. Also obtain some wide-range pH paper, and then return to your laboratory bench to make the following simple observations.

pH of Plasma

Test the pH of the plasma with wide-range pH paper. Record the pH observed. ______________________

Color and Clarity of Plasma

Hold the test tube up to a source of natural light. Note and record its color and degree of transparency. Is it clear, translucent, or opaque?

Color ______________________

Degree of transparency ______________________

Consistency

While wearing gloves, dip your finger and thumb into plasma, and then press them firmly together for a few seconds. Gently pull them apart. How would you describe the consistency of plasma (slippery, watery, sticky, granular)? Record your observations.

____

Activity 2

Examining the Formed Elements of Blood Microscopically

In this section, you will observe blood cells on an already prepared (purchased) blood slide or on a slide prepared from your own blood or blood provided by your instructor.

- If you are using the purchased blood slide, obtain a slide and begin your observations at step 6.
- If you are testing blood provided by a biological supply source or an animal hospital, obtain a tube of the supplied blood, disposable gloves, and the supplies listed in step 1, except for the lancets and alcohol swabs. After donning gloves, go to step 3b to begin your observations.
- If you are examining your own blood, you will perform all the steps described below *except* step 3b.

1. Obtain two glass slides, a glass stirring rod, dropper bottles of Wright's stain and distilled water, two or three lancets, cotton balls, and alcohol swabs. Bring this equipment to the laboratory bench. Clean the slides thoroughly and dry them.

2. Open the alcohol swab packet, and scrub your third or fourth finger with the swab. (Because the pricked finger may be a little sore later, it is better to prepare a finger on the non-dominant hand.) Swing your hand in a cone-shaped path for 10 to 15 seconds. This will dry the alcohol and cause your fingers to become filled with blood. Then, open the lancet packet and grasp the lancet by its blunt end. Quickly jab the pointed end into the prepared finger to produce a free flow of blood. It is *not* a good idea to squeeze or "milk" the finger, because this forces out tissue fluid as well as blood. If the blood is not flowing freely, make another puncture.

! Under no circumstances is a lancet to be used for more than one puncture. Dispose of the lancets in the designated disposal container immediately after use.

3a. With a cotton ball, wipe away the first drop of blood; then allow another large drop of blood to form. Touch the blood to one of the cleaned slides approximately 1.3 cm, or ½ inch, from the end. Then quickly (to prevent clotting) use the second slide to form a blood smear (**Figure 29.2**). When properly prepared, the blood smear is uniformly thin. If the blood smear appears streaked, the blood probably began to clot or coagulate before the smear was made, and another slide should be prepared. Continue at step 4.

3b. Dip a glass rod in the blood provided, and transfer a generous drop of blood to the end of a cleaned microscope slide. For the time being, lay the glass rod on a paper towel on the bench. Then, as described in step 3a (Figure 29.2), use the second slide to make your blood smear.

4. Allow the blood smear slide to air dry. When it is completely dry, it will look dull. Place it on a paper towel, and add 5 to 10 drops of Wright's stain. Count the number of drops of stain used. Allow the stain to remain on the slide for 3 to 4 minutes, and then add an equal number of drops of distilled water. Allow the water and Wright's stain mixture to remain on the slide for 4 or 5 minutes or until a metallic green film or scum is apparent on the fluid surface.

5. Rinse the slide with a stream of distilled water. Then flood it with distilled water, and allow it to lie flat until the slide becomes translucent and takes on a pink cast. Then stand the slide on its long edge on the paper towel, and allow it to dry completely. Once the slide is dry, you can begin your observations.

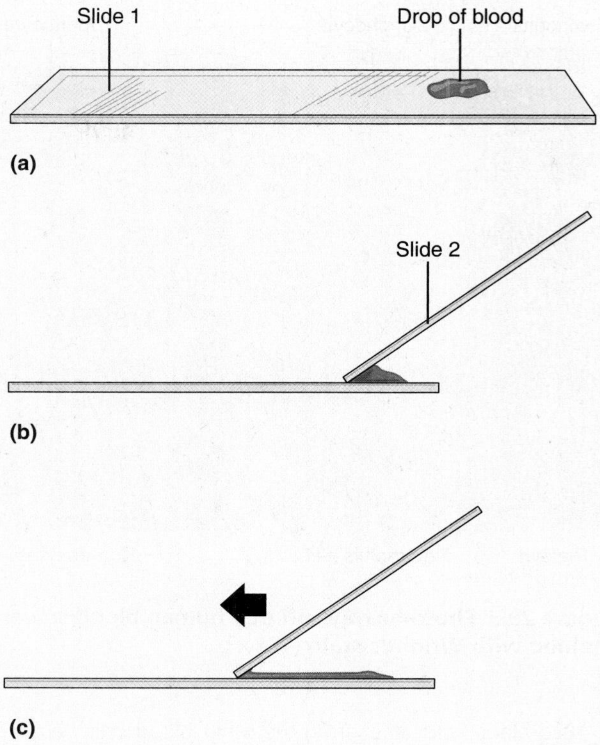

Figure 29.2 Procedure for making a blood smear. **(a)** Place a drop of blood on slide 1 approximately ½ inch from one end. **(b)** Hold slide 2 at a 30° to 40° angle to slide 1 (it should touch the drop of blood) and allow blood to spread along entire bottom edge of angled slide. **(c)** Smoothly advance slide 2 to end of slide 1 (blood should run out before reaching the end of slide 1). Then lift slide 2 away from slide 1 and place slide 1 on a paper towel. Dispose of slide 2 in the appropriate container.

6. Obtain a microscope and scan the slide under low power to find the area where the blood smear is the thinnest. After scanning the slide in low power to find the areas with the largest numbers of WBCs, read the following descriptions of cell types, and find each one in the art illustrating blood cell types (in Figure 29.1 and Table 29.1). (The formed elements are also shown in **Figure 29.3**, p. 420, and Figure 29.4.) Then, switch to the oil immersion lens, and observe the slide carefully to identify each cell type.

7. Set your prepared slide aside for use in Activity 3.

Erythrocytes

Erythrocytes, or red blood cells, which average 7.5 μm in diameter, vary in color from an orange-pink color to pale pink, depending on the effectiveness of the stain. They have a distinctive biconcave disc shape and appear paler in the center than at the edge (see Figure 29.3).

As you observe the slide, notice that the red blood cells are by far the most numerous blood cells seen in the field. Their number averages 4.5 million to 5.5 million cells per cubic millimeter of blood (for women and men, respectively).

Text continues on next page. →

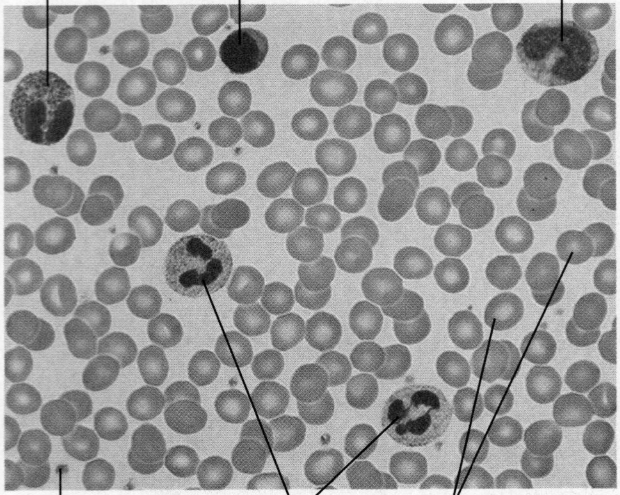

Figure 29.3 Photomicrograph of a human blood smear stained with Wright's stain (765×).

Red blood cells differ from the other blood cells because they are anucleate (lacking a nucleus) when mature and circulating in the blood. As a result, they are unable to reproduce or repair damage and have a limited life span of 100 to 120 days, after which they begin to fragment and are destroyed, mainly in the spleen.

In various anemias, the red blood cells may appear pale (an indication of decreased hemoglobin content) or may be nucleated (an indication that the bone marrow is turning out cells prematurely). +

29

Leukocytes

Leukocytes, or white blood cells, are nucleated cells that are formed in the bone marrow from the same stem cells *(hemocytoblast)* as red blood cells. They are much less numerous than the red blood cells, averaging from 4800 to 10,800 cells per cubic millimeter. The life span of leukocytes varies. They can survive for minutes or decades, depending on the type of leukocyte and tissue activity. Basically, white blood cells are protective, pathogen-destroying cells that are transported to all parts of the body in the blood or lymph. Important to their protective function is their ability to move in and out of blood vessels, a process called **diapedesis**, and to wander through body tissues by **amoeboid motion** to reach sites of inflammation or tissue destruction. They are classified into two major groups, depending on whether or not they contain conspicuous granules in their cytoplasm.

Granulocytes make up the first group. The granules in their cytoplasm stain differentially with Wright's stain, and they have peculiarly lobed nuclei, which often consist of expanded nuclear regions connected by thin strands of nucleoplasm. There are three types of granulocytes: **neutrophils**, **eosinophils**, and **basophils**.

The second group, **agranulocytes**, contains no *visible* cytoplasmic granules. Although found in the blood, they are much more abundant in lymphoid tissues. There are two types of agranulocytes: **lymphocytes** and **monocytes**. The specific characteristics of leukocytes are described in Table 29.1. Photomicrographs of the leukocytes illustrate their different appearances (**Figure 29.4**).

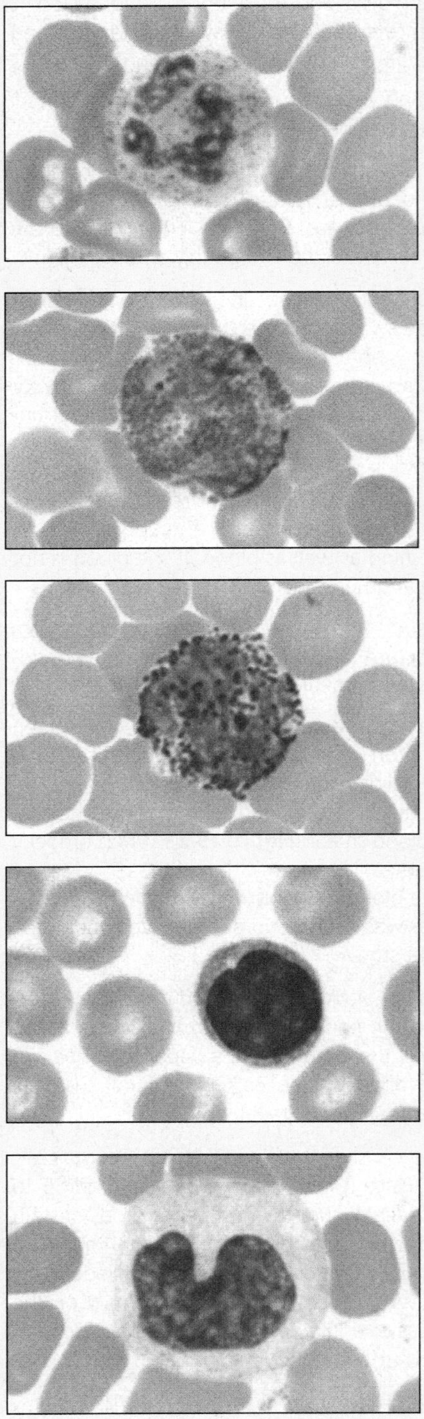

Figure 29.4 Leukocytes. In each case, the leukocytes are surrounded by erythrocytes (1330×, Wright's stain).

Instructors may assign this figure as an Art Labeling Activity using Mastering A&P™

Students are often asked to list the leukocytes in order from the most abundant to the least abundant. The following phrase may help you with this task: *N*ever *l*et *m*onkeys *e*at *b*ananas (neutrophils, lymphocytes, monocytes, eosinophils, basophils).

Platelets

Platelets are cell fragments of large multinucleate cells (**megakaryocytes**) formed in the bone marrow. They appear as darkly staining, irregularly shaped bodies interspersed among the blood cells (see Figure 29.3). The normal platelet count in blood ranges from 150,000 to 400,000 per cubic millimeter. Platelets are instrumental in the clotting process that occurs in plasma when blood vessels are ruptured.

After you have identified these cell types on your slide, observe charts and three-dimensional models of blood cells if these are available. Do not dispose of your slide, as you will use it later for the differential white blood cell count.

Hematologic Tests

When someone enters a hospital as a patient, several hematologic tests are routinely done to determine general level of health as well as the presence of pathological conditions. You will be conducting the most common of these tests in this exercise.

! Materials such as cotton balls, lancets, and alcohol swabs are used in nearly all of the following diagnostic tests. These supplies are at the general supply area and should be properly disposed of (glassware to the bleach bucket, lancets in a designated disposal container, and disposable items to the autoclave bag) immediately after use.

Other necessary supplies and equipment are at specific supply areas marked according to the test with which they are used. Since nearly all of the tests require a finger stick, if you will be using your own blood it might be wise to quickly read through the tests to determine in which instances more than one preparation can be done from the same finger stick. A little planning will save you the discomfort of multiple finger sticks.

An alternative to using blood obtained from the finger stick technique is using heparinized blood samples supplied by your instructor. The purpose of using heparinized tubes is to prevent the blood from clotting. Thus blood collected and stored in such tubes will be suitable for all tests except coagulation time testing.

Total White and Red Blood Cell Counts

A **total WBC count** or **total RBC count** determines the total number of that cell type per unit volume of blood. Total WBC and RBC counts are a routine part of any physical exam. Most clinical agencies use computers to conduct these counts. Total WBC and RBC counts will not be done here, but the importance of such counts (both normal and abnormal values) is briefly described below.

Total White Blood Cell Count

Since white blood cells are an important part of the body's defense system, it is essential to note any abnormalities in them.

Leukocytosis, a WBC count over 11,000 cells/mm^3, may indicate bacterial or viral infection, metabolic disease, hemorrhage, or poisoning by drugs or chemicals. A decrease in the white cell number below 4000/mm^3 **(leukopenia)** is usually due to exposure to certain chemicals or toxins, including anticancer agents. A person with leukopenia lacks the usual protective mechanisms. **Leukemia** is a malignant disorder of the lymphoid tissues characterized by uncontrolled proliferation of abnormal WBCs accompanied by a reduction in the number of RBCs and platelets. It is detectable not only by a total WBC count but also by a differential WBC count. +

Total Red Blood Cell Count

Since RBCs are absolutely necessary for oxygen transport, a doctor typically investigates any excessive change in their number immediately.

An increase in the number of RBCs **(polycythemia)** may result from bone marrow cancer or from living at high altitudes where less oxygen is available. A decrease in the number of RBCs results in anemia. The term **anemia** simply indicates a decreased oxygen-carrying capacity of blood that may result from a decrease in RBC number or size or a decreased hemoglobin content of the RBCs. A decrease in RBCs may result suddenly from hemorrhage or more gradually from conditions that destroy RBCs or hinder RBC production. +

Differential White Blood Cell Count

To obtain a **differential white blood cell count**, 100 WBCs are counted and classified according to type. Such a count is routine in a physical examination and in diagnosing illness, since any abnormality in percentages of WBC types may indicate a problem and the source of pathology.

29

Activity 3

Conducting a Differential WBC Count

1. Use the slide prepared for the identification of the blood cells in Activity 2 or a prepared slide provided by your instructor. Begin at the edge of the smear, and move the slide in a systematic manner on the microscope stage—either up and down or from side to side (as indicated in **Figure 29.5** on p. 422).

2. Record each type of white blood cell you observe by making a count in the first blank column of the **Activity 3 chart** on p. 422 (for example, 𝍸 || = 7 cells) until you have observed and recorded a total of 100 WBCs. Using the following equation, compute the percentage of each WBC type counted, and record the percentages on the **Hematologic Test Data Sheet** on the last page of the exercise, preceding the Review Sheet.

Text continues on next page. →

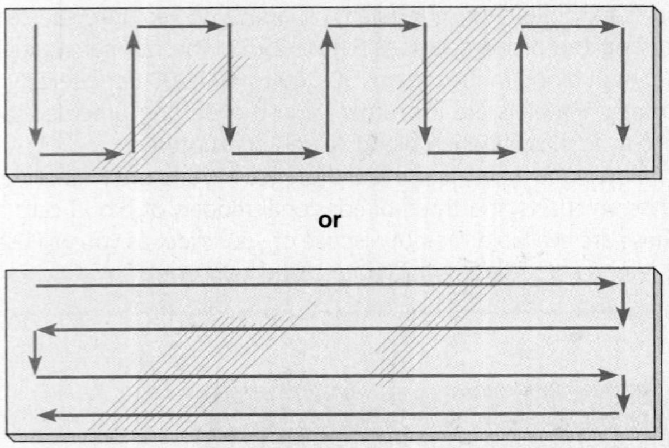

Figure 29.5 Alternative methods of moving the slide for a differential WBC count.

$$\text{Percent (\%)} = \frac{\text{\# observed}}{\text{Total \# counted}} \times 100$$

3. Select a slide marked "Unknown sample," record the slide number, and use the count chart below to conduct a differential count. Record the percentages on the data sheet (p. 428).

How does the differential count from the unknown sample slide compare to the normal percentages given for each type in Table 29.1?

____

____

Activity 3: Count of 100 WBCs		
	Number observed	
Cell type	Student blood smear	Unknown sample # ____
Neutrophils		
Eosinophils		
Basophils		
Lymphocytes		
Monocytes		

Using the text and other references, try to determine the blood pathology on the unknown slide. Defend your answer.

____

____

4. How does your differential white blood cell count compare to the percentages given in Table 29.1?

____

Hematocrit

The **hematocrit** is routinely determined when anemia is suspected. Centrifuging whole blood spins the formed elements to the bottom of the tube, with plasma forming the top layer (see Figure 29.1). Since the blood cell population is primarily RBCs, the hematocrit is generally considered equivalent to the RBC volume, and this is the only value reported. However, the relative percentage of WBCs can be differentiated, and both WBC and plasma volume will be reported here. Normal hematocrit values for the male and female, respectively, are 47.0 ± 5 and 42.0 ± 5.

29

Activity 4

Instructors may assign a related Pre-Lab Video Coaching Activity for Determining the Hematocrit using Mastering A&P™

Determining the Hematocrit

The hematocrit is determined by the micromethod, so only a drop of blood is needed. If possible (and the centrifuge allows), all members of the class should prepare their capillary tubes at the same time so the centrifuge can be run only once.

1. Obtain two heparinized capillary tubes, capillary tube sealer or modeling clay, a lancet, alcohol swabs, and some cotton balls.

2. If you are using your own blood, use an alcohol swab to cleanse a finger, prick the finger with a lancet, and allow the blood to flow freely. Wipe away the first few drops and, holding the red-line-marked end of the capillary tube to the blood drop, allow the tube to fill at least three-fourths full by capillary action (**Figure 29.6a**). If the blood is not flowing freely, the end of the capillary tube will not be completely submerged in the blood during filling, air will enter, and you will have to prepare another sample.

If you are using instructor-provided blood, simply immerse the red-marked end of the capillary tube in the blood sample and fill it three-quarters full as just described.

3. Plug the blood-containing end by pressing it into the capillary tube sealer or clay (Figure 29.6b). Prepare a second tube in the same manner.

4. Place the prepared tubes opposite one another in the radial grooves of the microhematocrit centrifuge with the sealed ends abutting the rubber gasket at the centrifuge periphery (Figure 29.6c). This loading procedure balances the centrifuge and prevents blood from spraying everywhere by centrifugal force. *Make a note of the numbers of the grooves your tubes are in.* When all the tubes have been loaded, make sure the centrifuge is properly balanced, and secure the centrifuge cover. Turn the centrifuge on, and set the timer for 4 or 5 minutes.

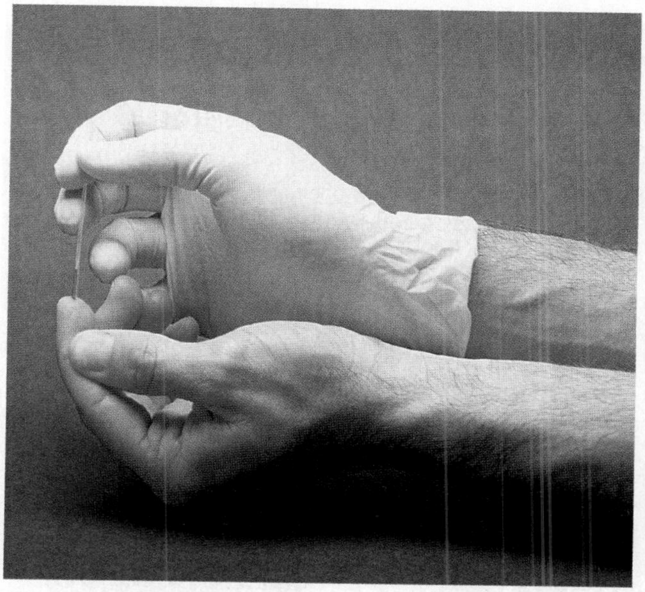

(a)

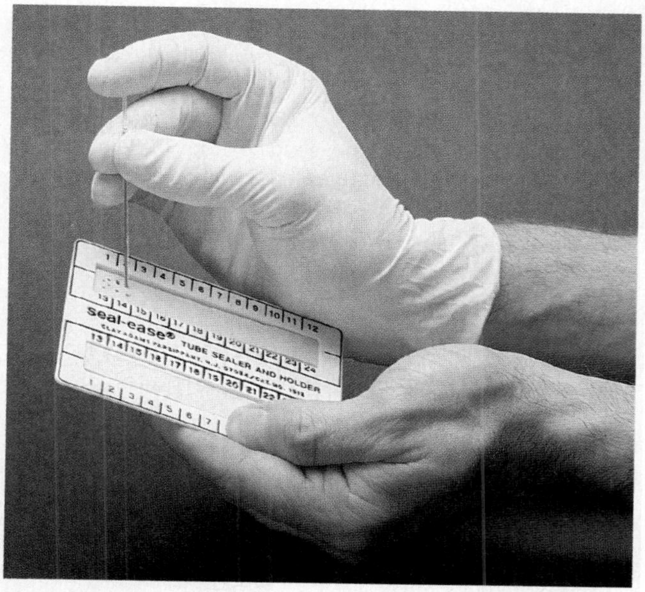

(b)

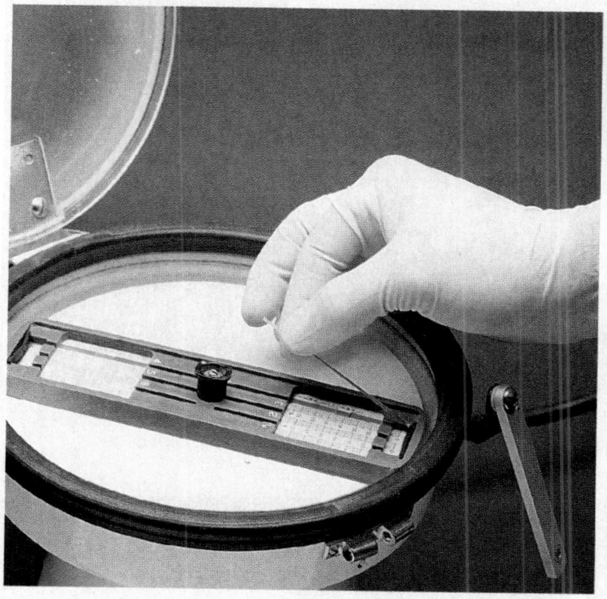

(c)

Figure 29.6 **Steps in a hematocrit determination.**
(a) Fill a heparinized capillary tube with blood.
(b) Plug the blood-containing end of the tube with clay.
(c) Place the tube in a microhematocrit centrifuge. (Centrifuge must be balanced.)

5. Determine the percentage of RBCs, WBCs, and plasma by using the microhematocrit reader. The RBCs are the bottom layer, the plasma is the top layer, and the WBCs are the buff-colored layer between the two. If the reader is not available, use a millimeter ruler to measure the length of the filled capillary tube occupied by each element, and compute its percentage by using the following formula:

$$\frac{\text{Height of the column composed of the element (mm)}}{\text{Height of the original column of whole blood (mm)}} \times 100$$

Record your calculations below and on the data sheet (p. 428).

% RBC ________ % WBC ________ % plasma ________

Usually WBCs constitute 1% of the total blood volume. How do your blood values compare to this figure and to the normal percentages for RBCs and plasma? (See Figure 29.1.)

____

As a rule, a hematocrit is considered a more accurate test than the total RBC count for determining the RBC composition of the blood. A hematocrit within the normal range generally indicates a normal RBC number, whereas an abnormally high or low hematocrit is cause for concern.

Hemoglobin Concentration

As noted earlier, a person can be anemic even with a normal RBC count. Since hemoglobin (Hb) is the RBC protein responsible for oxygen transport, perhaps the most accurate way of measuring the oxygen-carrying capacity of the blood is to determine its hemoglobin content. Oxygen, which combines reversibly with the heme (iron-containing portion) of the hemoglobin molecule, is picked up by the blood cells in the lungs and unloaded in the tissues. Thus, the more hemoglobin molecules the RBCs contain, the more oxygen they will be able to transport. Normal blood contains 12 to 18 g of hemoglobin per 100 ml of blood. Hemoglobin content in men is slightly higher (13 to 18 g) than in women (12 to 16 g).

Activity 5

Determining Hemoglobin Concentration

Several techniques have been developed to estimate the hemoglobin content of blood, ranging from the old, rather inaccurate Tallquist method to expensive hemoglobinometers, which are precisely calibrated and yield highly accurate results. Directions for both the Tallquist method and a hemoglobinometer are provided here.

Tallquist Method

1. Obtain a Tallquist hemoglobin scale, test paper, lancets, alcohol swabs, and cotton balls.

2. Use instructor-provided blood or prepare the finger as previously described. (For best results, make sure the alcohol evaporates before puncturing your finger.) Place one good-sized drop of blood on the special absorbent paper provided with the color scale. The blood stain should be larger than the holes on the color scale.

3. As soon as the blood has dried and loses its glossy appearance, match its color, under natural light, with the color standards by moving the specimen under the comparison scale so that the blood stain appears at all the various apertures. (Do not allow the blood to dry to a brown color, as this will result in an inaccurate reading.) Because the colors on the scale represent 1% variations in hemoglobin content, it may be necessary to estimate the percentage if the color of your blood sample is intermediate between two color standards.

4. On the data sheet (p. 428) record your results as the percentage of hemoglobin concentration and as grams per 100 ml of blood.

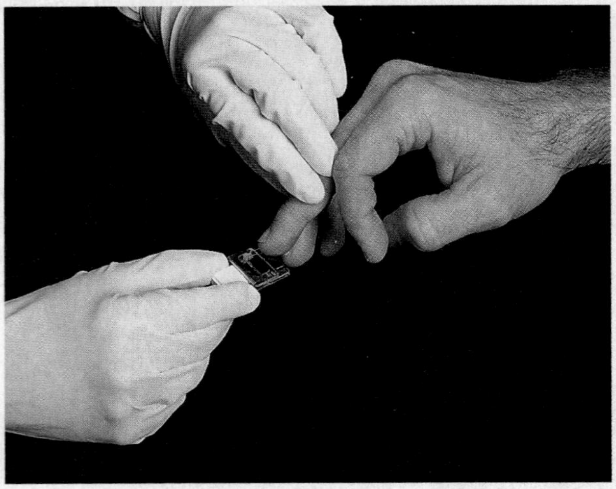

(a) A drop of blood is added to the moat plate of the blood chamber. The blood must flow freely.

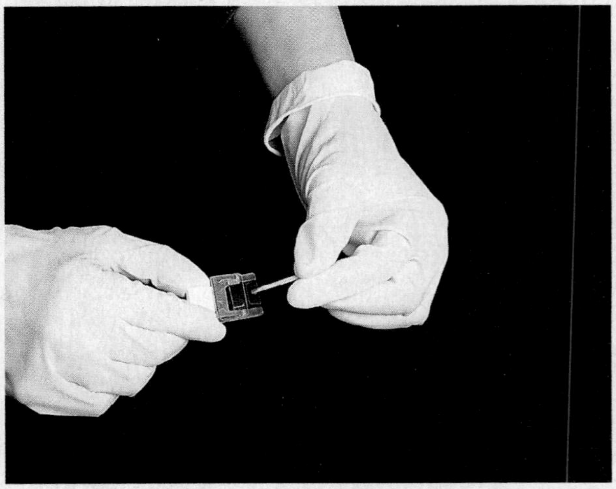

(b) The blood sample is hemolyzed with a wooden hemolysis applicator. Complete hemolysis requires 35 to 45 seconds.

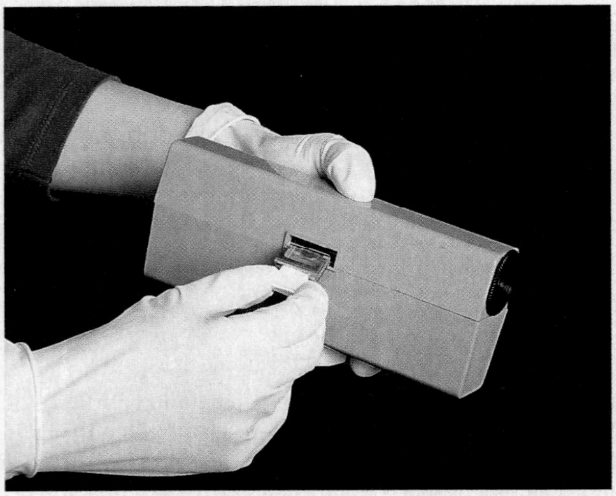

(c) The charged blood chamber is inserted into the slot on the side of the hemoglobinometer.

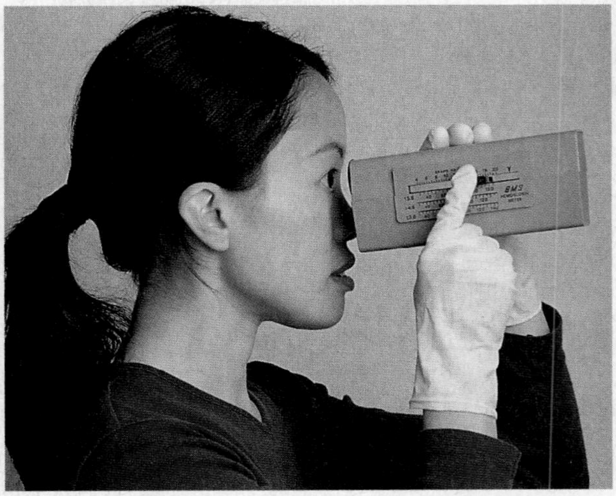

(d) The colors of the green split screen are found by moving the slide with the right index finger. When the two colors match in density, the grams/100 ml and % Hb are read on the scale.

Figure 29.7 Hemoglobin determination using a hemoglobinometer.

29

Hemoglobinometer Determination

1. Obtain a hemoglobinometer, hemolysis applicator, alcohol swab, and lens paper. Test the hemoglobinometer light source to make sure it is working; if not, request new batteries before proceeding and test it again.

2. Remove the blood chamber from the slot in the side of the hemoglobinometer, and disassemble the blood chamber by separating the glass plates from the metal clip. Notice as you do this that the larger glass plate has an H-shaped depression cut into it that acts as a moat to hold the blood, whereas the smaller glass piece is flat and serves as a coverslip.

3. Clean the glass plates with an alcohol swab, and then wipe them dry with lens paper. Hold the plates by their sides to prevent smearing during the wiping process.

4. Reassemble the blood chamber (remember: larger glass piece on the bottom with the moat up), but leave the moat plate about halfway out to provide adequate exposed surface to charge it with blood.

5. Obtain a drop of blood (from the provided sample or from your fingertip as before), and place it on the depressed area of the moat plate that is closest to you (**Figure 29.7a**).

6. Using the wooden hemolysis applicator, stir or agitate the blood to rupture (lyse) the RBCs (Figure 29.7b). This usually takes 35 to 45 seconds. Hemolysis is complete when the blood appears transparent rather than cloudy.

7. Push the blood-containing glass plate all the way into the metal clip, and then firmly insert the charged blood chamber back into the slot on the side of the instrument (Figure 29.7c).

8. Hold the hemoglobinometer in your left hand with your left thumb resting on the light switch located on the underside of the instrument. Look into the eyepiece and notice that there is a green area divided into two halves (a split field).

9. With the index finger of your right hand, slowly move the slide on the right side of the hemoglobinometer back and forth until the two halves of the green field match (Figure 29.7d).

10. Note and record on the data sheet (p. 428) the grams of Hb (hemoglobin)/100 ml of blood indicated on the uppermost scale by the index mark on the slide. Also record % Hb, indicated by one of the lower scales.

11. Disassemble the blood chamber again, and carefully place its parts (glass plates and clip) into a bleach-containing beaker.

Generally speaking, the relationship between the hematocrit and grams of hemoglobin per 100 ml of blood is 3:1—for example, a hematocrit of 36% with 12 g of Hb per 100 ml of blood is a ratio of 3:1. How do your values compare?

____

Record on the data sheet (p. 428) the value obtained from your data.

Coagulation Time

Hemostasis is a protective mechanism that is set into motion when a blood vessel breaks. Hemostasis responds rapidly to stop bleeding. During hemostasis, three events occur in the following order: vascular spasm, platelet plug formation, and coagulation (blood clotting). Platelet plug formation and coagulation are illustrated in **Figure 29.8** (p. 426).

Blood clotting, or **coagulation**, is a process that requires the interaction of many substances normally present in the plasma (clotting factors, or procoagulants) as well as some released by platelets and injured tissues. The injured tissues and platelets release **tissue factor (TF)** and **phosphatidylserine** (formerly known as **platelet factor 3**) respectively, which trigger the clotting mechanism, or cascade. Tissue factor and phosphatidylserine interact with other blood protein clotting factors and calcium ions to form **prothrombin activator**, which in turn converts **prothrombin** (present in plasma) to **thrombin**. Thrombin then acts enzymatically to polymerize (combine) the soluble **fibrinogen** proteins (present in plasma) into insoluble **fibrin**, which forms a meshwork of strands that traps the RBCs and forms the basis of the clot (Figure 29.8b). Normally, blood removed from the body clots within 2 to 6 minutes.

29

Activity 6

Determining Coagulation Time

1. Obtain a *nonheparinized* capillary tube, a timer (or watch), a lancet, cotton balls, a triangular file, and alcohol swabs.

2. Clean and prick the finger to produce a free flow of blood. Discard the lancet in the disposal container.

3. Place one end of the capillary tube in the blood drop, and hold the opposite end at a lower level to collect the sample.

4. Lay the capillary tube on a paper towel after collecting the sample.

Record the time. ______________________________

5. At 30-second intervals, make a small nick on the tube close to one end with the triangular file, and then carefully break the tube. Slowly separate the ends to see whether a gel-like thread of fibrin spans the gap. When this occurs, record below and on the data sheet (p. 428) the time for coagulation to occur. Are your results within the normal time range?

____

6. Put used supplies in the autoclave bag and broken capillary tubes into the sharps container.

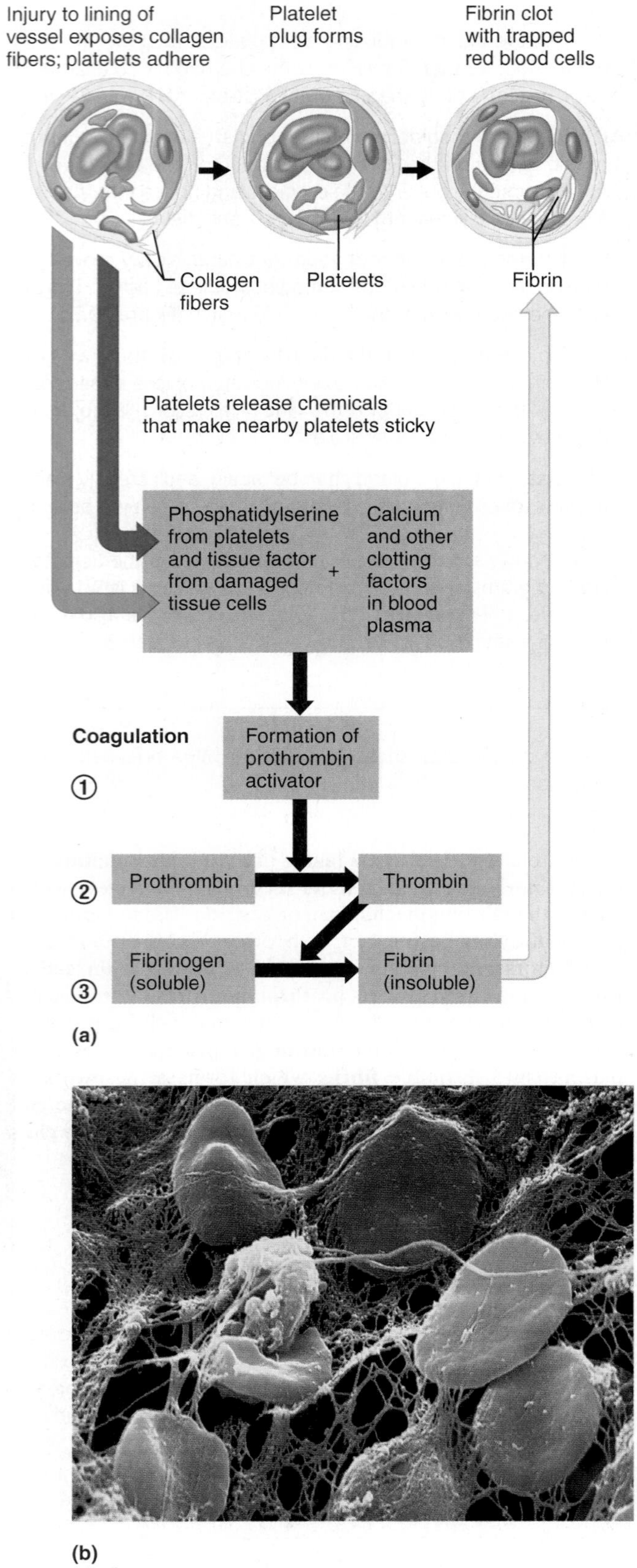

Figure 29.8 Events of platelet plug formation and coagulation. (a) Simple schematic of events. Steps numbered 1–3 represent the major events of coagulation. **(b)** Photomicrograph of RBCs trapped in a fibrin mesh (2700×).

Blood Typing

Blood typing is a system of blood classification based on the presence of specific glycoproteins on the outer surface of the RBC plasma membrane. Such proteins are called **antigens** or **agglutinogens** and are genetically determined. For ABO blood groups, these antigens are accompanied by plasma proteins, called **antibodies** or **agglutinins**. These antibodies act against RBCs carrying antigens that are not present on the person's own RBCs. If the donor blood type doesn't match, the recipient's antibodies react with the donor's blood antigens, causing the RBCs to clump, agglutinate, and eventually hemolyze. It is because of this phenomenon that a person's blood must be carefully typed before a whole blood or packed cell transfusion.

Several blood typing systems exist, based on the various possible antigens, but the factors routinely typed for are antigens of the ABO and Rh blood groups which are most commonly involved in transfusion reactions. The basis of the ABO typing is shown in Table 29.2.

Individuals whose red blood cells carry the Rh antigen are Rh positive (approximately 85% of the U.S. population); those lacking the antigen are Rh negative. Unlike ABO blood groups, the blood of neither Rh-positive (Rh^+) nor Rh-negative (Rh^-) individuals carries preformed anti-Rh antibodies. This is understandable in the case of the Rh-positive individual. However, Rh-negative persons who receive transfusions of Rh-positive blood become sensitized by the Rh antigens of the donor RBCs, and their systems begin to produce anti-Rh antibodies. On subsequent exposures to Rh-positive blood, typical transfusion reactions occur, resulting in the clumping and hemolysis of the donor blood cells.

Instructors may assign a related Pre-Lab Video Coaching Activity for Blood Typing using Mastering A&P™

Activity 7

Typing for ABO and Rh Blood Groups

Blood may be typed on microscope slides or using blood test cards. Each method is described in this activity. The artificial blood kit does not use any body fluids and produces results similar to but not identical to results for human blood.

Typing Blood Using Glass Slides

1. Obtain two clean microscope slides, a wax marking pencil, anti-A, anti-B, and anti-Rh typing sera, toothpicks, lancets, alcohol swabs, medicine dropper, and the Rh typing box.

2. Divide slide 1 into halves with the wax marking pencil. Label the lower left-hand corner "anti-A" and the lower right-hand corner "anti-B." Mark the bottom of slide 2 "anti-Rh."

3. Place one drop of anti-A serum on the *left* side of slide 1. Place one drop of anti-B serum on the *right* side of slide 1. Place one drop of anti-Rh serum in the center of slide 2.

4. If you are using your own blood, cleanse your finger with an alcohol swab, pierce the finger with a lancet, and wipe away the first drop of blood. Obtain 3 drops of freely flowing blood, placing one drop on each side of slide 1 and a drop on slide 2. Immediately dispose of the lancet in a designated disposal container.

   If using instructor-provided animal blood or red blood cells treated with EDTA (an anticoagulant), use a medicine dropper to place one drop of blood on each side of slide 1 and a drop of blood on slide 2.

Table 29.2 ABO Blood Typing

ABO blood type	Antigens present on RBC membranes	Antibodies present in plasma	% of U.S. population		
			White	Black	Asian
A	A	Anti-B	40	27	28
B	B	Anti-A	11	20	27
AB	A and B	None	4	4	5
O	Neither	Anti-A and anti-B	45	49	40

5. Quickly mix each blood-antiserum sample with a *fresh* toothpick. Then dispose of the toothpicks and used alcohol swab in the autoclave bag.

6. Place slide 2 on the Rh typing box and rock gently back and forth. (A slightly higher temperature is required for precise Rh typing than for ABO typing.)

7. After 2 minutes, observe all three blood samples for evidence of clumping. The agglutination that occurs in the positive test for the Rh factor is very fine and difficult to interpret; thus if there is any question, observe the slide under the microscope. Record your observations in the **Activity 7 chart**.

8. Interpret your ABO results (see the examples of each type) in **Figure 29.9**. If you observe clumping on slide 2, you are Rh positive. If not, you are Rh negative.

9. Record your blood type on the data sheet (p. 428).

10. Put the used slides in the bleach-containing bucket at the general supply area; put disposable supplies in the autoclave bag.

Activity 7: Blood Typing

Result	Observed (+)	Not observed (−)
Presence of clumping with anti-A		
Presence of clumping with anti-B		
Presence of clumping with anti-Rh		

Using Blood Typing Cards

1. Obtain a blood typing card marked A, B, and Rh, dropper bottles of anti-A serum, anti-B serum, and anti-Rh serum, toothpicks, lancets, and alcohol swabs.

2. Place a drop of anti-A serum in the spot marked anti-A, place a drop of anti-B serum on the spot marked anti-B, and place a drop of anti-Rh serum on the spot marked anti-Rh (or anti-D).

3. Carefully add a drop of blood to each of the spots marked "Blood" on the card. If you are using your own blood, refer to step 4 in the Activity 7 section Typing Blood Using Glass Slides. Immediately discard the lancet in the designated disposal container.

4. Using a new toothpick for each test, mix the blood sample with the antibody. Dispose of the toothpicks appropriately.

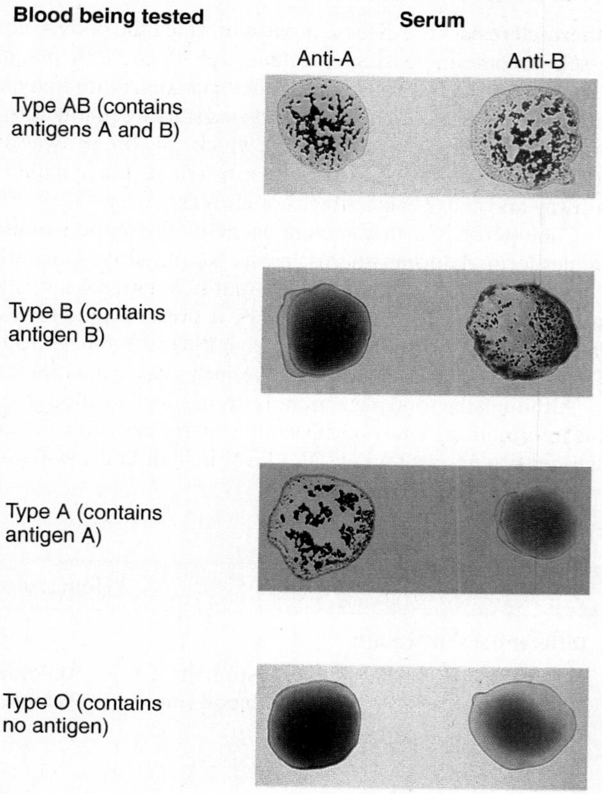

Figure 29.9 Blood typing of ABO blood types. When serum containing anti-A or anti-B antibodies (agglutinins) is added to a blood sample, agglutination will occur between the antibody and the corresponding antigen (agglutinogen A or B). As illustrated, agglutination occurs with both sera in blood group AB, with anti-B serum in blood group B, with anti-A serum in blood group A, and with neither serum in blood group O.

Instructors may assign this figure as an Art Labeling Activity using Mastering A&P™

5. Gently rock the card to allow the blood and antibodies to mix.

6. After 2 minutes, observe the card for evidence of clumping. The Rh clumping is very fine and may be difficult to observe. Record your observations in the Activity 7 chart. (Use Figure 29.9 to interpret your results.)

7. Record your blood type on the data sheet (p. 428), and discard the card in an autoclave bag.

29

Activity 8

Observing Demonstration Slides

Look at the slides of *macrocytic hypochromic anemia*, *microcytic hypochromic anemia*, *sickle cell anemia*, *lymphocytic leukemia* (chronic), and *eosinophilia* that have been put on demonstration by your instructor. Record your observations in the appropriate section of the Review Sheet. You can refer to your notes, the text, and other references later to respond to questions about the blood pathologies represented on the slides.

Cholesterol Concentration in Plasma

Atherosclerosis is the disease process in which the body's blood vessels become increasingly occluded, or blocked, by plaques. By narrowing the arteries, the plaques can contribute to hypertensive heart disease. They also serve as starting points for the formation of blood clots (thrombi), which may break away and block smaller vessels farther downstream in the circulatory pathway and cause heart attacks or strokes.

Cholesterol is a major component of the smooth muscle plaques formed during atherosclerosis. No physical examination of an adult is considered complete until cholesterol levels are assessed along with other risk factors. A normal value for total plasma cholesterol in adults ranges from 130 to 200 mg per 100 ml of plasma; you will use blood to make such a determination.

Although the total plasma cholesterol concentration is valuable information, it may be misleading, particularly if a person's high-density lipoprotein (HDL) level is high and low-density lipoprotein (LDL) level is relatively low. Cholesterol, being water insoluble, is transported in the blood complexed to lipoproteins. In general, cholesterol bound into HDLs is destined to be degraded by the liver and then eliminated from the body, whereas that forming part of the LDLs is "traveling" to the body's tissue cells. When LDL levels are excessive, cholesterol is deposited in the blood vessel walls; hence, LDLs are considered to carry the "bad" cholesterol.

Activity 9

Measuring Plasma Cholesterol Concentration

1. Go to the appropriate supply area, and obtain a cholesterol test card and color scale, lancet, and alcohol swab.
2. Clean your fingertip with the alcohol swab, allow it to dry, then prick it with a lancet. Place a drop of blood on the test area of the card. Put the lancet in the designated disposal container.
3. After 3 minutes, remove the blood sample strip from the card and discard in the autoclave bag.
4. Analyze the underlying test spot, using the included color scale. Record the cholesterol level below and on the **Hematologic Test Data Sheet**.

Total cholesterol level ____________ mg/dl

5. Before leaving the laboratory, use the spray bottle of bleach solution, and saturate a paper towel to thoroughly wash down your laboratory bench.

Hematologic Test Data Sheet

Differential WBC count:

WBC	Student blood smear	Unknown sample # ____
% neutrophils	________	________
% eosinophils	________	________
% basophils	________	________
% lymphocytes	________	________
% monocytes	________	________

Hematocrit:

RBC ______ % of blood volume

WBC ______ % of blood volume — not generally reported

Plasma ______ % of blood

Hemoglobin (Hb) content:

Tallquist method: ______ g/100 ml of blood; ______ % Hb

Hemoglobinometer (type:

______________________________)

____________ g/100 ml of blood; ____________ %Hb

Ratio (hematocrit to grams of Hb per 100 ml of blood):

____

Coagulation time: ____________

Blood typing:

ABO group ____________ Rh factor ____________

Total cholesterol level: ____________ **mg/dl of blood**

Instructors may assign a portion of the Review Sheet questions using Mastering A&P™

29 REVIEW SHEET
EXERCISE
Blood

Name ______________________ Lab Time/Date ______________________

Composition of Blood

1. What is the blood volume of an average-size adult male? __________ liters; an average adult female? __________ liters

2. What determines whether blood is bright red or a dull brick red? ______________________

____

3. Use the key to identify the cell type(s) or blood elements that fit the following descriptive statements. Some terms will be used more than once.

Key:

a. red blood cell	d. basophil	g. lymphocyte
b. megakaryocyte	e. monocyte	h. platelets
c. eosinophil	f. neutrophil	i. plasma

__________ 1. most numerous leukocyte

__________, __________, and __________ 2. granulocytes (3)

__________ 3. also called an erythrocyte; anucleate formed element

__________, __________, __________ 4. phagocytic leukocytes (3)

__________, __________ 5. agranulocytes

__________ 6. precursor cell of platelets

__________ 7. cell fragments

__________ 8. involved in destroying parasitic worms

__________ 9. releases histamine; promotes inflammation

__________ 10. produces antibodies

__________ 11. transports oxygen

__________ 12. primarily water, noncellular; the fluid matrix of blood

__________ 13. exits a blood vessel to develop into a macrophage

__________, __________, __________,

__________, __________ 14. the five types of white blood cells

4. Define *formed elements*. ______________________________

____

List the formed elements present in the blood. ______________________________

____

5. Describe the consistency and color of the plasma you observed in the laboratory. ______________________________

____

6. What is the average life span of a red blood cell? How does its anucleate condition affect this life span?

____

____

____

7. Identify the leukocytes shown in the photomicrographs below.

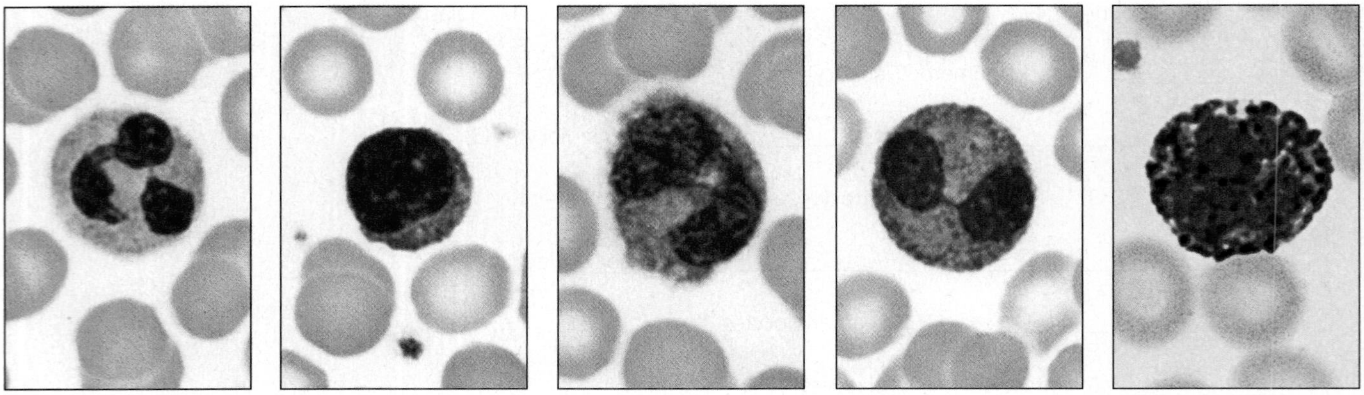

______________ ______________ ______________ ______________ ______________

8. Correctly identify the blood pathologies described in column A by matching them with selections from column B:

Column A		**Column B**
______________	1. abnormal increase in the number of WBCs	a. anemia
______________	2. abnormal increase in the number of RBCs	b. leukocytosis
______________	3. condition of too few RBCs or of RBCs with hemoglobin deficiencies	c. leukopenia
______________	4. abnormal decrease in the number of WBCs	d. polycythemia

Hematologic Tests

9. In the chart below, record information from the blood tests you read about or conducted. Complete the chart by recording values for healthy male adults and indicating the significance of high or low values for each test.

Test	Student test results	Normal values (healthy male adults)	Significance	
			High values	Low values
Total WBC count	No data			
Total RBC count	No data			
Hematocrit				
Hemoglobin determination				
Bleeding time	No data			
Coagulation time				

10. Why is a differential WBC count more valuable than a total WBC count when trying to determine the specific source of pathology? ______________________________

____

____

11. Discuss the effect of each of the following factors on RBC count. Consult an appropriate reference as necessary, and explain your reasoning.

long-term effect of athletic training (for example, running 4 to 5 miles per day over a period of 6 to 9 months):

____

____

a permanent move from sea level to a high-altitude area: ______________________________

____

____

12. Define *hematocrit*. ______________________________

13. If you had a high hematocrit, would you expect your hemoglobin determination to be high or low? ______________

Why? ______________________________

____

14. What is an anticoagulant? ____________________

Name two anticoagulants used in conducting the hematologic tests. ____________________

and ____________________

What is the body's natural anticoagulant? ____________________

15. If your blood agglutinates with anti-A but not anti-B sera, your ABO blood type would be ____________________.

To what ABO blood groups could you donate blood? ____________________

From which ABO donor types could you receive blood? ____________________

Which ABO blood type is most common? ____________________ Least common? ____________________

16. What blood type is theoretically considered the universal donor? ____________________ Why? ____________________

____

17. Assume the blood of two patients has been typed for ABO blood type.

Typing results
Mr. Adams:

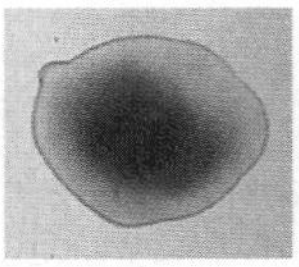

Blood drop and anti-A serum

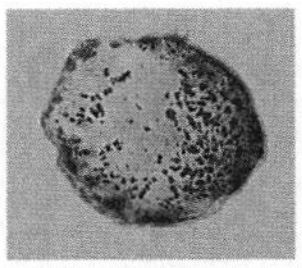

Blood drop and anti-B serum

Typing results
Mr. Calhoon:

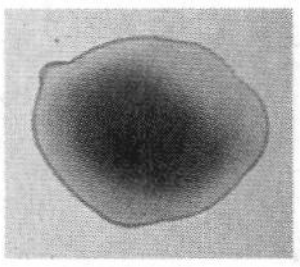

Blood drop and anti-A serum

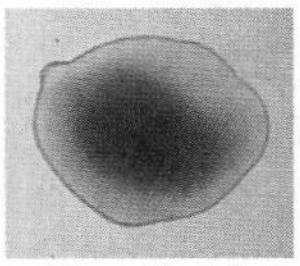

Blood drop and anti-B serum

On the basis of these results, Mr. Adams has type ____________________ blood, and Mr. Calhoon has type ____________________ blood.

18. Explain why an Rh-negative person does not have a transfusion reaction on the first exposure to Rh-positive blood but *does* have a reaction on the second exposure. ____________________

____

19. Record your observations of the five demonstration slides viewed.

a. Macrocytic hypochromic anemia: ____________________

b. Microcytic hypochromic anemia: ____________________

____

c. Sickle cell anemia: ____________________

d. Lymphocytic leukemia (chronic): ______________________________

____

e. Eosinophilia: ______________________________

Which of the slides above (a through e) corresponds with the following conditions?

______________ 1. iron-deficient diet

______________ 2. a type of bone marrow cancer

______________ 3. genetic defect that causes hemoglobin to become sharp/spiky

______________ 4. lack of vitamin B_{12}

______________ 5. a tapeworm infestation in the body

______________ 6. a bleeding ulcer

20. Provide the normal, or at least "desirable," range for plasma cholesterol concentration.

______________ mg/100 ml

21. ✚ Plasmapheresis is a procedure in which blood is removed, its plasma is separated from the formed elements, and the formed elements are returned to the patient or donor. Kidney transplants usually require that the donor and recipient have the same blood type. If plasmapheresis is administered to the patient before and after the transplant surgery, rejection of the kidney is unlikely to occur. Explain why.

____

____

22. ✚ Bleeding disorders are usually a result of thrombocytopenia, a deficiency of platelets. Considering the mechanism of hemostasis, explain why thrombocytopenia could lead to abnormal bleeding. ______________________________

____

____

EXERCISE 30

Anatomy of the Heart

Learning Outcomes

- Describe the location of the heart.
- Name and describe the covering and lining tissues of the heart.
- Name and locate the major anatomical areas and structures of the heart when provided with an appropriate model, image, or dissected sheep heart, and describe the function of each.
- Explain how the atrioventricular and semilunar valves operate.
- Distinguish blood vessels carrying oxygen-rich blood from those carrying carbon dioxide–rich blood, and describe the system used to color code them in images.
- Explain why the heart is called a double pump, and compare the pulmonary and systemic circuits.
- Trace the pathway of blood through the heart.
- Trace the functional blood supply of the heart, and name the associated blood vessels.
- Describe the histology of cardiac muscle, and state the importance of its intercalated discs and the spiral arrangement of its cells.

Go to Mastering A&P™ > Study Area to improve your performance in A&P Lab.

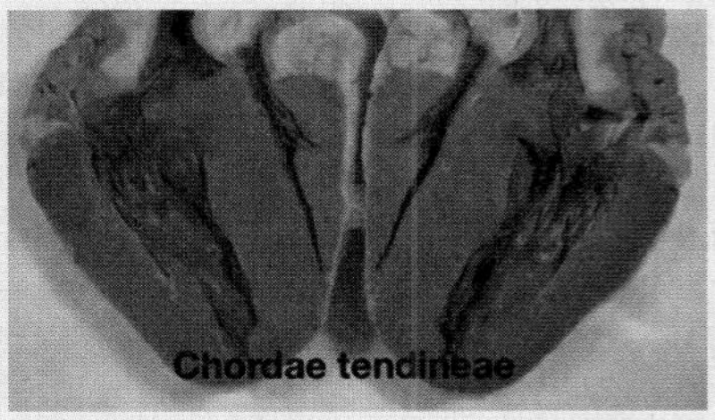

> Lab Tools >Bone and Dissection Videos > Sheep Heart

Instructors may assign new Building Vocabulary coaching activities, Pre-Lab Quiz questions, Art Labeling activities, related videos and coaching activities for the Sheep Heart dissection, Practice Anatomy Lab Practical questions (PAL), and more using the Mastering **A&P™** Item Library.

Pre-Lab Quiz

Instructors may assign these and other Pre-Lab Quiz questions using Mastering A&P™

1. The heart is enclosed in a double-walled sac called the:
   a. apex **b.** mediastinum **c.** pericardium **d.** thorax
2. The left ventricle discharges blood into the ___________, from which all systemic arteries of the body diverge to supply the body tissues.
   a. aorta
   b. pulmonary artery
   c. pulmonary vein
   d. vena cava
3. Circle the correct underlined term. The right atrioventricular valve, or <u>tricuspid valve</u> / <u>mitral valve</u>, prevents backflow into the right atrium when the right ventricle is contracting.
4. Circle the correct underlined term. The heart serves as a double pump. The <u>right</u> / <u>left</u> side serves as the pulmonary circuit pump, shunting carbon dioxide–rich blood to the lungs.
5. Circle the correct underlined term. In the heart, the <u>left</u> / <u>right</u> ventricle has thicker walls and a basically circular cavity shape.

Materials

- X-ray film of the human thorax; X-ray viewing box
- Three-dimensional heart model or laboratory chart showing heart anatomy
- Red and blue pencils
- Highlighter
- Three-dimensional models of cardiac and skeletal muscle
- Compound microscope
- Prepared slides of cardiac muscle (l.s.)
- Preserved or fresh sheep hearts, pericardial sacs intact (if possible)
- Dissecting instruments and tray
- Pointed glass rods or blunt probes
- Small plastic metric rulers
- Disposable gloves
- Container for disposal of organic debris

Text continues on next page. →

- Laboratory detergent
- Spray bottle with 10% household bleach solution

For instructions on animal dissections, see the dissection exercises (starting on p. 695) in the cat and fetal pig editions of this manual.

The major function of the **cardiovascular system** is transportation. Using blood as the transport vehicle, the system carries oxygen, nutrients, cell wastes, electrolytes, and many other substances vital to the body's homeostasis to and from the body cells. The system's propulsive force is the contracting heart, which can be compared to a muscular pump equipped with one-way valves. As the heart contracts, it forces blood into a closed system of large and small plumbing tubes (blood vessels) within which the blood is confined and circulated.

Gross Anatomy of the Human Heart

The **heart**, a cone-shaped organ approximately the size of a fist, is located within the mediastinum of the thorax. It is flanked laterally by the lungs, posteriorly by the vertebral column, and anteriorly by the sternum (**Figure 30.1**). Its more pointed **apex** extends slightly to the left and rests on the diaphragm, approximately at the level of the fifth intercostal space. Its broader **base**, from which the great vessels emerge, lies beneath the second rib and points toward the right shoulder.

☐ If an X-ray image of a human thorax is available, verify the relationships described above (otherwise, Figure 30.1 should suffice).

Figure 30.2 shows three views of the heart—anterior and posterior views and a frontal section.

The heart is enclosed within a double-walled sac called the pericardium. The loose-fitting superficial part of the sac is the **fibrous pericardium**. Deep to it is the *serous pericardium*, which lines the internal surface of the fibrous pericardium as the **parietal layer**. At the base of the heart, the parietal layer reflects back to cover the external surface of the heart as the **visceral layer**, or **epicardium**. The epicardium is an integral part of the heart wall. Serous fluid produced by these layers allows the heart to beat in a relatively frictionless environment.

The walls of the heart are composed of three layers:

- **Epicardium:** The outer layer, which is also the visceral pericardium.
- **Myocardium:** The middle layer and thickest layer, which is composed mainly of cardiac muscle. It is reinforced with dense fibrous connective tissue, the *cardiac skeleton*, which is thicker around the heart valves and at the base of the great vessels leaving the heart.
- **Endocardium:** The inner lining of the heart, which covers the heart valves and is continuous with the inner lining of the great vessels. It is composed of simple squamous epithelium resting on areolar connective tissue.

30

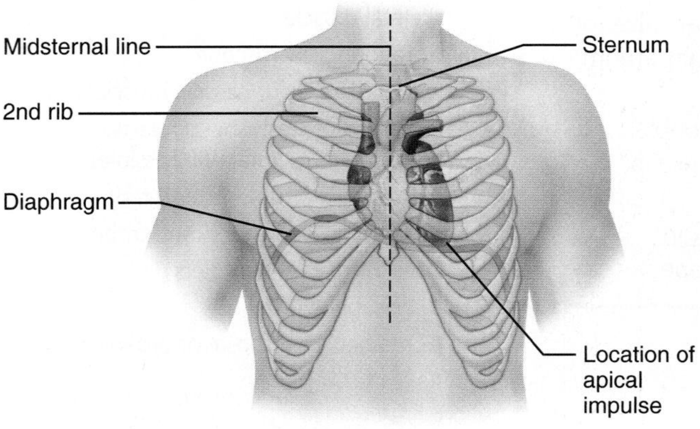

Figure 30.1 Location of the heart in the thorax. The apical impulse occurs where the heart's apex touches the chest wall.

Heart Chambers

The heart is divided into four chambers: two superior **atria** (singular: *atrium*) and two inferior **ventricles**. The septum that divides the heart longitudinally is referred to as the **interatrial septum** where it separates the atria, and the **interventricular septum** where it separates the ventricles. Functionally, the atria are receiving chambers and are relatively ineffective as pumps.

The inferior thick-walled ventricles, which form the bulk of the heart, are the discharging chambers. They force blood out of the heart into the large arteries that emerge from its base.

Heart Valves

Four valves enforce a one-way blood flow through the heart chambers. The **atrioventricular (AV) valves** are located between the atrium and the ventricle on the left and right side of the heart. The **semilunar (SL) valves** are located between a ventricle and a great vessel (**Figure 30.3**, p. 439).

- **Tricuspid valve:** The right AV valve has three flaplike cusps anchored to the **papillary muscles** of the ventricular wall by tiny, white collagenic cords called **chordae tendineae** (literally, "heart strings")
- **Mitral valve** (*bicuspid valve*): The left AV valve has two flaplike cusps anchored to the papillary muscles by chordae tendineae.

The AV valves are open and hang into the ventricles when blood is flowing into the atria and the ventricles are relaxed. When the ventricles contract, the blood in the ventricles is compressed, causing the AV valves to move superiorly and close the opening between the atrium and the ventricle. The chordae tendineae, pulled tight by the contracting papillary muscles, anchor the cusps in the closed position and prevent the backflow of blood from the ventricles into the atria. If unanchored, the cusps would move upward into the atria like an umbrella being turned inside out by a strong wind.

- **Pulmonary (SL) valve:** Has three pocketlike cusps located between the right ventricle and the pulmonary trunk.
- **Aortic (SL) valve:** Has three pocketlike cusps located between the left ventricle and the aorta.

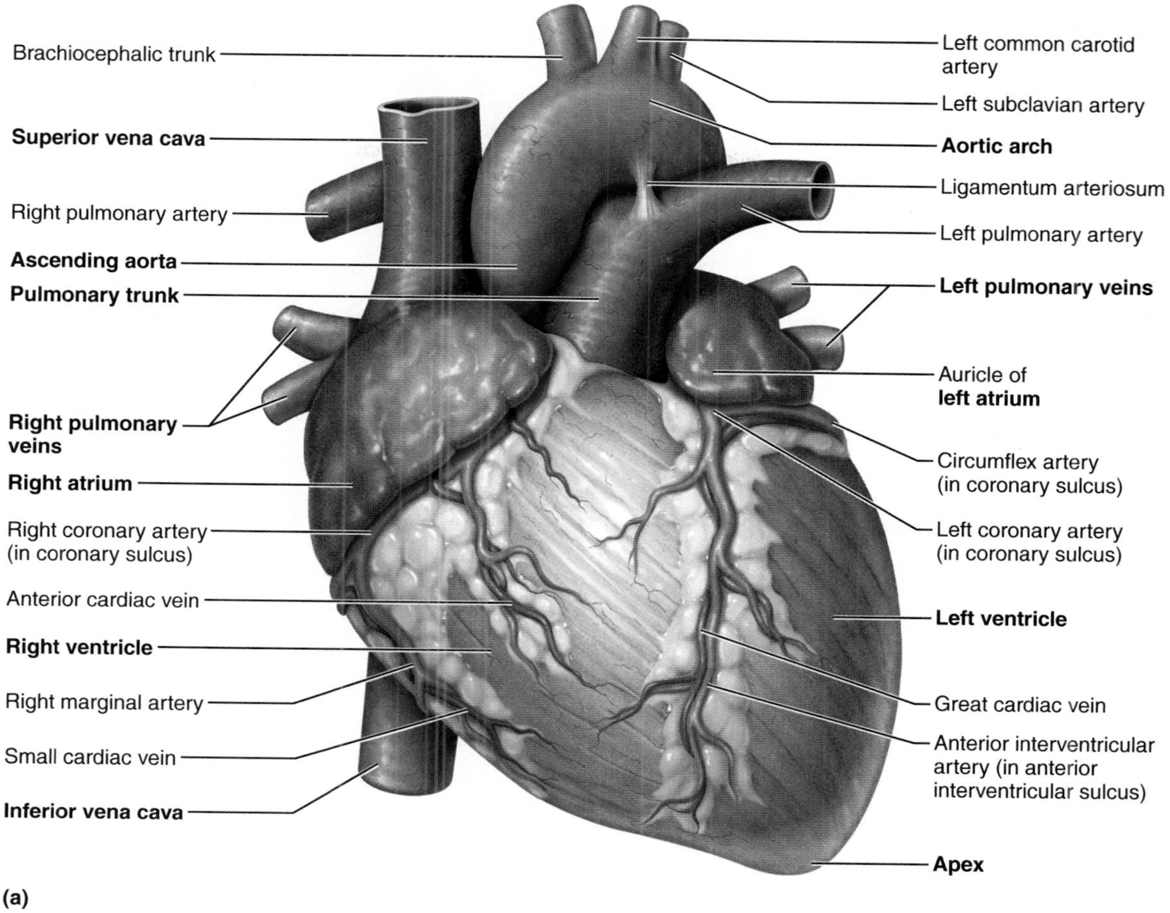

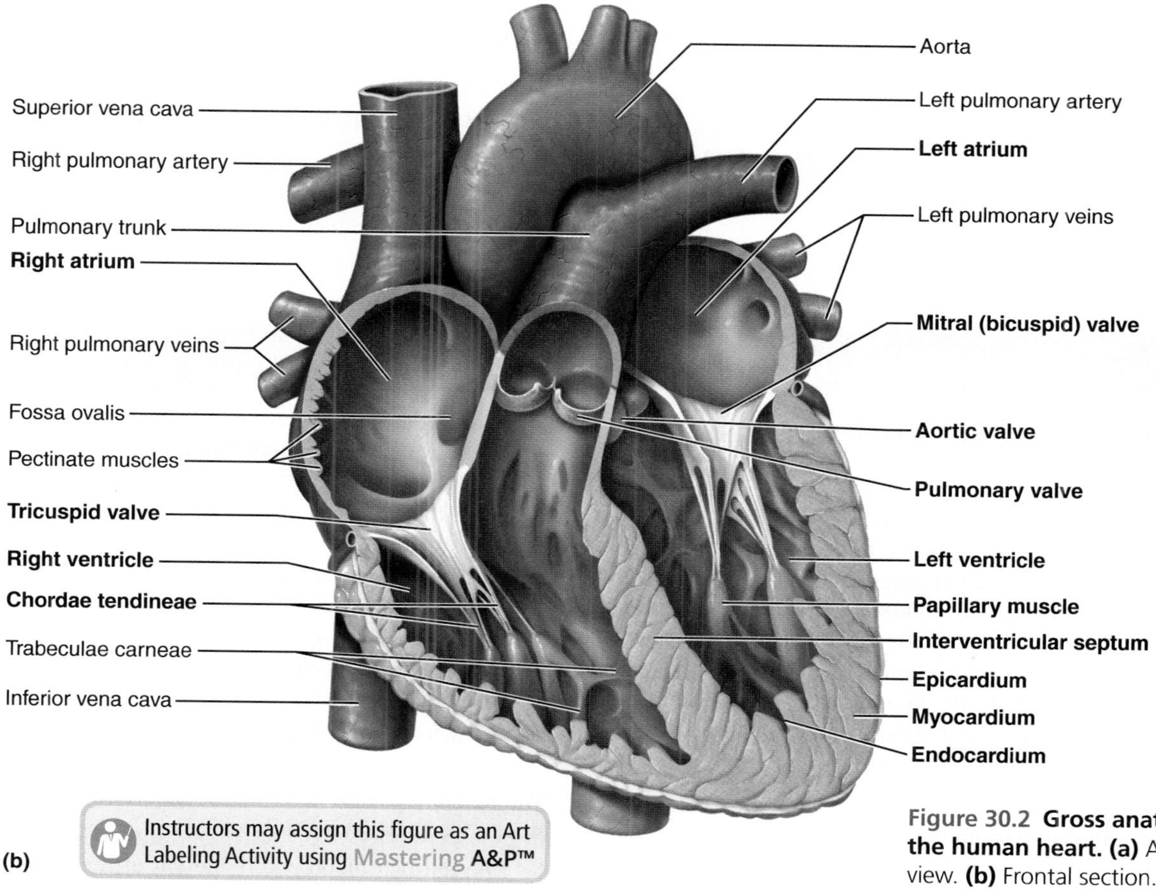

Instructors may assign this figure as an Art Labeling Activity using Mastering A&P™

Figure 30.2 **Gross anatomy of the human heart. (a)** Anterior view. **(b)** Frontal section.

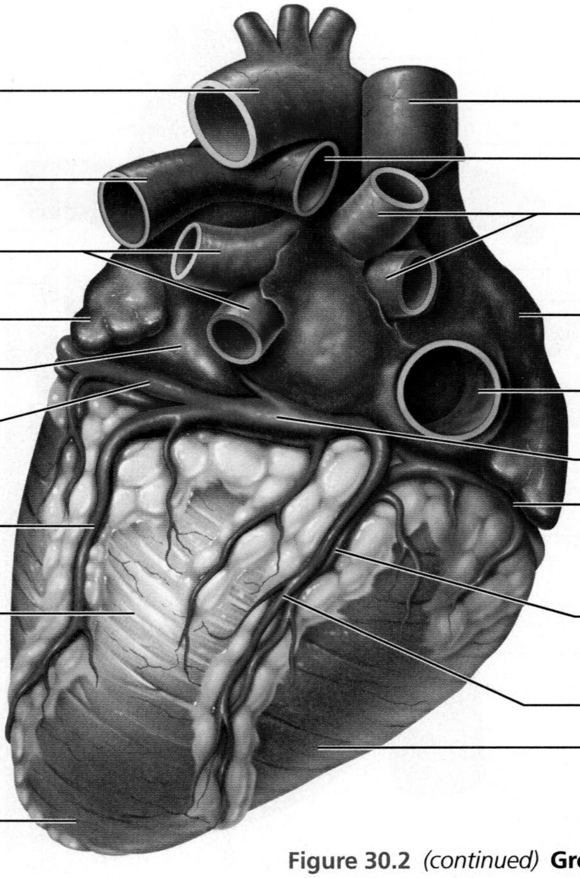

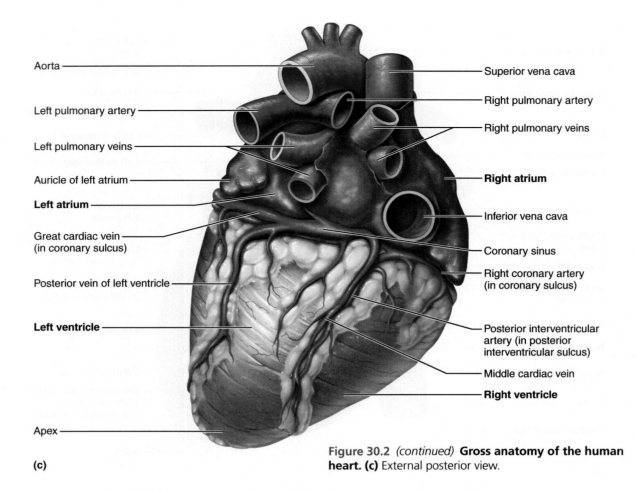

Figure 30.2 *(continued)* **Gross anatomy of the human heart. (c)** External posterior view.

The SL valves are open and flattened against the wall of the vessel when the contraction of the ventricles pushes blood into the great vessels. When the ventricles relax, blood flows backward toward the ventricle and the cusps fill with blood, closing the SL valves. This prevents the backflow of blood from the great vessels into the ventricles.

Activity 1

Using the Heart Model to Study Heart Anatomy

Locate in Figure 30.2 all the structures described above. Then, observe the human heart model and laboratory charts, and identify the same structures without referring to the figure.

30

Pulmonary, Systemic, and Coronary Circulations

Pulmonary and Systemic Circuits

The heart functions as a double pump:

- The right side of the heart pumps oxygen-poor blood entering its chambers to the lungs to unload carbon dioxide and to pick up oxygen. The blood vessels that carry blood to and from the lungs form the **pulmonary circuit**. The function of the pulmonary circuit is strictly to provide for gas exchange.
- The left side of the heart pumps oxygenated blood returning from the lungs to the body tissues. The blood vessels that carry blood to and from all body tissues form the **systemic circuit** (**Figure 30.4**, p. 440).

The following steps describe the blood flow through the right side of the heart (pulmonary circuit):

1. The right atrium receives oxygen-poor blood from the body via the venae cavae **(superior vena cava** and **inferior vena cava)** and the coronary sinus.
2. From the right atrium, blood flows through the tricuspid valve to the right ventricle.
3. From the right ventricle, blood flows through the pulmonary valve into the **pulmonary trunk**.
4. The pulmonary trunk branches into left and right **pulmonary arteries**, which carry blood to the lungs, where the blood unloads carbon dioxide and picks up oxygen.
5. Oxygen-rich blood returns to the heart via four pulmonary veins.

The remaining steps describe the blood flow through the left side of the heart (systemic circuit):

6. Oxygen-rich blood enters the left atrium via four pulmonary veins.

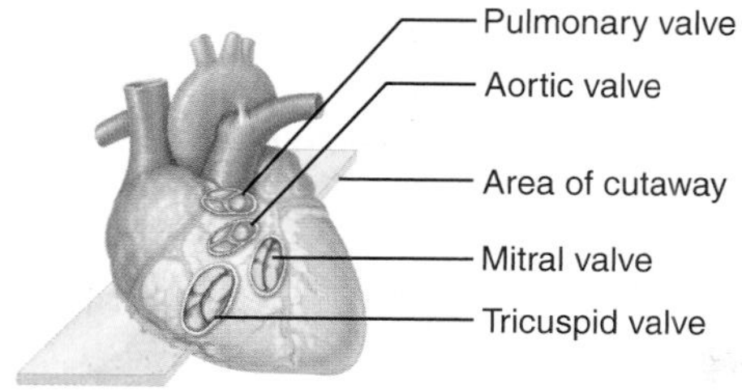

Figure 30.3 Heart valves. (a) Superior view of the two sets of heart valves (atria removed). **(b)** Photograph of the heart valves, superior view. **(c)** Photograph of the tricuspid valve. This inferior-to-superior view shows the valve as seen from the right ventricle. **(d)** Frontal section of the heart.

Instructors may assign this figure as an Art Labeling Activity using Mastering A&P™

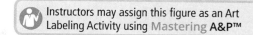

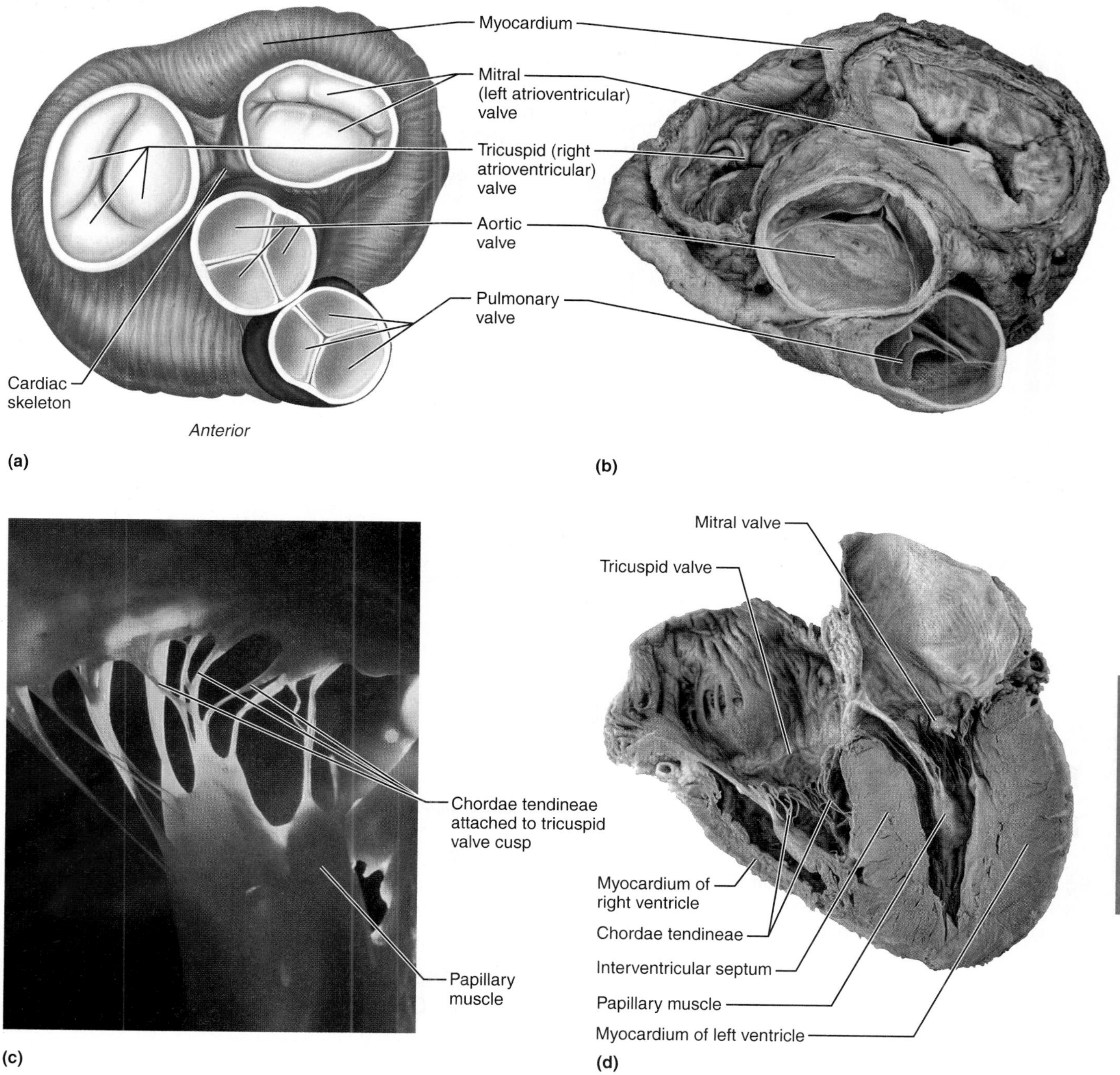

30

7. From the left atrium, blood flows through the mitral valve to the left ventricle.
8. From the left ventricle, blood flows through the aortic valve to the **aorta**.
9. Oxygen-rich blood is delivered to the body tissues by the systemic arteries.

Activity 2

Tracing the Path of Blood Through the Heart

Use colored pencils to trace the pathway of a red blood cell through the heart by adding arrows to the frontal section diagram (Figure 30.2b). Use red arrows for the oxygen-rich blood and blue arrows for the carbon dioxide–rich blood.

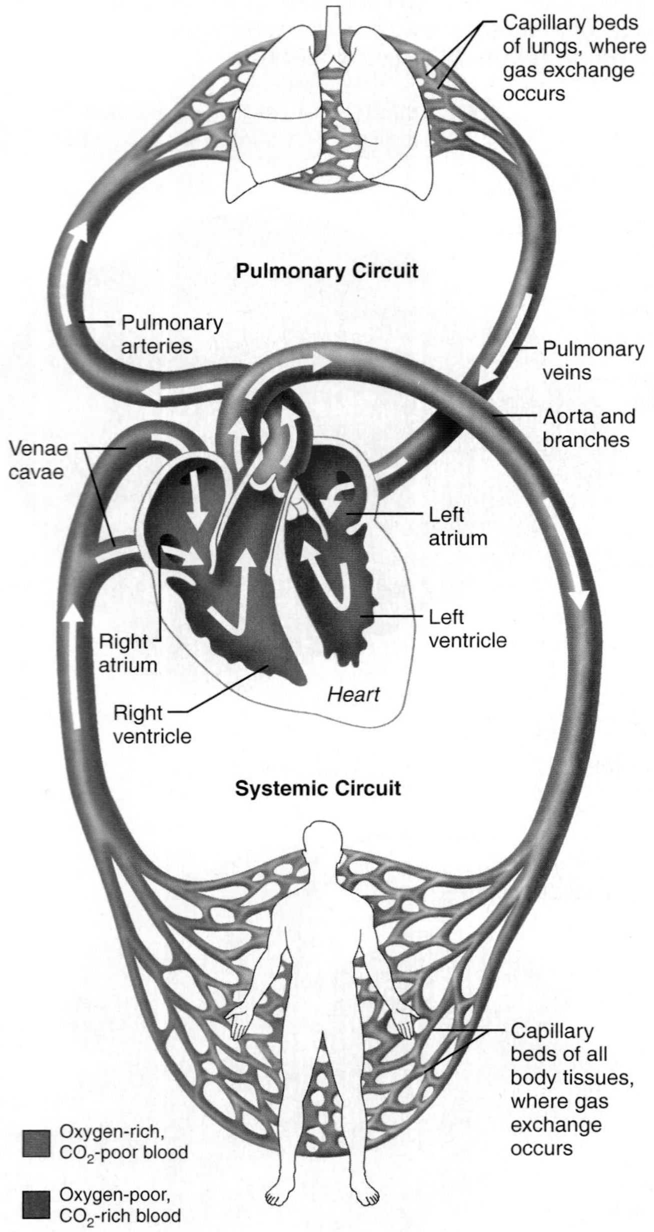

Figure 30.4 The systemic and pulmonary circuits. For simplicity, the actual number of two pulmonary arteries and four pulmonary veins has been reduced to one each in this art.

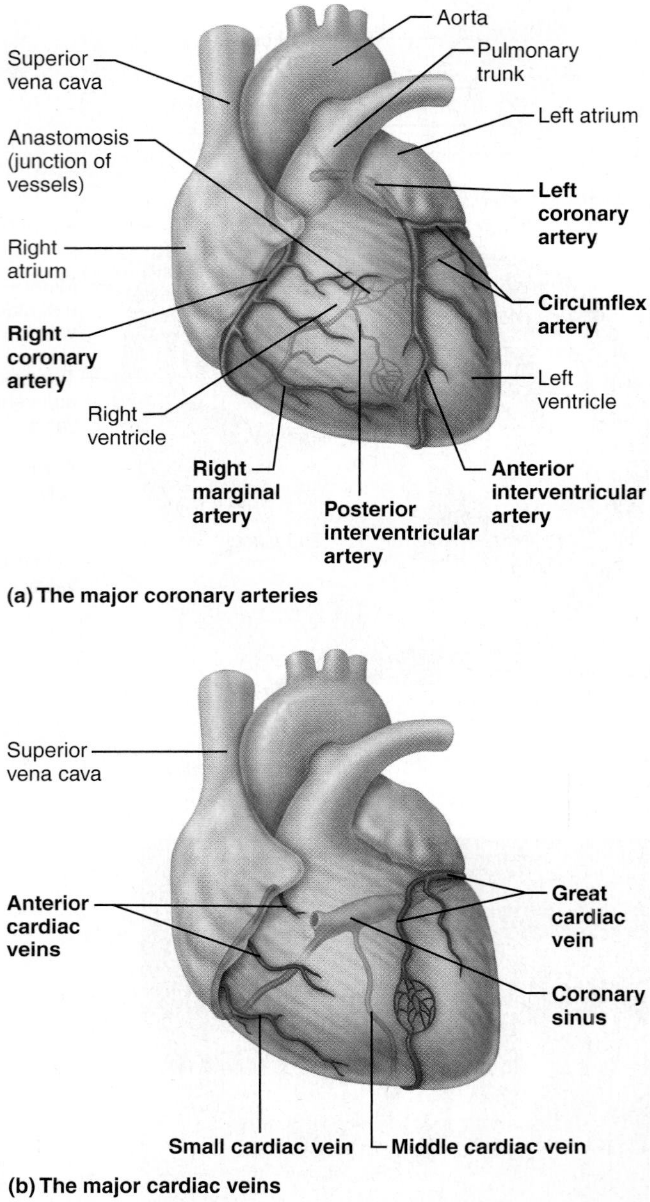

Figure 30.5 Coronary circulation.

Coronary Circulation

Even though the heart chambers are almost continually bathed with blood, this contained blood does not nourish the myocardium. The functional blood supply of the heart is provided by the coronary arteries (see Figure 30.2 and **Figure 30.5**). **Table 30.1** summarizes the arteries that supply blood to the heart and the veins that drain the blood.

Activity 3

Using the Heart Model to Study Cardiac Circulation

1. Obtain a highlighter and highlight all the cardiac blood vessels in Figure 30.2a and c. Note how arteries and veins travel together.

2. On a model of the heart, locate all the cardiac blood vessels shown in Figure 30.5. Use your finger to trace the pathway of blood from the right coronary artery to the lateral aspect of the right side of the heart and back to the right atrium. Name the arteries and veins along the pathway. Trace the pathway of blood from the left coronary artery to the anterior ventricular walls and back to the right atrium. Name the arteries and veins along the pathway. Note that there are multiple different pathways to distribute blood to these parts of the heart.

Microscopic Anatomy of Cardiac Muscle

The cardiac cells, crisscrossed by connective tissue fibers for strength, are arranged in spiral or figure-8-shaped bundles (**Figure 30.6**, overview diagram). When the heart contracts, its internal chambers become smaller, forcing the blood into the large arteries leaving the heart.

Activity 4

Examining Cardiac Muscle Tissue Anatomy

1. Observe the three-dimensional model of cardiac muscle, examining its branching cells and the areas where the cells interlock, the **intercalated discs**.
2. Compare the model of cardiac muscle to the model of skeletal muscle. Note the similarities and differences between the two kinds of muscle tissue.
3. Observe a longitudinal section of cardiac muscle under high power. Identify the nucleus, striations, intercalated discs, and sarcolemma of the individual cells, and compare your observations to Figure 30.6.

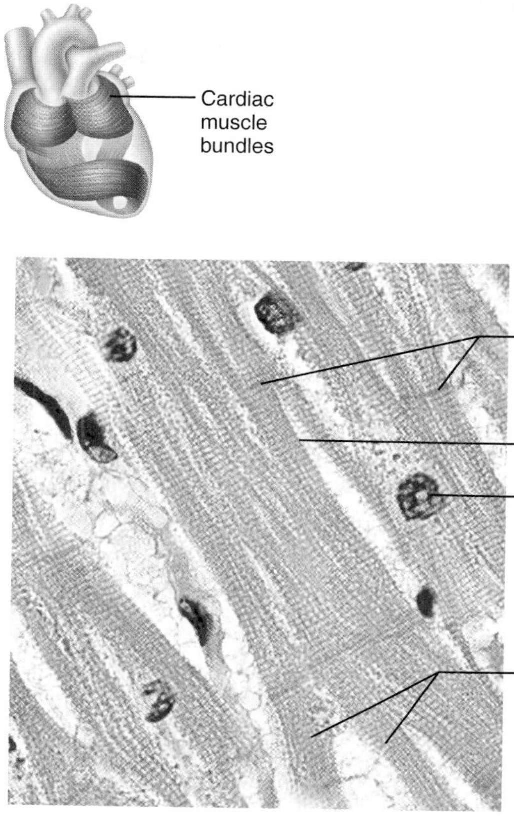

Figure 30.6 Photomicrograph of cardiac muscle (665×). The overview diagram illustrates the circular and spiral arrangement of cardiac muscle bundles in the myocardium of the heart.

Table 30.1 Coronary Circulation (Figure 30.5)

Arteries	Description	Areas supplied/branches
Right coronary artery (RCA)	Branches from the ascending aorta just above the aortic valve and encircles the heart in the coronary sulcus.	Its branches include the right marginal artery and the posterior interventricular artery.
Right marginal artery	Branches off the RCA and is located in the lateral portion of the right ventricle.	Supplies the lateral right side of the heart.
Posterior interventricular artery	Branches off the RCA and is located in the posterior interventricular sulcus.	Supplies the posterior walls of the ventricles and the posterior portion of the interventricular septum. Near the apex of the heart it merges (anastomoses) with the anterior interventricular artery.
Left coronary artery (LCA)	Branches from the ascending aorta and passes posterior to the pulmonary trunk.	Its branches include the anterior interventricular artery and the circumflex artery.
Anterior interventricular artery	Branches off the LCA and is located in the anterior interventricular sulcus. This artery is referred to clinically as the left anterior descending artery (LAD).	Supplies the anterior portion of the interventricular septum and the anterior walls of both ventricles.
Circumflex artery	Branches off the LCA: located in the coronary sulcus.	Supplies the left atrium and the posterior portion of the left ventricle.
Veins	**Description**	**Areas drained**
Great cardiac vein	Located in the anterior interventricular sulcus, parallel to the anterior interventricular artery.	Anterior portions of the right and left ventricles.
Middle cardiac vein	Located in the posterior interventricular sulcus, parallel to the posterior interventricular artery.	Posterior portions of the right and left ventricles.
Small cardiac vein	Located on the lateral right ventricle, parallel to the right marginal artery.	Lateral right ventricle.
Coronary sinus	Located in the coronary sulcus on the posterior surface of the heart; drains into the right atrium.	The entire heart; the great, middle and small cardiac veins all drain into the coronary sinus.
Anterior cardiac veins	Located on the anterior surface of the right atrium.	They drain directly into the right atrium.

30

DISSECTION

The Sheep Heart

Dissecting a sheep heart is a valuable exercise because it is similar in size and structure to the human heart. (Refer to **Figure 30.7** as you proceed with the dissection.)

1. Obtain a preserved sheep heart, a dissecting tray, dissecting instruments, a glass probe, a plastic ruler, and gloves.

2. Observe the texture of the fibrous pericardium. Also, note its point of attachment to the heart. Where is it attached?

___

3. If the fibrous pericardial sac is still intact, slit it open and cut it from its attachments. Observe the slippery parietal pericardium that lines the sac and the visceral pericardium (epicardium) that covers the heart wall. Using a sharp scalpel, carefully pull a little of the epicardium away from the myocardium. How do its position, thickness, and apposition to the heart differ from those of the parietal pericardium?

___

___

___

4. Examine the external surface of the heart. Notice the accumulation of adipose tissue, which in many cases marks the separation of the chambers and the location of the coronary arteries. Carefully scrape away some of the fat with a scalpel to expose the coronary blood vessels.

5. Identify the base and apex of the heart, and then identify the two wrinkled **auricles**, earlike flaps of tissue projecting from the atria. The balance of the heart muscle is ventricular tissue. To identify the left ventricle, compress the ventricles on each side of the longitudinal fissures carrying the coronary blood vessels. The side that feels thicker and more solid is the left ventricle. The right ventricle feels much thinner and somewhat flabby when compressed. This difference reflects the greater demand placed on the left ventricle, which must pump blood through the much longer systemic circuit, a pathway with much higher resistance than the pulmonary circuit served by the right ventricle. Hold the heart in its anatomical position (Figure 30.7a), with the anterior surface uppermost. In this position the left ventricle composes the entire apex and the left side of the heart.

6. Identify the pulmonary trunk and the aorta extending from the superior aspect of the heart. The pulmonary trunk is more anterior, and you may see its division into the right and left

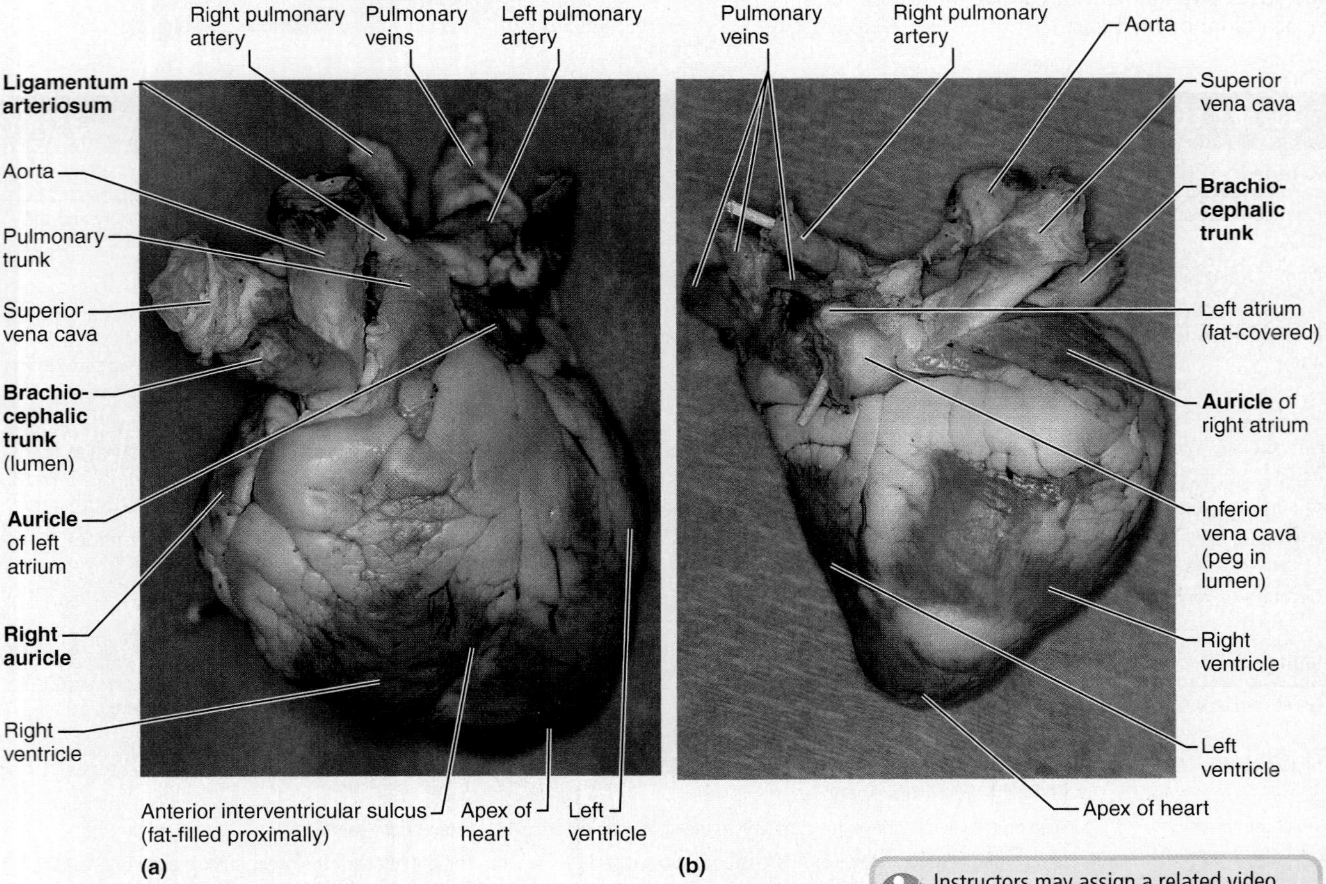

Figure 30.7 Anatomy of the sheep heart. (a) Anterior view. **(b)** Posterior view.

Instructors may assign a related video and coaching activity for the Sheep Heart dissection using Mastering A&P™

pulmonary arteries if it has not been cut too closely to the heart. The thicker-walled aorta, which branches almost immediately, is located just beneath the pulmonary trunk. The first observable branch of the sheep aorta, the **brachiocephalic trunk**, is identifiable unless the aorta has been cut immediately as it leaves the heart.

Gently pull on the aorta with your gloved fingers or forceps to stretch it. Repeat with the venae cavae.

Which vessel is easier to stretch? ____________________

How does the elasticity of each vessel relate to its ability to withstand pressure?

____

Carefully clear away some of the fat between the pulmonary trunk and the aorta to expose the **ligamentum arteriosum**, a cordlike remnant of the **ductus arteriosus**. (In the fetus, the ductus arteriosus allows blood to pass directly from the pulmonary trunk to the aorta, thus bypassing the nonfunctional fetal lungs.)

7. Cut through the wall of the aorta until you see the aortic valve. Identify the two openings to the coronary arteries just above the valve. Insert a probe into one of these holes to see if you can follow the course of a coronary artery across the heart.

8. Turn the heart to view its posterior surface (compare it to the view in Figure 30.7b). Notice that the right and left ventricles appear equal-sized in this view. Try to identify the four thin-walled pulmonary veins entering the left atrium. Identify the superior and inferior venae cavae entering the right atrium. Because of the way the heart is trimmed, the pulmonary veins and superior vena cava may be very short or missing. If possible, compare the approximate diameter of the superior vena cava with the diameter of the aorta.

Which is larger? ____________________

Which has thicker walls? ____________________

Why do you suppose these differences exist?

____

____

____

9. Insert a probe into the superior vena cava, through the right atrium, and out the inferior vena cava. Use scissors to cut along the probe so that you can view the interior of the right atrium. Observe the tricuspid valve.

How many cusps does it have? ____________________

10. Return to the pulmonary trunk and cut through its anterior wall until you can see the pulmonary valve (**Figure 30.8**, p. 444). How does its action differ from that of the tricuspid valve?

____

____

____

Extend the cut through the pulmonary trunk into the right ventricle. Cut down, around, and up through the tricuspid valve to make the cut continuous with the cut across the right atrium (see Figure 30.8).

11. Reflect the cut edges of the superior vena cava, right atrium, and right ventricle to obtain the view seen in the dissection photo (Figure 30.8). Observe the comblike ridges of muscle throughout most of the right atrium. This is called **pectinate muscle** (*pectin* = comb). Identify, on the ventral atrial wall, the large opening of the inferior vena cava, and follow it to its external opening with a probe. Notice that the atrial walls in the vicinity of the venae cavae are smooth and lack the roughened appearance (pectinate musculature) of the other regions of the atrial walls. Just below the inferior vena caval opening, identify the opening of the **coronary sinus**, which returns venous blood of the coronary circulation to the right atrium. Nearby, locate an oval depression, the **fossa ovalis**, in the interatrial septum. This depression marks the site of an opening in the fetal heart, the **foramen ovale**, which allows blood to pass from the right to the left atrium, thus bypassing the fetal lungs.

12. Identify the papillary muscles in the right ventricle, and follow their attached chordae tendineae to the cusps of the tricuspid valve. Notice the pitted and ridged appearance **(trabeculae carneae)** of the inner ventricular muscle.

13. Identify the **moderator band** (septomarginal band), a bundle of cardiac muscle fibers connecting the interventricular septum to anterior papillary muscles. It contains a branch of the atrioventricular bundle and helps coordinate contraction of the ventricle.

14. Make a longitudinal incision through the left atrium and continue it into the left ventricle. Notice how much thicker the myocardium of the left ventricle is than that of the right ventricle. Measure the thickness of right and left ventricular walls in millimeters and record the numbers.

____

How do your numbers compare with those of your classmates?

____

Text continues on next page. →

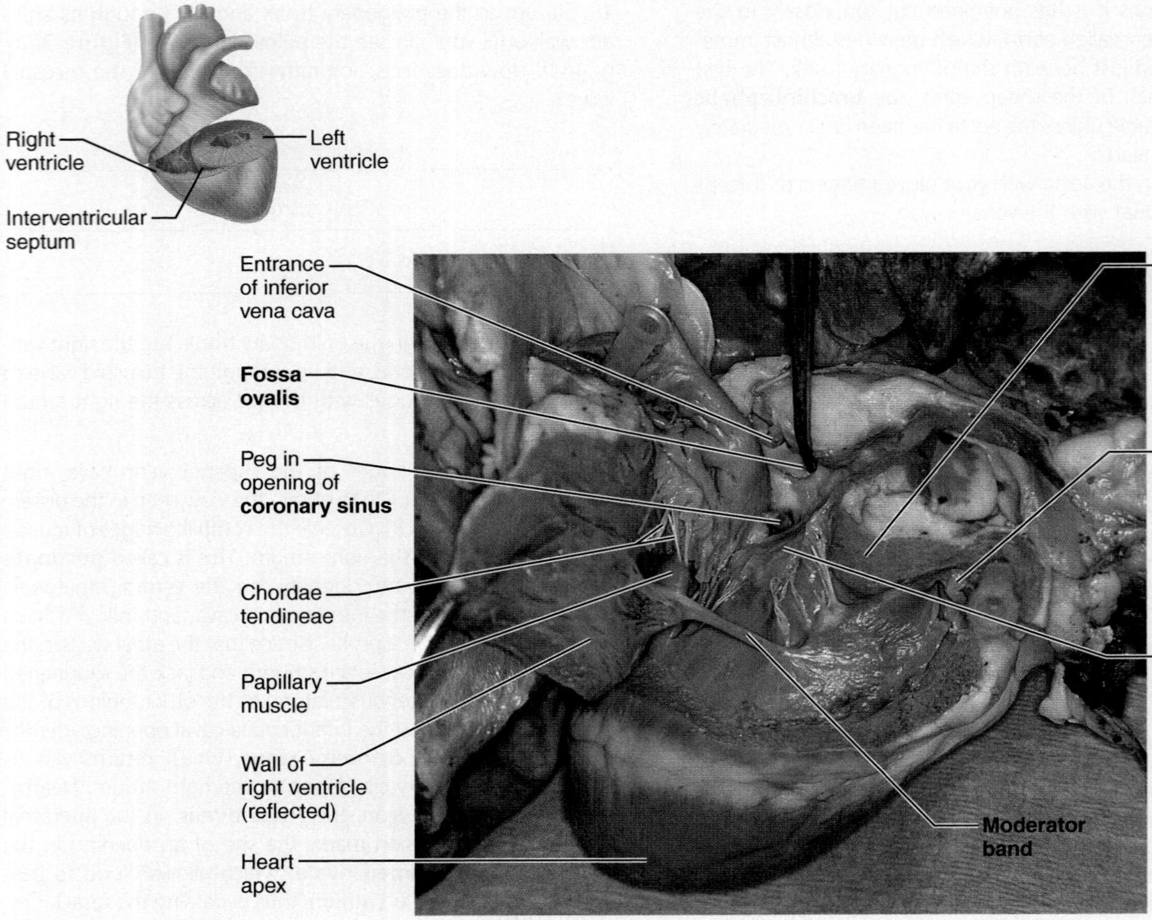

Figure 30.8 Right side of the sheep heart opened and reflected to reveal internal structures. Overview diagram illustrates the anatomical differences between the right and left ventricles. The left ventricle has thicker walls, and its cavity is basically circular. By contrast, the right ventricle cavity is crescent-shaped and wraps around the left ventricle.

Instructors may assign a related video and coaching activity for the Sheep Heart dissection using Mastering A&P™

30

Compare the *shape* of the left ventricular cavity to the shape of the right ventricular cavity. (See overview diagram in Figure 30.8.)

____

____

Are the papillary muscles and chordae tendineae observed in the right ventricle also present in the left ventricle? ________

Count the number of cusps in the mitral valve. How does this compare with the number seen in the tricuspid valve?

____

How do the sheep valves compare with their human counterparts?

____

15. Reflect the cut edges of the atrial wall, and attempt to locate the entry points of the pulmonary veins into the left atrium. Follow the pulmonary veins, if present, to the heart exterior with a probe. Notice how thin walled these vessels are.

16. Dispose of the organic debris in the designated container, clean the dissecting tray and instruments with detergent and water, and wash the lab bench with bleach solution before leaving the laboratory.

Instructors may assign a portion of the Review Sheet questions using Mastering A&P™

30 EXERCISE REVIEW SHEET
Anatomy of the Heart

Name ______________________ Lab Time/Date ______________________

Gross Anatomy of the Human Heart

1. An anterior view of the heart is shown here. Match each structure listed with the correct letter in the figure.

______ 1. right atrium

______ 2. right ventricle

______ 3. left atrium

______ 4. left ventricle

______ 5. superior vena cava

______ 6. ascending aorta

______ 7. aortic arch

______ 8. brachiocephalic trunk

______ 9. left common carotid artery

______ 10. left subclavian artery

______ 11. pulmonary trunk

______ 12. left pulmonary arteries

______ 13. ligamentum arteriosum

______ 14. left pulmonary veins

______ 15. right coronary artery

______ 16. circumflex artery

______ 17. anterior interventricular artery

______ 18. apex of heart

______ 19. great cardiac vein

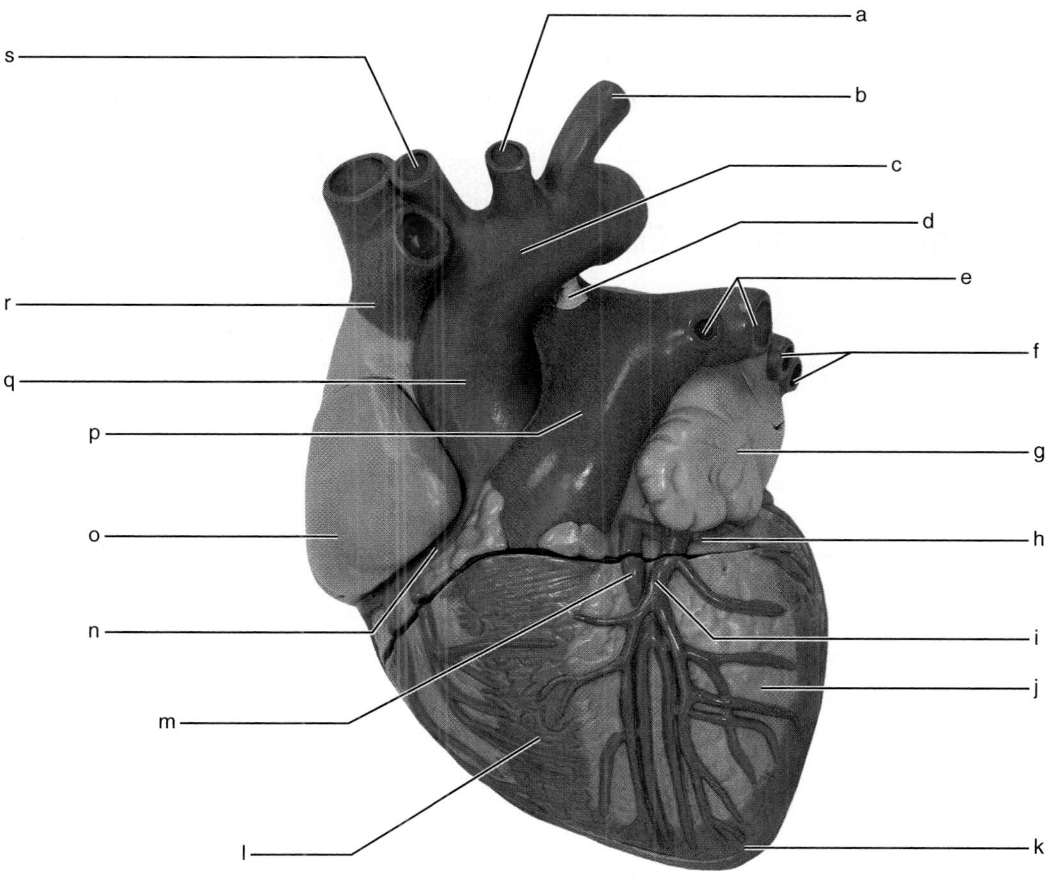

2. What is the function of the fluid that fills the pericardial sac? ______________________

____

3. Match the terms in the key to the descriptions provided below.

		Key:
______	1. location of the heart in the thorax	a. atria
______	2. tricuspid and mitral valves	b. atrioventricular valves
______	3. discharging chambers of the heart	c. coronary arteries
______	4. visceral pericardium	d. coronary sinus
______	5. receiving chambers of the heart	e. endocardium
______	6. layer composed of cardiac muscle	f. epicardium
______	7. provide nutrient blood to the heart muscle	g. mediastinum
______	8. lining of the heart chambers	h. myocardium
______	9. pulmonary and aortic valves	i. semilunar valves
______	10. drains blood into the right atrium	j. ventricles

4. Which valves are anchored by chordae tendineae? ______________________

____

5. Which valves close when the cusps fill with blood? ______________________

____

Pulmonary, Systemic, and Coronary Circulations

6. Describe the role of the pulmonary circuit. ______________________

____

____

7. Describe the role of the systemic circuit. ______________________

____

____

8. Name the three vessels that deliver oxygen-poor blood to the right atrium.

____

9. Starting with the right atrium, trace a drop of blood through the heart and lungs, naming the following structures: aorta, aortic valve, left atrium, left ventricle, mitral valve, pulmonary arteries, pulmonary capillaries, pulmonary valve, pulmonary trunk, pulmonary veins, right atrium, right ventricle, and tricuspid valve.

1. ______________________
2. ______________________
3. ______________________
4. ______________________
5. ______________________
6. ______________________
7. ______________________
8. ______________________
9. ______________________
10. ______________________
11. ______________________
12. ______________________
13. ______________________

Microscopic Anatomy of Cardiac Muscle

10. How would you distinguish the structure of cardiac muscle from that of skeletal muscle? ______________________

____

____

11. Add the following terms to the photograph of cardiac muscle below.

a. intercalated discs b. nucleus c. cardiac muscle cell

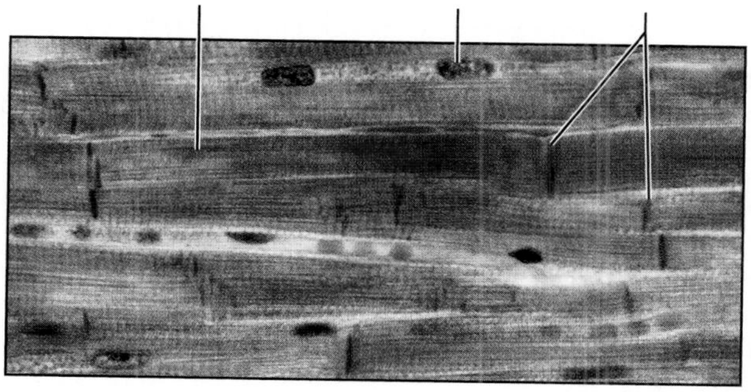

Describe the unique anatomical features of cardiac muscle. What role does the unique structure of cardiac muscle play in its function?

____

____

____

____

____

____

Dissection of the Sheep Heart

12. During the sheep heart dissection, you were asked initially to identify the right and left ventricles without cutting into the heart. During this procedure, what differences did you observe between the two chambers?

____

____

When you measured thickness of ventricular walls, was the right or left ventricle thicker? ______________________

Knowing that structure and function are related, how would you say this structural difference reflects the relative functions of these two heart chambers? ______________________

____

13. Semilunar valves prevent backflow into the ______________________; mitral and tricuspid valves prevent backflow into the ______________________. Using your own observations, explain how the operation of the semilunar valves differs from that of the atrioventricular valves. ______________________

____

____

____

14. Compare and contrast the structure of the mitral and tricuspid valves. ______________________

____

15. ✚ A proximal LAD lesion is a blockage in the left anterior descending artery, also known as the anterior interventricular artery. Explain why a heart attack caused by an obstruction of this artery is sometimes referred to as the "widow maker" heart attack. ______________________

____

____

16. ✚ Congestive heart failure (CHF) is an inability of the heart to pump sufficient blood to the body. One sign of CHF is excess fluid in the tissue spaces, known as edema. Describe the location of the edema if the left side of the heart fails, compared to the location of edema if the right side of the heart fails. ______________________

____

____

EXERCISE 31

Conduction System of the Heart and Electrocardiography

Learning Outcomes

- State the function of the intrinsic conduction system of the heart.
- List and identify the elements of the intrinsic conduction system, and describe how impulses are initiated and conducted through this system and the myocardium.
- Interpret the ECG in terms of depolarization and repolarization events occurring in the myocardium. Identify the P, QRS, and T waves on an ECG recording using an ECG recorder or BIOPAC®.
- Define *tachycardia*, *bradycardia*, and *fibrillation*.
- Calculate the heart rate, durations of the QRS complex, P-R interval, and Q-T interval from an ECG obtained during the laboratory period, and recognize normal values for the durations of these events.
- Describe and explain the changes in the ECG observed during experimental conditions such as exercise or breath holding.

Go to Mastering A&P™ > Study Area to improve your performance in A&P Lab.

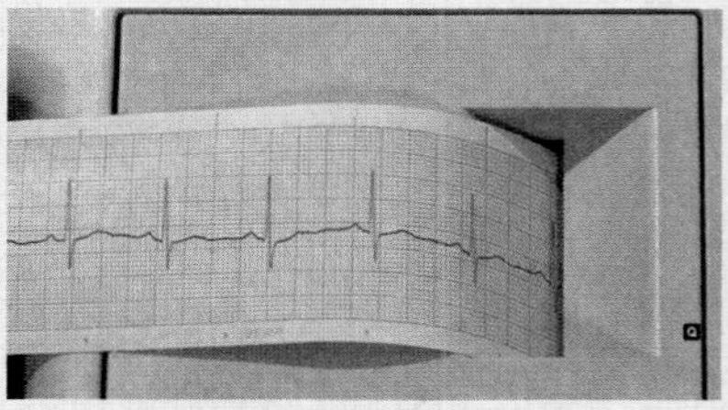

> Lab Tools > Pre-Lab Videos > ECG

Instructors may assign new Building Vocabulary coaching activities, Pre-Lab Quiz questions, Art Labeling activities, Pre-Lab Video Coaching Activities for ECG, Practice Anatomy Lab Practical questions (PAL), and more using the Mastering **A&P™** Item Library.

Pre-Lab Quiz

Instructors may assign these and other Pre-Lab Quiz questions using Mastering **A&P™**

1. Circle True or False. Cardiac muscle cells are electrically connected by gap junctions and behave as a single unit.
2. Because it sets the rate of depolarization for the normal heart, the ________ node is known as the pacemaker of the heart.
   a. atrioventricular **b.** Purkinje **c.** sinoatrial
3. In a typical ECG, the __________ wave signals the depolarization of the atria immediately before they contract.
   a. P **c.** R
   b. Q **d.** T
4. Circle the correct underlined term. A heart rate over 100 beats/minute is known as <u>tachycardia</u> / <u>bradycardia</u>.
5. Circle True or False. ECG can be used to calculate heart rate.

Materials

- ECG or BIOPAC® equipment:*

ECG recording apparatus, electrode paste, alcohol swabs, rubber straps or disposable electrodes

BIOPAC® BIOPAC® BSL System with BSL software version 4.0.1 and above (for Windows 10/8.x/7 or Mac OS X 10.9–10.12), data acquisition unit MP36/35 or MP45, PC or Mac computer, Biopac Student Lab electrode lead set, disposable vinyl electrodes.

Instructors using the MP36/35/30 data acquisition unit with BSL software versions earlier than 4.0.1 (for Windows or Mac) will need slightly different channel settings and collection strategies. Instructions for using the older data acquisition unit can be found on MasteringA&P.

- Cot or lab table; pillow (optional)
- Millimeter ruler

**Note: Instructions for using PowerLab® equipment can be found on MasteringA&P.*

Heart contraction results from a series of depolarization waves that travel through the heart preliminary to each beat. Because cardiac muscle cells are electrically connected by gap junctions, the entire myocardium behaves like a single unit, a **functional syncytium**.

The Intrinsic Conduction System

The ability of cardiac muscle to beat is intrinsic—it does not depend on impulses from the nervous system to initiate its contraction and will continue to contract rhythmically even if all nerve connections are severed. The **intrinsic conduction system** of the heart consists of **cardiac pacemaker cells**. The intrinsic conduction system ensures that heart muscle depolarizes in an orderly and sequential manner, from atria to ventricles, and that the heart beats as a coordinated unit.

The components of the intrinsic conduction system include the following (**Figure 31.1**):

- The **sinoatrial (SA) node**, located in the right atrium just inferior to the entrance to the superior vena cava
- The **atrioventricular (AV) node** in the lower atrial septum at the junction of the atria and ventricles
- The **AV bundle (bundle of His)** and right and left **bundle branches**, located in the interventricular septum
- The **subendocardial conducting network**, also called **Purkinje fibers**

Note that the atria and ventricles are separated from one another by a region of electrically inert connective tissue, so the depolarization wave can be transmitted to the ventricles only via the tract between the AV node and AV bundle. Thus, any damage to the AV node-bundle pathway partially or totally insulates the ventricles from the influence of the SA node.

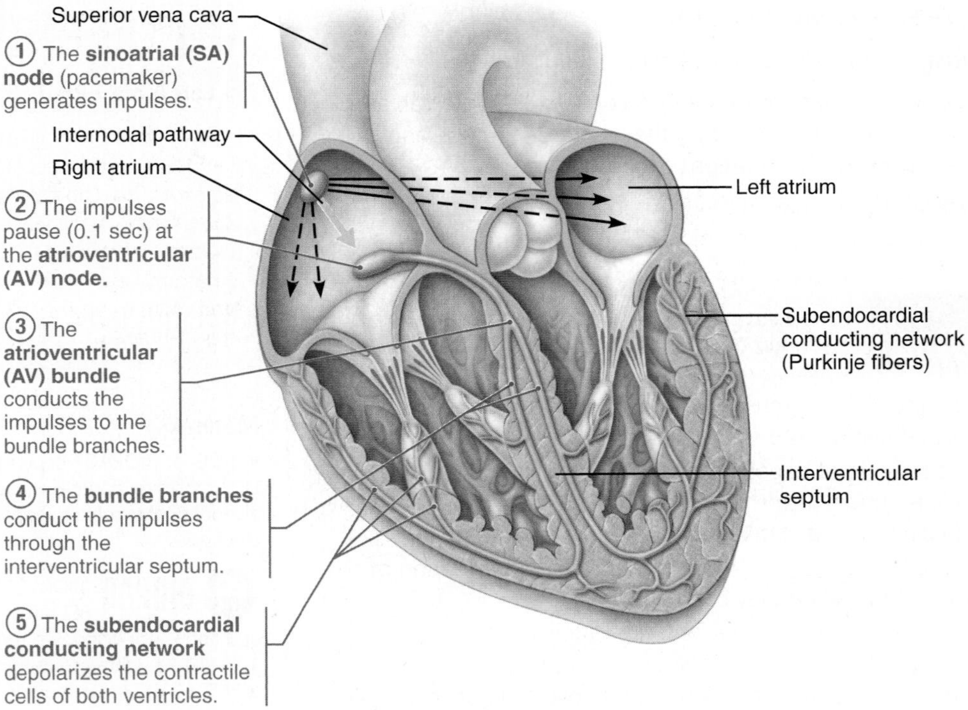

Figure 31.1 The intrinsic conduction system of the heart. Impulses travel through the heart in order (1) to (5) following the yellow pathway. Dashed-line arrows indicate transmission of the impulse from the SA node through the atria. Solid yellow arrow indicates transmission of the impulse from the SA node to the AV node via the internodal pathway.

Instructors may assign this figure as an Art Labeling Activity using Mastering A&P™

31

Electrocardiography

The conduction of impulses through the heart generates electrical currents that eventually spread throughout the body. These impulses can be detected on the body's surface and recorded with an instrument called an *electrocardiograph*. The graphic recording of the electrical changes occurring during the cardiac cycle is called an **electrocardiogram (ECG** or **EKG)** (**Figure 31.2**). A typical ECG has three recognizable waves, or *deflections*: the P wave, the QRS complex, and the T wave. For analysis, the ECG is divided into segments and intervals. A **segment** is a region between two waves. For example, the S-T segment is the region between the end of the S deflection and the start of the T wave. An **interval** is a region that contains a segment and one or more waves. For example, the Q-T interval includes the S-T segment as well as the QRS complex and the T wave. Boundaries for waves as well as some commonly measured segments and intervals are described in **Table 31.1**. The deflection waves of an ECG correlate to the depolarization and repolarization of the heart's chambers (**Figure 31.3**, p. 452).

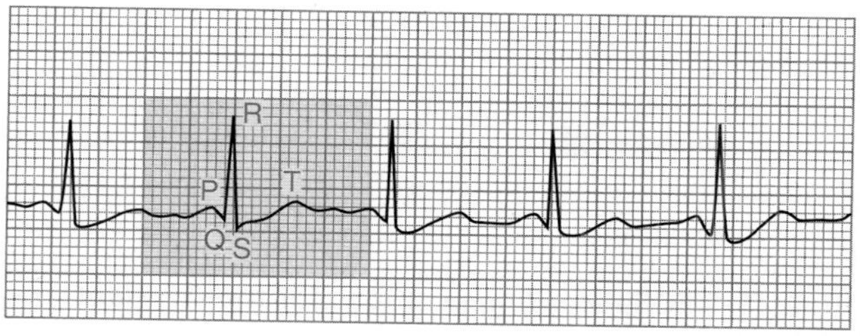

(a)

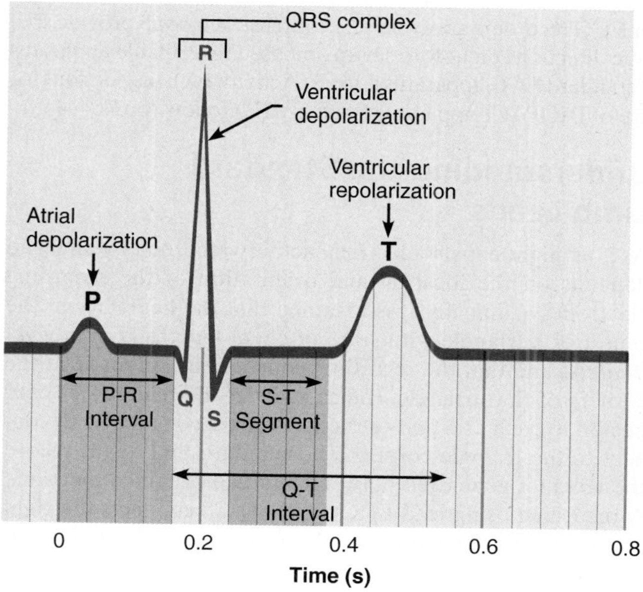

(b)

Figure 31.2 The normal electrocardiogram. (a) Regular sinus rhythm. **(b)** Waves, segments, and intervals of a normal ECG.

Instructors may assign this figure as an Art Labeling Activity using Mastering A&P™

Table 31.1 Boundaries of Each ECG Component

Feature	Boundaries
P wave	Start of P deflection to return to baseline
P-R interval	Start of P deflection to start of Q deflection
QRS complex	Start of Q deflection to S return to baseline
S-T segment	End of S deflection to start of T wave
Q-T interval	Start of Q deflection to end of T wave
T wave	Start of T deflection to return to baseline
T-P segment	End of T wave to start of next P wave
R-R interval	Peak of R wave to peak of next R wave

Abnormalities of the deflection waves and changes in the time intervals of the ECG are useful in detecting myocardial infarcts (heart attacks) or problems with the conduction system of the heart.

Table 31.2 summarizes some examples of abnormal electrocardiogram tracings and their possible clinical significance.

A heart rate over 100 beats/min is referred to as **tachycardia**; a rate below 60 beats/min is **bradycardia**. Although neither condition is pathological, prolonged tachycardia may progress to **fibrillation**, a condition of rapid uncoordinated heart contractions. Bradycardia in athletes is a positive finding; that is, it indicates an increased efficiency of cardiac functioning. Because *stroke volume* (the amount of blood ejected by a ventricle with each contraction) increases with physical conditioning, the heart can contract less often per minute and still meet circulatory demands.

Twelve standard leads are used to record an ECG for diagnostic purposes. Three of these are bipolar leads that measure the voltage difference between the arms, or an arm and a leg, and nine are unipolar leads. Together the 12 leads provide a fairly comprehensive picture of the electrical activity of the heart.

For this investigation, four electrodes are used (**Figure 31.4**, p. 452), and results are obtained from the three *standard limb leads* (also shown in Figure 31.4). Several types of physiographs

31

Table 31.2 Examples of Abnormal ECGs and Possible Clinical Significance

Finding	Possible clinical significance
Enlarged R wave	Enlarged ventricles.
Prolonged P-R interval	First-degree heart block. The signal from the SA node to the AV node is delayed longer than normal.
Prolonged Q-T interval (when compared to the R-R interval)	Increased risk of ventricular arrhythmias. This interval corresponds to the beginning of ventricular depolarization through ventricular repolarization.
S-T segment elevated from baseline	Myocardial infarction (heart attack).

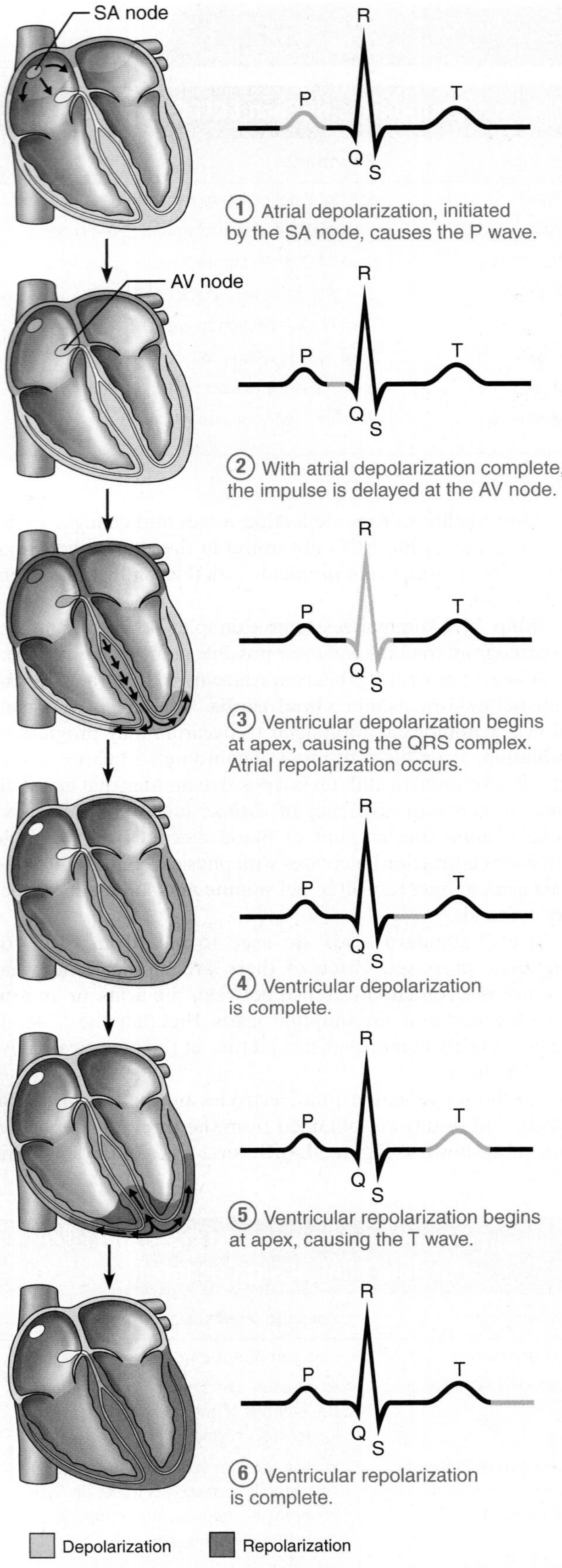

Figure 31.3 The sequence of depolarization and repolarization of the heart related to the deflection waves of an ECG tracing.

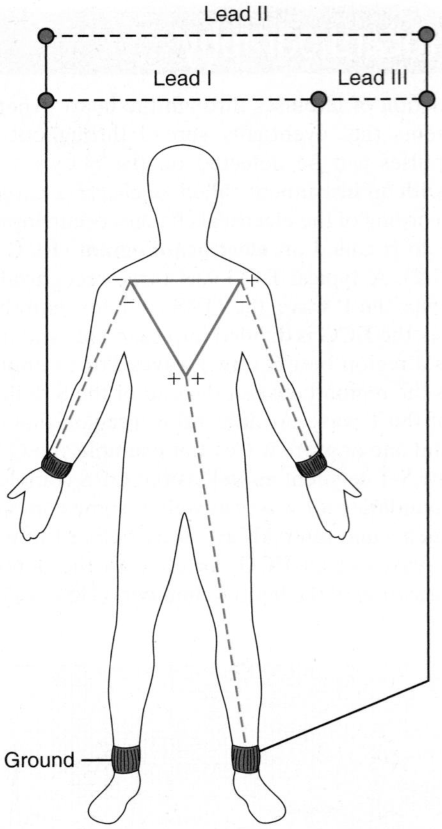

Figure 31.4 ECG recording positions for the standard limb leads.

or ECG recorders are available. Your instructor will provide specific directions on how to set up and use the available apparatus if standard ECG apparatus is used (Activity 1A). Instructions for use of BIOPAC® apparatus (Activity 1B) follow (pp. 454–458).

Understanding the Standard Limb Leads

As you might expect, electrical activity recorded by any lead depends on the location and orientation of the recording electrodes. Clinically, it is assumed that the heart lies in the center of a triangle with sides of equal lengths (*Einthoven's triangle*) and that the recording connections are made at the corners of that triangle. But in practice, the electrodes connected to each arm and to the left leg are considered to connect to the triangle corners. The standard limb leads record the voltages generated between any two of the connections. A recording using lead I (RA-LA), which connects the right arm (RA) and the left arm (LA), is most sensitive to electrical activity spreading horizontally across the heart. Lead II (RA-LL) and lead III (LA-LL) record activity along the vertical axis (from the base of the heart to its apex) but from different orientations. The significance of Einthoven's triangle is that the sum of the voltages of leads I and III equals that in lead II (Einthoven's law). Hence, if the voltages of two of the standard leads are recorded, that of the third lead can be determined mathematically.

Activity 1A

Instructors may assign a related Pre-Lab Video Coaching Activity using Mastering A&P™

Recording ECGs Using a Standard ECG Apparatus

Preparing the Subject

1. If using electrodes that require gel, first scrub the skin at the attachment site with an alcohol swab. Place the gel on four electrode plates, and position each electrode as follows. Attach an electrode to the anterior surface of each forearm, about 5 to 8 cm (2 to 3 in.) above the wrist, and secure them with rubber straps. In the same manner, attach an electrode to each leg, approximately 5 to 8 cm above the ankle. Disposable electrodes may be placed directly on the subject in the same areas.

2. Attach the appropriate tips of the patient cable to the electrodes. The cable leads are marked RA (right arm), LA (left arm), LL (left leg), and RL (right leg, the ground).

Making a Baseline Recording

1. Position the subject comfortably in a supine position on a cot or sitting relaxed on a laboratory chair.

2. Turn on the power switch, and adjust the sensitivity knob to 1. Set the paper speed to 25 mm/sec and the lead selector to the position corresponding to recording from lead I (RA-LA).

3. Set the control knob at the **RUN** position, and record the subject's at-rest ECG from lead I for 2 to 3 minutes or until the recording stabilizes. The subject should try to relax and not move unnecessarily, because the skeletal muscle action potentials will also be recorded.

4. Stop the recording and mark it "lead I."

5. Repeat the recording procedure for leads II (RA-LL) and III (LA-LL).

6. Each student should take a representative portion of one of the lead recordings and label the record with the name of the subject and the lead used. Identify and label the P, QRS, and T waves. The calculations you perform for your recording should be based on the following information: Because the paper speed was 25 mm/sec, each millimeter of paper corresponds to a time interval of 0.04 sec. Thus, if an interval requires 4 mm of paper, its duration is 4 mm $\times$ 0.04 sec/mm = 0.16 sec.

7. Calculate the heart rate. Obtain a millimeter ruler and measure the R to R interval. Enter this value into the following equation to find the time for one heartbeat.

_________ mm/beat $\times$ 0.04 sec/mm = _________ sec/beat

Now find the beats per minute, or heart rate, by using the figure just calculated for seconds per beat in the following equation:

60 sec/min ÷ (sec/beat) = beats/min

Measure the QRS complex, and calculate its duration.

____

Measure the Q-T interval, and calculate its duration.

____

Measure the P-R interval, and calculate its duration.

____

Are the calculated values within normal limits?

____

8. At the bottom of this page, attach sections of the ECG recordings from leads I through III. Make sure you indicate the paper speed, lead, and subject's name on each tracing. Also record the heart rate on the tracing.

Recording the ECG for Running in Place

1. Make sure the electrodes are securely attached to prevent electrode movement while recording the ECG.

2. Set the paper speed to 25 mm/sec, and prepare to make the recording using lead I.

3. Record the ECG while the subject is running in place for 3 minutes. Then have the subject sit down, but continue to record the ECG for an additional 4 minutes. *Mark the recording* at the end of the 3 minutes of running and at 1 minute after cessation of activity.

4. Stop the recording. Calculate the beats/min during the third minute of running, at 1 minute after exercise, and at 4 minutes after exercise. Record below:

__________ beats/min while running in place

__________ beats/min at 1 minute after exercise

__________ beats/min at 4 minutes after exercise

5. Compare this recording with the previous recording from lead I. Which intervals are shorter in the "running" recording?

____

6. Does the subject's heart rate return to resting level by 4 minutes after exercise?

____

Recording the ECG During Breath Holding

1. Position the subject comfortably in the sitting position.

2. Using lead I and a paper speed of 25 mm/sec, begin the recording. After approximately 10 seconds, instruct the subject to begin breath holding, and mark the record to indicate the onset of the 1-minute breath-holding interval.

Text continues on next page. →

3. Stop the recording after 1 minute, and remind the subject to breathe. Calculate the beats/minute during the 1-minute experimental (breath-holding) period.

Beats/min during breath holding: ______________________

4. Compare this recording with the lead I recording obtained under resting conditions.

What differences do you see? ______________________

____

Attempt to explain the physiological reason for the differences you have seen. (*Hint:* A good place to start might be to check hypoventilation or the role of the respiratory system in acid-base balance of the blood.)

____

____

____

____

Activity 1B

Electrocardiography Using BIOPAC®

In this activity, you will record the electrical activity of the heart under three different conditions: (1) while the subject is lying down, (2) after the subject sits up and breathes normally, and (3) after the subject has exercised and is breathing deeply.

In order to obtain a clear ECG, it is important that the subject:

- Remain still during the recording.
- Refrain from laughing or talking during the recording.
- When in the sitting position, keep arms and legs steady and relaxed.
- Remove metal watches and bracelets.

Setting Up the Equipment

1. Connect the BIOPAC® unit to the computer and turn the computer **ON**.

2. Make sure the BIOPAC® unit is **OFF**.

3. Plug in the equipment (as shown in **Figure 31.5**):
   - Electrode lead set—CH 1

4. Turn the BIOPAC® unit **ON**.

5. Place the three electrodes on the subject (as shown in **Figure 31.6**), and attach the electrode leads according to the colors indicated. Place two electrodes on the medial surface of each leg, 5 to 8 cm (2 to 3 in.) superior to the ankle. Place the other electrode on the right anterior forearm 5 to 8 cm above the wrist.

6. The subject should lie down and relax in a comfortable position with eyes closed. A chair or place to sit up should be available nearby.

7. Start the Biopac Student Lab program on the computer by double-clicking the icon on the desktop or by following your instructor's guidance.

8. Select lesson **L05-Electrocardiography (ECG) 1** from the menu, and click **OK**.

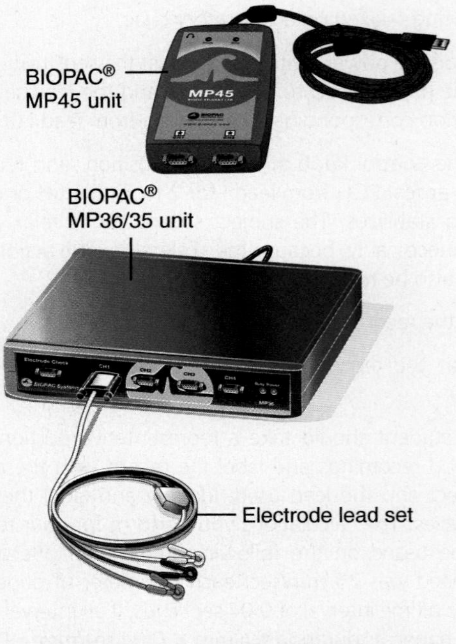

Figure 31.5 Setting up the BIOPAC® unit. Plug the electrode lead set into Channel 1. Leads are shown plugged into the MP36/35 unit.

9. Type in a filename that will save this subject's data on the computer hard drive. You may want to use the subject's last name followed by ECG-1 (for example, SmithECG-1), then click **OK**.

10. Because we are not recording all available lesson options, click the File Menu, choose **Lesson Preferences**. Choose **Do not calculate heart rate data** and click **OK**. Click the File Menu and choose **Lesson Preferences** again; select **Lesson Recordings** and click **OK**; click the box for **Deep breathing** to deselect it; and then click **OK**.

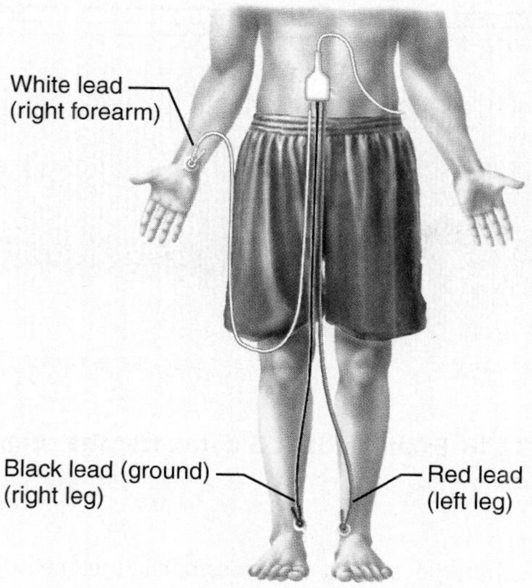

Figure 31.6 Placement of electrodes and the appropriate attachment of electrode leads by color.

Calibrating the Equipment

- Examine the electrodes and the electrode leads to be certain they are properly attached.

1. The subject must remain supine, still, and relaxed. With the subject in a still position, click **Calibrate**. This will initiate the process whereby the computer will automatically establish parameters to record the data.

2. The calibration procedure will stop automatically after 8 seconds.

3. Observe the recording of the calibration data, which should look similar to the data example (**Figure 31.7**).
   - If the data look very different, click **Redo Calibration** and repeat the steps above.
   - If the data look similar, proceed to the next section. **Don't** click **Done** until you have completed all 3 recordings.

Recording 1: Subject Supine

1. To prepare for the recording, remind the subject to remain still and relaxed while lying down.

2. Click **Continue** and when prepared, click **Record** and gather data for 20 seconds. At the end of 20 seconds, click **Suspend**.

3. Observe the data, which should look similar to the data example (**Figure 31.8**).
   - If the data look very different, click **Redo** and repeat the steps above. Be certain to check attachment of the electrodes and leads, and remind the subject not to move, talk, or laugh.
   - If the data look similar, move on to the next recording.

Recording 2: After Subject Sits Up, with Normal Breathing

1. Tell the subject to be ready to sit up in the designated location. With the exception of breathing, the subject should try

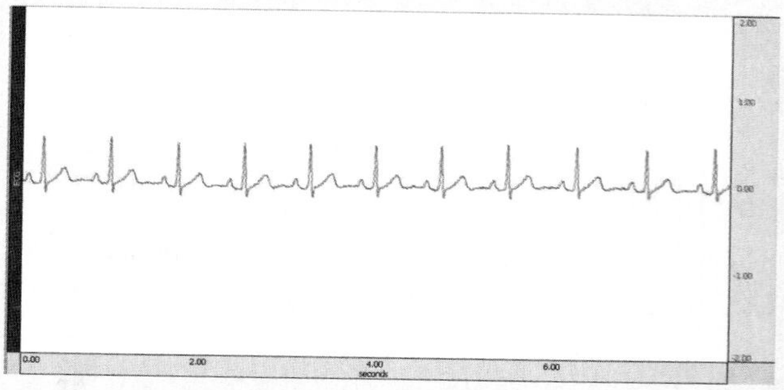

Figure 31.7 Example of calibration data.

to remain motionless after assuming the seated position. *If the subject moves too much during recording after sitting up, unwanted skeletal muscle artifacts will affect the recording.*

2. Click **Continue** and when prepared, instruct the subject to sit up. Immediately after the subject assumes a motionless state, click **Record**, and the data will begin recording.

3. At the end of 20 seconds, click **Suspend** to stop recording.

4. Observe the data, which should look similar to the data example (**Figure 31.9**, p. 456).
   - If the data look very different, have the subject lie down, then click **Redo**. Be certain to check attachment of the electrodes, then repeat steps 1–4 above. Do not click **Record** until the subject is motionless.
   - If the data look similar, move on to the next recording.

Recording 3: After Subject Exercises, with Deep Breathing

1. Click **Continue**, but **don't** click **Record** until after the subject has exercised. Remove the electrode pinch connectors from the electrodes on the subject.

2. Have the subject do a brief round of exercise, such as jumping jacks or running in place for 1 minute, in order to elevate the heart rate.

3. As quickly as possible after the exercise, have the subject resume a motionless, seated position. Reattach the pinch

Text continues on next page. →

31

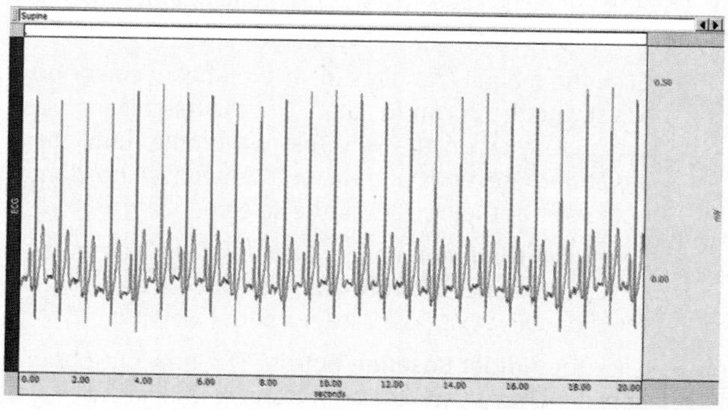

Figure 31.8 Example of ECG data while the subject is lying down.

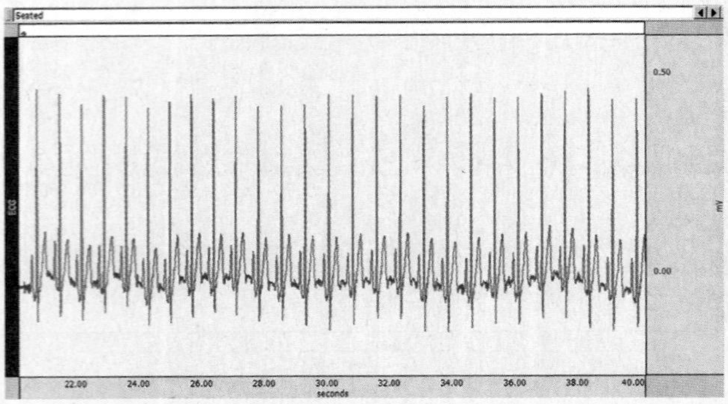

Figure 31.9 Example of ECG data after the subject sits up and breathes normally.

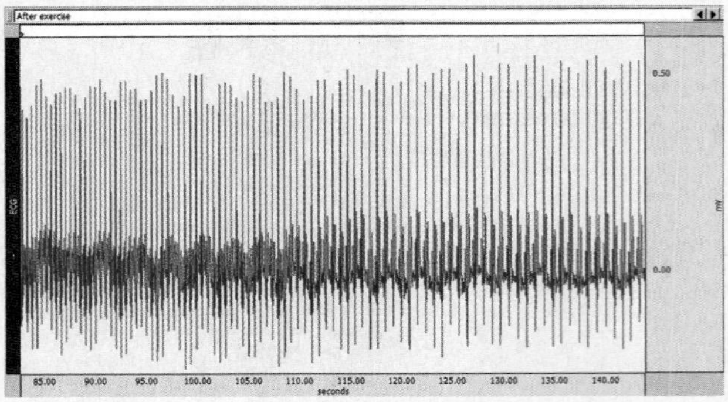

Figure 31.10 Example of ECG data after the subject exercises.

connectors. Once again, if the subject moves too much during recording, unwanted skeletal muscle artifacts will affect the data. After exercise, the subject is likely to be breathing deeply but otherwise should remain as still as possible.

4. Immediately after the subject assumes a motionless, seated state, click **Record**, and the data will begin recording. Record the ECG for 60 seconds in order to observe post-exercise recovery.

5. After 60 seconds, click **Suspend** to stop recording.

6. Observe the data, which should look similar to the data example (**Figure 31.10**).

- If the data look very different, click **Redo** and repeat the steps above. Be certain to check attachment of the electrodes and leads, and remember not to click **Record** until the subject is motionless.

7. When finished, click **Done** and then **Yes**. Remove the electrodes from the subject.

8. A pop-up window will appear. To record from another subject, select **Record from another subject** and return to step 5 under Setting Up the Equipment. If continuing to the Data Analysis section, select **Analyze current data file** and proceed to step 2 of the Data Analysis section.

Data Analysis

1. If just starting the BIOPAC® program to perform data analysis, enter **Review Saved Data** mode and choose the file with the subject's ECG data (for example, SmithECG-1).

2. Use the following tools to adjust the data in order to clearly view and analyze four consecutive cardiac cycles:

- Click the magnifying glass (near the I-beam cursor box) to activate the **Zoom** function. Use the magnifying glass cursor to click on the very first waveforms until there are about 4 seconds of data represented (see horizontal time scale at the bottom of the screen).
- Select the **Display** menu at the top of the screen, and click **Autoscale Waveforms** (or click Ctrl + Y). This function will adjust the data for better viewing.
- Click the **Adjust Baseline** button. Two new buttons will appear; simply click these buttons to move the waveforms **Up** or **Down** so they appear clearly in the center of the display window. Once they are centered, click **Exit**.

3. Note that the second and fourth channel/measurement boxes at the top of the screen are set to Delta T and bpm.

Channel	Measurement	Data
CH 1	Delta T	ECG
CH 1	BPM	ECG

Analysis of Recording 1: Subject Lying Down

1. Use the arrow cursor and click the I-beam cursor box for the "area selection" function.

2. First measure **Delta T** and **BPM** in Recording 1 (approximately seconds 0–20). Using the I-beam cursor, highlight from the peak of one R wave to the peak of the next R wave (as shown in **Figure 31.11**).

3. Observe that the computer automatically calculates the **Delta T** and **BPM** for the selected area. These measurements represent the following:

Delta T (difference in time): Computes the elapsed time between the beginning and end of the highlighted area

BPM (beats per minute): Computes the beats per minute when the area from the R wave of one cycle to the R wave of another cycle is highlighted

4. Record these data in the **Activity 1B: Recording 1 Samples chart** under R to R Sample 1 (round to the nearest 0.01 second and 0.1 beat per minute).

5. Using the I-beam cursor, highlight two other pairs of R to R areas in this recording. Record the data in the same chart under Samples 2 and 3.

6. Calculate the means of the data in this chart.

7. Next, use the **Zoom**, **Autoscale Waveforms**, and **Adjust Baseline** tools described in step 2 to focus in on one ECG waveform within Recording 1. (See the example in **Figure 31.12**).

8. Once a single ECG waveform is centered for analysis, click the I-beam cursor box to activate the "area selection" function.

9. Using the highlighting function and **Delta T** computation, measure the duration of every component of the ECG

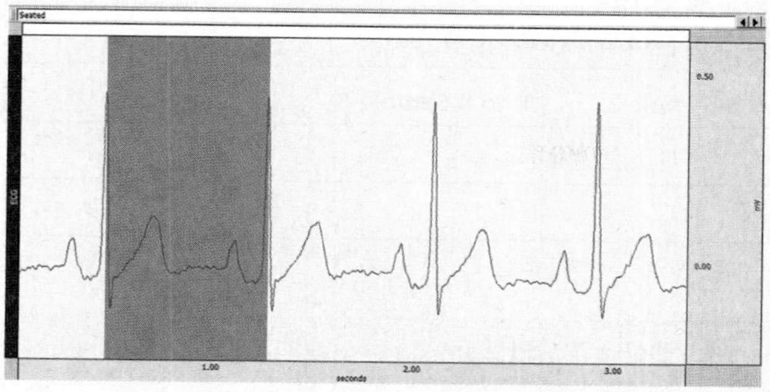

Figure 31.11 Example of highlighting from R wave to R wave.

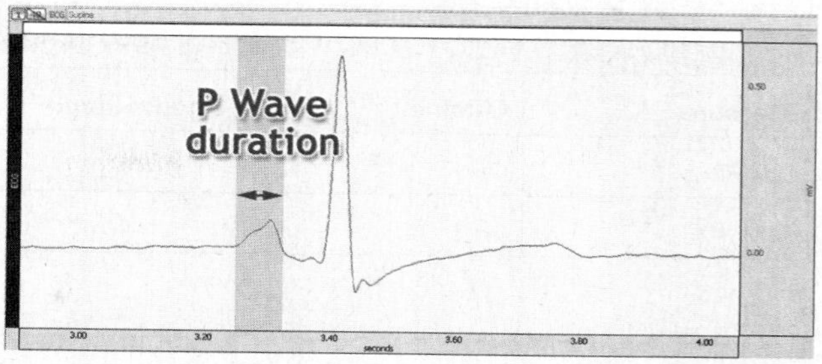

Figure 31.12 Example of a single ECG waveform with the first part of the P wave highlighted.

Activity 1B: Recording 1 Samples for Delta T and bpm					
Measure	**Channel**	**R to R Sample 1**	**R to R Sample 2**	**R to R Sample 3**	**Mean**
Delta T	CH 1				
BPM	CH 1				

waveform. (Refer to Figure 31.2b and Table 31.1 for guidance in highlighting each component.)

10. Highlight each component of one cycle. Observe the elapsed time, and record this data under Cycle 1 in the **Activity 1B: Recording 1 Elapsed Time chart**.

11. Scroll along the horizontal axis at the bottom of the data to view and analyze two additional cycles in Recording 1. Record the elapsed time for every component of Cycle 2 and Cycle 3 in the Recording 1 Elapsed Time chart.

12. In the same chart, calculate the means for the three cycles of data and record.

Activity 1B: Recording 1 Elapsed Time for ECG Components (seconds)				
Component	**Cycle 1**	**Cycle 2**	**Cycle 3**	**Mean**
P wave				
P-R interval				
QRS complex				
S-T segment				
Q-T interval				
T wave				
T-P segment				
R-R interval				

Analysis of Recording 2: Subject Sitting Up and Breathing Normally

1. Scroll along the horizontal time bar until you reach the data for Recording 2 (approximately seconds 20–40). A marker with "Seated" should denote the beginning of this data.

2. As in the analysis of Recording 1, use the I-beam tool to highlight and measure the **Delta T** and **BPM** between three different pairs of R waves in this recording, and record the data in the **Activity 1B: Recording 2 Samples chart** on p. 458.

Analysis of Recording 3: After Exercise with Deep Breathing

1. Scroll along the horizontal time bar until you reach the data for Recording 3 (approximately seconds 40–100). A marker with "After exercise" should denote the beginning of this data.

2. As before, use the I-beam tool to highlight and measure the **Delta T** and **BPM** between three pairs of R waves in this recording, and record the data in the **Activity 1B: Recording 3 Samples chart** on p. 458.

3. Using the instructions for steps 7–9 in the section Analysis of Recording 1, highlight and observe the elapsed time for each component of one cycle, and record these data under Cycle 1 in the **Activity 1B: Recording 3 Elapsed Time chart** on p. 458.

4. When finished, **Exit** from the file menu to quit.

Compare the average **Delta T** times and average **BPM** between the data in Recording 1 (lying down) and the data in Recording 3 (after exercise). Which is greater in each case?

____

____

Text continues on next page. →

Activity 1B: Recording 2 Samples for Delta T and bpm					
Measure	**Channel**	**R to R Sample 1**	**R to R Sample 2**	**R to R Sample 3**	**Mean**
Delta T	CH 1				
BPM	CH 1				

Activity 1B: Recording 3 Samples for Delta T and bpm					
Measure	**Channel**	**R to R Sample 1**	**R to R Sample 2**	**R to R Sample 3**	**Mean**
Delta T	CH 1				
BPM	CH 1				

What is the relationship between the elapsed time **(Delta T)** between R waves and the heart rate?

___

What event does the period between R waves correspond to?

___

Is there a change in heart rate when the subject makes the transition from lying down (Recording 1) to a sitting position (Recording 2)?

___

Examine the average duration of each of the ECG components in Recording 1 and the data in Recording 3. In the **Activity 1B: Average Duration chart**, record the average values for Recording 1 and the data for Recording 3. Draw a circle around those measures that fit within the normal range.

Compare the Q-T intervals in the data while the subject is at rest versus after exercise; this interval corresponds closely to the duration of contraction of the ventricles. Describe and explain any difference.

___

___

Compare the duration in the period from the end of each T wave to the next P wave while the subject is at rest versus after exercise. Describe and explain any difference.

___

___

A patient presents with a P-R interval three times longer than the normal duration. What might be the cause of this abnormality?

___

___

Activity 1B: Recording 3 Elapsed Time for ECG Components (seconds)	
Component	**Cycle 1**
P wave	
P-R interval	
QRS complex	
S-T segment	
Q-T interval	
T wave	
T-P segment	
R-R interval	

Activity 1B: Average Duration for ECG Components			
ECG component	**Normal duration (seconds)**	**Recording 1 (lying down)**	**Recording 3 (post-exercise)**
P wave	0.07–0.18		
P-R interval	0.12–0.20		
QRS complex	0.06–0.12		
S-T segment	<0.20		
Q-T interval	0.32–0.38		
T wave	0.10–0.25		
T-P segment	0–0.40		
R-R interval	varies		

Instructors may assign a portion of the Review Sheet questions using Mastering A&P™

REVIEW SHEET

EXERCISE 31

Conduction System of the Heart and Electrocardiography

Name ______________________ Lab Time/Date ______________________

The Intrinsic Conduction System

1. List the elements of the intrinsic conduction system in order, starting from the SA node.

SA node → ______________________ → ______________________ →

______________________ → ______________________

At what structure in the transmission sequence is the impulse temporarily delayed? ______________________

Why? ______________________

____

2. Even though cardiac muscle has an inherent ability to beat, the intrinsic conduction system plays a critical role in heart physiology.

What is that role? ______________________

____

Electrocardiography

3. Define *ECG*. ______________________

____

4. Draw an ECG wave form representing one heartbeat. Label the P wave, QRS complex, and T wave; the P-R interval; the S-T segment, and the Q-T interval.

5. Why does heart rate increase during running? ______________________

____

____

6. Match the terms below with their descriptions.

a. AV node	______	1.	the intrinsic conduction system structure that initiates atrial depolarization
b. interval	______	2.	a region on an ECG tracing that is between two waves but doesn't include a wave
c. P wave	______	3.	the deflection on the ECG that is a result of atrial depolarization
d. QRS complex	______	4.	the deflection on the ECG that is a result of ventricular depolarization
e. SA node	______	5.	the deflection on the ECG that is a result of ventricular repolarization
f. segment	______	6.	a region on an ECG tracing that includes a segment and at least one wave
g. T wave	______	7.	the intrinsic conduction system structure where the conduction of the impulse is delayed

7. Define the following terms.

1. *tachycardia*: ____________________

2. *bradycardia*: ____________________

3. *fibrillation*: ____________________

8. Abnormalities of heart valves can be detected more accurately by auscultation than by electrocardiography. Why is this so?

____

____

____

9. In patients with atrial fibrillation (AF), the atria generate as many as 500 action potentials per minute, causing the atria to spasm instead of contracting as a coordinated unit. A maze procedure involves making incisions in the atria and sewing them back together, resulting in the formation of scar tissue in the atria. If the scar tissue is electrically inert, how might this procedure prevent atrial fibrillation? ____________________

____

____

10. Long Q-T syndrome (LQTS) is characterized by a prolonged Q-T interval. When LQTS is symptomatic, patients experience episodes of fainting and arrhythmia. Describe the electrical events that occur during the Q-T interval. ____________________

____

____

32 EXERCISE Anatomy of Blood Vessels

Learning Outcomes

- Describe the tunics of blood vessel walls, and state the function of each layer.
- Correlate differences in artery, vein, and capillary structure with the functions of these vessels.
- Recognize a cross-sectional view of an artery and vein when provided with a microscopic view or appropriate image.
- List and identify the major arteries arising from the aorta, and indicate the body region supplied by each.
- Describe the cerebral arterial circle, and discuss its importance in the body.
- List and identify the major veins draining into the superior and inferior venae cavae, and indicate the body regions drained.
- Describe these special circulations in the body: pulmonary circulation, hepatic portal system, and fetal circulation, and discuss the important features of each.

Go to Mastering A&P™ > Study Area to improve your performance in A&P Lab.

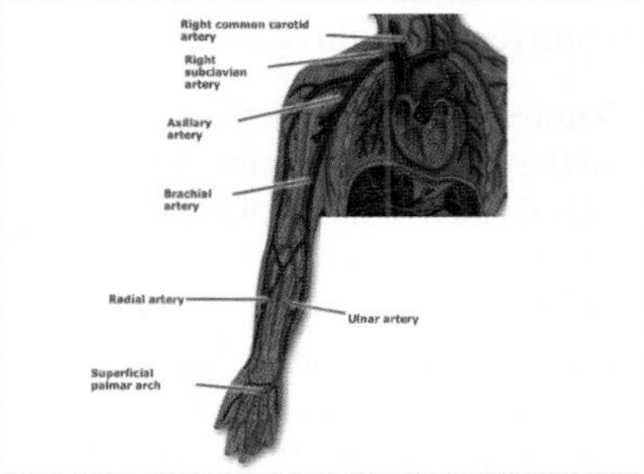

> Lab Tools > Practice Anatomy Lab > Anatomical Models

Instructors may assign new Building Vocabulary coaching activities, Pre-Lab Quiz questions, Art Labeling activities, Practice Anatomy Lab Practical questions (PAL), and more using the Mastering A&P™ Item Library.

Pre-Lab Quiz

Instructors may assign these and other Pre-Lab Quiz questions using Mastering A&P™

1. Circle the correct underlined term. Arteries / Veins drain tissues and return blood to the heart.
2. Circle True or False. Gas exchange takes place between tissue cells and blood through capillary walls.
3. Circle the correct underlined term. Veins draining the head and upper extremities empty into the superior / inferior vena cava.
4. The function of the _________ is to drain the digestive viscera and carry dissolved nutrients to the liver for processing.
   - **a.** fetal circulation
   - **b.** hepatic portal circulation
   - **c.** pulmonary circulation system
5. Circle the correct underlined term. In the developing fetus, the umbilical artery / vein carries blood rich in nutrients and oxygen to the fetus.

Materials

- Compound microscope
- Prepared microscope slides showing cross sections of an artery and vein
- Anatomical charts of human arteries and veins (or a three-dimensional model of the human circulatory system)
- Anatomical charts of the following specialized circulations: pulmonary circulation, hepatic portal circulation, fetal circulation, arterial supply of the brain (or a brain model showing this circulation)

For instructions on animal dissections, see the dissection exercises (starting on p. 695) in the cat and fetal pig editions of this manual.

Arteries, carrying blood away from the heart, and veins, which drain the tissues and return blood to the heart, function simply as conducting vessels or conduits. Only the tiny capillaries that connect the arterioles and venules and branch throughout the tissues directly serve the needs of the body's cells. It is through the capillary walls that exchanges are made between tissue cells and blood.

In this exercise you will examine the microscopic structure of blood vessels and identify the major arteries and veins of the systemic circulation and other special circulations.

Microscopic Structure of the Blood Vessels

Except for the microscopic capillaries, the walls of blood vessels are constructed of three coats, or *tunics* (**Figure 32.1**).

- **Tunica intima:** Lines the lumen of a vessel and is composed of the *endothelium*, subendothelial layer, and internal elastic membrane. The simple squamous cells of the endothelium fit closely together, forming an extremely smooth blood vessel lining that helps decrease resistance to blood flow.
- **Tunica media:** Middle coat, composed primarily of smooth muscle and elastin. The smooth muscle plays an active role in regulating the diameter of blood vessels, which in turn alters blood flow and blood pressure.
- **Tunica externa:** Outermost tunic, composed of areolar or fibrous connective tissue. Its function is basically supportive and protective. In larger vessels, the tunica externa contains a system of tiny blood vessels, the **vasa vasorum**.

In general, the walls of arteries are thicker than those of veins. The tunica media in particular tends to be much heavier and contains substantially more smooth muscle and elastic tissue. Arteries, which are closer to the pumping action of the heart, must be able to expand as an increased volume of blood is propelled into them during systole and then recoil passively as the blood flows off into the circulation during diastole. The anatomical differences between the different types of vessels reflect their functional differences. **Table 32.1** summarizes the structure and function of various blood vessels.

Valves in veins act to prevent backflow of blood in much the same manner as the semilunar valves of the heart. The skeletal

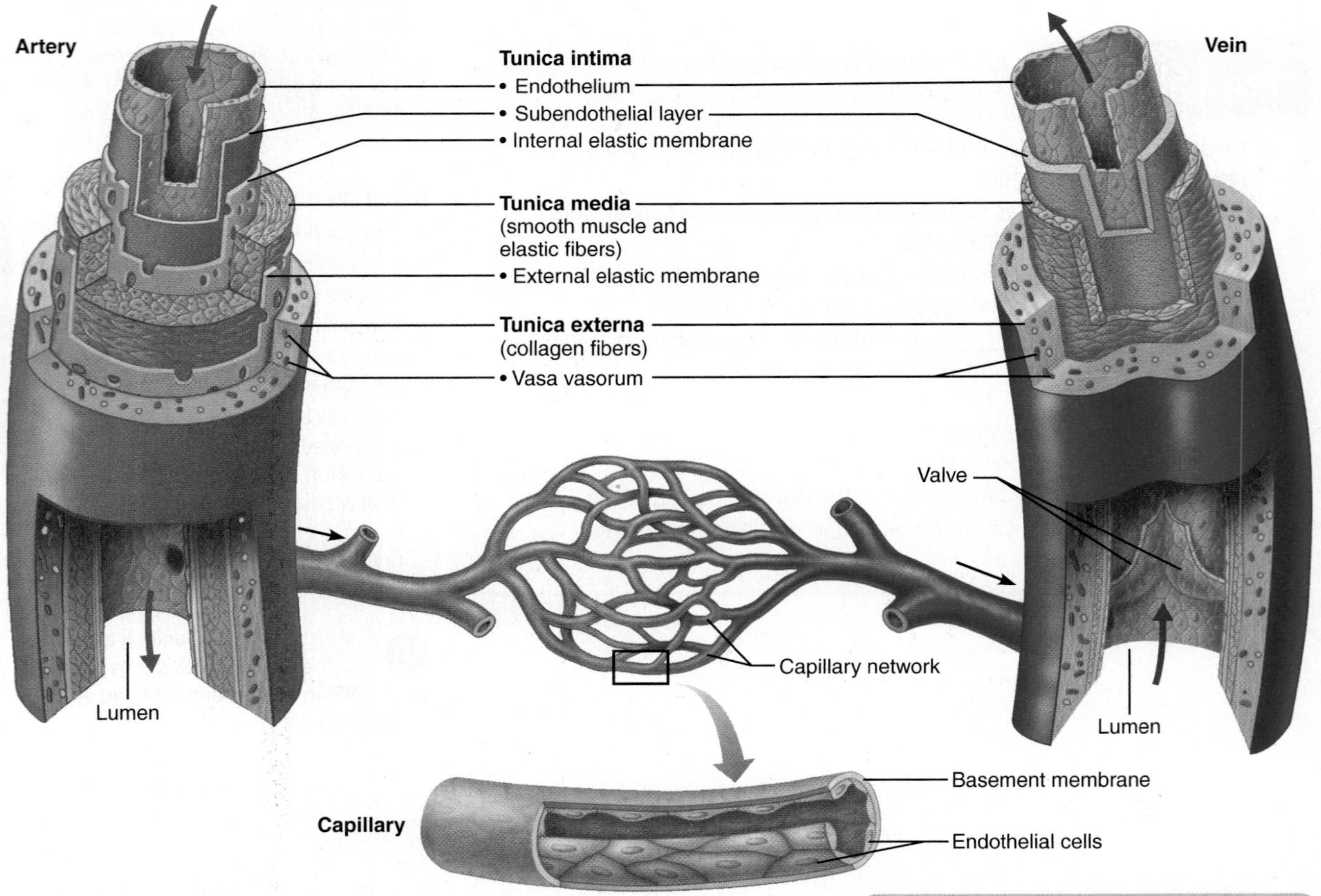

Figure 32.1 Generalized structure of arteries, veins, and capillaries.

Instructors may assign this figure as an Art Labeling Activity using Mastering A&P

Table 32.1 **Summary of Blood Vessel Anatomy and Physiology**

Type of vessel	Description	Average lumen diameter	Average wall thickness	Function
Elastic (conducting) arteries	Largest, most elastic arteries. Contain more elastic tissue than other arteries.	1.5 cm	1.0 mm	Act as a pressure reservoir, expanding and recoiling for continuous blood flow. Examples: aorta, brachiocephalic artery, and common carotid artery.
Muscular (distributing) arteries	Medium-sized arteries, accounting for most arteries found in the body. They have less elastic tissue and more smooth muscle than other arteries.	0.6 cm	1.0 mm	Better ability to constrict and less stretchable than elastic arteries. They distribute blood to specific areas of the body. Examples: brachial artery and radial artery.
Arterioles	Smallest arteries, with a very thin tunica externa and only a few layers of smooth muscle in the tunica media.	37 µm	6 µm	Blood flows from arterioles into a capillary bed. They play a role in regulating the blood flow to specific areas of the body.
Capillaries	Contain only a tunica intima.	9 µm	0.5 µm	Provide for the exchange of materials (gases, nutrients, etc.) between the blood and tissue cells.
Venules	Smallest veins. All tunics are very thin, with at most two layers of smooth muscle and no elastic tissue.	20 µm	1 µm	Drain capillary beds and merge to form veins.
Veins	Contain more fibrous tissue in the tunica externa than corresponding arteries. The tunica media is thinner, with a larger lumen than the corresponding artery.	0.5 cm	0.5 mm	Low-pressure vessels; return blood to the heart. Valves prevent the backflow of the blood.

muscle "pump" also promotes venous return; as the skeletal muscles surrounding the veins contract and relax, the blood is milked through the veins toward the heart. Anyone who has been standing relatively still for an extended time has experienced swelling in the ankles, caused by blood pooling in the feet during the period of muscle inactivity. Pressure changes that occur in the thorax during breathing also aid the return of blood to the heart.

☐ To demonstrate how efficiently venous valves prevent backflow of blood, perform the following simple experiment. Allow one hand to hang by your side until the blood vessels on the dorsal aspect become distended. Place two fingertips against one of the distended veins and, pressing firmly, move the superior finger proximally along the vein, and then release this finger. The vein will remain flattened and collapsed despite gravity. Then remove the distal fingertip, and observe the rapid filling of the vein.

Check the box when you have completed this task.

32

Activity 1

Examining the Microscopic Structure of Arteries and Veins

1. Obtain a slide showing a cross-sectional view of blood vessels and a microscope.

2. Scan the section to identify a thick-walled artery (use **Figure 32.2** as a guide). Very often, but not always, an arterial lumen will appear scalloped because its walls are constricted by the elastic tissue of the tunica media.

3. Identify a vein. Its lumen may be elongated or irregularly shaped and collapsed, and its walls will be considerably thinner. Notice the difference in the relative amount of elastic fibers in the tunica media of the two vessels. Also, note the thinness of the tunica intima layer, which is composed of flat squamous cells.

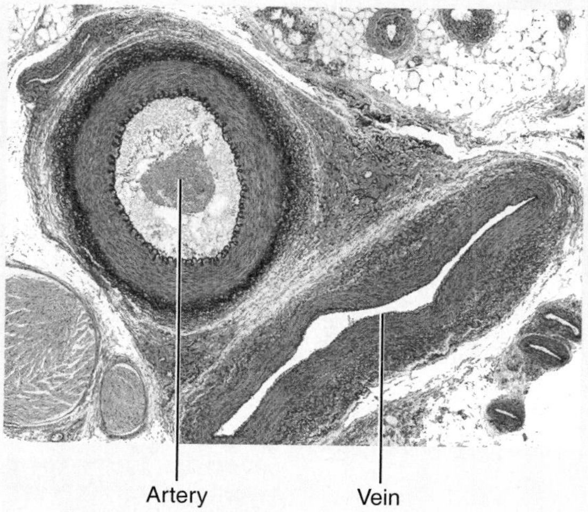

Figure 32.2 Photomicrograph of a muscular artery and the corresponding vein in cross section (40×).

Major Systemic Arteries of the Body

The **aorta** is the largest artery of the body. It has three main regions: ascending aorta, aortic arch, and descending aorta. Extending upward as the **ascending aorta** from the left ventricle, it arches posteriorly and to the left as the **aortic arch** and then courses downward as the **descending aorta** through the thoracic cavity. Called the **thoracic aorta** from T_5 to T_{12}, the descending aorta penetrates the diaphragm to enter the abdominal cavity just anterior to the vertebral column. As it enters the abdominal cavity, it becomes the **abdominal aorta**. The branches of the ascending aorta and aortic arch are summarized in **Table 32.2**, p. 466. Branches of the thoracic and abdominal aorta are summarized in Tables 32.3 and 32.4.

As you locate the arteries diagrammed on **Figure 32.3**, notice that in many cases, the name of the artery reflects the body region it travels through (axillary, subclavian, brachial, popliteal), the organ served (renal, hepatic), or the bone followed (tibial, femoral, radial, ulnar).

Aortic Arch

The **brachiocephalic** (literally, "arm-head") **trunk** is the first branch of the aortic arch (**Figure 32.4**). The other two major arteries branching off the aortic arch are the **left common carotid artery** and the **left subclavian artery**. The brachiocephalic trunk persists briefly before dividing into the **right common carotid artery** and the **right subclavian artery**.

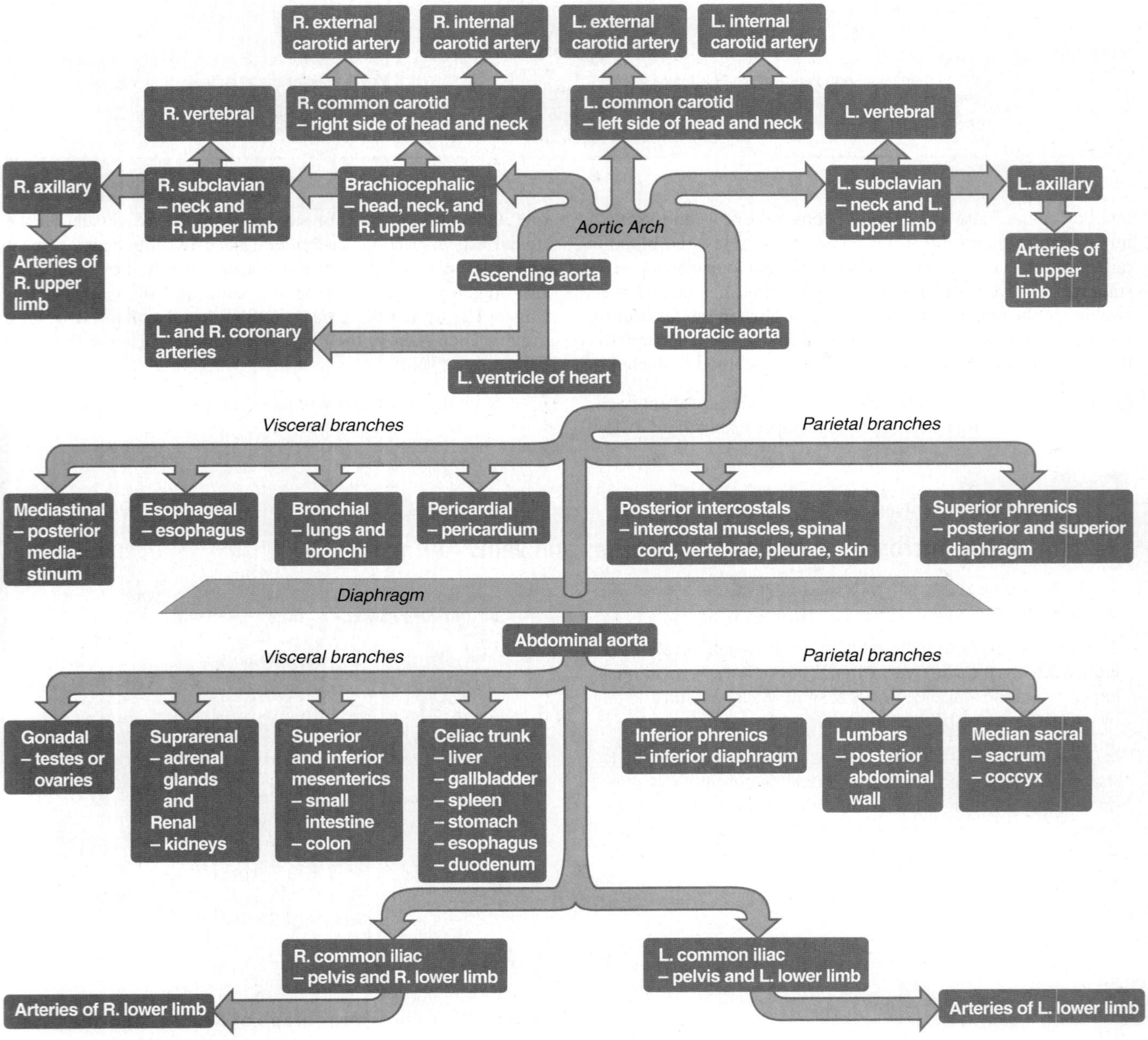

Figure 32.3 Schematic of the systemic arterial circulation. (R. = right, L. = left)

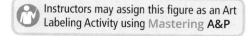

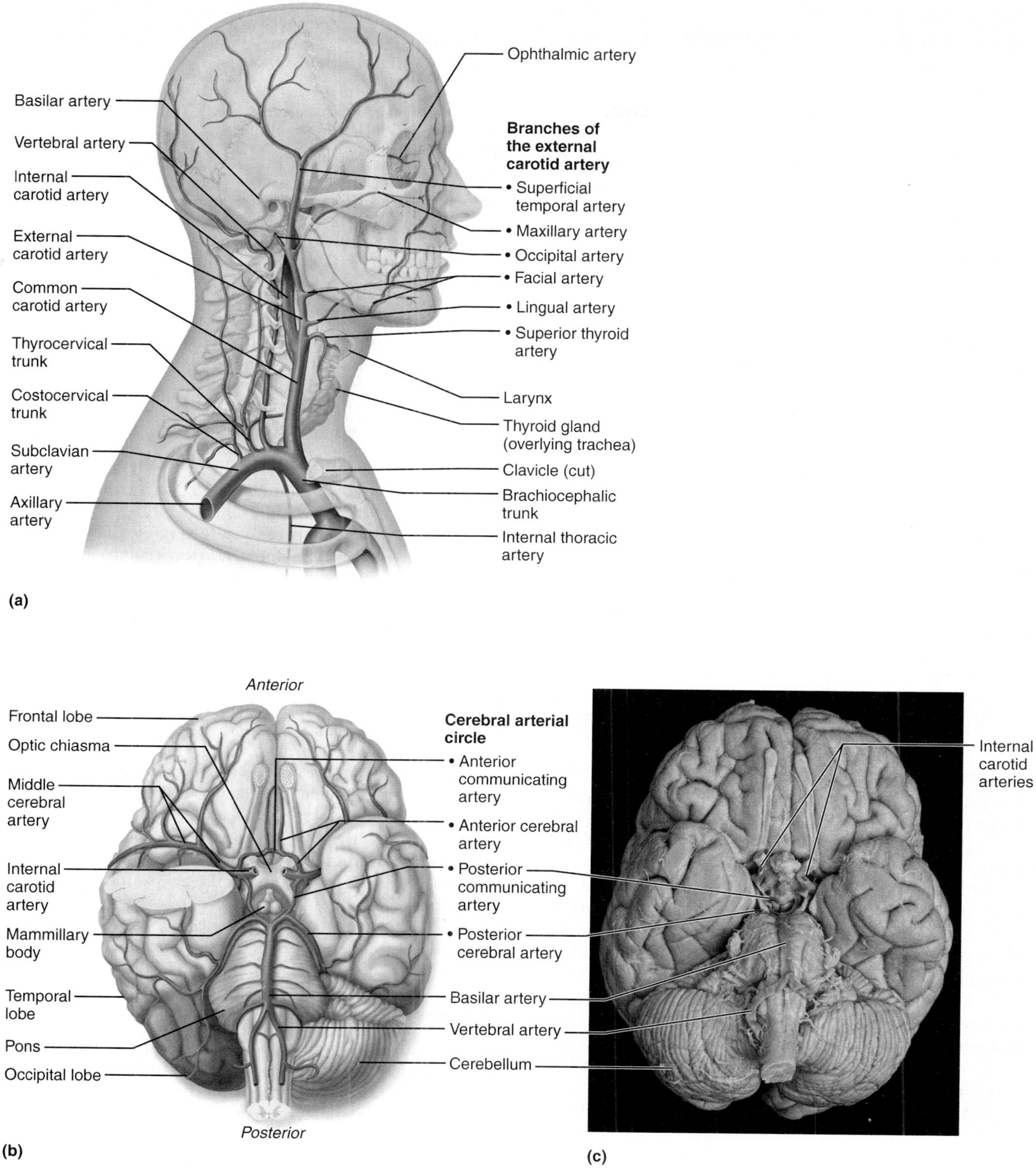

Figure 32.4 Arteries of the head, neck, and brain. (a) Right aspect. **(b)** Drawing of the cerebral arteries, inferior view. Right side of cerebellum and part of right temporal lobe have been removed. **(c)** Cerebral arterial circle (circle of Willis), inferior view of brain.

Instructors may assign this figure as an Art Labeling Activity using Mastering A&P

32

Table 32.2 The Aorta: Ascending Aorta and Aortic Arch (Figure 32.3)

Ascending aorta branches	Structures served
Right coronary artery	The myocardium of the heart (see Exercise 30)
Left coronary artery	The myocardium of the heart (see Exercise 30)
Aortic arch branches	**Structures served**
Brachiocephalic trunk (branches into right common carotid and right subclavian arteries)	Right common carotid artery – right side of the head and neck Right subclavian artery – right upper limb
Left common carotid artery	Left side of the head and neck
Left subclavian artery	Left upper limb

Arteries Serving the Head and Neck

The common carotid artery on each side divides to form an internal and an external carotid artery. The **internal carotid artery** serves the brain and gives rise to the **ophthalmic artery** that supplies orbital structures. The **external carotid artery** supplies the tissues external to the skull, largely via its **superficial temporal, maxillary, facial**, and **occipital** arterial branches. (Notice that several arteries are shown in the figure that are not described here.)

The right and left subclavian arteries each give off several branches to the head and neck. The first of these is the **vertebral artery**, which runs up the posterior neck to supply the cerebellum, part of the brain stem, and the posterior cerebral hemispheres. Issuing just lateral to the vertebral artery are the **thyrocervical trunk**, which mainly serves the thyroid gland and some scapular muscles, and the **costocervical trunk**, which supplies deep neck muscles and some of the upper intercostal muscles. In the armpit, the subclavian artery becomes the axillary artery, which serves the upper limb.

Arteries Serving the Brain

The brain is supplied by two pairs of arteries arising from the region of the aortic arch—the internal carotid arteries and the vertebral arteries. (Figure 32.4b is a diagram of the brain's arterial supply.)

32

Within the cranium, each internal carotid artery divides into **anterior** and **middle cerebral arteries**, which supply the bulk of the cerebrum. The right and left anterior cerebral arteries are connected by a short shunt called the **anterior communicating artery**. This shunt, along with shunts from each of the middle cerebral arteries, called the **posterior communicating arteries**, contribute to the formation of the **cerebral arterial circle** (*circle of Willis*), an arterial anastomosis at the base of the brain surrounding the pituitary gland and the optic chiasma.

The paired **vertebral arteries** diverge from the subclavian arteries and pass superiorly through the foramina of the transverse process of the cervical vertebrae to enter the skull through the foramen magnum. Within the skull, the vertebral arteries unite to form a single **basilar artery**, which continues superiorly along the anterior aspect of the brain stem, giving off branches to the pons, cerebellum, and inner ear. At the pons-midbrain border, the basilar artery divides to form the **posterior cerebral arteries**, which supply portions of the temporal and occipital lobes of the cerebrum and complete the cerebral arterial circle posteriorly.

The uniting of the blood supply of the internal carotid arteries and the vertebral arteries via the cerebral arterial circle is a protective device that provides alternate pathways for blood to reach the brain tissue in the case of arterial occlusion or impaired blood flow anywhere in the system.

Arteries Serving the Thorax and Upper Limbs

As the **axillary artery** runs through the axilla, it gives off several branches to the chest wall and shoulder girdle (**Figure 32.5**). These include the **thoracoacromial artery** (to shoulder and pectoral region), the **lateral thoracic artery** (lateral chest wall), the **subscapular artery** (to scapula and dorsal thorax), and the **anterior** and **posterior circumflex humeral arteries** (to the shoulder and the deltoid muscle). At the inferior edge of the teres major muscle, the axillary artery becomes the **brachial artery** as it enters the arm. The brachial artery gives off a major branch, the **deep artery of the arm**, and as it nears the elbow it gives off several small branches. At the elbow, the brachial artery divides into the **radial** and **ulnar arteries**, which follow the same-named bones to supply the forearm and hand.

The **internal thoracic arteries** that arise from the subclavian arteries supply the mammary glands, most of the thorax wall, and anterior intercostal structures via their **anterior intercostal artery** branches. The first two pairs of **posterior intercostal arteries** arise from the costocervical trunk, noted above. The more inferior pairs arise from the thoracic aorta. (Not shown in Figure 32.5 are the small arteries that serve the diaphragm [*phrenic arteries*], esophagus [*esophageal arteries*], bronchi [*bronchial arteries*], and other structures of the mediastinum [*mediastinal* and *pericardial arteries*].)

Thoracic Aorta

The thoracic aorta is the superior portion of the descending aorta (Figure 32.5). It begins where the aortic arch ends and ends just as it pierces the diaphragm. The main branches of the thoracic aorta are summarized in **Table 32.3**.

Abdominal Aorta

Although several small branches of the descending aorta serve the thorax, the more major branches of the descending aorta are those serving the abdominal organs and ultimately the lower limbs (**Figure 32.6**, pp. 468–469). Most of the branches of the abdominal aorta serve the abdominal organs. The major branches of the abdominal aorta are summarized in **Table 32.4** on p. 469.

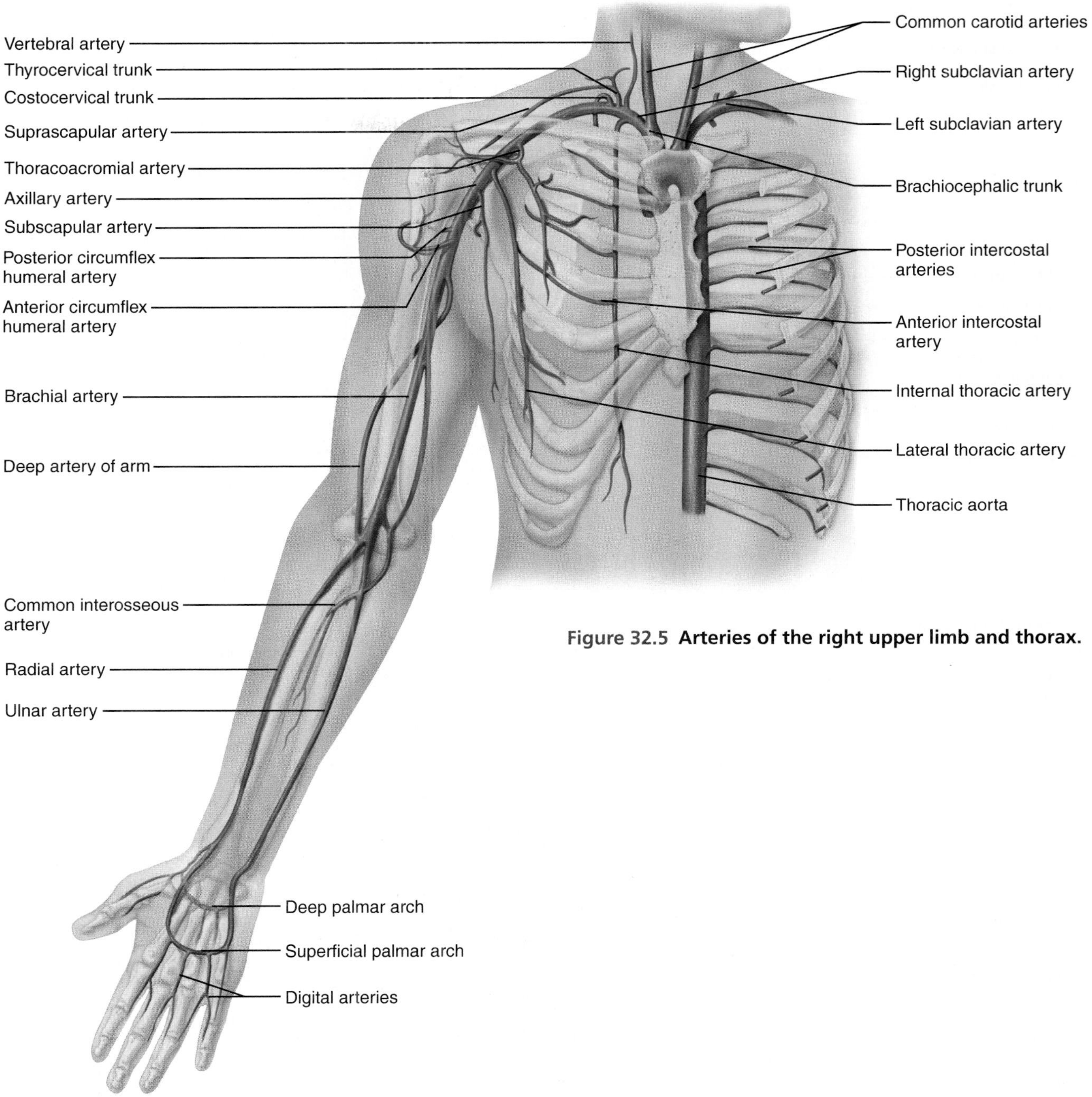

Figure 32.5 Arteries of the right upper limb and thorax.

Table 32.3 The Aorta: Thoracic Aorta (Figure 32.3) (Note that many of these arteries vary in number from person to person)

Visceral thoracic aorta branches	Structures served
Pericardial arteries	Pericardium, the serous membrane of the heart
Bronchial arteries	Bronchi, bronchioles, and lungs
Esophageal arteries	Esophagus
Mediastinal arteries	Posterior mediastinum
Parietal thoracic aorta branches	**Structures served**
Posterior intercostal arteries (inferior pairs)	Intercostal muscles, spinal cord, vertebrae, and skin
Subcostal arteries	Intercostal muscles, spinal cord, vertebrae, and skin
Superior phrenic arteries	Posterior, superior part of the diaphragm

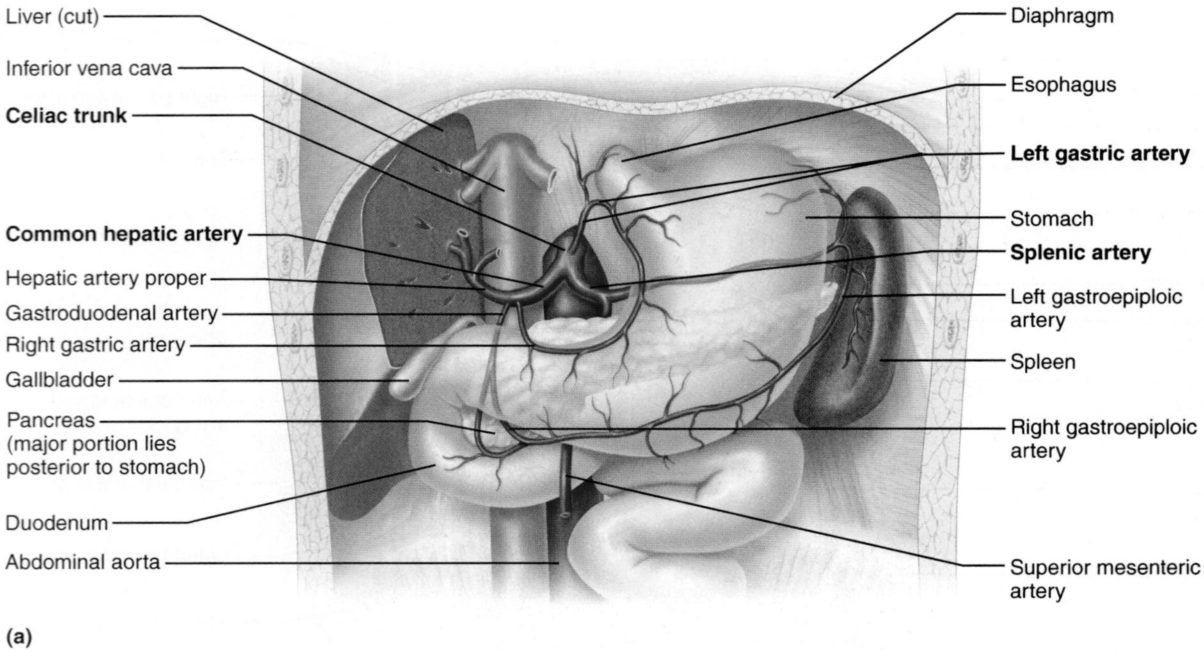

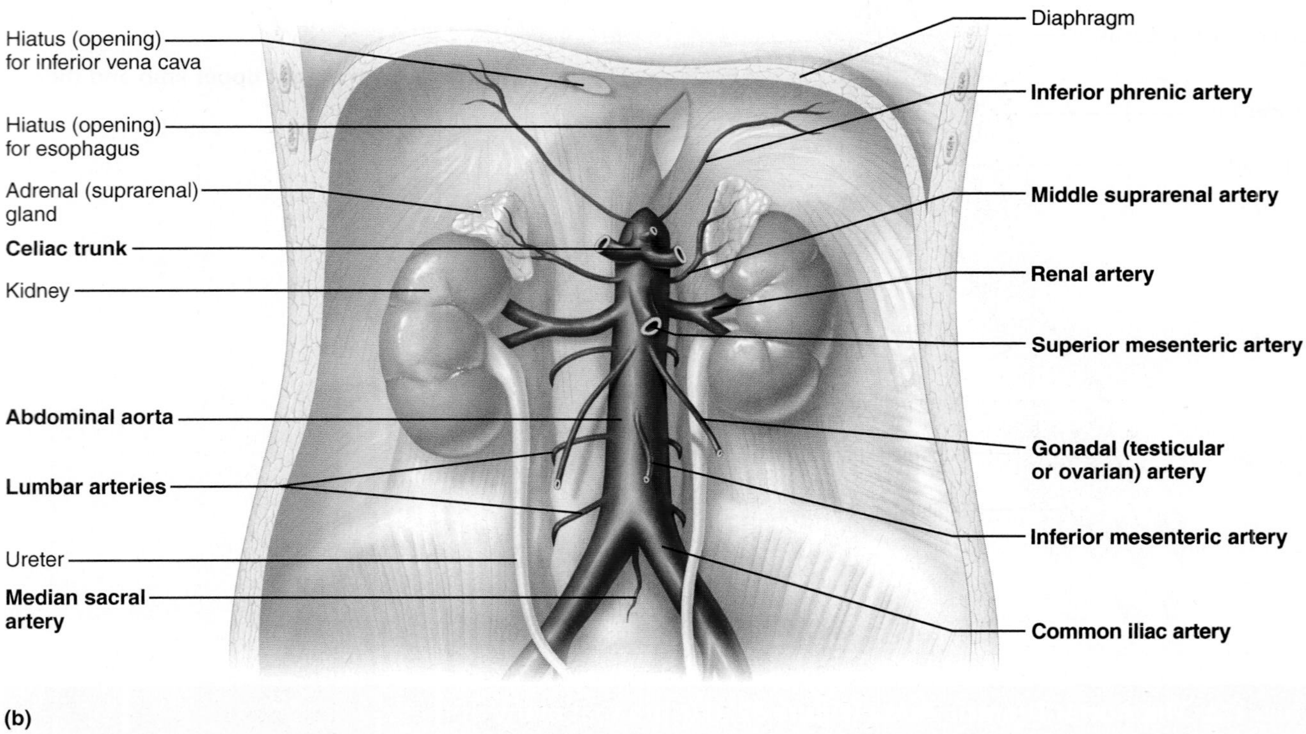

Figure 32.6 Arteries of the abdomen. (a) The celiac trunk and its major branches. **(b)** Major branches of the abdominal aorta.

32

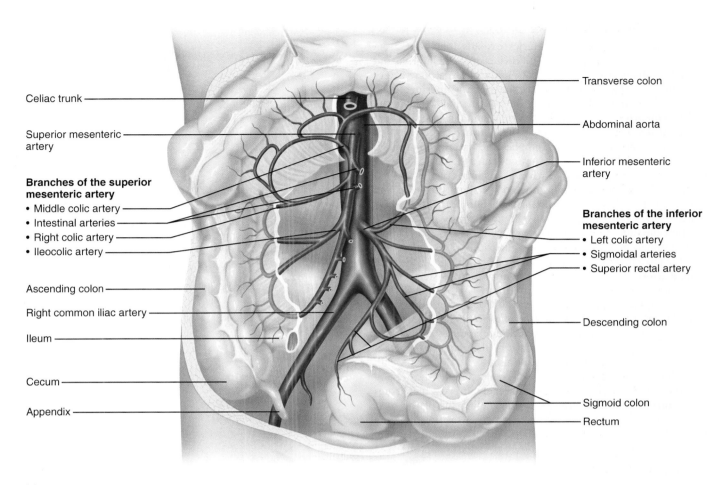

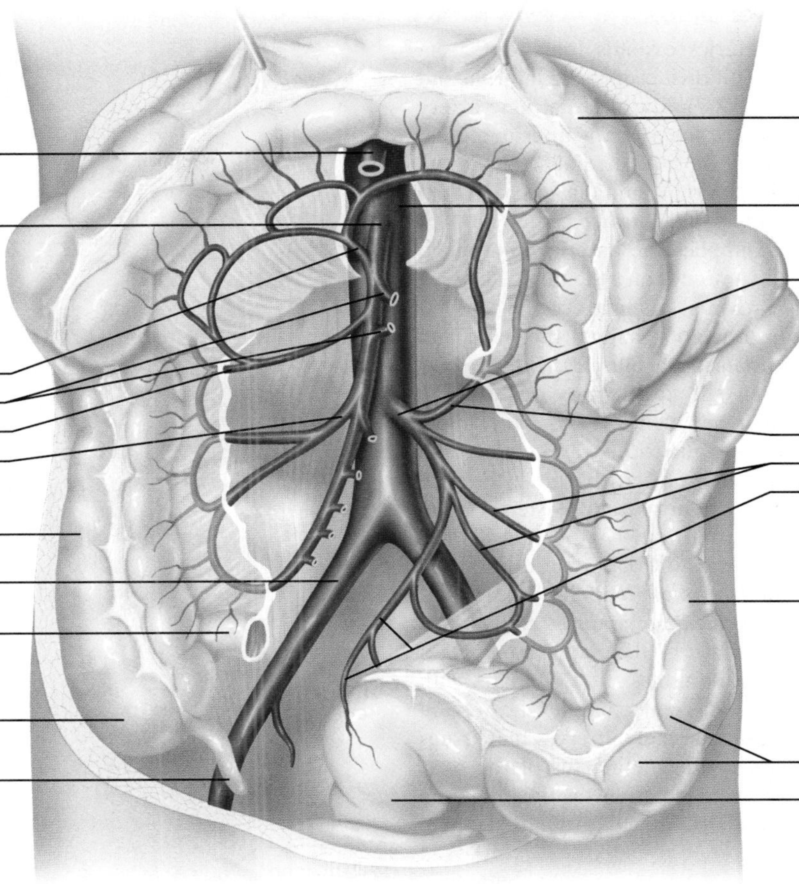

(c)

Figure 32.6 *(continued)* **(c)** Distribution of the superior and inferior mesenteric arteries, transverse colon pulled superiorly.

Table 32.4 The Aorta: Abdominal Aorta (Figure 32.6)

Branches	Structures served
Inferior phrenic arteries	Inferior surface of the diaphragm
Celiac trunk: left gastric artery	Stomach and esophagus
Celiac trunk: splenic artery	Branches to the spleen; short gastric arteries branch to the stomach; and the left gastroepiploic artery branches to the stomach
Celiac trunk: common hepatic artery	Branches into the hepatic artery proper (its branches serve the liver, gallbladder, and stomach) and the gastroduodenal artery (its branches serve the stomach, pancreas, and duodenum)
Superior mesenteric artery	Most of the small intestine and the first part of the large intestine
Middle suprarenal arteries	Adrenal glands that sit on top of the kidneys
Renal arteries	Kidneys
Gonadal arteries	Ovarian arteries (female) – ovaries Testicular arteries (male) – testes
Inferior mesenteric artery	Distal portion of the large intestine
Lumbar arteries	Posterior abdominal wall
Median sacral artery	Sacrum and coccyx
Common iliac arteries	The distal abdominal aorta splits to form the left and right common iliac arteries, which serve the pelvic organs, lower abdominal wall, and the lower limbs

Arteries Serving the Lower Limbs

Each of the common iliac arteries extends for about 5 cm (2 inches) into the pelvis before it divides into the internal and external iliac arteries (**Figure 32.7**). The **internal iliac artery** supplies the gluteal muscles via the **superior** and **inferior gluteal arteries**, and the adductor muscles of the medial thigh via the **obturator artery**, as well as the external genitalia and perineum (via the *internal pudendal artery*, not illustrated).

The **external iliac artery** supplies the anterior abdominal wall and the lower limb. As it continues into the thigh, its name changes to **femoral artery**. Proximal branches of the femoral artery, the **circumflex femoral arteries**, supply the head and neck of the femur and the hamstring muscles. The femoral artery gives off a deep branch, the **deep artery of the thigh** (also called the *deep femoral artery*), which is the main supply to the thigh muscles (hamstrings, quadriceps, and adductors). In the knee region, the femoral artery briefly becomes the **popliteal artery**; its subdivisions—the **anterior** and **posterior tibial arteries**—supply the leg, ankle, and foot. The posterior tibial, which supplies flexor muscles, gives off one main branch, the **fibular artery**, that serves the lateral calf (fibular muscles). It then divides into the **lateral** and **medial plantar arteries**, which supply blood to the plantar surface of the foot. The anterior tibial artery supplies the extensor muscles and terminates with the **dorsalis pedis artery**. The dorsalis pedis supplies the dorsum of the foot and continues on as the **arcuate artery**, which issues the **dorsal metatarsal arteries** to the metatarsus of the foot. The dorsalis pedis is often palpated in patients with circulation problems of the leg to determine the circulatory efficiency to the limb as a whole.

☐ Palpate your own dorsalis pedis artery.

Check the box when you have completed this task.

32

Activity 2

Locating Arteries on an Anatomical Chart or Model

Now that you have identified the arteries in Figures 32.3–32.7, attempt to locate and name them (without a reference) on a large anatomical chart or three-dimensional model of the vascular system.

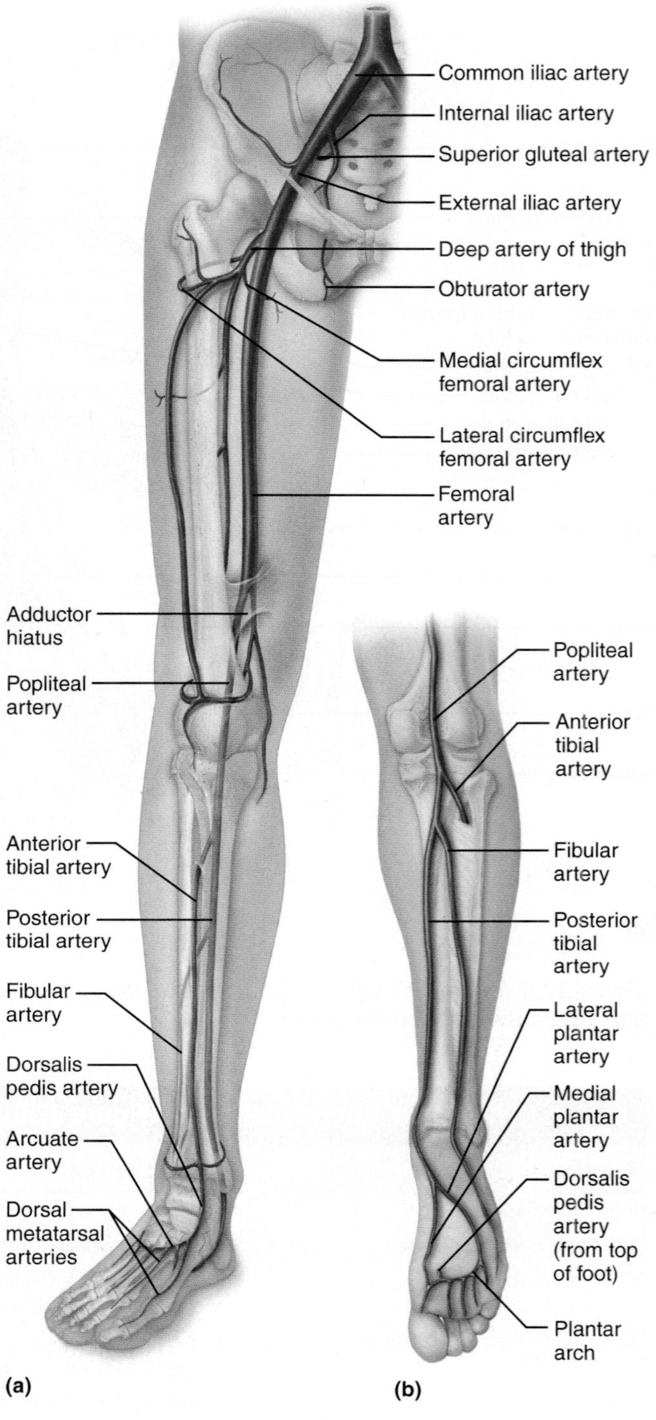

Figure 32.7 Arteries of the right pelvis and lower limb. **(a)** Anterior view. **(b)** Posterior view.

Major Systemic Veins of the Body

Arteries are generally located in deep, well-protected body areas. However, many veins follow a more superficial course and are often easily seen and palpated on the body surface. Most deep veins parallel the course of the major arteries, and in many cases the naming of the veins and arteries is identical except for the designation of the vessels as veins. Whereas the major systemic arteries branch off the aorta, the veins tend to converge on the venae cavae, which enter the right atrium of the heart. Veins draining the head and upper extremities empty into the **superior vena cava**, and those draining the lower body empty into the **inferior vena cava**. Figure 32.8, a schematic of the systemic veins and their relationship to the venae cavae, will get you started.

Veins Draining into the Inferior Vena Cava

The inferior vena cava, a much longer vessel than the superior vena cava, returns blood to the heart from all body regions

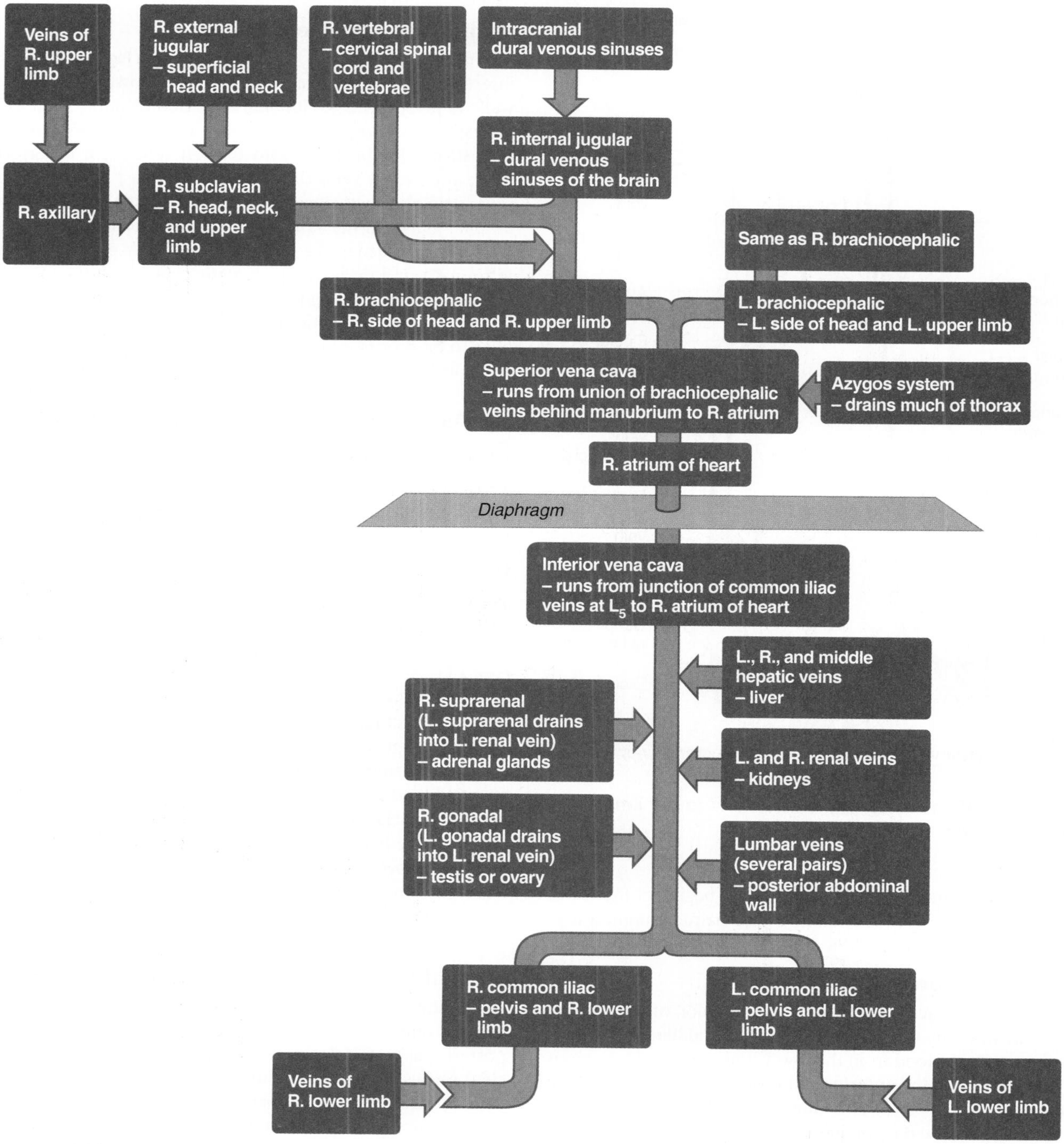

Figure 32.8 Schematic of systemic venous circulation.

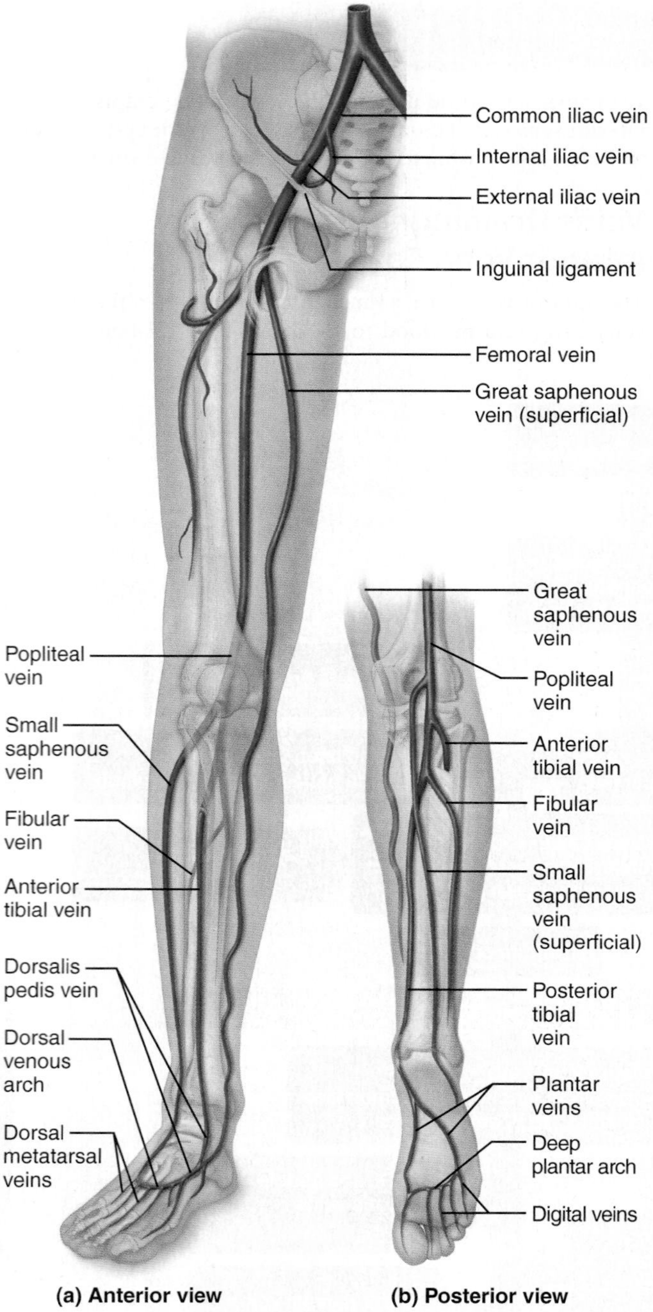

Figure 32.9 Veins of the right pelvis and lower limb. **(a)** Anterior view. **(b)** Posterior view.

below the diaphragm (see Figure 32.8). It begins in the lower abdominal region with the union of the paired **common iliac veins**, which drain venous blood from the legs and pelvis.

Veins of the Lower Limbs

Each common iliac vein is formed by the union of the **internal iliac vein**, draining the pelvis, and the **external iliac vein**, which receives venous blood from the lower limb (**Figure 32.9**). Veins of the leg include the **anterior** and **posterior tibial veins**, which serve the calf and foot. The anterior tibial vein is a superior continuation of the **dorsalis pedis vein** of the foot. The posterior tibial vein is formed by the union of the **medial** and **lateral plantar veins**, and ascends deep in the calf muscles. It receives the **fibular vein** in the calf and then joins with the anterior tibial vein at the knee to produce the **popliteal vein**, which crosses the back of the knee. The popliteal vein becomes the **femoral vein** in the thigh; the femoral vein in turn becomes the external iliac vein in the inguinal region.

The **great saphenous vein**, a superficial vein, is the longest vein in the body. Beginning in common with the **small saphenous vein** from the **dorsal venous arch**, it extends up the medial side of the leg, knee, and thigh to empty into the femoral vein. The small saphenous vein runs along the lateral aspect of the foot and through the calf muscle, which it drains, and then empties into the popliteal vein at the knee (Figure 32.9b).

Veins of the Abdomen

Moving superiorly in the abdominal cavity (**Figure 32.10**), the inferior vena cava receives blood from the posterior abdominal wall via several pairs of **lumbar veins**, and from the right ovary or testis via the **right gonadal vein**. (The **left gonadal [ovarian or testicular] vein** drains into the left renal vein superiorly.) The paired **renal veins** drain the kidneys. Just above the right renal vein, the **right suprarenal vein** (receiving blood from the adrenal gland on the same side) drains into the inferior vena cava, but its partner, the **left suprarenal vein**, empties into the left renal vein inferiorly. The **hepatic veins** drain the liver. The unpaired veins draining the digestive tract organs empty into a special vessel, the hepatic portal vein, which carries blood to the liver to be processed before it enters the systemic venous system. The hepatic portal system is discussed separately on p. 476.

Veins Draining into the Superior Vena Cava

Veins draining into the superior vena cava are named from the superior vena cava distally, *but remember that the flow of blood is in the opposite direction*.

Veins of the Head and Neck

The **right** and **left brachiocephalic veins** drain the head, neck, and upper extremities and unite to form the superior vena cava (**Figure 32.11**). Notice that although there is only one brachiocephalic artery, there are two brachiocephalic veins.

Branches of the brachiocephalic veins include the internal jugular, vertebral, and subclavian veins. The **internal jugular veins** are large veins that drain the superior sagittal sinus and other **dural sinuses** of the brain. As they run inferiorly, they receive blood from the head and neck via the **superficial temporal** and **facial veins**. The **vertebral veins** drain the posterior aspect of the head including the cervical vertebrae and spinal cord. The **subclavian veins** receive venous blood from the upper extremity. The **external jugular vein** joins the subclavian vein near its origin to return the venous drainage of the extracranial (superficial) tissues of the head and neck.

Veins of the Upper Limb and Thorax

As the subclavian vein passes through the axilla, it becomes the **axillary vein** and then the **brachial vein** as it courses along the posterior aspect of the humerus (**Figure 32.12**, p. 474). The brachial vein is formed by the union of the deep **radial** and **ulnar veins** of the forearm. The superficial venous drainage of the arm includes the **cephalic vein**, which courses along the lateral

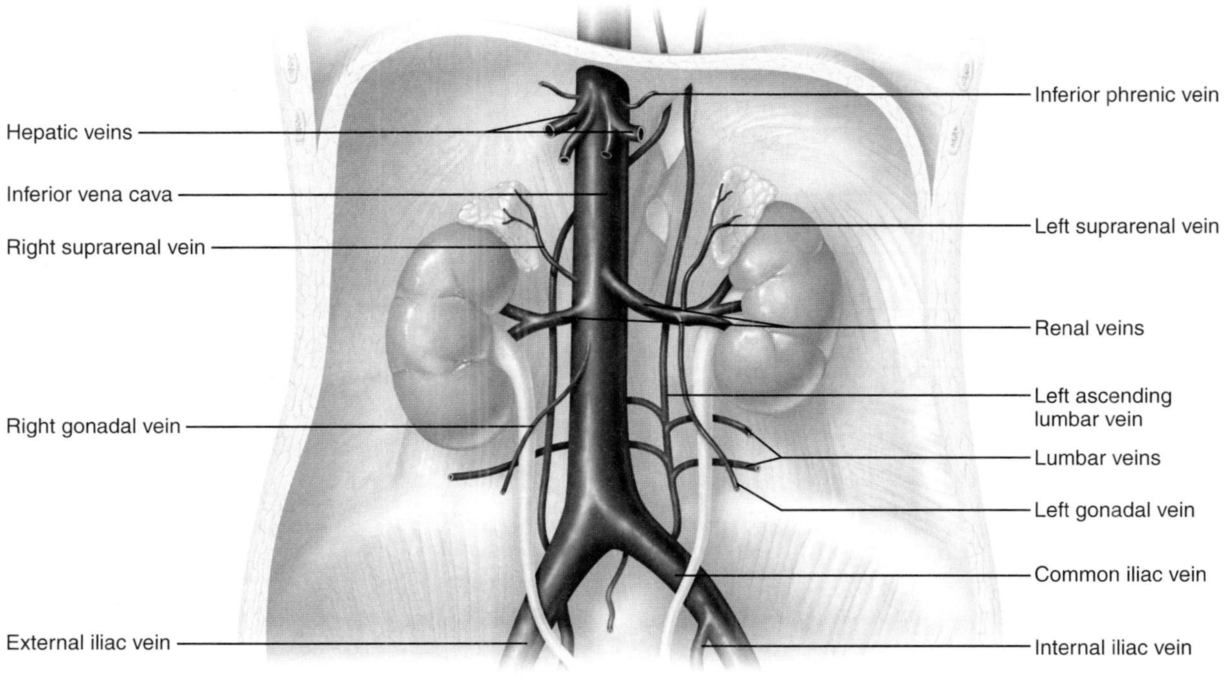

Figure 32.10 Venous drainage of abdominal organs not drained by the hepatic portal vein.

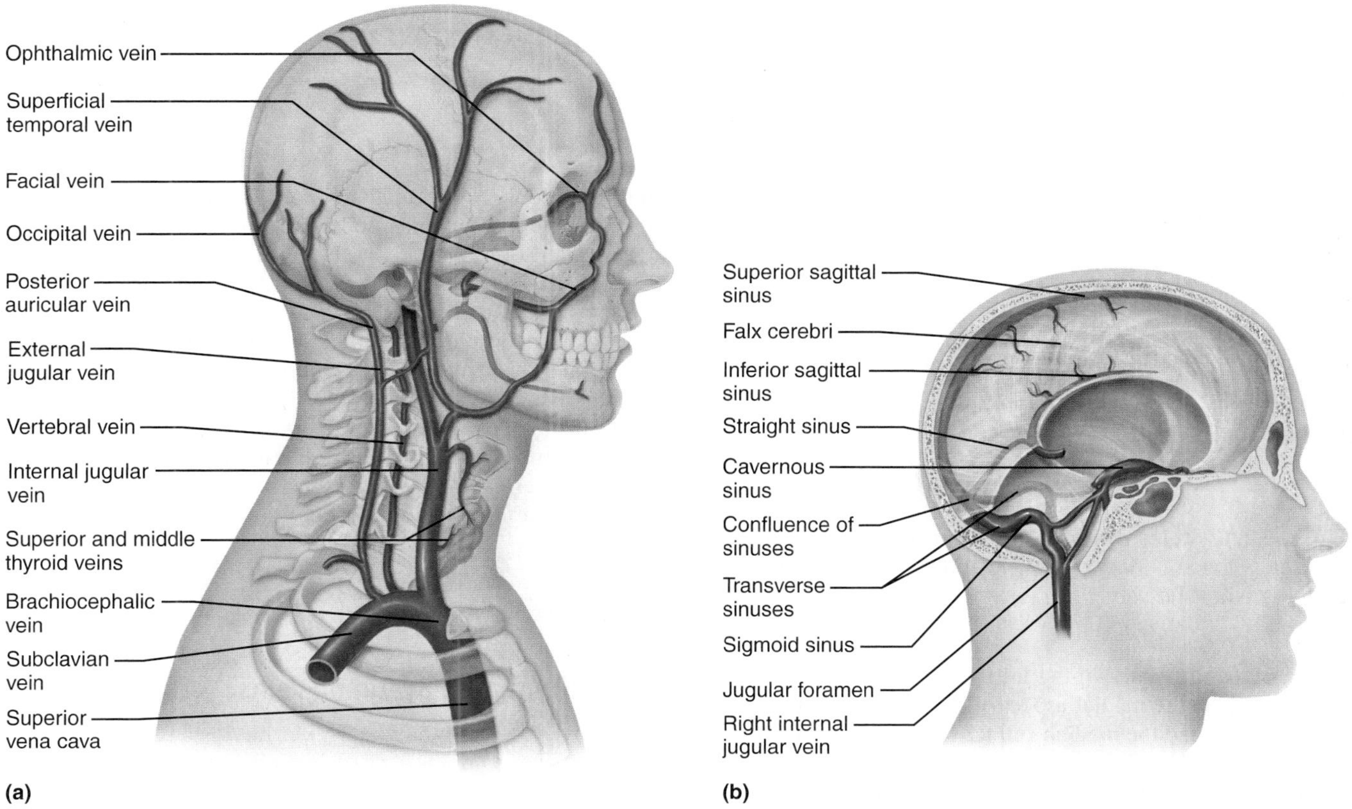

Figure 32.11 Venous drainage of the head, neck, and brain. (a) Veins of the head and neck, right superficial aspect. **(b)** Dural sinuses of the brain, right aspect.

32

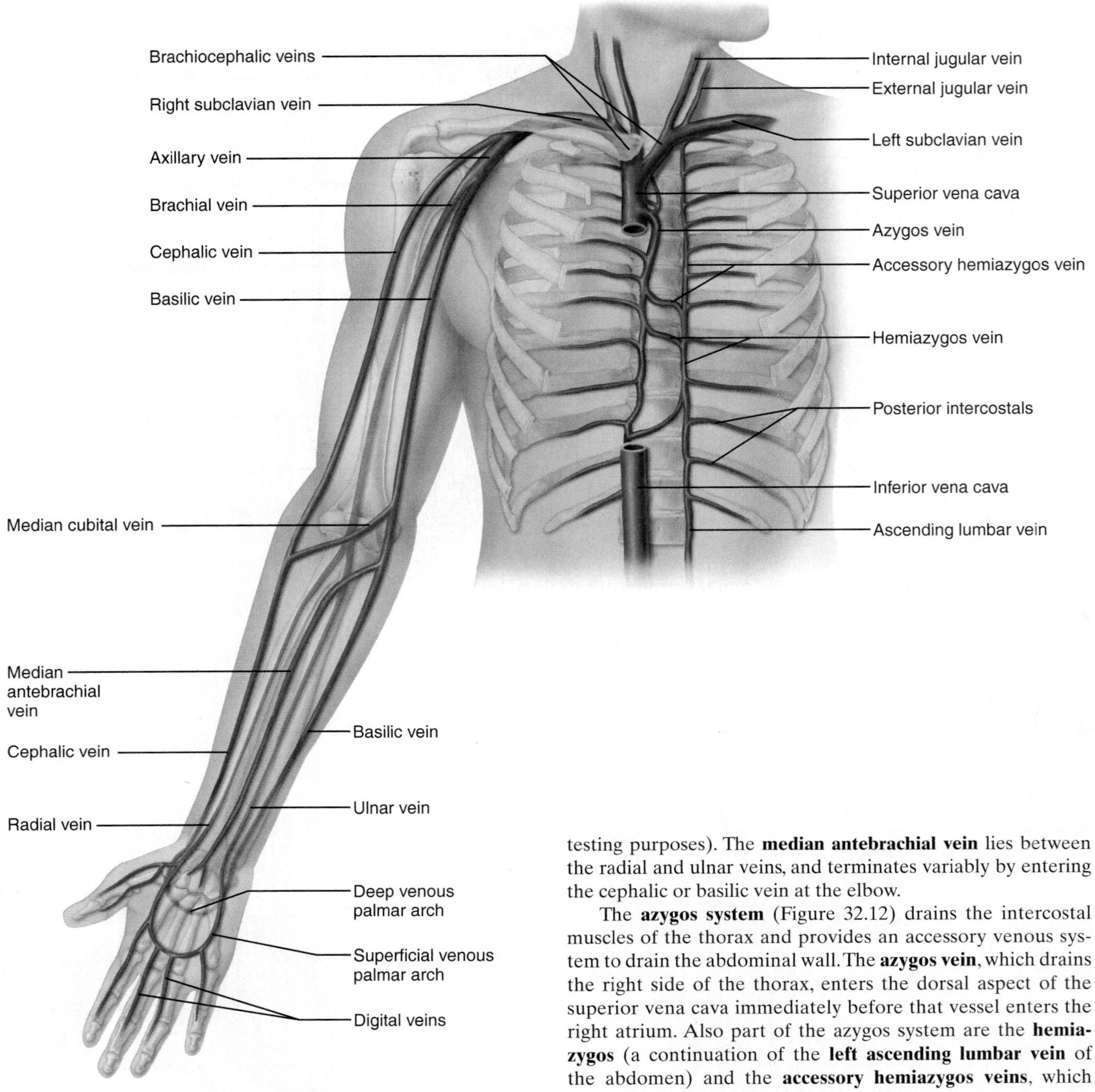

Figure 32.12 Veins of the thorax and right upper limb. For clarity, the abundant branching and anastomoses of these vessels are not shown.

aspect of the arm and empties into the axillary vein; the **basilic vein**, found on the medial aspect of the arm and entering the brachial vein; and the **median cubital vein**, which runs between the cephalic and basilic veins in the anterior aspect of the elbow (this vein is often the site of choice for removing blood for testing purposes). The **median antebrachial vein** lies between the radial and ulnar veins, and terminates variably by entering the cephalic or basilic vein at the elbow.

The **azygos system** (Figure 32.12) drains the intercostal muscles of the thorax and provides an accessory venous system to drain the abdominal wall. The **azygos vein**, which drains the right side of the thorax, enters the dorsal aspect of the superior vena cava immediately before that vessel enters the right atrium. Also part of the azygos system are the **hemiazygos** (a continuation of the **left ascending lumbar vein** of the abdomen) and the **accessory hemiazygos veins**, which together drain the left side of the thorax and empty into the azygos vein.

Activity 3

Identifying the Systemic Veins

Identify the important veins of the systemic circulation on the large anatomical chart or model without referring to the figures.

Special Circulations

Pulmonary Circulation

The pulmonary circulation (discussed previously in relation to heart anatomy on p. 438) differs in many ways from systemic circulation because it does not serve the metabolic needs of the body tissues with which it is associated (in this case, lung tissue). It functions instead to bring the blood into close contact with the alveoli of the lungs to permit gas exchanges that rid the blood of excess carbon dioxide and replenish its supply of vital oxygen. The arteries of the pulmonary circulation are structurally much like veins, and they create a low-pressure bed in the lungs. (If the arterial pressure in the systemic circulation is 120/80, the pressure in the pulmonary artery is likely to be approximately 24/8.) The functional blood supply of the lungs is provided by the **bronchial arteries** (not shown), which diverge from the thoracic portion of the descending aorta.

Pulmonary circulation begins with the large **pulmonary trunk**, which leaves the right ventricle and divides into the **right** and **left pulmonary arteries** about 5 cm (2 inches) above its origin. The right and left pulmonary arteries plunge into the lungs, where they subdivide into **lobar arteries** (three on the right and two on the left). The lobar arteries accompany the main bronchi into the lobes of the lungs and branch extensively within the lungs to form arterioles, which finally terminate in the capillary networks surrounding the alveolar sacs of the lungs. Diffusion of the respiratory gases occurs across the walls of the alveoli and **pulmonary capillaries**. The pulmonary capillary beds are drained by venules, which converge to form sequentially larger veins and finally the four **pulmonary veins** (two leaving each lung), which return the blood to the left atrium of the heart.

Activity 4

Identifying Vessels of the Pulmonary Circulation

Find the vessels of the pulmonary circulation in **Figure 32.13** and on an anatomical chart (if one is available).

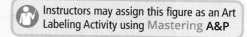

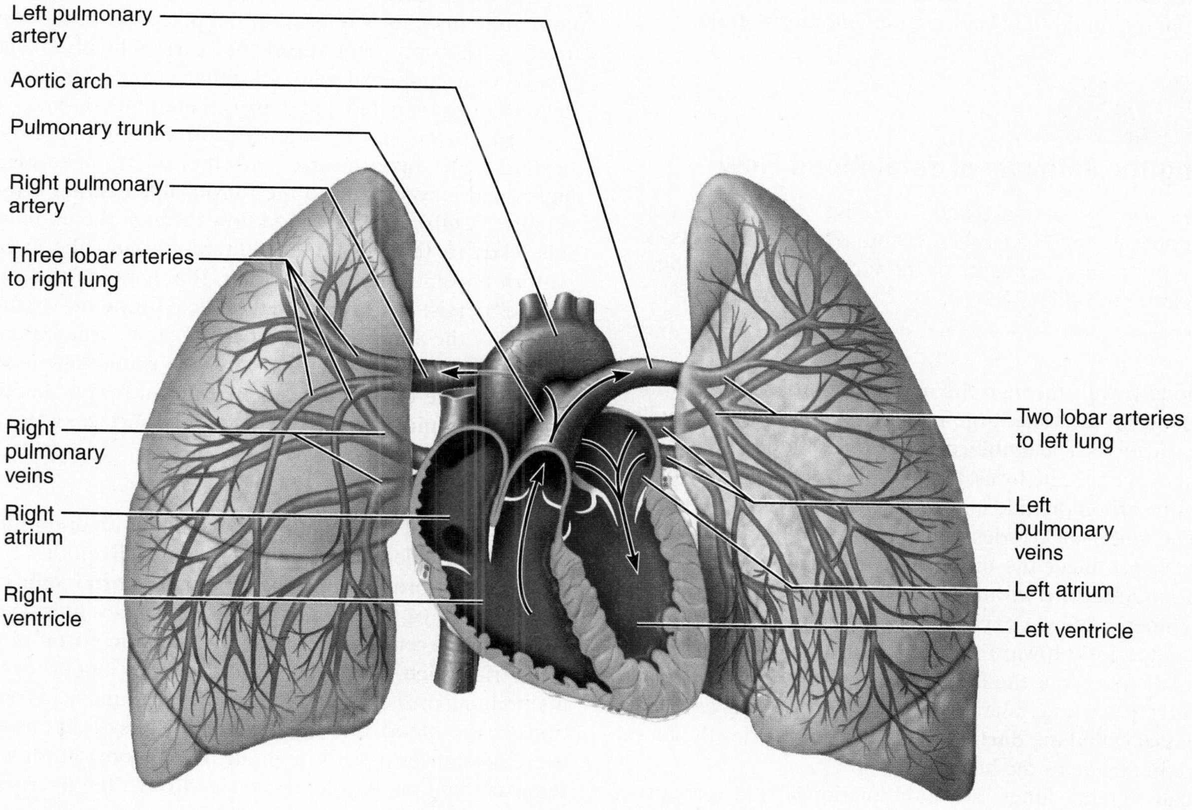

Figure 32.13 The pulmonary circulation. The pulmonary arterial system is shown in blue to indicate that the blood it carries is relatively oxygen-poor. The pulmonary venous drainage is shown in red to indicate that the blood it transports is oxygen-rich.

Instructors may assign this figure as an Art Labeling Activity using Mastering A&P

32

Group Challenge

Fix the Blood Trace

Several artery or vein sequences are listed below. Working in small groups, decide whether each sequence of blood vessels is correct or whether a blood vessel is missing. If correct, simply write "all correct." If incorrect, list the missing vessel, and draw an insertion mark ("v") on the arrow to indicate where the vessel would be located in the sequence. Refrain from using a figure or other reference to help with your decision. Instead, try to find the answers by working together. *Note:* The missing vessel will not be at the beginning or end of the sequence.

1. aortic arch → R. subclavian artery → R. axillary artery → R. brachial artery → R. radial artery → R. superficial palmar arch

____

2. abdominal aorta → R. common iliac artery → R. femoral artery → R. popliteal artery → R. anterior tibial artery → R. dorsalis pedis artery

____

3. ascending aorta → aortic arch → L. common carotid artery → L. internal carotid artery → L. anterior cerebral artery

____

4. R. median antebrachial vein → R. basilic vein → R. axillary vein → R. brachiocephalic vein → superior vena cava

____

Fetal Circulation

In a developing fetus, the lungs and digestive system are not yet functional, and all nutrient, excretory, and gaseous exchanges occur through the placenta (**Figure 32.14a**). Nutrients and oxygen move across placental barriers from the mother's blood into fetal blood, and carbon dioxide and other metabolic wastes move from the fetal blood supply to the mother's blood.

Activity 5

Tracing the Pathway of Fetal Blood Flow

Trace the pathway of fetal blood flow using Figure 32.14a and an anatomical chart (if available). Locate all the named vessels. Identify the named remnants of the foramen ovale and fetal vessels (refer to Figure 32.14b).

32

Fetal blood travels through the umbilical cord, which contains three blood vessels: one large umbilical vein and two smaller umbilical arteries. The **umbilical vein** carries blood rich in nutrients and oxygen to the fetus; the **umbilical arteries** carry blood laden with carbon dioxide and waste from the fetus to the placenta. The umbilical arteries, which transport blood away from the fetal heart, meet the umbilical vein at the *umbilicus* and wrap around the vein within the cord en route to their placental attachments. Newly oxygenated blood flows in the umbilical vein superiorly toward the fetal heart. Some of this blood perfuses the liver, but the larger proportion is ducted through the relatively nonfunctional liver to the inferior vena cava via a shunt vessel called the **ductus venosus**, which carries the blood to the right atrium of the heart.

Because fetal lungs are nonfunctional and collapsed, two shunting mechanisms ensure that blood almost entirely bypasses the lungs. Much of the blood entering the right atrium is shunted into the left atrium through the **foramen ovale**, a flaplike opening in the interatrial septum. The left ventricle then pumps the blood out the aorta to the systemic circulation. Blood that does enter the right ventricle and is pumped out of the pulmonary trunk encounters a second shunt, the **ductus arteriosus**, a short vessel connecting the pulmonary trunk and the aorta. Because the collapsed lungs present an extremely high-resistance pathway, blood more readily enters the systemic circulation through the ductus arteriosus.

The aorta carries blood to the tissues of the body; this blood ultimately finds its way back to the placenta via the umbilical arteries. The only fetal vessel that carries highly oxygenated blood is the umbilical vein. All other vessels contain varying degrees of oxygenated and deoxygenated blood.

At birth, or shortly after, the foramen ovale closes and becomes the **fossa ovalis**, and the ductus arteriosus collapses and is converted to the fibrous **ligamentum arteriosum** (Figure 32.14b). Lack of blood flow through the umbilical vessels leads to their eventual obliteration, and the circulatory pattern becomes that of the adult. Remnants of the umbilical arteries persist as the **median umbilical ligaments** on the inner surface of the anterior abdominal wall; the occluded umbilical vein becomes the **ligamentum teres** (or **round ligament** of the liver); and the ductus venosus becomes a fibrous band called the **ligamentum venosum** on the inferior surface of the liver.

Hepatic Portal Circulation

Blood vessels of the hepatic portal circulation drain the digestive viscera, spleen, and pancreas and deliver this blood to the liver for processing via the **hepatic portal vein** (formed by the union of the splenic and superior mesenteric veins). If a meal has recently been eaten, the hepatic portal blood will be nutrient-rich. The liver is the key body organ involved in maintaining proper sugar, fatty acid, and amino acid concentrations in the blood, and the hepatic portal system ensures that these substances pass through the liver before entering the systemic circulation. As blood travels through the liver sinusoids, some of the nutrients are removed to be stored or processed in various ways for release to the general circulation. At the same time, the hepatocytes are detoxifying alcohol and other possibly harmful chemicals present in the blood, and the liver's macrophages are removing bacteria and other debris from the passing blood. The liver in turn is drained by the hepatic veins that enter the inferior vena cava.

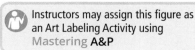

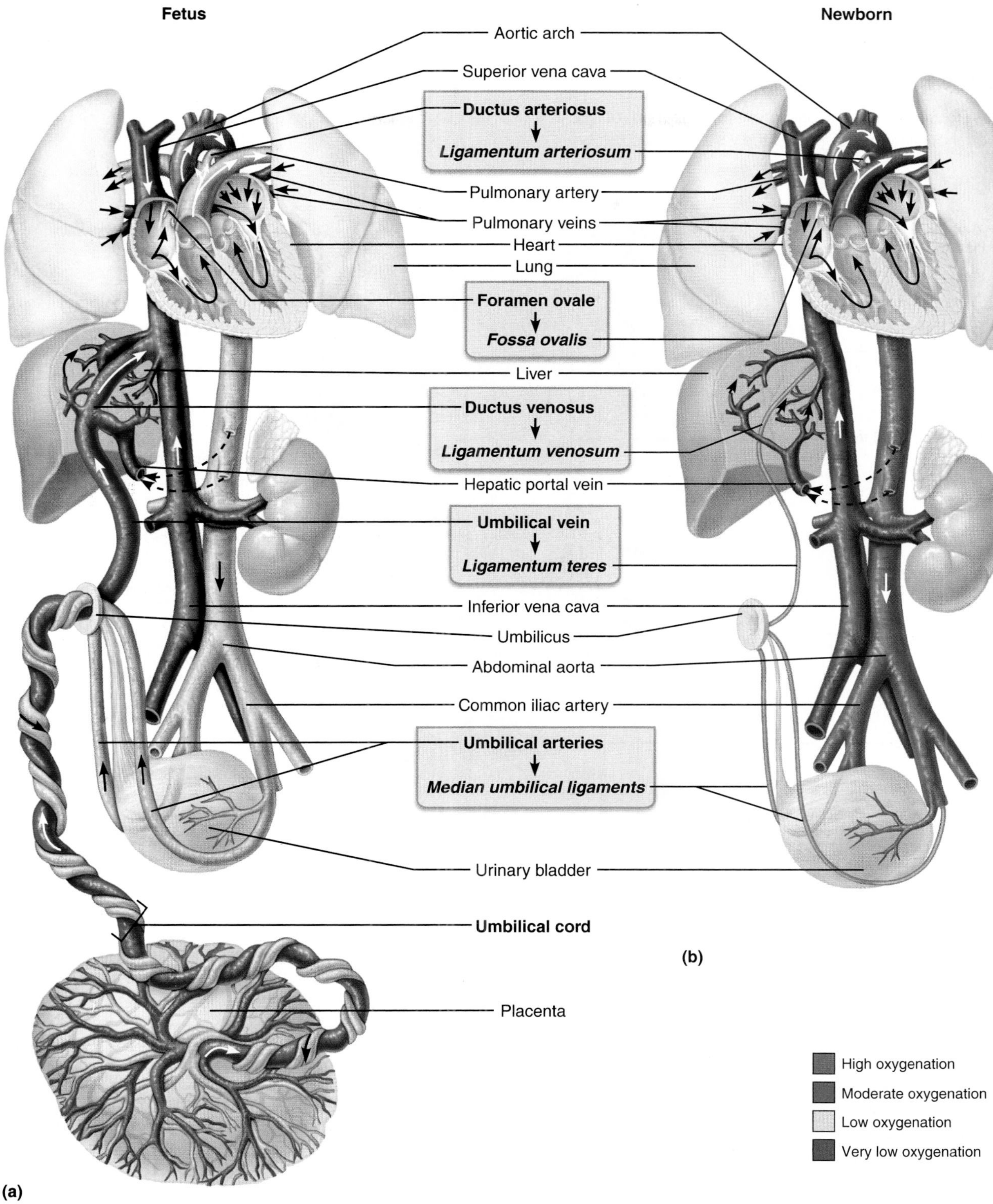

Figure 32.14 Circulation in the fetus and newborn. Arrows on blood vessels indicate direction of blood flow. Arrows in the blue boxes go from the fetal structure to what it becomes after birth (the postnatal structure). **(a)** Special adaptations for embryonic and fetal life. The umbilical vein (red) carries oxygen- and nutrient-rich blood from the placenta to the fetus. The umbilical arteries (pink) carry waste-laden blood from the fetus to the placenta. **(b)** Changes in the cardiovascular system at birth. The umbilical vessels are occluded, as are the liver and lung bypasses (ductus venosus and ductus arteriosus, and the foramen ovale).

Instructors may assign this figure as an Art Labeling Activity using Mastering A&P

Activity 6

Tracing the Hepatic Portal Circulation

Locate on **Figure 32.15** and on an anatomical chart of the hepatic portal circulation (if available), the vessels named below.

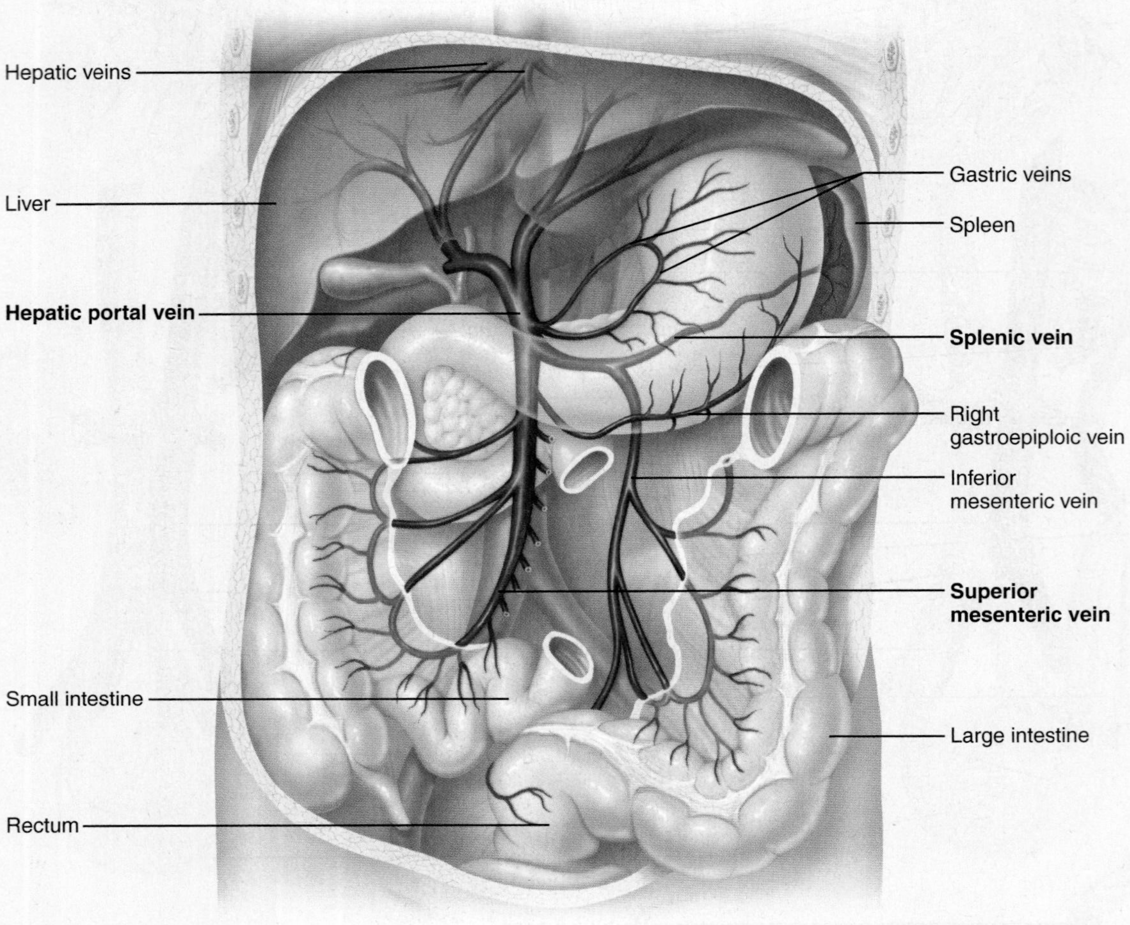

Figure 32.15 Hepatic portal circulation.

The **splenic vein** carries blood from the spleen, parts of the pancreas, and the stomach. The splenic vein unites with the **superior mesenteric vein** to form the hepatic portal vein. The superior mesenteric vein drains the small intestine, part of the large intestine, and the stomach. The **inferior mesenteric vein**, which drains the distal portion of the large intestine and rectum, empties into the splenic vein just before the splenic vein merges with the superior mesenteric vein.

For instructions on animal dissections, see the dissection exercises (starting on p. 695) in the cat and fetal pig editions of this manual.

32 REVIEW SHEET

EXERCISE

Anatomy of Blood Vessels

Instructors may assign a portion of the Review Sheet questions using Mastering A&P™

Name ______________________ Lab Time/Date ______________________

Microscopic Structure of the Blood Vessels

1. Cross-sectional views of an artery and of a vein are shown here. Identify each; on the lines to the sides, note the structural details that enabled you to make these identifications:

______________ (vessel type)

(a) ______________

(b) ______________

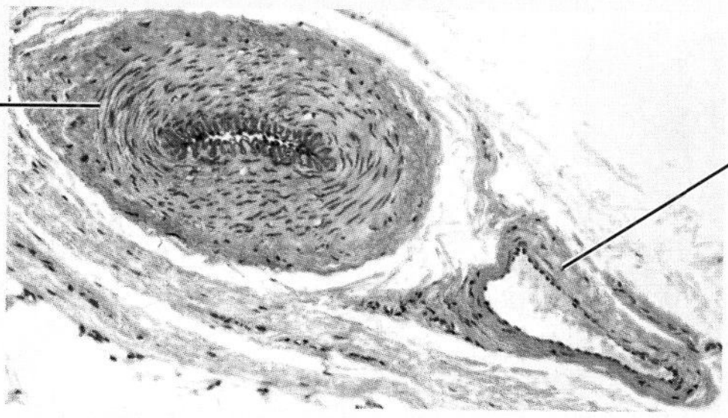

______________ (vessel type)

(a) ______________

(b) ______________

Now describe each tunic more fully by selecting its characteristics from the key below and placing the appropriate key letters on the answer lines.

Tunica intima __________ Tunica media __________ Tunica externa __________

Key:

a. innermost tunic
b. most superficial tunic
c. thin tunic of capillaries
d. regulates blood vessel diameter
e. contains smooth muscle and elastin
f. has a smooth surface to decrease resistance to blood flow

2. Describe the role that valves play in returning blood to the heart. ______________________

____

____

3. Name two events *occurring within the body* that aid in venous return.

______________________ and ______________________

4. Considering their functional differences, why do you think the walls of arteries are proportionately thicker than those of the corresponding veins? ______________________

____

____

Major Systemic Arteries and Veins of the Body

5. Use the key on the right to identify the arteries or veins described on the left.

______ 1. vessel that is paired in the venous system, but only a single vessel is present in the arterial system

______ 2. these arteries supply the myocardium

______, ______ 3. two paired arteries serving the brain

______ 4. vein that runs between the cephalic and basilic veins

______ 5. artery on the dorsum of the foot

______ 6. main artery that serves the thigh muscles

______ 7. supplies the diaphragm

______ 8. formed by the union of the radial and ulnar veins

______, ______ 9. two superficial veins of the arm

______ 10. artery serving the kidney

______ 11. veins draining the liver

______ 12. artery that supplies the distal half of the large intestine

______ 13. divides into the external and internal carotid arteries

______ 14. what the external iliac artery becomes on entry into the thigh

______ 15. drains blood from the spleen, pancreas, and part of the stomach

______ 16. supplies most of the small intestine

______ 17. join to form the inferior vena cava

______ 18. an arterial trunk that has three major branches, which run to the liver, spleen, and stomach

______ 19. major artery serving the tissues external to the skull

______, ______, ______, ______ 20. four veins serving the leg

______ 21. artery generally used to take the pulse at the wrist

Key:
- a. anterior tibial
- b. basilic
- c. brachial
- d. brachiocephalic
- e. celiac trunk
- f. cephalic
- g. common carotid
- h. common iliac
- i. coronary
- j. deep artery of the thigh
- k. dorsalis pedis
- l. external carotid
- m. femoral
- n. fibular
- o. great saphenous
- p. hepatic
- q. inferior mesenteric
- r. internal carotid
- s. median cubital
- t. phrenic
- u. posterior tibial
- v. radial
- w. renal
- x. splenic
- y. superior mesenteric
- z. vertebral

6. What is the function of the cerebral arterial circle?

____

7. The anterior and middle cerebral arteries arise from the ______________________ artery.

They serve the ______________________ of the brain.

8. Trace the pathway of a drop of blood from the aorta to the left occipital lobe of the brain, noting all structures through which it flows. ____

____

9. The human arterial and venous systems are diagrammed on this page and the next. Identify all indicated blood vessels.

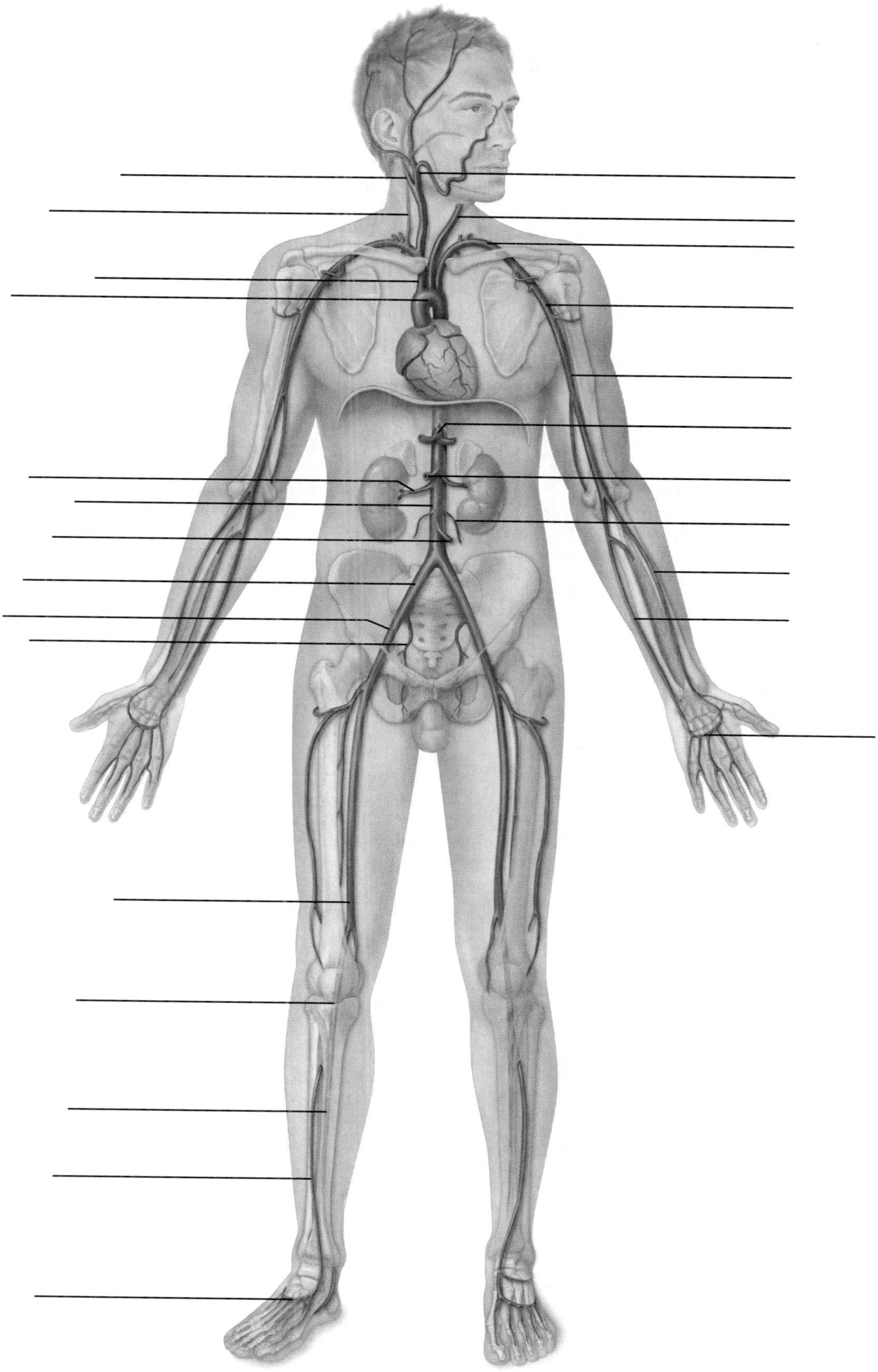

Arteries

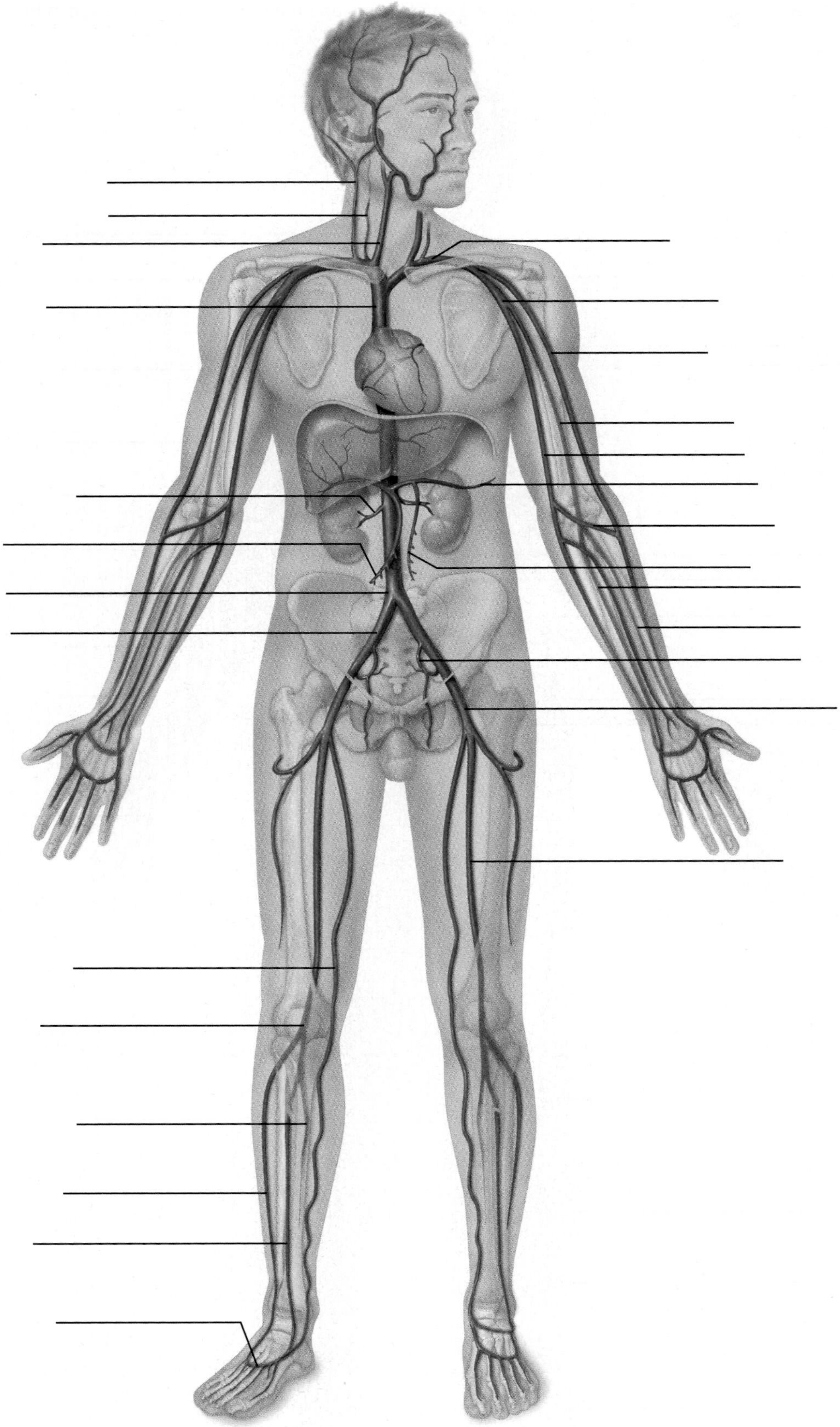

Veins

10. Trace the blood flow from the mitral valve to the tricuspid valve by way of the great toe: ______

______

______

______

______

Pulmonary Circulation

11. Trace the pathway of a carbon dioxide gas molecule in the blood from the inferior vena cava until it leaves the bloodstream. Name all structures (vessels, heart chambers, and others) passed through en route.

______

______

______

12. Most arteries of the adult body carry oxygen-rich blood, and the veins carry oxygen-poor blood.

How does this differ in the pulmonary arteries and veins? ______

______

13. How do the arteries of the pulmonary circulation differ structurally from the systemic arteries? What condition is indicated by this anatomical difference? ______

______

Fetal Circulation

14. For each of the following structures, first indicate its function in the fetus; then, note its fate (what happens to it or what it is converted to after birth). **Circle the blood vessel that carries the most oxygen-rich blood**.

Structure	Function in fetus	Fate and postnatal structure
Umbilical artery		
Umbilical vein		
Ductus venosus		
Ductus arteriosus		
Foramen ovale		

Hepatic Portal Circulation

15. Why is the blood that drains into the hepatic portal circulation nutrient-rich? ______________________________

____

16. Why is this blood carried to the liver before it enters the systemic circulation? ______________________________

____

17. The hepatic portal vein is formed by the union of the ____________________ and the ____________________.

The ____________________ vein carries blood from the ____________________, ____________________ and ____________________.

The ____________________ vein drains the ____________________, ____________________, and ____________________.

The ____________________ vein empties into the splenic vein and drains the ____________________ and ____________________.

18. Trace the flow of a drop of blood from the small intestine to the right atrium of the heart, noting all structures encountered

or passed through on the way. ____

____

19. ➕ A peripherally inserted central catheter (PICC) line involves the use of a long, thin tube to deliver medications or nutrients to a patient. For adult patients, it is usually inserted into the right cephalic vein. Trace the route that a medication would take to

reach the right atrium. List all vessels on the route. ______________________________

____

____

20. ➕ A patient with iliofemoral deep vein thrombosis (IFDVT) has a blood clot located in the femoral or external iliac vein. Such a patient is at risk of the clot traveling to the lungs, resulting in a pulmonary embolism. Trace the route of the clot from the femoral

vein to the pulmonary artery. List all vessels on the route. ______________________________

____

____

33 EXERCISE

Human Cardiovascular Physiology: Blood Pressure and Pulse Determinations

Learning Outcomes

- Define *systole*, *diastole*, and *cardiac cycle*.
- Indicate the normal length of the cardiac cycle, the relative pressure changes occurring within the atria and ventricles, the timing of valve closure, and the volume changes occurring in the ventricles during the cycle.
- Correlate the events of the ECG with the events of the cardiac cycle.
- Use the stethoscope to auscultate heart sounds, relate heart sounds to cardiac cycle events, and describe the clinical significance of heart murmurs.
- Demonstrate thoracic locations where the first and second heart sounds are most accurately auscultated.
- Define *pulse*, *pulse pressure*, *pulse deficit*, *blood pressure*, *sounds of Korotkoff*, and *mean arterial pressure (MAP)*.
- Determine a subject's apical and radial pulse.
- Determine a subject's blood pressure with a sphygmomanometer, and relate systolic and diastolic pressures to events of the cardiac cycle.
- Compare the value of venous pressure to systemic blood pressure, and describe how venous pressure is measured.
- Investigate the effects of exercise on blood pressure, pulse, and cardiovascular fitness.
- Indicate factors affecting blood flow and skin color.

Pre-Lab Quiz

Instructors may assign these and other Pre-Lab Quiz questions using Mastering A&P™

1. A graph illustrating the pressure and volume changes during one heartbeat is called the:
   a. blood pressure
   b. cardiac cycle
   c. conduction system of the heart
   d. electrical events of the heartbeat
2. Circle True or False. When ventricular systole begins, intraventricular pressure increases rapidly, closing the atrioventricular (AV) valves.
3. The term _________ refers to the alternating surges of pressure in an artery that occur with each contraction and relaxation of the left ventricle.
   a. diastole
   b. pulse
   c. murmur
   d. systole
4. Circle the correct underlined term. The pulse is most often taken at the lateral aspect of the wrist, above the thumb, by compressing the <u>popliteal</u> / <u>radial</u> artery.
5. In a blood pressure reading of 120/90, which number represents the *diastolic* pressure? ___________________

Go to Mastering A&P™ > Study Area to improve your performance in A&P Lab.

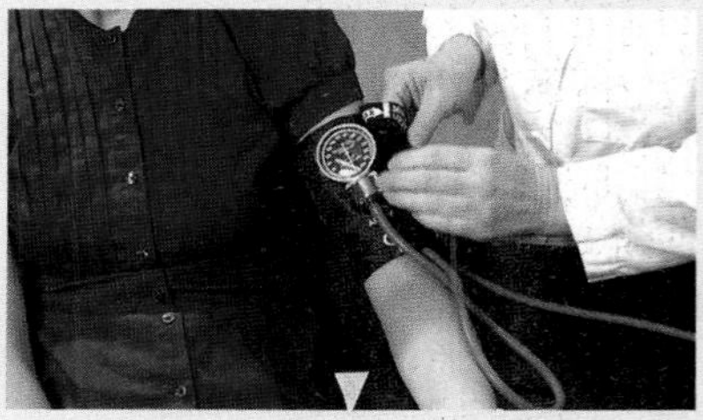

> Lab Tools > Pre-Lab Videos > Blood Pressure

Instructors may assign new Building Vocabulary coaching activities, Pre-Lab Quiz questions, Art Labeling activities, Pre-Lab Video Coaching Activities for Auscultating Heart Sounds, Measuring Pulses, and Measuring Blood Pressure, Practice Anatomy Lab Practical questions (PAL), PhysioEx activities, and more using the Mastering A&P™ Item Library.

Materials

- Recording of "Interpreting Heart Sounds" (if available on free loan from the local chapters of the American Heart Association) or any of the suitable Internet sites on heart sounds
- Stethoscope
- Alcohol swabs
- Watch (or clock) with second hand

BIOPAC® BSL System with BSL software version 4.0.1 and above (for Windows 10/8.x/7 or Mac OS X 10.9–10.12), data acquisition unit MP36/35, PC or Mac computer, BIOPAC pulse plethysmograph.

Text continues on next page. →

Instructors using the MP36/35/30 data acquisition unit with BSL software versions earlier than 4.0.1 (for Windows or Mac) will need slightly different channel settings and collection strategies. Instructions for using the older data acquisition unit can be found on MasteringA&P.

- Sphygmomanometer
- Felt marker
- Meter stick
- Cot (if available)
- Step stools (0.4 m [16 in.] and 0.5 m [20 in.] in height)
- Small basin suitable for the immersion of one hand
- Ice
- Laboratory thermometer

PEx PhysioEx™ 9.1 Computer Simulation Ex. 5 on p. PEx-75

Note: *Instructions for using PowerLab® equipment can be found on MasteringA&P.*

During this lab, we will conduct investigations of a few phenomena such as pulse, heart sounds, and blood pressure, all of which reflect the heart in action and the function of blood vessels. A discussion of the cardiac cycle will provide a basis for understanding and interpreting the various physiological measurements taken.

Cardiac Cycle

In a healthy heart, the two atria contract simultaneously. As they begin to relax, the ventricles contract simultaneously. According to general usage, the terms **systole** and **diastole** refer to events of ventricular contraction and relaxation, respectively. The **cardiac cycle** is equivalent to one complete heartbeat—during which both atria and ventricles contract and then relax. It is marked by a succession of changes in blood volume and pressure within the heart.

To better understand the events of the cardiac cycle, read the descriptions for a given phase in **Table 33.1**, and correlate each phase to the electrical events, heart sounds, pressure changes in the left side of the heart, and volume changes in the left side of the heart depicted in **Figure 33.1**. We begin with the heart in total relaxation, mid-to late diastole; however, you could start anywhere within the cycle.

Assuming the average heart beats approximately 75 beats per minute, the length of the cardiac cycle is about 0.8 second. Of this time period, atrial contraction occupies the first 0.1 second, which is followed by atrial relaxation and ventricular contraction for the next 0.3 second. The remaining 0.4 second is a period of total heart relaxation, the **quiescent period**.

Notice that two different types of events control the movement of blood through the heart: the alternate contraction and relaxation of the chambers, and the opening and closing of valves, which is entirely dependent on the pressure changes within the heart chambers.

Table 33.1 Events of the Cardiac Cycle

Phase	Description of events	Status of ventricles	Status of atria	Status of AV valves	Status of SL valves
① Ventricular filling (passive)	When atrial pressure is greater than ventricular pressure, the AV valves are forced open, and blood flows passively into the atria and on through to the ventricles.	Relaxed (diastole)	Relaxed (atrial diastole)	Opened	Closed
① Ventricular filling with atrial contraction	The atria contract to complete the filling of the ventricles. Ventricular diastole ends, and so the end diastolic volume (EDV) of the ventricles is achieved.	Relaxed (diastole)	Contracted (atrial systole)	Opened	Closed
② Isovolumetric contraction	The contraction of the ventricles begins, and ventricular pressure increases, closing the AV valves.	Contracted (systole)	Relaxed (atrial diastole)	Closed	Closed
③ Ventricular ejection	Ventricular pressure continues to rise; when the pressure in the ventricles exceeds the pressure in the great vessels exiting the heart, the SL valves open, and blood is ejected.	Contracted (systole)	Relaxed (atrial diastole)	Closed	Opened
④ Isovolumetric relaxation	The ventricles relax, decreasing the pressure in the ventricles; the decrease in pressure causes the SL valves to close. The **dicrotic notch** is the result of a pressure fluctuation that occurs when the aortic valve snaps shut.	Relaxed (diastole)	Relaxed (atrial diastole)	Closed	Closed

33

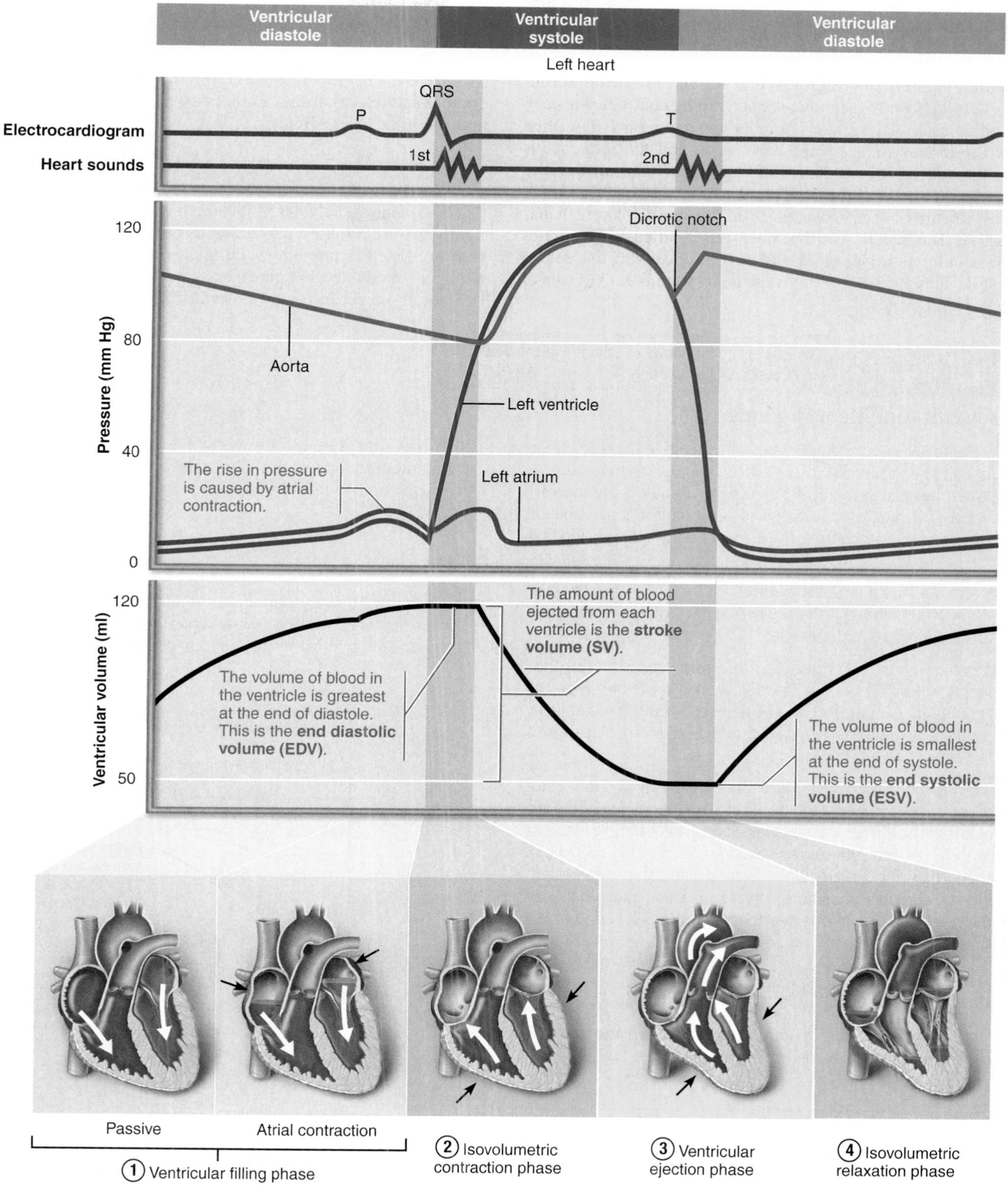

Figure 33.1 Summary of events occurring in the heart during the cardiac cycle.

Heart Sounds

Sounds heard in the cardiovascular system result from turbulent blood flow. Two distinct sounds can be heard during each cardiac cycle. These heart sounds are commonly described by the monosyllables "lub" and "dup"; and the sequence is designated lub-dup, pause, lub-dup, pause, and so on. The first heart sound (lub) is referred to as S_1 and is associated with closure of the AV valves at the beginning of ventricular systole. The second heart sound (dup), called S_2, occurs as the semilunar valves close and corresponds with the end of systole. Figure 33.1 correlates the heart sounds with events of the cardiac cycle.

☐ Listen to the recording "Interpreting Heart Sounds" or another suitable recording so that you may hear both normal and abnormal heart sounds.

Check the box when you have completed the task.

Abnormal heart sounds are called **murmurs** and often indicate valvular problems. In valves that do not close tightly, closure is followed by a swishing sound due to the backflow of blood (regurgitation). Distinct sounds, often described as high-pitched screeching, are associated with the tortuous flow of blood through constricted, or stenosed, valves. ✚

Activity 1

Instructors may assign a related Pre-Lab Video Coaching Activity using Mastering A&P

Auscultating Heart Sounds

In the following procedure, you will auscultate (listen to) your partner's heart sounds with a stethoscope. Several more sophisticated heart-sound amplification systems are on the market, and your instructor may prefer to use one if it is available. If so, directions for the use of this apparatus will be provided by the instructor.

1. Obtain a stethoscope and some alcohol swabs. Heart sounds are best auscultated if the subject's outer clothing is removed, so a male subject is preferable.

2. With an alcohol swab, clean the earpieces of the stethoscope. Allow the alcohol to dry. Notice that the earpieces are angled. For comfort and best auscultation, the earpieces should be angled in a *forward* direction when placed into the ears.

3. Don the stethoscope. Place the diaphragm of the stethoscope on your partner's thorax, just to the sternal side of the left nipple at the fifth intercostal space, and listen carefully for heart sounds. The first sound will be a longer, louder (more booming) sound than the second, which is short and sharp. After listening for a couple of minutes, try to time the pause between the second sound of one heartbeat and the first sound of the subsequent heartbeat.

How long is this interval? ______________ sec

How does it compare to the interval between the first and second sounds of a single heartbeat?

____

4. To differentiate individual valve sounds somewhat more precisely, auscultate the heart sounds over specific thoracic regions. (Refer to **Figure 33.2** for positioning of the stethoscope.)

Auscultation of AV Valves

As a rule, the mitral valve closes slightly before the tricuspid valve. You can hear the mitral valve more clearly if you place the stethoscope over the apex of the heart, which is at the fifth intercostal space, approximately in line with the middle region of the left clavicle. Listen to the heart sounds at this region; then move the stethoscope medially to the right margin of the sternum to auscultate the tricuspid valve. Can you detect the slight lag between the closure of the mitral and tricuspid valves?

____

There are normal variations in the site for "best" auscultation of the tricuspid valve. These range from the right sternal margin over the fifth intercostal space (depicted in Figure 33.2)

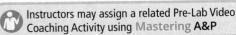

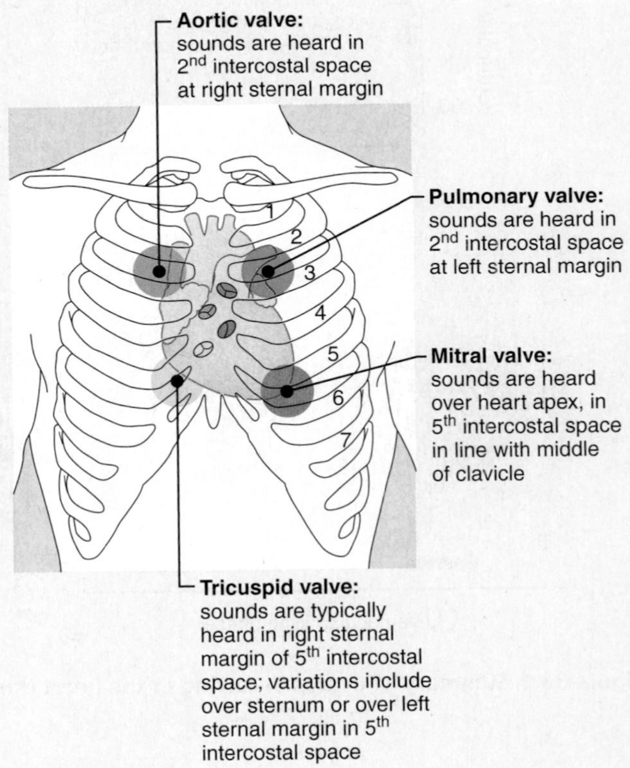

Figure 33.2 Areas of the thorax where heart sounds can best be detected.

to over the sternal body in the same plane, to the left sternal margin over the fifth intercostal space. If you have difficulty hearing closure of the tricuspid valve, try one of these other locations.

Auscultation of Semilunar Valves

Again there is a slight dyssynchrony of valve closure; the aortic valve normally snaps shut just ahead of the pulmonary valve. If the subject inhales deeply but gently, filling of the right ventricle (due to decreased intrapulmonary pressure) and closure of the pulmonary valve will be delayed slightly. The two sounds can therefore be heard more distinctly.

Position the stethoscope over the second intercostal space, just to the *right* of the sternum. The aortic valve is best heard at this position. As you listen, have your partner take a deep breath. Then move the stethoscope to the *left* side of the sternum in the same line, and auscultate the pulmonary valve. Listen carefully; try to hear the "split" between the closure of these two valves in the second heart sound.

Although at first it may seem a bit odd that the pulmonary valve issuing from the *right* heart is heard most clearly to the *left* of the sternum and the aortic valve of the left heart is best heard at the right sternal border, this is easily explained by reviewing heart anatomy.

The Pulse

The term **pulse** refers to the alternating surges of pressure (expansion and then recoil) in an artery that occur with each contraction and relaxation of the left ventricle. This difference between systolic and diastolic pressure is called the **pulse pressure**. (See p. 490.) Normally the pulse rate (pressure surges per minute) equals the heart rate (beats per minute), and the pulse averages 70 to 76 beats per minute in the resting state.

Parameters other than pulse rate are also useful clinically. You may also assess the regularity (or rhythmicity) of the pulse and its amplitude and/or tension—does the blood vessel expand and recoil (sometimes visibly) with the pressure waves? Can you feel it strongly, or is it difficult to detect? Is it regular like the ticking of a clock, or does it seem to skip beats?

Activity 2

Instructors may assign a related Pre-Lab Video Coaching Activity using Mastering A&P

Palpating Superficial Pulse Points

The pulse may be felt easily on any superficial artery when the artery is compressed over a bone or firm tissue. Palpate the following pulse or pressure points on your partner by placing the fingertips of the first two or three fingers of one hand over the artery. It helps to compress the artery firmly as you begin your palpation and then immediately ease up on the pressure slightly. In each case, notice the regularity of the pulse, and assess the degree of tension or amplitude. (**Figure 33.3** illustrates the superficial pulse points to be palpated.) Check off the boxes as you locate each pulse point.

- ☐ **Superficial temporal artery:** Anterior to the ear, in the temple region.
- ☐ **Facial artery:** Clench the teeth, and palpate the pulse just anterior to the masseter muscle on the mandible (in line with the corner of the mouth).
- ☐ **Common carotid artery:** At the side of the neck.
- ☐ **Brachial artery:** In the cubital fossa, at the point where it bifurcates into the radial and ulnar arteries.
- ☐ **Radial artery:** At the lateral aspect of the wrist, above the thumb.
- ☐ **Femoral artery:** In the groin.
- ☐ **Popliteal artery:** At the back of the knee.
- ☐ **Posterior tibial artery:** Just above the medial malleolus.
- ☐ **Dorsalis pedis artery:** On the dorsum of the foot.

Instructors may assign this figure as an Art Labeling Activity using Mastering A&P

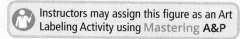

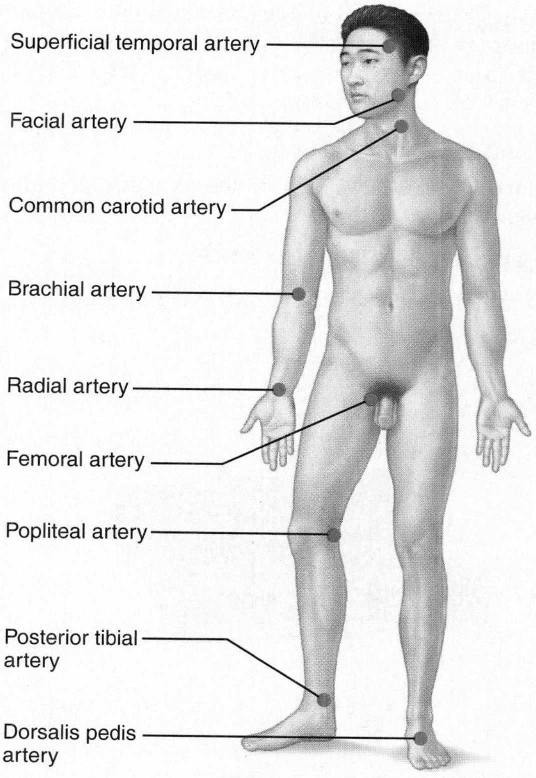

Figure 33.3 Body sites where the pulse is most easily palpated.

Text continues on next page. →

Which pulse point had the greatest amplitude?

____

Which had the least? ______________________________

Can you offer any explanation for this? ______________

____

____

____

Because of its easy accessibility, the pulse is most often taken on the radial artery. With your partner sitting quietly, practice counting the radial pulse for 1 minute. Make three counts, and average the results.

count 1 ____________ count 2 ____________

count 3 ____________ average ____________

Activity 3

Measuring Pulse Using BIOPAC®

Because of the elasticity of the arteries, blood pressure decreases and smooths out as blood moves farther away from the heart. A pulse, however, can still be felt in the fingers. A device called a plethysmograph or a piezoelectric pulse transducer can measure this pulse.

Setting Up the Equipment

33

1. Connect the BIOPAC® unit to the computer, and turn the computer **ON**.
2. Make sure the BIOPAC® unit is **OFF**.
3. Plug in the equipment (as shown in **Figure 33.4**).

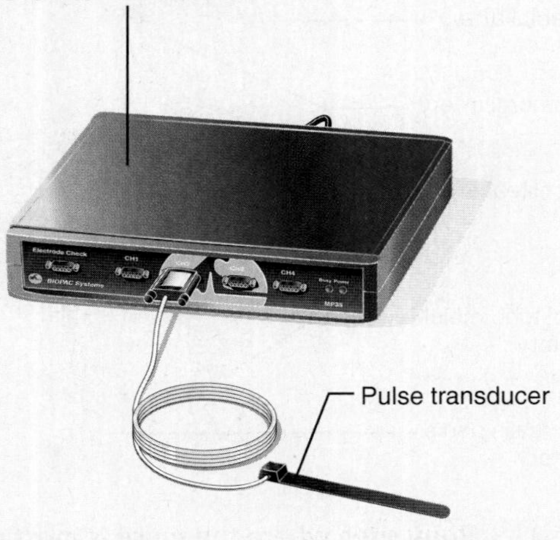

Figure 33.4 Setting up the BIOPAC®equipment. Plug the pulse tranducer into Channel 2.

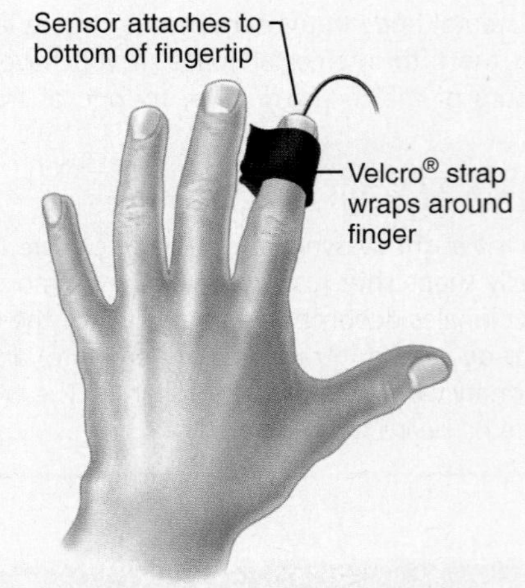

Figure 33.5 Placement of the pulse transducer around the tip of the index finger.

- Pulse transducer (plethysmograph)—CH 2

4. Turn the BIOPAC® unit **ON**.
5. Wrap the pulse transducer around the tip of the index finger (as shown in **Figure 33.5**). Wrap the Velcro® around the finger gently (if wrapped too tightly, it will reduce circulation to the finger and obscure the recording). *Do not wiggle the finger or move the plethysmograph cord during recording.*
6. Have the subject sit down, with the forearms supported and relaxed.
7. Start the Biopac Student Lab program on the computer by double-clicking the icon on the desktop or by following your instructor's guidance.
8. Select lesson **L07-ECG & Pulse** from the menu, and click **OK**.
9. Type in a filename that will save this subject's data on the computer hard drive. You may want to use the subject's last name followed by Pulse (for example, SmithPulse-1), then click **OK**.

Calibrating the Equipment

1. When the subject is relaxed, click **Calibrate**. Click **Ignore** when prompted for SS2L on Ch.1. The calibration will stop automatically after 8 seconds.
2. Observe the recording of the calibration data, which should look like the waveforms in the example (**Figure 33.6**). ECG data is not recorded in this activity.
   - If the data look very different, click **Redo Calibration** and repeat the steps above. If there is no signal, you may need to loosen the Velcro® on the finger and check all attachments.
   - If the data look similar, proceed to the next section.

Recording the Data

1. When the subject is ready, click **Record** to begin recording the pulse. After 30 seconds, click **Suspend**.
2. Observe the data, which should look similar to the pulse data example (**Figure 33.7**).

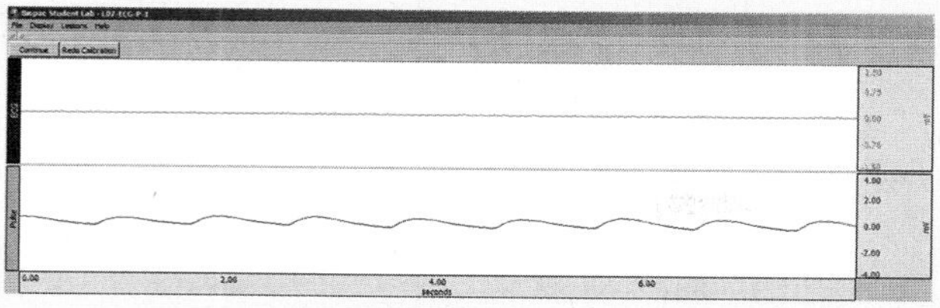

Figure 33.6 Example of waveforms during the calibration procedure.

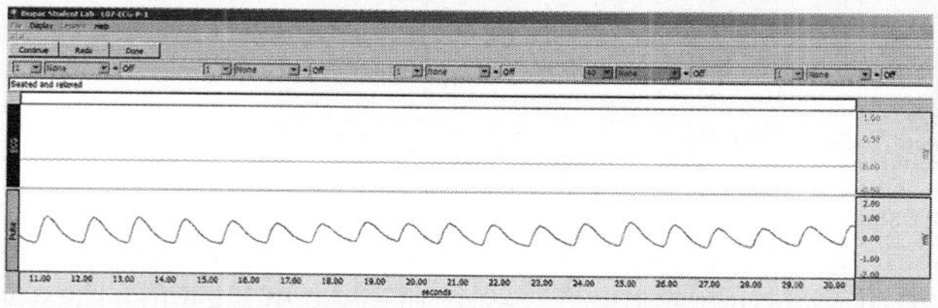

Figure 33.7 Example of waveforms during the recording of data.

- If the data look very different, click **Redo** and repeat the steps above.
- If the data look similar, go to the next step.

3. When you are finished, click **Done** and then click **Yes**. A pop-up window will appear; to record from another subject, select **Record from another subject** and return to step 5 under Setting Up the Equipment. If continuing to the Data Analysis section, select **Analyze current data file** and proceed to step 2 in the Data Analysis section.

Data Analysis

1. If just starting the BIOPAC® program to perform data analysis, enter **Review Saved Data** mode and choose the file with the subject's pulse data (for example, SmithPulse-1).

2. Observe that pulse data is in the lower scale.

3. To analyze the data, set up the first channel/measurement box at the top of the screen.

Channel	Measurement	Data
CH 40	Delta T	Pulse

Delta T: Measures the time elapsed in the selected area

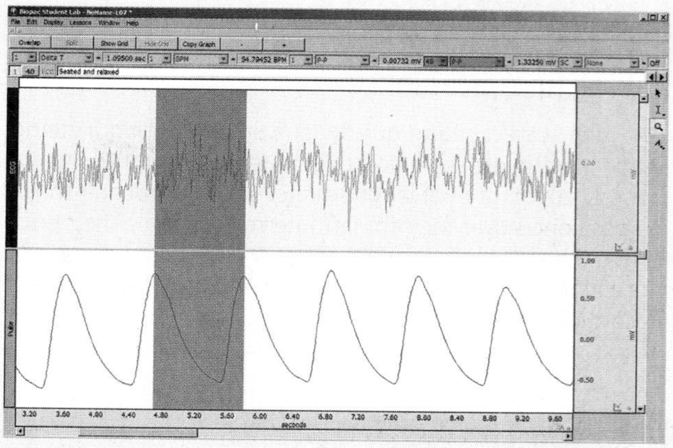

Figure 33.8 Using the I-beam cursor to highlight data for analysis.

4. Use the arrow cursor and click the I-beam cursor box on the lower right side of the screen to activate the "area selection" function. Using the activated I-beam cursor, highlight from the peak of one pulse to the peak of the next pulse (as shown in **Figure 33.8**).

Observe the elapsed time between heartbeats and record here

(to the nearest 0.01 second): ____________________

5. Calculate the beats per minute by inserting the elapsed time into this formula:

(1 beat/ ____________ sec) × (60 sec/min)

= ____________ beats/min

Optional Activity with BIOPAC® Pulse Measurement

To expand the experiment, you can measure the effects of heat and cold on pulse rate. To do this, you can submerge the subject's hand (the hand without the plethysmograph!) in hot and/or ice water for 2 minutes, and then record the pulse. Alternatively, you can investigate change in pulse rate after brief exercise such as jogging in place.

Apical-Radial Pulse

The correlation between the apical and radial pulse rates can be determined by simultaneously counting them. The **apical pulse** (actually the counting of heartbeats) may be slightly faster than the radial because of a slight lag in time as the blood rushes from the heart into the large arteries where it can be palpated. A difference between the values observed is referred to as a **pulse deficit**. A large difference may indicate cardiac impairment (a weakened heart that is unable to pump blood into the arterial tree to a normal extent), low cardiac output, or abnormal heart rhythms. Apical pulse counts are routinely ordered for patients with cardiac decompensation.

Activity 4

Taking an Apical Pulse

With the subject sitting quietly, one student, using a stethoscope, determines the apical pulse rate while another simultaneously counts the radial pulse rate. The stethoscope should be positioned over the fifth left intercostal space. The person taking the radial pulse should determine the starting point for the count and give the stop-count signal exactly 1 minute later. Record your values at right.

apical count ______________ beats/min

radial count ______________ pulses/min

pulse deficit ______________ pulses/min

Blood Pressure Determinations

Blood pressure (BP) is defined as the pressure the blood exerts against any unit area of the blood vessel walls, and it is generally measured in the arteries. Because the heart alternately contracts and relaxes, the resulting rhythmic flow of blood into the arteries causes the blood pressure to rise and fall during each beat. Thus you must take two blood pressure readings: the **systolic pressure**, which is the pressure in the arteries at the peak of ventricular contraction, and the **diastolic pressure**, which reflects the pressure during ventricular relaxation. Blood pressures are reported in millimeters of mercury (mm Hg), with the systolic pressure appearing first; 120/80 translates to 120 over 80, or a systolic pressure of 120 mm Hg and a diastolic pressure of 80 mm Hg. Normal blood pressure varies considerably from one person to another.

In this procedure, you will measure arterial pressure by indirect means and under various conditions. You will investigate and demonstrate factors affecting blood pressure and the rapidity of blood pressure changes.

Activity 5

Instructors may assign a related Pre-Lab Video Coaching Activity using Mastering A&P

Using a Sphygmomanometer to Measure Arterial Blood Pressure Indirectly

The **sphygmomanometer**, commonly called a *blood pressure cuff*, is an instrument used to obtain blood pressure readings by the auscultatory method (**Figure 33.9**). It consists of an inflatable cuff with an attached pressure gauge. The cuff is placed around the arm and inflated to a pressure higher than systolic pressure to occlude circulation to the forearm. As cuff pressure is gradually released, the examiner listens with a stethoscope for characteristic sounds called the **sounds of Korotkoff**, which indicate the resumption of blood flow into the forearm.

1. Work in pairs to obtain radial artery blood pressure readings. Obtain a felt marker, stethoscope, alcohol swabs, and a sphygmomanometer. Clean the earpieces of the stethoscope with the alcohol swabs, and check the cuff for the presence of trapped air by compressing it against the laboratory table. (A partially inflated cuff will produce erroneous measurements.)

2. The subject should sit in a comfortable position with one arm resting on the laboratory table (approximately at heart level if possible). Wrap the cuff around the subject's arm, just above the elbow, with the inflatable area on the medial arm surface. The cuff may be marked with an arrow; if so, the arrow should be positioned over the brachial artery (Figure 33.9). Secure the cuff by tucking the distal end under the wrapped portion or by bringing the Velcro® areas together.

3. Palpate the brachial pulse, and lightly mark its position with a felt pen. Don the stethoscope, and place its diaphragm over the pulse point.

! *Do not keep the cuff inflated for more than 1 minute.* If you have any trouble obtaining a reading within this time, deflate the cuff, wait 1 or 2 minutes, and try again. (A prolonged interference with blood pressure homeostasis can lead to fainting.)

4. Inflate the cuff to approximately 160 mm Hg pressure, and slowly release the pressure valve. Watch the pressure gauge as you listen carefully for the first soft thudding sounds of the blood spurting through the partially occluded artery. Mentally note this pressure (systolic pressure), and continue to release the cuff pressure. You will notice first an increase, then a muffling, of the sound. For the diastolic pressure, note the pressure at which the sound becomes muffled or disappears. Make two blood pressure determinations, and record your results below.

First trial:

systolic pressure ________

diastolic pressure ________

Second trial:

systolic pressure ________

diastolic pressure ________

5. Compute the **pulse pressure** for each trial. The pulse pressure is the difference between the systolic and diastolic pressures, and it indicates the amount of blood forced from the heart during systole, or the actual "working" pressure. A narrowed pulse pressure (less than 30 mm Hg) may be a signal of severe aortic stenosis, constrictive pericarditis, or tachycardia. A

33

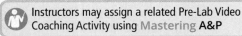

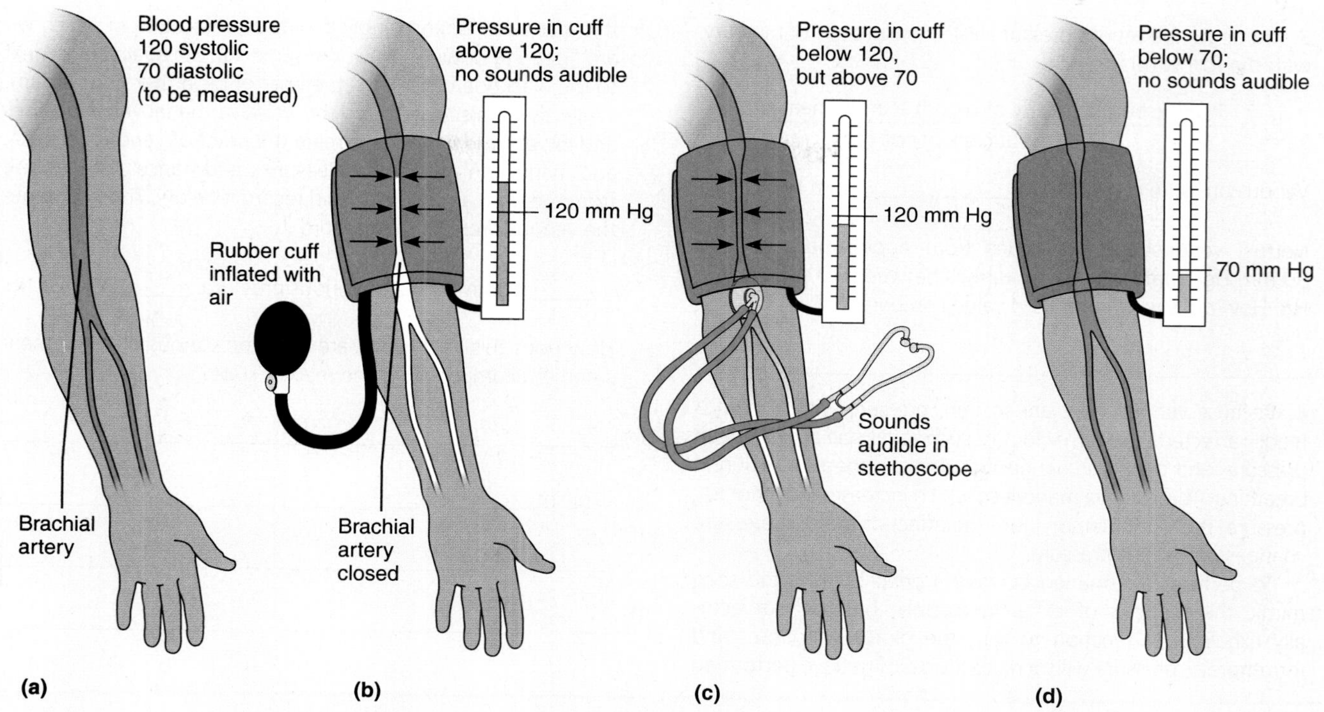

Figure 33.9 **Procedure for measuring blood pressure.** **(a)** The course of the brachial artery of the arm. Assume a blood pressure of 120/70. **(b)** The blood pressure cuff is wrapped snugly around the arm just above the elbow and inflated until blood flow into the forearm is stopped and no brachial pulse can be felt or heard. **(c)** Pressure in the cuff is gradually reduced while the examiner uses a stethoscope to listen (auscultate) for sounds (of Korotkoff) in the brachial artery. The pressure, read as the first soft tapping sounds are heard (the first point at which a small amount of blood is spurting through the constricted artery), is recorded as the systolic pressure. **(d)** As the pressure is reduced still further, the sounds become louder and more distinct, but when the artery is no longer restricted and blood flows freely, the sounds can no longer be heard. The pressure at which the sounds disappear is routinely recorded as the diastolic pressure.

widened pulse pressure (over 40 mm Hg) is common in hypertensive individuals.

Pulse pressure:

first trial _________ second trial _________

6. Compute the **mean arterial pressure (MAP)** for each trial using the following equation:

$$\text{MAP} = \text{diastolic pressure} + \frac{\text{pulse pressure}}{3}$$

first trial ___________ second trial ___________

33

Activity 6

Estimating Venous Pressure

It is not possible to measure venous pressure with the sphygmomanometer. The methods available for measuring it produce estimates at best, because venous pressures are so much lower than arterial pressures. The difference in pressure becomes obvious when these vessels are cut. If a vein is cut, the blood flows evenly from the cut. A lacerated artery produces rapid spurts of blood. In this activity, you will estimate venous pressures.

1. Obtain a meter stick, and ask your lab partner to stand with his or her right side toward the blackboard, arms hanging freely at the sides. On the board, mark the approximate level of the right atrium. (This will be just slightly higher than the point at which you ausculated the apical pulse.)

2. Observe the superficial veins on the dorsum of the right hand as the subject alternately raises and lowers it. Notice the collapsing and filling of the veins as internal pressures change. Have the subject repeat this action until you can determine the point at which the veins have just collapsed. Mark this hand level on the board. Then measure, in millimeters, the distance in the vertical plane from this point to the level of the right atrium (previously marked). Record this value. Distance of right arm from right atrium at point of venous collapse:

_____________ mm

Text continues on next page. →

3. Compute the venous pressure (P_V), in millimeters of mercury, with the following formula:

$$P_V = \frac{1.056 \text{ (specific gravity of blood)} \times \text{mm (measured)}}{13.6 \text{ (specific gravity of Hg)}}$$

Venous pressure computed: ______________ mm Hg

Normal venous pressure varies from approximately 30 to 90 mm Hg. That of the hand ranges between 30 and 40 mm Hg. How does your computed value compare?

____

4. Because venous walls are so thin, pressure within them is readily affected by external factors such as muscle activity, deep pressure, and pressure changes occurring in the thorax during breathing. The Valsalva maneuver, which increases intrathoracic pressure, is used to demonstrate the effect of thoracic pressure changes on venous pressure.

To perform this maneuver, take a deep breath, and then mimic the motions of exhaling forcibly, but without actually exhaling. In reaction to this, the glottis will close; and intrathoracic pressure will increase. (Most of us have performed this maneuver unknowingly in acts of defecation in which we are "bearing down.") Have the same subject again stand next to the blackboard mark for the level of his or her right atrium. While the subject performs the Valsalva maneuver and raises and lowers one hand, determine the point of venous collapse, and mark it on the board. Measure the distance of that mark from the right atrium level, and record it below. Then compute the venous pressure, and record it.

________ mm Venous pressure: ________ mm Hg

How does this value compare with the venous pressure measurement computed for the relaxed state?

____

Explain: ______________________________________

____

____

Activity 7

Observing the Effect of Various Factors on Blood Pressure and Heart Rate

Arterial blood pressure is directly proportional to cardiac output (CO, amount of blood pumped out of the left ventricle per unit time) and total peripheral resistance (TPR) to blood flow. The following equation shows this relationship:

$$BP = CO \times TPR$$

Total peripheral resistance is increased by blood vessel constriction (most importantly the arterioles), by an increase in blood viscosity, and by a loss of elasticity of the arteries (seen in arteriosclerosis). Any factor that increases either the cardiac output or the total peripheral resistance causes an almost immediate reflex rise in blood pressure. A close examination of these relationships reveals that many factors—age, weight, time of day, exercise, body position, emotional state, and various drugs, for example—alter blood pressure. The influence of a few of these factors is investigated here.

33

The following tests are done most efficiently if students work in groups of four: one student acts as the subject; two are examiners (one takes the radial pulse, and the other auscultates the brachial blood pressure); and a fourth student collects and records data. The sphygmomanometer cuff should be left on the subject's arm throughout the experiments (in a deflated state, of course) so that, at the proper times, the blood pressure can be taken quickly. In each case, take the measurements at least twice. For each of the following tests, students should formulate hypotheses, collect data, and write lab reports. (See Getting Started, on MasteringA&P.) Conclusions should be shared with the class.

Posture

To monitor circulatory adjustments to changes in position, take blood pressure and pulse measurements under the conditions noted in the **Activity 7 Posture chart**. Record your results on the chart.

Activity 7: Posture

	Trial 1		Trial 2	
	BP	**Pulse**	**BP**	**Pulse**
Sitting quietly				
Reclining (after 2 to 3 min)				
Immediately on standing from the reclining position ("at attention" stance)				
After standing for 3 min				

Activity 7: Exercise										
			Interval following test							
	Baseline		Immediately		1 min		2 min		3 min	
Harvard step test for 5 min at 30/min	**BP**	**P**	**BP**	**P**	**BP**	**P**	**BP**	**P**	**BP**	**P**
Well-conditioned individual	______	______	______	______	______	______	______	______	______	______
Poorly conditioned individual	______	______	______	______	______	______	______	______	______	______

Exercise

Blood pressure and pulse changes during and after exercise provide a good yardstick for measuring one's overall cardiovascular fitness. Although there are more sophisticated and more accurate tests that evaluate fitness according to a specific point system, the *Harvard step test* described here is a quick way to compare the relative fitness level of a group of people.

You will be working in groups of four, duties assigned as indicated above, except that student 4, in addition to recording the data, will act as the timer and call the cadence.

Any student with a known heart problem should refuse to participate as the subject.

All four students may participate as the subject in turn, if desired, but the bench stepping is to be performed *at least twice* in each group—once with a well-conditioned person acting as the subject, and once with a poorly conditioned subject.

Bench stepping is the following series of movements repeated sequentially:

1. Place one foot on the step.
2. Step up with the other foot so that both feet are on the platform. Straighten the legs and the back.
3. Step down with one foot.
4. Bring the other foot down.

The pace for the stepping will be set by the "timer" (student 4), who will repeat "Up-2-3-4, up-2-3-4" at such a pace that each "up-2-3-4" sequence takes 2 sec (30 cycles/min).

1. Student 4 should obtain the step (0.5 m [20-in.] tall for male subject or 0.4 m [16-in.] for a female subject) while baseline measurements are being obtained on the subject.
2. Once the baseline pulse and blood pressure measurements have been recorded on the **Activity 7 Exercise chart** (above), the subject is to stand quietly at attention for 2 minutes to allow his or her blood pressure to stabilize before beginning to step.
3. The subject is to perform the bench stepping for as long as possible, up to a maximum of 5 minutes, according to the cadence called by the timer. The subject must keep posture erect; the other group members should watch for and work against crouching. If the subject is unable to keep the pace for a span of 15 seconds, the test is to be terminated.
4. When the subject is stopped by the timer for crouching, stops voluntarily because he or she is unable to continue, or has completed 5 minutes of bench stepping, he or she is to sit down. Record the duration of exercise (in seconds), and measure the blood pressure and pulse immediately and thereafter at 1-minute intervals for 3 minutes post-exercise.

Duration of exercise: ______________ sec

5. Calculate the subject's *index of physical fitness* using the following formula:

$$\text{Index} = \frac{\text{duration of exercise in second} \times \text{mm } 100}{2 \times \text{sum of the three pulse counts in recovery}}$$

Interpret scores according to the following scale:

below 55	poor physical condition
55 to 62	low average
63 to 71	average
72 to 79	high average
80 to 89	good
90 and over	excellent

6. Record the test values on the Activity 7 Exercise chart above, and repeat the testing and recording procedure with the second subject.

When did you notice a greater elevation of blood pressure and pulse?

____

Explain: ______________________________________

____

Was there a sizable difference between the after-exercise values for well-conditioned and poorly conditioned individuals?

______________ Explain: _______________________

____

Did the diastolic pressure also increase? ______________

Text continues on next page. →

Activity 7: Stimulus							
Baseline		1 min		2 min		3 min	
BP	P	BP	P	BP	P	BP	P

Explain: ______________________________

____

A Noxious Sensory Stimulus (Cold)

Blood pressure can be affected by emotions and pain. This lability of blood pressure will be investigated through use of the **cold pressor test**, in which one hand will be immersed in unpleasantly (even painfully) cold water.

1. Measure the blood pressure and pulse of the subject as he or she sits quietly. Record these as the baseline values on the Activity 7 Stimulus chart (above).

2. Obtain a basin and thermometer, fill the basin with ice cubes, and add water. When the temperature of the ice bath has reached 5°C, immerse the subject's other hand (the non-cuffed limb) in the ice water. With the hand still immersed, take blood pressure and pulse readings at 1-minute intervals for a period of 3 minutes, and record the values on the chart.

How did the blood pressure change during cold exposure?

____

____

Was there any change in pulse? ______________________________

3. Subtract the respective baseline readings of systolic and diastolic blood pressure from the highest single reading of systolic and diastolic pressure obtained during cold immersion. (For example, if the highest experimental reading is 140/88 and the baseline reading is 120/70, then the differences in blood pressure would be systolic pressure, 20 mm Hg, and diastolic pressure, 18 mm Hg.) These differences are called the index of response. According to their index of response, subjects can be classified as follows:

Hyporeactors (stable blood pressure): Exhibit a rise of diastolic and/or systolic pressure ranging from 0 to 22 mm Hg or a drop in pressures

Hyperreactors (labile blood pressure): Exhibit a rise of 23 mm Hg or more in the diastolic and/or systolic blood pressure

Is the subject tested a hypo- or hyperreactor?

____

33

Skin Color as an Indicator of Local Circulatory Dynamics

Skin color reveals with surprising accuracy the state of the local circulation and allows inferences concerning the larger blood vessels and the circulation as a whole. The Activity 8 experiments on local circulation illustrate a number of factors that affect blood flow to the tissues.

Clinical expertise often depends upon good observation skills, accurate recording of data, and logical interpretation of the findings. One of the earliest compensatory responses of the body to impaired blood flow to the brain is constriction of cutaneous blood vessels. This reduces blood flow to the skin and diverts it into the circulatory mainstream to serve other, more vital tissues. As a result, the skin of the face and particularly of the extremities becomes pale, cold, and eventually moist with perspiration. Therefore, pale, cold, clammy skin should immediately prompt a suspicion that the circulation is dangerously inefficient.

Activity 8

Examining the Effect of Local Chemical and Physical Factors on Skin Color

The local blood supply to the skin (indeed, to any tissue) is influenced by (1) local metabolites, (2) oxygen supply, (3) local temperature, (4) autonomic nervous system impulses, (5) local vascular reflexes, (6) certain hormones, and (7) substances released by injured tissues. A number of these factors are examined in the simple experiments that follow. Each experiment should be conducted by students in groups of three or four. One student will act as the subject; the others will conduct the tests and make and record observations.

Vasodilation and Flushing of the Skin Due to Local Metabolites

1. Obtain a sphygmomanometer (blood pressure cuff) and stethoscope. You will also need a watch or clock with a second hand.

2. The subject should bare both arms by rolling up the sleeves as high as possible and then lay the forearms side by side on the bench top.

3. Observe the general color of the subject's forearm skin, and the normal contour and size of the veins. Notice whether skin color is bilaterally similar. Record your observations:

___

___

___

4. Apply the blood pressure cuff to one arm, and inflate it to 250 mm Hg. Keep it inflated for 1 minute. During this period, repeat the observations made above, and record the results:

___

___

5. Release the pressure in the cuff (leaving the deflated cuff in position), and again record the forearm skin color and the condition of the forearm veins. Make this observation immediately after deflation and then again 30 seconds later.

Immediately after deflation: ___

___

30 sec after deflation: ___

___

The above observations constitute your baseline information. Now conduct the following tests.

6. Instruct the subject to raise the cuffed arm above his or her head and to clench the fist as tightly as possible. While the hand and forearm muscles are tightly contracted, rapidly inflate the cuff to 240 mm Hg or more. This maneuver partially empties the hand and forearm of blood and stops most blood flow to the hand and forearm. Once the cuff has been inflated, the subject is to relax the fist and return the forearm to the bench top so that it can be compared to the other forearm.

7. Leave the cuff inflated for exactly 1 minute. During this interval, compare the skin color in the "ischemic" (blood-deprived) hand to that of the "normal" (noncuffed-limb) hand. Quickly release the pressure immediately after 1 minute.

What are the subjective effects (sensations felt by the subject, such as pain, cold, warmth, tingling, weakness) of stopping blood flow to the arm and hand for 1 minute? These sensations are "symptoms" of a change in function.

___

___

What are the objective effects (color of skin and condition of veins)?

___

___

How long does it take for the subject's ischemic hand to regain its normal color?

___

Effects of Venous Congestion

1. Again, but with a different subject, observe and record the appearance of the skin and veins on the forearms resting on the bench top. This time, pay particular attention to the color of the fingers, particularly the distal phalanges, and the nail beds. Record this information:

___

___

2. Wrap the blood pressure cuff around one of the subject's arms, and inflate it to 40 mm Hg. Maintain this pressure for 5 minutes. Make a record of the subjective and objective findings just before the 5 minutes are up, and then again immediately after release of the pressure at the end of 5 minutes.

Subjective (arm cuffed): ___

___

Objective (arm cuffed): ___

___

Subjective (pressure released): ___

___

Objective(pressurereleased): ___

___

3. With still another subject, conduct the following simple experiment: Raise one arm above the head, and let the other hang by the side for 1 minute. After 1 minute, quickly lay both arms on the bench top, and compare their color.

Color of raised arm: ___

Color of dependent arm: ___

Text continues on next page. →

From this and the two preceding observations, analyze the factors that determine tint of color (pink or blue) and intensity of skin color (deep pink or blue as opposed to light pink or blue). Record your conclusions.

____

____

____

____

____

Collateral Blood Flow

In some diseases, blood flow to an organ through one or more arteries may be completely and irreversibly obstructed. Fortunately, in most cases a given body area is supplied both by one main artery and by anastomosing channels connecting the main artery with one or more neighboring blood vessels. Consequently, an organ may remain viable even though its main arterial supply is occluded, as long as the **collateral vessels** are still functional.

The effectiveness of collateral blood flow in preventing ischemia can be easily demonstrated.

1. Check the subject's hands to be sure they are *warm* to the touch. If not, choose another subject, or warm the subject's hands in 35°C water for 10 minutes before beginning.

2. Palpate the subject's radial and ulnar arteries approximately 2.5 cm (1 in.) above the wrist flexure, and mark their locations with a felt marker.

3. Instruct the subject to supinate one forearm and to hold it in a partially flexed (about a 30° angle) position, with the elbow resting on the bench top.

4. Face the subject and grasp his or her forearm with both of your hands, the thumb and fingers of one hand compressing the marked radial artery and the thumb and fingers of the other hand compressing the ulnar artery. Maintain the pressure for 5 minutes, noticing the progression of the subject's hand to total ischemia.

5. At the end of 5 minutes, release the pressure abruptly. Record the subject's sensations, as well as the intensity and duration of the flush in the previously occluded hand. (Use the other hand as a baseline for comparison.)

____

____

____

6. Allow the subject to relax for 5 minutes; then repeat the maneuver, but this time *compress only the radial artery*. Record your observations.

____

____

How do the results of the first test differ from those of the second test with respect to color changes during compression and to the intensity and duration of reactive hyperemia (redness of the skin)?

____

____

____

____

7. Once again allow the subject to relax for 5 minutes. Then repeat the maneuver, *compressing only the ulnar artery*. Record your observations.

____

____

What can you conclude about the relative sizes of, and hand areas served by, the radial and ulnar arteries?

____

____

Effect of Mechanical Stimulation of Blood Vessels of the Skin

With moderate pressure, draw the blunt end of your pen across the skin of a subject's forearm. Wait 3 minutes to observe the effects, and then repeat with firmer pressure.

What changes in skin color do you observe with light-to-moderate pressure?

____

With heavy pressure? ______________________________

The redness, or *flare*, observed after mechanical stimulation of the skin results from a local inflammatory response promoted by chemical mediators released by injured tissues. These mediators stimulate increased blood flow into the area and leaking of fluid (from the capillaries) into the local tissues.

33

EXERCISE 33

REVIEW SHEET

Human Cardiovascular Physiology: Blood Pressure and Pulse Determinations

Name ______________________ Lab Time/Date ______________________

Cardiac Cycle

1. Using the grouped sets of terms to the right of the diagram, correctly identify each trace, valve closings and openings, and each time period of the cardiac cycle.

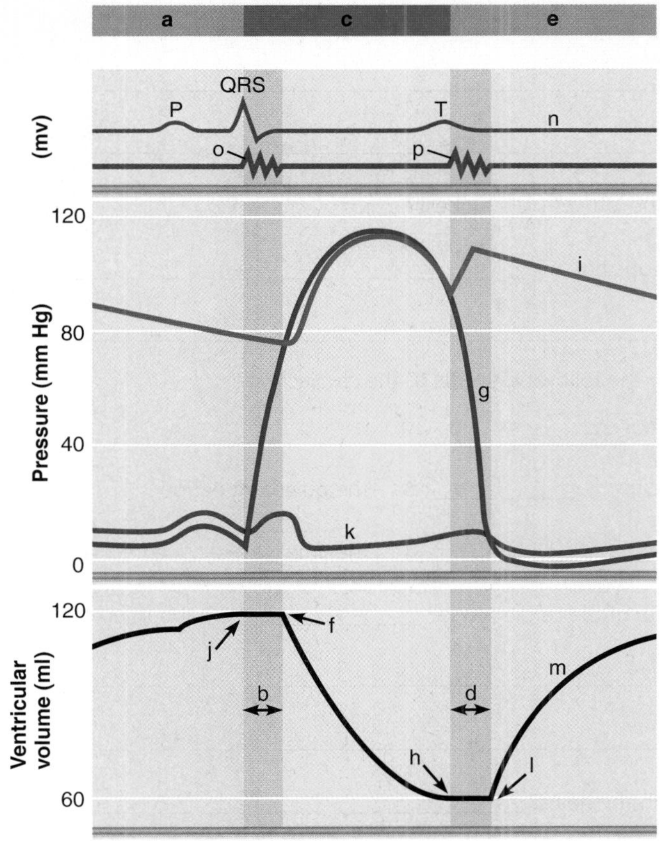

_______ 1. aortic pressure

_______ 2. atrial pressure (left)

_______ 3. ECG

_______ 4. first heart sound

_______ 5. second heart sound

_______ 6. ventricular pressure (left)

_______ 7. ventricular volume

_______ 8. aortic (semilunar) valve closes

_______ 9. aortic (semilunar) valve opens

_______ 10. AV and semilunar valves are closed (2 letters)

_______ 11. AV valve closes

_______ 12. AV valve opens

_______ 13. ventricular diastole (2 letters)

_______ 14. ventricular systole

2. Define the following terms.

systole: ____________________

diastole: ____________________

cardiac cycle: ____________________

3. Answer the following questions concerning events of the cardiac cycle.

During which phase of the cardiac cycle are the AV valves opened? ____________________

Describe the pressure difference that causes the AV valves to open. ____________________

During which phase of the cardiac cycle are the semilunar valves opened? ____________________

What event causes the semilunar valves to open? ____________________

Are both sets of valves closed during any part of the cycle? ____________________

If so, when? ____________________

Are both sets of valves open during any part of the cycle? ____________________

During which phase in the cardiac cycle is the pressure in the left ventricle highest? ____________________

What event results in the pressure deflection called the dicrotic notch? ____________________

____

4. Using the key below, indicate the time interval occupied by the following events of the cardiac cycle.

Key: a. 0.8 sec b. 0.4 sec c. 0.3 sec d. 0.1 sec

__________ 1. the length of the normal cardiac cycle

__________ 2. the time interval of atrial contraction

__________ 3. the quiescent period

__________ 4. the time interval of ventricular contraction

5. If an individual's heart rate is 80 beats/min, what is the length of the cardiac cycle? __________ What portion of the cardiac cycle decreases with a more rapid heart rate? ____________________

____

6. What two factors promote the movement of blood through the heart? ____________________

____________________ and ____________________

____

Heart Sounds

7. Complete the following statements.

The two monosyllables describing the heart sounds are _1_. The first heart sound is a result of closure of the _2_ valves, whereas the second is a result of closure of the _3_ valves. The heart chambers that have just been filled when you hear the first heart sound are the _4_, and the chambers that have just emptied are the _5_. Immediately after the second heart sound, both the _6_ and _7_ are filling with blood.

1. ______________________
2. ______________________
3. ______________________
4. ______________________
5. ______________________
6. ______________________
7. ______________________

8. As you listened to the heart sounds during the laboratory session, what differences in pitch, length, and amplitude (loudness) of the two sounds did you observe? ______________________

____

____

9. To auscultate most accurately, indicate where you would place your stethoscope for the following sounds:

closure of the tricuspid valve: ______________________

closure of the aortic valve: ______________________

apical heartbeat: ______________________

Which valve is heard most clearly when the apical heartbeat is auscultated? ______________________

10. No one expects you to be a full-fledged physician on such short notice, but on the basis of what you have learned about heart sounds, give an example of how abnormal sounds might be used to diagnose a heart problem.

____

____

____

The Pulse

11. Define *pulse*. ______________________

12. Describe the procedure used to take the pulse. ______________________

____

____

13. Identify the artery palpated at each of the pressure points listed.

at the wrist: ______________________

on the dorsum of the foot: ______________________

in front of the ear: ______________________

at the side of the neck: ______________________

14. When you were palpating the various pulse or pressure points, which appeared to have the greatest amplitude or tension?

__________ Why do you think this was so? __________

15. Assume someone has been injured in an auto accident and is hemorrhaging badly. What pressure point would you compress to help stop bleeding from each of the following areas?

the thigh: __________ the calf: __________

the forearm: __________ the thumb: __________

16. How could you tell by simple observation whether bleeding is arterial or venous? __________

17. You may sometimes observe a slight difference between the value obtained from an apical pulse (beats/min) and the value from an arterial pulse taken elsewhere on the body. What is this difference called?

Blood Pressure Determinations

18. Define *blood pressure*. __________

19. Identify the phase of the cardiac cycle to which each of the following apply.

systolic pressure: __________ diastolic pressure: __________

20. What is the name of the instrument used to compress the artery and record pressures in the auscultatory method of determining blood pressure? __________

21. What are the sounds of Korotkoff? __________

What causes the systolic sound? __________

What causes the disappearance of the sound? __________

22. Describe how blood pressure is measured using a sphygmomanometer. __________

23. Define *pulse pressure*. __________

Why is this measurement important? __________

24. Explain why *pulse pressure* is different from *pulse rate*. __________

25. How do venous pressures compare to arterial pressures? ______

Why? ______

26. What maneuver to increase the thoracic pressure illustrates the effect of external factors on venous pressure? ______

______ How is it performed? ______

______

27. What might an abnormal increase in venous pressure indicate? (Think!) ______

______

Observing the Effect of Various Factors on Blood Pressure and Heart Rate

28. What effect do the following have on blood pressure? (Indicate increase by ↑ and decrease by ↓.)

______ 1. increased diameter of the arterioles

______ 2. increased blood viscosity

______ 3. increased cardiac output

______ 4. hemorrhage

______ 5. arteriosclerosis

______ 6. increased pulse rate

29. In which position (sitting, reclining, or standing) is the blood pressure normally the highest?

______ The lowest? ______

What immediate changes in blood pressure did you observe when the subject stood up after being in the sitting or reclining position? ______

______

What changes in the blood vessels might account for the change? ______

______

______

After the subject stood for 3 minutes, what changes in blood pressure did you observe? ______

______

How do you account for this change? ______

30. What was the effect of exercise on blood pressure? ______

On pulse rate? ______

31. What effects of cold temperature did you observe on blood pressure in the laboratory? ______

______

What do you think the effect of heat would be? ______

Why? ______

32. Differentiate between a hypo- and a hyperreactor relative to the cold pressor test. ______________________

____

Skin Color as an Indicator of Local Circulatory Dynamics

33. Describe normal skin color and the appearance of the veins in the subject's forearm before any testing was conducted.

____

34. What changes occurred when the subject emptied the forearm of blood (by raising the arm and making a fist) and the flow was occluded with the cuff? ______________________

What changes occurred during venous congestion? ______________________

____

35. What is the importance of collateral blood supplies? ______________________

____

36. Explain the mechanism by which mechanical stimulation of the skin produced a flare. ______________________

____

37. + A left ventricular assist device (LVAD) is a medical device that is sometimes used as an alternative to heart transplant for patients who have a weakened left ventricle. Once implanted, the LVAD pumps blood from the left ventricle to the aorta. Name the phase of the cardiac cycle that this device assists:

______________________. Describe the status of the AV and SL valves during this phase. ______________________

38. + Fainting (syncope) during or shortly after urination (micturition) is common in men of advanced age. Although the phenomenon of micturition syncope is not well understood, it is known to be due to a sudden drop in blood pressure. Explain why dehydration would be an additional risk factor for micturition syncope. ______________________

____

____

EXERCISE 34

Frog Cardiovascular Physiology

Learning Outcomes

- Describe the properties of automaticity and rhythmicity as they apply to cardiac muscle.
- Discuss the anatomical differences between frog and human hearts.
- Compare the intrinsic rate of contraction of the "pacemaker" of the frog heart (sinus venosus) to that of the atria and ventricle.
- Define *extrasystole*, and explain when an extrasystole can be induced.
- Explain why it is important that cardiac muscle cannot be tetanized.
- Describe the effects of the following on heart rate: cold, heat, vagal stimulation, pilocarpine, atropine sulfate, epinephrine, digitalis, and potassium, sodium, and calcium ions.
- Define *ectopic pacemaker*, *vagal escape*, and *partial* and *total heart block*.
- Name the blood vessels associated with a capillary bed and describe microcirculation.
- Identify an arteriole, venule, and capillaries in a frog's web, and cite the differences between relative size of these vessels and the rate of blood flow through them.
- Discuss the effect of heat, cold, local irritation, and histamine on blood flow in capillaries, and explain how these responses help maintain homeostasis.

Pre-Lab Quiz

Instructors may assign these and other Pre-Lab Quiz questions using Mastering A&P™

1. Spontaneous depolarization-repolarization events occur in a regular and continuous manner in cardiac muscle, a property known as:
   a. automaticity
   b. rhythmicity
   c. synchronicity
2. Circle True or False. Heart rate can be modified by extrinsic impulses from the autonomic nerves.
3. Which chemical agent will you use to modify the frog heart rate?
   a. caffeine
   b. digitalis
   c. magnesium solution
   d. Ringer's solution
4. The ______________ nerve carries parasympathetic impulses to the heart.
   a. cardiac
   b. olfactory
   c. phrenic
   d. vagus
5. Circle the correct underlined term. The phenomenon of <u>vagal escape</u> / <u>heart block</u> occurs when the heart stops momentarily, and then begins to beat again.

Go to Mastering A&P™ > Study Area to improve your performance in A&P Lab.

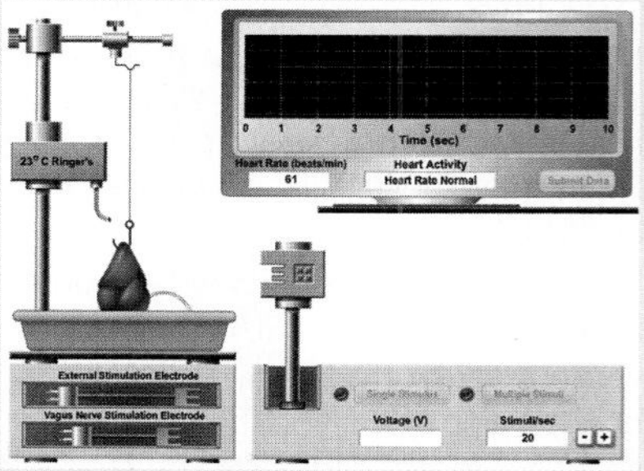

> Lab Tools > PhysioEx

Instructors may assign new Building Vocabulary coaching activities, Pre-Lab Quiz questions, Art Labeling activities, Practice Anatomy Lab Practical questions (PAL), PhysioEx activities, and more using the Mastering **A&P™** Item Library.

Materials

- Dissecting instruments and tray
- Disposable gloves
- Petri dishes
- Medicine dropper
- Millimeter ruler
- Disposal container for organic debris
- Frog Ringer's solutions (at room temperature, 5°C, and 32°C)
- Frogs*
- Thread

**Instructor will double-pith frogs as required for student experimentation.*

Text continues on next page. →

- Large rubber bands
- Fine common pins
- Frog board
- Cotton balls
- Physiograph or BIOPAC® equipment:

Physiograph (polygraph), physiograph paper and ink, force transducer, transducer cable, transducer stand, stimulator output extension cable, electrodes

BIOPAC® BIOPAC® BSL System with BSL software version 4.0.1 and above (for Windows 10/8.x/7) or (Mac OS X 10.9–10.12), data acquisition unit MP36/35, PC or Mac computer, BIOPAC® HDW100A tension adjuster or equivalent, BIOPAC® SS12LA force transducer with S-hook, small hook with thread, and transducer (or ring) stand. Optional: Ten gram weight for calibration

- Instructors using the MP36/35/30 data acquisition unit with BSL software versions earlier than 4.0.1 (for Windows or Mac) and will need slightly different channel settings and collection strategies. Instructions for using the older data acquisition unit can be found on MasteringA&P.
- Dropper bottles of freshly prepared solutions (using frog Ringer's solution as the solvent) of the following:

  2.5% pilocarpine

  5% atropine sulfate

  1% epinephrine

  2% digitalis

  2% calcium chloride ($CaCl_2$)

  0.7% sodium chloride (NaCl)

  5% potassium chloride (KCl)

  0.01% histamine

  0.01 *N* HCl
- Dissecting pins
- Paper towels
- Compound microscope

34

PEx PhysioEx™ 9.1 Computer Simulation Ex. 6 on p. PEx-93

Note: *Instructions for using PowerLab® equipment can be found on MasteringA&P.*

Investigations of human cardiovascular physiology are very interesting, but many areas obviously do not lend themselves to experimentation. However, investigation of frog cardiovascular physiology can be done and provides valuable data because the physiological mechanisms in these animals are similar, if not identical, to those in humans.

In this exercise, you will conduct cardiac investigations. In addition, you will observe the microcirculation in a frog's web and subject it to various chemical and thermal agents to demonstrate their influence on local blood flow.

Special Electrical Properties of Cardiac Muscle: Automaticity and Rhythmicity

Cardiac muscle differs from skeletal muscle both functionally and in its fine structure. Skeletal muscle must be electrically stimulated to contract. In contrast, heart muscle can and does depolarize spontaneously in the absence of external stimulation. This property, called **automaticity**, is due to plasma membranes that have reduced permeability to potassium ions but still allow sodium ions to slowly leak into the cells. This leakage causes the muscle cells to gradually depolarize until the action potential threshold is reached. Shortly thereafter, contraction occurs. Also, the spontaneous depolarization-repolarization events occur in a regular, continuous manner in cardiac muscle, a property called **rhythmicity**.

In the following experiment, you will observe these properties of cardiac muscle in vitro (that is, removed from the body). Work together in groups of three or four.

Activity 1

Investigating the Automaticity and Rhythmicity of Heart Muscle

1. Obtain a dissecting tray and instruments, disposable gloves, two petri dishes, frog Ringer's solution, a metric ruler, and a medicine dropper, and bring them to your laboratory bench.

2. Don the gloves, and then request and obtain a doubly pithed frog from your instructor. Quickly open the thoracic cavity and observe the heart rate in situ (at the site or within the body).

Record the heart rate: ________________ beats/min

3. Dissect out the heart and the gastrocnemius muscle of the calf, and place the removed organs in separate petri dishes containing frog Ringer's solution. (*Note:* The procedure for removing the gastrocnemius muscle is provided on pp. 235–236 in Exercise 14. The extreme care used in that procedure for the removal of the gastrocnemius muscle need not be exercised here.)

4. Observe the activity of the two organs for a few seconds.

Which is contracting? ________________________________

At what rate? ________________________________ beats/min

Is the contraction rhythmic? ________________________________

5. Sever the sinus venosus from the heart (**Figure 34.1**). The **sinus venosus** of the frog's heart corresponds to the SA node of the human heart.

Does the sinus venosus continue to beat? ________________

If not, lightly touch it with a probe to stimulate it. Record its rate of contraction.

Rate: ______________ beats/min

6. Sever the right atrium from the heart; then remove the left atrium. Does each atrium continue to beat?

__________________ Rate: __________________ beats/min

Does the ventricle continue to beat? ________________

Rate: ______________ beats/min

7. Notice that frogs have a single ventricle (Figure 34.1). Fragment the ventricle to determine how small the ventricular fragments must be before the automaticity of ventricular muscle is abolished. Measure these fragments and record their approximate size.

____________ mm × ____________ mm × ____________ mm

Which portion of the heart exhibited the most marked automaticity?

____

Which showed the least? ______________________

8. Properly dispose of the frog and heart fragments in the appropriate container before continuing.

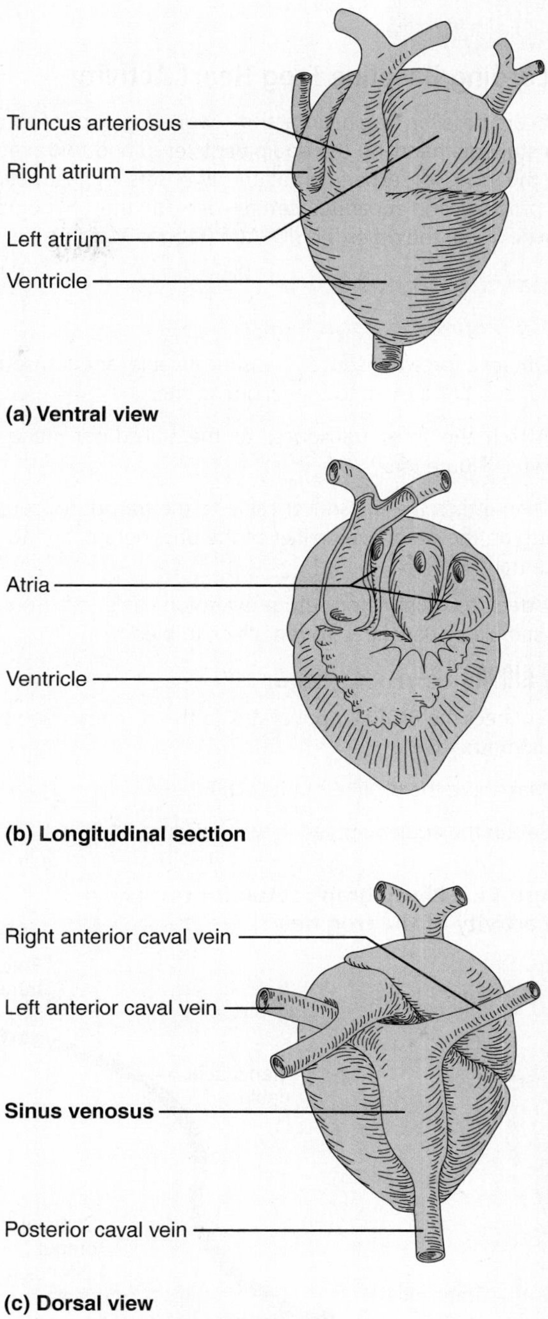

Figure 34.1 Anatomy of the frog heart.

34

Baseline Frog Heart Activity

The heart's effectiveness as a pump is dependent on both intrinsic (within the heart) and extrinsic (external to the heart) controls. In this activity, you will investigate some of these factors.

The nodal system, in which the "pacemaker" imposes its depolarization rate on the rest of the heart, is one intrinsic factor that influences the heart's pumping action. If its impulses fail to reach the ventricles (as in heart block), the ventricles continue to beat but at their own inherent rate, which is much slower than that usually imposed on them. Although heart contraction does not depend on nerve impulses, its rate can be modified by extrinsic impulses reaching the heart through the autonomic nerves. Additionally, cardiac activity is modified by various chemicals, hormones, ions, and metabolites. The effects of several of these chemical factors are examined in the next experimental series.

The frog heart has two atria and a single, incompletely divided ventricle (see Figure 34.1). The pacemaker is located in the sinus venosus, an enlarged region between the venae cavae and the right atrium.

Activity 2

Recording Baseline Frog Heart Activity

To record baseline frog heart activity, work in groups of four—two students handling the equipment setup and two preparing the frog for experimentation. Two sets of instructions are provided for apparatus setup—one for the physiograph (**Figure 34.2**), the other for BIOPAC® (**Figure 34.3**).

Apparatus Setups

Physiograph Apparatus Setup

1. Obtain a force transducer, transducer cable, and transducer stand, and bring them to the recording site.

2. Attach the force transducer to the transducer stand (as shown in Figure 34.2).

3. Then attach the transducer cable to the transducer coupler (input) on the channel amplifier of the physiograph and to the force transducer.

4. Attach the stimulator output extension cable to output on the stimulator panel (red to red, black to black).

BIOPAC® *Apparatus Setup*

1. Connect the BIOPAC® apparatus to the computer and turn the computer **ON**.

2. Make sure the BIOPAC® unit is **OFF**.

3. Set up the equipment (as shown in Figure 34.3).

4. Turn the BIOPAC® unit **ON**.

5. Launch the BIOPAC® Student Lab software by clicking the desktop icon or by following your instructor's guidance.

6. Open the Frog Heart template. Select the ***PRO* Lessons** tab in the startup screen and choose **A04 Frog Heart - Cardiac Rate & Contractility** and click **OK**.

7. Put the tension adjuster (BIOPAC® HDW100A, or equivalent) on the transducer stand, and attach the BIOPAC® SS12LA force transducer with the hook holes pointing down. Level the force transducer both horizontally and vertically.

8. Set the tension adjuster to approximately one-quarter of its full range. (***Note:*** *Do not firmly tighten any of the thumbscrews at this stage.*) Select a force range of 0 to 50 grams for this experiment.

9. Select and attach the small S-hook to the force transducer.

Optional: The force transducer can also be calibrated prior to preparing the frog by clicking the **Calibrate** button in the BSL graph template and following the prompts. A ten gram weight is required for calibration.

Figure 34.2 Physiograph setup for recording the activity of the frog heart.

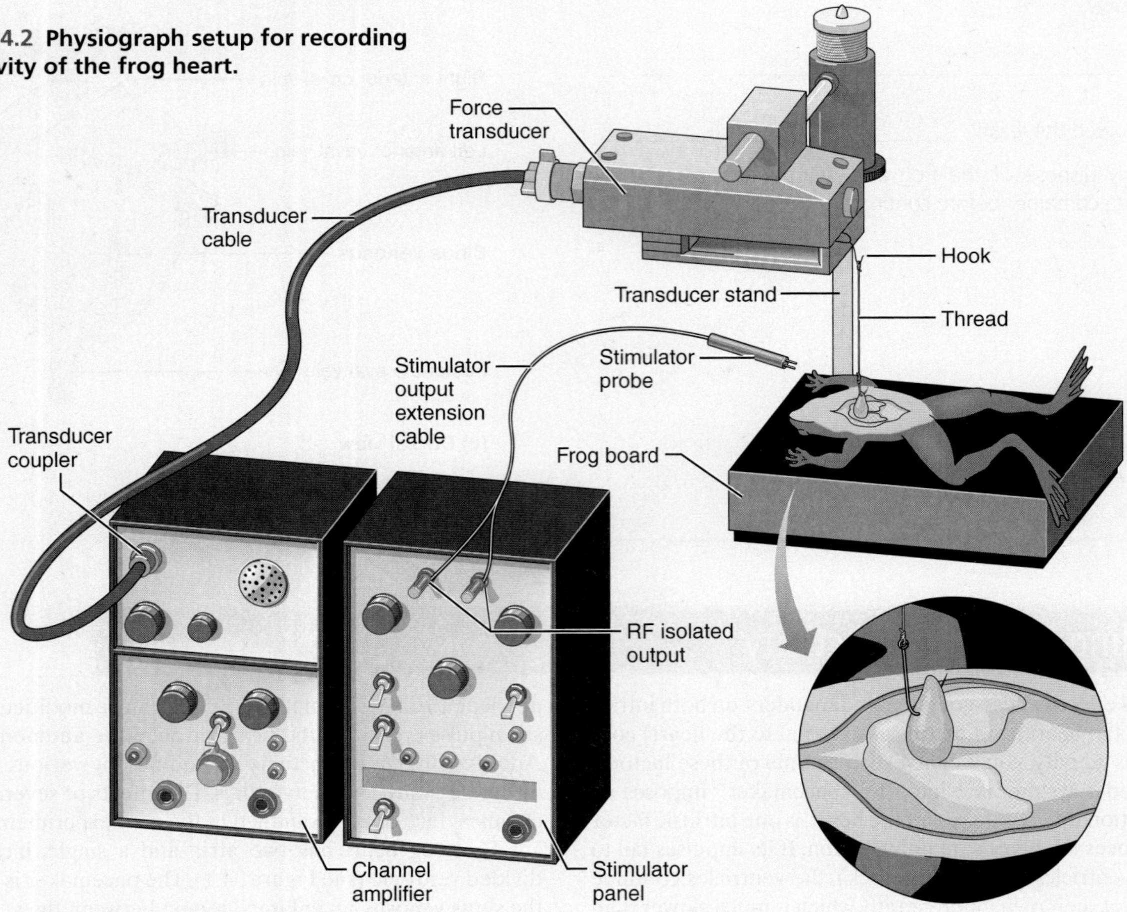

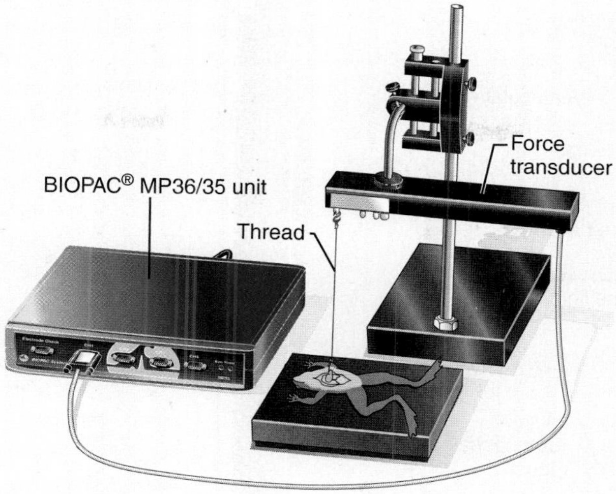

FIGURE 34.3 BIOPAC® setup for recording the activity of the frog heart. Plug the force transducer into channel 1. Transducer is shown plugged into the MP36/35 unit.

Preparation of the Frog

1. Obtain room-temperature frog Ringer's solution, a medicine dropper, dissecting instruments and tray, disposable gloves, fine common pins (physiograph) or small hook (BIOPAC®), cotton ball, frog board, large rubber bands, and some thread, and bring them to your bench.

2. Don the gloves, and obtain a doubly pithed frog from your instructor.

3. Make a longitudinal incision through the abdominal and thoracic walls with scissors, and then cut through the sternum to expose the heart.

4. Grasp the pericardial sac with forceps, and cut it open so that the beating heart can be observed.

5. Locate the vagus nerve, which runs down the lateral aspect of the neck and parallels the trachea and carotid artery. (In good light, it appears to be striated.) Slip an 18-inch length of thread under the vagus nerve so that it can later be lifted away from the surrounding tissues by the thread. Then place a Ringer's solution–soaked cotton ball over the nerve to keep it moistened until you are ready to stimulate it later in the procedure.

6. Using a medicine dropper, flush the heart with Ringer's solution. From this point on, the heart must be kept continually moistened with room-temperature Ringer's solution unless other solutions are being used for the experimentation.

7. Attach the frog to the frog board using large rubber bands.

Physiograph Frog Heart Preparation

1. Bend a common pin to a 90° angle, and tie to its head a thread 0.46–0.5 m (18–20 inches) long. Take care not to penetrate the ventricular chamber as you force the pin through the apex of the heart until the apex is well secured in the angle of the pin.

2. Tie the thread from the heart to the hook on the force transducer. Do not pull the thread too tightly. It should be taut enough to lift the heart apex upward, away from the thorax, but should *not* stretch the heart. Adjust the force transducer as necessary. (See Figure 34.2.)

BIOPAC® *Frog Heart Preparation*

1. Attach a small hook tied with thread to the frog heart, following the instructions in step 1 of the physiograph instructions above to insert it through the apex of the heart. Confirm that the prepared frog is firmly attached to the frog board, positioned below the ring stand with the line running vertically from the frog heart to the transducer.

2. Slide the tension adjuster/force transducer assembly down the ring stand until you can hang the loop loosely from the S-hook, then slide it back up until the line is taut but the heart muscle is not stretched (be careful not to tear the heart).

3. Position the tension adjuster and/or force transducer so that the top is level, with approximately 10 cm (4 inches) of line from the heart to the S-hook. Adjust the assembly so that the thread line runs vertically; for a true reflection of the muscle's contractile force, the muscle must not be pulled at an angle.

4. Use the tension adjuster knob to make the line taut and tighten all thumbscrews to secure positioning of the assembly. Let the setup sit for a minute, then recheck the tension to make sure nothing has slipped or stretched.

5. Data may be distorted if the transducer line is not pulling directly vertical from the frog heart to the S-hook. Once again, align the frog as described above and make sure that the heart is not twisting the thread. If it is twisting, you will need to *carefully* remove the hook and repeat the setup.

6. BIOPAC® calculates *rate* data, which always trails the actual rate by one cycle. Data collection is a sensitive process and may

Text continues on next page. →

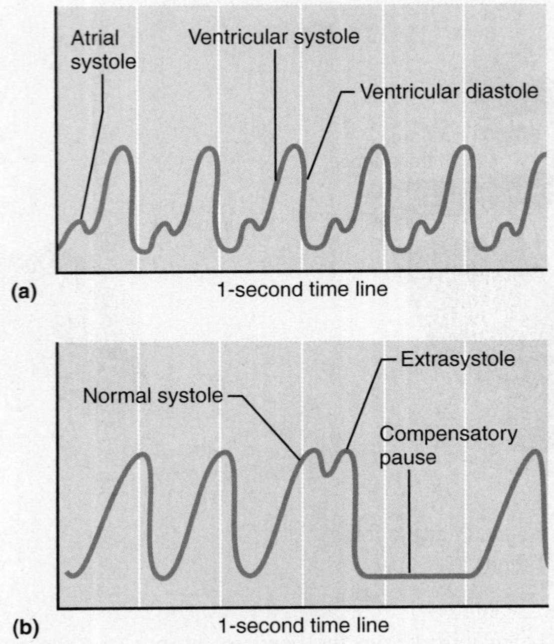

Figure 34.4 Physiograph recording of contractile activity of a frog heart. (a) Normal heartbeat. **(b)** Induction of an extrasystole.

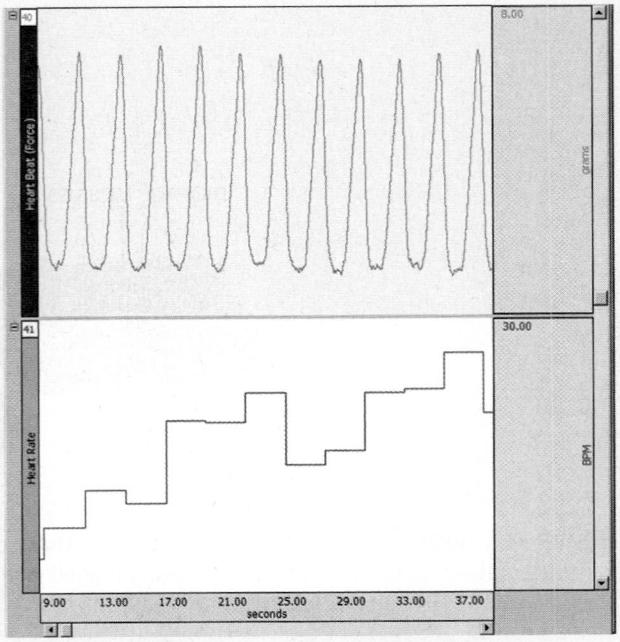

Figure 34.5 Example of baseline frog heart rate data.

display artifacts from table movement, heart movement (for instance, from breathing on the heart), and chemicals touching the heart. To get the best data, keep the experimental area stable, clean, and clear of obstructions. Hints for obtaining best data:

- Make sure the SS12LA force transducer is level on the horizontal and vertical planes.
- Set up the tension adjuster and force transducer in positions that minimize their movement when tension is applied. Keep the point of S-hook attachment as close as possible to the ring stand support.
- Position the tension adjuster so that you will not bump the cables or frog board when using the adjustment knob.
- Position and/or tape the force transducer cables where they will not be pulled or bumped easily.
- Make sure the frog board is on a stable surface.
- Make sure the frog is firmly attached to the frog board so it will not rise up when tension is applied.

Making the Baseline Recording

Using the Physiograph

1. Turn the amplifier on, and balance the apparatus according to instructions provided by your instructor. Set the paper speed at 0.5 cm/sec. Press the record and paper advance buttons.

2. Set the signal magnet or time marker at 1/sec.

3. Record 12 to 15 normal heartbeats. Be sure you can distinguish atrial and ventricular contractions (**Figure 34.4**). Then adjust the paper or scroll speed so that the peaks of ventricular contractions are approximately 2 cm apart. (Peaks indicate systole; troughs indicate diastole.) Pay attention to the relative force of heart contractions while recording.

Using BIOPAC®

1. Click **Start** to begin recording.

2. Observe at least five heart rate cycles, then click **Stop** to stop recording. Your data should look like that in the example (**Figure 34.5**).

3. Choose **Save** from the File menu, and type in a filename to save the recorded data. You may want to save by your team's name followed by FrogHeart-1 (for example, Smith-FrogHeart-1).

Analyzing the Baseline Data

Count the number of ventricular contractions per minute from your physiograph or BIOPAC® data, and record:

_______________ beats/min

Compute the A–V interval (period from the beginning of atrial contraction to the beginning of ventricular contraction).

_______________ sec

How do the two tracings compare in time?

____

____

Mark the atrial and ventricular systoles on the record. *Remember to keep the heart moistened with Ringer's solution*.

Activity 3

Investigating the Refractory Period of Cardiac Muscle Using the Physiograph

Repeated rapid stimuli can cause skeletal muscle to remain in a contracted state (as demonstrated in Exercise 14). In other words, the muscle can be tetanized. This is possible because of the relatively short refractory period of skeletal muscle. In this experiment, you will use the physiograph to investigate the refractory period of cardiac muscle and its response to stimulation.* During the procedure, one student should keep the stimulating electrodes in constant contact with the frog heart ventricle while another student operates the stimulator panel.

1. Using the physiograph, set the stimulator to deliver 20-V shocks of 2-msec duration, and begin recording.

2. Deliver single shocks at the beginning of ventricular contraction, the peak of ventricular contraction, and then later and later in the cardiac cycle.

3. Observe the recording for **extrasystoles**, which are extra beats that show up riding on the ventricular contraction peak. Also note the **compensatory pause**, which allows the heart to get back on schedule after an extrasystole. (See Figure 34.4b.)

During which portion of the cardiac cycle was it possible to induce an extrasystole?

____

____

4. Attempt to tetanize the heart by stimulating it at the rate of 20 to 30 impulses per second. What is the result?

____

____

Considering the function of the heart, why is it important that heart muscle cannot be tetanized?

____

____

*BIOPAC® users may investigate the refractory period of cardiac muscle by using PhysioEx Exercise 6.

Activity 4

Assessing Physical and Chemical Modifiers of Heart Rate

Now that you have observed normal frog heart activity, you will have an opportunity to investigate the effects of various factors that modify heart activity. In each case, record a few normal heartbeats before introducing the modifying factor. After removing the agent, allow the heart to return to its normal rate before continuing with the testing. On each record, indicate the point of introduction and removal of the modifying agent.

For each physical agent or solution that is applied:

If using the physiograph, increase the scroll or paper speed so that heartbeats appear as spikes 4 to 5 mm apart.

If using BIOPAC®, after applying each solution click the **Start** button. When the effect is observed, record the effect for five cycles, then click **Stop**. Choose **Save** from the File menu to save your data.

(*Note:* Repeat these steps for each physical agent or chemical solution that is being applied.)

Temperature

1. Obtain 5°C and 32°C frog Ringer's solutions and medicine droppers.

2. Bathe the heart with 5°C Ringer's solution, and continue to record until the recording indicates a change in cardiac activity and five cardiac cycles have been recorded.

3. Stop recording, pipette off the cold Ringer's solution, and flood the heart with room-temperature Ringer's solution.

4. Start recording again to determine the resumption of the normal heart rate. When this has been achieved, flood the heart with 32°C Ringer's solution, and again record five cardiac cycles after a change is noted.

5. Stop the recording, pipette off the warm Ringer's solution, and bathe the heart with room-temperature Ringer's solution once again.

What change occurred with the cold (5°C) Ringer's solution?

____

What change occurred with the warm (32°C) Ringer's solution?

____

Count the heart rate at the two temperatures, and record the data below.

__________ beats/min at 5°C; __________ beats/min at 32°C

Chemical Agents

Pilocarpine

Flood the heart with a 2.5% solution of pilocarpine. Record until you notice a change in the pattern of the ECG. Pipette off the excess pilocarpine solution, and proceed immediately to the next test. What happened when the heart was bathed in the pilocarpine solution?

____

Pilocarpine simulates the effect of parasympathetic (vagal) nerve stimulation by enhancing acetylcholine release; such drugs are called parasympathomimetic drugs.

Text continues on next page. →

34

Is pilocarpine an agonist or an antagonist of acetylcholine?

___

Atropine Sulfate

Apply a few drops of atropine sulfate to the frog's heart, and observe the recording. If no changes are observed within 2 minutes, apply a few more drops. When you observe a response, pipette off the excess atropine sulfate, and flood the heart with room-temperature Ringer's solution. What happens when the atropine sulfate is added?

___

___

Atropine is a drug that blocks the effect of the neurotransmitter acetylcholine, which is liberated by the parasympathetic nerve endings. Do your results accurately reflect this effect of atropine?

___

Is atropine an agonist or an antagonist of acetylcholine?

___

Epinephrine

Flood the frog heart with epinephrine solution, and continue to record until a change in heart activity is noted.

What are the results? ___

___

Which division of the autonomic nervous system does its effect mimic?

___

Digitalis

Pipette off the excess epinephrine solution, and rinse the heart with room-temperature Ringer's solution. Continue recording, and when the heart rate returns to baseline values, bathe it in digitalis solution. What is the effect of digitalis on the heart?

___

___

Digitalis is a drug commonly prescribed for heart patients with congestive heart failure. It slows heart rate, providing more time for venous return and decreasing the work of the weakened heart. These effects are thought to be due to inhibition of the sodium-potassium pump and enhancement of Ca^{2+} entry into the myocardial fibers.

Various Ions

To test the effect of various ions on the heart, apply the designated solution until you observe a change in heart rate or in strength of contraction. Pipette off the solution, flush with room-temperature Ringer's solution, and allow the heart to resume its normal rate before continuing. *Do not allow the heart to stop.* If the rate should decrease dramatically, flood the heart with room-temperature Ringer's solution.

Effect of Ca^{2+} (use 2% $CaCl_2$) ___

___

Effect of Na^{+} (use 0.7% NaCl) ___

___

Effect of K^{+} (use 5% KCl) ___

___

Potassium ion concentration is normally higher within cells than in the extracellular fluid. *Hyperkalemia* decreases the resting potential of plasma membranes, thus decreasing the force of heart contraction. In some cases, the conduction rate of the heart is so depressed that **ectopic pacemakers** (pacemakers appearing erratically and at abnormal sites in the heart muscle) appear in the ventricles, and fibrillation may occur. Was there any evidence of premature beats in the recording of potassium ion effects? ___

Was arrhythmia produced with any of the ions tested?

___ If so, which? ___

Vagus Nerve Stimulation

The vagus nerve carries parasympathetic impulses to the heart, which modify heart activity. If you are using the physiograph, you can test this by stimulating the vagus nerve.*

1. Remove the cotton placed over the vagus nerve. Using the previously tied thread, lift the nerve away from the tissues, and place the nerve on the stimulating electrodes.

2. Using a duration of 0.5 msec at a voltage of 1 mV, stimulate the nerve at a rate of 50/sec. Continue stimulation until the heart stops momentarily and then begins to beat again **(vagal escape)**. If no effect is observed, increase stimulus intensity and try again. If no effect is observed after a substantial increase in stimulus voltage, reexamine your "vagus nerve" to make sure that it is not simply strands of connective tissue.

3. Discontinue stimulation after you observe vagal escape, and flush the heart with room-temperature Ringer's solution until the normal heart rate resumes. What is the effect of vagal stimulation on heart rate?

___

*BIOPAC® users may observe the effects of vagal stimulation by using PhysioEx Exercise 6.

34

Intrinsic Conduction System Disturbance (Heart Block)

1. Moisten a 25-cm (10-inch) length of thread, and make a Stannius ligature (loop the thread around the heart at the junction of the atria and ventricle).

2. If using a physiograph, decrease the scroll or paper speed to achieve intervals of approximately 2 cm between the ventricular contractions, and record a few normal heartbeats.

3. Tighten the ligature in a stepwise manner while observing the atrial and ventricular contraction curves. As heart block occurs, the atria and ventricle will no longer show a 1:1 contraction ratio. Record a few beats each time you observe a different degree of heart block—a 2:1 ratio of atrial to ventricular contractions, 3:1, 4:1, and so on. As long as you can continue to count a whole-number ratio between the two chamber types, the heart is in **partial heart block**. When you can no longer count a whole number ratio, the heart is in **total**, or **complete, heart block**.

4. When total heart block occurs, release the ligature to see if the normal A–V rhythm is reestablished. What is the result?

____

____

5. Attach properly labeled recordings (or copies of the recordings) made during this procedure to the last page of this exercise for future reference.

6. Dispose of the frog remains and gloves in appropriate containers, and dismantle the experimental apparatus before continuing.

The Microcirculation and Local Blood Flow

The thin web of a frog's foot provides an opportunity to observe the flow of blood to, from, and within the capillary beds. The flow of blood through a capillary bed is called the **microcirculation**. In most body regions, a **terminal arteriole** branches into 10 to 20 true capillaries. The **true capillaries** are the exchange vessels that form the capillary bed. The true capillaries then drain into a **postcapillary venule** (Figure 34.6).

The total cross-sectional area of the capillaries in the body is much greater than that of the veins and arteries combined. Thus, the velocity of flow through the capillary beds is quite slow. Capillary flow is also intermittent, because if all capillary beds were filled with blood at the same time, there would be no blood at all in the large vessels. The flow of blood into the capillary beds is regulated by the diameter of the terminal arterioles, as well as by all of the arterioles upstream from it. The amount of blood flowing into the true capillaries of the bed is regulated chiefly by local chemical controls. Thus a capillary bed may be flooded with blood or almost entirely bypassed, depending on what is happening within the body or in a particular body region at any one time. You will investigate some of the local controls in the next group of experiments.

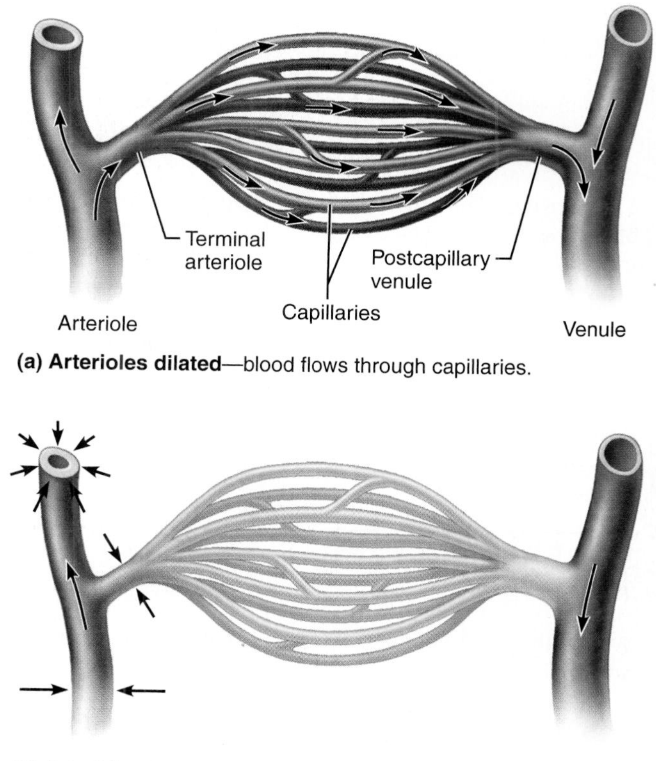

(a) Arterioles dilated—blood flows through capillaries.

(b) Arterioles constricted—no blood flows through capillaries.

Figure 34.6 Anatomy of a capillary typical bed.

Activity 5

Investigating the Effect of Various Factors on the Microcirculation

1. Obtain a frog board (with a hole at one end), dissecting pins, disposable gloves, frog Ringer's solution (room temperature, 5°C, and 32°C), 0.01 *N* HCl, 0.01% histamine solution, 1% epinephrine solution, a large rubber band, and some paper towels.

2. Put on the gloves, and obtain a frog (alive and hopping, *not* pithed). Moisten several paper towels with room-temperature

Text continues on next page. →

Ringer's solution, and wrap the frog's body securely with them. Leave one hind leg unsecured and extending beyond the paper cocoon.

3. Attach the frog to the frog board (or other supporting structure) with a large rubber band, and then carefully spread (but do not stretch) the web of the exposed hindfoot over the hole in the support. Have your partners hold the edges of the web firmly for viewing. Alternatively, secure the toes to the board with dissecting pins.

4. Obtain a compound microscope, and observe the web under low power to find a capillary bed. Focus on the vessels in high power. Keep the web moistened with Ringer's solution as you work. If the circulation seems to stop during your observations, massage the hind leg of the frog gently to restore blood flow.

5. Observe the red blood cells of the frog. Notice that unlike human RBCs, they are nucleated. Watch their movement through the smallest vessels—the capillaries. Do they move in single file, or do they flow through two or three cells abreast?

____

Are they flexible? __________ Explain. __________

Can you see any white blood cells in the capillaries?

__________ If so, which types? __________

6. Notice the relative speed (velocity) of blood flow through the blood vessels. Differentiate between the arterioles, which feed the capillary bed, and the venules, which drain it. This may be tricky, because images are reversed in the microscope. Thus, the vessel that appears to feed into the capillary bed will actually be draining it. You can distinguish between the vessels, however, if you consider that the flow is more pulsating and turbulent in the arterioles and smoother and steadier in the venules. How does the velocity of flow in the arterioles compare with that in the venules?

____

In the capillaries? __________

What is the relative difference in the diameter of the arterioles and capillaries?

____

Temperature

1. To investigate the effect of temperature on blood flow, flood the web with 5°C Ringer's solution two or three times to chill the entire area. Is a change in vessel diameter noticeable?

__________ Which vessels are affected? __________

How? __________

2. Blot the web gently with a paper towel, and then bathe the web with warm (32°C) Ringer's solution. Record your observations.

____

Inflammation

1. Pipette 0.01 *N* HCl onto the frog's web. Hydrochloric acid will act as an irritant and cause a localized inflammatory response. Is there an increase or decrease in the blood flow into the capillary bed following the application of HCl?

____

What purpose do these local changes serve during a localized inflammatory response?

____

____

____

2. Flush the web with room-temperature Ringer's solution and blot.

Histamine

1. Histamine, which is released in large amounts during allergic responses, causes extensive vasodilation. Investigate this effect by adding a few drops of histamine solution to the frog web. What happens?

____

____

How does this response compare to that produced by HCl?

____

2. Blot the web and flood with 32°C Ringer's solution as before. Now add a few drops of 1% epinephrine solution, and observe the web. What are epinephrine's effects on the blood vessels?

____

3. Return the dropper bottles to the supply area and the frog to the terrarium. Properly clean your work area before leaving the lab.

Instructors may assign a portion of the Review Sheet questions using Mastering A&P™

34 EXERCISE REVIEW SHEET
Frog Cardiovascular Physiology

Name ______________________________ Lab Time/Date ______________________

Special Electrical Properties of Cardiac Muscle: Automaticity and Rhythmicity

1. Define the following terms.

   automaticity: ______________________________

   rhythmicity: ______________________________

2. Describe the anatomical differences between frog and human hearts. ______________________________

   ____

   ____

3. Which region of the dissected frog heart had the highest intrinsic rate of contraction? ______________________________

   The greatest automaticity? ______________________________

   The greatest regularity or rhythmicity? ______________________________ How do these properties correlate with the duties of a pacemaker? ______________________________

   ____

   Is this region the pacemaker of the frog heart? ______________________________

   Which region had the lowest intrinsic rate of contraction? ______________________________

Investigating the Refractory Period of Cardiac Muscle

4. Define *extrasystole*. ______________________________

5. Respond to the following questions if you used a physiograph. ______________________________

   What was the effect of stimulation of the heart during ventricular contraction? ______________________________

   During ventricular relaxation (first portion)? ______________________________

   During the pause interval? ______________________________

   What does this indicate about the refractory period of cardiac muscle? ______________________________

Assessing Physical and Chemical Modifiers of Heart Rate

6. Describe the effect of thermal factors on the frog heart.

cold: ______________________ heat: ______________________

7. Once again refer to your recordings. Did the administration of the following produce any changes in force of contraction (shown by peaks of increasing or decreasing height)? If so, explain the mechanism.

epinephrine: ______________________

____

pilocarpine: ______________________

____

calcium ions: ______________________

____

8. Excessive amounts of each of the following ions would most likely interfere with normal heart activity. Note the type of changes caused in each case.

K^+: ______________________

Ca^{2+}: ______________________

Na^+: ______________________

9. Respond to the following questions if you used a physiograph. What was the effect of vagal stimulation on heart rate?

____

Which of the following factors cause the same (or very similar) heart rate–reducing effects: epinephrine, acetylcholine, atropine sulfate, pilocarpine, sympathetic nervous system activity, digitalis, potassium ions?

____

Which of the factors listed above would reverse or antagonize vagal effects? ______________________

____

10. What is vagal escape? ______________________

____

Why is vagal escape valuable in maintaining homeostasis? ______________________

____

11. How does the Stannius ligature used in the laboratory produce heart block? ______________________

____

____

12. Define *partial heart block*, and describe how it was recognized in the laboratory. ______________________

____

____

13. Define *total heart block*, and describe how it was recognized in the laboratory. ______________________

____

____

____

14. What do your heart block experiment results indicate about the spread of impulses from the atria to the ventricles?

____

____

Observing the Microcirculation Under Various Conditions

15. In what way are the red blood cells of the frog different from those of the human? ______________________

____

On the basis of this one factor, would you expect their life spans to be longer or shorter? ______________________

16. The following statements refer to your observation of one or more of the vessel types observed in the microcirculation in the frog's web. Characterize each statement by choosing the best response from the key.

Key: a. arterioles   b. venules   c. capillaries

__________ 1. smallest vessels observed

__________ 2. vessels in which blood flow is rapid, pulsating

__________ 3. vessels in which blood flow is least rapid

__________ 4. red blood cells pass through these vessels in single file

__________ 5. blood flow is smooth and steady

__________ 6. most numerous vessels

__________ 7. vessels that deliver blood to the capillary bed

__________ 8. vessels that serve the needs of the tissues via exchanges

__________ 9. vessels that drain the capillary beds

17. Discuss the effects of the following on blood vessel diameter (state specifically the blood vessels involved) and rate of blood flow. Then explain the importance of the observed reaction to the general well-being of the body.

local application of cold: ______________________________

____

local application of heat: ______________________________

____

inflammation (or application of HCl): ______________________________

____

histamine: ______________________________

____

18. Raynaud's syndrome is characterized by hyperactive constriction of blood vessels to the hands and feet in response to cold temperatures or emotional stress. If there is no underlying disease contributor, it is referred to as *primary* or *idiopathic* Raynaud's syndrome. Assuming that this disease follows a pattern similar to the normal physiological response to cold, which type of vessel would you expect to see constricting most, and why? ______________________________

____

____

Which branch of the autonomic nervous system controls these vessels? ______________________________

____

EXERCISE 35

The Lymphatic System and Immune Response

Learning Outcomes

- State the function of the lymphatic system, name its components, and compare its function to that of the blood vascular system.
- Describe the formation and composition of lymph, and discuss how it is transported through the lymphatic vessels.
- Relate immunological memory, specificity, and differentiation of self from nonself to immune function.
- Differentiate between the roles of B cells and T cells in the immune response.
- Describe the structure and function of lymph nodes, and indicate the location of T cells, B cells, and macrophages in a typical lymph node.
- Describe the major microanatomical features of the spleen and tonsils.
- Draw or describe the structure of the antibody monomer, and name the five immunoglobulin subclasses.
- Differentiate between antigen and antibody.
- Explain how the Ouchterlony test detects antigens by using the antigen-antibody reaction.

Go to Mastering A&P™ > Study Area to improve your performance in A&P Lab.

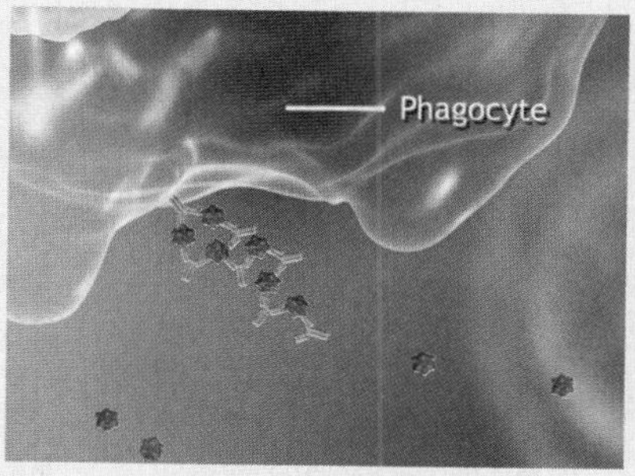

Instructors may assign new Building Vocabulary coaching activities, Pre-Lab Quiz questions, Art Labeling activities, Practice Anatomy Lab Practical questions (PAL), PhysioEx activities, and more using the Mastering A&P™ Item Library.

Pre-Lab Quiz

Instructors may assign these and other Pre-Lab Quiz questions using Mastering A&P™

1. Lymph is:
   a. excess blood that has escaped from veins
   b. excess tissue fluid that has leaked out of capillaries
   c. excess tissue fluid that has escaped from arteries
2. _________, which serve as filters for the lymphatic system, occur at various points along the lymphatic vessels.
   a. Glands **b.** Lymph nodes **c.** Valves
3. Circle True or False. The immune response is a systemic response that occurs when the body recognizes a substance as foreign and acts to destroy or neutralize it.
4. Three characteristics of the immune response are the ability to distinguish self from nonself, memory, and:
   a. autoimmunity **b.** specificity **c.** susceptibility
5. Circle the correct underlined term. T cells mediate humoral / cellular immunity because they destroy cells infected with viruses and certain bacteria and parasites.

Materials

- Large anatomical chart of the human lymphatic system
- Prepared slides of lymph node, spleen, and tonsil
- Compound microscope
- Wax marking pencil
- Petri dish containing simple saline agar
- Medicine dropper
- Dropper bottles of red and green food color
- Dropper bottles of goat antibody to horse serum albumin, goat antibody to bovine serum albumin, goat antibody to swine serum albumin, horse serum albumin diluted to 20% with physiological saline, unknown albumin sample diluted to 20% (prepared from horse, swine, and/or bovine albumin)
- Colored pencils

For instructions on animal dissections, see the dissection exercises (starting on p. 695) in the cat and fetal pig editions of this manual.

PEx PhysioEx™ 9.1 Computer Simulation Ex. 12 on p. PEx-177

The lymphatic system has two chief functions: (1) it transports tissue fluid that has leaked out of the vascular system and returns it to the blood vessels, and (2) it protects the body by removing foreign material such as bacteria from the lymphatic stream and by serving as a site for lymphocyte "policing" of body fluids and for lymphocyte multiplication.

The Lymphatic System

The **lymphatic system** consists of a network of lymphatic vessels (lymphatics), lymphoid tissue, lymph nodes, and a number of other lymphoid organs, such as the tonsils, thymus, and spleen. We will focus on the lymphatic vessels and lymph nodes

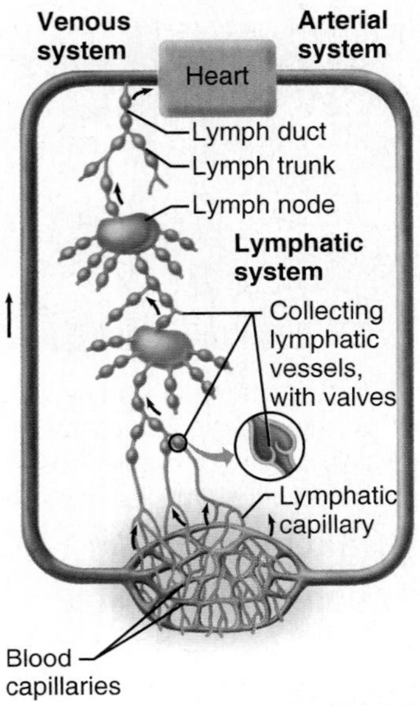

Figure 35.1 Lymphatic system. (a) Simplified scheme of the relationship of lymphatic vessels to blood vessels of the cardiovascular system. **(b)** Distribution of lymphatic vessels and lymph nodes. The green-shaded area represents body area drained by the right lymphatic duct. **(c)** Major veins in the superior thorax showing entry points of the thoracic and right lymphatic ducts. The major lymphatic trunks are also identified.

Instructors may assign this figure as an Art Labeling Activity using Mastering A&P

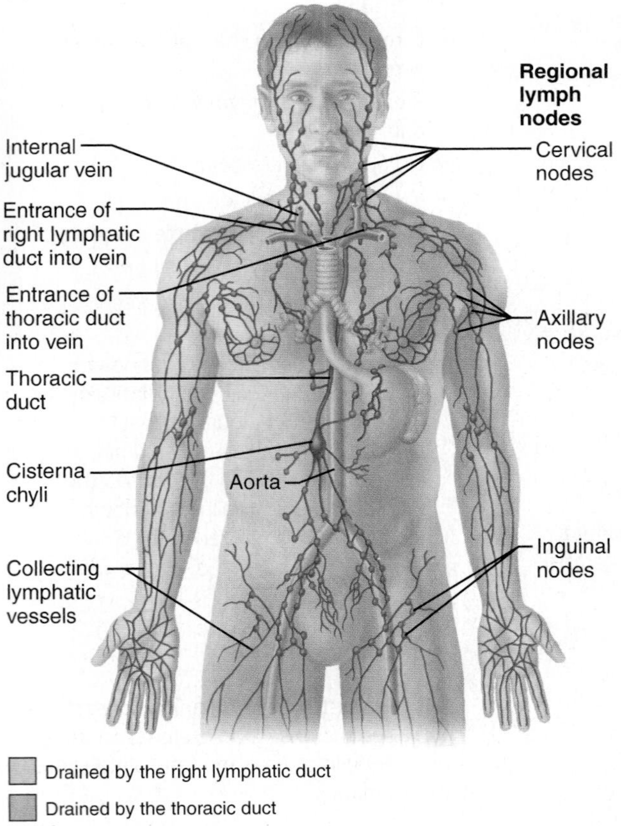

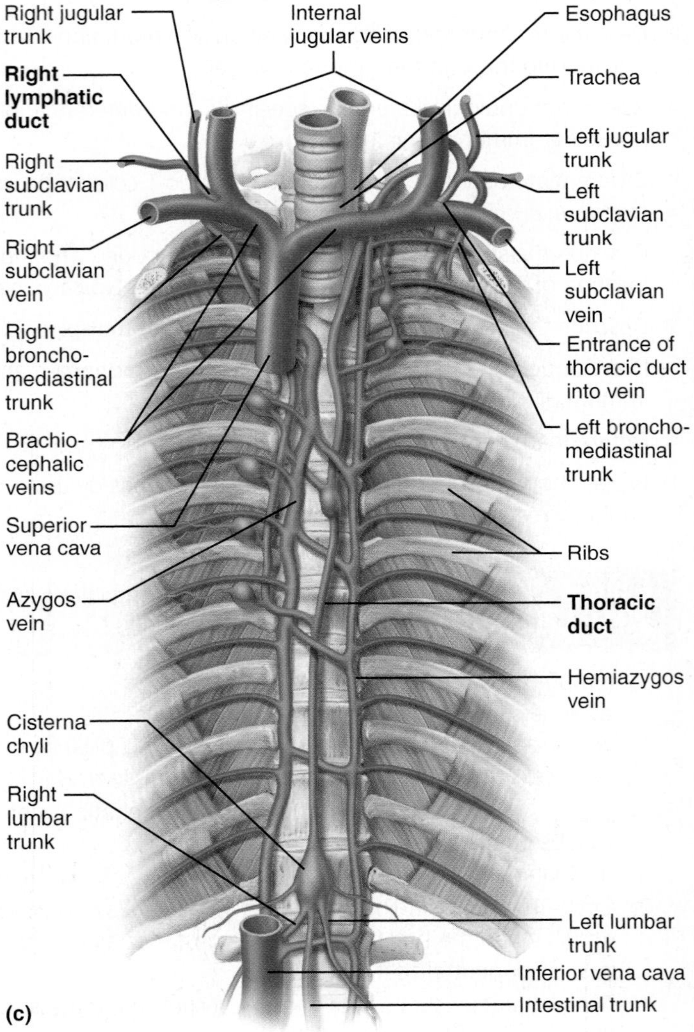

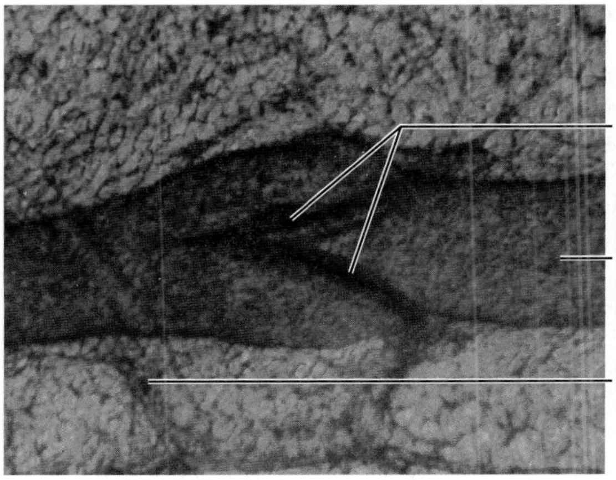

Figure 35.2 Collecting lymphatic vessel (700×).

in this section. The white blood cells that are the central actors in body immunity are described later in this exercise.

Distribution and Function of Lymphatic Vessels and Lymph Nodes

As blood circulates through the body, the hydrostatic and osmotic pressures operating at the capillary beds result in fluid outflow at the arterial end of the bed and in its return at the venous end. However, not all of the lost fluid is returned to the bloodstream by this mechanism. It is the microscopic, blind-ended **lymphatic capillaries** (**Figure 35.1a**), which branch through nearly all the tissues of the body, that pick up this leaked fluid and carry it through successively larger vessels—**collecting lymphatic vessels** to **lymphatic trunks**—until the lymph finally returns to the blood vascular system through one of the two large ducts in the thoracic region (Figure 35.1b). The **right lymphatic duct** drains lymph from the right upper extremity, head, and thorax delivered by the jugular, subclavian, and bronchomediastinal trunks. The large **thoracic duct** receives lymph from the rest of the body (see Figure 35.1c). The enlarged terminus of the thoracic duct is the **cisterna chyli**, which receives lymph from the digestive organs. In humans, both ducts empty the lymph into the venous circulation at the junction of the internal jugular vein and the subclavian vein, on their respective sides of the body. Notice that the lymphatic system, lacking both a contractile "heart" and arteries, is a one-way system; it carries lymph only toward the heart.

Like veins of the blood vascular system, the collecting lymphatic vessels have three tunics and are equipped with valves (**Figure 35.2**). However, lymphatics tend to be thinner-walled, to have *more* valves, and to anastomose (form branching networks) more than veins. Since the lymphatic system is a pumpless system, lymph transport depends largely on the milking action of the skeletal muscles and on pressure changes within the thorax that occur during breathing.

As lymph is transported, it filters through bean-shaped **lymph nodes**, which cluster along the lymphatic vessels of the body. There are hundreds of lymph nodes, but because they are usually embedded in connective tissue, they are not ordinarily seen. Particularly large collections of lymph nodes are found in the inguinal, axillary, and cervical regions of the body.

Activity 1

Identifying the Organs of the Lymphatic System

Study the large anatomical chart to observe the general plan of the lymphatic system. Notice the distribution of lymph nodes, various lymphatics, the lymphatic trunks, and the location of the right lymphatic duct, the thoracic duct, and the cisterna chyli.

For instructions on animal dissections, see the dissection exercises (starting on p. 695) in the cat and fetal pig editions of this manual.

The Immune Response

The collection of body defenses that protect us against microbes, cancer cells, and other foreign bodies is collectively known as the *immune response*. It consists of two main types of defense systems: the *innate (nonspecific) defenses* and the *adaptive (specific) defenses*.

- The innate defenses include surface barriers, such as the skin and mucous membranes, as well as internal defenses that include phagocytes, inflammation, and fever. We are born with these protective mechanisms.
- The adaptive defenses are referred to as specific defenses because the key players in this system have a "lock-and-key" recognition for foreign molecules.

Major Characteristics of the Adaptive Immune Response

Three characteristics of the adaptive immune response are (1) **memory**, (2) **specificity**, and (3) **self-tolerance**. The adaptive immune system's "memory" for previously encountered foreign antigens (the chicken pox virus, for example) is remarkably accurate and highly specific.

An almost limitless variety of things are *antigens*—that is, anything capable of provoking an immune response and reacting with the products of the response. Nearly all foreign proteins, many polysaccharides, bacteria and their toxins, viruses, mismatched RBCs, cancer cells, and many small molecules (haptens) can be antigenic. The cells that recognize antigens and initiate the adaptive immune response are lymphocytes, the second most numerous members of the leukocyte, or white blood cell (WBC), population. Each immunocompetent lymphocyte has receptors on its surface allowing it to bind with only one or a few very similar antigens, thus providing specificity.

As a rule, our own proteins are tolerated, a fact that reflects the ability of the immune system to distinguish our own tissues (self) from foreign antigens (nonself). Nevertheless, an inability to recognize self can and does occasionally happen, and the immune system then attacks the body's own tissues. This phenomenon is called *autoimmunity*. Autoimmune diseases include multiple sclerosis (MS), myasthenia gravis, Graves' disease, rheumatoid arthritis (RA), and diabetes mellitus.

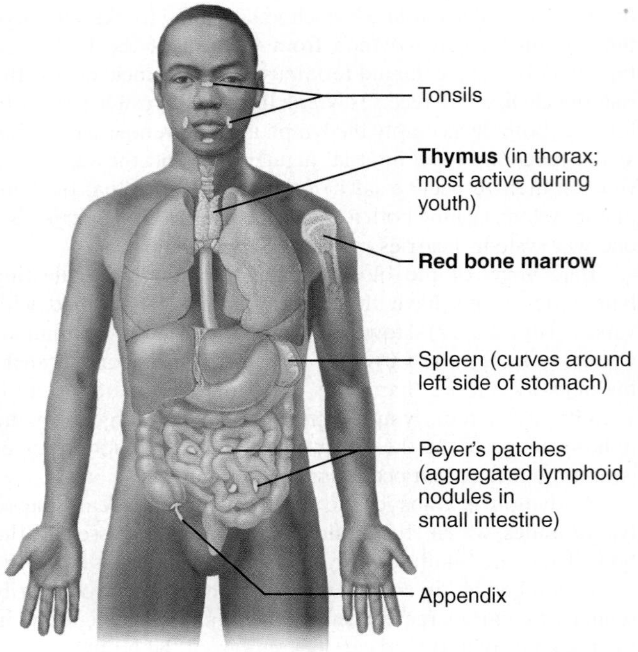

Figure 35.3 Lymphoid organs. Locations of the primary and secondary lymphoid organs. Primary lymphoid organs are in bold.

Instructors may assign this figure as an Art Labeling Activity using Mastering A&P

Organs, Cells, and Cell Interactions of the Immune Response

The immune system uses as part of its arsenal the **lymphoid organs** and **lymphoid tissues**, which include the thymus, lymph nodes, spleen, tonsils, appendix, and red bone marrow. Of these, the thymus and red bone marrow are considered to be the *primary lymphoid organs.* The others are *secondary lymphoid organs and tissues* (**Figure 35.3**).

35

The primary cells that provide for the adaptive immune response are the **B** and **T lymphocytes**, also known as **B** and **T cells**. The B and T cells originate in the red bone marrow. Each cell type must go through a maturation process whereby they become **immunocompetent** and **self-tolerant**. Immunocompetence involves the addition of receptors on the cell surface that recognize and bind to a specific antigen. Self-tolerance involves the cell's ability to distinguish self from nonself. B cells mature in the red bone marrow. T cells travel to the thymus for their maturation process.

After maturation, the B and T cells leave the bone marrow and thymus, respectively; enter the bloodstream; and travel to peripheral (secondary) lymphoid organs, where clonal selection occurs. **Clonal selection** is triggered when an antigen binds to the specific cell-surface receptors of a T or B cell. This event causes the lymphocyte to proliferate rapidly, forming a clone of like cells, all bearing the same antigen-specific receptors. Then, in the presence of certain regulatory signals, the members of the clone specialize, or differentiate—some form memory cells, and others become effector or regulatory cells. Upon subsequent meetings with the same antigen, the immune response proceeds considerably faster because the troops are already mobilized and awaiting further orders, so to speak. Additional characteristics of B and T cells are compared in **Table 35.1**.

Absence or failure of thymic differentiation of T lymphocytes results in a marked depression of both antibody and cell-mediated immune functions. Additionally, the observation that the thymus naturally shrinks with age has been correlated with the relatively immune-deficient status of elderly individuals. ✚

Activity 2

Studying the Microscopic Anatomy of a Lymph Node, the Spleen, and a Tonsil

1. Obtain a compound microscope and prepared slides of a lymph node, spleen, and a tonsil. As you examine the lymph node slide, notice the following anatomical features (depicted in **Figure 35.4**). The node is enclosed within a fibrous **capsule**, from which connective tissue septa (**trabeculae**) extend inward to divide the node into several compartments. Very fine strands of reticular connective tissue issue from the trabeculae, forming the stroma of the gland within which cells are found.

In the outer region of the node, the **cortex**, some of the cells are arranged in globular masses, referred to as germinal centers. The **germinal centers** contain rapidly dividing B cells. The rest of the cortical cells are primarily T cells that circulate continuously, moving from the blood into the node and then exiting from the node in the lymphatic stream.

In the internal portion of the lymph node, the **medulla**, the cells are arranged in cordlike fashion. Most of the medullary cells are macrophages. Macrophages are important not only

Table 35.1 Comparison of B and T Lymphocytes

Properties	B lymphocytes	T lymphocytes
Type of Immune Response	Humoral immunity (antibody-mediated immunity)	Cellular immunity (cell-mediated immunity)
Site of Maturation	Red bone marrow	Thymus
Effector Cells	Plasma cells (antibody-secreting cells)	Cytotoxic T cells, helper T cells, regulatory T cells
Memory Cell Formation	Yes	Yes
Functions	Plasma cells produce antibodies that inactivate antigen and tag antigen for destruction.	Cytotoxic T cells attack infected cells and tumor cells. Helper T cells activate B cells and other T cells.

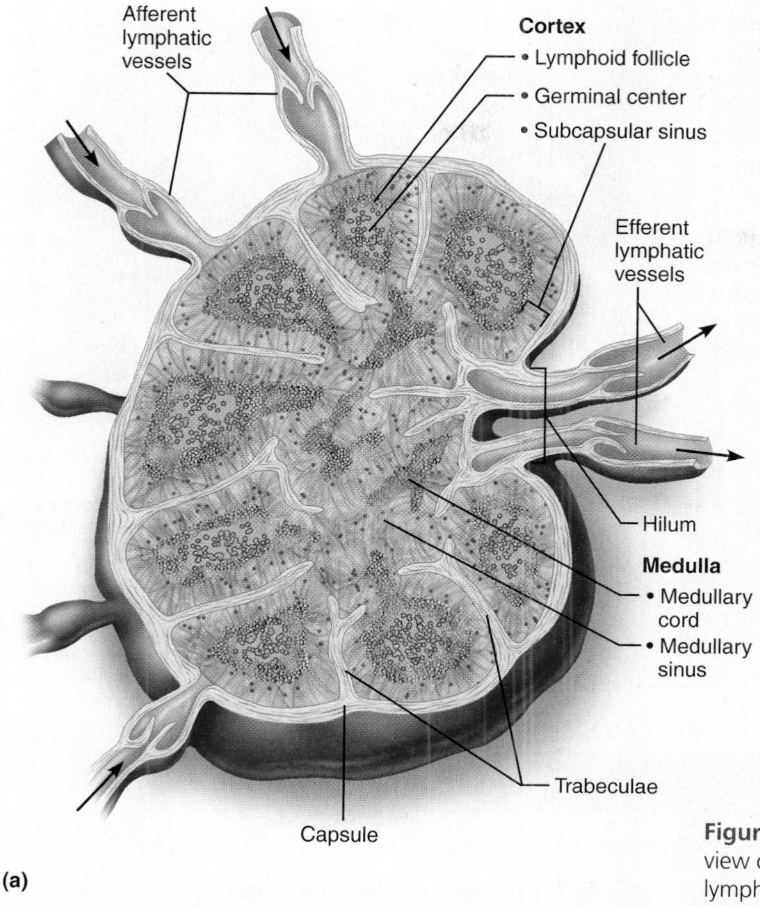

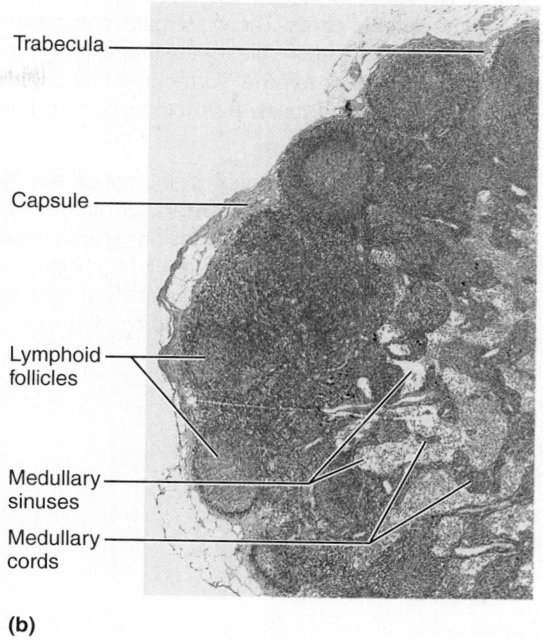

Instructors may assign this figure as an Art Labeling Activity using Mastering A&P

Figure 35.4 Structure of lymph node. (a) Longitudinal view of the internal structure of a lymph node and associated lymphatics. The arrows indicate the direction of the lymph flow. **(b)** Photomicrograph of part of a lymph node (20×).

for their phagocytic function but also because they play an essential role in "presenting" the antigens to the T cells.

Lymph enters the node through a number of *afferent vessels*, circulates through *lymph sinuses* within the node, and leaves the node through *efferent vessels* at the **hilum**. Since each node has fewer efferent than afferent vessels, the lymph flow stagnates somewhat within the node. This allows time for the generation of an immune response and for the macrophages to remove debris from the lymph before it reenters the blood vascular system.

2. As you observe the slide of the spleen, look for the areas of lymphocytes suspended in reticular fibers, the **white pulp**, clustered around central arteries (**Figure 35.5**). The remaining

Text continues on next page. →

35

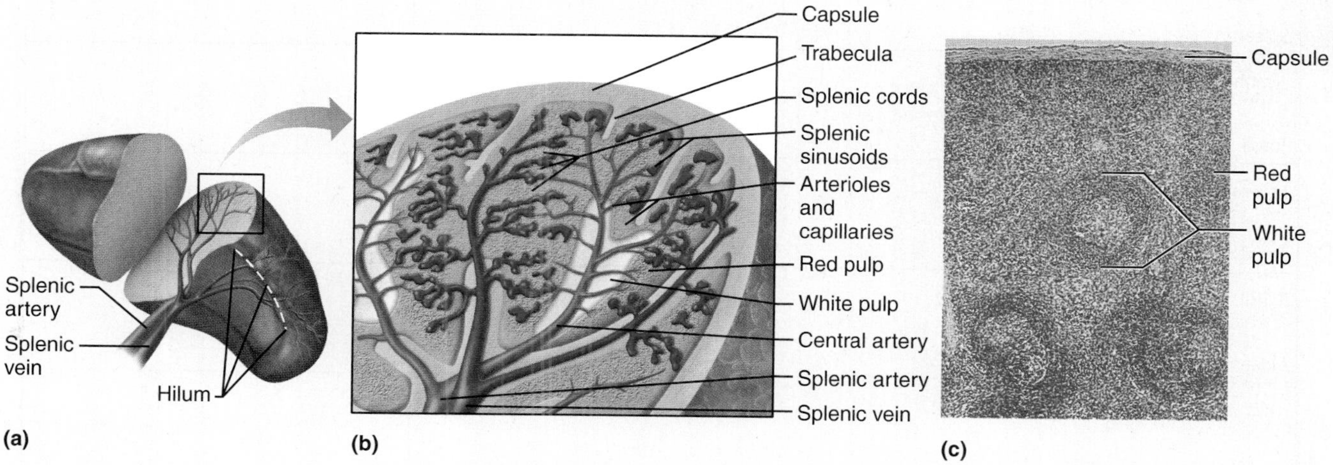

Figure 35.5 The spleen. (a) Gross structure. **(b)** Diagram of the histological structure. **(c)** Photomicrograph of spleen tissue showing white and red pulp regions (30×).

tissue in the spleen is the **red pulp**, which is composed of splenic sinusoids and areas of reticular tissue and macrophages called the **splenic cords**. The white pulp, composed primarily of lymphocytes, is responsible for the immune functions of the spleen. Macrophages remove worn-out red blood cells, debris, bacteria, viruses, and toxins from blood flowing through the sinuses of the red pulp.

3. As you examine the tonsil slide, notice the **lymphoid follicles** containing **germinal centers** surrounded by scattered lymphocytes. The characteristic **tonsillar crypts** (invaginations of the mucosal epithelium) of the tonsils trap bacteria and other foreign material (**Figure 35.6**). Eventually the bacteria work their way into the lymphoid tissue and are destroyed.

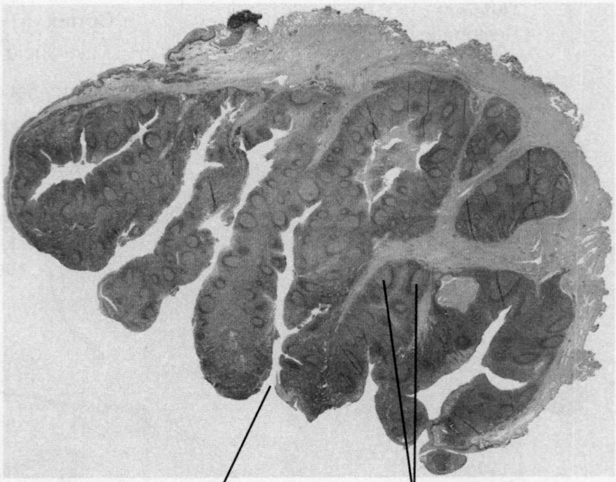

Figure 35.6 Histology of a palatine tonsil. The luminal surface is covered with epithelium that invaginates deeply to form crypts (10×).

Group Challenge

Compare and Contrast Lymphoid Organs and Tissues

Work in groups of three to discuss the characteristics of each lymphoid structure listed in the **Group Challenge chart** below. On a separate piece of paper, one student will record the characteristics for each structure for the group. The group will then consider each pair of structures listed in the chart, discuss the similarities and differences for each pair, and complete the chart based on consensus answers.

Use your textbook or another appropriate reference for comparing Peyer's patches and the thymus with other structures. Remember to consider both structural and functional similarities and differences.

Group Challenge: Comparing Lymphoid Structures		
Lymphoid pair	**Similarities**	**Differences**
Lymph node Spleen		
Lymph node Tonsil		
Peyer's patches Tonsils		
Tonsil Spleen		
Thymus Spleen		

35

Antibodies and Tests for Their Presence

Antibodies, or **immunoglobulins (Igs)**, are produced by effector B cells called *plasma cells*. They are a heterogeneous group of proteins that make up the general class of plasma proteins called **gamma globulins**. Antibodies are found not only in plasma but also (to greater or lesser extents) in all body secretions. Five major classes of immunoglobulins have been identified: IgM, IgG, IgD, IgA, and IgE. The immunoglobulin classes share a common basic structure but differ functionally and in their localization in the body.

All Igs are composed of one or more structural units called **antibody monomers**. A monomer consists of four protein chains bound together by disulfide bridges (**Figure 35.7**). Two of the chains are quite large and have a high molecular weight; these are the **heavy chains**. The other two chains are only half as long and have a low molecular weight. These are called **light chains**. The two heavy chains have a *constant (C) region*, in which the amino acid sequence is the same in a class of immunoglobulins, and a *variable (V) region*, in which the amino acid sequence varies considerably between antibodies. The same is true of the two light chains; each has a constant and a variable region.

The intact Ig molecule has a three-dimensional shape that is generally Y shaped. Together, the variable regions of the light and heavy chains in each "arm" construct one **antigen-binding site** uniquely shaped to "fit" a specific *antigenic determinant* (portion) of an antigen. Thus, each Ig monomer bears two identical sites that bind to identical antigenic determinants. Binding of the immunoglobulins to their complementary antigen(s) effectively immobilizes the antigens until they can be phagocytized or lysed by complement fixation.

The antigen-antibody reaction is used diagnostically in a variety of ways. One of the most familiar is blood typing. (See Exercise 29 for instructions on ABO and Rh blood typing.) The antigen-antibody test that we will use in this laboratory session is the Ouchterlony technique, which is used mainly for rapid screening of suspected antigens.

Ouchterlony Double-Gel Diffusion, an Immunological Technique

The Ouchterlony double-gel diffusion technique was developed in 1948 to detect the presence of particular antigens in sera or extracts. Antigens and antibodies are placed in wells in a gel and allowed to diffuse toward each other. If an antigen reacts with an antibody, a thin white line called a *precipitin line* forms. In the following activity, the double-gel diffusion technique will be used to identify antigens. Work in groups of no more than three.

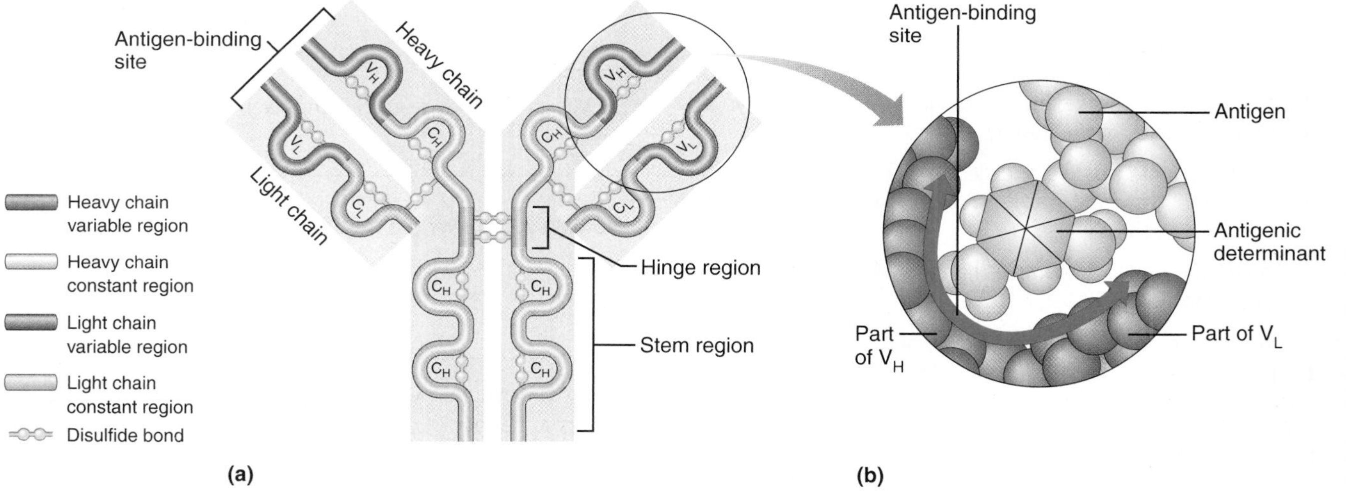

Figure 35.7 Antibody structure. (a) Schematic antibody structure consists of two heavy chains and two light chains connected by disulfide bonds. **(b)** Enlargement of an antigen-binding site of an immunoglobulin.

35

Activity 3

Using the Ouchterlony Technique to Identify Antigens

1. Obtain one each of the materials for conducting the Ouchterlony test: petri dish containing saline agar; medicine dropper; wax marking pencil; and dropper bottles of red and green food dye, horse serum albumin, an unknown serum albumin sample, and antibodies to horse, bovine, and swine albumin. Put your initials and the number of the unknown albumin sample used on the bottom of the petri dish near the edge.
2. Use the wax marking pencil and the template (**Figure 35.8**, p. 526) to divide the dish into three sections, and mark them I, II, and III.
3. Prepare sample wells (again using the template in Figure 35.8). Squeeze the medicine dropper bulb, and gently touch the tip to the surface of the agar. While releasing the bulb, push the tip down through the agar to the bottom of the dish. Lift the dropper vertically; this should leave a straight-walled well in the agar.
4. Repeat step 3 so that section I has two wells and sections II and III have four wells each.

Text continues on next page. →

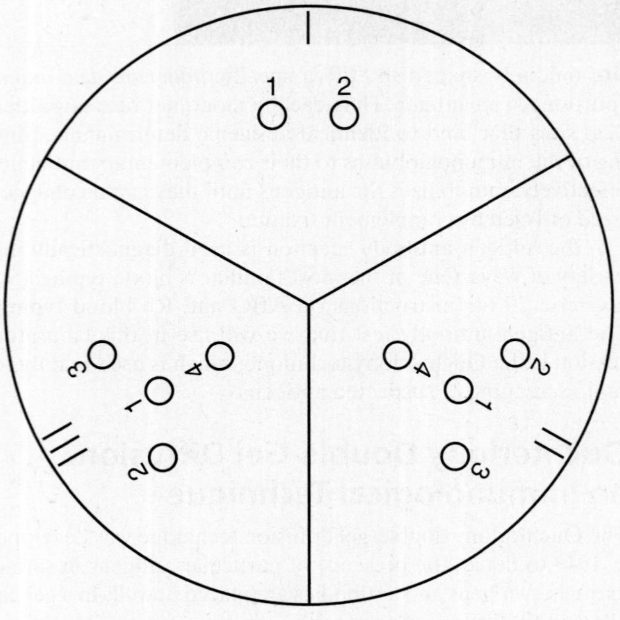

FIGURE 35.8 Template for well preparation for the Ouchterlony double-gel diffusion experiment.

5. To observe diffusion through the gel, nearly fill one well in section I with red dye and the other well with green dye. Be careful not to overfill the wells. Observe periodically for 30 to 45 minutes as the dyes diffuse through the agar. Draw your results as instructed in the Results section.

6. To demonstrate positive and negative results, fill the wells in section II as instructed (**Table 35.2**). A precipitin line should form only between wells 1 and 2.

7. To test the unknown sample, fill the wells in section III as instructed (**Table 35.3**).

8. Replace the cover on the petri dish, and incubate at room temperature for at least 16 hours. Make arrangements to observe the agar for precipitin lines after 16 hours. *The lines may begin to fade after 48 hours*. Draw the results as instructed in the following section, parts 1–3, indicating the location of all precipitin lines that form.

Results

1. For the demonstration of diffusion using dye, draw on the template (Figure 35.8) the appearance of section I of your dish after 30 to 45 minutes. Use colored pencils.

2. Draw section II on the template (Figure 35.8) as it appears after incubation, 16 to 48 hours. Be sure the wells are numbered.

Table 35.2 Section II

Well	Solution
1	Horse serum albumin
2	Goat anti–horse albumin
3	Goat anti–bovine albumin
4	Goat anti–swine albumin

Table 35.3 Section III

Well	Solution
1	Unknown # ______
2	Goat anti–horse albumin
3	Goat anti–bovine albumin
4	Goat anti–swine albumin

3. Draw section III on the template as it appears after incubation, 16 to 48 hours. Be sure the wells are numbered.

Unknown # ______

4. What evidence for diffusion did you observe in section I?

______

______

5. Is there any evidence of a precipitate in section I?

______

6. Which of the sera functioned as an antigen in section II?

______

7. Which antibody reacted with the antigen in section II?

______

How do you know? (Be specific about your observations.)

______

______

8. If swine albumin had been placed in well 1, what would you expect to happen? Explain.

______

______

9. If chicken albumin had been placed in well 1, what would you expect to happen? Explain.

______

______

10. What antigens were present in the unknown solution?

______

How do you know? (Be specific about your observations.)

______

______

EXERCISE 35

REVIEW SHEET
The Lymphatic System and Immune Response

Instructors may assign a portion of the Review Sheet questions using Mastering A&P™

Name ______________________ Lab Time/Date ______________________

The Lymphatic System

1. Match the terms below with the correct letters on the diagram.

_______ 1. appendix

_______ 2. axillary lymph nodes

_______ 3. cervical lymph nodes

_______ 4. cisterna chyli

_______ 5. inguinal lymph nodes

_______ 6. lymphatic vessels

_______ 7. Peyer's patches (in small intestine)

_______ 8. red bone marrow

_______ 9. right lymphatic duct

_______ 10. spleen

_______ 11. thoracic duct

_______ 12. thymus

_______ 13. tonsils

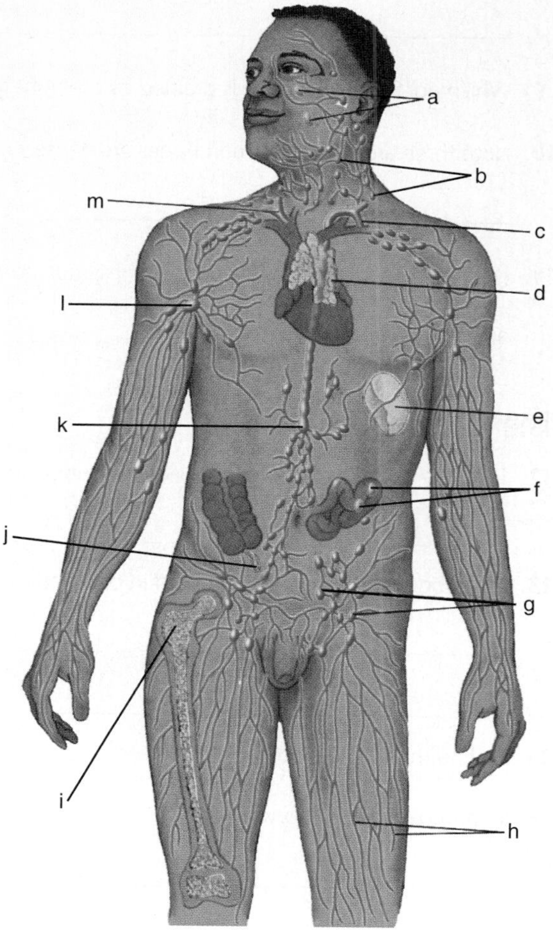

2. Explain why the lymphatic system is a one-way system, whereas the blood vascular system is a two-way system.

____

____

3. How do lymphatic vessels resemble veins? ______________________

How do lymphatic capillaries differ from blood capillaries? ______________________

____

4. What is the function of the lymphatic vessels? ______________________

____

5. What is lymph? ____________________

6. What factors are involved in the flow of lymphatic fluid? ____________________

____

7. What name is given to the terminal duct draining most of the body? ____________________

8. What is the cisterna chyli? ____________________

____

9. Which portion of the body is drained by the right lymphatic duct? ____________________

10. Note three areas where lymph nodes are densely clustered: ____________________,

____________________, and ____________________

11. What are the two major functions of the lymph nodes? ____________________

and ____________________

The Immune Response

12. Describe the effector cells involved in humoral immunity. ____________________

____

13. Describe the effector cells involved in cell-mediated immunity. What is the function of T cells in the immune response?

____

____

14. Define the following terms related to the operation of the immune system.

immunological memory: ____________________

____

specificity: ____________________

____

self-tolerance: ____________________

____

Studying the Microscopic Anatomy of a Lymph Node, the Spleen, and a Tonsil

15. In the diagram of a lymph node below, label the following: afferent lymphatic vessel, efferent lymphatic vessel, lymphoid follicle, trabeculae, subcapsular sinus, capsule, and hilum.

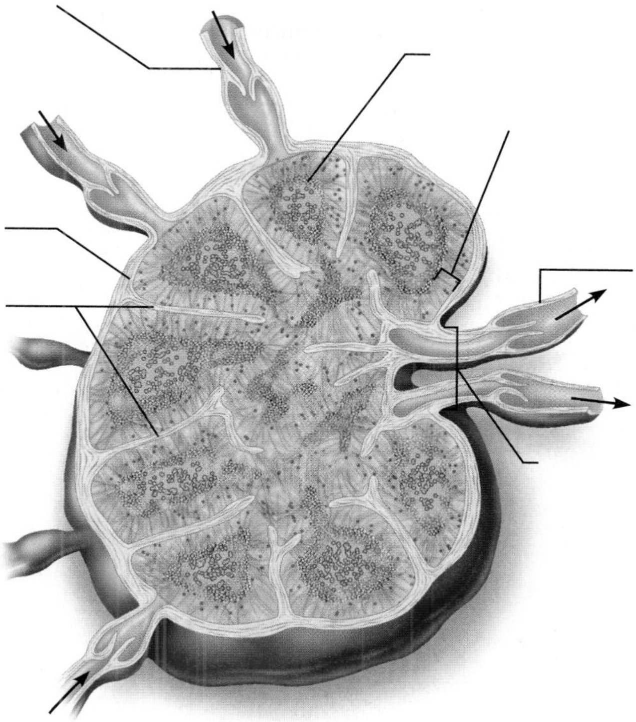

16. What structural characteristic ensures a *slow* flow of lymph through a lymph node? ______________________

____

Why is this desirable? ____

____

Antibodies and Tests for Their Presence

17. Distinguish between antigen and antibody. ______

18. Describe the four protein chains that make up the immunoglobulin monomer, and draw a typical monomer immunoglobulin in the box below. Label the variable regions and constant regions. ______

19. In the Ouchterlony test, what happened when the antibody to horse serum albumin mixed with horse serum albumin?

20. If the unknown antigen contained bovine and swine serum albumin, what would you expect to happen in the Ouchterlony test, and why? ______

21. ✚ Lymphedema is a condition characterized by insufficient movement of lymph in the lymphatic vessels. Fluid builds up in the tissues and in the lymphatic vessels of the limbs. Explain why exercise would have a positive effect on this condition.

22. ✚ Buboes are a key sign for diagnosing bubonic plague. Buboes are swollen lymph nodes that can become necrotic and turn black. Predict where on the body buboes would be most likely to develop in cases of the bubonic plague.

EXERCISE 36 Anatomy of the Respiratory System

Learning Outcomes

- State the major functions of the respiratory system.
- Define the following terms: *pulmonary ventilation*, *external respiration*, and *internal respiration*.
- Identify the major respiratory system structures on models or appropriate images, and describe the function of each.
- Describe the difference between the conducting and respiratory zones.
- Name the serous membrane that encloses each lung, and describe its structure.
- Demonstrate lung inflation in a fresh sheep pluck or preserved tissue specimen.
- Recognize the histologic structure of the trachea and lung tissue microscopically.

Go to Mastering A&P™ > Study Area to improve your performance in A&P Lab.

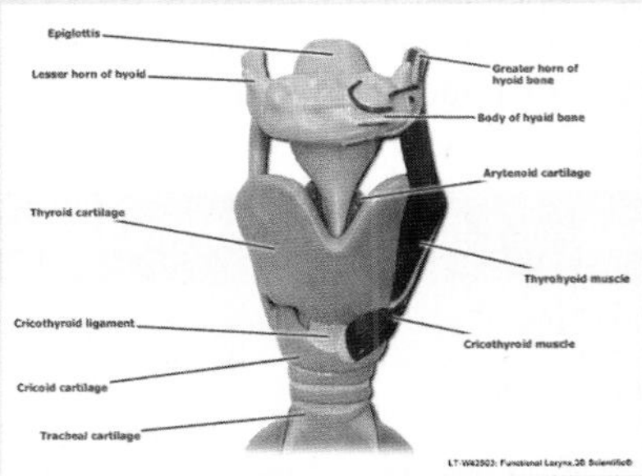

> Lab Tools > Practice Anatomy Lab > Anatomical Models

Instructors may assign new Building Vocabulary coaching activities, Pre-Lab Quiz questions, Art Labeling activities, Practice Anatomy Lab Practical questions (PAL), and more using the Mastering **A&P™** Item Library.

Pre-Lab Quiz

Instructors may assign these and other Pre-Lab Quiz questions using Mastering **A&P™**

1. The upper respiratory structures include the nose, the larynx, and the:
   a. epiglottis **c.** pharynx
   b. lungs **d.** trachea
2. Circle the correct underlined term. The <u>thyroid cartilage</u> / <u>arytenoid cartilage</u> is the largest and most prominent of the laryngeal cartilages.
3. Air flows from the larynx to the trachea and then enters the:
   a. left and right lungs **c.** pharynx
   b. left and right main bronchi **d.** segmental bronchi
4. Circle the correct underlined term. The lining of the trachea is: pseudostratified ciliated <u>columnar epithelium</u> / <u>transitional epithelium</u>, which propels dust particles, bacteria, and other debris away from the lungs.
5. __________, tiny balloonlike structures, are composed of a single thin layer of squamous epithelium. They are the main structural and functional units of the lung and the actual sites of gas exchange.

Materials

- Resin cast of the respiratory tree
- Human torso model
- Thoracic cavity structures model and/or chart of the respiratory system
- Larynx model (if available)
- Preserved inflatable lung preparation or sheep pluck fresh from the slaughterhouse
- Source of compressed air
- Dissecting tray
- Disposable gloves
- Disposable autoclave bag
- Prepared slides of the following (if available): trachea (cross section), lung tissue, both normal and pathological specimens
- Compound and stereomicroscopes

For instructions on animal dissections, see the dissection exercises (starting on p. 695) in the cat and fetal pig editions of this manual.

Body cells require an abundant and continuous supply of oxygen. The major role of the **respiratory system** is to supply the body with oxygen and dispose of carbon dioxide. To fulfill this role, at least four distinct processes, collectively referred to as **respiration**, must occur:

Pulmonary ventilation: The tidelike movement of air into and out of the lungs that allows the gases to be continuously changed and refreshed. Also more simply called *breathing*.

External respiration: The gas exchange between the blood and the air-filled chambers of the lungs.

Transport of respiratory gases: The transport of respiratory gases between the lungs and tissue cells of the body using blood as the transport vehicle.

Internal respiration: Exchange of gases between systemic blood and tissue cells.

Upper Respiratory System Structures

The upper respiratory system structures—the external nose, nasal cavity, pharynx, and paranasal sinuses—are summarized in **Table 36.1** and illustrated in **Figure 36.1**. As you read through the descriptions in the table, identify each structure in the figure. Note that different sources divide the upper and lower respiratory systems slightly differently.

Table 36.1 Structures of the Upper Respiratory System (Figure 36.1)

Structure	Description	Function
External Nose	Externally visible, its inferior surface has nostrils (nares). Supported by bone and cartilage and covered with skin.	The nostrils provide an entrance for air into the respiratory system.
Nasal Cavity (includes the structures listed below)	Lined with respiratory mucosa composed of pseudostratified ciliated columnar epithelium. The floor of the cavity is formed by the hard and soft palates.	Functions to filter, warm, and moisten incoming air; resonance chambers for voice production.
Nasal vestibule	Anterior portion of the nasal cavity; contains sebaceous and sweat glands and numerous hair follicles.	Filters coarse particles from the air.
Nasal septum	Formed by the vomer, perpendicular plate of the ethmoid bone, and septal cartilage.	Divides the nasal cavity into left and right sides.
Superior, middle, and inferior nasal conchae	Turbinates that project medially from the lateral walls of the cavity. Each concha has a corresponding meatus beneath it.	Increase the surface area of the mucosa, which enhances air turbulence and aids in trapping large particles in the mucus.
Posterior nasal apertures	Posterior openings of the nasal cavity.	Provide an exit for the air into the nasopharynx.
Pharynx (3 subdivisions: nasopharynx, oropharynx, and laryngopharynx—listed below)		
Nasopharynx	Superior portion of the pharynx located posterior to the nasal cavity; lined with pseudostratified ciliated columnar epithelium. The pharyngeal tonsil and openings of the pharyngotympanic tubes (surrounded by tubal tonsils) are located in this region.	Provides for the passage of air from the nasal cavity. Tonsils in the region provide protection against pathogens.
Oropharynx	Located posterior to the oral cavity and extends from the soft palate to the epiglottis; lined with stratified squamous epithelium. Its lateral walls contain the palatine tonsils. The lingual tonsils are in the anterior oropharynx at the base of the tongue.	Provides for the passage of air and swallowed food. Tonsils provide protection against pathogens.
Laryngopharynx	Extends from the epiglottis to the larynx; lined with stratified squamous epithelium. It diverges into respiratory and digestive branches.	Provides for the passage of air and swallowed food.
Pharyngotympanic Tube	Tube that opens into the lateral walls of the nasopharynx and connects the nasopharynx to the middle ear.	Allows the middle ear pressure to equalize with the atmospheric pressure.
Paranasal Sinuses	Surround the nasal cavity and are named for the bones in which they are located. Lined with pseudostratified ciliated columnar epithelium.	Act as resonance chambers for speech; warm and moisten incoming air.

Lower Respiratory System Structures

Anatomically, the lower respiratory structures include the larynx, trachea, bronchi, and lungs. The **larynx**, or voice box, attaches superiorly to the hyoid bone and contains a number of important structures summarized in **Table 36.2**, p. 534 and illustrated in **Figure 36.2**, p. 534. From the larynx, air enters the **trachea**, or windpipe, and travels down the trachea to the level of the *sternal angle*. There the passageway divides into the right and left **main (primary) bronchi** (**Figure 36.3**, p. 535). The right main bronchus is wider, shorter, and more vertical than the left, and foreign objects that enter the respiratory passageways are more likely to become stuck there.

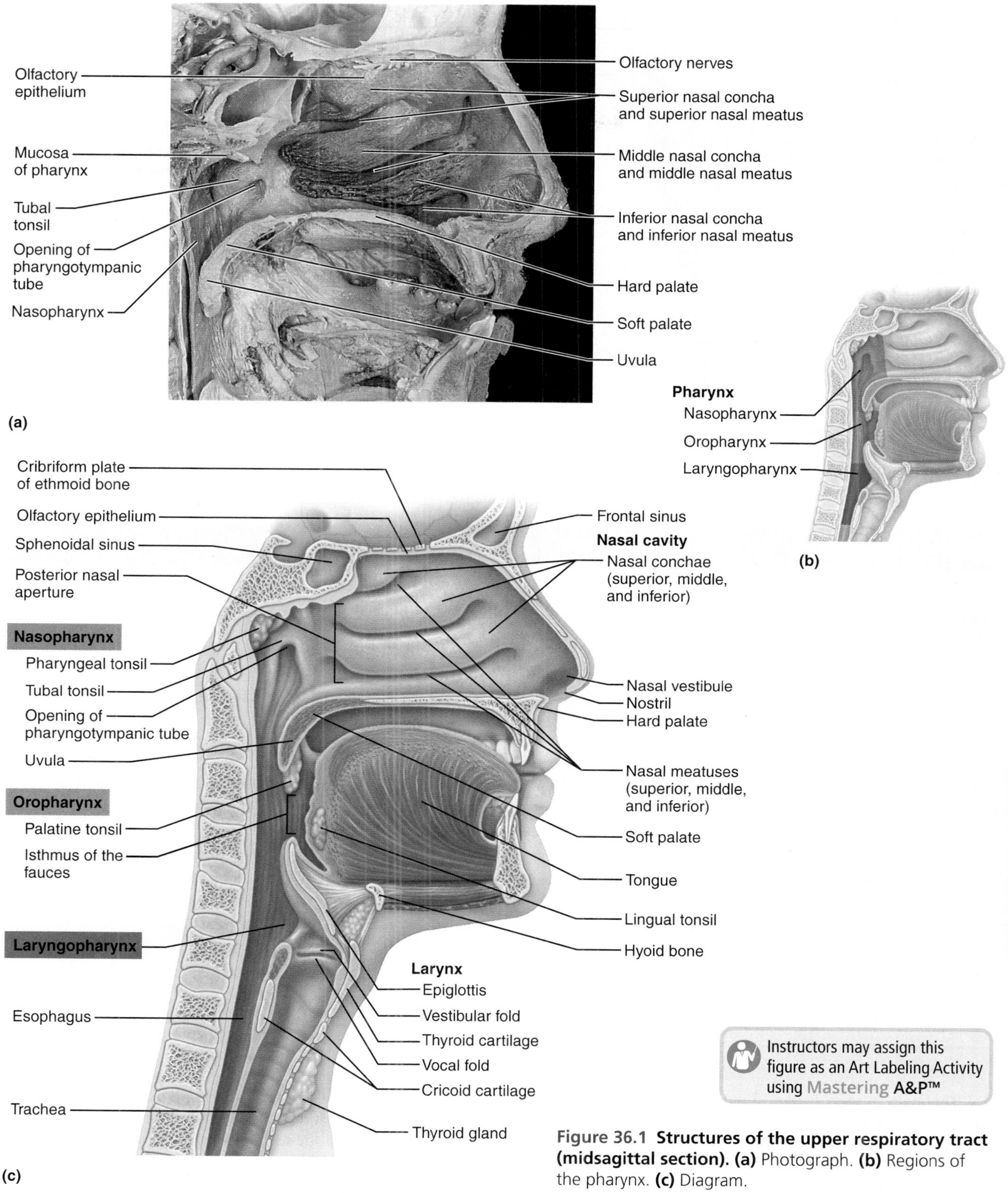

Instructors may assign this figure as an Art Labeling Activity using Mastering A&P™

Figure 36.1 Structures of the upper respiratory tract (midsagittal section). (a) Photograph. **(b)** Regions of the pharynx. **(c)** Diagram.

The trachea is lined with a ciliated, mucus-secreting, pseudostratified columnar epithelium. The cilia propel mucus laden with dust particles and other debris away from the lungs and toward the throat, where it can be expectorated or swallowed. The walls of the trachea are reinforced with C-shaped cartilaginous rings (see Figure 36.6, p. 537). These C-shaped cartilages serve a double function: The incomplete parts allow the esophagus to expand anteriorly when a large food bolus is swallowed. The solid portions reinforce the trachea walls to maintain its open passageway when pressure changes occur during breathing.

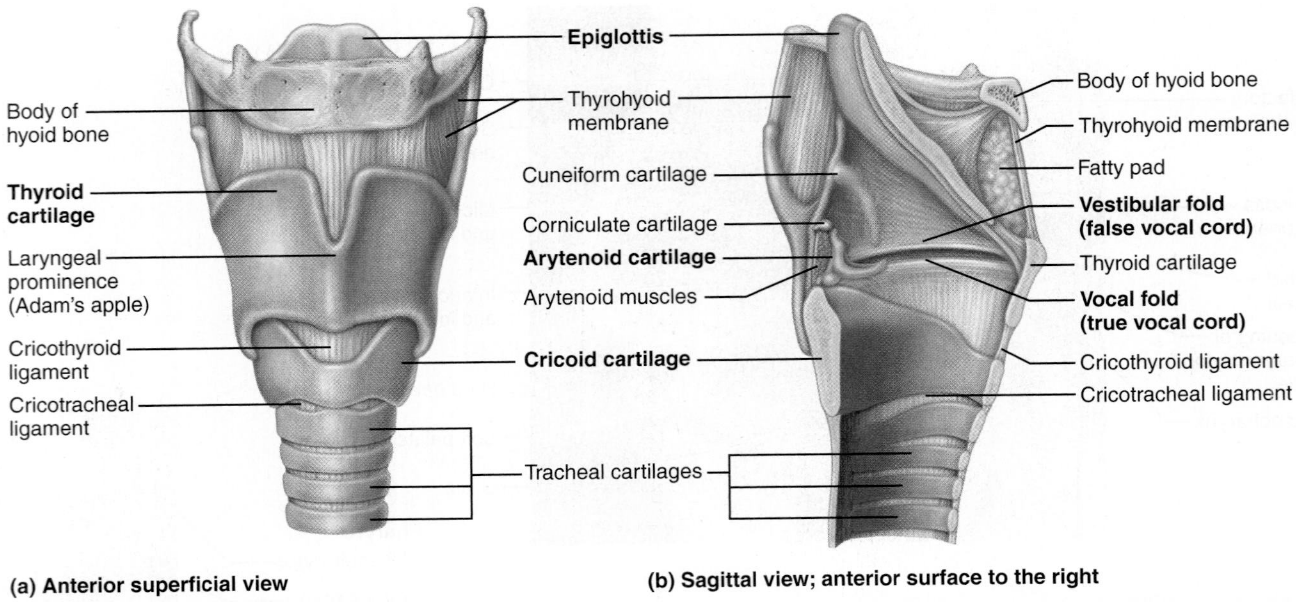

Figure 36.2 The larynx. Attached tracheal cartilages are also shown.

Instructors may assign this figure as an Art Labeling Activity using Mastering A&P™

Table 36.2 Structures of the Larynx (Figure 36.2)

Structure	Description	Function
Larynx (includes the structures listed below)	Tube connecting the laryngopharynx and the trachea. Nine cartilages are present. Epithelium superior to the vocal folds is stratified squamous. Epithelium inferior to the vocal folds is pseudostratified ciliated columnar.	Air passageway; prevents food from entering the lower respiratory tract. Responsible for voice production.
Thyroid cartilage	Large cartilage made up of hyaline cartilage. Its laryngeal prominence is commonly referred to as the Adam's apple.	Forms the framework of the larynx.
Cricoid cartilage	Single ring of hyaline cartilage located inferior to the thyroid cartilage and superior to the trachea.	Attaches the larynx to the trachea via the cricotracheal ligament.
Arytenoid cartilage	Paired pyramid-shaped hyaline cartilages.	Anchor the vocal folds (true vocal cords).
Corniculate cartilage	Paired small horn-shaped hyaline cartilages located atop the arytenoid cartilages.	Form part of the posterior wall of the larynx.
Cuneiform cartilage	Paired wedge-shaped hyaline cartilages.	Form the lateral aspect of the laryngeal wall.
Epiglottis	Single flap of elastic cartilage anchored to the inner rim of the thyroid cartilage.	"Guardian of the airways" forms a lid over the larynx during swallowing.
Vocal folds (true vocal cords)	Mucosal folds composed of mostly elastic fibers covered with mucous membrane; attached to the arytenoid cartilage.	Vibrate with expired air for sound production.
Vestibular folds (false vocal cords)	Located superior to the vocal folds. Mucosal folds similar in composition to the vocal folds.	Protect the vocal folds and help to close the glottis when we swallow.
Glottis	The vocal folds and the slitlike passageway between the vocal folds.	Plays a role in the Valsalva maneuver.

36

The main bronchi further divide into smaller branches, the **lobar (secondary) bronchi**. The lobar bronchi branch into the **segmental (tertiary) bronchi**, which divide repeatedly into smaller and smaller bronchi. Passages smaller than 1 mm in diameter are called **bronchioles**. Each bronchiole divides into many **terminal bronchioles**, which are less than 0.5 mm in diameter. Each terminal bronchiole branches into two or more **respiratory bronchioles** (Figure 36.3). All but the smallest branches have cartilaginous reinforcements in their walls. As the respiratory tubes get smaller and smaller, the relative amount of smooth muscle in their walls increases as the amount of cartilage declines and finally disappears. Additionally, the epithelium of the bronchioles changes from pseudostratified columnar to columnar and then to cuboidal in the terminal bronchioles and respiratory bronchioles. The continuous branching of the respiratory passageways in the lungs is often referred to as the **bronchial tree**.

- Observe a resin cast of respiratory passages if one is available.

The respiratory bronchioles subdivide into several **alveolar ducts**, which terminate in alveolar sacs that resemble clusters of grapes. **Alveoli**, tiny balloonlike expansions that each represent a single grape in the alveolar sac, are composed of a single thin layer of squamous epithelium overlying a basal lamina.

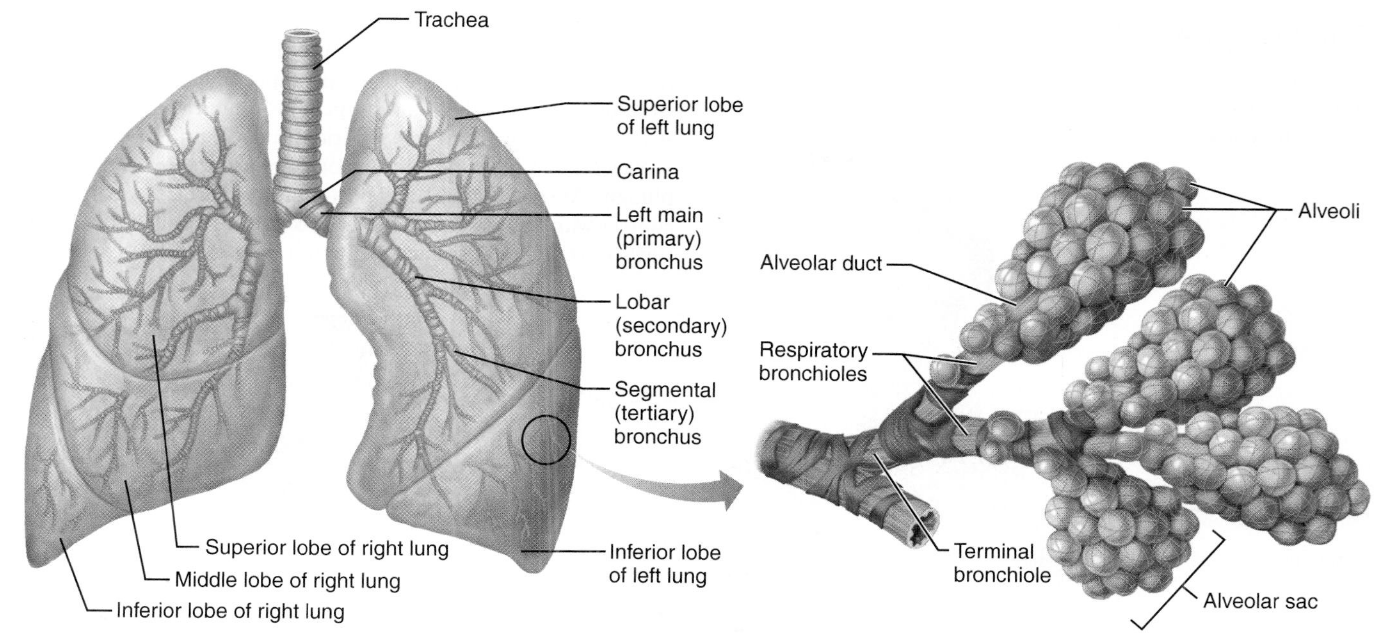

Figure 36.3 Structures of the lower respiratory tract.

Instructors may assign this figure as an Art Labeling Activity using Mastering A&P™

The external surfaces of the alveoli are covered with a network of pulmonary capillaries (**Figure 36.4**). Together, the alveolar and capillary walls and their fused basement membranes form the **respiratory membrane**, also called the *blood air barrier*.

Because gas exchanges occur across the respiratory membrane, the alveoli, alveolar ducts, and respiratory bronchioles are referred to collectively as **respiratory zone structures**. All other respiratory passageways (from the nasal cavity to the terminal bronchioles) simply serve as access or exit routes to and from these gas exchange chambers and are called **conducting zone structures** or *anatomical dead space*.

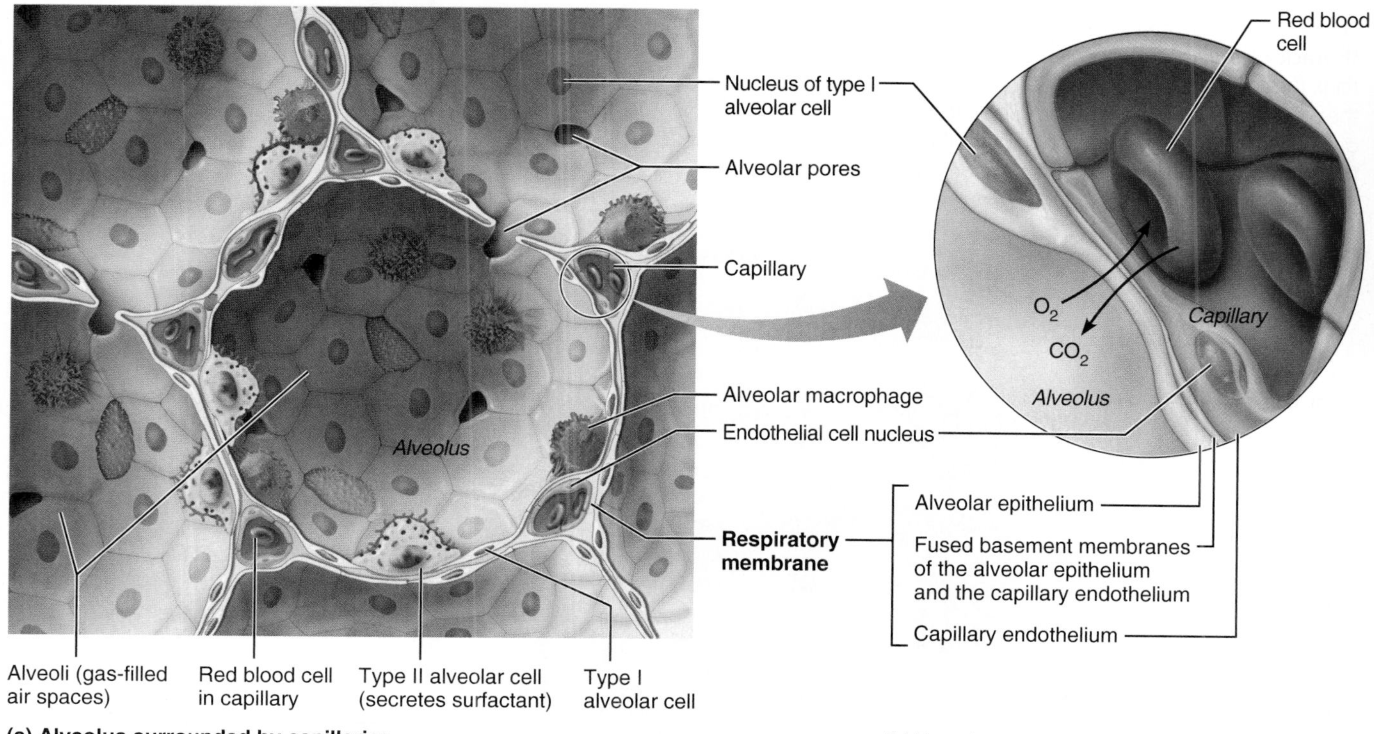

Figure 36.4 Diagram of the relationship between the alveoli and pulmonary capillaries involved in gas exchange.

Instructors may assign this figure as an Art Labeling Activity using Mastering A&P™

The Lungs and Pleurae

The paired lungs are soft, spongy organs that occupy the entire thoracic cavity except for the *mediastinum*, which houses the heart, bronchi, esophagus, and other organs (**Figure 36.5**). Each lung is connected to the mediastinum by a **root** containing its vascular and bronchial attachments. The structures of the root enter (or leave) the lung via a medial indentation called the **hilum**. All structures distal to the main bronchi are found within the lung. A lung's **apex**, the narrower superior aspect, lies just deep to the clavicle, and its **base**, the inferior concave surface, rests on the diaphragm. Anterior, lateral, and posterior lung surfaces are in close contact with the ribs and, hence, are collectively called the **costal surface**. The medial surface of the left lung exhibits a concavity

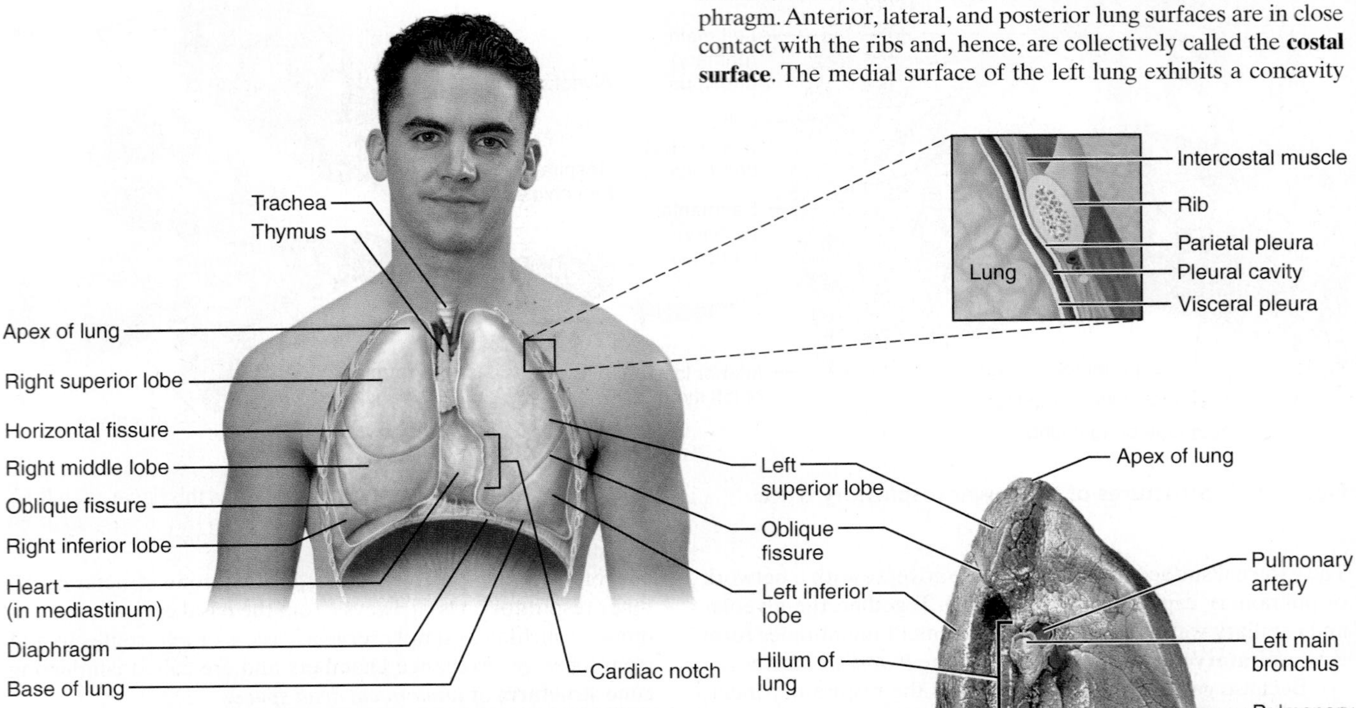

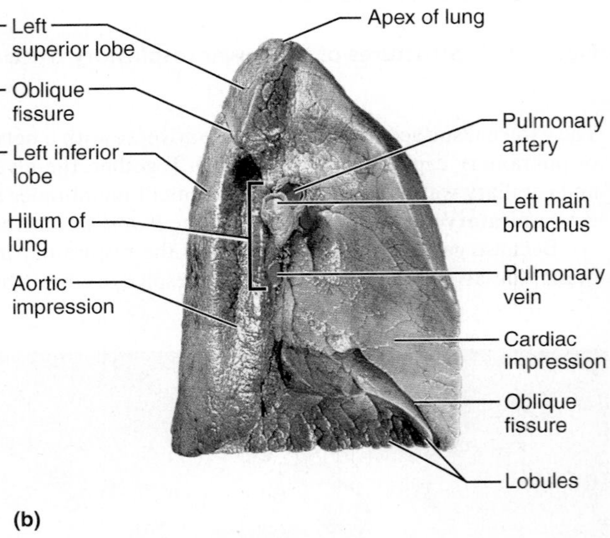

Instructors may assign this figure as an Art Labeling Activity using Mastering A&P™

Figure 36.5 Anatomical relationships of organs in the thoracic cavity. (a) Anterior view of the thoracic organs. The lungs flank the central mediastinum. The inset at upper right depicts the pleurae and the pleural cavity. **(b)** Photograph of medial aspect of left lung. **(c)** Transverse section through the superior part of the thorax, showing the lungs and the main organs in the mediastinum.

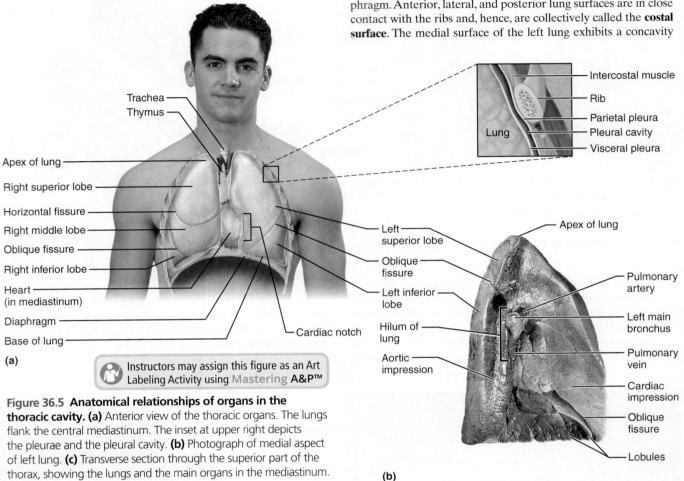

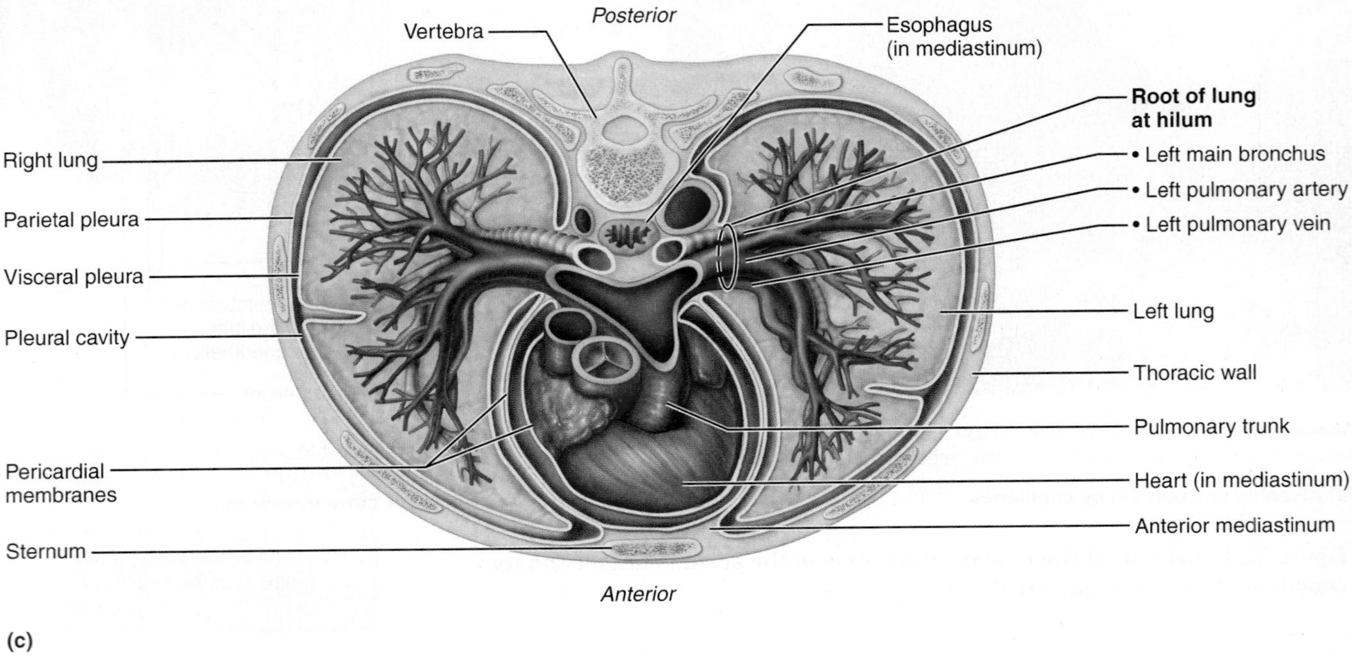

36

called the **cardiac notch**, which accommodates the heart where it extends left from the body midline. Fissures divide the lungs into a number of **lobes**—two in the left lung and three in the right. Other than the respiratory passageways and air spaces that make up the bulk of their volume, the lungs are mostly elastic connective tissue, which allows them to recoil passively during expiration.

Each lung is enclosed in a double-layered sac of serous membrane called the **pleura**. The outer layer, the **parietal pleura**, is attached to the thoracic walls and the **diaphragm**; the inner layer, covering the lung tissue, is the **visceral pleura**. The two pleural layers are separated by the **pleural cavity**, which is filled with a thin film of *pleural fluid*. Produced by the pleurae, this fluid allows the lungs to glide without friction over the thoracic wall during breathing.

Activity 1

Identifying Respiratory System Organs

Before proceeding, be sure to locate on the torso model, thoracic cavity structures model, larynx model, or an anatomical chart all the respiratory structures described—both upper and lower respiratory system organs.

For instructions on animal dissections, see the dissection exercises (starting on p. 695) in the cat and fetal pig editions of this manual.

Activity 2

Demonstrating Lung Inflation in a Sheep Pluck

A *sheep pluck* includes the larynx, trachea with attached lungs, the heart and pericardium, and portions of the major blood vessels found in the mediastinum.

Don disposable gloves, obtain a dissecting tray and a fresh sheep pluck (or a preserved pluck of another animal), and identify the lower respiratory system organs. Once you have completed your observations, insert a hose from an air compressor (vacuum pump) into the trachea, and allow air to flow alternately into and out of the lungs. Notice how the lungs inflate. This observation is educational in a preserved pluck, but it is a spectacular sight in a fresh one. When a fresh pluck used, the lung pluck changes color (becomes redder) as hemoglobin in trapped RBCs becomes loaded with oxygen.

Dispose of the gloves in the autoclave bag immediately after use.

Activity 3

Examining Prepared Slides of Trachea and Lung Tissue

1. Obtain a compound microscope and a slide of a cross section of the tracheal wall. Identify the smooth muscle layer, the hyaline cartilage supporting rings, and the pseudostratified ciliated epithelium (use **Figure 36.6** as a guide). Also try to identify a few goblet cells in the epithelium. (See Figure 6.3c and d on pp. 68–69.)

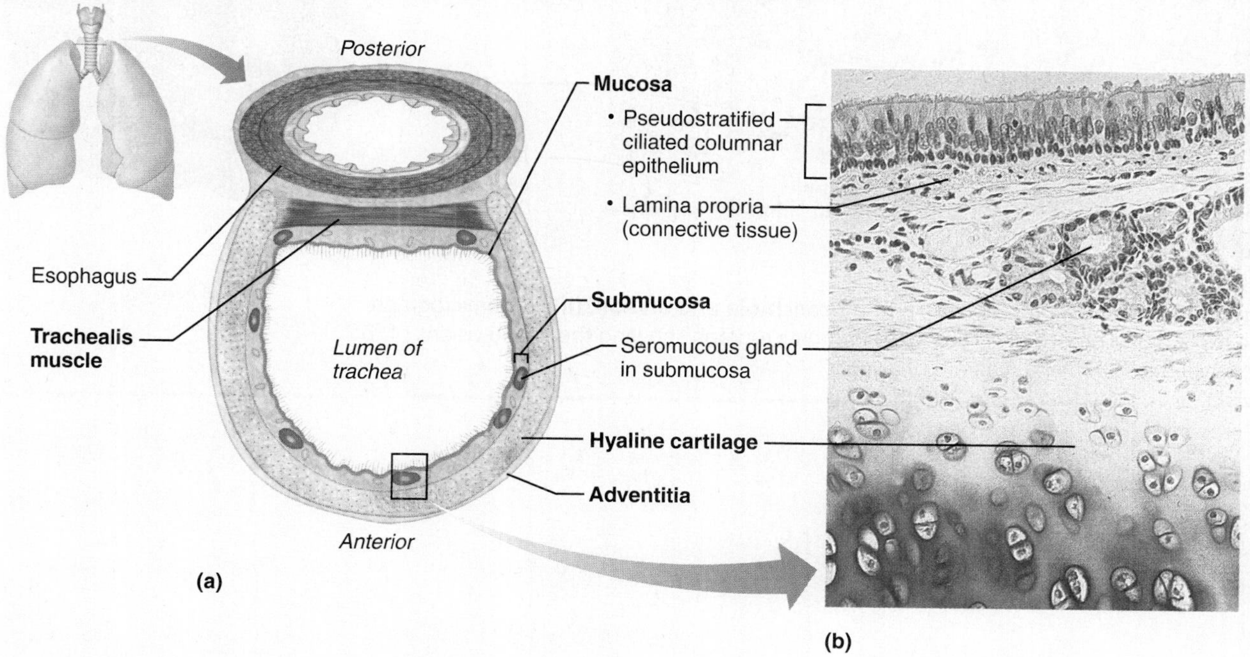

Figure 36.6 Tissue composition of the tracheal wall. (a) Cross-sectional view of the trachea. **(b)** Photomicrograph of a portion of the tracheal wall (125×).

Text continues on next page. →

2. Obtain a slide of lung tissue for examination. The alveolus is the main structural and functional unit of the lung and is the actual site of gas exchange. Identify a bronchiole (**Figure 36.7a**) and the simple squamous epithelium of the alveolar walls (Figure 36.7b).

3. Examine slides of pathological lung tissues, and compare them to the normal lung specimens. Record your observations in the review sheet at the end of this exercise.

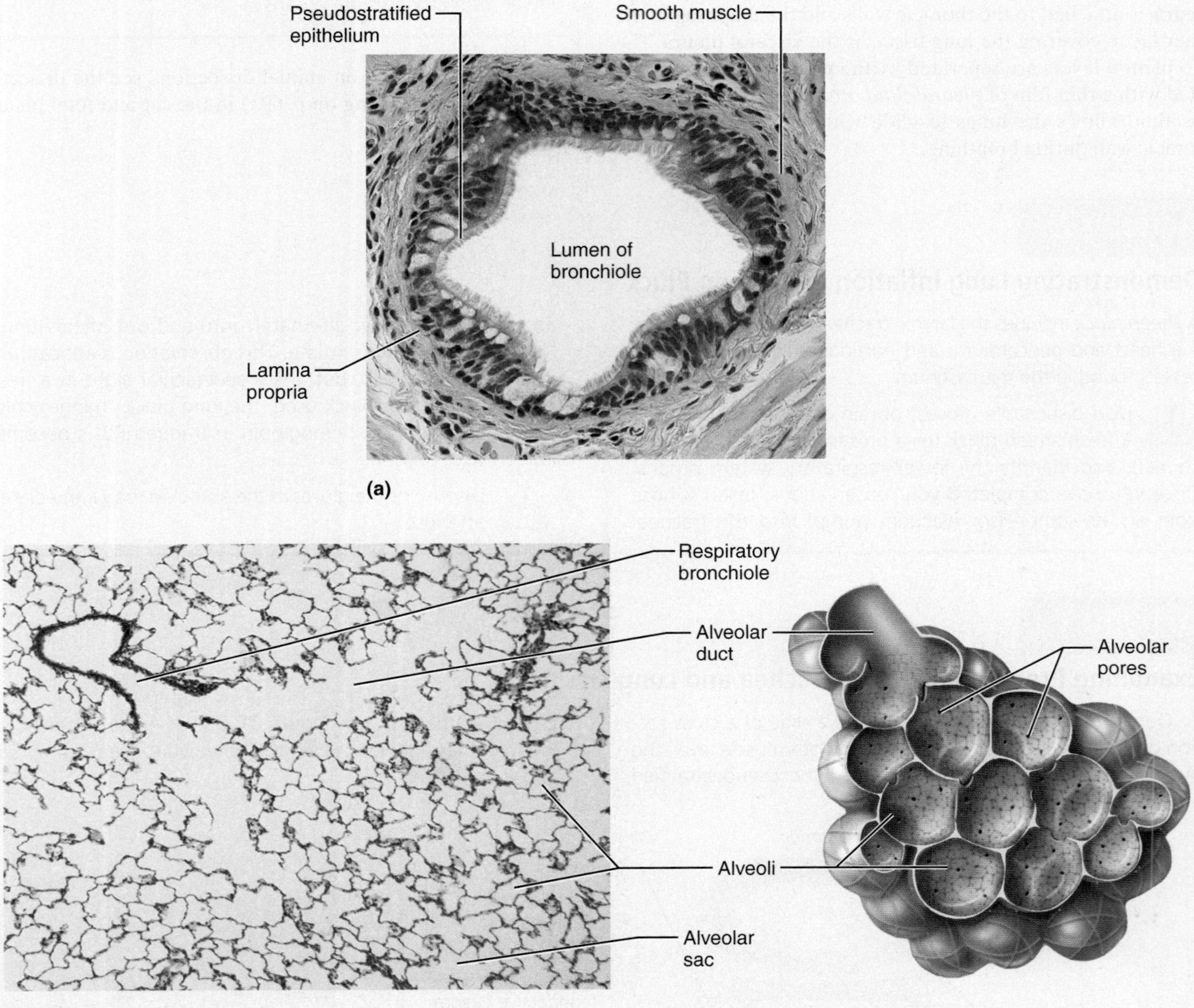

Figure 36.7 Microscopic structure of a bronchiole and alveoli. (a) Photomicrograph of a section of a bronchiole (180×). **(b)** Photomicrograph showing the final divisions of the bronchial tree (50×) and diagram of alveoli.

EXERCISE 36

REVIEW SHEET

Anatomy of the Respiratory System

Instructors may assign a portion of the Review Sheet questions using Mastering A&P™

Name ______________________ Lab Time/Date ______________________

Upper and Lower Respiratory System Structures

1. Complete the labeling of the model of the respiratory structures (sagittal section) shown below.

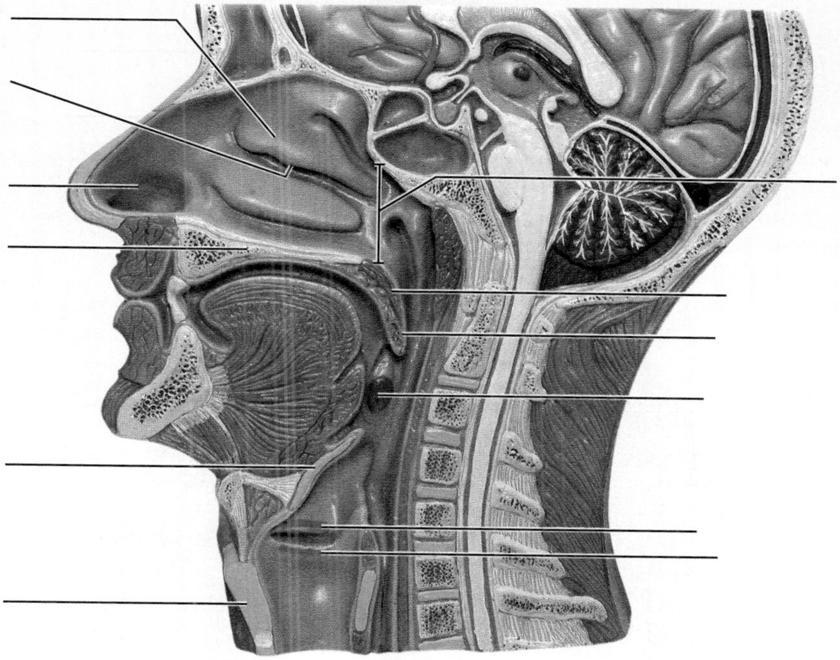

2. Two pairs of mucosal folds are found in the larynx. Which pair are the true vocal cords (superior or inferior)?

____

3. Name the specific cartilages in the larynx that correspond to the following descriptions.

forms the Adam's apple: ______________ shaped like a ring: ______________

a "lid" for the larynx: ______________ vocal cord attachment: ______________

4. Why is it important that the human trachea is reinforced with cartilaginous rings?

____

Why is it important that the rings are incomplete posteriorly?

____

5. What is the function of the pleural fluid? ______________________

____

____

6. Name two functions of the nasal conchae: ______________________________

and ______________________________

7. The following questions refer to the main bronchi.

Which is longer? ______________ Larger in diameter? ______________ More horizontal? ______________

Which more commonly traps a foreign object that has entered the respiratory passageways? ______________

8. Appropriately label all structures provided with leader lines on the model shown below.

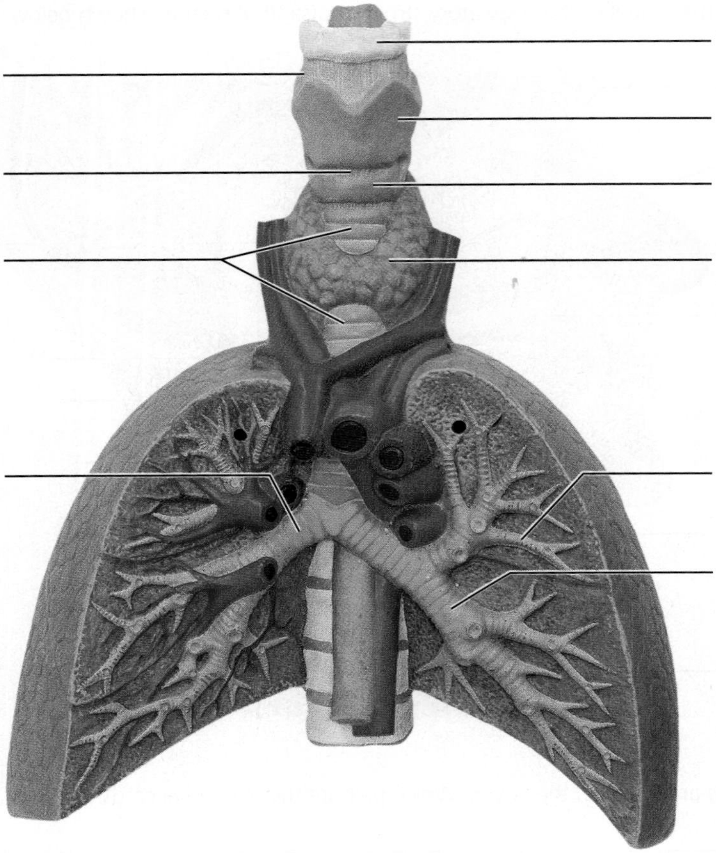

9. Trace a molecule of oxygen from the nostrils to the pulmonary capillaries of the lungs: Nostrils →

____

____

____

10. Match the terms in column B to the descriptions in column A.

Column A	Column B
________ 1. connects the larynx to the main bronchi	a. alveolus
________ 2. includes terminal and respiratory as subtypes	b. bronchiole
________ 3. food passageway posterior to the trachea	c. conchae
________ 4. covers the glottis during swallowing of food	d. epiglottis
________ 5. contains the vocal cords	e. esophagus
________ 6. indentation on the lung where the lung root structures enter and exit	f. hilum
________ 7. pleural layer lining the walls of the thorax	g. larynx
________ 8. site from which oxygen enters the pulmonary blood	h. palate
________ 9. connects the middle ear to the nasopharynx	i. pharyngotympanic tube
________ 10. pleural layer in contact with the surface of the lung	j. parietal pleura
________ 11. increases air turbulence in the nasal cavity	k. trachea
________ 12. separates the oral cavity from the nasal cavity	l. visceral pleura

11. What portions of the respiratory system are referred to as anatomical dead space? ________________________________

____

____

Why? ____

12. Define the following terms.

external respiration: ____

____

internal respiration: ____

____

Demonstrating Lung Inflation in a Sheep Pluck

13. Does the lung inflate part by part or as a whole, like a balloon? ________________________

14. What happened when the pressure was released? ________________________________

15. What type of tissue ensures this phenomenon? ________________________________

Examining Prepared Slides of Trachea and Lung Tissue

16. What structural characteristics of the alveoli make them an ideal site for the diffusion of gases?

___

17. If you observed pathological lung sections, record your observations. Also record how the tissue differed from normal lung tissue. Complete the table below using your answers.

Slide type	Observations	Comparison to normal lung tissue

18. ➕ Epiglottitis is a condition in which the epiglottis is inflamed. It is most often caused by a bacterial infection. Explain why this type of inflammation is life-threatening. ___

___

19. ➕ Pneumonia is an infectious disease in which fluid accumulates in the alveoli. Patients who are diagnosed with pneumonia are monitored for their oxygen saturation levels. Describe how pneumonia could affect the amount of oxygen in the blood. ___

___

37 EXERCISE Respiratory System Physiology

Learning Outcomes

- Define the following and provide volume figures if applicable:

inspiration	inspiratory reserve volume (IRV)
expiration	minute respiratory volume (MRV)
tidal volume (TV)	forced vital capacity (FVC)
vital capacity (VC)	forced expiratory volume (FEV_T)
expiratory reserve volume (ERV)	

- Explain the role of muscles and volume changes in the mechanical process of breathing.
- Describe bronchial and vesicular breathing sounds.
- Demonstrate proper usage of a spirometer or an airflow transducer and associated BIOPAC® equipment.
- Discuss the relative importance of various mechanical and chemical factors in producing respiratory variations.
- Explain the importance of the carbonic acid–bicarbonate buffer system in maintaining blood pH.

Go to Mastering A&P™ > Study Area to improve your performance in A&P Lab.

> Lab Tools > Pre-Lab Videos > Spirometry

Instructors may assign new Building Vocabulary coaching activities, Pre-Lab Quiz questions, Art Labeling activities, Pre-Lab Video Coaching Activities for Operating the Model Lung and Spirometry, Practice Anatomy Lab Practical questions (PAL), PhysioEx activities, and more using the Mastering **A&P™** Item Library.

Pre-Lab Quiz

Instructors may assign these and other Pre-Lab Quiz questions using Mastering **A&P™**

1. Which of the following processes does *not* occur during inspiration?
   a. diaphragm moves to a flattened position
   b. gas pressure inside the lungs is lowered
   c. inspiratory muscles relax
   d. size of thoracic cavity increases
2. During normal quiet breathing, about _____________ ml of air moves into and out of the lungs with each breath.
   a. 250 **c.** 1000
   b. 500 **d.** 2000
3. Circle True or False. The neural centers that control respiratory rhythm and maintain a rate of 12–18 respirations per minute are located in the medulla and thalamus.
4. The carbonic acid–bicarbonate buffer system stabilizes arterial blood pH at:
   a. 2.0 ± 1.00 **c.** 6.2 ± 0.07
   b. 7.4 ± 0.02 **d.** 9.5 ±1.15
5. Circle the correct underlined term. Acids / Bases released into the blood by the body cells tend to lower the pH of the blood and cause it to become acidic.

Materials

- Model lung (bell jar demonstrator)
- Tape measure with centimeter divisions (cloth or plastic)
- Stethoscope
- Alcohol swabs
- Spirometer or BIOPAC® equipment:

Spirometer, disposable cardboard mouthpieces, nose clips, table (on board) for recording class data, disposable autoclave bag, battery jar containing 70% ethanol solution

Text continues on next page. →

BIOPAC® BSL System with BSL software version 4.0.1 and above (for Windows 10/8.x/7 or Mac OS X 10.9–10.12), data acquisition unit MP36/35 or MP45, PC or Mac computer, BIOPAC® airflow transducer, BIOPAC® calibration syringe, disposable mouthpiece, nose clip, and bacteriological filter.

Instructors using the MP36/35/30 data acquisition unit with BSL software versions earlier than 4.0.1 (for Windows or Mac) will need slightly different channel settings and collection strategies. Instructions for using the older data acquisition unit can be found on MasteringA&P.

- Paper bag
- 0.05 *M* NaOH
- Phenol red in a dropper bottle
- 100-ml beakers
- Distilled water
- Straws
- Concentrated HCl and NaOH in dropper bottles
- 250- and 50-ml beakers
- Plastic wash bottles containing distilled water
- pH meter (standardized with buffer of pH 7)
- Buffer solution (pH 7)
- Graduated cylinder (100 ml)
- Glass stirring rod
- Animal plasma
- 0.01 *M* HCl in dropper bottles

PEx PhysioEx™ 9.1 Computer Simulation Ex. 7 on p. PEx-105

Note: *Instructions for using PowerLab® equipment can be found on MasteringA&P.*

37

The body's trillions of cells require O_2 and give off CO_2 as a waste the body must get rid of. The **respiratory system** provides the link with the external environment for both taking in O_2 and eliminating CO_2, but it doesn't work alone. The cardiovascular system via its contained blood provides the watery medium for transporting O_2 and CO_2 in the body. Let's look into how the respiratory system carries out its role.

Mechanics of Respiration

Pulmonary ventilation, or **breathing**, consists of two phases: **inspiration**, during which air is taken into the lungs, and **expiration**, during which air passes out of the lungs. As the inspiratory muscles (external intercostals and diaphragm) contract during inspiration, the size of the thoracic cavity increases. The diaphragm moves from its relaxed dome shape to a flattened position, increasing the vertical dimension. The external intercostals lift the rib cage, increasing the anterior-posterior and lateral dimensions (**Figure 37.1**). Because the lungs adhere to the thoracic walls like flypaper owing to the presence of serous fluid in the pleural cavity, the intrapulmonary volume (volume within the lungs) also increases, lowering the air (gas) pressure inside the lungs. The gases then expand to fill the available space, creating a partial vacuum that causes air to flow into the lungs—constituting the act of inspiration. During expiration, the inspiratory muscles relax, and the natural tendency of the elastic lung tissue to recoil decreases the intrathoracic and intrapulmonary volumes. As the gas molecules within the lungs are forced closer together, the intrapulmonary pressure rises to a point higher than atmospheric pressure. This causes gases to flow out of the lungs to equalize the pressure inside and outside the lungs—the act of expiration.

Activity 1

Instructors may assign a related Pre-Lab Video Coaching Activity using Mastering A&P™

Operating the Model Lung

Observe the bell jar model of the lungs, which demonstrates the principles involved in gas flows into and out of the lungs. It is a simple apparatus with a hard plastic dome-shaped container called a bell jar (representing the parietal pleura), the interior of the bell jar (representing the thoracic cavity), a rubber membrane (representing the diaphragm), two balloons (representing the lungs), and an inverted Y-shaped tube (representing the trachea and main bronchi).

1. Go to the demonstration area and work the model lung by moving the rubber diaphragm up and down. Notice the *relative* changes in balloon (lung) size as the volume of the thoracic cavity is alternately increased and decreased.

2. Check the appropriate columns in the chart concerning these observations in the review sheet at the end of this exercise.

3. A pneumothorax is a condition in which air has entered the pleural cavity, as with a puncture wound. Simulate a pneumothorax: Inflate the balloon lungs by pulling down on the diaphragm. Ask your lab partner to let air into the bottle–thoracic cavity by loosening the rubber stopper.

What happens to the balloon lungs?

____

4. After observing the operation of the model lung, conduct the following tests on your lab partner. Use the tape measure to determine chest circumference by placing the tape around the chest as high up under the armpits as possible. Record the measurements in centimeters in the appropriate space below for each of the conditions.

Quiet breathing:

Inspiration ______________ cm   Expiration ______________ cm

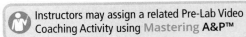

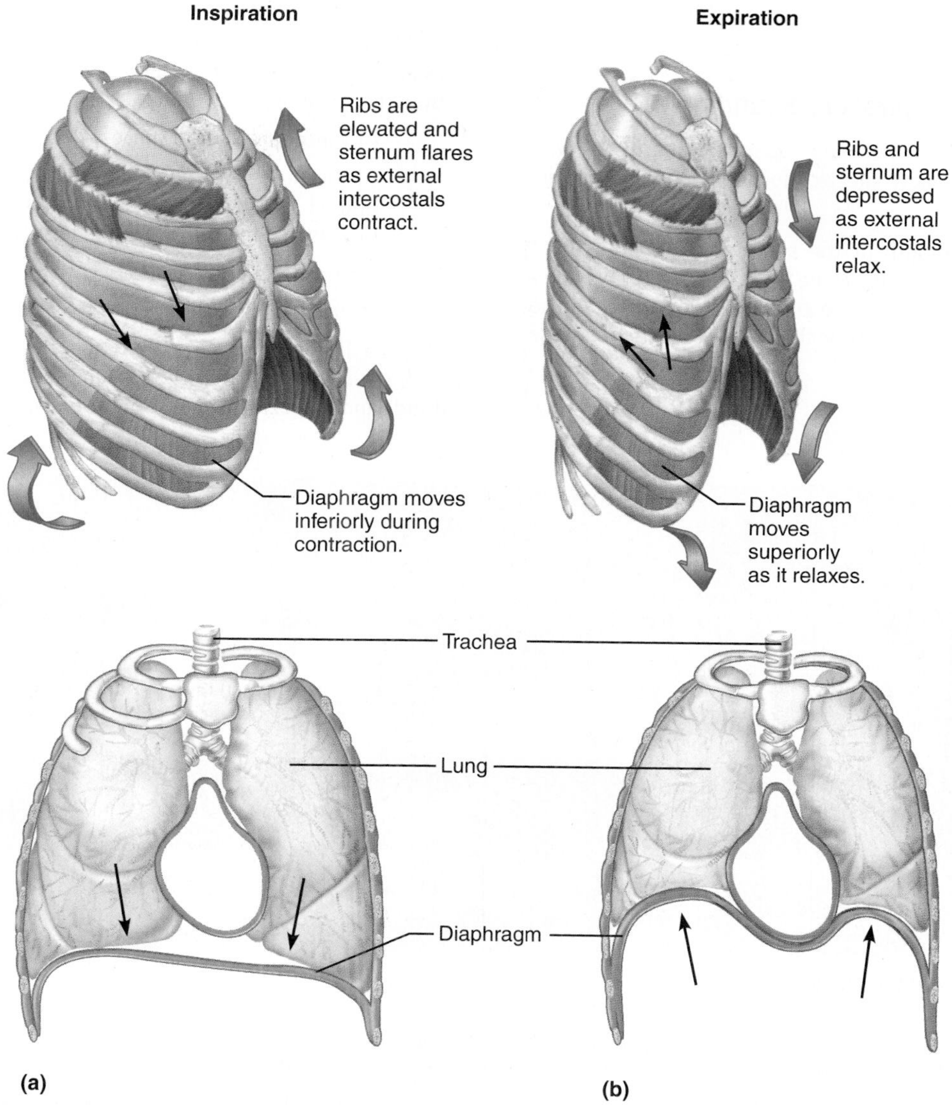

Figure 37.1 Rib cage and diaphragm positions during breathing. (a) At the end of a normal inspiration; ribs elevated, diaphragm contracted. **(b)** At the end of a normal expiration; ribs depressed, diaphragm relaxed.

Forced breathing:

Inspiration ____________ cm Expiration ____________ cm

Do the results coincide with what you expected on the basis of what you have learned thus far? ____________________

How does the structural relationship between the balloon-lungs and bottle–thoracic cavity differ from that seen in the human lungs and thoracic cavity?

____

____

____

Respiratory Sounds

As air flows in and out of the bronchial tree, it produces two characteristic sounds that can be auscultated with a stethoscope. The **bronchial sounds** are produced by air rushing through the large respiratory passageways (the trachea and the bronchi).

The second sound type, **vesicular breathing sounds**, apparently results from air filling the alveolar sacs and resembles the sound of a rustling of leaves.

Activity 2

Auscultating Respiratory Sounds

1. Obtain a stethoscope, and clean the earpieces with an alcohol swab. Allow the alcohol to dry before donning the stethoscope.

2. Place the diaphragm of the stethoscope on the throat of the test subject just below the larynx. Listen for bronchial sounds on inspiration and expiration. Move the stethoscope down toward the bronchi until you can no longer hear sounds.

3. Place the stethoscope over the following chest areas and listen for vesicular sounds during respiration (heard primarily during inspiration).
   - At various intercostal spaces
   - At the *triangle of auscultation* (a small depressed area of the back where the muscles fail to cover the rib cage; located just medial to the inferior part of the scapula)
   - Inferior to the clavicle

Diseased respiratory tissue, mucus, or pus can produce abnormal chest sounds such as rales (a rasping sound) and wheezing (a whistling sound). +

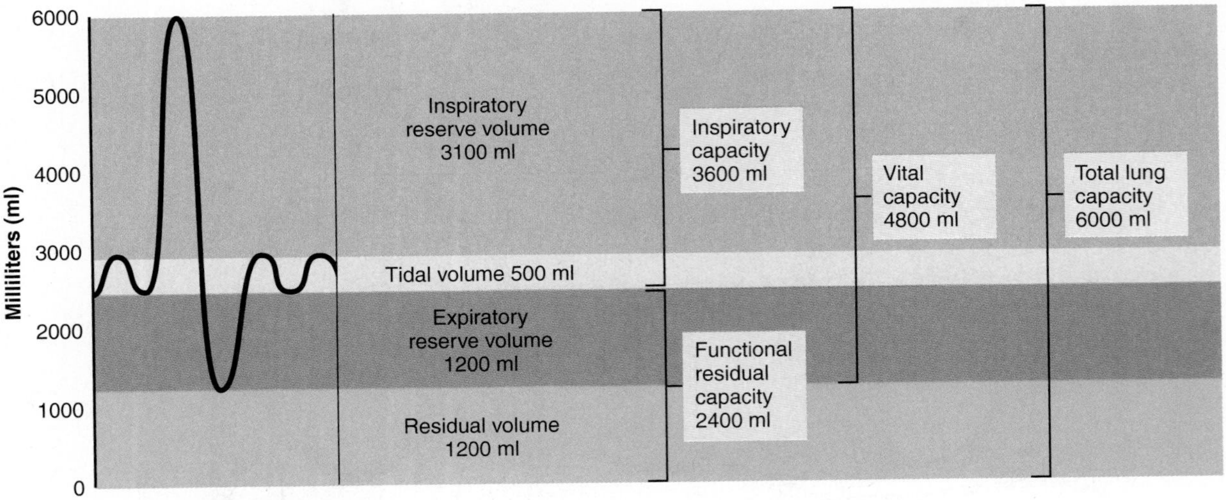

(a) Spirographic record for a male

	Measurement	Adult male average value	Adult female average value	Description
Respiratory volumes	Tidal volume (TV)	500 ml	500 ml	Amount of air inhaled or exhaled with each breath under resting conditions
	Inspiratory reserve volume (IRV)	3100 ml	1900 ml	Amount of air that can be forcefully inhaled after a normal tidal volume inspiration
	Expiratory reserve volume (ERV)	1200 ml	700 ml	Amount of air that can be forcefully exhaled after a normal tidal volume expiration
	Residual volume (RV)	1200 ml	1100 ml	Amount of air remaining in the lungs after a forced expiration
Respiratory capacities	Total lung capacity (TLC)	6000 ml	4200 ml	Maximum amount of air contained in lungs after a maximum inspiratory effort: TLC = TV + IRV + ERV + RV
	Vital capacity (VC)	4800 ml	3100 ml	Maximum amount of air that can be expired after a maximum inspiratory effort: VC = TV + IRV + ERV
	Inspiratory capacity (IC)	3600 ml	2400 ml	Maximum amount of air that can be inspired after a normal tidal volume expiration: IC = TV + IRV
	Functional residual capacity (FRC)	2400 ml	1800 ml	Volume of air remaining in the lungs after a normal tidal volume expiration: FRC = ERV + RV

(b) Summary of respiratory volumes and capacities for males and females

Figure 37.2 Respiratory volumes and capacities.

Respiratory Volumes and Capacities—Spirometry

A person's size, sex, age, and physical condition produce variations in respiratory volumes. Normal quiet breathing moves about 500 ml of air in and out of the lungs with each breath. As you have seen in the first activity, a person can usually forcibly inhale or exhale much more air than is exchanged in normal quiet breathing. The terms used for the measurable respiratory volumes and capacities are defined and illustrated with an idealized tracing in **Figure 37.2**.

Respiratory volumes can be measured, as in Activities 3 and 4, with an apparatus called a **spirometer**. There are two major types of spirometers, which give comparable results—the handheld dry, or wheel, spirometers (such as the Wright spirometer illustrated in **Figure 37.3**) and "wet" spirometers, such as the Phipps and Bird spirometer and the Collins spirometer (which is available in both recording and nonrecording varieties). The somewhat more sophisticated wet spirometer consists of a plastic or metal *bell* within a rectangular or cylindrical tank that air can be added to or removed from (**Figure 37.4**, p. 548).

In nonrecording spirometers, an indicator moves as air is *exhaled*, and only expired air volumes can be measured directly. By contrast, recording spirometers allow both inspired and expired gas volumes to be measured.

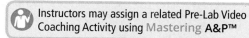

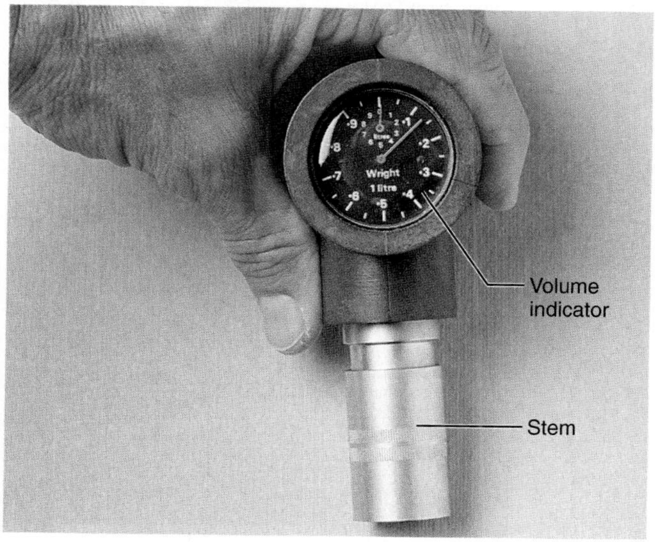

Figure 37.3 **The Wright handheld dry spirometer.** Reset to zero prior to each test.

Activity 3

Instructors may assign a related Pre-Lab Video Coaching Activity using Mastering A&P™

Measuring Respiratory Volumes Using Spirometers

The steps for using a nonrecording spirometer and a wet recording spirometer are given separately below.

Using a Nonrecording Spirometer

1. Before using the spirometer, count and record the subject's normal respiratory rate. The subject should face away from you as you make the count.

Respirations per minute: ________

Now identify the parts of the spirometer you will be using by comparing it to Figure 37.3 or 37.4a. Examine the spirometer volume indicator *before beginning* to make sure you know how to read the scale. Work in pairs, with one person acting as the subject while the other records the data of the volume determinations. *Reset the indicator to zero before beginning each trial.*

Obtain a disposable cardboard mouthpiece. Prior to inserting the cardboard mouthpiece, clean the valve assembly with an alcohol swab. Then insert the mouthpiece in the open end of the valve assembly (attached to the flexible tube) of the wet spirometer or over the fixed stem of the handheld dry spirometer. Before beginning, the subject should practice exhaling through the mouthpiece without exhaling through the nose, or prepare to use the nose clips. If you are using the handheld spirometer, make sure its dial faces upward so that the volumes can be easily read during the tests.

2. The subject should stand erect during testing. Conduct the test three times for each required measurement. Record the data where indicated in this section, and then find the average volume figure for that respiratory measurement. After you have completed the trials and computed the averages, enter the average values on the table prepared on the board for tabulation of class data,* and copy all averaged data onto the review sheet at the end of the exercise.

3. Measuring tidal volume (TV). The TV, or volume of air inhaled and exhaled with each quiet, normal respiration, is approximately 500 ml. To conduct the test, inhale a normal breath, and then exhale a normal breath of air into the spirometer mouthpiece. (Do not force the expiration!) Record the volume and repeat the test twice.

trial 1: __________ ml trial 2: __________ ml

trial 3: __________ ml average TV: __________ ml

***Note to the Instructor:** *The format of class data tabulation can be similar to that shown here. However, it would be interesting to divide the class into smokers and nonsmokers and then compare the mean average VC and ERV for each group. Such a comparison might help to determine whether smokers are handicapped in any way. It also might be a good opportunity for an informal discussion of the early warning signs of chronic bronchitis and emphysema, which are primarily smokers' diseases.*

Text continues on p. 549. →

37

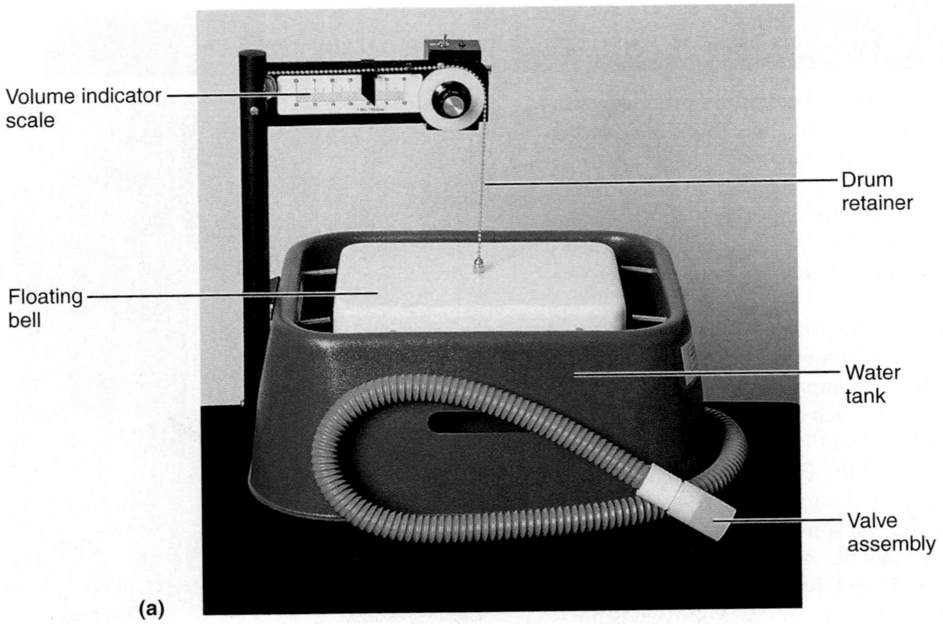

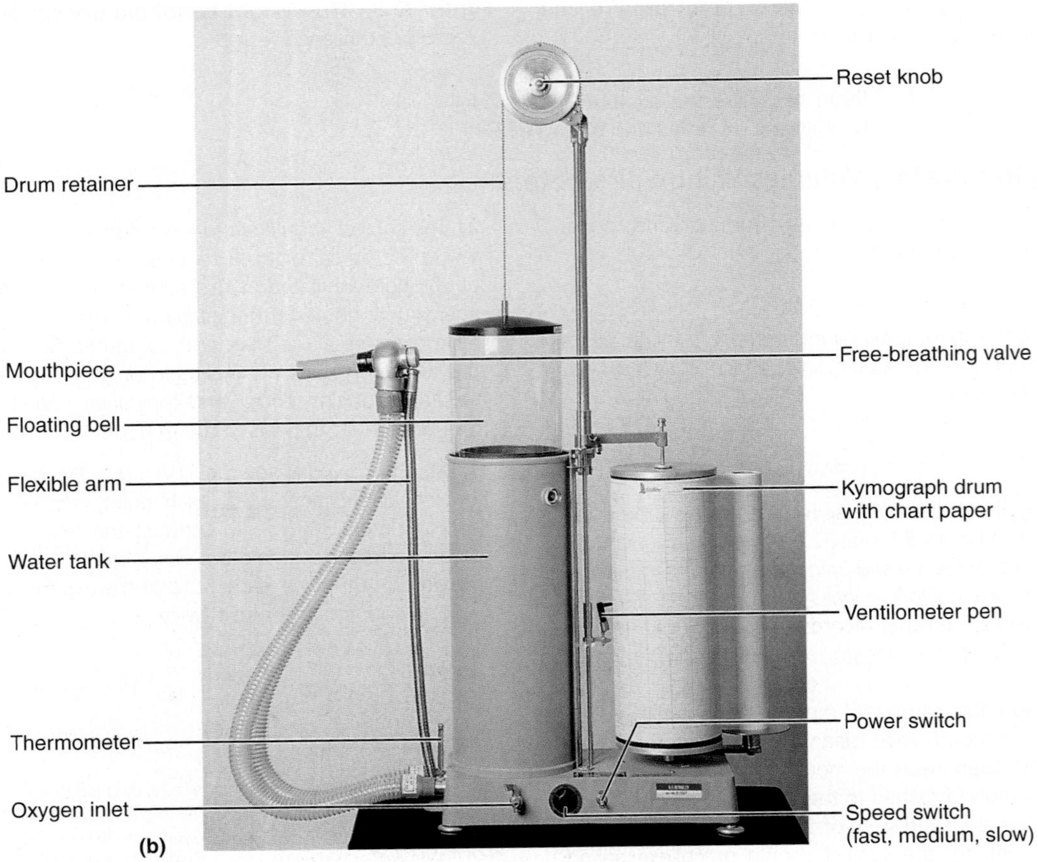

Figure 37.4 Wet spirometers. (a) The Phipps and Bird wet spirometer. **(b)** The Collins-9L wet recording spirometer.

4. Compute the subject's **minute respiratory volume (MRV)** using the following formula:

MRV = TV × respirations/min = ________ml/min

5. Measuring expiratory reserve volume (ERV). The ERV is the volume of air that can be forcibly exhaled after a normal expiration. Normally it ranges between 700 and 1200 ml.

Inhale and exhale normally two or three times, then insert the spirometer mouthpiece and exhale forcibly as much of the additional air as you can. Record your results, and repeat the test twice again.

trial 1: __________ ml trial 2: __________ ml

trial 3: __________ ml average ERV: __________ ml

ERV is dramatically reduced in conditions in which the elasticity of the lungs is decreased by a chronic obstructive pulmonary disease (COPD) such as **emphysema**. Since energy must be used to *deflate* the lungs in such conditions, expiration is physically exhausting to individuals suffering from COPD. ✚

6. Measuring vital capacity (VC). The VC, or total exchangeable air of the lungs (the sum of TV + IRV + ERV), normally ranges from 3100 ml to 4800 ml.

Breathe in and out normally two or three times, and then bend forward and exhale all the air possible. Then, as you raise yourself to the upright position, inhale as fully as possible. It is important to *strain* to inhale the maximum amount of air that you can. Quickly insert the mouthpiece, and exhale as forcibly as you can. Record your results and repeat the test twice again.

trial 1: __________ ml trial 2: __________ ml

trial 3: __________ ml average VC: __________ ml

7. The inspiratory reserve volume (IRV), or volume of air that can be forcibly inhaled following a normal inspiration, can now be computed using the average values obtained for TV, ERV, and VC and plugging them into the equation:

IRV = VC − (TV + ERV)

Record your calculated IRV: ______________________ ml

The normal IRV range is substantial, ranging from 1900 to 3100 ml. How does your calculated value compare?

____

Steps 8–10, which provide common directions for use of both nonrecording and recording spirometers, continue (on p. 551) after the wet recording spirometer directions.

Using a Wet Recording Spirometer

1. In preparation for recording, familiarize yourself with the spirometer by comparing it to Figure 37.4b.

2. Examine the chart paper, noting that its horizontal lines represent milliliter units. To apply the chart paper to the recording drum, first lift the drum retainer and then remove the kymograph drum. Wrap a sheet of chart paper around the drum, *making sure that the right edge overlaps the left*. Fasten it with tape, and then replace the kymograph drum and lower the drum retainer into its original position in the hole in the top of the drum.

3. Raise and lower the floating bell several times, noting as you do that the *ventilometer pen* moves up and down on the drum. This pen, which writes in black ink, will be used for recording and should be adjusted so that it records in the approximate middle of the chart paper. This adjustment is made by repositioning the floating bell using the *reset knob* on the metal pulley at the top of the spirometer apparatus. The other pen, the respirometer pen, which records in red ink, will not be used for these tests and should be moved away from the drum's recording surface.

4. Recording your normal respiratory rate. Clean the nose clips with an alcohol swab. While you wait for the alcohol to air dry, count and record your normal respiratory rate.

Respirations per minute: ______________

5. Recording tidal volume. After the alcohol has air dried, apply the nose clips to your nose. This will enforce mouth breathing.

Open the *free-breathing valve*. Insert a disposable cardboard mouthpiece into the end (valve assembly) of the breathing tube, and then insert the mouthpiece into your mouth. Practice breathing for several breaths to get used to the apparatus. At this time, you are still breathing room air.

Set the spirometer switch to **SLOW** (32 mm/min). Close the free-breathing valve, and breathe in a normal manner for 2 minutes to record your tidal volume—the amount of air inspired or expired with each normal respiratory cycle. This recording should show a regular pattern of inspiration-expiration spikes and should gradually move upward on the chart paper. (A downward slope indicates that there is an air leak somewhere in the system—most likely at the mouthpiece.) Notice that on an apparatus using a counterweighted pen (such as the Collins-9L Ventilometer shown in Figure 37.4b), inspirations are recorded by upstrokes and expirations are recorded by downstrokes.*

6. Recording vital capacity. To record your vital capacity, take the deepest possible inspiration you can and then exhale to the greatest extent possible—really *push* the air out. (The recording obtained should resemble that shown in Figure 37.5). Repeat the vital capacity measurement twice again. Then turn off the spirometer and remove the chart paper from the kymograph drum.

**If a Collins survey spirometer is used, the situation is exactly opposite: Upstrokes are expirations, and downstrokes are inspirations.*

Text continues on next page. →

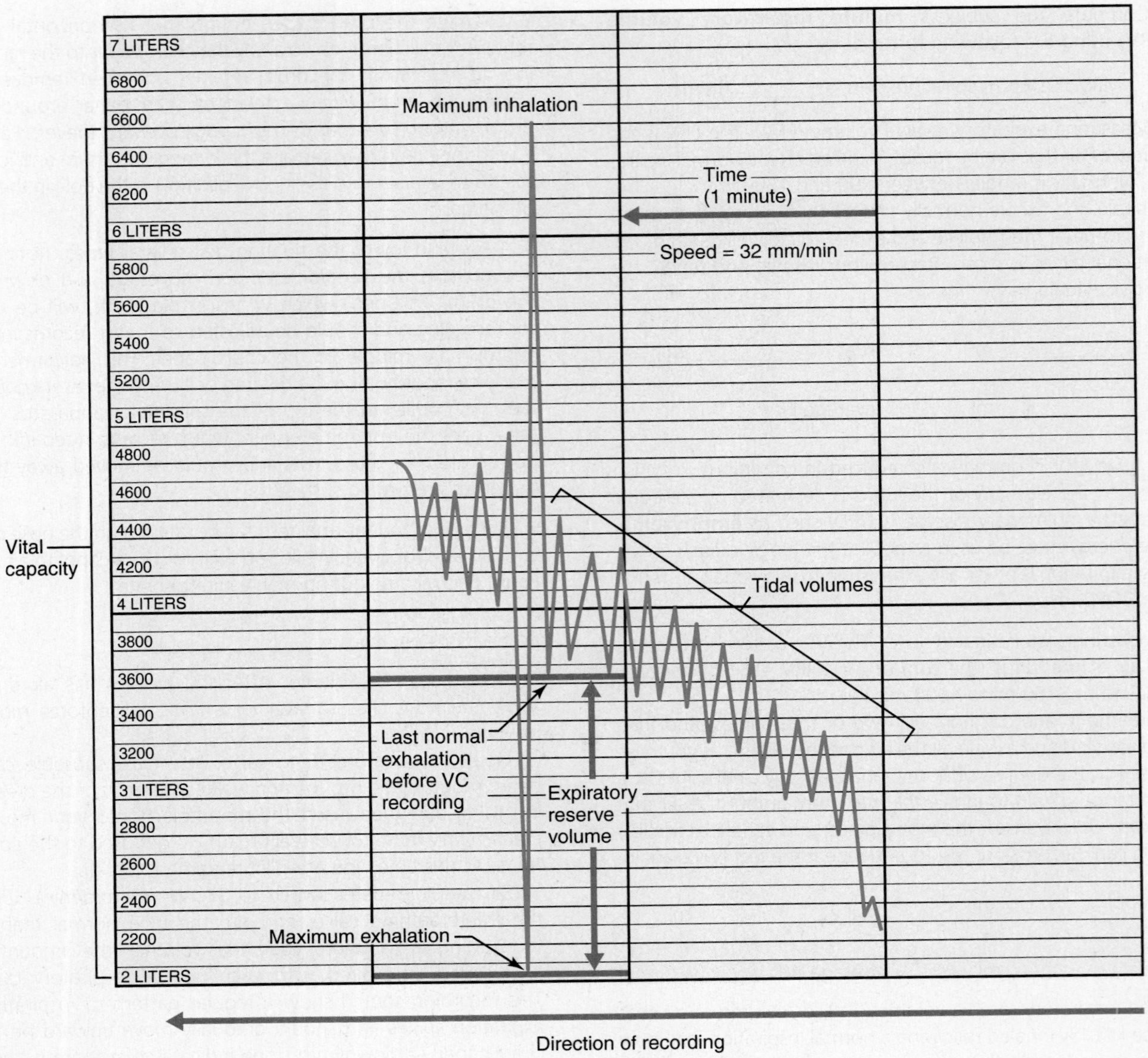

Figure 37.5 A typical spirometry recording of tidal volume, inspiratory capacity, expiratory reserve volume, and vital capacity. At a drum speed of 32 mm/min, each vertical column of the chart represents a time interval of 1 minute. (Note that downstrokes represent exhalations, and upstrokes represent inhalations.)

7. Determine and record your measured, averaged, and corrected respiratory volumes. Because the pressure and temperature inside the spirometer are influenced by room temperature and differ from those in the body, all measured values are to be multiplied by a **BTPS** (body temperature, atmospheric pressure, and water saturation) **factor**. At room temperature, the BTPS factor is typically 1.1 or very close to that value. Hence, you will multiply your average measured values by 1.1 to obtain your corrected respiratory volume values. Copy the averaged and corrected values onto the review sheet at the end of this exercise.

- Tidal volume (TV). Select a typical resting tidal breath recording. Subtract the millimeter value of the trough (exhalation) from the millimeter value of the peak (inspiration). Record this value below as *measured TV 1*. Select two other TV tracings to determine the TV values for the TV 2 and TV 3 measurements. Then, determine your average TV and multiply it by 1.1 to obtain the BTPS-corrected average TV value.

measured TV 1: _______ ml   average TV: _______ ml

measured TV 2: _______ ml   corrected average TV:

measured TV 3: _______ ml   _________ ml

Also compute your **minute respiratory volume (MRV)** using the following formula:

MRV = TV × respirations/min = ________ml/min

- Inspiratory capacity (IC). In the first vital capacity recording, find the expiratory trough immediately preceding the maximal inspiratory peak achieved during vital capacity determination. Subtract the milliliter value of that expiration from the value corresponding to the peak of the maximal inspiration that immediately follows. For example, according to our typical recording (**Figure 37.5**), these values would be

$$6600 - 3650 = 2950 \text{ ml}$$

Record your computed value and the results of the two subsequent tests on the appropriate lines below. Then calculate the measured and corrected inspiratory capacity averages, and record.

measured IC 1: _______ ml　　average IC: ________ ml

measured IC 2: _______ ml　　corrected average IC: ________ ml

measured IC 3: _______ ml

- Inspiratory reserve volume (IRV). Subtract the corrected average tidal volume from the corrected average for the inspiratory capacity and record below.

IRV = corrected average IC − corrected average TV

corrected average IRV: _________ ml

- Expiratory reserve volume (ERV). Subtract the number of milliliters corresponding to the trough of the maximal expiration obtained during the vital capacity recording from milliliters corresponding to the last *normal* expiration before the VC maneuver is performed. For example, according to our typical recording (Figure 37.5), these values would be

$$3650 \text{ ml} - 2050 \text{ ml} = 1600 \text{ ml}$$

Record your measured and averaged values (three trials) below.

measured ERV 1: _______ ml　　average ERV: _______ ml

measured ERV 2: _______ ml　　corrected average ERV:

measured ERV 3: _______ ml　　_______ ml

- Vital capacity (VC). Add your corrected values for ERV and IC to obtain the corrected average VC. Record below and on the review sheet at the end of this exercise.

corrected average VC: _____________ ml

Now continue with step 8 (below) whether you are following the procedure for the nonrecording or recording spirometer.

8. Figure out how closely your measured average vital capacity volume compares with the *predicted values* for someone your age, sex, and height. Obtain the predicted value either from the following equation or the appropriate table (see your instructor for the printed table). Notice that you will have to convert your height in inches to centimeters (cm) to find the corresponding value. This is easily done by multiplying your height in inches by 2.54.

Computed height: _____________ cm

$$\text{Male VC} = (0.052)\,H - (0.022)\,A - 3.60$$

$$\text{Female VC} = (0.041)\,H - (0.018)\,A - 2.69$$

Note: (VC) = vital capacity in liters, (*H*) = height in centimeters, and (*A*) = age in years.

Predicted VC (obtained from the equation or appropriate table):

___________ ml

Use the following equation to compute your VC as a percentage of the predicted VC value:

$$\% \text{ of predicted VC} = \left(\frac{\text{average VC}}{\text{predicted VC}}\right) \times 100$$

% predicted VC value: ___________%

9. Computing residual volume. A respiratory volume that cannot be experimentally demonstrated here is the residual volume (RV). RV is the amount of air remaining in the lungs after a maximal expiratory effort. The presence of residual air (usually about 1200 ml) that cannot be voluntarily flushed from the lungs is important because it allows gas exchange to go on continuously—even between breaths.

Although the residual volume cannot be measured directly, it can be approximated by using one of the following factors:

For ages 16–34　Factor = 0.250

For ages 35–49　Factor = 0.305

For ages 50–69　Factor = 0.445

Compute your predicted RV using the following equation:

$$\text{RV} = \text{VC} \times \text{factor}$$

! 10. Recording is finished for this subject. Before continuing with the next member of your group:

- Dispose of used cardboard mouthpieces in the autoclave bag.
- Swish the valve assembly (if removable) in the 70% ethanol solution, then rinse with tap water.
- Put a fresh mouthpiece into the valve assembly (or on the stem of the handheld spirometer). Using the procedures outlined above, measure and record the respiratory volumes for all members of your group.

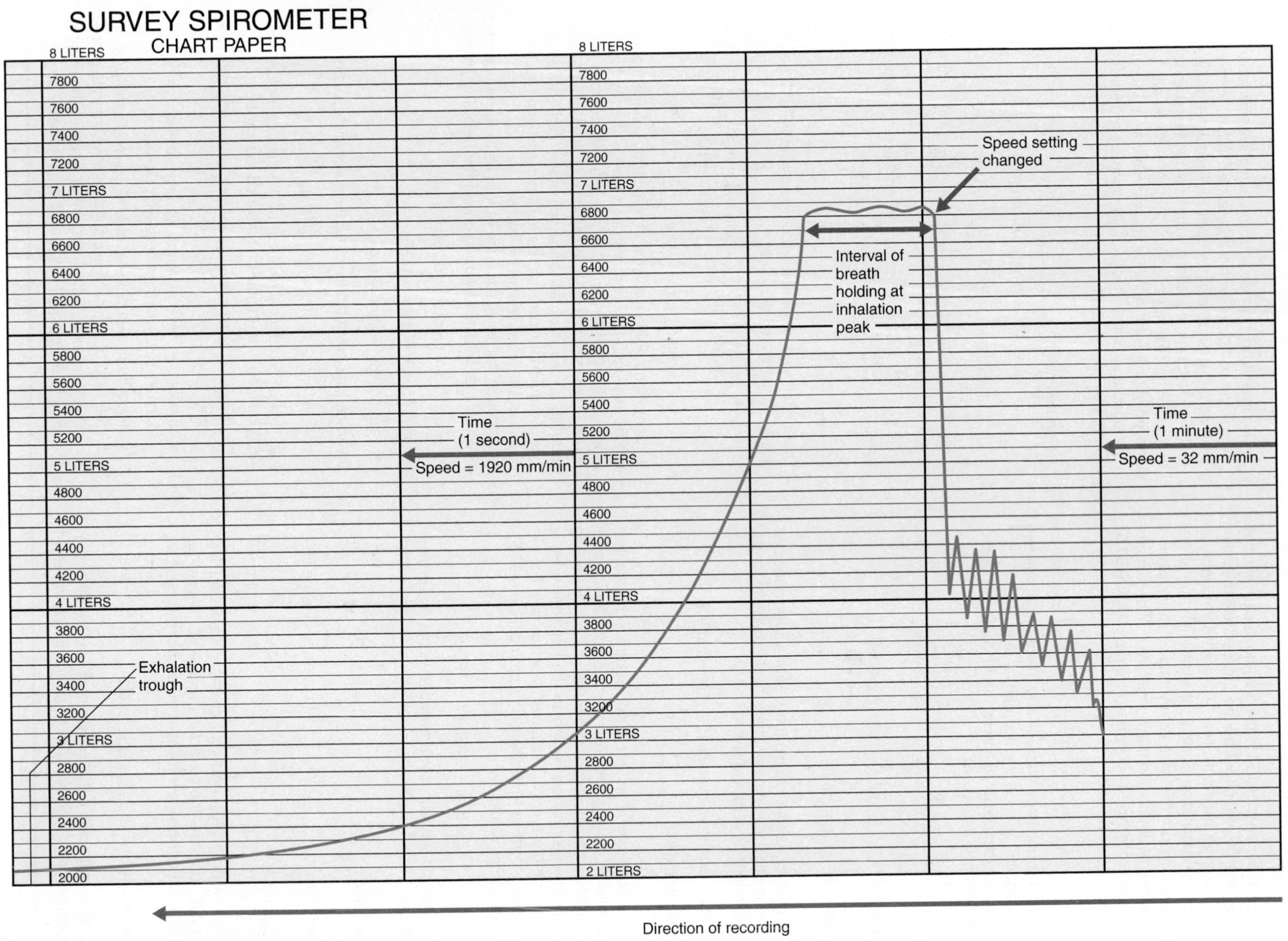

Figure 37.6 A recording of the forced vital capacity (FVC) and forced expiratory volume (FEV) or timed vital capacity test.

Forced Expiratory Volume (FEV_T) Measurement

Though not really diagnostic, pulmonary function tests can help the clinician distinguish between obstructive and restrictive pulmonary diseases. (In obstructive disorders, such as chronic bronchitis and asthma, airway resistance is increased, whereas in restrictive diseases, such as polio and tuberculosis, total lung capacity declines.) Two highly useful pulmonary function tests used for this purpose are the FVC and the FEV_T (**Figure 37.6**).

The **FVC** (forced vital capacity) measures the amount of gas expelled when the subject takes the deepest possible breath and then exhales forcefully and rapidly. This volume is reduced in those with restrictive pulmonary disease. The **FEV_T** (forced expiratory volume) involves the same basic testing procedure, but it specifically looks at the percentage of the vital capacity that is exhaled during specific time intervals of the FVC test. FEV_1, for instance, is the amount exhaled during the first second. Healthy individuals can expire 75% to 85% of their FVC in the first second. The FEV_1 is low in people with obstructive disease.

Activity 4

Measuring the FVC and FEV_1

Directions provided here for the FEV_T determination apply only to the recording spirometer.

1. Prepare to make your recording as described for the recording spirometer, steps 1–5 (on p. 549).

2. At a signal agreed upon by you and your lab partner, take the deepest inspiration possible and hold it for 1 to 2 seconds. As the inspiratory peak levels off, your partner is to change the drum speed to **FAST** (1920 mm/min) so that the distance between the vertical lines on the chart represents 1 second.

3. Once the drum speed is changed, exhale as much air as you can as rapidly and forcibly as possible.

4. When the tracing plateaus (bottoms out), stop recording and determine your FVC. Subtract the milliliter reading in the expiration trough (the bottom plateau) from the preceding inhalation peak (the top plateau). Record this value.

FVC: ______________ ml

5. Prepare to calculate the FEV_1. Draw a vertical line intersecting with the spirogram tracing at the precise point that exhalation began. Identify this line as *line 1*. From line 1, measure 32 mm horizontally to the left, and draw a second vertical line. Label this as *line 2*. The distance between the two lines represents 1 second, and the volume exhaled in the first second is read where line 2 intersects the spirogram tracing. Subtract that milliliter value from the milliliter value of the inhalation peak (at the intersection of line 1), to determine the volume of gas expired in the first second. According to the values given in the example (Figure 37.6), that figure would be 3400 ml (6800 ml – 3400 ml). Record your measured value below.

Milliliters of gas expired in second 1: ______________ ml

6. To compute the FEV_1 use the following equation:

$$FEV_1 = \frac{\text{volume expired in second 1}}{\text{FVC volume}} \times 100\%$$

Record your calculated value below and on the review sheet at the end of this exercise.

FEV_1: ________ % of FVC

Activity 5

Measuring Respiratory Volumes Using BIOPAC®

In this activity, you will measure respiratory volumes using the BIOPAC® airflow transducer. An example of these volumes is demonstrated in the computer-generated spirogram (**Figure 37.7**, p. 554). Since it is not possible to measure **residual volume (RV)** using the airflow transducer, assume that it is 1.0 liter for each subject, which is a reasonable estimation. Or enter a volume between 1 and 5 liters via Preferences. It is also important to estimate the **predicted vital capacity** of the subject for comparison to the measured value. A rough estimate of the vital capacity in liters (VC) of a subject can be calculated using the following formulas based on height in centimeters (*H*) and age in years (*A*).

$$\text{Male VC} = (0.052)H - (0.022)A - 3.60$$

$$\text{Female VC} = (0.041)H - (0.018)A - 2.69$$

Because many factors besides height and age influence vital capacity, it should be assumed that measured values up to 20% above or below the calculated predicted value are normal.

Setting Up the Equipment

1. Connect the BIOPAC® unit to the computer, and turn the computer **ON**.

2. Make sure the BIOPAC® unit is **OFF**.

3. Plug in the equipment (as shown in **Figure 37.8**, p. 554).
   - Airflow transducer—CH 1

4. Turn the BIOPAC® unit **ON**.

5. Place a *clean* bacteriological filter onto the end of the BIOPAC® calibration syringe (as shown in **Figure 37.9**, p. 554). Since the subject will be blowing through a filter, it is necessary to use a filter for calibration.

Text continues on next page. →

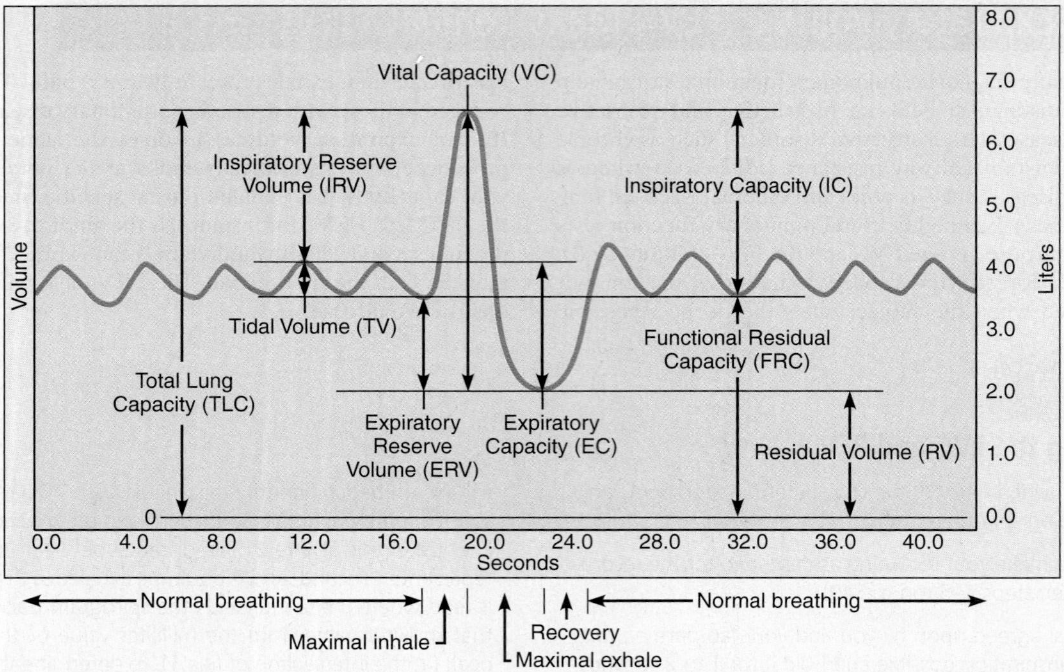

Figure 37.7 Example of a computer-generated spirogram.

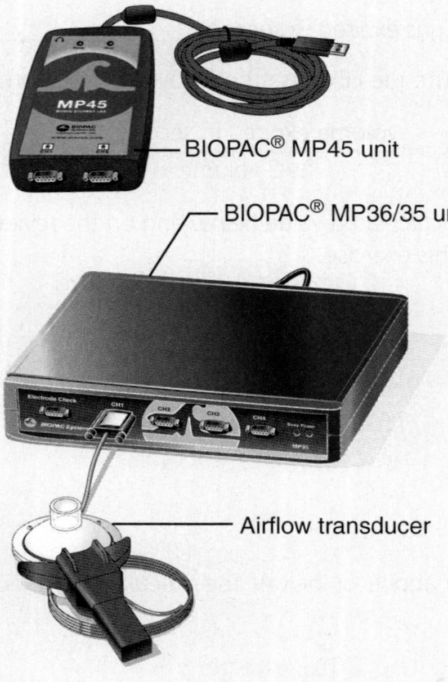

Figure 37.8 Setting up the BIOPAC® equipment. Plug the airflow transducer into Channel 1. Transducer is shown plugged into the MP36/35 unit.

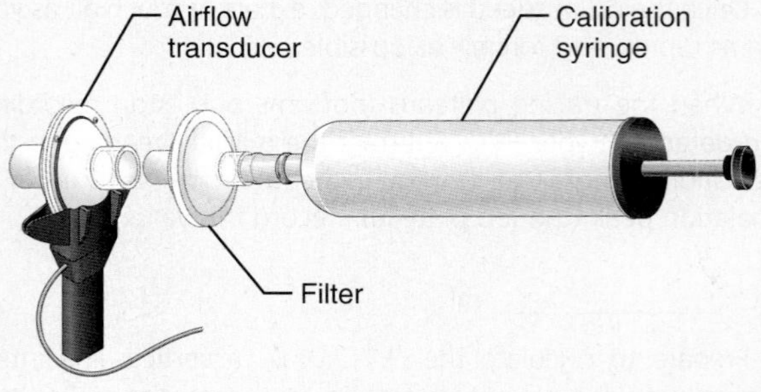

Figure 37.9 Placement of the calibration syringe and filter assembly onto the airflow transducer for calibration.

6. Start the Biopac Student Lab program on the computer by double-clicking the icon on the desktop or by following your instructor's guidance.

7. Select lesson **L12-Pulmonary Function-1** from the menu, and click **OK**.

8. Type in a filename that will save this subject's data on the computer hard drive. You may want to use the subject's last name followed by Pulmonary Function (PF)-1 (for example, SmithPF-1), then click **OK**.

At the start of the lesson, you have the option to record the subject's gender, age, and height. Domestic or metric units may be selected. These details are displayed in the Journal following the lesson.

Calibrating the Equipment

Two precautions must be followed:

- The airflow transducer is sensitive to gravity, so it must be held directly parallel to the ground during calibration and recording.
- Do not hold onto the airflow transducer when it is attached to the calibration syringe and filter assembly—the syringe tip is likely to break. (See **Figure 37.10** for the proper handling of the calibration assembly). The size of the calibration syringe can be altered via Preferences.

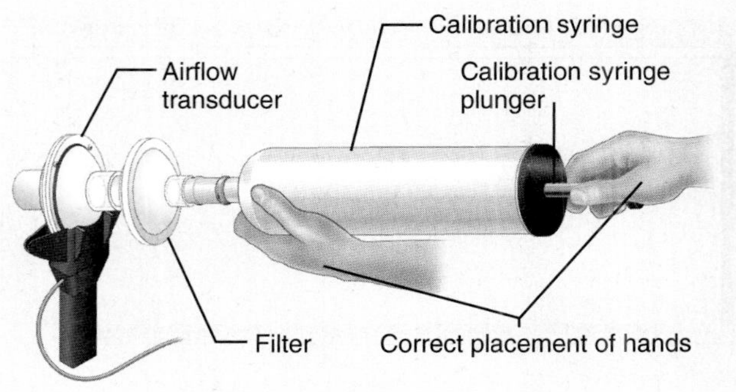

Figure 37.10 Proper handling of the calibration assembly.

1. Hold the airflow transducer upright and still, making sure no air is flowing through it, and click **Calibrate**. This part of the calibration will stop automatically after 8 seconds.

2. Verify that the airflow data is flat and centered. If necessary, click **Redo Calibration**. If airflow data is satisfactory, click **Continue**.

3. Insert the calibration syringe and filter assembly into the airflow transducer on the side labeled "Inlet." Make sure the calibration syringe plunger is pulled all the way out.

4. The final part of the calibration involves simulating five breathing cycles using the calibration syringe. A single cycle consists of:

- Pushing the plunger in (taking 1 second for this stroke)
- Waiting for 2 seconds
- Pulling the plunger out (taking 1 second for this stroke)
- Waiting 2 seconds

Remember to hold the airflow transducer directly parallel to the ground during calibration and recording.

5. When ready to perform this second stage of the calibration, click **Calibrate**. After you have completed five cycles, click **End Calibration**.

6. Observe the data, which should look similar to that in **Figure 37.11**.

- If the data look very different, click **Redo Calibration** and repeat the steps above.
- If the data look similar, gently remove the calibration syringe, leaving the air filter attached to the transducer. Proceed to the next section by clicking **Continue**.

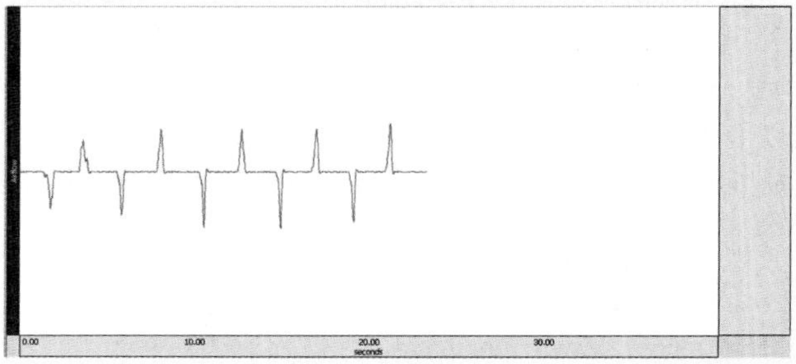

Figure 37.11 Example of calibration data.

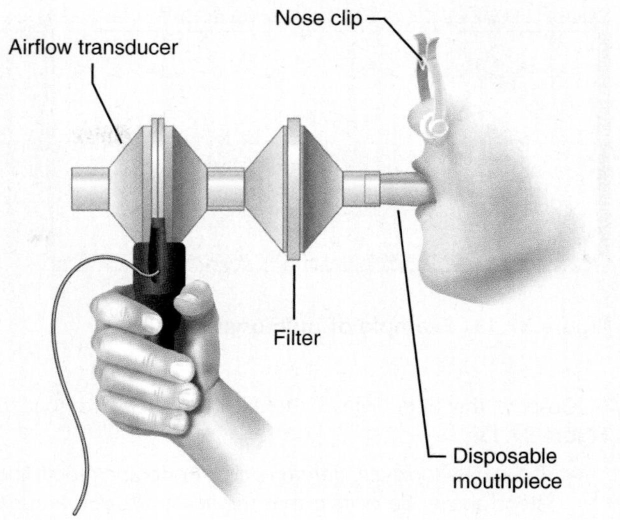

Figure 37.12 Proper equipment setup for recording data.

Recording the Data

Follow these procedures precisely, because the airflow transducer is very sensitive. Hints to obtain the best data:

- Always insert air filter on, and breathe through, the transducer side labeled **Inlet**.
- Keep the airflow transducer upright at all times.
- The subject should not look at the computer screen during the recording of data.
- The subject must keep a nose clip on throughout the experiment.

1. Insert a clean mouthpiece into the air filter that is already attached to the airflow transducer. *Be sure that the filter is attached to the **Inlet** side of the airflow transducer.*

2. Write the name of the subject on the mouthpiece and air filter. For safety purposes, each subject must use his or her own air filter and mouthpiece.

3. The subject should now place the nose clip on the nose (or hold the nose very tightly with finger pinch), wrap the lips tightly around the mouthpiece, and begin breathing normally through the airflow transducer (as shown in **Figure 37.12**).

4. When prepared, the subject will complete the following unbroken series with nose plugged and lips tightly sealed around the mouthpiece:

- Take five normal breaths (1 breath = inhale + exhale).
- Inhale as much air as possible.
- Exhale as much air as possible.
- Take five normal breaths.

5. When the subject is prepared to proceed, click **Record** on the first normal inhalation and proceed. When the subject finishes the last exhalation at the end of the series, click **Stop**.

Text continues on next page. →

37

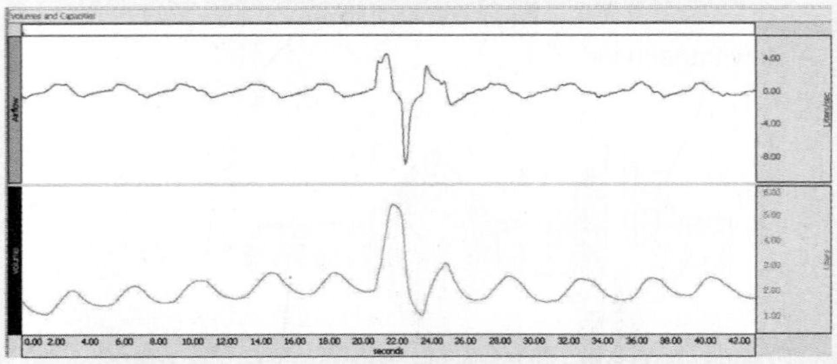

Figure 37.13 Example of pulmonary data.

6. Observe the data, which should look similar to that in **Figure 37.13**.
   - If the data look very different, click **Redo** and repeat the steps above. Be certain that the lips are sealed around the mouthpiece, the nose is completely plugged, and the transducer is upright.
   - If the data look similar, proceed to step 7.
7. When finished, click **Done**. A pop-up window will appear.
   - Click **Yes** if you are done and want to stop recording.
   - To record from another subject, select **Record from another Subject** and return to step 1 under Recording the Data. You will not need to redo the calibration procedure for the second subject.
   - If continuing to the Data Analysis section, select **Analyze current data file** and proceed to step 2 of the Data Analysis section.

Data Analysis

1. If just starting the BIOPAC® program to perform data analysis, enter **Review Saved Data** mode and choose the file with the subject's PF data (for example, SmithPF-1).
2. Observe how the channel numbers are designated: CH 1—Airflow; CH 2—Volume.
3. To set up the display for optimal viewing, hide CH 1—Airflow. To do this, hold down the Alt key (PC) or Option key (Mac) while using the cursor to click the Channel box 1 (the small box with a 1 at the upper left of the screen).
4. To analyze the data, set up the first pair of channel/ measurement boxes at the top of the screen by selecting the following channel and measurement type from the drop-down menu:

Channel	Measurement	Data
CH 2	P-P	Volume

5. Take two measures for an averaged TV calculation: Use the arrow cursor and click the I-beam cursor box on the lower right side of the screen to activate the "area selection" function. Using the activated I-beam cursor, highlight the inhalation of cycle 3 (as shown in **Figure 37.14**).

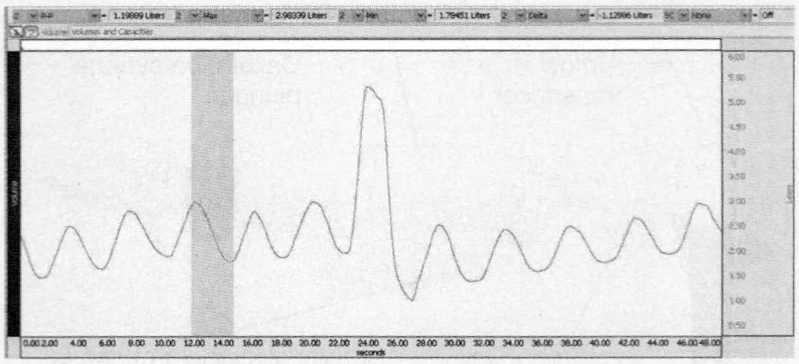

Figure 37.14 Highlighting data for the inhalation of the third breath.

Activity 5: Pulmonary Measurements	
Volumes	**Measurements (liters)**
Tidal volume (TV)	
Inspiratory reserve volume (IRV)	
Expiratory reserve volume (ERV)	
Vital capacity (VC)	
Residual volume (RV)	1.00 (assumed)

6. The computer automatically calculates the **P-P** value for the selected area. This measure is the difference between the highest and lowest values in the selected area. Note the value. Use the I-beam cursor to select the exhalation of cycle 3, and note the **P-P** value.
7. Calculate the average of the two **P-P** values. This represents the **tidal volume** (in liters). Record the value in the **Activity 5: Pulmonary Measurements chart** above.
8. Use the I-beam cursor to measure the IRV: Highlight from the peak of maximum inhalation to the peak of the last normal inhalation just before it (see Figure 37.7 for an example of IRV). Observe and record the Δ **(Delta)** value in the chart (to the nearest 0.01 liter).
9. Use the I-beam cursor to measure the ERV: Highlight from the trough of maximum exhalation to the trough of the last normal exhalation just before it (see Figure 37.7 for an example of ERV). Observe and record the Δ **(Delta)** value in the chart (to the nearest 0.01 liter).
10. Last, use the I-beam cursor to measure the VC: Highlight from the trough of maximum exhalation to the peak of maximum inhalation (see Figure 37.7 for an example of VC). Observe and record the **P-P** value in the chart (to the nearest 0.01 liter).
11. When finished, choose **File menu** and **Quit** to close the program.

Using the measured data, calculate the capacities listed in the **Activity 5: Calculated Pulmonary Capacities chart**.

Activity 5: Calculated Pulmonary Capacities		
Capacity	**Formula**	**Calculation (liters)**
Inspiratory capacity (IC)	= TV + IRV	
Functional residual capacity (FRC)	= ERV + RV	
Total lung capacity (TLC)	= TV + RV + IRV + ERV	

Use the formula in the introduction of this activity (p. 553) to calculate the predicted vital capacity of the subject based on height and age.

Predicted VC: __________ liters

How does the measured vital capacity compare to the predicted vital capacity?

____

____

____

Describe why height and weight might correspond with a subject's VC.

____

____

____

What other factors might influence the VC of a subject?

____

____

____

Factors Influencing Rate and Depth of Respiration

The neural centers that control respiratory rhythm and maintain a rate of 12 to 18 respirations/min are located in the medulla and pons. On occasion, input from the stretch receptors in the lungs (via the vagus nerve to the medulla) modifies the respiratory rate, as in cases of extreme overinflation of the lungs (Hering-Breuer reflex).

Death occurs when medullary centers are completely suppressed, as from an overdose of sleeping pills or gross overindulgence in alcohol, and respiration ceases completely. ✚

Although the neural centers initiate the basic rhythm of breathing, there is no question that physical phenomena such as talking, yawning, coughing, and exercise can modify the rate and depth of respiration. So, too, can chemical factors such as changes in oxygen or carbon dioxide concentrations in the blood or fluctuations in blood pH. This is especially important in initiating breathing in a newborn. The buildup of carbon dioxide in the blood triggers the baby's first breath. The experimental sequence in Activity 6 is designed to test the relative importance of various physical and chemical factors in the process of respiration.

37

Activity 6

Visualizing Respiratory Variations

In this activity, you will count the respiratory rate of the subject visually by observing the movement of the chest or abdomen.

1. Record quiet breathing for 1 minute with the subject in a sitting position.

Breaths per minute: ______________________________

2. Record the subject's breathing as he or she performs activities from the following list. Record your results on the review sheet at the end of this exercise.

talking	swallowing water
yawning	coughing
laughing	lying down
standing	running in place
doing a math problem (concentrating)	

3. Without recording, have the subject breathe normally for 2 minutes, then inhale deeply and hold his or her breath for as long as he or she can.

Breath-holding interval: ______________________ sec

As the subject exhales, record the recovery period (time to return to normal breathing—usually slightly over 1 minute):

Time of recovery period: ______________________ sec

Did the subject have the urge to inspire *or* expire during breath holding? ______________________

Without recording, repeat the above experiment, but this time exhale completely and forcefully *after* taking the deep breath.

Breath-holding interval ______________________ sec

Text continues on next page. →

Time of recovery period ______________________ sec

Did the subject have the urge to inspire *or* expire? ________

Explain the results. (*Hint*: The vagus nerve is the sensory nerve of the lungs and plays a role here.)

____

____

____

4. During the next task, a sensation of dizziness may develop. As the carbon dioxide is washed out of the blood by hyperventilation, the blood pH increases, leading to a decrease in blood pressure and reduced cerebral circulation.

! If you have a history of dizzy spells or a heart condition, do not perform this task.

The subject may experience a lack of desire to breathe after forced breathing is stopped. If the period of breathing cessation—apnea—is extended, cyanosis of the lips may occur.

Have the subject hyperventilate (breathe deeply and forcefully at the rate of 1 breath/4 sec) for about 30 sec.

Is the respiratory rate after hyperventilation faster *or* slower than during normal quiet breathing?

____

5. Repeat the hyperventilation step. After hyperventilation, the subject is to hold his or her breath as long as possible.

Breath-holding interval: ______________________

Can the breath be held for a longer or shorter time after hyperventilating?

____

6. Without recording, have the subject breathe into a paper bag for 3 minutes, then record breathing movements.

! *During the bag-breathing exercise, the subject's partner should watch the subject carefully.*

Is the breathing rate faster *or* slower than that recorded during normal quiet breathing?

____

After hyperventilating? ______________________________.

7. Run in place for 2 minutes, and then have your partner determine how long you can hold your breath.

Breath-holding interval: ______________________ sec

8. To prove that respiration has a marked effect on circulation, conduct the following test. Have your lab partner record the rate and relative force of your radial pulse before you begin.

Rate: ____________ beats/min Relative force: ____________

Inspire forcibly. Immediately close your mouth and nose to retain the inhaled air, and then make a forceful and prolonged expiration. Your lab partner should observe and record the condition of the blood vessels of your neck and face, and again immediately palpate the radial pulse.

Observations: ______________________________________

Radial pulse: ________ beats/min Relative force: ________

Explain the changes observed. _______________________

____

____

! Dispose of the paper bag in the autoclave bag. Observation of the test results should enable you to determine which chemical factor, carbon dioxide or oxygen, has the greatest effect on modifying the respiratory rate and depth.

37

Role of the Respiratory System in Acid-Base Balance of Blood

Blood pH must be relatively constant for the cells of the body to function optimally. The carbonic acid–bicarbonate buffer system of the blood is extremely important because it helps stabilize arterial blood pH at 7.4 ± 0.02.

When carbon dioxide diffuses into the blood from the tissue cells, much of it enters the red blood cells, where it combines with water to form carbonic acid (**Figure 37.15**):

$$H_2O + CO_2 \xrightarrow[\text{enzyme present in RBC}]{\text{carbonic anhydrase}} H_2CO_3 \text{ (carbonic acid)}$$

Some carbonic acid is also formed in the plasma, but that reaction is very slow because of the lack of the carbonic anhydrase enzyme. Shortly after it forms, carbonic acid dissociates to release bicarbonate (HCO_3^-) and hydrogen ions (H^+). The hydrogen ions that remain in the cells are neutralized when they combine with hemoglobin molecules. If they were not neutralized, the intracellular pH would become very acidic as H^+ ions accumulated. The bicarbonate ions diffuse out of the red blood cells into the plasma, where they become part of the carbonic acid–bicarbonate buffer system. As HCO_3^- follows its concentration gradient into the plasma, an electrical imbalance develops in the RBCs that draws Cl^- into them from the plasma. This exchange phenomenon is called the *chloride shift*.

Acids (more precisely, H^+) released into the blood by the body cells tend to lower the pH of the blood and to cause it to

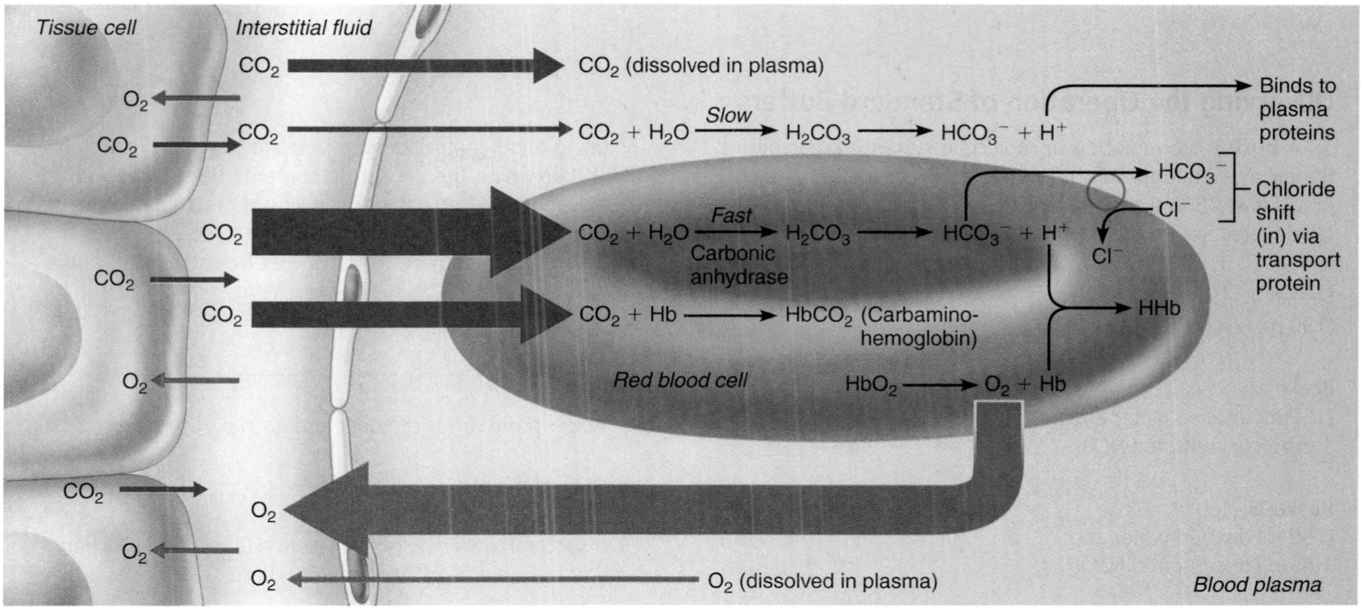

Figure 37.15 Oxygen release and carbon dioxide pickup at the tissues.

become acidic. On the other hand, basic substances that enter the blood tend to cause the blood to become more alkaline and the pH to rise. Both of these tendencies are resisted in large part by the carbonic acid–bicarbonate buffer system. If the H^+ concentration in the blood begins to increase, the H^+ ions combine with bicarbonate ions to form carbonic acid (a weak acid that does not tend to dissociate at physiological or acid pH) and are thus removed.

$$H^+ + HCO_3^- \rightarrow H_2CO_3$$

Likewise, as blood H^+ concentration drops below what is desirable and blood pH rises, H_2CO_3 dissociates to release bicarbonate ions and H^+ ions to the blood.

$$H_2CO_3 \rightarrow H^+ + HCO_3^-$$

The released H^+ lowers the pH again. The bicarbonate ions, being *weak* bases, are poorly functional under alkaline conditions and have little effect on blood pH unless and until blood pH drops toward acid levels.

In the case of excessively slow or shallow breathing (hypoventilation) or fast deep breathing (hyperventilation), the amount of carbonic acid in the blood can be greatly modified—increasing dramatically during hypoventilation and decreasing substantially during hyperventilation. In either situation, if the buffering ability of the blood is inadequate, respiratory acidosis or alkalosis can result. Therefore, maintaining the normal rate and depth of breathing is important for proper control of blood pH.

37

Activity 7

Demonstrating the Reaction Between Carbon Dioxide (in Exhaled Air) and Water

1. Fill a beaker with 100 ml of distilled water.
2. Add 5 ml of 0.05 *M* NaOH and five drops of phenol red. Phenol red is a pH indicator that turns yellow in acidic solutions.
3. Blow through a straw into the solution.

What do you observe?

____

____

What chemical reaction is taking place in the beaker?

____

____

4. Discard the straw in the autoclave bag.

Activity 8

Observing the Operation of Standard Buffers

1. A **buffer** is a molecule or molecular system that stabilizes the pH of a solution. To observe the action of a buffer system, obtain five 250-ml beakers and a wash bottle containing distilled water. Set up the following experimental samples:

Beaker 1:
(150 ml distilled water) pH ________

Beaker 2:
(150 ml distilled water and
1 drop concentrated HCl) pH ________

Beaker 3:
(150 ml distilled water and
1 drop concentrated NaOH) pH ________

Beaker 4:
(150 ml standard buffer solution
[pH 7] and 1 drop concentrated HCl) pH ________

Beaker 5:
(150 ml standard buffer solution
[pH 7] and 1 drop concentrated NaOH) pH ________

2. Using a pH meter standardized with a buffer solution of pH 7, determine the pH of the contents of each beaker and record above. After *each and every* pH recording, turn the pH meter switch to **STANDBY**, and rinse the electrodes thoroughly with a stream of distilled water from the wash bottle.

3. Add 3 more drops of concentrated HCl to beaker 4, stir, and record the pH: ________________

4. Add 3 more drops of concentrated NaOH to beaker 5, stir, and record the pH: ________________

How successful was the buffer solution in resisting pH changes when a strong acid (HCl) or a strong base (NaOH) was added?

____

Activity 9

Exploring the Operation of the Carbonic Acid–Bicarbonate Buffer System

To observe the ability of the carbonic acid–bicarbonate buffer system of blood to resist pH changes, perform the following simple experiment.

1. Obtain two small beakers (50 ml), animal plasma, graduated cylinder, glass stirring rod, and a dropper bottle of 0.01 *M* HCl. Using the pH meter standardized with the buffer solution of pH 7.0, measure the pH of the animal plasma. Use only enough plasma to allow immersion of the electrodes and measure the volume used carefully.

pH of the animal plasma: ________

2. Add 2 drops of the 0.01 *M* HCl solution to the plasma; stir and measure the pH again.

pH of plasma plus 2 drops of HCl: ________

3. Turn the pH meter switch to **STANDBY**, rinse the electrodes, and then immerse them in a quantity of distilled water (pH 7) exactly equal to the amount of animal plasma used. Measure the pH of the distilled water.

pH of distilled water: ________

4. Add 2 drops of 0.01 *M* HCl, swirl, and measure the pH again.

pH of distilled water plus the two drops of HCl: ________

Is the plasma a good buffer? ________

What component of the plasma carbonic acid–bicarbonate buffer system was acting to counteract a change in pH when HCl was added?

____

Instructors may assign a portion of the Review Sheet questions using Mastering A&P™

37 EXERCISE REVIEW SHEET Respiratory System Physiology

Name ______________________ Lab Time/Date ______________________

Mechanics of Respiration

1. For each of the following cases, check the column appropriate to your observations on the operation of the model lung.

	Diaphragm pushed up		Diaphragm pulled down	
Change	**Increased**	**Decreased**	**Increased**	**Decreased**
In internal volume of the bell jar (thoracic cavity)				
In internal pressure of the bell jar				
In the size of the balloons (lungs)				

2. Base your answers to the following on your observations in question 1.

Under what internal conditions of the thoracic cavity does air tend to flow into the lungs? ______________________

____

Under what internal conditions of the thoracic cavity does air tend to flow out of the lungs? Explain why this is so. ______________________

____

3. Activation of the diaphragm and the external intercostal muscles begins the inspiratory process. What effect does contraction of these muscles have on thoracic volume, and how is this accomplished? ______________________

____

____

4. What was the approximate increase in diameter of chest circumference during a quiet inspiration? ____________ cm

During forced inspiration? ____________ cm

5. What temporary physiological advantage is created by the substantial increase in chest circumference during forced inspiration? ______________________

____

Respiratory Sounds

6. Which of the respiratory sounds is heard during both inspiration and expiration? ______

Which is heard primarily during inspiration? ______

7. Where did you best hear the vesicular respiratory sounds? ______

Respiratory Volumes and Capacities—Spirometry or BIOPAC®

8. Write the respiratory volume term and the normal value that is described by the following statements.

Volume of air present in the lungs after a forceful expiration: ______

Volume of air that can be expired forcibly after a normal expiration: ______

Volume of air that is breathed in and out during a normal respiration: ______

Volume of air that can be inspired forcibly after a normal inspiration: ______

Volume of air corresponding to TV + IRV + ERV: ______

9. For the spirometer activities, record experimental respiratory volumes as determined in the laboratory. Corrected values and FEV_1 are for the recording spirometer only.

Average TV: ______ ml

Corrected value for TV: ______ ml

Average IRV: ______ ml

Corrected value for IRV: ______ ml

MRV: ______ ml/min

Average ERV: ______ ml

Corrected value for ERV: ______ ml

Average VC: ______ ml

Corrected value for VC: ______ ml

% predicted VC: ______ %

FEV_1: ______ % FVC

For the BIOPAC® activity, record the following experimental respiratory volumes as determined in the laboratory.

TV: ______ L

IRV: ______ L

ERV: ______ L

VC: ______ L

10. Explain how you would calculate the IRV assuming that you have measured VC, TV, and ERV. ______

______

11. How did your calculated vital capacity compare to your predicted vital capacity? Explain any differences you might have seen.

______

______

12. Describe the effect that age would have on vital capacity and why it would have this effect. ______

______

______

Factors Influencing Rate and Depth of Respiration

13. Where are the neural control centers of respiratory rhythm? ______________ and ______________

For questions 14–21, use your Activity 6 data.

14. In your data, what was the rate of quiet breathing?

Initial testing ______________ breaths/min

Test performed	**Observations (breaths per minute)**
Talking	
Yawning	
Laughing	
Standing	
Concentrating	
Swallowing water	
Coughing	
Lying down	
Running in place	

15. Record student data below.

Breath-holding interval after a deep inhalation: ______________ sec length of recovery period: ______________ sec

Breath-holding interval after a forceful expiration: ______________ sec length of recovery period: ______________ sec

After breathing quietly and taking a deep breath (which you held), was your urge to inspire *or* expire? ______________

After exhaling and then holding your breath, did you have a desire for inspiration *or* expiration? ______________

Explain these results. (*Hint*: What reflex is involved here?) ______________

____

____

16. Observations after hyperventilation: ______________

____

17. Breath-holding interval after hyperventilation: ______________ sec

Why does hyperventilation produce apnea or a reduced respiratory rate? ______________

____

18. Observations for rebreathing air: ______

______

Why does rebreathing air produce an increased respiratory rate? ______

______

19. What was the effect of running in place (exercise) on the duration of breath holding? ______

Explain this effect. ______

______

20. Record student data from the test illustrating the effect of respiration on circulation.

Radial pulse before beginning test: ______ /min

Radial pulse after testing: ______ /min

Relative pulse force before beginning test: ______

Relative force of radial pulse after testing: ______

Condition of neck and facial veins after testing: ______

Explain these data. ______

______

21. Do the following factors generally increase (indicate ↑) or decrease (indicate ↓) the respiratory rate and depth?

increase in blood CO_2: ______

increase in blood pH: ______

decrease in blood O_2: ______

decrease in blood pH: ______

Did it appear that CO_2 or O_2 had a more marked effect on modifying the respiratory rate? ______

22. Where are sensory receptors sensitive to changes in blood pressure located? ______

______

23. Where are sensory receptors sensitive to changes in O_2 levels in the blood located? ______

______

24. What is the primary factor that initiates breathing in a newborn infant? ______

______

25. Which, if any, of the measurable respiratory volumes would likely be increased in a person who is cardiovascularly fit, such as a runner or a swimmer?

______

______

Which, if any, of the measurable respiratory volumes would likely be decreased in a person who has smoked a lot for over 20 years?

______

______

26. Blood CO_2 levels and blood pH are related. When blood CO_2 levels increase, does the pH increase or decrease?

____________ Explain why. ____________

____

Role of the Respiratory System in Acid-Base Balance of Blood

27. Define *buffer*. ____________

28. How successful was the laboratory buffer (pH 7) in resisting changes in pH when the acid was added? ____________

____

When the base was added? ____________

How successful was the buffer in resisting changes in pH when the additional drops of the acid and base were added to the original samples? ____________

29. What buffer system operates in blood plasma? ____________

Which component of the buffer system resists a *drop* in pH? ____________ Which resists a *rise* in pH? ____________

30. Explain how the carbonic acid–bicarbonate buffer system of the blood operates. ____________

____

____

____

31. What happened when the carbon dioxide in exhaled air mixed with water? ____________

____

What role does exhalation of carbon dioxide play in maintaining relatively constant blood pH? ____________

____

____

32. ✚ Atelectasis is a collapsed lung. Explain how a pneumothorax might result in atelectasis and what should be done to restore the negative pressure of the pleural cavity. ____________

____

____

33. ✚ Pectus excavatum is a condition in which the anterior thoracic cage is caved inward because of abnormal development of the sternum and ribs. What effect would you expect this condition to have on vital capacity, and why? ____________

____

____

38 EXERCISE Anatomy of the Digestive System

Learning Outcomes

- State the overall function of the digestive system.
- Describe the general histologic structure of the alimentary canal wall, and identify the following structures on an appropriate image of the wall: mucosa, submucosa, muscularis externa, and serosa or adventitia.
- Identify on a model or image the organs of the alimentary canal, and name their subdivisions, if any.
- Describe the general function of each of the digestive system organs or structures.
- List and explain the specializations of the structure of the stomach and small intestine that contribute to their functional roles.
- Name and identify the accessory digestive organs, listing a function for each.
- Describe the anatomy of the generalized tooth, and name the human deciduous and permanent teeth.
- List the major enzymes or enzyme groups produced by the salivary glands, stomach, small intestine, and pancreas.
- Recognize microscopically or in an image the histologic structure of the following organs:

small intestine	tooth	liver
salivary glands	stomach	

Go to Mastering A&P™ > Study Area to improve your performance in A&P Lab.

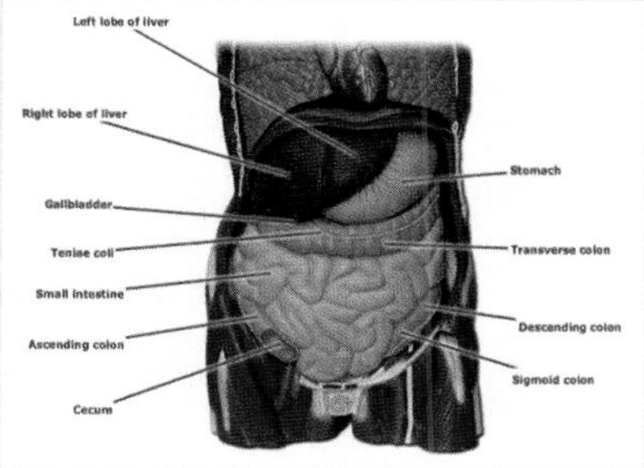

> Lab Tools > Practice Anatomy Lab > Anatomical Models

Instructors may assign new Building Vocabulary coaching activities, Pre-Lab Quiz questions, Art Labeling activities, Practice Anatomy Lab Practical questions (PAL), and more using the Mastering **A&P™** Item Library.

Materials

- Dissectible torso model
- Anatomical chart of the human digestive system
- Prepared slides of the liver and mixed salivary glands; of longitudinal sections of the esophagus-stomach junction and a tooth; and of cross sections of the stomach, duodenum, ileum, and large intestine
- Compound microscope
- Three-dimensional model of a villus (if available)
- Jaw model or human skull
- Three-dimensional model of liver lobules (if available)

For instructions on animal dissections, see the dissection exercises (starting on p. 695) in the cat and fetal pig editions of this manual.

Pre-Lab Quiz

Instructors may assign these and other Pre-Lab Quiz questions using Mastering **A&P™**

1. Circle the correct underlined term. Digestion / Absorption occurs when small molecules pass through epithelial cells into the blood for distribution to the body cells.
2. The ___________ abuts the lumen of the alimentary canal and consists of epithelium, lamina propria, and muscularis mucosae.
   a. mucosa **b.** serosa **c.** submucosa
3. Wavelike contractions of the digestive tract that propel food along are called:
   a. digestion **c.** ingestion
   b. elimination **d.** peristalsis
4. Circle the correct underlined term. The ascending colon / descending colon traverses down the left side of the abdominal cavity and becomes the sigmoid colon.
5. A tooth consists of two major regions, the crown and the:
   a. dentin **c.** gingiva
   b. enamel **d.** root

The **digestive system** provides the body with the nutrients, water, and electrolytes essential for health. The organs of this system ingest, digest, and absorb food and eliminate the undigested remains as feces.

The digestive system consists of a hollow tube extending from the mouth to the anus, into which various accessory organs empty their secretions (**Figure 38.1**). For ingested food to become available to the body cells, it must first be broken down into its smaller diffusible molecules—a process called **digestion**. The digested end products can then pass through the epithelial cells lining the tract into the blood for distribution to the body cells—a process termed **absorption**.

The organs of the digestive system are traditionally separated into two major groups: the **alimentary canal**, or **gastrointestinal (GI) tract**, and the **accessory digestive organs**. The alimentary canal consists of the mouth, pharynx, esophagus, stomach, and small and large intestines. The accessory structures include the teeth, which physically break down foods, and the salivary glands, gallbladder, liver, and pancreas, which secrete their products into the alimentary canal.

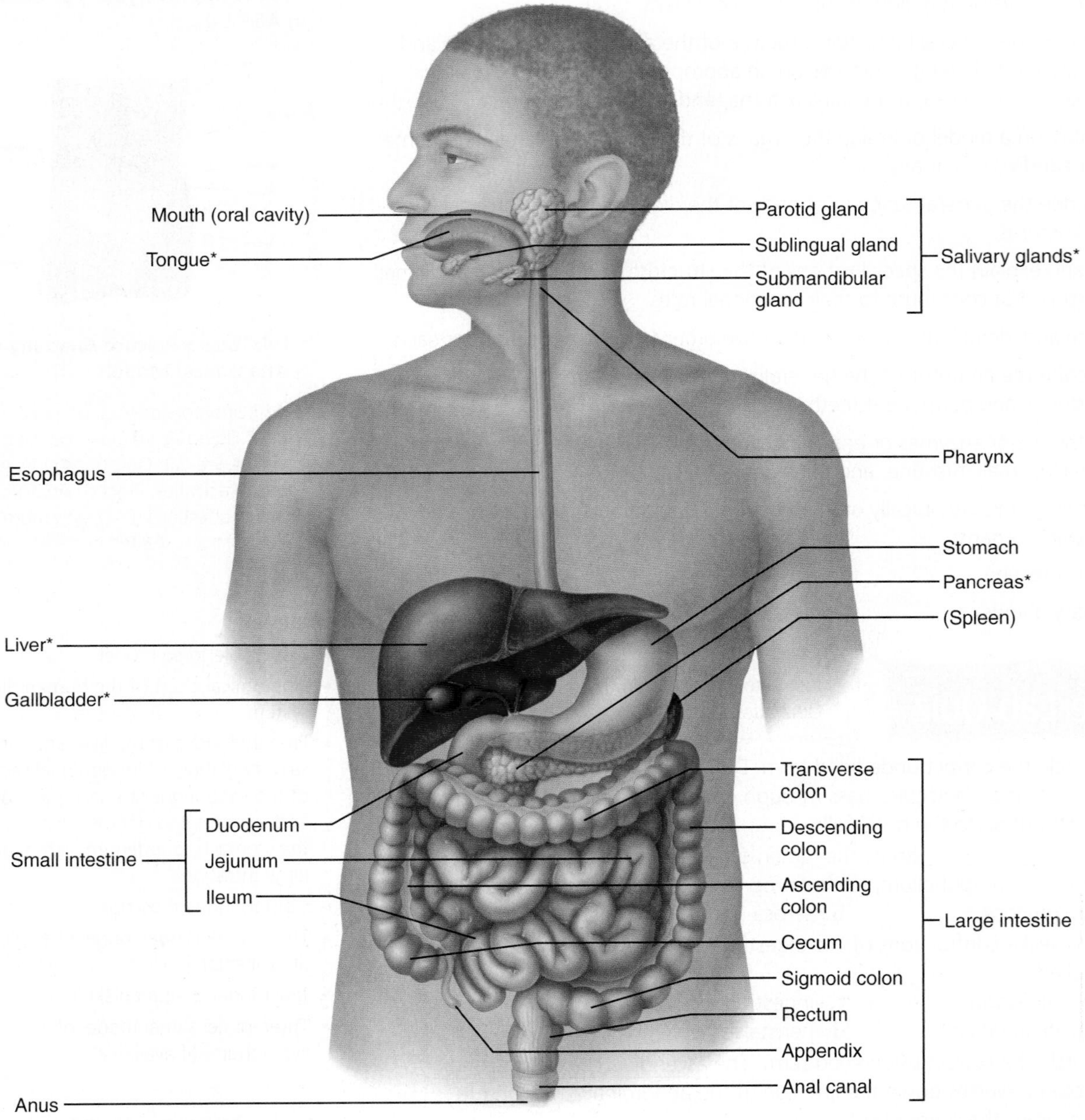

Figure 38.1 The human digestive system: alimentary tube and accessory organs. Organs marked with asterisks are accessory organs. Those without asterisks are alimentary canal organs (except the spleen, an organ of the lymphatic system).

Instructors may assign this figure as an Art Labeling Activity using Mastering A&P™

38

General Histological Plan of the Alimentary Canal

From the esophagus to the anal canal, the basic structure of the alimentary canal is similar. As we study individual parts of the alimentary canal, we will note how this basic plan is modified to provide the unique digestive functions of each subsequent organ.

Essentially, the alimentary canal wall has four basic layers or tunics. From the lumen outward, these are the *mucosa*, the *submucosa*, the *muscularis externa*, and either a *serosa* or *adventitia* (**Figure 38.2**). Each of these layers has a predominant tissue type and a specific function in the digestive process.

Table 38.1 on p. 571 summarizes the characteristics of the layers of the wall of the alimentary canal.

Organs of the Alimentary Canal

Activity 1

Identifying Alimentary Canal Organs

The sequential pathway and fate of food as it passes through the alimentary canal are described in the next sections. Identify each structure in Figure 38.1 and on the torso model or anatomical chart of the digestive system as you work.

Oral Cavity, or Mouth

Food enters the digestive tract through the **oral cavity**, or **mouth** (**Figure 38.3**, p. 570). Within this mucous membrane–lined cavity are the gums, teeth, tongue, and openings of the ducts of the salivary glands. The **lips (labia)** protect the anterior opening, the **oral orifice**. The **cheeks** form the mouth's lateral walls, and the **palate**, its roof. The anterior portion of the palate is referred to as the **hard palate** because the palatine processes of the maxillae and horizontal plates of the palatine bones underlie it. The posterior **soft palate** is a fibromuscular structure that is unsupported by bone. The **uvula**, a fingerlike projection of

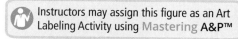

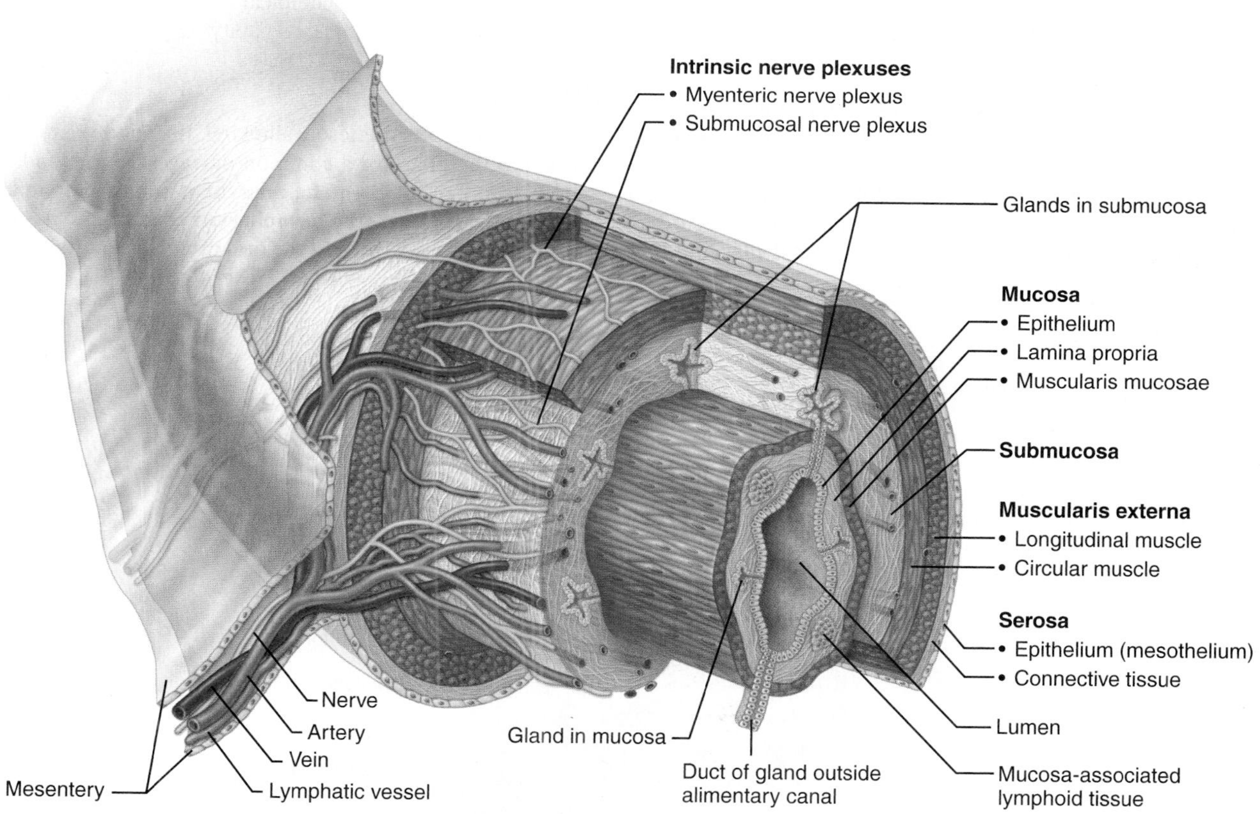

Figure 38.2 Basic structural pattern of the alimentary canal wall.

Instructors may assign this figure as an Art Labeling Activity using Mastering A&P™

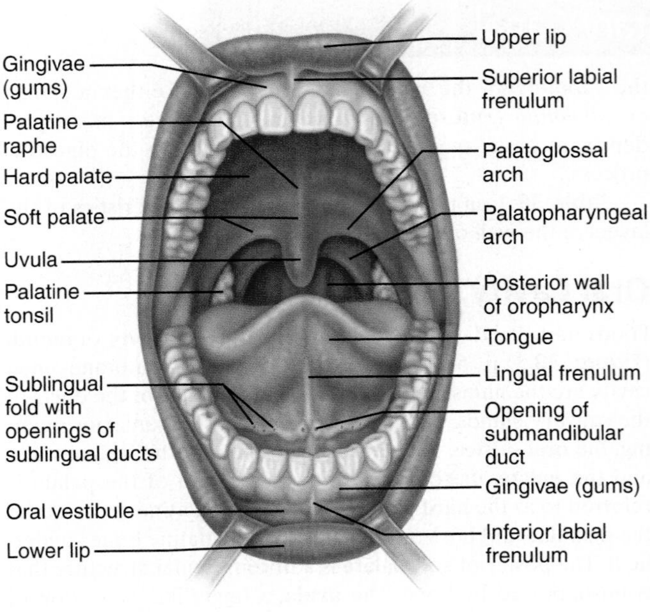

Figure 38.3 Anterior view of the oral cavity.

the soft palate, extends inferiorly from its posterior margin. The floor of the oral cavity is occupied by the muscular **tongue**, which is largely supported by the *mylohyoid muscle* (**Figure 38.4**) and attaches to the hyoid bone, mandible, styloid processes, and pharynx. A membrane called the **lingual frenulum** secures the inferior midline of the tongue to the floor of the mouth.

The space between the teeth and cheeks (or lips) is the **oral vestibule**; the area that lies within the teeth and gums is the **oral cavity proper**. (The teeth and gums are discussed in more detail on pp. 578–580.)

On each side of the mouth at its posterior end are masses of lymphoid tissue, the **palatine tonsils** (see Figure 38.3). Each lies in a concave area bounded anteriorly and posteriorly by membranes, the **palatoglossal arch** and the **palatopharyngeal arch**, respectively. Another mass of lymphoid tissue, the **lingual tonsil** (see Figure 38.4), covers the base of the tongue, posterior to the oral cavity proper. The tonsils, in common with other lymphoid tissues, are part of the body's defense system.

Very often in young children, the palatine tonsils become inflamed and enlarge, partially blocking the entrance to the pharynx posteriorly and making swallowing difficult and painful. This condition is called **tonsillitis**. ✚

Three pairs of salivary glands duct their secretion, saliva, into the oral cavity. One component of saliva, salivary amylase, begins the digestion of starchy foods within the oral cavity. (The salivary glands are discussed in more detail on p. 580.)

As food enters the mouth, it is mixed with saliva and masticated (chewed). The cheeks and lips help hold the food between the teeth during mastication, and the highly mobile tongue manipulates the food during chewing and initiates swallowing. Thus the mechanical and chemical breakdown of food begins before the food has left the oral cavity.

Pharynx

When the tongue initiates swallowing, the food passes posteriorly into the pharynx, a common passageway for food, fluid, and air (see Figure 38.4). The pharynx is subdivided anatomically into three parts—the **nasopharynx** (behind the nasal cavity), the **oropharynx** (behind the oral cavity, extending from the

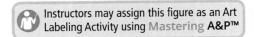

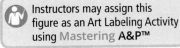

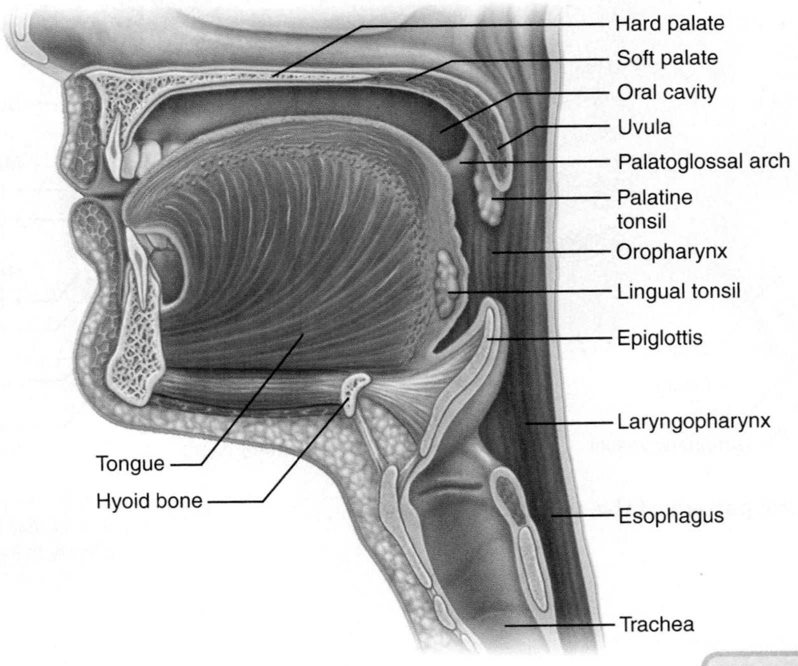

Figure 38.4 Sagittal view of the head showing oral cavity and pharynx.

Table 38.1 Alimentary Canal Wall Layers (Figure 38.2)

Layer	Subdivision of the layer	Tissue type	Major functions (generalized for the layer)
Mucosa	Epithelium	Stratified squamous epithelium in the mouth, oropharynx, laryngopharynx, esophagus, and anus; simple columnar epithelium in the remainder of the canal	Secretion of mucus, digestive enzymes, and hormones; absorption of end products into the blood; protection against infectious disease.
	Lamina propria	Areolar connective tissue with blood vessels; many lymphoid follicles, especially as tonsils and mucosa-associated lymphoid tissue (MALT)	
	Muscularis mucosae	A thin layer of smooth muscle	
Submucosa	N/A	Areolar and dense irregular connective tissue containing blood vessels, lymphatic vessels, and nerve fibers (submucosal nerve plexus)	Blood vessels absorb and transport nutrients. Elastic fibers help maintain the shape of each organ.
Muscularis externa	Circular layer Longitudinal layer	Inner layer of smooth muscle Outer layer of smooth muscle	Segmentation and peristalsis of digested food along the tract are regulated by the myenteric nerve plexus.
Serosa* (visceral peritoneum)	Connective tissue Epithelium (mesothelium)	Areolar connective tissue Simple squamous epithelium	Reduces friction as the digestive system organs slide across one another.

*Since the esophagus is outside the peritoneal cavity, the serosa is replaced by an adventitia made of aerolar connective tissue that binds the esophagus to surrounding tissues.

soft palate to the epiglottis), and the **laryngopharynx** (behind the larynx, extending from the epiglottis to the base of the larynx).

The walls of the pharynx consist largely of two layers of skeletal muscle: an inner layer of longitudinal muscle and an outer layer of circular constrictor muscles. Together these initiate wavelike contractions that propel the food inferiorly into the esophagus. The mucosa of the oropharynx and laryngopharynx, like that of the oral cavity, contains a protective stratified squamous epithelium.

Esophagus

The **esophagus** extends from the laryngopharynx through the diaphragm to the gastroesophageal sphincter in the superior aspect of the stomach. Approximately 25 cm long in humans, it is essentially a food passageway that conducts food to the stomach in a wavelike peristaltic motion. The esophagus has no digestive or absorptive function. The walls at its superior end contain skeletal muscle, which is replaced by smooth muscle in the area nearing the stomach. The **gastroesophageal sphincter**, a slight thickening of the smooth muscle layer at the esophagus-stomach junction, controls food passage into the stomach (Figure 38.6c).

Stomach

The **stomach** (**Figure 38.5**, p. 572) is primarily located in the upper left quadrant of the abdominopelvic cavity and is nearly hidden by the liver and diaphragm. The stomach is made up of several regions, summarized in **Table 38.2** on p. 573. *Mesentery* is the general term that refers to a double layer of peritoneum—a sheet of two serous membranes fused together—that extends from the digestive organs to the body wall. There are two mesenteries, the **greater omentum** and **lesser omentum**, that connect to the stomach. The lesser omentum extends from the liver to the **lesser curvature** of the stomach. The greater omentum extends from the **greater curvature** of the stomach, reflects downward, and covers most of the abdominal organs in an apronlike fashion. (Figure 38.7 on p. 575 illustrates the omenta as well as the other peritoneal attachments of the abdominal organs.)

The stomach is a temporary storage region for food as well as a site for mechanical and chemical breakdown of food. It contains a third (innermost) *obliquely* oriented layer of smooth muscle in its muscularis externa that allows it to churn, mix, and pummel the food, physically reducing it to smaller fragments. **Gastric glands** of the mucosa secrete hydrochloric acid (HCl) and hydrolytic enzymes. The *mucosal glands* also secrete a viscous mucus that helps prevent the stomach itself from being digested by the proteolytic enzymes. Most digestive activity occurs in the pyloric part of the stomach. After the food is processed in the stomach, it resembles a creamy mass called **chyme**, which enters the small intestine through the pyloric sphincter.

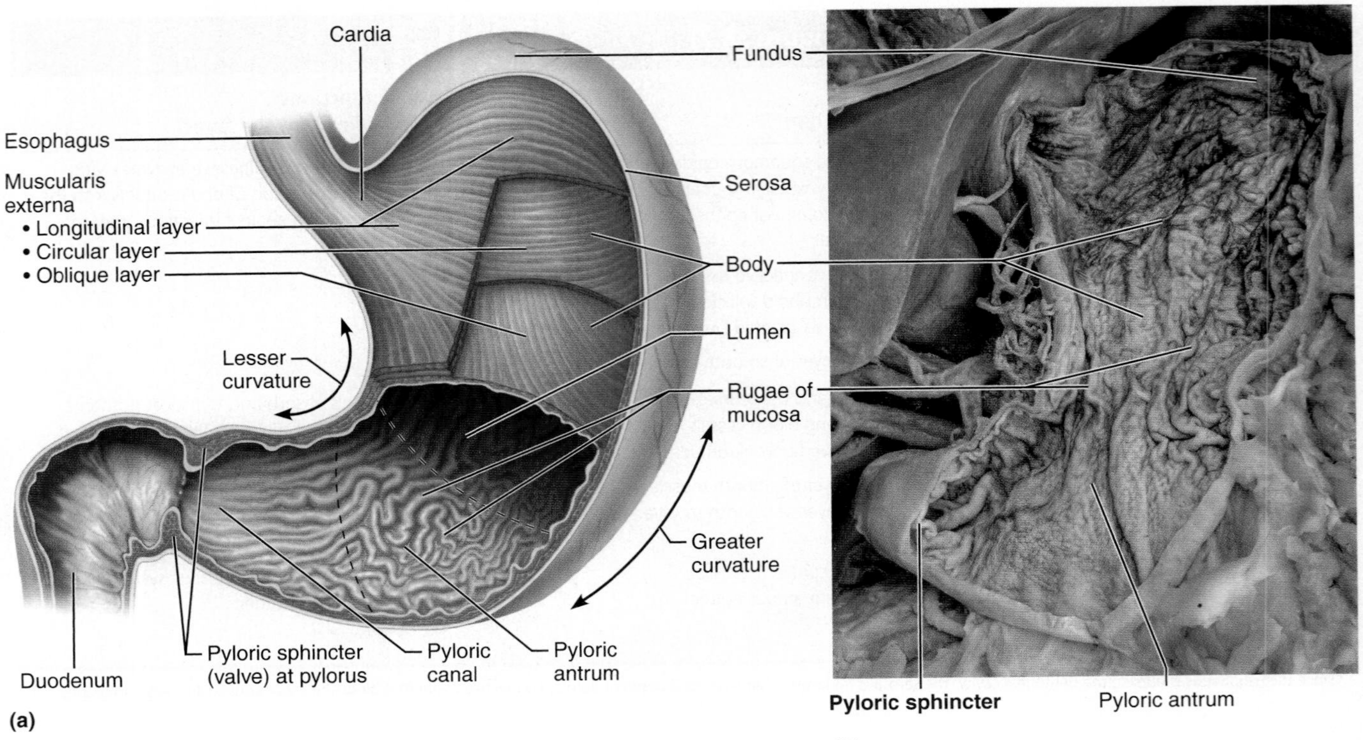

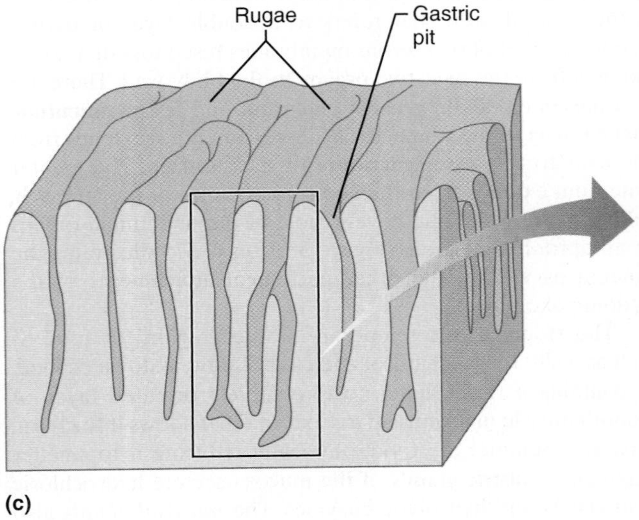

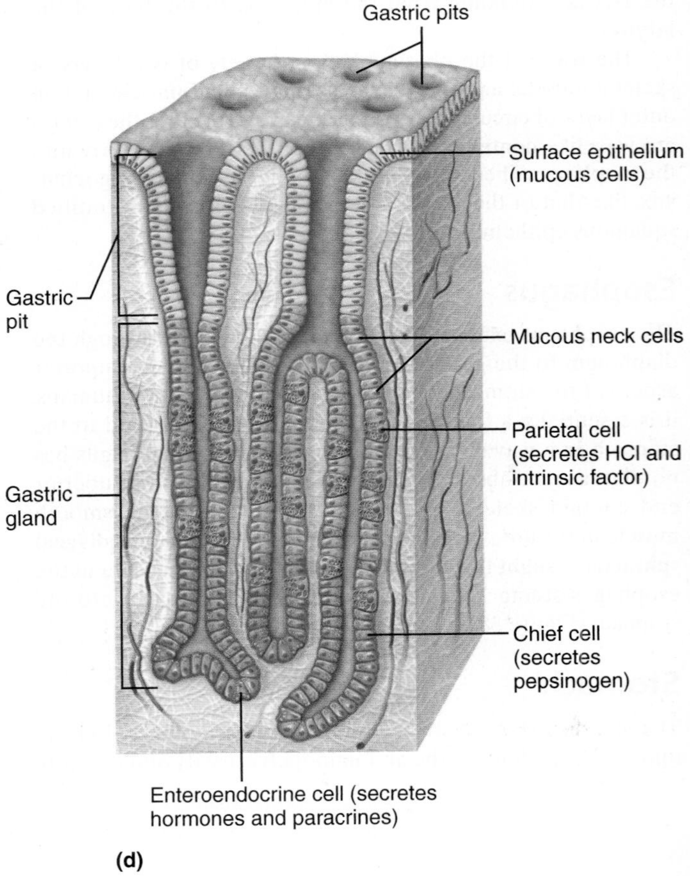

Figure 38.5 Anatomy of the stomach. (a) Gross internal and external anatomy. **(b)** Photograph of internal aspect of stomach. **(c, d)** Section of the stomach wall showing rugae and gastric pits.

Instructors may assign this figure as an Art Labeling Activity using Mastering A&P™

Table 38.2 Parts of the Stomach (Figure 38.5)

Structure	Description
Cardia (cardial part)	The area surrounding the cardial orifice through which food enters the stomach
Fundus	The dome-shaped area that is located superior and lateral to the cardia
Body	Midportion of the stomach and largest region
Pyloric part:	Funnel-shaped pouch that forms the distal stomach
Pyloric antrum	Wide superior portion of the pyloric part
Pyloric canal	Narrow tubelike portion of the pyloric part
Pylorus	Distal end of the pyloric part that is continuous with the small intestine
Pyloric sphincter	Valve that controls the emptying of the stomach into the small intestine

Activity 2

Studying the Histologic Structure of the Stomach and the Esophagus-Stomach Junction

1. **Stomach:** View the stomach slide first. Refer to **Figure 38.6a** on p. 574 as you scan the tissue under low power to locate the muscularis externa; then move to high power to more closely examine this layer. Try to pick out the three smooth muscle layers. How does the extra oblique layer of smooth muscle found in the stomach correlate with the stomach's churning movements?

____

____

____

Identify the gastric glands and the gastric pits (see Figures 38.5 and 38.6b). If the section is taken from the stomach fundus and is differentially stained, you can identify, in the gastric glands, the blue-staining **chief cells**, which produce pepsinogen, and the red-staining **parietal cells**, which secrete HCl and intrinsic factor. The enteroendocrine cells that release hormones and paracrines are indistinguishable. Draw a small section of the stomach wall, and label it appropriately.

2. **Esophagus-stomach junction:** Scan the slide under low power to locate the mucosal junction between the end of the esophagus and the beginning of the stomach. Draw a small section of the junction, and label it appropriately.

Compare your observations to Figure 38.6c. What is the functional importance of the epithelial differences seen in the two organs?

____

____

____

38

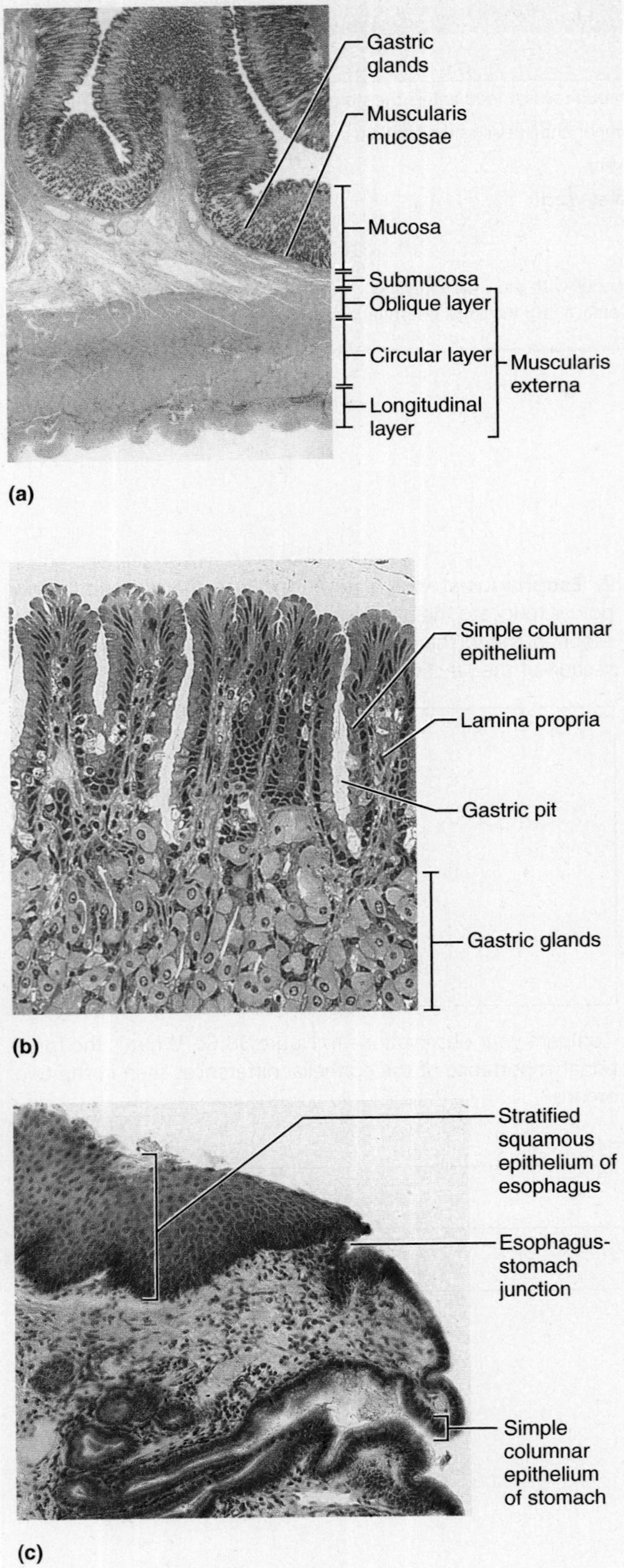

Figure 38.6 Histology of selected regions of the stomach and esophagus-stomach junction. **(a)** Stomach wall (12×). **(b)** Gastric pits and glands (130×). **(c)** Esophagus-stomach junction, longitudinal section (130×).

Small Intestine

The **small intestine** is a convoluted tube, 6 to 7 meters (about 20 feet) long in a cadaver but only about 2 m (6 feet) long during life because of its muscle tone. It extends from the pyloric sphincter to the ileocecal valve. The small intestine is suspended by a double layer of peritoneum, the fan-shaped **mesentery**, from the posterior abdominal wall (**Figure 38.7**), and it lies, framed laterally and superiorly by the large intestine, in the abdominal cavity. The small intestine has three subdivisions (see Figure 38.1):

1. The **duodenum** extends from the pyloric sphincter for about 25 cm (10 inches) and curves around the head of the pancreas; most of the duodenum lies in a retroperitoneal position.
2. The **jejunum**, continuous with the duodenum, extends for 2.5 m (about 8 feet). Most of the jejunum occupies the umbilical region of the abdominal cavity.
3. The **ileum**, the terminal portion of the small intestine, is about 3.6 m (12 feet) long and joins the large intestine at the **ileocecal valve.** It is located inferiorly and somewhat to the right in the abdominal cavity, but its major portion lies in the pubic region.

In the small intestine, enzymes from two sources complete the digestion process: **brush border enzymes**, which are hydrolytic enzymes bound to the microvilli of the columnar epithelial cells; and, more important, enzymes produced by the pancreas and ducted into the duodenum largely via the **main pancreatic duct**. Bile (formed in the liver) also enters the duodenum via the **bile duct** in the same area. At the duodenum, the ducts join to form the bulblike **hepatopancreatic ampulla** and empty their products into the duodenal lumen through the **major duodenal papilla**, an orifice controlled by a muscular valve called the **hepatopancreatic sphincter** (see Figure 38.15 on p. 581).

Nearly all nutrient absorption occurs in the small intestine, where three structural modifications increase the absorptive surface of the mucosa: the microvilli, villi, and circular folds (**Figure 38.8**, p. 576).

- **Microvilli:** Microscopic projections of the surface plasma membrane of the columnar epithelial lining cells of the mucosa.
- **Villi:** Fingerlike projections of the mucosa tunic that give it a velvety appearance and texture.
- **Circular folds:** Deep, permanent folds of the mucosa and submucosa layers that force chyme to spiral through the intestine, mixing it and slowing its progress. These structural modifications decrease in frequency and size toward the end of the small intestine. Any residue remaining undigested and unabsorbed at the terminus of the small intestine enters the large intestine through the ileocecal valve. The amount of lymphoid tissue in the submucosa of the small intestine (especially the aggregated lymphoid nodules called **Peyer's patches**, **Figure 38.9b**, p. 577) increases along the length of the small intestine and is very apparent in the ileum.

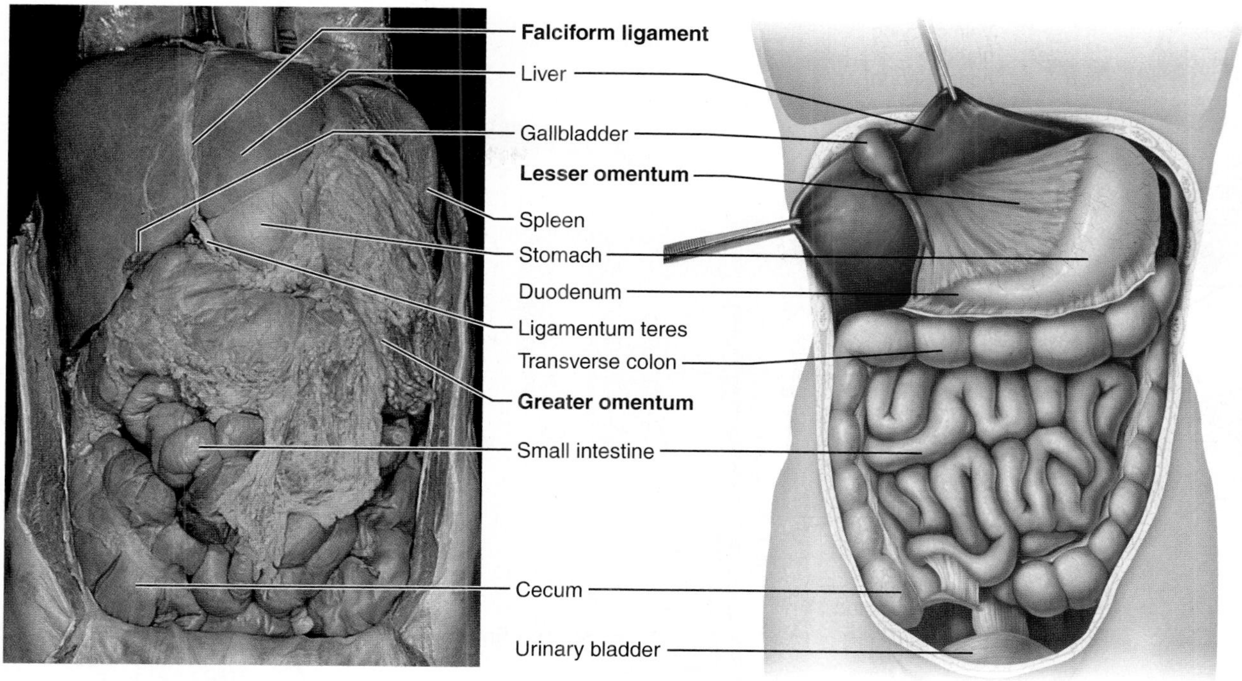

(a)

(b)

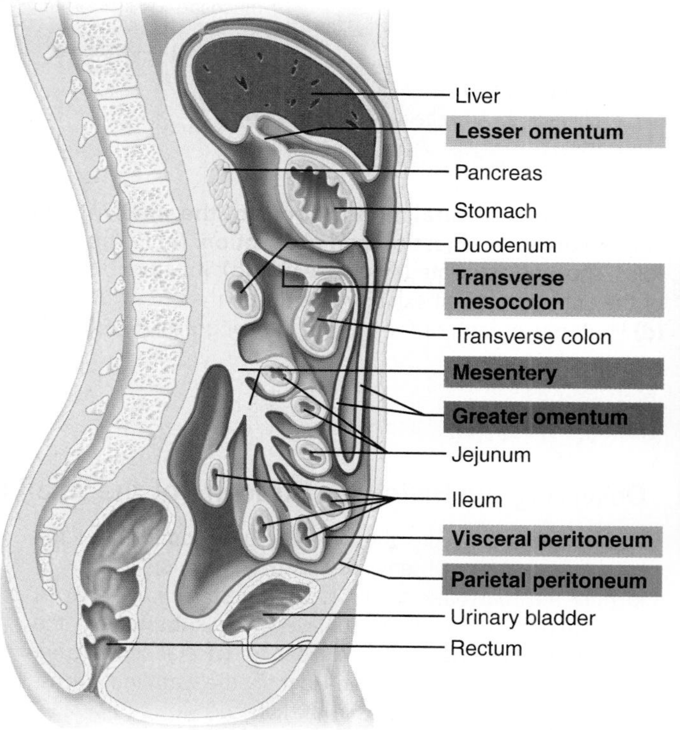

(c)

Instructors may assign this figure as an Art Labeling Activity using Mastering A&P™

Figure 38.7 Peritoneal attachments of the abdominal organs. Superficial anterior views of abdominal cavity: **(a)** photograph with the greater omentum in place and **(b)** diagram showing greater omentum removed and liver and gallbladder reflected superiorly. **(c)** Sagittal view of a male torso. Mesentery labels appear in colored boxes.

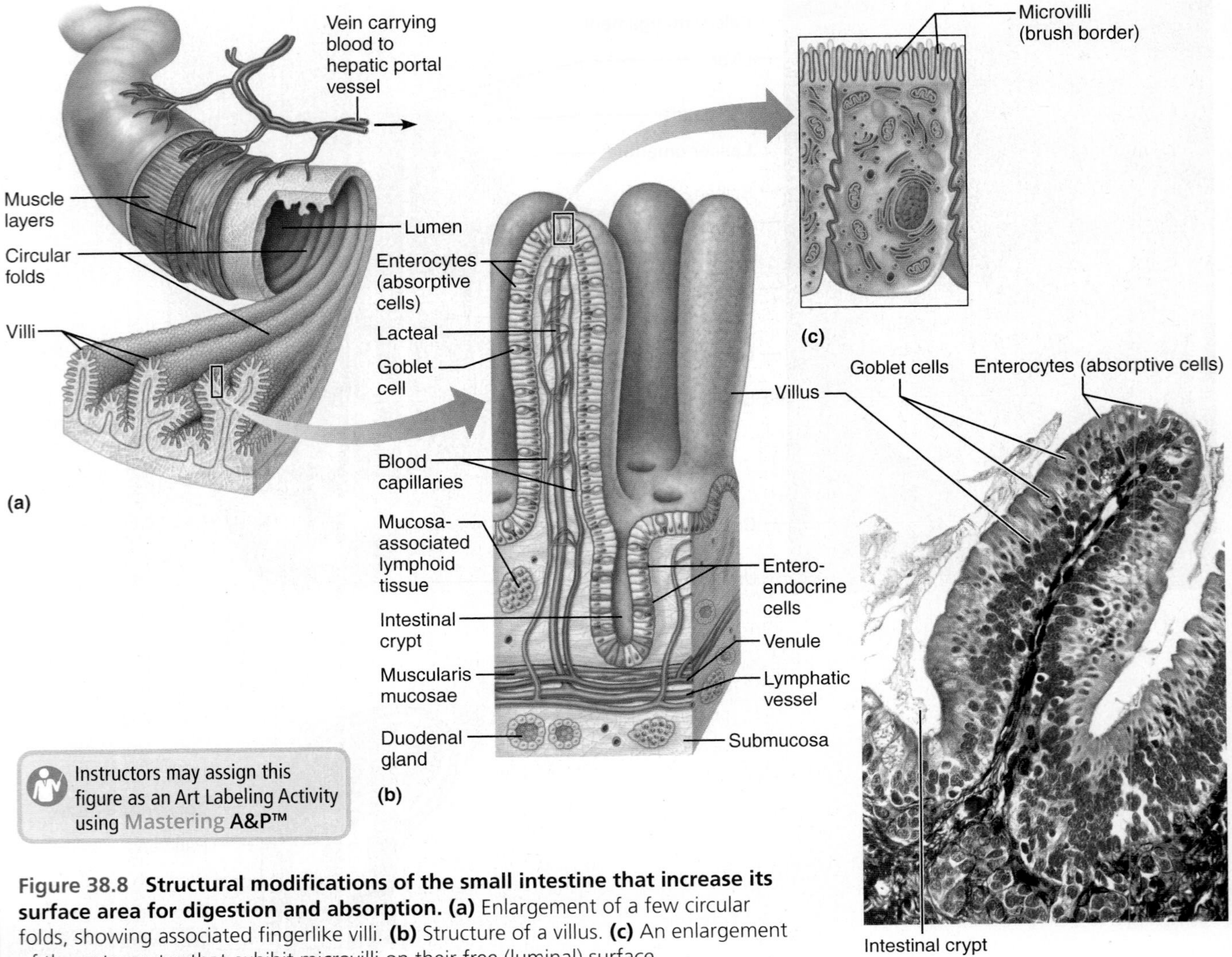

Instructors may assign this figure as an Art Labeling Activity using Mastering A&P™

Figure 38.8 Structural modifications of the small intestine that increase its surface area for digestion and absorption. (a) Enlargement of a few circular folds, showing associated fingerlike villi. **(b)** Structure of a villus. **(c)** An enlargement of the enterocytes that exhibit microvilli on their free (luminal) surface. **(d)** Photomicrograph of the mucosa showing villi (250×).

38

Activity 3

Observing the Histologic Structure of the Small Intestine

1. **Duodenum:** Secure the slide of the duodenum to the microscope stage. Observe the tissue under low power to identify the four basic tunics of the intestinal wall—that is, the **mucosa** and its three sublayers, the **submucosa**, the **muscularis externa**, and the **serosa**, or *visceral peritoneum*. Consult Figure 38.9a to help you identify the scattered mucus-producing **duodenal glands** in the submucosa.

What type of epithelium do you see here? ______________

____

Examine the large leaflike *villi*, which increase the surface area for absorption. Notice the scattered mucus-producing goblet cells in the epithelium of the villi. Note also the **intestinal crypts** (see also Figure 38.8), invaginated areas of the mucosa between the villi containing the cells that produce intestinal juice, a watery mucus-containing mixture that serves as a carrier fluid for absorption of nutrients from the chyme. Sketch and label a small section of the duodenal wall, showing all layers and villi.

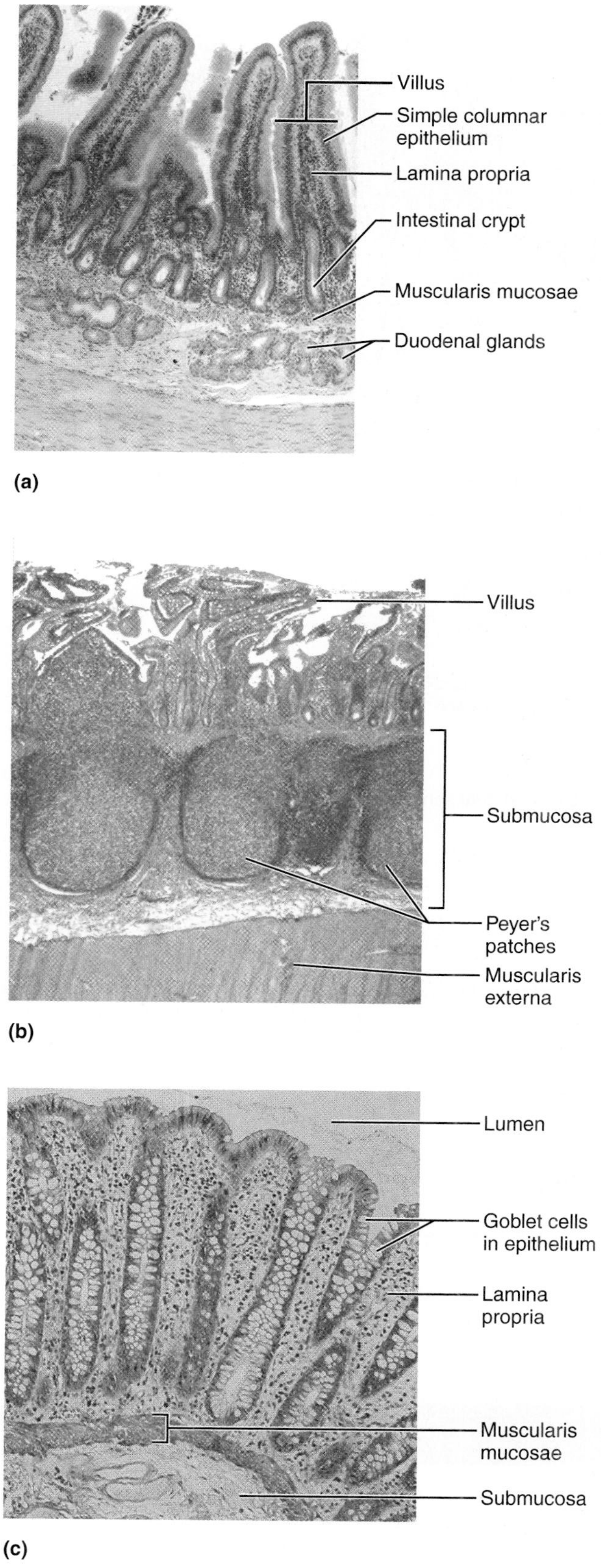

Figure 38.9 Histology of selected regions of the small and large intestines. Cross-sectional views. **(a)** Duodenum of the small intestine (95×). **(b)** Ileum of the small intestine (20×). **(c)** Large intestine (80×).

2. **Ileum:** The structure of the ileum resembles that of the duodenum, except that the villi are less elaborate (because most of the absorption has occurred by the time that chyme reaches the ileum). Secure a slide of the ileum to the microscope stage for viewing. Observe the villi, and identify the four layers of the wall and the large, generally spherical Peyer's patches (Figure 38.9b). What tissue type are Peyer's patches?

____

3. If a villus model is available, identify the following cells or regions before continuing: absorptive epithelium, goblet cells, lamina propria, the muscularis mucosae, capillary bed, and lacteal. If possible, also identify the intestinal crypts.

Large Intestine

The **large intestine** (**Figure 38.10**, p. 578) is about 1.5 m (5 feet) long and extends from the ileocecal valve to the anus. It encircles the small intestine on three sides and consists of the following subdivisions: **cecum**, **appendix**, **colon**, **rectum**, and **anal canal**.

The wormlike appendix, which hangs from the cecum, is a trouble spot in the large intestine. Since it is generally twisted, it provides an ideal location for bacteria to accumulate and multiply. Inflammation of the appendix, or appendicitis, is the result. +

The colon is divided into several distinct regions. The **ascending colon** travels up the right side of the abdominal cavity and makes a right-angle turn at the **right colic (hepatic) flexure** to cross the abdominal cavity as the **transverse colon**. It then turns at the **left colic (splenic) flexure** and continues down the left side of the abdominal cavity as the **descending colon**, where it takes an S-shaped course as the **sigmoid colon**. The sigmoid colon, rectum, and the anal canal lie in the pelvis anterior to the sacrum and thus are not considered abdominal cavity structures. Except for the transverse and sigmoid colons, the colon is retroperitoneal.

The anal canal terminates in the **anus**, the opening to the exterior of the body. The anal canal has two sphincters, a voluntary *external anal sphincter* composed of skeletal muscle, and an involuntary *internal anal sphincter* composed of smooth muscle. The sphincters are normally closed except during defecation, when undigested food and bacteria are eliminated from the body as feces.

In the large intestine, the longitudinal muscle layer of the muscularis externa is reduced to three longitudinal muscle bands called the **teniae coli**. Since these bands are shorter than the rest of the wall of the large intestine, they cause the wall to pucker into small pocketlike sacs called **haustra**. Fat-filled pouches of visceral peritoneum, called *epiploic appendages*, hang from the colon's surface.

The major function of the large intestine is to consolidate and propel the unusable fecal matter toward the anus and eliminate it from the body. While it does this task, it (1) provides a site where intestinal bacteria manufacture vitamins B and K; and (2) reclaims most of the remaining water from undigested food, thus conserving body water.

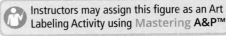

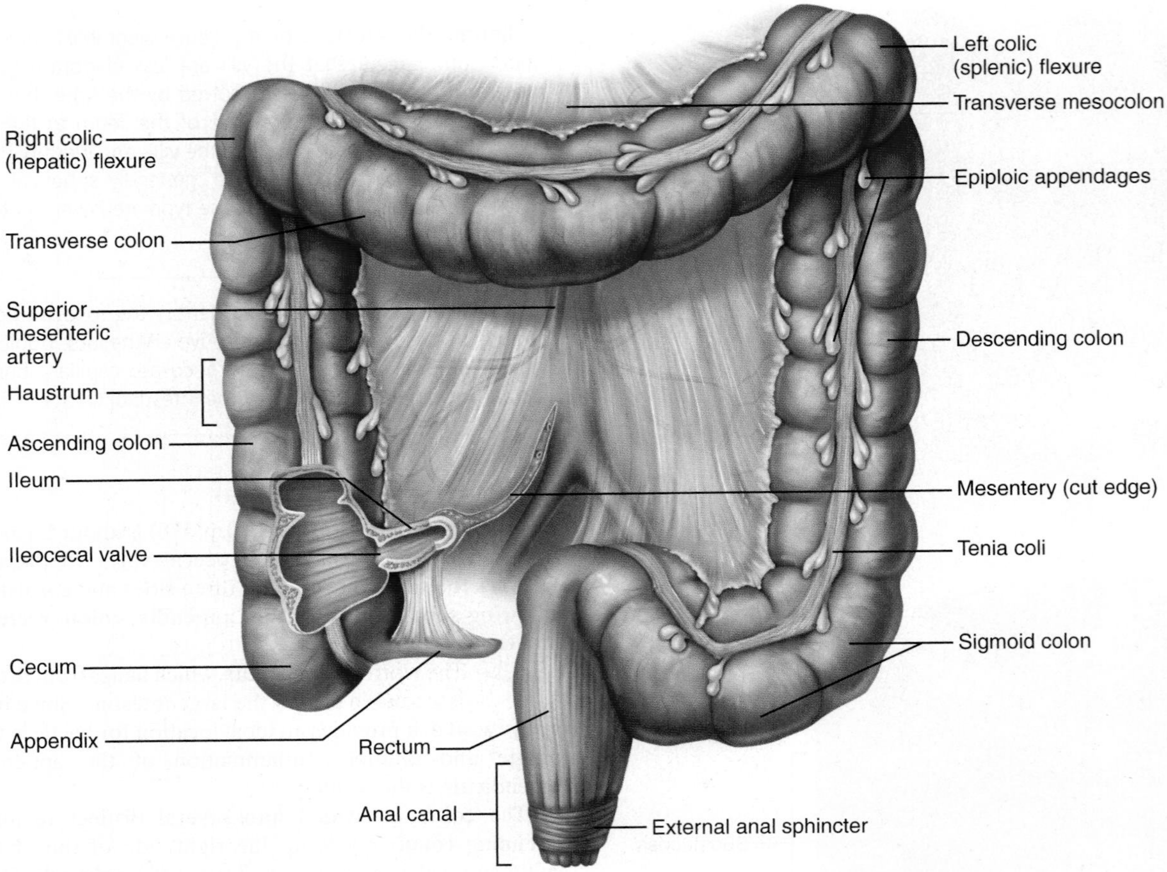

Figure 38.10 The large intestine. (Section of the cecum removed to show the ileocecal valve.)

Instructors may assign this figure as an Art Labeling Activity using Mastering **A&P™**

Watery stools, or **diarrhea**, result from any condition that rushes undigested food residue through the large intestine before it has had sufficient time to absorb the water. Conversely, when food residue remains in the large intestine for extended periods, excessive water is absorbed and the stool becomes hard and difficult to pass, causing **constipation**. +

38

Activity 4

Examining the Histologic Structure of the Large Intestine

Large intestine: Secure a slide of the large intestine to the microscope stage for viewing. Observe the numerous goblet cells in the epithelium (Figure 38.9c). Why do you think the large intestine produces so much mucus?

____

____

Accessory Digestive Organs

Teeth

By the age of 21, two sets of teeth have developed (**Figure 38.11**). The initial set, called the **deciduous** (or **milk**) **teeth**, normally appears between the ages of 6 months and 2½ years. The first of these to erupt are the lower central incisors. The child begins to shed the deciduous teeth around the age of 6, and a second set of teeth, the **permanent teeth**, gradually replaces them. As the deeper permanent teeth progressively enlarge and develop, the roots of the deciduous teeth are resorbed, leading to their final shedding.

Teeth are classified as **incisors**, **canines** *(eye teeth, cuspids)*, **premolars** *(bicuspids)*, and **molars**. The incisors are chisel shaped and exert a shearing action used in biting. Canines are cone-shaped teeth used for tearing food. The premolars have two *cusps* (grinding surfaces); the molars have broad crowns with rounded cusps specialized for the fine grinding of food.

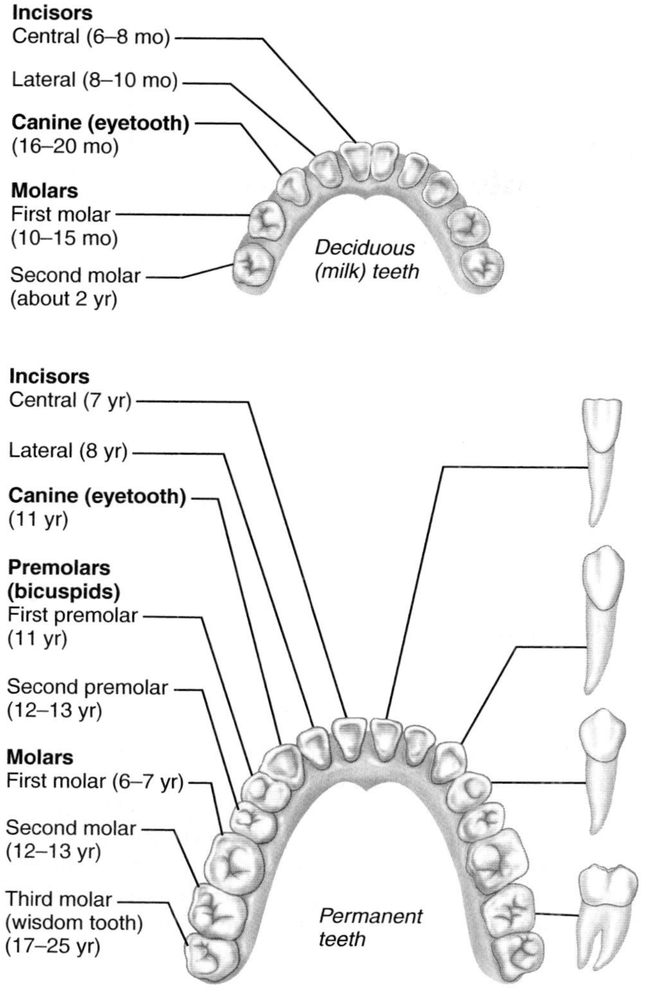

Figure 38.11 Human dentition. (Approximate time of teeth eruption shown in parentheses.)

Instructors may assign this figure as an Art Labeling Activity using Mastering **A&P™**

Dentition is described by means of a **dental formula**, which designates the numbers, types, and position of the teeth in one side of the jaw. Because tooth arrangement is bilaterally symmetrical, it is necessary to designate one only side of the jaw. The complete dental formula for the deciduous teeth from the medial aspect of each jaw and proceeding posteriorly is as follows:

$$\frac{\text{Upper teeth: 2 incisors, 1 canine, 0 premolars, 2 molars}}{\text{Lower teeth: 2 incisors, 1 canine, 0 premolars, 2 molars}} \times 2$$

This formula is generally abbreviated to read as follows:

$$\frac{2,1,0,2}{2,1,0,2} \times 2 \text{ (20 deciduous teeth)}$$

The permanent teeth are then described by the following dental formula:

$$\frac{2,1,2,3}{2,1,2,3} \times 2 \text{ (32 permanent teeth)}$$

Although 32 is designated as the normal number of permanent teeth, not everyone develops a full set. In many people, the third molars, commonly called *wisdom teeth*, never erupt.

Activity 5

Identifying Types of Teeth

Identify the four types of teeth (incisors, canines, premolars, and molars) on the jaw model or human skull.

A tooth consists of two major regions, the **crown** and the **root**. These two regions meet at the **neck** near the gum line. A longitudinal section made through a tooth shows the following basic anatomical plan (Figure 38.12). The crown is the superior portion of the tooth visible above the **gingiva**, or **gum**, which surrounds the tooth. The surface of the crown is covered by **enamel**. Enamel consists of 95% to 97% inorganic calcium salts and thus is heavily mineralized. The crevice between the end of the crown and the upper margin of the gingiva is referred to as the *gingival sulcus*.

That portion of the tooth embedded in the bone is the root. The outermost surface of the root is covered by **cement**, which is similar to bone in composition and less brittle than enamel. The cement attaches the tooth to the **periodontal ligament**, which holds the tooth in the tooth socket and exerts a cushioning effect. **Dentin**, which composes the bulk of the tooth, is the bonelike material interior to the enamel and cement.

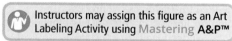

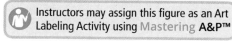

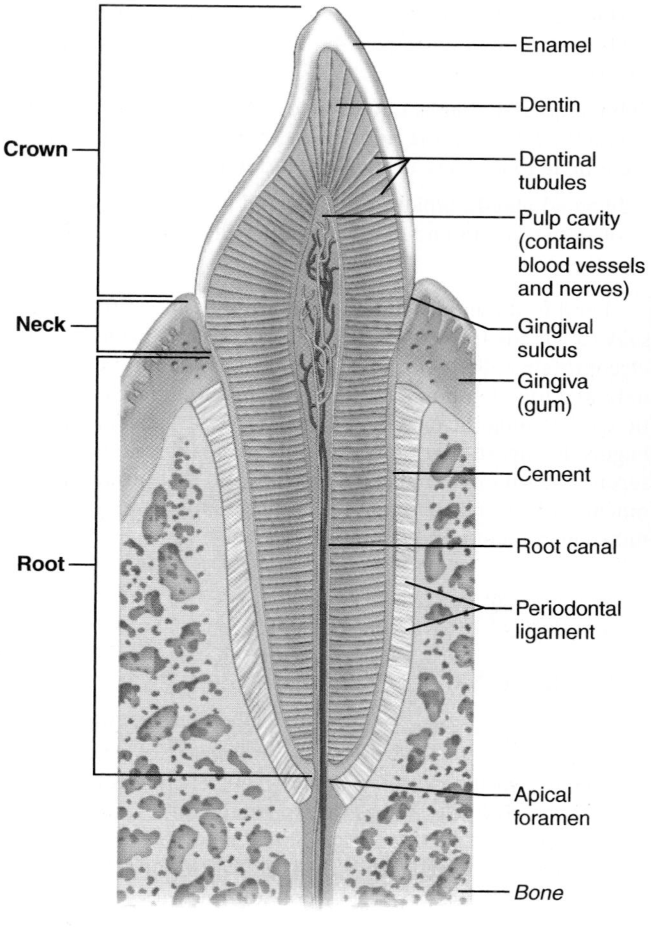

Figure 38.12 Longitudinal section of human canine tooth within its bony socket (alveolus).

Instructors may assign this figure as an Art Labeling Activity using Mastering **A&P™**

38

Dentin surrounds the **pulp cavity**, which is filled with pulp. **Pulp** is composed of connective tissue liberally supplied with blood vessels, nerves, and lymphatics that provide for tooth sensation and supplies nutrients to the tooth tissues. **Odontoblasts**, specialized cells in the outer margins of the pulp cavity, produce and maintain the dentin. Odontoblasts have slender processes that extend into the *dentinal tubules* of the dentin. The pulp cavity extends into distal portions of the root and becomes the **root canal**. An opening at the root apex, the **apical foramen**, provides a route of entry into the tooth for blood vessels, nerves, and other structures from the tissues beneath.

Activity 6

Studying Microscopic Tooth Anatomy

Observe a slide of a longitudinal section of a tooth, and compare your observations with the structures detailed in Figure 38.12. Identify as many of these structures as possible.

Salivary Glands

Three pairs of major **salivary glands** (see Figure 38.1) empty their secretions into the oral cavity.

Parotid glands: Large glands located anterior to the ear and ducting into the mouth over the second upper molar through the parotid duct.

Submandibular glands: Located along the medial aspect of the mandibular body in the floor of the mouth, and ducting under the tongue to the base of the lingual frenulum.

Sublingual glands: Small glands located most anteriorly in the floor of the mouth and emptying under the tongue via several small ducts.

Food in the mouth and mechanical pressure stimulate the salivary glands to secrete saliva. Saliva consists primarily of a glycoprotein called *mucin*, which moistens the food and helps to bind it together into a mass called a **bolus**, and a clear serous fluid containing the enzyme *salivary amylase*. Salivary amylase begins the digestion of starch. Parotid gland secretion is mainly serous; the submandibular is a mixed gland that produces both mucin and serous components; and the sublingual gland is a mixed gland that produces mostly mucin.

Activity 7

Examining Salivary Gland Tissue

Examine salivary gland tissue under low power and then high power to become familiar with the appearance of a glandular tissue. Notice the clustered arrangement of the cells around their ducts. The cells are basically triangular, with their pointed ends facing the duct opening. Differentiate between mucus-producing cells, which have a clear cytoplasm, and serous cells, which have granules in their cytoplasm. The serous cells often form *demilunes* (caps) around the more central mucous cells. (**Figure 38.13** may be helpful in this task.)

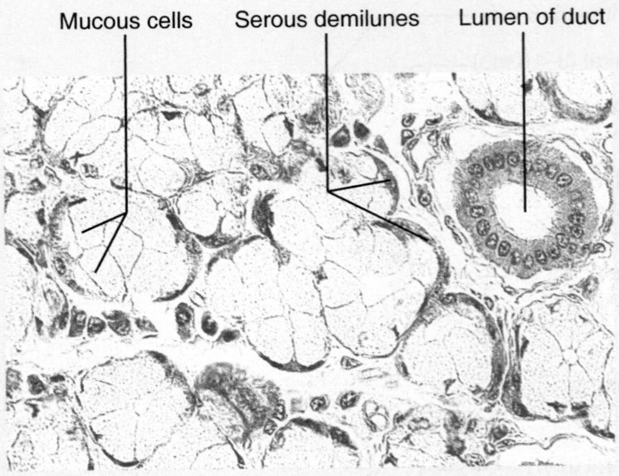

Figure 38.13 Histology of a mixed salivary gland. Sublingual gland (170×).

Liver and Gallbladder

The **liver** (see Figure 38.1), the largest gland in the body, is located inferior to the diaphragm, more to the right than the left side of the body. The human liver has four lobes and is suspended from the diaphragm and anterior abdominal wall by the **falciform ligament** (**Figure 38.14**).

The liver performs many metabolic roles. However, its digestive function is to produce bile, which leaves the liver through the **common hepatic duct** and then enters the duodenum through the **bile duct** (**Figure 38.15**). Bile has no enzymatic action but emulsifies fats, breaking up fat globules into small droplets. Without bile, very little fat digestion or absorption occurs.

When digestive activity is not occurring in the digestive tract, bile backs up into the **cystic duct** and enters the **gallbladder**, a small, green sac on the inferior surface of the liver. Bile is stored there until needed for the digestive process.

If the common hepatic or bile duct is blocked (for example, by wedged gallstones), bile is prevented from entering the small intestine, accumulates, and eventually backs up into the liver. This exerts pressure on the liver cells, and bile begins to enter the bloodstream. As the bile circulates through the body, the tissues become yellow, or jaundiced.

Blockage of the ducts is just one cause of jaundice. More often it results from actual liver problems such as **hepatitis**, (which is any inflammation of the liver,) or **cirrhosis**, a condition in which the liver is severely damaged and becomes hard and fibrous. +

As demonstrated by its highly organized anatomy, the liver (**Figure 38.16**, p. 582) is very important in the initial processing of the nutrient-rich blood draining the digestive organs. Its structural and functional units are called **lobules**. Each lobule is a basically hexagonal structure consisting of cordlike arrays of **hepatocytes**, or *liver cells*, which radiate outward from a central vein running upward in the longitudinal axis of the lobule. At each of the six corners of the lobule is a **portal triad**, so named because three basic structures are always present there: a *portal arteriole* (a branch of the *hepatic artery*, the functional blood supply of the liver), a *portal venule* (a branch of the *hepatic portal vein* carrying nutrient-rich blood from the digestive viscera),

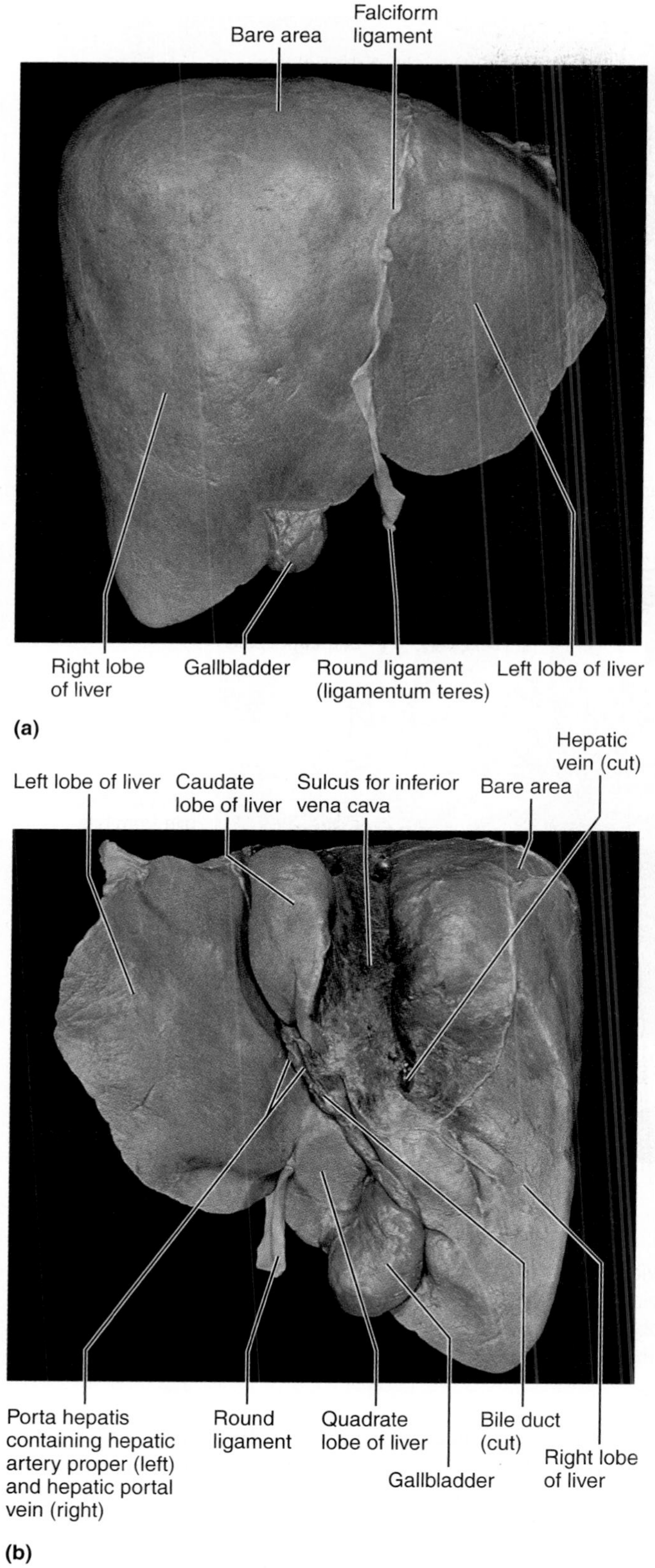

Figure 38.14 Gross anatomy of the human liver. **(a)** Anterior view. **(b)** Posteroinferior aspect. The four liver lobes are separated by a group of fissures in this view.

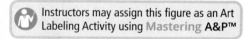

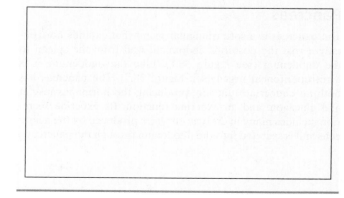

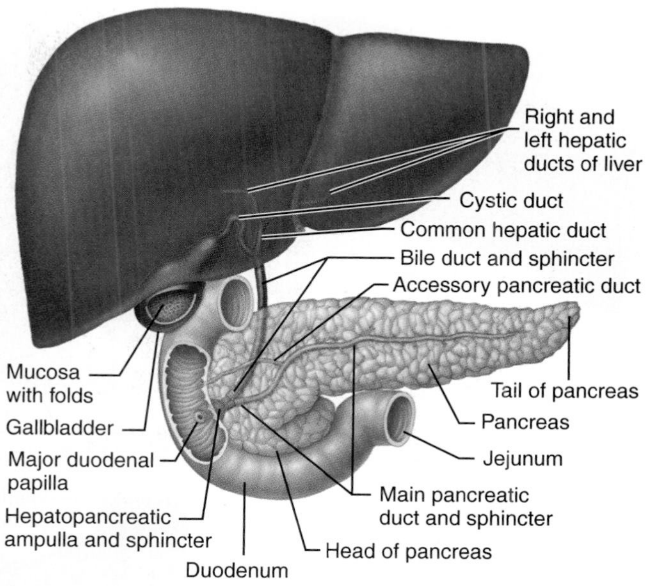

Figure 38.15 Ducts of accessory digestive organs.

Instructors may assign this figure as an Art Labeling Activity using Mastering A&P™

and a *bile duct*. Between the liver cells are blood-filled spaces, or **sinusoids**, through which blood from the hepatic portal vein and hepatic artery percolates. **Stellate macrophages**, special phagocytic cells also called **hepatic macrophages**, line the sinusoids and remove debris such as bacteria from the blood as it flows past, while the hepatocytes pick up oxygen and nutrients. The sinusoids empty into the central vein, and the blood ultimately drains from the liver via the *hepatic veins*.

Bile is continuously being made by the hepatocytes. It flows through tiny canals, the **bile canaliculi**, which run between adjacent cells toward the bile duct branches in the triad regions, where the bile eventually leaves the liver.

38

Activity 8

Examining the Histology of the Liver

Examine a slide of liver tissue and identify as many as possible of the structural features (see Figure 38.16). Also examine a three-dimensional model of liver lobules if this is available. Reproduce a small pie-shaped section of a liver lobule in the space below. Label the hepatocytes, the stellate macrophages, sinusoids, a portal triad, and a central vein.

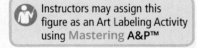

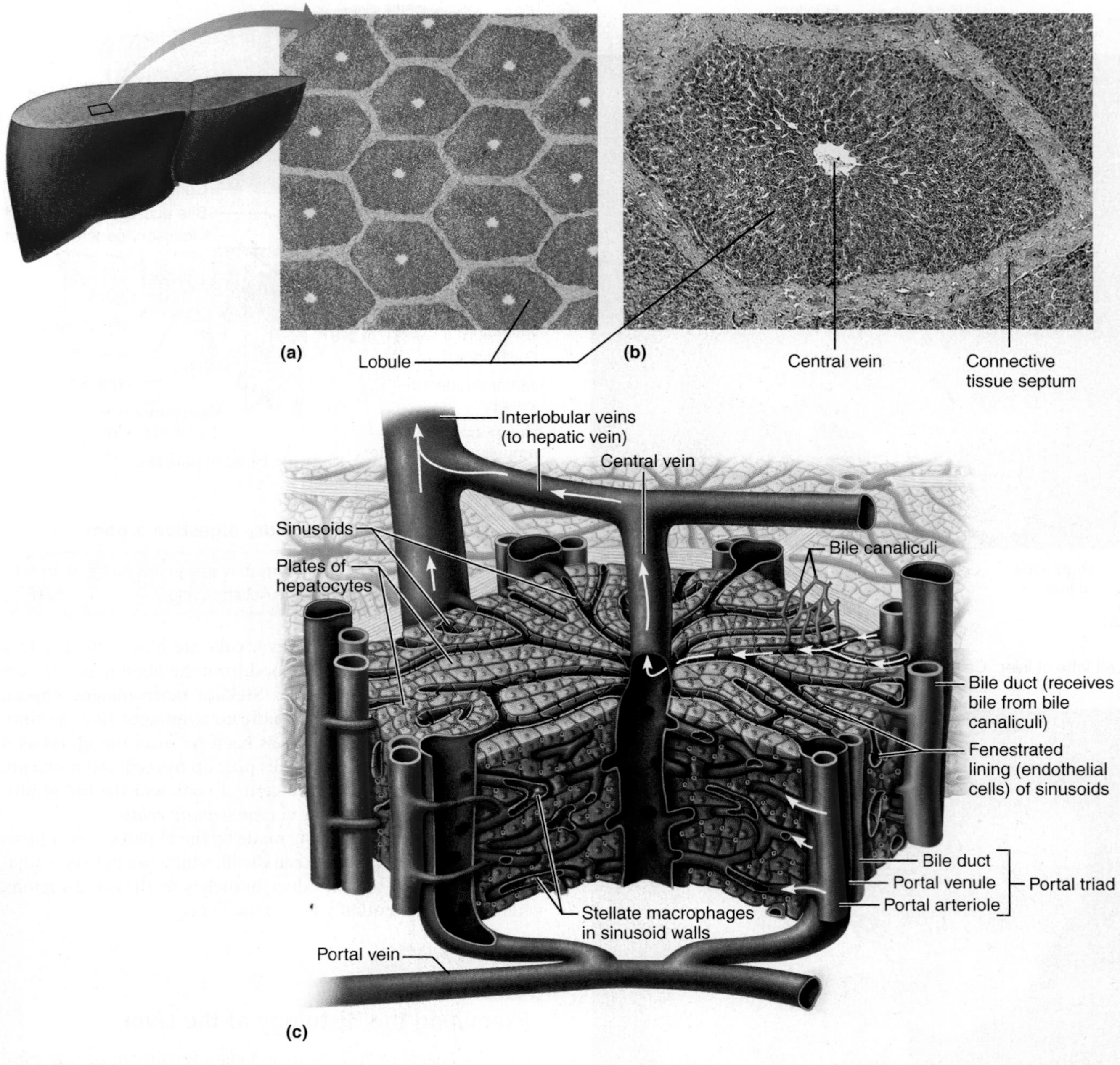

Figure 38.16 Microscopic anatomy of the liver. (a) Schematic view of the cut surface of the liver showing the hexagonal nature of its lobules. **(b)** Photomicrograph of one liver lobule (55×). **(c)** Enlarged three-dimensional diagram of one liver lobule. Arrows show direction of blood flow. Bile flows in the opposite direction toward the bile ducts.

Instructors may assign this figure as an Art Labeling Activity using Mastering A&P™

Pancreas

The **pancreas** is a soft, triangular gland that extends horizontally across the posterior abdominal wall from the spleen to the duodenum (see Figure 38.1). Like the duodenum, it is a retroperitoneal organ (see Figure 38.7). The pancreas has both an endocrine function, producing the hormones insulin and glucagon, and an exocrine function. Its exocrine secretion includes many hydrolytic enzymes produced by the acinar cells and is secreted into the duodenum through the pancreatic ducts. Pancreatic juice is very alkaline. Its high concentration of bicarbonate ion (HCO_3^-) neutralizes the acidic chyme entering the duodenum from the stomach, enabling the pancreatic and intestinal enzymes to operate at their optimal pH, which is slightly alkaline. (See Figure 27.3c.)

For instructions on animal dissections, see the dissection exercises (starting on p. 695) in the cat and fetal pig editions of this manual.

REVIEW SHEET

EXERCISE 38

Anatomy of the Digestive System

Instructors may assign a portion of the Review Sheet questions using Mastering A&P™

Name ______________________ Lab Time/Date ______________________

General Histological Plan of the Alimentary Canal

1. The general anatomical features of the alimentary canal are listed below. Fill in the table to complete the information.

Wall layer	Subdivisions of the layer	Major functions
Mucosa		
Submucosa	(Not applicable)	
Muscularis externa		
Serosa or adventitia	(Not applicable)	

Organs of the Alimentary Canal

2. The tubelike digestive system canal that extends from the mouth to the anus is known as the ______________ canal or the ______________ tract.

3. How is the muscularis externa of the stomach modified? ______________

How does this modification relate to the function of the stomach? ______________

4. What transition in epithelial type exists at the esophagus-stomach junction? ______________

How do the epithelia of these two organs relate to their specific functions?

____

5. Differentiate the colon from the large intestine. ______________

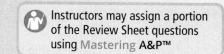

6. Match the items in column B with the descriptive statements in column A.

Column A

_________ 1. structure that suspends the small intestine from the posterior body wall

_________ 2. fingerlike extensions of the intestinal mucosa that increase the surface area for absorption

_________ 3. large collections of lymphoid tissue found in the submucosa of the small intestine

_________ 4. deep folds of the mucosa and submucosa that extend completely or partially around the circumference of the small intestine

_________ 5. mobile organ that manipulates food in the mouth and initiates swallowing

_________ 6. conduit for both air and food

_________ 7. food passageway that has no digestive/absorptive function

_________ 8. folds of the gastric mucosa

_________ 9. pocketlike sacs of the large intestine

_________ 10. projections of the plasma membrane of a mucosal epithelial cell

_________ 11. valve at the junction of the small and large intestines

_________ 12. primary region of nutrient absorption

_________ 13. membrane securing the tongue to the floor of the mouth

_________ 14. absorbs water and forms feces

_________ 15. area between the teeth and lips/cheeks

_________ 16. wormlike sac that outpockets from the cecum

_________ 17. initiates protein digestion

_________ 18. structure attached to the lesser curvature of the stomach

_________ 19. covers most of the abdominal organs like an apron

_________ 20. valve controlling food movement from the stomach into the duodenum

_________ 21. posterosuperior boundary of the oral cavity

_________ 22. region containing two sphincters through which feces are expelled from the body

_________ 23. bone-supported anterosuperior boundary of the oral cavity

Column B

a. anus

b. appendix

c. circular folds

d. esophagus

e. frenulum

f. greater omentum

g. hard palate

h. haustra

i. ileocecal valve

j. large intestine

k. lesser omentum

l. mesentery

m. microvilli

n. oral vestibule

o. Peyer's patches

p. pharynx

q. pyloric sphincter

r. rugae

s. small intestine

t. soft palate

u. stomach

v. tongue

w. villi

7. Correctly identify all organs depicted in the diagram below.

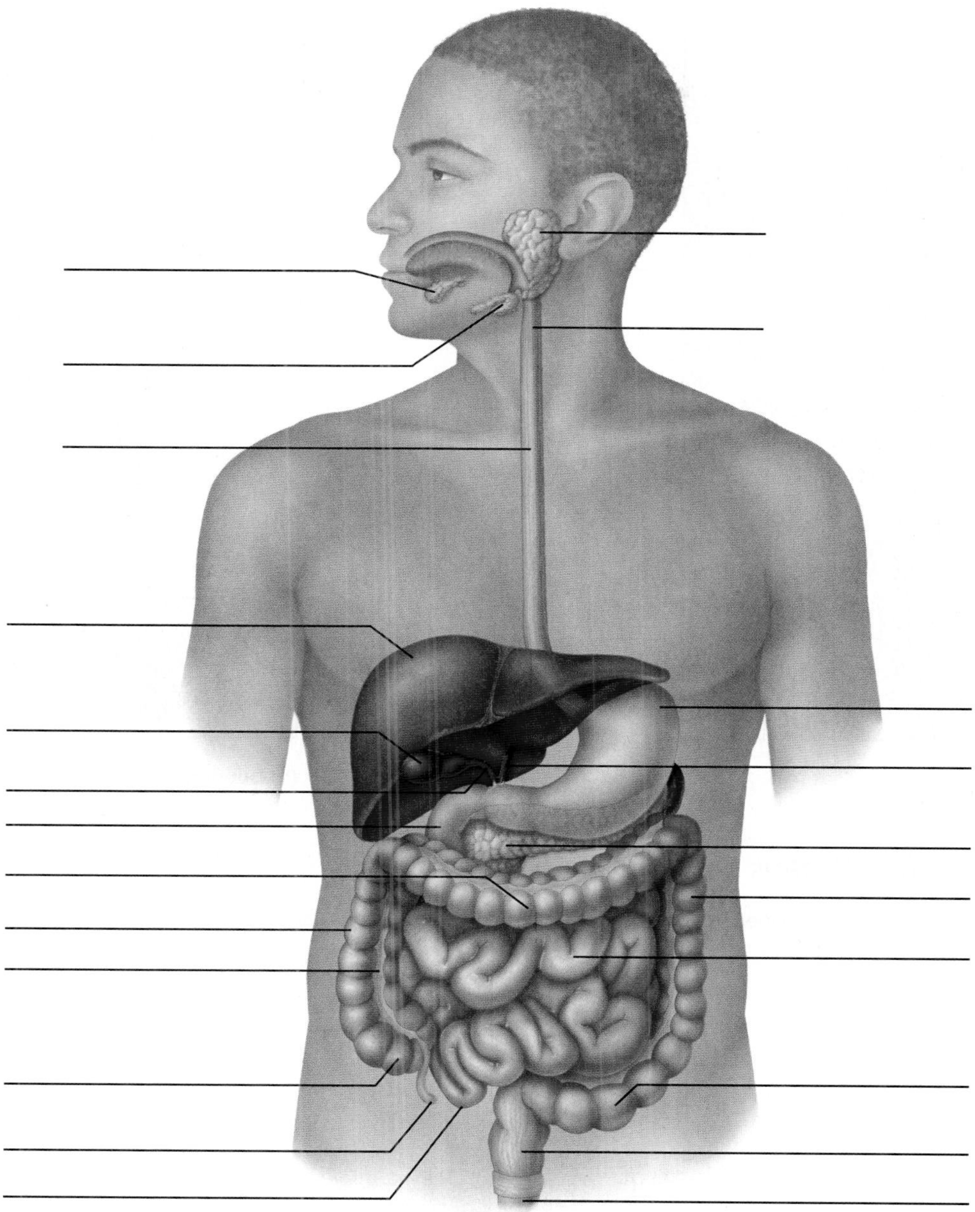

8. You have studied the histologic structure of a number of organs in this laboratory. The stomach and the duodenum are diagrammed below. Label the structures indicated by leader lines.

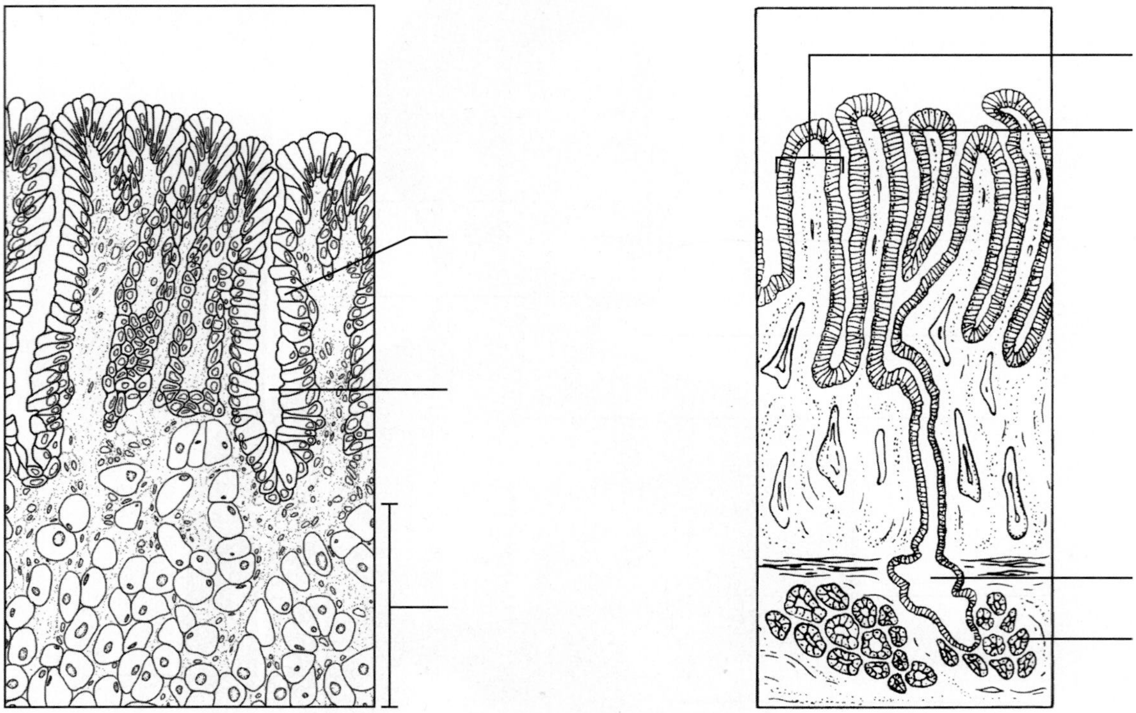

(a) Stomach

(b) Duodenum

Accessory Digestive Organs

9. Correctly label all structures provided with leader lines in the diagram of a molar below. (Note: Some of the terms in the key for question 10 may be helpful in this task.)

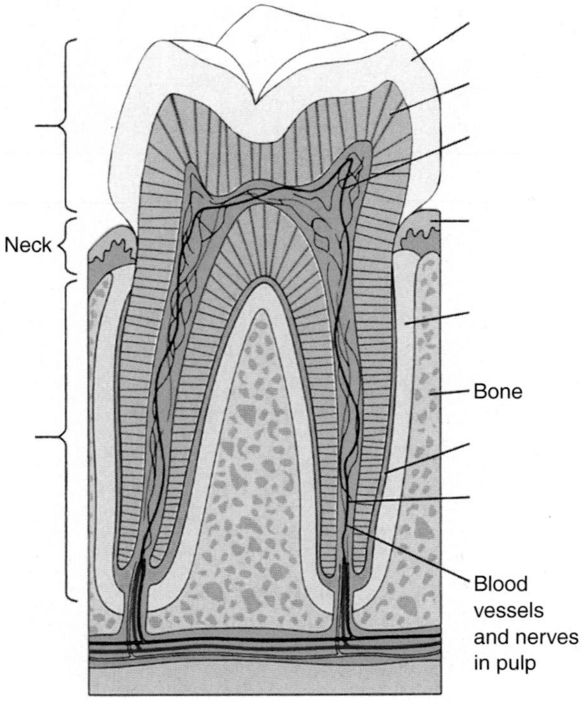

10. Use the key to identify each tooth area described below.

_________ 1. visible portion of the tooth	*Key:*	a. cement
_________ 2. material covering the tooth root		b. crown
_________ 3. hardest substance in the body		c. dentin
_________ 4. attaches the tooth to the tooth socket		d. enamel
_________ 5. portion of the tooth embedded in bone		e. gingival sulcus
_________ 6. forms the major portion of tooth structure; similar to bone		f. odontoblast
_________ 7. produces the dentin		g. periodontal ligament
_________ 8. site of blood vessels, nerves, and lymphatics		h. pulp
_________ 9. narrow gap between the crown and the gum		i. root

11. In the human, the number of deciduous teeth is _______________; the number of permanent teeth is _______________.

12. The dental formula for permanent teeth is $\frac{2,1,2,3}{2,1,2,3} \times 2$

Explain what this means. ____

____

What is the dental formula for the deciduous teeth? ____________ × ____________ (____________ deciduous teeth)

13. Which teeth are the "wisdom teeth"? ____

14. Various types of glands form a part of the alimentary canal wall or duct their secretions into it. Match the glands listed in column B with the function/locations described in column A.

Column A	Column B
_________ 1. produce(s) mucus; found in the submucosa of the small intestine	a. duodenal glands
_________ 2. produce(s) a product containing amylase that begins starch breakdown in the mouth	b. gastric glands
	c. intestinal crypts
_________ 3. produce(s) many enzymes and an alkaline fluid that is secreted into the duodenum	d. liver
_________ 4. produce(s) bile that it secretes into the duodenum via the bile duct	e. pancreas
_________ 5. produce(s) HCl and pepsinogen	f. salivary glands
_________ 6. found in the mucosa of the small intestine; produce(s) intestinal juice	

15. Which of the salivary glands produces a secretion that is mainly serous? ______________________________

16. What is the role of the gallbladder? ____

17. Name three structures that form a portal triad of the liver. ______________________________,

______________________________, and ______________________________

18. Where would you expect to find the stellate macrophages of the liver? ______________________________

What is their function? ______________________________

19. Why is the liver so dark red in the living animal? ______________________________

____

20. The pancreas has two major populations of secretory cells—those in the islets and the acinar cells. Which population serves the digestive process? ______________________________

21. ➕ Pyloric stenosis is a type of gastric outlet obstruction caused by a narrowing of the pyloric part of the stomach. It is most common in infants. Describe the clinical signs that you would expect to see with this condition. ______________________________

____

____

22. ➕ Surgical removal of the gallbladder is called a *cholecystectomy*. The presence of gallstones that block any of the ducts that carry bile is the usual reason for the surgery. Explain why the gallbladder is not an essential organ, and predict possible dietary changes that a patient might need to make post-cholecystectomy. ______________________________

____

____

EXERCISE 39 Digestive System Processes: Chemical and Physical

Learning Outcomes

- List the digestive system enzymes involved in the digestion of proteins, fats, and carbohydrates; state their site of origin; and summarize the conditions promoting their optimal functioning.
- Name the end products of protein, fat, and carbohydrate digestion.
- Define *enzyme*, *catalyst*, *control*, *substrate*, and *hydrolase*.
- Describe the different types of enzyme assays and the appropriate chemical tests to determine whether a particular foodstuff has been digested.
- Discuss the role of temperature and pH in the regulation of enzyme activity.
- State the function of bile in the digestive process.
- Explain why swallowing is both a voluntary and a reflex activity, and discuss the role of the tongue, larynx, and gastroesophageal sphincter in swallowing.
- Compare and contrast segmentation and peristalsis as mechanisms of mixing and propulsion in digestive tract organs.

Pre-Lab Quiz

Instructors may assign these and other Pre-Lab Quiz questions using Mastering A&P™

1. Circle the correct underlined term. Enzymes are catalysts / substrates that increase the rate of chemical reactions without becoming a part of the product.
2. One enzyme that you will be studying today, produced by the salivary glands and secreted into the mouth, hydrolyzes starch to maltose. It is ______________________.
3. Circle True or False. When you use iodine to test for starch, a color change to blue-black indicates a positive starch test.
4. The enzyme ______________, produced by the pancreas, is responsible for breaking down proteins.
   a. amylase **b.** kinase **c.** lipase **d.** trypsin
5. Circle True or False. Both smooth and skeletal muscles are involved in the propulsion of foodstuffs along the alimentary canal.

The food we eat must be processed so that its nutrients can reach the cells of our body. First the food is mechanically broken down into small particles, and then the particles are chemically (enzymatically) digested into the molecules that can be absorbed. Food digestion is a prerequisite to food absorption. (You have already studied mechanisms of passive and active absorption in Exercise 5 and/or PhysioEx™ Exercise 1. Before proceeding, review that material.)

Go to Mastering A&P™ > Study Area to improve your performance in A&P Lab.

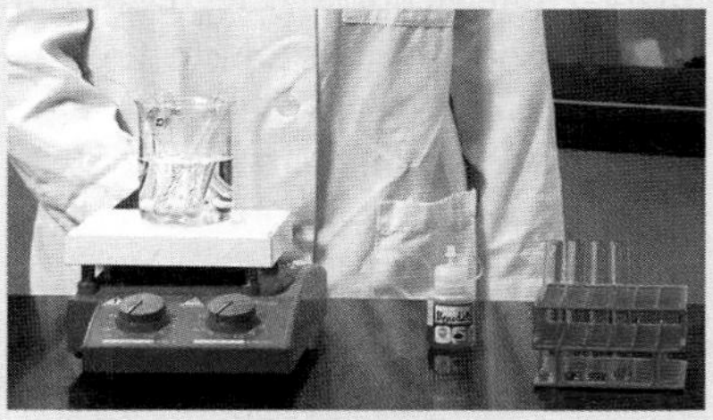

> Lab Tools > Pre-Lab Videos > Salivary Amylase

Instructors may assign new Building Vocabulary coaching activities, Pre-Lab Quiz questions, Art Labeling activities, Pre-Lab Video Coaching Activities for Salivary Amylase, Practice Anatomy Lab Practical questions (PAL), PhysioEx activities, and more using the Mastering A&P™ Item Library.

Materials

Part I: Enzyme Action

General Supply Area

- Hot plates
- 250-ml beakers
- Boiling chips
- Test tubes and test tube rack
- Wax markers
- Water bath set at 37°C (if not available, incubate at room temperature and double the time)
- Ice water bath
- Chart on board for recording class results

Activity 1: Starch Digestion

- Dropper bottle of distilled water

Text continues on next page. →

- Dropper bottles of the following:
  1% alpha-amylase solution*
  1% boiled starch solution, freshly prepared†
  1% maltose solution
  Lugol's iodine solution (IKI)
  Benedict's solution
- Spot plate

Activity 2: Protein Digestion

- Dropper bottles of 1% trypsin and 0.01% BAPNA solution

Activity 3: Bile Action and Fat Digestion

- Dropper bottles of 1% pancreatin solution, litmus cream (fresh cream to which powdered litmus is added to achieve a deep blue color), 0.1 *N* HCl, and vegetable oil
- Bile salts (sodium taurocholate)
- Parafilm® (small squares to cover the test tubes)

Part II: Physical Processes

Activity 5: Observing Digestive Movements

- Water pitcher
- Paper cups
- Stethoscope
- Alcohol swab
- Disposable autoclave bag
- Watch, clock, or timer

Activity 6: Video Viewing

- Television and VCR or DVD player for independent viewing of video by student
- *Interactive Physiology®*, Digestive System

PEx PhysioEx™ 9.1 Computer Simulation Ex. 8 on p. PEx-119

39

**The alpha-amylase must be a low-maltose preparation for good results.*

†Prepare by adding 1 g starch to 100 ml distilled water; boil and cool; add a pinch of salt (NaCl). Prepare fresh daily.

Digestion of Foodstuffs: Enzymatic Action

Enzymes are large protein molecules produced by body cells. They are biological **catalysts**, meaning that they increase the rate of a chemical reaction without themselves becoming part of the product. The digestive enzymes are hydrolytic enzymes, or **hydrolases**. Their **substrates**, or the molecules on which they act, are organic food molecules which they break down by adding water to the molecular bonds, thus cleaving the bonds between the chemical building blocks, or monomers.

Each enzyme hydrolyzes only one or a small group of substrate molecules, and specific environmental conditions are necessary for it to function optimally. Since digestive enzymes actually function outside the body cells in the digestive tract, their hydrolytic activity can also be studied in a test tube.

Figure 39.1 is a flowchart depicting the progressive digestion of carbohydrates, proteins, fats, and nucleic acids. It summarizes the specific enzymes involved, their site of formation, and their site of action. Acquaint yourself with the flowchart before beginning this experiment, and refer to it as necessary during the laboratory session.

General Instructions for Activities 1–3

Work in groups of four, with each group taking responsibility for setting up and conducting one of the following experiments. In each activity, you are directed to boil the contents of one or more test tubes. To do this, obtain a 250-ml beaker, boiling chips, and a hot plate from the general supply area. Place a few boiling chips into the beaker, add about 125 ml of water, and bring to a boil. Place the test tube for each specimen in the water for the number of minutes specified in the directions. You will also be using a 37°C bath and an ice water bath for parts of these experiments.

Upon completion of the experiments, each group should communicate its results to the rest of the class by recording them in a chart on the board. Each assay contains one or more **controls**, the specimens against which experimental samples are compared. All members of the class should observe the controls as well as the experimental results and be able to explain the tests used and the results observed and anticipated for each experiment.

Activity 1

Instructors may assign a related Pre-Lab Video Coaching Activity using Mastering **A&P™**

Assessing Starch Digestion by Salivary Amylase

1. From the general supply area, obtain a test tube rack, 10 test tubes, and a wax marking pencil. From the Activity 1 supply area, obtain a dropper bottle of distilled water and dropper bottles of maltose, amylase, and starch solutions.

2. In this experiment you will investigate the hydrolysis of starch to maltose by **salivary amylase**. You will need to be able to identify the presence of starch and maltose, the breakdown product of starch, to determine to what extent the enzymatic activity has occurred. Thus controls must be prepared to provide a known standard against which comparisons can be made. Starch decreases and sugar increases as hydrolysis occurs, according to the following formula:

$$\text{Starch} + \text{water} \xrightarrow{\text{amylase}} \text{maltose}$$

Text continues on p. 592. →

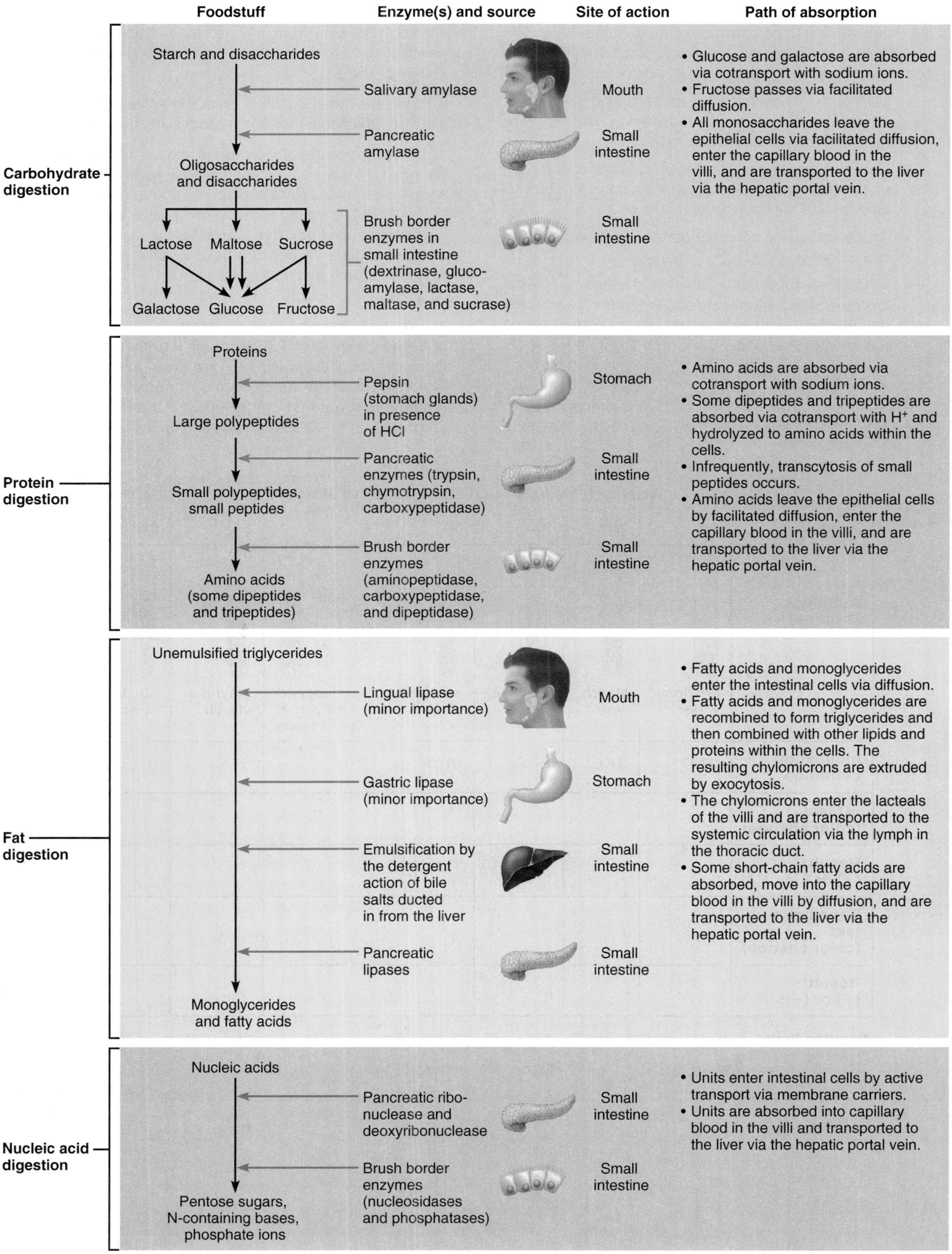

Figure 39.1 Flowchart of digestion and absorption of foodstuffs.

Two students should prepare the controls (tubes 1A to 3A) while the other two prepare the experimental samples (tubes 4A to 6A).

- Mark each tube with a wax pencil and load the tubes as indicated in the **Activity 1 chart** below, using 3 drops (gtt) of each indicated substance.
- Place all tubes in a rack in the 37°C water bath for approximately 1 hour. Shake the rack gently from time to time to keep the contents evenly mixed.
- At the end of the hour, perform the amylase assay described below.
- While these tubes are incubating, proceed to Physical Processes: Mechanisms of Food Propulsion and Mixing (p. 596). Be sure to monitor the time so as to complete this activity as needed.

Amylase Assay

1. After 1 hour, obtain a spot plate and dropper bottles of Lugol's iodine solution (for the IKI, or iodine, test) and Benedict's solution from the Activity 1 supply area. Set up your boiling water bath using a hot plate, boiling chips, and a 250-ml beaker.

2. While the water is heating, mark six depressions of the spot plate 1A–6A (*A* for amylase) for sample identification.

3. Using a pipet, transfer a drop of the sample from each of the tubes 1A–6A into the appropriately numbered spot. Into each sample drop, place a drop of Lugol's iodine (IKI) solution. A blue-black color indicates the presence of starch and is referred to as a **positive starch test**. If starch is not present, the mixture will not turn blue, which is referred to as a **negative starch test**. Record your results (+ for positive, – for negative) in the Activity 1 chart and on the board.

4. Into the remaining mixture in each tube, place 3 drops of Benedict's solution. Put each tube into the beaker of boiling water for about 5 minutes. If a green-to-orange precipitate forms, maltose is present; this is a **positive sugar test**. A **negative sugar test** is indicated by no color change. Record your results in the Activity 1 chart and on the board.

Activity 1: Salivary Amylase Digestion of Starch

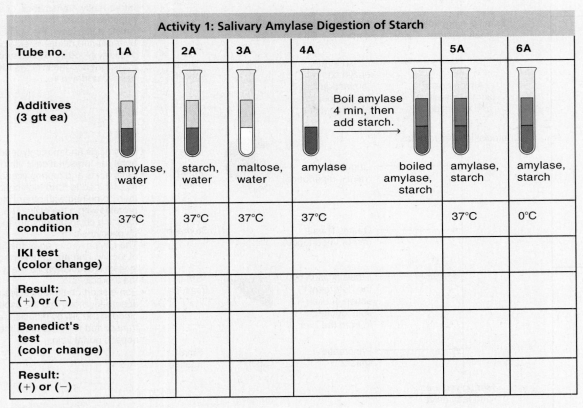

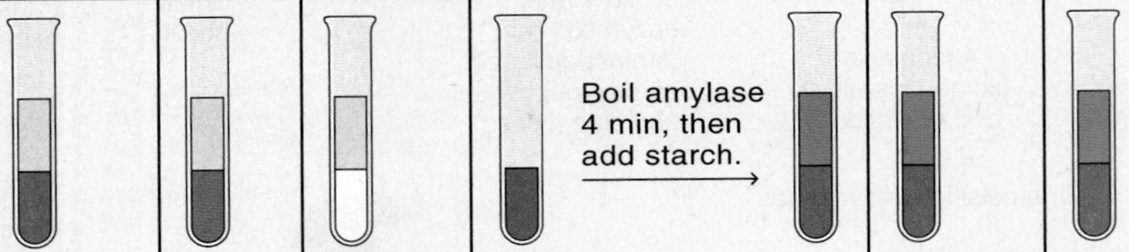

Tube no.	1A	2A	3A	4A	5A	6A
Additives (3 gtt ea)	amylase, water	starch, water	maltose, water	amylase; Boil amylase 4 min, then add starch. → boiled amylase, starch	amylase, starch	amylase, starch
Incubation condition	37°C	37°C	37°C	37°C	37°C	0°C
IKI test (color change)						
Result: (+) or (−)						
Benedict's test (color change)						
Result: (+) or (−)						

Additive key:

= Amylase = Starch = Maltose = Water

Protein Digestion by Trypsin

Trypsin, an enzyme produced by the pancreas, hydrolyzes proteins to small peptides. BAPNA (*N*-alpha-benzoyl-L-arginine-*p*-nitroanilide) is a synthetic trypsin substrate consisting of a dye covalently bound to an amino acid. Trypsin hydrolysis of BAPNA cleaves the dye molecule from the amino acid, causing the solution to change from colorless to bright yellow. The color change from clear to yellow is direct evidence of hydrolysis by trypsin.

Activity 2

Assessing Protein Digestion by Trypsin

1. From the general supply area, obtain five test tubes and a test tube rack, and from the Activity 2 supply area get a dropper bottle of trypsin and one of BAPNA. Bring these items to your bench.

2. Two students should prepare the controls (tubes 1T and 2T) while the other two prepare the experimental samples (tubes 3T to 5T).

- Mark each tube with a wax pencil, and load the tubes as indicated in the Activity 2 chart, using 3 drops (gtt) of each indicated substance.
- Place all tubes in a rack in the appropriate water bath for approximately 1 hour. Shake the rack occasionally to keep the contents well mixed.
- At the end of the hour, examine the tubes for the results of the trypsin assay (detailed below).

Trypsin Assay

Since BAPNA is a synthetic color-producing substrate, the presence of yellow color indicates a **positive hydrolysis test**; the dye molecule has been cleaved from the amino acid. If the sample mixture remains clear, a **negative hydrolysis test** has occurred.

Record the results in the Activity 2 chart and on the board.

Activity 2: Trypsin Digestion of Protein

Tube no.	1T	2T	3T	4T	5T
Additives (3 gtt ea)	trypsin, water	BAPNA, water	trypsin → Boil trypsin 4 min, then add BAPNA. → boiled trypsin, BAPNA	trypsin, BAPNA	trypsin, BAPNA
Incubation condition	37°C	37°C	37°C	37°C	0°C
Color change					
Result: (+) or (−)					

Additive key:
■ = Trypsin ■ = BAPNA □ = Water

Pancreatic Lipase Digestion of Fats and the Action of Bile

The treatment that fats and oils go through during digestion in the small intestine is a bit more complicated than that of carbohydrates or proteins—pretreatment with bile to physically emulsify the fats is required. Hence, two sets of reactions occur.

First:

$$\text{Fats/oils} \xrightarrow[\text{(emulsification)}]{\text{bile}} \text{minute fat/oil droplets}$$

Then:

$$\text{Fat/oil droplets} \xrightarrow[\text{(digestion)}]{\text{lipase}} \text{monoglycerides and fatty acids}$$

The term **pancreatin** describes the enzymatic product of the pancreas, which includes enzymes that digest proteins, carbohydrates, nucleic acids, and fats. It is used here to investigate the properties of **pancreatic lipase**, which hydrolyzes fats and oils to their component monoglycerides and two fatty acids.

Since fatty acids are organic acids, they acidify solutions, decreasing the pH. An easy way to recognize that digestion is ongoing or completed is to test pH. You will be using a pH indicator called *litmus blue* to follow these changes; it changes from blue to pink as the test tube contents become acidic.

Activity 3

Demonstrating the Emulsification Action of Bile and Assessing Fat Digestion by Lipase

1. From the general supply area, obtain nine test tubes and a test tube rack, plus one dropper bottle of each of the solutions in the Activity 3 supply area.

2. Although *bile*, a secretory product of the liver, is not an enzyme, it is important to fat digestion because of its emulsifying action. It physically breaks down large fat particles into smaller ones. Emulsified fats provide a larger surface area for enzymatic activity. To demonstrate the action of bile on fats, prepare two test tubes and mark them 1E and 2E (*E* for emulsified fats).

- To tube 1E, add 20 drops of water and 4 drops of vegetable oil.
- To tube 2E, add 20 drops of water, 4 drops of vegetable oil, and a pinch of bile salts.
- Cover each tube with a small square of Parafilm®, shake vigorously, and allow the tubes to stand at room temperature.

After 10 to 15 minutes, observe both tubes. If emulsification has not occurred, the oil will be floating on the surface of the water. If emulsification has occurred, the fat droplets will be suspended throughout the water, forming an emulsion.

In which tube has emulsification occurred? ______________

3. Two students should prepare the controls (1L and 2L, *L* for lipase) while the other two students in the group set up the experimental samples (3L to 5L, 4B, and 5B, where *B* is for bile), as illustrated in the **Activity 3 chart**.

- Mark each tube with a wax pencil and load the tubes using 5 drops (gtt) of each indicated solution.
- Place a pinch of bile salts in tubes 4B and 5B.
- Cover each tube with a small square of Parafilm®, and shake to mix the contents of the tube.
- Remove the Parafilm®, and place all tubes in a rack in the appropriate water bath for approximately 1 hour. Shake the test tube rack from time to time to keep the contents well mixed.
- At the end of the hour, perform the lipase assay below.

Lipase Assay

Fresh cream provides the fat substrate for this assay; add litmus powder to it to make litmus cream. The basis of this assay is a pH change that is detected by the litmus powder indicator. Alkaline or neutral solutions containing litmus are blue but will turn reddish in the presence of acid. If digestion occurs, the fatty acids produced will turn the litmus cream from blue to

Activity 3: Pancreatic Lipase Digestion of Fats

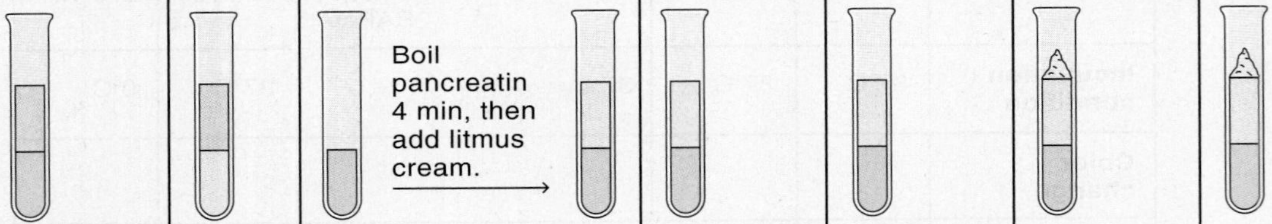

Tube no.	1L	2L	3L	4L	5L	4B	5B
Additives (5 gtt ea)	pancreatin, water	litmus cream, water	pancreatin → Boil pancreatin 4 min, then add litmus cream. → boiled pancreatin, litmus cream	pancreatin, litmus cream	pancreatin, litmus cream	pancreatin, litmus cream, bile salts	pancreatin, litmus cream, bile salts
Incubation condition	37°C	37°C	37°C	37°C	0°C	37°C	0°C
Color change							
Result: (+) or (−)							

Additive key:

= Pancreatin = Litmus cream = Water = Pinch bile salts

pink. Because the effect of hydrolysis by lipase is directly seen, additional assay reagents are not necessary.

1. To prepare a color control, add 0.1 *N* HCl drop by drop to tubes 1L and 2L (covering the tubes with a square of Parafilm® after each addition and shaking to mix) until the cream turns pink.

2. Record the color of the tubes in the Activity 3 chart and on the board.

Activity 4

Reporting Results and Conclusions

1. Share your results with the class as directed in the General Instructions (p. 590).

2. Suggest additional experiments, and carry out experiments if time permits.

3. Prepare a lab report for the experiments on digestion. (See Getting Started, on MasteringA&P.)

Group Challenge

Odd Enzyme Out

The following boxes each contain four digestive enzymes. One of the listed enzymes does not share a characteristic that the other three do. Working in groups of three, discuss the characteristics of the enzymes in each group. On a separate piece of paper, one student will record the characteristics for each enzyme for the group. Discuss the possible candidates for the "odd enzyme". Once the group has come to a consensus, circle the enzyme that doesn't belong with the others, and explain why it is singled out. Sometimes there may be multiple reasons why the enzyme doesn't belong with the others. Include as many as you can think of, but make sure it does not have the key characteristic shared by the other three.

1. Which is the "odd enzyme"?	**Why is it the odd one out?**
Trypsin Carboxypeptidase Pepsin Chymotrypsin	
2. Which is the "odd enzyme"?	**Why is it the odd one out?**
Lactase Pepsin Aminopeptidase Trypsin	
3. Which is the "odd enzyme"?	**Why is it the odd one out?**
Maltase Pancreatic lipase Nucleosidase Dipeptidase	
4. Which is the "odd enzyme"?	**Why is it the odd one out?**
Sucrase Dextrinase Glucoamylase Chymotrypsin	

Physical Processes: Mechanisms of Food Propulsion and Mixing

Although enzyme activity is a very important part of the overall digestion process, foods must also be processed physically (by chewing and churning) and moved by mechanical means along the tract if digestion and absorption are to be completed. Muscles are involved in producing the movements of foodstuffs along the gastrointestinal tract. Although we tend to think only of smooth muscles when visceral activities are involved, both skeletal and smooth muscles are involved in the physical processes. This fact is demonstrated by the simple activities that follow.

Deglutition (Swallowing)

Swallowing, or **deglutition**, is largely the result of skeletal muscle activity and occurs in two phases: *buccal* (mouth) and *pharyngeal-esophageal*. The buccal phase (**Figure 39.2** step ①) is voluntarily controlled and initiated by the tongue. Once begun, the process continues involuntarily in the pharynx and esophagus through peristalsis, resulting in the delivery of the swallowed contents to the stomach (Figure 39.2 steps ②–③).

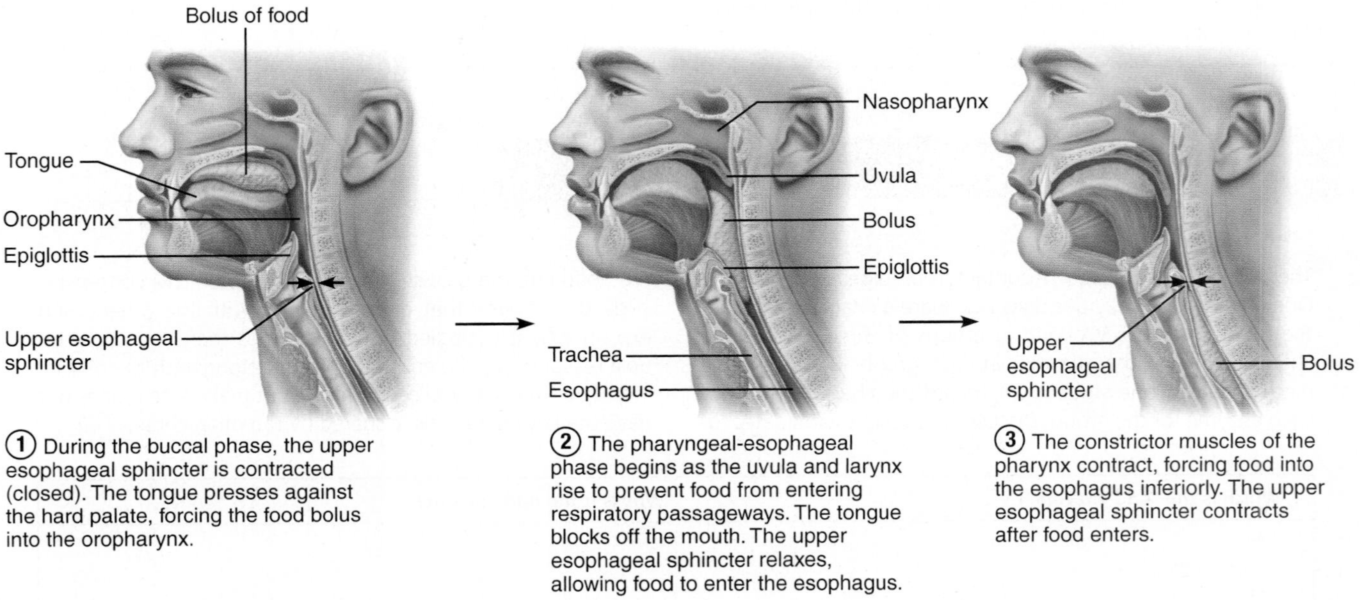
Bolus of food
Tongue
Oropharynx
Epiglottis
Upper esophageal sphincter
① During the buccal phase, the upper esophageal sphincter is contracted (closed). The tongue presses against the hard palate, forcing the food bolus into the oropharynx.
Nasopharynx
Uvula
Bolus
Epiglottis
Trachea
Esophagus
② The pharyngeal-esophageal phase begins as the uvula and larynx rise to prevent food from entering respiratory passageways. The tongue blocks off the mouth. The upper esophageal sphincter relaxes, allowing food to enter the esophagus.
Upper esophageal sphincter
Bolus
③ The constrictor muscles of the pharynx contract, forcing food into the esophagus inferiorly. The upper esophageal sphincter contracts after food enters.

Figure 39.2 Swallowing. The process of swallowing consists of voluntary (buccal) (step ①) and involuntary (pharyngeal-esophageal) phases (steps ②–③).

39

Activity 5

Observing Movements and Sounds of the Digestive System

1. Obtain a pitcher of water, a stethoscope, a paper cup, an alcohol swab, and an autoclave bag in preparation for making the following observations.

2. While swallowing a mouthful of water, consciously note the movement of your tongue during the process. Record your observations.

____

____

____

3. Repeat the swallowing process while your laboratory partner watches the externally visible movements of your larynx. (This movement is more obvious in a male, since males have a larger thyroid cartilage.) Record your observations.

____

____

What do these movements accomplish? ____________

____

____

4. Before donning the stethoscope, your lab partner should clean the earpieces with an alcohol swab. Then, he or she should place the diaphragm of the stethoscope over your abdominal wall, approximately 2.5 cm (1 inch) below the xiphoid process and slightly to the left, to listen for sounds as you again take two or three swallows of water. There should be two audible sounds—one when the water splashes against

the gastroesophageal sphincter, and the second when the peristaltic wave of the esophagus arrives at the sphincter and the sphincter opens, allowing water to gurgle into the stomach. Determine, as accurately as possible, the time interval between these two sounds, and record it below.

Interval between arrival of water at the sphincter and the

opening of the sphincter: __________ sec

This interval gives a fair indication of the time it takes for the peristaltic wave to travel down the 25 cm (10 inches) of the esophagus. (Actually the time interval is slightly less than it seems, because pressure causes the sphincter to relax before the peristaltic wave reaches it.)

! Dispose of the used paper cup in the autoclave bag.

Segmentation and Peristalsis

Although several types of movements occur in the digestive tract organs, peristalsis and segmentation are most important as mixing and propulsive mechanisms (**Figure 39.3**).

Peristaltic movements are the major means of propelling food through most of the alimentary canal. Essentially, they are waves of contraction followed by waves of relaxation that squeeze foodstuffs through the alimentary canal, and they are superimposed on segmental movements.

Segmental movements are local constrictions of the organ wall that occur rhythmically. They serve mainly to mix the foodstuffs with digestive juices and to increase the rate of absorption by continually moving different portions of the chyme over adjacent regions of the intestinal wall. However, segmentation is also an important means of food propulsion in the small intestine, and slow segmenting movements called *haustral contractions* are frequently seen in the large intestine.

Activity 6

Viewing Segmental and Peristaltic Movements

If a video showing some of the propulsive movements is available, go to a viewing station to view it before leaving the laboratory. Alternatively, use the *Interactive Physiology®* module on the Digestive System to observe gut motility.

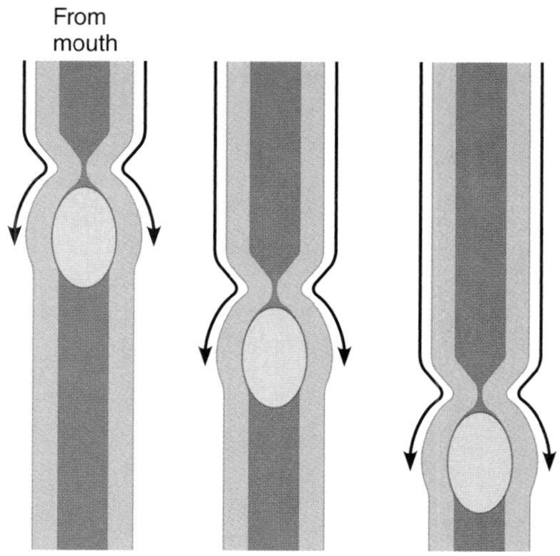

(a)

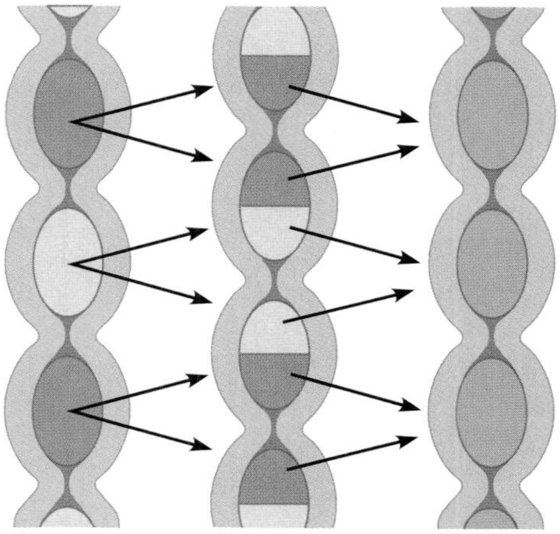

(b)

Figure 39.3 **Peristaltic and segmental movements of the digestive tract. (a)** Peristalsis: neighboring segments of the intestine alternately contract and relax, moving food along the tract. **(b)** Segmentation: single segments of intestine alternately contract and relax. Because inactive segments exist between active segments, food mixing occurs to a greater degree than food movement. Peristalsis is superimposed on segmentation movements.

39

Instructors may assign a portion of the Review Sheet questions using Mastering A&P™

REVIEW SHEET

EXERCISE 39

Digestive System Processes: Chemical and Physical

Name ______________________________ Lab Time/Date ______________________

Digestion of Foodstuffs: Enzymatic Action

1. Match the following definitions with the proper choices from the key.

Key: a. catalyst d. control c. enzyme d. substrate

__________ 1. substance on which a catalyst works

__________ 2. biologic catalyst; protein in nature

__________ 3. increases the rate of a chemical reaction without becoming part of the product

__________ 4. provides a standard of comparison for test results

2. Describe how an enzyme differs from a substrate. ______________________________

____

____

3. The enzymes of the digestive system are classified as hydrolases. What does this mean?

____

____

4. Fill in the following chart about the various digestive system enzymes encountered in this exercise.

Enzyme	Organ producing it	Site of action	Substrate(s)	Optimal pH
Salivary amylase				
Trypsin				
Lipase (pancreatic)				

5. Name the end products of digestion for the following macromolecules.

proteins: ______________________________

carbohydrates: ______________________________

fats: ______________________________

6. Explain why salivary amylase would not digest protein. ____

____

____

7. Explain why salivary amylase would be less active in the stomach than in the mouth. ______________________

____

8. In Activities 1 through 3, you used several indicators or tests in the laboratory to determine the presence or absence of certain substances. Choose the correct test or indicator from the key to correspond to the condition described below.

Key: a. Lugol's iodine (IKI) b. Benedict's solution c. litmus d. BAPNA

__________ 1. used to test for protein hydrolysis, which was indicated by a yellow color

__________ 2. used to test for the presence of starch, which was indicated by a blue-black color

__________ 3. used to test for the presence of fatty acids, which was evidenced by a color change from blue to pink

__________ 4. used to test for the presence of reducing sugars (maltose, sucrose, glucose) as indicated by a blue to green or orange color change

9. What conclusions can you draw when an experimental sample gives both a positive starch test and a positive maltose test after incubation? ____

Why was 37°C the optimal incubation temperature? ____

Why did very little, if any, starch digestion occur in test tube 4A? ____

____

When starch was incubated with amylase at 0°C, did you see any starch digestion? ______________________

Why or why not? ____

____

Assume you have said to a group of your peers that amylase is capable of starch hydrolysis to maltose. If you had not done control tube 1A, what objection to your statement could be raised? ____

____

What if you had not done tube 2A? ____

10. In the exercise concerning trypsin function, why was an enzyme assay such as Benedict's or Lugol's iodine (IKI), which test for the presence of a reaction product, not necessary? ____

Why was tube 1T necessary? ____

Why was tube 2T necessary? ____

Trypsin is a protease similar to pepsin, the protein-digesting enzyme in the stomach. Would trypsin work well in the stomach? __________ Why or why not? ____

11. In the procedure concerning pancreatic lipase digestion of fats and the action of bile salts, how did the appearance of tubes 1E and 2E differ? ______

______

Explain the reason for the difference. ______

______

Why did the litmus indicator change from blue to pink during fat hydrolysis? ______

______

Why is bile not considered an enzyme? ______

How did the tubes containing bile compare with those not containing bile? ______

What role does bile play in fat digestion? ______

______

12. The three-dimensional structure of a functional protein is altered by intense heat or nonphysiological pH even though peptide bonds may not break. Such inactivation is called denaturation, and denatured enzymes are nonfunctional. Explain why.

______

______

What specific experimental conditions resulted in denatured enzymes? ______

______

13. Pancreatic and intestinal enzymes operate optimally at a pH that is slightly alkaline, yet the chyme entering the duodenum from the stomach is very acid. How is the proper pH for the functioning of the pancreatic-intestinal enzymes ensured?

______

______

14. Assume you have been chewing a piece of bread for 5 or 6 minutes. How would you expect its taste to change during this interval? ______

Why? ______

15. Note the mechanism of absorption (passive or active transport) of the following food breakdown products, and indicate by a check mark (✓) whether the absorption would result in their movement into the blood capillaries or the lymphatic capillaries (lacteals).

Substance	**Mechanism of absorption**	**Blood**	**Lymph**
Monosaccharides			
Fatty acids and monoglycerides			
Amino acids			
Water			
Na^+, Cl^-, Ca^{2+}			

16. People on a strict diet to lose weight begin to metabolize stored fats at an accelerated rate. How does this condition affect blood pH? ______

17. Some of the digestive organs have groups of secretory cells that liberate hormones into the blood. These exert an effect on the digestive process by acting on other cells or structures and causing them to release digestive enzymes, expel bile, or increase the motility of the digestive tract. For each hormone below, note the organ producing the hormone and its effects on the digestive process. Include the target organs affected.

Hormone	Produced by	Target organ(s) and effects
Secretin		
Gastrin		
Cholecystokinin		

Physical Processes: Mechanisms of Food Propulsion and Mixing

18. Complete the following statements.

Swallowing, or _1_, occurs in two phases—the _2_ and _3_. One of these phases, the _4_ phase, is voluntary. During the voluntary phase, the _5_ is used to push the food into the back of the throat. During swallowing, the _6_ rises to ensure that its passageway is covered by the epiglottis so that the ingested substances don't enter the respiratory passageways. It is possible to swallow water while standing on your head because the water is carried along the esophagus involuntarily by the process of _7_. The pressure exerted by the foodstuffs on the _8_ sphincter causes it to open, allowing the foodstuffs to enter the stomach.

The two major types of propulsive movements that occur in the small intestine are _9_ and _10_. One of these movements, _11_, acts to continually mix the foods and to increase the absorption rate by moving different parts of the chyme mass over the intestinal mucosa, but it has less of a role in moving foods along the digestive tract.

1. ______
2. ______
3. ______
4. ______
5. ______
6. ______
7. ______
8. ______
9. ______
10. ______
11. ______

19. + Celiac disease is caused by a misdirected immune response to the protein gluten. The villi in the small intestine are damaged by the patient's own immune response. Enzyme supplements designed to digest gluten have proved to be ineffective. Hypothesize why these enzyme supplements would not be active in the stomach. ______

______

______

20. + Individuals with cystic fibrosis are plagued by increased production of mucus. This excess mucus has a variety of effects on the body, including decreased production of pancreatic enzymes and blockage of the pancreatic ducts that secrete enzymes.

Describe how impaired pancreatic enzyme secretion affects digestion. ______

______

______

EXERCISE 40

Anatomy of the Urinary System

Learning Outcomes

- List the functions of the urinary system.
- Identify, on a model or image, the urinary system organs, and state the general function of each.
- Compare the course and length of the urethra in males and females.
- Identify these regions of the dissected kidney (longitudinal section): hilum, cortex, medulla, medullary pyramids, major and minor calyces, pelvis, renal columns, and fibrous and perirenal fat capsules.
- Trace the blood supply of the kidney from the renal artery to the renal vein.
- Define *nephron*, and describe its anatomy.
- Define *glomerular filtration*, *tubular reabsorption*, and *tubular secretion*, and indicate the nephron areas involved in these processes.
- Define *micturition*, and explain the differences in the control of the internal and external urethral sphincters.
- Recognize the histologic structure of the kidney and ureter microscopically or in an image.

Go to Mastering A&P™ > Study Area to improve your performance in A&P Lab.

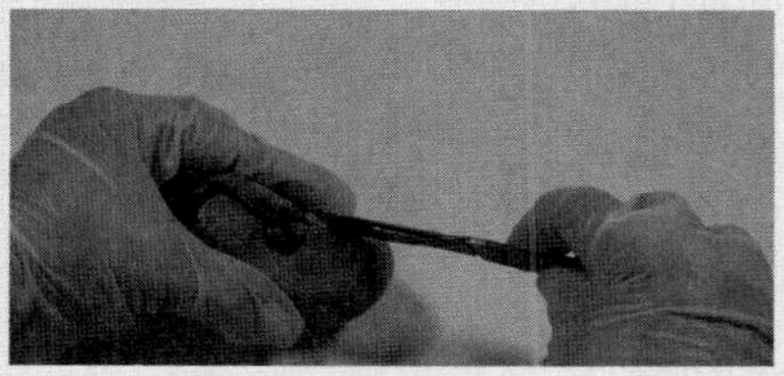

> Lab Tools > Bone and Dissection Videos > Sheep Kidney

Instructors may assign new Building Vocabulary coaching activities, Pre-Lab Quiz questions, Art Labeling activities, related videos and coaching activities for the Sheep Kidney dissection, Practice Anatomy Lab Practical questions (PAL), and more using the Mastering **A&P™** Item Library.

Pre-Lab Quiz

Instructors may assign these and other Pre-Lab Quiz questions using Mastering **A&P™**

1. Circle the correct underlined term. In its excretory role, the urinary system is primarily concerned with the removal of carbon-containing / nitrogenous wastes from the body.
2. The _________ perform(s) the excretory and homeostatic functions of the urinary system.
   a. kidneys
   b. ureters
   c. urinary bladder
   d. all of the above
3. This knot of coiled capillaries, found in the kidneys, forms the filtrate. It is the:
   a. arteriole
   b. glomerulus
   c. podocyte
   d. tubule
4. The section of the renal tubule closest to the glomerular capsule is the:
   a. collecting duct
   b. distal convoluted tubule
   c. nephron loop
   d. proximal convoluted tubule
5. Circle the correct underlined term. The afferent / efferent arteriole drains the glomerular capillary bed.

Materials

- Human dissectible torso model, three-dimensional model of the urinary system, and/or anatomical chart of the human urinary system
- Dissecting instruments and tray
- Pig or sheep kidney, doubly or triply injected
- Disposable gloves
- Three-dimensional models of the cut kidney and of a nephron (if available)
- Compound microscope
- Prepared slides of a longitudinal section of kidney and cross sections of the bladder
- "Sticky" notes

For instructions on animal dissections, see the dissection exercises (starting on p. 695) in the cat and fetal pig editions of this manual.

Metabolism of nutrients produces wastes, including carbon dioxide, nitrogenous wastes, and ammonia, that must be eliminated if the body's normal function is to continue. Excretory processes involve multiple organ systems, with the **urinary system** primarily responsible for the removal of nitrogenous wastes from the body. In addition to this excretory function, the kidney maintains the electrolyte, acid-base, and fluid balances of the blood and is thus a major homeostatic organ of the body.

To perform its functions, the kidney acts first as a blood filter and then as a filtrate processor. It allows toxins, metabolic wastes, and excess ions to leave the body in the urine, while retaining needed substances and returning them to the blood. Malfunction of the urinary system, particularly of the kidneys, leads to a failure in homeostasis which, unless corrected, is fatal.

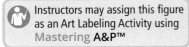

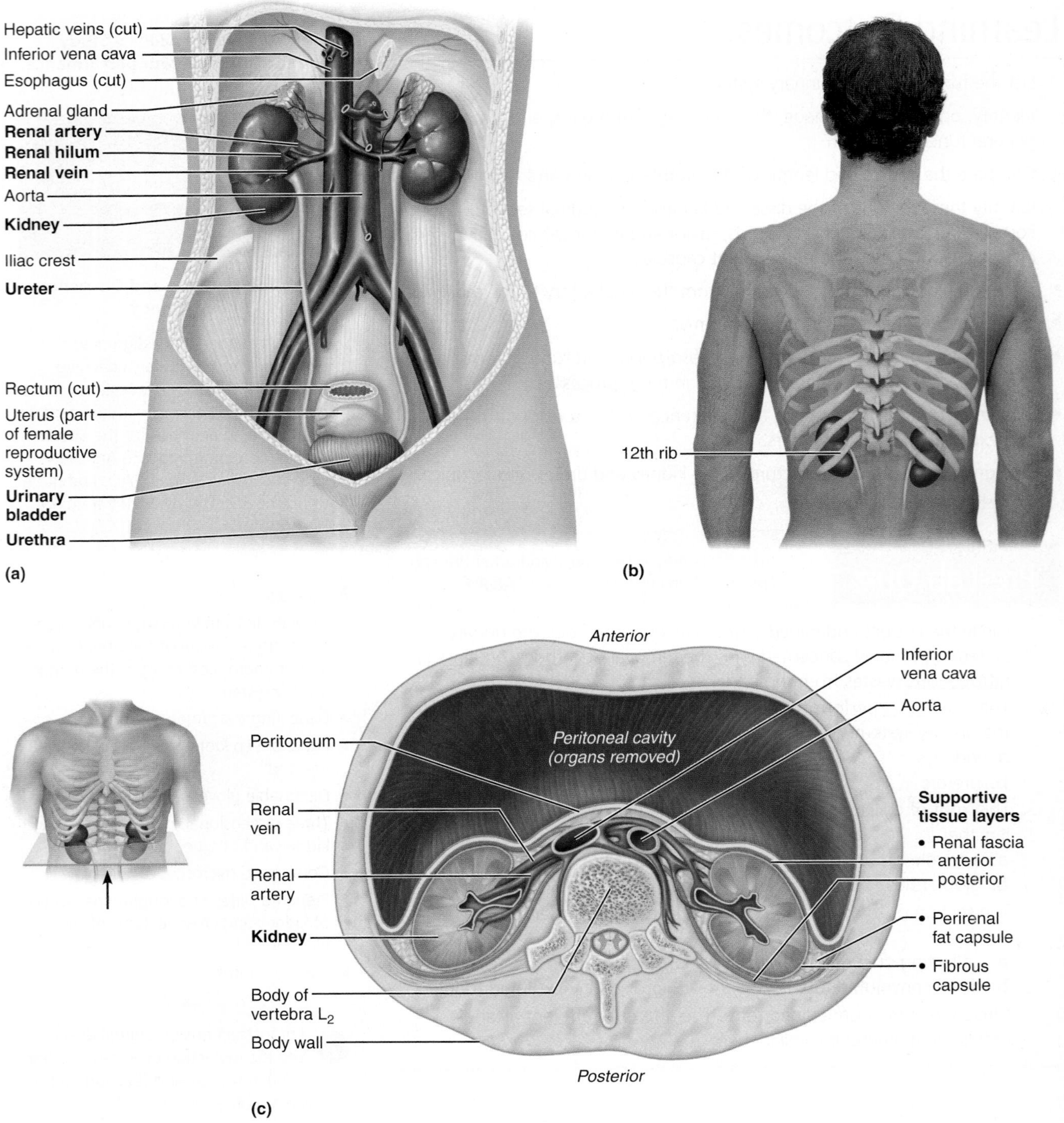

Figure 40.1 Organs of the urinary system. (a) Anterior view of the female urinary organs. Most unrelated abdominal organs have been removed. **(b)** Posterior in situ view showing the position of the kidneys relative to the twelfth ribs. **(c)** Cross section of the abdomen viewed from inferior direction. Note the retroperitoneal position and supportive tissue layers of the kidneys.

Instructors may assign this figure as an Art Labeling Activity using Mastering **A&P™**

Gross Anatomy of the Human Urinary System

The urinary system (**Figure 40.1**) consists of the paired kidneys and ureters and the single urinary bladder and urethra. The **kidneys** perform the functions described on p. 604 and manufacture urine in the process. The remaining organs of the system provide temporary storage reservoirs or transportation channels for urine.

Activity 1

Identifying Urinary System Organs

Examine the human torso model, a large anatomical chart, or a three-dimensional model of the urinary system to locate and study the anatomy and relationships of the urinary organs.

1. Locate the paired kidneys on the dorsal body wall in the superior lumbar region. Notice that they are not positioned at exactly the same level. Because it is crowded by the liver, the right kidney is slightly lower than the left kidney. Three layers of support tissue surround each kidney. Beginning with the innermost layer, they are (1) a transparent *fibrous capsule*, (2) a *perirenal fat capsule*, and (3) the fibrous *renal fascia* that holds the kidneys in place in a retroperitoneal position.

Text continues on next page →

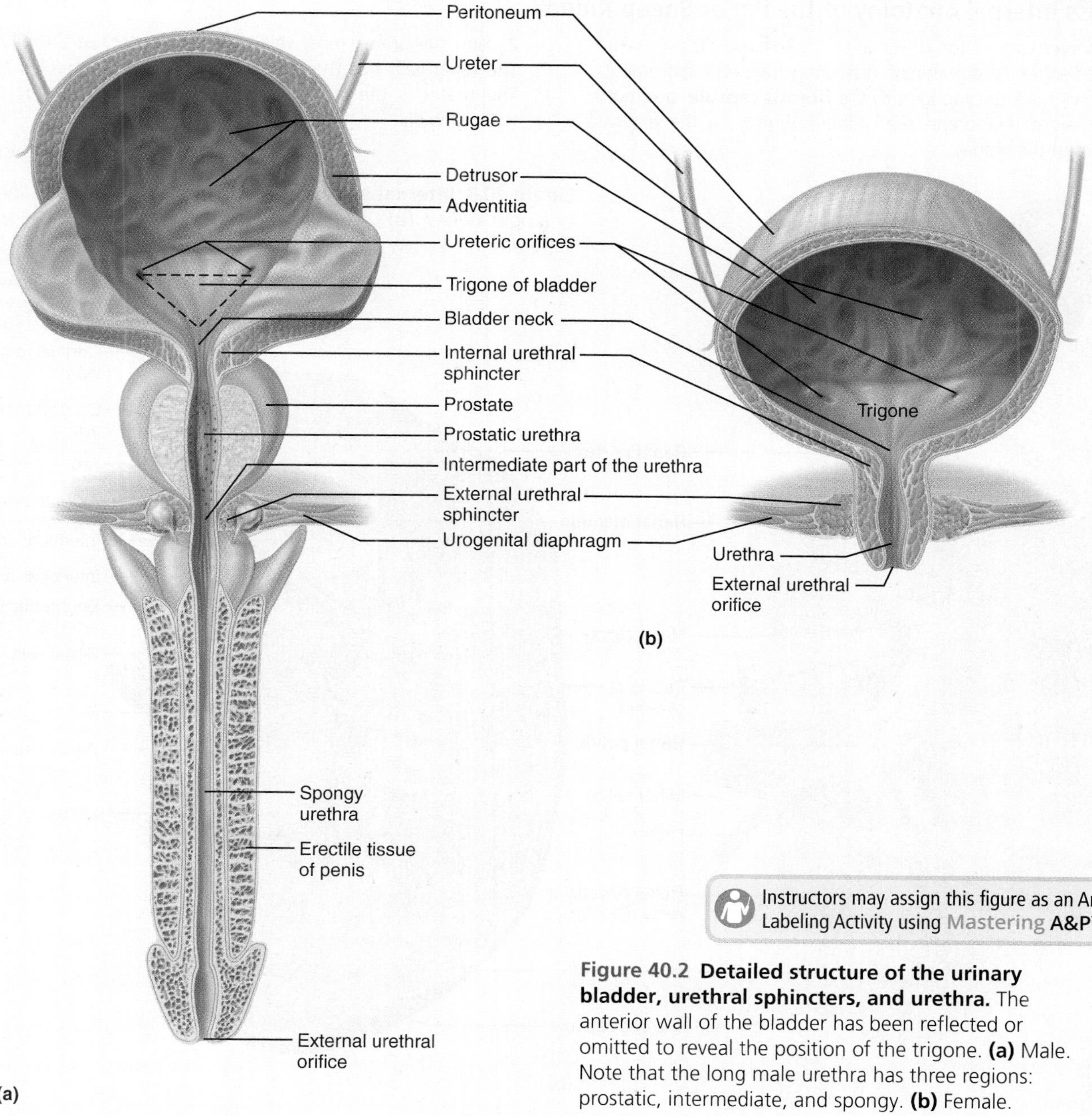

Instructors may assign this figure as an Art Labeling Activity using Mastering A&P™

Figure 40.2 Detailed structure of the urinary bladder, urethral sphincters, and urethra. The anterior wall of the bladder has been reflected or omitted to reveal the position of the trigone. **(a)** Male. Note that the long male urethra has three regions: prostatic, intermediate, and spongy. **(b)** Female.

2. Observe the **renal arteries** as they diverge from the descending aorta and plunge into the indented medial region, called the **hilum**, of each kidney. Note also the **renal veins**, which drain the kidneys, and the two **ureters**, which carry urine from the kidneys, moving it by peristalsis to the bladder for temporary storage.

3. Locate the **urinary bladder**, and observe the point of entry of the two ureters into this organ. Also locate the single **urethra**, which drains the bladder. The triangular region of the bladder delineated by the openings of the ureters and the urethra is referred to as the **trigone** (**Figure 40.2**, p. 605).

4. Follow the course of the urethra to the body exterior. In the male, it is approximately 20 cm (8 inches) long, travels the length of the **penis**, and opens at its tip. Its three named regions are the *prostatic urethra*, *intermediate part*, and *spongy urethra* (described in more detail in Exercise 42). The male urethra has a dual function: it carries urine to the body exterior, and it provides a passageway for semen ejaculation. Thus in the male, the urethra is part of both the urinary and reproductive systems. In females, the urethra is short, approximately 4 cm (1½ inches) long (see Figure 40.2). There are no common urinary-reproductive pathways in the female, and the female's urethra serves only to transport urine to the body exterior. Its external opening, the **external urethral orifice**, lies anterior to the vaginal opening.

Instructors may assign a related video and coaching activity for the Sheep Kidney Dissection

DISSECTION

Gross Internal Anatomy of the Pig or Sheep Kidney

1. In preparation for dissection, don gloves. Obtain a preserved sheep or pig kidney, dissecting tray, and instruments. Observe the kidney to identify the **fibrous capsule**, a smooth, transparent membrane that adheres tightly to the external aspect of the kidney.

2. Find the ureter, renal vein, and renal artery at the hilum. The renal vein has the thinnest wall and will be collapsed. The ureter is the largest of these structures and has the thickest wall.

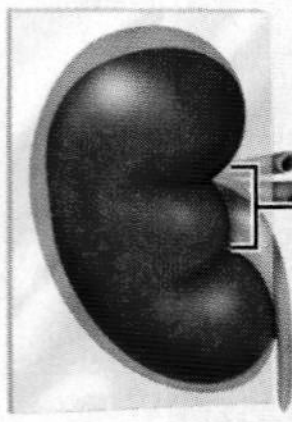

Figure 40.3 Internal anatomy of the kidney. (a) Photograph of a right kidney. **(b)** Diagram showing the larger blood vessels supplying the kidney tissue.

Instructors may assign this figure as an Art Labeling Activity using Mastering A&P™

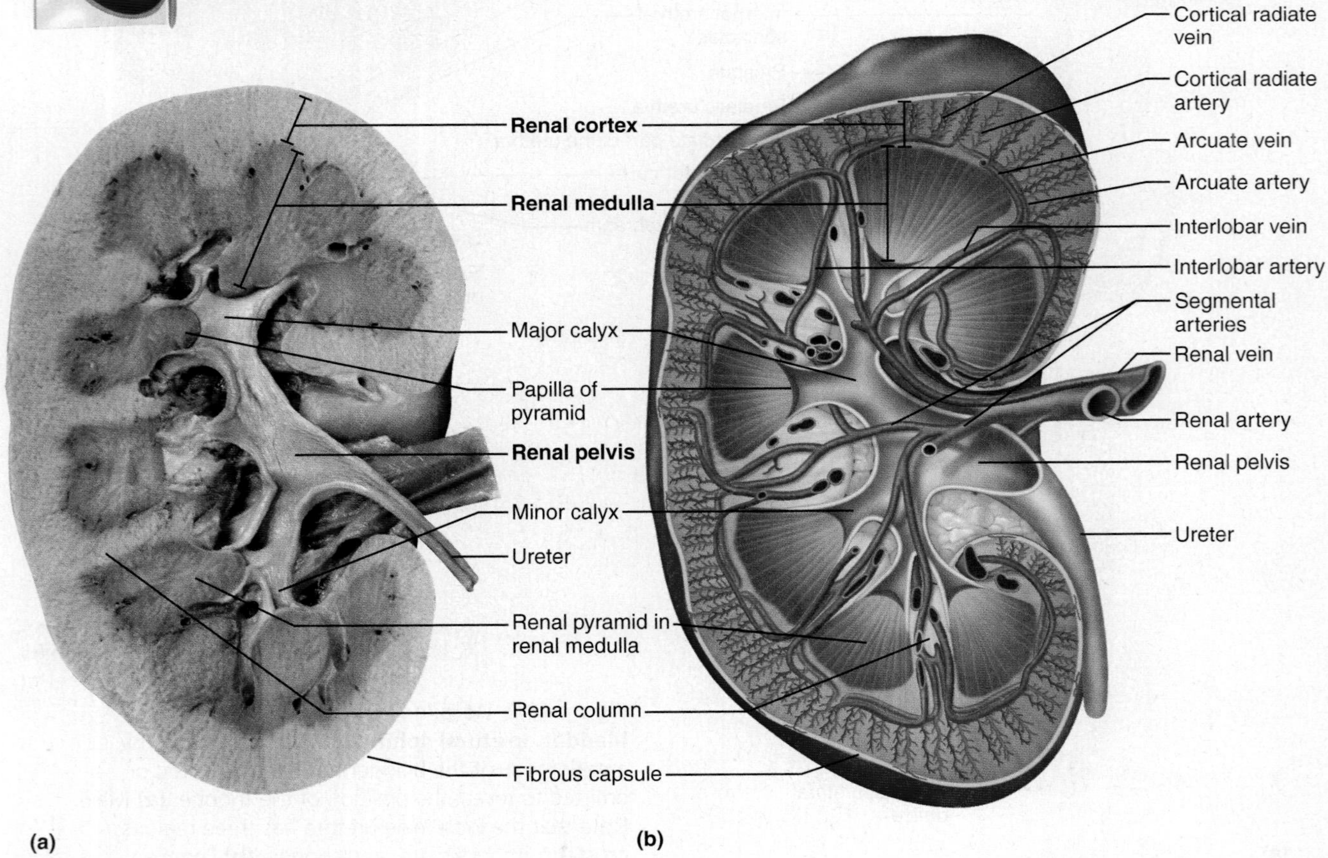

3. Section the kidney in the frontal plane through the longitudinal axis and locate the anatomical areas described and depicted in **Figure 40.3**.

Renal cortex: The superficial kidney region, which is lighter in color. If the kidney is doubly injected with latex, you will see a predominance of red and blue latex specks in this region indicating its rich blood supply.

Renal medulla: Deep to the cortex; a darker, reddish-brown color. The medulla is segregated into triangular regions that have a striped appearance—the **medullary**, or **renal pyramids**. The base of each pyramid faces toward the cortex. Its more pointed *papilla*, or *apex*, points to the innermost kidney region.

Renal columns: Areas of tissue that are more like the cortex in appearance, the columns dip inward between the pyramids, separating them.

Renal pelvis: Extending inward from the hilum; a relatively flat, basinlike cavity that is continuous with the **ureter**, which exits from the hilum region. Fingerlike extensions of the pelvis should be visible. The larger, or primary, extensions are called the **major calyces** (singular: *calyx*); subdivisions of the major calyces are the **minor calyces**. Notice that the minor calyces terminate in cuplike areas that enclose the papillae and collect urine draining from the pyramidal tips into the pelvis.

Approximately a fourth of the total blood flow of the body is delivered to the kidneys each minute by the large **renal arteries**. As a renal artery approaches the kidney, it breaks up into branches called **segmental arteries**, which enter the hilum. Each segmental artery, in turn, divides into several **interlobar arteries**, which ascend in the renal columns. At the cortex-medulla junction, the interlobar arteries branch into the **arcuate arteries**, which curve over the bases of the medullary pyramids. Small **cortical radiate arteries** branch off the arcuate arteries and ascend into the cortex, giving off the individual **afferent arterioles**, which lead to the **glomerulus**, a ball of capillaries found in the nephron. **Efferent arterioles** drain the glomerulus and feed into one of two capillary beds, either the **peritubular capillaries** or the **vasa recta**. Blood draining from the nephron capillary beds enters the **cortical radiate veins** and then drains through the **arcuate veins** and the **interlobar veins** to finally enter the **renal vein** in the pelvis region. There are no segmental veins.

Dispose of the kidney specimen as your instructor specifies.

Functional Microscopic Anatomy of the Kidney and Bladder

Kidney

Each kidney contains over a million **nephrons**, the structural and functional units responsible for filtering the blood and forming urine.

Each nephron consists of two major structures: a *renal corpuscle* and a *renal tubule*. The structures within the renal corpuscle and renal tubule are summarized individually in **Table 40.1** on p. 608 and illustrated in **Figure 40.4** on p. 609.

There are two kinds of nephrons, cortical and juxtamedullary (see Figure 40.4). **Cortical nephrons** are most numerous, making up about 85% of nephrons. They are located almost entirely within the renal cortex except for small parts of their nephron loops that dip into the renal medulla. The renal corpuscles of **juxtamedullary nephrons** are located deep in the cortex at the border with the medulla; their long nephron loops penetrate deeply into the medulla. Juxtamedullary nephrons play an important role in concentrating urine.

Urine formation is a result of three processes: *filtration*, *reabsorption*, and *secretion* (**Figure 40.5**, page 610). **Filtration**, the role of the glomerulus, is largely a passive process in which a portion of the blood passes from the glomerular capillary into the glomerular capsule. During **tubular reabsorption**, many of the filtrate components move through the tubule cells and return to the blood in the peritubular capillaries. Some of this reabsorption is passive, such as that of water, which passes by osmosis, but the reabsorption of most substances depends on active transport processes and is highly selective. Substances that are almost entirely reabsorbed from the filtrate include water, glucose, and amino acids. Various ions are selectively reabsorbed or allowed to go out in the urine according to what is required to maintain appropriate blood pH and electrolyte composition. Waste products including urea, creatinine, uric acid, and drug metabolites are reabsorbed to a much lesser degree or not at all. Most (75% to 80%) of tubular reabsorption occurs in the proximal convoluted tubule.

Tubular secretion is essentially the reverse process of tubular reabsorption. Substances such as hydrogen and potassium ions and creatinine move from the blood of the peritubular capillaries through the tubular cells into the filtrate to be disposed of in the urine.

The capillary vascular supply consists of three distinct capillary beds, the *glomerulus*, the *peritubular capillary bed*, and the *vasa recta*. Vessels leading to and from the glomerulus, the first capillary bed, are both arterioles: the **afferent arteriole** feeds the bed while the **efferent arteriole** drains it. The glomerular capillary bed is unique in the body. It is a high-pressure bed along its entire length. Its high pressure is a result of two major factors: (1) the bed is *fed and drained* by arterioles, and (2) the afferent feeder arteriole is larger in diameter than the efferent arteriole draining the bed. The high hydrostatic pressure created by these two anatomical features forces fluid and blood components smaller than proteins out of the glomerulus into the glomerular capsule. That is, it forms the filtrate that is processed by the nephron tubule.

The **peritubular capillary bed** arises from the efferent arteriole draining the glomerulus. This set of capillaries clings intimately to the renal tubule. The peritubular capillaries are *low-pressure*, porous capillaries adapted for absorption rather than filtration and readily take up the solutes and water reabsorbed from the filtrate by the tubule cells. Efferent arterioles that supply juxtaglomerular nephrons tend not to form peritubular capillaries. Instead they form long, straight, highly interconnected vessels called **vasa recta** that run parallel and close

Table 40.1 Structures of the Nephron (Figure 40.4)

Structure	Description	Epithelium	Function
Structures Within the Renal Corpuscle			
Glomerulus	A cluster of capillaries supplied by the afferent arteriole and drained by the efferent arteriole	Fenestrated endothelium (simple squamous)	Forms part of the filtration membrane
Visceral layer of the glomerular capsule	Podocytes that branch into foot processes	Simple squamous epithelium	Forms part of the filtration membrane. Spaces between the foot processes form filtration slits.
Parietal layer of the glomerular capsule	Outer impermeable wall of the glomerular capsule	Simple squamous epithelium	Forms the outside of the cuplike glomerular capsule. Plays no role in filtration.
Structures Within the Renal Tubule			
Proximal convoluted tubule (PCT)	Highly coiled first section of the renal tubule	Simple cuboidal with many microvilli and many mitochondria	Primary site of tubular reabsorption of water and solutes. Some secretion also occurs.
Descending limb of the nephron loop	First portion of the nephron loop	Simple cuboidal with some microvilli	Tubular reabsorption and secretion of water and solutes.
Descending thin limb of the nephron loop	A continuation of the descending limb	Simple squamous epithelium	Very permeable to water. Water is reabsorbed, but no solutes are reabsorbed.
Thick ascending limb of the nephron loop	In most nephrons, the ascending limb is thick	Cuboidal or low columnar, with very few aquaporins	Not permeable to water. Solutes are reabsorbed actively and passively.
Distal convoluted tubule (DCT)	Coiled distal portion of the tubule	Simple cuboidal with few microvilli but many mitochondria	Some reabsorption of water and solutes and secretion, which are regulated to meet the body's needs
Collecting duct	Receives filtrate from the DCT of multiple nephrons	Simple cuboidal epithelium with two specialized cell types: principal cells and intercalated cells	Some reabsorption and secretion to conserve body fluids, maintain blood pH, and regulate solute concentrations.

to the long nephron loops. The vasa recta is essential for the formation of concentrated urine.

Each nephron also has a **juxtaglomerular complex (JGC)** (**Figure 40.6**, p. 610) located where the most distal portion of the ascending limb of the nephron loop touches the afferent arteriole. Helping to form the JGC are (1) *granular cells* (also called *juxtaglomerular [JG] cells*) in the arteriole walls that sense blood pressure in the afferent arteriole, and (2) a group of columnar cells in the ascending limb of the nephron loop called the *macula densa* that monitors NaCl concentration in the filtrate. The role of the JGC is to regulate the rate of filtration and systemic blood pressure.

40

Activity 2

Studying Nephron Structure

1. Begin your study of nephron structure by identifying the glomerular capsule, proximal and distal convoluted tubule regions, and the nephron loop on a model of the nephron. Then, obtain a compound microscope and a prepared slide of kidney tissue to continue with the microscope study of the kidney.

2. Hold the longitudinal section of the kidney up to the light to identify cortical and medullary areas. Then secure the slide on the microscope stage, and scan the slide under low power.

3. Move the slide so that you can see the cortical area. Identify a glomerulus, which appears as a ball of tightly packed material containing many small nuclei (**Figure 40.7**, p. 611). It is usually surrounded by a vacant-appearing region corresponding to the space between the visceral and parietal layers of the glomerular capsule that surrounds it.

4. Notice that the renal tubules are cut at various angles. Try to differentiate between the fuzzy cuboidal epithelium of the proximal convoluted tubule, which has dense microvilli, and that of the distal convoluted tubule with sparse microvilli. Also identify the thin-walled nephron loop.

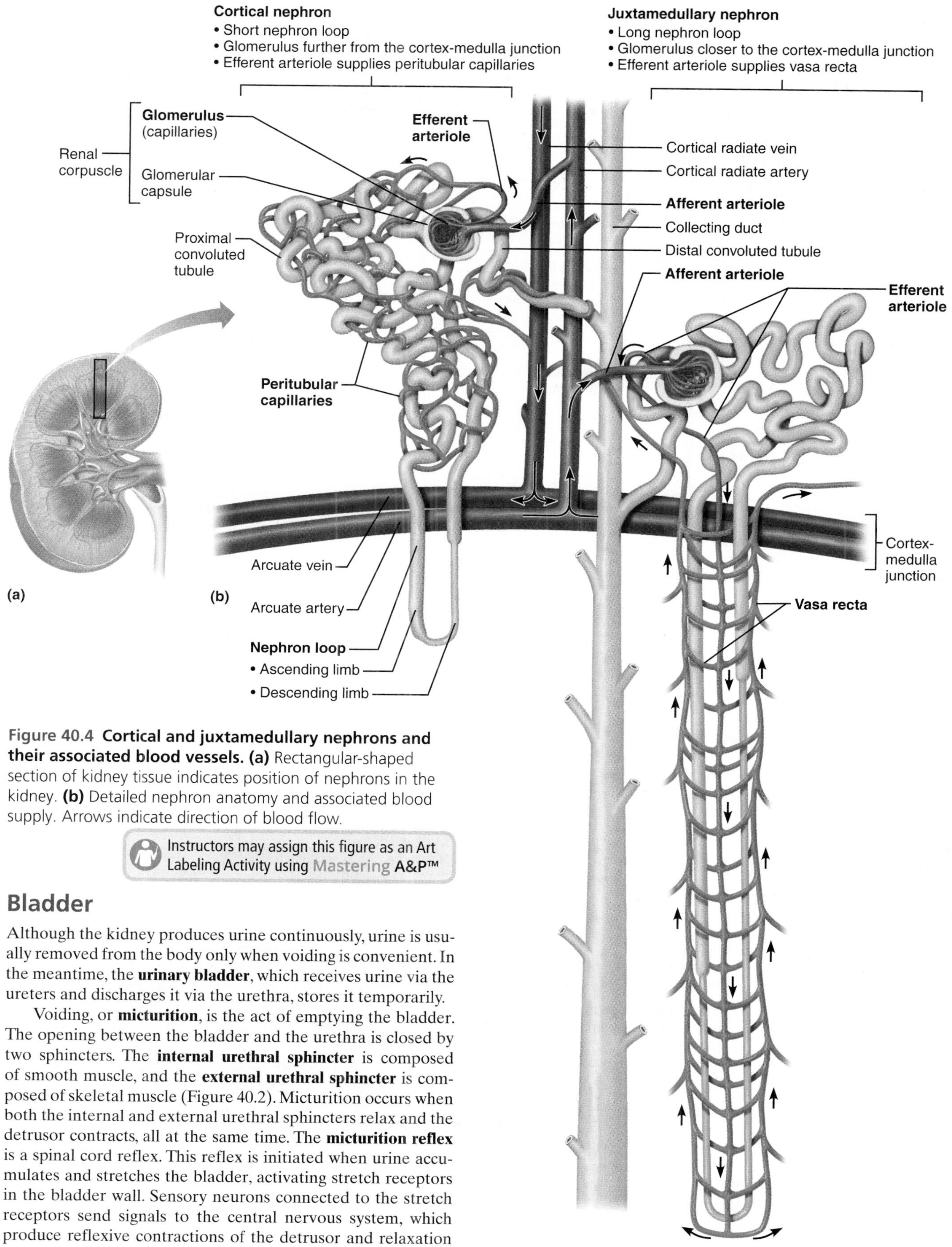

Figure 40.4 Cortical and juxtamedullary nephrons and their associated blood vessels. (a) Rectangular-shaped section of kidney tissue indicates position of nephrons in the kidney. **(b)** Detailed nephron anatomy and associated blood supply. Arrows indicate direction of blood flow.

Instructors may assign this figure as an Art Labeling Activity using Mastering A&P™

Bladder

Although the kidney produces urine continuously, urine is usually removed from the body only when voiding is convenient. In the meantime, the **urinary bladder**, which receives urine via the ureters and discharges it via the urethra, stores it temporarily.

Voiding, or **micturition**, is the act of emptying the bladder. The opening between the bladder and the urethra is closed by two sphincters. The **internal urethral sphincter** is composed of smooth muscle, and the **external urethral sphincter** is composed of skeletal muscle (Figure 40.2). Micturition occurs when both the internal and external urethral sphincters relax and the detrusor contracts, all at the same time. The **micturition reflex** is a spinal cord reflex. This reflex is initiated when urine accumulates and stretches the bladder, activating stretch receptors in the bladder wall. Sensory neurons connected to the stretch receptors send signals to the central nervous system, which produce reflexive contractions of the detrusor and relaxation

40

of the internal urethral sphincter through parasympathetic nervous system pathways. Somatic motor neurons leading to the external urethral sphincter are inhibited, causing the skeletal muscle to relax. With both sphincters open and the bladder contracting, urine is voided. Higher brain centers allow or inhibit the micturition reflex, depending on the convenience and desire to urinate. Inhibition of the micturition reflex relies in part on control of the external urethral sphincter.

Lack of voluntary control over the external urethral sphincter is referred to as **incontinence**. Incontinence is normal in children under 2 years old; in older children and adults, it can result from spinal cord injuries or urinary tract pathology.

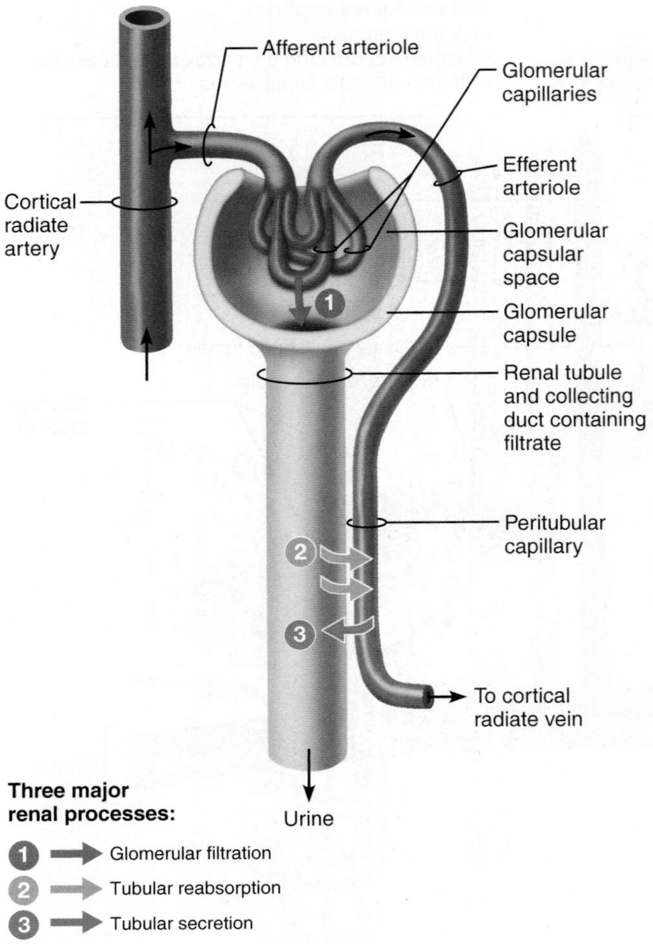

Instructors may assign this figure as an Art Labeling Activity using Mastering A&P™

Figure 40.5 A schematic, uncoiled nephron. A kidney actually has millions of nephrons acting in parallel. The three major renal processes by which the kidneys adjust the composition of plasma are depicted. Black arrows show the path of blood flow through the renal microcirculation.

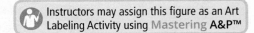

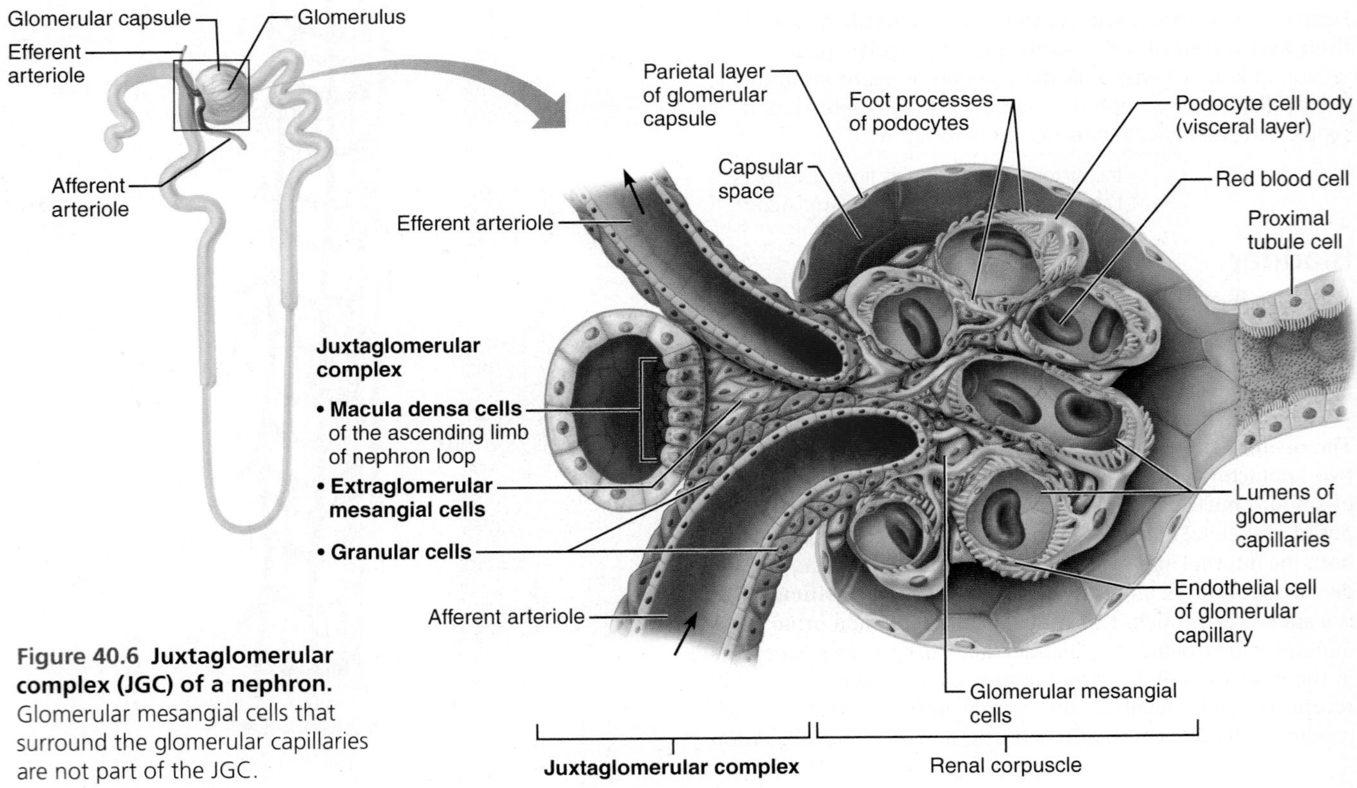

Figure 40.6 Juxtaglomerular complex (JGC) of a nephron. Glomerular mesangial cells that surround the glomerular capillaries are not part of the JGC.

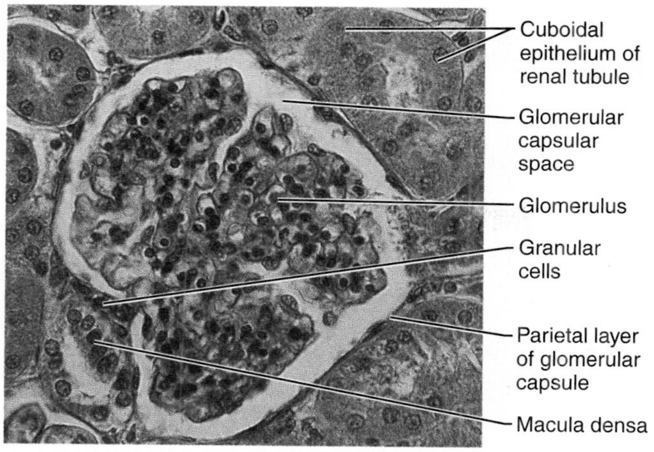

(a)

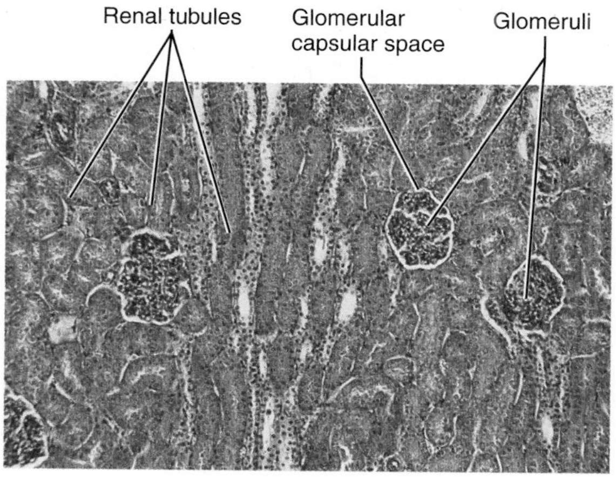

(b)

Figure 40.7 Microscopic structure of kidney tissue. **(a)** Detailed structure of the nephron (225×). **(b)** Low-power view of the renal cortex (67×).

Activity 3

Studying Bladder Structure

1. Return the kidney slide to the supply area, and obtain a slide of bladder tissue. Scan the bladder tissue. Identify its three layers: mucosa, muscular layer, and fibrous adventitia.

2. Study the highly specialized transitional epithelium of the mucosa. The plump, transitional epithelial cells have the ability to slide over one another, thus decreasing the thickness of the mucosa layer as the bladder fills and stretches to accommodate the increased urine volume. Depending on the degree of stretching of the bladder, the mucosa may be three to eight cell layers thick. (Compare the transitional epithelium of the mucosa to that shown in Figure 6.3h on p. 71).

3. Examine the heavy muscular wall (detrusor), which consists of three irregularly arranged muscular layers. The innermost and outermost muscle layers are arranged longitudinally; the middle layer is arranged circularly. Attempt to differentiate the three muscle layers.

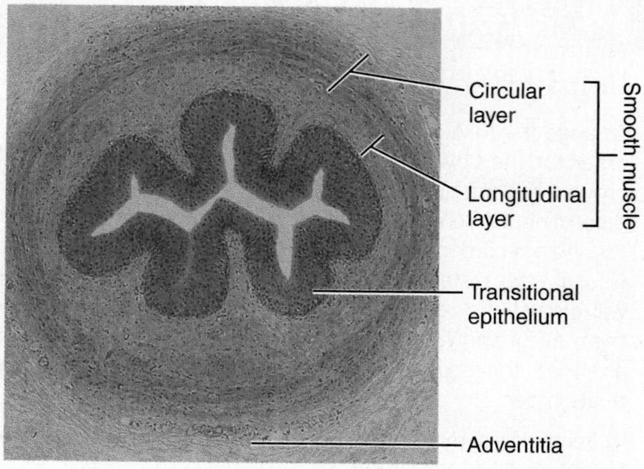

Figure 40.8 Structure of the ureter wall. Cross section of ureter (35×).

4. Draw a small section of the bladder wall, and label all regions or tissue areas.

5. Compare your sketch of the bladder wall to the structure of the ureter wall in **Figure 40.8**. How are the two organs similar histologically?

____

____

____

What is/are the most obvious differences?

____

____

For instructions on animal dissections, see the dissection exercises (starting on p. 695) in the cat and fetal pig editions of this manual.

Group Challenge

Urinary System Sequencing

Arrange the following sets of urinary structures in the correct order for the flow of urine, filtrate, or blood. Work in small groups, but refrain from using a figure or other reference to determine the sequence. Within your group, assign a facilitator and a recorder. The facilitator will list each term in a given set on a separate "sticky" note. All members of the group will discuss the correct order for the structures and arrange them accordingly. After all of the "sticky" notes have been arranged, the recorder will write down the terms in the appropriate order.

1. renal pelvis, minor calyx, renal papilla, urinary bladder, ureter, major calyx, and urethra

____

____

2. distal convoluted tubule, ascending limb of the nephron loop, glomerulus, collecting duct, descending limb of the nephron loop, proximal convoluted tubule, and glomerular capsule

____

____

3. segmental artery, afferent arteriole, cortical radiate artery, glomerulus, renal artery, interlobar artery, and arcuate artery ______________________________

____

4. arcuate vein, inferior vena cava, peritubular capillaries, renal vein, interlobar vein, cortical radiate vein, and efferent arteriole ______________________________

____

Instructors may assign a portion of the Review Sheet questions using Mastering A&P™

40 REVIEW SHEET

EXERCISE

Anatomy of the Urinary System

Name ______________________ Lab Time/Date ______________________

Gross Anatomy of the Human Urinary System

1. Complete the following statements.

The kidney is referred to as an excretory organ because it excretes __1__ wastes. It is also a major homeostatic organ because it maintains the electrolyte, __2__, and __3__ balance of the blood.

Urine is continuously formed by the structural and functional units of the kidneys, the __4__, and is routed down the __5__ by the mechanism of __6__ to a storage organ called the __7__. Eventually, the urine is conducted to the body __8__ by the urethra. In the male, the urethra is __9__ centimeters long and transports both urine and __10__. The female urethra is __11__ centimeters long and transports only urine.

Voiding or emptying the bladder is called __12__. Voiding has both voluntary and involuntary components. The voluntary sphincter is the __13__ sphincter, composed of skeletal muscle. An inability to control this sphincter is referred to as __14__.

1. ______________________
2. ______________________
3. ______________________
4. ______________________
5. ______________________
6. ______________________
7. ______________________
8. ______________________
9. ______________________
10. ______________________
11. ______________________
12. ______________________
13. ______________________
14. ______________________

2. Which layer of support tissue holds the kidneys in the retroperitoneal position?

____

3. Name the three structures that outline the triangular region of the bladder known as the trigone. ______________________

____

4. Label the photograph of the kidney model by selecting the letter for the correct structure from the key below.

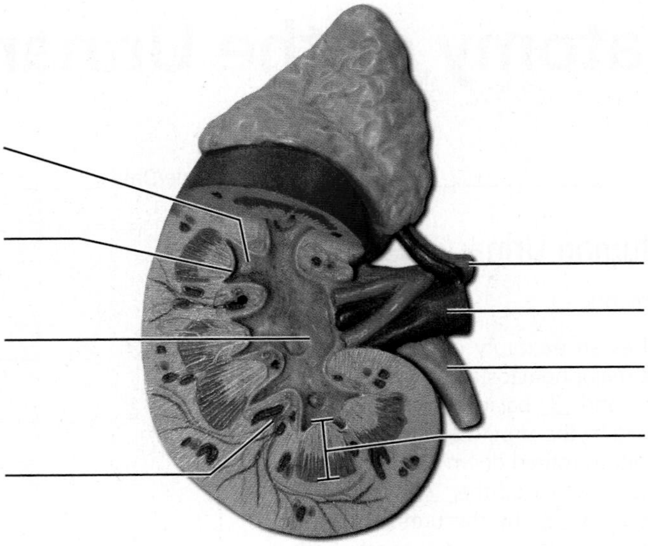

Key:

a.	minor calyx	d.	renal papilla	g.	renal vein
b.	renal artery	e.	renal pelvis	h.	ureter
c.	renal column	f.	renal pyramid		

Gross Internal Anatomy of the Pig or Sheep Kidney

5. Match the appropriate structure in column B to its description in column A.

Column A		Column B
____________	1. smooth membrane, tightly adherent to the kidney surface	a. cortex
____________	2. portion of the kidney containing mostly collecting ducts	b. fibrous capsule
____________	3. superficial region of kidney tissue	c. medulla
____________	4. basinlike area of the kidney, continuous with the ureter	d. minor calyx
____________	5. a cup-shaped extension of the pelvis that encircles the apex of a pyramid	a. renal column
____________	6. tissue running between the renal pyramids	f. renal pelvis

Functional Microscopic Anatomy of the Kidney and Bladder

6. Label the blood vessels and parts of the nephron by selecting the letter for the correct structure from the key below.

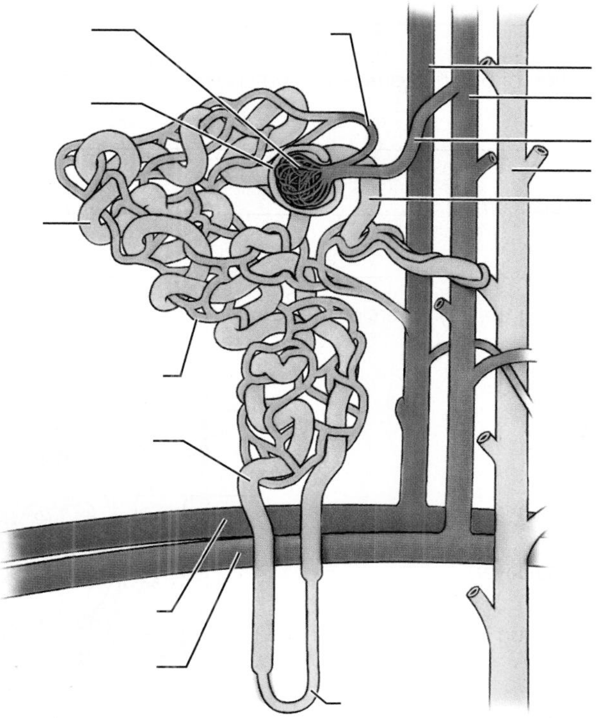

Key:

a. afferent arteriole
b. arcuate artery
c. arcuate vein
d. collecting duct
e. cortical radiate artery
f. cortical radiate vein
g. distal convoluted tubule
h. efferent arteriole
i. glomerular capsule
j. glomerulus
k. nephron loop—ascending limb
l. nephron loop—descending limb
m. peritubular capillary
n. proximal convoluted tubule

7. For each of the following descriptions of a structure, find the matching name in the question 6 key.

__________ 1. capillary specialized for filtration

__________ 2. capillary specialized for reabsorption

__________ 3. cuplike part of the renal corpuscle

__________ 4. location of macula densa

__________ 5. primary site of tubular reabsorption

__________ 6. receives urine from many nephrons

8. Explain *why* the glomerulus is such a high-pressure capillary bed. ______________________________

____

How does its high-pressure condition aid its function of filtrate formation? ______________________________

____

9. What structural modification of certain tubule cells enhances their ability to reabsorb substances from the filtrate?

____

10. Explain the mechanism of tubular secretion, and explain its importance in the urine formation process. ______

______

______

11. Compare and contrast the composition of blood plasma and glomerular filtrate. ______

______

12. Describe the role of the juxtaglomerular complex. ______

______

13. Label the drawing of the nephron using the key letters of the correct terms.

Key: a. granular cells

b. cuboidal epithelium

c. macula densa

d. glomerular capsule (parietal layer)

e. ascending limb of the nephron loop

f. glomerulus

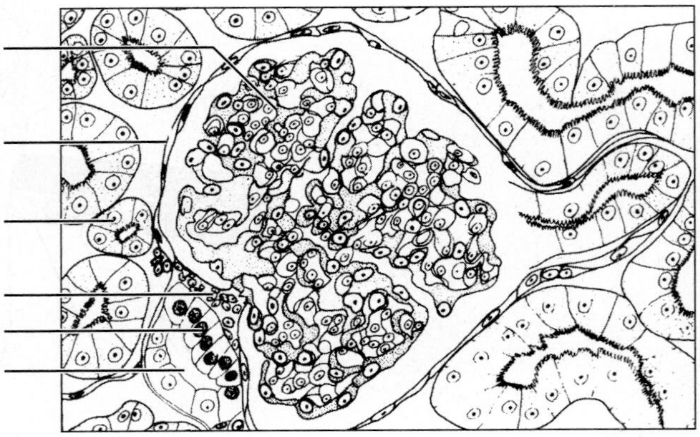

14. What is important functionally about the specialized epithelium (transitional epithelium) in the bladder?

______

______

15. ✚ A urinary tract infection (UTI) is an infection of any of the urinary tract structures: kidneys, ureter, bladder, or urethra. Considering the differences in the male and female anatomy and that the bacteria that cause UTIs are often found in the feces, explain why females are more likely to contract a UTI. ______

______

______

16. ✚ Acute glomerulonephritis (GN) is sudden inflammation of the glomeruli, most commonly a result of a streptococcal infection that causes the body to attack its own tissue. The damage to the filtration membrane increases permeability of the membrane for proteins and larger components. Describe the abnormal components of the urine for patients with GN. ______

______

EXERCISE 41
Urinalysis

Learning Outcomes

▶ List the physical characteristics of urine, and indicate the normal pH and specific gravity ranges.

▶ List substances that are normal urinary constituents.

▶ Conduct various urinalysis tests and procedures, and use them to determine the substances present in a urine specimen.

▶ Define the following urinary conditions:

calculi   albuminuria   hemoglobinuria
casts   glycosuria   ketonuria
           hematuria   pyuria

▶ Discuss the possible causes and implications of the conditions listed above.

Pre-Lab Quiz

 Instructors may assign these and other Pre-Lab Quiz questions using Mastering A&P™

1. Circle True or False. Glucose can usually be found in all normal urine.
2. _____, like other blood proteins, is/are too large to pass through the glomerular filtration membrane and is/are normally not found in urine.
   a. Albumin   c. Nitrates
   b. Chloride   d. Sulfate
3. Circle the correct underlined term. <u>Hematuria</u> / <u>Ketonuria</u>, the appearance of red blood cells in the urine, almost always indicates pathology of the urinary system.
4. Circle the correct underlined term. <u>Proteinuria</u> / <u>Pyuria</u>, the presence of white blood cells or pus in the urine, is consistent with inflammation of the urinary tract.
5. Circle the correct underlined term. <u>Casts</u> / <u>Calculi</u> are hardened cell fragments formed in the distal convoluted tubules and collecting ducts and flushed out of the urinary tract.

Blood composition depends on three major factors: diet, cellular metabolism, and urinary output. In 24 hours, the kidneys' 2 million nephrons filter 150 to 180 liters of blood plasma through their glomeruli into the tubules, where it is selectively processed by tubular reabsorption and secretion. In the same period, urinary output, which contains by-products of metabolism and excess ions, is 0.8 to 1.8 liters. In healthy people, the kidneys can maintain blood constancy despite wide variations in diet and metabolic activity.

Go to Mastering A&P™ > Study Area to improve your performance in A&P Lab.

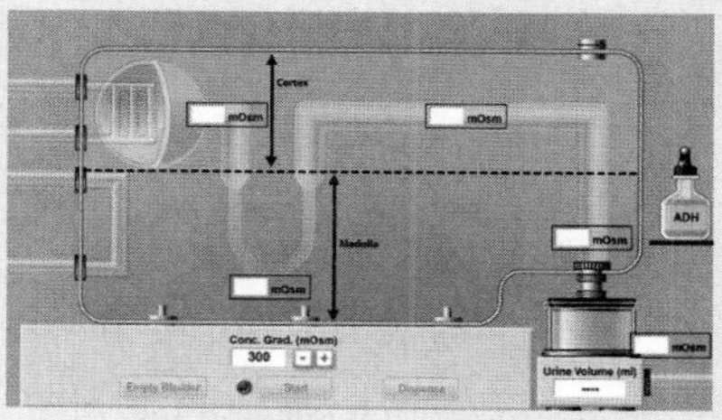

> Lab Tools > PhysioEx

Instructors may assign new Building Vocabulary coaching activities, Pre-Lab Quiz questions, Art Labeling activities, Pre-Lab Video Coaching Activities for Urinalysis, Practice Anatomy Lab Practical questions (PAL), PhysioEx activities, and more using the Mastering A&P™ Item Library.

Materials

▶ Disposable gloves
▶ Student urine samples collected at the beginning of the laboratory or "normal" artificial urine provided by the instructor*
▶ Numbered "pathological" urine specimens provided by the instructor*
▶ Wide-range pH paper
▶ Dipsticks: individual (Clinistix®, Ketostix®, Albustix®, Hemastix®) or combination (Chemstrip® or Multistix®)
▶ Urinometer
▶ Test tubes, test tube rack, and test tube holders
▶ 10-ml graduated cylinders

*Directions for making artificial urine are provided in the Instructor's Guide for this manual.

Text continues on next page. ➔

- Test reagents for sulfates: 10% barium chloride solution, dilute hydrochloric acid (HCl)
- Hot plate
- 500-ml beaker
- Test reagent for phosphates: dilute nitric acid (HNO_3), dilute ammonium molybdate
- Glass stirring rod
- Test reagent for chloride: 3.0% silver nitrate solution ($AgNO_3$), freshly prepared
- Clean microscope slide and coverslip
- Compound microscope
- Test reagent for urea: concentrated nitric acid in dropper bottles
- Test reagent for glucose: Clinitest® tablets; Clinitest® color chart
- Medicine droppers
- Timer (watch or clock with a second hand)
- Ictotest® reagent tablets and test mat
- Flasks and laboratory buckets containing 10% bleach solution
- Disposable autoclave bags
- *Demonstration*: Instructor-prepared specimen of urine sediment set up for microscopic analysis

PEx PhysioEx™ 9.1 Computer Simulation Ex. 9 on p. PEx-131.

41

Characteristics of Urine

Color and Transparency

Freshly voided urine is generally clear and pale yellow to amber in color. This normal yellow color is due to *urochrome*, a pigment metabolite that arises from the body's breakdown of hemoglobin and travels to the kidney as bilirubin or bile pigments. As a rule, color variations from pale yellow to deeper amber indicate the relative concentration of solutes to water in the urine. The greater the solute concentration, the deeper the color. Abnormal urine color may be due to certain foods, such as beets, various drugs, bile, or blood. Cloudy urine may indicate a urinary tract infection.

Odor

The odor of freshly voided urine is slightly aromatic, but bacterial action gives it an ammonia-like odor when left standing. Some drugs, vegetables (such as asparagus), and various disease processes (such as diabetes mellitus) alter the characteristic odor of urine. For example, the urine of a person with uncontrolled diabetes mellitus (and elevated levels of ketones) smells fruity or acetone-like.

pH

The pH of urine ranges from 4.5 to 8.0, but its average value is slightly acidic (usually around 6). Diet may markedly influence the pH of the urine. For example, a diet high in protein (meat, eggs, cheese) and whole wheat products increases the acidity of urine. Conversely, a vegetarian diet usually increases the alkalinity of the urine. A bacterial infection of the urinary tract may also cause the urine to become more alkaline.

Specific Gravity

Specific gravity is the relative weight of a specific volume of liquid compared with an equal volume of distilled water. The specific gravity of distilled water is 1.000, because 1 ml weighs 1 g. Since urine contains dissolved solutes, a given volume of urine weighs more than the same volume of water, and its customary specific gravity ranges from 1.001 to 1.030. Urine with a specific gravity of 1.001 contains few solutes and is considered very dilute. Dilute urine commonly results when a person drinks excessive amounts of water, uses diuretics, or suffers from diabetes insipidus or chronic renal failure. Conditions that produce urine with a high specific gravity include limited fluid intake, fever, diabetes mellitus, gonorrhea, and kidney inflammation, called *pyelonephritis*. If urine becomes excessively concentrated, some of the substances normally held in solution begin to precipitate or crystallize, forming **kidney stones**, or **renal calculi**.

Water is the largest component of urine, accounting for 95% of its volume. The second largest component of urine is urea. Nitrogenous wastes in the urine include urea, uric acid, and creatinine. *Urea* comes from the breakdown of proteins. *Uric acid* is a breakdown product from nucleic acids. *Creatinine* is a metabolite produced from the metabolism of creatine phosphate in muscle tissue.

Normal solute constituents of urine, in order of decreasing concentration, include urea; sodium, potassium, phosphate, and sulfate ions; creatinine; and uric acid. Much smaller but highly variable amounts of calcium, magnesium, and bicarbonate ions are also found in the urine. Abnormally high concentrations of any of these urinary constituents may indicate a pathological condition.

Abnormal Urinary Constituents

Abnormal urinary constituents are substances not normally present in the urine when the body is operating properly. +

When certain pathological conditions are present, urine composition often changes dramatically. **Table 41.1** identifies substances that are not normally found in the urine and describes their characteristics.

Casts

Any complete discussion of the varieties and implications of casts is beyond the scope of this exercise. However, because they always represent a pathological condition of the kidney or urinary tract, they should at least be mentioned. **Casts** are hardened cell fragments, usually cylindrical, which are formed in the distal convoluted tubules and collecting ducts and then flushed out of the urinary tract. Hyaline casts are formed from a mucoprotein secreted by tubule cells (Figure 41.1b, p. 622). These casts form when the filtrate flow rate is slow, the pH is low, or the salt concentration is high, all conditions that cause protein to denature. Red blood cell casts are typical in glomerulonephritis, as red blood cells leak through the filtration membrane and stick together in the tubules. White blood cell casts form when the kidney is inflamed, which is typically a result of pyelonephritis (a type of urinary tract infection) but sometimes occurs with glomerulonephritis. Degenerated renal tubule cells form granular casts (Figure 41.1b).

Activity 1

Instructors may assign a related Pre-Lab Video Coaching Activity using Mastering A&P

Analyzing Urine Samples

In this part of the exercise, you will use prepared dipsticks and perform chemical tests to determine the characteristics of normal urine as well as to identify abnormal urinary components. You will investigate two or more urine samples. The first, designated as the *standard urine specimen* in the **Activity 1 chart** (p. 621), will be either yours or a "standard" sample provided by your instructor. The second will be an unknown urine specimen provided by your instructor. Make the following determinations on both samples, and record your results by circling the appropriate item or description or by adding data to complete the chart. If you have more than one unknown sample, accurately identify each sample by number.

Text continues on next page. →

Table 41.1 Abnormal Urinary Constituents

Abnormal urinary constituent	Clinical term	Description	Possible conditions
Glucose	Glycosuria (glucosuria)	High blood sugar levels due to inadequate insulin levels; or can result when active transport mechanisms for glucose are exceeded temporarily	Pathological: uncontrolled diabetes mellitus Nonpathological: Excessive carbohydrate intake
Protein	Proteinuria (albuminuria)	Increased permeability of the glomerular filtration membrane (proteins are usually too large to pass through); albumin is the most abundant blood protein	Pathological: severe hypertension, glomerulonephritis, ingestion of poisons, bacterial toxins, kidney trauma Nonpathological: excessive physical exertion, pregnancy
Ketone bodies	Ketonuria	Excessive production of and accumulation intermediates of fat metabolism, which may result in acidosis	Uncontrolled diabetes mellitus, starvation, low-carbohydrate diets
Erythrocytes (RBCs)	Hematuria	Irritation of the urinary tract organs that results in bleeding; or a result of leakage of RBCs through a damaged filtration membrane	Bleeding in the tract: kidney stones, urinary tract tumors, trauma to urinary tract organs Damaged filtration membrane: glomerulonephritis
Hemoglobin	Hemoglobinuria	Fragmentation of erythrocytes, resulting in the release of hemoglobin into the plasma and subsequently into the filtrate	Hemolytic anemia, transfusion reactions, severe burns, poisonous snake bites, renal disease
Nitrites	Nitrituria	Results when gram-negative bacteria such as *E. coli* reduce nitrates to form nitrites	Urinary tract infections (UTIs)
Bile pigments	Bilirubinuria	Increased levels of bilirubin in the urine as a result of liver damage or blockage of the bile duct	Hepatitis, cirrhosis of the liver, gallstones
Leukocytes (WBCs)	Pyuria	Presence of WBCs or pus in the urine caused by inflammation of the urinary tract	Urinary tract infections (including pyelonephritis), gonorrhea

! *Obtain and wear disposable gloves throughout this laboratory session.* Although the instructor-provided urine samples are actually artificial urine (concocted in the laboratory to resemble real urine), you should still observe the techniques of safe handling of body fluids as part of your learning process. When you have completed the laboratory procedures: (1) dispose of the gloves, used pH paper strips, and dipsticks in the autoclave bag; (2) put used glassware in the bleach-containing laboratory bucket; (3) wash the lab bench down with 10% bleach solution.

Determination of the Physical Characteristics of Urine

1. Determine the color, transparency, and odor of your "standard" sample and one of the numbered pathological samples, and circle the appropriate descriptions in the Activity 1 chart.

2. Obtain a roll of wide-range pH paper to determine the pH of each sample. Use a fresh piece of paper for each test, and dip the strip into the urine to be tested two or three times before comparing the color obtained with the chart on the dispenser. Record your results in the chart. (If you will be using one of the combination dipsticks—Chemstrip® or Multistix®—you can use these dipsticks to determine pH.)

3. To determine specific gravity, obtain a urinometer cylinder and float. Mix the urine well, and fill the urinometer cylinder about two-thirds full with urine.

4. Examine the urinometer float to determine how to read its markings. In most cases, the scale has numbered lines separated by a series of unnumbered lines. The numbered lines give the reading for the first two decimal places. You must determine the third decimal place by reading the lower edge of the meniscus—the curved surface representing the urine-air junction—on the stem of the float.

5. Carefully lower the urinometer float into the urine. Make sure it is floating freely before attempting to take the reading. Record the specific gravity of both samples in the chart. Do not dispose of this urine if the samples that you have are less than 200 ml in volume because you will need to make several more determinations.

41

Determination of Inorganic Constituents in Urine

Sulfates

Using a 10-ml graduated cylinder, add 5 ml of urine to a test tube, and then add a few drops of dilute hydrochloric acid and 2 ml of 10% barium chloride solution. The appearance of a white precipitate (barium sulfate) indicates the presence of sulfates in the sample. Clean the graduated cylinder and the test tubes well after use. Record your results.

Phosphates

Obtain a hot plate and a 500-ml beaker. To prepare the hot water bath, half fill the beaker with tap water and heat it on the hot plate. Add 5 ml of urine to a test tube, and then add three or four drops of dilute nitric acid and 3 ml of ammonium molybdate. Mix well with a glass stirring rod, and then heat gently in a hot water bath. Formation of a yellow precipitate indicates the presence of phosphates in the sample. Record your results.

Chlorides

Place 5 ml of urine in a test tube, and add several drops of silver nitrate. The appearance of a white precipitate (silver chloride) is a positive test for chlorides. Record your results.

Nitrites

Use a combination dipstick to test for nitrites. Record your results.

Determination of Organic Constituents in Urine

Individual dipsticks or combination dipsticks (Chemstrip® or Multistix®) may be used for many of the tests in this section. If you are using combination dipsticks, be prepared to take the readings on several factors (pH, protein [albumin], glucose, ketones, blood/hemoglobin, leukocytes, urobilinogen, bilirubin, and nitrites) at the same time. Generally speaking, results for all of these tests may be read *during* the second minute after immersion, but readings taken after 2 minutes have passed should be considered invalid. Pay careful attention to the directions for method and time of immersion and disposal of excess urine from the strip, regardless of the dipstick used. Identify the dipsticks that you use in the chart. If you are testing your own urine and get an unanticipated result, it is helpful to know that most of the combination dipsticks produce false positive or negative results for certain solutes when the subject is taking vitamin C, aspirin, or certain drugs.

Urea

Put two drops of urine on a clean microscope slide and *carefully* add one drop of concentrated nitric acid to the urine. Slowly warm the mixture on a hot plate until it begins to dry at the edges, but do not allow it to boil or to evaporate to dryness. When the slide has cooled, examine the edges of the preparation under low power to identify the rhombic or hexagonal crystals of urea nitrate, which form when urea and nitric acid react chemically. Keep the light low for best contrast. Record your results.

Glucose

Use a combination dipstick or obtain a vial of Clinistix®, and conduct the dipstick test according to the instructions on the vial. Record your results in the Activity 1 chart.

Because the Clinitest® reagent is routinely used in clinical agencies for glucose determinations in pediatric patients, it is worthwhile to conduct this test as well. Obtain the Clinitest® tablets and the associated color chart. You will need a timer (watch or clock with a second hand) for this test. Using a medicine dropper, put 5 drops of urine into a test tube; then rinse the dropper and add 10 drops of water to the tube. Add a Clinitest® tablet. Wait 15 seconds and then compare the color obtained to the color chart. Record your results.

Protein

Use a combination dipstick or obtain the Albustix® dipsticks, and conduct the determinations as indicated on the vial. Record your results.

Ketones

Use a combination dipstick or obtain the Ketostix® dipsticks. Conduct the determinations as indicated on the vial. Record your results.

Blood/Hemoglobin

Test your urine samples for the presence of hemoglobin by using a Hemastix® dipstick or a combination dipstick according to the directions on the vial. Usually a short drying period is required before making the reading, so read the directions carefully. Record your results.

Bilirubin

Using a combination dipstick, determine if there is any bilirubin in your urine samples. Record your results.

Also conduct the Ictotest® for the presence of bilirubin. Using a medicine dropper, place one drop of urine in the center of one of the special test mats provided with the Ictotest® reagent tablets. Place one of the reagent tablets over the drop of urine, and then add two drops of water directly to the tablet. If the mixture turns purple when you add water, bilirubin is present. Record your results.

Leukocytes

Use a combination dipstick to test for leukocytes. Record your results.

Urobilinogen

Use a combination dipstick to test for urobilinogen. Record your results.

Clean up your area following the procedures described at the beginning of this activity.

Activity 1: Urinalysis Results

Observation or test	Normal values	Standard urine specimen	Unknown specimen (# _______)
Physical Characteristics			
Color	Pale yellow	Yellow: pale medium dark other ____________	Yellow: pale medium dark other ____________
Transparency	Clear	Clear Slightly cloudy Cloudy	Clear Slightly cloudy Cloudy
Odor	Aromatic	Describe: ________________	Describe: ________________
pH	4.5–8.0	________________	________________
Specific gravity	1.001–1.030	________________	________________
Inorganic Components			
Sulfates	Present	Present Absent	Present Absent
Phosphates	Present	Present Absent	Present Absent
Chlorides	Present	Present Absent	Present Absent
Nitrites	Absent	Present Absent	Present Absent
Dipstick: ________			
Organic Components			
Urea	Present	Present Absent	Present Absent
Glucose Dipstick: ________	Negative	Record results: ________________	Record results: ________________
Clinitest®	Negative	________________	________________
Protein Dipstick: ________	Negative	________________	________________
Ketone bodies Dipstick: ________	Negative	________________	________________
RBCs/hemoglobin Dipstick: ________	Negative	________________	________________
Bilirubin Dipstick: ________	Negative	________________	________________
Ictotest®	Negative (no color change)	Negative Positive (purple)	Negative Positive (purple)
Leukocytes	Absent	Present Absent	Present Absent
Dipstick: ________			
Urobilinogen	Present	Present Absent	Present Absent
Dipstick: ________			

Activity 2

Analyzing Urine Sediment Microscopically (Optional)

If your instructor so indicates, conduct a microscopic analysis of urine sediment in "real" urine. The urine sample to be analyzed microscopically has been centrifuged to spin the more dense urine components to the bottom of a tube, and some of the sediment has been mounted on a slide and stained with Sedi-Stain™ to make the components more visible.

Go to the demonstration microscope to conduct this study. Using the lowest light source possible, examine the slide under low power to determine whether you can see any common sediments (**Figure 41.1**).

Unorganized sediments: Chemical substances that form crystals or precipitate from solution; for example, calcium oxalates, carbonates, and phosphates; uric acid; ammonium ureates; and cholesterol. Also, if one has been taking antibiotics or certain drugs such as sulfa drugs, these may be detectable in the urine in crystalline form. Normal urine contains very small amounts of crystals, but conditions such as urinary retention or urinary tract infection may cause the appearance of much larger amounts. The high-power lens may be needed to view the various crystals, which tend to be much more minute than the organized cellular sediments.

Organized sediments: Include epithelial cells (rarely of any pathological significance), white blood cells, red blood cells, and casts. The presence of white blood cells, red blood cells, and casts other than trace amounts always indicates kidney pathology. Note that red blood cells, white blood cells, and epithelial cells can also form casts.

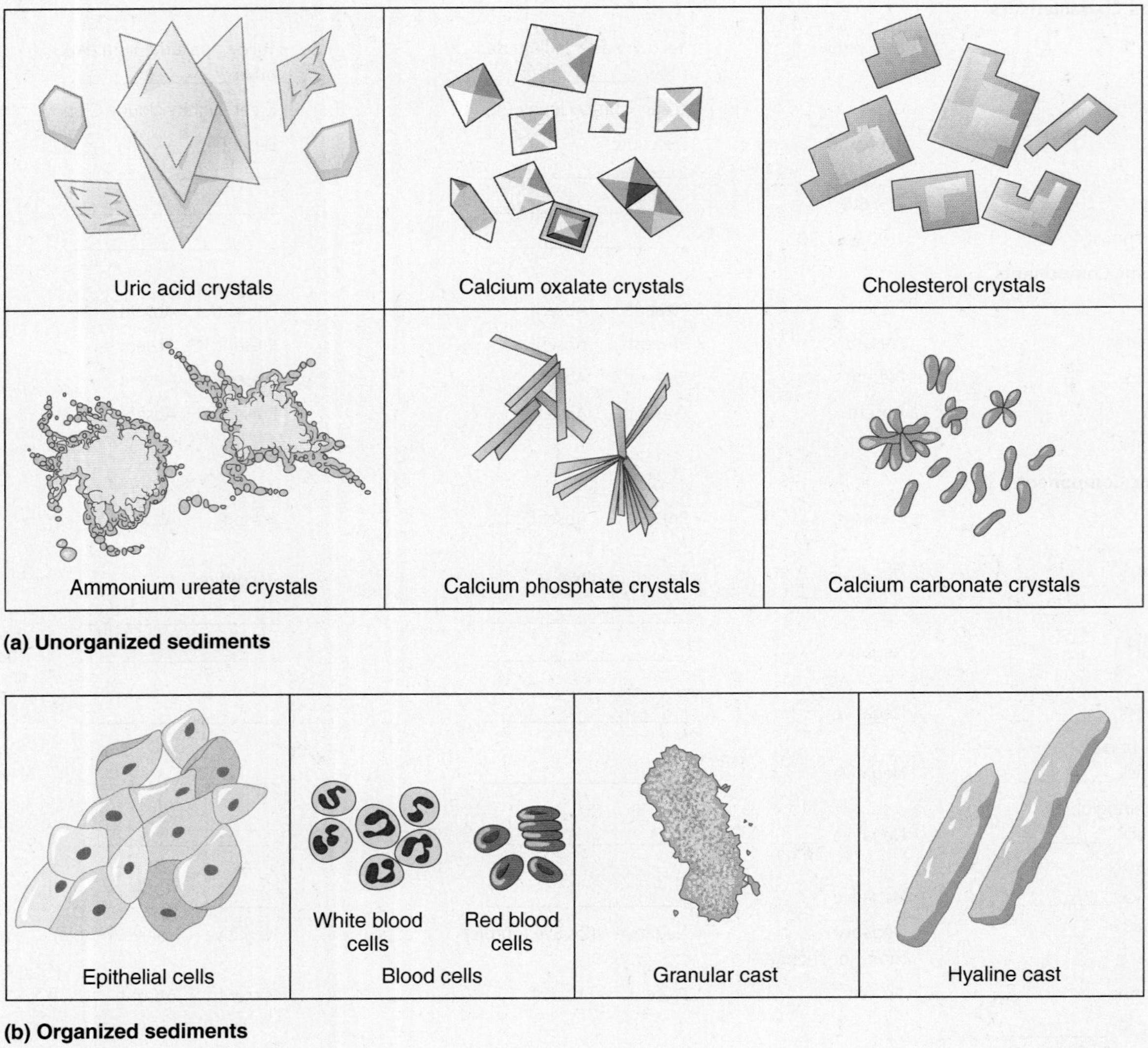

Figure 41.1 Examples of sediments.

Instructors may assign a portion of the Review Sheet questions using Mastering A&P™

41 EXERCISE REVIEW SHEET Urinalysis

Name ______________________ Lab Time/Date ______________________

Characteristics of Urine

1. What is the range for the volume of urine normally excreted in a 24-hour period? ______________________

2. Assuming normal conditions, note whether each of the following substances would be (a) in greater relative concentration in the urine than in the glomerular filtrate, (b) in lesser concentration in the urine than in the glomerular filtrate, or (c) absent from both the urine and the glomerular filtrate. Use an appropriate reference as needed.

________ 1. water	________ 6. amino acids	________ 11. uric acid
________ 2. phosphate ions	________ 7. glucose	________ 12. creatinine
________ 3. sulfate ions	________ 8. protein	________ 13. white blood cells
________ 4. potassium ions	________ 9. red blood cells	________ 14. nitrites
________ 5. sodium ions	________ 10. urea	

3. Explain why urinalysis is a routine part of any good physical examination. ______________________

4. What substance is responsible for the normal yellow color of urine? ______________________

5. Which has a greater specific gravity: 1 ml of urine or 1 ml of distilled water? ________ Explain your answer. ________

6. Explain the relationship between the color, specific gravity, and volume of urine. ______________________

Abnormal Urinary Constituents

7. Explain two reasons why glucose might be present in the urine.

8. Name two conditions associated with the presence of bile pigments in the urine.

9. Explain how a patient could have red blood cells in the urine even if the filtration membrane were still intact. ______________________

10. Several specific terms have been used to indicate the presence of abnormal urine constituents. Identify each of the abnormalities described below by inserting a term from the key at the right that names the condition.

		Key:
______________	1. presence of erythrocytes in the urine	a. glycosuria
______________	2. presence of hemoglobin in the urine	b. hematuria
______________	3. presence of glucose in the urine	c. hemoglobinuria
______________	4. presence of protein in the urine	d. ketonuria
______________	5. presence of ketone bodies in the urine	e. proteinuria
______________	6. presence of white blood cells in the urine	f. pyuria

11. What are renal calculi, and what conditions favor their formation? ______________________________

____

12. Glucose and protein are both normally absent in the urine, but the reason for their absence differs. Explain why glucose is normally absent. ______________________________

Explain why protein is normally absent. ______________________________

13. Name the three major nitrogenous wastes found in the urine. ______________________________,

______________________________, and ______________________________

14. Explain the difference between organized and unorganized sediments. ______________________________

____

____

15. Describe the effect that dehydration would have on the specific gravity of urine and why. ______________________________

____

16. ✚ Renal calculi contain crystallized calcium, magnesium, and uric acid. Explain why patients with hyperparathyroidism would be at risk for developing renal calculi. ______________________________

____

____

17. ✚ *Proteus mirabilis* produces the enzyme urease, which converts urea into ammonia. Explain why patients with a UTI caused by *Proteus mirabilis* would have a higher-than-normal urine pH. ______________________________

____

____

42 EXERCISE Anatomy of the Reproductive System

Learning Outcomes

- Discuss the general function of the reproductive system.
- Identify the structures of the male and female reproductive systems on an appropriate model or image, and list the general function of each.
- Define *semen*, state its composition, and name the organs involved in its production.
- Trace the pathway followed by a sperm from its site of formation to the external environment.
- Define *erection* and *ejaculation*.
- Define *gonad*, and name the gametes and endocrine products of the testes and ovaries, indicating the cell types or structures responsible for the production of each.
- Describe the microscopic structure of the penis, seminal glands, epididymis, uterine wall, and uterine tube, and relate structure to function.
- Explain the role of the fimbriae and ciliated epithelium of the uterine tubes in the movement of the egg from the ovary to the uterus.
- Identify the fundus, body, and cervical regions of the uterus.
- Define *endometrium*, *myometrium*, and *ovulation*.
- Describe the anatomy and discuss the reproduction-related function of female mammary glands.

Pre-Lab Quiz

Instructors may assign these and other Pre-Lab Quiz questions using Mastering A&P™

1. After sperm are produced, they enter the first part of the duct system, the:
   a. ductus deferens **c.** epididymis
   b. ejaculatory duct **d.** urethra
2. The prostate, seminal glands, and bulbo-urethral glands produce ____________, the liquid medium in which sperm leaves the body.
   a. seminal fluid **c.** urine
   b. testosterone **d.** water
3. Circle the correct underlined term. The interstitial endocrine cells / seminiferous tubules produce testosterone, the hormonal product of the testis.
4. The endocrine products of the ovaries are estrogen and:
   a. luteinizing hormone **c.** prolactin
   b. progesterone **d.** testosterone
5. Circle the correct underlined term. The labia majora / clitoris are/is homologous to the penis.

Go to Mastering A&P™ > Study Area to improve your performance in A&P Lab.

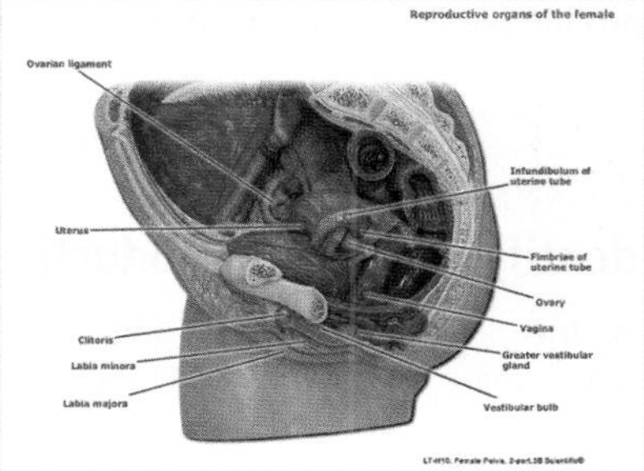

> Lab Tools > Practice Anatomy Lab > Anatomical Models

Instructors may assign new Building Vocabulary coaching activities, Pre-Lab Quiz questions, Art Labeling activities, Practice Anatomy Lab Practical questions (PAL), and more using the Mastering A&P™ Item Library.

Materials

- Three-dimensional models or large laboratory charts of the male and female reproductive tracts
- Prepared slides of cross sections of the penis, seminal glands, epididymis, uterus showing endometrium (secretory phase), and uterine tube
- Compound microscope

For instructions on animal dissections, see the dissection exercises (starting on p. 695) in the cat and fetal pig editions of this manual.

Most organ systems of the body function from the time they are formed to sustain the existing individual. However, the **reproductive system** begins its biological function, the production of offspring, at puberty.

The essential organs of reproduction are the **gonads**, the testes and the ovaries, which produce the sex cells, or **gametes**, and the sex hormones. The reproductive role of the male is to manufacture sperm and to deliver them to the female reproductive tract. The female, in turn, produces eggs. If the time is suitable, the combination of sperm and egg produces a fertilized egg, which is the first cell of a new individual. Once fertilization has occurred, the female uterus provides a nurturing, protective environment in which the embryo, later called the fetus, develops until birth.

Gross Anatomy of the Human Male Reproductive System

The primary reproductive organs of the male are the **testes**, which produce sperm and the male sex hormones. All other reproductive structures are ducts or sources of secretions, which aid in the safe delivery of the sperm to the body exterior or female reproductive tract.

Activity 1

Identifying Male Reproductive Organs

As the following organs and structures are described, locate them on **Figure 42.1** and then identify them on a three-dimensional model of the male reproductive system or on a large laboratory chart.

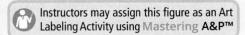

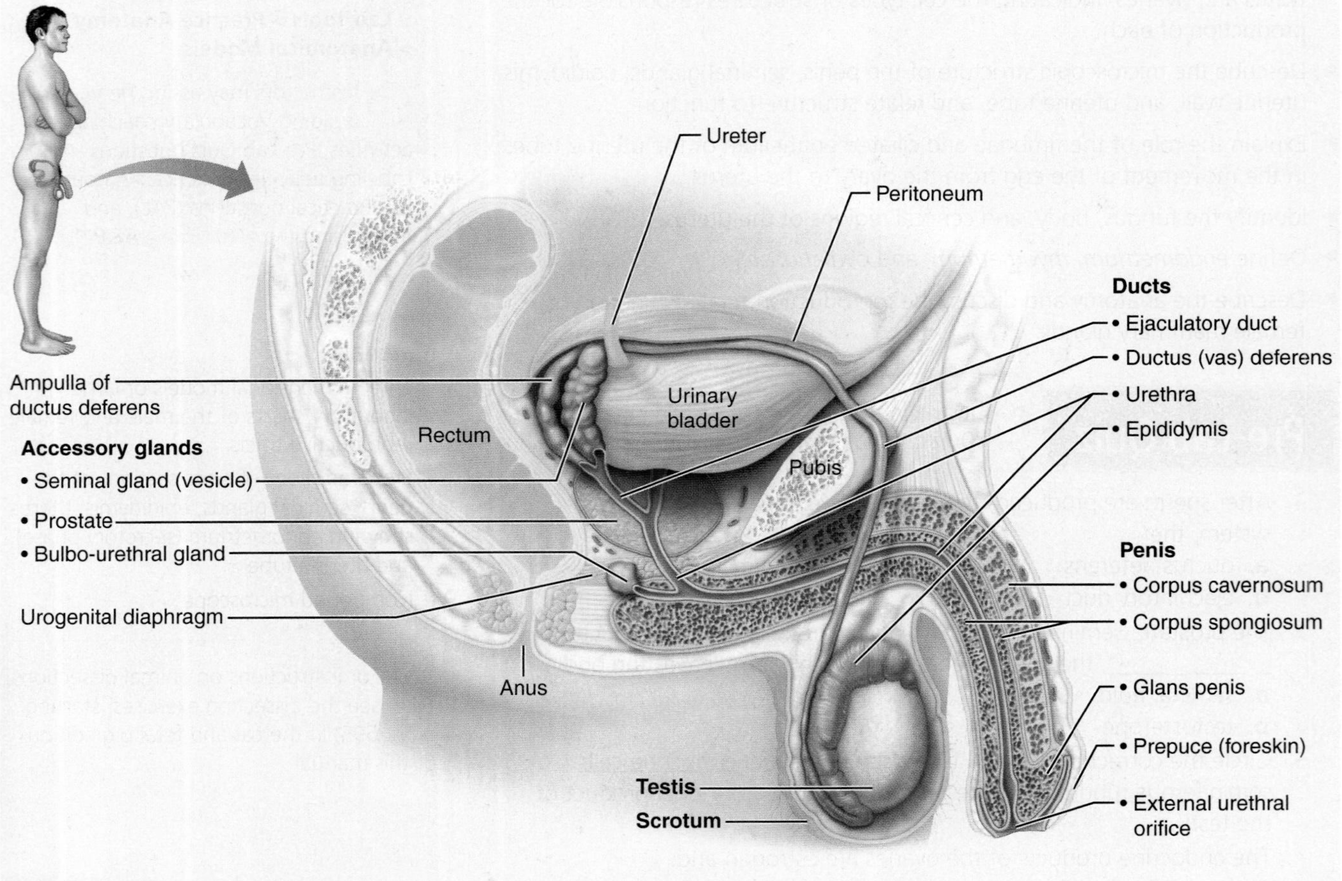

(a) Sagittal view

Figure 42.1 Reproductive organs of the human male. (a) Sagittal view.

Instructors may assign this figure as an Art Labeling Activity using Mastering A&P™

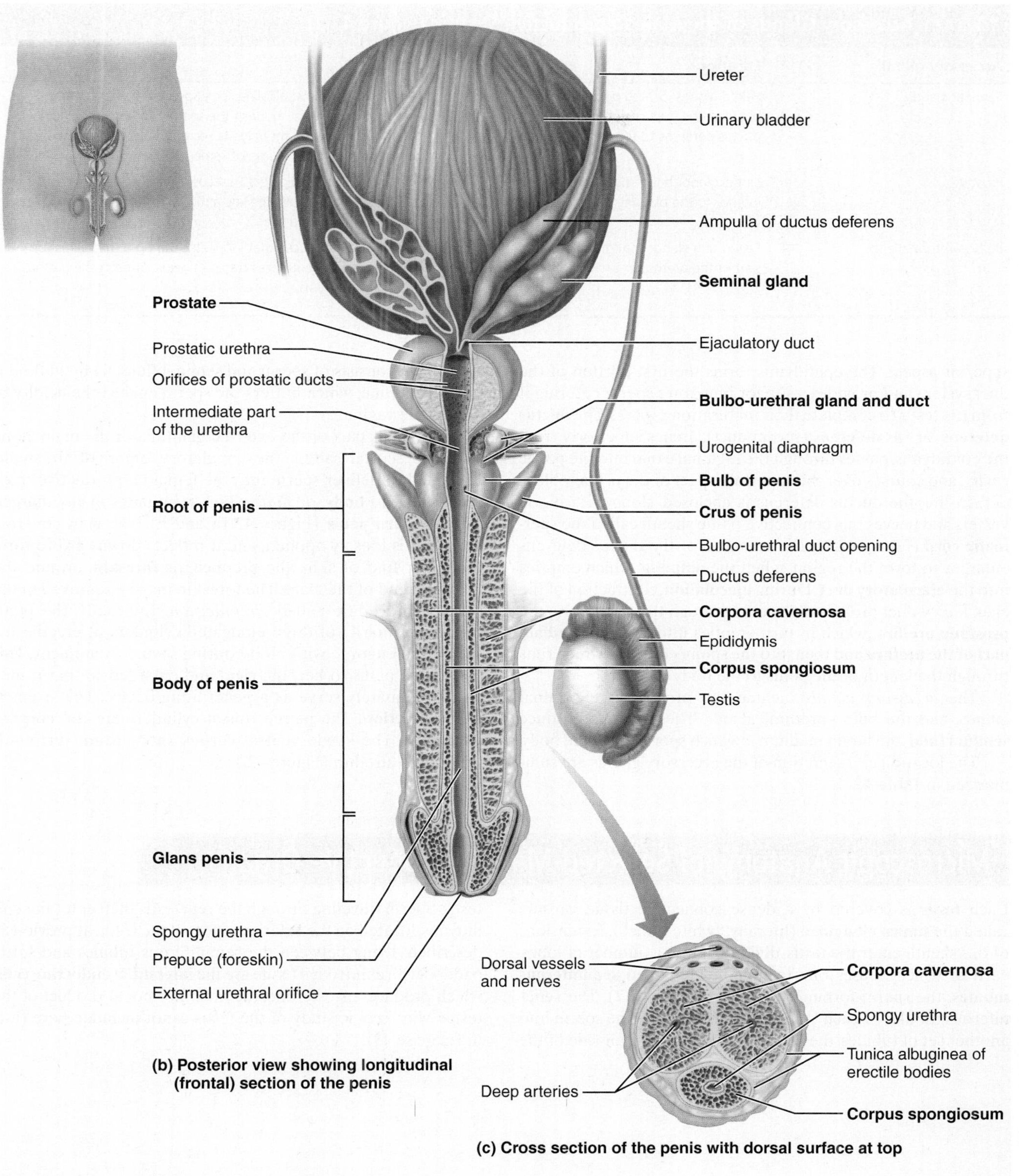

Figure 42.1 (*continued*) **(b)** Posterior view showing a longitudinal section of the penis. **(c)** Transverse section of the penis.

The paired oval testes lie in the **scrotum** outside the abdominopelvic cavity. The temperature there (approximately 94°F, or 34°C) is slightly lower than body temperature, a requirement for producing viable sperm.

The accessory structures forming the *duct system* are the epididymis, the ductus deferens, the ejaculatory duct, and the urethra. The **epididymis** is an elongated structure running up the posterior and lateral side of the testis and capping its

Table 42.1 Accessory Glands of the Male Reproductive System (Figure 42.1)

Accessory gland	Location	Secretion
Seminal glands	Paired glands located posterior to the urinary bladder. The duct of each gland merges with a ductus deferens to form the ejaculatory duct.	A thick, light yellow, alkaline secretion containing fructose and citric acid, which nourish the sperm, and prostaglandins for enhanced sperm motility. Its secretion has the largest contribution to the volume of semen.
Prostate	Single gland that encircles the prostatic urethra inferior to the bladder.	A milky, slightly acidic fluid that contains citric acid, several enzymes, and prostate-specific antigen (PSA). Its secretion plays a role in activating the sperm.
Bulbo-urethral glands	Paired tiny glands that drain into the intermediate part of the urethra.	A clear alkaline mucus that lubricates the tip of the penis for copulation and neutralizes traces of acidic urine in the urethra prior to ejaculation.

superior aspect. The epididymis forms the first portion of the duct system and provides a site for immature sperm entering it from the testis to complete their maturation process. The **ductus deferens**, or **vas deferens** (sperm duct), arches superiorly from the epididymis, passes through the inguinal canal into the pelvic cavity, and courses over the superior aspect of the urinary bladder. In life, the ductus deferens is enclosed along with blood vessels and nerves in a connective tissue sheath called the **spermatic cord** (Figure 42.2). The terminus of the ductus deferens enlarges to form the region called the **ampulla**, which empties into the **ejaculatory duct**. During **ejaculation**, contraction of the ejaculatory duct propels the sperm through the prostate to the **prostatic urethra**, which in turn empties into the **intermediate part of the urethra** and then into the **spongy urethra**, which runs through the length of the penis to the body exterior.

The *accessory glands* include the prostate, the seminal glands, and the bulbo-urethral glands. These glands produce **seminal fluid**, the liquid medium in which sperm leave the body.

The location and secretion of the accessory glands are summarized in **Table 42.1**.

Semen consists of sperm and seminal fluid. Seminal fluid is overall alkaline, which buffers the sperm against the acidity of the female vagina.

The **penis**, part of the external genitalia of the male along with the scrotal sac, is the copulatory organ of the male. Designed to deliver sperm into the female reproductive tract, it consists of a **body**, or *shaft*, which terminates in an enlarged tip, the **glans penis** (Figure 42.1a and b). The skin covering the penis is loosely applied, and it reflects downward to form a circular fold of skin, the **prepuce**, or **foreskin**, around the proximal end of the glans. The foreskin may be removed in the surgical procedure called *circumcision*. Internally, the penis consists primarily of three elongated cylinders of erectile tissue, which engorge with blood during sexual excitement. This causes the penis to become rigid and enlarged so that it may more adequately serve as a penetrating device. This event is called **erection**. The paired dorsal cylinders are the **corpora cavernosa**. The single ventral **corpus spongiosum** surrounds the spongy urethra (Figure 42.1c).

Microscopic Anatomy of Selected Male Reproductive Organs

Each **testis** is covered by a dense connective tissue capsule called the **tunica albuginea** (literally, "white tunic"). Extensions of this sheath enter the testis, dividing it into a number of lobes, each of which houses one to four highly coiled **seminiferous tubules**, the sperm-forming factories (**Figure 42.2**). The seminiferous tubules of each lobe converge to empty the sperm into another set of tubules, the **rete testis**, at the posterior side of the testis. Sperm traveling through the rete testis then enter the epididymis, located on the exterior aspect of the testis, as previously described. Lying between the seminiferous tubules and softly padded with connective tissue are the **interstitial endocrine cells**, which produce testosterone, the main hormonal product of the testis. Microscopic study of the testis is not included here (it is in Exercise 43).

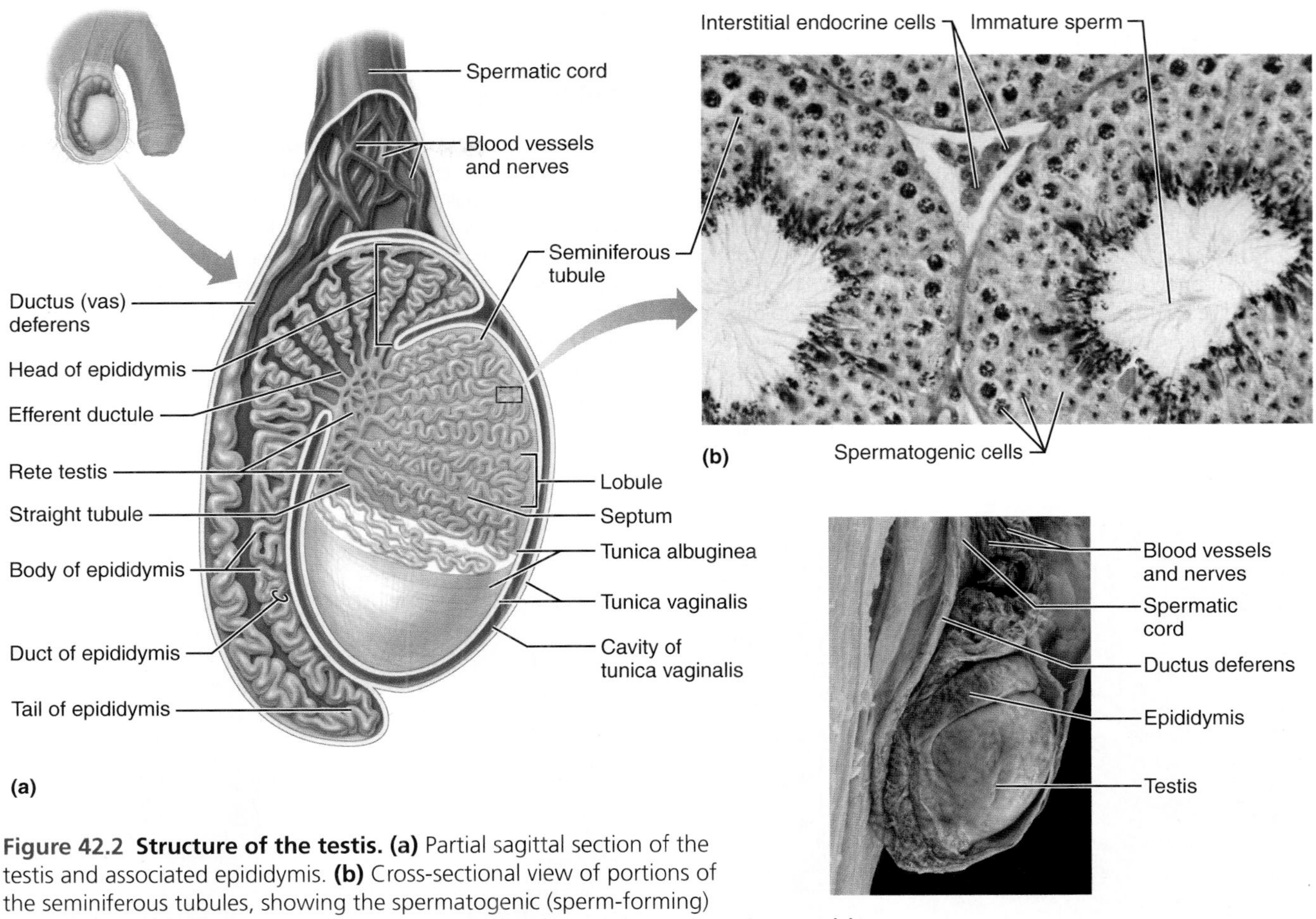

Figure 42.2 Structure of the testis. (a) Partial sagittal section of the testis and associated epididymis. **(b)** Cross-sectional view of portions of the seminiferous tubules, showing the spermatogenic (sperm-forming) cells, which make up the epithelium of the tubule walls, and the interstitial endocrine cells in the loose connective tissue between the tubules (200×). **(c)** External view of a testis from a cadaver; same orientation as in (a).

Instructors may assign this figure as an Art Labeling Activity using Mastering A&P™

Activity 2

Penis

Obtain a slide of a cross section of the penis. Scan the tissue under low power to identify the urethra and the cavernous bodies. Compare your observations to Figure 42.1c and **Figure 42.3**. Observe the lumen of the urethra carefully. What type of epithelium do you see?

____

Explain the function of this type of epithelium.

____

____

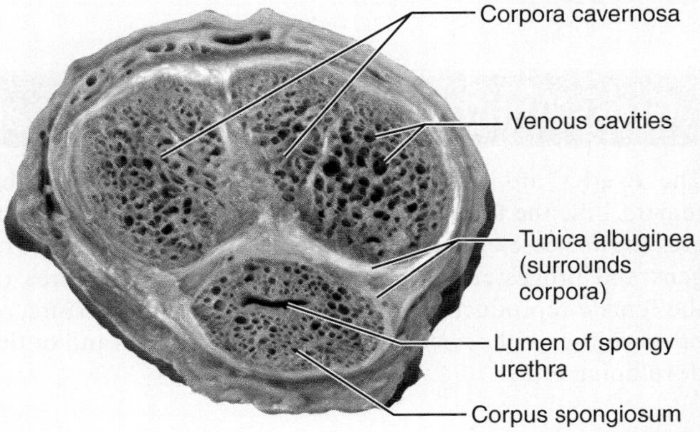

Figure 42.3 Transverse section of the penis (2×).

Activity 3

Seminal Gland

Obtain a slide showing a cross-sectional view of the seminal gland. Examine the slide at low magnification to get an overall view of the highly folded mucosa of this gland. Switch to higher magnification, and notice that the folds of the gland protrude into the lumen where they divide further, giving the lumen a honeycomb look (**Figure 42.4**). Notice that the loose connective tissue lamina propria is underlain by smooth muscle fibers—first a circular layer, and then a longitudinal layer.

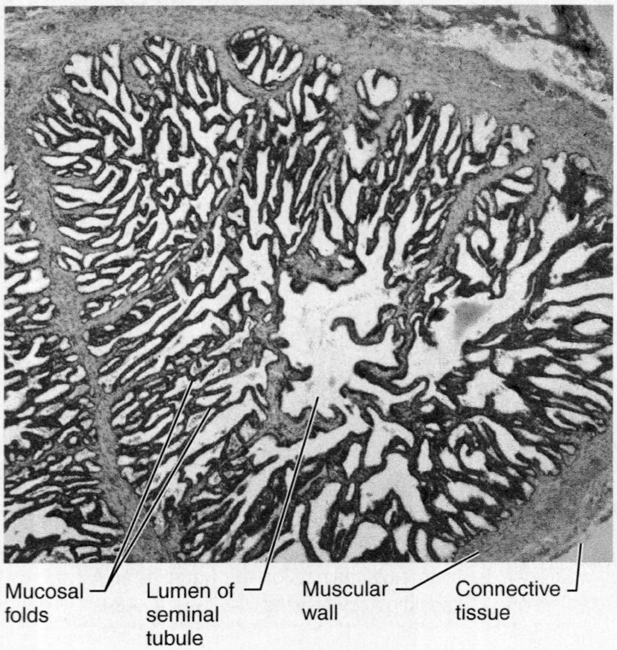

Figure 42.4 Cross-sectional view of a seminal gland with its elaborate network of mucosal folds. Glandular secretion is seen in the lumen (25×).

Activity 4

Epididymis

Obtain a slide of a cross section of the epididymis. Notice the abundant tubule cross sections resulting from the fact that the coiling epididymis tubule has been cut through many times in the specimen (**Figure 42.5**). Look for sperm in the lumen of the tubule. Examine the composition of the tubule wall carefully. Identify the *stereocilia* of the pseudostratified columnar epithelial lining. These nonmotile microvilli absorb excess fluid and pass nutrients to the sperm in the lumen. Now identify the smooth muscle layer. What do you think the function of the smooth muscle is?

____

____

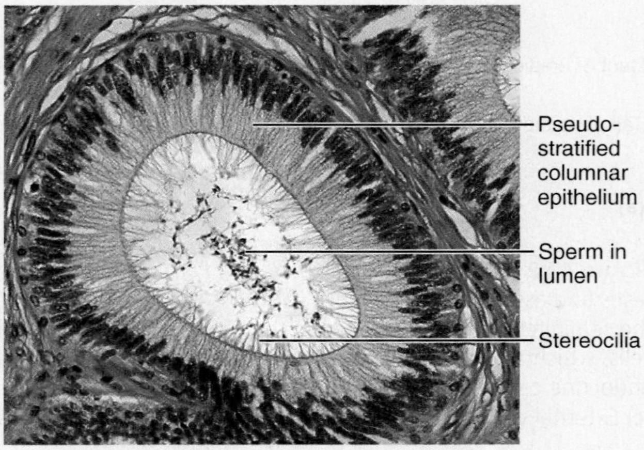

Figure 42.5 Cross section of epididymis (120×).

Gross Anatomy of the Human Female Reproductive System

42

The **ovaries** are the primary reproductive organs of the female. Like the testes of the male, the ovaries produce gametes (in this case eggs, or ova) and also sex hormones (estrogens and progesterone). The other accessory structures of the female reproductive system transport, house, nurture, or otherwise serve the needs of the reproductive cells and/or the developing fetus.

Activity 5

Identifying Female Reproductive Organs

As you read the descriptions of these structures, locate them in Figure 42.6 and Figure 42.7 and then on the female reproductive system model or large laboratory chart.

External Genitalia

The **external genitalia (vulva)** consist of the mons pubis, the labia majora and minora, the clitoris, the external urethral and vaginal orifices, the hymen, and the greater vestibular glands. **Table 42.2** summarizes the structures of the female external genitalia (**Figure 42.6**).

The diamond-shaped region between the anterior end of the labial folds, the ischial tuberosities laterally, and the anus posteriorly is called the **perineum**.

Internal Organs

The internal female organs include the vagina, uterus, uterine tubes, ovaries, and the ligaments and supporting structures that suspend these organs in the pelvic cavity (**Figure 42.7**, p. 632). The **vagina** extends for approximately 10 cm (4 inches) from the vestibule to the uterus superiorly. It serves as a copulatory

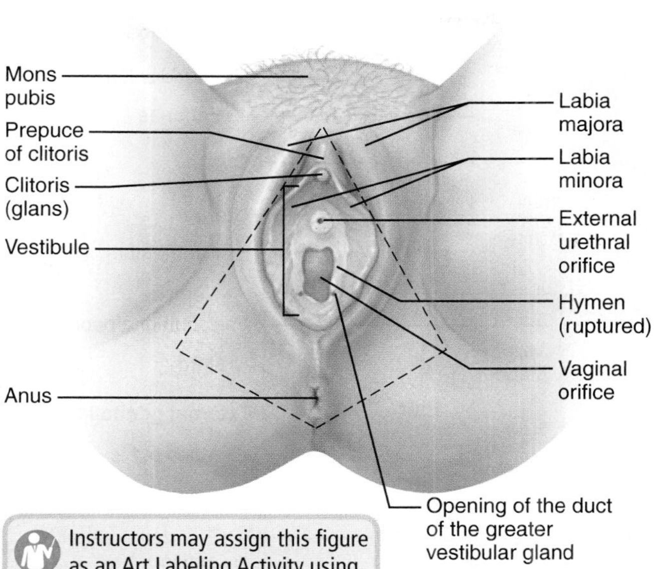

Instructors may assign this figure as an Art Labeling Activity using Mastering A&P™

Figure 42.6 External genitalia (vulva) of the human female. The region enclosed by dashed lines is the perineum.

organ because it receives the penis (and semen) during sexual intercourse. The vagina also provides a passageway for delivery of an infant and for menstrual flow. The pear-shaped **uterus**, situated between the bladder and the rectum, is a muscular organ with its narrow end, the **cervix**, directed inferiorly. The major portion of the uterus is referred to as the **body**; its superior rounded region above the entrance of the uterine tubes is called the **fundus**. A fertilized egg is implanted in the uterus, which houses the embryo or fetus during its development.

In some cases, the fertilized egg may implant in a uterine tube or even on the abdominal viscera, creating an **ectopic pregnancy**. Such implantations are usually unsuccessful and may even endanger the mother's life because the uterine tubes cannot accommodate the increasing size of the fetus. +

The **endometrium**, the thick mucosal lining of the uterus, has a superficial **functional layer**, or **stratum functionalis**, that sloughs off periodically (about every 28 days) in response to cyclic changes in the levels of ovarian hormones in the woman's blood. This sloughing-off process, which is accompanied by bleeding, is referred to as **menstruation**, or **menses**. The deeper **basal layer**, or **stratum basalis** (Figure 43.6b), forms a new functional layer after menstruation ends.

The **uterine**, or **fallopian**, **tubes** are about 10 cm (4 inches) long and extend from the ovaries in the peritoneal cavity to the superolateral region of the uterus. The distal ends of the tubes are funnel-shaped and have fingerlike projections called **fimbriae**. Unlike in the male duct system, there is no actual contact between the female gonad and the initial part of the female duct system—the uterine tube.

Because of this open passageway between the female reproductive organs and the peritoneal cavity, reproductive system infections, such as gonorrhea and other **sexually transmitted infections (STIs)**, also called *sexually transmitted diseases (STDs)*, can cause widespread inflammations of the pelvic viscera, a condition called **pelvic inflammatory disease (PID)**. +

The internal female organs are all retroperitoneal, except the ovaries. They are supported and suspended somewhat freely by ligamentous folds of peritoneum. The supporting structures for the uterus, uterine tubes, and ovaries are summarized in **Table 42.3** on p. 633.

Within the ovaries, the female gametes, or eggs, begin their development in saclike structures called *follicles*. The growing follicles also produce *estrogens*. When a developing egg has reached the appropriate stage of maturity, it is ejected from the ovary in an event called **ovulation**. The ruptured follicle is then converted to a second type of endocrine structure called a *corpus luteum*, which secretes progesterone and some estrogens.

The flattened almond-shaped ovaries lie adjacent to the uterine tubes but are not connected to them; consequently, an ovulated "egg," actually a secondary oocyte (see Exercise 43), enters the pelvic cavity. The waving fimbriae of the uterine tubes create fluid currents that, if successful, draw the egg into the lumen of the uterine tube. There the egg begins its passage to the uterus, propelled by the cilia of the tubule walls. The usual and most desirable site of fertilization is the uterine tube, because the journey to the uterus takes about 3 to 4 days and an egg is viable only for up to 24 hours after it is expelled from the ovary. Thus, sperm must swim upward through the vagina and uterus and into the uterine tubes to reach the egg. This must be an arduous journey, because they must swim against the downward current created by ciliary action—rather like swimming upstream!

Table 42.2 External Genitalia (Vulva) of the Human Female (Figure 42.6)

Structure	Description
Mons pubis	Rounded fatty eminence that cushions the pubic symphysis; covered with coarse pubic hair after puberty.
Labia majora (singular: *labium majus*)	Two elongated hair-covered skin folds that extend from the mons pubis. They contain sebaceous glands, apocrine glands, and adipose tissue. They are homologous to the scrotum.
Labia minora (singular: *labium minus*)	Two smaller folds located medial to the labia majora. They don't have hair or adipose tissue, but they do have many sebaceous glands.
Vestibule	Region located between the two labia minora. From anterior to posterior, it contains the clitoris, the external urethral orifice, and the vaginal orifice.
Clitoris	Small mass of erectile tissue located where the labia minora meet anteriorly. It is homologous to the penis.
Prepuce of the clitoris	Skin folds formed by the union of the labia minora; they serve to hood the clitoris.
External urethral orifice	Serves as the outlet for the urinary system. It has no reproductive function in the female.
Hymen	A thin fold of vascular mucous membrane that may partially cover the vaginal opening.
Greater vestibular glands	Pea-sized mucus-secreting glands located on either side of the hymen. They lubricate the distal end of the vagina during coitus. They are homologous to the bulbo-urethral glands of males.

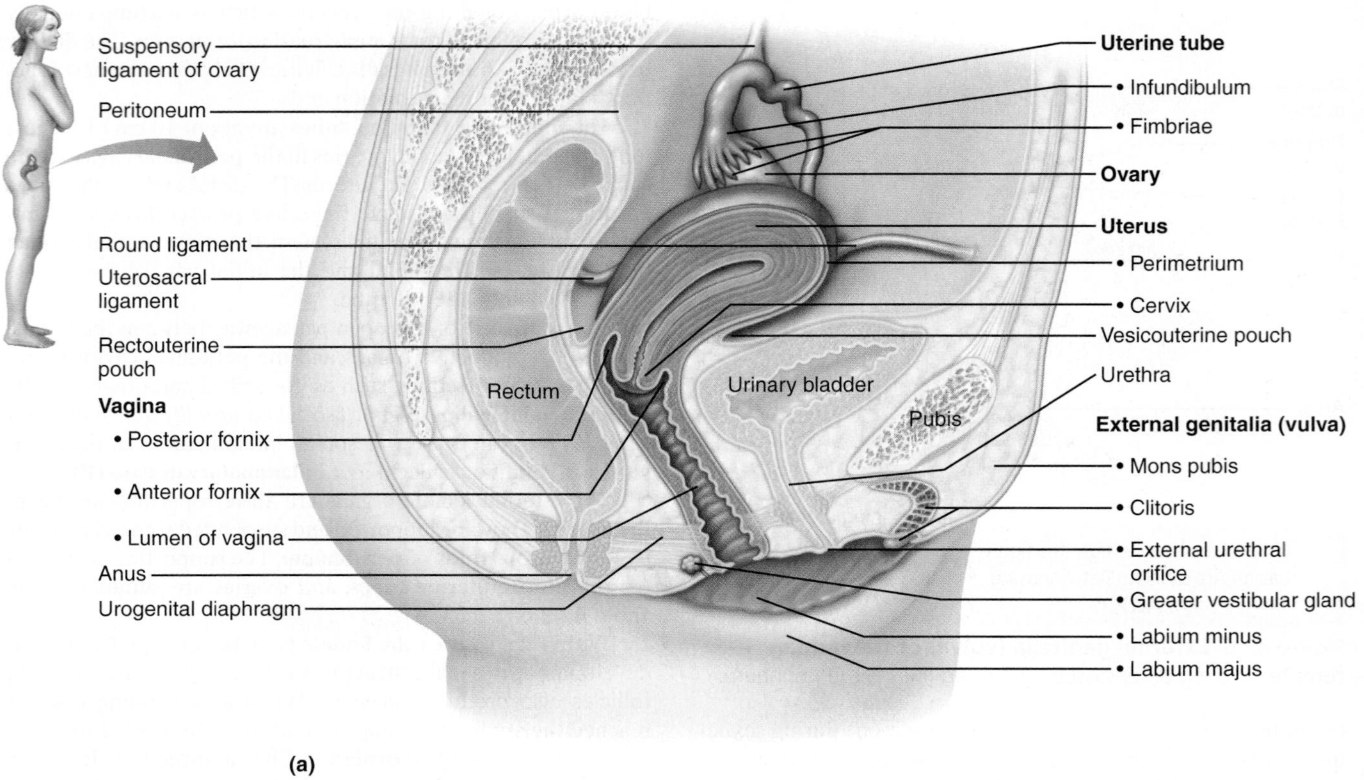

(a)

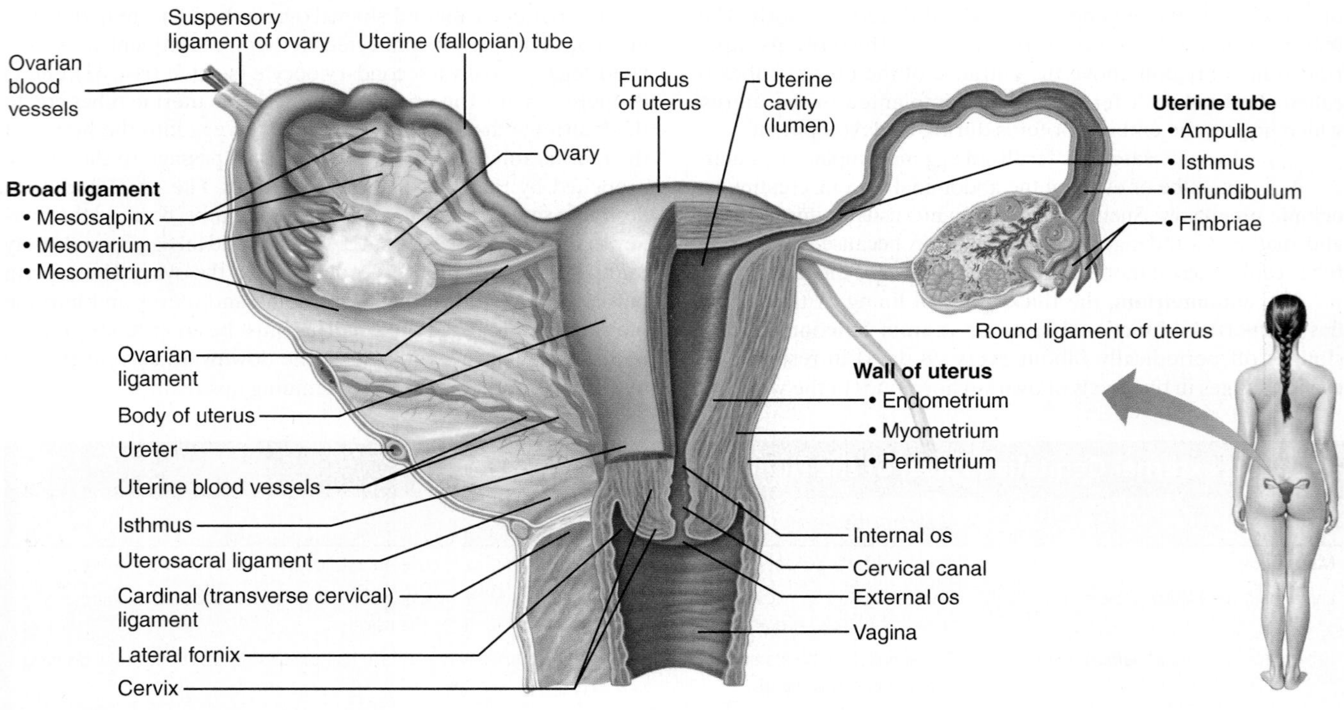

(b)

Figure 42.7 Internal reproductive organs of the human female. (a) Midsagittal section of the human female reproductive system. **(b)** Posterior view. The posterior walls of the vagina, uterus, and uterine tubes, as well as the broad ligament, have been removed on the right side to reveal the shape of the lumen of these organs.

Instructors may assign this figure as an Art Labeling Activity using Mastering A&P™

42

Table 42.3 Supporting Structures for the Uterus, Uterine Tubes, and Ovaries (Figure 42.7)

Structure	Description
Broad ligament	Fold of the peritoneum that drapes over the superior uterus to enclose the uterus and uterine tubes and anchors them to the lateral body walls
Mesometrium	Portion of the broad ligament that supports the uterus laterally (mesentery of the uterus)
Round ligaments	Anchor the uterus to the anterior pelvic wall by descending through the mesometrium and the inguinal canal; attach to the skin of one of the labia majora
Uterosacral ligaments	Secure the inferior uterus to the sacrum posteriorly
Cardinal (transverse cervical) ligaments	Connect the cervix and vagina to the pelvic wall laterally
Mesosalpinx	Portion of the broad ligament that anchors the uterine tube (mesentery of the uterine tube)
Mesovarium	Posterior fold of the broad ligament that supports the ovaries (mesentery of the ovaries)
Suspensory ligaments	A lateral continuation of the broad ligament that attaches the ovaries to the lateral pelvic wall
Ovarian ligaments	Anchors the ovaries to the uterus medially and is enclosed within the broad ligament

Microscopic Anatomy of Selected Female Reproductive Organs

Activity 6

Wall of the Uterus

Obtain a slide of a cross-sectional view of the uterine wall. Identify the three layers of the uterine wall—the endometrium, myometrium, and serosa. **Figure 42.8**, a photomicrograph that includes the secretory endometrium, will help with this study.

As you study the slide, notice that the bundles of smooth muscle are oriented in several different directions. What is the function of the **myometrium** (smooth muscle layer) during childbirth?

___

___

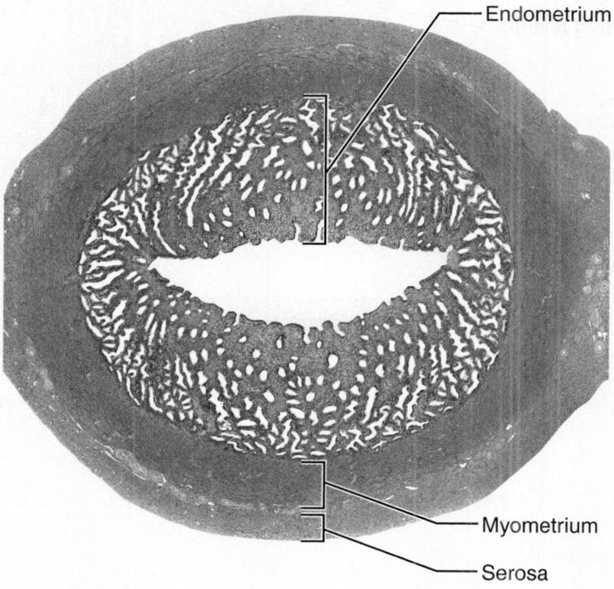

Figure 42.8 Cross-sectional view of the uterine wall. The mucosa is in the secretory phase. (3×).

Activity 7

Uterine Tube

Obtain a slide of a cross-sectional view of a uterine tube for examination. Notice that the mucosal folds nearly fill the tubule lumen (**Figure 42.9**). Then switch to high power to examine the ciliated secretory epithelium.

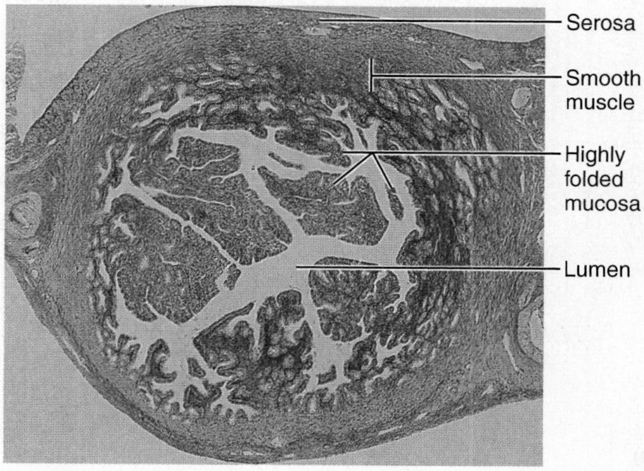

Figure 42.9 Cross-sectional view of the uterine tube (12×).

The Mammary Glands

The **mammary glands** exist within the breasts in both sexes, but they normally have a reproduction-related function only in females. Since the function of the mammary glands is to produce milk to nourish the newborn infant, their importance is more closely associated with events that occur when reproduction has already been accomplished. Periodic stimulation by the female sex hormones, especially estrogens, increases the size of the female mammary glands at puberty. During this period, the duct system becomes more elaborate, and fat is deposited—fat deposition is the more important contributor to increased breast size.

The rounded, skin-covered mammary glands lie anterior to the pectoral muscles of the thorax, attached to them by connective tissue. Slightly below the center of each breast is a pigmented area, the **areola**, which surrounds a centrally protruding **nipple** (**Figure 42.10**).

Internally each mammary gland consists of 15 to 25 **lobes** that radiate around the nipple and are separated by fibrous connective tissue and fat. Within each lobe are smaller chambers called **lobules**, containing the glandular **alveoli** that produce milk during lactation. The alveoli of each lobule pass the milk into a number of **lactiferous ducts**, which join to form an expanded storage chamber, the **lactiferous sinus**, as they approach the nipple. The sinuses open to the outside at the nipple. The description of mammary glands that we have just given applies only to nursing women or women in the last trimester of pregnancy.

For instructions on animal dissections, see the dissection exercises (starting on p. 695) in the cat and fetal pig editions of this manual.

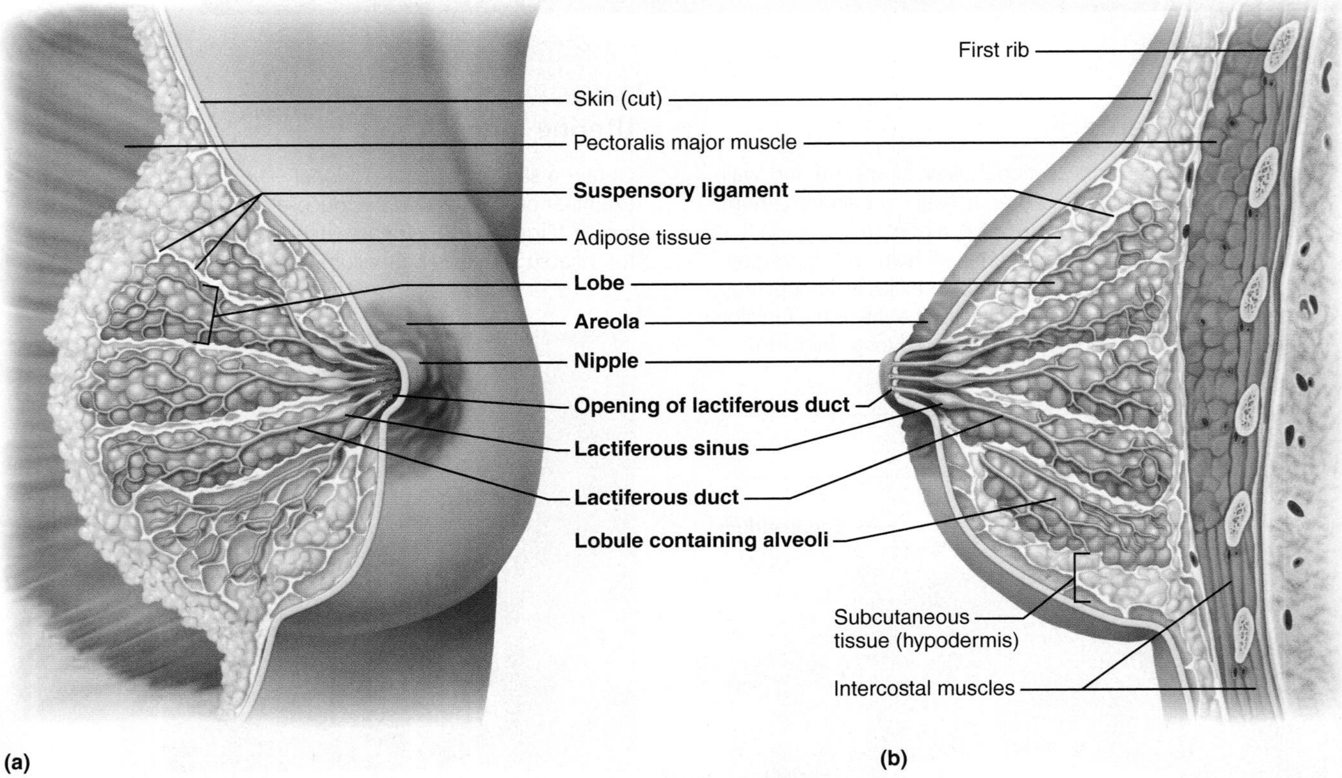

Figure 42.10 Anatomy of lactating mammary gland. (a) Anterior view of partially dissected breast. **(b)** Sagittal section of the breast.

Instructors may assign this figure as an Art Labeling Activity using Mastering **A&P™**

Instructors may assign a portion of the Review Sheet questions using Mastering A&P™

REVIEW SHEET

EXERCISE 42

Anatomy of the Reproductive System

Name ______________________________ Lab Time/Date ______________________________

Gross Anatomy of the Human Male Reproductive System

1. List the two main functions of the testis. ______________________________

and ______________________________

2. Identify all indicated structures or portions of structures on the photo of the model of the male reproductive system below.

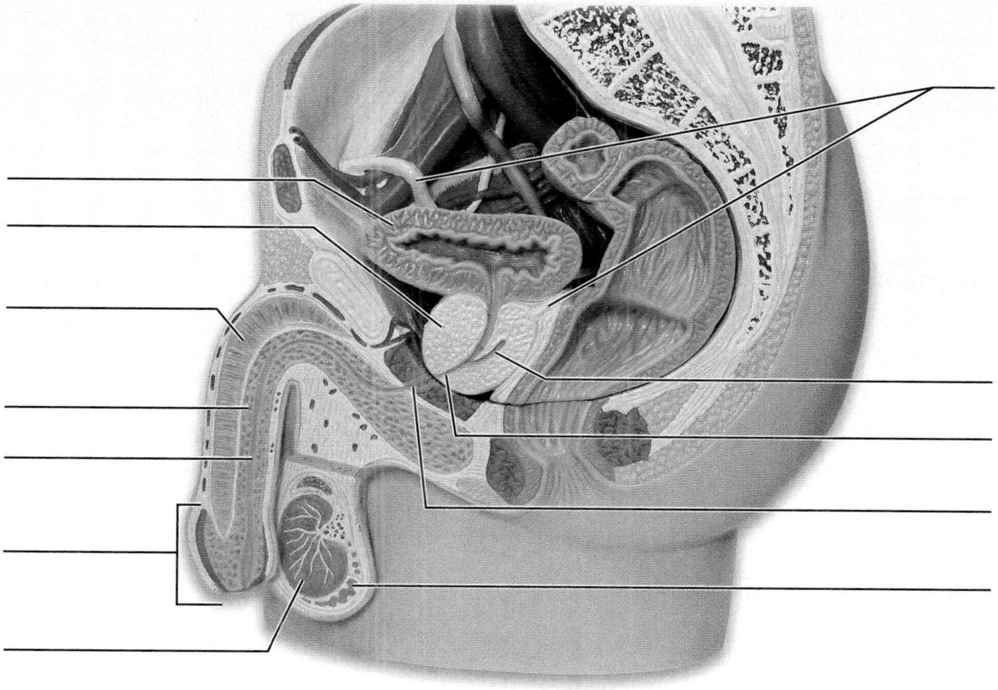

3. Why are the testes located in the scrotum rather than inside the ventral body cavity? ______________________________

____

4. An enlarged prostate is associated with a number of disorders, including prostatitis and prostate cancer. Describe the diagnostic exam used to detect an enlarged prostate. ______________________________

____

5. Name the two ducts that merge to form the ejaculatory duct. ______________________________

____

6. Match the terms in column B to the descriptive statements in column A.

Column A

_______________ 1. copulatory organ/penetrating device

_______________ 2. muscular passageway conveying sperm to the ejaculatory duct; in the spermatic cord

_______________ 3. distal urethra that transports both sperm and urine

_______________ 4. sperm maturation site

_______________ 5. location of the testis in adult males

_______________ 6. loose fold of skin encircling the glans penis

_______________ 7. portion of the urethra that is located in the urogenital diaphragm

_______________ 8. accessory gland that secretes fluid to cleanse the urethra prior to ejaculation

_______________ 9. accessory gland that secretes the largest contribution to semen

Column B

a. bulbo-urethral glands
b. ductus (vas) deferens
c. epididymis
d. intermediate part of the urethra
e. penis
f. prepuce
g. scrotum
h. seminal gland
i. spongy urethra

7. Describe the composition of semen, and name all structures contributing to its formation. _______________

____

____

8. Of what importance is the fact that seminal fluid is alkaline? _______________

____

9. What structures compose the spermatic cord? _______________

____

Where is it located? _______________

10. Using the following terms, trace the pathway of sperm from the testes to the urethra: rete testis, epididymis, seminiferous tubule, ductus deferens.

_______________ → _______________ → _______________ → _______________

Gross Anatomy of the Human Female Reproductive System

11. Name the structures composing the external genitalia, or vulva, of the female. _______________

____

____

12. On the photo of the model of the female reproductive system below, identify all indicated structures.

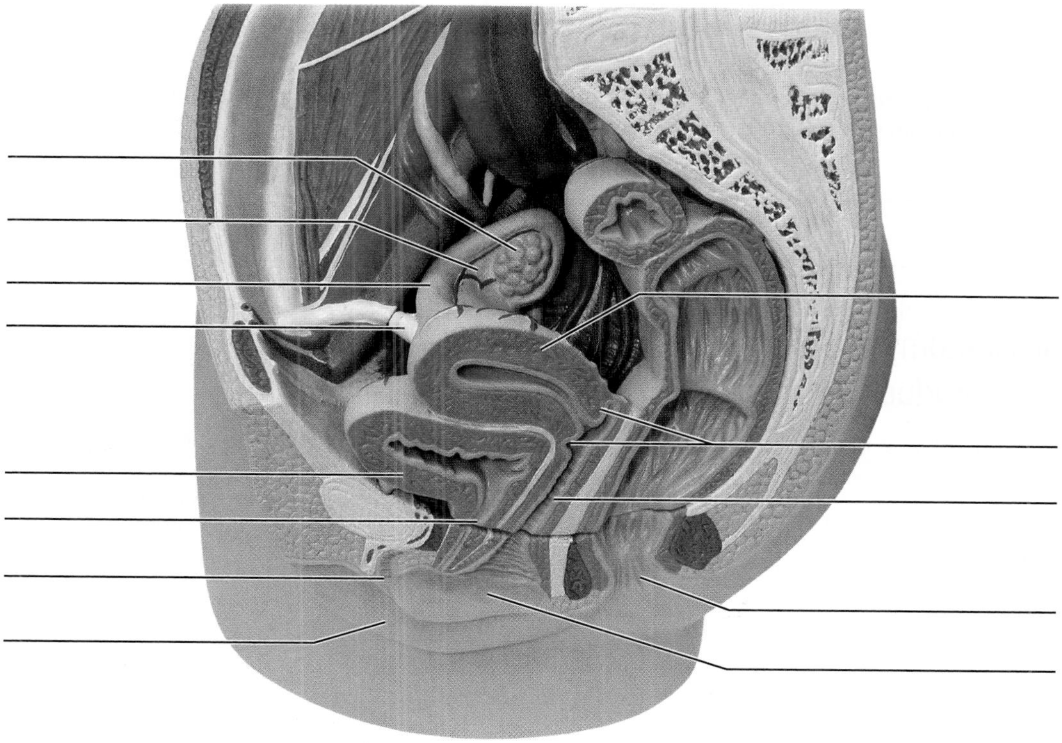

13. Identify the female reproductive system structures described below.

________________ 1. site of fetal development

________________ 2. copulatory canal

________________ 3. egg typically fertilized here

________________ 4. becomes erect during sexual excitement

________________ 5. glands homologous to the bulbo-urethral glands of males

________________ 6. partially closes the vaginal opening; a membrane

________________ 7. produces oocytes, estrogens, and progesterone

________________ 8. fingerlike ends of the uterine tube

14. Do any sperm enter the pelvic cavity of the female? Why or why not? ________________

____

15. What is an ectopic pregnancy, and how can it happen? ________________

____

____

16. Put the following vestibular-perineal structures in their proper order from the anterior to the posterior aspect: vaginal orifice, anus, external urethral opening, and clitoris.

Anterior limit: ______________ → ______________ → ______________ → ______________

17. Assume that a couple has just consummated the sex act and that the sperm have been deposited in the vagina. Trace the pathway of the sperm through the female reproductive tract.

____

18. Define *ovulation*. ____

Microscopic Anatomy of Selected Male and Female Reproductive Organs

19. The testis is divided into a number of lobes by connective tissue. Each of these lobes contains one to four ______________

______________________, which converge to empty sperm into another set of tubules called the

____.

20. What is the function of the cavernous bodies in the penis? ____

____

21. Name the three layers of the uterine wall from the inside out.

______________, ______________, ______________

Which of these is sloughed during menses? ____

Which contracts during childbirth? ____

22. Describe the epithelium found in the uterine tube. ____

____

23. Describe the arrangement of the layers of smooth muscle in the seminal gland. ____

____

24. What is the function of the stereocilia exhibited by the epithelial cells of the mucosa of the epididymis? ______________

____

25. On the diagram showing the sagittal section of the human testis, correctly identify all structures provided with leader lines.

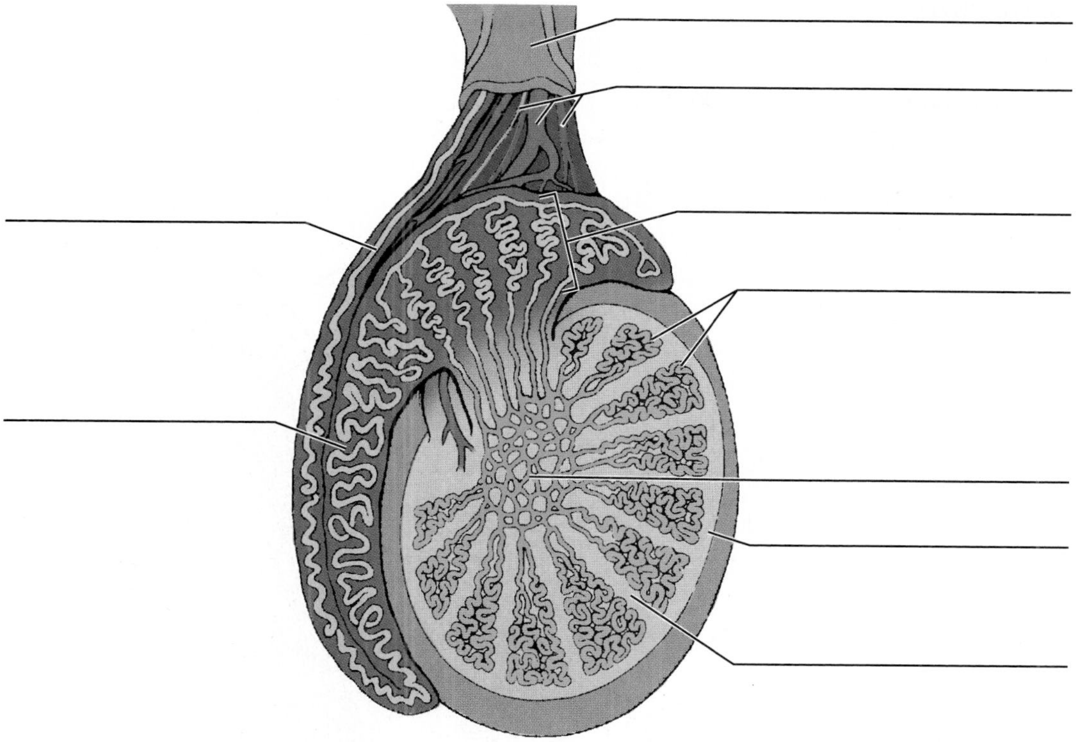

The Mammary Glands

26. Match the key term with the correct description.

________	glands that produce milk during lactation	a.	alveoli
________	subdivision of mammary lobes that contains alveoli	b.	areola
________	enlarged storage chamber for milk	c.	lactiferous duct
________	duct connecting alveoli to the storage chambers	d.	lactiferous sinus
________	pigmented area surrounding the nipple	e.	lobule
________	releases milk to the outside	f.	nipple

Key: a–f as listed above.

27. Using the key terms, correctly identify breast structures.

Key: a. adipose tissue

b. areola

c. lactiferous duct

d. lactiferous sinus

e. lobule containing alveoli

f. nipple

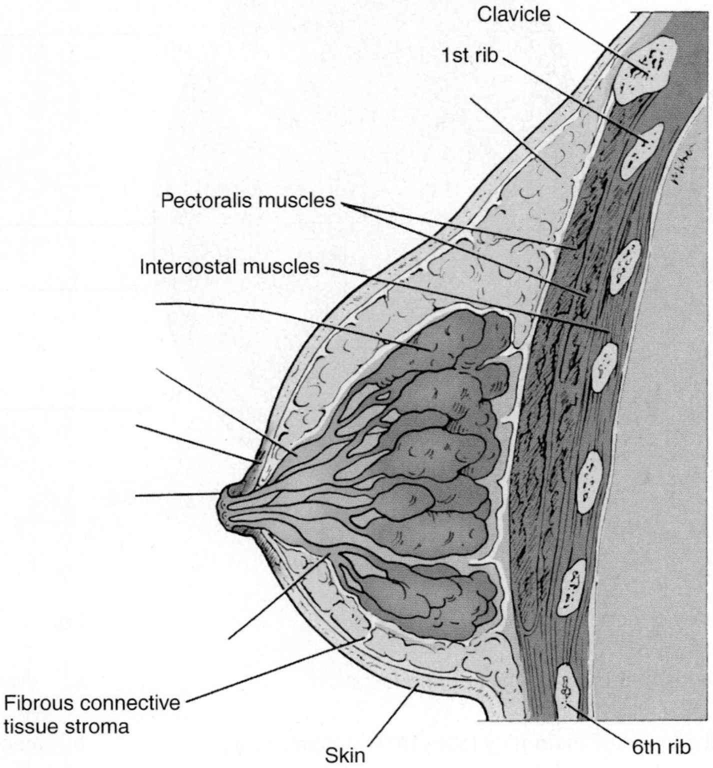

28. Cryptorchidism is failure of the testes to descend. Explain why this would cause sterility if not corrected. ______________________

____

____

29. Hysterectomy is a surgical removal of the uterus. It may or may not be accompanied by a salpingo-oophorectomy, removal of the uterine tubes and ovaries. Why would it be an advantage to leave the ovaries intact? ______________________

____

____

EXERCISE 43

Physiology of Reproduction: Gametogenesis and the Female Cycles

Learning Outcomes

- Define *meiosis*, *gametogenesis*, *oogenesis*, *spermatogenesis*, *synapsis*, *haploid*, *zygote*, and *diploid*.
- Cite similarities and differences between mitosis and meiosis.
- Describe the stages of spermatogenesis, and relate each to the cross-sectional structure of the seminiferous tubule.
- Define *spermiogenesis*, and relate the anatomy of sperm to their function.
- Describe the effects of FSH and LH on testicular function.
- Discuss the microscopic structure of the ovary; identify primary, secondary, and vesicular follicles, and the corpus luteum; list the hormones produced by the follicles and the corpus luteum.
- Relate the stages of oogenesis to follicle development in the ovary.
- Compare and contrast spermatogenesis and oogenesis.
- Discuss the effect of FSH and LH on the ovary, and describe the feedback relationship between anterior pituitary gonadotropins and ovarian hormones.
- List the phases of the menstrual cycle, and discuss the hormonal control of each.

Go to Mastering A&P™ > Study Area to improve your performance in A&P Lab.

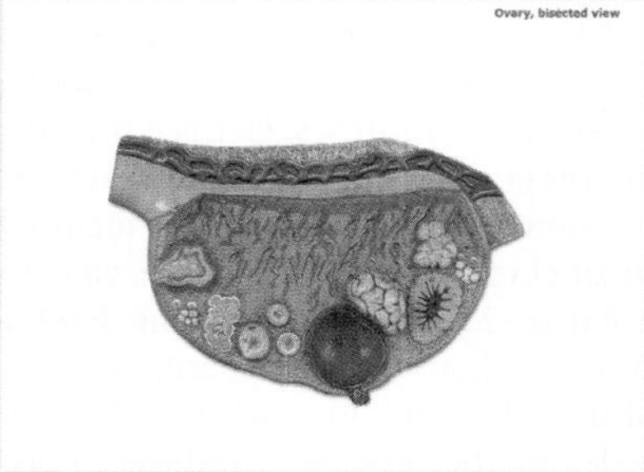

> Lab Tools > Practice Anatomy Lab > Anatomical Models

Instructors may assign new Building Vocabulary coaching activities, Pre-Lab Quiz questions, Art Labeling activities, Practice Anatomy Lab Practical questions (PAL), and more using the Mastering **A&P™** Item Library.

Pre-Lab Quiz

Instructors may assign these and other Pre-Lab Quiz questions using Mastering **A&P™**

1. Human gametes contain ___________ chromosomes.
   a. 13 **b.** 23 **c.** 36 **d.** 46
2. The end product of meiosis is:
   a. two diploid daughter cells
   b. two haploid daughter cells
   c. four diploid daughter cells
   d. four haploid daughter cells
3. ____________ extend inward from the periphery of the seminiferous tubule and provide nourishment to the spermatids as they begin their transformation into sperm.
   a. Interstitial endocrine cells
   b. Granulosa cells
   c. Sustentocytes
   d. Follicle cells
4. Circle the correct underlined term. The acrosome / midpiece of the sperm contains enzymes involved in the penetration of the egg.
5. Circle the correct underlined term. Within each ovary, the immature ovum develops in a saclike structure called a corpus / follicle.

Materials

- Three-dimensional models illustrating meiosis, spermatogenesis, and oogenesis
- Sets of "pop it" beads in two colors with magnetic centromeres, available in Chromosome Simulation Lab Activity from Ward's Natural Science
- Compound microscope
- Prepared slides of testis and human sperm
- *Demonstration*: microscopes set up to demonstrate the following stages of oogenesis in *Ascaris megalocephala*:

  Slide 1: Primary oocyte with fertilization membrane, sperm nucleus, and aligned tetrads apparent

  Slide 2: Formation of the first polar body

  Slide 3: Secondary oocyte with dyads aligned

  Slide 4: Formation of the ovum and second polar body

Text continues on next page. →

Slide 5: Fusion of the male and female pronuclei to form the fertilized egg

- Prepared slides of ovary and uterine endometrium (showing menstrual, proliferative, and secretory phases)

Human beings develop from the union of egg and sperm. Each of these gametes is a unique cell produced either in the ovary or testis. Unlike all other body cells, gametes have only half the normal chromosome number, and they are produced by a special type of nuclear division called meiosis.

Meiosis

The normal number of chromosomes in most human body cells is 46, the **diploid**, or **2*n***, chromosomal number. This number is made up of two sets of similar chromosomes, one set of 23 from each parent. Thus each body cell contains 23 pairs of similar chromosomes called **homologous chromosomes** or homologues (**Figure 43.1**). Each member of a homologous pair contains genes that code for the same traits.

Gametes contain only one member of each homologous pair of chromosomes. Therefore, each human gamete contains a total of 23 chromosomes, the **haploid**, or ***n***, chromosomal number. When egg and sperm fuse, they form a **zygote** that restores the diploid number of chromosomes. The zygote divides by the process of **mitosis** to produce the multicellular human body. Mitosis is the process by which most body cells divide. It produces two diploid daughter cells, each containing 46 chromosomes that are identical to those of the mother cell (see Exercise 4).

Gametogenesis is the process of gamete formation. It involves nuclear division by **meiosis**, which reduces the number of chromosomes by half. Before meiosis begins, the chromosomes in the *mother cells*, or stem cells, are replicated just as they are before mitosis. The identical copies remain together as *sister chromatids*. They are held together by a centromere, forming a structure called a **dyad** (Figure 43.1).

Two nuclear divisions, called meiosis I and meiosis II, occur during meiosis. Each has the same phases as mitosis—prophase, metaphase, anaphase, and telophase. Meiosis I is the *reduction division* of meiosis because the number of chromosomes is reduced from 2*n* to *n*. During prophase of meiosis I, the homologous chromosomes pair up in a process called **synapsis** that forms little groups of four chromatids, called **tetrads** (Figure 43.1). During synapsis, the free ends of adjacent maternal and paternal chromatids wrap around each other at one or more points, forming **chiasmata** (singular: **chiasma**). This event, **crossing over**, allows maternal and paternal chromosomes to exchange genetic material. The tetrads align randomly at the metaphase plate so that either the maternal or paternal chromosome may be on a given side of the plate. Then the two homologous chromosomes, each still composed of two sister chromatids, are pulled to opposite ends of the cell. At the end of meiosis I, each haploid daughter cell contains one member of each original homologous pair.

Meiosis II begins immediately without replication of the chromosomes. The dyads align on the metaphase plate, and the two sister chromatids are pulled apart, each now becoming a full chromosome. The net result of meiosis is four haploid daughter cells, each containing an equal share of chromosomes, 23 chromosomes. For this reason, meiosis II is sometimes called the **equational division of meiosis**. The events of crossing over and the random alignment of tetrads during meiosis I introduce great genetic variability. As a result, it is unlikely that any gamete is exactly like another.

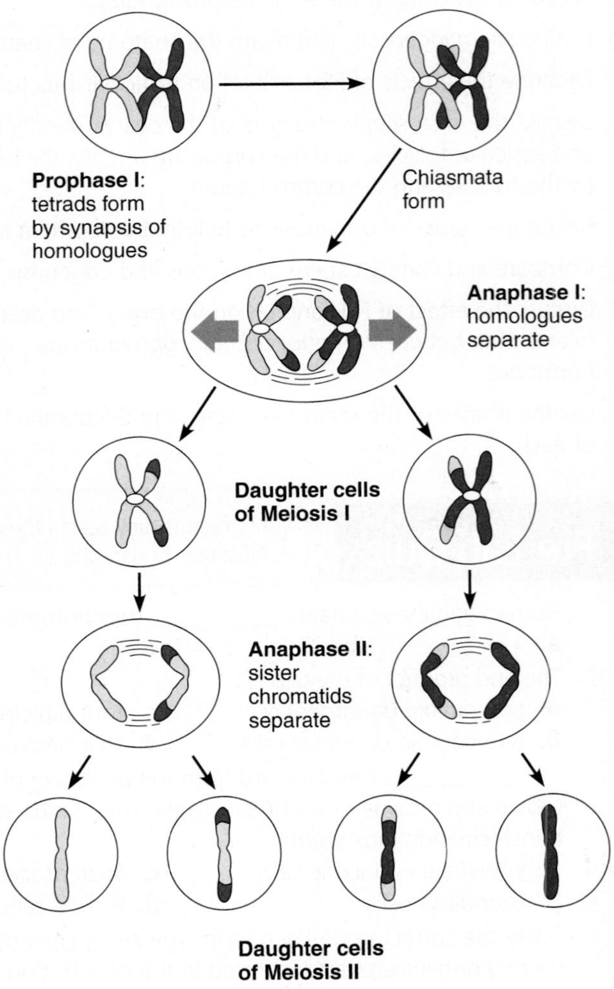

Figure 43.1 Summarized events of meiosis involving one pair of homologous chromosomes. Homologue from the male parent is purple; homologue from the female parent is pink.

Activity 1

Identifying Meiotic Phases and Structures

1. Obtain a model depicting the events of meiosis, and follow the sequence of events during meiosis I and II. Identify prophase, metaphase, anaphase, and telophase in each nuclear division. Also identify tetrads and chiasmata during meiosis I and dyads during meiosis II. Note ways in which the daughter cells resulting from meiosis I differ from the mother cell and how the gametes differ from both cell populations. Use the key on the model, your textbook, or an appropriate reference as necessary to aid you in these observations.

2. Using strings of colored "pop it" beads with magnetic centromeres, demonstrate the phases of meiosis, including crossing over, for a cell with a diploid (2*n*) number of 4. Use one bead color for the male chromosomes and another color for the female chromosomes.

3. Ask your instructor to verify the accuracy of your "creation" before returning the beads to the supply area.

Spermatogenesis

Human sperm production, or **spermatogenesis**, begins at puberty and continues without interruption throughout life. The average male ejaculate contains about a quarter billion sperm. Because only one sperm fertilizes an ovum, the perpetuation of the species will not be endangered by lack of sperm.

Spermatogenesis, the process of gametogenesis in males, occurs in the seminiferous tubules of the testes. The process of spermatogenesis is illustrated in **Figure 43.2**. The primitive stem cells, or **spermatogonia**, found at the tubule periphery, divide extensively to build up the stem cell line. Before puberty, all divisions are mitotic divisions that produce more spermatogonia. At puberty, however, under the influence of follicle-stimulating hormone (FSH) secreted by the anterior pituitary gland, each mitotic division of a spermatogonium produces one spermatogonium and one **primary spermatocyte**, which is destined to undergo meiosis. As meiosis occurs, the dividing cells approach the lumen of the tubule. Thus the progression of meiotic events can be followed from the tubule periphery to the lumen. It is important to recognize that **spermatids**, haploid cells that are the actual product of meiosis, are not functional gametes. They are nonmotile cells and have too much excess baggage to function well in a reproductive capacity. A subsequent process, called **spermiogenesis**, strips away the extraneous cytoplasm from the spermatid, converting it to a motile, streamlined **sperm**.

Instructors may assign this figure as an Art Labeling Activity using Mastering A&P™

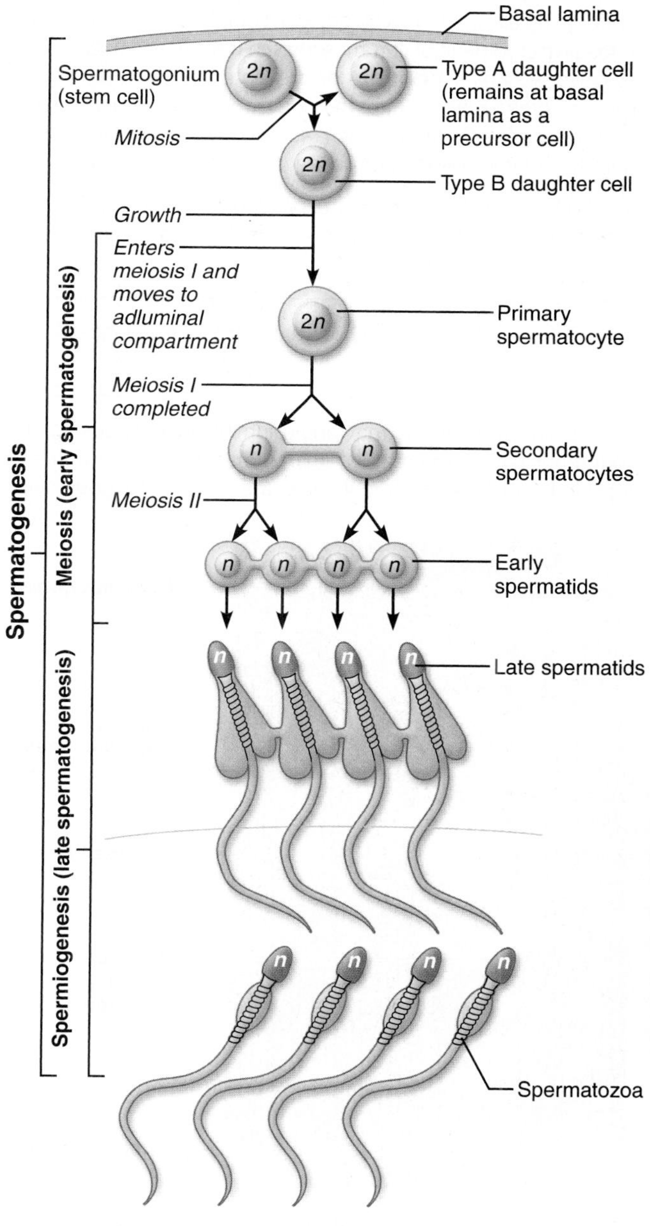

Figure 43.2 Spermatogenesis.

Activity 2

Examining Events of Spermatogenesis

1. Obtain a slide of the testis and a microscope. Examine the slide under low power to identify the cross-sectional views of the cut seminiferous tubules. Then rotate the high-power lens into position, and observe the wall of one of the cut tubules. **Figure 43.3** is a photomicrograph of a seminiferous tubule.

2. Scrutinize the cells at the periphery of the tubule. The cells in this area are the spermatogonia. About half of these will form primary spermatocytes, which begin meiosis. These are recognizable by their pale-staining nuclei with centrally located nucleoli. The remaining daughter cells resulting from mitotic divisions of spermatogonia stay at the tubule periphery to maintain the germ cell line.

3. Observe the cells in the middle of the tubule wall. There you should see a large number of spermatocytes that are obviously undergoing a nuclear division process. Look for coarse clumps of threadlike chromatin or chromosomes that have the appearance of coiled springs. Attempt to differentiate between the larger primary spermatocytes and the somewhat smaller secondary spermatocytes. Once formed, the secondary spermatocytes quickly undergo division and so are more difficult to find.

In which location would you expect to see cells containing tetrads, closer to the spermatogonia or closer to the lumen?

____

Would these cells be primary or secondary spermatocytes?

____

4. Examine the cells at the tubule lumen. Identify the small round-nucleated spermatids, many of which may appear lopsided and look as though they are starting to lose their cytoplasm. See if you can find a spermatid embedded in an elongated cell type—a **sustentocyte**, or *Sertoli cell*—which extends inward from the periphery of the tubule. The sustentocytes nourish the spermatids as they begin their transformation into sperm. Also in the adluminal area (area toward the lumen), locate immature sperm, which can be identified by their tails. The sperm develop directly from the spermatids by the loss of extraneous cytoplasm and the development of a propulsive tail.

5. Identify the **interstitial endocrine cells**, also called *Leydig cells*, lying external to and between the seminiferous tubules. Luteinizing hormone (LH) prompts these cells to produce testosterone, which acts synergistically with follicle-stimulating hormone (FSH) to stimulate sperm production. Both LH and FSH are named for their effects on the female gonad.

In the next stage of sperm development, spermiogenesis, all the superficial cytoplasm is sloughed off, and the remaining cell organelles are compacted into the three regions of the mature sperm. At the risk of oversimplifying, these anatomical regions are the *head*, the *midpiece*, and the *tail*, which correspond roughly to the activating and genetic region, the metabolic region (rich in mitochondria for ATP production), and the locomotor region (a typical flagellum powered by ATP), respectively. The mature sperm is a streamlined cell equipped with an organ of locomotion and a high rate of metabolism that enable it to move long distances quickly to get to the egg. It is a prime example of the correlation of form and function.

The pointed sperm head contains the DNA, or genetic material, of the chromosomes. Essentially it is the nucleus of the spermatid. Anterior to the nucleus is the **acrosome**, which contains enzymes necessary for penetration of the egg.

6. Obtain a prepared slide of human sperm, and view it with the oil immersion lens. Identify the head, acrosome, and tail regions of the sperm (**Figure 43.4**). Deformed sperm—for example, sperm with multiple heads or tails—are sometimes present in such preparations. Did you observe any?

______________ If so, describe them. ______________

____

7. Examine the model of spermatogenesis to identify the spermatogonia, the primary and secondary spermatocytes, the spermatids, and the functional sperm.

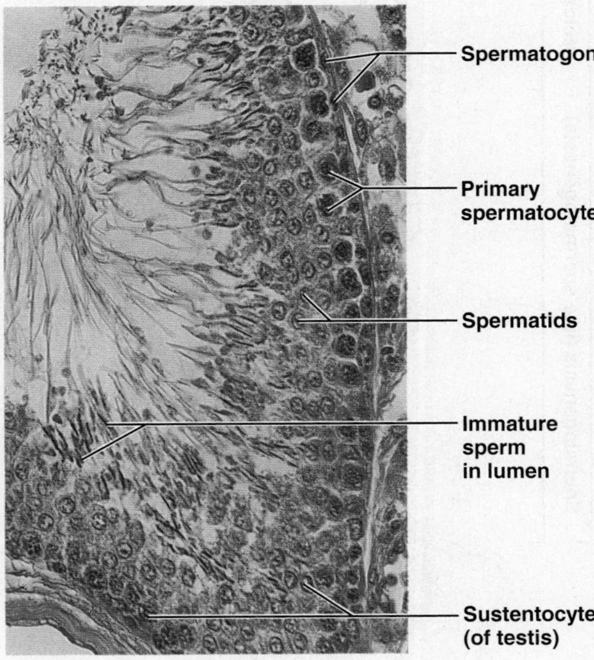

Figure 43.3 Micrograph of an active seminiferous tubule undergoing spermatogenesis (275×).

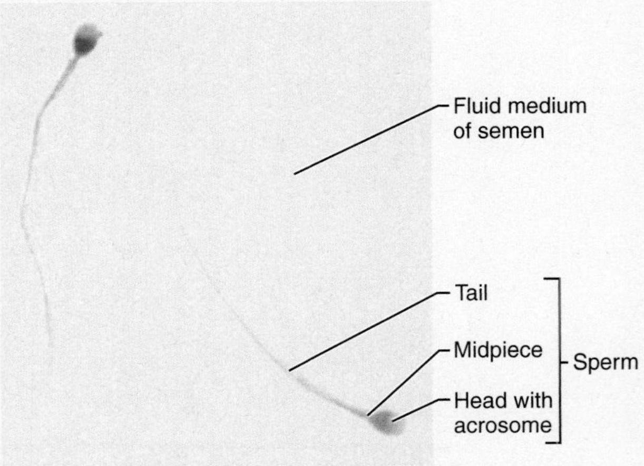

Figure 43.4 Sperm in semen. (1000×).

43

Demonstration of Oogenesis in *Ascaris* (Optional)

Oogenesis (the process of producing an egg) in mammals is difficult to demonstrate. However, the process may be studied rather easily in the transparent eggs of *Ascaris megalocephala*, an invertebrate roundworm parasite found in the intestine of mammals. Since its diploid chromosome number is 4, the chromosomes are easily counted.

Activity 3

Examining Meiotic Events Microscopically

Go to the demonstration area where the slides are set up, and make the following observations:

1. Scan the first demonstration slide to identify a *primary oocyte*, the cell type that begins the meiotic process. It will have what appears to be a relatively thick cell membrane; this is the *fertilization membrane* that the oocyte produces after sperm penetration. Find and study a primary oocyte that is undergoing meiosis I. Look for a barrel-shaped spindle with two tetrads (two groups of four beadlike chromosomes) in it. Most often the spindle is located at the periphery of the cell. The sperm nucleus may or may not be seen, depending on how the cell was cut.

2. Observe slide 2. Locate a cell in which half of each tetrad (a dyad) is being extruded from the cell surface into a smaller cell called the *first polar body*.

3. On slide 3, attempt to locate a *secondary oocyte* (a daughter cell produced during meiosis I) undergoing meiosis II.

   In this view, you should see two dyads (each with two beadlike chromosomes) on the spindle.

4. On slide 4, locate a cell in which the *second polar body* is being formed. In this case, both it and the ovum will now contain two chromosomes, the haploid number for *Ascaris*.

5. On slide 5, identify a *fertilized egg*, or a cell in which the sperm and ovum nuclei (actually *pronuclei*) are fusing to form a single nucleus containing four chromosomes.

Human Oogenesis and the Ovarian Cycle

Once the adult ovarian cycle is established, gonadotropic hormones produced by the anterior pituitary influence the development of ova in the ovaries and their cyclic production of female sex hormones. Within an ovary, each immature ovum develops within a saclike structure called a *follicle*, where it is encased by one or more layers of smaller cells. The surrounding cells are called **pre-granulosa cells** if one layer is present and **granulosa cells** when more than one layer is present.

The process of **oogenesis**, or female gamete formation, which occurs in the ovary, is similar to spermatogenesis occurring in the testis, but there are some important differences. Oogenesis begins with primitive stem cells called **oogonia**, located in the cortex of the ovaries of the developing female fetus (**Figure 43.5**, p. 646). During fetal development, the oogonia undergo mitosis thousands of times until their number reaches 2 million or more. They then become encapsulated by a single layer of squamouslike pre-granulosa cells and form the **primordial follicles** of the ovary. By the time the female child is born, most of her oogonia have increased in size and have become **primary oocytes**, which are in the prophase stage of meiosis I. Thus at birth, the female is presumed to have her lifetime supply of primary oocytes.

From birth until puberty, the primary oocytes are quiescent. Then, under the influence of FSH, one or sometimes more of the follicles begin to undergo maturation approximately every 28 days. The stages of maturation that follicles undergo are described in **Table 43.1** on p. 647 and illustrated in Figure 43.5 and **Figure 43.6** on p. 647.

In the female, meiosis produces only one functional gamete, in contrast to the four produced in the male. Another major difference is in the relative size and structure of the functional gametes. Sperm are tiny and equipped with tails for locomotion. They have few organelles and virtually no nutrient-containing cytoplasm; hence the nutrients contained in semen are essential to their survival. In contrast, the egg is a relatively large nonmotile cell, well stocked with cytoplasmic reserves that nourish the developing embryo until implantation can be accomplished. Essentially all the zygote's organelles are provided by the egg.

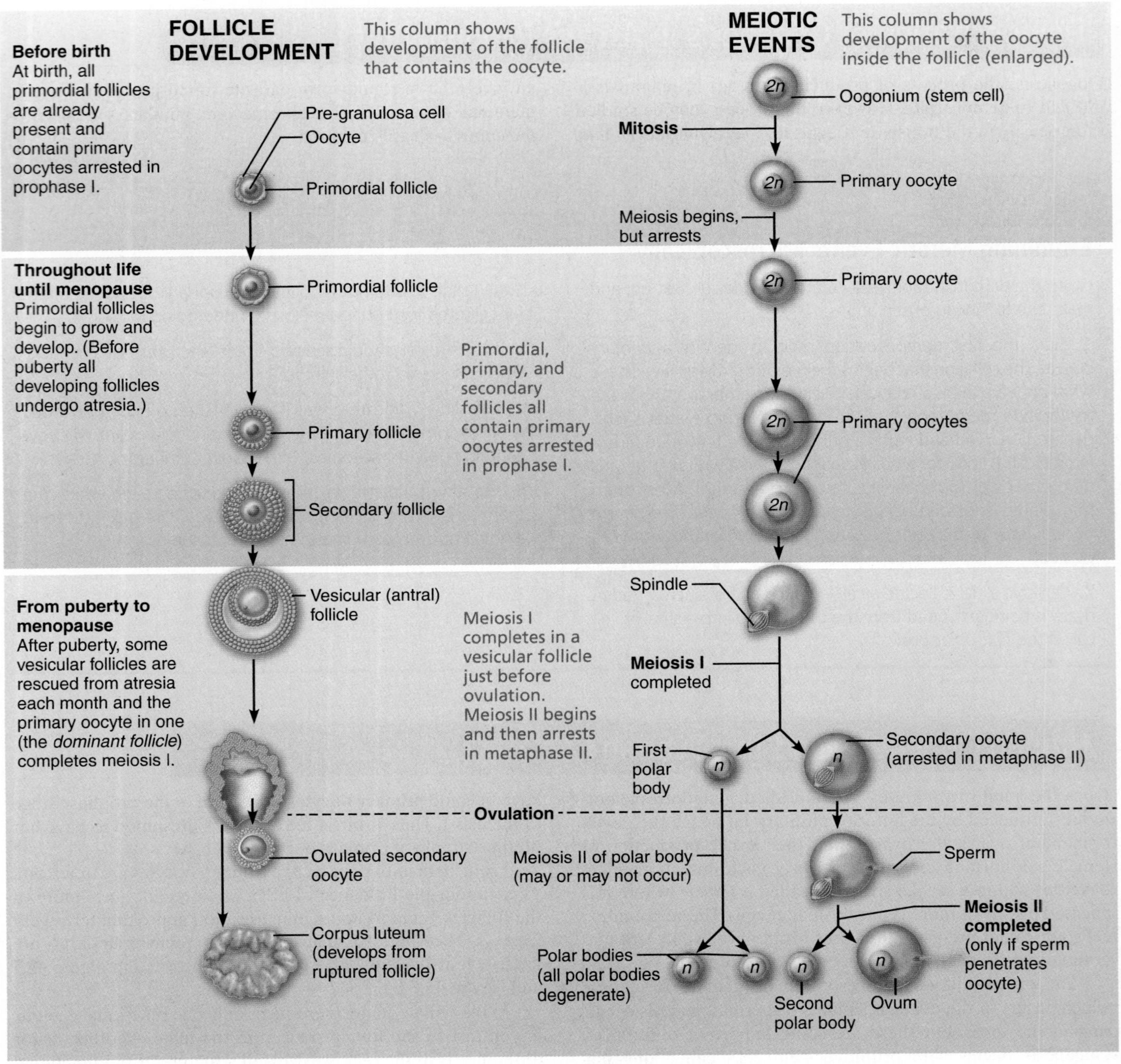

Figure 43.5 Events of oogenesis.

43

Activity 4

Examining Oogenesis in the Ovary

Because many different stages of ovarian development exist within the ovary at any one time, a single microscopic preparation will contain follicles at many different stages of development (Figure 43.6). Obtain a cross section of ovary tissue, and identify the following structures: germinal epithelium, primary follicle, late secondary follicle, vesicular follicle, and corpus luteum. Use Table 43.1 and Figure 43.6 to guide your observations.

Activity 5

Comparing and Contrasting Oogenesis and Spermatogenesis

Examine the model of oogenesis, and compare it with the spermatogenesis model. Note differences in the number, size, and structure of the functional gametes.

Table 43.1 Maturation Stages of Ovarian Follicles (Figures 43.5 and 43.6)

Follicle stage	Description
The follicles below (primordial through late secondary follicle) are primary oocytes arrested in prophase I.	
Primordial follicle	A primary oocyte surrounded by a single layer of squamous *pre-granulosa* cells. They are produced in the fetus and are present at birth.
Primary follicle	This stage begins at puberty. The squamous pre-granulosa cells become cuboidal and surround the primary oocyte.
Secondary follicle	The pre-granulosa cells divide and form a stratified cuboidal epithelium around the oocyte. The pre-granulosa cells are now called *granulosa cells*.
Late secondary follicle	The oocyte secretes a glycoprotein to form its extracellular membrane, the zona pellucida. This thick membrane is now surrounded by a layer of connective tissue and epithelial cells called the *theca folliculi*. Clear liquid has accumulated in between the granulosa cells.
The follicles below (vesicular and the ruptured follicle) are secondary oocytes arrested in metaphase II.	
Vesicular (antral) follicle	The liquid continues to accumulate between the granulosa cells and forms a large fluid-filled cavity called an *antrum*. There are now several layers of granulosa cells called the *corona radiata*. The primary oocyte completes the first meiotic division to produce a *secondary oocyte* and a *polar body*.
Ruptured follicle	The follicle ruptures, and the secondary oocyte is released with its corona radiata to begin its journey down the uterine tube. Note that meiosis II will be completed to produce an ovum only if fertilization occurs. If no fertilization occurs, the secondary oocyte degenerates.
The structures below (corpus luteum and corpus albicans) are remnants of the ruptured follicle.	
Corpus luteum	The "yellow body" formed by granulosa cells from the ruptured follicle. It secretes the steroid hormones estrogen and progesterone.
Corpus albicans	The "white body" that represents the further deterioration of the corpus luteum.

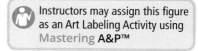

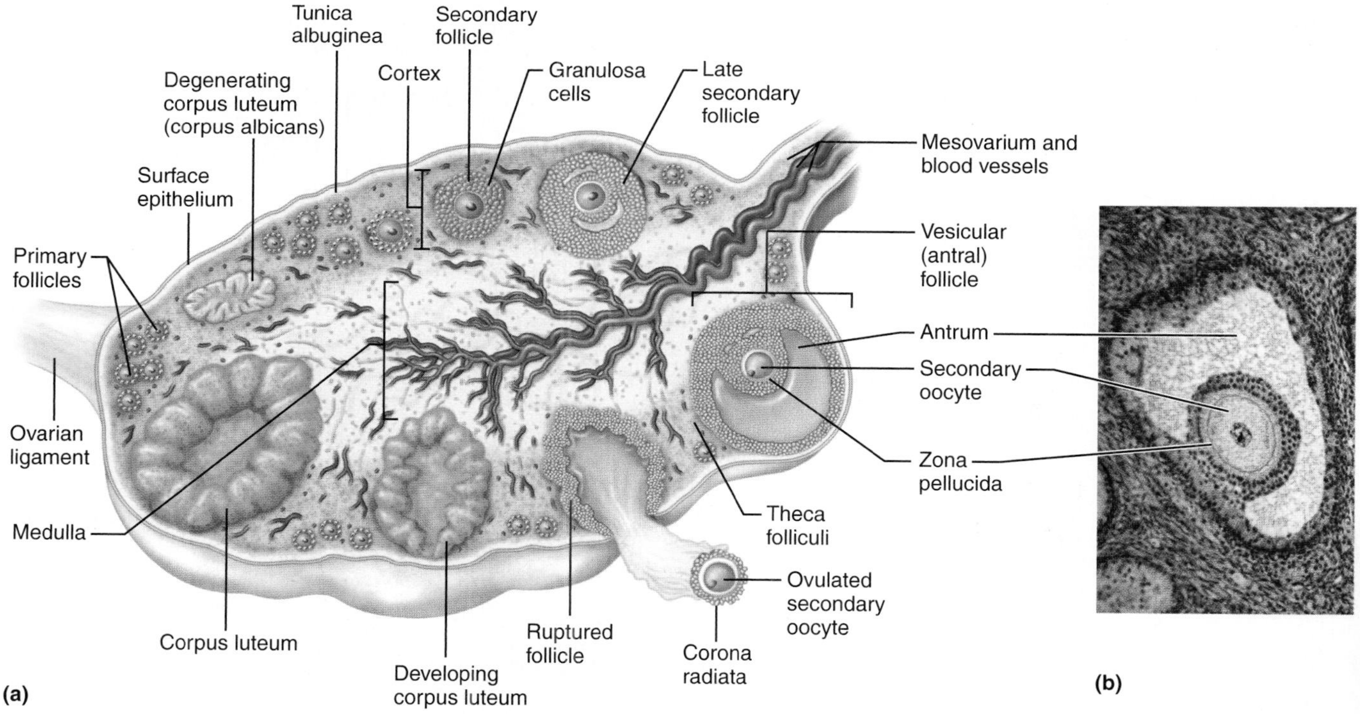

Figure 43.6 Anatomy of the human ovary. (a) The ovary has been sectioned to reveal the follicles in its interior. Note that not all the structures would appear in the ovary at the same time. **(b)** Photomicrograph of a vesicular (antral) follicle (75×).

Instructors may assign this figure as an Art Labeling Activity using Mastering A&P™

The Menstrual Cycle

The **uterine cycle**, or **menstrual cycle**, is hormonally controlled by estrogens and progesterone secreted by the ovary. It is normally divided into three stages: menstrual, proliferative, and secretory. Notice how the endometrial changes (**Figure 43.7**) correlate with hormonal and ovarian changes (**Figure 43.8**).

If fertilization has occurred, the embryo will produce a hormone much like LH, which will maintain the function of the corpus luteum. Otherwise, as the corpus luteum begins to deteriorate, lack of ovarian hormones in the blood causes blood vessels supplying the endometrium to kink and become spastic, setting the stage for menstruation to begin by the 28th day.

Although 28 days is a common length for the menstrual cycle (Figure 43.8d), its length is highly variable, sometimes as short as 21 days or as long as 38.

Activity 6

Observing Histological Changes in the Endometrium During the Menstrual Cycle

Obtain slides showing the menstrual, secretory, and proliferative phases of the endometrium of the uterus. Observe each carefully, comparing their relative thicknesses and vascularity. As you work, refer to the corresponding photomicrographs (Figure 43.7).

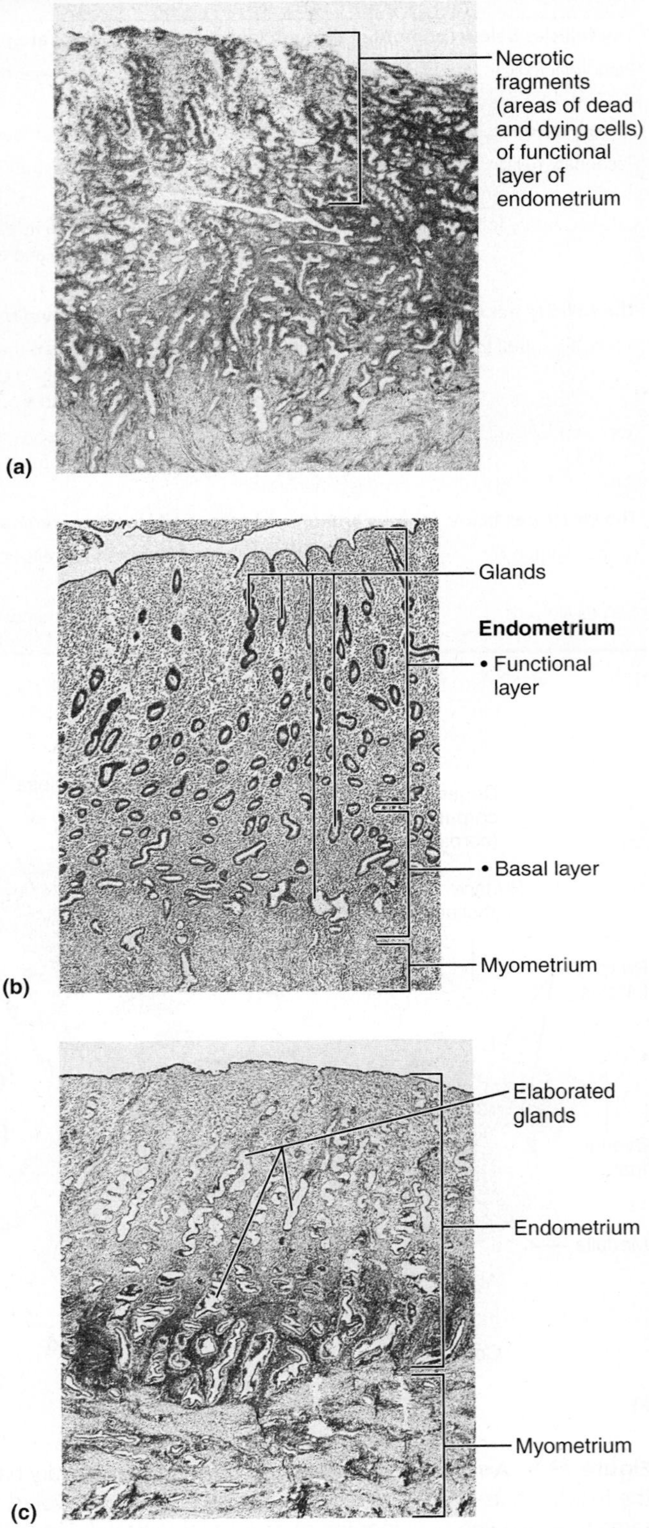

Figure 43.7 Endometrial changes during the menstrual cycle. (a) Onset of menstruation (8×). **(b)** Early proliferative phase (10×). **(c)** Early secretory phase (8×).

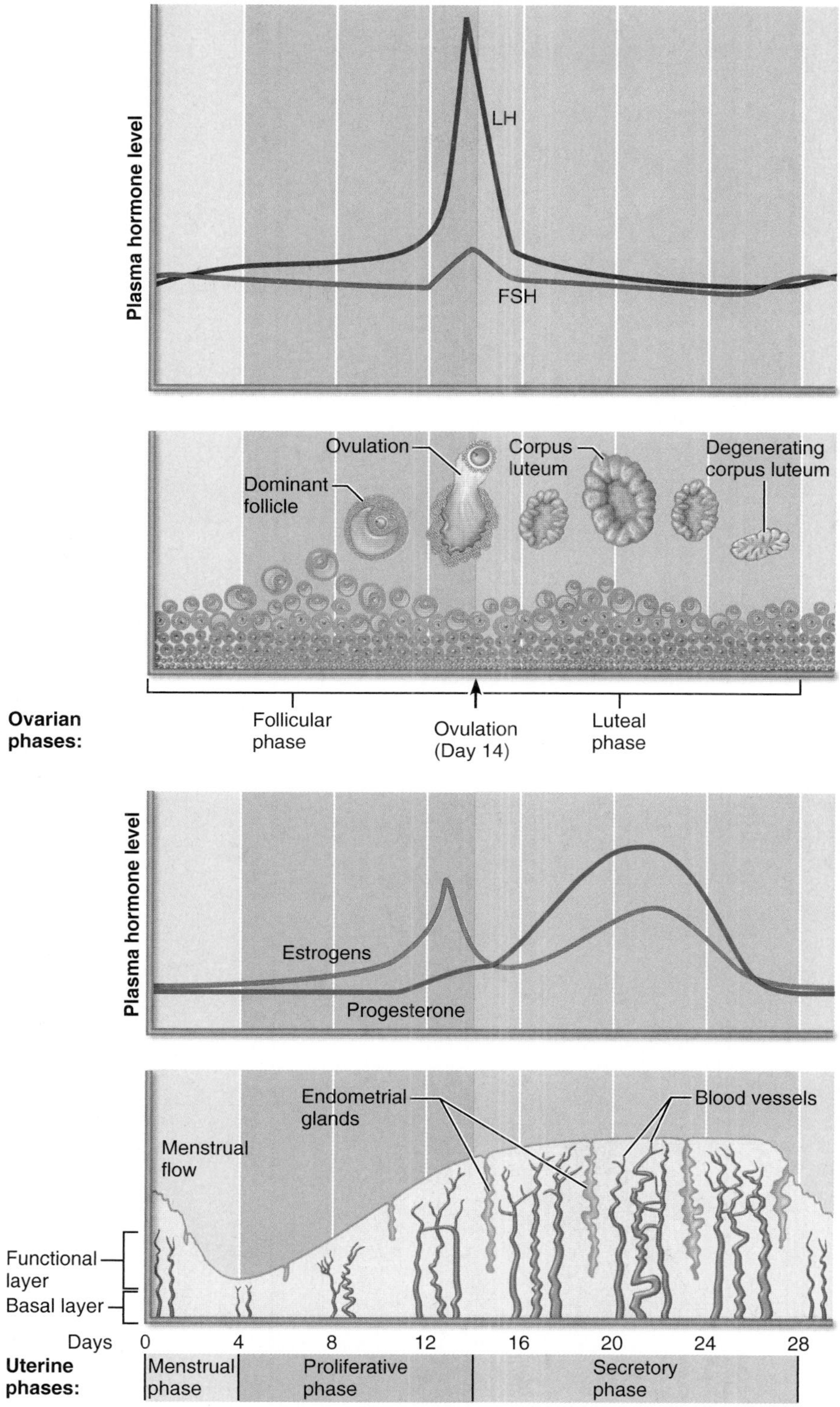

(a) Fluctuation of gonadotropin levels: Fluctuating levels of pituitary gonadotropins (follicle-stimulating hormone and luteinizing hormone) in the blood regulate the events of the ovarian cycle.

(b) Ovarian cycle: Structural changes in vesicular ovarian follicles and the corpus luteum are correlated with changes in the endometrium of the uterus during the uterine cycle (d). Recall that only vesicular follicles (in their antral phase) are hormone dependent. Primary and secondary follicles are not hormone dependent and develop in waves.

(c) Fluctuation of ovarian hormone levels: Fluctuating levels of ovarian hormones (estrogens and progesterone) cause the endometrial changes of the uterine cycle. The high estrogen levels are also responsible for the LH/FSH surge shown in (a).

(d) The three phases of the uterine cycle:
- Menstrual: The functional layer of the endometrium is shed.
- Proliferative: The functional layer of the endometrium is rebuilt.
- Secretory: Begins immediately after ovulation. Enrichment of the blood supply and glandular secretion of nutrients prepare the endometrium to receive an embryo.

Both the menstrual and proliferative phases occur before ovulation, and together they correspond to the follicular phase of the ovarian cycle. The secretory phase corresponds in time to the luteal phase of the ovarian cycle.

43

Figure 43.8 Correlation of anterior pituitary and ovarian hormones with structural changes in the ovary and uterus. The time bar applies to all parts of the figure. Note that the end of day 28 of the current cycle marks the beginning (0) of the first day of the next cycle.

REVIEW SHEET

EXERCISE 43

Instructors may assign a portion of the Review Sheet questions using Mastering A&P™

Physiology of Reproduction: Gametogenesis and the Female Cycles

Name ______________________ Lab Time/Date ______________________

Meiosis

1. The following statements refer to events occurring during mitosis and/or meiosis. For each statement, decide whether the event occurs in (a) mitosis only, (b) meiosis only, or (c) both mitosis and meiosis.

__________ 1. dyads are visible

__________ 2. tetrads are visible

__________ 3. product is two diploid daughter cells genetically identical to the mother cell

__________ 4. product is four haploid daughter cells quantitatively and qualitatively different from the mother cell

__________ 5. involves the phases prophase, metaphase, anaphase, and telophase

__________ 6. occurs throughout the body

__________ 7. occurs only in the ovaries and testes

__________ 8. provides cells for growth and repair

__________ 9. homologues synapse; chiasmata are seen

__________ 10. chromosomes are replicated before the division process begins

__________ 11. provides cells for production of offspring

__________ 12. consists of two consecutive nuclear divisions, without chromosomal replication occurring before the second division

2. Describe the process of synapsis. ______________________

____

3. How does crossing over introduce variability in the daughter cells? ______________________

____

4. Define *homologous chromosomes*. ______________________

____

Spermatogenesis

5. The cell types seen in the seminiferous tubules are listed in the key. Match the correct cell type or types with the descriptions given below.

Key: a. primary spermatocyte c. spermatogonium e. spermatid
b. secondary spermatocyte d. sustentocyte f. sperm

__________ 1. primitive stem cell

__________ 2. haploid (3 responses)

__________ 3. provides nutrients to developing sperm

__________ 4. product of meiosis II

__________ 5. product of spermiogenesis

__________ 6. product of meiosis I

6. Why are spermatids not considered functional gametes? __________

7. Differentiate *spermatogenesis* from *spermiogenesis*. __________

Oogenesis, the Ovarian Cycle, and the Menstrual Cycle

8. The sequence of events leading to gamete formation in the female begins during fetal development. By the time the child is born, all viable oogonia have been converted to __________.

In view of this fact, how does the total gamete potential of the female compare to that of the male?

9. The female gametes develop in structures called *follicles*. Describe the structure of a primordial follicle. __________

10. How are primary and vesicular follicles anatomically different? __________

11. What is a corpus luteum? __________

12. What are the hormones produced by the corpus luteum? __________

13. Use the key to identify the cell type you would expect to find in the following structures. The items in the key may be used once, more than once, or not at all.

Key: a. oogonium b. primary oocyte c. secondary oocyte d. ovum

_______ 1. forming part of the primary follicle in the ovary

_______ 2. in the uterine tube before fertilization

_______ 3. in the vesicular follicle of the ovary

_______ 4. in the uterine tube shortly after fertilization

_______ 5. primitive stem cell

14. In the following illustration of an ovary, label each leader line with the letter for the correct term from the key below.

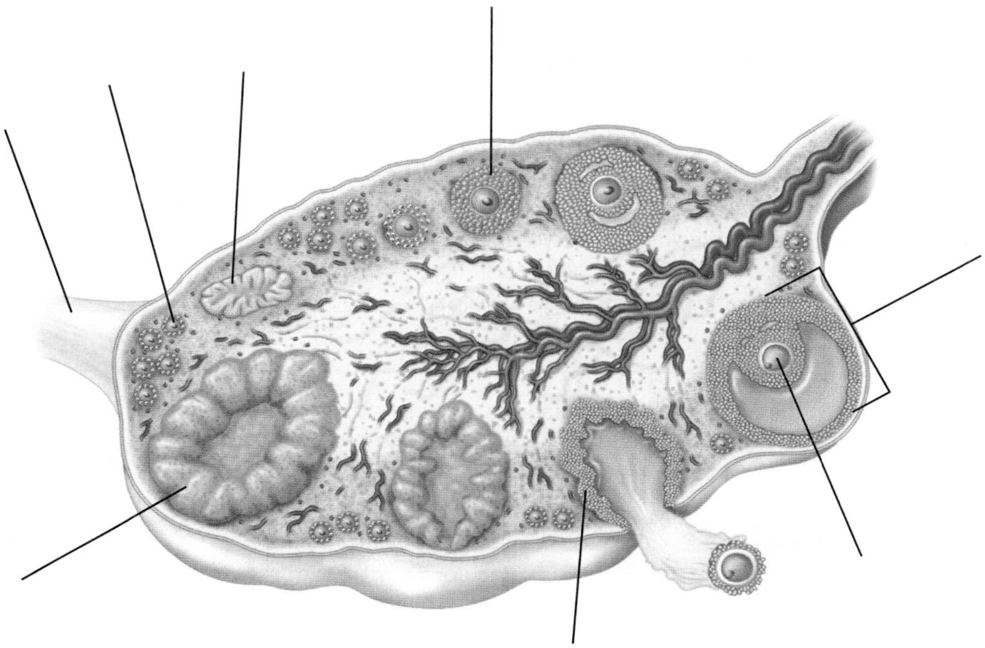

Key:

a. corpus albicans

b. corpus luteum

c. ovarian ligament

d. primary follicle

e. ruptured follicle

f. secondary follicle

g. secondary oocyte

h. vesicular (antral) follicle

(a)

15. The cellular products of spermatogenesis are four ________________; the final products of oogenesis are one ________________ and three ________________. What is the function of this unequal cytoplasmic division seen during oogenesis in the female? ________________________________

____

____

What is the fate of the polar bodies produced during oogenesis? ________________________________

Why? ________________________________

16. For each statement below dealing with plasma hormone levels during the female ovarian and menstrual cycles, decide whether the condition in column A is usually (a) greater than or (b) less than the condition in column B.

	Column A	**Column B**
________	1. amount of LH in the blood during menstruation	amount of LH in the blood at ovulation
________	2. amount of FSH in the blood on day 6 of the cycle	amount of FSH in the blood on day 20 of the cycle
________	3. amount of estrogen in the blood during menstruation	amount of estrogen in the blood at ovulation
________	4. amount of progesterone in the blood on day 14	amount of progesterone in the blood on day 23
________	5. amount of estrogen in the blood on day 10	amount of progesterone in the blood on day 10

17. What uterine tissue undergoes dramatic changes during the menstrual cycle? ________________________________

18. When during the female menstrual cycle would fertilization be most likely? Explain why. ________________________________

____

19. The menstrual cycle depends on events within the female ovary. The stages of the menstrual cycle are listed below. For each, note its approximate time span and the related events in the uterus; and then to the right, record the ovarian events occurring simultaneously. Pay particular attention to hormonal events.

Menstrual cycle stage	Uterine events	Ovarian events
Menstruation Days 1–5		
Proliferative Days 6–14		
Secretory Days 15–28		

20. Endometriosis occurs when fragments of endometrial tissue undergo retrograde (moving backward) menstruation, resulting in displaced tissue that often attaches to the peritoneum of the pelvic cavity. These fragments respond to hormonal changes, resulting in bouts of severe pain even after menstruation has ended. Explain how the female anatomy contributes to the ability of this tissue to relocate and attach to the peritoneum. ________________________________

____

21. Natural family planning, or fertility awareness, is a method that can be used to achieve or prevent pregnancy. It is based on the ability to predict ovulation. Measuring which hormone would be the best predictor for ovulation, and why?

____

____

EXERCISE 44 Survey of Embryonic Development

Learning Outcomes

- Define *fertilization* and *zygote*.
- Define and discuss the function of *cleavage* and *gastrulation*.
- Differentiate between the blastula and gastrula forms of the sea urchin and human using appropriate models or diagrams.
- Define *blastocyst*.
- Identify the following structures: embryoblast, trophoblast, chorionic villi, amnion, yolk sac, and allantois, and state the function of each.
- Describe the process and timing of implantation in the human.
- Define *decidua basalis* and *decidua capsularis*.
- Name the three primary germ layers, and list the organs or organ systems that arise from each.
- Describe the gross anatomy and general function of the human placenta.

Go to Mastering A&P™ > Study Area to improve your performance in A&P Lab.

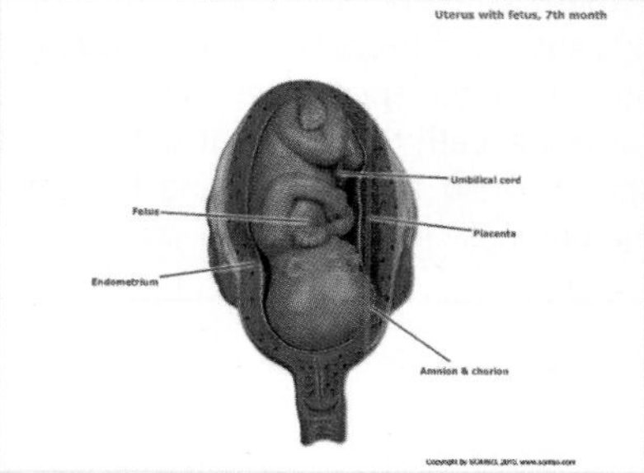

> Lab Tools > Practice Anatomy Lab > Anatomical Models

Instructors may assign new Building Vocabulary coaching activities, Pre-Lab Quiz questions, Art Labeling activities, Practice Anatomy Lab Practical questions (PAL), and more using the Mastering A&P™ Item Library.

Pre-Lab Quiz

Instructors may assign these and other Pre-Lab Quiz questions using Mastering A&P™

1. Circle the correct underlined term. The fertilized egg, or zygote / embryo, appears as a single cell surrounded by a fertilization membrane and a jellylike membrane.
2. Circle the correct underlined term. As a result of gastrulation, a three- / four-layered embryo forms, with each layer corresponding to a primary germ layer.
3. The ___________ implants in the uterine wall.
   a. zygote **b.** morula **c.** blastocyst **d.** gastrula
4. The ___________ gives rise to the epidermis of the skin and the nervous system.
   a. ectoderm **b.** endoderm **c.** mesoderm
5. By week 9 of development, the embryo is referred to as a:
   a. blastocyst **b.** blastomere **c.** fetus **d.** gastrula

Materials

- Prepared slides of sea urchin development (zygote through larval stages)
- Compound microscope
- Three-dimensional human development models or plaques (if available)
- *Demonstration*: Phases of human development in *Life Before Birth: Normal Fetal Development*, Second Edition (by Marjorie A. England, 1996, Mosby-Wolfe)
- Disposable gloves
- *Demonstration*: Pregnant cat, rat, or pig uterus with uterine wall dissected
- Three-dimensional model of pregnant human torso
- *Demonstration*: Fresh or formalin-preserved human placenta
- Prepared slide of placenta tissue

Early development in all animals involves three basic types of activities, which are integrated to ensure the formation of a viable offspring: (1) an increase in cell number and subsequent cell growth; (2) cellular specialization; and (3) morphogenesis, the formation of functioning organ systems. This exercise first provides a rather broad overview of the changes in structure that take place during embryonic development in sea urchins. The pattern of changes in this marine animal provides a basis of comparison with developmental events in the human.

Developmental Stages of Sea Urchins and Humans

Activity 1

Microscopic Study of Sea Urchin Development

1. Obtain a compound microscope and a set of slides depicting embryonic development of the sea urchin.

2. Observe the fertilized egg, or **zygote**, which appears as a single cell immediately surrounded by a jellylike membrane and a fertilization membrane. After an egg is penetrated by a sperm, the egg and the sperm nuclei fuse to form a single nucleus. This process is called **fertilization**. Within 2 to 5 minutes after sperm penetration, a fertilization membrane forms beneath the jelly coat to prevent the entry of additional sperm. Draw the zygote, and label the fertilization and jelly membranes.

Zygote

3. Observe the cleavage stages. Once fertilization has occurred, the zygote begins to divide, forming a mass of successively smaller and smaller cells, called **blastomeres**. This series of mitotic divisions without intervening growth periods is referred to as **cleavage**, and it results in a multicellular embryonic body. The cleavage stage of embryonic development provides a large number of building blocks (cells) with which to fashion the forming body.

As the division process continues, a solid ball of cells forms. At the 32-cell stage, it is called the **morula**, and the embryo resembles a berry in form. Then the cell mass hollows out to become the embryonic form called the **blastula**, which is a ball of cells surrounding a central cavity. The blastula is the final product of cleavage.

Identify and sketch the blastula stage of cleavage—a ball of cells with an apparently lighter center due to the presence of the central cavity.

Blastula

4. Identify the **early gastrula** form, which follows the blastula in the developmental sequence. The gastrula looks as though one end of the blastula had been indented or pushed into the central cavity, forming a two-layered embryo. In time, a third layer of cells appears between the initial two cell layers. Thus, as a result of **gastrulation**, a three-layered embryo forms, each layer corresponding to a **primary germ layer** from which all body tissues develop. The innermost layer, the **endoderm**, and the middle layer, the **mesoderm**, form the internal organs; the outermost layer, the **ectoderm**, forms the surface tissues of the body.

Draw a gastrula below. Label the ectoderm and endoderm. If you can see the third layer of cells, the mesoderm, budding off between the other two layers, label that also.

Gastrula

5. Gastrulation in the sea urchin is followed by the appearance of the free-swimming larval form, in which the three germ layers have differentiated into the various tissues and organs of the animal's body. If time allows, observe the larval form on the prepared slides.

Activity 2

Examining the Stages of Human Development

Use **Figure 44.1** and **Figure 44.2** on p. 658 and models or plaques of human development to identify the various stages of development and to respond to the questions posed below.

1. Observe the fertilized egg, or zygote, which appears as a single cell immediately surrounded by a jellylike *zona pellucida* and then a crown of granulosa cells (the *corona radiata*).

2. Next, observe the cleavage stages.

How is the human cleavage process similar to that in the sea urchin?

____

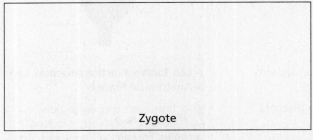

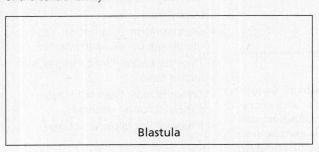

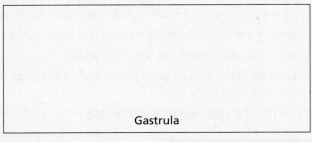

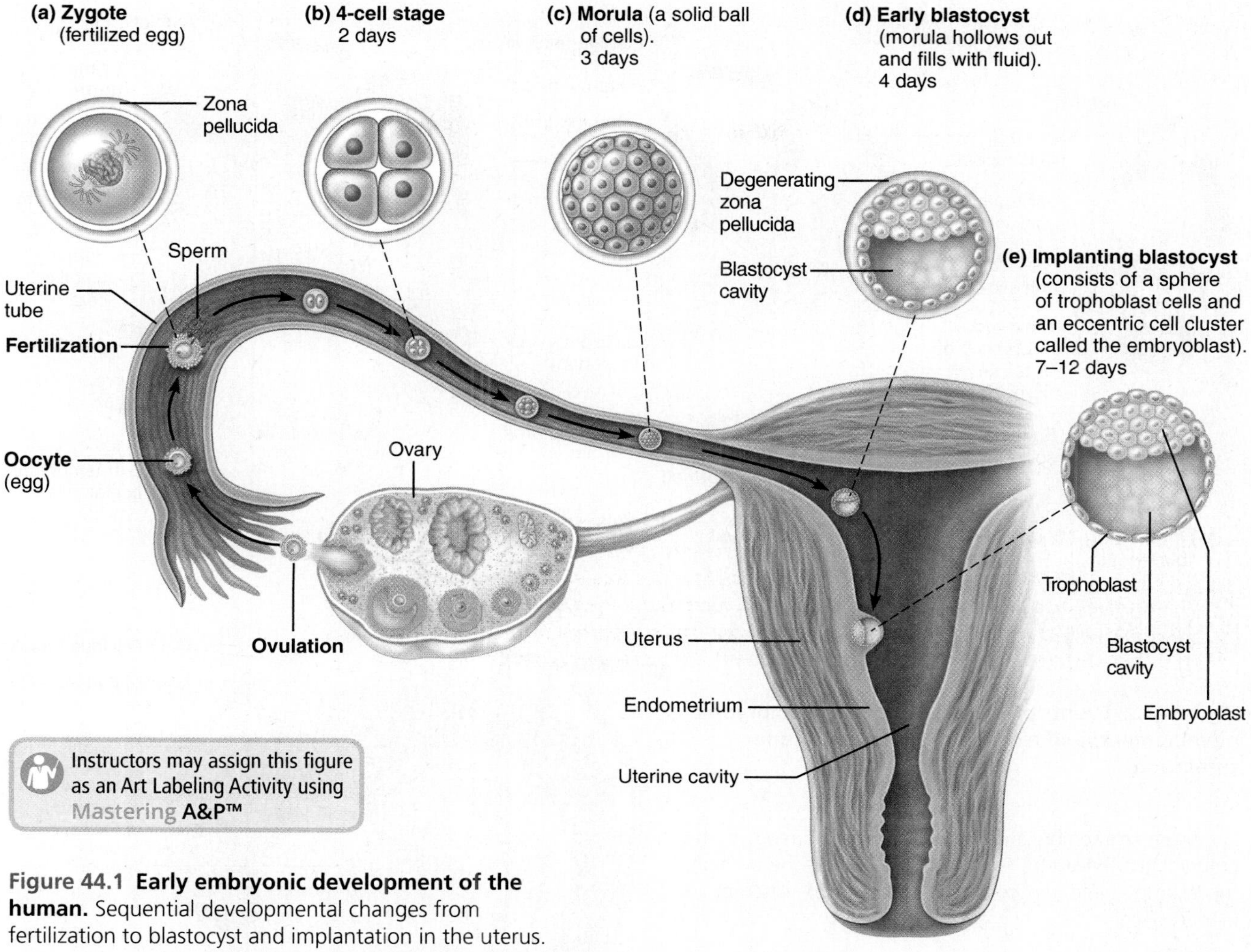

Instructors may assign this figure as an Art Labeling Activity using Mastering A&P™

Figure 44.1 Early embryonic development of the human. Sequential developmental changes from fertilization to blastocyst and implantation in the uterus.

Observe the blastula, the final product of cleavage, which is called the **blastocyst** in the human. The blastocyst is a fluid-filled sphere outlined by the **trophoblast**, a single layer of flattened cells enclosing the *embryoblast*. Unlike the sea urchin, only a portion of the blastocyst cells in the human contributes to the formation of the embryonic body—those seen at one side of the blastocyst (Figure 44.1e) forming the **embryoblast** (or inner cell mass). The trophoblast becomes an extraembryonic membrane called the **chorion**, which forms the fetal portion of the **placenta**. The embryoblast becomes the two-layered **embryonic disc**, which forms the embryo proper.

3. Observe the *implanting* blastocyst shown on the model or in the figures. By approximately day 7 after ovulation, a developing human embryo (blastocyst) is floating free in the uterine cavity. About that time, it adheres to the uterine wall over the embryoblast area, and implantation begins.

The trophoblast cells secrete enzymes that break down the endometrium to reach the blood vessels beneath, assisting in implantation. Implantation takes about 5 days. It is completed by day 12 after fertilization, and the embryoblast is imbedded in the endometrium. The portion of the endometrium that lies beneath the embryoblast and eventually lies beneath the embryo is the **decidua basalis**, and the portion of the endometrium that is on the luminal face of the uterus is the **decidua capsularis**.

By the time implantation has been completed, embryonic development has progressed to the **gastrula stage**, and the three primary germ layers are present and are beginning to differentiate (Figure 44.2). Within the next 6 weeks, virtually all of the body organ systems will have been laid down at least in rudimentary form by the germ layers. The three primary germ layers and their major derivatives are summarized in **Table 44.1** on p. 658.

By week 9 of development, the embryo is referred to as a **fetus**, and from this point on, the major activities are growth and tissue and organ specialization.

4. Again observe the blastocyst to follow the formation of the embryonic membranes and the placenta (Figure 44.2). Notice the villus extensions of the trophoblast. By the time implantation is complete, the trophoblast has differentiated into the chorion, and its large elaborate villi are lying in the blood-filled sinusoids in the uterine tissue. This composite of the decidua basalis and **chorionic villi** is called the **placenta** (Figure 44.3), and all exchanges to and from the embryo occur through the chorionic membranes.

Text continues on next page. →

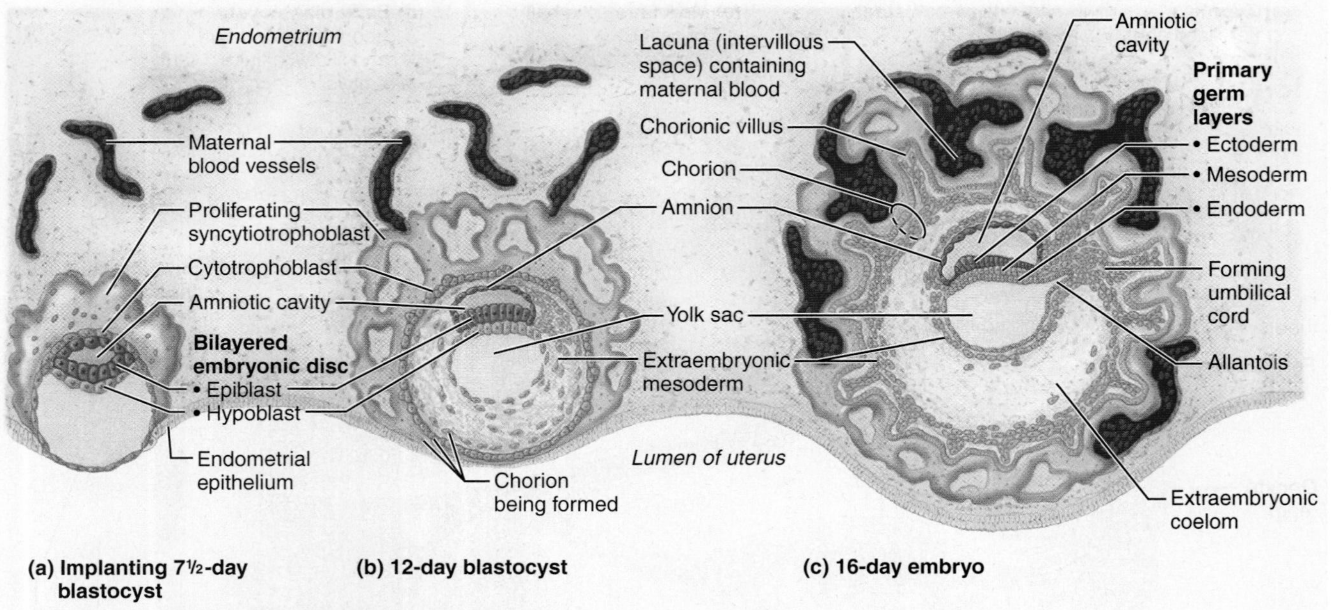

Instructors may assign this figure as an Art Labeling Activity using Mastering A&P™

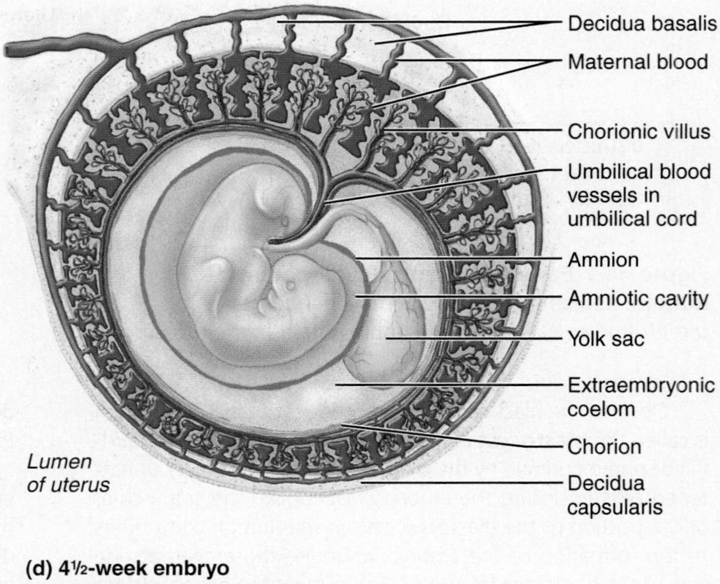

Figure 44.2 Events of placentation, early embryonic development, and extraembryonic membrane formation.

Three extraembryonic membranes (originating in the embryoblast) have also formed by this time—the amnion, the allantois, and the yolk sac (Figure 44.2). Attempt to identify each.

- The **amnion** encases the young embryonic body in a fluid-filled chamber that protects the embryo against mechanical trauma and temperature extremes and prevents adhesions during rapid embryonic growth.
- The **yolk sac** in humans has lost its original function, which was to pass nutrients to the embryo after digesting the yolk mass. The placenta has taken over that task; also, the human egg has very little yolk. However, the yolk sac is not totally useless. The embryo's first blood cells originate here,

44

Table 44.1 Major Derivatives of the Three Primary Germ Layers (Figure 44.2c)

Primary germ layer	Major derivatives
Ectoderm (outermost layer)	Epidermis and accessory skin structures; nervous system and organs of the special senses; epithelia of the oral cavity, the nasal cavity, and the anal canal; adrenal medulla, pituitary, and pineal glands
Mesoderm (middle layer)	Dermis; skeleton; skeletal, cardiac, and most smooth muscle; cartilage; blood vessels; kidneys and ureters; most lymphoid organs and tissues; internal reproductive organs; serous membranes; adrenal cortex
Endoderm (inner most layer)	Epithelial lining of the respiratory tract, GI tract, urinary tract, and reproductive tract; liver, gallbladder, and pancreas; thymus, parathyroid glands, and thyroid glands

and the primordial germ cells migrate from it into the embryo's body to seed the gonadal tissue. The yolk sac also forms part of the digestive tube.

- The **allantois**, which protrudes from the posterior end of the yolk sac, is also largely redundant in humans because of the placenta. In birds and reptiles, it is a repository for embryonic wastes. In humans, it is the structural basis on which the mesoderm migrates to form the **umbilical cord**, which attaches the embryo to the placenta.

5. Go to the demonstration area to view the photographic series, *Life Before Birth: Normal Fetal Development*. After viewing them, respond to the following questions.

What organs or organ systems appear *very* early in embryonic development?

____

____

Does development occur in a rostral to caudal (head to toe) direction, or vice versa?

____

Does development occur in a distal to proximal direction, or vice versa?

____

What is vernix caseosa? ______________________________

____

____

What is lanugo? ______________________________

____

____

In Utero Development

Activity 3

Identifying Fetal Structures

1. Put on disposable gloves. Go to the appropriate demonstration area and observe the fetuses in the Y-shaped animal (cat, rat, or pig) uterus. Identify the following fetal or fetal-related structures:

- **Placenta** (a composite structure formed from the uterine mucosa and the fetal chorion).

  Describe its appearance. ______________________________

  ____

  ____

  ____

- **Umbilical cord.** Describe its relationship to the placenta and fetus.

  ____

  ____

  ____

- **Amniotic sac.** Identify the transparent amnion surrounding a fetus. Open one amniotic sac, and note the amount, color, and consistency of the fluid.

  ____

  ____

  ____

  Remove a fetus, and observe the degree of development of the head, body, and extremities. Is the skin thick or thin?

  ____

2. Observe the model of a pregnant human torso. Identify the placenta. How does it differ in shape from the animal placenta observed?

____

____

____

Identify the umbilical cord. In what region of the uterus does implantation usually occur, as indicated by the position of the placenta?

____

____

What might be the consequence if it occurred lower?

____

____

____

44

Gross and Microscopic Anatomy of the Placenta

The placenta is a remarkable temporary organ. Composed of maternal and fetal tissues, it is responsible for providing nutrients and oxygen to the embryo and fetus while removing carbon dioxide and metabolic wastes.

Activity 4

Studying Placental Structure

1. Notice that the human placenta on display has two very different-appearing surfaces—one smooth and the other spongy, roughened, and torn-looking.

Which is the fetal side? ______________________

What is the basis for your conclusion? ______________________

____

____

Identify the umbilical cord. Within the cord, identify the umbilical vein and two umbilical arteries. What is the function of the umbilical vein?

____

The umbilical arteries? ______________________

____

2. Obtain a microscope slide of placental tissue. Observe the tissue carefully and identify the *intervillous spaces* (lacunae), which are blood filled in life (**Figure 44.3**). Identify the villi, and notice their rich vascular supply.

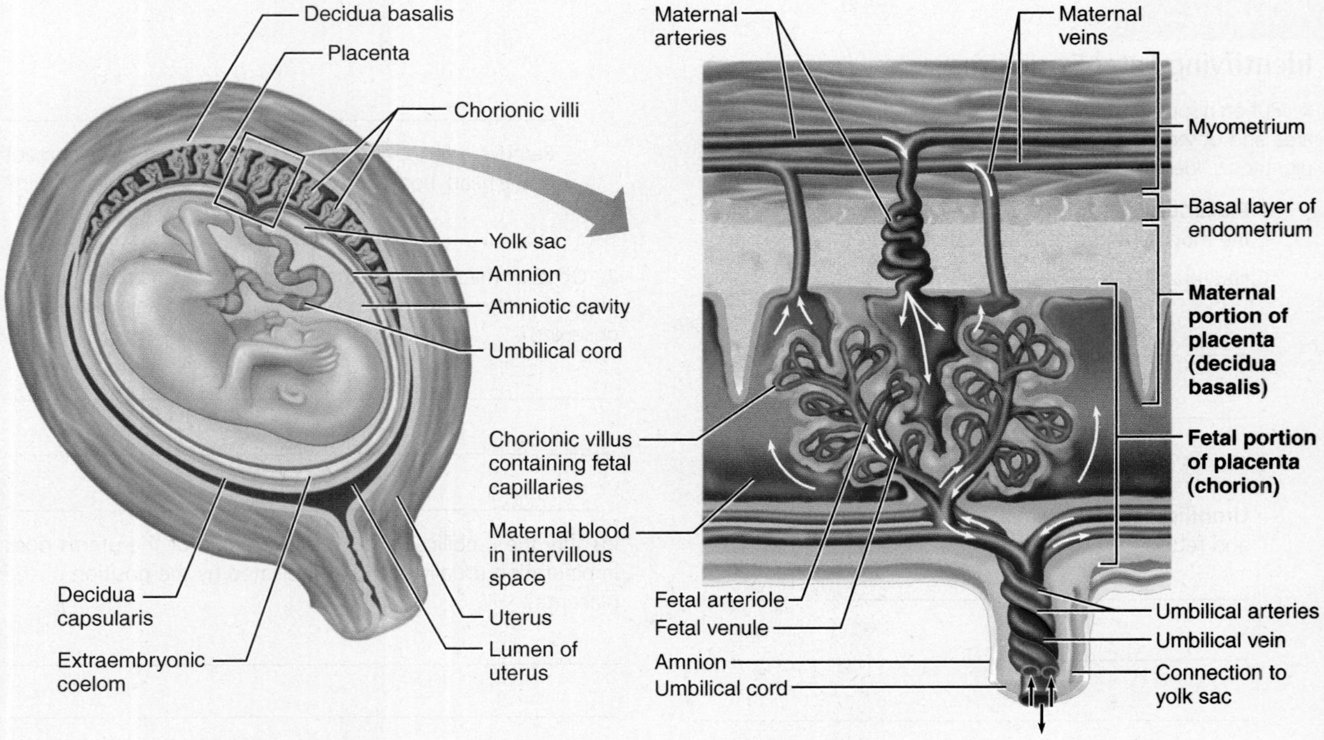

Figure 44.3 Diagrammatic representation of the structure of the placenta for a 13-week fetus.

Instructors may assign this figure as an Art Labeling Activity using Mastering **A&P™**

44

Instructors may assign a portion of the Review Sheet questions using Mastering A&P™

44 EXERCISE REVIEW SHEET Survey of Embryonic Development

Name ______________________ Lab Time/Date ______________________

Developmental Stages of Sea Urchins and Humans

1. Define *zygote*. ______________________

2. Describe how you were able to tell by observation when a sea urchin egg was fertilized. ______________________

____

3. Use the key choices to identify the embryonic stage or process described below.

Key: a. blastocyst (blastula in sea urchins) c. fertilization e. morula
b. cleavage d. gastrulation f. zygote

__________ 1. process of male and female pronuclei fusion

__________ 2. solid ball of embryonic cells

__________ 3. process of rapid mitotic cell division without intervening growth periods

__________ 4. cell resulting from combination of egg and sperm

__________ 5. process involving cell rearrangements to form the three primary germ layers

__________ 6. embryonic stage in which the embryo consists of a hollow ball of cells

4. What is the importance of cleavage in embryonic development? ______________________

____

How is cleavage different from mitotic cell division, which occurs later in life? ______________________

____

____

5. Which blastocyst derivatives or extraembryonic membranes have the following fates?

__________________ 1. forms the embryo proper

__________________ 2. becomes the extraembryonic membrane called the chorion

__________________ 3. produces the amnion, yolk sac, and allantois

__________________ 4. produces the primordial germ cells

__________________ 5. an extraembryonic membrane that provides the structural basis for the umbilical cord

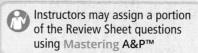

6. Using the letters on the diagram, correctly identify each of the following maternal or embryonic structures.

____________ amnion	____________ chorionic villi	____________ decidua capsularis	____________ forming umbilical cord
____________ chorion	____________ decidua basalis	____________ ectoderm	____________ mesoderm
		____________ endoderm	____________ uterine cavity

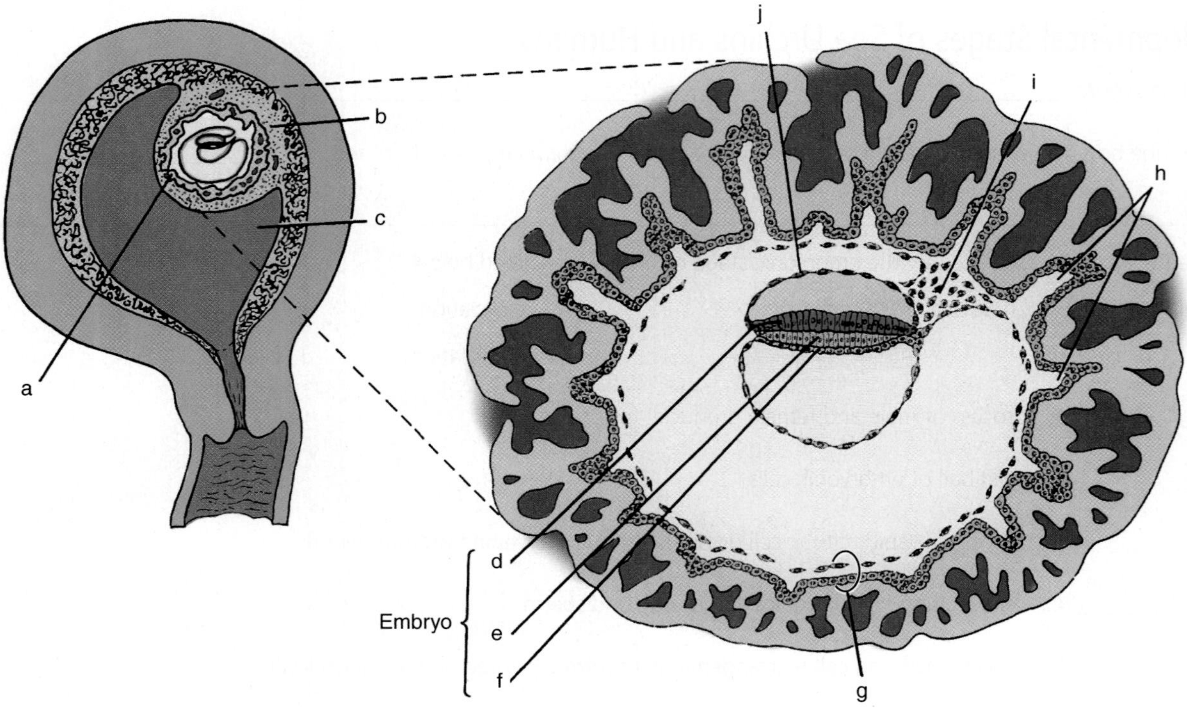

7. Explain the process and importance of gastrulation. ____

____

____

8. What is the function of the amnion and the amniotic fluid? ____________________________________

____

____

9. Describe the process of implantation, noting the role of the trophoblast cells. ______________________

____

____

10. How many days after fertilization is implantation generally completed? __________ What event in the female menstrual cycle ordinarily occurs just about this time if implantation does *not* occur? ________________________________

____

11. Referring to the illustrations and text of *Life Before Birth: Normal Fetal Development*, answer the following:

Which two organ systems are extensively developed in the *very young* embryo?

______________________________ and ______________________________

Describe the direction of development by circling the correct descriptions below:

proximal-distal distal-proximal caudal-rostral rostral-caudal

Does body control during infancy develop in the same directions? Think! Can an infant pick up a common pin (pincer grasp) or wave his arms earlier? Is arm-hand or leg-foot control achieved earlier?

____

12. Note whether each of the following organs or organ systems develops from the (a) ectoderm, (b) endoderm, or (c) mesoderm.

__________ 1. skeletal muscle __________ 4. respiratory mucosa __________ 7. nervous system

__________ 2. skeleton __________ 5. circulatory system __________ 8. serous membrane

__________ 3. lining of the GI tract __________ 6. epidermis of skin __________ 9. liver, pancreas

In Utero Development

13. Make the following comparisons between a human and the dissected structures of another pregnant mammal.

Comparison object	Human	Dissected animal
Shape of the placenta		
Shape of the uterus		

14. Where in the human uterus do implantation and placentation ordinarily occur? ______________________________

____

15. Describe the function(s) of the placenta. ______________________________

____

16. Which two extraembryonic membranes has the placenta more or less "put out of business"? ______________________________

____

17. When does the human embryo come to be called a fetus? ______________________________

18. What is the usual and most desirable fetal position in utero? ______________________________

Why is this the most desirable position? ______________________________

____

____

____

Gross and Microscopic Anatomy of the Placenta

19. Describe fully the gross structure of the human placenta as observed in the laboratory. ______________________________

____

____

20. What is the tissue origin of the placenta: fetal, maternal, or both? ______________________________

21. What placental barriers must be crossed to exchange materials? ______________________________

____

22. ✚ Sexually transmitted bacteria, including the causative agents of gonorrhea and chlamydia, can damage a variety of pelvic organs, leading to a condition called pelvic inflammatory disease (PID). PID often leads to scarring of the uterine tubes. Explain how this could result in infertility in women. ______________________________

____

____

23. ✚ PID is also a risk factor for ectopic pregnancies, in which a fertilized egg implants outside the uterus. Explain the connection between PID and ectopic pregnancies.

____

____

EXERCISE 45 Principles of Heredity

Learning Outcomes

- Define *allele*, *heterozygous*, *homozygous*, *dominance*, *recessiveness*, *genotype*, *phenotype*, and *incomplete dominance*.
- Work simple genetics problems using a Punnett square.
- State the basic laws of probability.
- Observe blood types as phenotypes, and determine their genotype basis.
- Separate variants of hemoglobin using agarose gel electrophoresis.

Go to Mastering A&P™ > Study Area to improve your performance in A&P Lab.

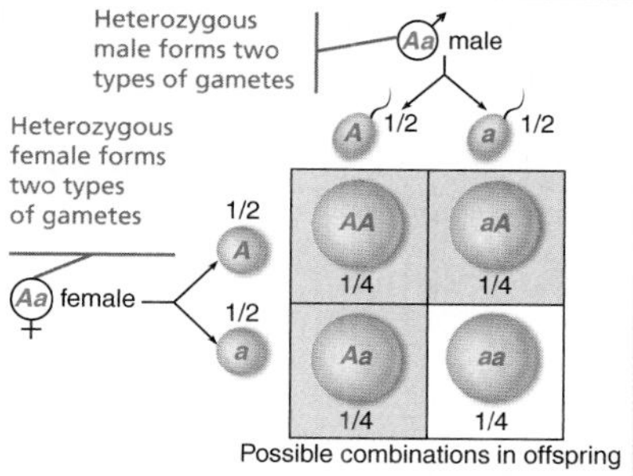

Instructors may assign new Building Vocabulary coaching activities, Pre-Lab Quiz questions, Art Labeling activities, and more using the Mastering A&P™ Item Library.

Pre-Lab Quiz

Instructors may assign these and other Pre-Lab Quiz questions using Mastering A&P™

1. Circle True or False. A heterozygous individual will have two of the same alleles in a chromosome pair.
2. The allele that is expressed and not masked is the ________ allele.
   a. dominant
   b. genotypic
   c. homozygous
   d. recessive
3. A condition known as ________ can result when heterozygous individuals exhibit a phenotype intermediate between homozygous individuals and both alleles are expressed in the offspring.
   a. incomplete dominance
   b. sex-linked inheritance
   c. total dominance
   d. total heterozygosity
4. Circle True or False. Males are more likely to inherit hemophilia because it is a result of receiving the sex-linked gene from the father.
5. Circle True or False. The inheritance of ABO blood type involves three possible alleles.

The field of genetics is bristling with excitement. Complex gene-splicing techniques have allowed researchers to precisely isolate genes coding for specific proteins and then to use those genes to harvest large amounts of specific proteins and even to cure some dreaded human diseases. At present, growth hormone, insulin, erythropoietin, and interferon produced by these genetic engineering techniques are available for clinical use, and the list is growing daily.

Introduction to the Language of Genetics

In humans all cells, except eggs and sperm, contain 46 chromosomes, that is, the diploid number. The diploid chromosomal number actually represents two complete (or nearly complete) sets of genetic instructions—one from the mother and the other from the father—or 23 pairs of *homologous chromosomes*.

Materials

- Pennies (for coin tossing)
- Scratch paper (for drawing Punnett squares)

Activity 5: Blood typing

- Anti-A and Anti-B sera
- Clean microscope slides
- Toothpicks
- Wax pencils
- Sterile lancets
- Alcohol swabs
- Beaker containing 10% bleach solution
- Disposable autoclave bag

Activity 6: Hemoglobin phenotyping

- Electrophoresis equipment and power supply
- 1.2% agarose gels
- 1X TBE (Tris-Borate/EDTA) buffer pH 8.4

Text continues on next page. →

- Micropipette or variable automicropipette with tips
- Marking pen
- Safety goggles (student-provided)
- Metric ruler
- Plastic baggies
- Disposable gloves
- Coomassie blue protein stain solution
- Coomassie blue de-stain solution
- Distilled water
- Staining tray
- 100-ml gradulated cylinder
- Hemoglobin samples dissolved in TBE solubilizing buffer with bromophenol blue: HbA, labeled A; HbS, labeled S; HbA + HbS, labeled AS; and unknown samples of each

Different molecular versions of genes coding for the same traits are called **alleles**. The alleles occur on the same location on homologous chromosomes and may be identical or different. When both alleles in a homologous chromosome pair are the same, the individual is **homozygous** for that trait. When the alleles are different, the individual is **heterozygous** for the given trait; and often only one of the alleles, called the **dominant gene**, exerts its effects. The allele with less potency, the **recessive gene**, is present but suppressed or masked. Whereas dominant genes, or alleles, exert their effects in both homozygous and heterozygous conditions, as a rule recessive alleles *must* be present in double dose—that is, the homozygous condition—to exert their influence.

An individual's actual genetic makeup, that is, whether the person is homozygous or heterozygous for the various alleles, is called the **genotype**. The expression of the genotype, for example, the presence or absence of a trait is referred to as a **phenotype**.

The complete story of heredity is much more complex than just outlined. In actuality, the expression of most traits (for example, eye color) is determined by multiple alleles or the interaction of several gene pairs. Simple dominant-recessive patterns of inheritance are rare. Many human traits previously thought to be expressed by a single gene displaying a dominant-recessive inheritance pattern (such as freckles, dimples, and the ability to roll the tongue) don't actually follow this pattern. The dominant-recessive pattern of inheritance is most clearly seen in disorders such as those listed in **Table 45.1**. Our emphasis here will be to investigate only the less complex aspects of genetics.

Table 45.1 Traits Determined by Simple Dominant-Recessive Inheritance

Phenotype due to expression of:	
Dominant genes (ZZ or Zz)	**Recessive genes (zz)**
Syndactyly (webbed digits)	Normal digits
Achondroplasia (heterozygous: dwarfism; homozygous: lethal)	Normal endochondral ossification
Huntington's disease	Absence of Huntington's disease
Normal skin pigmentation	Albinism
Absence of Tay-Sachs disease	Tay-Sachs disease
Absence of cystic fibrosis	Cystic fibrosis

Dominant-Recessive Inheritance

One of the best ways to master the terminology and learn the principles of heredity is to work out the solutions to some genetic crosses in much the same way Gregor Mendel did in his classic experiments on pea plants.

To work out the various simple monohybrid (one pair of alleles) crosses in this exercise, you will be given the genotype of the parents. You will then determine the possible genotypes of their offspring by using a grid called the *Punnett square*, and you will record the percentages of both genotype and phenotype. To illustrate the procedure, an example of one of Mendel's pea plant crosses is outlined next.

Alleles: *T* (determines *tallness*; dominant)

t (determines *dwarfism*; recessive)

Genotypes of parents: *TT* (♂) × *tt* (♀)

Phenotypes of parents: Tall × dwarf

To use the Punnett, or checkerboard, square, write the alleles (actually gametes) of one parent across the top and the gametes of the other parent down the left side. Then combine the gametes across and down to achieve all possible combinations (possible genotypes of their offspring), as follows:

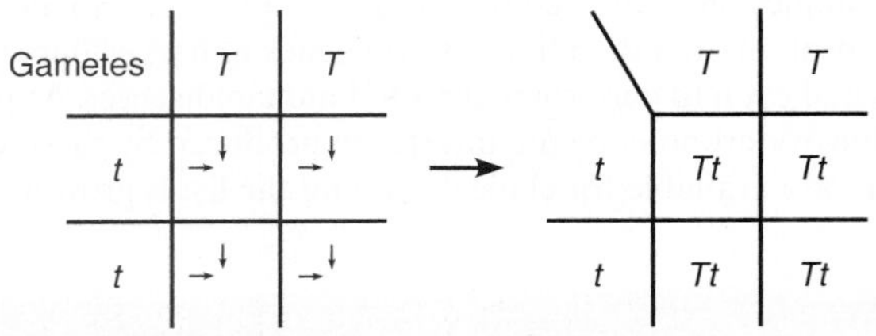

Results: Genotypes 100% *Tt* (all heterozygous)
Phenotypes 100% tall (because *T*, which determines tallness, is dominant, and all contain the *T* allele)

Activity 1

Working Out Crosses Involving Dominant and Recessive Genes

For each of the following crosses, draw your own Punnett square and use the technique outlined above to determine the genotypes and phenotypes of the pea plant offspring.

1. Genotypes of parents: *Tt* (♂) × *tt* (♀)

% of each genotype: ____________________

% of each phenotype: ________% tall, ________% dwarf

2. Genotypes of parents: *Tt* (♂) × *Tt* (♀)

% of each genotype: ____________________

% of each phenotype: ________% tall, ________% dwarf

3. Genotypes of parents: *TT* (♂) × *Tt* (♀)

% of each genotype: ____________________

% of each phenotype: ________% tall, ________% dwarf

Incomplete Dominance

The concepts of dominance and recessiveness are somewhat arbitrary and artificial in some instances because so-called dominant genes may be expressed differently in homozygous and heterozygous individuals. This produces a condition called **incomplete dominance**. In such cases, both alleles express themselves in the offspring. The crosses are worked out in the same manner as indicated previously, but heterozygous offspring exhibit a phenotype intermediate between that of the homozygous individuals. Some examples follow.

Activity 2

Working Out Crosses Involving Incomplete Dominance

1. The inheritance of flower color in snapdragons illustrates the principle of incomplete dominance. The geno-type *RR* is expressed as a red flower, *Rr* yields pink flowers, and *rr* produces white flowers. Work out the following crosses to determine the expected phenotypes and both genotypic and phenotypic percentages.

a. Genotypes of parents: *RR* × *rr*

% of each genotype: ____________________

% of each phenotype: ____________________

b. Genotypes of parents: *Rr* × *rr*

% of each genotype: ____________________

% of each phenotype: ____________________

____

c. Genotypes of parents: *Rr* × *Rr*

% of each genotype: ____________________

% of each phenotype: ____________________

____

2. In humans, the inheritance of sickle cell anemia/trait is determined by a single pair of alleles that exhibit incomplete dominance. Individuals homozygous for the sickling gene *(s)* have *sickle cell anemia*. In double dose *(ss)*, the sickling gene causes production of a very abnormal hemoglobin, which crystallizes and becomes sharp and spiky under conditions of oxygen deficit. Heterozygous individuals *(Ss)* have the *sickle cell trait*; they make both normal and sickling hemoglobin. Usually these individuals are healthy, but prolonged decreases in blood oxygen levels can lead to a sickle cell crisis. Individuals with the genotype *SS* form normal hemoglobin. Work out the following crosses:

a. Parental genotypes: *SS* × *ss*

% of each genotype: ____________________

% of each phenotype: ____________________

b. Parental genotypes: *ss* × *Ss*

% of each genotype: ____________________

% of each phenotype: ____________________

Sex-Linked Inheritance

A cell's chromosomes can be stained, photographed, and digitally rearranged to produce an image called a *karyotype*, which shows the complete human diploid chromosomal complement displayed in homologous pairs (**Figure 45.1**).

Of the 23 pairs of homologous chromosomes, 22 pairs are referred to as **autosomes**. The autosomes guide the expression of most body traits. The 23rd pair, the **sex chromosomes** (X and Y), determine the genetic sex of an individual (male = XY; female = XX). The Y sex chromosome is only about a third the size of the X sex chromosome, and it lacks many of the genes that are found on the X.

Inherited traits determined by genes on the sex chromosomes are said to be *sex-linked*, and genes present *only* on the X sex chromosome are said to be *X-linked*. Some examples of X-linked genes include those that determine normal color vision (or, conversely, color blindness, most commonly red-green color blindness), and normal clotting ability (as opposed to hemophilia). The alleles that determine red-green color blindness and hemophilia are recessive alleles. In females, *both* X chromosomes must carry the recessive alleles for a woman to express either of these conditions, and thus they tend to be infrequently seen. However, should a male receive an X-linked recessive allele for these conditions, he will exhibit the recessive phenotype because his Y chromosome does not contain alleles for that gene.

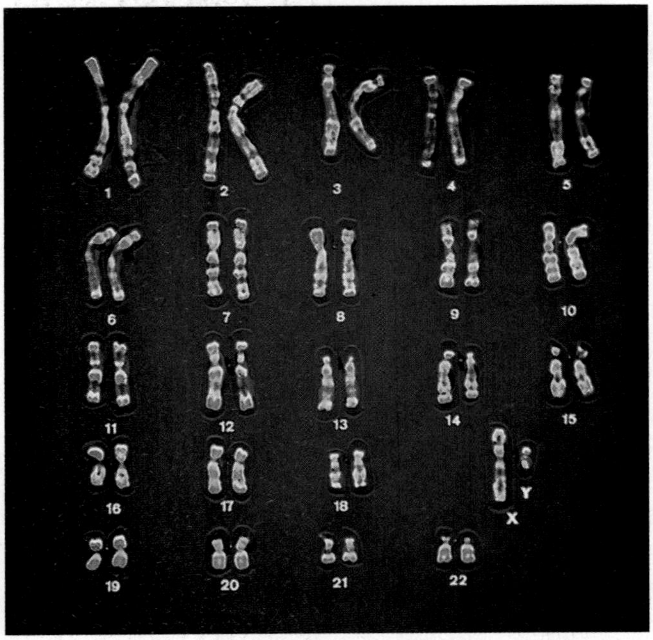

Figure 45.1 Karyotype (chromosomal complement) of human male. Each pair of homologous chromosomes is numbered except the sex chromosomes, which are identified by their letters, X and Y.

The critical point to understand about X-linked inheritance is the *absence* of male to male (that is, father to son) transmission of sex-linked genes. The X of the father *will* pass to each of his daughters but to none of his sons. Males always inherit sex-linked conditions from their mothers via the X chromosome.

Activity 3

Working Out Crosses Involving Sex-Linked Inheritance

1. A heterozygous woman carrying the recessive gene for red-green color blindness marries a man who is red-green color-blind. Assume the dominant gene is X^C (allele for normal color vision) and the recessive gene is X^c (determines red-green color blindness). The mother's genotype is X^CX^c, and the father's X^cY. Do a Punnett square to determine the answers to the following questions.

According to the laws of probability, what percentage of all their children will be red-green color-blind?

__________ %

What is the percentage of red-green color-blind individuals by sex?

__________% males; __________% females

What percentage of all children will be carriers? ______%

What is the sex of the carriers? ____________________

2. A heterozygous woman carrying the recessive gene for hemophilia marries a man who is not a hemophiliac. Assume the dominant gene is X^H and the recessive gene is X^h. The woman's genotype is X^HX^h, and her husband's genotype is X^HY. What is the potential percentage and sex of their offspring who will be hemophiliacs?

__________% males; __________% females

What percentage can be expected to lack the allele for hemophilia?

__________%

What is the anticipated sex and percentage of individuals who will be carriers for hemophilia?

__________%; __________sex

Probability

Parceling out of chromosomes to gametes during meiosis and the combination of egg and sperm are random events. Hence, the possibility that certain genomes will arise and be expressed is based on the laws of probability. The randomness of gene recombination from each parent determines individual uniqueness and explains why siblings, however similar, never have totally corresponding traits (unless, of course, they are identical twins). The Punnett square method that you have been using to work out the genetics problems actually provides information on the *probability* of the appearance of certain genotypes considering all possible events. Probability *(P)* is defined as:

$$P = \frac{\text{number of specific events or cases}}{\text{total number of events or cases}}$$

If an event is certain to happen, its probability is 1. If it happens one out of every two times, its probability is ½; if one out of four times, its probability is ¼, and so on.

When figuring the probability of separate events occurring together (or consecutively), the probability of each event must be multiplied together to get the final probability figure. For example, the probability of a penny coming up "heads" in each toss is ½ (because it has two sides—heads and tails). But the probability of a tossed penny coming up heads four times in a row is $\frac{1}{2} \times \frac{1}{2} \times \frac{1}{2} \times \frac{1}{2} = \frac{1}{16}$.

Activity 4

Exploring Probability

1. Obtain two pennies and perform the following simple experiment to explore the laws of probability.

a. Toss one penny into the air ten times, and record the number of heads (H) and tails (T) observed.

__________ heads __________ tails

Probability: __________/10 tails; __________/10 heads

b. Now simultaneously toss two pennies into the air for 24 tosses, and record the results of each toss below. In each case, report the probability in the lowest fractional terms.

# of HH: __________ Probability: __________

# of HT: __________ Probability: __________

# of TT: __________ Probability: __________

Does the first toss have any influence on the second?

____

c. Do a Punnett square using HT for the "alleles" of one "parental" coin and HT for the "alleles" of the other.

Probability of HH: __________

Probability of HT: __________

Probability of TT: __________

How closely do your coin-tossing results correlate with the percentages obtained from the Punnett square results?

____

____

2. Determine the probability of having a boy or girl offspring for each conception.

Parental genotypes: XY × XX

Probability of males: __________%

Probability of females: __________%

Multiple-Allele Inheritance

Some genes exhibit more than two allele forms, leading to a phenomenon called **multiple-allele inheritance**. Inheritance of the ABO blood type is based on the existence of three alleles designated as I^A, I^B, and *i*. Both I^A and I^B are dominant over *i*, but neither is dominant over the other. The I^A and I^B alleles are *codominant*. Thus the possession of I^A and I^B will yield type AB blood, whereas the possession of the I^A and *i* alleles will yield type A blood, and so on (as explained in Exercise 29). The four ABO blood groups, or phenotypes, are A, B, AB, and O. Their correlation to genotype is indicated in **Table 45.2**.

Table 45.2 Blood Groups

ABO blood group (phenotype)	Genotype
A	I^AI^A or I^Ai
B	I^BI^B or I^Bi
AB	I^AI^B
O	*ii*

45

Activity 5

Using Blood Type to Explore Phenotypes and Genotypes

An individual with a phenotype of blood type A could be homozygous with a genotype of I^AI^A or heterozygous with a genotype of I^AI^i. An individual with a phenotype of blood type AB must be heterozygous with a genotype of I^AI^B. Similarly, an individual with a phenotype of blood type O must be homozygous with a genotype of I^iI^i.

1. Determine the potential genotypes and phenotypes for the offspring of a mother who is heterozygous for blood type A and a father who is heterozygous for blood type B.

Genotype of the mother: ____________________

Genotype of the father: ____________________

Potential genotypes for the offspring: ____________________

Potential phenotypes for the offspring: ____________________

2. Determine the potential genotypes and phenotypes for the offspring of a mother with the phenotype of blood type O and a father with the phenotype of blood type AB.

Genotype of the mother: ____________________

Genotype of the father: ____________________

Potential genotypes for the offspring: ____________________

Potential phenotypes for the offspring: ____________________

3. A child's phenotype is blood type A, and the mother's phenotype is blood type O.

Determine the:

Potential genotype(s) for the child: ____________________

Potential genotype(s) for the father: ____________________

If you have previously typed your blood, record your phenotype and potential genotypes below. If not, type your blood following your instructor's instructions (see Exercise 29, pp. 426–427), and then enter your results below.

Dispose of any blood-soiled supplies by placing the glassware in the bleach-containing beaker and all other items in the autoclave bag.

Your blood type phenotype: ____________________

Your potential blood type genotype(s): ____________________

Hemoglobin Phenotype Identification Using Agarose Gel Electrophoresis

Agarose gel electrophoresis separates molecules based on size and charge. In the appropriate buffer with an alkaline pH, hemoglobin molecules will move toward the anode of the apparatus at different speeds, based on the number of negative charges on the molecules.

Sickle cell anemia and sickle cell trait are discussed in Activity 2 of this exercise. The beta chains of hemoglobin S (HbS) contain a base substitution where a valine replaces glutamic acid. As a result of the substitution, HbS has fewer negative charges than the predominant form of adult hemoglobin (HbA) and can be separated from HbA using agarose gel electrophoresis.

Activity 6

Using Agarose Gel Electrophoresis to Identify Normal Hemoglobin, Sickle Cell Anemia, and Sickle Cell Trait

1. You will need an electrophoresis unit and power supply, a 1.2% agarose gel with eight wells, 1X TBE buffer, micropipettes or a variable automatic micropipette (2–20 µl) with tips, samples of hemoglobin (marked A, AS, S, and unknown #__________) dissolved in TBE solubilizing buffer containing bromophenol blue, a marking pen, safety goggles, metric ruler, plastic bag, and disposable gloves.

2. Record the number of your unknown sample. __________

3. Place the agarose gel into the electrophoresis unit.

4. Using a micropipette, carefully add 15 µl of hemoglobin sample A to wells number 1 and 5, AS to wells 2 and 6, S to wells 3 and 7, and the unknown samples to wells 4 and 8.

5. Slowly add electrophoresis buffer until the gels are covered with about 0.25 cm of buffer.

6. The electrophoresis unit runs with high voltage. Do not attempt to open it while the power supply is attached. Close and lock the electrophoresis unit, and connect the unit to the power source, red to red and black to black.

7. Run the unit about 50 minutes at 120 volts until the bromophenol blue is about 0.25 cm from the anode.

8. Turn the power supply **OFF**. Disconnect the cables.

9. Open the electrophoresis unit, carefully remove the gel, and slide the gel into a plastic bag. You should be able to see the hemoglobin bands on the gel. Mark each of the bands on the gel with the marking pen.

Alternatively, the gels may be stained with Coomassie blue. Obtain a flask of Coomassie blue stain, a flask of destaining solution, a flask of distilled water, a staining tray, and a 100-ml graduated cylinder, and do the following:

a. Carefully remove the gel from the plastic plate, and place the gel into a staining dish.

b. Add about 30 ml of stain (enough stain to cover the gel), and be sure the agarose is not stuck to the dish.

c. Allow the gel to remain in the stain for at least an hour (more time might be necessary), and then remove the stain and pour into an appropriate waste container. Rinse the gel and dish with distilled water.

d. Add about 100 ml of de-staining solution. Change the solution after a day. If the background stain has been reduced enough to see the bands, place the staining dish over a light source, and observe the bands. If the stain is still too dark, repeat the de-staining process until the bands can be observed.

e. To store the gels, refrigerate in a bag with a small amount of de-staining solution or dry on a glass plate.

10. Draw the banding patterns for samples 1 through 8 in the figure for question 13 in the Review Sheet (p. 675). Based on the banding patterns of the known samples, what are the genotypes of your unknown samples? Record on the Review Sheet chart.

11. Rinse the electrophoresis unit in distilled or de-ionized water, and clean the glass plates with soap and water.

Instructors may assign a portion of the Review Sheet questions using Mastering A&P™

45 EXERCISE REVIEW SHEET
Principles of Heredity

Name ______________________ Lab Time/Date ______________________

Introduction to the Language of Genetics

1. Match the key choices with the definitions given below.

Key:

a. alleles	d. genotype	g. phenotype
b. autosomes	e. heterozygous	h. recessive
c. dominant	f. homozygous	i. sex chromosomes

__________ 1. the combination of alleles for a given gene

__________ 2. chromosomes that determine most body characteristics

__________ 3. situation in which an individual has identical alleles for a particular trait

__________ 4. genes not expressed unless they are present in homozygous condition

__________ 5. expression of a genetic trait

__________ 6. situation in which an individual has different alleles making up his or her genotype for a particular trait

__________ 7. genes for the same trait that may have different expressions

__________ 8. chromosomes that determine genetic sex

__________ 9. the more potent gene allele; masks the expression of the less potent allele

Dominant-Recessive Inheritance

2. In humans, syndactyly (webbed digits) is inherited by possession of a dominant allele *(S)*. If a man who is homozygous and has normal digits (*ss*) marries a woman who is heterozygous for syndactyly and has webbed digits, what percentage of their children would be expected to have webbed digits?

__________%

3. Cystic fibrosis is due to an abnormal recessive gene (*c*). Only homozygous recessive individuals exhibit this disease. If the parents are both heterozygous for cystic fibrosis (*Cc*), what percentage of their children will have cystic fibrosis?

__________%

Incomplete Dominance

4. Tail length on a bobcat is controlled by incomplete dominance. The alleles are *T* for normal tail length and *t* for tail-lessness.

   How would you describe the tails of heterozygous *(Tt)* bobcats? ______________________________

5. If a woman and a man who both carry the trait for sickle cell anemia have children, what percentage of the various phenotypes would you expect from a cross between these individuals?

   __________% with sickle cell anemia

   __________% with sickle cell trait

   __________% with normal hemoglobin

Sex-Linked Inheritance

6. What does it mean when someone says a particular characteristic is sex-linked? ______________________________

   ____

7. You are a male, and you have been told that hemophilia "runs in your genes." Whose ancestors, your mother's or your father's, should you investigate? ____________________ Why? ______________________________

   ____

   ____

8. An X^CX^c female marries an X^CY man. Do a Punnett square for this match.

   What is the probability of producing a red-green color-blind son? __________

   A red-green color-blind daughter? __________

   A daughter who is a carrier for the red-green color-blind allele? __________

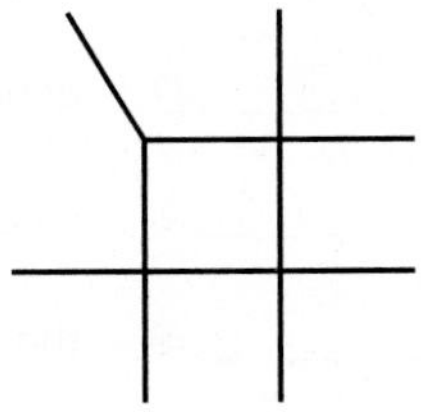

Probability

9. What is the probability of having three daughters in a row? ______________________________

10. What is the probability of having five sons in a row? ______________________________

Multiple-Allele Inheritance

11. If a child's phenotype is blood type A and mother also has the phenotype blood type A, list the potential genotypes for the father of the child. ______________________________

    ____

12. In the days before DNA fingerprinting, blood typing was used as a means of paternity testing. It was not possible to prove paternity, since many males would have the same blood type, but it was possible to exclude a man who couldn't possibly be the father based on blood typing. For a child with type O blood born to a mother with type A blood, is a man with blood type AB

possibly the father? Why or why not? ______________________________

____

What must the genotype of the mother be? ______________________________

Using Agarose Gel Electrophoresis to Identify Hemoglobin Phenotypes

13. Draw the banding patterns you obtained on the space below.

Sample	Well	Banding pattern
1. A	1. ☐	
2. AS	2. ☐	
3. S	3. ☐	
4. Unknown	4. ☐	
5. A	5. ☐	
6. AS	6. ☐	
7. S	7. ☐	
8. Unknown	8. ☐	

14. Why does sickle cell hemoglobin behave differently from normal hemoglobin during agarose gel electrophoresis?

____

____

15. ✚ A child is born with sickle cell disease. What can you conclude about the potential genotypes of the parents? ______________________________

____

____

16. ✚ A mother finds out that her son is red-green color-blind. She is not color-blind. What can you conclude about her genotype? Explain your answer.

____

____

EXERCISE 46 Surface Anatomy Roundup

Learning Outcomes

- Define *surface anatomy*, and explain why it is an important field of study; define *palpation*.
- Describe and palpate the major surface features of the cranium, face, and neck.
- Describe the easily palpated bony and muscular landmarks of the back, and locate the vertebral spines on the living body.
- List the bony surface landmarks of the thoracic cage, explain how they relate to the major soft organs of the thorax, and explain how to find the second to eleventh ribs.
- Name and palpate the important surface features on the anterior abdominal wall, and explain how to palpate a full bladder.
- Define and explain the following: *linea alba*, *umbilical hernia*, examination for an inguinal hernia, *linea semilunaris*, and *McBurney's point*.
- Locate and palpate the main surface features of the upper limb.
- Explain the significance of the cubital fossa, pulse points in the distal forearm, and the anatomical snuff box.
- Describe and palpate the surface landmarks of the lower limb.
- Explain exactly where to administer an injection in the gluteal region and in the other major sites of intramuscular injection.

Go to Mastering A&P™ > Study Area to improve your performance in A&P Lab.

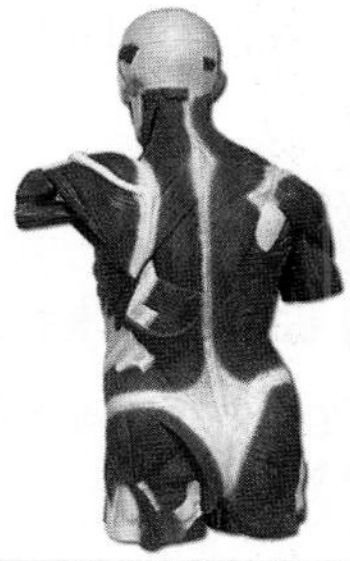

> Lab Tools > Practice Anatomy Lab > Anatomical Models

Instructors may assign new Building Vocabulary coaching activities, Pre-Lab Quiz questions, Art Labeling activities, Practice Anatomy Lab Practical questions (PAL), and more using the Mastering A&P™ Item Library.

Materials

- Articulated skeletons
- Three-dimensional models or charts of the skeletal muscles of the body
- Hand mirror
- Stethoscope
- Alcohol swabs
- Washable markers

Pre-Lab Quiz

Instructors may assign these and other Pre-Lab Quiz questions using Mastering A&P™

1. The epicranial aponeurosis binds to the subcutaneous tissue of the cranium to form the:
   a. mastoid process **c.** true scalp
   b. occipital protuberance **d.** xiphoid process
2. The ____________ is the most prominent neck muscle and also the neck's most important landmark.
   a. buccinator **c.** masseter
   b. epicranius **d.** sternocleidomastoid
3. The three boundaries of the ____________ are the trapezius medially, the latissimus dorsi inferiorly, and the scapula laterally.
   a. torso triangle **c.** triangle of back muscles
   b. triangle of ausculation **d.** triangle of McBurney
4. Circle True or False. With the exception of a full bladder, most internal pelvic organs are not easily palpated through the skin of the body surface.
5. On the dorsal surface of your hand is a grouping of superficial veins known as the ____________, which provides a site for drawing blood and inserting intravenous catheters.
   a. anatomical snuff box **c.** radial and ulnar veins
   b. dorsal venous network **d.** palmar arches

Surface anatomy is a valuable branch of anatomical and medical science. True to its name, **surface anatomy** does indeed study the *external surface* of the body, but more important, it also studies *internal* organs as they relate to external surface landmarks and as they are seen and felt through the skin. Feeling internal structures through the skin with the fingers is called **palpation** (literally, "touching").

Surface anatomy is living anatomy, better studied in live people than in cadavers. It can provide a great deal of information about the living skeleton (almost all bones can be palpated) and about the muscles and blood vessels that lie near the body surface. Furthermore, a skilled examiner can learn a good deal about the heart, lungs, and other deep organs by performing a surface assessment. Thus, surface anatomy serves as the basis of the standard physical examination. If you are planning a career in the health sciences or physical education, a study of surface anatomy will show you where to take pulses, where to insert tubes and needles, where to locate broken bones and inflamed muscles, and where to listen for the sounds of the lungs, heart, and intestines.

We will take a regional approach to surface anatomy, exploring the head first and proceeding to the trunk and the limbs. You will be observing and palpating your own body as you work through the exercise, because your body is the best learning tool of all. To aid your exploration of living anatomy, skeletons and muscle models or charts are provided around the lab so that you can review the bones and muscles you will encounter. Whenever possible, have a study partner assume the role of the subject for skin sites that you cannot reach on your own body.

Activity 1

Palpating Landmarks of the Head

The head (**Figure 46.1** and **Figure 46.2**) is divided into the cranium and the face.

Cranium

1. Run your fingers over the superior surface of your head. Notice that the underlying cranial bones lie very near the surface. Proceed to your forehead and palpate the **superciliary arches** (brow ridges) directly superior to your orbits (Figure 46.1).

2. Move your hand to the posterior surface of your skull, where you can feel the knoblike **external occipital protuberance**. Run your finger directly laterally from this projection to feel the ridgelike *superior nuchal line* on the occipital bone. This line, which marks the superior extent of the muscles of the posterior neck, serves as the boundary between the head and the neck. Now feel the prominent **mastoid process** on each side of the cranium just posterior to your ear.

3. The **frontal belly** of the epicranius (Figure 46.2) inserts superiorly onto the broad aponeurosis called the *epicranial aponeurosis* (Table 13.1, p. 198) that covers the superior surface of the cranium. This aponeurosis binds tightly to the overlying subcutaneous tissue and skin to form the true **scalp**. Push on your scalp, and confirm that it slides freely over the underlying cranial bones. Because the scalp is only loosely bound to the skull, people can easily be "scalped" (in industrial accidents, for example). The scalp is richly vascularized by a large number of arteries running through its subcutaneous tissue. Most arteries of the body constrict and close after they are cut or torn, but those in the scalp are unable to do so because they are held open by the dense connective tissue surrounding them.

What do these facts suggest about the amount of bleeding that accompanies scalp wounds?

____

____

Face

The surface of the face is divided into many different regions, including the *orbital*, *nasal*, *oral* (mouth), and *auricular* (ear) areas.

1. Trace a finger around the entire margin of the bony orbit. The **lacrimal fossa**, which contains the tear-gathering lacrimal sac, may be felt on the medial side of the eye socket.

2. Touch the most superior part of your nose, its **root**, which lies between the eyebrows (Figure 46.2). Just inferior to this, between your eyes, is the **bridge** of the nose formed by the nasal bones. Continue your finger's progress inferiorly along the nose's anterior margin, the **dorsum nasi**, to the tip of the nose, the **apex**. Place one finger in a nostril and another finger on the flared winglike **ala** that defines the nostril's lateral border.

3. Grasp your **auricle**, the shell-like part of the external ear that surrounds the opening of the **external acoustic meatus** (Figure 46.1). Now trace the ear's outer rim, or **helix**, to the **lobule** (earlobe) inferiorly. The lobule is easily pierced, and since it is not highly sensitive to pain, it provides a convenient place to hang an earring or obtain a drop of blood for clinical blood analysis. Next, place a finger on your temple just anterior

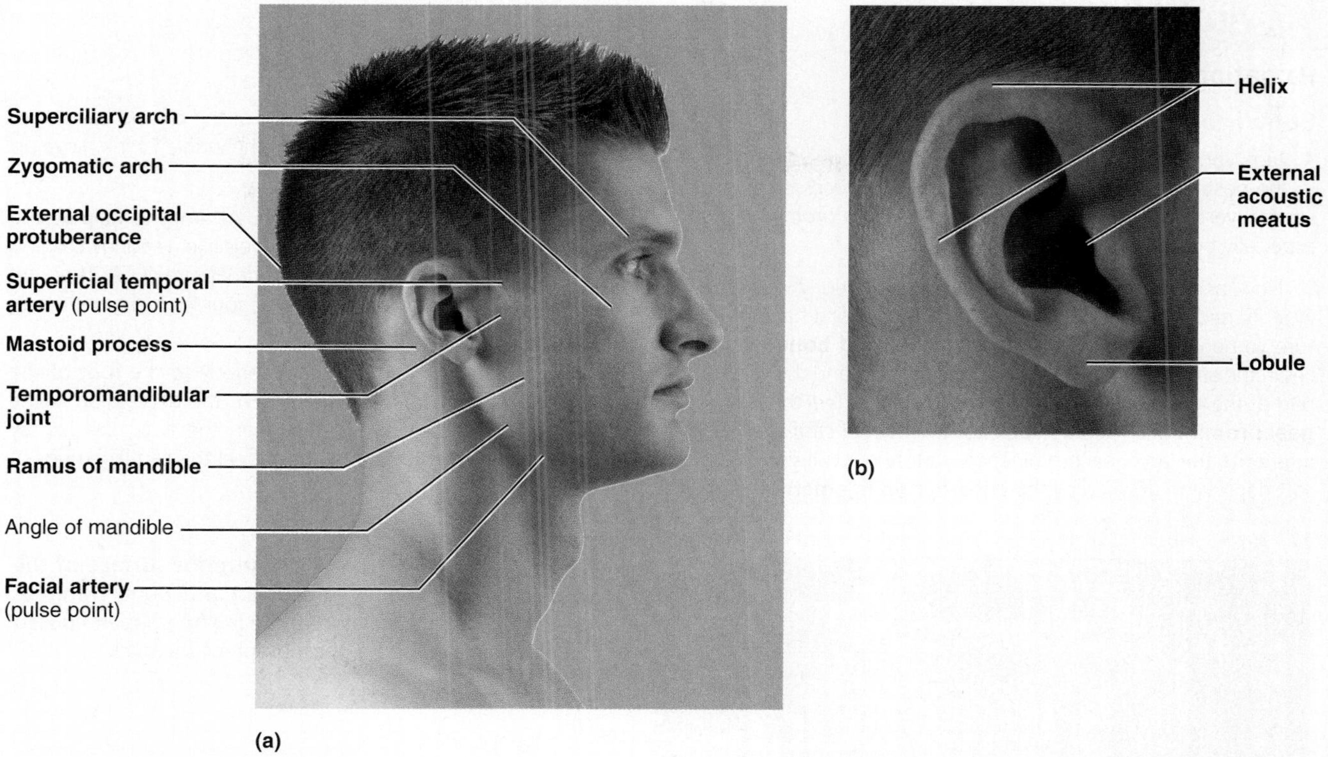

Figure 46.1 Surface anatomy of the head. (a) Lateral aspect. **(b)** Close-up of an auricle.

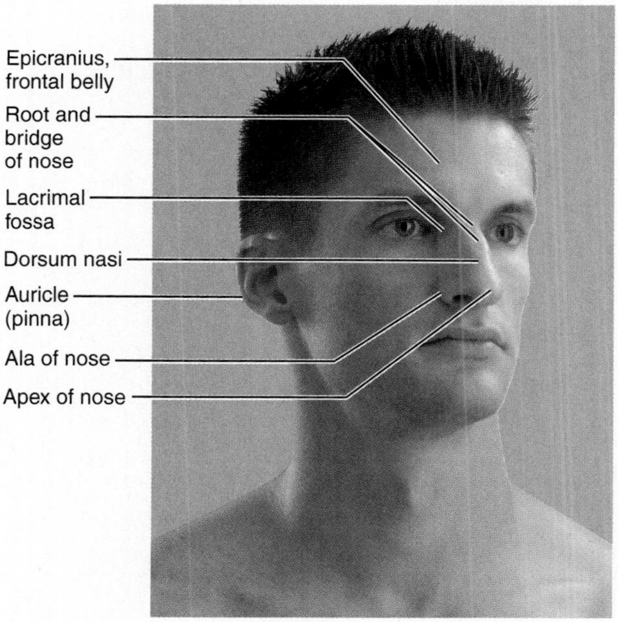

Figure 46.2 Surface structures of the face.

to the auricle. There, you may be able to feel the pulsations of the **superficial temporal artery**, which ascends to supply the scalp (Figure 46.1).

4. Run your hand anteriorly from your ear toward the orbit, and feel the **zygomatic arch** just deep to the skin. This bony arch is easily broken by blows to the face. Next, place your fingers on the skin of your face, and feel it bunch and stretch as you contort your face into smiles, frowns, and grimaces. You are now monitoring the action of several of the subcutaneous **muscles of facial expression** (Table 13.1, p. 198).

5. On your lower jaw, palpate the parts of the bony **mandible**: its anterior body and its posterior ascending **ramus**. Press on the skin over the mandibular ramus, and feel the **masseter muscle** bulge when you clench your teeth. Palpate the anterior border of the masseter, and trace it to the mandible's inferior margin. At this point, you will be able to detect the pulse of your **facial artery** (Figure 46.1). Finally, to feel the **temporomandibular joint**, place a finger directly anterior to the external acoustic meatus of your ear, and open and close your mouth several times. The bony structure you feel moving is the *condylar process of the mandible*.

Activity 2

Palpating Landmarks of the Neck

Bony Landmarks

1. Run your fingers inferiorly along the back of your neck, in the posterior midline, to feel the *spinous processes* of the cervical vertebrae. The spine of C_7, the *vertebra prominens*, is especially prominent.

2. Now, beginning at your chin, run a finger inferiorly along the anterior midline of your neck (**Figure 46.3**). The first hard structure you encounter will be the U-shaped **hyoid bone**, which lies in the angle between the floor of the mouth and the vertical part of the neck. Directly inferior to this, you will feel the **laryngeal prominence** (Adam's apple) of the thyroid cartilage. Just inferior to the laryngeal prominence, your finger will sink into a soft depression (formed by the **cricothyroid ligament**) before proceeding onto the rounded surface of the **cricoid cartilage**. Now swallow several times, and feel the whole larynx move up and down.

3. Continue inferiorly to the trachea. Attempt to palpate the *isthmus of the thyroid gland*, which feels like a spongy cushion over the second to fourth tracheal rings (Figure 46.3). Then, try to palpate the two soft lateral *lobes* of your thyroid gland along the sides of the trachea.

4. Move your finger all the way inferiorly to the root of the neck, and rest it in the **jugular notch**, the depression in the superior part of the sternum between the two clavicles. By pushing deeply at this point, you can feel the cartilage rings of the trachea.

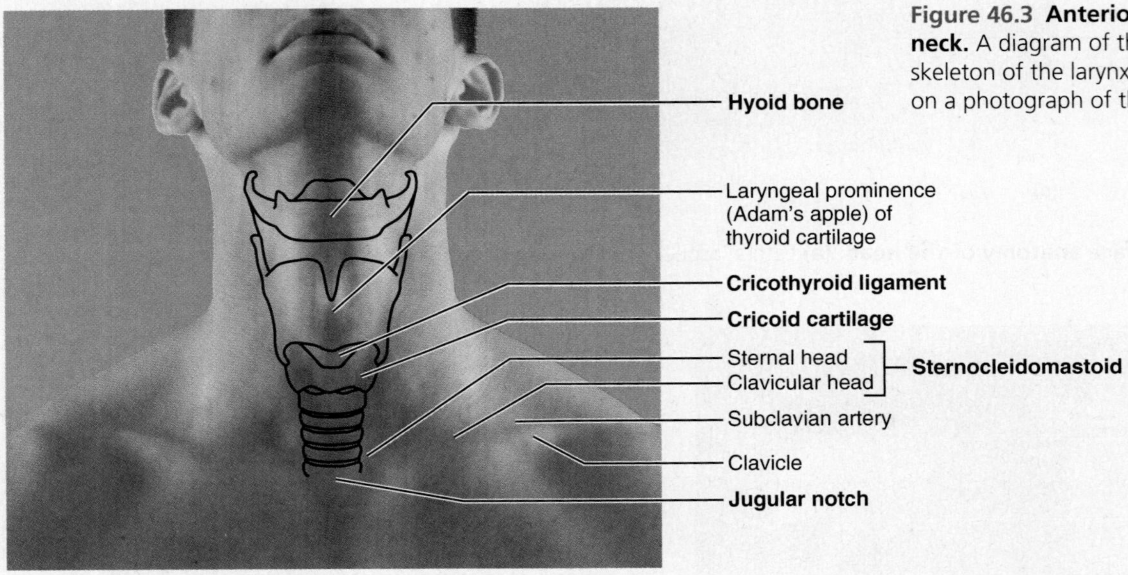

Figure 46.3 Anterior surface of the neck. A diagram of the underlying skeleton of the larynx is superimposed on a photograph of the neck.

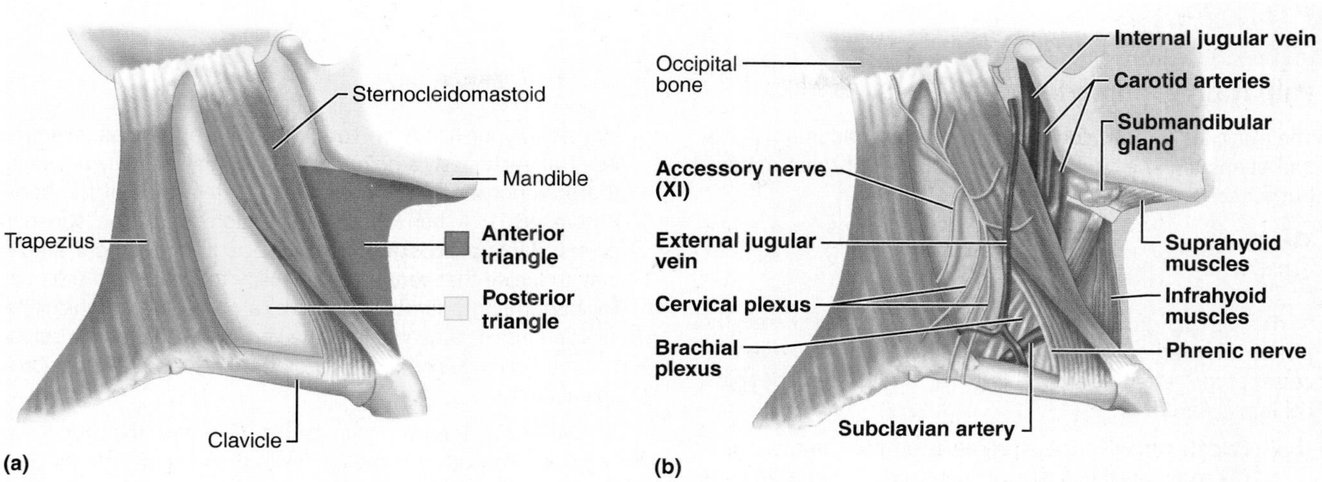

Figure 46.4 Anterior and posterior triangles of the neck. (a) Boundaries of the triangles. **(b)** Some contents of the triangles.

Muscles

The **sternocleidomastoid** is the most prominent muscle in the neck and the neck's most important surface landmark. You can best see and feel it when you turn your head to the side.

Obtain a hand mirror, hold it in front of your face, and turn your head sharply from right to left several times. You will be able to see both heads of this muscle, the **sternal head** medially and the **clavicular head** laterally (Figure 46.3). Several important structures lie beside or beneath the sternocleidomastoid:

- The *cervical lymph nodes* lie both superficial and deep to this muscle. Swollen cervical nodes provide evidence of infections or cancer of the head and neck.
- The *common carotid artery* and *internal jugular vein* lie just deep to the sternocleidomastoid, a relatively superficial location that exposes these vessels to danger in slashing wounds to the neck.
- Just lateral to the inferior part of the sternocleidomastoid is the large **subclavian artery** on its way to supply the upper limb. By pushing on the subclavian artery at this point, one can stop the bleeding from a wound anywhere in the associated limb.
- Just anterior to the sternocleidomastoid, superior to the level of your larynx, you can feel a carotid pulse—the pulsations of the **external carotid artery** (Figure 46.4).
- The *external jugular vein* descends vertically, just superficial to the sternocleidomastoid and deep to the skin (Figure 46.4b). To make this vein "appear" on your neck, stand before the mirror, and gently compress the skin superior to your clavicle with your fingers.

Triangles of the Neck

The sternocleidomastoid muscles divide each side of the neck into the posterior and anterior triangles (Figure 46.4a).

1. The **posterior triangle** is defined by the sternocleidomastoid anteriorly, the trapezius posteriorly, and the clavicle inferiorly. Palpate the borders of the posterior triangle.

   The **anterior triangle** is defined by the inferior margin of the mandible superiorly, the midline of the neck anteriorly, and the sternocleidomastoid posteriorly.

2. The contents of these two triangles include nerves, glands, blood vessels, and small muscles (Figure 46.4b). The posterior triangle contains the **accessory nerve** (cranial nerve XI), most of the **cervical plexus**, and the **phrenic nerve**. In the inferior part of the triangle are the **external jugular vein**, the trunks of the **brachial plexus**, and the **subclavian artery**. These structures are relatively superficial and are easily cut or injured by wounds to the neck.

   In the neck's anterior triangle, important structures include the **submandibular gland**, the **suprahyoid** and **infrahyoid muscles**, and parts of the **carotid arteries** and **jugular veins** that lie superior to the sternocleidomastoid.

   - Palpate your carotid pulse.

A wound to the posterior triangle of the neck can lead to long-term loss of sensation in the skin of the neck and shoulder, as well as partial paralysis of the sternocleidomastoid and trapezius muscles. Explain these effects. +

____

____

Activity 3

Palpating Landmarks of the Trunk

The trunk of the body consists of the thorax, abdomen, pelvis, and perineum. The *back* includes parts of all of these regions, but for convenience it is treated separately.

The Back

Bones

1. The vertical groove in the center of the back is called the **posterior median furrow** (**Figure 46.5**). The *spinous processes* of the vertebrae are visible in the furrow when the spinal column is flexed.

- Palpate a few of these processes on your partner's back (C_7 and T_1 are the most prominent and the easiest to find).
- Also palpate the posterior parts of some ribs, as well as the prominent **spine of the scapula** and the scapula's long **medial border**.

  The scapula lies superficial to ribs 2 to 7; its **inferior angle** is at the level of the spinous process of vertebra T_7. The medial end of the scapular spine lies opposite the T_3 spinous process.

2. Now feel the **iliac crests** (superior margins of the iliac bones) in your own lower back. You can find these crests effortlessly by resting your hands on your hips. Locate the most superior point of each crest, a point that lies roughly halfway between the posterior median furrow and the lateral side of the body (Figure 46.5). A horizontal line through these two superior points, the **supracristal line**, intersects L_4, providing a simple way to locate that vertebra. The ability to locate L_4 is essential for performing a *lumbar puncture*, a procedure in which the clinician inserts a needle into the vertebral canal of the spinal column directly superior or inferior to L_4 and withdraws cerebrospinal fluid.

3. The *sacrum* is easy to palpate just superior to the cleft in the buttocks. You can feel the *coccyx* in the extreme inferior part of that cleft, just posterior to the anus.

Muscles

The largest superficial muscles of the back are the **trapezius** superiorly and **latissimus dorsi** inferiorly (Figure 46.5). Furthermore, the deeper **erector spinae** muscles are very evident in the lower back, flanking the vertebral column like thick vertical cords.

1. Shrug your shoulders to feel the trapezius contracting just deep to the skin.

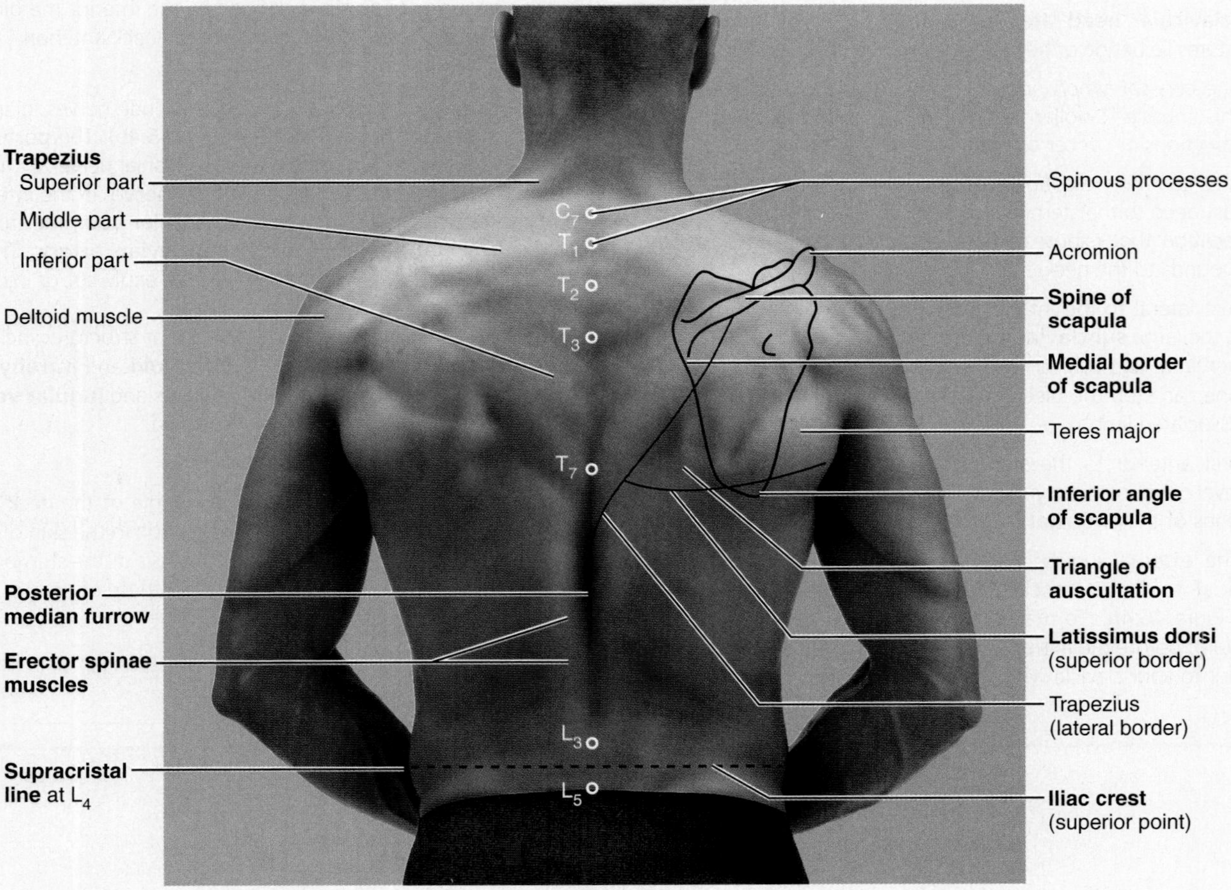

Figure 46.5 Surface anatomy of the back.

46

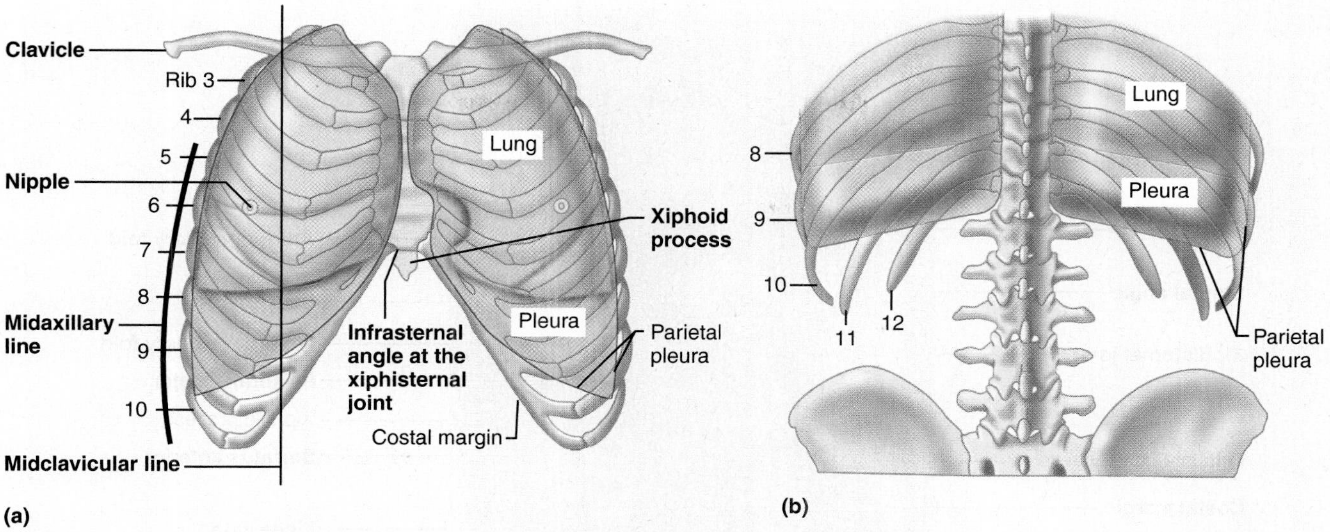

Figure 46.6 The bony rib cage as it relates to the underlying lungs and pleural cavities. Both the pleural cavities (blue) and the lungs (pink) are outlined. **(a)** Anterior view. **(b)** Posterior view.

2. Feel your partner's erector spinae muscles contract and bulge as he straightens his spine from a slightly bent-over position.

The superficial muscles of the back fail to cover a small area of the rib cage called the **triangle of auscultation** (Figure 46.5). This triangle lies just medial to the inferior part of the scapula. Its three boundaries are formed by the trapezius medially, the latissimus dorsi inferiorly, and the scapula laterally. The physician places a stethoscope over the skin of this triangle to listen for lung sounds (*auscultation* = listening). To hear the lungs clearly, the doctor first asks the patient to fold the arms together in front of the chest and then flex the trunk.

What do you think is the precise reason for having the patient take this action?

____

____

3. Have your partner assume the position just described. After cleaning the earpieces with an alcohol swab, use the stethoscope to auscultate the lung sounds. Compare the clarity of the lung sounds heard over the triangle of auscultation to that over other areas of the back.

The Thorax

Bones

1. Start exploring the anterior surface of your partner's bony *thoracic cage* (**Figure 46.6** and **Figure 46.7**, p. 684) by defining the extent of the *sternum*. Use a finger to trace the sternum's triangular *manubrium* inferior to the jugular notch, its flat *body*, and the tongue-shaped **xiphoid process**. Now palpate the ridgelike **sternal angle**, where the manubrium meets the body of the sternum. Locating the sternal angle is important because it directs you to the second ribs (which attach to it). Once you find the second rib, you can count down to identify every other rib in the thorax (except the first and sometimes the twelfth rib, which lie too deep to be palpated). The sternal angle is a highly reliable landmark—it is easy to locate, even in overweight people.

2. By locating the individual ribs, you can mentally "draw" a series of horizontal lines of "latitude" that you can use to map and locate the underlying visceral organs of the thoracic cavity. Such mapping also requires lines of "longitude," so let us construct some vertical lines on the wall of your partner's trunk. As he lifts an arm straight up in the air, extend a line inferiorly from the center of the axilla onto his lateral thoracic wall. This is the **midaxillary line** (Figure 46.6a). Now estimate the midpoint of his **clavicle**, and run a vertical line inferiorly from that point toward the groin. This is the **midclavicular line**, and it will pass about 1 cm medial to the nipple.

3. Next, feel along the V-shaped inferior edge of the rib cage, the **costal margin**. At the **infrasternal angle**, the superior angle of the costal margin, lies the **xiphisternal joint**. The heart lies on the diaphragm deep to the xiphisternal joint.

4. The thoracic cage provides many valuable landmarks for locating the vital organs of the thoracic and abdominal cavities. On the anterior thoracic wall, ribs 2–6 define the superior-to-inferior extent of the female breast, and the fourth intercostal space indicates the location of the **nipple** in men, children, and small-breasted women. The right costal margin runs across the anterior surface of the liver and gallbladder. Surgeons must be aware of the inferior margin of the *pleural cavities* because if they accidentally cut into one of these cavities, a lung collapses. The inferior pleural margin lies adjacent to vertebra T_{12} near the posterior midline (Figure 46.6b) and runs horizontally across the back to reach rib 10 at the midaxillary line. From there, the pleural margin ascends to rib 8 in the midclavicular line (Figure 46.6a) and to the level of the xiphisternal joint near the

Text continues on next page. →

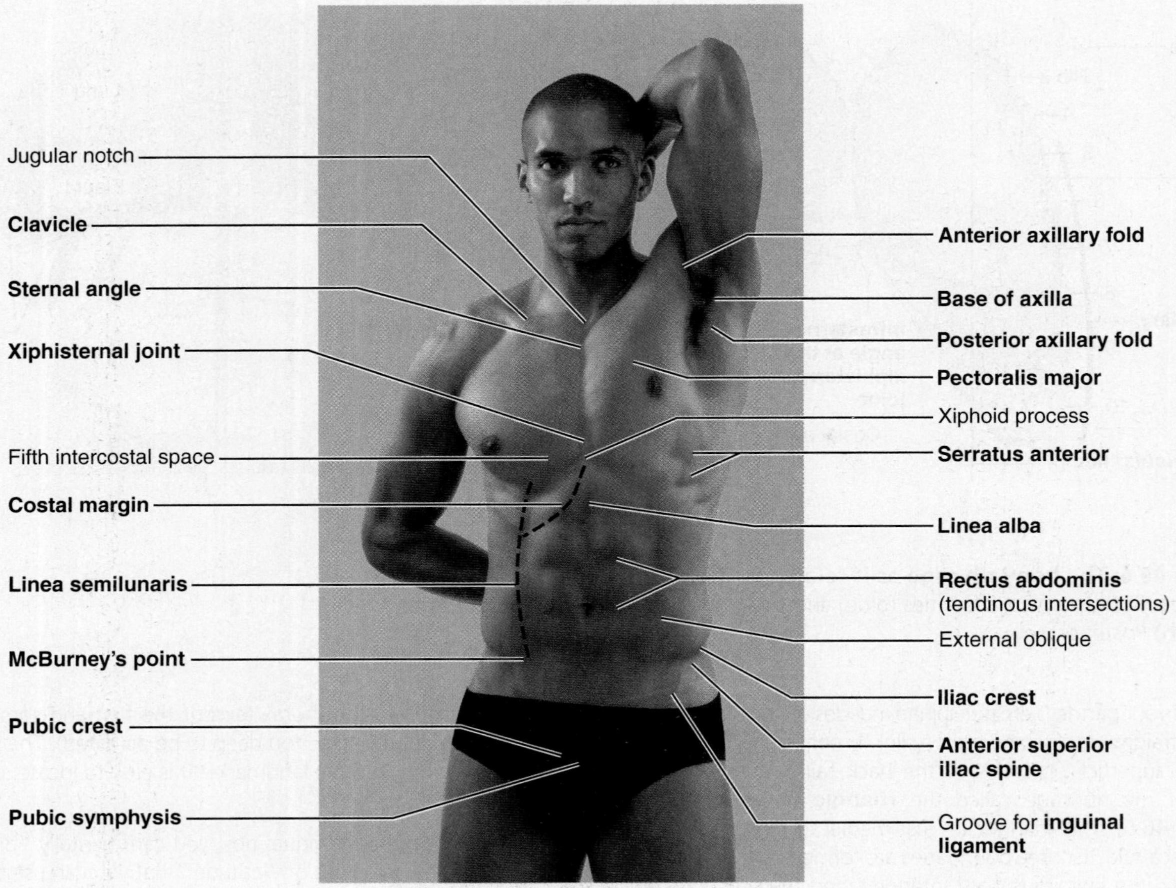

Figure 46.7 The anterior thorax and abdomen.

anterior midline. The *lungs* do not fill the inferior region of the pleural cavity. Instead, their inferior borders run at a level that is two ribs superior to the pleural margin, until they meet that margin near the xiphisternal joint.

5. Let's review the relationship of the *heart* to the thoracic cage. The superior right corner of the heart lies at the junction of the third rib and the sternum; the superior left corner lies at the second rib, near the sternum; the inferior left corner lies in the fifth intercostal space in the midclavicular line; and the inferior right corner lies at the sternal border of the sixth rib. You may wish to outline the heart on your chest or that of your lab partner by connecting the four corner points with a washable marker.

Muscles

The main superficial muscles of the anterior thoracic wall are the **pectoralis major** and the anterior slips of the **serratus anterior** (Figure 46.7).

- Palpate these two muscles on your chest. They both contract during push-ups, and you can confirm this by pushing yourself up from your desk with one arm while palpating the muscles with your opposite hand.

46

Activity 4

Palpating Landmarks of the Abdomen

Bony Landmarks

The anterior abdominal wall (Figure 46.7) extends inferiorly from the costal margin to an inferior boundary that is defined by several landmarks. Palpate these landmarks as they are described below.

1. **Iliac crest**. Locate the iliac crests by resting your hands on your hips.

2. **Anterior superior iliac spine**. Representing the most anterior point of the iliac crest, this spine is a prominent landmark. It can be palpated in everyone, even those who are overweight. Run your fingers anteriorly along the iliac crest to its end.

3. **Inguinal ligament**. The inguinal ligament, indicated by a groove on the skin of the groin, runs medially from the anterior superior iliac spine to the pubic tubercle of the pubis.

4. **Pubic crest.** You will have to press deeply to feel this crest on the pubis near the median **pubic symphysis**. The **pubic tubercle**, the most lateral point of the pubic crest, is easier to palpate, but you will still have to push deeply.

Inguinal hernias occur immediately superior to the inguinal ligament and may exit from a medial opening called the **superficial inguinal ring**. To locate this ring, one would palpate the pubic tubercle. An inguinal hernia in a male can be detected by pushing into the superficial inguinal ring (**Figure 46.8**). +

Muscles and Other Surface Features

The central landmark of the anterior abdominal wall is the *umbilicus* (navel). Running superiorly and inferiorly from the umbilicus is the **linea alba** ("white line"), represented in the skin of lean people by a vertical groove (Figure 46.7). The linea alba is a tendinous seam that extends from the xiphoid process to the pubic symphysis, just medial to the rectus abdominis muscles (Table 13.4, pp. 204–208). The linea alba is a favored site for surgical entry into the abdominal cavity because the surgeon can make a long cut through this line with no muscle damage and minimal bleeding.

Several kinds of hernias involve the umbilicus and the linea alba. In an **acquired umbilical hernia**, the linea alba weakens until intestinal coils push through it just superior to the navel. The herniated coils form a bulge just deep to the skin.

Another type of umbilical hernia is a **congenital umbilical hernia**, present in some infants: The umbilical hernia is seen as a cherry-sized bulge deep to the skin of the navel that enlarges whenever the baby cries. Congenital umbilical hernias are usually harmless, and most correct themselves automatically before the child's second birthday. +

1. **McBurney's point** is the spot on the anterior abdominal skin that lies directly superficial to the base of the appendix (Figure 46.7). It is located one-third of the way along a line between the right anterior superior iliac spine and the umbilicus. Try to find it on your body.

McBurney's point is often the place where the pain of appendicitis is experienced most acutely. Pain at McBurney's point after the pressure is removed (rebound tenderness) can indicate appendicitis. This is not a *precise* method of diagnosis, however.

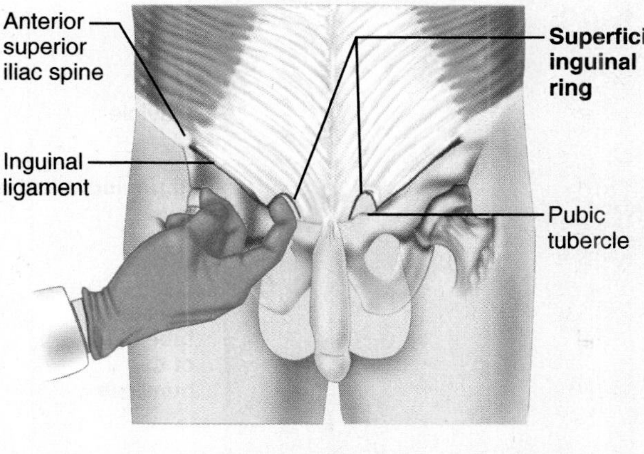

Figure 46.8 Clinical examination for an inguinal hernia in a male. The examiner palpates the patient's pubic tubercle, pushes superiorly to invaginate the scrotal skin into the superficial inguinal ring, and asks the patient to cough. If an inguinal hernia exists, it will push inferiorly and touch the examiner's fingertip.

2. Flanking the linea alba are the vertical straplike **rectus abdominis** muscles (Figure 46.7). Feel these muscles contract just deep to your skin as you do a bent-knee sit-up (or as you bend forward after leaning back in your chair). In the skin of lean people, the lateral margin of each rectus muscle makes a groove known as the **linea semilunaris** (half-moon line). On your right side, estimate where your linea semilunaris crosses the costal margin of the rib cage. The *gallbladder* lies just deep to this spot, so this is the standard point of incision for gallbladder surgery. In muscular people, three horizontal grooves can be seen in the skin covering the rectus abdominis. These grooves represent the **tendinous intersections**, fibrous bands that subdivide the rectus muscle. Because of these subdivisions, each rectus abdominis muscle presents four distinct bulges. Try to identify these insertions on yourself or your partner.

3. The only other major muscles that can be seen or felt through the anterior abdominal wall are the lateral **external obliques**. Feel these muscles contract as you cough, strain, or raise your intra-abdominal pressure in some other way.

4. The anterior abdominal wall can be divided into four quadrants (Figure 1.7, p. 9). A clinician listening to a patient's **bowel sounds** places the stethoscope over each of the four abdominal quadrants, one after another. Normal bowel sounds, which result as peristalsis moves air and fluid through the intestine, are high-pitched gurgles that occur every 5 to 15 seconds.

- Use the stethoscope to listen to your own or your partner's bowel sounds.

Abnormal bowel sounds can indicate intestinal disorders. Absence of bowel sounds indicates a halt in intestinal activity, which follows long-term obstruction of the intestine, surgical handling of the intestine, peritonitis, or other conditions. Loud tinkling or splashing sounds, by contrast, indicate an increase in intestinal activity. Such loud sounds may accompany gastroenteritis (inflammation and upset of the GI tract) or a partly obstructed intestine. +

The Pelvis and Perineum

The bony surface features of the *pelvis* are considered with the bony landmarks of the abdomen (p. 684) and the gluteal region (p. 689). Most *internal* pelvic organs are not palpable through the skin of the body surface. A full *bladder*, however, becomes firm and can be felt through the abdominal wall just superior to the pubic symphysis. A bladder that can be palpated more than a few centimeters above this symphysis is retaining urine and dangerously full, and it should be drained by catheterization.

Activity 5

Palpating Landmarks of the Upper Limb

Axilla

The **base of the axilla** is the groove in which the underarm hair grows (Figure 46.7). Deep to this base lie the axillary *lymph nodes* (which swell and can be palpated in breast cancer), the large *axillary vessels* serving the upper limb, and much of the brachial plexus. The base of the axilla forms a "valley" between two thick, rounded ridges, the **axillary folds**. Just anterior to the base, clutch your **anterior axillary fold**, formed by the pectoralis major muscle. Then grasp your **posterior axillary fold**. This fold is formed by the latissimus dorsi and teres major muscles of the back as they course toward their insertions on the humerus.

Shoulder

1. Again locate the prominent spine of the scapula posteriorly (Figure 46.5). Follow the spine to its lateral end, the flattened **acromion** on the shoulder's summit. Then, palpate the **clavicle** anteriorly, tracing this bone from the sternum to the shoulder (**Figure 46.9**). Notice the clavicle's curved shape.

2. Now locate the junction between the clavicle and the acromion on the superolateral surface of your shoulder, at the **acromioclavicular joint**. To find this joint, thrust your arm anteriorly repeatedly until you can palpate the precise point of pivoting action.

3. Next, place your fingers on the **greater tubercle** of the humerus. This is the most lateral bony landmark on the superior surface of the shoulder. It is covered by the thick **deltoid muscle**, which forms the rounded superior part of the shoulder. Intramuscular injections are often given into the deltoid, about 5 cm (2 inches) inferior to the greater tubercle (Figure 46.17a, p. 690).

Arm

Remember, according to anatomists, the arm runs only from the shoulder to the elbow, and not beyond.

1. In the arm, palpate the humerus along its entire length, especially along its medial and lateral sides.

2. Feel the **biceps brachii** muscle contract on your anterior arm when you flex your forearm against resistance. The medial boundary of the biceps is represented by the **medial bicipital furrow** (Figure 46.9). This groove contains the large *brachial artery*, and by pressing on it with your fingertips you can feel your *brachial pulse*. Recall that the brachial artery is the artery routinely used in measuring blood pressure with a sphygmomanometer.

3. All three heads of the **triceps brachii** muscle (lateral, long, and medial) are visible through the skin of a muscular person (**Figure 46.10**).

Elbow Region

1. In the distal part of your arm, near the elbow, palpate the two projections of the humerus, the **lateral** and **medial epicondyles** (Figures 46.9 and 46.10). Midway between the epicondyles, on the posterior side, feel the **olecranon**, which forms the point of the elbow.

2. Confirm that the two epicondyles and the olecranon all lie in the same horizontal line when the elbow is extended. If these three bony processes do not line up, the elbow is dislocated.

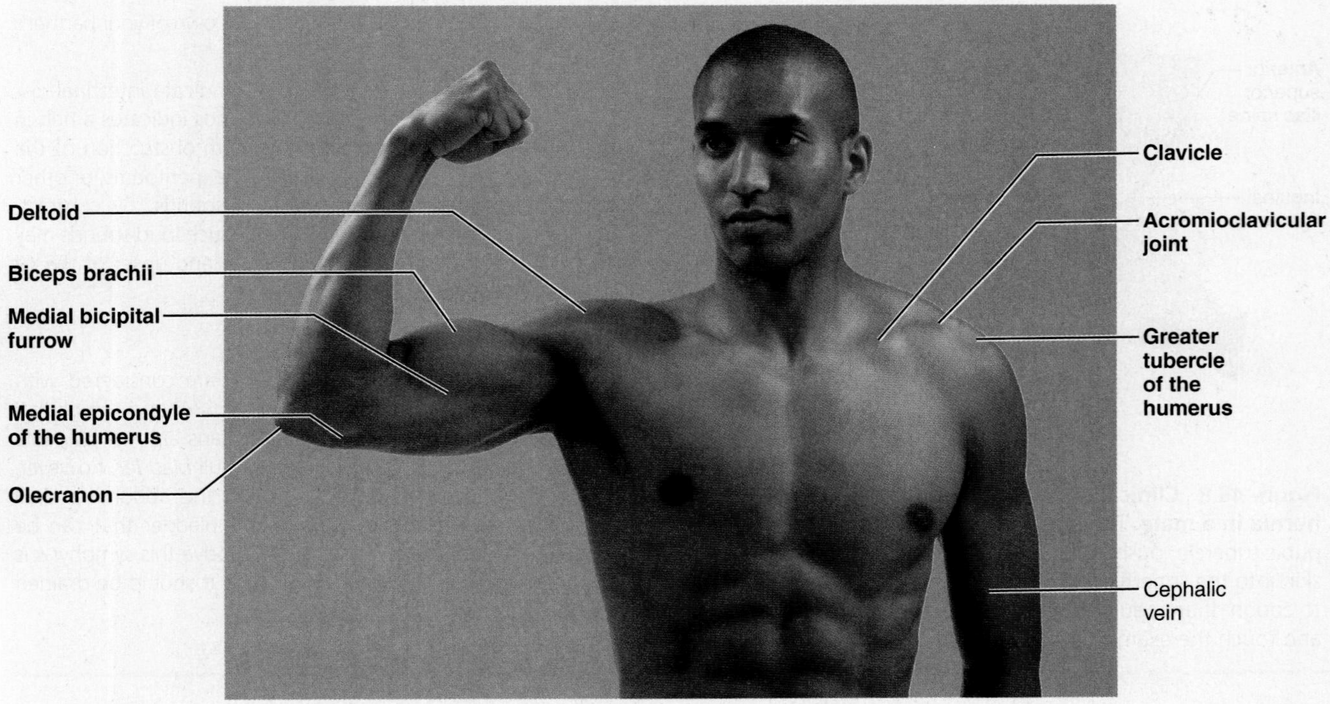

Figure 46.9 Shoulder and arm.

46

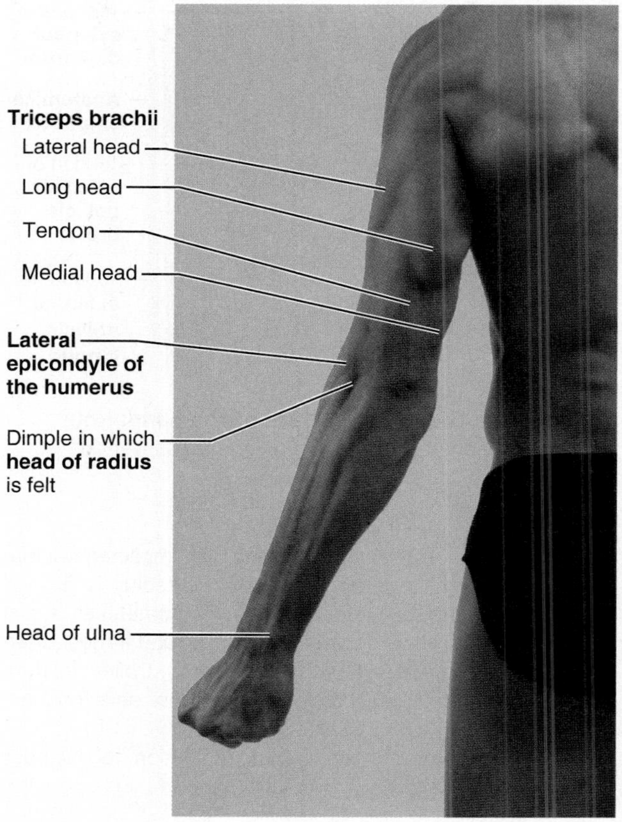

Figure 46.10 Surface anatomy of the upper limb, posterior view.

3. Now feel along the posterior surface of the medial epicondyle. You are palpating your ulnar nerve.

4. On the anterior surface of the elbow is a triangular depression called the **cubital fossa** (**Figure 46.11**). The triangle's superior *base* is formed by a horizontal line between the humeral epicondyles; its two inferior sides are defined by the **brachioradialis** and **pronator teres** muscles (Figure 46.11b). Try to define these boundaries on your own limb. To find the brachioradialis muscle, flex your forearm against resistance, and watch this muscle bulge through the skin of your lateral forearm. To feel your pronator teres contract, palpate the cubital fossa as you pronate your forearm against resistance. (Have your partner provide the resistance.)

Superficially, the cubital fossa contains the **median cubital vein** (Figure 46.11a). Clinicians often draw blood from this superficial vein and insert intravenous (IV) catheters into it to administer medications, transfused blood, and nutrient fluids. The large **brachial artery** lies just deep to the median cubital vein (Figure 46.11b), so a needle must be inserted into the vein from a shallow angle (almost parallel to the skin) to avoid puncturing the artery. Tendons and nerves are also found deep in the fossa (Figure 46.11b).

5. The median cubital vein interconnects the larger **cephalic** and **basilic veins** of the upper limb. These veins are visible through the skin of lean people (Figure 46.11a). Examine your arm to see if your cephalic and basilic veins are visible.

Forearm and Hand

The two parallel bones of the forearm are the medial *ulna* and the lateral *radius*.

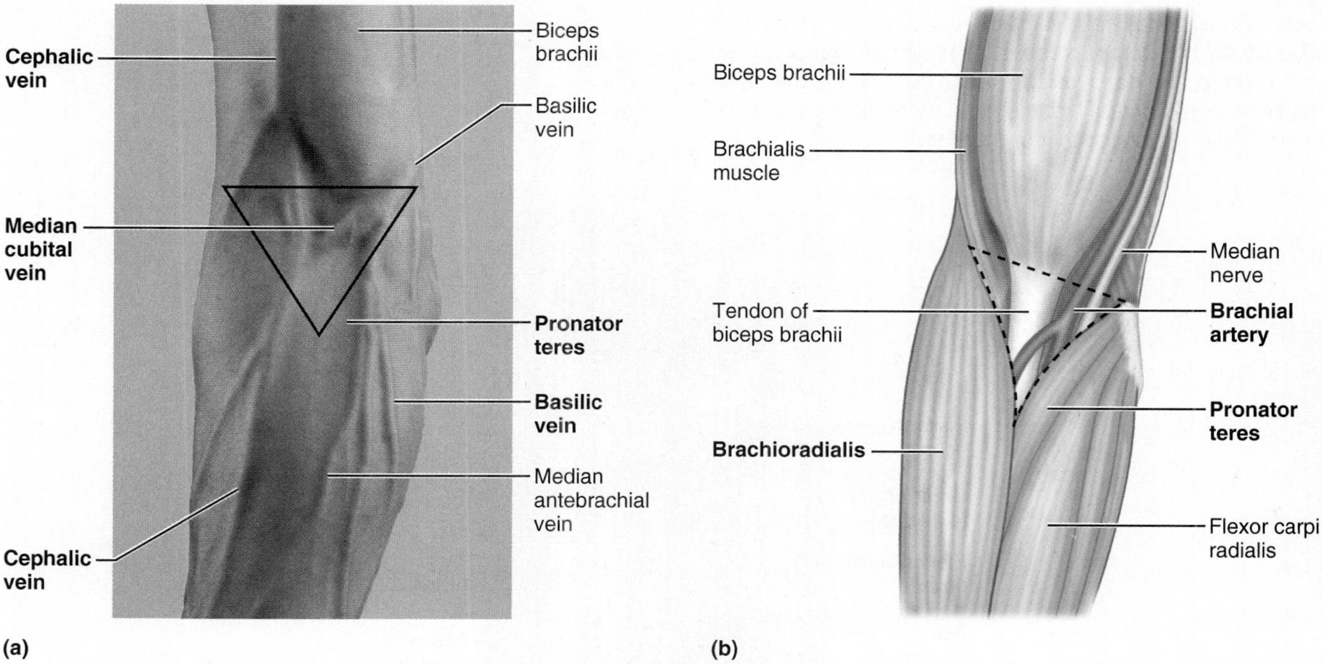

Figure 46.11 The cubital fossa on the anterior surface of the right elbow (outlined by the triangle). (a) Photograph. **(b)** Diagram of deeper structures in the fossa.

Text continues on next page. →

46

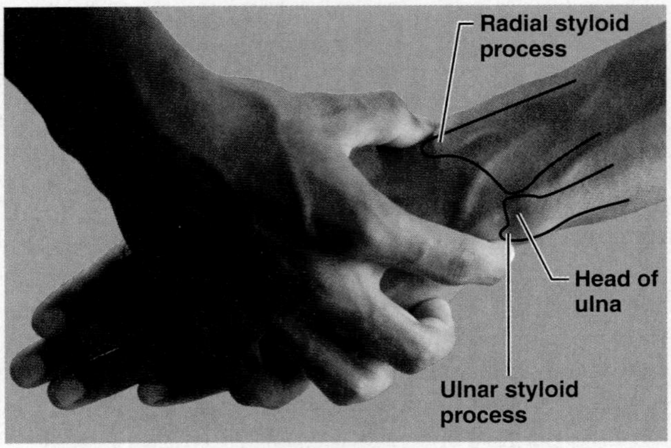

Figure 46.12 **A way to locate the ulnar and radial styloid processes.** The right hand is palpating the left hand in this picture. Note that the head of the ulna is not the same as the ulnar styloid process. The radial styloid process lies about 1 cm distal to the ulnar styloid process.

1. Feel the ulna along its entire length as a sharp ridge on the posterior forearm (confirm that this ridge runs inferiorly from the olecranon). As for the radius, you can feel its distal half, but most of its proximal half is covered by muscle. You can, however, feel the rotating **head** of the radius. To do this, extend your forearm, and note that a dimple forms on the posterior lateral surface of the elbow region (Figure 46.10). Press three fingers into this dimple, and rotate your free hand as if you were turning a doorknob. You will feel the head of the radius rotate as you perform this action.

2. Both the radius and ulna have a knoblike **styloid process** at their distal ends. Palpate these processes at the wrist (**Figure 46.12**). Do not confuse the ulnar styloid process with the conspicuous **head of the ulna**, from which the styloid process stems. Confirm that the radial styloid process lies about 1 cm (0.4 inch) distal to that of the ulna.

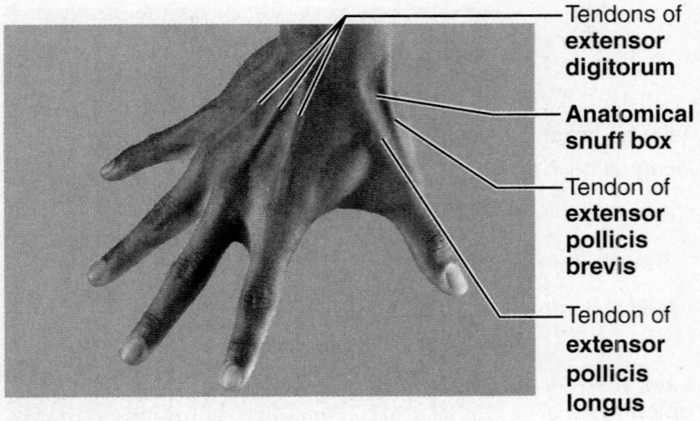

Figure 46.14 **The dorsal surface of the hand.** Note especially the anatomical snuff box and dorsal venous network.

Colles' fracture of the wrist is an impacted fracture in which the distal end of the radius is pushed proximally into the shaft of the radius. This sometimes occurs when someone falls on outstretched hands, and it most often happens to elderly women with osteoporosis. Colles' fracture bends the wrist into curves that resemble those on a fork. +

Can you deduce how physicians use palpation to diagnose Colles' fracture?

____

____

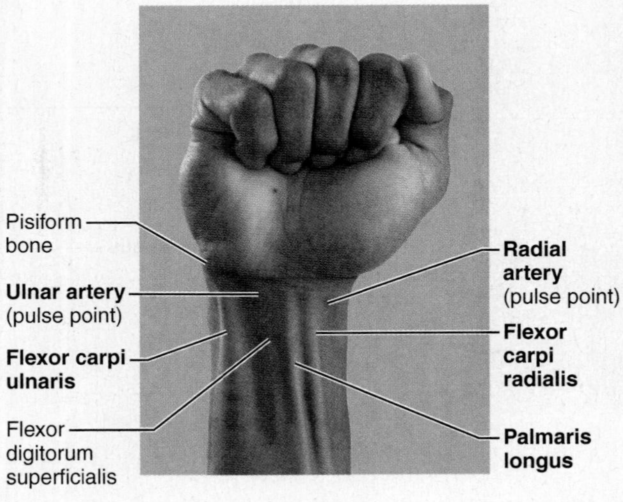

Figure 46.13 **The anterior surface of the distal forearm and fist.** The tendons of the flexor muscles guide the clinician to several sites for pulse taking.

46

3. Next, feel the major groups of muscles within your forearm. Flex your hand and fingers against resistance, and feel the anterior *flexor muscles* contract. Then extend your hand at the wrist, and feel the tightening of the posterior *extensor muscles*.

4. Near the wrist, the anterior surface of the forearm reveals many significant features (**Figure 46.13**). Flex your fist against resistance; the tendons of the main wrist flexors will bulge the skin of the distal forearm. The tendons of the **flexor carpi radialis** and **palmaris longus** muscles are most obvious. The palmaris longus, however, is absent from at least one arm in 30% of all people, so your forearm may exhibit just one prominent tendon instead of two. The **radial artery** lies just lateral to (on the thumb side of) the flexor carpi radialis tendon, where the pulse is easily detected (Figure 46.13). Feel your radial pulse here. The *median nerve*, which innervates the thumb, lies deep to the palmaris longus tendon. Finally, the **ulnar artery** lies on the medial side of the forearm, just lateral to the tendon of the **flexor carpi ulnaris**. Locate and feel your ulnar arterial pulse (Figure 46.13).

5. Extend your thumb and point it posteriorly to form a triangular depression in the base of the thumb on the back of your hand. This is the **anatomical snuff box** (**Figure 46.14**). Its two elevated borders are defined by the tendons of the thumb extensor muscles, **extensor pollicis brevis** and **extensor pollicis longus**. The radial artery runs within the snuff box, so this is another site for taking a radial pulse. The main bone on the floor of the snuff box is the scaphoid bone of the wrist, but the radial styloid process is also present here. If displaced by a bone fracture, the radial styloid process will be felt outside of the snuff box rather than within it. The "snuff box" took its name from the fact that people once put snuff (tobacco for sniffing) in this hollow before lifting it up to the nose.

6. On the dorsal surface of your hand, observe the superficial veins just deep to the skin. This is the **dorsal venous network**, which drains superiorly into the cephalic vein. This venous network provides a site for drawing blood and inserting intravenous catheters and is preferred over the median cubital vein for these purposes. Next, extend your hand and fingers, and observe the tendons of the **extensor digitorum** muscle.

7. The anterior surface of the hand also contains some features of interest (**Figure 46.15**). These features include the *epidermal ridges* (fingerprints) and many **flexion creases** in the skin. Grasp your **thenar eminence** (the bulge on the palm that contains the thumb muscles) and your **hypothenar eminence** (the bulge on the medial palm that contains muscles that move the little finger).

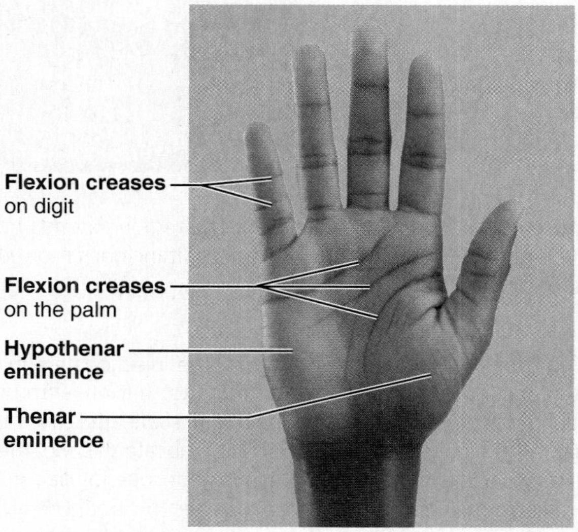

Figure 46.15 The palmar surface of the hand.

Activity 6

Palpating Landmarks of the Lower Limb

Gluteal Region

Dominating the gluteal region are the two *prominences* (cheeks) of the buttocks (**Figure 46.16**, p. 690). These are formed by subcutaneous fat and by the thick **gluteus maximus** muscles. The midline groove between the two prominences is called the **natal cleft** (*natal* = rump) or **gluteal cleft**. The inferior margin of each prominence is the horizontal **gluteal fold**, which roughly corresponds to the inferior margin of the gluteus maximus.

1. Try to palpate your **ischial tuberosity** just above the medial side of each gluteal fold (it will be easier to feel if you sit down or flex your thigh first). The ischial tuberosities are the robust inferior parts of the ischial bones, and they support the body's weight during sitting.

2. Next, palpate the **greater trochanter** of the femur on the lateral side of your hip. This trochanter lies just anterior to a hollow and about 10 cm (one hand's breadth, or 4 inches) inferior to the iliac crest. To confirm that you have found the greater trochanter, alternately flex and extend your thigh. Because this trochanter is the most superior point on the lateral femur, it moves with the femur as you perform this movement.

3. To palpate the sharp **posterior superior iliac spine**, locate your iliac crests again, and trace each to its most posterior point. You may have difficulty feeling this spine, but it is indicated by a distinct dimple in the skin that is easy to find. This dimple lies two to three finger breadths lateral to the midline of the back. The dimple also indicates the position of the *sacroiliac joint*, where the hip bone attaches to the sacrum of the spinal column. You can check *your* "dimples" out in the privacy of your home.

The gluteal region is a major site for administering intramuscular injections. When such injections are given, extreme care must be taken to avoid piercing the major nerve that lies just deep to the gluteus maximus muscle.

This thick *sciatic nerve* innervates much of the lower limb. Furthermore, the needle must avoid the gluteal nerves and gluteal blood vessels, which also lie deep to the gluteus maximus.

Text continues on next page. →

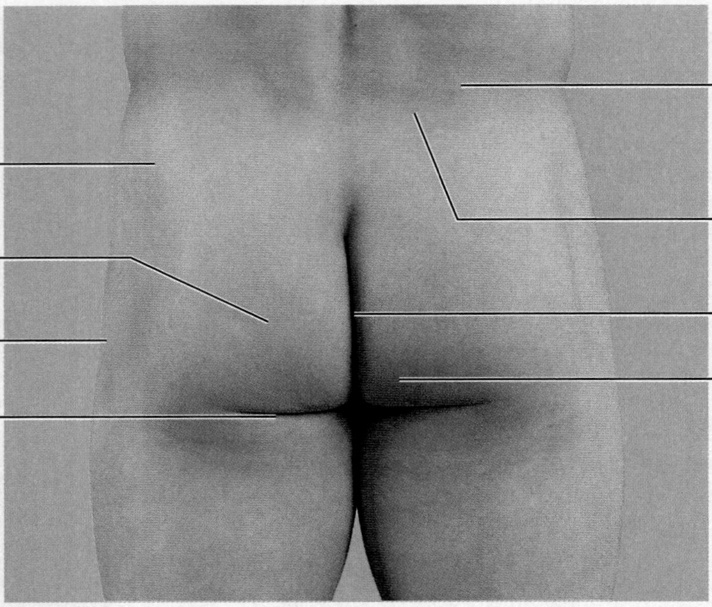

Figure 46.16 The gluteal region. The region extends from the iliac crests superiorly to the gluteal folds inferiorly. Therefore, it includes more than just the prominences of the buttock.

To avoid harming these structures, the injections are most often applied to the **gluteus *medius*** (not maximus) muscle superior to the cheeks of the buttocks, in a safe area called the **ventral gluteal site** (**Figure 46.17b**). To locate this site, mentally draw a line laterally from the posterior superior iliac spine (dimple) to the greater trochanter; the injection would be given 5 cm (2 inches) superior to the midpoint of that line. Another safe way to locate the ventral gluteal site is to approach the lateral side of the patient's left hip with your extended right hand (or the right hip with your left hand). Then, place your thumb on the anterior superior iliac spine and your index finger as far posteriorly on the iliac crest as it can reach. The heel of your hand comes to lie on the greater trochanter, and the needle is inserted in the angle of the V formed between your thumb and index finger about 4 cm (1.5 inches) inferior to the iliac crest.

Gluteal injections are not given to small children because their "safe area" is too small to locate with certainty and because the gluteal muscles are thin at this age. Instead, infants and toddlers receive intramuscular shots in the prominent **vastus lateralis** muscle of the thigh (Figure 46.17c).

Text continues on p. 692. →

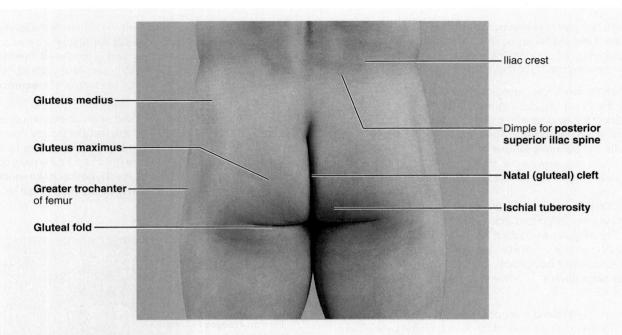

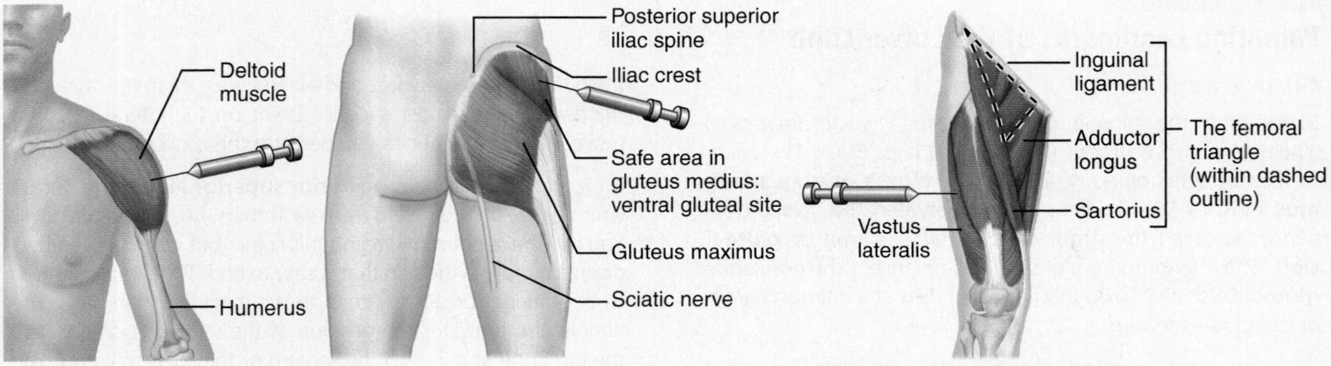

Figure 46.17 Three major sites of intramuscular injections. (a) Deltoid muscle of the arm (for injection volumes of less than 1 ml). **(b)** Ventral gluteal site (gluteus medius). **(c)** Vastus lateralis in the lateral thigh. The femoral triangle is also shown.

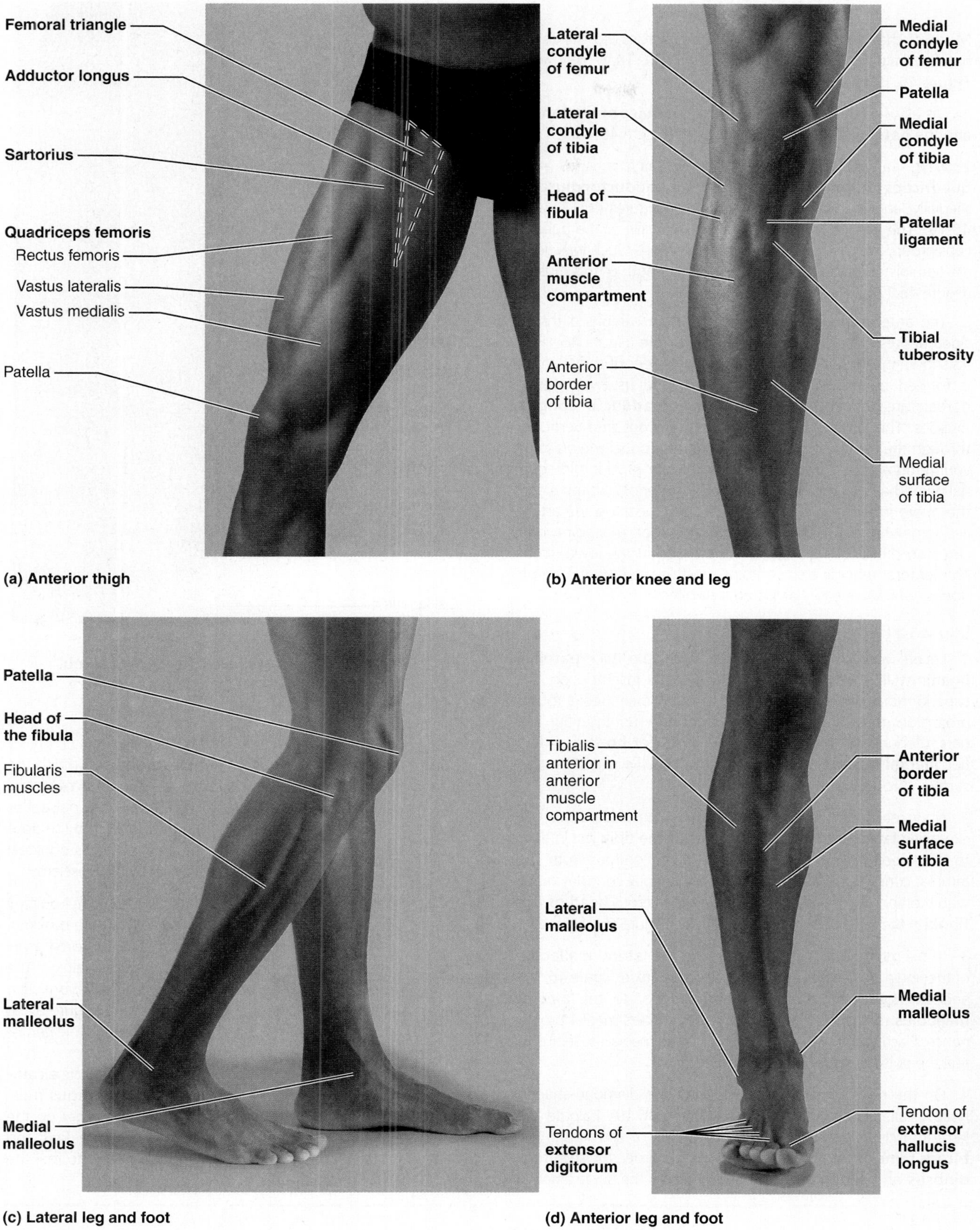

Figure 46.18 Anterior surface of the lower limb.

Thigh

Much of the femur is clothed by thick muscles, so the thigh has few palpable bony landmarks (**Figure 46.18**, p. 691, and **Figure 46.19**).

1. Distally, feel the **medial** and **lateral condyles of the femur** and the **patella** anterior to the condyles (Figure 46.18b).

2. Next, palpate your three groups of thigh muscles—the **quadriceps femoris muscles** anteriorly, the **adductor muscles** medially, and the **hamstrings** posteriorly (Figures 46.18a and 46.19). The **vastus lateralis**, the lateral muscle of the quadriceps group, is a site for intramuscular injections. Such injections are administered about halfway down the length of this muscle (Figure 46.17c).

3. The anterosuperior surface of the thigh exhibits a three-sided depression called the **femoral triangle** (Figure 46.18a). As shown in Figure 46.17c, the superior border of this triangle is formed by the **inguinal ligament**, and its two inferior borders are defined by the **sartorius** and **adductor longus** muscles. The large *femoral artery* and *vein* descend vertically through the center of the femoral triangle. To feel the pulse of your femoral artery, press inward just inferior to your midinguinal point (halfway between the anterior superior iliac spine and the pubic tubercle). Be sure to push hard, because the artery lies somewhat deep. By pressing very hard on this point, one can stop the bleeding from a hemorrhage in the lower limb. The femoral triangle also contains most of the *inguinal lymph nodes*, which are easily palpated if swollen.

Leg and Foot

1. Locate your patella again, then follow the thick **patellar ligament** inferiorly from the patella to its insertion on the superior tibia (Figure 46.18b and c). Here you can feel a rough projection, the **tibial tuberosity**. Continue running your fingers inferiorly along the tibia's sharp **anterior border** and its flat **medial surface**—bony landmarks that lie very near the surface throughout their length.

2. Now, return to the superior part of your leg, and palpate the expanded **lateral** and **medial condyles of the tibia** just inferior to the knee. You can distinguish the tibial condyles from the femoral condyles because you can feel the tibial condyles move with the tibia during knee flexion. Feel the bulbous **head of the fibula** in the superolateral region of the leg (Figure 46.18b and c).

3. In the most distal part of the leg, feel the **lateral malleolus** of the fibula as the lateral prominence of the ankle (Figure 46.18c and d). Notice that this lies slightly inferior to the **medial malleolus** of the tibia, which forms the ankle's medial prominence. Place your finger just posterior to the medial malleolus to feel the pulse of your *posterior tibial artery*.

46

4. On the posterior aspect of the knee is a diamond-shaped hollow called the **popliteal fossa** (Figure 46.19). Palpate the large muscles that define the four borders of this fossa: The **biceps femoris** forming the superolateral border, the **semitendinosus** and **semimembranosus** defining the superomedial border, and the two heads of the **gastrocnemius** forming the inferior border. The main vessels to the leg, the *popliteal artery* and *vein*, lie deep within this fossa. To feel a popliteal pulse, flex your leg at the knee and push your fingers firmly into the popliteal fossa. If a physician is unable to feel a patient's popliteal pulse, the femoral artery may be narrowed by atherosclerosis.

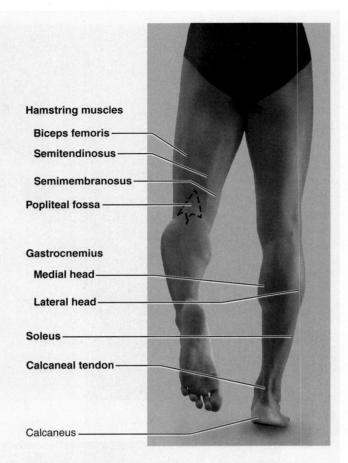

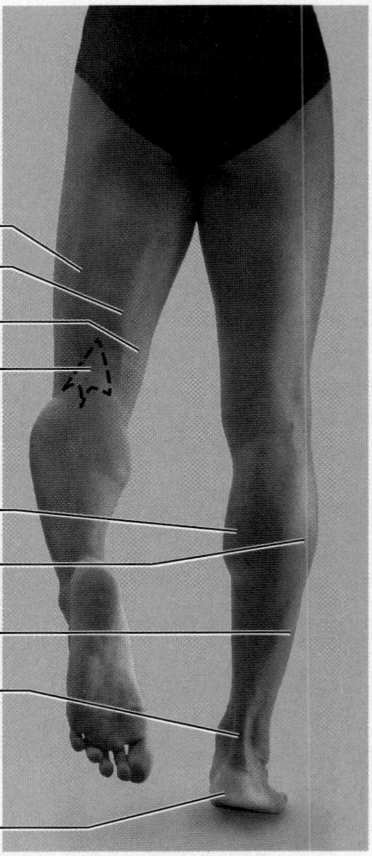

Figure 46.19 Posterior surface of the lower limb. Notice the diamond-shaped popliteal fossa posterior to the knee.

5. Observe the dorsum (superior surface) of your foot. You may see the superficial **dorsal venous arch** overlying the proximal part of the metatarsal bones (Figure 46.18d). This arch gives rise to both saphenous veins (the main superficial veins of the lower limb). Visible in lean people, the *great saphenous vein* ascends along the medial side of the entire limb (Figure 32.9, p. 472). The *small saphenous vein* ascends through the center of the calf.

As you extend your toes, observe the tendons of the **extensor digitorum longus** and **extensor hallucis longus** muscles on the dorsum of the foot. Finally, place a finger on the extreme proximal part of the space between the first and second metatarsal bones. Here you should be able to feel the pulse of the **dorsalis pedis artery**.

Instructors may assign a portion of the Review Sheet questions using Mastering A&P™

46 EXERCISE REVIEW SHEET
Surface Anatomy Roundup

Name ______________________ Lab Time/Date ______________________

______ **1.** A blow to the cheek is most likely to break what superficial bone or bone part? (a) superciliary arches, (b) mastoid process, (c) zygomatic arch, (d) ramus of the mandible

______ **2.** Rebound tenderness (a) occurs in appendicitis, (b) is whiplash of the neck, (c) is a sore foot from playing basketball, (d) occurs when the larynx falls back into place after swallowing.

______ **3.** The anatomical snuff box (a) is in the nose, (b) contains the radial styloid process, (c) is defined by tendons of the flexor carpi radialis and palmaris longus, (d) cannot really hold snuff.

______ **4.** Some landmarks on the body surface can be seen or felt, but others are abstractions that you must construct by drawing imaginary lines. Which of the following pairs of structures is abstract and invisible? (a) umbilicus and costal margin, (b) anterior superior iliac spine and natal cleft, (c) linea alba and linea semilunaris, (d) McBurney's point and midaxillary line, (e) lacrimal fossa and sternocleidomastoid

______ **5.** Many pelvic organs can be palpated by placing a finger in the rectum or the vagina, but only one pelvic organ is readily palpated through the skin. This is the (a) nonpregnant uterus, (b) prostate, (c) full bladder, (d) ovaries, (e) rectum.

______ **6.** Contributing to the posterior axillary fold is/are (a) pectoralis major, (b) latissimus dorsi, (c) trapezius, (d) infraspinatus, (e) pectoralis minor, (f) a and e.

______ **7.** Which of the following is *not* a pulse point? (a) anatomical snuff box, (b) inferior margin of mandible anterior to masseter muscle, (c) center of distal forearm at palmaris longus tendon, (d) medial bicipital furrow on arm, (e) dorsum of foot between the first two metatarsals

______ **8.** Which pair of ribs inserts on the sternum at the sternal angle? (a) first, (b) second, (c) third, (d) fourth, (e) fifth

______ **9.** The inferior angle of the scapula is at the same level as the spinous process of which vertebra? (a) C_5, (b) C_7, (c) T_3, (d) T_7, (e) L_4

______ **10.** An important bony landmark that can be recognized by a distinct dimple in the skin is the (a) posterior superior iliac spine, (b) ulnar styloid process, (c) shaft of the radius, (d) acromion.

______ **11.** A nurse missed a patient's median cubital vein while trying to withdraw blood and then inserted the needle far too deeply into the cubital fossa. This error could cause any of the following problems, *except* this one: (a) paralysis of the ulnar nerve, (b) paralysis of the median nerve, (c) bruising the insertion tendon of the biceps brachii muscle, (d) blood spurting from the brachial artery.

______ **12.** Which of these organs is almost impossible to study with surface anatomy techniques? (a) heart, (b) lungs, (c) brain, (d) nose

______ **13.** A preferred site for inserting an intravenous medication line into a blood vessel is the (a) medial bicipital furrow on arm, (b) external carotid artery, (c) dorsal venous network of hand, (d) popliteal fossa.

______ **14.** One listens for bowel sounds with a stethoscope placed (a) on the four quadrants of the abdominal wall; (b) in the triangle of auscultation; (c) in the right and left midaxillary line, just superior to the iliac crests; (d) inside the patient's bowels (intestines), on the tip of an endoscope.

______ **15.** A stab wound in the posterior triangle of the neck could damage any of the following structures *except* the (a) accessory nerve, (b) phrenic nerve, (c) external jugular vein, (d) external carotid artery.

16. What procedure requires locating the supracristal line? ______________________

____

What disease is this procedure used to detect? ______________________

17. Describe the procedure used to detect a full urinary bladder. ______________________

____

18. A patient is experiencing mastoiditis. Where would you expect the inflammation to be located? ______________________

____

DISSECTION EXERCISE 1

Dissection and Identification of Cat Muscles

Learning Outcomes

- Name and locate muscles on a dissected cat.
- Recognize similarities and differences between human and cat musculature.

Materials

- Disposable gloves or protective skin cream
- Safety glasses
- Preserved and injected cat (one for every two to four students)
- Dissecting instruments and tray
- Name tag and large plastic bag
- Paper towels
- Embalming fluid
- Organic debris container

Go to Mastering A&P™ > Study Area to improve your performance in A&P Lab.

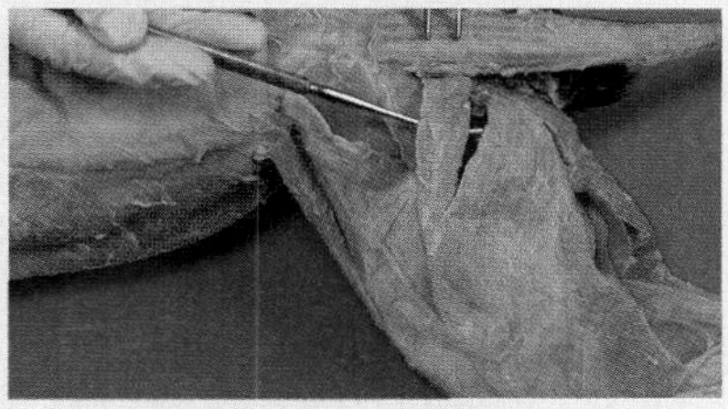

> Animations & Videos > Cat Dissection Videos > Superficial Muscles of the Trunk

Instructors may assign Cat Dissection Videos, Practice Anatomy Lab Practical questions (PAL) for the dissections, and more using the Mastering **A&P™** Item Library.

The skeletal muscles of all mammals are named in a similar fashion. However, some muscles that are separate in lower animals are fused in humans, and some muscles present in lower animals are absent in humans. In this exercise, you will dissect the cat musculature to enhance your knowledge of the human muscular system. Since the aim is to become familiar with the muscles of the human body, you should pay particular attention to the similarities between cat and human muscles. However, pertinent differences will be pointed out as you encounter them. Refer to a discussion of the anatomy of the human muscular system as you work (see Exercise 13).

When dissecting, wear safety glasses and a lab coat or apron over your clothes to prevent staining your clothes with embalming fluid. Also, read through this entire exercise before coming to the lab.

Activity 1

Preparing the Cat for Dissection

The preserved laboratory animals purchased for dissection have been embalmed with a solution that prevents deterioration of the tissues. The animals are generally delivered in plastic bags that contain a small amount of the embalming fluid. *Do not dispose of this fluid* when you remove the cat; the fluid prevents the cat from drying out. It is very important to keep the cat's tissues moist because you will probably use the same cat from now until the end of the course.

1. Don disposable gloves and safety glasses, and then obtain a cat, dissecting tray, dissecting instruments, and a name tag. Using a pencil, mark the name tag with the names of the members of your group, and set it aside. You will attach the name tag to the plastic bag at the end of the dissection so that you can identify your animal in subsequent laboratory sessions.

2. To begin removing the skin, place the cat ventral side down on the dissecting tray. Cutting away from yourself with a newly bladed scalpel, make a short, shallow incision in the midline of the neck, just to penetrate the skin. From this point on, use scissors. Continue to cut the length of the back to the sacrolumbar region, stopping at the tail (**Figure D1.1**, p. 696).

3. From the dorsal surface of the tail region, continue the incision around the tail, encircling the anus and genital organs. The skin will not be removed from this region.

4. Before you begin to remove the skin, check with your instructor. He or she may want you to skin only the right or left side of the cat. Beginning again at the dorsal tail region, make an incision through the skin down each hind leg to be skinned nearly to the ankle. Continue the cut completely around the ankle.

5. Return to the neck. Cut the skin around the circumference of the neck.

Text continues on next page. →

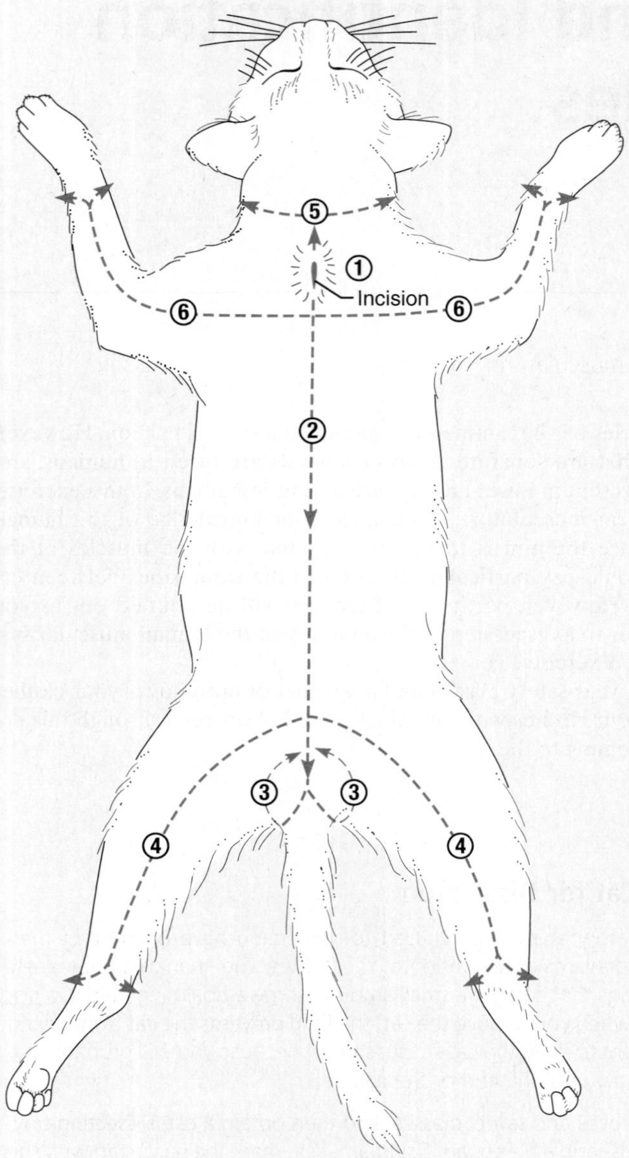

Figure D1.1 Incisions to be made in skinning a cat. Numbers indicate sequence.

6. Cut down each foreleg to be skinned to the wrist. Completely cut through the skin around the wrist (Figure D1.1).

7. Now free the skin from the loose connective tissue (superficial fascia) that binds it to the underlying structures. With one hand, grasp the skin on one side of the midline dorsal incision. Then, using your fingers or a blunt probe, break through the "cottony" connective tissue fibers to release the skin from the muscle beneath. Work toward the ventral surface and then toward the neck. As you pull the skin from the body, you should see small, white, cordlike structures extending from the skin to the muscles at fairly regular intervals. These are the cutaneous nerves that serve the skin. You will also see (particularly as you approach the ventral surface) that a thin layer of muscle fibers remains adhered to the skin. This is the **cutaneous maximus** muscle, which enables the cat to move its skin rather like our facial muscles allow us to express emotion. Where the cutaneous maximus fibers cling to those of the deeper muscles, carefully cut them free. Along the ventral surface of the trunk, notice the two lines of nipples associated with the mammary glands. These are more prominent in females, especially if they are pregnant or were recently lactating.

8. You will notice as you start to free the skin in the neck that it is more difficult to remove. Take extra care and time in this area. The large flat **platysma** muscle in the ventral neck region (a skin muscle like the cutaneous maximus) will remain attached to the skin. The skin will not be removed from the top of the head since the cat's head muscles are not sufficiently similar to human head muscles to merit study.

9. Complete the skinning process by freeing the skin from the forelimbs, the lower torso, and the hindlimbs in the same manner. The skin may be more difficult to remove as you approach the paws so you may need to take additional time in these areas to avoid damaging the underlying muscles and tendons. *Do not discard the skin.*

10. Inspect your skinned cat. Notice that it is difficult to see any cleavage lines between the muscles because of the overlying connective tissue, which is white or yellow. If time allows, carefully remove as much of the fat and fascia from the surface of the muscles as possible, using forceps or your fingers. The muscles, when exposed, look grainy or threadlike and are light brown. If you carry out this clearing process carefully and thoroughly, you will be ready to begin your identification of the superficial muscles.

11. If the muscle dissection exercises are to be done at a later laboratory session, follow the cleanup instructions noted in the box below. *Prepare your cat for storage in this way every time the cat is used.*

Preparing the Dissection Animal for Storage

1. To prevent the internal organs from drying out, dampen a layer of folded paper towels with embalming fluid, and wrap them snugly around the animal's torso. (Do not use *water-soaked* paper towels, which encourages mold growth.) Make sure the dissected areas are completely enveloped.

2. Return the animal's skin flaps to their normal position over the ventral cavity body organs.

3. Place the animal in a plastic storage bag. Add more embalming fluid if necessary, press out excess air, and securely close the bag with a rubber band or twine.

4. Make sure your name tag is securely attached, and place the animal in the designated storage container.

5. Clean all dissecting equipment with soapy water, rinse, and dry it for return to the storage area. Wash down the lab bench, and properly dispose of organic debris and your gloves before leaving the laboratory. Return safety glasses to the appropriate location.

Activity 2

Dissecting Neck and Trunk Muscles

The proper dissection of muscles involves careful separation of one muscle from another and transection of superficial muscles in order to study those lying deeper. In general, when you are instructed to **transect** a muscle, you should first completely free it from all adhering connective tissue and *then* cut through the belly (fleshiest part) of the muscle about halfway between its origin and insertion points. *Use caution when working around points of muscle origin or insertion, and do not remove the fascia associated with such attachments.*

As a rule, all the fibers of one muscle are held together by a connective tissue sheath (epimysium) and run in the same general direction. Before you begin dissection, observe your skinned cat. If you look carefully, you can see changes in the direction of the muscle fibers, which will help you to locate the muscle borders. Pulling in slightly different directions on two adjacent muscles will usually expose subtle white lines created by the connective tissue surrounding the muscles and allow you to find the normal cleavage line between them. After you have identified cleavage lines, *use a blunt probe* to break the connective tissue between muscles and to separate them. If the muscles separate as clean, distinct bundles, your procedure is probably correct. If they appear ragged or chewed up, you are probably tearing a muscle apart rather than separating it from adjacent muscles. Because of time considerations, in this exercise you will identify only the muscles that are most easily identified and separated out.

Anterior Neck Muscles

1. Examine the anterior neck surface of the cat, and identify the following superficial neck muscles. The *platysma* belongs in this group but was probably removed during the skinning process. Refer to **Figure D1.2** as you work. The **sternomastoid** muscle and the more lateral and deeper **cleidomastoid** muscle (not visible in Figure D1.2) are joined in humans to form the sternocleidomastoid. The large external jugular veins, which drain the head, should be obvious crossing the anterior aspect of these muscles. The **mylohyoid** muscle parallels the bottom aspect of the chin, and the **digastric** muscles form a V over the mylohyoid muscle. Although it is not one of the neck muscles, you can now identify the fleshy **masseter** muscle, which flanks the digastric muscle laterally. Finally, the **sternohyoid** is a narrow muscle between the mylohyoid (superiorly) and the inferior sternomastoid.

2. The deeper muscles of the anterior neck of the cat are small and straplike and hardly worth the effort of dissection. However, one of these deeper muscles can be seen with a minimum of extra effort. Transect the sternomastoid and sternohyoid muscles approximately at midbelly. Reflect the cut ends to reveal the bandlike **sternothyroid** muscle (not visible in Figure D1.2), which runs along the anterior surface of the throat just deep and lateral to the sternohyoid muscle. The cleidomastoid muscle, which lies deep to the sternomastoid, is also more easily identified now.

Text continues on p. 699. →

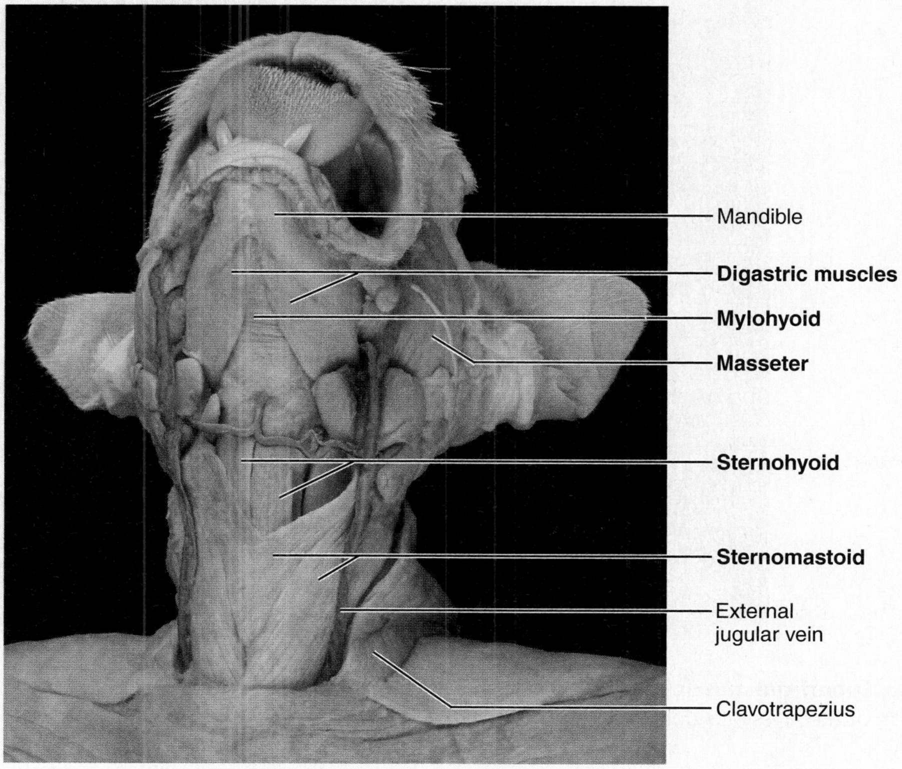

Figure D1.2 Superficial muscles of the anterior neck of the cat.

DISSECTION 1

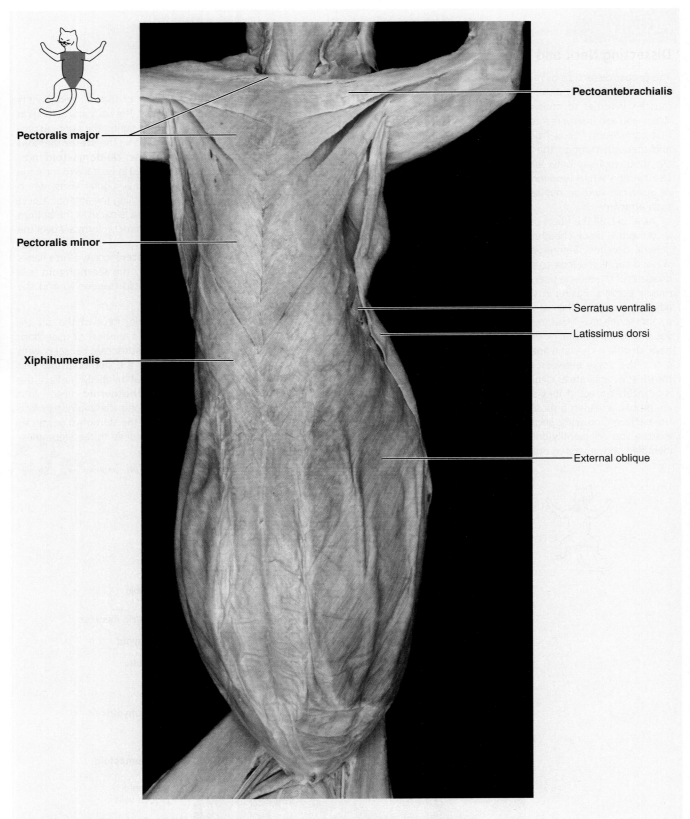

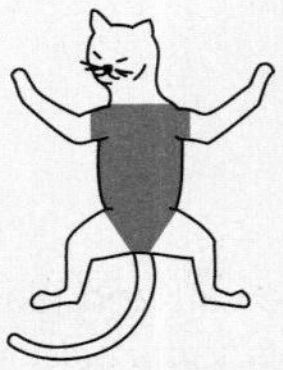

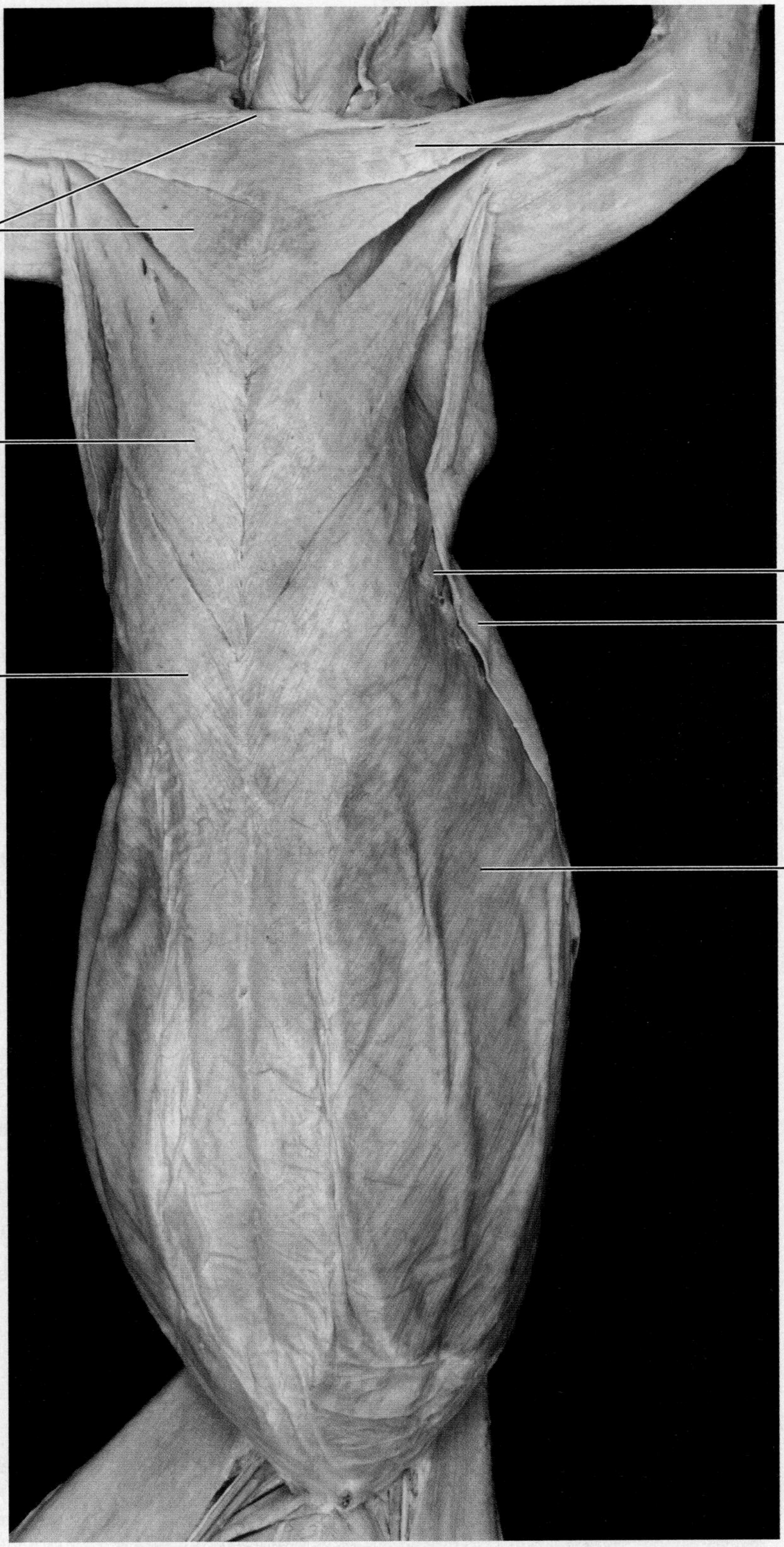

Figure D1.3 Superficial muscles of the thorax and abdominal wall, ventral view. Note location of the latissimus dorsi.

Superficial Chest Muscles

In the cat, the chest or pectoral muscles adduct the arm, just as they do in humans. However, humans have only two pectoral muscles, and cats have four—the pectoralis major, pectoralis minor, xiphihumeralis, and pectoantebrachialis (**Figure D1.3**). However, because of their relatively great degree of fusion, the cat's pectoral muscles appear to be a single muscle. The pectoral muscles are rather difficult to dissect and identify because they do not separate from one another easily.

The **pectoralis major** is 5 to 8 cm (2 to 3 inches) wide and can be seen arising on the manubrium, just inferior to the sternomastoid muscle of the neck, and running to the humerus. Its fibers run at right angles to the long axis of the cat's body.

The **pectoralis minor** lies beneath the pectoralis major and extends posterior to it on the abdominal surface. It originates on the sternum and inserts on the humerus. Its fibers run obliquely to the long axis of the body, which helps to distinguish it from the pectoralis major. Contrary to what its name implies, the pectoralis minor is a larger and thicker muscle than the pectoralis major.

The **xiphihumeralis** can be distinguished from the posterior edge of the pectoralis minor only by virtue of the fact that its origin is lower—on the xiphoid process of the sternum. Its fibers run parallel to and are fused with those of the pectoralis minor.

The **pectoantebrachialis** is a thin, straplike muscle, about 1.3 cm (½ inch) wide, lying over the pectoralis major. Notice that the pectoralis major is visible both anterior and posterior to the borders of the pectoantebrachialis. It originates from the manubrium, passes laterally over the pectoralis major, and merges with the muscles of the forelimb approximately halfway down the humerus. It has no homologue in humans.

Identify, free, and trace out the origin and insertion of the cat's chest muscles (refer to Figure D1.3).

Muscles of the Abdominal Wall

The superficial trunk muscles include those of the abdominal wall (Figure D1.3 and **Figure D1.4**, p. 700). Cat musculature in this area is quite similar in function to that of humans.

1. Complete the dissection of the more superficial anterior trunk muscles of the cat by identifying the origins and insertions of the muscles of the abdominal wall. Work carefully here. These muscles are very thin, and it is easy to miss their boundaries. Begin with the **rectus abdominis**, a long band of muscle approximately 2.5 cm (1 inch) wide running immediately lateral to the midline of the body on the abdominal surface. Humans have four transverse *tendinous intersections* in the rectus abdominis (see Figure 13.9a, p. 206), but they are absent or difficult to identify in the cat. Identify the **linea alba**, the longitudinal band of connective tissue that separates the rectus abdominis muscles. Note the relationship of the rectus abdominis to the other abdominal muscles and their fascia.

2. The **external oblique** is a sheet of muscle immediately beside the rectus abdominis (Figure D1.4). Carefully free and then transect the external oblique to reveal the anterior attachment of the rectus abdominis. Reflect the external oblique; observe the deeper **internal oblique** muscle. Notice which way the fibers run.

How does the fiber direction of the internal oblique compare to that of the external oblique?

____

____

3. Free and then transect the internal oblique muscle to reveal the fibers of the **transversus abdominis**, whose fibers run transversely across the abdomen.

Superficial Muscles of the Shoulder and the Dorsal Trunk and Neck

Dissect the superficial muscles of the dorsal surface of the trunk. Refer to **Figure D1.5** on p. 701.

1. Turn your cat on its ventral surface, and start your observations with the **trapezius group**. Humans have a single large *trapezius muscle*, but the cat has three separate muscles—the clavotrapezius, acromiotrapezius, and spinotrapezius—that together perform a similar function. The prefix in each case (clavo-, acromio-, and spino-) reveals the muscle's site of insertion. The **clavotrapezius**, the most anterior muscle of the group, is homologous to that part of the human trapezius that inserts into the clavicle. Slip a probe under this muscle, and follow it to its apparent origin.

Where does the clavotrapezius appear to originate?

____

Is this similar to its origin in humans? ____________________

The fibers of the clavotrapezius are continuous posteriorly with those of the clavicular part of the cat's deltoid muscle (clavodeltoid), and the two muscles work together to flex the humerus. Release the clavotrapezius muscle from adjoining muscles. The **acromiotrapezius** is a large, thin, nearly square muscle easily identified by its aponeurosis, which passes over the vertebral border of the scapula. It originates from the cervical and T_1 vertebrae and inserts into the scapular spine. The triangular **spinotrapezius** runs from the thoracic vertebrae to the scapular spine. This is the most posterior of the trapezius muscles in the cat. Now that you know where they are located, pull on the three trapezius muscles to mimic their action.

Do the trapezius muscles appear to have the same functions in cats as in humans?

____

2. The **levator scapulae ventralis**, a flat, straplike muscle, can be located in the triangle created by the division of the fibers of the clavotrapezius and acromiotrapezius. Its anterior fibers run underneath the clavotrapezius from its origin at

Text continues on next page. →

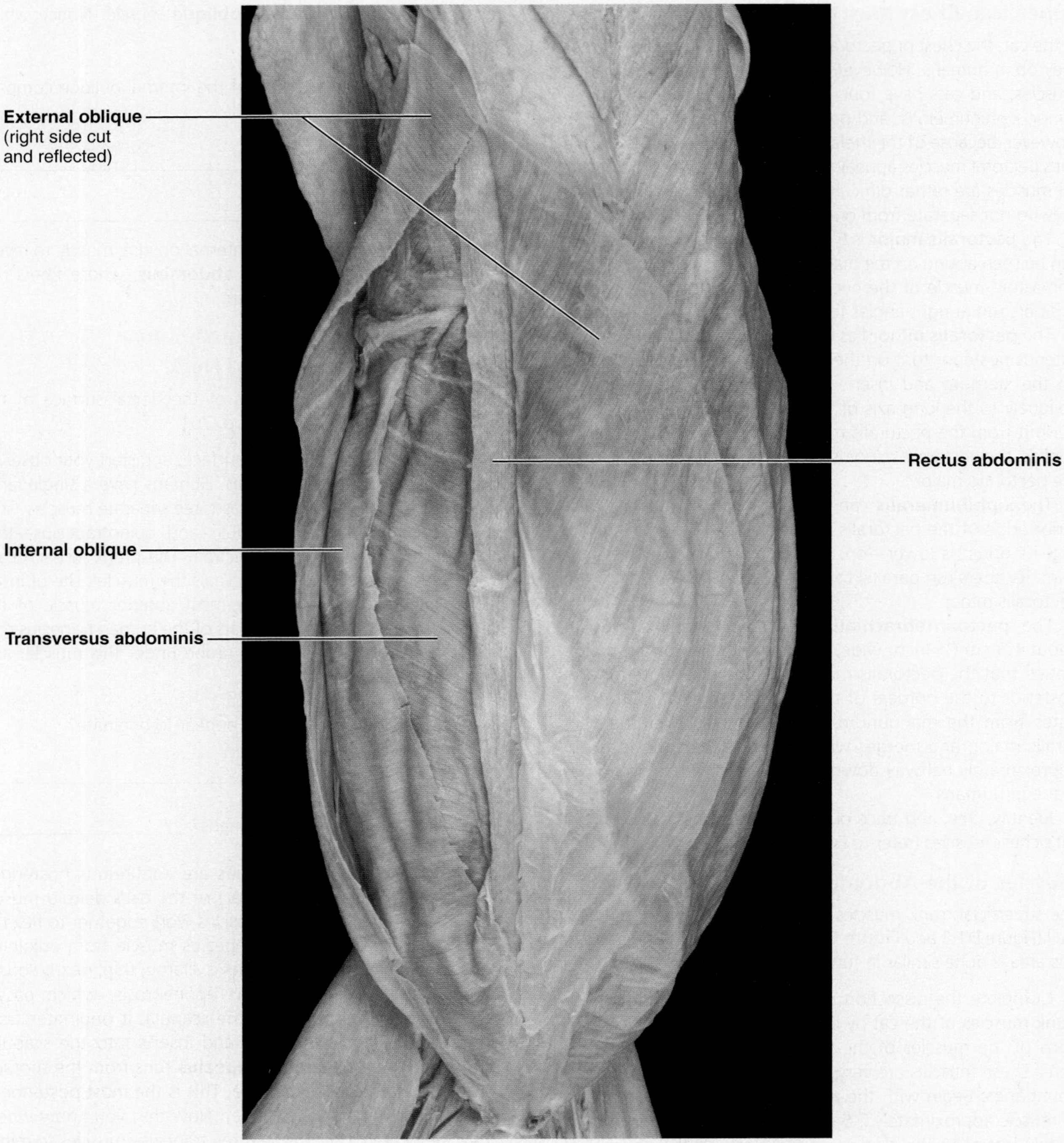

Figure D1.4 Muscles of the abdominal wall of the cat.

the occipital bone, and it inserts on the vertebral border of the scapula. In the cat it helps to hold the upper edges of the scapulae together and draws them toward the head.

What is the function of the levator scapulae in humans?

____

____

3. The **deltoid group**: Like the trapezius, the human *deltoid muscle* is represented by three separate muscles in the cat—the clavodeltoid, acromiodeltoid, and spinodeltoid. The **clavodeltoid** (also called the *clavobrachialis*), the most superficial muscle of the shoulder, is a continuation of the clavotrapezius below the clavicle, which is this muscle's point of origin (Figure D1.5). Follow its path down the forelimb to the point where it merges along a white line with the pectoantebrachialis. Separate it from the pectoantebrachialis, and then transect it and pull it back.

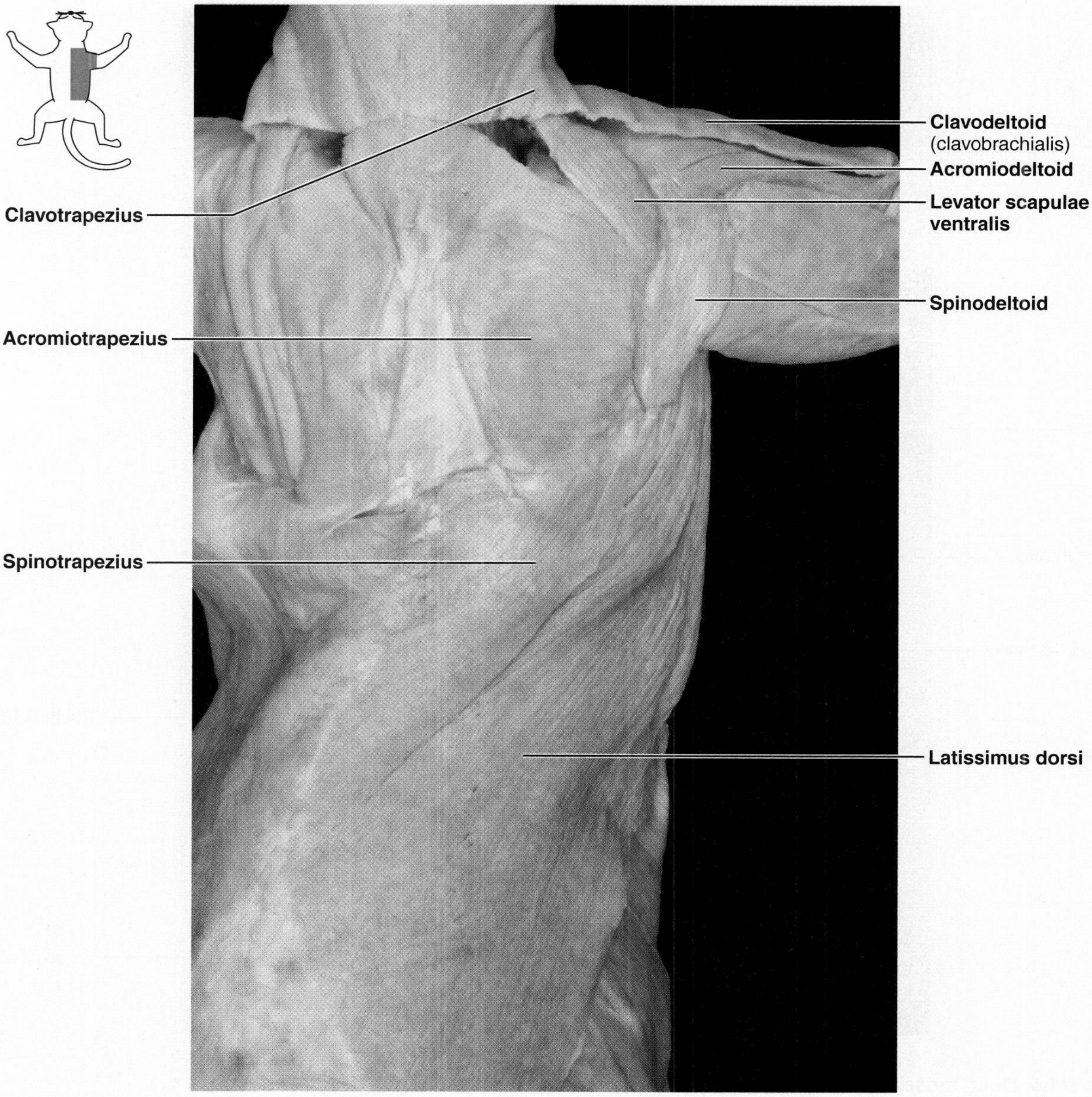

Figure D1.5 Superficial muscles of the anterodorsal aspect of the shoulder, trunk, and neck of the cat.

Where does the clavodeltoid insert? ______________________

What do you think the function of this muscle is?

____

____

The **acromiodeltoid** lies posterior to the clavodeltoid and runs over the top of the shoulder. This small triangular muscle originates on the acromion of the scapula. It inserts into the spinodeltoid (a muscle of similar size) posterior to it. The **spinodeltoid** is covered with fascia near the anterior end of the scapula. Its tendon extends under the acromiodeltoid muscle and inserts on the humerus. Notice that its fibers run obliquely to those of the acromiodeltoid. Like the human deltoid muscle, the acromiodeltoid and spinodeltoid muscles in the cat abduct and rotate the humerus.

4. The **latissimus dorsi** is a large, thick, flat muscle covering most of the lateral surface of the posterior trunk; it extends and adducts the arm. Its anterior edge is covered by the spinotrapezius and may appear ragged because it has been cut off from the cutaneous maximus muscle attached to the skin. As in humans, it inserts into the humerus. But before inserting, its fibers merge with the fibers of many other muscles, among them the xiphihumeralis of the pectoralis group.

Text continues on next page. →

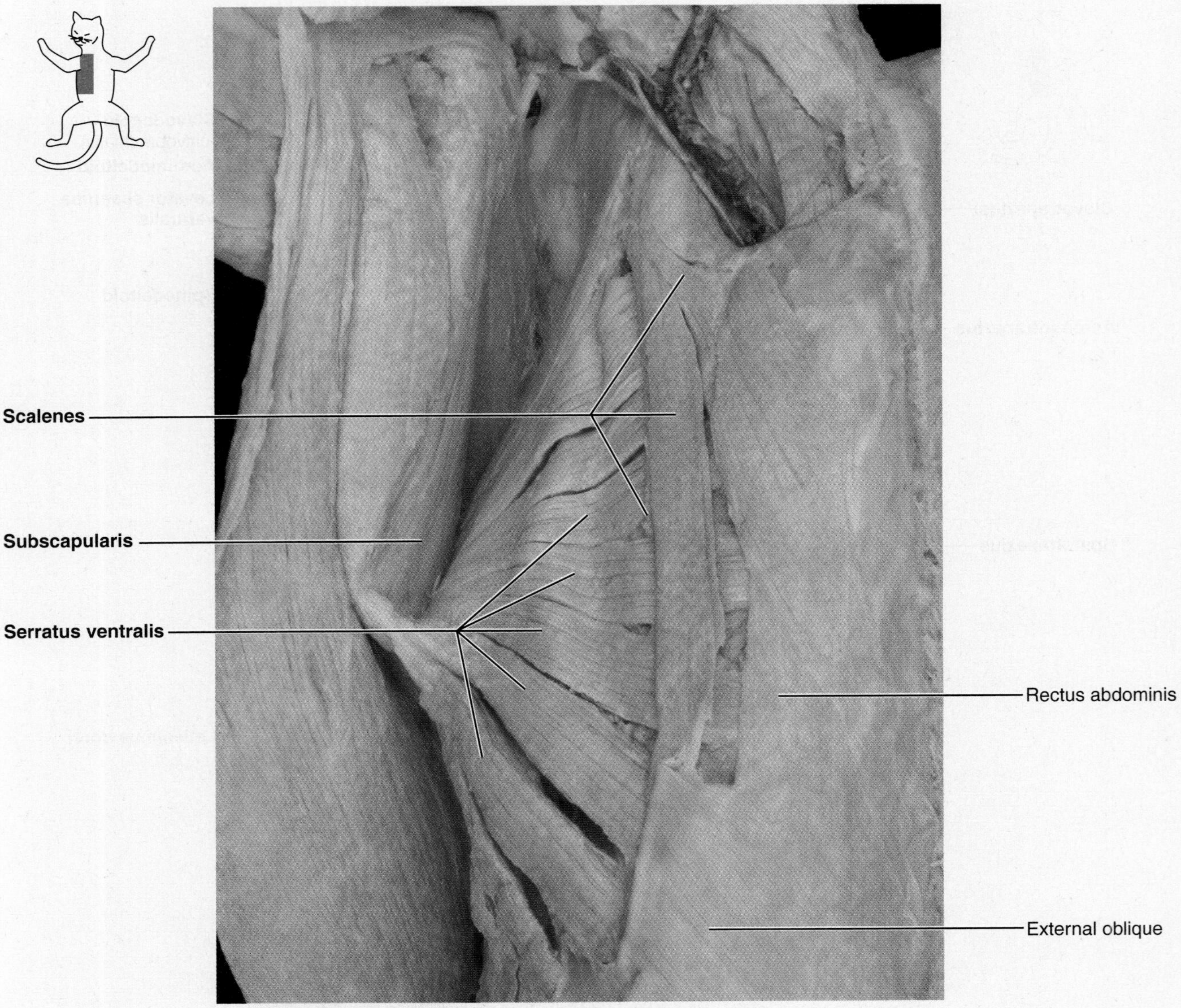

Figure D1.6 Deep muscles of the inferolateral thorax of the cat.

Deep Muscles of the Laterodorsal Trunk and Neck

1. In preparation for identifying deep muscles of the dorsal trunk, transect the latissimus dorsi, the muscles of the pectoralis group, and the spinotrapezius and reflect them back. Be careful not to damage the large brachial nerve plexus, which lies in the axillary space beneath the pectoralis group.

2. The **serratus ventralis** corresponds to two separate muscles in humans. The posterior portion is homologous to the *serratus anterior* of humans, arising deep to the pectoral muscles and covering the lateral surface of the rib cage. It is easily identified by its fingerlike muscular origins, which arise on the first 9 or 10 ribs. It inserts into the scapula (**Figure D1.6**). The anterior portion of the serratus ventralis, which arises from the cervical vertebrae, is homologous to the *levator scapulae* in humans; both pull the scapula toward the sternum. Trace this muscle to its insertion. In general, in the cat, this muscle pulls the scapula posteriorly and downward.

3. Reflect the upper limb to reveal the **subscapularis**, which occupies most of the ventral surface of the scapula (Figure D1.6). Humans have a homologous muscle.

4. Locate the anterior, posterior, and middle **scalene** muscles on the lateral surface of the cat's neck and trunk. The most prominent and longest of these muscles is the middle scalene, which lies between the anterior and posterior members. The scalenes originate on the ribs and run cephalad over the serratus ventralis to insert in common on the cervical vertebrae. These muscles draw the ribs anteriorly and bend the neck downward; thus they are homologous to the human scalene muscles, which elevate the ribs and flex the neck. (Notice that the difference is only one of position. Humans walk erect, but cats are quadrupeds.)

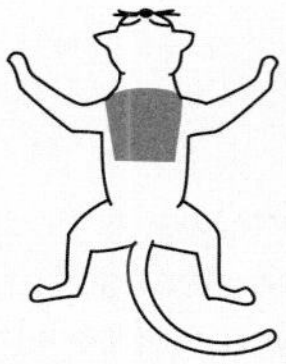

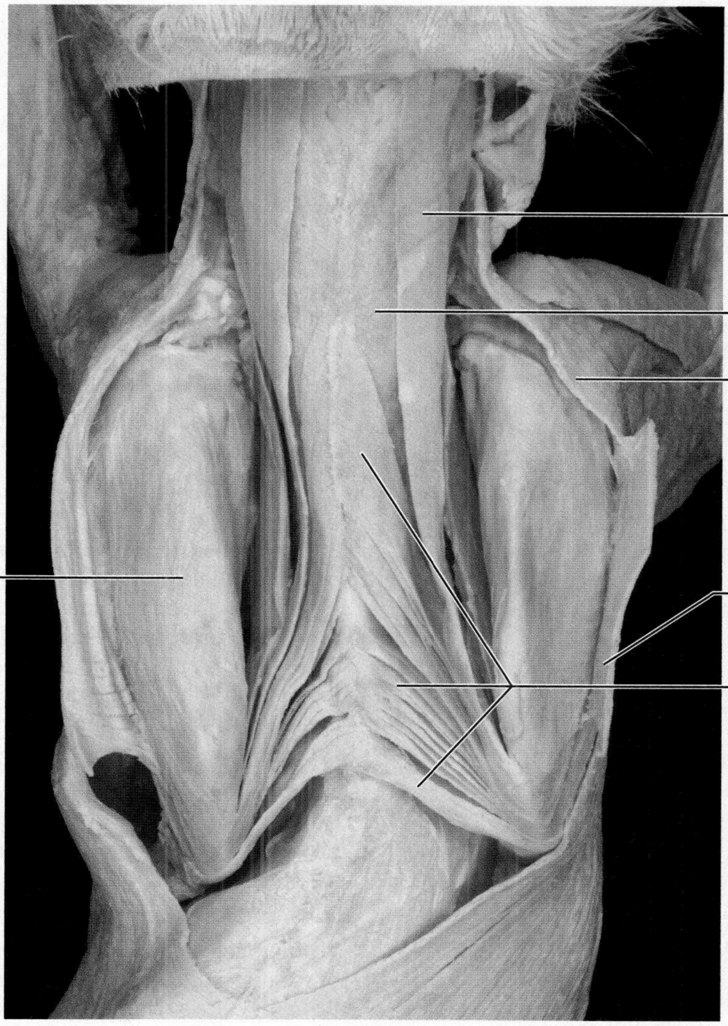

Figure D1.7 Deep muscles of the superior aspect of the dorsal thorax of the cat.

5. Reflect the flaps of the transected latissimus dorsi, spinodeltoid, acromiodeltoid, and levator scapulae ventralis. The **splenius** is a large flat muscle occupying most of the side of the neck close to the vertebrae (**Figure D1.7**). As in humans, it originates on the ligamentum nuchae at the back of the neck and inserts into the occipital bone. It raises the head.

6. To view the rhomboid muscles, lay the cat on its side and hold its forelegs together to spread the scapulae apart. The rhomboid muscles lie between the scapulae and beneath the acromiotrapezius. All the rhomboid muscles originate on the vertebrae and insert on the scapula. They hold the dorsal part of the scapula to the cat's back.

There are three rhomboids in the cat. The ribbonlike **rhomboid capitis**, the most anterolateral muscle of the group, has no counterpart in the human body. The **rhomboid minor**, located posterior to the rhomboid capitis, is much larger. The fibers of the rhomboid minor run transversely to those of the rhomboid capitis. The most posterior muscle of the group, the **rhomboid major**, is so closely fused to the rhomboid minor that many consider them to be one muscle—the **rhomboideus**, which is homologous to human *rhomboid muscles*.

7. The **supraspinatus** and **infraspinatus** muscles are similar to the same muscles in humans.The supraspinatus can be found under the acromiotrapezius, and the infraspinatus is deep to the spinotrapezius. Both originate on the lateral scapular surface and insert on the humerus.

DISSECTION 1

Activity 3

Dissecting Forelimb Muscles

Cat forelimb muscles fall into the same three categories as human upper limb muscles, but in this section the muscles of the entire forelimb are considered together (refer to **Figure D1.8**).

Muscles of the Lateral Surface

1. The **triceps brachii** muscle of the cat is easily identified if the cat is placed on its side. It is a large fleshy muscle covering the posterior aspect and much of the side of the humerus. As in humans, this muscle arises from three heads, which originate from the humerus and scapula and insert jointly into the olecranon of the ulna. Remove the fascia from the superior region of the lateral arm surface to identify the lateral and long heads of the triceps. The long head is approximately twice as long as the lateral head and lies medial to it on the posterior arm surface. The medial head can be exposed by transecting the lateral head and pulling it aside. Now pull on the triceps muscle.

How does the function of the triceps muscle compare in cats and in humans?

____

Anterior and distal to the medial head of the triceps is the tiny **anconeus** muscle (not visible in Figure D1.8), sometimes called the fourth head of the triceps muscle. Notice its darker color and the way it wraps the tip of the elbow.

2. The **brachialis** can be located anterior to the lateral head of the triceps muscle. Identify its origin on the humerus, and trace its course as it crosses the elbow and inserts on the ulna. It flexes the cat's foreleg.

Identifying the forearm muscles is difficult because of the tough fascia sheath that encases them, but give it a try.

3. Remove as much of the connective tissue as possible, and cut through the ligaments that secure the tendons at the wrist (transverse carpal ligaments) so that you will be able to follow the muscles to their insertions. Begin your identification of the forearm muscles at the lateral surface of the forearm. The muscles of this region look very similar and are difficult to identify accurately unless a definite order is followed. Thus you will begin with the most anterior muscles and proceed to the posterior aspect. Remember to check carefully the tendons of insertion to verify your muscle identifications.

4. The ribbonlike muscle on the lateral surface of the humerus is the **brachioradialis**. Observe how it passes down the forearm to insert on the radial styloid process. (If you did not remove the fascia very carefully, you may have also removed this muscle.)

5. The **extensor carpi radialis longus** has a broad origin and is larger than the brachioradialis. It extends down the anterior surface of the radius (Figure D1.8). Transect this muscle to view the **extensor carpi radialis brevis** (not shown in Figure D1.8), which is partially covered by and sometimes fused with the extensor carpi radialis longus. Both muscles have origins, insertions, and actions similar to their human counterparts.

6. You can see the entire **extensor digitorum communis** along the lateral surface of the forearm. Trace it to its four tendons, which insert on the second to fifth digits. This muscle extends these digits. The **extensor digitorum lateralis** (absent in humans) also extends the digits. This muscle lies immediately posterior to the extensor digitorum communis.

7. Follow the **extensor carpi ulnaris** from the lateral epicondyle of the humerus to the ulnar side of the fifth metacarpal. Often this muscle has a shiny tendon, which helps in its identification.

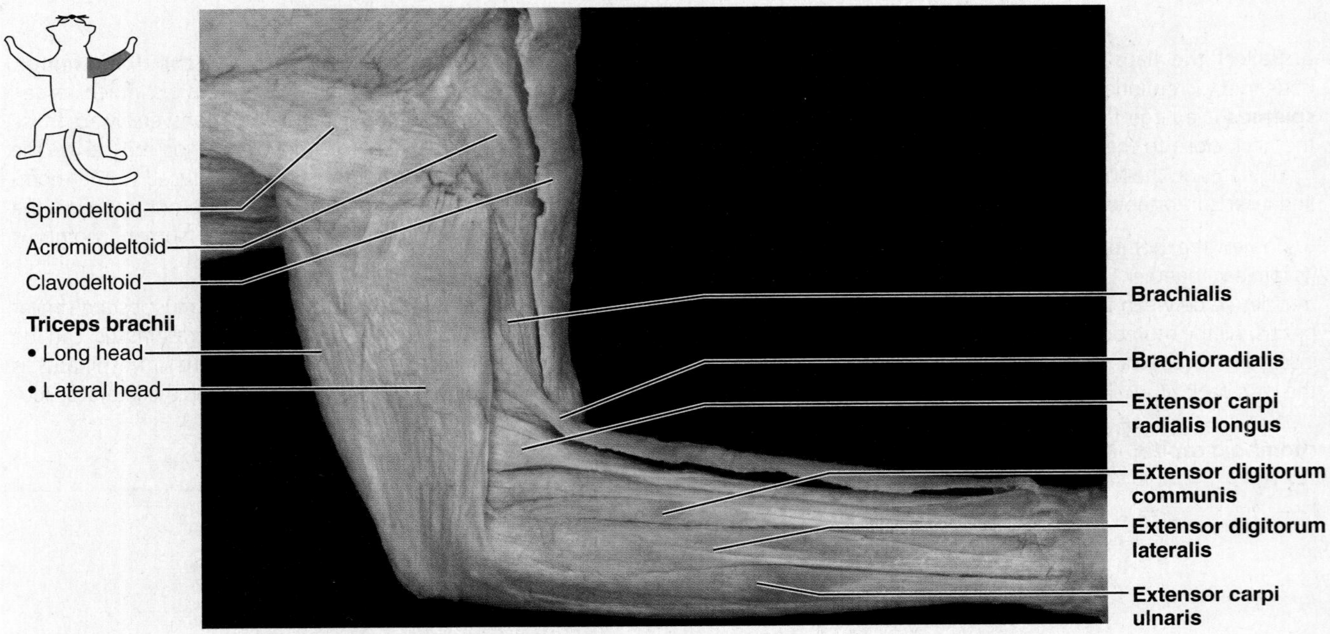

Figure D1.8 Lateral surface of the forelimb of the cat.

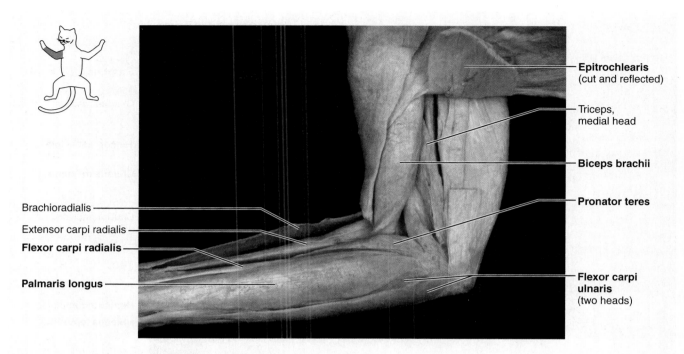

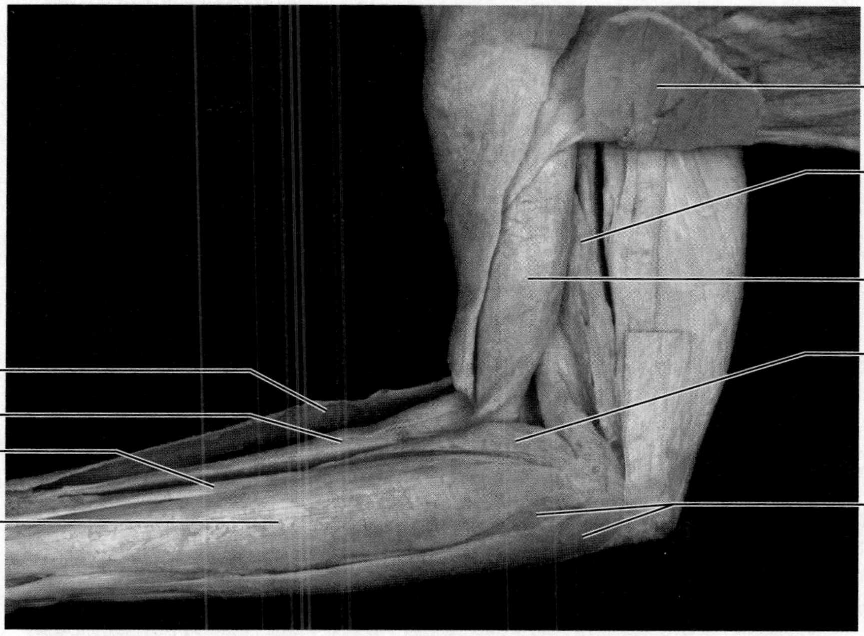

Figure D1.9 Medial surface of the forelimb of the cat.

Muscles of the Medial Surface

1. The **biceps brachii** (**Figure D1.9**) is a large spindle-shaped muscle medial to the brachialis on the anterior surface of the humerus. Pull back the cut ends of the pectoral muscles to get a good view of the biceps. This muscle is much more prominent in humans, but its origin, insertion, and action are very similar in cats and in humans. Follow the muscle to its origin.

Does the biceps have two heads in the cat? _______________

2. The broad, flat, exceedingly thin muscle on the posteromedial surface of the arm is the **epitrochlearis**. Its tendon originates from the fascia of the latissimus dorsi, and the muscle inserts into the olecranon of the ulna. This muscle extends the forearm of the cat; it is not found in humans.

3. The **coracobrachialis** (not illustrated) of the cat is insignificant (approximately 1.3 cm, or ½ inch, long) and can be seen as a very small muscle crossing the ventral aspect of the shoulder joint. It runs beneath the biceps brachii to insert on the humerus and has the same function as the human coracobrachialis.

4. Turn the cat so that the ventral forearm muscles (mostly flexors and pronators) can be observed (refer to Figure D1.9). As in humans, most of these muscles arise from the medial epicondyle of the humerus. The **pronator teres** runs from the medial epicondyle of the humerus and declines in size as it approaches its insertion on the radius. Do not bother to trace it to its insertion.

5. Like its human counterpart, the **flexor carpi radialis** runs from the medial epicondyle of the humerus to insert into the second and third metacarpals.

6. The large flat muscle in the center of the medial surface is the **palmaris longus**. Its origin on the medial epicondyle of the humerus abuts that of the pronator teres and is shared with the flexor carpi radialis. The palmaris longus extends down the forearm to terminate in four tendons on the digits. This muscle is proportionately larger in cats than in humans.

The **flexor carpi ulnaris** arises from a two-headed origin (medial epicondyle of the humerus and olecranon of the ulna). Its two bellies pass downward to the wrist, where they are united by a single tendon that inserts into the carpals of the wrist. As in humans, this muscle flexes the wrist.

Activity 4

Dissecting Hindlimb Muscles

Remove the fat and fascia from all thigh surfaces, but do not cut through or remove the **fascia lata** (or iliotibial band), which is a tough white aponeurosis covering the anterolateral surface of the thigh from the hip to the leg. If the cat is a male, the cordlike sperm duct will be embedded in the fat near the pubic symphysis. Carefully clear around, but not in, this region.

Posterolateral Hindlimb Muscles

1. Turn the cat on its ventral surface, and identify the following superficial muscles of the hip and thigh (refer to **Figure D1.10** on p. 706). Viewing the lateral aspect of the hindlimb, you will identify these muscles in sequence from the anterior to

Text continues on next page. →

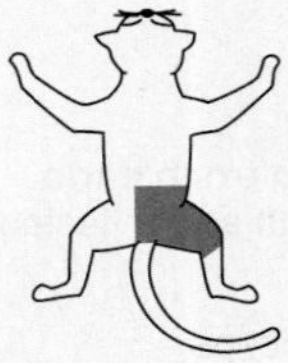

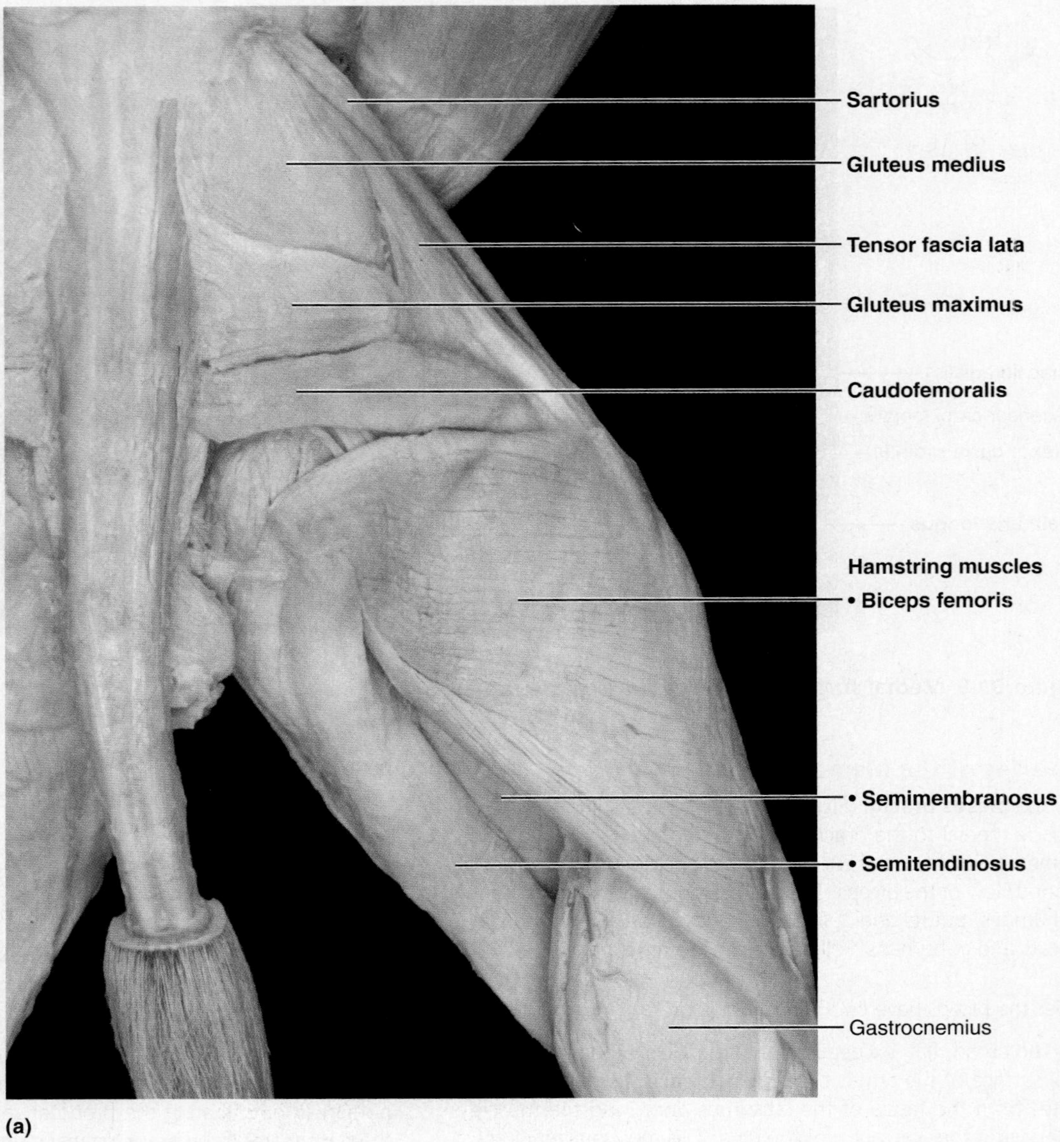

Figure D1.10 Muscles of the posterolateral thigh in the cat. (a) Superficial view.

the posterior aspects of the hip and thigh. Most anterior is the **sartorius**, seen in this view as a thin band (Figure D1.10a). Approximately 4 cm (1½ inches) wide, it extends around the lateral aspect of the thigh to the anterior surface, where the major portion of it lies (see Figure D1.12a). Free it from the adjacent muscles and pass a blunt probe under it to trace its origin and insertion. Homologous to the sartorius muscle in humans, it adducts and rotates the thigh, but in addition, the cat sartorius acts as a knee extensor. Transect this muscle.

2. The **tensor fascia lata** is posterior to the sartorius. It is wide at its superior end, where it originates on the iliac crest, and narrows as it approaches its insertion into the fascia lata, which runs to the proximal tibial region. Transect its superior end and pull it back to expose the **gluteus medius** lying beneath it. This is the largest of the gluteus muscles in the cat. It originates on the ilium and inserts on the greater trochanter of the femur. The gluteus medius overlays and obscures the gluteus minimus, pyriformis, and gemellus muscles, which will not be identified here.

3. The **gluteus maximus** is a small triangular hip muscle posterior to the superior end of the tensor fasciae latae and paralleling it. In humans the gluteus maximus is a large fleshy muscle forming most of the buttock mass. In the cat it is only about 1.3 cm (½ inch) wide and 5 cm (2 inches) long, and it is smaller than the gluteus medius. The gluteus maximus covers part of the gluteus medius as it extends from the sacral region to the end of the femur. It abducts the thigh.

4. Posterior to the gluteus maximus, identify the triangular **caudofemoralis**, which originates on the caudal vertebrae and inserts into the patella via an aponeurosis. There is no homologue to this muscle in humans; in cats it abducts the thigh and flexes the vertebral column.

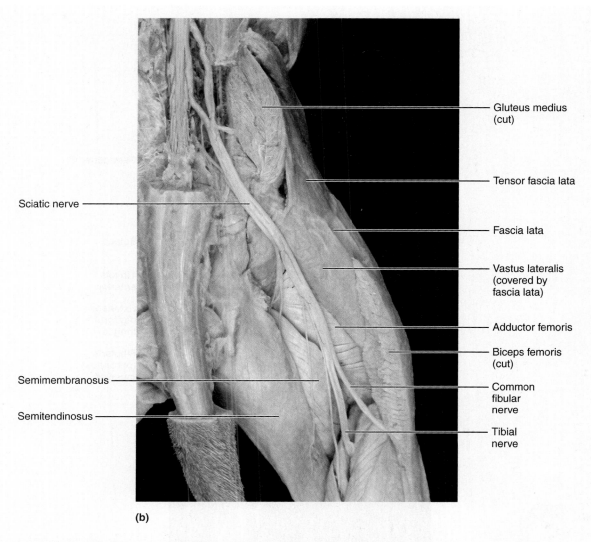

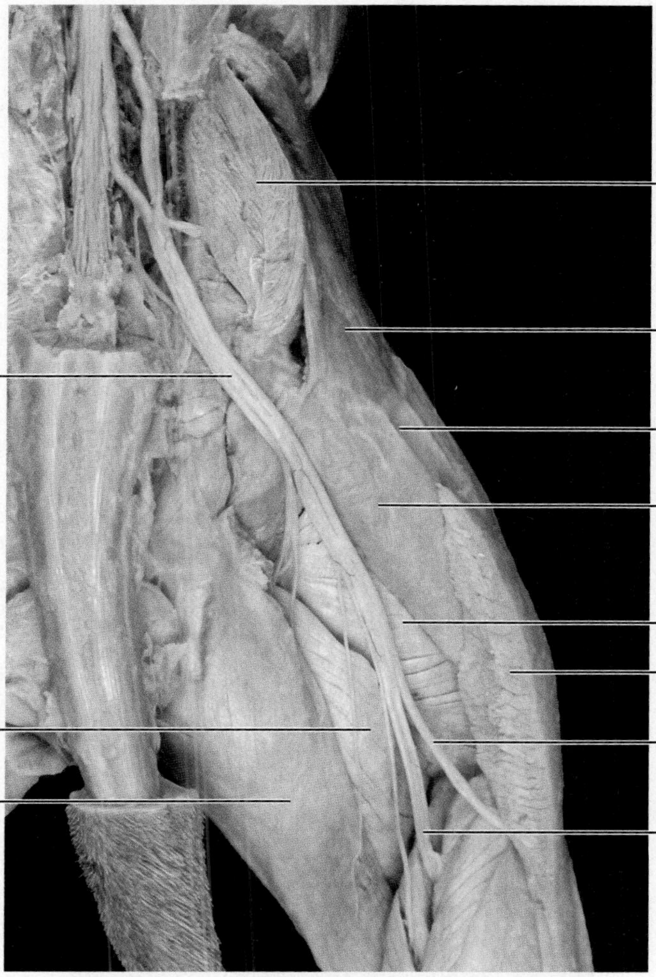

(b)

Figure D1.10 *(continued)* **(b)** Deep muscles.

5. The **hamstring muscles** of the hindlimb include the biceps femoris, the semitendinosus, and the semimembranosus muscles. The **biceps femoris** is a large, powerful muscle that covers about three-fourths of the posterolateral surface of the thigh. It is 4 cm (1½ inches) to 5 cm (2 inches) wide throughout its length. Trace it from its origin on the ischial tuberosity to its insertion on the tibia. Part of the **semitendinosus** can be seen beneath the posterior border of the biceps femoris. Transect and reflect the biceps muscle to reveal the whole length of the semitendinosus and the large sciatic nerve positioned under the biceps (Figure D1.10b). Contrary to what its name implies ("half-tendon"), this muscle is muscular and fleshy except at its insertion. It is uniformly about 2 cm (¾ inch) wide as it runs down the thigh from the ischial tuberosity to the medial side of the tibia. It flexes the knee. The **semimembranosus**, a large muscle lying medial to the semitendinosus and largely obscured by it, is best seen in an anterior view of the thigh (Figure D1.12b). If desired, however, the semitendinosus can be transected to view it from the posterior aspect. The semimembranosus is larger and broader than the semitendinosus. Like the other hamstrings, it originates on the ischial tuberosity and inserts on the medial epicondyle of the femur and the medial tibial surface.

How does the semimembranosus compare with its human homologue?

____

____

6. Remove the heavy fascia covering the lateral surface of the shank (leg). Moving from the posterior to the anterior aspect, identify the following muscles on the posterolateral shank (**Figure D1.11**, p. 708). First reflect the lower portion of the biceps femoris to see the origin of the **triceps surae**, the large composite muscle of the calf. Humans also have a triceps surae. The **gastrocnemius**, part of the triceps surae, is the largest muscle on the shank. As in humans, it has two heads

Text continues on next page. →

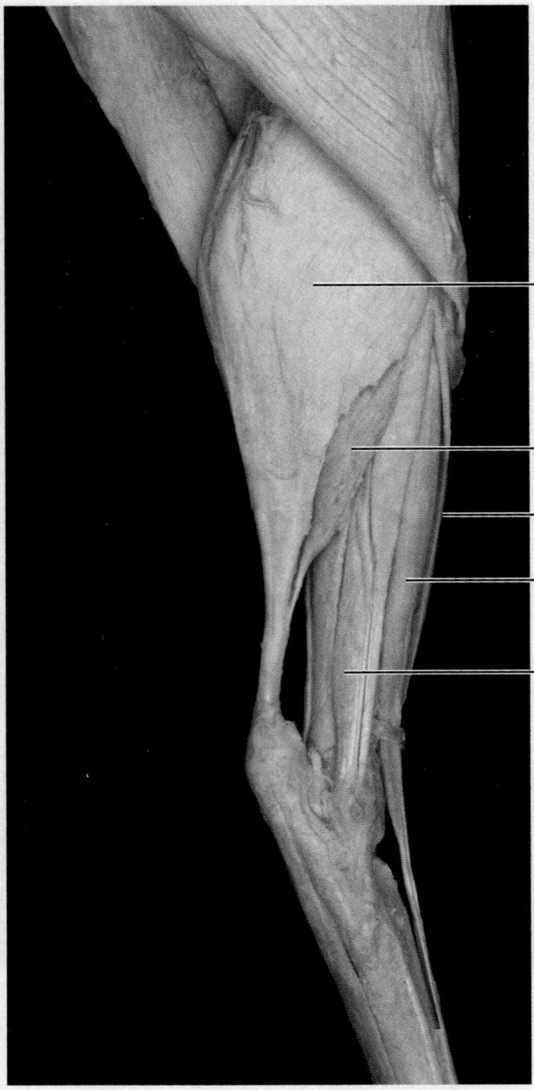

Figure D1.11 Superficial muscles of the posterolateral aspect of the shank (leg).

and inserts via the calcaneal tendon into the calcaneus. Run a probe beneath this muscle and then transect it to reveal the **soleus**, which is deep to the gastrocnemius.

7. Another important group of muscles in the leg is the **fibularis (peroneus) muscles**, which collectively appear as a slender, evenly shaped superficial muscle lying anterior to the triceps surae. Originating on the fibula and inserting on the digits and metatarsals, the fibularis muscles flex the foot.

8. The **extensor digitorum longus** lies anterior to the fibularis muscles. Its origin, insertion, and action in cats are similar to the homologous human muscle. The **tibialis anterior** is anterior to the extensor digitorum longus. The tibialis anterior is roughly triangular in cross section and heavier at its proximal end. Locate its origin on the proximal fibula and tibia and its insertion on the first metatarsal. You can see the sharp edge of the tibia at the anterior border of this muscle. As in humans, it is a foot flexor.

Anteromedial Hindlimb Muscles

1. Turn the cat onto its dorsal surface to identify the muscles of the anteromedial hindlimb (**Figure D1.12**). Note once again the straplike sartorius at the surface of the thigh, which you have already identified and transected. It originates on the ilium and inserts on the medial region of the tibia.

2. Reflect the cut ends of the sartorius to identify the **quadriceps** muscles. The most medial muscle of this group, the **vastus medialis**, lies just beneath the sartorius. Resting close to the femur, it arises from the ilium and inserts into the patellar ligament. The small spindle-shaped muscle anterior

Text continues on p. 710. →

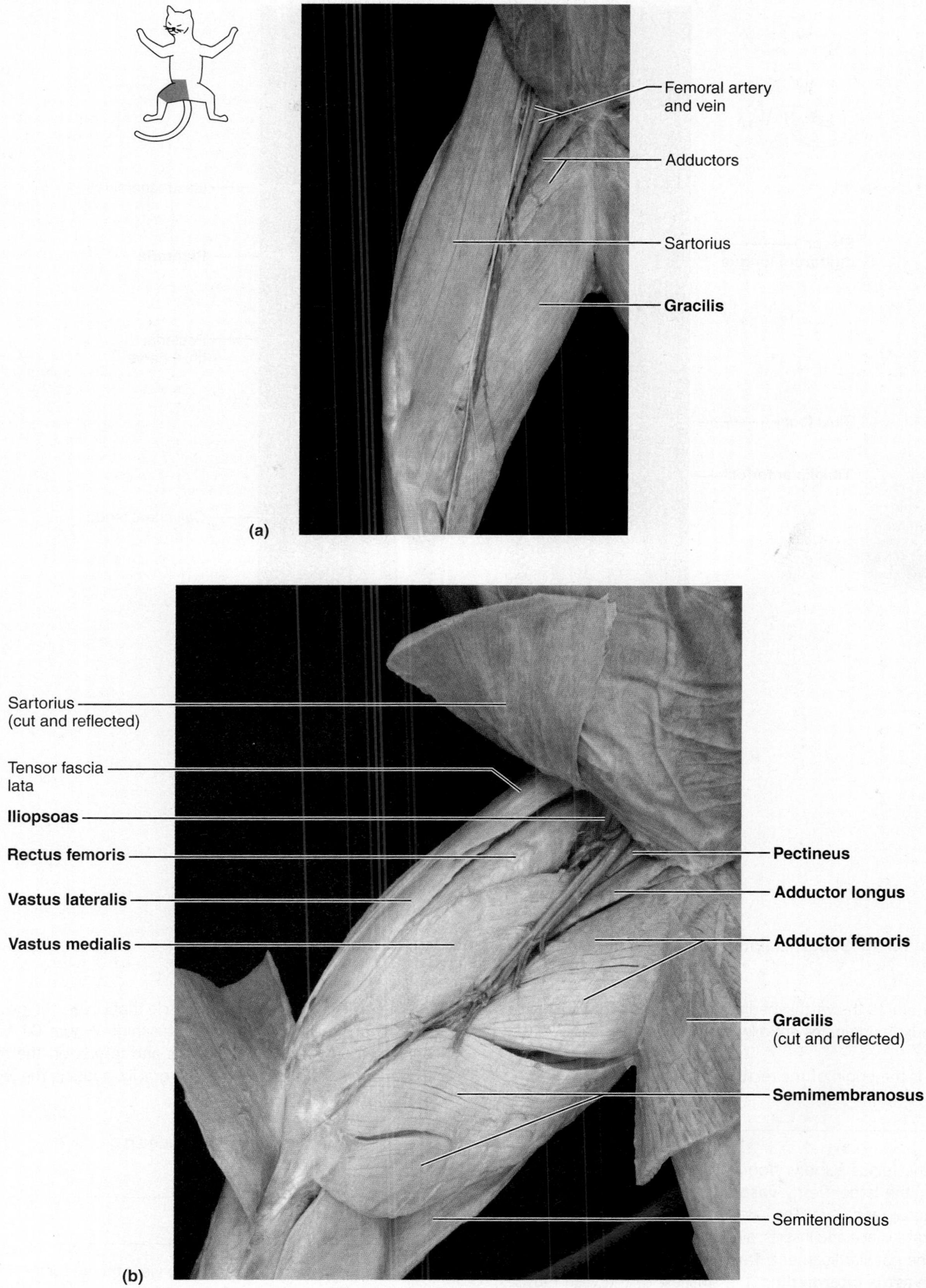

Figure D1.12 Superficial muscles of the anteromedial thigh. (a) Gracilis and sartorius are intact in this superficial view of the right thigh. **(b)** The gracilis and sartorius are transected and reflected to show deeper muscles.

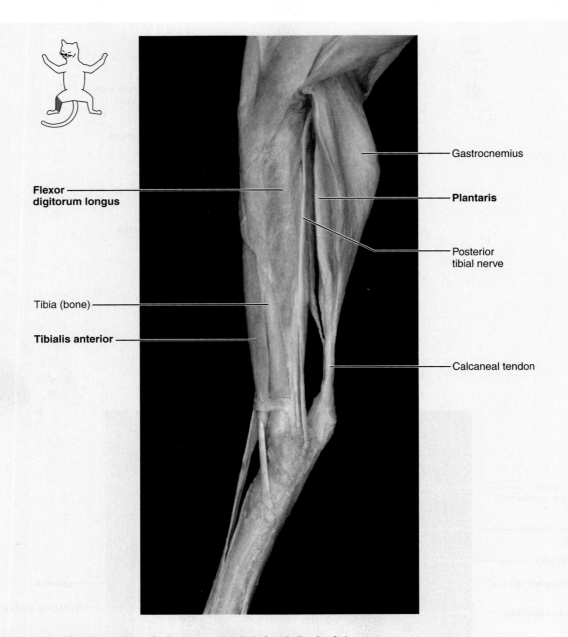

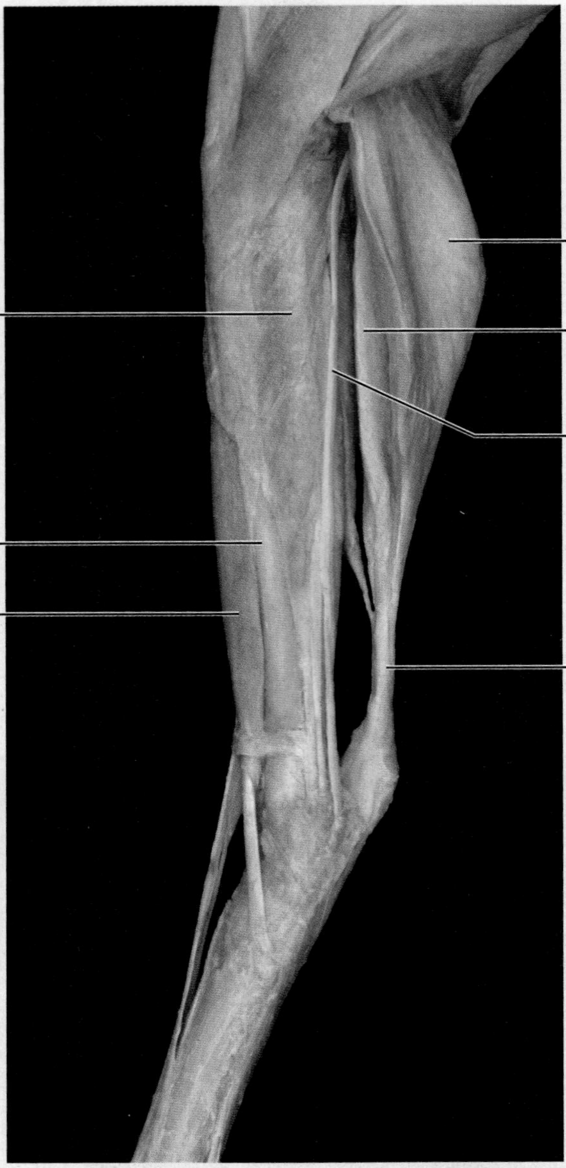

Figure D1.13 Superficial muscles of the anteromedial shank (leg) of the cat.

and lateral to the vastus medialis is the **rectus femoris**, In cats this muscle originates entirely from the femur.

What is the origin of the rectus femoris in humans?

____

Free the rectus femoris from the most lateral muscle of this group, the large, fleshy **vastus lateralis**, which lies deep to the tensor fascia lata. The vastus lateralis arises from the lateral femoral surface and inserts, along with the other vasti muscles, into the patellar ligament. Transect this muscle to identify the deep **vastus intermedius**, the smallest of the vasti muscles. It lies medial to the vastus lateralis and merges superiorly with the vastus medialis. The vastus intermedius is not shown in the figure.

3. The **gracilis** is a broad thin muscle that covers the posterior portion of the medial aspect of the thigh (Figure D1.12a). It originates on the pubic symphysis and inserts on the medial proximal tibial surface. In cats the gracilis adducts the leg and draws it posteriorly.

How does this compare with the human gracilis?

____

____

4. Free and transect the gracilis to view the adductor muscles deep to it.The **adductor femoris** is a large muscle that lies beneath the gracilis and abuts the semimembranosus medially. Its origin is the pubic ramus and the ischium, and its fibers pass

downward to insert on most of the length of the femoral shaft. The adductor femoris is homologous to the human *adductor magnus*, and *adductor brevis*. Its function is to extend the thigh after it has been drawn forward, and to adduct the thigh. A small muscle about 2.5 cm (1 inch) long—the **adductor longus**—touches the superior margin of the adductor femoris. It originates on the pubis and inserts on the proximal surface of the femur.

5. Before continuing your dissection, locate the **femoral triangle**, an important area bordered by the proximal edge of the sartorius and the adductor muscles. It is usually possible to identify the femoral artery (injected with red latex) and the femoral vein (injected with blue latex), which span the triangle (Figure D1.12a). (You will identify these vessels again in your study of the circulatory system.) If your instructor wishes you to identify the pectineus and iliopsoas, remove these vessels and go on to steps 6 and 7.

6. Examine the superolateral margin of the adductor longus to locate the small **pectineus**. It is sometimes covered by the gracilis (which you have cut and reflected). The pectineus, which originates on the pubis and inserts on the proximal end of the femur, is similar in all ways to its human homologue.

7. Just lateral to the pectineus you can see a small portion of the **iliopsoas**, a long and cylindrical muscle. Its origin is on the transverse processes of T_1 through T_{12} and the lumbar vertebrae, and it passes posteriorly toward the body wall to insert on the medial aspect of the proximal femur. The iliopsoas flexes and laterally rotates the thigh. It corresponds to the human iliopsoas.

8. Reidentify the gastrocnemius of the shank and then the **plantaris**, which is fused with the lateral head of the gastrocnemius (**Figure D1.13**). It originates from the lateral aspect of the femur and patella, and its tendon passes around the calcaneus to insert on the second phalanx. Working with the triceps surae, it flexes the digits and extends the foot.

9. Anterior to the plantaris is the **flexor digitorum longus**, a long, tapering muscle with two heads. It originates on the lateral surfaces of the proximal fibula and tibia and inserts via four tendons into the terminal phalanges. As in humans, it flexes the toes.

10. The **tibialis posterior** is a long, flat muscle lateral and deep to the flexor digitorum longus (not shown in Figure D1.13). It originates on the medial surface of the head of the fibula and the ventral tibia. It merges with a flat, shiny tendon to insert into the tarsals.

11. The **flexor hallucis longus** (also not illustrated) is a long muscle that lies lateral to the tibialis posterior. It originates from the posterior tibia and passes downward to the ankle. It is a uniformly broad muscle in the cat. As in humans, it is a flexor of the great toe.

12. Before you leave the laboratory, follow the boxed instructions to prepare your cat for storage and to clean the area (p. 696).

DISSECTION REVIEW

Many human muscles are modified from those of the cat (or any quadruped). The following questions refer to these differences.

1. How does the human trapezius muscle differ from the cat's?

____

2. How does the deltoid differ?

____

____

3. How does the biceps brachii differ?

____

____

4. How do the size and orientation of the human gluteus maximus muscle differ from that in the cat?

____

____

____

5. How does the origin of the rectus femoris in the cat differ from that in humans?

____

____

6. Explain how the tendinous intersections of the rectus abdominis of the cat differ from those found in humans.

____

7. Match each term in column B to its description in column A.

Column A	Column B
________ 1. to separate muscles	a. dissect
________ 2. to fold back a muscle	b. embalm
________ 3. to cut through a muscle	c. reflect
________ 4. to preserve tissue	d. transect

DISSECTION EXERCISE 2

Dissection of Cat Spinal Nerves

Learning Outcome

- Identify on a dissected animal the musculocutaneous, radial, median, and ulnar nerves of the forelimb and the femoral, saphenous, sciatic, common fibular (peroneal), and tibial nerves of the hindlimb.

Materials

- Disposable gloves
- Safety glasses
- Dissecting instruments and tray
- Animal specimen from previous dissection
- Embalming fluid
- Paper towels
- Organic debris container

Go to Mastering A&P™ > Study Area to improve your performance in A&P Lab.

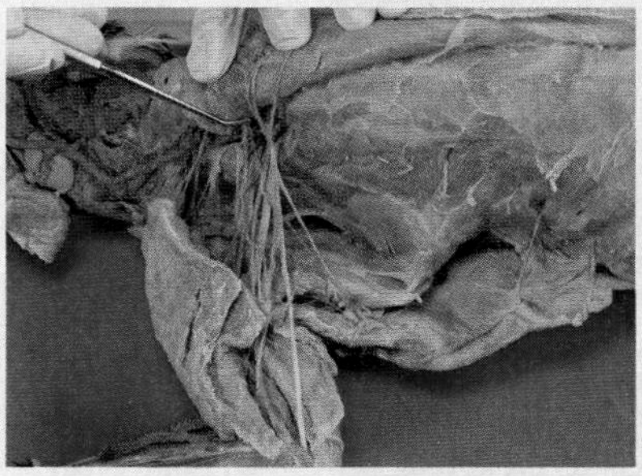

> Animations & Videos > Cat Dissection Videos > The Brachial Plexus

Instructors may assign Cat Dissection Videos, Practice Anatomy Lab Practical questions (PAL) for the dissections, and more using the Mastering **A&P™** Item Library.

The cat has 38 or 39 pairs of spinal nerves compared to 31 in humans. Of these, 8 are cervical, 13 thoracic, 7 lumbar, 3 sacral, and 7 or 8 caudal. A complete dissection of the cat's spinal nerves would be extraordinarily time-consuming and is not warranted in a basic anatomy and physiology course. However, it is desirable for you to have some dissection work to complement your study of the anatomical charts. Thus at this point you will carry out a partial dissection of the brachial plexus and lumbosacral plexus and identify some of the major nerves. Refer to a discussion of human spinal nerves as you work (see Exercise 19).

Activity 1

Dissecting Nerves of the Brachial Plexus

1. Don disposable gloves and safety glasses. Place your cat specimen on the dissecting tray, dorsal side down. Reflect the cut ends of the left pectoralis muscles to expose the large brachial plexus in the axillary region (**Figure D2.1**, p. 714). Use forceps to carefully clear away the connective tissue around the exposed nerves as far back toward their points of origin as possible.

2. The **musculocutaneous nerve** is the most superior nerve of this group. It splits into two subdivisions that run under the margins of the coracobrachialis and biceps brachii muscles. Trace its fibers into the ventral muscles of the arm it serves.

3. Locate the large **radial nerve** inferior to the musculocutaneous nerve. The radial nerve serves the dorsal muscles of the arm and forearm. Follow it into the three heads of the triceps brachii muscle.

4. In the cat, the **median nerve** is closely associated with the brachial artery and vein (Figure D2.1). It travels through the arm to supply the ventral muscles of the forearm (with the exception of the flexor carpi ulnaris and the ulnar head of the flexor digitorum profundus). It also innervates some of the intrinsic hand muscles, as in humans. Locate and follow it to the ventral forearm muscles.

5. The **ulnar nerve** is the most posterior of the large brachial plexus nerves. Follow it as it travels down the forelimb, passing over the medial epicondyle of the humerus, to supply the flexor carpi ulnaris, the ulnar head of the flexor digitorum profundus, and the hand muscles.

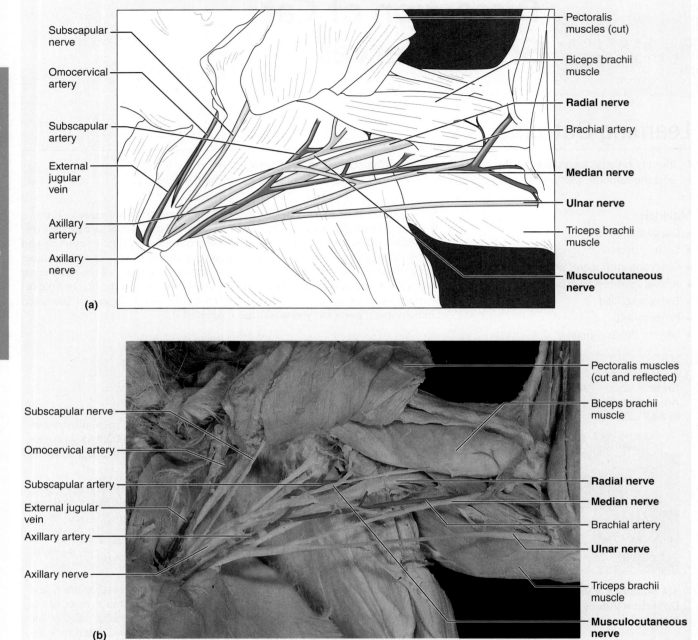

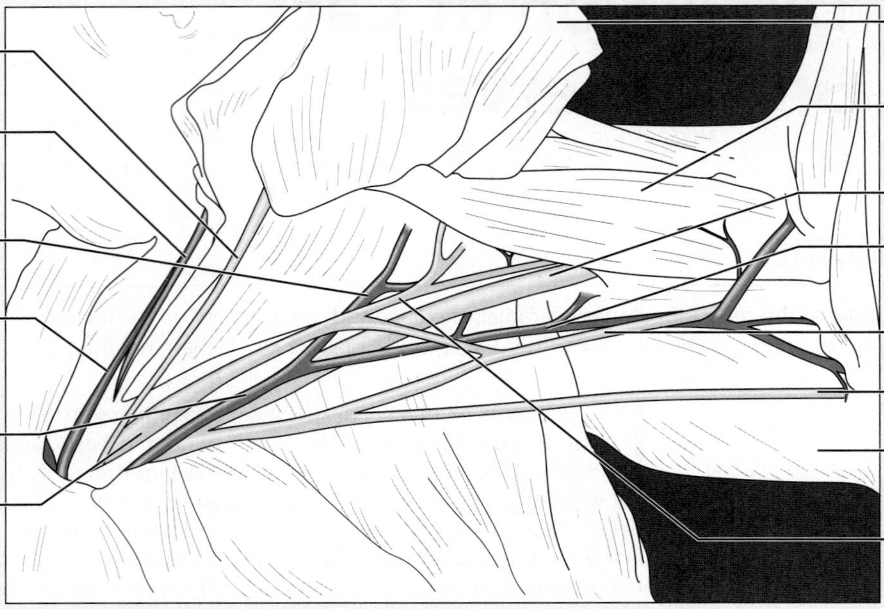

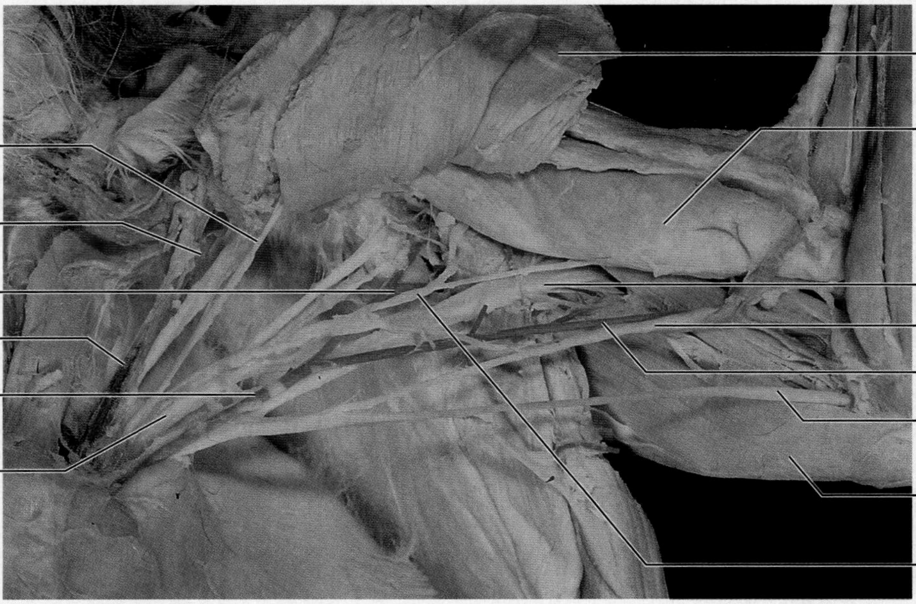

Figure D2.1 Brachial plexus and major blood vessels of the left forelimb of the cat, ventral aspect. (a) Diagram. **(b)** Photograph.

Activity 2

Dissecting Nerves of the Lumbosacral Plexus

1. To locate the **femoral nerve** arising from the lumbar plexus, first identify the *femoral triangle*, which is bordered by the sartorius and adductor muscles of the anterior thigh (**Figure D2.2**). The large femoral nerve travels through this region after emerging from the psoas major muscle in close association with the femoral artery and vein. Follow the nerve into the muscles and skin of the anterior thigh, which it supplies. Notice also its cutaneous branch in the cat, the **saphenous nerve**, which continues down the anterior medial surface of the thigh with the great saphenous artery and vein to supply the skin of the anterior shank and foot.

2. Turn the cat ventral side down so you can view the posterior aspect of the lower limb (**Figure D2.3**, p. 716). Reflect the

Text continues on p. 717. →

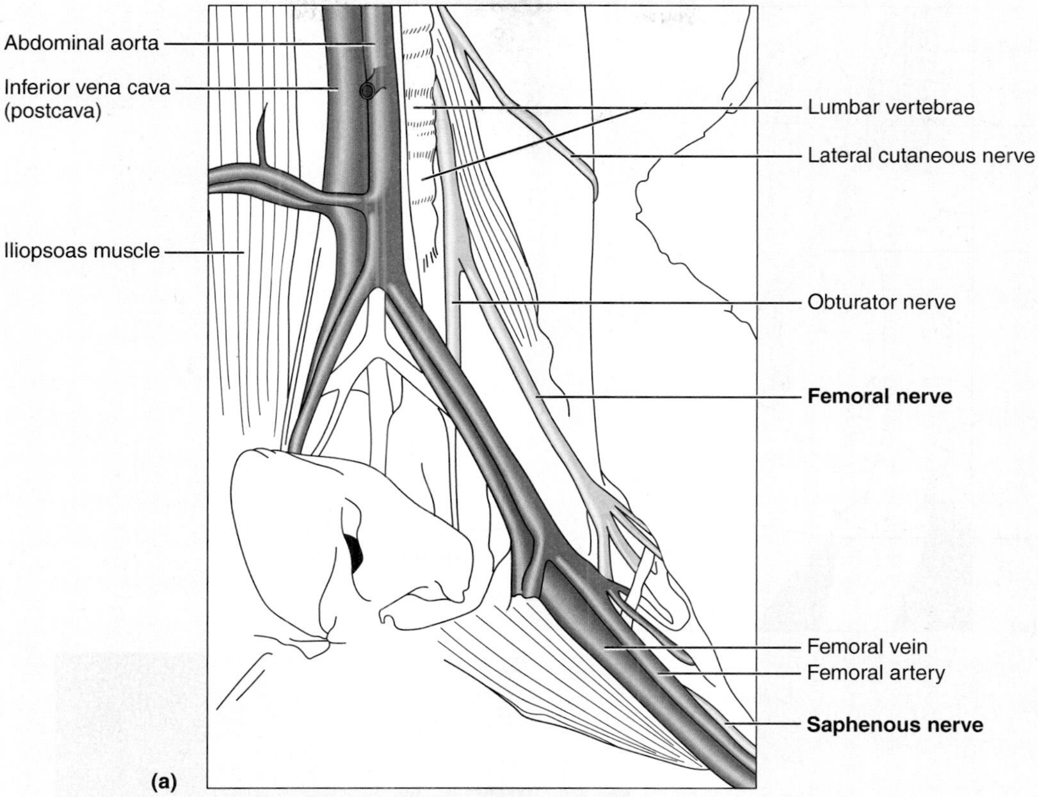

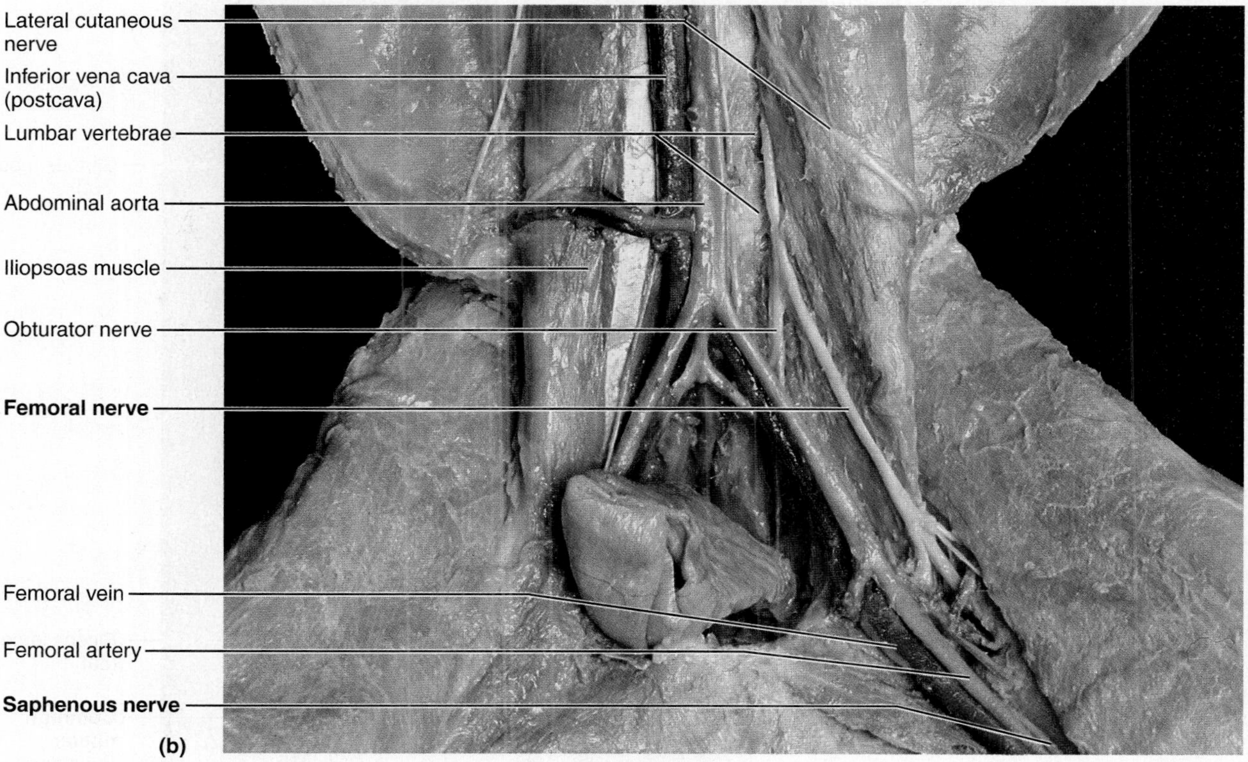

Figure D2.2 Lumbar plexus of the cat, ventral aspect. (a) Diagram. **(b)** Photograph.

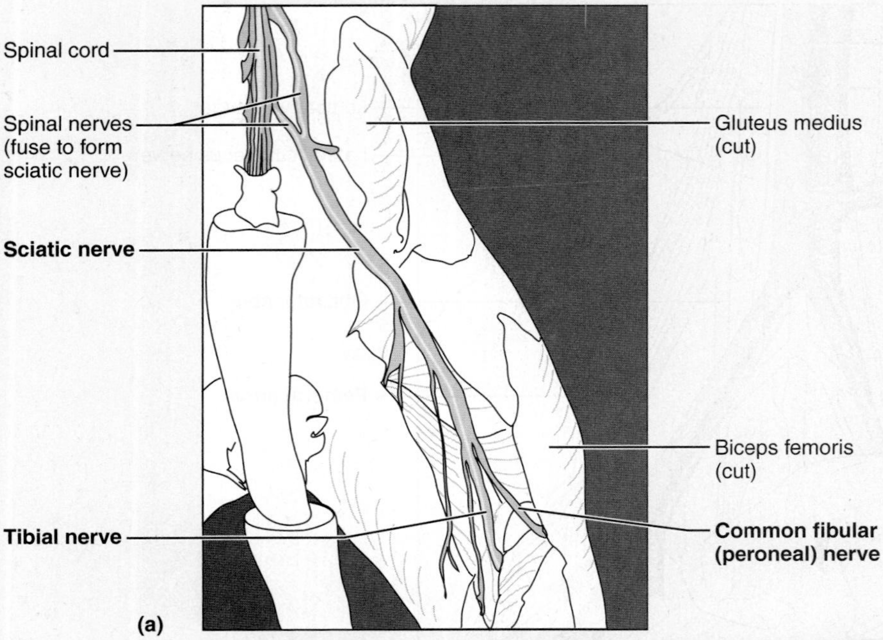

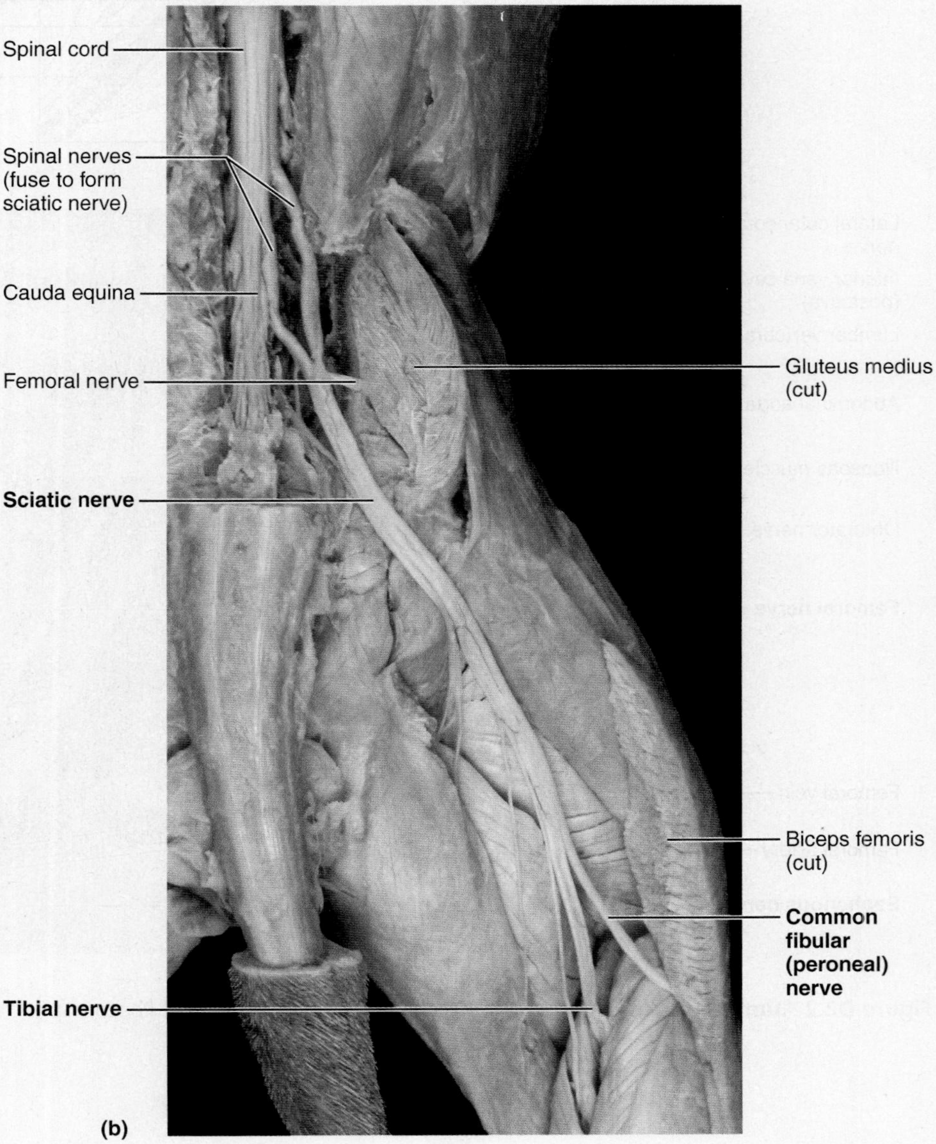

Figure D2.3 Sacral plexus of the cat, dorsal aspect. (a) Diagram. **(b)** Photograph.

ends of the transected biceps femoris muscle to view the large cordlike sciatic nerve. The **sciatic nerve** arises from the sacral plexus and serves the dorsal thigh muscles and all the muscles of the leg and foot. Follow the nerve as it travels down the posterior thigh lateral to the semimembranosus muscle. Note that just superior to the gastrocnemius muscle of the calf, it divides into its two major branches, which serve the leg.

3. Identify the **tibial nerve** medially and the **common fibular (peroneal) nerve**, which curves over the lateral surface of the gastrocnemius.

4. Before you leave the laboratory, follow the boxed instructions to prepare your cat for storage and to clean the area (p. 696).

DISSECTION REVIEW

1. From anterior to posterior, put in their proper order the nerves issuing from the brachial plexus (i.e., the median, musculocutaneous, radial, and ulnar nerves).

____

____

2. Which of the nerves named above serves most of the cat's forearm extensor muscles? ______________ Which serves the forearm flexors? ______________

3. Just superior to the gastrocnemius muscle, the sciatic nerve divides into its two main branches, the ______________ and ______________ nerves.

4. What name is given to the cutaneous nerve of the cat's thigh? ______________

DISSECTION EXERCISE 3

Identification of Selected Endocrine Organs of the Cat

Learning Outcomes

- ▶ Prepare the cat for observation by opening the ventral body cavity.
- ▶ Identify and name the major endocrine organs on a dissected cat.

Materials

- ▶ Disposable gloves
- ▶ Safety glasses
- ▶ Dissecting instruments and tray
- ▶ Animal specimen from previous dissections
- ▶ Bone cutters
- ▶ Embalming fluid
- ▶ Paper towels
- ▶ Organic debris container

Go to Mastering A&P™ > Study Area to improve your performance in A&P Lab.

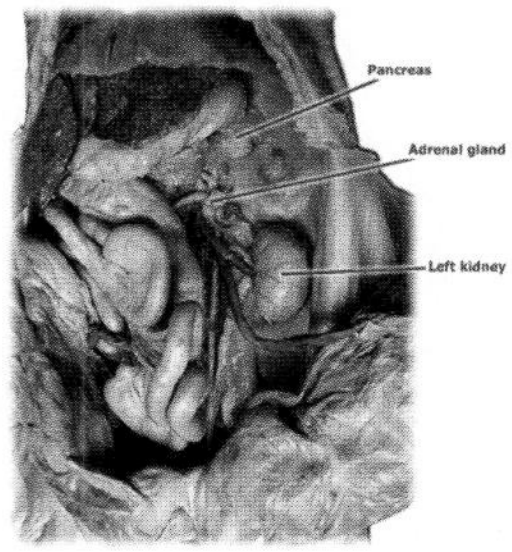

> Lab Tools > Practice Anatomy Lab > Cat > Endocrine System

Instructors may assign Cat Dissection Videos, Practice Anatomy Lab Practical questions (PAL) for the dissections, and more using the Mastering **A&P™** Item Library.

Activity 1

Opening the Ventral Body Cavity

1. Don gloves and safety glasses, and then obtain your dissection animal. Place the animal on the dissecting tray, ventral side up. Using scissors, make a longitudinal median incision through the ventral body wall. Begin your cut just superior to the midline of the pubis, and continue it anteriorly to the rib cage. Check the incision guide provided in **Figure D3.1** as you work.

2. Angle the scissors slightly (1.3 cm, or ½ inch) to the right or left of the sternum, and continue the cut through the rib cartilages (just lateral to the body midline),

Text continues on next page. →

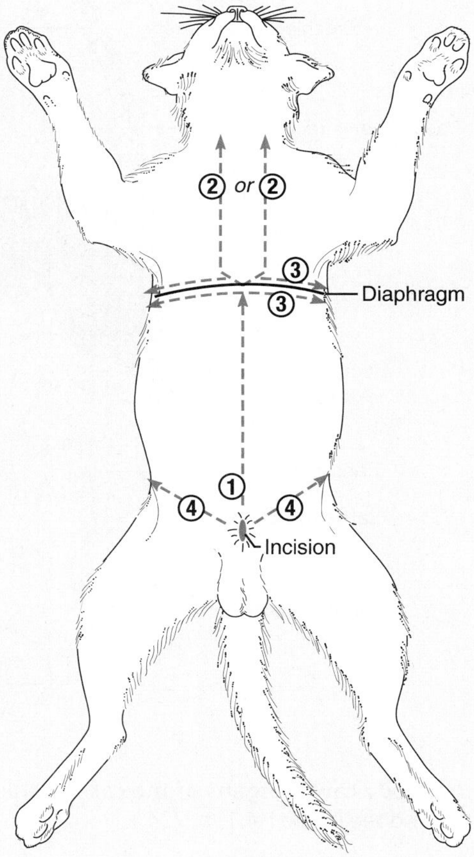

Figure D3.1 Incisions to be made in opening the ventral body cavity of a cat. Numbers indicate sequence.

to the base of the throat. Your instructor may have you use heavier bone cutters to cut through the rib cartilages.

3. Make two lateral cuts on both sides of the ventral body surface, anterior and posterior to the diaphragm, which separates the thoracic and abdominal parts of the ventral body cavity. *Leave the diaphragm intact*. Spread the thoracic walls laterally to expose the thoracic organs.

4. Make an angled lateral cut on each side of the median incision line just superior to the pubis, and spread the flaps to expose the abdominal cavity organs.

Activity 2

Identifying Organs

Figure D3.2 provides a general overview of the ventral body cavity organs of the cat. It is also useful to compare cat endocrine organs to those of the human (see Exercise 27). Since you will study the organ systems housed in the ventral body cavity in later units, the objective here is simply to identify the most important organs and those that will help you to locate the desired endocrine organs (marked *). Examine the schematic and photographic cat images showing the relative positioning of several of the animal's endocrine organs as you work (**Figure D3.3**).

Neck and Thoracic Cavity Organs

Trachea: The windpipe; runs down the midline of the throat and then divides just anterior to the lungs to form the bronchi, which plunge into the lungs on either side.

***Thyroid gland:** Its dark lobes straddle the trachea (Figure D3.3b). Thyroid hormones are the main hormones regulating the body's metabolic rate. In general, the metabolic rate of a species of animal is inversely proportional to its size.

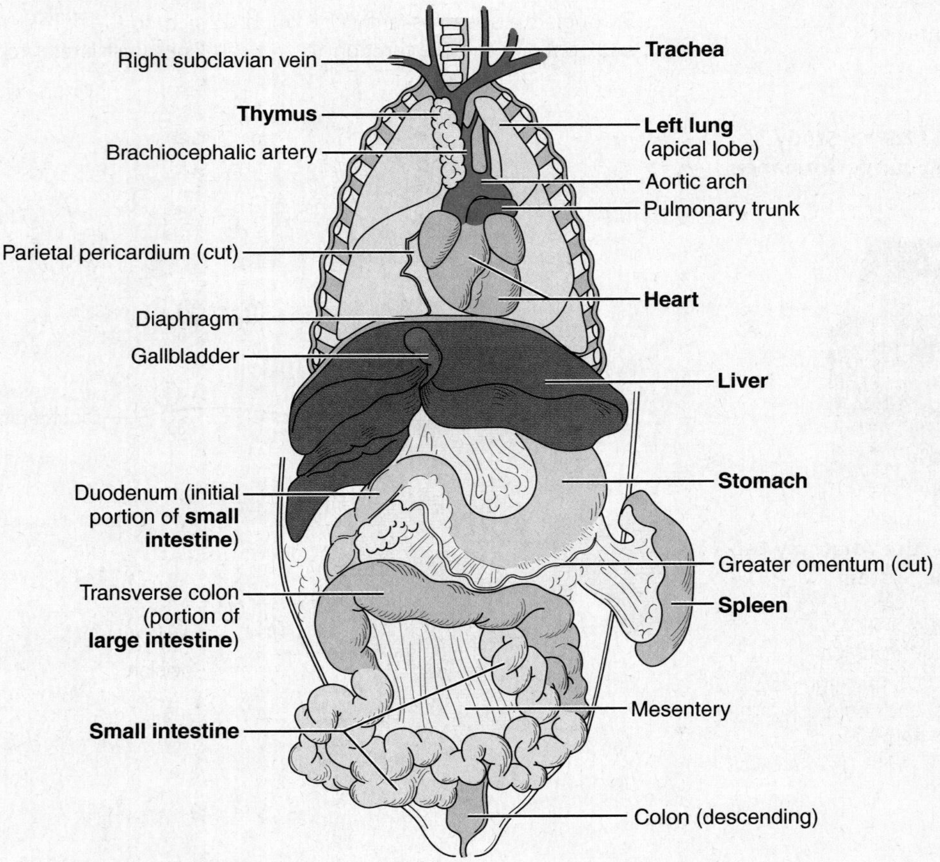

Figure D3.2 Ventral body cavity organs of the cat. Superficial view with greater omentum removed. (Also see Figure D4.1, p. 724.)

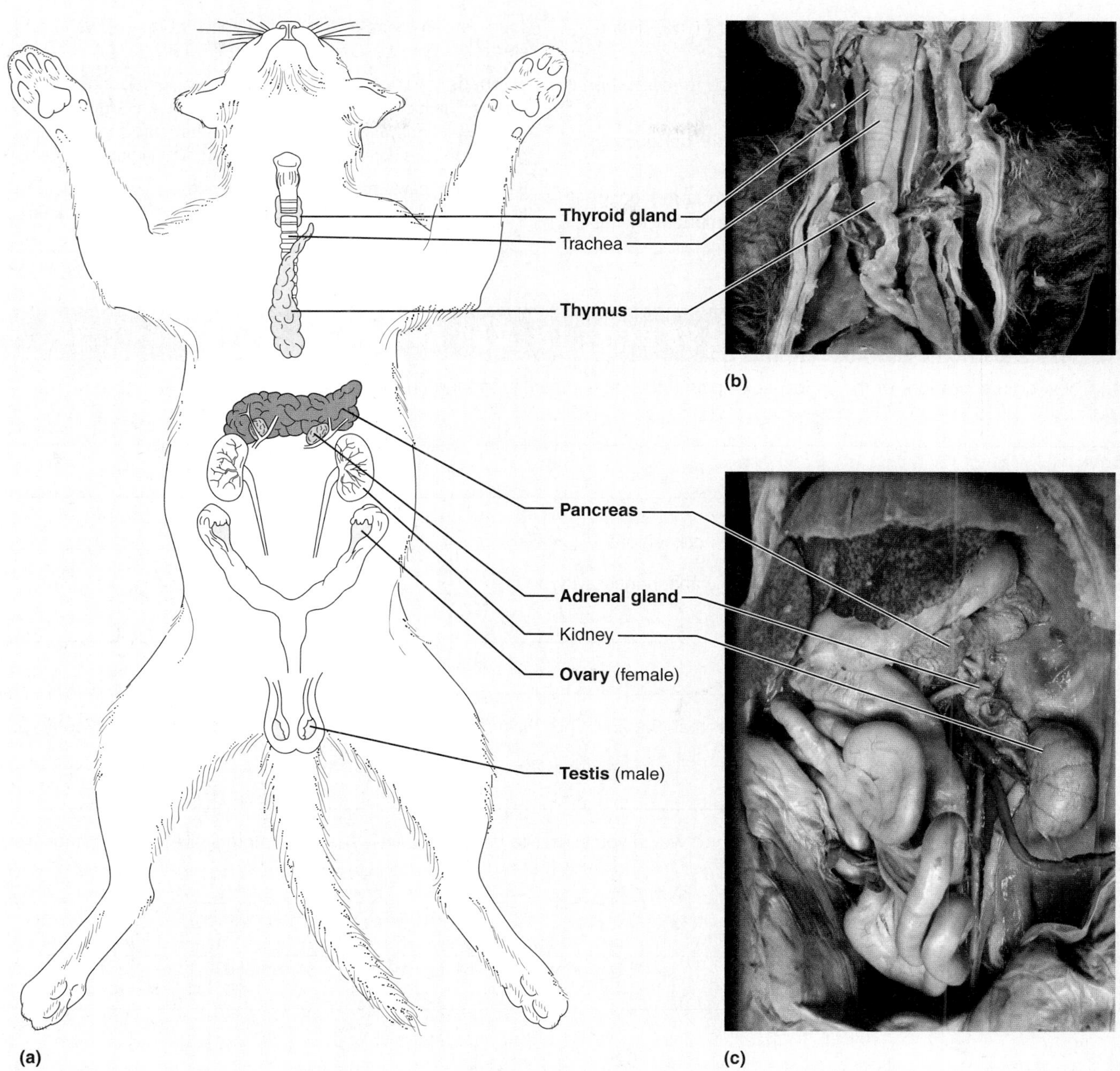

Figure D3.3 Endocrine organs in the cat. (a) Drawing. **(b)** and **(c)** Photographs.

***Thymus:** Glandular structure superior to and partly covering the heart (Figure D3.3b).The hormones of the thymus are intimately involved in programming the immune system. If you have a young cat, the thymus will be quite large. In old cats, most of this organ has been replaced by fat.

Heart: In the mediastinum enclosed by the pericardium.

Lungs: Paired organs flanking the heart.

Abdominopelvic Cavity Organs

Liver: Large multilobed organ lying under the umbrella of the diaphragm.

- Lift the drapelike, fat-infiltrated greater omentum covering the abdominal organs to expose the following organs:

Stomach: Dorsally located sac to the left side of the liver.

Spleen: Flattened brown organ curving around the lateral aspect of the stomach.

Small intestine: Tubelike organ continuing posteriorly from the stomach.

Large intestine: Takes a U-shaped course around the small intestine to terminate in the rectum.

- Lift the first section of the small intestine with your forceps to see the pancreas.

***Pancreas:** Diffuse gland located in delicate mesentery lying deep to and between the small intestine and stomach

Text continues on next page. →

(Figure D3.3c). This gland is extremely important in regulating blood sugar levels.

- Push the intestines to one side with a probe to reveal the deeper organs in the abdominal cavity.

Kidneys: Bean-shaped organs located toward the dorsal body wall surface and behind the peritoneum.

***Adrenal glands:** Visible above and medial to each kidney, these small glands produce corticosteroids important in the stress response and in preventing abnormalities of water and electrolyte balance in the body (Figure D3.3c).

***Gonads (ovaries or testes):** Sex organs producing sex hormones. The location of the gonads is illustrated in Figure D3.3a, but their identification is deferred until the reproductive system organs are considered (see Exercise 42 and Dissection Exercise 9).

Before you leave the laboratory, follow the boxed instructions to prepare your cat for storage and to clean the area (p. 696).

DISSECTION REVIEW

1. How do the locations of the endocrine organs in the cat compare with those in the human?

____

____

2. Name two endocrine organs located in the neck region: ______________________ and ______________________

3. Name three endocrine organs located in the abdominal cavity.

____

____

4. Given the assumption (not necessarily true) that human beings have more stress than cats, which endocrine organs would you expect to be relatively larger in humans?

____

5. Cats are smaller animals than humans. Which would you expect to have a (relatively speaking) more active thyroid gland—cats or humans? ______________________ Why? ______________________________________

____

____

DISSECTION EXERCISE 4

Dissection of the Blood Vessels of the Cat

Learning Outcomes

- Identify some of the most important blood vessels of the cat.
- Point out anatomical differences between the vascular system of the human and the cat.

Materials

- Disposable gloves
- Safety glasses
- Dissecting instruments and tray
- Animal specimen from previous dissections
- Bone cutters
- Scissors
- Embalming fluid
- Paper towels
- Organic debris container

Go to Mastering A&P™ > Study Area to improve your performance in A&P Lab.

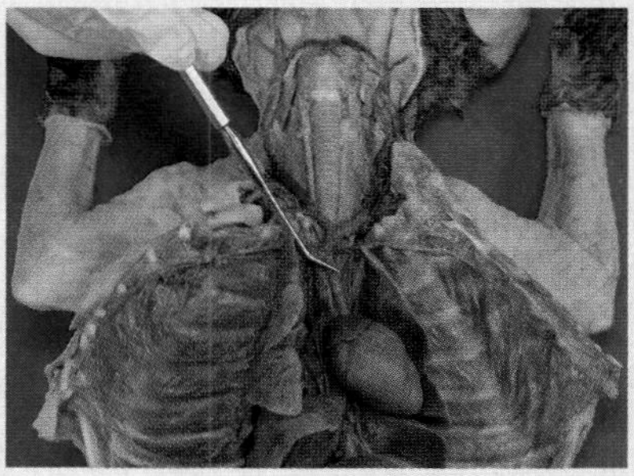

> Animations & Videos > Cat Dissection Videos > Blood Vessels of the Thorax

Instructors may assign Cat Dissection Videos, Practice Anatomy Lab Practical questions (PAL) for the dissections, and more using the Mastering A&P™ Item Library.

Activity 1

Opening the Ventral Body Cavity

If you have completed the identification of selected endocrine organs (Dissection Exercise 3), you have already opened your animal's ventral body cavity and identified many of its organs. In such a case, begin this exercise with Activity 3, Preparing to Identify the Blood Vessels (p. 724). Consult a discussion of the anatomy of the human blood vessels as you work (see Exercise 32).

If you have not already opened your animal's ventral body cavity, do so now by following your instructor's directions (see Dissection Exercise 3, Activity 1, p. 719).

Activity 2

Preliminary Organ Identification

A helpful prelude to identifying and tracing the blood supply of the various organs of the cat is a preliminary identification of ventral body cavity organs (**Figure D4.1**, p. 724). Since you will study the organ systems contained in the ventral cavity in later units, the objective here is simply to identify the most important organs. Locate and identify the following body cavity organs (refer to Figure D4.1).

Thoracic Cavity Organs

Heart: In the mediastinum enclosed by the pericardium.

Lungs: Flanking the heart.

Thymus: Superior to and partially covering the heart. (See Figure D3.2, p. 720.) The thymus is quite large in young cats but is largely replaced by fat as cats age.

Abdominal Cavity Organs

Liver: Posterior to the diaphragm.

- Lift the large, drapelike, fat-infiltrated greater omentum covering the abdominal organs to expose the following:

Stomach: Dorsally located and to the left of the liver.

Spleen: A flattened, brown organ curving around the lateral aspect of the stomach.

Small intestine: Continues posteriorly from the stomach.

Large intestine: Takes a U-shaped course around the small intestine and terminates in the rectum.

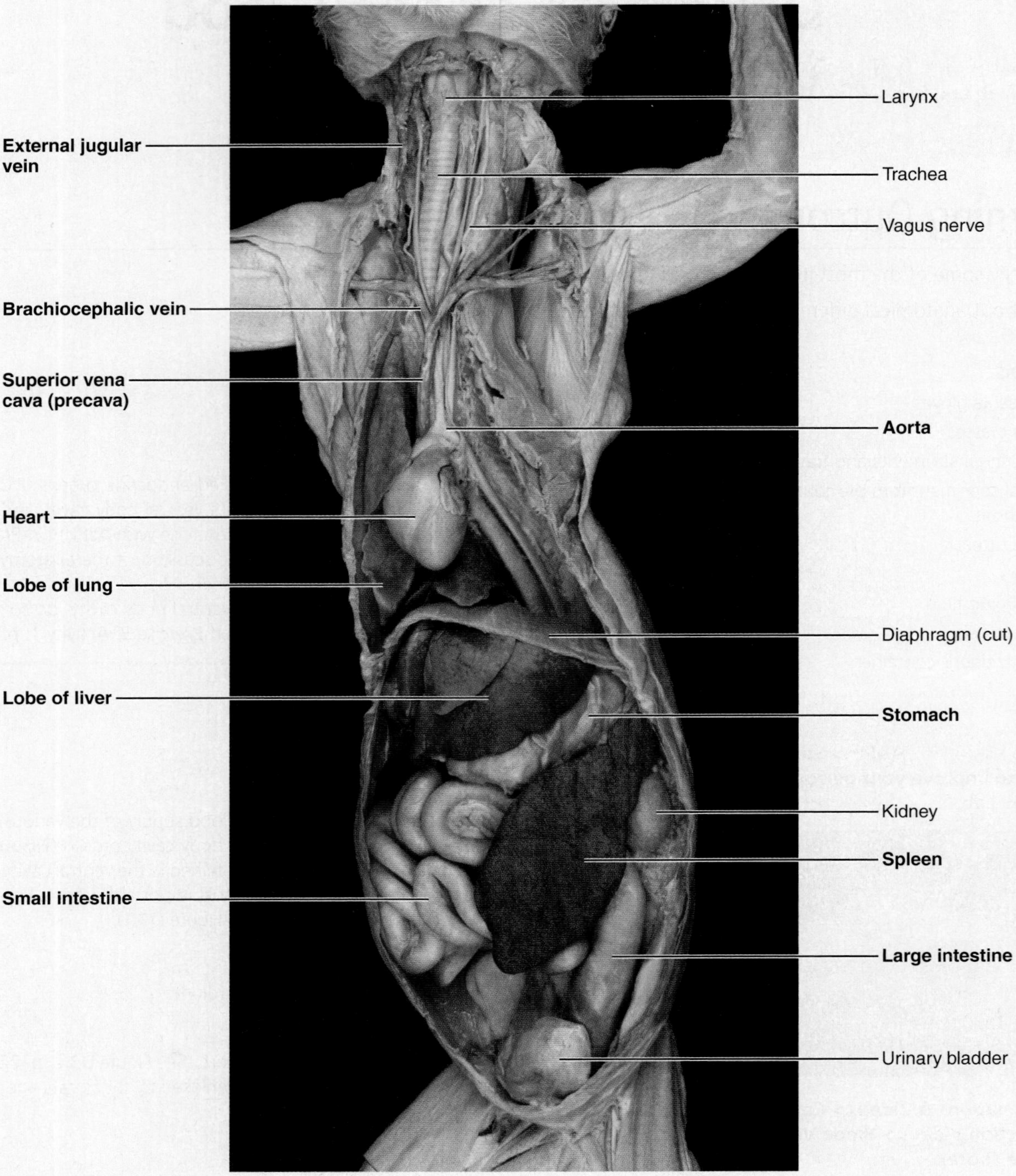

Figure D4.1 Ventral body cavity organs of the cat. (Greater omentum has been removed.)

Activity 3

Preparing to Identify the Blood Vessels

1. Don gloves and safety glasses, and then obtain your dissection animal. Place the animal on the dissection tray, ventral side up. Carefully clear away any thymus tissue or fat obscuring the heart and the large vessels associated with the heart. Before identifying the blood vessels, try to locate the *phrenic nerve* (from the cervical plexus), which innervates the diaphragm. The phrenic nerves lie ventral to the root of the lung on each side as they pass to the diaphragm. Also attempt to locate the *vagus*

nerve (cranial nerve X) passing laterally along the trachea and dorsal to the root of the lung.

2. Slit the parietal pericardium and reflect it superiorly. Then, cut it away from its heart attachments. Review the structures of the heart. Notice its pointed inferior end (apex) and its broader superior base. Identify the two *atria*, which appear darker than the inferior *ventricles*.

3. Identify the **aorta**, the largest artery in the body, issuing from the left ventricle. Also identify the *coronary arteries* in the sulcus on the ventral surface of the heart. As an aid to blood vessel identification, the arteries of laboratory dissection specimens are injected with red latex; the veins are injected with blue latex. Exceptions to this will be noted as they are encountered.

4. Identify the two large venae cavae—the **superior** and **inferior venae cavae**—entering the right atrium. The superior vena cava is the largest dark-colored vessel entering the base of the heart. These vessels are called the precava and postcava, respectively, in the cat. The caval veins drain the same relative body areas as in humans. Also identify the **pulmonary trunk** (usually injected with blue latex) extending anteriorly from the right ventricle. The right and left pulmonary arteries branch off of the pulmonary trunk. Trace the **pulmonary arteries** until they enter the lungs. Locate the **pulmonary veins** entering the left atrium and the ascending aorta arising from the left ventricle and running dorsal to the precava and to the left of the body midline.

Activity 4

Identifying the Arteries of the Cat

Begin your dissection of the arterial system of the cat. Refer to **Figure D4.2**, p. 726 and **Figure D4.3**, p. 727.

1. Reidentify the aorta as it emerges from the left ventricle. The first branches of the aorta are the **coronary arteries**, which supply the myocardium. The coronary arteries emerge from the base of the aorta and can be seen on the surface of the heart. Follow the aorta as it arches (aortic arch), and identify its major branches. In the cat, the aortic arch gives off two large vessels, the **brachiocephalic artery** and the **left subclavian artery**. The brachiocephalic artery has three major branches, the right subclavian artery and the right and left common carotid arteries. Note that in humans, the left common carotid artery directly branches off the aortic arch.

2. Follow the **right common carotid artery** along the right side of the trachea as it moves anteriorly, giving off branches to the neck muscles, thyroid gland, and trachea. At the level of the larynx, it branches to form the **external** and **internal carotid arteries**. The internal carotid is quite small in the cat and it may be difficult to locate. It may even be absent. The distribution of the carotid arteries parallels that in humans.

3. Follow the **right subclavian artery** laterally. It gives off four branches, the first being the tiny **vertebral artery**, which along with the internal carotid artery provides the arterial circulation of the brain. Other branches of the subclavian artery include the **costocervical trunk** (to the costal and cervical regions), the **thyrocervical trunk** (to the shoulder), and the **internal thoracic (mammary) artery** (serving the ventral thoracic wall). As the subclavian passes in front of the first rib it becomes the **axillary artery**. Its branches, which may be difficult to identify, supply the trunk and shoulder muscles. These are the **ventral thoracic artery** (to the pectoral muscles), the **long thoracic artery** (to pectoral muscles and latissimus dorsi), and the **subscapular artery** (to the trunk muscles). As the axillary artery enters the arm, it is called the **brachial artery**, and it travels with the median nerve down the length of the humerus. At the elbow, the brachial artery branches to produce the two major arteries serving the forearm and hand, the **radial** and **ulnar arteries**.

4. Return to the thorax, lift the left lung, and follow the course of the *descending aorta* through the thoracic cavity. The esophagus overlies it along its course. Notice the paired **intercostal arteries** that branch laterally from the aorta in the thoracic region.

5. Follow the aorta through the diaphragm into the abdominal cavity. Carefully pull the peritoneum away from its ventral surface and identify the following vessels:

Celiac trunk: The first branch diverging from the aorta immediately as it enters the abdominal cavity; supplies the stomach, liver, gallbladder, pancreas, and spleen. (Trace as many of its branches to these organs as possible.)

Superior mesenteric artery: Immediately posterior to the celiac trunk; supplies the small intestine and most of the large intestine. (Spread the mesentery of the small intestine to observe the branches of this artery as they run to supply the small intestine.)

Adrenolumbar arteries: Paired arteries diverging from the aorta slightly posterior to the superior mesenteric artery; supply the muscles of the body wall and adrenal glands.

Renal arteries: Paired arteries supplying the kidneys.

Gonadal arteries (testicular or ovarian): Paired arteries supplying the gonads.

Inferior mesenteric artery: An unpaired thin vessel arising from the ventral surface of the aorta posterior to the gonadal arteries; supplies the second half of the large intestine.

Iliolumbar arteries: Paired, rather large arteries that supply the body musculature in the iliolumbar region.

External iliac arteries: Paired arteries that continue through the body wall and pass under the inguinal ligament to the hindlimb.

6. After giving off the external iliac arteries, the aorta persists briefly and then divides into three arteries: the two **internal iliac arteries**, which supply the pelvic viscera, and the **median sacral artery**. As the median sacral artery enters the tail, it

Text continues on p. 728. →

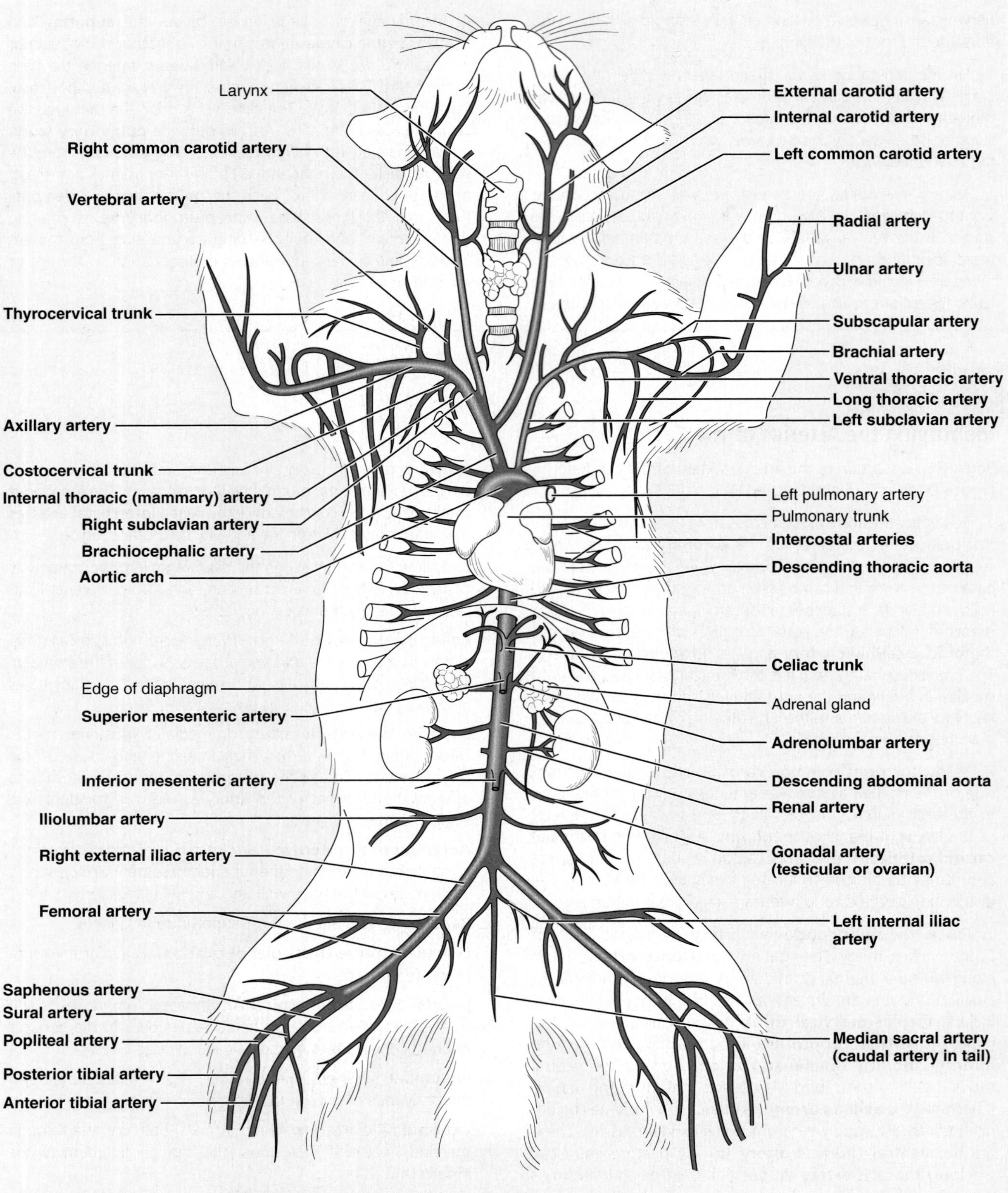

Larynx
Right common carotid artery
Vertebral artery
Thyrocervical trunk
Axillary artery
Costocervical trunk
Internal thoracic (mammary) artery
Right subclavian artery
Brachiocephalic artery
Aortic arch
Edge of diaphragm
Superior mesenteric artery
Inferior mesenteric artery
Iliolumbar artery
Right external iliac artery
Femoral artery
Saphenous artery
Sural artery
Popliteal artery
Posterior tibial artery
Anterior tibial artery
External carotid artery
Internal carotid artery
Left common carotid artery
Radial artery
Ulnar artery
Subscapular artery
Brachial artery
Ventral thoracic artery
Long thoracic artery
Left subclavian artery
Left pulmonary artery
Pulmonary trunk
Intercostal arteries
Descending thoracic aorta
Celiac trunk
Adrenal gland
Adrenolumbar artery
Descending abdominal aorta
Renal artery
Gonadal artery (testicular or ovarian)
Left internal iliac artery
Median sacral artery (caudal artery in tail)

Figure D4.2 Arterial system of the cat. (See also Figure D4.3.)

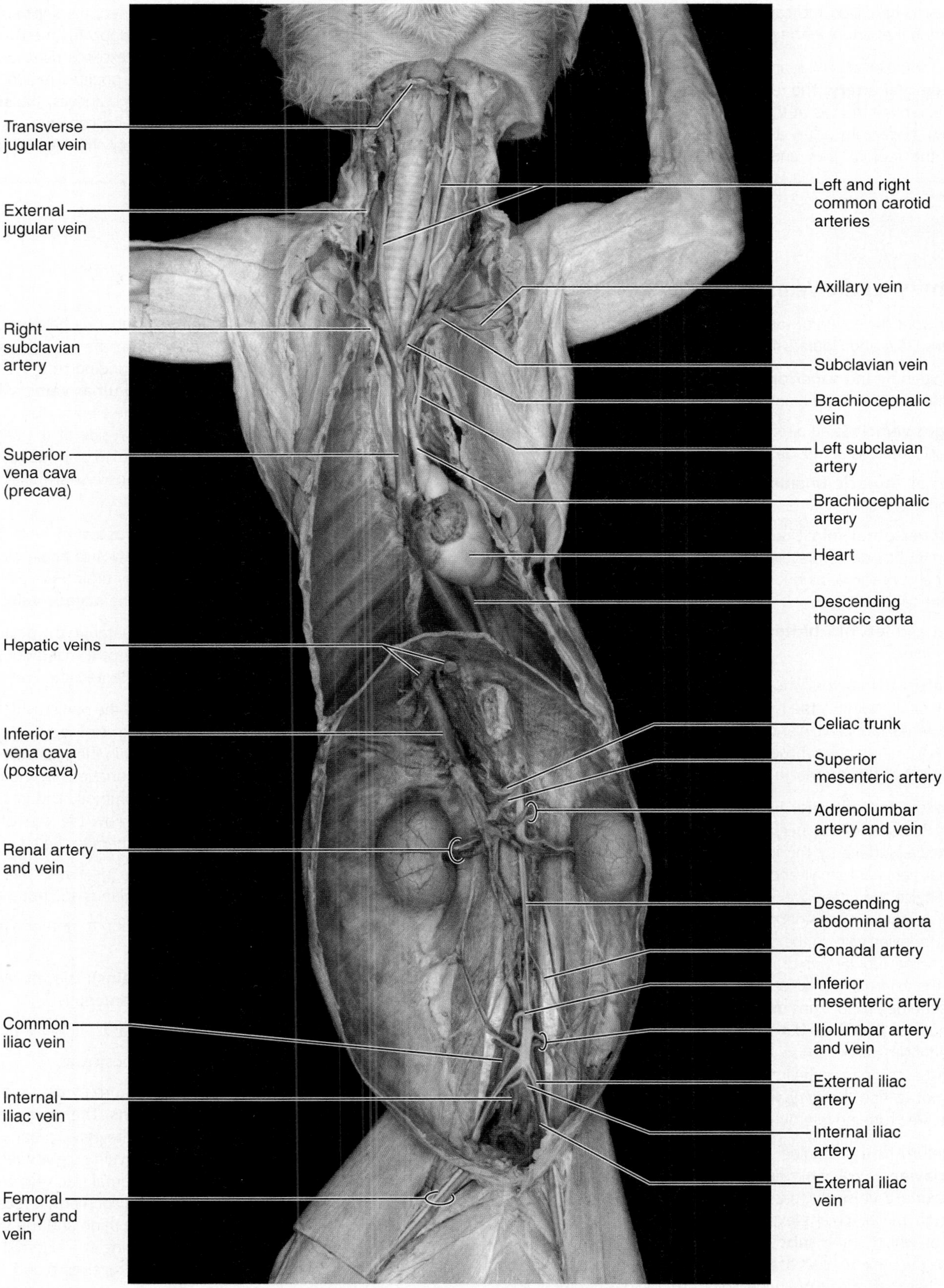
Transverse jugular vein
External jugular vein
Right subclavian artery
Superior vena cava (precava)
Hepatic veins
Inferior vena cava (postcava)
Renal artery and vein
Common iliac vein
Internal iliac vein
Femoral artery and vein
Left and right common carotid arteries
Axillary vein
Subclavian vein
Brachiocephalic vein
Left subclavian artery
Brachiocephalic artery
Heart
Descending thoracic aorta
Celiac trunk
Superior mesenteric artery
Adrenolumbar artery and vein
Descending abdominal aorta
Gonadal artery
Inferior mesenteric artery
Iliolumbar artery and vein
External iliac artery
Internal iliac artery
External iliac vein

Figure D4.3 Cat dissected to reveal major blood vessels.

comes to be called the **caudal artery**. Note that there is no common iliac artery in the cat.

7. Trace the external iliac artery into the thigh, where it becomes the **femoral artery**. The femoral artery is most easily identified in the *femoral triangle* at the medial surface of the upper thigh. Follow the femoral artery as it courses through the thigh (along with the femoral vein and nerve) and gives off branches to the thigh muscles. As you approach the knee, the **saphenous artery** branches off the femoral artery to supply the medial portion of the leg. The femoral artery then descends deep to the knee to become the **popliteal artery** in the popliteal region. The popliteal artery in turn gives off two main branches, the **sural artery** and the **posterior tibial artery**, and continues as the **anterior tibial artery**. These branches supply the leg and foot.

Activity 5

Identifying the Veins of the Cat

Begin your dissection of the venous system of the cat. Refer to **Figure D4.4** and Figure D4.3.

1. Reidentify the **superior vena cava (precava)** as it enters the right atrium. Trace it anteriorly to identify veins that enter it.

Azygos vein: Passing directly into its dorsal surface; drains the thoracic intercostal muscles.

Internal thoracic (mammary) veins: Drain the chest and abdominal walls.

Right vertebral vein: Drains the spinal cord and brain; usually enters right side of precava approximately at the level of the internal thoracic veins but may enter the brachiocephalic vein in your specimen.

Right and **left brachiocephalic veins:** Form the precava by their union.

2. Reflect the pectoral muscles, and trace the brachiocephalic vein laterally. Identify the two large veins that unite to form it—the external jugular vein and the subclavian vein. Notice that this differs from humans, whose brachiocephalic veins are formed by the union of the internal jugular and subclavian veins.

3. Follow the **external jugular vein** as it travels anteriorly along the side of the neck to the point where it is joined on its medial surface by the **internal jugular vein**. The internal jugular veins are small and may be difficult to identify in the cat. Notice the difference in cat and human jugular veins. The internal jugular is considerably larger in humans and drains into the subclavian vein. In the cat, the external jugular is larger, and the internal jugular vein drains into it. Several other vessels drain into the external jugular vein (transverse scapular vein draining the shoulder, facial veins draining the head, and others). These are not discussed here but are shown on the figure and may be traced if time allows. Also, identify the *common carotid artery*, since it accompanies the internal jugular vein in this region, and attempt to find the *sympathetic trunk*, which is located in the same area running lateral to the trachea.

4. Return to the shoulder region and follow the path of the **subclavian vein** as it moves laterally toward the arm. It becomes the **axillary vein** as it passes in front of the first rib and runs through the brachial plexus, giving off several branches, the first of which is the **subscapular vein**. The subscapular vein drains the proximal part of the arm and shoulder. The four other branches that receive drainage from the shoulder, pectoral, and latissimus dorsi muscles are shown in the figure but need not be identified in this dissection.

5. Follow the axillary vein into the arm, where it becomes the **brachial vein**. You can locate this vein on the medial side of the arm accompanying the brachial artery and nerve. Trace it to the point where it receives the **radial** and **ulnar veins** (which drain the forelimb) at the inner bend of the elbow. Also locate the superficial **cephalic vein** on the dorsal side of the arm. It communicates with the brachial vein via the median cubital vein in the elbow region and then enters the transverse scapular vein in the shoulder.

6. Reidentify the **inferior vena cava (postcava)**, and trace it to its passage through the diaphragm. Notice again as you follow its course that the intercostal veins drain into a much smaller vein lying dorsal to the postcava, the **azygos vein**.

7. Attempt to identify the **hepatic veins** entering the postcava from the liver. These may be seen if some of the anterior liver tissue is scraped away where the postcava enters the liver.

8. Displace the intestines to the left side of the body cavity, and proceed posteriorly to identify the following veins in order. All of these veins empty into the postcava and drain the organs served by the same-named arteries. In the cat, variations in the connections of the veins to be located are common, and in some cases the postcaval vein may be double below the level of the renal veins. If you observe deviations, call them to the attention of your instructor.

Adrenolumbar veins: From the adrenal glands and body wall.

Renal veins: From the kidneys (it is common to find two renal veins on the right side).

Gonadal veins (testicular or ovarian veins): The left vein of this venous pair enters the left renal vein anteriorly.

Iliolumbar veins: Drain muscles of the back.

Common iliac veins: Unite to form the postcava.

The common iliac veins are formed in turn by the union of the **internal iliac** and **external iliac veins**. The more medial internal iliac veins receive branches from the pelvic organs and gluteal region, whereas the external iliac vein receives venous drainage from the lower limb. As the external iliac vein enters the thigh by running beneath the inguinal ligament, it receives the **deep femoral vein**, which drains the thigh and the external genital region. Just inferior to that point, the external iliac vein becomes the **femoral vein**, which receives blood from the thigh, leg, and foot. Follow the femoral vein down the

Text continues on p. 731. →

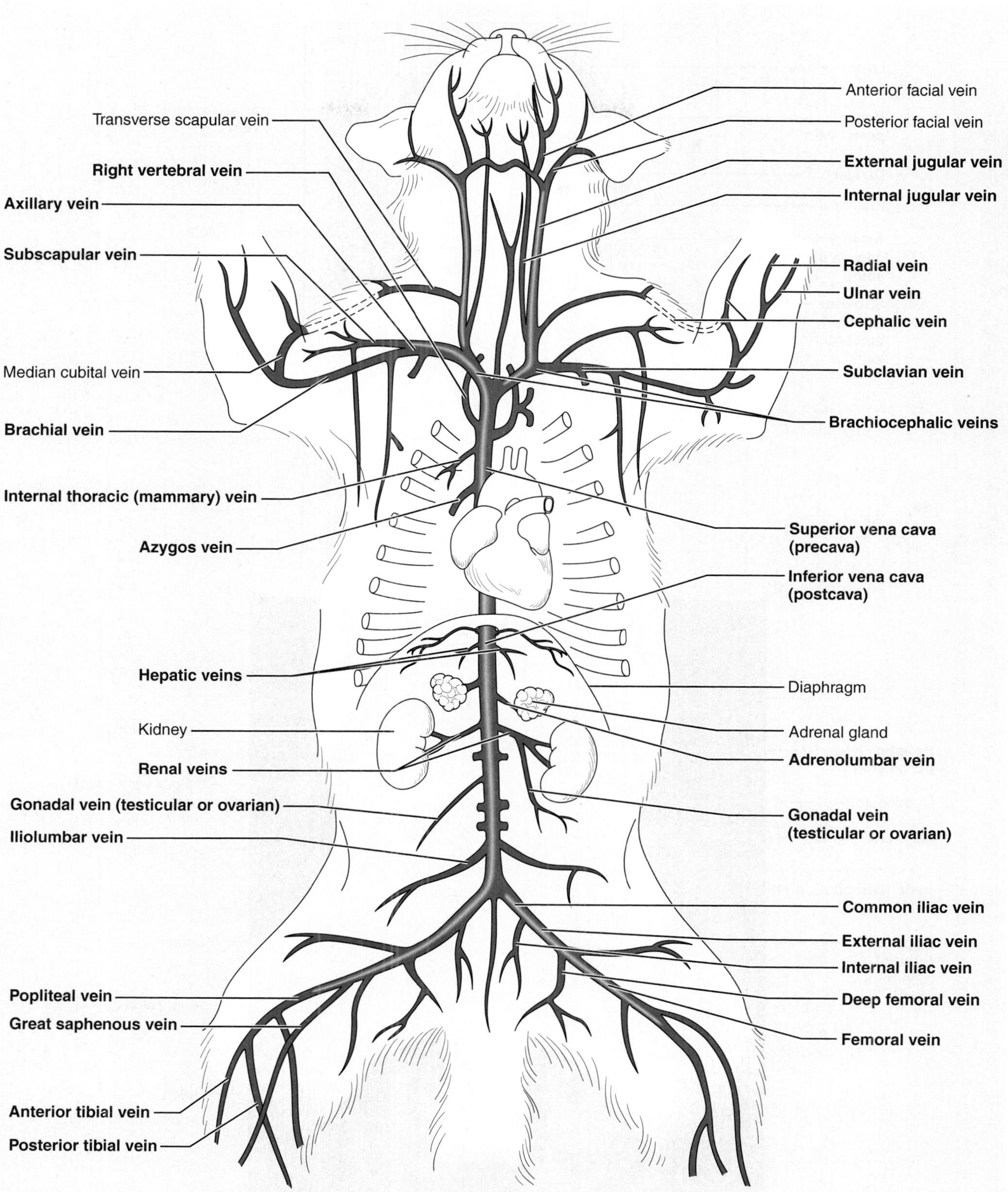

Figure D4.4 Venous system of the cat. See also Figure D4.3.

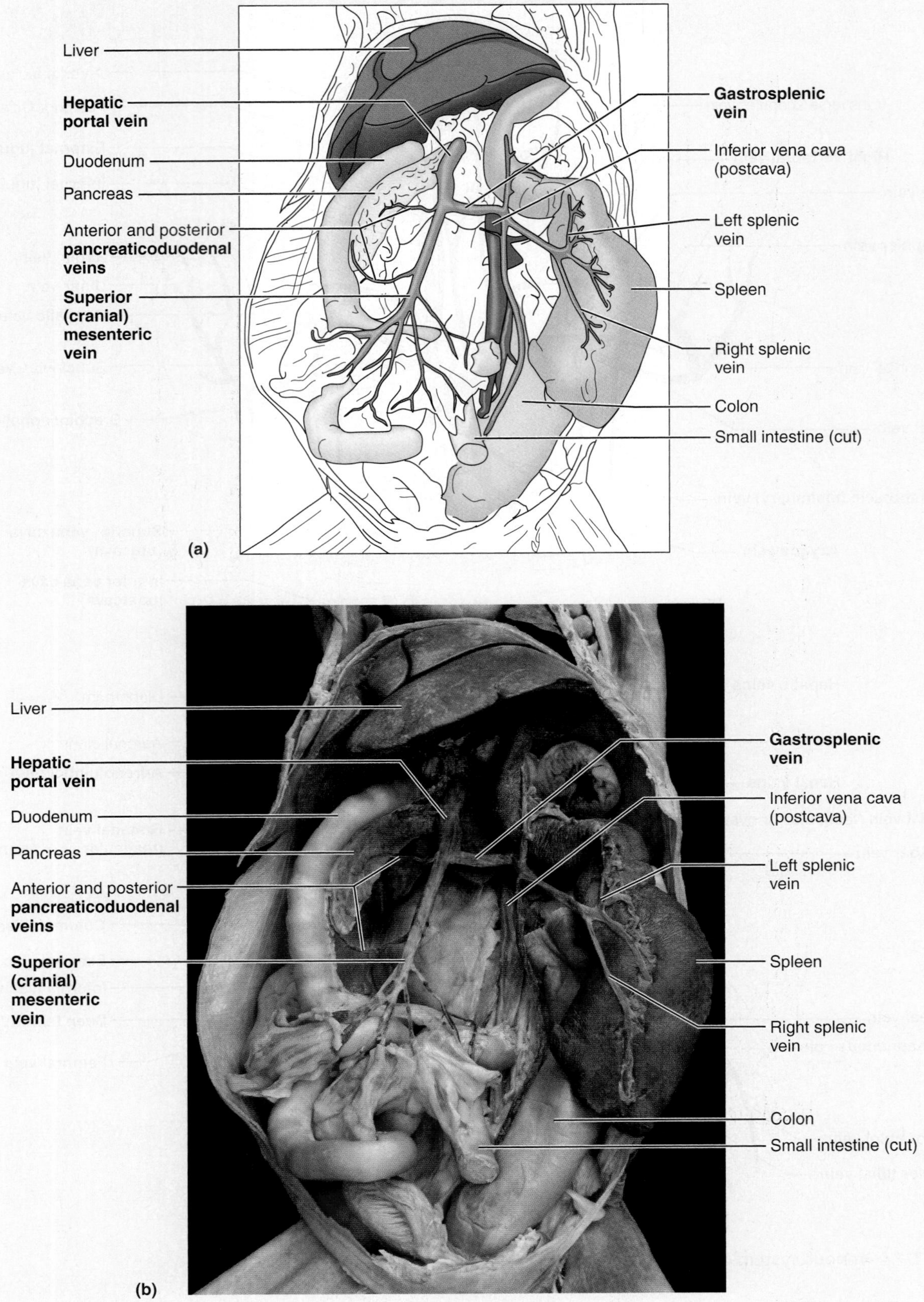

Figure D4.5 Hepatic portal circulation of the cat. (a) Diagram. **(b)** Photograph of hepatic portal system of the cat just posterior to the liver and pancreas. Intestines have been partially removed. The mesentery of the small intestine has been partially dissected to show the veins of the portal system.

thigh to identify the **great saphenous vein**, a superficial vein that travels up the inner aspect of the calf and across the inferior portion of the gracilis muscle (accompanied by the great saphenous artery and nerve) to enter the femoral vein. The femoral vein is formed by the union of this vein and the popliteal vein. The **popliteal vein** is located deep in the thigh beneath the semimembranosus and semitendinosus muscles in the popliteal space accompanying the popliteal artery. Trace the popliteal vein to its point of division into the **posterior** and **anterior tibial veins**, which drain the leg.

9. Trace the hepatic portal drainage system in your cat. Refer to **Figure D4.5**. Locate the **hepatic portal vein** by removing the peritoneum between the first portion of the small intestine and the liver. It appears brown because of coagulated blood, and it is unlikely that it or any of the vessels of this circulation contain latex. In the cat, the hepatic portal vein is formed by the union of the gastrosplenic and superior mesenteric veins. In the human, the hepatic portal vein is formed by the union of the splenic and superior mesenteric veins. If possible, locate the following vessels, which empty into the hepatic portal vein.

Gastrosplenic vein: Carries blood from the spleen and stomach; located dorsal to the stomach.

Superior (cranial) mesenteric vein: A large vein draining the small and large intestines and the pancreas.

Inferior (caudal) mesenteric vein (not shown): Parallels the course of the inferior mesenteric artery and empties into the superior mesenteric vein. In humans, this vessel merges with the splenic vein.

Pancreaticoduodenal veins (anterior and posterior): The anterior branch empties into the hepatic portal vein; the posterior branch empties into the superior mesenteric vein. In humans, both of these are branches of the superior mesenteric vein.

10. Before you leave the laboratory, follow the boxed instructions to prepare your cat for storage and to clean the area (p. 696).

DISSECTION REVIEW

1. What differences did you observe between the origins of the left common carotid arteries in the cat and in the human?

____

____

Between the origins of the internal and external iliac arteries?

____

____

____

2. How do the relative sizes of the external and internal jugular veins differ in the human and the cat?

____

____

3. In the cat the inferior vena cava is also called the ______________________,

and the superior vena cava is also referred to as the ______________________

4. Describe the location of the following blood vessels:

ascending aorta: ______________________

aortic arch: ______________________

descending thoracic aorta: ______________________

descending abdominal aorta: ______________________

DISSECTION EXERCISE 5

The Main Lymphatic Ducts of the Cat

Learning Outcome

▶ Compare and contrast lymphatic structures of the cat to those of a human.

Materials

- Disposable gloves
- Safety glasses
- Dissecting instruments and tray
- Animal specimen from previous dissections
- Embalming fluid
- Paper towels
- Organic debris container

Go to Mastering A&P™ > Study Area to improve your performance in A&P Lab.

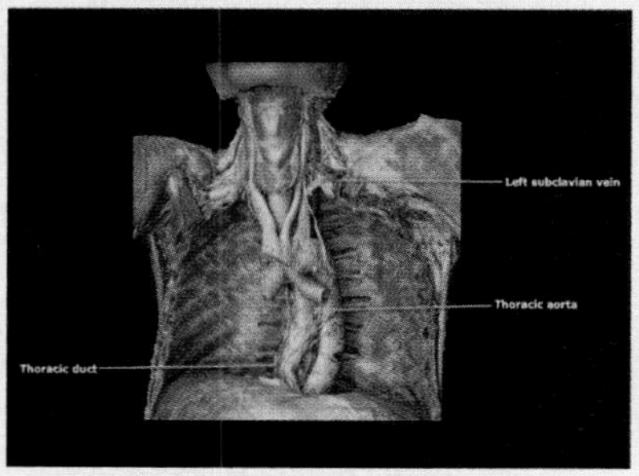

> Lab Tools > Practice Anatomy Lab > Human Cadaver > Lymphatic System

Instructors may assign Cat Dissection Videos, Practice Anatomy Lab Practical questions (PAL) for the dissections, and more using the Mastering A&P™ Item Library.

Activity 1

Identifying the Main Lymphatic Ducts of the Cat

1. Don disposable gloves and safety glasses. Obtain your cat and a dissecting tray and instruments. Because lymphatic vessels are extremely thin walled, it is difficult to locate them in a dissection unless the animal has been triply injected (with yellow or green latex for the lymphatic system). However, the large thoracic duct can be localized and identified. Compare your observations to those of the human lymphatic system (see Exercise 35).

2. Move the thoracic organs to the side to locate the **thoracic duct**. Typically it lies just to the left of the mid-dorsal line, abutting the dorsal aspect of the descending aorta. It is usually about the size of pencil lead and red-brown with a segmented or beaded appearance caused by the valves within it. Trace it anteriorly to the site where it passes behind the left brachiocephalic vein and then bends and enters the venous system at the junction of the left subclavian and external jugular veins. If the veins are well injected, some of the blue latex may have slipped past the valves and entered the first portion of the thoracic duct.

3. While in this region, also attempt to identify the short **right lymphatic duct** draining into the right subclavian vein, and notice the collection of lymph nodes in the axillary region.

4. If the cat is triply injected, trace the thoracic duct posteriorly to identify the **cisterna chyli**, a saclike enlargement. This structure, which receives fat-rich lymph from the intestine, begins at the level of the diaphragm and can be localized posterior to the left kidney.

5. Before you leave the laboratory, follow the boxed instructions to prepare your cat for storage and to clean the area (p. 696).

DISSECTION REVIEW

1. How does the cat's lymphatic drainage pattern compare to that of humans? ____________________

____

2. What is the role of each of the following? ____________________

   a. thoracic duct ____________________

   ____

   b. right lymphatic duct ____________________

   ____

   c. cisterna chyli ____________________

   ____

D I S S E C T I O N
E X E R C I S E

6 Dissection of the Respiratory System of the Cat

Learning Outcome

- Identify the major respiratory system organs in a dissected animal.

Materials

- Disposable gloves
- Safety glasses
- Dissecting instruments and tray
- Animal specimen from previous dissections
- Embalming fluid
- Paper towels
- Dissecting microscope
- Organic debris container

Go to Mastering A&P™ > Study Area to improve your performance in A&P Lab.

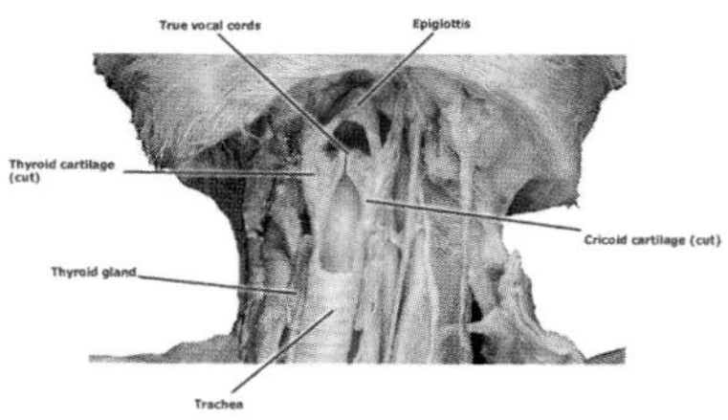

> Lab Tools > Practice Anatomy Lab > Cat > Respiratory System

Instructors may assign Cat Dissection Videos, Practice Anatomy Lab Practical questions (PAL) for the dissections, and more using the Mastering **A&P™** Item Library.

In this dissection exercise, you will be examining both the gross and the microscopic structure of respiratory system organs. Don disposable gloves and safety glasses, and then obtain your dissection animal and dissecting tray and instruments. Refer to a discussion of the human respiratory system as you work (see Exercise 37).

Activity 1

Identifying Organs of the Respiratory System

1. Examine the external nares, oral cavity, and oropharynx (**Figure D6.1**). Use a probe to demonstrate the continuity between the oropharynx and the nasopharynx above. The pharynx continues as the laryngopharynx, which lies immediately dorsal to the larynx (not shown).

2. After securing the animal to the dissecting tray dorsal surface down, expose the more distal respiratory structures by retracting the cut muscle and rib cage. Do not sever nerves and blood vessels located on either side of the trachea if these have not been studied. If you have not previously opened the thoracic cavity, make a medial longitudinal incision through the neck muscles and thoracic musculature to expose and view the thoracic organs (see Figure D3.1, p. 719).

Text continues on p. 737. →

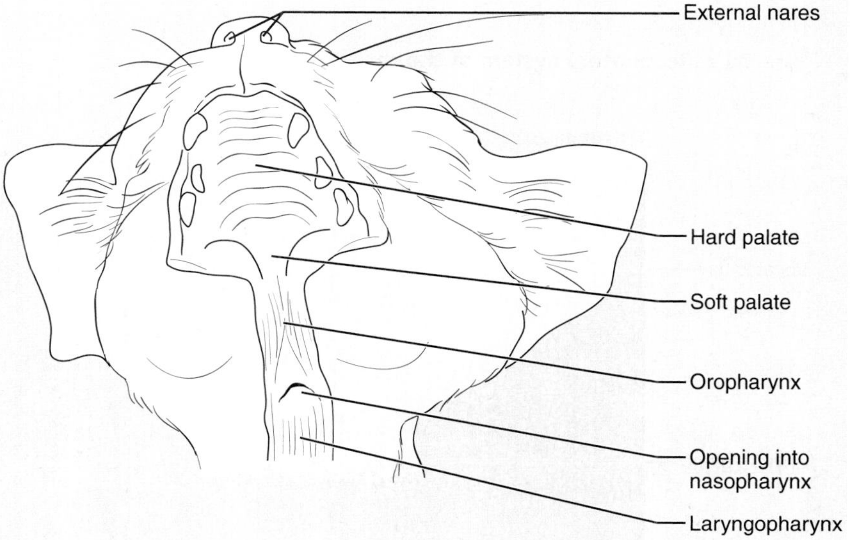

Figure D6.1 External nares, oral cavity, and pharynx of the cat. The larynx has been dissected free and reflected toward the thorax.

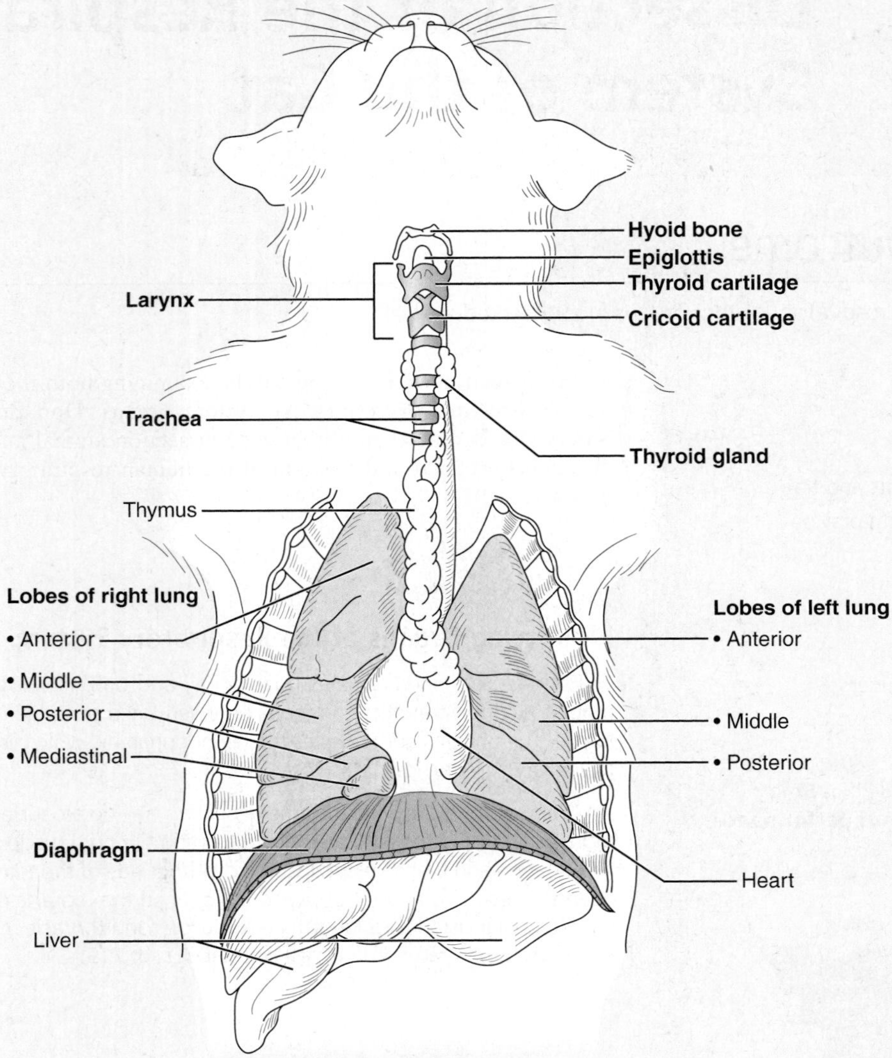

Figure D6.2 Respiratory system of the cat.

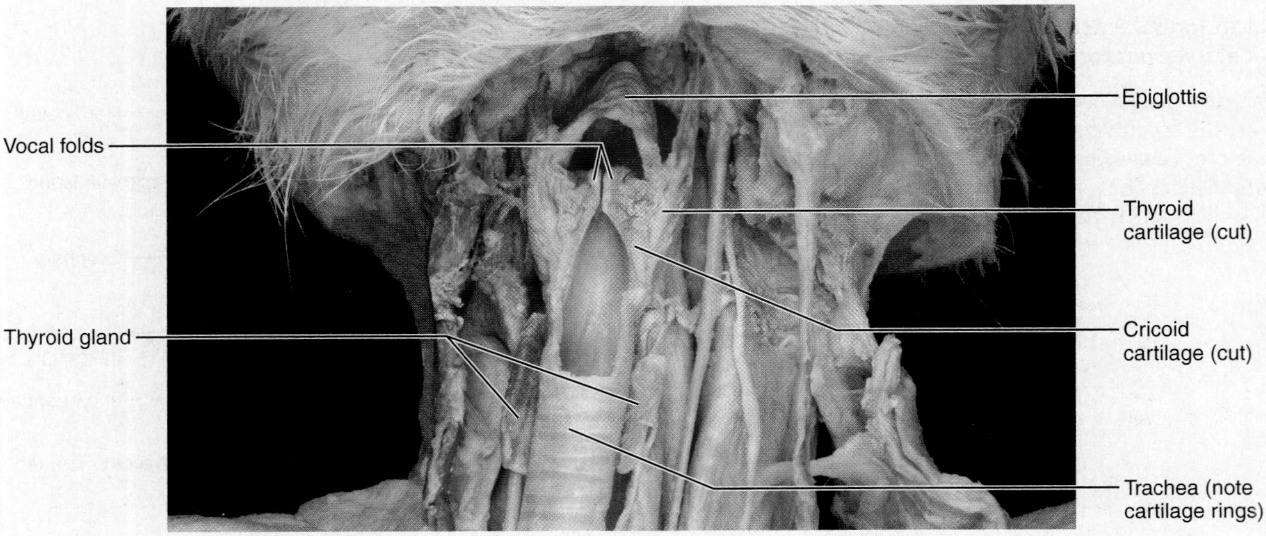

Figure D6.3 Larynx (opened), trachea, and thyroid gland.

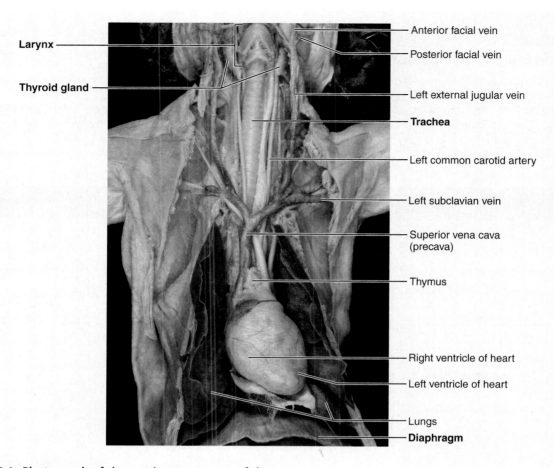

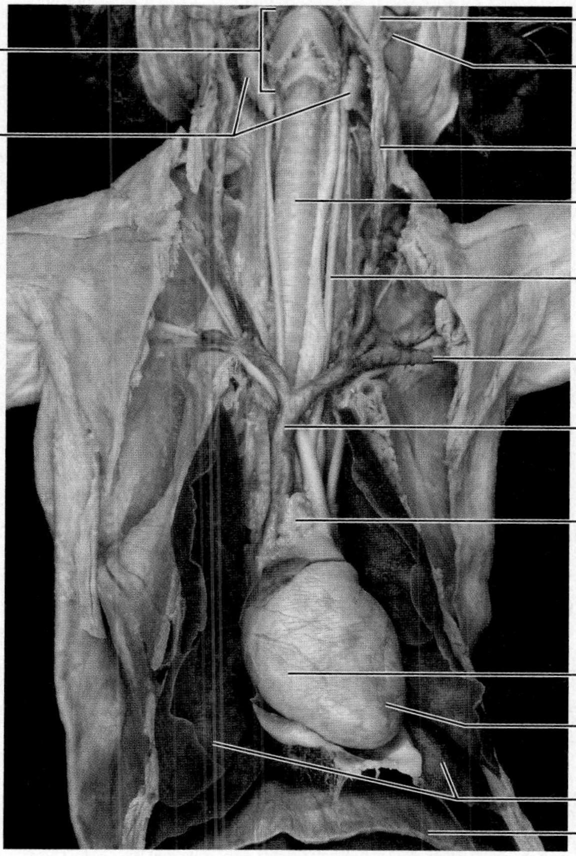

Figure D6.4 Photograph of the respiratory system of the cat.

3. Use **Figure D6.2**, **Figure D6.3**, and **Figure D6.4** to help you identify the structures named in steps 3 through 5. Examine the **trachea**, and determine by finger examination whether the cartilage rings are complete or incomplete posteriorly. Locate the **thyroid gland** inferior to the larynx on the trachea. Free the **larynx** from the attached muscle tissue for ease of examination. Identify the **thyroid** and **cricoid cartilages** and the flaplike **epiglottis**. Find the **hyoid bone**, located anterior to the larynx. Make a longitudinal incision through the ventral wall of the larynx, and locate the vocal and vestibular folds on the inner wall (Figure D6.3).

4. Locate the large *right* and *left common carotid arteries* and the *external jugular veins* on either side of the trachea. Also locate a conspicuous white band, the *vagus nerve*, which lies alongside the trachea, adjacent to the common carotid artery.

5. Examine the contents of the thoracic cavity (Figure D6.4). Follow the trachea as it bifurcates into two *main (primary) bronchi*, which plunge into the **lungs**. Note that there are two *pleural cavities* containing the lungs and that each lung is composed of many lobes. In humans there are three lobes in the right lung and two in the left. How does this compare to what is seen in the cat?

____

____

In the mediastinum, identify the *pericardial sac* (if it is still present) containing the heart. Examine the *pleura*, and note its exceptionally smooth texture.

6. Locate the **diaphragm** and the **phrenic nerve**. The phrenic nerve, clearly visible as a white "thread" running along the pericardium to the diaphragm, controls the activity of the diaphragm in breathing. Lift one lung, and find the esophagus beneath the parietal pleura. Follow it through the diaphragm to the stomach.

Activity 2

Observing Lung Tissue Microscopically

Make a longitudinal incision in the outer tissue of one lung lobe beginning at a main bronchus. Attempt to follow part of the respiratory tree from this point down into the smaller subdivisions. Carefully observe the cut lung tissue (under a dissecting microscope, if one is available), noting the richness of the vascular supply and the irregular or spongy texture of the lung.

Before you leave the laboratory, follow the boxed instructions to prepare your cat for storage and to clean the area (p. 696).

DISSECTION REVIEW

1. Are the cartilaginous rings in the cat trachea complete or incomplete?

___

2. Describe the appearance of the bronchial tree in the cat lung.

___

___

3. Describe the appearance of lung tissue under the dissection microscope.

___

___

DISSECTION EXERCISE 7

Dissection of the Digestive System of the Cat

Learning Outcomes

- Identify on a dissected animal the organs composing the alimentary canal, and name their subdivisions if any.
- Name and identify the accessory organs of digestion in the dissection animal.

Materials

- Disposable gloves
- Safety glasses
- Dissecting instruments and tray
- Animal specimen from previous dissections
- Bone cutters
- Hand lens
- Embalming fluid
- Paper towels
- Organic debris container

Go to Mastering A&P™ > Study Area to improve your performance in A&P Lab.

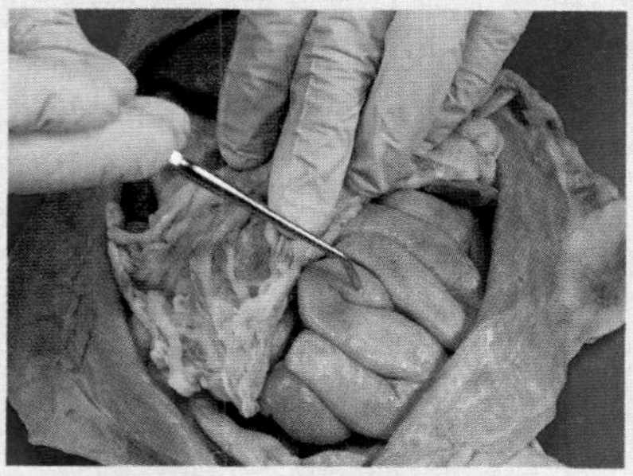

> Animations & Videos > Cat Dissection Videos > The Peritoneum

Instructors may assign Cat Dissection Videos, Practice Anatomy Lab Practical questions (PAL) for the dissections, and more using the Mastering **A&P™** Item Library.

Don gloves and safety glasses, and obtain your dissection animal. Secure it to the dissecting tray, dorsal surface down. Obtain all necessary dissecting instruments. If you have completed the dissection of the circulatory and respiratory systems, the abdominal cavity is already exposed and many of the digestive system structures have been previously identified. However, duplication of effort generally provides a good learning experience, so all of the digestive system structures will be traced and identified in this exercise. Refer to a discussion of the human digestive system as you work (see Exercise 38).

If the abdominal cavity has not been previously opened, make a midline incision from the rib cage to the pubic symphysis (**Figure D7.1**). Then make four lateral cuts—two parallel to the rib cage and two at the inferior margin of the abdominal

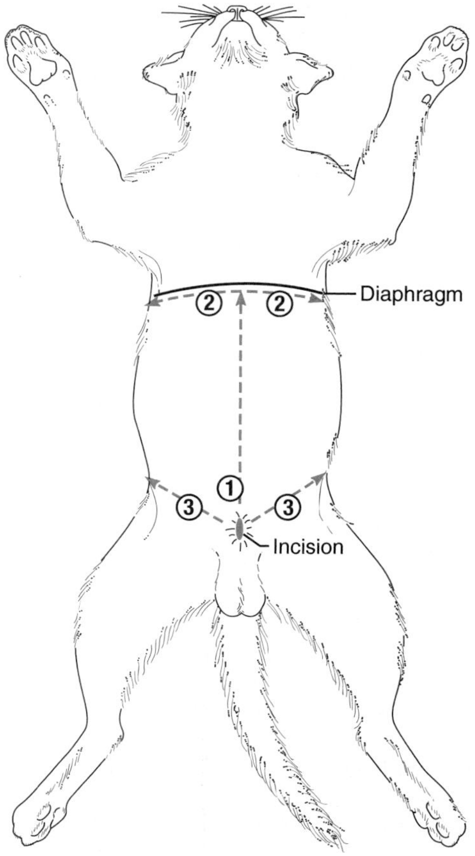

Figure D7.1 Incisions to be made in opening the ventral body cavity of the cat. Numbers indicate sequence.

cavity so that the abdominal wall can be reflected back while you examine the abdominal contents. Observe the shiny membrane lining the inner surface of the abdominal wall, which is the **parietal peritoneum**.

Activity 1

Identifying Alimentary Canal Organs

1. Locate the abdominal alimentary canal structures. Refer to **Figure D7.2**.

2. Identify the large reddish brown **liver** (**Figure D7.3** on p. 742) just beneath the diaphragm and the greater omentum, an apron of mesentery riddled with fat (not shown), that covers the abdominal contents.The greater omentum is attached to the greater curvature of the stomach; its immune cells and macrophages help to protect the abdominal cavity. Lift the greater omentum, noting its two-layered structure and attachments, and lay it to the side or remove it to make subsequent organ identifications easier. Does the liver of the cat have the same number of lobes as the human liver?

____

3. Lift the liver and examine its inferior surface to locate the **gallbladder**, a dark greenish sac embedded in the liver's ventral surface. Identify the **falciform ligament**, a delicate layer of mesentery separating the main lobes of the liver (right and left median lobes) and attaching the liver superiorly to the abdominal wall. Also identify the thickened area along the posterior edge of the falciform ligament, the *round ligament*, or *ligamentum teres*, a remnant of the umbilical vein of the fetus.

4. Displace the left lobes of the liver to expose the **stomach**. Identify the esophagus as it enters the stomach and the cardial part, fundus, body, and pyloric part of the stomach. What is the general shape of the stomach?

____

Locate the **lesser omentum**, the serous membrane attaching the lesser curvature of the stomach to the liver, and identify the large spleen curving around the greater curvature of the stomach.

Make an incision through the stomach wall to expose the inner surface of the stomach. When the stomach is empty, its mucosa has large folds called **rugae**. Can you see rugae? As the stomach fills, the rugae gradually disappear and are no longer visible. Identify the **pyloric sphincter** at the distal end of the stomach.

5. Lift the stomach and locate the **pancreas**, which appears as a grayish or brownish diffuse glandular mass in the mesentery. It extends from the vicinity of the spleen and greater curvature of the stomach and wraps around the duodenum. Attempt to find the **pancreatic duct** as it empties into the duodenum at a bulbous area referred to as the **hepatopancreatic ampulla** (refer to Figure D7.3). Tease away the fine connective tissue, locate the **bile duct** close to the pancreatic duct, and trace its path superiorly to the point where it diverges into the **cystic duct** (gallbladder duct) and the **common hepatic duct** (duct from the liver). Notice that the duodenum assumes a looped position.

6. Lift the **small intestine** to investigate the manner in which it is attached to the posterior body wall by the **mesentery**.

Observe the mesentery closely. What types of structures do you see in this double peritoneal fold?

____

Other than providing support for the intestine, what functions does the mesentery have?

____

____

Trace the path of the small intestine from its proximal (duodenal) end to its distal (ileal) end. Can you see any obvious differences in the external anatomy of the small intestine from one end to the other?

____

With a scalpel, slice open the distal portion of the ileum and flush out the inner surface with water. Feel the inner surface with your fingertip. How does it feel?

____

Use a hand lens to see whether you can see any **villi** and to locate the areas of lymphatic tissue called **Peyer's patches**, which appear as scattered white patches on the inner intestinal surface (see Figure 38.9, p. 577).

Return to the duodenal end of the small intestine. Make an incision into the duodenum. As before, flush the surface with water, and feel the inner surface. Does it feel any different from the ileal mucosa?

__________ If so, describe the difference. __________

____

Use the hand lens to observe the villi. What differences do you see in the villi in the two areas of the small intestine?

____

____

7. Make an incision into the junction between the ileum and cecum to locate the ileocecal valve (**Figure D7.4** on p. 742). Observe the **cecum**, the initial expanded part of the large intestine. You may have to remove lymph nodes from this area to observe it clearly. Does the cat have an appendix?

____

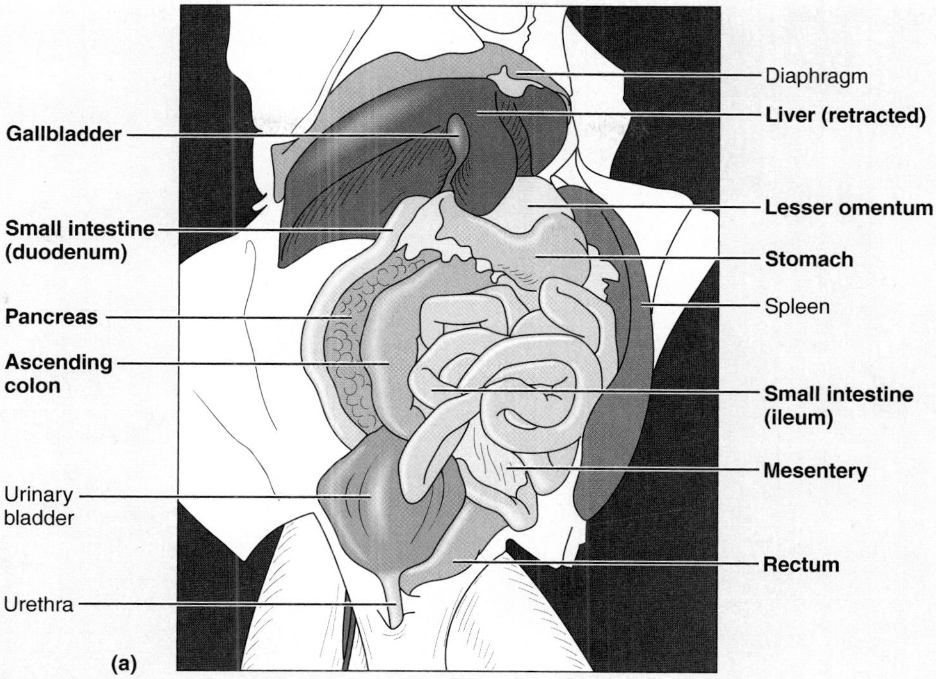

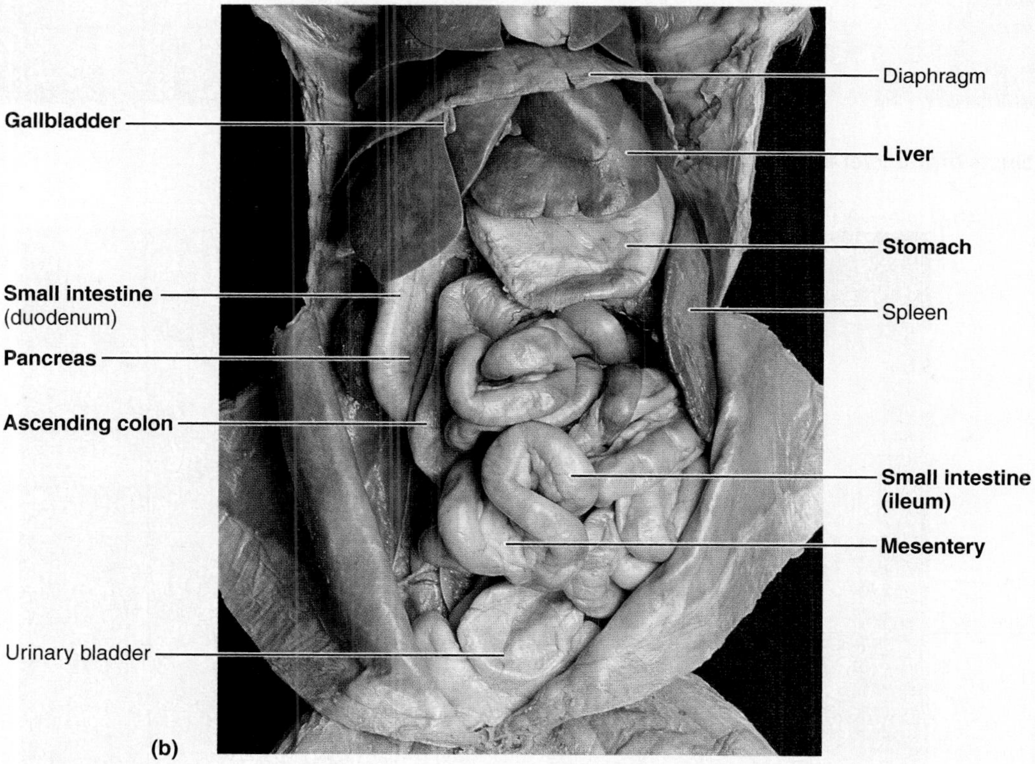

Figure D7.2 Digestive organs of the cat. (a) Diagram. **(b)** Photograph. The greater omentum has been cut from its attachment to the stomach.

8. Identify the short ascending, transverse, and descending portions of the **colon** and the **mesocolon**, a membrane that attaches the colon to the posterior body wall. Trace the descending colon to the **rectum**, which penetrates the body wall, and identify the **anus** on the exterior surface of the specimen.

Identify the two portions of the peritoneum, the parietal peritoneum lining the abdominal wall (identified previously) and the visceral peritoneum, which is the outermost layer of the wall of the abdominal organs.

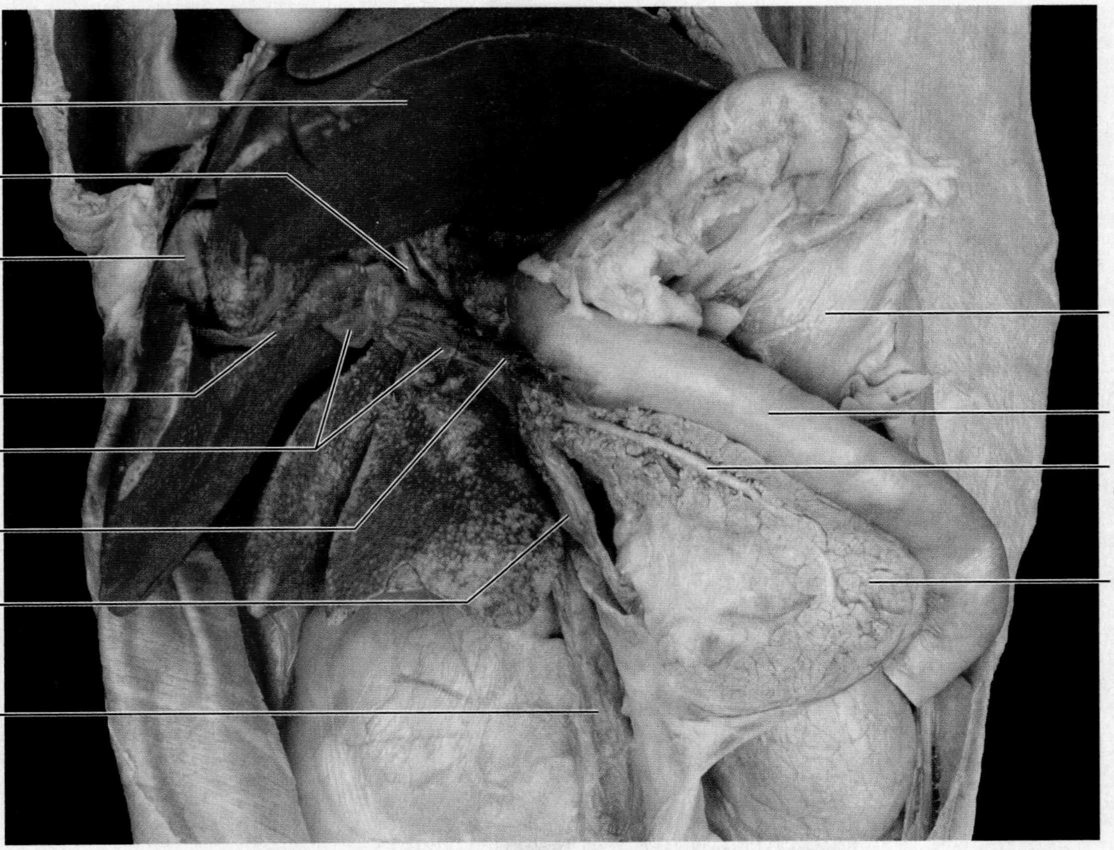

Figure D7.3 Ducts of the liver and pancreas.

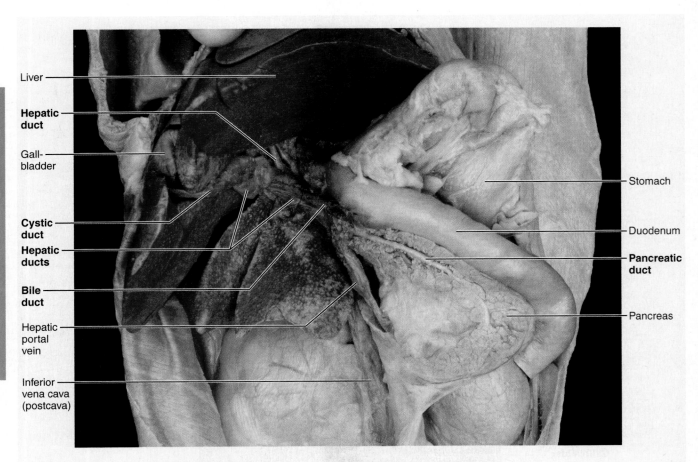

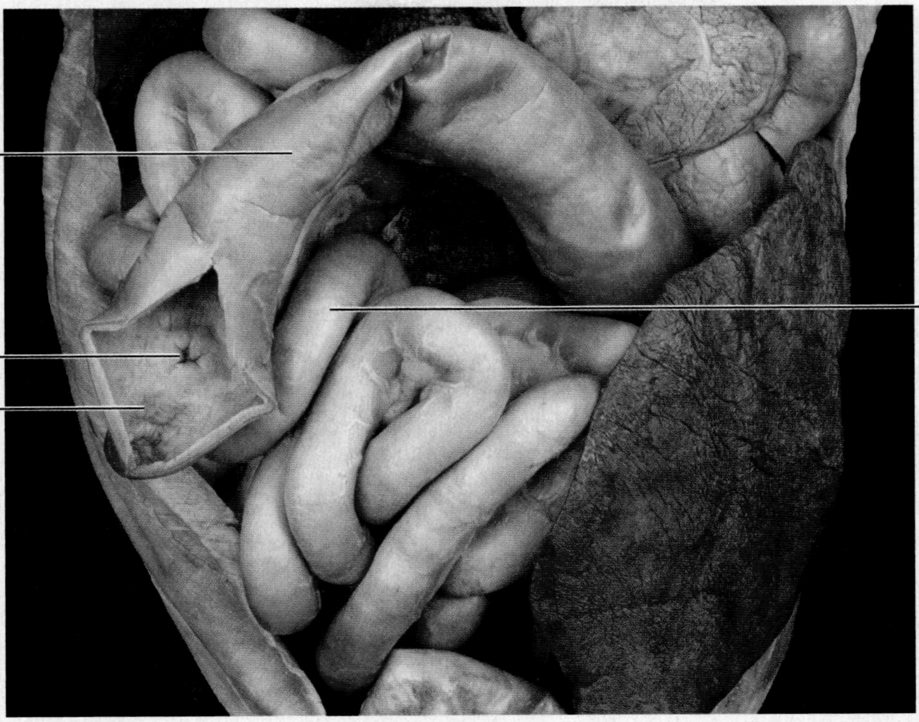

Figure D7.4 Ileocecal valve.

Activity 2

Exposing and Viewing the Salivary Glands and Oral Cavity Structures

1. To expose and identify the **salivary glands**, remove the skin from one side of the head and clear the connective tissue away from the angle of the jaw, below the ear, and superior to the masseter muscle. Many lymph nodes are in this area, and you should remove them if they obscure the salivary glands, which are lighter tan and lobular in texture. The cat possesses five pairs of salivary glands, but only those glands described in humans are easily localized and identified (**Figure D7.5**). Locate the **parotid gland** on the cheek just inferior to the ear. Follow its duct over the surface of the masseter muscle to the angle of the mouth.

Text continues on next page. →

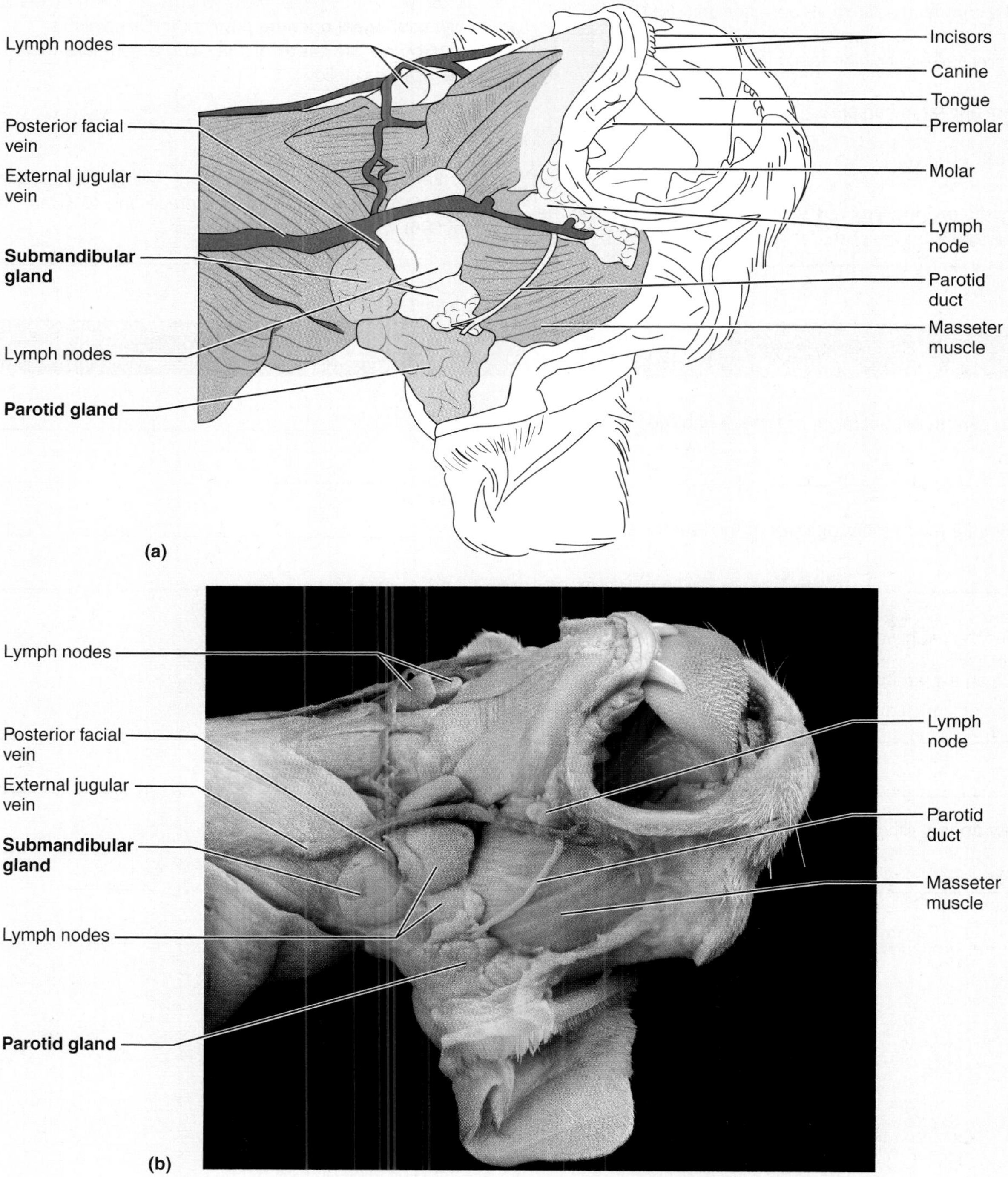

Figure D7.5 Salivary glands of the cat. (a) Diagram. **(b)** Photograph.

The **submandibular gland** is posterior to the parotid, near the angle of the jaw, and the **sublingual gland** (not shown in the figure) is just anterior to the submandibular gland within the lower jaw. The ducts of the submandibular and sublingual glands run deep and parallel to each other and empty on the side of the frenulum of the tongue. These need not be identified on the cat.

2. To expose and identify the structures of the oral cavity, cut through the mandibular angle with bone cutters to free the lower jaw from the maxilla.

Identify the **hard** and **soft palates**, and use a probe to trace the bony hard palate to its posterior limits (see Figure D6.1, p. 735). Note the transverse ridges, or *rugae*, on the hard palate, which play a role in holding food in place while chewing.

Do these appear in humans? ______________________

Does the cat have a uvula? ______________________

Identify the **oropharynx** at the rear of the oral cavity and the palatine tonsils on the posterior walls at the junction between the oral cavity and oropharynx. Identify the **tongue**, and rub your finger across its surface to feel the papillae. Some of the papillae, especially at the anterior end of the tongue, should feel sharp and bristly. These are the filiform papillae. What do you think their function is?

____

Locate the **lingual frenulum** attaching the tongue to the floor of the mouth. Trace the tongue posteriorly until you locate the **epiglottis**, the flap of tissue that covers the entrance to the respiratory passageway when swallowing occurs. Identify the **esophageal opening** posterior to the epiglottis.

Observe the **teeth** of the cat.The dental formula for the adult cat is as follows:

$$\frac{3,1,3,1}{3,1,2,1} \times 2 = 30$$

3. Before you leave the laboratory, follow the boxed instructions to prepare your cat for storage and to clean the area (p. 696).

D I S S E C T I O N 7

DISSECTION REVIEW

1. Compare the appearance of tongue papillae in cats and humans. ______________________

____

2. Compare the number of lobes of the liver in cats and humans. ______________________

____

3. Does the cat have a uvula? ______________ An appendix? ______________

4. Give an explanation for the different adult dental formulas in cats and humans.

____

____

5. How do the villi differ in the duodenum and the ileum? Explain.

____

____

DISSECTION EXERCISE 8

Dissection of the Urinary System of the Cat

Learning Outcome

- Identify the urinary system organs on a dissection specimen.

Materials

- Disposable gloves
- Safety glasses
- Dissecting instruments and tray
- Animal specimen from previous dissections
- Hand magnifying lens
- Embalming fluid
- Paper towel
- Organic debris container

Go to Mastering A&P™ > Study Area to improve your performance in A&P Lab.

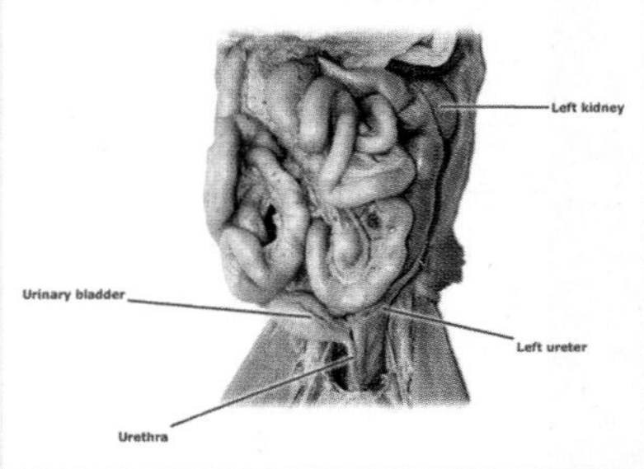

> Lab Tools > Practice Anatomy Lab > Cat > Urinary System

Instructors may assign Cat Dissection Videos, Practice Anatomy Lab Practical questions (PAL) for the dissections, and more using the Mastering **A&P™** Item Library.

The structures of the reproductive and urinary systems are often considered together as the *urogenital system*, since they have common embryological origins. However, in dissections of a specimen it is useful to study the systems separately. Here we will identify the structures of the urinary tract (**Figure D8.1**, p. 746, and **Figure D8.2**, p. 747), with only a few references to neighboring reproductive structures. (The anatomy of the reproductive system is studied in Dissection Exercise 9.) Refer to a discussion of the human urinary system as you work (see Exercise 40).

Activity 1

Identifying Organs of the Urinary System

1. Don gloves and safety glasses. Obtain your dissection specimen, and place it ventral side up on the dissecting tray. Reflect the abdominal viscera (in particular, the small intestine) to locate the kidneys high on the dorsal body wall (Figure D8.1). Note that the **kidneys** in the cat, as well as in the human, are retroperitoneal (behind the peritoneum). Carefully remove the peritoneum, and clear away the bed of fat that surrounds the kidneys. Then locate the adrenal glands that lie superiorly and medial to the kidneys.

2. Identify the **renal artery** (red latex injected), the **renal vein** (blue latex injected), and the ureter at the hilum region of the kidney. You may find two renal veins leaving one kidney in the cat, but not in humans.

3. To observe the gross internal anatomy of the kidney, slit the connective tissue *fibrous capsule* encasing a kidney, and peel it back. Make a midfrontal cut through the kidney, and examine one cut surface with a hand lens to identify the granular *cortex* and the central darker *medulla*, which will appear striated. Notice that the cat's renal medulla consists of just one pyramid as compared to the multipyramidal human kidney.

4. Trace the tubelike **ureters** to the **urinary bladder**, a smooth muscular sac located superior to the small intestine. If your cat is a female, be careful not to confuse the ureters with the uterine tubes, which lie superior to the bladder in the same general region (Figure D8.1). Observe the sites where the ureters enter the bladder. How would you describe the entrance point anatomically?

____

____

5. Cut through the bladder wall, and examine the region where the **urethra** exits to see if you can discern any evidence of the *internal urethral sphincter*.

6. If your cat is a male, identify the prostate (part of the male reproductive system), which encircles the urethra distal to the neck of the bladder (Figure D8.2). Notice that the urinary bladder is somewhat fixed in position by ligaments.

Text continues on p. 748. →

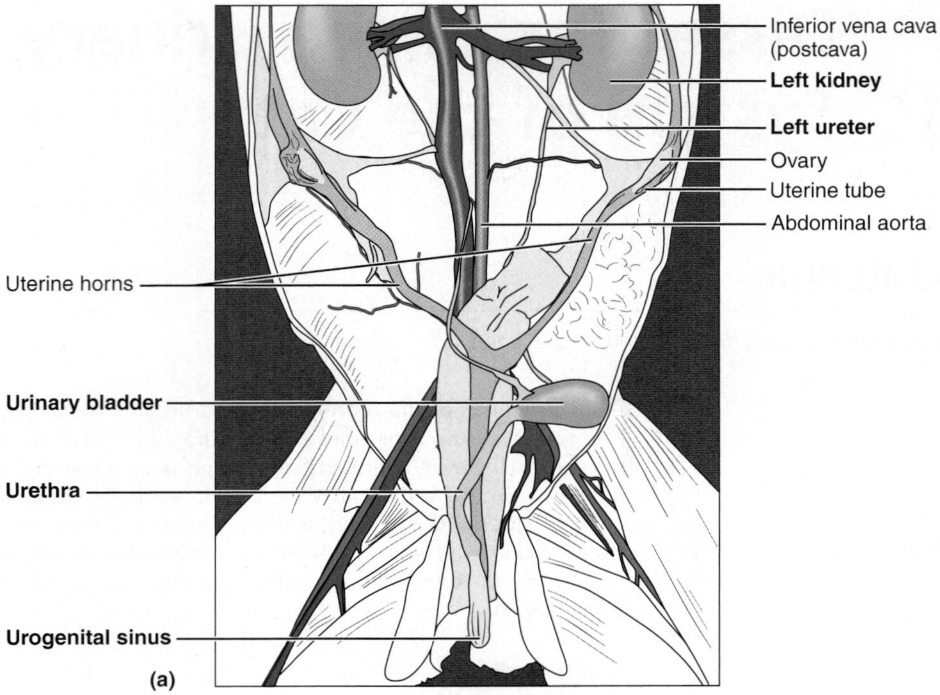

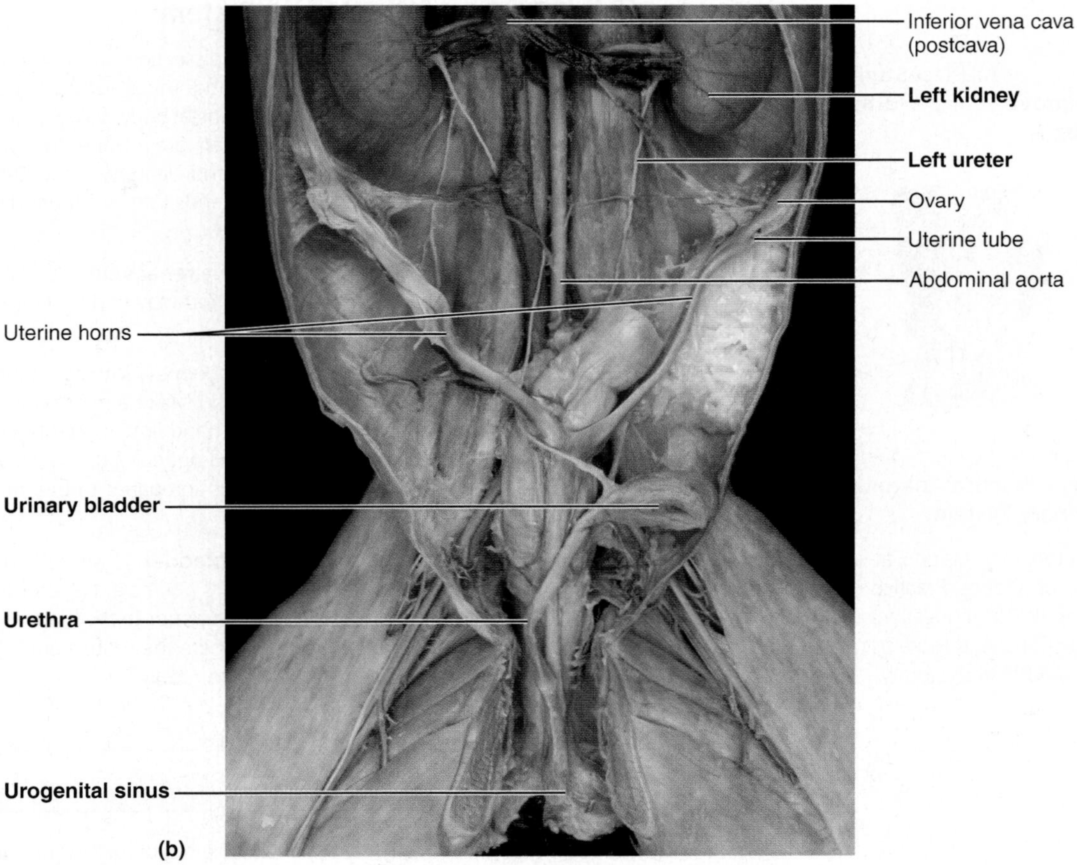

Figure D8.1 Urinary system of the female cat. (Reproductive structures are also indicated.) **(a)** Diagram. **(b)** Photograph of female urogenital system.

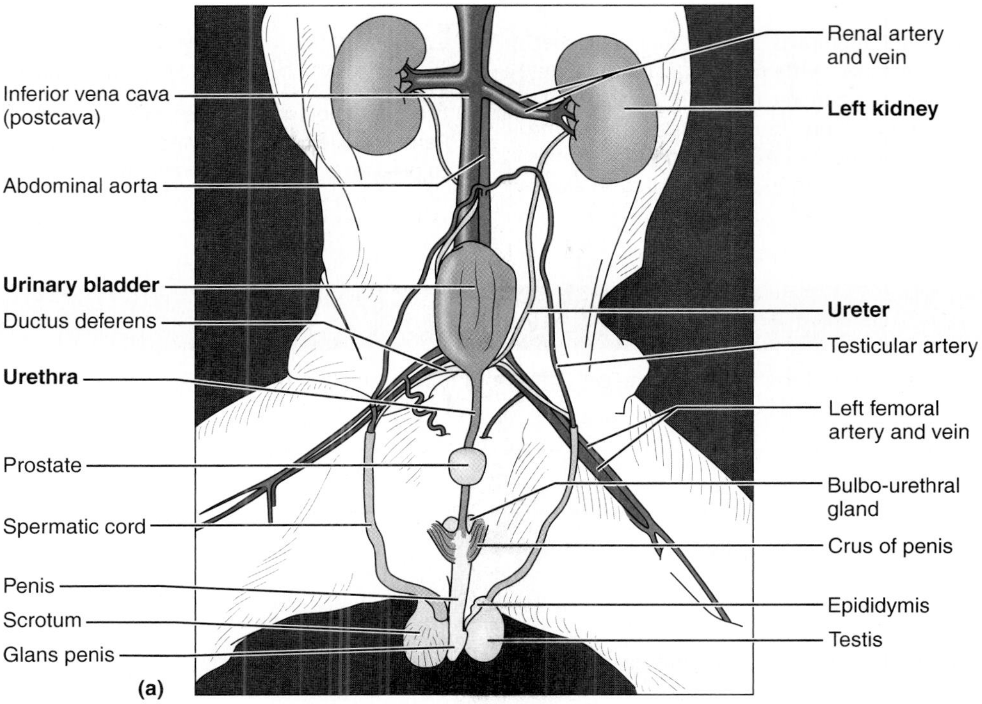

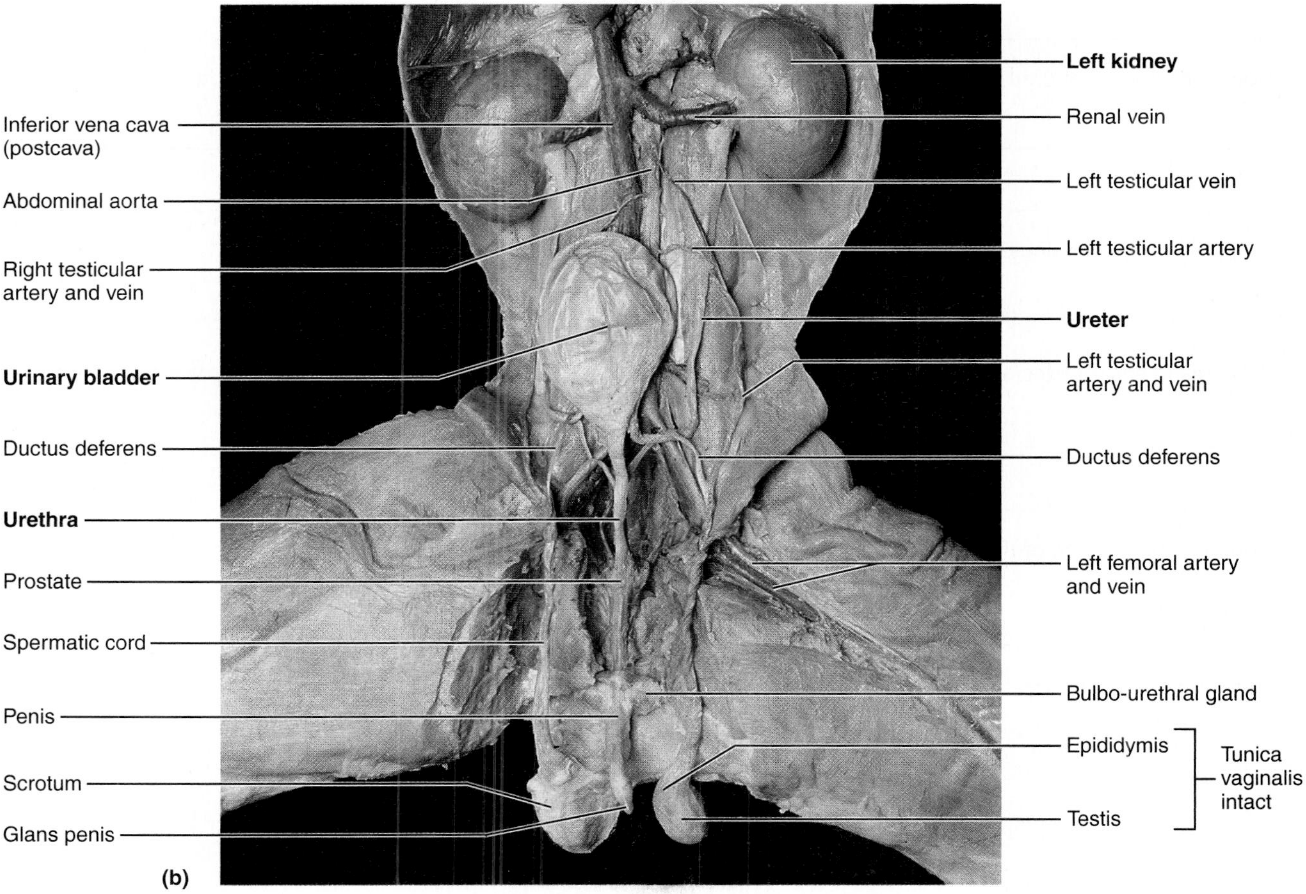

Figure D8.2 Urinary system of the male cat. (Reproductive structures are also indicated.) **(a)** Diagram. **(b)** Photograph of male urogenital system.

7. Using a probe, trace the urethra as it exits from the bladder. In the male, it enters the penis. In the female cat, it terminates in the **urogenital sinus**, a common chamber into which both the vagina and the urethra empty. In the human female, the vagina and the urethra have separate external openings. At this time, do not perform dissection to expose the urethra along its entire length, because you might damage the reproductive structures, which you may study in a separate exercise.

8. To complete this exercise, observe a cat of the opposite sex. Before you leave the laboratory, follow the boxed instructions to prepare your cat for storage and to clean the area (p. 696).

DISSECTION REVIEW

1. a. How does the position of the kidneys in the cat differ from their position in humans?

____

____

b. In what way is the position similar?

____

2. Distinguish between the functions of a ureter and those of the urethra.

____

____

3. How does the site of urethral emptying in the female cat differ from its termination point in the human female?

____

____

4. What is a urogenital sinus?

____

5. What gland encircles the neck of the bladder in the male? ______________ Is this part of the urinary system?

______________ What is its function? ______________________________

____

6. Compare the location of the adrenal glands in the cat to the location in humans.

____

____

DISSECTION EXERCISE 9

Dissection of the Reproductive System of the Cat

Learning Outcomes

- Identify the major reproductive structures of a male and a female dissection animal.
- Recognize and discuss pertinent differences between the reproductive structures of humans and the dissection animal.

Materials

- Disposable gloves
- Safety glasses
- Dissecting instruments and tray
- Animal specimen from previous dissections
- Bone cutters
- Small metric rulers (for female cats)
- Embalming fluid
- Paper towels
- Organic debris container

Go to Mastering A&P™ > Study Area to improve your performance in A&P Lab.

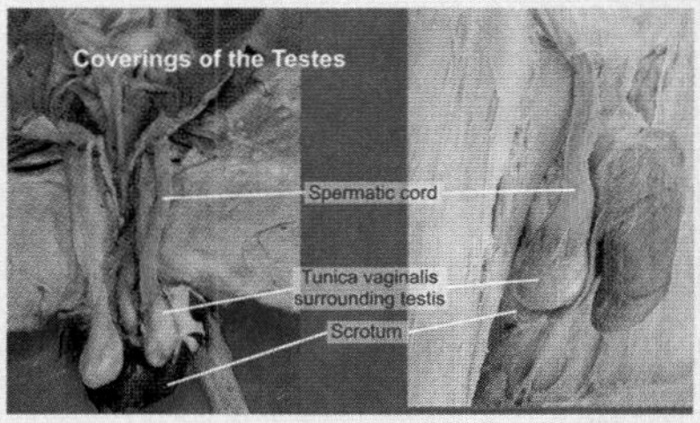

> Animations & Videos > Cat Dissection Videos > Male Reproductive Structures

Instructors may assign Cat Dissection Videos, Practice Anatomy Lab Practical questions (PAL) for the dissections, and more using the Mastering **A&P™** Item Library.

Don gloves and safety glasses. Obtain your cat, a dissecting tray, and the necessary dissecting instruments. After you have completed the study of the reproductive structures of your specimen, observe a cat of the opposite sex. The following instructions assume that the abdominal cavity has been opened in previous dissection exercises. Refer to a discussion of the human reproductive system as you work (see Exercise 42).

Activity 1

Identifying Organs of the Male Reproductive System

Identify the male reproductive structures. Refer to **Figure D9.1** on p. 750.

1. Identify the **penis**, and notice the prepuce covering the glans. Carefully cut through the skin overlying the penis to expose the cavernous tissue beneath, then cross section the penis to see the relative positioning of the three cavernous bodies.

2. Identify the **scrotum**, and then carefully make a shallow incision through the scrotum to expose the **testes**. Notice the abundant connective tissue stretching between the inner wall of the scrotum and testis surface, and note that the scrotum is divided internally.

3. Lateral to the medial aspect of the scrotal sac, locate the **spermatic cord**, which contains the testicular artery, vein, and nerve, as well as the ductus deferens, and follow it up through the inguinal canal into the abdominal cavity. It is not necessary to cut through the pubis; a slight tug on the spermatic cord in the scrotal sac region will reveal its position in the abdominal cavity. Carefully loosen the spermatic cord from the connective tissue investing it, and follow its course as it travels superiorly in the pelvic cavity. Then follow the **ductus deferens** as it loops over the ureter and then courses posterior to the bladder and enters the prostate. Using bone cutters, carefully cut through the pubic symphysis to follow the urethra.

4. Notice that the **prostate**, an enlarged whitish glandular mass abutting the urethra, is comparatively smaller in the cat than in the human, and it is more distal to the bladder. In the human, the prostate is immediately adjacent to the base of the bladder. Carefully slit open the prostate to follow the ductus deferens to the urethra. The male cat urethra, like that of the human, serves as both a urinary and a sperm duct. In the human, the ductus deferens is joined by the duct of the seminal gland to form the ejaculatory duct, which enters the prostate. Seminal glands are not present in the cat.

5. Trace the **urethra** to the proximal ends of the cavernous tissues of the penis, each of which is anchored to the ischium by a band of connective tissue called the

Text continues on p. 751. →

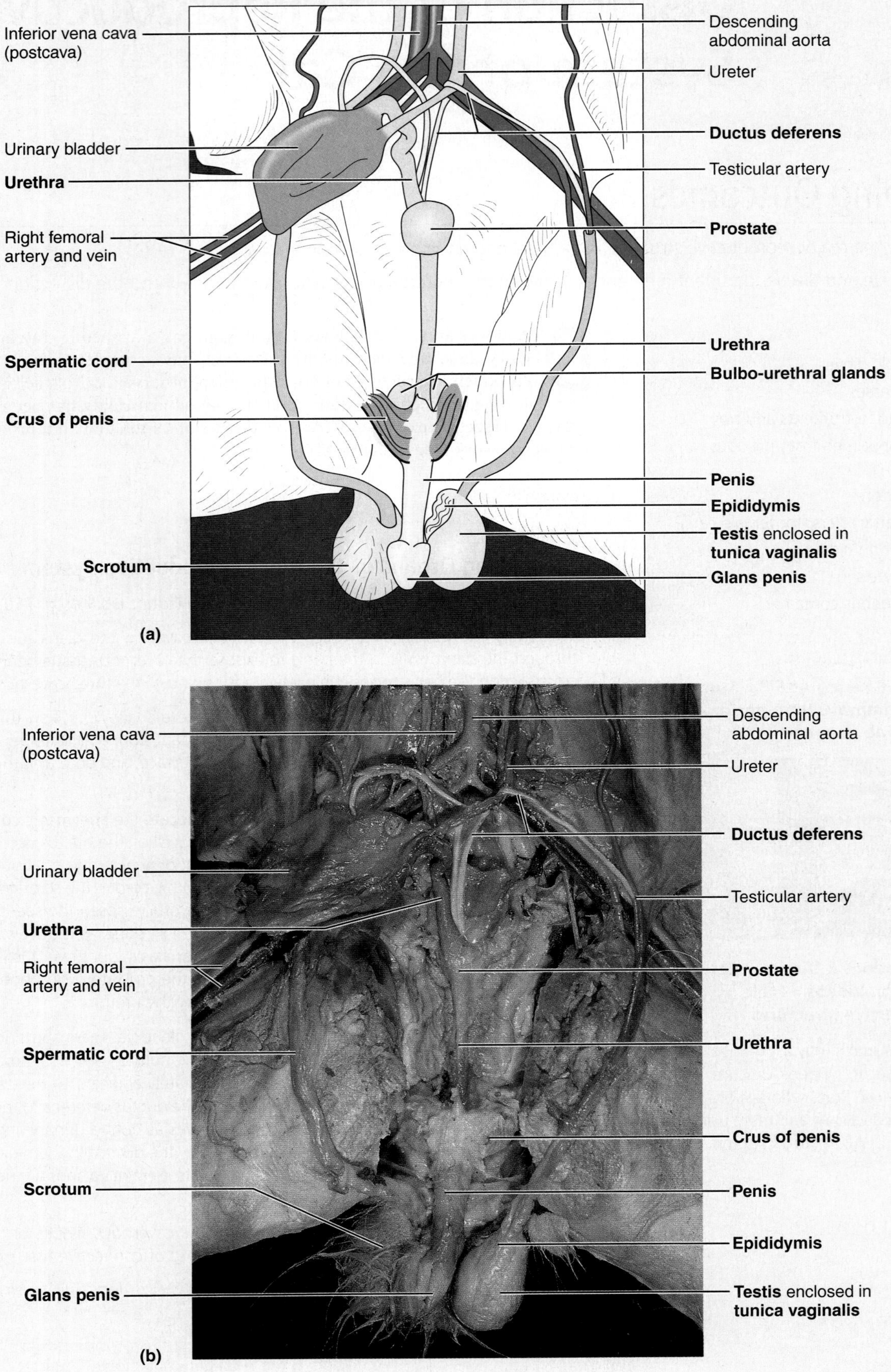

Figure D9.1 Reproductive system of the male cat. (a) Diagram. **(b)** Photograph.

crus of the penis. The crus is covered ventrally by the ischiocavernosus muscle, and the **bulbo-urethral gland** lies beneath it (Figure D9.1).

6. Once again, turn your attention to the testis. Cut it from its attachment to the spermatic cord, and carefully slit open the **tunica vaginalis** capsule enclosing it. Identify the **epididymis** running along one side of the testis. Make a longitudinal cut through the testis and epididymis. Can you see the tubular nature of the epididymis and the rete testis portion of the testis with the naked eye?

Activity 2

Identifying Organs of the Female Reproductive System

Identify the female reproductive structures. Refer to **Figure D9.2** on p. 752.

1. Unlike the pear-shaped simplex, or one-part, uterus of the human, the uterus of the cat is Y-shaped (bipartite, or bicornuate) and consists of a **uterine body** from which two **uterine horns** diverge. Such an enlarged uterus enables the animal to produce litters. Examine the abdominal cavity, and identify the bladder and the body of the uterus lying just dorsal to it.

2. Follow one of the uterine horns as it travels superiorly in the body cavity. Identify the thin mesentery (the *broad ligament*) that helps anchor it and the other reproductive structures to the body wall. Approximately halfway up the length of the uterine horn, it should be possible to identify the more important *round ligament*, a cord of connective tissue extending laterally and posteriorly from the uterine horn to the region of the body wall that would correspond to the inguinal region of the male.

3. Examine the **uterine tube** and **ovary** at the distal end of the uterine horn just caudal to the kidney. Observe how the funnel-shaped end of the uterine tube curves around the ovary. As in the human, the distal end of the tube is fimbriated, or fringed, and the tube is lined with ciliated epithelium. The uterine tubes of the cat are tiny and much shorter than in the human. Identify the **ovarian ligament**, a short, thick cord that extends from the uterus to the ovary and anchors the ovary to the body wall. Also observe the *ovarian artery* and *vein* passing through the mesentery to the ovary and uterine structures.

4. Return to the body of the uterus, and follow it caudally to the pelvis. Use bone cutters to cut through the pubic symphysis, cutting carefully so you do not damage the urethra deep to it. Expose the pelvic region by pressing the thighs dorsally. Follow the uterine body caudally to where it narrows to its sphincterlike cervix, which protrudes into the vagina. Note the point where the urethra draining the bladder and the **vagina** enter a common chamber, the **urogenital sinus.** How does this anatomical arrangement compare to that seen in the human female?

____

____

5. On the cat's exterior, observe the **vulva**, which is similar to the human vulva. Identify the slim **labia majora** surrounding the urogenital opening.

6. To determine the length of the vagina, which is difficult to ascertain by external inspection, slit through the vaginal wall just superior to the urogenital sinus, and cut toward the body of the uterus with scissors. Reflect the cut edges, and identify the muscular cervix of the uterus. Measure the distance between the urogenital sinus and the cervix. Approximately how long is the vagina of the cat?

____

7. To complete this exercise, observe a cat of the opposite sex.

8. Before you leave the laboratory, follow the boxed instructions to prepare your cat for storage and to clean the area (p. 696).

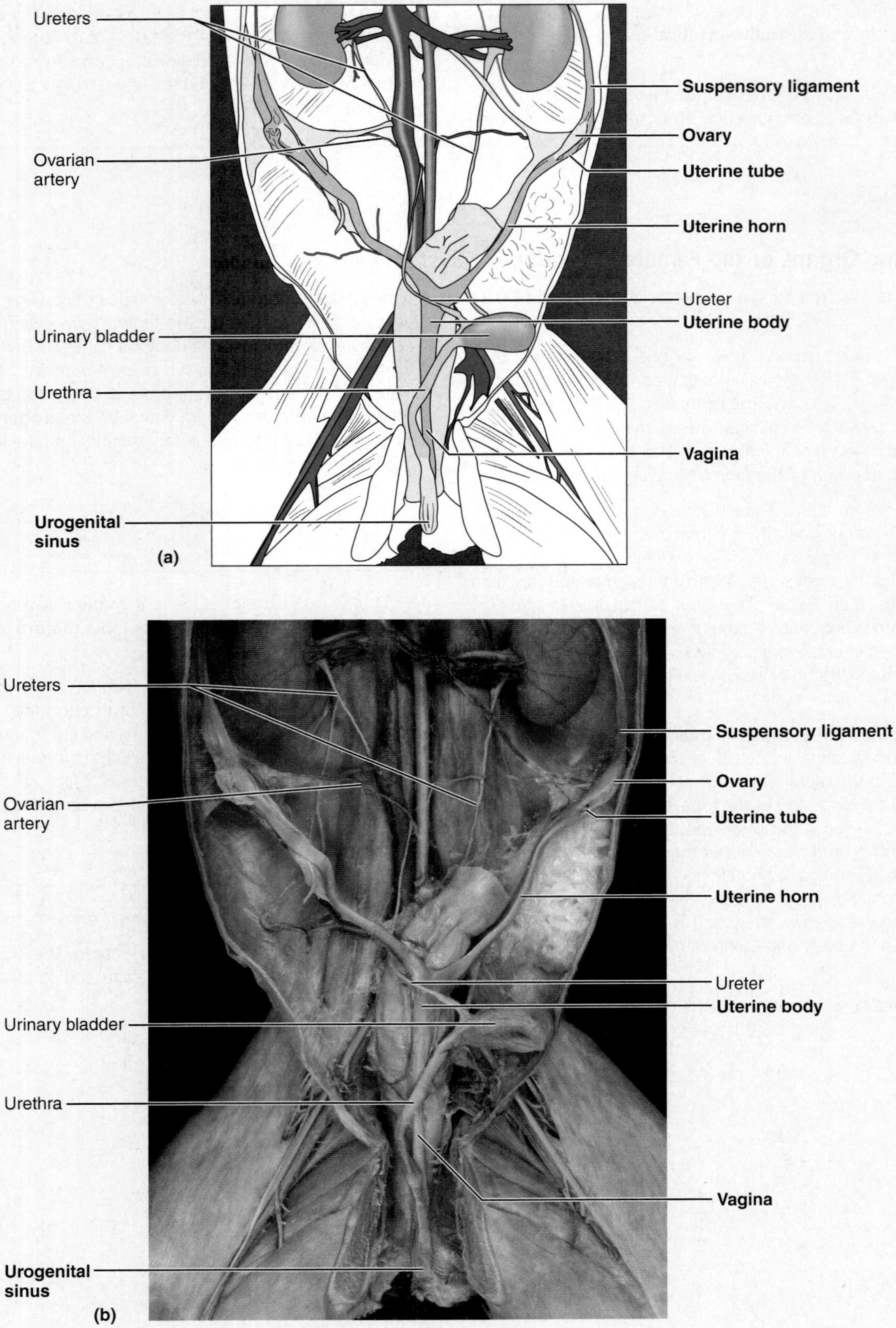

Figure D9.2 Reproductive system of the female cat. (a) Diagram. **(b)** Photograph.

DISSECTION REVIEW

1. The female cat has a ________________ uterus; that of the human female is ______________________________.

   Explain the difference in structure of these two uterine types. ______________________________

   ____

   ____

2. What reproductive advantage is conferred by the feline uterine type?

   ____

   ____

3. Cite differences noted between the cat and the human relative to the following structures:

   uterine tubes ______________________________

   ____

   site of entry of ductus deferens into the urethra ______________________________

   ____

   location of the prostate ______________________________

   ____

   seminal glands ______________________________

   urethral and vaginal openings in the female ______________________________

   ____

   ____

PhysioEx™ 9.1

PhysioEx™ 9.1 by
Peter Zao, North Idaho College
Timothy Stabler, Indiana University Northwest
Lori Smith, American River College
Andrew Lokuta, University of Wisconsin–Madison
Edwin Griff, University of Cincinnati

Cell Transport Mechanisms and Permeability

PRE-LAB QUIZ

1. Circle the correct underlined term: In active transport / passive transport processes, the cell must provide energy in the form of ATP to power the process.
2. The movement of particles from an area of greater concentration to an area of lesser concentration is:
   a. diffusion
   b. osmosis
   c. active transport
   d. kinetic energy
3. All of the following are true of active transport *except*:
   a. ATP is used to power active transport.
   b. Solutes are moving with their concentration gradient.
   c. It uses a membrane-bound carrier protein.
   d. It can only occur in certain animals.
4. Circle the correct underlined term: In Exercise 1, the dialysis tubing will mimic the nucleus / plasma membrane of a cell.
5. Circle the correct underlined term: The larger the molecular weight / concentration of a compound, the larger the pore size required for passive transport of that compound.

Exercise Overview

The molecular composition of the plasma membrane allows it to be selective about what passes through it. It allows nutrients and appropriate amounts of ions to enter the cell and keeps out undesirable substances. For that reason, we say the plasma membrane is **selectively permeable**. Valuable cell proteins and other substances are kept within the cell, and metabolic wastes pass to the exterior.

Transport through the plasma membrane occurs in two basic ways: either passively or actively. In **passive processes**, the transport process is driven by concentration or pressure differences (*gradients*) between the interior and exterior of the cell. In **active processes**, the cell provides energy (ATP) to power the transport.

Two key passive processes of membrane transport are **diffusion** and **filtration**. Diffusion is an important transport process for every cell in the body. **Simple diffusion** occurs without the assistance of membrane proteins, and **facilitated diffusion** requires a membrane-bound carrier protein that assists in the transport.

In both simple and facilitated diffusion, the substance being transported moves *with* (or *along* or *down*) the *concentration gradient* of the solute (from a region of its higher concentration to a region of its lower concentration). The process does not require energy from the cell. Instead, energy in the form of **kinetic energy** comes from the constant motion of the molecules. The movement of solutes continues until the solutes are evenly dispersed throughout the solution. At this point, the solution has reached **equilibrium**.

A special type of diffusion across a membrane is **osmosis**. In osmosis, water moves with its concentration gradient, from a higher concentration of water to a lower concentration of water. It moves in response to a higher concentration of solutes on the other side of a membrane.

In the body, the other key passive process, **filtration**, usually occurs only across capillary walls. Filtration depends upon a *pressure gradient* as its driving force. It is not a selective process. It is dependent upon the size of the pores in the filter.

The two key active processes (recall that active processes require energy) are **active transport** and **vesicular transport**. Like facilitated diffusion, active transport uses a membrane-bound carrier protein. Active transport differs from facilitated diffusion because the solutes move *against* their concentration gradient and because ATP is used to power the transport. Vesicular transport includes phagocytosis, endocytosis, pinocytosis, and exocytosis. These processes are not covered in this exercise. The activities in this exercise will explore the cell transport mechanisms individually.

ACTIVITY 1

Simulating Dialysis (Simple Diffusion)

OBJECTIVES

1. To understand that diffusion is a passive process dependent upon a solute concentration gradient.
2. To understand the relationship between molecular weight and molecular size.
3. To understand how solute concentration affects the rate of diffusion.
4. To understand how molecular weight affects the rate of diffusion.

Introduction

Recall that all molecules possess *kinetic energy* and are in constant motion. As molecules move about randomly at high speeds, they collide and bounce off one another, changing direction with each collision. For a given temperature, all matter has about the same average kinetic energy. Smaller molecules tend to move faster than larger molecules because kinetic energy is directly related to both mass and velocity ($KE = ½ mv^2$).

When a **concentration gradient** (difference in concentration) exists, the net effect of this random molecular movement is that the molecules eventually become evenly distributed throughout the environment—in other words, diffusion occurs. **Diffusion** is the movement of molecules from a region of their higher concentration to a region of their lower concentration. The driving force behind diffusion is the kinetic energy of the molecules themselves.

The diffusion of particles into and out of cells is modified by the plasma membrane, which is a physical barrier. In general, molecules diffuse passively through the plasma membrane if they are small enough to pass through its pores (and are aided by an electrical and/or concentration gradient) or if they can dissolve in the lipid portion of the membrane (as in the case of CO_2 and O_2). A membrane is called *selectively permeable*, *differentially permeable*, or *semipermeable* if it allows some solute particles (molecules) to pass but not others.

The diffusion of *solute particles* dissolved in water through a selectively permeable membrane is called **simple diffusion**. The diffusion of *water* through a differentially permeable membrane is called **osmosis**. Both simple diffusion and osmosis involve movement of a substance from an area of its higher concentration to an area of its lower concentration, that is, *with* (or *along* or *down*) its concentration gradient.

This activity provides information on the passage of water and solutes through selectively permeable membranes. You can apply what you learn to the study of transport mechanisms in living, membrane-bounded cells. The dialysis membranes used each have a different *molecular weight cutoff* (*MWCO*), indicated by the number below it. You can think of MWCO in terms of pore size: the larger the MWCO number, the larger the pores in the membrane. The molecular weight of a solute is the number of grams per mole, where a mole is the constant Avogadro's number 6.02×10^{23} molecules/mole. The larger the molecular weight, the larger the mass of the molecule. The term molecular mass is sometimes used instead of molecular weight.

EQUIPMENT USED The following equipment will be depicted on-screen: left and right beakers—used for diffusion of solutes; dialysis membranes with various molecular weight cutoffs (MWCOs).

Experiment Instructions

Go to the home page in the PhysioEx software and click **Exercise 1: Cell Transport Mechanisms and Permeability**. Click **Activity 1: Simulating Dialysis (Simple Diffusion)**, and take the online **Pre-lab Quiz** for Activity 1.

After you take the online Pre-lab Quiz, click the **Experiment** tab and begin the experiment. The experiment instructions are reprinted here for your reference. The opening screen for the experiment is shown below.

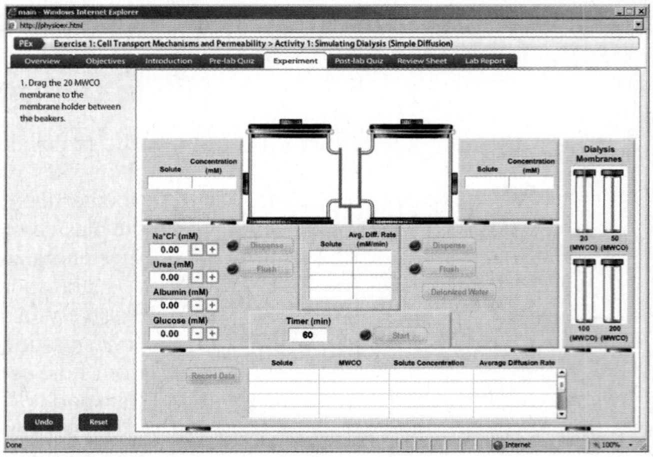

Windows Explorer 2007, Microsoft Corporation

1. Drag the 20 MWCO membrane to the membrane holder between the beakers.

2. Increase the Na^+Cl^- concentration to be dispensed to the left beaker to 9.00 m*M* by clicking the + button beside the Na^+Cl^- display. Click **Dispense** to fill the left beaker with 9.00 m*M* Na^+Cl^- solution.

3. Note that the concentration of Na^+Cl^- in the left beaker is displayed in the concentration window to the left of the beaker. Click **Deionized Water** and then click **Dispense** to fill the right beaker with deionized water.

4. After you start the run, the barrier between the beakers will descend, allowing the solutions in each beaker to have access to the dialysis membrane separating them. You will

be able to determine the amount of solute that passes through the membrane by observing the concentration display to the side of each beaker. A level above zero in Na^+Cl^- concentration in the right beaker indicates that Na^+ and Cl^- ions are diffusing from the left beaker into the right beaker through the selectively permeable dialysis membrane. Note that the timer is set to 60 minutes. The simulation compresses the 60-minute time period into 10 seconds of real time. Click **Start** to start the run and watch the concentration display to the side of each beaker for any activity.

5. Click **Record Data** to display your results in the grid (and record your results in Chart 1).

CHART 1 **Dialysis Results (average diffusion rate in m*M*/min)**

	Membrane MWCO			
Solute	**20**	**50**	**100**	**200**
Na^+Cl^-				
Urea				
Albumin				
Glucose				

PREDICT Question 1
The molecular weight of urea is 60.07. Do you think urea will diffuse through the 20 MWCO membrane?

____

6. Click **Flush** beneath each of the beakers to prepare for the next run.

7. Increase the urea concentration to be dispensed to the left beaker to 9.00 m*M* by clicking the + button beside the urea display. Click **Dispense** to fill the left beaker with 9.00 m*M* urea solution.

8. Click **Deionized Water** and then click **Dispense** to fill the right beaker with deionized water.

9. Click **Start** to start the run and watch the concentration display to the side of each beaker for any activity.

10. Click **Record Data** to display your results in the grid (and record your results in Chart 1).

11. Click the 20 MWCO membrane in the membrane holder to automatically return it to the membrane cabinet and then click **Flush** beneath each beaker to prepare for the next run.

12. Drag the 50 MWCO membrane to the membrane holder between the beakers. Increase the Na^+Cl^- concentration to be dispensed to the left beaker to 9.00 m*M*. Click **Dispense** to fill the left beaker with 9.00 m*M* Na^+Cl^- solution.

13. Click **Deionized Water** and then click **Dispense** to fill the right beaker with deionized water.

14. Click **Start** to start the run and watch the concentration display to the side of each beaker for any activity.

15. Click **Record Data** to display your results in the grid (and record your results in Chart 1).

16. Click **Flush** beneath each of the beakers to prepare for the next run.

17. Increase the Na^+Cl^- concentration to be dispensed to the left beaker to 18.00 m*M*. Click **Dispense** to fill the left beaker with 18.00 m*M* Na^+Cl^- solution.

18. Click **Deionized Water** and then click **Dispense** to fill the right beaker with deionized water.

19. Click **Start** to start the run and watch the concentration display to the side of each beaker for any activity.

20. Click **Record Data** to display your results in the grid (and record your results in Chart 1).

21. Click the 50 MWCO membrane in the membrane holder to automatically return it to the membrane cabinet and then click **Flush** beneath each beaker to prepare for the next run.

22. Drag the 100 MWCO membrane to the membrane holder between the beakers. Increase the Na^+Cl^- concentration to be dispensed to the left beaker to 9.00 m*M*. Click **Dispense** to fill the left beaker with 9.00 m*M* Na^+Cl^- solution.

23. Click **Deionized Water** and then click **Dispense** to fill the right beaker with deionized water.

24. Click **Start** to start the run and watch the concentration display to the side of each beaker for any activity.

25. Click **Record Data** to display your results in the grid (and record your results in Chart 1).

26. Click **Flush** beneath each of the beakers to prepare for the next run.

27. Increase the urea concentration to be dispensed to the left beaker to 9.00 m*M*. Click **Dispense** to fill the left beaker with 9.00 m*M* urea solution.

28. Click **Deionized Water** and then click **Dispense** to fill the right beaker with deionized water.

29. Click **Start** to start the run and watch the concentration display to the side of each beaker for any activity.

30. Click **Record Data** to display your results in the grid (and record your results in Chart 1).

31. Click the 100 MWCO membrane in the membrane holder to automatically return it to the membrane cabinet and then click **Flush** beneath each beaker to prepare for the next run.

PREDICT Question 2
Recall that glucose is a monosaccharide, albumin is a protein with 607 amino acids, and the average molecular weight of a single amino acid is 135 g/mole. Will glucose or albumin be able to diffuse through the 200 MWCO membrane?

____

32. Drag the 200 MWCO membrane to the membrane holder between the beakers. Increase the glucose concentration to be dispensed to the left beaker to 9.00 m*M*. Click **Dispense** to fill the left beaker with 9.00 m*M* glucose solution.

33. Click **Deionized Water** and then click **Dispense** to fill the right beaker with deionized water.

34. Click **Start** to start the run and watch the concentration display to the side of each beaker for any activity.

35. Click **Record Data** to display your results in the grid (and record your results in Chart 1).

36. Click **Flush** beneath each of the beakers to prepare for the next run.

37. Increase the albumin concentration to be dispensed to the left beaker to 9.00 m*M*. Click **Dispense** to fill the left beaker with 9.00 m*M* albumin solution.

38. Click **Deionized Water** and then click **Dispense** to fill the right beaker with deionized water.

39. Click **Start** to start the run and watch the concentration display to the side of each beaker for any activity.

40. Click **Record Data** to display your results in the grid (and record your results in Chart 1).

After you complete the experiment, take the online **Post-lab Quiz** for Activity 1.

Activity Questions

1. Did any solutes move through the 20 MWCO membrane? Why or why not?

____

____

2. Did Na^+Cl^- move through the 50 MWCO membrane?

____

____

3. Describe how the size of a molecule (molecular weight) affects its rate of diffusion.

____

____

4. What happened to the rate of diffusion when you increased the Na^+Cl^- solute concentration?

____

____

ACTIVITY 2

Simulated Facilitated Diffusion

OBJECTIVES

1. To understand that some solutes require a carrier protein to pass through a membrane because of size or solubility limitations.
2. To observe how the concentration of solutes affects the rate of facilitated diffusion.
3. To observe how the number of transport proteins affects the rate of facilitated diffusion.
4. To understand how transport proteins can become saturated.

Introduction

Some molecules are lipid insoluble or too large to pass through pores in the cell's plasma membrane. Instead, they pass through the membrane by a passive transport process called **facilitated diffusion**. For example, sugars, amino acids, and ions are transported by facilitated diffusion. In this form of transport, solutes combine with carrier-protein molecules in the membrane and are then transported *with* (or *along* or *down*) their concentration gradient. The carrier-protein molecules in the membrane might have to change shape slightly to accommodate the solute, but the cell does not have to expend the energy of ATP.

Because facilitated diffusion relies on carrier proteins, solute transport varies with the number of available carrier-protein molecules in the membrane. The carrier proteins can become saturated if too much solute is present and the maximum transport rate is reached. The carrier proteins are embedded in the plasma membrane and act like a shield, protecting the hydrophilic solute from the lipid portions of the membrane.

Facilitated diffusion typically occurs in one direction for a given solute. The greater the concentration difference between one side of the membrane and the other, the greater the rate of facilitated diffusion.

EQUIPMENT USED The following equipment will be depicted on-screen: left and right beakers—used for diffusion of solutes; dialysis membranes with various molecular weight cutoffs (MWCOs); membrane builder—used to build membranes with different numbers of glucose protein carriers.

Experiment Instructions

Go to the home page in the PhysioEx software and click **Exercise 1: Cell Transport Mechanisms and Permeability**. Click **Activity 2: Simulated Facilitated Diffusion**, and take the online **Pre-lab Quiz** for Activity 2.

After you take the online Pre-lab Quiz, click the **Experiment** tab and begin the experiment. The experiment instructions are reprinted here for your reference. The opening screen for the experiment is shown on the following page.

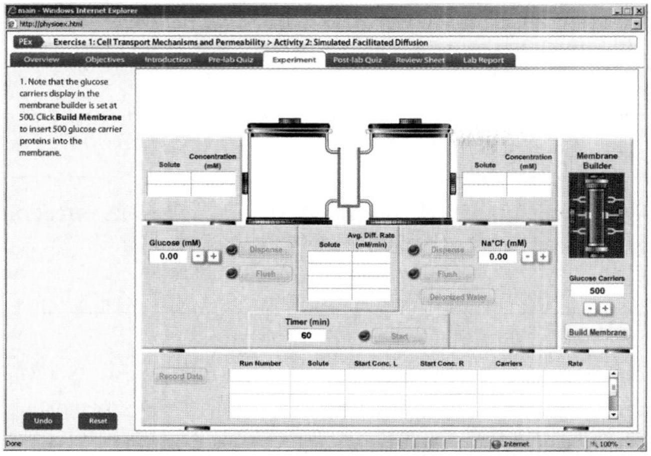

Windows Explorer 2007, Microsoft Corporation

1. Note that the glucose carriers display in the membrane builder is set at 500. Click **Build Membrane** to insert 500 glucose carrier proteins into the membrane.

2. Drag the membrane to the membrane holder between the beakers.

3. Increase the glucose concentration to be dispensed to the left beaker to 2.00 m*M* by clicking the + button beside the glucose display. Click **Dispense** to fill the left beaker with 2.00 m*M* glucose solution.

4. Note that the concentration of glucose in the left beaker is displayed in the concentration window to the left of the beaker. Click **Deionized Water** and then click **Dispense** to fill the right beaker with deionized water.

5. After you start the run, the barrier between the beakers will descend, allowing the solutions in each beaker to have access to the dialysis membrane separating them. You will be able to determine the amount of solute that passes through the membrane by observing the concentration display to the side of each beaker. A level above zero in glucose concentration in the right beaker indicates that glucose is diffusing from the left beaker into the right beaker through the selectively permeable dialysis membrane. Note that the timer is set to 60 minutes. The simulation compresses the 60-minute time period into 10 seconds of real time. Click **Start** to start the run and watch the concentration display to the side of each beaker for any activity.

6. Click **Record Data** to display your results in the grid (and record your results in Chart 2).

CHART 2 Facilitated Diffusion Results (glucose transport rate, m*M*/min)

	Number of glucose carrier proteins		
Glucose concentration	**500**	**700**	**100**
2 m*M*			
8 m*M*			
10 m*M*			
2 m*M* w/2.00 m*M* Na^+Cl^-			

7. Click **Flush** beneath each of the beakers to prepare for the next run.

8. Increase the glucose concentration to be dispensed to the left beaker to 8.00 m*M* by clicking the + button beside the glucose display. Click **Dispense** to fill the left beaker with 8.00 m*M* glucose solution.

9. Click **Deionized Water** and then click **Dispense** to fill the right beaker with deionized water.

10. Click **Start** to start the run and watch the concentration display to the side of each beaker for any activity.

11. Click **Record Data** to display your results in the grid (and record your results in Chart 2).

12. Click the membrane in the membrane holder to automatically return it to the membrane builder and then click **Flush** beneath each beaker to prepare for the next run.

? PREDICT Question 1
What effect do you think increasing the number of protein carriers will have on the glucose transport rate?

____

13. Increase the number of glucose carriers to 700 by clicking the + button beneath the glucose carriers display. Click **Build Membrane** to insert 700 glucose carrier proteins into the membrane.

14. Drag the membrane to the membrane holder between the beakers. Increase the glucose concentration to be dispensed to the left beaker to 2.00 m*M*. Click **Dispense** to fill the left beaker with 2.00 m*M* glucose solution.

15. Click **Deionized Water** and then click **Dispense** to fill the right beaker with deionized water.

16. Click **Start** to start the run and watch the concentration display to the side of each beaker for any activity.

17. Click **Record Data** to display your results in the grid (and record your results in Chart 2).

18. Click **Flush** beneath each of the beakers to prepare for the next run.

19. Increase the glucose concentration to be dispensed to the left beaker to 8.00 m*M*. Click **Dispense** to fill the left beaker with 8.00 m*M* glucose solution.

20. Click **Deionized Water** and then click **Dispense** to fill the right beaker with deionized water.

21. Click **Start** to start the run and watch the concentration display to the side of each beaker for any activity.

22. Click **Record Data** to display your results in the grid (and record your results in Chart 2).

23. Click the membrane in the membrane holder to automatically return it to the membrane builder and then click **Flush** beneath each beaker to prepare for the next run.

24. Decrease the number of glucose carriers to 100 by clicking the – button beneath the glucose carriers display. Click **Build Membrane** to insert 100 glucose carrier proteins into the membrane.

25. Drag the membrane to the membrane holder between the beakers. Increase the glucose concentration to be dispensed to the left beaker to 10.00 m*M*. Click **Dispense** to fill the left beaker with 10.00 m*M* glucose solution.

26. Click **Deionized Water** and then click **Dispense** to fill the right beaker with deionized water.

27. Click **Start** to start the run and watch the concentration display to the side of each beaker for any activity.

28. Click **Record Data** to display your results in the grid (and record your results in Chart 2).

29. Click the membrane in the membrane holder to automatically return it to the membrane builder and then click **Flush** beneath each beaker to prepare for the next run.

30. Increase the number of glucose carriers to 700. Click **Build Membrane** to insert 700 glucose carrier proteins into the membrane.

? PREDICT Question 2
What effect do you think adding Na^+Cl^- will have on the glucose transport rate?

____

31. Increase the glucose concentration to be dispensed to the left beaker to 2.00 m*M*. Click **Dispense** to fill the left beaker with 2.00 m*M* glucose solution.

32. Increase the Na^+Cl^- concentration to be dispensed to the right beaker to 2.00 m*M*. Click **Dispense** to fill the right beaker with 2.00 m*M* Na^+Cl^- solution.

33. Click **Start** to start the run and watch the concentration display to the side of each beaker for any activity.

34. Click **Record Data** to display the results in the grid (and record your results in Chart 2).

After you complete the experiment, take the online **Post-lab Quiz** for Activity 2.

Activity Questions

1. Are the solutes moving with or against their concentration gradient in facilitated diffusion?

____

____

____

2. What happened to the rate of facilitated diffusion when the number of carrier proteins was increased?

____

____

3. Explain why equilibrium was not reached with 10 m*M* glucose and 100 membrane carriers.

____

____

4. In the simulation you added Na^+Cl^- to test its effect on glucose diffusion. Explain why there was no effect.

____

____

ACTIVITY 3

Simulating Osmotic Pressure

OBJECTIVES

1. To explain how osmosis is a special type of diffusion.
2. To understand that osmosis is a passive process that depends upon the concentration gradient of water.
3. To explain how tonicity of a solution relates to changes in cell volume.
4. To understand conditions that affect osmotic pressure.

Introduction

A special form of diffusion, called **osmosis**, is the diffusion of water through a selectively permeable membrane. (A membrane is called *selectively permeable*, *differentially permeable*, or *semipermeable* if it allows some molecules to pass but not others.) Because water can pass through the pores of most membranes, it can move from one side of a membrane to the other relatively freely. Osmosis takes place whenever there is a difference in water concentration between the two sides of a membrane.

If we place distilled water on both sides of a membrane, *net* movement of water does not occur. Remember, however, that water molecules would still move between the two sides of the membrane. In such a situation, we would say that there is no *net* osmosis.

The concentration of water in a solution depends on the number of solute particles present. For this reason, increasing the solute concentration coincides with decreasing the water concentration. Because water moves down its concentration gradient (from an area of its higher concentration to an area of its lower concentration), it always moves *toward* the solution with the highest concentration of solutes. Similarly, solutes also move down their concentration gradients.

If we position a *fully* permeable membrane (permeable to solutes and water) between two solutions of differing concentrations, then all substances—solutes and water—diffuse freely, and an equilibrium will be reached between the two sides of the membrane. However, if we use a selectively permeable membrane that is impermeable to the solutes, then we have established a condition where water moves but solutes do not. Consequently, water moves toward the more concentrated solution, resulting in a *volume increase* on that side of the membrane.

By applying this concept to a closed system where volumes cannot change, we can predict that the *pressure* in the more concentrated solution will rise. The force that would need to be applied to oppose the osmosis in a closed system is the **osmotic pressure**. Osmotic pressure is measured in *millimeters of mercury* (*mm Hg*). In general, the more impermeable the solutes, the higher the osmotic pressure.

Osmotic changes can affect the volume of a cell when it is placed in various solutions. The concept of **tonicity** refers to the way a solution affects the volume of a cell. The tonicity of a solution tells us whether or not a cell will shrink or swell. If the concentration of impermeable solutes is the *same* inside and outside of the cell, the solution is **isotonic**. If there is a *higher* concentration of impermeable solutes *outside* the cell than in the cell's interior, the solution is **hypertonic**. Because the net movement of water would be out of the cell, the cell would *shrink* in a hypertonic solution. Conversely, if the concentration of impermeable solutes is *lower* outside of the cell than in the cell's interior, then the solution is **hypotonic**. The net movement of water would be into the cell, and the cell would *swell* and possibly burst.

EQUIPMENT USED The following equipment will be depicted on-screen: left and right beakers—used for diffusion of solutes; dialysis membranes with various molecular weight cutoffs (MWCOs).

Experiment Instructions

Go to the home page in the PhysioEx software and click **Exercise 1: Cell Transport Mechanisms and Permeability**. Click **Activity 3: Simulating Osmotic Pressure**, and take the online **Pre-lab Quiz** for Activity 3.

After you take the online Pre-lab Quiz, click the **Experiment** tab and begin the experiment. The experiment instructions are reprinted here for your reference. The opening screen for the experiment is shown below.

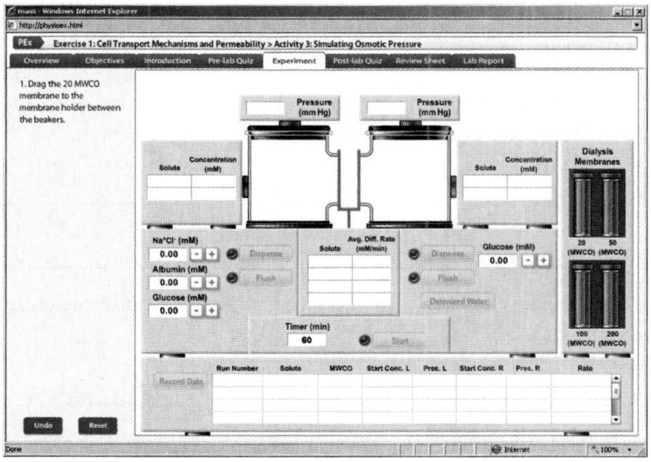

Windows Explorer 2007, Microsoft Corporation

1. Drag the 20 MWCO membrane to the membrane holder between the beakers.

2. Increase the Na^+Cl^- concentration to be dispensed to the left beaker to 5.00 m*M* by clicking the + button beside the Na^+Cl^- display. Click **Dispense** to fill the left beaker with 5.00 m*M* Na^+Cl^- solution.

3. Note that the concentration of Na^+Cl^- in the left beaker is displayed in the concentration window to the left of the beaker. Click **Deionized Water** and then click **Dispense** to fill the right beaker with deionized water.

4. After you start the run, the barrier between the beakers will descend, allowing the solutions in each beaker to have access to the dialysis membrane separating them. You can observe the changes in pressure in the two beakers by watching the pressure display above each beaker. You will also be able to determine the amount of solute that passes through the membrane by observing the concentration display to the side of each beaker. A level above zero in Na^+Cl^- concentration in the right beaker indicates that Na^+ and Cl^- ions are diffusing from the left beaker into the right beaker through the selectively permeable dialysis membrane. Note that the timer is set to 60 minutes. The simulation compresses the 60-minute time period into 10 seconds of real time. Click **Start** to start the run and watch the pressure display above each beaker for any activity.

5. Click **Record Data** to display your results in the grid (and record your results in Chart 3).

CHART 3 Osmosis Results

Solute	Membrane (MWCO)	Pressure on left (mm Hg)	Diffusion rate (m*M*/min)
Na^+Cl^-			
Na^+Cl^-			
Na^+Cl^-			
Glucose			
Glucose			
Glucose			
Albumin w/glucose			

6. Click **Flush** beneath each of the beakers to prepare for the next run.

7. Increase the Na^+Cl^- concentration to be dispensed to the left beaker to 10.00 m*M* by clicking the + button beside the Na^+Cl^- display. Click **Dispense** to fill the left beaker with 10.00 m*M* Na^+Cl^- solution.

8. Click **Deionized Water** and then click **Dispense** to fill the right beaker with deionized water.

PREDICT Question 1
What effect do you think increasing the Na^+Cl^- concentration will have?

____

9. Click **Start** to start the run and watch the pressure display above each beaker for any activity.

10. Click **Record Data** to display your results in the grid (and record your results in Chart 3).

11. Click the 20 MWCO membrane in the membrane holder to automatically return it to the membrane cabinet and then click **Flush** beneath each beaker to prepare for the next run.

12. Drag the 50 MWCO membrane to the membrane holder between the beakers. Increase the Na^+Cl^- concentration to be dispensed to the left beaker to 10.00 m*M*. Click **Dispense** to fill the left beaker with 10.00 m*M* Na^+Cl^- solution.

13. Click **Deionized Water** and then click **Dispense** to fill the right beaker with deionized water.

14. Click **Start** to start the run and watch the pressure display above each beaker for any activity.

15. Click **Record Data** to display your results in the grid (and record your results in Chart 3).

16. Click the 50 MWCO membrane in the membrane holder to automatically return it to the membrane cabinet and then click **Flush** beneath each beaker to prepare for the next run.

17. Drag the 100 MWCO membrane to the membrane holder between the beakers. Increase the glucose concentration to be dispensed to the left beaker to 8.00 m*M* by clicking the + button beside the glucose display beneath the left beaker. Click **Dispense** to fill the left beaker with 8.00 m*M* glucose solution.

18. Click **Deionized Water** and then click **Dispense** to fill the right beaker with deionized water.

19. Click **Start** to start the run and watch the pressure display above each beaker for any activity.

20. Click **Record Data** to display your results in the grid (and record your results in Chart 3).

21. Click **Flush** beneath each of the beakers to prepare for the next run.

22. Increase the glucose concentration to be dispensed to the left beaker to 8.00 m*M*. Click **Dispense** to fill the left beaker with 8.00 m*M* glucose solution.

23. Increase the glucose concentration to be dispensed to the right beaker to 8.00 m*M* by clicking the + button beside the glucose display beneath the right beaker. Click **Dispense** to fill the right beaker with 8.00 m*M* glucose solution.

24. Click **Start** to start the run and watch the pressure display above each beaker for any activity.

25. Click **Record Data** to display your results in the grid (and record your results in Chart 3).

26. Click the 100 MWCO membrane in the membrane holder to automatically return it to the membrane cabinet and then click **Flush** beneath each beaker to prepare for the next run.

27. Drag the 200 MWCO membrane to the membrane holder between the beakers. Increase the glucose concentration to be dispensed to the left beaker to 8.00 m*M*. Click **Dispense** to fill the left beaker with 8.00 m*M* glucose solution.

28. Click **Deionized Water** and then click **Dispense** to fill the right beaker with deionized water.

29. Click **Start** to start the run and watch the pressure display above each beaker for any activity.

30. Click **Record Data** to display your results in the grid (and record your results in Chart 3).

31. Click **Flush** beneath each of the beakers to prepare for the next run.

32. Increase the albumin concentration to be dispensed to the left beaker to 9.00 m*M*. Click **Dispense** to fill the left beaker with 9.00 m*M* albumin solution.

33. Increase the glucose concentration to be dispensed to the right beaker to 10.00 m*M*. Click **Dispense** to fill the right beaker with 10.00 m*M* glucose solution.

> ? PREDICT **Question 2**
> What do you think will be the pressure result of the current experimental conditions?
>
> ______________________________

34. Click **Start** to start the run and watch the pressure display above each beaker for any activity.

35. Click **Record Data** to display your results in the grid (and record your results in Chart 3).

After you complete the experiment, take the online **Post-lab Quiz** for Activity 3.

Activity Questions

1. Which membrane resulted in the greatest pressure with Na^+Cl^- as the solute? Why?

____

____

2. Explain what happens to the osmotic pressure with increasing solute concentration.

____

____

3. If the solutes are allowed to diffuse, is osmotic pressure generated?

____

____

4. If the solute concentrations are equal, is osmotic pressure generated? Why or why not?

____

____

ACTIVITY 4

Simulating Filtration

OBJECTIVES

1. To understand that filtration is a passive process dependent upon a pressure gradient.
2. To understand that filtration is not a selective process.
3. To explain that the size of the membrane pores will determine what passes through.
4. To explain the effect that increasing the hydrostatic pressure has on the filtration rate and how this correlates to events in the body.
5. To understand the relationship between molecular weight and molecular size.

Introduction

Filtration is the process by which water and solutes pass through a membrane (such as a dialysis membrane) from an area of higher hydrostatic (fluid) pressure into an area of lower hydrostatic pressure. Like diffusion, filtration is a passive process. For example, fluids and solutes filter out of the capillaries in the kidneys into the kidney tubules because blood pressure in the capillaries is greater than the fluid pressure in the tubules. So, if blood pressure increases, the rate of filtration increases.

Filtration is not a selective process. The amount of *filtrate*—the fluids and solutes that pass through the membrane—depends almost entirely on the *pressure gradient* (the difference in pressure between the solutions on the two sides of the membrane) and on the *size* of the *membrane pores*. Solutes that are too large to pass through are retained by the capillaries. These solutes usually include blood cells and proteins. Ions and smaller molecules, such as glucose and urea, can pass through.

In this activity the pore size is measured as a *molecular weight cutoff* (*MWCO*), which is indicated by the number below the filtration membrane. You can think of MWCO in terms of pore size: the larger the MWCO number, the larger the pores in the filtration membrane. The molecular weight of a solute is the number of grams per mole, where a mole is the constant Avogadro's number 6.02×10^{23} molecules/mole. You will also analyze the filtration membrane for the presence or absence of solutes that might be left sticking to the membrane.

EQUIPMENT USED The following equipment will be depicted on-screen: top and bottom beakers—used for filtration of solutes; dialysis membranes with various molecular weight cutoffs (MWCOs); membrane residue analysis station—used to analyze the filtration membrane.

Experiment Instructions

Go to the home page in the PhysioEx software and click **Exercise 1: Cell Transport Mechanisms and Permeability**. Click **Activity 4: Simulating Filtration** and take the online **Pre-lab Quiz** for Activity 4.

After you take the online Pre-lab Quiz, click the **Experiment** tab and begin the experiment. The experiment instructions are reprinted here for your reference. The opening screen for the experiment is shown above.

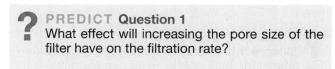

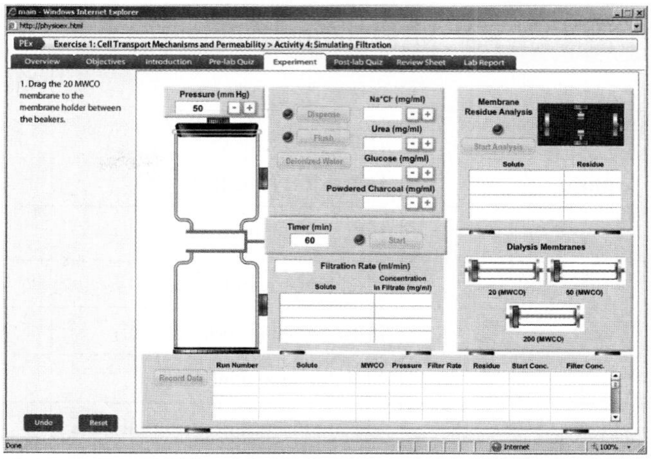

Windows Explorer 2007, Microsoft Corporation

1. Drag the 20 MWCO membrane to the membrane holder between the beakers.

2. Increase the concentration of Na^+Cl^-, urea, glucose, and powdered charcoal to be dispensed to 5.00 mg/ml by clicking the + button beside the display for each solute. Click **Dispense** to fill the top beaker.

3. After you start the run, the membrane holder below the top beaker retracts, and the solution will filter through the membrane into the beaker below. You will be able to determine whether solute particles are moving through the filtration membrane by observing the concentration displays beside the bottom beaker. A rise in detected solute concentration indicates that the solute particles are moving through the filtration membrane. Note that the pressure is set at 50 mm Hg and the timer is set to 60 minutes. The simulation compresses the 60-minute time period into 10 seconds of real time. Click **Start** to start the run and watch the concentration displays beside the bottom beaker for any activity.

4. Drag the 20 MWCO membrane to the holder in the membrane residue analysis unit. Click **Start Analysis** to begin analysis (and cleaning) of the membrane.

5. Click **Record Data** to display your results in the grid (and record your results in Chart 4, p. PEx-12).

6. Click the 20 MWCO membrane in the membrane holder to automatically return it to the membrane cabinet and then click **Flush** to prepare for the next run.

PREDICT Question 1
What effect will increasing the pore size of the filter have on the filtration rate?

____

7. Drag the 50 MWCO membrane to the membrane holder between the beakers. With the concentration of Na^+Cl^-, urea, glucose, and powdered charcoal still set to 5.00 mg/ml, click **Dispense** to fill the top beaker.

8. Click **Start** to start the run and watch the concentration displays beside the bottom beaker for any activity.

CHART 4 Filtration Results

		Membrane (MWCO)			
		20	50	200	200
Solute	**Filtration rate (ml/min)**				
Na^+Cl^-	Filter concentration (mg/ml)				
	Membrane residue				
Urea	Filter concentration (mg/ml)				
	Membrane residue				
Glucose	Filter concentration (mg/ml)				
	Membrane residue				
Powdered charcoal	Filter concentration (mg/ml)				
	Membrane residue				

9. Drag the 50 MWCO membrane to the holder in the membrane residue analysis unit. Click **Start Analysis** to begin analysis (and cleaning) of the membrane.

10. Click **Record Data** to display your results in the grid (and record your results in Chart 4).

11. Click the 50 MWCO membrane in the membrane holder to automatically return it to the membrane cabinet and then click **Flush** to prepare for the next run.

12. Drag the 200 MWCO membrane to the membrane holder between the beakers. With the concentration of Na^+Cl^-, urea, glucose, and powdered charcoal still set to 5.00 mg/ml, click **Dispense** to fill the top beaker.

13. Click **Start** to start the run and watch the concentration displays beside the bottom beaker for any activity.

14. Drag the 200 MWCO membrane to the holder in the membrane residue analysis unit. Click **Start Analysis** to begin analysis (and cleaning) of the membrane.

15. Click **Record Data** to display your results in the grid (and record your results in Chart 4).

16. Click the 200 MWCO membrane in the membrane holder to automatically return it to the membrane cabinet and then click **Flush** to prepare for the next run.

? PREDICT **Question 2**
What will happen if you increase the pressure above the beaker (the driving pressure)?

____

17. Increase the pressure to 100 mm Hg by clicking on the + button beside the pressure display above the top beaker.

18. Drag the 200 MWCO membrane to the membrane holder between the beakers. With the concentration of Na^+Cl^-, urea, glucose, and powdered charcoal still set to 5.00 mg/ml, click **Dispense** to fill the top beaker.

19. Click **Start** to start the run and watch the concentration displays beside the bottom beaker for any activity.

20. Drag the 200 MWCO membrane to the holder in the membrane residue analysis unit. Click **Start Analysis** to begin analysis (and cleaning) of the membrane.

21. Click **Record Data** to display your results in the grid (and record your results in Chart 4).

After you complete the Experiment, take the online **Post-lab Quiz** for Activity 4.

Activity Questions

1. Explain your results with the 20 MWCO filter. Why weren't any of the solutes present in the filtrate?

____

____

2. Describe two variables that affected the rate of filtration in your experiments.

____

____

3. Explain how you can increase the filtration rate through living membranes.

____

____

4. Judging from the filtration results, indicate which solute has the largest molecular weight.

____

ACTIVITY 5

Simulating Active Transport

OBJECTIVES

1. To understand that active transport requires cellular energy in the form of ATP.
2. To explain how the balance of sodium and potassium is maintained by the Na^+-K^+ pump, which moves both ions against their concentration gradients.
3. To understand coupled transport and be able to explain how the movement of sodium and potassium is independent of other solutes, such as glucose.

Introduction

Whenever a cell uses cellular energy (ATP) to move substances across its membrane, the process is an *active transport process*. Substances moved across cell membranes by an active transport process are generally unable to pass by diffusion. There are several reasons why a substance might not be able to pass through a membrane by diffusion: it might be too large to pass through the membrane pores, it might not be lipid soluble, or it might have to move *against*, rather than with, a concentration gradient.

In one type of active transport, substances move across the membrane by combining with a carrier-protein molecule. This kind of process resembles an enzyme-substrate interaction. ATP hydrolysis provides the driving force, and, in many cases, the substances move *against* concentration gradients or electrochemical gradients or both. The carrier proteins are commonly called **solute pumps**. Substances that are moved into cells by solute pumps include amino acids and some sugars. Both of these kinds of solutes are necessary for the life of the cell, but they are lipid insoluble and too large to pass through membrane pores.

In contrast, sodium ions (Na^+) are ejected from the cells by active transport. There is more Na^+ outside the cell than inside the cell, so Na^+ tends to remain in the cell unless actively transported out. In the body, the most common type of solute pump is the Na^+-K^+ (sodium-potassium) pump, which moves Na^+ and K^+ in opposite directions across cellular membranes. Three Na^+ ions are ejected from the cell for every two K^+ ions entering the cell. Note that there is more K^+ inside the cell than outside the cell, so K^+ tends to remain outside the cell unless actively transported in.

Membrane carrier proteins that move more than one substance, such as the Na^+-K^+ pump, participate in *coupled transport*. If the solutes move in the same direction, the carrier is a *symporter*. If the solutes move in opposite directions, the carrier is an *antiporter*. A carrier that transports only a single solute is a *uniporter*.

EQUIPMENT USED The following equipment will be depicted on-screen: Simulated cell inside a large beaker.

Experiment Instructions

Go to the home page in the PhysioEx software and click **Exercise 1: Cell Transport Mechanisms and Permeability**. Click **Activity 5: Simulating Active Transport** and take the online **Pre-lab Quiz** for Activity 5.

After you take the online Pre-lab Quiz, click the **Experiment** tab and begin the experiment. The experiment instructions are reprinted here for your reference. The opening screen for the experiment is shown below.

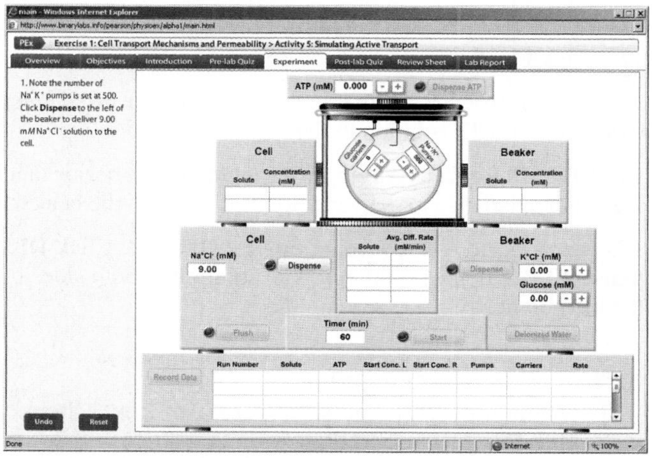

Windows Explorer 2007, Microsoft Corporation

1. Note the number of Na^+-K^+ pumps is set at 500. Click **Dispense** to the left of the beaker to deliver 9.00 m*M* Na^+Cl^- solution to the cell.

2. Increase the K^+Cl^- concentration to be delivered to the beaker to 6.00 m*M* by clicking the + button beside the K^+Cl^- display. Click **Dispense** to the right of the beaker to deliver 6.00 m*M* K^+Cl^- solution to the beaker.

3. Increase the ATP concentration to 1.00 m*M* by clicking the + button beside the ATP display above the beaker. Click **Dispense ATP** to deliver 1.00 m*M* ATP solution to both sides of the membrane.

4. After you start the run, the solutes will move across the cell membrane, simulating active transport. You will be able to determine the amount of solute that is transported across the membrane by observing the concentration displays on both sides of the beaker (the display on the left shows the concentrations inside the cell and the display on the right shows the concentrations inside the beaker). Note that the timer is set to 60 minutes. The simulation compresses the 60-minute time period into 10 seconds of real time. Click **Start** to start the run and watch the concentration displays on both sides of the beaker for any activity.

5. Click **Record Data** to display your results in the grid.

6. Click **Flush** to reset the beaker and simulated cell.

7. Click **Dispense** to the left of the beaker to deliver 9.00 m*M* Na^+Cl^- solution to the cell.

8. Increase the K^+Cl^- concentration to be delivered to the beaker to 6.00 m*M* by clicking the + button beside the K^+Cl^- display. Click **Dispense** to the right of the beaker to deliver 6.00 m*M* K^+Cl^- solution to the beaker.

9. Increase the ATP concentration to 3.00 m*M* by clicking the + button beside the ATP display above the beaker. Click **Dispense ATP** to deliver 3.00 m*M* ATP solution to both sides of the membrane.

10. Click **Start** to start the run and watch the concentration displays on both sides of the beaker for any activity.

11. Click **Record Data** to display your results in the grid.

12. Click **Flush** to reset the beaker and simulated cell.

13. Click **Dispense** to the left of the beaker to deliver 9.00 m*M* Na^+Cl^- solution to the cell.

14. Click **Deionized Water** to the right of the beaker and then click **Dispense** to deliver deionized water to the beaker.

15. Increase the ATP concentration to 3.00 m*M*. Click **Dispense ATP** to deliver 3.00 m*M* ATP solution to both sides of the membrane.

PREDICT Question 1
What do you think will result from these experimental conditions?

____

16. Click **Start** to start the run and watch the concentration displays on both sides of the beaker for any activity.

17. Click **Record Data** to display your results in the grid.

18. Click **Flush** to reset the beaker and simulated cell.

19. Increase the number of Na^+-K^+ pumps to 800 by clicking the + button beneath the Na^+-K^+ pump display. Click **Dispense** to the left of the beaker to deliver 9.00 m*M* Na^+Cl^- solution to the cell.

20. Increase the K^+Cl^- concentration to be delivered to the beaker to 6.00 m*M*. Click **Dispense** to the right of the beaker to deliver 6.00 m*M* K^+Cl^- solution to the beaker.

21. Increase the ATP concentration to 3.00 m*M*. Click **Dispense ATP** to deliver 3.00 m*M* ATP solution to both sides of the membrane.

22. Click **Start** to start the run and watch the concentration displays on both sides of the beaker for any activity.

23. Click **Record Data** to display your results in the grid.

24. Click **Flush** to reset the beaker and simulated cell.

25. With the number of Na^+-K^+ pumps still set to 800, increase the number of glucose carriers to 400 by clicking the + button beneath the glucose carriers display. Click **Dispense** to the left of the beaker to deliver 9.00 m*M* Na^+Cl^- solution to the cell.

PREDICT Question 2
Do you think the addition of glucose carriers will affect the transport of sodium or potassium?

____

26. Increase the K^+Cl^- concentration to be delivered to the beaker to 6.00 m*M*. Increase the glucose concentration to be delivered to the beaker to 10.00 m*M*. Click **Dispense** to the right of the beaker to deliver 6.00 m*M* K^+Cl^- and 10.00 m*M* glucose solution to the beaker.

27. Increase the ATP concentration to 3.00 m*M*. Click **Dispense ATP** to deliver 3.00 m*M* ATP solution to both sides of the membrane.

28. Click **Start** to start the run and watch the concentration displays on both sides of the beaker for any activity.

29. Click **Record Data** to display your results in the grid.

After you complete the experiment, take the online **Post-lab Quiz** for Activity 5.

Activity Questions

1. In the initial trial the number of Na^+-K^+ pumps is set to 500, the Na^+Cl^- concentration is set to 9.00 m*M*, the K^+Cl^- concentration is set to 6.00 m*M*, and the ATP concentration is set to 1.00 m*M*. Explain what happened and why. What would happen if no ATP had been dispensed?

____

____

2. Why was there no transport when you dispensed only Na^+Cl^-, even though ATP was present?

____

____

3. What happens to the rate of transport of Na^+ and K^+ when you increase the number of Na^+-K^+ pumps?

____

____

4. Explain why the Na^+ and K^+ transports were unaffected by the addition of glucose.

____

____

NAME ______________________

LAB TIME/DATE ______________________

Cell Transport Mechanisms and Permeability

ACTIVITY 1 Simulating Dialysis (Simple Diffusion)

1. Describe two variables that affect the rate of diffusion. ______________________

2. Why do you think the urea was not able to diffuse through the 20 MWCO membrane? How well did the results compare with your prediction? ______________________

3. Describe the results of the attempts to diffuse glucose and albumin through the 200 MWCO membrane. How well did the results compare with your prediction? ______________________

4. Put the following in order from smallest to largest molecular weight: glucose, sodium chloride, albumin, and urea. ______________________

ACTIVITY 2 Simulated Facilitated Diffusion

1. Explain one way in which facilitated diffusion is the same as simple diffusion and one way in which it differs. ______________________

2. The larger value obtained when more glucose carriers were present corresponds to an increase in the rate of glucose transport. Explain why the rate increased. How well did the results compare with your prediction? ______________________

3. Explain your prediction for the effect Na^+Cl^- might have on glucose transport. In other words, explain why you picked the choice that you did. How well did the results compare with your prediction? ______________________

ACTIVITY 3 Simulating Osmotic Pressure

1. Explain the effect that increasing the Na^+Cl^- concentration had on osmotic pressure and why it has this effect. How well did the results compare with your prediction? ______________________

2. Describe one way in which osmosis is similar to simple diffusion and one way in which it is different. ______________

____

3. Solutes are sometimes measured in milliosmoles. Explain the statement, "Water chases milliosmoles." ______________

____

4. The conditions were 9 m*M* albumin in the left beaker and 10 m*M* glucose in the right beaker with the 200 MWCO membrane in place. Explain the results. How well did the results compare with your prediction? ______________

____

ACTIVITY 4 Simulating Filtration

1. Explain in your own words why increasing the pore size increased the filtration rate. Use an analogy to support your statement. How well did the results compare with your prediction? ______________

____

2. Which solute did not appear in the filtrate using any of the membranes? Explain why. ______________

____

3. Why did increasing the pressure increase the filtration rate but not the concentration of solutes? How well did the results compare with your prediction? ______________

____

ACTIVITY 5 Simulating Active Transport

1. Describe the significance of using 9 m*M* sodium chloride inside the cell and 6 m*M* potassium chloride outside the cell, instead of other concentration ratios. ______________

____

2. Explain why there was no sodium transport even though ATP was present. How well did the results compare with your prediction? ______________

3. Explain why the addition of glucose carriers had no effect on sodium or potassium transport. How well did the results compare with your prediction? ______________

4. Do you think glucose is being actively transported or transported by facilitated diffusion in this experiment? Explain your answer. ______________

____

Skeletal Muscle Physiology

PRE-LAB QUIZ

1. Circle the correct underlined term: Tendons / Ligaments attach skeletal muscle to the periosteum of bones.
2. A motor unit consists of:
   a. a specialized region of the motor end plate
   b. a motor neuron and all of the muscle cells it innervates
   c. a motor fiber and a neuromuscular junction
   d. a muscle fiber's plasma membrane
3. During the ________________ phase of a muscle twitch, chemical changes—such as the release of calcium—are occurring intracellularly as the muscle prepares for contraction.
   a. latent
   b. excitation
   c. coupling
   d. relaxation
4. Circle the correct underlined term: An action potential in a motor neuron triggers the release of acetylcholine / magnesium.
5. Circle True or False: When a weak muscle contraction occurs, each motor unit still develops its maximum tension even though fewer motor units are activated.
6. Circle True or False: During wave summation, muscle twitches follow each other closely, resulting in a step-like increase in force, so that each successive twitch peaks slightly higher than the one before, resulting in a staircase effect.
7. The smallest stimulus required to induce an action potential in a muscle fiber's plasma membrane is known as the:
   a. maximal voltage
   b. stimulus voltage
   c. strong voltage
   d. threshold voltage
8. Muscles do not shorten even though they are actively contracting during a(n) ________________ contraction.
   a. isotonic
   b. isometric
   c. tetanic
   d. treppe

Exercise Overview

Humans make voluntary decisions to walk, talk, stand up, and sit down. Skeletal muscles, which are usually attached to the skeleton, make these actions possible (view Figure 2.1, p. PEx-18). Skeletal muscles characteristically span two joints and attach to the skeleton via **tendons**, which attach to the periosteum of a bone. Skeletal muscles are composed of hundreds to thousands of individual cells called **muscle fibers**, which produce **muscle tension** (also referred to as **muscle force**). Skeletal muscles are remarkable machines. They provide us with the manual dexterity to create magnificent works of art and can generate the brute force needed to lift a 45-kilogram sack of concrete.

When a skeletal muscle is isolated from an experimental animal and mounted on a **force transducer**, you can generate **muscle contractions** with controlled **electrical stimulation**. Importantly, the contractions of this isolated muscle are known to mimic those of working muscles in the body. That is, in vitro experiments reproduce in vivo functions. Therefore, the activities you perform in this exercise will give you valuable insight into skeletal muscle physiology.

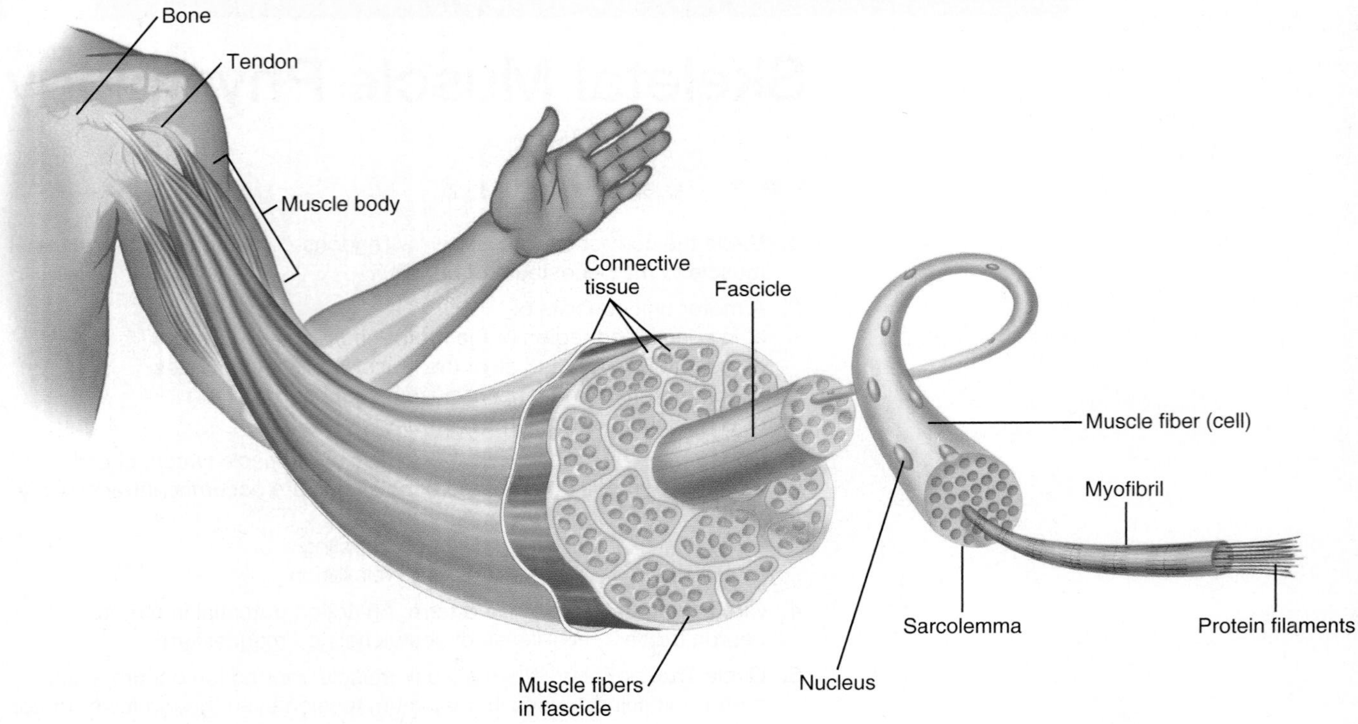

FIGURE 2.1 Structure of a skeletal muscle.

ACTIVITY 1

The Muscle Twitch and the Latent Period

OBJECTIVES

1. To understand the terms *excitation-contraction coupling, electrical stimulus, muscle twitch, latent period, contraction phase*, and *relaxation phase*.
2. To initiate muscle twitches with electrical stimuli of varying intensity.
3. To identify and measure the duration of the latent period.

Introduction

A **motor unit** consists of a **motor neuron** and all of the **muscle fibers** it innervates. The motor neuron and a muscle fiber intersect at the **neuromuscular junction**. Specifically, the neuromuscular junction is the location where the axon terminal of the neuron meets a specialized region of the muscle fiber's plasma membrane. This specialized region is called the **motor end plate**.

An action potential in a motor neuron triggers the release of acetylcholine from its terminal. Acetylcholine then diffuses onto the muscle fiber's plasma membrane (or **sarcolemma**) and binds to receptors in the motor end plate, initiating a change in ion permeability that results in a *graded depolarization* of the muscle plasma membrane (the end-plate potential). The events that occur at the neuromuscular junction lead to the **end-plate potential**. The end-plate potential triggers a series of events that results in the contraction of a muscle cell. This entire process is called **excitation-contraction coupling**.

You will be simulating excitation-contraction coupling in this and subsequent activities, but you will be using electrical pulses, rather than acetylcholine, to trigger action potentials. The pulses will be administered by an electrical stimulator that can be set for the precise voltage, frequency, and duration of shock desired. When applied to a muscle that has been surgically removed from an animal, a single electrical stimulus will result in a **muscle twitch**—the mechanical response to a single action potential. A muscle twitch has three phases: the *latent period*, the *contraction phase*, and the *relaxation phase*.

1. The **latent period** is the period of time that elapses between the generation of an action potential in a muscle cell and the start of muscle contraction. Although no force is generated during the latent period, chemical changes (including the release of calcium from the sarcoplasmic reticulum) occur intracellularly in preparation for contraction.

2. The **contraction phase** starts at the end of the latent period and ends when muscle tension peaks.

3. The **relaxation phase** is the period of time from peak tension until the end of the muscle contraction

EQUIPMENT USED The following equipment will be depicted on-screen: intact, viable skeletal muscle dissected off the leg of a frog; electrical stimulator—delivers the desired amount and duration of stimulating voltage to the muscle via electrodes resting on the muscle; mounting stand—includes a force transducer to measure the amount of force, or tension, developed by the muscle; oscilloscope—displays the stimulated muscle twitch and the amount of active, passive, and total force developed by the muscle.

Experiment Instructions

Go to the home page in the PhysioEx software and click **Exercise 2: Skeletal Muscle Physiology**. Click **Activity 1: The Muscle Twitch and the Latent Period**, and take the online **Pre-lab Quiz** for Activity 1.

After you take the online Pre-lab Quiz, click the **Experiment** tab and begin the experiment. The experiment instructions are reprinted here for your reference. The opening screen for the experiment is shown below.

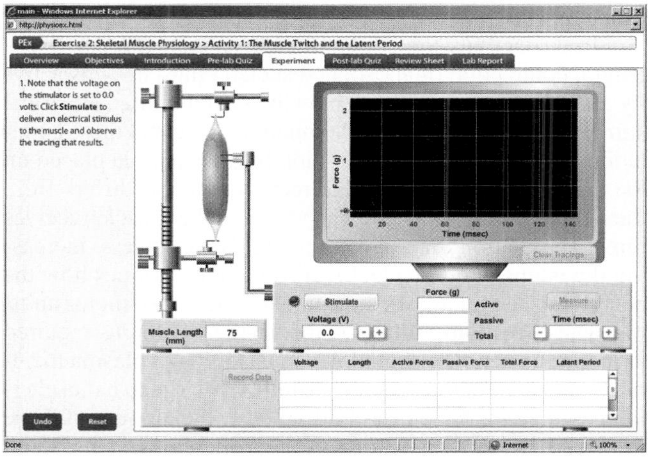

Windows Explorer 2007, Microsoft Corporation

1. Note that the voltage on the stimulator is set to 0.0 volts. Click **Stimulate** to deliver an electrical stimulus to the muscle and observe the tracing that results.

2. The tracing on the oscilloscope indicates active muscle force. Note whether any muscle force developed with the voltage set to zero. Click **Record Data** to display your results in the grid (and record your results in Chart 1).

CHART 1 Latent Period Results

Voltage	Active force (g)	Latent period (msec)

3. Increase the voltage to 3.0 volts by clicking the + button beside the voltage display.

4. Click **Stimulate** and observe the tracing that results.

5. Note the muscle force that developed. Click **Record Data** to display your results in the grid (and record your results in Chart 1).

6. Click **Clear Tracings** to remove the tracings from the oscilloscope.

7. Increase the voltage to 4.0 volts by clicking the + button beside the voltage display.

8. Click **Stimulate** and observe the tracing that results. Note that the trace starts at the left side of the screen and stays flat for a short period of time. Remember that the X-axis displays elapsed time in milliseconds. Also note how the force during the twitch also changes.

9. Click **Measure** on the stimulator. A thin, vertical yellow line appears at the far left side of the oscilloscope screen. To measure the length of the latent period, you measure the time between the application of the stimulus and the beginning of the first observable response (here, an increase in force). Click the + button beside the time display. You will see the vertical yellow line start to move across the screen. Watch what happens in the time (msec) display as the line moves across the screen. Keep clicking the + button until the yellow line reaches the point in the tracing where the graph stops being a flat line and begins to rise (this is the point at which muscle tension starts to develop). If the yellow line moves past the desired point, click the – button to move it backward.

When the yellow line is positioned correctly, click **Record Data** to display the latent period in the grid (and record your results in Chart 1).

10. Click **Clear Tracings** to remove the tracings from the oscilloscope.

? PREDICT Question 1
Will changes to the stimulus voltage alter the duration of the latent period? Explain.

____

11. You will now gradually increase the voltage to observe how changes to the stimulus voltage alter the duration of the latent period.

- Increase the voltage by 2.0 volts.
- Click **Stimulate** and observe the tracing that results.
- Click **Measure** on the stimulator and then click the + button until the yellow line reaches the point in the tracing where the graph stops being a flat line and begins to rise.
- Click **Record Data** (and record your results in Chart 1).

Repeat this step until you reach 10.0 volts.

After you complete the experiment, take the online **Post-lab Quiz** for Activity 1.

Activity Questions

1. Draw a graph that depicts a single skeletal muscle twitch, placing time on the X-axis and force on the Y-axis. Label the phases of this muscle twitch and describe what is happening in the muscle during each phase.

2. During the latent period of a skeletal muscle twitch, there is an apparent lack of muscle activity. Describe the electrical and chemical changes that occur in the muscle during this period.

____

____

____

ACTIVITY 2

The Effect of Stimulus Voltage on Skeletal Muscle Contraction

OBJECTIVES

1. To understand the terms *motor neuron*, *muscle twitch*, *motor unit*, *recruitment*, *stimulus voltage*, *threshold stimulus*, and *maximal stimulus*.
2. To understand how motor unit recruitment can increase the tension a whole muscle develops.
3. To identify a threshold stimulus voltage.
4. To observe the effect of increases in stimulus voltage on a whole muscle.
5. To understand how increasing stimulus voltage to an isolated muscle in an experiment mimics motor unit recruitment in the body.

Introduction

A skeletal muscle produces **tension** (also known as **muscle force**) when nervous or electrical stimulation is applied. The force generated by a whole muscle reflects the number of active **motor units** at a given moment. A strong muscle contraction implies that many motor units are activated, with each unit developing its maximal tension, or force. A weak muscle contraction implies that fewer motor units are activated, but each motor unit still develops its maximal tension. By increasing the number of active motor units, we can produce a steady increase in muscle force, a process called **motor unit recruitment**.

Regardless of the number of **motor units** activated, a single stimulated contraction of whole skeletal muscle is called a **muscle twitch**. A tracing of a muscle twitch is divided into three phases: the latent period, the contraction phase, and the relaxation phase. The latent period is a short period between the time of muscle stimulation and the beginning of a muscle response. Although no force is generated during this interval, chemical changes occur intracellularly in preparation for contraction (including the release of calcium from the sarcoplasmic reticulum). During the contraction phase, the myofilaments utilize the cross-bridge cycle and the muscle develops tension. Relaxation takes place when the contraction has ended and the muscle returns to its normal resting state and length.

In this activity you will stimulate an isometric, or fixed-length, contraction of an isolated skeletal muscle. This activity allows you to investigate how the strength of an electrical stimulus affects whole-muscle function. Note that these simulations involve indirect stimulation by an electrode placed on the surface of the muscle. Indirect stimulation differs from the situation in vivo, where each fiber in the muscle receives direct stimulation via a nerve ending. Nevertheless, increasing the intensity of the electrical stimulation mimics how the nervous system increases the number of activated motor units.

The **threshold voltage** is the smallest stimulus required to induce an action potential in a muscle fiber's plasma membrane, or sarcolemma. As the **stimulus voltage** to a muscle is increased beyond the threshold voltage, the amount of force produced by the whole muscle also increases. This result occurs because, as more voltage is delivered to the whole muscle, more muscle fibers are activated and, thus, the total force produced by the muscle increases. Maximal tension in the whole muscle occurs when all the muscle fibers have been activated by a sufficiently strong stimulus (referred to as the **maximal voltage**). Stimulation with voltages greater than the maximal voltage will not increase the force of contraction. This experiment is analogous to, and accurately mimics, muscle activity in vivo, where the recruitment of additional motor units increases the total muscle force produced. This phenomenon is called *motor unit recruitment*.

EQUIPMENT USED The following equipment will be depicted on-screen: intact, viable skeletal muscle dissected off the leg of a frog; electrical stimulator—delivers the desired amount and duration of stimulating voltage to the muscle via electrodes resting on the muscle; mounting stand—includes a force transducer to measure the amount of force, or tension, developed by the muscle; oscilloscope—displays the stimulated muscle twitch and the amount of active, passive, and total force developed by the muscle.

Experiment Instructions

Go to the home page in the PhysioEx software and click **Exercise 2: Skeletal Muscle Physiology**. Click **Activity 2: The Effect of Stimulus Voltage on Skeletal Muscle Contraction**, and take the online **Pre-lab Quiz** for Activity 2.

After you take the online Pre-lab Quiz, click the **Experiment** tab and begin the experiment. The experiment instructions are reprinted here for your reference. The opening screen for the experiment is shown on the following page.

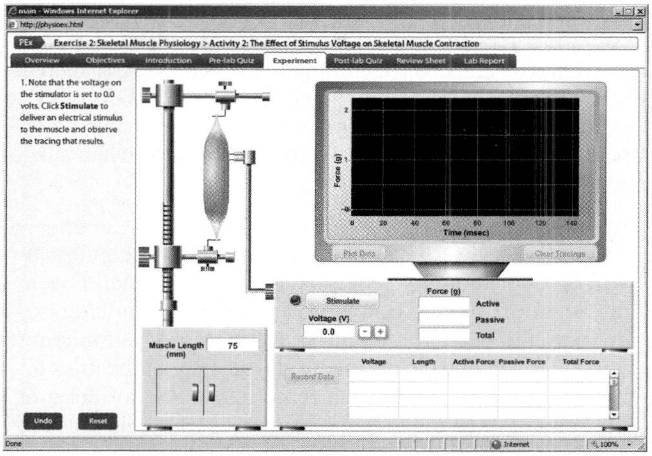

Windows Explorer 2007, Microsoft Corporation

1. Note that the voltage on the stimulator is set to 0.0 volts. Click **Stimulate** to deliver an electrical stimulus to the muscle and observe the tracing that results.

2. Note the active force display and then click **Record Data** to display your results in the grid (and record your results in Chart 2).

3. Increase the voltage to 0.2 volts by clicking the + button beside the voltage display. Click **Stimulate** to deliver an electrical stimulus to the muscle and observe the tracing that results.

4. Note the active force display and then click **Record Data** to display your results in the grid (and record your results in Chart 2).

5. You will now gradually increase the voltage and stimulate the muscle to determine the minimum voltage required to generate active force.

- Increase the voltage by 0.1 volts and then click **Stimulate**.
- If no active force is generated, increase the voltage by 0.1 volts and stimulate the muscle again. When active force is generated, click **Record Data** to display your results in the grid (and record your results in Chart 2).

6. Enter the threshold voltage for this experiment in the field below and then click **Submit** to record your answer in the lab report. ___________ volts

7. Click **Clear Tracings** to clear the tracings on the oscilloscope.

? PREDICT Question 1
As the stimulus voltage is increased from 1.0 volt up to 10 volts, what will happen to the amount of active force generated with each stimulus?

____

8. Increase the voltage on the stimulator to 1.0 volt and then click **Stimulate**.

9. Note the active force display and then click **Record Data** to display your results in the grid (and record your results in Chart 2).

CHART 2 Effect of Stimulus Voltage on Skeletal Muscle Contraction

Voltage	Active force (g)

10. You will now gradually increase the voltage and stimulate the muscle to determine the maximal voltage.

- Increase the voltage by 0.5 volts.
- Click **Stimulate** and observe the tracing that results.
- Note the active force display and then click **Record Data** to display your results in the grid (and record your results in Chart 2).

Repeat this step until you reach 10.0 volts.

11. Click **Plot Data** to view a summary of your data on a plotted grid. Click **Submit** to record your plot in the lab report.

12. Enter the maximal voltage for this experiment in the field below and then click **Submit** to record your answer in the lab report. __________ volts

After you complete the experiment, take the online **Post-lab Quiz** for Activity 2.

Activity Questions

1. For a single skeletal muscle twitch, explain the effect of increasing stimulus voltage.

____

____

2. How is this effect achieved in vivo?

____

____

ACTIVITY 3

The Effect of Stimulus Frequency on Skeletal Muscle Contraction

OBJECTIVES

1. To understand the terms *stimulus frequency*, *wave summation*, and *treppe*.
2. To observe the effect of an increasing stimulus frequency on the force developed by an isolated skeletal muscle.
3. To understand how increasing stimulus frequency to an isolated skeletal muscle induces the summation of twitch force.

Introduction

As demonstrated in Activity 2, increasing the stimulus voltage to an isolated skeletal muscle (up to a maximal value) results in an increase of force produced by the whole muscle. This experimental result is analogous to motor unit recruitment in the body. Importantly, this result relies on being able to increase the single stimulus intensity in the experiment. You will now explore another way to increase the force produced by an isolated skeletal muscle.

When a muscle first contracts, the force it is able to produce is less than the force it is able to produce with subsequent stimulations within a relatively short time span. **Treppe** is the progressive increase in force generated when a muscle is stimulated in succession, such that muscle twitches follow one another closely, with each successive twitch peaking slightly higher than the one before. This step-like increase in force is why treppe is also known as the staircase effect. For the first few twitches, each successive twitch produces slightly more force than the previous twitch as long as the muscle is allowed to fully relax between stimuli and the stimuli are delivered relatively close together.

When a skeletal muscle is stimulated repeatedly, such that the stimuli arrive one after another within a short period of time, muscle twitches can overlap with each other and result in a stronger muscle contraction than a stand-alone twitch. This phenomenon is known as wave summation. **Wave summation** occurs when muscle fibers that are developing tension are stimulated again before the fibers have relaxed. Thus, wave summation is achieved by increasing the **stimulus frequency**, or rate of stimulus delivery to the muscle. Wave summation occurs because the muscle fibers are already in a partially contracted state when subsequent stimuli are delivered.

EQUIPMENT USED The following equipment will be depicted on-screen: intact, viable skeletal muscle dissected off the leg of a frog; an electrical stimulator—delivers the desired amount and duration of stimulating voltage to the muscle via electrodes resting on the muscle; mounting stand—includes a force transducer to measure the amount of force, or tension, developed by the muscle; oscilloscope—displays the stimulated muscle twitch and the amount of active, passive, and total force developed by the muscle.

Experiment Instructions

Go to the home page in the PhysioEx software and click **Exercise 2: Skeletal Muscle Physiology**. Click **Activity 3: The Effect of Stimulus Frequency on Skeletal Muscle Contraction**, and take the online **Pre-lab Quiz** for Activity 3.

After you take the online Pre-lab Quiz, click the **Experiment** tab and begin the experiment. The experiment instructions are reprinted here for your reference. The opening screen for the experiment is shown below.

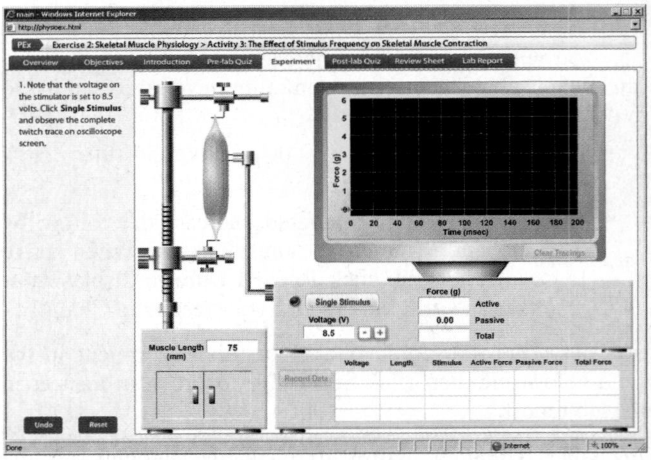

Windows Explorer 2007, Microsoft Corporation

1. Note that the voltage on the stimulator is set to 8.5 volts. Click **Single Stimulus** and observe the tracing that results on the oscilloscope.

2. Note the active force display and then click **Record Data** to display your results in the grid (and record your results in Chart 3).

3. Click **Single Stimulus** and allow the trace to rise and completely fall. *Immediately after* the trace has returned to baseline, click **Single Stimulus** again.

CHART 3 Effect of Stimulus Frequency on Skeletal Muscle Contraction

Voltage	Stimulus	Active force (g)

4. Note the active force for the second muscle twitch and click **Record Data** to display your results in the grid (and record your results in Chart 3).

5. You should have observed an increase in active force generated by the muscle with the immediate second stimulus. This increase demonstrates the phenomenon of treppe. Click **Clear Tracings** to clear the tracings on the oscilloscope.

6. You will now investigate the process of wave summation. Click **Single Stimulus** and watch the trace rise and begin to fall. *Before* the trace falls completely back to the baseline, click **Single Stimulus** again. (You can simply click **Single Stimulus** twice in quick succession in order to achieve this.)

7. Note the active force for the second muscle twitch and click **Record Data** to display your results in the grid (and record your results in Chart 3).

? PREDICT Question 1
As the stimulus frequency increases, what will happen to the muscle force generated with each successive stimulus? Will there be a limit to this response?

____

____

____

8. Now stimulate the muscle at a higher frequency by clicking **Single Stimulus** four times in rapid succession.

9. Note the active force display and then click **Record Data** to display your results in the grid (and record your results in Chart 3).

10. Click **Clear Tracings** to clear the tracings on the oscilloscope.

? PREDICT Question 2
In order to produce sustained muscle contractions with an active force value of 5.2 grams, do you think you need to increase the stimulus voltage?

____

11. Increase the voltage to 10.0 volts by clicking the + button beside the voltage display. After setting the voltage, click **Single Stimulus** four times in rapid succession.

12. Note the active force display and then click **Record Data** to display your results in the grid (and record your results in Chart 3).

13. Click **Clear Tracings** to clear the tracings on the oscilloscope.

14. Return the voltage to 8.5 volts by clicking the – button beside the voltage display. After setting the voltage, click **Single Stimulus** as many times as you can in rapid succession. Note the active force display. If you did not achieve an active force of 5.2 grams, click **Clear Tracings** and then click **Single Stimulus** even more rapidly. Repeat this step until you achieve an active force of 5.2 grams.

When you achieve an active force of 5.2 grams, click **Record Data** to display your results in the grid (and record your results in Chart 3).

After you complete the experiment, take the online **Post-lab Quiz** for Activity 3.

Activity Questions

1. Why is treppe also known as the staircase effect?

____

____

2. What changes are thought to occur in the skeletal muscle to allow treppe to be observed?

____

____

3. How does the frequency of stimulation affect the amount of force generated by a skeletal muscle?

____

____

4. Explain how wave summation is achieved in vivo.

____

____

ACTIVITY 4

Tetanus in Isolated Skeletal Muscle

OBJECTIVES

1. To understand the terms *stimulus frequency*, *unfused tetanus*, *fused tetanus*, and *maximal tetanic tension*.
2. To observe the effect of an increasing stimulus frequency on an isolated skeletal muscle.
3. To understand how increasing the stimulus frequency to an isolated skeletal muscle leads to unfused or fused tetanus.

Introduction

As demonstrated in Activity 3, increasing the **stimulus frequency** to an isolated skeletal muscle results in an increase in force produced by the whole muscle. Specifically, you observed that, if electrical stimuli are applied to a skeletal muscle in quick succession, the overlapping twitches generated more force with each successive stimulus. However, if stimuli continue to be applied frequently to a muscle over a prolonged period of time, the maximum possible muscle force from each stimulus will eventually reach a plateau—a state known as **unfused tetanus**. If stimuli are then applied with even greater frequency, the twitches will begin to fuse so that the peaks and valleys of each twitch become indistinguishable from one another—this state is known as **complete (fused) tetanus**. When the stimulus frequency reaches a value beyond which no further increases in force are generated by the muscle, the muscle has reached its **maximal tetanic tension**.

EQUIPMENT USED The following equipment will be depicted on-screen: intact, viable skeletal muscle dissected off the leg of a frog; electrical stimulator—delivers the desired amount and duration of stimulating voltage to the muscle via electrodes resting on the muscle; mounting stand—includes a force transducer to measure the amount of force, or tension, developed by the muscle; oscilloscope—displays the stimulated muscle twitch and the amount of active, passive, and total force developed by the muscle.

Experiment Instructions

Go to the home page in the PhysioEx software and click **Exercise 2: Skeletal Muscle Physiology**. Click **Activity 4: Tetanus in Isolated Skeletal Muscle** and take the online **Pre-lab Quiz** for Activity 4.

After you take the online Pre-lab Quiz, click the **Experiment** tab and begin the experiment. The experiment instructions are reprinted here for your reference. The opening screen for the experiment is shown above.

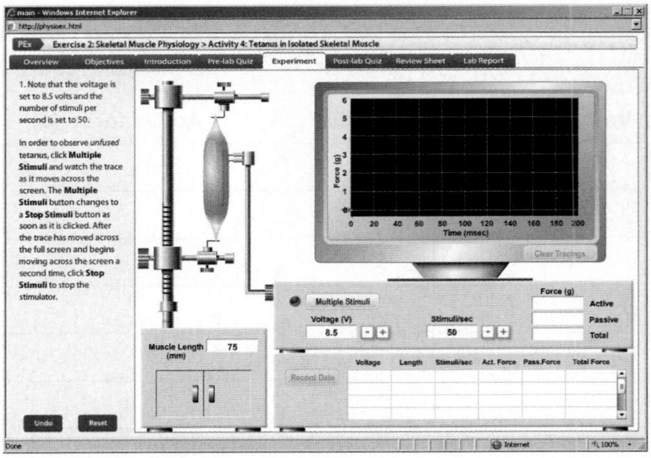

Windows Explorer 2007, Microsoft Corporation

1. Note that the voltage is set to 8.5 volts and the number of stimuli per second is set to 50. To observe *unfused* tetanus, click **Multiple Stimuli** and watch the trace as it moves across the screen. The **Multiple Stimuli** button changes to a **Stop Stimuli** button after it is clicked. After the trace has moved across the full screen and begins moving across the screen a second time, click **Stop Stimuli** to stop the stimulator.

2. Click **Record Data** to display your results in the grid (and record your results in Chart 4).

CHART 4 Tetanus in Isolated Skeletal Muscle

Stimuli/second	Active force (g)

? PREDICT Question 1
As the stimulus frequency increases further, what will happen to the muscle tension and twitch appearance with each successive stimulus? Will there be a limit to this response?

____

____

3. In order to observe *fused* tetanus, increase the stimuli/sec setting to 130 by clicking the + button beside the stimuli/sec display. Click **Multiple Stimuli** and observe the resulting trace. After the trace has moved across the full screen and begins moving across the screen a second time, click **Stop Stimuli**.

4. Note the fused tetanus and click **Record Data** to display your results in the grid (and record your results in Chart 4).

5. Click **Clear Tracings** to clear the oscilloscope screen.

6. Increase the stimuli/sec setting to 140 by clicking the + button beside the stimuli/sec display. Click **Multiple Stimuli** and observe the resulting trace. After the trace has moved across the full screen and begins moving across the screen a second time, click **Stop Stimuli**.

7. Note the fused tetanus and click **Record Data** to display your results in the grid (and record your results in Chart 4).

8. Click **Clear Tracings** to clear the oscilloscope screen.

9. You will now observe the effect of incremental increases in the number of stimuli per second above 140 stimuli per second.

- Increase the stimuli/sec setting by 2.
- Click **Multiple Stimuli** and observe the resulting trace. After the trace has moved across the full screen and begins moving across the screen a second time, click **Stop Stimuli**.
- Click **Record Data** to display your results in the grid (and record your results in Chart 4).
- Click **Clear Tracings** to clear the oscilloscope screen.

Repeat this step until you reach 150 stimuli per second.

After you complete the experiment, take the online **Post-lab Quiz** for Activity 4.

Activity Questions

1. Explain what you think is being summated in the skeletal muscle to allow a high stimulus frequency to induce a smooth, continuous skeletal muscle contraction.

____

____

2. Why do many toddlers receive a tetanus shot (and then subsequent booster shots, as needed, later in life)? How does the condition known as "lockjaw" relate to tetanus shots?

____

____

ACTIVITY 5

Fatigue in Isolated Skeletal Muscle

OBJECTIVES

1. To understand the terms *stimulus frequency*, *complete (fused) tetanus*, *fatigue*, and *rest period*.
2. To observe the development of skeletal muscle fatigue.
3. To understand how the length of intervening rest periods determines the onset of fatigue.

Introduction

As demonstrated in Activities 3 and 4, increasing the stimulus frequency to an isolated skeletal muscle induces an increase of force produced by the whole muscle. Specifically, if voltage stimuli are applied to a muscle frequently in quick succession, the skeletal muscle generates more force with each successive stimulus.

However, if stimuli continue to be applied frequently to a muscle over a prolonged period of time, the maximum force of each twitch eventually reaches a plateau—a state known as *unfused tetanus*. If stimuli are then applied with even greater frequency, the twitches begin to fuse so that the peaks and valleys of each twitch become indistinguishable from one another—this state is known as **complete (fused) tetanus**. When the **stimulus frequency** reaches a value beyond which no further increase in force is generated by the muscle, the muscle has reached its **maximal tetanic tension**.

In this activity you will observe the phenomena of skeletal muscle *fatigue*. Fatigue refers to a decline in a skeletal muscle's ability to maintain a constant level of force, or tension, after prolonged, repetitive stimulation. You will also demonstrate how intervening **rest periods** alter the onset of fatigue in skeletal muscle. The causes of fatigue are still being investigated and multiple molecular events are thought to be involved, though the accumulations of lactic acid, ADP, and P_i in muscles are thought to be the major factors causing fatigue in the case of high-intensity exercise.

Common definitions for **fatigue** are:

- The failure of a muscle fiber to produce tension because of previous contractile activity.
- A decline in the muscle's ability to maintain a constant force of contraction after prolonged, repetitive stimulation.

EQUIPMENT USED The following equipment will be depicted on-screen: intact, viable skeletal muscle dissected off the leg of a frog; electrical stimulator—delivers the desired amount and duration of stimulating voltage to the muscle via electrodes resting on the muscle; mounting stand—includes a force transducer to measure the amount of force, or tension, developed by the muscle; oscilloscope—displays the stimulated muscle twitch and the amount of active, passive, and total force developed by the muscle.

Experiment Instructions

Go to the home page in the PhysioEx software and click **Exercise 2: Skeletal Muscle Physiology**. Click **Activity 5: Fatigue in Isolated Skeletal Muscle**, and take the online **Pre-lab Quiz** for Activity 5.

After you take the online Pre-lab Quiz, click the **Experiment** tab and begin the experiment. The experiment instructions are reprinted here for your reference. The opening screen for the experiment is shown on the following page.

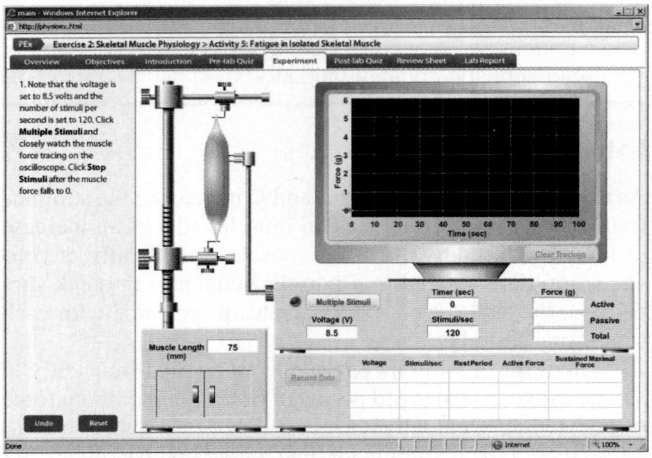

Windows Explorer 2007, Microsoft Corporation

1. Note that the voltage is set to 8.5 volts and the number of stimuli per second is set to 120. Click **Multiple Stimuli** and closely watch the muscle force tracing on the oscilloscope. Click **Stop Stimuli** after the muscle force falls to 0.

2. Click **Record Data** to display your results in the grid (and record your results in Chart 5).

CHART 5 Fatigue Results

Rest period (sec)	Active force (g)	Sustained maximal force (sec)

3. Click **Clear Tracings** to clear the oscilloscope screen.

? PREDICT Question 1
If the stimulator is briefly turned off for defined periods of time, what will happen to the length of time that the muscle is able to sustain maximal developed tension when the stimulator is turned on again?

4. To demonstrate the onset of fatigue after a variable rest period, you will be clicking the **Multiple Stimuli** button on and off three times. Read through the steps below before proceeding. Watch the timer closely to help you determine when to turn the stimulator back on.

- Click **Multiple Stimuli**.
- After the muscle force falls to 0, click **Stop Stimuli** to turn off the stimulator.
- Wait 10 seconds, then click **Multiple Stimuli** to turn the stimulator back on.
- Click **Stop Stimuli** after the muscle force falls to 0.
- Wait 20 seconds, then click **Multiple Stimuli** to turn the stimulator back on.
- Click **Stop Stimuli** after the muscle force falls to 0.

5. Click **Record Data** to display your results in the grid (and record your results in Chart 5).

After you complete the experiment, take the online **Post-lab Quiz** for Activity 5.

Activity Questions

1. What proposed mechanisms most likely explain why fatigue develops?

2. What would you recommend to an interested friend as the best ways to delay the onset of fatigue?

ACTIVITY 6

The Skeletal Muscle Length-Tension Relationship

OBJECTIVES

1. To understand the terms *isometric contraction*, *active force*, *passive force*, *total force*, and *length-tension relationship*.
2. To understand the effect that resting muscle length has on tension development when the muscle is maximally stimulated in an isometric experiment.
3. To explain the molecular basis of the skeletal muscle length-tension relationship.

Introduction

Skeletal muscle contractions are either isometric or isotonic. When a muscle attempts to move a load that is equal to the force generated by the muscle, the muscle contracts isometrically. During an **isometric** contraction, the muscle stays at a fixed length (*isometric* means "same length"). An example of isometric muscle contraction is when you stand in a doorway and push on the doorframe. The load that you are attempting to move (the doorframe) can easily equal the force generated by your muscles, so your muscles do not shorten even though they are actively contracting.

Isometric contractions are accomplished experimentally by keeping both ends of the muscle in a fixed position while electrically stimulating the muscle. Resting length (the length of the muscle before stimulation) is an important factor in determining the amount of force that a muscle can develop when stimulated. **Passive force** is generated by stretching the muscle and results from the elastic recoil of the tissue itself. This passive force is largely caused by the protein titin, which acts as a molecular bungee cord. **Active force** is generated when myosin thick filaments bind to actin thin filaments, thus engaging the cross bridge cycle and ATP hydrolysis. Think of the skeletal muscle as having two force properties: it exerts passive force when it is stretched (like a rubber band exerts passive force) and active force when it is stimulated. **Total force** is the sum of passive and active forces.

This activity allows you to set and hold constant the length of the isolated skeletal muscle and subsequently stimulate it with individual maximal voltage stimuli. A graph relating the three forces generated and the fixed length of the muscle will be automatically plotted after you stimulate the muscle. In muscle physiology this graph is known as the **isometric length-tension relationship**. The results of this simulation can be applied to human muscles to understand how optimum resting length will result in maximum force production.

To understand why muscle tissue behaves as it does, you must understand tension at the cellular level. If you have difficulty understanding the results of this activity, review the sliding filament model of muscle contraction. Think of the length-tension relationship in terms of those sarcomeres that are too short, those that are too long, and those that have the ideal amount of thick and thin filament overlap.

EQUIPMENT USED The following equipment will be depicted on-screen: intact, viable skeletal muscle dissected off the leg of a frog; electrical stimulator—delivers the desired amount and duration of stimulating voltage to the muscle via electrodes resting on the muscle; mounting stand—includes (1) a force transducer to measure the amount of force, or tension, developed by the muscle and (2) a gearing system that allows the hook through the muscle's lower tendon to be moved up or down, thus altering the fixed length of the muscle; oscilloscope—displays the stimulated muscle twitch and the amount of active, passive, and total force developed by the muscle.

Experiment Instructions

Go to the home page in the PhysioEx software and click **Exercise 2, Skeletal Muscle Physiology**. Click **Activity 6, The Skeletal Muscle Length-Tension Relationship**, and take the online **Pre-lab Quiz** for Activity 6.

After you take the online Pre-lab Quiz, click the **Experiment** tab and begin the experiment. The experiment instructions are reprinted here for your reference. The opening screen for the experiment is shown below.

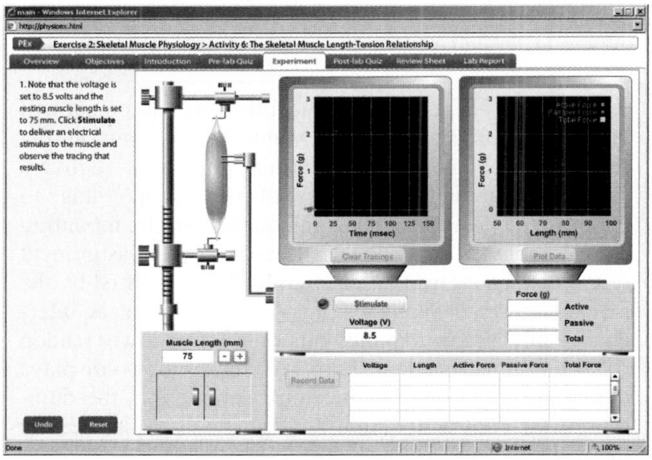

Windows Explorer 2007, Microsoft Corporation

1. Note that the voltage is set to 8.5 volts and the resting muscle length is set to 75 mm. Click **Stimulate** to deliver an electrical stimulus to the muscle and observe the tracing that results.

2. You should see a single muscle twitch tracing on the left oscilloscope display and three data points (representing active, passive, and total force generated during this twitch) plotted on the right display. The yellow box represents the total force, the red dot contained within the yellow box represents the active force, and the green square represents the passive force. Click **Record Data** to display your results in the grid (and record your results in Chart 6).

CHART 6 Skeletal Muscle Length-Tension Relationship

Length (mm)	Active force (g)	Passive force (g)	Total force (g)

? PREDICT Question 1
As the resting length of the muscle is changed, what will happen to the amount of total force the muscle generates during the stimulated twitch?

____

____

3. You will now gradually shorten the muscle to determine the effect of muscle length on active, passive, and total force.

- Shorten the muscle by 5 mm by clicking the – button beside the muscle length display.
- Click **Stimulate** to deliver an electrical stimulus to the muscle and note the values of the total, active, and passive forces relative to those observed at the original 75 mm.
- Click **Record Data** to display your results in the grid (and record your results in Chart 6).

Repeat these steps until you reach a muscle length of 50 mm.

4. Click **Clear Tracings** to clear the left oscilloscope display.

5. Lengthen the muscle to 80 mm by clicking the + button beside the muscle length display. Click **Stimulate** to deliver an electrical stimulus to the muscle and note the values of the total, active, and passive forces relative to those observed at the original 75 mm.

6. Click **Record Data** to display your results in the grid (and record your results in Chart 6).

7. You will now gradually lengthen the muscle to determine the effect of muscle length on active, passive, and total force.

- Lengthen the muscle by 10 mm by clicking the + button beside the muscle length display.
- Click **Stimulate** to deliver an electrical stimulus to the muscle and note the values of the total, active, and passive forces relative to those observed at the original 75 mm.
- Click **Record Data** to display your results in the grid(and record your results in Chart 6).

Repeat these steps until you reach a muscle length of 100 mm.

8. Click **Plot Data** to view a summary of your data on a plotted grid. Click **Submit** to record your plot in the lab report.

After you complete the experiment, take the online **Post-lab Quiz** for Activity 6.

Activity Questions

1. Explain what happens in the skeletal muscle sarcomere to result in the changes in active, passive, and total force when the resting muscle length is changed.

___

___

2. Explain the dip in the total force curve as the muscle was stretched to longer lengths. (Hint: Keep in mind that you are measuring the sum of active and passive forces.)

___

___

ACTIVITY 7

Isotonic Contractions and the Load-Velocity Relationship

OBJECTIVES

1. To understand the terms *isotonic concentric contraction*, *load*, *latent period*, *shortening velocity*, and *load-velocity relationship*.
2. To understand the effect that increasing load (that is, weight) has on an isolated skeletal muscle when the muscle is stimulated in an isotonic contraction experiment.
3. To understand the load-velocity relationship in isolated skeletal muscle.

Introduction

Skeletal muscle contractions can be described as either isometric or isotonic. When a muscle attempts to move an object (the **load**) that is equal in weight to the force generated by the muscle, the muscle is observed to contract isometrically. In an isometric contraction, the muscle stays at a fixed length (*isometric* means "same length").

During an **isotonic contraction**, the skeletal muscle length changes and, thus, the load moves a measurable distance. If the muscle length shortens as the load moves, the contraction is called an **isotonic *concentric* contraction.** An isotonic concentric contraction occurs when a muscle generates a force greater than the load attached to the muscle's end. In this type of contraction, there is a **latent period** during which there is a rise in muscle tension but no observable movement of the weight. After the muscle tension exceeds the weight of the load, an isotonic concentric contraction can begin. Thus, the latent period gets longer as the weight of the load gets larger. When the building muscle force exceeds the load, the muscle shortens and the weight moves. Eventually, the force of the muscle contraction will decrease as the muscle twitch begins the relaxation phase, and the load will therefore start to return to its original position.

An isotonic twitch is not an all-or-nothing event. If the load is increased, the muscle must generate more force to move it and the latent period will therefore get longer because it will take more time for the necessary force to be generated by the muscle. The speed of the contraction (muscle **shortening velocity**) also depends on the load that the muscle is attempting to move. Maximal shortening velocity is attained with minimal load attached to the muscle. Conversely, the heavier the load, the slower the muscle twitch. You can think of lifting an object from the floor as an example. A light object can be lifted quickly (high velocity), whereas a heavier object will be lifted with a slower velocity for a shorter duration.

In an isotonic muscle contraction experiment, one end of the muscle remains free (unlike in an isometric contraction experiment, where both ends of the muscle are held in a fixed position). Different weights (loads) can then be attached to the free end of the isolated muscle, while the other end is held in a fixed position by the force transducer. If the weight (the load) is less than the tension generated by the whole muscle, then the muscle will be able to lift it with a measurable distance, velocity, and duration. In this activity, you will change the weight (load) that the muscle will try to move as it shortens.

EQUIPMENT USED The following equipment will be depicted on-screen: intact, viable skeletal muscle dissected off the leg of a frog; electrical stimulator—delivers the desired amount and duration of stimulating voltage to the muscle via electrodes resting on the muscle; mounting stand—includes a ruler that allows a rapid measurement of the distance (cm) that the weight (load) is lifted by the isolated muscle; several weights (in grams)—can be interchangeably attached to the hook on the free lower tendon of the mounted skeletal muscle; oscilloscope—displays the stimulated isotonic concentric contraction, the duration of the contraction, and the distance that muscle lifts the weight (load).

Experiment Instructions

Go to the home page in the PhysioEx software and click **Exercise 2: Skeletal Muscle Physiology**. Click **Activity 7: Isotonic Contractions and the Load-Velocity Relationship**, and take the online **Pre-lab Quiz** for Activity 7.

After you take the online Pre-lab Quiz, click the **Experiment** tab and begin the experiment. The experiment instructions are reprinted here for your reference. The opening screen for the experiment is shown below.

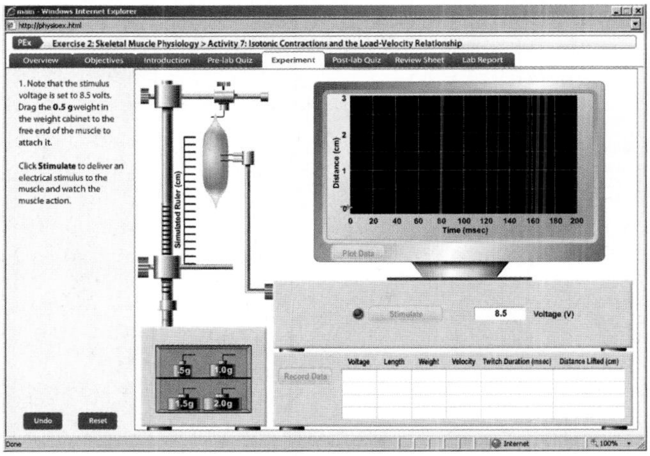

Windows Explorer 2007, Microsoft Corporation

1. Note that the stimulus voltage is set to 8.5 volts. Drag the 0.5-g weight in the weight cabinet to the free end of the muscle to attach it. Click **Stimulate** to deliver an electrical stimulus to the muscle and watch the muscle action.

2. Observe that, as the muscle shortens in length, it lifts the weight off the platform. The muscle then lengthens as it relaxes and lowers the weight back down to the platform. Click **Stimulate** again and try to watch both the muscle and the oscilloscope screen at the same time.

3. Click **Record Data** to display your results in the grid (and record your results in Chart 7).

CHART 7 Isotonic Contraction Results

Weight (g)	Velocity (cm/sec)	Twitch duration (msec)	Distance lifted (cm)

PREDICT Question 1
As the load on the muscle *increases*, what will happen to the latent period, the shortening velocity, the distance that the weight moved, and the contraction duration?

____

____

4. Remove the 0.5-g weight by dragging it back to the weight cabinet. Drag the 1.0-g weight to the free end of the muscle to attach it. Click **Stimulate** and observe the muscle and the oscilloscope screen.

5. Click **Record Data** to display your results in the grid (and record your results in Chart 7).

6. Remove the 1.0-g weight by dragging it back to the weight cabinet. Drag the 1.5-g weight to the free end of the muscle to attach it. Click **Stimulate** and observe the muscle and the oscilloscope screen.

7. Click **Record Data** to display your results in the grid (and record your results in Chart 7).

8. Remove the 1.5-g weight by dragging it back to the weight cabinet. Drag the 2.0-g weight to the free end of the muscle to attach it. Click **Stimulate** and observe the muscle and the oscilloscope screen.

9. Click **Record Data** to display your results in the grid (and record your results in Chart 7).

10. Click **Plot Data** to generate a muscle load-velocity relationship. Watch the display carefully as the program animates the development of a load-velocity relationship for the data you have collected. Click **Submit** to record your plot in the lab report.

After you complete the experiment, take the online **Post-lab Quiz** for Activity 7.

Activity Questions

1. Explain the relationship between the load attached to a skeletal muscle and the initial velocity of skeletal muscle shortening.

____

____

____

2. Explain why it will take you longer to perform ten repetitions lifting a 20-pound weight than it would to perform the same number of repetitions with a 5-pound weight.

____

____

____

NAME ______________________

LAB TIME/DATE ______________________

Skeletal Muscle Physiology

ACTIVITY 1 The Muscle Twitch and the Latent Period

1. Define the terms *skeletal muscle fiber*, *motor unit*, *skeletal muscle twitch*, *electrical stimulus*, and *latent period.* ______________________

2. What is the role of acetylcholine in a skeletal muscle contraction? ______________________

3. Describe the process of excitation-contraction coupling in skeletal muscle fibers. ______________________

4. Describe the three phases of a skeletal muscle twitch. ______________________

5. Does the duration of the latent period change with different stimulus voltages? How well did the results compare with your prediction? ______________________

6. At the threshold stimulus, do sodium ions start to move into or out of the cell to bring about the membrane depolarization? ______________________

ACTIVITY 2 The Effect of Stimulus Voltage on Skeletal Muscle Contraction

1. Describe the effect of increasing stimulus voltage on isolated skeletal muscle. Specifically, what happened to the muscle force generated with stronger electrical stimulations and why did this change occur? How well did the results compare with your prediction? ______________________

____

2. How is this change in whole-muscle force achieved in vivo? ______________________

____

3. What happened in the isolated skeletal muscle when the maximal voltage was applied? ______________________

____

ACTIVITY 3 The Effect of Stimulus Frequency on Skeletal Muscle Contraction

1. What is the difference between stimulus intensity and stimulus frequency? ______________________

2. In this experiment you observed the effect of stimulating the isolated skeletal muscle multiple times in a short period with complete relaxation between the stimuli. Describe the force of contraction with each subsequent stimulus. Are these results called treppe or wave summation? ______________________

____

3. How did the frequency of stimulation affect the amount of force generated by the isolated skeletal muscle when the frequency of stimulation was increased such that the muscle twitches did not fully relax between subsequent stimuli? Are these results called treppe or wave summation? How well did the results compare with your prediction? ______________________

____

____

____

4. To achieve an active force of 5.2 g, did you have to increase the stimulus voltage above 8.5 volts? If not, how did you achieve an active force of 5.2 g? How well did the results compare with your prediction? ______________________

____

____

5. Compare and contrast frequency-dependent wave summation with motor unit recruitment (previously observed by increasing the stimulus voltage). How are they similar? How was each achieved in the experiment? Explain how each is achieved in vivo. ______________________

____

ACTIVITY 4 Tetanus in Isolated Skeletal Muscle

1. Describe how increasing the stimulus frequency affected the force developed by the isolated whole skeletal muscle in this activity. How well did the results compare with your prediction? ______________________________

____

2. Indicate what type of force was developed by the isolated skeletal muscle in this activity at the following stimulus frequencies: at 50 stimuli/sec, at 140 stimuli/sec, and above 146 stimuli/sec. ______________________________

____

3. Beyond what stimulus frequency is there no further increase in the peak force? What is the muscle tension called at this frequency? ______________________________

ACTIVITY 5 Fatigue in Isolated Skeletal Muscle

1. When a skeletal muscle fatigues, what happens to the contractile force over time? ______________________________

____

2. What are some proposed causes of skeletal muscle fatigue? ______________________________

____

3. Turning the stimulator off allows a small measure of muscle recovery. Thus, the muscle will produce more force for a longer time period if the stimulator is briefly turned off than if the stimuli were allowed to continue without interruption. Explain why this might occur. How well did the results compare with your prediction? ______________________________

____

____

____

4. List a few ways that humans could delay the onset of fatigue when they are vigorously using their skeletal muscles. ______

____

ACTIVITY 6 The Skeletal Muscle Length-Tension Relationship

1. What happens to the amount of total force the muscle generates during the stimulated twitch? How well did the results compare with your prediction? ______________________________

____

____

2. What is the key variable in an isometric contraction of a skeletal muscle? ______________________________

3. Based on the unique arrangement of myosin and actin in skeletal muscle sarcomeres, explain why active force varies with changes in the muscle's resting length. ______

______

______

4. What skeletal muscle lengths generated passive force? (Provide a range.) ______

5. If you were curling a 7-kg dumbbell, when would your bicep muscles be contracting isometrically? ______

ACTIVITY 7 Isotonic Contractions and the Load-Velocity Relationship

1. If you were using your bicep muscles to curl a 7-kg dumbbell, when would your muscles be contracting isotonically?

______

2. Explain why the latent period became longer as the load became heavier in the experiment. How well did the results compare with your prediction? ______

______

______

3. Explain why the shortening velocity became slower as the load became heavier in this experiment. How well did the results compare with your prediction? ______

______

______

4. Describe how the shortening distance changed as the load became heavier in this experiment. How well did the results compare with your prediction? ______

______

______

5. Explain why it would take you longer to perform 10 repetitions lifting a 10-kg weight than it would to perform the same number of repetitions with a 5-kg weight. ______

______

6. Describe what would happen in the following experiment: A 2.5-g weight is attached to the end of the isolated whole skeletal muscle used in these experiments. Simultaneously, the muscle is maximally stimulated by 8.5 volts and the platform supporting the weight is removed. Will the muscle generate force? Will the muscle change length? What is the name for this type of contraction? ______

______

Neurophysiology of Nerve Impulses

PRE-LAB QUIZ

1. Circle the correct underlined term: Neurons / Neuroglial cells are capable of generating an electrical signal.
2. The _________ is the potential difference between the inside of a cell and the outside of a cell across the membrane.
   a. conductance
   b. resting membrane potential
   c. permeable potential
   d. active membrane potential
3. If the response to a stimulus is a change from a negative potential to a less negative potential, the change is called:
   a. repolarization
   b. stimulus recovery
   c. depolarization
   d. stimulus response
4. What type of channels open when the membrane depolarizes?
   a. voltage-gated potassium channels
   b. ligand-gated sodium channels
   c. voltage-gated sodium channels
   d. ligand-gated potassium channels
5. The region where the neurotransmitter is released from one neuron and binds to a receptor on a target cell is the:
   a. chemical synapse
   b. receptor potential
   c. postsynaptic synapse
   d. postsynaptic potential

Exercise Overview

The nervous system contains two general types of cells: **neurons** and neuroglia (or glial cells). This exercise focuses on neurons. Neurons respond to their local environment by generating an electrical signal. For example, sensory neurons in the nose generate a signal (called a **receptor potential**) when odor molecules interact with receptor proteins on the membrane of these olfactory sensory neurons. Thus, sensory neurons can respond directly to sensory stimuli. The receptor potential can trigger another electrical signal (called an **action potential**), which travels along the membrane of the sensory neuron's axon to the brain—you could say that the action potential is conducted to the brain.

The action potential causes the release of **chemical neurotransmitters** onto neurons in olfactory regions of the brain. These chemical neurotransmitters bind to receptor proteins on the membrane of these brain **interneurons**. In general, interneurons respond to chemical neurotransmitters released by other neurons. In the nose the odor molecules are sensed by sensory neurons. In the brain the odor is perceived by the activity of interneurons responding to neurotransmitters. Any resulting action or behavior is caused by the subsequent activity of **motor neurons**, which can stimulate muscles to contract (see Exercise 2).

In general each neuron has three functional regions for signal transmission: a receiving region, a conducting region, and an output region, or secretory region. Sensory neurons often have a receptive ending specialized to detect a specific sensory stimulus, such as odor, light, sound, or touch. The **cell body** and **dendrites** of interneurons receive stimulation by neurotransmitters at structures called **chemical synapses** and produce **synaptic potentials**. The conducting

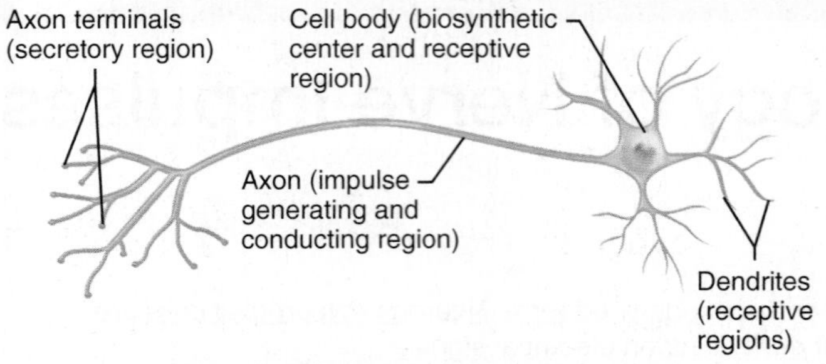

FIGURE 3.1 A neuron with functional areas identified.

region is usually an **axon**, which ends in an output region (the axon terminal) where neurotransmitter is released (view Figure 3.1).

Although the neuron is a single cell surrounded by a continuous plasma membrane, each region contains distinct membrane proteins that provide the basis for the functional differences. Thus, the receiving end has receptor proteins and proteins that generate the receptor potential, the conducting region has proteins that generate and conduct action potentials, and the output region has proteins to package and release neurotransmitters. Membrane proteins are found throughout the neuronal membrane—many of these proteins transport ions (see Exercise 1).

The signals generated and conducted by neurons are electrical. In ordinary household devices, electric current is carried by electrons. In biological systems, currents are carried by positively or negatively charged **ions**. Like charges repel each other and opposite charges attract. In general, ions cannot easily pass through the lipid bilayer of the plasma membrane and must pass through **ion channels** formed by integral membrane proteins. Some channels are usually open (leak channels) and others are gated, meaning that the channel can be in an open or closed configuration. Channels can also be selective for which ions are allowed to pass. For example, sodium channels are mostly permeable to sodium ions when open, and potassium channels are mostly permeable to potassium ions when open. The term **conductance** is often used to describe **permeability**. In general, ions will flow through an open channel from a region of higher concentration to a region of lower concentration (see Exercise 1). In this exercise you will explore some of these characteristics applied to neurons.

Although it is possible to measure the ionic currents through the membrane (even the currents passing through single ion channels), it is more common to measure the potential difference, or voltage, across the membrane. This membrane voltage is usually called the **membrane potential**, and the units are **millivolts (mV)**. One can think of the membrane as a battery, a device that separates and stores charge. A typical household battery has a positive and a negative pole so that when it is connected, for example through a lightbulb in a flashlight, current flows through the bulb. Similarly, the plasma membrane can store charge and has a relatively positive side and a relatively negative side. Thus, the membrane is said to be **polarized**. When these two sides (intracellular and extracellular) are connected through open ion channels, current in the form of ions can flow in or out across the membrane and thus change the membrane voltage.

ACTIVITY 1

The Resting Membrane Potential

OBJECTIVES

1. To define the term *resting membrane potential*.
2. To measure the resting membrane potential in different parts of a neuron.
3. To determine how the resting membrane potential depends on the concentrations of potassium and sodium.
4. To understand the ion conductances/ion channels involved in the resting membrane potential.

Introduction

The receptor potential, synaptic potentials, and action potentials are important signals in the nervous system. These potentials refer to changes in the membrane potential from its resting level. In this activity you will explore the nature of the resting potential. The **resting membrane potential** is really a potential difference between the inside of the cell (intracellular) and the outside of the cell (extracellular) across the membrane. It is a steady-state condition that depends on the resting permeability of the membrane to ions and on the intracellular and extracellular concentrations of those ions to which the membrane is permeable.

For many neurons, Na^+ and K^+ are the most important ions, and the concentrations of these ions are established by transport proteins, such as the Na^+-K^+ pump, so that the intracellular Na^+ concentration is low and the intracellular K^+ concentration is high. Inside a typical cell, the concentration of K^+ is ~150 m*M* and the concentration of Na^+ is ~5 m*M*. Outside a typical cell, the concentration of K^+ is ~5 m*M* and the concentration of Na^+ is ~150 m*M*. If the membrane is permeable to a particular ion, that ion will diffuse down its concentration gradient from a region of higher concentration to a region of lower concentration. In the generation of the resting membrane potential, K^+ ions diffuse out across the membrane, leaving behind a net negative charge—large anions that cannot cross the membrane.

The membrane potential can be measured with an amplifier. In the experiment the extracellular solution is connected to a ground (literally, the earth) which is defined as 0 mV. To record the voltage across the membrane, a microelectrode is inserted through the membrane without significantly damaging it. Typically, the microelectrode is made by pulling a thin glass pipette to a fine hollow point and filling the pulled pipette with a salt solution. The salt solution conducts electricity like a wire, and the glass insulates it. Only the tip of the microelectrode is inserted through the membrane, and the filled tip of the microelectrode makes electrical contact with the intracellular solution. A wire connects the microelectrode to the input of the amplifier so that the amplifier records the membrane potential, the voltage across the membrane between the intracellular and grounded extracellular solutions.

The membrane potential and the various signals can be observed on an oscilloscope. An electron beam is pulled up or down according to the voltage as it sweeps across a phosphorescent screen. Voltages below 0 mV are negative and voltages above 0 mV are positive. For this first activity, the time of the sweep is set for 1 second per division, and the sensitivity is set to 10 mV per division; a division is the distance between gridlines on the oscilloscope.

EQUIPMENT USED The following equipment will be depicted on-screen: neuron (in vitro)—a large, dissociated (or cultured) neuron; three extracellular solutions—control, high potassium, and low sodium; microelectrode—a probe with a very small tip that can impale a single neuron (In an actual wet lab, a microelectrode manipulator is used to position the microelectrode. For simplicity, the microelectrode manipulator will not be depicted in this activity.); microelectrode manipulator controller—controls movement of the manipulator; microelectrode amplifier—used to measure the voltage between the microelectrode and a reference; oscilloscope—used to observe voltage changes.

Experiment Instructions

Go to the home page in the PhysioEx software, and click **Exercise 3: Neurophysiology of Nerve Impulses**. Click **Activity 1: The Resting Membrane Potential**, and take the online **Pre-lab Quiz** for Activity 1.

After you take the online Pre-lab Quiz, click the **Experiment** tab and begin the experiment. The experiment instructions are reprinted here for your reference. The opening screen for the experiment is shown below.

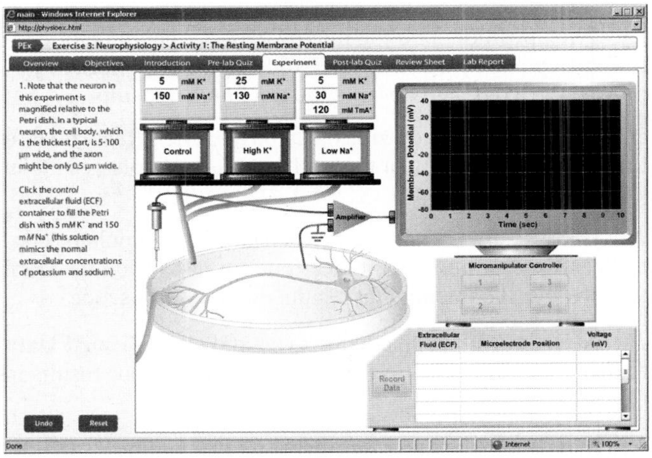

Windows Explorer 2007, Microsoft Corporation

1. Note that the neuron in this experiment is magnified relative to the petri dish. In a typical neuron, the cell body, which is the thickest part, is 5–100 μm wide, and the axon might be only 0.5 μm wide.

   Click the ***control*** **extracellular fluid (ECF)** container to fill the petri dish with 5 m*M* K^+ and 150 m*M* Na^+ (this solution mimics the normal extracellular concentrations of potassium and sodium).

2. Note that a reference electrode is already positioned in the petri dish. This reference electrode is connected to ground through the amplifier.

   Click position **1** on the microelectrode manipulator controller to position the microelectrode tip in the solution, just outside the cell body, and observe the tracing that results on the oscilloscope.

3. Note the oscilloscope tracing of the voltage outside the cell body and click **Record Data** to display your results in the grid (and record your results in Chart 1).

4. Click position **2** on the microelectrode manipulator controller to position the microelectrode tip just inside the cell body and observe the tracing that results.

5. Note the oscilloscope tracing of the voltage inside the cell body and click **Record Data** to display your results in the grid (and record your results in Chart 1). This is the resting membrane potential; that is, the potential difference between intracellular and extracellular membrane voltages. By convention, the extracellular resting membrane voltage is taken to be 0 mV.

6. Click position **3** on the microelectrode manipulator controller to position the microelectrode tip in the solution, just outside the axon, and observe the tracing that results.

7. Note the oscilloscope tracing of the voltage outside the axon and click **Record Data** to display your results in the grid (and record your results in Chart 1).

CHART 1 Resting Membrane Potential

Extracellular fluid (ECF)	Microelectrode position	Voltage (mV)

8. Click position **4** on the microelectrode manipulator controller to position the microelectrode tip just inside the axon and observe the tracing that results.

9. Note the oscilloscope tracing of the voltage inside the axon and click **Record Data** to display your results in the grid (and record your results in Chart 1).

> ? PREDICT **Question 1**
> Predict what will happen to the resting membrane potential if the extracellular K^+ concentration is increased.
>
> ______________________________
>
> ______________________________

10. You will now change the concentrations of the ions in the extracellular fluid to determine which ions contribute most to the separation of charge across the membrane. The extracellular potassium concentration is normally low, so you will first increase the extracellular potassium concentration.

In the high K^+ ECF solution the K^+ concentration has been increased fivefold, from 5 to 25 m*M*. To keep the number of positive charges in the extracellular solution constant, the Na^+ concentration has been reduced by 20 m*M*, from 150 to 130 m*M*. As you will see, this relatively small decrease in Na^+ will not by itself change the membrane potential. Note that in this activity, the generation of the action potential (which is covered in Activities 3–9) is blocked with a toxin. Click the **high K^+ ECF** container to change the solution in the petri dish to 25 m*M* K^+ and 130 m*M* Na^+.

11. Note the voltage inside the axon and click **Record Data** to display your results in the grid (and record your results in Chart 1).

12. Click position **3** on the microelectrode manipulator controller to position the microelectrode tip in the solution, just outside the axon, and observe the tracing that results.

13. Note the voltage outside the axon and click **Record Data** to display your results in the grid (and record your results in Chart 1).

14. Click position **1** on the microelectrode manipulator controller to position the microelectrode tip in the solution, just outside the cell body, and observe the tracing that results.

15. Note the voltage outside the cell body and click **Record Data** to display your results in the grid (and record your results in Chart 1).

16. Click position **2** on the microelectrode manipulator controller to position the microelectrode tip just inside the cell body and observe the tracing that results on the oscilloscope.

17. Note the voltage inside the cell body and click **Record Data** to display your results in the grid (and record your results in Chart 1).

18. Click the *control* ECF container to change back to the normal K^+ concentration and note the change in voltage inside the cell body.

19. You will now decrease the extracellular Na^+ concentration (the extracellular Na^+ concentration is normally high).

The extracellular sodium concentration in the low Na^+ solution has been decreased fivefold, from 150 m*M* to 30 m*M*. To keep the number of positive charges constant in the extracellular solution, the Na^+ has been replaced by the same amount of a large monovalent cation. Note that the extracellular Na^+ concentration, even in the low Na^+ ECF, is higher than the intracellular Na^+ concentration. Click the **low Na^+ ECF** container to change the solution in the petri dish to 5 m*M* K^+ and 30 m*M* Na^+.

20. Note the voltage inside the cell body and click **Record Data** to display your results in the grid (and record your results in Chart 1).

21. Click position **1** on the microelectrode manipulator controller to position the microelectrode tip in the solution, just outside the cell body, and observe the tracing that results.

22. Note the voltage outside the cell body and click **Record Data** to display your results in the grid (and record your results in Chart 1).

23. Click position **3** on the microelectrode manipulator controller to position the microelectrode tip in the solution, just outside the axon, and observe the tracing that results.

24. Note the voltage outside the axon and click **Record Data** to display your results in the grid (and record your results in Chart 1).

25. Click position **4** on the microelectrode manipulator controller to position the microelectrode tip just inside the axon and observe the tracing that results on the oscilloscope.

26. Note the voltage inside the axon and click **Record Data** to display your results in the grid (and record your results in Chart 1).

After you complete the experiment, take the online **Post-lab Quiz** for Activity 1.

Activity Questions

1. Explain why the resting membrane potential had the same value in the cell body and in the axon.

____

____

2. Describe what would happen to a resting membrane potential if the sodium-potassium transport pump was blocked.

____

____

3. Describe what would happen to a resting membrane potential if the concentration of large intracellular anions that are unable to cross the membrane is experimentally increased.

____

____

ACTIVITY 2

Receptor Potential

OBJECTIVES

1. To define the terms *sensory receptor*, *receptor potential*, *sensory transduction*, *stimulus modality*, and *depolarization*.
2. To determine the *adequate stimulus* for different sensory receptors.
3. To demonstrate that the receptor potential amplitude increases with stimulus intensity.

Introduction

The receiving end of a sensory neuron, the **sensory receptor**, has receptor proteins (as well as other membrane proteins) that can generate a signal called the **receptor potential** when the sensory neuron is stimulated by an appropriate, adequate stimulus. In this activity you will use the same recording instruments and microelectrode that you used in Activity 1. However, in this activity, you will record from the sensory receptor of three different sensory neurons and examine how these neurons respond to sensory stimuli of different modalities.

The sensory region will be shown disconnected from the rest of the neuron so that you can record the receptor potential in isolation. Similar results can sometimes be obtained by treating a whole neuron with chemicals that block the responses generated by the axon. The molecules localized to the sensory receptor ending are able to generate a receptor potential when an adequate stimulus is applied. The energy in the stimulus (for example, chemical, physical, or heat) is changed into an electrical response that involves the opening or closing of membrane ion channels. The general process that produces this change is called **sensory transduction**, which occurs at the receptor ending of the sensory neuron. Sensory transduction can be thought of as a type of signal transduction where the signal is the sensory stimulus.

You will observe that, with an appropriate stimulus, the amplitude of the receptor potential increases with stimulus intensity. Such a response is an example of a potential that is graded with stimulus intensity. These responses are sometimes referred to as *graded potentials*, or *local potentials*. Thus, the receptor potential is a graded, or local, potential. If the response (receptor potential) is a change in membrane potential from the negative resting potential to a less negative level, the membrane becomes less polarized and the change is called **depolarization**.

EQUIPMENT USED The following equipment will be depicted on-screen: three sensory receptors—Pacinian (lamellar) corpuscle, olfactory receptor, and free nerve ending; microelectrode—a probe with a very small tip that can impale a single neuron (In an actual wet lab, a microelectrode manipulator is used to position the microelectrodes. For simplicity, the microelectrode manipulator will not be depicted in this activity.); microelectrode amplifier—used to measure the voltage between the microelectrode and a reference; stimulator—used to select the stimulus modality (pressure, chemical, heat, or light) and intensity (low, moderate, or high); oscilloscope—used to observe voltage changes.

Experiment Instructions

Go to the home page in the PhysioEx software, and click **Exercise 3: Neurophysiology of Nerve Impulses**. Click **Activity 2: Receptor Potential**, and take the online **Pre-lab Quiz** for Activity 2.

After you take the online Pre-lab Quiz, click the **Experiment** tab and begin the experiment. The experiment instructions are reprinted here for your reference. The opening screen for the experiment is shown below.

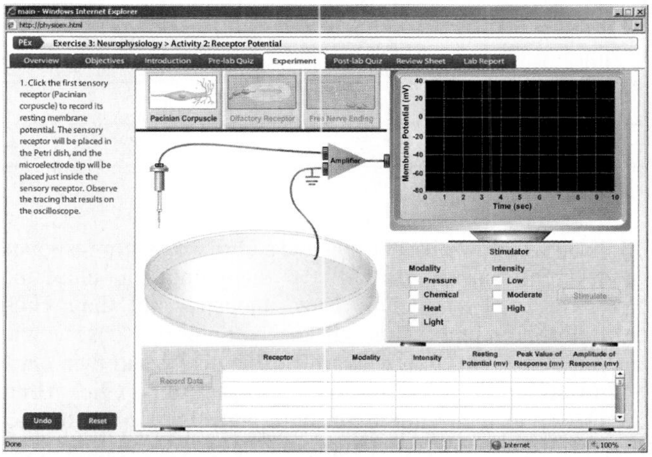

Windows Explorer 2007, Microsoft Corporation

1. Note that the timescale on the oscilloscope has been changed from 1 second per division to 10 milliseconds per division, so that you can observe the responses recorded in the sensory receptors more clearly. Click the first sensory receptor (Pacinian corpuscle) to record its resting membrane potential. The sensory receptor will be placed in the petri dish, and the microelectrode tip will be placed just inside the sensory receptor. Observe the tracing that results on the oscilloscope.

2. Note the voltage inside the sensory receptor and click **Record Data** to display your results in the grid (and record your results in Chart 2).

PREDICT Question 1
The adequate stimulus for a Pacinian corpuscle is pressure or vibration on the skin. For a Pacinian corpuscle, which modality will induce a receptor potential of the largest amplitude?

____

____

CHART 2 Receptor Potential

	Receptor potential (mV)		
Stimulus modality	**Pacinian (lamellar) corpuscle**	**Olfactory receptor**	**Free nerve ending**
None			
Pressure			
Low			
Moderate			
High			
Chemical			
Low			
Moderate			
High			
Heat			
Low			
Moderate			
High			
Light			
Low			
Moderate			
High			

3. You will now observe how the sensory receptor responds to different sensory stimuli. On the stimulator, click the **Pressure** modality. Click **Low** intensity and then click **Stimulate** to stimulate the sensory receptor and observe the tracing that results. Click **Moderate** intensity and then click **Stimulate** and observe the tracing that results. Click **High** intensity and then click **Stimulate** and observe the tracing that results. Click **Record Data** to display your results in the grid (and record your results in Chart 2).

4. On the stimulator, click the **Chemical** (odor) modality. Click **Low** intensity and then click **Stimulate** to stimulate the sensory receptor and observe the tracing that results. Click **Moderate** intensity and then click **Stimulate** and observe the tracing that results. Click **High** intensity and then click **Stimulate** and observe the tracing that results. Click **Record Data** to display your results in the grid (and record your results in Chart 2).

5. On the stimulator, click the **Heat** modality. Click **Low** intensity and then click **Stimulate** to stimulate the sensory receptor and observe the tracing that results. Click **Moderate** intensity and then click **Stimulate** and observe the tracing that results. Click **High** intensity and then click **Stimulate** and observe the tracing that results. Click **Record Data** to display your results in the grid (and record your results in Chart 2).

6. On the stimulator, click the **Light** modality. Click **Low** intensity and then click **Stimulate** to stimulate the sensory receptor and observe the tracing that results. Click **Moderate** intensity and then click **Stimulate** and observe the tracing that results. Click **High** intensity and then click **Stimulate** and observe the tracing that results. Click **Record Data** to display your results in the grid (and record your results in Chart 2).

? PREDICT Question 2
The adequate stimuli for olfactory receptors are chemicals, typically odorant molecules. For an olfactory receptor, which modality will induce a receptor potential of the largest amplitude?

____

7–12. Repeat steps 1–6 with the next sensory receptor: olfactory receptor.

13–18. Repeat steps 1–6 with the next sensory receptor: free nerve ending.

After you complete the experiment, take the online **Post-lab Quiz** for Activity 2.

Activity Questions

1. Are graded receptor potentials always depolarizing? Do graded receptor potentials always make it easier to induce action potentials?

____

____

2. Based on the definition of membrane depolarization in this activity, define membrane *hyperpolarization*.

____

____

3. What do you think is the adequate stimulus for sensory receptors in the ear? Can you think of a stimulus that would inappropriately activate the sensory receptors in the ear if the stimulus had enough intensity?

____

____

ACTIVITY 3

The Action Potential: Threshold

OBJECTIVES

1. To define the terms *action potential*, *nerve*, *axon hillock*, *trigger zone*, and *threshold*.
2. To predict how an increase in extracellular K^+ could trigger an action potential.

Introduction

In this activity you will explore changes in potential that occur in the axon. Axons are long, thin structures that conduct a signal called the **action potential**. A **nerve** is a bundle of axons.

Axons are typically studied in a nerve chamber. In this activity the axon will be draped over wires that make electrical contact with the axon and can therefore record the electrical activity in the axon. Because the axon is so thin, it is very difficult to insert an electrode across the membrane into the axon. However, some of the charge (ions) that crosses the membrane to generate the action potential can be recorded from outside the membrane (extracellular recording), as you will do in this activity. The molecular mechanisms underlying the action potential were explored more than 50 years ago with intracellular recording using the giant axons of the squid, which are about 1 millimeter in diameter.

In this activity the axon will be artificially disconnected from the cell body and dendrites. In a typical multipolar neuron (view Figure 3.1 in the Exercise Overview), the axon extends from the cell body at a region called the **axon hillock**. In a myelinated axon, this first region is called the initial segment. An action potential is usually initiated at the junction of the axon hillock and the initial segment; therefore, this region is also referred to as the **trigger zone**.

You will use an electrical stimulator to explore the properties of the action potential. Current passes from the stimulator to one of the stimulation wires, then across the axon, and then back to the stimulator through a second wire. This current will depolarize the axon. Normally, in a sensory neuron, the depolarizing receptor potential spreads passively to the axon hillock and produces the depolarization needed to evoke the action potential. Once an action potential is generated, it is regenerated down the membrane of the axon. In other words, the action potential is **propagated**, or *conducted*, down the axon (see Activity 6).

You will now generate an action potential at one end of the axon by stimulating it electrically and record the action potential that is propagated down the axon. The extracellular action potential that you record is similar to one that would be recorded across the membrane with an intracellular microelectrode, but much smaller. For simplicity, only one axon is depicted in this activity.

EQUIPMENT USED The following equipment will be depicted on-screen: nerve chamber; axon; oscilloscope—used to observe timing of stimuli and voltage changes in the axon; stimulator—used to set the stimulus voltage and to deliver pulses that depolarize the axon; stimulation wires (S); recording electrodes (wires R1 and R2)—used to record voltage changes in the axon. (The first set of recording electrodes, R1, is 2 centimeters from the stimulation wires, and the second set of recording electrodes, R2, is 2 centimeters from R1.)

Experiment Instructions

Go to the home page in the PhysioEx software, and click **Exercise 3: Neurophysiology of Nerve Impulses**. Click **Activity 3: The Action Potential: Threshold**, and take the online **Pre-lab Quiz** for Activity 3.

After you take the online Pre-lab Quiz, click the **Experiment** tab and begin the experiment. The experiment instructions are reprinted here for your reference. The opening screen for the experiment is shown below.

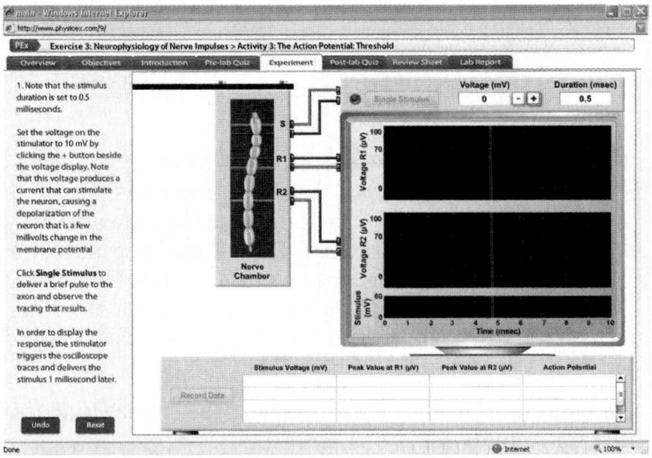

Windows Explorer 2007, Microsoft Corporation

1. Note that the stimulus duration is set to 0.5 milliseconds. Set the voltage on the stimulator to 10 mV by clicking the + button beside the voltage display. Note that this voltage produces a current that can stimulate the neuron, causing a depolarization of the neuron that is a change of a few millivolts in the membrane potential.

Click **Single Stimulus** to deliver a brief pulse to the axon and observe the tracing that results. In order to display the response, the stimulator triggers the oscilloscope traces and delivers the stimulus 1 millisecond later.

2. Note that the recording electrodes R1 and R2 record the extracellular voltage, rather than the actual membrane potential. The 10 mV depolarization at the site of stimulation only occurs locally at that site and is not recorded farther down the axon. At this initial stimulus voltage, there was no action potential. Click **Record Data** to display your results in the grid (and record your results in Chart 3, p. PEx-42).

CHART 3 Threshold			
Stimulus voltage (mV)	**Peak value at R1 (μV)**	**Peak value at R2 (μV)**	**Action potential**

3. You will increase the stimulus voltage until you observe an action potential at recording electrode 1 (R1). Increase the voltage by 10 mV by clicking the + button beside the voltage display and then click **Single Stimulus**. The voltage at which you first observe an action potential is the **threshold voltage**. Note that the action potential recorded extracellularly is quite small. Intracellularly, the membrane potential would change from −70 mV to about +30 mV. Click **Record Data** to display your results in the grid (and record your results in Chart 3).

? PREDICT **Question 1**
How will the action potential at R1 (or R2) change as you continue to increase the stimulus voltage?

____

4. You will now continue to observe the effects of incremental increases of the stimulus voltage. Increase the voltage by 10 mV by clicking the + button beside the voltage display and then click **Single Stimulus**. Repeat this step until you reach the maximum voltage the stimulator can deliver.

Repeat this step until you stimulate the axon at 50 mV and then click **Record Data** to display your results in the grid (and record your results in Chart 3).

After you complete the experiment, take the online **Post-lab Quiz** for Activity 3.

Activity Questions

1. Explain why the threshold voltage is not always the same value (between axons and within an axon).

____

____

2. Describe how the action potential is regenerated by local ion flux at each location on the axon.

____

____

3. Why doesn't the peak value of the action potential increase with stronger stimuli?

____

____

ACTIVITY 4

The Action Potential: Importance of Voltage-Gated Na^+ Channels

OBJECTIVES

1. To define the term *voltage-gated channel.*
2. To describe the effect of tetrodotoxin on the voltage-gated Na^+ channel.
3. To describe the effect of lidocaine on the voltage-gated Na^+ channel.
4. To examine the effects of tetrodotoxin and lidocaine on the action potential.
5. To predict the effect of lidocaine on pain perception and to predict the site of action in the sensory neurons (nociceptors) that sense pain.

Introduction

The action potential (as seen in Activity 3) is generated when voltage-gated sodium channels open in sufficient numbers. **Voltage-gated sodium channels** open when the membrane depolarizes. Each sodium channel that opens allows Na^+ ions to diffuse into the cell down their electrochemical gradient. When enough sodium channels open so that the amount of sodium ions that enters via these voltage-gated channels overcomes the leak of potassium ions (recall that the potassium leak via passive channels establishes and maintains the negative resting membrane potential), threshold for the action potential is reached, and an action potential is generated.

In this activity you will observe what happens when these voltage-gated sodium channels are blocked with chemicals. One such chemical is tetrodotoxin (TTX), a toxin found in puffer fish, which is extremely poisonous. Another such chemical is lidocaine, which is typically used to block pain in dentistry and minor surgery.

EQUIPMENT USED The following equipment will be depicted on-screen: nerve chamber; axon; oscilloscope—used to observe timing of stimuli and voltage changes in the axon; stimulator—used to set the stimulus voltage and the interval between stimuli and to deliver pulses that depolarize the axon; stimulation wires (S); recording electrodes (wires R1 and R2)—used to record voltage changes in the axon (The first set of recording electrodes, R1, is 2 centimeters from the stimulation wires, and the second set of recording electrodes, R2, is 2 centimeters from R1.); tetrodotoxin (TTX); lidocaine.

Experiment Instructions

Go to the home page in the PhysioEx software and click **Exercise 3: Neurophysiology of Nerve Impulses.** Click **Activity 4: The Action Potential: Importance of**

Voltage-Gated Na⁺ Channels, and take the online **Pre-lab Quiz** for Activity 4.

After you take the online Pre-lab Quiz, click the **Experiment** tab and begin the experiment. The experiment instructions are reprinted here for your reference. The opening screen for the experiment is shown below.

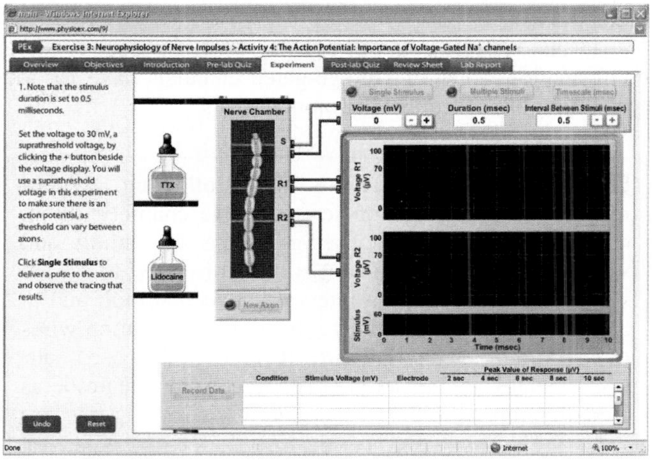

Windows Explorer 2007, Microsoft Corporation

1. Note that the stimulus duration is set to 0.5 milliseconds. Set the voltage to 30 mV, a suprathreshold voltage, by clicking the + button beside the voltage display. You will use a suprathreshold voltage in this experiment to make sure there is an action potential, as threshold can vary between axons. Click **Single Stimulus** to deliver a pulse to the axon and observe the tracing that results.

2. Enter the peak value of the response at R1 and R2 in the field below and then click **Submit** to record your answer in the lab report. _____________ μV

3. Click **Timescale** on the stimulator to change the timescale on the oscilloscope from milliseconds to seconds.

4. You will now deliver successive stimuli separated by 2.0-second intervals to observe what the control action potentials look like at this timescale. Set the interval between stimuli to 2.0 seconds by clicking the + button beside the "Interval between Stimuli" display. Click **Multiple Stimuli** to deliver pulses to the axon every 2 seconds. The stimuli will be stopped after 10 seconds.

5. Note the peak values of the responses at R1 and R2 and click **Record Data** to display your results in the grid (and record your results in Chart 4).

? PREDICT **Question 1**
If you apply TTX between recording electrodes R1 and R2, what effect will the TTX have on the action potentials at R1 and R2?

____

____

6. Drag the dropper cap of the TTX bottle to the axon between recording electrodes R1 and R2 to apply a drop of TTX to the axon.

7. Click **Multiple Stimuli** to deliver pulses to the axon every 2 seconds. The stimuli will be stopped after 10 seconds.

8. Note the peak values of the responses at R1 and R2 and click **Record Data** to display your results in the grid (and record your results in Chart 4).

9. Click **New Axon** to select a new axon. TTX is irreversible and there is no known antidote for TTX poisoning.

? PREDICT **Question 2**
If you apply lidocaine between recording electrodes R1 and R2, what effect will the lidocaine have on the action potentials at R1 and R2?

____

____

10. Drag the dropper cap of the lidocaine bottle to the axon between recording electrodes R1 and R2 to apply a drop of lidocaine to the axon.

11. Set the interval between stimuli to 2.0 seconds by clicking the + button beside the "Interval between Stimuli" display. Click **Multiple Stimuli** to deliver pulses to the axon every 2 seconds. The stimuli will be stopped after 10 seconds.

CHART 4 **Effects of Tetrodotoxin and Lidocaine**

			Peak value of response (μV)				
Condition	Stimulus voltage (mV)	Electrodes	2 sec	4 sec	6 sec	8 sec	10 sec

12. Note the peak values of the responses at R1 and R2. For simplicity, this experiment was performed on a single axon, where the action potential is an "all-or-none" event. If you had treated a bundle of axons (a nerve), each with a slightly different threshold and sensitivity to the drugs, you would likely see the peak values of the action potentials decrease more gradually as more and more axons were blocked. Click **Record Data** to display your results in the grid (and record your results in Chart 4).

After you complete the experiment, take the online **Post-lab Quiz** for Activity 4.

Activity Questions

1. If depolarizing membrane potentials open voltage-gated sodium channels, what closes them?

____

2. Why must a sushi chef go through years of training to prepare puffer fish for human consumption?

____

3. For action potential generation and propagation, are there any other cation channels that could substitute for the voltage-gated sodium channels if the sodium channels were blocked?

____

ACTIVITY 5

The Action Potential: Measuring Its Absolute and Relative Refractory Periods

OBJECTIVES

1. To define *inactivation* as it applies to a voltage-gated sodium channel.
2. To define the *absolute refractory period* and *relative refractory period* of an action potential.
3. To define the relationship between stimulus frequency and the generation of action potentials.

Introduction

Voltage-gated sodium channels in the plasma membrane of an excitable cell open when the membrane depolarizes. About 1–2 milliseconds later, these same channels inactivate, meaning they no longer allow sodium to go through the channel. These inactivated channels cannot be reopened by depolarization for an additional period of time (usually many milliseconds). Thus, during this time, fewer sodium channels can be opened. There are also voltage-gated potassium channels that open during the action potential. These potassium channels open more slowly. They contribute to the repolarization of the action potential from its peak, as more potassium flows out through this second type of potassium channel (recall there are also passive potassium channels that let potassium leak out, and these leak channels are always open). The flux through extra voltage-gated potassium channels opposes the depolarization of the membrane to threshold, and it also causes the membrane potential to become transiently more negative than the resting potential at the end of an action potential. This phase is called after-hyperpolarization, or the undershoot.

In this activity you will explore what consequences the conformation states of voltage-gated channels have for the generation of subsequent action potentials.

EQUIPMENT USED The following equipment will be depicted on-screen: nerve chamber; axon; oscilloscope—used to observe timing of stimuli and voltage changes in the axon; stimulator—used to set the stimulus voltage and the interval between stimuli and to deliver pulses that depolarize the axon; stimulation wires (S); recording electrode (wires R1)—used to record voltage changes in the axon. (The recording electrode is 2 centimeters from the stimulation wires.)

Experiment Instructions

Go to the home page in the PhysioEx software and click **Exercise 3: Neurophysiology of Nerve Impulses**. Click **Activity 5: The Action Potential: Measuring Its Absolute and Relative Refractory Periods**, and take the online **Pre-lab Quiz** for Activity 5.

After you take the online Pre-lab Quiz, click the **Experiment** tab and begin the experiment. The experiment instructions are reprinted here for your reference. The opening screen for the experiment is shown below.

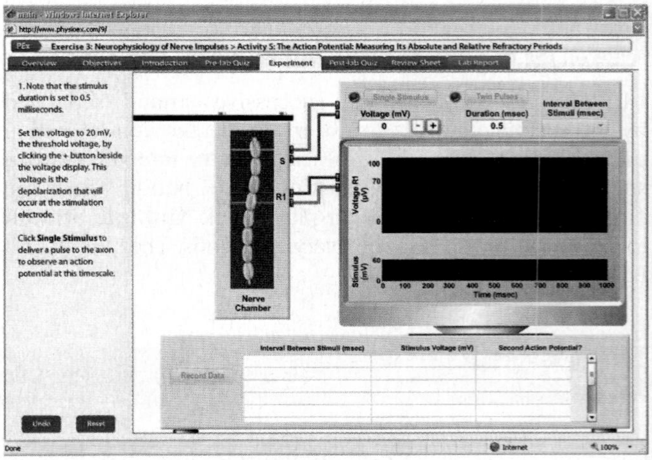

Windows Explorer 2007, Microsoft Corporation

1. Note that the stimulus duration is set to 0.5 milliseconds. Set the voltage to 20 mV, the threshold voltage, by clicking the + button beside the voltage display. This voltage is the depolarization that will occur at the stimulation electrode. Click **Single Stimulus** to deliver a pulse to observe an action potential at this timescale.

2. You will now deliver two successive stimuli separated by 250 milliseconds. Set the interval between stimuli to 250 milliseconds by selecting 250 in the "Interval between Stimuli"

pull-down menu. Click **Twin Pulses** to deliver two pulses to the axon and observe the tracing that results. Click **Record Data** to display your results in the grid (and record your results in Chart 5).

CHART 5 Absolute and Relative Refractory Periods

Interval between stimuli (msec)	Stimulus voltage (mV)	Second action potential?

3. Decrease the interval between stimuli to 125 milliseconds by selecting 125 in the "Interval between Stimuli" pull-down menu. Click **Twin Pulses** to deliver two pulses to the axon and observe the tracing that results. Click **Record Data** to display your results in the grid (and record your results in Chart 5).

4. Decrease the interval between stimuli to 60 milliseconds by selecting 60 in the "Interval between Stimuli" pull-down menu. Click **Twin Pulses** to deliver two pulses to the axon and observe the tracing that results.

Note that, at this stimulus interval, the second stimulus did not generate an action potential. Click **Record Data** to display your results in the grid (and record your results in Chart 5).

5. A second action potential can be generated at this stimulus interval, but the stimulus intensity must be increased. This interval is part of the relative refractory period, the time after an action potential when a second action potential can be generated if the stimulus intensity is increased.

Increase the stimulus intensity by 5 mV by clicking the + button beside the voltage display and then click **Twin Pulses** to deliver two pulses to the axon. Repeat this step until you generate a second action potential. After you generate a second action potential, click **Record Data** to display your results in the grid (and record your results in Chart 5)

? PREDICT **Question 1**
If you further decrease the interval between the stimuli, will the threshold for the second action potential change?

____

____

6. You will now decrease the interval until the second action potential fails again. (So that you can clearly observe two action potentials at the shorter interval between stimuli, the timescale on the oscilloscope has been set to 10 msec per division.) Decrease the interval between stimuli by 50% and then click **Twin Pulses** to deliver two pulses to the axon. When the second action potential fails, click **Record Data** to display your results in the grid (and record your results in Chart 5).

7. You will now increase the stimulus intensity until a second action potential is generated again. Increase the stimulus intensity by 5 mV by clicking the + button beside the voltage display and then click **Twin Pulses** to deliver two pulses to the axon. Repeat this step until you generate a second action potential. After you generate a second action potential, click **Record Data** to display your results in the grid (and record your results in Chart 5).

8. You will now determine the interval between stimuli at which a second action potential cannot be generated, no matter how intense the stimulus. Increase the stimulus intensity to 60 mV (the highest voltage on the stimulator). Decrease the interval between stimuli by 50% and then click **Twin Pulses** to deliver two pulses to the axon. Repeat this step until the second action potential fails.

The interval at which the second action potential fails is the **absolute refractory period**, the time after an action potential when the neuron cannot fire a second action potential, no matter how intense the stimulus. Click **Record Data** to display your results in the grid (and record your results in Chart 5).

After you complete the experiment, take the online **Post-lab Quiz** for Activity 5.

Activity Questions

1. Explain how the absolute refractory period ensures directionality of action potential propagation.

____

____

2. Some tissues (for example, cardiac muscle) have long absolute refractory periods. Why would this be beneficial?

____

____

3. What do you think is the benefit of a relative refractory period in an axon of a sensory neuron?

____

____

ACTIVITY 6

The Action Potential: Coding for Stimulus Intensity

OBJECTIVES

1. To observe the response of axons to longer periods of stimulation.
2. To examine the relationship between stimulus intensity and the frequency of action potentials.

Introduction

As seen in Activity 3, the action potential has a constant amplitude, regardless of the stimulus intensity—it is an "all-or-none" event. As seen in Activity 5, the absolute refractory period is the time after an action potential when the neuron cannot fire a second action potential, no matter how intense the stimulus, and the relative refractory period is the time after an action potential when a second action potential can be generated if the stimulus intensity is increased.

In this activity you will use these concepts to begin to explore how the axon codes the stimulus intensity as *frequency*, the number of events (in this case, action potentials) per unit time. To demonstrate this phenomenon you will use longer periods of stimulation that are more representative of real-life stimuli. For example, when you encounter an odor, the odor is normally present for seconds (or longer), unlike the very brief stimuli used in Activities 3–5. These longer stimuli allow the axon of the neuron to generate additional action potentials as soon as it has recovered from the first. As seen in Activity 5, the length of this recovery period changes depending on the stimulus intensity. For example, at threshold, a second action potential can occur only after the axon has recovered from the absolute refractory period and the entire relative refractory period.

We will not consider the phenomenon of adaptation, which is a decrease in the response amplitude that often occurs with prolonged stimuli. For example, with most odors, after many seconds, you no longer smell the odor, even though it is still present. This decrease in response is due to adaptation.

EQUIPMENT USED The following equipment will be depicted on-screen: nerve chamber; axon; oscilloscope—used to observe timing of stimuli and voltage changes in the axon; stimulator—used to set the voltage and duration of stimuli and to deliver pulses that depolarize the axon; stimulation wires (S); recording electrode (wires R1)—used to record voltage changes in the axon. (The recording electrode is 2 centimeters from the stimulation wires.)

Experiment Instructions

Go to the home page in the PhysioEx software and click **Exercise 3: Neurophysiology of Nerve Impulses**. Click **Activity 6: The Action Potential: Coding for Stimulus Intensity**, and take the online **Pre-lab Quiz** for Activity 6.

After you take the online Pre-lab Quiz, click the **Experiment** tab and begin the experiment. The experiment instructions are reprinted here for your reference. The opening screen for the experiment is shown below.

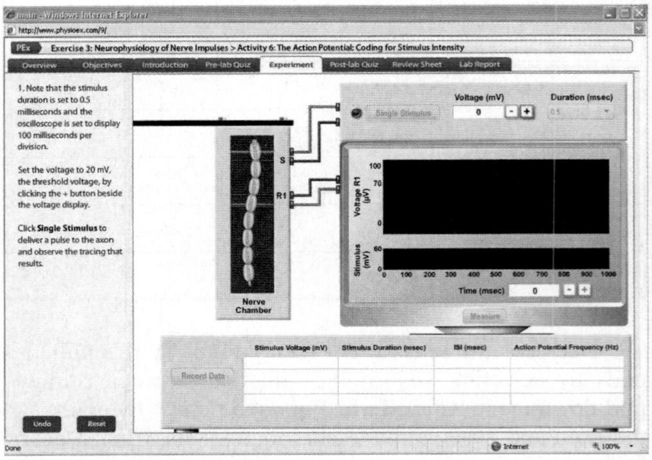

Windows Explorer 2007, Microsoft Corporation

1. Note that the stimulus duration is set to 0.5 milliseconds and the oscilloscope is set to display 100 milliseconds per division. Set the voltage to 20 mV, the threshold voltage, by clicking the + button beside the voltage display. Click **Single Stimulus** to deliver a pulse to the axon and observe the tracing that results.

2. Note how the action potential looks at this timescale and click **Record Data** to display your results in the grid (and record your results in Chart 6).

CHART 6 Frequency of Action Potentials

Stimulus voltage (mV)	Stimulus duration (msec)	ISI (msec)	Action potential frequency (Hz)

3. Increase the stimulus duration to 500 milliseconds by selecting 500 from the duration pull-down menu. Click **Single Stimulus** to deliver a pulse to the axon and observe the tracing that results. The stimulus is delivered after a delay of 100 milliseconds so that you can easily see the timing of the stimulus.

4. At the site of stimulation, the stimulus keeps the membrane of the axon at threshold for a long time, but this depolarization does not spread to the recording electrode. After one action potential has been generated and the axon has fully recovered from its absolute and relative refractory periods, the stimulus is still present to generate another action potential.

Measure the time (in milliseconds) between action potentials. This interval should be a bit longer than the relative refractory period (measured in Activity 5). Click **Measure** to help determine the time between action potentials. A thin, vertical yellow line appears at the far left side of the oscilloscope screen. You can move the line in 10-millisecond increments by clicking the + and − buttons beside the time display, which shows the time at the line. Click **Submit** to display your answer in the data table (and record your results in Chart 6).

5. The interval between action potentials is sometimes called the interspike interval (ISI). Action potentials are sometimes referred to as spikes because of their rapid time course. From the ISI, you can calculate the action potential frequency. The frequency is the reciprocal of the interval and is usually expressed in hertz (Hz), which is events (action potentials) per second. From the ISI you entered, calculate the frequency of action potentials with a prolonged (500 msec) threshold stimulus intensity. Frequency = 1/ISI. Click **Submit** to display your answer in the data table (and record your results in Chart 6).

6. A stimulus intensity of 30 mV was able to generate a second action potential toward the end of the relative refractory period in Activity 5. With this stronger stimulus, the second action potential can occur after a shorter time. Increase the stimulus intensity to 30 mV by clicking the + button beside the voltage display. Click **Single Stimulus** to deliver this stronger stimulus and observe the tracing that results.

7. Click **Submit** to display your answer in the data table (and record your results in Chart 6). Click **Measure** to help determine the time between action potentials. A thin, vertical yellow line appears at the far left side of the oscilloscope screen. You can move the line in 10-millisecond increments by clicking the + and − buttons beside the time display, which shows the time at the line.

8. From the ISI you entered, calculate the frequency of action potentials with a prolonged (500 msec) 30-mV stimulus intensity. Frequency = 1/ISI. Click **Submit** to display your answer in the data table (and record your results in Chart 6).

9. A stimulus intensity of 45 mV was able to generate a second action potential in the middle of the relative refractory period in Activity 5. With this even stronger stimulus, the second action potential can occur after an even shorter time. Increase the stimulus intensity to 45 mV.

? PREDICT Question 1
What effect will the increased stimulus intensity have on the frequency of action potentials?

____

____

10. Click **Single Stimulus** to deliver the stronger, 45-mV stimulus and observe the tracing that results.

11. Click **Submit** to display your answer in the data table (and record your results in Chart 6). Click **Measure** to help determine the time between action potentials. A thin, vertical yellow line appears at the far left side of the oscilloscope screen. You can move the line in 10-millisecond increments by clicking the + and − buttons beside the time display, which shows the time at the line.

12. From the ISI you entered, calculate the frequency of action potentials with a prolonged (500 msec) 45-mV stimulus intensity. Frequency = 1/ISI. Click **Submit** to display your answer in the data table (and record your results in Chart 6).

After you complete the experiment, take the online **Post-lab Quiz** for Activity 6.

Activity Questions

1. Compare the action potential frequency in a temperature-sensitive sensory neuron exposed to warm water and then hot water.

____

____

2. When a long-duration stimulus is applied, what two determinants of an action potential refractory period are being overcome?

____

____

3. Suggest several ways to pharmacologically overcome a neuron's refractory period and thereby increase the action potential frequency.

____

____

ACTIVITY 7

The Action Potential: Conduction Velocity

OBJECTIVES

1. To define and measure *conduction velocity* for an action potential.
2. To examine the effect of myelination on conduction velocity.
3. To examine the effect of axon diameter on conduction velocity.

Introduction

Once generated, the action potential is propagated, or conducted, down the axon. In other words, all-or-none action potentials are regenerated along the entire length of the axon. This propagation ensures that the amplitude of the action potential does not diminish as it is conducted along the axon. In some cases, such as the sensory neuron traveling from your toe to the spinal cord, the axon can be quite long (in this case, up to 1 meter). Propagation/conduction occurs because there are voltage-gated sodium and potassium channels located along the axon and because the large depolarization that constitutes the action potential (once generated at the trigger zone) easily brings the next region of the axon to threshold. The **conduction velocity** can be easily calculated by knowing both the distance the action potential travels and the amount of time it takes. Velocity has the units of distance per time, typically meters/second. An experimental stimulus artifact (see Activity 3) provides a convenient marker of the stimulus time because it travels very quickly (for our purposes, instantaneously) along the axon.

Several parameters influence the conduction velocity in an axon, including the axon diameter and the amount of myelination. **Myelination** refers to a special wrapping of the membrane from glial cells (or neuroglia) around the axon. In the central nervous system, oligodendrocytes are the glia that wrap around the axon. In the peripheral nervous system, the Schwann cells are the glia that wrap around the axon. Many glial cells along the axon contribute a myelin sheath, and the myelin sheaths are separated by gaps called nodes of Ranvier.

In this activity you will compare the conduction velocities of three axons: (1) a large-diameter, heavily myelinated axon, often called an A fiber (the terms axon and fiber are synonymous), (2) a medium-diameter, lightly myelinated axon (called the B fiber), and (3) a thin, unmyelinated fiber (called the C fiber). Examples of these axon types in the body include the axon of the sensory Pacinian corpuscle (an A fiber), the axon of both the olfactory sensory neuron and a free nerve ending (C fibers), and a visceral sensory fiber (a B fiber).

EQUIPMENT USED The following equipment will be depicted on-screen: nerve chamber; three axons—A fiber, B fiber, and C fiber; oscilloscope—used to observe timing of stimuli and voltage changes in the axon; stimulator—used to set the stimulus voltage and to deliver pulses that depolarize the axon; stimulation wires (S); recording electrodes (wires R1 and R2)—used to record voltage changes in the axon. (The first set of recording electrodes, R1, is 2 centimeters from the stimulation wires, and the second set of recording electrodes, R2, is 2 centimeters from R1.)

Experiment Instructions

Go to the home page in the PhysioEx software and click **Exercise 3: Neurophysiology of Nerve Impulses**. Click **Activity 7: The Action Potential: Conduction Velocity**, and take the online **Pre-lab Quiz** for Activity 7.

After you take the online Pre-lab Quiz, click the **Experiment** tab and begin the experiment. The experiment instructions are reprinted here for your reference. The opening screen for the experiment is shown below.

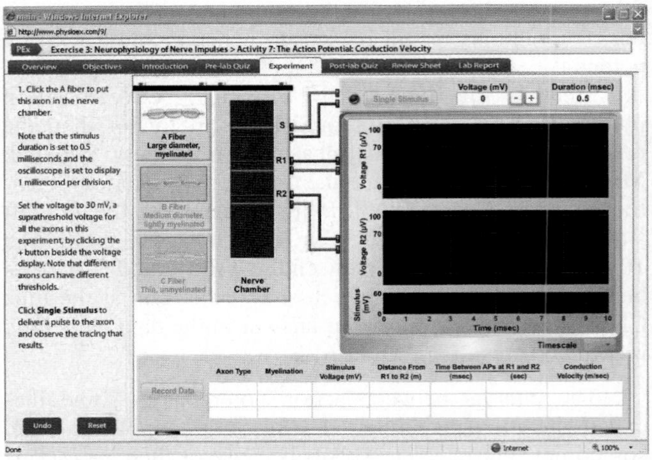

Windows Explorer 2007, Microsoft Corporation

1. Click the A fiber to put this axon in the nerve chamber. Note that the stimulus duration is set to 0.5 milliseconds and the oscilloscope is set to display 1 millisecond per division.

Set the voltage to 30 mV, a suprathreshold voltage for all the axons in this experiment, by clicking the + button beside the voltage display. Note that different axons can have different thresholds. Click **Single Stimulus** to deliver a pulse to the axon and observe the tracing that results.

2. Click **Record Data** to display your results in the grid (and record your results in Chart 7).

3. Note the difference in time between the action potential recorded at R1 and the action potential recorded at R2. The distance between these sets of recording electrodes is 10 centimeters (0.1 m). Convert the time from milliseconds to seconds and then click **Submit** to display your results in the grid (and record your results in Chart 7).

4. Calculate the conduction velocity in meters/second by dividing the distance between R1 and R2 (0.1 m) by the time it took for the action potential to travel from R1 to R2. Click

CHART 7 Conduction Velocity

Axon type	Myelination	Stimulus voltage (mV)	Distance from R1 to R2 (m)	Time between action potentials at R1 and R2		Conduction velocity (m/sec)
				(msec)	(sec)	

Submit to display your results in the grid (and record your results in Chart 7).

? PREDICT **Question 1**
How will the conduction velocity in the B fiber compare with that in the A fiber?

____

____

5. Click the **B fiber** to put this axon in the nerve chamber. Set the timescale on the oscilloscope to 10 milliseconds per division by selecting 10 in the timescale pull-down menu. Click **Single Stimulus** to deliver a pulse to the axon and observe the tracing that results.

6–8. Repeat steps 2–4 with the B fiber (and record your results in Chart 7).

? PREDICT **Question 2**
How will the conduction velocity in the C fiber compare with that in the B fiber?

____

____

9 Click the **C fiber** to put this axon in the nerve chamber. Set the timescale on the oscilloscope to 50 milliseconds per division by selecting 50 in the timescale pull-down menu. Click **Single Stimulus** to deliver a pulse to the axon and observe the tracing that results.

10–12. Repeat steps 2–4 with the C fiber (and record your results in Chart 7).

After you complete the experiment, take the online **Post-lab Quiz** for Activity 7.

Activity Questions

1. The squid utilizes a very large-diameter, unmyelinated axon to execute a rapid escape response when it perceives danger. How is this possible, given that the axon is unmyelinated?

____

____

2. When you burn your finger on a hot stove, you feel sharp, immediate pain, which later becomes slow, throbbing pain. These two types of pain are carried by different pain axons. Speculate on the axonal diameter and extent of myelination of these axons.

____

____

3. Why do humans possess a mixture of axons, some large-diameter, heavily myelinated axons and some small-diameter, relatively unmyelinated axons?

____

____

ACTIVITY 8

Chemical Synaptic Transmission and Neurotransmitter Release

OBJECTIVES

1. To define *neurotransmitter*, *chemical synapse*, *synaptic vesicle*, and *postsynaptic potential*.
2. To determine the role of calcium ions in neurotransmitter release.

Introduction

A major function of the nervous system is communication. The axon conducts the action potential from one place to another. Often, the axon has branches so that the action potential is conducted to several places at about the same time. At the end of each branch, there is a region called the axon terminal that is specialized to release packets of chemical neurotransmitters from small (~30-nm diameter) intracellular membrane-bound vesicles, called **synaptic vesicles**. **Neurotransmitters** are extracellular signal molecules that act on local targets as paracrine agents, on the neuron releasing the chemical as autocrine agents, and sometimes as hormones (endocrine agents) that reach their target(s) via the circulation. These chemicals are released by exocytosis and diffuse across a small extracellular space (called the synaptic gap, or synaptic cleft) to the target (most often the receiving end of another neuron or a muscle or gland). The neurotransmitter molecules often bind to membrane receptor proteins on the target, setting in motion a sequence of molecular events that can open or close membrane ion channels and cause the membrane potential in the target cell to change. This region where the neurotransmitter is released from one neuron and binds to a receptor on a target cell is called a **chemical synapse**, and the change in membrane potential of the target is called a synaptic potential, or **postsynaptic potential**.

In this activity you will explore some of the steps in neurotransmitter release from the axon terminal. Exocytosis of synaptic vesicles is normally triggered by an increase in calcium ions in the axon terminal. The calcium enters from outside the cell through membrane calcium channels that are opened by the depolarization of the action potential. The axon terminal has been greatly magnified in this activity so that you can visualize the release of neurotransmitter. Different from the other activities in this exercise, however, this procedure of directly seeing neurotransmitter release is not easily done in the lab; rather, neurotransmitter is usually detected by the postsynaptic potentials it triggers or by collecting and analyzing chemicals at the synapse after robust stimulation of the neurons.

EQUIPMENT USED The following equipment will be depicted on-screen: neuron (in vitro)—a large, dissociated (or cultured) neuron with magnified axon terminals; four extracellular solutions—control Ca^{2+}, no Ca^{2+}, low Ca^{2+}, and Mg^{2+}.

Experiment Instructions

Go to the home page in the PhysioEx software and click **Exercise 3: Neurophysiology of Nerve Impulses**. Click **Activity 8: Chemical Synaptic Transmission and Neurotransmitter Release**, and take the online **Pre-lab Quiz** for Activity 8.

After you take the online Pre-lab Quiz, click the **Experiment** tab and begin the experiment. The experiment instructions are reprinted here for your reference. The opening screen for the experiment is shown below.

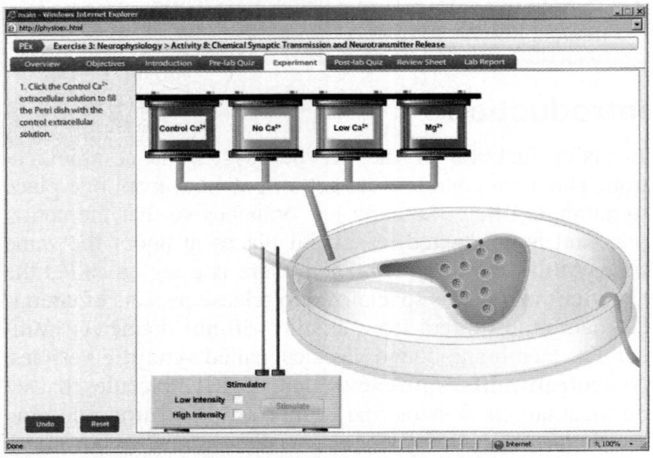

Windows Explorer 2007, Microsoft Corporation

1. Click the **control Ca^{2+}** extracellular solution to fill the petri dish with the control extracellular solution.

2. Click **Low Intensity** on the stimulator and then click **Stimulate** to stimulate the neuron (axon) with a threshold stimulus that generates a low frequency of action potentials. Observe the release of neurotransmitter.

3. Click **High Intensity** on the stimulator and then click **Stimulate** to stimulate the neuron with a longer, more intense stimulus to generate a burst of action potentials. Observe the release of neurotransmitter.

PREDICT Question 1
You have just observed that each action potential in a burst can trigger additional neurotransmitter release. If calcium ions are removed from the extracellular solution, what will happen to neurotransmitter release at the nerve terminal?

____

____

4–6. Repeat steps 1–3 with the *no Ca^{2+}* extracellular solution.

PREDICT Question 2
What will happen to the amount of neurotransmitter release when low amounts of calcium are added back to the extracellular solution?

____

____

7–9. Repeat steps 1–3 with the *low Ca^{2+}* extracellular solution.

PREDICT Question 3
What will happen to neurotransmitter release when magnesium is added to the extracellular solution?

____

____

10–12. Repeat steps 1–3 with the *Mg^{2+}* extracellular solution.

After you complete the experiment, take the online **Post-lab Quiz** for Activity 8.

Activity Questions

1. If you added more sodium to the extracellular solution, could the sodium substitute for the missing calcium?

____

____

2. How does botulinum toxin block synaptic transmission? Why is it used for cosmetic procedures?

____

____

ACTIVITY 9

The Action Potential: Putting It All Together

OBJECTIVES

1. To identify the functional areas (for example, the sensory ending, axon, and postsynaptic membrane) of a two-neuron circuit.
2. To predict and test the responses in each functional area to a very weak, subthreshold stimulus.
3. To predict and test the responses in each functional area to a moderate stimulus.
4. To predict and test the responses in each functional area to an intense stimulus.

Introduction

In the nervous system, sensory neurons respond to adequate sensory stimuli, generating action potentials in the axon if the stimulus is strong enough to reach threshold (the action potential is an "all-or-nothing" event). Via chemical synapses, these sensory neurons communicate with interneurons that process the information. Interneurons also communicate with motor neurons that stimulate muscles and glands, again, usually via chemical synapses.

After performing Activities 1–8, you should have a better understanding of how neurons function by generating changes from their resting membrane potential. If threshold is reached, an action potential is generated and propagated. If the stimulus is more intense, then action potentials are generated at a higher frequency, causing the release of more neurotransmitter at the next synapse. At an excitatory synapse the chemical neurotransmitter binds to receptors at the receiving end of the next cell (usually the cell body or dendrites of an interneuron), causing ion channels to open, resulting in a depolarization toward threshold for an action potential in the interneuron's axon. This depolarizing synaptic potential (called an excitatory postsynaptic potential) is graded in amplitude, depending on the amount of neurotransmitter and the number of channels that open. In the axon, the amplitude of this synaptic potential is coded as the frequency of action potentials. Neurotransmitters can also cause inhibition, which will not be covered in this activity.

In this activity you will stimulate a sensory neuron, predict the response of that cell and its target, and then test those predictions.

EQUIPMENT USED The following equipment will be depicted on-screen: neuron (in vitro)—a large, dissociated (or cultured) neuron; interneuron (in vitro)—a large, dissociated (or cultured) interneuron; microelectrodes—small probes with very small tips that can impale a single neuron (In an actual wet lab, a microelectrode manipulator is used to position the microelectrodes. For simplicity, the microelectrode manipulator will not be depicted in this activity.); microelectrode amplifier—used to measure the voltage between the microelectrodes and a reference; oscilloscope—used to observe the changes in voltage across the membrane of the neuron and interneuron; stimulator—used to set the stimulus intensity (low or high) and to deliver pulses to the neuron.

Experiment Instructions

Go to the home page in the PhysioEx software and click **Exercise 3: Neurophysiology of Nerve Impulses**. Click **Activity 9: The Action Potential: Putting It All Together**, and take the online **Pre-lab Quiz** for Activity 9.

After you take the online Pre-lab Quiz, click the **Experiment** tab and begin the experiment. The experiment instructions are reprinted here for your reference. The opening screen for the experiment is shown above.

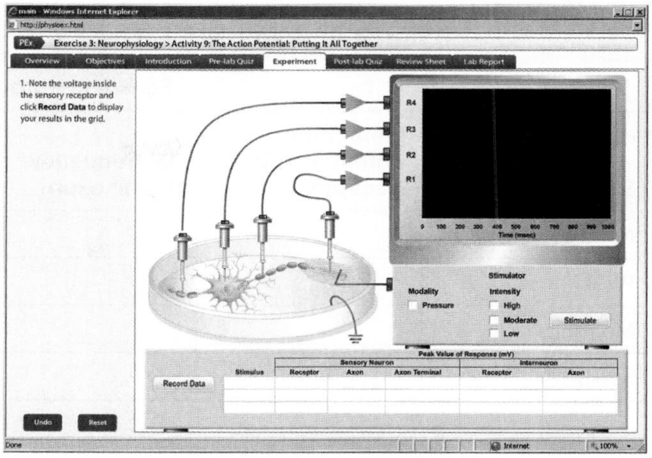

Windows Explorer 2007, Microsoft Corporation

1. Note the membrane potential at the sensory receptor and the receiving end of the interneuron and click **Record Data** to display your results in the grid (and record your results in Chart 9).

? PREDICT Question 1
What will happen if you apply a very weak, subthreshold stimulus to the sensory receptor?

____

____

2. Click **Very Weak** intensity on the stimulator and then click **Stimulate** to stimulate the receiving end of the sensory neuron and observe the tracing that results.

3. Click **Record Data** to display your results in the grid (and record your results in Chart 9). The stimulus lasts 500 msec.

? PREDICT Question 2
What will happen if you apply a moderate stimulus to the sensory receptor?

____

____

4. Click **Moderate** intensity on the stimulator and then click **Stimulate** to stimulate the sensory receptor and observe the tracing that results.

5. Click **Record Data** to display your results in the grid (and record your results in Chart 9).

? ? PREDICT Question 3
What will happen if you apply a strong stimulus to the sensory receptor?

____

____

CHART 9 Putting It All Together

	Sensory neuron			Interneuron	
Stimulus	**Membrane Potential (mV) Receptor**	**AP frequency (Hz) in axon**	**Vesicles released from axon terminal**	**Membrane potential (mV) receiving end**	**AP frequency (Hz) in axon**
None					
Weak					
Moderate					
Strong					

6. Click **Strong** intensity on the stimulator and then click **Stimulate** to stimulate the sensory receptor and observe the tracing that results.

7. Click **Record Data** to display your results in the grid (and record your results in Chart 9).

After you complete the experiment, take the online **Post-lab Quiz** for Activity 9.

Activity Questions

1. Why were the peak values of the action potentials at R2 and R4 the same when you applied a strong stimulus?

____

____

2. If the axons were unmyelinated, would the peak value of the action potential at R4 change relative to that at R2?

____

____

NAME ______________________

LAB TIME/DATE ______________________

Neurophysiology of Nerve Impulses

ACTIVITY 1 The Resting Membrane Potential

1. Explain why increasing extracellular K^+ reduces the net diffusion of K^+ out of the neuron through the K^+ leak channels.

____

2. Explain why increasing extracellular K^+ causes the membrane potential to change to a less negative value. How well did the results compare with your prediction? ______________________

____

____

3. Explain why a change in extracellular Na^+ did not alter the membrane potential in the resting neuron. ______________________

____

4. Discuss the relative permeability of the membrane to Na^+ and K^+ in a resting neuron. ______________________

____

____

5. Discuss how a change in Na^+ or K^+ conductance would affect the resting membrane potential. ______________________

____

ACTIVITY 2 Receptor Potential

1. Sensory neurons have a resting potential based on the efflux of potassium ions (as demonstrated in Activity 1). What passive channels are likely found in the membrane of the olfactory receptor, in the membrane of the Pacinian corpuscle, and in the membrane of the free nerve ending? ______________________

____

____

2. What is meant by the term *graded potential*? ______________________

____

____

3. Identify which of the stimulus modalities induced the largest amplitude receptor potential in the Pacinian corpuscle. How well did the results compare with your prediction? ____________________

____

4. Identify which of the stimulus modalities induced the largest-amplitude receptor potential in the olfactory receptors. How well did the results compare with your prediction? ____________________

____

5. The olfactory receptor also contains a membrane protein that recognizes isoamyl acetate and, via several other molecules, transduces the odor stimulus into a receptor potential. Does the Pacinian corpuscle likely have this isoamyl acetate receptor protein? Does the free nerve ending likely have this isoamyl acetate receptor protein? ____________________

____

6. What type of sensory neuron would likely respond to a green light? ____________________

____

____

ACTIVITY 3 The Action Potential: Threshold

1. Define the term *threshold* as it applies to an action potential. ____________________

____

____

2. What change in membrane potential (depolarization or hyperpolarization) triggers an action potential? ____________________

____

3. How did the action potential at R1 (or R2) change as you increased the stimulus voltage above the threshold voltage? How well did the results compare with your prediction? ____________________

____

4. An action potential is an "all-or-nothing" event. Explain what is meant by this phrase. ____________________

____

____

5. What part of a neuron was investigated in this activity? ____________________

____

____

ACTIVITY 4 The Action Potential: Importance of Voltage-Gated Na^+ Channels

1. What does TTX do to voltage-gated Na^+ channels? ______________________________

____

____

2. What does lidocaine do to voltage-gated Na^+ channels? How does the effect of lidocaine differ from the effect of TTX?

____

____

3. A nerve is a bundle of axons, and some nerves are less sensitive to lidocaine. If a nerve, rather than an axon, had been used in the lidocaine experiment, the responses recorded at R1 and R2 would be the sum of all the action potentials (called a compound action potential). Would the response at R2 after lidocaine application necessarily be zero? Why or why not? ____________

____

____

4. Why are fewer action potentials recorded at R2 when TTX is applied between R1 and R2? How well did the results compare with your prediction? ______________________________

____

____

5. Why are fewer action potentials recorded at R2 when lidocaine is applied between R1 and R2? How well did the results compare with your prediction? ______________________________

____

____

6. Pain-sensitive neurons (called nociceptors) conduct action potentials from the skin or teeth to sites in the brain involved in pain perception. Where should a dentist inject the lidocaine to block pain perception? ______________________________

____

____

ACTIVITY 5 The Action Potential: Measuring Its Absolute and Relative Refractory Periods

1. Define *inactivation* as it applies to a voltage-gated sodium channel. ______________________________

____

____

2. Define the *absolute refractory period.* ____________________

3. How did the threshold for the second action potential change as you further decreased the interval between the stimuli? How well did the results compare with your prediction? ____________________

4. Why is it harder to generate a second action potential during the relative refractory period? ____________________

ACTIVITY 6 The Action Potential: Coding for Stimulus Intensity

1. Why are multiple action potentials generated in response to a long stimulus that is above threshold? ____________________

2. Why does the frequency of action potentials increase when the stimulus intensity increases? How well did the results compare with your prediction? ____________________

3. How does threshold change during the relative refractory period? ____________________

4. What is the relationship between the interspike interval and the frequency of action potentials? ____________________

ACTIVITY 7 The Action Potential: Conduction Velocity

1. How did the conduction velocity in the B fiber compare with that in the A fiber? How well did the results compare with your prediction? ____________________

2. How did the conduction velocity in the C fiber compare with that in the B fiber? How well did the results compare with your prediction? ______________________________

3. What is the effect of axon diameter on conduction velocity? ______________________________

4. What is the effect of the amount of myelination on conduction velocity? ______________________________

5. Why did the time between the stimulation and the action potential at R1 differ for each axon? ______________________________

6. Why did you need to change the timescale on the oscilloscope for each axon? ______________________________

ACTIVITY 8 Chemical Synaptic Transmission and Neurotransmitter Release

1. When the stimulus intensity is increased, what changes: the number of synaptic vesicles released or the amount of neurotransmitter per vesicle? ______________________________

2. What happened to the amount of neurotransmitter release when you switched from the control extracellular fluid to the extracellular fluid with no Ca^{2+}? How well did the results compare with your prediction? ______________________________

3. What happened to the amount of neurotransmitter release when you switched from the extracellular fluid with no Ca^{2+} to the extracellular fluid with low Ca^{2+}? How well did the results compare with your prediction? ______________________________

4. How did neurotransmitter release in the Mg^{2+} extracellular fluid compare to that in the control extracellular fluid? How well did the result compare with your prediction? ______________________________

5. How does Mg^{2+} block the effect of extracellular calcium on neurotransmitter release? ______________________________

ACTIVITY 9 The Action Potential: Putting It All Together

1. Why is the resting membrane potential the same value in both the sensory neuron and the interneuron? ______________________________

2. Describe what happened when you applied a very weak stimulus to the sensory receptor. How well did the results compare with your prediction? ______________________________

3. Describe what happened when you applied a moderate stimulus to the sensory receptor. How well did the results compare with your prediction? ______________________________

4. Identify the type of membrane potential (graded receptor potential or action potential) that occurred at R1, R2, R3, and R4 when you applied a moderate stimulus. (View the response to the stimulus.) ______________________________

5. Describe what happened when you applied a strong stimulus to the sensory receptor. How well did the results compare with your prediction? ______________________________

PhysioEx 9.1

EXERCISE 4

Endocrine System Physiology

PRE-LAB QUIZ

1. Define *metabolism*. ______________________________
2. Circle the correct underlined term: Hormones are chemicals secreted from endocrine / exocrine glands.
3. The most important hormone for maintaining metabolism and body heat is:
   a. steroid hormone
   b. thyroxine
   c. thyroid-stimulating hormone
   d. adrenaline
4. Circle the correct underlined term: Negative feedback mechanisms / Positive feedback mechanisms ensure that if the body needs a hormone it will be produced until there is too much of it.
5. After menopause, the ovaries will stop producing and secreting:
   a. progesterone
   b. follicle-stimulating hormone
   c. estrogen
   d. androgen
6. ______________ is the hormone produced by the beta cells of the pancreas that allows our cells to absorb glucose from the bloodstream.
   a. Insulin c. Cortisol
   b. Glucagon d. Mellitol
7. Circle True or False: Cortisol is a hormone secreted by the adrenal medulla.
8. Circle the correct underlined term: Diabetes mellitus type 1 / type 2 results when the body is able to produce enough insulin but fails to respond to it.

Exercise Overview

In the human body the **endocrine system** (in addition to the nervous system) coordinates and integrates the functions of different physiological systems (view Figure 4.1, p. PEx-60). Thus, the endocrine system plays a critical role in maintaining **homeostasis**. This role begins with chemicals, called **hormones**, secreted from ductless **endocrine glands**, which are tissues that have an epithelial origin. Endocrine glands secrete hormones into the extracellular fluid compartments. More specifically, the blood usually carries hormones (sometimes attached to specific plasma proteins) to their **target cells**. Target cells can be very close to, or very far from, the source of the hormone.

Hormones bind to high-affinity **receptors** located on the target cell's surface, in its cytosol, or in its nucleus. These hormone receptors have remarkable **sensitivity**, as the hormone concentration in the blood can range from 10^{-9} to 10^{-12} molar! A hormone-receptor complex forms and can then exert a **biological action** through signal-transduction cascades and alteration of gene transcription at the target cell. The physiological response to hormones can vary from seconds to hours to days, depending on the chemical nature of the hormone and its receptor location in the target cell.

The chemical structure of the hormone is important in determining how it will interact with target cells. *Peptide* and *catecholamine hormones* are fast-acting hormones that attach to a plasma-membrane receptor and cause a second-messenger cascade in the cytoplasm of the target cell. For example, a chemical called cAMP (cyclic adenosine monophosphate) is synthesized from a molecule of ATP.

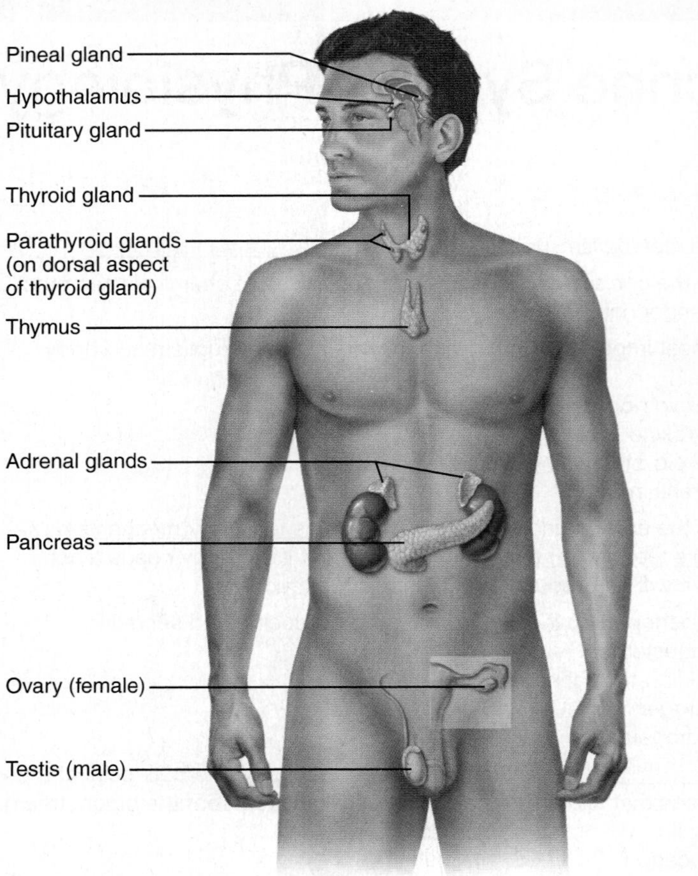

FIGURE 4.1 Selected endocrine organs of the body.

The synthesis of this chemical makes the cell more metabolically active and, therefore, more able to respond to a stimulus.

Steroid hormones and *thyroxine* (thyroid hormone) are slow-acting hormones that enter the target cell and interact with the nucleus to affect the transcription of various proteins that the cell can synthesize. The hormones enter the nucleus and attach at specific points on the DNA. Each attachment causes the production of a specific mRNA, which is then moved to the cytoplasm, where ribosomes can translate the mRNA into a protein.

Keep in mind that the organs of the endocrine system do not function independently. The activities of one endocrine gland are often coordinated with the activities of other glands. No one system functions independently of any other system. For this reason, we will be stressing feedback mechanisms and how we can use them to predict, explain, and understand hormone effects.

Given the powerful influence that hormones have on homeostasis, **negative feedback mechanisms** are important in regulating hormone secretion, synthesis, and effectiveness at target cells. Negative feedback ensures that if the body needs a particular hormone, that hormone will be produced until there is too much of it. When there is too much of the hormone, its release will be inhibited.

Rarely, the body regulates hormones via a *positive feedback mechanism.* The release of *oxytocin* from the posterior pituitary is one of these rare instances. Oxytocin is a hormone that causes the muscle layer of the uterus, called the *myometrium,* to contract during childbirth. This contraction of the myometrium causes additional oxytocin to be released, allowing stronger contractions. Unlike what happens in negative feedback mechanisms, the increase in circulating levels of oxytocin does not inhibit oxytocin secretion.

Many experimental methods can be used to study the functions of an endocrine gland. These methods include removing the gland from an animal and then injecting, implanting, or feeding glandular extracts into a normal animal or an animal deprived of the gland being studied. In this exercise you will use these methods to gain a deeper understanding of the *function* and *regulation* of some of the endocrine glands.

ACTIVITY 1

Metabolism and Thyroid Hormone

OBJECTIVES

1. To understand the terms *basal metabolic rate* (*BMR*), *thyroid-stimulating hormone* (*TSH*), *thyroxine*, *goiter*, *hypothyroidism*, *hyperthyroidism*, *thyroidectomized*, and *hypophysectomized.*
2. To observe how negative feedback mechanisms regulate hormone release.
3. To understand thyroxine's role in maintaining the basal metabolic rate.
4. To understand the effect of TSH on the basal metabolic rate.
5. To understand the role of the hypothalamus in regulating the secretion of thyroxine and TSH.

Introduction

Metabolism is the broad range of biochemical reactions occurring in the body. Metabolism includes *anabolism* and *catabolism.* Anabolism is the building up of small molecules into larger, more complex molecules via enzymatic reactions. Energy is stored in the chemical bonds formed when larger, more complex molecules are formed.

Catabolism is the breakdown of large, complex molecules into smaller molecules via enzymatic reactions. The breaking of chemical bonds in catabolism releases energy that the cell can use to perform various activities, such as forming ATP. The cell does not use all the energy released by bond breaking. Much of the energy is released as heat to maintain a fixed body temperature, especially in humans. Humans are *homeothermic* organisms that need to maintain a fixed body temperature to maintain the activity of the various metabolic pathways in the body.

The most important hormone for maintaining metabolism and body heat is **thyroxine** (thyroid hormone), also known as *tetraiodothyronine*, or T_4. Thyroxine is secreted by the thyroid gland, located in the neck.

The production of thyroxine is controlled by the pituitary gland, or hypophysis, which secretes **thyroid-stimulating hormone (TSH)**. The blood carries TSH to its target tissue, the thyroid gland. TSH causes the thyroid gland to increase in size and secrete thyroxine into the general circulation. If TSH levels are too high, the thyroid gland enlarges. The resulting glandular swelling in the neck is called a **goiter**.

The **hypothalamus** in the brain is also a vital participant in thyroxine and TSH production. It is a primary endocrine gland that secretes several hormones that affect the pituitary gland, or hypophysis, which is also located in the brain.

Thyrotropin-releasing hormone (TRH) is directly linked to thyroxine and TSH secretion. TRH from the hypothalamus stimulates the anterior pituitary to produce TSH, which then stimulates the thyroid to produce thyroxine.

These events are part of a classic negative feedback mechanism. When circulation levels of thyroxine are low, the hypothalamus secretes more TRH to stimulate the pituitary gland to secrete more TSH. The increase in TSH further stimulates the secretion of thyroxine from the thyroid gland. The increased levels of thyroxine will then influence the hypothalamus to reduce its production of TRH.

TRH travels from the hypothalamus to the pituitary gland via the **hypothalamic-pituitary portal system**. This specialized arrangement of blood vessels consists of a single **portal vein** that connects two capillary beds. The hypothalamic-pituitary portal system transports many other hormones from the hypothalamus to the pituitary gland. The hypothalamus primarily secretes *tropic* hormones, which stimulate the secretion of other hormones. TRH is an example of a tropic hormone because it stimulates the release of TSH from the pituitary gland. TSH itself is also an example of a tropic hormone because it stimulates production of thyroxine.

In this activity you will investigate the effects of thyroxine and TSH on a rat's metabolic rate. The metabolic rate will be indicated by the amount of oxygen the rat consumes per time per body mass. You will perform four experiments on three rats: a normal rat, a thyroidectomized rat (a rat whose thyroid gland has been surgically removed), and a hypophysectomized rat (a rat whose pituitary gland has been surgically removed). You will determine (1) the rat's basal metabolic rate, (2) its metabolic rate after it has been injected with thyroxine, (3) its metabolic rate after it has been injected with TSH, and (4) its metabolic rate after it has been injected with propylthiouracil, a drug that inhibits the production of thyroxine.

EQUIPMENT USED The following equipment will be depicted on-screen: three refillable syringes—used to inject the rats with propylthiouracil (a drug that inhibits the production of thyroxine by blocking the incorporation of iodine into the hormone precursor molecule), thyroid-stimulating hormone (TSH), and thyroxine; airtight, glass animal chamber—provides an isolated, sealed system in which to measure the amount of oxygen consumed by the rat in a specified amount of time (Opening the clamp on the left tube allows outside air into the chamber, and closing the clamp will create a closed, airtight system. The T-connector on the right tube allows you to connect the chamber to the manometer or to connect the fluid-filled manometer to the syringe filled with air.); soda lime (found at the bottom of the glass chamber)—absorbs the carbon dioxide given off by the rat; manometer— U-shaped tube containing fluid (As the rat consumes oxygen in the isolated, sealed system, this fluid will rise in the left side of the U-shaped tube and fall in the right side of the tube.); syringe—used to inject air into the tube and thus measure the amount of air that is needed to return the fluid columns in the manometer to their original levels; animal scale—used to measure body weight; three white rats—a *normal* rat, a *thyroidectomized* (Tx) rat (a rat whose thyroid gland has been surgically removed), and a *hypophysectomized* (Hypox) rat (a rat whose pituitary gland has been surgically removed).

Experiment Instructions

Go to the home page in the PhysioEx software and click **Exercise 4: Endocrine System Physiology.** Click **Activity 1: Metabolism and Thyroid Hormone**, and take the online **Pre-lab Quiz** for Activity 1.

After you take the online Pre-lab Quiz, click the **Experiment** tab and begin the experiment. The experiment instructions are reprinted here for your reference. The opening screen for the experiment is shown below.

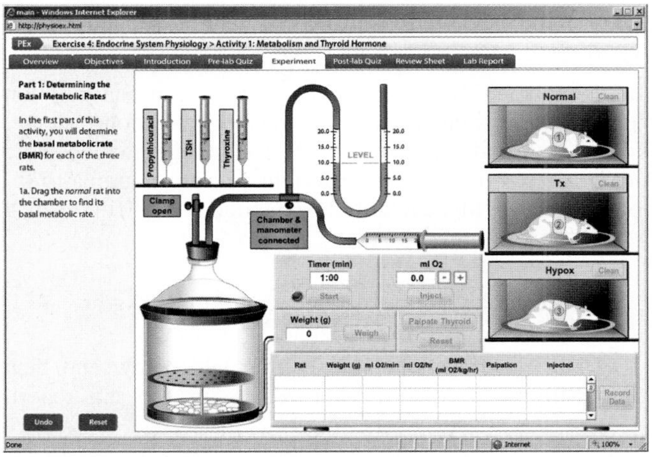

Windows Explorer 2007, Microsoft Corporation

Part 1: Determining the Basal Metabolic Rates

In the first part of this activity, you will determine the basal metabolic rate (BMR) for each of the three rats.

1a. Drag the *normal* rat into the chamber to find its BMR.

1b. Click **Weigh** to determine the rat's weight.

1c. Click the clamp on the left tube (top of the chamber) to close it. This will prevent any outside air from entering the chamber and ensure that the only oxygen the rat is breathing is the oxygen inside the closed system.

1d. Note that the timer is set to one minute. Click **Start** beneath the timer to measure the amount of oxygen consumed by the rat in one minute in the sealed chamber. Note what happens to the water levels in the manometer as time progresses.

1e. Click the T-connector knob to connect the manometer and syringe.

1f. Click the clamp on the left tube (top of the chamber) to open it so the rat can breathe outside air.

1g. Observe the difference between the level in the left and right arms of the manometer. Estimate the volume of O_2 that you will need to inject to make the levels equal by counting the divisions on both sides. This volume is equivalent to the amount of oxygen that the rat consumed during the minute in the sealed chamber. Click the + button under the ml O_2 display until you reach the estimated volume. Then click **Inject** and watch what happens to the fluid in the two arms. When the volume levels are equalized, the word "Level" will appear and stay on the screen.

- If you have not injected enough oxygen, the word "Level" will not appear. Click the + to increase the volume and then click **Inject** again.

- If you have injected too much oxygen, the word "Level" will flash and then disappear. Click the button to decrease the volume and then click **Inject** again. Click **Record Data** when the levels are equalized.

1h. Calculate the oxygen consumption per hour for this rat using the following equation:

$$\frac{\text{ml } O_2 \text{ consumed}}{1 \text{ minute}} \times \frac{60 \text{ minutes}}{1 \text{ hr}} = \text{ml } O_2/\text{hr}$$

Enter the oxygen consumption per hour in the field below and then click **Submit** to record your results in the lab report. _______________ ml O_2/hr

1i. Now that you have calculated the oxygen consumption per hour for this rat, you can calculate the metabolic rate per kilogram of body weight with the following equation (note that you need to convert the weight data from grams to kilograms to use this equation): Metabolic rate = (ml O_2/hr)/(weight in kg) = ml O_2/kg/hr.

$$\text{Metabolic rate} = \frac{\text{ml } O_2/\text{hr}}{\text{weight in kg}} = \text{ml } O_2/\text{kg/hr}$$

Enter the metabolic rate in the field below and then click **Submit** to record your results in the lab report. _______________ ml O_2/kg/hr

1j. Click **Palpate Thyroid** to manually check the size of the thyroid and, thus, whether a goiter is present. After reviewing the findings, click **Submit** to record your results in the lab report.

1k. Drag the rat from the chamber back to its cage and then click **Restore** (beneath **Palpate Thyroid**) to restore the apparatus to its initial state.

? PREDICT Question 1
Make a prediction about the basal metabolic rate (BMR) of the remaining rats compared with the BMR of the normal rat you just measured.

____

____

2a.–2k. Repeat steps 1a–1k for the *thyroidectomized (Tx)* rat.

3a.–3k. Repeat steps 1a–1k for the *hypophysectomized (Hypox)* rat.

? PREDICT Question 2
What do you think will happen to the metabolic rates of the rats after you inject them with thyroxine?

____

____

Part 2: Determining the Effect of Thyroxine on Metabolic Rate

In this part of the activity, you will investigate the effects of thyroxine injections on the metabolic rates of all three rats.

4a. Drag the syringe filled with *thyroxine* to the *normal* rat's hindquarters. Release the mouse button to inject thyroxine into the rat. (In this experiment, the effects of the injection are immediate. In a wet lab, you would have to inject the rats daily with thyroxine for 1–2 weeks).

4b. In this part of the activity, the rat's weight, the amount of oxygen consumed by the rat in one minute, the rat's oxygen consumption per hour, the rat's metabolic rate, and the result of the thyroid palpation will be generated automatically after you drag the rat into the chamber.

Drag the injected rat into the chamber and note the results (and record your results in Chart 1).

4c. Drag the rat from the chamber back to its cage and then click **Clean** to clear all traces of thyroxine from the rat and clean the syringe. (In this experiment, the thyroxine is removed instantly. In a wet lab, clearance would take weeks or require that a different rat be used.)

5a.–5c. Repeat steps 4a–4c with the *thyroidectomized (Tx)* rat (and record your results in Chart 1).

6a.–6c. Repeat steps 4a–4c with the *hypophysectomized (Hypox)* rat (and record your results in Chart 1).

? PREDICT Question 3
What do you think will happen to the metabolic rates of the rats after you inject them with TSH?

____

____

Part 3: Determining the Effect of TSH on Metabolic Rate

In this part of the activity, you will investigate the effects of TSH injections on the metabolic rates of all three rats.

7a. Drag the syringe filled with *TSH* to the *normal* rat's hindquarters. Release the mouse button to inject TSH into the rat. (In this experiment, the effects of the injection are immediate. In a wet lab, you would have to inject the rats daily with TSH for 1–2 weeks.)

7b. In this part of the activity, the rat's weight, the amount of oxygen consumed by the rat in one minute, the rat's oxygen consumption per hour, the rat's metabolic rate, and the result of the thyroid palpation will be generated automatically after you drag the rat into the chamber.

Drag the injected rat into the chamber and note the results (and record your results in Chart 1).

7c. Drag the rat from the chamber back to its cage and then click **Clean** to clear all traces of TSH from the rat and clean the syringe. (In this experiment, the TSH is removed instantly. In a wet lab, clearance would take weeks or require that a different rat be used.)

8a.–8c. Repeat steps 7a–7c with the *thyroidectomized (Tx)* rat (and record your results in Chart 1).

9a.–9c. Repeat steps 7a–7c with the *hypophysectomized (Hypox)* rat (and record your results in Chart 1).

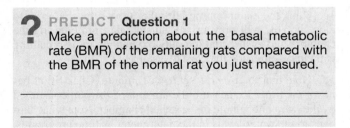

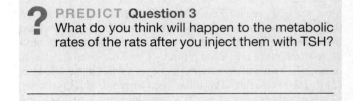

CHART 1 **Effects of Hormones on Metabolic Rate**

	Normal rat	Thyroidectomized rat	Hypophysectomized rat
Baseline			
Weight	______ grams	______ grams	______ grams
ml O_2 used in 1 minute	______ ml	______ ml	______ ml
ml O_2 used per hour	______ ml	______ ml	______ ml
Metabolic rate	______ ml O_2/kg/hr	______ ml O_2/kg/hr	______ ml O_2/kg/hr
Palpation results	______	______	______
With thyroxine			
Weight	______ grams	______ grams	______ grams
ml O_2 used in 1 minute	______ ml	______ ml	______ ml
ml O_2 used per hour	______ ml	______ ml	______ ml
Metabolic rate	______ ml O_2/kg/hr	______ ml O_2/kg/hr	______ ml O_2/kg/hr
Palpation results	______	______	______
With TSH			
Weight	______ grams	______ grams	______ grams
ml O_2 used in 1 minute	______ ml	______ ml	______ ml
ml O_2 used per hour	______ ml	______ ml	______ ml
Metabolic rate	______ ml O_2/kg/hr	______ ml O_2/kg/hr	______ ml O_2/kg/hr
Palpation results	______	______	______
With propylthiouracil			
Weight	______ grams	______ grams	______ grams
ml O_2 used in 1 minute	______ ml	______ ml	______ ml
ml O_2 used per hour	______ ml	______ ml	______ ml
Metabolic rate	______ ml O_2/kg/hr	______ ml O_2/kg/hr	______ ml O_2/kg/hr
Palpation results	______	______	______

? PREDICT Question 4
Propylthiouracil (PTU) is a drug that inhibits the production of thyroxine by blocking the attachment of iodine to tyrosine residues in the follicle cells of the thyroid gland (iodinated tyrosines are linked together to form thyroxine). What do you think will happen to the metabolic rates of the rats after you inject them with PTU?

____

____

Part 4: Determining the Effect of Propylthiouracil on Metabolic Rate

In this part of the activity, you will investigate the effects of propylthiouracil injections on the metabolic rates of all three rats.

10a. Drag the syringe filled with *propylthiouracil* to the *normal* rat's hindquarters. Release the mouse button to inject propylthiouracil into the rat. (In this experiment, the effects of the injection are immediate. In a wet lab, you would have to inject the rats daily with propylthiouracil for 1–2 weeks).

10b. In this part of the activity, the rat's weight, the amount of oxygen consumed by the rat in one minute, the rat's oxygen consumption per hour, the rat's metabolic rate, and the result of the thyroid palpation will be generated automatically after you drag the rat into the chamber.

Drag the injected rat into the chamber and note the results (and record your results in Chart 1).

10c. rag the rat from the chamber back to its cage and then click **Clean** to clear all traces of propylthiouracil from the rat and clean the syringe. (In this experiment, the propylthiouracil is removed instantly. In a wet lab, clearance would take weeks or require that a different rat be used.)

11a.–11c. Repeat steps 10a–10c with the *thyroidectomized (Tx)* rat (and record your results in Chart 1).

12a.–12c. Repeat steps 10a–10c with the *hypophysectomized (Hypox)* rat (and record your results in Chart 1).

After you complete the experiment, take the online **Post-lab Quiz** for Activity 1.

Activity Questions

1. Using a water-filled manometer, you observed the amount of oxygen consumed by rats in a sealed chamber. What happened to the carbon dioxide the rat produced while in the sealed chamber?

2. What would happen to the fluid levels of the manometer (and, thus, the results of the metabolism experiment) if the rats in the sealed chamber were engaged in physical activity (such as running in a wheel)?

3. Describe the role of the hypothalamus in the production of thyroxine.

4. What does it mean if a hormone is a *tropic* hormone?

5. How could you treat a thyroidectomized rat so that it functions like a "normal" rat? How would you verify that your treatments were safe and effective?

6. What is the role of the hypothalamus in the production of thyroid-stimulating hormone (TSH)?

7. How does thyrotropin-releasing hormone (TRH) travel from the hypothalamus to the pituitary gland?

8. Why didn't the administration of TSH have any effect on the metabolic rate of the thyroidectomized rat?

9. Why didn't the administration of propylthiouracil have any effect on the metabolic rate of either the thyroidectomized rat or the hypophysectomized rat?

10. Propylthiouracil inhibits the production of thyroxine by blocking the attachment of iodine to the amino acid tyrosine. What naturally occurring problem in some parts of the world does this drug mimic?

ACTIVITY 2

Plasma Glucose, Insulin, and Diabetes Mellitus

OBJECTIVES

1. To understand the use of the terms *insulin*, *type 1 diabetes mellitus*, *type 2 diabetes mellitus*, and *glucose standard curve.*
2. To understand how fasting plasma glucose levels are used to diagnose diabetes mellitus.
3. To understand the assay that is used to measure plasma glucose.

Introduction

Insulin is a hormone produced by the beta cells of the endocrine portion of the pancreas. This hormone is vital to the regulation of **plasma glucose** levels, or "blood sugar," because the hormone enables our cells to absorb glucose from the bloodstream. Glucose absorbed from the blood is either used as fuel for metabolism or stored as glycogen (also known as animal starch), which is most notable in liver and muscle cells. About 75% of glucose consumed during a meal is stored as glycogen. As humans do not feed continuously (we are considered "discontinuous feeders"), the production of glycogen from a meal ensures that a supply of glucose will be available for several hours after a meal.

Furthermore, the body has to maintain a certain level of plasma glucose to continuously serve nerve cells because these cell types use only glucose for metabolic fuel. When glucose levels in the plasma fall below a certain value, the alpha cells of the pancreas are stimulated to release the hormone **glucagon**. Glucagon stimulates the breakdown of stored glycogen into glucose, which is then released back into the blood.

When the pancreas does not produce enough insulin, **type 1 diabetes mellitus** results. When the pancreas produces sufficient insulin but the body fails to respond to it, **type 2 diabetes mellitus** results. In either case, glucose remains in the bloodstream, and the body's cells are unable to take it up to serve as the primary fuel for metabolism. The kidneys then filter the excess glucose out of the plasma. Because the reabsorption of filtered glucose involves a finite number of transporters in kidney tubule cells, some of the excess glucose is not reabsorbed into the circulation. Instead, it passes out of the body in urine (hence *sweet urine*, as the name **diabetes mellitus** suggests).

The inability of body cells to take up glucose from the blood also results in skeletal muscle cells undergoing protein catabolism to free up amino acids to be used in forming glucose in the liver. This action puts the body into a negative nitrogen balance from the resulting protein depletion and tissue wasting. Other associated problems include poor wound healing and poor resistance to infections.

This activity is divided into two parts. In Part 1, you will generate a **glucose standard curve**, which will be explained in the experiment. In Part 2, you will use the glucose standard curve to measure the fasting plasma glucose levels from several patients to diagnose the presence or absence of diabetes mellitus. A patient with FPG values greater than or equal to 126 mg/dl in two FPG tests is diagnosed with diabetes. FPG values between 110 and 126 mg/dl indicate impairment or borderline impairment of insulin-mediated glucose uptake by cells. FPG values less than 110 mg/dl are considered normal.

EQUIPMENT USED The following equipment will be depicted on-screen: deionized water—used to adjust the volume so that it is the same for each reaction; glucose standard; enzyme color reagent; barium hydroxide; heparin; blood samples from five patients; test tubes—used as reaction vessels for the various tests; test tube incubation unit—used to incubate, mix, and centrifuge the samples; spectrophotometer—used to measure the amount of light absorbed or transmitted by a pigmented solution.

Experiment Instructions

Go to the home page in the PhysioEx software and click **Exercise 4: Endocrine System Physiology.** Click **Activity 2: Plasma Glucose, Insulin, and Diabetes Mellitus**, and take the online **Pre-lab Quiz** for Activity 2.

After you take the online Pre-lab Quiz, click the **Experiment** tab and begin the experiment. The experiment instructions are reprinted here for your reference. The opening screen for the experiment is shown below.

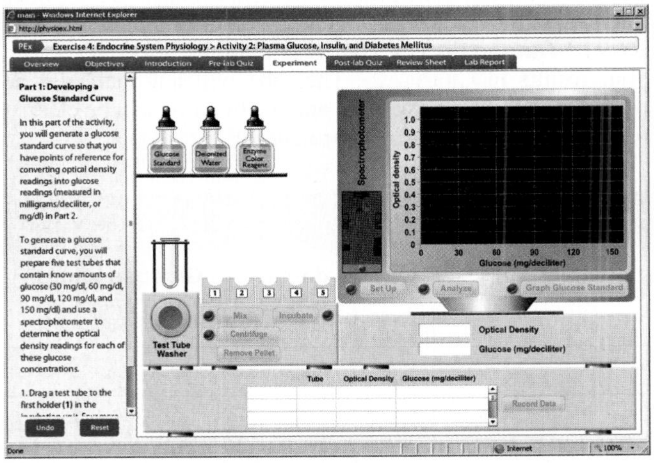

Windows Explorer 2007, Microsoft Corporation

Part 1: Developing a Glucose Standard Curve

In this part of the activity, you will generate a glucose standard curve so that you have points of reference for converting optical density readings into glucose readings (measured in milligrams/deciliter, or mg/dl) in Part 2.

To generate a glucose standard curve, you will prepare five test tubes that contain known amounts of glucose (30 mg/dl, 60 mg/dl, 90 mg/dl, 120 mg/dl, and 150 mg/dl) and use a spectrophotometer to determine the optical density readings for each of these glucose concentrations.

1. Drag a test tube to the first holder (**1**) in the incubation unit. Four more test tubes will automatically be placed in the incubation unit.

2. Drag the dropper cap of the glucose standard bottle to the first tube in the incubation unit to dispense one drop of glucose standard solution into the tube. The dropper will automatically move across and dispense glucose standard to the remaining tubes. Note that each tube receives one additional drop of glucose standard (tube 2 receives 2 drops, tube 3 receives 3 drops, tube 4 receives 4 drops, and tube 5 receives 5 drops).

3. Drag the dropper cap of the deionized water bottle to the first tube in the incubation unit to dispense four drops of deionized water into the tube. The dropper will automatically move across and dispense deionized water to the remaining tubes. Note that each tube receives one less drop of deionized water (tube 2 receives 3 drops, tube 3 receives 2 drops, tube 4 receives 1 drop, and tube 5 does not receive any drops).

4. Click **Mix** to mix the contents of the tubes.

5. Click **Centrifuge** to centrifuge the contents of the tubes. After the centrifugation process, the tubes will automatically rise.

6. Click **Remove Pellet** to remove any pellets formed during the centrifugation process. Pellets can contain reagent precipitates and debris from the laboratory environment.

7. Drag the dropper cap of the enzyme color reagent bottle to the first tube in the incubation unit to dispense five drops of enzyme color reagent into each tube.

8. Click **Incubate** to incubate the contents of the tubes. The incubation unit will gently agitate the test tube rack, evenly mixing the contents of all test tubes throughout the incubation.

9. Click **Set Up** on the spectrophotometer to warm up the instrument and get it ready for your sample readings.

10. Drag tube 1 to the spectrophotometer.

11. Click **Analyze** to analyze the sample. A data point will appear on the monitor to show the optical density and the glucose concentration of the sample. These values will also appear in the optical density and glucose displays.

12. Click **Record Data** to display your results in the grid (and record your results in Chart 2.1). The tube will automatically be placed in the test tube washer.

CHART 2.1 Glucose Standard Curve Results

Tube	Optical density	Glucose (mg/dl)
1		
2		
3		
4		
5		

13. You will now analyze the samples in the remaining tubes.

- Drag the next tube into the spectrophotometer.
- Click **Analyze** to analyze the sample. A data point will appear on the monitor to show the optical density and the glucose concentration of the sample. These values will also appear in the optical density and glucose displays.
- Click **Record Data** to display your results in the grid (and record your results in Chart 2.1). The tube will automatically be placed in the test tube washer.

Repeat this step until you analyze all five tubes.

14. Click **Graph Glucose Standard** to generate the glucose standard curve on the monitor. You will use this graph in Part 2.

? PREDICT Question 1
How would you measure the amount of plasma glucose in a patient sample?

____

____

Part 2: Measure Fasting Plasma Glucose Levels

In this part of the activity, you will use the glucose standard curve you generated in Part 1 to measure the fasting plasma glucose levels from five patients to diagnose the presence or absence of diabetes mellitus. Note the addition of two reagent bottles (barium hydroxide and heparin) and blood samples from the five patients. To undergo the fasting plasma glucose (FPG) test, patients must fast for a minimum of 8 hours prior to the blood draw.

A patient with FPG values greater than or equal to 126 mg/dl in two FPG tests is diagnosed with diabetes. FPG values between 110 and 126 mg/dl indicate impairment or borderline impairment of insulin-mediated glucose uptake by cells. FPG values less than 110 mg/dl are considered normal.

15. Drag a test tube to the first holder (**1**) in the incubation unit. Four more test tubes will automatically be placed in the incubation unit.

16. Drag the dropper cap of the first patient blood sample to the first tube in the incubation unit to dispense three drops of the sample. Three drops from each sample will automatically be dispensed into a separate tube.

17. Drag the dropper cap of the deionized water bottle to the first tube in the incubation unit to dispense five drops of deionized water into each tube.

18. Barium hydroxide dissolves and thus clears both proteins and cell membranes (so that clear glucose readings can be obtained). Drag the dropper cap of the barium hydroxide bottle to the first tube in the incubation unit to dispense five drops of barium hydroxide into each tube.

19. Drag the dropper cap of the heparin bottle to the first tube in the incubation unit to dispense a drop of heparin into each tube. Heparin prevents blood clots, which would interfere with clear glucose readings.

20. Click **Mix** to mix the contents of the tubes.

21. Click **Centrifuge** to centrifuge the contents of the tubes. After the centrifugation process, the tubes will automatically rise.

22. Click **Remove Pellet** to remove any pellets formed during the centrifugation process. Pellets can contain reagent precipitates and debris from the laboratory environment.

23. Drag the dropper cap of the enzyme color reagent bottle to the first tube in the incubation unit to dispense five drops of enzyme color reagent into each tube.

24. Click **Incubate** to incubate the contents of the tubes. The incubation unit will gently agitate the test tube rack, evenly mixing the contents of all test tubes throughout the incubation.

25. Click **Set Up** on the spectrophotometer to warm up the instrument and get it ready for your sample readings.

26. Click **Graph Glucose Standard** to display the glucose standard curve you generated in Part 1 on the monitor.

27. Drag tube 1 to the spectrophotometer.

28. Click **Analyze** to analyze the sample. A horizontal line will appear on the monitor to show the optical density of the sample. The optical density will also appear in the optical density display.

29. Drag the movable ruler (the vertical red line on the right side of the monitor) to the intersection of the horizontal yellow line (the optical density of the sample) and the glucose standard curve. Note the change in the glucose display as you move the line. The glucose concentration where the lines intersect is the fasting plasma glucose for this patient. Click **Record Data** to display your results in the grid (and record your results in Chart 2.2). The tube will automatically be placed in the test tube washer, and the monitor will be cleared (except for the glucose standard curve).

CHART 2.2 Fasting Plasma Glucose Results

Sample	Optical density	Glucose (mg/dl)
1		
2		
3		
4		
5		

30. You will now analyze the samples in the remaining tubes.

- Drag the next tube into the spectrophotometer.
- Click **Analyze** to analyze the sample. A data point will appear on the monitor to show the optical density and the glucose concentration of the sample. These values will also appear in the optical density and glucose displays.

- Click **Record Data** to display your results in the grid. The tube will automatically be placed in the test tube washer (and record your results in Chart 2.2).

 Repeat this step until you analyze all five tubes.

After you complete the experiment, take the online **Post-lab Quiz** for Activity 2.

Activity Questions

1. How would you know if your glucose standard curve was aberrant and thus inappropriate for patient diagnostics?

____

____

2. What are potential sources of variability when generating a glucose standard curve?

____

____

3. What recommendations would you make to a patient with fasting plasma glucose levels in the impaired/borderline-impaired range who was in the impaired/borderline-impaired range for the oral glucose tolerance test?

____

____

4. The amount of corn syrup in the American diet has been described as alarmingly high (especially in the foods that children eat). In the context of this activity, predict the likely trends in the fasting plasma glucose levels of our children as they mature.

____

____

ACTIVITY 3

Hormone Replacement Therapy

OBJECTIVES

1. To understand the terms *hormone replacement therapy*, *follicle-stimulating hormone (FSH)*, *estrogen*, *calcitonin*, *osteoporosis*, *ovariectomized*, and *T score*.
2. To understand how estrogen levels affect bone density.
3. To understand the potential benefits of hormone replacement therapy.

Introduction

Follicle-stimulating hormone (FSH) is an anterior pituitary peptide hormone that stimulates ovarian follicle growth. Developing ovarian follicles then produce and secrete a steroid hormone called **estrogen** into the plasma. Estrogen has numerous effects on the female body and homeostasis, including the stimulation of bone growth and protection against **osteoporosis** (a reduction in the quantity of bone characterized by decreased bone mass and increased susceptibility to fractures).

After menopause, the ovaries stop producing and secreting estrogen. One of the effects and potential health problems of menopause is a loss of bone density that can result in osteoporosis and bone fractures. For this reason, postmenopausal treatments to prevent osteoporosis often include hormone replacement therapy. Estrogen can be administered to increase bone density. Calcitonin (secreted by C cells in the thyroid gland) is another peptide hormone that can be administered to counteract the development of osteoporosis. Calcitonin inhibits osteoclast activity and stimulates calcium uptake and deposition in long bones.

In this activity you will use three **ovariectomized** rats that are no longer producing estrogen because their ovaries have been surgically removed. A **T score** is a quantitative measurement of the mineral content of bone, used as an indicator of the structural strength of the bone and as a screen for osteoporosis. The three rats were chosen because each has a baseline T score of 2.61, indicating osteoporosis. T scores are interpreted as follows: normal = +1 to −0.99; osteopenia (bone thinning) = −1.0 to −2.49; osteoporosis = −2.5 and below.

You will administer either estrogen therapy or calcitonin therapy to these rats, representing two types of **hormone replacement therapy**. The third rat will serve as an untreated control and receive daily injections of saline. The vertebral bone density (VBD) of each rat will be measured with dual X-ray absorptiometry (DXA) to obtain its T score after treatment.

EQUIPMENT USED The following equipment will be depicted on-screen: three ovariectomized rats (Note that if this were an actual wet lab, the ovariectomies would have been performed on the rats a month before the experiment to ensure that no residual hormones remained in the rats' systems.); saline; estrogen; calcitonin; reusable syringe—used to inject the rats; anesthesia—used to immobilize the rats for the X-ray scanning; dual X-ray absorptiometry bone-density scanner (DXA)—used to measure vertebral bone density of the rats.

Experiment Instructions

Go to the home page in the PhysioEx software and click **Exercise 4: Endocrine System Physiology.** Click **Activity 3: Hormone Replacement Therapy**, and take the online **Pre-lab Quiz** for Activity 3.

After you take the online Pre-lab Quiz, click the **Experiment** tab and begin the experiment. The experiment instructions are reprinted here for your reference. The opening screen for the experiment is shown on the following page.

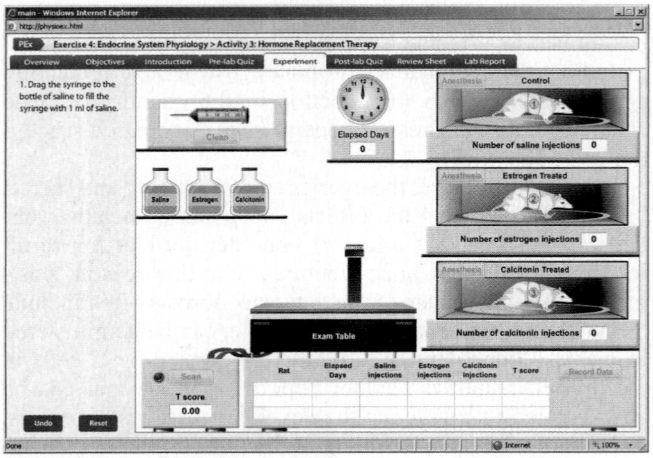

Windows Explorer 2007, Microsoft Corporation

1. Drag the syringe to the bottle of saline to fill the syringe with 1 ml of saline.

2. Drag the syringe to the *control* rat, placing the tip of the needle in the rat's lower abdominal area. Injections into this area are considered *intraperitoneal* and will quickly be circulated by the abdominal blood vessels.

3. Click **Clean** beneath the syringe holder to clean the syringe of all residues.

4. Drag the syringe to the bottle of estrogen to fill the syringe with 1 ml of estrogen.

5. Drag the syringe to the *estrogen-treated* rat, placing the tip of the needle in the rat's lower abdominal area.

6. Click **Clean** beneath the syringe holder to clean the syringe of all residues.

7. Drag the syringe to the bottle of calcitonin to fill the syringe with 1 ml of calcitonin.

8. Drag the syringe to the *calcitonin-treated* rat, placing the tip of the needle in the rat's lower abdominal area.

9. Click **Clean** beneath the syringe holder to clean the syringe of all residues.

10. Click the clock face to advance one day (24 hours).

11. Each rat must receive seven injections over the course of seven days (one injection per day). The remaining injections will be automated. Click the clock face to repeat the series of injections until you have injected each of the rats seven times.

PREDICT Question 1
What effect will the saline injections have on the control rat's vertebral bone density?

____

PREDICT Question 2
What effect will the estrogen injections have on the estrogen-treated rat's vertebral bone density?

____

PREDICT Question 3
What effect will the calcitonin injections have on the calcitonin-treated rat's vertebral bone density?

____

12. Click **Anesthesia** above the *control* rat's cage to immobilize the control rat with a gaseous anesthetic for X-ray scanning.

13. Drag the anesthetized rat to the exam table for X-ray scanning.

14. Click **Scan** to activate the scanner. The T score will appear in the T score display. Click **Record Data** to record your results in the grid (and record your results in Chart 3). The control rat will be automatically returned to its cage.

CHART 3 Hormone Replacement Therapy Results

Rat	T score

15. You will now obtain the T scores for the remaining rats. Perform these steps to obtain the T score for the *estrogen-treated* rat, then repeat these steps to obtain the T score for the *calcitonin-treated* rat.

- Click **Anesthesia** above the rat's cage to immobilize the rat with a gaseous anesthetic for X-ray scanning.
- Drag the anesthetized rat to the exam table for X-ray scanning.
- Click **Scan** to activate the scanner. The T score will appear in the T score display.
- Click **Record Data** to record your results in the grid (and record your results in Chart 3). The rat will be automatically returned to its cage.

After you complete the experiment, take the online **Post-lab Quiz** for Activity 3.

Activity Questions

1. Recently, hormone replacement therapy has been prominent in the popular press. Describe a hormone replacement therapy that you have seen in the news, and highlight its benefits, its potential risks, the reasons to continue and the reasons to discontinue its use.

____

____

2. In hormone replacement therapy, how is the hormone dose determined by the prescribing physician?

____

____

ACTIVITY 4

Measuring Cortisol and Adrenocorticotropic Hormone

OBJECTIVES

1. To understand the terms *cortisol, adrenocorticotropic hormone* (*ACTH*), *corticotropin-releasing hormone* (*CRH*), *Cushing's syndrome*, *iatrogenic*, *Cushing's disease*, and *Addison's disease.*
2. To understand how CRH controls ACTH secretion and ACTH controls cortisol secretion.
3. To understand how negative feedback mechanisms influence the levels of tropic CRH and ACTH.
4. To measure the blood levels of cortisol and ACTH in five patients and correlate these readings with symptoms and diagnoses.
5. To distinguish between Cushing's syndrome and Cushing's disease.

Introduction

Cortisol, a hormone secreted by the *adrenal cortex*, is important in the body's response to many kinds of stress. Cortisol release is stimulated by **adrenocorticotropic hormone (ACTH)**, a tropic hormone released by the anterior pituitary. A *tropic* hormone stimulates the secretion of another hormone. ACTH release, in turn, is stimulated by **corticotropin-releasing hormone (CRH)**, a tropic hormone from the hypothalamus. Increased levels of cortisol negatively feed back to inhibit the release of both ACTH and CRH.

Increased cortisol in the blood, or *hypercortisolism*, is referred to as **Cushing's syndrome** if the increase is caused by an adrenal gland tumor. Cushing's syndrome can also be **iatrogenic** (that is, physician induced). For example, physician-induced Cushing's syndrome can occur when glucocorticoid hormones, such as prednisone, are administered to treat rheumatoid arthritis, asthma, or lupus. Cushing's syndrome is often referred to as "steroid diabetes" because it results in hyperglycemia. In contrast, **Cushing's disease** is hypercortisolism caused by an anterior pituitary tumor. People with Cushing's disease exhibit increased levels of ACTH and cortisol.

Decreased cortisol in the blood, or *hypocortisolism*, can occur because of adrenal insufficiency. In primary adrenal insufficiency, also known as **Addison's disease**, the low cortisol is directly caused by gradual destruction of the adrenal cortex and ACTH levels are typically elevated as a compensatory effect. Secondary adrenal insufficiency also results in low levels of cortisol, usually caused by damage to the anterior pituitary. Therefore, the levels of ACTH are also low in secondary adrenal insufficiency.

As you can see, a variety of endocrine disorders can be related to both high and low levels of cortisol and ACTH. Table 4.1 summarizes these endocrine disorders.

TABLE 4.1 Cortisol and ACTH Disorders

	Cortisol level	ACTH level
Cushing's syndrome (primary hypercortisolism)	High	Low
Iatrogenic Cushing's syndrome	High	Low
Cushing's disease (secondary hypercortisolism)	High	High
Addison's disease (primary adrenal insufficiency)	Low	High
Secondary adrenal insufficiency (hypopituitarism)	Low	Low

EQUIPMENT USED The following equipment will be depicted on-screen: plasma samples from five patients; HPLC (high-performance liquid chromatography) column—used to quantitatively measure the amount of cortisol and ACTH in the patient samples; HPLC detector—provides the hormone concentration in the patient sample; reusable syringe—used to inject the patient samples into the HPLC injection port; HPLC injection port—used to inject the patient samples into the HPLC column.

Experiment Instructions

Go to the home page in the PhysioEx software and click **Exercise 4: Endocrine System Physiology**. Click **Activity 4: Measuring Cortisol and Adrenocorticotropic Hormone**, and take the online **Pre-lab Quiz** for Activity 4.

After you take the online Pre-lab Quiz, click the **Experiment** tab and begin the experiment. The experiment instructions are reprinted here for your reference. The opening screen for the experiment is shown below.

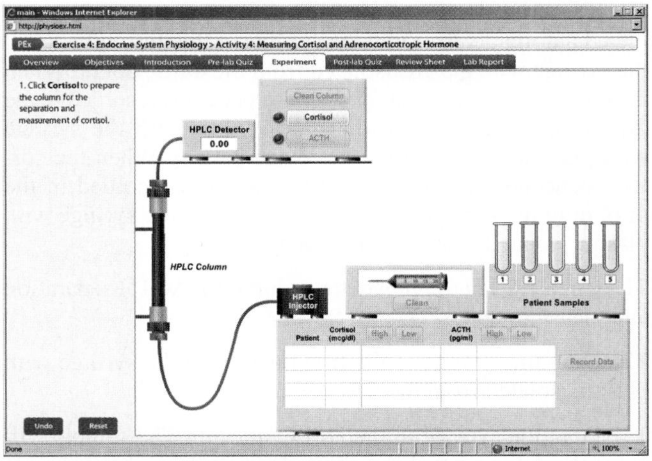

Windows Explorer 2007, Microsoft Corporation

1. Click **Cortisol** to prepare the column for the separation and measurement of cortisol.

2. Drag the syringe to the first tube to fill the syringe with plasma isolated from the first patient.

3. Drag the syringe to the HPLC injector. The sample will enter the tubing and flow through the column. The cortisol concentration in the patient sample will appear in the HPLC detector display.

4. Click **Record Data** to display your results in the grid (and record your results in Chart 4).

CHART 4 Measurement of Cortisol

Patient	Cortisol (mcg/dl)	Cortisol level	ACTH (pg/ml)	ACTH level
1				
2				
3				
4				
5				

5. Click **Clean** beneath the syringe to prepare it for the next sample. Click **Clean Column** to remove residual cortisol from the column.

6. Drag the syringe to the second tube to fill the syringe with plasma isolated from the second patient.

7. Drag the syringe to the HPLC injector. The sample will enter the tubing and flow through the column. The cortisol concentration in the patient sample will appear in the HPLC detector display.

8. Click **Record Data** to display your results in the grid (and record your results in Chart 4).

9. Click **Clean** beneath the syringe to prepare it for the next sample. Click **Clean Column** to remove residual cortisol from the column.

10. The procedure for the remaining samples will be completed automatically. Drag the syringe to the third tube to fill the syringe with plasma isolated from the third patient. When the cortisol concentration for the third patient is recorded in the grid, drag the syringe to the fourth tube to fill the syringe with plasma isolated from the fourth patient. When the cortisol concentration for the fourth patient is recorded in the grid, drag the syringe to the fifth tube to fill the syringe with plasma isolated from the fifth patient.

11. Click **ACTH** to prepare the column for ACTH separation and measurement.

12. Drag the syringe to the first tube to fill the syringe with plasma isolated from the first patient.

13. Drag the syringe to the HPLC injector. The sample will enter the tubing and flow through the column. The ACTH concentration in the patient sample will appear in the HPLC detector display.

14. Click **Record Data** to display your results in the grid.

15. Click **Clean** beneath the syringe to prepare it for the next sample. Click **Clean Column** to remove residual ACTH from the column.

16. Drag the syringe to the second tube to fill the syringe with plasma isolated from the second patient.

17. Drag the syringe to the HPLC injector. The sample will enter the tubing and flow through the column. The ACTH concentration in the patient sample will appear in the HPLC detector display.

18. Click **Record Data** to display your results in the grid (and record your results in Chart 4).

19. Click **Clean** beneath the syringe to prepare it for the next sample. Click **Clean Column** to remove residual ACTH from the column.

20. The procedure for the remaining samples will be completed automatically. Drag the syringe to the third tube to fill the syringe with plasma isolated from the third patient. When the ACTH concentration for the third patient is recorded in the grid, drag the syringe to the fourth tube to fill the syringe with plasma isolated from the fourth patient. When the ACTH concentration for the fourth patient is recorded in the grid, drag the syringe to the fifth tube to fill the syringe with plasma isolated from the fifth patient.

21. Indicate whether the cortisol and ACTH concentrations (levels) for each patient are high or low using the breakpoints shown in Table 4.2. Click the row of the patient and then click **High** or **Low** next to cortisol and ACTH.

TABLE 4.2 Abnormal Morning Cortisol and ACTH Levels

ACTH level	High	Low
Cortisol	≥23 mcg/dl	<5 mcg/dl
ACTH	≥80 pg/ml	<20 pg/ml

Note: 1 mcg = 1 μg = 1 microgram

After you complete the experiment, take the online **Post-lab Quiz** for Activity 4.

Activity Questions

1. Discuss the benefits and drawbacks of giving glucocorticoids to young children that have significant allergy-induced asthma.

____

____

2. Explain the difference between Cushing's syndrome and Cushing's disease.

____

____

NAME ______________________________

LAB TIME/DATE ______________________

Endocrine System Physiology

ACTIVITY 1 Metabolism and Thyroid Hormone

Part 1

1. Which rat had the fastest basal metabolic rate (BMR)? ______________________________

2. Why did the metabolic rates differ between the normal rat and the surgically altered rats? How well did the results compare with your prediction? ______________________________

3. If an animal has been thyroidectomized, what hormone(s) would be missing in its blood? ______________________________

4. If an animal has been hypophysectomized, what effect would you expect to see in the hormone levels in its body? ______________________________

Part 2

5. What was the effect of thyroxine injections on the normal rat's BMR? ______________________________

6. What was the effect of thyroxine injections on the thyroidectomized rat's BMR? How does the BMR in this case compare with the normal rat's BMR? Was the dose of thyroxine in the syringe too large, too small, or just right? ______________________________

7. What was the effect of thyroxine injections on the hypophysectomized rat's BMR? How does the BMR in this case compare with the normal rat's BMR? Was the dose of thyroxine in the syringe too large, too small, or just right? ______________

____

____

Part 3

8. What was the effect of thyroid-stimulating hormone (TSH) injections on the normal rat's BMR? ______________

____

____

9. What was the effect of TSH injections on the thyroidectomized rat's BMR? How does the BMR in this case compare with the normal rat's BMR? Why was this effect observed? ______________

____

10. What was the effect of TSH injections on the hypophysectomized rat's BMR? How does the BMR in this case compare with the normal rat's BMR? Was the dose of TSH in the syringe too large, too small, or just right? ______________

____

____

Part 4

11. What was the effect of propylthiouracil (PTU) injections on the normal rat's BMR? Why did this rat develop a palpable goiter?

____

____

12. What was the effect of PTU injections on the thyroidectomized rat's BMR? How does the BMR in this case compare with the normal rat's BMR? Why was this effect observed? ______________

____

13. What was the effect of PTU injections on the hypophysectomized rat's BMR? How does the BMR in this case compare with the normal rat's BMR? Why was this effect observed? ______________

____

____

ACTIVITY 2 Plasma Glucose, Insulin, and Diabetes Mellitus

1. What is a glucose standard curve, and why did you need to obtain one for this experiment? Did you correctly predict how you would measure the amount of plasma glucose in a patient sample using the glucose standard curve? ______________

____

____

2. Which patient(s) had glucose reading(s) in the diabetic range? Can you say with certainty whether each of these patients has type 1 or type 2 diabetes? Why or why not? ______________

____

____

3. Describe the diagnosis for patient 3, who was also pregnant at the time of this assay. ______________

____

____

4. Which patient(s) had normal glucose reading(s)? ______________

____

5. What are some lifestyle choices these patients with normal plasma glucose readings might recommend to the borderline impaired patients? ______________

____

____

ACTIVITY 3 Hormone Replacement Therapy

1. Why were ovariectomized rats used in this experiment? How does the fact that the rats are ovariectomized explain their baseline T scores? ______________

____

____

2. What effect did the administration of saline injections have on the control rat? How well did the results compare with your prediction? ______________

____

____

3. What effect did the administration of estrogen injections have on the estrogen-treated rat? How well did the results compare with your prediction? ____

____

____

4. What effect did the administration of calcitonin injections have on the calcitonin-treated rat? How well did the results compare with your prediction? ____

____

____

5. What are some health risks that postmenopausal women must consider when contemplating estrogen hormone replacement therapy? ____

____

____

ACTIVITY 4 Measuring Cortisol and Adrenocorticotropic Hormone

1. Which patient would most likely be diagnosed with Cushing's disease? Why? ______________________________

____

____

2. Which two patients have hormone levels characteristic of Cushing's syndrome? ______________________________

____

____

3. Patient 2 is being treated for rheumatoid arthritis with prednisone. How does this information change the diagnosis? ______

____

____

4. Which patient would most likely be diagnosed with Addison's disease? Why? ______________________________

____

____

Cardiovascular Dynamics

PRE-LAB QUIZ

1. Circle the correct underlined term: Flow rate in a pipe is directly / inversely proportional to the pipe's resistance.
2. Total blood flow is proportional to cardiac output, the:
   a. amount of blood the heart pumps per minute
   b. amount of blood the heart pumps in a day
   c. amount of blood moving through the entire body area in a set time
   d. amount of blood added to the flow rate
3. Circle True or False: The longer the blood vessel length, the greater the resistance.
4. Blood vessel radius is controlled by:
   a. skeletal muscle contraction outside the vessel walls
   b. smooth muscle contraction or relaxation within the blood vessels
   c. skeletal muscle relaxation outside of the blood vessels
   d. percentage of red blood cells as compared to blood volume
5. Circle the correct underlined term: The more viscous a fluid, the more / less resistance it has to flow.
6. The volume remaining in the ventricles at the end of diastole, immediately before cardiac contraction, is called ________.
   a. stroke volume
   b. end systolic volume
   c. residual cardiac volume
   d. end diastolic volume
7. Circle the correct underlined term: Systole / Diastole occurs as the heart chambers relax and fill with blood.
8. In the human body, the heart beats about _____ strokes per minute.
9. The degree to which the ventricles are stretched by the end diastolic volume is known as ____________.
10. A condition that can lead to heart disease, ______________, is a result of plaque buildup in the arteries.
   a. hypotension
   b. sickle cell anemia
   c. atherosclerosis
   d. polycythemia

Exercise Overview

The cardiovascular system is composed of a pump—the heart—and blood vessels that distribute blood containing oxygen and nutrients to every cell of the body. The principles governing blood flow are the same physical laws that apply to the flow of liquid through a system of pipes. For example, one very basic law in fluid mechanics is that the flow rate of a liquid through a pipe is directly proportional to the difference between the pressures at the two ends of the pipe (the **pressure gradient**) and inversely proportional to the pipe's **resistance** (a measure of the degree to which the pipe hinders, or resists, the flow of the liquid).

$$\text{Flow} = \text{pressure gradient/resistance} = \Delta P/R$$

This basic law also applies to blood flow. The "liquid" is blood, and the "pipes" are blood vessels. The pressure gradient is the difference between the pressure in arteries and the pressure in veins that results when blood is pumped into arteries. Blood flow rate is directly proportional to the pressure gradient and inversely proportional to resistance.

Blood flow is the amount of blood moving through a body area or the entire cardiovascular system in a given amount of time. Total blood flow is proportional to **cardiac output** (the amount of blood the heart is able to pump per minute). Blood flow to specific body areas can vary dramatically in a given time period. Organs differ in their requirements from moment to moment, and blood vessels have different-sized diameters in their lumen (opening) to regulate local blood flow to various areas in response to the tissues' immediate needs. Consequently, blood flow can increase to some areas and decrease to other areas at the same time.

Resistance is a measure of the degree to which the blood vessel hinders, or resists, the flow of blood. The main factors that affect resistance are (1) blood vessel *radius*, (2) blood vessel *length*, and (3) blood *viscosity*.

Radius

The smaller the blood vessel radius, the greater the resistance, because of frictional drag between the blood and the vessel walls. Contraction of smooth muscle of the blood vessel, or **vasoconstriction**, results in a decrease in the blood vessel radius. Lipid deposits can also cause the radius of an artery to decrease, preventing blood from reaching the coronary tissue, which frequently leads to a heart attack. Alternately, relaxation of smooth muscle of the blood vessel, or **vasodilation**, causes an increase in the blood vessel radius. Blood vessel radius is the single most important factor in determining blood flow resistance.

Length

The longer the vessel length, the greater the resistance—again, because of friction between the blood and vessel walls. The length of a person's blood vessels change only as a person grows. Otherwise, the length generally remains constant.

Viscosity

Viscosity is blood "thickness," determined primarily by **hematocrit**—the fractional contribution of red blood cells to total blood volume. The higher the hematocrit, the greater the viscosity. Under most physiological conditions, hematocrit does not vary much and blood viscosity remains more or less constant.

The Effect of Blood Pressure and Vessel Resistance on Blood Flow

Blood flow is directly proportional to blood pressure because the pressure difference (ΔP) between the two ends of a vessel is the driving force for blood flow. Peripheral resistance is the friction that opposes blood flow through a blood vessel. This relationship is represented in the following equation:

$$\text{Blood flow (ml/min)} = \frac{\Delta P}{\text{peripheral resistance}}$$

Three factors that contribute to peripheral resistance are blood viscosity (η), blood vessel length (L), and the radius of the blood vessel (r). These relationships are expressed in the following equation:

$$\text{Peripheral resistance} = \frac{8L\eta}{\pi r^4}$$

From this equation you can see that the viscosity of the blood and the length of the blood vessel are directly proportional to peripheral resistance. The peripheral resistance is inversely proportional to the fourth power of the vessel radius. If you combine the two equations, you get the following result:

$$\text{Blood flow (ml/min)} = \frac{\Delta P \pi r^4}{8L\eta}$$

From this combination you can see that blood flow is directly proportional to the fourth power of vessel radius, which means that small changes in vessel radius result in dramatic changes in blood flow.

ACTIVITY 1

Studying the Effect of Blood Vessel Radius on Blood Flow Rate

OBJECTIVES

1. To understand how blood vessel radius affects blood flow rate.
2. To understand how vessel radius is changed in the body.
3. To understand how to interpret a graph of blood vessel radius versus blood flow rate.

Introduction

Controlling **blood vessel radius** (one-half of the diameter) is the principal method of controlling blood flow. Controlling blood vessel radius is accomplished by contracting or relaxing the smooth muscle within the blood vessel walls (vasoconstriction or vasodilation).

To understand why radius has such a pronounced effect on blood flow, consider the physical relationship between blood and the vessel wall. Blood in direct contact with the vessel wall flows relatively slowly because of the friction, or drag, between the blood and the lining of the vessel. In contrast, blood in the center of the vessel flows more freely because it is not rubbing against the vessel wall. The free-flowing blood in the middle of the vessel is called the **laminar flow**. Now picture a fully constricted (small-radius) vessel and a fully dilated (large-radius) vessel. In the fully constricted vessel, proportionately more blood is in contact with the vessel wall and there is less laminar flow, significantly impeding the rate of blood flow in the fully constricted vessel relative to that in the fully dilated vessel.

In this activity you will study the effect of blood vessel radius on blood flow. The experiment includes two glass beakers and a tube connecting them. Imagine that the left beaker is your heart, the tube is an artery, and the right beaker is a destination in your body, such as another organ.

EQUIPMENT USED The following equipment will be depicted on-screen: left beaker—simulates blood flowing from the heart; flow tube between the left and right beaker—simulates an artery; right beaker—simulates another organ (for example, the biceps brachii muscle).

Experiment Instructions

Go to the home page in the PhysioEx software and click **Exercise 5: Cardiovascular Dynamics**. Click **Activity 1: Studying the Effect of Blood Vessel Radius on Blood Flow Rate**, and take the online **Pre-lab Quiz** for Activity 1.

After you take the online Pre-lab Quiz, click the **Experiment** tab and begin the experiment. The experiment instructions are reprinted here for your reference. The opening screen for the experiment is shown below.

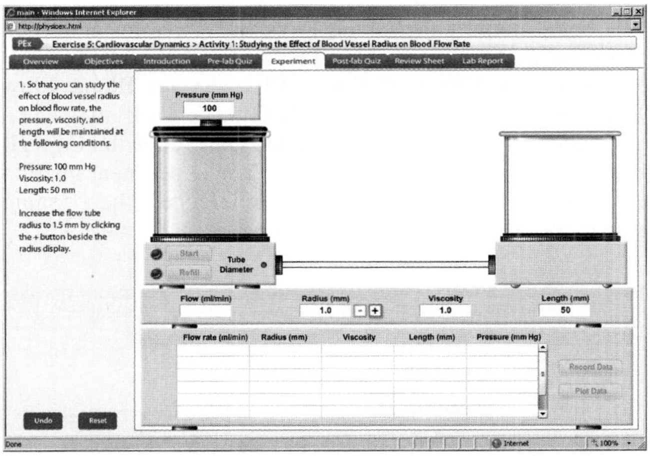

Windows Explorer 2007, Microsoft Corporation

1. So that you can study the effect of blood vessel radius on blood flow rate, the pressure, viscosity, and length will be maintained at the following conditions:

Pressure: 100 mm Hg

Viscosity: 1.0

Length: 50 mm

Increase the flow tube radius to 1.5 mm by clicking the + button beside the radius display.

2. Click **Start** and then watch the fluid move into the right beaker. (Fluid moves slowly under some conditions—be patient!) Pressure propels fluid from the left beaker to the right beaker through the flow tube. The flow rate is shown in the flow rate display after the left beaker has finished draining.

3. Click **Record Data** to display your results in the grid (and record your results in Chart 1).

4. Click **Refill** to replenish the left beaker.

? PREDICT Question 1
What do you think will happen to the flow rate if the radius is increased by 0.5 mm?

____

CHART 1 Effect of Blood Vessel Radius on Blood Flow Rate

Flow (ml/min)	Radius (mm)

5. Increase the flow tube radius to 2.0 mm by clicking the + button beside the radius display. Click **Start** and watch the fluid move into the right beaker.

6. Click **Record Data** to display your results in the grid (and record your results in Chart 1).

7. Click **Refill** to replenish the left beaker.

8. You will now observe the effect of incremental increases in flow tube radius.

- Increase the flow tube radius by 0.5 mm.
- Click **Start** and then watch the fluid move into the right beaker.
- Click **Record Data** to display your results in the grid (and record your results in Chart 1).
- Click **Refill** to replenish the left beaker.

Repeat this step until you reach a flow tube radius of 5.0 mm.

? PREDICT Question 2
Do you think a graph plotted with radius on the X-axis and flow rate on the Y-axis will be linear (a straight line)?

____

9. Click **Plot Data** to view a summary of your data on a plotted grid. Radius will be displayed on the X-axis and flow rate will be displayed on the Y-axis. Click **Submit** to record your plot in the lab report.

After you complete the experiment, take the online **Post-lab Quiz** for Activity 1.

Activity Questions

1. Describe the relationship between vessel radius and blood flow rate.

____

____

2. In this activity you altered the radius of the flow tube by clicking the + and − buttons. Explain how and why the radius of blood vessels is altered in the human body.

3. Describe the appearance of your plot of blood vessel radius versus blood flow rate and relate the plot to the relationship between these two variables.

4. Describe an advantage of slower blood velocity in some areas of the body, for example, in the capillaries of our fingers.

ACTIVITY 2

Studying the Effect of Blood Viscosity on Blood Flow Rate

OBJECTIVES

1. To understand how blood viscosity affects blood flow rate.
2. To list the components in the blood that contribute to blood viscosity.
3. To explain conditions that might lead to viscosity changes in the blood.
4. To understand how to interpret a graph of viscosity versus blood flow.

Introduction

Viscosity is the thickness, or "stickiness," of a fluid. The more viscous a fluid, the more resistance to flow. Therefore, the flow rate will be slower for a more viscous solution. For example, consider how much more slowly maple syrup pours out of a container than milk does.

The viscosity of blood is due to the presence of plasma proteins and formed elements, which include white blood cells (leukocytes), red blood cells (erythrocytes), and platelets (thrombocytes). Formed elements and plasma proteins in the blood slide past one another, increasing the resistance to flow. With a viscosity of 3–5, blood is much more viscous than water (usually given a viscosity value of 1).

A body in homeostatic balance has a relatively stable blood consistency. Nevertheless, it is useful to examine the effects of blood viscosity on blood flow to predict what might occur in the human cardiovascular system when homeostatic imbalances occur. Factors such as dehydration and altered blood cell numbers do alter blood viscosity. For example, polycythemia is a condition in which excess red blood cells are present, and certain types of anemia result in fewer red blood cells. Increasing the number of red blood cells increases blood viscosity, and decreasing the number of red blood cells decreases blood viscosity.

In this activity you will examine the effects of blood viscosity on blood flow rate. The experiment includes two glass beakers and a tube connecting them. Imagine that the left beaker is your heart, the tube is an artery, and the right beaker is a destination in your body, such as another organ.

EQUIPMENT USED The following equipment will be depicted on-screen: left beaker—simulates blood flowing from the heart; flow tube between the left and right beaker—simulates an artery; right beaker—simulates another organ (for example, the biceps brachii muscle).

Experiment Instructions

Go to the home page in the PhysioEx software and click **Exercise 5: Cardiovascular Dynamics**. Click **Activity 2: Studying the Effect of Blood Viscosity on Blood Flow Rate**, and take the online **Pre-lab Quiz** for Activity 2.

After you take the online Pre-lab Quiz, click the **Experiment** tab and begin the experiment. The experiment instructions are reprinted here for your reference. The opening screen for the experiment is shown below.

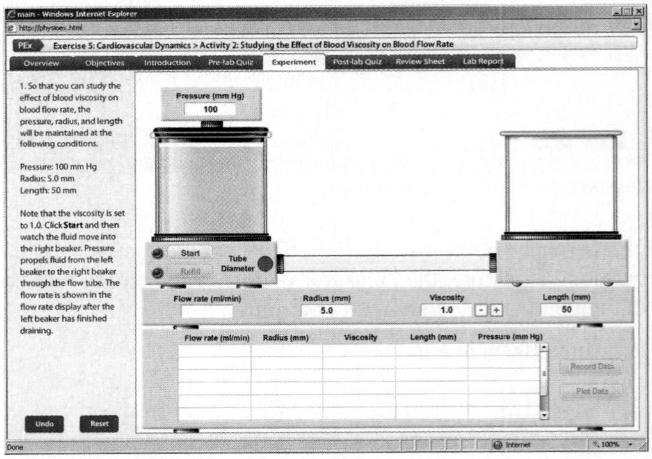

Windows Explorer 2007, Microsoft Corporation

1. So that you can study the effect of blood viscosity on blood flow rate, the pressure, radius, and length will be maintained at the following conditions:

Pressure: 100 mm Hg

Radius: 5.0 mm

Length: 50 mm

Note that the viscosity is set to 1.0. Click **Start** and then watch the fluid move into the right beaker. Pressure propels fluid from the left beaker to the right beaker through the flow tube. The flow rate is shown in the flow rate display after the left beaker has finished draining.

2. Click **Record Data** to display your results in the grid (and record your results in Chart 2).

CHART 2 Effect of Blood Viscosity on Blood Flow Rate

Flow (ml/min)	Viscosity

3. Click **Refill** to replenish the left beaker.

? PREDICT **Question 1**
What effect do you think increasing the viscosity will have on the fluid flow rate?

4. Increase the fluid viscosity to 2.0 by clicking the + button beside the viscosity display. Click **Start** and then watch the fluid move into the right beaker.

5. Click **Record Data** to display your results in the grid (and record your results in Chart 2).

6. Click **Refill** to replenish the left beaker.

7. You will now observe the effect of incremental increases in viscosity.
   - Increase the viscosity by 1.0.
   - Click **Start** and then watch the fluid move into the right beaker.
   - Click **Record Data** to display your results in the grid (and record your results in Chart 2).
   - Click **Refill** to replenish the left beaker.

   Repeat this step until you reach a viscosity of 8.0.

8. Click **Plot Data** to view a summary of your data on a plotted grid. Viscosity will be displayed on the X-axis and flow rate will be displayed on the Y-axis. Click **Submit** to record your plot in the lab report.

After you complete the experiment, take the online **Post-lab Quiz** for Activity 2.

Activity Questions

1. Describe the effect on blood flow rate when blood viscosity was increased.

2. Explain why the relationship between viscosity and blood flow rate is inversely proportional.

3. What might happen to blood flow if you increased the number of blood cells?

ACTIVITY 3

Studying the Effect of Blood Vessel Length on Blood Flow Rate

OBJECTIVES

1. To understand how blood vessel length affects blood flow rate.
2. To explain conditions that can lead to blood vessel length changes in the body.
3. To compare the effect of blood vessel length changes with the effect of blood vessel radius changes on blood flow rate.

Introduction

Blood vessel lengths increase as we grow to maturity. The longer the vessel, the greater the resistance to blood flow through the blood vessel because there is a larger surface area in contact with the blood cells. Therefore, when blood vessel length increases, friction increases. Our blood vessel lengths stay fairly constant in adulthood, unless we gain or lose weight. If we gain weight, blood vessel lengths can increase, and if we lose weight, blood vessel lengths can decrease.

In this activity you will study the physical relationship between blood vessel length and blood flow. Specifically, you will study how blood flow changes in blood vessels of constant radius but different lengths. The experiment includes two glass beakers and a tube connecting them. Imagine that the left beaker is your heart, the tube is an artery, and the right beaker is a destination in your body, such as another organ.

EQUIPMENT USED The following equipment will be depicted on-screen: left beaker—simulates blood flowing from the heart; flow tube between the left and right beaker—simulates an artery; right beaker—simulates another organ (for example, the biceps brachii muscle).

Experiment Instructions

Go to the home page in the PhysioEx software and click **Exercise 5: Cardiovascular Dynamics**. Click **Activity 3: Studying the Effect of Blood Vessel Length on Blood Flow Rate**, and take the online **Pre-lab Quiz** for Activity 3.

After you take the online Pre-lab Quiz, click the **Experiment** tab and begin the experiment. The experiment instructions are reprinted here for your reference. The opening screen for the experiment is shown on the following page.

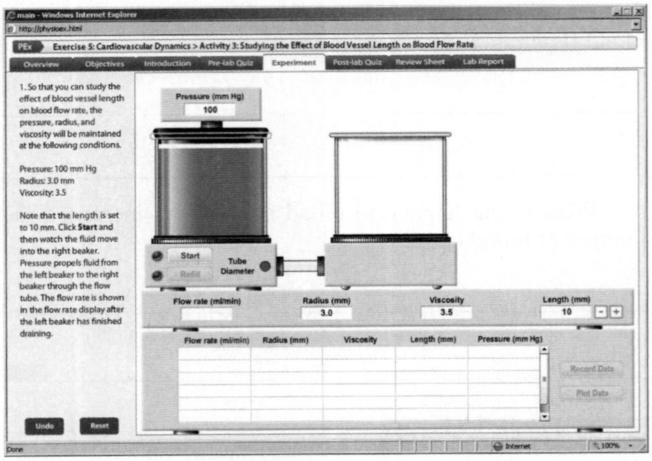

Windows Explorer 2007, Microsoft Corporation

1. So that you can study the effect of blood vessel length on blood flow rate, the pressure, radius, and viscosity will be maintained at the following conditions:

Pressure: 100 mm Hg

Radius: 3.0 mm

Viscosity: 3.5

Note that the length is set to 10 mm. Click **Start** and then watch the fluid move into the right beaker. Pressure propels fluid from the left beaker to the right beaker through the flow tube. The flow rate is shown in the flow rate display after the left beaker has finished draining.

2. Click **Record Data** to display your results in the grid (and record your results in Chart 3).

CHART 3 Effect of Blood Vessel Length on Blood Flow Rate

Flow (ml/min)	Flow Tube length (mm)

3. Click **Refill** to replenish the left beaker.

PREDICT Question 1
What effect do you think increasing the flow tube length will have on the fluid flow rate?

____

4. Increase the flow tube length to 15 mm by clicking the + button beside the length display. Click **Start** and then watch the fluid move into the right beaker.

5. Click **Record Data** to display your results in the grid (and record your results in Chart 3).

6. Click **Refill** to replenish the left beaker.

7. You will now observe the effect of incremental increases in flow tube length.

- Increase the flow tube length by 5 mm.
- Click **Start** and then watch the fluid move into the right beaker.
- Click **Record Data** to display your results in the grid (and record your results in Chart 3).
- Click **Refill** to replenish the left beaker.

Repeat this step until you reach a flow tube length of 40 mm.

8. Click **Plot Data** to view a summary of your data on a plotted grid. Length will be displayed on the X-axis and flow rate will be displayed on the Y-axis. Click **Submit** to record your plot in the lab report.

After you complete the experiment, take the online **Post-lab Quiz** for Activity 3.

Activity Questions

1. Is the relationship between blood vessel length and blood flow rate directly proportional or inversely proportional? Why?

____

____

2. Which of the following can vary in size more quickly: blood vessel diameter or blood vessel length?

____

____

3. Describe what happens to resistance when blood vessel length increases.

____

____

ACTIVITY 4

Studying the Effect of Blood Pressure on Blood Flow Rate

OBJECTIVES

1. To understand how blood pressure affects blood flow rate.
2. To understand what structure produces blood pressure in the human body.
3. To compare the plot generated for pressure versus blood flow to those generated for radius, viscosity, and length.

Introduction

The pressure difference between the two ends of a blood vessel is the driving force behind blood flow. This pressure difference is referred to as a pressure gradient. In the cardiovascular system, the force of contraction of the heart provides the initial pressure and vascular resistance contributes to the pressure gradient. If the heart changes its force of contraction, the blood vessels need to be able to respond to the change in force. Large arteries close to the heart have more elastic tissue in their tunics in order to accommodate these changes.

In this activity you will look at the effect of pressure changes on blood flow (recall from the blood flow equation that a change in blood flow is directly proportional to the pressure gradient). The experiment includes two glass beakers and a tube connecting them. Imagine that the left beaker is your heart, the tube is an artery, and the right beaker is a destination in your body, such as another organ.

EQUIPMENT USED The following equipment will be depicted on-screen: left beaker—simulates blood flowing from the heart; flow tube between the left and right beaker—simulates an artery; right beaker—simulates another organ (for example, the biceps brachii muscle).

Experiment Instructions

Go to the home page in the PhysioEx software and click **Exercise 5: Cardiovascular Dynamics**. Click **Activity 4: Studying the Effect of Blood Pressure on Blood Flow Rate**, and take the online **Pre-lab Quiz** for Activity 4.

After you take the online Pre-lab Quiz, click the **Experiment** tab and begin the experiment. The experiment instructions are reprinted here for your reference. The opening screen for the experiment is shown below.

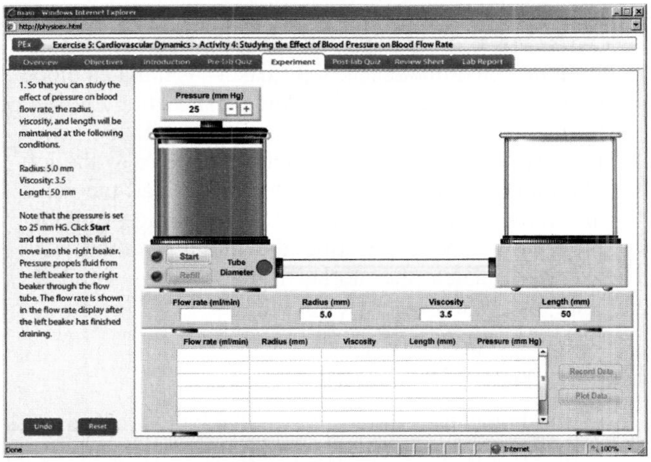

Windows Explorer 2007, Microsoft Corporation

1. So that you can study the effect of pressure on blood flow rate, the radius, viscosity, and length will be maintained at the following conditions:

Radius: 5.0 mm

Viscosity: 3.5

Length: 50 mm

Note that the pressure is set to 25 mm Hg. Click **Start** and then watch the fluid move into the right beaker. Pressure propels fluid from the left beaker to the right beaker through the flow tube. The flow rate is shown in the flow rate display after the left beaker has finished draining.

2. Click **Record Data** to record your results in the grid (and record your results in Chart 4).

CHART 4 Effect of Blood Pressure on Blood Flow Rate

Flow (ml/min)	Pressure (mm Hg)

3. Click **Refill** to replenish the left beaker.

PREDICT Question 1
What effect do you think increasing the pressure will have on the fluid flow rate?

____

4. Increase the pressure to 50 mm Hg by clicking the + button beside the pressure display. Click **Start** and then watch the fluid move into the right beaker.

5. Click **Record Data** to record your results in the grid (and record your results in Chart 4).

6. Click **Refill** to replenish the left beaker.

7. You will now observe the effect of incremental increases in pressure.

- Increase the pressure by 25 mm Hg.
- Click **Start** and then watch the fluid move into the right beaker.
- Click **Record Data** to display your results in the grid (and record your results in Chart 4).
- Click **Refill** to replenish the left beaker.

Repeat this step until you reach a pressure of 200 mm Hg.

PREDICT Question 2
Do you think a graph plotted with pressure on the X-axis and flow rate on the Y-axis will be linear (a straight line)?

____

8. Click **Plot Data** to view a summary of your data on a plotted grid. Pressure will be displayed on the X-axis and flow rate will be displayed on the Y-axis. Click **Submit** to record your plot in the lab report.

After you complete the experiment, take the online **Post-lab Quiz** for Activity 4.

Activity Questions

1. How does increasing the driving pressure affect the blood flow rate?

____

____

2. Is the relationship between blood pressure and blood flow rate directly proportional or inversely proportional? Why?

____

____

3. How does the cardiovascular system increase pressure?

____

____

4. Although changing blood pressure can be used to alter the blood flow rate, this approach causes problems if it continues indefinitely. Explain why.

____

____

ACTIVITY 5

Studying the Effect of Blood Vessel Radius on Pump Activity

OBJECTIVES

1. To understand the terms *systole* and *diastole*.
2. To predict how a change in blood vessel radius will affect flow rate.
3. To predict how a change in blood vessel radius will affect heart rate.
4. To observe the compensatory mechanisms for maintaining blood pressure.

Introduction

In the human body, the heart beats approximately 70 strokes each minute. Each heartbeat consists of a filling interval, when blood moves into the chambers of the heart, and an ejection period, when blood is actively pumped into the aorta and the pulmonary trunk.

The pumping activity of the heart can be described in terms of the phases of the cardiac cycle. Heart chambers fill during **diastole** (relaxation of the heart) and pump blood out during **systole** (contraction of the heart). As you can imagine, the length of time the heart is relaxed is one factor that determines the amount of blood within the heart at the end of the filling interval. Up to a point, increasing ventricular filling time results in a corresponding increase in ventricular volume. The volume in the ventricles at the end of diastole, just before cardiac contraction, is called the **end diastolic volume**, or **EDV**. The volume ejected by a single ventricular contraction is the **stroke volume**, and the volume remaining in the ventricle after contraction is the **end systolic volume**, or **ESV**.

The human heart is a complex, four-chambered organ consisting of two individual pumps (the right and left sides). The right side of the heart pumps blood through the lungs into the left side of the heart. The left side of the heart, in turn, delivers blood to the systems of the body. Blood then returns to the right side of the heart to complete the circuit.

Recall that cardiac output **(CO)** is equal to blood flow. To determine CO, you multiply heart rate (HR) by stroke volume (SV): $CO = HR \times SV$. From the equation for flow ($flow = \Delta P/R$), you can determine the equation for blood pressure: $\Delta P = flow \times R$. Substituting CO in the equation for flow, you get: $\Delta P = HR \times SV \times R$.

Therefore, to maintain blood pressure, the cardiovascular system can alter heart rate, stroke volume, or resistance. For example, if resistance decreases, heart rate can increase to maintain the pressure difference.

In this activity you will explore the operation of a simple, one-chambered pump and apply the physical concepts in the experiment to the operation of either of the two pumps of the human heart. The stroke volume and the difference in pressure will remain constant. You will explore the effect that a change in resistance has on heart rate and the compensatory mechanisms that the cardiovascular system uses to maintain blood pressure.

EQUIPMENT USED The following equipment will be depicted on-screen: left beaker—simulates blood coming from the lungs; flow tube connecting the left beaker and the pump—simulates the pulmonary veins; pump—simulates the left ventricle (the valve to the left of the pump simulates the bicuspid valve, and the valve to the right of the pump simulates the aortic semilunar valve); flow tube connecting the pump and the right beaker—simulates the aorta; right beaker—simulates blood going to the systemic circuit.

Experiment Instructions

Go to the home page in the PhysioEx software and click **Exercise 5: Cardiovascular Dynamics**. Click **Activity 5: Compensation: Studying the Effect of Blood Vessel Radius on Pump Activity** and take the online **Pre-lab Quiz** for Activity 5.

After you take the online Pre-lab Quiz, click the **Experiment** tab and begin the experiment. The experiment instructions are reprinted here for your reference. The opening screen for the experiment is shown on the following page.

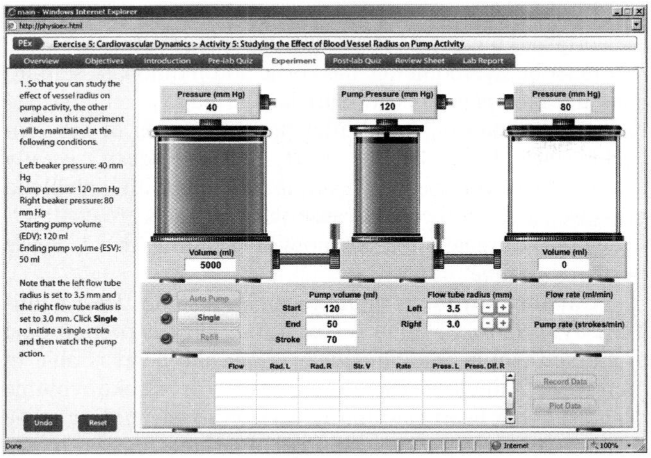

Windows Explorer 2007, Microsoft Corporation

1. So that you can study the effect of vessel radius on pump activity, the other variables in this experiment will be maintained at the following conditions:

Left beaker pressure: 40 mm Hg

Pump pressure: 120 mm Hg

Right beaker pressure: 80 mm Hg

Starting pump volume (EDV): 120 ml

Ending pump volume (ESV): 50 ml

Note that the left flow tube radius is set to 3.5 mm and the right flow tube radius is set to 3.0 mm. Click **Single** to initiate a single stroke and then watch the pump action.

2. Click **Auto Pump** to initiate 10 strokes and then watch the pump action. The flow rate is shown in the flow rate display and the pump rate is shown in the pump rate display after the left beaker has finished draining.

3. Click **Record Data** to display your results in the grid (and record your results in Chart 5).

CHART 5 Effect of Blood Vessel Radius on Pump Activity

Flow rate (ml/min)	Right radius (mm)	Pump rate (strokes/min)

4. Click **Refill** to replenish the left beaker.

PREDICT Question 1
If you increase the flow tube radius, what will happen to the pump rate to maintain constant pressure?

____

5. Increase the right flow tube radius to 3.5 mm by clicking the + button beside the right flow tube radius display. Click **Auto Pump** to initiate 10 strokes and then watch the pump action.

6. Click **Record Data** to display your results in the grid (and record your results in Chart 5).

7. Click **Refill** to replenish the left beaker.

8. You will now observe the effect of incremental increases in the right flow tube radius.

- Increase the right flow tube radius by 0.5 mm.
- Click **Auto Pump** to initiate 10 strokes and then watch the pump action.
- Click **Record Data** to display your results in the grid (and record your results in Chart 5).
- Click **Refill** to replenish the left beaker.

Repeat this step until you reach a right flow tube radius of 5.0 mm.

9. Click **Plot Data** to view a summary of your data on a plotted grid. Right flow tube radius will be displayed on the X-axis and flow rate will be displayed on the Y-axis. Click **Submit** to record your plot in the lab report.

After you complete the experiment, take the online **Post-lab Quiz** for Activity 5.

Activity Questions

1. Describe the position of the pump during diastole.

____

____

2. Describe the position of the pump during systole.

____

____

3. Describe what happened to the flow rate when the blood vessel radius was increased.

____

____

4. Explain what happened to the resistance and the pump rate to maintain pressure when the radius was increased.

____

____

ACTIVITY 6

Studying the Effect of Stroke Volume on Pump Activity

OBJECTIVES

1. To understand the effect a change in venous return has on stroke volume.
2. To explain how stroke volume is changed in the heart.
3. To explain the Frank-Starling law of the heart.
4. To define *preload*, *contractility*, and *afterload*.
5. To distinguish between intrinsic and extrinsic control of contractility of the heart.
6. To explore how heart rate and stroke volume contribute to cardiac output and blood flow.

Introduction

In a normal individual, 60% of the blood contained within the heart is ejected from the heart during ventricular systole, leaving 40% of the blood behind. The blood ejected by the heart—the **stroke volume**—is the difference between the **end diastolic volume (EDV)**, the volume in the ventricles at the end of diastole, just before cardiac contraction, and **end systolic volume (ESV)**, the volume remaining in the ventricle after contraction. That is, stroke volume = EDV − ESV. Many factors affect stroke volume, the most important of which include *preload*, *contractility*, and *afterload*. We will look at these defining factors and how they relate to stroke volume.

The Frank-Starling law of the heart states that, when more than the normal volume of blood is returned to the heart by the venous system, the heart muscle will be stretched, resulting in a more forceful contraction of the ventricles. This, in turn, will cause more than normal blood to be ejected by the heart, raising the stroke volume. The degree to which the ventricles are stretched by the end diastolic volume (EDV) is referred to as the **preload**. Thus, the preload results from the amount of ventricular filling between strokes, or the magnitude of the EDV. Ventricular filling could increase when the heart rate is slow because there will be more time for the ventricles to fill. Exercise increases venous return and, therefore, EDV. Factors such as severe blood loss and dehydration decrease venous return and EDV.

The **contractility** of the heart refers to strength of the cardiac muscle contraction (usually the ventricles) and its ability to generate force. A number of extrinsic mechanisms, including the sympathetic nervous system and hormones, control the force of cardiac muscle contraction, but they are not the focus of this activity. The focus of this activity will be the intrinsic controls of contractility (those that reside entirely within the heart). When the end diastolic volume increases, the cardiac muscle fibers of the ventricles stretch and lengthen. As the length of the cardiac sarcomere increases, so does the force of contraction. Cardiac muscle, like skeletal muscle, demonstrates a **length-tension relationship**. At rest, cardiac muscles are at a less than optimum overlap length for maximum tension production in the healthy heart. Therefore, when the heart experiences an increase in stretch with an increase in venous return and, therefore, EDV, it can respond by increasing the force of contraction, yielding a corresponding increase in stroke volume.

Afterload is the back pressure generated by the blood in the aorta and the pulmonary trunk. Afterload is the threshold that must be overcome for the aortic and pulmonary semilunar valves to open. This pressure is referred to as an *after*load because the load is placed after the contraction of the ventricles starts. In the healthy heart, afterload doesn't greatly change stroke volume. However, individuals with high blood pressure can be affected because the ventricles are contracting against a greater pressure, possibly resulting in a decrease in stroke volume.

Cardiac output is equal to the heart rate (HR) multiplied by the stroke volume. Total blood flow is proportional to cardiac output (the amount of blood the heart is able to pump per minute). Therefore, when the stroke volume decreases, the heart rate must increase to maintain cardiac output. Conversely, when the stroke volume increases, the heart rate must decrease to maintain cardiac output.

Even though our simple pump in this experiment does not work exactly like the human heart, you can apply the concepts presented to basic cardiac function. In this activity you will examine how the activity of the pump is affected by changing the starting (EDV) and ending volumes (ESV).

EQUIPMENT USED The following equipment will be depicted on-screen: left beaker—simulates blood coming from the lungs; flow tube connecting the left beaker and the pump—simulates the pulmonary veins; pump—simulates the left ventricle (the valve to the left of the pump simulates the bicuspid valve, and the valve to the right of the pump simulates the aortic semilunar valve); flow tube connecting the pump and the right beaker—simulates the aorta; right beaker—simulates blood going to the systemic circuit.

Experiment Instructions

Go to the home page in the PhysioEx software and click **Exercise 5: Cardiovascular Dynamics**. Click **Activity 6: Studying the Effect of Stroke Volume on Pump Activity** and take the online **Pre-lab Quiz** for Activity 6.

After you take the online Pre-lab Quiz, click the **Experiment** tab and begin the experiment. The experiment instructions are reprinted here for your reference. The opening screen for the experiment is shown below.

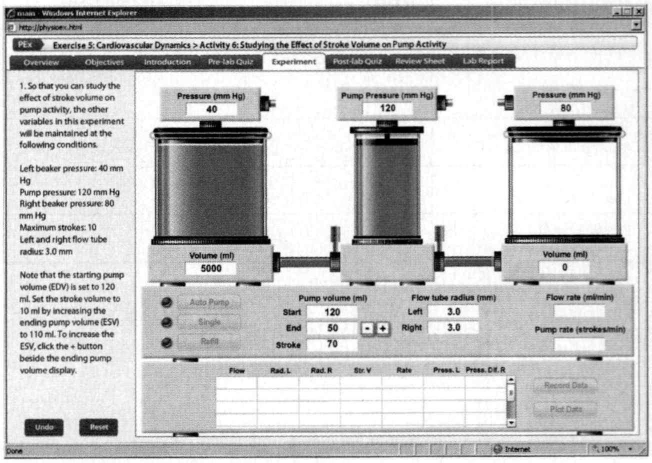

Windows Explorer 2007, Microsoft Corporation

1. So that you can study the effect of stroke volume on pump activity, the other variables in this experiment will be maintained at the following conditions:

Left beaker pressure: 40 mm Hg

Pump pressure: 120 mm Hg

Right beaker pressure: 80 mm Hg

Maximum strokes: 10

Left and right flow tube radius: 3.0 mm

Note that the starting pump volume (EDV) is set to 120 ml. Set the stroke volume to 10 ml by increasing the ending pump volume (ESV) to 110 ml. To increase the ESV, click the + button beside the ending pump volume display.

2. Click **Auto Pump** to initiate 10 strokes and then watch the pump action. The flow rate is shown in the flow rate display and the pump rate is shown in the pump rate display after the left beaker has finished draining.

3. Click **Record Data** to display your results in the grid (and record your results in Chart 6).

CHART 6 Effect of Stroke Volume on Pump Activity

Flow rate (ml/min)	Stroke volume (ml)	Pump rate (strokes/min)

4. Click **Refill** to replenish the left beaker.

PREDICT Question 1
If the pump rate is analogous to the heart rate, what do you think will happen to the rate when you increase the stroke volume?

____

5. Increase the stroke volume to 20 ml by decreasing the ESV. To decrease the ending pump volume, click the − button beside the ending pump volume display.

6. Click **Auto Pump** to initiate 10 strokes and then watch the pump action.

7. Click **Record Data** to display your results in the grid (and record your results in Chart 6).

8. Click **Refill** to replenish the left beaker.

9. You will now observe the effect of incremental increases in the stroke volume.

- Increase the stroke volume by 10 ml by decreasing the ending pump volume (ESV).
- Click **Auto Pump** to initiate 10 strokes and then watch the pump action.
- Click **Record Data** to display your results in the grid (and record your results in Chart 6).
- Click **Replenish** to refill the left beaker.

Repeat this step until you reach a stroke volume of 60 ml.

10. Increase the stroke volume by 20 ml by decreasing the ending pump volume (ESV). Click **Auto Pump** to initiate 10 strokes and then watch the pump action.

11. Click **Record Data** to display your results in the grid (and record your results in Chart 6).

12. Click **Refill** to replenish the left beaker.

13. Increase the stroke volume by 20 ml by decreasing the ending pump volume (ESV). Click **Auto Pump** to initiate 10 strokes and then watch the pump action.

14. Click **Record Data** to display your results in the grid (and record your results in Chart 6).

15. Click **Plot Data** to view a summary of your data on a plotted grid. Stroke volume will be displayed on the X-axis and flow rate will be displayed on the Y-axis. Click **Submit** to record your plot in the lab report.

After you complete the Experiment, take the online **Post-lab Quiz** for Activity 6.

Activity Questions

1. Describe how the heart responds to an increase in end diastolic volume (include the terms *preload* and *contractility* in your explanation).

____

____

2. Explain what happened to the pump rate when the stroke volume increased. Why?

____

____

3. Judging from the simulation results, explain why an athlete's resting heart rate might be lower than that of an average person.

____

____

ACTIVITY 7

Compensation in Pathological Cardiovascular Conditions

OBJECTIVES

1. To understand how aortic stenosis affects flow of blood through the heart.
2. To explain ways in which the cardiovascular system might compensate for changes in peripheral resistance.
3. To understand how the heart compensates for changes in afterload.
4. To explain how valves affect the flow of blood through the heart.

Introduction

If a blood vessel is compromised, your cardiovascular system can compensate to some degree. Aortic valve stenosis is a condition where there is a partial blockage of the aortic semilunar valve, increasing resistance to blood flow and left ventricular **afterload**. Therefore, the pressure that must be reached to open the aortic valve increases. The heart could compensate for a change in afterload by increasing contractility, the force of contraction. Increasing contractility will increase cardiac output by increasing stroke volume. To increase contractility, the myocardium becomes thicker. Athletes similarly improve their hearts through cardiovascular conditioning. That is, the thickness of the myocardium increases in diseased hearts with aortic valve stenosis and in athletes' hearts (though the chamber volume increases in athletes' hearts and decreases in diseased hearts).

Valves are important in the heart because they ensure that blood flows in one direction through the heart. The valves in the activity will ensure that blood moves in a single direction. Because the right flow tube represents the aorta (which is actually on the left side of the heart), decreasing the right flow tube radius simulates stenosis, or narrowing of the aortic valve.

Plaques in the arteries, known as **atherosclerosis**, can similarly cause an increase in resistance. An increase in peripheral resistance results in a decreased flow rate. Atherosclerosis is a type of **arteriosclerosis** in which the arteries have lost their elasticity. Atherosclerosis is one of the conditions that leads to heart disease.

In this activity you will test three different compensation mechanisms and predict which mechanism will make the best improvement in flow rate. The three mechanisms include (1) increasing the left flow tube radius (that is, increasing preload), (2) increasing the pump's pressure (that is, increasing contractility), and (3) decreasing the pressure in the right beaker (that is, decreasing afterload).

EQUIPMENT USED The following equipment will be depicted on-screen: left beaker—simulates blood coming from the lungs; flow tube connecting the left beaker and the pump—simulates the pulmonary veins; pump—simulates the left ventricle (the valve to the left of the pump simulates the bicuspid valve, and the valve to the right of the pump simulates the aortic semilunar valve); flow tube connecting the pump and the right beaker—simulates the aorta; right beaker—simulates blood going to the systemic circuit.

Experiment Instructions

Go to the home page in the PhysioEx software and click **Exercise 5: Cardiovascular Dynamics**. Click **Activity 7: Compensation in Pathological Cardiovascular Conditions** and take the online **Pre-lab Quiz** for Activity 7.

After you take the online Pre-lab Quiz, click the **Experiment** tab and begin the experiment. The experiment instructions are reprinted here for your reference. The opening screen for the experiment is shown below.

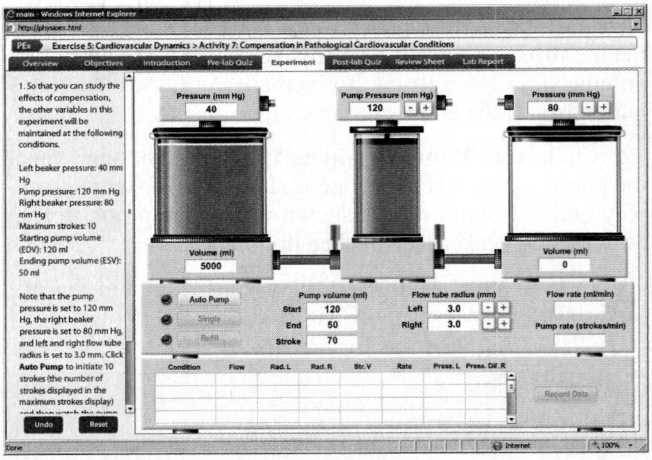

Windows Explorer 2007, Microsoft Corporation

1. So that you can study the effects of compensation, the other variables in this experiment will be maintained at the following conditions:

Left beaker pressure: 40 mm Hg

Maximum strokes: 10

Starting pump volume (EDV): 120 ml

Ending pump volume (ESV): 50 ml

Note that the pump pressure is set to 120 mm Hg, the right beaker pressure is set to 80 mm Hg, and left and right flow tube radius is set to 3.0 mm. Click **Auto Pump** to initiate 10 strokes (the number of strokes displayed in the maximum strokes display) and then watch the pump action. The flow rate is shown in the flow rate display and the pump rate is shown in the pump rate display after the left beaker has finished draining.

2. Click **Record Data** to display your results in the grid (and record your results in Chart 7). This will be your baseline, or "normal," data point for flow rate.

3. Click **Refill** to replenish the left beaker.

4. Decrease the right flow tube radius to 2.5 mm by clicking the − button beside the right flow tube radius display. Click **Auto Pump** to initiate 10 strokes and then watch the pump action.

5. Click **Record Data** to display your results in the grid (and record your results in Chart 7).

6. Click **Refill** to replenish the left beaker.

CHART 7 Compensation Results

Condition	Flow rate (ml/min)	Left radius (mm)	Right radius (mm)	Pump rate (strokes/min)	Pump pressure (mm Hg)	Right beaker pressure (mm Hg)

PREDICT Question 1
You will now test three mechanisms to compensate for the decrease in flow rate caused by the decreased flow tube radius. Which mechanism do you think will have the greatest compensatory effect?

___

7. Increase the left flow tube radius to 3.5 mm by clicking the + button beside the left flow tube radius display. Click **Auto Pump** to initiate 10 strokes and then watch the pump action.

8. Click **Record Data** to display your results in the grid (and record your results in Chart 7).

9. Click **Refill** to replenish the left beaker.

10. Increase the left flow tube radius to 4.0 mm. Click **Auto Pump** to initiate 10 strokes and then watch the pump action.

11. Click **Record Data** to display your results in the grid (and record your results in Chart 7).

12. Click **Refill** to replenish the left beaker.

13. Increase the left flow tube radius to 4.5 mm. Click **Auto Pump** to initiate 10 strokes and then watch the pump action.

14. Click **Record Data** to display your results in the grid (and record your results in Chart 7).

15. Click **Refill** to replenish the left beaker.

16. Decrease the left flow tube radius to 3.0 mm by clicking the − button beside the left flow tube radius display and increase the pump pressure to 130 mm Hg by clicking the + button beside the pump pressure display. Click **Auto Pump** to initiate 10 strokes and then watch the pump action.

17. Click **Record Data** to display your results in the grid (and record your results in Chart 7).

18. Click **Refill** to replenish the left beaker.

19. Increase the pump pressure to 140 mm Hg by clicking the + button beside the pump pressure display. Click **Auto Pump** to initiate 10 strokes and then watch the pump action.

20. Click **Record Data** to display your results in the grid (and record your results in Chart 7).

21. Click **Refill** to replenish the left beaker.

22. Increase the pump pressure to 150 mm Hg. Click **Auto Pump** to initiate 10 strokes and then watch the pump action.

23. Click **Record Data** to display your results in the grid (and record your results in Chart 7).

24. Click **Refill** to replenish the left beaker.

25. Decrease the pump pressure to 120 mm Hg by clicking the − button beside the pump pressure display and decrease the right (destination) beaker pressure to 70 mm Hg by clicking the − button beside the right beaker pressure display. Click **Auto Pump** to initiate 10 strokes and then watch the pump action.

26. Click **Record Data** to display your results in the grid (and record your results in Chart 7).

27. Click **Refill** to replenish the left beaker.

28. Decrease the right (destination) beaker pressure to 60 mm Hg by clicking the − button beside the right beaker pressure display. Click **Auto Pump** to initiate 10 strokes and then watch the pump action.

29. Click **Record Data** to display your results in the grid (and record your results in Chart 7).

30. Click **Refill** to replenish the left beaker.

31. Decrease the right (destination) beaker pressure to 50 mm Hg. Click **Auto Pump** to initiate 10 strokes and then watch the pump action.

32. Click **Record Data** to display your results in the grid (and record your results in Chart 7).

33. Click **Refill** to replenish the left beaker.

> **?** PREDICT **Question 2**
> What do you think will happen if the pump pressure and the beaker pressure are the same?
>
> ______________________________

34. Increase the right (destination) beaker pressure to 120 mm Hg by clicking the + button beside the right beaker pressure display. Click **Auto Pump** to initiate 10 strokes and then watch the pump action.

After you complete the experiment, take the online **Post-lab Quiz** for Activity 7.

Activity Questions

1. Explain why a thicker myocardium is seen in both the athlete's heart and the diseased heart.

____

____

2. Describe what the term *afterload* means.

____

____

3. Explain which mechanism in the simulation had the greatest compensatory effect.

____

____

4. Describe the mechanism used in the human heart to compensate for aortic stenosis.

____

____

NAME ______________________________

LAB TIME/DATE ______________________

Cardiovascular Dynamics

ACTIVITY 1 Studying the Effect of Blood Vessel Radius on Blood Flow Rate

1. Explain how the body establishes a pressure gradient for fluid flow. ______________________________

____

2. Explain the effect that the flow tube radius change had on flow rate. How well did the results compare with your prediction?

____

____

3. Describe the effect that radius changes have on the laminar flow of a fluid. ______________________________

____

4. Why do you think the plot was not linear? (Hint: Look at the relationship of the variables in the equation.) How well did the results compare with your prediction? ______________________________

____

ACTIVITY 2 Studying the Effect of Blood Viscosity on Blood Flow Rate

1. Describe the components in the blood that affect viscosity. ______________________________

____

2. Explain the effect that the viscosity change had on flow rate. How well did the results compare with your prediction?

____

____

3. Describe the graph of flow versus viscosity. ______________________________

____

4. Discuss the effect that polycythemia would have on viscosity and on blood flow. ______________________________

____

____

ACTIVITY 3 Studying the Effect of Blood Vessel Length on Blood Flow Rate

1. Which is more likely to occur, a change in blood vessel radius or a change in blood vessel length? Explain why.

____

____

2. Explain the effect that the change in blood vessel length had on flow rate. How well did the results compare with your prediction? _______________

____

3. Explain why you think blood vessel radius can have a larger effect on the body than changes in blood vessel length (use the blood flow equation). _______________

____

4. Describe the effect that obesity would have on blood flow and why. _______________

____

ACTIVITY 4 Studying the Effect of Blood Pressure on Blood Flow Rate

1. Explain the effect that pressure changes had on flow rate. How well did the results compare with your prediction?

____

2. How does the plot differ from the plots for tube radius, viscosity, and tube length? How well did the results compare with your prediction? _______________

____

3. Explain why pressure changes are not the best way to control blood flow. _______________

____

4. Use your data to calculate the increase in flow rate in ml/min/mm Hg. _______________

____

ACTIVITY 5 Studying the Effect of Blood Vessel Radius on Pump Activity

1. Explain the effect of increasing the right flow tube radius on the flow rate, resistance, and pump rate. _______________

____

2. Describe what the left and right beakers in the experiment correspond to in the human heart. ______________________

____

3. Briefly describe how the human heart could compensate for flow rate changes to maintain blood pressure. ____________

____

ACTIVITY 6 Studying the Effect of Stroke Volume on Pump Activity

1. Describe the Frank-Starling law in the heart. ______________________________________

____

2. Explain what happened to the pump rate when you increased the stroke volume. Why do you think this occurred? How well did the results compare with your prediction? ______________________________________

____

3. Describe how the heart alters stroke volume. ______________________________________

____

4. Describe the intrinsic factors that control stroke volume. ______________________________

____

ACTIVITY 7 Compensation in Pathological Cardiovascular Conditions

1. Explain how the heart could compensate for changes in peripheral resistance. ______________________

____

2. Which mechanism had the greatest compensatory effect? How well did the results compare with your prediction? _______

____

3. Explain what happened when the pump pressure and the beaker pressure were the same. How well did the results compare with your prediction? ____

____

4. Explain whether it would be better to adjust heart rate or blood vessel diameter to achieve blood flow changes at a local level (for example, in just the digestive system). ______________________________________

____

Cardiovascular Physiology

PRE-LAB QUIZ

1. Circle True or False: Cardiac muscle and some types of smooth muscle have the ability to contract without any external stimuli.
2. The total cardiac action potential lasts 250–300 milliseconds and has __________ phases.
   a. two
   b. three
   c. four
   d. five
3. The __________ nerve carries parasympathetic signals to the heart.
   a. phrenic
   b. vagus
   c. splanchnic
   d. visceral
4. Circle True or False: Stimulation of the parasympathetic nervous system increases the heart rate and the force of contraction of the heart.
5. What is *vagal escape*?
   ____
6. The frog heart is different from the human heart in that it contains:
   a. one atrium and two ventricles
   b. two atria and a single, incompletely divided ventricle
   c. two atria and two ventricles (it is not different)
7. The __________ node has the fastest rate of depolarization and determines the heart rate.
   a. atrioventricular
   b. atriosinus
   c. sinoatrial
   d. ventriculosino
8. Circle the correct underlined term: Humans are able to maintain an external body temperature of 35.8–38.2°C in spite of environmental conditions and are referred to as poikilotherms / homeotherms.
9. Circle True or False: Sympathetic nerve fibers release epinephrine and acetylcholine at their cardiac synapses.
10. The resting cell membrane in cardiac muscle cells favors the movement of __________ ions.
    a. potassium
    b. calcium
    c. sodium
    d. magnesium

Exercise Overview

Cardiac muscle and some types of smooth muscle contract spontaneously, without any external stimuli. Skeletal muscle is unique in that it requires depolarizing signals from the nervous system to contract. The heart's ability to trigger its own contractions is called **autorhythmicity**.

If you isolate cardiac pacemaker muscle cells, place them into cell culture, and observe them under a microscope, you can see the cells contract. Autorhythmicity occurs because the plasma membrane in cardiac pacemaker muscle cells has reduced permeability to potassium ions but still allows sodium and calcium ions to slowly leak into the cells. This leakage causes the muscle cells to slowly depolarize until the action potential threshold is reached and L-type calcium channels open, allowing Ca^{2+} entry from the extracellular fluid. Shortly thereafter, contraction of the remaining cardiac muscle occurs prior

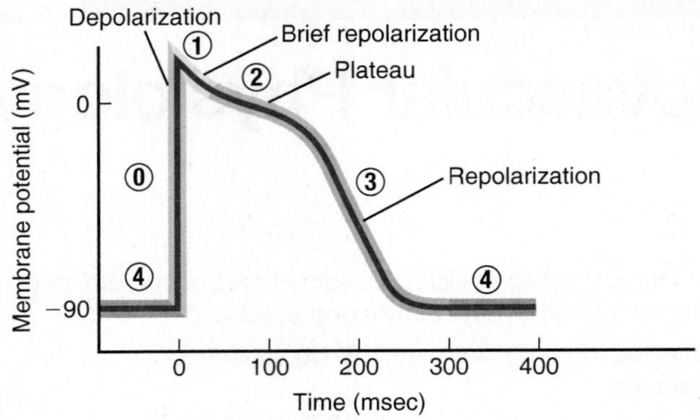

FIGURE 6.1 The cardiac action potential.

to potassium-dependent repolarization. The spontaneous depolarization-repolarization events occur in a regular and continuous manner in cardiac pacemaker muscle cells, leading to **cardiac action potentials** in the majority of cardiac muscle.

There are five main phases of membrane polarization in a cardiac action potential (view Figure 6.1).

- **Phase 0** is similar to depolarization in the neuronal action potential. Depolarization causes voltage-gated sodium channels in the cell membrane to open, increasing the flow of sodium ions into the cell and increasing the membrane potential.
- In **phase 1**, the open sodium channels begin to inactivate, decreasing the flow of sodium ions into the cell and causing the membrane potential to fall slightly. At the same time, voltage-gated potassium channels close and voltage-gated calcium channels open. The subsequent decrease in the flow of potassium out of the cell and increase in the flow of calcium into the cell act to depolarize the membrane and curb the fall in membrane potential caused by the inactivation of sodium channels.
- In **phase 2**, known as the **plateau phase**, the membrane remains in a depolarized state. Potassium channels stay closed, and long-lasting (L-type) calcium channels stay open. This plateau lasts about 0.2 seconds, or 200 milliseconds.
- In **phase 3**, the membrane potential gradually falls to more negative values when a second set of potassium channels that began opening in phases 1 and 2 allows significant amounts of potassium to flow out of the cell. The falling membrane potential causes calcium channels to close, reducing the flow of calcium into the cell and repolarizing the membrane until the resting potential is reached.
- In **phase 4**, the resting membrane potential is again established in cardiac muscle cells and is maintained until the next depolarization arrives from neighboring cardiac pacemaker cells.

The total cardiac action potential lasts 250–300 milliseconds.

ACTIVITY 1

Investigating the Refractory Period of Cardiac Muscle

OBJECTIVES

1. To observe the autorhythmicity of the heart.
2. To understand the phases of the cardiac action potential.
3. To induce extrasystoles and observe them on the oscilloscope tracing of contractile activity in the isolated, intact frog heart.
4. To relate the presence or absence of wave summation and tetanus in cardiac muscle to the refractory period of the cardiac action potential.

Introduction

Recall that **wave summation** occurs when a skeletal muscle is stimulated with such frequency that muscle twitches overlap and result in a stronger contraction than a single muscle twitch. When the stimulations are frequent enough, the muscle reaches a state of fused tetanus, during which the individual muscle twitches cannot be distinguished. Tetanus occurs in skeletal muscle because skeletal muscle has a relatively short **absolute refractory period** (a period during which action potentials cannot be generated no matter how strong the stimulus).

Unlike skeletal muscle, cardiac muscle has a relatively long refractory period and is thus incapable of wave summation. In fact, cardiac muscle is incapable of reacting to *any* stimulus before approximately the middle of phase 3, and will not respond to a normal cardiac stimulus before phase 4. The period of time between the beginning of the cardiac action potential and the approximate middle of phase 3 is the **absolute refractory period**. The period of time between the absolute refractory period and phase 4 is the **relative refractory period**. The total refractory period of cardiac muscle is 200–250 milliseconds—almost as long as the contraction of the cardiac muscle.

In this activity you will use external stimulation to better understand the refractory period of cardiac muscle. You will use a frog heart, which is anatomically similar to the human heart. The frog heart has two atria and a single, incompletely divided ventricle.

EQUIPMENT USED The following equipment will be depicted on-screen: oscilloscope display—displays the contractile activity from the frog heart; electrical stimulator—used to apply electrical shocks to the frog heart; electrode holder—locks electrodes in place for stimulation; external stimulation electrode; apparatus for sustaining an isolated frog heart—includes 23°C Ringer's solution; frog heart.

Experiment Instructions

Go to the home page in the PhysioEx software and click **Exercise 6: Cardiovascular Physiology**. Click **Activity 1: Investigating the Refractory Period of Cardiac Muscle**, and take the online **Pre-lab Quiz** for Activity 1.

After you take the online Pre-lab Quiz, click the **Experiment** tab and begin the experiment. The experiment

instructions are reprinted here for your reference. The opening screen for the experiment is shown below.

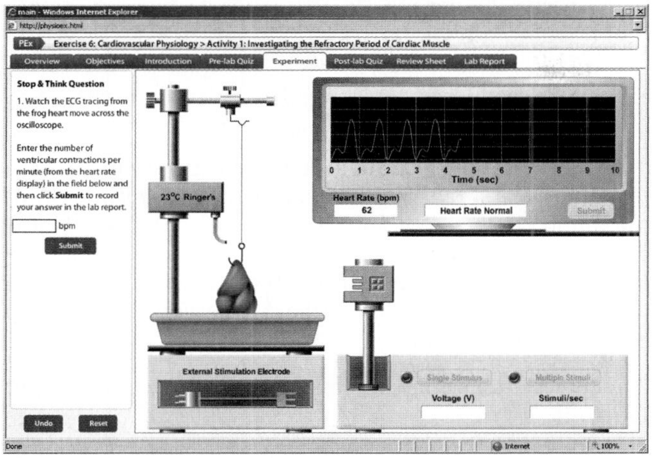

Windows Explorer 2007, Microsoft Corporation

1. Watch the contractile activity from the frog heart on the oscilloscope. Enter the number of ventricular contractions per minute (from the heart rate display) in the field below and then click **Submit** to record your answer in the lab report.

______ beats/min

2. Drag the external stimulation electrode to the electrode holder to the right of the frog heart. The electrode will touch the ventricular muscle tissue.

> **?** PREDICT **Question 1**
> When you increase the frequency of the stimulation, what do you think will happen to the amplitude (height) of the ventricular systole wave?
>
> ____

3. Deliver single shocks in succession by clicking **Single Stimulus** rapidly. You might need to practice to acquire the correct technique. You should see a "doublet," or double peak, which contains an **extrasystole**, or extra contraction of the ventricle, and then a compensatory pause, which allows the heart to get back on schedule after the extrasystole. When you see a doublet, click **Submit** to record the tracing in the lab report.

> **?** PREDICT **Question 2**
> If you deliver multiple stimuli (20 stimuli per second) to the heart, what do you think will happen?
>
> ____

4. Click **Multiple Stimuli** to deliver electrical shocks to the heart at a rate of 20 stimuli/sec. The **Multiple Stimuli** button changes to a **Stop Stimuli** button as soon as it is clicked. Observe the effects of stimulation on the contractile activity and, after a few seconds, click **Stop Stimuli** to stop the stimuli.

After you complete the experiment, take the online **Post-lab Quiz** for Activity 1.

Activity Questions

1. Describe how the frog heart and human heart differ anatomically.

____

____

2. What does an extrasystole correspond to? How did you induce an extrasystole?

____

____

3. Explain why it is important that wave summation and tetanus do not occur in the cardiac muscle.

____

____

ACTIVITY 2

Examining the Effect of Vagus Nerve Stimulation

OBJECTIVES

1. To understand the role that the sympathetic and parasympathetic nervous systems have on heart activity.
2. To explain the consequences of vagal stimulation and vagal escape.
3. To explain the functionality of the sinoatrial node.

Introduction

The autonomic nervous system has two branches: the **sympathetic** nervous system ("fight or flight") and **parasympathetic** nervous system ("resting and digesting"). At rest both the sympathetic and parasympathetic nervous systems are working but the parasympathetic branch is more active. The sympathetic nervous system becomes more active when needed, for example, during exercise and when confronting danger.

Both the parasympathetic and sympathetic nervous systems supply nerve impulses to the heart. Stimulation of the sympathetic nervous system increases the rate and force of contraction of the heart. Stimulation of the parasympathetic nervous system decreases the heart rate without directly changing the force of contraction. The vagus nerve (cranial nerve X) carries the signal to the heart. If stimulation of the vagus nerve (vagal stimulation) is excessive, the heart will stop beating. After a short time, the ventricles will begin to beat again. The resumption of the heartbeat is referred to as **vagal escape** and can be the result of sympathetic reflexes or initiation of a rhythm by the Purkinje fibers.

The **sinoatrial node (SA node)** is a cluster of autorhythmic cardiac cells found in the right atrial wall in the human heart. The SA node has the fastest rate of spontaneous depolarization, and, for that reason, it determines the heart rate and is therefore referred to as the heart's "**pacemaker**." In the absence of parasympathetic stimulation, sympathetic stimulation, and hormonal controls, the SA node generates action potentials 100 times per minute.

EQUIPMENT USED The following equipment will be depicted on-screen: oscilloscope display—displays the contractile activity from the frog heart; electrical stimulator—used to apply electrical shocks to the frog heart; electrode holder—locks electrodes in place for stimulation; vagus nerve stimulation electrode; apparatus for sustaining an isolated, intact frog heart—includes 23°C Ringer's solution; frog heart with vagus nerve (thin, white strand to the right).

Experiment Instructions

Go to the home page in the PhysioEx software and click **Exercise 6: Cardiovascular Physiology**. Click **Activity 2: Examining the Effect of Vagus Nerve Stimulation**, and take the online **Pre-lab Quiz** for Activity 2.

After you take the online Pre-lab Quiz, click the **Experiment** tab and begin the experiment. The experiment instructions are reprinted here for your reference. The opening screen for the experiment is shown below.

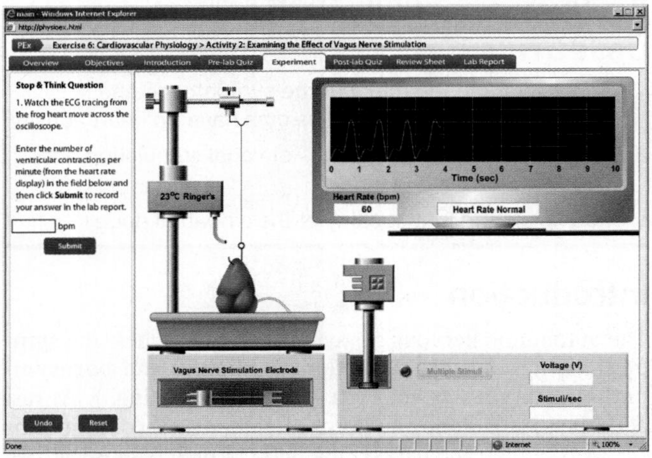

Windows Explorer 2007, Microsoft Corporation

1. Watch the contractile activity from the frog heart on the oscilloscope. Enter the number of ventricular contractions per minute (from the heart rate display) in the field below and then click **Submit** to record your answer in the lab report.

_____ beats/min

2. Drag the vagus nerve stimulation electrode to the electrode holder to the right of the heart. Note that, when the electrode locks in place, the vagus nerve is draped over the electrode. Stimuli will go directly to the vagus nerve and indirectly to the heart.

3. Enter the number of ventricular contractions per minute (from the heart rate display) in the field below and then click **Submit** to record your answer in the lab report.

_____ beats/min

PREDICT Question 1
What do you think will happen if you apply multiple stimuli to the heart by indirectly stimulating the vagus nerve?

4. Click **Multiple Stimuli** to deliver electrical shocks to the vagus nerve at a rate of 50 stimuli/sec. The **Multiple Stimuli** button changes to a **Stop Stimuli** button as soon as it is clicked. Observe the effects of stimulation on the contractile activity and, after waiting at least 20 seconds (the tracing will make two full sweeps across the oscilloscope), click **Stop Stimuli** to stop the stimuli.

After you complete the experiment, take the online **Post-lab Quiz** for Activity 2.

Activity Questions

1. Describe how stimulation of the vagus nerves affects the heart rate.

2. How does the sympathetic nervous system affect heart rate and the force of contraction?

3. Describe the mechanism of vagal escape.

4. What would happen to the heart rate if the vagus nerve were cut?

ACTIVITY 3

Examining the Effect of Temperature on Heart Rate

OBJECTIVES

1. To define the terms *hyperthermia* and *hypothermia*.
2. To contrast the terms *homeothermic* and *poikilothermic*.
3. To understand the effect that temperature has on the frog heart.
4. To understand the effect that temperature could have on the human heart.

Introduction

Humans are **homeothermic**, which means that the human body maintains an internal body temperature within the 35.8–38.2°C range even though the external temperature is changing. When the external temperature is elevated, the hypothalamus is signaled to activate heat-releasing mechanisms, such as sweating and vasodilation, to maintain the body's internal temperature. During extreme external temperature conditions, the body might not be able to maintain homeostasis and either **hyperthermia** (elevated body temperature) or **hypothermia** (low body temperature) could result. In contrast, the frog is a **poikilothermic** animal. Its internal body temperature changes depending on the temperature of its external environment because it lacks internal homeostatic regulatory mechanisms.

Ringer's solution, also known as Ringer's irrigation, consists of essential electrolytes (chloride, sodium, potassium, calcium, and magnesium) in a physiological solution and is required to keep the isolated, intact heart viable. In this activity you will explore the effect of temperature on heart rate using a Ringer's solution incubated at different temperatures.

EQUIPMENT USED The following equipment will be depicted on-screen: oscilloscope display—displays the contractile activity from the frog heart; electrical stimulator—used to apply electrical shocks to the frog heart; electrode holder—locks electrodes in place for stimulation; external stimulation electrode; apparatus for sustaining an isolated, intact frog heart—includes 5°C, 23°C, and 32°C Ringer's solution; frog heart.

Experiment Instructions

Go to the home page in the PhysioEx software and click **Exercise 6: Cardiovascular Physiology**. Click **Activity 3: Examining the Effect of Temperature on Heart Rate**, and take the online **Pre-lab Quiz** for Activity 3.

After you take the online Pre-lab Quiz, click the **Experiment** tab and begin the experiment. The experiment instructions are reprinted here for your reference. The opening screen for the experiment is shown below.

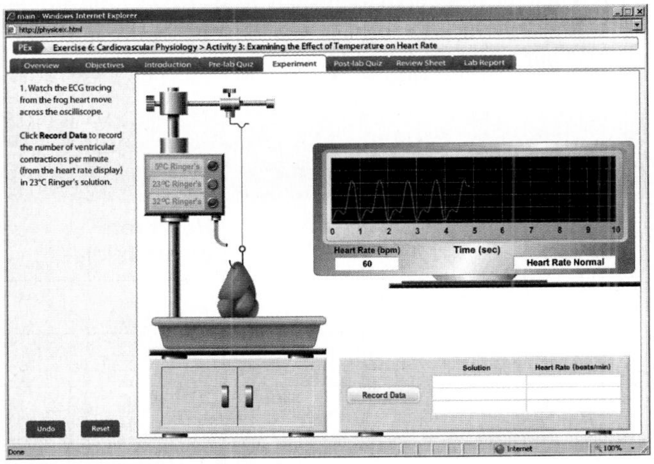

Windows Explorer 2007, Microsoft Corporation

1. Watch the contractile activity from the frog heart on the oscilloscope. Click **Record Data** to record the number of ventricular contractions per minute (from the heart rate display) in 23°C Ringer's solution.

PREDICT Question 1
What effect will decreasing the temperature of the Ringer's solution have on the heart rate of the frog?

____

2. Click **5°C Ringer's** to observe the effects of lowering the temperature.

3. When the heart activity display reads *Heart Rate Stable*, click **Record Data** to display your results in the grid (and record your results in Chart 3).

CHART 3 Effect of Temperature on Heart Rate

Solution	Heart rate (beats/min)

4. Click **23°C Ringer's** to bathe the heart and return it to room temperature. When the heart activity display reads *Heart Rate Normal*, you can proceed.

PREDICT Question 2
What effect will increasing the temperature of the Ringer's solution have on the heart rate of the frog?

____

5. Click **32°C Ringer's** to observe the effects of increasing the temperature.

6. When the heart activity display reads *Heart Rate Stable*, click **Record Data** to display your results in the grid (and record your results in Chart 3).

After you complete the experiment, take the online **Post-lab Quiz** for Activity 3.

Activity Questions

1. Explain the importance of Ringer's solution (essential electrolytes in physiological saline) in maintaining the autorhythmicity of the heart.

____

____

2. Describe the effect of lower temperature on heart rate.

___

___

3. Explain the effect that fever would have on heart rate. Explain why.

___

___

ACTIVITY 4

Examining the Effects of Chemical Modifiers on Heart Rate

OBJECTIVES

1. To distinguish between cholinergic and adrenergic modifiers of heart rate.
2. To define agonist and antagonist modifiers of heart rate.
3. To observe the effects of epinephrine, pilocarpine, atropine, and digitalis on heart rate.
4. To relate chemical modifiers of the heart rate to sympathetic and parasympathetic activation.

Introduction

Although the heart does not need external stimulation to beat, it can be affected by extrinsic controls, most notably the autonomic nervous system. The sympathetic nervous system is activated in times of "fight or flight," and sympathetic nerve fibers release **norepinephrine** (also known as **noradrenaline**) and **epinephrine** (also known as **adrenaline**) at their cardiac synapses.

Norepinephrine and epinephrine increase the frequency of action potentials by binding to β_1 adrenergic receptors embedded in the plasma membrane of **sinoatrial (SA) node** (pacemaker) cells. Working through a cAMP second-messenger mechanism, binding of the ligand opens sodium and calcium channels, increasing the rate of depolarization and shortening the period of repolarization, thus increasing the heart rate.

The parasympathetic nervous system, our "resting and digesting branch," usually dominates, and parasympathetic nerve fibers release **acetylcholine** at their cardiac synapses. Acetylcholine decreases the frequency of action potentials by binding to muscarinic cholinergic receptors embedded in the plasma membrane of the SA node cells. Acetylcholine indirectly opens potassium channels and closes calcium and sodium channels, decreasing the rate of depolarization and, thus, decreasing heart rate.

Chemical modifiers that inhibit, mimic, or enhance the action of acetylcholine in the body are labeled **cholinergic**. Chemical modifiers that inhibit, mimic, or enhance the action of epinephrine in the body are **adrenergic**. If the modifier works in the same fashion as the neurotransmitter (acetylcholine or norepinephrine), it is an **agonist**. If the modifier works in opposition to the neurotransmitter, it is an **antagonist**. In this activity you will explore the effects of pilocarpine, atropine, epinephrine, and digitalis on heart rate.

EQUIPMENT USED The following equipment will be depicted on-screen: oscilloscope display—displays the contractile activity; apparatus for sustaining an isolated intact frog heart—includes 23°C Ringer's solution; pilocarpine; atropine; epinephrine; digitalis; frog heart.

Experiment Instructions

Go to the home page in the PhysioEx software and click **Exercise 6: Cardiovascular Physiology**. Click **Activity 4: Examining the Effects of Chemical Modifiers on Heart Rate**, and take the online **Pre-lab Quiz** for Activity 4.

After you take the online Pre-lab Quiz, click the **Experiment** tab and begin the experiment. The experiment instructions are reprinted here for your reference. The opening screen for the experiment is shown below.

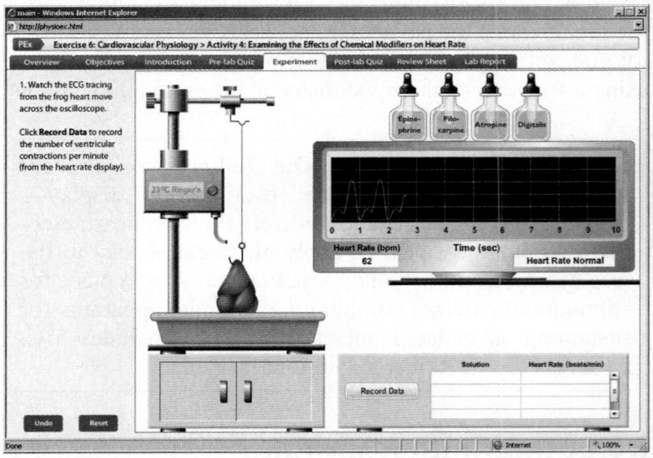

Windows Explorer 2007, Microsoft Corporation

1. Watch the contractile activity from the frog heart on the oscilloscope. Click **Record Data** to record the number of ventricular contractions per minute (from the heart rate display) and record your results in Chart 4.

2. Drag the dropper cap of the epinephrine bottle to the frog heart to release epinephrine onto the heart.

3. Observe the contractile activity and the heart activity display. When the heart activity display reads *Heart Rate Stable*, click **Record Data** to display your results in the grid (and record your results in Chart 4).

CHART 4 Effects of Chemical Modifiers on Heart Rate

Solution	Heart rate (beats/min)

4. Click **23°C Ringer's** (room temperature) to bathe the heart and flush out the epinephrine. When the heart activity display reads *Heart Rate Normal*, you can proceed.

? PREDICT **Question 1**
Pilocarpine is a cholinergic drug, an acetylcholine agonist. Predict the effect that pilocarpine will have on heart rate.

____

5. Drag the dropper cap of the pilocarpine bottle to the frog heart to release pilocarpine onto the heart.

6. Observe the contractile activity and the heart activity display. When the heart activity display reads *Heart Rate Stable*, click **Record Data** to display your results in the grid (and record your results in Chart 4).

7. Click **23°C Ringer's** (room temperature) to bathe the heart and flush out the pilocarpine. When the heart activity display reads *Heart Rate Normal*, you can proceed.

? PREDICT **Question 2**
Atropine is another cholinergic drug, an acetylcholine antagonist. Predict the effect that atropine will have on heart rate.

____

8. Drag the dropper cap of the atropine bottle to the frog heart to release atropine onto the heart.

9. Observe the contractile activity and the heart activity display. When the heart activity display reads *Heart Rate Stable*, click **Record Data** to display your results in the grid (and record your results in Chart 4).

10. Click **23°C Ringer's** (room temperature) to bathe the heart and flush out the atropine. When the heart activity display reads *Heart Rate Normal*, you can proceed.

11. Drag the dropper cap of the digitalis bottle to the frog heart to release digitalis onto the heart.

12. Observe the contractile activity and the heart activity display. When the heart activity display reads *Heart Rate Stable*, click **Record Data** to display your results in the grid (and record your results in Chart 4).

After you complete the experiment, take the online **Post-lab Quiz** for Activity 4.

Activity Questions

1. Define *agonist* and *antagonist*. Clearly distinguish between the two and give examples used in this activity.

____

____

2. Describe the effect of epinephrine on heart rate and force of contraction.

____

____

3. What is the effect of atropine on heart rate?

____

____

4. Describe the effect of digitalis on heart rate and force of contraction.

____

____

ACTIVITY 5

Examining the Effects of Various Ions on Heart Rate

OBJECTIVES

1. To understand the movement of ions that occurs during the cardiac action potential.
2. To describe the potential effect of potassium, sodium, and calcium ions on heart rate.
3. To explain how calcium channel blockers might be used pharmaceutically to treat heart patients.
4. To define the terms *inotropic* and *chronotropic*.

Introduction

In cardiac muscle cells, action potentials are caused by changes in permeability to ions due to the opening and closing of ion channels. The permeability changes that occur for the cardiac muscle cell involve potassium, sodium, and calcium ions. The concentration of potassium is greater inside the cardiac muscle cell than outside the cell. Sodium and calcium are present in larger quantities outside the cell than inside the cell.

The resting cell membrane favors the movement of potassium more than sodium or calcium. Therefore, the resting membrane potential of cardiac cells is determined mainly by the ratio of extracellular and intracellular concentrations of potassium. View Table 6.1on p. PEx-100 for a summary of the phases of the cardiac action potential and ion movement during each phase.

Calcium channel blockers are used to treat high blood pressure and abnormal heart rates. They block the movement of calcium through its channels throughout all phases of the cardiac action potentials. Consequently, because less calcium gets through, both the rate of depolarization and the force of the contraction are reduced. Modifiers that affect heart rate are **chronotropic**, and modifiers that affect the force of contraction are **inotropic**. Modifiers that lower heart rate are negative chronotropic, and modifiers that increase heart rate are positive chronotropic. The same adjectives describe inotropic modifiers. Therefore, negative inotropic drugs decrease the force of contraction of the heart and positive inotropic drugs increase the force of contraction of the heart.

TABLE 6.1

Phase of cardiac action potential	Ion movement
Phase 0 (rapid depolarization)	Sodium moves in
Phase 1 (small repolarization)	Sodium movement decreases
Phase 2 (plateau)	Potassium movement out decreases Calcium moves in
Phase 3 (repolarization)	Potassium moves out Calcium movement decreases
Phase 4 (resting potential)	Potassium moves out Little sodium or calcium moves in

EQUIPMENT USED The following equipment will be depicted on-screen: oscilloscope display—displays the contractile activity from the frog heart; apparatus for sustaining frog heart—includes 23°C Ringer's solution; calcium ions; sodium ions; potassium ions; frog heart.

Experiment Instructions

Go to the home page in the PhysioEx software and click **Exercise 6: Cardiovascular Physiology**. Click **Activity 5: Examining the Effects of Various Ions on Heart Rate**, and take the online **Pre-lab Quiz** for Activity 5.

After you take the online Pre-lab Quiz, click the **Experiment** tab and begin the experiment. The experiment instructions are reprinted here for your reference. The opening screen for the experiment is shown below.

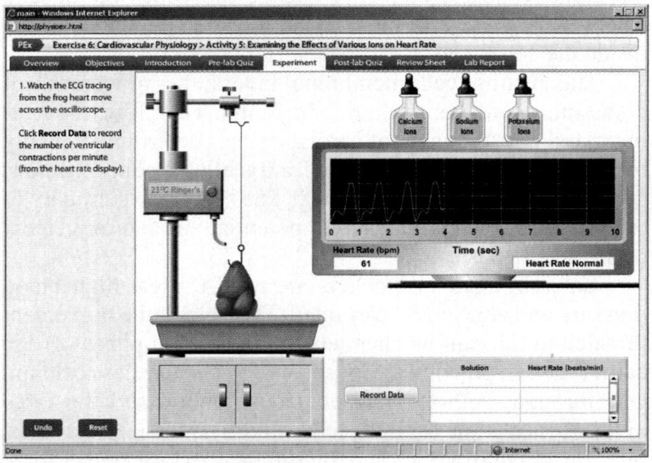

Windows Explorer 2007, Microsoft Corporation

1. Watch the contractile activity from the frog heart move on the oscilloscope. Click **Record Data** to record the number of ventricular contractions per minute (from the heart rate display).

? PREDICT Question 1
Because calcium channel blockers are negative chronotropic and negative inotropic, what effect do you think increasing the concentration of calcium will have on heart rate?

____

2. Drag the dropper cap of the calcium ions bottle to the frog heart to release calcium ions onto the heart. Note the change in heart rate after you drop the calcium ions onto the heart.

3. When the heart activity display reads *Heart Rate Stable*, click **Record Data** to display your results in the grid (and record your results in Chart 5).

CHART 5 Effects of Various Ions on Heart Rate

Solution	Heart rate (beats/min)

4. **Click 23°C Ringer's** (room temperature) to bathe the heart and flush out the calcium. When the heart activity display reads *Heart Rate Normal*, you can proceed.

5. Drag the dropper cap of the sodium ions bottle to the frog heart to release sodium ions onto the heart. Note the immediate change in the heart rate and the change in heart rate over time after you drop the sodium ions onto the heart.

6. After waiting at least 20 seconds (the tracing will make two full sweeps across the oscilloscope), click **Record Data** to display your results in the grid (and record your results in Chart 5).

7. Click **23°C Ringer's** (room temperature) to bathe the heart and flush out the sodium. When the heart activity display reads *Heart Rate Normal*, you can proceed.

? PREDICT Question 2
Excess potassium outside of the cardiac cell decreases the resting potential of the plasma membrane, thus decreasing the force of contraction. What effect (if any) do you think it will *initially* have on heart rate?

____

8. Drag the dropper cap of the potassium ions bottle to the frog heart to release potassium ions onto the heart. Note the immediate change in heart rate and the change in heart rate over time after you drop the potassium ions onto the heart.

9. After waiting at least 20 seconds (the tracing will make two full sweeps across the oscilloscope), click **Record Data** to display your results in the grid (and record your results in Chart 5).

After you complete the experiment, take the online **Post-lab Quiz** for Activity 5.

Activity Questions

1. Define chronotropic and inotropic effects on the heart.

____

____

2. Describe the effect of adding calcium ions to the frog heart.

____

____

3. Calcium channel blockers are often used to treat high blood pressure. Explain how their effects would benefit individuals with high blood pressure.

____

____

4. Describe the initial effect of adding potassium ions to the frog heart.

____

____

NAME ______________________

LAB TIME/DATE ______________

Cardiovascular Physiology

ACTIVITY 1 Investigating the Refractory Period of Cardiac Muscle

1. Explain why the larger waves seen on the oscilloscope represent ventricular contraction.

____

____

2. Explain why the amplitude of the wave did not change when you increased the frequency of the stimulation. (Hint: Relate your response to the refractory period of the cardiac action potential.) How well did the results compare with your prediction?

____

____

3. Why is it only possible to induce an extrasystole during relaxation? ______________________

____

4. Explain why wave summation and tetanus are not possible in cardiac muscle tissue. How well did the results compare with your prediction? ______________________

____

ACTIVITY 2 Examining the Effect of Vagus Nerve Stimulation

1. Explain the effect that extreme vagus nerve stimulation had on the heart. How well did the results compare with your prediction?

____

____

2. Explain two ways that the heart can overcome excessive vagal stimulation. ______________________

____

3. Describe how the sympathetic and parasympathetic nervous systems work together to regulate heart rate. ______________________

____

4. What do you think would happen to the heart rate if the vagus nerve was cut? ______________________

____

ACTIVITY 3 Examining the Effect of Temperature on Heart Rate

1. Explain the effect that decreasing the temperature had on the frog heart. How do you think the human heart would respond? How well did the results compare with your prediction? ________________

2. Describe why Ringer's solution is required to maintain heart contractions. ________________

3. Explain the effect that increasing the temperature had on the frog heart. How do you think the human heart would respond? How well did the results compare with your prediction? ________________

ACTIVITY 4 Examining the Effects of Chemical Modifiers on Heart Rate

1. Describe the effect that pilocarpine had on the heart and why it had this effect. How well did the results compare with your prediction? ________________

2. Atropine is an acetylcholine antagonist. Does atropine inhibit or enhance the effects of acetylcholine? Describe your results and how they correlate with how the drug works. How well did the results compare with your prediction? ________________

3. Describe the benefits of administering digitalis. ________________

4. Distinguish between cholinergic and adrenergic chemical modifiers. Include examples of each in your discussion. ________________

ACTIVITY 5 Examining the Effects of Various Ions on Heart Rate

1. Describe the effect that increasing the calcium ions had on the heart. How well did the results compare with your prediction? ________________

2. Describe the effect that increasing the potassium ions initially had on the heart in this activity. Relate this to the resting membrane potential of the cardiac muscle cell. How well did the results compare with your prediction? ________________

3. Describe how calcium channel blockers are used to treat patients and why. ________________

Respiratory System Mechanics

PRE-LAB QUIZ

1. Circle the correct underlined term: Blood enriched with carbon dioxide / oxygen returns to the heart from the body tissues.
2. Which of the following occurs during expiration?
   a. external intercostal muscles contract
   b. diaphragm contracts
   c. diaphragm relaxes
   d. abdominal wall muscles contract
3. Circle True or False: When the diaphragm contracts, the volume in the thoracic cavity decreases.
4. The volume of air remaining in the lungs after a forceful and complete respiration is known as:
   a. tidal volume
   b. vital capacity
   c. residual volume
   d. reserve volume
5. Circle the correct underlined term: Obstructive / Restrictive lung diseases reduce both respiratory volumes and capacities.
6. Circle True or False: During an acute asthma attack, there is a significant loss of elastic recoil as the disease destroys the alveoli of the lungs.
7. Circle the correct underlined term: Total lung capacity / Vital capacity is the maximum amount of air that can be inspired and then expired after maximal effort.
8. The actual site of gas exchange in the lungs occurs in the:
   a. alveoli
   b. diaphragm
   c. external intercostals
   d. internal intercostals
9. Circle the correct underlined term: A spirometer / inhaler is a device that can measure the volume of air inspired and expired in a specific period of time.
10. Circle True or False: During heavy exercise, both the rate of breathing and the residual volume will increase to their maximum limits.

Exercise Overview

The physiological function of the respiratory system is essential to life. If problems develop in most other physiological systems, we can survive for some time without addressing them. But if a persistent problem develops within the respiratory system (or the circulatory system), death can occur in minutes.

The primary role of the respiratory system is to distribute oxygen to, and remove carbon dioxide from, *all* the cells of the body. The respiratory system works together with the circulatory system to achieve this. **Respiration** includes **ventilation**, or the movement of air into and out of the lungs (breathing), and the transport (via blood) of oxygen and carbon dioxide between the lungs and body cells (view Figure 7.1, p. PEx-106). The heart pumps deoxygenated blood to pulmonary capillaries, where gas exchange occurs between blood and **alveoli** (air sacs in the lungs), thus oxygenating the blood. The heart then pumps the oxygenated blood to body tissues, where oxygen is used for cell metabolism. At the same time, carbon dioxide (a waste product of metabolism) from body tissues diffuses into the blood. This carbon dioxide–enriched, oxygen-reduced blood then returns to the heart, completing the circuit.

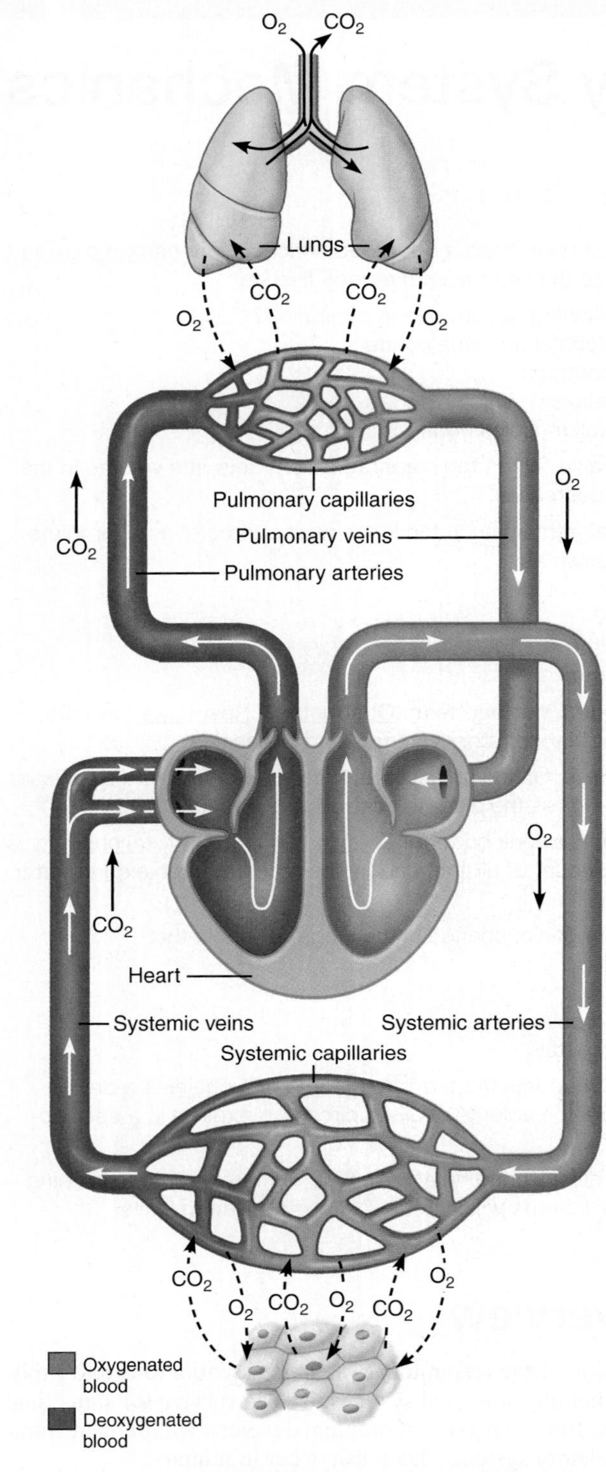

FIGURE 7.1 Relationship between external respiration and internal respiration.

Ventilation is the result of skeletal muscle contraction. When the **diaphragm**—a dome-shaped muscle that divides the thoracic and abdominal cavities—and the **external intercostal muscles** contract, the volume in the thoracic cavity increases. This increase in thoracic volume reduces the pressure in the thoracic cavity, allowing atmospheric gas to enter the lungs (a process called **inspiration**). When the diaphragm and the external intercostals relax, the pressure in the thoracic cavity increases as the volume decreases, forcing air out of the lungs (a process called **expiration**). Inspiration is considered an *active* process because muscle contraction requires the use of ATP, whereas expiration is usually considered a *passive* process because the muscles relax, rather than contract. When a person is running, however, expiration becomes an active process, resulting from the contraction of **internal intercostal muscles** and **abdominal muscles**. In this case, both inspiration and expiration are considered *active* processes because muscle contraction is needed for both.

The amount of air that flows into and out of the lungs in 1 minute is the pulmonary **minute ventilation**, which is calculated by multiplying the **frequency of breathing** by the volume of each breath (the **tidal volume**). Ventilation must be regulated at all times to maintain oxygen in arterial blood and carbon dioxide in venous blood at their normal levels—that is, at their normal **partial pressures**. The *partial pressure* of a gas is the proportion of pressure that the gas exerts in a mixture. For example, in the atmosphere at sea level, the total pressure is 760 mm Hg. Oxygen makes up 21% of the total atmosphere and, therefore, has a partial pressure (P_{O_2}) of 160 mm Hg (760 mm Hg $\times$ 0.21).

Oxygen and carbon dioxide diffuse down their partial pressure gradients, from high partial pressures to low partial pressures. Oxygen diffuses from the alveoli of the lungs into the blood, where it can dissolve in plasma and attach to hemoglobin, and then diffuses from the blood into the tissues. Carbon dioxide (produced by the metabolic reactions of the tissues) diffuses from the tissues into the blood and then diffuses from the blood into the alveoli for export from the body.

In this exercise you will investigate the basic mechanics and regulation of the respiratory system. The concepts you will explore with a simulated lung will help you understand the operation of the human respiratory system in better detail.

ACTIVITY 1

Measuring Respiratory Volumes and Calculating Capacities

OBJECTIVES

1. To understand the use of the terms *ventilation, inspiration, expiration, diaphragm, external intercostals, internal intercostals, abdominal-wall muscles, expiratory reserve volume (ERV), forced vital capacity (FVC), tidal volume (TV), inspiratory reserve volume (IRV), residual volume (RV), and forced expiratory volume in one second (FEV_1).*
2. To understand the roles of skeletal muscles in the mechanics of breathing.
3. To understand the volume and pressure changes in the thoracic cavity during ventilation of the lungs.
4. To understand the effects of airway radius and, thus, resistance on airflow.

Introduction

The two phases of **ventilation**, or breathing, are (1) **inspiration**, during which air is taken into the lungs, and (2) **expiration**, during which air is expelled from the lungs. Inspiration occurs as the **external intercostal muscles** and the **diaphragm** contract. The diaphragm, normally a dome-shaped muscle, flattens as it moves inferiorly while the external intercostal muscles, situated between the ribs, lift the rib cage. These cooperative actions increase the thoracic volume. Air rushes into the lungs because this increase in thoracic volume creates a partial vacuum.

During quiet expiration, the inspiratory muscles relax, causing the diaphragm to rise superiorly and the chest wall to move inward. Thus, the **thorax** returns to its normal shape because of the elastic properties of the lung and thoracic wall. As in a deflating balloon, the pressure in the lungs rises, forcing air out of the lungs and airways. Although expiration is normally a *passive* process, **abdominal-wall muscles** and the **internal intercostal muscles** can also contract during expiration to force additional air from the lungs. Such forced expiration occurs, for example, when you exercise, blow up a balloon, cough, or sneeze.

Normal, quiet breathing moves about 500 ml (0.5 liter) of air (the **tidal volume**) into and out of the lungs with each breath, but this amount can vary due to a person's size, sex, age, physical condition, and immediate respiratory needs. In this activity you will measure the following respiratory volumes (the values given for the normal adult male and female are approximate).

Tidal volume (TV): Amount of air inspired and then expired with each breath under resting conditions (500 ml)

Inspiratory reserve volume (IRV): Amount of air that can be forcefully inspired after a normal tidal volume inspiration (male, 3100 ml; female, 1900 ml)

Expiratory reserve volume (ERV): Amount of air that can be forcefully expired after a normal tidal volume expiration (male, 1200 ml; female, 700 ml)

Residual volume (RV): Amount of air remaining in the lungs after forceful and complete expiration (male, 1200 ml; female, 1100 ml)

Respiratory capacities are calculated from the respiratory volumes. In this activity you will calculate the following respiratory capacities.

Total lung capacity (TLC): Maximum amount of air contained in lungs after a maximum inspiratory effort: TLC = TV + IRV + ERV + RV (male, 6000 ml; female, 4200 ml)

Vital capacity (VC): Maximum amount of air that can be inspired and then expired with maximal effort: VC = TV + IRV + ERV (male, 4800 ml; female 3100 ml)

You will also perform two pulmonary function tests in this activity.

Forced vital capacity (FVC): Amount of air that can be expelled when the subject takes the deepest possible inspiration and forcefully expires as completely and rapidly as possible

Forced expiratory volume (FEV_1): Measures the percentage of the vital capacity that is expired during 1 second of the FVC test (normally 75%–85% of the vital capacity)

EQUIPMENT USED The following equipment will be depicted on-screen: simulated human lungs suspended in a glass bell jar; rubber diaphragm—used to seal the jar and change the volume and, thus, pressure in the jar (As the diaphragm moves inferiorly, the volume in the bell jar increases and the pressure drops slightly, creating a partial vacuum in the bell jar. This partial vacuum causes air to be sucked into the tube at the top of the bell jar and then into the simulated lungs. As the diaphragm moves up, the decreasing volume and rising pressure within the bell jar forces air out of the lungs.); adjustable airflow tube—connects the lungs to the atmosphere; oscilloscope; three different breathing patterns: normal tidal volumes, expiratory reserve volume (ERV), and forced vital capacity (FVC).

Experiment Instructions

Go to the home page in the PhysioEx software and click **Exercise 7: Respiratory System Mechanics**. Click **Activity 1: Measuring Respiratory Volumes and Calculating Capacities**, and take the online **Pre-lab Quiz** for Activity 1.

After you take the online Pre-lab Quiz, click the **Experiment** tab and begin the experiment. The experiment instructions are reprinted here for your reference. The opening screen for the experiment is shown below.

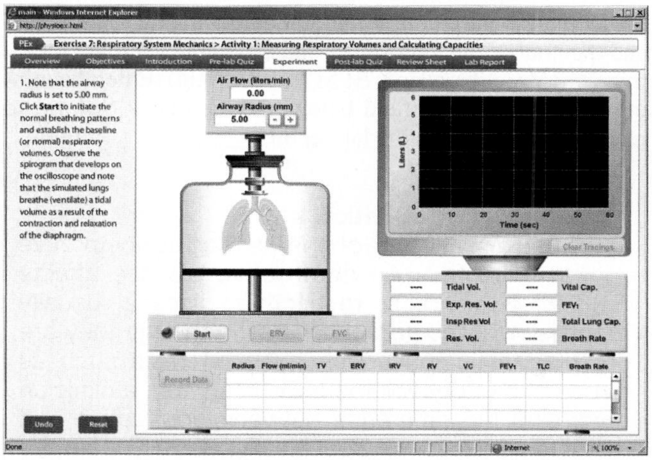

Windows Explorer 2007, Microsoft Corporation

1. Note that the airway radius is set to 5.00 mm. Click **Start** to initiate the normal breathing patterns and establish the baseline (or normal) respiratory volumes. Observe the spirogram that develops on the oscilloscope and note that the simulated lungs breathe (ventilate) a tidal volume as a result of the contraction and relaxation of the diaphragm.

2. Click **Record Data** to display your results in the grid (and record your results in Chart 1, p. PEx-108).

3. Click **Clear Tracings** to clear the spirogram on the oscilloscope.

4. You will now complete the measurement of respiratory volumes and determine the respiratory capacities. First, click **Start** to initiate the normal breathing pattern. After 10 seconds, click **ERV**. Wait another 10 seconds and then click **FVC** to

CHART 1 Respiratory Volumes and Capacities

Radius (mm)	Flow (ml/min)	TV (ml)	ERV (ml)	IRV (ml)	RV (ml)	VC (ml)	FEV_1 (ml)	TLC (ml)

complete the measurement of respiratory volumes. When you click ERV, the program will simulate forced expiration using the contraction of the internal intercostal muscles and abdominal-wall muscles. When you click FVC, the lungs will first inspire maximally and then expire fully to demonstrate forced vital capacity.

5. Note that, in addition to the tidal volume, the expiratory reserve volume, inspiratory reserve volume, and residual volume were measured. The vital capacity and total lung capacity were calculated from those volumes. Click **Record Data** to display your results in the grid (and record your results in Chart 1).

6. Minute ventilation is the amount of air that flows into and then out of the lungs in a minute. Minute ventilation (ml/min) = TV (ml/breath) × BPM (breaths/min). Enter the minute ventilation in the field below and then click **Submit** to record your answer in the lab report. __________ ml/min

? PREDICT Question 1
Lung diseases are often classified as obstructive or restrictive. An **obstructive** disease affects *airflow*, and a **restrictive** disease usually reduces *volumes and capacities*. Although they are not diagnostic, pulmonary function tests such as forced expiratory volume (FEV_1) can help a clinician determine the difference between obstructive and restrictive diseases. Specifically, an FEV_1 is the forced volume expired in 1 second.

In obstructive diseases such as chronic bronchitis and asthma, airway radius is decreased. Thus, FEV_1 will:

____.

7. You will now explore what effect changing the airway radius has on pulmonary function. Decrease the airway radius to 4.50 mm by clicking the – button beneath the airway radius display.

8. Click **Start** to initiate the normal breathing pattern. After 10 seconds, click **ERV**. Wait another 10 seconds and then click **FVC**. The FEV_1 will appear in the FEV_1 display beneath the oscilloscope.

9. Click **Record Data** to display your results in the grid (and record your results in Chart 1).

10. You will now gradually decrease the airway radius.

- Decrease the airway radius by 0.50 mm by clicking the – button beneath the airway radius display.
- Click **Start** to initiate the normal breathing pattern. After 10 seconds, click **ERV**. Wait another 10 seconds and then click **FVC**. The FEV_1 will appear in the FEV_1 display beneath the oscilloscope.
- Click **Record Data** to display your results in the grid (and record your results in Chart 1).

Repeat this step until you reach an airway radius of 3.00 mm.

11. A useful way to express FEV_1 is as a percentage of the forced vital capacity (FVC). Using the FEV_1 and FVC values from the data grid, calculate the FEV_1 (%) by dividing the FEV_1 volume by the FVC volume (in this case, the VC is equal to the FVC) and multiply by 100%. Enter the FEV_1 (%) for an airway radius of 5.0 mm in the field below and then click **Submit** to record your answer in the lab report.

FEV_1 (%) for an airway radius of 5.0 (mm): __________

12. Enter the FEV_1 (%) for an airway radius of 3.00 mm in the field below and then click **Submit** to record your answer in the lab report.

FEV_1 (%) for an airway radius of 3.00 (mm): __________

After you complete the experiment, take the online **Post-lab Quiz** for Activity 1.

Activity Questions

1. When you forcefully exhale your entire expiratory reserve volume, any air remaining in your lungs is called the residual volume (RV). Why is it impossible to further exhale the RV (that is, *where* is this air volume trapped, and *why* is it trapped)?

____

____

2. How do you measure a person's RV in a laboratory?

____

____

3. Draw a spirogram that depicts a person's volumes and capacities before and during a significant cough.

ACTIVITY 2

Comparative Spirometry

OBJECTIVES

1. To understand the terms *spirometry, spirogram, emphysema, asthma, inhaler, moderate exercise, heavy exercise, tidal volume (TV), expiratory reserve volume (ERV), inspiratory reserve volume (IRV), residual volume (RV), vital capacity (VC), total lung capacity (TLC), forced vital capacity (FVC)*, and *forced expiratory volume in one second (FEV_1)*.
2. To observe and compare spirograms collected from resting, healthy patients to those taken from an emphysema patient.
3. To observe and compare spirograms collected from resting, healthy patients to those taken from a patient suffering an acute asthma attack.
4. To observe and compare the spirogram collected from an asthmatic patient *while* suffering an acute asthma attack to that taken after the patient uses an inhaler for relief.
5. To observe and compare spirograms collected from volunteers engaged in moderate exercise and heavy exercise.

Introduction

In this activity you will explore the changes to normal respiratory volumes and capacities when pathophysiology develops and during aerobic exercise by recruiting volunteers to breathe into a water-filled spirometer. The spirometer is a device that measures the volume of air inspired and expired by the lungs over a specified period of time. Several lung capacities and flow rates can be calculated from this data to assess pulmonary function. With your knowledge of respiratory mechanics, you can predict, document, and explain changes to the volumes and capacities in each state.

Emphysema breathing: With emphysema, there is a significant loss of elastic recoil in the lung tissue. This loss of elastic recoil occurs as the disease destroys the walls of the alveoli. Airway resistance is also increased as the lung tissue in general becomes more flimsy and exerts less anchoring on the surrounding airways. Thus, the lung becomes overly compliant and expands easily. Conversely, a great effort is required to expire because the lungs can no longer passively recoil and deflate. Each expiration requires a noticeable and exhausting muscular effort, and a person with emphysema expires slowly.

Acute asthma attack breathing: During an acute asthma attack, bronchiole smooth muscle spasms and, thus, the airways become constricted (that is, reduced in diameter). They also become clogged with thick mucus secretions. These changes lead to significantly increased airway resistance.

Underlying these symptoms is an airway inflammatory response brought on by triggers such as allergens (for example, dust and pollen), extreme temperature changes, and even exercise. Like with emphysema, the airways collapse and pinch closed before a forced expiration is completed. Thus, the volumes and peak flow rates are significantly reduced during an asthma attack. Unlike with emphysema, the elastic recoil is not diminished in an acute asthma attack.

When an acute asthma attack occurs, many people seek to relieve symptoms with an inhaler, which atomizes the medication and allows for direct application onto the afflicted airways. Usually, the medication includes a smooth muscle relaxant (for example, a β_2 agonist or an acetylcholine antagonist) that relieves the bronchospasms and induces bronchiole dilation. The medication can also contain an anti-inflammatory agent, such as a corticosteroid, that suppresses the inflammatory response. The use of the inhaler reduces airway resistance.

Breathing during exercise: During *moderate* aerobic exercise, the human body has an increased metabolic demand, which is met, in part, by changes in respiration. Specifically, both the rate of breathing and the tidal volume increase. These two respiratory variables do not increase by the same amount. The increase in the tidal volume is greater than the increase in the rate of breathing. During *heavy* exercise, further changes in respiration are required to meet the extreme metabolic demands of the body. In this case both the rate of breathing and the tidal volume increase to their maximum tolerable limits.

EQUIPMENT USED The following equipment will be depicted on-screen: a classic water-filled spirometer with an attached rotating drum that records the analog spirogram in real time; breathing patterns from a variety of patients: unforced breathing and forced vital capacity for a "normal" patient, a patient with emphysema, and a patient with asthma (during an attack and after using an inhaler); and the breathing patterns from a patient during moderate and heavy exercise.

Experiment Instructions

Go to the home page in the PhysioEx software and click **Exercise 7: Respiratory System Mechanics**. Click **Activity 2: Comparative Spirometry**, and take the online **Pre-lab Quiz** for Activity 2.

After you take the online Pre-lab Quiz, click the **Experiment** tab and begin the experiment. The experiment

instructions are reprinted here for your reference. The opening screen for the experiment is shown below.

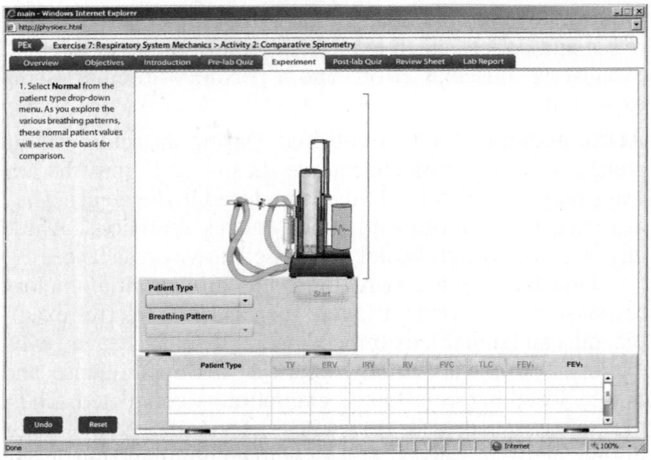

Windows Explorer 2007, Microsoft Corporation

1. Select **Normal** from the patient type drop-down menu. As you explore the various breathing patterns, these normal patient values will serve as the basis for comparison.

2. Select **Unforced Breathing** from the breathing pattern drop-down menu.

3. Click **Start** to record the patient's unforced breathing pattern and watch as the drum starts turning and the spirogram develops on the paper rolling off the drum.

4. Note the volume levels (in milliliters) on the Y-axis of the spirogram. When half the screen is filled with unforced tidal volumes and the spirogram has paused, select **Forced Vital Capacity** from the breathing pattern drop-down menu.

5. Click **Start** to record the patient's forced vital capacity. The spirogram ends as the paper rolls to the right edge of the screen.

6. Click on each of the buttons in the data recorder to measure respiratory volumes and capacities. Start with tidal volume (TV) and work your way to the right. When you measure each volume or capacity, (1) a bracket appears on the spirogram to indicate where that measurement originates and (2) the value (in milliliters) displays in the grid. After you complete all the measurements, the FEV_1 (%) ratio will automatically be calculated. The FEV_1 (%) = (FEV_1/FVC) × 100%. Record your results in Chart 2.

PREDICT Question 1
With emphysema, there is a significant loss of elastic recoil in the lung tissue and a noticeable, exhausting muscular effort is required for each expiration. Inspiration actually becomes easier because the lung is now overly compliant. What lung values will change (from those of the normal patient) in the spirogram when the patient with emphysema is selected?

____

7. Select **Emphysema** from the patient type drop-down menu.

8. Select **Unforced Breathing** from the breathing pattern drop-down menu.

9. Click **Start** to record the patient's unforced breathing pattern and watch as the drum starts turning and the spirogram develops on the paper rolling off the drum.

10. Note the volume levels on the Y-axis of the spirogram. When half the screen is filled with unforced tidal volumes and the spirogram has paused, select **Forced Vital Capacity** from the breathing pattern drop-down menu.

11. Click **Start** to record the patient's forced vital capacity. The spirogram ends as the paper rolls to the right edge of the screen.

12. Click on each of the buttons in the data recorder to measure respiratory volumes and capacities. Start with tidal volume (TV) and work your way to the right. Record your results in Chart 2.

CHART 2 Spirometry Results

Patient type	TV (ml)	ERV (ml)	IRV (ml)	RV (ml)	FVC (ml)	TLC (ml)	FEV_1 (ml)	FEV_1 (%)
Normal								
Emphysema								
Acute asthma attack								
Plus inhaler								
Moderate exercise								
Heavy exercise								

PREDICT Question 2
During an acute asthma attack, airway resistance is significantly increased by (1) increased thick mucous secretions and (2) airway smooth muscle spasms. What lung values will change (from those of the normal patient) in the spirogram for a patient suffering an acute asthma attack?

____

13. Select **Acute Asthma Attack** from the patient type drop-down menu.

14. Select **Unforced Breathing** from the breathing pattern drop-down menu.

15. Click **Start** to record the patient's unforced breathing pattern and watch as the drum starts turning and the spirogram develops on the paper rolling off the drum.

16. Note the volume levels on the Y-axis of the spirogram. When half the screen is filled with unforced tidal volumes and the spirogram has paused, select **Forced Vital Capacity** from the breathing pattern drop-down menu.

17. Click **Start** to record the patient's forced vital capacity. The spirogram ends as the paper rolls to the right edge of the screen.

18. Click on each of the buttons in the data recorder to measure respiratory volumes and capacities. Start with tidal volume (TV) and work your way to the right. Record your results in Chart 2.

PREDICT Question 3
When an acute asthma attack occurs, many people seek relief from the increased airway resistance by using an inhaler. This device atomizes the medication and induces bronchiole dilation (though it can also contain an anti-inflammatory agent). What lung values will change *back* to those of the normal patient in the spirogram after the asthma patient uses an inhaler?

____

19. Select **Plus Inhaler** from the patient type drop-down menu.

20. Select **Unforced Breathing** from the breathing pattern drop-down menu.

21. Click **Start** to record the patient's unforced breathing pattern and watch as the drum starts turning and the spirogram develops on the paper rolling off the drum.

22. Note the volume levels on the Y-axis of the spirogram. When half the screen is filled with unforced tidal volumes and the spirogram has paused, select **Forced Vital Capacity** from the breathing pattern drop-down menu.

23. Click **Start** to record the patient's forced vital capacity. The spirogram ends as the paper rolls to the right edge of the screen.

24. Click on each of the buttons in the data recorder to measure respiratory volumes and capacities. Start with tidal volume (TV) and work your way to the right. Record your results in Chart 2.

PREDICT Question 4
During moderate aerobic exercise, the human body will change its respiratory cycle in order to meet increased metabolic demands. During heavy exercise, further changes in respiration are required to meet the extreme metabolic demands of the body. Which lung value will change more during moderate exercise, the ERV or the IRV?

____

25. Select **Moderate Exercise** from the patient type drop-down menu. Note that the selection of a breathing pattern is not applicable because our central nervous system automatically adjusts and maintains the depth and frequency of breathing to meet the increased metabolic demands while we exercise. We do not normally alter this pattern with conscious intervention.

26. Click **Start** to record the patient's breathing pattern and watch as the drum starts turning and the spirogram develops on the paper rolling off the drum.

27. Click on each of the buttons in the data recorder to measure respiratory volumes and capacities. Start with tidal volume (TV) and work your way to the right. *ND* indicates this measurement or calculation was not done. Record your results in Chart 2.

28. Select **Heavy Exercise** from the patient type drop-down menu.

29. Click **Start** to record the patient's breathing pattern and watch as the drum starts turning and the spirogram develops on the paper rolling off the drum.

30. Click on each of the buttons in the data recorder to measure respiratory volumes and capacities. Start with tidal volume (TV) and work your way to the right. Record your results in Chart 2.

After you complete the experiment, take the online **Post-lab Quiz** for Activity 2.

Activity Questions

1. Why is residual volume (RV) above normal in a patient with emphysema?

____

____

2. Why did the asthmatic patient's inhaler medication fail to return all volumes and capacities to normal values right away?

____

____

3. Looking at the spirograms generated in this activity, state an easy way to determine whether a person's exercising effort is moderate or heavy.

____

____

ACTIVITY 3

Effect of Surfactant and Intrapleural Pressure on Respiration

OBJECTIVES

1. To understand the terms *surfactant*, *surface tension*, *intrapleural space*, *intrapleural pressure*, *pneumothorax*, and *atelectasis*.
2. To understand the effect of surfactant on surface tension and lung function.
3. To understand how negative intrapleural pressure prevents lung collapse.

Introduction

At any gas-liquid boundary, the molecules of the liquid are attracted more strongly to each other than they are to the gas molecules. This unequal attraction produces tension at the liquid surface, called **surface tension**. Because surface tension resists any force that tends to increase surface area of the gas-liquid boundary, it acts to decrease the size of hollow spaces, such as the alveoli, or microscopic air spaces within the lungs.

If the film lining the air spaces in the lung were pure water, it would be very difficult, if not impossible, to inflate the lungs. However, the aqueous film covering the alveolar surfaces contains **surfactant**, a detergent-like mixture of lipids and proteins that decreases surface tension by reducing the attraction of water molecules to each other. You will explore the importance of surfactant in this activity.

Between breaths, the pressure in the pleural cavity, the **intrapleural pressure**, is less than the pressure in the alveoli. Two forces cause this negative pressure condition: (1) the tendency of the lung to recoil because of its elastic properties and the surface tension of the alveolar fluid and (2) the tendency of the compressed chest wall to recoil and expand outward. These two forces pull the lungs away from the thoracic wall, creating a partial vacuum in the pleural cavity.

Because the pressure in the intrapleural space is lower than atmospheric pressure, any opening created in the pleural membranes equalizes the intrapleural pressure with atmospheric pressure by allowing air to enter the pleural cavity, a condition called **pneumothorax**. A pneumothorax can then lead to lung collapse, a condition called **atelectasis**. In this activity, the **intrapleural space** is the space between the wall of the glass bell jar and the outer wall of the lung it contains.

EQUIPMENT USED The following equipment will be depicted on-screen: simulated human lungs suspended in a glass bell jar; rubber diaphragm—used to seal the jar and change the volume and, thus, pressure in the jar (As the diaphragm moves inferiorly, the volume in the bell jar increases and the pressure drops slightly, creating a partial vacuum in the bell jar. This partial vacuum causes air to be sucked into the tube at the top of the bell jar and then into the simulated lungs. As the diaphragm moves up, the decreasing volume and rising pressure within the bell jar forces air out of the lungs.); valve—allows intrapleural pressure in the left side of the bell jar to equalize with atmospheric pressure; surfactant—amphipathic lipids (dipalmitoylphosphatidylcholine, phosphatidylglycerol, and palmitic acid) and short, synthetic peptides in a mixture that mimics the surfactant found in human lungs (surfactant molecules reduce surface tension in alveoli by adsorbing to the air-water interface, with their hydrophilic parts in the water and their hydrophobic parts facing toward the air); oscilloscope.

Experiment Instructions

Go to the home page in the PhysioEx software and click **Exercise 7: Respiratory System Mechanics**. Click **Activity 3: Effect of Surfactant and Intrapleural Pressure on Respiration**, and take the online **Pre-lab Quiz** for Activity 3.

After you take the online Pre-lab Quiz, click the **Experiment** tab and begin the experiment. The experiment instructions are reprinted here for your reference. The opening screen for the experiment is shown below.

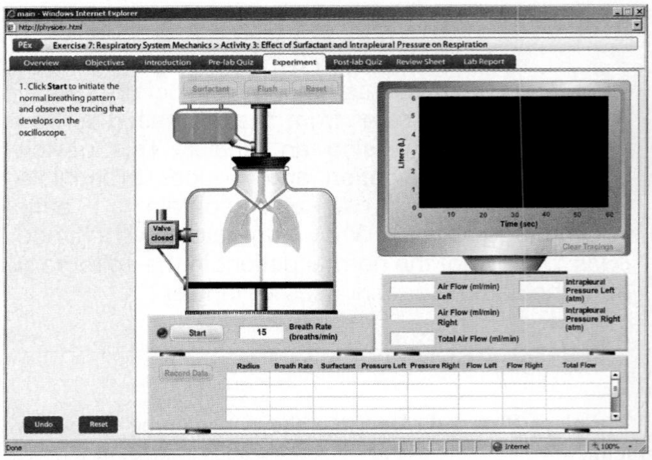

Windows Explorer 2007, Microsoft Corporation

1. Click **Start** to initiate the normal breathing pattern and observe the tracing that develops on the oscilloscope.

2. Click **Record Data** to display your results in the grid (and record your results in Chart 3). This data represents breathing in the absence of surfactant.

3. Click **Surfactant** twice to dispense two aliquots of the synthetic lipids and peptides onto the interior lining of the lungs.

4. Click **Start** to initiate breathing in the presence of surfactant and observe the tracing that develops.

5. Click **Record Data** to display your results in the grid (and record your results in Chart 3).

CHART 3 Effect of Surfactant and Intrapleural Pressure on Respiration

Surfactant	Intrapleural pressure left (atm)	Intrapleural pressure right (atm)	Airflow left (ml/min)	Airflow right (ml/min)	Total airflow (ml/min)

PREDICT Question 1
What effect will adding more surfactant have on these lungs?

____

6. Click **Surfactant** twice to dispense two more aliquots of the synthetic lipids and proteins onto the interior lining of the lungs.

7. Click **Start** to initiate breathing in the presence of additional surfactant and observe the tracing that develops.

8. Click **Record Data** to display your results in the grid (and record your results in Chart 3).

9. Click **Clear Tracings** to clear the tracing on the oscilloscope.

10. Click **Flush** to clear the lungs of surfactant from the previous run.

11. Click **Start** to initiate breathing and observe the tracing that develops. Notice the negative pressure condition displayed below the oscilloscope when the lungs inflate.

12. Click **Record Data** to display your results in the grid (and record your results in Chart 3).

13. Click the valve on the left side of the glass bell jar to open it.

14. Click **Start** to initiate breathing and observe the tracing that develops.

15. Click **Record Data** to display your results in the grid (and record your results in Chart 3).

PREDICT Question 2
What will happen to the collapsed lung in the left side of the glass bell jar if you close the valve?

____

16. Click the valve on the left side of the glass bell jar to close it.

17. Click **Start** to initiate breathing and observe the tracing that develops.

18. Click **Record Data** to display your results in the grid (and record your results in Chart 3).

19. Click the **Reset** button above the glass bell jar to draw the air out of the intrapleural space and return the lung to its normal resting condition.

20. Click **Start** to initiate breathing and observe the tracing that develops.

21. Click **Record Data** to display your results in the grid (and record your results in Chart 3).

After you complete the experiment, take the online **Post-lab Quiz** for Activity 3.

Activity Questions

1. Why is normal quiet breathing so difficult for premature infants?

____

____

2. Why does a pneumothorax frequently lead to atelectasis?

____

____

PhysioEx™ 9.1

REVIEW SHEET

EXERCISE 7

NAME ______________________

LAB TIME/DATE ______________________

Respiratory System Mechanics

ACTIVITY 1 Measuring Respiratory Volumes and Calculating Capacities

1. What would be an example of an everyday respiratory event the ERV button simulates? ______________________

2. What additional skeletal muscles are utilized in an ERV activity?

____

3. What was the FEV_1 (%) at the initial radius of 5.00 mm?

____

4. What happened to the FEV_1 (%) as the radius of the airways decreased? How well did the results compare with your prediction?

____

5. Explain why the results from the experiment suggest that there is an obstructive, rather than a restrictive, pulmonary problem.

____

ACTIVITY 2 Comparative Spirometry

1. What lung values changed (from those of the normal patient) in the spirogram when the patient with emphysema was selected? Why did these values change as they did? How well did the results compare with your prediction?

____

____

2. Which of these two parameters changed more for the patient with emphysema, the FVC or the FEV_1? ______________________

3. What lung values changed (from those of the normal patient) in the spirogram when the patient experiencing an acute asthma attack was selected? Why did these values change as they did? How well did the results compare with your prediction?

____

____

____

4. How is having an acute asthma attack similar to having emphysema? How is it different? ______________________

____

5. Describe the effect that the inhaler medication had on the asthmatic patient. Did all the spirogram values return to "normal"? Why do you think some values did not return all the way to normal? How well did the results compare with your prediction?

6. How much of an increase in FEV_1 do you think is required for it to be considered significantly improved by the medication?

7. With moderate aerobic exercise, which changed more from normal breathing, the ERV or the IRV? How well did the results compare with your prediction?

8. Compare the breathing rates during normal breathing, moderate exercise, and heavy exercise.

ACTIVITY 3 Effect of Surfactant and Intrapleural Pressure on Respiration

1. What effect does the addition of surfactant have on the airflow? How well did the results compare with your prediction?

2. Why does surfactant affect airflow in this manner?

3. What effect did opening the valve have on the left lung? Why does this happen?

4. What effect on the collapsed lung in the left side of the glass bell jar did you observe when you closed the valve? How well did the results compare with your prediction?

5. What emergency medical condition does opening the left valve simulate?

6. In the last part of this activity, you clicked the Reset button to draw the air out of the intrapleural space and return the lung to its normal resting condition. What emergency procedure would be used to achieve this result if these were the lungs in a living person? ______________________________

____

____

7. What do you think would happen when the valve is opened if the two lungs were in a single large cavity rather than separate cavities? ______________________________

____

Chemical and Physical Processes of Digestion

PRE-LAB QUIZ

1. Circle the correct underlined term: The liver / stomach produces pepsin which, in the presence of hydrochloric acid, digests protein.
2. ______________ is a hydrolytic enzyme that breaks starch down to maltose.
   a. Pepsin
   b. Salivary amylase
   c. Pancreatic lipase
   d. Bile
3. A ______________ is made in order to compare a known standard to an experimental standard.
4. Circle True or False: A positive IKI test for starch will yield a bright orange to green color.
5. An enzyme has a pocket or ______________, which the substrate or substrates must fit into temporarily in order for catalysis to occur.
   a. reagent
   b. hydrolase
   c. active site
   d. polysaccharide
6. During digestion, the chief cells of the stomach secrete ____________, which is responsible for the digestion of protein.
   a. peptidase
   b. amylase
   c. lipase
   d. pepsin
7. Circle the correct underlined term: Positive / Negative controls are used to determine whether there are any contaminating substances in the reagents used in an experiment.
8. Circle True or False: At room temperature, both fats and oils are liquid and are soluble in water.
9. A solution containing fatty acids formed by lipase activity will have a ______________ than a solution without such fatty acid production.
   a. lower pH
   b. higher pH
   c. lower temperature
   d. higher temperature

Exercise Overview

The **digestive system**, also called the gastrointestinal system, consists of the digestive tract (also called the gastrointestinal tract, or GI tract) and accessory glands that secrete enzymes and fluids needed for digestion. The digestive tract includes the mouth, pharynx, esophagus, stomach, small intestine, colon, rectum, and anus. The major functions of the digestive system are to ingest food, to break food down to its simplest components, to extract nutrients from these components for absorption into the body, and to eliminate wastes.

Most of the food we consume cannot be absorbed into our bloodstream without first being broken down into smaller subunits. **Digestion** is the process of

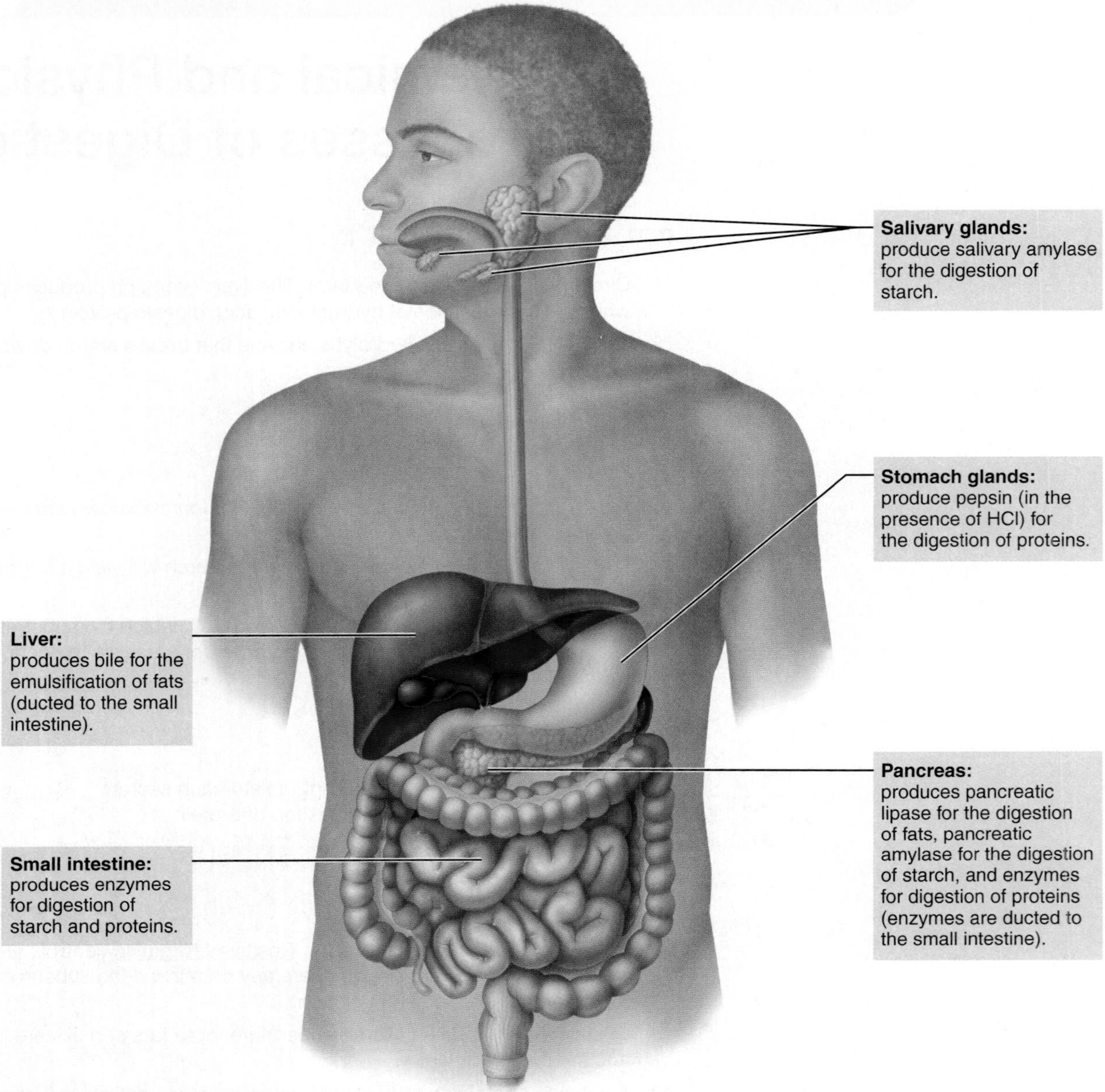

FIGURE 8.1 The human digestive system. A few sites of chemical digestion and the organs that produce the enzymes of chemical digestion.

breaking down food molecules into smaller molecules with the aid of enzymes in the digestive tract. **Enzymes** are large protein molecules produced by body cells. They are biological catalysts that increase the rate of a chemical reaction without becoming part of the product. The digestive enzymes are hydrolytic enzymes, or **hydrolases**, which break down organic food molecules, or **substrates**, by adding water to the molecular bonds, thus cleaving the bonds between the subunits, or monomers.

A hydrolytic enzyme is highly specific in its action. Each enzyme hydrolyzes one substrate molecule or, at most, a small group of substrate molecules. Specific environmental conditions are necessary for an enzyme to function optimally. For example, in extreme environments, such as high temperature, an enzyme can unravel, or denature, because of the effect that temperature has on the three-dimensional structure of the protein.

Because digestive enzymes actually function outside the body cells in the digestive tract lumen, their hydrolytic activity can also be studied in vitro in a test tube. Such in vitro studies provide a convenient laboratory environment for investigating the effect of various factors on enzymatic activity. View Figure 8.1 for an overview of chemical digestion sites in the body.

ACTIVITY 1

Assessing Starch Digestion by Salivary Amylase

OBJECTIVES

1. To explain how enzyme activity can be assessed with enzyme assays: the IKI assay and the Benedict's assay.
2. To define *enzyme*, *catalyst*, *hydrolase*, *substrate*, and *control*.
3. To understand the specificity of amylase action.
4. To name the end products of carbohydrate digestion.
5. To perform the appropriate chemical tests to determine whether digestion of a particular food has occurred.
6. To discuss the possible effect of temperature and pH on amylase activity.

Introduction

In this activity you will investigate the hydrolysis of starch to maltose by **salivary amylase**, the enzyme produced by the salivary glands and secreted into the mouth. For you to be able to detect whether or not enzymatic action has occurred, you need to be able to identify the presence of the substrate and the product to determine to what extent hydrolysis has occurred. Thus, **controls** must be prepared to provide a known standard against which comparisons can be made. With positive controls, all of the required substances are included and a positive result is expected. Sometimes negative controls are included. With negative controls, a negative result is expected. Negative results with negative controls validate the experiment. Negative controls are used to determine whether there are any contaminating substances in the reagents. So, when a positive result is produced but a negative result is expected, one or more contaminating substances are present to cause the change.

With amylase activity, starch decreases and maltose increases as digestion proceeds according to the following equation.

$$\text{Starch} + \text{water} \xrightarrow{\text{amylase}} \text{maltose}$$

Because the chemical changes that occur as starch is digested to maltose cannot be seen by the naked eye, you need to conduct an **enzyme assay**, the chemical method of detecting the presence of digested substances. You will perform two enzyme assays on each sample. The IKI assay detects the presence of starch, and the Benedict's assay tests for the presence of reducing sugars, such as glucose or maltose, which are the digestion products of starch. Normally a caramel-colored solution, IKI turns blue-black in the presence of starch. Benedict's reagent is a bright blue solution that changes to green to orange to reddish brown with increasing amounts of maltose. It is important to understand that enzyme assays only indicate the presence or absence of substances. It is up to you to analyze the results of the experiments to decide whether enzymatic hydrolysis has occurred.

EQUIPMENT USED The following equipment will be depicted on-screen: amylase—an enzyme that digests starch; starch—a complex carbohydrate substrate; maltose—a disaccharide substrate; pH buffers—solutions used to adjust the pH of the solution; deionized water—used to adjust the volume so that it is the same for each reaction; test tubes—used as reaction vessels for the various tests; incubators—used for temperature treatments (boiling, freezing, and 37°C incubation); IKI—found in the assay cabinet; used to detect the presence of starch; Benedict's reagent—found in the assay cabinet; used to detect the products of starch digestion (this includes the reducing sugars maltose and glucose).

Experiment Instructions

Go to the home page in the PhysioEx software and click **Exercise 8: Chemical and Physical Processes of Digestion**. Click **Activity 1: Assessing Starch Digestion by Salivary Amylase**, and take the online **Pre-lab Quiz** for Activity 1.

After you take the online Pre-lab Quiz, click the **Experiment** tab and begin the experiment. The experiment instructions are reprinted here for your reference. The opening screen for the experiment is shown below.

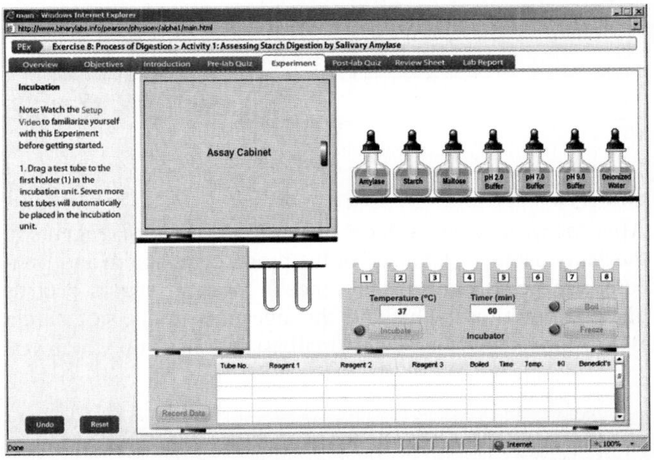

Windows Explorer 2007, Microsoft Corporation

Incubation

1. Drag a test tube to the first holder **(1)** in the incubation unit. Seven more test tubes will automatically be placed in the incubation unit.

2. Add the substances indicated below to tubes 1 through 7.

Tube 1: amylase, starch, pH 7.0 buffer

Tube 2: amylase, starch, pH 7.0 buffer

Tube 3: amylase, starch, pH 7.0 buffer

Tube 4: amylase, deionized water, pH 7.0 buffer

Tube 5: deionized water, starch, pH 7.0 buffer

Tube 6: deionized water, maltose, pH 7.0 buffer

Tube 7: amylase, starch, pH 2.0 buffer

Tube 8: amylase, starch, pH 9.0 buffer

To add a substance to a test tube, drag the dropper cap of the bottle on the solutions shelf to the top of the test tube.

3. Click the number **(1)** under the first test tube. The tube will descend into the incubation unit. All other tubes should remain in the raised position.

4. Click **Boil** to boil tube 1. After boiling for a few moments, the tube will automatically rise.

5. Click the number **(2)** under the second test tube. The tube will descend into the incubation unit. All other tubes should remain in the raised position.

6. Click **Freeze** to freeze tube 2. After freezing for a few moments, the tube will automatically rise.

7. Click **Incubate** to start the run. Note that the incubation temperature is set at 37°C and the timer is set at 60 min. The incubation unit will gently agitate the test tube rack, evenly mixing the contents of all test tubes throughout the incubation. The simulation compresses the 60-minute time period into 10 seconds of real time, so what would be a 60-minute incubation in real time will take only 10 seconds in the simulation. When the incubation time elapses, the test tube rack will automatically rise, and the doors to the assay cabinet will open.

? PREDICT **Question 1**
What effect do you think boiling and freezing will have on the activity of the amylase enzyme?

____

____

Assays

After the assay cabinet doors open, notice the two reagents in the assay cabinet. IKI tests for the presence of starch and Benedict's reagent detects the presence of reducing sugars, such as glucose or maltose, which are the digestion products of starch. Below the reagents are eight small assay tubes into which you will dispense a small amount of test solution from the incubated samples in the incubation unit, plus a drop of IKI.

8. Drag the first tube in the incubation unit to the first small assay tube on the left side of the assay cabinet to decant approximately half of the contents in the test tube into the assay tube. The decanting step will automatically repeat for the remaining tubes in the incubation unit.

9. Drag the IKI dropper cap to the first assay tube to dispense a drop of IKI into the assay tube. The dropper will automatically move across and dispense IKI to the remaining tubes.

10. Inspect the tubes for color change. A blue-black color indicates a positive starch test. If starch is not present, the mixture will look like diluted IKI, a negative starch test. Intermediate starch amounts result in a pale-gray color. Click **Record Data** to display your results in the grid (and record your results in Chart 1).

11. Drag the Benedict's reagent dropper cap to the test tube in the first holder **(1)** in the incubation unit to dispense five drops of Benedict's reagent into the tube. The dropper will automatically move across and dispense Benedict's reagent to the remaining tubes.

12. Click **Boil**. The entire tube rack will descend into the incubation unit and automatically boil the tube contents for a few moments.

13. Inspect the tubes for color change. A green-to-reddish color indicates that a reducing sugar is present; this is a positive sugar test. An orange-colored sample contains more sugar than a green sample. A reddish-brown color indicates even more sugar. A negative sugar test is indicated by no color change from the original bright blue. Click **Record Data** to display your results in the grid (and record your results in Chart 1).

After you complete the experiment, take the online **Post-lab Quiz** for Activity 1.

CHART 1 **Salivary Amylase Digestion of Starch**

Tube No.	1	2	3	4	5	6	7	8
Additives	Amylase Starch pH 7.0 buffer	Amylase Starch pH 7.0 buffer	Amylase Starch pH 7.0 buffer	Amylase Deionized water pH 7.0 buffer	Deionized water Starch pH 7.0 buffer	Deionized water Maltose pH 7.0 buffer	Amylase Starch pH 2.0 buffer	Amylase Starch pH 9.0 buffer
Incubation condition	Boil first, then incubate at 37°C for 60 minutes	Freeze first, then incubate at 37°C for 60 minutes	37°C 60 minutes	37°C 60 minutes	37°C 60 minutes	37°C 60 minutes	37°C 60 minutes	37°C 60 minutes
IKI test								
Benedict's test								

Activity Questions

1. Describe the effect that boiling had on the activity of amylase. Why did boiling have this effect? How does the effect of freezing differ from the effect of boiling?

____

____

2. What is the purpose for including tube 3 and what can you conclude from the result?

____

____

3. Describe how you determined the optimal pH for amylase activity.

____

____

4. Judging from what you learned in this activity, suggest a reason why salivary amylase would be much less active in the stomach.

____

____

ACTIVITY 2

Exploring Amylase Substrate Specificity

OBJECTIVES

1. Explain how hydrolytic enzyme activity can be assessed with the IKI assay and the Benedict's assay.
2. Understand the specificity that enzymes have for their substrate.
3. Understand the difference between the substrates starch and cellulose.
4. Explain what would be the substrate specificity of peptidase.
5. Explain how bacteria might aid in digestion.

Introduction

In this activity you will investigate the specificity that enzymes have for their substrates. To do this you will hydrolyze starch to maltose and maltotriose using **salivary amylase**, the enzyme produced by the salivary glands and secreted into the mouth. To detect whether or not enzymatic action has occurred, you need to be able to identify the presence of the substrate and the product to determine to what extent hydrolysis has occurred. The **substrate** is the substance that the enzyme acts on. The enzyme has a pocket called the **active site**, which the substrate or substrates must fit into temporarily for catalysis to occur. The substrate is often held in the active site by non-covalent bonds (weak bonds), such as ionic bonds and hydrogen bonds.

With amylase activity, starch decreases and sugar increases as digestion proceeds according to the following equation.

$$\text{Starch} + \text{water} \xrightarrow{\text{amylase}} \text{maltose} + \text{maltotriose} + \text{starch}$$

Because the chemical changes that occur as starch is digested to maltose cannot be seen by the naked eye, you need to conduct an **enzyme assay**, the chemical method of detecting the presence of digested substances. You will perform two enzyme assays on each sample. The IKI assay detects the presence of starch or cellulose and the Benedict's assay tests for the presence of reducing sugars, such as glucose or maltose, which are the digestion products of starch. Normally a caramel-colored solution, IKI turns blue-black in the presence of starch or cellulose. Benedict's reagent is a bright blue solution that changes to green to orange to reddish brown with increasing amounts of maltose. It is important to understand that enzyme assays only indicate the presence or absence of substances. It is up to you to analyze the results of the experiments to decide whether enzymatic hydrolysis has occurred.

Starch is a polysaccharide found in plants, where it is used to store energy. Plants also have the polysaccharide **cellulose**, which provides rigidity to their cell walls. Both polysaccharides are polymers of glucose, but the glucose molecules are linked differently. You will be testing salivary amylase to determine whether it digests cellulose. Also, you will investigate to see whether a bacterial suspension can digest cellulose and whether **peptidase**, a pancreatic enzyme that digests peptides, can break down starch.

EQUIPMENT USED The following equipment will be depicted on-screen: amylase—an enzyme that digests starch; starch—a polysaccharide; pH 7.0 buffer—a solution used to set the pH of the test tube solution; deionized water—used to adjust the test tube solution volume so it is the same for each reaction; glucose—a reducing sugar that is the monosaccharide subunit of both starch and cellulose; cellulose—a complex carbohydrate found in the cell wall of plants; peptidase—a pancreatic enzyme that breaks down peptides; bacteria—a suspension of live bacteria; test tubes—used as reaction vessels for the various tests; incubators—used for temperature treatments (37°C incubation); IKI—found in the assay cabinet; used to detect the presence of starch or cellulose; Benedict's reagent—found in the assay cabinet; used to detect the products of starch and cellulose digestion.

Experiment Instructions

Go to the home page in the PhysioEx software and click **Exercise 8: Chemical and Physical Processes of Digestion**. Click **Activity 2: Exploring Amylase Substrate Specificity**, and take the online **Pre-lab Quiz** for Activity 2.

After you take the online Pre-lab Quiz, click the **Experiment** tab and begin the experiment. The experiment instructions are reprinted here for your reference. The opening screen for the experiment is shown on the following page.

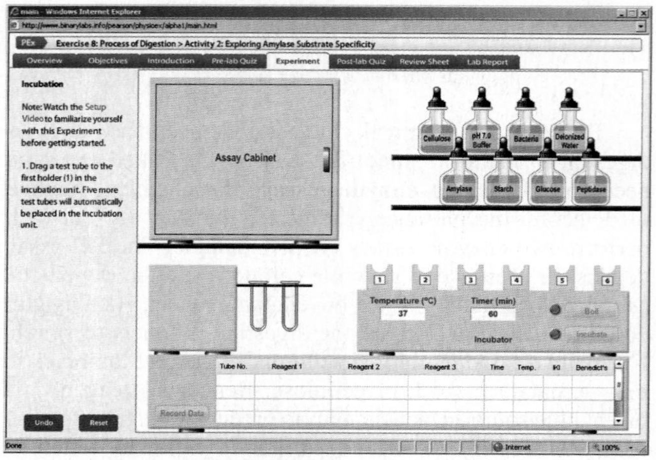

Windows Explorer 2007, Microsoft Corporation

Incubation

1. Drag a test tube to the first holder (**1**) in the incubation unit. Five more test tubes will automatically be placed in the incubation unit.

2. Add the substances indicated below to tubes 1 through 6.

Tube 1: amylase, starch, pH 7.0 buffer

Tube 2: amylase, glucose, pH 7.0 buffer

Tube 3: amylase, cellulose, pH 7.0 buffer

Tube 4: cellulose, pH 7.0 buffer, deionized water

Tube 5: peptidase, starch, pH 7.0 buffer

Tube 6: bacteria, cellulose, pH 7.0 buffer

To add a substance to a test tube, drag the dropper cap of the bottle on the solutions shelf to the test tube.

3. Click **Incubate** to start the run. Note that the incubation temperature is set at 37°C and the timer is set at 60 min. The incubation unit will gently agitate the test tube rack, evenly mixing the contents of all test tubes throughout the incubation. The simulation compresses the 60-minute time period into 10 seconds of real time, so what would be a 60-minute incubation in real life will take only 10 seconds in the simulation. When the incubation time elapses, the test tube rack will automatically rise, and the doors to the assay cabinet will open.

PREDICT Question 1
Do you think test tube 3 will show a positive Benedict's test?

____

____

Assays

After the assay cabinet doors open, notice the two reagents in the assay cabinet. IKI tests for the presence of starch and Benedict's reagent detects the presence of reducing sugars, such as glucose or maltose, which are the digestion products of starch. Below the reagents are seven small assay tubes into which you will dispense a portion of the incubated samples, plus a drop of IKI.

4. Drag the first tube in the incubation unit to the first small assay tube on the left side of the assay cabinet to decant approximately half of the contents in the test tube into the assay tube. The decanting step will automatically repeat for the remaining tubes in the incubation unit.

5. Drag the IKI dropper cap to the first assay tube to dispense a drop of IKI into the assay tube. The dropper will automatically dispense IKI into the remaining tubes.

6. Inspect the tubes for color change. A blue-black color indicates a positive starch test. If starch is not present, the mixture will look like diluted IKI, a negative starch test. Intermediate starch amounts result in a pale-gray color. Click **Record Data** to display your results in the grid (and record your results in Chart 2).

7. Drag the Benedict's reagent dropper cap to the test tube in the first holder (**1**) in the incubation unit to dispense five drops of Benedict's reagent into the tube. The dropper will automatically move across and dispense Benedict's reagent to the remaining tubes.

8. Click **Boil**. The entire tube rack will descend into the incubation unit and automatically boil the tube contents for a few moments.

9. Inspect the tubes for color change. A green-to-reddish color indicates that a reducing sugar is present; this is a positive sugar test. An orange-colored sample contains more sugar than a green sample. A reddish-brown color indicates even more sugar. A negative sugar test is indicated by no

CHART 2 Enzyme Digestion of Starch and Cellulose

Tube No.	1	2	3	4	5	6
Additives	Amylase Starch pH 7.0 buffer	Amylase Glucose pH 7.0 buffer	Amylase Cellulose pH 7.0 buffer	Deionized water Cellulose pH 7.0 buffer	Peptidase Starch pH 7.0 buffer	Bacteria Cellulose pH 7.0 buffer
Incubation condition	37°C 60 minutes	37°C 60 minutes	37°C 60 minutes	37°C 60 minutes	37°C 60 minutes	37°C 60 minutes
IKI test						
Benedict's test						

color change from the original bright blue. Click **Record Data** to display your results in the grid (and record your results in Chart 2).

After you complete the Experiment, take the online **Post-lab Quiz** for Activity 2.

Activity Questions

1. Does amylase use cellulose as a substrate?

2. What effect did the addition of bacteria have on the digestion of cellulose?

3. What effect did the addition of peptidase to the starch have? Why?

4. What is the smallest subunit into which starch can be broken down?

ACTIVITY 3

Assessing Pepsin Digestion of Protein

OBJECTIVES

1. Explain how the enzyme activity of pepsin can be assessed with the BAPNA assay.
2. Identify the substrate specificity of pepsin.
3. Discuss the effects of temperature and pH on pepsin activity.
4. Understand the pH specificity of enzyme activity and how it relates to human physiology.

Introduction

In this activity, you will explore the digestion of protein (**peptides**). Peptides are two or more **amino acids** linked together by a peptide bond. A peptide chain containing 10 to 100 amino acids is typically called a **polypeptide**. **Proteins** can consist of a large peptide chain (more than 100 amino acids) or even multiple peptide chains.

During digestion, **chief cells** of the stomach glands secrete a protein-digesting enzyme called **pepsin**. Pepsin **hydrolyzes** peptide bonds. This activity breaks up ingested proteins and polypeptides into smaller peptide chains and free amino acids. In this activity, you will use **BAPNA** as a **substrate** to assess pepsin activity. BAPNA is a synthetic "peptide" that releases a yellow dye **product** when hydrolyzed. BAPNA solutions turn yellow in the presence of an active peptidase, such as pepsin, but otherwise remain colorless.

To quantify the pepsin activity in each test solution, you will use a **spectrophotometer** to measure the amount of yellow dye produced. A spectrophotometer shines light through the sample and then measures how much light is absorbed. The fraction of light absorbed is expressed as the sample's **optical density**. Yellow solutions, where BAPNA has been hydrolyzed, will have optical densities greater than zero. The greater the optical density, the more hydrolysis has occurred. Colorless solutions, in contrast, do not absorb light and will have an optical density near zero.

Some negative controls are included in this activity. With negative controls, a negative result is expected. Negative results with negative controls validate the experiment. Negative controls are used to determine whether there are any contaminating substances in the reagents. So, when a positive result is produced but a negative result is expected, one or more contaminating substances are present to cause the change.

EQUIPMENT USED The following equipment will be depicted on-screen: pepsin—an enzyme that digests peptides; BAPNA—a synthetic "peptide"; pH buffers—solutions used to set the pH of the test tube solution; deionized water—used to adjust the test tube solution volume so it is the same for each reaction; test tubes—used as reaction vessels for the various tests; incubators—used for temperature treatments (boiling and 37°C incubation); spectrophotometer—found in the assay cabinet; used to measure the optical density of solutions.

Experiment Instructions

Go to the home page in the PhysioEx software and click **Exercise 8: Chemical and Physical Processes of Digestion.** Click **Activity 3: Assessing Pepsin Digestion of Protein**, and take the online **Pre-lab Quiz** for Activity 3.

After you take the online Pre-lab Quiz, click the **Experiment** tab and begin the experiment. The experiment instructions are reprinted here for your reference. The opening screen for the experiment is shown below.

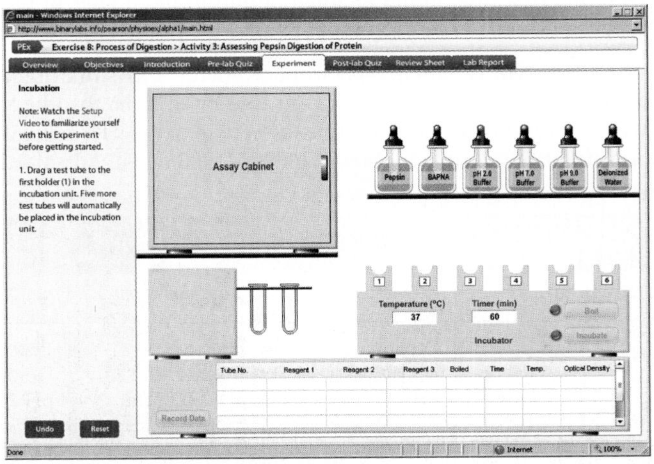

Windows Explorer 2007, Microsoft Corporation

Incubation

1. Drag a test tube to the first holder (**1**) in the incubation unit. Five more test tubes will automatically be placed in the incubation unit.

2. Add the substances indicated below to tubes 1 through 6.

Tube 1: pepsin, BAPNA, pH 2.0 buffer

Tube 2: pepsin, BAPNA, pH 2.0 buffer

Tube 3: pepsin, deionized water, pH 2.0 buffer

Tube 4: deionized water, BAPNA, pH 2.0 buffer

Tube 5: pepsin, BAPNA, pH 7.0 buffer

Tube 6: pepsin, BAPNA, pH 9.0 buffer

To add a substance to a test tube, drag the dropper cap of the bottle on the solutions shelf to the test tube.

3. Click the number **(1)** under the first test tube. The tube will descend into the incubation unit. All other tubes should remain in the raised position.

4. Click **Boil** to boil tube 1. After boiling for a few moments, the tube will automatically rise.

5. Click **Incubate** to start the run. Note that the incubation temperature is set at 37°C and the timer is set at 60 min. The incubation unit will gently agitate the test tube rack, evenly mixing the contents of all test tubes throughout the incubation. The simulation compresses the 60-minute time period into 10 seconds of real time, so what would be a 60-minute incubation in real life will take only 10 seconds in the simulation. When the incubation time elapses, the test tube rack will automatically rise, and the doors to the assay cabinet will open. The spectrophotometer is in the assay cabinet.

? PREDICT Question 1
At which pH do you think pepsin will have the highest activity?

____

____

Assays

6. You will now use the spectrophotometer to measure how much yellow dye was liberated from BAPNA hydrolysis. Drag the first tube in the incubation unit to the holder in the spectrophotometer to drop the tube into the holder.

7. Click **Analyze**. The spectrophotometer will shine light through the solution to measure the amount of light absorbed, which it reports as the solution's optical density. The optical density of the sample is shown in the optical density display.

8. Click **Record Data** to display your results in the grid (and record your results in Chart 3).

9. Drag the tube to its original position in the incubation unit.

10. Analyze the remaining five tubes by repeating the following steps for each tube.

- Drag the tube to the holder in the spectrophotometer to drop the tube into the holder.
- Click **Analyze**.
- Drag the tube to its original position in the incubation unit.

After you have analyzed all five tubes, click **Record Data** to display your results in the grid (and record your results in Chart 3).

After you complete the experiment, take the online **Post-lab Quiz** for Activity 3.

Activity Questions

1. Describe the significance of the optimum pH for pepsin observed in the simulation and the secretion of pepsin by the chief cells of the gastric glands.

____

____

2. Would pepsin be active in the mouth? Explain your answer.

____

____

3. What are the subunit products of peptide digestion?

____

____

4. Describe the reason for including control tube 4.

____

____

CHART 3 Pepsin Digestion of Protein

Tube No.	**1**	**2**	**3**	**4**	**5**	**6**
Additives	Pepsin BAPNA pH 2.0 buffer	Pepsin BAPNA pH 2.0 buffer	Pepsin Deionized water pH 2.0 buffer	Deionized water BAPNA pH 2.0 buffer	Pepsin BAPNA pH 7.0 buffer	Pepsin BAPNA pH 9.0 buffer
Incubation condition	Boil first, then incubate at 37°C for 60 minutes	37°C 60 minutes	37°C 60 minutes	37°C 60 minutes	37°C 60 minutes	37°C 60 minutes
Optical density						

ACTIVITY 4

Assessing Lipase Digestion of Fat

OBJECTIVES

1. Explain how the enzyme activity of pancreatic lipase can be assessed with a pH-based measurement.
2. Identify the hydrolysis products of fat digestion.
3. Understand the role that bile plays in fat digestion.
4. Understand the significance of pH specificity of lipase activity and how it relates to human physiology.
5. Discuss the difficulty of using pH to measure digestion when comparing the activity of lipase at various pHs.

Introduction

Fats and oils belong to a diverse class of molecules called lipids. **Triglycerides**, a type of lipid, make up both fats and oils. At room temperature, fats are solid and oils are liquid. Both are poorly soluble in water. This insolubility of triglycerides presents a challenge during digestion because they tend to clump together, leaving only the surface molecules exposed to **lipase** enzymes. To overcome this difficulty, **bile salts** are secreted into the small intestine during digestion to physically emulsify lipids. Bile salts act like a detergent, separating the lipid clumps and increasing the surface area accessible to lipase enzymes.

As a result, two reactions must occur. First,

$$\text{Triglyceride clumps} \xrightarrow[\text{(emulsification)}]{\text{bile}} \text{minute triglyceride droplets}$$

Then,

$$\text{Triglyceride} \xrightarrow{\text{lipase}} \text{monoglyceride} + \text{two fatty acids}$$

Lipase hydrolyzes each triglyceride to a monoglyceride and two fatty acids. In addition to the **pancreatic lipase** secreted into the small intestine, **lingual lipase** and **gastric lipase** are also secreted. Even though bile salts are not secreted in the mouth or the stomach, small amounts of lipids are digested by these other lipases.

Because some of the end products of fat digestion are acidic (that is, fatty acids), lipase activity can be easily measured by monitoring the solution's **pH**. A solution containing fatty acids liberated by lipase activity will have a lower pH than a solution without such fatty acid production. You will record pH in this activity with a **pH meter**.

EQUIPMENT USED The following equipment will be depicted on-screen: lipase—an enzyme that digests triglycerides; vegetable oil—a mixture of triglycerides; bile salts—a solution that physically separates fats into smaller droplets; pH buffers—solutions used to set the pH of the test tube solution; deionized water—used to adjust the test tube solution volume so it is the same for each reaction; test tubes—used as reaction vessels for the various tests; incubators—used for temperature treatments (boiling and 37°C incubation); pH meter—found in the assay cabinet; used to measure pH.

Experiment Instructions

Go to the home page in the PhysioEx software and click **Exercise 8: Chemical and Physical Processes of Digestion**. Click **Activity 4: Assessing Lipase Digestion of Fat**, and take the online **Pre-lab Quiz** for Activity 4.

After you take the online Pre-lab Quiz, click the **Experiment** tab and begin the experiment. The experiment instructions are reprinted here for your reference. The opening screen for the experiment is shown below.

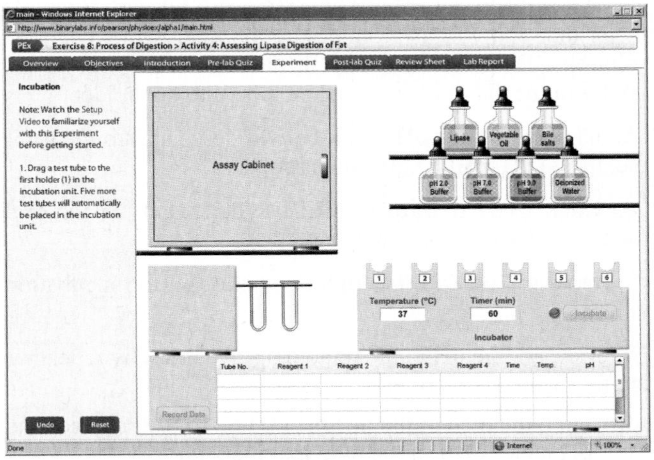

Windows Explorer 2007, Microsoft Corporation

Incubation

1. Drag a test tube to the first holder **(1)** in the incubation unit. Five more test tubes will automatically be placed in the incubation unit.

2. Add the substances indicated below to tubes 1 through 6.

Tube 1: lipase, vegetable oil, bile salts, pH 7.0 buffer

Tube 2: lipase, vegetable oil, deionized water, pH 7.0 buffer

Tube 3: lipase, deionized water, bile salts, pH 9.0 buffer

Tube 4: deionized water, vegetable oil, bile salts, pH 7.0 buffer

Tube 5: lipase, vegetable oil, bile salts, pH 2.0 buffer

Tube 6: lipase, vegetable oil, bile salts, pH 9.0 buffer

To add a substance to a test tube, drag the dropper cap of the bottle on the solutions shelf to the test tube.

3. Click **Incubate** to start the run. Note that the incubation temperature is set at 37°C and the timer is set at 60 min. The incubation unit will gently agitate the test tube rack, evenly mixing the contents of all test tubes throughout the incubation. The simulation compresses the 60-minute time period into 10 seconds of real time, so what would be a 60-minute incubation in real life will take only 10 seconds in the simulation. When the incubation time elapses, the test tube rack will automatically rise, and the doors to the assay cabinet will open.

? PREDICT Question 1
Which tube do you think will have the highest lipase activity?

____

____

Assays

4. After the assay cabinet doors open, you will see a pH meter that you will use to measure the final pH of your test solutions. Drag the first tube in the incubation unit to the holder in the pH meter to drop the tube into the holder.

5. Click **Measure pH**. A probe will descend into the sample, take a pH reading, and then retract.

6. Click **Record Data** to display your results in the grid (and record your results in Chart 4).

7. Drag the tube to its original position in the incubation unit.

8. Measure the pH in the remaining five tubes by repeating the following steps for each tube.

- Drag the tube in the incubation unit to the holder in the pH meter to drop the tube into the holder.
- Click **Measure pH.**
- Drag the tube to its original position in the incubation unit.

After you have measured the pH in all five tubes, click **Record Data** to display your results in the grid (and record your results in Chart 4).

After you complete the experiment, take the online **Post-lab Quiz** for Activity 4.

Activity Questions

1. Describe how lipase activity is measured in the simulation.

____

____

2. Can you determine if fat hydrolysis occurred in tube 5? Why or why not?

____

____

3. Would pancreatic lipase be active in the mouth? Why or why not?

____

____

4. Describe the physical separation of fats by bile salts.

____

____

CHART 4 Pancreatic Lipase Digestion of Triglycerides and the Action of Bile

Tube No.	1	2	3	4	5	6
Additives	Lipase Vegetable oil Bile salts pH 7.0 buffer	Lipase Vegetable oil Deionized water pH 7.0 buffer	Lipase Deionized water Bile salts pH 9.0 buffer	Deionized water Vegetable oil Bile salts pH 7.0 buffer	Lipase Vegetable oil Bile salts pH 2.0 buffer	Lipase Vegetable oil Bile salts pH 9.0 buffer
Incubation condition	37°C 60 minutes	37°C 60 minutes	37°C 60 minutes	37°C 60 minutes	37°C 60 minutes	37°C 60 minutes
pH						

NAME ______________________

LAB TIME/DATE ______________________

Chemical and Physical Processes of Digestion

ACTIVITY 1 Assessing Starch Digestion by Salivary Amylase

1. List the substrate and the subunit product of amylase. ______________________

2. What effect did boiling and freezing have on enzyme activity? Why? How well did the results compare with your prediction? ______________________

3. At what pH was the amylase most active? Describe the significance of this result. ______________________

4. Briefly describe the need for controls and give an example used in this activity. ______________________

5. Describe the significance of using a 37°C incubation temperature to test salivary amylase activity. ______________________

ACTIVITY 2 Exploring Amylase Substrate Specificity

1. Describe why the results in tube 1 and tube 2 are the same. ______________________

2. Describe the result in tube 3. How well did the results compare with your prediction? ______________________

3. Describe the usual substrate for peptidase. ______________________

4. Explain how bacteria can aid in digestion. ______________________

ACTIVITY 3 Assessing Pepsin Digestion of Protein

1. Describe the effect that boiling had on pepsin and how you could tell that it had that effect. ______

   Was your prediction correct about the optimal pH for pepsin activity? Discuss the physiological correlation behind your results.

   ______

2. What do you think would happen if you reduced the incubation time to 30 minutes for tube 5? ______

ACTIVITY 4 Assessing Lipase Digestion of Fat

1. Explain why you can't fully test the lipase activity in tube 5. ______

2. Which tube had the highest lipase activity? How well did the results compare with your prediction? Discuss possible reasons why it may or may not have matched. ______

3. Explain why pancreatic lipase would be active in both the mouth and the intestine. ______

4. Describe the process of bile emulsification of lipids and how it improves lipase activity. ______

Renal System Physiology

PRE-LAB QUIZ

1. All of the following are functions of the kidney *except:*
   a. regulate the body's acid-base balance
   b. regulate the body's electrolyte balance
   c. regulate plasma volume
   d. regulate smooth muscle function of bladder
2. Circle the correct underlined term: During urine formation, the process of tubular secretion / tubular reabsorption leaves behind mostly salt water and wastes after moving most of the filtrate back into the blood.
3. Circle True or False: The glomerular filtration rate in humans is such that in 24 hours, approximately 180 liters of filtrate is produced.
4. The __________ is a capillary knot that filters fluid from the blood into the renal tubule.
   a. glomerulus
   b. distal convoluted tubule
   c. proximal convoluted tubule
   d. loop of Henle
5. Circle the correct underlined term: Starling forces / Tubular secretions are responsible for driving protein-free fluid out of the glomerular capillaries and into Bowman's capsule.
6. What hormone is responsible for increasing the water permeability of the collecting duct, allowing water to flow to areas of higher solute concentration?
   a. aldosterone
   b. antidiuretic hormone
   c. diuretic hormone
   d. thyroxine
7. Circle the correct underlined term: Aldosterone / Thyroid hormone acts on the distal convoluted tubule to cause sodium to be reabsorbed and potassium to be excreted.
8. Under what conditions would excess glucose be eliminated in the urine?

   ____

Exercise Overview

The **kidney** is *both* an excretory and a regulatory organ. By filtering the water and solutes in the blood, the kidneys are able to *excrete* excess water, waste products, and even foreign materials from the body. However, the kidneys also *regulate* (1) plasma osmolarity (the concentration of a solution expressed as osmoles of solute per liter of solvent), (2) plasma volume, (3) the body's acid-base balance, and (4) the body's electrolyte balance. All these activities are extremely important for maintaining homeostasis in the body.

The paired kidneys are located between the posterior abdominal wall and the abdominal peritoneum. The right kidney is slightly lower than the left kidney. Each human kidney contains approximately one million **nephrons**, the functional units of the kidney.

Each nephron is composed of a **renal corpuscle** and a **renal tubule**. The renal corpuscle consists of a "ball" of capillaries, called the *glomerulus*, which is enclosed by a fluid-filled capsule, called *Bowman's capsule*, or the glomerular capsule. An **afferent arteriole** supplies blood to the glomerulus. As blood flows through the glomerular capillaries, protein-free plasma filters into the Bowman's capsule, a process called **glomerular filtration**. An **efferent arteriole** then drains the glomerulus of the remaining blood (view Figure 9.1, p. PEx-132).

The filtrate flows from Bowman's capsule into the start of the renal tubule, called the **proximal convoluted tubule**, then into the **loop of Henle**, a U-shaped

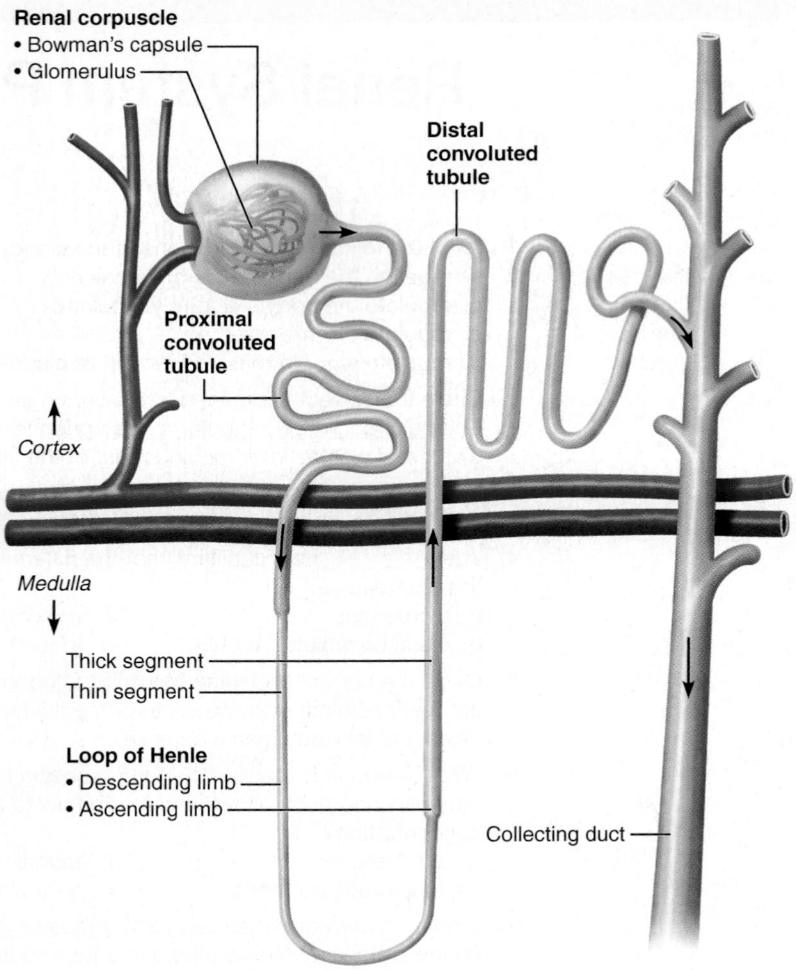

FIGURE 9.1 Location and structure of nephrons.

hairpin loop, and, finally, into the **distal convoluted tubule** before emptying into a **collecting duct**. From the collecting duct, the filtrate flows into, and collects in, the minor calyces.

The nephron performs three important functions that process blood into filtrate and urine: (1) glomerular filtration, (2) tubular reabsorption, and (3) tubular secretion. **Glomerular filtration** is a passive process in which fluid passes from the lumen of the glomerular capillary into the glomerular capsule of the renal tubule. **Tubular reabsorption** moves most of the filtrate back into the blood, leaving mainly salt water and the wastes in the lumen of the tubule. Some of the desirable, or needed, solutes are actively reabsorbed, and others move passively from the lumen of the tubule into the interstitial spaces. **Tubular secretion** is essentially the reverse of tubular reabsorption and is a process by which the kidneys can rid the blood of additional unwanted substances, such as creatinine and ammonia.

The reabsorbed solutes and water that move into the interstitial space between the nephrons need to be returned to the blood, or the kidneys will rapidly swell like balloons. The **peritubular capillaries** surrounding the renal tubule reclaim the reabsorbed substances and return them to general circulation. Peritubular capillaries arise from the efferent arteriole exiting the glomerulus and empty into the renal veins leaving the kidney.

ACTIVITY 1

The Effect of Arteriole Radius on Glomerular Filtration

OBJECTIVES

1. To understand the terms *nephron*, *glomerulus*, *glomerular capillaries*, *renal tubule*, *filtrate*, *Bowman's capsule*, *renal corpuscle*, *afferent arteriole*, *efferent arteriole*, *glomerular capillary pressure*, and *glomerular filtration rate*.
2. To understand how changes in afferent arteriole radius impact glomerular capillary pressure and filtration.
3. To understand how changes in efferent arteriole radius impact glomerular capillary pressure and filtration.

Introduction

Each of the million **nephrons** in each kidney contains two major parts: (1) a tubular component, the **renal tubule**, and (2) a vascular component, the **renal corpuscle** (view Figure 9.1). The **glomerulus** is a tangled capillary knot that filters fluid from the blood into the lumen of the renal tubule. The function of the renal tubule is to process the filtered fluid, also called the **filtrate**. The beginning of the renal tubule is an enlarged end called **Bowman's capsule** (or the glomerular capsule), which surrounds the glomerulus and serves to funnel the filtrate into the rest of the renal tubule. Collectively, the glomerulus and Bowman's capsule are called the renal corpuscle.

Two arterioles are associated with each glomerulus: an **afferent arteriole** feeds the **glomerular capillary** bed and an **efferent arteriole** drains it. These arterioles are responsible for blood flow through the glomerulus. The diameter of the efferent arteriole is smaller than the diameter of the afferent arteriole, restricting blood flow out of the glomerulus. Consequently, the pressure in the glomerular capillaries forces fluid through the endothelium of the capillaries into the lumen of the surrounding Bowman's capsule. In essence, everything in the blood except for the blood cells (red and white) and plasma proteins is filtered through the glomerular wall. From the Bowman's capsule, the filtrate moves into the rest of the renal tubule for processing. The job of the tubule is to reabsorb all the beneficial substances from its lumen and allow the wastes to travel down the tubule for elimination from the body.

During glomerular filtration, blood enters the glomerulus from the afferent arteriole and protein-free plasma flows from the blood across the walls of the glomerular capillaries and into the Bowman's capsule. The **glomerular filtration rate** is an index of kidney function. In humans, the filtration rate ranges from 80 to 140 ml/min, so that, in 24 hours, as much as 180 liters of filtrate is produced by the glomeruli. The filtrate formed is devoid of cellular debris, is essentially protein free, and contains a concentration of salts and organic molecules similar to that in blood.

The glomerular filtration rate can be altered by changing arteriole resistance or arteriole hydrostatic pressure. In this activity, you will explore the effect of arteriole radius on glomerular capillary pressure and filtration in a single nephron. You can apply the concepts you learn by studying a single nephron to understand the function of the kidney as a whole.

EQUIPMENT USED The following equipment will be depicted on-screen: source beaker for blood (first beaker on left side of screen)—simulates blood flow and pressure (mm Hg) from general circulation to the nephron; drain beaker for blood (second beaker on left side of screen)—simulates the renal vein; flow tube with adjustable radius—simulates the afferent arteriole and connects the blood supply to the glomerular capillaries; second flow tube with adjustable radius—simulates the efferent arteriole and drains the glomerular capillaries into the peritubular capillaries, which ultimately drain into the renal vein (drain beaker); simulated nephron (The filtrate forms in Bowman's capsule, flows through the renal tubule—the tubular components—and empties into a collecting duct, which in turn drains into the urinary bladder.); nephron tank; glomerulus—"ball" of capillaries that forms part of the filtration membrane; glomerular (Bowman's) capsule—forms part of the filtration membrane and a capsular space where the filtrate initially forms; proximal convoluted tubule; loop of Henle; distal convoluted tubule; collecting duct; drain beaker for filtrate (beaker on right side of screen)—simulates the urinary bladder.

Experiment Instructions

Go to the home page in the PhysioEx software and click **Exercise 9: Renal System Physiology**. Click **Activity 1: The Effect of Arteriole Radius on Glomerular Filtration**, and take the online **Pre-lab Quiz** for Activity 1.

After you take the online Pre-lab Quiz, click the **Experiment** tab and begin the experiment. The experiment instructions are reprinted here for your reference. The opening screen for the experiment is shown below.

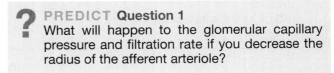

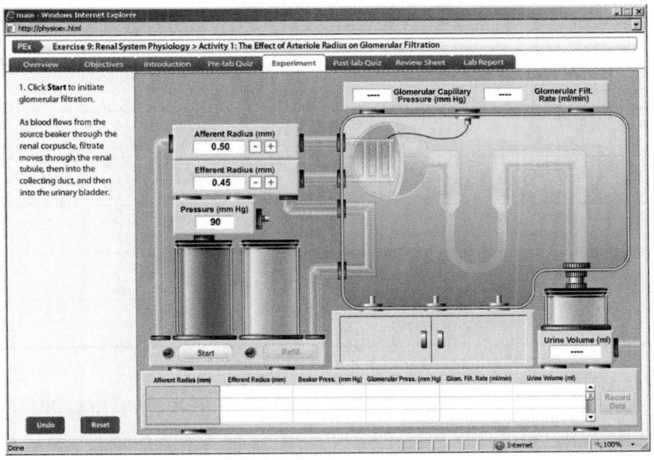

Windows Explorer 2007, Microsoft Corporation

1. Click **Start** to initiate glomerular filtration. As blood flows from the source beaker through the renal corpuscle, filtrate moves through the renal tubule, then into the collecting duct, and then into the urinary bladder.

2. The glomerular capillary pressure display shows the hydrostatic blood pressure in the glomerular capillaries that promotes filtration, and the filtration rate display shows the flow rate of the fluid moving from the lumen of the glomerular capillaries into the lumen of Bowman's capsule. Click **Record Data** to display your results in the grid (and record your results in Chart 1, p. PEx-134).

3. Click **Refill** to replenish the source beaker and prepare the nephron for the next run.

PREDICT Question 1
What will happen to the glomerular capillary pressure and filtration rate if you decrease the radius of the afferent arteriole?

____

4. Decrease the radius of the afferent arteriole to 0.45 mm by clicking the − button beside the afferent radius display. Click **Start** to initiate glomerular filtration.

5. Note the glomerular capillary pressure and glomerular filtration rate displays and click **Record Data** to display your results in the grid (and record your results in Chart 1).

6. Click **Refill** to replenish the source beaker and prepare the nephron for the next run.

CHART 1 **Effect of Arteriole Radius on Glomerular Filtration**

Afferent arteriole radius (mm)	Efferent arteriole radius (mm)	Glomerular capillary pressure (mm Hg)	Glomerular filtration rate (ml/min)

7. You will now observe the effect of incremental decreases in the radius of the afferent arteriole.

- Decrease the radius of the afferent arteriole by 0.05 mm by clicking the − button beside the afferent radius display.
- Click **Start** to initiate glomerular filtration.
- Note the glomerular capillary pressure and glomerular filtration rate displays and click **Record Data** to display your results in the grid (and record your results in Chart 1).
- Click **Refill** to replenish the source beaker and prepare the nephron for the next run.

Repeat this step until you reach an afferent arteriole radius of 0.35 mm.

> **PREDICT Question 2**
> What will happen to the glomerular capillary pressure and filtration rate if you increase the radius of the afferent arteriole?
>
> __________________________________

8. Increase the radius of the afferent arteriole to 0.55 mm by clicking the + button beside the afferent radius display. Click **Start** to initiate glomerular filtration.

9. Note the glomerular capillary pressure and glomerular filtration rate displays and click **Record Data** to display your results in the grid (and record your results in Chart 1).

10. Click **Refill** to replenish the source beaker and prepare the nephron for the next run.

11. Increase the radius of the afferent arteriole to 0.60 mm. Click **Start** to initiate glomerular filtration.

12. Note the glomerular capillary pressure and glomerular filtration rate displays and click **Record Data** to display your results in the grid (and record your results in Chart 1).

13. Click **Refill** to replenish the source beaker and prepare the nephron for the next run.

> **PREDICT Question 3**
> What will happen to the glomerular capillary pressure and filtration rate if you decrease the radius of the efferent arteriole?
>
> __________________________________

14. Decrease the radius of the afferent arteriole to 0.50 mm by clicking the − button beside the afferent radius display. Click **Start** to initiate glomerular filtration.

15. Note the glomerular capillary pressure and glomerular filtration rate displays and click **Record Data** to display your results in the grid (and record your results in Chart 1).

16. Click **Refill** to replenish the source beaker and prepare the nephron for the next run.

17. You will now observe the effect of incremental decreases in the radius of the efferent arteriole.

- Decrease the radius of the efferent arteriole by 0.05 mm by clicking the − button beside the efferent radius display.
- Click **Start** to initiate glomerular filtration.

- Note the glomerular capillary pressure and glomerular filtration rate displays and click **Record Data** to display your results in the grid (and record your results in Chart 1).
- Click **Refill** to replenish the source beaker and prepare the nephron for the next run.

Repeat this step until you reach an efferent arteriole radius of 0.30 mm.

After you complete the experiment, take the online **Post-lab Quiz** for Activity 1.

Activity Questions

1. Activation of sympathetic nerves that innervate the kidney leads to a decreased urine production. Knowing that fact, what do you think the sympathetic nerves do to the afferent arteriole?

____

____

2. How is this effect of the sympathetic nervous system beneficial? Could this effect become harmful if it goes on too long?

____

____

ACTIVITY 2

The Effect of Pressure on Glomerular Filtration

OBJECTIVES

1. To understand the terms *glomerulus*, *glomerular capillaries*, *renal tubule*, *filtrate*, *Starling forces*, *Bowman's capsule*, *renal corpuscle*, *afferent arteriole*, *efferent arteriole*, *glomerular capillary pressure*, and *glomerular filtration rate*.
2. To understand how changes in glomerular capillary pressure affect glomerular filtration rate.
3. To understand how changes in renal tubule pressure affect glomerular filtration rate.

Introduction

Cellular metabolism produces a complex mixture of waste products that must be eliminated from the body. This excretory function is performed by a combination of organs, most importantly, the paired kidneys. Each kidney consists of approximately one million nephrons, which carry out three crucial processes: (1) glomerular filtration, (2) tubular reabsorption, and (3) tubular secretion.

Both the blood pressure in the **glomerular capillaries** and the **filtrate** pressure in the **renal tubule** can have a significant impact on the **glomerular filtration rate**. During glomerular filtration, blood enters the **glomerulus** from the **afferent arteriole**. **Starling forces** (hydrostatic and osmotic pressure gradients) drive protein-free fluid between the blood in the glomerular capillaries and the filtrate in **Bowman's capsule**. The glomerular filtration rate is an index of kidney function. In humans, the filtration rate ranges from 80 to 140 ml/min, so that, in 24 hours, as much as 180 liters of filtrate is produced by the glomerular capillaries. The filtrate formed is devoid of blood cells, is essentially protein free, and contains a concentration of salts and organic molecules similar to that in blood.

Approximately 20% of the blood that enters the glomerular capillaries is normally filtered into Bowman's capsule, where it is then referred to as filtrate. The unusually high hydrostatic blood pressure in the glomerular capillaries promotes this filtration. Thus, the glomerular filtration rate can be altered by changing the afferent arteriole resistance (and, therefore, the hydrostatic pressure). In this activity you will explore the effect of blood pressure on the glomerular filtration rate in a single nephron. You can apply the concepts you learn by studying a single nephron to understand the function of the kidney as a whole.

EQUIPMENT USED The following equipment will be depicted on-screen: left source beaker (first beaker on left side of screen)—simulates blood flow and pressure (mm Hg) from general circulation to the nephron; drain beaker for blood (second beaker on left side of screen)—simulates the renal vein; flow tube with adjustable radius—simulates the afferent arteriole and connects the blood supply to the glomerular capillaries; second flow tube with adjustable radius—simulates the efferent arteriole and drains the glomerular capillaries into the peritubular capillaries, which ultimately drain into the renal vein (drain beaker); simulated nephron (The filtrate forms in Bowman's capsule, flows through the renal tubule—the tubular components—and empties into a collecting duct, which in turn drains into the urinary bladder.); nephron tank; glomerulus—"ball" of capillaries that forms part of the filtration membrane; glomerular (Bowman's) capsule—forms part of the filtration membrane and a capsular space where the filtrate initially forms; proximal convoluted tubule; loop of Henle; distal convoluted tubule; collecting duct; one-way valve between end of collecting tube (duct) and urinary bladder—used to restrict the flow of filtrate into the urinary bladder, increasing the volume and pressure in the renal tubule; drain beaker for filtrate (beaker on right side of screen)—simulates the urinary bladder.

Experiment Instructions

Go to the home page in the PhysioEx software and click **Exercise 9: Renal System Physiology**. Click **Activity 2: The Effect of Pressure on Glomerular Filtration**, and take the online **Pre-lab Quiz** for Activity 2.

After you take the online Pre-lab Quiz, click the **Experiment** tab and begin the experiment. The experiment instructions are reprinted here for your reference. The opening screen for the experiment is shown on the following page.

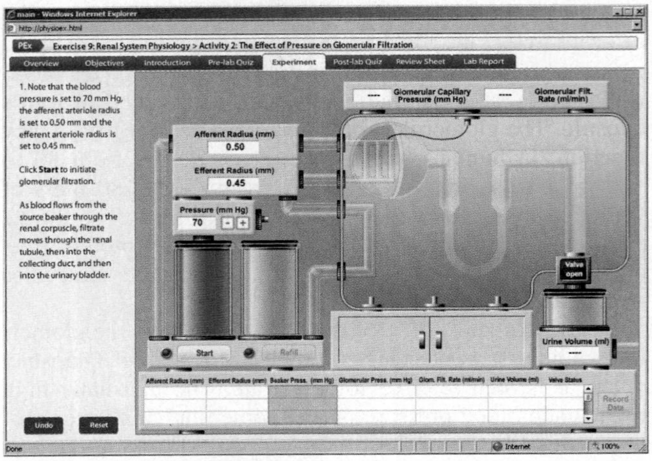

Windows Explorer 2007, Microsoft Corporation

1. Note that the blood pressure is set to 70 mm Hg, the afferent arteriole radius is set to 0.50 mm, and the efferent arteriole radius is set to 0.45 mm. Click **Start** to initiate glomerular filtration. As blood flows from the source beaker through the renal corpuscle, filtrate moves through the renal tubule, then into the collecting duct, and then into the urinary bladder.

2. The glomerular capillary pressure display shows the hydrostatic blood pressure in the glomerular capillaries that promotes filtration, and the filtration rate display shows the flow rate of the fluid moving from the lumen of the glomerular capillaries into the lumen of Bowman's capsule. Click **Record Data** to display your results in the grid (and record your results in Chart 2.)

3. Click **Refill** to replenish the source beaker and prepare the nephron for the next run.

? PREDICT Question 1
What will happen to the glomerular capillary pressure and filtration rate if you increase the blood pressure in the left source beaker?

____

4. Increase the blood pressure to 80 mm Hg by clicking the + button beside the pressure display. Click **Start** to initiate glomerular filtration.

5. Note the glomerular capillary pressure and glomerular filtration rate displays and click **Record Data** to display your results in the grid (and record your results in Chart 2).

6. Click **Refill** to replenish the source beaker and prepare the nephron for the next run.

7. You will now observe the effect of further incremental increases in blood pressure.

- Increase the blood pressure by 10 mm Hg by clicking the + button beside the pressure display.
- Click **Start** to initiate glomerular filtration.
- Note the glomerular capillary pressure and glomerular filtration rate displays and click **Record Data** to display your results in the grid (and record your results in Chart 2).
- Click **Refill** to replenish the source beaker and prepare the nephron for the next run.

Repeat this step until you reach a blood pressure of 100 mm Hg.

? PREDICT Question 2
What will happen to the filtrate pressure in Bowman's capsule (not directly measured in this experiment) and the filtration rate if you close the one-way valve between the collecting duct and the urinary bladder?

____

8. Note that the valve between the collecting duct and the urinary bladder is open. Decrease the blood pressure to 70 mm Hg by clicking the button beside the pressure display. Click **Start** to initiate glomerular filtration.

CHART 2 Effect of Pressure on Glomerular Filtration

Blood pressure (mm Hg)	Valve (open or closed)	Glomerular capillary pressure (mm Hg)	Glomerular filtration rate (ml/min)	Urine volume (ml)

9. Note the glomerular capillary pressure and glomerular filtration rate displays and click **Record Data** to display your results in the grid (and record your results in Chart 2).

10. Click **Refill** to replenish the source beaker and prepare the nephron for the next run.

11. Click the valve between the collecting duct and the urinary bladder to close it. Click **Start** to initiate glomerular filtration.

12. Note the glomerular capillary pressure and glomerular filtration rate displays and click **Record Data** to display your results in the grid (and record your results in Chart 2).

13. Click **Refill** to replenish the source beaker and prepare the nephron for the next run.

14. Increase the blood pressure to 100 mm Hg. Click **Start** to initiate glomerular filtration.

15. Note the glomerular capillary pressure and glomerular filtration rate displays and click **Record Data** to display your results in the grid (and record your results in Chart 2).

16. Click **Refill** to replenish the source beaker and prepare the nephron for the next run.

17. Click the valve between the collecting duct and the urinary bladder to open it. Click **Start** to initiate glomerular filtration.

18. Note the glomerular capillary pressure and glomerular filtration rate displays and click **Record Data** to display your results in the grid (and record your results in Chart 2).

After you complete the experiment, take the online **Post-lab Quiz** for Activity 2.

Activity Questions

1. Judging from the results in this laboratory activity, what *should be* the effect of blood pressure on glomerular filtration?

____

____

2. Persistent high blood pressure with inadequate glomerular filtration is now a frequent problem in Western cultures. Using the concepts in this activity, explain this health problem.

____

____

ACTIVITY 3

Renal Response to Altered Blood Pressure

OBJECTIVES

1. To understand the terms *nephron*, *renal tubule*, *filtrate*, *Bowman's capsule*, *blood pressure*, *afferent arteriole*, *efferent arteriole*, *glomerulus*, *glomerular filtration rate*, and *glomerular capillary pressure*.
2. To understand how blood pressure affects glomerular capillary pressure and glomerular filtration.
3. To observe which is more effective: changes in afferent or efferent arteriole radius when changes in blood pressure occur.

Introduction

In humans approximately 180 liters of filtrate flows into the **renal tubules** every day. As demonstrated in Activity 2, the **blood pressure** supplying the **nephron** can have a substantial impact on the **glomerular capillary pressure** and **glomerular filtration**. However, under most circumstances, glomerular capillary pressure and glomerular filtration remain relatively constant despite changes in blood pressure because the nephron has the capacity to alter its **afferent** and **efferent arteriole** radii.

During glomerular filtration, blood enters the **glomerulus** from the afferent arteriole. **Starling forces** (primarily hydrostatic pressure gradients) drive protein-free fluid out of the glomerular capillaries and into **Bowman's capsule**. Importantly for our body's homeostasis, a relatively constant glomerular filtration rate of 125 ml/min is maintained despite a wide range of blood pressures that occur throughout the day for an average human.

Activities 1 and 2 explored the independent effects of arteriole radii and blood pressure on glomerular capillary pressure and glomerular filtration. In the human body, these effects occur simultaneously. Therefore, in this activity, you will alter both variables to explore their combined effects on glomerular filtration and observe how changes in one variable can compensate for changes in the other to maintain an adequate glomerular filtration rate.

EQUIPMENT USED The following equipment will be depicted on-screen: left source beaker (first beaker on left side of screen)—simulates blood flow and pressure (mm Hg) from general circulation to the nephron; drain beaker for blood (second beaker on left side of screen)—simulates the renal vein; flow tube with adjustable radius—simulates the afferent arteriole and connects the blood supply to the glomerular capillaries; second flow tube with adjustable radius—simulates the efferent arteriole and drains the glomerular capillaries into the peritubular capillaries, which ultimately drain into the renal vein (drain beaker); simulated nephron (The filtrate forms in Bowman's capsule, flows through the renal tubule—the tubular components—and empties into a collecting duct, which in turn drains into the urinary bladder.); nephron tank; glomerulus—"ball" of capillaries that forms part of the filtration membrane; glomerular (Bowman's) capsule—forms part of the filtration membrane and a capsular space where the filtrate initially forms; proximal convoluted tubule; loop of Henle; distal convoluted tubule; collecting duct; one-way valve between end of collecting tube (duct) and urinary bladder—used to restrict the flow of filtrate into the urinary bladder, increasing the volume and pressure in the renal tubule; drain beaker for filtrate (beaker on right side of screen)—simulates the urinary bladder.

Experiment Instructions

Go to the home page in the PhysioEx software and click **Exercise 9: Renal System Physiology**. Click **Activity 3: Renal Response to Altered Blood Pressure**, and take the online **Pre-lab Quiz** for Activity 3.

After you take the online Pre-lab Quiz, click the **Experiment** tab and begin the experiment. The experiment instructions are reprinted here for your reference. The opening screen for the experiment is shown below.

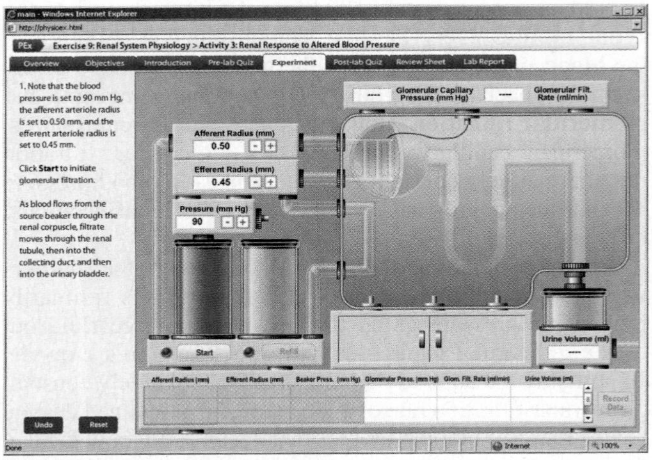

Windows Explorer 2007, Microsoft Corporation

1. Note that the blood pressure is set to 90 mm Hg, the afferent arteriole radius is set to 0.50 mm, and the efferent arteriole radius is set to 0.45 mm. Click **Start** to initiate glomerular filtration. As blood flows from the source beaker through the renal corpuscle, filtrate moves through the renal tubule, then into the collecting duct, and then into the urinary bladder.

2. The glomerular capillary pressure display shows the hydrostatic blood pressure in the glomerular capillaries that promotes filtration, and the filtration rate display shows the flow rate of the fluid moving from the lumen of the glomerular capillaries into the lumen of Bowman's capsule. Click **Record Data** to display your results in the grid (and record your results in Chart 3).

3. Click **Refill** to replenish the source beaker and prepare the nephron for the next run.

4. You will now observe how the nephron might operate to keep the glomerular filtration rate relatively constant despite a large drop in blood pressure. Decrease the blood pressure to 70 mm Hg by clicking the − button beside the pressure display. Click **Start** to initiate glomerular filtration.

5. Note the glomerular capillary pressure and glomerular filtration rate displays and click **Record Data** to display your results in the grid (and record your results in Chart 3).

6. Click **Refill** to replenish the source beaker and prepare the nephron for the next run.

7. Increase the afferent arteriole radius to 0.60 mm by clicking the + button beside the afferent radius display. Click **Start** to initiate glomerular filtration.

8. Note the glomerular capillary pressure and glomerular filtration rate displays and click **Record Data** to display your results in the grid (and record your results in Chart 3).

9. Click **Refill** to replenish the source beaker and prepare the nephron for the next run.

10. Return the afferent arteriole radius to 0.50 mm by clicking the − button beside the afferent radius display and decrease the efferent radius to 0.35 mm by clicking the button beside the efferent radius display. Click **Start** to initiate glomerular filtration.

11. Note the glomerular capillary pressure and glomerular filtration rate displays and click **Record Data** to display your results in the grid (and record your results in Chart 3).

12. Click **Refill** to replenish the source beaker and prepare the nephron for the next run.

PREDICT Question 1
What will happen to the glomerular capillary pressure and glomerular filtration rate if both of these arteriole radii changes are implemented simultaneously with the low blood pressure condition?

____

CHART 3 Renal Response to Altered Blood Pressure

Afferent arteriole radius (mm)	Efferent arteriole radius (mm)	Blood pressure (mm Hg)	Glomerular capillary pressure (mm Hg)	Glomerular filtration rate (ml/min)

13. Set the afferent arteriole radius to 0.60 mm and keep the efferent arteriole radius at 0.35 mm. Click **Start** to initiate glomerular filtration.

14. Note the glomerular capillary pressure and glomerular filtration rate displays and click **Record Data** to display your results in the grid (and record your results in Chart 3).

After you complete the experiment, take the online **Post-lab Quiz** for Activity 3.

Activity Questions

1. How could an increased urine volume be viewed as beneficial to the body?

___

___

2. Diuretics are frequently given to people with persistent high blood pressure. Why?

___

___

ACTIVITY 4

Solute Gradients and Their Impact on Urine Concentration

OBJECTIVES

1. To understand the terms *antidiuretic hormone (ADH)*, *reabsorption*, *loop of Henle*, *collecting duct*, *tubule lumen*, *interstitial space*, and *peritubular capillaries*.
2. To explain the process of water reabsorption in specific regions of the nephron.
3. To understand the role of ADH in water reabsorption by the nephron.
4. To describe how the kidneys can produce urine that is four times more concentrated than the blood.

Introduction

As filtrate moves through the tubules of a nephron, solutes and water move *from* the **tubule lumen** *into* the **interstitial spaces** of the nephron. This movement of solutes and water relies on the total solute concentration gradient in the interstitial spaces surrounding the tubule lumen. The interstitial fluid is comprised mostly of NaCl and urea. When the nephron is permeable to solutes or water, equilibrium will be reached between the interstitial fluid and the tubular fluid contents.

Antidiuretic hormone (ADH) increases the water permeability of the **collecting duct**, allowing water to flow to areas of higher solute concentration, from the tubule lumen into the surrounding interstitial spaces. **Reabsorption** describes this movement of filtered solutes and water from the lumen of the renal tubules back into the plasma. The reabsorbed solutes and water that move into the interstitial space need to be returned to the blood, or the kidneys will rapidly swell like balloons. The **peritubular capillaries** surrounding the renal tubule reclaim the reabsorbed substances and return them to general circulation. Peritubular capillaries arise from the efferent arteriole exiting the glomerulus and empty into the renal veins leaving the kidney.

Without reabsorption, we would excrete the solutes and water that our bodies need to maintain homeostasis. In this activity you will examine the process of passive reabsorption that occurs while filtrate travels through a nephron and urine is formed. While completing the experiment, assume that when ADH is present, the conditions favor the formation of the most concentrated urine possible.

EQUIPMENT USED The following equipment will be depicted on-screen: simulated nephron surrounded by interstitial space between the nephron and peritubular capillaries (Reabsorbed solutes, such as glucose, will move from the lumen of the tubule into the interstitial space, and then into the peritubular capillaries that branch out from the efferent arteriole.); drain beaker for filtrate—simulates the urinary bladder; antidiuretic hormone (ADH).

Experiment Instructions

Go to the home page in the PhysioEx software and click **Exercise 9: Renal System Physiology**. Click **Activity 4: Solute Gradients and Their Impact on Urine Concentration**, and take the online **Pre-lab Quiz** for Activity 4.

After you take the online Pre-lab Quiz, click the **Experiment** tab and begin the experiment. The experiment instructions are reprinted here for your reference. The opening screen for the experiment is shown below.

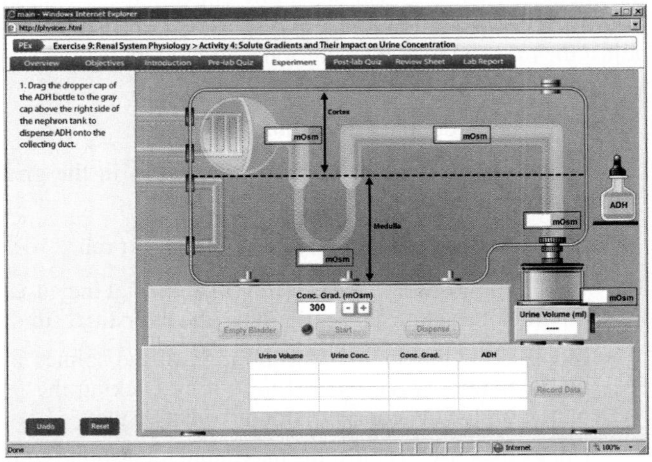

Windows Explorer 2007, Microsoft Corporation

1. Drag the dropper cap of the ADH bottle to the gray cap above the right side of the nephron tank to dispense ADH onto the collecting duct.

2. Click **Dispense** beneath the concentration gradient display to adjust the maximum total solute concentration in the interstitial fluid to 300 mOsm. Because the blood solute concentration is also 300 mOsm, there is no osmotic difference between the lumen of the tubule and the surrounding interstitial fluid.

3. Click **Start** to initiate filtration. Filtrate will flow through the nephron, and solutes and water will move out of the tubules into the interstitial space. Fluid will also move

back into the peritubular capillaries, thus completing the process of reabsorption.

4. Click **Record Data** to display your results in the grid (and record your results in Chart 4).

CHART 4 Solute Gradients and Their Impact on Urine Concentration

Urine volume (ml)	Urine concentration (mOsm)	Concentration gradient (mOsm)

5. Click **Empty Bladder** to prepare for the next run.

> ? PREDICT **Question 1**
> What will happen to the urine volume and concentration as the solute gradient in the interstitial space is increased?
>
> ______________________________

6. Increase the maximum concentration of the solutes in the interstitial space to 600 mOsm by clicking the + button beside the concentration gradient display. Click **Dispense** to adjust the maximum total solute concentration in the interstitial fluid.

7. Click **Start** to initiate filtration.

8. Click **Record Data** to display your results in the grid (and record your results in Chart 4).

9. Click **Empty Bladder** to prepare for the next run.

10. You will now observe the effect of incremental increases in maximum total solute concentration in the interstitial fluid.

- Increase the maximum concentration of the solutes in the interstitial space by 300 mOsm by clicking the + button beside the concentration gradient display.
- Click **Dispense** to adjust the maximum total solute concentration in the interstitial fluid.
- Click **Start** to initiate filtration.
- Click **Record Data** to display your results in the grid (and record your results in Chart 4).
- Click **Empty Bladder** to prepare for the next run.

Repeat this step until you reach the maximum total solute concentration in the interstitial fluid of 1200 mOsm.

After you complete the experiment, take the online **Post-lab Quiz** for Activity 4.

Activity Questions

1. From what you learned in this activity, speculate on ways that desert rats are able to concentrate their urine significantly more than humans.

____

____

2. Judging from this activity, what would be a reasonable mechanism for diuretics?

____

____

ACTIVITY 5

Reabsorption of Glucose via Carrier Proteins

OBJECTIVES

1. To understand the terms *reabsorption, carrier proteins, apical membrane, secondary active transport, facilitated diffusion,* and *basolateral membrane*.
2. To understand the role that glucose carrier proteins play in removing glucose from the filtrate.
3. To understand the concept of a glucose carrier transport maximum and why glucose is not normally present in the urine.

Introduction

Reabsorption is the movement of filtered solutes and water from the lumen of the renal tubules back into the plasma. Without reabsorption, we would excrete the solutes and water that our bodies require for homeostasis.

Glucose is not very large and is therefore easily filtered out of the plasma into Bowman's capsule as part of the filtrate. To ensure that glucose is reabsorbed into the body so that it can fuel cellular metabolism, glucose **carrier proteins** are present in the proximal tubule cells of the nephron. There are a finite number of these glucose carriers in each renal tubule cell. Therefore, if too much glucose is present in the filtrate, it will not all be reabsorbed and glucose will be inappropriately excreted into the urine.

Glucose is first absorbed by **secondary active transport** at the **apical membrane** of proximal tubule cells and then it leaves the tubule cell via **facilitated diffusion** along the **basolateral membrane**. Both types of carrier proteins that transport these molecules across the tubule membranes are transmembrane proteins. Because carrier proteins are needed to move glucose from the lumen of the nephron into the interstitial spaces, there is a limit to the amount of glucose that can be reabsorbed. When all glucose carriers are bound with the glucose they are transporting, excess glucose in the filtrate is eliminated in urine.

In this activity, you will examine the effect of varying the number of glucose transport proteins in the *proximal convoluted tubule*. It is important to note that, normally, the number of glucose carriers is constant in a human kidney and that it is the plasma glucose that varies during the day. Plasma glucose will be held constant in this activity, and the number of glucose carriers will be varied.

EQUIPMENT USED The following equipment will be depicted on-screen: simulated nephron surrounded by interstitial space between the nephron and peritubular capillaries (Reabsorbed solutes, such as glucose, will move from the lumen of the tubule into the interstitial space, and then into the peritubular capillaries that branch out from the efferent arteriole.); drain beaker for filtrate—simulates the urinary bladder; glucose carrier protein control box—used to adjust the number of glucose carriers that will be inserted into the proximal tubule.

Experiment Instructions

Go to the home page in the PhysioEx software and click **Exercise 9: Renal System Physiology**. Click **Activity 5: Reabsorption of Glucose via Carrier Proteins**, and take the online **Pre-lab Quiz** for Activity 5.

After you take the online Pre-lab Quiz, click the **Experiment** tab and begin the experiment. The experiment instructions are reprinted here for your reference. The opening screen for the experiment is shown below.

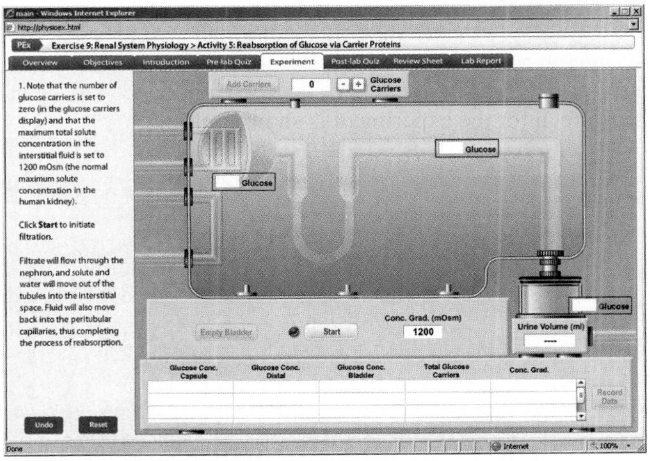

Windows Explorer 2007, Microsoft Corporation

1. Note that the number of glucose carriers is set to zero (in the glucose carriers display) and that the maximum total solute concentration in the interstitial fluid is set to 1200 mOsm (the normal maximum solute concentration in the human kidney). Click **Start** to initiate filtration. Filtrate will flow through the nephron, and solute and water will move out of the tubules into the interstitial space. Fluid will also move back into the peritubular capillaries, thus completing the process of reabsorption.

2. Click **Record Data** to display your results in the grid (and record your results in Chart 5). The concentrations of glucose in Bowman's capsule, the distal convoluted tubule, and the urinary bladder will be displayed in the grid.

CHART 5 Reabsorption of Glucose via Carrier Proteins

Glucose concentration (m*M*)			
Bowman's capsule	Distal convoluted tubule	Urinary bladder	Glucose carriers

3. Click **Empty Bladder** to prepare the nephron for the next run.

PREDICT Question 1
What will happen to the glucose concentration in the urinary bladder as glucose carriers are added to the proximal tubule?

____

4. Increase the number of glucose carriers to 100 (an arbitrary number) by clicking the + button beside the glucose carriers display. Click **Add Carriers** to insert the specified number of glucose carrier proteins per unit area into the membrane of the proximal tubule.

5. Click **Start** to initiate filtration.

6. Click **Record Data** to display your results in the grid (and record your results in Chart 5).

7. Click **Empty Bladder** to prepare the nephron for the next run.

8. You will now observe the effect of incremental increases in the number of glucose carriers.

- Increase the number of glucose carriers by 100 by clicking the + button beside the glucose carriers display.
- Click **Add Carriers** to insert the specified number of glucose carrier proteins per unit area into the membrane of the proximal tubule.
- Click **Start** to initiate filtration.
- Click **Record Data** to display your results in the grid (and record your results in Chart 5).
- Click **Empty Bladder** to prepare the nephron for the next run.

Repeat this step until you have inserted 400 glucose carrier proteins per unit area into the membrane of the proximal tubule.

After you complete the experiment, take the online **Post-lab Quiz** for Activity 5.

Activity Questions

1. Why would your family physician at the turn of the twentieth century taste your urine?

____

____

ACTIVITY 6

The Effect of Hormones on Urine Formation

OBJECTIVES

1. To understand the terms *antidiuretic hormone (ADH)*, *aldosterone*, *reabsorption*, *loop of Henle*, *distal convoluted tubule*, *collecting duct*, *tubule lumen*, and *interstitial space*.
2. To understand how the hormones aldosterone and ADH affect renal processes in a human kidney.
3. To understand the role of ADH in water reabsorption by the nephron.
4. To understand the role of aldosterone in solute reabsorption and secretion by the nephron.

Introduction

The concentration and volume of urine excreted by our kidneys will change depending on what our body needs for homeostasis. For example, if a person consumes a large quantity of water, the excess water will be eliminated as a large volume of dilute urine. On the other hand, when dehydration occurs, there is a clear benefit in being able to produce a small volume of concentrated urine to retain water. Activity 4 demonstrated how the total solute concentration gradient in the interstitial spaces surrounding the tubule lumen makes it possible to excrete concentrated urine.

Aldosterone is a hormone produced by the adrenal cortex under the control of the body's *renin-angiotensin system*. A decrease in blood pressure is detected by cells in the afferent arteriole, triggering the release of renin. Renin acts as a proteolytic enzyme, causing angiotensinogen to be converted into angiotensin I. Endothelial cells throughout the body possess a *converting enzyme* that converts angiotensin I into angiotensin II. Angiotensin II signals the adrenal cortex to secrete aldosterone. Aldosterone acts on the distal convoluted tubule cells in the nephron to promote the reabsorption of sodium from filtrate *into* the body and the secretion of potassium *from* the body. This electrolyte shift, coupled with the addition of **antidiuretic hormone (ADH)**, also causes more water to be reabsorbed into the blood, resulting in increased blood pressure.

ADH is manufactured by the hypothalamus and stored in the posterior pituitary gland. ADH levels are influenced by the osmolality of body fluids and the volume and pressure of the cardiovascular system. A 1% change in body osmolality will cause this hormone to be secreted. The primary action of this hormone is to increase the permeability of the collecting duct to water so that more water is reabsorbed into the body by inserting aquaporins, or water channels, in the apical membrane. Without this water reabsorption, the body would quickly dehydrate.

Thus, our kidneys tightly regulate the amount of water and solutes excreted to maintain water balance in the body. If water intake is down, or if there has been a fluid loss from the body, the kidneys work to conserve water by making the urine very hyperosmotic (having a relatively high solute concentration) to the blood. If there has been a large intake of fluid, the urine is more hypo-osmotic. In the normal individual, urine osmolarity varies from 50 to 1200 milliosmoles/kg of water.

EQUIPMENT USED The following equipment will be depicted on-screen: simulated nephron surrounded by interstitial space between the nephron and peritubular capillaries (Reabsorbed solutes, such as glucose, will move from the lumen of the tubule into the interstitial space, and then into the peritubular capillaries that branch out from the efferent arteriole.); drain beaker for filtrate—simulates the urinary bladder; aldosterone; antidiuretic hormone (ADH).

Experiment Instructions

Go to the home page in the PhysioEx software and click **Exercise 9: Renal System Physiology**. Click **Activity 6: The Effect of Hormones on Urine Formation**, and take the online **Pre-lab Quiz** for Activity 6.

After you take the online Pre-lab Quiz, click the **Experiment** tab and begin the experiment. The experiment instructions are reprinted here for your reference. The opening screen for the experiment is shown below.

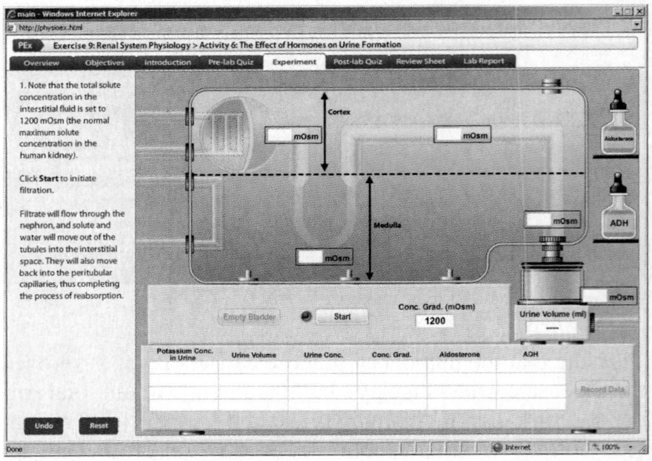

Windows Explorer 2007, Microsoft Corporation

1. Note that the total solute concentration in the interstitial fluid is set to 1200 mOsm (the normal maximum solute concentration in the human kidney). Click **Start** to initiate filtration. Filtrate will flow through the nephron, and solute and water will move out of the tubules into the interstitial space. They will also move back into the peritubular capillaries, thus completing the process of reabsorption.

2. Click **Record Data** to display your results in the grid (and record your results in Chart 6). You will use this baseline data to compare the conditions of the filtrate and urine volume in the presence of the hormones aldosterone and ADH.

CHART 6	The Effect of Hormones on Urine Formation			
Potassium concentration (mM)	Urine volume (ml)	Urine concentration (mOsm)	Aldosterone	ADH

3. Click **Empty Bladder** to prepare the nephron for the next run.

PREDICT Question 1
What will happen to the urine volume (compared with baseline) when aldosterone is added to the distal tubule?

____

4. Drag the dropper cap of the aldosterone bottle to the gray cap above the right side of the nephron tank to dispense aldosterone into the tank surrounding the distal tubule and the collecting duct.

5. Click **Start** to initiate filtration.

6. Click **Record Data** to display your results in the grid (and record your results in Chart 6).

7. Click **Empty Bladder** to prepare the nephron for the next run.

8. Drag the dropper cap of the ADH bottle to the gray cap at the top right side of the nephron tank to dispense ADH into the tank surrounding the distal tubule and the collecting duct.

PREDICT Question 2
What will happen to the urine volume (compared with baseline) when ADH is added to the collecting duct?

____

9. Click **Start** to initiate filtration.

10. Click **Record Data** to display your results in the grid (and record your results in Chart 6).

11. Click **Empty Bladder** to prepare the nephron for the next run.

PREDICT Question 3
What will happen to the urine volume and the urine concentration (compared with baseline) in the presence of both aldosterone and ADH?

____

12. Drag the dropper cap of the aldosterone bottle and then the dropper cap of the ADH bottle to the gray cap above the right side of the nephron tank to dispense aldosterone and ADH into the tank surrounding the distal convoluted tubule and the collecting duct.

13. Click **Start** to initiate filtration.

14. Click **Record Data** to display your results in the grid (and record your results in Chart 6).

After you complete the experiment, take the online **Post-lab Quiz** for Activity 6.

Activity Questions

1. Why does ethanol consumption lead to a dramatic increase in urine production?

____

____

2. Why do angiotensin converting enzyme (ACE) inhibitors given to people with hypertension lead to increased urine production?

____

____

NAME ______________________

LAB TIME/DATE ______________________

Renal System Physiology

ACTIVITY 1 The Effect of Arteriole Radius on Glomerular Filtration

1. What are two primary functions of the kidney? ______________________

____

2. What are the components of the renal corpuscle? ______________________

____

3. Starting at the renal corpuscle, list the components of the renal tubule as they are encountered by filtrate. ______________________

____

____

4. Describe the effect of decreasing the afferent arteriole radius on glomerular capillary pressure and filtration rate. How well did the results compare with your prediction? ______________________

____

5. Describe the effect of increasing the afferent arteriole radius on glomerular capillary pressure and filtration rate. How well did the results compare with your prediction? ______________________

____

6. Describe the effect of decreasing the efferent arteriole radius on glomerular capillary pressure and filtration rate. How well did the results compare with your prediction? ______________________

____

7. Describe the effect of increasing the efferent radius on glomerular capillary pressure and filtration rate. ______________________

____

ACTIVITY 2 The Effect of Pressure on Glomerular Filtration

1. As blood pressure increased, what happened to the glomerular capillary pressure and the glomerular filtration rate? How well did the results compare with your prediction? ______________________

____

2. Compare the urine volume in your baseline data with the urine volume as you increased the blood pressure. How did the urine volume change? ______

______

3. How could the change in urine volume with the increase in blood pressure be viewed as being beneficial to the body?

______

______

4. When the one-way valve between the collecting duct and the urinary bladder was closed, what happened to the filtrate pressure in Bowman's capsule (this is not directly measured in this experiment) and the glomerular filtration rate? How well did the results compare with your prediction? ______

______

5. How did increasing the blood pressure alter the results when the valve was closed? ______

______

ACTIVITY 3 Renal Response to Altered Blood Pressure

1. List the several mechanisms you have explored that change the glomerular filtration rate. How does each mechanism specifically alter the glomerular filtration rate? ______

______

2. Describe and explain what happened to the glomerular capillary pressure and glomerular filtration rate when *both* arteriole radii changes were implemented simultaneously with the low blood pressure condition. How well did the results compare with your prediction? ______

______

______

3. How could you adjust the afferent or efferent radius to compensate for the effect of reduced blood pressure on the glomerular filtration rate? ______

______

4. Which arteriole radius adjustment was more effective at compensating for the effect of low blood pressure on the glomerular filtration rate? Explain why you think this difference occurs. ______

______

5. In the body, how does a nephron maintain a near-constant glomerular filtration rate despite a constantly fluctuating blood pressure? ______

______

ACTIVITY 4 Solute Gradients and Their Impact on Urine Concentration

1. What happened to the urine concentration as the solute concentration in the interstitial space was increased? How well did the results compare to your prediction? ______

______

______

2. What happened to the volume of urine as the solute concentration in the interstitial space was increased? How well did the results compare to your prediction? ______

______

3. What do you think would happen to urine volume if you did not add ADH to the collecting duct? ______

______

______

4. Is most of the tubule filtrate reabsorbed into the body or excreted in urine? Explain. ______

______

______

5. Can the reabsorption of solutes influence water reabsorption from the tubule fluid? Explain. ______

______

ACTIVITY 5 Reabsorption of Glucose via Carrier Proteins

1. What happens to the concentration of glucose in the urinary bladder as the number of glucose carriers increases? ______

______

______

2. What types of transport are utilized during glucose reabsorption and where do they occur? ______

______

3. Why does the glucose concentration in the urinary bladder become zero in these experiments? ______

______

4. A person with type 1 diabetes cannot make insulin in the pancreas, and a person with untreated type 2 diabetes does not respond to the insulin that is made in the pancreas. In either case, why would you expect to find glucose in the person's urine?

______

______

ACTIVITY 6 The Effect of Hormones on Urine Formation

1. How did the addition of aldosterone affect urine volume (compared with baseline)? Can the reabsorption of solutes influence water reabsorption in the nephron? Explain. How well did the results compare with your prediction? ______

2. How did the addition of ADH affect urine volume (compared with baseline)? How well did the results compare with your prediction? Why did the addition of ADH also affect the concentration of potassium in the urine (compared with baseline)?

3. What is the principal determinant for the release of aldosterone from the adrenal cortex? ______

4. How did the addition of both aldosterone and ADH affect urine volume (compared with baseline)? How well did the results compare with your prediction? ______

5. What is the principal determinant for the release of ADH from the posterior pituitary gland? Does ADH favor the formation of dilute or concentrated urine? Explain why. ______

6. Which hormone (aldosterone or ADH) has the greater effect on urine volume? Why? ______

7. If ADH is not available, can the urine concentration still vary? Explain your answer. ______

8. Consider this situation: you want to reabsorb sodium ions but you do not want to increase the volume of the blood by reabsorbing large amounts of water from the filtrate. Assuming that aldosterone and ADH are both present, how would you adjust the hormones to accomplish the task? ______

Acid-Base Balance

PRE-LAB QUIZ

1. Circle the correct term: A(n) acid / base is often found in the form of a hydroxyl ion (OH^-) or a bicarbonate ion (HCO_3^-) and binds the H^+.
2. A condition that can result from traveling to high altitudes or hyperventilation, __________ is caused by having too little carbon dioxide in the blood.
   a. respiratory acidosis
   b. respiratory arrest
   c. respiratory alkalosis
   d. renal acidosis
3. Which of the following is *not* a chemical buffering system used by the body to maintain pH homeostasis?
   a. bicarbonate
   b. sodium
   c. phosphate
   d. protein
4. Circle True or False: Blood and tissue fluids normally have a pH range of between 6.9 and 7.8.
5. Hypoventilation will result in elevated levels of which gas in the blood? __________________
6. Circle the correct underlined term: Vomiting, constipation, and ingestion of antacids and bicarbonate are all possible causes of metabolic alkalosis / metabolic acidosis
7. Fever, stress, or the ingestion of food will all cause the rate of cell metabolism to ______________.
   a. increase b. decrease c. remain the same
8. Circle True or False: Both the renal system and the respiratory system compensate for metabolic acidosis and alkalosis by excreting bicarbonate ions.
9. The body's primary method of compensating for conditions of respiratory acidosis or respiratory alkalosis is ____________________.

Exercise Overview

pH denotes the hydrogen ion concentration, $[H^+]$, in a solution (such as body fluids). The reciprocal relationship between pH and $[H^+]$ is defined by the following equation.

$$pH = \log(1/[H^+])$$

Because the relationship is reciprocal, $[H^+]$ is higher at *lower* pH values (indicating higher acid levels) and lower at *higher* pH values (indicating lower acid levels).

The pH of a body's fluid is also referred to as its **acid-base balance**. An **acid** is a substance that releases H^+ in solution. A **base**, often a hydroxyl ion (OH^-) or bicarbonate ion (HCO_3^-), is a substance that binds, or buffers, the H^+. A **strong acid** completely dissociates in solution, releasing all of its hydrogen ions and, thus, lowering the solution's pH. A **weak acid** dissociates incompletely and does not release all of its hydrogen ions in solution, producing a lesser effect on the solution's pH. A **strong base** has a strong tendency to bind to H^+, raising the solution's pH. A **weak base** binds less of the H^+, producing a lesser effect on the solution's pH.

The pH of body fluids is very tightly regulated. Blood and tissue fluids normally have a pH between 7.35 and 7.45. Under pathological conditions, blood pH as low as 6.9 or as high as 7.8 has been recorded, but a higher or lower pH cannot sustain human life. The narrow range from 7.35 to 7.45 is remarkable when you consider the vast number of biochemical reactions that take place in the body. The human body normally produces a large amount of H^+ as the result of metabolic processes; ingested acids; and the products of fat, sugar, and amino acid metabolism. The regulation of a relatively constant internal pH is one of the major physiological functions of the body's organ systems.

To maintain pH homeostasis, the body utilizes both *chemical* and *physiological* buffering systems. Chemical buffers are composed of a mixture of weak acids and weak bases. They help regulate the body's pH levels by binding H^+ and removing it from solution as its concentration begins to rise or by releasing H^+ into solution as its concentration begins to fall. The body's three major chemical buffering systems are the *bicarbonate*, *phosphate*, and *protein buffer systems*. We will not focus on chemical buffering systems in this exercise, but keep in mind that chemical buffers are the fastest form of compensation and can return pH to normal within a fraction of a second.

The body's two major physiological buffering systems are the **renal system** and the **respiratory system**. The renal system is the slower of the two, taking hours to days to do its work. The respiratory system usually works within minutes, but cannot handle the amount of pH change that the renal system can. These physiological buffer systems help regulate body pH by controlling the output of acids, bases, or carbon dioxide (CO_2) from the body. For example, if there is too much acid in the body, the renal system may respond by excreting more H^+ from the body in urine. Similarly, if there is too much carbon dioxide in the blood, the respiratory system may respond by increasing ventilation to expel the excess carbon dioxide. Carbon dioxide levels have a direct effect on pH because the addition of carbon dioxide to the blood results in the generation of more H^+. The following equation shows what happens when carbon dioxide combines with water in the blood, producing carbonic acid.

$$H_2O + CO_2 \rightleftarrows \underset{\text{carbonic acid}}{H_2CO_3} \rightleftarrows H^+ + \underset{\text{bicarbonate ion}}{HCO_3^-}$$

ACTIVITY 1

Hyperventilation

OBJECTIVES

1. To introduce pH homeostasis in the body.
2. To understand the normal ranges for pH and P_{CO_2}.
3. To recognize respiratory alkalosis and its causes.
4. To interpret an oscilloscope tracing for hyperventilation and compare it with a tracing for normal breathing.

Introduction

Acid-base imbalances can have respiratory and metabolic causes. When diagnosing these disorders, two key signs are evaluated: the pH and the partial pressure of carbon dioxide in the blood (**P_{CO_2}**). The normal range for pH is between 7.35 and 7.45, and the normal range for P_{CO_2} is between 35 and 45 mm Hg. When the pH falls below 7.35, the body is said to be in a state of **acidosis**. When the pH rises above 7.45, the body is said to be in a state of **alkalosis**.

Respiratory alkalosis is the condition of too little carbon dioxide in the blood. Respiratory alkalosis commonly results from traveling to high altitude (where the air contains less oxygen) or hyperventilation, which can be brought on by fever, panic attack, or anxiety. Hyperventilation, defined as an increase in the rate and depth of breathing, removes carbon dioxide from the blood faster than it is being produced by the cells of the body, reducing the amount of H^+ in the blood and, thus, increasing the blood's pH. The following equation shows the shift in the equilibrium that results in the increase in blood pH due to less carbon dioxide in the blood.

$$H_2O + \mathbf{CO_2} \leftarrow \underset{\text{carbonic acid}}{H_2CO_3} \leftarrow \mathbf{H^+} + \underset{\text{bicarbonate ion}}{HCO_3^-}$$

The renal system can compensate for alkalosis by retaining H^+ and excreting bicarbonate ions to lower the blood pH levels back to the normal range.

EQUIPMENT USED The following equipment will be depicted on-screen: simulated lung chamber; pH meter; oscilloscope; two breathing patterns: normal and hyperventilation.

Experiment Instructions

Go to the home page in the PhysioEx software and click **Exercise 10: Acid-Base Balance**. Click **Activity 1: Hyperventilation**, and take the online **Pre-lab Quiz** for Activity 1.

After you take the online Pre-lab Quiz, click the **Experiment** tab and begin the experiment. The experiment instructions are reprinted here for your reference. The opening screen for the experiment is shown below.

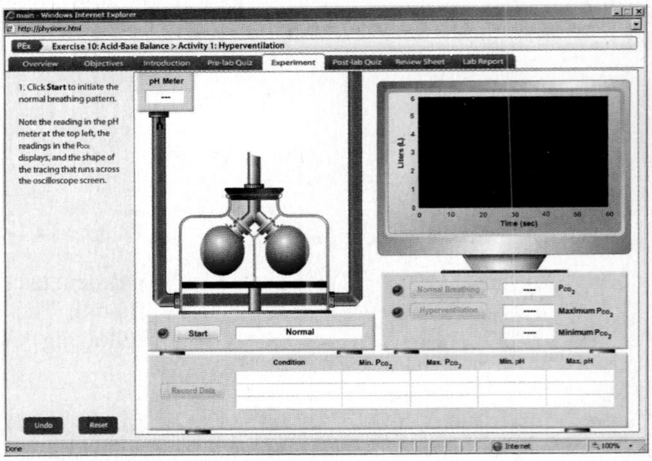

Windows Explorer 2007, Microsoft Corporation

1. Click **Start** to initiate the normal breathing pattern. Note the reading in the pH meter at the top left, the readings in the P_{CO_2} displays, and the shape of the tracing that runs across the oscilloscope screen.

2. Click **Record Data** to display your results in the grid (and record your results in Chart 1).

PREDICT Question 1
What do you think will happen to the pH and P_{CO_2} levels with hyperventilation?

____

CHART 1 Hyperventilation Breathing Patterns				
Condition	Minimum P_{CO_2}	Maximum P_{CO_2}	Minimum pH	Maximum pH

3. Click **Start** to initiate the normal breathing pattern. After the normal breathing tracing runs for 10 seconds, click **Hyperventilation** to initiate the hyperventilation breathing pattern. Note the reading in the pH meter at the top left, the readings in the P_{CO_2} displays, and the shape of the tracing that runs across the oscilloscope screen.

4. Click **Record Data** to display your results in the grid (and record your results in Chart 1).

5. Click **Start** to initiate the normal breathing pattern. After the normal breathing tracing runs for 10 seconds, click **Hyperventilation** to initiate the hyperventilation breathing pattern. After the hyperventilation tracing runs for 10 seconds, click **Normal Breathing** to return to the normal breathing pattern. Note the reading in the pH meter at the top left, the readings in the P_{CO_2} displays, and the shape of the tracing that runs across the oscilloscope screen.

6. Click **Record Data** to display your results in the grid (and record your results in Chart 1).

After you complete the experiment, take the online **Post-lab Quiz** for Activity 1.

Activity Questions

1. At what pH range is the body considered to be in a state of respiratory alkalosis?

____

____

2. How can the body compensate for respiratory alkalosis?

____

____

3. How did the tidal volume change with hyperventilation?

____

____

4. What might cause a person to hyperventilate?

____

____

ACTIVITY 2

Rebreathing

OBJECTIVES

1. To understand how rebreathing can simulate hypoventilation.
2. To observe the results of respiratory acidosis.
3. To describe the causes of respiratory acidosis.

Introduction

The body is said to be in a state of **acidosis** when the pH of the blood falls below 7.35 (although a pH of 7.35 is technically not acidic). Respiratory acidosis is the result of impaired respiration, or *hypoventilation*, which leads to the accumulation of too much carbon dioxide in the blood. The causes of impaired respiration include airway obstruction, depression of the respiratory center in the brain stem, lung disease (such as emphysema and chronic bronchitis), and drug overdose.

Recall that carbon dioxide contributes to the formation of carbonic acid when it combines with water through a reversible reaction catalyzed by carbonic anhydrase. The carbonic acid then dissociates into hydrogen ions and bicarbonate ions. Because hypoventilation results in elevated carbon dioxide levels in the blood, the equilibrium shifts, the H^+ levels increase, and the pH value of the blood decreases.

$$H_2O + \mathbf{CO_2} \rightarrow \underset{\text{carbonic acid}}{H_2CO_3} \rightarrow \mathbf{H^+} + \underset{\text{bicarbonate ion}}{HCO_3^-}$$

Rebreathing is the action of breathing in air that was just expelled from the lungs. Rebreathing results in the accumulation of carbon dioxide in the blood. Breathing into a paper bag is an example of rebreathing. (Note that breathing into a paper bag can deplete the body of oxygen and is therefore not the best therapy for hyperventilation because it can mask other life-threatening emergencies, such as a heart attack or asthma.) In this activity, you will observe what happens to pH and carbon dioxide levels in the blood during rebreathing. In the body, the kidneys regulate the acid-base balance by altering the amount of H^+ and HCO_3^- excreted in the urine.

EQUIPMENT USED The following equipment will be depicted on-screen: simulated lung chamber; pH meter; oscilloscope; two breathing patterns: normal and rebreathing.

Experiment Instructions

Go to the home page in the PhysioEx software and click **Exercise 10: Acid-Base Balance**. Click **Activity 2: Rebreathing**, and take the online **Pre-lab Quiz** for Activity 2.

After you take the online Pre-lab Quiz, click the **Experiment** tab and begin the experiment. The experiment instructions are reprinted here for your reference. The opening screen for the experiment is shown below.

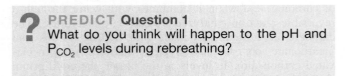

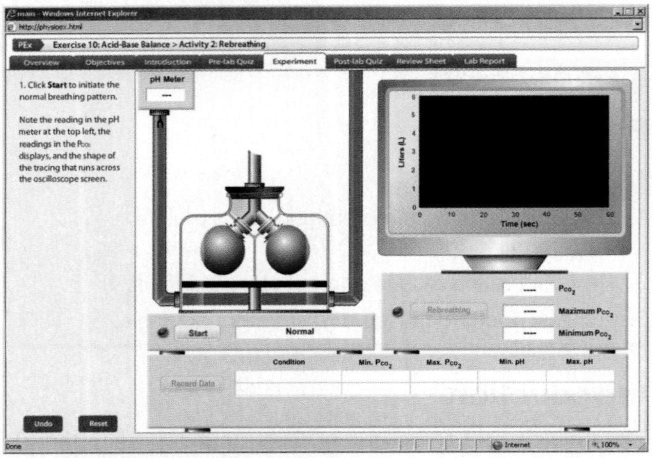

Windows Explorer 2007, Microsoft Corporation

1. Click **Start** to initiate the normal breathing pattern. Note the reading in the pH meter at the top left, the readings in the P_{CO_2} displays, and the shape of the tracing that runs across the oscilloscope screen.

2. Click **Record Data** to display your results in the grid (and record your results in Chart 2).

> **? PREDICT Question 1**
> What do you think will happen to the pH and P_{CO_2} levels during rebreathing?

3. Click **Start** to initiate the normal breathing pattern. After the normal breathing tracing runs for 10 seconds, click **Rebreathing** to initiate the rebreathing pattern. Note the reading in the pH meter at the top left, the readings in the P_{CO_2} displays, and the shape of the tracing that runs across the oscilloscope screen.

4. Click **Record Data** to display your results in the grid (and record your results in Chart 2).

After you complete the experiment, take the online **Post-lab Quiz** for Activity 2.

Activity Questions

1. Did the pH level of the blood change at all with rebreathing? If so, how did it change?

2. What happens to the pH level of the blood when there is too much carbon dioxide remaining in the blood?

3. How did the tidal volumes change with rebreathing?

4. Describe two ways in which too much carbon dioxide might remain in the blood.

ACTIVITY 3

Renal Responses to Respiratory Acidosis and Respiratory Alkalosis

OBJECTIVES

1. To understand renal compensation mechanisms for respiratory acidosis and respiratory alkalosis.
2. To explore the functional unit of the kidneys that responds to acid-base balance.
3. To observe the changes in ion concentrations that occur with renal compensation.

Introduction

The kidneys play a major role in maintaining fluid, electrolyte, and acid-base balance in the body's internal environment. By regulating the amount of water lost in the urine, the kidneys defend the body against excessive hydration or dehydration. By regulating the acidity of urine and the rate of electrolyte excretion, the kidneys maintain plasma pH and electrolyte levels within normal limits.

CHART 2 Normal Breathing Patterns

Condition	Minimum P_{CO_2}	Maximum P_{CO_2}	Minimum pH	Maximum pH

Renal compensation is the body's primary method of compensating for conditions of respiratory acidosis or respiratory alkalosis. The kidneys regulate the acid-base balance by altering the amount of H^+ and HCO_3^- excreted in the urine. If we revisit the equation for the dissociation of carbonic acid, a weak acid, we see that the conservation of bicarbonate ion (base) has the same net effect as the loss of acid, H^+.

$$H_2O + CO_2 \rightleftarrows \underset{\text{carbonic acid}}{H_2CO_3} \rightleftarrows H^+ + \underset{\text{bicarbonate ion}}{HCO_3^-}$$

In this activity you will examine how the renal system compensates for respiratory acidosis or respiratory alkalosis. Respiratory acidosis is generally caused by the accumulation of carbon dioxide in the blood from hypoventilation, but it can also be caused by rebreathing. Acidosis results in a lower-than-normal blood pH. Respiratory alkalosis is caused by a depletion of carbon dioxide, often caused by an episode of hyperventilation, and results in an elevated blood pH.

You will primarily be working with the variable P_{CO_2}. Recall that the normal range for pH is between 7.35 and 7.45 and the normal range for P_{CO_2} is between 35 and 45 mm Hg. You will observe how increases and decreases in P_{CO_2} affect the levels of H^+ and HCO_3^- that the kidneys excrete in urine. The functional unit for adjusting the plasma composition is the **nephron**. Remember that although the renal system can partially compensate for pH imbalances with a respiratory cause, the kidneys cannot fully compensate if respirations have not returned to normal because the carbon dioxide levels will still be abnormal.

EQUIPMENT USED The following equipment will be depicted on-screen: source beaker for blood (first beaker on left side of screen); drain beaker for blood (second beaker on left side of screen); simulated nephron (The filtrate forms in Bowman's capsule and flows through the renal tubule—the tubular components, and empties into a collecting duct which, in turn drains into the urinary bladder.); nephron tank; glomerulus—"ball" of capillaries that forms part of the filtration membrane; glomerular (Bowman's) capsule—forms part of the filtration membrane and a capsular space where the filtrate initially forms; proximal convoluted tubule; loop of Henle; distal convoluted tubule; collecting duct; drain beaker for filtrate (beaker on right side of screen)—simulates the urinary bladder.

Experiment Instructions

Go to the home page in the PhysioEx software and click **Exercise 10: Acid-Base Balance**. Click **Activity 3: Renal Responses to Respiratory Acidosis and Respiratory Alkalosis** and take the online **Pre-lab Quiz** for Activity 3.

After you take the online Pre-lab Quiz, click the **Experiment** tab and begin the experiment. The experiment instructions are reprinted here for your reference. The opening screen for the experiment is shown above.

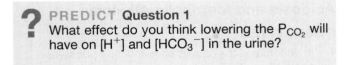

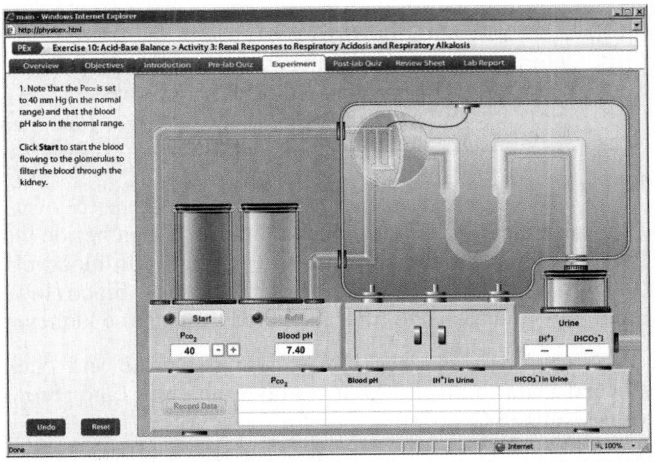

Windows Explorer 2007, Microsoft Corporation

1. Note that the P_{CO_2} is set to 40 mm Hg (in the normal range) and that the blood pH is also in the normal range. Click **Start** to start the blood flowing to the glomerulus to filter the blood through the kidney.

2. Note the $[H^+]$ and $[HCO_3^-]$ in the urine and click **Record Data** to display your results in the grid (and record your results in Chart 3).

CHART 3 **Renal Responses to Respiratory Acidosis and Respiratory Alkalosis**

P_{CO_2}	Blood pH	$[H^+]$ in urine	$[HCO_3^-]$ in urine

3. Click **Refill** to replenish the source beaker.

? PREDICT Question 1
What effect do you think lowering the P_{CO_2} will have on $[H^+]$ and $[HCO_3^-]$ in the urine?

____

4. Lower the P_{CO_2} to 30 by clicking the − button beside the P_{CO_2} display. Note the corresponding increase in blood pH (above the normal range). Click **Start** to start the blood flowing to the glomerulus to filter the blood through the kidney.

5. Note the $[H^+]$ and $[HCO_3^-]$ in the urine and click **Record Data** to display your results in the grid (and record your results in Chart 3).

6. Click **Refill** to replenish the source beaker.

? PREDICT Question 2
What effect do you think raising the P_{CO_2} will have on $[H^+]$ and $[HCO_3^-]$ in the urine?

____

7. Raise the P_{CO_2} to 60 by clicking the + button beside the P_{CO_2} display. Note the corresponding decrease in blood pH (below the normal range). Click **Start** to start the blood flowing to the glomerulus to filter the blood through the kidney.

8. Note the $[H^+]$ and $[HCO_3^-]$ in the urine and click **Record Data** to display your results in the grid (and record your results in Chart 3).

After you complete the experiment, take the online **Post-lab Quiz** for Activity 3.

Activity Questions

1. Describe how the kidneys respond to respiratory acidosis.

____

____

2. What P_{CO_2} corresponded to respiratory acidosis?

____

3. Describe how the kidneys respond to respiratory alkalosis.

____

____

4. What P_{CO_2} corresponded to respiratory alkalosis?

____

ACTIVITY 4

Respiratory Responses to Metabolic Acidosis and Metabolic Alkalosis

OBJECTIVES

1. To understand the causes of metabolic acidosis and metabolic alkalosis.
2. To observe the physiological changes that occur with an increase and decrease in metabolic rate.
3. To explain how the respiratory system compensates for metabolic acidosis and alkalosis.

Introduction

Conditions of acidosis and alkalosis that do not have respiratory causes are termed *metabolic acidosis and metabolic alkalosis*. **Metabolic acidosis** is characterized by low plasma HCO_3^- and pH. The causes of metabolic acidosis include:

- **Ketoacidosis**, a buildup of keto acids that can result from diabetes mellitus
- **Salicylate poisoning**, a toxic condition resulting from ingestion of too much aspirin or oil of wintergreen (a substance often found in laboratories)
- The ingestion of too much alcohol, which metabolizes into acetic acid
- Diarrhea, which results in the loss of bicarbonate with the elimination of intestinal contents
- Strenuous exercise, which can cause a buildup of lactic acid from anaerobic muscle metabolism

Metabolic alkalosis is characterized by elevated plasma HCO_3^- and pH. The causes of metabolic alkalosis include:

- Ingestion of alkali, such as antacids or bicarbonate
- Vomiting, which can result in the loss of too much H^+
- Constipation, which may result in significant reabsorption of HCO_3^-

Increases or decreases in the body's normal metabolic rate can also result in metabolic acidosis or alkalosis. Recall that carbon dioxide—a waste product of metabolism—mixes with water in plasma to form carbonic acid, which in turn forms H^+.

$$H_2O + CO_2 \rightleftarrows \underset{\text{carbonic acid}}{H_2CO_3} \rightleftarrows H^+ + \underset{\text{bicarbonate ion}}{HCO_3^-}$$

An increase in the normal metabolic rate causes more carbon dioxide to form as a metabolic waste product, resulting in the formation of more H^+ and, therefore, lower plasma pH, potentially causing acidosis. Other acids that are also normal metabolic waste products (such as ketone bodies and phosphoric, uric, and lactic acids) would likewise accumulate with an increase in metabolic rate.

Conversely, a decrease in the normal metabolic rate causes less carbon dioxide to form as a metabolic waste product, resulting in the formation of less H^+ and, therefore, higher plasma pH, potentially causing alkalosis. Many factors can affect the rate of cell metabolism. For example, fever, stress, or the ingestion of food all cause the rate of cell metabolism to *increase*. Conversely, a fall in body temperature or a decrease in food intake causes the rate of cell metabolism to *decrease*.

The respiratory system compensates for metabolic acidosis or alkalosis by expelling or retaining carbon dioxide in the blood. During metabolic acidosis, respiration increases to expel carbon dioxide from the blood, thus decreasing $[H^+]$ and raising the pH. During metabolic alkalosis, respiration decreases to promote the accumulation of carbon dioxide in the blood, thus increasing $[H^+]$ and decreasing the pH.

The renal system also compensates for metabolic acidosis and alkalosis by conserving or excreting bicarbonate ions. Nevertheless, in this activity, you will focus on respiratory compensation of metabolic acidosis and alkalosis.

EQUIPMENT USED The following equipment will be depicted on-screen: simulated heart pump; simulated lung chamber; oscilloscope.

Experiment Instructions

Go to the home page in the PhysioEx software and click **Exercise 10: Acid-Base Balance**. Click **Activity 4: Respiratory Responses to Metabolic Acidosis and Metabolic Alkalosis**, and take the online **Pre-lab Quiz** for Activity 4.

After you take the online Pre-lab Quiz, click the **Experiment** tab and begin the experiment. The experiment instructions are reprinted here for your reference. The opening screen for the experiment is shown below.

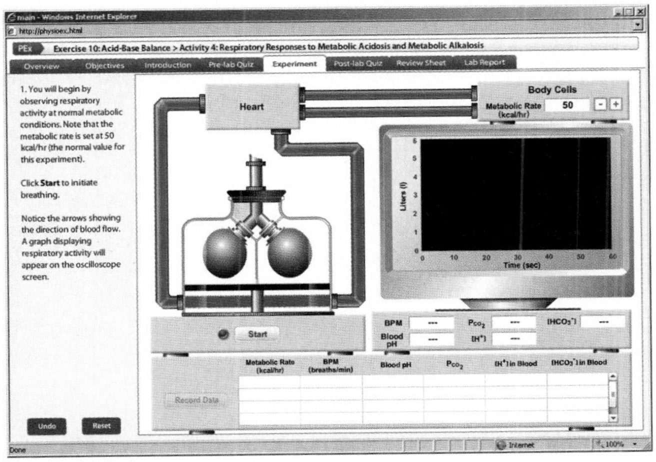

Windows Explorer 2007, Microsoft Corporation

1. You will begin by observing respiratory activity at normal metabolic conditions. Note that the metabolic rate is set at 50 kcal/hr (the normal value for this experiment). Click **Start** to initiate breathing and blood flow. Notice the arrows showing the direction of blood flow. A graph displaying respiratory activity will appear on the oscilloscope screen.

2. Note the data in the displays below the oscilloscope screen and click **Record Data** to display your results in the grid (and record your results in Chart 4).

3. Increase the metabolic rate to 60 kcal/hr by clicking the + button beside the metabolic rate display. Click **Start** to initiate breathing and blood flow.

4. Note the data in the displays below the oscilloscope screen and click **Record Data** to display your results in the grid (and record your results in Chart 4).

5. Click **Clear Tracings** to clear the tracings on the oscilloscope.

? PREDICT Question 1
What do you think will happen when the metabolic rate is increased to 80 kcal/hr?

____

6. Increase the metabolic rate to 80 kcal/hr by clicking the + button beside the metabolic rate display. Click **Start** to initiate breathing and blood flow.

7. Note the data in the displays below the oscilloscope screen and click **Record Data** to display your results in the grid (and record your results in Chart 4).

8. Click **Clear Tracings** to clear the tracings on the oscilloscope.

9. Decrease the metabolic rate to 40 kcal/hr by clicking the − button beside the metabolic rate display. Click **Start** to initiate breathing and blood flow.

10. Note the data in the displays below the oscilloscope screen and click **Record Data** to display your results in the grid (and record your results in Chart 4).

11. Click **Clear Tracings** to clear the tracings on the oscilloscope.

? PREDICT Question 2
What do you think will happen when the metabolic rate is decreased to 20 kcal/hr?

____

CHART 4 Respiratory Responses to Metabolic Acidosis and Metabolic Alkalosis

Metabolic rate	BPM (breaths/min)	Blood pH	P_{CO_2}	$[H^+]$ in blood	$[HCO_3^-]$ in blood

12. Decrease the metabolic rate to 20 kcal/hr by clicking the button beside the metabolic rate display. Click **Start** to initiate breathing and blood flow.

13. Note the data in the displays below the oscilloscope screen and click **Record Data** to display your results in the grid (and record your results in Chart 4).

After you complete the experiment, take the online **Post-lab Quiz** for Activity 4.

Activity Questions

1. Describe what happens to carbon dioxide and pH with increased metabolism.

____

____

2. Describe the respiratory response to metabolic acidosis.

____

____

3. When the respiratory system compensates for the metabolic acidosis, does the pH increase or decrease in value?

____

____

4. Describe the respiratory response to metabolic alkalosis.

____

____

Acid-Base Balance

NAME ______________________

LAB TIME/DATE ______________________

ACTIVITY 1 Hyperventilation

1. Describe the normal ranges for pH and carbon dioxide in the blood. ______________________

2. Describe what happened to the pH and the carbon dioxide levels with hyperventilation. How well did the results compare with your prediction? ______________________

3. Explain how returning to normal breathing after hyperventilation differed from hyperventilation without returning to normal breathing. ______________________

4. Describe some possible causes of respiratory alkalosis. ______________________

ACTIVITY 2 Rebreathing

1. Describe what happened to the pH and the carbon dioxide levels during rebreathing. How well did the results compare with your prediction? ______________________

2. Describe some possible causes of respiratory acidosis. ______________________

3. Explain how the renal system would compensate for respiratory acidosis. ______________________________

____

____

ACTIVITY 3 Renal Responses to Respiratory Acidosis and Respiratory Alkalosis

1. Describe what happened to the concentration of ions in the urine when the P_{CO_2} was lowered. How well did the results compare with your prediction? ______________________________

____

____

2. What condition was simulated when the P_{CO_2} was lowered? ______________________________

____

____

3. Describe what happened to the concentration of ions in the urine when the P_{CO_2} was raised. How well did the results compare with your prediction? ______________________________

____

____

4. What condition was simulated when the P_{CO_2} was raised? ______________________________

____

____

ACTIVITY 4 Respiratory Responses to Metabolic Acidosis and Metabolic Alkalosis

1. Describe what happened to the blood pH when the metabolic rate was increased to 80 kcal/hr. What body system was compensating? How well did the results compare with your prediction? ______________________________

____

____

2. List and describe some possible causes of metabolic acidosis. ______________________________

____

____

3. Describe what happened to the blood pH when the metabolic rate was decreased to 20 kcal/hr. What body system was compensating? How well did the results compare with your prediction? ______________________________

____

____

4. List and describe some possible causes of metabolic alkalosis. ______________________________

____

____

Blood Analysis

PRE-LAB QUIZ

1. The percentage of erythrocytes in a sample of whole blood is measured by the:
   a. hemoglobin
   b. hematocrit
   c. ABO blood type
   d. erythropoietin
2. Circle the correct underlined term: The protein that transports oxygen from the lungs to the cells of the body is hemoglobin / hematocrit.
3. A lower-than-normal hematocrit is known as __________, in which insufficient oxygen is transported to the body cells.
   a. polycythemia
   b. hemophilia
   c. hemochromatosis
   d. anemia
4. The erythrocyte sedimentation rate (ESR) can be used to follow the progression of all of the following diseases or conditions *except*:
   a. anemia
   b. rheumatoid arthritis
   c. acute appendicitis (within 24 hours)
   d. myocardial infarction
5. Circle the correct underlined term: A rouleaux formation / hemoglobinometer will be used to compare a standard color value to an experimental sample to determine hemoglobin content.
6. Circle the correct underlined term: A person with type AB / O blood has two recessive alleles and has neither type A nor type B antigen.
7. Circle True or False: Blood transfusion reactions are of little consequence and occur when the recipient has antibodies that react with the antigens present on the transfused cells.

Exercise Overview

Blood transports soluble substances to and from all cells of the body. Laboratory analysis of our blood can reveal important information about how well this function is being achieved. The five activities in this exercise simulate common laboratory tests performed on blood: (1) *hematocrit* determination, (2) *erythrocyte sedimentation rate*, (3) *hemoglobin* determination, (4) *blood typing*, and (5) total *cholesterol* determination.

Hematocrit refers to the percentage of red blood cells (RBCs), or erythrocytes, in a sample of whole blood. A hematocrit of 48 means that 48% of the volume of blood consists of RBCs. RBCs transport oxygen to the cells of the body. Therefore, the higher the hematocrit, the more RBCs are present in the blood and the greater the oxygen-carrying potential of the blood. Males usually have higher hematocrit levels than females because males have higher levels of testosterone. In addition to promoting the male sex characteristics, testosterone is responsible for stimulating the release of erythropoietin from the kidneys. Erythropoietin (EPO) is a hormone that stimulates the synthesis of RBCs. Therefore, higher levels of testosterone lead to more EPO secretion and, thus, higher hematocrit levels.

The **erythrocyte sedimentation rate (ESR)** measures the settling of RBCs in a vertical, stationary tube of blood during one hour. In a healthy individual, RBCs do not settle very much in an hour. In some disease conditions, increased production of fibrinogen and immunoglobulins causes the RBCs to clump together, stack up, and form a column (called a *rouleaux formation*). RBCs in a rouleaux formation are heavier and settle faster (that is, they display an increase in the sedimentation rate.)

Hemoglobin (Hb), a protein found in RBCs, is necessary for the transport of oxygen from the lungs to the cells of the body. Four polypeptide chains of amino acids comprise the globin part of the molecule. Each polypeptide chain has a heme unit—a group of atoms that includes an atom of iron to which a molecule of oxygen binds. Each polypeptide chain, if it folds correctly, can bind a molecule of oxygen. Therefore, each hemoglobin molecule can carry four molecules of oxygen. Oxygen combined with hemoglobin forms oxyhemoglobin, which has a bright red color.

All of the cells in the human body, including RBCs, are surrounded by a plasma membrane that contains genetically determined glycoproteins, called antigens. On RBC membranes, there are certain antigens, called **agglutinogens**, that determine a person's blood type. Blood typing is used to identify the **ABO blood groups**, which are determined by the presence or absence of two antigens: **type A** and **type B**. Because these antigens are genetically determined, a person has two copies (alleles) of the gene for these antigens, one copy from each parent.

Cholesterol is a lipid substance that is essential for life—it is an important component of all cell membranes and is the base molecule of steroid hormones, vitamin D, and bile salts. Cholesterol is produced in the human liver and is present in some foods of animal origin, such as milk, meat, and eggs. Because cholesterol is a hydrophobic lipid, it needs to be wrapped in protein packages, called **lipoproteins**, to travel in the blood (which is mostly water) from the liver and digestive organs to the cells of the body.

ACTIVITY 1

Hematocrit Determination

OBJECTIVES

1. To understand the terms *hematocrit*, *red blood cells*, *hemoglobin*, *buffy coat*, *anemia*, and *polycythemia*.
2. To understand how the hematocrit (packed red blood cell volume) is determined.
3. To understand the implications of elevated or decreased hematocrit.
4. To understand the importance of proper disposal of laboratory material that comes in contact with blood.

Introduction

Hematocrit refers to the percentage of **red blood cells (RBCs)**, or erythrocytes, in a sample of whole blood. A hematocrit of 48 means that 48% of the volume of blood consists of RBCs. RBCs transport oxygen to the cells of the body. Therefore, the higher the hematocrit, the more RBCs are present in the blood and the higher the oxygen-carrying potential of the blood. Hematocrit values are determined by spinning a microcapillary tube filled with a sample of whole blood in a special microhematocrit centrifuge. This procedure separates the blood cells from the blood plasma. A **buffy coat** layer of white blood cells (WBCs) appears as a thin, white layer *between* the heavier RBC layer and the lighter, yellow plasma.

The hematocrit is determined after centrifuging by measuring the height of the RBC layer (in millimeters) and dividing that by the height of the total blood sample (in millimeters). This calculation gives the percentage of the total blood volume consisting of RBCs. The average hematocrit for males is 42–52%, and the average hematocrit for females is 37–47%. A lower-than-normal hematocrit indicates **anemia**, and a higher-than-normal hematocrit indicates **polycythemia**.

Anemia is a condition in which insufficient oxygen is transported to the body's cells. There are many possible causes for anemia, including inadequate numbers of RBCs, a decreased amount of the oxygen-carrying pigment **hemoglobin** in the RBCs, and abnormally shaped hemoglobin. The heme portion of a hemoglobin molecule contains an atom of iron to which a molecule of oxygen can bind. If adequate iron is not available, the body cannot manufacture hemoglobin, resulting in the condition *iron-deficiency anemia. Aplastic anemia* results from the failure of the bone marrow to produce adequate red blood cell numbers. *Sickle cell anemia* is an inherited condition in which the protein portion of hemoglobin molecules folds incorrectly when oxygen levels are low. As a result, oxygen molecules cannot bind to the misshapen hemoglobin, the RBCs develop a sickle shape, and anemia results. Regardless of the underlying cause, anemia causes a reduction in the blood's ability to transport oxygen to the cells of the body.

Polycythemia refers to an increase in RBCs, resulting in a higher-than-normal hematocrit. There are many possible causes of polycythemia, including living at high altitudes, strenuous athletic training, and tumors in the bone marrow. In this activity you will simulate the blood test used to determine hematocrit.

EQUIPMENT USED The following equipment will be depicted on-screen: six heparinized capillary tubes (heparin keeps blood from clotting); blood samples from six individuals: sample 1: a healthy male living in Boston, sample 2: a healthy female living in Boston, sample 3: a healthy male living in Denver, sample 4: a healthy female living in Denver, sample 5: a male with aplastic anemia, sample 6: a female with iron-deficiency anemia; capillary tube sealer—a clay material (shown as an orange-yellow substance) used to seal the capillary tubes on one end so the blood sample can be centrifuged without having the blood spray out of the tube; microhematocrit centrifuge—used to centrifuge the samples (rotates at 14,500 revolutions per minute); metric ruler; biohazardous waste disposal—used to properly dispose of equipment that comes in contact with blood.

Experiment Instructions

Go to the home page in the PhysioEx software and click **Exercise 11: Blood Analysis**. Click **Activity 1: Hematocrit Determination**, and take the online **Pre-lab Quiz** for Activity 1.

After you take the online Pre-lab Quiz, click the **Experiment** tab and begin the experiment. The experiment

instructions are reprinted here for your reference. The opening screen for the experiment is shown below.

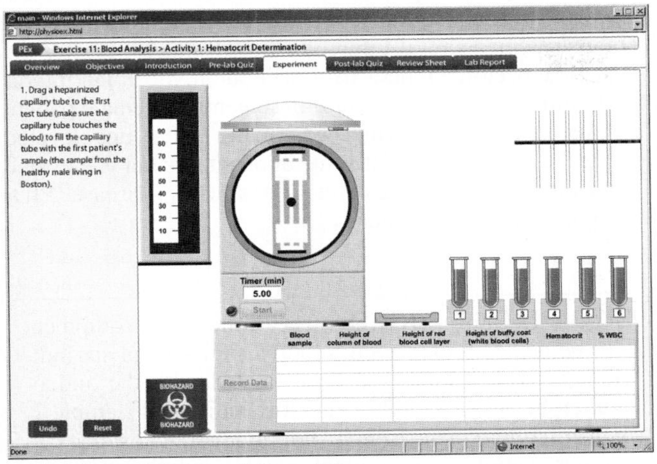

Windows Explorer 2007, Microsoft Corporation

1. Drag a heparinized capillary tube to the first test tube (make sure the capillary tube touches the blood) to fill the capillary tube with the first patient's sample (the sample from the healthy male living in Boston).

2. Drag the capillary tube containing sample 1 to the container of capillary tube sealer to seal one end of the tube.

3. Drag the capillary tube to the microhematocrit centrifuge. The remaining samples will automatically be prepared for centrifugation.

4. Note that the timer is set to 5 minutes. Click **Start** to centrifuge the samples for 5 minutes at 14,500 revolutions per minute. The simulation compresses the 5-minute time period into 5 seconds of real time.

5. Drag capillary tube 1 from the centrifuge to the metric ruler to measure the height of the column of blood and the height of each layer.

6. Click **Record Data** to display your results in the grid (and record your results in Chart 1).

7. Drag capillary tube 1 to the biohazardous waste disposal.

? PREDICT **Question 1**
Predict how the hematocrits of the patients living in Denver, Colorado (approximately one mile above sea level), will compare with the hematocrit levels of the patients living in Boston, Massachusetts (at sea level).

____

8. You will now measure the column and layer heights of the remaining samples.

- Drag the next capillary tube from the centrifuge to the metric ruler.
- Click **Record data** to display your results in the grid (and record your results in Chart 1). The tube will automatically be placed in the biohazardous waste disposal.

Repeat this step for each of the remaining samples.

After you complete the experiment, take the online **Post-lab Quiz** for Activity 1.

CHART 1 Hematocrit Determination

	Total height of column of blood (mm)	Height of red blood cell layer (mm)	Height of buffy coat (mm)	Hematocrit	% WBC
Sample 1 (healthy male living in Boston)					
Sample 2 (healthy female living in Boston)					
Sample 3 (healthy male living in Denver)					
Sample 4 (healthy female living in Denver)					
Sample 5 (male with aplastic anemia)					
Sample 6 (female with iron-deficiency anemia)					

Activity Questions

1. How do you calculate the hematocrit after you centrifuge the total blood sample? What does the result of this calculation indicate?

2. What is the significance of the "buffy coat" after you centrifuge the total blood sample?

3. As noted in the Exercise Overview, the average hematocrit for males is 42–52%, the average hematocrit for females is 37–47%, and erythropoietin is a hormone that is responsible for the synthesis of RBCs. Given this information, explain how a female could have a consistent hematocrit of 48, large, well-defined skeletal muscles, and an abnormally deep voice.

ACTIVITY 2

Erythrocyte Sedimentation Rate

OBJECTIVES

5. To understand *erythrocyte sedimentation rate (ESR), red blood cells (RBCs)*, and *rouleaux formation*.
6. To learn how to perform an erythrocyte sedimentation rate blood test.
7. To understand the results (and their implications) from an erythrocyte sedimentation rate blood test.
8. To understand the importance of proper disposal of laboratory material that comes in contact with blood.

Introduction

The **erythrocyte sedimentation rate (ESR)** measures the settling of **red blood cells (RBCs)** in a vertical, stationary tube of whole blood during one hour. In a healthy individual, red blood cells do not settle very much in an hour. In some disease conditions, increased production of fibrinogen and immunoglobulins cause the RBCs to clump together, stack up, and form a dark red column (called a **rouleaux formation**). RBCs in a rouleaux formation are heavier and settle faster (that is, they exhibit an increase in the settling rate).

The ESR is neither very specific nor diagnostic, but it can be used to follow the progression of certain diseases, including sickle cell anemia, some cancers, and inflammatory diseases, such as rheumatoid arthritis. When the disease worsens, the ESR increases. When the disease improves, the ESR decreases.

The ESR can be elevated in iron-deficiency anemia, and menstruating females sometimes develop anemia and show an increase in ESR. The ESR can also be used to evaluate a patient with chest pains because the ESR is elevated in established myocardial infarction (heart attack) but normal in angina pectoris (chest pain without myocardial infarction). Similarly, it can be useful in screening a female patient with severe abdominal pains because the ESR is not elevated within the first 24 hours of acute appendicitis but is elevated in the early stage of acute pelvic inflammatory disease (PID) or ruptured ectopic pregnancy.

EQUIPMENT USED The following equipment will be depicted on-screen: blood samples from six individuals (each sample has been treated with the anticoagulant heparin): sample 1: healthy individual, sample 2: menstruating female, sample 3: individual with sickle cell anemia, sample 4: individual with iron-deficiency anemia, sample 5: individual suffering a myocardial infarction, sample 6: individual with angina pectoris; sodium citrate—used to bind with calcium and prevent the blood samples from clotting so they can be easily poured into the narrow sedimentation rate tubes; test tubes—used as reaction vessels for the tests; sedimentation tubes (contained in cabinet); magnifying chamber—used to help read the millimeter markings on the sedimentation tubes; biohazardous waste disposal—used to properly dispose of equipment that comes in contact with blood.

Experiment Instructions

Go to the home page in the PhysioEx software and click **Exercise 11: Blood Analysis**. Click **Activity 2: Erythrocyte Sedimentation Rate**, and take the online **Pre-lab Quiz** for Activity 2.

After you take the online Pre-lab Quiz, click the **Experiment** tab and begin the experiment. The experiment instructions are reprinted here for your reference. The opening screen for the experiment is shown below.

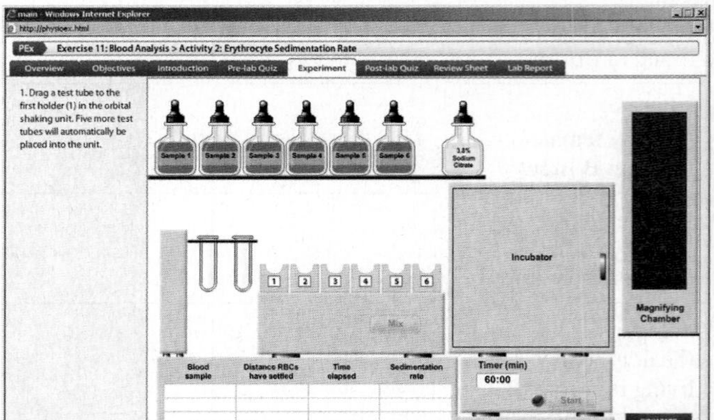

Windows Explorer 2007, Microsoft Corporation

1. Drag a test tube to the first holder (1) in the orbital shaking unit. Five more test tubes will automatically be placed into the unit.

2. Drag the dropper cap of the sample 1 bottle (the sample from the healthy individual) to the first test tube (1) in the orbital shaking unit to dispense one milliliter of blood into the tube. The remaining five samples will be automatically dispensed.

3. Drag the dropper cap of the 3.8% sodium citrate bottle to the first test tube to dispense 0.5 milliliters of sodium citrate into each of the tubes.

4. Click **Mix** to mix the samples.

5. Drag the first test tube to the first sedimentation tube in the incubator to pour the contents of the test tube into the sedimentation tube.

6. Drag the now empty test tube to the biohazardous waste disposal. The contents of the remaining test tubes will automatically be poured into the sedimentation tubes, and the empty tubes will automatically be placed in the biohazardous waste disposal.

7. Note that the timer is set to 60 minutes. Click **Start** to incubate the sedimentation tubes for 60 minutes. The simulation compresses the 60-minute time period into 6 seconds of real time.

8. Drag the first sedimentation tube to the magnifying chamber to examine the tube. The tube is marked in millimeters (the distance between two marks is 5 mm).

9. Click **Record Data** to display your results in the grid (and record your results in Chart 2).

10. Drag the sedimentation tube to the biohazardous waste disposal.

? PREDICT **Question 1**
How will the sedimentation rate for sample 6 (unhealthy individual) compare with the sedimentation rate for sample 1 (healthy individual)?

____

11. You will now measure the sedimentation rate for the remaining samples.

- Drag the next sedimentation tube to the magnifying chamber to examine the tube.
- Click **Record Data** to display your results in the grid (and record your results in Chart 2). The tube will automatically be placed in the biohazardous waste disposal.

Repeat this step for each of the remaining samples.

After you complete the experiment, take the online **Post-lab Quiz** for Activity 2.

Activity Questions

1. Why is ESR useful, even though it is neither specific nor sensitive?

____

____

2. Describe the physical process underlying an accelerated erythrocyte sedimentation rate.

____

____

ACTIVITY 3

Hemoglobin Determination

OBJECTIVES

9. To understand the terms *hemoglobin (Hb)*, *anemia*, *heme*, *oxyhemoglobin*, and *hemoglobinometer*.
10. To learn how to determine the amount of hemoglobin in a blood sample.
11. To understand the results and their implications when examining the amounts of hemoglobin present in a blood sample.
12. To understand the importance of proper disposal of laboratory material that comes in contact with blood.

CHART 2 Erythrocyte Sedimentation Rate

Blood sample	Distance RBCs have settled (mm)	Elapsed time	Sedimentation rate
Sample 1 (healthy individual)			
Sample 2 (menstruating female)			
Sample 3 (individual with sickle cell anemia)			
Sample 4 (individual with iron-deficiency anemia)			
Sample 5 (individual suffering a myocardial infarction)			
Sample 6 (individual with angina pectoris)			

Introduction

Hemoglobin (Hb), a protein found in red blood cells, is necessary for the transport of oxygen from the lungs to the cells of the body. Four polypeptide chains of amino acids comprise the globin part of the molecule. Each polypeptide chain has a **heme** unit—a group of atoms that includes an atom of iron to which a molecule of oxygen binds. Each polypeptide chain, if it folds correctly, can bind a molecule of oxygen. Therefore, each hemoglobin molecule can carry four molecules of oxygen. Oxygen combined with hemoglobin forms **oxyhemoglobin**, which has a bright-red color. Anemia results when insufficient oxygen is carried in the blood.

A quantitative hemoglobin measurement is used to determine the classification and possible causes of anemia and also gives useful information on some other disease conditions. For example, a person can have anemia with a normal red blood cell count if there is inadequate hemoglobin in the red blood cells. Normal blood contains an average of 12–18 grams of hemoglobin per 100 milliliters of blood. A healthy male has 13.5–18 g/100 ml and a healthy female has 12–16 g/100 ml. Hemoglobin levels increase in patients with polycythemia, congestive heart failure, and chronic obstructive pulmonary disease (COPD). Hemoglobin levels also increase when dwelling at high altitudes. Hemoglobin levels decrease in patients with anemia, hyperthyroidism, cirrhosis of the liver, renal disease, systemic lupus erythematosus, and severe hemorrhage.

The hemoglobin level of a blood sample is determined by stirring the blood with a wooden stick to rupture, or lyse, the red blood cells. The color intensity of the hemolyzed blood reflects the amount of hemoglobin present. A **hemoglobinometer** transmits green light through the hemolyzed blood sample and then compares the amount of light that passes through the sample to standard color intensities to determine the hemoglobin content of the sample.

EQUIPMENT USED The following equipment will be depicted on-screen: blood samples from five individuals: sample 1: healthy male, sample 2: healthy female, sample 3: female with iron-deficiency anemia, sample 4: male with polycythemia, sample 5: female Olympic athlete; hemolysis sticks—used to stir the blood samples to lyse the red blood cells, thereby releasing their hemoglobin; blood chamber dispenser—used to dispense a blood chamber slide with a depression for the blood sample; hemoglobinometer—used to analyze the hemoglobin level in each sample; biohazardous waste disposal—used to properly dispose of equipment that comes in contact with blood.

Experiment Instructions

Go to the home page in the PhysioEx software and click **Exercise 11: Blood Analysis**. Click **Activity 3: Hemoglobin Determination**, and take the online **Pre-lab Quiz** for Activity 3.

After you take the online Pre-lab Quiz, click the **Experiment** tab and begin the experiment. The experiment instructions are reprinted here for your reference. The opening screen for the experiment is shown below.

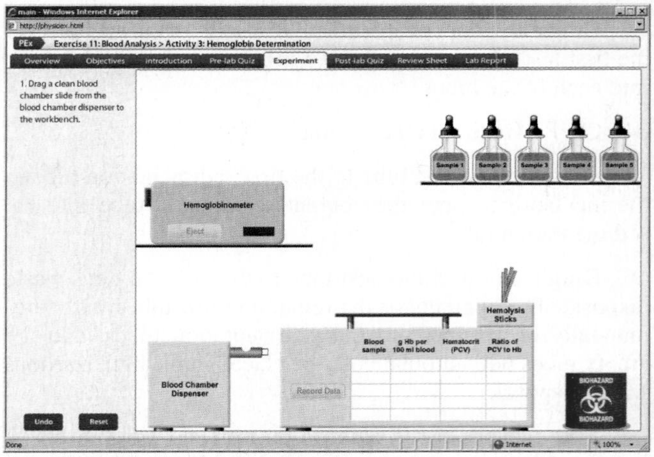

Windows Explorer 2007, Microsoft Corporation

1. Drag a clean blood chamber slide from the blood chamber dispenser to the workbench.

2. Drag the bottle cap from the sample 1 bottle (the sample from the healthy male) to the depression in the blood chamber slide to dispense a drop of blood into the depression.

3. Drag a hemolysis stick to the drop of blood in the chamber to stir the blood sample for 45 seconds, lysing the red blood cells and releasing their hemoglobin.

4. Drag the hemolysis stick to the biohazardous waste disposal.

5. Drag the blood chamber slide to the dark rectangular slot on the hemoglobinometer to analyze the sample. After you insert the blood chamber slide into the hemoglobinometer, you will see a blowup of the inside of the hemoglobinometer.

6. The left half of the circular field shows the intensity of green light transmitted by blood sample 1. The right half of the circular field shows the intensity of green light for known levels of hemoglobin present in blood. Drag the lever on the right side of the hemoglobinometer down until the shade of green in the right half of the field matches the shade of green in the left half of the field and then click **Record Data** to display your results in the grid (and record your results in Chart 3).

7. Click **Eject** to remove the blood chamber slide from the hemoglobinometer.

8. Drag the blood chamber slide from the hemoglobinometer to the biohazardous waste disposal.

? PREDICT Question 1
How will the hemoglobin levels for the female Olympic athlete (sample 5) compare with the hemoglobin levels for the healthy female (sample 2)?

____

CHART 3 Hemoglobin Determination

Blood sample	Hb in grams per 100 ml of blood	Hematocrit (PCV)	Ratio of PCV to Hb
Sample 1 (healthy male)		48	
Sample 2 (healthy female)		44	
Sample 3 (female with iron-deficiency anemia)		40	
Sample 4 (male with polycythemia)		60	
Sample 5 (female Olympic athlete)		60	

9. You will now measure the hemoglobin levels for each of the remaining samples.

- Drag a blood chamber slide to the workbench.
- Drag the bottle cap from the next sample bottle to the depression in the slide.
- Drag a hemolysis stick to the drop of blood in the chamber (after stirring the sample, the hemolysis stick will automatically be placed in the biohazardous waste disposal).
- Drag the blood chamber slide to the dark rectangular slot on the hemoglobinometer.
- Drag the lever on the right side of the hemoglobinometer down until the shade of green in the right half of the field matches the shade of green in the left half of the field and then click **Record Data** to display your results in the grid (and record your results in Chart 3).
- Click **Eject** to remove the blood chamber slide from the hemoglobinometer (the slide will automatically be placed in the biohazardous waste disposal).

Repeat this step until you analyze all five samples.

After you complete the experiment, take the online **Post-lab Quiz** for Activity 3.

Activity Questions

1. As mentioned in the introduction to this activity, hemoglobin levels increase for people living at high altitudes. Given that the atmospheric pressure of oxygen significantly declines as you ascend to higher elevations, why do you think hemoglobin levels would increase for those living at high altitudes?

____

____

2. Just by looking at the color of a freshly drawn blood sample, how could you distinguish between blood that is well oxygenated and blood that is poorly oxygenated?

____

____

ACTIVITY 4

Blood Typing

OBJECTIVES

1. To understand the terms *antigens*, *agglutinogens*, *ABO antigens*, *Rh antigens*, and *agglutinins*.
2. To learn how to perform a blood-typing assay.
3. To understand the results and their implications when examining agglutination reactions.
4. To understand the importance of proper disposal of laboratory material that comes in contact with blood.

Introduction

All of the cells in the human body, including red blood cells, are surrounded by a plasma membrane that contains genetically determined glycoproteins, called **antigens**. On red blood cell membranes, there are certain antigens, called **agglutinogens**, that determine a person's blood type. If a blood transfusion recipient has antibodies (called **agglutinins**) that react with the antigens present on the transfused cells, the red blood cells will become clumped together, or agglutinated, and then lysed, resulting in a potentially life-threatening blood transfusion reaction. It is therefore important to determine an individual's blood type before performing blood transfusions to avoid mixing incompatible blood. Although many different antigens are present on red blood cell membranes, the **ABO** and **Rh antigens** cause the most vigorous and potentially fatal transfusion reactions.

The ABO blood groups are determined by the presence or absence of two antigens: type A and type B. Because these antigens are genetically determined, a person has two copies (alleles) of the gene for these proteins, one copy from each parent. The presence of these antigens is due to a dominant allele, and their absence is due to a recessive allele.

- A person with type A blood can have two alleles for the type A antigen or one allele for the type A antigen and one allele for the absence of either the type A or type B antigen.
- A person with type B blood can have two alleles for the type B antigen or one allele for the type B antigen and one allele for the absence of either the type A or type B antigen.
- A person with type AB blood has one allele for the type A antigen and one allele for the type B antigen.
- A person with type O blood has two recessive alleles and has neither the type A nor type B antigen.

TABLE 11.1 ABO Blood Types

Blood type	Antigens on RBCs	Antibodies present in plasma
A	A	anti-B
B	B	anti-A
AB	A and B	none
O	none	anti-A and anti-B

Antibodies against the A and B antigens are found preformed in the blood plasma. A person has antibodies only for the antigens not on his or her red blood cells, so a person with type A blood will have anti-B antibodies. View Table 11.1 for a summary of the antigens on red blood cells and the antibodies in the plasma for each blood type.

The Rh factor is another genetically determined protein that can be present on red blood cell membranes. Approximately 85% of the population is Rh positive (Rh^+), and their red blood cells have this protein on their surface. Antibodies against the Rh factor are not found preformed in the plasma. They are produced by an Rh negative (Rh^-) individual only after exposure to blood cells from someone who is Rh^+. Such exposure can occur during pregnancy when Rh^+ blood cells from the baby cross the placenta and expose the mother to the antigen.

To determine an individual's blood type, drops of an individual's blood sample are mixed separately with antiserum containing antibodies to either type A antigens, type B antigens, or Rh antigens. An agglutination reaction (showing clumping) indicates the presence of the agglutinogen.

EQUIPMENT USED The following equipment will be depicted on-screen: blood samples from six individuals with different blood types; anti-A serum (blue bottle), anti-B serum (yellow bottle), and anti-Rh serum (white bottle), containing antibodies to the A antigen, B antigen, and Rh antigen, respectively; blood-typing slide dispenser; color-coded stirring sticks—used to mix the blood sample and the serum (blue: used with anti-A serum, yellow: used with the anti-B serum, white: used with the anti-Rh serum); light box—used to view the blood type samples; biohazardous waste disposal—used to properly dispose of equipment that comes in contact with blood.

Experiment Instructions

Go to the home page in the PhysioEx software and click **Exercise 11, Blood Analysis**. Click **Activity 4, Blood Typing**, and take the online **Pre-lab Quiz** for Activity 4.

After you take the online Pre-lab Quiz, click the **Experiment** tab and begin the experiment. The experiment instructions are reprinted here for your reference. The opening screen for the experiment is shown above.

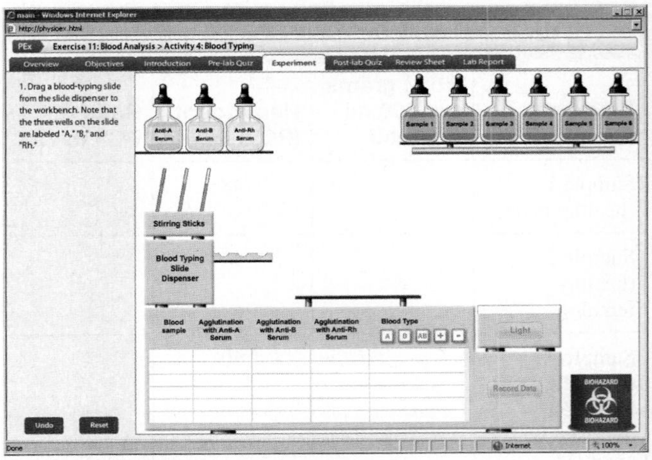

Windows Explorer 2007, Microsoft Corporation

1. Drag a blood-typing slide from the slide dispenser to the workbench. Note that the three wells on the slide are labeled "A," "B," and "Rh."

2. Drag the dropper cap of the sample 1 bottle to well A on the blood-typing slide to dispense a drop of blood into each well.

3. Drag the dropper cap of the anti-A serum bottle to well A on the blood-typing slide to dispense a drop of anti-A serum into the well.

4. Drag the dropper cap of the anti-B serum bottle to well B on the blood-typing slide to dispense a drop of anti-B serum into the well.

5. Drag the dropper cap of the anti-Rh serum bottle to well Rh on the blood-typing slide to dispense a drop of anti-Rh serum into the well.

6. Drag a blue-tipped stirring stick to well A to mix the blood and anti-A serum.

7. Drag the stirring stick to the biohazardous waste disposal.

8. Drag a yellow-tipped stirring stick to well B to mix the blood and anti-B serum.

9. Drag the stirring stick to the biohazardous waste disposal.

10. Drag a white-tipped stirring stick to well Rh to mix the blood and anti-Rh serum.

11. Drag the stirring stick to the biohazardous waste disposal.

12. Drag the blood-typing slide to the light box and then click **Light** to analyze the slide.

13. Under each of the wells, click **Positive** if agglutination occurred (the sample shows clumping) or click **Negative** if agglutination did not occur (the sample looks smooth).

CHART 4 Blood Typing Results

Blood sample	Agglutination with anti-A serum	Agglutination with anti-B serum	Agglutination with anti-Rh serum	Blood type
1				
2				
3				
4				
5				
6				

14. Click **Record Data** to display your results in the grid (and record your results in Chart 4).

15. Drag the blood-typing slide to the biohazardous waste disposal.

? PREDICT Question 1
If the patient's blood type is AB^-, what would be the appearance of the A, B, and Rh samples?

____

16. You will now analyze the remaining samples.

- Drag a blood-typing slide from the slide dispenser to the workbench. The next sample will be added to each well on the slide, the appropriate antiserum will be added to each well, the sample and antisera will be mixed, and the slide will be placed in the light box.
- Under each of the wells, click **Positive** if agglutination occurred (the sample shows clumping) or click **Negative** if agglutination did not occur (the sample looks smooth).
- Click **Record Data** to display your results in the grid (and record your results in Chart 4).

Repeat this step until you analyze all six samples.

17. You will now indicate the blood type for each sample and indicate whether the sample is Rh positive or Rh negative.

- Click the row for the sample in the grid (and record your results in Chart 4).
- Click A, B, AB, or O above the blood type column to indicate the blood type.
- Click the − button or the + button above the blood type column to indicate whether the sample is Rh negative or Rh positive.

Repeat this step for all six samples. Record your results in Chart 4.

After you complete the experiment, take the online **Post-lab Quiz** for Activity 4.

Activity Questions

1. Antibodies against the A and B antigens are found in the plasma, and a person has antibodies only for the antigens that are not present on their red blood cells. Using this information, list the antigens found on red blood cells and the antibodies in the plasma for blood types 1) AB−, 2) O+, 3) B−, and 4) A+.

____

____

2. If an individual receives a bone marrow transplant from someone with a different ABO blood type, what happens to the recipient's ABO blood type?

____

____

ACTIVITY 5

Blood Cholesterol

OBJECTIVES

5. To understand the terms *cholesterol*, *lipoproteins*, *low-density lipoprotein (LDL)*, *hypocholesterolemia*, *hypercholesterolemia*, and *atherosclerosis*.
6. To learn how to test for total blood cholesterol using a colorimetric assay.
7. To understand the results and their implications when examining total blood cholesterol.
8. To understand the importance of proper disposal of laboratory material that comes in contact with blood.

Introduction

Cholesterol is a lipid substance that is essential for life—it is an important component of all cell membranes and is the base molecule of steroid hormones, vitamin D, and bile salts. Cholesterol is produced in the human liver and is present in some foods of animal origin, such as milk, meat, and eggs. Because cholesterol is a water-insoluble lipid, it needs to be wrapped in protein packages, called **lipoproteins**, to travel in the blood (which is mostly water) from the liver and digestive organs to the cells of the body.

One type of lipoprotein package, called **low-density lipoprotein (LDL)**, has been identified as a potential source of damage to the interior of arteries. LDLs can contribute to **atherosclerosis**, the buildup of plaque, in these blood vessels.

A total blood cholesterol determination does not measure the level of LDLs, but it does provide valuable information about the total amount of cholesterol in the blood.

Less than 200 milligrams of total cholesterol per deciliter of blood is considered desirable. Between 200 and 239 mg/dl is considered borderline high cholesterol. Over 240 mg/dl is considered high blood cholesterol **(hypercholesterolemia)** and is associated with an increased risk of cardiovascular disease. Abnormally low blood cholesterol levels (total cholesterol lower than 100 mg/dl) can also suggest a problem. Low levels may indicate hyperthyroidism (overactive thyroid gland), liver disease, inadequate absorption of nutrients from the intestine, or malnutrition. Other reports link **hypocholesterolemia** (low blood cholesterol) to depression, anxiety, and mood disturbances, which are thought to be controlled by the level of available serotonin, a neurotransmitter. There is evidence of a relationship between low levels of blood cholesterol and low levels of serotonin in the brain.

In this test for total blood cholesterol, a sample of blood is mixed with enzymes that produce a colored reaction with cholesterol. The intensity of the color indicates the amount of cholesterol present. The cholesterol tester compares the color of the sample to the colors of known levels of cholesterol (standard values).

EQUIPMENT USED The following equipment will be depicted on-screen: lancets—sharp, needlelike instruments used to prick the finger to obtain a drop of blood; four patients (represented by an extended finger); alcohol wipes—used to cleanse the patient's fingertip before it is punctured with the lancet; color wheel—divided into shades of green that correspond to total cholesterol levels; cholesterol strips—contain chemicals that convert, by a series of reactions, the cholesterol in the blood sample into a green-colored solution; biohazardous waste disposal—used to properly dispose of equipment that comes in contact with blood.

Experiment Instructions

Go to the home page in the PhysioEx software and click **Exercise 11: Blood Analysis**. Click **Activity 5: Blood Cholesterol**, and take the online **Pre-lab Quiz** for Activity 5.

After you take the online Pre-lab Quiz, click the **Experiment** tab and begin the experiment. The experiment instructions are reprinted here for your reference. The opening screen for the experiment is shown below.

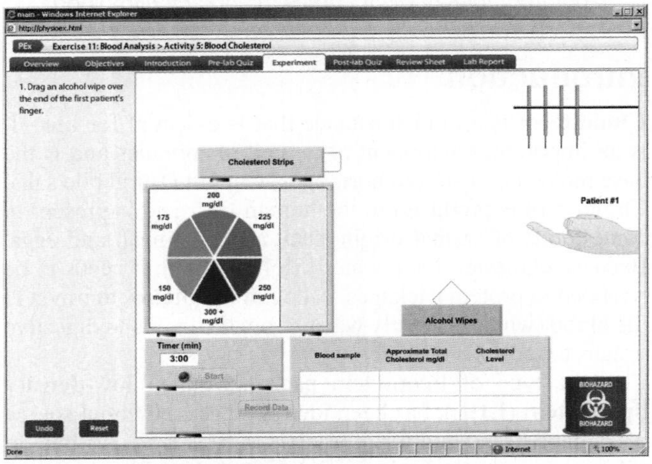

Windows Explorer 2007, Microsoft Corporation

1. Drag an alcohol wipe over the end of the first patient's finger.

2. Drag the alcohol wipe to the biohazardous waste disposal.

3. Drag a lancet to the tip of the patient's finger to prick the finger and obtain a drop of blood.

4. Drag the lancet to the biohazardous waste disposal.

5. Drag a cholesterol strip to the finger to transfer a drop of blood from the patient's finger to the strip.

6. Drag the cholesterol strip to the rectangular box to the right of the color wheel.

7. Click **Start** to start the timer. It takes three minutes for the chemicals in the cholesterol strip to react with the blood. The simulation compresses the 3-minute time period into 3 seconds of real time.

8. Click the color on the color wheel that most closely matches the color on the cholesterol strip.

9. Click **Record Data** to display your results in the grid (and record your results in Chart 5).

CHART 5 Total Cholesterol Determination

Blood sample	Approximate total cholesterol (mg/dl)	Cholesterol level
1		
2		
3		
4		

10. Drag the cholesterol test strip to the biohazardous waste disposal.

PREDICT Question 1
Patient 4 prefers to cook all his meat in lard or bacon grease. Knowing this dietary preference, you anticipate his total cholesterol level to be:

____

11. You will now test the total cholesterol levels for the remaining patients.

- Drag an alcohol wipe over the end of the patient's finger. The alcohol wipe will automatically be placed in the biohazardous waste disposal.
- Drag a lancet to the tip of the patient's finger to prick the finger and obtain a drop of blood. The lancet will automatically be placed in the biohazardous waste disposal.
- Drag a cholesterol strip to the finger to transfer a drop of blood from the patient's finger to the strip.

- Drag the cholesterol strip to the rectangular box to the right of the color wheel. The timer will automatically run for three minutes to allow the chemicals in the cholesterol strip to react with the blood.
- Click the color on the color wheel that most closely matches the color on the cholesterol strip.
- Click **Record Data** to display your results in the grid (and record your results in Chart 5). The cholesterol strip will automatically be placed in the biohazardous waste disposal.

Repeat this step until you determine the total cholesterol levels for all four patients.

After you complete the experiment, take the online **Post-lab Quiz** for Activity 5.

Activity Questions

1. Why do cholesterol plaques occur in arteries and not veins?

____

____

2. Phytosterols can alter absorption of certain molecules by the intestinal tract. Why would they be a beneficial dietary supplement for people with high LDL levels?

____

____

Blood Analysis

NAME ______________________________

LAB TIME/DATE ______________________

ACTIVITY 1 Hematocrit Determination

1. List the hematocrits for the healthy male (sample 1) and female (sample 2) living in Boston (at sea level) and indicate whether they are normal or whether they indicate anemia or polycythemia.

2. Describe the difference between the hematocrits for the male and female living in Boston. Why does this difference between the sexes exist?

3. List the hematocrits for the healthy male and female living in Denver (approximately one mile above sea level) and indicate whether they are normal or whether they indicate anemia or polycythemia.

4. How did the hematocrit levels of the Denver residents differ from those of the Boston residents? Why? How well did the results compare with your prediction?

5. Describe how the kidneys respond to a chronic decrease in oxygen and what effect this has on hematocrit levels.

6. List the hematocrit for the male with aplastic anemia (sample 5) and indicate whether it is normal or abnormal. Explain your response.

7. List the hematocrit for the female with iron-deficiency anemia (sample 6) and indicate whether it is normal or abnormal. Explain your response.

____

____

ACTIVITY 2 Erythrocyte Sedimentation Rate

1. Describe the effect that sickle cell anemia has on the sedimentation rate (sample 3). Why do you think that it has this effect?

____

____

2. How did the sedimentation rate for the menstruating female (sample 2) compare with the sedimentation rate for the healthy individual (sample 1)? Why do you think this occurs?

____

____

3. How did the sedimentation rate for the individual with angina pectoris (sample 6) compare with the sedimentation rate for the healthy individual (sample 1)? Why? How well did the results compare with your prediction?

____

____

4. What effect does iron-deficiency anemia (sample 4) have on the sedimentation rate?

____

____

5. Compare the sedimentation rate for the individual suffering a myocardial infarction (sample 5) with the sedimentation rate for the individual with angina pectoris (sample 6). Explain how you might use this data to monitor heart conditions.

____

____

ACTIVITY 3 Hemoglobin Determination

1. Is the male with polycythemia (sample 4) deficient in hemoglobin? Why?

____

____

2. How did the hemoglobin levels for the female Olympic athlete (sample 5) compare with the hemoglobin levels for the healthy female (sample 2)? Is either person *deficient* in hemoglobin? How well did the results compare with your prediction?

____

____

3. List conditions in which hemoglobin levels would be expected to decrease. Provide reasons for the change when possible.

____

____

4. List conditions in which hemoglobin levels would be expected to increase. Provide reasons for the change when possible.

____

____

5. Describe the ratio of hematocrit to hemoglobin for the healthy male (sample 1) and female (sample 2). (A normal ratio of hematocrit to grams of hemoglobin is approximately 3:1.) Discuss any differences between the two individuals.

____

____

6. Describe the ratio of hematocrit to hemoglobin for the female with iron-deficiency anemia (sample 3) and the female Olympic athlete (sample 5). (A normal ratio of hematocrit to grams of hemoglobin is approximately 3:1.) Discuss any differences between the two individuals.

____

____

ACTIVITY 4 Blood Typing

1. How did the appearance of the A, B, and Rh samples for the patient with AB^- blood type compare with your prediction?

____

____

2. Which blood sample contained the rarest blood type?

____

3. Which blood sample contained the universal donor?

____

4. Which blood sample contained the universal recipient?

____

5. Which blood sample did not agglutinate with any of the antibodies tested? Why?

____

____

6. What antibodies would be found in the plasma of blood sample 1? ______________________

____

7. When transfusing an individual with blood that is compatible but not the same type, it is important to separate packed cells from the plasma and administer only the packed cells. Why do you think this is done? (Hint: Think about what is *in plasma* versus what is *on RBCs*.)

____

____

8. List the blood samples in this activity that represent people who could donate blood to a person with type B^+ blood.

____

____

ACTIVITY 5 Blood Cholesterol

1. Which patient(s) had desirable cholesterol level(s)?

____

2. Which patient(s) had elevated cholesterol level(s)?

____

3. Describe the risks for the patient(s) you identified in question 2.

____

____

4. Was the cholesterol level for patient 4 low, desirable, or high? How well did the results compare with your prediction? What advice about diet and exercise would you give to this patient? Why?

____

____

5. Describe some reasons why a patient might have abnormally low blood cholesterol.

____

____

PhysioEx™ 9.1

EXERCISE 12

Serological Testing

PRE-LAB QUIZ

1. Circle True or False: Antibodies include proteins, polysaccharides, and various small molecules that stimulate antigen production.
2. You will use the ______________ technique to detect an infectious agent, *Chlamydia trachomatis*, by using labeled antibodies to detect the presence of the antigen.
   a. blotting technique
   b. Ouchterlony technique
   c. fluorescent antibody technique
   d. serology antigen technique
3. Circle True or False: In the life cycle of *Chlamydia trachomatis*, the infectious cell type is the large, less-dense reticulate body.
4. The ______________ is designed to determine if antigens are identical, related, or unrelated.
   a. Ouchterlony technique
   b. blotting technique
   c. fluorescent antibody technique
   d. serous antigen technique
5. Circle True or False: A typical antigen will always have only one site that allows for binding of antibodies or epitope.
6. ______________ occurs when a patient goes from testing negative for a specific antibody to testing positive for the same antibody.
   a. Death
   b. Seroconversion
   c. Precipitin
   d. Bleeding
7. Circle the correct underlined term: The ELISA assay / fluorescent antibody technique is considered enzyme linked because an enzyme is chemically linked to an antibody in this test.
8. Circle the correct underlined term: Gel electrophoresis / A constant region is a technique that uses electrical current to separate proteins based on their size and charge.
9. Circle True or False: In testing for HIV, the Western blotting technique is used as a confirmation of positive ELISA assay results because ELISA is prone to false positive results.

Exercise Overview

Immunology, the study of the immune system, focuses on chemical interactions that are difficult to observe. A number of chemical techniques have been developed to visually represent antibodies and antigens in the **serum**, the fluid portion of the blood with the clotting factors removed. The study and use of these techniques is referred to as **serology**. These techniques are performed in vitro, outside of the body, and are primarily used as diagnostic tools to detect disease. Other applications include pregnancy testing and drug testing. These immunological techniques depend upon the principle that an antibody binds only to specific, corresponding antigens. The tests are relatively expensive to perform, so these activities will allow you to perform them without the sometimes cost-prohibitive supplies.

Antigens and Antibodies

The word **antigen** is derived from two words: *anti*body and *gen*erator. Antigens do not produce antibodies, but early scientists noted that when antigens were present, antibodies appeared. Plasma cells actually produce antibodies.

Antigens include proteins, polysaccharides, and various small molecules that stimulate antibody production. Antigens are often molecules that are described as **nonself**, or foreign to the body. There are also self-antigens that act as identifier tags, such as the proteins found on the surface of red blood cells. Most often,

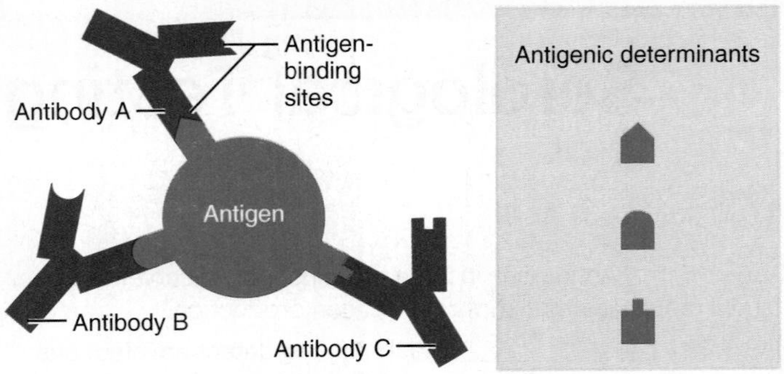

FIGURE 12.1 Antigen-antibody interaction with antigenic determinants.

antigens are a portion of an infectious agent, such as a bacterium or a virus, and the body produces antibodies in response to the presence of the infectious agent.

Antigens are often large and have multiple antigenic sites—locations that can bind to antibodies. We refer to these sites as **antigenic determinants**, or **epitopes**. The antibody has a corresponding antigen-binding site that has a "lock-and-key" recognition for the antigenic determinant on the antigen (view Figure 12.1). All of the simulated tests presented in this exercise take advantage of antigen-antibody specificity. These tests include direct fluorescent antibody technique, Ouchterlony technique, ELISA (enzyme-linked immunosorbent assay), and Western blotting technique.

Nonspecific Binding

The lock-and-key recognition that antigen and antibody have for each other is much like the specificity that an enzyme and its substrate have for one another. However, with antigen and antibody, **nonspecific binding** sometimes occurs. For this reason you will perform a number of washing steps in this exercise to remove any nonspecific binding.

Positive and Negative Controls

You will also use **positive** and **negative controls** to ensure that the test is working accurately. Positive controls include a substance that is known to react positively, thus giving you a standard against which to base your results. Negative controls include substances that should not react. A positive result with a negative control is a "false positive," which would invalidate all other results. Likewise, a negative result with a positive control is a "false negative," which would also invalidate your results.

ACTIVITY 1

Using Direct Fluorescent Antibody Technique to Test for Chlamydia

OBJECTIVES

1. To understand how fluorescent antibodies can be used diagnostically to detect the presence of a specific antigen.
2. To observe how to test for the sexually transmitted disease chlamydia.
3. To distinguish between antigens and antibodies.
4. To understand the terms *epitope* and *antigenic determinant*.
5. To observe nonspecific binding that can result between antigen and antibody.

Introduction

The direct fluorescent antibody technique uses antibodies to directly detect the presence of antigen. A fluorescent dye molecule attached to these antibodies acts as a visual signal for a positive result. This technique is typically used to test for antigens from infectious agents, such as bacteria or viruses. In this activity you will test for the presence of *Chlamydia trachomatis* (a bacterium that invades the cells of its host) using fluorescently labeled antibodies to detect the presence of the antigen and, therefore, the bacterium. *Chlamydia trachomatis* is an important infectious agent because it causes the sexually transmitted disease **chlamydia**. Left untreated, chlamydia can lead to sterility in men and women.

Chlamydia trachomatis is an obligate, intracellular bacterium, which means that it can only survive inside a host cell. The life cycle of the bacterium has two cellular types. The infectious cell type is the small, dense **elementary body**, which is capable of attaching to the host cell. The **reticulate body** is a larger, less-dense cell, which divides actively once inside the host cell. The reticulate body is also referred to as the vegetative form. The life cycle of *Chlamydia* begins when the elementary body enters the host cell and continues as the elementary body changes inside the host cell into a reticulate body. The reticulate body divides into more reticulate bodies and converts back to the elementary body form for release to infect other cells.

In this activity you will test three patient samples and two control samples for the *Chlamydia* infection. An epithelial scraping from the male urethra or from the cervix of the uterus is performed to collect squamous cells from the surface. The elementary bodies are measured by reacting antigen-specific antibodies to infected cells. The fluorescent dye attached to the antigen-specific antibodies makes the complex detectable. The sample is viewed with a fluorescent microscope. The presence of ten or more elementary bodies in a field of view with a diameter of 5 millimeters is considered a positive result. The elementary bodies will be stained green inside red host cells.

EQUIPMENT USED The following equipment will be depicted on-screen: five samples: patient A, patient B, patient C, a positive control, and a negative control; incubator; fluorescent microscope; 95% ethyl alcohol—used for fixing the sample to the microscope slide; chlamydia fluorescent antibody (Chlamydia FA)—antibodies specific for the *Chlamydia* antigen with a fluorescent dye attached; fluorescent antibody mounting media (FA mounting)—used to mount the prepared sample to the slide when ready for viewing under the microscope; phosphate buffered saline (PBS)—used to wash off excess antibodies and prevent nonspecific binding of the antigen and antibody; fluorescent antibody buffer (FA buffer)—used to remove excess ethyl alcohol; petri dishes—used for incubation of the slides to keep them moist; microscope slides—an incubation vessel where the antigen and antibody react; cotton-tipped applicators—used for application and mixing of the antibodies with the samples; filter paper—used to keep the samples moist in the petri dishes; biohazardous waste disposal.

Experiment Instructions

Go to the home page in the PhysioEx software and click **Exercise 12: Serological Testing**. Click **Activity 1: Using Direct Fluorescent Antibody Technique to Test for Chlamydia**, and take the online **Pre-lab Quiz** for Activity 1.

After you take the online Pre-lab Quiz, click the **Experiment** tab and begin the experiment. The experiment instructions are reprinted here for your reference. The opening screen for the experiment is shown below.

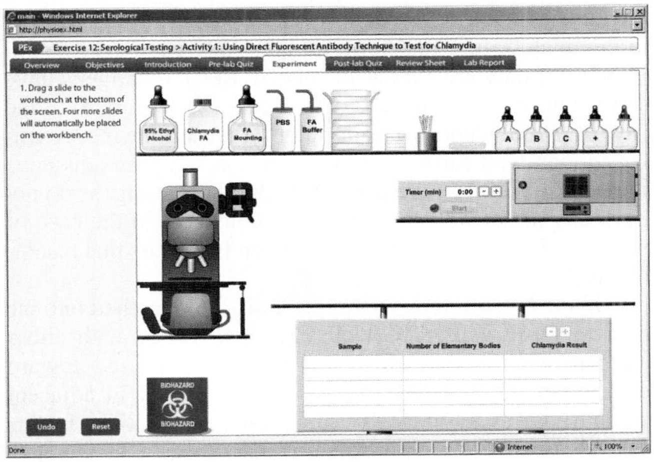

Windows Explorer 2007, Microsoft Corporation

1. Drag a slide to the workbench at the bottom of the screen. Four more slides will automatically be placed on the workbench.

2. The patient samples have been suspended in a small amount of buffer and placed in dropper bottles for ease of dispensing. Drag the dropper cap of the patient A sample bottle to the first slide on the workbench to dispense a drop of the sample onto the slide. A drop from each sample will be placed on a separate slide.

3. Drag the dropper cap of the 95% ethyl alcohol bottle to the first slide on the workbench to dispense three drops of ethyl alcohol onto each slide.

4. Set the timer to 5 minutes by clicking the + button beside the timer display. Click **Start** to start the timer and allow the ethyl alcohol to fix the sample to the slide and prevent the sample from being washed off in the subsequent washing steps. The simulation compresses the 5-minute time period into 5 seconds of real time.

5. Drag the fluorescent antibody (FA) buffer squirt bottle to the first slide to rinse all five slides and remove excess ethyl alcohol.

6. Drag an applicator stick to the chlamydia fluorescent antibody (FA) bottle to soak its cotton tip with antibodies that are specific for *Chlamydia* and labeled with a fluorescent tag.

7. Drag the applicator stick to the first slide to apply the chlamydia fluorescent antibody. Separate applicator sticks will automatically be soaked in chlamydia fluorescent antibody and applied to each slide. Each applicator will automatically be placed in the biohazardous waste disposal.

8. Drag a petri dish to the workbench. A piece of filter paper will be placed into the petri dish. The filter paper has been moistened with fluorescent antibody buffer to keep the samples from drying out during incubation. Four more petri dishes (and filter paper moistened with fluorescent antibody buffer) will automatically be placed on the workbench.

9. Drag the first slide into the first petri dish. The remaining four slides will automatically be placed into the remaining petri dishes and all five petri dishes will be loaded into the incubator.

10. Set the timer to 20 minutes by clicking the + button next to the timer display. Click **Start** to incubate the samples at 25°C. During incubation the antibodies will react with the corresponding antigens if they are present in the sample. The petri dishes will automatically be removed from the incubator when the time is complete. The simulation compresses the 20-minute incubation time period into 10 seconds of real time.

11. Drag the phosphate buffered saline (PBS) squirt bottle to the first petri dish to wash off excess antibodies and prevent nonspecific binding of the antigen and antibody. The timer will count down 10 minutes for a thorough washing.

12. Click the first petri dish to open the dish and remove the slide. The slides will automatically be removed from the remaining petri dishes.

13. Drag the first petri dish to the biohazardous waste disposal. The remaining petri dishes will automatically be placed in the biohazardous waste disposal.

14. Drag the dropper cap of the fluorescent antibody (FA) mounting media to the first slide to dispense a drop of mounting media onto each slide to mount the sample to the slide.

15. Drag the first slide (patient A) to the fluorescent microscope. Count the number of elementary bodies you see through the microscope (recall that elementary bodies stain green). Click **Submit** to display your results in the grid (and record your results in Chart 1). After you click **Submit**, the slide will automatically be placed in the biohazardous waste disposal.

CHART 1 Direct Fluorescent Antibody Technique Results

Sample	Number of elementary bodies	Chlamydia result
Patient A		
Patient B		
Patient C		
Positive control		
Negative control		

16. Repeat step 15 for Patient B.

17. Repeat step 15 for Patient C.

18. Repeat step 15 for the Positive Control.

19. Repeat step 15 for the Negative Control.

20. You will now indicate whether each sample is negative or positive for *Chlamydia*. Click the row for the sample in the grid and then click the − button or the + button above the Chlamydia result column to indicate whether the sample is negative or positive for *Chlamydia*. Repeat this step for all five samples. Record your results in Chart 1.

After you complete the experiment, take the online **Post-lab Quiz** for Activity 1.

Activity Questions

1. With this technique, is the antigen or antibody found on the patient sample? Explain how you know this.

____

____

2. Explain the difference between an antigen and an epitope (antigenic determinant).

____

____

3. When a sample has a small number of elementary bodies but not enough to be a positive result, there appears to have been some nonspecific binding that was not removed by the washing steps. Which sample displayed this property?

____

____

ACTIVITY 2

Comparing Samples with Ouchterlony Double Diffusion

OBJECTIVES

1. To observe the precipitation reaction between antigen and antibody.
2. To distinguish between *epitope* and *antigen*.
3. To understand the specificity that antibodies have for their epitopes.
4. To observe how related proteins might share epitopes in common.

Introduction

The Ouchterlony technique is also known as double diffusion. In this technique antigen and antibody diffuse toward each other in a semisolid medium made up of clear, clarified agar. When the antigen and antibody are in optimal proportions, cross-linking of the antigen and antibody occurs, forming an insoluble precipitate, called a **precipitin line**. These lines can then be used to visually identify similarities between antigens. If optimum proportions have not been met—for example, if there is excess antigen or excess antibody—then no visible precipitate will form. This technique provides easily visible evidence of the binding between antigen and antibody, and sophisticated equipment is not needed to observe the antigen-antibody reaction.

The Ouchterlony technique is designed to determine whether antigens are identical, related, or unrelated. Antigens have **identity** if they are identical. Identical antigens have all their antigenic determinants, or epitopes, in common. In the case of identity, precipitin lines diffuse into each other to completely fuse and form an arc. Antigens have **partial identity** if they are similar or related. Related antigens have some, but not all, antigenic determinants in common. In the case of partial identity, a spur pointing toward the more similar antigen well forms in addition to the arc. Antigens have **non-identity** if they are unrelated. Unrelated antigens do not have any antigenic determinants in common. In the case of non-identity, the lines intersect to form two spurs that resemble an X.

In the Ouchterlony technique, holes are punched into the agar to form wells. The wells are then loaded with either antigen or antibody, which are allowed to diffuse toward each other. Often, the same antigen is placed in adjacent wells to assess the purity of an antigen preparation. In this case a smooth arc with no spurs should be seen, as the antigens are identical. Multiple antibodies can also be placed in a center well. The antibodies will diffuse out in all directions and react with the antigens that are placed in the surrounding wells.

In this activity you will use human and bovine (from cows) albumin as the antigens, and the antibodies will be made in goats against albumin from either humans or cows. The goals are to identify an unknown antigen and to observe the patterns produced by the various relationships: identity, partial identity, and non-identity.

EQUIPMENT USED The following equipment will be depicted on-screen: goat anti–human albumin (Goat A-H)—an antiserum containing antibodies produced by goats against human albumin; goat anti–bovine albumin (Goat A-B)—an antiserum containing antibodies produced by goats against bovine (cow) albumin; bovine serum albumin (BSA); human serum albumin (HSA); unknown antigen; petri dishes filled with clear agar; well cutter.

Experiment Instructions

Go to the home page in the PhysioEx software and click **Exercise 12: Serological Testing**. Click **Activity 2: Comparing Samples with Ouchterlony Double Diffusion**: and take the online **Pre-lab Quiz** for Activity 2.

After you take the online Pre-lab Quiz, click the **Experiment** tab and begin the experiment. The experiment instructions are reprinted here for your reference. The opening screen for the experiment is shown on the following page.

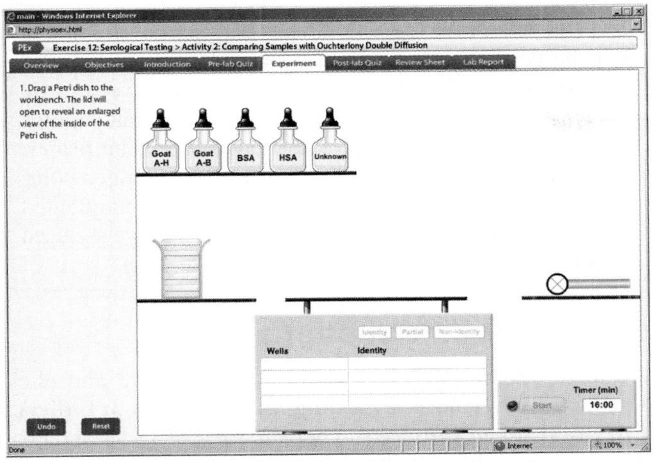

Windows Explorer 2007, Microsoft Corporation

1. Drag a petri dish to the workbench. The lid will open to reveal an enlarged view of the inside of the petri dish.

2. Drag the well cutter to the middle of the enlarged view of the petri dish to punch a hole in the agar in the middle of the petri dish. Drag the well cutter to the upper left, upper right, lower left, and lower right of the petri dish to punch four more holes in the agar. After you punch all five wells into the agar, the wells will be labeled 1–5.

3. Drag the dropper cap of the goat anti–human albumin (Goat A-H) bottle to well 1 to fill it with a sample.

4. Drag the dropper cap of the goat anti–bovine albumin (Goat A-B) bottle to well 1 to fill it with a sample.

5. Drag the dropper cap of the bovine serum albumin (BSA) bottle to well 2 to fill it with a sample.

6. Drag the dropper cap of the bovine serum albumin (BSA) bottle to well 3 to fill it with a sample.

7. Drag the dropper cap of the human serum albumin (HSA) bottle to well 4 to fill it with a sample.

8. Drag the dropper cap of the unknown antigen bottle to well 5 to fill it with a sample.

? PREDICT **Question 1**
How do you think human serum albumin and bovine serum albumin will compare?

____

9. Note that the timer is set to 16 hours. Click **Start** to start the timer. The antigen and antibodies will diffuse toward each other and form a precipitate, detected as a precipitin line. The simulation compresses the 16-hour time period into 10 seconds of real time.

10. You will now examine the precipitin lines that formed and indicate the relationship between each pair of antigens. Click the row for the wells containing the antigens in the grid and then click **Identity, Partial**, or **Non-Identity** above the identity column to indicate whether the antigens have identity, partial identity, or non-identity. Repeat this step for the four pairs of antigens. Record your results in Chart 2.

CHART 2 Ouchterlony Double Diffusion Results

Wells	Identity
2 and 5	
2 and 3	
3 and 4	
4 and 5	

After you complete the Experiment, take the online **Post-lab Quiz** for Activity 2.

Activity Questions

1. Which type of identity was present between the samples in this activity? Describe this type of identity.

____

____

2. Describe the importance of what you place in the center well.

____

____

3. Why do you think it is important for the agar to be clear and clarified?

____

____

4. Describe the role that albumin plays in the blood.

____

____

ACTIVITY 3

Indirect Enzyme-Linked Immunosorbent Assay (ELISA)

OBJECTIVES

1. To understand how the enzyme-linked immunosorbent assay (ELISA) is used as a diagnostic test.
2. To distinguish between the direct and the indirect ELISA.
3. To describe the basic structure of antibodies.
4. To define *seroconversion*.
5. To understand how the indirect ELISA is used to detect antibodies against HIV.

Introduction

The **enzyme-linked immunosorbent assay (ELISA)** is used to test for the presence of an antigen or antibody. The assay is considered enzyme linked because an enzyme is chemically linked to an antibody in both the direct and indirect versions of the test. Immunosorbent refers to the fact that either antigens or antibodies are being adsorbed (stuck) to plastic. If the test is designed to detect an antigen or antigens, it is a **direct ELISA** because it is directly looking for the foreign substance. An **indirect ELISA** is designed to detect antibodies that the patient has made against the antigen. A positive result with the indirect ELISA requires **seroconversion**. Seroconversion occurs when a patient goes from testing negative for a specific antibody to testing positive for the same antibody.

In the direct ELISA, a 96-well microtiter plate is coated with homologous antibodies made against the antigen of interest. The number of wells makes it easy to test many samples at the same time. The patient serum sample is added to the plate to test for the presence of the antigen that binds to the antibody coating on the plate. ELISA takes advantage of the fact that protein sticks well to plastic. A secondary antibody is added to the plate after the patient serum sample is added. If the antigen is present, a "sandwich" of antibody, antigen, and secondary antibody will form. The secondary antibody is chemically linked to an enzyme. When the substrate is added, the enzyme converts the substrate from a colorless compound to a colored compound. The amount of color produced will be proportional to the amount of antigen binding to the antibodies and thus indicates whether the patient is positive for the antigen. If the antigen is not present, the secondary (enzyme-linked) antibodies will be rinsed away with the washing steps and the substrate will not be converted and will remain colorless. A common use of the direct ELISA is a home pregnancy test, which detects human chorionic gonadotropin (hCG), a hormone present in the urine of pregnant women.

In the indirect ELISA, a 96-well microtiter plate is coated with antigens. The patient serum sample is added to test for the presence of antibodies that bind to the antigens on the plate. The secondary antibody that is added has an enzyme linked to it that binds to the **constant region** of the primary antibody if it is present in the patient sample. The constant region of an antibody has the same sequence of amino acids within a class of antibodies (for example, all IgG antibodies have the same constant region). The **variable region** of an antibody provides the diversity of antibodies and is the site to which the antigen binds. The configuration that forms in the indirect ELISA is antigen, primary antibody, and secondary antibody. Just as in the direct ELISA, the addition of substrate is used to determine whether the sample is positive for the presence of antibody.

In this activity you will use the indirect ELISA to test for the presence of antibodies made against human immunodeficiency virus (HIV). You will use positive and negative controls to verify the results. You will note that an indeterminate result can be obtained if there is not enough color produced to warrant a positive result. The cause of an indeterminate result could be either nonspecific binding or that the individual has been recently infected and has not yet produced enough antibodies for a positive result. In either case, the individual would be retested.

EQUIPMENT USED The following equipment will be depicted on-screen: five samples in the samples cabinet: patient A, patient B, patient C, a positive control, and a negative control; 96-well microtiter plate; multichannel pipettor; 100-μl pipettor; microtiter plate reader; pipettor tip dispenser; washing buffer; HIV antigen solution; developing buffer—secondary antibody conjugated with an enzyme; substrate solution; paper towels—used for blotting; biohazardous waste disposal.

Experiment Instructions

Go to the home page in the PhysioEx software and click **Exercise 12: Serological Testing**. Click **Activity 3: Indirect Enzyme-Linked Immunosorbent Assay (ELISA)**, and take the online **Pre-lab Quiz** for Activity 3.

After you take the online Pre-lab Quiz, click the **Experiment** tab and begin the experiment. The experiment instructions are reprinted here for your reference. The opening screen for the experiment is shown below.

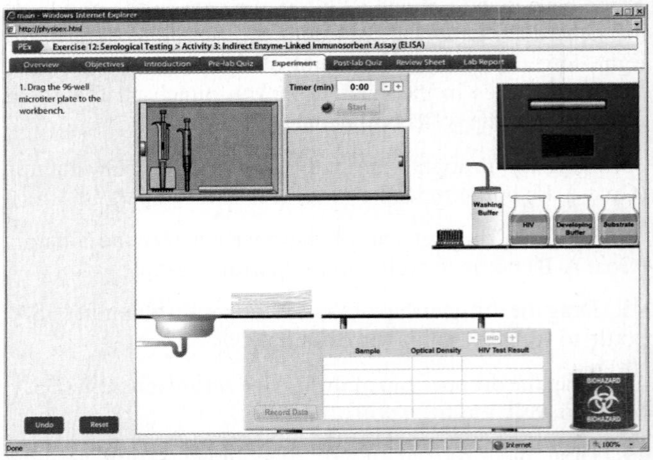

Windows Explorer 2007, Microsoft Corporation

1. Drag the 96-well microtiter plate to the workbench.

2. Drag the multichannel pipettor to the pipette tip dispenser to insert the tips.

3. Drag the multichannel pipettor to the HIV antigens bottle to draw the antigen solution into the tips.

4. Drag the multichannel pipettor directly over the microtiter plate to dispense the liquid into the wells in one row of the plate.

5. Drag the multichannel pipettor to the biohazardous waste disposal for removal and disposal of the tips.

6. Set the timer to 14 hours by clicking the + button beside the timer display. This incubation time allows the antigens to stick to the plastic wells of the microtiter plate. Click **Start** to start the timer. The simulation compresses the 14-hour time period into 10 seconds of real time.

7. Drag the washing buffer squeeze bottle to the microtiter plate to remove excess antigens that are not adsorbed (stuck) to the plate.

8. Drag the microtiter plate to the sink to dump the contents of the tray into the sink to remove the washing buffer and excess antigens that are not stuck to the plastic.

9. Drag the microtiter plate to the paper towels. The plate will be pressed to the surface of the paper towels to remove the remaining liquid from the wells. In a typical ELISA, you would perform multiple washing steps to reduce any nonspecific binding. The number of washing steps in this simulation has been reduced for simplicity.

10. Drag the 100-μl pipettor to the tip dispenser to place a tip onto the pipettor.

11. Drag the 100-μl pipettor to the test tube containing the positive control sample (+) to draw the sample into the tip.

12. Drag the 100-μl pipettor to the microtiter plate to dispense the sample into the wells of the plate. The tip will automatically be removed and disposed of in the biohazardous waste disposal. Each of the remaining samples will automatically be dispensed into plate.

13. Set the timer to 1 hour by clicking the + button beside the timer display. This incubation time allows the antigens stuck to the plastic to bind to the antibodies present in the sample. Click **Start** to start the timer. The simulation compresses the 1-hour time period into 10 seconds of real time.

14. Drag the washing buffer squeeze bottle to the microtiter plate to wash off excess antibodies and prevent nonspecific binding of the antigen and antibody.

15. Drag the microtiter plate to the sink to dump washing buffer and unbound antibodies into the sink.

16. Drag the microtiter plate to the paper towels. The plate will be pressed to the surface of the paper towels to remove the remaining liquid from the wells.

17. Drag the multichannel pipettor to the pipette tip dispenser to insert the tips.

18. Drag the multichannel pipettor to the developing buffer bottle to draw the developing buffer into the tips. The developing buffer contains the conjugated secondary antibody.

19. Drag the multichannel pipettor to the microtiter plate to dispense the solution into the wells. The tips will automatically be removed and disposed of in the biohazardous waste disposal.

20. Set the timer to 1 hour and then click **Start** to start the timer and allow the conjugated secondary antibody to bind to the primary antibody if it is present in the sample.

21. Drag the washing buffer squeeze bottle to the microtiter plate to remove any nonspecific binding that occurred.

22. Drag the microtiter plate to the sink to dump the contents of the tray into the sink.

23. Drag the microtiter plate to the paper towels. The plate will be pressed to the surface of the paper towels to remove the remaining liquid from the wells.

24. Drag the multichannel pipettor to the pipette tip dispenser to insert the tips.

25. Drag the multichannel pipettor to the substrate bottle to draw the substrate into the tips.

26. Drag the multichannel pipettor to the microtiter plate to dispense the solution into the wells. The tips will automatically be removed and disposed of in the biohazardous waste disposal.

27. An enlargement of the wells will appear. The development will progress over time. To determine the optical density for each sample (the samples are in the first row, from top to bottom, of the microtiter plate):

- Click the well and the optical density will appear in the window of the microtiter plate reader.
- Click **Record Data** to display your results in the grid (and record your results in Chart 3).

CHART 3 Indirect ELISA Results

Sample	Optical density	HIV test result
Patient A		
Patient B		
Patient C		
Positive control		
Negative control		

28. You will now indicate whether the result for each sample is negative, indeterminate, or positive for HIV.

- A result of <0.300 is read as negative for HIV-1.
- A result of 0.300–0.499 is read as indeterminate (need to retest).
- A result of >0.500 is read as positive for HIV-1.

Click the row for the sample in the grid and then click the − button, **IND**, or the + button above the HIV test result column to indicate whether the result for the sample is positive, indeterminate, or negative for HIV. Repeat this step for all five samples. Record your results in Chart 3.

After you complete the Experiment, take the online **Post-lab Quiz** for Activity 3.

Activity Questions

1. Describe how you can tell that this test is the indirect ELISA rather than the direct ELISA.

____

____

2. Describe what the secondary antibody binds to in this activity and why.

____

____

3. Define *seroconversion*. How can you tell that a sample has seroconverted?

____

____

ACTIVITY 4

Western Blotting Technique

OBJECTIVES

1. To compare the Western blotting technique to the ELISA.
2. To observe the use of the Western blotting technique to test for HIV.
3. To distinguish between antigens and antibodies.

Introduction

Southern blotting was developed by Ed Southern in 1975 to identify DNA. A variation of this technique, developed to identify RNA, was named Northern blotting, thus continuing the directional theme. Western blotting, another variation that identifies proteins, is named by the same convention.

Western blotting uses an electrical current to separate proteins on the basis of their size and charge. This technique uses **gel electrophoresis** to separate the proteins in a gel matrix. Because the resulting gel is fragile and would be difficult to use in further tests, the proteins are then transferred to a **nitrocellulose membrane**. The original Western blotting technique used blotting (diffusion) to transfer the proteins, but electricity is also used now for the transfer of the proteins to nitrocellulose strips. These strips are commercially available, eliminating the need for the electrophoresis and transfer equipment. In this activity you will begin the procedure after the HIV (human immunodeficiency virus) antigens have already been transferred to nitrocellulose and cut into strips.

Western blotting is also known as **immunoblotting** because the proteins that are transferred, or blotted, onto the membrane are later treated with antibodies—the same procedure used in the **indirect enzyme-linked immunosorbent assay (ELISA)**. The ELISA is considered enzyme linked because an enzyme is chemically linked to an antibody in both the direct and indirect versions of the test. Immunosorbent refers to the fact that either antigens or antibodies are being adsorbed (stuck) to plastic. If the test is designed to detect an antigen or antigens, it is a **direct ELISA** because it is directly looking for the foreign substance. An **indirect ELISA** is designed to detect antibodies that the patient has made against the antigen.

Similar to the secondary antibodies used in the indirect ELISA technique, the secondary antibodies in the Western blot have an enzyme attached to them, allowing for the use of color to detect a particular protein. The secondary antibody binds to the constant region of the primary antibody found in the patient's sample. The main difference between these techniques is that the ELISA technique uses a well that corresponds to a mixture of antigens, and the Western blot has a discrete protein band that represents the specific antigen that the antibody is recognizing. Like HIV, Lyme disease can also be detected with the Western blot technique.

The initial test for HIV is the ELISA, which is less expensive and easier to perform than the Western blot. The Western blot is used as a confirmatory test after a positive ELISA because the ELISA is prone to false-positive results. The bands from a positive Western blot are from antibodies binding to specific proteins and glycoproteins from the human immunodeficiency virus. A positive result from the Western blot is determined by the presence of particular protein bands (view Table 12.1).

EQUIPMENT USED The following equipment will be depicted on-screen: washing buffer; developing buffer—secondary antibody conjugated with an enzyme; substrate solution; five samples in the samples cabinet: patient A, patient B, patient C, positive control, and negative control; rocking apparatus; nitrocellulose strips; troughs; tray; biohazardous waste disposal.

Experiment Instructions

Go to the home page in the PhysioEx software and click **Exercise 12: Serological Testing**. Click **Activity 4: Western Blotting Technique**, and take the online **Pre-lab Quiz** for Activity 4.

After you take the online Pre-lab Quiz, click the **Experiment** tab and begin the experiment. The experiment instructions are reprinted here for your reference. The opening screen for the experiment is shown below.

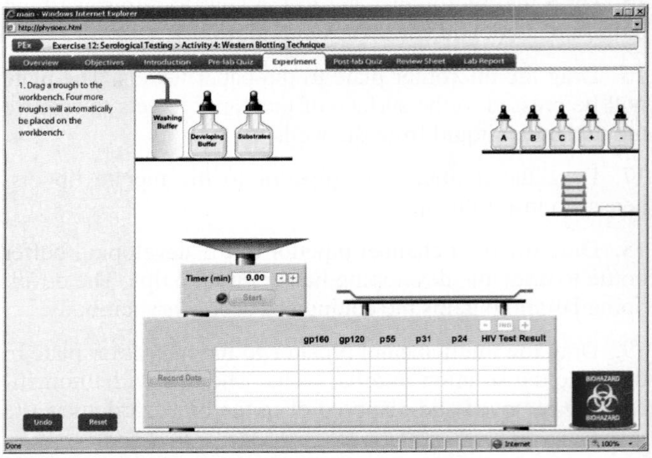

Windows Explorer 2007, Microsoft Corporation

1. Drag a trough to the tray on the workbench. Four more troughs will automatically be placed on the tray.

2. Click the stack of nitrocellulose strips to place a nitrocellulose strip in each trough.

3. Drag the dropper cap of the patient A sample bottle to the first trough to dispense the antiserum from patient A to the nitrocellulose strip. A drop of antiserum for each patient will be dispensed into a separate trough.

4. Drag the tray holding the five troughs to the rocking apparatus.

5. Set the timer to 60 minutes by clicking the + button beside the timer display. Click **Start** to gently rock the samples and allow the antibodies to react with the antigens bound to the nitrocellulose. The tray will automatically be returned to the workbench and each trough will be drained into the biohazardous waste disposal when the time is complete. The simulation compresses the 60-minute time period into 10 seconds of real time.

6. Drag the washing buffer squirt bottle to the first trough to dispense washing buffer in each trough. Each trough will be automatically drained into the biohazardous waste container. The washing step removes any nonspecific binding of antibodies that occurred.

7. Drag the dropper cap of the developing buffer bottle to the first trough to dispense developing buffer to each trough.

8. Drag the tray holding the five troughs to the rocking apparatus.

9. Set the timer to 60 minutes by clicking the + button beside the timer display. Click **Start** to gently rock the samples and allow the antibodies to react with the antibodies bound to the nitrocellulose. The tray will automatically be returned to the workbench and each trough will be drained into the biohazardous waste disposal when the time is complete.

10. Drag the washing buffer squirt bottle to the first trough to add washing buffer to each trough. Each trough will be automatically drained into the biohazardous waste container. The washing step removes any nonspecific binding of secondary conjugated antibodies. Excess secondary conjugated antibodies could react erroneously with the substrate and give a false-positive result.

11. Drag the dropper cap of the substrates bottle to the first trough to dispense the substrates (tetramethyl benzidine and hydrogen peroxide) into each trough. The substrates are the chemicals that are being changed by the enzyme that is linked to the antibody.

12. Drag the tray holding the five troughs to the rocking apparatus.

13. Set the timer to 10 minutes by clicking the + button beside the timer display. Click **Start** to gently rock the samples and allow the enzyme to react with the substrates. The tray will automatically be returned to the workbench when the time is complete. The simulation compresses the 10-minute time period into 10 seconds of real time.

14. To determine the antigens present for each sample:

- Click the nitrocellulose strip inside the trough to visualize the results.
- Click **Record Data** to display your results in the grid (and record your results in Chart 4). The bands present on the nitrocellulose strip represent the antibodies present in the sample that have reacted with the antigens (bands) on the strip (view Table 12.1).

Repeat this step for all five samples.

TABLE 12.1 HIV Antigens

Abbreviation	Description
gp160	Glycoprotein 160, a viral envelope precursor
gp120	Glycoprotein 120, a viral envelope protein that binds to CD4
p55	A precursor to the viral core protein p24
gp41	A final envelope glycoprotein
p31	Reverse transcriptase
p24	A viral core protein

15. You will now indicate whether the result for each sample is negative, indeterminate, or positive for HIV. The criteria for reporting a positive result varies slightly from agency to agency. The Centers for Disease Control and Prevention recommend the following criteria:

- If no bands are present, the result is negative.
- If bands are present but they do not match the criteria for a positive result, the result is indeterminate. Patients whose results are deemed indeterminate after multiple tests should be monitored and tested again at a later date.
- If either p31 or p24 is present *and* gp160 or gp120 is present, the result is positive. Click the row for the sample in the grid and then click the − button, **IND**, or the + button above the HIV test result column to indicate whether the result for the sample is positive, indeterminate, or negative for HIV.

Repeat this step for all five samples. Record your results in Chart 4.

After you complete the experiment, take the online **Post-lab Quiz** for Activity 4.

CHART 4 Western Blot Results

Sample	gp160	gp120	p55	p31	p24	HIV test result
Patient A						
Patient B						
Patient C						
Positive control						
Negative control						

Activity Questions

1. Describe how gel electrophoresis is used to separate proteins.

2. In a patient sample that is positive for HIV, would antibodies or antigens be present when using the Western blot technique? How do you know?

NAME ______________________

LAB TIME/DATE ______________________

Serological Testing

ACTIVITY 1 Using Direct Fluorescent Antibody Technique to Test for Chlamydia

1. Describe the importance of the washing steps in the direct antibody fluorescence test. ______________________

2. Explain where the epitope (antigenic determinant) is located. ______________________

3. Describe how a positive result is detected in this serological test. ______________________

4. How would the results be affected if a negative control gave a positive result? ______________________

ACTIVITY 2 Comparing Samples with Ouchterlony Double Diffusion

1. Describe how you were able to determine what antigen is in the unknown well. ______________________

2. Why does the precipitin line form? ______________________

3. Did you think human serum albumin and bovine serum albumin would have epitopes in common? How well did the results compare with your prediction? ____

ACTIVITY 3 Indirect Enzyme-Linked Immunosorbent Assay (ELISA)

1. Describe how the direct and indirect ELISA are different. ____

2. Discuss why a patient might test indeterminate. ____

3. How would your results have been affected if your negative control had given an indeterminate result? ____

4. Briefly describe the basic structure of antibodies. ____

ACTIVITY 4 Western Blotting Technique

1. Describe why the HIV Western blot is a more specific test than the indirect ELISA for HIV. ____

2. Explain the procedure for a patient with an indeterminate HIV Western blot result. ____

3. Briefly describe how the nitrocellulose strips were prepared before the patient samples were added to them. ____

4. Describe the importance of the washing steps in the procedure. ____

Credits

Illustrations

All illustrations are by Imagineering STA Media Services, except as noted below.

Scissors icon: Vectorpro/Shutterstock.

Button icon: justone/Shutterstock.

Video (arrow/camera) icons: Yuriy Vlasenko/Shutterstock.

Exercise 1 1.1: Imagineering STA Media Services/Precision Graphics. 1.2, 1.4: Precisions Graphics. 1.9: Source: Adapted from Marieb, Elaine N.; Mallatt, Jon B.; Wilhelm, Patricia Brady, *Human Anatomy*, 5e, F1.10, © 2008. Reprinted and Electronically reproduced by permission of Pearson Education, Inc., Upper Saddle River, New Jersey.

Exercise 3 3.2, 3.3, 3.4, 3.A3: Precision Graphics.

Exercise 4 4.3, Table 4.1.3, 4.RS1: Tomo Narashima. 4.2, Table 4.1.4–5: Imagineering STA Media Services/ Precision Graphics.

Exercise 7 7.1, 7.2, 7.6: Electronic Publishing Services, Inc. 7.5a: Steve Downing/Pearson Education.

Exercise 9 9.1–9.3, 9.6, 9.7, 9.9, 9.21: Nadine Sokol. 9.20c: Pearson Education.

Exercise 11 11.7d: Karen Krabbenhoft / Pearson Education.

Exercise 12 12.4: Imagineering STA Media Services/Precision Graphics. 12.5: Electronic Publishing Services, Inc.

Exercise 14 14.1–14.3: Precision Graphics. 14.7–14.9, 14.13–14.18, 14.U1: Biopac Systems.

Exercise 15 15.1: Imagineering STA Media Services/Precision Graphics.

Exercise 16 16.2: Precision Graphics.

Exercise 17 17.1, 17.5, 17.7–17.11, 17.RS01, 17.RS04: Electronic Publishing Services, Inc.

Exercise 18 18.4–18.7: Biopac Systems.

Exercise 19 19.1, 19.3, 19.RS01: Electronic Publishing Services, Inc.

Exercise 20 20.7–20.13: Biopac Systems.

Exercise 21 21.1: Electronic Publishing Services, Inc. 21.8, 21.9: Biopac Systems.

Exercise 23 23.1, 23.3, 23.4, 23.RS01: Electronic Publishing Services, Inc.

Exercise 24 24.1: Shirley Bortoli. 24.4, 24.5: Precision Graphics.

Exercise 25 25.1: Electronic Publishing Services, Inc./Precision Graphics. 25.2, 25.3, 25.7, 25.9: Electronic Publishing Services, Inc.

Exercise 26 26.1, 26.2: Electronic Publishing Services, Inc.

Exercise 29 29.2, 29.5: Precision Graphics.

Exercise 30 30.1: Electronic Publishing Services, Inc./Precision Graphics. 30.2, 30.3, 30.6: Electronic Publishing Services, Inc.

Exercise 31 31.1: Electronic Publishing Services, Inc. 31.2, 31.4: Precision Graphics. 31.7–31.12: Biopac Systems.

Exercise 32 32.3–32.13, 32.15: Electronic Publishing Services, Inc.

Exercise 33 33.2: Precision Graphics. 33.6–33.9: Biopac Systems.

Exercise 34 34.1, 34.2, 34.4: Precision Graphics. 34.5: Biopac Systems. 34.6: Electronic Publishing Services, Inc.

Exercise 35 35.8: Precision Graphics.

Exercise 36 36.1–36.5, 36.7: Electronic Publishing Services, Inc.

Exercise 37 37.5, 37.6: Precision Graphics. 37.11, 37.13, 37.14: Biopac Systems.

Exercise 38 38.1–38.5, 38.7, 38.8, 38.10, 38.15, 38.RS02, 38.RS03: Electronic Publishing Services, Inc. 38.16: Electronic Publishing Services, Inc./Precision Graphics.

Exercise 39 39.A1–A3: Precision Graphics.

Exercise 40 40.1, 40.2, 40.6: Electronic Publishing Services, Inc.

Exercise 41 41.1: Precision Graphics.

Exercise 42 42.1, 42.2, 42.7: Electronic Publishing Services, Inc.

Exercise 43 43.1: Precision Graphics. 43.2: Electronic Publishing Services, Inc.

Exercise 44 44.1: Electronic Publishing Services, Inc.

Cat Dissection Exercises 1.1, 3.1, 3.3a, 7.1: Precision Graphics. 2.1a, 2.2a, 3.2, 4.2, 4.4, 6.2, 7.2a, 8.2a, 9.1a: Kristin Mount.

Fetal Pig Dissection Exercises

1.3–1.8, 2.2, 3.1, 3.2, 4.1–4.5, 5.2, 6.1, 6.2, 7.1, 7.2, 8.1, 8.2: Kristin Mount.

PhysioEx Exercises All illustrations by BinaryLabs, Inc. except as noted 2.1: Precision Graphics/Source: Adapted from Stanfield, Principles of Human Physiology, 4e. F12.1, © 2011. Reprinted and Electronically reproduced by permission of Pearson Education, Inc. Upper Saddle River, NJ. 3.1: Imagineering STA Media Services. 4.1: Electronic Publishing Services, Inc. 6.1: Precision Graphics/Source: Adapted from Stanfield, *Principles of Human Physiology*, 4e. F13.13, © 2011. Reprinted and Electronically reproduced by permission of Pearson Education, Inc. Upper Saddle River, NJ. 7.1: Precision Graphics/ Source: Adapted from Stanfield, *Principles of Human Physiology*, 4e. F16.1, © 2011. Reprinted and Electronically reproduced by permission of Pearson Education, Inc. Upper Saddle River, NJ. 8.1: Electronic Publishing Services, Inc. 9.1: Electronic Publishing Services, Inc. 12.1: Imagineering STA Media Services.

Photographs

Visual Walkthrough Leif Saul/3B Scientific/Pearson Education, Inc. Karen Krabbenhoft/Pearson Education, Inc. Nik Merkulov/Shutterstock.

Exercise 1 1.3.1, 1.8a: John Wilson White/Pearson Education, Inc. 1.3a: CNRI/Science Source. 1.3b: Scott Camazine/Science Source. 1.3c: James Cavallini/Science Source.

Exercise 2 2.1, 2.2, 2.3a, 2.4a, 2.5b, c: Elena Dorfman/Pearson Education, Inc. 2.3b, 2.4b, 2.5a, 2.6: From *A Stereoscopic Atlas of Human Anatomy* by David L. Bassett, M.D. 2.RS1: Arcady/Shutterstock.

Exercise 3 3.1, 3.RS1: Vereshchagin Dmitry/Shutterstock. 3.5: Victor P. Eroschenko/Pearson Education, Inc.

Exercise 4 4.1: David M. Phillips/ Science Source. 4.5, 4.RS2: William Karkow/Pearson Education, Inc.

Exercise 5 5.2: Richard Megna/ Fundamental Photographs, NYC. 5.4: David M. Phillips/Science Source.

Exercise 6 6.3a, 6.3d, 6.3g, 6.5a–b, 6.5e, 6.5g, 6.5l, 6.7b–c: William Karkow/ Pearson Education, Inc. 6.3b–c, 6.3f, 6.3h, 6.5d, 6.5j–k: Allen Bell/Pearson Education, Inc. 6.3e, 6.5c, 6.5f, 6.5i, 6.7a: Nina Zanetti/Pearson Education, Inc. 6.5h: Steve Downing/Pearson Education, Inc. 6.6: Biophoto assoc./Science Source.

Exercise 7 7.2a, 7.5b: William Karkow/ Pearson Education, Inc. 7.6: Lisa Lee/ Pearson Education, Inc. 7.RS1–2: Somso. 7.RS3: Steve Downing/Pearson Education, Inc.

Exercise 8 8.4c: William Karkow/ Pearson Education, Inc. 8.5: Deborah Vaughan, PhD. 8.RS11: Steve Downing/ Pearson Education.

Exercise 9 9.1b, 9.2b, 9.4a–b, 9.21c–d, 9.RS1–2: Larry DeLay/Pearson Education, Inc. 9.3c, 9.5: Michael Wiley/Univ. of Toronto/Imagineering/Pearson Education, Inc. 9.10: Elena Dorfman/Pearson Education, Inc. 9.19b: Karen Krabbenhoft/ PAL 3.0/Pearson Education, Inc.

Exercise 10 10.5b: James Cavallini/Science Source. Table 10.4: Larry DeLay/Pearson Education. 10.RS3–6, 11, 13–14: Creative Digital Visions, LLC/Pearson Education. 10.RS6, 15,17: Karen Krabbenhoft/Pearson Education.

Exercise 11 11.5, 11.RS9: John Wilson White/Pearson Education, Inc. 11.6c: From *A Stereoscopic Atlas of Human Anatomy* by David L. Bassett, M.D. 11.8a: Mark Neilsen/Pearson Education, Inc. 11.8c: Karen Krabbenhoft/PAL 3.0/Pearson Education, Inc. 11.RS7: Karen Krabbenhoft / Pearson Education.

Exercise 12 12.1a: Marian Rice 12.3, 12.6: Victor P. Eroschenko/Pearson Education, Inc. 12.4: William Karkow/Pearson Education, Inc. 12.RS: Don W. Fawcett/Science Source.

Exercise 13 13.4b, 13.10, 13.12f, 13.14b: Karen Krabbenhoft/PAL 3.0/Pearson Education, Inc. 13.5b: Creative Digital Visions/Pearson Education, Inc. 13.9b: William Karkow/Pearson Education, Inc. 13.RS: 3B Scientific GmbH.

Exercise 15 15.2b, 15.6b: William Karkow/Pearson Education, Inc. 15.4, 15.6c: Nina Zanetti/Pearson Education, Inc. 15.3: Don W. Fawcett/Science Source. 15.6a: Sercomi/Science Source. 15.8b: Karen Krabbenhoft/Pearson Education, Inc.

Exercise 17 17.2c: Electronic Publishing Services, Inc. 17.3.2, 17.4b, 17.5b.2, 17.6, 17.10: Karen Krabbenhoft/PAL 3.0/Pearson Education, Inc. 17.7c: From *A Stereoscopic Atlas of Human Anatomy* by David L. Bassett, M.D. 17.11a, 17.11c: Sharon Cummings/Pearson Education, Inc. 17.12–17.14: Elena Dorfman/Pearson Education, Inc. 17.RS1: Karen Krabbenhoft/PAL 3.0/Pearson Education, Inc. 17.RS2: 3B Scientific GmbH.

Exercise 18 18.1a: RGB Ventures/SuperStock/Phanie/Alamy.

Exercise 19 19.2b–d, 19.8b: Karen Krabbenhoft/Pearson Education, Inc. 19.5: Lisa Lee/Pearson Education, Inc.

Exercise 21 21.4–21.6: Richard Tauber/Pearson Education, Inc.

Exercise 22 22.1b: Lisa Lee / Pearson Education. 22.1c–d, 22.2b: Victor P. Eroschenko/Pearson Education, Inc.

Exercise 23 23.4b: Lisa Lee/Pearson Education, Inc. 23.5: Elena Dorfman/Pearson Education, Inc. 23.6b: Stephen Spector/Pearson Education, Inc.

Exercise 24 24.6: Thalerngsak Mongkolsin/Shutterstock.

Exercise 25 25.4: Victor P. Eroschenko/Pearson Education, Inc. 25.6: Richard Tauber/Pearson Education, Inc. 25.8: Michael G. Wood. 25.RS: 3B Scientific GmbH.

Exercise 26 26.1b, 26.3: Victor P. Eroschenko/Pearson Education, Inc. 26.2d: Steve Downing/Pearson Education, Inc.

Exercise 27 27.3a, 27.3f: William Karkow/Pearson Education, Inc. 27.3b, 27.3d: Victor P. Eroschenko/Pearson Education, Inc. 27.3c, 27.RS1c: Lisa Lee/Pearson Education, Inc. 27.RS1a: Karen Krabbenhoft/Pearson Education, Inc. 27.3e, 27.RS01b: Lisa Lee / Pearson Education.

Exercise 29 29.3: William Karkow/Pearson Education, Inc. 29.4: Nina Zanetti/Pearson Education, Inc. 29.6, 29.7: Elena Dorfman/Pearson Education, Inc. 29.8b: Eye of Science/Science Source. 29.9, 29.RS: Jack Scanlan/Pearson Education, Inc.

Exercise 30 30.7, 30.8: Wally Cash/Pearson Education, Inc. 30.3d: Karen Krabbenhoft/PAL 3.0/Pearson Education, Inc. 30.3c: Philippe Plailly/Look at Sciences/Science Source. 30.6: William Karkow/Pearson Education, Inc. 30.RS1: 3B Scientific GmbH. 30.RS2: Steve Downing/Pearson Education, Inc.

Exercise 32 32.4c: Karen Krabbenhoft/Pearson Education, Inc. 32.RS1: Ed Reschke/Photolibrary/Getty Images.

Exercise 33 33.2: Lisa Lee / Pearson Education.

Exercise 35 35.2: Eric V. Grave/Science Source. 35.4b, 35.6: William Karkow/Pearson Education, Inc. 35.5c: Victor P. Eroschenko/Pearson Education, Inc.

Exercise 36 36.1a, 36.5b: From *A Stereoscopic Atlas of Human Anatomy* by David L. Bassett, M.D. 36.5a: Richard Tauber/Pearson Education, Inc. 36.6b: Victor P. Eroschenko/Pearson Education, Inc. 36.7a: William Karkow/Pearson Education, Inc. 36.7b: Lisa Lee/Pearson Education, Inc.

Exercise 37 37.3, 37.4: Elena Dorfman/Pearson Education, Inc.

Exercise 38 38.5b: Karen Krabbenhoft/PAL 3.0/Pearson Education, Inc. 38.6a, 38.9a: Nina Zanetti/Pearson Education, Inc. 38.6b, 38.13: Victor P. Eroschenko/Pearson Education, Inc. 38.6c: Lisa Lee/Pearson Education, Inc. 38.7a, 38.14: From *A Stereoscopic Atlas of Human Anatomy* by David L. Bassett, M.D. 38.8d: Steve Gschmeissner/Science Source. 38.9b: Steve Downing/Pearson Education, Inc. 38.9c: William Karkow/Pearson Education, Inc. 38.16b: M. I. Walker/Science Source.

Exercise 40 40.1b: Richard Tauber/Pearson Education, Inc. 40.3a: Karen Krabbenhoft/PAL 3.0/Pearson Education, Inc. 40.7, 40.8: Victor P. Eroschenko/Pearson Education, Inc. 40.RS1: 3B Scientific GmbH.

Exercise 42 42.2b, 42.5, 42.9: Victor P. Eroschenko/Pearson Education, Inc. 42.4: Roger C. Wagner. 42.8: Vetpathologist/Shutterstock. 42.RS1: 3B Scientific GmbH. 42.2c: Winston Charles Poulton / Pearson Education. 42.3: Karen Krabbenhoft / Pearson Education. 42.RS01, 42.RS02: Leif Saul / Pearson Education.

Exercise 43 43.3: Lisa Lee / Pearson Education. 43.4: William Karkow/Pearson Education, Inc. 43.6b: Steve Downing/Pearson Education, Inc. 43.7: Victor P. Eroschenko/Pearson Education, Inc.

Exercise 45 45.1: CNRI/Science Source.

Exercise 46 46.1–46.3, 46.5, 46.7, 46.9–46.11a, 46.12–46.16, 46.18, 46.19: John Wilson White/Pearson Education, Inc.

Cat Dissection Exercises 1.2–1.13, 2.3b, 4.1, 4.3, 4.5b, 6.3, 6.4, 7.3–7.5b, 8.1b, 9.2b: Shawn Miller (dissection) and Mark Nielsen (photography), Pearson Education. 2.1b, 9.1b: Paul Waring, Pearson Education. 2.2b, 7.2b, 8.2b: Elena Dorfman, Pearson Education. 3.3b,c: Yvonne Baptiste-Szymanski, Pearson Education.

Fetal Pig Dissection Exercises 1.1, 1.2: Jack Scanlan, Pearson Education. 1.3b–1.8b, 2.1, 4.1b–4.4b, 5.1, 5.2b, 6.1b, 6.2b, 7.1b, 7.2b, 8.1b. 8.2b: Elena Dorfman, Pearson Education. 3.2b,c: Charles J. Venglarik, Pearson Education.

Index

NOTE: Page numbers in **boldface** indicate a definition. A *t* following a page number indicates tabular material and an *f* indicates an illustration.